# STANDARD HANDBOOK FOR ELECTRICAL ENGINEERS

# ABOUT THE EDITORS

**Surya Santoso, Ph.D.**, is a professor of electrical and computer engineering in the Cockrell School of Engineering at the University of Texas at Austin. He is co-author of *Electrical Power Systems Quality*, co-editor of *Handbook of Electric Power Calculations*, and author of *Fundamentals of Electric Power Quality*. He is an IEEE Fellow.

**H. Wayne Beaty** is the former managing editor of *Electric Light & Power*, co-editor of *Handbook of Electric Power Calculations*, and co-author of *Electric Power Systems Quality*.

# STANDARD HANDBOOK FOR ELECTRICAL ENGINEERS

Surya Santoso, Ph.D. Editor

H. Wayne Beaty Editor

SEVENTEENTH EDITION

New York Chicago San Francisco Athens London
Madrid Mexico City Milan New Delhi
Singapore Sydney Toronto

**Library of Congress Control Number: 2017957636**



**Standard Handbook for Electrical Engineers, Seventeenth Edition**



2 3 4 5 6 7 QVS 23 22 21 20 19

ISBN 978-1-259-64258-6
MHID 1-259-64258-5

This book is printed on acid-free paper.

**Sponsoring Editor**
Michael McCabe

**Editing Supervisor**
Stephen M. Smith

**Production Supervisor**
Pamela A. Pelton

**Project Manager**
Kritika Kaushik, Cenveo® Publisher Services

**Copy Editors**
Surendra Shivam and Dipti Barthwal,
Cenveo Publisher Services

**Proofreaders**
Amitha Karkera and Raama Thenarasu

**Indexer**
Ariel Tuplano

**Art Director, Cover**
Jeff Weeks

**Composition**
Cenveo Publisher Services

# CONTENTS

# CONTRIBUTORS

**Erik Abromitis** *Generator Engineering, Siemens Energy, Inc.* (SEC. 14)

**Héctor J. Altuve Ferrer** *Distinguished Engineer and Dean, Schweitzer Engineering Laboratories, Inc. (SEL) University, Pullman, Washington* (SEC. 19)

**Kehinde Awodele** *University of Cape Town, Cape Town, South Africa* (SEC. 4)

**Ross Baldick** *Professor, University of Texas at Austin* (SEC. 21)

**Emilio Barocio** *Professor, Graduate Studies Program in Electrical Engineering, Universidad de Guadalajara, Guadalajara, Mexico* (SEC. 20)

**Simon Bartlett** *Professor, University of Queensland, Australia* (SEC. 18)

**Earle C. (Rusty) Bascom III** *Principal Engineer, Electrical Consulting Engineers, P.C.* (SEC. 5)

**H. Wayne Beaty** *Editor,* Standard Handbook for Electrical Engineers (SEC. 1)

**George W. Becker** *Senior Substation Engineer, POWER Engineers, Inc., Fort Mill, South Carolina* (SEC. 11)

**Stephen Boroczky** *Principal Engineer, Grid Systems, Australian Energy Market Operator (AEMO), Sydney, NSW, Australia* (SEC. 4)

**Mike Bryson** *Vice President of Operations, PJM Interconnection, United States* (SEC. 18)

**Spencer Burks** *Lower Colorado River Authority, Austin, Texas* (SEC. 4)

**Sayonsom Chanda** *Research Engineer, Idaho National Lab, Idaho Falls, Idaho* (SEC. 8)

**Hong Chen** *Senior Lead Engineer, PJM Interconnection, United States* (SEC. 18)

**Zhe Chen** *Professor of Electrical Engineering, Department of Energy Technology, Aalborg University, Aalborg, Denmark* (SEC. 9)

**Allen L. Clapp** *President, Clapp Research Associates, P.C.* (SEC. 5)

**Jose R. Daconti** *Senior Staff Consultant, Siemens Power Technologies International* (SEC. 5)

**Steven Darnell** *Principal Engineer, Systems Performance and Commercial, Australian Energy Market Operator (AEMO), Brisbane, QLD, Australia* (SEC. 4)

**Francisco de León** *Associate Professor, Department of Electrical and Computer Engineering, NYU Tandon School of Engineering, New York University, Brooklyn, New York* (SEC. 17)

**A. M. DiGioia, Jr.** *President, DiGioia Gray and Associates* (SEC. 5)

**Dale A. Douglass** *Principal Engineer, Douglass Power Consulting, LLC* (SEC. 5)

**Samuel A. Drinkut** *Generator Engineering, Siemens Energy, Inc.* (SEC. 14)

**Roger C. Dugan** *Senior Technical Executive, Electric Power Research Institute, Knoxville, Tennessee* (SEC. 22)

**Franklin T. Emery** *Generator Engineering, Siemens Energy, Inc.* (SEC. 14)

**Jim Eyer** *Principal and Senior Analyst, E&I Consulting, Oakland, California* (SEC. 10)

**Komla A. Folly** *University of Cape Town, Cape Town, South Africa* (SEC. 4)

**Pavlos S. Georgilakis** *National Technical University of Athens (NTUA), Athens, Greece* (SEC. 13)

**Jay Giri** *Director, GE Grid Software Solutions, United States* (SEC. 18)

**I. S. Grant** *Manager, TVA* (SEC. 5)

**Nikos Hatziargyriou** *Professor of Power Systems, School of Electrical and Computer Engineering, National Technical University of Athens, Athens, Greece, and CEO, Hellenic Electricity Distribution Network Operator SA (HEDNO)* (SECS. 8, 9)

**Alan Honecker** *Senior Manager, NEM Real Time Operations, Australian Energy Market Operator (AEMO), Sydney, NSW, Australia* (SEC. 4)

**Tianqi Hong** *Research Fellow, Department of Electrical and Computer Engineering, NYU Tandon School of Engineering, New York University, Brooklyn, New York* (SEC. 17)

**Alex Q. Huang** *Dula D. Cockrell Centennial Chair in Engineering, University of Texas at Austin* (SEC. 16)

**David Infield** *Professor of Renewable Energy Technologies, Department of Electronic and Electrical Engineering, University of Strathclyde, Glasgow, United Kingdom* (SEC. 9)

**Dan M. Ionel** *Professor and L. Stanley Pigman Chair in Power, Department of Electrical and Computer Engineering, University of Kentucky* (SECS. 14, 15)

**David S. Johnson** *Consultant, Pittsburgh, Pennsylvania* (SEC. 12)

**Haresh Kamath** *Senior Program Manager, Energy Storage and Distributed Generation, Electric Power Research Institute, Palo Alto, California* (SEC. 10)

**Leandro Kapolo** *NamPower, Windhoek, Namibia* (SEC. 4)

**Harold Kirkham** *Staff Scientist, Pacific Northwest National Laboratory, Richland, Washington* (SEC. 2)

**Kristian Koellner** *Lower Colorado River Authority, Austin, Texas* (SEC. 4)

**Martin Kopa** *ESKOM, Johannesburg, South Africa* (SEC. 4)

**Benjamin Kroposki** *Director, Power Systems Engineering Center, National Renewable Energy Laboratory, Golden, Colorado* (SEC. 10)

**K. V. N. Pawan Kumar** *Senior Engineer, POSOCO, India* (SEC. 18)

**Jianwei Liu** *Senior Lead Engineer, PJM Interconnection, United States* (SEC. 18)

**Kenneth Long** *Engineering Manager, Transmission & Distribution, Stantec Consulting Ltd., Portland, Oregon* (SEC. 12)

**Otto L. Lynch** *Vice President, Power Line Systems, Inc., Madison, Wisconsin* (SEC. 5)

**Om P. Malik** *Professor Emeritus, Department of Electrical and Computer Engineering, University of Calgary, Calgary, Alberta, Canada* (SECS. 14, 15)

**Robert Margolis** *Principal Energy Analyst, National Renewable Energy Laboratory, Golden, Colorado* (SEC. 10)

**Juan A. Martinez-Velasco** *Professor, Universitat Politècnica de Catalunya, Spain* (SEC. 24)

**Nhlanhla Mbuli** *ESKOM and University of Johannesburg, Johannesburg, South Africa* (SEC. 4)

**Mark F. McGranaghan** *Vice President, Electric Power Research Institute, Knoxville, Tennessee* (SEC. 22)

**Mark McVey** *Principal Engineer, Dominion, Richmond, Virginia* (SEC. 12)

**Mark Mehos** *Program Manager, Concentrating Solar Power, National Renewable Energy Laboratory, Golden, Colorado* (SEC. 10)

**Arturo R. Messina** *Professor, Graduate Studies Program in Electrical Engineering, Center for Research and Advanced Studies, Guadalajara, Mexico* (SEC. 20)

**Marco W. Migliaro** *President and CEO, IEEE Industry Standards and Technology Organization (IEEE-ISTO)* (SEC. 25)

**Federico Milano** *Professor, University College Dublin, Ireland* (SEC. 18)

**Sukumar Mishra** *Professor, Indian Institute of Technology, Delhi, India* (SEC. 20)

**Osama A. Mohammed** *Professor, Department of Electrical and Computer Engineering, Florida International University* (SEC. 14)

**John D. Mozer** *Professional Engineer, Retired* (SEC. 5)

**S. R. Narasimhan** *Additional General Manager, POSOCO, India* (SEC. 18)

**Jeffrey H. Nelson** *Manager, Project Development, Transmission, Tennessee Valley Authority, Chattanooga, Tennessee* (SEC. 12)

**Adam C. Newman** *Senior Director, Business Development and Alliance Management, IEEE Standards Association* (SEC. 25)

**Sarma Nuthalapati** *Principal EMS Network Applications Engineer, PEAK Reliability, Vancouver, Washington* (SEC. 4)

**Oladiran Obadina** *Electric Reliability Council of Texas (ERCOT), Austin, Texas* (SEC. 4)

**Teruo Ohno** *TEPCO Research Institute, Tokyo Electric Power Holdings, Inc., Japan* (SEC. 4)

**T. W. Olsen** *Retired Former Manager, Technology, Energy Management Division, Medium Voltage and Systems, Siemens Industry, Inc., Wendell, North Carolina* (SEC. 12)

**Patrick Panciatici** *Scientific Advisor, RTE France, France* (SEC. 18)

**Adam Peard** *Area Manager—System Analysis and Solutions, Network Planning, Western Power, Perth, WA, Australia* (SEC. 4)

**Vandana Rallabandi** *Research Engineer, Department of Electrical and Computer Engineering, University of Kentucky* (SECS. 14, 15)

**Shantha Ranatunga** *Specialist, Systems Performance and Commercial, Australian Energy Market Operator (AEMO), Brisbane, QLD, Australia* (SEC. 4)

**Ashhar Raza** *Research Fellow, Department of Electrical and Computer Engineering, NYU Tandon School of Engineering, New York University, Brooklyn, New York* (SEC. 17)

**Daniel Ruiz-Vega** *Professor, Graduate Program in Electrical Engineering, Instituto Politécnico Nacional, Mexico* (SEC. 20)

**Surya Santoso** *Professor, Department of Electrical and Computer Engineering, University of Texas at Austin* (SECS. 7, 22)

**S. C. Saxena** *Deputy General Manager, POSOCO, India* (SEC. 18)

**Nilanjan Senroy** *Associate Professor, Indian Institute of Technology, Delhi, India* (SEC. 20)

**Hamid R. Sharifnia** *Manager of Engineering, Stantec Consulting Ltd., Portland, Oregon* (SEC. 12)

**Xu She** *Lead Electrical Engineer, GE Global Research* (SEC. 16)

**S. K. Soonee** *Adviser, POSOCO, India* (SEC. 18)

**Anurag K. Srivastava** *Associate Professor, School of Electrical Engineering and Computer Science, Washington State University, Pullman, Washington* (SEC. 8)

**J. R. Stewart** *Consultant* (SEC. 5)

**Kai Sun** *Associate Professor, University of Tennessee, Knoxville, Tennessee* (SEC. 20)

**Resmi Surendran** *Senior Manager, Wholesale Market Operations and Analysis, Electric Reliability Council of Texas, Taylor, Texas* (SEC. 21)

**Simon Tam** *Manager of Transmission Operations, PJM Interconnection, United States* (SEC. 18)

**Narges Taran** *Research Engineer, Department of Electrical and Computer Engineering, University of Kentucky* (SECS. 14, 15)

**Mini Shaji Thomas** *Director, National Institute of Technology, Tiruchirappalli, India* (SEC. 4)

**Héctor Volskis** *Operador Nacional do Sistema Elétrico (ONS), Brazil* (SEC. 4)

**Michael W. Wactor** *Technical Director, Corporate Product Development, Powell Industries, Inc., Houston, Texas* (SEC. 12)

**Rahul Walawalkar** *President and Managing Director, Customized Energy Solutions India Pvt. Ltd., and Executive Director, India Energy Storage Alliance, Pune, India* (SEC. 10)

**Jianhui Wang** *Department of Electrical Engineering, Southern Methodist University, Dallas, Texas, and Energy Systems Division, Argonne National Laboratory, Argonne, Illinois* (SEC. 8)

**Xuanyuan Sharon Wang** *Jibei Electric Power Company, State Grid Corporation of China, Beijing, China* (SEC. 4)

**Diane Watkins** *Manager, Substation Field Engineering, Xcel Energy, Denver, Colorado* (SEC. 11)

**Jian Zhou** *Director, East China Grid, China* (SEC. 18)

# PREFACE

Electrical engineering is one of the largest and most diverse fields of science and engineering. In its early days in the late nineteenth century, electrical engineering dealt with the study and application of electric power, telephony, and radiotelegraphy. Pioneers of these nascent fields include Thomas Edison, Alexander Graham Bell, and Guglielmo Marconi. Driven by continuous innovation and two world wars, electrical engineering grew rapidly. By the early twenty-first century, it covered electronics, computing and information technology, integrated circuits and embedded systems, nanotechnology, electronic materials, and synthetic biology. This list will certainly continue to expand in the coming decades.

The first edition of the *Standard Handbook for Electrical Engineers* was written and compiled by "A Staff of Specialists" and published by the McGraw Publishing Company in 1907. Continuing its 100-plus years of legacy, this Handbook focuses on one particular branch of electrical engineering: electric power and its applications. The topics in the Handbook encompass the full spectrum of electric power engineering and include generation, transmission, distribution, operation, system protection, switchgear, power components, and electricity markets.

Since the publication of the Sixteenth Edition of this Handbook, several changes have taken place that have had an impact on the science and technology of electric power engineering. The new and significantly revised sections in the Seventeenth Edition are as follows.

The "Measurement and Instrumentation" section has been completely revised and updated to include the concept of and expression of uncertainty in measurement, digital measurement techniques, and power, energy, and phasor measurements, as well as the measurement of component values such as resistance, inductance, and capacitance.

"Interconnected Power Grids" is a new section that provides a comprehensive overview of network structures, transmission and distribution service providers, and electricity markets for interconnected power grids around the world, that is, in Australia, Brazil, China, India, Japan, North America, and Africa.

"Smart Grids and Microgrids" is a new section that introduces the ins and outs of microgrids and smart grids. Recent advances in computation, communication, controllable loads, and automation allow distribution circuits to operate in a complete island. Smart grids take advantage of various digital technologies to improve the system operation beyond that of a traditional grid.

Renewable energy, both wind and solar, has been integrated into transmission and distribution grids in a greater proportion in the past decades. Energy storage has become a key technology enabler of renewable integration and ancillary grid services. As a result, two new sections, "Wind Power Generation" and "Solar Power Generation and Energy Storage," have been added to the Handbook.

Power transformers are an essential apparatus in power transmission and delivery. "Power Transformers" is dedicated to covering a range of topics in this area, including the characteristics, types, design, insulation, and operation of power transformers.

Materials on prime movers and electric machines have been reorganized and consolidated into two sections, "Electric Machines: Generators" and "Electric Machines: Motors and Drives." Among the topics included in these sections are prime movers, dc and ac generators and motors, special-purpose electric machines, and drives.

"Power System Analysis" is a new section that provides fundamental knowledge necessary for analyzing power systems in steady state. It covers complex power, per-unit system, sequence impedance, power flow, and short-circuit analysis.

"Power System Operations" is another new section that discusses how interconnected power systems operate efficiently. Major topics in this section include power balance, frequency control, transmission operation and security, and energy management and outage management systems. The section also presents international perspectives on power system operation in the United States, Europe, China, Australia, and India.

Protective elements and systems are the first line of defense in detecting short-circuit faults and abnormal operating conditions. The main functions of such systems include disconnecting the protected elements and facilitating restoration. "Power System Protection," a new section of the Handbook, provides a comprehensive introduction to power system protection. It covers numerous protection techniques and their applications to equipment protection such as buses, lines, transformers, and generators.

Power system stability studies evaluate the ability of a power system to return to its stable operating condition without losing synchronism following a system disturbance. A new section on "Power System Stability and Control" has been added to the Handbook. It discusses small-signal and transient stability as well as the impact of wind and solar generation on system stability.

Electricity markets are artificial constructs designed to realize the objective of providing dependable electrical power at the lowest cost of production. Electricity markets are essential for the modern operation and management of interconnected power grids. A new section on "Electricity Markets" has been added to the Handbook, and it covers the philosophy and principles, characteristics and building blocks, and design and implementation of electricity markets.

The editors and contributors expect that this updated edition of the Handbook will continue to serve the professional careers of its readers. As with previous editions of the Handbook, the Seventeenth Edition contains, in a single volume, all major electric power topics and aims to be accurate and comprehensive in its technical treatment and to be of use in engineering practice and application as well as in study and preparation for such practice.

*Surya Santoso*

# ACKNOWLEDGMENTS

I would like to express my sincere gratitude to the contributors for their painstaking and immeasurable efforts in reviewing, updating, and authoring materials presented in the Seventeenth Edition of this Handbook. The contributors consist of more than 100 world-expert academics and practicing engineers in electric power and its various subspecialties. This Seventeenth Edition would not have been possible without their dedication and commitment. It is a privilege to work with such talented people.

The materials compiled and presented in this Handbook have undergone continuous revision and refinement to reflect advances in the field of electric power and its applications. Grateful acknowledgment is given to each author who contributed to this Handbook since its Thirteenth Edition (1993):

Donald G. Fink and Barry N. Taylor on "Units, Symbols, Constants, Definitions, and Conversion Factors"

Donald G. Fink, Gerald Fitzpatrick, Norman Belecki, George Burns, Forest Harris, B. W. Mangum, and Martin R. Riley on "Measurement and Instrumentation"

Glenn Davidson, Philip Mason Opsal, Donald J. Barta, T. W. Dakin, Charles A. Harper, Duane E. Lyon, Charles B. Rawlins, James Stubbins, John Tanaka, Anthony L. Von Holle, Kenneth L. Latimer, E. J. Croop, and Robert W. Bohl on "Properties of Materials"

J. A. Williams, H. Brian White, L. O. Barthold, J. A. Moran, and D. D. Wilson on "Alternating-Current Power Transmission"

Ram Adapa, Michael P. Bahrman, P. F. Albrecht, G. D. Breuer, K. Clark, R. C. Degeneff, H. J. Fielder, C. W. Flairty, D. W. Houghtaling, E. T. Jauch, J. J. LaForest, E. V. Larsen, J. C. McIver, F. Nozari, R. L. Rofini, H. M. Schneider, J. D. Stickler, J. Urbanek, and L. E. Zafanella on "Direct-Current Power Transmission"

Allen L. Clapp, Daniel J. Ward, Cheryl A. Warren, James L. Burke, and Walter J. Ros on "Power Distribution"

John Randolph, Philip C. Bolin, Allen L. Clapp, W. Bruce Dietzman, Joseph Basilesco, Rusko Matulic, and Philip R. Nannery on "Substations"

Craig A. Colopy, Carey J. Cook, Jon Hilgenkamp, Christopher McCarthy, Douglas M. Staszesky, Robert B. Hardin, Robert J. Landman, Kelly A. Shaw, Robert A. Brown, Ramsis S. Girgis, Louis C. Grove, James H. Harlow, Robert E. Kleeb, and Carl M. Pandza on "Switchgear and Power System Components"

O. A. Mohammed, Thomas W. Nehl, E. H. Myers, Erik Abromitis, Samuel A. Drinkut, Franklin T. Emery, John D. Amos, Aleksandar Prole, Lon W. Montgomery, James L. Kirtley, Jr., R. E. Appleyard, L. T. Rosenberg, William H. Day, Donald H. Hall, Lawrence R. Mizen, and Roy P. Allen on "Electric Machines: Generators"

Om P. Malik, Kenneth C. Cornelius, John H. Dulas, Alexander Kusko, Kelly A. Shaw, and Syed M. Peeran on "Electric Machines: Motors and Drives"

Amit Kumar Jain, Raja Ayyanar, P. Wood, L. Gyugyi, Jerome B. Brewster, T. M. Heinrich, R. M. Oates, B. R. Pelley, and Donald Galler on "Power Electronics"

John Adams, Hassan Bevrani, Math H. J. Bollen, Gustavo Brunello, Rujiroj Leelaruji, Christopher McCarthy, Yasunori Mitani, Sarma Nuthalapati, Oladiran Obadina, Paulo F. Ribeiro, Hesham

Shaalan, Douglas M. Staszesky, George R. Stoll, Resmi Surendran, Luigi Vanfretti, Masayuki Watanabe, Christa Lorber, James R. Latimer, Bruce F. Wollenberg, W. A. Elmore, and Jalal Gohari on "Power System Operations"

John B. Dagenhart on "Power Quality and Reliability"

A. P. (Sakis) Meliopoulos on "Lightning and Overvoltage Protection"

James V. Mitsche, M. M. Adibi, J. D. Cypert, and T. Q. Zhang on "Computer Applications in the Electric Power Industry"

I would like to thank H. Wayne Beaty, who has served as co-editor of this Handbook with Donald G. Fink since 1978 and later was editor from 2000 to 2013. He joins an esteemed lineage of previous editors of this Handbook: Frank F. Fowle (1915–1933), Archer E. Knowlton (1941–1957), and Donald G. Fink (1968–1987). In addition, I would like to thank my doctoral students, Harsha V. Padullaparti, Naveen Ganta, Piyapath Siratarnsophon, Suma Jothibasu, David Rosewater, and Quan Nguyen, for examining and reviewing a number of sections from the Handbook, as well as Dr. Grazia Todeschini of Swansea University, U.K., and Michael McCabe of McGraw-Hill for their input and encouragement throughout the process. Special appreciation goes to Kritika Kaushik and her team at Cenveo Publisher Services for their patience in fulfilling my requests, meticulous adherence to perfection, and untiring effort in editing and typesetting the manuscript.

Finally, one cannot overlook the support so generously given by his family during this enormous endeavor. When one lifts up his eyes and surveys the mountains, he asks "Where does my help come from?" Like the Psalmist's, his help comes from the LORD, the Maker of heaven and earth!

*Surya Santoso*

# 1

# UNITS, SYMBOLS, CONSTANTS, DEFINITIONS, AND CONVERSION FACTORS

H. Wayne Beaty

*Editor,* Standard Handbook for Electrical Engineers; *Senior Member, Institute of Electrical and Electronics Engineers; technical assistance provided by David B. Newell, Staff Scientist, National Institute of Standards and Technology, and Chair, CODATA Task Group on Fundamental Constants*



## 1.1 THE SI UNITS

The units of the quantities most commonly used in electrical engineering (volts, amperes, watts, ohms, etc.) are those of the metric system. They are embodied in the International System of Units (*Système International d'Unités*, abbreviated *SI*). The SI units are used throughout this handbook, in accordance with the established practice of electrical engineering publications throughout the world. Other units, notably the cgs (centimeter-gram-second) units, may have been used in citations in the earlier literature. The cgs electrical units are listed in Table 1-9 with conversion factors to the SI units.

The SI electrical units are based on the mksa (meter-kilogram-second-ampere) system. They have been adopted by the standardization bodies of the world, including the International Electrotechnical Commission (IEC), the American National Standards Institute (ANSI), and the Standards Board of the Institute of Electrical and Electronics Engineers (IEEE).

## 1.2 CGPM BASE QUANTITIES

Seven quantities have been adopted by the General Conference on Weights and Measures (CGPM[a]) as *base quantities*, that is, quantities that are not derived from other quantities. The base quantities are length, mass, time, electric current, thermodynamic temperature, amount of substance, and luminous intensity. Table 1-1 lists these quantities, the name of the SI unit for each, and the standard letter symbol by which each is expressed in the International System (SI).

**TABLE 1-1** SI Base Units

| Quantity | Unit | Symbol |
|---|---|---|
| Length | meter | m |
| Mass | kilogram | kg |
| Time | second | s |
| Electric current | ampere | A |
| Thermodynamic temperature* | kelvin | K |
| Amount of substance | mole | mol |
| Luminous intensity | candela | cd |

*Celsius temperature is, in general, expressed in degrees Celsius (symbol °C).

The units of the base quantities have been defined by the CGPM as follows:

*Meter.* The length of the path traveled by light in vacuum during a time interval of 1/299 792 458 of a second (CGPM).

*Kilogram.* The unit of mass; it is equal to the mass of the international prototype of the kilogram (CGPM).

NOTE: The prototype is a platinum-iridium cylinder maintained at the International Bureau of Weights and Measures, near Paris. The kilogram is approximately equal to the mass of 1000 cubic centimeters of water at its temperature of maximum density.

*Second.* The duration of 9 192 631 770 periods of the radiation corresponding to the transition between the two hyperfine levels of the ground state of the cesium 133 atoms (CGPM).

*Ampere.* The constant current that if maintained in two straight parallel conductors of infinite length, of negligible circular cross section, and placed 1 meter apart in vacuum would produce between these conductors a force equal to $2 \times 10^{-7}$ newton per meter of length (CGPM).

*Kelvin.* The unit of thermodynamic temperature is the fraction 1/273.16 of the thermodynamic temperature of the triple point of water (CGPM).

NOTE: The zero of the Celsius scale (the freezing point of water) is defined as 0.01 K below the triple point, that is, 273.15 K. See Table 1-27.

*Mole.* That amount of substance of a system that contains as many elementary entities as there are atoms in 0.012 kilogram of carbon-12 (CGPM).

NOTE: When the mole is used, the elementary entities must be specified. They may be atoms, molecules, ions, electrons, other particles, or specified groups of such particles.

*Candela.* The luminous intensity, in a given direction, of a source that emits monochromatic radiation of frequency $540 \times 10^{12}$ Hz and that has a radiant intensity in that direction of 1/683 watt per steradian (CGPM).

NOTE: Until January 1, 1948, the generally accepted unit of luminous intensity was the *international candle.* The difference between the candela and the international candle is so small that only measurements of high precision are affected. The use of the term *candle* is deprecated.

[a]From the initials of its French name, *Conference Générale des Poids et Mesures.*

## 1.3 SUPPLEMENTARY SI UNITS

Two additional SI units, numerics which are considered as dimensionless derived units (see Sec. 1.4), are the radian and the steradian, for the quantities plane angle and solid angle, respectively. Table 1-2 lists these quantities and their units and symbols. The supplementary units are defined as follows:

*Radian.* The plane angle between two radii of a circle that cut off on the circumference an arc equal in length to the radius (CGPM).

*Steradian.* The solid angle which, having its vertex in the center of a sphere, cuts off an area of the surface of the sphere equal to that of a square with sides equal to the radius of the sphere (CGPM).

**TABLE 1-2** SI Supplementary Units

| Quantity | Unit | Symbol |
|---|---|---|
| Plane angle | radian | rad |
| Solid angle | steradian | sr |

## 1.4 DERIVED SI UNITS

Most of the quantities and units used in electrical engineering fall in the category of SI derived units, that is, units which can be completely defined in terms of the base and supplementary quantities described above. Table 1-3 lists the principal electrical quantities in the SI system and shows their equivalents in terms of the base and supplementary units. The definitions of these quantities, as they appear in the *IEEE Standard Dictionary of Electrical and Electronics Terms* (ANSI/IEEE Std 100-1988), are

*Hertz.* The unit of frequency 1 cycle per second.

*Newton.* The force that will impart an acceleration of 1 meter per second to a mass of 1 kilogram.

**TABLE 1-3** SI Derived Units in Electrical Engineering

| | SI unit | | | |
|---|---|---|---|---|
| Quantity | Name | Symbol | Expression in terms of other units | Expression in terms of SI base units |
| Frequency (of a periodic phenomenon) | hertz | Hz | 1/s | $s^{-1}$ |
| Force | newton | N | | $m \cdot kg \cdot s^{-2}$ |
| Pressure, stress | pascal | Pa | $N/m^2$ | $m^{-1} \cdot kg \cdot s^{-2}$ |
| Energy, work, quantity of heat | joule | J | $N \cdot m$ | $m^2 \cdot kg \cdot s^{-2}$ |
| Power, radiant flux | watt | W | J/s | $m^2 \cdot kg \cdot s^{-3}$ |
| Quantity of electricity, electric charge | coulomb | C | $A \cdot s$ | $s \cdot A$ |
| Potential difference, electric potential, electromotive force | volt | V | W/A | $m^2 \cdot kg \cdot s^{-3} \cdot A^{-1}$ |
| Electric capacitance | farad | F | C/V | $m^{-2} \cdot kg^{-1} \cdot s^4 \cdot A^2$ |
| Electric resistance | ohm | Ω | V/A | $m^2 \cdot kg \cdot s^{-3} \cdot A^{-2}$ |
| Conductance | siemens | S | A/V | $m^{-2} \cdot kg^{-1} \cdot s^3 \cdot A^2$ |
| Magnetic flux | weber | Wb | $V \cdot s$ | $m^2 \cdot kg \cdot s^{-2} \cdot A^{-1}$ |
| Magnetic flux density | tesla | T | $Wb/m^2$ | $kg \cdot s^{-2} \cdot A^{-1}$ |
| Celsius temperature | degree Celsius | °C | K | |
| Inductance | henry | H | Wb/A | $m^2 \cdot kg \cdot s^{-2} \cdot A^{-2}$ |
| Luminous flux | lumen | lm | | $cd \cdot sr$* |
| Illuminance | lux | lx | $lm/m^2$ | $m^{-2} \cdot cd \cdot sr$* |
| Activity (of radionuclides) | becquerel | Bq | 1/s | $s^{-1}$ |
| Absorbed dose | gray | Gy | J/kg | $m^2 \cdot s^{-2}$ |
| Dose equivalent | sievert | Sv | J/kg | $m^2 \cdot s^{-2}$ |

*In this expression, the steradian (sr) is treated as a base unit. See Table 1-2.

*Pascal.* The pressure exerted by a force of 1 newton uniformly distributed on a surface of 1 square meter.

*Joule.* The work done by a force of 1 newton acting through a distance of 1 meter.

*Watt.* The power required to do work at the rate of 1 joule per second.

*Coulomb.* The quantity of electric charge that passes any cross section of a conductor in 1 second when the current is maintained constant at 1 ampere.

*Volt.* The potential difference between two points of a conducting wire carrying a constant current of 1 ampere, when the power dissipated between these points is 1 watt.

*Farad.* The capacitance of a capacitor in which a charge of 1 coulomb produces 1 volt potential difference between its terminals.

*Ohm.* The resistance of a conductor such that a constant current of 1 ampere in it produces a voltage of 1 volt between its ends.

*Siemens (mho).* The conductance of a conductor such that a constant voltage of 1 volt between its ends produces a current of 1 ampere in it.

*Weber.* The magnetic flux which decreases to zero when linked with a single turn induces in the turn a voltage whose time integral is 1 volt-second.

*Tesla.* The magnetic induction equal to 1 weber per square meter.

*Henry.* The inductance for which the induced voltage in volts is numerically equal to the rate of change of current in amperes per second.

*Lumen.* The flux through a unit solid angle (steradian) from a uniform point source of 1 candela; the flux on a unit surface all points of which are at a unit distance from a uniform point source of 1 candela.

*Lux.* The illumination on a surface of 1 square meter on which there is uniformly distributed a flux of 1 lumen; the illumination produced at a surface all points of which are 1 meter away from a uniform point source of 1 candela.

Table 1-4 lists other quantities and the SI derived unit names and symbols useful in engineering applications. Table 1-5 lists additional quantities and the SI derived units and symbols used in mechanics, heat, and electricity.

**TABLE 1-4** Examples of SI Derived Units of General Application in Engineering

| Quantity | SI unit | |
|---|---|---|
| | Name | Symbol |
| Angular velocity | radian per second | rad/s |
| Angular acceleration | radian per second squared | $rad/s^2$ |
| Radiant intensity | watt per steradian | W/sr |
| Radiance | watt per square meter steradian | $W \cdot m^{-2} \cdot sr^{-1}$ |
| Area | square meter | $m^2$ |
| Volume | cubic meter | $m^3$ |
| Velocity | meter per second | m/s |
| Acceleration | meter per second squared | $m/s^2$ |
| Wavenumber | 1 per meter | $m^{-1}$ |
| Density, mass | kilogram per cubic meter | $kg/m^3$ |
| Concentration (of amount of substance) | mole per cubic meter | $mol/m^3$ |
| Specific volume | cubic meter per kilogram | $m^3/kg$ |
| Luminance | candela per square meter | $cd/m^2$ |

**TABLE 1-5** Examples of SI Derived Units Used in Mechanics, Heat, and Electricity

| Quantity | SI unit | | |
|---|---|---|---|
| | Name | Symbol | Expression in terms of SI base units |
| Viscosity, dynamic | pascal second | Pa · s | $m^{-1} \cdot kg \cdot s^{-1}$ |
| Moment of force | newton meter | N · m | $m^2 \cdot kg \cdot s^{-2}$ |
| Surface tension | newton per meter | N/m | $kg \cdot s^{-2}$ |
| Heat flux density, irradiance | watt per square meter | $W/m^2$ | $kg \cdot s^{-3}$ |
| Heat capacity | joule per kelvin | J/K | $m^2 \cdot kg \cdot s^{-2} \cdot K^{-1}$ |
| Specific heat capacity, specific entropy | joule per kilogram kelvin | J/(kg · K) | $m^2 \cdot s^{-2} \cdot K^{-1}$ |
| Specific energy | joule per kilogram | J/kg | $m^2 \cdot s^{-2}$ |
| Thermal conductivity | watt per meter kelvin | W/(m · K) | $m \cdot kg \cdot s^{-3} \cdot K^{-1}$ |
| Energy density | joule per cubic meter | $J/m^3$ | $m^{-1} \cdot kg \cdot s^{-2}$ |
| Electric field strength | volt per meter | V/m | $m \cdot kg \cdot s^{-3} \cdot A^{-1}$ |
| Electric charge density | coulomb per cubic meter | $C/m^3$ | $m^{-3} \cdot s \cdot A$ |
| Electric flux density | coulomb per square meter | $C/m^2$ | $m^{-2} \cdot s \cdot A$ |
| Permittivity | farad per meter | F/m | $m^{-3} \cdot kg^{-1} \cdot s^4 \cdot A^2$ |
| Current density | ampere per square meter | $A/m^2$ | |
| Magnetic field strength | ampere per meter | A/m | |
| Permeability | henry per meter | H/m | $m \cdot kg \cdot s^{-2} \cdot A^{-2}$ |
| Molar energy | joule per mole | J/mol | $m^2 \cdot kg \cdot s^{-2} \cdot mol^{-1}$ |
| Molar entropy, molar heat capacity | joule per mole kelvin | J/(mol · K) | $m^2 \cdot kg \cdot s^{-2} \cdot K^{-1} mol^{-1}$ |

## 1.5 SI DECIMAL PREFIXES

All SI units may have affixed to them standard prefixes which multiply the indicated quantity by a power of 10. Table 1-6 lists the standard prefixes and their symbols. A substantial part of the extensive range ($10^{36}$) covered by these prefixes is in common use in electrical engineering (e.g., gigawatt, gigahertz, nanosecond, and picofarad). The practice of compounding a prefix (e.g., micromicrofarad) is deprecated (the correct term is picofarad).

## 1.6 USAGE OF SI UNITS, SYMBOLS, AND PREFIXES

Care must be exercised in using the SI symbols and prefixes to follow exactly the capital-letter and lowercase-letter usage prescribed in Tables 1-1 through 1-8, inclusive. Otherwise, serious confusion may occur. For example, pA is the SI symbol for $10^{-12}$ of the SI unit for electric current (picoampere), while Pa is the SI symbol for pressure (the pascal).

**TABLE 1-6** SI Prefixes Expressing Decimal Factors

| Factor | Prefix | Symbol | Factor | Prefix | Symbol |
|---|---|---|---|---|---|
| $10^{18}$ | exa | E | $10^{-1}$ | deci | d |
| $10^{15}$ | peta | P | $10^{-2}$ | centi | c |
| $10^{12}$ | tera | T | $10^{-3}$ | milli | m |
| $10^9$ | giga | G | $10^{-6}$ | micro | $\mu$ |
| $10^6$ | mega | M | $10^{-9}$ | nano | n |
| $10^3$ | kilo | k | $10^{-12}$ | pico | p |
| $10^2$ | hecto | h | $10^{-15}$ | femto | f |
| $10^1$ | deka | da | $10^{-18}$ | atto | a |

The spelled-out names of the SI units (e.g., volt, ampere, watt) are not capitalized. The SI letter symbols are capitalized only when the name of the unit stands for or is directly derived from the name of a person. Examples are V for volt, after Italian physicist Alessandro Volta (1745–1827); A for ampere, after French physicist André-Marie Ampère (1775–1836); and W for watt, after Scottish engineer James Watt (1736–1819). The letter symbols serve the function of abbreviations, but they are used without periods.

It will be noted from Tables 1-1, 1-3, and 1-5 that with the exception of the ampere, all the SI electrical quantities and units are derived from the SI base and supplementary units or from other SI derived units. Thus, many of the short names of SI units may be expressed in compound form embracing the SI units from which they are derived. Examples are the volt per ampere for the ohm, the joule per second for the watt, the ampere-second for the coulomb, and the watt-second for the joule. Such compound usage is permissible, but in engineering publications, the short names are customarily used.

Use of the SI prefixes with non-SI units is not recommended; the only exception stated in IEEE Standard 268 is the microinch. Non-SI units, which are related to the metric system but are not decimal multiples of the SI units such as the calorie, torr, and kilogram-force, are specially to be avoided.

**TABLE 1-7** Time and Angle Units Used in the SI System (Not Decimally Related to the SI Units)

| Name | Symbol | Value in SI unit |
|---|---|---|
| minute | min | 1 min = 60 s |
| hour | h | 1 h = 60 min = 3 600 s |
| day | d | 1 d = 24 h = 86 400 s |
| degree | ° | 1° = (π/180) rad |
| minute | ′ | 1′ = (1/60)° = (π/10 800) rad |
| second | ″ | 1″ = (1/60)′ = (π/648 000) rad |

A particular problem arises with the universally used units of time (minute, hour, day, year, etc.) that are nondecimal multiples of the second. Table 1-7 lists these and their equivalents in seconds, as well as their standard symbols (see also Table 1-19). The watthour (Wh) is a case in point; it is equal to 3600 joules. The kilowatthour (kWh) is equal to 3 600 000 joules or 3.6 megajoules (MJ). In the mid-1980s, the use of the kilowatthour persisted widely, although eventually it was expected to be replaced by the megajoule, with the conversion factor 3.6 megajoules per kilowatthour. Other aspects in the usage of the SI system are the subject of the following recommendations published by the IEEE:

*Frequency.* The CGPM has adopted the name *hertz* for the unit of frequency, but cycle per second is widely used. Although cycle per second is technically correct, the name *hertz* is preferred because of the widespread use of cycle alone as a unit of frequency. Use of cycle in place of cycle per second, or kilocycle in place of kilocycle per second, etc., is incorrect.

*Magnetic Flux Density.* The CGPM has adopted the name *tesla* for the SI unit of magnetic flux density. The name *gamma* shall not be used for the unit nanotesla.

*Temperature Scale.* In 1948, the CGPM abandoned *centigrade* as the name of the temperature scale. The corresponding scale is now properly named the *Celsius scale*, and further use of *centigrade* for this purpose is deprecated.

*Luminous Intensity.* The SI unit of luminous intensity has been given the name *candela*, and further use of the old name *candle* is deprecated. Use of the term *candle-power*, either as the name of a quantity or as the name of a unit, is deprecated.

*Luminous Flux Density.* The common British-American unit of luminous flux density is the lumen per square foot. The name *footcandle*, which has been used for this unit in the United States, is deprecated.

*Micrometer and Micron.* The names *micron* for micrometer and *millimicron* for nanometer are deprecated.

*Gigaelectronvolt (GeV).* Because billion means a thousand million in the United States but a million million in most other countries, its use should be avoided in technical writing. The term *billion electronvolts* is deprecated; use gigaelectronvolts instead.

*British-American Units.* In principle, the number of British-American units in use should be reduced as rapidly as possible. Quantities are not to be expressed in mixed units. For example, mass should be expressed as 12.75 lb, rather than 12 lb or 12 oz. As a start toward implementing this recommendation, the following should be abandoned:

1. British thermal unit (for conversion factors, see Table 1-25).
2. horsepower (see Table 1-26).
3. Rankine temperature scale (see Table 1-27).
4. U.S. dry quart, U.S. liquid quart, and U.K. (Imperial) quart, together with their various multiples and subdivisions. If it is absolutely necessary to express volume in British-American units, the cubic inch or cubic foot should be used (for conversion factors, see Table 1-17).
5. footlambert. If it is absolutely necessary to express luminance in British-American units, the candela per square foot or lumen per steradian square foot should be used (see Table 1-28A).
6. inch of mercury (see Table 1-23C).

## 1.7 OTHER SI UNITS

Table 1-8 lists units used in the SI system whose values are not derived from the base quantities but from experiment. The definitions of these units, given in the *IEEE Standard Dictionary* (ANSI/IEEE Std 100-1988) are

*Electronvolt.* The kinetic energy acquired by an electron in passing through a potential difference of 1 volt in vacuum.

NOTE: The electronvolt is equal to $1.60218 \times 10^{-19}$ joule, approximately (see Table 1-25B).

*Unified Atomic Mass Unit.* The fraction ½ of the mass of an atom of the nuclide $^{12}$C.

NOTE: u is equal to $1.660\ 54 \times 10^{-27}$ kg, approximately.

**TABLE 1-8** Units Used with the SI System Whose Values Are Obtained Experimentally

| Name | Symbol |
|---|---|
| electronvolt | eV |
| unified atomic mass unit | u |
| astronomical unit* | |
| parsec | pc |

*The astronomical unit does not have an international symbol. AU is customarily used in English, UA in French.

*Astronomical Unit.* The length of the radius of the unperturbed circular orbit of a body of negligible mass moving around the sun with a sidereal angular velocity of 0.017 202 098 950 radian per day of 86 400 ephemeris seconds.

NOTE: The International Astronomical Union has adopted a value for 1 AU equal to $1.496 \times 10^{11}$ meters (see Table 1-15C).

*Parsec.* The distance at which 1 astronomical unit subtends an angle of 1 second of arc. 1 pc = 206 264.8 AU = $30\ 857 \times 10^{12}$ m, approximately (see Table 1-15C).

## 1.8 CGS SYSTEMS OF UNITS

The units most commonly used in physics and electrical science, from their establishment in 1873 until their virtual abandonment in 1948, are based on the centimeter-gram-second (cgs) electromagnetic and electrostatic systems. They have been used primarily in theoretical work, as contrasted with the SI units (and their "practical unit" predecessors, see Sec. 1.9) used in engineering. Table 1-9 lists the principal cgs electrical quantities and their units, symbols, and equivalent values in SI units. Use of these units in electrical engineering publications has been officially deprecated by the IEEE since 1966.

**TABLE 1-9** CGS Units and Equivalents

| Quantity | Name | Symbol | Correspondence with SI unit |
|---|---|---|---|
| | Electromagnetic system | | |
| Current | abampere | abA | = 10 amperes (exactly) |
| Voltage | abvolt | abV | = $10^{-8}$ volt (exactly) |
| Capacitance | abfarad | abF | = $10^{9}$ farads (exactly) |
| Inductance | abhenry | abH | = $10^{-9}$ henry (exactly) |
| Resistance | abohm | abΩ | = $10^{-9}$ ohm (exactly) |
| Magnetic flux | maxwell | Mx | = $10^{-8}$ weber (exactly) |
| Magnetic field strength | oersted | Oe | = 79.577 4 amperes per meter |
| Magnetic flux density | gauss | G | = $10^{-4}$ tesla (exactly) |
| Magnetomotive force | gilbert | Gb | = 0.795 774 ampere |
| | Electrostatic system | | |
| Current | statampere | statA | = 3.335 641 × $10^{-10}$ ampere |
| Voltage | statvolt | statV | = 299.792 46 volts |
| Capacitance | statfarad | statF | = 1.112 650 × $10^{-12}$ farad |
| Inductance | stathenry | statH | = 8.987 554 × $10^{11}$ henrys |
| Resistance | statohm | statΩ | = 8.987 554 × $10^{11}$ ohms |
| | Mechanical units | | |
| | (equally applicable to the electrostatic and electromagnetic systems) | | |
| Work/energy | erg | erg | = $10^{-7}$ joule (exactly) |
| Force | dyne | dyn | = $10^{-5}$ newton (exactly) |

The cgs units have not been used to any great extent in electrical engineering, since many of the units are of inconvenient size compared with quantities used in practice. For example, the cgs electromagnetic unit of capacitance is the gigafarad.

## 1.9 PRACTICAL UNITS (ISU)

The shortcomings of the cgs systems were overcome by adopting the volt, ampere, ohm, farad, coulomb, henry, joule, and watt as "practical units," each being an exact decimal multiple of the corresponding electromagnetic cgs unit (see Table 1-9). From 1908 to 1948, the practical electrical units were embodied in the International System Units (ISU, not to be confused with the SI units). During these years, precise formulation of the units in terms of mass, length, and time was impractical because of imprecision in the measurements of the three basic quantities. As an alternative, the units were standardized by comparison with apparatus, called *prototype standards*. By 1948, advances in the measurement of the basic quantities permitted precise standardization by reference to the definitions of the basic units, and the International System Units were officially abandoned in favor of the absolute units. These in turn were supplanted by the SI units which came into force in 1950.

## 1.10 DEFINITIONS OF ELECTRICAL QUANTITIES

The following definitions are based on the principal meanings listed in the *IEEE Standard Dictionary* (ANSI/IEEE Std 100-1988), which should be consulted for extended meanings, compound terms, and related definitions. The United States Standard Symbols (ANSI/IEEE Std 260, IEEE Std 280) for these quantities are shown in parentheses (see also Tables 1-10 and 1-11). Electrical units used in the United States prior to 1969, with SI equivalents, are listed in Table 1-29.

*Admittance (Y).* An admittance of a linear constant-parameter system is the ratio of the phasor equivalent of the steady-state sine-wave current or current-like quantity (response) to the phasor equivalent of the corresponding voltage or voltage-like quantity (driving force).

*Capacitance (C).* Capacitance is that property of a system of conductors and dielectrics which permits the storage of electrically separated charges when potential differences exist between the conductors. Its value is expressed as the ratio of an electric charge to a potential difference.

*Coupling Coefficient (k).* Coefficient of coupling (used only in the case of resistive, capacitive, and inductive coupling) is the ratio of the mutual impedance of the coupling to the square root of the product of the self-impedances of similar elements in the two circuit loops considered. Unless otherwise specified, coefficient of coupling refers to inductive coupling, in which case $k = M/(L_1L_2)^{1/2}$, where $M$ is the mutual inductance, $L_1$ the self-inductance of one loop, and $L_2$ the self-inductance of the other.

*Conductance (G)*

1. The conductance of an element, device, branch, network, or system is the factor by which the mean-square voltage must be multiplied to give the corresponding power lost by dissipation as heat or as other permanent radiation or as electromagnetic energy from the circuit.
2. Conductance is the real part of admittance.

*Conductivity ($\gamma$).* The conductivity of a material is a factor such that the conduction current density is equal to the electric field strength in the material multiplied by the conductivity.

*Current (I).* Current is a generic term used when there is no danger of ambiguity to refer to any one or more of the currents described below. (For example, in the expression "the current in a simple series circuit," the word *current* refers to the conduction current in the wire of the inductor and to the displacement current between the plates of the capacitor.)

*Conduction Current.* The conduction current through any surface is the integral of the normal component of the conduction current density over that surface.

*Displacement Current.* The displacement current through any surface is the integral of the normal component of the displacement current density over that surface.

*Current Density (J).* Current density is a generic term used when there is no danger of ambiguity to refer either to conduction current density or to displacement current density or to both.

*Displacement Current Density.* The displacement current density at any point in an electric field is (in the International System) the time rate of change of the electric-flux-density vector at that point.

*Conduction Current Density.* The electric conduction current density at any point at which there is a motion of electric charge is a vector quantity whose direction is that of the flow of positive charge at this point, and whose magnitude is the limit of the time rate of flow of net (positive) charge across a small plane area perpendicular to the motion, divided by this area, as the area taken approaches zero in a macroscopic sense, so as to always include this point. The flow of charge may result from the movement of free electrons or ions but is not in general, except in microscopic studies, taken to include motions of charges resulting from the polarization of the dielectric.

*Damping Coefficient ($\delta$).* If $F$ is a function of time given by

$$F = A \exp(-\delta t) \sin(2\pi t/T)$$

then $\delta$ is the damping coefficient.

*Elastance (S).* Elastance is the reciprocal of capacitance.

*Electric Charge, Quantity of Electricity (Q).* Electric charge is a fundamentally assumed concept required by the existence of forces measurable experimentally. It has two forms known as positive and negative. The electric charge on (or in) a body or within a closed surface is the excess of one form of electricity over the other.

*Electric Constant, Permittivity of Vacuum ($\Gamma_e$).* The electric constant pertinent to any system of units is the scalar which in that system relates the electric flux density $D$ in vacuum, to $E$, the electric field strength ($D = \Gamma_e E$). It also relates the mechanical force between two charges in vacuum to their magnitudes and separation. Thus, in the equation $F = \Gamma_r Q_1 Q_2 / 4\pi\Gamma_e r^2$, the force $F$ between charges $Q_1$ and $Q_2$ separated by a distance $r\Gamma_e$ is the electric constant, and $\Gamma_r$ is a dimensionless factor which is unity in a rationalized system and $4\pi$ in an unrationalized system.

NOTE: In the cgs electrostatic system, $\Gamma_e$ is assigned measure unity and the dimension "numeric." In the cgs electromagnetic system, the measure of $\Gamma_e$ is that of $1/c^2$, and the dimension is $[L^{-2}T^2]$. In the International System, the measure of $\Gamma_e$ is $10^7/4\pi c^2$, and the dimension is $[L^{-3}M^{-1}T^4I^2]$. Here, $c$ is the speed of light expressed in the appropriate system of units (see Table 1-12).

*Electric Field Strength (**E**).* The electric field strength at a given point in an electric field is the vector limit of the quotient of the force that a small stationary charge at that point will experience, by virtue of its charge, as the charge approaches zero.

*Electric Flux ($\Psi$).* The electric flux through a surface is the surface integral of the normal component of the electric flux density over the surface.

*Electric Flux Density, Electric Displacement (**D**).* The electric flux density is a quantity related to the charge displaced within a dielectric by application of an electric field. Electric flux density at any point in an isotropic dielectric is a vector which has the same direction as the electric field strength, and a magnitude equal to the product of the electric field strength and the permittivity $\epsilon$. In a nonisotropic medium, $\epsilon$ may be represented by a tensor and $\mathbf{D}$ is not necessarily parallel to $\mathbf{E}$.

*Electric Polarization (**P**).* The electric polarization is the vector quantity defined by the equation $\mathbf{P} = (\mathbf{D} - \Gamma_e \mathbf{E})/\Gamma_r$, where $\mathbf{D}$ is the electric flux density, $\Gamma_e$ is the electric constant, $\mathbf{E}$ is the electric field strength, and $\Gamma_r$ is a coefficient that is set equal to unity in a rationalized system and to $4\pi$ in an unrationalized system.

*Electric Susceptibility ($\chi_e$).* Electric susceptibility is the quantity defined by $\chi_e = (\epsilon_r - 1)/\Gamma_r$, where $\epsilon_r$ is the relative permittivity and $\Gamma_r$ is a coefficient that is set equal to unity in a rationalized system and to $4\pi$ in an unrationalized system.

*Electrization ($E_i$).* The electrization is the electric polarization divided by the electric constant of the system of units used.

*Electrostatic Potential (V).* The electrostatic potential at any point is the potential difference between that point and an agreed-on reference point, usually the point at infinity.

*Electrostatic Potential Difference (V).* The electrostatic potential difference between two points is the scalar-product line integral of the electric field strength along any path from one point to the other in an electric field, resulting from a static distribution of electric charge.

*Impedance (Z).* An impedance of a linear constant-parameter system is the ratio of the phasor equivalent of a steady-state sine-wave voltage or voltage-like quantity (driving force) to the phasor equivalent of a steady-state sine-wave current or current-like quantity (response). In electromagnetic radiation, electric field strength is considered the driving force and magnetic field strength the response. In mechanical systems, mechanical force is always considered as a driving force and velocity as a response. In a general sense, the dimension (and unit) of impedance in a given application may be whatever results from the ratio of the dimensions of the quantity chosen as the driving force to the dimensions of the quantity chosen as the response. However, in the types of systems cited above, any deviation from the usual convention should be noted.

*Mutual Impedance.* Mutual impedance between two loops (meshes) is the factor by which the phasor equivalent of the steady-state sine-wave current in one loop must be multiplied to give the phasor equivalent of the steady-state sine-wave voltage in the other loop caused by the current in the first loop.

*Self-impedance.* Self-impedance of a loop (mesh) is the impedance of a passive loop with all other loops of the open-circuited network.

*Transfer Impedance.* A transfer impedance is the impedance obtained when the response is determined at a point other than that at which the driving force is applied.

NOTE: In the case of an electric circuit, the response may be determined in any branch except that which contains the driving force.

*Logarithmic Decrement* ($\Lambda$). If $F$ is a function of time given by

$$F = A \exp(-\delta t) \sin(2\pi t/T)$$

then the logarithmic decrement $\Lambda = T\delta$.

*Magnetic Constant, Permeability of Vacuum* ($\Gamma_m$). The magnetic constant pertinent to any system of units is the scalar which in that system relates the mechanical force between two currents in vacuum to their magnitudes and geometric configurations. For example, the equation for the force $F$ on a length $l$ of two parallel straight conductors of infinite length and negligible circular cross section, carrying constant currents $I_1$ and $I_2$ and separated by a distance $r$ in vacuum, is $F = \Gamma_m \Gamma_r I_{12} l/2\pi r$, where $\Gamma_m$ is the magnetic constant and $\Gamma_r$ is a coefficient set equal to unity in a rationalized system and to $4\pi$ in an unrationalized system.

NOTE: In the cgs electromagnetic system, $\Gamma_m$ is assigned the magnitude unity and the dimension "numeric." In the cgs electrostatic system, the magnitude of $\Gamma_m$ is that of $1/c^2$, and the dimension is $[L^{-2}T^2]$. In the International System, $\Gamma_m$ is assigned the magnitude $4\pi \times 10^{-7}$ and has the dimension $[LMT^{-2}I^{-2}]$.

*Magnetic Field Strength* (***H***). Magnetic field strength is that vector point function whose curl is the current density and which is proportional to magnetic flux density in regions free of magnetized matter.

*Magnetic Flux* ($\Phi$). The magnetic flux through a surface is the surface integral of the normal component of the magnetic flux density over the surface.

*Magnetic Flux Density, Magnetic Induction* (***B***). Magnetic flux density is that vector quantity which produces a torque on a plane current loop in accordance with the relation $\boldsymbol{T} = IA\boldsymbol{n} \times \boldsymbol{B}$, where $\boldsymbol{n}$ is the positive normal to the loop and $A$ is its area. The concept of flux density is extended to a point inside a solid body by defining the flux density at such a point as that which would be measured in a thin disk-shaped cavity in the body centered at that point, the axis of the cavity being in the direction of the flux density.

*Magnetic Moment* (***m***). The magnetic moment of a magnetized body is the volume integral of the magnetization. The magnetic moment of a loop carrying current $I$ is $\boldsymbol{m} = (1/2)\int \boldsymbol{r} \times d\boldsymbol{r}$, where $\boldsymbol{r}$ is the radius vector from an arbitrary origin to a point on the loop, and where the path of integration is taken around the entire loop.

NOTE: The magnitude of the moment of a plane current loop is $IA$, where $A$ is the area of the loop. The reference direction for the current in the loop indicates a clockwise rotation when the observer is looking through the loop in the direction of the positive normal.

*Magnetic Polarization, Intrinsic Magnetic Flux Density* ($\boldsymbol{J}$, $\boldsymbol{B}_i$). The magnetic polarization is the vector quantity defined by the equation $\boldsymbol{J} = (\boldsymbol{B} - \Gamma_m \boldsymbol{H})/\Gamma_r$, where $\boldsymbol{B}$ is the magnetic flux density, $\Gamma_m$ is the magnetic constant, $\boldsymbol{H}$ is the magnetic field strength, and $\Gamma_r$ is a coefficient that is set equal to unity in a rationalized system and to $4\pi$ in an unrationalized system.

*Magnetic Susceptibility* ($\chi_m$). Magnetic susceptibility is the quantity defined by $\chi_m = (\mu_r - 1)/\Gamma_r$, where $\mu_r$ is the relative permeability and $\Gamma_r$ is a coefficient that is set equal to unity in a rationalized system and to $4\pi$ in an unrationalized system.

*Magnetic Vector Potential* (***A***). The magnetic vector potential is a vector point function characterized by the relation that its curl is equal to the magnetic flux density and its divergence vanishes.

*Magnetization* ($\boldsymbol{M}$, $\boldsymbol{H}_i$). The magnetization is the magnetic polarization divided by the magnetic constant of the system of units used.

*Magnetomotive Force* ($\boldsymbol{F}_m$). The magnetomotive force acting in any closed path in a magnetic field is the line integral of the magnetic field strength around the path.

*Mutual Inductance (M).* The mutual inductance between two loops (meshes) in a circuit is the quotient of the flux linkage produced in one loop divided by the current in another loop, which induces the flux linkage.

*Permeability.* Permeability is a general term used to express various relationships between magnetic flux density and magnetic field strength. These relationships are either (1) *absolute permeability* ($\mu$), which in general is the quotient of a change in magnetic flux density divided by the corresponding change in magnetic field strength, or (2) *relative permeability* ($\mu_r$), which is the ratio of the absolute permeability to the magnetic constant.

*Permeance ($P_m$).* Permeance is the reciprocal of reluctance.

*Permittivity, Capacitivity ($\epsilon$).* The permittivity of a homogeneous, isotropic dielectric, in any system of units, is the product of its relative permittivity and the electric constant appropriate to that system of units.

*Relative Permittivity, Relative Capacitivity, Dielectric Constant ($\epsilon_r$).* The relative permittivity of any homogeneous isotropic material is the ratio of the capacitance of a given configuration of electrodes with the material as a dielectric to the capacitance of the same electrode configuration with a vacuum as the dielectric constant. Experimentally, vacuum must be replaced by the material at all points where it makes a significant change in the capacitance.

*Power (P).* Power is the time rate of transferring or transforming energy. *Electric power* is the time rate of flow of electrical energy. The *instantaneous electric power* at a single terminal pair is equal to the product of the instantaneous voltage multiplied by the instantaneous current. If both voltage and current are periodic in time, the time average of the instantaneous power, taken over an integral number of periods, is the *active power,* usually called simply the *power* when there is no danger of confusion.

If the voltage and current are sinusoidal functions of time, the product of the rms value of the voltage and the rms value of the current is called the *apparent power;* the product of the rms value of the voltage and the rms value of the in-phase component of the current is the *active power;* and the product of the rms value of the voltage and the rms value of the quadrature component of the current is called the *reactive power.*

The SI unit of instantaneous power and active power is the watt. The germane unit for apparent power is the voltampere and for reactive power it is the var.

*Power Factor ($F_p$).* Power factor is the ratio of active power to apparent power.

*Q.* Q, sometimes called *quality factor,* is that measure of the quality of a component, network, system, or medium considered as an energy storage unit in the steady state with sinusoidal driving force which is given by

$$Q = \frac{2\pi \times (\text{maximum energy in storage})}{\text{energy dissipated per cycle of the driving force}}$$

NOTE: For single components such as inductors and capacitors, the $Q$ at any frequency is the ratio of the equivalent series reactance to resistance, or of the equivalent shunt susceptance to conductance. For networks that contain several elements and for distributed parameter systems, the $Q$ is generally evaluated at a frequency of resonance. The *nonloaded* Q of a system is the value of Q obtained when only the incidental dissipation of the system elements is present. The *loaded* Q of a system is the value Q obtained when the system is coupled to a device that dissipates energy. The "period" in the expression for Q is that of the driving force, not that of energy storage, which is usually half of that of the driving force.

*Reactance (X).* Reactance is the imaginary part of impedance.

*Reluctance ($R_m$).* Reluctance is the ratio of the magnetomotive force in a magnetic circuit to the magnetic flux through any cross section of the magnetic circuit.

*Reluctivity ($\nu$).* Reluctivity is the reciprocal of permeability.

*Resistance (R)*

1. The resistance of an element, device, branch, network, or system is the factor by which the mean-square conduction current must be multiplied to give the corresponding power lost by dissipation as heat or as other permanent radiation or as electromagnetic energy from the circuit.
2. Resistance is the real part of impedance.

*Resistivity (ρ).* The resistivity of a material is a factor such that the conduction current density is equal to the electric field strength in the material divided by the resistivity.

*Self-inductance (L)*

1. Self-inductance is the quotient of the flux linkage of a circuit divided by the current in that same circuit which induces the flux linkage. If $v$ = voltage induced, $v = d(Li)/dt$.
2. Self-inductance is the factor $L$ in the $\frac{1}{2}Li^2$ if the latter gives the energy stored in the magnetic field as a result of the current $i$.

NOTE: Definitions 1 and 2 are not equivalent except when $L$ is constant. In all other cases, the definition being used must be specified. The two definitions are restricted to relatively slow changes in $i$, that is, to low frequencies, but by analogy with the definitions, equivalent inductances often may be evolved in high-frequency applications such as resonators and waveguide equivalent circuits. Such "inductances," when used, must be specified. The two definitions are restricted to cases in which the branches are small in physical size when compared with a wavelength, whatever the frequency. Thus, in the case of a uniform 2-wire transmission line it may be necessary even at low frequencies to consider the parameters as "distributed" rather than to have one inductance for the entire line.

*Susceptance (B).* Susceptance is the imaginary part of admittance.

*Transfer Function (H).* A transfer function is that function of frequency which is the ratio of a phasor output to a phasor input in a linear system.

*Transfer Ratio (H).* A transfer ratio is a dimensionless transfer function.

*Voltage, Electromotive Force (V).* The voltage along a specified path in an electric field is the dot product line integral of the electric field strength along this path. As defined, here voltage is synonymous with potential difference only in an electrostatic field.

## 1.11 DEFINITIONS OF QUANTITIES OF RADIATION AND LIGHT

The following definitions are based on the principal meanings listed in the *IEEE Standard Dictionary* (ANSI/IEEE Std 100-2000), which should be consulted for extended meanings, compound terms, and related definitions. The symbols shown in parentheses are from Table 1-10.

*Candlepower.* Candlepower is luminous intensity expressed in candelas (term deprecated by IEEE).

*Emissivity, Total Emissivity (ϵ).* The total emissivity of an element of surface of a temperature radiator is the ratio of its radiant flux density (radiant exitance) to that of a blackbody at the same temperature.

*Spectral Emissivity, ϵ(λ).* The spectral emissivity of an element of surface of a temperature radiator at any wavelength is the ratio of its radiant flux density per unit wavelength interval (spectral radiant exitance) at that wavelength to that of a blackbody at the same temperature.

*Light.* For the purposes of illuminating engineering, light is visually evaluated radiant energy.

NOTE 1: Light is psychophysical, neither purely physical nor purely psychological. Light is not synonymous with radiant energy, however restricted, nor is it merely sensation. In a general nonspecialized sense, light is the aspect of radiant energy of which a human observer is aware through the stimulation of the retina of the eye.

NOTE 2: Radiant energy outside the visible portion of the spectrum must not be discussed using the quantities and units of light; it is nonsense to refer to "ultraviolet light" or to express infrared flux in lumens.

*Luminance (Photometric Brightness) (L).* Luminance in a direction, at a point on the surface of a source, or of a receiver, or on any other real or virtual surface is the quotient of the luminous flux ($\Phi$) leaving, passing through, or arriving at a surface element surrounding the point, propagated in directions defined by an elementary cone containing the given direction, divided by the product of the solid angle of the cone ($d\omega$) and the area of the orthogonal projection of the surface element on a plane perpendicular to the given direction ($dA \cos \theta$). $L = d^2\Phi/[d\omega\,(da \cos \theta)] = dI/(dA \cos \theta)$. In the defining equation, $\theta$ is the angle between the direction of observation and the normal to the surface.

In common usage, the term *brightness* usually refers to the intensity of sensation which results from viewing surfaces or spaces from which light comes to the eye. This sensation is determined in part by the definitely measurable luminance defined above and in part by conditions of observation such as the state of adaptation of the eye. In much of the literature, the term *brightness,* used alone, refers to both luminance and sensation. The context usually indicates which meaning is intended.

*Luminous Efficacy of Radiant Flux.* The luminous efficacy of radiant flux is the quotient of the total luminous flux divided by the total radiant flux. It is expressed in lumens per watt.

*Spectral Luminous Efficacy of Radiant Flux, K(λ).* Spectral luminous efficacy of radiant flux is the quotient of the luminous flux at a given wavelength divided by the radiant flux at the wavelength. It is expressed in lumens per watt.

*Spectral Luminous Efficiency of Radiant Flux.* Spectral luminous efficiency of radiant flux is the ratio of the luminous efficacy for a given wavelength to the value at the wavelength of maximum luminous efficacy. It is a numeric.

NOTE: The term *spectral luminous efficiency* replaces the previously used terms *relative luminosity* and *relative luminosity factor.*

*Luminous Flux (Φ).* Luminous flux is the time rate of flow of light.

*Luminous Flux Density at a Surface.* Luminous flux density at a surface is luminous flux per unit area of the surface. In referring to flux incident on a surface, this is called *illumination (E).* The preferred term for luminous flux *leaving* a surface is *luminous exitance (M),* which has been called *luminous emittance.*

*Luminous Intensity (I).* The luminous intensity of a source of light in a given direction is the luminous flux proceeding from the source per unit solid angle in the direction considered ($I = d\Phi/d\omega$).

*Quantity of Light (Q).* Quantity of light (luminous energy) is the product of the luminous flux by the time it is maintained, that is, it is the time integral of luminous flux.

*Radiance (L).* Radiance in a direction, at a point on the surface, of a source, or of a receiver, or on any other real or virtual surface is the quotient of the radiant flux ($P$) leaving, passing through, or arriving at a surface element surrounding the point, and propagated in directions defined by an elementary cone containing the given direction, divided by the product of the solid angle of the cone ($d\omega$) and the area of the orthogonal projection of the surface element on a plane perpendicular to the given direction ($dA \cos \theta$). $L = d^2P/d\omega(dA \cos \theta) = dI/(dA \cos \theta)$. In the defining equation, $\theta$ is the angle between the normal to the element of the source and the direction of observation.

*Radiant Density (w).* Radiant density is radiant energy per unit volume.

*Radiant Energy (W).* Radiant energy is energy traveling in the form of electromagnetic waves.

*Radiant Flux Density at a Surface.* Radiant flux density at a surface is radiant flux per unit area of the surface. When referring to radiant flux incident on a surface, this is called *irradiance (E).* The preferred term for radiant flux *leaving* a surface is *radiant exitance (M),* which has been called *radiant emittance.*

*Radiant Intensity (I).* The radiant intensity of a source in a given direction is the radiant flux proceeding from the source per unit solid angle in the direction considered ($I = dP/d\omega$).

*Radiant Power, Radiant Flux (P).* Radiant flux is the time rate of flow of radiant energy.

## 1.12 LETTER SYMBOLS

Tables 1-10 and 1-11 list the United States Standard letter symbols for quantities and units (ANSI Std Y10.5, ANSI/IEEE Std 260). A *quantity symbol* is a single letter (e.g., $I$ for electric current) specified as to general form of type and modified by one or more subscripts or superscripts when appropriate. A *unit symbol* is a letter or group of letters (e.g., cm for centimeter), or in a few cases, a special sign, that may be used in the place of the name of the unit.

Symbols for quantities are printed in *italic* type, while symbols for units are printed in roman type. Subscripts and superscripts that are letter symbols for quantities or for indices are printed in roman type as follows:

| | |
|---|---|
| $C_p$ | heat capacity at constant pressure $p$ |
| $a_{ij}$, $a_{45}$ | matrix elements |
| $I_i$, $I_o$ | input current, output current |

For indicating the vector character of a quantity, ***boldface italic*** type is used (e.g., ***F*** for force). Ordinary italic type is used to represent the magnitude of a vector quantity.

The product of two quantities is indicated by writing $ab$. The quotient may be indicated by writing

$$\frac{a}{b}, \qquad a/b, \qquad \text{or} \qquad ab^{-1}$$

If more than one solidus (/) is required in any algebraic term, parentheses must be inserted to remove any ambiguity. Thus, one may write $(a/b)/c$ or $a/bc$, but not $a/b/c$.

Unit symbols are written in lowercase letters, except for the first letter when the name of the unit is derived from a proper name, and except for a very few that are not formed from letters. When a compound unit is formed by multiplication of two or more other units, its symbol consists of the symbols for the separate units joined by a raised dot (e.g., N · m for newton = meter). The dot may be omitted in the case of familiar compounds such as watthour (Wh) if no confusion would result. Hyphens should not be used in symbols for compound units. Positive and negative exponents may be used with the symbols for units.

When a symbol representing a unit that has a prefix (see Sec. 1.5) carries an exponent, this indicates that the multiple (or submultiple) unit is raised to the power expressed by the exponent.

*Examples:*

$$2\ \mathrm{cm}^3 = 2(\mathrm{cm})^3 = 2(10^{-2}\ \mathrm{m})^3 = 2 \cdot 10^{-6}\ \mathrm{m}^3$$

$$1\ \mathrm{ms}^{-1} = 1(\mathrm{ms})^{-1} = 1(10^{-3}\ \mathrm{s})^{-1} = 10^3\ \mathrm{s}^{-1}$$

*Phasor Quantities,* represented by complex numbers or complex time-varying functions, are extensively used in certain branches of electrical engineering. The following notation and typography are standard:

| | Notation | Remarks |
|---|---|---|
| Complex quantity | $Z$ | $Z = \lvert Z\rvert \exp(\mathrm{j}\phi)$<br>$Z = \mathrm{Re}\, Z + \mathrm{j}\, \mathrm{Im}\, Z$ |
| Real part | Re $Z$, $Z'$ | |
| Imaginary part | Im $Z$, $Z''$ | |
| Conjugate complex quantity | $Z^*$ | $Z^* = \mathrm{Re}\, Z - \mathrm{j}\, \mathrm{Im}\, Z$ |
| Modulus of $Z$ | $\lvert Z\rvert$ | |
| Phase of $Z$, Argument of $Z$ | arg $Z$ | $\arg Z = \phi$ |

**TABLE 1-10** Standard Symbols for Quantities

| Quantity | Quantity symbol | Unit based on International System | Remarks |
|---|---|---|---|
| Space and time: | | | |
| Angle, plane | $\alpha, \beta, \gamma, \theta, \phi, \psi$ | radian | Other Greek letters are permitted where no conflict results. |
| Angle, solid | $\Omega \cdots \omega$ | steradian | |
| Length | $l$ | meter | |
| Breadth, width | $b$ | meter | |
| Height | $h$ | meter | |
| Thickness | $d$, $\delta$ | meter | |
| Radius | $r$ | meter | |
| Diameter | $d$ | meter | |
| Length of path line segment | $s$ | meter | |
| Wavelength | $\lambda$ | meter | |
| Wave number | $\sigma \cdots \tilde{\nu}$ | reciprocal meter | $\sigma = 1/\lambda$ The symbol $\tilde{\nu}$ is used in spectroscopy. |
| Circular wave number Angular wave number | $k$ | radian per meter | $k = 2\pi/\lambda$ |
| Area | $A \cdots S$ | square meter | |
| Volume | $V$, $v$ | cubic meter | |
| Time | $t$ | second | |
| Period | $T$ | second | |
| Time constant | $\tau \cdots T$ | second | |
| Frequency | $f \cdots \nu$ | second | |
| Speed of rotation Rotational frequency | $n$ | revolution per second | |
| Angular frequency | $\omega$ | radian per second | $\omega = 2\pi f$ |
| Angular velocity | $\omega$ | radian per second | |
| Complex (angular) frequency Oscillation constant | $p \cdots s$ | reciprocal second | $p = -\delta + j\omega$ |
| Angular acceleration | $\alpha$ | radian per second squared | |
| Velocity | $v$ | meter per second | |
| Speed of propagation of electromagnetic waves | $c$ | meter per second | In vacuum, $c_0$ |
| Acceleration (linear) | $a$ | meter per second squared | |
| Acceleration of free fall Gravitational acceleration | $g$ | meter per second squared | |
| Damping coefficient | $\delta$ | neper per second | |
| Logarithmic decrement | $\Lambda$ | (numeric) | |
| Attenuation coefficient | $\alpha$ | neper per meter | |
| Phase coefficient | $\beta$ | radian per meter | |
| Propagation coefficient | $\gamma$ | reciprocal meter | $\gamma = \alpha + j\beta$ |
| Mechanics: | | | |
| Mass | $m$ | kilogram | |
| (Mass) density | $\rho$ | kilogram per cubic meter | Mass divided by volume |
| Momentum | $p$ | kilogram meter per second | |
| Moment of inertia | $I$, $J$ | kilogram meter squared | |

**TABLE 1-10** Standard Symbols for Quantities (*Continued*)

| Quantity | Quantity symbol | Unit based on International System | Remarks |
|---|---|---|---|
| Force | $F$ | newton | |
| Weight | $W$ | newton | Varies with acceleration of free fall |
| Weight density | $\gamma$ | newton per cubic meter | Weight divided by volume |
| Moment of force | $M$ | newton meter | |
| Torque | $T \cdots M$ | newton meter | |
| Pressure | $p$ | newton per square meter | The SI name *pascal* has been adopted for this unit. |
| Normal stress | $\sigma$ | newton per square meter | |
| Shear stress | $\tau$ | newton per square meter | |
| Stress tensor | $\sigma$ | newton per square meter | |
| Linear strain | $\varepsilon$ | (numeric) | |
| Shear strain | $\gamma$ | (numeric) | |
| Strain tensor | $\varepsilon$ | (numeric) | |
| Volume strain | $\theta$ | (numeric) | |
| Poisson's ratio | $\mu, \nu$ | (numeric) | Lateral contraction divided by elongation |
| Young's modulus<br>Modulus of elasticity | $E$ | newton per square meter | $E = \sigma/\varepsilon$ |
| Shear modulus<br>Modulus of rigidity | $G$ | newton per square meter | $G = \tau/\gamma$ |
| Bulk modulus | $K$ | newton per square meter | $K = -p/\theta$ |
| Work | $W$ | joule | |
| Energy | $E, W$ | joule | $U$ is recommended in thermodynamics for internal energy and for blackbody radiation. |
| Energy (volume) density | $w$ | joule per cubic meter | |
| Power | $P$ | watt | |
| Efficiency | $\eta$ | (numeric) | |
| Heat: | | | |
| Thermodynamic temperature | $T \cdots \Theta$ | kelvin | |
| Temperature<br>Customary temperature | $t \cdots \theta$ | degree Celsius | The word *centigrade* has been abandoned as the name of a temperature scale. |
| Heat | $Q$ | joule | |
| Internal energy | $U$ | joule | |
| Heat flow rate | $\Phi \cdots q$ | watt | Heat crossing a surface divided by time |
| Temperature coefficient | $\alpha$ | reciprocal kelvin | |
| Thermal diffusivity | $\alpha$ | square meter per second | |
| Thermal conductivity | $\lambda \cdots k$ | watt per meter kelvin | |
| Thermal conductance | $G_\theta$ | watt per kelvin | |
| Thermal resistivity | $\rho_\theta$ | meter kelvin per watt | |
| Thermal resistance | $R_\theta$ | kelvin per watt | |
| Thermal capacitance<br>Heat capacity | $C_\theta$ | joule per kelvin | |
| Thermal impedance | $Z_\theta$ | kelvin per watt | |
| Specific heat capacity | $c$ | joule per kelvin kilogram | Heat capacity divided by mass |
| Entropy | $S$ | joule per kelvin | |
| Specific entropy | $s$ | joule per kelvin kilogram | Entropy divided by mass |
| Enthalpy | $H$ | joule | |
| Radiation and light: | | | |
| Radiant intensity | $I \cdots I_e$ | watt per steradian | |
| Radiant power<br>Radiant flux | $P, \Phi \cdots \Phi_e$ | watt | |

(*Continued*)

**TABLE 1-10** Standard Symbols for Quantities (*Continued*)

| Quantity | Quantity symbol | Unit based on International System | Remarks |
|---|---|---|---|
| Radiant energy | $W, Q \cdots Q_e$ | joule | The symbol $U$ is used for the special case of blackbody radiant energy. |
| Radiance | $L \cdots L_e$ | watt per steradian square meter | |
| Radiant exitance | $M \cdots M_e$ | watt per square meter | |
| Irradiance | $E \cdots E_e$ | watt per square meter | |
| Luminous intensity | $I \cdots I_v$ | candela | |
| Luminous flux | $\Phi \cdots \Phi_v$ | lumen | |
| Quantity of light | $Q \cdots Q_v$ | lumen second | |
| Luminance | $L \cdots L_v$ | candela per square meter | |
| Luminous exitance | $M \cdots M_v$ | lumen per square meter | |
| Illuminance<br>Illumination | $E \cdots E_v$ | lux | |
| Luminous efficacy† | $K(\lambda)$ | lumen per watt | |
| Total luminous efficacy | $K, K_t$ | lumen per watt | |
| Refractive index<br>Index of refraction | $n$ | (numeric) | |
| Emissivity† | $\epsilon(\lambda)$ | (numeric) | |
| Total emissivity | $\epsilon, \epsilon_t$ | (numeric) | |
| Absorptance† | $\alpha(\lambda)$ | (numeric) | |
| Transmittance† | $\tau(\lambda)$ | (numeric) | |
| Reflectance† | $\rho(\lambda)$ | (numeric) | |
| Fields and circuits: | | | |
| Electric charge<br>Quantity of electricity | $Q$ | coulomb | |
| Linear density of charge | $\lambda$ | coulomb per meter | |
| Surface density of charge | $\sigma$ | coulomb per square meter | |
| Volume density of charge | $\rho$ | coulomb per cubic meter | |
| Electric field strength | $E \cdots K$ | volt per meter | |
| Electrostatic potential<br>Potential difference | $V \cdots \phi$ | volt | |
| Retarded scalar potential | $V_r$ | volt | |
| Voltage<br>Electromotive force | $V, E \cdots U$ | volt | |
| Electric flux | $\Psi$ | coulomb | |
| Electric flux density<br>(Electric) displacement | $D$ | coulomb per square meter | |
| Capacitivity<br>Permittivity<br>Absolute permittivity | $\epsilon$ | farad per meter | Of vacuum, $\varepsilon_v$ |
| Relative capacitivity<br>Relative permittivity<br>Dielectric constant | $\epsilon_r, \kappa$ | (numeric) | |
| Complex relative capacitivity | $\epsilon_r^*, \kappa^*$ | (numeric) | $\epsilon_r^* = \epsilon'_r - j\epsilon''_r$ |
| Complex relative permittivity | | | $\epsilon'_r$ is positive for lossy materials. The complex absolute permittivity $\epsilon^*$ is defined in analogous fashion. |
| Complex dielectric constant | | | |

**TABLE 1-10** Standard Symbols for Quantities (*Continued*)

| Quantity | Quantity symbol | Unit based on International System | Remarks | |
|---|---|---|---|---|
| Electric susceptibility | $\chi_e \cdots \epsilon_i$ | (numeric) | $\chi_e = \epsilon_r - 1$ | MKSA |
| Electrization | $E_i \cdots K_i$ | volt per meter | $E_i = (D/\Gamma_e) - E$ | MKSA |
| Electric polarization | $P$ | coulomb per square meter | $P = D - \Gamma_e E$ | MKSA |
| Electric dipole moment | $p$ | coulomb meter | | |
| (Electric) current | $I$ | ampere | | |
| Current density | $J \cdots S$ | ampere per square meter | | |
| Linear current density | $A \cdots \alpha$ | ampere per meter | Current divided by the breadth of the conducting sheet | |
| Magnetic field strength | $H$ | ampere per meter | | |
| Magnetic (scalar) potential<br>Magnetic potential difference | $U, U_m$ | ampere | | |
| Magnetomotive force | $F, F_m \cdots \mathscr{F}$ | ampere | | |
| Magnetic flux | $\Phi$ | weber | | |
| Magnetic flux density<br>Magnetic induction | $B$ | tesla | | |
| Magnetic flux linkage | $\Lambda$ | weber | | |
| (Magnetic) vector potential | $A$ | weber per meter | | |
| Retarded (magnetic) vector potential | $A_r$ | weber per meter | | |
| Permeability<br>Absolute permeability | $\mu$ | henry per meter | Of vacuum, $\mu_v$ | |
| Relative permeability | $\mu_r$ | (numeric) | | |
| Initial (relative) permeability | $\mu_o$ | (numeric) | | |
| Complex relative permeability | $\mu_r^*$ | (numeric) | $\mu_r^* = \mu'_r - j\mu''_r$<br>$\mu''_r$ is positive for lossy materials. The complex absolute permeability $\mu^*$ is defined in analogous fashion. | |
| Magnetic susceptibility | $\chi_m \cdots \mu_i$ | (numeric) | $\chi_m = \mu_r - 1$ | MKSA |
| Reluctivity | $\nu$ | meter per henry | $\nu = 1/\mu$ | |
| Magnetization | $H_i, M$ | ampere per meter | $H_i = (B/\Gamma_m) - H$ | MKSA |
| Magnetic polarization<br>Intrinsic magnetic flux density | $J, B_i$ | tesla | $J = B - \Gamma_m H$ | MKSA |
| Magnetic (area) moment | $m$ | ampere meter squared | The vector product $\boldsymbol{m} \times \boldsymbol{B}$ is equal to the torque. | |
| Capacitance | $C$ | farad | | |
| Elastance | $S$ | reciprocal farad | $S = 1/C$ | |
| (Self-) inductance | $L$ | henry | | |
| Reciprocal inductance | $\Gamma$ | reciprocal henry | | |
| Mutual inductance | $L_{ij}, M_{ij}$ | henry | If only a single mutual inductance is involved, $M$ may be used without subscripts. | |
| Coupling coefficient | $k \cdots \kappa$ | (numeric) | $k = L_{ij}(L_i L_j)^{-1/2}$ | |
| Leakage coefficient | $\sigma$ | (numeric) | $\sigma = 1 - k^2$ | |
| Number of turns (in a winding) | $N, n$ | (numeric) | | |
| Number of phases | $m$ | (numeric) | | |
| Turns ratio | $n \cdots n_*$ | (numeric) | | |

(*Continued*)

**TABLE 1-10** Standard Symbols for Quantities (*Continued*)

| Quantity | Quantity symbol | Unit based on International System | Remarks |
|---|---|---|---|
| Transformer ratio | $a$ | (numeric) | Square root of the ratio of secondary to primary self-inductance. Where the coefficient of coupling is high, $a \approx n_*$. |
| Resistance | $R$ | ohm | |
| Resistivity<br>Volume resistivity | $\rho$ | ohm meter | |
| Conductance | $G$ | siemens | $G = \text{Re } Y$ |
| Conductivity | $\gamma, \sigma$ | siemens per meter | $\gamma = 1/\rho$<br>The symbol $\sigma$ is used in field theory, as $\gamma$ is used for the propagation coefficient. |
| Reluctance | $R, R_m \cdots \mathcal{R}$ | reciprocal henry | Magnetic potential difference divided by magnetic flux |
| Permeance | $P, P_m \cdots \mathcal{P}$ | henry | $P_m = 1/R_m$ |
| Impedance | $Z$ | ohm | |
| Reactance | $X$ | ohm | |
| Capacitive reactance | $X_C$ | ohm | For a pure capacitance, $X_C = -1/\omega C$ |
| Inductive reactance | $X_L$ | ohm | For a pure capacitance, $X_L = \omega L$ |
| Quality factor | $Q$ | (numeric) | See $Q$ in Sec. 1.10. |
| Admittance | $Y$ | siemens | $Y = 1/Z = G + jB$ |
| Susceptance | $B$ | siemens | $B = \text{Im } Y$ |
| Loss angle | $\delta$ | radian | $\delta = (R/\|X\|)$ |
| Active power | $P$ | watt | |
| Reactive power | $Q \cdots P_q$ | var | |
| Apparent power | $S \cdots P_s$ | voltampere | |
| Power factor | $\cos \phi \cdots F_p$ | (numeric) | |
| Reactive factor | $\sin \phi \cdots F_q$ | (numeric) | |
| Input power | $P_i$ | watt | |
| Output power | $P_o$ | watt | |
| Poynting vector | $S$ | watt per square meter | |
| Characteristic impedance<br>Surge impedance | $Z_o$ | ohm | |
| Intrinsic impedance of a medium | $\eta$ | ohm | |
| Voltage standing-wave ratio | $S$ | (numeric) | |
| Resonance frequency | $f_r$ | hertz | |
| Critical frequency<br>Cutoff frequency | $f_c$ | hertz | |
| Resonance angular frequency | $\omega_r$ | radian per second | |
| Critical angular frequency<br>Cutoff angular frequency | $\omega_c$ | radian per second | |
| Resonance wavelength | $\lambda_r$ | meter | |
| Critical wavelength<br>Cutoff wavelength | $\lambda_c$ | meter | |
| Wavelength in a guide | $\lambda_g$ | meter | |
| Hysteresis coefficient | $k_h$ | (numeric) | |
| Eddy-current coefficient | $k_e$ | (numeric) | |
| Phase angle<br>Phase difference | $\phi, \theta$ | radian | |

†($\lambda$) is not part of the basic symbol but indicates that the quantity is a function of wavelength.

**TABLE 1-11** Standard Symbols for Units

| Unit | Symbol | Notes |
|---|---|---|
| ampere | A | SI unit of electric current |
| ampere (turn) | A | SI unit of magnetomotive force |
| ampere-hour | Ah | Also A · h |
| ampere per meter | A/m | SI unit of magnetic field strength |
| angstrom | Å | 1 Å = $10^{-10}$ m. Deprecated. |
| atmosphere, standard | atm | 1 atm = 101 325 Pa. Deprecated. |
| atmosphere, technical | at | 1 at = 1 kgf/cm$^2$. Deprecated. |
| atomic mass unit (unified) | u | The (unified) atomic mass unit is defined as one-twelfth of the mass of an atom of the $^{12}$C nuclide. Use of the old atomic mass (amu), defined by reference to oxygen, is deprecated. |
| atto | a | SI prefix for $10^{-18}$ |
| attoampere | aA | |
| bar | bar | 1 bar = 100 kPa. Use of the bar is strongly discouraged, except for limited use in meteorology. |
| barn | b | 1 b = $10^{-28}$ m$^2$ |
| barrel | bbl | 1 bbl = 42 gal$_{US}$ = 158.99 L |
| barrel per day | bbl/d | This is the standard barrel used for petroleum, etc. A different standard barrel is used for fruits, vegetables, and dry commodities. |
| baud | Bd | In telecommunications, a unit of signaling speed equal to one element per second. The signaling speed in bauds is equal to the reciprocal of the signal element length in seconds. |
| bel | B | |
| becquerel | Bq | SI unit of activity of a radionuclide |
| billion electronvolts | GeV | The name *gigaelectronvolt* is preferred for this unit. |
| bit | b | In information theory, the bit is a unit of information content equal to the information content of a message, the *a priori* probability of which is one-half.<br>In computer science, the bit is a unit of storage capacity. The capacity, in bits, of a storage device is the logarithm to the base two of the number of possible states of the device. |
| bit per second | b/s | |
| British thermal unit | Btu | |
| calorie (International Table calorie) | cal$_{IT}$ | 1 cal$_{IT}$ = 4.1868 J. Deprecated. |
| calorie (thermochemical calorie) | cal | 1 cal = 4.1840 J. Deprecated. |
| candela | cd | SI unit of luminous intensity |
| candela per square inch | cd/in$^2$ | Use of the SI unit, cd/m$^2$, is preferred. |
| candela per square meter | cd/m$^2$ | SI unit of luminance. The name *nit* is sometimes used for this unit. |
| candle | cd | The unit of luminous intensity has been given the name *candela;* use of the name *candle* for this unit is deprecated. |
| centi | c | SI prefix for $10^{-2}$ |
| centimeter | cm | |
| centipoise | cP | 1 cP = mPa · s. The name centipoise is deprecated. |
| centistokes | cSt | 1 cSt = 1 mm$^2$/s. The name centistokes is deprecated. |
| circular mil | cmil | 1 cmil = $(\pi/4) \cdot 10^{-6}$ in$^2$ |
| coulomb | C | SI unit of electric charge |
| cubic centimeter | cm$^3$ | |
| cubic foot | ft$^3$ | |
| cubic foot per minute | ft$^3$/min | |
| cubic foot per second | ft$^3$/s | |
| cubic inch | in$^3$ | |
| cubic meter | m$^3$ | |
| cubic meter per second | m$^3$/s | |
| cubic yard | yd$^3$ | |
| curie | Ci | A unit of activity of radionuclide. Use of the SI unit, the becquerel, is preferred, 1 Ci = 3.7 × $10^{10}$ Bq. |
| cycle | c | |

*(Continued)*

**TABLE 1-11** Standard Symbols for Units (*Continued*)

| Unit | Symbol | Notes |
|---|---|---|
| cycle per second | Hz, c/s | See hertz. The name *hertz* is internationally accepted for this unit; the symbol Hz is preferred to c/s. |
| darcy | D | 1 D = 1 cP (cm/s) (cm/atm) = 0.986 923 $\mu$m$^2$. A unit of permeability of a porous medium. By traditional definition, a permeability of one darcy will permit a flow of 1 cm$^3$/s of fluid of 1 cP viscosity through an area of 1 cm$^2$ under a pressure gradient of 1 atm/cm. For nonprecision work, 1 D may be taken equal to 1 $\mu$m$^2$ and 1 mD equal to 0.001 $\mu$m$^2$. Deprecated. |
| day | d | |
| deci | d | SI prefix for $10^{-1}$ |
| decibel | dB | |
| degree (plane angle) | . . . ° | |
| degree (temperature): | | |
| degree Celsius | °C | SI unit of Celsius temperature. The degree Celsius is a special name for the kelvin, for use in expressing Celsius temperatures or temperature intervals. |
| degree Fahrenheit | °F | Note that the symbols for °C, °F, and °R comprise two elements, written with no space between the ° and the letter that follows. The two elements that make the complete symbol are not to be separated. |
| degree Kelvin | | See kelvin |
| degree Rankine | °R | |
| deka | da | SI prefix for 10 |
| dyne | dyn | Deprecated. |
| electronvolt | eV | |
| erg | erg | Deprecated. |
| exa | E | SI prefix for $10^{18}$ |
| farad | F | SI unit of capacitance |
| femto | f | SI prefix for $10^{-15}$ |
| femtometer | fm | |
| foot | ft | |
| conventional foot of water | ftH$_2$O | 1 ftH$_2$O = 2989.1 Pa (ISO) |
| foot per minute | ft/min | |
| foot per second | ft/s | |
| foot per second squared | ft/s$^2$ | |
| foot pound-force | ft · lbf | |
| footcandle | fc | 1 fc = 1 lm/ft$^2$. The name *lumen per square foot* is also used for this unit. Use of the SI unit of illuminance, the lux (lumen per square meter), is preferred. |
| footlambert | fL | 1 fL = (1/$\pi$) cd/ft$^2$. A unit of luminance. One lumen per square foot leaves a surface whose luminance is one footlambert in all directions within a hemisphere. Use of the SI unit, the candela per square meter, is preferred. |
| gal | Gal | 1 Gal = 1 cm/s$^2$. Deprecated. |
| gallon | gal | 1 gal$_{UK}$ = 4.5461 L<br>1 gal$_{US}$ = 231 in$^3$ = 3.7854 L |
| gauss | G | The gauss is the electromagnetic CGS unit of magnetic flux density. Deprecated. |
| giga | G | SI prefix for $10^9$ |
| gigaelectronvolt | GeV | |
| gigahertz | GHz | |
| gilbert | Gb | The gilbert is the electromagnetic CGS unit of magnetomotive force. Deprecated. |
| grain | gr | |
| gram | g | |
| gram per cubic centimeter | g/cm$^3$ | |
| gray | Gy | SI unit of absorbed dose in the field of radiation dosimetry |

**TABLE 1-11** Standard Symbols for Units (*Continued*)

| Unit | Symbol | Notes |
|---|---|---|
| hecto | h | SI prefix for $10^2$ |
| henry | H | SI unit of inductance |
| hertz | Hz | SI unit of frequency |
| horsepower | hp | The horsepower is an anachronism in science and technology. Use of the SI unit of power, the watt, is preferred. |
| hour | h | |
| inch | in | |
| conventional inch of mercury | inHg | 1 inHg = 3386.4 Pa (ISO) |
| conventional inch of water | $inH_2O$ | 1 $inH_2O$ = 249.09 Pa (ISO) |
| inch per second | in/s | |
| joule | J | SI unit of energy, work, quantity of heat |
| joule per kelvin | J/K | SI unit of heat capacity and entropy |
| kelvin | K | In 1967, the CGPM gave the name *kelvin* to the SI unit of temperature which had formerly been called *degree kelvin* and assigned it the symbol K (without the symbol °). |
| kilo | k | SI prefix for $10^3$ |
| kilogauss | kG | Deprecated. |
| kilogram | kg | SI unit of mass |
| kilogram-force | kgf | Deprecated. In some countries, the name kilopond (kp) has been used for this unit. |
| kilohertz | kHz | |
| kilohm | kΩ | |
| kilometer | km | |
| kilometer per hour | km/h | |
| kilopound-force | klbf | Kilopound-force should not be misinterpreted as kilopond (see kilogram-force). |
| kilovar | kvar | |
| kilovolt | kV | |
| kilovoltampere | kVA | |
| kilowatt | kW | |
| kilowatthour | kWh | Also kW · h |
| knot | kn | 1kn = 1 nmi/h |
| lambert | L | 1 L = $(1/\pi)$ cd/cm$^2$. A CGS unit of luminance. One lumen per square centimeter leaves a surface whose luminance is one lambert in all directions within a hemisphere. Deprecated. |
| liter | L | 1 L = $10^{-3}$ m$^3$. The letter symbol 1 has been adopted for *liter* by the CGPM, and it is recommended in a number of international standards. In 1978, the CIPM accepted L as an alternative symbol. Because of frequent confusion with the numeral 1 the letter symbol 1 is no longer recommended for U.S. use. The script letter ℓ, which had been proposed, is not recommended as a symbol for liter. |
| liter per second | L/s | |
| lumen | lm | SI unit of luminous flux |
| lumen per square foot | lm/ft$^2$ | A unit of illuminance and also a unit of luminous exitance. Use of the SI unit, lumen per square meter, is preferred. |
| lumen per square meter | lm/m$^2$ | SI unit of luminous exitance |
| lumen per watt | lm/W | SI unit of luminous efficacy |
| lumen second | lm · s | SI unit of quantity of light |
| lux | lx | 1 lx = 1 lm/m$^2$. SI unit of illuminance |
| maxwell | Mx | The maxwell is the electromagnetic CGS unit of magnetic flux. Deprecated. |
| mega | M | SI prefix for $10^6$ |
| megaelectronvolt | MeV | |
| megahertz | MHz | |
| megohm | MΩ | |
| meter | m | SI unit of length |

(*Continued*)

**TABLE 1-11** Standard Symbols for Units (*Continued*)

| Unit | Symbol | Notes |
|---|---|---|
| metric ton | t | 1 t = 1000 kg. The name *tonne* is used in some countries for this unit, but use of this name in the U.S. is deprecated. |
| mho | mho | Formerly used as the name of the siemens (S). |
| micro | $\mu$ | SI prefix for $10^{-6}$ |
| microampere | $\mu$A | |
| microfarad | $\mu$F | |
| microgram | $\mu$g | |
| microhenry | $\mu$H | |
| microinch | $\mu$in | |
| microliter | $\mu$L | See note for *liter.* |
| micrometer | $\mu$m | |
| micron | $\mu$m | Deprecated. Use micrometer. |
| microsecond | $\mu$s | |
| microwatt | $\mu$W | |
| mil | mil | 1 mil = 0.001 in |
| mile (statute) | mi | 1 mi = 5280 ft |
| miles per hour | mi/h | Although use of mph as an abbreviation is common, it should not be used as a symbol. |
| milli | m | SI prefix for $10^{-3}$ |
| milliampere | mA | |
| millibar | mbar | Use of the bar is strongly discouraged, except for limited use in meteorology. |
| milligram | mg | |
| millihenry | mH | |
| milliliter | mL | See note for *liter.* |
| millimeter | mm | |
| conventional millimeter of mercury | mmHg | 1 mmHg = 133.322 Pa. Deprecated. |
| millimicron | nm | Use of the name *millimicron* for the nanometer is deprecated. |
| millipascal second | mPa · s | SI unit-multiple of dynamic viscosity |
| millisecond | ms | |
| millivolt | mV | |
| milliwatt | mW | |
| minute (plane angle) | . . . ′ | |
| minute (time) | min | Time may also be designated by means of superscripts as in the following example: $9^h46^m30^s$. |
| mole | mol | SI unit of amount of substance |
| month | mo | |
| nano | n | SI prefix for $10^{-9}$ |
| nanoampere | nA | |
| nanofarad | nF | |
| nanometer | nm | |
| nanosecond | ns | |
| nautical mile | nmi | 1 nmi = 1852 m |
| neper | Np | |
| newton | N | SI unit of force |
| newton meter | N · m | |
| newton per square meter | N/m$^2$ | SI unit of pressure or stress, see pascal. |
| nit | nt | 1 nt = 1 cd/m$^2$ The name *nit* is sometimes given to the SI unit of luminance, the candela per square meter. |
| oersted | Oe | The oersted is the electromagnetic CGS unit of magnetic field strength. Deprecated. |
| ohm | Ω | SI unit of resistance |
| ounce (avoirdupois) | oz | |
| pascal | Pa | 1 Pa = 1 N/m$^2$ SI unit of pressure or stress |

**TABLE 1-11** Standard Symbols for Units (*Continued*)

| Unit | Symbol | Notes |
|---|---|---|
| pascal second | Pa · s | SI unit of dynamic viscosity |
| peta | P | SI prefix for $10^{15}$ |
| phot | ph | 1 ph = lm/cm$^2$<br>CGS unit of illuminance. Deprecated. |
| pico | p | SI prefix for $10^{-12}$ |
| picofarad | pF | |
| picowatt | pW | |
| pint | pt | 1 pt (U.K.) = 0.568 26 L<br>1 pt (U.S. dry) = 0.550 61 L<br>1 pt (U.S. liquid) = 0.473 18 L |
| poise | P | Deprecated. |
| pound | lb | |
| pound per cubic foot | lb/ft$^3$ | |
| pound-force | lbf | |
| pound-force foot | lbf · ft | |
| pound-force per square foot | lbf/ft$^2$ | |
| pound-force per square inch | lbf/in$^2$ | Although use of the abbreviation psi is common, it should not be used as a symbol. |
| poundal | pdl | |
| quart | qt | 1 qt (U.K.) = 1.136 5 L<br>1 qt (U.S. dry) = 1.101 2 L<br>1 qt (U.S. liquid) = 0.946 35 L |
| rad | rd | A unit of absorbed dose in the field of radiation dosimetry. Use of the SI unit, the gray, is preferred. 1 rd = 0.01 Gy. |
| radian | rad | SI unit of plane angle |
| rem | rem | A unit of dose equivalent in the field of radiation dosimetry. Use of the SI unit, the sievert, is preferred. 1 rem = 0.01 Sv. |
| revolution per minute | r/min | Although use of rpm as an abbreviation is common, it should not be used as a symbol. |
| revolution per second | r/s | |
| roentgen | R | A unit of exposure in the field of radiation dosimetry |
| second (plane angle) | . . . ″ | |
| second (time) | s | SI unit of time |
| siemens | S | 1 S = 1 $\Omega^{-1}$<br>SI unit of conductance. The name mho has been used for this unit in the U.S. |
| sievert | Sv | SI unit of dose equivalent in the field of radiation dosimetry. Name adopted by the CIPM in 1978. |
| slug | slug | 1 slug = 14.593 9 kg |
| square foot | ft$^2$ | |
| square inch | in$^2$ | |
| square meter | m$^2$ | |
| square meter per second | m$^2$/s | SI unit of kinematic viscosity |
| square millimeter per second | mm$^2$/s | SI unit-multiple of kinematic viscosity |
| square yard | yd$^2$ | |
| steradian | sr | SI unit of solid angle |
| stilb | sb | 1 sb = 1 cd/cm$^2$<br>A CGS unit of luminance. Deprecated. |
| stokes | St | Deprecated. |
| tera | T | SI prefix for $10^{12}$ |
| tesla | T | 1 T = 1 N/(A · m) = 1 Wb/m$^2$. SI unit of magnetic flux density (magnetic induction). |
| therm | thm | 1 thm = 100 000 Btu |
| ton (short) | ton | 1 ton = 2000 lb |
| ton, metic | t | 1 t = 1000 kg. The name *tonne* is used in some countries for this unit, but use of this name in the U.S. is deprecated. |

(*Continued*)

**TABLE 1-11** Standard Symbols for Units (*Continued*)

| Unit | Symbol | Notes |
|---|---|---|
| (unified) atomic mass unit | u | The (unified) atomic mass unit is defined as one-twelfth of the mass of an atom of the $^{12}C$ nuclide. Use of the old atomic mass unit (amu), defined by reference to oxygen, is deprecated. |
| var | var | IEC name and symbol for the SI unit of reactive power |
| volt | V | SI unit of voltage |
| volt per meter | V/m | SI unit of electric field strength |
| voltampere | VA | IEC name and symbol for the SI unit of apparent power |
| watt | W | SI unit of power |
| watt per meter kelvin | W/(m · K) | SI unit of thermal conductivity |
| watt per steradian | W/sr | SI unit of radiant intensity |
| watt per steradian square meter | W/(sr · m$^2$) | SI unit of radiance |
| watthour | Wh | |
| weber | Wb | Wb = V · s<br>SI unit of magnetic flux |
| yard | yd | |
| year | a | In the English language, generally yr. |

**TABLE 1-12** Fundamental Constants of Physics and Chemistry
[An abbreviated list of the CODATA recommended values based on the 2014 adjustment]

| Quantity | Symbol | Numerical value | Unit | Relative std. uncert. $u_r$ |
|---|---|---|---|---|
| speed of light in vacuum | $c, c_0$ | 299 792 458 | m s$^{-1}$ | exact |
| magnetic constant | $\mu_0$ | $4\pi \times 10^{-7}$ = 12.566 370 614... × $10^{-7}$ | N A$^{-2}$ | exact |
| electric constant $1/\mu_0 c^2$ | $\epsilon_0$ | 8.854 187 817... × $10^{-12}$ | F m$^{-1}$ | exact |
| characteristic impedance of vacuum $\mu_0 c$ | $Z_0$ | 376.730 313 461... | Ω | exact |
| Newtonian constant of gravitation | $G$ | 6.674 08(31) × $10^{-11}$ | m$^3$ kg$^{-1}$ s$^{-2}$ | 4.7 × $10^{-5}$ |
| Planck constant | $h$ | 6.626 070 040(81) × $10^{-34}$ | J s | 1.2 × $10^{-8}$ |
| $h/2\pi$ | $\hbar$ | 1.054 571 800(13) × $10^{-34}$ | J s | 1.2 × $10^{-8}$ |
| elementary charge | $e$ | 1.602 176 620 8(98) × $10^{-19}$ | C | 6.1 × $10^{-9}$ |
| magnetic flux quantum $h/2e$ | $\Phi_0$ | 2.067 833 831(13) × $10^{-15}$ | Wb | 6.1 × $10^{-9}$ |
| conductance quantum $2e^2/h$ | $G_0$ | 7.748 091 731 0(18) × $10^{-5}$ | S | 2.3 × $10^{-10}$ |
| electron mass | $m_e$ | 9.109 383 56(11) × $10^{-31}$ | kg | 1.2 × $10^{-8}$ |
| proton mass | $m_p$ | 1.672 621 898(21) × $10^{-27}$ | kg | 1.2 × $10^{-8}$ |
| proton-electron mass ratio | $m_p/m_e$ | 1836.152 673 89(17) | | 9.5 × $10^{-11}$ |
| fine-structure constant $e^2/4\pi\epsilon_0\hbar c$ | $\alpha$ | 7.297 352 566 4(17) × $10^{-3}$ | | 2.3 × $10^{-10}$ |
| inverse fine-structure constant | $\alpha^{-1}$ | 137.035 999 139(31) | | 2.3 × $10^{-10}$ |
| Rydberg constant $\alpha^2 m_e c/2h$ | $R_\infty$ | 10 973 731.568 508(65) | m$^{-1}$ | 5.9 × $10^{-12}$ |
| Avogadro constant | $N_A, L$ | 6.022 140 857(74) × $10^{23}$ | mol$^{-1}$ | 1.2 × $10^{-8}$ |
| Faraday constant $N_A e$ | $F$ | 96 485.332 89(59) | C mol$^{-1}$ | 6.2 × $10^{-9}$ |
| molar gas constant | $R$ | 8.314 459 8(48) | J mol$^{-1}$ K$^{-1}$ | 5.7 × $10^{-7}$ |
| Boltzmann constant $R/N_A$ | $k$ | 1.380 6485 2(79) × $10^{-23}$ | J K$^{-1}$ | 5.7 × $10^{-7}$ |
| Stefan-Boltzmann constant $(\pi^2/60)k^4/\hbar^3 c^2$ | $\sigma$ | 5.670 367(13) × $10^{-8}$ | W m$^{-2}$ K$^{-4}$ | 2.3 × $10^{-6}$ |
| | | Non-SI units accepted for use with the SI | | |
| electron volt (e/C) J | eV | 1.602 176 620 8(98) × $10^{-19}$ | J | 6.1 × $10^{-9}$ |
| (unified) atomic mass unit $\frac{1}{12}m(^{12}C)$ | u | 1.660 539 040(20) × $10^{-27}$ | kg | 1.2 × $10^{-8}$ |

*Source:* CODATA recommended values of the fundamental physical constants: 2014; Peter J. Mohr, David B. Newell, and Barry N. Taylor; Rev. Mod. Phys. 88, 035009, 73 pages (2016).

## 1.13 GRAPHIC SYMBOLS

An extensive list of standard graphic symbols for electrical engineering has been compiled in IEEE Standard 315 (ANSI Y32.2). Those concerned with the preparation of circuit diagrams and graphic layouts should conform to these standard symbols to avoid confusion with earlier, nonstandard forms.

## 1.14 PHYSICAL CONSTANTS

Table 1-12 provides an abbreviated list of the CODATA recommended values of the fundamental constants of physics and chemistry based on the 2014 adjustment. A complete list of recommended values can be found in the reference paper given at the bottom of Table 1-12. Alternatively, they may also be found at physics.nist.gov/constants. Table 1-13 lists the values of some energy equivalents.

**TABLE 1-13** The Values of Some Energy Equivalents
[Derived from the relations $E = mc^2 = hc/\lambda = h\nu = kT$, and based on the 2014 CODATA adjustment of the values of the constants; 1 eV = $(e/\mathrm{C})$ J, 1 u = $m_u = \frac{1}{2} m(^{12}\mathrm{C}) = 10^{-3}$ kg mol$^{-1}/N_A$, and $E_h = 2R_\infty hc = \alpha^2 m_e c^2$ is the Hartree energy (hartree).]

| | Relevant unit | | | |
|---|---|---|---|---|
| | J | kg | $\mathrm{m}^{-1}$ | Hz |
| 1 J | (1 J) = 1 J | (1 J)/$c^2$ = 1.112 650 056... × $10^{-17}$ kg | (1 J)/$hc$ = 5.034 116 651(62) × $10^{24}$ $\mathrm{m}^{-1}$ | (1 J)/$h$ = 1.509 190 205(19) × $10^{33}$ Hz |
| 1 kg | (1 kg)$c^2$ = 8.987 551 787... × $10^{16}$ J | (1 kg) = 1 kg | (1 kg)$c/h$ = 4.524 438 411(56) × $10^{41}$ $\mathrm{m}^{-1}$ | (1 kg)$c^2/h$ = 1.356 392 512(17) × $10^{50}$ Hz |
| 1 $\mathrm{m}^{-1}$ | (1 $\mathrm{m}^{-1}$)$hc$ = 1.986 445 824(24) × $10^{-25}$ J | (1 $\mathrm{m}^{-1}$)$h/c$ = 2.210 219 057(27) × $10^{-42}$ kg | (1 $\mathrm{m}^{-1}$) = 1$\mathrm{m}^{-1}$ | (1 $\mathrm{m}^{-1}$)$c$ = 299 792 458 Hz |
| 1 Hz | (1 Hz)$h$ = 6.626 070 040(81) × $10^{-34}$ J | (1 Hz)$h/c^2$ = 7.372 497 201(91) × $10^{-51}$ kg | (1 Hz)/$c$ = 3.335 640 951... × $10^{-9}$ $\mathrm{m}^{-1}$ | (1 Hz) = 1 Hz |
| 1 K | (1 K)$k$ = 1.380 648 52(79) × $10^{-23}$ J | (1 K)$k/c^2$ = 1.536 178 65(88) × $10^{-40}$ kg | (1 K)$k/hc$ = 69.503 457(40) $\mathrm{m}^{-1}$ | (1 K)$k/h$ = 2.083 661 2(12) × $10^{10}$ Hz |
| 1 eV | (1 eV) = 1.602 176 620 8(98) × $10^{-19}$ J | (1 eV)/$c^2$ = 1.782 661 907(11) × $10^{-36}$ kg | (1 eV)/$hc$ = 8.065 544 005(50) × $10^{5}$ $\mathrm{m}^{-1}$ | (1 eV)/$h$ = 2.417 989 262(15) × $10^{14}$ Hz |
| 1 u | (1 u)$c^2$ = 1.492 418 062(18) × $10^{-10}$ J | (1 u) = 1.660 539 040(20) × $10^{-27}$ kg | (1 u)$c/h$ = 7.513 006 616 6(34) × $10^{14}$ $\mathrm{m}^{-1}$ | (1 u)$c^2/h$ = 2.252 342 720 6(10) × $10^{23}$ Hz |
| 1 $E_h$ | (1 $E_h$) = 4.359 744 650(54) × $10^{-18}$ J | (1 $E_h$)/$c^2$ = 4.850 870 129(60) × $10^{-35}$ kg | (1 $E_h$)/$hc$ = 2.194 746 313 702(13) × $10^{7}$ $\mathrm{m}^{-1}$ | (1 $E_h$)/$h$ = 6.579 683 920 711(39) × $10^{15}$ Hz |

| | Relevant unit | | | |
|---|---|---|---|---|
| | K | eV | u | $E_h$ |
| 1 J | (1 J)/$k$ = 7.242 973 1(42) × $10^{22}$ K | (1 J) = 6.241 509 126(38) ×$10^{18}$ eV | (1 J)/$c^2$ = 6.700 535 363(82) × $10^{9}$ u | (1 J) = 2.293 712 317(28) × $10^{17}$ $E_h$ |
| 1 kg | (1 kg)$c^2/k$ = 6.509 659 5(37) × $10^{39}$ K | (1 kg)$c^2$ = 5.609 588 650(34) × $10^{35}$ eV | (1 kg) = 6.022 140 857(74) × $10^{26}$ u | (1 kg)$c^2$ = 2.061 485 823(25) × $10^{34}$ $E_h$ |
| 1 $\mathrm{m}^{-1}$ | (1 $\mathrm{m}^{-1}$)$hc/k$ = 1.438 777 36(83) × $10^{-2}$ K | (1 $\mathrm{m}^{-1}$)$hc$ = 1.239 841 973 9(76) × $10^{-6}$ eV | (1 $\mathrm{m}^{-1}$)$h/c$ = 1.331 025 049 00(61) × $10^{-15}$ u | (1 $\mathrm{m}^{-1}$)$hc$ = 4.556 335 252 767(27) × $10^{-8}$ $E_h$ |
| 1 Hz | (1 Hz)$h/k$ = 4.799 244 7(28) × $10^{-11}$ K | (1 Hz)$h$ = 4.135 667 662(25) × $10^{-15}$ eV | (1 Hz)$h/c^2$ = 4.439 821 661 6(20) × $10^{-24}$ u | (1 Hz)$h$ = 1.519 829 846 008 8(90) × $10^{-16}$ $E_h$ |
| 1 K | (1 K) = 1 K | (1 K)$k$ = 8.617 330 3(50) × $10^{-5}$ eV | (1 K)$k/c^2$ = 9.251 084 2(53) × $10^{-14}$ u | (1 K)$k$ = 3.166 810 5(18) × $10^{-6}$ $E_h$ |
| 1 eV | (1 eV)/$k$ = 1.160 452 21(67) × $10^{4}$ K | (1 eV) = 1 eV | (1 eV)/$c^2$ = 1.073 544 110 5(66) × $10^{-9}$ u | (1 eV) = 3.674 932 248(23) × $10^{-2}$ $E_h$ |
| 1 u | (1 u)$c^2/k$ = 1.080 954 38(62) × $10^{13}$ K | (1 u)$c^2$ = 931.494 095 4(57) × $10^{6}$ eV | (1 u) = 1 u | (1 u)$c^2$ = 3.423 177 690 2(16) × $10^{7}$ $E_h$ |
| 1 $E_h$ | (1 $E_h$)/$k$ = 3.157 751 3(18) × $10^{5}$ K | (1 $E_h$) = 27.211 386 02(17) eV | (1 $E_h$)/$c^2$ = 2.921 262 319 7(13) × $10^{-8}$ u | (1 $E_h$) = 1 $E_h$ |

## 1.15 NUMERICAL VALUES

Extensive use is made in electrical engineering of the constants $\pi$ and $\epsilon$ and of the numbers 2 and 10, the latter in logarithmic units and number systems. Table 1-14 lists functions of these numbers to 9 or 10 significant digits. In most engineering applications (except those involving the difference of large, nearly equal numbers), five significant digits suffice. The use of the listed values in computations with electronic hand calculators will suffice in most cases to produce results more than adequate for engineering work.

**TABLE 1-14** Numerical Values Used in Electrical Engineering

Functions of $\pi$:

$\pi = 3.141\ 592\ 654$
$1/\pi = 0.318\ 309\ 886$
$\pi^2 = 9.869\ 604\ 404$
$\sqrt{\pi} = 1.772\ 453\ 851$
$\pi/180° = 0.017\ 453\ 293$ (= radians per degree)
$180°/\pi = 57.295\ 779\ 51$ (= degrees per radian)

Functions of $\epsilon$:

$\epsilon = 2.718\ 281\ 828$
$1/\epsilon = 0.367\ 879\ 441$
$1 - 1/\epsilon = 0.632\ 120\ 559$
$\epsilon^2 = 7.389\ 056\ 096$
$\sqrt{\epsilon} = 1.648\ 721\ 271$

Logarithms to the base 10:

$\log_{10} \pi = 0.497\ 149\ 873$
$\log_{10} \epsilon = 0.434\ 294\ 482$
$\log_{10} 2 = 0.301\ 029\ 996$
$\log_{10} x = (\ln x)(0.434\ 294\ 482) = (\log_2 x)(0.301\ 029\ 996)$

Natural logarithms (to the base $\epsilon$):

$\ln \pi = 1.144\ 729\ 886$
$\ln 2 = 0.693\ 147\ 181$
$\ln 10 = 2.302\ 585\ 093$
$\ln x = (\log_{10} x)(2.302\ 585\ 093) = (\log_2 x)(0.693\ 147\ 181)$

Logarithms to the base 2:

$\log_2 \pi = 1.651\ 496\ 130$
$\log_2 \epsilon = 1.442\ 695\ 042$
$\log_2 10 = 3.321\ 928\ 096$
$\log_2 x = (\log_{10} x)(3.321\ 928\ 096) = (\ln x)(1.442\ 695\ 042)$

Powers of 2:

$2^5 = 32$
$2^{10} = 1024$
$2^{15} = 32{,}768$
$2^{20} = 1{,}048{,}576$
$2^{25} = 33{,}554{,}432$
$2^{30} = 1{,}073{,}741{,}824$
$2^{40} = 1.099\ 511\ 628 \times 10^{12}$
$2^{50} = 1.125\ 899\ 907 \times 10^{15}$
$2^{100} = 1.267\ 650\ 601 \times 10^{30}$

Logarithmic units:

| Power ratio | Current or voltage ratio | Decibels* | Nepers† |
|---|---|---|---|
| 1 | 1 | 0 | 0 |
| 2 | 1.414 214 | 3.010 300 | 0.346 574 |
| 3 | 1.732 051 | 4.771 213 | 0.549 306 |
| 4 | 2 | 6.020 600 | 0.693 147 |
| 5 | 2.236 068 | 6.989 700 | 0.804 719 |
| 10 | 3.162 278 | 10 | 1.151 293 |
| 15 | 3.872 983 | 11.760 913 | 1.354 025 |

(*Continued*)

**TABLE 1-14** Numerical Values Used in Electrical Engineering (*Continued*)

Values of $2^{(2N)}$:

| Value of $N$ | Value of $2^{(2N)}$ |
|---|---|
| 1 | 4 |
| 2 | 16 |
| 3 | 256 |
| 4 | 65,536 |
| 5 | 4,294,967,296 |
| 6 | $1.844\ 674\ 407 \times 10^{19}$ |
| 7 | $3.402\ 823\ 668 \times 10^{38}$ |
| 8 | $1.157\ 920\ 892 \times 10^{77}$ |
| 9 | $1.340\ 780\ 792 \times 10^{154}$ |
| 10 | $1.797\ 693\ 132 \times 10^{308}$ |

*The decibel is defined for power ratios only. It may be applied to current or voltage ratios only when the resistances through which the currents flow or across which the voltages are applied are equal.

†The neper is defined for current and voltage ratios only. It may be applied to power ratios only when the respective resistances are equal.

## 1.16 CONVERSION FACTORS

The increasing use of the metric system in British and American practice has generated a need for extensive tables of multiplying factors to facilitate conversions from and to the SI units. Tables 1-15 through 1-28 list these conversion factors.

| Table | Quantity | SI unit | Subtabulation | Basis of grouping |
|---|---|---|---|---|
| 1-15 | Length | meter | 1-15A | Units decimally related to one meter |
| | | | 1-15B | Units less than one meter |
| | | | 1-15C | Units greater than one meter |
| | | | 1-15D | Other length units |
| 1-16 | Area | square meter | 1-16A | Units decimally related to one square meter |
| | | | 1-16B | Nonmetric area units |
| | | | 1-16C | Other area units |
| 1-17 | Volume/capacity | cubic meter | 1-17A | Units decimally related to one cubic meter |
| | | | 1-17B | Nonmetric volume units |
| | | | 1-17C | U.S. liquid capacity measures |
| | | | 1-17D | British liquid capacity measures |
| | | | 1-17E | U.S. and U.K. dry capacity measures |
| | | | 1-17F | Other volume and capacity units |
| 1-18 | Mass | kilogram | 1-18A | Units decimally related to one kilogram |
| | | | 1-18B | Less than one pound-mass |
| | | | 1-18C | One pound-mass and greater |
| | | | 1-18D | Other mass units |
| 1-19 | Time | second | 1-19A | One second and less |
| | | | 1-19B | One second and greater |
| | | | 1-19C | Other time units |
| 1-20 | Velocity | meter per second | | |
| 1-21 | Density | kilogram per cubic meter | 1-21A | Units decimally related to one kilogram per cubic meter |
| | | | 1-21B | Nonmetric density units |
| | | | 1-21C | Other density units |
| 1-22 | Force | newton | | |
| 1-23 | Pressure | pascal | 1-23A | Units decimally related to one pascal |
| | | | 1-23B | Units decimally related to one kilogram-force per square meter |
| | | | 1-23C | Units expressed as heights of liquid |
| | | | 1-23D | Nonmetric pressure units |

(*Continued*)

(*Continued*)

| | | | | |
|---|---|---|---|---|
| 1-24 | Torque/bending moment | newton meter | | |
| 1-25 | Energy/work | joule | 1-25A | Units decimally related to one joule |
| | | | 1-25B | Units less than 10 joules |
| | | | 1-25C | Units greater than 10 joules |
| 1-26 | Power | watt | 1-26A | Units decimally related to one watt |
| | | | 1-26B | Nonmetric power units |
| 1-27 | Temperature | kelvin | | |
| 1-28 | Light | candela per square meter | 1-28A | Luminance units |
| | | lux | 1-28B | Illuminance units |

*Statements of Equivalence.* To avoid ambiguity, the conversion tables have been arranged in the form of statements of equivalence, that is, each unit listed at the left-hand edge of each table is stated to be equivalent to a multiple or fraction of each of the units to the right in the table. For example, the uppermost line of Table 1-15B represents the following statements:

**Column 2.** 1 meter is equal to 1.093 613 30 yards
**Column 3.** 1 meter is equal to 3.280 839 89 feet
**Column 4.** 1 meter is equal to 39.370 078 7 inches
**Column 5.** 1 meter is equal to $3.937\ 007\ 87 \times 10^4$ mils
**Column 6.** 1 meter is equal to $3.937\ 007\ 87 \times 10^7$ microinches

This table contains similar statements relating the meter, yard, foot, inch, mil, and microinch to each other, that is, conversion factors between the non-SI units as well as to and from the SI unit are given. In all, these tables contain over 1700 such statements. Exact conversion factors are indicated in **boldface** type.

*Tabulation Groups.* To produce tables that can be contained on individual pages of the handbook, units of a given quantity have been arranged in separate subtabulations identified by capital letters. Each such subtabulation represents a group of units related to each other decimally, by magnitude or by usage. Each subtabulation contains the SI unit,[b] so equivalent values can be found between units that are tabulated in separate tables. For example, to obtain equivalence between pounds per cubic foot and tonnes per cubic meter, we read from the fourth line of Table 1-21B:

1 pound per cubic foot is equal to 16.018 463 4 kilograms per cubic meter

From the first line of Table 1-21A, we find:

1 kilogram per cubic meter is equal to 0.001 metric ton per cubic meter

Hence,

1 pound per cubic foot is equal to 16.018 463 4 kilograms per cubic meter
= 0.016 018 463 4 metric ton per cubic meter

[b]In Tables 1-17C, 1-17D, 1-17E, and 1-18B, a decimal submultiple of the SI unit (the liter and gram, respectively) is listed because it is most commonly used in conjunction with the other units in the respective tables. The procedure for linking the subtables is unchanged.

**TABLE 1-15** Length Conversion Factors

(Exact conversions are shown in **boldface** type. Repeating decimals are underlined.) The SI unit of length is the meter.

A. Length units decimally related to one meter

| | Meters (m) | Kilometers (km) | Decimeters (dm) | Centimeters (cm) | Millimeters (mm) | Micrometers ($\mu$m) | Nanometers (nm) | Ångströms (Å) |
|---|---|---|---|---|---|---|---|---|
| 1 meter = | **1** | **0.001** | **10** | **100** | **1 000** | **1 000 000** | **$10^9$** | **$10^{10}$** |
| 1 kilometer = | **1 000** | **1** | **10 000** | **100 000** | **1 000 000** | **$10^9$** | **$10^{12}$** | **$10^{13}$** |
| 1 decimeter = | **0.1** | **0.000 1** | **1** | **10** | **100** | **100 000** | **$10^8$** | **$10^8$** |
| 1 centimeter = | **0.01** | **0.000 01** | **0.1** | **1** | **10** | **10 000** | **$10^7$** | **$10^8$** |
| 1 millimeter = | **0.001** | **$10^{-6}$** | **0.01** | **0.1** | **1** | **1 000** | **1 000 000** | **$10^7$** |
| 1 micrometer (micron) = | **$10^{-6}$** | **$10^{-9}$** | **0.000 01** | **0.000 1** | **0.001** | **1** | **1 000** | **10 000** |
| 1 nanometer = | **$10^{-9}$** | **$10^{-12}$** | **$10^{-8}$** | **$10^{-7}$** | **$10^{-6}$** | **0.001** | **1** | **10** |
| 1 ångström = | **$10^{-10}$** | **$10^{-13}$** | **$10^{-9}$** | **$10^{-8}$** | **$10^{-7}$** | **0.000 1** | **0.1** | **1** |

B. Nonmetric length units less than one meter

| | Meters (m) | Yards (yd) | Feet (ft) | Inches (in) | Mils (mil) | Microinches ($\mu$in) |
|---|---|---|---|---|---|---|
| 1 meter = | **1** | 1.093 613 30 | 3.280 839 89 | 39.370 078 7 | $3.937\,007\,87 \times 10^4$ | $3.937\,007\,87 \times 10^7$ |
| 1 yard = | **0.914 4** | **1** | **3** | **36** | **36 000** | **$3.6 \times 10^7$** |
| 1 foot = | **0.304 8** | **1/3** = $0.333\,\underline{3}$ | **1** | **12** | **12 000** | **$1.2 \times 10^7$** |
| 1 inch = | **0.025 4** | **1/36** = $0.027\,\underline{7}$ | **1/12** = $0.083\,\underline{3}$ | **1** | **1 000** | **1 000 000** |
| 1 mil = | **$2.54 \times 10^{-5}$** | $2.77\underline{7} \times 10^{-5}$ | $8.33\underline{3} \times 10^{-5}$ | **0.001** | **1** | **1 000** |
| 1 microinch = | **$2.54 \times 10^{-8}$** | $2.77\underline{7} \times 10^{-8}$ | $8.33\underline{3} \times 10^{-8}$ | **$10^{-8}$** | **0.001** | **1** |

C. Nonmetric length units greater than one meter (with equivalents in feet)

| | Meters (m) | Rods (rd) | Statute miles (mi) | Nautical miles (nmi) | Astronomical units (AU) | Parsecs (pc) | Feet (ft) |
|---|---|---|---|---|---|---|---|
| 1 meter = | **1** | 0.198 838 78 | $6.213\,711\,92 \times 10^{-4}$ | $5.399\,568\,04 \times 10^{-4}$ | $6.684\,491\,98 \times 10^{-12}$ | $3.240\,733\,17 \times 10^{-17}$ | 3.280 839 89 |
| 1 rod = | **5.029 2** | **1** | **0.003 125** | $2.715\,550\,76 \times 10^{-3}$ | $3.361\,764\,71 \times 10^{-11}$ | $1.629\,829\,53 \times 10^{-16}$ | **16.5** |
| 1 statute mile = | **1 609.344** | **320** | **1** | 0.868 976 24 | $1.075\,764\,71 \times 10^{-8}$ | $5.215\,454\,50 \times 10^{-14}$ | **5 280** |
| 1 nautical mile = | **1 852** | 368.249 423 | 1.150 779 45 | **1** | $1.237\,967\,91 \times 10^{-8}$ | $6.001\,837\,80 \times 10^{-14}$ | 6 076.115 48 |
| 1 astronomical unit* = | **$1.496 \times 10^{11}$** | $2.974\,628\,17 \times 10^{10}$ | 92 957 130.3 | 80 777 537.8 | **1** | $4.848\,136\,82 \times 10^{-6}$ | $4.908\,136\,48 \times 10^{11}$ |
| 1 parsec = | $3.085\,721\,50 \times 10^{16}$ | $6.135\,611\,02 \times 10^{15}$ | $1.917\,378\,44 \times 10^{13}$ | $1.666\,156\,32 \times 10^{13}$ | 206 264.806 | **1** | $1.012\,375\,82 \times 10^{17}$ |
| 1 foot = | **0.304 8** | $0.060\,\underline{606}$ | $1.893\,\underline{939} \times 10^{-4}$ | $1.645\,788\,33 \times 10^{-4}$ | $2.037\,433\,16 \times ^{-12}$ | $9.877\,754\,72 \times 10^{-18}$ | **1** |

*(Continued)*

**TABLE 1-15** Length Conversion Factors (*Continued*)

(Exact conversions are shown in **boldface** type. Repeating decimals are <u>underlined</u>.) The SI unit of length is the meter.

| D. Other length units |
|---|
| 1 cable = **720** feet = **219.456** meters |
| 1 cable (U.K.) = **608** feet = **185.318 4** meters |
| 1 chain (engineers') = **100** feet = **30.48** meters |
| 1 chain (surveyors') = **66** feet = **20.116 8** meters |
| 1 fathom = **6** feet = **1.828 8** meters |
| 1 fermi = **1** femtometer = $\mathbf{10^{-15}}$ meter |
| 1 foot (U.S. Survey) = 0.304 800 6 meter |
| 1 furlong = **660** feet = **201.168** meters |
| 1 hand = **4** inches = **0.101 6** meter |
| 1 league (international nautical) = 3 nautical miles = **5 556** meters |
| 1 league (statute) = **3** statute miles = **4 828.032** meters |
| 1 league (U.K. nautical) = 5 559.552 meters |
| 1 light-year = 9.460 895 2 × $10^{15}$ meters (= distance traveled by light in vacuum in one sidereal year) |
| 1 link (engineers') = **1** foot = **0.304 8** meter |
| 1 link (surveyors') = **7.92** inches = **0.201 168** meter |
| 1 micron = **1** micrometer = $\mathbf{10^{-6}}$ meter |
| 1 millimicron = **1** nanometer = $\mathbf{10^{-9}}$ meter |
| 1 myriameter = **10 000** meters |
| 1 nautical mile (U.K.) = 1 853.184 meters |
| 1 pale = **1** rod = **5.029 2** meters |
| 1 perch (linear) = **1** rod = **5.029 2** meters |
| 1 pica = **1/6** inch (approx.) = 4.217 518 × $10^{-3}$ meter |
| 1 point = **1/72** inch (approx.) = 3.514 598 × $10^{-4}$ meter |
| 1 span = **9** inches = **0.228 6** meter |

*As defined by the International Astronomical Union.

**TABLE 1-16** Area Conversion Factors

(Exact conversions are shown in **boldface** type. Repeating decimals are underlined.) The SI unit of area is the square meter.

A. Area units decimally related to one square meter

| | Square meters $(m)^2$ | Square kilometers $(km)^2$ | Hectares (square hectometers) $(hm)^2$ | Square centimeters $(cm)^2$ | Square millimeters $(mm)^2$ | Square micrometers $(\mu m)^2$ | Barns (b) |
|---|---|---|---|---|---|---|---|
| 1 square meter = | **1** | **$10^{-6}$** | **0.000 1** | **10 000** | **1 000 000** | **$10^{12}$** | **$10^{28}$** |
| 1 square kilometer = | **1 000 000** | **1** | **100** | **$10^{10}$** | **$10^{12}$** | **$10^{18}$** | **$10^{34}$** |
| 1 hectare = | **10 000** | **0.01** | **1** | **$10^{8}$** | **$10^{10}$** | **$10^{16}$** | **$10^{32}$** |
| 1 square centimeter = | **0.000 1** | **$10^{-10}$** | **$10^{-8}$** | **1** | **100** | **$10^{8}$** | **$10^{24}$** |
| 1 square millimeter = | **$10^{-6}$** | **$10^{-12}$** | **$10^{-10}$** | **0.01** | **1** | **$10^{6}$** | **$10^{22}$** |
| 1 square micrometer = | **$10^{-12}$** | **$10^{-18}$** | **$10^{-16}$** | **$10^{-8}$** | **$10^{-6}$** | **1** | **$10^{16}$** |
| 1 barn = | **$10^{-28}$** | **$10^{-34}$** | **$10^{-32}$** | **$10^{-24}$** | **$10^{-22}$** | **$10^{-16}$** | **1** |

B. Nonmetric area units (with square meter equivalents)

| | Square meters $(m)^2$ | Square statute miles $(mi)^2$ | Acres (acre) | Square rods $(rd)^2$ | Square yards $(yd)^2$ | Square feet $(ft)^2$ | Square inches $(in)^2$ | Circular mils (cmil) |
|---|---|---|---|---|---|---|---|---|
| 1 square meter = | **1** | $3.861\ 021\ 59 \times 10^{-7}$ | $2.471\ 053\ 82 \times 10^{-4}$ | $3.953\ 686\ 10 \times 10^{-2}$ | 1.195 990 05 | 10.763 910 4 | 1 550.003 10 | $1.973\ 525\ 24 \times 10^{9}$ |
| 1 square statute mile = | 2 589 988.1 | **1** | **640** | **102 400** | **3 097 600** | **27 878 400** | $4.014\ 489\ 60 \times 10^{9}$ | $5.111\ 406\ 91 \times 10^{15}$ |
| 1 acre = | 4 046.856 11 | **1/640 = 0.001 562 5** | **1** | **160** | **4 840** | **43 560** | **6 272 640** | $7.986\ 573\ 30 \times 10^{12}$ |
| 1 square rod = | 25.292 852 6 | **$9.765\ 625 \times 10^{-6}$** | **1/160 – 0.006 25** | **1** | **30.25** | **272.25** | **39 204** | $4.991\ 608\ 31 \times 10^{10}$ |
| 1 square yard = | **0.836 127 36** | $3.228\ 305\ 79 \times 10^{-7}$ | $2.066\ 115\ 70 \times 10^{-4}$ | $3.305\ 785\ 12 \times 10^{-2}$ | **1** | **9** | **1 296** | $1.650\ 118\ 45 \times 10^{9}$ |
| 1 square foot = | **0.092 903 04** | $3.587\ 006\ 43 \times 10^{-8}$ | $2.295\ 684\ 11 \times 10^{-5}$ | $3.673\ 094\ 58 \times 10^{-3}$ | **1/9** = 0.111 111 | **1** | **144** | $1.833\ 464\ 95 \times 10^{8}$ |
| 1 square inch = | **$6.451\ 6 \times 10^{-4}$** | $2.490\ 976\ 69 \times 10^{-10}$ | $1.594\ 225\ 08 \times 10^{-7}$ | $2.550\ 760\ 13 \times 10^{-5}$ | $7.716\ 049\ 38 \times 10^{-4}$ | **1/144** = 0.006 944 44 | **1** | $1.273\ 239\ 55 \times 10^{6}$ |
| 1 circular mil = | $5.067\ 074\ 79 \times 10^{-10}$ | $1.956\ 408\ 51 \times 10^{-16}$ | $1.252\ 101\ 45 \times 10^{-13}$ | $2.003\ 362\ 32 \times 10^{-11}$ | $6.060\ 171\ 01 \times 10^{-10}$ | $5.454\ 153\ 91 \times 10^{-9}$ | $7.853\ 981\ 63 \times 10^{-7}$ | **1** |

Exact conversions are:

1 acre = **4 046.856 422 4** square meters

1 square mile = **2 589 988.110 336** square meters

C. Other area units

1 are = **100** square meters

1 centiare (centare) = **1** square meter

1 perch (area) = **1** square rod = **30.25** square yards = 25.292 852 6 square meters

1 rod = **40** square rods = 1 011.714 11 square meters

1 section = **1** square statute mile = 2 589 988.1 square meters

1 township = **36** square statute miles = 93 239 572 square meters

**TABLE 1-17** Volume and Capacity Conversion Factors

(Exact conversions are shown in **boldface** type. Repeating decimals are underlined.) The SI unit of volume is the cubic meter.

A. Volume units decimally related to one cubic meter

| | Cubic meters (steres) $(m)^3$ | Cubic decimeters $(dm)^3$ | Cubic centimeters $(cm)^3$ | Liters (L) | Centiliters (cL) | Milliliters (mL) | Microliters ($\mu$L) |
|---|---|---|---|---|---|---|---|
| 1 cubic meter = | **1** | **1 000** | **1 000 000** | **1 000** | **100 000** | **1 000 000** | **$10^9$** |
| 1 cubic decimeter = | **0.001** | **1** | **1 000** | **1** | **100** | **1 000** | **1 000 000** |
| 1 cubic centimeter = | **0.000 001** | **0.001** | **1** | **0.001** | **0.1** | **1** | **1 000** |
| 1 liter = | **0.001** | **1** | **1 000** | **1** | **100** | **1 000** | **1 000 000** |
| 1 centiliter = | **0.000 01** | **0.01** | **10** | **0.01** | **1** | **10** | **10 000** |
| 1 milliliter = | **0.000 001** | **0.001** | **1** | **0.001** | **0.1** | **1** | **1 000** |
| 1 microliter = | **$10^{-9}$** | **0.000 001** | **0.001** | **0.000 001** | **0.000 1** | **0.001** | **1** |

B. Nonmetric volume units (with cubic meter and liter equivalents)

| | Cubic meters (steres) $(m)^3$ | Liters (L) | Cubic inches $(in)^3$ | Cubic feet $(ft)^3$ | Cubic yards $(yd)^3$ | Barrels (U.S.) (bbl) | Acre-Feet (acre-ft) | Cubic miles $(mi)^3$ |
|---|---|---|---|---|---|---|---|---|
| 1 cubic meter = | **1** | **1 000** | 6.102 374 41 × $10^4$ | 35.314 666 | 1.307 950 62 | 6.289 810 97 | 8.107 131 94 × $10^{-4}$ | 2.399 127 59 × $10^{-10}$ |
| 1 liter = | **0.001** | **1** | 61.023 744 1 | 0.035 314 66 | 1.307 950 62 × $10^{-3}$ | 6.289 810 97 × $10^{-3}$ | 8.107 131 93 × $10^{-7}$ | 2.399 127 59 × $10^{-13}$ |
| 1 cubic inch = | **1.638 706 4 × $10^{-5}$** | **1.638 706 4 × $10^{-2}$** | **1** | **1/1 728** = 5.787 037 03 × $10^{-4}$ | **1/46 656** = 2.143 347 05 × $10^{-5}$ | 1.030 715 32 × $10^{-4}$ | 1.328 520 90 × $10^{-8}$ | 3.931 465 73 × $10^{-15}$ |
| 1 cubic foot = | 2.831 684 66 × $10^{-2}$ | **28.316 846 592** | **1 728** | **1** | **1/27** = 0.037 037 | 0.178 107 61 | **1/43 560** = 2.295 684 11 × $10^{-5}$ | 6.793 572 78 × $10^{-12}$ |
| 1 cubic yard = | 0.764 554 86 | 764.554 858 | **46 656** | **27** | **1** | 4.808 905 38 | 6.198 347 11 × $10^{-4}$ | 1.834 264 65 × $10^{-10}$ |
| 1 barrel (U.S.) = | 0.158 987 29 | 158.987 294 | **9 702** | 5.614 583 33 | 0.207 947 53 | **1** | 1.288 930 98 × $10^{-4}$ | 3.814 308 05 × $10^{-11}$ |
| 1 acre-foot = | 1.233 481 84 | 1.233 481 84 × $10^6$ | 7.527 168 00 × $10^7$ | **43 560** | 1 613 333 33 | 7 758.367 34 | **1** | 2.959 280 30 × $10^{-7}$ |
| 1 cubic mile = | 4.168 181 83 × $10^9$ | 4.168 181 83 × $10^{12}$ | 2.543 580 61 × $10^{14}$ | **1.471 979 52 × $10^{11}$** | **5.451 776 × $10^9$** | 26.217 074 9 × $10^9$ | **3 379 200** | **1** |

C. United States liquid capacity measures (with liter equivalents)

| | Liters (L) | Gallons (U.S. gal) | Quarts (U.S. qt) | Pints (U.S. pt) | Gills (U.S. gi) | Fluid ounces (U.S. floz) | Fluidrams (U.S. fldr) | Minims (U.S. minim) |
|---|---|---|---|---|---|---|---|---|
| 1 liter = | **1** | 0.264 172 05 | 1.056 688 | 2.113 376 | 8.453 506 | 33.814 023 | 270.512 18 | 16 230.73 |
| 1 gallon, U.S. = | 3.785 411 8 | **1** | **4** | **8** | **32** | **128** | **1 024** | **61 440** |
| 1 quart, U.S. = | **0.946 352 946** | **1/4 = 0.25** | **1** | **2** | **8** | **32** | **256** | **15 360** |
| 1 pint, U.S. = | 0.473 176 5 | **1/8 = 0.125** | **1/2 = 0.5** | **1** | **4** | **16** | **128** | **7 680** |
| 1 gill, U.S. = | 0.118 294 1 | **1/32 = 0.031 25** | **1/8 = 0.125** | **1/4 = 0.25** | **1** | **4** | **32** | **1 920** |
| 1 fluid ounce, U.S. = | 2.957 353 × $10^{-2}$ | **1/128 = 0.007 812 5** | **1/32 = 0.031 25** | 1/16 = 0.062 5 | **1/4 = 0.25** | **1** | **8** | **480** |
| 1 fluidram, U.S. = | 3.696 691 2 × $10^{-3}$ | **1/102 4 = 9.765 625 × $10^{-4}$** | **1/256 = 3.906 25 × $10^{-3}$** | **1/128 = 0.007 812 5** | **1/32 = 0.031 25** | **1/8 = 0.125** | **1** | **60** |
| 1 minim, U.S. = | 6.161 152 × $10^{-5}$ | **1/61 440 =** 1.627 604 $\underline{16}$ × $10^{-5}$ | **1/15 360 =** 6.510 416 $\underline{66}$ × $10^{-5}$ | **1/7 680 =** 1.302 083 3$\underline{3}$ × $10^{-4}$ | **1/1 920 =** 5.208 333 $\underline{3}$ × $10^{-4}$ | **1/480 =** 2.083 333 $\underline{3}$ × $10^{-3}$ | **1/60 =** 0.016 666 $\underline{6}$ | **1** |

D. British Imperial liquid capacity measures (with liter equivalents)

| | Liters (L) | Gallons (U.K. gal) | Quarts (U.K. qt) | Pints (U.K. pt) | Gills (U.K. gi) | Fluid ounces (U.K. floz) | Fluidrams (U.K. fldr) | Minims (U.K. minim) |
|---|---|---|---|---|---|---|---|---|
| 1 liter = | **1** | 0.219 969 2 | 0.879 876 6 | 1.759 753 | 7.039 018 | 35.195 06 | 281.560 5 | 16 893.63 |
| 1 gallon, U.K. = | 4.546 092 | **1** | **4** | **8** | **32** | **160** | **1 280** | **76 800** |
| 1 quart, U.K. = | 1.136 523 | **1/4 = 0.25** | **1** | **2** | **8** | **40** | **320** | **19 200** |
| 1 pint, U.K. = | 0.568 261 5 | **1/8 = 0.125** | **1/2 = 0.5** | **1** | **4** | **20** | **160** | **9 600** |
| 1 gill, U.K. = | 0.142 065 4 | **1/32 = 0.031 25** | **1/8 = 0.125** | **1/4 = 0.25** | **1** | **5** | **40** | **2 400** |
| 1 fluid ounce, U.K. = | 2.841 307 × $10^{-2}$ | **1/160 = 0.006 25** | **1/40 = 0.025** | **1/20 = 0.05** | **1/5 = 0.2** | **1** | **8** | **480** |
| 1 fluidram, U.K. = | 3.551 634 × $10^{-3}$ | **1/1 280 = 7.812 5 × $10^{-4}$** | **1/320 = 0.003 125** | **1/160 = 0.006 25** | **1/40 = 0.025** | **1/8 = 0.125** | **1** | **60** |
| 1 minim, U.K. = | 5.919 391 × $10^{-5}$ | **1/76 800 =** 1.302 083 3$\underline{3}$ × $10^{-5}$ | **1/19 200 =** 5.208 333 3$\underline{3}$ × $10^{-5}$ | **1/9 600 =** 1.041 666 6$\underline{6}$ × $10^{-4}$ | **1/2 400 =** 4.166 666 6$\underline{6}$ × $10^{-4}$ | **1/480 =** 2.083 333 3$\underline{3}$ × $10^{-3}$ | **1/60 =** 0.016 666 $\underline{66}$ | **1** |

(*Continued*)

**TABLE 1-17** Volume and Capacity Conversion Factors (*Continued*)
(Exact conversions are shown in **boldface** type. Repeating decimals are underlined.) The SI unit of volume is the cubic meter.

E. United States and British dry capacity measures (with liter equivalents)

| | | U.S. dry measures | | | | British dry measures | | | |
|---|---|---|---|---|---|---|---|---|---|
| | Liters (L) | Bushels (U.S. bu) | Pecks (U.S. peck) | Quarts (U.S. qt) | Pints (U.S. pt) | Bushels (U.K. bu) | Pecks (U.K. peck) | Quarts (U.K. qt) | Pints (U.K. pt) |
| 1 liter = | **1** | 0.028 377 59 | 0.113 510 37 | 0.908 082 99 | 1.816 165 98 | 0.027 496 1 | 0.109 984 6 | 0.879 876 6 | 1.759 753 4 |
| 1 bushel, U.S. = | 35.239 070 | **1** | **4** | **32** | **64** | 0.968 938 7 | 3.875 754 9 | 31.006 04 | 62.012 08 |
| 1 peck, U.S. = | 8.809 767 5 | **1/4 = 0.25** | **1** | **8** | **16** | 0.242 234 7 | 0.968 938 7 | 7.751 509 | 15.503 02 |
| 1 quart, U.S. = | 1.101 220 9 | **1/32 = 0.031 25** | **1/8 = 0.125** | **1** | **2** | 0.030 279 34 | 0.121 117 3 | 0.968 938 7 | 1.937 878 |
| 1 pint, U.S. = | 0.550 610 5 | **1/64 = 0.015 625** | **1/16 = 0.062 5** | **1/2 = 0.5** | **1** | 0.015 139 67 | 0.060 558 67 | 0.484 469 3 | 0.968 938 7 |
| 1 bushel, U.K. = | 36.368 73 | 1.032 057 | 4.128 228 | 33.025 82 | 66.051 65 | **1** | **4** | **32** | **64** |
| 1 peck, U.K. = | 9.092 182 | 0.258 014 3 | 1.032 057 | 8.256 456 | 16.512 91 | **1/4 = 0.25** | **1** | **8** | **16** |
| 1 quart, U.K. = | 1.136 523 | 0.032 251 78 | 0.129 007 1 | 1.200 950 | 2.401 900 | **1/32 = 0.031 25** | **1/8 = 0.125** | **1** | **2** |
| 1 pint, U.K. = | 0.568 261 4 | 0.016 125 89 | 0.064 503 6 | 0.516 028 4 | 1.032 057 | **1/64 = 0.015 625** | **1/64 = 0.062 5** | **1/2 = 0.5** | **1** |

Exact conversion: 1 dry pint, U.S. = **33.600 312 5** cubic inches

F. Other volume and capacity units

1 barrel, U.S. (used for petroleum, etc.) = **42** gallons = 0.158.987 296 cubic meter
1 barrel ("old barrel") = **31.5** gallons = 0.119 240 cubic meter
1 board foot = **144** cubic inches = $2.359\ 737 \times 10^{-3}$ cubic meter
1 cord = **128** cubic feet = 3.624 556 cubic meters
1 cord foot = **16** cubic feet = 0.453 070 cubic meter
1 cup = **8** fluid ounces, U.S. = $2.365\ 882 \times 10^{-4}$ cubic meter
1 gallon (Canadian, liquid) = $4.546\ 090 \times 10^{-3}$ cubic meter
1 perch (volume) = 24.75 cubic feet = 0.700 842 cubic meter
1 stere = **1** cubic meter
1 tablespoon = **0.5** fluid ounce, U.S. = $1.478\ 677 \times 10^{-5}$ cubic meter
1 teaspoon = **1/6** fluid ounce, U.S. = $4.928\ 922 \times 10^{-6}$ cubic meter
1 ton (register ton) = **100** cubic feet = 2.831 684 66 cubic meters

**TABLE 1-18** Mass Conversion Factors

(Exact conversions are shown in **boldface** type. Repeating decimals are underlined.) The SI unit of mass is the kilogram.

A. Mass units decimally related to one kilogram

| | Kilograms (kg) | Tonnes (metric tons) | Grams (g) | Decigrams (dg) | Centigrams (cg) | Milligrams (mg) | Micrograms ($\mu$g) |
|---|---|---|---|---|---|---|---|
| 1 kilogram = | **1** | **0.001** | **1 000** | **10 000** | **100 000** | **1 000 000** | **$10^9$** |
| 1 tonne = | **1 000** | **1** | **1 000 000** | **$10^7$** | **$10^8$** | **$10^9$** | **$10^{12}$** |
| 1 gram = | **0.001** | **0.000 001** | **1** | **10** | **100** | **1 000** | **1 000 000** |
| 1 decigram = | **0.000 1** | **$10^{-7}$** | **0.1** | **1** | **10** | **100** | **100 000** |
| 1 centigram = | **0.000 01** | **$10^{-8}$** | **0.01** | **0.1** | **1** | **10** | **10 000** |
| 1 milligram = | **0.000 001** | **$10^{-9}$** | **0.001** | **0.01** | **0.1** | **1** | **1 000** |
| 1 microgram = | **$10^{-9}$** | **$10^{-12}$** | **0.000 001** | **0.000 01** | **0.000 1** | **0.001** | **1** |

B. Nonmetric mass units less than one pound-mass (with gram equivalents)

| | Grams (g) | Avoirdupois ounces-mass ($oz_m$, avdp) | Troy ounces-mass ($oz_m$, troy) | Avoirdupois drams (dr avdp) | Apothecary drams (dr apoth) | Pennyweights (dwt) | Grains (grain) | Scruples (scruple) |
|---|---|---|---|---|---|---|---|---|
| 1 gram = | **1** | 0.035 273 962 | 0.032 150 747 | 0.564 383 39 | 0.257 205 97 | 0.643 014 93 | 15.432 358 4 | 0.771 617 92 |
| 1 avdp ounce-mass = | 28.349 523 1 | **1** | 0.911 458 33 | **16** | 7.291 666 66 | 18.227 166 7 | **437.5** | **21.875** |
| 1 troy ounce-mass = | 31.103 476 8 | 1.097 142 86 | **1** | 17.554 285 7 | **8** | **20** | **480** | **24** |
| 1 avdp dram = | 1.771 845 20 | **1/16 = 0.062 5** | 0.056 966 15 | **1** | 0.455 729 17 | 1.139 322 92 | **27.343 75** | **1.367 187 5** |
| 1 apothecary dram = | 3.887 934 58 | 0.137 142 857 | **1/8 = 0.125** | 2.194 285 70 | **1** | **2.5** | **60** | **3** |
| 1 pennyweight = | 1.555 173 83 | 0.054 863 162 | **1/20 = 0.05** | 0.877 714 28 | **1/2.5 = 0.4** | **1** | **24** | **1.2** |
| 1 grain = | **0.064 798 91** | **1/437.5** = 2.285 714 29 $\times 10^{-3}$ | **1/480** = 0.002 083 333 | 3.657 142 85 $\times 10^{-2}$ | **1/60** = 0.016 666 66 | **1/24** = 0.041 666 66 | **1** | **0.05** |
| 1 scrople = | 1.295 078 20 | 4.571 428 58 $\times 10^{-2}$ | **1/24** = 0.041 666 66 | 0.731 428 57 | **1/3** = 0.333 333 33 | **5/6** = 0.833 333 33 | **20** | **1** |

*(Continued)*

**TABLE 1-18** Mass Conversion Factors (*Continued*)
(Exact conversions are shown in **boldface** type. Repeating decimals are underlined.) The SI unit of mass is the kilogram.

| C. Nonmetric mass units of one pound-mass and greater (with kilogram equivalents) | | | | | | | | |
|---|---|---|---|---|---|---|---|---|
| | Kilograms (kg) | Long tons (long ton) | Short tons (short ton) | Long hundredweights (long cwt) | Short hundredweights (short cwt) | Slugs (slug) | Avoirdupois pounds-mass ($lb_m$, avdp) | Troy pounds-mass ($lb_m$, troy) |
| 1 kilogram = | **1** | 9.842 065 28 × $10^{-1}$ | 1.102 311 31 × $10^{-3}$ | 1.968 411 31 × $10^{-2}$ | 2.204 622 62 × $10^{-2}$ | 0.068 521 77 | 2.204 622 62 | 2.679 228 89 |
| 1 long ton = | 1 016.046 9 | **1** | **1.12** | **20** | **22.4** | 69.621 329 | **2 240** | $2722.222\ 2\underline{22}$ |
| 1 short ton = | **907.184 74** | **200/224** = 0.892 857 14 | **1** | **4 000/224** = 17.857 142 9 | **20** | 62.161 901 | **2 000** | 2 430.555 55 |
| 1 long hundredweight = | 50.802 345 4 | **0.05** | **0.056** | **1** | **1.12** | 3.481 066 4 | **112** | $136.111\ 11\underline{1}$ |
| 1 short hundredweight = | **45.359 237** | **10/224** = 0.044 642 86 | **0.05** | **100/112** = 0.892 857 14 | **1** | 3.108 095 0 | **100** | $121.527\ 7\underline{77}$ |
| 1 slug = | 14.593 903 | 0.014 363 41 | 0.016 087 02 | 0.287 268 3 | 0.321 740 5 | **1** | 32.174 05 | 39.100 406 |
| 1 avdp pound-mass = | **0.453 592 37** | **1/2 240** = 4.464 285 71 × $10^{-1}$ | **0.000 5** | **1/1 12** = 8.928 571 43 × $10^{-3}$ | **0.01** | 3.108 095 0 × $10^{-2}$ | **1** | $1.215\ 277\ 7\underline{77}$ |
| 1 troy pound-mass = | 0.373 241 72 | 3.673 469 37 × $10^{-1}$ | 4.114 285 70 × $10^{-1}$ | 7.346 938 79 × $10^{-3}$ | 8.228 571 45 × $10^{-3}$ | 0.025 575 18 | 0.822 857 14 | **1** |

Exact conversions: 1 long ton = **1 016.046 908 8** kilograms
1 troy pound-mass = **0.373 241 721 6** kilogram

D. Other mass units

1 assay ton = 29.166 667 grams
1 carat (metric) = **200** milligrams
1 carat (troy weight) = **3$^1/_6$** grains = 205.196 55 milligrams
1 myriagram = **10** kilograms
1 quintal = **100** kilograms
1 stone = **14** pounds, avdp = **6.350 293 18** kilograms

**TABLE 1-19** Time Conversion Factors

(Exact conversions are shown in **boldface** type. Repeating decimals are underlined.) The SI unit of time is the second.

A. Time units of one second and less

| | Seconds (s) | Milliseconds (ms) | Microseconds ($\mu$s) | | Picoseconds (ps) |
|---|---|---|---|---|---|
| 1 second = | **1** | **1 000** | **1 000 000** | $\mathbf{10^{9}}$ | $\mathbf{10^{12}}$ |
| 1 millisecond = | **0.001** | **1** | **1 000** | **1 000 000** | $\mathbf{10^{9}}$ |
| 1 microsecond = | **0.000 001** | **0.001** | **1** | **1 000** | **1 000 000** |
| 1 nanosecond = | $\mathbf{10^{-9}}$ | **0.000 001** | **0.001** | **1** | **1 000** |
| 1 picosecond = | $\mathbf{10^{-12}}$ | $\mathbf{10^{-9}}$ | **0.000 001** | **0.001** | **1** |

B. Time units of one second and greater

| | Mean solar seconds (s) | Mean solar minutes (min) | Mean solar hours (h) | Mean solar days (d) | Mean solar weeks (w) | Calendar (Gregorian) year (yr) |
|---|---|---|---|---|---|---|
| 1 second = | **1** | **1/60** = 0.016 666 6 | **1/3 600** = 0.000 277 7 | **1/86 400** = $1.157\ 407\ 407 \times 10^{-5}$ | **1/604 800** = $1.653\ 439\ 15 \times 10^{-6}$ | $3.168\ 873\ 85 \times 10^{-8}$ |
| 1 minute = | **60** | **1** | **1/60** = 0.016 666 6 | **1/1 440** = 0.000 694 44 | **1/10 080** = $9.920\ 634\ 92 \times 10^{-5}$ | $1.901\ 324\ 31 \times 10^{-6}$ |
| 1 hour = | **3 600** | **60** | **1** | **1/24** = 0.041 666 6 | **1/168** = $5.952\ 380\ 95 \times 10^{-3}$ | $1.140\ 794\ 50 \times 10^{-4}$ |
| 1 day = | **86 400** | **1 440** | **24** | **1** | **1/7 = 0.142 857 14** | $2.737\ 907\ 00 \times 10^{-3}$ |
| 1 week = | **604 800** | **10 080** | **168** | **7** | **1** | $1.916\ 534\ 90 \times 10^{-2}$ |
| **1 calendar year = (Gregorian)** | 31 556 952 | 525 949.2 | 8 765.82 | 365.242 5 | 52.117 5 | **1** |

NOTES: The conventional calendar year of 365 days can be used in rough calculations only; the modern calendar is based on the Gregorian year of 365.2425 mean solar days, the value chosen by Pope Gregory XIII in 1582. This value requires that a leap-year day be introduced every four years as February 29, except that centennial years (1900, 2000, etc.) are leap years only when divisible by 400. The remaining difference between the Gregorian year and the tropical year (see below) introduces an error of 1 day in 3300 years.

The tropical year is the interval between successive vernal equinoxes and has been defined by the International Astronomical Union for noon of January 1, 1900 as 31 556 925.974 7 seconds = 365.242 198 79 mean solar days. The tropical year decreases by approximately 5.3 milliseconds per year. The sidereal year is the interval between successive returns of the sun to the direction of the same star. Sidereal time units, given in Table 1-18C, are used primarily in astronomy. The SI second, defined by the atomic process of the cesium atom, is equal to the mean solar second within the limits of their definition.

C. Other time units

1 decade = **10** Gregorian years
1 fortnight = **14** days = **1 209 600** seconds
1 century = **100** Gregorian years
1 millennium = **1 000** Gregorian years
1 sidereal year = 366.256 4 sidereal days = 31 558 149.8 seconds
1 sidereal day = 86 164.091 seconds
1 sidereal hour = 3 590.170 seconds
1 sidereal minute = 59.836 17 seconds
1 sidereal second = 0.997 269 6 second
1 shake = $\mathbf{10^{-8}}$ seconds

**TABLE 1-20** Velocity Conversion Factors

The SI unit of velocity is the meter per second.

| | Meters per second (m/s) | Kilometers per hour (km/h) | Statute miles per hour (mi/h) | Knots (kn) | Feet per minute (ft/min) | Feet per second (ft/s) | Inches per second (in/s) |
|---|---|---|---|---|---|---|---|
| 1 meter per second = | **1** | **3.6** | 2.236 936 29 | 1.943 844 49 | 196.850 394 | 3.280 839 89 | 39.370 078 7 |
| 1 kilometer per hour = | **1/3.6** = 0.277 777 | **1** | 0.621 371 19 | 0.539 956 80 | 54.680 664 9 | 0.911 344 42 | 10.936 133 0 |
| 1 statute mile per hour = | **0.447 04** | **1.609 344** | **1** | 0.868 976 24 | **88** | **88/60** = 1.466 666 | **88/5 = 17.6** |
| 1 knot = | 0.514 444 | **1.852** | 1.150 779 45 | **1** | 101.268 592 | 1.687 780 99 | 20.253 718 4 |
| 1 foot per minute = | **0.005 08** | **0.018 288** | 0.011 363 | $9.874\ 730\ 01 \times 10^{-3}$ | **1** | **1/60** = 0.016 666 | **1/5 = 0.2** |
| 1 foot per second = | **0.304 8** | **1.097 28** | 0.681 818 | 0.592 483 80 | **60** | **1** | **12** |
| 1 inch per second = | **0.025 4** | **0.091 44** | 0.056 818 | 0.049 373 65 | **5** | **1/12** = 0.083 333 | **1** |

NOTE: The velocity of light in vacuum, $c$ = 299 792 458 meters per second = 670 616 629 statute miles per hour
= 186 282.397 statute miles per second
= 0.983 571 056 feet per nanosecond

Other velocity units

1 foot per hour = $8.466\ 667 \times 10^{-5}$ meter per second
1 statute mile per minute = **26.822 4** meters per second
1 statute mile per second = **1 609.344** meters per second

**TABLE 1-21** Density Conversion Factors

(Exact conversions are shown in **boldface** type. Repeating decimals are underlined.) The SI unit of density is the kilogram per cubic meter.

A. Density units decimally related to one kilogram per cubic meter

| | Kilograms per cubic meter ($kg/m^3$) | Tonnes per cubic meter ($t/m^3$) | Grams per cubic meter ($g/m^3$) | Grams per liter (g/L) | Milligrams per liter (mg/L) | Micrograms per milliliter ($\mu g/mL$) |
|---|---|---|---|---|---|---|
| 1 kilogram per cubic meter = | **1** | **0.001** | **1 000** | **1** | **1 000** | **1 000** |
| 1 tonne per cubic meter = | **1 000** | **1** | **1 000 000** | **1 000** | **1 000 000** | **1 000 000** |
| 1 gram per cubic meter = | **0.001** | **0.000 001** | **1** | **0.001** | **1** | **1** |
| 1 gram per liter = | **1** | **0.001** | **1 000** | **1** | **1 000** | **1 000** |
| 1 milligram per liter = | **0.001** | **0.000 001** | **1** | **0.001** | **1** | **1** |
| 1 microgram per milliliter = | **0.001** | **0.000 001** | **1** | **0.001** | **1** | **1** |

B. Nonmetric density units (with kilogram per cubic meter equivalents)

| | Kilograms per cubic meter ($kg/m^3$) | Short tons per cubic mile (short $tons/mi^3$) | Avoirdupois pounds per acrefoot (lb avdp/acre-ft) | Avoirdupois pounds per cubic foot (lb $avdp/ft^3$) | Avoirdupois pounds per cubic inch (lb $avdp/in^3$) | Avoirdupois ounces per U.S. quart (oz advp/U.S. qt) | Avoirdupois drams per U.S. fluid ounce (dr advp/U.S. floz) | Grains per U.S. fluid ounce (grain/U.S. floz) |
|---|---|---|---|---|---|---|---|---|
| 1 kilogram per cubic meter = | **1** | 4 594 934 | 2 719.362 0 | $6.242\,796\,1 \times 10^{-2}$ | $3.612\,729\,20 \times 10^{-5}$ | $3.338\,161\,6 \times 10^{-2}$ | $1.669\,080\,82 \times 10^{-2}$ | 0.456 389 28 |
| 1 short ton per cubic mile = | $2.176\,451\,9 \times 10^{-7}$ | **1** | $5.918\,560\,5 \times 10^{-4}$ | $1.358\,714\,5 \times 10^{-8}$ | $7.862\,931\,3 \times 10^{-12}$ | $7.265\,348\,2 \times 10^{-9}$ | $3.632\,674\,1 \times 10^{-9}$ | $9.933\,0931\,1 \times 10^{-8}$ |
| 1 avdp pound per acrefoot = | $3.677\,333\,2 \times 10^{-4}$ | 1 689.600 0 | **1** | $2.295\,684\,1 \times 10^{-5}$ | $1.328\,520\,9 \times 10^{-8}$ | $1.227\,553\,2 \times 10^{-5}$ | $6.137\,766\,2 \times 10^{-6}$ | $1.678\,295\,5 \times 10^{-4}$ |
| 1 avdp pound per cubic foot = | 16.018 463 4 | 73 598 976 | **43 560** | **1** | **1/1 728** = $5.787\,037\,03 \times 10^{-4}$ | 0.534 722 2 | 0.267 361 1 | 7.310 655 0 |
| 1 avdp pound per cubic inch = | 27 679.905 | $1.271\,790\,4 \times 10^{11}$ | **75 271 680** | **1 728** | **1** | **924** | **462** | 12 632.812 |
| 1 avdp ounce per U.S. quart = | 29.956 608 | $1.376\,395\,5 \times 10^{8}$ | 81 462.86 | 1.870 130 0 | $1.082\,251\,1 \times 10^{-3}$ | **1** | **0.5** | 13.671 874 |
| 1 avdp dram per U.S. fluid ounce = | 59.913 216 | $2.752\,793\,0 \times 10^{8}$ | 162 925.72 | 3.740 259 8 | $2.164\,502\,3 \times 10^{-3}$ | **2** | **1** | 27.343 748 |
| 1 grain per U.S. fluid ounce = | 2.191 111 9 | 10 067 357 | 5 958.426 3 | 0.136 786 65 | $7.915\,894\,0 \times 10^{-5}$ | 0.073 142 86 | 0.036 571 43 | **1** |

C. Other density units

1 grain per gallon, U.S. = 17.118 06 grams per cubic meter
1 gram per cubic centimeter = **1 000** kilograms per cubic meter
1 avdp ounce per gallon, U.S. = 7.489 152 kilograms per cubic meter
1 avdp ounce per cubic inch = 1 729.994 kilograms per cubic meter
1 avdp pound per gallon, U.S. = 119.826 4 kilograms per cubic meter
1 slug per cubic foot = 515.379 kilograms per cubic meter
1 long ton per cubic yard = 1 328.939 kilograms per cubic meter

**TABLE 1-22** Force Conversion Factors

(Exact conversions are shown in **boldface** type. Repeating decimals are underlined.) The SI unit of force is the newton (N).

| | Newtons (N) | Kips (kip) | Slugs-force ($slug_f$) | Kilograms-force (kilopond) ($kg_f$) | Avoirdupois pounds-force ($lb_f$ avdp) | Avoirdupois ounces-force ($oz_f$ advp) | Poundals (pdl) | Dynes (dyn) |
|---|---|---|---|---|---|---|---|---|
| 1 newton = | **1** | 2.248 089 43 × $10^{-4}$ | 6.987 275 24 × $10^{-3}$ | 0.101 971 62 | 0.224 808 94 | 3.596 943 09 | 7.233 014 2 | **100 000** |
| 1 kip = | 444 8.221 62 | **1** | 31.080 949 | 453.592 370 | **1 000** | **16 000** | 32 174.05 | 444 822 162 |
| 1 slug-force = | 143.117 305 | 0.032 174 05 | **1** | 14.593 903 | 32.174 05 | 514 784 80 | 1 035.169 5 | 14 311 730 |
| 1 kilogram force (kilopond) = | **9.806 650** | 2.204 622 62 × $10^{-3}$ | 6.852 176 3 × $10^{-2}$ | **1** | 2.204 622 62 | 35.273 961 9 | 70 931 638 4 | **980 665** |
| 1 avdp pound force = | 4.448 221 62 | **0.001** | 3.108 094 88 × $10^{-2}$ | 0.453 592 37 | **1** | **16** | 32.174 05 | 444 822.162 |
| 1 avdp ounce force = | 0.278 013 85 | **1/16 000 = 0.000 062 5** | 1.942 559 30 × $10^{-3}$ | 2.834 952 3 × $10^{-2}$ | **1/16 = 0.062 5** | **1** | 2.010 878 03 | 27 801.385 |
| 1 poundal = | 0.138 254 95 | 3.108 094 9 × $10^{-5}$ | 9.660 253 9 × $10^{-4}$ | 0.140 980 81 | 0.031 080 95 | 0.497 295 18 | **1** | 13 825.495 |
| 1 dyne = | **0.000 01** | 2.248 089 43 × $10^{-8}$ | 6.987 275 24 × $10^{-8}$ | 1.019 716 21 × $10^{-6}$ | 2.248 089 43 × $10^{-6}$ | 3.596 943 10 × $10^{-5}$ | 7.233 014 2 × $10^{-5}$ | **1** |

The exact conversion is 1 avdp pound-force = **4.448 221 615 260 5** newtons.

**TABLE 1-23** Pressure/Stress Conversion Factors

(Exact conversions are shown in **boldface** type. Repeating decimals are <u>underlined</u>.) The SI unit of pressure or stress is the pascal (Pa).

| A. Pressure units decimally related to one pascal | | | | | |
|---|---|---|---|---|---|
| | Pascals (Pa) | Bars (bar) | Decibars (dbar) | Millibars (mbar) | Dynes per square centimeter (dyn/cm²) |
| 1 pascal = | **1** | **0.000 01** | **0.000 1** | **0.01** | **10** |
| 1 bar = | **100 000** | **1** | **10** | **1 000** | **1 000 000** |
| 1 decibar = | **10 000** | **0.1** | **1** | **100** | **100 000** |
| 1 millibar = | **100** | **0.001** | **0.01** | **1** | **1 000** |
| 1 dyne per square centimeter = | **0.1** | **0.000 001** | **0.000 01** | **0.001** | **1** |

| B. Pressure units decimally related to one kilogram-force per square meter (with pascal equivalents) | | | | | |
|---|---|---|---|---|---|
| | Kilograms-force per square meter ($kg_f/m^2$) | Kilograms-force per square centimeter ($kg_f/cm^2$) | Kilograms-force per square millimeter ($kg_f/mm^2$) | Grams-force per square centimeter ($g_f/cm^2$) | Pascals (Pa) |
| 1 kilogram-force per square meter = | **1** | **0.000 1** | **0.000 001** | **0.1** | **9.806 65** |
| 1 kilogram-force per square centimeter = | **10 000** | **1** | **0.01** | **1 000** | **98 066.5** |
| 1 kilogram-force per square millimeter = | **1 000 000** | **100** | **1** | **100 000** | **9 806 650** |
| 1 gram-force per square centimeter = | **10** | **0.001** | **0.000 01** | **1** | **98.066 5** |
| 1 pascal = | 0.101 971 62 | $1.019\,7162 \times 10^{-5}$ | $1.019\,716\,2 \times 10^{-7}$ | $1.019\,716\,2 \times 10^{-2}$ | 1 |

NOTE: **1** atmosphere (technical) = 1 kilogram-force per square centimeter = **98 066.5** pascals.

*(Continued)*

**TABLE 1-23** Pressure/Stress Conversion Factors (*Continued*)
(Exact conversions are shown in **boldface** type. Repeating decimals are underlined.) The SI unit of pressure or stress is the pascal (Pa).

C. Pressure units expressed as heights of liquid (with pascal equivalents)

| | Millimeters of mercury at 0°C (mmHg, 0°C) | Centimeters of mercury at 60°C (cmHg, 60°C) | Inches of mercury at 32°F (inHg, 32°F) | Inches of mercury at 60°F (inHg, 60°F) | Centimeters of water at 4°C ($cmH_2O$, 4°C) | Inches of water at 60°F ($inH_2O$, 60°F) | Feet of water at 39.2°F ($ftH_2O$, 39.2°F) | Pascals (Pa) |
|---|---|---|---|---|---|---|---|---|
| 1 millimeter of mercury, 0°C = | **1** | 0.100 282 | 0.039 370 1 | 0.039 481 3 | 1.359 548 | 0.535 775 6 | 0.044 604 6 | 133.322 4 |
| 1 centimeter of mercury, 60°C = | 9.971 830 | **1** | 0.392 591 9 | 0.393 700 8 | 13.557 18 | 5.342 664 | 0.444 789 5 | 1 329.468 |
| 1 inch of mercury, 32°F = | **25.4** | 2.547 175 | **1** | 1.002 824 8 | 34.532 52 | 13.608 70 | 1.132 957 | 3 386.389 |
| 1 inch of mercury, 60°C = | 25.328 45 | **2.54** | 0.997 183 1 | **1** | 34.435 25 | 13.570 37 | 1.129 765 | 3 376.85 |
| 1 centimeter of water, 4°C = | 0.735 539 | 0.073 762 | 0.028 958 | 0.029 040 0 | **1** | 0.394 083 8 | 0.032 808 4 | 98.063 8 |
| 1 inch of water, 60°F = | 1.866 453 | 0.187 173 | 0.073 482 | 0.073 690 0 | 2.537 531 | **1** | 0.083 252 4 | 248.840 |
| 1 foot of water, 39.2°F = | 22.419 2 | 2.248 254 | 0.882 646 | 0.885 139 | 30.479 98 | 12.011 67 | **1** | 2 988.98 |
| 1 pascal = | $7.500\ 615 \times 10^{-3}$ | $7.521\ 806 \times 10^{-4}$ | $2.952\ 998 \times 10^{-4}$ | $2.961\ 34 \times 10^{-4}$ | $1.019\ 74 \times 10^{-2}$ | $4.018\ 65 \times 10^{-3}$ | $3.345\ 62 \times 10^{-4}$ | **1** |

NOTE: 1 torr = 1 millimeter of mercury at 0°C = 133.322 4 pascals.

D. Nonmetric pressure units (with pascal equivalents)

| | Atmospheres (atm) | Avoirdupois pounds-force per square inch ($lb/in^2$) | Avoirdupois pounds-force per square foot ($lb_f/ft^2$, avdp) | Poundals per square foot ($pdl/ft^2$) | Pascals (Pa) |
|---|---|---|---|---|---|
| 1 atmosphere = | **1** | 14.695 95 | 2 116.217 | 68 087.24 | **101 325** |
| 1 avdp pound-force per square inch = | $6.804\ 60 \times 10^{-2}$ | **1** | **144** | 4 633.063 | 6 894.757 |
| 1 avdp pound-force per square foot = | $4.725\ 414 \times 10^{-4}$ | **1/144** = 0.006 944 | **1** | 32.174 05 | 47.880 26 |
| 1 poundal per square foot = | $1.468\ 704 \times 10^{-5}$ | $2.158\ 399 \times 10^{-4}$ | 0.031 080 9 | **1** | 1.488 164 |
| 1 pascal = | $9.869\ 233 \times 10^{-6}$ | $1.450\ 377 \times 10^{-4}$ | 0.020 885 4 | 0.671 968 9 | **1** |

NOTE: 1 normal atmosphere = 760 torr = **101 325** pascals.

**TABLE 1-24** Torque/Bending Moment Conversion Factors

(Exact conversions are shown in **boldface** type. Repeating decimals are <u>underlined</u>.) The SI unit of torque is the newton-meter (N · m).

| | Newton-meters (N · m) | Kilogram-force-meters ($kg_f \cdot m$) | Avoirdupois pound-force-feet ($lb_f \cdot ft$, avdp) | Avoirdupois pound-force-inches ($lb_f \cdot in$, avdp) | Avoirdupois ounce-force-inches ($oz_f \cdot in$, avdp) | Dyne-centimeters (dyne · cm) |
|---|---|---|---|---|---|---|
| 1 newton-meter = | **1** | 0.101 971 6 | 0.737 562 1 | 8.850 748 1 | 141.611 9 | **10 000 000** |
| 1 kilogram-force-meter = | **9.806 65** | **1** | 7.233 013 | 86.796 16 | 1 388.739 | **98 066 500** |
| 1 avdp pound-force-foot = | 1.355 818 | 0.138 255 0 | **1** | **12** | **192** | 13 558 180 |
| 1 avdp pound-force-inch = | 0.112 984 8 | $1.152\ 124 \times 10^{-2}$ | **1/12** = 0.083 33<u>3</u> | **1** | **16** | 1 129 848 |
| 1 avdp ounce-force-inch = | $7.061\ 552 \times 10^{-3}$ | $7.200\ 779 \times 10^{-4}$ | **1/192** = 0.005 208 <u>3</u> | **1/16 = 0.062 5** | **1** | 70 615.52 |
| 1 dyne-centimeter = | $\mathbf{10^{-7}}$ | $1.017\ 716 \times 10^{-8}$ | $7.375\ 621 \times 10^{-8}$ | $8.850\ 748 \times 10^{-7}$ | $1.416\ 119 \times 10^{-5}$ | **1** |

**TABLE 1-25** Energy/Work Conversion Factors

(Exact conversions are shown in **boldface** type. Repeating decimals are underlined.) The SI unit of energy and work is the joule (J).

A. Energy/work units decimally related to one joule

| | Joules (J) | Megajoules (MJ) | Kilojoules (kJ) | Millijoules (mJ) | Microjoules (μJ) | Ergs (erg) |
|---|---|---|---|---|---|---|
| 1 joule = | **1** | **0.000 001** | **0.001** | **1 000** | **1 000 000** | $\mathbf{10^{7}}$ |
| 1 megajoule = | **1 000 000** | **1** | **1 000** | $\mathbf{10^{9}}$ | $\mathbf{10^{12}}$ | $\mathbf{10^{13}}$ |
| 1 kilojoule = | **1 000** | **0.001** | **1** | **1 000 000** | $\mathbf{10^{9}}$ | $\mathbf{10^{10}}$ |
| 1 millijoule = | **0.001** | $\mathbf{10^{-9}}$ | $\mathbf{10^{-6}}$ | **1** | **1 000** | **10 000** |
| 1 microjoule = | **0.000 001** | $\mathbf{10^{-12}}$ | $\mathbf{10^{-9}}$ | **0.001** | **1** | **10** |
| 1 erg = | $\mathbf{10^{-7}}$ | $\mathbf{10^{-13}}$ | $\mathbf{10^{-10}}$ | **0.000 1** | **0.1** | **1** |

NOTE: 1 watt-second = 1 joule.

B. Energy/work units less than ten joules (with joule equivalents)

| | Joules (J) | Foot-poundals (ft · pdl) | Foot-pounds-force (ft · $lb_f$) | Calories (International Table) (cal, IT) | Calories (thermochemical) (cal, thermo) | Electronvolts (eV) |
|---|---|---|---|---|---|---|
| 1 joule = | **1** | 23.730 36 | 0.737 562 1 | 0.238 845 9 | 0.239 005 7 | $6.241\ 46 \times 10^{18}$ |
| 1 foot-poundal = | $4.214\ 011 \times 10^{-2}$ | **1** | $3.108\ 095 \times 10^{-2}$ | $1.006\ 499 \times 10^{-2}$ | $1.007\ 173 \times 10^{-2}$ | $2.630\ 16 \times 10^{17}$ |
| 1 foot-pound-force = | 1.355 818 | 32.174 05 | **1** | 0.323 831 6 | 0.324 048 3 | $8.462\ 28 \times 10^{18}$ |
| 1 calorie (Int. Tab.) = | **4.186 8** | 99.854 27 | 3.088 025 | **1** | 1.000 669 | $2.613\ 17 \times 10^{19}$ |
| 1 calorie (thermo) = | **4.184** | 99.287 83 | 3.085 960 | 0.999 331 2 | **1** | $2.611\ 43 \times 10^{19}$ |
| 1 electronvolt = | $1.602\ 19 \times 10^{-18}$ | $3.802\ 05 \times 10^{-18}$ | $1.181\ 71 \times 10^{-19}$ | $3.826\ 77 \times 10^{-20}$ | $3.829\ 33 \times 10^{-20}$ | **1** |

C. Energy/work units greater than ten joules (with joule equivalents)

| | Joules (J) | British thermal units, International Table (Btu, IT) | British thermal units, thermochemical (Btu, thermo) | Kilowatthours (kWh) | Horsepower-hours, electrical (hp · h, elec) | Kilocalories, International Table (kcal, IT) | Kilocalories, thermochemical (kcal, thermo) |
|---|---|---|---|---|---|---|---|
| 1 joule = | **1** | $9.478\ 170 \times 10^{-4}$ | $9.484\ 516\ 5 \times 10^{-4}$ | $\mathbf{1/(3.6 \times 10^{6})}$ $\mathbf{2.77\underline{7} \times 10^{-7}}$ | $3.723\ 562 \times 10^{-7}$ | $2.388\ 459 \times 10^{-4}$ | $2.390\ 057\ 4 \times 10^{-4}$ |
| 1 British thermal unit, Int. Tab. = | 1 055.056 | **1** | 1.000 669 | $2.930\ 711\ 1 \times 10^{-4}$ | $3.928\ 567 \times 10^{-4}$ | 0.251 995 8 | 0.252 164 4 |
| 1 British thermal unit (thermo) = | 1 054.35 | 0.999 331 | **1** | $2.928\ 745 \times 10^{-4}$ | $03.925\ 938 \times 10^{-4}$ | 0.251 827 2 | 0.251 995 7 |
| 1 kilowatthour = | **3 600 000** | 3 412.141 | 3 414.426 | **1** | **1/0.746** = 1.340 482 6 | 859.845 2 | 860.420 7 |
| 1 horsepower hour, electrical = | **2 685 600** | 2 545.457 | 2 547.162 | **0.746** | **1** | 641.444 5 | 641.873 8 |
| 1 kilocalorie, Int. Tab. = | **4 186.8** | 3.968 320 | 3.970 977 | **0.001 163** | $1.558\ 981 \times 10^{-3}$ | **1** | 1.000 669 |
| 1 kilocalorie, thermochemical = | **4 184** | 3.965 666 | 3.968 322 | 0.001 162 <u>2</u> | $1.557\ 938\ 6 \times 10^{-3}$ | 0.999 331 | **1** |

The exact conversion is 1 British thermal unit, International Table = **1 055.055 852 62** joules.

**TABLE 1-26** Power Conversion Factors

(Exact conversions are shown in **boldface** type. Repeating decimals are underlined.) The SI unit of power is the watt (W).

A. Power units decimally related to one watt

| | Watts (W) | Megawatts (MW) | Kilowatts (kW) | Milliwatts (mW) | Microwatts (μW) | Picowatts (pW) | Ergs per second (ergs/s) |
|---|---|---|---|---|---|---|---|
| 1 watt = | **1** | **0.000 001** | **0.001** | **1 000** | **1 000 000** | $\mathbf{10^{9}}$ | $\mathbf{10^{7}}$ |
| 1 megawatt = | **1 000 000** | **1** | **1 000** | $\mathbf{10^{9}}$ | $\mathbf{10^{12}}$ | $\mathbf{10^{15}}$ | $\mathbf{10^{13}}$ |
| 1 kilowatt = | **1 000** | **0.001** | **1** | **1 000 000** | $\mathbf{10^{9}}$ | $\mathbf{10^{12}}$ | $\mathbf{10^{10}}$ |
| 1 milliwatt = | **0.001** | $\mathbf{10^{-9}}$ | **0.000 001** | **1** | **1 000** | **1 000 000** | **10 000** |
| 1 microwatt = | **0.000 001** | $\mathbf{10^{-12}}$ | $\mathbf{10^{-9}}$ | **0.001** | **1** | **1 000** | **10** |
| 1 picowatt = | $\mathbf{10^{-9}}$ | $\mathbf{10^{-15}}$ | $\mathbf{10^{-12}}$ | **0.000 001** | **0.001** | **1** | **0.01** |
| 1 erg per second = | $\mathbf{10^{-7}}$ | $\mathbf{10^{-13}}$ | $\mathbf{10^{-10}}$ | **0.000 1** | **0.1** | **100** | **1** |

NOTE: 1 watt = 1 joule per second (J/s).

B. Nonmetric power units (with watt equivalents)

| | British thermal units (International Table) per hour (Btu/hr, IT) | British thermal units (thermochemical) per minute (Btu/min, thermo) | Avoirdupois foot-pounds-force per second (ft · $lb_f$/s avdp) | Kilocalories per minute (thermochemical) (kcal/min, thermo) | Kilocalories per second (International Table) (kcal/s, IT) | Horsepower (electrical) (hp, elec) | Horsepower (mechanical) (hp, mech) | Watts (W) |
|---|---|---|---|---|---|---|---|---|
| 1 British thermal unit (Int. Tab.) per hour = | **1** | 0.016 677 8 | 0.216 158 1 | $4.202\,740\,5 \times 10^{-3}$ | $6.999\,883\,1 \times 10^{-5}$ | $3.928\,567\,0 \times 10^{-4}$ | $3.930\,148\,0 \times 10^{-4}$ | 0.293 071 1 |
| 1 British thermal unit (thermo) per minute = | 59.959 853 | **1** | 12.960 810 | 0.251 995 7 | $4.197\,119\,5 \times 10^{-3}$ | 0.023 555 6 | 0.023 565 1 | 17.572 50 |
| 1 foot-pound-force per second = | 4.626 242 6 | 0.077 155 7 | **1** | 0.019 442 9 | $3.238\,315\,7 \times 10^{-4}$ | $1.817\,450\,4 \times 10^{-3}$ | **1/550** = $1.818\,18\underline{1\,8} \times 10^{-3}$ | 1.355 818 |
| 1 kilocalorie per minute (thermo) = | 237.939 98 | 3.968 321 7 | 51.432 665 | **1** | 0.016 655 5 | 0.093 476 3 | 0.093 513 9 | 69.733 333 |
| 1 kilocalorie per second (Int. Tab.) = | 14 285.953 | 238.258 64 | 3 088.025 1 | 60.040 153 | **1** | 5.612 332 4 | 5.614 591 1 | **4 186.800** |
| 1 horsepower (electrical) = | 2 545.457 4 | 42.452 696 | 550.221 34 | 10.697 898 | 0.178 179 0 | **1** | 1.000 402 4 | **746** |
| 1 horsepower (mechanical) = | 2 544.433 4 | 42.435 618 | **550** | 10.693 593 | 0.178 107 4 | 0.999 597 7 | **1** | 745.699 9 |
| 1 watt = | 3.412 141 3 | 0.056 907 1 | 0.737.562 1 | 0.014 340 3 | $2.388\,459\,0 \times 10^{-4}$ | **1/746** = $1.340\,482\,6 \times 10^{-3}$ | $1.341\,022\,0 \times 10^{-3}$ | **1** |

NOTE: The horsepower (mechanical) is defined as a power equal to **550** foot-pounds-force per second.
Other units of horsepower are:
1 horsepower (boiler) = 9 809.50 watts
1 horsepower (metric) = 735.499 watts
1 horsepower (water) = 746.043 watts
1 horsepower (U.K.) = 745.70 watts
1 ton (refrigeration) = 3 516.8 watts

**TABLE 1-27** Temperature Conversions
(Conversions in **boldface** type are exact. Continuing decimals are underlined.)

| Celsius (°C) °C = 5(°F–32)/9 | Fahrenheit (°F) °F = [9(C°)/5] + 32 | Absolute (K) K = °C + 273.15 |
|---|---|---|
| **–273.15** | **–459.67** | **0** |
| **–200** | **–328** | **73.15** |
| **–180** | **–292** | **93.15** |
| **–160** | **–256** | **113.15** |
| **–140** | **–220** | **133.15** |
| **–120** | **–184** | **153.15** |
| **–100** | **–148** | **173.15** |
| **–80** | **–112** | **193.15** |
| **–60** | **–76** | **213.15** |
| **–40** | **–40** | **233.15** |
| **–20** | **–4** | **253.15** |
| –17.77 | 0 | 255.372 |
| **0** | **32** | **273.15** |
| **5** | **41** | **278.15** |
| **10** | **50** | **283.15** |
| **15** | **59** | **288.15** |
| **20** | **68** | **293.15** |
| **25** | **77** | **298.15** |
| **30** | **86** | **303.15** |
| **35** | **95** | **308.15** |
| **40** | **104** | **313.15** |
| **45** | **113** | **318.15** |
| **50** | **122** | **323.15** |
| **55** | **131** | **328.15** |
| **60** | **140** | **333.15** |
| **65** | **149** | **338.15** |
| **70** | **158** | **343.15** |
| **75** | **167** | **348.15** |
| **80** | **176** | **353.15** |
| **85** | **185** | **358.15** |
| **90** | **194** | **363.15** |
| **95** | **203** | **368.15** |
| **100** | **212** | **373.15** |
| **105** | **221** | **378.15** |
| **110** | **230** | **383.15** |
| **115** | **239** | **378.15** |
| **120** | **248** | **393.15** |
| **140** | **284** | **413.15** |
| **160** | **320** | **433.15** |
| **180** | **356** | **453.15** |
| **200** | **392** | **473.15** |
| **250** | **482** | **523.15** |
| **300** | **572** | **573.15** |
| **350** | **662** | **623.15** |
| **400** | **752** | **673.15** |
| **450** | **842** | **723.15** |
| **500** | **932** | **773.15** |
| **1 000** | **1 832** | **1 273.15** |
| **5 000** | **9 032** | **5 273.15** |
| **10 000** | **18 032** | **10 273.15** |

NOTE: Temperature in kelvins equals temperature in degrees Rankine divided by 1.8 [K = °R/**1.8**].

**TABLE 1-28** Light Conversion Factors
(Exact conversions are shown in **boldface** type. Repeating decimals are underlined.)

A. Luminance units. The SI unit of luminance is the candela per square meter ($cd/m^2$).

| | Candelas per square meter ($cd/m^2$) | Candelas per square foot ($cd/ft^2$) | Candelas per square inch ($cd/in^2$) | Apostilbs (asb) | Stilbs (sb) | Lamberts (L) | Footlamberts (fL) |
|---|---|---|---|---|---|---|---|
| 1 candela per square meter = | **1** | **0.092 903 04** | **6.451 6** $\times$ **10**$^{-4}$ | $\pi$ = 3.141 592 65 | **0.000 1** | **(0.000 1)** $\pi$ = 3.141 592 65 $\times 10^{-4}$ | 0.291 863 51 |
| 1 candela per square foot = | 10.763 910 4 | **1** | **1/144** = 0.006 944 44 | 33.815 821 8 | 1.076 391 04 $\times 10^{-3}$ | 3.381 582 18 $\times 10^{-3}$ | $\pi$ = 3.141 592 65 |
| 1 candela per square inch = | 1 550.003 1 | **144** | **1** | 4 869.478 4 | 0.155 000 31 | 0.486 947 84 | 452.389 342 |
| 1 apostilb = | **1/$\pi$=** 0.318 309 89 | 0.029 571 96 | 2.053 608 06 $\times 10^{-4}$ | **1** | 3.183 098 86 $\times 10^{-5}$ | **0.000 1** | **0.092 903 04** |
| 1 stilb = | **10 000** | **929.030 4** | **6.451 6** | 31 415.926 5 | **1** | $\pi$ = 3.141 592 65 | 2 918.635 |
| 1 lambert = | **10 000/$\pi$=** 3 183.098 86 | 295.719 561 | 2.053 608 06 | **10 000** | **1/$\pi$=** 0.318 309 89 | **1** | **929.030 4** |
| 1 footlambert = | 3.426 259 1 | **1/$\pi$=** 0.318 309 89 | 2.210 485 32 $\times 10^{-3}$ | 10.763 910 4 | 3.426 259 1 $\times 10^{-4}$ | 1.076 391 03 $\times 10^{-3}$ | **1** |

NOTE: 1 nit (nt) = 1 candela per square meter ($cd/m^2$).
1 stilb (sb) = 1 candela per square centimeter ($cd/cm^2$).

B. Illuminance units. The SI unit of illuminance is the lux (lux).

| | Luxes (lx) | Phots (ph) | Footcandles (fc) | Lumens per square inch ($lm/in^2$) |
|---|---|---|---|---|
| 1 lux = | **1** | **0.000 1** | **0.092 903 04** | **6.451 6** $\times$ **10**$^{-4}$ |
| 1 phot = | **10 000** | **1** | **929.030 4** | **6.451 6** |
| 1 footcandle = | 10.763 910 4 | 1.076 391 04 $\times 10^{-3}$ | **1** | **1/144** = 0.006 944 44 |
| 1 lumen per square inch = | 1 550.003 1 | 0.155 000 31 | **144** | **1** |

NOTE: 1 lux (lux) = 1 lumen per square meter ($lm/m^2$).
1 phot (ph) = 1 lumen per square centimeter ($lm/cm^2$).
1 footcandle (fc) = 1 lumen per square foot ($lm/ft^2$).

*Use of Conversion Factors.* Conversion factors are multipliers used to convert a quantity expressed in a particular unit *(given unit)* to the same quantity expressed in another unit *(desired unit)*. To perform such conversions, the *given unit* is found at the left-hand edge of the conversion table, and the *desired unit* is found at the top of the same table. Suppose, for example, the quantity 1000 feet is to be converted to meters. The given unit, foot, is found in the left-hand edge of the third line of Table 1-15B. The desired unit, meter, is found at the top of the first column in that table. The conversion factor (**0.304 8**, exactly) is located to the right of the given unit and below the desired unit. The given quantity, 1000 feet, is multiplied by the conversion factor to obtain the equivalent length in meters, that is, 1000 feet is 1000 × 0.304 8 = 304.8 meters.

The general rule is: Find the given unit at the left side of the table in which it appears and the desired unit at the top of the same table; note the conversion factor to the right of the given unit and below the desired unit. Multiply the quantity expressed in the given unit by the conversion factor to find the quantity expressed in the desired unit.

Listings of conversion factors are often arranged as follows:

| *To convert from* | *To* | *Multiply by* |
|---|---|---|
| (Given unit) | (Desired unit) | (Conversion factor) |

The equivalences listed in the accompanying conversion tables can be cast in this form by placing the given unit (at the left of each table) under "To convert from," the desired units (at the top of the table) under "To," and the conversion factor, found to the right and below these units, under "Multiply by."

*Use of Two Tables to Find Conversion Factors.* When the given and desired units do not appear in the same table, the conversion factor between them is found in two steps. The *given unit* is selected at the left-hand edge of the table in which it appears, and an *intermediate conversion factor*, applicable to the SI unit shown at the top of the same table, is recorded. The *desired unit* is then found at the top of another table in which it appears, and another *intermediate conversion factor*, applicable to the SI unit at the left-hand edge of that table, is recorded. The conversion factor between the given and desired units is the product of these two intermediate conversion factors.

**TABLE 1-29** U.S. Electrical Units Used Prior to 1969, with SI Equivalents

| A. Legal units in the U.S. prior to January 1948 | |
|---|---|
| 1 ampere (US-INT) | = 0.999 843 ampere (SI) |
| 1 coulomb (US-INT) | = 0.999 843 coulomb (SI) |
| 1 farad (US-INT) | = 0.999 505 farad (SI) |
| 1 henry (US-INT) | = 1.000 495 henry (SI) |
| 1 joule (US-INT) | = 1.000 182 joule (SI) |
| 1 ohm (US-INT) | = 1.000 495 ohm (SI) |
| 1 volt (US-INT) | = 1.000 338 volt (SI) |
| 1 watt (US-INT) | = 1.000 182 watt (SI) |
| **B. Legal units in the U.S. from January 1948 to January 1969** | |
| 1 ampere (US-48) | = 1.000 008 ampere (SI) |
| 1 coulomb (US-48) | = 1.000 008 coulomb (SI) |
| 1 farad (US-48) | = 0.999 505 farad (SI) |
| 1 henry (US-48) | = 1.000 495 henry (SI) |
| 1 joule (US-48) | = 1.000 017 joule (SI) |
| 1 ohm (US-48) | = 1.000 495 ohm (SI) |
| 1 volt (US-48) | = 1.000 008 volt (SI) |
| 1 watt (US-48) | = 1.000 017 watt (SI) |

For example, it is required to convert 100 cubic feet to the equivalent quantity in cubic centimeters. The given quantity (cubic feet) is found in the fourth line at the left of Table 1-17B. Its intermediate conversion factor with respect to the SI unit is found below the cubic meters to be $2.831\ 684\ 66 \times 10^{-2}$. The desired quantity (cubic centimeters) is found at the top of the third column in Table 1-17A. Its intermediate conversion factor with respect to the SI unit, found under the cubic centimeters and to the right of the cubic meters, is 1 000 000. The conversion factor between cubic feet and cubic centimeters is the product of these two intermediate conversion factors, that is, 1 cubic foot is equal to $2.831\ 684\ 66 \times 10^{-2} \times 1\ 000\ 000 = 28\ 316.846\ 6$ cubic centimeters. The conversion from 100 cubic feet to cubic centimeters then yields $100 \times 28\ 316.846\ 6 = 2\ 831\ 684.66$ cubic centimeters.

*Conversion of Electrical Units.* Since the electrical units in current use are confined to the International System, conversions to or from non-SI units are fortunately not required in modern practice. Conversions to and from the older cgs units, when required, can be performed using the conversions shown in Table 1-9. Slight differences from the SI units occur in the electrical units legally recognized in the United States prior to 1969. These differences involve amounts smaller than that customarily significant in engineering; they are listed in Table 1-29.

## 1.17 BIBLIOGRAPHY

### 1.17.1 Standards

ANSI/IEEE Std 268; Metric Practice. New York, Institute of Electrical and Electronics Engineers.

Graphic Symbols for Electrical and Electronics Diagrams, IEEE Std 315-1975 (also published as ANSI Std Y32.2-1975). New York, Institute of Electrical and Electronics Engineers.

IEEE Standard Letter Symbols for Units of Measurement, ANSI/IEEE Std 260.1-2004. New York, Institute of Electrical and Electronics Engineers, ANSI Letter Symbols Units of Measurements (SI Units, Customary Inch-Pound Units, and Certain Other Units).

Letter Symbols for Quantities Used in Electrical Science and Electrical Engineering; ANSI Std Y10.5. Also published as IEEE Std 280; New York, Institute of Electrical and Electronics Engineers.

SI Units and Recommendations for the Use of Their Multiples and of Certain Other Units; International Standards ISO-1000 (E). Available in the United States from ANSI. New York, American National Standards Institute. Also identified as IEEE Std 322 and ANSI Z210.1.

### 1.17.2 Collections of Units and Conversion Factors

*Encyclopaedia Britannica* (see under "Weights and Measures"). Chicago, Encyclopaedia Britannica, Inc.

*McGraw-Hill Encyclopedia of Science and Technology* (see entries by name of quantity or unit and vol. 20 under "Scientific Notation"). New York, McGraw-Hill.

Mohr, Peter J. and Barry N. Taylor, *CODATA*: 2014; Recommended Values of the Fundamental Physical Constants; *Reviews of Modern Physics*, July-September 2016, vol. 88, pp. 1–73, http://www.physics.nist.gov/constants.

National Institute of Standards and Technology Units of Weight and Measure—International (Metric) and U.S. Customary; *NIST Misc. Publ.* 286. Washington, Government Printing Office.

The Introduction of the IAU System of Astronomical Constants into the Astronomical Ephemeris and into the American Ephemeris and Nautical Almanac (Supplement to the American Ephemeris 1968). Washington, United States Naval Observatory, 1966.

The Use of SI Units (The Metric System in the United Kingdom), PD 5686. London, British Standards Institution. See also British Std 350, Part 2, and PD 6203 Supplement 1.

*The World Book Encyclopedia* (see under "Weights and Measures"). Chicago, Field Enterprises Educational Corporation.

World Weights and Measures, Handbook for Statisticians, Statistical Papers, Series M, No. 21, Publication Sales No. 66, XVII, 3. New York, United Nations Publishing Service.

### 1.17.3 Books and Papers

Brownridge, D. R.: *Metric in Minutes.* Belmont, California, Professional Publications, Inc., 1994.

Cornelius, P., de Groot, W., and Vermeulen, R.: Quantity Equations, Rationalization and Change of Number of Fundamental Quantities (in three parts); *Appl. Sci. Res.*, 1965, vol. B12, pp. 1, 235, 248.

*IEEE Standard Dictionary of Electrical and Electronics Terms*, ANSI/IEEE Std 100-2000. New York, Institute of Electrical and Electronics Engineers, 2000.

Page, C. H.: Physical Entities and Mathematical Representation; *J. Res. Natl. Bur. Standards*, October–December 1961, vol. 65B, pp. 227–235.

Silsbee, F. B.: Systems of Electrical Units; *J. Res. Natl. Bur. Standards*, April–June 1962, vol. 66C, pp. 137–178.

Young, L.: *Systems of Units in Electricity and Magnetism.* Edinburgh, Oliver & Boyd Ltd., 1969.

# 2

# MEASUREMENT AND INSTRUMENTATION

Harold Kirkham

*Staff Scientist, Pacific Northwest National Laboratory, Richland, Washington*

## 2.1 INTRODUCTION

Since the last edition of this handbook was published, the field of measurement has gone through a veritable revolution, because of the advances in digital technology and the reduction in cost of digital hardware. The approach taken in this edition is to acknowledge the improved performance and ease of use that the digital instrument provides, and to the greatest extent possible to relegate analog measurements to a safe distance.

Metrologists anticipate further advances in measurement because the digital revolution continues, in its own quiet way. Modern digital instruments are capable of accomplishing more than their analog predecessors, and as they advance, the gap between old and new will widen.

But ultimately, measurement is an application-driven endeavor. The observer cannot be separated from the application. As instrumentation advances, it will be increasingly important for the observer to understand his or her role in the process, to a depth that has not been common before now. In fact, as measurements become seemingly easier to make, their meaning should be increasingly scrutinized.

One particular quantity provides a striking example. The quantity "reactive power" is very useful in power engineering. It is founded on the model of real power, which is defined for a single-phase circuit by the equation $P = VI\cos\Phi$, where $\Phi$ is the angle between the voltage and the current. Reactive power, which does no real work, is given by $P = VI\sin\Phi$, a version of the same equation with cosine replaced by sine. However, these equations are based on the assumption of a sine-wave model for all the quantities involved. In the real world, perfect sine-waves are rare. If the waves are distorted, the digital instrument has a multitude of ways to handle the harmonics. The user may have to make a choice! The topic of reactive power has been the subject of a sometimes bitter debate, yet resolution has seemingly evaded those involved. We will consider the matter in Sec. 2.7.

## 2.2 WHAT IS MEASUREMENT?

When we make a measurement, we are doing a special kind of data compression. We are saying, in effect, that everything we want to know about this signal, or that component, is contained in the result of the measurement we are going to make. Given a voltage, for example, we might want to know the amplitude, or the frequency. That means that the very notion of a measurement assumes that we know, even before we make the measurement, something about the thing we are going to assign a value to. We know, for example, whether the quantity before us on the bench is characterized by a voltage or a mass, and we use different instruments for these measurements.

If the quantity is electrical, we have more detailed expectations. In power engineering, we know the current is usually represented as a signal that is either alternating sinusoidally or is steady, for example. We have in our minds, in other words, a conceptual model of the thing to be measured, and we expect the thing being measured to have a magnitude that can be expressed in terms of that model.

The act of measuring is one that uses some aspect of the physical world to obtain values for that conceptual model. These days, we are comfortable that a model is a mathematical thing. Some examples are shown in Table 2-1.

It is a lot easier to acknowledge the existence of an equation defining what is being measured in the case of a digital instrument than an analog one, but it was true even in the analog instrument. The stationary indicating pointer indicated that two forces were in balance: that equality was the sign that the equation was solved.

For some measurements, nonelectrical experiments can be done to show the validity of a result. For example, power and energy are quantities whose electrical measurements are verifiable by calorimetry.

Much stronger assumptions are made about some things being characterized. In the case of a signal assumed to represent direct current, or a sinusoidal quantity called a phasor, for example, the form of the equation is fixed by the assumption that the signal indeed has the given character. It follows that if the signal does not have that character, the instrument may not give an accurate answer. Therefore, it is up to the user to be sure that the instrument settings are appropriate for the signal.

**TABLE 2-1** Examples of Measurement Equations

| Measurand | | |
|---|---|---|
| Name | Equation | Comment |
| Direct current | $I_c = \langle i(t) \rangle$ | Result is an average over some time |
| Phasor | $x(t) = X_m \cos\{\omega t + \varphi\}$ | Result is values for $X_m$, $\omega$ and $\varphi$, for all time |
| Changing phasor | $x(t) = X_m \cos\left\{\left(\omega' + \frac{C'_\varphi}{2}\right)t + \varphi' + \frac{C'_\omega}{2}t^2\right\} \Bigg\vert -\frac{w}{2} < t < \frac{w}{2}$ | Result is a value for $X_m$, $\omega_{app}$, $\varphi$ and $C'_\omega$, where $\omega_{app}$ is the apparent frequency $\omega' + \frac{C'_\varphi}{2}$, which includes the effect of changing phase due to frequency difference. Values apply only during observation window |
| Resistance | $R = \langle v(t) \rangle / \langle i(t) \rangle$ | Assumes dc, averaged over some time |
| rms voltage | $v_{rms} = \sqrt{\frac{1}{T}\int_0^T [v(t)]^2\, dt}$ | Assumes a periodic function with known period $T$. Connects AC and DC measurements |
| Power | $P = \frac{1}{kT}\int_\tau^{\tau_1 + kT} p\, dt$ | Assumes instantaneous power $p$ is periodic with known period $T$ |
| Energy | $E = \int_{t_1}^{t_2} P\, dt$ | No assumption of periodicity. The basis of the energy (billing) meter |
| Power factor | Various | The familiar equations that define power factor and reactive power, PF = $VI \cos \Phi$ and $S^2 = P^2 + Q^2$ apply only in the absence of harmonics. There is no agreement on the form otherwise |
| Reactive power | Various | |

The same is true for a changing phasor. Mathematically, the domain of the sine and cosine is from −1 to +1 in amplitude and $-\infty$ to $+\infty$ in time. Within this domain, the amplitude, frequency, and phase do not change. If they are changing, the equation can be modified and the results of the measurement applied only to the time window of observation used for the measurement. (In the example given, the frequency and phase are shown as changing during a window of width $w$.)

A seemingly arbitrary set of measurements contains some things that are familiar to power engineers: power factor, and reactive power. The problem with these quantities is that there are several ways to define the quantity when it contains harmonics, and they do not give the same result. A study by NIST, for example, identified 10 different ways power factor was being calculated in commercial instruments (Nelson, 2011). However, there is no experiment that can be done to declare any one definition right and the others wrong. The observer is free to use whichever definition best suits the needs of the moment.

The output of the instrument is called the result of a measurement, or usually just a result. A metrologist might refer to the number as the declared value. It is the value of a parameter that describes some aspect of the model, such as a power level or a voltage amplitude or its frequency. In other words, the result tells us something about the model, based on the real world. For example, if the quantity being measured is the voltage on a wall outlet, the result of a measurement might be a statement such as "the ac voltage of this signal is 121 V," perhaps with the unspoken addition "assuming it is a sine wave."

It happens that it is not possible to know the exact value of a physical quantity being measured. But we can learn something of it via observation (that is, measurement). A combination of adverse factors (discussed below) adds up to a situation where we cannot say the results are exact. The best we can do in measuring is to put bounds on the specification of the result. The factors include inadequacies of the model, noise on the signal, and artifacts of the A/D converter (or the bearings of a moving coil instrument). That 121-V signal should be described by not one but three numbers, as 121 V ±1 V with a confidence of 95%. In other words, if the measurement were repeated many times (and if the signal did not change), we would expect 95% of the results to be within 1 V of 121 V.

### 2.2.1 Uncertainty

That ±1-V figure is called the *uncertainty*. The statement of the uncertainty is considered to be an essential part of the statement of the result of a measurement. Usually, the statement is made with a confidence level of 95%, which is almost the same as saying $2\sigma$.[a] If the confidence number is omitted, it is fair to assume it is 95%. But it should be borne in mind that there is always one chance in 20 that the reading is wrong by a bigger number than the stated uncertainty.

Why is there uncertainty? First, though it may be counterintuitive: for most measurements the "true" value of something being measured is a useful and at the same time misleading concept. True value brings to mind something that is inherent in the thing being measured, and has no connection with the observer. The result of a measurement is in contrast to that: it is a value ascribed by an instrument to a parameter of a model that has been selected by the observer. Some aspect of the physical world is used to adjust this parameter, but the number that comes out is not anything fundamental to the thing being measured.

Measurements yield only information about the parameters of a model. The results can be used to make inferences about the physical world, and that is all. But the concept of true value is well-entrenched in the mind, and we acknowledge that it can be useful.

If a measurement is repeated, and the results show multiple different values, the treatment of the differences is usually based on the notion that these results are scattered around the true value. In fact, the causes of the scatter, while they are not always known, are divided by metrologists into two categories, one of which can be thought of as "averaging to zero." These are called type A uncertainties. Type B uncertainties do not have that feature: no matter how many times a measurement is made, and no matter what kind of statistical processing is done, type B uncertainties do not get smaller.

In general, the uncertainties can be combined in certain (rigorous) ways so that statement of the result of repeated measurements should be of the form given above:

$$Y = y \pm u \text{ with a confidence level of } z\%$$

Here $Y$ is the result, $y$ is the measured (and processed) value, and $u$ is the uncertainty. The confidence level $z$ is something that is part of the rigorous processing of the measurements. What the statement does is it establishes a range of possible values, usually assumed to have a Gaussian distribution.

The new and rigorous approach to uncertainty can be dated to recommendation INC-1 of the *Bureau International des Poids et Mesures* (BIPM) in 1980 which stated, among other things, that

> The uncertainty in the result of a measurement generally consists of several components which may be grouped into two categories according to the way in which their numerical value is estimated.
>
> Type A. Those which are evaluated by statistical methods
>
> Type B. Those which are evaluated by other means
>
> There is not always a simple correspondence between the classification into categories A or B and the previously used classification into "random" and "systematic" uncertainties. The term "systematic uncertainty" can be misleading and should be avoided.

The work was converted into a Standard with the publication (in 1993)[b] of the *ISO Guide to Expression of Uncertainty in Measurement*, often referred to as GUM.

Note that the effect of type A uncertainties can be reduced by repeated measurement and appropriate data processing. The effect of type B uncertainties is not reduced by repeated data taking. With care, the effect of type A uncertainties can be made small compared to that of type B uncertainties. There is clearly a limit to the benefit of multiple readings. It is often (but not always)

[a]A level of 99% confidence corresponds to $2.58\sigma$.

[b]Some corrections were made, and the document was reissued in 1998. The 1998 version is the current version.

true that most measurement systems are dominated by type B uncertainties. (That, too, may seem counterintuitive.)

Let us focus for a moment on type B uncertainties. Remember, these are where the evaluation of the uncertainty is not based on repeated measurements and applying statistics. The uncertainties are evaluated using engineering judgment and looking at all the relevant information available, which may include

- Previous measurement data
- Experience with, or knowledge of, the behavior and property of relevant instruments
- Specifications of measuring equipment
- Data on uncertainties of standards
- Calibration reports
- Uncertainties assigned to reference data taken from handbooks

It is therefore wise to think about all these things when expressing the result of a measurement.

A third type of uncertainty is called *definitional uncertainty*. In theory, a perfect definition of the thing to be measured requires an infinitely large amount of information. All practical definitions are therefore less than perfect. Minimizing the definitional uncertainty is of importance in the metrology laboratory, where considerable effort is put into making sure it is as small as it can be made. It is safe enough to leave that matter to the metrology laboratory, where it affects their statement of uncertainties on the things being calibrated.

For the practical engineer or technician making a measurement, definitional uncertainty comes down to making sure that the instrument in use is designed for the job at hand, and that its settings are appropriate. If the instrument is given a highly distorted waveform, but was calibrated on the assumption of a sine wave, the reading may be quite unexpected, and quite meaningless to the needs of the observer. It is simple to connect the instrument and take the reading, but it is crucial to be sure you are getting what you want from it.

In summary, measurement is a process that bridges the physical world and the conceptual, and it cannot do so with perfection. For the observer, measurement is a way to get the real world to "hold still" for a while, so the observer can know something about it. Few things are truly constant over time, and a measurement provides a way for someone to keep track of the changes. The height of a growing child, for example, may be recorded by dated marks on a wall. In general, once a measurement is made, the result of that measurement is a fixed entity. It may be written down, scratched on a wall, or recorded in a database. But it does not change after the measurement is made.

The thing that was measured, of course, is not subject to that constraint. Since a guide such as this is aimed at *applying* the science of measurement, we will remember that as we proceed from this fundamental introduction. Mostly, we expect the thing we measure to change practically as soon as we finish measuring it.

### 2.2.2 Model of the Process of Measurement

We have already noted that the instrument we use in making a measurement depends on what we are measuring. For the purposes of electrical measurements, this section mostly assumes that the instrument is digital. That makes it more straightforward to understand the processes taking place. The reader interested in finding out more about analog instrumentation will have no difficulty finding information: earlier versions of this handbook would provide a starting place.

For the purposes of characterizing electrical quantities, measurement is a three-step process:

1. Scaling and preparation of a signal
2. Obtaining of samples of that signal, expressed as multiples of a voltage reference
3. Digital processing and conversion to conventional units

***Scaling and Preparation.*** First, the quantity to be measured is selected and (external to the instrument) stepped up or down to a level suitable for the instrument to be used. In the world of electrical power, the voltage may be high—possibly very high—and a voltage signal must be scaled down greatly. The current may also be high, and a signal representing it has both to be scaled down and isolated from the high voltage.

The resulting signal becomes the thing known as the *realized quantity*, the quantity made real for the purposes of measurement. It may represent the voltage on a bus, or the current through a breaker; either way, it is a scaled version of some other quantity, and the scale factor is known (and assumed constant within some specified limit).

***Sampling.*** Second, the signal is sampled and converted by an A/D converter to a set of values known accurately in terms of a voltage reference in the A/D step. The comparison stage, in a digital instrument, is thus a comparison with a voltage reference.[c] In a modern digital instrument, this step occurs as soon as possible inside the instrument. There is a minimum amount of analog processing.

***Digital Processing.*** Third, an algorithm of some sort, based on a mathematical model selected by the observer and built into the instrument, is applied to the signal samples. The processing may include filtering, rectifying, phase detecting, rms calculations, or Fourier transforms. The result of the calculations, the *declared value* of the instrument, is converted to the appropriate units and made available to the observer.

As Sec. 1 has demonstrated, for electrical quantities the values of the units are well known and defined by physical values that are thought to have been constant since the beginning of time: the quantum physical standards.[d] In a digital instrument, the signal is digitized by A/D converters that compare the voltage being observed with a reference whose stability is well established, such as a band gap.[e] The output of the A/D converter is thus expressed in terms of a very stable reference.

***Generic Measurement System.*** Altogether then, a generic measurement system consists of a transducer that converts the input quantity to a voltage, an A/D converter that samples the input signal and produces values in terms of an accurately known reference, and an algorithm that calculates the desired quantity in terms of the proper units. Figure 2-1 shows the elements of a generic measurement system.

Figure 2-1 shows the "noisy" physical domain on the left, and an A/D converter, triggered by a time reference, putting the signal from the continuous-time domain into the discrete-time domain. In the discrete-time domain the effect of noise is minimized. In any measurement system, it is noise that establishes the lower limit of the size of signal that can be measured, and therefore the dynamic range of the measurement. The limits on the conversion process are the jitter on the time reference, and noise (and drift) on the analog voltage reference. These matters are well understood: a basic reference is Horowitz and Hill (1989). It is also true that in the digital domain one may regard the signal as "pure" information, something in the conceptual domain. It is in this domain that the syntax of mathematics rules. (The amount of mathematics that you can do in the world of wires and transformer is not zero, but it is rather limited. Nevertheless, many very useful instruments take advantage of some "physical mathematics.")

In what follows, we will see how these fundamental ideas about measurement are applied to make instruments to measure the various electrical parameters. We will look at the instruments themselves, and the systems they are part of. Many instrumentation systems have external networks

[c]In the days of analog instruments, the second and third step had a fuzzy boundary. The comparison was implemented as a balance of forces, and the measurement result obtained only when the needle stopped moving. The mathematical model was built into that balance along with the comparison. The conversion to conventional units was then done by suitably scaling the position of the pointer.

[d]The unit of mass is still (in 2017) an exception to this statement. But the kilogram will become a quantum standard within a few years.

[e]The reader may sometimes see definitions of "measurement" that are based on comparing the thing being measured with a reference of the same kind of thing. A weight balance is of this kind. The equivalent in electrical measurements is the bridge circuit, in which physical quantities are adjusted to balance one another. That is an example of a network that can be regarded in some way as external to the measuring system. Many elegant circuits have been devised so that the measuring system need indicate only the balance condition. In this section, we regard such things as examples of the first step of the measurement process, scaling, and preparation.

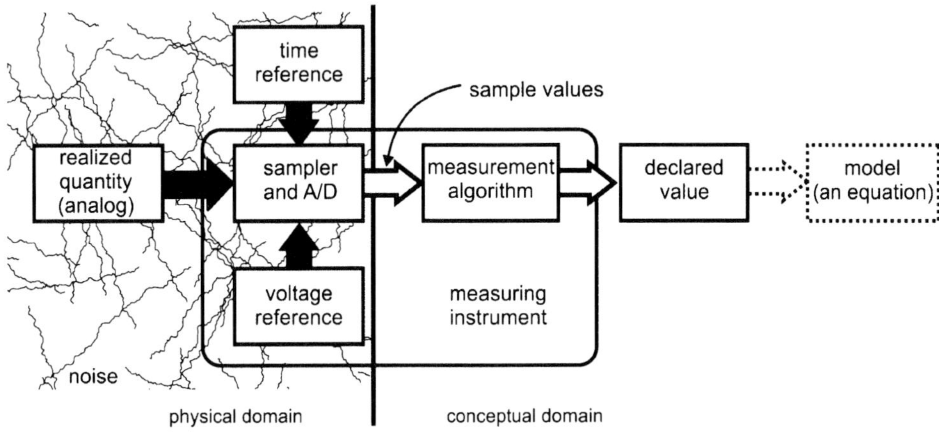


**FIGURE 2-1** Major elements of a measurement system.

serving a number of purposes. They may step the incoming signal down in terms of its energy level, they may isolate the signal form a high voltage, they may be arranged so that the measuring instrument is comparing two similar quantities, as in a bridge circuit, or they may be serving to reduce the effects of such things as wiring capacitance. But first, we will look at the calibration of instruments.

## 2.3 CALIBRATION

Calibration is a process that enables the user to have confidence in the accuracy of the result of the measurement. It is part of a process that allows measurements to be traced to standards, ultimately to the standards for the units discussed in Sec. 1. Since all physical components are subject to change with time, it is important that an instrument be calibrated before it is used. Just how long a time can elapse between calibration and use depends on the expected rates at which crucial elements of the instrument change. There is a time known as MTBOOT (mean time between out of tolerance) that sets the calibration interval. One of the most important things the user of an instrument can do is to use only an instrument with an up-to-date calibration. It seems self-evident that an instrument that is not within its calibration interval might give erroneous results.

The calibration process consists of two parts:

1. Comparison
2. Adjustment

First, a comparison is made between what might be called a high-quality item such as a reference standard, and the unit under test. The outcome of this step is that a difference is measured. This difference is called an error. In other words, it is the difference between two numbers that are known, unlike an uncertainty, which cannot be estimated so simply.

Second, taking account of the relative uncertainties, a correction factor (or a series of such factors) is calculated that should be applied to the unit under test during a regular measurement to yield a corrected result. In an analog instrument, it might be a potentiometer used to trim a value in a gain- or phase-controlling circuit. In a digital instrument, it might be a table of correction factors applicable under various circumstances. The correction will have the effect of reducing the measurement error, making the system being calibrated as accurate as it may be.

There are several ways to do the comparison of the first step. Which one is used is determined by what is being calibrated.

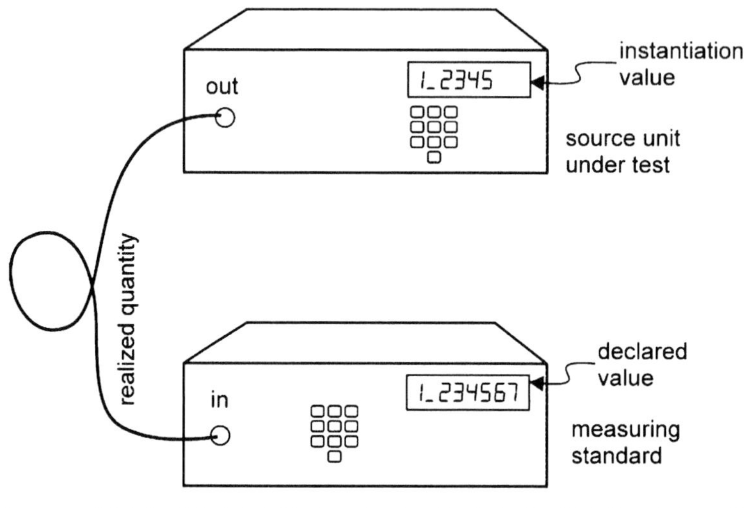


**FIGURE 2-2** Calibrating a source.

### 2.3.1 Reference Measuring System

If the unit being calibrated is a signal generator, or a voltage source, it can be calibrated by connecting it to a high-quality measuring instrument, as shown in Fig. 2-2.[f] In the figures, the "instantiation value" is the value set by the metrologist, here on a controlled source that is being used for comparison. While the word instantiation came from the world of computers and is not common in metrology, it is preferred here as a way to identify a value that is known and controlled, but not constant from one test to another. That use allows "reference" to be used to describe a value thought to be constant for the life of the equipment, such as the voltage on a band-gap device.

### 2.3.2 Reference Source

If the thing being calibrated is itself a measuring instrument, a high-quality source is needed for calibration, as in Fig. 2-3. Again, the metrologist is responsible for setting the instantiation value.

### 2.3.3 No Stable Source Available

It may be that there is no sufficiently stable source to allow the calibration of a measuring device that is expected to have low uncertainty. In this case, use is made of a source of more modest quality that is external to both the calibrating artifact and the unit under test, as in Fig. 2-4. The instantiation value is no longer something that is precisely known. The error is therefore the difference between the declared value of measuring standard and the unit under test.

In Fig. 2-4, the source is depicted as an ancient signal generator. It looks as if it has vacuum tubes inside that would cause the frequency drift as they warm up! The idea here is that the meter being calibrated and the measuring standard should agree, even if the stimulus value changes as the calibration proceeds. The notion of the drifting signal generator is an exaggeration, of course.

[f] In the figures that follow, we indicate the instrument with the higher accuracy by an increased display resolution.

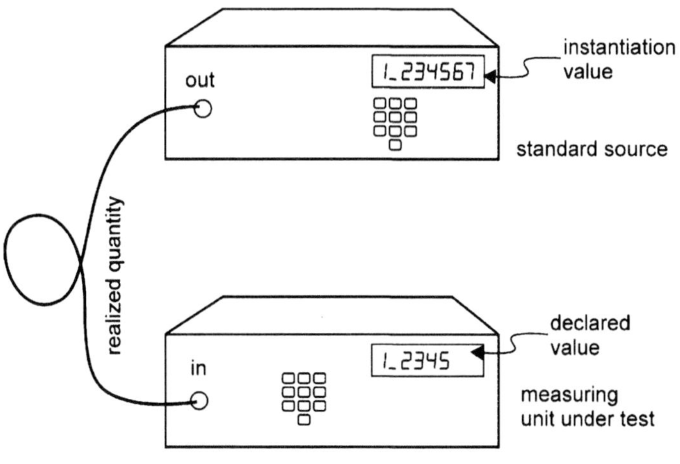


**FIGURE 2-3** Calibrating a measuring device.

This is a very common way to calibrate an instrument, even if the source is good. Current transformers are usually calibrated this way: it is thus not necessary to control the large primary current with great accuracy.

### 2.3.4 External Networks in Calibration

We have seen that in power system measurements, part of the instrumentation scheme is often external to the instrument itself. The various instrument transformers are examples of this. Systems are also used in calibration which may be regarded as an adaptation of the method of Fig. 2-4, in that the effect of the external network is to produce a zero to be measured. Various signals, and the components themselves, are used to do some arithmetic to change the quantities submitted to the measuring instrument. It is usually not very complicated mathematics that is implemented, but much ingenuity

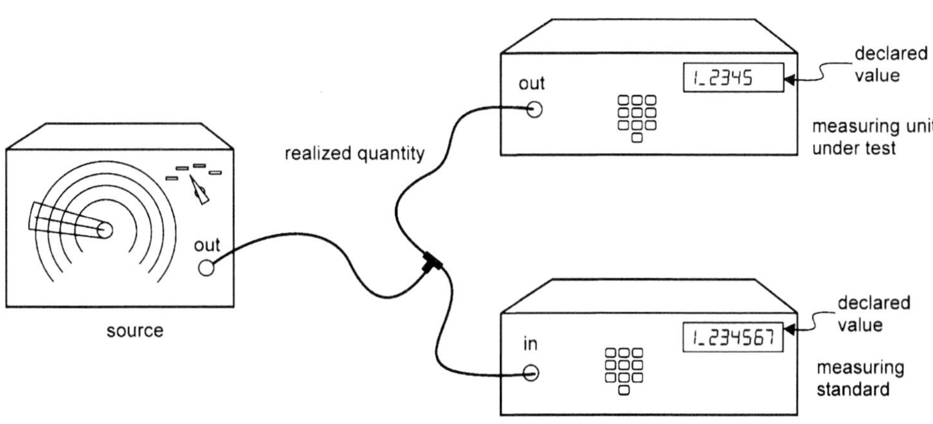


**FIGURE 2-4** Calibration using (relatively) low-quality source.

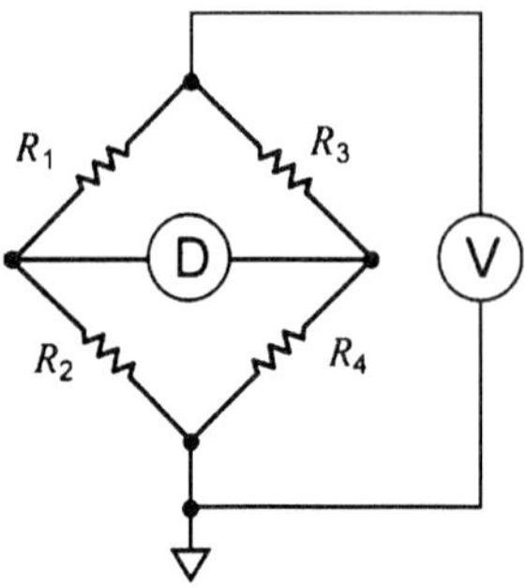


**FIGURE 2-5** Basic arrangement known as Wheatstone bridge.

has been shown over the years, and many of the methods are quite elegant. In some, two signals are made equal by means of a network of some kind: adjustments are made *to some element of the network* to make their difference zero. There is no declared value comparison to produce an error number per se, because the values of the adjusted elements give the measurements result. Automated systems are capable of performing the required calculations, and they do display the end result.

Such methods have their origin in the metrology laboratory, and these days that is mostly where they would be encountered. It is worth examining the fundamentals, even if the practitioner in the field will encounter such things only occasionally. Much progress has been made because of these networks.

***Wheatstone Bridge.*** The most familiar version of such a calibration is probably the Wheatstone bridge. The basic arrangement is shown in Fig. 2-5.

In this configuration, no matter the voltage $V$, the fraction of that voltage that appears at the junction between $R_1$ and $R_2$ is given by $R_2/(R_1 + R_2)$. Similarly, the voltage that appears at the junction of $R_3$ and $R_4$ is given, as a fraction of the total voltage, by $R_4/(R_3 + R_4)$. When there is no difference between these voltages, either the voltage $V$ has become zero or the bridge (said to be "balanced") is in a state in which

$$\frac{R_2}{R_1 + R_2} = \frac{R_4}{R_3 + R_4}$$

whence

$$R_2(R_3 + R_4) = R_4(R_1 + R_2)$$

$$R_2R_3 + R_2R_4 = R_4R_1 + R_4R_2$$

$$\frac{R_2R_3}{R_4} + R_2 = R_1 + R_2$$

and so finally

$$\frac{R_2R_3}{R_4} = R_1$$

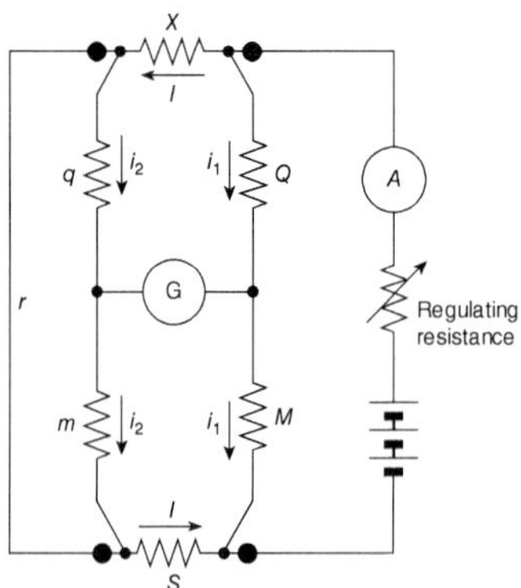


**FIGURE 2-6** Kelvin double bridge.

***Kelvin Double Bridge.*** The Kelvin double bridge, shown in Fig. 2-6, is an adaptation of the Wheatstone bridge specifically aimed at low resistance measurements. The problem with making measurements of low-value resistances is that the resistance of the contacts may be so large as to produce a large error in the result. The bridge is most easily understood when examined in form shown in Fig. 2-6.

In the figure, the circuit that carries most of the current is shown as the heavy lines. The large black dots represent the current-carrying connections to the resistors $X$ and $S$. The small black dots represent the potential connections to these resistors. Here $X$ is the unknown and $S$ is a standard resistance. They are connected by a wire that may have a small resistance $r$.

The resistances $Q$, $q$, $M$, and $m$ are physical things, but their value can include contributions due to contact resistance and wire resistance. In use, $Q/M$ is made equal to $q/m$, without taking these small error sources into account (as they cannot easily be measured).

If the low-value elements $m$, $q$, and $r$ are zero, it is evident that the bridge will balance when ($X/S = Q/M$), from which $X$ can be calculated. But there is more to it than that.

The advantages of the double bridge are first that (as in any bridge) it is not necessary to know the current at which the measurement is being made, at least not particularly accurately. Since the current may be large, however, in view of the low values of resistance to be measured, it is customary to provide a means of regulating it, as shown in the figure. Second, it is not necessary to know the errors due to contact resistance in order to eradicate their effect.

As far as the first advantage is concerned, the same idea applies to many measurements. For example, in the calibration of a current transformer, the same current is passed through the reference CT and the unknown CT, and it is not necessary to know the current precisely in order to calibrate the unknown CT.

As for the second advantage, overcoming load resistance and contact resistance, note that at balance the current in the detector $G$ is zero. Therefore, the current in $q$ is the same as the current in $m$, and therefore the current $I$ is the same in the unknown and the standard, regardless of contact resistance and regardless of the resistance $r$ of the wire connecting the two.

Again because the detector has no voltage across it, it is clear that

$$\text{Voltage drop across } Q = \text{voltage drop across } X + \text{voltage drop across } q$$

That is

$$I_1 Q = IX + i_2 q$$

and in the same way

$$i_1 M = IS + i_2 m$$

The resistances $q$ and $m$ are in series across the wire resistance $r$, so that the current $I$ divides so that the current $i_2$ is given by

$$I_2 Q = \frac{r}{r+q+m} I$$

Substituting this value of $i_2$ in the earlier equations gives

$$I_1 Q = IX + \frac{rq}{r+q+m} I$$

And

$$I_1 M = IS + \frac{rm}{r+q+m} I$$

The actual value of $i_1$ can be cancelled by division:

$$\frac{Q}{M} = \frac{X + \dfrac{rq}{r+q+m}}{S + \dfrac{rm}{r+q+m}}$$

from which

$$MX = QS + \frac{Qmr}{r+q+m} - \frac{Mqr}{r+q+m}$$

and so

$$X=\frac{QS}{M}+\frac{r}{r+q+m}\left(\frac{Qm}{M}-q\right)$$

or

$$X=\frac{QS}{M}+\frac{mr}{r+q+m}\left(\frac{Q}{M}-\frac{q}{m}\right)$$

Now it becomes clear that the second term can be made very small by making the resistance of the wire $r$ as small as possible, and by making the ratio $Q/M$ as nearly equal to $q/m$ as possible. If the term is made vanishingly small, we end up with the usual bridge equation

$$X=\frac{QS}{M}$$

However, the story does not end here. Theoretically, the process of balancing the bridge would involve changing the values of two ratios: $Q/M$ and $q/m$. In practice, the resistors that give these ratios are usually not changed.

What is done instead is this. The values are chosen to give an approximate balance—it is assumed that a fairly good idea of the value of the unknown is available. Then, a variable resistance $x$ is placed across $X$ to produce a modified value $X'$. At balance,

$$X'=\frac{QS}{M}$$

and we know from the circuit that

$$\frac{1}{X'}=\frac{1}{X}+\frac{1}{x}$$

so that the value of $X$ can be obtained easily enough. It is also perfectly possible to shunt the standard resistor $S$, in the event the unknown $X$ is smaller than the standard.

There are a couple of advantages to this approach. First, the value of only one resistance $x$ needs to be adjusted to give balance, and second, the variable resistance is a nonsmall value component whose resistance need not be known with any great precision.

Because errors due to contact and lead resistances are eliminated, the Kelvin double bridge is one of the best methods for the precise measurement of low resistances. Commercial versions were produced for many years.

Other methods based on external networks of this general type are the current comparator and the transformer ratio arm bridge. These are devices that are essentially passive, and the comparison is done in the magnetic field of a shielded transformer core. With care, uncertainties of a few parts per million are achievable. These (and other bridge-type networks) are generally analyzed on the assumption that the components are linear and the waveforms sinusoidal. At balance, a transformer in such a bridge configuration has zero flux in the core, so it is operating at its most linear point. Examples of its use are given in this chapter in connection with calibrating current and voltage transformers.

The point of all this is to note that there is combining of information in the external network that achieves some specific objective, and results in the need to know only a zero of voltage.

The method is remembered today in what is known as a four-terminal resistance.[g] Such a thing is quite commonly used even outside of standards laboratories. Current is injected by one pair of terminals, and the voltage measured by another pair. The resistance of the source leads is relatively unimportant, provided the value is low enough to permit the source to drive the current through the resistance being measured. See Fig. 2-7.

The resistance of the sense leads is also relatively unimportant, as the current in them is very small (assuming the voltmeter has high resistance). It is possible with this kind of measurement that the

[g]This technique also goes by the name of the Kelvin connection method.

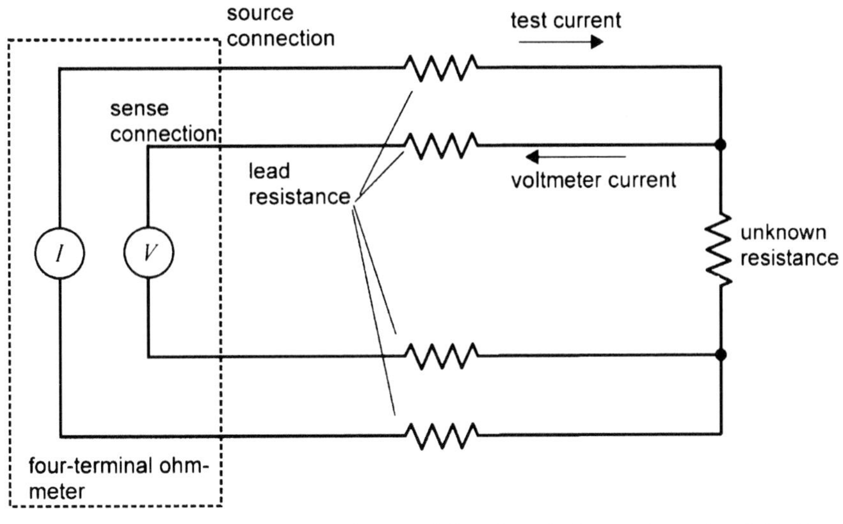


**FIGURE 2-7** Four-terminal resistance measurement.

biggest source of error is the thermoelectric effect, a small voltage that develops across the junction of two dissimilar metals. (The effect is the basis of the thermocouple.) If the various connections cannot be maintained at the same temperature, the effect can be calculated and reduced by reversing polarities.

It should be pointed out that there is an assumption made that the resistance is constant during the measurement and that the resistance is linear, and not a function of the instantaneous value of the applied voltage. Those assumptions mean that the usual expression for the resistance in terms of the instantaneous applied voltage and current $R = \langle v(t)\rangle / \langle i(t)\rangle$ does not depend on applying any particular value of voltage.

Figure 2-8 shows the appearance of two representative small implementations that might be used in an electronic system for current measurement. In the metrology laboratory, the four-terminal resistor is likely the size of a coffee cup, kept (and used) immersed in a bath of oil.

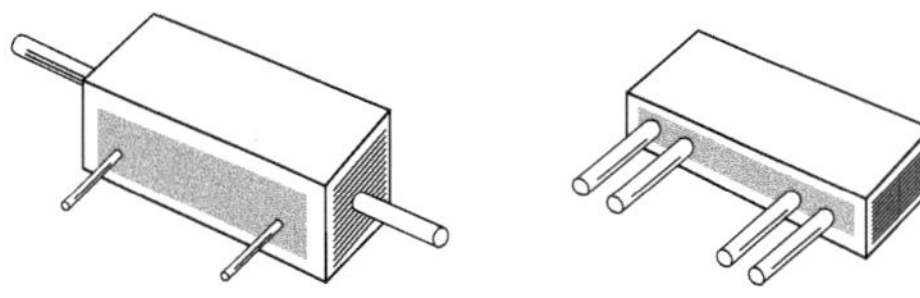

**FIGURE 2-8** Small four-terminal resistors.

### 2.3.5 Propagating Uncertainties

Usually, more than one device or element is involved in a measurement, and each contributes its own uncertainty to the uncertainty of the overall result. Consider the situation where some parameter $P$ is measured. Its value is a function of several things that are involved in the measurement. We can write the collection in a general way as

$$P = f(E_1 + E_2 + \dots E_n)$$

The combination might be a voltage from a voltage transformer with known uncertainty $E_1$ is applied to a voltage divider with known uncertainty $E_2$ and, in the end, an instrument with uncertainty $E_n$. If the first term of the Taylor series expansion is taken we get what is called a first-order approximation for the change $\delta P$ in the parameter $P$ that result from changes $\Delta E_i$ in the uncertainties $E_i$:

$$\delta P = \left(\frac{\partial P}{\partial E_1}\right)\Delta E_1 + \left(\frac{\partial P}{\partial E_2}\right)\Delta E_2 + \dots \left(\frac{\partial P}{\partial E_n}\right)\Delta E_n$$

The partial derivative $\left(\frac{\partial P}{\partial E_1}\right)\Delta E_1$ are evaluated at the expected operating point.[h] If the uncertainties are assumed to be uncorrelated, this expression yields the familiar square root of the sum of the squares:

$$\delta P = \sqrt{\left(\frac{\partial P}{\partial E_1}\Delta E_1\right)^2 + \left(\frac{\partial P}{\partial E_2}\Delta E_2\right)^2 + \ldots \left(\frac{\partial P}{\partial E_n}\Delta E_n\right)^2}$$

We can use the Wheatstone bridge to illustrate this. Three of the resistors of the Wheatstone bridge are known, and the unknown can be found in terms of their values. That means that three of the resistors control the overall uncertainty of the measurement. It can be seen easily enough that the combined uncertainty of $R_1$ is simply the combination of the uncertainties of the other three resistors. On the assumption that these uncertainties are uncorrelated, they would be added geometrically, as

$$U_{R_1} = \sqrt{U_{R_2}^2 + U_{R_3}^2 + U_{R_4}^2}$$

where $U$ is the uncertainty of the component indicated by the subscript. Such an expression is a simple example of propagating uncertainties, but it should be remembered that it assumes that the uncertainties are uncorrelated. It would not apply to a set of resistors that were all temperature-dependent, for example.

The voltage applied does not come into the equation. Further, the voltage across the detector is required only to be adjusted to zero, so the uncertainty of the voltmeter is required to be low only at the zero point. The voltmeter would ideally have a very high resistance, but at balance there is no voltage across its resistance, so no current flows even if the resistance is not very high. The balance condition is not altered, only the sensitivity as balance is approached and at balance.

### 2.3.6 Test Uncertainty Ratio

In all the calibration methods we have looked at so far, the error is found by a comparison of two numbers. The numbers are known: they are the declared value of two instruments. It is assumed that the uncertainties of the instruments have also been estimated.

For one instrument to be used to calibrate another, it is normally required that the uncertainties of the reference be lower than the thing being calibrated. To overcome the problem of assigning "blame" for the uncertainty, it used to be a rule of thumb that the reference should have 10 times lower uncertainty than the thing being tested. The ratio is called the test uncertainty ratio (TUR). Though some metrologists do adopt that rule, it is generally considered rather conservative. If we assume that the uncertainties are not correlated, most laboratories require a TUR of 4:1. This is in keeping with NSI/NCSL document Z540-1-1994 which states: "The laboratory shall ensure that calibration uncertainties are sufficiently small so that the adequacy of the measurement is not affected." It also states "Collective uncertainty of the measurement standards shall not exceed 25% of the acceptable tolerance (e.g., manufacturer specifications)."

There are no guarantees, but if this rule is adopted, the result of the calibration is statistically likely to result in a calibration that ensures the device being calibrated is performing within its stated uncertainty.

[h]It is assumed that the partial derivatives exist at this point.

## 2.4 INSTRUMENTATION

The instrumentation system that we will consider here is a digital one. The capability of digital processing greatly exceeds the earlier analog systems. Such systems can make measurements with greater speed and accuracy, make measurements that could not be implemented in analog form, and often have the ability for some degree of self-checking.

For most such instruments in power system use, there is an analog front end, however, that scales and isolates the high-energy signals involved. We described this as scaling and preparation earlier. We will consider this topic next.

### 2.4.1 Instrument Transformers

The material that follows is a brief summary of information on conventional instrument transformers. For more extensive information, consult IEEE Standard C57.13, Requirement for Instrument Transformers; American National Standards Institute; American National Standard C12, Code for Electricity Metering; *Electrical Meterman's Handbook*, Edison Electric Institute; manufacturer's literature; and textbooks on electrical measurements.

*AC range extension* beyond the reasonable capability of indicating instruments is accomplished with instrument transformers, since the use of heavy-current shunts and high-voltage multipliers would be dangerous to the operator and prohibitive both in cost and power consumption. Instrument transformers are also used to isolate instruments from power lines and to permit instrument circuits to be grounded.

Instrument transformers are broadly classified in two general types: (1) dry type, having molded insulation (sometimes only varnish-impregnated paper or cloth) usually intended for indoor installation, although large numbers of modern transformers have molded insulation suitable for outdoor operation on circuits up to 15 kV to ground; and (2) liquid-filled types in steel tanks with high-voltage primary terminals, intended for installation on circuits above 15 kV. They are further classified according to accuracy: (1) metering transformers having highest accuracy, usually at relatively low burdens; and (2) relaying and control transformers which in general have higher burden capacity and lower accuracy but larger dynamic range. Instrument transformers are routinely available for rated primary voltages up to 765 kV.

The current circuits of instruments and meters normally have very low impedance, and current transformers must be designed for operation into such a low-impedance secondary burden. The insulation from the primary to secondary of the transformer must be adequate to withstand line-to-ground voltage, since the connected instruments are usually at ground potential. Normal design is for operation with a rated secondary current of either 1 A or 5 A, and the input current may be as high as many thousand amperes.

The potential circuits of instruments are of relatively high impedance, and voltage transformers are designed for operation into a somewhat high-impedance secondary burden. In the usual design, the rated secondary voltage is 120 V. For temporary use, the user should confirm that the overall instrument burden is *low* enough for the proper operation of the instrument transformer. A digital instrument, on its own, may not be.

***Current Transformers.*** Current transformers are also classified according to their mechanical structure:

1. Wound primary, having more than one turn through the core window
2. Through type, wherein the circuit conductor (cable or busbar) is passed through the window
3. Bar type, having a bar, rod, or tube mounted in the window
4. Bushing type, intended for mounting on the insulating bushing of a power transformer or circuit breaker

The nominal current ratio of such a transformer is the inverse of the turns ratio, but for accurate current measurement, the actual ratio is determined under loading corresponding to use conditions.

The ratio errors and phase angle errors are categorized by ANSI/IEEE standard C57.13 and IEC standards.

The overload capacity of station-type current transformers and the mechanical strength of the winding and core structure must be high to withstand possible short circuits on the line. Various compensation schemes are used in many transformers to retain ratio accuracy up to several times rated current.

The secondary circuit of a CT *must never be opened* while the transformer is excited by primary current. High voltages are induced which may be hazardous both to insulation and to personnel. The accuracy of the transformer will be adversely affected.

***Voltage Transformers.*** Voltage transformers (potential transformers) are connected between the lines whose potential difference is to be determined and are used to step the voltage down (usually to 120 V) and to supply the voltage circuits of the connected instrument burden. Their basic construction is similar to that of a power transformer operating at the same input voltage, except that they are designed for optimal performance with the high-impedance secondary loads of the connected instruments. The core is operated at high flux density, and the insulation must be appropriate to the line-to-ground voltage.

With the development of higher transmission-line voltages (345, 500, and 765 kV), the *coupling-capacitor voltage transformer* (CCVT) has come into use for metering purposes to replace the conventional voltage transformer, which, at these voltages, is bulkier and more costly. The metering CCVT, shown in Fig. 2-9, consists of a modular capacitive divider which reduces the line voltage $V_1$ to a voltage $V_2$ (10–20 kV), with a series-resonant inductor to tune out the high impedance and make available energy transfer across the divider to operate the voltage transformer which further reduces the voltage to *VM*, the metering level. Required metering accuracy may be 0.3% or better, though they should be recalibrated periodically because the capacitance of the divider may change with use.[i]

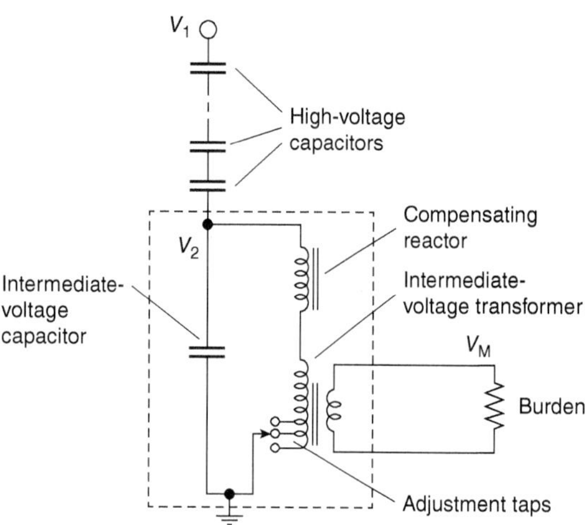


**FIGURE 2-9** CCVT internal arrangement.

*Standard burdens* and standard accuracy requirements for instrument transformers are given in American National Standard C57.13.

*Accuracy.* Most well-designed instrument transformers have sufficient accuracy for metering purposes. Where higher accuracy is required, see App. D of ANSI C12, The Code for Electricity Metering.

***Nonconventional Instrument Transformers.*** The purpose of the instrument transformer is to produce a small-scale copy, faithful in scale factor and phase angle, of the voltage or current being measured. The copy is isolated from the high voltage and the high energy levels of the power system. When instrument transformers were first introduced, around 1900, they had an added attraction: they could be made to furnish enough energy to operate an electromechanical relay. It is during this era that the standard values of 5 A and 120 V arose. A further advantage became evident: these

[i]Like certain kinds of smaller capacitor, the elements used in CCVTs sometimes develop a short circuit. The energy available in the power system rapidly vaporizes a small amount of the capacitor, and there is no lasting short. However, the capacitance has been changed. Over time, this effect can result in the device no longer meeting its specified accuracy.

relays required so much energy to operate them that they were unlikely to be operated by any stray (induced) current or voltage.

The modern electronic relay does not require such a high value of energy to operate. Other means of delivering the necessary low-energy signal have been developed. Collectively, we could call these methods "nonconventional instrument transformers," but some of them are not even transformers.

A current in a wire is associated with a magnetic field (in the wire itself and in the space surrounding it) and with a voltage drop in the series resistance of the wire. Both of these effects can be used for current measurement.

The magnetic field can be measured in several ways:

- Force (the ordinary analog ammeter is essentially a force measuring device that compares the force generated by the field with the strength of a spring)
- Magnetoresistance (materials exist whose resistance is a function of the applied magnetic field)
- Hall effect (a voltage is generated in a material in proportion to the applied field and the magnitude of a control current)
- Magnetostriction (materials exist whose physical size is a function of the applied magnetic field)
- Faraday effect (rotation of the plane of polarization of light in an optical material)

In addition to the magnetic field effects, the voltage drop across a resistor can be measured easily enough.

Note that all of these methods will work for dc as well as ac.

These various methods of transduction from magnetic field to some other parameter do not give isolation, however, except for the optical method, and that has received a good deal of commercial attention. A 1994 review paper (Sensors, 1994) categorized the optical technologies into five types. The interested reader is directed to that paper for further details on optical measurement methods.

The field of optical current measurement has not grown large, possibly because of the high development costs of the technology. However, some optical schemes are in use, though they may not be apparent to the casual observer. A series capacitor protection scheme of one manufacturer makes extensive use of what the review paper called type 1: a conventional CT with an optical readout. Further, optical "CTs" are a favored method of measuring the current in dc power lines.

A further development of the CT is worthy of mention even though it is not a widely used commercial device. It is sometimes called the "active CT." Precision active methods for measuring current originated at the National Research Council of Canada, the Canadian national metrology institute (NMI). The goal was to improve the accuracy of the transformer. In the end, even clamp-on transformers with openable cores were shown to be practicable with high accuracy. Active measurement is based on the fact that the errors in a current transformer come from the current needed to magnetize the core. Because of that, not all the primary current contributes to secondary current. The problem is solved by nulling the core magnetization from a separate energy source. Consider the configuration shown in Fig. 2-10.

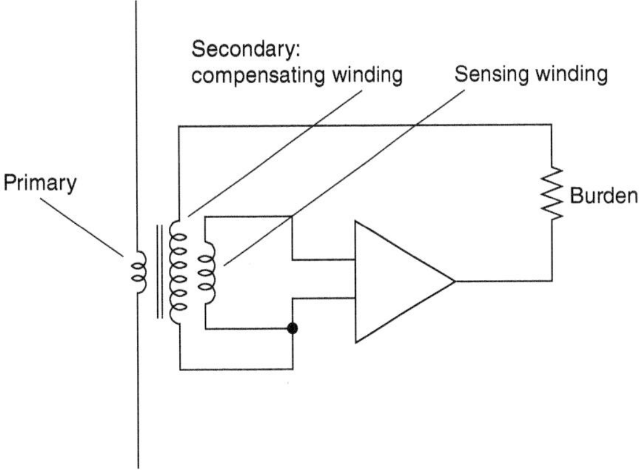

**FIGURE 2-10** Active current transformer.

In addition to the primary turns and the secondary turns, a sensing winding is wound on the transformer core. It is connected to an operational amplifier that is part of a feedback circuit. The input connections of an op-amp are essentially a short circuit, so the amplifier becomes a perfect burden (zero resistance load) for the newly created CT formed by the primary winding and the added sensing winding. The winding is connected so that a current from the op-amp cancels the flux that gave rise to the op-amp input signal. The flux in the core is driven to zero. The secondary ampere-turns, using energy from the op-amp, exactly compensate the primary ampere-turns.

So long as the flux in the core is zero, the compensation current is a perfect copy of the primary current, scaled by the turns ratio of the primary winding and the compensation winding. While the figure shows the burden connected to the amplifier output, the same current may also be observed at the grounded end of the compensation circuit.[j]

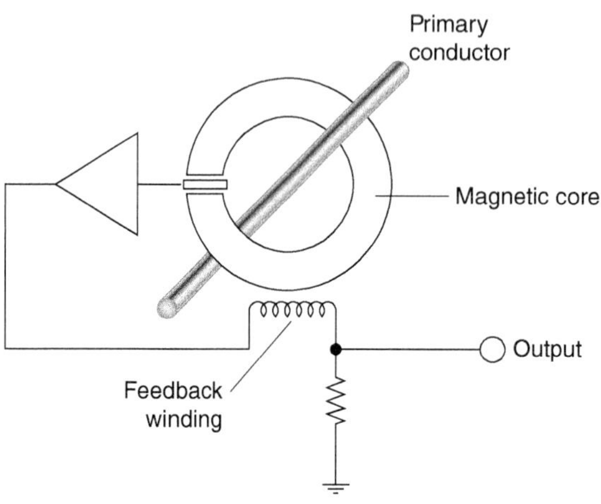


**FIGURE 2-11** Closed-loop Hall-effect current transformer.

Another adaptation of the feedback method is the closed-loop Hall sensor, shown in Fig. 2-11. Systems built on this principle are commercially available. The difference between this scheme and the active CT method is that the core flux is sensed by a Hall device, instead of a transformer winding. That means the sensor can operate down to dc.

Unlike the situation with an ordinary Hall effect current measurement, where the magnetic circuit functions as a concentrator, the output of the closed-loop version is substantially independent of the properties of the Hall crystal.

***Current Transformer Calibration.*** To calibrate a current transformer at a large current, it has been considered difficult to control the current with precision enough to calibrate a device intended to have (say) 0.3% uncertainty. (See Sec. 2.3.6 for further information.) An adaptation of the system shown in Fig. 2-4 has been used for many years, resulting in a scheme in which the primary current need be known only approximately. An external network used to make the equality visible between a reference device and a CT under test. Consider the setup shown in Fig. 2-12.

The current is generated by a source shown at the top of the diagram. It is passed through the primary winding of the CT being tested, and the primary winding of a device known as a current comparator. This is specially wound transformer that is designed to operate under conditions of ampere-turn balance; that is, the net ampere-turns in the core is designed to be zero. The secondary circuit is arranged so that the current from the CT under test cancels the ampere-turns from the primary. This cancellation is detected by a special winding connected to the detector, which should indicate zero. Of course, since the CT under test is not perfect, there will be a small remaining current in the current comparator. The in-phase and quadrature components of this current are cancelled by adding a small current to a compensation winding, by means of an adjustable resistor and an adjustable capacitor. When the detector indicates zero, there is no flux in the current comparator, which means that it is operating as a perfect transformer. The ampere-turns contributions are *exactly* given by the number of turns on the various windings, so provided these have been counted exactly, the

[j]For improved accuracy, and laboratory use, the method is modified to form what are known as multistage transformers. In a multistage transformer there are two or more cores, arranged concentrically. The sensing winding is on its own core, and it sees only the defects of the working core. It is possible to have a separate secondary circuit rated for the usual 5 A, if desired. The inner core of a two-stage CT represents the *error* in the working core. Therefore, even if this signal is not compensated with perfect accuracy, the uncertainty of the measurement can still be significantly reduced. For example, if a CT has an uncertainty of 1% without compensation, and compensation circuit has a 1% uncertainty, the overall measurement uncertainty is reduced to 1% of 1%, or 0.01%. Much better performance than this is usual in multistage CTs.

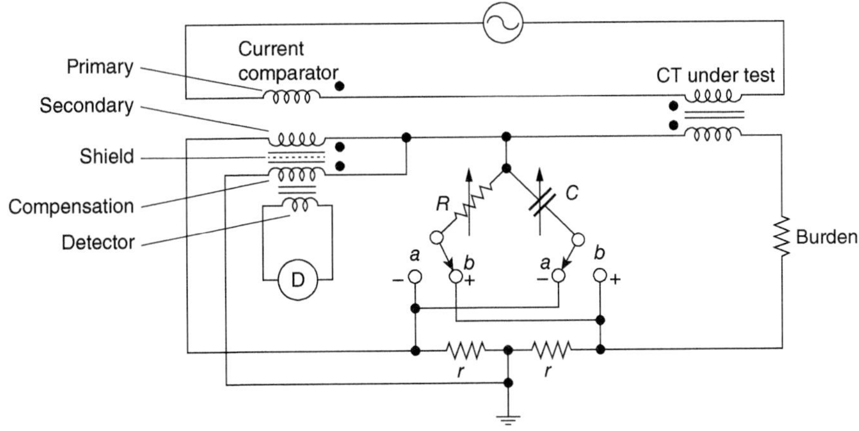


**FIGURE 2-12** Current transformer calibration using current comparator.

ratio is known exactly. The errors of the CT being tested can be determined to an uncertainty of a few parts per million by this method. The error can be written $\epsilon \pm \left(\frac{r}{R} + j\omega rC\right)$, so the ratio error and the phase angle error are separate. There are variations on this theme that give greater flexibility of use, but the general idea is the same. For more details, see the relevant standard: IEEE C57.13.

***Voltage Transformer Calibration.*** The current comparator can be adapted to use in calibrating voltage transformers (Fig. 2-13). The primary and secondary windings of the VT are connected to separate windings of the current comparator, via well-characterized capacitors. The high-voltage side is generally connected via a gas capacitor of 50 to 100 pF capacitance, and the low voltage capacitor may be 1 nF.

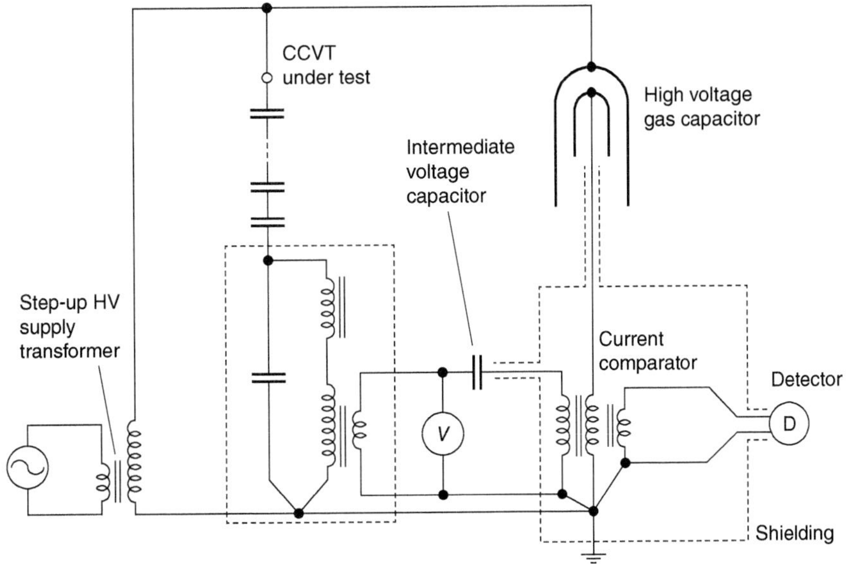


**FIGURE 2-13** Voltage transformer calibration using current comparator.

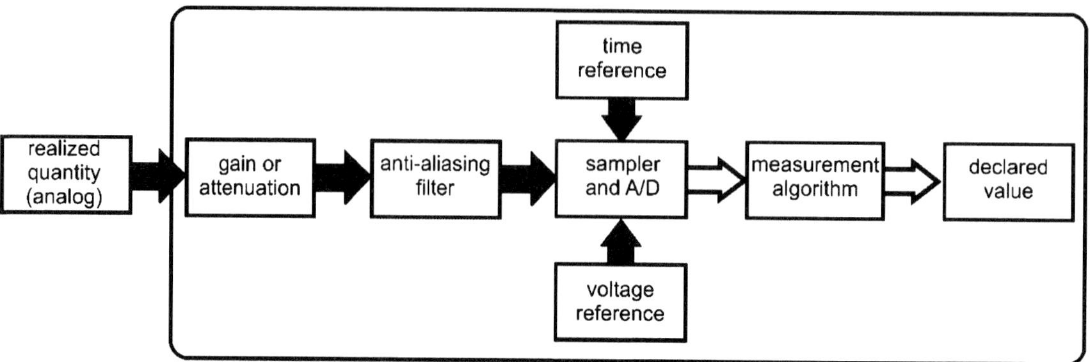


**FIGURE 2-14** Generic digital measurement system.

Since the ratio of the current comparator is known, and the ratio of the capacitors is known, the ratio of the transformer can be calculated.

### 2.4.2 Sampling

Modern instrumentation systems are digital. To make them work, the right sort of signal has to be fed to the algorithms that execute the calculations that implement the measurements. The process begins with an analog quantity and ends with a result (or declared value) as in Fig. 2-14. Central to the process is sampling.

Since the real world is analog, the input to the measurement system is a continuous-time analog signal. An original input quantity (perhaps pressure) is converted to an electrical quantity (e.g., voltage) by a suitably designed transducer. We are concerned here with the electrical signal that enters the instrument.

The signal is sampled periodically to produce a discrete-time digital signal that is processed by whatever measurement algorithm is being used before being output as the measurement result, or declared value.

The details of what takes place in the instrument depend on the requirements on the measurement, particularly the bandwidth of the signal and the accuracy of the result. We will look next at a few of these details.

***Antialiasing.*** The idea of sampling at a certain minimum rate arose in the development of communications. In order to reproduce a certain signal that was digitized and sent over a communications link, the signal had to be sampled at a rate at least twice that of the highest frequency component of the signal. As applied to measurements, the way of expressing that is that the signal should be bandlimited to a frequency corresponding to half the sample rate. That frequency is called the Nyquist frequency, after Harry Nyquist, who worked on sampling theory at Bell Labs.[k] If that is not done, and if the signal being measured includes energy above the Nyquist frequency, some strange things happen. The energy appears to exist *within* the band. This effect is called *aliasing*. Figure 2-15 shows a situation where the signal contains energy *only* above the Nyquist frequency.

[k]Those who are sticklers for detail may wish to know that the theorem that proved this was first developed by an English mathematician named Whittaker, and was later proved by Shannon (1984), also at Bell Labs. Nyquist himself seems not to have been involved.

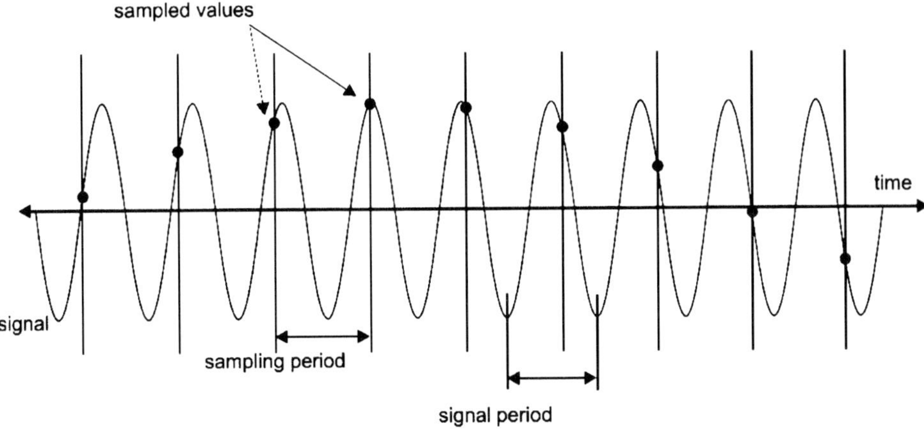


**FIGURE 2-15** Sampling more slowly than the Nyquist frequency.

It may be that some particular signal contains no energy above the Nyquist frequency. But the instrument designer cannot know that in advance, so a filter—an antialiasing filter—is always put ahead of the A/D converter. This filter is in the part of the signal path that is analog, and it may be the last analog filter that the signal sees, as other filtering is done in the digital domain.

It is worth noting that the advances in digital technology are such that it is sometimes easier to sample at a frequency considerably higher than might be considered necessary from knowledge of the expected signals. Such *oversampling* simplifies the design of the antialiasing filter, as it may not need a particularly steep roll-off so long as it is reasonably expected that it will contain little or no energy at the (much higher) cutoff frequency.[l]

***A/D Converter.*** The A/D converter is expected to convert the instantaneous voltage at its input into a digital representation. Since the conversion process takes some finite time, it is usual to have the analog voltage sampled at the right instant and held as a constant analog value for digital conversion. So there really are two parts to the process, the hold and the conversion. Holding is done by putting the voltage onto a capacitor, a process that can be done very rapidly.

The A/D conversion process involves a trade-off for the instrument designer. The faster conversion technologies are associated with lower resolution, and vice versa. A delta-sigma converter might be able to furnish 24-bit resolution, but is limited to some tens of kilosamples per second (kS/s).[m] A flash converter can convert video signals at above 100 MS/s, but with resolution limited to 8 bits. Between these extremes is the successive approximation register (SAR) that can convert (say) a 16-bit output at rates up to a few MS/s.

The process of A/D conversion is associated with its own uncertainties and errors. These are easy to visualize in the analog domain. The conversion may be slightly nonlinear, and it may have a small zero offset and a gain error. The worst-case values of all of these are part of the specification of the converter, and it can be assumed that the user of the instrument does not need to dig into the topic.

[l]The Nyquist criterion is based on an assumption that all that is known about the signal is contained within the signal itself. That assumption is reasonable when one is thinking of a communication system that has to transmit a replica of some signal that is unknowable in advance. However, that is not often the case in measurement. The observer usually has a fair idea of what to expect of the signal or component being measured. Because of that additional knowledge, the Nyquist criterion can sometimes safely be violated. Called compressive sensing or sparse sampling, it is a topic beyond the scope of the present handbook.

[m]The reader may wonder why anyone would need a 24-bit converter. Its accuracy would seem to be the sort of level attained in a metrology laboratory: better than 1 in $10^7$. The need lies usually not so much in total accuracy as in dynamic range. When the signal is likely to have a very large dynamic range (as in the case of the ground accelerations due to an earthquake, for example), the 24-bit converter can be very useful.

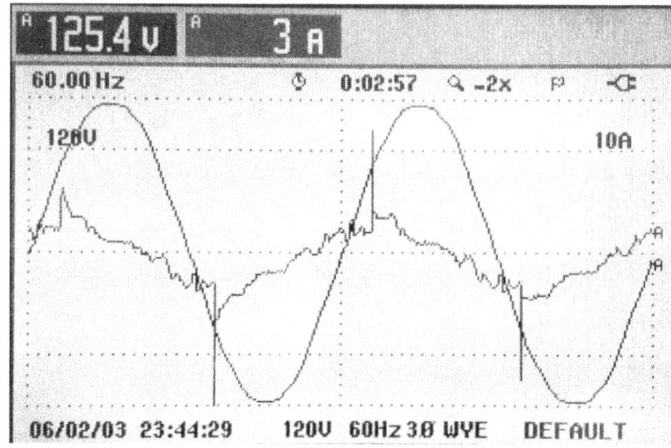


**FIGURE 2-16** Power analyzer screen showing LED lamp current.

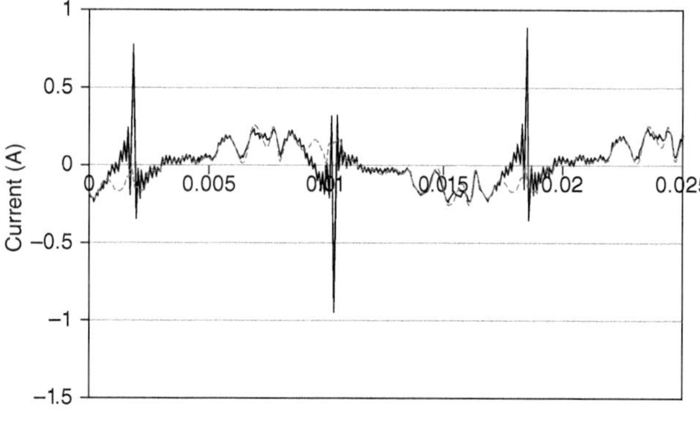


**FIGURE 2-17** Reconstructed current waveform.

***Filtering in the Measurement Algorithm.*** At the beginning of the section, we described measurement as a process whereby the parameters of some chosen model are adjusted so that the result matches the input signal. Except in rare cases, there is more than just measurement going on, there is filtering. The signal may contain energy at frequencies that are of no interest and would affect the result of the measurement, for example. Or it may be just a matter of filtering out distortion or noise to reduce its effect on accuracy.

There are two main types of digital filter: the finite impulse response (FIR) filter and the infinite impulse response (IIR) filter. Each is capable of a level of performance that could not be achieved with analog filters (e.g., in terms of the flatness of the passband or the steepness of the attenuation). For the user, the difference is seen in the time response.

This is worth knowing about, since the effect is sometimes seen indirectly in the instrument output. In the measurement of the current into an LED lamp, for example, the power analyzer showed that the lamp was drawing most of its current in a 10-A pulse lasting a very short period of time, as in Fig. 2-16.

The oscilloscope showed that the fast "spike" was in fact a pulse some microseconds long, followed by a short period of ringing. However, when the signal was reconstructed from the reported frequency/phase spectrum of a different analyzer, the spike assumed the form shown in Fig. 2-17.

The reconstruction shows a broadened spike, with what appears to be ringing that both *precedes* and follows the actual current. Such an effect is characteristic of the IIR filter—what we are seeing here is an artifact of our reconstruction, and is not a good representation of the signal. The reader is cautioned not to trust such waveforms unless they have been verified by a different system.

## 2.5 USE OF INSTRUMENTS

A number of general precautions should be observed in electrical measurements, and sources of error should be avoided, as detailed below:

1. The uncertainties of the instruments, any reference standards, and the methods used should be known so that appropriate choice of these measuring things may be made. It should be noted that instrument *accuracy classes* state the accuracy obtained during calibration in the laboratory. Operation of an instrument over a prolonged period, may cause errors due to material incompatibilities, or resistance changes in instrument elements may occur because of long-term heating under load. The accuracy figure given in the owner manual is therefore to be regarded as applicable only if the device in question has a valid calibration certificate.

2. In the days of analog instrumentation, the *average of several readings* was regarded as better than one. With digital instruments, repeated measurements may be worth doing as a way of checking. But be aware that merely writing down or logging successive readings rarely counts as repeating a measurement. The sources of error that repetition is aimed at averaging out might be contact resistance, or local temperature: error sources such as these will not have changed. It is extremely unlikely that anything in the digital instrument will have changed in a way that would lead to error-reduction by averaging.
3. The *range* of the measuring instrument should be suited to the application. Voltage transformers, wattmeters, and watthour meters should be operated near to rated voltage for best performance.[n] Care should be taken to avoid both momentary and sustained overloads.
4. *Magnetic or electric fields* may be large enough to interfere with the low-energy inputs of some instruments, producing errors in the results of the measurement. Alternating fields may induce electromotive forces (emfs) in loops formed in connections or even the internal wiring some high impedance systems. It is always some sort of noise that limits the smallest observable signal. Effort spent reducing the effect of sources external to the instrument may be worthwhile, as sometimes this electrical noise may even obscure the desired reading.

   The effects of stray alternating fields on may be controlled by careful grounding and shielding. Static fields are rarely much trouble. It also should be noted that Zener-diode-based references are affected by magnetic fields. This may alter the performance of some digital meters, though most use band gap references that are not so sensitive.
5. In measurements involving high resistances and small currents, *leakage paths* across insulating components of the measuring arrangement should be eliminated if they shunt portions of the measuring circuit. This is done by providing a guard circuit to intercept current in such shunt paths or to keep points at the same potential between which there might otherwise be unwanted currents. More information is given below in the subsection Guarding.
6. Variations in *ambient temperature* or temperature rise from self-heating under load may cause errors in resistive shunts. If the temperature coefficient and the instrument temperature are known (perhaps measured by a non-contact instrument), readings can be corrected where precision requirements justify it. Where measurements involve extremely small potential differences, thermal emfs resulting from temperature differences between junctions of dissimilar metals may produce errors; heat from the observer's hand or heat generated by the friction of a sliding contact may cause such effects.
7. *Phase-defect angles* in resistors, inductors, or capacitors and in instruments and instrument transformers must be taken into account in many ac measurements, particularly those near-zero power factor.
8. Large *potential differences* are to be avoided between the windings of an instrument or between its windings and frame. Electrostatic forces and capacitive currents may produce reading errors, and very large potential difference may result in insulating breakdown. Instruments should be connected in the ground leg of a circuit where feasible. When an instrument must be at a high potential, its case must be adequately insulated from ground and connected to the line in which the instrument circuit is connected, or the instrument should be enclosed in a screen that is connected to the line. Such an arrangement may involve shock hazard to the operator, and proper safety precautions must be taken.
9. *Electrostatic charges* and consequent disturbance to readings may result from a variety of causes in a high-impedance measurement. Low-level measurements in very dry weather may be seriously affected by charges on the clothing of the observer; it is almost impossible to find an indoor location that does not have some static electric field. Grounding and shielding are called for.

[n]While it may have been obvious that using an analog meter near to its full scale improved the resolution, it is no less true for digital instruments. Some manufacturers present the user a curve showing the effect of range-changing on accuracy: the best results are obtained near the top of each range.

The performance of certain insulators is adversely affected by high humidity. Some kinds of fiberglass, even if used indoors, have increased surface leakage when it is raining outside. More information is given below in the subsection Bulk and Surface Measurements.

## 2.6 POWER MEASUREMENT

It is well-known that the power $P$ in an electrical circuit is given by the volts times the amps. If the current is dc, the actual calculation may be $P=v(t)\times i(t)$ where the sampled values are averaged over some observation window. If the voltage is alternating, the instantaneous power $p=v(t)\times i(t)$ can be integrated over some number of cycles $k$ to give the power $P=\frac{1}{kT}\int_{\tau}^{\tau_1+kT} p\,dt$. The result is, of course, the average over the interval. The response of such systems may be as high as several kHz. Measurements can often be made to higher frequencies with somewhat reduced accuracy. Such an arrangement is well-adapted to the measurement of power in situations where current or voltage waveforms are badly distorted.

*Correction for wattmeter power consumption* may be important when the power measured is small. When the wattmeter is connected directly to the circuit (without the interposition of instrument transformers), the instrument reading will include the power consumed in the element connected next to the load being measured. If the instrument loss cannot be neglected, it is usually better to connect the voltage circuit next to the load and include its power consumption rather than that of the current circuit, since it is generally more nearly constant and is more easily calculated.

A theorem established by Blondel (1894) shows that the minimum number of wattmeters in a circuit is one less than the number of wires. Thus, a *three-phase three-wire circuit* requires two wattmeters connected as shown in Fig. 2-18; total power is the sum of the two readings under all conditions of load and power factor. If the load is balanced, at unity power factor each instrument will read half the load; at 50% power factor one instrument reads all the load and the other reading is zero; at less than 50% power factor one reading will be negative.

*Three-phase four-wire circuits* require three wattmeters as shown in Fig. 2-19. Total power is the sum of the three readings under all conditions of load and power factor. A three-phase Y system with a grounded neutral is the equivalent of a four-wire system and requires the use of three wattmeters. If the load is known to be balanced, one wattmeter can be used with its current circuit in series with one conductor and the voltage circuit connected between that conductor and the neutral. Total power is three times the wattmeter reading in this instance.

*Corrections for instrument transformers* in the measurement of power are of two kinds. *Ratio* errors, resulting from deviations of the actual ratio from its nominal, may be obtained from a calibration curve showing true ratio at the instrument burden imposed on the transformer and for the

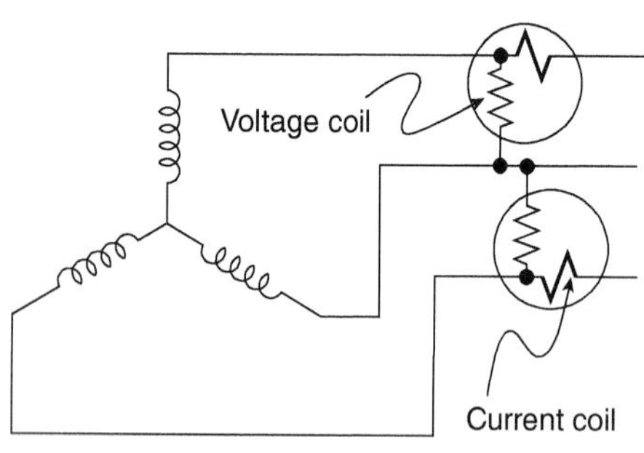


**FIGURE 2-18** Power in three-phase, three-wire circuit, two wattmeters.

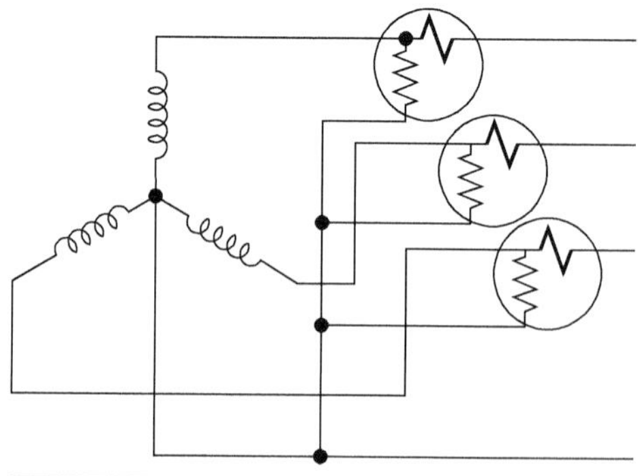

**FIGURE 2-19** Power in three-phase, four-wire circuit, three wattmeters.

current or voltage of the measurement. The effect of *phase-angle* changes introduced by instrument transformers is modification in the angle between the current and the voltage applied to the wattmeter; the resulting error depends on the power factor of the circuit and may be positive or negative depending on phase relations, as shown in the table below.

An earlier edition of this handbook showed a formula for finding the power in the circuit driving the instrument transformers in terms of the ratio errors of the transformers and the ratio of something labeled the *true* power factor. (The italics was in the original.) This true power factor was defined as $\cos\Phi = \cos(\Phi_{app} \pm\alpha \pm \beta \pm \gamma)$, where $\Phi_{app}$ is the phase angle of the secondary circuit, $\alpha$ is the angle of the wattmeter's voltage circuit, $\beta$ is the phase angle of the current transformer, and $\gamma$ is the phase angle of the voltage transformer. This is not the same as the apparent power factor $\cos\Phi_{app}$ (obtained by dividing the wattmeter reading and by the measured voltamps). With $K_c$ and $K_v$ as the true ratios of the current and voltage transformers, respectively, then the actual power in the circuit can be found from

$$\text{Main-circuit power} = K_c K_v \frac{\cos\Phi}{\cos\Phi_{app}} \times \text{wattmeter reading}$$

These angles—$\alpha$, $\beta$, and $\gamma$—are given positive signs when they act to decrease and negative when they act to increase the phase angle between instrument current and voltage with respect to that of the circuit. This is so because a decreased phase angle gives too large a reading and requires a negative correction (and vice versa), as shown in the following table of signs. The reader is cautioned that such corrections as these are likely accurate only for undistorted waves.

| | Sign to be used for phase angle | | | | | |
|---|---|---|---|---|---|---|
| | $\alpha$ wattmeter | | $\beta$ current transf. | | $\gamma$ voltage transf. | |
| Line power factor | Lead* | Lag | Lead | Lag | Lead | Lag |
| Lead | + | – | – | + | + | – |
| Lag | – | + | + | – | – | + |

*In general, $\alpha$ will be leading only when the inductance of the potential coil has been overcompensated with capacitance.

## 2.7 REACTIVE POWER

The term "power factor" arose in the days when the voltages and currents on the power system had relatively low levels of harmonics. At that time *reactive power* (reactive voltamperes, or vars) used to be measured by a wattmeter with its current coil in series with the circuit and the current in its voltage element in quadrature with the circuit voltage. With low levels of harmonics, that method of measurement of reactive power was surely adequate. However, even at the time the term was first used (Fleming, 1891), it was recognized that the straightforward relationship and seeming Pythagorean relationship between apparent power, real power, and reactive power was valid only for undistorted voltages and currents.

Now that we have power systems that have more harmonics, it becomes evident that there is no one definition that we can call correct for the quantity we know as reactive power. With the advent of digital instrumentation, the calculation of something with the label reactive power can be done according to any of several definitions that all give the same result if the signals are sinusoidal. Some instruments actually allow the user to select a method. IEEE Std 1459-2010 proposes a method that retains the Pythagorean relationship, but in spite of the fact that it has the weight of IEEE behind it, it should not be considered authoritative on the matter.

In particular, two definitions have been argued over for some while. An early definition was due to Budeanu (1927), who wrote a definition for reactive power that was a parallel with real power in the way it dealt with harmonics. Real power is defined for a sinusoidal wave as the product of the voltage and the current, times the cosine of the angle between them: $P = V \times I \times \cos\Phi$. That relationship is independent of the frequency, so it can be generalized and written for a distorted wave modeled as the sum of harmonics: $P = \Sigma_h V_h \times I_h \times \cos\Phi_h$, where the subscript $h$ signifies the harmonic number. Budeanu defined the reactive power simply by using the sin instead of the cosine: $P = \Sigma_h V_h \times I_h \times \sin\Phi_h$.

As it happens, this form of the definition does not give the anticipated Pythagorean relationship $S^2 = P^2 + Q^2$. Now, this relation holds for sinusoidal quantities, but there is no reason at all that it should be expected for distorted waves. Nevertheless, evidently having lost sight of the fact that there was no particular merit to retaining the Pythagorean relationship (because $P$ and $Q$ are scalar quantities), several scientists devoted time and energy to the quest. The reader is therefore cautioned that measurement results for reactive power in situations with distorted waveforms may not always agree with one another. Unlike the situation for power, where an experiment that involves heat measurement can be designed to validate the definition of power, no such experiment can exist for reactive power. One definition is as good (or as bad) as another.

## 2.8 POWER-FACTOR MEASUREMENT

The *power factor of a single-phase circuit* is the ratio of the real power in watts, as measured with a wattmeter, to the apparent power in voltamperes, obtained as the product of the voltage and current. When the waveform is sinusoidal (and only then), the power factor is also equal to the cosine of the phase angle.

If the signals have harmonics, the power factor is still defined as $PF = P/S$ but both terms on the right can be expanded to include the harmonics. According to IEEE Std 1459-2010, the equation becomes

$$PF = \frac{P}{S} = \frac{P_1 + P_H}{\sqrt{S_1^2 + S_N^2}} = \frac{P_1 / S_1[1 + (P_H / P_1)]}{\sqrt{1 + (1 + S_N / S_1)^2}}$$

where $S_N = \sqrt{(THD_I)^2 + (THD_V)^2}$, the subscripts being current and voltage. The reader may observe that this definition, depending as it does on knowledge of the total harmonic distortion in the signal, would require a very special kind of instrument to make the measurement. And the reader is also reminded to be careful to understand the *meaning* of such a measurement.

The *power factor of a polyphase circuit* which is balanced is the same as that of the individual phases. When the phases are not balanced, the IEEE standard defines the positive phase sequence power factor the same as for single phase, except that just the positive phase sequence numbers are used.

*Power-factor meters,* which indicate the power factor of a circuit directly, are made both as portable and as switchboard types. They may not read accurately if the load is unbalanced or distorted. It would be wise the check the owner manual.

## 2.9 ENERGY MEASUREMENTS

The subject of metering electric power and energy is extensively covered in the American National Standard C21, Code for Electricity Metering, American National Standards Institute. It covers definitions, circuit theory, performance standards for new meters, test methods, and installation standards for watthour meters, demand meters, pulse recorders, instrument transformers, and auxiliary devices. Further detailed information may be found in the *Handbook for Electric Metermen,* Edison Electric Institute.

The practical unit of electrical energy is the watthour, which is the energy expended in one hour when the power (or rate of expenditure) is 1 W. Energy is measured in watthours (or kilowatthours) by means of a *watthour meter*. A watthour meter has until recently been a motor mechanism in which a rotor element revolves at a speed proportional to power flow and drives a registering device on which energy consumption is integrated. Meters for continuous current are usually of the mercury-motor type, whereas those for alternating current utilize the principle of the induction motor.

***Polyphase Meter Connections.*** Obviously, it is extremely important that the various circuits of a polyphase meter be properly connected. If, for example, the current-coil connections are interchanged and the line power factor is 50%, the meter will run at the normal 100% power-factor speed, thus giving an error of 100%.

A test for correct connections is as follows: If the line power factor is over 50%, rotation will always be forward when the potential or the current circuit of either element is disconnected, but in one case the speed will be less than in the other. If the power factor is less than 50%, the rotation in one case will be backward. When it is not known whether the power factor is less or greater than 50%, this may be determined by disconnecting one element and noting the speed produced by the remaining element. Then change the voltage connection of the remaining element from the middle wire to the other outside wire and again note the speed. If the power factor is over 50%, the speed will be different in the two cases but in the same direction. If the power factor is less than 50%, the rotation will be in opposite directions in the two cases.

When instrument transformers are used, care must be exercised in determining correct connections; if terminals of similar instantaneous polarity have been marked on both current and voltage transformers, these connections can be verified and the usual test made to determine power factor. If the polarities have not been marked, or if the identities of instrument transformer leads have been lost in a conduit, the correct connections can still be established, but the procedure is more lengthy.

***Use of Instrument Transformers with Watthour Meters.*** When the capacity of the circuit is over 200 A, instrument current transformers are generally used to step down the current to 5 A. If the voltage is over 480 V, current transformers are almost invariably employed, irrespective of the magnitude of the current, in order to insulate the meter from the line; in such cases, voltage transformers are also used to reduce the voltage to 120 V. Transformer polarity markings must be observed for correct registration. The ratio and phase-angle errors of these transformers must be taken into account where high accuracy is important, as in the case of a large installation. These errors can be largely compensated for by adjusting the meter speed.

***Reactive Voltampere-Hour (Var-Hour) Meters.*** Reactive voltampere-hour (var-hour) meters are generally ordinary watthour meters in which the current coil is inserted in series with the load in the usual manner while the voltage coil is arranged to receive a voltage in quadrature with the load voltage. In two-phase circuits, this is easily accomplished by using two meters as in power measurements, with the current coils connected directly in series with those of the "active" meters but with the voltage coils connected across the quadrature phases. Evidently, if the meters are connected to rotate forward for an inductive load, they will rotate backward for capacitive loads. For three-phase three-wire circuits and three-phase four-wire circuits, phase-shifting transformers are used normally and complex connections result.

***Errors of var-Hour Meters.*** The two- and three-phase arrangements described above give correct values of reactive energy when the voltages and currents are balanced. The two-phase arrangement still gives correct values for unbalanced currents but will be in error if the voltages are unbalanced. Both three-phase arrangements give erroneous readings for unbalanced currents or voltages; an autotransformer arrangement usually will show less error for a given condition of unbalance than the simple arrangement with interchanged potential coils.

*Total var-hours,* or "apparent energy" expended in a load, is of interest to engineers because it determines the heating of generating, transmitting, and distributing equipment and hence their rating and investment cost. The apparent energy may be computed if the power factor is constant, from the observed watthours $P$ and the observed reactive var-hours $Q$; thus var-hours. This method may be greatly in error when the power factor is not constant; the computed value is always too small.

A number of devices have been offered for the direct measurement of the apparent energy. In one class (*a*) are those in which the meter power factor is made more or less equal to the line power factor. This is accomplished automatically (in the Angus meter) by inserting a movable member in the voltage coil pole structure which shifts the resulting flux as line power factor changes. In others, autotransformers are used with the voltage elements to give a power factor in the meter close to expected line power factor. By using three such pairs of autotransformers and three complete polyphase watthour-meter elements operating on a single register, with the record determined by the meter running at the highest speed, an accuracy of about 1% is achieved, with power factors ranging from unity down to 40%. In the other class (*b*), vector addition of active and reactive energies is accomplished either by electromagnetic means or by electromechanical means, many of them very ingenious. But the result obtained with the use of modern watthour and var-hour meters are generally adequate for most purposes.

The *accuracy* of a watthour meter is the percentage of the total energy passed through a meter which is registered by the dials. The watthours indicated by the meter in a given time are noted, while the actual watts are simultaneously measured with standard instruments. Because of the time required to get an accurate reading from the register, it is customary to count revolutions of the rotating element instead of the register. The accuracy of the gear-train ratio between the rotating element and the first dial of the register can be determined by count. Since the energy represented by one revolution, or the *watthour constant,* has been assigned by the manufacturer and marked on the meter, the indicated watthours will be $K_h R$, where $K_h$ is the watthour constant and $R$ the number of revolutions.

***Reference Standards.*** *Reference standards* for dc meter tests in the laboratory may be ammeters and voltmeters, in portable or laboratory-standard types, or potentiometers; in ac meter tests, use is made of indicating wattmeters and a time reference standard such as a stopwatch, clock, or tuning-fork or crystal-controlled oscillator together with an electronic digital counter. A more common reference is a standard watthour meter, which is started and stopped automatically by light pulsing through the anticreep holes of the meter under test.

The *portable standard* watthour meter (often called *rotating standard*) method of watthour-meter testing is most often used because only one observer is required and it is more accurate with fluctuating loads. Rotating standards are watthour meters similar to regular meters, except that they are made with extra care, are usually provided with more than one current and one voltage range, and are portable. A pointer, attached directly to the shaft, moves over a dial divided into 100 parts so that fractions of a revolution are easily read. Such a standard meter is used by connecting it to measure the same energy as is being measured by the meter to be tested; the comparison is made by the "switch" method, in which the register only (in dc standards) or the entire moving element (in ac standards) is started at the beginning of a revolution of the meter under test, by means of a suitable switch, and stopped at the end of a given number of revolutions. The accuracy is determined by direct comparison of the number of whole revolutions of the meter under test with the revolutions (whole and fractional) of the standard. Another method of measuring speed of rotation in the laboratory is to use a tiny mirror on the rotating member which reflects a beam of light into a photoelectric cell; the resulting impulses may be recorded on a chronograph or used to define the period of operation of a synchronous electric clock, etc.

*Watthour meters used with instrument transformers* are usually checked as secondary meters; that is, the meter is removed from the transformer secondary circuits (current transformers must first be short-circuited) and checked as a 5-A 120-V meter in the usual manner. The meter accuracy is adjusted so that when the known corrections for ratio and phase-angle errors of the current and potential transformers have been applied, the combined accuracy will be as close to 100% as possible, at all load currents and power factors. An overall check is seldom required both because of the difficulty and because of the decreased accuracy as compared with the secondary check.

*General precautions* to be observed in testing watthour meters are as follows: (1) The test period should always be sufficiently long and a sufficiently large number of independent readings should be taken to ensure the desired accuracy. (2) Capacity of the standards should be so chosen that readings will be taken at reasonably high percentages of their capacity in order to make observational or scale errors as small as possible. (3) Where indicating instruments are used on a fluctuating load, their average deflections should be estimated in such a manner as to include the time of duration of each deflection as well as the magnitude. (4) Instruments should be so connected that neither the standards nor the meter being tested is measuring the voltage-circuit loss of the other, that the same voltage is impressed on both, and that the same load current passes through both. (5) When the meter under test has not been previously in circuit, sufficient time should be allowed for the temperature of the voltage circuit to become constant. (6) Guard against the effect of stray fields by locating the standards and arranging the temporary test wiring in a judicious manner.

***Meter Constants.*** The following definitions of various meter constants are taken from the Code for Electricity Metering, 6th ed., ANSI C12.

*Register constant* $K_r$ is the factor by which the register reading must be multiplied in order to provide proper consideration of the register or gear ratio and of the instrument-transformer ratios to obtain the registration in the desired units.

*Register ratio* $R_r$ is the number of revolutions of the first gear of the register, for one revolution of the first dial pointer.

*Watthour constant* $K_h$ is the registration expressed in watthours corresponding to one revolution of the rotor. (When a meter is used with instrument transformers, the watthour constant is expressed in terms of primary watthours. For a secondary test of such a meter, the constant is the primary watthour constant, divided by the product of the nominal ratios of transformation.)

*Test current* of a watthour meter is the current marked on the nameplate by the manufacturer (identified as TA on meters manufactured since 1960) and is the current in amperes which is used as the basis for adjusting and determining the percentage registration of a watthour meter at heavy and light loads.

*Percentage registration* of a meter is the ratio of the actual registration of the meter to the true value of the quantity measured in a given time, expressed as a percentage. Percentage registration is also sometimes referred to as the *accuracy* or *percentage accuracy* of a meter. The value of one revolution having been established by the manufacturer in the design of the meter, meter watthours = $K_h \times R$, where $K_h$ is the watthour constant and $R$ is the number of revolutions of rotor in $S$ seconds. The corresponding power in meter watts is $P_m = (3600 \times R \times K_h)/S$. Hence, multiplying by 100 to convert to terms of percentage registration (accuracy),

$$\text{Percentage registration} = \frac{K_h \times R \times 3600 \times 100}{PS}$$

where $P$ is true watts. This is the basic formula for watthour meters in terms of true watt reference.

***Average Percentage Registration (Accuracy) of Watthour Meters.*** The Code for Electricity Metering makes the following statement under the heading, "Methods of Determination":

> The percentage registration of a watthour meter is, in general, different at light load than at heavy load, and may have still other values at intermediate loads. The determination of the average percentage registration of a watthour meter is not a simple matter as it involves the characteristics of the meter and the loading. Various methods are used to determine one figure which represents the average percentage registration, the method being prescribed by commissions in many cases. Two methods of determining the average percentage registration (commonly called "average accuracy" or "final average accuracy") are in common use: *Method 1.* Average percentage registration is the weighted average of the percentage registration at light load (LL) and at heavy load (HL), giving the heavy-load registration a weight of 4. *Method 2.* Average percentage registration is the average of the percentage registration at light load (LL) and at heavy load (HL).

***In-Service Performance Tests.*** *In-service performance tests,* as specified in the Code for Electricity Metering, ANSI C12, shall be made in accordance with a *periodic test schedule,* except that self-contained single-phase meters, self-contained polyphase meters, and three-wire network meters also may be tested under either of two other systems, provided that all meters are tested under the same system. These systems are the *variable interval plan* and the *statistical sampling plan.*

The chief characteristic of the *periodic-internal system* is that a fixed percentage of the meters in service shall be tested annually. In the test intervals specified below, the word *years* means calendar years. The periods stated are recommended test intervals. There may be situations in which individual meters, groups of meters, or types of meters should be tested more frequently. In addition, because of the complexity of installations using instrument transformers and the importance of large loads, more frequent inspection and test of such installations may be desirable. In general, periodic test schedules should be as follows:

1. Meters with surge-proof magnets and without demand registers or pulse initiators—16 years.
2. Meters without surge-proof magnets and without demand registers or pulse initiators—8 years.

The chief weaknesses of the preceding periodic test schedule are that it fails to recognize the differences in accuracy characteristics of various types of meters as a result of technical advance in meter design and construction, and fails to provide incentives for maintenance and modernization programs.

The *variable interval plan* provides for the division of meters into homogeneous groups and the establishment of a testing rate for each group based on the results of in-service performance tests made on meters longest in service without test. The maximum test rate recommended is 25% per year. The minimum test rate is recommended for the testing of a sufficient number of meters to provide adequate data to determine the test rate for the succeeding year. The provisions of the variable interval plan recognize the difference between various meter types and encourage adequate meter maintenance and replacement programs. See Sec. 8.1.8.5 of ANSI C12 for details of operation of this plan.

The *statistical sampling program* included is purposely not limited to a specific method, since it is recognized that there are many acceptable ways of achieving good results. The general provisions of the statistical sampling program provide for the division of meters into homogeneous groups, the annual selection and testing of a random sample of meters of each group, and the evaluation of the test results. The program provides for accelerated testing, maintenance, or replacement if the analysis of the sample test data indicates that a group of meters does not meet the performance criteria. See Sec. 8.1.8.6 of ANSI C12 for details of the operation of this program.

***Ampere-Hour Meters.*** Ampere-hour meters measure only electrical quantity, that is, coulombs or ampere-hours, and therefore, where they are used in the measurement of electrical energy, the potential is assumed to remain constant at a "declared" value, and the meter is calibrated or adjusted accordingly.

*Ampere-hour or volt-hour meters for alternating current* are not practical but ampere-squared-hour or volt-squared-hour meters are readily built in the form of the induction watthour meter. Ampere-hours or volt-hours are then obtainable by extracting the square root of the registered quantities.

***Maximum-Demand Meters.*** Some methods of selling energy involve the maximum amount which is taken by the customer in any period of a prescribed length, that is, the maximum demand. Many types of meters for measuring this demand have been developed, but space permits only a brief description of a few. There are two general classes of demand meters in common use: (1) integrated-demand meters and (2) thermal, logarithmic, or lagged-demand meters. Both have the same function, which is to meter energy in such a way that the registered value is a measure of the load as it affects the heating (and therefore the load-carrying capacity) of the electrical equipment.

***Integrated-Demand Meters.*** Integrated-demand meters consist of an integrating meter element (kWh or kvarh) driving a mechanism in which a timing device returns the demand actuator to zero at the end of each timing interval, leaving the maximum demand indicated on a passive pointer, display, or chart, which in turn is manually reset to zero at each reading period, generally 1 month. Such demand mechanisms operate on what is known as the block-interval principle. There are three types of block-interval demand registers: (1) the indicating type, in which the maximum demand obtained between each reading period is indicated on a scale or numeric display, (2) the cumulative type, in which the accumulated total of maximum demand during the preceding periods is indicated during the period after the device has been reset and before it is again reset, that is, the maximum demand for any one period is equal or proportional to the difference between the accumulated readings before and after reset, and (3) the multiple-pointer form, in which the demand is obtained by reading the position of the multiple pointers relative to their scale markings. The multiple pointers are resettable to zero.

Another form of demand meter, usually in a separate housing from its associated watthour meter, is the recording type, in which the demand is transferred as a permanent record onto a tape by printing, punching, or magnetic means or onto a circular or strip chart. A special form of tape recording for demand metering that has come into wide use in recent years is the pulse recorder, in which pulses from a pulse initiator in the watthour meter are recorded on magnetic tape or punched paper tape in a form usable for machine translation by digital-data-processing techniques. Advantages of this system are its great flexibility, freedom from the operating difficulties inherent in inked charts, and freedom from many of the personal errors of manual reading and interpretation of charts.

***Thermal, Logarithmic, or Lagged-Demand Meters.*** These are devices in which the indication of the maximum demand is subject to a characteristic time lag by either mechanical or thermal means. The indication is often designed to follow the exponential heating curve of electrical equipment. Such a response, inherent in thermal meters, averages on a logarithmic and continuous basis, which means that more recent loads are heavily weighted but that, as time passes, their effect decreases. The time characteristics for the lagged meter are defined as the nominal time required for 90% of the final indication with a constant load suddenly applied.

***Concordance of Demand Meters and Registers.*** The measurement of demand may be obtained with meters and registers having various operating principles and employing various means of recording or indicating the demand. On a constant load of sufficient duration, accurate demand meters and registers of both classifications will give the same value of maximum demand, within the limits of tolerance specified. On varying loads, the values given by accurate meters and registers of different classifications may differ because of the different underlying principles of the meters themselves. In commercial practice, the demand of an installation or a system is given with acceptable accuracy by the record or indication of any accurate demand meter or register of acceptable type.

## 2.10 MEASURING COMPONENT VALUES

It used to be that in the laboratory, resistances, inductances, and capacitances were measured by bridges, as described briefly in Sec. 2.3.4. These days, digital multimeters are frequently used to measure all these parameters.

### 2.10.1 Resistance Measurements

Even relatively inexpensive multimeters can make low-power measurements of resistance values between a few ohms and a hundred megohms or so. The resolution of such instruments varies from 1% of full scale to a part per million of full scale. These meters generally use a constant-current source with a known current controlled by comparing the voltage drop on an internal "standard" resistor to the voltage produced by a Zener diode. The current is set at such a level as to make the meter direct-reading in terms of the displayed voltage; that is, the number displayed by the meter reflects the voltage drop across the resistor, but the decimal point is moved and the scale descriptor is displayed as appropriate.

Multimeters often use three or more fixed currents and several voltage ranges to produce seven or more decade ranges with the full-scale reading from 1.4 to 3.9 times the range. For example, on the 1000-Ω range, full scale may be 3999.999 Ω. Power dissipated in the measured resistor generally does not exceed 30 mW and reaches that level only in the lowest ranges where resistors are usually designed to handle many times that power. The most accurate multimeters have a resolution of 1 to 10 ppm of range on all ranges above the 10-Ω range. Their sensitivity, linearity, and short-term stability make it possible to compare nominally equal resistors by substitution with an uncertainty two to three times the least count of the meter. This permits their use in making very accurate measurements, up to 10 ppm, or resistors whose values are close to those of standards at hand.

Many less expensive multimeters have only two leads or terminals to use to make measurements. In those cases, the leads from the meter to the resistor to be measured become part of the measured resistance. For low resistances, the lead resistance must be measured and subtracted out, or zeroed out.

### 2.10.2 Capacitance Measurements

The measurement of capacitance, like the measurement of resistance, is now routinely done by digital multimeters. Some measure by passing the same current through the capacitance to be measured and a reference resistor, and computing the impedance of the unknown from the ratio of the voltage drops across it and the reference resistor. This is a phasor calculation done using a 90° phase reference that can be generated using digital synthesis techniques.

Many automated bridges are intended for testing of precision components over a broad range of frequencies and with programmable direct current or voltage biases. Their accuracies range from a few percent at high frequencies to 0.01% or better at audio frequencies.

### 2.10.3 Inductance Measurements

Inductance measurement methods at power and audio frequency are (1) null methods employing bridges, or (2) deflection methods in which the inductance is computed from the measured values of the signals involved.

Bridges for inductance measurements can assume a variety of forms, depending on available components and reference standards, magnitude and time constant of the inductance to be measured, and a variety of other factors. In a four-arm bridge similar to the Wheatstone network, an inductance can be (1) compared with another inductance in an *adjacent* arm with two resistors forming the "ratio" arms or (2) measured in terms of a combination of resistance and capacitance in the *opposite* arm with two resistors as the "product" arms. It is generally better, where possible, to measure inductance in terms of capacitance and resistance rather than by comparison with another inductance because the problems of stray fields and coupling between bridge components are more easily avoided.

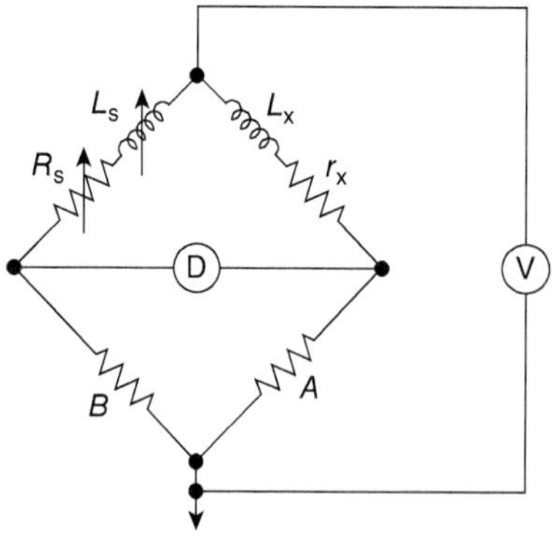

The basic circuits will be described for a few bridges which can be used to measure inductance, but a more detailed reference than this handbook (Golding, 1938; Harris, 1952; Horowitz and Hill, 1989) should be consulted for details.

In the balance equations which will be stated below, the inductance, $L$ or $M$, will be expressed in henrys, the resistance $R$ in ohms, capacitance in farads, and $\omega$ is $2\pi$ × frequency in hertz. The time constant of an inductor is $L/R$; its storage factor $Q$ is $\omega L/R$.

*Inductance comparison* is accomplished in the simple Wheatstone network shown in Fig. 2-20, in which $A$ and $B$ are resistive ratio arms, $L_x$ and $r_x$ represent the inductor and the associated resistance being measured, and $L_s$


FIGURE 2-20 Inductance bridge.

and $r_s$ are the reference inductor and the associated resistance (including that of the inductor itself) required to make the time constants of the two inductive arms equal. At balance,

$$\frac{A}{B}=\frac{L_x}{L_s}=\frac{r_x}{r_s}$$

An inductometer may be used to achieve balance, together with an adjustable resistance in the same bridge arm, as indicated in the diagram. If only a fixed-value standard inductor is available, balance can be secured by varying one of the ratio arms, but there also must be an adjustable resistance in series with $L_x$ or $L_s$ to balance the time constants of the inductive arms. Care must be taken to ensure that there is no inductive coupling between $L_s$ and $L_x$, since this would lead to a measurement error.

The *Maxwell-Wien bridge* for the determination of inductance in terms of capacitance and resistance is shown in Fig. 2-21. The balance equations are $L_x = ASC$ and $r_x = AS/B$. This bridge is widely used for accurate inductance measurements. It is most easily balanced by adjustments of capacitor $C$ and resistor $B$; these elements are in quadrature, and therefore their adjustments do not interact.

*Anderson's bridge,* shown in Fig. 2-22, can be used for measurement over a wide range of inductances with reasonable values of $R$ and $C$. Its balance equations are $L_x = CAS\ (1 + R/S + R/B)$ and $r_x = AS/B$. Balance adjustments are best made with $R$ and $r_x$. This bridge also has been used to measure the residuals of resistors, a substitution method being employed in which the unknown and a loop of resistance wire with calculable residuals are substituted in turn into the $L$ arm. If $A$ and $B$ are equal and if the resistances of the unknown and the calculable loop are matched, the residuals in the various bridge arms do not enter the final calculation, except the residual of $\Delta r_x$, the change in $r_x$ between balances. The elimination of bridge-arm residuals from the *exact* balance equations is characteristic of substitution methods, and quite generally, residuals or corrections to the arms that are unchanged between the balances do not have to be taken into account in the final calculation when the difference is small between the substituted quantities.

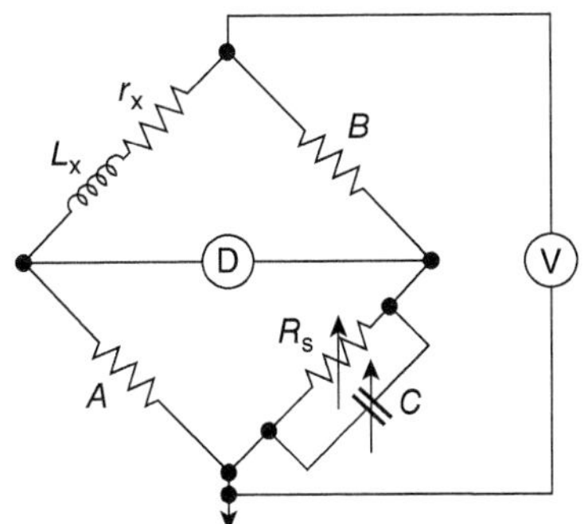


**FIGURE 2-21** Maxwell-Wien inductance-capacitance bridge.

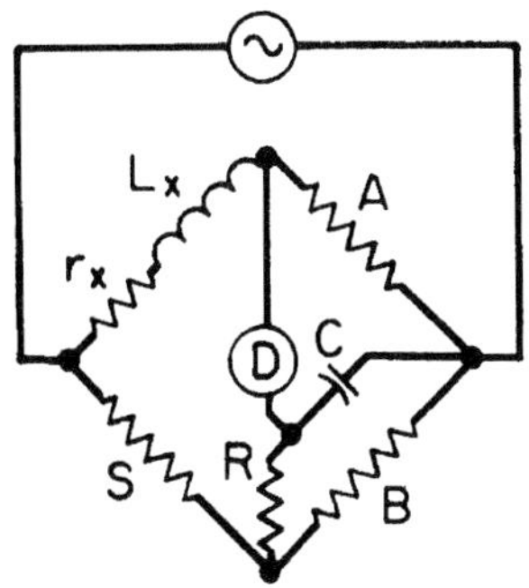


**FIGURE 2-22** Anderson's bridge.

*Owen's bridge,* shown in Fig. 2-23, can be used to measure a wide range of inductance with a standard capacitor $C_b$ of fixed value, by varying the resistance arms $S$ and $A$. In operation, the resistance $S$ and capacitor $C_b(r_b)$ are usually fixed, balance being secured by successive adjustments of $A$ and $R$. At balance,

$$r_x + R = (C_b/C_a)S + \omega L_x \omega C_b r_b$$

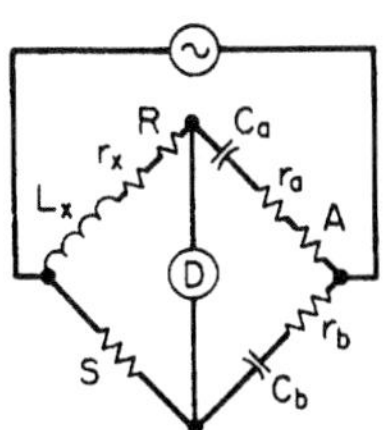


**FIGURE 2-23** Owen's inductance-capacitance bridge.

and $L_x(1 + \tan \delta_b \tan \delta_x) = C_b S(A + r_a)$. If $C_b(r_b)$ is a loss-free air capacitor so that $r_b = 0$ and $\tan \delta_b = 0$, $r_x = (C_b/C_z)S' - R$ and $L_x = C_b S(A + r_a)$. This is a bridge which is much used for examining the properties of magnetic materials; inductance may be measured with direct current superposed. With a low-reactance blocking capacitor in series with the detector and another in series with the source, a dc supply may be connected across the test inductance without current resulting in any other branch of the network; a high-reactance, low-resistance "choke" coil should be connected in series with the dc source.

*Mutual inductance* can be measured readily if an adjustable standard of proper range is available. Connections are made so the range of measurement is limited to values that can be read with the desired precision. Care should be taken in arranging the circuit to avoid coupling between the mutual inductors.

*Iron-cored inductors* vary in value with frequency and current, so measurements must be made at known current and frequency; bridge methods can, of course, be adapted to this measurement, care being exercised to ensure that the current capacities of the various bridge components are not exceeded. In such a case, the waveform of the voltage drop across the circuit branch containing the inductor may not be sinusoidal, whereas that across the other side of the bridge, containing linear resistances and reactances, may be undistorted. Generally, a tuned detector should be used.

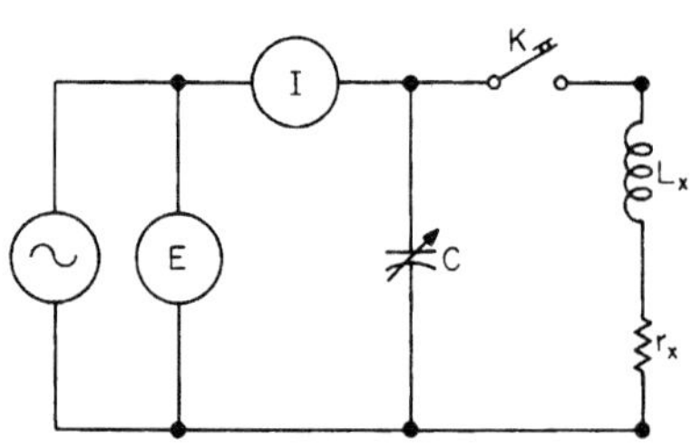


**FIGURE 2-24** Reactance-impedance method of measuring inductance.

A *reactance-impedance method,* suitable for high frequencies, is indicated in Fig. 2-24. The capacitor $C$ is adjusted until the same current is indicated by the ammeter with switch $K$ open or closed. (The applied voltage must be constant.) Then, $L_x = (1/2\omega^2 C)$ H if $C$ is in farads and the frequency is $f = \omega/2\pi$. The waveform must be practically sinusoidal and the ammeter of negligible impedance. This method may be used to measure the effective inductance of choke coils with superposed direct current.

## 2.11 FREQUENCY MEASUREMENT

Mathematically, the word frequency as applied to a sinusoidal function applies for all time, from minus infinity to plus infinity. Therefore, for any given system, the result of a single measurement at some time would apply at all other times, too. Obviously, this definition is of little use in the electric power system. While the speed of the generators is very nearly constant, it is not *exactly* constant at all times. Therefore, while the word frequency is still used, it is, in fact, not defined for this situation.

Most engineers think of the frequency as the derivative of the phase. If the signal being observed is represented by a sinusoid $x(t) = A\cos(\omega t + \varphi)$, that derivative relationship is true. The phase is $\psi(t) = \omega t + \varphi$, and the derivative is $\omega$. Power engineers tend to think of $\varphi$ as the phase, however. Thus, since that is not a function of time in the equation of the sinusoid, its derivative is zero. Evidently, this is not the right way to think of frequency or phase. (The problem faced by power engineers in this regard is explored later in Sec. 2.13.3.)

Electronic counters are widely used for frequency measurements. They work by counting the number of cycles of an input signal in a very accurately known time interval. The time interval is based on the output of an internal standard oscillator (clock) that may derive its time from a high-quality source such as WWV or GPS. Most counters of laboratory quality also can be used to measure time intervals, and the ratio of the frequencies of two input signals.

One could think of counting the zero crossings (or cycles) of the waveform as a reasonable way of learning the frequency. However, to obtain a resolution of 1 mHz on a power system operating at 60 Hz would require counting for about 17 seconds. Measurement devices that do this are readily available. One could think of the result as describing the *average* frequency over the interval.

Measurement of frequency with greater accuracy in shorter times has been a goal of engineering for some time (Law, 1933). With the advent of the phasor measurement unit, solutions have been found. See Sec. 2.13.3 for more information.

## 2.12 "UNUSUAL" MEASUREMENTS

### 2.12.1 High Voltage and Resistance

The measurement of high resistances is of interest in insulator studies. (An insulator is, after all, just a resistor with a very high value.) Most often, such resistances are measured at high voltage, partly because they will be used at high voltage, and partly because the thermal noise they produce increases with the absolute temperature and the resistance: $v_{rms} = \sqrt{4\, k_B\, T\, R\, \Delta f}$, where $k_B$ is Boltzmann's constant, $T$ the absolute temperature, $R$ the resistance, and $\Delta f$ the bandwidth. A higher voltage puts the measurement further above the noise floor.

There are several challenges to face in making the measurement. The measurements tend to be slow to make, because the large resistance values combine with stray capacitance to produce large time constants, and they tend to be sensitive to stray coupling. With dc measurements, there is a possibility that there is charge trapped in the insulator, and that can affect the result of a direct-reading method or a bridge. Further, the current that flows in a very large-value resistance such as an insulator may flow over the surface or through the bulk material, and it may be important to separate these two components of the current.

***Bulk and Surface Measurements.*** Suppose that a cylindrical sample of material is connected to apparatus to measure the voltage across it and the current through it, as shown in Fig. 2-25. (The figure shows both a cross-section and a plan view.) There will in fact be two components to the current in the external circuit, one through the bulk of the material and one over its surface. In general, the surface current is dependent on the condition of the external atmosphere and the surface itself. If the material is an insulator, with very high bulk resistance, the surface current may be large enough to dominate the total.

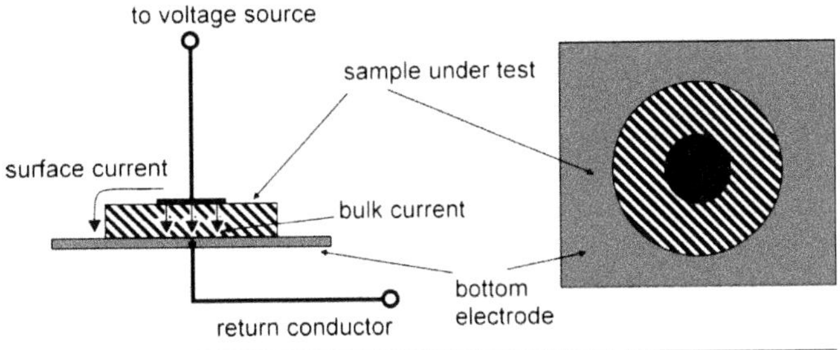


**FIGURE 2-25** Measurement of sample resistance.

***Guarding.*** A technique of *guarding* is customarily used to separate the two current components. Suppose that an additional electrode is included in the setup. Since the sample is circular, the additional electrode is annular, and surrounds the original upper electrode. Most commonly, the guard electrode is an electrode that surrounds the active electrode, as seen in Fig. 2-26.

In Fig. 2-26, the guard electrode and the center electrode are at the same potential (since there is no voltage drop across a current measurement device), so that there is no surface current (horizontal in the sectional view). It follows that the resistance calculated from these voltage and current

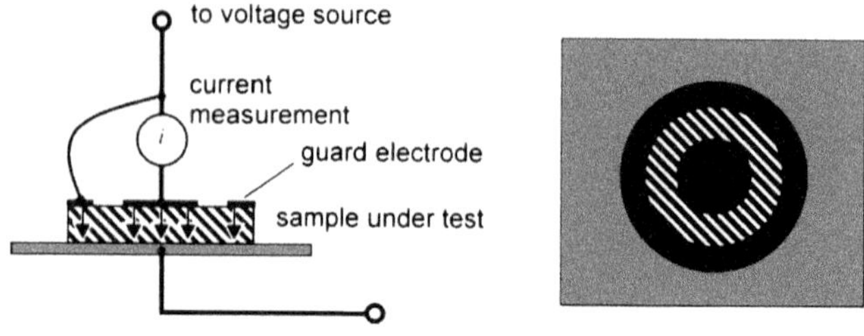


FIGURE 2-26 Measurement of bulk resistance.

measurements must be the bulk resistance. If the annular space between the center (active) electrode and the guard is small, the current injection into the material is uniform (vertical in the sectional view), and the resistance value calculated from the measurement results is accurate.

If the current is instead measured only from the guard electrode, the setup measures the surface current only, and while there is still a bulk current, its value does not affect the result. For better accuracy, the annular space should be larger than the optimum for bulk resistance measurements, however. The setup shown in Fig. 2-27 allows the calculation of surface resistance.

Note that some current still flows through the bulk of the material, but since this current is not included in the resistance calculation, its value does not affect the result. The lower electrode, since it carries the unmonitored current, could be considered a guard electrode in this configuration.

***Dielectric Loss.*** *Dielectric loss,* which occurs in cables and insulating bushings used at high voltages, represents an undesirable absorption of available energy and, more important, a restriction on the capacity of cables and insulating structures used in high-voltage power transmission. The problem of measuring the power consumed in these insulators is quite special, since their power factor is extremely low and the usual wattmeter techniques of power measurement are not applicable.

The Schering bridge, shown in Fig. 2-28, is an adaptation of the Wheatstone configuration that has been used to measure the loss angles of high-voltage power cables and high-voltage insulators. For this purpose, the supply voltage is connected as shown, and a ground connection is made at the junction of the two resistive branches. The impedances of the capacitive branches are much higher than the resistive ones, so the balance adjustments may be made close to ground potential.

Figure 2-28 shows the basic circuit of the bridge, as described by Schering and Semm in 1920. The balance equations are $C_x = C_s\ R_1/R_2$ and $\tan\delta_x = \omega R_1 P$, where $C_x$ is the cable or bushing whose losses

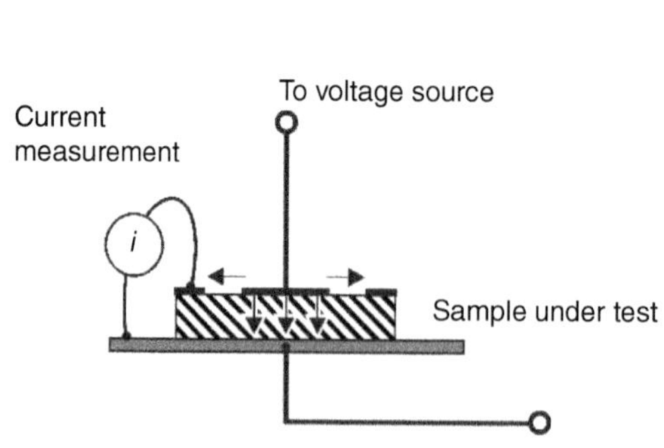


FIGURE 2-27 Measurement of surface resistance.

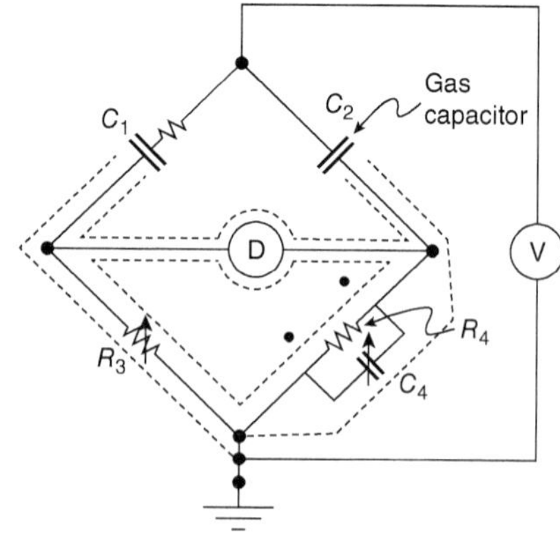


FIGURE 2-28 Schering and Semm's bridge for measuring dielectric loss.

are to be determined, $C_s$ is a loss-free high-voltage air-dielectric capacitor, $R_1$ and $R_2$ are noninductive resistors, and $P$ is an adjustable low-voltage capacitor having negligible loss.

***Ground Resistivity.*** Grounding systems are needed for a variety of reasons in power systems. For example, a connection to ground may be made for single-pole operation of a dc line that is normally operated as a bipole, or to ground a surge arrestor, a substation, or a transmission tower. Broadly speaking, ground connections are important in power systems, and challenging to make.

When a grounding system is planned, the local ground conditions are first evaluated. That means that the resistivity is measured at the planned site. Because of the possibility of induced voltages and currents, such measurements can be hazardous, and a test plan based on sound safety rules should be regarded as essential. This part of the handbook will not teach the whole complicated topic of ground measurements. The intent in this subsection is to explain the basic principles. For details on the many methods and precautions, the reader is referred to IEEE Std 81 and IEE Std 81.2.

The matter is made complicated by the fact that generally speaking, the ground itself is not homogeneous, nor are its properties stationary. The resistivity is affected by salt content, moisture content, and temperature. The distances that the electrode wires have to pass leaves them open to induced voltages and currents. (The effects of these measurement results can be reduced by filtering, but they may still be hazardous to the people involved.) Further, there may be conducting objects in the vicinity (buried or visible) that can lead to errors in the results of the measurements.

There are several ways the resistivity can be measured. One is similar to the four-wire resistance measurement described in the subsection Kelvin Double Bridge. Two probes are inserted into the ground, and a current passed between them. Two additional probes are inserted into the ground between the first two (and along a line joining them) and the voltage between these is measured with a high-impedance voltmeter.

The ratio of voltage to current obviously is the units of resistance. Suppose that the value measured is $R$. From this number can be calculated the ground resistivity according to formulas that take account of the geometry of the situation. If the probes are evenly spaced with separation $a$, and all driven to the same depth, and if the depth does not exceed 10% of the separation, the resistivity can be found from $\rho = 2\pi aR$. The arrangement is shown in Fig. 2-29.

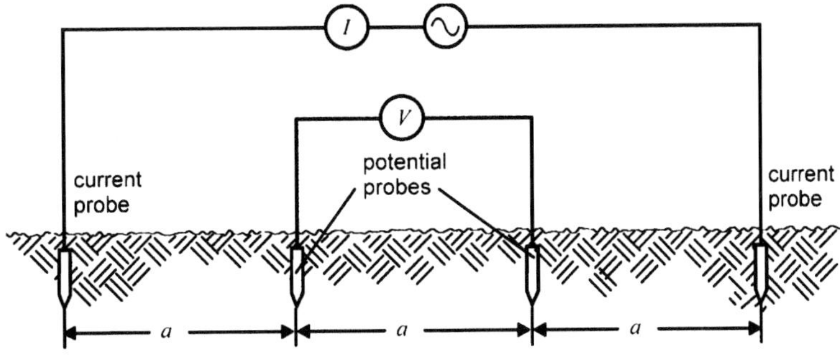


**FIGURE 2-29** Four-point, equal spacing.

Measurements of this kind normally require many repetitions, to reduce the effect of local variations in the soil or layer being measured. A simplification that requires only two of the probes to be measured is obtained when the inner probes are moved closer together. In this case, if the depth of the probes is $b$, and this is small compared to the probe separations $c$ and $d$, the resistivity can be obtained from $\rho = \dfrac{\pi c(c+d)R}{d}$. This configuration is shown in Fig. 2-30.

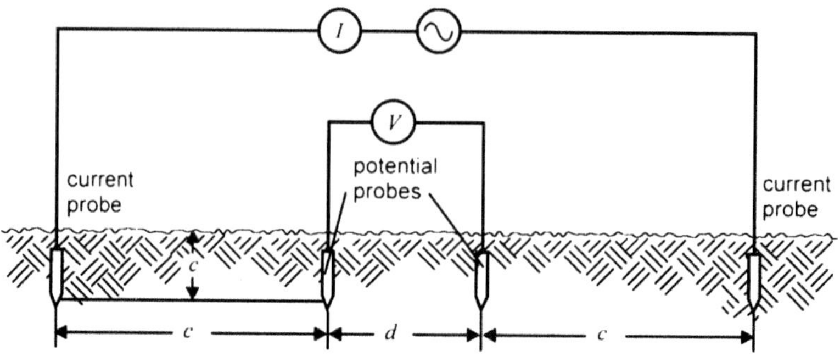


**FIGURE 2-30** Four-point, unequally spaced.

The resistivity is supposed to be applicable at a depth given by what is effectively the radius of the equipment, $c + d/2$. That allows the method to give results applicable to greater depth for greater dimensions, and it may then be possible to "see" the existence of nonuniform layers in the ground.

For greater appreciation of the complexities and nuances involved, the reader is urged to consult the IEEE standards.

***Ground Resistance.*** There are times when it is necessary to know the impedance to ground of an existing grounding system. (For observations on measuring the ground resistivity, see the previous subsection.) The difficulty of making a measurement of grounding resistance or impedance is that it is not known in advance what path an injected current will take. The resistance measured is not that of a small component with two or four connections, in which current goes in one end and comes out the other. Instead, current will be injected into an electrode inserted into the ground, and it will emerge from another some considerable distance away. The electrodes may be so far apart that they may have to be monitored by different personnel. The current may be visualized as spreading out from a source electrode and converging on the sink electrode. But how wide the path is and how deep into the earth the current goes depends on many factors that are generally not known before the measurement is made.

It is possible to use ac or dc to make the measurements, but if dc is used it is advisable to reverse the polarity periodically to reduce the effect of electrochemical action in the ground. Measurements at low frequency are often preferred since the current flowing in the event of a fault will be at power frequency. Strictly, it is the impedance value that is needed, but the word resistance is most commonly used because the impedance is largely resistive.

In the "fall-of-potential" method, a current is injected into the ground electrode, forming a circuit of large size, as the other ground electrode is a distance $D$ away. This distance should be at least five times further away from the middle of the ground electrode structure than its largest dimension. In a manner similar to that used in the four-terminal measurement of resistance (see the subsection Kelvin Double Bridge), the voltage is measured *inside* the circuit formed by the current injection electrodes. It is usual to have the distance from the ground electrode under test to the potential probe that is in the direction of the current probe be located at 62% of the distance to the current probe, as this is the location that would yield the appropriate resistance if the ground were uniform (Curdts, 1958). Figure 2-31 shows the arrangement.

It is important to check that the probe locations are free from any unwanted effects caused by buried structure. To do this, the location of the two potential probes is moved, and the impedance values calculated and plotted as a function of the distances. A representative result is shown in Fig. 2-32.

Both potential electrodes show a region of calculated impedance, where the value is relatively constant. According to the standard (IEEE Std 81-2012) the value in this region is taken as the ground electrode resistance to ground.

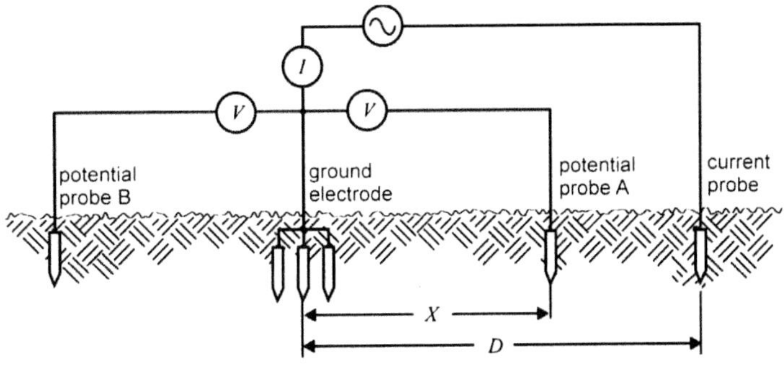


**FIGURE 2-31** Fall-of-potential connections.

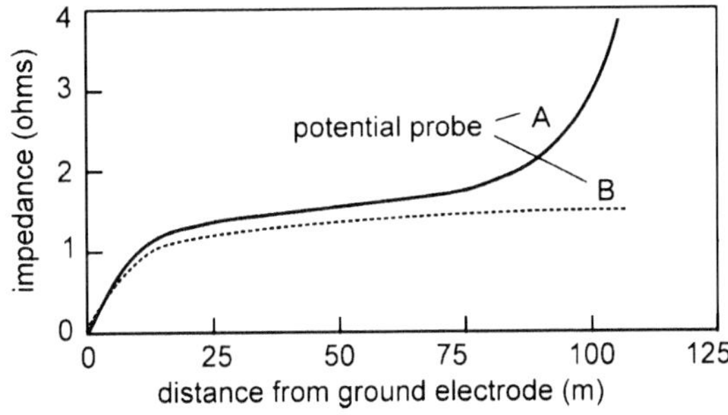


**FIGURE 2-32** Impedance as a function of spacing for fall-of-potential method.

There are other methods to make the measurement of the resistance to ground, but the fall-of-potential method has the advantage that the resistance to ground of the other electrodes is relatively unimportant.

## 2.13 PHASOR MEASUREMENT

The voltages and currents in the power system are customarily described by sinusoidal functions such as $x(t) = A\cos(\omega t + \varphi)$, where the $x$ could stand for the voltage or the current, the $A$ for the amplitude, and $\omega$ for the angular frequency. The time is given by $t$, and $\varphi$ is discussed below. Historically, the three parameters that describe this function were measured separately. The amplitude could be measured by a voltmeter or an ammeter, and the frequency by a frequency meter as described in Sec. 2.11. However, since the phase angle is so important in understanding the power flowing a power system, measuring the angle between one part of the system and another has been a subject of research for some time. Late in the 20th century, a method of measuring the angle between some part of the power system and a reference signal that could be derived from a precise knowledge of the time was demonstrated, and devices knows as phasor measurement units (PMUs) became commercially available. Since all PMUs use the same reference signal, the angle between any two points in the power system can therefore be calculated by subtraction.

The PMU is given signals from instrumentation transformers from which it derives the voltage and the current magnitudes and their angles, and reports these values as they were found in sequential observation windows. The PMU also furnishes values for the frequency and for the rate of change of frequency.

The applicable standards, discussed below, classify PMUs as P class and M class. P class responds more rapidly than M class, and may have an observation window of just two cycles. PMUs report the results of their measurements at times set by standard. The rates are linked to the nominal power frequency, and to UTC. PMUs differ from previous instrumentation in that (1) they report their measurement results simultaneously across the entire power system and at much higher rate than earlier instruments (rates may be as high as 60 times per second) and (2) the angle is included in the measured results.

### 2.13.1 Magnitude

The magnitude numbers reported by the PMU are the usual values for $A$ in the cosine equation above. The PMU may also perform the phase sequence calculations and report the positive sequence number. The amplitude is assumed constant across the window, and no rate of change is measured or reported.

### 2.13.2 Angle

It is well known that the power flow across a transmission line is a function of the voltages at the ends and the sine of the angle between those voltages. This relationship is used in the solution of the power system load flow, but until fairly recently there was no system to routinely measure that angle. That problem was solved by the advent of the PMU.

Angle is one of those things that has to be measured with respect to something. Rather than measure the angles from some point in the power system to some others, the PMU measures the angle between the signal at its input and an artificially created reference at exactly the nominal power system frequency of 50 or 60 Hz. This reference is synchronized to UTC. Typically a sinusoidal signal is not actually generated, but the timing of it is given by a pulse derived from a GPS receiver in the PMU. (It may also be that multiple PMUs in a substation get their time signal distributed from a single receiver.) From PMU data, the angle between any two points of interest in the power system can be obtained by subtraction.

The power system is only rarely at exactly the nominal frequency, though it is almost always very close to it. Since it is well known that the phase angle between two signals at different frequencies is not defined, it is fair to ask how the PMU can measure the angle between the fixed-frequency reference and the varying-frequency power system. The explanation depends on understanding what is meant by the word "phase."

Power engineers represent the voltage and currents they generate by a sinusoid. To them, the $\varphi$ in the equation $x(t) = A\cos(\omega t + \varphi)$ is the phase. Strictly speaking, that is true only because they have adopted the convention (which most of them have forgotten because of the convenient application of phasor diagrams) that the complete argument of the cosine is the phase $\psi(t) = \omega t + \varphi$, the value at $t = 0$ is what power engineers call the phase. To say that the phase angle between two signals of different frequency is not defined is to recall that time-dependent term $\omega t$ in the whole argument. Unless the complete time-dependent history of the two signals is known, the relative phase of two signals with different $\omega$ is, indeed, not defined.

However, within the short observation window of the PMU, the whole argument of the cosine can be measured, even if its history is not known. Since the times of every sample of the signal are quite accurately known, the value for the $\psi$ for any given measurement window is measurable. That is to say, for the reference frequency, we have a series of samples of a wave described by $x_r(t) = A_r \cos(\omega_r t)$ and for the signal we have samples of $x_s(t) = A_s \cos(\omega_s t + \varphi)$. The reference wave is synchronized to UTC by defining it to reach a positive maximum at the second tick of UTC. Since the frequency is defined as exactly the nominal value, there is effectively a zero of time at the peak of every reference every cycle.

The details of how the angle is measured in a commercial PMU are proprietary, but one may imagine a solution that uses a quadrature phase detector on the two signals, or one that uses the

Fourier transform method (Phadke and Thorp, 2008). [Other methods are also possible (Kirkham and Riepnieks, 2016) shows that the problem can be solved as a fitting problem.]

It should be pointed out that because the complete history of the frequency is not recorded by the PMU, the phase value reported is the principal value, that is, the value closest to the local zero. As time passes, this value sometimes goes from slightly less than 180° to slightly more than −180°. The user who needs a trend will have to "unwrap" the reported values (Venkatasubramanian, 2016) in order to avoid a sudden discontinuity in numbers.

If a phase reference is defined at some point in time, such as midnight, the total phase of the power system at any later time is a measure of the time error that would be accumulated on a synchronous clock. There have been discussions of the possibility of adding such a feature to the PMU, but it is not presently being done.

### 2.13.3 Frequency

To measure the frequency by counting zero crossings has long been the traditional way because frequency is the time-dependent term in the cosine equation that describes many audio and radio signals. It is also known that the equation has a domain of time from minus infinity to plus infinity, and that the values of the coefficients in the equation are fixed for all this time. That is not the situation with the power system, because the frequency changes (slightly) with every momentary imbalance between load and generation.

The achievement of the PMU in measuring frequency is quite remarkable. Under steady conditions, the PMU achieves a resolution of 1 mHz with a set of signal samples occupying just a few cycles. The responses of different PMUs to a sudden change in frequency are not always the same, however. The cause is almost certainly due to differences in signal filtering, a situation whose resolution requires some reworking of the applicable standard. Presently, therefore, some users of PMU data interested in data during transient events resort to using successive phase angle reports to calculate their own value for frequency.

### 2.13.4 Rate of Change of Frequency

When it was realized that the power system frequency was not quite constant, it was thought that it would be useful to know the rate at which it was changing. (Strictly speaking, of course, the label "frequency" is inappropriate, as that term is reserved for a nonchanging parameter, whose rate of change is therefore defined as zero.) The first standard for the PMU (IEEE 1344-1995) levied a requirement to furnish a value for this quantity, but it was not until the third standard (C37.118.1-2011) was issued that there was a requirement for its accuracy, 0.01 Hz/s for P class, less accuracy for M class. It was soon found that PMUs were not able to meet these requirements, and they were "relaxed" in an Amendment (IEEE C37.118.1a-2014). The difficulty of making this measurement arises because the problem is "ill-conditioned," and it remains to be seen where the technology will go in an attempt to resolve the matter.

## 2.14 BIBLIOGRAPHY

Allen, E., Kosterev, D., and Pourbeik, P. (n.d.). Validation of Power System Models. *Proc IEEE PES General Meeting*, 2010.

Berrisford, A. J. (2012). "Smart meters should be smarter," *IEEE Power and Energy Society General Meeting*, San Diego, CA, July 2012.

Blondel, A. (1894). Measurements of the Energy of Polyphase Currents. *Proc International Electrical*, pp. 112–117.

Budeanu, C. (1927). *Puisances reacives et fictives*. Bucharest: Instytut Romain de l'Energie.

C37.118.1. (2014). IEEE Standard for Synchrophasor Measurements for Power Systems. Amendment 1: Modification of Selected Performance Requirements, IEEE.

Curdts, E. (1958). Some of the Fundamental Aspects of Ground Resistance Measurements. *AIEE,* vol. 77(5), pp. 760–767.

Fleming, J. (1891). On Some Effects of Alternating-Current Flow in Circuits Having Capacity and Self-Induction. *Journal IEE,* vol. 20(94), pp. 362–408.

Fleming, J. (1892). Experimental Researches on Alternating Current Transformers. *Journal IEE,* vol. XXI, pp. 594–686.

Golding, E. (1938). *Electrical Measurements and Measuring Instruments,* 2nd ed., Pitman and Sons, London.

Harris, F. (1952). *Electrical Measurements,* Wiley, New York.

Horowitz, P. and Hill, W. (1989). *The Art of Electronics.* Cambridge University Press, Cambridge, UK.

IEEE Standard for Synchrophasor Measurements for Power Systems. (Dec. 2011). *IEEE Std.C37.118.1-2011 (Revision of IEEE Std.C37.118-2005),* pp. 1–61.

Kirkham, H. and Riepnieks, A. (2016). Students' Simple Method for Determining the Parameters of an AC Signal. *Proc RTUCon 2016.* IEEE, Riga, Latvia.

Law, R. (1933). A New Radiofrequency Phase Meter. *Review of Scientific Instruments,* vol. 4, no. 10, pp. 537–539.

Metrology, W. G. (2008). *GUM: Evaluation of Measurement Data—Guide to the Expression of Uncertainty in Measurement.* BIPM, Paris.

NCSL International 174 Writing Committee (ASC Z540). (2006). *ANSIINCSL Z540.3-2006: Requirements for the Calibration of Measuring and Test Equipment.* American National Standards Institute.

Nelson, T. (2011). *Definitions for Calculations of VA, VAh, VAR, and VARh for Poly-Phase Electricity Meters, NEMA C12.24 TR-2011.* NEMA.

Phadke, A. and Thorp, J. (2008). *Synchronized Phasor Measurements and Their Applications.* Springer, New York.

Sensors, W. G. (1994). Optical Current Transducers for Power Systems: A Review. *IEEE Transactions of Power Delivery,* vol. 9(4), Oct. 1994, pp. 1778–1788.

Shannon, C. (1948). A Mathematical Theory of Communication. *The Bell System Technical Journal,* vol. 27, pp. 379–423, pp. 623–656.

van Deursen, A. and van Waes, J. (2002). Mitigation of Ground Loop Effects in High-Voltage. *IEEE Transactions on Instrumentation and Measurement,* vol. 51(3), pp. 480–486.

Venkatasubramanian, V. (2016). Real-Time Strategies for Unwrapping of Synchrophasor Phase Angles. *IEEE Transactins on Power Systems.,* Vol. 31, No. 6, Nov. 2016, pp. 5033–5041.

# 3 PROPERTIES OF MATERIALS



## 3.1 CONDUCTOR MATERIALS

### 3.1.1 General Properties

***Conducting Materials.*** A *conductor of electricity* is any substance or material which will afford continuous passage to an electric current when subjected to a difference of electric potential. The greater the density of current for a given potential difference, the more efficient the conductor is said to be. Virtually, all substances in solid or liquid state possess the property of electric conductivity in some degree, but certain substances are relatively efficient conductors, while others are almost totally devoid of this property. The metals, for example, are the best conductors, while many other substances, such as metal oxides and salts, minerals, and fibrous materials, are relatively poor conductors, but their conductivity is beneficially affected by the absorption of moisture. Some of the less-efficient conducting materials such as carbon and certain metal alloys, as well as the efficient conductors such as copper and aluminum, have very useful applications in the electrical arts.


Grateful acknowledgment is given to past contributors to this section: Glenn Davidson, Philip Mason Opsal, Donald J. Barta, T. W Dakin, Charles A. Harper, Duane E. Lyon, Charles B. Rawlins, James Stubbins, and John Tanaka.

Certain other substances possess so little conductivity that they are classed as nonconductors, a better term being insulators or dielectrics. In general, all materials which are used commercially for conducting electricity for any purpose are classed as conductors.

***Definition of Conductor.*** A *conductor* is a body so constructed from conducting material that it may be used as a carrier of electric current. In ordinary engineering usage, a conductor is a material of relatively high conductivity.

***Types of Conductors.*** In general, a conductor consists of a solid wire or a multiplicity of wires stranded together, made of a conducting material and used either bare or insulated. Only bare conductors are considered in this subsection. Usually the conductor is made of copper or aluminum, but for applications requiring higher strength, such as overhead transmission lines, various composite constructions are used. For conductors having very low conductivity and used as resistor materials, a group of special alloys is available.

***Definition of Circuit.*** An *electric circuit* is the path of an electric current, or more specifically, it is a conducting part or a system of parts through which an electric current is intended to flow. Electric circuits in general possess four fundamental electrical properties, consisting of resistance, inductance, capacitance, and leakage conductance. That portion of a circuit which is represented by its conductors will also possess these four properties, but only two of them are related to the properties of the conductor considered by itself. Capacitance and leakage conductance depend in part on the external dimensions of the conductors and their distances from one another and from other conducting bodies, and in part on the dielectric properties of the materials employed for insulating purposes. The inductance is a function of the magnetic field established by the current in a conductor, but this field as a whole is divisible into two parts, one being wholly external to the conductor and the other being wholly within the conductor; only the latter portion can be regarded as corresponding to the magnetic properties of the conductor material. The resistance is strictly a property of the conductor itself. Both the resistance and the internal inductance of conductors change in effective values when the current changes with great rapidity as in the case of high-frequency alternating currents; this is termed the *skin effect*.

In certain cases, conductors are subjected to various mechanical stresses. Consequently, their weight, tensile strength, and elastic properties require consideration in all applications of this character. Conductor materials as a class are affected by changes in temperature and by the conditions of mechanical stress to which they are subjected in service. They are also affected by the nature of the mechanical working and the heat treatment which they receive in the course of manufacture or fabrication into finished products.

## 3.1.2 Metal Properties

***Specific Gravity and Density.*** Specific gravity is the ratio of mass of any material to that of the same volume of water at 4°C. Density is the unit weight of material expressed as pounds per cubic inch, grams per cubic centimeter, etc., at some reference temperature, usually 20°C. For all practical purposes, the numerical values of specific gravity and density are the same, expressed in g/cm$^3$.

*Density and Weight of Copper.* Pure copper, rolled, forged, or drawn and then annealed, has a density of 8.89 g/cm$^3$ at 20°C or 8.90 g/cm$^3$ at 0°C. Samples of high-conductivity copper usually will vary from 8.87 to 8.91 and occasionally from 8.83 to 8.94. Variations in density may be caused by microscopic flaws or seams or the presence of scale or some other defect; the presence of 0.03% oxygen will cause a reduction of about 0.01 in density. Hard-drawn copper has about 0.02% less density than annealed copper, on average, but for practical purposes the difference is negligible.

The international standard of density, 8.89 at 20°C, corresponds to a weight of 0.32117 lb/in$^3$ or $3.0270 \times 10^{-6}$ lb/(cmil)(ft) or $15.982 \times 10^{-3}$ lb/(cmil)(mile). Multiplying either of the last two figures by the square of the diameter of the wire in mils will produce the total weight of wire in pounds per foot or per mile, respectively.

*Copper Alloys.* Density and weight of copper alloys vary with the composition. For hard-drawn wire covered by ASTM Specification B105, the density of alloys 85 to 20 is 8.89 g/cm$^3$ (0.32117 lb/in$^3$) at 20°C; alloy 15 is 8.54 (0.30853); alloys 13 and 8.5 is 8.78 (0.31720).

*Copper-Clad Steel.* Density and weight of copper-clad steel wire and cable is a mean between the density of copper and the density of steel, which can be calculated readily when the relative volumes or cross sections of copper and steel are known. For practical purposes, a value of 8.15 g/cm$^3$ (0.29444 lb/in$^3$) at 20°C is used. See Table 3-1 for weight, breaking strength, and dc resistance.

*Aluminum Wire.* Density and weight of aluminum wire (commercially hard-drawn) is 2.705 g/cm$^3$ (0.0975 lb/in$^3$) at 20°C. Physical characteristics are shown in Fig. 3-1. The density of electrolytically refined aluminum (99.97% Al) and of hard-drawn wire of the same purity is 2.698 at 20°C. With less pure material there is an appreciable decrease in density on cold working. Annealed metal having a density of 2.702 will have a density of about 2.700 when in the hard-drawn or fully cold-worked conditions (see NBS Circ. 346, pp. 68 and 69).

*Aluminum-Clad Wire.* Density and weight of aluminum-clad wire is a mean between the density of aluminum and the density of steel, which can be calculated readily when the relative volumes or cross sections of aluminum and steel are known. For practical purposes, a value of 6.59 g/cm$^3$ (0.23808 lb/in$^3$) at 20°C is used.

*Aluminum Alloys.* Density and weight of aluminum alloys vary with type and composition. For hard-drawn aluminum alloy wire 5005-H19 and 6201-T81, a value of 2.703 g/cm$^3$ (0.09765 lb/in$^3$) at 20°C is used.

*Pure Iron and Galvanized Steel Wire.* Density and weight of pure iron is 7.90 g/cm$^3$ [$2.690 \times 10^{-6}$ lb/(cmil)(ft)] at 20°C. Density and weight of galvanized steel wire (EBB, BB, HTL-85, HTL-135, and HTL-195) with Class A weight of zinc coating are 7.83 g/cm$^3$ (0.283 lb/in$^3$) at 20°C, with Class B are 7.80 g/cm$^3$ (0.282 lb/in$^3$), and with Class C are 7.78 g/cm$^3$ (0.281 lb/in$^3$). The weight, breaking strength, and dc resistance values of galvanized-steel wire are shown in Fig. 3-2. The physical characteristics of smaller size galvanized-steel strand are shown in Fig. 3-3.

**Percent Conductivity.** It is very common to rate the conductivity of a conductor in terms of its percentage ratio to the conductivity of chemically pure metal of the same kind as the conductor is primarily constituted or in ratio to the conductivity of the international copper standard. Both forms of the conductivity ratio are useful for various purposes. This ratio also can be expressed in two different terms, one where the conductor cross sections are equal and therefore termed the *volume-conductivity ratio* and the other where the conductor masses are equal and therefore termed the *mass-conductivity ratio.*

**International Annealed Copper Standard.** The International Annealed Copper Standard (IACS) is the internationally accepted value for the resistivity of annealed copper of 100% conductivity. This standard is expressed in terms of mass resistivity as 0.15328 $\Omega \cdot$ g/m$^2$, or the resistance of a uniform round wire 1 m long weighing 1 g at the standard temperature of 20°C. Equivalent expressions of the annealed copper standard in various units of mass resistivity and volume resistivity are as follows:

| | |
|---|---|
| 0.15328 | $\Omega \cdot$ g/m$^2$ |
| 875.20 | $\Omega \cdot$ lb/mi$^2$ |
| 1.7241 | $\mu\Omega \cdot$ cm |
| 0.67879 | $\mu\Omega \cdot$ in at 20°C |
| 10.371 | $\Omega \cdot$ cmil/ft |
| 0.017241 | $\Omega \cdot$ mm$^2$/m |

**TABLE 3-1** Copper-Clad Steel Wire and Cable—Weight, Breaking Strength, DC Resistance (Based on ASTM Specifications B227 and B228)

| Conductor size,* AWG or in | Conductor stranding: No. of wires | Conductor stranding: Wire size, AWG | Conductor diam., in | Conductor area: Cir mils | Conductor area: Sq in | Conductor weight, lb: Per 1,000 ft | Conductor weight, lb: Per mile | Breaking strength, min., lb: High strength, Conductivity IACS 40% | Breaking strength, min., lb: High strength, Conductivity IACS 30% | Breaking strength, min., lb: Extra-high strength, Conductivity IACS 30% | DC resistance at 20°C (68°F), ohms per 1,000 ft: Conductivity IACS 40% | DC resistance at 20°C (68°F), ohms per 1,000 ft: Conductivity IACS 30% |
|---|---|---|---|---|---|---|---|---|---|---|---|---|
| | | | | | | Solid (B227) | | | | | | |
| 4 | . . . | . . . | 0.2043 | 41,740 | 0.03278 | 115.8 | 611.6 | 3,541 | 3,934 | 4,672 | 0.6337 | 0.8447 |
| 5 | . . . | . . . | 0.1819 | 33,090 | 0.02599 | 91.86 | 485.0 | 2,938 | 3,250 | 3,913 | 0.7990 | 1.065 |
| 0.165 | . . . | . . . | 0.1650 | 27,230 | 0.02138 | 75.55 | 398.9 | 2,523 | 2,780 | 3,368 | 0.9715 | 1.295 |
| 6 | . . . | . . . | 0.1620 | 26,240 | 0.02061 | 72.85 | 384.6 | 2,433 | 2,680 | 3,247 | 1.008 | 1.343 |
| 7 | . . . | . . . | 0.1443 | 20,820 | 0.01635 | 57.77 | 305.0 | 2,011 | 2,207 | 2,681 | 1.270 | 1.694 |
| 8 | . . . | . . . | 0.1285 | 16,510 | 0.01297 | 45.81 | 241.9 | 1,660 | 1,815 | 2,204 | 1.602 | 2.136 |
| 0.128 | . . . | . . . | 0.1280 | 16,380 | 0.01287 | 45.47 | 240.1 | 1,647 | 1,802 | 2,188 | 1.614 | 2.152 |
| 9 | . . . | . . . | 0.1144 | 13,090 | 0.01028 | 36.33 | 191.8 | 1,368 | 1,491 | 1,790 | 2.020 | 2.693 |
| 0.104 | . . . | . . . | 0.1040 | 10,820 | 0.008495 | 30.01 | 158.5 | 1,177 | 1,283 | 1,487 | 2.445 | 3.260 |
| 10 | . . . | . . . | 0.1019 | 10,380 | 0.008155 | 28.81 | 152.1 | 1,130 | 1,231 | 1,460 | 2.547 | 3.396 |
| 12 | . . . | . . . | 0.0808 | 6,530 | 0.005129 | 18.12 | 95.68 | 785 | . . . | . . . | 4.051 | . . . |
| 0.080 | . . . | . . . | 0.0800 | 6,400 | 0.005027 | 17.76 | 93.77 | 770 | . . . | 900 | 4.133 | 5.509 |

| | | | | | | | | | | | | |
|---|---|---|---|---|---|---|---|---|---|---|---|---|
| Stranded (B228) | | | | | | | | | | | | |
| 7/8 | 19 | 5 | 0.910 | 628,900 | 0.4940 | 1,770 | 9,344 | 50,240 | 55,570 | 66,910 | 0.04264 | 0.05685 |
| 13/16 | 19 | 6 | 0.810 | 498,800 | 0.3917 | 1,403 | 7,410 | 41,600 | 45,830 | 55,530 | 0.05377 | 0.07168 |
| 23/32 | 19 | 7 | 0.721 | 395,500 | 0.3107 | 1,113 | 5,877 | 34,390 | 37,740 | 45,850 | 0.06780 | 0.09039 |
| 21/32 | 19 | 8 | 0.642 | 313,700 | 0.2464 | 882.7 | 4,660 | 28,380 | 31,040 | 37,690 | 0.08550 | 0.1140 |
| 9/16 | 19 | 9 | 0.572 | 248,800 | 0.1954 | 700.0 | 3,696 | 23,390 | 25,500 | 30,610 | 0.1078 | 0.1437 |
| 5/8 | 7 | 4 | 0.613 | 292,200 | 0.2295 | 818.9 | 4,324 | 22,310 | 24,780 | 29,430 | 0.09143 | 0.1219 |
| 9/16 | 7 | 5 | 0.546 | 231,700 | 0.1820 | 649.4 | 3,429 | 18,510 | 20,470 | 24,650 | 0.1153 | 0.1537 |
| 1/2 | 7 | 6 | 0.486 | 183,800 | 0.1443 | 515.0 | 2,719 | 15,330 | 16,890 | 20,460 | 0.1454 | 0.1938 |
| 7/16 | 7 | 7 | 0.433 | 145,700 | 0.1145 | 408.4 | 2,157 | 12,670 | 13,910 | 16,890 | 0.1833 | 0.2444 |
| 3/8 | 7 | 8 | 0.385 | 115,600 | 0.09077 | 323.9 | 1,710 | 10,460 | 11,440 | 13,890 | 0.2312 | 0.3081 |
| 11/32 | 7 | 9 | 0.343 | 91,650 | 0.07198 | 256.9 | 1,356 | 8,616 | 9,393 | 11,280 | 0.2915 | 0.3886 |
| 5/16 | 7 | 10 | 0.306 | 72,680 | 0.05708 | 203.7 | 1,076 | 7,121 | 7,758 | 9,196 | 0.3676 | 0.4900 |
| . . . | 3 | 5 | 0.392 | 99,310 | 0.07800 | 277.8 | 1,467 | 8,373 | 9,262 | 11,860 | 0.2685 | 0.3579 |
| . . . | 3 | 6 | 0.349 | 78,750 | 0.06185 | 220.3 | 1,163 | 6,934 | 7,639 | 9,754 | 0.3385 | 0.4513 |
| . . . | 3 | 7 | 0.311 | 62,450 | 0.04905 | 174.7 | 922.4 | 5,732 | 6,291 | 7,922 | 0.4269 | 0.5691 |
| . . . | 3 | 8 | 0.277 | 49,530 | 0.03890 | 138.5 | 731.5 | 4,730 | 5,174 | 6,282 | 0.5383 | 0.7176 |
| . . . | 3 | 9 | 0.247 | 39,280 | 0.03085 | 109.9 | 580.1 | 3,898 | 4,250 | 5,129 | 0.6788 | 0.9049 |
| . . . | 3 | 10 | 0.220 | 31,150 | 0.02446 | 87.13 | 460.0 | 3,221 | 3,509 | 4,160 | 0.8559 | 1.141 |
| . . . | 3 | 12 | 0.174 | 19,590 | 0.01539 | 54.80 | 289.3 | 2,236 | . . . | . . . | 1.361 | . . . |

***Note:*** 1 in = 2.54 cm; 1 in$^2$ = 64.5 cm$^2$; 1 ft = 0.3048 m; 1 mi = 1.61 km; 1 lb = 0.4536 kg.
*To determine copper equivalent of copper-clad steel conductor, multiply circular-mil area by percent conductivity expressed as a decimal.

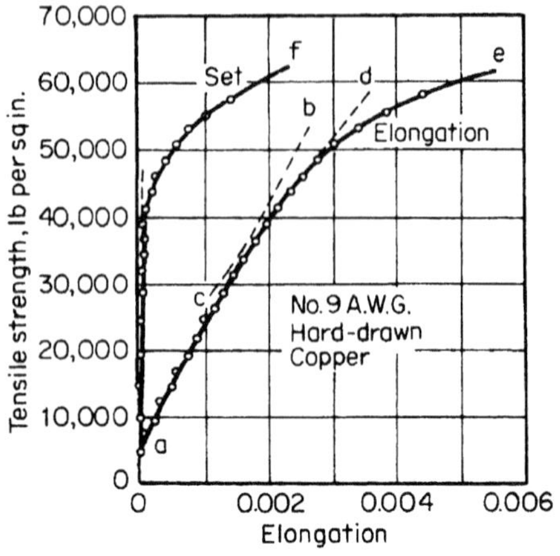


**FIGURE 3-1** Stress-strain curves of No. 9 AWG hard-drawn copper wire. (Watertown Arsenal test.)

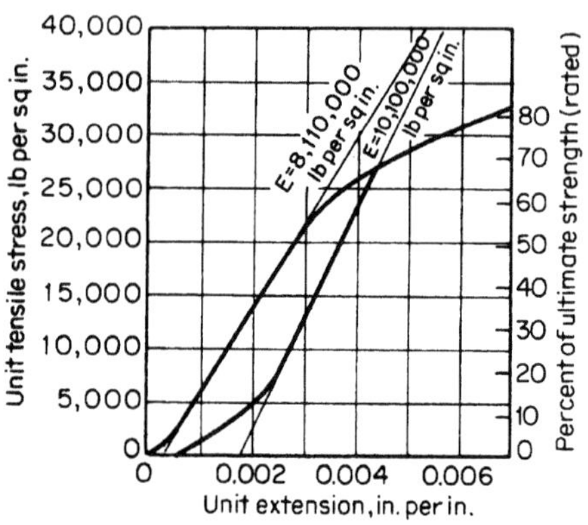


**FIGURE 3-2** Repeated stress-strain curve, 795,000 cmil ACSR; 54 × 0.1212 aluminum strands, 7 × 0.1212 steel strands.

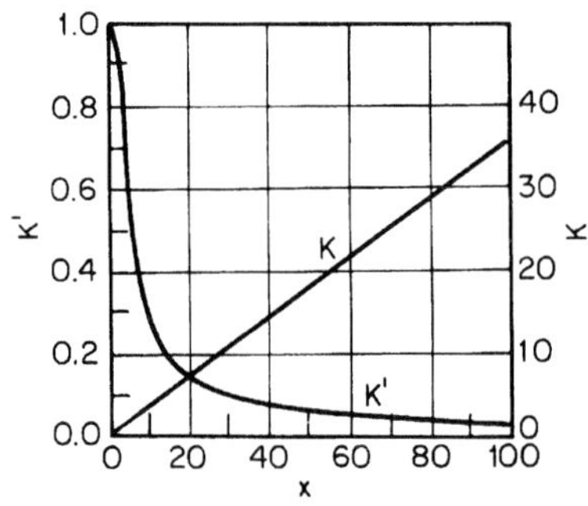


**FIGURE 3-3** *K* and *K'* for values of *x* from 0 to 100.

The preceding values are the equivalent of 1/58 $\Omega \cdot mm^2/m$, so the volume conductivity can be expressed as 58 S · $mm^2/m$ at 20°C.

***Conductivity of Conductor Materials.*** Conductivity of conductor materials varies with chemical composition and processing. The physical characteristics of aluminum cable are shown in Table 3-2.

***Electrical Resistivity.*** *Electrical resistivity* is a measure of the resistance of a unit quantity of a given material. It may be expressed in terms of either mass or volume; mathematically,

Mass resistivity: $$\delta = \frac{Rm}{l^2} \tag{3-1}$$

Volume resistivity: $$\rho = \frac{RA}{l} \tag{3-2}$$

where *R* is resistance, *m* is mass, *A* is cross-sectional area, and *l* is length.

Electrical resistivity of conductor materials varies with chemical composition and processing.

**TABLE 3-2** Aluminum Conductor—Physical Characteristics 1350-H19 Classes AA and A

| Cable code word | Conductor size | | Current-carrying capacity* A | Stranding | | Conductor diam, in | Rated strength, lb | Nominal weight, lb† | |
|---|---|---|---|---|---|---|---|---|---|
| | cmils or AWG | in² | | Class | No. and diam of wires, in | | | Per 1000 ft | Per mile |
| Peachbell | 6 | 0.0206 | 95 | A | 7 × 0.0612 | 0.184 | 563 | 24.6 | 130 |
| Rose | 4 | 0.0328 | 130 | A | 7 × 0.0772 | 0.232 | 881 | 39.2 | 207 |
| Iris | 2 | 0.0522 | 175 | AA, A | 7 × 0.0974 | 0.292 | 1,350 | 62.3 | 329 |
| Pansy | 1 | 0.0657 | 200 | AA, A | 7 × 0.1093 | 0.328 | 1,640 | 78.5 | 414 |
| Poppy | 1/0 | 0.0829 | 235 | AA, A | 7 × 0.1228 | 0.368 | 1,990 | 99.1 | 523 |
| Aster | 2/0 | 0.1045 | 270 | AA, A | 7 × 0.1379 | 0.414 | 2,510 | 124.9 | 659 |
| Phlox | 3/0 | 0.1317 | 315 | AA, A | 7 × 0.1548 | 0.464 | 3,040 | 157.5 | 832 |
| Oxlip | 4/0 | 0.1663 | 365 | AA, A | 7 × 0.1739 | 0.522 | 3,830 | 198.7 | 1,049 |
| Sneezewort | 250,000 | 0.1964 | 405 | AA | 7 × 0.1890 | 0.567 | 4,520 | 234.7 | 1,239 |
| Valerian | 250,000 | 0.1963 | 405 | A | 19 × 0.1147 | 0.574 | 4,660 | 234.6 | 1,239 |
| Daisy | 266,800 | 0.2097 | 420 | AA | 7 × 0.1953 | 0.586 | 4,830 | 250.6 | 1,323 |
| Laurel | 266,800 | 0.2095 | 425 | A | 19 × 0.1185 | 0.593 | 4,970 | 250.4 | 1,322 |
| Peony | 300,000 | 0.2358 | 455 | A | 19 × 0.1257 | 0.629 | 5,480 | 281.8 | 1,488 |
| Tulip | 336,400 | 0.2644 | 495 | A | 19 × 0.1331 | 0.666 | 6,150 | 316.0 | 1,668 |
| Daffodil | 350,000 | 0.2748 | 506 | A | 19 × 0.1357 | 0.679 | 6,390 | 328.4 | 1,734 |
| Canna | 397,500 | 0.3124 | 550 | AA, A | 19 × 0.1447 | 0.724 | 7,110 | 373.4 | 1,972 |
| Goldentuft | 450,000 | 0.3534 | 545 | AA | 19 × 0.1539 | 0.770 | 7,890 | 422.4 | 2,230 |
| Cosmos | 477,000 | 0.3744 | 615 | AA | 19 × 0.1584 | 0.793 | 8,360 | 447.5 | 2,363 |
| Syringa | 477,000 | 0.3743 | 615 | A | 37 × 0.1135 | 0.795 | 8,690 | 447.4 | 2,362 |
| Zinnia | 500,000 | 0.3926 | 635 | AA | 19 × 0.1622 | 0.811 | 8,760 | 469.2 | 2,477 |
| Hyacinth | 500,000 | 0.3924 | 635 | A | 37 × 0.1162 | 0.813 | 9,110 | 469.0 | 2,476 |
| Dahlia | 556,500 | 0.4369 | 680 | AA | 19 × 0.1711 | 0.856 | 9,750 | 522.1 | 2,757 |
| Mistletoe | 556,500 | 0.4368 | 680 | AA, A | 37 × 0.1226 | 0.858 | 9,940 | 522.0 | 2,756 |
| Meadowsweet | 600,000 | 0.4709 | 715 | AA, A | 37 × 0.1273 | 0.891 | 10,700 | 562.8 | 2,972 |
| Orchid | 636,000 | 0.4995 | 745 | AA, A | 37 × 0.1311 | 0.918 | 11,400 | 596.9 | 3,152 |
| Heuchera | 650,000 | 0.5102 | 755 | AA | 37 × 0.1325 | 0.928 | 11,600 | 609.8 | 3,320 |
| Verbena | 700,000 | 0.5494 | 790 | AA | 37 × 0.1375 | 0.963 | 12,500 | 656.6 | 3,467 |
| Flag | 700,000 | 0.5495 | 790 | A | 61 × 0.1071 | 0.964 | 12,900 | 656.8 | 3,468 |
| Violet | 715,500 | 0.5622 | 800 | AA | 37 × 0.1391 | 0.974 | 12,800 | 672.0 | 3,548 |
| Nasturtium | 715,500 | 0.5619 | 800 | A | 61 × 0.1083 | 0.975 | 13,100 | 671.6 | 3,546 |
| Petunia | 750,000 | 0.5892 | 825 | AA | 37 × 0.1424 | 0.997 | 13,100 | 704.3 | 3,719 |
| Cattail | 750,000 | 0.5892 | 825 | A | 61 × 0.1109 | 0.998 | 13,500 | 704.2 | 3,718 |
| Arbutus | 795,000 | 0.6245 | 855 | AA | 37 × 0.1466 | 1.026 | 13,900 | 746.4 | 3,941 |
| Lilac | 795,000 | 0.6248 | 855 | A | 61 × 0.1142 | 1.028 | 14,300 | 746.7 | 3,943 |
| Cockscomb | 900,000 | 0.7072 | 925 | AA | 37 × 0.1560 | 1.092 | 15,400 | 845.2 | 4,463 |
| Snapdragon | 900,000 | 0.7072 | 925 | A | 61 × 0.1215 | 1.094 | 15,900 | 845.3 | 4,463 |
| Magnolia | 954,000 | 0.7495 | 960 | AA | 37 × 0.1606 | 1.124 | 16,400 | 895.8 | 4,730 |
| Goldenrod | 954,000 | 0.7498 | 960 | A | 61 × 0.1251 | 1.126 | 16,900 | 896.1 | 4,731 |
| Hawkweed | 1,000,000 | 0.7854 | 990 | AA | 37 × 0.1644 | 1.151 | 17,200 | 938.7 | 4,956 |
| Camellia | 1,000,000 | 0.7849 | 990 | A | 61 × 0.1280 | 1.152 | 17,700 | 938.2 | 4,954 |
| Bluebell | 1,033,500 | 0.8124 | 1,015 | AA | 37 × 0.1672 | 1.170 | 17,700 | 970.9 | 5,126 |
| Larkspur | 1,033,500 | 0.8122 | 1,015 | A | 61 × 0.1302 | 1.172 | 18,300 | 970.6 | 5,125 |
| Marigold | 1,113,000 | 0.8744 | 1,040 | AA, A | 61 × 0.1351 | 1.216 | 19,700 | 1,045 | 5,518 |
| Hawthorn | 1,192,500 | 0.9363 | 1,085 | AA, A | 61 × 0.1398 | 1.258 | 21,100 | 1,119 | 5,908 |
| Narcissus | 1,272,000 | 0.999 | 1,130 | AA, A | 61 × 0.1444 | 1.300 | 22,000 | 1,194 | 6,304 |
| Columbine | 1,351,500 | 1.062 | 1,175 | AA, A | 61 × 0.1489 | 1.340 | 23,400 | 1,269 | 6,700 |
| Carnation | 1,431,000 | 1.124 | 1,220 | AA, A | 61 × 0.1532 | 1.379 | 24,300 | 1,344 | 7,096 |
| Gladiolus | 1,510,500 | 1.187 | 1,265 | AA, A | 61 × 0.1574 | 1.417 | 25,600 | 1,419 | 7,492 |
| Coreopsis | 1,590,000 | 1.250 | 1,305 | AA | 61 × 0.1615 | 1.454 | 27,000 | 1,493 | 7,883 |
| Jessamine | 1,750,000 | 1.375 | 1,385 | AA | 61 × 0.1694 | 1.525 | 29,700 | 1,643 | 8,675 |

*(Continued)*

**TABLE 3-2** Aluminum Conductor—Physical Characteristics 1350-H19 Classes AA and A (*Continued*)

| Cable code word | Conductor size | | Current-carrying capacity* A | Stranding | | Conductor diam, in | Rated strength, lb | Nominal weight, lb† | |
|---|---|---|---|---|---|---|---|---|---|
| | cmils or AWG | in² | | Class | No. and diam of wires, in | | | Per 1000 ft | Per mile |
| Cowslip | 2,000,000 | 1.570 | 1,500 | A | 91 × 0.1482 | 1.630 | 34,200 | 1,876 | 9,911 |
| Sagebrush | 2,250,000 | 1.766 | 1,600 | A | 91 × 0.1572 | 1.729 | 37,700 | 2,132 | 11,257 |
| Lupine | 2,250,000 | 1.962 | 1,700 | A | 91 × 0.1657 | 1.823 | 41,800 | 2,368 | 12,503 |
| Bitterroot | 2,750,000 | 2.159 | 1,795 | A | 91 × 0.1738 | 1.912 | 46,100 | 2,606 | 13,760 |
| Trillium | 3,000,000 | 2.356 | 1,885 | A | 127 × 0.1537 | 1.996 | 50,300 | 2,844 | 15,016 |
| Bluebonnet | 3,500,000 | 2.749 | 2,035 | A | 127 × 0.1660 | 2.158 | 58,700 | 3,350 | 17,688 |

*Class of stranding.* The class of stranding must be specified on all orders. Class AA stranding is usually specified for bare conductors used on overhead lines. Class A stranding is usually specified for conductors to be covered with weather-resistant (weatherproof) materials and for bare conductors where greater flexibility than afforded by Class AA is required.

*Lay.* The direction of lay of the outside layer of wires with Class AA and Class A stranding will be right hand unless otherwise specified.

***Note:*** 1 in² = 64.5 cm²; 1 in = 2.54 cm; 1 lb = 0.4536 kg; 1 ft = 0.3048 m; 1 mi = 1.61 km.

*Ampacity for conductor temperature rise of 40°C over 40°C ambient with a 2 ft/s crosswind and an emissivity factor of 0.5 without sun.

†Nominal conductor weights are based on ASTM standard stranding increments. Actual weights will vary with lay lengths. Invoicing will be based on actual weights.

***Effects of Temperature Changes.*** Within the temperature ranges of ordinary service there is no appreciable change in the properties of conductor materials, except in electrical resistance and physical dimensions. The change in resistance with change in temperature is sufficient to require consideration in many engineering calculations. The change in physical dimensions with change in temperature is also important in certain cases, such as in overhead spans and in large units of apparatus or equipment.

***Temperature Coefficient of Resistance.*** Over moderate ranges of temperature, such as 100°C, the change of resistance is usually proportional to the change of temperature. Resistivity is always expressed at a standard temperature, usually 20°C (68°F). In general, if $R_{t1}$ is the resistance at a temperature $t_1$ and $\alpha_{t1}$ is the temperature coefficient at that temperature, the resistance at some other temperature $t_2$ is expressed by the formula

$$R_{t_2} = R_{t_1}[1+\alpha_{t_1}(t_2 - t_1)] \tag{3-3}$$

Over wide ranges of temperature, the linear relationship of this formula is usually not applicable, and the formula then becomes a series involving higher powers of $t$, which is unwieldy for ordinary use. Equation (3-3) takes no account of the change in dimensions with change in temperature and therefore applies to the case of conductors of constant mass, usually met in engineering work.

When the temperature of reference $t_1$ is changed to some other value, the coefficient also changes. Upon assuming the general linear relationship between resistance and temperature previously mentioned, the new coefficient at any temperature $t$ within the linear range is expressed

$$\alpha_t = \frac{1}{(1/\alpha_{t_1})+(t-t_1)} \tag{3-4}$$

The reciprocal of $\alpha$ is termed the *inferred absolute zero of temperature.*

The *coefficient for copper of less than standard (or 100%) conductivity* is proportional to the actual conductivity, expressed as a decimal percentage. Thus, if $n$ is the percentage conductivity

(95% = 0.95), the temperature coefficient will be $\alpha_t' = n\alpha_t$, where $\alpha_t$ is the coefficient of the annealed copper standard.

The coefficients are computed from the formula

$$\alpha_t = \frac{1}{[1/n(0.00393)] + (t_1 - 20)} \tag{3-5}$$

*Copper Alloys and Copper-Clad Steel Wire.* Temperature-resistance coefficients for copper alloys usually can be approximated by multiplying the corresponding coefficient for copper (100% IACS) by the alloy conductivity expressed as a decimal. For some complex alloys, however, this relation does not hold even approximately, and suitable values should be obtained from the supplier. The temperature-resistance coefficient for copper-clad steel wire is 0.00378/°C at 20°C.

*Aluminum Cable Steel-Reinforced (ACSR).* ACSR conductors are widely used in the electric utility industry. The physical characteristics of an ACSR Conductor is shown in Table 3-3.

*Aluminum-Alloy Wires and Aluminum-Clad Wire.* Temperature-resistance coefficients for aluminum-alloy wires are for 5005-H19, 0.00353/°C, and for 6201-T81, 0.00347/°C at 20°C. Temperature-resistance coefficient for aluminum-clad wire is 0.0036/°C at 20°C.

*Galvanized Steel Wire and Galvanized Steel Strand.* The characteristics (weight, breaking strength, and dc resistance) of this wire and strand are shown in Tables 3-4 and 3-5.

*Typical Composite Conductors.* Temperature-resistance coefficients for typical composite conductors are as follows:

| Type | Approximate temperature coefficient per °C at 20°C |
|---|---|
| Copper–copper-clad steel | 0.00381 |
| ACSR (aluminum-steel) | 0.00403 |
| Aluminum–aluminum alloy | 0.00394 |
| Aluminum–aluminum-clad steel | 0.00396 |

***Reduction of Observations to Standard Temperature.*** A table of convenient corrections and factors for reducing resistivity and resistance to standard temperature, 20°C, will be found in Copper Wire Tables, *NBS Handbook* 100.

***Resistivity-Temperature Constant.*** The *change of resistivity per degree* may be readily calculated, taking account of the expansion of the metal with rise of temperature. The proportional relation between temperature coefficient and conductivity may be put in the following convenient form for reducing *resistivity* from one temperature to another. *The change of resistivity of copper per degree Celsius is a constant, independent of the temperature of reference and of the sample of copper. This "resistivity-temperature constant" may be taken, for general purposes, as* 0.00060 Ω *(meter, gram), or* 0.0068 $\mu\Omega \cdot cm$.

Details of the calculation of the resistivity-temperature constant will be found in Copper Wire Tables, *NBS Handbook* 100; also see this reference for expressions for the temperature coefficients of resistivity and their derivation.

**TABLE 3-3** ACSR Conductor—Physical Characteristics

| ACSR | | | | | | | | | | | | | Rated strength, lb | | | |
|---|---|---|---|---|---|---|---|---|---|---|---|---|---|---|---|---|
| | Cross section | | | | Stranding No. and diam. of strand, in | | Diameter, in | | Nominal weight, lb[†] | | | Zinc-coated core | | | |
| | Aluminum | | Total, | | | | | | Per 1000 ft | | | | | | |
| Code word | cmils or AWG | in² | in² | Current-carrying capacity,* A | Aluminum | Steel | Complete cond. | Steel core | Total | A1 | Steel | Standard weight coating | Class B coating | Class C coating | Aluminum-coated core |
| Turkey | 6 | 0.0206 | 0.0240 | 95 | 6 × 0.0661 | 1 × 0.0661 | 0.198 | 0.0661 | 36.1 | 24.5 | 11.6 | 1,190 | 1,160 | 1,120 | 1,120 |
| Swan | 4 | 0.0328 | 0.0382 | 130 | 6 × 0.0834 | 1 × 0.0834 | 0.250 | 0.0834 | 57.4 | 39.0 | 18.4 | 1,860 | 1,810 | 1,760 | 1,760 |
| Swanate | 4 | 0.0328 | 0.0411 | 130 | 7 × 0.0772 | 1 × 0.1029 | 0.257 | 0.1029 | 67.0 | 39.0 | 28.0 | 2,360 | 2,280 | 2,200 | 2,160 |
| Sparrow | 2 | 0.0522 | 0.0608 | 175 | 6 × 0.1052 | 1 × 0.1052 | 0.316 | 0.1052 | 91.3 | 62.0 | 29.3 | 2,850 | 2,760 | 2,680 | 2,640 |
| Sparate | 2 | 0.0522 | 0.0654 | 175 | 7 × 0.0974 | 1 × 0.1299 | 0.325 | 0.1299 | 106.7 | 62.0 | 44.7 | 3,640 | 3,510 | 3,390 | 3,260 |
| Robin | 1 | 0.0657 | 0.0767 | 200 | 6 × 0.1181 | 1 × 0.1181 | 0.355 | 0.1182 | 115.1 | 78.2 | 36.9 | 3,550 | 3,450 | 3,340 | 3,290 |
| Raven | 1/0 | 0.0830 | 0.0968 | 230 | 6 × 0.1327 | 1 × 0.1327 | 0.398 | 0.1327 | 145.3 | 98.7 | 46.6 | 4,380 | 4,250 | 4,120 | 3,980 |
| Quail | 2/0 | 0.1046 | 0.1221 | 265 | 6 × 0.1490 | 1 × 0.1490 | 0.447 | 0.1490 | 183.2 | 124.4 | 58.8 | 5,310 | 5,130 | 5,050 | 4,720 |
| Pigeon | 3/0 | 0.1317 | 0.1537 | 310 | 6 × 0.1672 | 1 × 0.1672 | 0.502 | 0.1672 | 230.8 | 156.7 | 74.1 | 6,620 | 6,410 | 6,300 | 5,880 |
| Penguin | 4/0 | 0.1662 | 0.1939 | 350 | 6 × 0.1878 | 1 × 0.1878 | 0.563 | 0.1878 | 291.1 | 197.7 | 93.4 | 8,350 | 8,080 | 7,950 | 7,420 |
| Waxwing | 266,800 | 0.2094 | 0.2210 | 430 | 18 × 0.1217 | 1 × 0.1217 | 0.609 | 0.1217 | 289.5 | 250.3 | 39.2 | 6,880 | 6,770 | 6,650 | 6,540 |
| Owl | 266,800 | 0.2096 | 0.2368 | 410 | 6 × 0.2109 | 7 × 0.0703 | 0.633 | 0.2109 | 342.4 | 250.5 | 91.9 | 9,680 | 9,420 | 9,160 | 9,160 |
| Partridge | 266,800 | 0.2095 | 0.2436 | 440 | 26 × 0.1013 | 7 × 0.0788 | 0.642 | 0.2364 | 367.3 | 251.7 | 115.6 | 11,300 | 11,000 | 10,600 | 10,640 |
| Merlin | 336,400 | 0.2642 | 0.2789 | 500 | 18 × 0.1367 | 1 × 0.1367 | 0.684 | 0.1367 | 365.2 | 315.7 | 49.5 | 8,680 | 8,540 | 8,400 | 8,260 |
| Linnet | 336,400 | 0.2640 | 0.3070 | 510 | 26 × 0.1137 | 7 × 0.0884 | 0.720 | 0.2652 | 462.5 | 317.0 | 145.5 | 14,100 | 13,700 | 13,300 | 13,300 |
| Oriole | 336,400 | 0.2642 | 0.3259 | 515 | 30 × 0.1059 | 7 × 0.1059 | 0.741 | 0.3177 | 527.1 | 318.1 | 209.0 | 17,300 | 16,700 | 16,200 | 15,900 |
| Chickadee | 397,500 | 0.3121 | 0.3295 | 555 | 18 × 0.1486 | 1 × 0.1486 | 0.743 | 0.1486 | 431.6 | 373.1 | 58.5 | 9,940 | 9,780 | 9,690 | 9,530 |
| Brant | 397,500 | 0.3122 | 0.3527 | 565 | 24 × 0.1287 | 7 × 0.0858 | 0.772 | 0.2574 | 512.1 | 375.0 | 137.1 | 14,600 | 14,300 | 13,900 | 13,900 |
| Ibis | 397,500 | 0.3119 | 0.3627 | 570 | 26 × 0.1236 | 7 × 0.0961 | 0.783 | 0.2883 | 546.6 | 374.7 | 171.9 | 16,300 | 15,800 | 15,300 | 15,100 |
| Lark | 397,500 | 0.3121 | 0.3849 | 575 | 30 × 0.1151 | 7 × 0.1151 | 0.806 | 0.3453 | 622.7 | 375.8 | 246.9 | 20,300 | 19,600 | 18,900 | 18,600 |
| Pelican | 477,000 | 0.3747 | 0.3955 | 625 | 18 × 0.1628 | 1 × 0.1628 | 0.814 | 0.1628 | 518.0 | 447.8 | 70.2 | 11,800 | 11,600 | 11,500 | 11,100 |
| Flicker | 477,000 | 0.3747 | 0.4233 | 635 | 24 × 0.1410 | 1 × 0.1628 | 0.846 | 0.2820 | 614.6 | 450.1 | 164.5 | 17,200 | 16,700 | 16,200 | 16,000 |
| Hawk | 477,000 | 0.3744 | 0.4354 | 640 | 26 × 0.1354 | 7 × 0.1053 | 0.858 | 0.3159 | 656.0 | 449.6 | 206.4 | 19,500 | 18,900 | 18,400 | 18,100 |
| Hen | 477,000 | 0.3747 | 0.4621 | 645 | 30 × 0.1261 | 7 × 0.1261 | 0.883 | 0.3783 | 747.4 | 451.1 | 296.3 | 23,800 | 23,000 | 22,100 | 21,300 |
| Osprey | 556,500 | 0.4369 | 0.4612 | 690 | 18 × 0.1758 | 1 × 0.1758 | 0.879 | 0.1758 | 604.1 | 522.2 | 81.9 | 13,700 | 13,500 | 13,400 | 12,900 |
| Parakeet | 556,500 | 0.4372 | 0.4938 | 700 | 24 × 0.1523 | 7 × 0.1015 | 0.914 | 0.3045 | 716.9 | 525.1 | 191.8 | 19,800 | 19,300 | 18,700 | 18,500 |

*Ampacity for conductor temperature rise of 40°C over 40°C ambient with a 2-ft/s crosswind and an emissivity factor of 0.5.

[†]Nominal conductor weights are based on ASTM standard stranding increments. Actual weight will vary with standard tolerances for wire diameters and lay length. Invoicing will be based on actual weight.

**TABLE 3-4** Galvanized-Steel Wire—Weight, Breaking Strength, DC Resistance (ASTM Specifications A111 and A326)

| Conductor size, BWG | Conductor diam., in | Conductor area, $in^2$ | Weight at 20°C* (68°F), lb/mi | Breaking strength, min., lb | | | | | DC resistance at 20°C (68°F), max., Ω/mi | | | | |
|---|---|---|---|---|---|---|---|---|---|---|---|---|---|
| | | | | Grade EBB† | Grade BB† | Grade 85 | Grade 135 | Grade 195 | Grade EBB | Grade BB | Grade 85 | Grade 135 | Grade 195 |
| 4 | 0.238 | 0.04449 | 797 | 2,028 | 2,270 | ... | ... | ... | 6.27 | 7.02 | | | |
| 6 | 0.203 | 0.03237 | 580 | 1,475 | 1,650 | ... | ... | ... | 8.62 | 9.65 | | | |
| 8 | 0.165 | 0.02138 | 383 | 975 | 1,090 | ... | ... | ... | 13.0 | 14.6 | | | |
| 9 | 0.148 | 0.01720 | 308 | 785 | 880 | 1,462 | ... | ... | 16.2 | 18.2 | 18.8 | | |
| 10 | 0.134 | 0.01410 | 253 | 645 | 720 | 1,199 | ... | ... | 19.8 | 22.2 | 22.9 | | |
| 11 | 0.120 | 0.01131 | 203 | 515 | 575 | ... | ... | ... | 24.7 | 27.6 | | | |
| 12 | 0.109 | 0.009331 | 167 | 425 | 475 | 793 | 1,213 | 1,800 | 29.9 | 33.5 | 34.7 | 38.9 | 38.9 |
| 14 | 0.083 | 0.005411 | 97.0 | 247 | 275 | 460 | ... | ... | 51.6 | 57.7 | 59.8 | | |

***Note:*** 1 in = 2.54 cm; 1 $in^2$ = 64.5 $cm^2$; 1 lb = 0.4536 kg; 1 mi = 1.61 km.
*Density = 7.83 g per cu cm at 20°C.
†ASTM designation: Extra Best Best (EBB), Best Best (BB).

**TABLE 3-5** Galvanized-Steel Strand—Dimensions, Weight, Breaking Strength (ASTM Specifications A363, A475)

| Strand diameter, in | | Stranding | | | | Breaking strength, minimum, lb | | | | | | | |
|---|---|---|---|---|---|---|---|---|---|---|---|---|---|
| | | | | | | Utilities grade* | | | | | | | |
| Nominal | Actual | No. of wires | Diameter of coated wires, in | Strand area, sq in | Strand weight, lb per 1000 ft | 1 | 2 | 3 | 4 | Common | Siemens-Martin | High strength | Extra-high strength |
| 1¼ | 1.253 | 37 | 0.179 | 0.9311 | 3248 | . . . | . . . | . . . | . . . | 44,600 | 73,000 | 113,600 | 162,200 |
| 1⅛ | 1.127 | 37 | 0.161 | 0.7533 | 2691 | . . . | . . . | . . . | . . . | 36,000 | 58,900 | 91,600 | 130,800 |
| 1 | 1.001 | 37 | 0.143 | 0.5942 | 2057 | . . . | . . . | . . . | . . . | 28,300 | 46,200 | 71,900 | 102,700 |
| 1 | 1.000 | 19 | 0.200 | 0.5969 | 2073 | . . . | . . . | . . . | . . . | 28,700 | 47,000 | 73,200 | 104,500 |
| ⅞ | 0.885 | 19 | 0.177 | 0.4675 | 1581 | . . . | . . . | . . . | . . . | 21,900 | 35,900 | 55,800 | 79,700 |
| ¾ | 0.750 | 19 | 0.150 | 0.3358 | 1155 | . . . | . . . | . . . | . . . | 16,000 | 26,200 | 40,800 | 58,300 |
| ⅝ | 0.625 | 19 | 0.125 | 0.2332 | 796 | . . . | . . . | . . . | . . . | 11,000 | 18,100 | 28,100 | 40,200 |
| ⅝ | 0.621 | 7 | 0.207 | 0.2356 | 813 | . . . | . . . | . . . | . . . | 11,600 | 19,100 | 29,600 | 42,400 |
| 9/16 | 0.565 | 19 | 0.113 | 0.1905 | 637 | . . . | . . . | . . . | . . . | 9,640 | 16,100 | 24,100 | 33,700 |
| 9/16 | 0.564 | 7 | 0.188 | 0.1943 | 671 | . . . | . . . | . . . | . . . | 9,600 | 15,700 | 24,500 | 35,000 |
| ½ | 0.500 | 19 | 0.100 | 0.1492 | 504 | . . . | . . . | . . . | . . . | 7,620 | 12,700 | 19,100 | 26,700 |
| ½ | 0.495 | 7 | 0.165 | 0.1497 | 517 | . . . | . . . | . . . | **25,000** | **7,400** | **12,100** | **18,800** | 26,900 |
| 7/16 | 0.435 | 7 | 0.145 | 0.1156 | 399 | . . . | . . . | . . . | **18,000** | **5,700** | **9,350** | **14,500** | 20,800 |
| ⅜ | 0.360 | 7 | 0.120 | 0.07917 | 273 | . . . | . . . | . . . | **11,500** | **4,250** | **6,950** | 10,800 | 15,400 |
| ⅜ | 0.356 | 3 | 0.165 | 0.06415 | 220.3 | . . . | . . . | **8500** | | | | | |
| 5/16 | 0.327 | 7 | 0.109 | 0.06532 | 225 | **6000** | | | | | | | |
| 5/16 | 0.312 | 7 | 0.104 | 0.05946 | 205 | . . . | . . . | . . . | . . . | **3,200** | 5,350 | **8,000** | 11,200 |
| 5/16 | 0.312 | 3 | 0.145 | 0.04954 | 170.6 | . . . | . . . | **6500** | | | | | |
| 9/32 | 0.279 | 7 | 0.093 | 0.04755 | 164 | **4600** | . . . | . . . | . . . | 2,570 | 4,250 | 6,400 | 8,950 |
| ¼ | 0.240 | 7 | 0.080 | 0.03519 | 121 | . . . | . . . | . . . | . . . | **1,900** | **3,150** | 4,750 | 6,650 |
| ¼ | 0.259 | 3 | 0.120 | 0.03393 | 116.7 | . . . | **3150** | **4500** | | | | | |
| 7/32 | 0.216 | 7 | 0.072 | 0.02850 | 98.3 | . . . | . . . | . . . | . . . | 1,540 | 2,560 | 3,850 | 5,400 |
| 3/16 | 0.195 | 7 | 0.065 | 0.02323 | 80.3 | **2400** | | | | | | | |
| 3/16 | 0.186 | 7 | 0.062 | 0.02113 | 72.9 | . . . | . . . | . . . | . . . | **1,150** | 1,900 | 2,850 | 3,990 |
| 5/32 | 0.156 | 7 | 0.052 | 0.01487 | 51.3 | . . . | . . . | . . . | . . . | **870** | 1,470 | 2,140 | 2,940 |
| ⅛ | 0.123 | 7 | 0.041 | 0.00924 | 31.8 | . . . | . . . | . . . | . . . | **540** | 910 | 1,330 | 1,830 |
| Elongation in 24 in | | | | | | | | % | | | | | |
| . . . | . . . | . . . | . . . | . . . | . . . | 10 | 8 | 5 | 4 | 10 | 8 | 5 | 4 |

***Note:*** Sizes and grades in bold-faced type are those most commonly used and readily available.
1 in = 2.54 cm; 1 in² = 64.5 cm²; 1 lb = 0.4536 kg; 1 ft = 0.3048 m.
*Used principally by communication and power and light industries.

***Temperature Coefficient of Expansion.*** *Temperature coefficient of expansion* (linear) of pure metals over a range of several hundred degrees is not a linear function of the temperature but is well expressed by a quadratic equation

$$\frac{L_{t_2}}{L_{t_1}} = 1 + [\alpha(t_2 - t_1) + \beta(t_2 - t_1)^2] \tag{3-6}$$

Over the temperature ranges for ordinary engineering work (usually 0 to 100°C), the coefficient can be taken as a constant (assumed linear relationship) and a simplified formula employed

$$L_{t_2} = L_{t_1}[1 + \alpha_{t_1}(t_2 - t_1)] \tag{3-7}$$

Changes in linear dimensions, superficial area, and volume take place in most materials with changes in temperature. In the case of linear conductors, only the change in length is ordinarily important.

The coefficient for changes in superficial area is approximately twice the coefficient of linear expansion for relatively small changes in temperature. Similarly, the volume coefficient is 3 times the linear coefficient, with similar limitations.

***Specific Heat.*** Specific heat of electrolytic tough pitch copper is 0.092 cal/(g)(°C) at 20°C (see *NBS Circ.* 73). Specific heat of aluminum is 0.226 cal/(g)(°C) at room temperature (see *NBS Circ.* C447, Mechanical Properties of Metals and Alloys). Specific heat of iron (wrought) or very soft steel from 0 to 100°C is 0.114 cal/(g)(°C); the true specific heat of iron at 0°C is 0.1075 cal/(g)(°C) (see *International Critical Tables,* vol. II, p. 518; also ASM, *Metals Handbook*).

***Thermal Conductivity of Electrolytic Tough Pitch Copper.*** Thermal conductivity of electrolytic tough pitch copper at 20°C is 0.934 cal/(cm$^2$)(cm)(s)(°C), adjusted to correspond to an electrical conductivity of 101% (see *NBS Circ.* 73).

***Thermal-Electrical Conductivity Relation of Copper.*** The Wiedemann-Franz-Lorenz law, which states that the ratio of the thermal and electrical conductivities at a given temperature is independent of the nature of the conductor, holds closely for copper. The ratio $K/\lambda T$ (where $K$ = thermal conductivity, $\lambda$ = electrical conductivity, $T$ = absolute temperature) for copper is 5.45 at 20°C.

***Thermal Conductivity***

*Copper Alloys.*

| | Thermal conductivity (volumetric) at 20°C | |
|---|---|---|
| ASTM alloy (Spec. B105) | Btu per sq ft per ft per h per °F | Cal per sq cm per cm per sec per °C |
| 8.5 | 31 | 0.13 |
| 15 | 50 | 0.21 |
| 30 | 84 | 0.35 |
| 55 | 135 | 0.56 |
| 80 | 199 | 0.82 |
| 85 | 208 | 0.86 |

*Aluminum.* The determination made by the Bureau of Standards at 50°C for aluminum of 99.66% purity is 0.52 cal/(cm$^2$)(cm)(s)(°C) (*Circ.* 346; also see *Smithsonian Physical Tables and International Critical Tables*).

*Iron.* Thermal conductivity of iron (mean) from 0 to 100°C is 0.143 cal/(cm$^2$)(cm)(s)(°C); with increase of carbon and manganese content, it tends to decrease and may reach a figure of approximately 0.095 with about 1% carbon, or only about half of that figure if the steel is hardened by water quenching (see *International Critical Tables*, vol. II, p. 518).

***Copper.*** Copper is a highly malleable and ductile metal of reddish color. It can be cast, forged, rolled, drawn, and machined. Mechanical working hardens it, but annealing will restore it to the soft state. The density varies slightly with the physical state, 8.9 being an average value. It melts at 1083°C (1981°F) and in the molten state has a sea-green color. When heated to a very high temperature, it vaporizes and burns with a characteristic green flame. Copper readily alloys with many other metals. In ordinary atmospheres it is not subject to appreciable corrosion. Its electrical conductivity is very sensitive to the presence of slight impurities in the metal.

Copper, when exposed to ordinary atmospheres, becomes oxidized, turning to a black color, but the oxide coating is protective, and the oxidizing process is not progressive. When exposed to moist air containing carbon dioxide, it becomes coated with green basic carbonate, which is also protective. At temperatures above 180°C it oxidizes in dry air. In the presence of ammonia it is readily oxidized in air, and it is also affected by sulfur dioxide. Copper is not readily attacked at high temperatures below the melting point by hydrogen, nitrogen, carbon monoxide, carbon dioxide, or steam. Molten copper readily absorbs oxygen, hydrogen, carbon monoxide, and sulfur dioxide, but on cooling, the occluded gases are liberated to a great extent, tending to produce blowholes or porous castings. Copper in the presence of air does not dissolve in dilute hydrochloric or sulfuric acid but is readily attacked by dilute nitric acid. It is also corroded slowly by saline solutions and seawater.

Commercial grades of copper in the United States are electrolytic, oxygen-free, Lake, fire-refined, and casting. *Electrolytic copper* is that which has been electrolytically refined from blister, converter, black, or Lake copper. *Oxygen-free copper* is produced by special manufacturing processes which prevent the absorption of oxygen during the melting and casting operations or by removing the oxygen by reducing agents. It is used for conductors subjected to reducing gases at elevated temperature, where reaction with the included oxygen would lead to the development of cracks in the metal. *Lake copper* is electrolytically or fire-refined from Lake Superior native copper ores and is of two grades, low resistance and high resistance. *Fire-refined copper* is a lower-purity grade intended for alloying or for fabrication into products for mechanical purposes; it is not intended for electrical purposes. *Casting copper* is the grade of lowest purity and may consist of furnace-refined copper, rejected metal not up to grade, or melted scrap; it is exclusively a foundry copper.

***Hardening and Heat-Treatment of Copper.*** There are but two well-recognized methods for hardening copper, one is by mechanically working it, and the other is by the addition of an alloying element. The properties of copper are not affected by a rapid cooling after annealing or rolling, as are those of steel and certain copper alloys.

***Annealing of Copper.*** Cold-worked copper is softened by annealing, with decrease of tensile strength and increase of ductility. In the case of pure copper hardened by cold reduction of area to one-third of its initial area, this softening takes place with maximum rapidity between 200 and 325°C. However, this temperature range is affected in general by the extent of previous cold reduction and the presence of impurities. The greater the previous cold reduction, the lower is the range of softening temperatures. The effect of iron, nickel, cobalt, silver, cadmium, tin, antimony, and tellurium is to lower the conductivity and raise the annealing range of pure copper in varying degrees.

| Commercial grade | ASTM Designation | Copper content, minimum % |
|---|---|---|
| Electrolytic | B5 | 99.900 |
| Oxygen-free electrolytic | B170 | 99.95 |
| Lake, low resistance | B4 | 99.900 |
| Lake, high resistance | B4 | 99.900 |
| Fire-refined | B216 | 99.88 |
| Casting | B119 | 98 |

***Alloying of Copper.*** Elements that are soluble in moderate amounts in a solid solution of copper, such as manganese, nickel, zinc, tin, and aluminum, generally harden it and diminish its ductility but improve its rolling and working properties. Elements that are but slightly soluble, such as bismuth and lead, do not harden it but diminish both the ductility and the toughness and impair its hot-working properties. Small additions (up to 1.5%) of manganese, phosphorus, or tin increase the tensile strength and hardness of cold-rolled copper.

*Brass* is usually a binary alloy of copper and zinc, but brasses are seldom employed as electrical conductors, since they have relatively low conductivity through comparatively high tensile strength. In general, brass is not suitable for use when exposed to the weather, owing to the difficulty from stress-corrosion cracking; the higher the zinc content, the more pronounced this becomes.

*Bronze* in its simplest form is a binary alloy of copper and tin in which the latter element is the hardening and strengthening agent. This material is rather old in the arts and has been used to some extent for electrical conductors for past many years, especially abroad. Modern bronzes are frequently ternary alloys, containing as the third constituent such elements as phosphorus, silicon, manganese, zinc, aluminum, or cadmium; in such cases, the third element is usually given in the name of the alloy, as in phosphor bronze or silicon bronze. Certain bronzes are quaternary alloys, or contain two other elements in addition to copper and tin.

In bronzes for use as electrical conductors, the content of tin and other metals is usually less than in bronzes for structural or mechanical applications, where physical properties and resistance to corrosion are the governing considerations. High resistance to atmospheric corrosion is always an important consideration in selecting bronze conductors for overhead service.

***Commercial Grades of Bronze.*** Various bronzes have been developed for use as conductors, and these are now covered by ASTM Specification B105. They all have been designed to provide conductors having high resistance to corrosion and tensile strengths greater than hard-drawn copper conductors. The standard specification covers 10 grades of bronze, designated by numbers according to their conductivities. Bronze fittings are generally used to terminate copper wires or ground wires to steel structures, or to connect buried copper conductors to form a ground grid for electrical safety grounding. Bronze fittings are generally bolted to conductors to make the electrical connection.

***Copper-Beryllium Alloy.*** *Copper-beryllium alloy* containing 0.4% of beryllium may have an electrical conductivity of 48% and a tensile strength (in 0.128-in wire) of 86,000 lb/in$^2$. A content of 0.9% of beryllium may give a conductivity of 28% and a tensile strength of 122,000 lb/in$^2$. The effect of this element in strengthening copper is about 10 times as great as that of tin.

***Copper-Clad Steel Wire.*** *Copper-clad steel wire* has been manufactured by a number of different methods. The general object sought in the manufacture of such wires is the combination of the high conductivity of copper with the high strength and toughness of iron or steel. The principal manufacturing processes now in commercial use are (*a*) coating a steel billet with a special flux, placing it in a vertical mold closed at the bottom, heating the billet and mold to yellow heat, and then casting molten copper around the billet, after which it is hot-rolled to rods and cold-drawn to wire, and (*b*) electroplating a dense coating of copper on a steel rod and then cold drawing to wire.

***Aluminum.*** *Aluminum* is a ductile metal, silver-white in color, which can be readily worked by rolling, drawing, spinning, extruding, and forging. Its specific gravity is 2.703. Pure aluminum melts at 660°C (1220°F). Aluminum has relatively high thermal and electrical conductivities. The metal is always covered with a thin, invisible film of oxide which is impermeable and protective in character. Aluminum, therefore, shows stability and long life under ordinary atmospheric exposure.

Exposure to atmospheres high in hydrogen sulfide or sulfur dioxide does not cause severe attack of aluminum at ordinary temperatures, and for this reason, aluminum or its alloys can be used in atmospheres which would be rapidly corrosive to many other metals.

Aluminum parts should, as a rule, not be exposed to salt solutions while in electrical contact with copper, brass, nickel, tin, or steel parts, since galvanic attack of the aluminum is likely to occur. Contact with cadmium in such solutions results in no appreciable acceleration in attack on the aluminum, while contact with zinc (or zinc-coated steel as long as the coating is intact) is generally beneficial, since the zinc is attacked selectively and it cathodically protects adjacent areas of the aluminum.

Most organic acids and their water solutions have little or no effect on aluminum at room temperature, although oxalic acid is an exception and is corrosive. Concentrated nitric acid (about 80% by weight) and fuming sulfuric acid can be handled in aluminum containers. However, more dilute solutions of these acids are more active. All but the most dilute (less than 0.1%) solutions of hydrochloric and hydrofluoric acids have a rapid etching action on aluminum.

Solutions of the strong alkalies, potassium, or sodium hydroxides dissolve aluminum rapidly. However, ammonium hydroxide and many of the strong organic bases have little action on aluminum and are successfully used in contact with it (see *NBS Circ.* 346).

Aluminum in the presence of water and limited air or oxygen rapidly converts into aluminum hydroxide, a whitish powder.

Commercial grades of aluminum in the United States are designated by their purity, such as 99.99, 99.6, 99.2, 99.0%. Electrical conductor alloy aluminum 1350, having a purity of approximately 99.5% and a minimum conductivity of 61.0% IACS, is used for conductor purposes. Specified physical properties are obtained by closely controlling the kind and amount of certain impurities.

***Annealing of Aluminum.*** Cold-worked aluminum is softened by annealing, with decrease of tensile strength and increase of ductility. The annealing temperature range is affected in general by the extent of previous cold reduction and the presence of impurities. The greater the previous cold reduction, the lower is the range of softening temperatures.

***Alloying of Aluminum.*** Aluminum can be alloyed with a variety of other elements, with a consequent increase in strength and hardness. With certain alloys, the strength can be further increased by suitable heat treatment. The alloying elements most generally used are copper, silicon, manganese, magnesium, chromium, and zinc. Some of the aluminum alloys, particularly those containing one or more of the following elements—copper, magnesium, silicon, and zinc—in various combinations, are susceptible to heat treatment.

Pure aluminum, even in the hard-worked condition, is a relatively weak metal for construction purposes. Strengthening for castings is obtained by alloying elements. The alloys most suitable for cold rolling seldom contain less than 90% to 95% aluminum. By alloying, working, and heat treatment, it is possible to produce tensile strengths ranging from 8500 lb/in$^2$ for pure annealed aluminum up to 82,000 lb/in$^2$ for special wrought heat-treated alloy, with densities ranging from 2.65 to 3.00. Alloys of aluminum are principally alloys 5005 and 6201 covered by ASTM Specifications B396 and B398.

In recent years, high-temperature-tolerant alloys have been developed that allow aluminum to be operated at high temperatures without losing strength. Zirconium is used as the alloying element. These alloys are manufactured under IEC Standard 62004, which specifies four alloys. Collectively they are referred to as "thermal resistant aluminum alloys." These alloys have densities at 20°C of 2.703 g/cm$^3$ and a mass temperature coefficient of resistivity of between 0.00360 and 0.0040/°C. These alloys, when combined with a steel core in an ACSR style construction, can operate continuously at temperatures from 150 to 230°C, and in emergencies (up to 400 h cumulatively) from 150 to 310°C depending on the alloy.

*Aluminum-clad steel wires* have a relatively heavy layer of aluminum surrounding and bonded to the high-strength steel core. The aluminum layer can be formed by compacting and sintering a layer of aluminum powder over a steel rod, by electroplating a dense coating of aluminum on a steel rod, or by extruding a coating of aluminum on a steel rod and then cold drawing to wire.

***Silicon.*** *Silicon* is a light metal having a specific gravity of approximately 2.34. There is lack of accurate data on the pure metal because its mechanical brittleness bars it from most industrial uses. However, it is very resistant to atmospheric corrosion and to attack by many chemical reagents. Silicon is of fundamental importance in the steel industry, but for this purpose it is obtained in the form of ferrosilicon, which is a coarse granulated or broken product. It is very useful as an alloying element in steel for electrical sheets and substantially increases the electrical resistivity, and thereby reduces the core losses. Silicon is peculiar among metals in the respect that its temperature coefficient of resistance may change sign in some temperature ranges, the exact behavior varying with the impurities.

***Beryllium.*** *Beryllium* is a light metal having a specific gravity of approximately 1.84, or nearly the same as magnesium. It is normally hard and brittle and difficult to fabricate. Copper is materially strengthened by the addition of small amounts of beryllium, without very serious loss of electrical conductivity. The principal use for this metal appears to be as an alloying element with other metals such as aluminum and copper. Beryllium is also toxic. Reference should be made to Material Safety Data Sheets for precautions in handling.

***Sodium.*** *Sodium* is a soft, bright, silvery metal obtained commercially by the electrolysis of absolutely dry fused sodium chloride. It is the most abundant of the alkali group of metals, is extremely reactive, and is never found free in nature. It oxidizes readily and rapidly in air. In the presence of water (it is so light that it floats) it may ignite spontaneously, decomposing the water with evolution of hydrogen and formation of sodium hydroxide. This can be explosive. Sodium should be handled with respect, since it can be dangerous when handled improperly. It melts at 97.8°C, below the boiling point of water and in the same range as many fuse metal alloys. Sodium is approximately one-tenth as heavy as copper and has roughly three-eighths the conductivity; hence 1 lb of sodium is about equal electrically to 3½ lb of copper.

## 3.1.3 Conductor Properties

***Definitions of Electrical Conductors***

*Wire.* A rod or filament of drawn or rolled metal whose length is great in comparison with the major axis of its cross section. The definition restricts the term to what would ordinarily be understood by the term solid wire. In the definition, the word slender is used in the sense that the length is great in comparison with the diameter. If a wire is covered with insulation, it is properly called an insulated wire, while primarily the term wire refers to the metal; nevertheless, when the context shows that the wire is insulated, the term wire will be understood to include the insulation.

*Conductor.* A wire or combination of wires not insulated from one another, suitable for carrying an electric current. The term conductor is not to include a combination of conductors insulated from one another, which would be suitable for carrying several different electric currents. Rolled conductors (such as bus bars) are, of course, conductors but are not considered under the terminology here given.

*Stranded Conductor.* A conductor composed of a group of wires, usually twisted, or any combination of groups of wires. The wires in a stranded conductor are usually twisted or braided together.

*Cable.* A stranded conductor (single-conductor cable) or a combination of conductors insulated from one another (multiple-conductor cable). The component conductors of the second kind of cable may be either solid or stranded, and this kind of cable may or may not have a common insulating covering. The first kind of cable is a single conductor, while the second kind is a group of several conductors. The term cable is applied by some manufacturers to a solid wire heavily insulated and lead covered; this usage arises from the manner of the insulation, but such a conductor is not included under this definition of cable. The term cable is a general one, and in practice, it is usually applied only to the larger sizes. A small cable is called a stranded wire or a cord, both of which are defined below. Cables may be bare or insulated, and the latter may be armored with lead or with steel wires or bands.

*Strand.* One of the wires of any stranded conductor.

*Stranded Wire.* A group of small wires used as a single wire. A wire has been defined as a slender rod or filament of drawn metal. If such a filament is subdivided into several smaller filaments or strands and is used as a single wire, it is called stranded wire. There is no sharp dividing line of size between a stranded wire and a cable. If used as a wire, for example, in winding inductance coils or magnets, it is called a stranded wire and not a cable. If it is substantially insulated, it is called a cord, defined below.

*Cord.* A small cable, very flexible and substantially insulated to withstand wear. There is no sharp dividing line in respect to size between a cord and a cable, and likewise no sharp dividing line in respect to the character of insulation between a cord and a stranded wire. Usually the insulation of a cord contains rubber.

*Concentric Strand.* A strand composed of a central core surrounded by one or more layers of helically laid wires or groups of wires.

*Concentric-Lay Conductor.* Conductor constructed with a central core surrounded by one or more layers of helically laid wires.

*Rope-Lay Conductor.* Conductor constructed of a bunch-stranded or a concentric-stranded member or members, as a central core, around which are laid one or more helical layers of such members.

*N-Conductor Cable.* A combination of N conductors insulated from one another. It is not intended that the name as given here actually be used. One would instead speak of a "3-conductor cable," a "12-conductor cable," etc. In referring to the general case, one may speak of a "multiple-conductor cable."

*N-Conductor Concentric Cable.* A cable composed of an insulated central conducting core with N-1 tubular-stranded conductors laid over it concentrically and separated by layers of insulation. This kind of cable usually has only two or three conductors. Such cables are used in carrying alternating currents. The remark on the expression "N conductor" given for the preceding definition applies here also. (Additional definitions can be found in ASTM B354.)

***Wire Sizes.*** Wire sizes have been for many years indicated in commercial practice almost entirely by gage numbers, especially in America and England. This practice is accompanied by some confusion because numerous gages are in common use. The most commonly used gage for electrical wires, in America, is the *American wire gage.* The most commonly used gage for steel wires is the *Birmingham wire gage.*

There is no legal standard wire gage in this country, although a gage for sheets was adopted by Congress in 1893. In England, there is a legal standard known as the *Standard wire gage.* In Germany, France, Austria, Italy, and other continental countries, practically no wire gage is used, but wire sizes are specified directly in millimeters. This system is sometimes called the *millimeter wire gage.* The wire sizes used in France, however, are based to some extent on the old Paris gage (*jauge de Paris de 1857*) (for a history of wire gages, see *NBS Handbook* 100, Copper Wire Tables; also see *Circ.* 67, Wire Gages, 1918).

There is a tendency to *abandon gage numbers* entirely and specify wire sizes by the *diameter in mils* (thousandths of an inch). This practice holds particularly in writing specifications and has the great advantages of being both simple and explicit. A number of wire manufacturers also encourage this practice, and it was definitely adopted by the U.S. Navy Department in 1911.

*Mil* is a term universally employed in this country to measure wire diameters and is a unit of length equal to one-thousandth of an inch. *Circular mil* is a term universally used to define cross-sectional areas, being a unit of area equal to the area of a circle 1 mil in diameter. Such a circle, however, has an area of 0.7854 (or $\pi/4$) mil$^2$. Thus a wire 10 mils in diameter has a cross-sectional area of 100 cmils or 78.54 mils$^2$. Hence, a cmil equals 0.7854 mil$^2$.

*American wire gage,* also known as the *Brown & Sharpe gage,* was devised in 1857 by J. R. Brown. It is usually abbreviated AWG. This gage has the property, in common with a number of other gages, that its sizes represent approximately the successive steps in the process of wire drawing. Also, like many other gages, its numbers are retrogressive, a larger number denoting a smaller wire, corresponding to the operations of drawing. These gage numbers are not arbitrarily chosen, as in many gages, but follow the mathematical law upon which the gage is founded.

*Basis of the AWG* is a simple mathematical law. The gage is formed by the specification of two diameters and the law that a given number of intermediate diameters are formed by geometric progression. Thus, the diameter of No. 0000 is defined as 0.4600 in and of No. 36 as 0.0050 in. There are

38 sizes between these two; hence the ratio of any diameter to the diameter of the next greater number is given by this expression

$$\sqrt[39]{\frac{0.4600}{0.0050}} = \sqrt[39]{92} = 1.1229322 \tag{3-8}$$

The square of this ratio = 1.2610. The sixth power of the ratio, that is, the ratio of any diameter to the diameter of the sixth greater number = 2.0050. The fact that this ratio is so nearly 2 is the basis of numerous useful relations or shortcuts in wire computations.

There are a number of approximate rules applicable to the AWG which are useful to remember:

1. An increase of three gage numbers (e.g., from No. 10 to 7) doubles the area and weight and consequently halves the dc resistance.
2. An increase of six gage numbers (e.g., from No. 10 to 4) doubles the diameter.
3. An increase of 10 gage numbers (e.g., from No. 10 to 1/0) multiplies the area and weight by 10 and divides the resistance by 10.
4. A No. 10 wire has a diameter of about 0.10 in, an area of about 10,000 cmils, and (for standard annealed copper at 20°C) a resistance of approximately 1.0 Ω/1000 ft.
5. The weight of No. 2 copper wire is very close to 200 lb/1000 ft (90 kg/304.8 m).

*Steel wire gage,* also known originally as the *Washburn & Moen gage* and later as the *American Steel & Wire Co.'s gage,* was established by Ichabod Washburn in 1830. This gage, with a number of its sizes rounded off to thousandths of an inch, is also known as the *Roebling gage.* It is used exclusively for steel wire and is frequently employed in wire mills.

*Birmingham wire gage,* also known as *Stubs' wire gage* and *Stubs' iron wire gage,* is said to have been established early in the eighteenth century in England, where it was long in use. This gage was used to designate the Stubs soft-wire sizes and should not be confused with Stubs' steel-wire gage. The numbers of the Birmingham gage were based on the reductions of size made in practice by drawing wire from rolled rod. Thus, a wire rod was called "No. 0," "first drawing No. 1," and so on. The gradations of size in this gage are not regular, as will appear from its graph. This gage is generally in commercial use in the United States for iron and steel wires.

*Standard wire gage,* which more properly should be designated *(British) Standard wire gage,* is the legal standard of Great Britain for all wires adopted in 1883. It is also known as the *New British Standard gage,* the *English legal standard gage,* and the *Imperial wire gage.* It was constructed by so modifying the Birmingham gage that the differences between consecutive sizes become more regular. This gage is largely used in England but never has been used extensively in America.

*Old English wire gage,* also known as the *London wire gage,* differs very little from the Birmingham gage. Formerly it was used to some extent for brass and copper wires but is now nearly obsolete.

*Millimeter wire gage,* also known as the *metric wire gage,* is based on giving progressive numbers to the progressive sizes, calling 0.1 mm diameter "No. 1," 0.2 mm "No. 2," etc.

***Conductor-Size Designation.*** America uses, for sizes up to 4/0, mil, decimals of an inch, or AWG numbers for solid conductors and AWG numbers or circular mils for stranded conductors; for sizes larger than 4/0, circular mils are used throughout. Other countries ordinarily use square millimeter area.

Conductor-size conversion can be accomplished from the following relation:

$$\text{cmils} = \text{in}^2 \times 1{,}273{,}200 = \text{mm}^2 \times 1973.5 \tag{3-9}$$

Measurement of wire diameters may be accomplished in many ways but most commonly by means of a micrometer caliper. Stranded cables are usually measured by means of a circumference tape calibrated directly in diameter readings.

***Stranded Conductors.*** Stranded conductors are used generally because of their increased flexibility and consequent ease in handling. The greater the number of wires in any given cross section, the greater will be the flexibility of the finished conductor. Most conductors above 4/0 AWG in size are stranded. Generally, in a given concentric-lay stranded conductor, all wires are of the same size and the same material, although special conductors are available embodying wires of different sizes and materials. The former will be found in some insulated cables and the latter in overhead stranded conductors combining high-conductivity and high-strength wires.

The flexibility of any given size of strand obviously increases as the total number of wires increases. It is a common practice to increase the total number of wires as the strand diameter increases in order to provide reasonable flexibility in handling. So-called flexible concentric strands for use in insulated cables have about one to two more layers of wires than the standard type of strand for ordinary use.

***Number of Wires in Standard Conductors.*** Where all the strands are the same diameter (natural stranding), each successive layer in a concentrically stranded conductor contains six more wires than the preceding one. The total number of wires in a conductor is

For 1-wire core constructions (1, 7, 19, etc.),

$$N = 3n(n+1)+1 \tag{3-10}$$

For 3-wire core constructions (3, 12, etc.),

$$N = 3n(n+2)+3 \tag{3-11}$$

where $n$ is number of layers over core, which is not counted as a layer.

Wire size in stranded conductors is

$$d = \sqrt{\frac{A}{N}} \tag{3-12}$$

where $A$ is total conductor area in circular mils, and $N$ is total number of wires.

Copper cables are manufactured usually to certain cross-sectional sizes specified in total circular mils or by gage numbers in AWG. This necessarily requires individual wires drawn to certain prescribed diameters, which are different as a rule from normal sizes in AWG.

Diameter of stranded conductors (circumscribing circle) is

$$D = d(2n+k) \tag{3-13}$$

where $d$ is diameter of individual wire, $n$ is number of layers over core, which is not counted as a layer, $k$ is 1 for constructions having 1-wire core (1, 7, 19, etc.), and 2.155 for constructions having 3-wire core (3, 12, etc.).

For standard concentric-lay stranded conductors, the following rule gives a simple method of determining the outside diameter of a stranded conductor from the known diameter of a solid wire of the same cross-sectional area: *To obtain the diameter of concentric-lay stranded conductor, multiply the diameter of the solid wire of the same cross-sectional area by the appropriate factor as follows:*

| Number of wires | Factor | Number of wires | Factor |
|---|---|---|---|
| 3 | 1.244 | 91 | 1.153 |
| 7 | 1.134 | 127 | 1.154 |
| 12 | 1.199 | 169 | 1.154 |
| 19 | 1.147 | 217 | 1.154 |
| 37 | 1.151 | 271 | 1.154 |
| 61 | 1.152 | | |

Area of stranded conductors is

$$A = Nd^2 \text{cmils} = \tfrac{1}{4}\pi N d^2 \times 10^{-6} \text{in}^2 \tag{3-14}$$

where $N$ is total number of wires, and $d$ is individual wire diameter in mils.

***Effects of Stranding.*** All wires in a stranded conductor except the core wire form continuous helices of slightly greater length than the axis or core. This causes a slight increase in weight and electrical resistance and slight decrease in tensile strength and sometimes affects the internal inductance, as compared theoretically with a conductor of equal dimensions but composed of straight wires parallel with the axis.

***Lay or Pitch.*** The axial length of one complete turn, or helix, of a wire in a stranded conductor is sometimes termed the *lay* or *pitch*. This is often expressed as the *pitch ratio*, which is the ratio of the length of the helix to its *pitch diameter* (diameter of the helix at the centerline of any individual wire or strand equals the outside diameter of the helix minus the thickness of one wire or strand). If there are several layers, the pitch expressed as an axial length may increase with each additional layer, but when expressed as the ratio of axial length to pitch diameter of helix, it is usually the same for all layers, or nearly so. In commercial practice, the pitch is commonly expressed as the ratio of axial length to outside diameter of helix, but this is an arbitrary designation made for convenience of usage. The *pitch angle* is shown in Fig. 3-4, where *ac* represents the axis of the stranded conductor and $l$ is the axial length of one complete turn or helix, *ab* is the length of any individual wire $l + \Delta l$ in one complete turn, and *bc* is equal to the circumference of a circle corresponding to the pitch diameter $d$ of the helix. The angle *bac*, or $\theta$, is the pitch angle, and the pitch ratio is expressed by $p = l/d$. There is no standard pitch ratio used by manufacturers generally, since it has been found desirable to vary this depending on the type of service for which the conductor is intended. Applicable lay lengths generally are included in industry specifications covering the various stranded conductors. For bare overhead conductors, a representative commercial value for pitch length is 13.5 times the outside diameter of each layer of strands.

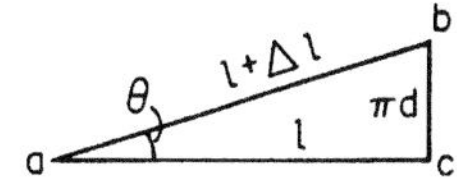


**FIGURE 3-4** Pitch angle in concentric-lay cable.

***Direction of Lay.*** The direction of lay is the lateral direction in which the individual wires of a cable run over the top of the cable as they recede from an observer looking along the axis. *Right-hand lay* recedes from the observer in clockwise rotation or like a right-hand screw thread; *left-hand lay* is the opposite. The outer layer of a cable is ordinarily applied with a right-hand lay for bare overhead conductors and left-hand lay for insulated conductors, although the opposite lay can be used if desired.

***Increase in Weight due to Stranding.*** Referring to Fig. 3-4, the increase in weight of the spiral members in a cable is proportional to the increase in length

$$\frac{l+\Delta l}{l} = \sec\theta = \sqrt{1+\tan^2\theta}$$

$$= \sqrt{1+\frac{\pi^2}{p^2}} = 1+\frac{1\pi^2}{2p^2} - \frac{1}{8}\left(\frac{\pi^2}{p^2}\right)^2 + \cdots \tag{3-15}$$

As a first approximation this ratio equals $1 + 0.5(\pi^2/p^2)$, and a pitch of 15.7 produces a ratio of 1.02. This correction factor should be computed separately for each layer if the pitch $p$ varies from layer to layer.

***Increase in Resistance due to Stranding.*** If it were true that no current flows from wire to wire through their lineal contacts, the proportional increase in the total resistance would be the same as the proportional increase in total weight. If all the wires were in perfect and complete contact with each other, the total resistance would decrease in the same proportion that the total weight increases, owing to the slightly increased normal cross section of the cable as a whole. The contact resistances are normally sufficient to make the actual increase in total resistance nearly as much, proportionately, as the increase in total weight, and for practical purposes they are usually assumed to be the same.

***Decrease in Strength due to Stranding.*** When a concentric-lay cable is subjected to mechanical tension, the spiral members tend to tighten around those layers under them and thus produce internal compression, gripping the inner layers and the core. Consequently, the individual wires, taken as a whole, do not behave as they would if they were true linear conductors acting independently. Furthermore, the individual wires are never exactly alike in diameter or in strength or in elastic properties. For these reasons, there is ordinarily a loss of about 4% to 11% in total tensile efficiency, depending on the number of layers. This reduction tends to increase as the pitch ratio decreases. Actual tensile tests on cables furnish the most dependable data on their ultimate strength.

*Tensile efficiency of a stranded conductor* is the ratio of its breaking strength to the sum of the tensile strengths of all its individual wires. Concentric-lay cables of 12 to 16 pitch ratio have a normal tensile efficiency of approximately 90%; rope-lay cables, approximately 80%. In the United States, ASTM standards specify the reduction factors based on the number of wires and number of layers in the conductor.

***Preformed Cable.*** This type of cable is made by preforming each individual wire (except the core) into a spiral of such length and curvature that the wire will fit naturally into its normal position in the cable instead of being forced into that shape under the usual tension in the stranding machine. This method has the advantage in cable made of the stiffer grades of wire that the individual wires do not tend to spread or untwist if the strand is cut in two without first binding the ends on each side of the cut.

***Weight.*** A uniform cylindrical conductor of diameter $d$, length $l$, and density $\delta$ has a total weight expressed by the formula

$$W = \delta l \frac{\pi d^2}{4} \tag{3-16}$$

The weight of any conductor is commonly expressed in pounds per unit of length, such as 1 ft, 1000 ft, or 1 mi. The weight of stranded conductors can be calculated using Eq. (3-16), but allowance must be made for increase in weight due to stranding. Rope-lay stranding has greater increase in weight because of the multiple stranding operations.

***Breaking Strength.*** The maximum load that a conductor attains when tested in tension to rupture.

***Total Elongation at Rupture.*** When a sample of any material is tested under tension until it ruptures, measurement is usually made of the total elongation in a certain initial test length. In certain kinds of testing, the initial test length has been standardized, but in every case, the total elongation at rupture should be referred to the initial test length of the sample on which it was measured. Such elongation is usually expressed in percentage of original unstressed length and is a general index of the ductility of the material. Elongation is determined on solid conductors or on individual wires before stranding; it is rarely determined on stranded conductors.

***Elasticity.*** All materials are deformed in greater or lesser degree under application of mechanical stress. Such deformation may be either of two kinds, known, respectively, as *elastic deformation* and *permanent deformation*. When a material is subjected to stress and undergoes deformation but resumes its original shape and dimensions when the stress is removed, the deformation is said to be *elastic*. If the stress is so great that the material fails to resume its original dimensions when the stress is removed, the permanent change in dimensions is termed *permanent deformation* or *set*. In general, the stress at which appreciable permanent deformation begins is termed the *working elastic limit*. Below this limit of stress the behavior of the material is said to be *elastic*, and in general, the deformation is proportional to the stress.

***Stress and Strain.*** The *stress* in a material under load, as in simple tension or compression, is defined as the total load divided by the area of cross section normal to the direction of the load, assuming the load to be uniformly distributed over this cross section. It is commonly expressed in pounds per square inch. The *strain* in a material under load is defined as the total deformation measured in the direction of the stress, divided by the total unstressed length in which the measured deformation occurs, or the deformation per unit length. It is expressed as a decimal ratio or numeric.

In order to show the complete behavior of any given conductor under tension, it is customary to make a graph in terms of loading or stress as the ordinates and elongation or strain as the abscissas. Such graphs or curves are useful in determining the elastic limit and the yield point if the loading is carried to the point of rupture. Graphs showing the relationship between stress and strain in a material tested to failure are termed *load-deformation* or *stress-strain curves.*

*Hooke's law* consists of the simple statement that the stress is proportional to the strain. It obviously implies a condition of perfect elasticity, which is true only for stresses less than the elastic limit.

***Stress-Strain Curves.*** A typical stress-strain diagram of hard-drawn copper wire is shown in Fig. 3-1, which represents No. 9 AWG. The curve *ae* is the actual stress-strain curve; *ab* represents the portion which corresponds to true elasticity, or for which Hooke's law holds rigorously; *cd* is the tangent *ae* which fixes the Johnson elastic limit; and the curve *af* represents the set, or permanent elongation due to flow of the metal under stress, being the difference between *ab* and *ae.* A typical stress-strain diagram of hard-drawn aluminum wire, based on data furnished by the Aluminum Company of America, is shown in Fig. 3-5.

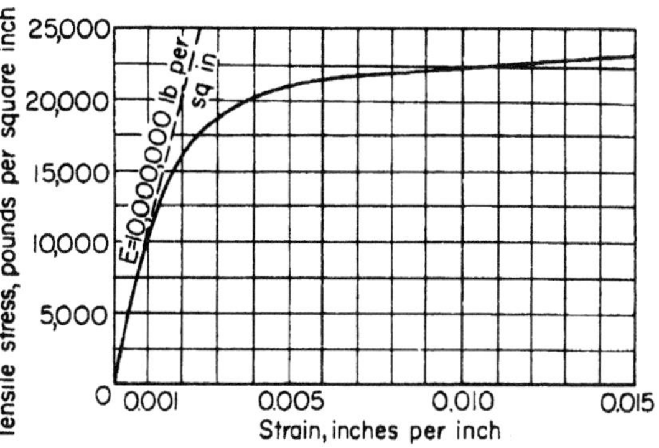


**FIGURE 3-5** Typical stress-strain curve of hard drawn aluminum wire.

***Modulus (or Coefficient) of Elasticity.*** *Modulus (or coefficient) of elasticity* is the ratio of internal stress to the corresponding strain or deformation. It is a characteristic of each material, form (shape or structure), and type of stressing. For deformations involving changes in both volume and shape, special coefficients are used. For conductors under axial tension, the ratio of stress to strain is called *Young's modulus.*

If $F$ is the total force or load acting uniformly on the cross section $A$, the stress is $F/A$. If this magnitude of stress causes an elongation $e$ in an original length $l$, the strain is $e/l$. Young's modulus is then expressed

$$M = \frac{Fl}{Ae} \tag{3-17}$$

If a material were capable of sustaining an elastic elongation sufficient to make $e$ equal to $l$, or such that the elongated length is double the original length, the stress required to produce this result would equal the modulus. This modulus is very useful in computing the sags of overhead conductor spans under loads of various kinds. It is usually expressed in pounds per square inch.

Stranding usually lowers the Young's modulus somewhat, rope-lay stranding to a greater extent than concentric-lay stranding. When a new cable is subjected initially to tension and the loading is carried up to the maximum working stress, there is an apparent elongation which is greater than the subsequent elongation under the same loading. This is apparently due to the removal of a very slight slackness in the individual wires, causing them to fit closely together and adjust themselves to the conditions of tension in the strand, and to the radial pressure the outer layers over inner layers, causing the strands to deform slightly at the crossover points, effectively reducing the diameter of the helix of that layer. When a new cable is loaded to the working limit, unloaded, and then reloaded, the value of Young's modulus determined on initial loading may be on the order of one-half to two-thirds of its true value on reloading. The latter figure should approach within a few percent of the modulus determined by test on individual straight wires of the same material.

For those applications where elastic stretching under tension needs consideration, the stress-strain curve should be determined by test, with the precaution not to prestress the cable before test unless it will be prestressed when installed in service. Commercially used values of Young's modulus for conductors are given in Table 3-6.

***Young's Modulus for ACSR.*** The permanent modulus of ACSR depends on the proportions of steel and aluminum in the cable and on the distribution of stress between aluminum and steel. This latter condition depends on temperature, tension, and previous maximum loadings. Because of the interchange of stress between the steel and the aluminum caused by changes of tension and temperature, computer programs are ordinarily used for sag-tension calculations.

Because ACSR is a composite cable made of aluminum and steel wires, additional phenomena occur which are not found in tests of cable composed of a single material. As shown in Fig. 3-2, the

**TABLE 3-6** Young's Moduli for Conductors

| Conductor | Young's modulus,* lb/in² | | Reference |
|---|---|---|---|
| | Final† | Virtual initial‡ | |
| Copper wire, hard-drawn | $17.0 \cdot 10^6$ | $14.5 \cdot 10^6$ | Copper Wire Engineering Assoc. |
| Copper wire, medium hard-drawn | $16.0 \cdot 10^6$ | $14.0 \cdot 10^6$ | Anaconda Wire and Cable Co. |
| Copper cable, hard-drawn, 3 and 12 wire | $17.0 \cdot 10^6$ | $14.0 \cdot 10^6$ | Copper Wire Engineering Assoc. |
| Copper cable, hard-drawn, 7 and 19 wire | $17.0 \cdot 10^6$ | $14.5 \cdot 10^6$ | Copper Wire Engineering Assoc. |
| Copper cable, medium hard-drawn | $15.5 \cdot 10^6$ | $14.0 \cdot 10^6$ | Anaconda Wire and Cable Co. |
| Bronze wire, alloy 15 | $14.0 \cdot 10^6$ | $13.0 \cdot 10^6$ | Anaconda Wire and Cable Co. |
| Bronze wire, other alloys | $16.0 \cdot 10^6$ | $14.0 \cdot 10^6$ | Anaconda Wire and Cable Co. |
| Bronze cable, alloy 15 | $13.0 \cdot 10^6$ | $12.0 \cdot 10^6$ | Anaconda Wire and Cable Co. |
| Bronze cable, other alloys | $16.0 \cdot 10^6$ | $14.0 \cdot 10^6$ | Anaconda Wire and Cable Co. |
| Copper-clad steel wire | $24.0 \cdot 10^6$ | $22.0 \cdot 10^6$ | Copperweld Steel Co. |
| Copper-clad steel cable | $23.0 \cdot 10^6$ | $20.5 \cdot 10^6$ | Copperweld Steel Co. |
| Copper–copper-clad steel cable, type E | $19.5 \cdot 10^6$ | $17.0 \cdot 10^6$ | Copperweld Steel Co. |
| Copper–copper-clad steel cable, type EK | $18.5 \cdot 10^6$ | $16.0 \cdot 10^6$ | Copperweld Steel Co. |
| Copper–copper-clad steel cable, type F | $18.0 \cdot 10^6$ | $15.5 \cdot 10^6$ | Copperweld Steel Co. |
| Copper–copper-clad steel cable, type 2A to 6A | $19.0 \cdot 10^6$ | $16.5 \cdot 10^6$ | Copper Wire Engineering Assoc. |
| Aluminum wire | $10.0 \cdot 10^6$ | | Reynolds Metals Co. |
| Aluminum cable | $9.1 \cdot 10^6$ | $7.3 \cdot 10^6$ | Reynolds Metals Co. |
| Aluminum-alloy wire | $10.0 \cdot 10^6$ | | Reynolds Metals Co. |
| Aluminum-alloy cable | $9.1 \cdot 10^6$ | $7.3 \cdot 10^6$ | Reynolds Metals Co. |
| Aluminum-steel cable, aluminum wire | $7.2\text{–}9.0 \cdot 10^6$ | | Aluminum Co. of America |
| Aluminum-steel cable, steel wire | $26.0\text{–}29.0 \cdot 10^6$ | | Aluminum Co. of America |
| Aluminum-clad steel wire | $23.5 \cdot 10^6$ | $22.0 \cdot 10^6$ | Copperweld Steel Co. |
| Aluminum-clad steel cable | $23.0 \cdot 10^6$ | $21.5 \cdot 10^6$ | Copperweld Steel Co. |
| Aluminum-clad steel–aluminum cable: | | | |
| AWAC 5/2 | $13.5 \cdot 10^6$ | $12.0 \cdot 10^6$ | Copperweld Steel Co. |
| AWAC 4/3 | $15.5 \cdot 10^6$ | $14.0 \cdot 10^6$ | Copperweld Steel Co. |
| AWAC 3/4 | $17.5 \cdot 10^6$ | $16.0 \cdot 10^6$ | Copperweld Steel Co. |
| AWAC 2/5 | $19.0 \cdot 10^6$ | $18.0 \cdot 10^6$ | Copperweld Steel Co. |
| Galvanized-steel wire, Class A coating | $28.5 \cdot 10^6$ | | Indiana Steel & Wire Co. |
| Galvanized-steel cable, Class A coating | $27.0 \cdot 10^6$ | | Indiana Steel & Wire Co. |

*Note:* 1 lb/in² = 6.895 kPa.

*For stranded cables the moduli are usually less than for solid wire and vary with number and arrangement of strands, tightness of stranding, and length of lay. Also, during initial application of stress, the stress-strain relation follows a curve throughout the upper part of the range of stress commonly used in transmission-line design.

†Final modulus is the ratio of stress to strain (slope of the curve) obtained after fully prestressing the conductor. It is used in calculating design or final sags and tensions.

‡Virtual initial modulus is the ratio of stress to strain (slope of the curve) obtained during initial sustained loading of new conductor. It is used in calculating initial or stringing sags and tensions.

part of the curve obtained in the second stress cycle contains a comparatively large "foot" at its base, which is caused by the difference in extension at the elastic limits of the aluminum and steel.

***Elastic Limit.*** This is variously defined as the limit of stress beyond which permanent deformation occurs or the stress limit beyond which Hooke's law ceases to apply or the limit beyond which the stresses are not proportional to the strains or the *proportional limit*. In some materials, the elastic limit occurs at a point which is readily determined, but in others it is quite difficult to determine because the stress-strain curve deviates from a straight line but very slightly at first, and the point of departure from true linear relationship between stress and strain is somewhat indeterminate.

Dean J. B. Johnson of the University of Wisconsin, well-known authority on materials of construction, proposed the use of an arbitrary determination referred to frequently as the *Johnson definition of elastic limit*. This proposal, which has been quite largely used, was that an *apparent elastic limit* be employed, defined as that point on the stress-strain curve at which the rate of deformation is 50% greater than at the origin. The apparent elastic limit thus defined is a practical value, which is suitable for engineering purposes because it involves negligible permanent elongation.

The *Johnson elastic limit* is that point on the stress-strain curve at which the natural tangent is equal to 1.5 times the tangent of the angle of the straight or linear portion of the curve, with respect to the axis of ordinates, or $Y$ axis.

***Yield Point.*** In many materials, a point is reached on the stress-strain diagram at which there is a marked increase in strain or elongation without an increase in stress or load. The point at which this occurs is termed the *yield point*. It is usually quite noticeable in ductile materials but may be scarcely perceptible or possibly not present at all in certain hard-drawn materials such as hard-drawn copper.

***Prestressed Conductors.*** In the case of some materials, especially those of considerable ductility, which tend to show permanent elongation or "drawing" under loads just above the initial elastic limit, it is possible to raise the working elastic limit by loading them to stresses somewhat above the elastic limit as found on initial loading. After such loading, or prestressing, the material will behave according to Hooke's law at all loads less than the new elastic limit. This applies not only to many ductile materials, such as soft or annealed copper wire, but also to cables or stranded conductors, in which there is a slight inherent slack or looseness of the individual wires that can be removed only under actual loading. It is sometimes the practice, when erecting such conductors for service, to prestress them to the working elastic limit or safe maximum working stress and then reduce the stress to the proper value for installation at the stringing temperature without wind or ice.

***Resistance.*** *Resistance* is the property of an electric circuit or of any body that may be used as part of an electric circuit which determines for a given current the average rate at which electrical energy is converted into heat. The term is properly applied only when the rate of conversion is proportional to the square of the current and is then equal to the power conversion divided by the square of the current. A uniform cylindrical conductor of diameter $d$, length $l$, and *volume resistivity* $\rho$ has a total resistance to continuous currents expressed by the formula

$$R = \frac{\rho l}{\pi d^2/4} \tag{3-18}$$

The resistance of any conductor is commonly expressed in ohms per unit of length, such as 1 ft, 1000 ft, or 1 mi. When used for conducting alternating currents, the effective resistance may be higher than the dc resistance defined above. In the latter case, it is a common practice to apply the proper factor, or ratio of effective ac resistance to dc resistance, sometimes termed the *skin-effect resistance ratio*. This ratio may be determined by test, or it may be calculated if the necessary data are available.

***Magnetic Permeability.*** *Magnetic permeability* applies to a field in which the flux is uniformly distributed over a cross section normal to its direction or to a sufficiently small cross section of a nonuniform field so that the distribution can be assumed as substantially uniform. In the case of a cylindrical conductor, the magnetomotive force (mmf) due to the current flowing in the conductor varies from zero at the center or axis to a maximum at the periphery or surface of the conductor and sets up a flux in circular paths concentric with the axis and perpendicular to it but of nonuniform distribution between the axis and the periphery. If the permeability is nonlinear with respect to the mmf, as is usually true with magnetic materials, there is no correct single value of permeability which fits the conditions, although an apparent or equivalent average value can be determined. In the case of other forms of cross section, the distribution is still more complex, and the equivalent permeability may be difficult or impossible to determine except by test.

***Internal Inductance.*** A uniform cylindrical conductor of nonmagnetic material, or of unit permeability, has a constant magnitude of internal inductance per unit length, independent of the conductor diameter. This is commonly expressed in microhenrys or millihenrys per unit of length, such as 1 ft, 1000 ft, or 1 mi. When the conductor material possesses magnetic susceptibility, and when the magnetic permeability $\mu$ is constant and therefore independent of the current strength, the internal inductance is expressed in absolute units by the formula

$$L = \frac{\mu l}{2} \tag{3-19}$$

In most cases, $\mu$ is not constant but is a function of the current strength. When this is true, there is an effective permeability, one-half of which ($\mu/2$) expresses the inductance per centimeter of length, but this figure of permeability is virtually the ratio of the effective inductance of the conductor of susceptible material to the inductance of a conductor of material which has a permeability of unity. When used for conducting alternating currents, the effective inductance may be less than the inductance with direct current; this is also a direct consequence of the same skin effect which results in an increase of effective resistance with alternating currents, but the overall effect is usually included in the figure of effective permeability. It is usually the practice to determine the effective internal inductance by test, but it may be calculated if the necessary data are available.

If $L'$ is the effective inductance of a linear conductor to sinusoidal alternating current of a given frequency, then

$$L' = L_1 + K'L_2 \tag{3-20}$$

where $L_1$ is external portion of inductance, $L_2$ is internal portion (due to the magnetic field within the conductor), and $K'$ is determined from Table 3-2 in terms of $x$. Thus, the total effective inductance per unit length of conductor is

$$L' = 2\ln\frac{d}{a} + K' - \frac{\mu}{2} \tag{3-21}$$

The inductance is here expressed in abhenrys per centimeter of conductor, in a linear circuit; $a$ is the radius of the conductor, and $d$ is the separation between the conductor and its return conductor, expressed in the same units.

***Skin Effect.*** *Skin effect* is a phenomenon which occurs in conductors carrying currents whose intensity varies rapidly from instant to instant but does not occur with continuous currents. It arises from the fact that elements or filaments of variable current at different points in the cross section of a conductor do not encounter equal components of inductance, but the central or axial filament meets the maximum inductance, and in general the inductance offered to other filaments of current decreases as the distance of the filament from the axis increases, becoming a minimum at the surface or periphery of the conductor. This, in turn, tends to produce unequal current density over the cross section as a whole; the density is a minimum at the axis and a maximum at the periphery. Such distribution of the current density produces an increase in effective resistance and a decrease in effective internal inductance; the former is of more practical importance than the latter. In the case of large copper and aluminum conductors at commercial power frequencies and in the case of most conductors at carrier and radio frequencies, the increase in resistance should be considered.

***Skin-Effect Ratios.*** If $R'$ is the effective resistance of a linear cylindrical conductor to sinusoidal alternating current of given frequency and $R$ is the true resistance with continuous current, then

$$R' = KR \quad \text{ohms} \tag{3-22}$$

where $K$ is determined from Table 3-7 in terms of $x$. The value of $x$ is given by

$$x = 2\pi a\sqrt{\frac{2f\mu}{\rho}} \tag{3-23}$$

where $a$ is the radius of the conductor in centimeters, $f$ is the frequency in cycles per second, $\mu$ is the relative magnetic permeability of the conductor (here assumed to be constant), and $\rho$ is the resistivity in abohm-centimeters (abohm = $10^{-9}$ Ω). Relative permeability is the ratio of the magnetic permeability of a substance to the permeability of a vacuum. For non-magnetic materials such as copper or aluminum, the magnetic permeability can be taken as 1.

For practical calculation, Eq. (3-23) can be written

$$x = 0.063598\sqrt{\frac{f\mu}{R}} \tag{3-24}$$

where $R$ is dc resistance at operating temperature in ohms per mile.

**TABLE 3-7** Skin-Effect Ratios

| $x$ | $K$ | $K'$ | $x$ | $K$ | $K'$ | $x$ | $K$ | $K'$ | $x$ | $K$ | $K'$ |
|---|---|---|---|---|---|---|---|---|---|---|---|
| 0.0 | 1.00000 | 1.00000 | 2.9 | 1.28644 | 0.86012 | 6.6 | 2.60313 | 0.42389 | 17.0 | 6.26817 | 0.16614 |
| 0.1 | 1.00000 | 1.00000 | 3.0 | 1.31809 | 0.84517 | 6.8 | 2.67312 | 0.41171 | 18.0 | 6.62129 | 0.15694 |
| 0.2 | 1.00001 | 1.00000 | 3.1 | 1.35102 | 0.82975 | 7.0 | 2.74319 | 0.40021 | 19.0 | 6.97446 | 0.14870 |
| 0.3 | 1.00004 | 0.99998 | 3.2 | 1.38504 | 0.81397 | 7.2 | 2.81334 | 0.38933 | 20.0 | 7.32767 | 0.14128 |
| 0.4 | 1.00013 | 0.99993 | 3.3 | 1.41999 | 0.79794 | 7.4 | 2.88355 | 0.37902 | 21.0 | 7.68091 | 0.13456 |
| 0.5 | 1.00032 | 0.99984 | 3.4 | 1.45570 | 0.78175 | 7.6 | 2.95380 | 0.36923 | 22.0 | 8.03418 | 0.12846 |
| 0.6 | 1.00067 | 0.99966 | 3.5 | 1.49202 | 0.76550 | 7.8 | 3.02411 | 0.35992 | 23.0 | 8.38748 | 0.12288 |
| 0.7 | 1.00124 | 0.99937 | 3.6 | 1.52879 | 0.74929 | 8.0 | 3.09445 | 0.35107 | 24.0 | 8.74079 | 0.11777 |
| 0.8 | 1.00212 | 0.99894 | 3.7 | 1.56587 | 0.73320 | 8.2 | 3.16480 | 0.34263 | 25.0 | 9.09412 | 0.11307 |
| 0.9 | 1.00340 | 0.99830 | 3.8 | 1.60314 | 0.71729 | 8.4 | 3.23518 | 0.33460 | 26.0 | 9.44748 | 0.10872 |
| 1.0 | 1.00519 | 0.99741 | 3.9 | 1.64051 | 0.70165 | 8.6 | 3.30557 | 0.32692 | 28.0 | 10.15422 | 0.10096 |
| 1.1 | 1.00758 | 0.99621 | 4.0 | 1.67787 | 0.68632 | 8.8 | 3.37597 | 0.31958 | 30.0 | 10.86101 | 0.09424 |
| 1.2 | 1.01071 | 0.99465 | 4.1 | 1.71516 | 0.67135 | 9.0 | 3.44638 | 0.31257 | 32.0 | 11.56785 | 0.08835 |
| 1.3 | 1.01470 | 0.99266 | 4.2 | 1.75233 | 0.65677 | 9.2 | 3.51680 | 0.30585 | 34.0 | 12.27471 | 0.08316 |
| 1.4 | 1.01969 | 0.99017 | 4.3 | 1.78933 | 0.64262 | 9.4 | 3.58723 | 0.29941 | 36.0 | 12.98160 | 0.07854 |
| 1.5 | 1.02582 | 0.98711 | 4.4 | 1.82614 | 0.62890 | 9.6 | 3.65766 | 0.29324 | 38.0 | 13.68852 | 0.07441 |
| 1.6 | 1.03323 | 0.98342 | 4.5 | 1.86275 | 0.61563 | 9.8 | 3.72812 | 0.28731 | 40.0 | 14.39545 | 0.07069 |
| 1.7 | 1.04205 | 0.97904 | 4.6 | 1.89914 | 0.60281 | 10.0 | 3.79857 | 0.28162 | 42.0 | 15.10240 | 0.06733 |
| 1.8 | 1.05240 | 0.97390 | 4.7 | 1.93533 | 0.59044 | 10.5 | 3.97477 | 0.26832 | 44.0 | 15.80936 | 0.06427 |
| 1.9 | 1.06440 | 0.96795 | 4.8 | 1.97131 | 0.57852 | 11.0 | 4.15100 | 0.25622 | 46.0 | 16.51634 | 0.06148 |
| 2.0 | 1.07816 | 0.96113 | 4.9 | 2.00710 | 0.56703 | 11.5 | 4.32727 | 0.24516 | 48.0 | 17.22333 | 0.05892 |
| 2.1 | 1.09375 | 0.95343 | 5.0 | 2.04272 | 0.55597 | 12.0 | 4.50358 | 0.23501 | 50.0 | 17.93032 | 0.05656 |
| 2.2 | 1.11126 | 0.94482 | 5.2 | 2.11353 | 0.53506 | 12.5 | 4.67993 | 0.22567 | 60.0 | 21.46541 | 0.04713 |
| 2.3 | 1.13069 | 0.93527 | 5.4 | 2.18389 | 0.51566 | 13.0 | 4.85631 | 0.21703 | 70.0 | 25.00063 | 0.04040 |
| 2.4 | 1.15207 | 0.92482 | 5.6 | 2.25393 | 0.49764 | 13.5 | 5.03272 | 0.20903 | 80.0 | 28.53593 | 0.03535 |
| 2.5 | 1.17538 | 0.91347 | 5.8 | 2.32380 | 0.48086 | 14.0 | 5.20915 | 0.20160 | 90.0 | 32.07127 | 0.03142 |
| 2.6 | 1.20056 | 0.90126 | 6.0 | 2.39359 | 0.46521 | 14.5 | 5.38560 | 0.19468 | 100.0 | 35.60666 | 0.02828 |
| 2.7 | 1.22753 | 0.88825 | 6.2 | 2.46338 | 0.45056 | 15.0 | 5.56208 | 0.18822 | ∞ | ∞ | 0 |
| 2.8 | 1.25620 | 0.87451 | 6.4 | 2.53321 | 0.43682 | 16.0 | 5.91509 | 0.17649 | | | |

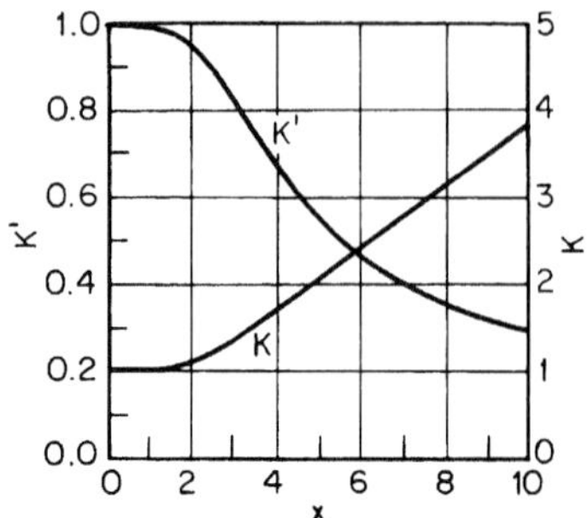


**FIGURE 3-6** *K* and *K′* for values of *x* from 0 to 10.

Values of *K* and *K′* in terms of *x* are shown in Table 3-7 and Figs. 3-3 and 3-6 (see *NBS Circ.* 74, pp. 309–311, for additional tables, and *Sci. Paper* 374).

Value of $\mu$ for nonmagnetic materials (copper, aluminum, etc.) is 1; for magnetic materials, it varies widely with composition, processing, current density, etc., and should be determined by test in each case.

***Alternating-Current Resistance.*** For small conductors at power frequencies, the frequency has a negligible effect, and dc resistance values can be used. For large conductors, frequency must be taken into account in addition to temperature effects. To do this, first calculate the dc resistance at the operating temperature, then determine the skin-effect ratio *K*, and finally determine the ac resistance at operating temperature.

AC resistance for copper conductors *not* in close proximity can be obtained from the skin-effect ratios given in Tables 3-7 and 3-8.

***AC Resistance for Aluminum Conductors.*** The increase in resistance and decrease in internal inductance of cylindrical aluminum conductors can be determined from data. It is not the same as for copper conductors of equal diameter but is slightly less because of the higher volume resistivity of aluminum.

***AC Resistance for ACSR.*** In the case of ACSR conductors, the steel core is of relatively high resistivity, and therefore its conductance is usually neglected in computing the total resistance of such strands. The effective permeability of the grade of steel employed in the core is also relatively small. It is approximately correct to assume that such a conductor is hollow and consists exclusively of its aluminum wires; in this case, the laws of skin effect in tubular conductors will be applicable. Conductors having a single layer of aluminum wires over the steel core have higher ac/dc ratios than those having multiple layers of aluminum wires.

***Inductive Reactance.*** Present practice is to consider inductive reactance as split into two components: (1) that due to flux within a radius of 1 ft including the internal reactance within the conductor of radius *r* and (2) that due to flux between 1 ft radius and the equivalent conductor spacing *Ds* or geometric mean distance (GMD). The fundamental inductance formula is

$$L = 2\ln\frac{D_s}{r} + \frac{\mu}{2} \quad \text{abH/(cm)(conductor)} \tag{3-25}$$

This can be rewritten

$$L = 2\ln\frac{D_s}{1} + 2\ln\frac{1}{r} + \frac{\mu}{2} \tag{3-26}$$

where the term 2 ln (*Ds*/1) represents inductance due to flux between 1 ft radius and the equivalent conductor spacing, and 2 ln (1/*r*) + ($\mu$/2) represents the inductance due to flux within 1 ft radius [2 ln (1/*r*) represents inductance due to flux between conductor surface and 1 ft radius, and $\mu/2$ represents internal inductance due to flux within the conductor].

By definition, *geometric mean radius* (GMR) of a conductor is the radius of an infinitely thin tube having the same internal inductance as the conductor. Therefore,

$$L = 2\ln\frac{D_s}{1} + 2\ln\frac{1}{\text{GMR}} \tag{3-27}$$

**TABLE 3-8** Skin-Effect Ratios—Copper Conductors *Not* in Close Proximity

| Conductor size, Mcm | Skin-effect ratio ***K*** at 60 cycles and 65°C (149°F) | | | | | | | | | | | | | | | |
|---|---|---|---|---|---|---|---|---|---|---|---|---|---|---|---|---|
| | | | Inside conductor diameter, in | | | | | | | | | | | | | |
| | 0* | | 0.25 | | 0.50 | | 0.75 | | 1.00 | | 1.25 | | 1.30 | | 2.00 | |
| | Outside diameter, in | ***K*** | Outside diameter, in | ***K*** | Outside diameter, in | ***K*** | Outside diameter, in | ***K*** | Outside diameter, in | ***K*** | Outside diameter, in | ***K*** | Outside diameter, in | ***K*** | Outside diameter, in | ***K*** |
| 3000 | 1.998 | 1.439 | 2.02 | 1.39 | 2.08 | 1.36 | 2.15 | 1.29 | 2.27 | 1.23 | 2.39 | 1.19 | 2.54 | 1.15 | 2.87 | 1.08 |
| 2500 | 1.825 | 1.336 | 1.87 | 1.28 | 1.91 | 1.24 | 2.00 | 1.20 | 2.12 | 1.16 | 2.25 | 1.12 | 2.40 | 1.09 | 2.75 | 1.05 |
| 2000 | 1.631 | 1.239 | 1.67 | 1.20 | 1.72 | 1.17 | 1.80 | 1.12 | 1.94 | 1.09 | 2.09 | 1.06 | 2.25 | 1.05 | 2.61 | 1.02 |
| 1500 | 1.412 | 1.145 | 1.45 | 1.12 | 1.52 | 1.09 | 1.63 | 1.06 | 1.75 | 1.04 | 1.91 | 1.03 | 2.07 | 1.02 | 2.47 | 1.01 |
| 1000 | 1.152 | 1.068 | 1.19 | 1.05 | 1.25 | 1.03 | 1.39 | 1.02 | 1.53 | 1.01 | 1.72 | 1.01 | | | | |
| 800 | 1.031 | 1.046 | 1.07 | 1.04 | 1.16 | 1.02 | 1.28 | 1.01 | 1.45 | 1.01 | | | | | | |
| 600 | 0.893 | 1.026 | 0.94 | 1.02 | 1.04 | 1.01 | | | | | | | | | | |
| 500 | 0.814 | 1.018 | 0.86 | 1.01 | 0.97 | 1.01 | | | | | | | | | | |
| 400 | 0.728 | 1.012 | 0.78 | 1.01 | | | | | | | | | | | | |
| 300 | 0.630 | 1.006 | | | | | | | | | | | | | | |

***Note:*** 1 in = 2.54 cm.

*For standard concentric-stranded conductors (i.e., inside diameter = 0).

Since inductive reactance = $2\pi fL$, for practical calculation Eq. (3-27) can be written

$$X = 0.004657 f \log \frac{D_s}{1} + 0.004657 f \log \frac{1}{\text{GMR}} \quad \Omega(\text{mi})(\text{conductor}) \tag{3-28}$$

In the conductor tables in this section, inductive reactance is calculated from Eq. (3-28), considering that

$$X = x_a + x_d \tag{3-29}$$

Inductive reactance for conductors using steel varies in a manner similar to ac resistance.

***Capacitive Reactance.*** The capacitive reactance can be considered in two parts also, giving

$$X = \frac{4.099}{f} \log \frac{D_s}{1} + \frac{4.099}{f} \log \frac{1}{r} \quad \text{M}\Omega(\text{mi})(\text{conductor}) \tag{3-30}$$

In the conductor tables in this section, capacitive reactance is calculated from Eq. (3-30), it being considered that

$$X' = x_a' + x_d' \tag{3-31}$$

It is important to note that in capacitance calculations the conductor radius used is the actual physical radius of the conductor.

***Capacitive Susceptance***

$$B = \frac{1}{x_a' + x_a'} \quad \mu\text{S}(\text{mi})(\text{conductor}) \tag{3-32}$$

***Charging Current***

$$I_C = eB \times 10^{-3} \quad \text{A}/(\text{mi})(\text{conductor}) \tag{3-33}$$

where $e$ is voltage to neutral in kilovolts.

***Bus Conductors.*** *Bus conductors* require that greater attention be given to certain physical and electrical characteristics of the metals than is usually necessary in designing line conductors. These characteristics are current-carrying capacity, emissivity, skin effect, expansion, and mechanical deflection. To obtain the most satisfactory and economical designs for bus bars in power stations and substations, where they are used extensively, consideration must be given to choice not only of material but also of shape. Both copper and aluminum are used for bus bars, and in certain outdoor substations, steel has proved satisfactory. The most common bus bar form for carrying heavy current, especially indoors, is flat copper bar. Bus bars in the form of angles, channels, and tubing have been developed for heavy currents and, because of better distribution of the conducting material, make more efficient use of the metal both electrically and mechanically. All such designs are based on the need for proper current-carrying capacity without excess bus bar temperatures and on the necessity for adequate mechanical strength.

***Hollow (Expanded) Conductors.*** Hollow (expanded) conductors were designed for use on high-voltage transmission lines where in order to reduce corona loss, it is desirable to increase the outside diameter without increasing the area beyond that needed for maximum line economy. Not only is the initial corona voltage considerably higher than for conventional conductors of equal cross section, but the current-carrying capacity for a given temperature rise is also greater because of the larger surface area available for cooling and the better disposition of the metal with respect to skin effect when carrying alternating currents. Expanded conductors have been used in the past, but are seldom specified today. They were air expanded or used a paper filler.

*Air-expanded ACSR* is a conductor whose diameter has been increased by aluminum skeletal wires between the steel core and the outer layers of aluminum strands creating air spaces. A conductor having the necessary diameter to minimize corona effects on lines operating above 300 kV will, many times, have more metal than is economical if the conductor is made conventionally.

***Composite Conductors.*** *Composite conductors* are those made up of usually two different types of wire having differing characteristics. They are generally designed for a ratio of physical and electrical characteristics different from those found in homogeneous materials. *Aluminum conductors, steel reinforced (ACSR)* and *aluminum conductors, aluminum alloy reinforced (ACAR)* are types commonly used in overhead transmission and distribution lines.

Cables of this type are particularly adaptable to long-span construction or other service conditions requiring more than average strength combined with liberal conductance. They lend themselves readily to economical, dependable use on transmission lines, rural distribution lines, railroad electrification, river crossings, and many kinds of special construction.

*Self-damping ACSR* conductors are used to limit aeolian vibration to a safe level regardless of conductor tension or span length. They are concentrically stranded conductors composed of two layers of trapezoidal-shaped wires or two layers of trapezoidal-shaped wires and one layer of round wires of 1350 (EC) alloy with a high-strength, coated steel core. The trapezoidal wire layers are self-supporting, and separated by gaps from adjacent layers (Fig. 3-7). Impact between layers during aeolian vibration causes damping action.

**FIGURE 3-7** Self-damping ACSR conductor.

*ACSR/TW* is similar to self-damping ACSR in its use of trapezoidal-shaped wires, but does not have the annular gaps between layers. ACSR/TW has a smaller diameter and smoother surface than conventional round-wire ACSR of the same area, and thus may have reduced wind loading.

*Twisted pair* (T-2 is a trademark) *conductors* are fabricated by twisting two conventional conductors together with a pitch of about 9 ft (2.7 m). Twisted pair conductors are designed to resist aeolian vibration caused by laminar winds blowing across conductors. The pair of conductors presents a constantly changing diameter to the wind (from one diameter to two diameters). This constantly changing diameter prevents the formation of von Karman vortices that produce vibration. Severity of wind-induced galloping when the conductor is coated with ice is reduced because an ice profile that is uniform along the conductor length cannot form on the variable profile presented by the conductor.

*Aluminum Conductor Steel Supported.* *Steel-supported aluminum conductors (SSAC)* are similar to conventional ACSR but employ an aluminum alloy in the annealed condition. The annealed aluminum has increased electrical conductivity, and the conductor has improved sag-tension characteristics for high-temperature service.

## 3.1.4 Fusible Metals and Alloys

*Fusible alloys* having melting points in the range from about 60 to 200°C are made principally of bismuth, cadmium, lead, and tin in various proportions. Many of these alloys have been known under the names of their inventors (see index of alloys in *International Critical Tables*, vol. 2).

*Fuse metals* for electric fuses of the open-link enclosed and expulsion types are ordinarily made of some low-fusible alloy; aluminum also is used to some extent. The resistance of the fuse causes dissipation of energy, liberation of heat, and rise of temperature. Sufficient current obviously will melt the fuse, and thus open the circuit if the resulting arc is self-extinguishing. Metals which volatilize readily in the heat of the arc are to be preferred to those which leave a residue of globules of hot metal. The rating of any fuse depends critically on its shape, dimensions, mounting, enclosure, and any other factors which affect its heat-dissipating capacity.

Fusing currents of different kinds of wire were investigated by W. H. Preece, who developed the formula

$$I = ad^{3/2} \tag{3-34}$$

where $I$ is fusing current in amperes, $d$ is diameter of the wire in inches, and $a$ is a constant depending on the material. He found the following values for $a$:

| | | | |
|---|---|---|---|
| Copper | 10,244 | Iron | 3,148 |
| Aluminum | 7,585 | Tin | 1,642 |
| Platinum | 5,172 | Alloy (2Pb-1Sn) | 1,318 |
| German silver | 5,230 | Lead | 1,379 |
| Platinoid | 4,750 | | |

Although this formula has been used to a considerable extent in the past, it gives values that usually are erroneous in practice because it is based on the assumption that all heat loss is due to radiation. A formula of the general type

$$I = kd^{n} \tag{3-35}$$

can be used with accuracy if $k$ and $n$ are known for the particular case (material, wire size, installation conditions, etc.).

Fusing current-time for copper conductors and connections may be determined by an equation developed by I. M. Onderdonk

$$33\left(\frac{I}{A}\right)^2 S = \log\left(\frac{T_m - T_a}{234 + T_a} + 1\right) \tag{3-36}$$

$$I = A\sqrt{\frac{\log\left(\frac{T_m - T_a}{234 + T_a} + 1\right)}{33S}} \tag{3-37}$$

where $I$ is current in amperes, $A$ is conductor area in circular mils, $S$ is time current applied in seconds, $T_m$ is melting point of copper in degrees Celsius, and $T_a$ is ambient temperature in degrees Celsius.

### 3.1.5 Miscellaneous Metals and Alloys

***Contact Metals.*** *Contact metals* may be grouped into three general classifications:

*Hard metals,* which have high melting points, for example, tungsten and molybdenum. Contacts of these metals are employed usually where operations are continuous or very frequent and current has nominal value of 5 to 10 A. Hardness to withstand mechanical wear and high melting point to resist arc erosion and welding are their outstanding advantages. A tendency to form high-resistance oxides is a disadvantage, but this can be overcome by several methods, such as using high-contact force, a hammering or wiping action, and a properly balanced electric circuit.

*Highly conductive* metals, of which silver is the best for both electric current and heat. Its disadvantages are softness and a tendency to pit and transfer. In sulfurous atmosphere, a resistant sulfide surface will form on silver, which results in high contact-surface resistance. These disadvantages are overcome usually by alloying.

*Noncorroding* metals, which for the most part consist of the noble metals, such as gold and the platinum group. Contacts of these metals are used on sensitive devices, employing extremely light pressures or low current in which clean contact surfaces are essential. Because most of these metals are soft, they are usually alloyed.

The metals commonly used are tungsten, molybdenum, platinum, palladium, gold, silver, and their alloys. Alloying materials are copper, nickel, cadmium, iron, and the rarer metals such as iridium and ruthenium. Some are prepared by powder metallurgy.

***Tungsten.*** *Tungsten* (W) is a hard, dense, slow-wearing metal, a good thermal and electrical conductor, characterized by its high melting point and freedom from sticking or welding. It is manufactured in several grades having various grain sizes.

***Molybdenum.*** *Molybdenum* (Mo) has contact characteristics about midway between tungsten and fine silver. It often replaces either metal where greater wear resistance than that of silver or lower contact-surface resistance than that of tungsten is desired.

***Platinum.*** *Platinum* (Pt) is one of the most stable of all metals under the combined action of corrosion and electrical erosion. It has a high melting point and does not corrode and surfaces remain clean and low in resistance under most adverse atmospheric and electrical conditions. *Platinum alloys* of iridium (Ir), ruthenium (Ru), silver (Ag), or other metals are used to increase hardness and resistance to wear.

***Palladium.*** *Palladium* (Pd) has many of the properties of platinum and is frequently used as an alternate for platinum and its alloys. *Palladium alloys* of silver (Ag), ruthenium (Ru), nickel (Ni), and other metals are used to increase hardness and resistance to wear.

***Gold.*** *Gold* (Au) is similar to platinum in corrosion resistance but has a much lower melting point. Gold and its alloys are ductile and easily formed into a variety of shapes. Because of its softness, it is usually alloyed. *Gold alloys* of silver (Ag) and other metals are used to impart hardness and improve resistance to mechanical wear and electrical erosion.

***Silver.*** *Silver* (Ag) has the highest thermal and electrical conductivity (110%, IACS) of any metal. It has low contact-surface resistance, since its oxide decomposes at approximately 300°F. It is available commercially in three grades:

| | Typical composition, % | |
|---|---|---|
| Grade | Silver | Copper |
| Fine silver | 99.95+ | |
| Sterling silver | 92.5 | 7.5 |
| Coin silver | 90 | 10 |

*Fine* silver is used extensively under low contact pressure where sensitivity and low contact-surface resistance are essential or where the circuit is operated infrequently.

*Sterling and coin* silvers are harder than fine silver and resist transfer at low voltage (6 to 8 V) better than fine silver. Since their contact-surface resistance is greater than that of fine silver, higher contact-closing forces should be used.

*Silver alloys* of copper (Cu), nickel (Ni), cadmium (Cd), iron (Fe), carbon (C), tungsten (W), molybdenum (Mo), and other metals are used to improve hardness, resistance to wear and arc erosion, and for special applications.

***Selenium.*** *Selenium* is a nonmetallic element chemically resembling sulfur and tellurium and occurs in several allotropic forms varying in specific gravity from 4.3 to 4.8. It melts at 217°C and boils at 690°C. At 0°C, it has a resistivity of approximately 60,000 $\Omega \cdot$ cm. The dielectric constant ranges from 6.1 to 7.4. It has the peculiar property that its resistivity decreases on exposure to light; the resistivity in darkness may be anywhere from 5 to 200 times the resistivity under exposure to light.

### 3.1.6 Bibliography

*Aluminum Electrical Conductor Handbook,* 3d ed. The Aluminum Association, Waldorf, Md., 1989.

*Current Temperature Characteristics of Aluminum Conductors.* Alcoa Conductor Products Company, Pittsburgh, Pa., 1965.

*Electric Utility Engineering Reference Book,* vol. 3. Westinghouse Electric Corporation, East Pittsburgh, Pa., 1959.

Johnson, J. B. 1925. *Materials of Construction.* John Wiley & Sons, Inc., New York.

*Metals Handbook.* 1998. Materials Park, Ohio, ASM International.

*Overload and Fault Current Limitations of Bare Aluminum Conductors.* Alcoa Conductor Products Company, Pittsburgh, Pa., 1965.

*Resistance and Reactance of Aluminum Conductors.* Alcoa Conductor Products Company, Pittsburgh, Pa., 1965.

Smith, C. S. 1930. "Thermal Conductivity of Copper Alloys: I. Copper-Zinc Alloy." *Trans. AIMME, Inst. Metals Div.,* vol. 89, p. 84.

Stauffacher, E. R. 1928. "Short-Time Current-Carrying Capacity of Copper Wires." *Gen. Elec. Rev.,* June.

*Underground Systems Reference Book.* Edison Electric Institute, Transmission and Distribution Committee, Washington, D.C., 1957.

## 3.2 MAGNETIC MATERIALS

### 3.2.1 Definitions

The following definitions of terms relating to magnetic materials and the properties and testing of these materials have been selected from ASTM Standard. Terms primarily related to magnetostatics are indicated by the symbol * and those related to magnetodynamics are indicated by the symbol **. General (nonrestricted) terms are not marked.

***AC Excitation N1I/l1. The ratio of the rms ampere-turns of exciting current in the primary winding of an inductor to the effective length of the magnetic path.*

***Active (Real) Power P.* The product of the rms current $I$ in an electric circuit, the rms voltage $E$ across the circuit, and the cosine of the angular phase difference $\theta$ between the current and the voltage.

$$P = EI\cos\theta \tag{3-38}$$

NOTE: The portion of the active power that is expended in a magnetic core is the total core loss $Pc$.

*Aging, Magnetic.* The change in the magnetic properties of a material resulting from metallurgical change. This term applies whether the change results from a continued normal or a specified accelerated aging condition.

NOTE: This term implies a deterioration of the magnetic properties of magnetic materials for electronic and electrical applications, unless otherwise specified.

*Ampere-turn.* Unit of magnetomotive force in the rationalized mksa system. One ampere-turn equals $4\pi/10$, or 1.257 gilberts.

*Ampere-turn per Meter.* Unit of magnetizing force (magnetic field strength) in the rationalized mksa system. One ampere-turn per meter is $4\pi \times 10^{-3}$, or 0.01257 oersted.

*Anisotropic Material.* A material in which the magnetic properties differ in various directions.

*Antiferromagnetic Material.* A feebly magnetic material in which almost equal magnetic moments are lined up antiparallel to each other. Its susceptibility increases as the temperature is raised until a critical (Neél) temperature is reached; above this temperature the material becomes paramagnetic.

***Apparent Power* $P_a$. The product (volt-amperes) of the rms exciting current and the applied rms terminal voltage in an electric circuit containing inductive impedance. The components of this impedance due to the winding will be linear, while the components due to the magnetic core will be nonlinear.

***Apparent Power; Specific,* $P_{a(B,f)}$. The value of the apparent power divided by the active mass of the specimen (volt-amperes per unit mass) taken at a specified maximum value of cyclically varying induction $B$ and at a specified frequency $f$.

**Coercive Force* $H_c$. The (dc) magnetizing force at which the magnetic induction is zero when the material is in a symmetrically cyclically magnetized condition.

**Coercive Force, Intrinsic,* $H_{ci}$. The (dc) magnetizing force at which the intrinsic induction is zero when the material is in a symmetrically cyclically magnetized condition.

**Coercivity Hcs.* The maximum value of coercive force.

***Core Loss; Specific,* $P_{c(B,f)}$. The active power (watts) expended per unit mass of magnetic material in which there is a cyclically varying induction of a specified maximum value $B$ at a specified frequency $f$.

***Core Loss (Total)* $P_c$. The active power (watts) expended in a magnetic circuit in which there is a cyclically alternating induction.

NOTE: Measurements of core loss are normally made with sinusoidally alternating induction, or the results are corrected for deviations from the sinusoidal condition.

*Curie Temperature* $T_c$. The temperature above which a ferromagnetic material becomes paramagnetic.

**Demagnetization Curve.* That portion of a normal (dc) hysteresis loop which lies in the second or fourth quadrant, that is, between the residual induction point $B_r$ and the coercive force point $H_c$. Points on this curve are designated by the coordinates $B_d$ and $H_d$.

*Diamagnetic Material.* A material whose relative permeability is less than unity.

NOTE: The intrinsic induction $B_i$, is oppositely directed to the applied magnetizing force $H$.

*Domains, Ferromagnetic.* Magnetized regions, either macroscopic or microscopic in size, within ferromagnetic materials. Each domain per se is magnetized to intrinsic saturation at all times, and this saturation induction is unidirectional within the domain.

***Eddy-Current Loss, Normal,* $P_e$. That portion of the core loss which is due to induced currents circulating in the magnetic material subject to an *SCM* excitation.

**Energy Product* $B_dH_d$. The product of the coordinate values of any point on a demagnetization curve.

**Energy-Product Curve, Magnetic.* The curve obtained by plotting the product of the corresponding coordinates $B_d$ and $H_d$ of points on the demagnetization curve as abscissa against the induction $B_d$ as ordinates.

NOTE 1: The maximum value of the energy product $(B_dH_d)_m$ corresponds to the maximum value of the external energy.

NOTE 2: The demagnetization curve is plotted to the left of the vertical axis and usually the energy-product curve to the right.

***Exciting Power, rms, $P_z$.* The product of the rms exciting current and the rms voltage induced in the exciting (primary) winding on a magnetic core.

NOTE: This is the apparent volt-amperes required for the excitation of the magnetic core only. When the core has a secondary winding, the induced primary voltage is obtained from the measured open-circuit secondary voltage multiplied by the appropriate turns ratio.

***Exciting Power, Specific $P_{z(B,f)}$.* The value of the rms exciting power divided by the active mass of the specimen (volt-amperes/unit mass) taken at a specified maximum value of cyclically varying induction B and at specified frequency *f*.

*Ferrimagnetic Material.* A material in which unequal magnetic moments are lined up antiparallel to each other. Permeabilities are of the same order of magnitude as those of ferromagnetic materials, but are lower than they would be if all atomic moments were parallel and in the same direction. Under ordinary conditions, the magnetic characteristics of ferrimagnetic materials are quite similar to those of ferromagnetic materials.

*Ferromagnetic Material.* A material that, in general, exhibits the phenomena of hysteresis and saturation, and whose permeability is dependent on the magnetizing force.

*Gauss (Plural Gausses).* The unit of magnetic induction in the cgs electromagnetic system. The gauss is equal to 1 maxwell per square centimeter or $10^{-4}$ T. See *magnetic induction (flux density).*

*Gilbert.* The unit of magnetomotive force in the cgs electromagnetic system. The gilbert is a magnetomotive force of $10/4\pi$ ampere-turns. See *magnetomotive force.*

**Hysteresis Loop, Intrinsic.* A hysteresis loop obtained with a ferromagnetic material by plotting (usually to rectangular coordinates) corresponding dc values of intrinsic induction $B_i$ for ordinates and magnetizing force $H$ for abscissas.

**Hysteresis Loop, Normal.* A closed curve obtained with a ferromagnetic material by plotting (usually to rectangular coordinates) corresponding dc values of magnetic induction B for ordinates and magnetizing force $H$ for abscissas when the material is passing through a complete cycle between equal definite limits of either magnetizing force $\pm H_m$ or magnetic induction $\pm B_m$. In general, the normal hysteresis loop has mirror symmetry with respect to the origin of the $B$ and $H$ axes, but this may not be true for special materials.

**Hysteresis-Loop Loss $W_h$.* The energy expended in a single slow excursion around a normal hysteresis loop is given by the following equation:

$$W_h = \int \frac{HdB}{4\pi} \quad \text{ergs} \tag{3-39}$$

where the integrated area enclosed by the loop is measured in gauss-oersteds.

***Hysteresis Loss, Normal, $P_h$.*

1. The power expended in a ferromagnetic material, as a result of hysteresis, when the material is subjected to an *SCM* excitation.
2. The energy loss/cycle in a magnetic material as a result of magnetic hysteresis when the induction is cyclic (but not necessarily periodic).

*Hysteresis, Magnetic.* The property of a ferromagnetic material exhibited by the lack of correspondence between the changes in induction resulting from increasing magnetizing force from decreasing magnetizing force.

*Induction B.* See *magnetic induction (flux density).*

**Induction, Intrinsic, $B_i$.* The vector difference between the magnetic induction in a magnetic material and the magnetic induction that would exist in a vacuum under the influence of the same magnetizing force. This is expressed by the equation

$$B_i = B - \Gamma_m H \tag{3-40}$$

NOTE: In the cgs-em system, $B_i/4\pi$ is often called *magnetic polarization.*

*Induction Maximum*

*1. $B_m$—the maximum value of $B$ in a hysteresis loop. The tip of this loop has the magnetostatic coordinates $H_m$, $B_m$, which exist simultaneously.

**2. $B_{max}$—the maximum value of induction, in a flux-current loop.

NOTE: In a flux-current loop, the magnetodynamic values $B_{max}$ and $H_{max}$ do not exist simultaneously; $B_{max}$ occurs later than $H_{max}$.

**Induction, Normal, B.* The maximum induction in a magnetic material that is in a symmetrically cyclically magnetized condition.

NOTE: Normal induction is a magnetostatic parameter usually measured by hallistic methods.

**Induction, Remanent, $B_d$.* The magnetic induction that remains in a magnetic circuit after the removal of an applied magnetomotive force.

NOTE: If there are no air gaps or other inhomogeneities in the magnetic circuit, the remanent induction $B_r$ will equal the residual induction $B_r$; if air gaps or other inhomogeneities are present, $B_d$ will be less than $B_r$.

**Induction, Residual, $B_r$.* The magnetic induction corresponding to zero magnetizing force in a magnetic material that is in a symmetrically cyclically magnetized condition.

**Induction, Saturation, $B_r$.* The maximum intrinsic induction possible in a material.

**Induction Curve, Intrinsic (Ferric).* A curve of a previously demagnetized specimen depicting the relation between intrinsic induction and corresponding ascending values of magnetizing force. This curve starts at the origin of the $B_i$ and $H$ axes.

**Induction Curve, Normal.* A curve of a previously demagnetized specimen depicting the relation between normal induction and corresponding ascending values of magnetizing force. This curve starts at the origin of the $B$ and $H$ axes.

*Isotropic Material.* Material in which the magnetic properties are the same for all directions.

*Magnetic Circuit.* A region at whose surface the magnetic induction is tangential.

NOTE: A practical magnetic circuit is the region containing the flux of practical interest, such as the core of a transformer. It may consist of ferromagnetic material with or without air gaps or other feebly magnetic materials such as porcelain and brass.

*Magnetic Constant (Permeability of Space) $\Gamma_m$.* The dimensional scalar factor that relates the mechanical force between two currents to their intensities and geometrical configurations. That is,

$$dF = \Gamma_m I_1 I_2 dl_1 \times \frac{dl_2 \times r_1}{nr^2} \tag{3-41}$$

where $\Gamma_m$ = magnetic constant when the element of force $dF$ of a current element $I_1\, dl_1$ on another current element $I_2\, dl_2$ is at a distance $r$
$r_1$ = unit vector in the direction from $dl_1$ to $dl_2$
$n$ = dimensionless factor, the symbol $n$ is unity in unrationalized systems and $4\pi$ in rationalized systems

NOTE 1: The numerical values of $\Gamma_m$ depend on the system of units employed. In the cgs-em system, $\Gamma_m = 1$; in the rationalized mksa system, $\Gamma_m = 4\pi \times 10^{-7}$ h/m.

NOTE 2: The magnetic constant expresses the ratio of magnetic induction to the corresponding magnetizing force at any point in a vacuum and therefore is sometimes called the permeability of space $\mu_r$.

NOTE 3: The magnetic constant times the relative permeability is equal to the absolute permeability:

$$\mu_{\text{abs}} = \Gamma_m \mu_r \tag{3-42}$$

*Magnetic Field Strength H.* See *magnetizing force.*

*Magnetic Flux $\phi$.* The product of the magnetic induction B and the area of a surface (or cross section) A when the magnetic induction B is uniformly distributed and normal to the plane of the surface.

$$\phi = BA \tag{3-43}$$

where $\phi$ = magnetic flux
$B$ = magnetic induction
$A$ = area of the surface

NOTE 1: If the magnetic induction is not uniformly distributed over the surface, the flux $\phi$ is the surface integral of the normal component of $B$ over the area:

$$\phi = \iint_s B dA \tag{3-44}$$

NOTE 2: Magnetic flux is scalar and has no direction.

*Magnetic Flux Density B.* See *magnetic induction (flux density).*

*Magnetic Induction (Flux Density) B.* That magnetic vector quantity which at any point in a magnetic field is measured either by the mechanical force experienced by an element of electric current at the point, or by the electromotive force induced in an elementary loop during any change in flux linkages with the loop at the point.

NOTE 1: If the magnetic induction $B$ is uniformly distributed and normal to a surface or cross section, then the magnetic induction is

$$B = \phi / A \tag{3-45}$$

where $B$ = magnetic induction
$\phi$ = total flux
$A$ = area

NOTE 2: $B_{in}$ is the instantaneous value of the magnetic induction and $B_m$ is the maximum value of the magnetic induction.

*Magnetizing Force (Magnetic Field Strength) H.* That magnetic vector quantity at a point in a magnetic field which measures the ability of electric currents or magnetized bodies to produce magnetic induction at the given point.

NOTE 1: The magnetizing force $H$ may be calculated from the current and the geometry of certain magnetizing circuits. For example, in the center of a uniformly wound long solenoid,

$$H = C(NI/l) \tag{3-46}$$

where $H$ = magnetizing force
$C$ = constant whose value depends on the system of units
$N$ = number of turns
$I$ = current
$l$ = axial length of the coil

If $I$ is expressed in amperes and $l$ is expressed in centimeters, then $C = 4\pi/10$ in order to obtain $H$ in the cgs = em unit, the oersted. If $I$ is expressed in amperes and $l$ is expressed in meters, then $C = 1$ in order to obtain $H$ in the mksa unit, ampere-turn per meter.

NOTE 2: The magnetizing force $H$ at a point in air may be calculated from the measured value of induction at the point by dividing this value by the magnetic constant $\Gamma_m$.

***Magnetizing Force, AC.* Three different values of dynamic magnetizing force parameters are in common use:

1. $H_L$—an assumed peak value computed in terms of peak magnetizing current (considered to be sinusoidal).
2. $H_x$—an assumed peak value computed in terms of measured rms exciting current (considered to be sinusoidal).
3. $H_p$—computed in terms of a measured peak value of exciting current, and thus equal to the value $H'_{max}$.

***Magnetodynamic.* The magnetic condition when the values of magnetizing force and induction vary, usually periodically and repetitively, between two extreme limits.

*Magnetomotive Force F.* The line integral of the magnetizing force around any flux loop in space.

$$F = \oint H\,dl \tag{3-47}$$

where $F$ = magnetomotive force
$H$ = magnetizing force
$dl$ = unit length along the loop

NOTE: The magnetomotive force is proportional to the net current linked with any closed loop of flux or closed path

$$F = CNI \tag{3-48}$$

where $F$ = magnetomotive force
$N$ = number of turns linked with the loop
$I$ = current in amperes
$C$ = constant whose value depends on the system of units. In the cgs system, $C = 4\pi/10$. In the mksa system, $C = 1$

**Magnetostatic.* The magnetic condition when the values of magnetizing force and induction are considered to remain invariant with time during the period of measurement. This is often referred to as a dc (direct-current) condition.

*Magnetostriction.* Changes in dimensions of a body resulting from magnetization.

*Maxwell.* The unit of magnetic flux in the cgs electromagnetic system. One maxwell equals $10^{-8}$ weber. See *magnetic flux.*

$$e = -N\frac{d\phi}{dt} \times 10^{-8} \tag{3-49}$$

where $e$ = induced instantaneous emf volts
$d\phi/dt$ = time rate of change of flux, maxwells per second
$N$ = number of turns surrounding the flux, assuming each turn is linked with all the flux

*Oersted.* The unit of magnetizing force (magnetic field strength) in the cgs electromagnetic system. One oersted equals a magnetomotive force of 1 gilbert/cm of flux path. One oersted equals $100/4\pi$ or 79.58 ampere-turns per meter. See *magnetizing force (magnetic field strength).*

*Paramagnetic Material.* A material having a relative permeability which is slightly greater than unity, and which is practically independent of the magnetizing force.

***Permeability, AC.* A generic term used to express various dynamic relationships between magnetic induction $B$ and magnetizing force $H$ for magnetic material subjected to a cyclic excitation by alternating or pulsating current. The values of ac permeability obtained for a given material depend fundamentally on the excursion limits of dynamic excitation and induction, the method and conditions of measurement, and also on such factors as resistivity, thickness of laminations, frequency of excitation, etc.

NOTE: The numerical value for any permeability is meaningless unless the corresponding $B$ or $H$ excitation level is specified. For incremental permeabilities, not only the corresponding dc $B$ or $H$ excitation level must be specified but also the dynamic excursion limits of dynamic excitation range ($\Delta B$ or $\Delta H$).

AC permeabilities in common use for magnetic testing are

1. ***Impedance (rms) permeability* $\mu_z$. The ratio of the measured peak value of magnetic induction to the value of the apparent magnetizing force $H_z$ calculated from the measured rms value of the exciting current, for a material in the *SCM* condition.

NOTE: The value of the current used to compute $H_z$ is obtained by multiplying the measured value of rms exciting current by 1.414. This assumes that the total exciting current is magnetizing current and is sinusoidal.

2. ***Inductance permeability* $\mu_L$. For a material in an *SCM* condition, the permeability is evaluated from the measured inductive component of the electric circuit representing the magnetic specimen. This circuit is assumed to be composed of paralleled linear inductive and resistive elements $\omega L_1$ and $R_1$.
3. ***Peak permeability* $\mu_p$. The ratio of the measured peak value of magnetic induction to the peak value of the magnetizing force $H_p$, calculated from the measured peak value of the exciting current, for a material in the *SCM* condition.

Other ac permeabilities are:

4. *Ideal permeability* $\mu_a$. The ratio of the magnetic induction to the corresponding magnetizing force after the material has been simultaneously subjected to a value of ac magnetizing force approaching saturation (of approximate sine waveform) superimposed on a given dc magnetizing force, and the ac magnetizing force has thereafter been gradually reduced to zero. The resulting ideal permeability is thus a function of the dc magnetizing force used.

NOTE: Ideal permeability, sometimes called anhysteretic permeability, is principally significant to feebly magnetic material and to the Rayleigh range of soft magnetic material.

5. ***Impedance, permeability, incremental,* $\mu_{\Delta z}$. Impedance permeability $\mu_z$ obtained when an ac excitation is superimposed on a dc excitation, *CM* condition.
6. ***Inductance permeability, incremental,* $\mu_{\Delta L}$. Inductance permeability $\mu_L$ obtained when an ac excitation is superimposed on a dc excitation, *CM* condition.
7. ***Initial dynamic permeability* $\mu_{0d}$. The limiting value of inductance permeability $\mu_L$ reached in a ferromagnetic core when, under *SCM* excitation, the magnetizing current has been progressively and gradually reduced from a comparatively high value to zero value.

NOTE: This same value, $\mu_{0d}$, is also equal to the initial values of both impedance permeability $\mu_x$ and peak permeability $\mu_p$.

8. ***Instantaneous permeability (coincident with* $B_{max}$*)* $\mu_t$. With *SCM* excitation, the ratio of the maximum induction $B_{max}$ to the instantaneous magnetizing force $H_t$, which is the value of apparent magnetizing force $H'$ determined at the instant when $B$ reaches a maximum.
9. ***Peak permeability, incremental,* $\mu_{\Delta p}$. Peak permeability $\mu_p$ obtained when an ac excitation is superimposed on dc excitation, *CM* condition.

**Permeability, DC.* Permeability is a general term used to express relationships between magnetic induction B and magnetizing force H under various conditions of magnetic excitation. These relationships are either (1) absolute permeability, which in general is the quotient of a change in magnetic induction divided by the corresponding change in magnetizing force, or (2) relative permeability, which is the ratio of the absolute permeability to the magnetic constant $\Gamma_m$.

NOTE 1: The magnetic constant $\Gamma_m$ is a scalar quantity differing in value and uniquely determined by each electromagnetic system of units. In the unrationalized cgs system, $\Gamma_m$ is 1 gauss/oersted and in the mksa rationalized system $\Gamma_m = 4\pi \times 10^{-7}$ H/m.

NOTE 2: Relative permeability is a pure number which is the same in all unit systems. The value and dimension of absolute permeability depend on the system of units employed.

NOTE 3: For any ferromagnetic material, permeability is a function of the degree of magnetization. However, initial permeability $\mu_0$ and maximum permeability $\mu_m$ are unique values for a given specimen under specified conditions.

NOTE 4: Except for initial permeability $\mu_0$, a numerical value for any of the dc permeabilities is meaningless unless the corresponding $B$ or $H$ excitation level is specified.

NOTE 5: For the incremental permeabilities $\mu_\Delta$ and $\mu_{\Delta i}$, a numerical value is meaningless unless both the corresponding values of mean excitation level ($B$ or $H$) and the excursion range ($\Delta B$ or $\Delta H$) are specified.

The following dc permeabilities are frequently used in magnetostatic measurements primarily concerned with the testing of materials destined for use with permanent or dc excited magnets:

1. **Absolute permeability* $\mu_{\text{abs}}$. The sum of the magnetic constant and the intrinsic permeability. It is also equal to the product of the magnetic constant and the relative permeability.

$$\mu_{\text{abs}} = \Gamma_m + \mu_i = \Gamma_m \mu_r \tag{3-50}$$

2. **Differential permeability* $\mu_d$. The absolute value of the slope of the hysteresis loop at any point, or the slope of the normal magnetizing curve at any point.
3. **Effective circuit permeability* $\mu_{\text{eff}}$. When a magnetic circuit consists of two or more components, each individually homogeneous throughout but having different permeability values, the effective (overall) permeability of the circuit is that value computed in terms of the total magnetomotive force, the total resulting flux, and the geometry of the circuit.

NOTE: For a symmetrical series circuit in which each component has the same cross-sectional area, reluctance values add directly, giving

$$\mu_{\text{eff}} = \frac{l_1 + l_2 + l_3 + \cdots}{l_1/\mu_1 + l_2/\mu_2 + l_3/\mu_3 + \cdots} \tag{3-51}$$

For a symmetrical parallel circuit in which each component has the same flux path length, permeance values add directly, giving

$$\mu_{\text{eff}} = \frac{\mu_1 A_1 + \mu_2 A_2 + \mu_3 A_3 + \cdots}{A_1 + A_2 + A_3 + \cdots} \tag{3-52}$$

4. **Incremental intrinsic permeability* $\mu_{\Delta i}$. The ratio of the change in intrinsic induction to the corresponding change in magnetizing force when the mean induction differs from zero.
5. **Incremental permeability* $\mu_\Delta$. The ratio of a change in magnetic induction to the corresponding change in magnetizing force when the mean induction differs from zero. It equals the slope of a straight line joining the excursion limits of an incremental hysteresis loop.

NOTE: When the change in $H$ is reduced to zero, the incremental permeability $\mu_\Delta$ becomes the reversible permeability $\mu_{\text{rev}}$.

6. **Initial permeability* $\mu_0$. The limiting value approached by the normal permeability as the applied magnetizing force $H$ is reduced to zero. The permeability is equal to the slope of the normal induction curve at the origin of linear $B$ and $H$ axes.
7. **Intrinsic permeability* $\mu_i$. The ratio of intrinsic induction to the corresponding magnetizing force.
8. **Maximum permeability* $\mu_m$. The value of normal permeability for a given material where a straight line from the origin of linear $B$ and $H$ axes becomes tangent to the normal induction curve.
9. **Normal permeability* $\mu$ *(without subscript)*. The ratio of the normal induction to the corresponding magnetizing force. It is equal to the slope of a straight line joining the extrusion limits of a normal hysteresis loop, or the slope of a straight line joining any point ($H_m$, $B_m$) on the normal induction curve to the origin of the linear $B$ and $H$ axes.
10. **Relative permeability* $\mu_r$. The ratio of the absolute permeability of a material to the magnetic constant $\Gamma_m$ giving a pure numeric parameter.

NOTE: In the cgs-em system of units, the relative permeability is numerically the same as the absolute permeability.

11. *Reversible permeability* $\mu_{rev}$. The limit of the incremental permeability as the change in magnetizing force approaches zero.
12. *Space permeability* $\mu_o$. The permeability of space (vacuum), identical with the magnetic constant $\Gamma_m$.

***Reactive Power (Quadrature Power)* $P_q$. The product of the rms current in an electric circuit, the rms voltage across the circuit, and the sine of the angular phase difference between the current and the voltage.

$$P_q = EI \sin\theta \tag{3-53}$$

where $P_q$ = reactive power, vars
$E$ = voltage, volts
$I$ = current, amperes
$\theta$ = angular phase by which $E$ leads $I$

NOTE: The reactive power supplied to a magnetic core having an *SCM* excitation is the product of the magnetizing current and the voltage induced in the exciting winding.

**Remanence* $B_{dm}$. The maximum value of the remanent induction for a given geometry of the magnetic circuit.

NOTE: If there are no air gaps or other inhomogeneities in the magnetic circuit, the remanence $B_{dm}$ is equal to the retentivity $B_{rs}$; if air gaps or other inhomogeneities are present, $B_{dm}$ will be less than $B_{rs}$.

**Retentivity* $B_{rs}$. That property of a magnetic material which is measured by its maximum value of the residual induction.

NOTE: Retentivity is usually associated with saturation induction.

*Symmetrically Cyclically Magnetized Condition, SCM.* A magnetic material is in an SCM condition when, under the influence of a magnetizing force that varies cyclically between two equal positive and negative limits, its successive hysteresis loops or flux-current loops are both identical and symmetrical with respect to the origin of the axes.

*Tesla.* The unit of magnetic induction in the mksa (Giorgi) system. The tesla is equal to 1 Wb/m$^2$ or 10$^4$ gausses.

*Var.* The unit of reactive (quadrature) power in the mksa (Giorgi) and the practical systems.

*Volt-Ampere.* The unit of apparent power in the mksa (Giorgi) and the practical systems.

*Watt.* The unit of active power in the mksa (Giorgi) and the practical systems. One watt is a power of 1 J/s.

*Weber.* The unit of magnetic flux in the mksa and in the practical system. The weber is the magnetic flux whose decrease to zero when linked with a single turn induces in the turn a voltage whose time integral is 1 v/s. One weber equals 10$^8$ maxwells. See *magnetic flux*.

### 3.2.2 Magnetic Properties and Their Application

The relative importance of the various magnetic properties of a magnetic material varies from one application to another. In general, properties of interest may include normal induction, hysteresis, dc permeability, ac permeability, core loss, and exciting power. It should be noted that there are various means of expressing ac permeability. The choice depends primarily on the ultimate use.

Techniques for the magnetic testing of many magnetic materials are described in the ASTM standards. The magnetic and electric circuits employed in magnetic testing of a specimen are as free as possible from any unfavorable design factors which would prevent the measured magnetic data from being representative of the inherent magnetic properties of the specimen. The flux "direction" in the specimen is normally specified, since most magnetic materials are magnetically anisotropic. In most ac magnetic tests, the waveform of the flux is required to be sinusoidal.

As a result of the existence of unfavorable conditions, such as those listed and described below, the performance of a magnetic material in a magnetic device can be greatly deteriorated from that which would be expected from magnetic testing of the material. Allowances for these conditions, if present, must be made during the design of the device if the performance of the device is to be correctly predicted.

*Leakage.* A principal difficulty in the design of many magnetic circuits is due to the lack of a practicable material which will act as an insulator with respect to magnetic flux. This results in magnetic flux seldom being completely confined to the desired magnetic circuit. Estimates of leakage flux for a particular design may be made based on experience and/or experimentation.

*Flux Direction.* Some magnetic materials have a very pronounced directionality in their magnetic properties. Failure to utilize these materials in their preferred directions results in impaired magnetic properties.

*Fabrication.* Stresses introduced into magnetic materials by the various fabricating techniques often adversely affect the magnetic properties of the materials. This occurs particularly in materials having high permeability. Stresses may be eliminated by a suitable stress-relief anneal after fabrication of the material to final shape.

*Joints.* Joints in an electromagnetic core may cause a large increase in total excitation requirements. In some cores operated on ac, core loss may also be increased.

*Waveform.* When a sinusoidal voltage is applied to an electromagnetic core, the resulting magnetic flux is not necessarily sinusoidal in waveform, especially at high inductions. Any harmonics in the flux waveform cause increases in core loss and required excitation power.

*Flux Distribution.* If the maximum and minimum lengths of the magnetic path in an electromagnetic core differ too much, the flux density may be appreciably greater at the inside of the core structure than at the outside. For cores operated on ac, this can cause the waveform of the flux at the extremes of the core structure to be distorted even when the total flux waveform is sinusoidal.

### 3.2.3 Types of Magnetism

Any substance may be classified into one of the following categories according to the type of magnetic behavior it exhibits:

1. Diamagnetic
2. Paramagnetic
3. Antiferromagnetic
4. Ferromagnetic
5. Ferrimagnetic

Substances that fall into the first three categories are so weakly magnetic that they are commonly thought of as *nonmagnetic*. In contrast, ferromagnetic and ferrimagnetic substances are strongly magnetic and are thereby of interest as *magnetic materials*. The magnetic behavior of any ferromagnetic or ferrimagnetic material is a result of its spontaneously magnetized magnetic domain structure and is characterized by a nonlinear normal induction curve, hysteresis, and saturation.

The pure elements which are ferromagnetic are iron, nickel, cobalt, and some of the rare earths. Ferromagnetic materials of value to industry for their magnetic properties are almost invariably alloys of the metallic ferromagnetic elements with one another and/or with other elements.

Ferrimagnetism occurs mainly in the ferrites, which are chemical compounds having ferric oxide ($Fe_2O_3$) as a component. In recent years, some of the magnetic ferrites have become very important in certain magnetic applications. The magnetic ferrites saturate magnetically at lower inductions than do the great majority of metallic ferromagnetic materials. However, the electrical resistivities of ferrites are at least several orders of magnitude greater than those of metals.

***Commercial Magnetic Materials.*** *Commercial magnetic materials* are generally divided into two main groups, each composed of ferromagnetic and ferrimagnetic substances:

1. Magnetically "soft" materials
2. Magnetically "hard" materials

The distinguishing characteristic of "soft" magnetic materials is high permeability. These materials are employed as core materials in the magnetic circuits of electromagnetic equipment. "Hard" magnetic materials are characterized by a high maximum magnetic energy product $BH_{max}$. These materials are employed as permanent magnets to provide a constant magnetic field when it is inconvenient or uneconomical to produce the field by electromagnetic means.

## 3.2.4 "Soft" Magnetic Materials

A wide variety of "soft" magnetic materials have been developed to meet the many different requirements imposed on magnetic cores for modern electrical apparatus and electronic devices. The various soft magnetic materials will be considered under three classifications:

1. Materials for solid cores
2. Materials for laminated cores
3. Materials for special purposes

## 3.2.5 Materials for Solid Cores

These materials are used in dc applications such as yokes of dc dynamos, rotors of synchronous dynamos, and cores of dc electromagnets and relays. Proper annealing of these materials improves their magnetic properties. The principal magnetic requirements for the solid-core materials are high saturation, high permeability at relatively high inductions, and at times, low coercive force.

*Wrought iron* is a ferrous material, aggregated from a solidifying mass of pasty particles of highly refined metallic iron, into which is incorporated, without subsequent fusion, a minutely and uniformly distributed quantity of slag. The better types of wrought iron are known as *Norway iron* and *Swedish iron* and are widely used in relays after being annealed to reduce coercive force and to minimize magnetic aging.

*Cast irons* are irons which contain carbon in excess of the amount which can be retained in solid solution in austenite at the eutectic temperature. The minimum carbon content is about 2%, while the practical maximum carbon content is about 4.5%. Cast iron was used in the yokes of dc dynamos in the early days of such machines.

*Gray cast iron* is a cast iron in which graphite is present in the form of flakes. It has very poor magnetic properties, inferior mechanical properties, and practically no ductility. It does lend itself well to the casting of complex shapes and is readily machinable.

*Malleable cast iron* is a cast iron in which the graphite is present as temper carbon nodules. It is magnetically better than gray cast iron.

*Ductile (nodular) cast iron* is a cast iron with the graphite essentially spheroidal in shape. It is magnetically better than gray cast iron. Ductile cast iron has the good castability and machinability of gray cast iron together with much greater strength, ductility, and shock resistance.

## 3.2.6 Carbon Steels

Carbon steels may contain from less than 0.1% carbon to more than 1% carbon. The magnetic properties of a carbon steel are greatly influenced by the carbon content and the disposition of the carbon. Low-carbon steels (less than 0.2% carbon) have magnetic properties which are similar to those of wrought iron and far superior to those of any of the cast irons.

*Wrought carbon steels* are widely used as solid-core materials. The low-carbon types are preferred in most applications.

*Cast carbon steels* replaced cast iron many years ago as the material used in the yokes of dc machines, but have since largely been supplanted in this application by wrought (hot-rolled) carbon-steel plates of welding quality.

## 3.2.7 Materials for Laminated Cores

The materials most widely employed in wound or stacked cores in electromagnetic devices operated at the commercial power frequencies (50 and 60 Hz) are the electrical steels and the specially processed carbon steels designated as *magnetic lamination steels.* The principal magnetic requirements for these materials are low core loss, high permeability, and high saturation. ASTM publishes standard specifications for these materials. On a tonnage basis, production of these materials far exceeds that of any other magnetic material.

*Electrical steels* are flat-rolled low-carbon silicon-iron alloys. Since applications for electrical steels lie mainly in energy-loss-limited equipment, the core losses of electrical steels are normally guaranteed by the producers. The general category of electrical steels may be divided into classifications of (1) nonoriented materials and (2) grain-oriented materials.

Electrical steels are usually graded by high-induction core loss. Both ASTM and AISI have established and published designation systems for electrical steels based on core loss.

The ASTM core loss type designation consists of six or seven characters. The first two characters are 100 times the nominal thickness of the material in millimeters. The third character is a code letter which designates the class of the material and specifies the sampling and testing practices. The last three or four characters are 100 times the maximum permissible core loss in watts per pound at a specified test frequency and induction.

The AISI designation system has been discontinued but is still widely used. The AISI type designation for a grade consisted of the letter M followed by a number. The letter M stood for magnetic material, and the number was approximately equal to 10 times the maximum permissible core loss in watts per pound for 0.014-in material at 15 kG, 60 Hz in 1947.

*Nonoriented electrical steels* have approximately the same magnetic properties in all directions in the plane of the material (see Figs. 3-8 and 3-9). The common application is in punched laminations for large and small rotating machines and for small transformers. Today, nonoriented materials are always cold-rolled to final thickness. Hot rolling to final thickness is no longer practiced. Nonoriented materials are available in both fully processed and semiprocessed conditions.

Fully processed nonoriented materials have their magnetic properties completely developed by the producer. Stresses introduced into these materials during fabrication of magnetic cores must be relieved by annealing to achieve optimal magnetic properties in the cores. In many applications, however, the degradation of the magnetic properties during fabrication is slight and/or can be tolerated, and the stress-relief anneal is omitted. Fully processed nonoriented materials contain up to about 3.5% silicon. Additionally, a small amount (about 0.5%) of aluminum is usually present. The common thicknesses are 0.014, 0.0185, and 0.025 in.

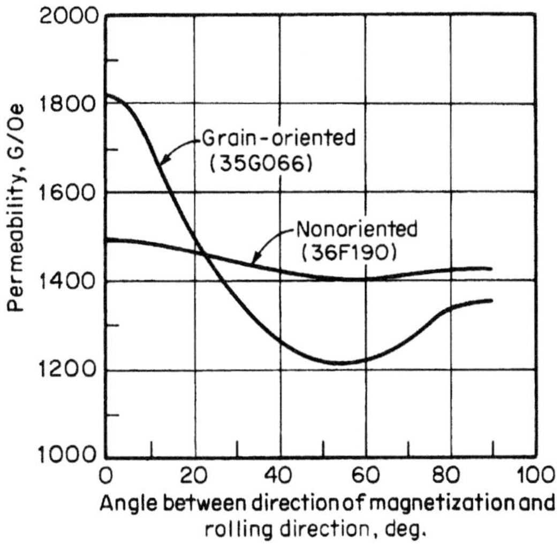


**FIGURE 3-8** Effect of direction of magnetization on normal permeability at 10 Oe of fully processed electrical steels.

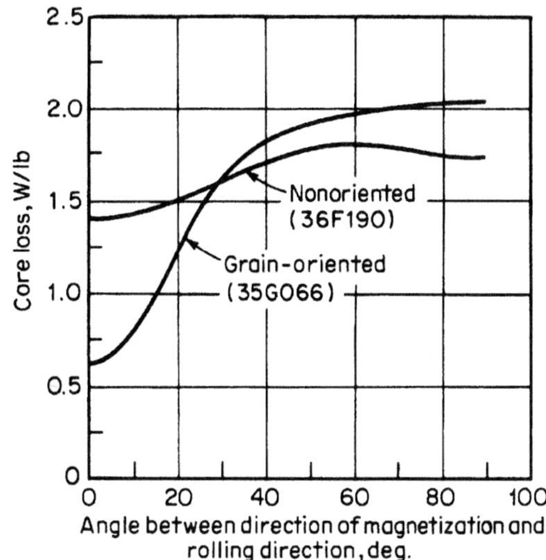


**FIGURE 3-9** Effect of direction of magnetization on core loss at 15 kG, 60 Hz or fully processed electrical steel.

Semiprocessed nonoriented materials do not have their inherent magnetic properties completely developed by the producer and must be annealed properly to achieve both decarburization and grain growth. These materials are used primarily in high-volume production of small laminations and cores which would require stress-relief annealing if made from fully processed material. Semiprocessed nonoriented materials contain up to about 3% silicon. Additionally, a small amount (about 0.5%) of aluminum is usually present. The carbon content may be as high as 0.05% but should be reduced to 0.005% or less by the required anneal. The common thicknesses of semiprocessed nonoriented materials are 0.0185 and 0.025 in.

*Grain-oriented electrical steels* have a pronounced directionality in their magnetic properties (Figs. 3-8 and 3-9). This directionality is a result of the "cube-on-edge" crystal structure achieved by proper composition and processing. Grain-oriented materials are employed most effectively in magnetic cores in which the flux path lies entirely or predominantly in the rolling direction of the material. The common application is in cores of power and distribution transformers for electric utilities.

Grain-oriented materials are produced in a fully processed condition, either unflattened or thermally flattened, in thicknesses of 0.0090, 0.0106, 0.0118, and 0.0138 in. Unflattened material has appreciable coil set or curvature. It is used principally in making spirally wound or formed cores. These cores must be stress-relief annealed to relieve fabrication stresses. Thermally flattened material is employed principally in making sheared or stamped laminations. Annealing of the laminations to remove both residual stresses from the thermal-flattening and fabrication stresses is usually recommended. However, special thermally flattened materials are available which do not require annealing when used in the form of wide flat laminations.

Two types of grain-oriented electrical steels are currently being produced commercially. The regular type, which was introduced many years ago, contains about 3.15% silicon and has grains about 3 mm in diameter. The high-permeability type, which was introduced more recently, contains about 2.9% silicon and has grains about 8 mm in diameter. In comparison with the regular type, the high-permeability type has better core loss and permeability at high inductions.

Some characteristics and applications for electrical steels are shown in Table 3-9.

*Surface insulation* of the surfaces of electrical steels is needed to limit the interlaminar core losses of magnetic cores made of electrical steels. Numerous surface insulations have been developed to meet the requirements of various applications. The various types of surface insulations have been classified by AISI.

**TABLE 3-9** Some Characteristics and Typical Applications for Specific Types of Electrical Steels

| ASTM type | Some characteristics | Typical applications |
|---|---|---|
| | Oriented types | |
| 23G048 through 35G066 or 27H076 through 35H094 or 27P066 through 35P076 | Highly directional magnetic properties due to grain orientation. Very low core loss and high permeability in rolling direction. | Highest-efficiency power and distribution transformers with lower weight per kVA. Large generators and power transformers. |
| | Nonoriented types | |
| 36F145 and 47F168 | Lowest core loss, conventional grades. Excellent permeability at low inductions. | Small power transformers and rotating machines of high efficiency. |
| 36F158 through 64F225 or 47S178 and 64S194 | Low core loss, good permeability at low and intermediate inductions. | High-reactance cores, generators, stators of high-efficiency rotating equipment. |
| 36F190 through 64F270 or 47S188 through 64S260 | Good core loss, good permeabilty at all inductions, and low exciting current. Good stamping properties. | Small generators, high-efficiency, continuous duty rotating ac and dc machines. Small motors, ballasts, and relays. |
| 47F290 through 64F600 or 47S250 through 64S350 | Ductile, good stamping properties, good permeabilty at high inductions. | |

*Annealing* of laminations or cores made from electrical steels is performed to accomplish either stress relief in fully processed material or decarburization and grain growth in semiprocessed material. Both batch-type annealing furnaces and continuous annealing furnaces are employed. The former is best suited for low-volume or varied production, while the latter is best suited for high-volume production.

Stress-relief annealing is performed at a soak temperature in the range from 730 to 845°C. The soak time need be no longer than that required for the charge to reach soak temperature. The heating and cooling rates must be slow enough so that excessive thermal gradients in the material are avoided. The annealing atmosphere and other annealing conditions must be such that chemical contamination of the material is avoided.

Annealing for decarburization and grain growth is performed at a soak temperature in the range from 760 to 870°C. Atmospheres of hydrogen or partially combusted natural gas and containing water vapor are often used. The soak time required for decarburization depends not only on the temperature and atmosphere but also on the dimensions of the laminations or cores being annealed. If the dimensions are large, long soak times may be required.

*Magnetic lamination steels* are cold-rolled low-carbon steels intended for magnetic applications, primarily at power frequencies. The magnetic properties of magnetic lamination steels are not normally guaranteed and are generally inferior to those of electric steels. However, magnetic lamination steels are frequently used as core materials in small electrical devices, especially when the cost of the core material is a more important consideration than the magnetic performance.

Usually, but not always, stamped laminations or assembled core structures made from magnetic lamination steels are given a decarburizing anneal to enhance the magnetic properties. Optimal magnetic properties are obtained when the carbon content is reduced to 0.005% or less from its initial value, which may approach 0.1%. The soak temperature of the anneal is in the range from 730 to 790°C. The atmosphere most often used at the present time is partially combusted natural gas with a suitable dew point. Soak time depends to a considerable degree on the dimensions of the laminations or core structures being annealed.

Three types of magnetic lamination steels are produced. Type 1 is usually made to a controlled chemical composition and is furnished in the full-hard or annealed condition without guaranteed magnetic properties. Type 2 is made to a controlled chemical composition, given special processing, and furnished in the annealed condition without guaranteed magnetic properties. After a suitable anneal, the magnetic properties of Type 2 are superior to those of Type 1. Type 2S is similar to Type 2, but the core loss is guaranteed.

### 3.2.8 Materials for Special Purposes

For certain applications of soft or nonretentive materials, special alloys and other materials have been developed, which, after proper fabrication and heat treatment, have superior properties in certain ranges of magnetization. Several of these alloys and materials will be described.

***Nickel-Iron Alloys.*** Nickel alloyed with iron in various proportions produces a series of alloys with a wide range of magnetic properties. With 30% nickel, the alloy is practically nonmagnetic and has a resistivity of 86 $\mu\Omega$/cm. With 78% nickel, the alloy, properly heat-treated, has very high permeability. These effects are shown in Figs. 3-10 and 3-11. Many variations of this series have been developed for special purposes. Table 3-10 lists some of the more important commercial types of nickel-iron alloys, with their approximate properties. These alloys are all very sensitive to heat treatment, so their properties are largely influenced thereby.

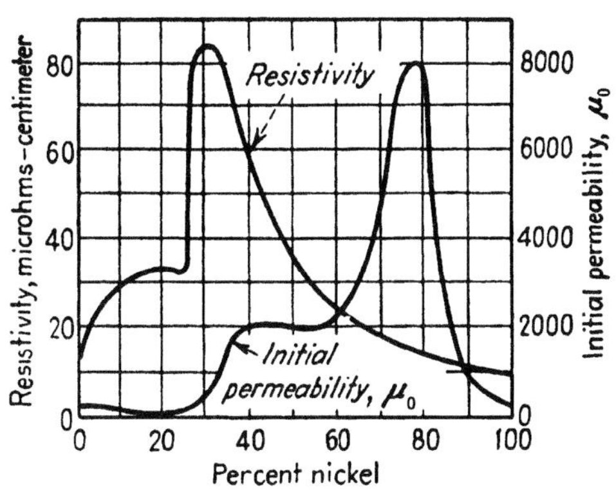


**FIGURE 3-10** Electrical resistivity and initial permeability of iron-nickel alloys with various nickel contents.

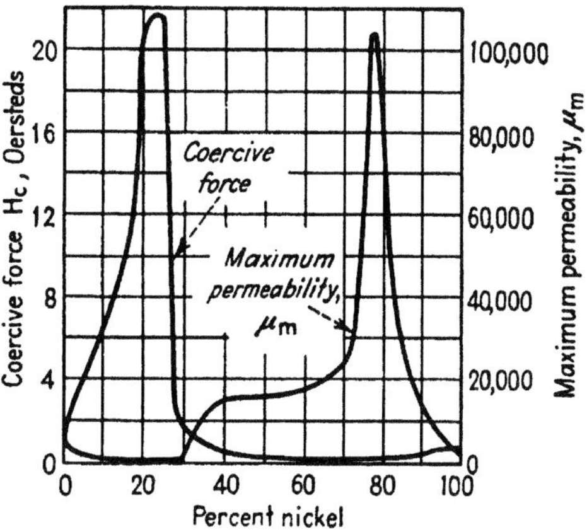


**FIGURE 3-11** Maximum permeability and coercive force of iron-nickel alloys with various nickel contents.

**TABLE 3-10** Special-Purpose Materials

| Name | Approximate composition, % | Saturation, G | Maximum permeability | Coercivity (from saturation), Oe | Initial permeability | Resistivity, $\mu\Omega \cdot$ cm |
|---|---|---|---|---|---|---|
| 78 Permalloy | 78.5 Ni | 10,500 | 70,000 | −0.05 | 8,000 | 16 |
| Mo Permalloy | 79 Ni, 4.0 Mo | 8,000 | 90,000 | −0.05 | 20,000 | 55 |
| Supermalloy | 79 Ni, 5 Mo | 7,900 | 900,000 | −0.002 | 100,000 | 60 |
| 48% nickel-iron | 48 Ni | 16,000 | 60,000 | −0.06 | 5,000 | 45 |
| Monimax | 47 Ni, 3 Mo | 14,500 | 35,000 | −0.10 | 2,000 | 80 |
| Sinimax | 43 Ni, 3 Si | 11,000 | 35,000 | −0.10 | 3,000 | 85 |
| Mumetal | 77 Ni, 5 Cu, 2 Cr | 6,500 | 85,000 | −0.05 | 20,000 | 60 |
| Deltamax | 50 Ni | 15,500 | 85,000 | −0.10 | | 45 |

***Permalloy.*** This is a term applied to a number of nickel-iron alloys developed by the Bell Laboratories, each specified by a prefix number indicating the nickel content. The term is usually associated with the 78.5% nickel-iron alloys, the important properties of which are high permeability and low hysteresis loss in relatively low magnetizing fields. These properties are obtained by a unique heat treatment consisting of a high-temperature anneal, preferably in hydrogen, with slow cooling followed by rapid cooling from about 625°C. The alloy is very sensitive to mechanical strain, so it is desirable to heat-treat the alloy in its final form. The addition of 3.8% chromium or molybdenum increases the resistivity from 16 to 65 and 55 $\mu\Omega \cdot$ cm, respectively, without seriously impairing the magnetic quality. In fact, low-density permeabilities are better with these additions. These alloys have found their principal application as a material for the continuous loading of submarine cables and in loading coils for landlines.

By special long-time high temperature treatments, maximum permeability values greater than 1 million have been obtained. The double treatment required by the 78% Permalloy is most effective when the strip is thin, say, under 10 mils. For greater thicknesses, the quick cooling from 625°C is not uniform throughout the section, and loss of quality results.

A 48% nickel-iron was developed for applications requiring a moderately high-permeability alloy with higher saturation density than 78 Permalloy. The same general composition is marketed under many names, such as Hyperm 50, Hipernik, Audiolloy, Allegheny Electric Metal, 4750, and Carpenter 49 alloy. Annealing is recommended after all mechanical operations are completed. These alloys have found extensive use in radio, radar, instrument, and magnetic-amplifier components.

*Deltamax.* By the use of special techniques of cold reduction and annealing, the 48% nickel-iron alloy develops directional properties resulting in high permeability and a square hysteresis loop in the rolling direction. A similar product is sold under the name of Orthonic. For optimal properties, these materials are rapidly cooled after a 2-h anneal in pure hydrogen at 1100°C. They are generally used in wound cores of thin tape for applications such as pulse transformers and magnetic amplifiers.

***Iron-Nickel-Copper-Chromium.*** The addition of copper and chromium to high-nickel-iron alloys has the effect of raising the permeability at low flux density. Alloys of this type are marketed under the names of Mumetal, 1040 alloy, and Hymu 80. For optimal properties, they are annealed after cutting and forming for 4 h at 1100°C in pure hydrogen and cooled slowly. Important applications are as magnetic shielding for instruments and electronic equipment and as cores in magnetic amplifiers.

***Constant-Permeability Alloys.*** *Constant-permeability alloys* having a moderate permeability, which is quite constant over a considerable range of flux densities, are desirable for use in circuits in which waveform distortion must be kept at a minimum. Isoperm and Conpernik are two alloys of this type. They are nickel-iron alloys containing 40% to 55% nickel which have been severely cold-worked. Perminvar is the name given to a series of cobalt-nickel-iron alloys (e.g., 50% nickel, 25% cobalt, 25% iron) which also exhibit this characteristic of constant permeability over a low (~800 G) density range. When magnetized to higher flux densities, they give a double loop constricted at the origin so as to give no measurable remanence or coercive force. The characteristics of the alloys in this group vary greatly with the chemical content and the heat treatment. A sample containing approximately 45 Ni, 25 Co, and 30 Fe, baked for 24 h at 425°C and slowly cooled, had hysteresis losses as follows: At 100 G, $214 \times 10^{-4}$ erg/(cm$^3$)(cycle); at 1003 G, 15.27 ergs; at 1604 G, 163 ergs; at 4950 G, 1736 ergs; and at 13,810 G, 4430 ergs. Over the range of flux densities in which the permeability is constant (from 0 to 600 G), the hysteresis loss is very small, or on the order of the foregoing figure for 100 G. The resistivity of the sample was 19.63 $\mu\Omega \cdot$ cm.

***Monel.*** *Monel metal* is an alloy of 67% nickel, 28% copper, and 5% other metals. It is slightly magnetic below 95°C.

***Iron-Cobalt Alloys.*** The addition of cobalt to iron has the effect of raising the saturation intensity of iron up to about 36% cobalt ($Fe_2Co$). This alloy is useful for pole pieces of electromagnets and for any application where high magnetic intensity is desired. It is workable hot but quite brittle cold. *Hyperco* contains approximately 1/3 Co, 2/3 Fe, plus 1% to 2% "added element." Total core loss is about 2.5 W/lb at 15 kG and 0.010 in thick. It is available as hot-rolled sheet, cold-rolled strip, plates, and forgings. The 50% cobalt-iron alloy Permendur has a high permeability in fields up to 50 Oe and, with about 2% vanadium added, can be cold-rolled.

***Iron-Silicon Aluminum Alloys.*** Aluminum in small percentages, usually under 0.5%, is a valuable addition to the iron-silicon alloy. Its principal function appears to be as a deoxidizer. Masumoto has investigated soft magnetic alloys containing much higher percentages of aluminum and found several that have high permeabilities and low hysteresis losses. Certain compositions have very low magnetostriction and anisotropy, high initial permeability, and high electrical resistivity. An alloy of 9.6% silicon and 6% aluminum with iron has better low-flux-density properties than the Permalloys. However, poor ductility has limited these alloys to dc applications in cast configurations or in insulated pressed-powder cores for high-frequency uses. These alloys are commonly known as Sendust. The material has been prepared in sheet form by special processes.

***Temperature-Sensitive Alloys.*** Inasmuch as the Curie point of metal may be moved up or down the temperature scale by the addition of other elements, it is possible to select alloys which lose their ferromagnetism at almost any desired temperature up to 1115°C, the change point in cobalt. Iron-based alloys are ordinarily used to obtain the highest possible permeability at points below the Curie temperature. Nickel, manganese, chromium, and silicon are the most effective alloy elements for this purpose, and most alloys made for temperature-control applications, such as instruments, reactors, and transformers, use one or more of these. The Carpenter Temperature Compensator 30 is a nickel-copper-iron alloy which loses its magnetism at 55°C and is used for temperature compensation in meters.

***Heusler's Alloys.*** *Heusler's alloys* are ferromagnetic alloys composed of "nonmagnetic" elements. Copper, manganese, and aluminum are frequently used as the alloying elements. The saturation induction is about one-third that of pure iron.

## 3.2.9 High-Frequency Materials Applications

Magnetic materials used in reactors, transformers, inductors, and switch-mode devices are selected on the basis of magnetic induction, permeability, and associated material power losses at the design frequency. Control of eddy currents becomes of primary importance to reduce losses and minimize skin effect produced by eddy-current shielding. This is accomplished by the use of high-permeability alloys in the form of wound cores of thin tape, or compressed, insulated powder iron alloy cores, or sintered ferrite cores.

Typically, the thin magnetic strip material is used in applications where operating frequencies range from 400 Hz to 20 kHz. Power conditioning equipment frequently operates at 10 kHz and up, and the magnetic materials used are compressed, powdered iron-alloy cores or sintered ferrite cores. Power losses in magnetic materials are of great concern, especially so when operated at high frequencies.

***3% Silicon-Iron Alloys.*** 3% *Silicon-iron alloys* for high-frequency use are available in an insulated 0.001- to 0.006-in-thick strip that exhibits high effective permeability and low losses at relatively high flux densities. This alloy, as well as other rolled-to-strip soft magnetic alloys, is used to make laminated magnetic cores by various methods, including (1) the wound-core approach for winding toroids and C and E cores, (2) stamped or sheared-to-length laminations for laid-up transformers, and (3) stamped laminations of various configurations (rings E, I, F, L, DU, etc.) for assembly into transformer cores. Laminated core materials usually are annealed after all fabricating and stamping operations have been completed in order to develop the desired magnetic properties of the material. Subsequent forming, bending, or machining may impair the magnetic characteristics developed by the anneal.

***Amorphous Metal Alloys.*** *Amorphous metal alloys* are made using a new technology which produces a thin (0.001 to 0.003 in) ribbon from rapidly quenched molten metal. The alloy solidifies before the atoms have a chance to segregate or crystallize, resulting in a glasslike atomic structure material of high electrical resistivity, 125 to 130 $\mu\Omega \cdot$ cm. A range of magnetic properties may be developed in these materials by using different alloying elements. Amorphous metal alloys may be used in the same high-frequency applications as the cast, rolled-to-strip, silicon-iron, and nickel-iron alloys.

***Nickel-Iron Powder Cores.*** *Nickel-iron powder cores* are made of insulated alloy powder, which is compressed to shape and heat-treated. The alloy composition most widely used is 2-81 Permalloy powder composed of 2% molybdenum, 81% nickel, and balance iron. Another less widely used powder, Sendust, is made of 7% to 13% silicon, 4% to 7% aluminum, and balance iron. Prior to pressing, the powder particles are thinly coated with an inorganic, high-temperature insulation which can withstand the high compacting pressures and the high-temperature (650°C) hydrogen atmosphere anneal. The insulation of the particles lowers eddy-current loss and provides a distributed air gap which can be controlled to provide cores in a range of permeabilities. The 2-81 Permalloy cores are commercially available in permeability ranges of 14 to 300, and Sendust cores have permeabilities ranging from 10 to 140.

These types of nickel-iron powder cores find use in applications where inductance must remain relatively constant when the magnetic component experiences changes in dc current or temperature. Additional stability over temperature can also be achieved by the addition of low-Curie-temperature powder materials to neutralize the naturally positive permeability-temperature coefficient of the alloy powder. Some applications are in telephone loading coils or filter chokes for power conditioning equipment where output voltage ripple must be minimized. Other uses are for pulse transformers and switch-mode power supplies where low power losses are desired. Operating frequencies can range from 1.0 kHz for 300 permeability materials to 500 kHz for the 14 permeability materials.

***Powdered-Iron Cores.*** *Powdered-iron cores* are manufactured from various types of iron powders whose particle sizes range from 2 to 100 $\mu$m. The particles are electrically insulated from one another using special insulating materials. The insulated powder is blended with phenolic or epoxy binders and a mold-release agent. The powder is then dry-pressed in a variety of shapes including toroids, E cores, threaded tuning cores, cups, sleeves, slugs, bobbins, and other special shapes. A low-temperature bake of the pressed product produces a solid component in which the insulated particles provide a built-in air gap, reducing eddy-current losses, increasing electrical Q, and thus allowing higher operating frequencies. The use of different iron powder blends and insulation systems provides a range of permeability, from 4 to 90, for use over the frequency spectrum of 50 Hz to 250 MHz. Applications include high-frequency transformers, tuning coils, variable inductors, rf chokes, and noise suppressors for power supply and power control circuits.

***Ferrite Cores.*** *Ferrite cores* are molded from a mixture of metallic oxide powders such that certain iron atoms in the cubic crystal of magnetite (ferrous ferrite) are replaced by other metal atoms, such as Mn and Zn, to form manganese zinc ferrite, or by Ni and Zn to form nickel zinc ferrite. Manganese zinc ferrite is the material most commercially available and is used in devices operating below 1.5 MHz. Nickel zinc ferrites are used mainly for filter applications above that frequency. They resemble ceramic materials in production processes and physical properties. The electrical resistivities correspond to those of semiconductors, being at least 1 million times those of metals. Magnetic permeability $\mu_0$ may be as high as 10,000. The Curie point is quite low, however, in the range 100 to 300°C. Saturation flux density is generally below 5000 G (Fig. 3-12). Ferrite materials are available in several compositions which, through processing, can improve one or two magnetic parameters (magnetic induction, permeability, low hysteresis loss, Curie temperature) at the expense of the other parameters. The materials are fabricated into shapes such as toroids; E, U, and I cores; beads; and self-shielding pot cores. Ferrite cores find use in filter applications up to 1.0 MHz, high-frequency power transformers operating at 10 to 100 kHz, pulse transformer delay lines, adjustable-air-gap inductors, recording heads, and filters used in high-frequency electronic circuits.

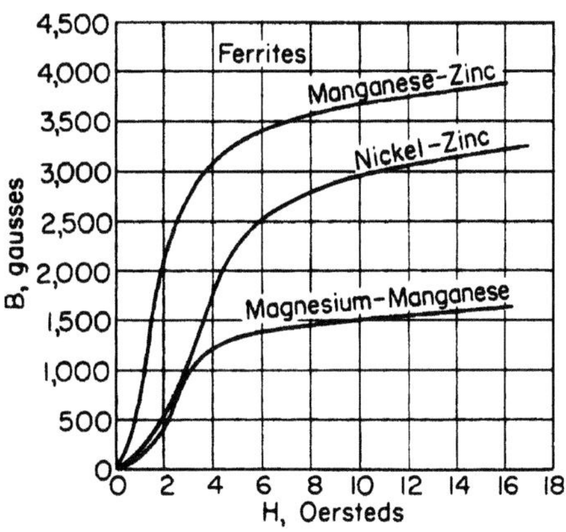


**FIGURE 3-12** Typical normal-induction curves for *soft* ferrites.

***Permanent-Magnet Materials.*** *Permanent-magnet materials* that are commercially available may be grouped into five classes as follows:

1. Quench-hardened alloys
2. Precipitation-hardened cast alloys
3. Ceramic materials
4. Powder compacts and elongated single-domain materials
5. Ductile alloys

## 3.2.10 Quench-Hardened Alloys

Early permanent magnets were made of low-carbon steel (1% C) that was hardened by heat treatment. Later developments saw improvements in the magnetic properties through the use of alloying elements of tungsten, chromium, and cobalt. The chrome steels are less expensive than the cobalt steels, and both find use in hysteresis clutch and motor applications.

***Ceramic Magnet Material.*** *Ceramic magnet material* usage is increasing yearly because of improved magnetic properties and the high cost of cobalt used in metallic alloy magnets. The basic raw material used in these magnets is iron oxide in combination with either strontium carbonate or barium carbonate. The iron oxide and carbonate mixture is calcined, and then the aggregate is ball-milled to a particle size of about 1.0 $\mu$m. The material is compacted in dies using the dry powder or a water-based slurry of the powder. High pressures are needed to press the parts to shape. In some ceramic grades, a magnetic field is applied during pressing to orient the material in order to obtain a preferred magnetic orientation. Parts are sintered at high temperatures and ground to finished size using diamond grinding wheels with suitable coolants. Ceramic magnets are hard and brittle, exhibit high electrical resistivities, and have lower densities than cast magnet alloys.

Made in the form of rings, blocks, and arcs, ceramic magnets find use in applications for loudspeakers, dc motors, microwave oven magnetron tubes, traveling wave tubes, holding magnets, chip collectors, and magnetic separator units. Ceramic magnet arcs find wide use in the auto industry in engine coolant pumps, heating-cooling fan motors, and window lift motors. As with other magnets, they are normally supplied nonmagnetized and are magnetized in the end-use structure using magnetizing fields of the order of 10,000 Oe to saturate the magnet. The brittleness of the material necessitates proper design of the magnet support structure so as not to impart mechanical stress to the magnet.

***Rare Earth Cobalt Magnets.*** *Rare earth cobalt magnets* have the highest energy product and coercivity of any commercially available magnetic material. Magnets are produced by powder metallurgy techniques from alloys of cobalt (65% to 77%), rare earth metals (23% to 35%), and sometimes copper and iron. The rare earth metal used is usually samarium, but other metals used are praseodymium, cerium, yttrium, neodymium, lathanum, and a rare earth metal mixture called *misch metal.* The rare earth alloy is ground to a fine particle size (1 to 10 $\mu$m), and the powder is then die-compacted in a strong magnetic field. The part is then sintered and abrasive-ground to finish tolerances.

Although this material uses comparatively expensive raw materials, the high value of coercive force (5500 to 9500 Oe) leads to small magnet size and good temperature stability. These magnets find use in miniature electronic devices such as motors, printers, electron beam focusing assemblies, magnetic bearings, and traveling wave tubes. Plastic-bonded rare earth magnets are also being made, but the magnetic value of the energy product is only a fraction of the sintered product.

***Ductile Alloys.*** *Ductile alloys* include the materials Cunife, Vicalloy, Remalloy, chromium-cobalt-iron (Cr-Co-Fe), and in a limited sense, manganese-aluminum-carbon (Mn-Al-C). They are sufficiently ductile and malleable to be drawn, forged, or rolled into wire or strip forms. A final heat treatment after forming develops the magnetic properties. Cunife has a directional magnetism developed as a result of cold working and finds wide use in meters and automotive speedometers. Vicalloy has been used as a high-quality and high-performance magnetic recording tape and in hysteresis clutch applications. Remalloy has been used extensively in telephone receivers but is now being replaced by a newer, less costly magnetic material.

New permanent-magnet materials that are now being produced are the Cr-Co-Fe alloy and the Mn-Al-C alloy. The Cr-Co-Fe alloy family contains 20% to 35% chromium and from 5% to 25% cobalt. This alloy is unique among permanent-magnet alloys due to its good hot and cold ductility, machinability, and excellent magnetic properties. The heat treatment of the alloy involves a rapid cooling from approximately 1200°C to a spinoidal decomposition phase occurring at about 600°C. The magnetic phase developed in the spinoidal decomposition process may be oriented by a heat treatment in a magnetic field, or the material may be magnetically oriented by "deformation aging" as would be accomplished in a wire-drawing operation. The magnetic properties that can be developed are comparable with those of Alnico 5 and are superior to those of the other ductile alloys, Cunife, Vicalloy, and Remalloy. Western Electric has introduced a Cr-Co-Fe alloy which replaces Remalloy in the production of telephone receiver magnets and at a lower cost due to reduced cobalt.

***The Mn-Al-C Alloy.*** *The Mn-Al-C alloy* achieves permanent-magnetic properties (*Br*, 5500 G; *Hc*, 2300 Oe; Mg · Oe energy product, 5 Mg · Oe) when mechanical deformation of the alloy takes place at a temperature of about 720°C. Mechanical deformation may be performed by warm extrusion. Magnet size is limited by the amount of deformation needed to develop and orient the magnetic phase in the alloy. The alloying elements are inexpensive, but the tooling and equipment needed in the deformation process is expensive and may be a factor in the economical production of this magnet alloy. Magnets of this alloy would find use in loudspeakers, motor applications, and microwave oven magnetron tubes. The low density, 5.1 g/cm$^3$, is desirable for motors where reduced inertia and weight savings are important. The low Curie temperature, 320°C, limits the use of this alloy to applications where the ambient temperature is less than 125°C.

***Permanent-Magnet Design.*** *Permanent-magnet design* involves the calculation of magnet area and magnet length to produce a specific magnetic flux density across a known gap, usually with the magnet having the smallest possible volume. Designs are developed from magnet material hysteresis loop data of the second quadrant, commonly called *demagnetization curves.*

Other considerations are the operating temperature of the magnetic assembly, magnet weight, and cost. Also, care should be exercised in the calculation of any steel return path cross section to ensure that it is adequate to carry the flux output of the magnet. Table 3-11 illustrates the range of magnetic characteristics that may be considered in the design. Detailed magnetic and material specifications may be obtained from the magnet manufacturer.

**TABLE 3-11** Comparison of Magnetic and Physical Properties of Selected Commercial Materials

| | Alnico 5 | Alnico 9 | Ferrite | Co5R |
|---|---|---|---|---|
| $B_r$, G | 12800 | 10500 | 4100 | 9500 |
| $H_c$, Oe | 640 | 1500 | 2900 | 6500 |
| $BdHd$, Mg · Oe | 5.5 | 9.0 | 4.0 | 22.0 |
| Curie point, °C | 850 | 815 | 470 | 740 |
| Temperature coefficient, %/°C | 0.02 | 0.02 | 0.19 | 0.03 |
| Density, g/cm$^3$ | 7.3 | 7.3 | 4.9 | 8.6 |
| Energy/unit weight | 0.8 | 1.2 | 0.8 | 2.6 |

### 3.2.11 Bibliography

AIP. (1976). *Magnetism and Magnetic Materials.* Ed. by J. J. Becker and G. H. Lander. American Institute of Physics, New York.

Anderson, J. C. (1968). *Magnetism and Magnetic Materials.* Chapman & Hall, London.

ASTM. (1970). *Direct-Current Magnetic Measurements for Soft Magnetic Materials.* Committee A-6 on Magnetic Properties, American Society for Testing and Materials, West Conshohocken, Pa.

Bradley, F. N. (1971). *Materials for Magnetic Functions.* Hayden Book Co., New York.

Brailsford, F. (1960). *Magnetic Materials.* Wiley, New York.

Brailsford, F. (1968). *An Introduction to the Magnetic Properties of Materials.* Longmans, London.

Chen, Chih-Wen. (1977). *Magnetism and Metallurgy of Soft Magnetic Materials.* North-Holland, Amsterdam.

Connolly, T. F., and Copenhaver, E. D. 1972. *Bibliography of Magnetic Materials and Tabulation of Magnetic Transition Temperatures.* IFI/Plenum, New York.

Cullity, B. D. (1972). *Introduction to Magnetic Materials.* Addison-Wesley, Reading, Mass.

Heck, C. (1974). *Magnetic Materials and Their Application.* Translated from the German by Stuart S. Hill. Butterworths, London.

Nussbaum, A. (1967). *Electronic and Magnetic Behavior of Materials.* Prentice-Hall, Englewood Cliffs, N.J.

Schieber, M. M. (1967). *Experimental Magnetochemistry: Nonmetallic Magnetic Materials.* North Holland, Amsterdam.

Sittig, M. (1970). *Magnetic Material.* Noyes Data Corp., Park Ridge, N.J.

Stapleton, R. E. (1968). *Magnetic and Electrical Materials Capable of Operating in the 800-1600° F Temperature Range.* American Society for Testing and Materials, West Conshohocken, Pa.

Thompson, J. E. (1968). *The Magnetic Properties of Materials.* CRC Press; Cleveland.

## 3.3 INSULATING MATERIALS

### 3.3.1 General Properties

***Electrical Insulation and Dielectric Defined.*** Electrical insulation is a medium or a material which, when placed between conductors at different potentials, permits only a small or negligible current in phase with the applied voltage to flow through it. The term *dielectric* is almost synonymous with electrical insulation, which can be considered the applied dielectric. A perfect dielectric passes no conduction current but only capacitive charging current between conductors. Only a vacuum at low stresses between uncontaminated metal surfaces satisfies this condition.

The range of resistivities of substances which can be considered insulators is from greater than $10^{20}$ Ω · cm downward to the vicinity of $10^6$ Ω · cm, depending on the application and voltage stress. There is no sharp boundary defined between low-resistance insulators and semiconductors. If the voltage stress is low and there is little concern about the level of current flow (other than that which would heat and destroy the insulation), relatively low-resistance insulation can be tolerated.

***Circuit Analogy of a Dielectric or Insulation.*** Any dielectric or electrical insulation can be considered as equivalent to a combination of capacitors and resistors which will duplicate the current-voltage behavior at a particular frequency or time of voltage application. In the case of some dielectrics, simple linear capacitors and resistors do not adequately represent the behavior. Rather, resistors and capacitors with particular nonlinear voltage-current or voltage-charge relations must be postulated to duplicate the dielectric current-voltage characteristic.

The simplest circuit representation of a dielectric is a parallel capacitor and resistor, as shown in Fig. 3-13 for $R_S = 0$. The perfect dielectric would be simply a capacitor. Another representation of a dielectric is a series-connected capacitor and resistor as in Fig. 3-13 for $R_p = \infty$, while still another involves both $R_S$ and $R_p$.

The ac dielectric behavior is indicated by the phase diagram (Fig. 3-14). The perfect dielectric capacitor has a current which leads the voltage by 90°, but the imperfect dielectric has a current which leads the voltage by less than 90°. The dielectric phase angle is $\theta$, and the difference, $90° - \theta = \delta$, is the loss angle. Most measurements of dielectrics give directly the tangent of the loss angle tan $\delta$ (known as the *dissipation factor*) and the capacitance $C$. In Fig. 3-13, if $R_p = \infty$, the series $R_s - C$ has a $\tan \delta = 2\pi f C_S R_s$, and if $R_s = 0$, the parallel $R_p - C$ has a $\tan \delta = \frac{1}{2}\pi f C_p R_p$.

The ac power or heat loss in the dielectric is $V^2 2\pi f C \tan \delta$ watts, or $VI \sin \delta$ watts, where $\sin \delta$ is known as the power factor, $V$ is the applied voltage, $I$ is the total current through the dielectric, and $f$ is the frequency. From this it can be seen that the equivalent parallel conductance of the dielectric $\sigma$ (the inverse of the equivalent parallel resistance $\rho$) is $2\pi f C \tan \delta$. The ac conductivity is

$$\sigma = (5/9) f \epsilon' \tan\delta \times 10^{-12} \Omega^{-1}\text{cm}^{-1} = 1/\rho \qquad \text{(3-54)}$$

where $\epsilon'$ is the permittivity (or relative dielectric constant) and $f$ is the frequency. (The IEEE now recommends the symbol $\epsilon'$ for the dielectric constant relative to a vacuum. The literature on dielectrics and insulation also has used $\kappa$ [kappa] for this dimensionless quantity or $\epsilon_r'$. In some places, $\epsilon'$ has been used to indicate the absolute dielectric constant, which is the product of the relative dielectric constant and the dielectric constant of a vacuum $\epsilon_0$, which is equal to $8.85 \times 10^{-12}$ F/m.) $\kappa_0$ also has been used to represent the dielectric constant of a vacuum. While the ac conductivity theoretically increases in proportion to the frequency, in practice, it will depart from this proportionality insofar as $\epsilon'$ and tan $\delta$ change with frequency.

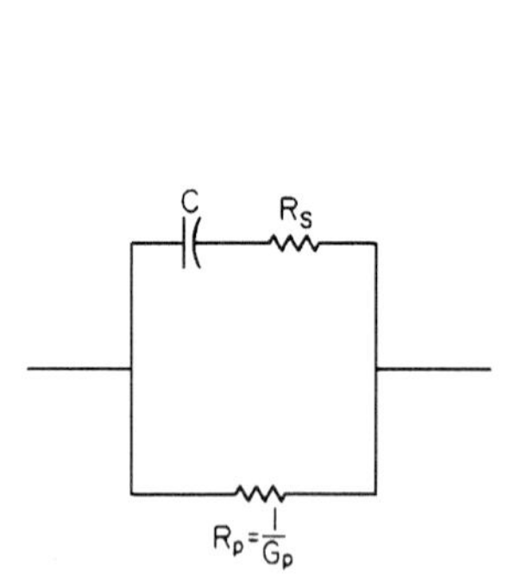


**FIGURE 3-13** Equivalent circuit of a dielectric.

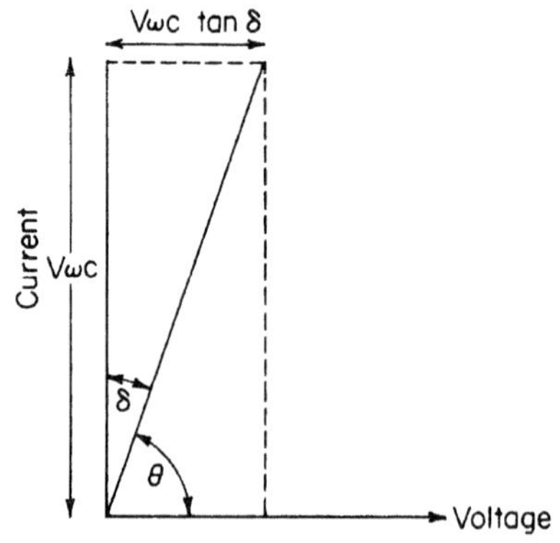


**FIGURE 3-14** Current-voltage phase relation in a dielectric.

***Capacitance and Permittivity or Dielectric Constant.*** The capacitance between plane electrodes in a vacuum (with fringing neglected) is

$$C = \epsilon' \epsilon_0 A/t = 0.0884 \times 10^{-12} A/t \qquad \text{farads} \tag{3-55}$$

where $\epsilon_0$ is the dielectric constant of a vacuum, $A$ is the area in square centimeters, and $t$ is the spacing of the plates in centimeters. $\epsilon_0$ is $0.225 \times 10^{-12}$ F/in when $A$ and $t$ are expressed in inch units.

When a dielectric material fills the volume between the electrodes, the capacitance is higher by virtue of the charges within the molecules and atoms of the material, which attract more charge to the capacitor planes for the same applied voltage. The capacitance with the dielectric between the electrodes is

$$C = \epsilon' \epsilon_0 A/t \tag{3-56}$$

where $\epsilon'$ is the relative dielectric constant of the material. The capacitance relations for several other commonly occurring situations are

Coaxial conductors: $$C = \frac{2\pi\epsilon'\epsilon_0 L}{\ln(r_2/r_1)} \qquad \text{farads} \tag{3-57}$$

Concentric spheres: $$C = \frac{4\pi\epsilon'\epsilon_0 r_1 r_2}{r_2 - r_1} \qquad \text{farads} \tag{3-58}$$

Parallel cylindrical conductors: $$C = \frac{2\pi\epsilon'\epsilon_0 L}{\cosh^{-1}(D/2r)} \qquad \text{farads} \tag{3-59}$$

In these equations, $L$ is the length of the conductors, $r_2$ and $r_1$ are the outer and inner radii, and $D$ is the separation between centers of the parallel conductors with radii $r$. For dimensions in centimeters, $\epsilon_0$ is 0.0884 F/cm.

The value of $\epsilon'$ depends on the number of atoms or molecules per unit volume and the ability of each to be polarized (i.e., to have a net displacement of their charge in the direction of the applied voltage stress). Values of $\epsilon'$ range from unity for vacuum to slightly greater than unity for gases at atmospheric pressure, 2 to 8 for common insulating solids and liquids, 35 for ethyl alcohol and 91 for pure water, and 1000 to 10,000 for titanate ceramics (see Table 3-12 for typical values).

The relative dielectric constant of materials is not constant with temperature, frequency, and many other conditions and is more appropriately called the *dielectric permittivity*. Refer to the volume by Smyth (1955) for a discussion of the relation of $\epsilon'$ to molecular structure and to von Hippel (1954) and other tables of dielectric materials from the MIT Laboratory for Insulation Research. The permittivity of many liquids has been tabulated in *NBS Circ.* 514. The *Handbook of Chemistry and Physics* (Chemical Rubber Publishing Co.) also lists values for a number of plastics and other materials.

The permittivity of many plastics, ceramics, and glasses varies with the composition, which is frequently variable in nominally identical materials. In the case of some plastics, it varies with degree of cure and in the case of ceramics with the firing conditions. Plasticizers often have a profound effect in raising the permittivity of plastic compositions.

There is a force of attraction between the plates of a capacitor having an applied voltage. The stored energy is ½ $CV^2$ J. The force equals the derivative of this energy with respect to the plate separation: (½) $\epsilon'\epsilon_0 E^2 \times 10^2$ N/cm² or (½) $\epsilon'\epsilon_0 E^2 \times 10$ bar, where $E$ is the electric field in volts per centimeter. The force increases proportionally to the capacitance or permittivity. This leads to a force of attraction of dielectrics into an electric field, that is, a net force which tends to move them toward a region of high field. If two dielectrics are present, the one with higher permittivity will displace the one with lower permittivity in the higher-field region. For example, air bubbles in a liquid are repelled from high-field regions. Correspondingly, elongated dielectric bodies are rotated into the direction of the electric field. In general, if the voltage on a dielectric system is maintained constant, the dielectrics move (if they are able) to create a higher capacitance.

**TABLE 3-12** Dielectric Permittivity (Relative Dielectric Constant), $\epsilon'$

| | *k* | | *k* |
|---|---|---|---|
| Inorganic crystalline | | Polymer resins | |
| NaCl, dry crystal | 5.5 | Nonpolar resins | |
| $CaCO_2$ (av) | 9.15 | Polyethylene | 2.3 |
| $Al_2O_2$ | 10.0 | Polystyrene | 2.5–2.6 |
| MgO | 8.2 | Polypropylene | 2.2 |
| BN | 4.15 | Polytetrafluoroethylene | 2.0 |
| $TiO_2$ (av) | 100 | Polar resins | |
| $BaTiO_2$ crystal | 4,100 | Polyvinyl chloride (rigid) | 3.2–3.6 |
| Muscovite mica | 7.0–7.3 | Polyvinyl acetate | 3.2 |
| Fluorophlogopite (synthetic mica) | 6.3 | Polyvinyl fluoride | 8.5 |
| | | Nylon | 4.0–4.6 |
| Ceramics | | Polyethylene terephthalate | 3.25 |
| Alumina | 8.1–9.5 | Cellulose cotton fiber (dry) | 5.4 |
| Steatite | 5.5–7.0 | Cellulose Kraft fiber (dry) | 5.9 |
| Forsterite | 6.2–6.3 | Cellulose cellophane (dry) | 6.6 |
| Aluminum silicate | 4.8 | Cellulose triacetate | 4.7 |
| Typical high-tension porcelain | 6.0–8.0 | Tricyanoethyl cellulose | 15.2 |
| Titanates | 50–10,000 | Epoxy resins unfilled | 3.0–4.5 |
| Beryl | 4.5 | Methylmethacrylate | 3.6 |
| Zirconia | 8.0–10.5 | Polyvinyl acetate | 3.7–3.8 |
| Magnesia | 8.2 | Polycarbonate | 2.9–3.0 |
| Glass-bonded mica | 6.4–9.2 | Phenolics (cellulose-filled) | 4–15 |
| | | Phenolics (glass-filled) | 5–7 |
| Glasses | | Phenolics (mica-filled) | 4.7–7.5 |
| Fused silica | 3.8 | Silicones (glass-filled) | 3.1–4.5 |
| Corning 7740 (common laboratory Pyrex) | 5.1 | | |

***Resistance and Resistivity of Dielectrics and Insulation.*** The measured resistance $R$ of insulation depends on the geometry of the specimen or system measured, which for a parallel-plate arrangement is

$$R = \rho t / A \qquad \text{ohms} \tag{3-60}$$

where $t$ is the insulation thickness in centimeters, $A$ is the area in square centimeters, and $\rho$ is the dielectric resistivity in ohm-centimeters. If $t$ and $A$ vary from place to place, the effective "insulation resistance" will be determined by the effective integral of the $t/A$ ratio over all the area under stress, on the assumption that the material resistivity $\rho$ does not change. If the material is not homogeneous and materials of different resistivities appear in parallel, the system can be treated as parallel resistors: $R = R_a R_b/(R_a + R_b)$. In this case, the lower-resistivity material usually controls the overall behavior. But if materials of different resistivities appear in series in the electric field, the higher-resistivity material generally will control the current, and a majority of the voltage will appear across it, as in the case of series resistors.

The resistance of dielectrics and insulation is usually time-dependent and (for the same reason) frequency-dependent. The dc behavior of dielectrics under stress is an extension of the low-frequency behavior. The ac and dc resistance and permittivity can, in principle, be related for comparable times and frequencies.

Current flow in dielectrics can be divided into parts: (*a*) the true dc current, which is constant with time and would flow indefinitely, is associated with a transport of charge from one electrode into the dielectric, through the dielectric, and out into the other electrode, and (*b*) the polarization or absorption current, which involves, not charge flow through the interface between the dielectric and the electrode, but rather the displacement of charge within the dielectric. This is illustrated in

Fig. 3-15, where it is shown that the displaced or absorbed charge is responsible for a reverse current when the voltage is removed.

Polarization current results from any of the various forms of limited charge displacement which can occur in the dielectric. The displacement occurring first (within less than nanoseconds) is the electronic and intramolecular charged atom displacement responsible for the very high frequency permittivity. The next slower displacement is the rotation of dipolar molecules and groups which are relatively free to move. The displacement most commonly observed in dc measurements, that is, currents changing in times of the order of seconds and minutes, is due to the very slow rotation of dipolar molecules and ions moving up to internal barriers in the material or at the conductor surfaces. When those slower displacement polarizations occur, the dielectric constant declines with increasing frequency and approaches the square of the optical refractive index $\eta^2$ at optical frequencies.

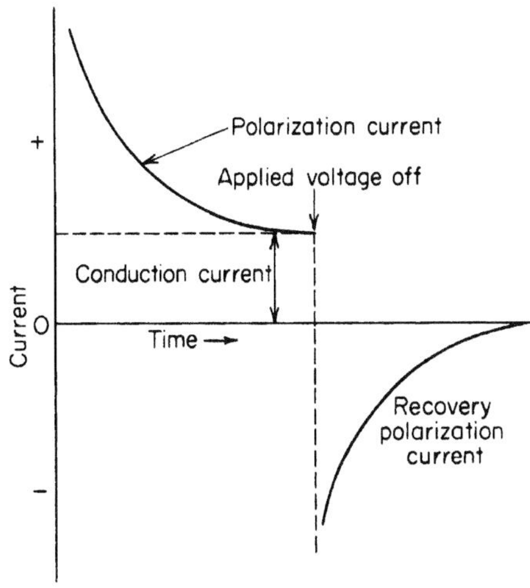


**FIGURE 3-15** Typical dc dielectric current behavior.

In composite dielectrics (material with relatively lower resistance intermingled with a material of relatively higher resistance), a large interfacial or Maxwell-Wagner type of polarization can occur. A circuit model of such a situation can be represented by placing two of the circuits of Fig. 3-13 in series and making the parallel resistance of one much lower than the other. To get the effect, it is necessary that the time constant $R_pC$ be different for each material.

A simple model of the polarization current predicts an exponential decline of the current with time: $I_p = Ae^{-\alpha t}$, similar to the charging of a capacitor through a resistor. Composite materials are likely to have many different time constants, $\alpha = 1/RC$, superimposed. It is found empirically that the polarization or absorption current decreases inversely as a simple negative exponent of the time

$$I = At^{-n} \tag{3-61}$$

The ratio of the current at 1 min to that at 10 min has been called the *polarization index* and is used to indicate the quality of composite machine insulation. A low polarization index associated with a low resistance sometimes indicates parallel current leakage paths through or over the surface of insulation (e.g., in adsorbed water films).

The level of the conduction current which flows essentially continuously through insulation is an indication of the level of the ionic concentration and mobility in the material. Frequently, as with salt in water, the ions are provided by dissolved, absorbed, or included impurity electrolytes in the material rather than by the material itself. Purifying the material will therefore often raise the resistivity. If it is liquid, purification can be done with adsorbent clays or ion-exchange resins.

The conductivity of ions in an insulation is given by the equation

$$\sigma = \mu ec \qquad \Omega^{-1}\cdot\text{cm}^{-1} \tag{3-62}$$

where $\mu$ is the ion mobility, $e$ is the ionic charge in coulombs, and $c$ is the ionic concentration per cubic centimeter. The mobility, expressed in centimeters per second-volt per centimeter, decreases inversely with the effective internal viscosity and is very low for hard resins, but it increases with temperature and with softness of the resin or fluidity of liquids. The ionic conductivity also varies widely with material purity. Among the polymers and resins, nonpolar resins such as polyethylene are likely to have high resistivities, on the order of $10^{16}$ or greater, since they do not readily dissolve or dissociate ionic impurities. Harder or crystalline polar resins have higher resistivity than do similar softer resins of similar dielectric constant and purity. Resins and liquids of higher dielectric constant usually have higher conductivities because they dissolve ionic impurities better, and the impurities dissociate to ions much more readily in a higher dielectric constant medium. Ceramics and glasses have lower resistivity if they contain alkali ions (sodium and potassium), since these ions are highly mobile.

Water is particularly effective in decreasing the resistivity by increasing the ionic concentration and mobility of materials, on the surface as well as internally. Water associates with impurity ions or ionizable constituents within or on the surface or interfaces. It helps to dissociate the ions by virtue of its high dielectric constant and provides a local environment of greater mobility, particularly as surface water films.

The ionic conductivity $\sigma$, exclusive of polarization effects, can be expected to increase exponentially with temperature according to the relation

$$\sigma = \sigma_0 e^{-B/T} \tag{3-63}$$

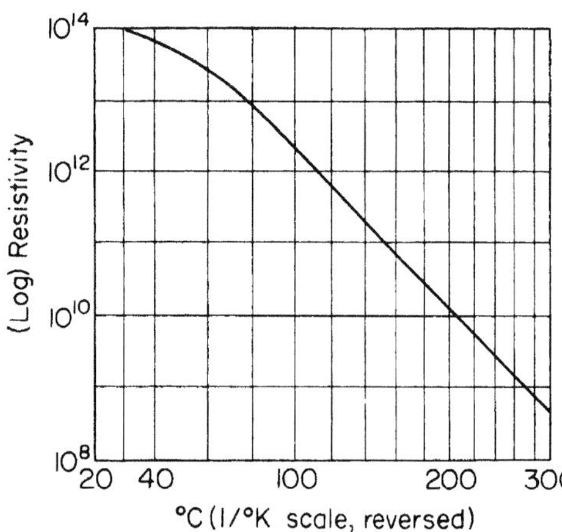


**FIGURE 3-16** Typical dielectric resistivity-temperature dependence. (*Corning.*)

where $T$ is the Kelvin temperature and $\sigma_0$ and $B$ are constants. This relation, log $\sigma$ versus $1/T$, is shown in Fig. 3-16. It is often observed that at lower temperatures, where the resistivity is higher, the resistivity tends to be lower than the extrapolated higher temperature line would predict. There are at least two possible reasons for this: the effect of adsorbed moisture and the contribution of a very slowly decaying polarization current.

***Variation of Dielectric Properties with Frequency.*** The permittivity of dielectrics invariably tends downward with increasing frequency, owing to the inability of the polarizing charges to move with sufficient speed to follow the increasing rate of alternations of the electric field. This is indicated in Fig. 3-17. The sharper decline in permittivity is known as a *dispersion region.* At the lower frequencies, the ionic-interface polarization declines first; next, the molecular dipolar polarizations decline. With some polar polymers, two or more dipolar dispersion regions may occur owing to different parts of the molecular rotation.

Figure 3-17 is typical of polymers and liquids but not of glasses and ceramics. Glasses, ceramics, and inorganic crystals usually have much flatter permittivity-frequency curves, similar to that shown for the nonpolar polymer, but at a higher level, owing to their atom-ion displacement polarization, which can follow the electric field usually up to infrared frequencies.

The dissipation factor–frequency curve indicates the effect of ionic migration conduction at low frequency. It shows a maximum at a frequency corresponding to the permittivity dispersion region. This maximum is usually associated with a molecular dipolar rotation and occurs when the rotational mobility is such that the molecular rotation can just keep up with frequency of the applied field. Here it has its maximum movement in phase with the voltage, thus contributing to conduction current. At lower frequencies, the molecule dipole can rotate faster than the field and contributes more to permittivity. At higher frequencies it cannot move fast enough. Such a dispersion region can also occur because of ionic migration and interface polarization if the interfaces are closely spaced and if the frequency and mobility have the required values.

The frequency region where the dipolar dispersion occurs depends on the rotational mobility. In mobile, low-viscosity liquids, it is in the 100- to 10,000-MHz range. In viscous liquids, it occurs in the region of 1 to 100 MHz. In soft polymers it may occur in the audio-frequency range, and with

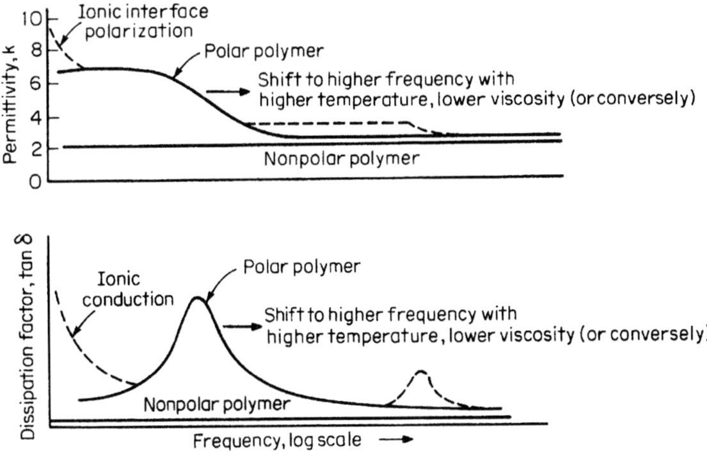


**FIGURE 3-17** Typical variation in dielectric properties with frequency.

hard polymers it is likely to be at very low frequency (indistinguishable from dc properties). Since the viscosity is affected by the temperature, increased temperature shifts the dispersion to higher frequencies.

***Variation of Dielectric Properties with Temperature.*** The trend in ac permittivity and conductivity, as measured by the dissipation factor, is controlled by the increasing ionic migrational and dipolar molecular rotational mobility with increasing temperature. This curve, which is indicated in Fig. 3-18, is in most respects a mirror image of the frequency trend shown in Fig. 3-17, since the two effects are interrelated.

The permittivity-dispersion and dissipation factor maximum region occurs below room temperature for viscous liquids and still lower for mobile liquids. In fact, mobile liquids may crystallize

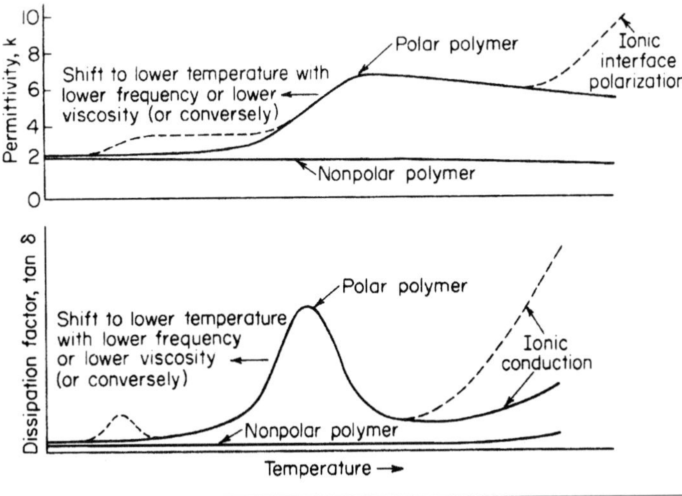


**FIGURE 3-18** Typical variation in dielectric properties with temperature.

before they would show dispersion, except at high frequencies. With polymers, the dissipation factor maxima is likely to occur, at power frequencies, at a temperature close to a softening-point or internal second-order transition-point temperature. Dielectric dispersion and mechanical modulus dispersion usually can be correlated at the same temperature for comparable frequencies.

***Composite Dielectrics.*** The dielectric properties of composite dielectrics are generally a weighted average of the individual component properties, unless there is interaction, such as dissolving (as opposed to intermixing) of one material in another, or chemical reaction of one with another. Interfaces created by the mixing present a special factor, which often can lead to a higher dissipation factor and lower resistivity as a result of moisture and/or impurity concentration at the interface.

The ac properties of sheets of two dielectrics of dielectric constant $k_1$ and $k_2$ and of thickness $t_1$ and $t_2$ placed in series are related to the properties of the individual materials by the series of capacitance and impedance relation

$$C = \frac{k_0 k_1 k_2 A}{k_1 t_2 + k_2 t_1} \tag{3-64}$$

$$\tan\delta = \frac{(t_1/t_2)\epsilon_2' \tan\delta_1 + \epsilon_1' \tan\delta_2}{\epsilon_1' + \epsilon_2'(t_1/t_2)} \tag{3-65}$$

Similarly, the properties of two dielectrics in parallel are

$$C = \epsilon_0 \left( \frac{\epsilon_1' A_1}{t_1} + \frac{\epsilon_2' A_2}{t_2} \right) \tag{3-66}$$

$$\tan\delta = \frac{t_2 \epsilon_1' A_1 \tan\delta_1 + t_1 \epsilon_2' A_2 \tan\delta_2}{t_2 \epsilon_1' A_1 + t_1 \epsilon_2' A_2} \tag{3-67}$$

With steady dc voltages, the resistivities control the current. With equal-area layer dielectrics in series,

$$R = R_1 + R_2 = \frac{1}{A}(\rho_1 t_1 + \rho_2 t_2) \tag{3-68}$$

When the dielectrics are in parallel and of equal thickness $t$,

$$R = \frac{R_1 R_2}{R_1 + R_2} = \frac{\rho_1 \rho_2 t}{\rho_1 A_2 + \rho_2 A_1} \tag{3-69}$$

***Potential Distribution in Dielectrics.*** The maximum potential gradient in dielectrics is of critical significance insofar as the breakdown is concerned, since breakdown or corona is usually initiated at the region of highest gradient. In a uniform-field arrangement of conductors or electrodes, the maximum gradient is simply the applied voltage divided by the minimum spacing. In divergent fields, the gradient must be obtained by calculation (which is possible for some simple arrangements) or by field mapping.

A common situation is the coaxial geometry with inner and outer radii $R_1$ and $R_2$. The gradient at radius $r$ (centimeters) with voltage $V$ applied is given by the equation

$$E = \frac{V}{r \ln(R_2/R_1)} \quad \text{V/cm} \tag{3-70}$$

The gradient is a maximum at $r = R_1$.

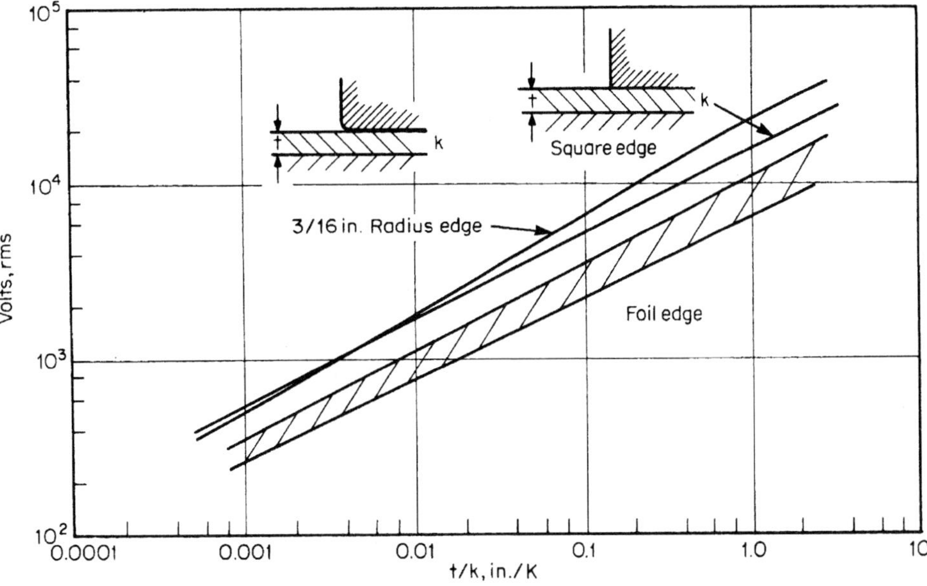


**FIGURE 3-19** Corona threshold voltage at conductor edges in air as a function of insulation thickness.

When different dielectrics appear in series, the greater stress with ac fields is on the material having the lower dielectric constant. This material will frequently break down first unless its dielectric strength is much higher.

$$\frac{E_1}{E_2} = \frac{\epsilon_2'}{\epsilon_1'} \quad \text{and} \quad E_1 = \frac{V}{t_1 + t_2\epsilon_1'/\epsilon_2'} \tag{3-71}$$

The effect of the insulation thickness and dielectric constant (as well as the sharpness of the conductor edge) to create sufficient electric stress for local air breakdown (partial discharges) is shown in Fig. 3-19. With dc fields, the stress distributes according to the resistivities of the materials, the higher stress being on the higher-resistivity material.

***Dielectric Strength.*** This is defined by the ASA as the maximum potential gradient that the material can withstand without rupture. Practically, the strength is often reported as the breakdown voltage divided by the thickness between electrodes, regardless of electrode stress concentration.

Breakdown appears to require not only sufficient electric stress but also a certain minimum amount of energy. It is a property which varies with many factors such as thickness of the specimen, size and shape of electrodes used in applying stress, form or distribution of the field of electric stress in the material, frequency of the applied voltage, rate and duration of voltage application, fatigue with repeated voltage applications, temperature, moisture content, and possible chemical changes under stress.

The practical dielectric strength is decreased by defects in the material, such as cracks, and included conducting particles and gas cavities. As will be shown in more detail in later subsections on gases and liquids, the dielectric strength is quite adversely affected by conducting particles.

To state the dielectric strength correctly, the size and shape of specimen, method of test, temperature, manner of applying voltage, and other attendant conditions should be particularized as definitely as possible.

ASTM standard methods of dielectric strength testing should be used for making comparison tests of materials, but the levels of dielectric strength measured in such tests should not be expected to apply in service for long times. It is best to test an insulation in the same configuration in which it would be used. Also, the possible decline in dielectric strength during long-time exposure to the service environment, thermal aging, and partial discharges (corona), if they exist at the applied service voltage, should be considered. ASTM has thermal life test methods for assessing the long-time endurance of some forms of insulation such as sheet insulation, wire enamel, and others. There are IEEE thermal life tests for some systems such as random wound motor coils.

The dielectric strength varies as the time and manner of voltage application. With unidirectional pulses of voltage, having rise times of less than a few microseconds, there is a time lag of breakdown, which results in an apparent higher strength for very short pulses. In testing sheet insulation in mineral oil, usually a higher strength for pulses of slow rise time and somewhat higher strength for dc voltages is observed.

The trend in breakdown voltage with time is typical of many solid insulation systems.

With ac voltages, the apparent strength declines steadily with time as a result of partial discharges (in the ambient medium at the conductor or electrode edge). These penetrate the solid insulation. The discharges result from breakdown of the gas or liquid prior to the breakdown of the solid. Mica in particular, as well as other inorganic materials, is more resistant to such discharges. Organic resins should be used with caution where the ac voltage gradient is high and partial discharges (corona) may be present. Since the presence of partial discharges on insulation is so important to the long-time voltage endurance, their detection and measurement have become very important quality control and design tools. If discharges continuously strike the insulation within internal cavities or on the surface, the time to failure usually varies inversely as the applied frequency, since the number of discharges per unit time increases almost in direct proportion to the frequency. But in some cases, ambient conditions prevent continuous discharges.

When organic resin insulation is fabricated to avoid partial discharges using conductors or electrodes intimately bonded to the insulation, as in extruded polyethylene cables with a plastic semiconducting interface between the resin and the coaxial inner and outer metal conductors, respectively, the voltage endurance is greatly extended. Imperfections, however, in this "semicon"-resin interface, or at conducting particle inclusions in the resin, can lead to local discharges and the development of "electrical tree" growth. Vacuum impregnating and casting electrodes or conductors into resin also tend to avoid cavities and surface discharges and greatly improve the voltage endurance at high stresses.

The dc strength of solid insulation is usually higher and declines much less with time than the ac strength, since corona discharges are infrequent.

The dielectric strength is much higher where surface discharges are avoided and when the electric field is uniform. This can be achieved with solid materials by recessing spherical cavities into the material and using conducting paint electrodes.

The "intrinsic" electric strength of solid materials measured in uniform fields, avoiding surface discharges, ranges from levels on the order of 0.5 to 1 MV/cm for alkali halide crystals, which are about the lowest, upward to somewhat more than 10 MV/cm. Polymers and some inorganic materials, such as mica and aluminum oxide, have strengths of 2 to 20 MV/cm for thin films. The strength decreases with increasing thickness and with temperature above a critical temperature (which is usually from 1 to 100°C), below which the strength has a level value or a moderate increase with increasing temperature. Below the critical temperature, the breakdown is believed to be strictly electronic in nature and is constant or increases slightly with temperature. Above this temperature, it declines owing to dielectric thermal heating.

The breakdown voltage of thin insulation materials containing defects, which give the minimum breakdown voltage, declines as the area under stress increases. The effect of area on the strength can be estimated from the standard deviation $S$ of tests on smaller areas by applying minimum value statistics: $V_1 - V_2 = 1.497\, S \log(A_1/A_2)$, where $V_1$ and $V_2$ are the breakdown voltages of areas $A_1$ and $A_2$.

If the ac or dc conductivity of a dielectric is high or the frequency is high, breakdown can occur as a result of dielectric heating, which raises the temperature of the material sufficiently to cause melting or decomposition, formation of gas, etc. This effect can be detected by measuring the conductivity as a function of applied electric stress. If the conductivity rises with time, with constant voltage, and at constant ambient temperature, this is evidence of an internal dielectric heating. If the heat transfer to the electrodes and ambient surroundings is adequate, the internal temperature eventually may stabilize, but if this heat transfer is inadequate, the temperature will rise until breakdown occurs. The criterion of this sort of breakdown is the heat balance between dielectric heat input and loss to the surroundings.

The dielectric heat input is given by the equation

$$\sigma E^2 = (5/9\epsilon' f \tan\delta \times 10^{-12})E^2 \qquad \text{W/cm}^3 \tag{3-72}$$

where $E$ is the field in volts per centimeter. When this quantity is on the order of 0.1 or greater, dielectric heating can be a problem. It is much more likely to occur with thick insulation and at elevated temperatures.

***Water Penetration.*** Water penetration into electrical insulation also degrades the dielectric strength by several mechanisms. The effect of water to increase the insulation conductivity contributes thereby to a decreased dielectric strength, probably by a thermal breakdown mechanism. Another effect noticed recently, particularly in polyethylene cables, is the development of "water" or "electrochemical trees." Water (and/or a similar high dielectric constant chemical) can diffuse through polyethylene and collect at tiny hygroscopic inclusion sites, where the water or chemical is adsorbed. Then the electric field causes an expansion and growth of the adsorbed water or chemical in the electric field direction. This may completely bridge the insulation or possibly increase the local electric stress at the site so as to produce an electric tree and eventual breakdown.

***Ionizing Radiation.*** Ionizing radiation, as from nuclear sources, may degrade insulation dielectric strength and integrity by causing polymer chain scission, and cracking of some plastics, as well as gas bubbles in liquids. Also, the conductivity levels in solids and liquids are increased.

***Arc Tracking of Insulation.*** High-current arc discharges between conductors across the surface of organic resin insulation may carbonize the material and produce a conducting track. In the presence of surface water films, formed from rain or condensation, etc., small arc discharges form between interrupted parts of the water film, which is fairly conducting, and conducting tracks grow progressively across the surface, eventually bridging between conductors and causing complete breakdown. Materials vary widely in their resistance to tracking, and there are a variety of dry and wet tests for this property. With proper fillers, some organic resins can be made essentially nontracking. Some resins such as polymethyl methacrylate and polymethylene oxide burst into flame under arcing conditions.

***Thermal Aging.*** Organic resinous insulating materials in particular are subject in varying degrees to deterioration due to thermal aging, which is a chemical process involving decomposition or modification of the material to such an extent that it may no longer function adequately as the intended insulation. The aging effects are usually accelerated by increased temperature, and this characteristic is used to make accelerated tests to failure or to an extent of deterioration considered dangerous. Such tests are made at appreciably higher than normal operating temperatures, if the expected life is to be several years or more, since useful accelerated tests reasonably should be completed in less than a year.

Frequently, other environmental factors influence the life in addition to the temperature. These include presence or absence of oxygen, moisture, and electrolysis. Mechanical and electrical stress may reduce the life by setting a required level of performance at which the insulation must perform. If this level is high, less deterioration of the insulation is required to reach this level.

Sometimes a complete apparatus is life-tested, as well as smaller specimens involving only one insulation material or a simple combination of these in a simple model. New tests are being devised continually, but there has been some standardization of tests by the IEEE and ASTM and internationally by the IEC.

It is important to note that frequently materials are assigned temperature ratings based on tests of the material alone. Often that material, combined with others in an apparatus or system, will perform satisfactorily at appreciably higher temperatures. Conversely, because of incompatibility with other materials, it may not perform at as high a temperature as it would alone. For this reason, it is considered desirable to make functional operating tests on complete systems. These can also be accelerated at elevated temperatures and environmental exposure conditions such as humidification, vibration, cold-temperature cycling, etc., introduced intermittently. The basis for temperature rating of apparatus and materials is discussed thoroughly in *IEEE Std* 1. Tests for determining ratings are described in *IEEE Std* 98, 99, and 101.

***Application of Electrical Insulation.*** In applying an insulating material, it is necessary to consider not only the electrical requirements but also the mechanical and environmental conditions of the application. Mechanical failure often leads to electrical failure, and mechanical failure is frequently the primary cause for failure of an aged insulation.

The initial properties of an insulation are frequently more than adequate for the application, but the effects of aging and environment may degrade the insulation rapidly to the point of failure. Thus, the thermal and environmental stability should be considered of equal importance. The effects of moisture and surface dirt contamination should be particularly considered, if these are likely to occur.

***Bibliography***

Bartnikas, R., and Eichhorn, R. M. (eds). 1983. *Electrical Properties of Solid Insulating Materials: Molecular Structure and Electrical Behavior.* ASTM, West Conshohocken, Pa.

Bartnikas, R., and McMahon, E. J. 1979. *Engineering Dielectrics: Corona Measurement and Interpretation.* ASTM, West Conshohocken, Pa.

Clark, F. M. 1962. *Insulating Materials for Design and Engineering Practice.* Wiley, New York.

Dakin, T. W. 1948. "Electrical Insulation Deterioration Treated as a Chemical Rate Phenomenon." *AIEE Trans.* vol. 67, p. 113.

Dakin, T. W., Philofsky, H. M., and Divens, W. C. 1954. "Effect of Electrical Discharges on the Breakdown 01 Solid Insulation. *AIEE Trans.*, vol. 73-1, pp. 155–162.

Mandelcorn, L., and Sommerman, G. M. L. 1963. "Electrical Tracking Resistance of Polymers." *AIEE Trans.*, vol. 70-III, pp. 69–74.

Mason, J. H. 1959. *Progress in Dielectrics.* Heywood & Co., London.

Peek, F. W. 1929. *Dielectric Phenomena in High-Voltage Engineering.* McGraw-Hill, New York.

Roth, A. 1959. *Hochspannungstechnik.* Springer-Verlag, Vienna.

Smyth, C. P. 1955. *Dielectric Behavior and Structure.* McGraw-Hill, New York.

Studniarz, S. A., and Dakin, T. W. 1982. "The Voltage Endurance of Cast Epoxy Resin, Part II." Paper presented at the IEEE International Symposium on Electrical Insulation, New York.

von Hippel, A. 1954. *Dielectric Materials and Applications.* Massachusetts Institute of Technology, Cambridge, Mass.

von Hippel, A. 1954. *Dielectrics and Waves.* Wiley, New York.

Weber, K. H., and Endicott, H. S. 1956. "Area Effect and Its External Basis for the Electric Breakdown of Transformer Oil." *AIEE Trans.*, vol. 5-III, p. 371.

Whitehead, S. 1951. *Dielectric Breakdown of Solids.* Oxford University Press, New York.

## 3.3.2 Insulating Gases

***General Properties of Gases.*** A gas is a highly compressible dielectric medium, usually of low conductivity and with a dielectric constant only a little greater than unity, except at high pressures. In high electric fields, the gas may become conducting as a result of impact ionization of the gas molecules by electrons accelerated by the field and by secondary processes which produce partial breakdown (corona) or complete breakdown. Conditions which ionize the gas molecules, such as very high temperatures and ionizing radiation (ultraviolet rays, x-rays, gamma rays, high-velocity electrons, and ions such as alpha particles), will also produce some conduction in a gas.

The gas density $d$ (grams per liter) increases with pressure $p$ (torrs or millimeters of mercury) and gram-molecular weight $M$ and decreases inversely with the absolute temperature $T$ (degrees Celsius + 273) according to the relation

$$d = \frac{M}{22.4}\frac{p}{760}\frac{273}{T} \quad g/L \tag{3-73}$$

The preceding relation is exact for ideal gases but is only approximately correct for most common gases.

If the gas is a vapor in equilibrium with a liquid or solid, the pressure will be the vapor pressure of the liquid or solid. The logarithm of the pressure varies as $-\Delta H/RT$, where $\Delta H$ is the heat of vaporization in calories per mole and $R$ is the molar gas constant, 1.98 cal/(mol)(°C). This relation also applies to all common atmospheric gases at low temperatures, below the points where they liquify.

### *Dielectric Properties at Low Electric Fields*

*Dielectric Constant.* The dielectric constant k of gases is a function of the molecular electrical polarizability and the gas density. It is independent of magnetic and electric fields except when a significant number of ions is present.

*Conduction.* The conductivity of a pure molecular gas at moderate electric stress and moderate temperature can be assumed, in the absence of any ionizing effect such as ionizing radiation, to be practically zero. Ionizing radiation induces conduction in the gas to a significant extent, depending on the amount absorbed and the volume of gas under stress. The energy of the radiation must exceed, directly or indirectly, the ionization energy of the gas molecules and thus produce an ion pair (usually an electron and positive ion). The threshold ionization energy is on the order of 10 to 25 electronvolts (eV)/molecule for common gases (10.86 eV for methyl alcohol, 12.2 for oxygen, 15.5 for nitrogen, and 24.5 for helium). Only very short wavelength ultraviolet light is effective directly in photoionization, since 10 eV corresponds to a photon of ultraviolet with a wavelength of 1240 Å. Since the photoelectric work function of metal surfaces is much lower (2 to 6 eV; e.g., copper about 4 eV), the longer-wavelength ultraviolet commonly present is effective in ejecting electrons from a negative conductors surface. Such cathode-ejected electrons give the gas apparent conductivity.

High-energy radiation from nuclear disintegration is a common source of ionization in gases. Nuclear sources usually produce gamma rays on the order of $10^6$ eV energy. Only a small amount is absorbed in passing through a low-density gas. A flux of 1 R/h produces ion pairs corresponding to a saturation current (segment *ab* of Fig. 3-20) of $0.925 \times 10^{-13}$ A/cm$^3$ of air at 1 atm pressure if all the ions formed are collected at the electrodes. The effect is proportional to the flux and the gas density.

At a voltage stress below about 100 V/cm, some of the ions formed will recombine before being collected, and the current will be correspondingly less (segment *oa* of Fig. 3-20). Higher stresses do not increase the current if all the ions formed are collected. A very small current, on the order of $10^{-21}$ A/cm$^3$ of air, is attributable to cosmic rays and residual natural radioactivity.

Electrons (beta rays) produce much more ionization per path length than gamma rays, because they are slowed down by collisions and lose their energy more quickly. Correspondingly, the slower alpha particles (positive helium nuclei) produce a very dense ionization in air over a short range. For example, a 3-million-eV (MeV) alpha particle has a range in air of 1.7 cm and creates a total of

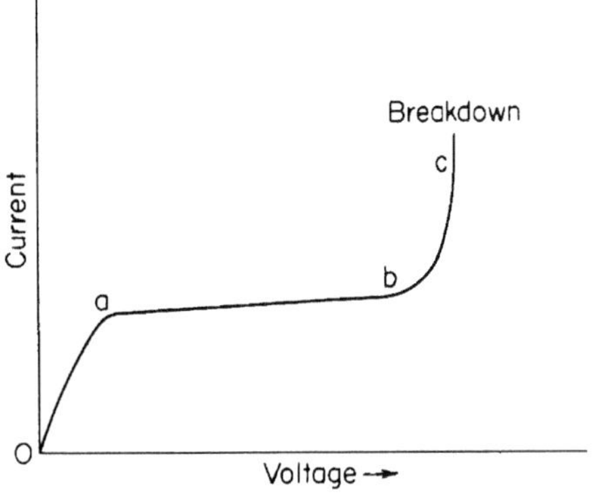


**FIGURE 3-20** Current-voltage behavior of a lightly ionized gas.

$6.8 \times 10^5$ ion pairs. A beta particle (an electron) of the same energy creates only 40 ion pairs per centimeter and has a range of 13 m in air.

It should be noted that ionizing radiation of significant levels has only a small effect on gas dielectric strength. For example, the ionization current produced by a corona discharge from a needle point is typically much higher than that produced by a radiation flux of significant level, $10^{11}$ gamma photons per square centimeter.

At temperatures increasing above 600°C, it has been shown that thermionic electron emissions from negative conductor surfaces produce significant currents compared with levels typical of electrical insulation.

Since the rate of production of ions by the various sources mentioned above is limited, the current in the gas does not follow Ohm's law, unless the rate of collection of the ions at the electrodes is small compared with the rate of production of these ions, as in the initial part of segment *oa* in Fig. 3-20.

### Dielectric Breakdown

*Uniform Fields.* The dielectric breakdown of gases is a result of an exponential multiplication of free electrons induced by the field. It is generally assumed that the initiation of breakdown requires only one electron. However, if only a few electrons are present prior to breakdown, it is not easily possible to measure the trend of current shown in Fig. 3-20. If the breakdown is completed between metal electrodes, the spark develops extremely rapidly into an arc, involving copious emission of electrons from the cathode metal and, if the necessary current flow is permitted, vaporization of metal from the electrodes. Table 3-13 gives the dielectric strength of typical gases.

**TABLE 3-13** Relative Dielectric Strengths of Gases (0.1-in gap)

| | | | |
|---|---|---|---|
| Air | 0.95 | $CF_4$ | 1.1 |
| $N_2$ | 1.0 | $C_2F_6$ | 1.9 |
| $CO_2$ | 0.90 | $C_3F_8$ | 2.3 |
| $H_2$ | 0.57 | $C_4F_8$ cyclic | 2.8 |
| A | 0.28 | $CF_2Cl_2$ | 2.4 |
| Ne | 0.13 | $C_2F_5Cl$ | 2.6 |
| He | 0.14 | $C_2F_4Cl_2$ | 3.3 |
| $SF_6$ | 2.3–2.5 | | |

In uniform electric fields, breakdown occurs at a critical voltage which is a function of the product of the pressure $p$ and spacing $d$ (Paschen's law).

It would be more accurate to consider the gas density-spacing product, since the dielectric strength varies with the temperature only as the latter affects the gas density. It will be noted that the electric field at breakdown decreases as the spacing increases. This is typical of all gases and is due to the fact that a minimum amount of multiplication of electrons must occur before breakdown occurs. A single electron accelerated by the field creates an avalanche which grows exponentially as $e^{\alpha}x$, where $x$ is the distance and $\alpha$ is the Townsend ionization coefficient (electrons formed by collision per centimeter), which increases rapidly with electric field. At small spacings, $\alpha$ and the field must be higher for sufficient multiplication. In divergent electric fields or large spacings, it has been found that when the integral *int*$\alpha_E\, dx$ increases to about 18.4 ($10^8$ electrons), sufficient space charge develops to produce a streamer type of breakdown. It seems to be apparent that the final step in gas breakdown before arc development is the development of a branched filamentary streamer which proceeds more easily from the positive electrode toward the negative electrode.

***Relative Dielectric Strengths of Gases.*** The relative dielectric strength, with few exceptions, tends upward with increasing molecular weight. There are a number of factors other than molecular or atomic size which influence the retarding effect on electrons. These include ability to absorb electron energy on collision and trap electrons to form negative ions. The noble atomic gases (helium, argon, neon, etc.) are poorest in these respects and have the lowest dielectric strengths. Table 3-13 gives the relative dielectric strengths of a variety of gases at 1 atm pressure at a $p \, . \, d$ value of 1 atm × 0.25 cm. The relative strengths vary with the $p \, . \, d$ value, as well as gap geometry, and particularly in divergent fields where corona begins before breakdown. It is best to consult specific references with regard to divergent field breakdown values.

***Corona and Breakdown in Nonuniform Fields between Conductors.*** In nonuniform fields, when the ratio of spacing to conductor radius of curvature is about 3 or less, breakdown occurs without prior corona. The breakdown voltage is controlled by the integral of the Townsend ionization coefficient $\alpha$ across the gap. At larger ratios of spacing to radius of curvature, corona discharge occurs at voltage levels below complete gap breakdown.

Corona in air at atmospheric pressure occurs before breakdown when the ratio of outer to inner radius of coaxial electrodes exceeds 2.72 or where the ratio of gap to sphere radius between spheres exceeds 2.04. These discharges project some distance from the small-radii conductor but do not continue out into the weaker electric field region until a higher voltage level is reached.

Such partial breakdowns are often characterized by rapid pulses of current and radio noise. With some conductors at intermediate voltages between onset and complete breakdown, they blend into a pulseless glow discharge around the conductor. When corona occurs before breakdown, it creates an ion space charge around the conductor, which modifies the electric field, reducing the stress at sharp conductor points in the intermediate voltage range. At higher voltages, streamers break out of the space-charge region and cross the gap.

The surface voltage stress at which corona begins increases above that for uniform field breakdown stress, since the field to initiate breakdown must extend over a finite distance. An empirical relation developed by Peek is useful for expressing the maximum surface stress for corona onset in air for several geometries of radius $r$ cm:

$$\text{For concentric cylinders}: E = 31\delta\left(1+\frac{0.308}{\sqrt{\delta r}}\right) \quad \text{kV/cm} \tag{3-74}$$

$$\text{For parallel wires}: E = 29.8\delta\left(1+\frac{0.301}{\sqrt{\delta r}}\right) \quad \text{kV/cm} \tag{3-75}$$

$$\text{For spheres}: E = 27.2\delta\left(1+\frac{0.54}{\sqrt{\delta r}}\right) \quad \text{kV/cm} \tag{3-76}$$

where $\delta$ is the density of air relative to that at 25°C and 1 atm pressure.

***Corona Discharges on Insulator Surfaces.*** It has been shown by a number of investigators that the discharge-threshold voltage stress on or between insulator surfaces is the same as between metal electrodes. Thus, the threshold voltage for such discharges can be calculated from the series dielectric-capacitance relation for internal gaps of simple shapes, such as plane and coaxial gaps, insulated conductor surfaces, and hollow spherical cavities.

The corona-initiating voltage at a conductor edge on a solid barrier depends on the electric stress concentration and generally on the ratio of the barrier thickness to its dielectric constant, except with low surface resistance. Any absorbed water or conducting film raises the corona threshold voltage by reducing stress concentration at the conductor edge on the surface.

It is sometimes possible to overvolt such gaps considerably prior to the first discharge, and the offset voltage may be below the proper voltage due to surface-charge concentration. With ac voltages, pulse discharges occur regularly back and forth each half cycle, but with dc voltage, the first discharge deposits a surface charge on the insulator surface which must leak away before another discharge can occur. Thus, corona on or between insulator surfaces is very intermittent with steady dc voltages, but discharges occur when the voltage is raised or lowered.

***Flashover on Solid Surfaces in Gases.*** As has been mentioned in the previous subsection on partial discharges, the breakdown in gases is influenced by the presence of solid insulation between conductors. This insulation increases the electric stress in the gas. A particular case of this is the complete breakdown between conductors across or around solid insulator surfaces. This can occur when the conductors are on the same side of the insulation or on opposite sides. A significant reduction in flashover voltage can occur whenever a significant part of the electric field passes through the insulation. The reduction is influenced by the percentage of electric flux which passes through the solid insulation and the dielectric constant of the insulation.

***Bibliography***

Cobine, J. D. (1958). *Gaseous Conductors.* Dover Publications, New York.

Curran, S. C., and Craggs, J. D. (1949). *Counting Tubes.* Butterworth Scientific Publications, London.

Dakin, T. W. (1974). "Breakdown of Gases in Uniform Fields—Paschen Curves for Nitrogen, Air and SF6." *Electra*, vol. 92, p. 67.

Dakin, T. W., and Berg, D. (1962). "Theory of Gas Breakdown," in *Progress in Dielectrics* 4. Academic Press, New York.

Dakin, T. W., and Mullen, G. A. (1972). "Continuous Recording of Outdoor Insulator Surface Conductance." *IEEE Trans. Electrical Insulations*, vol. EI-7, pp. 169–175.

Hall, H. C., and Russek, R. M. (1954). "Discharge Inception and Extinction in Dielectric Voids." *IEEE Proc.* vol. 101-2, p. 47.

Loeb, L. B. (1956). "Electrical Breakdown of Gases," in *Encyclopedia of Physics,* vol. 22. Springer-Verlag, Berlin.

MacDougall, F. H. (1952). *Physical Chemistry.* Macmillan, New York.

Meek, J. M., and Craggs, J. D. (1953). *Electrical Breakdown a/Gases.* Oxford University Press, New York.

Narbut, P., Berg, D., Works, C. N., and Dakin, T. W. (1964). "Factors Controlling Electric Strength of Gaseous Insulation." *AIEE Trans.*, vol. 78, no. 3, pp. 59–74.

Peek, F. W. (1929). *Dielectric Phenomena High-Voltage Engineering.* McGraw-Hill, New York.

Smyth, C. (1949). *Dielectric Behavior and Structure.* McGraw-Hill, New York.

Trump, J. G., Cloud, R. W., Mann, J. G., & Hanson, E. P. (1950). Influence of Electrodes on DC Breakdown in Gases at High Pressure. *Electrical Engineering*, vol. 69(11), pp. 961–964.

Winkelnkemper, H., Krasucki, Z., Gerhold, J., and Dakin, T. W. (1977). "Breakdown of Gases in Uniform Fields, Paschen Curves for Hydrogen, Carbon Dioxide, and Helium." *Electra*, vol. 92, p. 67.

## 3.3.3 Insulating Oils and Liquids

***General Considerations.*** Typical insulating liquids are natural or synthetic organic compounds and frequently consist of mixtures of essentially isomeric compounds with some range of molecular weight. The mixture of very similar but not exactly the same molecules, with a range of molecular size and with chain and branched hydrocarbons, prevents crystallization and results in a low freezing point, together with a relatively high boiling point. Typical insulating liquids have permittivities (dielectric constants) of 2 to 7 and a wide range of conductivities depending on their purity. The dc conductivity in these liquids is usually due to dissolved impurities, which are ionized by dissociation. Higher ionized impurity and conductivity levels occur in liquids having higher permittivities and lower viscosities.

The function of insulating liquids is to provide electrical insulation and heat transfer. As insulation, the liquid is used to displace air in the system and provide a medium of high electric strength to fill pores, cracks, and gaps in insulation systems. It is usually necessary to fill and impregnate systems with liquid under vacuum so that all air bubbles are eliminated. If air is completely displaced in all high-electric-field regions, the corona threshold voltage and breakdown voltage for the system are greatly increased. The viscosity selected for a liquid insulation is often a compromise to provide the best balance between electrical insulation and heat transfer and other limitations such as flammability, solidification at low temperatures, and pressure development at high temperatures in sealed systems.

The most commonly used insulating liquids are natural hydrocarbon mineral oils refined to give low conductivity and selected viscosity and vapor-pressure levels for transformer, circuit-breaker, and cable applications. A number of synthetic fluids are also used for particular applications where the higher cost above that of mineral oil is warranted by the requirements of the application or by the improved performance in relation to the apparatus design.

***Mineral Insulating Oils.*** Mineral insulating oils are hydrocarbons (compounds of hydrogen and carbon) refined from crude petroleum deposits from the ground. They consist partly of aliphatic compounds with the general formula $C_nH_{2n} + 2$ and $C_nH_{2n}$, comprising a mixture of straight- and branched-chain and cyclic or partially cyclic compounds. Many oils also contain a sizable fraction of aromatic compounds related to benzene, naphthalene, and derivatives of these with aliphatic side chains. The ratio of aromatic to aliphatic components depends on the source of the oil and its refining treatment. The percentage of aromatics is of importance to the gas-absorption or evaluation characteristics under electrical discharges and to the oxidation characteristics.

The important physical properties of a mineral oil (as for other insulating liquids as well) are listed in Table 3-14 for three types of mineral oils. In addition to these properties, mineral oils which are exposed to air in their application have distinctive oxidation characteristics which vary with type of oil and additives and associated materials.

Many manufacturers now approve the use of any of several brands of mineral insulating oil in their apparatus provided that they meet their specifications which are similar to ASTM D1040 (values from which are tabulated in Table 3-14). Low values of dielectric strength may indicate water or dirt contamination. A high neutralization number will indicate acidity, developed very possibly from oxidation, particularly if the oil has been used already. Presence of sulfur is likely to lead to corrosion of metals in the oil.

The solubility of gases and water in mineral oil is of importance in regard to its function in apparatus. Solubility is proportional to the partial pressure of the gas above the oil

$$S = S_0(p/p_0) \tag{3-77}$$

where $S$ is the amount dissolved at pressure $p$ if the solubility is expressed as the amount $S_0$ dissolved at pressure $p_0$.

The solubility is frequently expressed in volume percent of the oil. Values for solubility of some common gases in transformer oil at atmospheric pressure (760 torr) and 25°C are air 10.8%, nitrogen 9.0%, oxygen 14.5%, carbon dioxide 99.0%, hydrogen 7%, and methane 30% by volume. The solubilities of all the gases, except $CO_2$, increase slightly with increasing temperature. Water is dissolved in new transformer oil to the extent of about 60 to 80 ppm at 100% relative humidity and 25°C.

**TABLE 3-14** Characteristic Properties of Insulating Liquids

| | Mineral oil | | |
|---|---|---|---|
| Type of liquid | Transformer | Cable and capacitor | Solid cable |
| Specific gravity | 0.88 | 0.885 | 0.93 |
| Viscosity, Saybolt sec at 37.8°C | 57–59 | 0.100 | 100 |
| Flash point, °C | 135 | 165 | 235 |
| Fire point, °C | 148 | 185 | 280 |
| Pour point, °C | −45 | −45 | −5 |
| Specific heat | 0.425 | 0.412 | ........... |
| Coefficient of expansion | 0.00070 | ........... | 0.00075 |
| Thermal conductivity, cal/(cm) (s) (°C) | 0.39 | ........... | ........... |
| Dielectric strength,* kV~ | 30 | ........... | ........... |
| Permittivity at 25°C | 2.2 | ........... | ........... |
| Resistivity, Ω · cm × $10^{12}$ | 1–10 | 50–100 | 1–10 |

*ASTM D877.

The amount dissolved is proportional to the relative humidity. Solubility of water increases with oxidation of the oil and the addition of polar impurities, with which the water becomes associated. Larger quantities of water can be suspended in the oil as fine droplets.

***Dielectric Properties of Mineral Oils.*** The permittivity of mineral insulating oils is low, since they are essentially nonpolar, containing only a few molecules with electric dipole moments. Some oils possess a minor fraction of polar constituents, which have not been identified. These contribute a dipolar character to the dielectric properties at low temperature and/or high frequency. A typical permittivity for American transformer oil at 60 Hz is 2.19 at 25°C, declining almost linearly to 2.11 at 100°C. At low temperatures and high frequencies, values of permittivity as high as 2.85 have been noted in oils with a relatively high level of polar constituents.

The dc conductivity levels of mineral oils range from about $10^{-15}\ \Omega^{-1} \cdot \text{cm}^{-1}$ for pure new oils up to $10^{-12}\ \Omega^{-1} \cdot \text{cm}^{-1}$ for contaminated used oils. This conductivity is due to dissociated impurity ions or ions developed by oil oxidation. It increases approximately exponentially with temperature about 1 decade in 80°C.

Alternating-current dissipation-factor values are nearly proportional to the dc conductivity $10^{-13}\ \Omega^{-1}$, corresponding to a tan $\delta$ of 0.008. If no electrode polarization or interfacial polarization effects at solid barrier surface are present, the dc conductivity $\sigma$ should be related to the ac conductivity (tan $\delta$) by $\sigma = 5/9\epsilon' f \tan\delta \times 10^{-12}$, where $\epsilon'$ is the dielectric permittivity (Table 3-12) and $f$ is the frequency.

Corona or partial breakdown can occur in mineral oil, as with any liquid or gas, when the electric stress is locally very high and complete breakdown is limited by a solid barrier or large oil gap (as with a needle point in a large gap). Such discharges produce hydrogen and methane gas, and sometimes carbon with larger discharges. Dissolved air is also sometimes released by the discharge. If the gas bubbles formed are not ejected away from the high field, they will reduce the subsequent discharge threshold voltage to as much as 80%. The resistance of insulating oils to partial discharges is measured by two ASTM gassing tests: D2298 (Merrill test) and D2300 (modified Pirelli test). These tests measure the amount of decomposition gas evolved under specified conditions of exposure to partial discharges. A minimum amount of gas is, of course, preferred, particularly in applications for cables or capacitors. In fact, conventional mineral oils are inadequate in this respect for application in modern 60-Hz power capacitor designs.

***Deterioration of Oil.*** Deterioration of oil in apparatus partially open or "breathing" is subject to air oxidation. This leads to acidity and sludge. There is no correlation between the amount of acid and the likelihood of sludging or the amount of sludge. Sludge clogs the ducts, reduces the heat transfer, and accelerates the rate of deterioration. ASTM tests for oxidation of oils are D1904, D1934, D1313, and D1314. Copper and lead and certain other metals accelerate the oxidation of mineral oils. Oils are considerably more stable in nitrogen atmospheres.

Inhibitors are now commonly added both to new and to used oils to delay the oxidation. Ditertiary butyl paracresol (DBPC) is the inhibitor most commonly used at present.

***Servicing, Filtering, and Treating.*** Oil in service is usually maintained by testing for acidity, dielectric strength, inhibitor content, interfacial tension, neutralization number, peroxide number, pour point, power factor, refractive index and specific optical dispersion, resistivity, saponification, sludge, corrosive sulfur, viscosity, and water content, as outlined in ASTM D117. These properties indicate various types of contamination or deterioration which might affect the operation of the insulating oil.

Depending on the voltage rating of the apparatus, the oil is maintained above 16 to 22 kV (ASTM test D877). The usual contaminants are water, sludge, acids, and in circuit-breaker oils, carbon. The centrifuge is best suited for removing large quantities of water, heavier solid particles, etc. The blotter filter press is used for the removal of minute quantities of water, fine carbon, etc. In another method, after removing the larger particles, the oil is heated and sprayed into a vacuum chamber, where the water and volatile acids are removed. Sludge and very fine solids are then taken out by a blotter filter press. All units are assembled together so that the process is completed in a single pass. Some work has been done in reclaiming oil by treating it to reduce acidity. One process is similar to the later stages in refining. Another treatment uses activated alumina, Fuller's earth, or silica gel.

It has been found that analysis of the dissolved gas in oil or above the oil in oil-insulated transformers and cables is a good diagnostic tool to detect electrical faults, particularly, or deterioration, generally. For example, continuing or intermittent partial discharges produce hydrogen and low-molecular-weight hydrocarbons such as methane, ethane, and ethylene which accumulate in the oil and can be measured accurately to assess the magnitude of the fault. Higher-current arc faults produce acetylene in addition to $H_2$ and other low-molecular-weight hydrocarbons. Thermal deterioration of cellulosic or paper insulation is indicated by elevated concentrations of CO and $CO_2$ in the oil.

***Synthetic Liquid Insulation.*** Synthetic chlorinated diphenyl and chlorinated benzene liquids (askarels) have been used widely from the mid-1930s up to the mid-1970s and are still in service in many power capacitors and transformers, where they were adopted for their nonflammability as well as good electrical characteristics. Since the mid-1970s, their use has been banned in most countries due to their alleged toxicity and resistance to biodegradation in the environment. Now, when apparatus containing these fluids, which are commonly referred to as *PCBs*, are taken out of service, environmental regulations in the United States require that the fluid not be released into the environment. Waste fluid should be incinerated at high temperature with HCl reactive absorbent scrubbers in the stack, since this acid gas is a product of the combustion.

New synthetic fluids have been developed and are now widely applied in power capacitors where the electrical stresses are very high. These fluids include aromatic (containing benzene rings) hydrocarbons, some of which have excellent resistance to partial discharges. They are not fire-resistant, however.

Very high boiling, low-vapor-pressure, high-flash-point (>300°C) hydrocarbon oils are being tried for power transformers with some fire resistance. Methods for assessing the risk of fire with such liquids, as well as with silicones, are still being debated.

Perchlorethylene (tetrachloroethylene), a nonpolar liquid, is now in use in sealed medium-power transformers, where nonflammability is required. With a boiling point at atmospheric pressure of 121°C, this fluid is completely nonflammable. It is also widely used in dry cleaning. Other important classes of synthetic insulating fluids are discussed in the following subsections.

***Fluorocarbon Liquids.*** A number of nonpolar nonflammable perfluorinated aliphatic compounds, in which the hydrogen has been completely replaced by fluorine, are available with different ranges of viscosity and boiling point from below room temperature to more than 200°C. These compounds have low permittivities (near 2.0) and very low conductivity. They are inert chemically and have low solubilities for most other materials. The chemical formula for these compounds is one of the following: $C_nF_{2n}$, $C_nF_{2n+2}$, or $C_nF_{2n}O$. The presence of the oxygen in the latter formula does not seem to reduce the stability. These compounds have been used for filling electronic apparatus and large transformers to give high heat-transfer rates together with high dielectric strength. The vapors of these liquids also have high dielectric strengths.

***Silicone Fluids.*** These fluids, chemically formed from Si—O chains with organic (usually methyl) side groups, have a high thermal stability, low temperature coefficient of viscosity, low dielectric losses, and high dielectric strength. They can be obtained with various levels of viscosity and correlated vapor pressures. Rated service temperatures extend from −65 to 200°C, some having short-time capability up to 300°C. Their permittivity is about 2.6 to 2.7, declining with increasing temperature. These fluids have a tendency to form heavier carbon tracks than other insulating liquids when breakdown occurs. They cannot be considered fireproof but will reduce the risk of fire due to their low vapor pressure.

***Ester Fluids.*** There are a few applications, mostly for capacitors, where organic ester compounds are used. These liquids have a somewhat higher permittivity, in the range of about 4 to 7, depending on the ratio of ester groups to hydrocarbon chain lengths. Their conductivities are generally somewhat higher than those of the other insulating liquids discussed here. The compounds are easily subject to hydrolysis with water to form acids and alcohols and should be kept dry, particularly if the temperature is raised. Their thermal stability is poor. Specifically, dibutyl sebacate has been used in high-frequency capacitors and castor oil in energy-storage capacitors.

***Bibliography***

Berberich, L. J. 1947. "Oxidation Inhibitors in Electrical Insulating Oils." *ASTM Bulletin 149*, pp. 65–73. ASTM, West Conshohocken, Pa.

Berberich, L. J., Blodgett, R. B., and Bartlett, S. C. 1961. "Influence of Gaseous Electric Discharge on Hydrocarbon Oils." *AIEE Trans.*, vol. 80, p. 528.

Clark, F. M. 1962. *Insulating Materials for Design and Engineering Practices.* Wiley, New York.

Dakin, T. W., Studniarz, S. A., and Hummert, G. T. 1972. "Annual Report, NRC-NAS Conference on Electrical Insulation and Breakdown." NRC, Washington D.C.

Gruse, W. A., and Stevens, D. R. 1960. *Chemical Technology of Petroleum.* McGraw-Hill, New York.

Kaufman, R. B., Shimanski, E. J., and MacFadyen, K. W. 1955. "Gas and Moisture Equilibrium in Transformer Oils." *AIEE Trans.*, vol. 74, no. 3, p. 312.

Mandelcorn, L., Dakin, T. W., Miller, R. L., and Mercier, G. 1979. "High-Voltage Power Capacitor Dielectrics: Recent Developments," in *Proceedings of IEEE Conference*, publication no. 79CH1510-7-EI. IEEE, New York.

Peek, F. W. 1929. *Dielectric Phenomena in High-Voltage Engineering.* McGraw-Hill, New York.

Rohlfs, A. F., and Turner, F. J. 1956. "Correlation between the Breakdown Strength of Large Oil Gaps and Oil Quality Gauges." *AIEE Trans.*, vol. 75, no. 3, pp. 45–51.

Roth, A. 1959. *Hochspannungstechnik.* Springer-Verlag, Vienna.

Weber, K. H., and Endicott, H. S. 1956. "Area Effect and Its External Basis for the Electric Breakdown of Transformer Oil." *AIEE Trans.*, vol. 75, no. 3, p. 371.

## 3.3.4 Insulated Conductors

Insulated conductors vary from those carrying only a few volts to those carrying thousands of volts. They range from low-voltage bell wire with conductor gage of 22 to 24 to power cables with conductors of 2000 kcmil or 1013 mm$^2$ in cross-sectional area. The conductors can be round, rectangular, braided, or stranded. They can be of aluminum or copper. The insulation can be thin as in magnet wire or thick as in underground or marine cables. The insulation system can vary with functional application. It can be extruded or taped. It can be thermoplastic or thermoset. It can be a polymer in combination with cotton or glass cloth. There can be several different layers with different functional roles. Some of the applications for insulated conductors are communications, control, bell, building, hookup, fixture, appliance, and motor lead. The insulation technology for magnet wire and for power cables has been studied extensively because of the severe stresses seen by these insulation systems.

***Flexible Cords.*** Flexible cords and cables cover appliance and lamp cords, extension cords for home or industrial use, elevator traveling cables, decorative-lighting wires and cords, mobile home wiring, and wiring for appliances that get hot (e.g., hot plates, irons, cooking appliances). The requirements for these cables vary a great deal with application. They must be engineered to be water-resistant, impact-resistant, temperature-tolerant, flex-tolerant, linearly strong, and flame-resistant and have good electrical insulation characteristics.

***Magnet Wire Insulation.*** The term *magnet wire* includes an extremely broad range of sizes of both round and rectangular conductors used in electrical apparatus. Common round-wire sizes for copper are AWG No. 42 (0.0025 in) to AWG No. 8 (0.1285 in). A significant volume of aluminum magnet wire is produced in the size range of AWG No. 4 to AWG No. 26. Ultrafine sizes of round wire, used in very small devices, range as low as AWG No. 60 for copper and AWG No. 52 for aluminum.

Approximately 20 different "enamels" are used commercially at present in insulating magnet wire. Magnet wire insulations are high in electrical, physical, and thermal performance and best in space factor. The most widely used polymers for film-insulated magnet wire are based on polyvinyl acetals, polyesters, polyamideimides, polyimides, polyamides, and polyurethanes. Many magnet wire constructions use different layers of these polymer types to achieve the best combination of properties. The most commonly used magnet wire is NEMA MW-35C, Class 200, which is constructed with a polyester basecoat and a polyamideimide topcoat. Polyurethanes are employed where ease of solderability without solvent or mechanical striping is required. The thermal class of polyurethane insulations has been increased up to Class 155 and even Class 180. Magnet wire products also are produced with fabric layers (fiberglass or Dacron-fiberglass) served over bare or conventional film-insulated magnet wire. Self-bonding magnet wire is produced with a thermoplastic cement as the outer layer, which can be heat-activated to bond the wires together.

***Power Cables.*** Insulated power cables are used extensively in underground residential distribution. There has been extensive replacement of PILC, or paper in lead cable, with extruded polymer-insulated cables. Although PILC is still dominant for underground transmission cables, extruded polymeric cables are also beginning to be used for these high-voltage applications. Typical cable sizes with the cross section of the conductor are shown in the Table 3-15.

**TABLE 3-15** Typical Cable Sizes

| Cable size | Conductor cross section, $mm^2$ |
|---|---|
| AWG 2 | 33.6 |
| AWG 1 | 42.4 |
| AWG 1/0 | 53.5 |
| AWG 2/0 | 67.4 |
| AWG 3/0 | 85.0 |
| AWG 4/0 | 107.2 |
| 500 kcmil | 253.5 |
| 750 kcmil | 379.5 |
| 1000 kcmil | 507.0 |
| 2000 kcmil | 1013.0 |

Typically, a cable rated at 15 kV will have insulation of wall thickness 175 mil (4.45 mm); one rated at 35 kV will have a wall thickness of 345 mil (8.76 mm); one rated at 69 kV will have insulation thickness of 650 mil (16.5 mm); and a 138-kV cable will have insulation of wall thickness 850 mil (21.6 mm). A cable construction includes the conductor shield, insulation, and insulation shield. In addition, most cables these days have a jacket to diminish moisture penetration into the insulation.

The conductor shield is a semiconductive material applied to the conductor to smooth out the stress. Since the conductors, especially the stranded conductors, have "bumps" that can enhance the field, the role of the semiconductor is to present an even voltage stress to the insulation. The insulation shield fulfills a similar role on the outer surface of the insulation. Grit, or especially metal particles, can be sites where breakdown begins. A clean interface and a semiconductive material prevent such sites from forming. The formulation of the conductor shield and the insulation shield is different. The formulation also depends on the insulating material used. A number of different materials have been used as the matrix material for semiconductive shields. These include low-density polyethylene (LDPE), ethylene–ethyl acrylate (EEA), ethylene–vinyl acetate (EVA), ethylene–propylene rubber (EPR), ethylene–propylene diene monomer (EPDM), butyl rubber, and various proprietary formulations. These materials, in themselves, are not conducting. They are made conducting by loading the polymer with carbon.

There are two insulations in use for power cables. One is cross-linked polyethylene (XLPE) and the other is ethylene–propylene rubber (EPR). These insulating materials will be described in greater detail in the following paragraphs.

Most of the cables being installed in the latter part of the 1990s are jacketed. The jacket provides protection against oil, grease, and chemicals. However, the primary role played by the jacket is to slow down the ingress of moisture, since moisture in the presence of an electric field causes the insulation to degrade by a process called *treeing*. One of the materials used extensively as a jacket material is linear low-density polyethylene (LLDPE). Jackets are approximately 50 mil (1.27 mm) thick. The discussion thus far has not described the chemistry of each of these insulating materials. The terms *thermoset* and *thermoplastic* are used without explanation. Material names such as PE, PTFE, PVC, and silicones are used without characterizing the chemistry or structure.

A *thermoplastic* resin is one with a melting point. With rising temperature, a thermoplastic resin first undergoes a glass-transition temperature ($T_g$) and then a crystalline melting point ($T_m$). Below the glass-transition temperature, a polymer is rigid and exhibits properties associated with the crystalline state. Above the glass-transition temperature, the material becomes plastic and viscous, and the material starts to slowly approach the structure of the liquid state. The glass-transition state can be detected by plotting the dielectric constant, refractive index, specific heat, coefficient of expansion, or electrical conductivity as a function of temperature.

There is one characteristic slope below the glass-transition state and another steeper slope above the glass-transition temperature. Approximate values for $T_g$ and $T_m$ for polyethylene are −128 and 115°C, and for polystyrene they are 80 and 240°C. It is difficult to give exact values for a given generic polymer. This is so because the exact value will depend a great deal on the variation in the character of a particular polymer, with all the variations being grouped together and called by a common name. For example, for polyethylene, the molecular weight (the degree of polymerization) of the resin, the degree of branching, and the size or length of the branches will affect both $T_g$ and $T_m$. With polyvinyl chloride, the steroregularity, copolymerization, and plasticization all will affect $T_g$ and $T_m$.

A *thermoset* resin does not exhibit a visible melting point. An epoxy or a phenolic resin has a three-dimensional network structure. The three-dimensional structure results in a rigid framework that cannot be made fluid without breaking a large number of bonds. The thermoplastic resins, on the other hand, are linear. They might be thought of as strands of spaghetti. The strands can slip by one another and can be fluid. The analogy to a bowl of spaghetti can be used in understanding how a thermoset material can be formed by cross-linking a thermoplastic polymer such as polyethylene. The cross-linking reaction forms bonds between the linear strands of polyethylene to form a three-dimensional structure. To visualize the cross-linking, an analogy that can be used is that of the bowl of spaghetti left in a refrigerator overnight. Once the strands of spaghetti stick together, the mass is no longer fluid. The mass can be taken out of the bowl, and it will retain the shape of the bowl. The only way to fluidize this spaghetti is to break most or all the bonds formed between the individual strands.

Some of the insulation materials used for insulated conductors are polyethylene (PE), ethylene–propylene rubber (EPR), polyvinyl chloride (PVC), fluorinated ethylene propylene (FEP), ethylene chlorotrifluoroethylene, polytetrafluoroethylene (PTFE), butyl rubber, neoprene, nitrile–butadiene rubber (NBR), latex, polyamide, and polyimide.

Polyethylene is made by polymerizing ethylene, a gas with a boiling point of −104°C. A reaction carried out at high temperature (up to 250°C) and high pressure (between 1000 and 3000 atm) produces low-density polyethylene. The reason for the low density is that the short and long branches on the long chains prevent the chains from packing efficiently into a crystalline mass. The use of Ziegler-Natta catalysts results in high-density polyethylene. The use of the catalyst results in less branching and thereby a polymer that can pack more efficiently into crystalline domains. Recently, shape-selective catalysts have become available that produce polyethylene polymers that can be made with designer properties. Even though polyethylene consists of chains of carbons, the properties can vary depending on molecular weight and molecular shape. Polyethylene sold for insulating purposes has only small amounts of additives. There is always some antioxidant. For cross-linked polyethylene, the residues of the cross-linking agent are present. Additives to inhibit treeing are added.

Ethylene–propylene rubber is a copolymer made from ethylene and propylene. The physical properties of the neat polymer are such that it is not useful unless compounded. The finished compounded product has as much as 40% to 50% filler content. Fillers consist of clays, calcium carbonate, barium sulfate, or various types of silica. In addition to the filler, EPR is compounded with plasticizer, antioxidants, flame retardants, process aids, ion scavengers, coupling agents, a curing coagent, and a curative.

Polyvinyl chloride is a polymer made from vinyl chloride, a gas boiling at −14°C. It is partially syndiotactic; that is, the stereochemistry of the carbons on which the chlorines are attached is more or less alternating. By being only partially syndiotactic, the crystallinity is low. However, the polymer is still fairly rigid, and for use where flexibility is desired, the polymer must be plasticized.

Dibutylphthalate is often used as a plasticizer. In addition to plasticizers, PVC contains heat and light stabilizers. Oxides, hydroxides, or fatty acid salts of lead, barium, tin, or cadmium are typical stabilizers.

Polytetrafluoroethylene (or Teflon) is a polymer made from tetrafluoroethylene, a nontoxic gas boiling at −76°C. It is a linear polymer consisting of chains made of $CF_2$ units. Its crystallinity is quite high, and its crystalline melting point is 327°C. It is resistant to almost all reagents, even up to the boiling point of the reagent. It is attacked only by molten alkali metals or the alkali metal dissolved in liquid ammonia. Polytetrafluoroethylene exhibits excellent electrical properties. It has a low dielectric constant and a low loss factor. These electrical properties do not change even when the polymer is kept at 250°C for long periods of time.

Fluorinated ethylene propylene is a copolymer made from tetrafluoroethylene and hexafluoropropylene. It compares in toughness, chemical inertness, and heat stability to polytetrafluoroethylene (PTFE).

Polychlorotrifluoroethylene has performance properties that are surpassed only by PTFE and FEP. The crystalline melting point is 218°C, as compared with 327°C for PTFE. It retains useful properties to 150°C, as opposed to 250°C for PTFE. The advantage for polychlorotrifluoroethylene is that its melt viscosity is so low enough that molding and extrusion become more feasible than for PTFE and FEP.

Polyamides or nylons are long-chain linear polymers made by molecules linked by amide linkages. Nylon 66 is made from hexamethylene diamine and adipic acid. Nylon 66 exhibits high strength, elasticity, toughness, and abrasion resistance. Nylon 6 is made from caprolactam, a cyclic amide. To form a polymer, the caprolactam opens and the amine group and carboxylic acid group form intermolecular amide links rather than the intramolecular amide link in the cyclic compound.

Polyimides are polymers connected by imide bonds. An amide is formed when the OH group of a carboxylic acid is replaced by the NH of an amine. An imide is a related structure formed when the noncarbonyl oxygen of an acid anhydride is replaced by a nitrogen of an amine. A polyimide is usually formed from an aromatic diamine and an aromatic dianhydride. The aromatic nature of the polyimide imparts thermal stability.

Rubbers used for electrical insulation can be either natural rubber or one of the synthetic rubbers. Natural rubber is obtained from the latex of different plants. The primary commercial source is the tree *Hevea brasiliensis.* Natural rubber is an isoprenoid compound wherein the isoprene (2-methyl-1,3-butadiene) is the unit of a high-molecular-weight polymer with a degree of polymerization of around 5000. Rubber without processing is too gummy to be of practical use. It is vulcanized (cross-linked) by reaction with sulfur. Natural rubber is flexible and elastic and exhibits good electrical characteristics.

Butyl rubbers are synthetic rubbers made by copolymerizing isobutylene (2-methyl-1-propene) with a small amount of isoprene. The purpose of isoprene is to introduce a double bond into the polymer chain so that it can be cross-linked. Butyl rubbers are mostly amorphous, with crystallization taking place on stretching. They are characterized by showing a low permeability to gases, thus making them the material of choice for inner tubes of automobile tires. They are reasonably resistant to oxidative aging. Butyl rubbers have good electrical properties.

*Polychloroprene* or *neoprene* is a generic term for polymers or copolymers of chloroprene (2-chloro-1,3-butadiene). Neoprene is an excellent rubber with good oil resistance. It has resistance to oxidative degradation, and is stable at high temperatures. Its properties are such that it would make excellent automobile tires, but the cost of the polymer makes it noncompetitive for this market. Its desirable properties are exploited for wire and cable insulations.

Nitrile rubbers are polymers of butadiene and acrylonitrile. Nitrile rubbers are used where oil resistance is needed. The degree of oil resistance varies with acrylonitrile content of the copolymer. With 18% acrylonitrile content, the oil resistance is only fair. With 40% acrylonitrile content, the oil resistance is excellent. The oil resistance is characterized by retention of low swelling, good tensile strength, and good abrasion resistance after being immersed in gasoline or oil. Nitrile rubbers can be used in contact with water or antifreeze. For use in wire insulation where oil resistance is needed, nitrile rubber is slightly better than neoprene.

**TABLE 3-16** Thermal Conductivity of Materials Commonly Used for Electrical Design

| Material | Thermal conductivity W/(in)(°C) | Thermal conductivity Btu/(h)(ft)(°F) |
|---|---|---|
| Silver | 10.6 | 241 |
| Copper | 9.6 | 220 |
| Eutectic bond | 7.50 | 171.23 |
| Gold | 7.5 | 171 |
| Aluminum | 5.5 | 125 |
| Beryllia 95% | 3.9 | 90.0 |
| Molybdenum | 3.7 | 84 |
| Cadmium | 2.3 | 53 |
| Nickel | 2.29 | 52.02 |
| Silicon | 2.13 | 48.55 |
| Palladium | 1.79 | 40.46 |
| Platinum | 1.75 | 39.88 |
| Chromium | 1.75 | 39.88 |
| Tin | 1.63 | 36.99 |
| Steel | 1.22 | 27.85 |
| Solder (60–40) | 0.91 | 20.78 |
| Lead | 0.83 | 18.9 |
| Alumina 95% | 0.66 | 15.0 |
| Kovar | 0.49 | 11.1 |
| Epoxy resin, BeO-filled | 0.088 | 2.00 |
| Silicone RTV, BeO-filled | 0.066 | 1.5 |
| Quartz | 0.05 | 1.41 |
| Silicon dioxide | 0.035 | 0.799 |
| Borosilicate glass | 0.026 | 0.59 |
| Glass frit | 0.024 | 0.569 |
| Conductive epoxy | 0.020 | 0.457 |
| Sylgard resin | 0.009 | 0.21 |
| Epoxy glass laminate | 0.007 | 0.17 |
| Doryl cement | 0.007 | 0.17 |
| Epoxy resin, unfilled | 0.004 | 0.10 |
| Silicone RTV, BeO-filled | 0.004 | 0.10 |
| Air | | 0.016 |

**TABLE 3-17** Thermal-Conductivity Conversion Factors

| From | To: $\frac{(cal)(cm)}{(s)(cm^2)(°C)}$ | $\frac{(W)(cm)}{(cm^2)(°C)}$ | $\frac{(W)(in)}{(in^2)(°C)}$ | $\frac{(Btu)(ft)}{(h)(ft^2)(°F)}$ |
|---|---|---|---|---|
| $\frac{(cal)(cm)}{(s)(cm^2)(°C)}$ | 1 | 4.18 | 10.62 | 241.9 |
| $\frac{(W)(cm)}{(cm^2)(°C)}$ | $2.39 \times 10^{-1}$ | 1 | 2.54 | 57.8 |
| $\frac{(W)(in)}{(in^2)(°C)}$ | $9.43 \times 10^{-2}$ | $3.93 \times 10^{-1}$ | 1 | 22.83 |
| $\frac{(Btu)(ft)}{(h)(ft^2)(°F)}$ | $4.13 \times 10^{-3}$ | $1.73 \times 10^{-2}$ | $4.38 \times 10^{-2}$ | 1 |

**Bibliography**

Billmeyer, F. W., Jr. 1962. *Textbook of Polymer Science*, 2nd ed. Wiley-Interscience, New York.

National Electrical Code. 1996. The National Fire Protection Agency, Quincy, Mass.

NEMA Standard MW 1000-1997. NEMA, Rosslyn, Va.

UL Standard for Safety, UL-62. Northbrook, Ill: Underwriters Laboratory Inc.

ASTM D2307. ASTM, Philadelphia, Pa.

## 3.3.5 Thermal Conductivity of Electrical Insulating Materials

One of the general characteristics of electrical insulating materials is that they are also good thermal insulating materials. This is true, in varying degrees, for the entire spectrum of insulating materials, including air, fluids, plastics, glasses, and ceramics. While the thermal insulating properties of electrical insulating materials are not especially important for electrical and electronic designs which are not heat sensitive, modern designs are increasingly heat sensitive. This is often because higher power levels are being dissipated from smaller part volumes, thus tending to raise the temperature of critical elements of the product design. This results in several adverse effects, including degradation of electrical performance and degradation of many insulating materials, especially insulating papers and plastics. The net result is reduced life and/or reduced reliability of the electrical or electronic part. To maximize life and reliability, much effort has been devoted to data and guidelines for gaining the highest possible thermal conductivity, consistent with optimization of product design limitations such as fabrication, cost, and environmental stresses. This subsection will present data and guidelines which will be useful to electrical and electronic designers in selection of electrical insulating materials for best meeting thermal design requirements. Also, methods of determining thermal conductivity $K$ will be described.

***Basic Thermal-Conductivity Data.*** The thermal-conductivity values for a range of materials commonly used in electrical design are shown in Table 3-16. These data show the ranking of the range of materials, both conductors and insulating materials, from high to low. The magnitude of the differences in conductor and plastic thermal-conductivity values can be seen. Note that one ceramic, 95% beryllia, has a higher thermal-conductivity value than some metals—thus making beryllia highly considered for high-heat-dissipating designs which allow its use. Thermal conductivity is variously reported in many different units, and convenient conversions are shown in Table 3-17.

Values of thermal conductivity do not change drastically up to 100°C or higher, and hence only a single value is usually given for plastics. For higher-temperature applications, such as with ceramics, the temperature effect should be considered. In addition to bulk insulating materials, insulating coatings are frequently used.

**Bibliography**

Harper, C. A. (1995). *Electronic Packaging and Interconnecting Handbook*, 2nd ed. McGraw-Hill, New York.

Harper, C. A. (1996). *Handbook of Hybrid Microelectronics*, 2nd ed. McGraw-Hill, New York.

Harper, C. A. (1996). *Handbook of Plastics and Elastomers*, 3rd ed. McGraw-Hill, New York.

Sergent, J. E., and Drum, A. (1997). *Thermal Management Handbook*. McGraw-Hill, New York.

# 4

# INTERCONNECTED POWER GRIDS

Sarma Nuthalapati

*Principal EMS Network Applications Engineer, PEAK Reliability, Vancouver, Washington*

Stephen Boroczky

*Principal Engineer, Grid Systems, Australian Energy Market Operator (AEMO), Sydney, NSW, Australia*

Steven Darnell

*Principal Engineer, Systems Performance and Commercial, Australian Energy Market Operator (AEMO), Brisbane, QLD, Australia*

Alan Honecker

*Senior Manager, NEM Real Time Operations, Australian Energy Market Operator (AEMO), Sydney, NSW, Australia*

Adam Peard

*Area Manager—System Analysis and Solutions, Network Planning, Western Power, Perth, WA, Australia*

Shantha Ranatunga

*Specialist, Systems Performance and Commercial, Australian Energy Market Operator (AEMO), Brisbane, QLD, Australia*

Héctor Volskis

*Operador Nacional do Sistema Elétrico (ONS), Brazil*

Xuanyuan Sharon Wang

*Jibei Electric Power Company, State Grid Corporation of China, Beijing, China*

## Mini Shaji Thomas

*Director, National Institute of Technology, Tiruchirappalli, India*

## Teruo Ohno

*TEPCO Research Institute, Tokyo Electric Power Holdings, Inc., Japan*

## Spencer Burks

*Lower Colorado River Authority, Austin, Texas*

## Kristian Koellner

*Lower Colorado River Authority, Austin, Texas*

## Komla A. Folly

*University of Cape Town, Cape Town, South Africa*

## Kehinde Awodele

*University of Cape Town, Cape Town, South Africa*

## Leandro Kapolo

*NamPower, Windhoek, Namibia*

## Nhlanhla Mbuli

*ESKOM and University of Johannesburg, Johannesburg, South Africa*

## Martin Kopa

*ESKOM, Johannesburg, South Africa*

## Oladiran Obadina

*Electric Reliability Council of Texas (ERCOT), Austin, Texas*

## 4.1 INTRODUCTION

*BY SARMA NUTHALAPATI*

This section provides an overview of interconnected power grids. Any typical power grid consists of various components as shown in Fig. 4-1. It consists of a generation system that has generators generating power, a transmission system that transmits power at high voltages to different load centers, and a distribution system that distributes power to different loads. Figure 4-2 shows a typical layout of a distribution system for a city. It shows how power from high voltage network (transmission system) is being converted into medium voltage level network through extra high voltage (EHV) substations and further distributed to different regions of the city through medium voltage network, which is finally distributed to various loads through distribution substations and distribution feeders shown in Fig. 4-3.

Different parts of the world have different practices of connecting the power grid and managing it. This section provides details about the interconnected grid across different parts of the world and gives readers a glimpse of various features of any interconnected power grid. Each subsection discusses the structure, planning, and operation aspects of the respective grid, as well as a future perspective.

## 4.2 INTERCONNECTED POWER GRIDS IN AUSTRALIA

*BY STEPHEN BOROCZKY, STEVEN DARNELL, ALAN HONECKER, ADAM PEARD, AND SHANTHA RANATUNGA*

### 4.2.1 Introduction

Australia's power systems have developed from a number of independent regional systems, which evolved as population and industry developed in dispersed coastal areas of Australia.

Subsequent interconnection has led to two main coastal interconnected systems: the national grid, where the National Electricity Market (NEM) operates in the east, and the South West Interconnected System (SWIS), where the Wholesale Electricity Market (WEM) operates in the west. Both of these systems operate at 50 Hz but at various high voltages that reflect their independent regional origins [1-6]. Figure 4-4 depicts the extent of the interconnected power systems in Australia.

A number of smaller transmission systems and numerous isolated power systems service the more remote regions of Australia.

### 4.2.2 System Statistics

From the demand data in Table 4-1, it can be surmised that regional maximum demand has not been increasing over the last 5 to 10 years. Coupled with the loss of some industrial load, this is largely due to the increasing penetration of "behind the meter" rooftop solar photovoltaic (PV) installations.

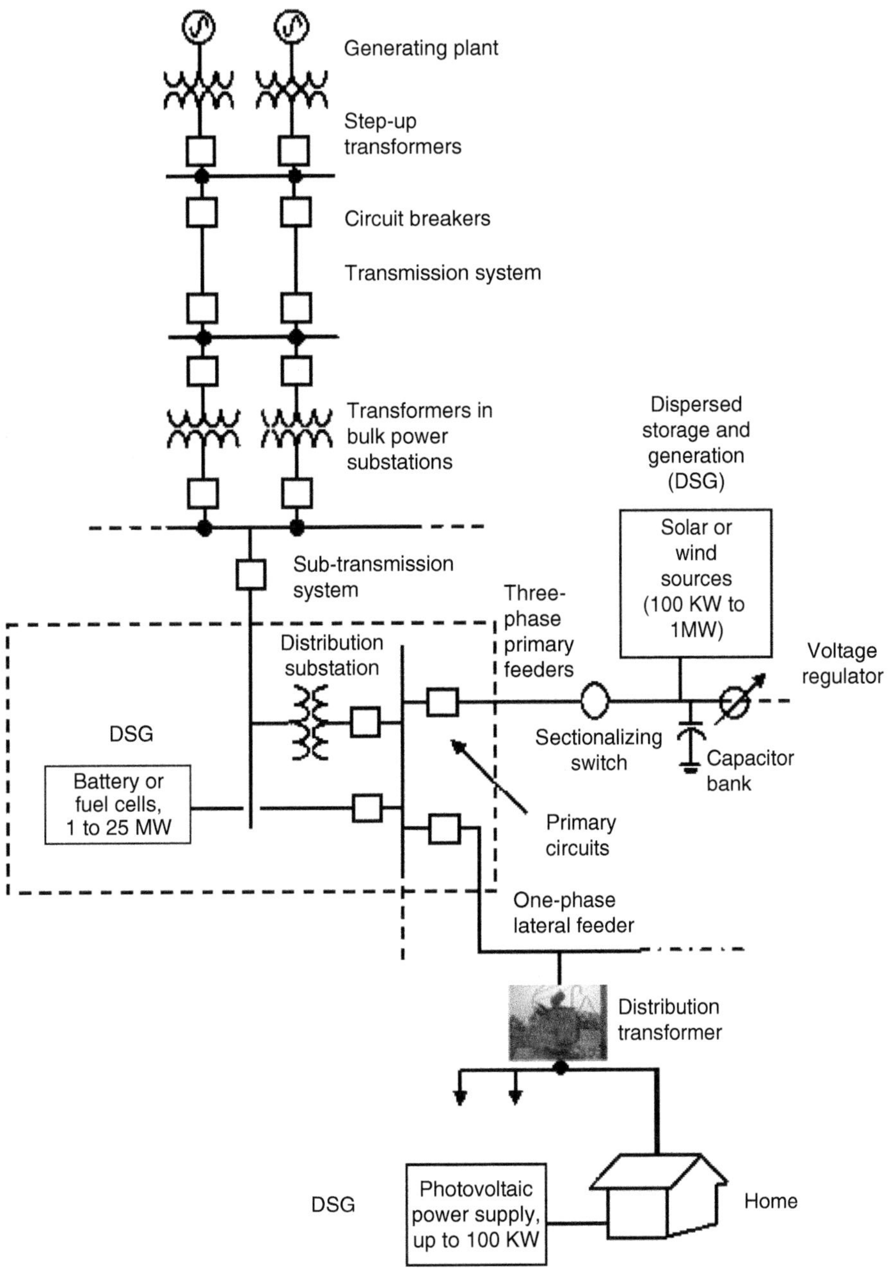


**FIGURE 4-1** Typical components of a power grid.

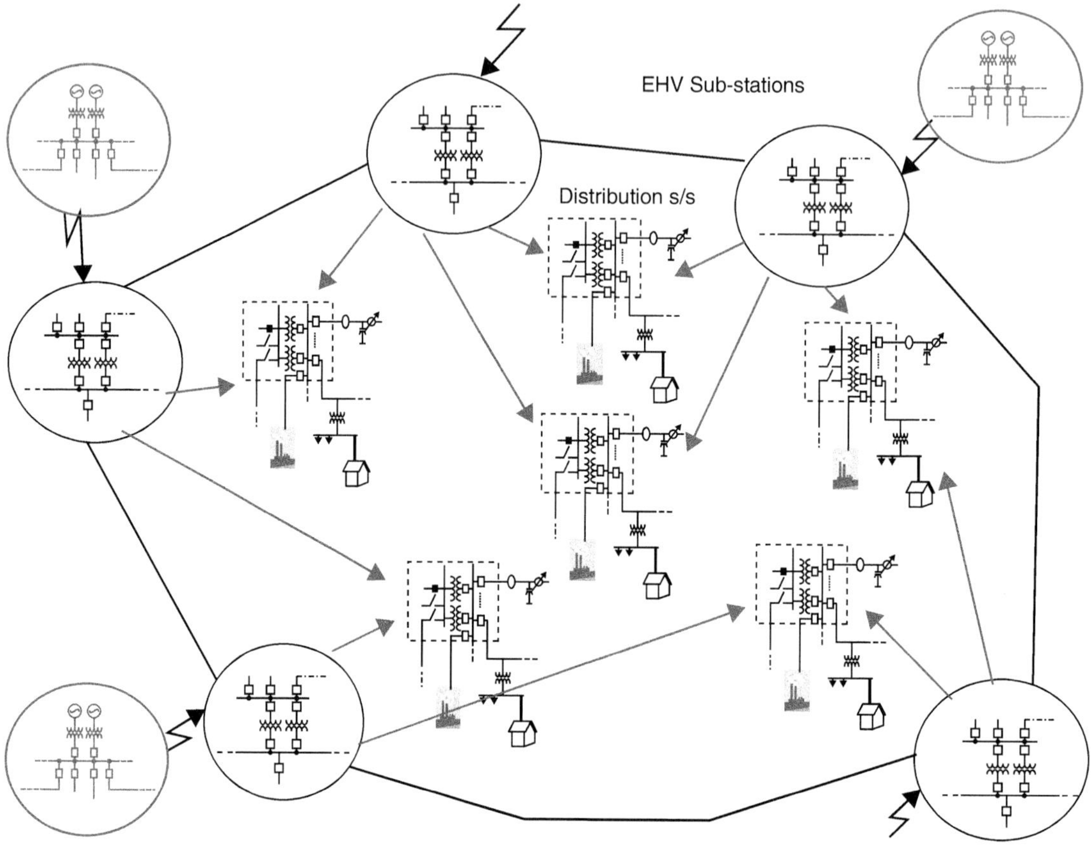


**FIGURE 4-2** Interconnected power grid in a typical city.

This has resulted in recent stagnant or even negative demand growth. Residential rooftop PV has grown to such an extent that the NEM, particularly in South Australia, has the highest percentage of households with PV installations in the world [17].

The energy mix shown in Fig. 4-5 shows the dominance of coal as an energy source in Australia, with currently roughly 75% of energy supplied by large coal fired power stations but nevertheless represents a significant reduction from 1990 levels. This scene is set to change further with Australia's commitment to the 2015 Paris 21st Conference of Parties emission abatement targets to reduce carbon emissions by 26% to 28% below 2005 levels by 2030 [8,9].

### 4.2.3 National Electricity Market (NEM)

The national grid, operated as the NEM, is the largest interconnected system in Australia and represents the power system that spans eastern and southern states of Australia. It consists of an interconnection of state and regional based power systems, interconnected by AC and DC interconnectors. The AC transmission system predominantly runs along the coast stretching from Cairns in Far North Queensland through NSW and Victoria to Port Lincoln in South Australia for more than 4000 km,

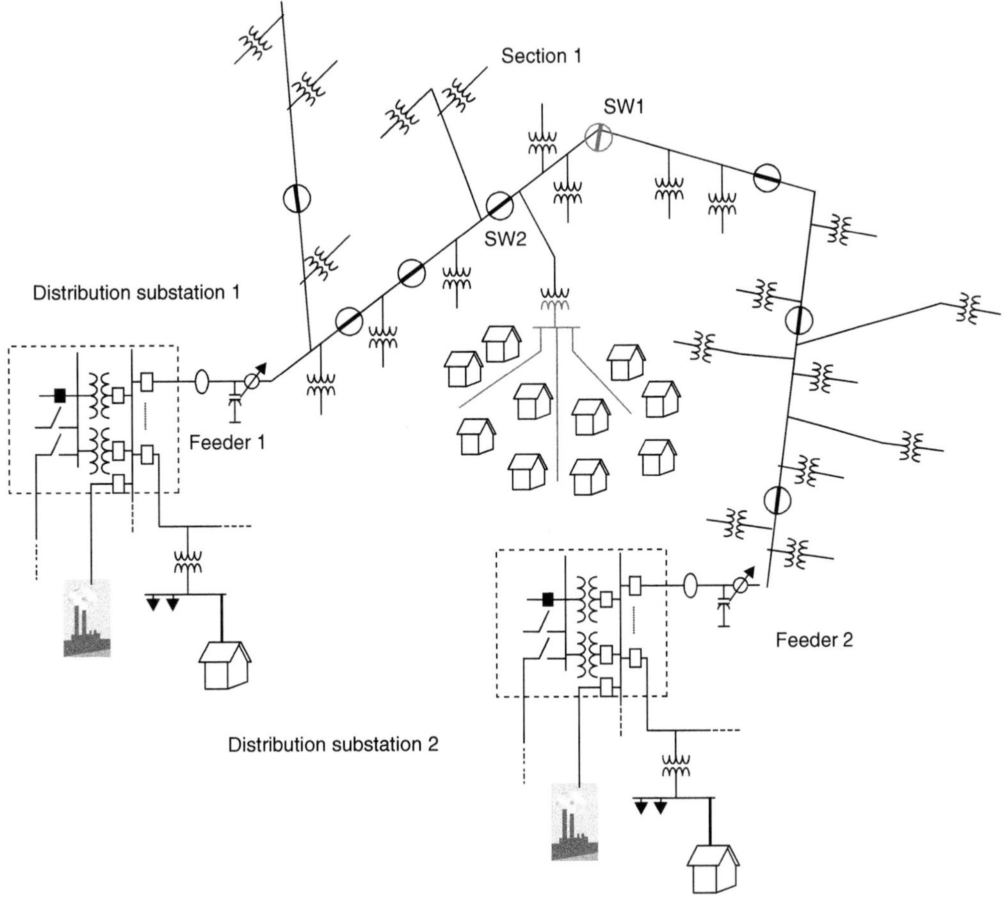


**FIGURE 4-3** Typical layout of a distribution feeder. SW1/SW2: Normally open/Normally closed sectionalizing switch on a distribution feeder.

along with a DC interconnection to the island of Tasmania. As such it is a loosely meshed, long, thin network that represents one of the longest interconnected power systems in the world. This presents its own unique operational challenges.

### *Governing Bodies and Participants*

*Australian Competition and Consumer Commission.* The Australian Competition and Consumer Commission (ACCC) role is to protect, strengthen, and supplement Australian markets by enforcement of the *Competitions and Consumer Act 2010.* With respect to the NEM, it works with the Australian Energy Regulator (AER) to support fair trading, to promote competition and economic efficiency in the Australian markets and to remedy market failure [1].

*Australian Energy Regulator.* AER regulates the energy markets and networks under national energy market legislation and rules. It monitors compliance and enforces the rules under which the energy markets operate.

**TABLE 4-1** Australian Regional Based Data—2016 [1,4,7,8]

| State or territory | Interconnected system | Maximum demand* (MW) (year obtained) | System operator/ market operator | Transmission Network Service Provider (TNSP) | Dominant transmission voltages (kV) | Distribution Network Service Provider (DNSP) |
|---|---|---|---|---|---|---|
| Queensland | NEM | 9154 (2015) | AEMO | PowerLink | 275, 132, 110 | Energex, Ergon Energy |
| NSW and ACT | NEM | 14744 (2011) | AEMO | TransGrid | 500, 330, 132 | AusGrid, Endeavour Energy, Essential Energy, ActewAGL |
| Victoria | NEM | 10576 (2009) | AEMO | Ausnet Services, AEMO† | 500, 220, 66 | CitiPower and PowerCor, Jemena, United Energy, Ausnet Services |
| Tasmania | NEM | 1790 (2008) | AEMO | TasNetworks | 220, 110 | TasNetworks |
| South Australia | NEM | 3399 (2011) | AEMO | ElectraNet SA | 275, 132 | SA Power Networks |
| Western Australia‡ | SWIS | 4286 (2016 | AEMO | Western Power | 330, 220, 132 | Western Power |
| Northern Territory | None | 360§ (2016) | Power and Water Corporation | Power and Water Corporation | 132, 66 | Power and Water Corporation |

*Demand is measure of the total electrical power requirement met by generating units.
†AEMO has a TNSP responsibility of planning of the Victorian transmission system.
‡Western Australia represents the South West Interconnected System (SWIS) only.
§Total demand for the combined regulated networks in Northern Territory.

*Australian Energy Market Commission.* The Australian Energy Market Commission makes and amends the National Electricity Rules that underpin the NEM. It conducts independent reviews and provides advice to governments on the development of electricity markets.

*Australian Energy Market Operator.* The Australian Energy Market Operator (AEMO) is responsible for the day-to-day operation of the NEM interconnected power system as well as a number of energy markets including the NEM wholesale electricity market. It is the independent system operator and the independent market operator of the NEM. Some of its operational responsibilities include:

- The secure operation of the NEM interconnected power system.
- Frequency control of the two DC connected NEM islands.
- Operation of the NEM and the various ancillary service markets.

To fulfil its responsibilities in the NEM, AEMO operates the power system from two geographically separated control centers on the East coast of Australia. They are operated in a "co-primary" fashion, where both control centers are operated as primary control centers. Responsibilities are either duplicated or shared between the two control centers. Some functions that cannot be shared, such as automatic generation control (AGC), are initiated from an active site but can be transferred at any time. The two control centers are resourced so that either can seamlessly assume responsibility for the entire NEM if needed.

Other AEMO functions include the operation of:

- Wholesale Electricity Market (WEM) in Western Australia, including all power system security functions

**FIGURE 4-4** Map of Australia depicting the national grid in the East and South, and the SWIS in the South West corner of Western Australia. http://www.aemo.com.au/aemo/apps/visualisations/map.html.

- Energy Retail markets in the NEM
- Various other Gas markets in Australia

In its role as the national transmission planner, AEMO provides the long term strategic view of the development of the national transmission grid in the NEM. This culminates in the regular publication of reports such as the National Transmission Network Development Plan [3], the Electricity Statement of Opportunities (ESOO) [2] and the National Electricity Forecasting Report [4].

*Transmission Network Service Providers.* Transmission Network Service Providers (TNSP) are generally state based transmission asset owners. They maintain, control and operate the assets in their portfolio. Transmission assets are usually operated under the direction of AEMO.

As the asset owner, the TNSP owns and maintains all the supervisory and control equipment, as well as communication equipment associated with the operation of their assets. While each TNSP is somewhat unique, they all maintain a primary control center geographically located within their asset area with a "hot standby" control center in a physically different location.

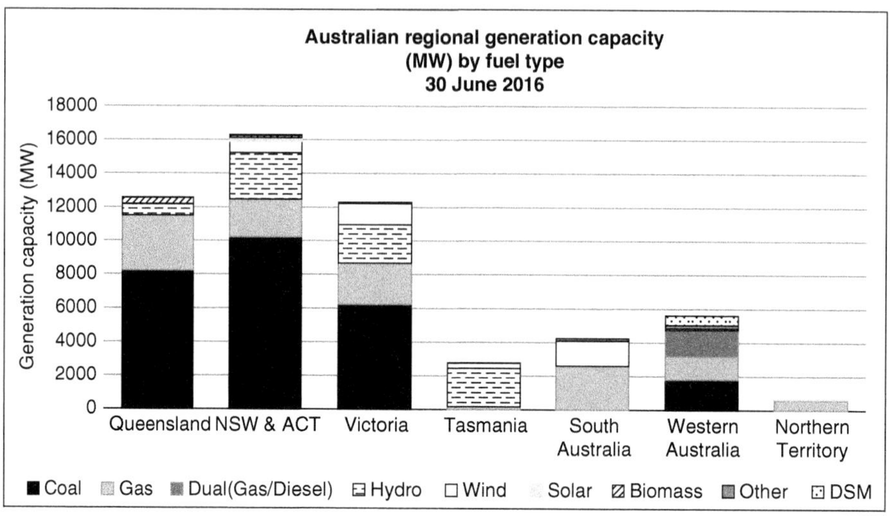


**FIGURE 4-5** Australian state and territory generation and Demand Side Management (DSM) capacity by fuel type [2,7,9,14].

In planning their network, the TNSP will identify and assess emerging network limitations. An annual planning review identifies these constraints, as well as any aging infrastructure, and proposes options to address them. The annual planning review process may involve joint planning with Distribution Network Service Providers (DNSP), neighboring TNSPs and with AEMO to determine an efficient plan which not only considers network options but also options such as generation or demand side management [10,11].

Before any transmission augmentation can proceed, feasibility studies must be undertaken using the AER's "Regulatory Investment Test for Transmission" framework to identify the most economically efficient option that addresses these limitations [12].

*Distribution Network Service Providers.* Distribution Network Service Providers (DNSP) are responsible for the reticulation of electricity from the bulk supply points to the consumer. They maintain and manage the secure operation of the distribution assets in their portfolio. The DNSP will generally have both a primary and a backup control center to manage their distribution networks.

*NEM Market Generators.* Market Generators are power station asset owners or operators that participate as market players in the NEM. The National Electricity Rules state that generators of capacity larger than 30 MW should be registered as either Scheduled Generators or Semi-Scheduled Generators to be dispatched by AEMO's central dispatch process.[a] They own or operate all the equipment associated with the generation of power, including any generator transformers that connect the generation asset to a high voltage bus in the transmission network. Scheduled generation will respond to AGC signals to follow their dispatch target and to provide frequency response for the interconnected system. They will also respond to requests from AEMO to manage the transmission voltage levels, either by adjusting their generator transformer tap position or by modifying generator excitation.

***Operating the National Electricity Market.*** The central dispatch process is integral to the operation of the NEM and is the normal mechanism through which AEMO has to adjust power flows

[a]National Electricity Rules [5], Rule 2.2.

in the NEM. All MW control actions are initiated by manipulation of constraints in this security-constrained dispatch process that is determined every 5 minutes. The interregional interconnectors remain the weak points and determine many of the constraints in this dispatch process. Table 4-2 summarizes the nature of these interconnector limitations.

*Frequency Control.* Dispatch instructions for scheduled generators are mostly via automatic generation control (AGC). Participants can follow dispatch instructions by AGC or manual control at their discretion. Renewable or intermittent dispatch is by semi-scheduled dispatch systems. Semi-scheduled dispatch means dispatch is binding only when constrained. Dispatch targets are derived from centralized wind and solar forecasting systems—the Australian Wind Energy Forecasting System (AWEFS) and a similar solar forecasting ASEFS.

Frequency control is via eight separate minor markets known as frequency control ancillary services (FCAS). These eight FCAS services are raise and lower for regulation, 6-second, 60-second, and 5-minute response. Regulation is dispatched constantly on a 4-second cycle by AGC and seeks to correct for the 5-minute dispatch error and maintain system frequency at 50 Hz. The other FCAS services respond locally to frequency disturbances and seek to contain, stabilize and recover to normal frequency tolerance bands in response to contingency.

As Tasmania is an isolated AC island, only connected to the mainland by a single HVDC cable, AEMO operates Tasmania as a separate AGC area from the mainland. AEMO dispatches the whole interconnected system, both market and operational in its entirety, so AGC is not complicated by the need to deploy tie line controls.

*Power System Security.* The nature of the NEM means that it can present some unique operational challenges. While, static constraints, such as thermal line or transformer limitations and equipment voltage limitations need to be addressed, all too often the limiting constraint is due to dynamic phenomena, such as transient stability, oscillatory stability, or voltage stability. As a result, accurate assessment of these dynamic limitations is critical to operate the interconnected system within its technical envelope.

**TABLE 4-2** Summary of NEM Interconnector Characteristics [15,16]

| Interconnector | Type | Connected regions | Nominal rating | Type of limiting constraints |
|---|---|---|---|---|
| Queensland—NSW Interconnector (QNI) | AC | Queensland—New South Wales | 1078 MW South, 500 MW North (nominal) | Thermal, voltage stability, oscillatory stability |
| Terranora interconnector via DirectLink | AC interconnector in series with Voltage Source Converter HVDC | Queensland—New South Wales | 224 MW | Thermal and voltage stability in nearby region |
| NSW—Victoria | AC | New South Wales—Victoria | 1700 MW | Thermal, transient stability, voltage stability |
| Heywood interconnector | AC | Victoria—South Australia | 650 MW in either direction (design—currently testing) | Thermal, transient, voltage and oscillatory stability |
| MurrayLink | Voltage Source Converter HVDC | Victoria—South Australia | 220 MW | Can influence voltage stability on Victorian end, Local AC network, Thermal |
| Basslink | Commutating HVDC with Submarine Cable | Victoria—Tasmania | 478 MW South,* 594 MW (short term) North | Thermal limitations in Tasmanian which are mitigated by advanced Special Protection schemes |

*Basslink MW flows as measured at the inverter end.

The NEM market solution deploys constraints to respect static limitations and to estimate the impact of these dynamic limitations. Real-time dynamic assessment tools that simulate and estimate transient and voltage security limits are then used to give confidence that the system is being operated securely.

By its nature, the system is susceptible to oscillatory instability and depends on active damping controls of power system stabilizers and SVC power oscillation dampers to maintain stable and secure operation. Real-time measurement of system damping completes the picture with the assessment of oscillatory stability issues. If issues are detected, additional constraints can be added to the market solution to achieve secure operation. By monitoring the damping of the interconnected power system, AEMO will constrain interconnector flow should poor damping be detected or the damping monitor be unavailable.

AEMO is also in the process of automating the voltage control of the main transmission system to maintain system voltages within their respective normal and short-term post-contingent limits. An optimization engine determines a reactive dispatch based on violations and forecast voltage profiles and sends control requests electronically, via SCADA or other means, to the reactive plant operators that include both TNSP's and market generators.

***NEM Market Structure.*** The NEM is an energy only zonal pricing market. The five pricing zones are generally the coastal regions of five states: Queensland, New South Wales, Victoria, South Australia, and Tasmania.

Spot prices in the NEM are set at nodes in the five pricing zones at 5-minute intervals. Associated generation is likewise dispatched at 5-minute intervals by a security constrained dispatch algorithm. These five regional reference prices (RRP) are defined as the marginal cost of energy at the reference node in each zone. Static loss factors modify the RRP at connection points.

The market is formed by generator offers on the supply side, and demand forecasts on the demand side. Settlement is in 30-minute intervals and is derived from the average of the associated 5 minutes prices. Spot prices are capped at AUD$14,000 and AUD$1000 and in general range between average around AUD$20-AUD$80. Financial markets trade spot market derivatives (exchange and bilateral contracts) that enable participants to manage spot market risk.

Forecast market outcomes cover a 2-day rolling window (pre-dispatch). Pre-dispatch is nonbinding and for information only. It forecasts at 5-minute resolution for 1 hour and at 30-minute resolution for the remainder of the 2-day period.

Reserve management is over a 2-year period and forecasts capacity reserve (Projected Assessment of System Adequacy—PASA) and like pre-dispatch is for information only. PASA forecasts at a half-hourly resolution over 6 days and at a daily resolution over the remainder of the 2-year forecast. The ESOO assesses supply adequacy by running hourly Monte Carlo simulations for 10 years to help stakeholders assess opportunities in the NEM.

***Operational Challenges.*** One of the emerging operational challenges facing Australia is a result of the changing nature of the generation mix. This is most acute in South Australia where high inertia synchronous generation is being displaced largely by low-inertia non-synchronous generation in the form of renewable energy sources (RESs) and distributed rooftop PV.

While no issues have been identified under system normal conditions, there may nevertheless be extreme operating conditions where issues could become apparent unless preventive measures are taken. Conditions of high renewables, high distributed PV generation, and low inertia, coupled with the risk of islanding can lead to:

- High Rate of Change of Frequency (RoCoF) in South Australia and subsequent impact on equipment.
- Insufficient frequency control ancillary service to control frequency.
- Reduced effectiveness of Under Frequency Load Shedding schemes to abate extreme power system events.
- Reduced fault levels, reducing the effectiveness of protection equipment to detect and clear faults.

The industry is examining these issues and looking at ways to address them [6].

The NEM has been in operation for about 20 years and one of its defining features is the dominance of large coal-fired generation—currently 75% of energy supplied, was as high as 90%. In recent times new investment has been mainly in intermittent generation—wind and solar—guided by government incentives. In the near future one of the main challenges for the governing institutions of the NEM will be to manage a transition from large coal-fired generation to smaller distributed intermittent generation.

### 4.2.4 South West Interconnected System (SWIS)

The SWIS represents the largest interconnected power system in Western Australia. While its geographical footprint of around 260,000 km$^2$ is only a small part of Western Australia, it services the majority of the population in the state.

The market operating in the SWIS is referred to as the Wholeseale Electricity Market (WEM). It has both scheduled and non-scheduled generation as well as a number of demand-side management programs that participate in this market.

Some of the defining characteristics of the SWIS include predominant coal and gas generation in the south, gas and wind generation in the north, and wind and gas in the east, along with large industrial loads that appear at the extents of the system. Its maximum demand is forecast to grow at only 1.4% per annum. [14].

***Governing Bodies***

*Minister for Energy (WA).* The Minister for Energy established the initial WEM rules and approves proposed changes.

*Public Utilities Office.* The Public Utilities Office provides advice to the Minister for Energy and administers emergency plans.

*Economic Regulation Authority.* The Economic Regulation Authority monitors compliance to the WEM rules, conducts market surveillance to ensure no abuse of market power and provides financial oversight.

The Economic Regulation Authority regulates Western Power's network, which is the only regulated Network in Western Australia. This includes acting as the authority over the Technical Rules[b] which detail the technical requirements to be met by Western Power and by users who connect facilities to the transmission and distribution systems which make up the Western Power Network.

*Market Advisory Committee.* The Market Advisory Committee is designed to provide advice to AEMO on various aspects of market design and operation and includes members representing all types of market participants (generators, customers, network operators, system operator as well as AEMO).

*Australian Energy Market Operator.* AEMO in its role in Western Australia has taken over the day-to-day operation of the WEM and the System Management functions of operating the power system, dispatch, system security, and system reliability.

*Western Power.* Western Power is the Transmission Network Operator and the Distribution Network Operator for the SWIS. They own and operate the bulk of the transmission and distribution assets within the SWIS.

***Operating the SWIS.*** The connection arrangements and dispatch of generation in the WEM is fundamentally based on an unconstrained dispatch philosophy.[c] The WEM operates on a 30-minute trading interval using less sophisticated central dispatch processes than the NEM. To manage power system security issues that arise from time to time during outage conditions, such as thermal or stability limitations, operators typically need to adjust generation dispatch manually, rather than relying on a central dispatch engine that automates the economic dispatch of generation, subject to constraints on the system.

[b] https://www.erawa.com.au/electricity/electricity-access/western-power-network/technical-rules/technical-rules

[c] When all transmission network elements are in service.

Five ancillary services are provided in the WEM:

- Load Following (LFAS), or regulation, is provided by automatic generation control (AGC) that constantly adjusts generation to maintain system frequency at 50 Hz.
- Spinning Reserve (SRAS) provides raise contingency services to stabilize and recover normal frequency in response to generation loss contingencies.
- Load Rejection Reserve (LRRAS) provides lower contingency services to stabilize and recover normal frequency in response to load loss contingencies.
- System restart services (SRS) provide black start capability, and
- Dispatch support services (DSS) cover out-of-merit costs to manage constraints in the system.

While dispatch is essentially an unconstrained process, there are a number of emerging constraints that need to be managed.

The Eastern Goldfields, in the east of the SWIS, can present considerable operational challenges. It is connected to the rest of the SWIS by a single 650 km 220 kV transmission line and with the relatively low inertia of the local generation and minimal reactive reserve; it can present voltage, transient, and oscillatory stability issues. A number of special control schemes are in place to control these stability issues in order to maximize the transfer capability. A number of damping monitors have also been installed to identify oscillatory stability issues in real-time and to improve power system security in the area.

There are also a number of other emerging voltage stability, thermal, and capacity constraints that have been identified in the SWIS. During system normal conditions, with all transmission elements in service, these limitations should not require any changes in the merit order dispatch plan for generation. During outage conditions the limitations can be more onerous and operators sometimes adjust dispatch to ensure system security and reliability requirements are maintained.

***Wholesale Electricity Market Structure.*** Western Australia's Wholesale Electricity Market (WEM) is combination of an energy market and a capacity market.

A Reserve Capacity Mechanism is designed to ensure that there is sufficient installed capacity (including both generation and demand side management options) to meet the expected peak demands for the year including any minimum reserve margins. Capacity payments are paid to capacity providers and are funded by customers.

Long-term bilateral contracts between market participants can be for energy or capacity and are off-market settlement. As the market operator, AEMO's only interest in these contracts is that the bilateral energy transactions need to be scheduled the day ahead.

The Short Term Energy Market (STEM) is an energy-only forward market that facilitates trading around the bilateral contract positions. Its primary purpose is to facilitate economic energy trade between market participants.

The combination of bilateral contracts and the STEM results in the day-ahead "Net Contract Positions." A Balancing Market then determines a common balancing price to account for the difference between these net contract positions and the actual real-time outcomes that meet system demand on the day.

Frequency control services are provided by a combination of administered procurement and pricing mechanisms for contingency services as well as a market for frequency regulation service.

***Wholesale Electricity Market (WEM) Reform.*** The WEM is currently undergoing market reform with the objectives of reducing electricity costs without compromising safe and reliable supply, reducing government risks, and to attract private investment in the energy market.

Some of the areas being reviewed include:

- Replacement of the unconstrained dispatch model with a constrained dispatch model, allowing system limitations to be represented as constraints in the dispatch.

- Reserve capacity auctions.
- Ex-ante pricing.

The challenge for the future of the WEM will be in implementing these reforms and in delivering the reform objectives.

### 4.2.5 References

[1] State of the Energy Market 2015, *Australian Energy Regulator,* https://www.aer.gov.au/publications/state-of-the-energy-market-reports/state-of-the-energy-market-2015.

[2] Electricity Statement of Opportunities for The National Electricity Market, *Australian Energy Market Operator*, August 2016, https://www.aemo.com.au/Electricity/National-Electricity-Market-NEM/Planning-and-forecasting/NEM-Electricity-Statement-of-Opportunities.

[3] National Transmission Network Development Plan for the National Electricity Market, *Australian Energy Market Operator,* November 2015, http://www.aemo.com.au/Electricity/National-Electricity-Market-NEM/Planning-and-forecasting/National-Transmission-Network-Development-Plan.

[4] National Electricity Forecasting Report for the National Electricity Market, *Australian Energy Market Operator,* June 2016, https://www.aemo.com.au/Electricity/National-Electricity-Market-NEM/Planning-and-forecasting/National-Electricity-Forecasting-Report.

[5] National Electricity Rules, *Australian Energy Market Commission,* http://www.aemc.gov.au/Energy-Rules/National-electricity-rules/Current-Rules.

[6] Future Power System Security Program, *Australian Energy Market Operator,* August 2016, https://www.aemo.com.au/Electricity/National-Electricity-Market-NEM/Security-and-reliability/FPSSP-Reports-and-Analysis.

[7] Annual Planning Report 2015/16, *Western Power,* https://www.westernpower.com.au/media/1619/annual-planning-report-2015-16.pdf.

[8] Network Management Plan 2013/14 to 2018/19, *Power and Water Corporation,* January 2015 and January 2016, https://www.powerwater.com.au/__data/assets/pdf_file/0020/64226/Network_Management_Plan_2013-14_to_2018-19.pdf.

[9] 2014/15 Annual Report Powering the NT, *Territory Generation,* http://territorygeneration.com.au/news_and_publications/news/2015/territory_generation_annual_report_2014–15.

[10] Victorian Electricity Planning Approach, *Australian Energy Market Operator,* June 2016, https://www.aemo.com.au/Electricity/National-Electricity-Market-NEM/Planning-and-forecasting/Victorian-transmission-network-service-provider-role.

[11] New South Wales Annual Planning Report, *TransGrid,* 2016, https://www.transgrid.com.au/news-views/news/2015/Pages/2015-Transmission-Annual-Planning-Report-released.aspx.

[12] Regulatory Investment Test for Transmission Application Guidelines, *Australian Energy Regulator,* June 2010, https://www.aer.gov.au/networks-pipelines/guidelines-schemes-models-reviews/regulatory-investment-test-for-transmission-rit-t-and-application-guidelines-2010.

[13] Wholesale Electricity Market Design Summary, *Independent Market Operator (now AEMO),* 2012, https://www.aemo.com.au/Electricity/-/media/F12B82DEB2484DD0848FD5C277DB5CA8.ashx.

[14] Deferred 2015 Electricity Statement of Opportunities for the Wholesale Electricity Market, *Australian Energy Market Operator*, June 2016, https://www.aemo.com.au/-/media/Files/Electricity/WEM/Planning_and_Forecasting/ESOO/2015/Deferred-2015-Electricity-Statement-of-Opportunities-for-the-WEM.ashx.

[15] Interconnector Capabilities for the National Electricity Market, *Australian Energy Market Operator,* September 2015, http://www.aemo.com.au/-/media/Files/PDF/Interconnector-Capabilities-v2.pdf

[16] Heywood Interconnector: Overview of Upgrade and Current Status, *Australian Energy Market Operator*, August 2015, https://www.aemo.com.au/media/Files/Other/planning/The20Heywood20Interconnector20UpgradeUpdate202015.pdf

## 4.3 INTERCONNECTED POWER GRID IN BRAZIL

*BY HÉCTOR VOLSKIS*

### 4.3.1 Introduction

Brazil spans a large part of the South American continent. The distance of the far ends of the Brazilian territory (from North to South, and from East to West) is about 3900 km.

Electricity started in Brazil in the end of 19th century. After the Second World War started the idea to create an interconnected power system considering the Brazilian characteristics: many hydro generations with reservoirs (sources) faraway from great consuming centers (loads) and a rainfall that allows coordinating the use of the reservoirs to maximize managing energy/water.

Brazilian Interconnect Power System-BIPS covers two-thirds of the Brazilian Territory (5 million km$^2$) and hydro generation is dominant (70% of the installed capacity).

### 4.3.2 Structure

Today BIPS attends near 97% of the country's electricity consumption with a large transmission network that includes over 120,000 km of 230, 345, 440, 500, 525, and 765 kV AC transmission lines, two 600 kV HVDC transmission lines and more than 550 substations (Fig. 4-6).

For operational purposes, BIPS is divided into four interconnected regions—South, Southeast/Midwest, North and Northeast. The BIPS is characterized with a dominant hydroelectric power generation (amount to more than 70% of the total installed capacity of 140,000 MW and more than 90% of the total energy production), and long distance power transfers from generation parks to load centers. The hydro generation parks are formed by plants in cascade formation located along 12 major hydrographic basins all over the country and many of them are not close to the major load centers in the South and Southeast region.

Rainfall and the resulting inflow patterns are distinct among regions, and vary significantly over the year for each region, as well between dry and wet years.

Solar and wind power generation start to grow: Wind has 9390 MW installed capacity and Solar 27 MW (2016—Banco de Informações de Geração—ANEEL—Brazilian Electricity Regulatory Agency).

International asynchronous interconnections with Argentina (Garabi—2000 MW) and Uruguay (Melo—500 MW) are established through HVDC frequency converters.

### 4.3.3 Planning

Since 1996, Brazil has been struggling to redesign its energy sector, giving opportunity to private companies to invest and be responsible for the energy supply in the country. Four important organs compose the division of energy policy in Brazil. The CNPE (National Council for Energy Policy) is responsible for advising the government about the right policies and the right decisions about promoting the conscious use of energy resources in the country. The MME (Ministry of Mines and Energy) implements the political decisions taken by the CNPE. It is also responsible for defining preventive actions of security of the energy distribution systems in case of imbalances between supply and demand. The National Agency of Electric Energy (ANEEL) regulates and supervises the energy distribution systems. Finally, the Power Research Company (EPE) manages the research in energy sector in areas such as oil, natural gas, coal, renewable energy resources, and energy efficiency.

The interconnected system of production and transmission of electric power in Brazil is a large hydrothermal system, with a strong predominance of hydro plants and multiple owners. Only 1.7% of the country's electricity production capacity is out of the BIPS, in small isolated systems located mainly in the Amazonian region (source: ONS, 2017 PEN). In December 2016, the installed capacity in Brazil reached a total of 142,042 MW, of which 101,598 MW in hydro power plants (including small plants) and 29,950 MW in thermal power plants and 9,611 MW in wind plants (ONS, 2017 PEN).

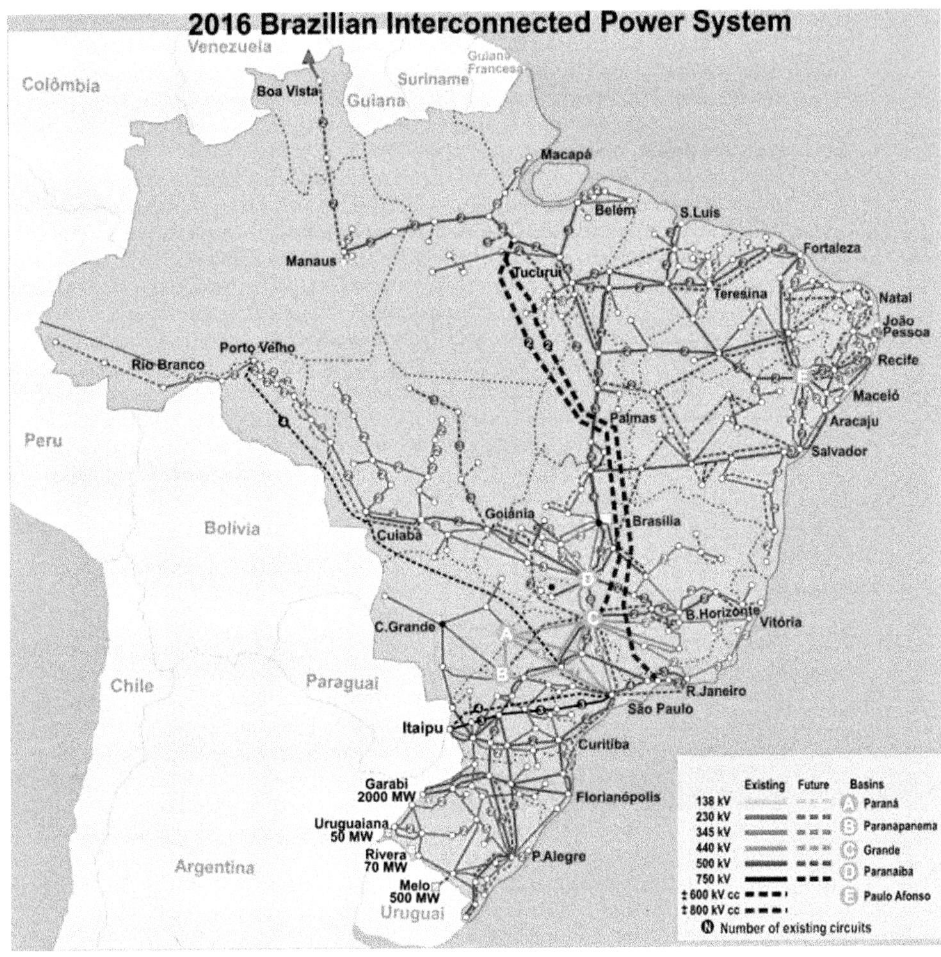


**FIGURE 4-6** Brazilian Interconnected Power System (BIPS).
*Source*: ONS Annual Report.

Due to the magnitude of the system with large reservoirs spread over large geographic areas, any decision implies different spatial and temporal consequences, making the problem quite complex. Then, there is a relationship between the decision-making at any stage and its future consequences. If in the present, the option is to use lots of water for power generation, system reservoirs levels will be lower, so if a period of low inflows occurs, the deficit risk regarding demanded electric power supply will increase, which will drive to the necessity to operate the thermal power plants, increasing operation costs. Likewise, if in the present it is chosen to generate thermal energy in order to store hydraulic energy and if in the future a period of high flows occurs, system power spillage will be necessary, which leads to a more expensive and unnecessary operation. Therefore, it is necessary that the BIPS operation is preceded by a planning, in addition, the coordination of the operation of the reservoir system of the power sector, in conjunction with the operation of thermal power plants complementation allows the best use of the natural flows, avoiding the waste of water and excessive fuel costs. This coordination is done within the so-called Operation Planning of the Interconnected Power System, currently performed by ONS, Brazilian Independent System Operator.

Today this planning is made in three steps, and in each of them the mathematical models used have different planning horizons, discretization of time and degree of detailing in the representation of the generating units and operational constraints. These models are linked through coupling, at the end of their horizons, of allocation policies of hydro and thermal resources produced by the model of the previous step, forming a chain of steps that comes from medium-term to the short-term including more details at each step.

At the top of the chain is the medium-term planning, where the stochastic optimization model, gets the allocation policy of hydro and thermal resources of minimum cost for each month considering a time horizon of 5 to 10 years. The hydroelectric plant is represented in an aggregated way in four equivalent reservoirs of energy, representing the subsystems of the south, southeast, northeast, and north. Next, in the short-term planning, also of stochastic optimization, determines a scheduling for each system's power plant for the weeks of the following month and for the next month. At the base of the chain, is the daily programming, calculate the generation dispatch for each half hour of the following day. In this planning, the main objective is to minimize the expected value of the operation cost (thermal generation spending and penalties for not meeting demand) over the planning horizon, taking into account physical constraints and system reliability. However, in the planning, one must consider a lot of activities related to the multiple use of water in reservoirs in conjunction with the generation dispatch and multi period optimization of reservoirs. It is highlighted water withdrawals for other uses and flood control.

### 4.3.4 Operation and Control

The Operador Nacional do Sistema Elétrico (ONS) is a nonprofitable company responsible to operate the BIPS. In Brazil there is only one Independent System Operator. ONS was created in 1998 with a mandate for performing a centralized cost-based scheduling and dispatch of BIPS after the deregulation of Brazil electricity market took effect in late 90s. It was created to substitute the previous cooperative structure and collegiate entities for operation coordination, which had shared utility responsibilities. The new model institutes ONS as the Brazilian Independent System Operator (ISO) in charge of the over 90,000 MW National Interconnected Power System. Today, there are 110 transmission companies, 170 generators and 95 distribution companies and high voltage consumers participating in the Brazilian electricity market.

A hierarchical structure of control centers is used by ONS to operate in a global and integrated way the "Operation Network." This network is the union of "Basic Network" (230 kV and upper voltage levels), the "Complementary Network" (facilities which impact the Basic Network) and the integrated power plants. ONS existing control center structure (see Fig. 4-7) is composed by:

CNOS—Nation System Operation Center. The higher level one hierarchical is responsible for coordination, supervision, and control of the basic and complementary network.

COSR—Regional System Operation Centers: Four centers owned by ONS (South located in Florianópolis, South-East in Rio de Janeiro, North located in Brasília and North-East in Recife). Responsible for coordination, supervision and control of the regional/local basic and complementary network, control of generation dispatch of independent power producers and command and dispatch execution of power plants under AGC.

One of ONS main operational task is to help realize the economic gains through inter-regional power transfers to take advantage of seasonal rainfall and water flow differences in each of its operating regions. This is realized through optimization of hydro resources utilization and hydro-thermal coordination. The result has a direct impact on the overall operating cost of the system.

On the other hand, for a system of this proportion, disturbances due to significant generation and load unbalances may cause excessive variations in the system frequency, tension collapse situations, and even system separation of certain parts of the BIPS network and loss of important load centers.

The studies of the dynamic behavior of BIPS have also shown that inter-area low-frequency electromechanical oscillations (0.3 to 0.8 Hz), usually well damped, could in some disturbances spread with severe consequences.

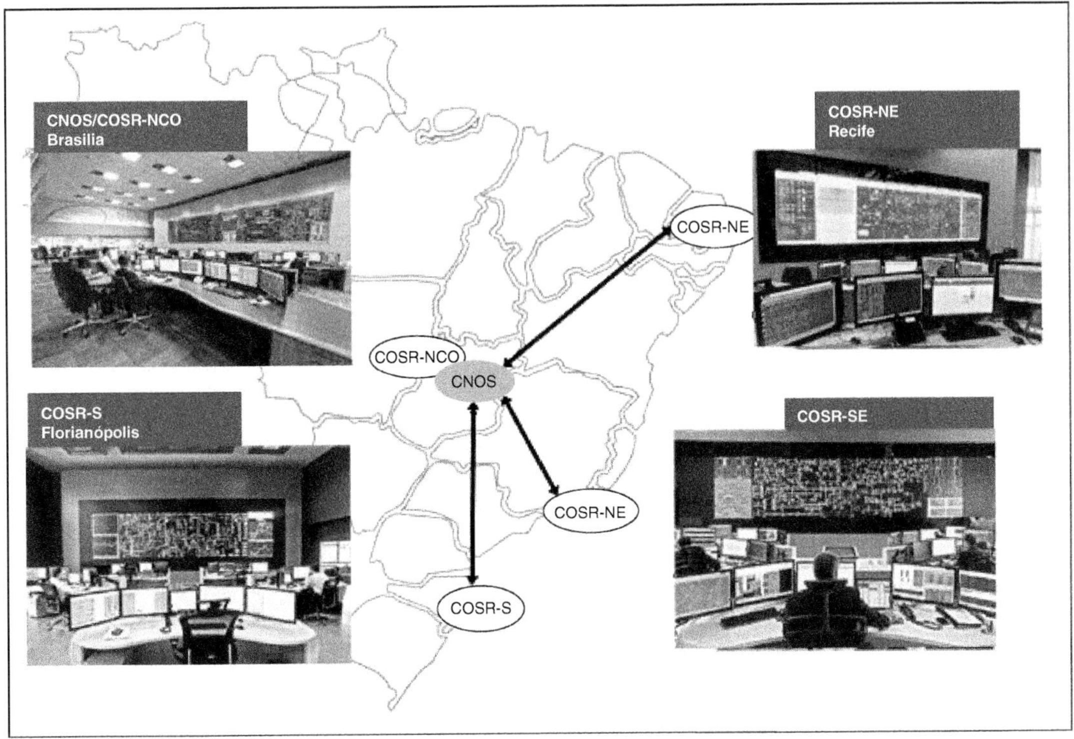


**FIGURE 4-7** ONS control centers.
*Source:* ONS Annual Report.

To avoid such situations, ONS has deployed hundreds of System Integrity Protection Schemes (SIPS) that will take pre-determined actions, such as load shedding or generator tripping, in the event of predefined system contingencies, such as losing one or more circuits of a major transmission path.

The economic and reliable operation of BIPS must also accommodate the needs of a deregulated electricity market in Brazil.

The main operation challenge of BIPS for ONS thus is how to achieve optimal hydro resource utilization while ensuring a reliable system operation within the constraints of physical limits and market operation regulations.

### 4.3.5 Smart Grid Initiatives

Since 2008, there has been a growing interest in smart energy technologies among Latin American countries, with Brazil leading the way. In 2010, almost all Brazilian electric utilities started to study Smart Grid in order to prepare them on this technology and to strategically direct their investments in new infrastructure and Research and Development (R&D) projects toward the modernization of own electric system (Fig. 4-8).

The following issues might be considered as motivating factors for Smart Grid implementation in Brazil: (i) Reduction of non-technical losses; (ii) Increase of the operational efficiency; (iii) Expansion and automatization of the electric power system with standardized smart technologies; (iv) Increase the system and power quality, especially for industries and high-tech based companies.

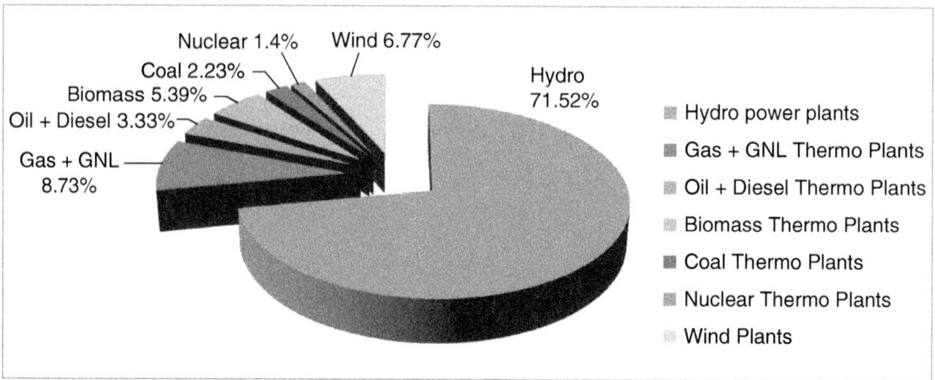


**FIGURE 4-8** Brazil's domestic electricity supply in 2016.
*Source*: Brazilian Energy Balance 2017, Energy Research Agency.

Regarding the barriers for Smart Grid implementation in Brazil, most of them are the same as other countries. They are principally: (i) Market uncertainty and lack of policies on market structure and rules; (ii) Low public awareness and engagement; (ii) Interoperability and scalability assurance; (iv) Revenue uncertainty due to lack of regulatory definitions. In addition to these barriers, it is worth mentioning other particular issues in Brazil: (i) The electric power grid in Brazil is very large and it requires a huge amount of investment; (ii) In Brazil there are large and low density rural and remote areas.

All-important entities are working to solve barriers allowing Smart Grid implementation in Brazil. We have CNPE, MME, ANEEL, EPE, ONS, and Brazilian Research Institutes like CEPEL.

Currently, Brazil gets around 83.7% of its electricity from renewables—made up of the hydro, biomass, and wind segments of the pie chart below.

Brazil's energy mix consists of 40% renewables—the hydro, firewood, and charcoal—and sugar cane products. Figure 4-9 shows pie chart of segments of and other renewables segments of the pie chart below.

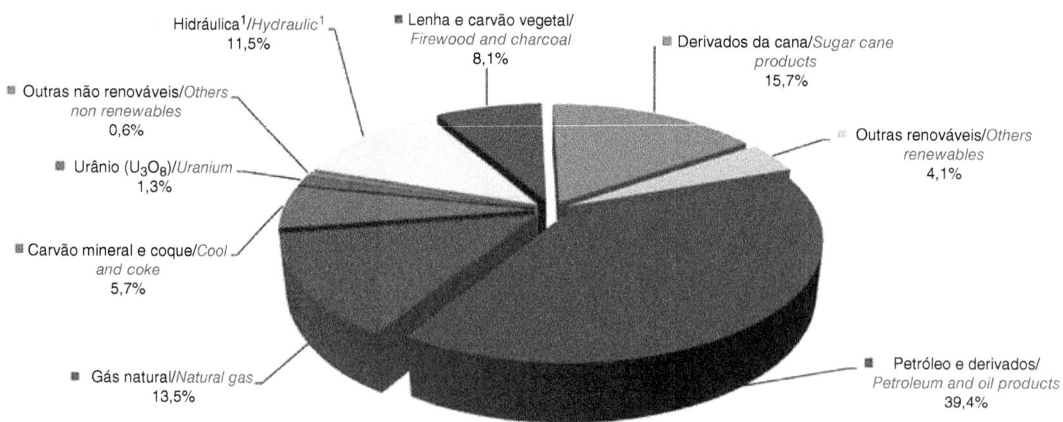


**FIGURE 4-9** Brazil's domestic energy supply in 2014.
*Source*: Brazilian Energy Balance 2015, Energy Research Agency.

### 4.3.6 Conclusion

The BIPS is very large (Brazil is a continental country when comparing with some country) and it requires a huge amount of investment. Nowadays, hydro generation corresponds a 67% of total generation capacity. Important generation sources are far away from the big load centers. To operating the BIPS, Brazil has only one Independent System Operator named ONS. The short term planning and operation of BIPS is responsibility of ONS. One of ONS main operational task is to help realize the economic gains through inter-regional power transfers to take advantage of seasonal rainfall and water flow differences in each of its operating regions. This is realized through optimization of hydro resources utilization and hydro-thermal coordination. The result has a direct impact on the overall operating cost of the system.

Since 2008, there has been a growing interest in smart energy technologies among Latin American countries, with Brazil leading the way. Almost all Brazilian electric utilities started to study Smart Grid in order to prepare them on this technology and to strategically direct their investments in new infrastructure and R&D projects toward the modernization of own electric system.

Some barriers for Smart Grid implementation in Brazil: Market uncertainty; Lack of policies on market structure and rules; Low public awareness and engagement; Interoperability and scalability assurance; Revenue uncertainty due to lack of regulatory definitions.

All-important entities are working to solve barriers allowing Smart Grid implementation in Brazil.

### 4.3.7 References

[1] Brazilian Interconnected Power System (BIPS) in numbers; http://www.ons.org.br/pt/paginas/sobre-o-sin/o-sistema-em-numeros.

[2] RA 2015 ONS Annual report and financial statements; http://www.ons.org.br/download/biblioteca_virtual/relatorios_anuais/RAONS_2015/html/home-eng.html/.

[3] Brazilian Energy Balance 2015, Energy Research Agency; https://ben.epe.gov.br/downloads/Relatorio_Final_BEN_2015.pdf/.

## 4.4 INTERCONNECTED POWER GRID IN CHINA

*BY XUANYUAN SHARON WANG*

### 4.4.1 Introduction

Being the world's most populous country with a rapidly growing economy, China is the world's largest energy producer and consumer. China's interconnected power systems are featured by being large in size and high in voltage grades, coupled with diversified generation resources and fast growing demand. In addition, China developed ultra-high-voltage (UHV) grids to transmit massive power from energy centers in remote areas to load centers in big cities due to reverse distribution of primary energy and electricity demand. This subsection outlines China's power grids in various aspects to provide readers with a brief review of the system.

***History and How It Evolved.*** China's power grid experienced a long process of development during its quest to keep up with generation increase and load growing. This process is characterized by four stages that entail evolvement of voltage levels and interconnections.

1. Stage One: The development of provincial grids (prior to the 1970s). The first 220 kV line was built in 1954 with an objective of exporting power from Fengman power plant. Later on in the

1960s and 1970s, the connection of isolated city grids commenced and it ended up forming 220 kV provincial grids that covered their respective provincial territories.

2. Stage Two: The development of inter-provincial grids (1970s to 1980s). In 1972, 330 kV lines were adopted with an aim of exporting power from Liu Jiaxia hydro-power plant. In 1981, 500 kV lines commenced operation of transmitting power from Pingdingshan to Wuhan. By the end of the 1980s, seven regional power grids were formed to cover Northeast China, North China, Central China, East China, Northwest China, Sichuan and Chongqiong, and South China. The 500 kV transmission lines became the backbone structure in most regional power grids except for in the Northwest region where 330 kV network prevailed.
3. Stage Three: The development of inter-regional grids (1989 to 2009). In September 1989, the ±500 kV DC transmission project commenced operation between Central and East China. This transmission project had a capacity of 1200 MW and it became a big achievement in China's power system toward inter-regional grids. In addition, construction of transmission lines for Three Gorges project made active progresses, which played a key role in inter-regional connections for optimal resource allocation.
4. Stage Four: The development of UHV grids (2009 till now). China has launched a research and development of UHV technology and its grid applications since 2004. The progress had been positive and on January 6, 2009, the first 1000 kV UHV AC power line was put into operation. This was followed by an establishment of another two UHV AC lines and four UHV DC lines. Additionally, four UHVAC lines and six UHVDC lines are under construction, totaling China's UHV power lines up to over 30,000 km. Moreover, the Zhundong-Wannan ±1100 kV UHVDC transmission project currently under construction is considered as the world's highest voltage level, largest capacity (12,000 MW), and longest distance (3324 km) in transmission systems. China aims at building a strong and smart power grid that comprises of the UHV transmission system as the network backbone and maintains coordinated interactions in multi-level grid operations.

### 4.4.2 Structure

China's power system is composed of seven regional grids—Northeast China, North China, East China, Central China, Northwest China, Southwest China, and South China (Fig. 4-10). The Northeast China grid supplies power to Heilongjiang, Jilin, Liaoning and East Inner Mongolia,

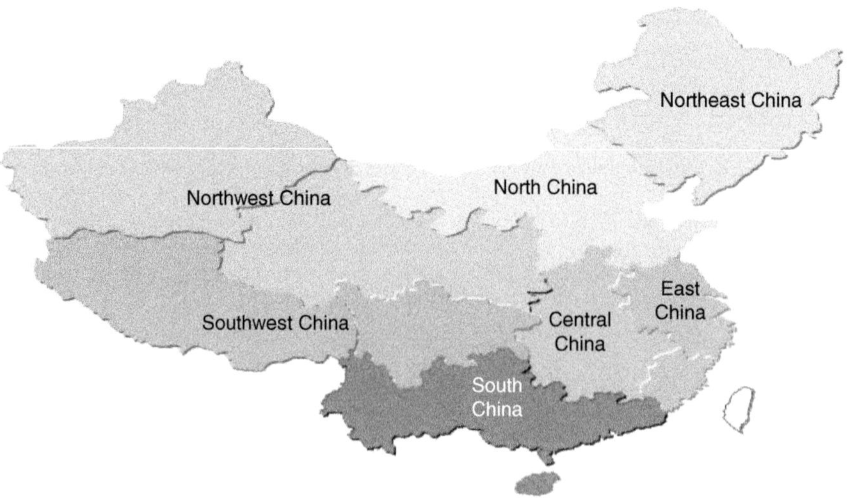


**FIGURE 4-10** Regional power grids in China.

with generation capacity of 96 GW. The North China grid supplies power to Beijing, Tianjin, Hebei, Shanxi, Shangdong, and West Inner Mongolia, with generation capacity of 341 GW. The East China grid covers Shanghai, Jiangsu, Zhejiang, Anhui, and Fujian provinces, with generation capacity of 301 GW. The Central China grid covers Henan, Hubei, Hunan, and Jiangxi, with generation capacity of 194 GW. The Northwest China grid covers Shaanxi, Gansu, Qinghai, and Xinjiang, with generation capacity of 198 GW. The Southwest China covers Sichuan, Chongqing, and Tibet, with generation capacity of 109 GW. The South grid has a territory covering Guangdong, Guangxi, Yunnan, Guizhou, and Hainan, with generation capacity of 268 GW. In 2015, China's installed capacity amounted to 1507 GW and electricity production was 5600 TWh, both being the world's largest.

***Voltage Levels.*** The continued growth in system capacity and electricity demand has led to needs for higher transmission capacity and hence voltage levels of transmission lines are gradually raised. Introduction of a higher voltage level was normally timed with the point when a large new generation plant was integrated to the system and typically took 20 to 30 years in China. At present, China has two series of voltage levels in AC transmission lines, 1000/500/220/110(66)/35/10/0.4 kV and 750/330(220)/110/35/10/0.4 kV. DC transmission voltage grades include ±500(±400), ±660, and ±800 kV. 1000 kV UHV AC and ±800 kV UHV DC transmission lines already commenced operation. Additionally, ±1100 kV UHV AC transmission lines are under construction.

## 4.4.3 Interconnection of Different Regions

Regional connections are accelerated with the advancement of UHV technologies. As of 2015, central China was connected synchronously to North China through a 1000 kV UHV AC line and to Southwest through 500 kV AC lines. Asynchronous DC connections was developed between Northeast China and North China through a DC tie, between North China and Northwest China through ±660 kV lines, between Central China and East China through ±500 kV lines, between Central China and Northwest China through ±800 kV lines and a DC tie, between Central China and South China through ±500 kV lines, between Southwest China and East China through ±800 kV lines, between Southwest China and Northwest China through ±500 kV lines, between Northwest China and Tibet through ±400 kV lines.

Based on inter-regional connections, six synchronous power grids have thus far been developed in China to realize nationwide connection with the exception of Taiwan. Six synchronous grids are Northeast China, North-Central-Southwest China, East China, Northwest China, South China, and Tibet Interconnection.

## 4.4.4 Planning

Aligned with the national economic and social development planning process, China's electricity planning cycle takes 5 years. It is a coordinated effort that jointly plans generation, transmission, and distribution in regards to capacity, construction timeline, and system integration. The National Energy Administration (NEA) issued Power System Planning Policy in 2016 as a guideline for electricity planning. The policy outlines that National and Provincial Energy Administrations are accountable for overall coordination of electricity planning process.

Utility companies are responsible for providing information, running simulations, making proposals, and reviewing the consolidated plans. Electric power planning institutes are obliged with the responsibility of individual plan studies and the final plan consolidation. The national electricity plan focuses on large power plants, inter-provincial transmission projects, and lines above 500 kV. Provincial electricity plans focus on generation and transmission projects that are not included in the national plan as well as distribution projects. The national plan is published by the NEA, whereas provincial ones are published by each province after being approved by the NEA for an overall consistency.

The process of planning usually commences 2 years ahead and published in the first half of the first year during its 5-year implementation period. If there are unpredictable factors affecting implementation of the plans, revision processes may commence in the second year.

***Grid Development Plan.*** A grid development plan is an essential component in the entire electricity planning. This is because a robust transmission grid is critical to implementation of national energy strategies as well as enhancement of the system's reliability and economics. The 13th 5-year grid development plan published by State Grid of China (SGCC) outlines that consolidation of current wide-area synchronous grids need to be accelerated for optimal operations. By the end of 2020, four interconnections are expected to form. They are three power exporting synchronous grids that are Southwest, Northwest, and Northeast grids, and one power importing synchronous grid comprised of Central, North, and East China grids. The plan published by China Southern Power Grid (CSG) anticipates transporting power from the west to the east using DC transmission lines. In addition, current five provincial grids in south China will gradually merge into two synchronous grids with reasonable scale, clear structure and relative independence.

***Load Growth and Generation Mix***

*Electricity Demand.* Since the 1980s, China's electricity consumption has maintained a high growth momentum with an average annual growth rate of 7.8% between 1980 and 2000 and 10% between 2000 and 2015. In 2015, China's total electricity consumption reached 5.6 TkWh.

Despite that total power demand is the largest in the world, China's consumption per capita is much lower compared to developed countries. For instance, China's electricity consumption per capita was 4138 kWh in 2015, which is equivalent to that of the United States in the early 1960s. Based on anticipated improvement of Chinese people's living standards and more electrical appliances at households, a rising trend is expected. China's electricity consumption per capita is projected to be 5691 kWh in 2020, which will be equivalent to that of the United States in the mid-1960s, Japan in the mid-1980s, and the United Kingdom in the mid-1990s.

As China is still at a stage of late industrialization and rapid urbanization, economic and social development will maintain its growing trend at the foreseeable future. More so, the proportion of electricity at end-energy use will continue to increase in order to meet the need for low-carbon-emission economic and social development. Therefore, electricity demand is projected to rise. Further predictions show that for periods of 2016 to 2020 and 2020 to 2030, China's total electricity consumption grow by 6.4% to 7.4% and 3.0% annually to reach 8.1 to 8.6 TkWh and 11.9 TkWh, respectively; peak load grow by 7.3% to 8.3% and 3.4% annually to reach 1340 to 1400 GW and 1920 GW, respectively.

*Generation Mix.* China's installed power capacity grew annually at an average rate of 8.2% from 1980 to 2000 and 10.9% from 2000 to 2015. At the end of 2015, China's total installed capacity reached 1507 GW, consisting of 989 GW of thermal generation (65.7%), 320 GW of hydropower (21.2%), 130 GW of wind power (8.5%), 41 GW of solar power (2.8%), and 27 GW of nuclear power (1.8%). Installed capacity of Hydropower, wind, and solar became the largest in the world.

China's generation mix will be continuously adjusted and optimized toward a clean and green structure to achieve low-carbon development goals. According to the National Energy Development Strategy and Action Plan, China predicts that non-fossil energy will account for 15% and 20% of primary energy consumption by 2020 and 2030. It is also predicted that installed capacity of wind power and solar power will continue to grow rapidly, while thermal power will decrease gradually. According to planned generation mix, China's installed capacity will reach 2070 GW in total by 2020, including 1120 GW coal generation, 240 GW wind generation, 150 GW solar generation, 347 GW hydro generation. This will help in increasing renewable capacity to over 800 GW, accounting for 39.3% in total generation mix.

*Renewables.* China has identified clean energy development as a key solution towards a low-carbon society so that it can achieve sustainable growth by being less dependent on fossil fuels.

Since 2005, the average annual growth rate of renewable capacity is 15.1%. The main contributors for renewable capacity were wind at 58.2% and solar at 89.6%.

Despite that China is a world leader in installed capacity of wind, solar, and hydropower, the country faces uneven renewable energy distribution. This is because hydro resource is abundant in the southwest and onshore wind resource is rich in the north, the northeast, and the northwest. Solar resource is abundant in the northwest, Tibet, and Inner Mongolia. This resulted in highly concentrated development of wind and solar power in resource bases located in remote areas that have less population and are far from load centers. More so, there are less fuel types in generation mix in north, northeast, and northwest China. Power supplies such as pumped storage and gas units have capabilities of operating and adjusting quickly, but they are less than 2% in total installed capacity. The situation is even worse in winter as most coal plants are expected to operate at certain levels to supply heat so they end up failing to pick up fluctuation in wind and solar power. These facts make issues associated with integration and operation of large-scale wind and solar generation in China become significant. One proposed solution is to accelerate development of UHV grids as a transmission corridor for long-distance power transfer and nation-wide resource allocation.

It is expected that wind generation will continue to be developed in large scale and still locate in north, northeast, and northwest China, whereas Solar power will be either centrally developed in Qinghai, Gansu, and other west provinces or as distributed generation in Jiangsu, Zhejiang, and other central provinces.

## 4.4.5 Operation and Control

***Control Center.*** China's power sector's reform in 2002 unbundled generation from vertically integrated power industry while grid companies still kept businesses in transmission, distribution, retail, as well as the responsibility of operating the system.

Dispatch centers hence belong to grid companies as an internal division accountable for operating the grid to achieve its highest security, reliability, and economical efficiency. The national dispatch center, China's only national level power dispatch center, affiliated with SGCC, has the highest authority in system operations and is primarily responsible for operating UHV grids, inter-regional power transmission lines, and large power plants deployed across regions. Six SGCC regional dispatch centers plus one CSG dispatch center are responsible for operating inter-provincial lines, transmission grids of 500 kV and above that are not under the national center's oversight, and large power plants deployed across provinces. Thirty-three provincial centers are responsible for operating 220 (330) kV grids and power plants managed by provinces. Over four hundred prefectural centers as well as over 1600 county centers take the responsibility of operating grids at 110 kV and below as well as a few small-size local generators.

***Hierarchy and Inter-Control Center Operation.*** China's power system is a hybrid grid composed of both AC and DC systems, characterized by its complex structure and large size. To improve system controllability and management effectiveness, control centers at five different levels all follow the concept of coordinated operation and hierarchical management to ensure reliable and secure operation of the system. Being the highest in the structure, the national control center has the authority to give orders to subordinate centers with respect to system operations and control. Power exchanges across regions are scheduled by the national dispatch center and implemented by regional centers. Inter-provincial power exchanges are planned by regional centers and correspondent provinces dispatch generation accordingly.

***Transaction and Settlements.*** At present, the majority of electricity is purchased by grid companies from power plants and sold to end users. The annual amount of electricity a conventional power plant produces is determined by Development and Reform Commissions through allocation processes based on average minimum-run hours as well as principles of energy save and emission reduction. Utility companies then formulate generators' operating schedules and issue dispatching orders accordingly. Renewable units are not set by annual generating amount; instead, as long as grids can maintain safe and reliable operation, they would run to the full extent. Power trades exist, though not much, mainly in forms of power purchase between big customers and generators as well as generation rights trades, through bilateral or centralized pool markets.

Inter-regional and Inter-provincial power transactions are primarily mid to long term, determined by planning processes to implement national energy strategies on transferring power from the west to the east and from the north to the south. When power shortage occurs or there is residual capacity to purchase cost-effective power, short-term trades can be made between areas.

Utility companies are responsible for settlements and billing of all power transactions. Currently, electricity pricing is set by the government on most power transactions except for market-based trades such as power purchase between big customers and generators.

In 2015, the issue of No. 9 document by Chinese government started a new round of reform in China's electricity market and power industry. It mandates the establishment of relatively dependent power exchange centers, opens up retail sectors, encourages big customers to purchase electricity directly from power plants, and promotes development of wholesale markets. The business and profit model of utility companies will be changed from being the single buyer and the single seller in the market to ones who provide transmission and distribution service and charge service fees under regulator's supervision. It is also expected that generators' price and schedules be more market based and less regulated.

### 4.4.6 Regulatory Bodies

In China, regulation of power sectors is jointly carried out by the National Development and Reform Commission (NDRC), the National Energy Administration (NEA), and the State Owned Assets Supervision and Administration Commission (SASAC).

The NDRC is a macroeconomic management agency under the State Council, which has a broad administrative and planning control over the Chinese economy. In the power sector, the NDRC is responsible to formulate plans for the development of China's energy industry as well as guide and promote industry restructuring. It also regulates electricity tariff and generators' annual schedules.

The NEA, overseen by NDRC, was a consolidation of the former NEA and the State Electricity Regulatory Commission in 2013 to strengthen the integrated administration of energy industry in concert with the NDRC. Some of NEA's responsibilities include drafting laws and regulations concerning the supervision and administration of energy development, supervising and regulating electricity market as well as safety and reliability of electricity production, supervising and examining power rates, setting prices for ancillary services, taking actions during any electricity emergencies, organizing or participating in investigation of safety mishaps during electricity production.

In China, most large generation and utility companies are state-owned enterprises and therefore are supervised and regulated by the SASAC. The SASAC, a special commission directly under the State Council, was founded in 2003 through the consolidation of various other industry-specific ministries. It is responsible for ensuring efficient management of state-owned assets, appointing top executives and approving any mergers or sales of stock or assets, as well as drafting laws related to state-owned enterprises.

### 4.4.7 Future Grid Initiatives

***Smart Grid Initiatives.*** Smart grid in China refers to a modernized power grid that is supported by a UHV grid as its backbone, features coordinated development of grids at different levels, and covers various power segments including generation, transmission, distribution, consumption, and dispatch. It integrates modern telecommunication, automatic control, decision support, and advanced power technologies and it is characterized by being informative, automatic and interactive. Smart grid is capable of friendly integration of renewable resources as well as interaction with users, with smart response and system self-recovery capabilities that enable it to substantially improve the safety, reliability, and operational efficiency of the power system.

***Generation.*** A vital goal in smart grid development is its capability to integrate large-scale renewable energy generation and resolve challenges associated with variability and uncertainty of intermittent resources. This requires adoption of forecasting tools that possess high accuracy for wind and solar resources, proper grid-connection plans as well as optimal and coordinated control mechanism on integrating multiple types of renewable generation. A national pilot project commenced in 2015 in Zhangjiakou and it specialized in co-operation of wind, solar, energy storage, and transmission systems. The project, being the largest in scale, launched 450 MW wind generation, 100 MW solar generation, 20 MW energy storage, and a 220 kV smart substation. The project is successful in integration of large-scale renewables using latest technologies and demonstrates economic and reliable co-operation of a system that comprises multiple models of wind turbines, large scale power control equipment of PV system, and chemical energy storage of various sizes and types.

***Transmission.*** Online condition monitoring and real-time diagnosis for transmission equipment are fundamentals in building smart transmission. SGCC has completed installation of master stations for condition monitoring master stations of transmission and transformation equipment in 26 provinces (autonomous regions and municipalities) located in its operating area by the end of 2014. This accomplished online condition monitoring of 4263 transmission lines as well as transmission and transformation equipment in 3597 substations.

***Transformation.*** Smart substation is a substantial support to elevate the overall intelligence level of power system. It is achieved by station digitalization and design compactness as well as equipment and business integration. By the end of 2014, SGCC has built 1527 smart substations, out of which 1135 substations are at 110 (66) kV, 344 substations are at 220 kV, 29 substations are at 500 kV, 12 substations are at 330 kV, and 7 stations are at 750 kV.

***Distribution.*** The development of smart distribution made significant breakthroughs in the fields of distribution system self-recovery control, distribution terminal intelligentization, and distributed generation connection. By the end of 2014, SGCC has built smart distribution grids covering core centers of 78 cities and operating distribution automation systems for over 10,000 lines at 10 kV. Through this enhancement, distribution grids were improved in operational controllability and system reliability by demonstrating time reduction of unplanned outages, scope limitation of fault impacts, and reliability improvement of supplies.

***Consumption.*** China conducted a series of projects in areas of smart meters, power consumption information acquisition, interactive marketing, demand-side management, user-side distributed generation, electric vehicle charging/swapping facilities, power quality monitoring, and power optical fiber to home. By the end of 2014, SGCC has built a power consumption information acquisition system containing 240 million smart meters, realizing remote automatic meter reading, self-service recharging, real-time consumption monitoring, line loss monitoring, and orderly load shedding management. Power optical fibers, a successful integration of power cables and optical fibers, were introduced into 470,000 households to provide end-users with not only electricity but internet, telecom, radio, TV signals, and other value-added services. Twenty-eight smart communities were built in Beijing, Shanghai, and other locations providing service platforms covering 287,000 households. The electric vehicle battery charging/swapping networks were built and it accumulated installation of 24,000 charging piles and 618 charging/swap stations.

***Dispatch.*** Intelligent dispatch focuses on conducting proactive and intelligent monitoring, analyses, early warning, decision-making support, and self-recovery control to ensure efficient utilization of renewable energy while maintaining safe and reliable operation of power grids. By the end of 2014, the smart grid operation system developed by SGCC has been implemented at 33 provincial dispatch centers and 5 regional dispatch centers, sharing 890,000 real-time data from 7011 plants or substations to achieve panoramic operation monitoring for lines at 220 kV and above. It also integrated phasor measurement unit (PMU) data from 2451 substations or power plants to provide dynamic perception of faults in lines at 500 kV and above. Additionally, alarm information of lines at 500 kV and above were shared among all control centers by deploying smart alarm function at both control centers and substations.

***Microgrids.*** As an effective way to utilize local renewable distributed generation, microgrids create sustainable, reliable, and more cost-effective energy system in conjunction with conventional grids and have gained great attention in China with rapid growth in renewables. In July 2015, the NEA issued Guidance on Promoting Renewable Microgrids Demonstration Projects Development, as a policy to encourage building microgrids with quantitative goals. By the end of 2015, the number of China's microgrid demonstration projects in operation reached a total of 56.

Existing microgrids in China are mainly supplied by solar and wind resources, either connected or disconnected from centralized grids. The application of microgrids is categorized into three scenarios: remote areas, islands, and cities. Microgrids in remote areas focus on rural electrification in distant regions such as Tibet, Qinhai, Xinjiang, and Inner Mongolia with very low population density and abundant renewable resources. Microgrids in islands alleviate the situation where habitants highly rely on diesel for electricity but are limited to diesel's supply shortage as well as high cost. Microgrids in cities incorporate distributed renewables to provide end customers with a cleaner and more diversified power supplies as well as proven economic benefits.

## 4.4.8 Conclusion

The existing China's power grid evolved based on the need for societal and economic development as well as national energy strategies, driven by technology innovation and advancement. It provides a fundamental infrastructure for secure, clean, efficient, and sustainable development in the national energy sector. For years to come, China's power grid is envisioned as a grid utilizing UHV transmission lines in long-distance massive clean electricity transfer from energy bases to load centers. It is also anticipated to be a robust, widely interconnected, highly intelligent, accessible and interactive, secure and reliable, as well as cost effective system combined with both AC and DC technologies.

In addition, China proposed to build global energy interconnection to facilitate efforts in meeting global energy demand with clean and green energy at the United Nation Development summit in New York in 2015. It suggested connecting power grids globally through smart grids via UHV networks as its backbone and clean energy as main resources. Regardless of whether or not this would happen or when it would happen, the Chinese do hope for a cleaner, reliable, and sustainable energy system for the benefit of all mankind.

## 4.4.9 References


[1] Liu, Z., *Global Energy Interconnection*. Amsterdam: Elsevier, 2015.

[2] Liu, Z., *Ultra-High Voltage AC/DC Grids*. Amsterdam: Elsevier, 2014.

[3] Liu, Z., *Electric Power and Energy in China*. Beijing: China Electric Power Press, 2012.

[4] Zhao, Z., *Development and Prospect of China Power Grids*, in Electric Power, Vol. 37, No. 1, 2004, pp. 6–11.

[5] Zhang, Q. et al., *Review and Outlook for World's Large Power Grids*, Liu, Z. and Shu,Y., Eds., Beijing: China Electric Power Press, 2016.

[6] Shu, Y., *Promote the Scientific Development of Wind and Other Renewables*, in China Energy News, Sept. 10th 2012, pp. 12.

[7] Liu, Z., *Development of Global Energy Interconnection for an Era of Sustainable Society*, Keynote speech at Global Energy Interconnection Summit, Beijing, March 2016.

[8] Zheng, B., *Development of Network Interconnection in China*, in Power System Technology, Vol. 27, No. 2, 2003, pp. 30–33.

[9] Zhou, X., Chen, S., Lu, Z., *Review and Prospect for Power System Development and Related Technologies: A Concept of Three-Generation Power Systems*, Vol. 33, No. 22, Proceedings of the CSEE, 2013.

[10] Zhang, Y., *Analysis on the Development Strategies of the UHV Grid in China*, in Proceedings of the CSEE, Vol. 29, No. 22, 2009, pp. 1–7.

[11] Prospect for 13th Five-Year Blue Print in Power, in Rural Power Management, Vol. 1, pp. 1, 2016.

[12] Electric Power Law of the People's Republic of China, 1995, 2015 Amendment.

[13] NEA, *Power System Planning Policy*, May 2016.

[14] State Council of the People's Republic of China, *Action Plan of Energy Development Strategy*, June 2014.

[15] State Council of the People's Republic of China, *Deepening the Reform on Power Restructuring (9th document)*, March 2015.

[16] State Council of the People's Republic of China, *Power Industry Restructuring Plan (5th document)*, Feb. 2002.

[17] Ministry of Commerce of the People's Republic of China, *Renewable Energy Law of the People's Republic of China*, December 20, 2013.

[18] China Electricity Council, Data and Statistics of China Power Industry, http://english.cec.org.cn/No.110.index.htm.

[19] National Development and Reform Commission, http://www.sdpc.gov.cn/.

[20] National Energy Administration, http://www.nea.gov.cn/.

[21] State-Owned Assets Supervision and Administration Commission of the State Council, http://www.sasac.gov.cn/.

[22] State Grid of China Corporation, http://www.sgcc.com.cn/.

[23] China Southern Power Grid Company Limited, http://www.csg.cn/.

## 4.5 INTERCONNECTED POWER GRID IN INDIA

*BY MINI SHAJI THOMAS*

### 4.5.1 Introduction

The Indian Power Sector is among the largest in the world and has been showing tremendous growth in the last few decades in terms of installed capacity, transmission capability, interconnections, transmission voltages, and integration of renewable sources.

The history of Electric Power in India tracks down to July 24, 1879 with the demonstration of electric light in Calcutta (Kolkata). With the success of this, further demonstrations were conducted in Bombay (Mumbai) in 1882 at Crawford Market and Bombay Electric Supply and Tramways Company (BEST) set up a generating station in 1905. Darjeeling Municipality set up the first Hydroelectric Installation in India near a tea estate at Sidrapong. On August 5, 1905 the first electric street light in Asia was lit in Bangalore and the first electric train ran between Kurla and Victoria Terminus in Bombay in 1925.

Indian Power Sector was concentrated in and around a few urban areas at the time of independence in 1947. In the 50s, huge river valleys projects came up and some limited inter-connected systems were set up to provide power to nearby population and efforts were made to set up projects for irrigation, agriculture and flood control. In the 60s there were huge developments in the power sector such as increase in generating unit size, transmission voltage and interconnection as there was rapid industrialization happening and power grids at the state level started to evolve.

The 70s and 80s saw steady increase in the transmission voltage, establishment of thermal power station in response to rapid urbanization. The 90s saw the development in HVDC System for bulk power supply over large distances for inter-regional power transfer and back-to-back connections [5]. The important milestones in the development of Indian Power Sector are given in Table 4-3.

**TABLE 4-3** Important Milestones in the Development of Indian Power Sector

| | |
|---|---|
| 1897–98 | First hydro (130 kW) Darjeeling/thermal (1 MW) in Calcutta by CESC. |
| 1910 | Indian Electricity Act 1910 enacted to regulate supply by the Licensees to the consumers. |
| 1948 | Indian Electricity (Supply) Act 1948 (ES Act). Formation of State Electricity Boards with full powers to control generation, distribution, and utilization of electricity within their respective states and Central Electricity Authority (CEA) for planning and development of power system. |
| 1964 | Five Regional Electricity Boards (REBs) formed by the Government of India to ensure integrated grid operation and regional cooperation on power. |
| 1976 | Creation of Central Generating Companies of NTPC, NHPC, NPC, NLC, and NEEPCO. |
| 1991 | ES Act 1948 amended for the formation of private Generating companies. 100% foreign investment in power sector without any export obligations. |
| 1998 | Electricity Regulatory Commission Act 1998 enacted, formation of Central Electricity Regulatory Commission (CERC) and State Electricity Regulatory Commissions (SERC). Regulatory power of the State governments transferred to SERC. |
| 1998 | Act amended to provide for of Central Transmission Utility (CTU) and State Transmission Utilities (STU). |
| 2000 | Indian Electricity Grid Code (IEGC) and Availability Based Tariff (ABT). |
| 2003 | Electricity Act 2003 enacted by the Parliament. |
| 2004 | Open Access allowed. |
| 2007 | Power Markets emerged. |
| 2009 | Unscheduled Interchange (UI) Regulations introduced. |
| 2010 | Renewable energy Certificate, sharing of Inter-State Transmission Charges and Losses. |

## 4.5.2 Institutional Set Up

The Institutional set up of the Indian Power Sector is given in Fig. 4-11. In India, Electric Power is a concurrent subject governed by both Central and State Governments. The overall monitoring and control is by the Ministry of Power, Government of India. The set up also includes public sector enterprises and central generating companies such as National Thermal Power Corporation (NTPC), National Hydro Electric Power Corporation (NHPC), and Nuclear Power Corporation (NPC).

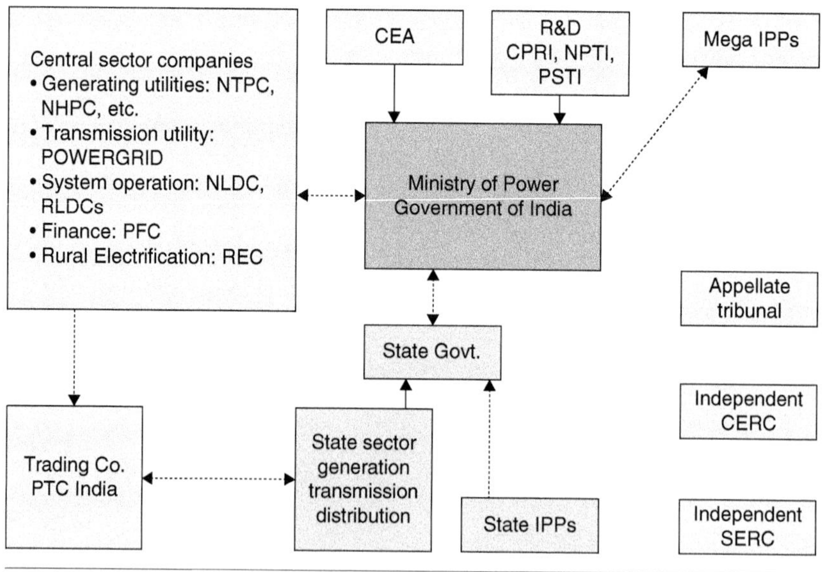


**FIGURE 4-11** Institutional set up of Indian Power Sector.

The public sector transmission company is Power Grid Corporation of India and the Grid Control Operations are managed by the newly created Power System Operations Corporation (POSOCO). The research Institutions created by the Central Government include Central Power Research Institute (CPRI), National Power Training Institute (NPTI), and Power System Training Institute (PSTI) for capacity building. Each state in India has its own Ministry of Power managing the generation, transmission, and distribution of power in the state and also to liaison with central entities.

The Central Electricity Authority (CEA) advises the Ministry of Power on technical, financial, and economic matters. It prepares the National Power Plan, conducts load surveys, and comes out with the planning criteria to be followed regarding major generation and transmission infrastructure in the country.

Power Finance Corporation (PFC) is a central sector entity for funding of power projects across the country. Rural Electrification Corporation (REC) is set up exclusively for the electrification of villages.

India has an independent regulatory authority for electricity, the Central Electricity Regulatory Commission (CERC), set up under the Electricity Regulatory Commission's Act 1998 as an independent statutory body with quasi-judicial powers. CERC regulates traffic related matter and interstate bulk sale of power, advises the Central Government on the formulation of tariff policy frames the guidelines regarding tariff and promotes competition and efficiency in the power sector. There are State Electricity Regulatory Commissions that regulate the activities at the state level.

### 4.5.3 Makeup and Size

Indian Power Grid is one of the very large power grids in the world with an installed capacity of 305,554 MW as on August 31, 2016. Each state in India has its own generation transmission and distribution of electric power. Indian Electricity Generation is mostly coal based (69%) with renewable Generation (14%) edging over hydro Power (14%) recently and the rest contributed by other sources. Hydropower's share has declined steeply from the mid-1960s, when it was over 45%, which dropped to 26% in 2005 and now to 14%.

Figure 4-12 gives a comprehensive picture of the growth in installed capacity from thermal and renewable energy till now.

### 4.5.4 Voltage Levels

India has been leading the efforts in increased transmission voltage among the countries in the world. The transmission voltage was 220 kV in the early 70s and 400 kV was introduced in 1977 in India. In 2000, 765 kV transmission lines were introduced, and 2014–15 saw the testing of 1200 kV substation at Bina in the state of Madhya Pradesh. In 2016 a test line of 1200 kV was set up successfully by a consortium of public and private companies under the leadership of Power Grid Corporation of India. This is the first 1200 kV line in the world to be operational and the transmission world is looking towards India for innovations like this. In high voltage DC (HVDC) scenario, 500 kV HVDC was tested in 1990 and 800 kV HVDC line was tested in 2012. The transmission utility "Powergrid" operates about 131,728 circuit km of transmission lines at 800/765 kV, 400 kV, 220 kV and 132 kV EHVAC and +500 kV HVDC levels and 213 substations. The transformation capacity is about 266,163 MVA as on 31st August 2016. This gigantic transmission network, spread over length and breadth of the country, is consistently maintained at an availability of over 99%.

The distribution voltages range from 66 kV, 33 kV, and 11 kV in India and the household supply is at 400 V for three phase and 230 V for single phase. Indian power sector operates at a frequency of 50 Hz.

### 4.5.5 Interconnection of Different Regions

India has a fair share of resources required for electricity generation, however, the distribution of natural resources (coal, water resources, wind, etc.) is spatial, with coal reserves in the eastern region, water resources in the northern Himalayas and northeast region and wind energy abundant in the

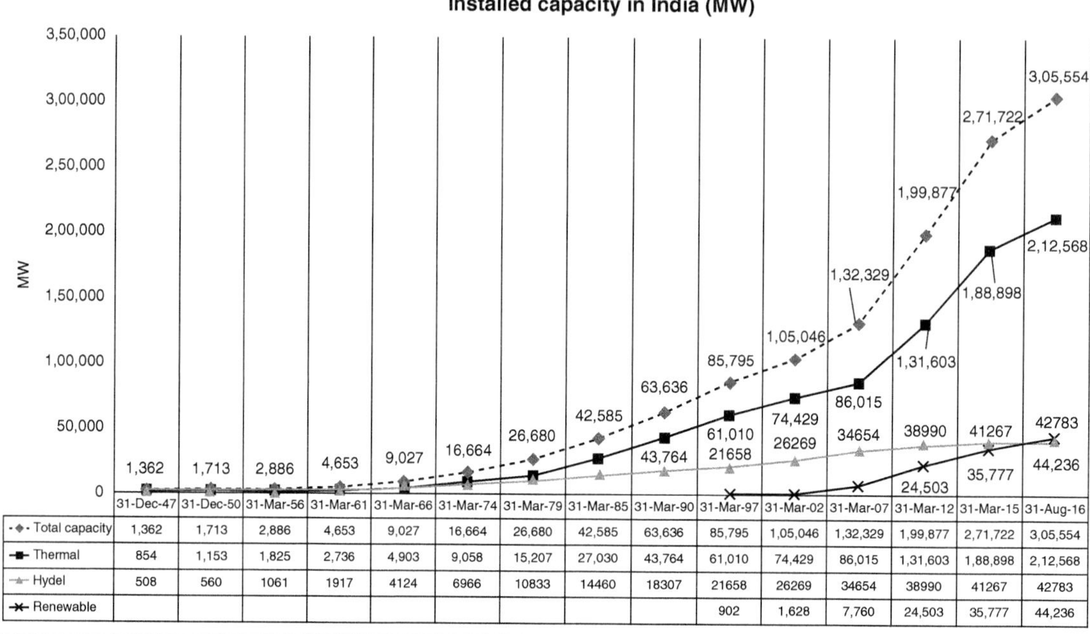


| | 31-Dec-47 | 31-Dec-50 | 31-Mar-56 | 31-Mar-61 | 31-Mar-66 | 31-Mar-74 | 31-Mar-79 | 31-Mar-85 | 31-Mar-90 | 31-Mar-97 | 31-Mar-02 | 31-Mar-07 | 31-Mar-12 | 31-Mar-15 | 31-Aug-16 |
|---|---|---|---|---|---|---|---|---|---|---|---|---|---|---|---|
| Total capacity | 1,362 | 1,713 | 2,886 | 4,653 | 9,027 | 16,664 | 26,680 | 42,585 | 63,636 | 85,795 | 1,05,046 | 1,32,329 | 1,99,877 | 2,71,722 | 3,05,554 |
| Thermal | 854 | 1,153 | 1,825 | 2,736 | 4,903 | 9,058 | 15,207 | 27,030 | 43,764 | 61,010 | 74,429 | 86,015 | 1,31,603 | 1,88,898 | 2,12,568 |
| Hydel | 508 | 560 | 1061 | 1917 | 4124 | 6966 | 10833 | 14460 | 18307 | 21658 | 26269 | 34654 | 38990 | 41267 | 42783 |
| Renewable | | | | | | | | | | 902 | 1,628 | 7,760 | 24,503 | 35,777 | 44,236 |

**FIGURE 4-12** The growth of installed capacity in India from Independence in 1947 till now.

coastal regions. Hence, some of the states have abundance of power where as other regions and states have acute shortage of power.

Hence, there was a need to set up central power generating agencies for coal and hydro generation coupled with high voltage transmission network crisscrossing the entire country. This led to the setting up of large thermal and hydro generating station where the raw material were available in the 70s and 80s and wiring of high voltage transmission lines across the country.

The 29 states and 7 union territories of India are distributed in 5 regions—Northern, Southern, Eastern, Western, and Northeastern (NE) regions. Figure 4-13 shows a comprehensive picture of the five regions with the characteristics of each region. As mentioned earlier the northern region with eight states and two union territories is a deficit region, always drawing power from the east and northeast region which are high in generation capacity and low in loads.

In the 70s and 80s, the states used to generate electricity, transmit and distribute within the state. Although the five regional electricity boards were formed as early as 1964, the pooling of resources within the region among states started in the 80s. The five regional grids operated with five frequencies in the 80s and early 90s. In October, 1991 the east and the northeast were synchronized, and in March, 2003 the western region was synchronized with east and northeast, forming the central grid, thus reducing the frequencies to three in the north, central, and southern grids. In August, 2006 the northern region was synchronized with the central grid forming the "New Grid."

In December, 2013 the southern grid was synchronized with the New Grid fulfilling the dream of "One Nation One Grid One Frequency," thus creating the All India grid with a single frequency, catapulting it to the status of very large power system.

India is also leading the efforts for the formation of the SAARC grid interconnecting Bhutan, Nepal, Sri Lanka, and Bangladesh.

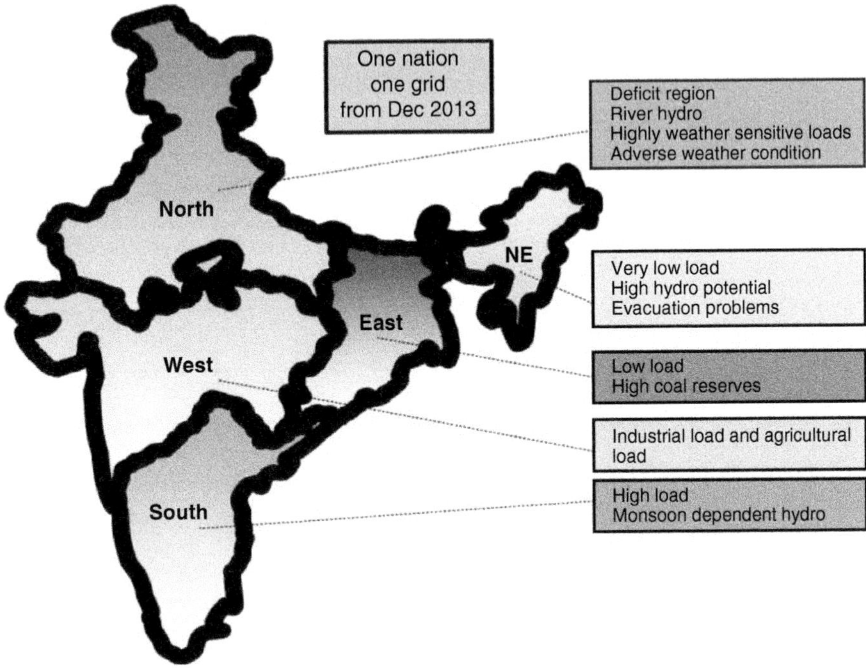


**FIGURE 4-13** The five Regions in India with their characteristics.

### 4.5.6 Renewables

Till recently India was adding large-scale conventional power plants which were not enough to meet the ever increasing demand for electric power. Currently India has a sizable footprint of renewable energy at 45 GW installed capacity, mainly wind at 27.7 GW, solar at 8 GW and the rest contributed by small hydro, biopower, and other sources. Refer to Fig. 4-12 for an overview of the growth.

The Government of India is aiming to deploy and integrate 175 GW of Renewable Energy by 2022, the 75th year of Indian independence. The constituents being: (1) solar (utility-scale, distributed, off-grid/mini-grid—100 GW); (2) wind (utility-scale—60 GW); (3) small hydro (5 GW); (4) bioenergy (10 GW). This is feasible now as the solar and wind technologies are commercially viable. The Jawaharlal Nehru National Solar Mission aims to generate 100,000 MW of solar power by 2022, creating a positive environment among investors keen to tap into India's potential.

### 4.5.7 Operation and Control

India has one of the most modern and well-established transmission coordination and operations in the world, monitoring and controlling the transmission of power, especially interregional transfer and pricing issues.

The setup of the transmission control centers are given in Fig. 4-14 where the National Load Despatch Center (NLDC), New Delhi, started functioning in 2009, whereas the five regional load dispatch centers (RLDCs) began operation in early 90s. There are five RLDCs, as shown in Fig. 4-14, which coordinates the activities of the states under each region. The state load dispatch

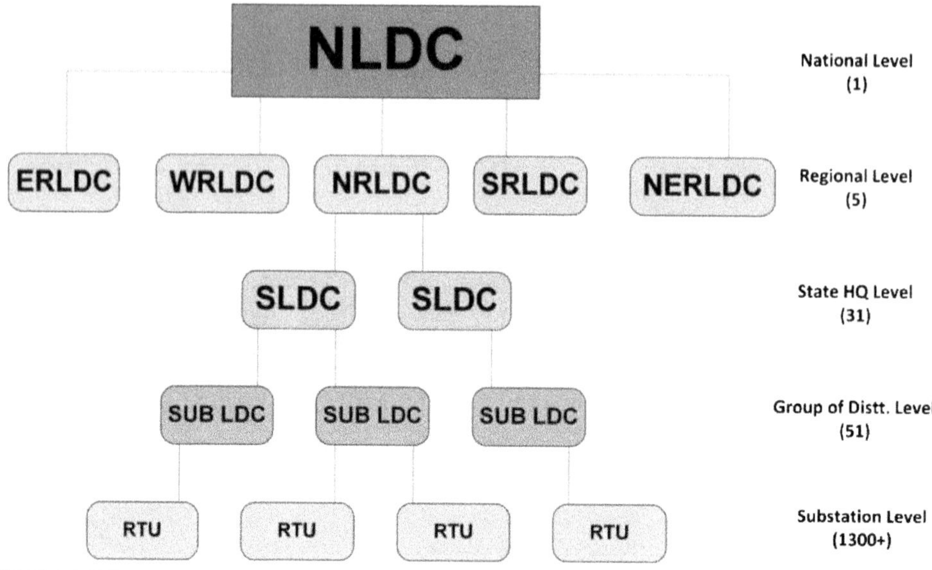


**FIGURE 4-14** Transmission control center organizational diagram.

centers (SLDCs) are connected to the RLDCs. For example, the southern regional load dispatch center (SRLDC) monitors the states of Andhra Pradesh, Tamil Nadu, Karnataka, Kerala, and Puducherry and is connected to the respective SLDCs. The SLDCs, depending on the size of the state, have one or more sub-load dispatch centers which are connected to the major generating stations and substations. Hence the LDCs will have complete information regarding the generation, and flow of power. The RTUs at generating stations and substations of central utilities report directly to RLDCs.

RLDCs and SLDCs are apex bodies and all constituents have to comply with the directions of NLDC for ensuring integrated operation of the inter-state transmission systems, to achieve maximum efficiency and economy.

Implementation of Indian Electricity Grid Code (IEGC) w.e.f. Feburary 1, 2000 was a milestone in the history of Indian power sector, which puts obligations on various players in the grid for maintaining security of the inter-state transmission system. It brings set of rules to be followed by all utilities connected to the inter-state transmission system. The regional grids are proposed to be operated as loose power pools and strict control of tie line/generation schedule is not envisaged. Incentives/disincentives to give signals for correct grid operation are built in features of the Availability Based Tariff (ABT). Unscheduled interchanges are billed at a frequency linked rate which varies linearly from 0 at 50.5 Hz, the main suppliers of bulk power are the central sector and state owned power stations.

### 4.5.8 Unified Real-Time Dynamic State Measurement (URTDSM) Scheme

The power grids are expected to operate closer to their limits in order to maximize utilization of the network and India is no exception. The role of the system operator and the decisions taken have become very critical, who completely rely on the information available to them in real time.

The existing SCADA or EMS systems acquire analog and digital information such as voltage, frequency, active and reactive power flows and circuit breaker status through RTUs or IEDs spread throughout the system. This information is updated once every 2 seconds for status points and

10 seconds for analog values. Lack of a coordinated accurate time stamp for recorded data makes any reconstruction of a timeline difficult and is time consuming.

Wide area monitoring through high-speed communication helps in securing the system in minimum amount of time, as the scan times are to the tune of 25 to 120 scans per second and time stamped angular separation of nodes are available using this technology, which will greatly enhance the information availability to the operator. Synchrophasor technology comprise of Phasor Measurement Units (PMUs), Phasor Data Concentrators (PDCs), Historian, communication network, real time visualization, and offline toolboxes.

India has embarked on an ambitious project of installing PMUs and PDCs to help the system operation with better visualization and analytical tools for appropriate action during a contingency. This unified real-time dynamic state measurement (URTDSM) scheme requires a large number of PMUs and PDCs to be deployed. In the first phase, placement of 1186 PMUs at all lines in HVDC terminal stations, 400 kV and above voltage level substations, generating stations stepped up at 220 kV level and above where fiberoptic cable along with communication equipment is either existing or being implemented. Placement of 27 Nodal PDC at strategic substation, 25 Master PDC at SLDCs, 5 Super PDC at RLDCs, 2 PDCs at Main and Backup NLDC, with 16 Remote consoles are underway in this phase at the moment. Development of Analytics for various applications using PMU data is also being taken up in parallel. In phase 2, placement of balance 483 PMUs and respective equipment at similar locations along with provision of fiberoptic connectivity and communication equipment will be completed.

### 4.5.9 Smart Grid Initiatives in India [10]

Government of India has taken a number of initiatives for transforming the existing grid to a smarter one. The most prominent initiatives being:

a. National Smart Grid Mission (NSGM)
b. Integrated power development scheme
c. Indian Smart Grid Taskforce
d. Smart city and AMRUT projects
e. India Smart Grid Forum

***National Smart Grid Mission.*** GOI has established a NSGM in power sector to plan and monitor implementation of policies and programs related to smart grid activities in India in March 2015. Ministry of Power (MoP), Ministry of New and Renewable Energy (MNRE), Ministry of Urban Development (MoUD), and Ministry of Heavy Industry (MoHI) are associated with this mission. Since Smart Grid is a dynamic and evolving concept due to constant technological innovations, the objectives, structure and functions of NSGM are such that it allows sufficient freedom and flexibility of operations without needing to refer the matter to different ministers/agencies frequently. NSGM is a three-tier body consisting Governing Council, Empowered Committee, and Technical committee. Smart Grid Knowledge Center (SGKC) is developed by Powergrid Corporation of India Limited, the transmission utility of India with funding from MoP. SGKC acts as a resource center for providing technical support to the mission in all technical matters, including development of technical workforce, capacity building, and outreach, suggesting curriculum changes in technical education, etc.

***Integrated Power Development Scheme (for Sub-Transmission and Distribution System).*** The Government of India has implemented Integrated Power Development Scheme (IPDS) with an objective to cover the works relating to strengthening of sub-transmission and distribution system, including provisioning of solar panels, metering of distribution transformers, feeder, consumers in urban areas, and IT enablement of distribution sector in 2014. All the distribution companies (private and state power departments) are provided financial assistance to enhance and modernize the infrastructure. The DISCOMS assess the need of strengthening the urban distribution networks and has formulated bankable Detailed Project Reports (DPRs) which is recommended by existing Distribution Reforms Committee (DRC) at state level. Earlier, in an effort to move toward Smart grid, GOI had launched Accelerated Power Development Program (APDP) in 2000 to 2001 and Accelerated Power Development and Reforms Program (APDRP) during 2002 to 2003. APDRP scheme was restructured as a Central Sector Scheme and renamed as Restructured Accelerated Power Development and Reforms Program (RAPDRP) to overcome the shortcomings of APDRP. The program mainly considers on reduction of sustained loss, establishment of reliable and automated systems, collection of accurate data, adoption of Information Technology, etc. for demonstrable performance. The States/Utilities have constituted the State Electricity Regulatory Commission so that it can achieve the following target of AT&C loss reduction at utility level: (a) utilities having AT&C loss above 30%—reduction by 3% per year; (b) utilities having AT&C loss below 30%—reduction by 1.5% per year. It commits a time frame for introduction of measures for better accountability at all levels in the project area. It is heartening to note that most of the distribution companies have been able to achieve the targets well ahead of time. Case studies related to Integrated Power Development Scheme and RAPDRP are cited later in this section in the case study subsection.

***Indian Smart Grid Taskforce.*** India Smart Grid Task Force (ISGTF) was set up in September, 2010 under the aegis of the Ministry of Power to serve as Government's focal point for activities related to "Smart Grid" including a road map for implementation of Smart Grids in India. The main functions of ISGTF are to ensure awareness, coordination, and integration of the diverse activities related to Smart grid technologies, practices and services for Smart Grid Research and Development, coordinate and integrate other relevant intergovernmental activities, collaborate on interoperability framework, review and validate recommendations from India Smart Grid Forum, etc.

***Smart City and AMRUT Projects.*** A smart city is an urban center of the future which is ideally environmentally green, safe, and efficient. It owns advanced technical and infrastructural assist to provide high quality of life with economical growth. ISO37120 standards are drafted for framework and guidelines of smart city. Indian central government has launched three mega projects 100 Smart Cities project, AMRUT City projects, and Housing Scheme for all for transforming urban India.

***India Smart Grid Forum.*** India Smart Grid Forum (ISGF) is a non-profit voluntary consortium of public and private stakeholders with the prime objective of accelerating development of Smart Grid technologies in the Indian Power Sector. In 2010, ISGF was set up to provide a platform through which different wings such as utilities, industry, academia, and other stakeholders could participate in the development of Indian smart grid systems by giving their relevant inputs to the government's decision-making. The aim of the Forum is to comfort the Indian power sector to use Smart Grid technologies in an efficient, cost-effective, new and scalable manner by bringing all the main stakeholders and technologies together. ISGF leverage the global experience and standards by coordinating and cooperating with significant global and Indian bodies.

***Smart Grid Pilot Projects.*** Ministry of Power has allotted 14 Smart Grid Pilot Projects to be implemented by state-owned Distribution Utilities. The projects incorporate automated metering infrastructure (AMI) for residential and industrial consumer, different portals such as community Portal, Consumer Portal, employee portal, etc. and data analytics for decision making and support. These projects cover or focus on automated metering infrastructure, outage management system, peak load management, power quality management, etc. Various benefits are envisaged such as aggregate technical and commercial losses are reduced, peak load consumption is reduced, reduction in failure unforeseen outages and recovery time, reduction in billing, meter reading, maintenance cost, etc.

The smart grid projects in India amalgamate power sector with communication sector and manage energy consumption, peak load shifting, supply demand management system, improves grid stability, etc. These projects open up the markets, introduce flexible tariff system and empower end-consumer. A saturated rollout is needed to be achieved for the capital investment. Enhancement in perceptive of consumer behavior gives a very positive impact on the smart grid concept and has increased competitive environment leading to open access market. Meeting regulatory requirements and expectation is another challenge. Proper communication for data collection and management of these enormous and complex data are needed to be accounted. The deployment of smart grid and smart city can help Indian power system to achieve significant goals.

### 4.5.10 References

[1] Thomas, Mini S. and McDonald, John D.: "*Power system SCADA and Smart Grids*," CRC Press, Taylor and Francis, USA, April 2015.

[2] Agrawal, V. K., Porwal, R. K., Kumar, Rajesh, Vivek, P., Muthukumar, T., Deployment of System Protection Schemes for enhancing reliability of power system: Operational experience of wide area SPS in Northern Regional Power System in India, in *Power and Energy Systems (ICPS), 2011 International Conference on*, 2011, pp. 1–6.

[3] Mukhopadhyay, S., Dube, S. K., and Soonee, S. K., "Development of power market in India," *2006 IEEE Power Engineering Society General Meeting*, 2006, USA.

[4] NITI Ayog: Report of the Expert Group on 175 GW renewable energy by 2022, 2015. http://niti.gov.in/writereaddata/files/writereaddata/files/document_publication/report-175-GW-RE.pdf.

[5] Ministry of Power: "*National Smart Grid Mission*," March 2015. Available at: http://powermin.nic.in/upload/pdf/National_Smart_Grid_Mission_OM.pdf.

[6] Ministry of Power: "*Integrated Power Development Scheme*," December, 2014. Available at: http://powermin.nic.in/upload/pdf/Integrated_Power_Development_Scheme.pdf.

[7] Ministry of Power: "*Guidelines for the Re-structured APDRP during XI Plan*," December 2008. Available at: http://powermin.nic.in/upload/pdf/Guidelines_APDR-P_XI_Plan.pdf.

[8] India smart grid forum: "*Smart Grid Bulletin*," Vol. 1, issue *9, Sept.*, 2014. Available at: http://indiasmartgrid.org/en/Lists/newsletter/Attachments/16/ISGF%20Smart%20Grid%20Bulletin%20-%20Issue%209%20(September%202014).pdf.

[9] Thomas, Mini S.: '*Smart Cities: An Indian Perspective*', IEEE smart grid newsletter, May 2015.

[10] Thomas, Mini S.: "Smart Grid Initiatives in India" IET Engineering & Technology Reference, 2016, 6 pp. DOI: 10.1049/etr.2015.0070, Online ISSN 2056-4007.

## 4.6 INTERCONNECTED POWER GRID IN JAPAN

*BY TERUO OHNO*

### 4.6.1 Introduction

There are 10 vertically integrated electric power companies that own generation, transmission, distribution, and retail in Japan. Among these 10 utilities, power systems of 9 utilities are interconnected mainly for the following purposes (Table 4-4):

- To share generation reserve
- To stabilize frequency
- Power interchange to reduce generation cost under the normal condition
- Power interchange for the power system security under the emergency condition
- Power provision from power stations co-developed by multiple utilities

Figure 4-15 shows the interconnection of power systems in Japan. The interconnection between the eastern and western part of Japan is not a synchronous connection as they use different frequencies. The eastern part of Japan has been using a frequency of 50 Hz since 1896 when the Tokyo Dento Company introduced three-phase AC generators (two units × 265 kW) to its Asakusa Power Station from the German manufacturer AEG (Allgemeine Elektricitäts-Gesellschaft AG). In contrast, the western part of Japan has been using a frequency of 60 Hz since the same year when Osaka Dento Company introduced AC generators (four units × 150 kW) to its Saiwai-cho Power Station from the U.S. manufacturer GE.

**TABLE 4-4** Major Events for the Interconnection of Power Systems in Japan

| Year | Events |
|---|---|
| 1959 | Tohoku EPCO (Electric Power Company) was interconnected to TEPCO (Tokyo EPCO) at 275 kV to send power from newly-developed large hydro power plant, Tagokura Power Station and Honna Power Station. |
| 1960 | Chubu EPCO was interconnected to Kansai EPCO at 275 kV. |
| 1962 | Chugoku EPCO was interconnected to Kyushu EPCO at 220 kV. The interconnection line, Shin-Kanmon Kansen, was upgraded for the 220 kV operation.<br>Chugoku EPCO was interconnected to Shikoku EPCO at 220 kV. |
| 1964 | Hokuriku EPCO was interconnected to Kansai EPCO at 275 kV.<br>All utilities are interconnected on the 60 Hz side by this connection. |
| 1965 | TEPCO was interconnected to Chubu EPCO at 275 kV at Sakuma Frequency Converter Station. This was the first interconnection between the 50 Hz system and the 60 Hz system. |
| 1979 | Hokkaido EPCO was interconnected to Tohoku EPCO through ±125 kV HVDC lines.<br>All utilities are interconnected on the 50 Hz side by this connection. |
| 1995 | The interconnection between Tohoku EPCO and TEPCO was upgraded by the commissioning of new 500 kV interconnection lines. |
| 2000 | The interconnection between Kansai EPCO and Shikoku EPCO was upgraded by the commissioning of new ±250 kV HVDC lines, Kii-Suido HVDC lines. |

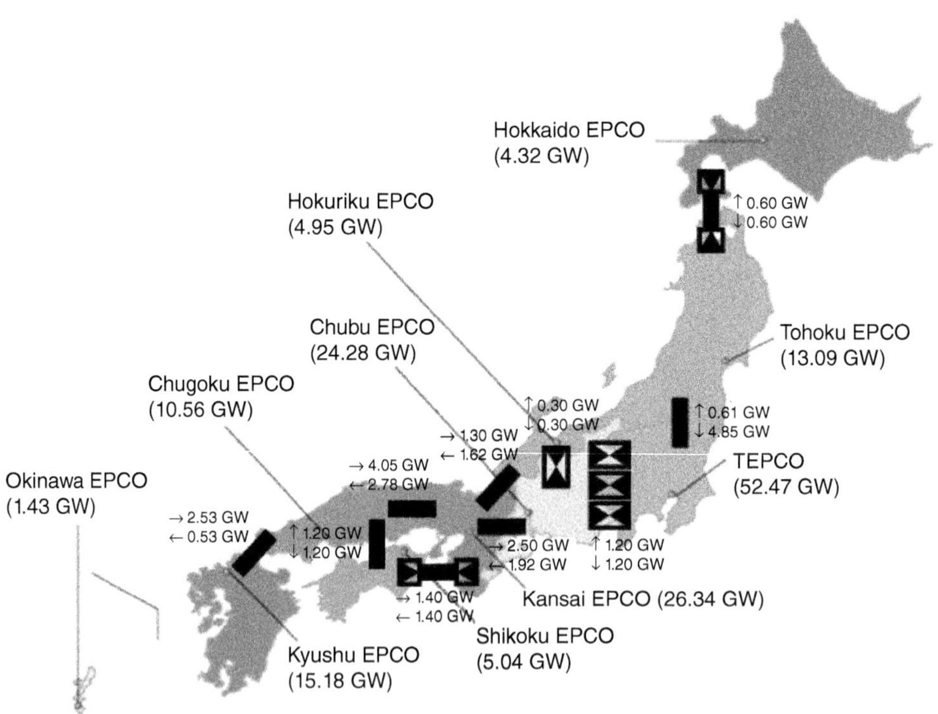


*Note*: Values in parentheses are averages of first three highest forecasted demands in each area in 2016, measured at transmission line ends of power stations. Values by the interconnection are transmission capacities of the interconnection during daytime on weekdays in August 2016.

**FIGURE 4-15** Interconnected power systems in Japan.

The unification of two frequencies has been studied since the 1940s, but the enormous required cost has almost prohibited the unification. The study was recently conducted after the Great East Japan Earthquake in 2011 when rolling blackouts occurred due to the shortage of generation. The study found that the required cost for the unification is approximately 100,000 million USD (1 USD = 100 JPY) for utilities to replace their generation, transmission, and distribution facilities. Additional cost will be necessary for the customer side also to replace their facilities.

All utilities in the eastern part of Japan, Hokkaido EPCO, Tohoku EPCO, and TEPCO, use a frequency of 50 Hz, but Hokkaido EPCO is not synchronously connected to the other two utilities. Also, 1 of 10 utilities, Okinawa EPCO, is not interconnected to the other nine utilities due to its distance.

### 4.6.2 Paradigm Shift after Great East Japan Earthquake

Until recently, it is a prime responsibility of each vertically integrated utility to have enough generation to supply demand in each area. Even though power systems of nine utilities are interconnected, the interconnection was for utilities to help, not to compete, each other. Therefore, the capacity of the interconnection was limited to the minimum required level, considering the cost to build the interconnection. In addition, the number of the AC connection between utilities is normally limited to one, and DC connections were built to enhance the interconnection capacity when necessary. Thanks to this practice, there has not been loop flow between utilities or cascading outages to neighboring utilities.

Other reasons which has been limiting the interconnection capacity include:

- Because of the longitudinal shape of Japan, areas of nine utilities are aligned in series. Since the backbone of Japan is mostly mountainous, it is geographically difficult to build the meshed interconnection.
- The bulk power system in Japan tends to carry large power due to the difficulty to build transmission lines. It is typical for these lines to carry two to three times of SIL (Surge Impedance Loading) by the reactive power support from shunt capacitors. When the power flow through the interconnection is increased, it will, in many cases, lead to the transient instability. To solve the transient stability problem, it is often necessary to upgrade the bulk power system inside utilities.

The Great East Japan Earthquake in 2011 and the shortage of supply after the earthquake have significantly changed the concept of the interconnection explained above. The following roles are now expected to the interconnection:

- Improve the supply reliability under extreme events, such as the Great East Japan Earthquake, by the enhancement of the power interchange.
- Reduce the electricity tariff by the introduction of the higher competition through the interconnection. Realize the EDC (economic dispatching control) with all generators in Japan, reducing a chance of market split.
- Accommodate higher penetration of RES.

In line with this new expectations, a new HVDC line between Hokkaido EPCO and Tohoku EPCO and another new HVDC line between TEPCO and Chubu EPCO are under construction.

To facilitate the extended roles of the interconnection, the Organization for Cross-regional Coordination of Transmission Operators, Japan (OCCTO) was established and has started its operation since April 1st, 2015. The main roles of OCCTO for the planning and operation of the interconnected power system are:

- Aggregate the electricity supply-demand plan and the network development plan submitted by electricity companies.

- Enhance the interconnection capacity including the frequency converter stations between the eastern and western power system.
- Cross-regionally operate the interconnected power system.
- Under the normal condition, cross-regionally control the electricity supply-demand balance and frequency, which are controlled by transmission and distribution companies in each area.
- Under emergency conditions, instruct the adjustment of the cross-regional power interchange and the increase of generation.

### 4.6.3 Planning of Interconnection Lines

OCCTO has established its Network Codes in April 2015 [1]. The Network Codes have been amended three times since then, and the latest version has been effective since July 11, 2016. The Network Codes define the planning and operation of the transmission network, including interconnection lines.

Based on the Network Codes, OCCTO initiates the cross-regional network development planning process by a proposal from OCCTO itself or electric power suppliers or by a request from the government. The proposal from OCCTO itself is divided into two categories: a proposal to enable the stable supply and a proposal to facilitate the cross-regional trade.

To enable the stable supply, OCCTO proposes the start of the cross-regional network development planning process under the following conditions:

- Unplanned outages of generators led to a power interruption even with the cross-regional power interchange was adjusted to the total transfer capacity.
- The development of interconnection lines is found to be necessary for the stable supply as a study of extreme contingencies and disasters.

Additionally, the start of the cross-regional network development planning process is proposed by OCCTO to facilitate the cross-regional trade, for example, under the following conditions:

- The available transfer capacity of interconnection was equal to or less than 5% of the total transfer capacity for the duration which was equal to or longer than 20% of total time of the past 1 year.
- In the annual usage plan of interconnection lines, the available transfer capacity of interconnection is equal to or less than 5% of the total transfer capacity for the duration which is equal to or longer than 20% of total time of the annual usage plan.
- In the long-term usage plan of interconnection lines, the available transfer capacity of interconnection is 10% or less for three or more fiscal years.

The cost of the development of interconnection lines is borne by beneficiaries. For example, when the interconnection line helps the stable supply by increasing the possible cross-regional power interchange in case of a lack of supply due to a disaster, the cost of the development of the interconnection line is borne by transmission and distribution companies whose stable supply is enhanced by the interconnection line. The cost is eventually shared by electricity users in the service area of the transmission and distribution companies.

### 4.6.4 Future Outlook

After the Great East Japan Earthquake, the future of the energy and electricity supply has been discussed throughout Japan. In 2015, as a result of the discussions, the Japanese government established the energy mix in 2030, considering 3E+S: energy security, economic efficiency, environment, and safety. According to the energy mix, the RES will account for 22% to 24% of the electricity supply in 2030 as shown in Fig. 4-16, while the RES was approximately 10% of the electricity supply in 2015 (Table 4-5).

**TABLE 4-5** Integration of PVs in Japan

| Area (EPCO) | Connected PVs (GW)* | Lowest peak demand (GW)† | Percentage (%) |
|---|---|---|---|
| Hokkaido | 1.02 | 3.08 | 33.1 |
| Tohoku | 2.67 | 7.91 | 33.8 |
| Hokuriku | 0.61 | 2.52 | 24.2 |
| Chugoku | 3.02 | 5.54 | 54.5 |
| Shikoku | 1.84 | 2.65 | 69.6 |
| Kyushu | 6.24 | 7.88 | 79.2 |

*As of the end of July 2016, except for Hokuriku (August 5, 2016) and Chugoku (August 26, 2016).

†At 1 pm on May 12, 2014, except for Hokkaido (at noon on May 26, 2014) and Shikoku (at noon on May 12, 2014).

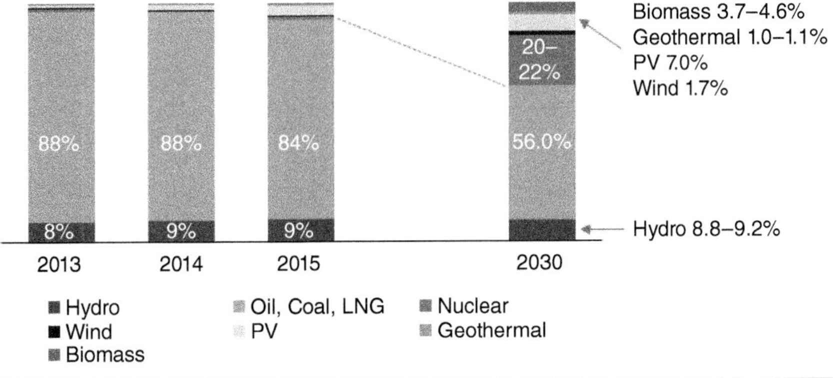


**FIGURE 4-16** Energy mix in Japan in 2030.

Due to the recent increase in PVs, however, the supply-demand operation is already an issue in utilities who are experiencing relatively high penetration of PVs for their demand levels. To solve the issue and achieve the target energy mix in 2030, it is important to use the interconnection more effectively and further facilitate the cross-regional operation. For example, the transaction of regulating power through the interconnection is considered to accommodate more RES.

In addition to RES, the smart grid and smart community business is also growing with the help of the public funding. These businesses aim to extend across the interconnection in the future, and the interconnection is expected to accommodate the transaction. The role of the interconnection will become wider and more important in the future.

## 4.6.5 Reference

[1] Organization for Cross-regional Coordination of Transmission Operators: "Network Codes," April 2015. Available on the Web: https://www.occto.or.jp/en/companies/guideline/files/network_codes_20160928.pdf.

# 4.7 INTERCONNECTED POWER GRID IN NORTH AMERICA

*BY SARMA NUTHALAPATI, SPENCER BURKS, AND KRISTIAN KOELLNER*

## 4.7.1 Introduction

The first power grid in North America was placed into service in the 1880s in New York City where a small DC grid was created to feed electric lighting in Manhattan [1]. In the decades to follow, small generating plants were constructed in city centers to light up those towns. In the decades to follow, these smaller grids began to interconnect, particularly during the First World War, and they continued to grow through the Second World War [2,3]. During the 1930s, efforts were made to provide power to rural areas and connect remote hydroelectric resources to load centers, and entities such as the Tennessee Valley Authority (TVA), Bonneville Power Administration (BPA), and the Lower Colorado River Authority (LCRA) were created.

As electricity use continued to grow, larger capacity generators and transmission lines were constructed. In 1978, the Public Utilities Regulatory Policies Act was passed which encouraged the creation of power markets for non-utility power producers, moving power grids towards privatization.

In response to the Northeast blackout in August 2003 that affected over 50 million people in the United States and Canada, The Energy Policy Act of 2005 was passed giving the Federal Energy Regulatory Commission (FERC) the authority to enforce reliability standards for the bulk electric system in the United States [4]. FERC then certified the North American Electric Reliability Corporation (NERC) as the "electric reliability organization" which now further develops and enforces these reliability standards in the United States, Canada, and part of the Baja California peninsula in Mexico.

Today, there are four main interconnections in North America which have evolved and continue to operate asynchronously at 60 Hz.

## 4.7.2 Structure

The four main interconnections in North America (Fig. 4-17):

- Western Interconnection—One of the two largest interconnections in North America. It stretches from the Western Canadian Provinces of British Columbia and Alberta through the Western states of the United States and down to Baja California.
- Eastern Interconnection—The other of the two largest interconnections in North America. It includes central and eastern provinces in Canada and central and eastern states in the United States, but excludes most of Texas and Quebec.
- The Electric Reliability Council of Texas (ERCOT)—This smaller interconnection consists of approximately 85% of the electric load in the state of Texas with DC interchanges between the Eastern and Western Interconnections and part of Mexico [5].
- Quebec Interconnection—wholly operated by Hydro Quebec in the Canadian Province of Quebec with DC interchanges with New Brunswick, Ontario, and the U.S. Northeast [6].

Each interconnection consists of an electric power transmission system which allows for the bulk transfer of electricity between power generation sites, such as power plants, and load-serving substations. These substations are then the starting point of the power distribution system which connects to all end-use entities such as residential homes and businesses.

## 4.7.3 Voltage Levels

Voltage levels in North America typically range from 69 to 765 kV for transmission and 2.4 to 35 kV for distribution. Standard residential service is 120 V/240 V in single phase, while some industrial services can be 480 V to 4.16 kV three phases (Fig. 4-18).

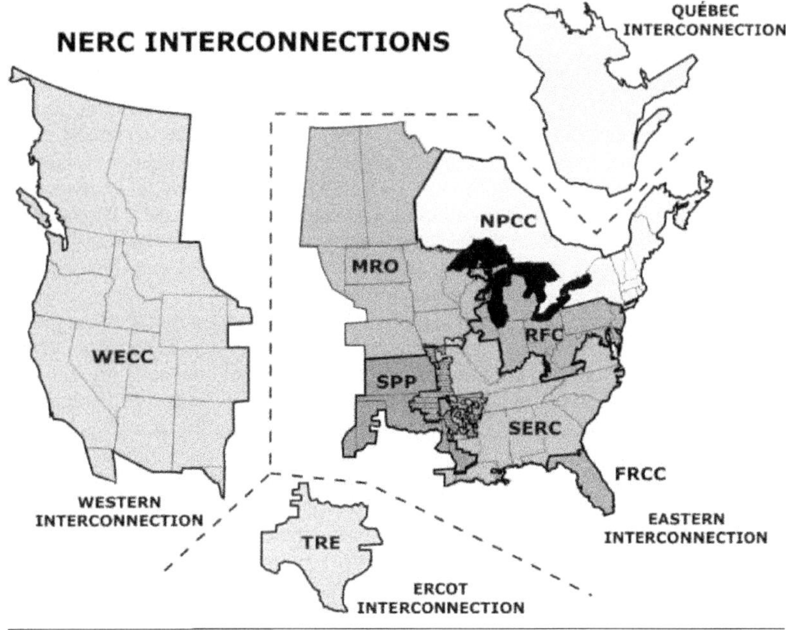


**FIGURE 4-17** Interconnections in North America.
*Source*: www.nerc.com.

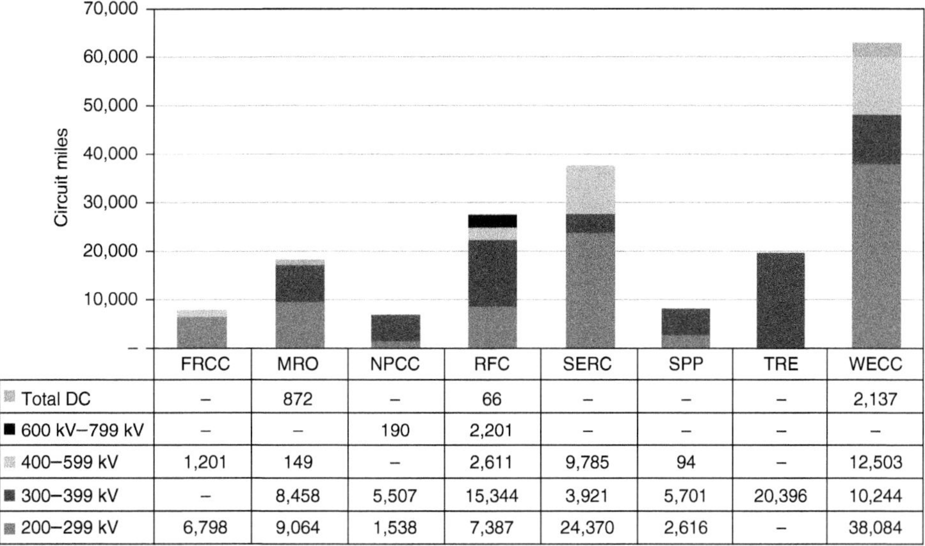


| | FRCC | MRO | NPCC | RFC | SERC | SPP | TRE | WECC |
|---|---|---|---|---|---|---|---|---|
| Total DC | – | 872 | – | 66 | – | – | – | 2,137 |
| 600 kV–799 kV | – | – | 190 | 2,201 | – | – | – | – |
| 400–599 kV | 1,201 | 149 | – | 2,611 | 9,785 | 94 | – | 12,503 |
| 300–399 kV | – | 8,458 | 5,507 | 15,344 | 3,921 | 5,701 | 20,396 | 10,244 |
| 200–299 kV | 6,798 | 9,064 | 1,538 | 7,387 | 24,370 | 2,616 | – | 38,084 |

**FIGURE 4-18** Existing transmission as of last of 2014.
*Source*: *Developed by DOE from NERC (2015b) http://www.nerc.com/pa/RAPA/tads/Pages/default.aspx.*

### 4.7.4 Functional Tiers of the Power Grid

In North America, interconnections are typically split into three separate categories—transmission, generation, and distribution. Transmission (generally above 60 kV) consists of interconnected transmission owners and operations mixed together to form the larger interconnection. Local control centers operates their independent transmission networks in coordination with a larger operator for the interconnection. Generation consists of independent generation stations operated on a unit by unit basis. They bid into power markets to supply power and are dispatched in the most economical way possible while maintaining grid reliability requirements. Distribution (generally below 60 kV) begins at load-serving substations connected to the transmission system and provides lower voltage paths from the substations to each individual residential, commercial, and industrial customer.

The most basic task of grid operators is to manage appropriate power flows and voltage levels. On the transmission system, bulk power is transferred along transmission lines from generation sources to load-serving distribution substations. Transmission conductors have thermal limits which cannot be exceeded without damage to the conductor or causing the conductor to sag beyond safe clearance levels. In many locations, autotransformers are used to step up or down voltage levels to efficiently transport the power flows across a further distance. These autotransformers also have thermal limits which cannot be exceeded without damaging the insulation on the windings inside the transformer. Damage to insulation increases aging of the transformer until it eventually fails or needs to be replaced.

For the most part, electricity demand is not controllable and power generation must match demand in real-time. Grid operators use centralized control systems to dispatch generation units to control these power flows. Generators are dispatched in the most economical way possible without violating thermal ratings on equipment. Good utility practice dictates that the generation units should be dispatched in a manner such that the power grid can lose any single element without causing any thermal limits to be exceeded, i.e. "N-1 secure". Phase-shifting transformers may be installed to give operators additional control over real power flows across key transmissions elements. In some circumstances, dynamic stability issues are identified and are factored in to generation dispatch in addition to the thermal limits.

Another consideration with power flows is voltage angle separation. Since transmission elements are primarily inductive by nature, power flows lead to differences in voltage angle across large geographical areas. If the difference becomes too large, system separation can occur and result in wide-area blackouts. Grid operators actively monitor voltage angle conditions to ensure that this does not occur. By controlling regional generation dispatch, power flows can be altered to minimize the angle separation across the transmission system.

Grid operators control voltage on the transmission system using both static and dynamic reactive control devices. The primary source of reactive power are steam-based thermal generation units operating synchronous machines with controllable excitation systems. These units stabilize voltage and maintain a reactive margin to respond to dynamic grid events such as line faults and generator trips. Modern renewable energy generators are also able to function as reactive resources when their energy source is available, but they accomplish this using specialized control systems since they are typically coupled to the grid using power electronics instead of synchronous machines. Since the locations of generators in a power system are not always where the reactive resources are needed the most, static reactive devices such as capacitor or reactor banks are installed and switched as needed. For grid locations which either require constant precision or are highly sensitive to varying power flows, many flexible reactive devices are installed and controlled through power electronics in the form of Static VAR Compensators (SVCs) or, alternatively, a Static Compensator (STATCOM).

### 4.7.5 System Protection

An important issue in operating the transmission system is managing system faults and equipment failures which result in short-circuit conditions. During these events, voltage is suppressed across the system, sometimes across hundreds of miles, and large currents flow from generators and inductive motors. It is imperative that faulted elements are isolated from the power system as quickly as possible.

Suppressed voltages, if not resolved quickly, can lead to motor stall—a condition where connected inductive motors experience a locked rotor condition. Motor stall will increase the reactive demand on the system and further suppress voltage until there is a wide area voltage collapse. Fault currents increase loading on transmission elements and can cause permanent damage to transmission lines, induce transformer insulation failures, and cause generators to trip offline. The largest risk of poor system protection is a dynamic instability event which results in cascading loss of load, generation, and transmission elements. Power quality during faults is another issue affecting customers such as industrial plants with sensitive equipment. A lengthy fault and voltage sag can result in halting production, loss of product, and even safety concerns at these facilities.

Isolation of faults is achieved using elaborate relaying schemes connected to circuit breakers which identify and switch out faulted elements. The primary relay scheme utilized at the transmission level are distance relays which are configured using known system impedances to determine the distance to the fault based on current and voltage measurements. If the fault is located within a defined zone, then the relay will trip accordingly. Due to the inherent variance in these measurements due to changing system configurations and differences in fault impedances, and the need for redundancy, relays are configured with multiple zones which are time coordinated to provide overlapping coverage and allow for the correct element to be isolated. In modern relay schemes, high-speed pilot communication between relays allows the protection scheme to reduce fault clearing times by validating fault location and eliminating the need for time-based coordination. As a secondary measure, inverse time over current relays are installed to detect fault currents on phase or neutral conductors. Improper relay coordination or the failure of relaying and communication equipment can lead to equipment damage, over-tripping, cascading outages, or uncontrolled generator separation.

The majority of faults are phase to ground, and common causes of faults include lightning, animal contact, failed structures and insulators, transformer insulation failures, and foreign contact on exposed conductors.

Differential schemes are also used for protection of transmission elements. Differential schemes rely on Kirchhoff's current law (KCL) and monitor all current inputs and outputs in a given zone, and if these values do not sum to zero then circuit breakers isolate the element. The benefit of differential schemes is that they operate quickly and do not require any coordination with other relaying equipment since the location of the fault is already known.

### 4.7.6 HVDC

In North America, High Voltage Direct Current (HVDC) is commonly used to allow for power transfers across long distances or between the asynchronous interconnections. Long distance HVDC lines can be used to connect generation-rich regions such as hydropower in Northern Oregon to load centers such as in Southern California at lower cost and with more controllability than AC alternatives. Cost benefits arise from reduced structure and conductor requirements and smaller Right-of-Way requirements. In cases of generation shortage or arbitrage pricing between interconnections, the HVDC ties act as generation sources and sinks in respective transmission systems.

HVDC ties exist between the Eastern, Western, and ERCOT interconnections. The smaller ERCOT and Quebec interconnections have additional HVDC ties with Mexico and the Eastern Interconnection, respectively.

### 4.7.7 Distribution System

The majority of distribution system are built with three-phase radial circuits called feeders. The feeders are routed through cities and rural areas at primary voltages between 2.4 kV and 35 kV. Individual three-phase and single phase circuits are tapped off of this circuit and provide a primary voltage source near customer property. Smaller transformers are then attached to these lines to provide secondary voltage to customers and include a low voltage service line attaching to the meter for individual premises. The secondary voltage of these service lines typically ranges from 120/240 V for homes to 277/480 V for larger commercial or industrial customers. However, some customers will take primary voltage service and manage their own secondary system. These customers are typically large industrial customers with either sensitive load requiring special power quality equipment or of a large enough scale to create cost savings relative to typical utility rates.

Characteristics of these distribution systems vary based on the nature of load they serve. Urban areas have higher load density that reduces the distance between load-serving substations to approximately 3 to 5 miles apart. This results in shorter feeders and allows for many ties between feeders. These ties allow for load to be transferred between feeder circuits during distribution outages. In some regions, distribution automation systems are installed to coordinate the restoration of large sections of load in response to faulted sections of line. Some commercial and industrial customers who require high reliability such as universities, hospitals, and sports stadiums will be configured with multiple distribution feeds and equipment to switch between the feeds if any outage occurs at any point of time.

Rural areas have much lower load density and feeders tend to extend in length. Some rural feeders are 20 to 30 miles in length without the ability to move load during outage conditions. Reliability can be relatively worse in these areas due to the large amount of line exposure and lack of alternative feeders. Higher distribution primary voltage (22 to 35 kV) is more common in these areas due to voltage drop effects.

Some downtown areas with extremely high residential or commercial density will have highly redundant distribution systems known as distribution networks. These system are configured with multiple feeders from multiple substation power transformers for each individual premise. This allows for one or more distribution faults to occur without interrupting customer service. Downtown networks are typically installed in duct banks under city streets and in vaults located inside of large skyscrapers. The vaults enclose distribution transformers which step down from primary to secondary voltage level to be distributed throughout the building. If the build is too tall to effectively distribute power at secondary voltage, then multiple vaults will be installed many stories above ground level with primary conductors running vertically through the building. Customer outages are extremely uncommon in these networked distribution systems, but higher fault duty inside of small electrical vaults raises concerns over arc-flash safety.

Voltage is controlled on the distribution system to maintain voltage at each customer's meter. Substation power transformers are typically equipment with either a load tap changer or stand-alone regulators which adjust the voltage on the low-side of the transformer. This shields the distribution system from any steady-state voltage fluctuations on the transmission system and accounts for varying load levels.

The majority of customers load consists of inductive motor load which has high reactive power demands. It is common for distribution entities to install capacitor banks along feeders to improve the load power factor as seen by the transmission system and "flatten" the voltage profile along a feeder. Longer feeders, such as in rural areas, may rely upon these capacitors or downstream regulators to maintain proper voltage to the meter. Modern day distribution control centers will monitor the voltage along these feeders and switch capacitor banks in and out of service to correct both voltage and power factor. Power factor correction also reduces overall losses in the distribution system which falls under some utilities' Volt-VAR Optimization (VVO) programs for energy efficiency.

In a radial system, coordination and fault location is simplified. Protection begins inside the substation with feeder circuit breakers which monitor for and isolate faults down the line. For longer feeders, distribution reclosers are installed on the line such that a fault at the end of the feeder can be isolated without causing an outage to the entire feeder. Taps off of the main feeder circuit and secondary transformers are fused to help minimize the outage impact and aid in locating a fault in the distribution system. Modern distribution control centers will monitor smart meters to determine which upstream protective element operated and dispatch personnel to the site as quickly as possible. Reclosing schemes are common on feeder circuit breakers and reclosers with an initial fast trip time and reclose cycle to attempt to clear temporary faults before fuses have time to operate. This is known as a "fuse-saving" scheme.

Communication equipment is typically not required for distribution relaying and inverse time-over-current relays are coordinated to allow downstream devices to operate first. Communication equipment is used for locations with automatic fault isolation and restoration as discussed previously or in some locations with large distribution generation (DG).

DG—the installation of small generators on the distribution system—is becoming increasingly common in North America. Residential solar installations have a minimal impact on the distribution system infrastructure, but some larger DG installation may require special considerations. Large installations (greater than 5 MW) may be served via an "express feeder" which is a dedicated distribution feed for the DG. Industrial plants with on-site generation are beginning to take advantage of some market programs to generate on the grid when electricity prices are high and available generation on the transmission system is scarce and system protection must be able to correctly isolate faults with power flows in multiple directions.

### 4.7.8 Generation

In North America, utility scale power generation is predominantly steam based thermal generation using coal, natural gas, or nuclear fission as heat sources. The steam is then converted into electric power using synchronous machines which convert mechanical energy to electrical energy. Coal has historically been the dominant form for power generation until recent times where natural gas has begun taking its place due to availability, economics and emissions.

Renewable energy in the form of hydroelectric power has been in place in North America for decades and remains the largest source of renewable electric power in the United States. Wind and solar have been steadily growing as electricity sources over the last decade due to changes in technology and governmental subsidies. Biomass and geothermal energy sources also exist, although at a smaller penetration level (Fig. 4-19).

Grid level storage also exists at the utility scale. Pumped hydroelectric storage, such as TVA's Raccoon Mountain facility where water is pumped into a lake which can be released through hydrogeneration units at any time, is the main source of storage. Thermal, compressed air, battery, and flywheel based storage also play significant roles (Fig. 4-20).

### 4.7.9 Planning

***Distribution Planning.*** Planning for the power grid occurs in three inter-related stages—distribution, transmission, and generation—with the most basic form being the distribution system by distribution planning engineers. System expansion projects are either linked to specific new customers, such as industrial plants and commercial or residential developments, or regional changes in system demand.

For new customer projects, builders and developers work directly with distribution utilities to provide service to new premises. Larger customer connections are routed through distribution planning groups for study. These distribution planners maintain computer-based models of the distribution system to perform load flow studies to assess if any system upgrades are required to support the new load. These loads and system upgrades are then incorporated into 1- to 5-year planning models that also factor in year-over-year aggregate increases in load. These increases are typically determined utilizing recorded feeder-level data or aggregating meter data. Planning engineers then make system upgrade recommendations to support changes to system load and reliability criteria. Overall increases in demand may lead to the installation of new distribution substations, new or larger substation transformers, new feeders, or upgrades to existing feeders. As new feeders and substations are installed, existing feeders are split into segments with normally open switches to connect the feeders together. This allows for utilities to move load between feeders in the case of localized power outages. Reducing the overall length of feeders also has the benefit of reducing the frequency of outage occurrences due to reduced exposure to typical causes of faults and minimizing the number of outage premises (Fig. 4-21).

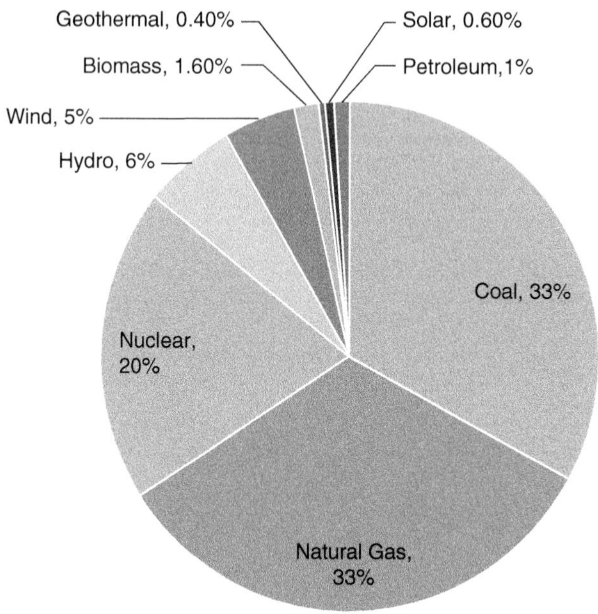


**FIGURE 4-19** U.S. electricity generation by energy source, U.S. Energy Information Administration, 2015. https://www.eia.gov/tools/faqs/faq.cfm?id=427&t=3

***Transmission Planning.*** The next level of planning is for the transmission system. Since the transmission system is an interconnected network, in comparison to radial distribution feeders, multiple transmission owners coordinate within their respective interconnections to develop common transmission models.

Common transmission models are used for long-term transmission planning, usually ranging from 1 to 10 years. Distribution planning entities will submit their forecasted load data based upon

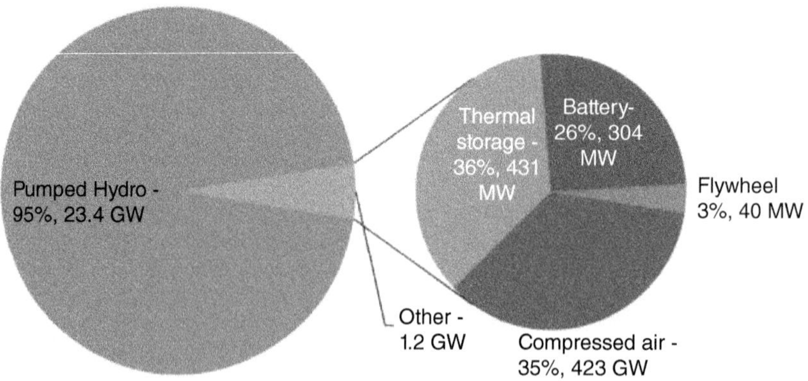


**FIGURE 4-20** Rate power of U.S. grid storage projects, Department of Energy, 2013. http://energy.gov/sites/prod/files/2014/09/f18/Grid%20Energy%20Storage%20December%202013.pdf.

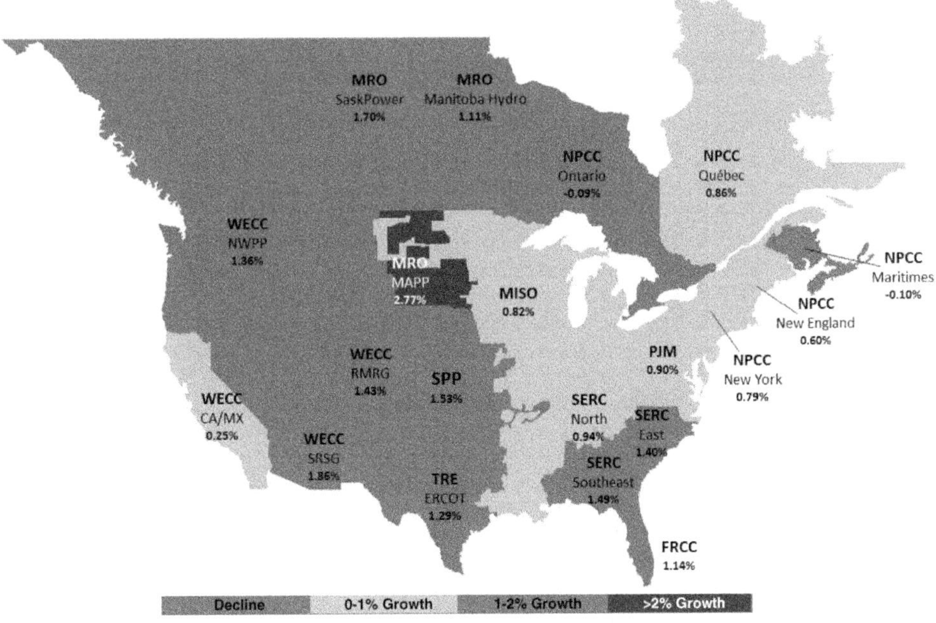


**FIGURE 4-21** Ten-year compound growth rate, NERC 2014.
*Source*: http://www.nerc.com/pa/RAPA/ra/Reliability%20Assessments%20DL/2014LTRA_ERATTA.pdf.

known connection projects and forecasted regional growth for use in these transmission models. Known generation interconnection additions and retirements are also factored into these models. Individual transmission entities and regional transmission entities will recommend projects based on established system security requirements and economic options are chosen. These projects may consist of, for example, installing larger conductor on existing transmission lines, increasing overall autotransformer capacity at a specific substation, upgrading terminal equipment on transmission elements to increase capacity of existing assets, the construction of new transmission lines and substations, or converting select transmission elements to a higher voltage level.

The construction of new transmission elements is a much more involved process for transmission than distribution. At distribution levels, small utility easements along roads are usually sufficient for installation of new facilities. In contrast, transmission facilities are much larger and frequently invoke the use of eminent domain. This requires transmission owners to prove the necessity of such a project with affected parties and regional Public Utility Commissions (PUCs). Once the necessity is determined, transmission planners propose multiple routes and options to affected property owners. Ultimately, a route is decided upon and approved by the PUCs. This process becomes more challenging for projects crossing state boundaries where multiple PUCs are involved.

Similarly to distribution, transmission planners also perform studies for new load serving substations and generation interconnections. These studies are run to assess if there will be system security issues which arise from the new addition and to recommend any system expansion projects or improvements if required. For generator interconnections, transmission planners may also study and recommend projects to ensure that generators will be able to operate at an economic output with constraints from the transmission system.

***Generation Planning.*** Generation planning differs by region, with some regions operating with both energy and capacity markets and others with energy only markets.

Energy only markets (e.g. ERCOT), as the name suggests, only purchase energy and energy services needed to operate the grid. Energy and "ancillary" energy services are sold in real-time or in day-ahead markets. The economics behind these markets allow for "surge pricing" when demand is high and supply is low. The assumption is that the long-term equilibrium state of prices will adjust such they support fixed costs associated with the construction of power plants.

Capacity markets differ in that they create an avenue for regional planning entities to influence new generation construction. In large, new generators must bid into an auction to win the right to install generation capacity. The winners of the auction are then able to recover costs associated with construction of a new plant equal to their bid. Requests for bids are based upon forecasted demand levels, adjustments in existing generation supply, and the need to maintain a predetermined reserve margin to ensure system security. In the United States there are organized capacity markets in four ISOs: the PJM Interconnection; the New York ISO; ISO-New England; and the Midcontinent ISO [7].

Both energy and capacity markets typically have mechanisms which reward generation in some locations instead of others. The higher-valued locations are typically closer to system load have fewer transmission constraints for new generation additions, and benefit more for ancillary energy services. Examples of ancillary energy services include frequency regulation (maintaining 60 Hz), spinning reserves (in the case of generation trips), fast-frequency response (similar to spinning reserves but addresses dynamic grid responses), and black start units (which are able to quickly and reliably start in the extreme case of a wide-area blackout). Both markets also have mechanisms which allow demand response technologies to function as viable substitutes for additional generation capacity.

***Other Aspects of Grid Planning.*** Grid planning considers additional factors beyond those discussed above, including many aspects guided by NERC reliability standards, regional criteria, and utility's individual criteria. Power angle and voltage stability limits are also determined by engineers and projects are recommended to mitigate the threat of system angle separation and wide-area voltage collapses. Planning to limit the available fault duty (i.e. short circuit level) for energized equipment is also a consideration.

Frequency response is also monitored to ensure that generation markets allow for the operation of the power grid without leading to cascading generation or load outages due to transient frequency responses improperly damped. Frequency response is a function of net rotating inertia on the power grid at any given point in time. After a generator trip occurs, inertial energy is transferred to the grid to arrest the decaying frequency until a new steady-state equilibrium is reached. If frequency dips too low, motor loads can stall and generation protection schemes or underfrequency load shed (UFLS) can actuate. As a larger proportion of generation made up of solid-state coupled RESs such as wind and solar are connected, the monitoring of frequency response becomes increasingly important.

Due to the importance of dynamic planning criteria in preventing low-risk high-impact events, model validation has become increasingly important. Measurement equipment with high sampling rates are used for this purpose. A technology known as Phasor Measurement Units (PMUs) have been deployed across North America in recent years to gather data necessary to create dynamic planning models. In the Western Interconnection, grid operators even use a 1400-MW "braking resistor" at the BPA's Chief Joseph substation to manually induce a controlled, transient event which can be used to record the system response across a wide area [8].

Integration of renewable resources and grid storage are another concern. Due to the intermittency of renewable resources such as wind and solar, planning entities will study an interconnection's ability to respond to events such as decreases in wind energy and forecast the amount of installed capacity which can be relied upon over peak demand conditions. Another facet of renewables is building infrastructure to support it. Large transmission investments were made decades ago to support hydroelectric power and more recently to support wind energy. These projects connect generation resources which are located far away from load centers. The Competitive Renewable Energy Zone (CREZ) facilities in the ERCOT interconnection is an example of such a project [9].

Another consideration is Sub-Synchronous Resonance (SSR). This electromechanical phenomena occurs when transmission series capacitors alter system impedance such that a resonance condition exists between the transmission system and turbine generators. When conditions are unfavorable, a turbine can oscillate at a frequency less than 60 Hz causing stress to turbine generators which can quickly lead to shaft damage. Planning engineers must study locations where SSR can potentially occur and take prevention measures.

A more unexpected consideration for planning Engineers in North America is driven by space weather. A Coronal Mass Ejection (CME) by the sun can create a phenomenon known as a Geo-Magnetic Disturbance (GMD) where large currents flowing through Earth's atmosphere induce a quasi-DC current flow in the transmission system. The current can lead to excessive heating, high reactive power consumption, and harmonic injection by autotransformers due to half-cycle saturation of these non-linear inductors. The tripping of shunt capacitors combined with increased reactive power demand can lead to grid situations where a fault can lead to a wide-area outage. In March 1989, a severe geomagnetic storm actually caused a massive blackout to Hydro-Quebec's power grid [10]. The event is of particular concern for grid operators in Canada and Northern states which are impacted the most. Planning entities will perform studies to determine the impact of GMD and develop measures to prevent any potential grid outages.

### 4.7.10 References


[1] "Edison's Miracle of Light," Public Broadcasting Station (PBS); http://www.pbs.org/wgbh/americanexperience/features/introduction/light-introduction/.

[2] "Transmission Issues and Power Exchanges in Texas," Harold L. Hughes, Public Utility Commission of Texas; http://oaktrust.library.tamu.edu/bitstream/handle/1969.1/92173/ESL-IE-92-04-09.pdf.

[3] "Energy In Brief," U.S. Energy Information Administration; http://www.eia.gov/energy_in_brief/article/power_grid.cfm.

[4] "The Public Utility Regulatory Policies Act," The National Museum of American History; http://americanhistory.si.edu/powering/past/history4.htm.

[5] "ERCOT Quick Facts, April 2013," ERCOT ISO; http://www.ercot.com/content/news/presentations/2013/ERCOT_Quick_Facts_Apr%202013.pdf.

[6] "Transmission System Overview," Hydro-Quebec; http://www.hydroquebec.com/transenergie/en/reseau-bref.html.

[7] "Marginal Success—Capacity Markets in the U.S.," Platt's; https://www.platts.com/IM.Platts.Content%5Caboutplatts%5Cmediacenter%5Cpdf%5Cinsightdec13_uspower.pdf.

[8] "A Tutorial on Detection and Characterization of Special Behavior in Large Electric Power Systems," Pacific Northwest National Laboratory; http://www.pnl.gov/main/publications/external/technical_reports/PNNL-14655.pdf.

[9] "CREZ Transmission Lines," The Texas Tribune; https://www.texastribune.org/tribpedia/crez-transmission-lines/about/.

[10] "Mach 1989, Quebec experienced a blackout caused by a solar storm," Hydro-Quebec; http://www.hydroquebec.com/learning/notions-de-base/tempete-mars-1989.html.


## 4.8 INTERCONNECTED POWER GRID IN SOUTHERN AFRICAN COUNTRIES

*BY KOMLA A. FOLLY, KEHINDE AWODELE, LEANDRO KAPOLO, NHLANHLA MBULI, MARTIN KOPA, AND OLADIRAN OBADINA*

### 4.8.1 Introduction

Africa is blessed with energy sources (both renewable and non-renewable) vast enough to meet all energy needs. In the last few decades, progress has been made with respect to energy development in the SADC region, with several strategic plans being enacted. However, in terms of electricity access, SADC region lags behind other regional economic communities in Africa. More than two-third of the population in the SADC region do not have access to electricity. The situation in the rural areas is worst, on average about 95% of the people living in rural areas do not have access to electricity. Although SADC member states have abundant energy resources (both renewable and non-renewable), they are unable to effectively tap into these resources. As a result, energy production and consumption throughout the region are unevenly distributed. There is an unprecedented opportunity in the SADC member states to explore aspects of smart grid concepts that could improve the reliability, security and efficiency of the electricity power network that will contribute to the region's economic and environmental health.

## 4.8.2 Evolution of Southern Africa Power Pool

***Historic Development.*** In 1980, southern African countries that had attained political independence and majority rule, otherwise known as Front-Line States, convened an "Economic Summit of the Majority-Ruled States of Southern Africa" in Lusaka, Zambia. That summit created the Southern Africa Development Coordination Conference (SADCC). Over a decade later, at the 1992 summit held in Windhoek, capital of newly independent Namibia, the SADCC Heads of State and Government signed a Treaty transforming SADCC into Southern African Development Community (SADC).

The event was a culmination of a process, which had fostered the experience of working together and a sense of regional identity. Figure 4-22 shows the historic development of the power system interconnection in SADC.

The idea of power trade within southern Africa stretches back as far as 1906 when the Victoria Falls Power Company was registered in Southern Rhodesia "to harness the Victoria Falls and supply electricity to the mining industry on the Witwatersrand" in South Africa. This vision could not be realized at the time because the technology to transmit power over long distances did not exist. The transmission of power from the Zambezi to the Witwatersrand had to wait until the 1970s, when Cahora Bassa was built. Down South, abundant coal resources in South Africa provided power to the mines in the Witwatersrand.

*The Cahora Bassa—SA Link.* In the late 1960s, the Portuguese colonial regime in Mozambique began investigations into the development of a major power complex downstream of Kariba, at

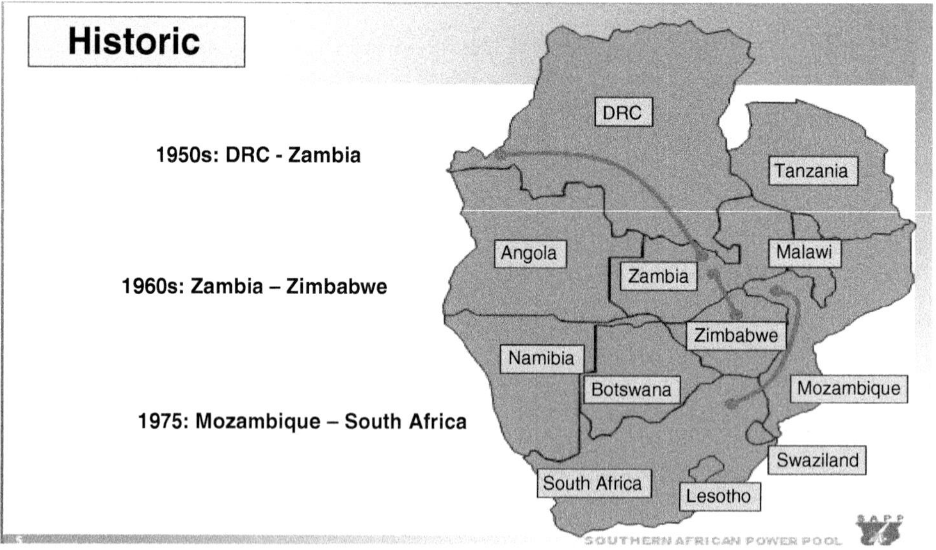


**FIGURE 4-22** Historic development of the Power Interconnection in SADC [1].

Cahora Bassa on the Zambezi. The only market large enough for the proposed 2075 MW station was South Africa. Eskom and other South African companies were heavily involved in the planning and execution of the project, which was completed over the period 1969 to 1978. By the time that the last turbines were commissioned, Mozambique was embroiled in a civil war and the 1360 km HVDC power line was put out of operation by sabotage attacks in 1981. The line was only restored to full operation 17 years later in 1998.

*Development of Kariba.* In the Federation of Rhodesia and Nyasaland, the Kariba hydroelectric scheme was developed to supply the Northern Rhodesian (Zambian) Copperbelt and the mines and industry in Southern Rhodesia (Zimbabwe).

The Kariba scheme was the first significant grid interconnection involving the construction of the Kariba Dam and 666 MW power station complex on the south bank (1955 to 1959) and the installation of 330 kV power lines providing a high-voltage backbone to the electricity systems of the two countries sharing the Zambezi River border on which the dam was built. Following the break-up of the Federation of Rhodesia and Nyasaland in 1963 and Zambia's independence in 1964, the 900 MW Kafue Gorge power station and the 600 MW Kariba North power stations were developed with support from the World Bank and friendly countries that supported Zambia's desire for self-sufficiency. By the late 1970s it was the Southern Rhodesia that now depended on its northern neighbor for power imports.

***The Central African Power Corridor (CAPCO).*** On November 25, 1963, the Government of Southern Rhodesia (today Zimbabwe) and the government of Northern Rhodesia (today Zambia) signed an agreement related to the Central African Power Corporation (CAPCO) at Salisbury (today known as Harare). The aim was to see that there is an integrated system for the control of the generation of electric power and its transmission in the territories of the said governments, which was at that time, the responsibility of the Federal Power Board. CAPCO was to continue to operate and fully develop a transmission system from Kariba connecting both the Zambian and the Zimbabwean networks consisting of 330 kV overhead lines under the joint ownership and control of the two governments. This agreement was put into operation from 1964 and it was carried on until after the independence of both countries, first the Northern Rhodesia when it became Zambia in 1964 and later the Southern Rhodesia when it became Zimbabwe in 1980.

Having survived the liberation struggle that put Zambia and the Southern Rhodesia on different sides, the Central African Power Corporation (CAPCO), was dissolved in 1987 and re-constituted as the Zambezi River Authority (ZRA), which is focused on the maintenance of the Kariba Dam complex and the regulation of the shared water resources of the Zambezi. CAPCO's generation and transmission functions in the respective countries were taken over by the national power utilities, ZESA in Zimbabwe and ZESCO in Zambia.

***Development of the South African Network.*** Further South, in 1909 the Earl of Selborne, Governor-General of South Africa, established a Power Companies Commission to enquire into the desirability of the establishment of large electric power companies in the Transvaal. The commission recommended that: "Since the supply of electric power leads to the establishment of a virtual monopoly in a commodity which has become practically a necessity of modern civilisation, it should, while being left as far as possible to private enterprise, at the same time be placed under government control and subjected to regulations which shall secure the equitable supply of power, the public safety and public interests generally" [2].

This commission further recommended that the electricity supply industry remain in private hands mainly because of the need to attract foreign investment in industry in South Africa and also because the need for state capital for growth meant that the government was simply not in a position to finance the construction of a major power company.

The Transvaal Power Act of 1910 enabled the VFTPC (Victoria Falls and Transvaal Power Company Ltd) and the Rand Mines Power Supply Company to obtain licenses to construct new power systems. The Act ended the fragmented and uncontrolled development of the power transmission industry but not of the distribution industry.

The Electricity Act of 1922 concluded the work done by Dr Merz [3], repealed the Transvaal Power Act of 1910 and was the first electricity Act to apply to the Union of South Africa as a whole. The first chapter of the Act provided for the establishment of a commission (to be known as the Electricity Supply Commission). On March 6, 1923, the birth of the Electricity Supply Commission (Escom as it later became known) was announced with the following notice in the

> The Electricity Supply Commission was established as a body corporate in law and had responsibility inter alia for the establishment, acquisition, maintenance and working of undertakings for an efficient supply of electricity; the investigation of new or additional facilities to supply electricity within an area; and the co-ordination and co-operation of existing undertakings to stimulate the provision, whenever required, of a cheap and abundant supply of electricity.
>
> Government Gazette [3]

The 1922 Electricity Act resulted in further centralization of the electricity industry and greater government control and ownership. Private ownership was not rejected but it became subject to more control. By 1948, Escom negotiated a take-over of VFTPC for 14.5 million pounds and this provided Escom with a well-established power system able to meet the demands of the Rand undertaking. The Electricity Act of 1958 was replaced in its entirety by a new Electricity Act of 1987 where Escom was renamed Eskom. Eskom had jurisdiction over tariff levels while the Electricity Control Board had jurisdiction over tariff structure.

On April 1, 1995, a new regulatory authority, the National Electricity Regulator (NER), was established (in terms of the Electricity Act of 1987 as amended) as successor to the Electricity Control Board. The main objective of the NER was to control the electricity supply industry in terms of the Act. Its main regulatory areas were pricing and tariffs, licensing, customer complaints, and dispute resolution as well as quality of service and supply.

***Southern African Power Pool (SAPP).*** The Southern African Power Pool (SAPP) is a regional utility grouping that was created by SADC member states to harness and create a platform for the regional power utilities to trade electricity amongst themselves while also improving the security of electricity supply. The SAPP member countries and the interconnected grid are shown in Fig. 4-23.

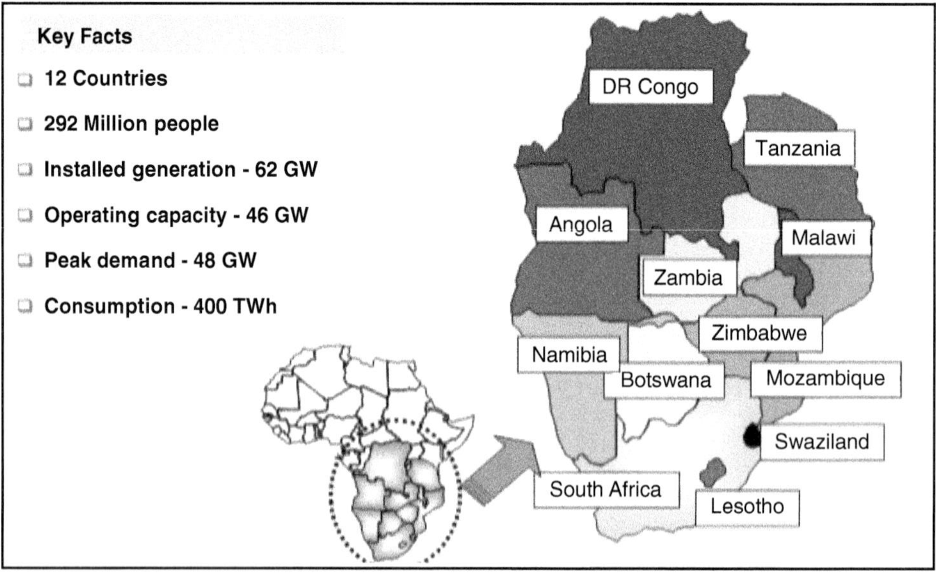


**FIGURE 4-23** SAPP members countries [4].

At present, SAPP comprises all 12 SADC member countries in the subcontinent (the other SADC members are the island states of Madagascar, Mauritius, and Seychelles). Nine of these are operating members, that is, countries that are part of the interconnected grid, which carries around 97% of the energy produced by SAPP countries.

The formal process of establishing a power pool started with the establishment of the Southern African Development Coordination Conference (SADCC) in the early 1980s.

The drive toward greater regional cooperation in Southern Africa received an unlikely boost from the extreme drought in 1991 to 1992, which severely affected hydropower production in the Zambezi basin, leading to economically and socially disruptive load shedding in Zimbabwe and Zambia. This is the event that led to the first tripartite agreement between Zimbabwe, Zambia, and the Democratic Republic of the Congo (DRC) to source power from DRC for supply to Zimbabwe. The drought also expedited plans for the interconnection projects to connect Zimbabwe to the South African grid and to the Cahora Bassa power station. The interconnections allowed the drought-resistant coal-fired capacity of South Africa to provide backup for this and future droughts. The dismantling of apartheid in South Africa in 1994 removed the political constraints on South Africa's participation in regional activities.

SAPP itself was the culmination of efforts at coordinated energy development undertaken as part of the political goal of regional integration of the Southern African Development Community (SADC). Given the dominant role of South Africa in the power market, it was not possible to make much progress as long as South Africa was not a member of SADC. By 1995, later at the SADC summit held in August, it then became possible to formally establish the power pool. The founding agreements of SAPP were inspired by bilateral and multilateral agreements that were already in existence.

SAPP was then created through a treaty by those governments as a body, and because it deals with energy infrastructures and operations thereof, it was then placed under the SADC's Directorate of Infrastructure and Services (DIS).

The initial benefits of SAPP include increase in system reliability coordination of Northern Hydro with South African coal and the provision of a forum for regional solutions to electric energy problems (capacity building). The expected future benefits are the development and the facilitation of the regional spot energy market that will attract investment in generation.

### 4.8.3 Structure

***Makeup and Size.*** By 2016, the membership of SAPP has grown to include Operating (OP) members, Non-Operating (NP) members, Observers (OB), Independent Power Producers (IPPs) and Independent Transmission Companies (ITCs) as allowed in the revised governance documents. Table 4-6 shows the list of SAPP members as of 2016. The SAPP regional grouping capacity can roughly be divided into two distinct parts, namely, the hydro-potential north and the thermal potential south as shown in Fig. 4-24. This allows the southern part of the region to take advantage of cheap hydro energy generated in the north during the rainy season, with the net flow reversed during the dry season when the thermal south would export energy to its northern neighbors. The bulk of the power is generated from coal, concentrated in South Africa's nothern provinces, eastern Botswana, and western Zimbabwe. South Africa also has a nuclear power plant in the Western Cape and a hydropower plant in the Drakensburg Mountains.

The generation in the rest of the SADC countries is predominantly hydro-based, with power stations being located in the Zambezi Basin countries of Zambia, Zimbabwe, Mozambique, Malawi; at Inga in the Democratic Republic of Congo; the Kwanza Basin in central Angola; the Kunene Basin in the northern Namibia; and also in Tanzania. The initial operational statistics gave the following generation mix for SAPP: 74.3% coal, 20.1% hydro, 4% nuclear, and 1.6% diesel and gas at the onset of SAPP (1995); and over time this has changed to accommodate new technologies in the mix.

***SAPP Energy Resources.*** By 2016 the current generation mix in SAPP is 62.05% coal, 21.02% hydro, 4.38% diesel and gas, 4.03% wind, 3.01% nuclear, 2.94% solar PV, 1.51% open cycle gas turbine (OCGT), 0.97% concentrated solar power (CSP), 0.07% Biomass, and 0.03% landfill as

**TABLE 4-6** Details of the Utilities that Form the Membership of SAPP [5]

| No | Full name of utility | Status | Abbreviation | Country |
|---|---|---|---|---|
| 1 | Botswana Power Corporation | OP | BPC | Botswana |
| 2 | Electricidade de Mocambique | OP | EDM | Mozambique |
| 3 | Electricity Supply Corporation of Malawi | NP | ESCOM | Malawi |
| 4 | Empresa Nacional de Electridade | NP | ENE | Angola |
| 5 | ESKOM | OP | Eskom | South Africa |
| 6 | Lesotho Electricity Corporation | OP | LEC | Lesotho |
| 7 | NAMPOWER | OP | NamPower | Namibia |
| 8 | Societe Nationale d'Electricite | OP | SNEL | DRC |
| 9 | Swaziland Electricity Corporation | OP | SEC | Swaziland |
| 10 | Tanzania Electricity Company Ltd | NP | TANESCO | Tanzania |
| 11 | ZESCO Limited | OP | ZESCO | Zambia |
| 12 | Zimbabwe Electricity Authority | OP | ZESA | Zimbabwe |
| 13 | Copperbelt Hydro Power Corporation | ITC | CEC | Zambia |
| 14 | Lusemfwa Hydro Power Station | IPP | LHPS | Zambia |
| 15 | Hidroelectrica de Cahora Bassa | OB | HCB | Mozambique |
| 16 | Zimbabwe Electricity Authority | OB | MOTRACO | Mozambique |

IPP = independent power producer; ITC = Independent Transmission Company; NOP = non-operating member; OB = observer; OP = operating member.

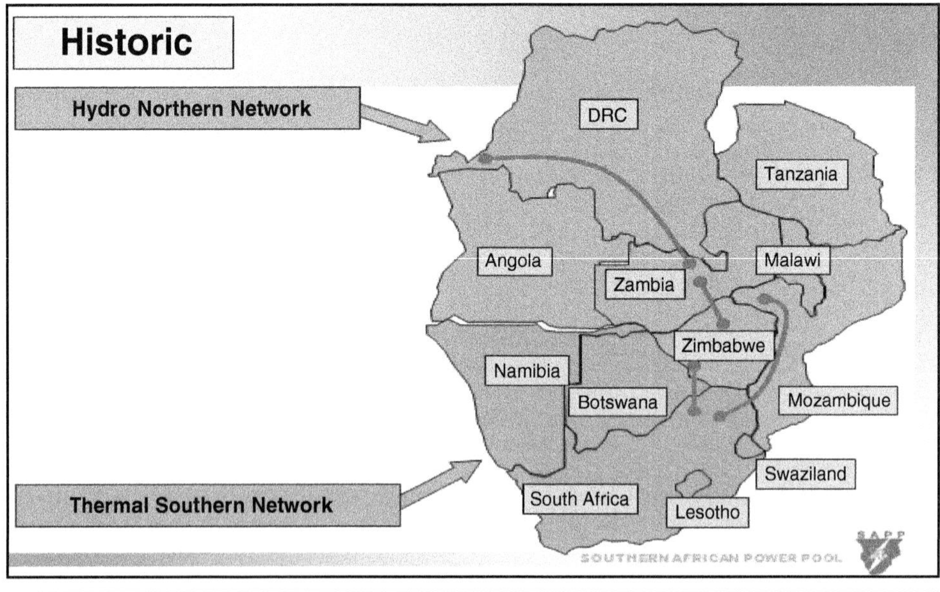


**FIGURE 4-24** Source diversity of SAPP [1].

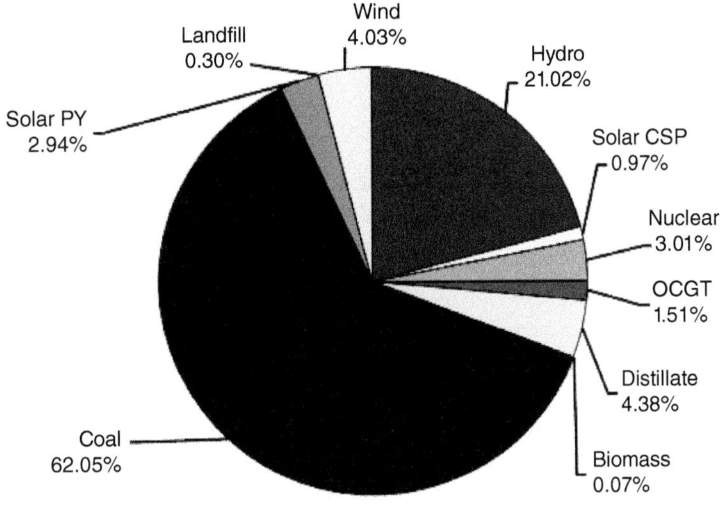


**FIGURE 4-25** Current generation mix in SAPP—2016 [6].

shown in Fig. 4-25. The coal generation is predominantly in the South (South Africa, Botswana, and Zimbabwe) and the hydropower in the North in the Zambezi Basin (Zambia, Zimbabwe, Mozambique, and Malawi), DRC and Cunene (Angola and Namibia). The nuclear power station (the only one in Africa) is in the Western Cape (in South Africa), which is far from the coal-fired power plants in the northern and eastern provinces. Most of the diesel power plants are for small isolated rural networks.

Cheap electricity in the region is brought about by the relatively cheap running cost of power stations in the thermal southern region, where there is an abundance of coal and the power stations are virtually built on top of coal fields, cutting down on transportation logistics. In the hydro northern region, the two major perennial rivers systems, namely, the Zambezi and the Congo Rivers, provide abundant free water as a driver of hydro power stations. Similar to the case of the thermal South's coal power stations, most of the north's hydro power stations have also been built between the 1960s and 1980s, most of which have been paid off. Until about 2007–8, the region had relatively cheap resources and a large operating margin. This hampered any new investment in the generation of power plants. While the supply side remained relatively constant, the demand side continued to grow until 2007–8 when the demand started to outstrip the supply side, causing wide spread load shedding activities in the region. This, for the first time, threatened security of supply throughout the region.

***Current Transmission System in SAPP.*** The footprint of the SAPP regional grid is presented in Fig. 4-26, with the existing and planned transmission interconnectors between the utilities of member countries shown. Although the 220 kV, 275 kV, and 330 kV transmission voltages are used in the SAPP system, the main transmission voltage is 400 kV. There has also been the use of 765 kV technology since the 1980s.

The 765 kV voltage was introduced in the South African networks and a number of lines at this voltage have been in operation. These lines were built in the corridor between Gauteng and the Western Cape Provinces which are located over 1500 km apart. Gauteng Province is located very close to the generation pool in Mpumalanga, where the majority of coal reserves is located. On the other hand, the Western Cape has a significant load, which is much higher than the generation located in the area, and the only mechanism of supporting the load is to transport power generated from Mpumalanga via the lines linked to the Gauteng Province. Among the constraints of power transfer in this scenario is voltage stability, and commissioning lines at 765 kV provided the solution.

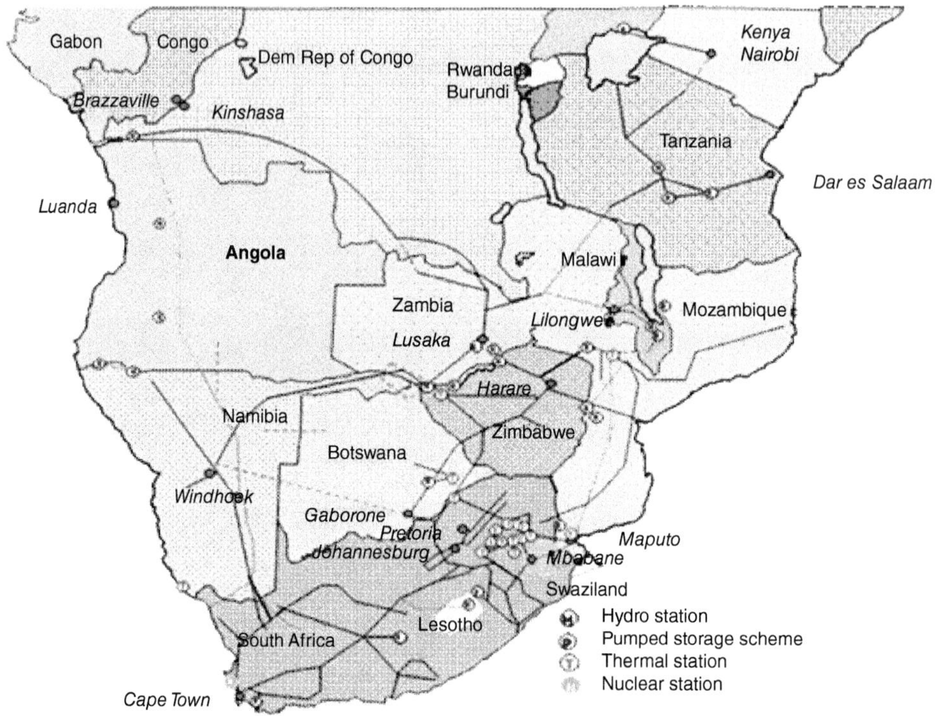


**FIGURE 4-26** Geographic Layout of the Interconnected Grid of Southern Africa [7].

There have also been advances in the utilization of Flexible AC Transmission System (FACTS) devices and examples of these are discussed below.

In the Namibian power system, SVCs are at Auas substation to solve 50 Hz parallel resonance concerns [7] and Omburu substation to deal with voltage control problems in the network [8]. In the Eskom grid, numerous SVCs are also operational. Again due to the centralized nature of generation, that is, being concentrated in the coal pool in Mpumalanga, transmission lines to load centers tend to be long, posing all sorts of power transfer challenges. The Eskom SVCs are currently in operation mainly as solutions to potential overvoltage problems following rejection of major loads in the remote load centers and to enhance transfer of power to far-flung areas.

There has been some experience in the SAPP network high voltage direct current (HVDC) transmission systems. Transferring power between Mozambique and South Africa, is the Apollo Songo ±533 kV, 1920 MW HVDC link. To tap the unutilized hydropower potential in Mozambique and to provide for the increasing load in South Africa in the 1970s, Cahora Bassa dam was developed and the link was conceived to transmit power generated at this facility over a distance in excess of 1400 km to Apollo Converter Station in Pretoria, South Africa. Due to internal strife within Mozambique, the line could not be operated until 1998 when it was rehabilitated and brought back into operation. In the DRC, one of the longest HVDC transmission lines, the Inga Kolwezi ±560 kV, 560 MW HVDC scheme is in operation. It was planned for commissioning in the early 1970s, but due to wars in that country it could only be brought into operation in 1982. It transmits power over a distance of 1700 km from Inga Falls to the mining load district of Katanga [9]. The Caprivi Link, connecting Zambezi converter station in Namibia's Caprivi region near the Zambian border and Gerus converter station in central Namibia is the only VSC-HVDC scheme in Southern Africa [10]. It connects the ac networks of Zambia, where hydro power is generated, and Namibia via a 950 km overhead line at ±350 kV dc.

### 4.8.4 Operation and Control

***SAPP Control Areas.*** A Control Area is defined in the Southern African Power Pool (SAPP) as: "an electrical system with borders defined by points of the interconnection and capable of maintaining continuous balance between the generation under its control, the consumption of electricity in the control area and the scheduled interchanges with other control areas" [12].

As per the SAPP operating guidelines, the criteria for control area operation is that the control equipment of each control area shall be designed and operated to enable the Control Area Operator to continuously meet its System and Interconnection control obligations and measure its performance. Control Area Operators are required to offer Control Area Services and Regulating Reserves for the secure control and operation of the interconnected system. A System Operator or Electricity Supply Enterprise that does not meet the criteria for a Control Area Operator must be hosted by a Control Area [13].

The SAPP currently has three Control Areas namely [14]:

- ZESCO, which covers Zambia and the Democratic Republic of Congo;
- ZESA, which covers Zimbabwe and northern Mozambique; and,
- Eskom, which covers South Africa, Botswana, Namibia, Swaziland, Lesotho, and southern Mozambique.

The Electricity Supply Enterprises within a Control Area shall contract with the Control Area Operator for all applicable Control Area Services and required Regulating Reserves. Some of the Control Area services include system control, that is, tie line control and frequency control, energy interchange transaction scheduling and accounting, and inadvertent energy management [13].

***SAPP System Power Frequency Control.*** Power frequency is common across interconnected systems, so all customers are affected when the frequency deviates too far from its nominal value. Frequency is one common factor in an interconnected power system and, for Southern Africa, the nominal frequency is 50 Hz. The rate of frequency change depends on the inertias of rotating machines connected to the network and the difference between supply and demand. The frequency can be varied either by changing the generator's MW or real power output or by changing the customers' MW demand [14]. Control area operators are required to perform primary frequency control in their own areas depending on the amount of generation that they have on line. In primary frequency control, the generators in their control areas act directly in response to the actual frequency using the decentralized frequency control approach.

The three control area operators, Eskom, ZESA, and ZESCO, perform secondary control using the centralized frequency control approach where generators and loads change their output on instruction from a central coordinator located at their national control centers (NCCs). Secondary frequency control is performed either manually or via automatic generating control (AGC). AGC is a centralized control loop that coordinates the generators and its main function is to restore the system frequency to the nominal value or to the agreed dead band. The agreed dead band for frequency control in SAPP is 50 Hz $\pm$ 0.15 Hz.

***Operational Constraints.*** The main challenges observed on the SAPP interconnected system under steady state conditions, are those of power transfer during trading transactions being limited by thermal, voltage and voltage collapse constraints. This is due to transmission capacity constraints, especially while considering loss of major interconnecting transmission lines between the various utilities. Figure 4-27 shows Eskom tie lines that interconnect with SAPP [4].

Because of the nature of the power system and various changes in generation and loads, there are always power oscillations in the system. These oscillations are mainly due to the energy transfer between the rotating masses of the machines on the power systems interconnected by weak transmission lines. When these oscillations are excited by system disturbances, they grow to amplitudes that can cause undesirable effects on the system and could damage power system plant and equipment. A number of studies that have been conducted on the SAPP interconnected system, indicated that

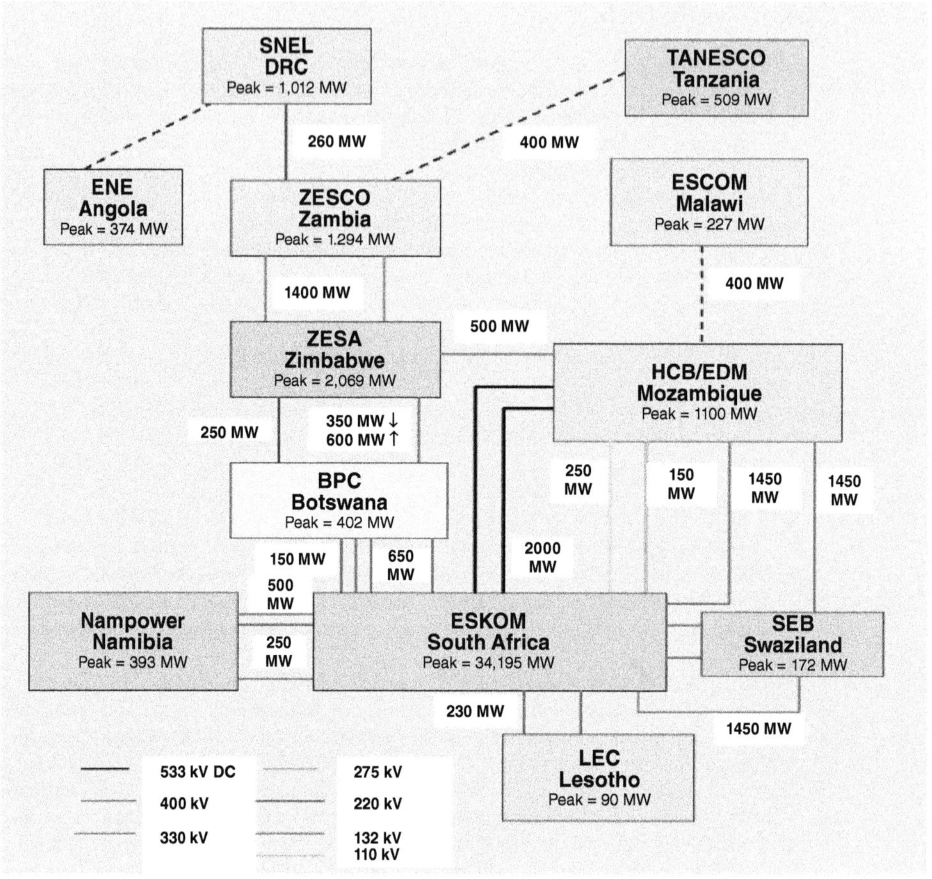


**FIGURE 4-27** Eskom tie lines interconnecting with SAPP [4].

there are undamped 0.3 Hz inter-area power oscillations between the predominantly hydro system in the North (Cahora Bassa, ZESA, ZESCO, and SNEL) and the predominantly thermal system in the South (BPC and Eskom).

These inter-area power oscillations are a characteristic of the relatively weak link between the two "systems." When triggered, the oscillations are poorly damped especially when the ±175 MVar static var compensator installed at Insukamini substation along this link is out of service. Other studies conducted on the Eskom system have revealed that there are also 0.6 Hz inter-area oscillations between the Mpumalanga generation pool and the generators in the Cape network within South Africa. A number of power system stabilizers on the Eskom network have been tuned to damp out the inter-area oscillations on the Eskom system.

***Implementation of the Wide Area Monitoring System in the South African Power System.*** Power oscillations are a growing concern among power system operators worldwide. The stability of these oscillations is of vital concern, and is a prerequisite for secure system operation. Wide Area Measurement Systems (WAMS) are increasingly used to monitor and improve these oscillations that are observed on the power systems. A phasor-based WAMS is a network of fast synchronized measurements of voltage and current phasors (synchrophasors) that enables users to monitor the angular stability and dynamics of a power system. Continuous monitoring enables operators to perform corrective actions promptly, before an issue escalates and presents a risk to the integrity of the system. In this way, the reliability and security of a power network is improved.

Most system operators are faced with challenges in operating modern power system networks because of capacity constraints, reserve shortages, and high penetration of intermittent renewable generation. Insufficient real-time information regarding congestion on transmission corridors and stressed equipment may result in conservative load shedding to save the network from collapsing. The Eskom System Operator identified the need to enhance the situational awareness of controllers at the national control center of the power system by implementing synchronized phasor measurements technology to improve the power system reliability and operational security during normal and highly stressed operating conditions. Eskom plans to have 50 PMUs in the future. The utility company already has 15 PMUs in operation and 7 are being commissioned. Planned installation rate is eight per annum. Already, a number of system disturbances on the interconnected system have been effectively and efficiently analyzed using the information obtained from the WAMS system.

***Energy Trading in SAPP.*** Following the creation of SAPP, a coordination center (CC) for SAPP was established in February 2000 with offices in Harare, Zimbabwe. The center would help develop a spot market for electricity in the region and manage the transformation of the power pool from a cooperative pool to a competitive one with open markets for electricity. Some on the functions of the SAPP coordination center are to: implement SAPP objectives; provide a focal point for SAPP activities; facilitate the STEM; monitor the operations of SAPP transactions between the members; and to carry out technical studies on the power pool to evaluate the impact of future projects on the operation of the pool.

*Bilateral and Multilateral Contracts.* The trading arrangements between members have continued to operate predominantly under the pre-SAPP–type bilateral and multilateral contracts. Therefore, SAPP continues to go through a transition and is migrating from a cooperative pool to a competitive pool. A cooperative pool uses cost based trade whereas a competitive pool uses bid-based trade. SAPP's focus has thus been to introduce a short-term energy market (STEM) to facilitate the trading of surplus energy not committed under existing contracts.

*Short-Term Energy Market (STEM).* The STEM that was developed and used over the period 2001 to 2007 is a notable achievement, even though the amounts involved were always a small proportion of the region's total annual energy consumption, which is about 300,000 GWh. STEM will be a firm energy market and the only commodity that will be traded is energy. Energy will be sold through offers and bids by daily, weekly, or monthly contracts. The offers and bids will be matched at the coordination center and the results of successful bidders will be published on the billboard. It is hoped that STEM would be a precursor to the full spot market in the region.

Long term bilateral agreements between participants will be given priority for transmission on the SAPP inter-connectors.

*Day-Ahead Market.* STEM has been replaced by a fully competitive day-ahead market (DAM), but most of the electricity trade in the region would continue to be via long-term bilateral contracts. DAM has been in operation since 2008 and it is a step in achieving a full energy trading on a SPOT market. The DAM is also a firm energy market where hourly energy contracts for each of the 24 hours of the following day, or a future day may be traded. It also caters for block bids for periods as specified by the SAPP Market Operator (MO).

*New Trading Platform Development.* By 2016 SAPP was busy concluding the development of the new trading platform (the SAPP-NTP); this new trading platform is to include day-ahead market (DAM), month- and week-ahead markets, as well as intraday market (IDM) that is an hour-ahead market. It is internet based with the server accessible to configured participants in their respective utilities.

*Recent Load and Generation Status.* The peak demands in the various member utilities and the total demand of the SAPP network for 2014 are summarized in Table 4-7. The load in the region reached a peak of 45.6 GW, with Eskom peak load reaching 35.9 GW, that is, 78% of the total peak [5].

**TABLE 4-7** Peak load in the Southern Interconnected System for year 2014 [5]

| Utility name | Peak demand (MW) |
|---|---|
| SNEL | 1040 |
| TANESCO | 890 |
| ESCOM | 278 |
| ZESCO | 1681 |
| ENE | 1073 |
| ZESA | 1671 |
| HCB/EdM | 1606 |
| BPC | 578 |
| NamPower | 611 |
| ESKOM | 35896 |
| SEC | 205 |
| LEC | 129 |
| **TOTAL** | **45658** |

### 4.8.5 Renewable Energy in Southern Africa

***Renewable Energy Potential.*** Renewable energy resources (RESs) such as hydro, solar, wind, biomass, geothermal, and tidal waves abound in different countries in Southern Africa [15]. These potentials have been exploited in varying degrees in the different countries, and RESs now account for about 23.5% of total electricity generation with hydro being the major source [16]. The use of biodiesel and bioethanol for transport is established in Malawi and Zimbabwe while other countries like Angola, Mozambique, South Africa, Swaziland, and Zambia are establishing mandates for blending biodiesel and ethanol with fossil fuels [16]. In Mauritius, electricity generation from Bagasse has been considerably exploited, increasing from 70 GWh/year in 1992 to 360 GWh/year in 2002 [17,18]. Namibia, South Africa, and Zambia have geothermal energy sites whose potentials have been investigated [17]. Tanzania also has some geothermal sites [15]. Table 4-8 shows the renewable energy capacity of the Southern African countries.

**TABLE 4-8** Renewable Energy Capacity in Southern African Countries, 2014 [16]

| | Technology type | | | | | | | | | |
|---|---|---|---|---|---|---|---|---|---|---|
| Country | Large-scale hydro | Medium-scale hydro | Small-scale hydro | Pumped storage | Solar PV | Onshore wind | Biomass/ waste | Bio-gas | Total | % Change 2000 to 2014 |
| Angola | 861 | 16 | 1 | 0 | 0 | 0 | 0 | 0 | 878 | 225 |
| Botswana | 0 | 0 | 0 | 0 | 1 | 0 | 0 | 0 | 1 | 100 |
| DRC | 2360 | 50 | 6 | 0 | 0 | 0 | 0 | 0 | 2416 | 1 |
| Lesotho | 72 | 3 | 2 | 0 | 0 | 0 | 0 | 0 | 77 | 0 |
| Madagascar | 130 | 34 | 1 | 0 | 3 | 1 | 0 | 0 | 169 | 55 |
| Malawi | 346 | 4 | 1 | 0 | 1 | 0 | 17 | 0 | 369 | 21 |
| Mauritius | 42 | 17 | 2 | 0 | 18 | 1 | 271 | 0 | 351 | 32 |
| Mozambique | 2182 | 3 | 1 | 0 | 1 | 0 | 0 | 0 | 2187 | 0 |
| Namibia | 332 | 0 | 0 | 0 | 5 | 0 | 0 | 0 | 337 | 35 |
| Seychelles | 0 | 0 | 0 | 0 | 0 | 6 | 0 | 0 | 6 | 600 |
| South Africa | 653 | 30 | 3 | 0590 | 922 | 570 | 242 | 13 | 4023 | 60 |
| Swaziland | 55 | 6 | 2 | 0 | 0 | 0 | 75 | 0 | 138 | 48 |
| Tanzania | 553 | 14 | 6 | 0 | 11 | 0 | 62 | 0 | 646 | 8 |
| Zambia | 2244 | 11 | 2 | 0 | 2 | 0 | 43 | 0 | 2302 | 26 |
| Zimbabwe | 680 | 6 | 2 | 0 | 5 | 0 | 97 | 0 | 790 | 6 |
| **SADC** | 10510 | 194 | 29 | 1590 | 969 | 578 | 807 | 13 | 14690 | 26 |

***Barriers to Widespread Use of Renewable Energy in the Region.*** In spite of the benefits of renewable energy and its potentials in the region, there are some barriers to its widespread use which include: absence of legal and regulatory framework in most countries, poor institutional framework, lack of coordination and linkage in renewable energy technology (RET) programs, price distortion, high initial capital, weak dissemination strategies, and lack of expertise [17]. Some of these issues are however receiving attention and the initial capital cost of some of the technologies such as solar PV is fast reducing as indicated in [19].

In recent years, some efforts have been made to improve the deployment of renewable energy in the SADC region through the Renewable Energy Support Programme (RESP) and the SADC Renewable Energy Strategy and Action Plan (RESAP). The region has plans to develop renewable energy resources (hydro, wind, solar, etc.). The development of Renewable energy resources has been driven primarily by electricity supply shortages in several key countries, accentuated by the absence of new investments in grid electricity generating capacity [16,17]. Other incentives for the development include the changing economics of wind and solar energy, emergence of new policy concepts such as Feed-in-Tariffs (FITs), auctioning of power supply to Independent Power Producers (IPPs), net metering and Renewable Energy Certificates (RECs). To expedite renewable energy development and energy efficiency, and reduce dependence on fossil fuels, SADC member states are developing their own targets and policies. A set of target activities was established in the SADC Energy Protocol (1996). For the period 2004 to 2018, a target of 70% access to modern energy sources by rural communities was set in the 2003 SADC Regional Indicative Strategic Development Plan (RISDP) [16]. This framework forms the basis for the development and operation of the renewable energy policies within the region but does not give any specific targets for member states to aim at, at the national level [17].

***Harmonization.*** To promote the widespread use of renewable energy technology in the region, a harmonized subregional framework for the renewable energy sector was developed, which comprised standardization and policy alignment in order to narrow the differences in the legal and regulatory issues, standards, regulations, and codes of practice of the different countries, promote the development and widespread utilization of new and renewable energy in the region, and promote regional trade in RETs [17]. In pursuance of renewable energy development and energy efficiency initiatives in the region, the SADC energy ministers approved in principle the formation of a SADC center for Renewable Energy and Energy Efficiency (SACREEE) with Namibia as host country [16].

*Renewable Energy Generation in South Africa* There have been major strides in the area of renewable energy generation, especially in relation to the Renewable Energy Independent Power Producer Procurement Program (REIPPPP) in South Africa. The country is a signatory to the United Nations Framework Convention of Climate Change (UNFCC) and is committed to sustainable development and reducing greenhouse emissions. With these ends in consideration, and with the desire to increase the dwindling reserves after 2010, the REIPPPP was initiated to source renewable energy generation from the private sector. The sourcing of generators is done in a sequence of bid windows, with Ministerial determination stipulating the type and amount of renewable generation to be procured. Bid rounds 1, 2, 3, 3.5, and 4 of the REIPPP have been conducted, but allocations for only rounds 1 to 3.5 had been concluded, with the total number projects awarded equaling 64, and having an aggregated capacity of 3193.5 MW.

The breakdown of capacity and the number of projects awarded, both by fuel type, is summarized in Fig. 4-28, where it is shown that the program has aimed at procuring renewable energy from a wide range of energy sources. The majority of projects awarded by the end of 2014 were in the areas of wind and solar photovoltaic generation.

### 4.8.6 Future Outlook

***Smart Grid Initiatives in SADC Countries.*** Smart grid is a broad concept that covers the entire electricity supply chain and is characterized by the use of technologies to intelligently integrate the generation, transmission, and distribution of electricity. Some of the benefits of smart grid include optimized asset utilization and efficiency, increased integration of clean and renewable energy,

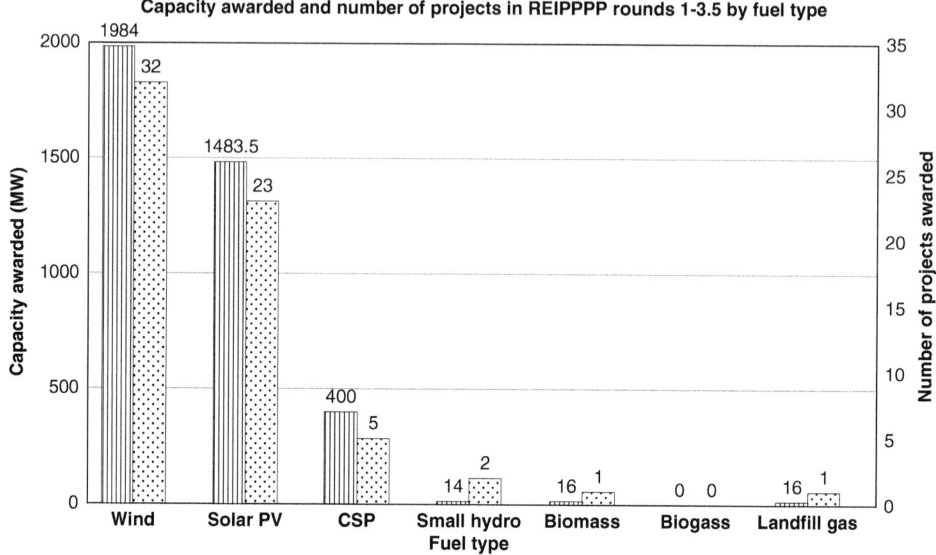


**FIGURE 4-28** Generation procured from private producers at the end of 2014 by various fuel types and amount [4].

better integration of distributed generation (DG) and micro-grid/mini-grid, improved security, enabling active participation by consumers in demand response that allows a two-way communication between the utilities and customers, grid resilience, etc. [20]. Smart grid is critical for future economic growth and social development of SADC region and will be required to meet the goals of the Africa Power Vision (APV) which according to [21] is, "A long-term plan for increasing access to reliable and affordable energy by using Africa's diversified energy resources in a coherent and well balanced manner, consistent with Agenda 2063, Africa's new transformation strategy. APV primarily seeks to drive and rapidly accelerate the implementation of critical energy projects in Africa under the Programme for Infrastructure Development in Africa (PIDA)."

Efforts to promote the concepts of smart grid are gaining momentum in several SADC countries with South Africa taking the lead. For example, in South Africa, the National Energy Development Institute (SANEDI) in partnership with industry took the initiative to create the South African Smart Grid Initiative (SASGI) in 2012 with the objectives of developing a Smart Grid vision for South Africa, providing policy and creating a framework and a platform for smart grid deployment [22]. The South African (state-owned) power utility Eskom, and the various municipalities have started experimenting with several technologies of smart grid. Eskom is looking at rolling out advanced metering infrastructure (AMI) to improve their billing system, the revenue and the demand-side management as well as to improve the demand response and the efficiency of the system [23].

*Energy Efficiency.* Energy efficiency forms a significant part of smart grid. For Africa and SADC region, smart grid will come from the application of intelligent energy technology to optimize the use of generation resources and the delivery of power. In this context, several SADC member states and utilities are taking advantage of energy efficiency as a complement to renewable energy to reduce demand and delay the requirement for new generation capacity. Some of the main energy efficiency initiatives in SADC regions are: (a) the use of compact fluorescent lamps (CFLs) or light-emitting diodes (LEDs); (b) the adoption of solar water systems and hot water load control (i.e., remotely turning off of conventional water heaters during peak periods); (c) the use of demand side management (DSM) that could be linked to demand response. All the SADC member states, except two

(Madagascar and Seychelles) have adopted the CFLs energy efficient approach where incandescent light bulbs are exchanged for CFLs [16].

*Challenges in Implementing Smart Grid Technologies in Southern Africa.* The urgency of speeding up the implementation of smart grid technologies in Southern Africa cannot be overstated. However, there are still significant challenges that must be overcome to deploy smart grids at the scale they are needed. Some of these challenges are technical, legal and regulatory, financial, and educational [24]. For example, ageing and outdated infrastructure will require a major overhaul and augmentation to support smart grids; old legacy systems cannot always be retrofitted with new technologies and early retirement of equipment may become an issue; many regulatory policies are old and outdated to deal with the consequences of smart grids; there is a lack of human skills and the "know how" required to deal with highly sophisticated equipment; greater public engagement and participation is lacking; significant investments are required to purchase the new technologies envisioned for communicating information between the end users, electricity service providers, and to modernize the ageing transmission and distribution infrastructure. There is a need for governments to establish clear and consistent policies that will facilitate innovative investment in the smart grid [24]. Also, greater public engagement (in particular, customers and consumers) will be required.

***Mini-Grids/Micro-Grids Potential in SADC Region.*** Currently, the electricity access rate in most SADC countries is very low, and the majority of the population does not have access to modern fuels like natural gas, kerosene, or propane. They still rely on traditional use of biomass for cooking. This is hazardous to health and inefficient [25]. The region is faced with electricity shortage due to lack of enough generation capacity and ageing transmission and distribution infrastructure. As the demand for electricity grows, the countries in the SADC region will require alternative energy solutions based on the more flexible and decentralized (and off-grid) systems such as distributed generators, micro-grids and/or mini-grids rather than the legacy centralized power grid.

In most SADC countries, extending the grids to rural areas is often not financially viable. With the vast range of natural and renewable resources, the opportunities of micro-grid in the SADC region are enormous. Some international and regional bodies such as the African Development Bank's Sustainable Energy Fund for Africa (SEFA) and the Regional Electricity Regulators Association of Southern Africa (RERA) are promoting the growth of mini-grids in sub-Saharan Africa (including Southern Africa) to unlock the region's potential for clean energy and increase energy access in isolated communities. RERA, backed by the Africa-EU Renewable Energy Cooperation Program (RECP), managed by the EU Energy Initiative Partnership Dialogue Facility (EUEI-PDF) has developed guidelines to assist countries in the SADC region to create a framework for attracting investment in mini-grids [26].

Alongside traditional and renewable generation, micro-grids or mini-grids (i.e., large scale micro-grids) offer the most attractive options for increasing access to electricity for the majority in remote and low income communities. These are localized grouping of electricity generation (renewable and non-renewable), energy storage, energy control and conversion, energy monitoring and management and load management tools which can operate while connected to traditional electricity grid or function independently. For some, micro-grid holds the promise of becoming a basic building block in the implementation of the next generation smart infrastructure [27].

In the last few years, renewable energy based micro-grids have received increasing attention in SADC region due to the values they can offer. Most of the micro-grids are solar PV based, although other renewable sources such as wind, geothermal, biomass, etc. have also been used. RESs such as solar, wind, etc. have enormous potentials in contributing to Southern Africa's electricity portfolio [28].

Tanzania is leading in the development of mini-grids, most of which are Solar PV or mini-hydro, through the innovative standardized Power Purchase agreement (PPA) using an avoided-cost feed-in-tariff (FIT). The government of Tanzania established the rural Energy Agency (REA) to focus on off-grid and renewable energy projects in Tanzania. According to the 2012 Tanzania Power Master Plan, the Tanzanian government is targeting 30% connectivity by 2015. Earlier in 2016, Jumeme

Rural Power Supply Ltd. unveiled their hybrid solar/mini-grids. There are plans to develop and implement solar-hybrid mini-grids in rural growth centers in Tanzania [29].

Several mini-grid initiatives and renewable energy projects are currently underway in other SADC member states such as Zimbabwe, Zambia, Swaziland, South Africa, Seychelles, Namibia, Mozambique, Mauritius, Malawi, Madagascar, Lesotho, DRC, Botswana and Angola [30]. The developments of mini-grids in those countries are undertaken by private companies, regional concessionaries or large International non-governmental organizations (NGOs) [16].

### *Future Perspective on Grid Development*

*Generation Projects.* Major projects are committed for commissioning in the SAPP network in the next 5 years. The total capacity committed is 24,062 MW. Table 4-9 breaks down the contributions by various countries to the regional aggregate. About 11,274 MW, 47% of the planned capacity, will be for projects in South Africa. In relation to the installed capacity of 61.8 GW at the end of 2014, the total commissioned capacity represents an increase in installed capacity of close to 20% that can be considered as significant.

If one looks at South Africa, there are ambitious plans for generation capacity expansions beyond 2019. In fact, the Integrated Resource Plan (IRP) [26] of the country states that the total installed generation capacity is planned to be 85,241 MW by 2030, representing an additional generation capacity of about 41,704 MW to the 2014 scenario, a virtual doubling of generation capacity of the network. The breakdown of this planned capacity by fuel type is shown in Fig. 4-29.

Coal generation is expected to dominate the new capacity to be created with 41,704 MW (48.19%) planned for installation. Plans for adding to the fleet of gas turbines entails installing 9170 MW (10.76%) and 1896 MW (2.22%) of OCGT and CCGT, respectively.

The plan also includes a significant 11,400 MW (13.37%) of possible nuclear power plants. RESs also feature strongly in the plan, with 11,800 MW (13.84%) of wind and 600 MW of CSP. Pumped storage capacity at 2192 MW (3.42%) also features in the plans.

*The Future of SAPP.* SAPP is striving to have a fully integrated, robust grid with a generation mix which can handle the climatic, technical, and economic conditions of the subcontinent. There is a drive to enhance and increase rural electrification projects as well as harnessing new/renewable energy resources for the region. There is also a drive to interconnect the non-operating members into the pool to enhance trade and ensure security of supply. Figure 4-30 shows the future SAPP pool inter-connector projects.

**TABLE 4-9** Planned Generation Capacity Additions in SAPP for the Period 2015 to 2019 [4].

| Country | 2015 | 2016 | 2017 | 2018 | 2019 | Total (MW) |
|---|---|---|---|---|---|---|
| Angola | — | 1280 | 2271 | — | — | 3551 |
| Botswana | — | — | — | 300 | — | 300 |
| DRC | 430 | — | 150 | — | — | 580 |
| Lesotho | — | — | — | — | — | 0 |
| Malawi | — | — | — | 74 | 300 | 374 |
| Mozambique | 205 | 40 | — | 600 | — | 845 |
| Namibia | — | 15 | — | — | 800 | 815 |
| South Africa | 1828 | 3462 | 3032 | 1476 | 1476 | 11274 |
| Swaziland | — | — | — | — | 12 | 12 |
| Tanzania | 150 | — | 500 | 1140 | 300 | 2090 |
| Zambia | 135 | — | 300 | 101 | 1090 | 1626 |
| Zimbabwe | 15 | — | 120 | 1200 | 1260 | 2595 |
| **TOTAL** | **2763** | **4797** | **6373** | **4891** | **5238** | **24062** |

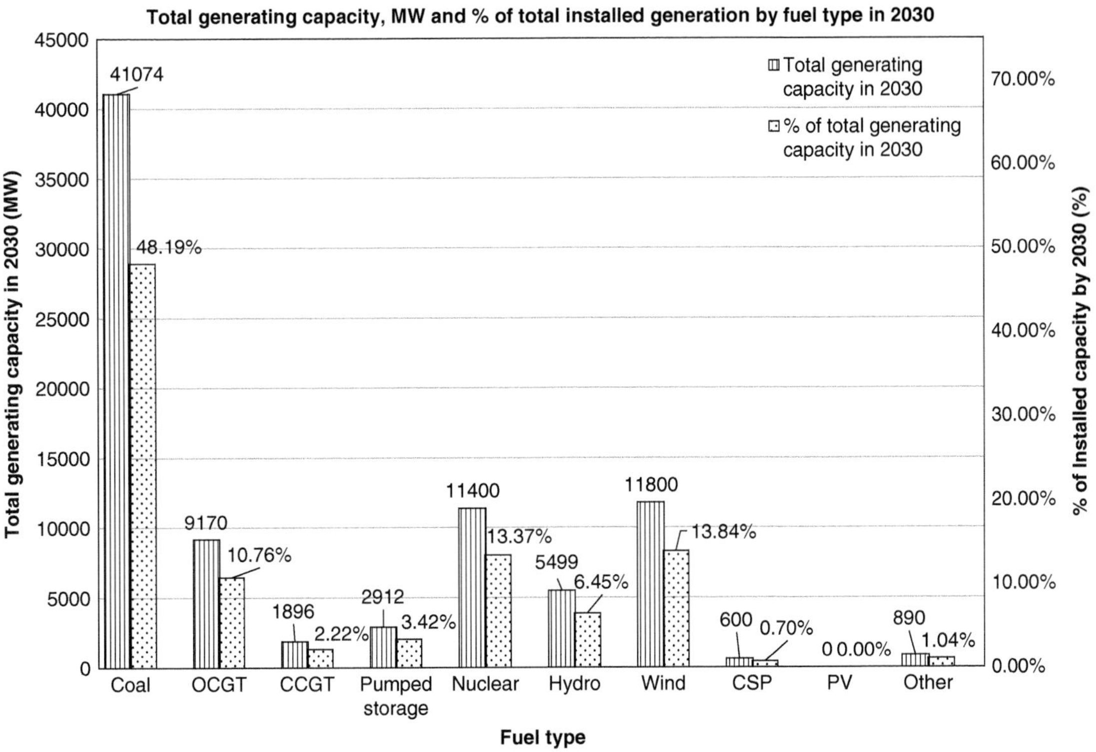

**Total generating capacity, MW and % of total installed generation by fuel type in 2030**

| Fuel type | Total generating capacity in 2030 (MW) | % of total generating capacity in 2030 |
|---|---|---|
| Coal | 41074 | 48.19% |
| OCGT | 9170 | 10.76% |
| CCGT | 1896 | 2.22% |
| Pumped storage | 2912 | 3.42% |
| Nuclear | 11400 | 13.37% |
| Hydro | 5499 | 6.45% |
| Wind | 11800 | 13.84% |
| CSP | 600 | 0.70% |
| PV | 0 | 0.00% |
| Other | 890 | 1.04% |

**FIGURE 4-29** Status of generation capacity in South Africa in 2030 (Adapted from [26]).

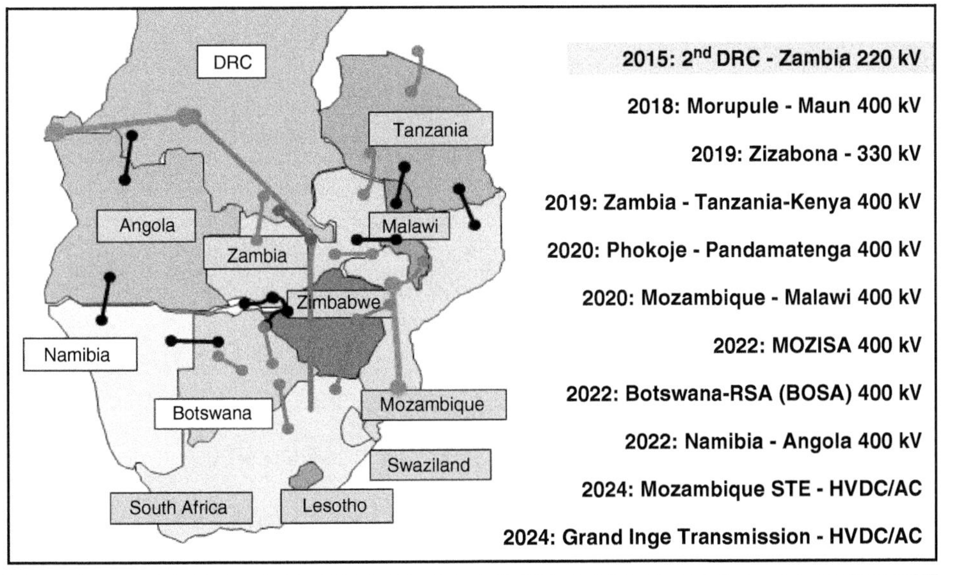


**FIGURE 4-30** Future SAPP pool inter-connector projects [5].

### 4.8.7 Conclusion

A lack of access to electricity is a fundamental brake on development in many parts of Africa. It is clear that economic development in SADC region cannot be achieved without affordable energy, and it cannot be sustained unless the energy is reliable and secure. It is expected that SADC state members will continue to take advantage of recent smart grid concepts and technologies to meet the growing demand for electricity and greatly improve the efficiency, reliability, and security of the electricity supply. There is a need for governments, research institutions, industry, the financial sector, and international organizations to work together to achieve the "smart grid vision" of SADC region and through that the African Power Vision. SADC governments must promote broad deployment of energy efficiency, deal effectively with carbon capture and storage and put the right policies in place.

### 4.8.8 References


[1] SAPP, "Southern Africa Power Pool Annual Report," SAPP Coordination Centre Office, 1998.

[2] Mountain, B. "Towards a Pricing Strategy for the South African Electricity Supply and Distribution Industry, page 64," in *MSc Dissertation, University of Cape Town, Cape Town*, Cape Town, 1994.

[3] South Rhodesian Government, "Agreement Relating to the Central African Power Corporation," South Rhodesian Government Gazette Extra Ordinary, Vol. 41, No 51, 25 November, Signed at Salisbury, 1963.

[4] SAPP, "Southern Africa Power Pool Annual Report," Southern Africa Power Pool Coordination Centre, 2015.

[5] SAPP, "Southern Africa Power Pool Annual Report," Southern Africa Power Pool Coordination Centre Office, 2014.

[6] Vajeth, O. "Introduction to SAPP," Southern Africa Power Pool, 2016.

[7] SAPP Grid, "Southern Africa Power Pool," 2016. [Online]. Available: http://www.sapp.co.zw/sappgrid.html. [Accessed 1 August 2016].

[8] Hammad, A., Bosshoff, S., Van der Merwe, W. C., Van Dysk, C., Otto, W., and Kleywnstuber, U. H. E., "SVC for Mitigating 50 Hz Resonance of a Long 400 kV AC Interconnection," Singapore, 1999.

[9] Halonen, M., Rudin, S., Thorvaldsson, B., Kleyenstruber, U., Boshoff, S., and Van der Merwe, C., "SVC for Resonance Control in Nampower Electrical Power System," Vancouver, BC, 2001.

[10] ABB, "Inga-Kolwezi," 2014. [Online]. Available: http://new.abb.com/systems/hvdc/references/inga-Kolwezi. [Accessed 1 August 2016].

[11] Magg, T. G., Manchen, M., Krige, E., Kandjii, E., Paisson, R., and Wasborg, J., "Comparison between Energized System and Real-Time Simulator Testing, CIGRE 2012," 2012. [Online]. Available: https://library.e.abb.com/public/..../Caprivi-link-HVDC-Interconnector-Comparison-between-energized-System.pdf. [Accessed 1 August 2016].

[12] SAPP, "Operating Guidelines," Southern Africa Power Pool, 1996.

[13] SAPP, "Operating Guidelines," Southern Africa Power Pool, 2013a.

[14] SAPP, "Power Quality in Electrical Power Systems," Southern Africa Power Pool, 2013b.

[15] IRENA, "Southern Africa Power Pool: Planning and Prospects for Renewable Energy," IRENA (International Renewable Energy Agency), 2013c.

[16] Stiles, G. and Murove, C., "SADC Renewable Energy and Energy Efficiency Status Report," Renewable energy Policy Network for the 21st Century (REN21), Paris, 2015.

[17] ECA, "Sustainable Energy: A Framework for New and Renewable Energy in Southern Africa ECA/SA/TPUB/2005/6," Economic Commission for Africa (ECA), Southern Africa Office, 2006.

[18] CEB, "Integrated Electricity Plan (2003–2012)," Central Electricity Board, Mauritius, Curepipe, Mauritius, 2003.

[19] IRENA, "Renewable Power generation Costs for 2014," IRENA (International Renewable Energy Agency), 2015.

[20] USDoE, "What is the Smart Grid? Office of Electricity Delivery and Energy Reliability," 2016. [Online]. Available: http://www.smartgrid.gov/the_smart_grid/smart_grid.html. [Accessed 17 August 2016].

[21] NEPAD, "Africa Power Vision Concept note & Implementation Plan: From Vision to Action," January 2015. [Online]. Available: http://www.un.org/esa/ffd/wp-content/.../NEPAD-APV-Exec-Sum-ENG.pdf. [Accessed 2 August 2016].

[22] SASGI, "South Africa Smart Grid Initiative," 2012. [Online]. Available: http://www.sasgi.org.za/about-sasgi/. [Accessed 2 August 2016].

[23] Engineering News, "South Africa Opts for Incremental Smart-grid Migration," 20 March 2012. [Online]. Available: http://www.engineeringnews.co.za/article/business-needs-crucial-for-eskom-smart-grid-migration. [Accessed 4 August 2016].

[24] Folly, K. A., "Challenges in Implementing Smart Grid technologies in Africa," Africa Utility Week: Cape Town, 2013.

[25] Waagsaether, K., "Overview of the Energy Picture for SADc Countries, with a Focus on Renewable Energy," Southern African Faith Communities Development Institute(SAFCEI): http://safcei.org/product/sadc-energy-report/, 2014.

[26] SADoE, "Integrated Resource Plan 2010–2030," Department of Energy, 2010.

[27] Ainah, P. K. and Folly, K. A., "Development of Micro-Grid in Sub-Saharan Africa: An Overview," *International Review of Electrical Engineering (IRE),* Vol. 10, no. 6, pp. 633–645, 2015.

[28] Folly, K. A., "Wind Energy Integration into the South African Grid: Prospects and Challenges," in Wind Energy: Developments, Potential and Challenges, pp. 93–120, Nova Publisher, USA, 2016.

[29] Walker, M., "HOMER Energy Asssists with Mini-Grid Solar Project in Tanzania, Homer microgrid," 2015. [Online]. Available: http://microgridnews.com/homer-energy-assists-with-mini-grid-solar-project-in-tanzania/. [Accessed 10 July 2016].

[30] Renewable Energy World, "Microgrids Seen as Answer for 620 Million African without Power," 2015. [Online]. Available: http://www.renewableenergyworld.com/articles/2015/11/minigrids-seen-as-answer-for-620-million-africans-without-power.html. [Accessed 3 July 2016].

# 5 ALTERNATING-CURRENT POWER TRANSMISSION

Jose R. Daconti

*Senior Staff Consultant, Siemens Power Technologies International; Senior Member, IEEE; Distinguished Member, CIGRE*

Allen L. Clapp

*President, Clapp Research Associates, P.C.; Life Member, IEEE; Senior Member, ASCE*

A. M. DiGioia, Jr.

*President, DiGioia Gray and Associates; Fellow, ASCE; Member, IEEE*

Dale A. Douglass

*Principal Engineer, Douglass Power Consulting, LLC; Fellow, IEEE*

I. S. Grant

*Manager, TVA; Fellow, IEEE*

Otto L. Lynch

*Vice President, Power Line Systems, Inc., Madison, Wisconsin*

John D. Mozer

*Professional Engineer, Retired; Life Member and SEI Fellow, ASCE*

J. R. Stewart

*Consultant; Fellow, IEEE*

Earle C. (Rusty) Bascom III

*Principal Engineer, Electrical Consulting Engineers, P.C.; Senior Member, IEEE*

## 5.1 OVERHEAD AC POWER TRANSMISSION

*BY JOSE R. DACONTI, ALLEN L. CLAPP, A. M. DIGIOIA, JR., DALE A. DOUGLASS, I. S. GRANT, OTTO L. LYNCH, JOHN D. MOZER, AND J. R. STEWART*

Overhead transmission of electric power remains one of the most important elements of today's electric power system. Transmission systems deliver power from generating plants to industrial sites and to substations from which distribution systems supply residential and commercial service. Those transmission systems also interconnect electric utilities, permitting power exchange when it is of economic advantage and to assist one another when generating plants are out of service because of damage or routine repairs. Total investment in transmission and substations is approximately 10% of the investment in generation.

Since the beginning of the electrical industry, research has been directed toward higher and higher voltages for transmission. As systems have grown, higher-voltage systems have rarely displaced existing systems, but have instead overlayed them. Economics have typically dictated that an overlay voltage should be between 2 and 3 times the voltage of the system it is reinforcing. Thus, it is common to see, for example, one system using lines rated 115, 230, and 500 kilovolts (kV). The highest ac voltage in commercial use is 765 kV although 1100 kV lines have seen limited use in Japan and Russia. Research and test lines have explored voltages as high as 1500 kV, but it is unlikely that, in the foreseeable future, use will be made of voltages higher than those already in service. This plateau in growth is due to a corresponding plateau in the size of generators and power plants, more homogeneity in the geographic pattern of power plants and loads, and adverse public reaction to overhead lines. Recognizing this plateau, some focus has been placed on making intermediate voltage lines more compact. Important advances in design of transmission structures as well as in the components used in line construction, particularly insulators, were made during the mid-1980s to mid-1990s. Current research promises some further improvements in lines of existing voltage including uprating and new designs for HVDC.

### 5.1.1 Transmission Systems

The fundamental purpose of the electric utility transmission system is to transmit power from generating units to the distribution system that ultimately supplies the loads. This objective is served by transmission lines that connect the generators into the transmission network, interconnect various areas of the transmission network, interconnect one electric utility with another, or deliver the electrical power from various areas within the transmission network to the distribution substations. Transmission system design is the selection of the necessary lines and equipment which will deliver the required power and quality of service for the lowest overall average cost over the service life. The system must also be capable of expansion with minimum changes to existing facilities.

Electrical design of ac systems involves (1) power flow requirements; (2) system stability and dynamic performance; (3) selection of voltage level; (4) voltage and reactive power flow control; (5) conductor selection; (6) losses; (7) corona-related performance (radio, audible, and television noise); (8) electromagnetic field effects; (9) insulation and overvoltage design; (10) switching arrangements; (11) circuit-breaker duties; and (12) protective relaying.

Mechanical design includes (1) sag and tension calculations; (2) conductor composition; (3) conductor spacing (minimum spacing to be determined under electrical design); (4) types of insulators; and (5) selection of conductor hardware.

Structural design includes (1) selection of the type of structures to be used, (2) mechanical loading calculations, (3) foundations, and (4) guys and anchors.

Miscellaneous features of transmission-line design are (1) line location, (2) acquisition of right-of-way, (3) profiling, (4) locating structures, (5) inductive coordination (considers line location and electrical calculations), (6) means of communication, and (7) seismic factors.

### 5.1.2 Voltage Levels

Standard transmission voltages are established in the United States by the American National Standards Institute (ANSI). There is no clear delineation between distribution, subtransmission, and transmission voltage levels. In some systems, 69 kV may be a transmission voltage while in other systems it is classified as distribution, depending on function. Table 5-1 shows the standard voltages listed in ANSI Standards C84 and C92.2, all of which are in use at present.

The nominal system voltages of 345, 500, and 765 kV from Table 5-1 are classified as extrahigh voltages (EHV). They are used extensively in the United States and in certain other parts of the world. In addition, 400-kV EHV transmission is used, principally in Europe. EHV is used for the transmission of large blocks of power and for longer distances than would be economically feasible at the lower voltages. EHV may be used also for interconnections between systems or superimposed on large power-system networks to transfer large blocks of power from one area to another.

One voltage level above 800 kV, namely, 1100 kV nominal (1200 kV maximum), is presently standardized. This level is not widely used, although sufficient research and development have been completed to prove technical practicability.[1–3]

**TABLE 5-1** Standard System Voltages, kV

| Rating | | Rating | |
|---|---|---|---|
| Nominal | Maximum | Nominal | Maximum |
| 34.5 | 36.5 | 230 | 242 |
| 46 | 48.3 | 345 | 362 |
| 69 | 72.5 | 500 | 550 |
| 115 | 121 | 765 | 800 |
| 138 | 145 | 1100 | 1200 |
| 161 | 169 | | |

### 5.1.3 Electrical Properties of Conductors

***Positive-Sequence Resistance and Reactances.*** The conductors most commonly used for transmission lines have been aluminum conductor steel-reinforced (ACSR), all-aluminum conductor (AAC), all-aluminum alloy conductor (AAAC), and aluminum conductor alloy-reinforced (ACAR), but conductors able to operate at higher temperatures such as ACSS are available for a modest price premium and are becoming more common. Research is progressing on new high-temperature ceramic-cored conductors. Tables of the electrical characteristics of the most commonly used ACSR conductors are in Sec. 3. Characteristics of other conductors can be found in conductor handbooks or manufacturers' literature and websites.

The per mile resistance, inductive reactance, and capacitive reactance can be determined from the data in the tables of Sec. 3 and the spacing factors $X_d$ and $X_d'$.

The positive-sequence resistance is listed as the 60-Hz value at 50°C. The expression for inductive reactance per mile is

$$X_L = 0.004657 f \log \frac{D}{\text{GMR}} \tag{5-1}$$

where $D$ = equivalent spacing in feet, GMR = geometric mean radius in feet as given in the conductor tables of Sec. 3, and $f$ = frequency in hertz. GMR for ACSR conductor is given at 60 Hz. However, 60-Hz values of GMR can be used at other commercial power-system frequencies with small error. $X_L$ also can be expressed as

$$X_L = X_a + X_d = 0.004657 f \log \frac{1}{\text{GMR}} + 0.004657 f \log D \tag{5-2}$$

When the spacing is 1 ft, $X_d$ becomes zero. Thus $X_a$ is frequently called the "one-foot" inductive reactance. The expression for capacitive shunt reactance per mile is:

$$X_c = \frac{4.099 \times 10^6}{f} \log \frac{D}{r_c} \tag{5-3}$$

where $r_c$ is the conductor radius in feet. $X_c$ can also be expressed as

$$X_c = X_a' + X_d'$$

where

$$X_a' = \frac{4.099 \times 10^6}{f} \log \frac{1}{r_c} \tag{5-4}$$

and

$$X'_d = \frac{4.099 \times 10^6}{f} \log D \qquad (5\text{-}5)$$

*Bundle conductors* consist of two or more conductors per phase mechanically and electrically connected and supported by an insulator assembly. The positive-sequence resistance is, to a first approximation, the 60-Hz, 50°C values in Sec. 3 tables divided by the number of conductors per phase. General formulas for the inductance and capacitance of bundle conductors are

$$L_\phi = \frac{1}{n}\left[0.74113 \log \frac{r_c}{\text{GMR}} + 0.74113 \log \frac{24(S_{gm})^n}{d(M_{gm})^{n-1}}\right] \quad \text{mH/mi} \qquad (5\text{-}6)$$

From Eq. (5-6) inductive reactance is found to be

$$X_L = \frac{1}{n}\left[K + 0.004657 f \log \frac{24(S_{gm})^n}{d(M_{gm})^{n-1}}\right] \quad \Omega/\text{mi at 60 Hz} \qquad (5\text{-}7)$$

and the capacitance is

$$C_\phi = \frac{0.03883n}{\log[24(S_{gm})^n / d(M_{gm})^{n-1}]} \quad \mu\text{F/mi} \qquad (5\text{-}8)$$

In the above, $n$ = number of conductors per phase (bundle); $d$ = diameter of conductor in inches; $S_{gm}$ = geometric mean distance between conductors of different phases in feet, found by taking the mean distance from all conductors of one phase to all conductors of the other phases; $M_{gm}$ = geometric mean distance in feet between the $n$ conductors of one phase; $K$ = internal conductor reactance defined as

$$K = 0.004657 f \log \frac{r_c}{\text{GMR}} \quad \Omega/\text{mi} \qquad (5\text{-}9)$$

**TABLE 5-2** Equivalent Reactances

| Bundle | $X_{aeq}$ | $X'_{aeq}$ |
|---|---|---|
| 2 conductors | $\frac{1}{2}(X_a - X_s)$ | $\frac{1}{2}(X'_a - X'_s)$ |
| 3 conductors | $\frac{1}{3}(X_a - 2X_s)$ | $\frac{1}{3}(X'_a - 2X'_s)$ |
| 4 conductors | $\frac{1}{4}(X_a - 3X_s)$ | $\frac{1}{4}(X'_a - 3X'_s)$ |

The inductive series reactance and capacitive shunt reactances for bundled conductors can also be found by using the $X_a + X_d$ method, by determining the equivalent $X_a$ and $X'_a$ of the conductor bundle. The expressions for the equivalents are given in Table 5-2. These expressions are for three-conductor bundles on equilateral spacing and for four-conductor bundles on square spacing. The subscript $s$ indicates the spacing of the conductors within the bundle in feet. Values for $X_a$ and $X'_a$ are in the conductor tables in Sec. 3. Values for $X_s$ and $X'_s$ are from the same formulas as $X_d$ and $X'_d$.

$$X_s = 0.004657 f \log s \qquad (5\text{-}10)$$

$$X'_s = \frac{4.099 \times 10^8}{f} \log s \qquad (5\text{-}11)$$

where $s$ is in feet and $f$ is frequency in hertz. Equation (5-11) is correct for a ratio of spacing $s$ to conductor radius $r$ of 5 or more.

**TABLE 5-3** Typical Transmission-Line Impedance*

| Voltage, kV | $R_1$ | $X_{L1}$ | $X_{C1}$ | $R_0$ | $X_{L0}$ | $X_{C0}$ | $X_0/X_1$ |
|---|---|---|---|---|---|---|---|
| 69 | 0.280 | 0.709 | 0.166 | 0.687 | 2.74 | 0.315 | 3.86 |
| 115 | 0.119 | 0.723 | 0.169 | 0.625 | 2.45 | 0.265 | 3.39 |
| 230 | 0.100 | 0.777 | 0.182 | 0.591 | 2.26 | 0.275 | 2.91 |
| 345 | 0.060 | 0.590 | 0.138 | 0.551 | 1.99 | 0.208 | 3.37 |
| 500 | 0.028 | 0.543 | 0.127 | 0.463 | 1.90 | 0.198 | 3.50 |
| 765 | 0.019 | 0.548 | 0.128 | 0.428 | 1.77 | 0.185 | 3.23 |

*$R_1$, $X_{L1}$, $R_0$, $X_{L0}$ are in ohms per mile; $X_{C0}$, $X_{C1}$ are in megohm-miles.
*Note:* 1 mi = 1.61 km.

The value of $X_{aeq}$ is added to $X_d$ (the spacing factor, which is determined for the mean spacing between the conductors of the different phases). $X'_{aeq}$ and $X'_d$ are handled in a like manner.

***Zero-Sequence Impedances.*** When earth-return currents due to faults or other causes are to be calculated, negative- and zero-sequence impedances must be determined in addition to positive-sequence quantities. Negative-sequence quantities are the same as the positive-sequence values for transmission lines. Precise determination of the zero-sequence quantities is difficult because of the variability of the earth-return path.

Calculation of zero-sequence impedance parameters is far more complex than for positive-sequence quantities, being a function of conductor size, spacing, relative position of conductors with respect to overhead ground wires, electrical characteristics of overhead ground wires, and the resistivity of the earth-return circuit. Reference 4 includes a detailed analysis of zero-sequence parameters, which are normally calculated using digital computer programs.

Table 5-3 lists representative values of positive- and zero-sequence impedances for different voltage transmission lines with shield wires. Zero-sequence reactance increases for unshielded lines.

***Nominal-π Representation.*** Transmission lines can be represented by nominal $\pi$ as in Fig. 5-1, in which half the capacitive susceptance, in siemens, is connected at each end of the line. The nominal-$\pi$ representation is used in digital computer studies involving lines of moderate length (usually under 100 mi).

***Nominal-T Representation.*** The nominal-T representation of a transmission line is shown in Fig. 5-2. The total line susceptance *b*, in siemens, is concentrated at *A*, the midpoint of the line.

***ABCD Parameters.*** These line parameters (general circuit constants) are defined by the equations

$$E_s = AE_r - BI_r \tag{5-12}$$

$$I_s = CE_r - DI_r \tag{5-13}$$

For a short line (under 100 mi) if $Z_1 = R + j\omega L$ and $Z_2 = 2/jb$ (refer to the nominal-$\pi$ line of Fig. 5-1)

$$A = D = \frac{Z_1 + Z_2}{Z_2} \tag{5-14}$$

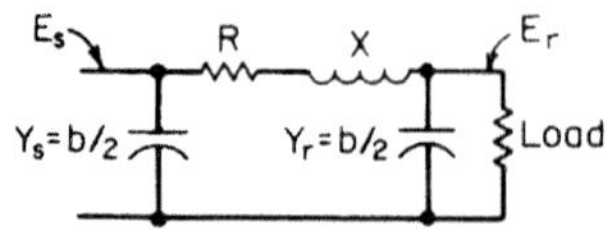


**FIGURE 5-1** Nominal-$\pi$ line.

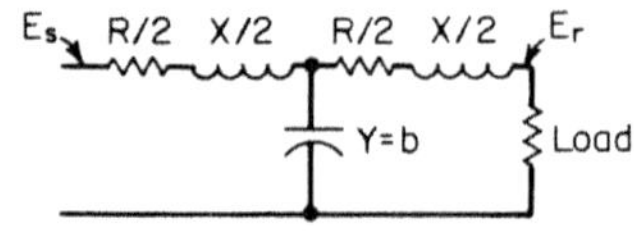


**FIGURE 5-2** Nominal-*T* line.

$$B = Z_1 l \tag{5-15}$$

$$C = \left( \frac{Z_1 + 2Z_2}{Z_2^2} \right) / l \tag{5-16}$$

For longer lines where $l$ is the length of the line

$$A = D = \cosh(\gamma l) \tag{5-17}$$

$$B = Z_c \sinh(\gamma l) \tag{5-18}$$

$$C = \frac{\sinh(\gamma l)}{Z_c} \tag{5-19}$$

where

$$\gamma = \sqrt{(R + j\omega L)(j\omega C)} \tag{5-20}$$

and

$$Z_c = \sqrt{\frac{R + j\omega L}{j\omega C}} \tag{5-21}$$

and $R$, $L$, and $C$ are line resistance, inductance, and capacitance per mile.

Formulas for *ABCD* constants for various circuit configurations are given in Table 5-4.

**TABLE 5-4** Formulas for Generalized Circuit Constants

| No. | Type of network | Diagram | Equivalent constants $A_t$ | $B_t$ | $C_E$ | $D_t$ |
|---|---|---|---|---|---|---|
| 1 | Series impedance | $E_s$ Z $E_r$ | 1 | $Z$ | O | 1 |
| 2 | Shunt admittance | $E_{sn}$ Y $E_{rn}$ | 1 | O | Y | 1 |
| 3 | Uniform line | $E_s$ ABC $E_r$ | $A$ | B | C | $A$ |
| 4 | Two uniform lines | $E_s$ $A_2B_2C_2$ $A_1B_1C_1$ $E_r$ | $A_1A_2 + C_1B_2$ | $B_1A_2 + A_1B_2$ | $A_1C_2 + A_2C_1$ | $A_1A_2 + B_1C_2$ |
| 5 | Two nonuniform lines or networks | $E_s$ $A_2B_2C_2D_2$ $A_1B_1C_1D_1$ $E_r$ | $A_1A_2 + C_1B_2$ | $B_1A_2 + D_1B_2$ | $A_1C_2 + D_2C_1$ | $D_1D_2 + B_1C_2$ |
| 6 | General network and sending transformer impedance | $E_s$ $Z_{TS}$ ABCD $E_r$ | $A + CZ_{TS}$ | $B + DZ_{TS}$ | C | D |
| 7 | General network and receiving transformer impedance | $E_s$ ABCD $Z_{TR}$ $E_r$ | $A$ | $B + AZ_{TR}$ | C | $D + CZ_{TR}$ |
| 8 | Two networks in parallel | $E_s$ $A_1B_1C_1D_1$ $A_2B_2C_2D_2$ $E_r$ | $\frac{A_1B_2 + A_2B_1}{B_1 + B_2}$ | $\frac{B_1B_2}{B_1 + B_2}$ | $C_1 + C_2 + \frac{(A_1 - A_2)(D_2 - D_1)}{B_1 + B_2}$ | $\frac{D_1B_2 + D_2B_1}{B_1 + B_2}$ |

***Note:*** All constants in this table are complex quantities; $A = a_1 + ja_2$ and $D = d_1 + jd_2$ are numerical values, $B = b_1 + jb_2$ = ohms, and $C = c_1 + jc_2$ = siemens. As a check on calculations of *ABCD* constants, note that $AD - BC = 1$.

**TABLE 5-5** SIL of Typical Transmission Lines

| System kV | $Z_s$, Ω | SIL, MW |
|---|---|---|
| | Overhead lines | |
| 230 | 367 | 144 |
| 345 | 300 | 400 |
| 500 | 285 | 880 |
| 765 | 280 | 2090 |
| 1200 | 250 | 5760 |
| | Cables | |
| 230 | 38 | 1390 |
| 345 | 25 | 4760 |

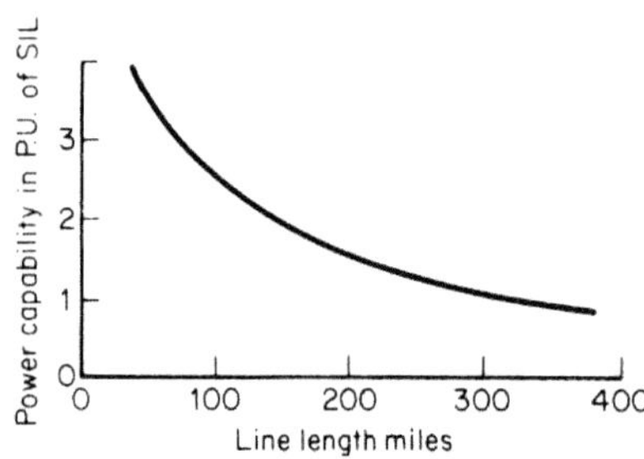


**FIGURE 5-3** Overhead line loading in terms of SIL.

***Surge Impedance Loading.*** The surge impedance of a transmission line is the characteristic impedance with resistance set equal to zero (i.e., $R$ is assumed small compared to $j\omega L$ of Eq. 5-21).

$$Z_s = \sqrt{\frac{L}{C}} \tag{5-22}$$

The power which flows in a lossless transmission line terminated in a resistive load equal to the line's surge impedance is denoted as the surge impedance loading (SIL) of the line. Under these conditions, the receiving end voltage $E_R$ equals the sending end voltage $E_S$ in the magnitude, but lags $E_S$ by an angle $\delta$ corresponding to the travel time of the line. For a three-phase line

$$\text{SIL} = \frac{(E_{L\text{-}L})^2}{Z_S} \tag{5-23}$$

Since $Z_s$ has no reactive component, there is no reactive power in the line, $Q_S = Q_R = 0$. This indicates that for SIL the reactive losses in the line inductance are exactly offset by reactive power supplied by the shunt capacitance or $I^2\omega L = E^2\omega C$.

SIL is a useful measure of transmission-line capability even for practical lines with resistance, as it indicates a loading where the line's reactive requirements are small. For power transfer significantly above SIL, shunt capacitors may be needed to minimize voltage drop along the line, while for transfer significantly below SIL, shunt reactors may be needed.

SILs for typical transmission lines are given in Table 5-5. Cables normally have current ratings (ampacity) considerably below SIL, while overhead line current ratings may be either greater than or less than SIL. Figure 5-3 presents illustrative overhead line loadability as a function of line length and SIL.

Although Fig. 5-3 is illustrative only of loading limits, it is a useful estimating tool. Long lines tend to be stability-limited and have a lower loading limit than shorter lines, which tend to be voltage-drop- or conductor-ampacity-limited.

## 5.1.4 Electrical Environmental Effects

***Corona and Field Effects.*** There are two categories of electrical environmental effects of power transmission lines. Corona effects are those caused by electrical stresses at the conductor surface which result in air ionization ("corona") and include radio, television, and audible noise. Field effects are those caused by induction to objects in proximity to the line. While the generic term is electromagnetic effects, within the electric power industry the fields are divided into two types: electric-field effects and magnetic-field effects. Electric fields, related to the voltage of the line, are the primary cause of induction to vehicles, buildings, and objects of comparable size. Magnetic fields, related to the currents in the line, are the primary cause of induction to long objects, such as fences and pipelines.

***Assessment Criteria.*** In an electrical environmental analysis, it is important to determine the proper criteria for assessment of the impact. For example, the audible noise criterion in a commercial or industrial area would be inappropriate in a quiet residential neighborhood.[5] Likewise, ground-level electric field criteria on a parking lot would be different from that in terrain inaccessible by motor vehicles. For audible noise, the only concern is annoyance, but for electric fields, safety, annoyance, and perception levels all may have to be considered.

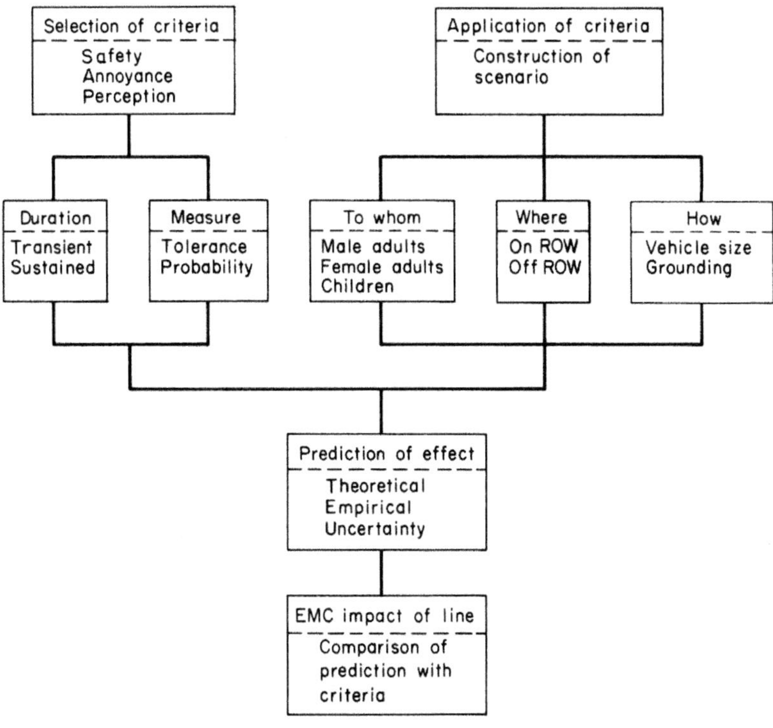


**FIGURE 5-4** Factors affecting transmission line EMC for shock effects.

Probability of exposure is also an important criterion. The impact of radio noise in arid locations is different from that in places with considerable rainfall. Since different people have different perception and annoyance thresholds, statistical evaluations are necessary, recognizing that some percentage of people will find a generally accepted noise level annoying. Because of the combination of worst-case events which are normally assumed in an electrical environmental analysis, the overall probability of annoyance is usually considerably smaller than initially presumed.

A predictive model is necessary to calculate the expected effect. Depending on the specific effect, it may be an empirical formula or may be quite sophisticated. However, it is only by calculating the effect and comparing it with specified criteria that the overall impact can be assessed. This is illustrated by Fig. 5-4,[6] which is a flowchart of the analysis procedure for an example case of electric-field-induced shock.

***Audible Noise.*** Corona-produced audible noise during foul weather, particularly during or following rain, can be an important design parameter for high-voltage ac transmission lines. Audible noise has two components, a random noise component and a low-frequency hum, each produced by different physical mechanisms. While the hum component is closely correlated with corona loss on the line, the random noise is not. Of these two, the most frequent cause of annoyance is the random noise, and it is this which is calculated and compared with acceptance criteria.

Analyses to predict levels of audible noise consider $A$-weighted sound level [dB($A$)] during rain, including

$L_{50}$, which is the level exceeded 50% of the time during rain (considering all rain storms over a period of time, usually 1 year).

$L_5$, which is the level exceeded 5% of the time during rain.

Average, which is the average level of noise expected during rain. (This is usually close to the $L_{50}$ value and is sometimes called "wet-conductor" noise.)

Heavy rain, which is the level expected during heavy rain. (This usually is representative of laboratory artificial rain tests but is assumed representative of the $L_5$ level.)

Reference 7 compares audible noise formulas, which have been developed throughout the world. One formula for both $L_5$ and $L_{50}$ values is given by

$g$ = Average-maximum surface gradient of conductor or conductor bundle, kV/cm

$n$ = Number of subconductors in a phase (or pole) bundle

$d$ = Diameter of subconductors, cm

$D$ = Distance from line to point at which noise level is to be calculated, m

SL = $A$-weighted sound level of the noise produced by the line, dB($A$)

$N_p$ = number of phases

AN = $A$-weighted sound level of the noise produced by one phase of the line, dB($A$)

$\text{AN}_0$ = A reference $A$-weighted sound level, dB($A$)

$K_1, K_2, K_3, K_4$ = Constant coefficients

Application = All line geometries

Noise measure = $L_5$ rain and $L_{50}$ rain

Range of validity = 230–1500 kV, $1 \le n \le 16$, $2 \le d \le 6$

For each phase, the $L_5$ noise level is given by

$$\text{AN}_5 = \frac{-665}{g} + 20 \log n + 44 \log d - 10 \log D - 0.02\, D + \text{AN}_0 + K_1 + K_2 \qquad \text{(5-24)}$$

with

$$\begin{aligned} \text{AN}_0 &= 75.2 && \text{for } n < 3 \\ &= 67.9 && \text{for } n \ge 3 \\ K_1 &= 7.5 && \text{for } n = 1 \\ &= 2.6 && \text{for } n = 2 \\ &= 0 && \text{for } n \ge 3 \\ K_2 &= 0 && \text{for } n < 3 \\ &= 22.9(n-1)\frac{d}{B} && \text{for } n \ge 3 \end{aligned}$$

where $B$ is the bundle diameter, cm.

The $L_{50}$ level for each phase is obtained from

$$\text{AN}_{50} = \text{AN}_5 - \Delta A \qquad \text{(5-25)}$$

where

$$\begin{aligned} \Delta A &= 14.2\frac{g_c}{g} - 8.2 && \text{for } n < 3 \\ &= 14.2\frac{g_c}{g} - 10.4 - 8\left[(n-1)\frac{d}{B}\right] && \text{for } n > 3 \end{aligned}$$

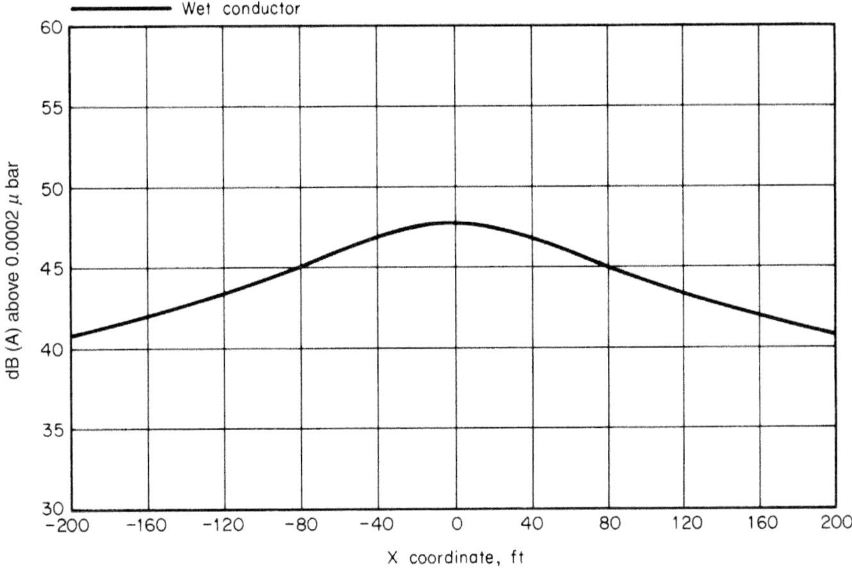


**FIGURE 5-5** Audible noise profile at ground level for a transmission line.

and

$$g_c = 24.4(d^{-0.24}) \qquad \text{for } n \le 8$$

$$= 24.4(d^{-0.24}) - 0.25(n-8) \qquad \text{for } n > 8$$

$$\text{SL} = 10 \log \sum_{i=1}^{Np} 10^{AN_i/10} \tag{5-26}$$

Figure 5-5 illustrates a typical presentation of audible noise calculations. The profile, in this case for a representative 500-kV line and wet conductors, quantifies the level of noise in dB($A$) greater than 0.002 $\mu$bar as a function of distance from the centerline of the structure. From this method of presentation, analysis of maximum levels as well as effect on width of right-of-way can be analyzed. Similarly, design variables such as conductor size, spacing, and configuration; height of conductors; and weather variations can be considered.

Figure 5-6[3,8] quantifies experience with transmission-line audible noise complaints. These occur mostly during wet-conductor conditions and low ambient noise, such as after rain or during fog. During heavy-rain conditions, the noise of the rain masks the line noise. Other factors during heavy rain, such as closed windows, combine to make this condition less likely to result in complaints even though the noise is louder. In the absence of

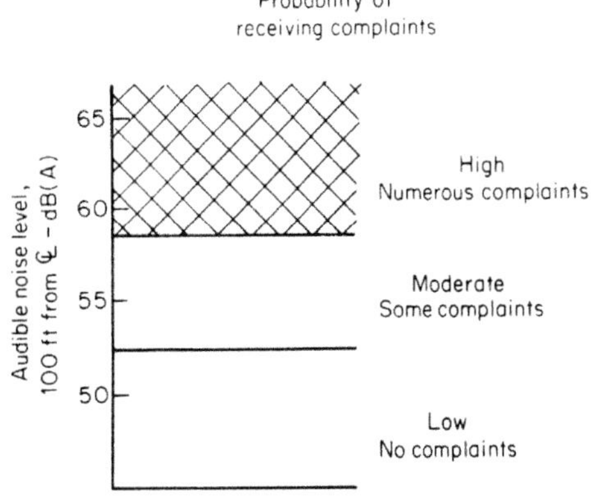


**FIGURE 5-6** Audible noise compliance guidelines.

local noise regulations, comparison of calculated $L_{50}$ or average audible noise with Fig. 5-6 gives a reasonable preliminary evaluation of the possibility of audible noise annoyance. When measurements are to be taken to confirm ambient noise or line noise, care must be taken to follow proper procedures.[9]

***Radio and Television Noise.*** Electromagnetic interference from overhead power lines is caused by two phenomena: complete electrical discharges across small gaps (microsparks) and partial electrical discharges (corona). Gap-type sources occur at insulators, line hardware, and defective equipment and are a construction and maintenance problem rather than a design consideration. They are responsible for about 90% of radio noise complaints and can be located and eliminated as they occur.[10] Conductor and hardware corona is considered during the design phase. On a properly designed line, conductor corona noise rarely results in television interference complaints except perhaps in weak signal fringe areas.

The specification of "corona-free" hardware is important to eliminate electromagnetic interference from conductor support hardware, and is especially important as lines are constructed with closer spacings and resulting higher electric fields on the hardware. Conductor clamps and other fittings, which were formerly acceptable at traditional phase spacings, may not be adequate for compact lines.

For ac lines, radio and television noise are functions of the weather. Fair-weather noise may be significant and varies with the season, wind velocity, and barometric pressure.

Two families of computation methods are available for radio noise: those based on conductor laboratory tests and analytical propagation theory (semianalytical methods) and those based on an empirical formula using data from long-term tests on operating lines (comparative methods).

The comparison method[11] is useful for conventional geometries and designs:

$$\mathrm{RI} = -150.4 + 120 \log g + 40 \log d + 20 \log \frac{h}{D^2} + 10[1 - (\log 10f)^2] \tag{5-27}$$

where $g$ = average maximum surface gradient of conductor or conductor-bundle, kV peak/cm
$d$ = subconductor diameter, mm
$h$ = height of phase, m
$D$ = radial distance to observer, m
$f$ = frequency, MHz
RI = fair-weather radio noise, dB

RI is calculated for each phase and the maximum value is used as the RI of the line. Average foul-weather RI levels are assumed to be 17 dB above fair weather, and heavy-rain RI 24 dB above fair weather. Other methods are described in Ref. 3.

As with audible noise, the most useful data presentation is the level of radio noise as a function of distance from the centerline of the structure. An illustrative example for a specific 500-kV line is shown in Fig. 5-7.

There are no generally accepted RI limits in the United States, because of the impossibility of setting universal criteria for all land use and local conditions.[12] A Canadian standard exists for RI limits and is a useful guide.[13]

Two quantities are required to set criteria for evaluation of radio noise. These are the level of signal strength in the line vicinity and an appropriate signal/noise ratio. This latter ratio is typically assumed to be 24 to 26 dB at the edge of the right-of-way. Primary signal strengths may be 54 dB above 1 μV (0.5 mV/m) in rural areas to 88 dB or more in cities.

Prediction of television noise is not as advanced as that of radio noise, primarily because of the limited number of actual cases of conductor corona television interference. As with radio noise, most television interference complaints result from microsparks which can be located and eliminated as they occur. These are not generally a design consideration. In the few cases where corona-caused television noise has occurred in foul weather, it has often been possible to remedy the situation by an improvement in the receiving antennas rather than changes to the transmission-line design. References 3 and 14 contain some work on prediction and evaluation of TVI.

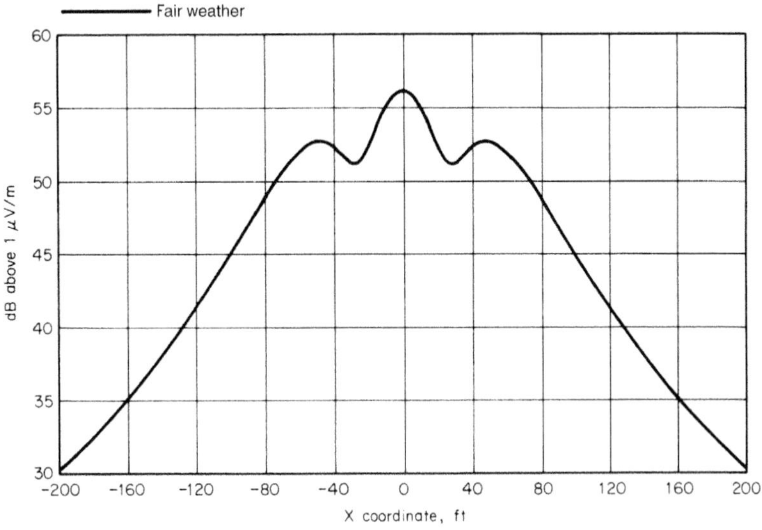


**FIGURE 5-7** Radio noise profile at ground level for transmission line.

***Gaseous Oxidants.*** Gaseous oxidants can be produced by corona activity in air and, in sufficient concentrations, may produce adverse effects on flora and fauna. The most important oxidants are ozone ($O_3$) and oxides of nitrogen (mainly NO and $NO_2$), where ozone is the major constituent.

Federal standards limit photochemical oxidants to 0.12 part per million for a maximum of 1-h concentration not to be exceeded more than once per year. Some states have more restrictive regulation; for example, the Minnesota Pollution Control Agency standards are for 0.07 ppm by volume (130 μg/m$^3$). Ozone can be detected by smell at minimum concentrations of 0.01 to 0.15 ppm.

Analytic studies and field measurements have been conducted on both operating and test lines.[15-22] The highest calculated value for 1-mi/h wind parallel to the line was 0.019 ppm maximum ground-level concentration. Measurements have indicated that transmission-line contribution to gaseous oxidants cannot be detected within statistical limits of significance and accuracy. With instrumentation capable of detecting 0.002 ppm, the transmission-line contribution was indistinguishable from ambient.

Thus, gaseous oxidants are not a concern with respect to electric power transmission lines.

***Ground-Level Electric Fields.*** Ground-level electric field effects of overhead power transmission lines relate to the possibility of exposure to electric discharges from objects in the field of the line. These may be steady currents or spark discharges. Other areas which have received attention are the possibility of fuel ignition and interference with wearers of prosthetic devices (e.g., pacemakers).[23]

It is appropriate to consider unlikely conditions when setting and applying electric-field safety criteria because of possible consequences; thus statistical considerations are necessary. Annoyance criteria need not be as stringent and mitigating factors can be considered.

*Electric-Field Calculations.* The resultant electric fields in proximity to a transmission line are the superposition of the fields due to the three-phase conductors. The conducting earth must be represented by image charges located below the conductors at a depth equal to the conductor height.

For example, consider the three-conductor line of Fig. 5-8. The effect of earth can be represented by replacing the earth with image conductors as shown in Fig. 5-8. At 60 Hz and for typical values

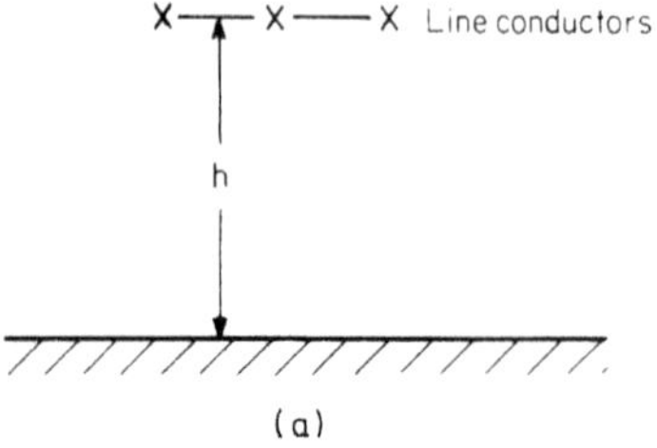


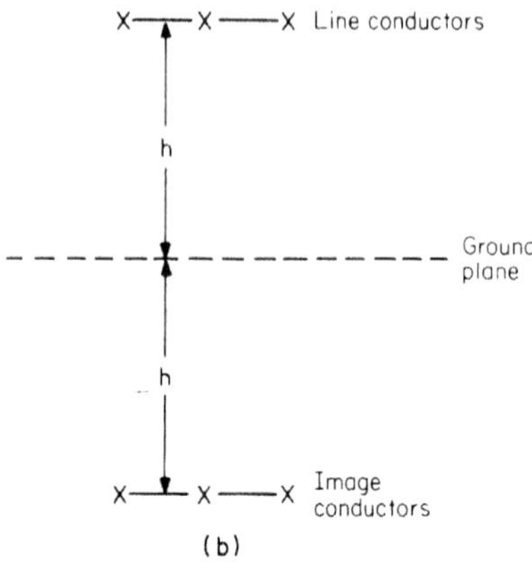


**FIGURE 5-8** Representation of conducting earth: (*a*) earth; (*b*) image.

of earth resistivity, the relaxation time of the earth (the time required for charges to redistribute themselves due to an externally applied field) is so small compared to the power frequency wave that for each instant of time the charge is distributed on the earth's surface as in the static condition (i.e., the earth appears to be a perfect conductor).

The electric fields surrounding the transmission line are a function of the instantaneous charges on the line. Usually, however, the charges are not known, but the voltages to ground of the different conductors are. Since the charge $Q$ on each conductor is a function of the voltage on all conductors, an $n \times n$ capacitance matrix results, where $n$ is the number of conductors, according to the formula

$$[Q] = [C][V] \tag{5-28}$$

which, for a three-conductor configuration (ignoring shield wires), is

$$Q_1 = C_{11}V_1 + C_{12}V_2 + C_{13}V_3 \tag{5-29}$$

$$Q_2 = C_{21}V_1 + C_{22}V_2 + C_{23}V_3 \tag{5-30}$$

$$Q_3 = C_{31}V_1 + C_{32}V_2 + C_{33}V_3 \tag{5-31}$$

The off-diagonal (mutual) capacitance terms significantly affect the final results. The individual terms of the capacitance matrix are computed by

$$C_{nm} = \left.\frac{Q_n}{V_m}\right| \text{all other voltages} = 0 \tag{5-32}$$

where $n$ and $m$ are conductors.

The potential coefficient matrix is, however, more amenable to computation and is defined by

$$[V] = [P][Q] \tag{5-33}$$

whose individual terms are given by

$$P_{nm} = \left.\frac{V_n}{Q_m}\right| \text{all other charges} = 0 \tag{5-34}$$

This is an open-circuit matrix where the individual terms can be computed by assuming a charge at one conductor and calculating the voltage at the prescribed location assuming all the other conductors nonexistent (open-circuited). For a single conductor of radius $r$ and a height $h$ above the earth, the self-potential coefficient is given by

$$P_{nm} = \frac{1}{2\pi\varepsilon_o}\ln\frac{2h}{r} \tag{5-35}$$

For two conductors $n$ and $m$ where $d_{nm}$ is the distance between them, and $d_{nm'}$ is the distance between conductor $n$ and the image of conductor $m$, the mutual potential coefficient is given by

$$P_{nm} = \frac{1}{2\pi\varepsilon_o} \ln \frac{d_{nm'}}{d_{nm}} \tag{5-36}$$

This potential coefficient matrix can be calculated and inverted to yield the capacitance matrix:

$$[C] = [P]^{-1} \tag{5-37}$$

This capacitance matrix allows the calculation of the charges on the individual conductors for the given initial voltage distribution according to Eqs. (5-29) through (5-31). Once these charges are obtained, the desired electric fields can be determined.

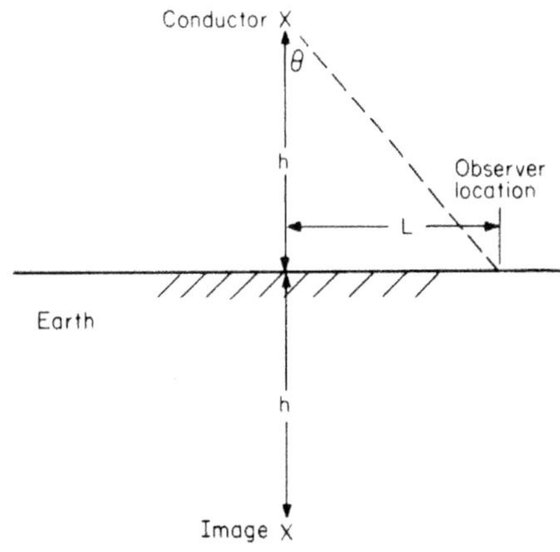


**FIGURE 5-9** Single conductor.

For the single conductor and observer location of Fig. 5-9, the ground-level electric field is determined from

$$E = \frac{Q_l}{2\pi\varepsilon_o r} \tag{5-38}$$

The distance from the conductor to the observer is

$$r = \sqrt{h^2 + L^2} \tag{5-39}$$

Thus

$$E = \frac{Q_l}{2\pi\varepsilon_o \sqrt{h^2 + L^2}} \tag{5-40}$$

$Q$ must be determined from $[Q] = [C][V]$. For a single conductor this equation reduces to

$$Q_l = P^{-1}V = \frac{1}{\frac{1}{(2\pi\varepsilon_o)} \ln (2h/r)} V \tag{5-41}$$

For a multiconductor configuration, $Q$ would come from the full matrix calculation.

$E$ is radially directed from the line charge. The vertical component is

$$|E| \cos\theta = \frac{Q_l}{2\pi\varepsilon_o \sqrt{h^2 + L^2}} \frac{h}{\sqrt{h^2 + L^2}} = \frac{Q_l}{2\pi\varepsilon_o} \frac{h}{h^2 + L^2} \tag{5-42}$$

The vertical component of the electric field at ground level because of the image is equal to the field from the conductor, since the image is the geometric mirror image and has the opposite sign charge. Thus, the total ground-level field is given by

$$E = \frac{Q_l}{\pi\varepsilon_o} \frac{h}{h^2 + L^2} \tag{5-43}$$

At ground level, the horizontal components of the electric fields of the conductor and its image cancel and the resultant field is purely vertical.

For a three-phase line, the fields of the three conductors and their images are computed separately and added.

For fields extremely close to the line conductors, care must be taken to represent the local effects properly. For example, the surface field around the conductor is not uniform. For a bundled conductor, it is more nearly represented by a sinusoid. Farther from the conductors, a GMR representation will suffice.

For a bundle of diameter $D$ with $n$ conductors of radius $r$, the GMR is given by

$$\text{GMR} = \frac{D}{2}\sqrt[n]{\frac{2nr}{D}} \tag{5-44}$$

Replacing the conductor radius with the bundle GMR gives the appropriate representation.

Figure 5-10 illustrates a representative electric-field profile, in kV rms per meter, from the centerline of the structure. This presentation clearly illustrates the maximum field, the location of the maximum, and the effect on right-of-way width considerations. Sensitivity to various parameters can also be quickly evaluated.

*Criteria for Evaluation.* The effects of electric fields on humans is due to discharges from objects insulated from ground; typically vehicles, buildings, and fences which become electrically charged by induction from the line. Table 5-6 summarizes effects on humans, ranging from no perception through severe shock and possible ventricular fibrillation.[24]

Criteria for spark discharges are expressed in terms of stored charge or stored energy on the charged object. Levels for perception in adult males are of the order of 0.12 mJ, while experience indicates that approximately 2 mJ results in an annoying spark. Safety is seldom of concern, since approximately 25 J is required for injury, a value beyond that expected on objects beneath transmission lines.

Deno's work, using test data, relates short-circuit current to the undisturbed electric field for objects insulated from ground.[23] Initial calculations assume the worst possible combination of circumstances; no leakage path to ground exists for the object, complete grounding of the person involved, steady contact, and orientation of the vehicle parallel to the line. Table 5-7 lists sample criteria and electric fields needed to meet them for three sample vehicles.

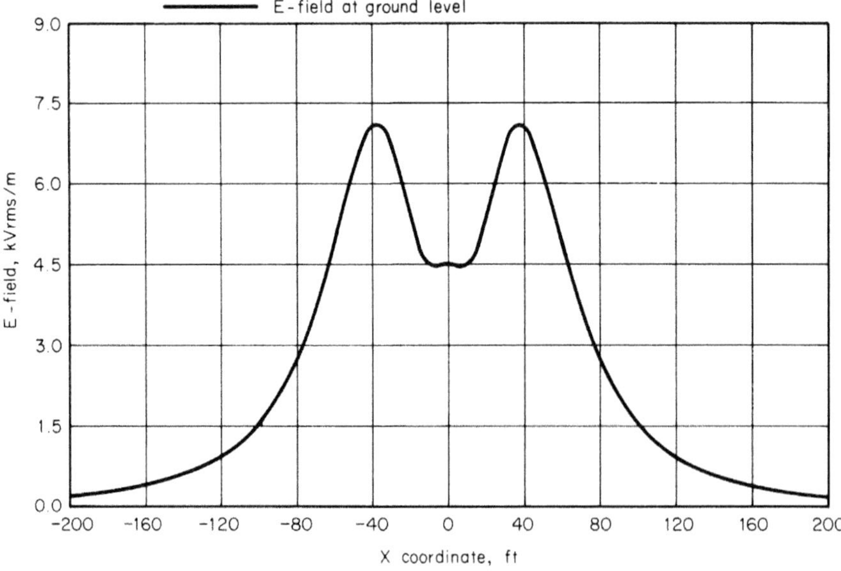


**FIGURE 5-10** Electric-field profile at ground level for transmission line.

**TABLE 5-6** Threshold Levels for 60-Hz Contact Currents

| rms current, mA | Threshold reaction and/or sensation |
|---|---|
| | Perception |
| 0.09 | Touch perception for 1% of women |
| 0.13 | Touch perception for 1% of men |
| 0.24 | Touch perception for 50% of women |
| 0.33 | Grip perception for 1% of women |
| 0.36 | Touch perception for 50% of men |
| 0.49 | Grip perception for 1% of men |
| 0.73 | Grip perception for 50% of women |
| 1.10 | Grip perception for 50% of men |
| | Startle |
| 2.2 | Estimated borderline hazardous reaction, 50% probability for women (arm contact) |
| 3.2 | Estimated borderline hazardous reaction, 50% probability for women (pinched contacts) |
| | Let-go |
| 4.5 | Estimated let-go for 0.5% of children |
| 6.0 | Let-go for 0.5% of women |
| 9.0 | Let-go for 0.5% of men |
| 10.5 | Let-go for 50% of women |
| 16.0 | Let-go for 50% of men |
| | Respiratory tetanus |
| 15 | Breathing difficult for 50% of women |
| 23 | Breathing difficult for 50% of men |
| | Fibrillation |
| 35 | Estimated 3-s fibrillating current for 0.5% of 20-kg (44-lb) children |
| 100 | Estimated 3-s fibrillating current for 0.5% of 70-kg (150-lb) adults |
| | Established standards |
| 0.50 | ANSI standard for maximum leakage (portable appliance) |
| 0.75 | ANSI standard for maximum leakage (installed appliance) |
| 5.0 | NESC recommended limit for induced current under transmission line |

**TABLE 5-7** Limiting Electric Field for Given Criteria, kV/m

| | | Sample vehicles | | |
|---|---|---|---|---|
| | | Autos, pickups | Farm vehicles | Buses, trailer trucks |
| | Sample criteria | A | B | C |
| Safety | 5 mA | 22.32 | 10.86 | 6.33 |
| | 25 J | 259.00 | 159.00 | 106.50 |
| Annoyance | 2 mA | 8.92 | 4.35 | 2.50 |
| | 2 mJ | 2.37 | 1.41 | 0.95 |
| Perception | 1.1 mA | 4.91 | 2.39 | 1.39 |
| | 0.12 mJ | 0.58 | 0.35 | 0.23 |

**TABLE 5-8** Likely Range of Maximum Vertical Electric Field for Various Voltage Transmission Lines

| Line voltage, kV | Near-ground vertical electric field, kV/m |
|---|---|
| 765 | 8–13 |
| 500 | 5–9 |
| 345 | 4–6 |
| 230 | 2–3.5 |
| 161 | 2–3 |
| 138 | 2–3 |
| 115 | 1–2 |
| 69 | 1–1.5 |

High voltages may develop due to electric-field coupling, but the available short-circuit current is small (i.e., high-impedance source); thus calculations are based on a Norton equivalent and the short-circuit current. A relatively high resistance ground is sufficient to reduce electric-field-coupled voltage.

Table 5-8 lists maximum electric fields on the right-of-way under lines of different voltage classes. The fields attenuate rapidly with distance from the line and are usually much lower at the right-of-way edge.

***Fuel Ignition.*** Theoretical calculations indicate that if several unlikely conditions exist simultaneously, a spark could release sufficient energy to ignite gasoline vapors. These conditions include a perfectly grounded person refueling a car perfectly insulated from ground with a metal can while the car is parked directly under a line. The spark would have to occur in the precise location of optimum fuel-air mixture. Research[3,25] confirms the low probability of accidental fuel ignition under actual conditions.

No confirmed cases of accidental ignition under transmission lines exist, confirming the low probability of these factors occurring simultaneously. Because of the consequences of a gasoline fire, some electric utilities advise that gasoline-fueled vehicles not be refueled near a line of 500 kV or above. If refueling were necessary, the vehicle could be grounded or the can connected to the vehicle to prevent sparks.

***Ground-Level Magnetic Fields.*** Magnetic-field coupling affects objects which parallel the line for a distance, such as fences and pipelines, and is generally negligible for vehicle- or building-sized objects. As opposed to electric-field coupling, magnetic-field coupling is a low-voltage, low-impedance source with relatively high short-circuit currents. Single grounds are ineffective in preventing magnetically coupled voltages and multiple low-resistance grounds are needed. The resistance of the person touching a fence or pipeline is the dominant current-limiting impedance in the equivalent electrical circuit.[26] Calculations are based on a "longitudinal electromotive force" approach and are described in Refs. 27 to 29.

A consideration in the calculation of magnetic fields, which is different from the electric-field calculation, concerns the images. A perfectly conducting earth can be assumed for the electric-field problem, even for realistic values of earth resistivity. The assumption of a transmission line in free space (no earth at all) gives a closer approximation to the ground-level magnetic fields than does the assumption of a perfectly conducting earth for measurements near the line. At distances beyond 100 m, the effect of earth becomes increasingly more significant. The effect of conducting earth is frequently treated by use of an image conductor located at a greater depth in the earth than the conductors are above the earth. Distances of several hundred meters are commonly used for this image depth, according to the relation $D = 660\sqrt{\rho/f}$ meters where $\rho$ is the earth resistivity in ohm-meters and $f$ is the frequency. Magnetic-field calculations are given in Ref. 9, including the use of Carson's terms to evaluate the effects of imperfectly conducting earth.

It is normally adequate to consider conductors in free space without images. For the conductor of Fig. 5-9 without its image

$$B = \frac{\mu_o I}{2\pi r} = \frac{\mu_o I}{2\pi\sqrt{h^2 + L^2}} \tag{5-45}$$

This is then separated into vertical and horizontal components by multiplying by sin $\theta$ and cos $\theta$. In general, both components must be retained. For a three-phase line, all conductors must be computed. Horizontal and vertical components of $B$ from the three conductors must then be combined individually as phasors, considering the angles of the different currents. The combined horizontal and vertical components in general have different angles, causing their resultant to trace an ellipse in time. Single-axis magnetic-field meters with the sensing coil oriented for a maximum reading give the magnitude of the major axis of the field ellipse. A three-axis meter of the type presently used for data logging responds to the square root of the sum of the squares of the three field components (the "resultant" field). The resultant field can be as much as 41% greater than the major axis of the field ellipse for circularly polarized fields of the type which result from symmetrical conductor configurations.[30]

In the same manner, image currents at some assumed depth can be computed and their fields included. The use of matrix calculations allows inclusion of ground wires and bundled conductors as is the case of electric fields.

With both electric and magnetic fields it is essential to follow proper measurement procedures[30] for comparison with calculations.

For electric fields it is important that the field not be perturbed by the presence of the operator or other nearby objects. For both electric and magnetic fields, it is necessary to accurately know the conductor positions, the conductor height, the distance to the observer, and the line operating conditions (voltage and current). Magnetic-field measurements frequently differ from calculations for a number of reasons beyond errors in distance and clearance measurement:

1. Line current is continually varying, so in general it is not as well known as line voltage. In addition to uncertainty concerning the current magnitude at the time of the field measurement, line current unbalance in both magnitude and phase angle can be important. Unbalance has an increasingly significant effect on the magnetic field, the farther one moves from the line. Spot measurements, especially in homes and near distribution lines, are of limited usefulness to characterize exposure. For this reason, it is often advisable to statistically characterize the magnetic field. A statistical description of the field over time can be developed from measurements or calculations which assume balanced currents. It is also sometimes useful to develop a statistical distribution for a specific current level and an assumed maximum unbalance.
2. Related to current unbalance is circulating current in the shield wires, return currents in the earth, and currents in nearby pipes. These currents may cause significant differences between calculation and measurement.
3. The difference between single- and three-axis instruments has been described above. Two operators with different instruments can determine different answers based on the principles of measurement.
4. In nonuniform fields, such as around appliances, the size of the sensing coil and presence or absence of ferromagnetic core material will affect the reading of instruments equally well calibrated in a uniform field. Calibration must be made in a calibrating coil sufficiently large that the field is uniform over the area of the sensing coil, yet not so large that other nearby currents do not affect the field.
5. Harmonic currents have different effects depending on the frequency response of the instrument. Some instruments have a response linearly increasing with frequency, some are flat with frequency, and others have bandpass filters of different waveshapes.

### 5.1.5 Line Insulation

***Requirements.*** The electrical operating performance of a transmission line depends primarily on the insulation. An insulator not only must have sufficient mechanical strength to support the greatest loads of ice and wind that may be reasonably expected, with an ample margin, but must be so designed as to withstand severe mechanical abuse, lightning, and power arcs without mechanically failing. It must prevent a flashover for practically any power-frequency operating condition and many transient voltage conditions, under any conditions of humidity, temperature, rain, or snow, and with such accumulations of dirt, salt, and other contaminants that are not periodically washed off by rains.[31]

***Insulator Materials.*** The majority of present insulators are made of glazed porcelain. Porcelain is a ceramic product obtained by the high-temperature vitrification of clay, finely ground feldspar, and silica. Insulators of high-grade electrical porcelain of the proper chemical composition free from laminations, holes, and cooling stresses have been available for many years.

The insulator glaze seals the porcelain surface and is usually dark brown, but other colors such as gray and blue are used. Porcelain insulators for transmission may be disks, posts, or long-rod types.

Porcelain insulators have been used at all transmission line voltages and, if correctly manufactured and applied, have high reliability.

A typical porcelain disk insulator is shown in Fig. 5-11.

Glass insulators have been used on a significant proportion of transmission lines. These are made from toughened glass, and are usually clear and colorless or light green. For transmission voltages they are available only as disk types. Most glass disk insulators will shatter when damaged, but without mechanically releasing the conductor. This provides a simple method of inspection.

Synthetic insulators, originally pioneered by the General Electric Company in 1963 for high-voltage transmission lines,[32] and more recently introduced by several manufacturers, are finding increasing acceptance. Most consist of a fiberglass rod covered by weather sheds of skirts of polymer (silicon rubber, polytetrafluoroethylene, cycloaliphatic resin, etc.)[33] as shown in Fig. 5-12. Other types include a cast polymer concrete called Polysil R[34] and a coreless type with alternating metal and insulating sections.[35]

Improvements in design and manufacture in recent years have made synthetic insulators increasingly attractive since their strength-to-weight ratio is significantly higher than that of porcelain and can result in reduced tower costs, especially on EHV and UHV transmission lines.

These insulators are usually manufactured as long-rod or post types. The light weight of most designs and resistance to damage aids construction. In addition, their performance under contaminated conditions may be significantly better than that of porcelain.[36]

Use of synthetic insulators on transmission lines is relatively recent and a few questions are still under study, in particular the lifetime behavior of insulating shed materials under contaminated conditions. It has been found necessary to use grading rings on some types at higher voltages to prevent damage to the sheds, and a very small number of insulators have experienced "brittle fractures," in which the fiberglass core breaks close to an end fitting. Despite these problems it appears that reliable synthetic insulators are presently available.

***Insulator Design.*** Transmission insulators may be strings of disks (either cap and pin or ball and socket), long-rods, or line posts. Posts are only infrequently applied above 230 kV.

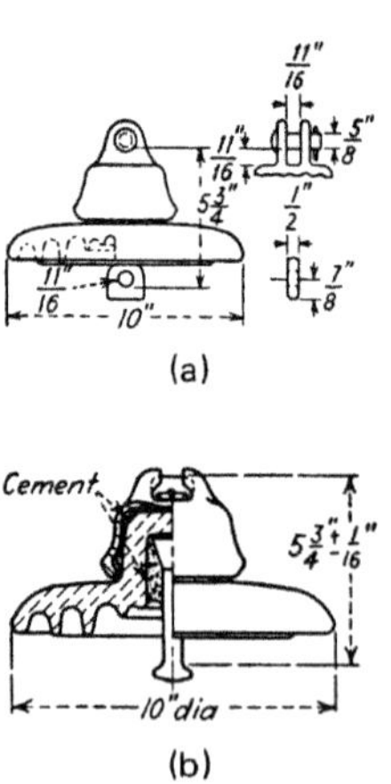


**FIGURE 5-11** Typical porcelain disk insulator: (*a*) clevis type; (*b*) ball-and-socket type. (*Locke Insulators Inc.*)

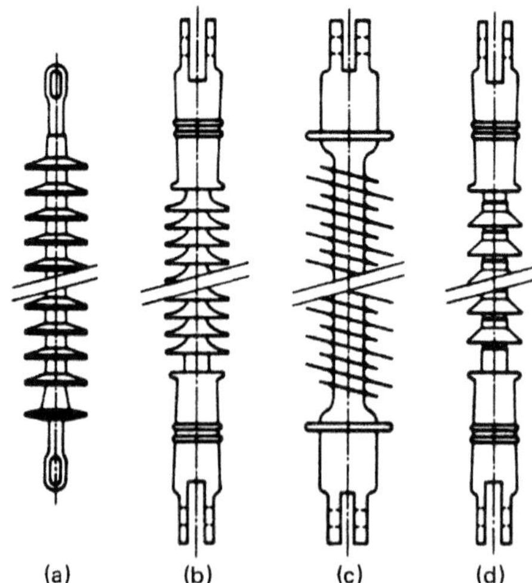


**FIGURE 5-12** Typical nonceramic insulators.

Present suspension insulators conform to ANSI Standard C29.2, and standards have been established for 15,000-, 25,000-, 36,000-, and 50,000-lb ratings. It is common practice to use a factor of safety of 2 for the maximum mechanical stress applied to porcelain or glass insulators. For fiberglass-core insulators it is more common for the manufacturer to supply a recommended maximum working load.

Each manufacturer supplies catalogs which provide a physical description of the insulator's mechanical characteristics, wet and dry 60-Hz flashover strength, and positive- and negative-impulse ($1.2 \times 50\ \mu s$) critical (50%) flashover strength. Switching surge performance ($250 \times 3000\ \mu s$) is usually not supplied. In clean conditions most insulators of equivalent dimensions have very similar performance.

Suspension insulator strings, that is, insulators used to support the conductor weight at a suspension or tangent structure, may be in I (vertical) or V configurations. The V configuration is used to prevent conductor movement and resultant clearance reductions at the structure. At dead-end or tension structures the insulators must also support the conductor tension, and it is not uncommon for these tension strings to be given a slightly higher flashover strength (e.g., by adding disks) to reduce the likelihood of a flashover that might lead to insulator string mechanical failure. Two or more strings of insulators in parallel can be used on suspension and tension strings to provide higher mechanical strength if required.

The electrical strength of line insulation may be determined by power frequency, switching surge, or lightning performance requirements. At different line voltages, different parameters tend to dominate. Table 5-9 shows typical line insulation levels and the controlling parameter. In compacted or uprated designs, considerably fewer insulators than these have been successfully used.[37,38]

Detailed descriptions of insulation design for electrical performance for different conditions, line voltages, and line types are available[39–41] from a number of studies.

**TABLE 5-9** Typical Line Insulation

| Line voltage, kV | No. of standard disks | Controlling parameter (typical) |
|---|---|---|
| 115 | 7–9 | Lightning or contamination |
| 138 | 7–10 | Lightning or contamination |
| 230 | 11–12 | Lightning or contamination |
| 345 | 16–18 | Lightning, switching surge, or contamination |
| 500 | 24–26 | Lightning, switching surge, or contamination |
| 765 | 30–37 | Switching surge or contamination |

***Insulator Standards.*** The NEMA Publication *High Voltage Insulator Standards*, and AIEE Standard 41 have been combined in ANSI C29.1 through C29.9. Standard C29.1 covers all electrical and mechanical tests for all types of insulators. The standards for the various insulators covering flashover voltages; wet, dry, and impulse; radio influence; leakage distance; standard dimensions; and mechanical-strength characteristics are as follows: Ceramic C29.2, suspension; C29.3, spool; C29.4, strain; C29.5, low- and medium-voltage pin; C29.6, high-voltage pin; C29.7, high-voltage line post; C29.8, apparatus pin; C29.9, apparatus post, C29.12 and C29.13, nonceramic suspension; C29.17 and C29.18, nonceramic line post. These standards should be consulted when specifying or purchasing insulators.

***Line Insulation Design***

*Power-Frequency Design.* The criteria for power-frequency design is usually that flashover shall not occur for normal operating conditions, including reduced clearances to the structure from high wind. A typical wind-design limit is the 50- or 100-year return period wind, that is, a wind velocity which occurs only once in 50 or 100 years. This velocity is obtained from local wind records and may be typically 80 to 100 mi/h. Maximum operating voltages are designed by ANSI C84 and C92 standards and are 5% or 10% above the nominal value.

In clean conditions, power-frequency voltage is not a controlling parameter for insulator design (as distinct from air-gap clearance). However, even in quite lightly contaminated conditions it may become so.

Design for contamination is usually expressed as inches of creepage per kilovolt, where the creepage distance is the length of the shortest path for a current over the insulator surface and ranges up to

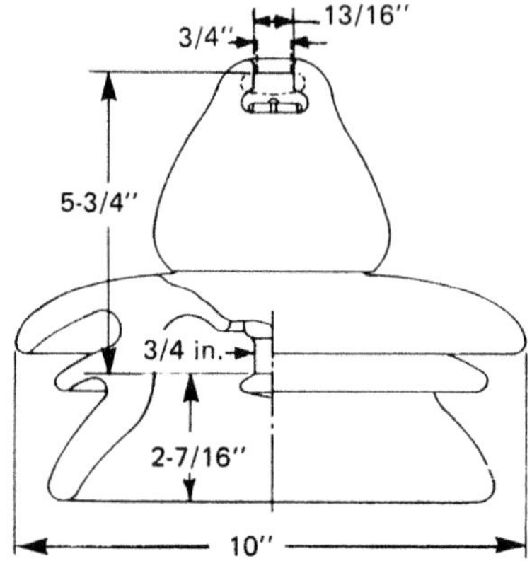


**FIGURE 5-13** Typical fog-type disk insulator.

2 in/kV or more for heavy contamination. Standard insulator disks (10 × 5¾ in) have a typical creepage length of 11.5 in per disk. To avoid very long insulator strings for contamination, disks with additional creepage distance are made. The creepage can be extended by use of lengthened skirts and deeper grooves in the underside. Fog-type disks have up to 21.5 in of creepage per 13½ × 8-in units. A typical fog-type insulator is illustrated in Fig. 5-13.

In extremely contaminated conditions, insulation with extended creepage may not be enough. In these cases insulator washing or the use of a silicone or petroleum grease coating (replaced at regular intervals) may be used.

Table 5-10 provides a simplified indication of creepage distance as a function of contamination,[39] and Fig. 5-14 shows guidelines from the IEEE application guide.[40]

For nonceramic insulation the same approach is used, except that subject to manufacturer's recommendations, a reduction in creepage distance up to 30% may be possible. This is due to the physical behavior of the nonceramic insulating material in moist conditions.

Another approach that has sometimes been used to combat contamination effects is the semiconductive glaze insulator. The semiconducting glaze allows a small but definite power-frequency current to flow over the surface. The insulator does not improve the standard test values, such as wet and dry power-frequency flashover and short-time impulse flashover, although it may have some value under switching surge conditions.

The glaze has a surface resistivity of about 10 MΩ per square. This is achieved by special formulations of materials involving, at the present stage of development, the use of tin-antimony additive to a more normal glaze composition. The presence of this small leakage current, of the order of 1 to

**TABLE 5-10** Insulation Requirements for Contamination: Provisional EHV Line Insulation Design Table for Various Contamination Conditions
Standard 5⁵/₃₄ × 10-in vertical insulator units

| Contamination | | Equivalent amount | Provisional design values | | |
|---|---|---|---|---|---|
| | | | Leakage distance | Average kV rms | |
| Class | Types | NaCl, mg/cm² | in/kV rms line to ground | Per in axial length | Per unit |
| A | Clean atmosphere—rural and forest regions; no industrial contamination | 0–0.03 | Insulation requirements not set by contamination | | |
| B | Slight atmospheric contamination; suburbs of large industrial regions; railways; frequent washing rains | 0.04 | 1.04 | 2.0 | 11.5 |
| C | Moderate contamination containing soluble salts up to 5%; furnaces, dust from metallurgical plants, mine dust, fly ash, fertilizer dust in small quantities | 0.06 | 1.31 | 1.6 | 9.1 |
| D | Severe contamination containing 15% or more of soluble salts; dust from aluminum and chemical works, cement plants, heavy agricultural fertilizing, fly ash with high salt or sulfur content | 0.12 | 1.74 | 1.2 | 6.9 |
| E | Salt precipitation—seaside regions, salt marshes | 0.30 | 2.11 | 1.0 | 5.7 |

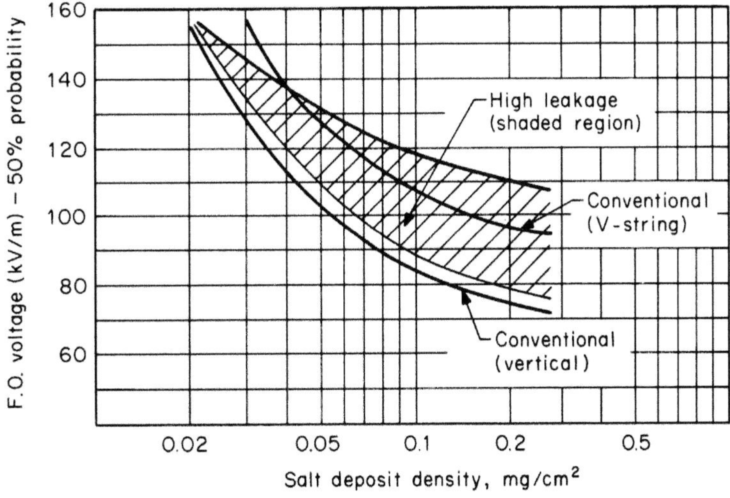


**FIGURE 5-14** Power frequency withstand voltage of contaminated suspension insulators in fog expressed in kV/m of connection length (spacing).

2 mA for suspension insulators, but which can be several times that value for large porcelains (such as are used in high-voltage bushings) has three effects:

1. Linearization of the voltage distribution over the insulator or string of insulators. This aids greatly in improving the performance of the insulator with respect to corona disturbance and RIV performance, plus having some benefits under dry and clean conditions.
2. Heating of the insulator. This occurs because of the power loss associated with the leakage current flow to a temperature which is usually about 5°C over the ambient air conditions. The heating effect enables the insulator to remain dry during conditions of fog or mist. This eliminates the majority of contaminated-insulator flashovers which occur when accumulated contamination becomes damp. This damp contamination condition is the most usual cause for contaminated-insulator flashover because most contaminants are more electrically conducting when damp or wet.
3. The elimination of "dry banding," which is recognized as another major cause of flashover of standard insulators when contaminated. This occurs when the insulator has been thoroughly wetted, such as in a rain storm which wets but does not thoroughly clean the contamination from the insulator's surface. Under these conditions, dry bands will form as the standard insulator dries, and arcs strike across the dry-band area. These arcs can progress until flashover of the entire insulator occurs. With a semiconducting insulator, the relatively low resistance of the glaze shunts the dry-band area as the insulator dries and prevents the striking of the small power-frequency arcs.

The improved performance possible with semiconducting insulators has been proved in the laboratory and field,[42–46] but, because of the energy losses associated with the inherent leakage current, they are not widely used.

In some severe contamination areas, the problem has been effectively attacked by the use of silicone grease coatings. The unique amoebic action of a thick layer of silicone grease on an insulating surface is such as to envelop conducting solid particles which are said to "load" up the silicone grease to the saturation point, at which time the "used" silicone grease is removed and replaced with new silicone grease. In severe contamination areas, the greasing and degreasing cycles may be required every few months; in less severe contamination areas, the cycle may be a year or more depending on experience acquired. In this manner, the time between insulator cleanings can be greatly extended, thus making for substantial savings. Once the silicone coating is used, the coatings must usually be wiped off and replaced manually, as necessary. Among the manufacturers of silicone grease are the General Electric Company and the Dow-Corning Corporation.

For the cleaning operation to remove contamination from the insulator surface, many contaminants such as salt deposits and water-soluble conducting liquids can be successfully removed by hot-line washing, using high-pressure water and insulated nozzles and hoses. Another method is "dry cleaning" by the use of an abrasive powder such as a limestone mixture or biodegradable plastic pellets, discharged at high pressure through hose and nozzle on the insulating surface. In many cases either hot-line washing or dry cleaning alone is sufficient to cope with the rate of accumulation encountered with the particular contaminant. An exception is substantially conducting materials, which take a chemical "set" after exposure to water, such as cement dust, some forms of gypsum, or asbestos, which often must first be manually chipped off or scrubbed off the insulating surface and then covered with silicone grease as previously described.

It should be emphasized that these problems may be very severe or even nonexistent, due to the variability of contamination exposure, which in turn depends on the chemical and electrical nature of the contaminant, prevailing wind direction, persistence of fog, smog, or other weather factors.

To monitor buildup of contaminants, some utilities collect data at the site to warn operating departments of an impending flashover, so as to promptly implement contamination-combative procedures.

*Switching Surge Design.* Operation of a circuit breaker on a transmission line can cause transient overvoltages, although flashovers due to such switching surges are rare in lines below 500 kV. If the breaker is opening, this may be due to restrikes across the breaker contacts as they separate, although restriking has been nearly eliminated with present breaker technology. If the breaker is closing, the cause may be unequal voltages on each side of the breaker, including the effect of residual charge on the line from a recent deenergization. The crest magnitudes of switching surges are normally defined in per unit of nominal power-frequency-crest phase-to-ground voltage. For example, on a 138-kV line (145 kV maximum), the per unit value is 118 kV. Typical switching surges range from 1 to as high as 4 or 5 per unit, and the varying characteristics of breaker operations provide a distribution of surge magnitudes which is often modeled as a truncated gaussian distribution.

The criterion for switching surge design is usually that flashover shall not occur for most or all switching events. Several design methods have been used, including

1. The maximum expected surge is determined, for example, from a transient network analyzer (TNA) or digital study, and the line insulation is designed to withstand that surge.
2. Rather than the maximum surge, a surge value corresponding to a statistical level is used, typically the 2% value (i.e., the crest value determined from the statistical distribution of surge crests, such that the level will be exceeded by only 2% of all surges).
3. Rather than design insulation to withstand a maximum surge, a statistical approach is used to design for a low number of flashovers per switching event. Typical levels are one flashover per 100 or 1000 breaker operations. This often results in a more economical design than either of the withstand approaches above.
4. By modeling the statistical distribution of switching surge crests, the distribution of insulator flashover with voltage, and the statistical distribution of weather that can be obtained from local weather stations, a probabilistic design can be prepared using a relatively simple computer program based on the allowable flashover rate. Typical procedures, data, and examples for such calculations are provided in several publications.[47,48]

*Impulse Surge Design.* Impulse surges on a line are caused by lightning strokes to or near the line. At transmission insulation levels, only strokes that directly intercept the line are capable of causing flashovers.

A number of methods of calculating transmission-line lightning performance are described in Section 23. A computer program for this simplified calculation method is available from the *IEEE WG on Transmission Line Lightning Performance,* and more sophisticated programs for evaluation of multicircuit lines are available from a number of sources.

It is unusual for line insulation to be determined by lightning performance alone. More typically, insulation is determined by other requirements and the lightning performance is then verified. If this performance is unsatisfactory, it is often more efficient to change other design parameters such as shield wires or grounding than to add insulation.

Other methods of improving lightning performance have included addition of surge arresters at relatively frequent intervals along a line, and on double-circuit lines the use of unbalanced insulation so one circuit will flash over first and protect the other. Use of line arresters is most beneficial in regions of high ground resistance. Use of unbalanced insulation can improve the performance of the circuit with the highest insulation, but at the detriment of overall line performance.

*Phase-to-Phase Insulation.* The controlling paths for flashovers on most presently installed transmission lines are phase-to-ground, since there are usually grounded structure components between phases. However, for some new designs, such as the Chainette,[49] and compact lines the controlling path may be phase-to-phase air gaps or even phase-to-phase insulators.

Design methods for phase-to-phase insulation are essentially the same as for phase-to-ground insulation. Until recently, there was lack of knowledge of conductor clearance at midspan under various dynamic loading conditions, and lack of phase-to-phase switching surge data. Research studies sponsored by EPRI have now provided adequate design information on both topics.[41,47,48]

***Protective and Grading Devices.*** Damage to insulators from heavy arcs was a serious maintenance problem in the past, and several devices were developed to ensure that an arc would stay clear of the insulator string. Subsequent improvements in the use of overhead ground wires and fast relaying have reduced the likelihood of insulator damage to the point that arc protection devices are now rarely used in the United States.

Earlier protective measures consisted of attaching small horns to the clamp, but it was found that horns with a large spread both at the top of the insulator and at the clamp were required to be effective. Under lightning impulse the arc tends to cascade the string, and tests show that the gap between horns should be considerably less than the length of the insulator string. Protection by arcing horns thus resulted in either a reduced flashover voltage or an increase in the number of units and length of the string. In any event, flashover persisted as a power arc until the line tripped out. For these reasons arcing horns have not been used in the United States for many years, although they are fairly common in Europe.

The arcing ring or grading shield is mainly for the purpose of improving the voltage distribution over the insulator string, and its effectiveness is due to the more uniform field. Protection of the insulator is not, therefore, dependent on simply providing a shorter arcing path, as is the case with horns. Efficient rings are rather large in diameter and, for suspension strings, clearances to the structure should be at least as great as from ring to ring. These considerations have made this device generally unattractive for modern construction. Grading rings are now used only at very high voltages for special applications, or with nonceramic insulators. Corona shields help improve the voltage distribution at the line ends of insulator strings.

### 5.1.6 Line and Structure Location

***Preparation for Construction.*** The cost of preparing for transmission-line construction is a considerable part of the total costs—under some conditions as much as 25%. Right-of-way and clearing are more or less fixed by local conditions, but the cost of surveys, accompanying maps, profiles, and engineering layout is to some extent governed by judgment. Many times in the past the overall costs have been increased by right-of-way difficulties and by delays in receiving proper materials because of inadequate preparations. The engineering work, properly carried out, makes it possible to obtain the right-of-way and complete the clearing well in advance of construction and to purchase every item of material and deliver it to the correct location.

The work of locating and laying out a line does not require great refinement, but careful planning is essential. With inexperienced surveyors or drafters, it must be assumed that errors will be made, and every possible device must be used to discover these errors before construction is started.

***Location.*** The general character of the line location should be determined because it has a definite bearing on the type of design. In extreme cases, such as difficult mountainous sections or in highly developed areas near cities, this may be a determining factor in the selection of the conductor and type of structures.

With today's importance of the transmission grid, all transmission lines must be routed so that accessibility for inspection and repairs can be made quickly. Line location is a matter of judgment and requires a person of wide general experience capable of correctly weighing the divergent requirements for inexpensive and available right-of-way, low construction costs, and convenience in maintenance. In mountainous country or in thickly populated areas, it is generally not advisable to attempt a direct route or try to locate on long tangents. Small angles of a few degrees cost little more and add little to the length of line. Most designs provide suspension structures for line angles of 5° to 15° which are not excessively costly, although additional right-of-way costs should be considered for guyed structures. It is also a good idea to avoid high, exposed ridges to afford protection against both wind and lightning.

Following a general reconnaissance by ground and air, for which 10 to 20 days per 100 mi should be allowed for traditional surveying methods, and the assembling of all available maps and information, control points can be established for a general route or areas selected for more detailed study which may prove to be determining factors in the location of the line.

With this preliminary work completed, the major difficulties should have been determined. The policy as to such matters as right-of-way condemnation, electrical environmental assessments, telephone coordination, navigable-stream crossing, air routes, airports, and crossings with other utilities must be decided as definitely as possible.

Preliminary specifications should be issued before the final survey is started. These should include (1) outline drawings of the various structures with the important dimensions; (2) conductor sag curves and a sag template; (3) the maximum spans and angles for each type of structure; and (4) the requirements for right-of-way and clearing. Estimated costs are valuable, especially comparative costs of the various types of structure. With this information the field engineer can often, in a difficult section, choose the location best suited to the design.

Aerial maps can often be secured at much less cost than preliminary surveys, and in highly developed areas may be used to advantage for completely laying out the line without sending surveyors into the area until after the right-of-way has been secured. In today's digital world, many lines can be routed without any additional aerial reconnaissance.

Photographs taken at approximately ½ mi to the inch give sufficient detail for most work. Such maps can be photographically enlarged about four times for special detail. With a ½-mi-to-the-inch scale, the route of the line can be determined within a width of about 3 mi and sufficient landmarks located on a fairly accurate map to serve as a guide for flying the line. For modern digital photograph, 6-in pixels with 200 DPI imagery usually suffices for these photographs.

*Location Survey.* The actual survey party can typically be divided into four divisions, each of which can complete at least a mile a day in average weather and country. Their operations may be carried out separately or nearly concurrently by allowing a full week's separation between successive operations and transferring personnel as needed.

The work falls naturally into the following: (1) an alignment party, choosing the exact location and cutting out the line; (2) a staking party, driving stakes at 100-ft stations and locating all obstructions; (3) a level party, taking elevations and side slopes; and (4) a property and topography party, locating property lines.

A field drafting force located at a convenient point for receiving field notes can complete the final plan and profile drawings as fast as the survey can be made.

The method of procedure and size of survey organization depend on the character of the country, the length and type of line, the experienced personnel available, and the schedule which must be maintained. In level, sparsely populated country, satisfactory but incomplete property surveys and profiles have been made during an open dry winter for a wood H-frame line 50 mi in length in approximately 4 months' time, with the personnel averaging a crew of eight and an engineer. Modern survey methods such as light distance and ranging (LiDAR) and modern line design program will greatly reduce this amount of time and manpower required.

On a development involving the construction of several hundred miles of steel-tower line, the survey for a 65-mi line in rather difficult country, including 25 mi of inaccessible mountainous country, was completed with property maps and profiles in the form for permanent records in 2 months' time with a crew of about 20 and a locating engineer.

*Purchase.* Generally, right-of-way is not purchased in fee, but a perpetual easement is secured in which the owner grants the necessary rights to construct and operate the line but retains ownership and use of the land. The width of the right-of-way may be stated as a definite width or in general terms, but the easement must provide for (1) a means of access to each structure; (2) permission to erect all structures and guys; (3) all trees and brush to be cleared over a specified width for erection; (4) the removal of trees, which would not safely clear the conductor if the conductor were to swing out under maximum wind or which would not safely clear the conductor if they were to fall; and (5) the removal of buildings, lumber piles, haystacks, etc., which constitute a fire hazard. One of the major causes of serious line outages is the neglect to adhere strictly to conservative rules for clearing.

***Structure Spotting.*** The efficient location of structures on the profile is an important component of line design. Structures of appropriate height and strength must be located to provide adequate conductor ground clearance and minimum cost. In the past, most tower spotting has been done manually, using templates, but several computer programs have been available for a number of years for the same purpose.

*Manual Tower Spotting.* A celluloid template, shaped to the form of the suspended conductor, is used to scale the distance from the conductor to the ground and to adjust structure locations and heights to (1) provide proper clearance to the ground; (2) equalize spans; and (3) grade the line (Fig. 5-15).

The template has been traditionally cut as a parabola on the maximum sag (usually at the maximum operating temperature of 100°C or higher) of the anticipated ruling span for the line and it should be extended by computing the sag as proportional to the square of the span for spans both shorter and longer than the ruling span. By extending the template to a span of several thousand feet, clearances may be scaled on steep hillsides. The form of the template is based on the fact that, at the time when the conductor is erected, the horizontal tensions must be equal in all spans of every length, both level and inclined, if the insulators hang plumb at the stringing temperature. The template, therefore, must be cut to a catenary or, as has been traditionally done in the past, approximately, a parabolic approximation. The parabolic approximation is accurate to within about one-half of 1% for sags up to 5% of the span, which is well within the necessary refinement for traditional designs with considerable clearance buffers, but may not be accurate enough for lines being re-rated today.

Since vertical ground clearances are being established, the maximum operating temperature (100°C and higher) no-wind curve is often used in the template. This temperature is required by the *National Electrical Safety Code* ANSI C1-2012 for clearance above ground, rails, water, buildings, signs, tanks, etc., as the minimum temperature to be used. Thus, it is appropriate in most sections of the United States for neutrals, guys, communication cables on messengers, and similar items that are not expected to have significant heating from line losses. However, this temperature is not appropriate in many areas in the southwest United States for these items and is not appropriate for high-current distribution and transmission phase conductors—maximum operating temperature is required.

IEEE Std. 738 contains information on calculating both steady-state and transient conductor temperatures. Wind speed affects conductor temperature—the lower the wind speed, the lower will be the cooling effect and the higher will be the temperature of the conductor. Note that, in some locales (such as near a seacoast), the highest conductor temperature may occur on less than the hottest day, since the inland area heats up more on the hottest days and creates a thermally driven wind bringing cool air in from the coast. Thus, a series of templates may be required for the same conductor (or cable) for different expected current loadings or different areas of the system. Special conditions may call for additional clearance checks. For example, if it is known that a line will have high temperature rise because of an emergency load current, conductor clearance should be checked for the estimated emergency conductor temperature rating. Glaze ice and wet snow loadings can also create excessive sag of conductors. Such occurrences may not normally be considered in line design, since when they occur, the line may be taken out of service until the ice or snow drops. However, lines that are subject to high ice loads will experience permanent deformation of the wire due to these loads and this should be considered in the original design. On existing lines where this was not considered in the original design, the wires may exhibit additional sag over the design sags and the wires may be closer to the ground and other obstacles than anticipated (see http://www.nerc.com/fileUploads/FilePressReleases/PR%20Facilityh%Ratings$2007@20Oct%2010.pdf).

The template must be used subject to a "Creep" and "Load" correction for the conductors. Creep is a nonelastic conductor stretch which continues for the life of the line, with the rate of elongation decreasing with time. For example, the creep elongation during the first 6 months is equal to that of the next 9½ years. All conductors of all materials are subject to creep, but conductors with aluminum are especially sensitive to creep. The conductor manufacturers should always be consulted for these values. "Sag-Tension Calculation Methods for Overhead Lines" by CIGRE Task Force B2.12.3 discusses creep, load, and how the final sags and tensions are calculated and is available at http://www.cigre.org.

Precise values for creep are impossible to determine, since they vary with both temperature and tension, which are continuously varying during the life of the line. For example, it is found that a 1000-ft span of 954,000-cmil 48/7 ACSR when subjected to a constant tension of approximately 18% of its ultimate strength at a temperature of 60°F will have a sag increase in 1 day of approximately 5.5 in; in 10 days, 13 in; in 1 year, 27 in; in 10 years, 44 in; and in 30 years, 52 in.

Unless it is known that the line will have a life of less than 10 years, no less than 10 years' creep should be allowed for in the design. It is possible to prestress the creep out of small conductors, but for large conductors this requires time and special tensioning facilities not normally available. Also the time lost in constructing an EHV line will more than pay for the extra structure height required to compensate for the creep. Prestressing changes the modulus of elasticity, and this new modulus should be used in the design.

Precise values for load are also difficult to determine as the maximum load that the line will experience cannot be accurately predicted. If the amount of ice and/or wind are over-predicted, the conductor will never reach its design sag and money will have been overspent making the structure taller than it needed to be. However, if the amount of ice and/or wind are under-predicted, the conductor will sag lower than designed and there will be unanticipated clearance violations resulting in violations of code requirements and/or costly line flashovers.

The vertical weight supported at any structure is the weight of the length of conductor between low points of the sag in the two adjacent spans. For bare-conductor weights, this distance between low points can be scaled by using a template of the sag at any desired temperature. The maximum weight under loaded conditions should be scaled from a template made for the loaded sags. For most problems, the horizontal distance may be taken as equal to the conductor length. Distances to the low point of the sag may be computed by Eq. (5-58).

*Uplift.* On steep inclined spans the low point may fall beyond the lower support; this indicates that the conductor in the uphill span exerts a negative or upward pull on the lower tower. The amount of this upward pull is equal to the weight of the conductor from the lower tower to the low point in the sag. Should the upward pull of the uphill span be greater than the downward load of the next adjacent span, actual uplift would be caused, and the conductor would tend to swing clear of the tower.

It is important that abrupt changes in elevation of the structures should not occur, so that the conductor will not tend to swing clear of any structure even at low temperatures. This condition would be indicated if the 0°F curve of the template can be adjusted to hang free of the center support and just touch the adjacent supports on either side. In northern states it would be well to add a curve to the template for the below-zero temperatures experienced.

*Insulator Swing.* The uplift condition should not even be approached in laying out suspension insulator construction; that is, each tower should carry a considerable weight of conductor. The minimum weight that should be allowed on any structure may be logically determined by finding the transverse angle to which the insulator string may swing without reducing the clearance from the conductor to the structure too greatly. Also, the ratio of vertical weight to horizontal wind load should be limited to avoid insulator swing beyond this angle. The maximum wind is usually assumed at a temperature of 60°F. The wind pressure, measured in pounds per square foot, to be used in swing calculations is a matter of judgment and depends on local conditions. Under high-wind conditions it is reasonable to require somewhat less than normal clearances. Generally a clearance corresponding to about 75% of the flashover value of the insulator is adequate. The insulator will swing in the direction of the resultant of the vertical and horizontal forces acting on the insulator string as shown in Fig. 5-15.

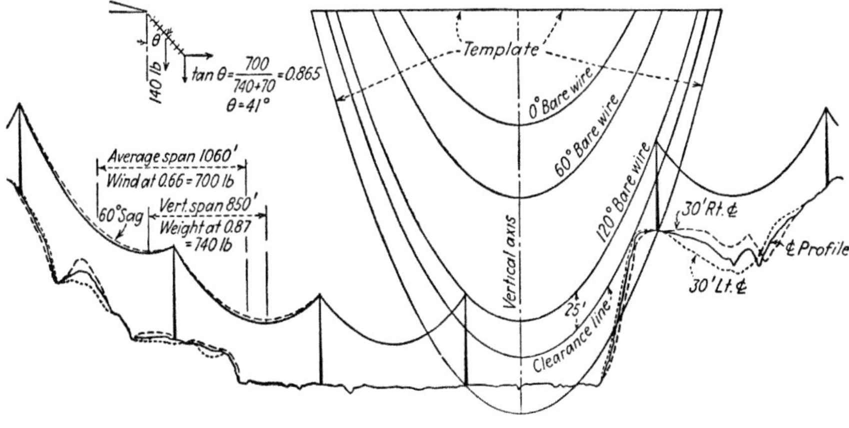


**FIGURE 5-15** Sag template determines clearances of a suspended conductor from the ground.

*Long Spans.* Rough country may necessitate spans considerably longer than contemplated in the design and may involve a number of factors including (1) proper clearance between conductors, (2) excessive tensions under maximum load, and (3) structures adequate to carry the additional loads.

Safe horizontal clearance between conductors is often based on the National Electrical Safety Code (NESC) formula, in which the spacing $a$ in inches is given as proportional to the square root of sag; $s$ is in inches.

$$a = 0.3 \text{ in/kV} + 8\sqrt{\frac{s}{12}} \tag{5-46}$$

This relation was developed for, and is useful on, comparatively short span lines of the smaller conductors and for voltages up to 69 kV; but for very long spans and heavy conductors, the formula results in spacings considerably larger than have proved satisfactory. It also results in spacings that are questionably small for very light conductors on long spans. Percy H. Thomas proposed an empirical formula[69] which takes into account the weight of the conductor and its diameter, requiring less spacing for heavy conductors and a greater spacing for small conductors by the ratio of diameter $D$ in inches to weight $w$ in pounds per foot ($D/w$) as a means of determining the required conductor spacing for the average span of the line. The factor $C$ in Eq. (5-71) includes an allowance to permit the standard spacing to be used on somewhat longer spans than average construction. The same formula, however, may be used to examine the spacings which have been successfully used on maximum spans and a value for $C$ selected from experience for determining the safe spacing required for an occasional unusually long span.

Excessive tensions on very long spans may be avoided by dead-ending at both ends and computing such a stringing sag as will result in the same maximum tension as elsewhere in the line. Such a span will be found to have considerably greater stringing sag and lower stringing tension than the normal span. Sag curves or charts are often prepared giving the sag for dead-end spans of various lengths such that the maximum tension under loaded conditions will be the same.

Dead-end construction is costly, and consideration should be given to avoiding this additional expense. It is common practice to permit spans up to double the average span without dead ends, although spans of this length may require additional spacing between wires. A careful examination of some trial figures on the sags and tensions developed in a long span will often indicate how great a span may be carried on suspension structures. The maximum loaded tension which would occur in a long span, if this span were dead-ended and sagged to the same stringing tension as the rest of the line, compared with the maximum tension for normal span lengths, is a good indication of the necessity for dead-end construction.

In case a number of long spans are encountered in a line or section of line, it may prove more economical to reduce the tension in the entire section to the long-span values and accept an increase in sag and corresponding reduction in span length in order to avoid dead ends.

Once all the structures have been spotted on the profile to meet the above requirements, the actual ruling span of each dead-end section should be calculated and compared to the original assumed ruling span used. If the actual ruling span of a section is more than 5% different than the assumed ruling span, that section should again be re-spotted using a new template with the revised ruling span. This process should be repeated perpetually on the line until all actual ruling spans are with 5% of the ruling span used to spot the structures.

***Computerized Line Design.***[50–52] In a line of any significant length there are a very large number of possible structure location sequences which meet the requirement for minimum electrical clearances yet also meet the maximum load limits of the chosen structure family. With considerable design experience, it is possible to select a reasonably economical structure spotting solution, but no manual structure spotting method can explore all the possibilities nor find the lowest-cost solution.

In the past 25 years, computer programs have become available to explore all possible structure spotting combinations, selecting the least cost available. In addition to exploring minimum-cost structure spotting combinations for new lines, these computer programs also allow the user to explore uprating alternatives including rerating, reconductoring, inserting structures, raising structures and attachment points, and retensioning the existing conductors. With the advent of more and more powerful personal computers and easier-to-use graphical interfaces, these programs are easily applied even to relatively small line designs. Such programs are particularly attractive when modern digital surveying methods such as LiDAR, which obtains complete three-dimensional terrain data, existing structure locations, heights, and catenaries, can be used to develop complete three-dimensional models of the transmission corridor. These models allow for differential pole and tower body/leg extensions to be selected, as well as guying requirements due to uneven terrain. These models not only enable highly accurate line designs, but can be used for vegetation analysis to find growing and falling tree infractions which can trip a line and lead to serious grid interruptions.

Digital data collection and analysis allows the line designer to explore a number of design aspects that were simply impossible to do manually. For example, Fig. 5-16 shows the result of a series of lowest-cost numerical tower spotting calculations made to explore the effects of conductor type (all-aluminum conductor, low-steel 45/7 ACSR, and high-steel 54/7 ACSR) and conductor stringing tension expressed as a percent of rated breaking strength (RBS). Each data point represents an optimized tower spotting calculation. It's interesting to note that the lowest-cost solution is the weakest conductor at a modest tension level.

In addition to the line design and optimization benefits, the computer programs also provide highly detailed and accurate plan and profile drawings and other important construction documentation such as stringing charts, offset clipping reports, and constructions staking reports. Accurate Bill of Materials can be developed virtually eliminating the errors, omissions, and conservative assumption discussed in the earlier part of this subsection that are normally associated with manual line-design methods.

## 5.1.7 Mechanical Design of Overhead Spans

***Catenary Calculations for Stranded Conductors.*** The energized conductors of transmission and distribution lines must be placed in a manner that limits the opportunity for contact by people or equipment. Overhead conductors, however, elongate with time, temperature, and tension, thereby changing their original positions after installation. Despite the effects of weather and loading on a line, the conductors must remain at safe distances from buildings, objects, and people or vehicles passing beneath the line at all times. To ensure this safety, the shape of the terrain along the right-of-way, the height and lateral position of the conductor between support

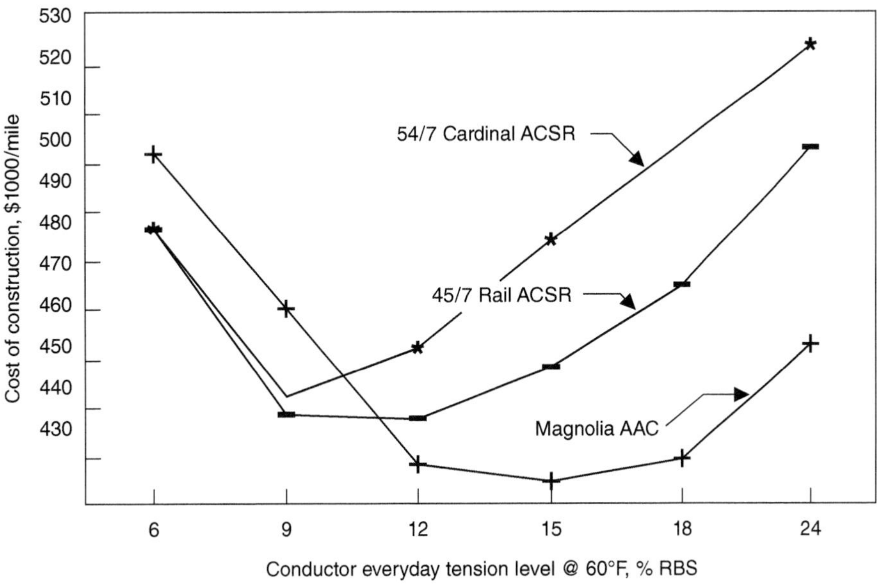


**FIGURE 5-16** Cost of construction versus conductor tensions for 1200-ft (366 m) wind span.

points, and the position of the conductor between support points under all wind, ice, and temperature conditions must be known.

Bare overhead transmission or distribution conductors are typically flexible and uniform in weight along their lengths. Because of these characteristics, they take the form of a catenary between support points. The shape of the catenary[53,54] changes with conductor temperature, ice and wind loading, and time. To ensure adequate vertical and horizontal clearance under all weather and electrical loadings, and to ensure that the breaking strength of the conductor is not exceeded, the behavior of the conductor catenary under all conditions must be incorporated into the line design. The required prediction of the future behavior of the conductor are determined through calculations commonly referred to as *sag-tension calculations,* which predict the behavior of conductors according to recommended tension limits under varying loading conditions. These tension limits specify certain percentages of the conductor's rated breaking strength that is not to be exceeded on installation or during the life of the line. These conditions, along with the elastic and permanent elongation properties of the conductor, provide the basis for determining the amount of resulting sag during installation and long-term operation of the line.

Accurately determined initial sag limits are essential in the line design process. Final sags and tensions depend on initial installed sags and tensions and on proper handling during installation. The final sag shape of conductors is used to select support point heights and span lengths so that the minimum clearances will be maintained over the life of the line. If the conductor is damaged or the initial sags are incorrect, the line clearances may be violated or the conductor may break during heavy ice or wind loadings.

***Sag and Tension in Level Spans.*** A bare stranded overhead conductor is normally held clear of objects, people, and other conductors by attachment to insulators on supporting structures at each end of the span. The elevation differences between the supporting structures affect the shape of the conductor catenary. The catenary's shape has a distinct effect on the sag and tension of the conductor, which can be determined using well-defined mathematical equations.

The shape of a catenary is a function of the conductor weight per unit length $w$, the horizontal component of tension, $H$, the span length $S$, and the sag of the conductor $D$. Conductor sag and span length are illustrated in Fig. 5-17 for a level span.

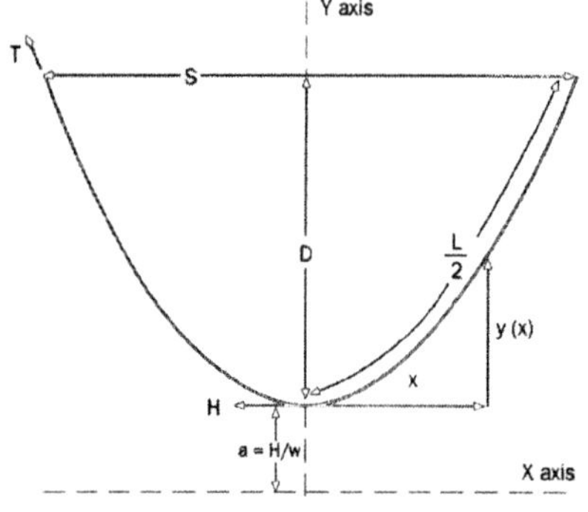


**FIGURE 5-17** The catenary curve for level spans.

The exact catenary equation uses hyperbolic functions. Relative to the low point of the catenary curve shown in Fig. 5-17, the height of the conductor $y(x)$ above this low point is given by the following equation:

$$y(x) = \frac{H}{w}\left[\cosh\left(\frac{wx}{H}\right) - 1\right] \cong \frac{wx^2}{2H} \tag{5-47}$$

Note that $x$ is positive in either direction from the low point of the catenary. The expression to the right is an approximate parabolic equation based on a MacLaurin series expansion of the hyperbolic cosine.

For a level span, the low point is in the center and the sag $D$ is found by substituting $x = S/2$ in the preceding equations. The exact catenary and approximate parabolic equations for sag become the following:

$$D = \frac{H}{w}\left[\cosh\left(\frac{wS}{2H}\right) - 1\right] \cong \frac{wS^2}{8H} \tag{5-48}$$

The ratio $H/w$ which appears in all of the preceding equations is commonly referred to as the *catenary constant*. An increase in the catenary constant causes the catenary curve to become shallower and the sag to decrease. Although it varies with conductor temperature, ice and wind loading, and time, the catenary constant typically has a value in the range of several thousand feet for most transmission-line catenaries.

The approximate, or parabolic, expression is sufficiently accurate as long as the sag is less than 5% of the span length. As an example, consider a 1000-ft (304.8-m) span of Drake ACSR conductor with a per unit weight of 1.096 lb/ft (15.99 N/m) installed at a tension of 4500 lb (20.016 kN). The catenary constant $H/w$ is 4106 ft (1251.8 m). The calculated sag is 30.48 ft (9.293 m) and 30.44 ft (9.280 m) using the hyperbolic and approximate parabolic equations, respectively. For this case where the sag-to-span ratio is 3.4%, the difference in calculated sag between the hyperbolic and parabolic equations is 0.48 in (1.3 cm).

The horizontal component of tension $H$ is equal to the conductor tension at the point in the catenary where the conductor slope is horizontal. For a level span, this is the midpoint of the span. At the ends of the level span, the conductor tension $T$ is equal to the horizontal component plus the conductor weight per unit length $w$ multiplied by the sag $D$, as shown in the following:

$$T = H + wD \tag{5-49}$$

Given the conditions in the preceding example calculation for a 1000-ft (304.8-m) level span of ACSR Drake, the tension at the attachment points $T$ exceeds the 4500-lb (20.016-N) horizontal component of tension $H$ by only 36 lb (162 N), a difference of only 0.8%.

This shows that the use of horizontal tension $H$ and parabolic equations for the catenary are adequate for typical transmission spans and sags. However, there is little reason to use either approximation in numerical methods or computer programs.

***Conductor Length.*** Application of calculus to the catenary equation allows the calculation of the conductor length $L(x)$ measured along the conductor from the low point of the catenary in either direction.

The equation for catenary length between the supports is

$$L(x) = \frac{H}{w}\sinh\left(\frac{wx}{H}\right) \cong x\left(1 + \frac{x^2w^2}{6H^2}\right) \tag{5-50}$$

For a level span, the conductor length corresponding to $x = S/2$, is half of the total conductor length $L$; thus

$$L = \left(\frac{2H}{w}\right)\sinh\left(\frac{Sw}{2H}\right) \cong S\left(1+\frac{S^2w^2}{24H^2}\right) \tag{5-51}$$

The parabolic equation for conductor length can also be expressed as a function of sag $D$ by substitution of the sag parabolic equation [Eq. (5-48)]:

$$L = S + \frac{8D^2}{3S} \tag{5-52}$$

***Sag and tension in inclined spans*** may be analyzed using essentially the same equations that were used for level spans. The catenary equation for the conductor height above the low point in the span is the same. However, the span is considered to consist of two separate sections, one to the right of the low point and the other to the left as shown in Fig. 5-18. The shape of the catenary relative to the low point is unaffected by the difference in suspension point elevation (span inclination). In each direction from the low point, the conductor elevation $y(x)$ relative to the low point is given by Eq. (5-47):

$$y(x) = \frac{H}{w}\left[\cosh\left(\frac{wx}{H}\right) - 1\right] \approx \frac{wx^2}{2H} \tag{5-53}$$

Note that $x$ is considered positive in either direction from the low point.

The horizontal distance $x_L$ from the left support point to the low point in the catenary is

$$x_L = \frac{S}{2}\left(1+\frac{h}{4D}\right) \tag{5-54}$$

The horizontal distance $x_R$ from the right support point to the low point of the catenary is

$$x_R = \frac{S}{2}\left(1-\frac{h}{4D}\right) \tag{5-55}$$

where $S$ = horizontal distance between support points
$h$ = vertical distance between support points
$S_1$ = straight-line distance between support points
$D$ = sag measured vertically from a line through the points of conductor support to a line tangent to the conductor

(as shown in Fig. 5-18). The midpoint sag $D$ is approximately equal to the sag in a horizontal span, with a length equal to the inclined span $S_1$.

Knowing the horizontal distance from the low point to the support point in each direction, we can apply the preceding equations for $y(x)$, $L$, $D$, and $T$ to each side of the inclined span. The total conductor length $L$ in the inclined span is equal to the sum of the lengths in the $x_R$ and $x_L$ subspan sections:

$$L \approx S + (x_L^3 + x_R^3)\left(\frac{w^2}{6H^2}\right) \tag{5-56}$$

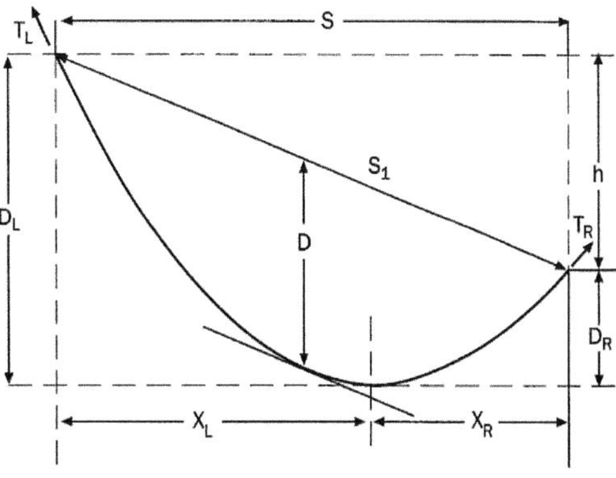


**FIGURE 5-18** Inclined catenary span.

In each subspan, the sag is relative to the corresponding support point elevation

$$D_R = \frac{wx_R^2}{2H}, \qquad D_L = \frac{wx_L^2}{2H} \tag{5-57}$$

or in terms of sag $D$ and the vertical distance between support points

$$D_R = D\left(1 - \frac{h}{4D}\right)^2, \qquad D_L = D\left(1 + \frac{h}{4D}\right)^2 \tag{5-58}$$

and the maximum tension is

$$T_R = H + wD_R, \qquad T_L = H + wD_L \tag{5-59}$$

or in terms of upper and lower support points:

$$T_u = T_1 + wh \tag{5-60}$$

where $D_R$ = sag in right subspan section
$D_L$ = sag in left subspan section
$T_R$ = tension in right subspan section
$T_L$ = tension in left subspan section
$T_u$ = tension in conductor at upper support
$T_1$ = tension in conductor at lower support

The horizontal conductor tension is equal at both supports. The vertical component of conductor tension is greater at the upper support, and the resultant tension, $T_u$, is also greater.

***Conductor and Structure Loads.*** When the conductors are exposed to ice and/or wind, effective conductor loading per unit length increases over that of the bare conductor weight per unit length. During occasions of heavy ice and/or wind load, the conductor catenary tension increases dramatically along with the loads on angle and deadend structures. Both the conductors and support structures can fail unless these conditions are considered in the design of the overhead transmission line.

*Ice Loading.* Ice loading of overhead conductors may take several physical forms (glace ice, rime ice, wet snow). The impact of lower-density ice formation is usually considered in the design of line sections at high altitudes. The formation of ice on overhead conductors has the following influence on line design:

- Ice loads determine the maximum vertical conductor loads that structures and foundations must withstand.
- In combination with simultaneous wind loads, ice loads also determine the maximum transverse loads on structures.
- In regions of heavy ice loads, the maximum sags and the permanent increase in sag with time (difference between initial and final sags) may be due to ice loadings.

Ice loads for use in designing overhead lines are normally derived on the basis of past experience, code requirements, state regulations and analysis of historical weather data. Mean recurrence intervals for extreme ice loadings are a function of local conditions along various routings. The impact of varying assumptions concerning ice loading can be investigated with design software.

*The National Electrical Safety Code (NESC)*[55] is the regulatory document that specifies the required ice and wind loading conditions for the design of overhead transmission lines and supporting structures in the United States. The NESC specifies three loading conditions denoted as (1) *district loading*, (2) *extreme wind loading*, and (3) *extreme ice with concurrent wind loading*.

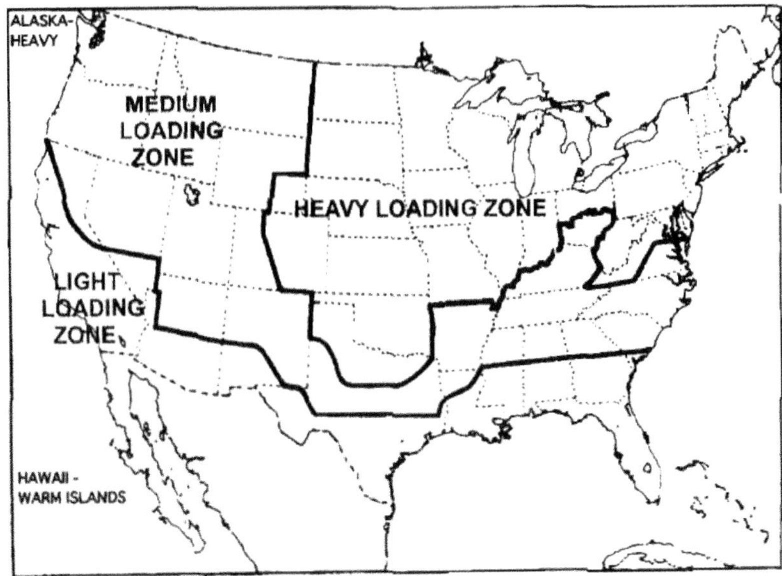


The Warm Island Loading District includes American Samoa, Guam, Hawaii, Puerto Rico, Virgin Islands, and other islands located from 0° to 25° latitude, north or south.

**FIGURE 5-19** NESC loading districts map. (*Reprinted with permission from the IEEE.*)[55]

*District Loading.* In this case, the United States is divided into four loading districts shown on map in Fig. 5-19. The specified radial ice thickness and horizontal wind pressure for each of these zones are given in Table 5-11. This table also gives the conductor temperature which is to be used along with the ice thickness and wind pressure in each of these zones. The ice is assumed to be a

**TABLE 5-11** Ice Thickness, Wind Pressures, Temperatures and Added Load Constants (from NESC)[55]

| | Loading districts (see Fig. 5-19) | | | | | | |
|---|---|---|---|---|---|---|---|
| | | | | Warm islands located at 0° to 25° latitude* | | | |
| | Heavy | Medium | Light | Altitudes sea level to 9000 ft | Altitudes above 9000 ft | Extreme wind loading | Extreme ice loading with concurrent wind |
| Radial thickness of ice (in) | 0.5 | 0.25 | 0 | 0 | 0.25 | 0 | See Fig. 5-19 |
| Horizontal wind pressure (lb/ft$^2$) | 4 | 4 | 9 | 9 | 4 | See Fig. 5-19 | See Fig. 5-19 |
| Temperature (°F) | 0 | +15 | +30 | +50 | +15 | +60 | +15 |
| Constant K to be added to the resultant load (lb/ft) | 0.3 | 0.2 | 0.05 | 0.05 | 0.2 | 0 | 0 |

*Islands located at 0° to 25° latitude include American Samoa, Guam, Hawaii, Puerto Rico and the Virgin Islands.

uniformly thick glaze ice coated around each conductor having a density of 57 lb/ft$^3$. The wind is assumed to acting in a horizontal direction perpendicular to the plane of each sagged conductor.

Therefore, the corresponding wind and ice loading acting on each conductor are calculated to be

$$\text{Conductor wind load (lb/ft): } w_w = p \times (D_c + 2t)/12 \qquad (5\text{-}61)$$

$$\text{Conductor ice load (lb/ft): } w_i = 1.244 \times t\,(D_c + t) \qquad (5\text{-}62)$$

where $p$ is the wind pressure in lb/ft$^2$, $t$ is the radial ice thickness in inches for the appropriate district loading from Table 5-11 and $D_c$ is the bare conductor diameter in inches. If the conductor is subject to both wind and ice loading, the resultant load acting on the conductor becomes

$$\text{Resultant conductor load (lb/ft): } w_R = \sqrt{w_w^2 + (w_c + w_i)^2} \qquad (5\text{-}63)$$

in which $w_c$ is the conductor weight in lb/ft.

For district loadings, the NESC also requires that an additional loading $K$ (in lb/ft) listed in Table 5-11 for each loading district be added to the resultant ice and wind loading when performing sag and tension calculations. Therefore, the total resultant conductor load $w$ for district loadings is

$$w = w_R + K \qquad (5\text{-}64)$$

The wind load on the supporting structures is calculated by the following equation:

$$\text{Structure wind load (lb): } W_s = p \times C_f \times A \qquad (5\text{-}65)$$

where $p$ is the wind pressure from Table 5-11 for the appropriate loading district, $C_f$ is the structure force coefficient from Tables 5-14 or 5-15, and $A$ is the projected area on the windward side of the structure.

*Extreme Wind Loadings.* For overhead lines and/or supporting structures that are more than 60 ft above the ground, the NESC[55] requires that they also be designed for extreme wind loadings. For this loading case, the ice thickness $t$ is zero, and the wind pressure $p$ (in lb/ft$^2$) used in Eq. (5-61) to calculate the conductor horizontal wind loading is given by the following formula:

$$p = 0.00256 \times V^2 \times k_z \times G_{RF} \times C_f \qquad (5\text{-}66)$$

in which $V$ is the basic wind speed in mph given on the maps in Fig. 5-20, $k_z$ is the velocity pressure exposure coefficient given in Table 5-12, $G_{RF}$ is the gust response factor given in Table 5-13 and $C_f$ is the force coefficients given in Tables 5-14 and 5-15. The force coefficient $C_f$ for stranded conductors can vary significantly depending on wind speed and stranding and, therefore, is usually assumed to be 1.0. The vertical conductor load is simply the weight of the conductor $w_c$ and the resultant conductor load is given by Eq. (5-64) with an ice load $w_i$ equal to zero. The $K$ loading assumed in the district loading case is *not* added to on the resultant conductor load for this loading case.

The basic wind speed map in Fig. 5-20 is taken from ASCE Standard 7-05,[56] and these wind speeds are the 50-year return period 3-s gust wind speeds for the contiguous United States measured at 10 m above ground in open terrain. This is the terrain condition that should be assumed for the design of overhead transmission lines. The NESC[55] and ASCE 7-05[56] give more detailed basic wind speed maps for Alaska, and for the hurricane zones of the Gulf of Mexico and the southeastern, mid-Atlantic and north-Atlantic U.S. coastlines.

The effective height $h$ used in Tables 5-12 and 5-13 for determining the $k_z$ and $G_{RF}$ values is the distance above ground level to the center of pressure of the conductors or structure. For the conductors, this center of pressure can be approximated as the average attachment height of the conductors to the support structure insulators minus one-third the average sag of the conductors. For support structures with total heights of 200 ft or less, the effective height $h$ can be approximated as two-thirds the average height of the support structures. For structures taller than 200 ft, the values of $k_z$ should be varied over the height of the structure to represent the increase in the wind speed with height above ground.

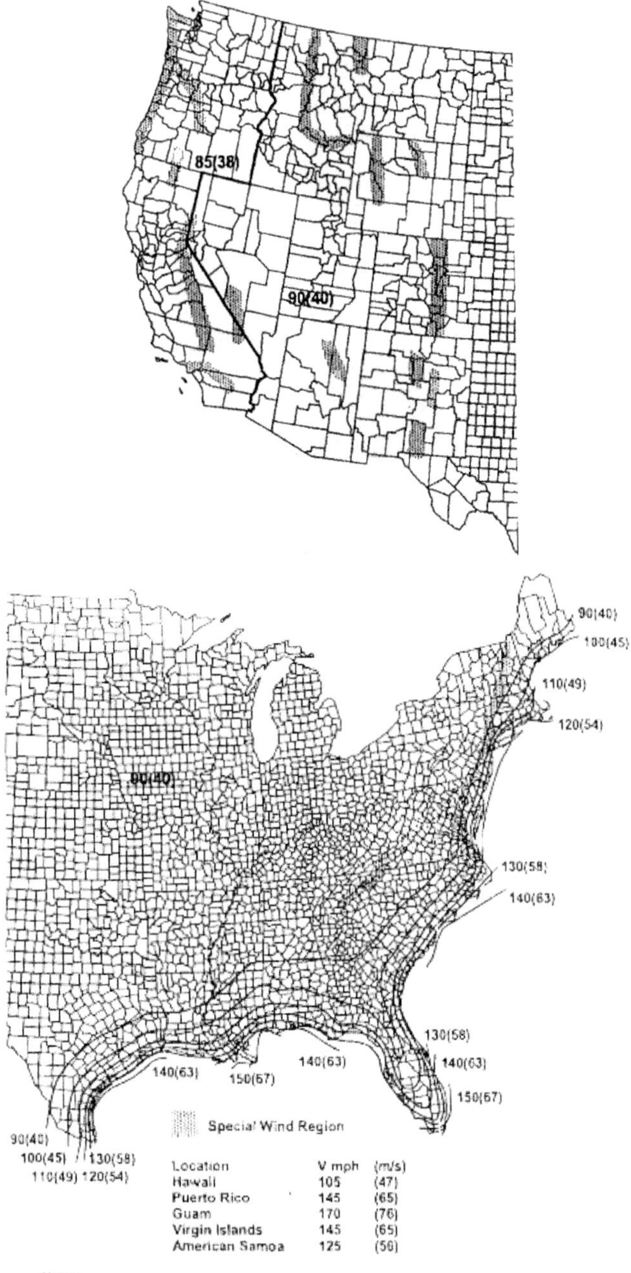


**FIGURE 5-20** Basic wind speeds for the contiguous United States. *(Reprinted with permission from ASCE 7-05.)*[56]

**TABLE 5-12** Velocity Pressure Exposure Coeffcient, $k_z$ (from NESC)[55]

| Height $h$ (ft) | $k_z$ (structure) | $k_z$ (conductors at specified height on the structure, and component) |
|---|---|---|
| 0–33 | 0.9 | 1.0 |
| >33–50 | 1.0 | 1.1 |
| >50–80 | 1.1 | 1.2 |
| >80–115 | 1.2 | 1.3 |
| >115–165 | 1.3 | 1.4 |
| >165–250 | 1.4 | 1.5 |
| >250 | Use formulas in NESC[55] | Use formulas in NESC[55] |

**TABLE 5.13** Structure and Conductor Gust Response Factors, $G_{RF}$ (from NESC)[55]

| | | Conductor $G_{RF}$ for span length, L (ft) | | | | | | |
|---|---|---|---|---|---|---|---|---|
| Height $h$ (ft) | Structure $G_{RF}$ | -250 | 250–500 | >500–750 | >750–1000 | >1000–1500 | >1500–2000 | >2000 |
| 0–33 | 1.02 | 0.93 | 0.86 | 0.79 | 0.75 | 0.73 | 0.69 | * |
| >33–50 | 0.97 | 0.88 | 0.82 | 0.76 | 0.72 | 0.70 | 0.67 | * |
| >50–80 | 0.93 | 0.86 | 0.80 | 0.75 | 0.71 | 0.69 | 0.66 | * |
| >80–115 | 0.89 | 0.83 | 0.78 | 0.73 | 0.70 | 0.68 | 0.65 | * |
| >115–165 | 0.86 | 0.82 | 0.77 | 0.72 | 0.69 | 0.67 | 0.64 | * |
| >165–250 | 0.83 | 0.80 | 0.75 | 0.71 | 0.68 | 0.66 | 0.63 | * |
| >250 | * | * | * | * | * | * | * | * |

*For heights greater than 250 ft and/or spans greater than 2000 ft, use the formulas given in the NESC[55]

**TABLE 5-14** Force Coefficients $C_f$ for Pole Structures and Members for Different Shapes (from ASCE Manual 74)[57]

| Pole or member shape | Force coefficient $C_f$ |
|---|---|
| Circular | 0.9 |
| Square or rectangle | 2.0 |
| 6-Sided polygonal | 1.4 |
| 8-Sided polygonal | 1.4 |
| 12-Sided polygonal | 1.0 |
| 16-Sided polygonal | 0.9 |

**TABLE 5-15** Force Coefficients $C_f$ for Normal Wind on Latticed Towers Having Flat-Sided Members (from ASCE Manual 74)[57]

| Solidity ratio, $\phi$ | Force coefficient $C_f^*$ | |
|---|---|---|
| | Square-section towers | Triangular-section towers |
| <0.025 | 4.0 | 3.6 |
| 0. 025–0.44 | 4.1–5.2$\phi$ | 3.7–4.5$\phi$ |
| 0.45–0.69 | 1.8 | 1.7 |
| 0.70–1.00 | 1.3 + 0.7$\phi$ | 1.0 + $\phi$ |

*$C_f$ values account for both the windward and leeward faces, including shielding of the leeward face. If the tower has round-section members, these values for $C_f$ can be modified by the correction factors in Table 5-16.

**TABLE 5-16** Correction Factors for Normal Wind on Latticed Towers Having Round-Section Members (from ASCE Manual 74)[57]

| Solidity ratio, $\phi$ | Correction factor |
|---|---|
| <0.30 | 0.67 |
| 0.30–0.79 | 0.67$\phi$ + 0.47 |
| 0.80–1.00 | 1.00 |

*Extreme Ice with Concurrent Wind Loadings.* For this loading case, the extreme ice thickness and the concurrent wind speed are given on the maps in Fig. 5-21, which are taken from ASCE Standard 7-05.[56] These maps give the 50-year return period uniform ice thicknesses due to freezing rain (glaze ice) along with the concurrent 3-s gust wind speeds for the contiguous United States. The NESC[55] and ASCE Standard 7-05[56] include detailed maps of this loading condition for Alaska, as well as for the regions denoted by the arrows on the maps in Fig. 5-21.

For this loading case, the vertical ice loading on the conductor is calculated by Eq. (5-62) and the concurrent horizontal wind loading on the conductor is calculated by Eq. (5-61) with the wind pressure $p$ determined from Eq. (5-66). The resultant conductor loading is calculated by Eq. (5-63).

*Structure Load and Strength Factors.* To enhance the safety of the support structures and components of an overhead line system, the NESC[55] specifies *load factors* (equal to or greater than 1.0) that are applied to the loads acting on the supporting structures and components; and *strength factors* (equal to or less than 1.0) that are applied to the calculated or specified strengths of the various support components. Table 5-17 gives the NESC load factors to be applied to loads acting on the supporting structures of *Grade B* overhead line construction, which is the highest grade of construction specified in the NESC and represents most high-voltage overhead lines. Where vertical loads significantly reduce the stress in a supporting structure member, a vertical load factor of 1.0 should be used for the design of such member. In all loading cases, the member shall be designed for the worst case loading.

Table 5-18 gives the NESC strength factors for structures and components in grade B line construction. The strength factors in this table apply to the design of new construction. For structures and component materials that may experience deterioration with time, consideration of the effects of this deterioration on strength shall be considered in assigning the appropriate strength factors.

The load and strength factors discussed above apply to pole structures as well as tower structures, but do not apply to insulators (see Sec. 27 of the NESC[55] for insulator loading and strength requirements).

***Conductor Tension Limits.*** The NESC recommends limits on the tension of an overhead conductor as a percentage of the conductor's rated breaking strength. The tension limits are 60%, under

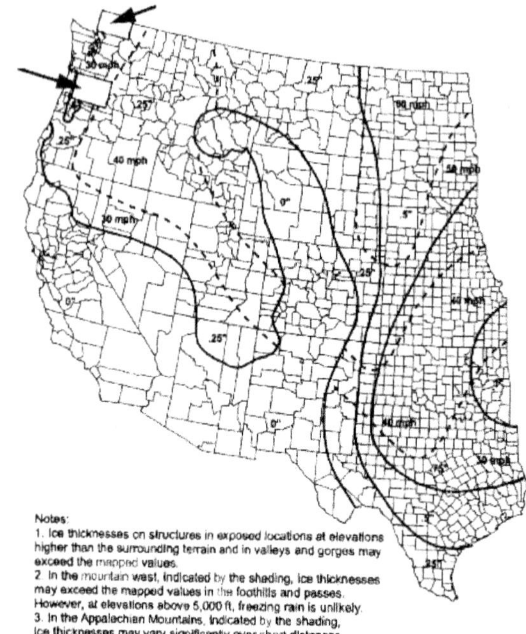


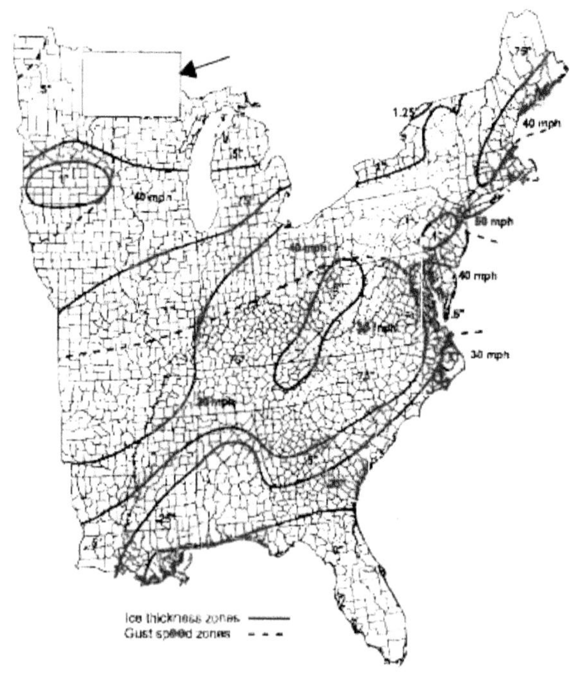


50-YEAR MEAN RECURRENCE INTERVAL UNIFORM ICE THICKNESSES DUE TO FREEZING RAIN WITH CONCURRENT 3-SECOND GUST SPEEDS: CONTIGUOUS 48 STATES.

**FIGURE 5-21** Uniform ice thickness with concurrent wind. (*Reprinted with permission from ASCE 7-05.*)[56]

**TABLE 5-17** Load Factors for Wires, Structures, Cross-arms, Support Hardware, Guys, Foundations and Anchors to be Used with the Strength Factors of Table 5.18 (From NESC)[55]

| Loading case | Load factors for grade B construction | |
|---|---|---|
| Nesc district loading | Vertical loads* | 1.50 |
| | Transverse loads | |
| | Wind load | 2.50 |
| | Wire tension | 1.65 |
| | Longitudinal loads | |
| | In general | 1.10 |
| | At deadends | 1.65 |
| Extreme wind loading | Wind load | 1.00 |
| | All other loads | 1.00 |
| Extreme ice loading with concurrent wind | Ice and wind load | 1.00 |
| | All other loads | 1.00 |

*Where vertical loads significantly reduce the stress in a structure member, a vertical load factor of 1.0 should be used for the design of such member. Such member shall be designed for the worst-case loading.

**TABLE 5-18** Strength Factors for Structures, Crossarms, Braces, Support Hardware, Guys, Foundations and Anchors to Be Used with the Load Factors of Table 5.17 (from NESC)[55]

| Loading case | Strength factors for grade B construction | |
|---|---|---|
| Strength factors for use with NESC district loading | Metal and prestressed-concrete structures, crossarms and braces | 1.00 |
| | Wood and reinforced-concrete structures, crossarms and braces | 0.65 |
| | Fiber-reinforced polymer structures, crossarms and braces | 1.00 |
| | Support hardware | 1.00 |
| | Guy wire | 0.90 |
| | Guy anchor and foundation | 1.00 |
| Strength factors for use with extreme wind loading and extreme ice with concurrent wind loading | Metal and prestressed-concrete structures, crossarms and braces | 1.00 |
| | Wood and reinforced-concrete structures, crossarms and braces | 0.75 |
| | Fiber-reinforced polymer structures, crossarms and braces | 1.00 |
| | Support hardware | 1.00 |
| | Guy wire | 0.90 |
| | Guy anchor and Foundation | 1.00 |

district ice and wind loading, of Fig. 5-19 80% under extreme wind loading of Fig. 5-20 and extreme ice and concurrent wind loading of Fig. 5-21).

The aeolian vibration limits of the NESC are 35% initial unloaded (when installed) and 25% final unloaded (after undergoing the specified loadings for an appreciable period and inelastic deformation has occurred as a result).

Traditionally, the NESC specified a temperature of 60°F for the aeolian vibration checks. The 2012 NESC requires use of the same temperatures as required for the 60% maximum tension check for the applicable loading district (0, 15, 30, or 50°F).

An exception allows use of a higher temperature (not to exceed 60°F) if vibration control measures are taken or an engineering study, manufacturer's recommendation, or experience indicates that aeolian vibration is unlikely.

It is common, however, for lower unloaded tension limits to be used. Except in areas experiencing severe ice loading, it is not unusual to find tension limits of 60% maximum, 25% unloaded initial, and 15% unloaded final. This set of specifications could easily result in an actual maximum tension on the order of only 35% to 40%, an initial tension of 20%, and a final unloaded tension level of 15%. In this case, the 15% tension limit is said to govern.

Transmission-line conductors are seldom covered with ice, and winds on the conductor are usually much lower than those used in maximum load calculations. Under such everyday conditions, tension limits are specified to limit aeolian vibration to safe levels. Even with everyday lower tension levels of 15% to 20%, it is assumed that vibration control devices will be used in those sections of the line which are subject to severe vibration. Aeolian vibration levels, and thus appropriate unloaded tension limits, vary with the type of conductor, the terrain, the span length, and the use of dampers. Special conductors such as ACSS, SDC, and VR, which exhibit high self-damping properties, may be installed to the full code limits, if desired.

***Experimental Stress-Strain Curves.*** Sag-tension calculations are normally done by computer and allow the user to enter many different loading and conductor temperature conditions. Both initial and final conditions are calculated, and multiple tension constraints can be specified. The complex stress-strain behavior of conductors can be modeled by computer algorithms, including temperature, elastic and plastic effects.

Stress-strain curves for bare overhead conductors include a minimum of an initial curve and a final curve over a range of elongations from 0% to 0.45%. For conductors consisting of two materials, an initial and final curve for each is included. Creep curves for various lengths of time are typically included as well.

Overhead conductors are not purely elastic. They stretch with tension, but when the tension is reduced to zero, they do not return to their initial length. Thus, conductors are plastic; the change in conductor length cannot be expressed with a simple linear equation. The permanent length increase that occurs in overhead conductors yields the difference in initial and final sag-tension data found in most computer programs.

Figure 5-22 shows a typical stress-strain curve[58] for a 26/7 ACSR conductor; the curve is valid for conductor sizes ranging from 266.8 to 795 kcmil. A 795-kcmil 26/7 ACSR Drake conductor has a breaking strength of 31,500 lb (14,000 kg) and an area of 0.7264 in$^2$ (46.9 mm$^2$) so that it fails at an average stress of 43,000 lb/in$^2$ (30 kg/mm$^2$). The stress-strain curve illustrates that at a stress equal to 50% of the conductor's breaking strength (21,500 lb/in$^2$), the elongation is less than 0.3%. This translates to an elongation of 1.8 ft (0.55 m) in a 600-ft (180-m) span.

Note that the component curves for the steel core and the aluminum-stranded outer layers are separated. This separation allows for changes in the relative curve locations as the temperature of the conductor changes.

Figure 5-23 is a stress-strain curve[58] for an all-aluminum 37-strand conductor ranging in size from 250 to 1033.5 kcmil. Because the conductor is made entirely of aluminum, there is only one initial curve and one final curve.

***Permanent Conductor Elongation due to High Tensions.*** Once a conductor has been installed to an initial tension, it can elongate further. Such elongation results from two phenomena: permanent elongation due to high-tension levels resulting from ice and wind loads and creep elongation under everyday tension levels. These types of conductor elongation are discussed in the following subsections. Both Figs. 5-22 and 5-23 indicate that when the conductor is initially installed, it elongates nonlinearly. If the conductor tension increases to a relatively high level under ice and wind loading, the conductor will elongate. When the wind and ice loads abate, the conductor elongation will reduce along a curve parallel to the final curve, but will never return to its original length.

For example, refer to Fig. 5-23 and assume that a newly strung 795-kcmil, 37-strand Arbutus AAC has an everyday tension of 2780 lb. The conductor area is 0.6245 in$^2$, so the everyday stress is 4450 lb/in$^2$ and the elongation is 0.062%. Following an extremely heavy ice and wind load event, assume that the conductor stress reaches 18,000 lb/in$^2$. When the conductor tension decreases back to everyday levels, the conductor elongation will be permanently increased by more than 0.2%. Also the sag under everyday conditions will be correspondingly greater, and the tension will be less. In most computer sag-tension methods, final sag-tension values are calculated for such permanent elongation due to heavy loading conditions.

***Permanent Elongation at Everyday Tensions (Creep).*** Conductors permanently elongate under tension even if the tension level never exceeds everyday levels. This permanent elongation caused by everyday tension levels is called *creep*.[58] Creep can be determined by long-term laboratory creep

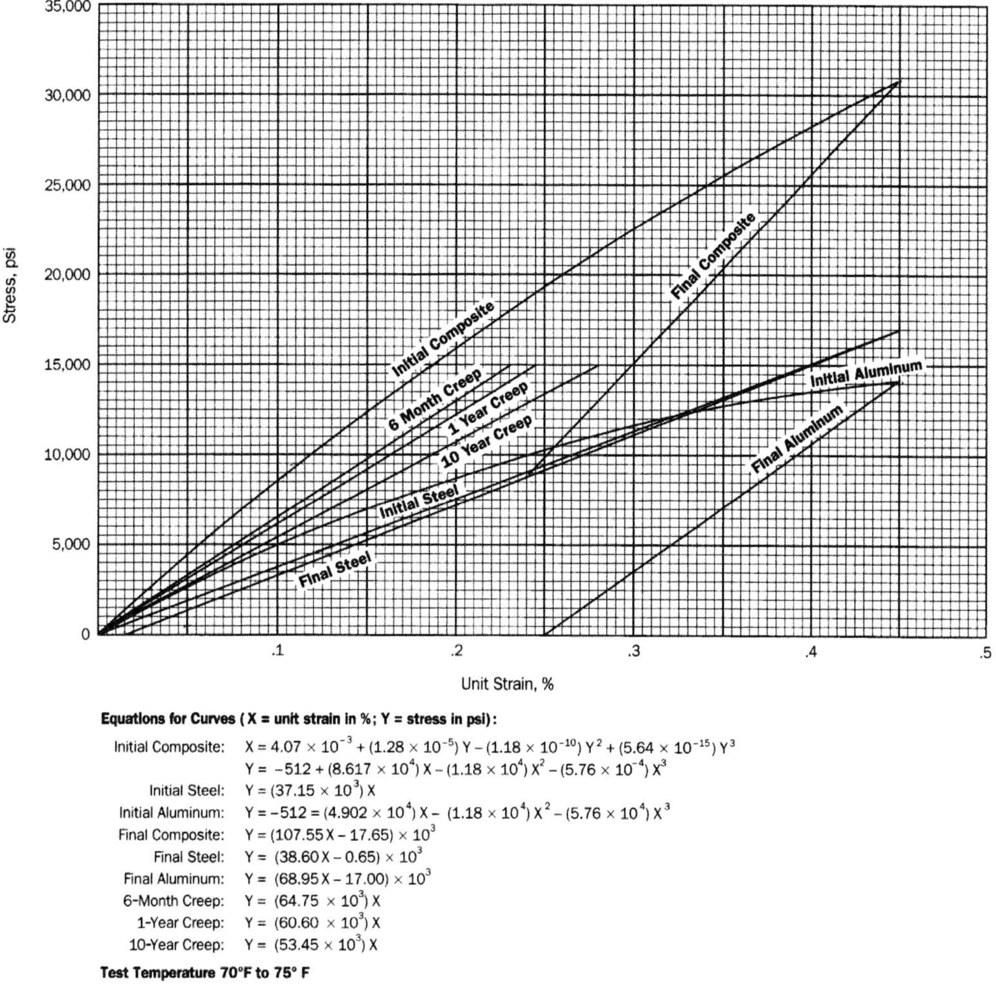


**FIGURE 5-22** Stress-strain curves for 26/7-strand ACSR.

tests. The results of the tests are used to generate creep-versus-time curves. On the stress-strain graphs, creep curves are often shown for 6-month, 1-year, and 10-year periods. Figure 5-23 shows these typical creep curves for a 37-strand 250- to 1033.5-kcmil AAC. In Fig. 5-23, assume that the conductor tension remains constant at the initial stress of 4450 lb/in$^2$. At the intersection of this stress level and the initial elongation curve, 6-month, 1-year, and 10-year creep curves, the conductor elongation from the initial elongation of 0.062% increases to 0.11%, 0.12%, and 0.15%, respectively. Because of creep elongation, the resulting final sags are greater and the conductor tension is less than the initial values.

Creep elongation in aluminum conductors is quite predictable as a function of time and obeys a simple exponential relationship. Thus, the permanent elongation due to creep at everyday tension can be found for any period of time after initial installation. Creep elongation of copper and steel strands is much less and is normally ignored. Permanent increase in conductor length due to heavy load occurrences cannot be predicted at the time a line is built. The reason for this unpredictability is

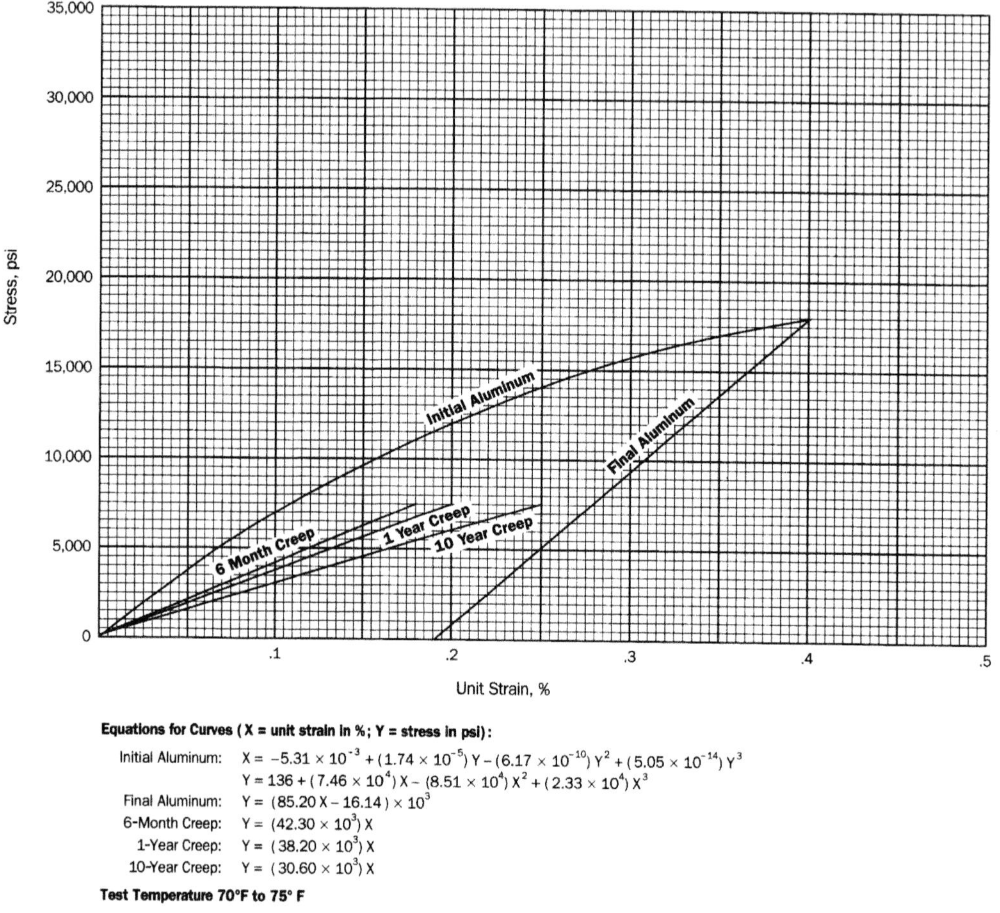


**FIGURE 5-23** Stress-strain curves for 37-strand AAC.

that the occurrence of heavy ice and wind loads is random. A heavy ice storm may occur the day after the line is built or may never occur over the life of the line.

***Sag-Tension Tables.*** To illustrate the result of typical sag-tension calculations, Table 5-19 present initial and final sag-tension data for a 795-kcmil 26/7 ACSR Drake AAC conductor in NESC light and heavy loading areas for a span of 600 ft. Typical tension constraints of 15% final unloaded at 60°F, 25% initial unloaded at 60°F, and 60% initial at maximum loading are used. The calculations for this table were performed with the Alcoa SAG10™ program, version 1.1.

***Initial versus Final Sags and Tensions.*** Rather than calculated as a function of time, most sag-tension calculations are based on initial and final loading conditions. Initial sags and tensions are simply the sags and tensions at the time the line is built. Final sags and tensions are calculated assuming that (1) the specified ice and wind loading has occurred and (2) the conductor has experienced 10 years of creep elongation at a conductor temperature of 60°F at the user-specified initial tension.

With most sag-tension calculation methods, final sags are calculated for both heavy ice/wind loads and for creep elongation. The final sag-tension values reported to the user are those with the greatest increase in sag.

**TABLE 5-19** Sag and Tension Data for 795 kcmil 26/7 ACSR Drake Conductor
Span = 600 ft; NESC heavy loading district; creep is not a factor.

| | | | | | Final | | Initial | |
|---|---|---|---|---|---|---|---|---|
| Temperature, °F | Ice, in | Wind, lb/ft² | k, lb/ft | Resultant weight, lb/ft | Sag, ft | Tension, lb | Sag, ft | Tension, lb |
| 0 | 0.50 | 4.00 | 0.30 | 2.509 | 11.14 | 10153 | 11.14 | 10153 |
| 32 | 0.50 | 0.00 | 0.00 | 2.094 | 44.54 | 8185 | 11.09 | 8512 |
| −20 | 0.00 | 0.00 | 0.00 | 1.094 | 6.68 | 7372 | 6.27 | 7855 |
| 0 | 0.00 | 0.00 | 0.00 | 1.094 | 7.56 | 6517 | 6.89 | 7147 |
| 30 | 0.00 | 0.00 | 0.00 | 1.094 | 8.98 | 5490 | 7.95 | 6197 |
| 60 | 0.00 | 0.00 | 0.00 | 1.094 | 10.44 | 4725 | 9.12 | 5402 |
| 90 | 0.00 | 0.00 | 0.00 | 1.094 | 11.87 | 4157 | 10.36 | 4759 |
| 120 | 0.00 | 0.00 | 0.00 | 1.094 | 13.24 | 3727 | 11.61 | 4248 |
| 167 | 0.00 | 0.00 | 0.00 | 1.094 | 14.29 | 3456 | 13.53 | 3649 |
| 212 | 0.00 | 0.00 | 0.00 | 1.094 | 15.24 | 3241 | 15.24 | 3241 |

### 5.1.8 Mechanical Interaction of Suspension Spans

Transmission lines are normally designed in line sections with each end of the section terminated by a strain structure that allows no longitudinal movement of the conductor.[54] Structures within each line section are typically suspension structures that support the conductor vertically but allow free movement of the conductor attachment point either longitudinally or transversely.

***Tension Differences for Adjacent Dead-End Spans.*** Table 5-20 contains initial and final sag and tension data for a 700-ft (213-m) and a 1000-ft (305-m) dead-end span with an ACSR Drake conductor that was initially installed with the same 6300-lb tension limits at 60°F. Note that the differences between final tensions at 60°F is 260 lb, which is due entirely to the difference in span length. Even the initial tension (equal at 60°F) differs by approximately 880 lb at −20°F and 610 lb at 167°F.

***Tension Equalization by Suspension Insulators.*** At a typical suspension structure, the conductor is supported vertically by a suspension insulator assembly, but allowed to move freely in the direction of the conductor axis. This conductor movement is possible because of insulator swing along the conductor axis. Changes in conductor tension between spans, caused by changes in temperature, load, and time are normally equalized by insulator swing, eliminating horizontal tension differences across suspension structures.

***Ruling-Span Approximation.*** The sag and tension for a series of suspension spans in a line section can be found using the ruling-span concept.[53,54] The ruling-span (RS) for the line section is defined by the following equation:

$$\mathrm{RS} = \sqrt{\frac{S_1^3 + S_2^3 + \cdots + S_n^3}{S_1 + S_2 + \cdots + S_n}}$$

where RS = ruling span for the line section containing $n$ suspension spans
$S_1$ = span length of first suspension span
$S_2$ = span length of second suspension span
$S_n$ = span length of $n$th suspension span

Alternatively, a generally satisfactory method for estimating the ruling span is to take the sum of the average suspension span length plus two-thirds of the difference between the maximum span

**TABLE 5-20** Tension Differences in Adjacent Dead-End Spans for 795-kcmil 26/7 ACSR Drake Conductor

| | | | | | Final | | Initial | |
|---|---|---|---|---|---|---|---|---|
| Temperature, °F | Ice, in | Wind, lb/ft² | *k*, lb/ft | Resultant weight, lb/ft | Sag, ft | Tension, lb | Sag, ft | Tension, lb |
| Span = 700 ft; NESC heavy loading district; area = 0.7264 in²; creep *is* a factor. | | | | | | | | |
| 0 | 0.50 | 4.00 | 0.30 | 2.509 | 13.61 | 11318 | 13.55 | 11361 |
| 32 | 0.50 | 0.00 | 0.00 | 2.094 | 13.93 | 9224 | 13.33 | 9643 |
| −20 | 0.00 | 0.00 | 0.00 | 1.094 | 8.22 | 8161 | 7.60 | 8824 |
| 0 | 0.00 | 0.00 | 0.00 | 1.094 | 9.19 | 7301 | 8.26 | 8115 |
| 30 | 0.00 | 0.00 | 0.00 | 1.094 | 10.75 | 6242 | 9.39 | 7142 |
| 60 | 0.00 | 0.00 | 0.00 | 1.094 | 12.36 | 5429 | 10.65 | 6300 |
| 90 | 0.00 | 0.00 | 0.00 | 1.094 | 13.96 | 4809 | 11.99 | 5596 |
| 120 | 0.00 | 0.00 | 0.00 | 1.094 | 15.52 | 4330 | 13.37 | 5020 |
| 167 | 0.00 | 0.00 | 0.00 | 1.094 | 16.97 | 3960 | 15.53 | 4326 |
| 212 | 0.00 | 0.00 | 0.00 | 1.094 | 18.04 | 3728 | 17.52 | 3837 |
| Span = 1000 ft; area = 0.7264 in²; NESC heavy loading district; creep is *not* a factor. | | | | | | | | |
| 0 | 0.50 | 4.00 | 0.30 | 2.509 | 25.98 | 12116 | 25.98 | 12116 |
| 32 | 0.50 | 0.00 | 0.00 | 2.094 | 26.30 | 9990 | 25.53 | 10290 |
| −20 | 0.00 | 0.00 | 0.00 | 1.094 | 18.72 | 7318 | 17.25 | 7940 |
| 0 | 0.00 | 0.00 | 0.00 | 1.094 | 20.09 | 6821 | 18.34 | 7469 |
| 30 | 0.00 | 0.00 | 0.00 | 1.094 | 22.13 | 6197 | 20.04 | 6840 |
| 60 | 0.00 | 0.00 | 0.00 | 1.094 | 24.11 | 5689 | 21.76 | 6300* |
| 90 | 0.00 | 0.00 | 0.00 | 1.094 | 26.04 | 5271 | 23.49 | 5839 |
| 120 | 0.00 | 0.00 | 0.00 | 1.094 | 30.14 | 4923 | 27.82 | 5444 |
| 167 | 0.00 | 0.00 | 0.00 | 1.094 | 30.14 | 4559 | 27.82 | 4935 |
| 212 | 0.00 | 0.00 | 0.00 | 1.094 | 31.47 | 4369 | 30.24 | 4544 |

*Design condition.

and the average span. However, some judgment must be exercised in using this method because a large difference between the average and maximum span may cause a substantial error in the ruling span value.

As discussed earlier, suspension spans are supported by suspension insulators that are free to move in the direction of the conductor axis. This freedom of movement allows the tension in each suspension span to be equal to that calculated for the ruling span. This assumption is valid for the suspension spans and ruling span under the same conditions of temperature and load, for both initial and final sags. For level spans, sag in each suspension span is given by the parabolic sag equation

$$D_i = \frac{w(S_i^2)}{8H_{RS}} \tag{5-67}$$

where $D_i$ = sag in the $i$th span
$S_i$ = span length of the $i$th span
$H_{RS}$ = horizontal tension from ruling span sag-tension calculations

Suspension spans vary in length, although typically not over a large range. Conductor temperature during sagging varies over a range considerably smaller than that used for line design purposes.

If the sag in any suspension span exceeds approximately 5% of the span length, a correction factor should be added to the sag obtained from Eq. (5-67), or the sag should be calculated using the catenary method presented in Eq. (5-69). This correction factor may be calculated as follows:

$$\text{Correction} = D^2\left(\frac{w}{8H}\right) \tag{5-68}$$

where $D$ = sag obtained from parabolic equation
$w$ = weight of conductor, lb/ft
$H$ = horizontal tension, lb

The catenary equation for calculating the sag in a suspension or stringing span is

$$\text{Sag} = \frac{H}{w}\left[\cosh\frac{Sw}{2H} - 1\right] \tag{5-69}$$

where $S$ = span length, ft
$H$ = horizontal tension, lb
$w$ = resultant weight, lb/ft

***Stringing Sag Tables.*** Conductors are typically installed in line section lengths consisting of multiple spans. The conductor is pulled from the conductor reel at a point near one strain structure, progressing through travelers attached to each suspension structure to a point near the next strain structure. After stringing, the conductor tension is increased until the sag in one or more suspension spans reaches the appropriate stringing sags according to the ruling span for the line section. The calculation of stringing sags is based on the preceding sag equation.

Table 5-22 shows a typical stringing sag table for a 600-ft ruling span of ACSR Drake with suspension spans ranging from 400 to 700 ft and conductor temperatures of 20 to 100°F. All the values in this stringing table have been calculated using the parabolic sag equation with ruling-span initial tensions shown in Table 5-21.

***Line Design Sag-Tension Parameters.*** In laying out a transmission line, the first step is to survey the route and create a plan-profile of the selected right-of-way. The plan-profile drawings serve an

**TABLE 5-21** Sag and Tension Data for 795-kcmil 26/7 ACSR Drake 600-ft Ruling Span
NESC heavy loading district; area = 0.7264 in$^2$; creep is *not* a factor.

| | | | | | Final | | Initial | |
|---|---|---|---|---|---|---|---|---|
| Temperature, °F | Ice, in | Wind, lb/ft$^2$ | $k$, lb/ft | Resultant weight, lb/ft | Sag, ft | Tension, lb | Sag, ft | Tension, lb |
| 0 | 0.50 | 4.00 | 0.30 | 2.509 | 11.14 | 10153 | 11.14 | 10153 |
| 32 | 0.50 | 0.00 | 0.00 | 2.094 | 44.54 | 8185 | 11.09 | 8512 |
| −20 | 0.00 | 0.00 | 0.00 | 1.094 | 6.68 | 7372 | 6.27 | 7855 |
| 0 | 0.00 | 0.00 | 0.00 | 1.094 | 7.56 | 6517 | 6.89 | 7147 |
| 30 | 0.00 | 0.00 | 0.00 | 1.094 | 8.98 | 5490 | 7.95 | 6197 |
| 60 | 0.00 | 0.00 | 0.00 | 1.094 | 10.44 | 4725 | 9.12 | 5402 |
| 90 | 0.00 | 0.00 | 0.00 | 1.094 | 11.87 | 4157 | 10.36 | 4759 |
| 120 | 0.00 | 0.00 | 0.00 | 1.094 | 13.24 | 3727 | 11.61 | 4248 |
| 167 | 0.00 | 0.00 | 0.00 | 1.094 | 14.29 | 3456 | 13.53 | 3649 |
| 212 | 0.00 | 0.00 | 0.00 | 1.094 | 15.24 | 3241 | 15.24 | 3241 |

**TABLE 5-22** Stringing Sag Table for 795 kcmil-26/7 ACSR Drake 600-ft Ruling Span
Controlling design condition; 15% RBS at 60°F; no ice or wind, final; NESC heavy loading district.

| Horizontal tension, lb | 6493 | 6193 | 5910 | 5645 | 5397 | 5166 | 4952 | 4753 | 4569 |
|---|---|---|---|---|---|---|---|---|---|
| Temperature, °F | 20 | 30 | 40 | 50 | 60 | 70 | 80 | 90 | 100 |
| Spans | | | | | Sag, ft-in | | | | |
| 400 | 3-4 | 3-6 | 3-8 | 3-11 | 4-1 | 4-3 | 4-5 | 4-7 | 4-9 |
| 410 | 3-6 | 3-9 | 3-11 | 4-1 | 4-3 | 4-5 | 4-8 | 4-10 | 5-0 |
| 420 | 3-9 | 3-11 | 4-1 | 4-3 | 4-6 | 4-8 | 4-10 | 5-1 | 5-3 |
| 430 | 3-11 | 4-1 | 4-3 | 4-6 | 4-8 | 4-11 | 5-1 | 5-4 | 5-6 |
| 440 | 4-1 | 4-3 | 4-6 | 4-8 | 4-11 | 5-2 | 5-4 | 5-7 | 5-10 |
| 450 | 4-3 | 4-6 | 4-8 | 4-11 | 5-2 | 5-4 | 5-7 | 5-10 | 6-1 |
| 460 | 4-5 | 4-8 | 4-11 | 5-2 | 5-4 | 5-7 | 5-10 | 6-1 | 6-4 |
| 470 | 4-8 | 4-11 | 5-1 | 5-4 | 5-7 | 5-10 | 6-1 | 6-4 | 6-7 |
| 480 | 4-10 | 5-1 | 5-4 | 5-7 | 5-10 | 6-1 | 6-4 | 6-8 | 6-11 |
| 490 | 5-1 | 5-4 | 5-7 | 5-10 | 6-1 | 6-4 | 6-8 | 6-11 | 7-2 |
| 500 | 5-3 | 5-6 | 5-9 | 6-1 | 6-4 | 6-7 | 6-11 | 7-2 | 7-6 |
| 510 | 5-6 | 5-9 | 6-0 | 6-4 | 6-7 | 6-11 | 7-2 | 7-6 | 7-9 |
| 520 | 5-8 | 6-0 | 6-3 | 6-7 | 6-10 | 7-2 | 7-6 | 7-9 | 8-1 |
| 530 | 5-11 | 6-2 | 6-6 | 6-10 | 7-1 | 7-5 | 7-9 | 8-1 | 8-5 |
| 540 | 6-2 | 6-5 | 6-9 | 7-1 | 7-5 | 7-9 | 8-1 | 8-5 | 8-9 |
| 550 | 6-4 | 6-8 | 7-0 | 7-4 | 7-8 | 8-0 | 8-4 | 8-8 | 9-1 |
| 560 | 6-7 | 6-11 | 7-3 | 7-7 | 7-11 | 8-4 | 8-0 | 9-0 | 9-5 |
| 570 | 6-10 | 7-2 | 7-6 | 7-10 | 8-3 | 8-7 | 9-0 | 9-4 | 9-9 |
| 580 | 7-1 | 7-5 | 7-9 | 8-2 | 8-6 | 8-11 | 9-4 | 9-8 | 10-1 |
| 590 | 7-4 | 7-8 | 8-1 | 8-5 | 8-10 | 9-3 | 9-7 | 10-0 | 10-5 |
| 600 | 7-7 | 7-11 | 8-4 | 8-9 | 9-1 | 9-6 | 9-11 | 10-4 | 10-9 |
| 610 | 7-10 | 8-3 | 8-7 | 9-0 | 9-5 | 9-10 | 10-3 | 10-9 | 11-2 |
| 620 | 8-1 | 8-6 | 8-11 | 9-4 | 9-9 | 10-2 | 10-7 | 11-1 | 11-6 |
| 630 | 8-4 | 8-9 | 9-2 | 9-7 | 10-1 | 10-6 | 11-0 | 11-5 | 11-11 |
| 640 | 8-8 | 9-1 | 9-6 | 9-11 | 10-5 | 10-10 | 11-4 | 11-9 | 12-3 |
| 650 | 8-11 | 9-4 | 9-9 | 10-3 | 10-9 | 11-2 | 11-8 | 12-2 | 12-8 |
| 660 | 9-2 | 9-7 | 10-1 | 10-7 | 11-1 | 11-6 | 12-0 | 12-6 | 13-1 |
| 670 | 9-5 | 9-11 | 10-5 | 10-11 | 11-5 | 11-11 | 12-5 | 12-11 | 13-5 |
| 680 | 9-9 | 10-3 | 10-8 | 11-2 | 11-9 | 12-3 | 12-9 | 13-4 | 13-10 |
| 690 | 10-0 | 10-6 | 11-0 | 11-6 | 12-1 | 12-7 | 13-2 | 13-8 | 14-3 |
| 700 | 10-4 | 10-10 | 11-4 | 11-11 | 12-5 | 13-0 | 13-6 | 14-1 | 14-8 |

important function in linking together the various stages involved in the design and construction of the line. These drawings, based on the route survey, show the location and elevation of all natural and artificial obstacles to be traversed by, or adjacent to, the proposed line. These plan-profiles are drawn to scale and provide the basis for tower spotting and line design work.

Once the plan-profile is completed, one or more estimated ruling spans for the line may be selected. On the basis of these estimated ruling spans and the maximum design tensions, sag-tension data may be calculated providing initial and final sag values. From these data, sag templates may be constructed to the same scale as the plan-profile for each ruling span, and used to graphically spot structures.

***Catenary Constants.*** The sag in a ruling span is equal to the weight per unit length $w$ times the span length $S$ squared, divided by 8 times the horizontal component of the conductor tension $H$. The ratio of conductor horizontal tension $H$ to weight per unit length $w$ is the catenary constant $H/w$. For a ruling-span sag-tension calculation using eight loading conditions, a total of 16 catenary values could be defined, one for initial and one for final tensions under each loading condition.

Catenary constants can be defined for each loading condition of interest and are used in locating structures. Some typical uses of catenary constants are to avoid overloading, ensure that ground clearance is sufficient at all points along the right-of-way, and minimize blowout or uplift under cold-weather conditions. To do this, the following catenary constants are typically found: (1) the maximum line temperature; (2) heavy ice and wind loading; (3) wind blowout; and (4) minimum conductor temperature. Under any of these loading conditions, the catenary constant allows sag calculation at any point within the span.

***Wind and Weight Span.*** The maximum wind span of any structure is equal to the distance measured from center to center of the two adjacent spans supported by the structure. The wind span is used to determine the maximum horizontal force a structure must withstand under high-wind conditions. The wind span is not dependent on conductor sag or tension, only on the horizontal span length.

The weight span of a structure is a measure of the maximum vertical force a structure must withstand. The weight span is equal to conductor weight per unit length times the horizontal distance between the low points of sag of the two adjacent spans. The maximum weight span for a structure is dependent on the design ice and wind loading condition. When the elevations of adjacent structures are the same, the wind and weight spans are equal.

***Uplift at Suspension Structures.*** Conductor uplift, shown in Fig. 5-24, occurs when the weight span of a structure is negative. On steeply inclined spans, the low point of sag may fall beyond the lower support. This indicates that the conductor in the uphill span is exerting a negative, or upward, force on the lower tower. The amount of this upward force is equal to the weight of the conductor from the lower tower to the low point in the sag. If the upward pull of the uphill span is greater than the downward load of the next adjacent span, actual uplift will occur and the conductor will swing free of the tower. This usually occurs under minimum temperature conditions and must be dealt with by adding weights to the insulator suspension string or using a strain structure.

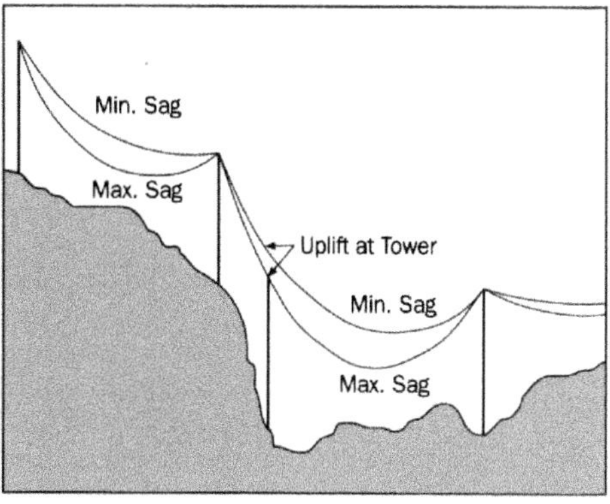


**FIGURE 5-24** Illustration of conductor uplift.

***Tower Spotting.*** Given sufficiently detailed plan-profile drawings, structure heights, wind/weight spans, catenary constants, and minimum ground clearances, structure locations can be chosen such that ground clearance is maintained and structure loads are acceptable. Tower spotting can be performed using a sag template, plan-profile drawing, and structure heights, or numerically, by using one of several commercial computer programs. Tower spotting and optimization of line design parameters are discussed in Sec. 5.1.6.

***Unbalanced Ice Loads.***[59,60] The jump of the conductor resulting from ice dropping off one span of an ice-covered line has been the cause of many serious outages on long-span lines where conductors are arranged in the same vertical plane. The vertical spacing required to prevent "ice-jump" trouble may be estimated by static calculations of the differential sag of two vertically adjacent conductors assuming one conductor has ice and the other has no ice. Normally, sufficient clearance is provided to accommodate this sag difference, including a factor for sag error with a margin for switching surge withstand.

Utility practice, based on historically satisfactory field performance, is typically based on the following criteria for vertical phase-to-phase clearance due to ice:

A maximum sag error of 6 in is assumed.

The upper conductor is assumed to be subject to maximum ice load, typically 1 in for short ruling spans, 0.5 in for unusually long spans.

The lower conductor is assumed to be completely free of ice.

Clearance sufficient to withstand the maximum switching surge is to be provided.

The calculation is performed using normal sag-tension procedures. For certain line designs (e.g., compact lines with post insulators), dynamic techniques are possible.[41,47] However, the trouble has been practically cleared up by horizontally offsetting the conductors from 18 in to 3 ft on medium-voltage lines. The conductor jumps in practically a vertical plane, and this should be true if no wind is blowing, since then all forces and reactions are in a vertical direction.

***Wind-Induced Motion of Overhead Conductors.*** In addition to ordinary "blowout" of overhead conductors (i.e., the swinging motion of the conductor due to the normal component of wind), there are three types of cyclic wind-induced motion that can be a source of damage to structures or conductors or that can result in sufficient reduction in electrical clearance to cause flashover. The categories of wind-induced cyclic motion are eolian vibration, galloping, and wake-induced oscillation.

Eolian vibration can occur when conductors are exposed to a steady low-velocity wind. If the amplitude of such vibration is sufficient, it can result in strand fatigue and/or fatigue of conductor accessories. The amplitude of vibration can be reduced by reducing the conductor tension, adding damping by using dampers (or clamps with damping characteristics), or by the use of special conductors which either provide more damping than standard conductors or are shaped so as to prevent resonance between the tensioned conductor span and the wind-induced vibration force.

Galloping is normally confined to conductors with a coating of glaze ice over at least part of their circumference and thus is not a problem in those areas where ice storms do not occur. It may be controlled by the use of various accessories attached to the conductor in the span to change mechanical and/or damping characteristics. Specially shaped conductors or conductor accessories which alter the iced conductor's aerodynamic characteristics, particularly those that increase aerodynamic damping, are also effective. The amplitude of galloping motions can be reduced by the use of higher conductor tensions and evidence suggests higher tensions can also reduce the possibility of occurrence. Galloping and eolian vibration occur in both single and bundled conductors.

Wake-induced oscillation is limited to lines having bundled conductors and results from aerodynamic forces on the downstream conductor of the bundle as it moves in and out of the wake of the upstream conductor. Wake-induced oscillation is controlled by maintaining sufficiently large conductor spacing in the bundle, unequal subspan lengths, and tilting the bundles.

*Eolian Vibration.* As wind blows across a conductor, vortices are shed from the top and bottom of the conductor. The vortex shedding is accompanied by a varying pressure on the top and bottom of the conductor that encourages cyclic vibration of the conductor perpendicular to the direction of wind flow. The frequency at which this alternating pressure occurs is given by the expression

$$f = 3.26 \times \frac{U}{d} \tag{5-70}$$

where $U$ = wind speed, mi/h
$d$ = conductor diameter, in
$f$ = frequency, Hz

For a 1.0-in-diameter conductor exposed to a 10-mi/h wind, the vortex shedding force oscillates at 32.6 Hz. To develop significant amplitudes, there must be a resonance between this oscillating wind force and the vibrating catenary (conductor). The fundamental frequency of vibration of the suspended conductor is in the range of 0.1 to 1.0 Hz. Therefore, the eolian vibration force will be unlikely to excite a fundamental span mode. This is verified by actual conductor performance where significant amplitudes are usually observed for frequencies in the range of 10 to 100 Hz. Practical wind speeds cause vortex shedding forces of greater than 10 Hz, eliminating frequencies below this level, and frequencies above 100 Hz are not present because of the rapid increase in conductor self-damping for these higher frequencies.

The maximum alternating stress resulting in strand fatigue normally occurs at the conductor clamp. The stress is related to the amplitude of conductor vibration and is the amplitude normally measured by field recording devices. Stress and amplitude of vibration can be related by analytical means such as the Poffenberger-Swart formula.[61] The amplitude of eolian vibration is fixed by the balance of energy input from the wind-induced vortex shedding forces and the energy loss due to conductor, accessory, and structure damping. The addition of dampers to the conductor has been established as an effective means of control.[62] Special conductors such as SDC and SSAC[63] have also been shown effective in reducing the strand stress levels.

Another effective means of limiting vibration fatigue problems is to increase the self-damping of standard conductors by reducing tension. As a practical approximation, stringing conductors to a final unloaded tension of 15% or less at the minimum seasonal average temperature (usually 0 to 30°F) will prevent vibration fatigue problems. Higher tensions are routinely used in areas where the line is parallel to existing lines and the higher tension on the existing line has not resulted in problems. The use of vibration dampers or special antivibration conductors can also allow the use of higher tension levels.

As with single conductors, bundled conductors are subjected to eolian vibration. However, the interaction of conductors in the bundle due to slightly different tensions and increased damping from spacers results in lower vibration levels for bundles than for single conductors in the same wind exposure.

*Galloping.* Both bundled and single conductors are subject to galloping during or after glaze ice storms. Galloping oscillations occur at frequencies near the fundamental span mode or its second or third harmonic (0.1 to 1.0 Hz) and exhibit maximum amplitudes as large as the conductor sag. While there has been extensive debate concerning the galloping mechanism, and considerable experimental and analytical study, it appears that there presently exist a number of control methods that are effective in reducing the amplitude and incidence of galloping motion. In-span hardware, such as the "detuning pendulum" developed by EPRI,[64] and the "wind damper" developed by Richardson,[65] are effective for existing spans where galloping occurs. The T2 conductor,[66] developed by Kaiser, and several other hardware devices are available to control galloping in new lines.

In contrast, control methods such as sleet melting by use of high current levels appear to be almost totally ineffective in stopping galloping and can result in annealing damage to the conductor.

*Wake-Induced Oscillation.* Bundled conductors are subject to wake-induced oscillations with amplitudes and frequencies typically between that of eolian vibration and galloping. The frequencies of oscillation are normally in the range of 1 to 10 Hz, and the amplitudes are in the range of 10 conductor diameters. The modes in which such vibration occurs are considerably more complex than the modes exhibited during either galloping or the almost invisible eolian vibrations. The source of wind energy for wake-induced oscillation is, as the name suggests, the wake from the windward conductor of the bundle which causes the motion of the downwind conductor.

There are three basic approaches to the control of wake-induced oscillation.[67] Two involve reducing the input of wind energy, and the third involves detuning the mechanical bundle system to prevent resonance. The methods based on reducing wind energy input to the bundle are bundle tilting and bundle sizing. By tilting the bundle to angles of 20° or more, the downwind conductors are moved to the edge of the upwind conductor's wake and the energy input is reduced. By keeping the subconductor spacing to the order of 20 times the conductor diameter, the wind energy input to the windward conductor is reduced by being moved to a wake region of reduced intensity. The third commonly used method to control or eliminate wake-induced oscillations is to stagger the length or simply to shorten the average subspan length. This method does not control those oscillations where the bundle moves as a rigid body and is somewhat dependent on the mechanical characteristics of the spacers.

In comparison to the damage that can result from eolian vibration or galloping, field reports of wake-induced oscillation damage are usually of a minor nature, primarily conductor abrasion from clashing and spacer breakage, neither of which normally results in system outages.

## 5.1.9 Supporting Structures

***Types of Supporting Structure.*** Numerous types of structure are used for supporting transmission-line conductors, for example, self-supporting steel towers, guyed steel towers, self-supporting aluminum towers, guyed aluminum towers, self-supporting steel poles, flexible and semiflexible steel towers and poles, rope suspension, wood poles, wood H frames, and concrete poles. The type of supporting structure to use depends on such factors as the location of the line, importance of the line, desired life of the line, money available for initial investment, cost of maintenance, and availability of material. Because of the wide conductor spacing required for electrical clearances and insulation, the high tensile stresses used in conductors and ground cables to pull these cables up to a sag which will keep the heights of the structures within reason, the long spans necessary for crossing ravines in mountainous country, and the reliance to be placed on a major trunk line, lines exceeding 345 kV are frequently built of self-supporting steel towers although guyed and rope-suspension structures are increasingly applied. A line built with self-supporting steel towers is very satisfactory in all respects, as it requires less inspection and has a maximum life with minimum maintenance costs. However, high-strength aluminum-alloy towers are available, and their use is on the increase. They have the advantage of better resistance to corrosive atmospheres than steel.[68] The structural configurations and design details are the same as with steel, with the added problem of greater deflections when stresses are applied owing to the lower modulus of elasticity of aluminum. The effect of long-time creep of aluminum is yet to be determined. Self-supporting steel poles are frequently used in congested districts where right-of-way is limited and short spans are necessary. The advent of EHV has brought a great variety of new structural configurations. Details of some of these have been published. *Electrical World,* Nov. 15, 1965, pp. 95–118, contains outline drawings of 35 towers and six wood-pole H-frame structures as applied to EHV, as well as a tabulation of specification items of EHV lines in the United States and Canada. The *Transmission Line Reference Book, 345 kV and Above,* 2d ed., 1982, published by EPRI,[3] contains details of a broad spectrum of 345- through 800-kV structures.

*Wood poles* are used extensively where they are readily available. Medium- and lower-voltage lines can be built economically with such poles fitted with either steel or wood crossarms. *Wood H frames* composed of two poles tied together at the top with wood or steel crossarms have been successfully used for the higher-voltage lines up to 345 kV. To take full advantage of the transverse strength, such poles can be braced internally for at least a portion of their height with wood X bracing.

*Concrete poles* have been used in some parts of the world where timber is scarce and where the ingredients for making concrete are readily obtainable. Another advantage is that they are impervious to insect damage and other forms of decay prevalent with wood structures in tropical or subtropical climates. They are generally cast in units, by using standard forms, and transported to the site, although they may be manufactured where used. Concrete poles should always have sufficient prestressed steel reinforcement to take care of the bending stresses due to wind loads, pulls from cables, and the like, in addition to being designed as columns under vertical loads. In all structures conductor configuration and the effect of various forces which may act upon them must be taken into account.

***Conductor Spacing and Clearances***

*Horizontal Configuration.* The minimum spacing of conductors on structures where post-type or V-string insulators are used on medium-length spans will generally depend on the least separation that can be used at midspan without the conductors approaching too closely under adverse wind- or ice-loading conditions.

With suspension insulators a different problem exists, as the swing of the insulator string has to be considered and clearances to the structure determined. This will generally give conductors a spacing at the supports which will be greater than the required midspan separation. One typical rule is to calculate the swing of the insulator string, both with the wind on the bare conductor and the wind on the ice-coated conductor with the corresponding vertical loads acting at the point of conductor suspension, to determine which condition gives the maximum deflection. The vertical loads should be taken on a length of span which is two-thirds the span for the horizontal loads. This will allow a certain amount of leeway in using a standard height structure at a location where the ground is lower than at the two adjacent structures. After the length of the insulator string has been determined electrically and the angle of insulator swing calculated, a normal electrical clearance is established to the structure from the deflected position of the conductor, which, when applied to the three conductors in their relative positions, will determine the necessary horizontal separation of the phases at the supports. This separation should then be checked to see whether it is sufficient for the midspan separation required. Midspan separations that will not be subject to flashover if the conductors begin to swing out of step are usually inherent on high-voltage lines owing to the clearances required at the structures. On very long spans and on the longer spans of low-voltage lines, these spacings may be insufficient. Thomas[69] proposed a horizontal-spacing formula for the determination of safe midspan spacings in windy territory where gusts and strong eddies might cause wires to start swinging at different periods

$$\delta = \frac{CdD}{w} + A + \frac{L}{2} \tag{5-71}$$

in which $\delta$ = horizontal spacing in feet; $C$ = an experience factor discussed later; $d$ = percent sag of the condition to be studied; $D$ = overall diameter of the conductor; $w$ = conductor weight, in pounds per foot, used in calculating $d$; $A$ = arcing distance of the line voltage (1 ft/110 kV); and $L$ = length, in feet, of the swinging portion of the insulator string. Thomas proposed an experience factor of 4 for copper and 3.5 for ACSR.[69] It has since been found that, in areas not subject to frequent violent winds, values of $C$ as low as 1 will provide safe midspan spacings. Thomas was doubtful whether as to the added $L/2$ distance is necessary, since insulators seldom swing out of step. This doubt seems to have been justified.

*Vertical Configuration.* Where the conductors are arranged in vertical configuration, the same electrical clearances will apply for the same voltage as for horizontal configuration, but it may be necessary to increase the vertical separation somewhat to prevent the conductors from coming together or approaching too closely at the center of the span when unequal ice-loading conditions occur or the ice falls off a lower conductor first.

In Fig. 5-25, $\theta$ = angle of insulator swing from vertical, $H$ = horizontal span, $V$ = vertical span, $w$ = weight of conductor with or without ice load per lineal foot, and $w_i$ = weight of insulator string including hardware. Then

$$\tan\theta = \frac{Hw_e}{Vw + w_i/2} \tag{5-72}$$

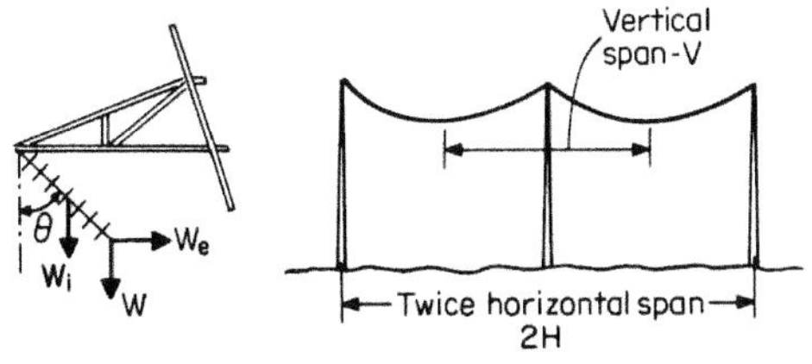


**FIGURE 5-25** Determination of suspension insulator swing.

Ground wires, if used, are located above the conductors for lightning protection and in such a position that there is no danger of contact with the conductors at midspan. As ground wires are generally strung with less sag than the conductor cables, ample clearance at midspan is readily obtainable.

These considerations, taken together with the maximum vertical sag to be used and the height required for the conductors above ground level, will determine the height and width of the supporting structure. Extensions can be used where the terrain requires a higher structure than normal.

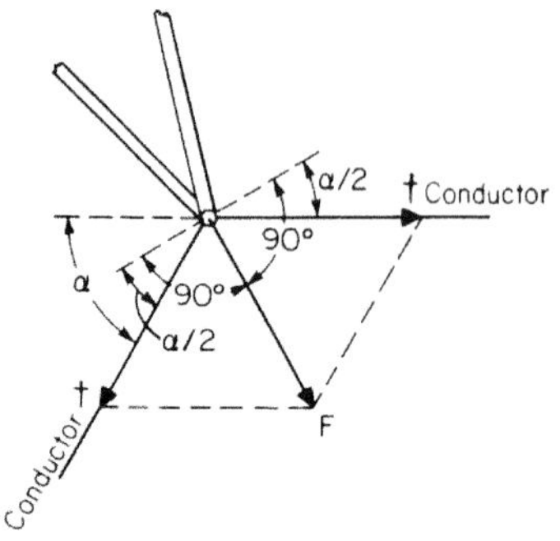


**FIGURE 5-26** Determination of transverse forces.

***Transverse Forces on Support Structures.*** Transverse forces acting on towers or poles are due to wind on the conductors and ground cables (and ice coating if in ice districts), wind on the structures, and horizontal components of the tensions in the cables at angle turns in the line (Fig. 5-26).

The stress due to an angle in the line is computed by finding the resultant force produced by the wires in the two adjacent spans. For example, in Fig. 5-26, if the change in the direction of the line is the angle $a$ and the stresses $t$ in the adjacent spans are equal to each other, the resultant force

$$F = 2t \sin\frac{a}{2} \tag{5-73}$$

Table 5-23 gives the resultant force $F$ due to a tension $t$ of 1000 lb in each conductor of two adjacent spans. The resultant force due to each conductor may be thus computed and the moments about the ground line determined. These moments may be added to those produced by the wind pressure to find the maximum stress.

**TABLE 5-23** Resultant Force due to Equal Tensions of 1000 lb in Adjacent Spans

| Angle | | | Angle | | |
|---|---|---|---|---|---|
| $a$ deg | $a/2$ deg | Resultant $F$, lb | $a$ deg | $a/2$ deg | Resultant $F$, lb |
| 10 | 5 | 174.4 | 70 | 35 | 1147.2 |
| 20 | 10 | 347.2 | 80 | 40 | 1285.6 |
| 30 | 15 | 517.6 | 90 | 45 | 1414.2 |
| 40 | 20 | 684.0 | 100 | 50 | 1532.0 |
| 50 | 25 | 845.2 | 110 | 55 | 1638.4 |
| 60 | 30 | 1000.0 | 120 | 60 | 1732.0 |

*Note:* 1 lb = 4.448 N.

In applying wind loads to the structure, the appropriate force coefficients, exposure coefficients, and gust response factors should be used.

***Longitudinal Forces on Support Structures.*** Longitudinal forces acting on towers or poles are due mainly to the maximum tension which is assumed to exist in the conductor and ground wires if broken. Ordinarily, especially with suspension insulator strings, these tensions are balanced in the adjacent spans; but if a conductor breaks, a distinct force is produced along the line due to unbalanced tension. If the break occurs on a conductor at the end of a crossarm, there is, in addition to the longitudinal force, a torsional force introduced which must be resisted by the structure. Wind acting in the direction of the line is not ordinarily a factor, as the maximum tension in the conductor is produced when the wind is blowing transverse to the line. As to the reduced stress which occurs in a span from the breaking of a conductor with the suspension insulator string deflecting in the direction of the line, the best practice is to ignore this reduction in tension, as the force due to breaking may cause an impact which more than offsets the reduction in tension. Special release clamps were devised for use on the Plymouth Meeting–Siegfried line of the Philadelphia Electric Company so that, if an insulator string deflected to an angle of 20° in the direction of the line, the clamping mechanism would release the pressure with only the friction in the saddle holding the conductor. This reduced the tension in the conductor considerably, and by assuming a low value for the tension in the conductor due to a break, a more economical structure was obtained.

***Vertical Forces on Support Structures.*** Vertical forces acting on towers or poles are those caused by the weight of that portion of the conductors, plus ice loading if any, which is supported by the structure in question. In addition, there are the weights due to insulators and accessories and the weight of the structure itself. If a structure is located in a valley, there may actually be uplift on it, if the vertical components of the tensions in the conductors exceed the downward loads.

***Combined Forces on Support Structures.*** In determining the maximum forces acting on towers or poles, it is necessary to combine the transverse forces, longitudinal forces (including torsion), and vertical forces so that they act simultaneously. Several different combinations of loading conditions may be desirable, as follows:

1. A condition with all conductors intact and the full transverse and vertical forces acting. These forces should correspond to the appropriate extreme wind and ice loadings and the NESC district loadings. The NESC also specifies overload capacity factors which must be applied to the transverse, vertical, and longitudinal loadings to provide adequate strength of the support structures. These factors depend on the grade of construction and on the type of structure. ASCE Manual 74[57] specifies load factors that can be applied to the extreme wind and ice loading cases to increase the reliability for important or long transmission lines.
2. A condition with all conductors intact, except the number it is desired to assume broken, with the transverse and vertical forces computed for each particular conductor, according to whether it is assumed broken. The longitudinal forces due to broken conductors must be combined with the transverse and vertical forces at all points of support where the conductors are assumed broken. It is customary, when more than one conductor is assumed broken, to consider all breaks in the same span and at the supports which will produce the maximum overturning moment, the maximum torque, or a combination of both.
3. A condition in some localities where extra-heavy vertical loads, caused by an unusually large formation of ice on the conductors, may occur. These loads are combined with the weight of the structure.
4. A condition with vertical loads acting upward at the conductor supports.

NOTE: It is not customary to combine transverse and longitudinal loads with the loads specified under items 3 and 4.

Other factors may enter into the determination of the maximum forces acting on supporting structures in special cases, such as the horizontal and vertical components of tensions in guys and the addition of pole-top transformers, switches, and working platforms.[70]

The proper number of conductors to assume broken is a debatable question and depends upon what margin of safety is desired and the amount of money it is desired to invest for this security. Generally speaking, the minimum number of conductors to assume broken for tangent suspension single-circuit towers should be either one ground wire or any one conductor, and for double-circuit towers either one ground wire and one conductor or any two conductors on the same side of the tower and in the same span, by using the different cable supports for application of the forces to determine the maximum stress in each member of the tower. Anchor or dead-end towers should be able to withstand all or any number of conductors and ground wires broken. Generally, the condition of broken conductors and ground wires on one side of the tower will produce greater stresses in the web members than if all the conductors and ground wires are considered broken, owing to the unbalanced torsional forces existing when only the conductors and ground wires on one side of the tower are broken.

***Types of Metal Structure.*** Structures may support single, double, or multiple circuits. The first two types are generally used for transmission lines except in congested areas where right-of-way is very expensive and it is desired to transmit large blocks of power over one line. In such a case three or more circuits may be supported by the structures.

*Self-Supporting or Rigid Structures.* On both single- and double-circuit tower lines of any considerable length, at least three kinds of towers are required for economic reasons:

1. A tangent suspension tower which can be used for normal spans where no angles in the line occur (Figs. 5-27 and 5-28).
2. An angle suspension tower which can be used for normal spans with a small-angle turn in the line or with longer spans on tangents.
3. An angle tower which can be used for normal spans with a large-angle turn in the line, with extralong spans on tangent, or as a full dead-end tower for anchoring. Insulators may be either suspended or in the strain position.

Very often it is desirable to introduce a fourth kind of tower with insulators always in the strain position to take care of exceptionally large-angle turns in the line; in extremely long spans on tangent; and also, where required, as a full dead-end tower. When this type of tower is provided, the tower listed in item 3 may be of lighter construction and not used for dead-end purposes.

Double-circuit towers with the vertical configuration of conductors, as used on different lines, are very much alike in appearance, generally being square in cross section. It is customary to locate the middle conductors outside the upper and lower conductors.

With single-circuit towers and the conductors arranged in horizontal configuration, a different problem arises which has resulted in the design of special patented structures for the higher-voltage lines with wide conductor spacing. The shape of these towers has been developed with a view to minimizing the weight of steel required in the superstructure and also reducing the size of footings by minimizing the effect of torsion. The more common types are the Blaw-Knox tower (Fig. 5-27), or

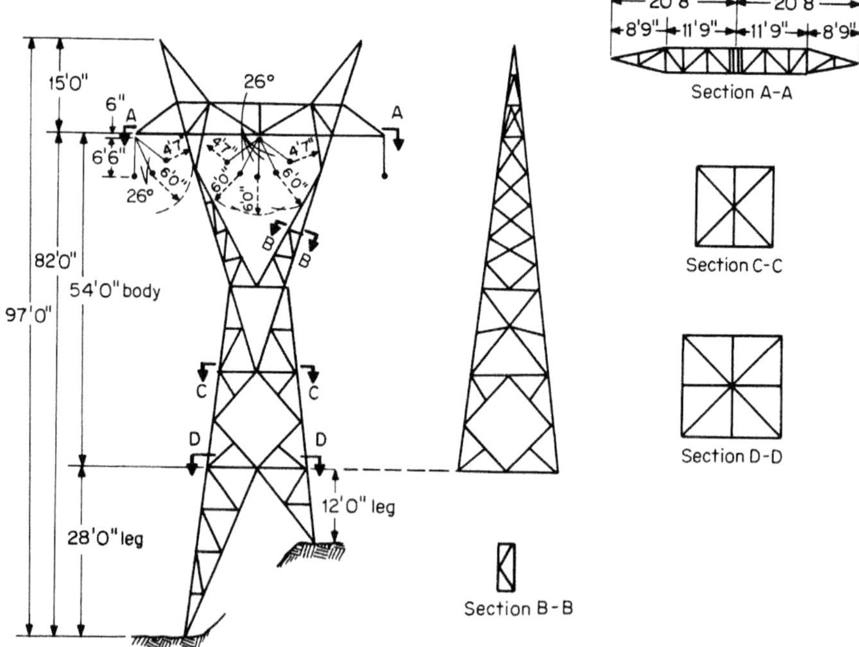


**FIGURE 5-27** Tennessee Valley Authority 161-kV single-circuit tangent suspension corset-type tower. (*Designed by Blaw-Knox Co.*)

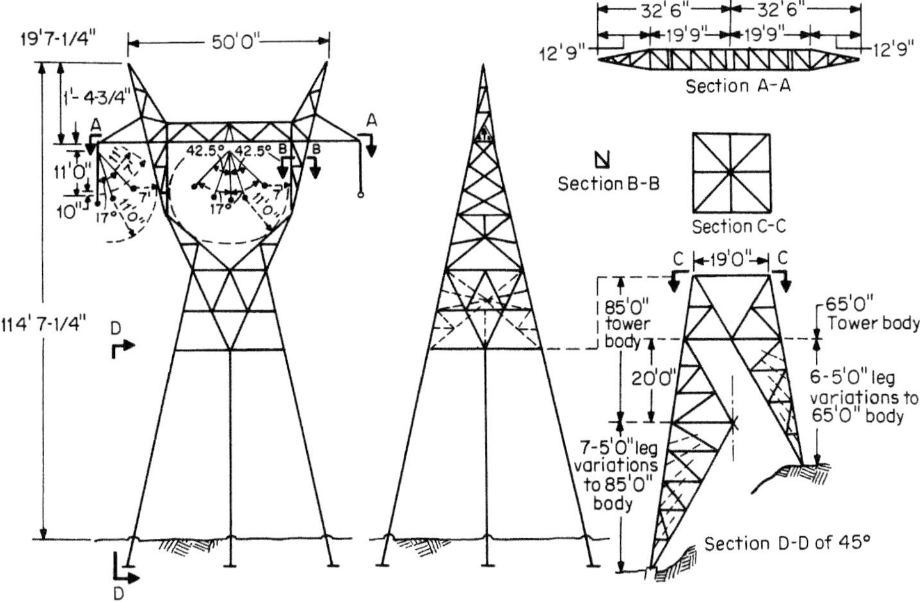


**FIGURE 5-28** City of Los Angeles 287-kV tangent suspension rotated-type tower. (*Designed by American Bridge Co.*)

corseted type, as originally used on the Plymouth Meeting–Conowingo line of the Philadelphia Electric Company; and the American Bridge Company's rotated tower (Fig. 5-28), used on the first Hoover Dam–Los Angeles line (this line has since been uprated to 500 kV) and also by the Bonneville system and on lines of the Tennessee Valley Authority. Either of these types serves the purpose for which it was intended. The theory behind the rotated tower is that the greatest overturning moment is caused by a combination of the transverse forces and longitudinal forces, due to broken conductors, acting simultaneously, which produces a resultant force acting at an angle of approximately 45° with the direction of the line. In this case the whole four tower legs are resisting the overturning moment, thereby reducing foundation loads and consequently costs. Under normal conditions of loading, with only the transverse forces acting, the legs on the diagonal separation will take care of the overturning moment. Obviously the greatest advantage of the rotated type over the nonrotated type is on tangent towers and towers used for small-angle turns in the line when the transverse and longitudinal forces are approximately equal.

Figure 5-29*a* shows a TVA 500-kV conventional-design tangent self-supporting tower for a bundle-conductor line having three 971,600-cmil ACSR conductors per phase. The overhead ground-wire clamps are suspended and insulated from the tower by means of distribution-type guy strain insulators. The overhead ground wires are composed of seven strands of No. 9 Alumoweld and are used for carrier-current communication channels.[71] Each ground-wire insulator is provided with a spill-over gap to protect it during lightning discharges.

It is interesting to compare Figs. 5-29*a* and *b*. Both show 500-kV towers, but Fig. 5-29*b* is designed for a narrower right-of-way. The wind side swing of the conductors in a span is half the sag at 30° side swing, and this is common to both towers. Therefore, the saving in right-of-way for Fig. 5-29*b* is 40 ft plus 7 ft, 7 in, less 30 ft, 3 in, or 17 ft, 4 in on each side, or a total of approximately 35 ft.

Figure 5-30 shows a light-suspension 500-kV single-circuit tower typically used by the Bonneville Power Administration (BPA), supporting three 1,192,500-cmil ACSR Bunting conductors per phase and two 7-strand No. 8 Alumoweld overhead ground wires. BPA uses continuous overhead ground wires throughout its entire 500-kV network except on single-circuit lines west of the Cascade

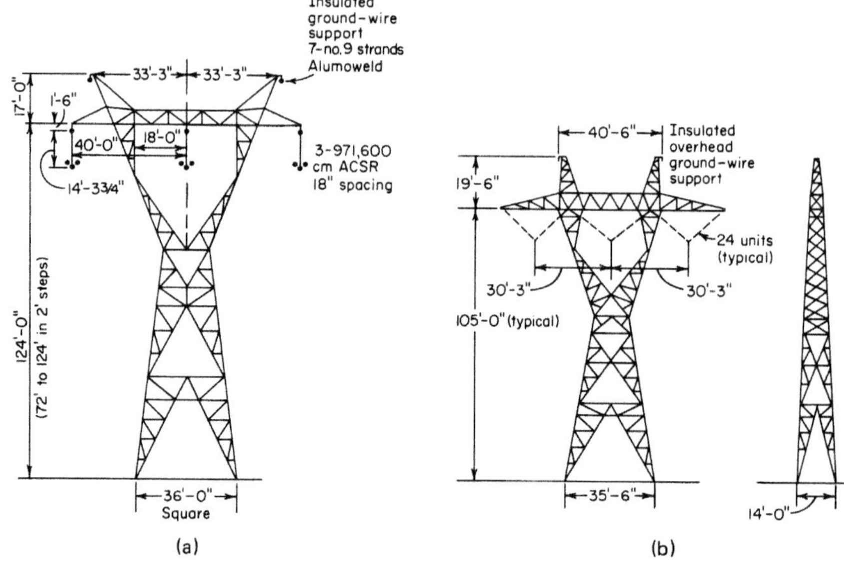


**FIGURE 5-29** (*a*) A 500-kV tangent self-supporting tower (*Tennessee Valley Authority*); (*b*) 500 kV semiflexible steel tangent tower (*Arkansas Power & Light Co.*).

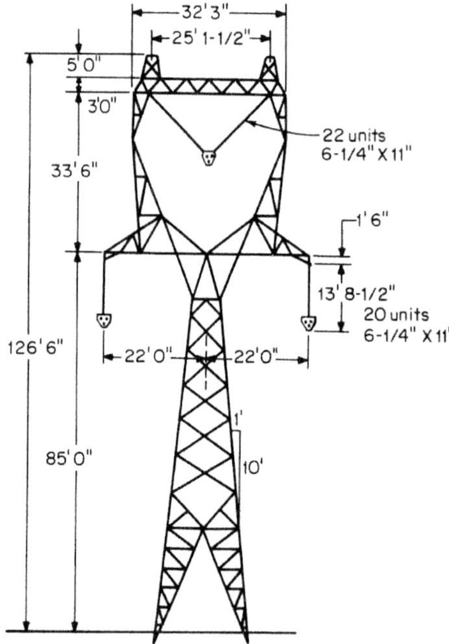


**FIGURE 5-30** Suspension-type 500-kV tower. (*Bonneville Power Administration.*)

Mountains. In the latter case, overhead ground wires extend 1 mi out from the substations. Typically, BPA 500-kV lines are designed to withstand 100-mi/h winds and solid ice coatings up to 1½ in.

A steel suspension self-supporting tower used by Hydro-Quebec for 735-kV Manicouagan lines is shown in Fig. 5-31. Line conductors consist of a four-conductor bundle per phase, where each conductor is a 1,028,000-cmil ACSR insulated with 33 insulator units (5¾ × 10 in) per phase. This type of tower was used on the first stages of the Manicouagan project since September 1965, and in subsequent stages of the same project.

A unique structure is the 765-kV self-supporting steel tower used by American Electric Power Company (Fig. 5-32). This tower, weighing from 44,000 to 66,500 lb, including grillage foundation, was designed by American Bridge Company for erection in parts, if desired, by a Skycrane helicopter. Like AEP's 765-kV V tower shown in Fig. 5-32, there are 30 insulator disks (5¾ × 10 in) per leg of V strings in the outside phases and 32 insulator units in the middle phase. Also, like the tower shown in Fig. 5-32, this tower is designed to meet the same special

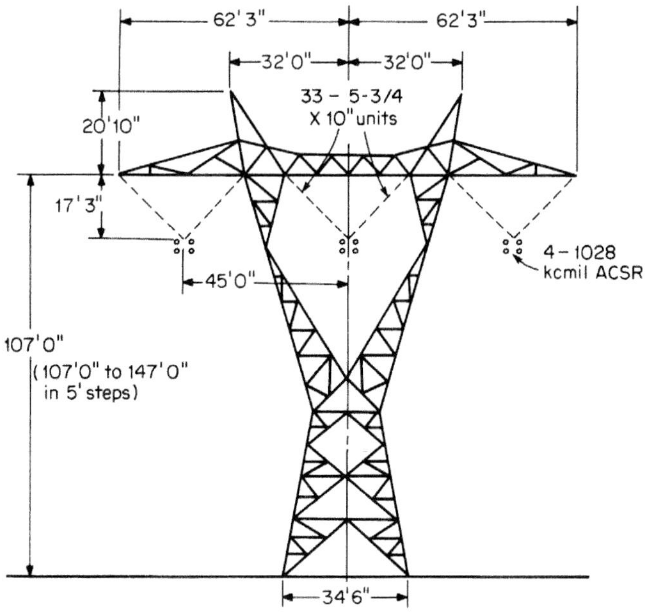


**FIGURE 5-31** Steel suspension self-supporting tower for 735-kV Manicouagan lines. (*Hydro-Quebec.*)

AEP loading criteria already described. Two overhead ground wires provide a 15° shielding angle to the outside four-conductor bundles.

*Semiflexible Structures.* Such structures have been used to some extent for the voltages under EHV. This type of tower has a narrow base in the direction of the line. The ground wires are strung tightly to take up unbalanced loads due to broken conductors and form part of the structural system. In case a conductor breaks, the unbalanced load will be taken up by the ground wires and transmitted by them to the next anchor tower.

With the advent of EHV and bundle conductors, semiflexible self-supporting towers are receiving more consideration, and some are being used. With the heavy bundle conductors, the breaking of one conductor is not serious, and the breaking of all conductors of a phase is practically nonexistent. Possible causes are airplanes and tornadoes, which no practical tower could withstand. Figure 5-29*b* shows such a tower as used on the 500-kV system of the Arkansas Power and Light Company. The overhead ground wires are insulated from the towers as they are in Fig. 14-29*a* for communication purposes. Figure 5-33*a* shows a steel-saving semiflexible tower used by the Pacific Gas and Electric Company. Note the X guying used between tower legs to obtain the required lateral strength.

*Guyed Towers.* Such towers overcome the weakness of semiflexible towers in line with the line. They can be used for single-conductor lines or for any other service. Figure 5-33*b* shows a guyed steel tower used by the Pacific Gas and Electric Company in mountainous country. A feature of the tower is that the legs do not have to be of equal length. This tower has the same internal X guying as the tower of Fig. 5-33*a*. The self-supporting feature of the tower of Fig. 5-33*a* is replaced by four guys in the direction of the line and with an increase in strength.

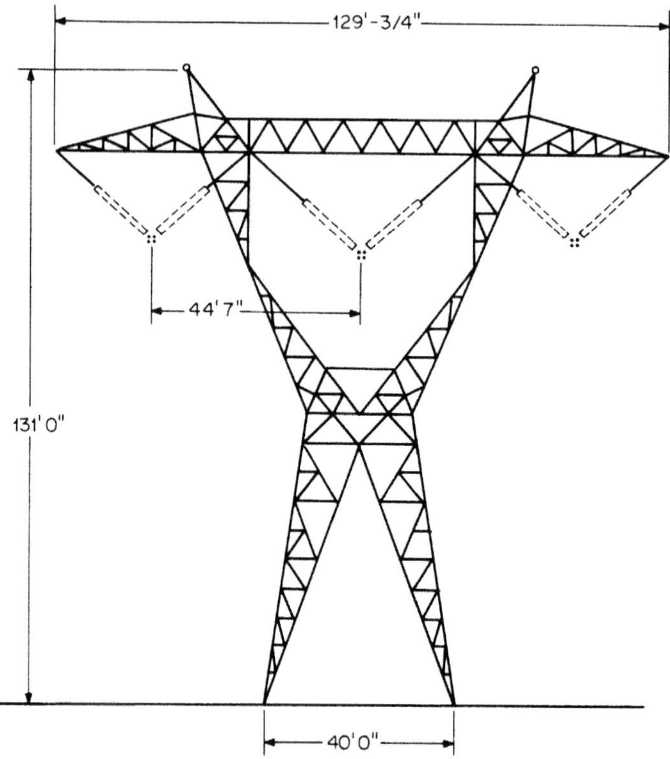


**FIGURE 5-32** Steel 765-kV self-supporting suspension tower. (*American Electric Power Co.*)

Figure 5-34 shows a Kaiser aluminum guyed-V 345-kV tower used by the American Electric Power Company. Weighing from 3350 to 5400 lb (1510 to 2450 kg), this tower was erected by using a helicopter to "tilt up" the assembled tower by pivoting about a special hinge at the center foundation. This allows the use of a helicopter with a lifting capacity smaller than the weight of the tower since most of the tower weight is supported by the foundation while it is being tilted up. There are 15 insulator units (5¾ × 10 in) per leg of the V strings on this tower.

Figure 5-35 shows a Kaiser aluminum guyed-V 765-kV tower also used by American Electric Power Company. There are 30 insulator units (5¾ × 10 in) per leg of the V strings in the outside phases and 32 insulator units in the middle phase.

Each of these towers has been designed to withstand special AEP loading criteria which include 100-mi/h winds with no ice, 50-mi/h winds with 1 in of ice, and 1¼ in of ice with no wind, in addition to the NESC loading requirements.

*Tubular Steel Poles.* These poles are being used on city streets and in congested areas where a wide right-of-way cannot be gained. They have been used for voltages up to and including 345 kV. Vertical configuration of conductors is used for all high-voltage lines. Insulators may be side post or suspension[72] on cantilever arms or a combination[73] of the two. Figure 5-36 shows a 230-kV pole used on a line of the Arizona Public Service Company in Phoenix. These poles are of tubular steel in three sections with telescoping joints. The poles are tapered, with a diameter of 24 in at the base and 10.8 in at the top. The mast arms are 8 ft long, of tubular steel, with brackets bolted to the poles with two ¾-in through bolts. The poles are spaced approximately 300 ft apart. Insulator side swing is

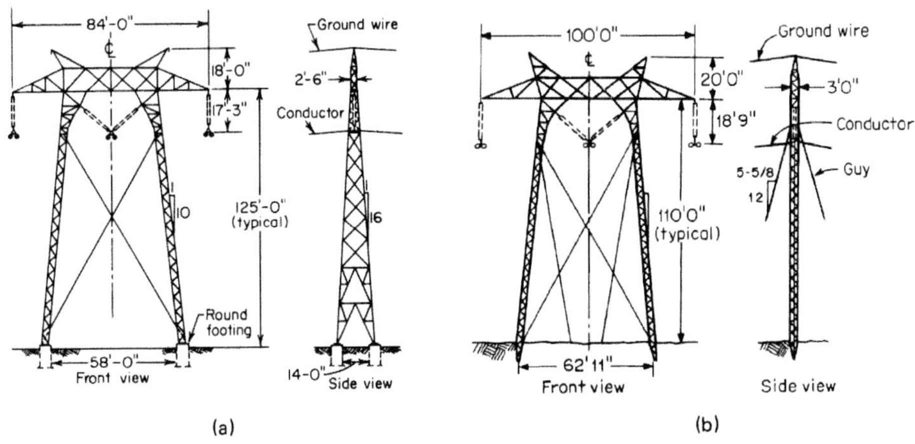


**FIGURE 5-33** (*a*) A 500-kV steel tower used in valley areas (*Pacific Gas and Electric Co.*); (*b*) a 500-kV steel tower used in mountainous areas (*Pacific Gas & Electric Co.*).

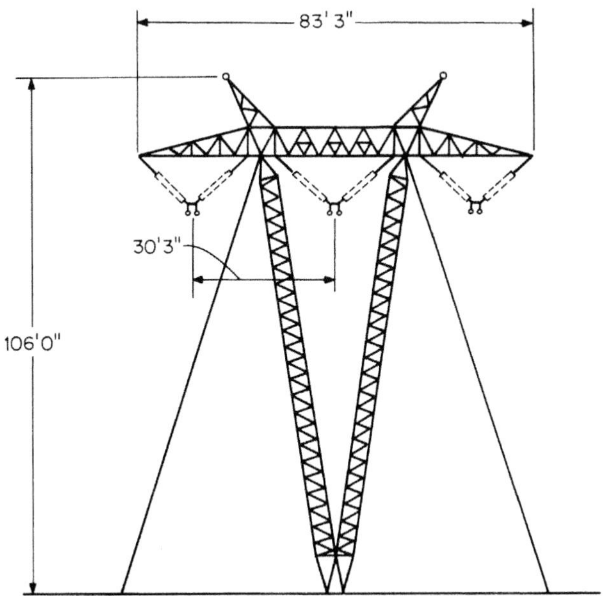


**FIGURE 5-34** A 345-kV guyed-V aluminum suspension tower. (*American Electric Power Co.*)

reduced by a 200-lb combined hold-down weight and corona shield. The poles present a pleasing appearance and have elicited no objections even with a line installed on each side of a 60-ft street. The poles, side arms, and accessories were furnished by the Union Metal Manufacturing Company of Canton, Ohio.

The New Orleans Public Service Company 230-kV line[73] is of similar construction but is designed for hurricane-force winds. The poles are 12-sided, elliptical, high-strength steel, with the short diameter,

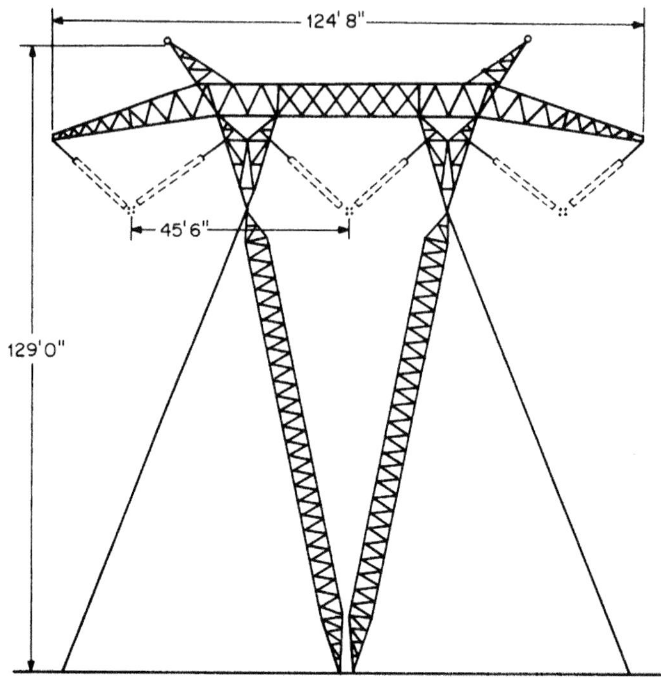


**FIGURE 5-35** A 765-kV guyed-V aluminum tower. (*American Electric Power Co.*)

which is in line with the line, 75% of the long diameter. The insulators are a combination of 12 suspension insulators and a swivel-ended strut (side-post) insulator equal to 12 suspension insulators, to prevent side swing of the suspension insulators. Some poles have side-post 230-kV insulators only. The poles have no base for bolting to a foundation but do have baseplates and are set in concrete in holes 25 ft deep. The holes are made by driving 32-in-diameter steel casings to a depth somewhat deeper than 25 ft and cleaning them out.

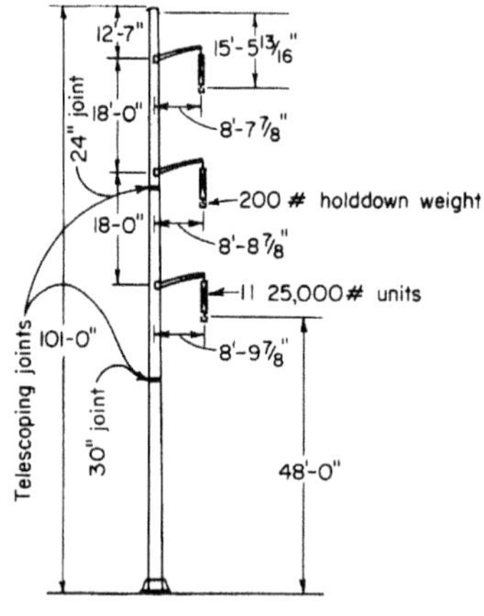


**FIGURE 5-36** A 230-kV steel pole. (*Arizona Public Service Co.*)

*Special Structures.* Transposition towers require special structures where it is not expedient to make the transpositions on a standard tower by the use of special crossarms. Long spans over rivers and bays and crossings over important highways and trunkline railroads frequently require towers which either are much higher than normal or must have a larger factor of safety against collapse. Anchor towers near substations, towers for mounting switches, and towers for turning 90° angles also may come under this classification. Such special structures are designed to suit local requirements and are subject to regulations of the U.S. Army Engineer Corps in the case of navigable-river crossings, to state public-utility commissions or other bodies for highway crossings, and to the particular regulations of railroads which are concerned.

***Stresses in Structures—Design.*** Stresses in towers can be computed analytically by the historic graphic methods or by use of one of the available computer programs for tower analysis and design. Most of the design procedures assume the foundations are rigid. In actual practice, with towers set in the ground, an uneven settlement of the foundations may produce excess stresses which must be considered and taken care of in the overload factor provided against failure. Some flexible lattice-type towers and most pole-type structures will undergo sufficiently large horizontal displacements such that the nonlinear stresses due to the vertical loads acting in conjunction with these displacements (the so-called $P$-$\Delta$ effect) must be considered in the design.

Tension members may be designed for their full net area of section with bolt holes deducted, and compression members may be designed in accordance with the strength formulas given in ASCE Manual 52.[74]

For short panels in narrow-faced towers, it is economical to use a tension and compression system of web bracing without horizontal struts, as the stresses in the corner posts will be reduced considerably, with torsional stresses eliminated entirely. The bracing should make an angle of 30° to 45° with the horizontal, with all web members in any one panel of the same size to divide the loads equally. Where a tension system of diagonal bracing is used with horizontal struts taking the compression, the angle of the bracing may be increased to 50° or more from the horizontal. Long, unsupported lengths of main members can be broken up by the use of redundant members. The lower panels of towers should be so designed that variable-length leg extensions, which are interchangeable, can be employed to take care of sloping ground. Square extensions of any desired length may be used where towers higher than standard are necessary, with variable-length leg extensions fitted to the bottom of the square extensions if required.

***Structure Tests.*** When an important transmission line is built using structures of new design, at least one type of structure (generally the tangent suspension tower or the type which is to be used most frequently) should be tested, first with the working loads as specified for the design of the tower applied, and finally with the ultimate loads which the tower is expected to withstand, applied. Steel poles should always be tested on rigid foundations with the transverse, longitudinal, and vertical loads applied simultaneously to introduce all direct bending and torsional stresses. Crossarms should be tested for the additional torsional stresses introduced, where pin insulators are used, and combined with the longitudinal loads and the heaviest vertical loads specified. Equivalent concentrated loads may be used in some cases to avoid applying a multiplicity of small loads at different points, which would cause delay in shifting loads, but care should be taken to see that all combinations of loads or individual loads which will produce the maximum stress in each member are applied. After the structure has successfully withstood all the specified loads, a destruction test is desirable to determine the overload factor. This can generally be made with the test loads which cause the maximum stresses in the greatest number of members on a tower in place, by increasing the transverse loads indefinitely until failure occurs. After a test is completed, members of a tower should be examined for elongation of bolt holes, straightness, etc.

Towers should be tested with the protective coating which is to be used in service on the steel, and the foundations should be the same as those for which the towers are designed. If it is impossible to test towers on earth foundations, they may be tested on rigid foundations, but a test on rigid foundations will undoubtedly show a greater overload factor than may be expected in service.

## 5.1.10 Line Accessories (Lines under EHV)

***Suspension Clamps.*** These designs are fairly well standardized for the usual conductors. Simple, light, well-designed clamps in both malleable iron and forged steel are available for almost any conductor. The seat and clamping surfaces should be smooth, without any projections or sharp bends, and should be formed to support the conductor on long, easy curves and at the comparatively sharp bends formed at horizontal and vertical angles. Heavy, complicated clamps, unless very carefully designed, are generally avoided to allow as much freedom as possible at the support. For the same reason, care is exercised to avoid rigid connections of any kind.

***Trunnion-Type Clamps.*** These are designed to give an almost completely flexible connection by supporting the clamp on a pivot, approximately on the axis of the conductor (Fig. 5-37). Thus any vibration of the conductor tends to be transmitted through the clamp, eliminating much of the heavy binding stresses caused by a fixed support.

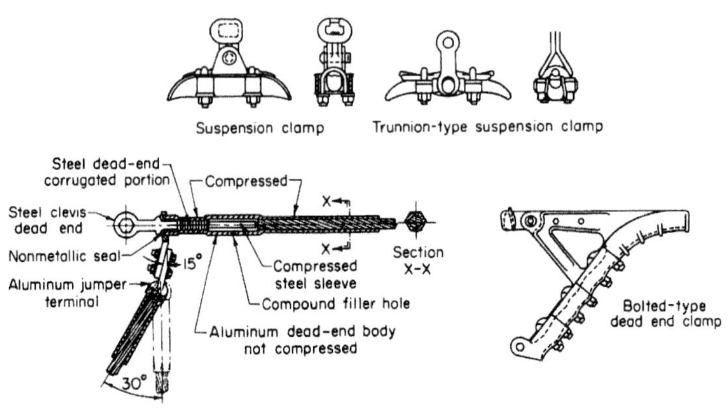


**FIGURE 5-37** Conductor clamps.

The suspension clamp is intended primarily to support the weight of the conductor and to prevent any longitudinal movement from accidental unequal tensions in adjacent spans. It is generally considered desirable but not always essential that the suspension clamp hold the conductor in case of a break. For large conductors under heavy tensions it is difficult to design a light, flexible connection that will not slip under such a contingency.

***Slip, or Releasing, Clamps.*** Several especially heavy lines have been designed on the proposition that, since suspension clamps could not reasonably be secured that would positively hold the conductor, a clamp should be used that would hold under all ordinary conditions but would slip at something like one-half the maximum conductor tension in case of a break. This arrangement justified a considerable reduction in the exceedingly large longitudinal design loads on the towers and resulted in a considerable saving in tower and foundation costs. Several designs of slip clamps and releasing clamps have been used.

***Dead-End Clamps.*** These clamps are of the bolted type and are available for practically all copper and aluminum conductors. However, for the larger ACSR conductors the compression-type dead-end clamp is generally used (see Fig. 5-37). This is very similar to the compression splice used on ACSR.

The dead end for the steel core, which may have a clevis or an eye-type end, is pressed on after the aluminum sleeve has been slipped out over the conductor. The aluminum sleeve is then slipped back over the steel sleeve until the aluminum body makes contact with the shoulder of the steel sleeve. The electrical connection tongue on the aluminum body is aligned with the clevis or eye of the steel-core dead end as required, after which the aluminum body is filled with the nonoxidizing compound furnished with the body and the body is compressed. Similar pressed-on dead ends are available for copper, Copperweld, and other conductors. Several manufacturers furnish ACSR dead ends in all sizes required.

***Armor Rods.*** These rods are quite generally used on ACSR lines as a protection against fatigue of the aluminum strands from vibration. Armor rods consist of a bundle of aluminum rods, somewhat larger in diameter than the strands of the conductor, laid parallel to the length of the conductor and arranged to form a complete covering. These are spirally twisted by a tool to lie approximately parallel with the lay of the strands in the cable and are clamped in place at each end. The suspension clamp is attached at the center, with the armor extending 2 or 3 ft on each side. The bending stresses caused by vibration are reduced by the increased diameter and area of metal and distributed over a longer section of conductor.

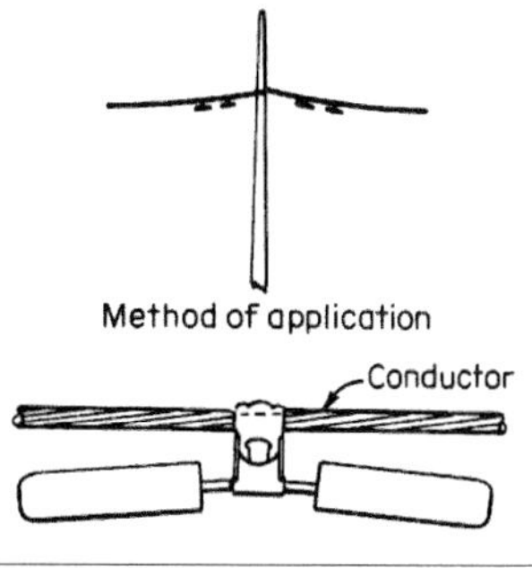


**FIGURE 5-38** Vibration damper.

***Vibration Dampers.*** The Stockbridge damper, as well as several other designs, are devices for damping vibration out of the entire span. Such dampers have been used on ACSR, copper, and steel conductors and ground wires, as illustrated in Fig. 5-38. The cause of conductor vibration and the action of the Stockbridge damper are outlined in the paragraph on wind-inclined motion.

*Overhead Ground-Wire Vibration.* Overhead ground wires are especially subject to vibration; in fact, most steel ground wires will often be found in rather irregular vibration of small amplitude which generally does not appear to have any ill effects. Ground-wire attachments should be made with at least as great care as given the conductor clamps. Rigid clamps have been almost entirely abandoned in favor of a suspension clamp similar to that used on the conductor and attached by links or shackles so as to give a perfectly flexible connection.

***Hardware.*** Many items of hardware have become fairly well standardized. The dimensions of the eye of eyebolts, the length of thread on various-length bolts, end links, and hardware for suspension insulators are quite uniform. It is usually possible to obtain about identical stock material from a number of manufacturers. Many other items such as shackles, guy clamps, and crossarm braces are furnished in such a wide variety that considerable care is required to choose the most commonly used but suitable stock items. Much expense and confusion in both construction and maintenance are saved by limiting the number of hardware items.

***Insulating Braces and Guys.*** With the use of wood to increase the impulse insulation to decrease the line's sensitivity to lightning flashovers, steel crossarm braces have been replaced with wood on a number of lines. Connections are made by pressed-steel fittings. The use of a 48-in wood brace in place of steel, for additional wood insulation, is roughly the equivalent in lightning-flashover strength of adding one suspension unit to the insulation. The effect on 60-Hz flashover is, however, negligible.

To obtain equal wood insulation at guyed structures to what may be obtained on unguyed construction requires long wood insulators in the guys. These guy insulators are quite efficient because of the high tensile strength of clear wood; an ultimate strength of 6000 lb/in$^2$ on the net section is conservative. A 2- × 2-in fir insulator will develop the full strength of a $^3/_8$-in Siemens-Martin guy strand. The design of the connection to the pressed-steel fitting requires only that several bolts of insufficient diameter be used to give the necessary bearing area between the wood and the shank of the bolt. The bolts should be placed alternately through the face and side of the stick to prevent splitting.

Reinforced fiberglass is receiving increased favor as guy-wire insulators in place of wood, and as crossarm braces in place of wood or steel. Impulse flashover voltages of fiberglass line hardware can be supplied by the manufacturers.

***Guys.*** The various grades of guy strand are almost universally furnished in accordance with ASTM specifications. The ultimate strength for each size and grade is given in Sec. 4. The so-called double-galvanized is generally used. Common guy strand is not ordinarily employed in transmission construction, as the best-quality galvanizing is not furnished in this grade.

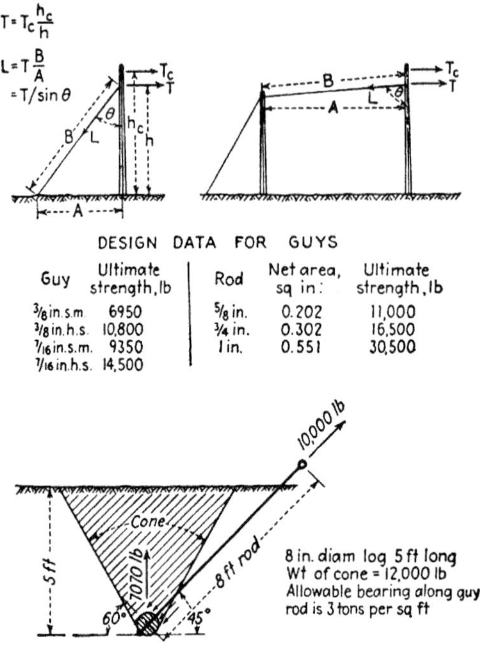


**FIGURE 5-39** General guy and log anchor-strength calculations.

Siemens-Martin strand is most commonly used for the lighter lines and high-strength strands for heavy construction.

More than one size of guy strand is not economical for a line, and often the same size may be used for several designs. The 3/8-in size, in either Siemens-Martin or high-strength grade, is most generally used both for guys and for overhead ground wires.

In the usual wood-pole construction great refinement is not required in designing guys. Usually it is sufficient to determine the number of guys, of the size and quality to be used on the line, required to support the load, an additional guy being employed for any fractional part. In transmission construction a safety factor of 2 is general for guys, although this may be somewhat reduced.

A common problem in guy design is illustrated in Fig. 5-39. The ratio of the guy load $L$ to the conductor load $T$ is the same as the ratio of the length of the guy $B$ to the distance $A$. The length $B$ is readily determined from a sketch drawn to scale.

$$L = T\frac{B}{A} \quad \text{or} \quad L = \frac{T}{\sin\theta} \tag{5-74}$$

If the conductor load $T_c$ is above the point of the attachment of the guy, then

$$T = \frac{T_c h_c}{h} \tag{5-75}$$

## 5.1.11 Foundations

***Effect of Foundation Rotational Displacement.*** Displacement criteria for tower foundations and for single-shaft foundations are discussed below. When foundations rotate and move vertical loads into eccentric (off center) positions, the bending moments due to eccentric vertical loading will increase and add to bending moments due to wind and wire tension forces. Appropriate consideration of increases in bending moments due to foundation rotation is required when determining applicable displacement criteria.

***Lattice-Tower Foundation Loads and Displacement Criteria.*** Lattice-tower foundation loads consist of vertical tension (uplift) or compression forces and horizontal shear forces. For tangent and small-line-angle towers, the vertical loads on a foundation may be either uplift or compression. For terminal and line angle towers, the foundations on one side may always be loaded in uplift while the other side may always be loaded in compression. The distribution of horizontal forces between the foundations of a lattice tower vary with the bracing of the structure. A typical free-body diagram of foundation loads is shown in Fig. 5-40.

When the foundations of a tower displace and the geometric relationship of the tower to its foundations remains the same, any increase in load due to this displacement will have a minimal effect on the tower and its foundation. However, foundation movements which change the geometric relationship between the tower and its foundations will redistribute the loads in the tower members and

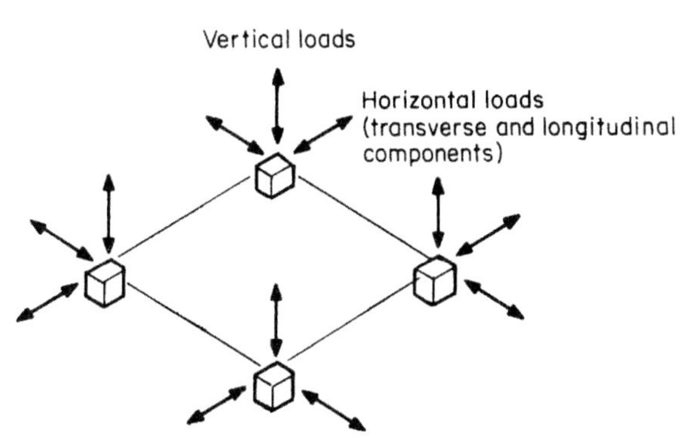


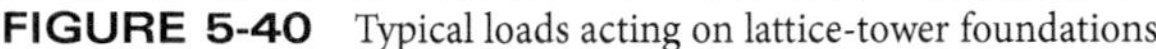

**FIGURE 5-40** Typical loads acting on lattice-tower foundations.

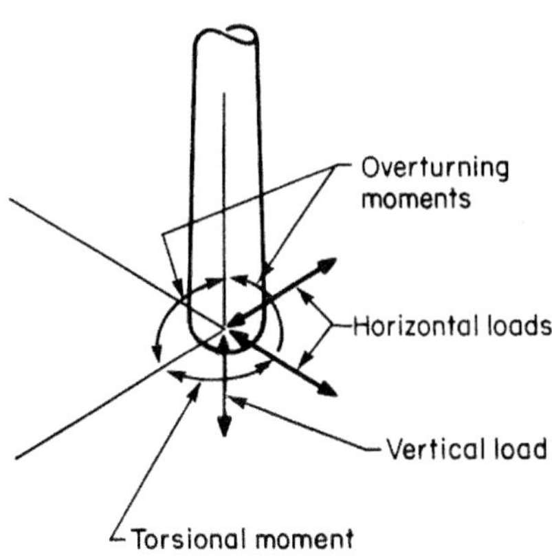


**FIGURE 5-41** Typical loads acting on foundations for single-shaft structures.

foundations. This will usually cause greater reactions on the foundation that moves least relative to the tower, which in turn will tend to equalize this differential displacement.

Presently, the effects of differential foundation movements are normally not included in tower design. Several options are available should the engineer decide to consider differential foundation displacements in the tower design. These options include designing the foundations to satisfy performance criteria which will not cause significant secondary loads on the tower, or design the tower to withstand specified differential foundation movements.

***Single-Shaft Foundation Loads and Displacement Criteria.*** Single-pole structures have one foundation so that differential foundation movement is precluded. The foundation reactions consist of a large overturning moment and usually relatively small horizontal, vertical, or torsional loads. Figure 5-41 presents a free-body diagram of the loads.

For single-shaft structures, the foundation movement of concern is the angular rotation of the shaft in the vertical plane and horizontal displacement of the top of the foundation. When these displacements have been determined, the displacement of the conductors can be computed. Under high-wind loading, a corresponding deflection of the conductors perpendicular to the transmission line can be permitted. Accordingly, a large ground-line displacement of the foundation could also be permitted. As a result of foundation rotation, the clearance between the conductors and the structure would be decreased only for structures with single-string insulators. The midspan ground clearance and the change in line angle would also decrease a negligible amount.

In establishing displacement criteria for single-shaft-structure foundations, consideration should be given to how much total, as well as permanent, displacement can be permitted. In some cases, large permanent displacements might be aesthetically unacceptable and replumbing of the structures and/or their foundations may be required. In establishing displacement criteria, the cost of replumbing should be compared to the cost of a foundation that is more resistant to displacement.

For terminal and large-line-angle structures, large foundation deflections parallel to the conductor may be intolerable. For these structures, excessive deflections may reduce the conductor-to-ground clearance or affect the load capacity of adjacent structures. There are also problems in the stringing and sagging of conductors if the deflections are excessive. These problems are usually resolved by designing a more deflection-resistant foundation, construction methods, or use of permanent guys.

***Framed Structure Foundation Loads and Displacement Criteria.*** These structures are dependent in part for their stability on one or more of their joints resisting moment. The foundation reactions are dependent on which joints can resist moment and the relative stiffness of the members. Figures 5-42 and 5-43 present free-body diagrams of four- and two-legged framed structures. If the bases of structures are assumed as pins or universal joints, then the moments acting on the foundations will theoretically be zero.

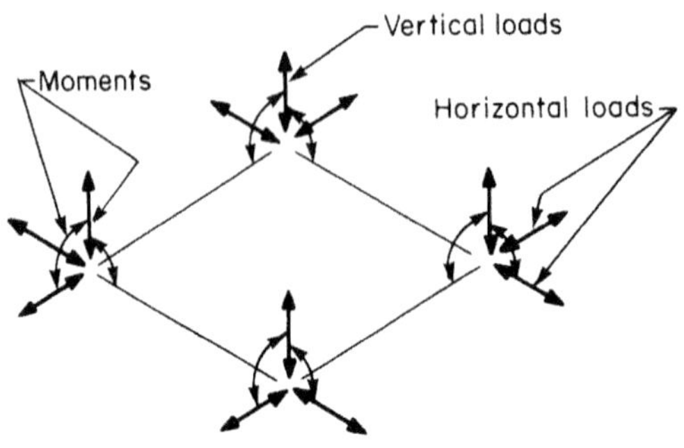


FIGURE 5-42 Typical loads acting on foundations for four-legged framed structures.

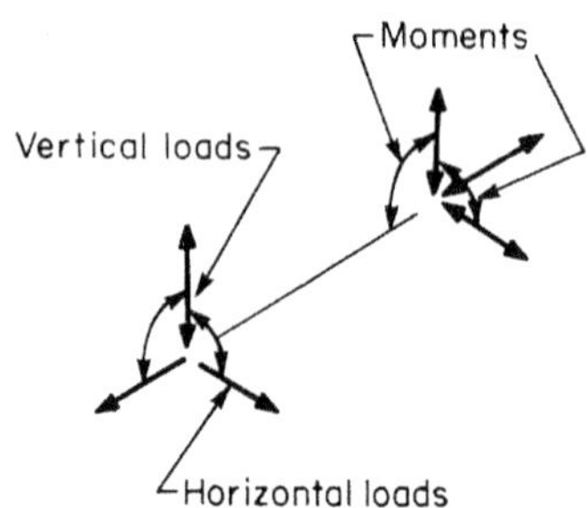


FIGURE 5-43 Typical loads acting on foundations for two-legged framed structures.

Many different types of two-legged, H-framed structures are in use in transmission lines. This has been particularly true since visual impact has become of greater concern.

The H-frame structure is particularly applicable for wood, tubular steel, and concrete poles. The crossarm may be pin-connected to the poles. These structures may be unbraced, braced, or internally guyed as shown in Fig. 5-44.

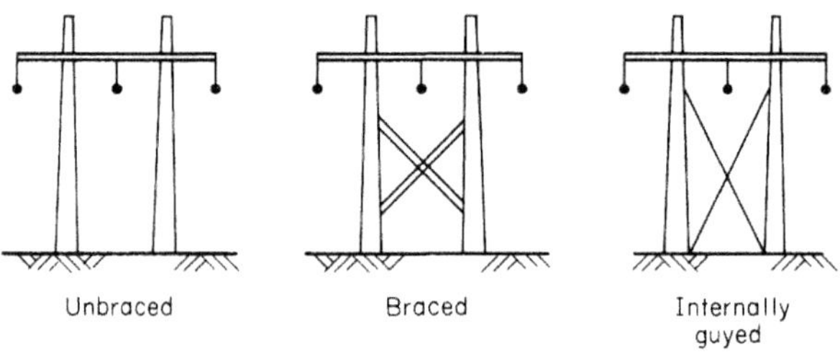


FIGURE 5-44 Typical H-frame structures.

As with lattice towers, past practice has not usually included the influence of foundation displacement and rotation in H-frame structure design. Significant foundation movements will redistribute the frame and foundation loads. The foundations can be designed to experience movements which will not produce significant secondary stresses, or the structure can be designed to a predetermined maximum allowable foundation displacement and rotation.

### *Typical HV and EHV Guyed Structures*

*Guyed Portal Structures.* The design and utilization of guyed-mast transmission structures (guyed towers) has evolved very rapidly since the mid-1960s or so, and the subject warrants discussion. A high-strength wire stranding used in tension and a latticed mast in compression with limited bending are two of the most efficient structural components, a point not lost on designers of transmission lines in northern Europe during the early days of transmission of electricity.

The most popular form was the guyed portal tower (Fig. 5-45), in which two tripods, each essentially consisting of a mast and two guys, were held apart at the tops by a rigid crossarm. The arrangement permitted the use of two mast footings (compression with a little shear load) and if the structure was not too tall, the four guys could be brought to two anchors, each doing double duty. The structure also offered the possibility of easy ground assembly and erection by one-piece rotation using a gin pole or A frame. Portal tower application was to a large extent limited to flat terrain and relatively short towers (short spans or low voltages) as rough terrain required masts of different lengths, which

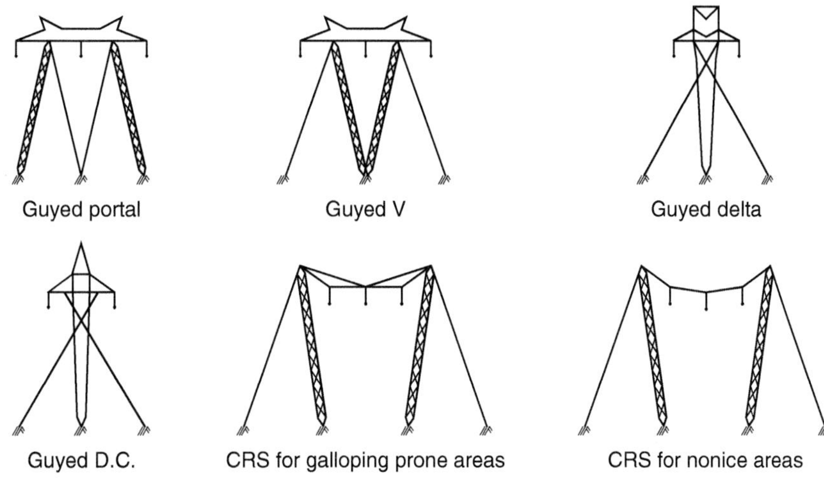


**FIGURE 5-45** Typical externally guyed structures.

complicated the usual erection methods, and tall structures made common guy anchors an impossibility or else greatly increased the tower stresses under transverse loading.

*Guyed-V Tower.* In 1957, the guyed V tower was developed[75] so that the guyed-mast principle would be more suitable for rough terrain; the guyed V consisted of the same two tripods moved laterally but was now held together by the crossarm.

Introduction of the guyed V helped promote the initial and serious use of helicopters for line construction for the first guyed-V towers, which were designed in aluminum. The very light weight of both the aluminum material and guyed-mast concept permitted transport and erection of a complete structure with the limited capacity of the helicopters then commercially available. Guyed-V tower use expanded rapidly with thousands of kilometers of lines built at up to the 500-kV level, fabricated in both steel and aluminum as the economies of the V principle were proving to be sufficient to justify use even when helicopter erection was not warranted.

Application expanded into the 750-kV class in the United States, Brazil, and Canada, and nominal spans were consistently in the 450- to 500-m range as the low cost of the extra height of taller towers (two masts and some guy wire) were extending the optimum envelope.

Single-mast guyed towers such as the delta were developed that offered compaction benefits and were also easy to use on rough terrain as they also require only one compression footing; the guy lengths are cut as required to fit the terrain. The guyed-V tower became the most widely used tower at the higher voltages, although all forms of guyed-mast structures were usually restricted to open terrain and areas where widespread guys were not considered to be hazards to the use of large farm equipment or where land occupancy was not a problem.

*Cross-Rope Suspension (CRS) or Chainette.* The next step in guyed-mast development followed from the failure of a 750-kV class tower due to material defect, a failure in a remote area where construction had been by a large mobile crane that could move only on the frozen winter roads. The replacement guyed-V structure was too heavy for available helicopter lift, and the repair was greatly delayed until crane access came with winter weather.

As voltages increase, all towers tend to become top-heavy, and the guyed-V or guyed-$\Delta$ are no exceptions. The urge to dispense with the crossarm led to the development of the *cross-rope suspension* (CRS) tower or *Chainette*, as it is referred to in Quebec, Canada.

The CRS concept was actually derived from the successful CRS system built in the mountains of British Columbia, Canada in 1955[76,77] and uses one or more wire ropes suspended between two guyed masts to replace the crossarm and support the insulator strings. Even at 1000 kV, the individual masts are well within the capacity of modest low cost helicopters, and thus the problems of initial construction and emergency replacement are easily solved, with crane or helicopter. The initial design of the CRS[78,79] made use of the six-part suspension assembly because it was feared that galloping of a single phase might find the support point forces transferred through the rope system to the other phases and thus promote widespread and damaging activity. The CRS found application on the third, fourth, and fifth lines of the 735-kV James Bay system in Quebec and on a section of 500-kV lines in Oregon, in the United States.

The next applications and development came about in South Africa at 400 kV and in the Argentine at 500 kV, where a single cross rope is used because of the absence of any significant icing and thus no fear of galloping.[77]

All CRS lines must use a special construction or spacer rope extending between the tops of the masts; the initial use is to position the mast tops before the conductors are in place to provide tension to the system and subsequently to provide means of access to the phases by use of a wheeled ladder for both stringing and sagging work and for maintenance.

The CRS type of construction has a few negative aspects as it requires space at the tower sites (an open area that can be farmed), but the positive points are many. Both single- or multiple-rope CRS systems permit reduced phase spacing with increased SIL values and limited only by gradient effects or fear of wind clashing. On the 1150 km of the 500 kV line in the Argentine studies showed that clashings were a negligible threat and the reduced phase spacings available by the CRS increased SIL values and reduced compensation costs by many millions of dollars. All wire rope components including the guys are precut to length and can be tested to working loads, and length adjustment is included in only one of the guys. The weight of structural steel in a tower will be about 50% of that of a comparable guyed-V or guyed-Δ tower. With overhead wires attached directly to the tops of the masts, the lightning protection is about as good as possible and the extra relaxation provided by the suspended wire assembly added to the insulator strings ensures that cascading potential is negligible.

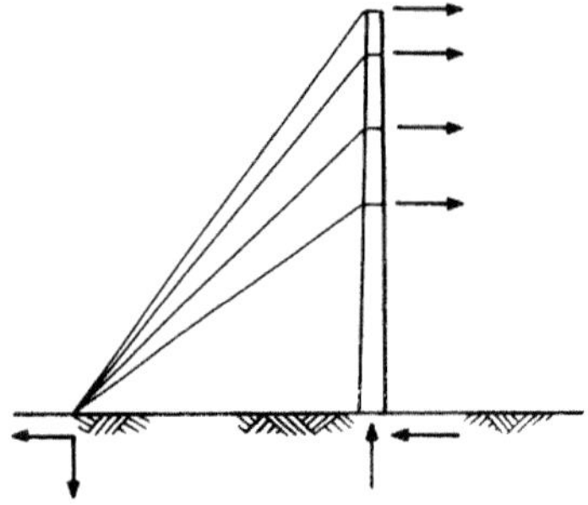

**FIGURE 5-46** Single-shaft externally guyed structure.

*Guying Rigid Structures.* The guyed structures of the preceding subsection consisted of structural components pivoted or free to rotate at the connections to the foundations. Thus, the guy loads are in all cases determined by simple static analysis. The addition of guys to structures that are fixed or rigidly attached to the foundations (Fig. 5-46) produce arrangements that are indeterminate to varying degrees and for some, the analysis of strength can be quite complex and performance will be dependent on introducing and maintaining precise levels of guy pretension.

The problems arise because of the large strains or stretch that guys develop in order to resist the loads and the varying degrees of rigidity of the structures that they are guying. Guyed-pole structures, as shown in Fig. 5-46, if they are wood and single-pole or the very common H-frame structure, are readily guyed, since the wood components are relatively flexible and allow the guys to develop load before the poles deflect enough to fail.

However, guyed-metal-pole structures that are fixed at their bases become more difficult to assess and analysis is usually needed to confirm both the distribution of loads between the guys and the poles themselves and also to set the pretension needed in the guys to ensure that the deflections or distortions of the components remain within limits. The design engineer must be conscious of the fact that deflections of poles can introduce large additional bending stresses caused by the $P$-$\Delta$ effect—the pole compression acting on the moment arm of the bow of the pole.

The most complex situation arises from attempts to guy rigid lattice structures (Fig. 5-47). In order to resist a load applied to the structure, the guys will normally stretch so much that the rigid structure will already have failed. Successful guying in such a manner requires either oversized

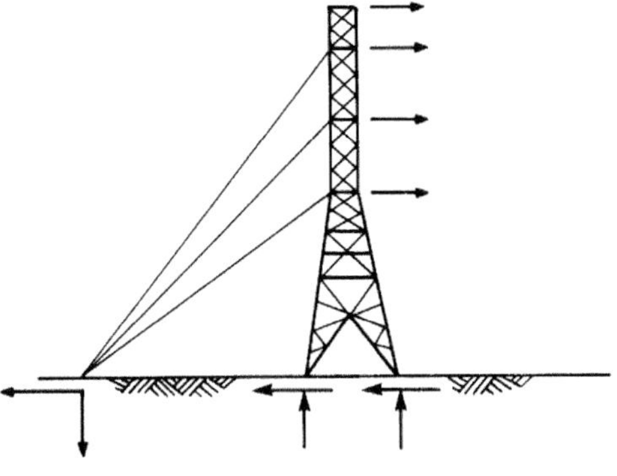

**FIGURE 5-47** Externally guyed lattice tower.

guys to limit the stretch of the guys or very precise levels of pretension, set and maintained to distribute the loads in desired manner between the structure and the guys.

There is one condition under which guying of a rigidly framed structure is readily done and that is when the purpose is to restrict the movement of the structure on failure. Longitudinal guying of rigid structures or even of H frames is a method of creating stop towers to limit the movement of slack and thus stop a cascading of structure if one is under way. Transverse guys are sometimes applied outside of heavy angle structures to restrict their failure mode and thus limit the amount of slack that can be introduced into the line system if the tower failure occurs under an extreme ice loading.

***Foundation Types.*** A wide variety of foundation types can be used with self-supporting or guyed lattice, framed, and single-shaft structures. They include the following:

| Lattice tower | Framed and single-shaft structures |
|---|---|
| Steel grillages | Concrete spread foundations |
| Concrete spread foundation | Drilled shafts |
| Rock foundation | Direct embedment |
| Drilled shafts | |

*Steel Grillages.* Figure 5-48 shows three typical types of steel grillages. Figure 5-48*a* is a pyramid arrangement in which the leg stub is connected to four smaller stubs which in turn are connected to the grillage at the base. The advantage of this type of construction is that the pyramid can transfer the horizontal shear load down to the grillage base by truss action. However, the pyramid arrangement does not permit much flexibility for adjusting the assembly, if needed. In addition, it is difficult to compact the backfill inside the pyramid.

Figure 5-48*b* shows a grillage foundation which has the single leg stub carried directly to the grillage base. The horizontal shear is transferred through shear members that engage the passive lateral resistance of the adjacent compacted soil.

Figure 5-48*c* also has the single leg stub carried directly to the grillage base. This type of grillage foundation has a leg reinforcer which increases the area for mobilizing passive soil pressure as well as increasing the leg strength. The shear is transferred to the soil via the leg and reinforcer and resisted by passive soil pressure. The base grillage of these three typical foundations consists of steel beams, angles, or channels which transfer the compressive or uplift load to the soil.

The advantages of steel grillage foundations are that they can be purchased with the tower steel and concrete is not required at the site. The disadvantage is that these foundations usually must be designed before any soil borings are obtained and may have to be enlarged by pouring a concrete base around the grillage if actual soil conditions are not as good as those assumed in the original design. In addition, large grillages are difficult to set with required accuracy. The placement and compaction of the backfill material are critical to the actual load-carrying capacity and load-displacement characteristics of the foundation.

*Concrete Spread Foundations.* This type of foundation consists of a base mat and a square or round pier. It is constructed of reinforced concrete. There are several variations as indicated in Fig. 5-49. The stub angle can be bent and the pier and mat centered. The mat can be located so that the projection from the stub angle intersects the centroid of the mat or the pier itself can be battered to the tower leg slope.

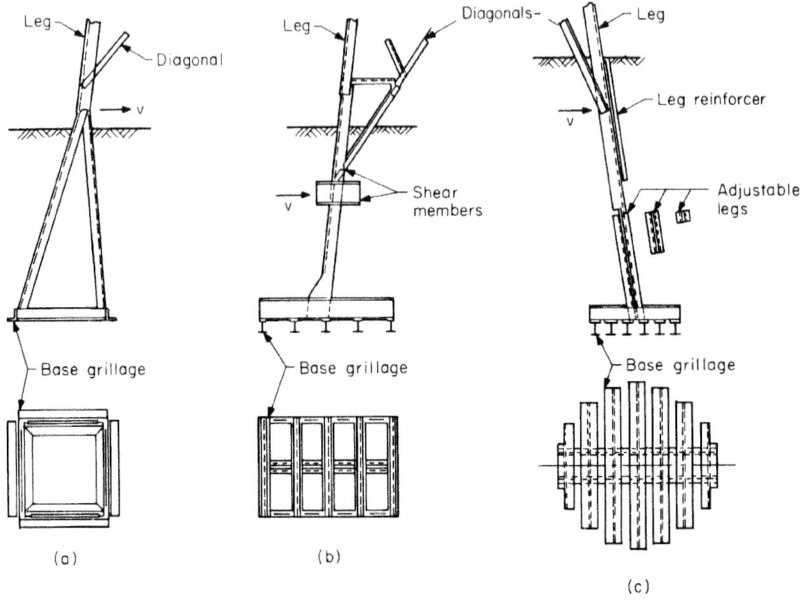


**FIGURE 5-48** Typical steel grillage foundations.

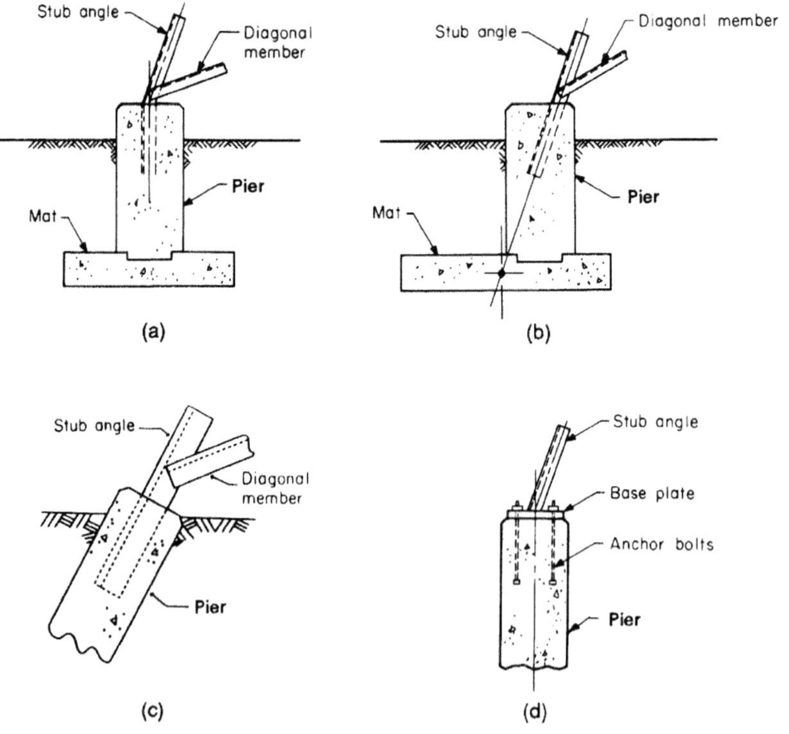


**FIGURE 5-49** Typical concrete spread foundations (reinforcing not shown).

The stub angle is embedded in the top of the pier so that the upper exposed section can be spliced directly to the main tower leg and diagonals. The stub angle should be of adequate size to resist the axial loads transmitted from the main leg and diagonals plus any secondary bending moment from the horizontal shear, if applicable. The stub angle must be embedded in the concrete to a sufficient length to transmit the load to the concrete. Bolted clip angles or welded stud shear connectors may be added to the lower end of the stub to reduce its length. Anchor bolts can also be used in lieu of the direct embedment stub angle as shown in Fig. 5-49*d*.

*Rock Foundations.* Many areas of the United States have bedrock either exposed at the ground surface or covered with a thin mantle of soil. Relatively simple, economical, and efficient rock foundations may be installed where this type of terrain is encountered. A rock foundation can be designed to resist both uplift and compression loads plus horizontal shear and, in some structure applications, bending moments. Where suitable bedrock is encountered at the surface or close to the surface, a rock foundation, as shown in Fig. 5-50, can be installed. Bolted clip angles or welded stud shear connectors may be added to the lower end of the stud to reduce its length.

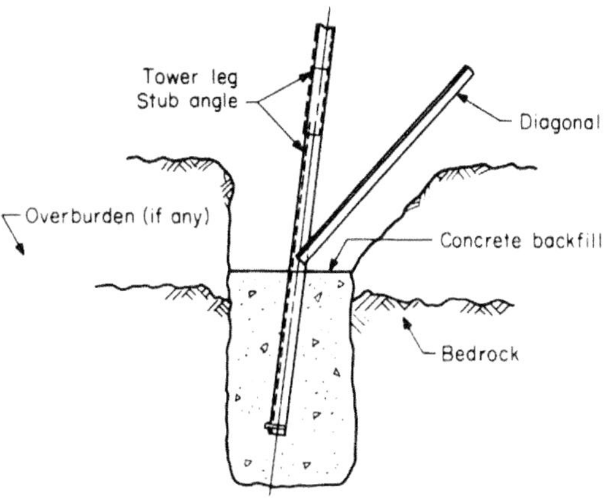


**FIGURE 5-50** Rock foundations.

*Drilled Shafts.* The drilled concrete shaft is the most common type of foundation presently being used to support lattice towers, framed structures, and single shafts. Drilled concrete shafts are constructed by power augering a circular excavation, placing the reinforcing steel and pouring concrete to form a shaft foundation. Lattice towers are attached by embedment of a stub angle or through the use of baseplates and anchor bolts. Framed structures and single shafts are attached through the use of baseplates and anchor bolts.

Drilled shafts can be constructed in a wide variety of soil and rock types. However, the construction of drilled concrete shafts may encounter problems under certain soil conditions. For example, granular soils may collapse into the excavation before concrete can be poured. In soft, cohesive soils, squeezing or shear failure of the soil can occur, producing a reduced diameter or the excavation may become completely obstructed before the concrete is placed. This soil movement in the excavation can result in ground-surface settlement. Thus, casing and/or drilling mud may be required in granular and soft cohesive soils to maintain an open excavation.

*Direct Embedment.* Direct embedment refers to wood, steel, or concrete pole foundations (both single-shaft and H-framed structures) constructed by power augering a circular excavation in the ground, inserting the pole directly into the excavation, and backfilling the void between the pole and the sides of the excavation. Thus, the pole acts as its own foundation by transferring loads to the in situ soil via the backfill. This technique has been traditionally used for wood-pole foundations and has recently been employed for metal- and concrete-pole foundations.

The quality of backfill, method of placement, and degree of compaction strongly influence the stiffness and strength of a direct embedment foundation. Corrosion of an embedded metal pole is also an important consideration. It should be noted that the presence of granular or soft, cohesive soils may cause the same construction problems for direct embedment foundations as for drilled concrete-shaft foundations.

***Subsurface Investigations.*** The technical requirement of assuring a safe and cost-effective foundation design for transmission structures requires a thorough knowledge of the subsurface conditions along the right-of-way (ROW). The intent of this subsection is to provide a guide for performing an adequate subsurface investigation for the design of transmission-line structure foundations.

When designing a foundation, the engineer should be concerned with the following factors: (1) the ultimate load-bearing capacity of the subsurface material, and (2) the allowable displacements of the foundation. In the case of grillages and directly embedded poles, the quality of available backfill materials is also a concern. Hence, the objectives of a subsurface investigation are to determine the stratigraphy, physical characteristics, and engineering properties (particularly the strength and deformation characteristics) of the soil or rock underlying a given site.

To determine the most cost-effective foundation, it is necessary to consider the engineering and physical properties of the subsurface materials; construction costs; the construction aspects of a particular foundation type and how they are influenced by such factors as groundwater elevation, safety requirements, contractor capability and experience, and environmental constraints.

The scope of a subsurface investigation will vary depending on foundation loads, type of structure, and probable foundation types, types of subsurface materials, and previous knowledge of subsurface conditions along the line route. It is necessary to use engineering judgment when considering the scope of the subsurface investigation. A detailed outline for developing a cost-effective investigation is given in an EPRI report.[80]

***Design of Spread Foundations.*** Spread foundations are used to support lattice towers. The design of spread foundations for transmission towers must consider both the direction (uplift or compression) and orientation (inclination and eccentricity) of the applied loads. The foundation must be designed to prevent excessive displacement or shear (bearing capacity or uplift) failure of the support soil. A detailed presentation of estimating the uplift and compression capacity of spread foundations is given in an EPRI report.[80]

***Design of Drilled Shaft Foundations.*** Drilled shaft foundations are used to support lattice-tower, framed, and single-shaft structures. This type of foundation supports vertical compression loads through a combination of side shear and end bearing and supports vertical uplift loads by side shear. Lateral loads and overturning moments are supported by lateral resistance of the soil and/or rock in which the shaft is embedded plus the vertical shearing resistance on the perimeter of the shaft, and the horizontal shear on the base and the base moment.

*Compression and Uplift Capacity.* Methods for computing the compression and uplift capacity of drilled shaft foundations are given in an EPRI report.[81]

*Lateral Load Capacity.* The response of a drilled shaft to lateral loads is the result of complex interactions between the shaft and the soil and/or rock in which it is embedded. A common method of modeling this interaction is called the subgrade modulus approach. Reference 82 provides a detailed explanation of a method for determining the lateral capacity of drilled shafts. A computer program, MFAD (Moment Foundation Analysis and Design), originally called PADLL, which was developed as part of the EPRI research project eliminates the simplifying assumptions associated with prior models.

*Direct Embedment.* The response of direct embedment foundations in compression, uplift, and lateral loads is similar to that of drilled concrete shafts. Most of the analytical techniques used in drilled shaft design are relevant to direct embedment design. The principal differences between direct embedment foundations and drilled concrete shaft foundations are: (1) the backfill which intervenes between the pole and the in situ soil, and (2) the stiffness of the embedded structure shaft relative to that of a drilled concrete shaft. Drilled shafts transfer loads directly to the in situ soil. A detailed presentation for the analysis and design of directly embedded poles is given in an EPRI report.[83] Results from 12 full-scale tests on single-pole direct embedment foundations are given in a related report.[84]

*Construction Considerations.* Factors affecting, and problems associated with, the construction of drilled shaft foundations can be divided into two general areas: (1) geotechnical factors influencing construction, and (2) construction-related problems. A common geotechnical occurrence is the erosion (sloughing, caving in) of loose granular soil layers, mainly below the groundwater level. Another geotechnical factor influencing the overall capacity of the foundation is the release of stresses due to excavation, especially when the hole is left open for a long period of time (say, one or more days, depending on soil conditions). Some construction-related problems are associated with the use of drilling mud (slurry method), which tends to leave an undesirable film (up to various inches of thickness) of soft material adhered to the walls of the hole. Another frequently occurring problem is the perturbance and remolding generated by the use of casing. Also, problems arise when special geometry is requested by the designer for the drilled shaft (under-ream, shear rings, etc.). In all of these cases, in general, good communication is required between field personnel and designers to permit early detection of these conditions.

***Design of Anchors and Anchor Foundations.*** An anchor is a device which will provide resistance to an upward (tensile) force transferred to the anchor by a guy wire or structure leg member. An anchor may be a steel plate, wooden log, or concrete slab buried in the ground, a deformed bar or a steel cable grouted into a hole drilled into either soil or rock, or one of several manufactured anchors which are either drilled or rotated into the ground. Anchorage may also be provided by vertical or battered drilled shafts or piles. Typical types of anchors are shown in Fig. 5-51.

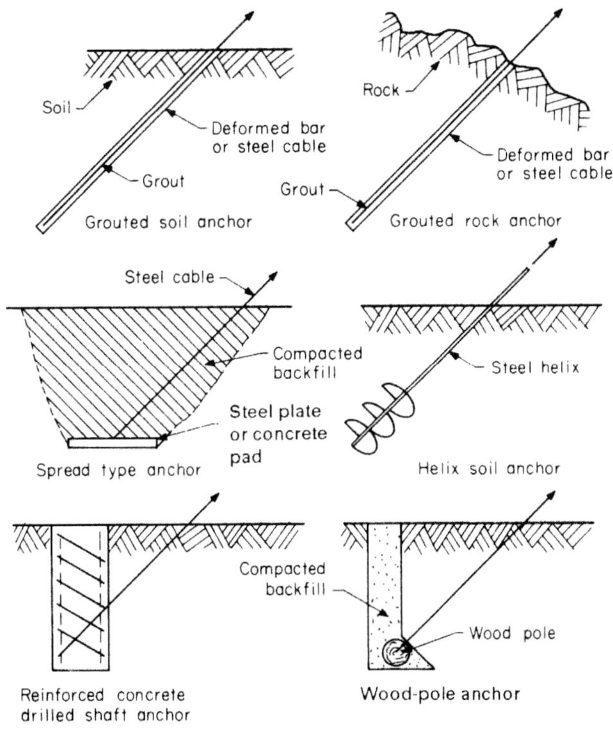


**FIGURE 5-51** Typical anchors.

Anchors may be classified as either deadman or prestressed. Deadman anchors are defined as those anchors which are not loaded until the structure is loaded. Prestressed anchors are loaded to specified load levels during installation of the anchor. An advantage of a prestressed anchor is that most of the initial strains of the anchorage system have been removed before the structural load is applied. Therefore, the full capacity of the anchor can be attained at very small deformation (movements in soil of less than ¼ in are typical). Another advantage of prestressed anchors is that they are proof-loaded to their design load at the time of installation. Disadvantages of prestressed anchors are that they are generally more expensive than deadman anchors and they should not be used in compressible soils.

Another advantage of prestressed anchors is that shallow anchors may obtain additional strength by the increased effective stress created by the influence of the bearing plate on the soils adjacent to the anchors.

Deadman anchors may include any of the systems shown in Fig. 5-51. Initial strains in deadman anchors may be reduced by as much as 50% by prestressing them to their design load at the time of installation.

*Anchor Application.* Anchors are used to permanently support guyed structures, as well as to temporarily support other structure types during erection and stringing. The legs of lattice towers can be anchored directly by rock anchors or helix-type anchors. The uplift capacity of spread foundations may be increased through the use of anchors as shown in Fig. 5-52. Guys and anchors are also extensively used to terminate wire loads on wood structures and to increase wood structure capacity for high transverse loading. At intermediate structure locations, guys and anchors may be utilized to provide additional longitudinal strength. Anchors can be used to increase the load capacity of existing foundations.

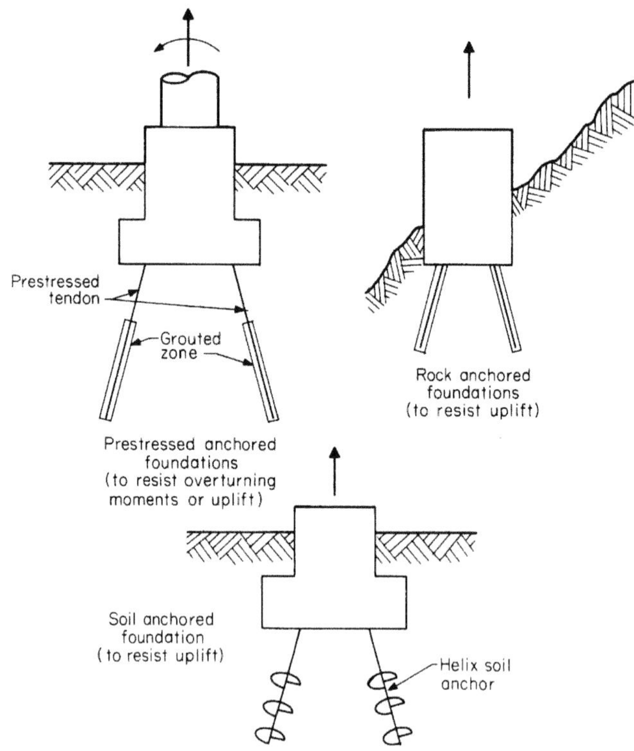


**FIGURE 5-52** Typical anchored spread foundations.

*Design.* The design of an anchor depends on a knowledge of the peak and residual shear strength properties of the soil or rock in which it is embedded. In rock, it is also important to know the degree and depth of any weathering which may have occurred, together with the orientation and spacing of joints and foliation. In addition, an understanding of the load characteristics and the structure deflection tolerance combined with the guy cable elongation is important in selecting and designing the type of anchor. Anchor pullout tests are often conducted to confirm design assumptions where prior experience is lacking.

References 80, 85, and 86 provide detailed information on the design of anchors.

## 5.1.12 Overhead Line Uprating and Upgrading

Performing a transmission line uprating can be very attractive in terms of getting smaller costs and shorter leadtimes when compared to building a new line. Besides that, it can postpone the need of new lines, reduce congestion costs and avoid unnecessary load shedding during contingencies. However, before starting the uprating task, it is important to evaluate some feasibility issues, as well as to choose the most appropriate type of uprating (Thermal Uprating or Voltage Uprating) for a specific transmission line. Each type of uprating requires different kinds of previous analysis and shall be done by its own methods. Sometimes a transmission line uprating also requires a line upgrading. Transmission line upgrading is related to physical modifications to the line.[87,88]

### *Uprating Feasibility Issues*

*Technical Feasibility.* For this kind of analysis it is important to consider at least the following points:

- *System load requirements.* It is important to evaluate for how long the uprated/upgraded line will satisfy the load requirements.
- *Assessment of current conditions and life expectancy of transmission line materials.* It is important to make this kind of evaluation for the main transmission line components, such as towers, foundations, conductors, insulators, and hardware.
- *Potential margins for uprating/upgrading.* It is important to check electrical clearances, mechanical strengths, ROW width, as well as the possibility of compliance with the requirements of safety codes (e.g., NESC), regulatory bodies and government agencies (e.g., navigable streams, public lands, air lanes).
- *Utility considerations.* Sometimes electric utilities are not authorized to take the transmission line out of service to perform the necessary uprate/upgrade services. In these cases it is important to check if the mentioned services can be done with the line in service.

*Economical/Financial Feasibility.* For this kind of analysis it is important to consider at least the following points:

- Uprating/upgrading costs vs. new line costs. It is important to remember that technical analysis of old lines usually requires data gathering and this can be very expensive and time consuming. Besides that, it is necessary to estimate what will be the need of the uprated line in terms of additional ROW. Other costs that can be relevant are related to construction (material and labor), maintenance and operation of the uprated line. Environmental costs are usually higher for new lines.
- Uprating/upgrading costs vs. uprating/upgrading benefits.

*Environmental Feasibility.* For this kind of analysis it is important to consider at least the following points:

- *Environmental considerations.* Usually not so critical when compared to new lines. However, it may be necessary to deal with historical societies, environmental groups, concerned neighbors, and so forth.

- Right-of-way easements. If significant changes will be made to the original line, it is necessary to check the validity of the previous ROW terms of use. It can be difficult to get licensing for the modified line. It is also important to check the existence of ROW encroachments and line crossings that would be unacceptable by the uprated line.

***Thermal Uprating***

*Effectiveness.* This kind of uprating can be a good option when the line loading is limited by thermal constraints and the line has margin in terms of maximum allowable conductor temperature.

*Previous Analysis to Perform.* Before proceeding with a transmission line thermal uprating it is necessary to analyze maximum allowable conductor temperature, conductor-to-ground clearance, magnetic fields, ROW issues, and sometimes structural strengths.

*Usual Techniques.* Some of the techniques used to perform transmission line thermal uprating are described as follows:

- Performing dynamic thermal rating monitoring
- Raising the thermal limit imposed to the line by some inexpensive substation equipment (e.g., disconnect switches)
- Raising the thermal limit by making similar the thermal limits of all line sections
- Keeping appropriate conductor-to-ground clearances while increasing the maximum allowable conductor temperature
- Bundling the original line conductor with another one, or replacing the line conductor by a more conductive one, to increase the line current-carrying capacity

Some of these techniques have large structural impacts. Performing a transmission line thermal uprating can bring some impact to the substation equipment.

***Voltage Uprating.*** This kind of transmission line uprating can result in a much higher rating increase than thermal uprating. Besides that, transferring the same amount of power in a higher voltage level reduces the line current, and consequently, line losses and voltage drops. However, voltage uprating is typically more expensive than thermal uprating due to the need of also uprating the voltage class of the terminal substations equipment.

Voltage uprating may be limited by the requirements of the National Electrical Safety Code (NESC) ANSI C2-2012 for increased ground clearances, conductor-to-conductor clearances, or conductor-to-tower clearances at the higher voltage. If any of these clearances is a limitation at the higher voltage, a better option may be to reconductor the line with one of the newer high-strength, low-sag conductors that offers greater current carrying capacity and has attributes compatible with the original line design.

*Effectiveness.* This kind of uprating can be a good option when: the line loading is limited by voltage drop or stability considerations; the line has margins in terms of electrical clearances; the uprating can be done with minimal line modifications or it will be applied to several circuits simultaneously; or the line design criteria can be relaxed.

*Previous Analysis to Perform.* Before proceeding with a transmission line voltage uprating it is necessary to analyze tower clearances, conductor-to-ground clearance, corona performance, electric fields, ROW issues, and sometimes structural strengths.

*Some Usual Voltage Uprating Techniques.* Some of the techniques used to perform transmission line voltage uprating are described as follows:

- Addition of insulator units to the transmission line insulator strings
- Replacement of standard insulator units by polymeric or antifog units

- Application of strut insulators (or V strings) to prevent swinging of suspension strings
- Keeping appropriate conductor-to-ground clearances while increasing the transmission line operating voltage
  - Retensioning the existing conductors
  - Performing sag adjustments (cutting out conductor lengths, sliding conductor clamps)
  - Increasing the conductor height at the attachment support (converting suspension strings to pseudo dead-end strings)
  - Increasing the attachment support height
  - Raising towers
  - Moving towers
  - Inserting additional towers
  - Performing terrain contouring (rural areas)
- Bundling the original line conductor with another one, or replacing the line conductor by a bigger one, to assure a good corona performance
- Performing line compaction
- Converting a 3-phase double-circuit line to a 6-phase single-circuit line
- Converting a low voltage double-circuit line to a high-voltage single-circuit line
- Converting HVAC lines to HVDC lines

Some of these techniques have large structural impacts.

### 5.1.13 References

1. Barnes, H. C., and Thoren, B., The AEP-ASEA UHV Test Station and Line, IEEE Conf., Paper C73-319-1, presented at IEEE Summer Power Meeting, 1973.
2. Development of Ultra-high Voltage Transmission, Bonneville Power Administration, Portland, OR, July 1974.
3. *Transmission Line Reference Book, 345 kV and Above*, 2nd ed., Electric Power Research Institute, Palo Alto, CA, 1982.
4. Clarke, E., *Circuit Analysis of A-C Power Systems*, Wiley, New York, 1943.
5. Keast, D. N., Assessing the Impact of Audible Noise from AC Transmission Lines: A Proposed Method, *IEEE Trans. Power Appar. Syst.*, May/June 1980, vol. PAS-99, no. 3, p. 1021.
6. Clayton, R. E., and Stewart, J. R., Transmission Line Electromagnetic Compatibility, 1975 IEEE Electromagnetic Compatibility Symposium Record, IEEE publication 75CH1002-5EMC.
7. A Comparison of Methods for Calculating Audible Noise of High Voltage Transmission Lines, IEEE Task Force Report, *IEEE Trans. Power Appar. Syst.*, Oct. 1982, vol. PAS-101, no. 10, p. 4290.
8. Perry, D. E., An Analysis of Transmission Line Audible Noise Based upon Field and Three-Phase Test Line Measurements, *IEEE Trans. Power Appar. Syst.*, May/June 1972, vol. PAS-91, p. 857.
9. Measurement of Audible Noise from Transmission Lines, IEEE Task Force Report, *IEEE Trans. Power Appar. Syst.*, March 1981, vol. PAS-100, no. 3, p. 1442.
10. Loftness, M., *AC Power Interference Manual*, Percival Publishing, Tumwater, WA.
11. Chartier, V. L., et al., Investigation of Corona and Field Effects of AC/DC Hybrid Transmission Lines, *IEEE Trans. Power Appar. Syst.*, Jan. 1981, vol. PAS-100, no. 1, p. 72.
12. Review of Technical Considerations on Limits to Interference from Power Lines and Stations, IEEE Committee Report, *IEEE Trans. Power Appar. Syst.*, Jan./Feb. 1980, vol. PAS-99, no. 1, p. 365.
13. *Tolerable Limits and Methods of Measurement of Electromagnetic Interference from Alternating Current High Voltage Power Systems*, CSA Standard C108.3.1-1975, Canadian Standards Association.
14. *Human Response to Interference with TV Picture Quality*, Report EL-1587, Project 68-4, Electric Power Research Institute, Palo Alto, CA, 1980.
15. Scherer, H. N., Jr., Ware, B. J., and Shih, C. H., Gaseous Effluents Due to EHV Transmission Line Corona, IEEE paper T 72 550-2, *IEEE Trans. Power Appar. Syst.*, May/June 1973, vol. PAS-92, no. 3.

16. Frydman, M., Levy, A., and Miller, S. E., Oxidant Measurements in the Vicinity of Energized 765 kV Lines, IEEE paper T 72 551-0, *IEEE Trans. Power Appar. Syst.*, May/June 1973, vol. PAS-92, no. 3.

17. Fern, W. J., and Brabets, R. I., Field Investigation of Ozone Adjacent to High Voltage Transmission Lines, IEEE paper T 74 057-6, *IEEE Trans. Power Appar. Syst.*, Sept./Oct. 1974, vol. PAS-93, no. 5.

18. Frydman, M., and Shih, C. H., Effects of the Environment on Oxidants Production in AC Corona, IEEE paper T 73 407-4, *IEEE Trans. Power Appar. Syst.*, Jan./Feb. 1974, vol. PAS-93, no. 1.

19. Roach, J. F., Chartier, V. L., and Dietrich, F. M., Experimental Oxidant Production Rates for EHV Transmission Lines and Theoretical Estimates of Ozone Concentrations Near Operating Lines, IEEE paper T 73 414-0, *IEEE Trans. Power Appar. Syst.*, March/April 1974, vol. PAS-93, no. 2.

20. Abel, W. A., Comparison of Ozone Instrumentation, IEEE paper A 78 166-7, abstract in *IEEE Trans. Power Appar. Syst.*, July/Aug. 1978, vol. PAS-97, no. 4, p. 1009.

21. Roach, J. F., et al., Ozone Concentration Measurements on the C-Line at the Apple Grove 750 kV Project and Theoretical Estimates of Ozone Concentrations Near 765 kV Lines of Normal Design, *IEEE Trans. Power Appar. Syst.*, July/Aug. 1978, vol. PAS-97, no. 4, p. 1392.

22. Sebo, S. A., et al., Examination of Ozone Emanating from EHV Transmission Line Corona Discharges, *IEEE Trans. Power Appar. Syst.*, March/April 1976, vol. PAS-95, no. 2.

23. *The Electrostatic and Electromagnetic Effects of AC Transmission Lines*, IEEE Tutorial 79 EH 0145-3 PWR, 1979.

24. Electric and Magnetic Field Coupling from High Voltage AC Power Transmission Lines-Classification of Short-Term Effects on People, IEEE Committee paper, *IEEE Trans. Power Appar. Syst.*, Nov./Dec. 1978, vol. PAS-97, no. 6, p. 2243.

25. Chiu, M. C., Fuel Ignition by High Voltage Capacitive Discharges, report JHU PPSET-18, March 1983, John Hopkins University Applied Physics Laboratory, Laurel, MD.

26. Hamaam, M. S., and Baishiki, R. S., A Range of Body Impedance Values for Low Voltage, Low Source Impedance Systems of 60 Hz; *IEEE Trans. Power Appar. Syst.*, May 1983, vol. PAS-102, no. 5, p. 1097.

27. Jaffa, K. C., Magnetic Field Induction from Overhead Transmission and Distribution Power Lines on Parallel Fences, *IEEE Trans. Power Appar. Syst.*, April 1981, vol. PAS-100, no. 4, p. 1624.

28. Dabkowski, J., The Calculation of Magnetic Coupling from Overhead Transmission Lines, *IEEE Trans. Power Appar. Syst.*, Aug. 1981, vol. PAS-100, no. 8, p. 3850.

29. Taylor, R. J., Hazard Analysis for Magnetic Induction from Electric Transmission Lines, report JHU PPSE T-23, March 1982, John Hopkins University Applied Physics Laboratory, Laurel, MD.

30. *Procedures for Measurement of Power Frequency Electric and Magnetic Fields from AC Power Lines.* IEEE Standard 644, IEEE, New York, 1994.

31. Kaminski, J., Jr., Long Time Mechanical and Electrical Strength in Suspended Insulators, *Trans. AIEE*, Aug. 1963, p. 446.

32. Nicholas, F. S., and Vose, F. C., A Polymer Insulator for High Voltage Transmission Lines, *Elec. Eng.*, vol. 82, 1963.

33. Abilgaard, E. H., Bauer, E. A., et al., Composite Longrod Insulators and Their Influence on the Design of Overhead Lines, CIGRE paper 22-03, 1976.

34. Development of Polymer Bonded Silica, (Polysil) for Electrical Applications, EPRI Report EL488, May 1977.

35. The Metapol Insulator: Dulmison (Australia) Inc. Catalog.

36. Karady, G., and Lamontagne, G., Electrical and Contamination Performance of Synthetic Insulators for 735 kV Transmission Lines, IEEE paper A76 502-5, presented at IEEE PES Summer Meeting, Portland, OR, July 18–23, 1976.

37. Broschat, M., Transmission Line Uprating 115 kV to 230 kV, Report on Operating Performance, *IEEE Trans.*, March/April 1972, pp. 545–548.

38. Update Line to 345 kV on same ROW, *Electr. World*, Nov. 15, 1973, pp. 66–67.

39. *EHV Transmission Line Reference Book*, EEI, 1968.

40. IEEE Working Group on Insulator Contamination, Application Guide for Insulators in a Contaminated Environment, *IEEE Trans. Power Appar. Syst.*, Sept./Oct. 1979, pp. 1676–1690.

41. *Transmission Line Reference Book, 115-138 kV Compact Line Design*, EPRI Publication, 1978, Palo Alto, CA.

42. Moran, J. H., The Effect of Cold Switch-on on Semi-conducting Glazed Insulators, IEEE paper C74 071-7, presented at IEEE Power Meeting, New York, Jan. 1974.

43. Moran, J. H., and Powell, D. G., A Possible Solution to the Insulator Contamination Problem, IEEE paper 71CP41, presented at IEEE Power Meeting, New York, Jan. 1971.

44. Falter, S. L., and Powell, D. G., Radio Influence Voltage Characteristics of Transmission Line Assemblies, Using Semi-conducting Glazed Insulators, IEEE paper C73 416-5 presented at IEEE Summer Power Meeting, 1973, Vancouver, B.C.

45. Fukui, H., Naito, K., Irie, T., and Komoto, I., A Practical Study on Application of Semi-conducting Glaze Insulators to Transmission Line, IEEE paper T74 073-3, presented at IEEE Winter Power Meeting, New York, Jan. 1974.

46. Nigol, O., Reichman, J., and Rosenblatt, G., Development of New Semi-conductive Glaze Insulators, Paper T73 420-7, *IEEE Trans. Power Appar. Syst.*, March/April 1974, vol. PAS-93, pp. 614–622.

47. Bundled Circuit Design for 115-138 kV Compact Transmission Lines, EPRI Report EL 1314 (2 vols.), Feb. 1980.

48. Phase to Phase Switching Surge Design, EPRI Report EL 1550, Sept. 1980.

49. Souchereau et al., Validation of a Chainette Tower for a 735 kV Line, CIGRE paper 22-04, 1978.

50. Converti, V., Hyland, E. J., and Tickle, D. E., Optimized Transmission Tower Spotting on Digital Computer, *AIEE CP60-1201*, Oct. 1960.

51. Peyrot, A. H., Peyrot, E. M., and Carton, T., Interaction and Integration in Power Line Design, *IEEE Computer Appl. Power*, 1992, vol. 5, no. 4.

52. Carton, T., and Peyrot, A. H., Computer-Aided Structural and Geometric Design of Power Lines, *IEEE Trans. Power Syst.*, Feb. 1992, vol. 7, no. 1, pp. 438–443.

53. Ehrenburg, D. O., Transmission Line Catenary Calculations, AIEE paper, Committee on Power Transmission & Distribution, July 1935.

54. Winkelman, P. F., Sag-Tension Computations and Field Measurements of Bonneville Power Administration, AIEE paper 59-900, June 1959.

55. *National Electrical Safety Code*, ANSI C2-2012, Published by IEEE Standards Association, New York, NY, 2012.

56. SEI/ASCE Standard 7-05, *Minimum Design Loads for Buildings and Other Structures*, American Society of Civil Engineering, New York, NY, 2005.

57. ASCE Manual 74: *Guidelines for Electrical Transmission Line Structural Loading*, 3rd ed., American Society of Civil Engineers, New York, 2010.

58. *Aluminum Association, Stress-Strain-Creep Curves for Aluminum Overhead Electrical Conductors*, publication SSCC-723613, The Aluminum Association., Inc., Waldorf, MD, 1997.

59. Greisser, V. H., Effects of Ice Loading on Transmission Lines, *Trans. AIEE*, 1913, vol. 32, p. 1829.

60. Healy, E. S., and Wright, J. A., Unbalanced Conductor Tensions, *Trans. AIEE*, 1926, p. 1064.

61. Poffenberger, J. C., and Swart, R. L., Differential Displacement and Dynamic Conductor Strain, *IEEE Trans.*, vol. PAS-84, 1965, pp. 281–289.

62. *Transmission Line Reference Book—Wind-Induced Conductor Motion*, Electric Power Research Institute, Palo Alto, CA, 1979, chap. 3.4.

63. Ibid., chap. 3.5.

64. Nigol, O., and Havard, D. G., Control of Torsionally-Induced Conductor Galloping with Detuning Pendulums, IEEE paper A78 125-7, Jan. 1978.

65. Richardson, A. S., Design and Performance of an Aerodynamic Anti-Galloping Device, IEEE Conf. paper C68 670-PWR, June 1968.

66. Douglass, D. A., and Roche, J. B., Anti-Galloping Potential of a New Twisted Conductor Design, *Proc. Canadian Electrical Association Int. Symp. Overhead Conductor Dynamics*, June 1981, pp. 83–98.

67. *Transmission Line Reference Book—Wind-Induced Conductor Motion*, chap. 5.

68. Sellers, A. H., and Williams, J. E., All-Aluminum Transmission Tower Line, *Trans. AIEE*, June 1961, p. 169.

69. Thomas, P. H., Formula for Minimum Horizontal Spacing, *Trans. AIEE*, 1928, vol. 47, p. 1323.

70. Farr, F. W., Ferguson, C. M., McMurtrie, N. J., Steiner, J. R., White, H. B., and Zobel, E. S., A Guide to Transmission Structure Design Loadings, *Trans. IEEE Power Group*, Nov. 1964, p. 1073.

71. Farmer, G. E., The Use of Insulated Ground Wires on a Transmission Line for Communication Purposes, *IEEE Trans. Power Appar. Syst.*, Dec. 1963, p. 884.
72. Ramthun, M. K., Pitzel, B. H., and Campbell, D. W., Stream-Lined 230-kV Transmission Passes Overhead in City's Streets, *Electr. World*, June 29, 1964, p. 94.
73. Stumpf, M. W., and Mouton, R. A., 12-Sided Single Poles Carry 760 MVA Capacity Line (New Orleans, LA), *Electr. World*, Nov. 16, 1964, p. 94.
74. ASCE Manual 52, *Guide for the Design of Steel Transmission Towers*, 2d ed., American Society of Civil Engineers, New York, 1988.
75. White, H. B., Design of Chute des Passes 345-kV Transmission Line, AIEE paper 60-72 WGM, 1960.
76. White, H. B., Cross Suspension System, Kemano Kitimat Transmission Line, AIEE paper CP 58-432 WGM, 1958.
77. Ritky, F., and White, H. B., Unique Suspension System Conquers Rugged Terrain, *T & D World*, Aug. 1997, vol. 49, no. 8.
78. White, H. B., Structural System for the James Bay Transmission Lines: Hydro Quebec Symposium on EHV and UHV Alternating Current, IREQ, 1973.
79. Lecomte, D., and Meyere, P., Evolution of the Design for the 735-kV Transmission Lines of Hydro-Quebec, CIGRE paper 22-08, 1980.
80. Cornell University, *Transmission Line Structure Foundations for Uplift/Compression Loadings*, Electric Power Research Institute Report EL-2870, Palo Alto, CA, Feb. 1983.
81. Cornell University, *Critical Evaluation of Design Methods for Foundations under Axial Uplift and Compression Loads*, Electric Power Research Institute, Report EL-3771, Palo Alto, CA, Nov. 1984.
82. GAI Consultants, Inc., *Laterally Loaded Drilled Pier Research*, vols. 1 and 2, Electric Power Research Institute, Report EL-2197, Palo Alto, CA, Jan. 1982.
83. GAI Consultants, Inc., *Direct Embedment Foundation Research*, Electric Power Research Institute, Report EL-6309, Palo Alto, CA, April 1989.
84. GAI Consultants, Inc., *Direct Embedment Foundation Research, Load Test Summaries*, Electric Power Research Institute, Report EL-6849, Palo Alto, CA, June 1990.
85. Goldberg, D. T., Jaworski, W. E., and Gordon, M. D., *Lateral Support System and Underpinning*, vol. I, *Design and Construction*, prepared for Federal Highway Administration, U.S. Department of Commerce Publication PB-257 210, April 1, 1976.
86. Post-Tensioning Institute., *Post-Tensioning Manual*, 5th ed., Phoenix, AZ, 1990.
87. Rural Utilities Services Bulletin 1724E-203, Guide for Upgrading RUS Transmission Lines, 1994.
88. Southwire Company, Overhead Conductor Manual, 1994.


## 5.2 UNDERGROUND POWER TRANSMISSION

*BY EARLE C. (RUSTY) BASCOM III*

### 5.2.1 Cable Applications

Traditionally, underground cable systems have been installed in major urban areas where overhead lines are not practical, there is congested access to existing utility substations, locations such as airport approaches because of safety issues, water crossings where overhead lines are not feasible, and for aesthetics. Material and installation costs for cables are much more costly than overhead lines, although rights-of-way and permitting costs often make the underground cables competitive alternatives to overhead lines. Environmental permitting can be a long process for new transmission projects, and underground alternatives often see fewer challenges although construction of underground along existing public utility corridors—mainly road ways—can be disruptive and require careful traffic management plans. "Hybrid" lines using segments of overhead and underground have become common for new projects, but matching power transfer capacity of cables to lines is a challenge.[1]

The power cable industry evolved considerably early in the 21st century with oil-impregnated paper insulation (pipe-type cable in North America and self-contained fluid filled cable worldwide) being displaced by extruded-dielectric insulated cables—principally cross-linked polyethylene (XLPE)—that are now commercially used up to 500 kV. New high-pressure pipe-type (HPPT) cables are rarely installed now due to concerns about the dielectric liquids, higher maintenance, and only one available supplier, but there are a few thousand circuit miles that continue to operate; retrofitting cable pipes with extruded cable is of interest. As with pipe-type, new self-contained liquid-filled (SCLF) cable installations are now rare and manufacturing of these types has diminished significantly. Mass-impregnated nondraining (MIND) paper cables are used for most long high voltage dc (HVDC) circuits, but polymer-insulated extruded cables are now becoming more common from several manufacturers for both land and underwater applications. Compressed-gas insulated transmission systems are now uncommon for new installations and only used for special applications—typically short lengths and high-power transfers within substations.

### 5.2.2 Cable System Considerations and Types

***Cable Integration into Utility System.*** Planning and operating considerations for underground cables are different from those for overhead lines.[2] Special attention should be paid to properly representing the cables for utility system analyses, capacitance effects including charging currents, reactor application, system restoration, inductance effects including load sharing, surge impedance loading, insulation coordination, system insulation requirements, and losses. As compared to overhead lines, positive sequence impedances of cables are much lower so modeling of cables and lines in power systems requires appropriate selection of electrical parameters. Cables are largely capacitive as compared to overhead lines and power transformers that are inductive; these characteristics affect reactive compensation and switching studies. For comparative purposes, Table 5-24 lists several characteristics of typical overhead lines and underground cables for 345 kV transmission.

Overhead lines may be limited by thermal (sag) limits, stability limits or voltage limits. Because of their very high surge impedance loading limit and charging current, ac cables are always thermally limited due to capacitive charging current. Cables have lower ratings than overhead lines so matching underground capacity to overhead often requires more than one cable per phase. Since cable ratings and cable circuit impedance are almost always lower, load flow studies should be carefully evaluated when integrating long cable lines in parallel or networked with overhead lines.

**TABLE 5-24** Typical Electrical Characteristics, 345-kV Overhead Line and Underground Cables

| Parameter | Overhead line | Underground XLPE | Underground HPFF (PPP) |
|---|---|---|---|
| Shunt capacitance, μF/mi | 0.020 | 0.28 | 0.48 |
| Series inductance, mH/mi | 1.5 | 0.94 | 0.61 |
| Series Reactance, Ω/mi | 0.57 | 0.36 | 0.28 |
| Charging current, A/mi | 1.5 | 21.4 | 37.8 |
| Dielectric loss, kW/mi | 0+ | 4.3 | 5.5 |
| Reactive power, MVA/mi | 0.9 | 12.8 | 23.7 |
| Surge impedance, Ω | 285 | 28.6 | 18.4 |
| Surge impedance loading limit, MW | 418 | 4166 | 7129 |

### 5.2.3 Extruded-Dielectric Systems

Extruded-dielectric (XD) systems sometimes called solid-dielectric cables, provide a simpler, often lower-cost alternative to the paper-insulated cables that have historically been used for transmission cables. Present XD cable insulation technology uses cross-linked polyethylene (XLPE) and ethylene-propylene-rubber (EPR); EPR-insulated cables are used up to 138 kV, and generally with lower rating requirements, due to dielectric loss and charging current limits and constraints on available maximum conductor sizes. XLPE-insulated cable is the most common and widely available insulation type used for transmission cables and is commercially available up to 500 kV. The absence of dielectric fluid greatly simplifies the ancillary equipment and accessory complexity, and removes concerns about fluid leaks. XD cables also have lower capacitance than paper-insulated cables, simplifying their integration into the utility system. Although the first XD cables were installed in the United States in the 1960s, they did not find extensive use until the mid-1980s. XD cables are installed in conduits with splice vaults (manholes) or directly buried; at transmission voltages, separate vaults are typically used for each three-phase circuit to allow safe maintenance and repair while other cables in the same trench remain energized.

Extruded-dielectric cables consist of a copper or aluminum conductor, conductor semiconducting shield to ensure a smooth electrical profile in the insulation, an extruded polymeric insulation, an extruded insulation shield, and typically a metallic radial moisture barrier consisting of copper or aluminum foil laminate, corrugated copper, aluminum or stainless steel, or extruded lead, possibly with concentric shield wires, and a polymeric (usually polyethylene) jacket (Fig. 5-53). Copper or aluminum foil with additional shield wires has become common; lead is now seldom used except on submarine projects due to that metal's increasing cost and perception of environmental impact.

Extruded cable is sensitive to manufacturing processes, so exacting manufacturing processes and quality assurance programs are used throughout including steps to minimize contaminants in the handling and application of the insulation materials and semiconducting shields (sometimes called

**FIGURE 5-53** Example XLPE transmission cable with segmented copper conductor with copper foil laminate sheath and copper concentric shield wires.

"screens") between the insulation and other cable layers, as well as factory acceptance and field-commissioning tests. Early use of XLPE insulation showed sensitivity to moisture so modern manufacturing processes use dry curing and cooling of the XLPE insulation with cable designs that use radial moisture barriers to prevent moisture ingress into the insulation and water-swellable tapes and compounds to limit affected cable that has sustained external mechanical damage.

Sheath bonding is an important design consideration and requires more careful consideration than at distribution voltage levels. If both ends of the sheath are grounded as is common with distribution, conductor current induces currents in the sheaths, causing heat losses that de-rate the cable by 10% to 35%. Alternative sheath bonding modes are used to manage these losses. One method is to ground the metallic sheath/screen at only one end of a section (i.e., "single point bonding") while leaving the other end ungrounded; the elimination of the continuous electrical path avoids circulating currents and the 10% to 35% rating loss but requires consideration for controlling the resulting induced voltage (approximately 165 V/1000 A of phase current/kilometer of single-point bonded length). Another method of bonding involves transposing sheath connections (i.e., "cross-bonding"), usually of three nearly equal-length installation sections, that cancels the circulating current and also reduces the induced voltage.[3] Single point bonding and cross bonding connections are done through link boxes at joints and near terminations. Figure 5-54 shows sheath cross-bonding of a single power cable circuit.

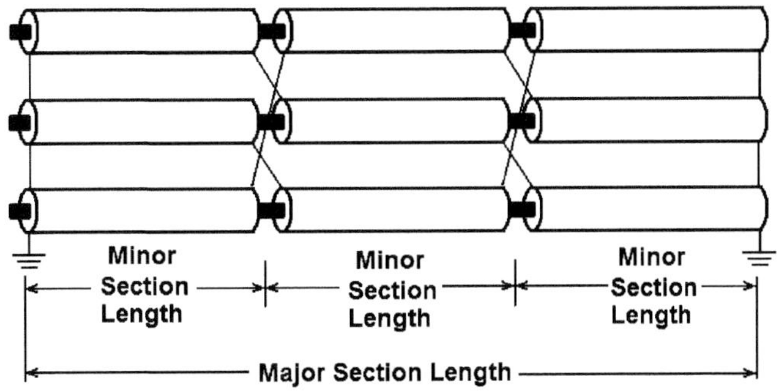


**FIGURE 5-54** Conceptual configuration of sheath cross bonding.

Extruded-dielectric transmission cables are typically installed individually, either directly buried or in duct. In some cases, the three individual phases are installed at one time in a plastic or steel pipe as part of a "trenchless" installation, in a manner resembling a pipe-type cable installation, and retrofitting of pipe-type cable pipes with extruded cables increasingly is being considered. Section lengths range from 300 to 1000 m, depending upon the installation mode, reel shipping lengths, and allowable sheath voltages, but lengths up to 7500 ft for transmission cables have been installed.

Modern cable joints are constructed using pre-molded splices (Fig. 5.55). As cable conductor sizes and higher-voltage cables have been used, methods to constrain and manage thermal mechanical movement are evaluated in detail for large XD cables in conduits and splice vaults.

Terminations have either porcelain or polymer housings, and some designs have a small amount of dielectric liquid (e.g., silicone oil) or gas ($SF_6$) in the termination.

### 5.2.4 High-Pressure Fluid-Filled Systems

Pipe-type cable had been used extensively in the United States since 1932 and in a few other areas around the world until the 1990s using paper-insulated technology that dates to the late 1800s. As utilities increasingly moved towards XLPE cable systems, utility use of pipe-type cable has dropped significantly, and manufacturing capabilities have diminished as a result. As of the writing of this handbook, only one manufacturer of pipe-type cable remains worldwide and has considered cutting back on capacity.

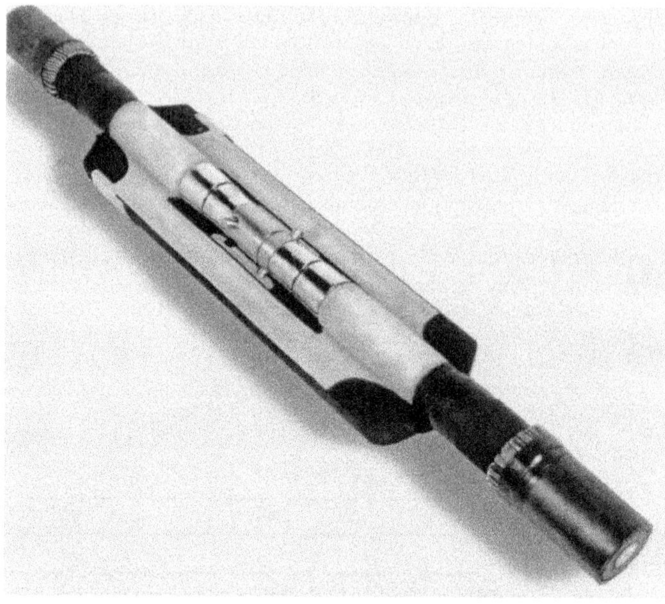

**FIGURE 5-55** "Click-fit" premolded joint. (*Courtesy of Prysmian.*)

**FIGURE 5-56** Pipe-type cable.

Pipe-type cable consists of a welded, coated, cathodically protected steel pipe, typically 6.625 to 10.75 in outer diameter in which mass-impregnated cables are pulled (Fig. 5-56). The cables themselves consist of a copper (occasionally aluminum) conductor, conductor shield, taped insulation (kraft paper or laminated paper-polypropylene), insulation shield, outer shielding, and skid wire to prevent cable damage as it is pulled into the pipe. Pipe cable installation sections between splices are typically 1500 to 3000 ft (500 to 1000 m) but have been up to 8000 ft for directionally drilled installations, and terminations are used for transitions in substations and risers. The pipe is pressurized to a nominal pressure of 200 psig (1.4 MPa) with either a dielectric liquid using a shipping-contained size pumping plant or dry nitrogen gas using a small (about the size of a telephone booth) gas cabinet consisting of a couple of nitrogen gas bottles and pressure regulator. Cathodic protection is used on the cable pipe as additional protection for the pipe corrosion coating.

### 5.2.5 Self-Contained Liquid-Filled (SCLF) Systems

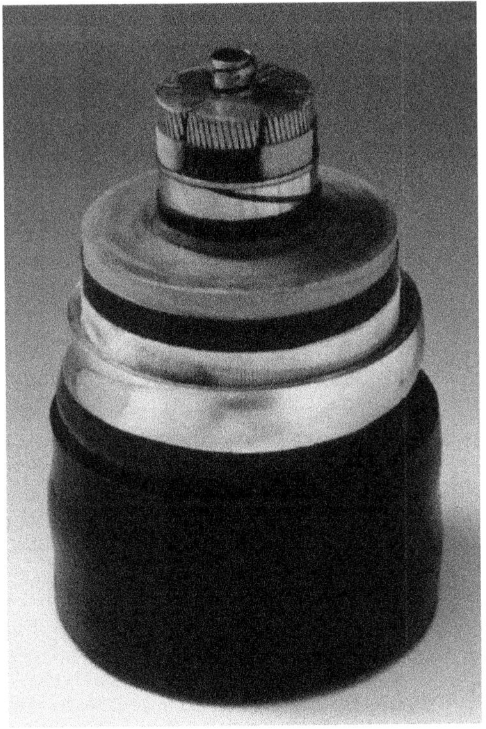

FIGURE 5-57 Self-contained cable. (*Courtesy of Prysmian.*)

This system type was developed in the 1920s using similar insulation technology as pipe-type cable, and was the worldwide standard outside the United States through the 1980s. There are many miles in operation at 345 to 525 kV, for both land and submarine installations, although new installations are uncommon as XLPE cable is now used more extensively. The self-contained cables (Fig. 5-57) have hollow core (providing a longitudinal feeding path for the dielectric liquid) conductors of copper or aluminum, taped kraft paper or laminated paper polypropylene insulation, taped semiconductive layers, and a sheath of reinforced lead, corrugated copper or aluminum for a moisture barrier, to carry fault current and to contain positive internal pressure, and an insulated plastic jacket. Submarine cables may have additional concentric shield wires, galvanized steel, copper or aluminum armor wires, and polypropylene yarn for bedding.

### 5.2.6 Direct Current Cables

DC cables are most often used for long water crossings and their shore ends, though there are numerous land installations with additional projects planned or under construction. Charging current and dielectric losses make ac cables unsuitable for lines more than 30 to 50 mi long (50 to 80 km)—depending upon voltage level and insulation type, but dc cables avoid these losses while reducing the number of cables (3 to 2) for each circuit, often with a much higher transmission capacity. The required dc to ac convertor stations limit widespread use of HVDC. HVDC cables were traditionally made with mass-impregnated paper insulation but polymeric HVDC cables are now common. For both insulation types, the conductors are usually concentric stranded, and operating temperature is limited to 50 to 70°C because the insulation is resistively graded rather than capacitively graded like in ac cables. HVDC cables have similar features as ac cables (insulation thickness related to voltage, radial moisture barrier, and insulating jacket).

### 5.2.7 Other Special Cables

Other types of cables used for power transmission are available although used for very specialized applications and with limited use. Compressed gas-insulated transmission lines (GITL) consist of concentric tubular aluminum bus—an outer shield and inner conductor—with sulfur hexafluoride ($SF_6$) gas between for electrical insulation, generally, for short sections within substations and can provide very high capacity. The aluminum bus is not buried because of corrosion concerns.

Superconducting cables are also an emerging technology using specialized conductor that is typically cooled by liquid nitrogen (i.e., 77 K) and has the benefit of no resistance (although reactance is still present in ac systems). Cold and warm dielectric options have been under development, although the present cryogenic cooling plant technology typically limits circuit lengths to a few thousand feet so superconducting cables are currently used only for short installations.

### 5.2.8 Cable Capacity Ratings

Cable capacity ratings—often called "ampacity"—is the current-carrying capacity of the circuit, although cable ratings in general may be listed in megavolt-amperes (MVA). Normal or continuous ratings are those that may be applied to the cable circuit for an extended or indefinite period of time without degrading the life of the cable system. Transient or emergency ratings are permissible temporary, typically 15 min to 300 h, applications of load that will result in temperature above normal operating temperatures and may degrade life. Ratings may be developed and cataloged ("book" ratings) or can be determined based upon measurements of various parameters in real time ("dynamic" or "real-time" ratings).

Rating calculations for insulated power cables are most often evaluated for buried cables, and the focus of most rating standards are based on various buried installation configurations, though insulated power cables may also be installed in air or placed at the bottom of a body of water; buried cables are almost always the most limiting configuration. Cable ratings for ac cables are most often placed on a conductor temperature limit associated with limiting damage to the cable insulation immediately adjacent to the conductor, while dc cables may be limited by an outer insulation temperature. The conductor temperature is determined by heat generated within the cable—electrical losses ($I^2R$) and dielectric losses (ac cables only)—which passes in a radial direction to ambient earth through various thermal resistance layers in the cable, duct (when present), trench and surrounding soil and caused by mutual heating effects from other nearby cables. The temperature of the conductor can be determined from some form of the following thermal equivalent circuit (see Fig. 5-58 for XD cables).

For ampacity calculations, current-dependent heat losses vary with respect to time, usually modeled on a daily (24-h) load cycle, while dielectric losses in ac cables are present as long as the cable is energized. Rating procedures include populating the heat sources and thermal resistances shown in Fig. 5-58, paying attention to issues such as insulation type and material characteristics. A unique characteristic of buried power cables is the relatively long thermal time constant (30 to 150 h, depending on system type) compared to all other power cable equipment; this means that although cables may be limiting during normal ratings, they could have a much higher emergency rating capability. Pipe-type cables may have higher ratings by virtue of removing the dielectric liquid from the pipe, cooling it through some form of heat exchanger, and then reintroducing it to the pipe while also circulating the liquid; some utilities implement this operating method. A description of various calculation procedures for the multitude of configuration and operating modes is beyond the scope of this reference. The reader is referred to Refs. 4 and 5 for additional information, with support from other Refs. 6 to 10).

An important consideration for rating of buried cables is the environment in which the cables are installed. The ability for heat to move through the soil is characterized by the soil's thermal resistivity that typically varies from 0.5 to 4.0 K · m/W and is dependent on moisture content, material, mixture,

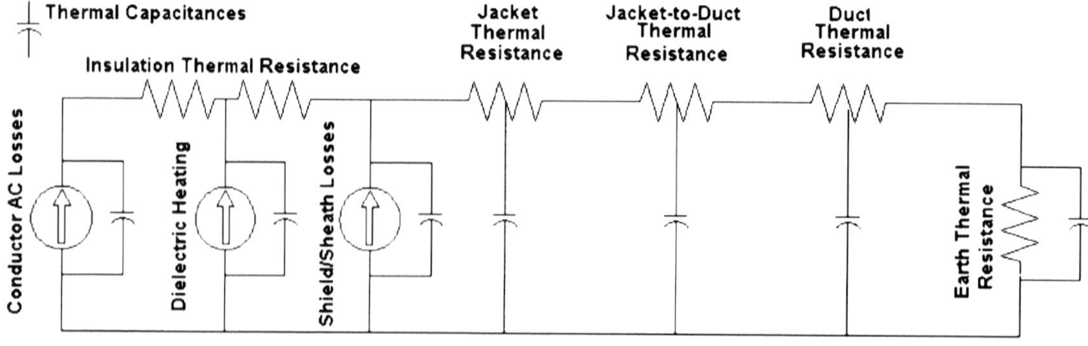


**FIGURE 5-58** Equivalent thermal circuit for XD cable installed in conduit.

size of particles, and density. Cable design includes carefully selecting the materials to be put back in the cable trench—called "thermal backfill"—and can consist of granular materials (e.g., "thermal sand") or flowable fill materials (e.g., "low thermal-resistivity concrete" or "fluidized thermal backfill"). The earth thermal resistance accounts for up to 75% of the total thermal resistance from conductor to ambient, so knowledge of the native soil characteristics and selection of the thermal backfill are important design parameters. Methods for measuring and intepretting native soil and special thermal backfill materials are well documented.[11]

### 5.2.9 Cable Uprating and Dynamic Ratings

Cable uprating and upgrading are techniques that are applied to existing cable circuits to get more power through existing infrastructure without building new lines. Many of the basic strategies used for uprating are important to consider for basic ampacity when a new circuit is installed but are also evaluated after a circuit is already in service, often in an effort to defer multi-million dollar new installation projects for years. In addition, rights-of-way are sometimes difficult to obtain even for underground circuits so getting more capacity through existing lines becomes critically important. Underground cables are especially amenable to uprating, principally because of the very large thermal time constant of the cable/earth system. Also, many older circuits were designed and rated using conservative assumptions; these assumptions may be reviewed carefully, often revealing additional circuit capacity.

Utilities have successfully implemented many approaches for uprating.[12] Typical uprating approaches are summarized below:

- *Characterize the thermal circuit:* In most cases, design engineers properly make conservative assumptions when rating the circuit. A careful audit of route as-built drawings, route thermal survey of soil and backfill thermal resistivity, load shape, etc. will provide accurate data and can often permit re-rating the circuit 5% to 10% higher.
- *Perform loading analysis:* Detailed analysis of loading for the preceding weeks, cable construction data, and trench cross-section information can permit refining the ampacity model, better selection of daily load/loss factors, often resulting in significant ampacity increases. Conservative assumptions about load/loss factors can also be refined.
- *Continuous temperature monitoring:* Some utilities have thermocouples or installed fibers that may be used to check for hot spots or have continuous monitoring through substation-installed remote terminal units[13], or distributed temperature sensing (DTS) equipment can be used to identify hot spots and infer effective soil thermal resistivity along the cable route. Continuous temperature monitoring with DTS equipment provides comprehensive coverage for the entire circuit and can be used during periods of high loading to better evaluate the assumptions used for ampacity calculations. Some utilities combine DTS with real-time ratings. A DTS system for continuous temperature monitoring is shown in Fig. 5-59.
- *Mitigate hot spots:* Localized hot spots—a few meters of cable—along a cable route may limit the overall capacity of the circuit. These locations may result from excess overburden, poor thermal resistivity soils or trench backfills, poor soil compaction, cable congestion (many circuits in one area causing mutual heating), etc. These areas may be mitigated, or oversized cable could be installed just for the affected area. Pipe cables can implement fluid circulation or oscillation to mitigate localized hot spots.
- *Reconductor:* Technically called "upgrading," substantial ampacity increases can be achieved for pipe-type cables or XD/SCFF cables in duct by removing the present cable and installing a larger-conductor cable, conductor with lower losses or a lower loss insulation (XLPE instead of EPR, PPP instead of kraft paper). Voltage upgrades might also be used providing a power increase proportional to the new voltage increase.
- *Add forced cooling:* For high-pressure fluid-filled (HPFF) cables, it is possible to remove the liquid from the cable pipe, pass it through a heat exchanger, and send it back into the pipe through a separate liquid line. Although installation and operating costs are high, forced cooling can add 40%

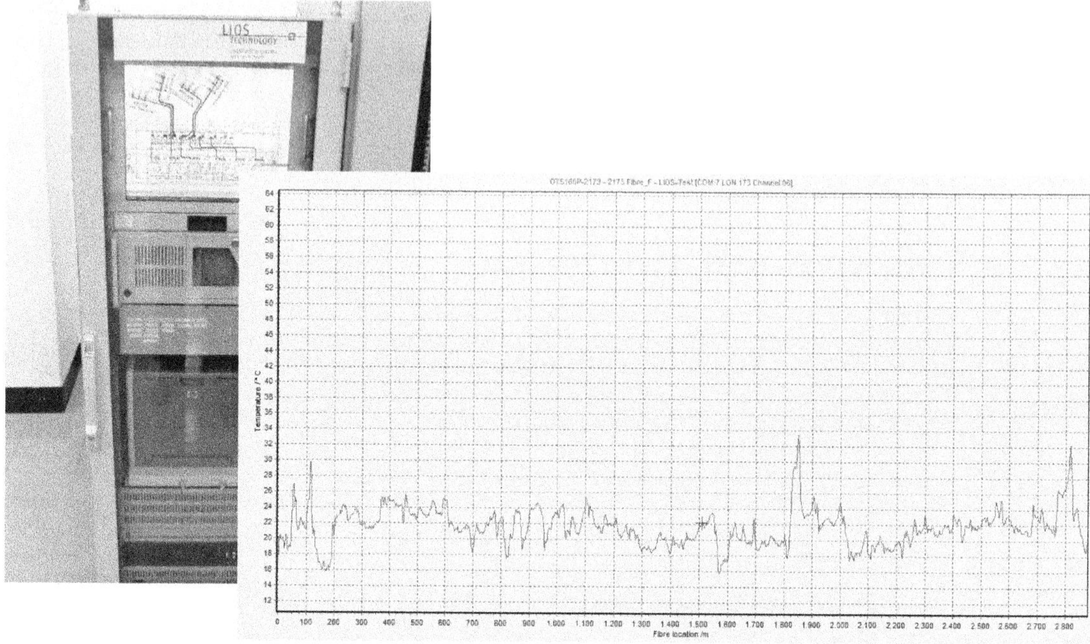


**FIGURE 5-59** Permanently installed DTS system located in a utility substation control building. (*Photo courtesy of LIOS Technology Inc.*)

to the capacity of an existing cable system. Forced cooling could also be used with XD/SCFF cables using parallel water pipes, but this is uncommon.

- *Dynamic rating:* Many cables operate below their thermal limits much of the time, and their conductor temperatures are much lower than allowable values. Therefore, the utility is not taking full advantage of the available capacity because the cable could carry more current without exceeding allowable temperatures. In many cases, the cables can even operate above traditional maximum temperatures without unduly decreasing cable life. Dynamic rating systems—essentially a computer monitoring parameters such as load and ambient temperature in real time and reporting rating updates regularly for use by operators—make this uprating methodology possible and generally economical.

### 5.2.10 Soil Thermal Properties and Controlled Backfill

The earth thermal resistance portion of the cable equivalent thermal circuit (see Fig. 5.58) accounts for the greatest percentage of the total thermal resistance for buried cables—often more than half the total. Native soil thermal resistivity can vary by an order of magnitude along a cable route, and it can vary by a factor of two to three at one location as a function of seasonal moisture content. Accurately characterizing thermal resistivity for soil samples, developing thermal dry-out curves, and choosing the proper value to use in ampacity calculations is an important part of cable system design.

Soil testing is often done in the field (*in situ*) or laboratory (using samples) to determine the thermal characteristics of the soil in which the cables will be installed. A thermal property analyzer (TPA) is used for these types of measurements (Fig. 5-60). For a thermal resistance test, a thermal needle consisting of a heater and thermistor are placed into the soil sample (Fig. 5-61). During the test, a constant heat output is applied to the thermal needle while the change in temperature is recorded; the slope of the recorded time-temperature curve is proportional to the soil thermal resistivity. This information is then used for ampacity calculations and to design the backfill materials to place in the cable trench.

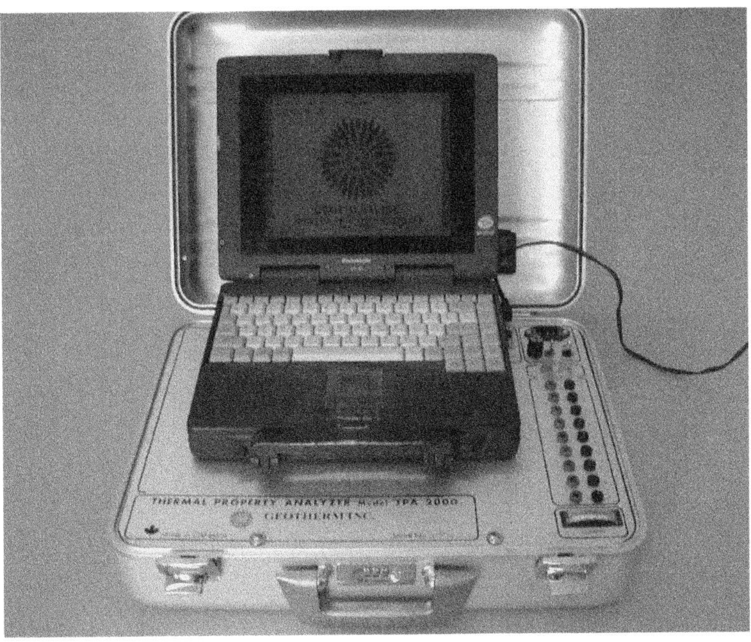

**FIGURE 5-60** Thermal property analyzer or "TPA." (*Courtesy of Geotherm, Inc.*)

In the laboratory, thermal dry-out curves may also be prepared that show the effect of soil moisture content on thermal resistivity. Generally, as soils have increased moisture content, the thermal resistivity decreases (Fig. 5-62). Since cables produce heat, some drying should be considered when evaluating ampacities.

Significant ampacity increases can be achieved at reasonable cost by placing a controlled backfill with a low, stable thermal resistivity around the cables, ducts, or pipes. A good controlled backfill is characterized by low thermal resistivity, less than 1 K · m/W, when completely dry, and a fairly flat curve of thermal resistivity versus moisture content. The preferred backfill material should be designed for each project using locally available materials, when possible, to provide the best installation and thermal characteristics at a reasonable cost. Well-graded sands or limestone screenings have been used since the 1950s, but many utilities have been installing Fluidized Thermal Backfill (FTB™), which is a low thermal resistivity, free-flowing engineered material consisting of natural mineral aggregates, sands, cement, water, and a fluidizer (Fig. 5-63). It is delivered in ready-mix concrete trucks, and flows throughout the trench so there is no need for compaction.

**FIGURE 5-61** Thermal probe being inserted into an auger during geotechnical testing.

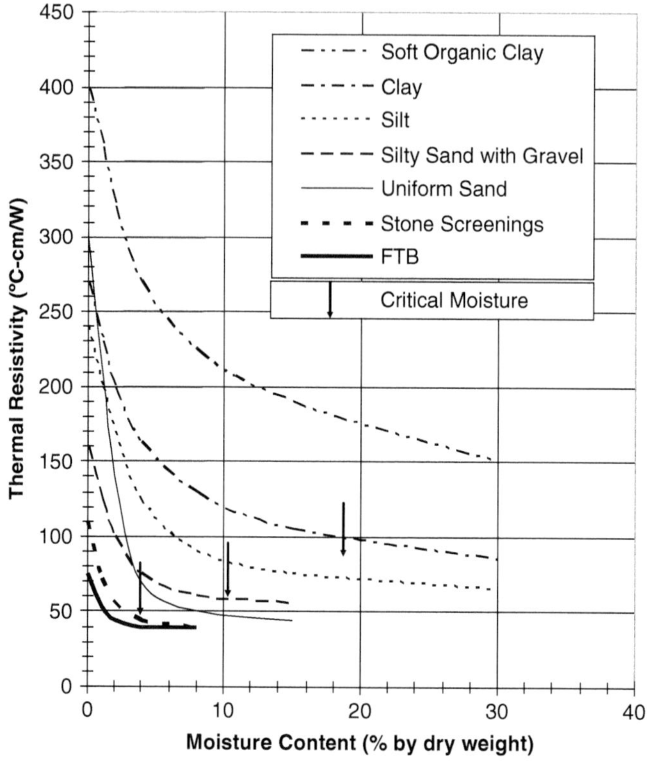


**FIGURE 5-62** Soil thermal dry-out curves.

**FIGURE 5-63** FTB being installed in a cable trench.

Segments of cable systems are often installed with trenchless methods such as pipe jacking or horizontal directional drilling. Some of the trenchless installations use a steel, concrete, or polymer casing in which conduits or pipes are installed. The annular space between the inner conduits or pipes and the casing can be filled with a low-thermal resistivity grout to improve ampacity similar to using special backfill with a normal open-trench installation. Factors such as grout pumping pressures and mix design components for thermal performance are balanced to meet the requirements of a given project.

### 5.2.11 Electrical Characteristics

***Calculation of Electrical Parameters.*** Electrical characteristics for cables are calculated using various established methods. Resistance, capacitance, and inductance can be calculated using the industry standard procedures for cable ampacity.[7] Sequence impedances for cables can be approximated using various sources.[14,15] Rigorous calculations for pipe cables are difficult because of the presence of the steel (ferromagnetic) pipe; Ref. 16 has traditionally been considered the best approach available for pipe-type, though it is known to be in error in many cases because of the great variability in line pipe magnetic permeability. For typical single-core (XD, SCFF) transmission cable installations, a detailed procedure exists[17] but does not easily lend itself to hand calculations. Induced sheath voltages may be calculated using the appropriate IEEE standard.[4] Induced voltages are an important consideration for single-point bonded and cross-bonded cable systems and should be evaluated when a circuit is out of service for maintenance and a parallel cable circuit will remain energized and carrying load. Single-point bonded and cross-bonded cable systems should have the appropriate link boxes and sheath voltage limiters to manage impulse voltages that can occur during faults. Ground continuity conductors are used with single-point bonded systems and should be properly sized for expected fault current levels and clearing times.

Although surge impedance loading is never an issue for cables (unlike overhead lines) because cables ratings are always constrained by thermal limits, system studies sometimes require the surge impedance for modeling power systems. This can be calculated using Eq. (5-76), where $f$ is the power frequency, $\varepsilon$ is the dielectric constant of the insulation, $D_i$ is the diameter of the cable over the insulation, and $D_c$ is the diameter of the cable conductor.

$$Z_S = \frac{f}{\sqrt{\varepsilon}} \ln\left(\frac{D_i}{D_c}\right) \tag{5-76}$$

Fault current capability of a cable can be an issue, particularly in dense power systems. The fault current capability of underground cables can be calculated assuming adiabatic conditions using Eq. (5-77), where conductor area is in square millimeters, and time is in seconds (i.e., at 60 Hz, 30 cycle clearing time will be 0.5 s). $T_1$ is the temperature of the sheath or conductor prior to the fault in °C, $T$ and $k$ are material parameters for conductor or sheath material (Table 5-25), and $T_2$ is the allowable maximum temperature for the insulation or jacket material (whichever is lower) in contact

**TABLE 5-25** Constants to Calculate Short Circuit Current Capability

| Conductor/sheath material | Constant, k J/Ω · m | Inferred temperature of zero resistance, $T$ °C |
|---|---|---|
| Copper | 50237 | −234.5 |
| Aluminum | 21144 | −228.0 |
| Lead | 1641 | −236.5 |

with the conductor or sheath. Conductor cross-sectional area is specified in square millimeters, and time is specified in seconds.

$$I_{SC} = (\text{Conductor Area})\sqrt{\frac{k}{\text{time}}\ln\left(\frac{T_2 - T}{T_1 - T}\right)} \tag{5-77}$$

Though polyethylene jackets are often used, many manufacturers allow higher shield/sheath temperatures than 150°C (some up to 250°C) based on experimental evaluation and the very conservative adiabatic assumptions (Table 5-26).

**TABLE 5-26** Polymer Temperature Limits for Fault Current Calculations

| Insulation/Jacket Material | Temperature, $T_2$ °C |
|---|---|
| High-density polyethylene (HDPE) | 150 |
| Cross-linked polyethylene (XLPE) | 250 |
| Polyvinyl chloride (PVC) | 150 |
| Ethylene propylene rubber (EPR) | 250 |
| Impregnated paper | 150 |
| Laminated paper-polypropylene | 150 |

## 5.2.12 Magnetic Fields

Magnetic fields are generated by underground cables as a result of current flowing in the conductors of the cables. While magnetic fields of underground cables are generally less than those of comparable overhead lines, the closer proximity at typical installation depths may result in greater field levels directly above the cables. The intensity of magnetic fields is a function of conductor and shield currents, phase spacing, and distance from the source and can be calculated for ac cables using the Biot-Savart law as follows:

$$B = \frac{\mu_0 I}{4\pi}\int \frac{dS \times r}{r^2} = \frac{\mu_0 I}{2\pi r} \tag{5-78}$$

with $\mu_0$ as the permeability of free space ($4\pi \times 10^{-7}$ Wb/m$^2$), $I$ is the conductor current in amperes, and $r$ is the distance from the conductor in meters. Magnetic fields are reported in micro-Tesla (μT) or milli-Gauss (mG); one Tesla is $10^4$ Gauss, so 1 mG is equal to 0.1 μT.

Magnetic fields generated by three-phase cable circuits must account for the individual cable phase current magnitude and phase angle and that the magnetic field is actually the resultant of the minor and major (real and imaginary) axes of the rotating elliptical magnetic field phasor. This can be done easily with the use of a computer for cables that do not have ferromagnetic elements such as the steel armor on submarine cables, the metal casing often used with directionally drilled cables, or the steel pipe around pipe-type cables. Cables systems with ferromagnetic components should be modeled using finite element software to account for the nonlinear nature of the problems and the variability in the permeability of steel with field intensity. CIGRE has developed procedures to calculate magnetic fields for cables with[18] (HPFF) and without[19] (XD, SCFF) ferromagnetic components; these methods use empirical relationships to account for the shielding effects of the iron-based shielding.

The importance of magnetic field management is a topic that varies depending largely on public perception of the issue. Some utilities have investigated various magnetic field management methods to find the best solution for a particular application.[20,21] Mitigation methods include consideration for cable phase arrangements, increased burial depth, placement of ground continuity conductors and shield wires, and possible application of magnetic shielding using steel plates or pipes or exotic high-permeability shielding materials.

### 5.2.13 Installation

***XD and SCLF Cables.*** Installation of single core transmission cables (SCLF and XD) requires that the cables either be directly buried or installed in conduits—although they are sometimes installed in troughs, mounted on the inside of tunnels or laid in a submarine environment. Direct-buried cable systems usually have better ampacity performance and lower installed costs although most systems are installed in conduits to limit the length of open trench at any one time and to avoid the high costs associated with accessing the cables later for repair or replacement. Burial depths depend on local requirements but are typically on the order of 1 m cover over the top cable. Joint bay locations for direct buried cables or manholes for conduit systems are dictated by cable reel lengths (typically limited to 3300 ft or 1000 m for large cables), local restrictions on the permissible length of open trench, maximum allowable pulling tensions or by maximum permitted sheath voltages. Pulling tensions are a consideration for cables installed in conduits or pipes. When cables are to be installed in conduits or pipes, it is important to consider the maximum pulling tension and sidewall bearing pressure forces that may be encountered during the installation.[10] Fluid reservoirs for SCLF cables are spaced according to elevation changes or maximum transient hydraulic pressures during load cycling. Generally, each substation end has a set of reservoirs, although they may be placed in manholes as needed. Reference 22 gives details of a typical SCLF cable installation.

For conduit systems, the ducts are installed first and the cables pulled in and splicing done later (Fig. 5-64). This type of installation has benefits for urban environments where lengths of open trench must be limited and surface restoration must be done promptly, as well as avoiding the need to do future excavations to make a repair. Many utilities install spare conduits so that additional cables or other utilities can be installed along the same conduit run.

Direct buried installations require that the trench be opened for an entire pull section, with the placement of special thermal backfill staged prior to cable placement and after to completely fill the trench. Direct burial is less expensive than duct banks but complicates repair and requires long

**FIGURE 5-64** Duct bank being backfilled.

sections of trench to remain open during installation when electrical and civil activities have to be carefully coordinated.

***HPFF Cables.*** Pipe-type cables require much more complex installation procedures and specialized equipment than extruded-dielectric cables.[23,24] The footprint of the pipe and pipe-type manholes is generally smaller than an equivalent extruded cable circuit, but the pipe sections must be welded together, x-rayed, vacuum and pressure tested, etc. The pipe is backfilled and surface restored before the three cables are brought to the site and pulled in together. Installed sections between splices are typically 500 to 1000 m in length but can be up to 2500 m. Cable is delivered to the site on large sealed steel reels and placed in special trailers (Fig. 5-65 or separate trailers for each reel), lagging is removed from the reels, the three phases are brought together into a single pulling yoke, and the cable is pulled to the adjacent manhole with winch after calculations of tensions are performed. After cable is installed, splices and terminations are built, each requiring 4 to 7 days for a three-phase splice; environmental protection is used during this work.

**FIGURE 5-65** Reel trailer.

Pressurizing plants for high-pressure liquid-filled (HPLF) cables are generally assembled in weatherproof enclosures at the manufacturing facility, shipped to the site and set in place on foundations (which may include a moat), where electrical and hydraulic connections are made. Before they are energized, the cables must be pressurized very slowly to prevent insulation damage and ensure that any minor amounts of gas in the cable are dissolved in the dielectric liquid.

## 5.2.14 Special Considerations

***Submarine Installations and Water Crossings.*** Short water crossings, less than 2.4 km, can be accomplished by horizontal directional drilling in many cases ("plowing in" of cables is seldom done with new installations because difficulties in obtaining permits, etc.), and the cable system can be pulled into the casing pipe or directly into the bore in some instances; cable installations can be on the order of 2400 m.

Traditional cut-and-cover installations are suitable where water bottom conditions and environmental considerations permit. Extruded-dielectric, pipe-type, and self-contained cables can be suitable for these short crossings.

Longer water crossings can be installed by laying the cable on the water bottom or embedding in the water bottom by trenching, jetting, or plowing for mechanical protection. The cable laying ship or barge, with special navigation and positioning equipment, pays off cable from a large turntable and which is placed at the selected position using the appropriate laying equipment (Fig. 5-66). The cable laying ship has the capacity to hold several miles of cable, depending on the size, which minimizes (or fully eliminates) the number of splices that must be installed in the water. Detailed studies and evaluation are required for submarine cables due to the complexity and many factors affecting these types of projects.[25] Both ac and dc submarine cables may be installed using similar methods.

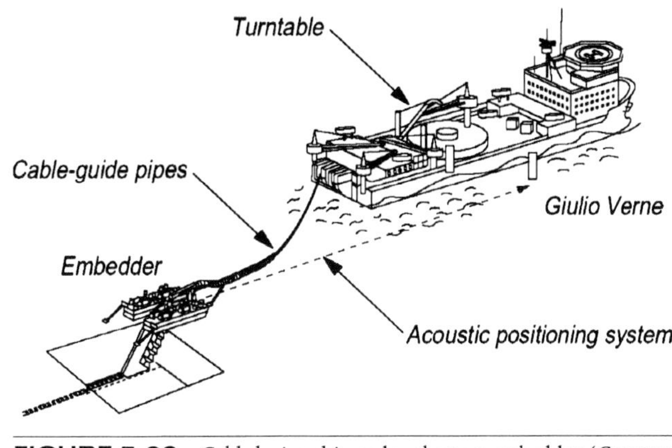


**FIGURE 5-66** Cable laying ship and sea bottom embedder. (*Courtesy of Prysmian.*)

***Horizontal Directional Drilling.*** Horizontal directional drilling (HDD) is an installation method for use under rivers, major highways, and other locations where open-cut trenching is not feasible. Drilling unit capabilities have increased and costs have decreased as the technology has matured, lowering the perceived risk of these types of installations. Installations of on the order of 7500 ft (2.3 km) have been done.[23,26] HDD installations are sometimes done under streets to avoid other utilities, minimize traffic disruption, and minimize the amount of contaminated soil that may be encountered.

Careful design work,[27] including characterization of the materials along the drill path, must be performed so the proper drilling equipment can be selected to ensure a successful installation without loss of drilling fluid to the surface. Detailed thermal analysis is required to properly account for the 10 to 30 m burial depth. The utility must decide whether to install a casing pipe—either plastic or steel—which will require a larger bore and which may decrease ampacity but which will help ensure successful conduit and cable installation. Entry and exit angle requirements of 8° to 25° can require a large setback for deep borings. Figure 5-67 shows a guided boring rig for a river crossing. Staging areas can be large, especially since casing pipe and ducts or cable pipe should be made-up beforehand and installed without stopping to limit the chances that the drilling mud may begin to set up or the borehole collapses.

Since the cable will be inaccessible for repairs, utilities should consider stocking a section of spare cable that is the full length of the bore. For pipe-type cables, or three XD cables installed in a single duct or pipe, the spare cable must be three times the length of the bore. In the case of extruded cables, some utilities have installed a fourth cable in a spare duct, to speed reenergization in the event of cable failure.

**FIGURE 5-67** Guided boring equipment for a river crossing.

In addition to horizontal directional drilling, other "trenchless" methods including pipe-jacking and micro-tunneling are used, usually for shorter crossings up to 400 ft (120 m). With these installation methods, pits are excavated on either side of a crossing and a casing is driven through a borehole that is cut by an auger or laborers using tools. Although changes in elevation can occur, pipe-jacking and micro-tunneling do not provide the ability to steer; these methods often permit a smaller equipment footprint on either side of the crossing and can be done by a wider range of contractors.

## 5.2.15 Accessories

***General.*** Although accessory costs are usually a small percentage of the total project cost, proper selection, and application of accessories are extremely important for long-term reliability of underground cable systems. Historically, outages due to accessories are more frequent than those due to problems in the cable itself, so minimizing splices is a general design preference.

***Extruded Dielectric Cable Accessories.*** Current technology for extruded splices is to use premolded joints, which became common in the early 1990s. In cross-bonded or single-point bonded systems, the splice must accommodate an insulator in the outer shielding to electrically isolate the shields/sheaths of adjacent cable sections and bring out connections that are connected to bonding cables and link boxes.

Terminations are typically slip-on, some with a dielectric liquid or $SF_6$ gas ("dry type") to fill any voids in the high-stress regions. The area above liquid-filled terminations includes allowances to accommodate fluid expansion. Both porcelain and polymer housings are available, with the general trend toward polymer (composite) bushings because of lower weight and better resistance to external mechanical damage. Figure 5-68 shows 69 kV polymer terminations. Terminations are available for air installation, installation in single-phase $SF_6$ bus, and installation in three-phase $SF_6$ bus.

Link boxes permit connections among the sheaths for grounding and cross-bonding and facilitate a means to isolate the metallic shield/sheath to perform dc jacket tests; the links must be removed when the cable jacket is tested electrically as part of routine maintenance. A sheath voltage limiter—resembling a 3.5- to 12-kV distribution-class lightning surge arrester—is often placed at the non-grounded end of a cable sheath to limit overvoltages which might damage the jacket.

***HPFF Cable Accessories.*** HPFF cable splices are normally individual hand-taped assemblies, which are bound together using an aluminum "spider" to maintain spacing and reduce flexing. They are contained in a 3- or 5-piece welded steel joint casing, which can be 40 cm diameter and 3.5 m long. Figure 5-69 shows a pipe-type cable splice with aluminum spiders in place prior to welding on the steel casing sleeves. Trifurcating splices provide the transition from three cables in a steel pipe, to individual cables in stainless-steel pipes rising up to a termination. Special "stop joints" are installed on high-pressure liquid-filled cables to minimize fluid loss in the event of a major leak, and to permit maintenance on the cable without draining large amounts of liquid.

Terminations (also called potheads) have paper rolls to help grade the electrical stress and may contain capacitor stacks to help grade electrical stress. Terminations are filled with dielectric oil or nitrogen gas, respectively, for fluid-filled or gas-filled systems. Pressure plants are installed to accommodate liquid-filled cable demand and expulsion under all loading conditions and include a large (8000 to 40,000 L) tank, pumps, valves, and alarms, while gas-filled systems are simplified gas regulators and alarms that may be valved off because the nitrogen gas is compressible.

**FIGURE 5-68** 69-kV polymer terminations during installation.

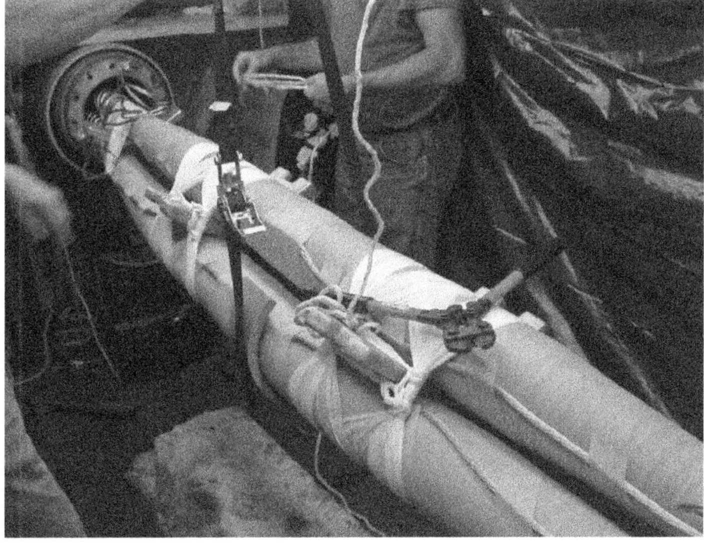

**FIGURE 5-69** Pipe-type cable splice being assembled.

**FIGURE 5-70** Isolator surge protector for pipe-type cable systems.

Cathodic protection is applied to protect the steel pipe in case the corrosion coating is damaged. The system consists of a rectifier to apply the −0.85V to protect the pipe and, on older pipe cable systems, a polarization cell to block dc while allowing a low impedance path for ac fault currents. Many utilities are replacing polarization cells with a solid-state device called an isolator surge protector (ISP) that performs the same function but is lower maintenance. Figure 5-70 shows an ISP.

***SCLF Cables Accessories.*** Normal joints are single-phase assemblies that have a special connector or ferrule to permit liquid flow through the conductor core. They are spliced in the same manner as a pipe-type splice with hand-applied paper tapes and have a nonmagnetic lead, copper, or aluminum housing. Insulators are placed in the housing to permit cross-bonding of the cable sheaths. Stop joints are placed every 1200 to 3000 m to isolate the liquid-supply sections. These joints may have to be placed closer together in areas with substantial elevation changes.

Terminations are very similar to those for pipe-type cables. Trifurcating joints are generally not required since almost all transmission-voltage self-contained cables are single-conductor design.

Liquid reservoirs are spaced along the circuit. Distances depend upon elevation changes, cable core diameter, and maximum allowable pressures. Typical spacings are 1500 to 5000 m. These reservoirs have alarms to indicate low liquid level or pressure. These alarms must be run back to the substation, usually in a separate communication duct, to notify operators of cable system problems.

## 5.2.16 Manufacturing

Manufacturing of transmission cable should be done using quality materials and an environment free of contaminants. Extruded dielectrics, particularly those with cross-linked polyethylene (XLPE) insulation, have been used increasingly in recent years as the manufacturing technology to use this type of insulation up to 500 kV has developed. Impregnated-paper insulation technology is still available but diminishing and there is currently only one pipe-type cable manufacturer remaining.

Manufacturing requires that copper or aluminum wire be drawn and wrapped together for the required conductor cross section. Depending on the size of the conductor, the conductor may be stranded or formed in segments that must be laid together. Extruded cables have a semiconductive layer of material applied over the conductor for a smooth, low electrical stress interface with the overlying insulation. Paper cables use tapes with carbon black for stress control.

Insulation is applied over the conductor by drawing the conductor through an appropriate type of machine. XD cables have insulation applied using a catenary, horizontal, or vertical extrusion machine. A "triple extrusion head" is the modern approach to applying extruded insulation where the conductor screen (shield), insulation, and insulation screen (shield) are applied in one process. For paper-insulated cables, paper tapes are helically applied dry until sufficient insulation has been built up for the operating voltage, then the completed cable core is heated under vacuum to remove moisture before dielectric liquid is applied in an impregnation tank.

Extruded cables are heated and degassed for a period for several days before a concentric wire shield, shield tapes and/or metallic sheath and insulating jacket are applied. Pipe-type cables are wrapped with a Mylar tape intercalated with a metallic shielding tape before applying helical skid wires.

Completed cable is placed on shipping reels. Extruded cable ends are sealed and lagged. Pipe cable reels are sealed and blanketed with dry nitrogen to limit moisture ingress. Self-contained cables are connected to sealed fluid reservoirs. Figure 5-71 shows extruded cables on shipping reels prior to final packaging.

**FIGURE 5-71** XLPE cable reels in a factory prior to shipping.

### 5.2.17 Operation and Maintenance

Operation and maintenance requirements of underground transmission cables are small compared with most electrical power equipment and are essentially confined to the dielectric fluid handling systems for pipe-type and self-contained cables, corrosion protection systems for pipe-type cables, and the pipe coating. Extruded cables have a protective jacket that should be tested periodically. By far, pipe-type cables have the highest maintenance requirements of any cable system type.

Extruded and self-contained transmission cables have an outer jacket typically constructed of polyethylene or polyvinyl chloride. The jacket is periodically tested to verify that there is no damage. Damage to the jacket may result in localized corrosion and unexpected circulating currents in the sheath that will cause $I^2R$ losses in the sheath, lowering the effective ampacity or overheating the cable. If the cables are cross bonded, the link boxes should be checked to be sure they remain dry, that the sheath voltage limiters are not damaged, and there is good ground continuity for the local earthing (Fig. 5-72).

All self-contained and pipe-type terminations and some extruded terminations contain small amounts of dielectric liquid (200 to 600 L). The terminations should be inspected to be sure there are no fluid leaks. In areas with high levels of air pollution, such as near the ocean, near unpaved roads

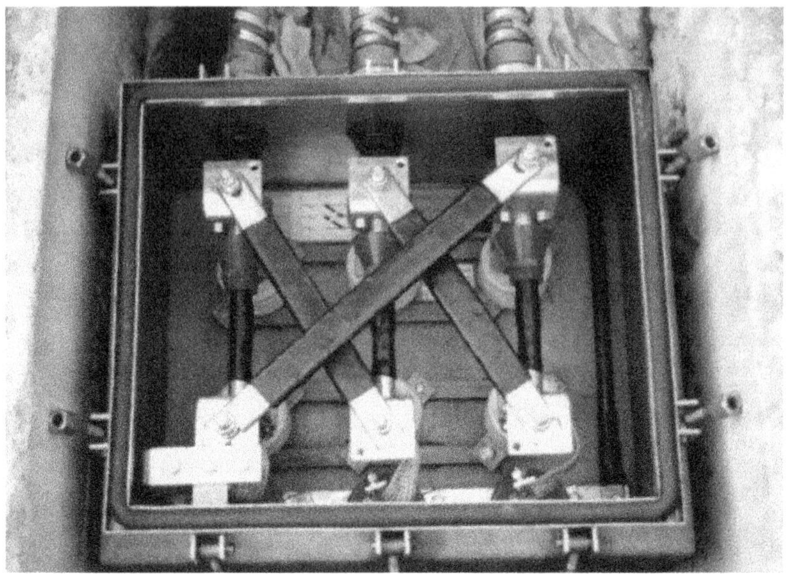

**FIGURE 5-72** Cross-bonding box with links installed.

or in industrial areas, it is important that the outside surface of the terminations be cleaned to avoid surface tracking. Porcelain terminations should be free of cracks or other mechanical damage.

Dissolved gas-in-oil analysis is a useful tool for pipe type and self-contained cables to determine the condition of a cable by examining the quantities and ratios of certain gases in the dielectric liquid. Field dissipation factor measurement can also be performed.[16] For both dissolved gas analysis and dissipation factor measurements, trending test results over time may offer additional insight into the condition of the cable systems. Utilities perform tests once every year or two, or more often if investigating a specific issue. Joints in locations susceptible to thermomechanical bending should be x-rayed every few years. Semiannual or quarterly dew-point readings should be taken on $SF_6$ gas-insulated bus.

A few utilities have experienced problems with dielectric fluid leaks in pipe-type cables caused by a combination of several factors including coating failure, cathodic protection problems, and stray currents in the earth. Monitoring of cathodic protection system performance and periodic testing of the pipe coating can limit the possibility of leaks, and leak detection and location using perfluorocarbon tracers or leak detection wire buried along the cable pipe have been used with success.

## 5.2.18 Fault Location

Although they are very uncommon, cable failures do occur and can be difficult to locate as compared to overhead lines where it is possible to visually inspect the entire line route. Locating cable faults requires the application of two types of techniques: terminal methods and tracer methods. Terminal methods are applied from one or the other end of the faulted cable and give an approximate location of a failure. Terminal methods include cable radar, electrical bridge techniques, and time domain reflectometry (cable radar). Tracer methods involve putting some type of signal on the faulted cable and then detecting changes in the signal when in close proximity to the fault location. Tracer techniques include tone sets, earth gradient, capacitive discharge ("thumping"), and magnetic pulse detection. Many times, the location of a failure is approximated with a terminal technique and then followed by a tracer technique to precisely locate the failure. The extent of fault location required depends on the installation of the cables. For example, if an entire manhole-to-manhole section will be replaced, fault locating only needs to localize the failure to the approximate area between manholes.

Pipe-type cable failures are particularly challenging. If fluid circulation is used on an HPFF cable, circulation should be stopped immediately once there is an indication of a failure to avoid contaminating the dielectric fluid. Mechanical damage or severe electrical faults can also cause fluid leaks that can be significant challenges. Since visible damage will generally not be apparent once the detection method has located the apparent fault location, the cable pipe must be x-rayed to determine if the exact location of the cable failure has been found. The liquid in the cable pipe must then be frozen on either side of the failure before the pipe can be cut and a repair initiated. These complications are one of the reasons extruded cables have generally supplanted pipe type. Fault location methods and strategies are summarized in Refs. 18 and 28.

## 5.2.19 Corrosion

Corrosion of underground cables can result from two phenomena: galvanic and electrolytic corrosion. Galvanic corrosion results from the electrochemical interaction of two or more materials, such as zinc and copper, with the level of galvanic action depending on the galvanic potential of the two materials, including materials in soils and seawater. The steel cable pipe of pipe-type cables or the metallic armor of submarine cables may experience galvanic or electrolytic corrosion. Galvanic corrosion can be prevented by isolating the cable or pipe from other materials by using a protective coating on the pipe, armor wires or outside the cable sheath. Pipe-type cables are often connected to a cathodic protection system where a negative voltage (relative to ground) is applied to the pipe using a rectifier and an isolator surge protector (or on older systems, a polarization cell) to provide a low-impedance path to ground for fault currents. Sacrificial anodes can also provide localized protection in the event the pipe-coating is damaged.

Electrolytic corrosion results from an externally imposed current leaving a conductive element where an anode zone is created. This might occur on the armor of a submarine cable or where holidays have developed in a pipe coating. Protecting a cable from electrolytic corrosion can be difficult because it is generally not possible to eliminate the source of the stray current, which causes electrolytic corrosion. Instead, Impressed Cathodic Current Protection (ICCP) is used. In this scheme, anode beds are connected to the positive terminal of a rectifier, and the cable armor, for example, is connected to the negative terminal. Anode beds are placed along the cable to produce an imposed current on the cable armor. The level of stray currents will designate the current density which must be produced by the ICCP system and anode beds in order to protect the cable[19] and is selected to counteract the corrosion currents that would otherwise impact the cable system.

## 5.2.20 Testing

There are various tests performed on underground cable systems during design (prequalification), after manufacture (factory acceptance tests including routine and sample tests), before and after installation (commissioning tests), and as part of maintenance. Various IEEE, IEC, AEIC, and ICEA standards have been developed that address this subject.

A manufacturer often performs a long-term "type test" prior to commercial production to prove the integrity of a cable and its accessories. Some purchasers of a cable system may also require that a type test be performed as part of their cable order, when a large cable system will be supplied, or when a new cable design or set of accessories is being used with a cable for the first time.

As part of manufacturing, routine and sample tests are performed and verified by the purchaser's representative to confirm that the cable was manufactured according to the designated specifications. Accessories also undergo routine tests and some purchasers choose to observe these tests or at least review completed test reports. Routine tests may include partial discharge (PD) testing, particularly on extruded cable systems, as well as capacitance, conductor resistance, dissipation factor ("tan $\delta$"), voltage withstand, and jacket integrity tests on extruded cables. Pipe-type cables on shipping reels may have the dew point and nitrogen pressure tested prior to shipping, and self-contained cables may have a check of dielectric fluid pressure rise to be sure there is no air in the system. Some of these tests may be performed just prior to cable installation after cable reels are on site to be sure the cable reels were not damaged during shipping.

After laying tests (commissioning tests) may include checks of conductor resistance, cable phasing, jacket integrity tests (on extruded or self-contained cables), an ac voltage test, a dc high potential ("hi-pot") test on paper insulated cables (only), as well as tests on dielectric fluids before filling cable pipes or pressure reservoirs. XLPE-insulated cables generally are subjected to an ac voltage withstand test at rated voltage (called a "soak" test) without carrying load or may be subjected to some higher voltage if appropriate test equipment is available, such as a variable-frequency resonant test set that can apply 1.4× to 1.7× rated voltage. This type of test set has some advantages including very low fault current energy to apply to a fault in the event a problem is detected during the test. Very-low-frequency (VLF) test sets may be employed for testing XLPE cables although they have limited availability at the higher transmission voltages.

Tests during operation and maintenance include periodic jacket integrity tests, partial discharge (PD), dissipation factor measurements (mainly on paper-insulated cables), and the various methods used to locate cable faults as described earlier. Dielectric fluid testing is performed on paper cable to check for dissolved gases (dissolved gas-in-oil analysis or "DGA"), which is often a good indicator of various problems with a cable system.

## 5.2.21 Future Developments

Research and development work on cable systems themselves has waned as the industry entered the 21st century. Most work has been on reducing the insulation thickness of XLPE cables so that upgrades of older self-contained cables are possible and also to retrofit XLPE cable into existing pipe-type cable pipes. Methods to facilitate field testing of cables and evaluation of thermal-mechanical movement of large-diameter extruded cables are being researched. Pipe-type cable development has essentially stopped, and manufacturing of this cable type is likely to diminish as the industry further moves to XLPE technology. Superconducting cables are also seeing continued development, but additional work remains for expanded use. The most significant power cable industry development has been the design and manufacturing of polymeric HVDC cables by most XLPE manufacturers. Installation contractors are seeking to install cable for longer sections using trenchless technology. As the early applications of XLPE transmission cables are approaching 30 years of operation, cable life assessment of extruded cable systems is becoming a topic of interest as utilities see their circuits age.

## 5.2.22 References


1. Bascom, E. C. III, Antoniello, V. D., "Underground Power Cable Considerations: Alternatives to Overhead," *Minnesota Power Systems Conference*, November 1–3, 2011.
2. Stewart, J. R., Williams, J. A., Wilson, D. D., "System Implications of Underground Cable Systems," *IEEE T&D Conference*, Anaheim, CA, September 15–19, 1986.
3. *IEEE Guide for the Application of Sheath-Bonding Methods for Single-Conductor Cables and the Calculation of Induced Voltages and Currents in Cable Sheaths*, ANSI/IEEE Std 575-2012.
4. "Calculation of the Continuous Current Rating of Cables (100% Load Factor)," *International Electrotechnical Commission*, Publication 60287.
5. Neher, J. H., McGrath, M. H., "The Calculation of Temperature Rise and Load Capability of Cable Systems," *Transactions of AIEE, Power Apparatus Systems*, vol. 76, pp. 752–772, October 1957.
6. Buckweitz, M. D., Pennell, D. B., "Forced Cooling of UG Lines," *Transmission & Distribution*, vol. 28, No. 4, pp. 51–58, April 1976.

7. "The Calculation of Continuous Ratings for Forced Cooled Cables," *CIGRE WG-08, Study Committee No. 21, Electra*, No. 66, pp. 59–84.
8. Purnhagen, D. W., "Designer's Handbook for Forced-Cooled High-Pressure Oil-Filled Pipe-Type Cable Systems," EPRI Report No. EL-3624, July 1984.
9. "The Calculation of Continuous Ratings for Forced Cooled Cables," *CIGRE WG-08, Study Committee No. 21, Electra*, No. 66, pp. 59–84.
10. Rifenburg, R. C., "Pipe Line Design for Pipe-Type Feeders," *Transactions of the AIEE*, pp. 1275–1288, December 1953.
11. Bascom, E. C. III, Patel, N., Parmar, D., "Thermal Environment Design Considerations for Ampacity of Buried Power Cables," *Proceedings of the IEEE Transmission & Distribution Conference*; Paper 14TD0241, Chicago, Illinois, April 14–17, 2014.
12. Bascom, E. C. III, "Underground Cable Uprating and Upgrading Tutorial," *Transactions of IEEE PES Transmission & Distribution Conference*, Paper 03TD0362 (Panel Session), Dallas, Texas, September 7–12, 2003.
13. Bascom, E. C. III, Clairmont, B., "Considerations for Advanced Temperature Monitoring of Underground Power Cables," *Proceedings of the IEEE Transmission & Distribution Conference*; Paper 14TD0212, Chicago, Illinois, April 14–17, 2014.
14. *Underground Transmission Systems Reference Book*, Electric Power Research Institute, TR-101670, 1992.
15. Neher, J. H., "The Phase Sequence Impedance of Pipe-Type Cables," *IEEE Transaction Power Apparatus & Systems*, vol. 83, pp. 795–804, August 1964.
16. Cooper, J. H., MacPhail, G. A., Cherukupalli, S., et al., "Insulation Condition Tests on Two Underground Transmission Cable Circuits at BC Hydro," Transactions of Doble Engineering Conference, Spring 1996.
17. Lewis, W. A., Allen, G. D., "Symmetrical-Component Circuit Constants and Neutral Circulating Currents for Concentric-Neutral Underground Distribution Cables," IEEE PAS, vol. PAS-97, No. 1, January/February 1978.
18. Bascom, E. C. III, von Herrmann, M. J., Zhao, T., "Power Cable Failure Modes and Fault Location Methods, Practices and Strategies," *Proceedings of the IEEE Transmission & Distribution Conference*; Paper 14TD0049, Chicago, Illinois, April 14–17, 2014.
19. Bascom, E. C. III, et al., "Construction Features and Environmental Factors Influencing Corrosion on a Self-Contained Fluid-Filled Submarine Cable Circuit in Long Island Sound," *IEEE Transactions on Power Delivery*, PE-034-PWRD-0-06-1997, January 1997.
20. Bascom, E. C. III, Cooper, J. H., Banker, W., Piteo, R., Regan, A. M., Boggs, S. A., "Magnetic Field Management Considerations for Underground Cable Duct Bank," *Transactions of IEEE PES Transmission & Distribution Conference*, Paper 05TD0399, Dallas, Texas, May 21–26, 2006.
21. Bascom, E. C. III, et al., "Considerations and Approach to Mitigating Underground Cable Magnetic Fields," *CIGRÉ-SEE Conference*, Paper #32, March 2011.
22. Kozak,S., Corbett, J. T., Bender, F. J., "Features of the New 138 kV Self-Contained Oil-Filled Cable System for Detroit Edison," *IEEE Transactions Power Apparatus & Systems*, vol. PAS-94, pp. 949–958, May/June 1975.
23. Bascom, E. C. III, Rezutko, J., "Novel Installation of a 138-kV Pipe-Type Cable System under Water Using Horizontal Directional Drilling," *Transactions of the IEEE Transmission & Distribution Conference & Exposition*, Chicago, April 14–17, 2014.
24. Hatcher, C. T., Gillette, R. W., Burrell, R. W., "345-kV Underground Transmission on the Consolidated Edison Company of New York System," *IEEE Transactions on Power Apparatus and Systems*, vol. 85, no. 4, pp. 353–360, April 1966.
25. IEEE Guide for the Planning, Design, Installation and Repair of Submarine Power Cable Systems, 1120-2004.
26. Bascom, E. C. III, Rezutko, J. "FPL Relocates Cable Circuits," *T & D World Magazine*, vol. 61, No. 4, pp. 24–32, April 2009.
27. Bascom, E. C. III, Williams, J., Kwilinski, M., "Technical Considerations for Applying Trenchless Technology Methods to Underground Power Cables," *Proceedings of the IEEE Transmission & Distribution Conference*; Paper TD2016-000133, Dallas, Texas, May 2–5, 2016.
28. Bascom, E. C. III, Von Dollen, D. W., Ng, H. W., "Computerized Underground Cable Fault Location Expertise," *IEEE T & D Conference*, Chicago, 1994.

# 6

# DIRECT-CURRENT POWER TRANSMISSION




Grateful acknowledgment is given to past contributors to this section: Ram Adapa and Michael P. Bahrman.

## 6.1 INTRODUCTION

High-voltage direct-current (HVDC) transmission is widely recognized as being advantageous for long-distance, bulk-power delivery, asynchronous interconnections, and long submarine cable crossings. Recent applications include integration of renewable energy sources such as wind and solar. HVDC lines and cables are less expensive and have lower losses than those for 3-phase ac transmission.

Typical HVDC lines utilize a bipolar configuration with two independent poles and are comparable to a double circuit ac line. Because of their controllability HVDC links offer firm capacity without limitation due to network congestion or loop flow on parallel paths. Higher power transfers are possible over longer distances with fewer lines with HVDC transmission than with ac transmission. Higher power transfers are possible without distance limitation to HVDC cable systems using fewer cables than with ac cable systems which require capacitive charging current.

HVDC systems became practical and commercially viable with the advent of high-voltage mercury-arc valves in the 1950s. Solid-state thyristor valves were introduced in the late 1960s, leading to simpler converter designs with lower operation and maintenance expenses and improved availability. In the late 1990s a number of newer converter technologies were introduced permitting wider use of HVDC transmission in applications, which might not otherwise be considered. A list of HVDC projects currently in operation or under construction is given in Table 6-1.

**TABLE 6-1** HVDC Project List*

| Name of HVDC system | Year commissioned/ upgraded/ retired | Nominal capacity (MW) | DC voltage (kV) | B-B line/ cable (km) | Location |
|---|---|---|---|---|---|
| Under Construction | | | | | |
| ESTLINK | 2006 | 350 | 150 | 106 | Estonia-Finland |
| BASSLINK | 2005 | 500 | 400 | 360 | Australia |
| NORNED | 2007 | 600 | 500 | 580 | Norway-Netherlands |
| THREE GORGES-SHANGHAI | 2007 | 3000 | ±500 | 900 | China |
| NEPTUNE | 2007 | 600 | 500 | 102 | U.S.A. |
| MISSION | 2007 | 150 | ±21 | B-B | U.S.A. |
| Operational | | | | | |
| VANCOUVER 1 | 1968 | 312 | +260 | 74 | Canada |
| VOLGOGRAD-DONBASS | 1962 | 720 | ±400 | 470 | Russia |
| SAKUMA | 1965/1993 | 300 | 2 × 125 | B-B | Japan |
| NEW ZEALAND HYBRID | 1965/92 | 1240 | +270/−350 | 612 | New Zealand |
| PACIFIC INTERTIE | 1970/84/89/02 | 3100 | ±500 | 1361 | U.S.A. |
| NELSON RIVER 1 | 1973/93 | 1854 | +463/−500 | 890 | Canada |
| GOTLAND HVDC LIGHT | 1999 | 50 | ±60 | 70 | Sweden |
| DIRECTLINK | 2000 | 3 × 60 | ±80 | 59 | Australia |
| MURRAYLINK | 2002 | 200 | ±150 | 176 | Australia |
| CROSS SOUND | 2002 | 330 | ±150 | 40 | U.S.A. |
| TROLL | 2004 | 2 × 40 | ±60 | 70 | Norway |
| EEL RIVER | 1972 | 320 | 2 × 80 | B-B | Canada |
| VANCOUVER 2 | 1977 | 370 | −280 | 74 | Canada |
| DAVID A. HAMIL | 1977 | 100 | 50 | B-B | U.S.A. |
| SHIN-SHINANO 1 | 1977 | 300 | 125 | B-B | Japan |
| SQUARE BUTTE | 1977 | 500 | ±250 | 749 | U.S.A. |
| CAHORA-BASSA | 1978 | 1920 | ±533 | 1420 | Mozambique-South Africa |
| C.U. | 1979 | 1128 | ±411 | 702 | U.S.A. |
| ACARAY | 1981 | 50 | 26 | B-B | Paraguay |
| INGA-SHABA | 1982 | 560 | ±500 | 1700 | Zaire |

**TABLE 6-1** HVDC Project List* (*Continued*)

| Name of HVDC system | Year commissioned/ upgraded/ retired | Nominal capacity (MW) | DC voltage (kV) | B-B line/ cable (km) | Location |
|---|---|---|---|---|---|
| EDDY COUNTRY | 1983 | 200 | 82 | B-B | U.S.A. |
| CHATEAUGUAY | 1984 | 2 × 500 | 2 × 140 | B-B | Canada |
| BLACKWATER | 1985 | 200 | 57 | B-B | U.S.A. |
| HIGHGATE | 1985 | 200 | 56 | B-B | U.S.A. |
| MADAWASKA | 1985 | 350 | 140 | B-B | Canada |
| MILES CITY | 1985 | 200 | 82 | B-B | U.S.A. |
| OKLAUNION | 1985 | 220 | 82 | B-B | U.S.A. |
| BROKEN HILL | 1986 | 40 | 2 × 17 (±8.33) | B-B | Australia |
| CROSS CHANNEL BP 1+2 | 1986 | 2000 | ±270 | 71 | France-U.K. |
| IPP (INTERMOUNTAIN) | 1986 | 1920 | ±500 | 784 | U.S.A. |
| ITAIPU 1 | 1986 | 3150 | ±600 | 796 | Brazil |
| ITAIPU 2 | 1987 | 3150 | ±600 | 796 | Brazil |
| URUGUAIANAI | 1987 | 54 | 18 | B-B | Brazil-Uruguay |
| VIRGINIA SMITH | 1987 | 200 | 50 | B-B | U.S.A. |
| FENNO-SKAN | 1989 | 572 | 400 | 234 | Finland-Sweden |
| McNEILL | 1989 | 150 | 42 | B-B | Canada |
| SILERU-BARSOOR | 1989 | 100 | ±200 | 196 | India |
| VINDHYACHAL | 1989 | 500 | 2 × 69.7 | B-B | India |
| RIHAND-DELHI | 1992 | 1500 | ±500 | 814 | India |
| SHIN-SHINANO 2 | 1992 | 300 | 125 | B-B | Japan |
| BALTIC CABLE | 1994 | 600 | 450 | 255 | Sweden-Germany |
| KONTEK | 1995 | 600 | 400 | 171 | Denmark-Germany |
| WELSH | 1995 | 600 | 162 | B-B | U.S.A. |
| CHANDRAPUR-RAMAGUNDUM | 1997 | 1000 | 2 × 205 | B-B | India |
| CHANDRAPUR-PADGHE | 1998 | 1500 | ±500 | 736 | India |
| HAENAM-CHEJU | 1998 | 300 | ±180 | 101 | South Korea |
| LEYTE-LUZON | 1998 | 440 | 350 | 443 | Philippines |
| VIZAG 1 | 1998 | 500 | 205 | B-B | India |
| MINAMI-FUKUMITZU | 1999 | 300 | 125 | B-B | Japan |
| KII CHANNEL | 2000 | 1400 | ±250 | 102 | Japan |
| SWEPOL LINK | 2000 | 600 | 450 | 230 | Sweden-Poland |
| GRITA | 2001 | 500 | 400 | 313 | Greece-Italy |
| HIGASHI-SHIMIZU | 2001 | 300 | 125 | B-B | Japan |
| MOYLE INTERCONNECTOR | 2001 | 2 × 250 | 2 × 250 | 64 | Scotland-N.Ireland |
| TIAN-GUANG | 2001 | 1800 | ±500 | 960 | China |
| THAILAND-MALAYSIA | 2001 | 600 | ±300 | 110 | Thailand-Malaysia |
| EAST-SOUTH INTERCONNECTOR | 2003 | 2000 | ±500 | 1400 | India |
| RAPID CITY TIE | 2003 | 2 × 100 | ±13 | B-B | U.S.A. |
| THREE GORGES CHANGZHOU | 2003 | 3000 | ±500 | 890 | China |
| GUI-GUANG | 2004 | 3000 | ±500 | 936 | China |
| THREE GORGES-GUANGDONG | 2004 | 3000 | ±500 | 900 | China |
| LAMAR | 2005 | 211 | ±63 | B-B | U.S.A. |
| VIZAG 2 | 2005 | 500 | ±88 | B-B | India |
| KONTI-SKAN 1 AND 2 | 1965/88/2005 | 740 | ±285 | 150 | Denmark-Sweden |
| SACOI | 1967/85/93 | 300 | ±200 | 385 | Italy-Corsica-Sardinia |
| SKAGERRAK 1-3 | 1976/77/93 | 1050 | 250/350 | 240 | Norway-Denmark |
| NELSON RIVER 2 | 1978/85 | 2000 | ±500 | 940 | Canada |
| HOKKAIDO-HONSHU | 1979/80/93 | 600 | ±250 | 167 | Japan |
| VYBORG | 1981/82/84/02 | 4 × 355 | 1 × 170 (±85) | B-B | Russia-Finland |
| GOTLAND II-III | 1983/87 | 260 | 150 | 98 | Sweden |
| QUEBEC-NEW ENGLAND | 1986/90/92 | 2250 | ±500 | 1500 | Canada-U.S.A. |

(*Continued*)

**TABLE 6-1** HVDC Project List* (*Continued*)

| Name of HVDC system | Year commissioned/ upgraded/ retired | Nominal capacity (MW) | DC voltage (kV) | B-B line/ cable (km) | Location |
|---|---|---|---|---|---|
| GESHA | 1989/90 | 1200 | ±500 | 1046 | China |
| GARABI 1&2 | 2000/02 | 2000 | ±70 | B-B | Argentina-Brazil |
| RIVERA | | 70 | | B-B | Uruguay |
| SASARAM | 2002 | 500 | 205 | B-B | India |
| Retired | | | | | |
| KINGSNORTH | 1972/1987 | 640 | | 82 | England |
| DUERNROHR 1 | 1983/1997 | 550 | 145 | B-B | Austria-Czech |
| ETZENRIHT | 1993/1997 | 600 | 160 | B-B | Germany-Czech |
| VIENNA SOUTH-EAST | 1993/1997 | 600 | 145 | B-B | Austria-Hungary |

*For the most recent list, contact Ms. Robyn L. Koropatnick, Teshmont Consultants LP, 1190 Waverly Street, Winnipeg, MB R3T 0P4 Canada. E-mail: rkoropatnick@teshmont.com.

## 6.2 APPLICATIONS

The significant increase in HVDC transmission can be attributed to one or more of the following reasons:

***Economical.*** HVDC transmission systems often provide a more economical alternative to ac transmission for long-distance, bulk-power delivery from remote resources such as hydroelectric developments, mine-mouth power plants, or generation from large-scale wind farms. Whenever long-distance transmission is discussed, the concept of "breakeven distance" frequently arises. This is where the savings in line costs and lower capitalized cost of losses offsets the higher converter station costs. A bipolar HVDC line uses only two insulated sets of conductors rather than three. This results in narrower right-of-way (ROW), smaller transmission towers, and lower line losses than with ac lines of comparable capacity. A rough approximation of the savings in line construction is 30%. Although breakeven distance is influenced by the costs of ROW and line construction with a typical value of 500 km, the concept itself is misleading because in many cases more ac lines are needed to deliver the same power over the same distance due to system stability limitations. Furthermore, the long-distance ac lines usually require intermediate switching stations and reactive power compensation. For example, the generator outlet transmission alternative for the ±250 kV, 500 MW Square Butte Project was two 345 kV series-compensated ac transmission lines. Similarly, the ±500 kV, 1600 MW Intermountain Power Project (IPP) ac alternative comprised two 500 kV ac lines. The IPP takes advantage of the double circuit nature of the bipolar line and includes a 100% short-term and 50% continuous monopolar overload. The first 6000 MW stage of the transmission for the Three Gorges Project in China would have required 5 × 500 kV ac lines as opposed to 2 × (±500) kV, 3000 MW bipolar HVDC lines (Fig. 6-1).

For underground or submarine cable systems there is considerable savings in installed cable costs and cost of losses with HVDC transmission. Depending on the power level to be transmitted, these savings can offset the higher converter station costs at distances of 40 km or more. Furthermore, there is a rapid drop-off in cable capacity with ac transmission over distance due to the reactive component of charging current. Although this can be compensated by intermediate shunt compensation for underground cables, it is not practical to do so for submarine cables. For a given cable conductor area, the line losses with HVDC cables, can be less than half those of ac cables. This is due to more conductors, reactive component of current, skin effect, and induced currents in the cable sheath and armor.

***Functional.*** The controllability and asynchronous nature of HVDC transmission provide a number of advantages for certain transmission applications. HVDC transmission capacity is firm and

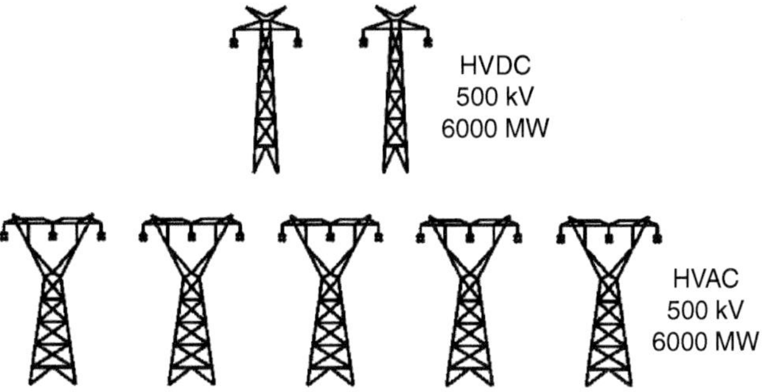


**FIGURE 6-1** HVDC and EHV ac alternatives for first stage of three Gorges outlet transmission.

utilization usually runs higher due to its controllability. This is because congestion or loop flow on parallel transmission paths does not result in scheduled curtailments for transmission loading relief.

With a cable system, unequal loadings or risk of postcontingency overloads often results in use of a series-connected phase-shifting transformer. These potential problems do not exist with a controlled HVDC cable system.

With HVDC transmission systems, interconnections can be made between asynchronous networks for more economic or reliable operation. The asynchronous interconnection allows interconnections of mutual benefit but provides a buffer between the two systems. Often these interconnections use back-to-back converters with no transmission line. The asynchronous links act as an effective "firewall" against propagation of cascading outages in one network from passing to another network. Many asynchronous interconnections exist in North America between the eastern and western interconnected systems, between the Electric Reliability Council of Texas (ERCOT) and its neighbors, that is, Mexico, Southwest Power Pool (SPP), and the western interconnect, and between Quebec and its neighbors, that is, New England and the Maritimes. The August 2003 northeast blackout provides an example of the firewall against cascading outages provided by asynchronous interconnections. As the outage propagated around the lower Great Lakes and through Ontario and New York, it stopped at the asynchronous interface with Quebec. Quebec was unaffected, the weak ac interconnections between New York and New England tripped, but the HVDC links from Quebec continued to deliver power to New England.

***Environmental.*** HVDC allows delivery of more power over fewer lines with narrower ROW. This is especially important in trying to access diverse resources in remote locations where lines may pass through environmentally sensitive or scenic areas. There is no induction or alternating electromagnetic fields from HVDC transmission. There is no physical restriction limiting the distance for underground cables. Underground cables can be used on shared ROW with other utilities without impacting reliability concerns over use of common corridors. Lower cable losses improves efficiency and results in less heating in the earth.

## 6.3 HVDC FUNDAMENTALS

### 6.3.1 Converter Behavior and Equations

Conventional HVDC transmission schemes utilize line-commutated, current-source converters. Such converters require a synchronous voltage source in order to operate. The basic building block used for HVDC conversion is the 3-phase, full-wave bridge referred to as a 6-pulse or Graetz bridge (Fig. 6-2). The term 6-pulse is due to the characteristic harmonic ripple in the dc output voltage,

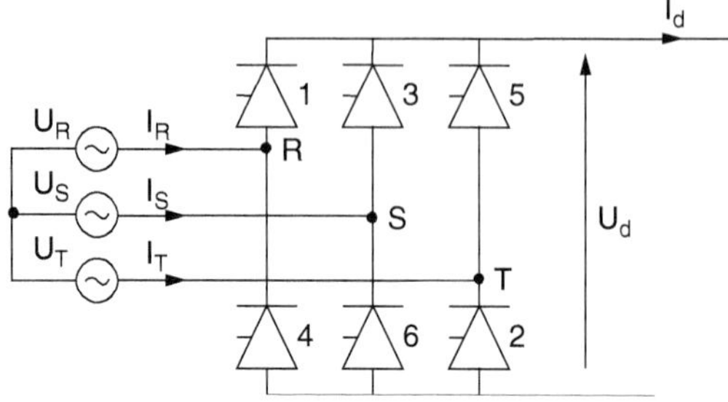


**FIGURE 6-2** Six-pulse bridge.

which is at multiples of 6 times the fundamental frequency. Each 6-pulse bridge is comprised of 6 controlled switching elements or thyristor valves. Each valve comprises a number of series-connected thyristors to achieve the desired dc voltage rating.

Converter dc output voltage is controlled by means of a delayed firing angle. Valve switching is synchronized to the ac source voltages via a phase-locked loop. The bridge is coupled to the ac bus via a converter transformer. Commutation of converter currents from one phase to another results in a converter voltage drop. Converter voltage drop is proportional to transformer reactance and current level $I_d$, resulting in a reduction of the dc voltage level $U_d$, due to commutation overlap $u$.

A set of equations has been derived to calculate $U_d$ as a function of the phase voltages, the commutation reactance $X_t$, dc current level $I_d$, and the delay angle $\alpha$. For rectifier operation converter polarity is positive, whereas for inverter operation it is negative bucking the direction of direct-current flow. Equations describing inverter operation use extinction angle $\gamma$.

The direct voltage across the 6-pulse bridge is calculated by Eq. (6-1) for rectifier operation and Eq. (6-2) for inverter operation.

$$\frac{U_{dR}}{2} = U_{diOR} \cdot \left[ cos\alpha - (d_{xR} + d_{rR}) \frac{I_d}{I_{dN}} \frac{U_{diORN}}{U_{diOR}} \right] - U_T \tag{6-1}$$

$$\frac{U_{dI}}{2} = U_{diOI} \cdot \left[ cos\gamma - (d_{xI} - d_{rI}) \frac{I_d}{I_{dN}} \frac{U_{diOIN}}{U_{diOI}} \right] + U_T \tag{6-2}$$

The nominal relative inductive direct voltage drop is defined by Eq. (6-3), where $X_t$ is the commutation reactance which includes the converter transformer reactance and any other reactances in the commutation circuit.

$$d_{xN} = \frac{3}{\pi} \frac{X_t \cdot I_{dN}}{U_{diON}} \tag{6-3}$$

The relative resistive direct voltage drop is defined by Eq. (6-4) where $P_{cu}$ is the transformer and smoothing reactor load losses and $R_{th}$ is current dependent voltage drop over the thyristors. The factor 2 is due to the fact that there are always two valves conducting at the same time.

$$d_r = \frac{P_{cu}}{U_{diON} \cdot I_{dN}} + \frac{2 \cdot R_{th} \cdot I_{dN}}{U_{diON}} \tag{6-4}$$

The overlap angle for the rectifier and inverter are described by Eqs. (6-5) and (6-6), respectively.

$$\cos(\alpha + \mu_R) = \cos\alpha - 2 \cdot d_{xNR} \frac{I_d U_{diONR}}{I_{dN} U_{diOR}} \tag{6-5}$$

$$\cos(\gamma + \mu_I) = \cos\gamma - 2 \cdot d_{xNI} \frac{I_d U_{diONI}}{I_{dN} U_{diOI}} \tag{6-6}$$

The reactive power consumption for a 12-pulse converter (two 6-pulse converters with 30° shift in valve voltages) connected in series is calculated with Eq. (6-7).

$$Q_d = 2 \cdot \chi \cdot I_d \cdot U_{diO} \tag{6-7}$$

where $\chi$ is the overlap function described by Eq. (6-8) for rectified operation and Eq. (6-9) for inverter operation.

$$\chi = \frac{1}{4} \cdot \frac{2 \cdot \mu + \sin 2\alpha - \sin 2(\alpha + \mu)}{\cos\alpha - \cos(\alpha + \mu)} \tag{6-8}$$

$$\chi = \frac{1}{4} \cdot \frac{2 \cdot \mu + \sin 2\gamma - \sin 2(\gamma + \mu)}{\cos\gamma - \cos(\gamma + \mu)} \tag{6-9}$$

The relationship between the no-load phase-phase ac voltage on the valve side and the ideal no-load direct voltage is shown in Eq. (6-10). The rms value of the rated ac current on the valve side of the converter transformer is shown in Eq. (6-11). The total rated MVA of the 3-phase transformer group feeding the 6-pulse converter bridge is according to Eq. (6-12).

$$U_{vo} = \frac{U_{diO}}{\sqrt{2}} \cdot \frac{\pi}{3} \tag{6-10}$$

$$I_{vN} = \sqrt{\frac{2}{3}} \cdot I_{dN} \tag{6-11}$$

$$S_N = \sqrt{3} \cdot U_{vN} \cdot I_{vN} = \frac{\pi}{3} \cdot U_{dION} \cdot I_{dN} \tag{6-12}$$

Figure 6-3 illustrates the commutation process and its effect on valve currents and dc voltage due to delay angle and overlap. The solid upper envelope of the phase voltages is the voltage top of the bridge

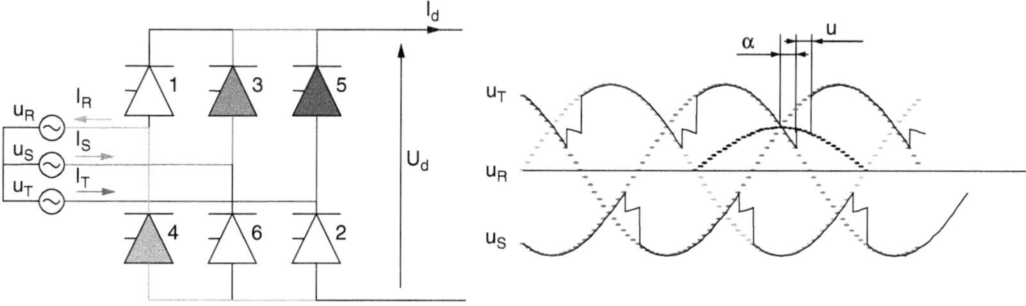


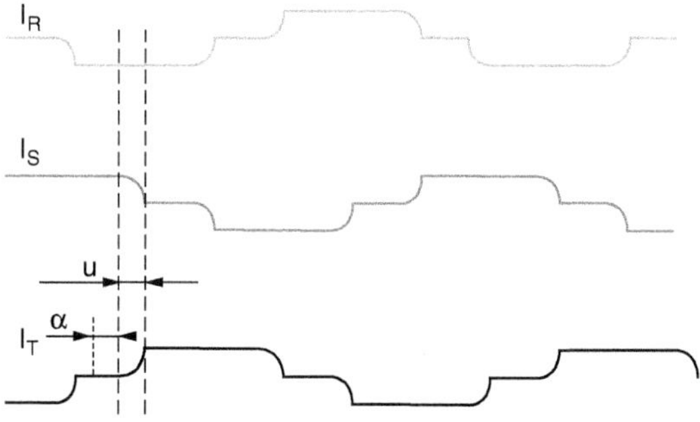


**FIGURE 6-3** Six-pulse bridge commutation with delay angle and overlap.

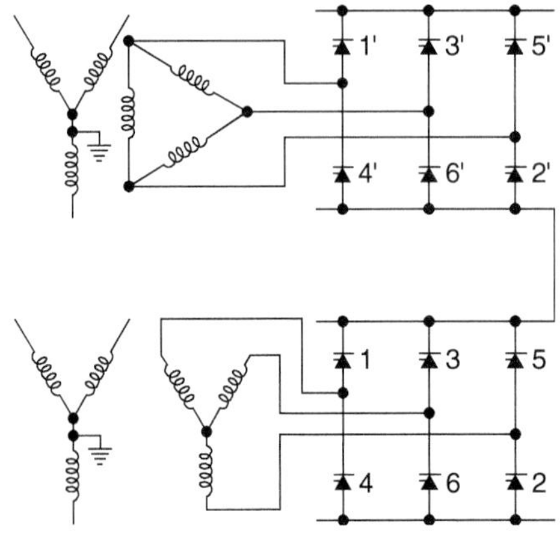


FIGURE 6-4 Twelve-pulse bridge.

with common valve cathodes, while the lower solid envelope is the voltage at the bottom of the bridge with the common valve anodes. The differential voltage across the bridge is the dc voltage $U_d$. The effect of the delay angle and commutation overlap on the dc voltage is evident. During commutation two valves in the same half bridge conduct simultaneously and the instantaneous voltage is half their sum.

The 6-pulse converter bridge can be used in rectifier operation with positive output voltage, $0 > \alpha < 90°$, converting ac to dc or in inverter operation with an output voltage that is negative with respect to the direction of dc current flow, $90 > \alpha < 180°$. By connecting two converters in series at opposite ends of a transmission line, one controlling dc voltage and the other controlling dc current, dc power transmission is achieved. The characteristic current harmonics ($f = 6n \pm 1$) are filtered on the ac side and the characteristic voltage harmonics ($f = 6n$) are filtered on the dc side to meet voltage distortion and telephone interference requirements.

The dc terminals of two 6-pulse bridges with ac voltage sources phase displaced by 30° can be connected in series for 12-pulse operation. In 12-pulse operation, the characteristic current and voltage harmonics have frequencies of $12n \pm 1$ and $12n$, respectively. The 30° phase displacement can easily be achieved by feeding one bridge through a transformer with a wye-connected secondary and the other transformer through a delta-connected secondary (Fig. 6-4). Most modern HVDC transmission schemes utilize 12-pulse converters to reduce the additional harmonic filtering requirements required for 6-pulse operation, for example, fifth and seventh on the ac side and sixth on the dc side. This is because although these harmonic currents still flow through the valves and the transformer windings, they are 180° out of phase and cancel out on the primary side.

### 6.3.2 Station Layout and System Configuration

A simplified single-line diagram for one pole with a 12-pulse converter is shown in Fig. 6-5. A photo and a CAD drawing of a monopolar converter station are shown in Figs. 6-6 and 6-7, respectively.

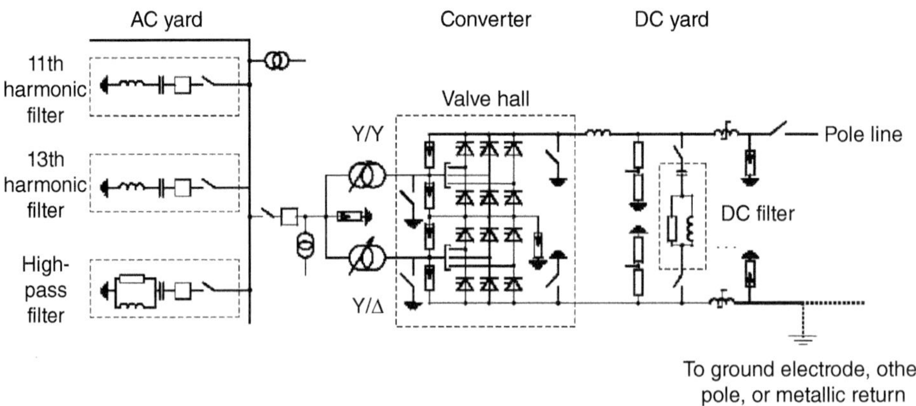


FIGURE 6-5 Simplified single-line diagram for monopole.

**FIGURE 6-6** Monopolar converter station.

An HVDC converter station comprises the following major subsystems:

- Thyristor valves
- Converter transformers
- AC harmonic filters
- DC harmonic filters
- Valve cooling
- Control and protection
- Auxiliary power
- Valve hall building

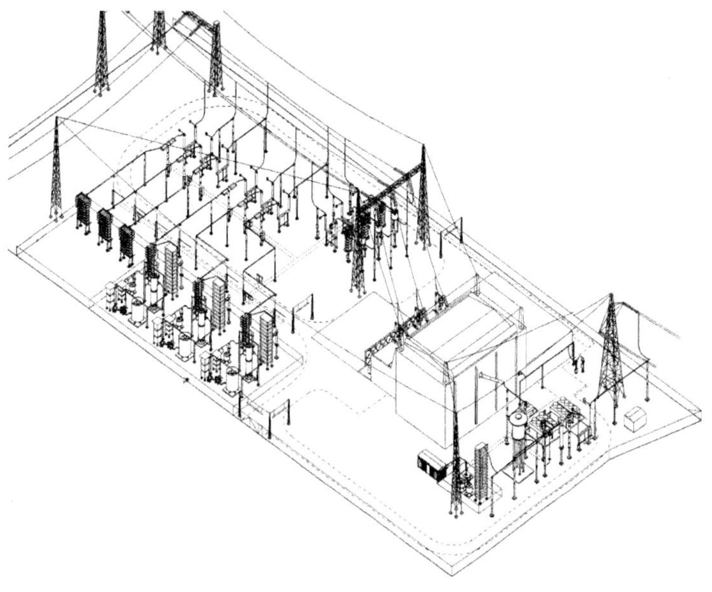

**FIGURE 6-7** CAD drawing of monopolar converter station.

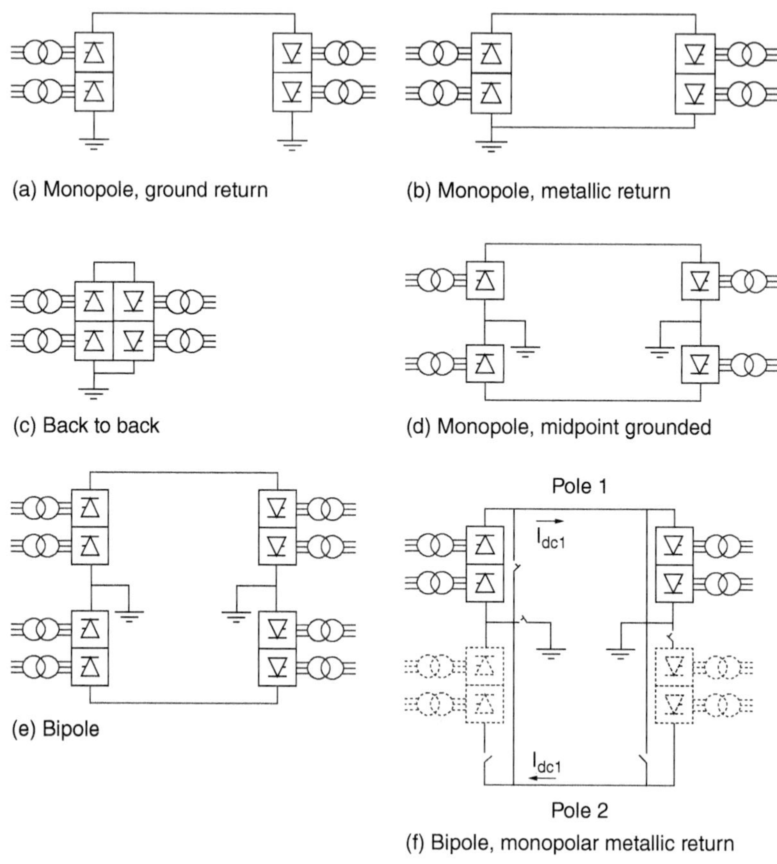


**FIGURE 6-8** HVDC operating configurations/modes.

The converter station layout depends on a number of factors such as the station configuration, that is, monopolar (Fig. 6-8), bipolar (Fig. 6-9) or back-to-back asynchronous tie (Fig. 6-10), valve design, ac system interconnection, filtering requirements, reactive power compensation requirements, land availability, and the local environment. In most cases, the thyristor valves are air-insulated, water-cooled, and enclosed in a converter building often referred to as a valve hall. For back-to-back ties

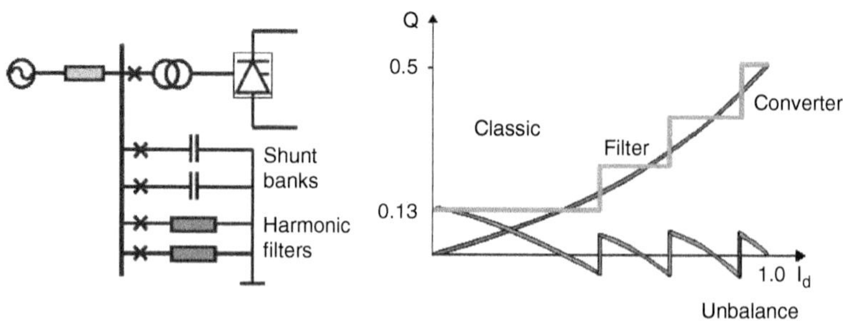


**FIGURE 6-9** Reactive power balance.

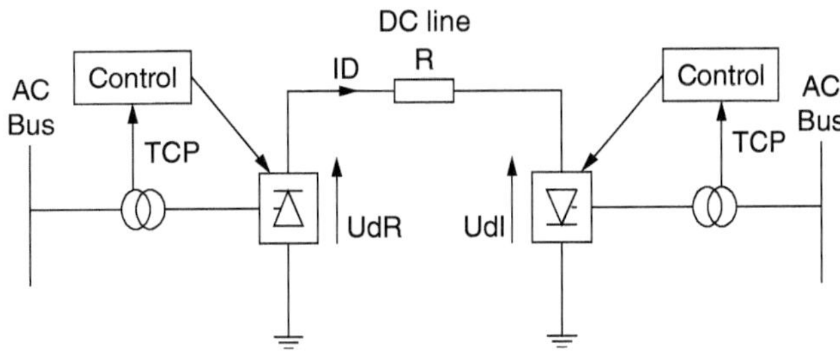


**FIGURE 6-10** HVDC control system.

with their characteristically low dc voltage, thyristor valves can be housed in prefabricated electrical enclosures in which case a valve hall is not required.

To obtain a more compact station design and reduce the number of insulated high-voltage wall bushings, converter transformers are often placed adjacent to the valve hall with valve winding bushings protruding through the building walls for connection to the valves. Double or quadruple valve structures housing valve modules are used within the valve hall. Valve arresters are located immediately adjacent to the valves. Indoor motor-operated grounding switches are used for personnel safety during maintenance. Closed loop valve cooling systems are used to circulate the cooling medium through the indoor thyristor valves with heat transfer to dry coolers or evaporative cooling towers located outdoors.

Monopolar systems with ground return are the simplest and least expensive systems for moderate power transfers since only two converters and one insulated cable or line conductor is required. Such systems are commonly used with low voltage electrode lines and sea electrodes to carry the return current in submarine cable crossings.

In some areas conditions are not conducive to monopolar earth or sea return. This could be the case in heavily congested areas, fresh water cable crossings, or areas with high earth resistivities. In such cases a metallic neutral or low voltage cable is used for the return path and the dc circuit uses a simple local ground reference.

Back-to-back stations are used for interconnection of asynchronous networks and use ac lines to connect on either side. In such systems, power transfer is limited by the relative capacities of the adjacent ac systems at the point of coupling.

As an economic alternative to a monopolar system with metallic return, the midpoint of a 12-pulse converter can be connected to earth directly or through an impedance and two half voltage cables or line conductors can be used. The converter is only operated in 12-pulse mode, so there is no earth current.

The most common configuration for modern overhead HVDC transmission lines is bipolar with a single 12-pulse converter for each pole at each terminal. This gives two independent dc circuits each capable of half capacity. For normal balanced operation there is no earth current. Monopolar earth return operation, often with overload capacity, can be used during outages of the opposite pole.

Earth return operation can be minimized during monopolar outages by using the opposite pole line for metallic return via pole/converter bypass switches at each end. This requires a metallic-return transfer breaker in the ground electrode line at one of the dc terminals to commutate the current from the relatively low resistance of the earth into that of the dc line conductor. Metallic return operation capability is provided for most dc transmission systems. This is not only effective during converter outages but also during line insulation failures where the remaining insulation strength is adequate to withstand the low resistive voltage drop in the metallic return path.

### 6.3.3 Reactive Power Compensation

As shown by Eqs. (6-7) through (6-9) in Sec. 6.3.1, HVDC conversion with line-commutated converters demands reactive power from the ac network at each HVDC terminal. The reactive power demand is a function of the firing angle in rectifier operation and extinction angle in inverter operation, the direct current and the overlap angle. The overlap angle is a function of the ac commutating voltage, the commutation reactance, and the dc current. As a rough approximation nominal reactive power demand at each terminal is about half the active power transfer.

The total reactive power produced by all the ac harmonic filters at each terminal is usually in the range of 30% to 40% of the converter rating. The filters therefore provide most of the reactive power compensation to meet the converter reactive power demand. The remaining reactive power necessary at the higher power levels can be provided from shunt capacitor banks, synchronous condensers, and static var compensators or nearby generation. Any reactive power mismatch must be provided or absorbed by the local ac system. Figure 6-9 shows the reactive power demand of a converter station, the reactive power from the filters, and the reactive power exchange with the ac network as a function of power transfer.

With weaker ac networks, that is, networks where the 3-phase symmetrical short circuit capacity is low compared to the rating of the dc converter station, various system constraints impact the reactive power compensation. With weaker systems, the size of the reactive power compensation elements may need to be reduced due to the voltage change on switching and the allowable reactive power exchange with the ac network. This may mean that filter banks may have to be subdivided with smaller branches. Sometimes, the minimum filtering requirements, for example, those at low power, exceed the reactive power demand of the converters, and shunt reactors are also required to absorb the excess vars from the filters.

### 6.3.4 Control and Operation of HVDC Links

The fundamental objectives of an HVDC control system are:

- To control basic system quantities such as dc line current, dc voltage, and transmitted power accurately and with sufficient speed of response
- To maintain adequate commutation margin in inverter operation so that the valves can recover their forward blocking capability after conduction before their voltage polarity reverses
- To control higher level quantities such as frequency in isolated mode or provide power oscillation damping to help stabilize the ac network
- To compensate for loss of a pole, a generator, or ac transmission circuit by rapid readjustment of power
- To ensure stable operation with reliable commutation in the presence of system disturbances
- To minimize system losses and reactive power consumption
- Ensure proper operation with fast and stable recoveries during system faults and disturbances

With HVDC transmission one terminal sets the dc voltage level, while the other regulates the dc current by controlling its output voltage relative to that maintained by the voltage-setting terminal. Since the dc line resistance is low, large changes in current and hence power can be made with relatively small changes in firing angle. Two independent methods exist for controlling the converter dc output voltage. These are (1) by changing the ratio between the direct voltage and the ac voltage by varying the delay angle $\alpha$ or (2) by changing the converter ac voltage via load tap changers (LTC) on the converter transformer. Although the former method is rapid, the latter method is slow due to the limited speed of response of the LTC. Use of high delay angles to achieve a larger dynamic range, however, increases the converter reactive power consumption. To minimize the reactive power demand while still providing adequate dynamic control range and commutation margin, the LTC is used at the rectifier terminal to keep the delay angle within its desired steady-state range, for example, 13° to 18°, and at the inverter to keep the extinction angle $\gamma$ within its desired range, for example, 17° to 20°, if the angle is used for dc voltage control or maintain rated dc voltage if operating in minimum commutation margin control mode.

Cooperation between the two terminals allows for efficient operation and provides for backup control modes for abrupt changes to the system voltages during disturbances. The converter control system at each terminal provides a static control characteristic. The intersection of the static control characteristics at the rectifier and inverter terminals determines the operating point. With the rectifier operating in constant current control and the inverter in constant angle control, as shown in Fig. 6-11, presents a stable operating point.

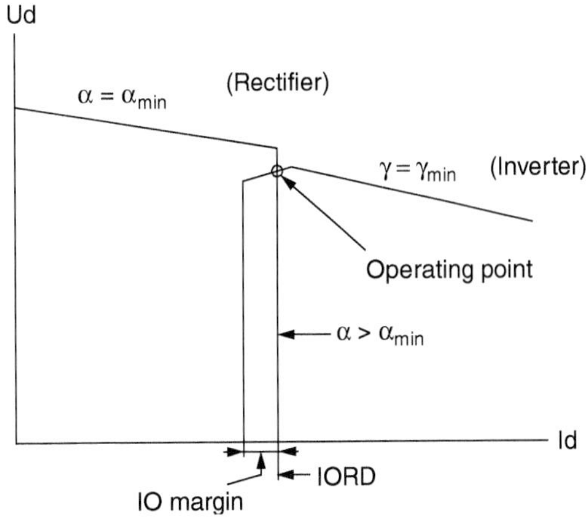


**FIGURE 6-11** Static operating characteristics.

Each converter terminal is equipped with a closed loop current control or current control amplifier (CCA) as shown in Fig. 6-12. The backup current regulator at the inverter comes into effect when the rectifier ac voltage is suddenly reduced, forcing the rectifier characteristic down resulting in a new operating point with the rectifier minimum firing angle setting the dc voltage and the inverter current order setting the current. This shift in operating point is referred to a mode shift. A dc voltage regulator may also be used with or without current compounding to achieve a positive slope at the inverter with minimum extinction angle or commutation margin as a backup. A mode shift can also occur for a sudden increase in inverter ac voltage if operating in constant extinction angle control.

Other control functions are needed to synchronize the valve firing to the ac system commutation voltages, to clear and recover from dc line faults, to translate the alpha orders to firing pulses and distribute them to the high-voltage valves, to minimize the reactive power consumption and achieve stable recoveries from large signal disturbances and faults in the ac network. Figure 6-13 shows these basic functions in the converter firing control (CFC).

The current order $I_o$ is received from the pole power control. If the dc voltage is very low as during faults, the current order is limited by the voltage-dependent current order limiter, VDCOL. The alpha firing order is then limited as to its minimum and maximum value and minimum valve firing voltage (UMIN) in the converter firing control. Alpha min is used in inverter operation to prevent firing in rectifier operation. Minimum commutation margin control is used in inverter operation to maintain the minimum voltage time area to ensure successful recovery of forward blocking capability after valve conduction.

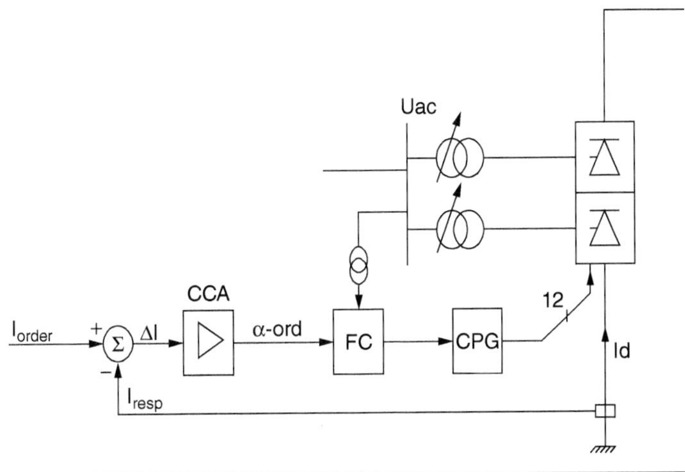


**FIGURE 6-12** Closed loop current control system.

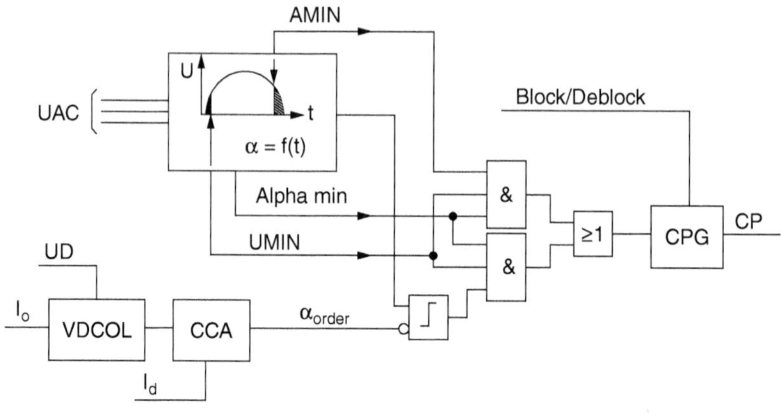


**FIGURE 6-13** Converter firing control.

Figure 6-14 shows the static characteristics of the rectifier and inverter with addition of the VDCOL. The VDCOL acts to limit the dc current order below its normal set point if the dc current is above its break point and the dc voltage is lower than its break point. Taking into account dynamic performance, the current limitation is very fast acting during decreasing voltage due to faults, while the recovery is slower upon system voltage recovery depending on ac system strength or ability to deliver reactive power to the converter during recovery.

The fundamental control functions described in the previous paragraphs are applied at the pole level and are independent of those on the other pole in a bipolar system. Coordination of the current orders between the terminals is required during ramping of the dc power during schedule changes. This is done during normal operation with secure communications between the terminals. Backup control strategies have been developed for communications outages. In a bipolar system, a master control is used for coordinated schedule changes and calculation of the current orders for each pole. The master control is used for compensation for loss of a pole by doubling the current order on the remaining pole subject to the equipment ratings. Figure 6-15 shows the current order coordination between the two terminals. For bipolar operation, the voltage fed to the power controller is the bipolar voltage

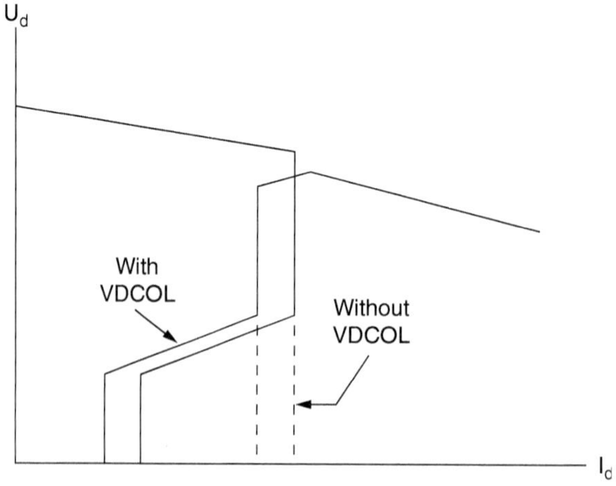


**FIGURE 6-14** $U_d$-$I_d$ characteristics with VDCOL.

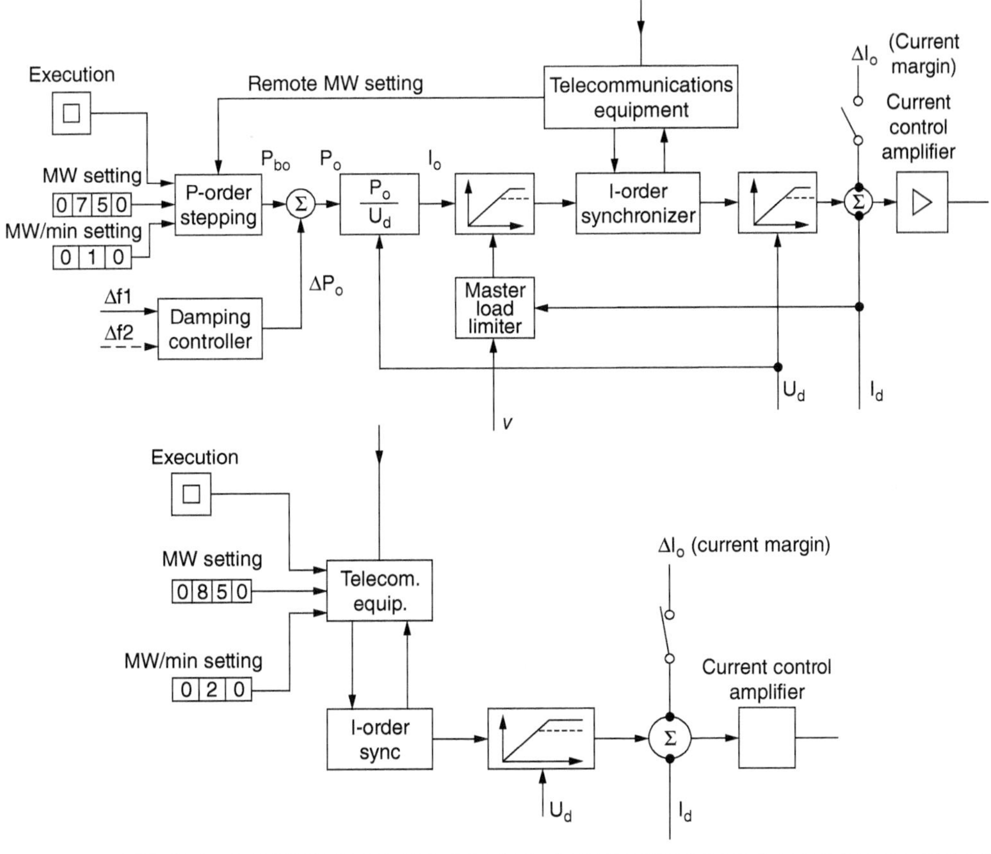


**FIGURE 6-15** Master power control and current order synchronization.

assuring equal current orders to each pole. Upon loss of a pole this voltage is cut in half. Normally, the master control is intentionally slow being only used for schedule changes. For loss of a pole, however, its response time is fast. The master control can also handle supplemental control functions such as power oscillation damping and frequency control. Synchronization of the current order is such that the current margin is maintained.

### 6.3.5 Multiterminal Operation

The same control principles used for two-terminal operation can be applied to multiterminal operation with one terminal being assigned to voltage control, while the other terminals control their respective dc current orders (Fig. 6-16). The master control must also ensure that the sum of the rectifier current orders equals the sum of the inverter current orders on a per pole basis during all operating conditions. If one of the terminals is limited or tripped, the residual mismatch is allocated among the remaining stations according to prioritized distribution factors to ensure that Kirchoff's law is met. If the tripped station is the voltage setting terminal (VST), one of the remaining stations must be assigned to voltage control. The same method for clearing dc line faults, force retard of the rectifier(s) to invert off the dc current, can be used along with fast-acting pole-isolating switches which in turn can be used to isolate a faulty terminal without using special purpose dc breakers.

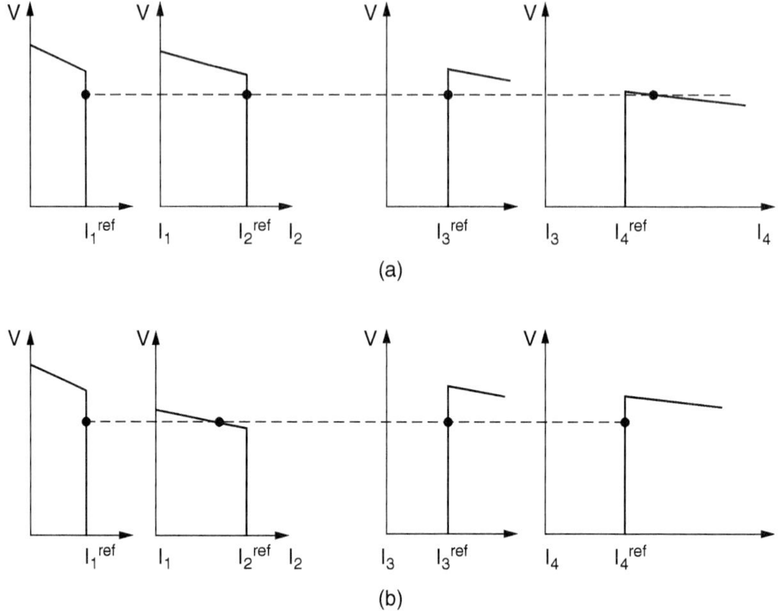


**FIGURE 6-16** Static characteristics for 4-terminal HVDC system illustrating mode shift from inverter 4 (upper set) to Rectifier 2 (lower set) due to depressed ac voltage at Rectifier 2.

### 6.3.6 Economics and Efficiency

The following factors influence the optimum solution for HVDC transmission systems:

- Power transfer requirements
- Transmission distance
- Capitalized cost of losses
- System configuration, that is, bipolar, monopolar, back-to-back OVHD line or cable system
- System connection voltages
- Relative system strength
- Reactive compensation requirements
- Environmental conditions
- Future expandability
- Transformer transport limitations

There is an economy of scale for HVDC transmission. It would cost less per kilowatt to transfer 3000 MW a distance of 800 km at ± 500 kV than it would to transfer 1000 MW. It would cost less per kilowatt to transfer 600 MW over a monopolar submarine cable system than it would to transfer the same power on a 2-pole cable system with each pole rated at half the capacity. A 550-MW back-to-back asynchronous link would cost less per kilowatt than a 150-MW link.

HVDC applications at locations with relatively low short circuit capacities typically cost more per kilowatt due to constraints on reactive power compensation and dynamic overvoltage mitigation measures. A typical terminal cost breakdown of an HVDC transmission system for an OVHD line is shown in Fig. 6-17.

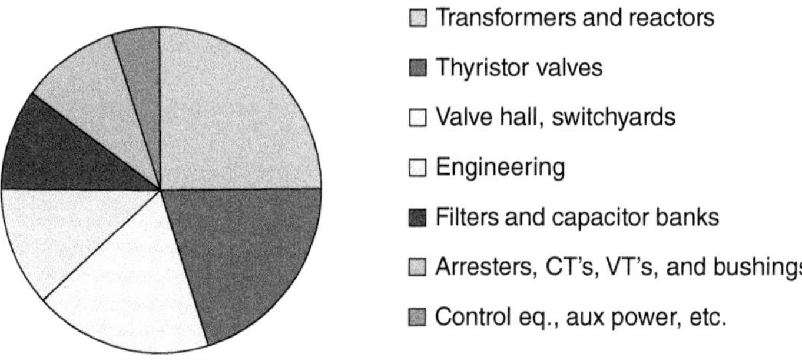


**FIGURE 6-17** Terminal cost.

## 6.4 ALTERNATIVE CONFIGURATIONS

### 6.4.1 Capacitor-Commutated Converters

Converters with series capacitors connected between the valves and the transformers were introduced in the late 1990s for weak-system back-to-back applications. These converters are referred to as capacitor-commutated converters (CCC). The series capacitor provides some of the converter reactive power compensation requirements automatically with load current and provides part of the commutation voltage improving voltage stability. The overvoltage protection of the series capacitors is simple since the fault currents are limited by the impedance of the converter transformers. The CCC configuration allows higher power ratings in areas where the ac network is close to its voltage stability limit. The asynchronous Garabi interconnection between Brazil and Argentina consists of $4 \times 550$ MW parallel CCC links. The Rapid City Tie between the eastern and western interconnected systems consists of $2 \times 100$ MW parallel CCC links (Fig. 6-18). Both installations use a modular design with converter valves located within prefabricated electrical enclosures.

### 6.4.2 Grid Power Flow Controller

A variation of the line-commutated design using a single 6-pulse converter has been used for a small back-to-back tie application. The term grid power flow controller (GPFC) has been used to describe this system design. By using a 6-pulse converter, there is no need for a second transformer secondary

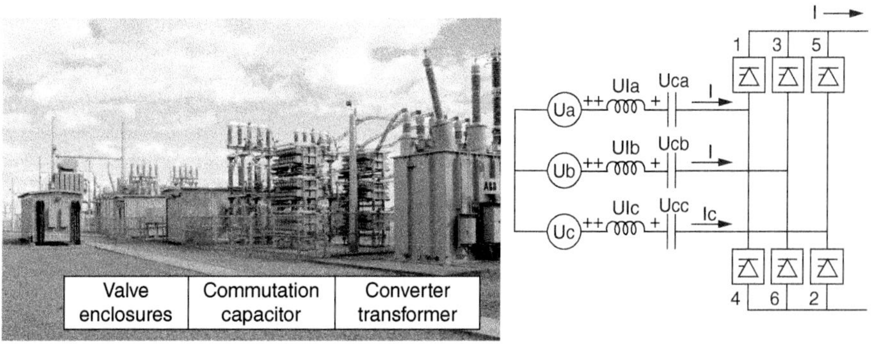


**FIGURE 6-18** Rapid City Tie with modular $2 \times 100$ MW capacitor commutated converters.

connection to obtain the requisite 30° phase displacement for 12-pulse operation. More ac harmonic filtering in the form of fifth and seventh branches is required, however. By using a 6-pulse converter and connecting the filters on the valve side, a simpler transformer connection can be utilized for matching the system voltage and blocking zero-sequence currents from flowing into the ac network. The ungrounded system has a large zero-sequence third order harmonic voltage component, however, appearing on the ungrounded neutrals and on the dc pole voltages, which increases the insulation levels. Despite using only one 6-pulse converter, the same number of series-connected thyristors is needed for the same dc voltage level.

### 6.4.3 Variable Frequency Transformer (VFT)

A technology that competes with HVDC for small capacity back-to-back ties in the 100 MW range was introduced in the early 2000s. A variable frequency transformer (VFT) is a machine rotating at the slip frequency between the two networks with high current between the rotor and stator passing through slip rings. The angle of the rotor is positioned to achieve a scheduled power flow by means of dc drives. The machine is connected to the network via step-up transformers. The reactive power demands of the VFT must be supplied by mechanically switched capacitor banks. Power control is slow due to having to move the inertia of the rotor, so it cannot respond quickly to a trip of generation on one of the isolated networks, for example. It cannot respond rapidly to variations in frequency or phase angle in the network so there will be inadvertent flow for fast variations. The VFT and its transformers provide an impedance, albeit a high one of around 40%, between the two networks. Therefore, the VFT will act as a voltage divider for faults in the network. This means that reactive power will be drained from one network due to a fault in the other. Losses of the VFT are higher than those for conventional HVDC.

## 6.5 STATION DESIGN AND EQUIPMENT

### 6.5.1 Thyristor Valves

For HVDC conversion, the thyristor valve must perform the following functions:

- Sequentially connect selected ac phases to the dc system per control pulses
- Conduct high current with low forward drop
- Block high voltages in both the forward and reverse directions
- Controllable and self monitoring
- Even voltage distribution and current turn-on
- Damp switching transients
- Fault tolerant and robust
- Accommodate cooling medium in high-voltage environment

Thyristor valves are built up of series-connected thyristor modules and saturable reactors to limit valve turn-on di/dt. Each module contains a number of series-connected thyristors mounted on heat sinks. Each thyristor level is paralleled by an RC network for even voltage distribution and damping of commutation overshoots. Voltage measurement across each thyristor level is provided for thyristor monitoring, forward protection, and recovery protection.

Each thyristor is coupled to the valve firing control at ground potential by means of two fiber optic links, one to carry valve trigger pulses to the thryistor gate circuit and the other for thyristor monitoring. Two types of thyristor triggering are used, electrically triggered thyristors (ETT) and light-triggered thyristors (LTT). Both triggering methods require voltage measurement at each thyristor level for monitoring and protection. ETT derives energy for gating from the RC damping circuit and

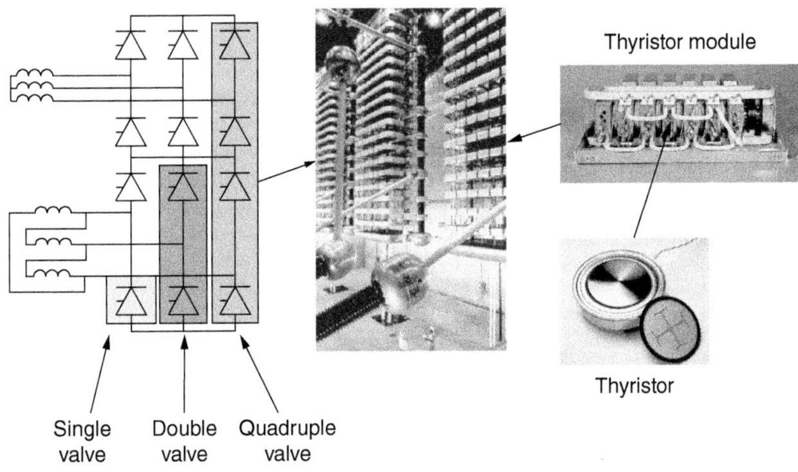


**FIGURE 6-19** 12-Pulse quadruple thyristor valve arrangement.

gating is initiated by trigger pulses generated by light-emitting diodes. LTT thyristors have an optical turning-on region integrated on the thyristor wafer itself and use higher-power trigger pulses provided by laser diodes. Each thyristor level is equipped with forward protection which gates the thyristor on if the forward blocking voltage becomes too high due to, for example, absence of a trigger pulse. In inverter operation, during the thyristor recovery time after conduction, the forward protection level can be temporarily lowered. This is called recovery protection. ETT permits recovery protection to be implemented independently at the individual thyristor level (Fig. 6-19).

## 6.5.2 Converter Transformers

Converter transformers are the link between the ac and dc systems. They provide isolation between the two systems, preventing dc voltage and current from reaching the ac system. They also provide the phase displacement necessary for 12-pulse operation through wye- and delta-valve winding connections. Converter transformers have regulating windings with load-tap changers to maintain the ac voltage and converter firing angle within a narrow band across the entire converter operating range. Converter transformer impedance also limits the valve short-circuit levels to within their handling capability. As shown by Eq. 6-12, the 3-phase rating of the converter transformer for a 6-pulse bridge is proportional to $U_{diON}$ and $I_{dN}$.

Converter transformer losses are those due to the fundamental frequency of load current plus those due to harmonics. The insulation design for converter transformers must take into account the direct voltage stresses superimposed on the normal ac voltage stresses. The ac stresses distribute as it would in a capacitive network while the dc voltage stresses distribute as according to a resistive network.

Transformer design depends on the bridge rating and type of converter connection and takes into account spare parts requirements and transport restrictions. For a small back-to-back, for example, a 3-phase bank with double secondary (wye and delta) may be used, that is, nine windings on a single core structure in a common tank for each 12-pulse converter bridge. For larger converters, three, single-phase transformers with double secondary windings may be used for each 12-pulse bridge. For the largest converter ratings where there may be some transport limitations, single-phase, two-winding transformers may be used, that is, six transformers per 12-pulse bridge (Fig. 6-20).

**FIGURE 6-20** Single-phase, three-winding converter transformer for a 3100 MW bipole.

### 6.5.3 Smoothing Reactor

A smoothing reactor is connected in series with the converter on the dc side to reduce the harmonic ripple in the dc current as well as reduce transient currents during faults. The smoothing reactor also protects the converter valves from voltage surges coming in on the dc line. The dc smoothing reactor together with shunt-connected dc filters serve to limit telephone interference disturbing currents from flowing on the dc line. Most smoothing reactors are air-core, naturally air-cooled.

### 6.5.4 AC Filters

Converters inject harmonic currents into the ac network. AC filters are used to prevent these harmonic currents from flowing into the ac network impedance causing voltage distortion and induced telephone interference in the audible frequency range. AC filters provide a low-impedance path to ground at the harmonic frequencies. The ac filter comprises high-voltage capacitor banks and lower-voltage reactors, resistors, and capacitors, which together form a circuit tuned to the characteristic harmonic(s). The lower-order filters are single- or double-tuned, band-pass filters, while the higher harmonics are often taken care of by high-pass filters (Fig. 6-21).

AC harmonic filter design involves calculating the harmonic currents generated and estimating harmonic impedance characteristics of the ac network across the whole range of operating conditions and tolerances. A filter design is then developed to meet the required performance requirements. Filter components are then rated with an adequate margin for the particular application.

The most common filter performance criteria are individual and total harmonic voltage distortion, $D_T$ and $D_h$, and weighted telephone interference factor (TIF), calculated as follows:

$$D_h = 100 \times V_h / V_1$$

$$D_T = \left( \sum_{h=2}^{49} D_h^2 \right)^{1/2}$$

$$\text{TIF} = \left[ \sum_{h=2}^{49} \left( F_h \cdot \frac{V_h}{V_1} \right)^2 \right]^{1/2}$$

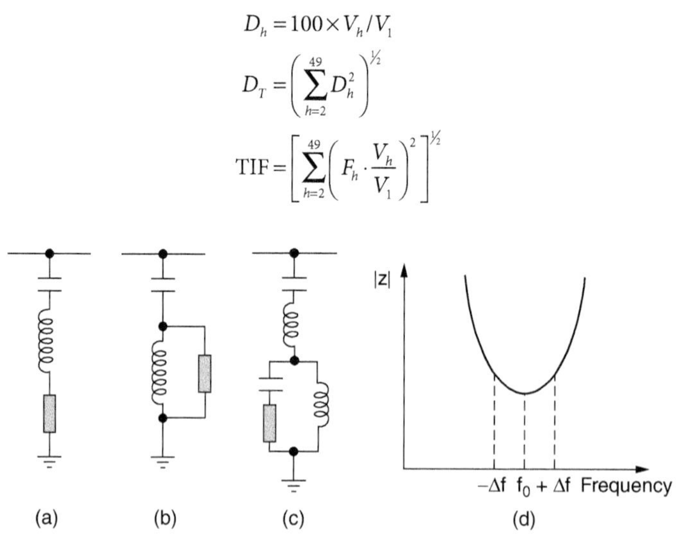


**FIGURE 6-21** (*a*) Bandpass filter, (*b*) highpass filter, (*c*) double bandpass filter, (*d*) impedance vs. frequency.

### 6.5.5 DC Filters

Filters are required on the dc side for dc to limit interference with communication circuits, which are inductively coupled to the dc line, for example, parallel telephone lines. The design criterion for dc harmonic filters is a function of relating to the flow of harmonic currents at any point along the dc line to the interference with adjacent telephone lines. Significant parameters are the relative location of telephone lines with respect to the dc line, their shielding, the presence of any ground wires, and the earth's resistivity. This criterion is typically expressed as equivalent disturbing current $I_{eq}$. Disturbance levels are lower in normal balanced bipolar mode, due to cancellation effects, than in monopolar mode.

DC filter design must take into account the entire dc network with all harmonic sources and operating modes. DC harmonic filters consist of band-pass and high-pass filters connected in shunt outside the smoothing reactor. Many modern HVDC links use a single 12th harmonic band-pass filter on each pole with active filtering for the higher order harmonics (Fig. 6-22). Active filtering consists of measuring the actual dc-side harmonics from the converter and counter-injecting the same amount with opposite polarity.

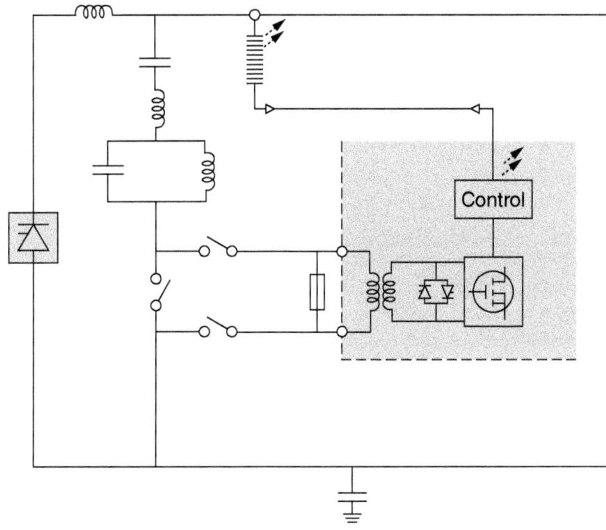


**FIGURE 6-22** Active dc harmonic filter.

### 6.5.6 Power Line Carrier (PLC) Filters

Commutation in HVDC converters discharges stray capacitances and generates electrical noise at the lower end of the power line carrier spectrum (PLC), that is, strongest at 30 to 70 kHz. This noise may pass onto the interconnecting ac and dc lines. Where low-level carriers exist at the lower end of the PLC spectrum, filters may be required.

### 6.5.7 Valve Cooling System

Thyristor valves must be cooled to avoid too high thyristor junction temperatures and to dissipate heat from the valve damping circuits and reactors. Valve cooling is accomplished by a deionized water loop circulating via insulated tuning to the individual thyristor heat sinks. Waste heat is passed to outdoor liquid-to-air coolers. Redundant variable speed pumps and coolers fed from redundant power supplies are used for reliability, availability, and ease of maintenance (Fig. 6-23).

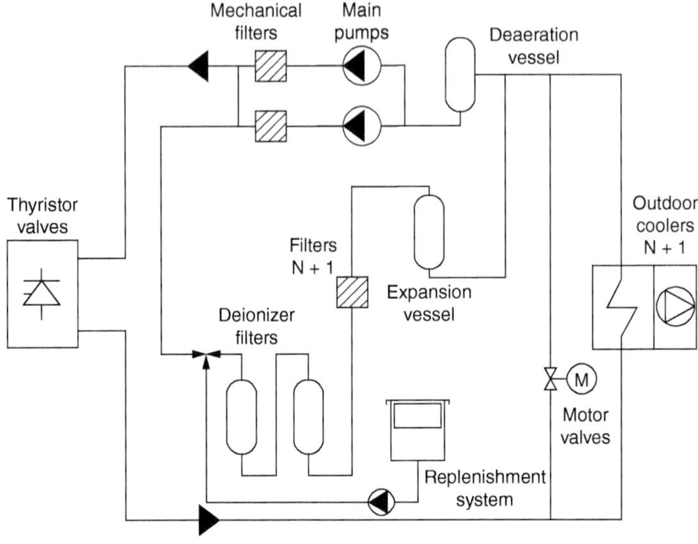


**FIGURE 6-23** Closed-loop water cooling system.

### 6.5.8 Reliability and Availability

To meet high levels of reliability and availability plus facilitate ease of maintenance, redundancy is commonly used in HVDC converter station design. Typical guaranteed unavailability values are 0.5% for forced outages and 1.0% for scheduled outages.

Redundant series-connected thyristor levels are used in the valves. The failure mode is short circuit of the thyristor, so operation can continue until a convenient time for restoring full redundancy. Redundant cooling pumps and cooler units are used. Use of redundant control and protection systems is often used. For major main circuit components, spare parts are provided at site to minimize the time for replacement.

## 6.6 VOLTAGE SOURCE CONVERTER–BASED HVDC TRANSMISSION

### 6.6.1 System Characteristics

Conventional HVDC transmission employs line-commutated, current-source converters with thyristor valves. These converters require a relatively strong synchronous voltage source in order to commutate. The conversion process demands reactive power from filters, shunt banks, or series capacitors, which are an integral part of the converter station. Any surplus or deficit in reactive power must be accommodated by the ac system. This difference in reactive power needs to be kept within a given band to keep the ac voltage within the desired tolerance. The weaker the system or the further away from generation, the tighter the reactive power exchange must be to stay within the desired voltage tolerance.

HVDC transmission using voltage source converters (VSCs) with pulse-width modulation (PWM) was introduced as HVDC Light® in the late 1990s by ABB. These VSC-based systems are force-commutated with insulated-gate bipolar transistor (IGBT) valves and solid-dielectric, extruded HVDC cables (Table 6-2).

**TABLE 6-2** HVDC VSC Projects Listing

| Project | Year commissioned | Power rating, MW | DC voltage, kV | Cable, km | Location |
|---|---|---|---|---|---|
| Hellsjon | 1997 | 3 | ± 10 | 10 | Sweden |
| Gotland Light | 1999 | 50 | ± 80 | 70 | Sweden |
| Direct Link | 2000 | 3 × 60 | ± 80 | 65 | Australia |
| Tjaerborg | 2000 | 7.2 | ± 9 | 4.4 | Denmark |
| Cross Sound Cable | 2002 | 330 | ± 150 | 40 | United States |
| Murraylink | 2002 | 200 | ± 150 | 180 | Australia |
| Troll Offshore | 2005 | 2 × 42 | ± 60 | 70 | Norway |
| Estlink | 2006 | 350 | ± 150 | 105 | Estonia/Finland |

HVDC transmission and reactive power compensation with VSC technology has certain attributes which can be beneficial to overall system performance. VSC converter technology can rapidly control both active and reactive power independently of one another. Reactive power can also be controlled at each terminal independent of the dc transmission voltage level. This control capability gives total flexibility to place converters anywhere in the ac network since there is no restriction on minimum network short-circuit capacity. Forced commutation with VSC even permits black start, that is, the converter can be used to synthesize a balanced set of 3-phase voltages like a virtual synchronous generator. The dynamic support of the ac voltage at each converter terminal improves the voltage stability and increases the transfer capability of the sending and receiving end ac systems.

### 6.6.2 Applications

The aforementioned attributes of VSC-based HVDC transmission makes it especially suitable in certain applications. These applications are summarized as follows:

***Underground Cable.*** HVDC cable systems do not face the distance limitations or suffer the higher losses of ac cable systems. Therefore, long-distance HVDC cable transmission is possible. Extruded HVDC cables are lighter, more flexible, and easier to splice than the mass-impregnated, oil-paper cables (MIND) used for conventional HVDC transmission, thus making them more conducive for land cable applications where transport limitations can drive up costs. The lower cost cable installations made possible by the extruded HVDC cables makes long-distance underground transmission economically feasible for use in areas with ROW constraints.

***Power Supply to Isolated Load.*** Forced-commutation, dynamic voltage control, and black-start capability allow VSC HVDC transmission to serve isolated loads on islands over long-distance submarine cables without any need for running expensive local generation.

***Offshore.*** The VSC transmission is compact and can feed production or transportation loads on offshore oil or gas platforms from shore. This can eliminate the need for more expensive, less efficient, or higher emission offshore power production. The VSC converters can operate at variable frequency to more efficiently drive large compressor or pumping loads using high-voltage motors.

***Asynchronous Interconnections.*** Interconnections between asynchronous networks are often at their periphery where the networks tend to be weak relative to the desired power transfer. The dynamic voltage support and improved voltage stability offered by VSC-based converters permits higher power transfers without as much need for ac system reinforcement. The VSC converters do not suffer commutation failures allowing fast recoveries from nearby ac faults. Economic power schedules, which reverse power direction, can be made without any restrictions since there is no minimum power or current restrictions.

***Urban Infeed.*** Power supply for large cities depends on local generation and power import capability. Local generation is often older and less efficient than newer units located remotely. Often, however, the older, less-efficient units located near the city center must be dispatched out-of-merit because they must be run for reliable voltage support or inadequate transmission. New transmission into large cities is difficult to site due to ROW and land-use constraints. Compact VSC-based underground transmission circuits can be placed on existing dual-use ROW to bring in power as well as provide voltage support, allowing a more economical power supply without compromising reliability. The receiving terminal acts like a virtual generator delivering power and voltage regulation. Stations are compact and housed mainly indoors making siting in urban areas somewhat easier.

***Outlet Transmission for Large-Scale Wind Generation.*** Large remote wind generation arrays require a collector system, reactive power support, and outlet transmission. Transmission for wind generation must often traverse scenic or environmentally sensitive areas or bodies of water. The VSC-based HVDC transmission allows efficient use of long-distance land or submarine cables and provides reactive support to the wind generation complex.

***Multiterminal Systems.*** The VSC HVDC transmission reverses power through reversal of current direction rather than polarity. This makes it easier to reverse power at an intermediate tap independently of the main power flow direction since voltage polarity reversal is not required. Conventional HVDC transmission requires switching for converter opposite pole connection or polarity reversal.

## 6.6.3 VSC Station Configuration and Design

HVDC transmission systems based on VSC converter technology are configured as shown in Fig. 6-24.

The transmission circuit consists of a bipolar two-wire HVDC system with converters connected pole-to-pole. The dc capacitors are used to provide a dc voltage source. The dc capacitors are grounded at their electrical center point to establish the earth reference potential for the transmission system. There is no earth return operation. The converters are coupled to the ac system through ac phase reactors and power transformers. Harmonic filters are located between the phase reactors and

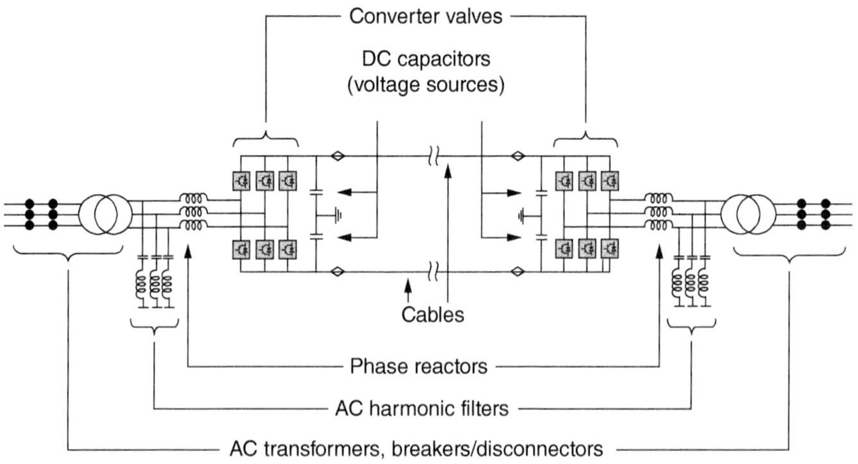


**FIGURE 6-24** VSC-based HVDC.

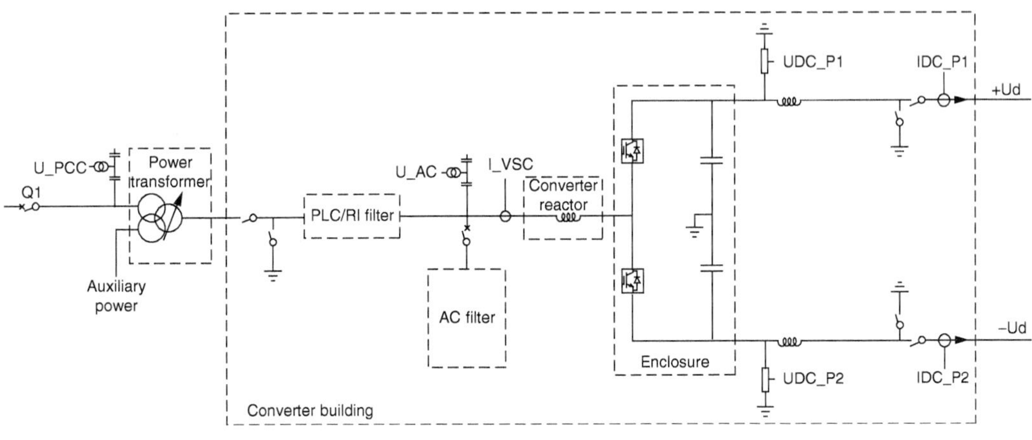


**FIGURE 6-25** Simplified SLD for VSC station.

power transformers. Therefore, the transformers are exposed to no dc voltage stresses or harmonics loading allowing use of ordinary power transformers.

A simplified single line diagram for a two-level VSC converter station is shown in Fig. 6-25. Principal station components are described in the following paragraphs.

***Power Transformer.*** The transformer is an ordinary single- or 3-phase power transformer with load tap changer. The secondary voltage, that is, the filter bus voltage, can be controlled with the tap changer to achieve the maximum active and reactive power, both consumption and generation, from the converter. The tap changer is located on the secondary side, which has the largest voltage swing, and also to ensure that the ratio between the line winding and a possible tertiary winding is fixed. The current in the transformer windings contains hardly any harmonics and is not exposed to any dc voltage. In order to maximize the active power transfer, the converter can generate a low frequency zero-sequence voltage (<0.2 pu), which is blocked by the ungrounded transformer secondary winding.

The transformer may be provided with a tertiary winding to feed the station auxiliary power system.

***Converter Reactors.*** The converter reactor is installed in series in each phase and is one of the key components in a voltage source converter to permit continuous and independent control of active and reactive power.

The main purposes of the converter reactors are to:

- Provide low-pass filtering of the PWM pattern to give the desired fundamental frequency voltage. The converter generates harmonics related to the switching frequency. The harmonic currents are blocked by the converter reactor and the harmonic content on the ac bus voltage is reduced by an ac filter.
- Provide active and reactive power control. The fundamental frequency voltage across the reactor defines the power flow (both active and reactive) between the ac and dc sides. Refer to typical P-Q diagram (Fig. 6-28) and active and reactive power definitions.
- Limit the short-circuit currents.

***DC-Capacitors.*** The primary objective of the valve dc side capacitor is to provide a low-inductance path for the turn-off switching currents and provide energy storage. The capacitor also reduces the harmonic ripple on the direct voltage. Disturbances in the system (e.g., ac faults) will cause dc voltage variations. The ability to limit these voltage variations depends on the size of the dc side capacitor. Since the dc capacitors are used indoors, dry capacitors are used.

***AC-Filters.*** Voltage source converters can be operated with different control schemes most of which use pulse width modulation to control the ratio between dc and ac side fundamental frequency voltage. Looking at the ac voltages on the converter side of the reactor, the voltage to ground consists of a square wave as indicated by Fig. 6-3. Connection of a large voltage source converter to a transmission or distribution system requires ac filters to remove the high-frequency components from introducing distortion or interference into the network. This is achieved by means of the converter reactor and the ac filters. The harmonics generated by VSC converters with PWM are higher in frequency than those from conventional HVDC converters. Therefore, smaller filter components can be used to meet performance requirements without large fundamental frequency reactive power generation. This makes the VSC converters better suited to weak-system applications.

The distorted waveform of the converter terminal voltage can be described as a series of harmonic voltages

$$E = \sum_h E_h \cos(h\Omega_1 t + \alpha_h)$$

where $E_h$ is the $h$th harmonic EMF. The magnitude of the harmonic EMFs will, naturally, vary with the dc voltage, the switching frequency (or pulse number) of the converter, etc. It will also depend on the chosen PWM control method and topology of the converter. For example, a converter can use sinusoidal PWM with third harmonic injection, that is, when a third harmonic is added on the fundamental frequency modulator to increase the power rating of the converter, or some form of harmonic cancellation such as optimized pulse width modulation, OPWM, can be used. Higher level converters can also be used to switch between a higher number of dc voltage levels, for example, a three-level converter can switch between the positive, zero, and negative dc voltage level. In a typical VSC scheme, ac filters contain two- or three-tuned or high-pass filter branches, which can be either grounded or ungrounded.

***DC Filters.*** For VSC converters in combination with extruded dc cables, the filtering on the dc side by the converter dc capacitor and the line smoothing reactor on the dc side is considered to give sufficient suppression of harmonics. However, under certain circumstances, if the dc cable route shares the same right of way or runs close by telephone circuits, railroad signaling wires, or similar, there is a possibility of exposure to harmonic interference from the cable. Under these circumstances and for conditions where a local preventive measure is not feasible, for example, improving the shielding of subscriber wires, the communications company should be consulted for permissible interference limits. A typical requirement can be expressed as an equivalent weighted residual current fed into the cable pair at each station. The current is calculated as

$$I_{eq} = (1/P_{800}) \times \sqrt{\sum_h (P_{hf_1} \times I_h)^2}$$

where $I_{eq}$ = weighted, 800 Hz equivalent disturbing current
$I_h$ = vector sum of harmonic currents in cable pair conductors and screens at harmonic $h$
$P_{hf_1}$ = weighting at the frequency of $h$ times the fundamental frequency

***High-Frequency (HF) Filters.*** In voltage source converters, the necessarily high *dv/dt* in the switching of valves means that the high-frequency (HF) noise generation is significantly higher than for conventional HVDC converters. To prevent this HF noise spreading from the converter to the connected power grids, particular attention is given to the design of the valves, to the shielding of the housings, and to ensuring proper HF grounding connections.

***IGBT Valves.*** The insulated gate bipolar transistor (IGBT) valves used in VSC converters are compromised of series-connected IGBT positions. The IGBT is a hybrid device exhibiting the low forward drop of a bipolar transistor as a conducting device (Fig. 6-26). Instead of the regular current-controlled base, the IGBT has a voltage-controlled capacitive gate, as in the MOSFET device.

A complete IGBT position consists of an IGBT, an antiparallel diode, a gate unit, a voltage divider, and a water-cooled heat sink. Each gate unit includes gate-driving circuits, surveillance circuits, and optical interface. The gate-driving electronics control the gate voltage and current at turn-on and turn-off, to achieve optimal turn-on and turn-off processes of the IGBT.

To be able to switch voltages higher than the rated voltage of one IGBT, many positions are connected in series in each valve similar to thyristors in conventional HVDC valves. All IGBTs must turn on and off at exactly the same moment, to achieve an evenly distributed voltage across the valve. Higher currents are handled by paralleling IGBT components or press packs.

**FIGURE 6-26** IGBT valve stacks with corona shields.

## 6.6.4 Converter Control

The fundamental frequency base apparent power of the converter measured at the filter bus between phase reactor and the ac harmonic filters along with its active and reactive power components are defined by the following equations. Voltage and current phasors used in these equations are according to Fig. 6-27.

$$S_b = P + jQ = \sqrt{3} \times U_F \times I_R$$

$$P = \frac{U_F \times U_C \times \sin\delta}{\omega L}$$

$$Q = \frac{U_F \times (U_F - U_C \times \cos\delta)}{\omega L}$$

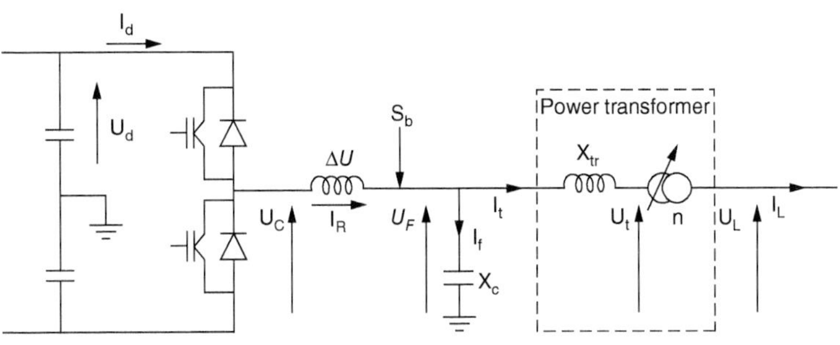


**FIGURE 6-27** Voltage source converter.

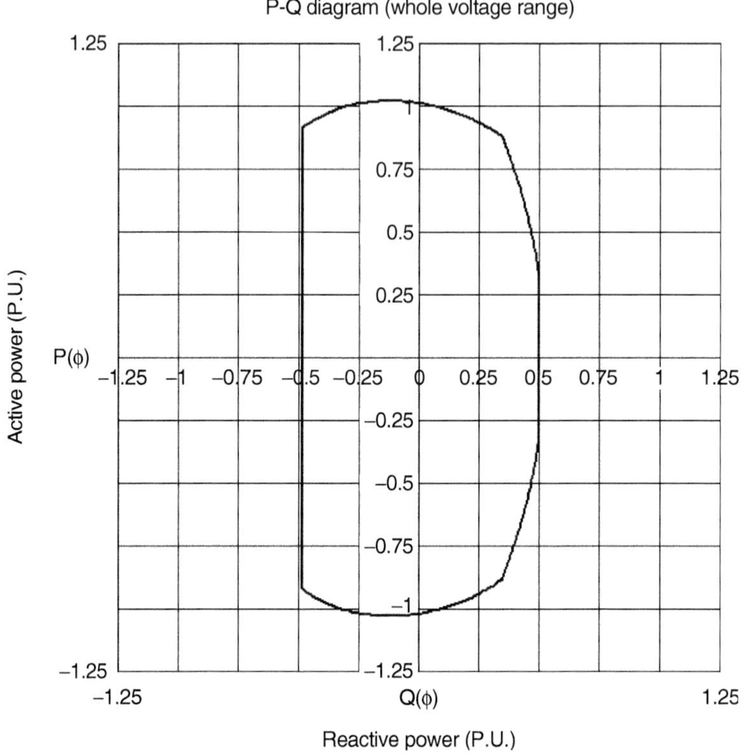


**FIGURE 6-28** VSC station net P-Q characteristics with practical limitations.

The inductance of the converter phase reactor is represented by $L$, and the phase angle between the filter voltage $U_F$ and converter voltage $U_C$ is represented by $\delta$.

The equations illustrate that the power can be controlled by changing the phase angle of the converter voltage with respect to the filter bus voltage, whereas the reactive power can be controlled by changing the magnitude of the converter voltage with respect to the filter bus voltage. Controlling these two aspects of the converter voltage operation in all four quadrants is possible as illustrated in the converter P-Q characteristics shown in Fig. 6-28. This means that the converter can be operated in the middle of its reactive power range near unity power factor to maintain dynamic reactive power reserve for contingency voltage support. It also means that the power transfer can be changed rapidly without altering the reactive power exchange with the ac network or waiting for switching of shunt compensation.

Being able to independently control ac voltage magnitude and phase relative to the system voltage allows use of separate active and reactive power control loops for HVDC system regulation.

The active power control loop can be set to control either the active power or the dc side voltage. In a dc link, one station will then be selected to control the active power while the other must be set to control the dc side voltage. The reactive power control loop can be set to control either the reactive power or the ac side voltage. Either of these two modes can be selected independently at either end of the dc link (Fig. 6-29).

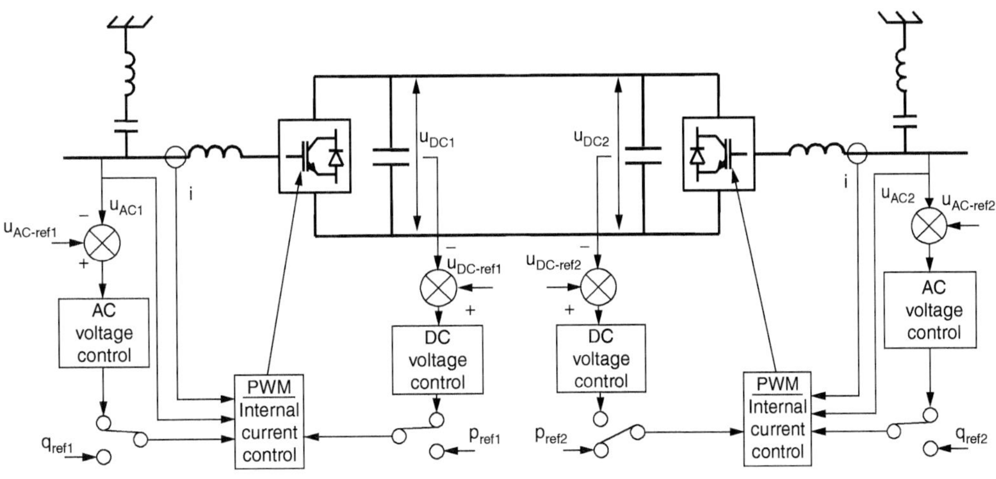


**FIGURE 6-29** VSC-based HVDC system control.

## 6.6.5 Pulse-Width Modulation (PWM) and Harmonic Generation

Pulse width modulation (PWM) of voltage source converters enables independent control of active and reactive power at a constant HVDC voltage using simple two-level converter topology as shown in Fig. 6-30.

A two-level VSC converter can synthesize a balanced set of 3-phase ac converter voltages by injecting either the positive or negative dc voltage on the converter side of the phase reactor. By varying the duration of the positive or negative voltage injections, a sinusoidal voltage with fundamental component at the system frequency can be created. Various PWM switching patterns can be used to minimize harmonics and lower converter switching losses. A PWM pattern with harmonic cancellation or optimized PWM and its harmonic content is shown in Fig. 6-31.

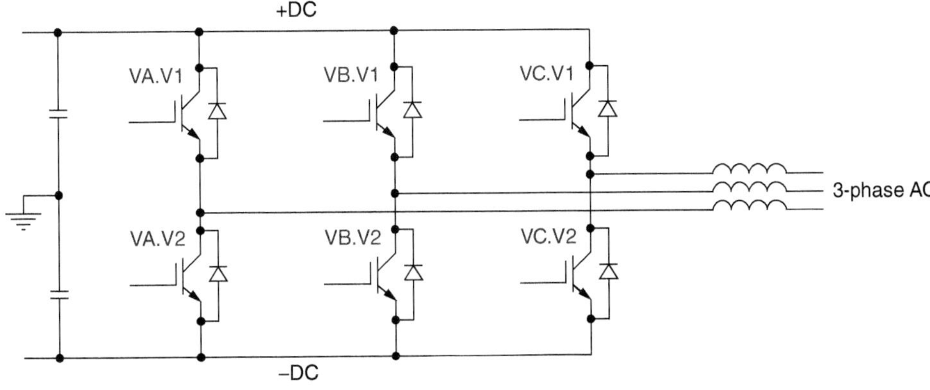


**FIGURE 6-30** VSC two-level converter topology.

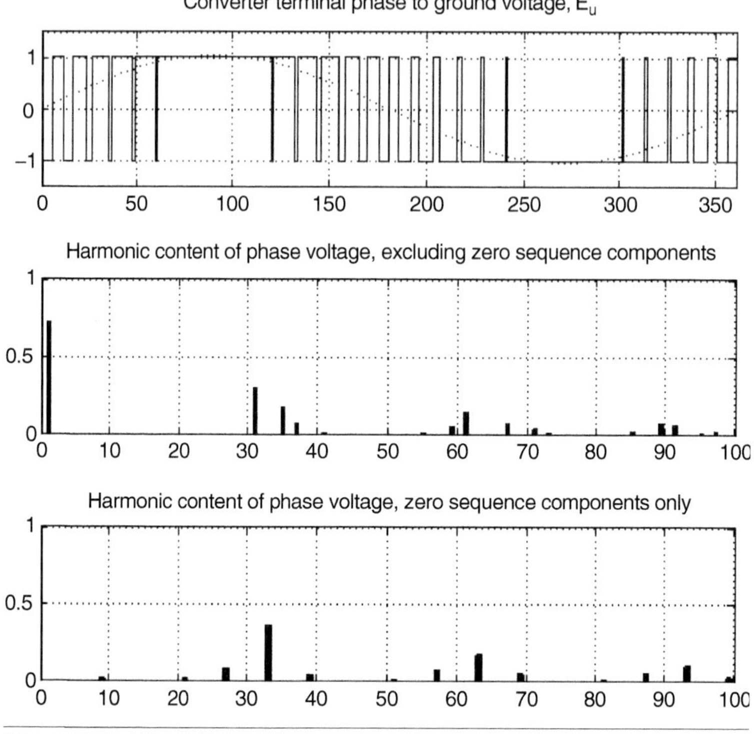


**FIGURE 6-31** PWM with harmonic cancellation for two-level VSC.

## 6.7 OVERHEAD LINES AND CABLES

### 6.7.1 Overhead Transmission Lines

General design criteria for transmission lines can be grouped in the following five categories:

- Power transmission capability
- Power losses
- Insulation coordination
- Corona and field effects
- Mechanical loading

Power transmission capacity is limited by the conductor sag and thermal capacity of the line for the ambient conditions. Emergency loading limits are sometimes used taking into account local conditions and increased sag. These factors affect both EHV ac and HVDC lines. Permissible power transfer levels on EHV ac lines are also affected by surge impedance loading, reactive power compensation, voltage profile, contingency reserve, and stability limits. Transmission on HVDC lines is not limited to reactive power constraints. The HVDC lines cannot become overloaded since the power flow is controlled, therefore contingency reserve is not usually required.

Power losses are due to resistive losses and corona losses. For a given ampacity, resistive losses are lower for an HVDC line than an EHV ac line since the same current is flowing in two sets of conductors in a bipolar dc line compared to three conductors for a 3-phase ac line. Furthermore,

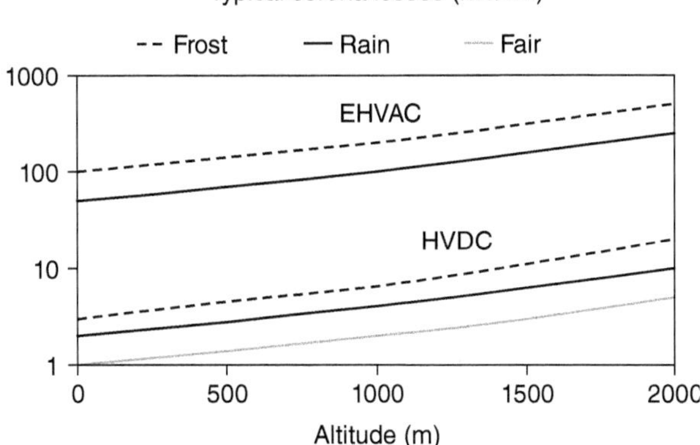


**FIGURE 6-32** Foul weather corona loss comparison of EHVAC and HVDC lines as a function of altitude.

the ac resistance is somewhat higher due to skin effect. Although corona losses for EHV ac lines are about the same as those for HVDC lines during fair weather conditions, they increase much more during foul weather conditions, for example, rain, frost, or snow (Fig. 6-32). This means that larger conductor bundles are needed for EHV ac. Dimensions of corona rings are less critical with HVDC. Due to the lower corona levels with HVDC lines, especially during foul weather, fewer bundled conductors are required to meet given requirements on audible noise (AN) or radio interference (RI).

Air clearance requirements are significantly lower for HVDC lines than for EHV ac lines but are more sensitive to altitude effects. Switching surges are significantly lower for HVDC lines than for EHV ac lines. Switching overvoltages govern the clearances for EHV ac lines whereas lightning overvoltages govern the clearances for HVDC lines.

Insulators made of conventional or composite materials can be used for HVDC. The dc operating voltage grading across the insulator string is resistive rather than capacitive. The lower clearance requirements on insulator string length together with the resistive voltage grading make insulator creepage distance more important for HVDC insulators, especially in areas prone to atmospheric pollution. The frequency and intensity of rain are also an important factor since rain washes away accumulated deposits periodically more so on the top surfaces. Additional insulator creepage distance can be achieved with larger sheds, longer skirts, or longer string lengths. A creepage distance of 2.8 cm/kV for lightly contaminated areas can be considered typical. Special considerations exist for insulator cap-and-pin design and choice of materials due to potential for external leakage currents. Collector rings can be used to trap contaminants mitigating uneven deposition along the insulator surface in polluted areas.

There is no electromagnetic induction from HVDC lines. There is an essential difference in acceptance level for dc fields than for ac fields with higher levels for static fields. The International Commission on Nonionizing Radiation Protection (ICNIRP) places a guideline of 40 μT on the maximum static electromagnetic field for continuous exposure to the general public. This compares to the nominal earth magnetic field of 50 μT. The dc magnetic field is very small for two conductors with current flowing in opposite directions at distances several multiples of the conductor spacing.

The HVDC line towers must bear less static and dynamic loading than EHV ac towers due to fewer conductors and insulators. The ROW requirements are narrower with HVDC. In areas where ROW widths are constrained, vertical configurations require less tower height. Balanced structure loading for vertical configurations can be achieved by use of "portal" structures with pole conductors passing through the center of the structure suspended with V-strings.

### 6.7.2 Underground and Submarine Cables

The HVDC is attractive for higher power transfers over longer distances due to the absence of charging currents and reactive power losses. Fewer cables are needed than for a 3-phase ac circuit. Furthermore, since there is no induction effect with HVDC, cable sheaths do not need to carry the same currents and steel armor can be used for stronger submarine cables.

In ac cables, stress created by the electrical field is distributed in inverse proportion to the capacitance of the cable dielectric. This results in the highest stresses close to the conductor. In dc cables, voltage distribution is determined by insulation resistance and space charges and is dependent on temperature. At higher conductor to sheath temperature gradient, the stress may become highest near the sheath.

Two types of cables are in common use for HVDC transmission: mass-impregnated, nondraining paper-insulated solid cables (MIND), and extruded polymer cables for lower voltage VSC applications (Figs. 6-33 and 6-34). Figure 6-35 shows voltage waveforms for transformer secondary winding, thyristor valve voltage, and dc voltage inside the smoothing reactor. With conventional HVDC, power reversal is achieved by voltage polarity reversal of the cable and 12-pulse harmonic voltages can be imposed on the cable insulation depending on the dc filter design. Figure 6-36 shows phase reactor voltage, valve voltage, and direct voltage for a VSC converter. With VSC transmission, the voltage polarity is constant regardless of transmission direction and switching transients are absorbed by the dc capacitor.

### 6.7.3 Ground Electrodes

Ground and sea return operation has been used for HVDC transmission to decrease investment costs and lower losses in monopolar submarine cable systems and as a temporary return path for pole outages in bipolar systems. Electrode design always ensures safe step potentials, but other important design factors must be taken into account.

Continuous earth return operation is not always possible due to local soil and geological conditions. With typical earth characteristics, return current penetrates deep within the earth and earth surface potential gradients are low and fall off rapidly with distance from the electrode. In cases with shallow, high-resistivity underlying bedrock, however, the current tends to flow more in the surface

**FIGURE 6-33** MIND cables for deep sea applications with conventional HVDC.

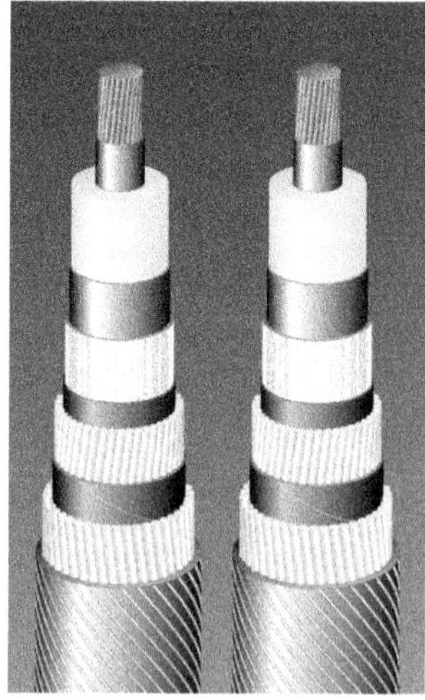

**FIGURE 6-34** Extruded polymer cables for deep sea applications with VSC-based HVDC.

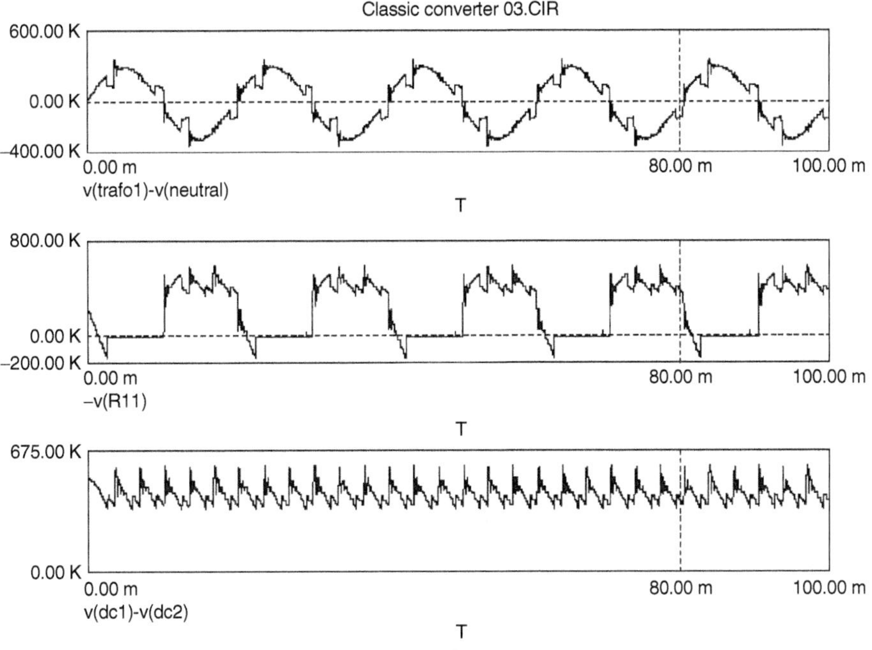


**FIGURE 6-35** Voltages for conventional HVDC transmission. Top trace—converter transformer voltage; Middle trace—valve voltage; Bottom trace—direct voltage (inside smoothing reactor).

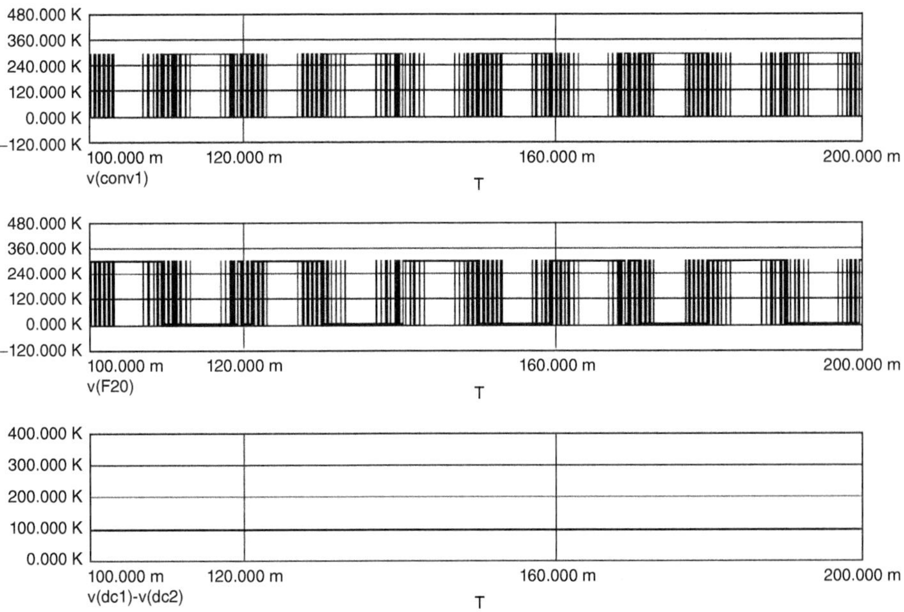


**FIGURE 6-36** Voltages for VSC-based HVDC transmission. Top trace—phase reactor voltage; Middle trace—valve voltage; Bottom trace—direct voltage.

layer and the potential gradient extends further from the electrode site. If other conducting underground utilities, such as pipelines, traverse the potential gradient near the electrode, there is risk of stray current pickup and discharge. Over a long period of time, stray current discharge could cause localized corrosion. Corrosion mitigation methods, such as controlled cathodic protection systems, insulating flanges, or sacrificial anodes, can be used or the ampere-hours for earth return operation can be limited through use of metallic return.

## 6.8 ULTRAHIGH-VOLTAGE DIRECT-CURRENT (UHVDC) TRANSMISSION

Most long-distance HVDC transmission systems with power levels above 1000 MW are at a bipolar voltage level of ± 500 kV. Voltage level for the 2 × 3150 MW Itaipu HVDC transmission system in Brazil has been operating at ± 600 kV since the mid-1980s. Transmission voltages of ± 600 kV to ± 800 kV are classified as UHVDC. Higher-power transfers can be achieved over longer distances with lower losses by increasing the dc voltage level into the UHVDC range. A considerable body of work is ongoing in this area for potential applications in China, India, and North America. The controllability and the mechanical and electrical characteristics of UHVDC lines make them in many respects more favorable for long-distance bulk power transmission than UHVAC lines. Figures 6-37 to 6-39 and Table 6-3 compare differences between UHVAC and UHVDC transmission lines.

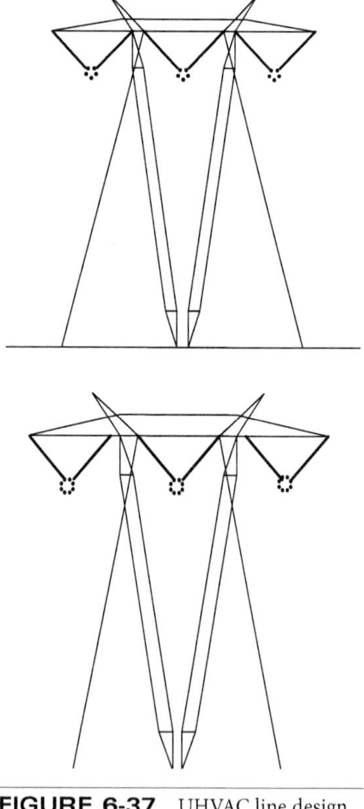

**FIGURE 6-37** UHVAC line design.

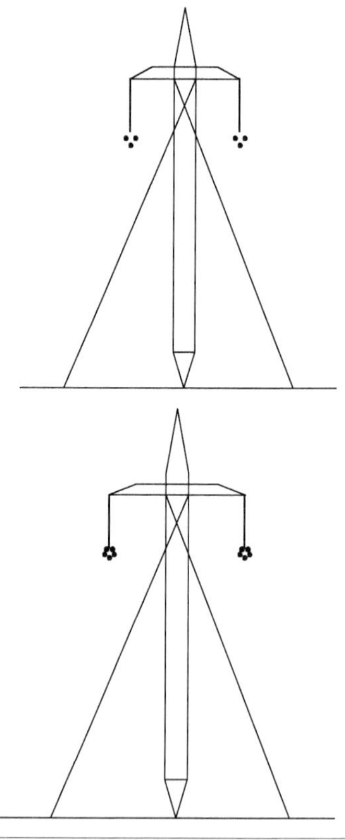

**FIGURE 6-38** UHVDC line design.

**TABLE 6-3** Comparison of Number of Lines for Given Power Transfer with UHVAC and UHVDC

| | | Cond. diam. | Thermal limit (line) | Thermal limit (s/s) | SIL | 1.5 × SIL | Required no. of lines | |
|---|---|---|---|---|---|---|---|---|
| | kv | mm | GW | GW | GW | GW | 8 GW | 12 GW |
| EHVAC | 800 | 5 × 35 | 7.5 | 5.5 | 2.5 | 3.8 | 4 | 5 |
| | 1000 | 8 × 35 | 15.0 | 6.9 | 4.3 | 6.5 | 3 | 3 |
| HVDC | ±600 | 3 × 50 | 8.0 | 5.8 | n.a. | n.a. | 2 | 3 |
| | ±800 | 5 × 50 | 17.7 | 5.8 | n.a. | n.a. | 2 | 3 |

**FIGURE 6-39** 800 kV EHV ac and ± 600 kV dc.

## 6.8.1 Modular Multilevel Voltage Source Converters

Existing 2- or 3-level converters using Pulse Width Modulation (PWM) have to switch the full dc voltage in large steps. This causes high harmonic distortion and high transient stresses resulting in HF noise.

With the new multilevel approach individual module capacitors are uniformly distributed throughout the topology and each level is individually controlled to generate a small voltage step. In this way, each module within the multilevel converter is a discrete voltage source in itself with a local capacitor to define its voltage step without creating ripple voltage distortion across the converter's other phases. By incrementally controlling each step, an almost sinusoidal voltage is generated at the ac outputs of the "multi-valves." The principle of operation of multilevel converter is shown in Fig. 6-40 and the modular multilevel converter topology is shown in Fig. 6-41.

Trans Bay HVDC Cable built by Siemens uses HVDC PLUS system with the modular multilevel converters. It is a 53-mile undersea high-voltage direct-current (HVDC) transmission link between San Francisco's City Center electrical power grid and a Pacific Gas & Electric substation near Pittsburg, California which is shown in Fig. 6-42. The HVDC PLUS system will transmit up to 400 MW at a direct-current (dc) voltage of 200 kV and is the first installation using its HVDC PLUS technology.

Major HVDC manufacturers—ABB, Siemens, and ALSTOM—are developing modular multilevel converter technology which will become standard for all future VSC-based HVDC projects.

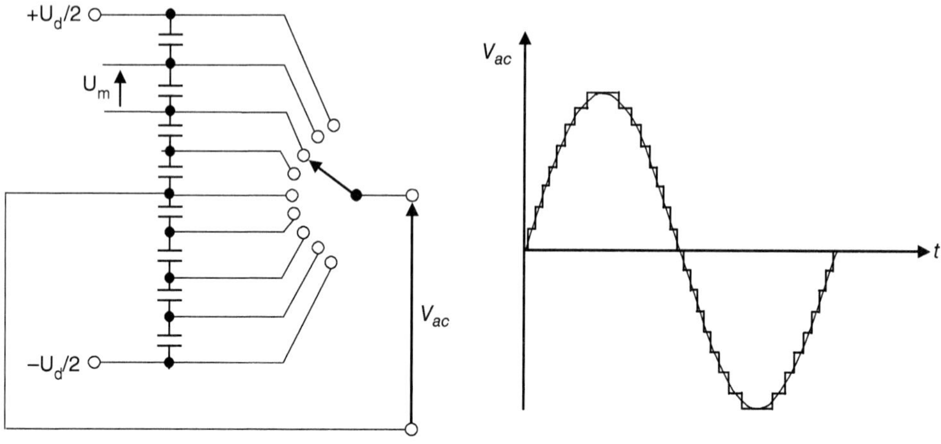


**FIGURE 6-40** Principle of operation of multilevel converter.

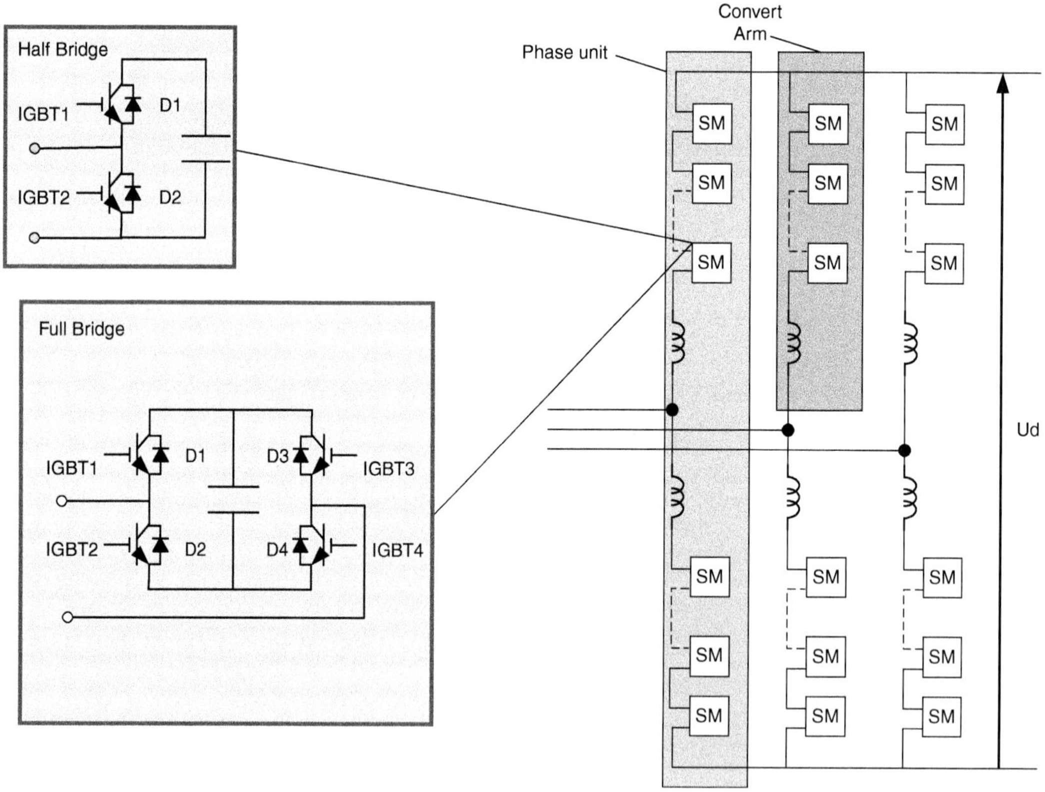

**FIGURE 6-41** Modular multilevel converter topology.

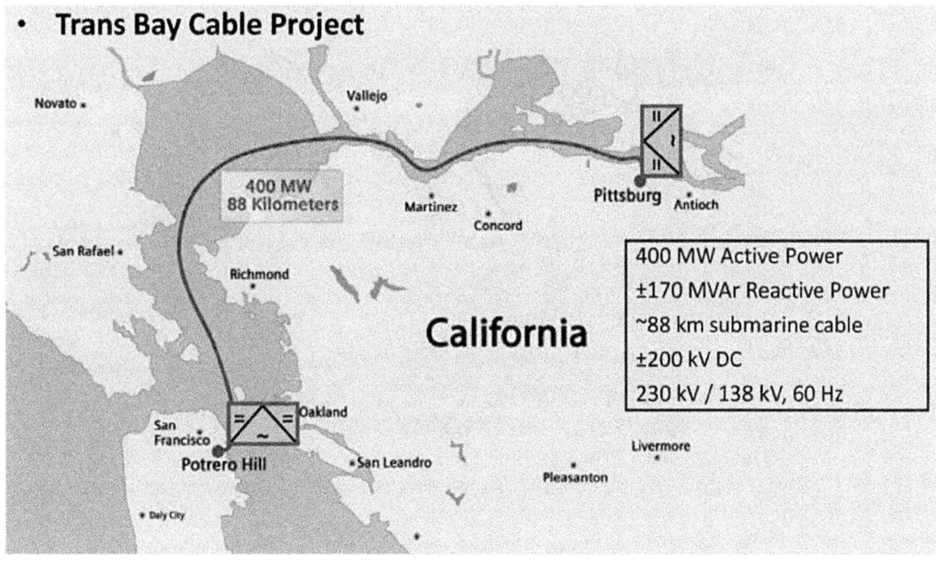


**FIGURE 6-42** Trans Bay Cable VSC DC cable.

## 6.9 UHVDC TRANSMISSION

In many parts of the world, especially countries like China, India, Southern Africa, and Brazil, the demand for power is growing and there is a need for transmitting 6000 MW to 8000 MW with UHVDC lines over 1000 to 2000 km using 800 kV or higher.

The first 800 kV UHV DC system, operated by the China Southern Power Grid Co. in Guangzhou, was built by Siemens, and commenced commercial operation in 2010. It allows China to tap more hydropower instead of adding new coal plants. The $CO_2$ emissions offset amounts to a whopping 33 million tons at the Yunnan-Guangdong project alone. The Yunnan–Guangdong UHVDC transmission system connects Chuxiong in the Yunnan to Suidong, Zengcheng in Guangdong, China. The transmission system has a transmission capacity of 5000 MW and a rated dc current of 3125 A. It has a total length of 1418 km (881 mi). It transmits electricity from the hydropower plants in the Yunnan to the Guangdong, including cities of Guangzhou and Shenzhen which is shown in Fig. 6-43.

Powergrid Corporation of India Ltd. selected ABB to deliver the world's first multi-terminal ±800 kV UHVDC transmission link. The link is known as North East–Agra UHVDC.

The link comprises four terminals located at three converter stations with a 33% continuous overload rating, and the power transmission system will thus have the possibility to convert 8000 MW—which is the largest HVDC transmission ever built. The North East–Agra UHVDC link will have a record 8000 MW converter capacity, including a 2000 MW redundancy, to transmit clean hydroelectric power from the North-Eastern and Eastern region of India to the city of Agra across a distance of 1728 km which is shown in Fig. 6-44.

More UHVDC projects at 800 kV and 1000 kV levels are being planned in both China and India.

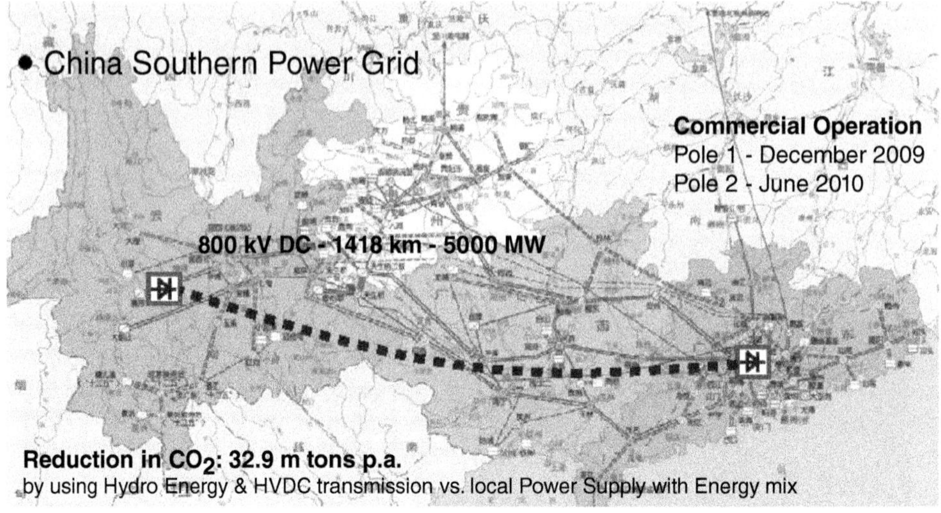


**FIGURE 6-43** Yunnan–Guangdong 800 kV UHVDC link in China.

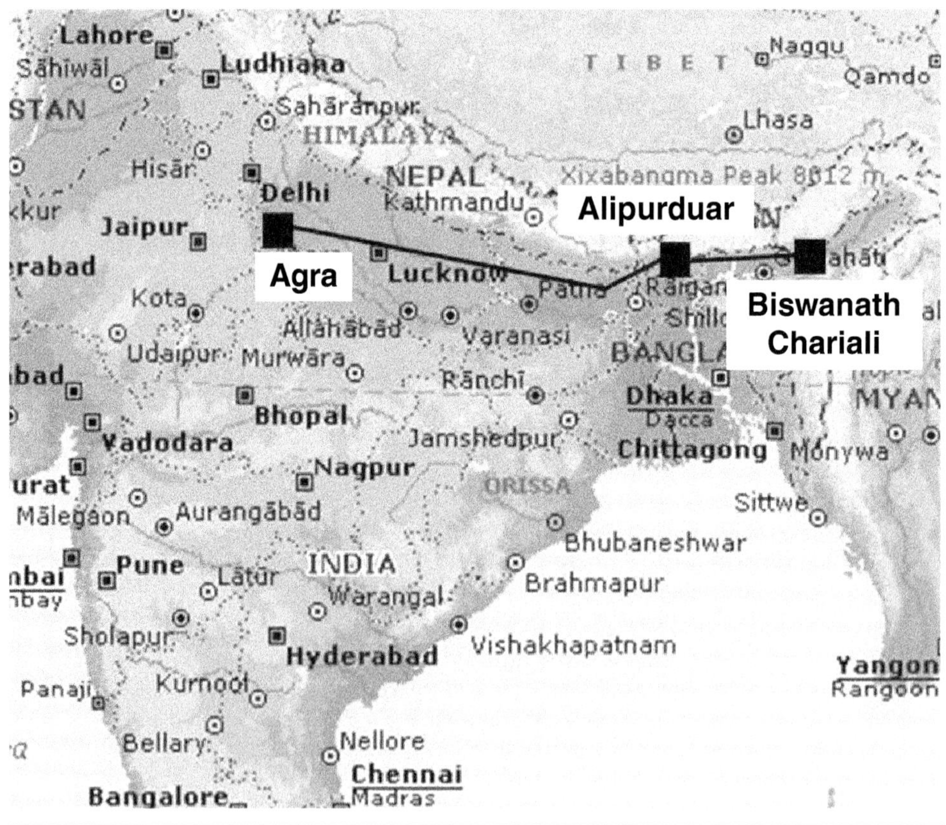


**FIGURE 6-44** North East–Agra UHVDC link in India.

## 6.10 BIBLIOGRAPHY


Adamson, C., and Hingorani, N.G.: *High Voltage Direct Current Power Transmission.* London, Garraway, 1960.

Arkell, C.A., Larsen, K.B., Dellby, B., and Luoini, G.: "Comparison of AC and DC Underground and Submarine Cable Transmission Systems," *CIGRE Symposium S 09-87 on AC/DC Transmission Interactions and Comparisons,* Paper No. 500-02, Boston, 1987.

Asplund, G., Eriksson, K., Jiang, H., Lindberg, J., Palsson, R., and Svensson, K.: "DC Transmission Based on Voltage Source Converters," CIGRE, Paris, 1998.

Asplund, G., Stromberg, H., Blidberg, I., Saksvik, O., and Loof, G.: "Outdoor Thyristor Valve for HVDC," *Proc. IEEE/Royal Institute of Technology Stockholm Power Tech: Power Electronics,* June 18–22, 1995.

Axelsson, U., Holm, A., Liljegren, C., Aberg, M., Eriksson, K., and Tollerz, O.: "The Gotland HVDC Light Project—Experience from Trial and Commercial Operation," CIRED Conf., Amsterdam, The Netherlands, June 18–21, 2001.

Bahrman, M.P., Larsen, E.V., Patel, H.S., and Piwko, R.J.: "Experience with HVDC-Turbine-Generator Torsional Interaction at Square Butte," *IEEE Trans. Power Appar. Syst.,* Vol. PAS-99, May-June 1980, pp. 966–975.

Carlsson, L., and Flisberg, G.: "Recent Classic HVDC Development," IEEE/PES T&D 2002 Latin American Conf., Sao Paulo, Brazil, March 18–22, 2002.

Carlsson, L., and Persson, A.: "New Technologies in HVDC Converter Design," *Proc. IEE Sixth International Conference on AC and DC Transmission,* April 29–May 3, 1996, pp. 387–392.

CIGRE Joint Working Group 13/14-08: "Circuit Breakers for Meshed Multiterminal HVDC Systems," Part I, "DC Side Substation Switching Under Normal and Fault Conditions," *Electra (CIGRE),* No. 163, December 1995, pp. 98–122; Part II, "Switching of Transmission Lines in Meshed HVDC Systems," *Electra (CIGRE),* No. 164, February 1996, pp. 63–82.

CIGRE Working Group 14-05: "Guide for Planning DC Links Terminating at AC System Locations Having Low Short Circuit Capacities," Part I, "AC/DC Interaction Phenomena," *CIGRE Publication* No. 68, 1992.

CIGRE SC 14: "Economic Assessment of HVDC Links," CIGRE Publ. 186.

Ekstrom, A., and Liss, G.: "A Refined HVDC Control System," *IEEE Trans. Power Systems,* Vol. 89, 1970, pp. 723–732.

Fletcher, D.E., and Patterson, N.A.: "The Equivalent Disturbing Current Method for DC Transmission Line Inductive Coordination Studies and DC Filter Performance Specification," *Proc. International Conf. on DC Power Transmission,* Montreal, June 4–8, 1984, pp. 198–204.

*Handbook for Insulation Coordination of High Voltage DC Converter Stations,* EPRI Report EL5414, Electric Power Research Institute, Palo Alto, CA, 1987.

Holmberg, P., Jonsson, T., and Lagerkvist, M: "Properties of Capacitor Commutated Converters in Long HVDC Cable Transmission," EPE'97, Trondheim, Norway.

*HVDC Ground Electrode Design,* EPRI Research Project 1467-1, Report EL-2020, International Engineering Co., Inc.

Jardini, J.A., Reis, L.B., Campos Barros, J.G., and Frontin, S.O.: "HVDC Transmission for Voltages above 600 kV: Evalualtion of DC System and Converter Station Major Equipment," *Proc. IEEE/IREQ International Conf. on DC Power Transmission,* Montreal, 1984, pp. 71–78.

Kanngiesser, K.W., Bowles, J.P., Ekstrom, A., Reeve, J., and Rumpf, E.: "HVdc Multiterminal Systems," *CIGRE,* Paper 14-08 (ELECTRA.), August 1974.

Karlsson, T., and Liss, G.: "HVDC Transmission with Extremely Long DC Cables, Control Strategies," IEEE/KTH Stockholm Power Tech Conf, Stockolm, Sweden, June 1995.

Kimbark, E.W.: *Direct Current Transmission.* New York, Wiley-Interscience, 1971.

Krishnayya, P.C.S.: "Important Characteristics of Thyristors of Valves for HVDC and Static Var Compensators," *CIGRE Conf. Proc.,* Paper 14-10, 1984.

Lahtinen, M.: "Connection of Harmonic Producing Installations in AC High Voltage Networks with Particular Reference to HVDC: Guide for Limiting Interference Caused by Harmonic Currents with Special Attention for Telecommunications Systems," *Electra (CIGRE),* No. 159, April 1995, pp. 26–48.

McCallum, D., Moreau, G., Primeau, J., Bahrman, M.P., Ekehov, B., and Soulier, D.: "Multiterminal Integration of the Nicolet Converter Station into the Quebec-New Enland Phase II HVDC Transmission System," *CIGRE Proceedings,* Paper 14-103, Paris, 1994.

Railing, B.D., Miller, J.J., Moreau, G., Bard, P., Ronstrom, L., and Lindberg, J.: "Cross Sound Cable Project Second Generation VSC Technology for HVDC," CIGRE, Paris, France, 2004.

Sakshaug, E.C., Kresge, J.S., and Miske, S.A.: "Arrester Protection of High Voltage DC Transmission System Converter Terminals," IEEE Transactions on Power Apparatus and Systems, vol. PAS-90, no. 4, pp. 1555–1562, July 1971.

Uhlmann, E.: *Power Transmission by Direct Current.* Berlin, Springer-Verlag, 1975.

Weimers, L.: "Bulk Power Transmission at Extra High Voltages. A Comparison between Transmission Lines for HVDC at Voltages above 600 kV DC and 800 kV AC," CEPSI 2004, Shanghai, China, Oct. 17–22, 2004.

Zhang, W., Isalsson, A.J., and Ekstrom, A.: "Analysis on the Control Principle of the Active DC Filter in the Lindome Converter Station of the Konti-Skan HVDC Link," IEEE Transactions on Power Systems, vol. 13, no. 2, pp. 374–381, May 1998.

# 7

# POWER DISTRIBUTION

Surya Santoso

*Professor, Department of Electrical and Computer Engineering, University of Texas at Austin; Fellow, IEEE*



Grateful acknowledgment is given to past contributors to this section: James J. Burke, Allen L. Clapp, Walter J. Ros, Daniel J. Ward, and Cheryl A. Warren.

## 7.1 OVERVIEW OF POWER DISTRIBUTION AND PRACTICE

Broadly speaking, *distribution* includes all parts of an electric utility system between bulk power sources and the consumers' service-entrance equipments. Some electric utility distribution engineers, however, use a more limited definition of distribution as that portion of the utility system between the distribution substations and the consumers' service-entrance equipment. In general, a typical distribution system consists of (1) subtransmission circuits with voltage ratings usually between 69 and 138 kV which deliver energy to the distribution substations, (2) distribution substations which convert the energy to a lower *primary system* voltage for local distribution and usually include facilities for voltage regulation of the primary voltage, (3) primary circuits or *feeders*, usually operating in the range of 4.16 to 34.5 kV and supplying the load in a well-defined geographic area, (4) distribution transformers with ratings from 10 to 2500 kVA which may be installed on poles or grade-level pads or in underground vaults near the consumers and transform the primary voltages to utilization voltages, (5) secondary circuits at utilization voltage which carry the energy from the distribution transformer along the street or rear-lot lines, and (6) service drops which deliver the energy from the secondary to the user's service-entrance equipment. Figures 7-1 and 7-2 depict the component parts of a typical distribution system.

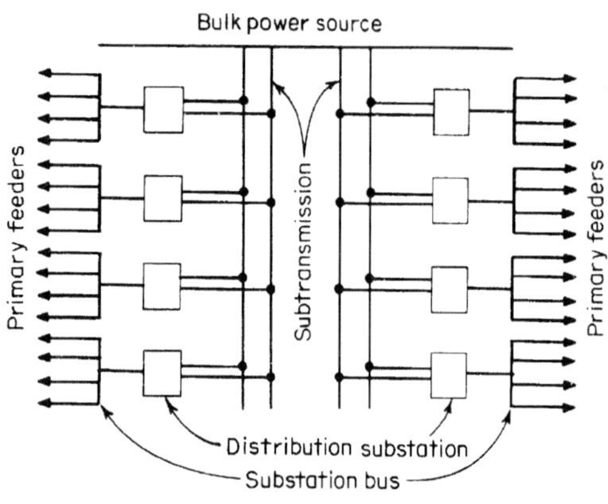


**FIGURE 7-1** Typical distribution system.

Distribution investment constitutes 50% of the capital investment of a typical electric utility system. Recent trends away from generation expansion at many utilities have put increased emphasis on distribution system development.

The function of distribution is to receive electric power from large, bulk sources and to distribute it to consumers at voltage levels and with degrees of reliability that are appropriate to the various types of users.

For single-phase residential users, American National Standard Institute (ANSI) C84.1-2016 defines *Voltage Range A* as 114/228 V to 126/252 V at the user's service entrance and 110/220 V to 126/252 V at the point of utilization. This allows for voltage drop in the consumer's system. Nominal voltage is 120/240 V. Within Range A utilization voltage, utilization equipment is designed and rated to give fully satisfactory performance.

As a practical matter, voltages above and below Range A do occur occasionally; however, ANSI C84.1 specifies that these conditions shall be limited in extent, frequency, and duration. When they do occur, corrective measures shall be undertaken within a reasonable time to improve voltages to meet Range A requirements.

Rapid dips in voltage which cause incandescent-lamp "flicker" should be limited to 4% or 6% when they occur infrequently and 3% or 4% when they occur several times per hour. Frequent dips, such as those caused by elevators and industrial equipment, should be limited to 1½% or 2%.

Reliability of service can be described by factors such as frequency and duration of service interruptions. While short and infrequent interruptions may be tolerated by residential and small commercial users, even a short interruption can be costly in the case of many industrial processes and can be dangerous in the case of hospitals and public buildings. For such sensitive loads, special measures are often taken to ensure an especially high level of reliability, such as redundancy in supply circuits and/or supply equipment. Certain computer loads may be sensitive not only to interruptions

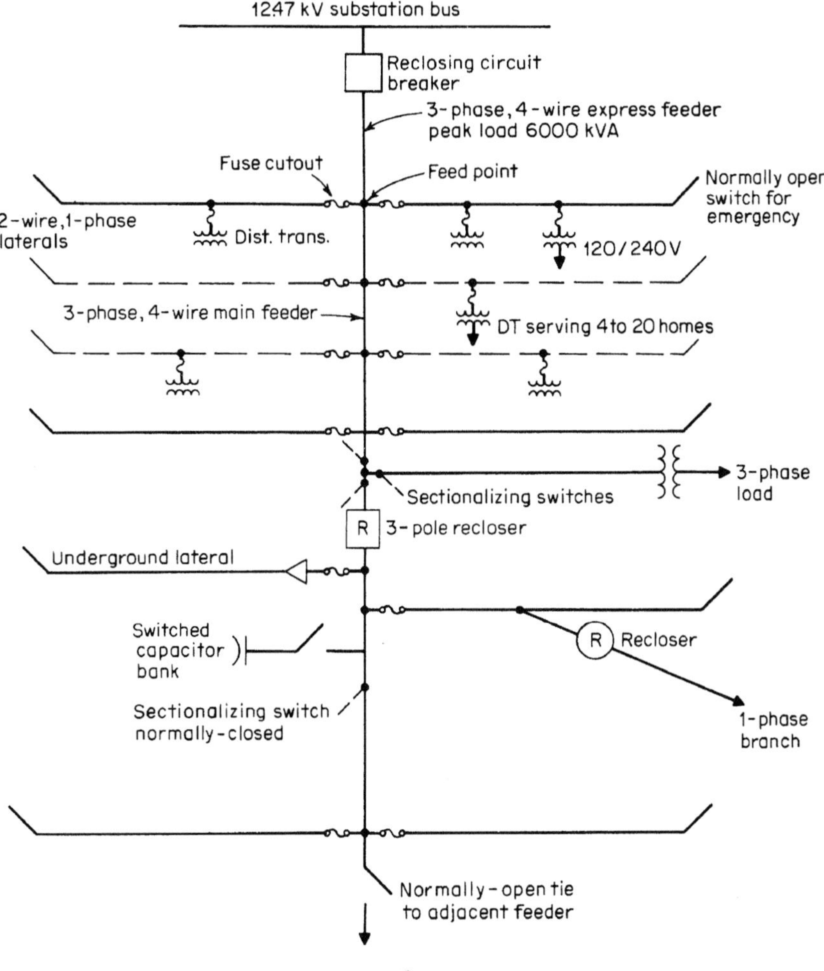


**FIGURE 7-2** One-line diagram of typical primary distribution feeder.

but even to severe voltage dips and may require special power-supply systems which are virtually uninterruptible.

From a system-planning and design point of view, the optimal choice of subtransmission voltage and system arrangement is closely interrelated with distribution substation size and with the primary distribution voltage level. At any given time, the most economical arrangement is achieved when the sum of the subtransmission, substation, and primary feeder costs to serve an area is a minimum over the life of the facilities. In practice, the number, size, and availability of bulk supply sources for feeding the subtransmission may be significant factors as well.

A distribution system should be designed so that anticipated load growth can be served at minimum expense. This flexibility is needed to handle load growth in existing areas as well as load growth in new areas of development.

Overhead and underground distribution systems are both used in large metropolitan areas. In the past in smaller towns and in the less-congested areas of larger cities, overhead distribution was almost universally used; the cost of underground distribution for residential areas was several times that of overhead. During the past 25 to 30 years, the cost of underground residential distribution (URD) has been reduced drastically through the development of low-cost, solid-dielectric cables suitable for direct burial, mass production of pad-mounted distribution transformers and accessories, mechanized cable-installation methods, etc. The cost of a typical URD system for a new residential subdivision is about 50% greater than that of an overhead system in many areas; in others, there is little or no differential due to local land conditions. As a result, some utilities will justifiably have some type of extra charge for underground. With the increased public interest in improving the appearance of residential areas and the declining cost of URD, the growth of URD has been extremely rapid. Today, perhaps as much as 70% of new residential construction is served underground. A number of states have enacted legislation making underground distribution mandatory for new residential subdivisions.

Rural service has been extended to most farmers and rural dwellers through the efforts of utilities, cooperatives, and government agencies. Rural construction must be of the least-expensive type consistent with durability and reliability because there may be only a few users per mile of line. Historically, rural construction has been overhead, but the advent of cable-plowing techniques has made underground economically competitive with overhead in some parts of the country, and a growing amount of rural distribution is being installed underground.

Higher primary voltages of 24.9Y/14.4 and 34.5Y/19.92 kV are continuing to grow in usage, although primary voltages in the 15-kV class predominate. The 5-kV class continues to decline in usage. Surveys indicate that in recent years approximately 78% of the overhead and underground line additions are at 15 kV, 11% are at 25 kV, and 7.5% are at 35 kV.

Generally, when a higher distribution voltage is initiated, it is built in new, rapidly growing load areas. The economic advantage of the higher voltages usually is not great enough to justify massive conversions of existing lower-voltage facilities to the higher level. The lower-voltage areas are contained and gradually compressed over a period of years as determined by economics, obsolescence, and convenience. Virtually, all modern primary systems serving residential and small commercial and small industrial loads are 4-wire, multigrounded, common-neutral systems.

## 7.2 DISTRIBUTION-SYSTEM AUTOMATION

Distribution automation (DA), a system to monitor and control the distribution system in real time, was gradually introduced in the 1970s more as a concept than a fully developed plan. Unlike the introduction of EMS, where utilities readily saw the benefits of automatic generation control and economic dispatch and adopted the technology, utilities were much more cautious in their approach to distribution automation.

Early distribution automation projects were undertaken by a handful of utilities. The technology was changing and evolving so much so that DA was being touted as an amorphous system capable of covering any imaginable function under the sun. A 1984 EPRI project, Guidelines for Evaluating Distribution Automation, focused attention on what functions could be automated and what value could be attached to those functions. A positive result of this project is that it got people thinking about what functions mattered most. However, it was a little bit ahead of its time in that there wasn't much standardization in systems employed for DA and one couldn't simply select functions of interest and expect to obtain a system that could be built for the total value of the functions selected. Then too, the choice of the communications systems (e.g., telephone, fiber optics, radio, carrier, etc.) proved to be a barrier to widespread implementation.

At the substation level, equipment loadings became an early focus, and asset management became a desired function for DA systems. In addition, the ability to trip distribution circuit breakers and transfer load between substations was commonplace as SCADA was added and this represented the extent of distribution automation to many companies.

Volt/var control, that is, controlling the combination of load tap changers (LTC) or voltage regulators and switched capacitor banks within a substation, was a function many companies incorporated with DA. With adoption of microprocessor relays and fault distance relaying, some incorporated the output information from fault distance relays and diagnostic alarms from various subsystems to be part of the DA package.

Moving outside the substation, controlling automated circuit tie switches was prompted by reliability considerations. Having SCADA links to other reclosers, particularly the ones with microprocessor controls, enabled more ability to remotely control field switching and achieve more rapid restoration of service.

In recent years, with the dropping prices of renewable energy and the growing awareness toward the environmental impact of fossil fuels, distribution utilities are facing increased levels of distributed energy resource (DER) integration. The distributed and variable nature of some of these DER creates challenges for a classical distribution grid built for centralized generation and predictable passive loads. While the impacts of the DER at low penetration levels can be ignored, as the levels rise, additional measures are necessary to manage the DER without adversely impacting the distribution grid operation practices. Distributed energy resource management system (DERMS) is a module of DA developed recently to support the DER management. The DERMS is a software solution that offers control and flexibility to effectively manage the DER such as solar and wind generation, energy storage, demand-response resources, electric vehicles, and combined heat and power facilities that supply or consume energy. The DERMS software provides the grid operators a set of tools to monitor, optimize, and dispatch DERs in real time to meet grid and market needs. As on date, many pilot DERMS projects have been started in the United States by utilities in states with lots of DER to handle such as Hawaii and California. The industrial giants like Siemens, ABB, and General Electric are offering DERMS products.

## 7.3 CLASSIFICATION AND APPLICATION OF DISTRIBUTION SYSTEMS

Distribution systems may be classified according to

- Voltage—120 V, 12,470 V, 34,500 V, etc.
- Scheme of connection—radial, loop, network, multiple, and series
- Loads—residential, small light and power, large light and power, street lighting, railways, etc.
- Number of conductors—2-wire, 3-wire, 4-wire, etc.
- Type of construction—overhead or underground
- Number of phases—single-phase, 2-phase, or 3-phase; and as to frequency: 25, 60 Hz, etc.

***Application of Systems.*** In American practice, alternating-current (ac) 60-Hz systems are almost universally used for electric power distribution. These systems comprise the most economical method of power distribution, owing in large measure to the ease of transforming voltages to levels appropriate to the various parts of the system. These transformations are accomplished by means of reliable and economical transformers. By proper system design and the application of overvoltage and overcurrent protective equipment, voltage levels and service reliability can be matched to almost any consumer requirement.

Single-phase residential loads generally are supplied by simple radial systems at 120/240 V. The ultimate in service reliability is provided in densely loaded business/commercial areas by means of grid-type secondary-network systems at 208Y/120 V or by "spot" networks, usually at 480Y/277 V. Secondary-network systems are used in about 90% of the cities in this country having a population of 100,000 or more and in more than one-third of all cities with populations between 25,000 and 100,000.

Where secondary-network systems do not supply sufficiently reliable service for critical loads, emergency generators and/or batteries are sometimes provided together with automatic switching equipment so that service can be maintained to the critical loads in the event that the normal utility supply is interrupted. Such loads are found in hospitals, computer centers, key industrial processes, etc.

Single-phase residential loads are almost universally supplied through 120/240-V, 3-wire, single-phase services. Large appliances, such as ranges, water heaters, and clothes dryers, are served at 240 V. Lighting, small appliances, and convenience outlets are supplied at 120 V.

An exception to the preceding comments occurs when the dwelling unit is in a distributed secondary-network area served at 280Y/120 V. In this case, large appliances are supplied at 208 V and small appliances at 120 V.

Three-phase, 4-wire, multigrounded, common-neutral primary systems, such as 12.47Y/7.2 kV, 24.9Y/14.4 kV, and 34.5Y/19.92 kV, are used almost exclusively. The fourth wire of these Y-connected systems is the neutral for both the primary and the secondary systems. It is grounded at many locations. Single-phase loads are served by distribution transformers, the primary windings of which are connected between a phase conductor and the neutral. Three-phase loads can be supplied by 3-phase distribution transformers or by single-phase transformers connected to form a 3-phase bank. Primary systems in the 15-kV class are most commonly used, but the higher voltages are gaining acceptance. Figure 7-2 illustrates a typical radial primary feeder.

The 4-wire system is particularly economic for URD systems because each primary lateral or branch circuit consists of only one insulated phase conductor and the bare, uninsulated neutral rather than two insulated conductors. Also, only one primary fuse is required at each transformer and one surge arrester in overhead installations.

Three-phase, 3-wire primary systems are not widely used for public distribution, except in California. They can be used to supply single-phase loads by means of distribution transformers having primary winding connected between 2-phase conductors. Single-phase primary laterals consist of two insulated phase conductors; each single-phase distribution transformer requires two fuses and two surge arresters (where used). Three-phase loads are served through 3-phase distribution transformers or appropriate 3-phase banks. Two-phase systems are rarely used today.

## 7.4 CALCULATION OF VOLTAGE REGULATION AND $I^2R$ LOSS

When a circuit supplies current to a load, it experiences a drop in voltage and a dissipation of energy in the form of heat. In dc circuits, voltage drop is equal to current in amperes multiplied by the resistance of the conductors, $V = IR$. In ac circuits, voltage drop is a function of load current and power factor and the resistance and reactance of the conductors. Heating is caused by conductor losses; for both dc and ac circuits they are computed as the square of current multiplied by conductor resistance in ohms. Watts = $I^2R$, or kW = $I^2R/1000$. Capacitance can usually be neglected for calculation in distribution circuits because its effect on voltage drop is negligible for the circuit lengths and operating voltages used. In circuit design, a conductor size should be selected so that it will carry the required load within specified voltage-drop limits and will have an optimized value of installed cost and cost of losses. Today, a conductor size meeting these criteria will operate well within safe operating temperature limits. In some cases, short-circuit current requirements will dictate the minimum conductor size.

Percent voltage drop or percent regulation is the ratio of voltage drop in a circuit to voltage delivered by the circuit, multiplied by 100 to convert to percent. For example, if the drop between a transformer and the last customer is 10 V and the voltage delivered to the customer is 240, the percent voltage drop is $10/240 \times 100 = 4.17\%$. Often the nominal or rated voltage is used as the denominator because the exact value of delivered voltage is seldom known.

Percent $I^2R$ or percent conductor loss of a circuit is the ratio of the circuit $I^2R$ or conductor loss, in kilowatts, to the kilowatts delivered by the circuit (multiplied by 100 to convert to percent). For example, assume a 240-V single-phase circuit consisting of 1000 ft of two No. 4/0 copper cables supplies a load of 100 A at unity power factor.

$$I^2R = 100^2 \times 2 \times 0.0512 = 1024 \text{ W} = 1.024 \text{ kW}$$
$$\text{Load delivered} = 240 \times 100 = 24{,}000 \text{ W} = 24 \text{ kW}$$
$$\% I^2R \text{ loss} = 1.024/24 \times 100 = 4.26\%$$

Direct-current voltage drop is easily calculated by multiplying load amperes $I$ by ohmic resistance $R$ of the conductors through which the current flows (see Sec. 4 for ohmic resistance of various conductors).

*Example.* A 500-ft dc circuit of two 4/0 copper cables carries 200 A. What is the voltage drop? Resistance of 1000 ft of 4/0 copper cable is 0.0512 Ω.

$$\text{Drop} = IR = 200 \times 0.0512 = 10.24 \text{ V}$$

If 240 is the delivered voltage,

$$\% \text{ regulation} = 10.24/240 \times 100 = 4.26\%$$

$I^2R$ or conductor loss in dc or ac circuits is calculated by multiplying the square of the current in amperes by ohmic resistance of the conductors through which the current flows. The result is in watts.

In dc circuits, percent voltage drop and percent conductor loss are identical.

$$\% \text{ voltage drop} = IR/V \times 100$$
$$\% I^2R = I^2R/VI \times 100 = IR/V \times 100$$

In ac circuits, the ratio of percent conductor loss to percent voltage regulation is given approximately by the following approximate formula:

$$\frac{\% I^2R \text{ loss}}{\% \text{ voltage drop}} = \frac{\cos\phi}{\cos\theta \cos(\phi - \theta)} \tag{7-1}$$

where $\theta$ = power-factor angle and $\phi$ = impedance angle; that is, $\tan \phi = X/R$.

Table 7-1 gives voltage drop in volts per 100,000 A · ft for 2-wire dc circuits for a number of conductor sizes. Ampere-feet is the product of the number of amperes of current flowing and the distance in feet between the sending and receiving terminals multiplied by 2 to take into account the drop in both the outgoing and return conductors; or the feet can be considered to be the total number of conductor feet, outgoing and return.

Table 7-1 also gives the voltage drop for 3-wire circuits when serving balanced loads, where the term "feet" is taken to mean twice the number of feet between sending and receiving terminals.

**TABLE 7-1** Voltage Drop in Volts per 100,000 A · ft, 2-Wire DC Circuits (Loop)

| Conductor size, AWG or kcmil | | Volts drop per 100,000 A · ft, 90° copper temp |
|---|---|---|
| Copper | Approx. equivalent aluminum | |
| 6 | 4 | 102.8 |
| 4 | 2 | 64.6 |
| 2 | 1/0 | 40.7 |
| 1/0 | 3/0 | 25.6 |
| 2/0 | 4/0 | 20.3 |
| 4/0 | 336 | 12.8 |
| 350 | 556 | 7.71 |
| 500 | 795 | 5.39 |
| 1000 | | 2.70 |
| 1500 | | 1.80 |
| 2000 | | 1.35 |

***Note:*** 1 ft = 0.3048 m.

*Example 1.* What is the voltage drop and percent voltage drop when 200 A dc flows 1500 ft one way through a 2-wire, 120-V, 556-kcmil aluminum circuit? First determine ampere-feet factor as 100 × 1500/100,000 = 1.5. From Table 7-1, the voltage drop is 7.71 V per 100,000 A · ft. This value multiplied by the 1.5 factor gives the total voltage drop = 1.5 × 7.71 = 11.6 V. The percent voltage drop = 11.6 × 100/120 = 9.64%. The percent conductor loss also is 9.64%, which is equivalent to 120 × 100 × 0.0954 = 1.16 kW.

*Example 2.* A mine 1 mile from a motor-generator station must receive 100 kW dc at not less than 575 V. Maximum voltage of the generator is 600 V. What conductor size should be used?

$$\text{Max. current} = \frac{100{,}000\ \text{W}}{575\ \text{V}} = 173.9\ \text{A}$$

$$\text{Loop ft} = 2 \times 5280 = 10{,}560\ \text{ft}$$

$$\frac{\text{A} \cdot \text{ft}}{100{,}000} = \frac{173.9 \times 10{,}560}{100{,}000} = 18.36$$

$$18.36 \times \text{voltage drop per } 100{,}000\ \text{A} \cdot \text{ft from Table 7-1} = 25\ \text{V}$$

Therefore, voltage drop per 100,000 A · ft = 25/18.36 = 1.36. From Table 7-1, the copper conductor size corresponding to 1.36 V/100,000 A · ft is 2000 kcmil copper.

***Calculating Voltage Drop in AC Circuits.*** The voltage drop per mile in each round wire of 3-phase 60-Hz line with equilateral spacing $D$ inches between centers or in each wire of a single-phase line $D$ inches between centers is

$$\tilde{V}\ \text{drop} = \tilde{I}R + j\tilde{I}\left(0.2794 \log \frac{D}{r} + 0.03034\mu\right) \quad \text{volts in phasor form} \qquad (7\text{-}2)$$

where $\tilde{I}$ is in phasor amperes, $R$ is the 60-Hz resistance of the wire per mile, Ω, log is the log to base 10, $r$ is the radius of round wire, in, and $\mu$ is the permeability of the wire (unity for nonmagnetic materials such as copper or aluminum). $j$ in Eq. (7-2) denotes an angle of 90°; $+j$ means 90° leading, $-j$ means 90° lagging. Thus, the expression for phasor current lagging the reference voltage is $\tilde{I} = I_x - jI_y = I\sqrt{\theta°}$ with reference to a conveniently chosen horizontal axis of reference—usually sending- or receiving-end voltage. The symbol ~ over $I$ or $V$ indicates phasor values. Voltage drops determined in this manner are also phasors and are with respect to the reference axis.

When wire is stranded, an equivalent radius must be used for $r$ in Eq. (7-2). $r = 0.528\sqrt{A}$ for 7 strands, $r = 0.5585\sqrt{A}$ for 19 strands, $r = 0.5675\sqrt{A}$ for 37 strands, where $r$ = equivalent radius, in, and $A$ = area of metal, in$^2$.

Frequency is 60 Hz for the constants in parentheses in Eq. (7-2), which gives reactance $X$ in ohms per mile. For 25 Hz, multiply by 25/60. The equation is sometimes written

$$\tilde{V}\ \text{drop per mile} = \tilde{I}(R + jX) = \tilde{I}\tilde{Z} \qquad \text{volts in phasor form} \qquad (7\text{-}3)$$

where $I$ is in phasor amperes and $Z = Z/\phi \cdot \Omega$/mi at 60 Hz.

Three unsymmetrically spaced wires $a$, $b$, and $c$ of a 3-phase circuit with correct transpositions can have voltage drop in each wire calculated by Eq. (7-2) by substituting for $D$ the geometric mean of the three interaxial distances.

$$D = \sqrt[3]{D_{ab} D_{bc} D_{ca}}$$

***The Phasor Method.*** In Eq. (7-3), $I$ is in vector amperes,

$$\tilde{I} = I_x - jI_y = I\angle\theta$$

where $\theta$ is the angle that the current lags (or leads) the voltage. The sending-end voltage is usually chosen as the axis, or phasor, of reference in drawing the phasor diagram. For example, consider

Fig. 7-3, where sending voltage $V_s = V_s \angle\theta°$, load current $I = I\angle\theta°$, circuit impedance $\tilde{Z} = Z = \angle\theta° = R + jX$, and load voltage $\tilde{V}_L = \tilde{V}_s - \tilde{I}\tilde{Z}$ (all phasors). The symbol $\angle$ is used for positive angles, assuming that the counterclockwise direction from the phasor or reference is positive and the clockwise directions negative. Assume that $V_s = 230/0°$, $\tilde{I} = 50\angle{-36.87°}$, $\tilde{Z} = 0.2\angle 71.57°$, and $\tilde{Z} = R + jX$. Thus

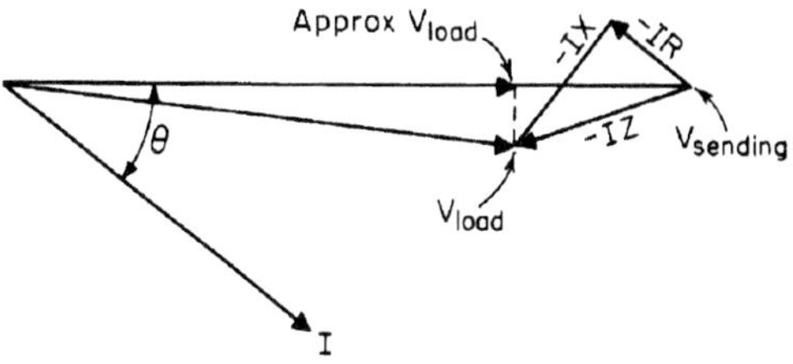


**FIGURE 7-3** Phasor diagram showing voltage relationships.

$$
\begin{aligned}
\tilde{V}_L &= 230/0° - 50\angle 36.87° \times 0.2\angle 71.57° \\
&= 230\angle\theta° - 10\angle 34.70° \\
&= 230 - 10 \cos 34.70° - j\, 10 \sin 34.70° \\
&= 230 - 8.22 - j\, 5.69 = 221.78 - j\, 5.69 \\
&= 221.78 \text{ (very nearly)}
\end{aligned}
$$

Neglecting the term $-j$ 5.69 simplifies the final calculation and gives the load voltage within a fraction of 1% of the precise result. This method is sufficiently accurate for practically all distribution engineering calculations and can be thought of as

$$
V \text{ drop} = IR\cos\theta + IX\sin\theta = IZ\cos(\phi - \theta) \tag{7-4}
$$

where $I$ and $Z$ are absolute magnitudes, not phasor quantities, $\phi$ is the impedance angle, and $\theta$ is the power-factor angle by which the current lags (or leads) the voltage. Calculating the drop in the above example by this method,

$$
\begin{aligned}
V \text{ drop} &= 50 \times 0.2 \times \cos 71.57° \times \cos 36.87° \\
&\quad + 50 \times 0.2 \times \sin 71.57° \times \sin 36.87° \\
&= 2.53 + 5.69 = 8.22 \text{ V}
\end{aligned}
$$

or

$$
\begin{aligned}
V \text{ drop} &= IZ\cos(\phi - \theta) = 50 \times 0.2 \times \cos(71.57° - 36.87°) \\
&= 10\cos(34.7°) = 10 \times 0.822 = 8.22 \text{ V}
\end{aligned}
$$

Impedance $Z$ can be visualized as the hypotenuse of a right triangle in which the base is the resistance $R$ and the altitude is the reactance $X$. In phasor form, $\tilde{Z} = R \pm jX$, where the positive sign is used for inductive reactance and the negative sign for capacitive reactance. Impedance also can be expressed as $\tilde{Z} = Z\angle\phi°$, where $Z$ is the absolute magnitude and $\phi$ is the angle between $\tilde{Z}$ and $R$ in Fig. 7-4. This angle is an absolute value in that it has no relationship to the axis of reference in a phasor diagram, as do voltage and current. Alternating current causes a voltage drop in resistance which is in time phase with the current and in inductive reactance a drop which leads the current by 90 electrical degrees, assuming the positive direction for measurement of angles is counterclockwise. Or conversely, the current in an inductive reactance lags the voltage drop by 90°.

***Impedance Values.*** Tables are available which give 60-Hz impedance values in ohms per 1000 ft for common sizes of wire and cable. The values can be expressed in the form $\tilde{Z} = R + jX$, which can be converted to the form $Z\angle\phi°$, if desired. The latter form is convenient to use in voltage-drop calculations when the current is expressed as $I\angle\phi°$.

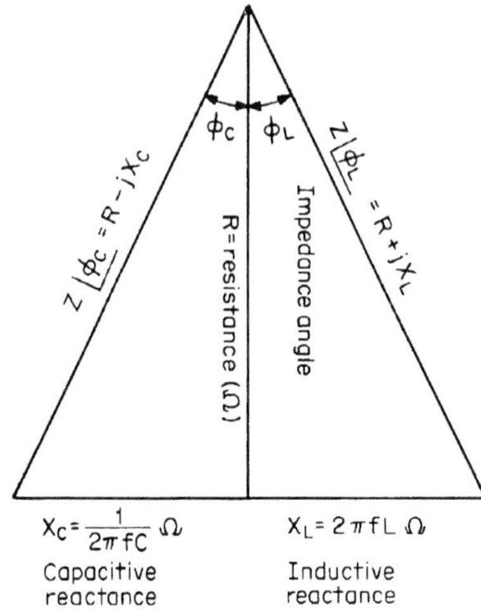


**FIGURE 7-4** Impedance diagrams for series connection of resistance and reactance ($L$ = inductance, in henrys; $C$ = capacitance, in farads; $f$ = frequency, in hertz).

***Power Factor.*** In typical distribution loads, the current lags the voltage, as shown in Fig. 7-3, where $\theta$ is shown as the angle between current and sending voltage and cos $\theta$ is referred to as the *power factor* of the circuit. In a purely resistive circuit, the current and voltage are in phase; consequently, the power factor is 1.0 or unity. In a purely inductive circuit, the voltage and current are out of phase by 90 electrical degrees, resulting in a power factor of zero. In a circuit consisting of a resistance in series with a reactance of equal ohmic value ($\phi = 45°$), $\theta = \pm 45°$ also. Thus, the power factor is cos 45° = 0.707, or 70.7%.

In a single-phase ac circuit, the load in kW can be expressed as

$$\text{kW} = EI\cos\theta \tag{7-5}$$

where $E$ = magnitude of rms line-to-neutral voltage, kV
$I$ = magnitude of current, rms amperes
$\theta$ = electrical angle between phasor voltage and current

From Eq. (7-5), it is obvious that the magnitude of the current for a given voltage and kilowatt load depends on the power factor, or

$$I = \text{kW}/(E\cos\theta) \tag{7-6}$$

The corresponding equations for balanced 3-phase circuits are

$$\text{kW} = \sqrt{3}\,EI\cos\theta \tag{7-7}$$

and

$$I = \text{kW}/(\sqrt{3}\,E\cos\theta) \tag{7-8}$$

where the symbols are as specified above, and $\theta$ is measured as the angle between the line-to-neutral voltage of a given phase and the current in that phase.

*Example.* Given a load of 500 kW at 80% power factor (lagging), 7.2 kV circuit voltage, 60-Hz, single-phase circuit using 1/0 aluminum conductor spaced 30 in on centers. The load is located 1 mi from the substation. What is the voltage drop? From tables on conductor characteristics,

$$r = 0.185\ \Omega/1000\ \text{ft}$$

$$x = 0.124\ \Omega/1000\ \text{ft}$$

Therefore,

$$R + jX = 5.28\,(0.185 + j\,0.124) = 0.9769 + j\,0.6547\ \Omega$$

From Eq. (7-6),

$$I = \frac{\text{kW}}{E\cos\theta} = \frac{500}{7.2\times0.8} = 86.81\ \text{A}$$

$$E = 7.2\angle\theta°$$

$$\cos\theta = 0.80$$

$$\theta = 36.87°$$

and $\sin\theta = 0.60$

From Eq. (7-4),[a]

$$\text{Voltage drop} = 2(IR\cos\theta + IX\sin\theta) = (86.81\times0.9769\times0.8 + 86.81\times0.6547\times0.6)$$
$$= 2(67.84 + 34.10) = 203.88 \text{ V}$$

***Calculation of 3-Phase Line Drops with Balanced Loads.*** In 3-phase circuits with balanced loads on each phase, the line-to-neutral voltage drop is merely the product of the phase current and the conductor impedance as determined from standard tables. There is no return current with balanced 3-phase loads. Thus, the line-to-line voltage drop is $\sqrt{3}$ times the line-to-neutral drop, or

$$V_{\text{drop } L\text{-}L} = \sqrt{3}(IR\cos\theta + IX\sin\theta) \tag{7-9}$$

For example, assume that the circuit of the preceding example now is a 3-phase 12.47-kV circuit 1 mi long with the same 1/0 aluminum conductors at an equivalent spacing of 30 in and a load of $3\times500 = 1500$ kW at 0.8 pf lagging. What is the line-to-line voltage drop? $R$ and $X$ are the same values as previously; that is, $R + jX = 0.9769 + j\,0.6547\ \Omega$.

The current per phase from Eq. (7-7) is

$$I = \frac{\text{kW}}{\sqrt{3}\,E\cos\theta} = \frac{1500}{\sqrt{3}\times12.47\times0.8} = 86.81 \text{ A}$$

as before,

$$V_{\text{drop } L\text{-}L} = \sqrt{3}(IR\cos\theta + IX\sin\theta)$$
$$= \sqrt{3}(86.81\times0.9769\times0.8 + 86.81\times0.6547\times0.6)$$
$$= 117.51 + 59.06 = 176.57 \text{ V (approx.)}$$

***Calculation of Voltage Drop in Unbalanced Unsymmetrical Circuits.*** If there are $n$ different wires $a, b, c, d, \cdots, n$ carrying currents $I_a, I_b, I_c, \cdots, I_n$, respectively, whether 2-, 3-phase, the voltage drop in wire $a$ per mile at 60 Hz is

$$I_aR_a + j\left[0.2794\left(I_a\log\frac{1}{r} + I_b\log\frac{1}{D_{ab}} + I_c\log\frac{1}{D_{ac}} + \cdots + I_n\log\frac{1}{D_{an}}\right) + 0.03034\mu I_a\right] \quad \text{volts in phasor form} \tag{7-10}$$

where currents are in phasor amperes, $R_a$ is 60-Hz ohmic resistance of conductor $a$ per mile, $r$ is equivalent radius, in inches, of conductor $a$, $D_{ab}$, $D_{ac}$, and $D_{an}$ are distances, in inches, between centers of conductors $a$ and $b$, $a$ and $c$, and $a$ and $n$, and $u$ is the permeability of conductor $a$ (unity for nonmagnetic material). To get the drop in $b$, replace all $a$'s by $b$'s and all $b$'s by $a$'s in Eq. (7-10); similarly, to get the drop in $c$, interchange $a$'s and $c$'s; likewise for $n$. For 25 Hz, multiply that part of Eq. (7-10) which is in brackets by 25/60. Equation (7-10) gives voltage drop for any degree of load unbalance, power factor, or conductor arrangements. In using this formula, calculations are made easier by choosing voltage to neutral as the reference axis.

***Approximate Method of Calculating Voltage Drop in Unbalanced, Unsymmetrical Circuits.*** Equation (7-10) requires laborious calculations and is used only when exact results are necessary. Voltage drops sufficiently accurate for engineering purposes can be calculated by using an equivalent

[a]The factor of 2 is used for a single-phase system to represent the impedance of the outgoing conductor and the return conductor.

impedance for each conductor. The reactance component of the equivalent impedance is computed from a spacing $D$ equal to the geometric means of the interaxial distances of the other conductors to the conductor being considered. For instance, if there are four conductors $a$, $b$, $c$, and $n$ for conductor $a$, $D=\sqrt[3]{D_{ab}, D_{ac}, D_{an}}$; for conductor $b$, $D=\sqrt[3]{D_{ab}, D_{bc}, D_{bn}}$.

***Phasor and Connection Diagrams.*** Phasor and connection diagrams are drawn in computing voltage drops in unbalanced circuits. Figure 7-5 shows an unbalanced 4-wire 3-phase 4160Y/2400-V circuit with assumed loads, power factors, and equivalent line impedances. Phase-to-neutral drops between source and load are given by the following, using one of the many possible voltage-notation conventions:

$$\begin{aligned} V_{na} - V_{n'a'} &= I_a Z_a + I_n Z_n \\ V_{nb} - V_{n'b'} &= I_b Z_b + I_n Z_n \\ V_{nc} - V_{n'c'} &= I_c Z_c + I_n Z_n \end{aligned} \tag{7-11}$$

Phase-to-phase drops between source and load are given by the following:

$$\begin{aligned} V_{ba} - V_{b'a'} &= I_a Z_a - I_b Z_b \\ V_{ac} - V_{a'c'} &= I_c Z_c - I_a Z_a \\ V_{cb} - V_{c'b'} &= I_b Z_b - I_c Z_c \end{aligned} \tag{7-12}$$

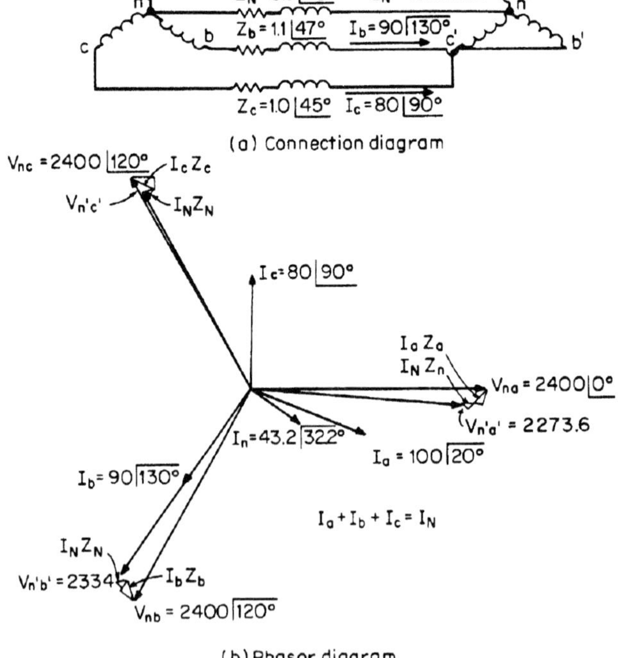


**FIGURE 7-5** Connections and phasor diagrams for unbalanced loads and unsymmetrical circuit.

In computing line-to-neutral drop in phase *a*, it is convenient to choose $V_{na}$ as the axis of reference.

$$V_{na} - V_{n'a'} = I_a Z_a + I_n Z_n = (100\angle 20^\circ)(1.2\angle 49^\circ) + X(43.2\angle 32.2^\circ)(0.5\angle 40^\circ)$$
$$= 120\angle 29^\circ + 21.6\angle 78^\circ = 126.4 + j61.9$$

Load voltage $V_{n'a'} = 2400 - 126.4 - j61.9 = 2273.6$ V (very nearly)

Likewise, in computing line-to-neutral drop in phase *b*, it is convenient to choose $V_{nb}$ as the axis of reference. The phasor diagram of Fig. 7-5 must be rotated in a counterclockwise direction 120°; then $I_b = 90\angle 10^\circ$ and $I_n = 43.2\angle 87.8^\circ$.

$$V_{nb} - V_{n'b'} = I_b Z_b + I_n Z_n = (90\angle 10^\circ(1.1\angle 47^\circ) + (43.2\angle 87.8^\circ)(0.5\angle 40^\circ) = 65.8 + j76.6$$

Load voltage $V_{n'b'} = 2400 - 65.8 - j76.6 = 2334.2$ V (very nearly)

Drop in the neutral conductor of a 4-wire 3-phase circuit or a 3-wire 2-phase circuit makes resultant drop on the more heavily loaded phases greater than it would be for the same current under balanced conditions. Likewise, net drop is less on more lightly loaded phases than for the same current when balanced.

***Distributed Loads, Voltage Drop, and $I^2R$ Loss.*** Voltage drop and conductor power losses resulting from a concentrated load on a distribution line can be calculated easily as shown in earlier parts of this section. However, distribution circuit loads are generally considered to be distributed—often, but not always, uniformly. Distributed load may be considered as effectively concentrated at one point along the circuit to calculate total voltage drop and at another point to calculate conductor $I^2R$ losses in the conductor. If the load is uniformly distributed along the feeder, the total voltage drop can be calculated by assuming that the entire load is concentrated at the midpoint of the circuit, and the total $I^2R$ losses can be calculated by assuming that the load is concentrated at a point one-third the total distance from the source.

However, if there is a superimposed through load beyond the given feeder section, this method of calculation becomes cumbersome. It is possible to develop a single precise equivalent circuit for both the voltage-drop and loss calculations. Figure 7-6 shows the load representation and equivalent for uniformly distributed loads. Equivalents also can be developed for other types of distribution. Figure 7-6 shows the equivalent circuit of two-thirds of the total load concentrated at three-quarters of the total distance from the source.

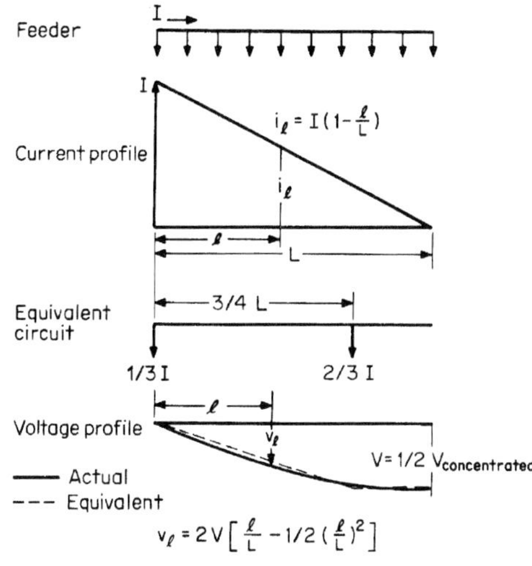


**FIGURE 7-6** Uniformly distributed loads.

## 7.5 THE SUBTRANSMISSION SYSTEM

***Definition.*** *Subtransmission* is that part of the utility system which supplies distribution substations from bulk power sources, such as large transmission substations or generating stations. In turn, the distribution substations supply primary distribution systems. Subtransmission has many of the characteristics of both transmission and distribution in that it moves relatively large amounts of power from one point to another, like transmission, and at the same time it provides area coverage, like distribution.

In some utility systems, transmission and subtransmission voltages are identical; in other systems, subtransmission is a separate and distinct voltage level (or levels). This is easy to account for because in the evolutionary development of utility systems, today's transmission voltage naturally tends to become tomorrow's subtransmission voltage, just as today's subtransmission voltage tends to become tomorrow's primary distribution voltage.

Because of the wide range of voltages used in subtransmission, and because of the wide variation in geographic conditions and local ordinances, subtransmission circuits are sometimes built on pole lines on city streets, or on tower lines on private rights-of-way, or in underground cables.

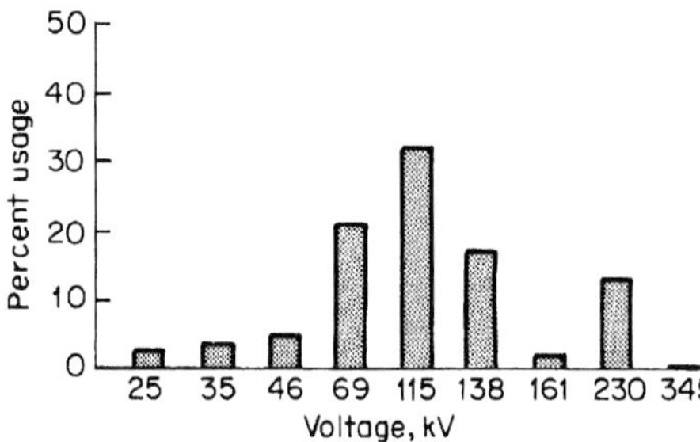


**FIGURE 7-7** Use of distribution substation high-voltage rating.

***Voltages.*** Voltages of subtransmission circuits range from 12 to 345 kV, but today the levels of 69, 115, and 138 kV are most common. The use of the higher voltages is expanding rapidly as higher primary voltages are receiving increased usage. Current practice as indicated by an informal utility survey is shown in Fig. 7-7; 115 and 138 kV together comprise about half the usage, 69 kV about 20%; 230 kV usage is becoming substantial, reflecting the growing use of 25- and 34.5-kV primary distribution.

Conductors of ACSR or aluminum generally have supplanted copper in overhead construction, and aluminum conductors are being used increasingly in cables.

***Voltage Regulation of Subtransmission.*** The size of conductors used in subtransmission systems is determined by (1) magnitude and power factor of the load, (2) emergency loading requirements, (3) distance that the load must be carried, (4) operating voltage, (5) permissible voltage drop under normal and emergency loading, and (6) optimal economic balance between installed cost of the conductor and cost of losses. Table 7-2 gives the line-to-neutral voltage drops per 100,000 A · ft for common cable and overhead conductor sizes and representative power factors for 34.5- and 69-kV subtransmission. Values in the table are based on the approximate formula (Eq. 7-4)

$$V_{\text{drop}} = IR \cos \theta + IX \sin \theta = IZ \cos (\phi - \theta)$$

where $R$, $X$, and $Z$ are 60-Hz resistance, reactance, and impedance in ohms per 1000 ft of a single conductor, $\theta$ is the power-factor angle in electrical degrees, and $\phi$ is the impedance angle, $\tan^{-1}(X/R)$.

***Examples of How to Use Table 7-2.*** Determine the voltage drop when a 3-phase 20,000-kVA load at 95% power factor is carried 10 mi over an overhead 69-kV circuit with No. 2/0 ACSR conductor. Assuming the receiving-end voltage to be 69 kV, the current is

$$I = \frac{\text{kVA}}{\sqrt{3}E} = \frac{20{,}000}{\sqrt{3} \times 69} = 167.35 \text{ A}$$

Circuit feet are

$$10 \times 5280 = 52{,}800 \text{ ft}$$

Thus

$$\frac{\text{A} \cdot \text{ft}}{100{,}000} = \frac{167.35 \times 52{,}800}{100{,}000} = 88.36$$

**TABLE 7-2** Voltage Drops per 100,000 A · ft* for 3-Phase, 60-Hz, 34.5- and 69-kV Subtransmission

| | Voltage class | | | | | | | | | | |
|---|---|---|---|---|---|---|---|---|---|---|---|
| | 34.5 kV | | | | | 69 kV | | | | | Approx. amp. capacity for air moving at 2 ft/s |
| | Lagging power factor | | | | | | | | | | |
| Conductor size | 0.7 | 0.8 | 0.9 | 0.95 | 1.00 | 0.7 | 0.8 | 0.9 | 0.95 | 1.00 | |
| | Underground subtransmission† | | | | | | | | | | |
| Aluminum: | | | | | | | | | | | |
| No. 1/0 | 18.3 | 19.9 | 21.1 | 21.5 | 21.0 | | | | | | |
| No. 2/0 | 15.4 | 16.5 | 17.4 | 17.6 | 16.9 | | | | | | |
| No. 4/0 | 10.7 | 11.2 | 11.5 | 11.4 | 10.5 | | | | | | |
| 350 kcmil | 7.69 | 7.84 | 7.77 | 7.55 | 6.50 | 8.04 | 8.10 | 7.92 | 7.62 | 6.38 | |
| 500 kcmil | 6.15 | 6.12 | 5.88 | 5.59 | 4.50 | 6.53 | 6.43 | 6.10 | 5.74 | 4.48 | |
| 750 kcmil | 4.96 | 4.80 | 4.44 | 4.10 | 3.00 | 5.25 | 5.05 | 4.63 | 4.23 | 3.01 | |
| 1000 kcmil | 4.32 | 4.12 | 3.73 | 3.37 | 2.30 | 4.69 | 4.44 | 3.96 | 3.55 | 2.32 | |
| | Overhead subtransmission‡ | | | | | | | | | | |
| ACSR: | | | | | | | | | | | |
| No. 4 | 42.9 | 45.5 | 47.3 | 47.5 | 44.7 | 43.6 | 46.1 | 47.7 | 47.8 | 44.7 | 120 |
| No. 2 | 31.5 | 32.5 | 32.7 | 32.1 | 28.4 | 32.2 | 33.1 | 33.1 | 32.4 | 28.4 | 165 |
| No. 1/0 | 24.1 | 24.1 | 23.2 | 22.1 | 18.0 | 24.8 | 24.7 | 23.7 | 22.4 | 18.0 | 225 |
| No. 2/0 | 21.6 | 21.2 | 20.1 | 18.8 | 14.6 | 22.3 | 21.8 | 20.5 | 19.1 | 14.6 | 260 |
| No. 4/0 | 17.3 | 16.6 | 15.1 | 13.8 | 9.66 | 18.0 | 17.2 | 15.5 | 14.1 | 9.66 | 355 |
| 336.4 kcmil | 12.7 | 11.8 | 10.4 | 9.13 | 5.57 | 13.4 | 12.4 | 10.8 | 9.44 | 5.57 | 480 |
| 477 kcmil | 11.2 | 10.3 | 8.72 | 7.44 | 3.92 | 12.0 | 10.9 | 9.15 | 7.75 | 3.92 | 605 |
| 795 kcmil | 9.73 | 8.68 | 7.06 | 5.78 | 2.37 | 10.4 | 9.28 | 7.49 | 6.09 | 2.37 | 850 |

***Note:*** 1 in = 25.4 mm; 1 in$^2$ = 645 mm$^2$; 1 ft = 0.3048 m. Regulation of copper conductors can be estimated with reasonable accuracy as that of aluminum conductors two sizes larger. For ampacities of cables, see Tables 7-22 and 7-23.

*Values in the table give the difference in absolute value between sending-end and receiving-end line-to-neutral voltages of a balanced 3-phase circuit.

†Underground cable impedances are based on 90°C conductor temperature with close triangular spacing of cables using typical solid-dielectric insulation, 100% insulation level, single conductor, shielded and jacketed.

‡Overhead conductor impedances are based on 50°C conductor temperature, ACSR construction, 600 A/in$^2$ density with 60-in equivalent spacing for 35 kV and 90 in for 69 kV.

From the overhead portion of Table 7-2, the voltage drop per 100,000 A · ft at 95% power factor for a No. 2/0 ACSR conductor is 19.1 V. Therefore, the total voltage drop for the example is $88.36 \times 19.1 = 1687.68$ V line-to-neutral. Since normal line-to-neutral voltage is $69/\sqrt{3} = 39.838$ kV, or 39,838 V, the percent voltage drop is $1687.68 \times 100/39{,}838 = 4.24\%$.

Assuming that permissible voltage drop is the limiting factor, what overhead ACSR conductor size should be used to supply a load of 40,000 kVA at 95% power factor and receiving-end voltage of 69 kV with a permissible drop of 5% and 8 mi between sending and receiving ends?

$$\text{Current} = \frac{40{,}000}{\sqrt{3} \times 69} = 334.71 \text{ A}$$

$$\text{Circuit feet} = 8 \times 5280 = 42{,}240 \text{ ft}$$

$$\frac{\text{A} \cdot \text{ft}}{100{,}000} = \frac{334.71 \times 42{,}240}{100{,}000} = 141.38$$

The permissible voltage drop is $0.05 \times 69{,}000/\sqrt{3} = 1991.92$ V line-to-neutral. The corresponding permissible voltage drop per 100,000 A · ft is

$$\frac{1991.92}{141.38} = 14.1 \text{ V}/100{,}000 \text{ A} \cdot \text{ft}$$

From Table 7-2 it is seen that this corresponds approximately to No. 4/0 ACSR.

***Subtransmission System Patterns.*** A wide variety of subtransmission system designs are in use, varying from simple radial systems to systems similar to networks. The radial system is not generally used because most utilities today plan their subtransmission-distribution substation systems so that one major contingency such as outage of a subtransmission circuit or failure of a distribution substation transformer will not result in loss of load—or at least the loss of load will be of short duration while automatic switching operations take place. Thus, loop and multiple circuit patterns predominate. Figures 7-8 and 7-9 illustrate the basic nature of these two patterns. The loop pattern implies that a single circuit originating at one bulk power source "loops" through several substations before terminating at another bulk source or even at the original source. Reinforcing ties, as indicated by the dotted connection, are used when the number of substations exceeds some predetermined level.

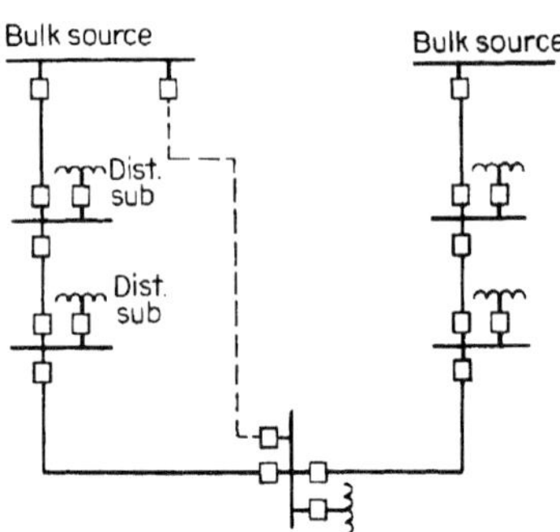


**FIGURE 7-8** Loop pattern.

Multiple circuit pattern implies the use of two or more circuits which are tapped at each substation, as illustrated in Fig. 7-9. The circuits may be radial or may terminate in a second bulk power source. Many variations of the two basic patterns are found. From a recent informal survey of approximately 50 major utilities, it appears that the two patterns are about equally used.

A vast majority of today's subtransmission is of overhead construction, much of it built on city streets as contrasted with private rights of way. However, appearance and environmental considerations, difficulty in obtaining substation sites and rights of way, and rapid growth of underground distribution are certain to exert continuing pressure on the undergrounding of subtransmission. Even with the use of direct-buried, solid-dielectric cables, the cost of underground subtransmission is many times the cost of overhead circuits, particularly where the overhead subtransmission can be built on city streets.

Thus, a requirement to build future subtransmission underground would have major impact on the balance of overall subtransmission-substation-primary distribution costs. It undoubtedly would

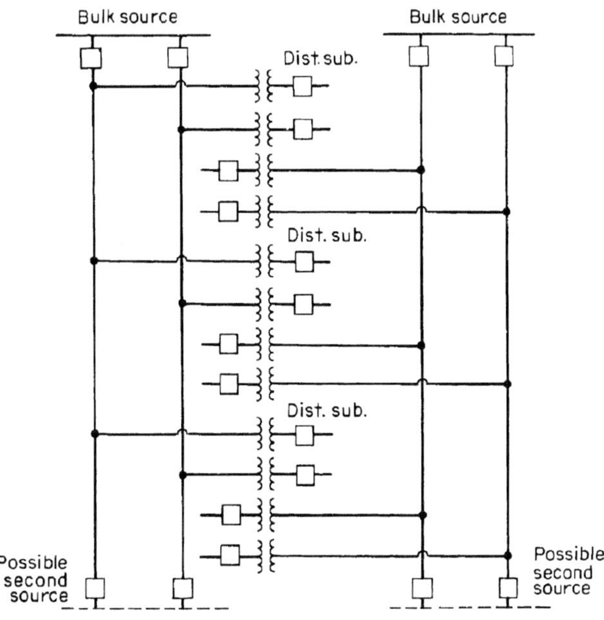


**FIGURE 7-9** Multiple pattern.

focus attention on minimizing the amount of subtransmission circuitry needed to cover the load area, which in turn would favor

Fewer, larger substations

Loop subtransmission pattern rather than multiple parallel circuits

Depending on load density in this area, it *could* favor

Higher primary voltage

Higher subtransmission voltage

Changes in either subtransmission or primary voltage levels are major decisions which require study in depth and ultimately the commitment of large financial resources.

## 7.6 PRIMARY DISTRIBUTION SYSTEMS

The primary distribution system takes energy from the low-voltage bus of distribution substations and delivers it to the primary windings of distribution transformers.

***Overhead Primary Systems.*** Typically, overhead primary distribution systems have been operated as radial circuits (normally open loops) from the substation outward. Figure 7-2 shows schematically a typical primary feeder in a predominantly residential area; an overhead 12.470Y/7.2-kV wye system is used for illustrative and functional purposes, but underground systems will be discussed later.

The main feeder backbone usually is a 3-phase 4-wire circuit from which the single-phase lateral or branch circuits are tapped through fuse cutouts to protect the system from faults on the lateral circuits. The single-phase lateral circuits consist of 1-phase conductor and the neutral. Distribution transformers are connected between the phase and the neutral; in this case they would have a rating of 7200 V.

Utilities use automatic reclosing feeder breakers and line reclosers to minimize service interruptions. However, serious problems involving the main will cause an outage to some or all of the feeder until line crews can locate the problem and manually operate pole-top disconnecting switches appropriately to isolate the problem and to pick up as much load as possible from adjacent feeders. Switches of this kind usually are found in both the main and lateral circuits, as indicated in Fig. 7-2. Also, it is often possible to make and to remove connections while the system is energized through the use of hot-line tools, hot-line clamps, insulated bucket trucks, etc.

Generally, this approach has provided an acceptable level of service because overhead system troubles are relatively easy to locate, and repair times are short. However, when the entire primary system is installed underground, while the frequency of serious trouble is expected to be lower than in overhead systems, it is likely that the time involved in pinpointing the location and making repairs will be much longer than in overhead systems.

***Underground System.*** While a relatively small percentage of new general-purpose feeders is being installed totally underground, the trend is growing and is expected to continue to grow. Since it is difficult to accomplish many maintenance and operating functions on an underground system while it is "hot," or energized, in contrast to overhead-system practices, specific provisions must be made in the system design to incorporate needed sectionalizing and overcurrent protective equipment.

The main feeder plan shown in Fig. 7-10 is reasonably typical of present practice on underground systems supplying basically residential and small commercial loads. Note that the main feeders are operated radially, but with normally open ties to adjacent main feeders. The main feeder switches usually are 3-phase, 600-A, manually operated load-break switches. The single-phase and 3-phase lateral circuits also are operated as normally open loops.

Switching in the 200-A circuits can be accomplished by means of either load-break switches or separable, insulated cable connectors. Usually, two main feeder switches are grouped along with the lateral circuit switching and protective equipment into one piece of pad-mounted equipment.

The primary feeders supplying secondary-network systems in metropolitan areas usually are radial 3-wire circuits consisting of 3/c cables in underground duct lines. The 3-phase network transformers are T-tapped to the primary feeders.

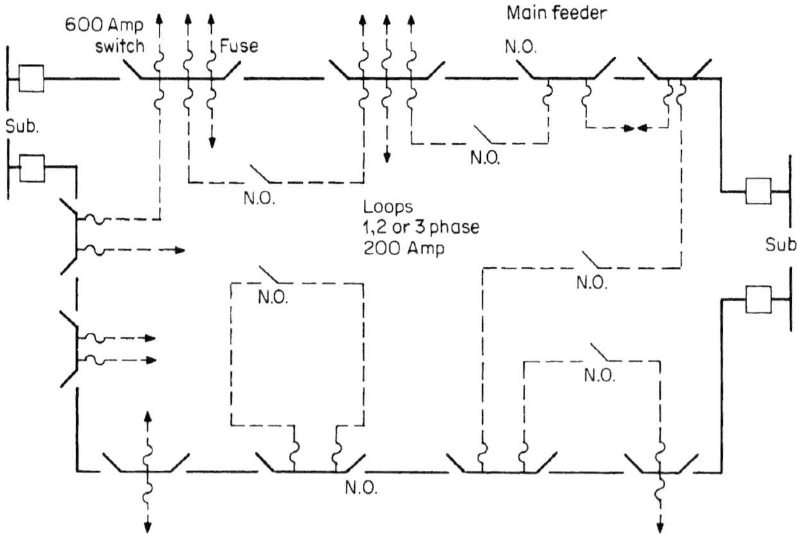


**FIGURE 7-10** Typical main-feeder underground circuit. (All switches closed unless shown otherwise.)

***Automation.*** With increasing emphasis on reliability of service, a definite trend is under way to make greater use of protective and sectionalizing equipment in the primary system in order to minimize the number of customers involved in an outage and to reduce the outage time. Proposed schemes run the gamut from manually operated devices to automatic devices remotely controlled from distribution centers. The remote-controlled schemes vary from some type of supervisory control to computer-controlled systems with built-in logic to cope quickly with the various problems which may arise.

***Primary-Distribution-System Voltage Levels.*** Since World War II, the 15-kV distribution class has become firmly entrenched and today represents 60% to 80% of all primary distribution activity. Very little expansion of lower-voltage systems is taking place. There is a trend, however, toward increasing usage of primary voltage levels above the 15-kV class. This trend has an impact on substation and subtransmission practices as well because higher primary voltages almost axiomatically lead to larger substations and higher subtransmission voltages.

The two principal voltages above 15 kV are 24.49Y/14.4 kV and 34.5Y/19.92 kV. New line additions at these voltage levels now average more than 20% of those at 15 kV.

To achieve economy, the higher primary voltages also require heavier feeder loadings which could imply reduced service reliability because more customers are affected by primary faults. Greater use of automatic switching and protective equipment can do much toward preserving a level of reliability to which the public has become accustomed. This is another reason that most observers believe that an increased amount of automation is inevitable in our distribution systems.

For example, a typical 12.47-kV feeder serves a normal peak load on the order of 6000 to 7000 kVA. On this basis, the probable peak loading of a fully developed 34.5-kV feeder would be expected to be in the neighborhood of 18,000 to 20,000 kVA.

Why go to high-voltage distribution (HVD)? Most of today's systems in the 15-kV class are not voltage-drop-limited, and cost of higher-voltage laterals and associated equipment needed to cover the load area is greater. The major economic advantages are

1. Larger (and fewer) substations
2. Fewer circuits
3. Possibility of eliminating a system voltage-transformation level where the new primary voltage is the former subtransmission level

Other advantages of HVD which are difficult to evaluate in dollars are

1. Reduced losses in early stages of development
2. Reduced voltage regulation
3. Greater distance or area coverage
4. Fewer circuits per route (reduced congestion)
5. Fewer circuit positions at substations
6. Fewer substation sites
7. Greater flexibility in supplying large spot loads

Some of the disadvantages of HVD have been

1. Cost of equipment
2. Reliability due to increased exposure
3. Higher equipment failure rates
4. Operability

***Conductor Sizes.*** The conductor sizes used in overhead primaries generally range from No. 2 AWG to 795 kcmil. ACSR and aluminum conductors have almost entirely displaced copper for new construction. Aerial cable is used occasionally for primary conductors in special situations where clearances are too close for open-wire construction or where adequate tree trimming is not practical. The type of construction more frequently used consists of covered conductors (nonshielded) supported from the messenger by insulating spacers of plastic or ceramic material. The conductor insulation, usually a solid dielectric such as polyethylene, has a thickness of about 150 mils for a 15-kV class circuit and is capable of supporting momentary contacts with tree branches, birds, and animals without puncturing. This type of construction is commonly referred to as *spacer cable*.

The conductor sizes most commonly used in underground primary distribution vary from No. 4 AWG to 1000 kcmil. Four-wire main feeders may employ 3- or 4-conductor cables, but single-conductor concentric-neutral cables are more popular for this purpose. The latter usually employ crosslinked polyethylene insulation, and often have a concentric neutral of one-half or one-third of the main conductor cross-sectional area.

The smaller-sized cables used in lateral circuits of URD systems are nearly always single-conductor, concentric-neutral, crosslinked polyethylene-insulated, and usually directly buried in the earth. Insulation thickness is on the order of 175 mils for 15-kV-class cables and 345 mils for 35-kV class with 100% insulation level.

Stranded or solid aluminum conductors have virtually supplanted copper for new construction, except where existing duct sizes are restrictive. With the solid-dielectric construction, in order to limit voltage gradient at the surface of the conductor within acceptable limits, a minimum conductor size of No. 2 AWG is common for 15-kV-class cables, and No. 1/0 AWG for 35-kV class.

***Voltage Regulation of Primary Distribution.*** Table 7-3 can be used to determine the voltage drop of an existing circuit when the load data are known or to determine minimum conductor size required to meet a given voltage-drop limit. Data are given for various underground-cable and overhead-conductor configurations for 12.47 and 34.5 kV.

*Example.* What is the voltage drop for a 34.5-kV overhead circuit 3 mi long using 4/0 aluminum conductor and carrying a balanced 3-phase load of 15,000 kVA at 90% power factor: The current is $15{,}000/\sqrt{3} \times 34.5 = 251$ A. The circuit feet are $3 \times 5280 = 15{,}840$ ft. Thus $\text{A} \cdot \text{ft}/100{,}000 = 251 \times 15{,}840/100{,}000 = 39.758$. From Table 7-3, the appropriate voltage drop per 100,000 A · ft is 14.0 V line-to-neutral. Therefore, the total voltage drop for the example is

$$39.758 \times 14.0 = 556.6 \text{ V line-to-neutral}$$

Since normal line-to-neutral voltage is $34{,}500\sqrt{3} = 19{,}920$ V, the percent voltage drop is

$$556.6 \times 100/19{,}920 = 2.79\%$$

*Example.* What is the minimum aluminum conductor size to carry 6000 kVA at 90% power factor of balanced 3-phase load over a 2-mi, 12.47Y/7.2-kV feeder with no more than a 3% voltage drop? Load current is $6000/\sqrt{3} \times 12.47 = 277.8$ A. Circuit feet $= 2 \times 5280 = 10{,}560$ ft. Thus

$$\frac{\text{A} \cdot \text{ft}}{100{,}000} = \frac{277.8 \times 10{,}560}{100{,}000} = 29.34$$

$$\text{Permissible voltage drop} = 0.03 \times \frac{12{,}470}{\sqrt{3}} = 216 \text{ V}$$

The corresponding drop per 100,000 A · ft is $216/29.34 = 7.36$ V, line-to-neutral. From Table 7-3, this value falls between 477 and 795 kcmil, so that the latter size would be chosen.

***Loading.*** Loading of primary feeders varies greatly depending on primary voltage, load density, emergency loading requirements, etc. Typical peak loads on 15-kV class feeders are 6 to 7000 kVA. Peak loads on 25- and 35-kV class, fully developed feeders probably will be proportionally greater in the future, assuming that appropriate measures can be taken to maintain acceptable reliability of service.

**TABLE 7-3** Line-to-Neutral Voltage Drops per 100,000 A·ft* for 12.47Y/7.2 and 34.5Y/19.92 kV and Balanced 3-Phase Loads

| | Voltage class | | | | | | | | | | |
|---|---|---|---|---|---|---|---|---|---|---|---|
| | 12.47Y/7.2 kV | | | | | 34.5Y/19.92 kV | | | | | Approx. amp. capacity for air moving at 2 ft/s |
| | Lagging power factor | | | | | | | | | | |
| Conductor size | 0.7 | 0.8 | 0.9 | 0.95 | 1.00 | 0.7 | 0.8 | 0.9 | 0.95 | 1.00 | |
| | Underground primary | | | | | | | | | | |
| Aluminum: | | | | | | | | | | | |
| Concentric neutral—direct buried, cross-linked polyethylene, conductor 70°C, neutral 60°C, earth resistivity 90 Ω·cm³, triplex configuration, full installation | | | | | | | | | | | |
| No. 1/0 | 17.1 | 18.5 | 19.8 | 20.2 | 19.8 | 17.6 | 19.0 | 20.1 | 20.4 | 19.8 | |
| No. 2/0 | 14.1 | 15.1 | 16.0 | 16.3 | 15.7 | 14.6 | 15.6 | 16.3 | 16.5 | 15.7 | |
| No. 4/0 | 9.82 | 10.4 | 10.7 | 10.7 | 9.96 | 10.3 | 10.8 | 11.0 | 10.9 | 9.95 | |
| 350 kcmil | 7.01 | 7.19 | 7.17 | 7.00 | 6.11 | 7.37 | 7.49 | 7.39 | 7.16 | 6.11 | |
| 500 kcmil | 5.66 | 5.69 | 5.55 | 5.31 | 4.40 | 6.04 | 6.00 | 5.76 | 5.47 | 4.40 | |
| 750 kcmil | 4.63 | 4.55 | 4.30 | 4.03 | 3.12 | 4.95 | 4.82 | 4.49 | 4.16 | 3.11 | |
| 1000 kcmil | 4.10 | 3.98 | 3.69 | 3.41 | 2.52 | 4.37 | 4.20 | 3.85 | 3.52 | 2.51 | |
| Single conductor shielded and jacked, cross-lined polyethylene, conductor 70°C, unigrounded shield, triplex configuration, full insulation | | | | | | | | | | | |
| 350 kcmil | 7.29 | 7.49 | 7.51 | 7.35 | 6.47 | 7.55 | 7.72 | 7.67 | 7.47 | 6.47 | |
| 500 kcmil | 5.78 | 5.82 | 5.67 | 5.45 | 4.54 | 6.08 | 6.07 | 5.86 | 5.58 | 4.54 | |
| 750 kcmil | 4.64 | 4.54 | 4.26 | 3.97 | 3.02 | 4.88 | 4.74 | 4.41 | 4.08 | 3.02 | |
| 1000 kcmil | 4.02 | 3.85 | 3.52 | 3.21 | 2.26 | 4.23 | 4.03 | 3.65 | 3.31 | 2.26 | |
| | Overhead primary† | | | | | | | | | | |
| No. 4 | 42.3 | 45.4 | 47.8 | 48.5 | 46.6 | 43.4 | 46.3 | 48.5 | 49.0 | 46.6 | 115 |
| No. 2 | 29.8 | 31.2 | 32.0 | 31.9 | 29.3 | 30.9 | 32.2 | 32.7 | 32.4 | 29.3 | 160 |
| No. 1/0 | 21.8 | 22.2 | 22.1 | 21.5 | 18.5 | 23.0 | 23.2 | 22.8 | 22.0 | 18.5 | 215 |
| No. 2/0 | 19.0 | 19.1 | 18.6 | 17.8 | 14.7 | 20.1 | 20.0 | 19.3 | 18.3 | 14.7 | 250 |
| No. 4/0 | 14.7 | 14.3 | 13.3 | 12.4 | 9.20 | 15.9 | 15.3 | 14.0 | 12.7 | 9.20 | 340 |
| 336.4 kcmil | 11.8 | 11.2 | 9.97 | 8.91 | 5.80 | 13.0 | 12.1 | 10.7 | 9.41 | 5.80 | 465 |
| 477 kcmil | 10.4 | 9.58 | 8.27 | 7.18 | 4.10 | 11.5 | 10.5 | 8.97 | 7.68 | 4.10 | 590 |
| 795 kcmil | 8.22 | 7.92 | 6.52 | 5.40 | 2.40 | 9.96 | 8.88 | 7.22 | 5.90 | 2.40 | 820 |

***Note:*** 1 in = 25.4 mm; 1 ft = 0.3048 m. For ampacities of cables, see Tables 7-23 and 7-24. Regulation of copper for overhead conductors can be estimated with reasonable accuracy the same as that of aluminum conductors two sizes larger. For single-phase overhead primaries, the voltage drop is approximately two times the 3-phase values given in the table. For underground single-phase primaries in concentric-neutral, direct-buried cables, see subsection on URD systems. Cables are 15- and 35-kV classes, respectively.

*Values in the table give the difference in absolute value between sending-end and receiving-end line-to-neutral voltages of a balanced 3-phase circuit, in volts.

†Overhead conductor impedances are based on 50°C conductor temperature, aluminum conductor with 30-in equivalent spacing for 12.47Y kV and 60-in for 34.5Y kV.

***Voltage Drop.*** Voltage drop in the primary feeder is an important factor in system design; however, it is only one of the many voltage-drop considerations involved in determining the range of voltages delivered to the customers' service entrances. American National Standard, "Voltage Ratings for Electric Power Systems and Equipment (60-Hz)," ANSI C84.1, defines in detail the voltage ranges which should be observed. Outside the distribution substation, voltage drops occur in the primary system, the distribution transformer, the secondary system, the service drop, and in the users' wiring systems as well. Remedial measures, such as voltage regulators and shunt capacitor banks, can be used to counteract or reduce the voltage drop due to load flow.

A traditional rough rule of thumb has been to allow a voltage drop of about 3% in the primary of urban and suburban systems at time of peak load. Actually, with typical load densities and primary systems of 15-kV class or higher, it is very probable that economic system designs have a primary voltage drop smaller than 3%.

In rural systems which are typified by long lines and light load densities, primary voltage drops may be somewhat larger. This is offset somewhat by the absence of secondaries in serving individual farms; however, the service drops often are longer than in urban systems. The design objective, of course, is to keep delivered voltage to all customers in an acceptable and satisfactory range.

## 7.7 THE COMMON-NEUTRAL SYSTEM

The 4-wire, multigrounded, common-neutral distribution system now is used almost exclusively because of the economic and operating advantages it offers. Usually, the windings of the substation transformers serving the primary system are wye-connected, and the neutral point is solidly grounded. Occasionally, a small amount of impedance is connected between the transformer neutral and ground in order to limit line-to-ground short-circuit currents on the primary system to a predetermined value. The neutral circuit must be a continuous metallic path along the primary routes of the feeder and to every user location. Where primary and secondary systems are both present, the same conductor is used as the "common" neutral for both systems. The neutral is grounded at each distribution transformer, at frequent intervals where no transformers are connected, and to metallic water pipes or driven grounds at each user's service entrance. The neutral carries a portion of the unbalanced or residual load currents for both the primary and secondary systems. The remainder of this current flows in the earth and/or the water system. For typical conditions, it is estimated that about one-half the return current flows in the neutral conductor, although the division can vary widely depending on earth resistivity and the relative routing of the electric and water systems. Figure 7-11 is a schematic representation of a common-neutral system.

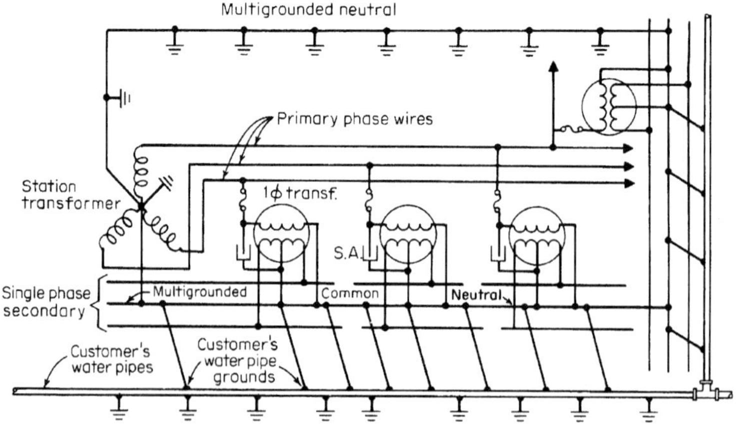


**FIGURE 7-11** Common-neutral methods of distribution.

***Grounding of Neutral.*** Rules related to grounding on the utility system neutral are given in the National Electrical Safety Code (NESC), ANSI C2, and regulations governing the grounding of the neutral on users' premises are stated in the National Electrical Code (NEC), NFPA 70. In brief, the secondary neutral is grounded at every service through a metallic water-piping system and through "made electrode grounds" such as other underground metal systems, building steel, or driven ground electrodes. The increasing use of nonmetallic water piping and insulating couplings on metal water systems is requiring the use of other grounding means. The secondary neutral also is grounded at the distribution transformer, usually by means of driven grounds. Although it is often general practice to install a metal butt plate or a wire butt wrap on poles to help in grounding the system neutral and other equipment, the NESC requires two such devices to equal one *made electrode*; as a result, neither can be used to satisfy the NESC requirement for a direct earth ground with a made electrode at each transformer or other arrester location.

The resistance to ground of a typical metallic water-piping system usually is less than 3 Ω. When made electrode grounds are used, they should have a resistance of not more than 25 Ω. Many utilities strive for lower values such as 5, 10, or 15 Ω.

Where there is no secondary neutral as such and no distribution transformers, the primary neutral should be grounded at intervals of not less than 1000 ft. Many utilities require grounding at smaller spacing, such as 500 ft; to meet the NESC requirements for a multigrounded neutral, there must be a minimum of the equivalent of four made electrodes in *each* mile. In URD systems, the primary circuits usually are in direct-buried, concentric neutral cable, so that excellent grounding is obtained.

The neutral must have a continuous metallic path between the substation and users' services. No disconnecting devices should be installed in the common neutral. In no case should the earth or buried metallic-piping systems be used as the only path for the return of normal load current.

***Size of Primary Neutral.*** On single-phase primary circuits (phase and neutral), the neutral conductor should be large enough to carry almost as much current as the phase conductor. Often the same neutral conductor size is used for both, or the neutral has "100%" conductivity.

In 3-phase primary circuits carrying reasonably balanced load, the neutral conductor can be considerably smaller than the phase conductors; 50% conductivity is not uncommon; some utilities specify size of neutral conductor, such as No. 1/0 aluminum, regardless of the size of the phase wires.

Secondary-system neutral conductors are often the same size as the phase conductors where open-wire construction is used. Where triplexed construction is used, the neutral frequently has a reduced cross section.

***4-Wire vs. 3-Wire Systems.*** The 4-wire, common-neutral primary system has many advantages over 3-wire systems.

1. Single-phase branch circuits, or laterals, consist of one insulated phase conductor and the neutral, rather than two insulated phase conductors. The economic advantage is very great in underground systems.
2. On overhead systems, only one lightning arrester is required at each single-phase distribution transformer, rather than two.
3. Only one primary bushing or cable termination is needed on each single-phase distribution transformer, rather than two. In the case of underground systems where the primary "loops through" each distribution transformer, two primary cable terminations or connectors are needed, rather than four.
4. Only one fuse or fuse cutout is needed in the primary of each single-phase distribution transformer. Not only is this a substantial economic advantage, but a short circuit in the primary of the transformer is interrupted positively by the action of a single fuse, and primary voltage is thereby removed from the transformer. In the case of the 3-wire system with the distribution transformer connected phase-to-phase, a second fuse must operate to remove primary voltage and the fault. There may be appreciable time between operation of the two fuses during which fault current continues to flow and abnormal voltages may be experienced by the user.
5. Single-phase primary lateral circuits can be protected by a single fuse cutout, rather than two. Line-to-ground short circuits are promptly cleared by operation of one fuse and voltage removed

from the branch circuit. In a 3-wire system (assumed grounded at the substation), single-phase lateral protection, if used, would require two fuse cutouts; a line-to-ground fault would blow only one fuse, leaving all the distribution transformers on that circuit excited at only 58% of normal as long as the faulted phase remains grounded. Under these conditions users' equipment would be exposed to abnormally low voltage. The ability to fuse lateral circuits contributes substantially to reliability of service, since a major amount of the total circuit exposure comprises the primary laterals in residential areas.

***Common-Neutral and Telephone Circuits.*** Usually, no problems are encountered in the joint use of poles for overhead distribution circuits and telephone circuits, particularly when the telephone circuits are in cable, as is now common practice. Also, in underground residential circuits, power cables and telephone cables often are installed in the same trench with no intentional physical separation of the power and communication facilities, that is, "random lay." Where separate grounding electrodes are employed for supply and communication facilities at customer's premises, the electrodes shall be bonded together with not less than No. 6 AWG copper wire.

## 7.8 VOLTAGE CONTROL

***System Voltage Levels and Voltage Ranges.*** Since about 1900, there have been several recommendations for certain voltages as standard or preferred for primary and secondary distribution systems, as well as for higher-voltage systems. The latest listing of standard system voltages is American National Standards Institute (ANSI) Standard C84.1-2016, "Voltage Ratings for Electric Power Systems and Equipment (60 Hz)." This standard was formulated by both utilities and manufacturers, and its recommendations are followed by both segments of the industry. Observance of this standard enables the utilities and manufacturers to work in harmony. In many states, ANSI C84 is the basis for rulings of the regulatory commission as far as voltage requirements are concerned.

This standard designates certain standard nominal voltages, including 120/240 V single-phase, 480Y/277 V, 12,470Y/7200 V, as well as the higher primary voltages, 24,940Y/14,400 V and 34,500Y/19,920 V, and others.

Using the nominal 120/240-V system as an example, the standard designates two different ranges of voltage, range A and range B. Range A service voltage specifies that a utility supply system be so designed and operated that most service voltages are within the limits specified, for example, 114/228 and 126/252 V. The occurrence of service voltages outside these limits is to be infrequent.

With the typical voltage drops between the service entrance and the points of utilization, the utilization equipment is designed and rated to give fully satisfactory performance within range A.

Range B service voltage includes voltages above and below range A that necessarily result from practical design and operating conditions on supply or user systems. These conditions are limited in extent, frequency, and duration. When they occur, corrective measures should be undertaken within a reasonable time to improve voltages to meet range A requirements.

Insofar as practicable, utilization equipment is designed to give acceptable performance within range B. The design and operating bogey of the utilities is to provide service voltage to all customers at all times within range A limits.

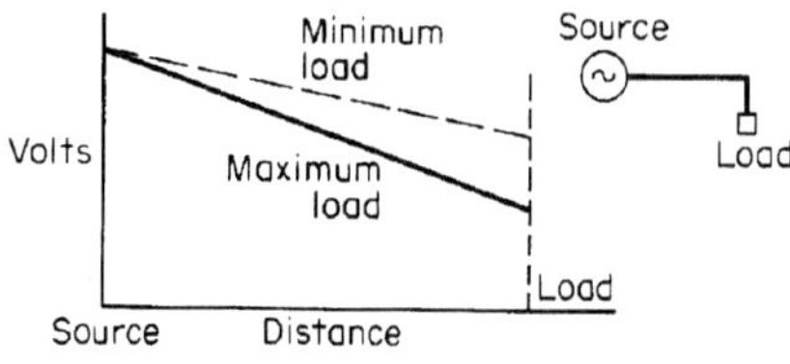


**FIGURE 7-12** Voltage profile for concentrated load.

***Voltage Profiles.*** It is usually convenient to discuss distribution-feeder-voltage regulation in terms of voltage *profiles* of the feeder, because the voltages are everywhere different on the feeder. A profile is simply a graph of feeder-voltage magnitude versus location on the feeder. For a simple case of one load at the end of the feeder (assuming uniform conductor), the one-line diagram and profile are as shown in Fig. 7-12.

The profile is a straight line between source and the load, and the voltage regulation at any point between is proportional to the distance from the source. It may be, as shown by the dashed-line profile, that minimum load is not zero, in which case the voltage variation is less than the calculated regulation, since regulation is usually calculated on the voltage difference between no-load and full-load conditions. If additional loads are distributed along the feeder, the profile becomes a broken line, and if the load is uniformly distributed, the profile becomes a smooth curve, as shown in Fig. 7-13.

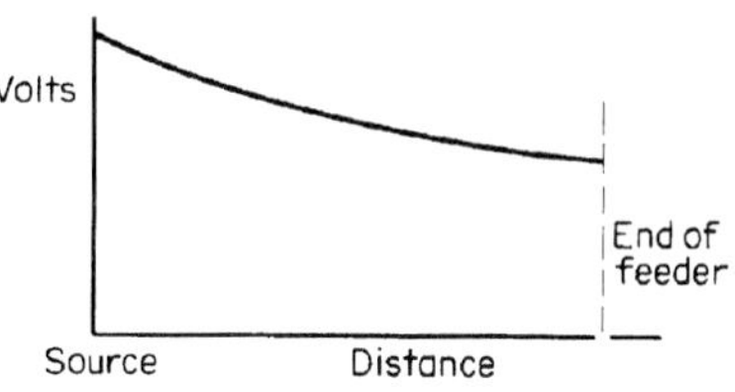


**FIGURE 7-13** Voltage profile for distributed load.

The shape of the profile is of less consequence than knowing the extremes, because there are generally customers connected at all points on the feeder, and no customer's voltage should be too high or too low. Since most feeders neither supply a single load nor are uniformly loaded, it usually is necessary to calculate the voltage profile on a piece-by-piece basis, representing the loads and feeder configurations as accurately as the situation warrants.

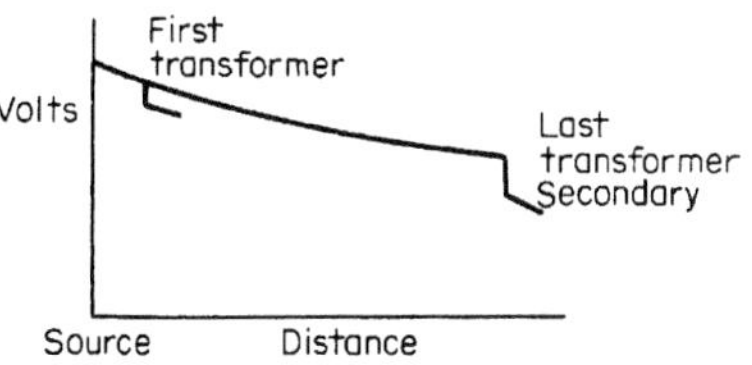


**FIGURE 7-14** Additional regulation due to transformer and secondary.

In addition to the distribution-feeder-voltage profile, there is additional regulation in the distribution transformer and its secondaries and services. This additional regulation can be added to the profile as shown in Fig. 7-14. For protection of the first customer on the feeder 0 from possible overvoltage, it is usual to assume only a partially loaded transformer rather than one at full load.

It is now possible to establish a limiting band of voltage within which all customers must lie for satisfactory service, usually range A. In turn, this also will establish the maximum permissible difference between the full-load and light-load primary voltage. The problem of holding the right voltage at each customer location at all times may be visualized by referring to Fig. 7-15.

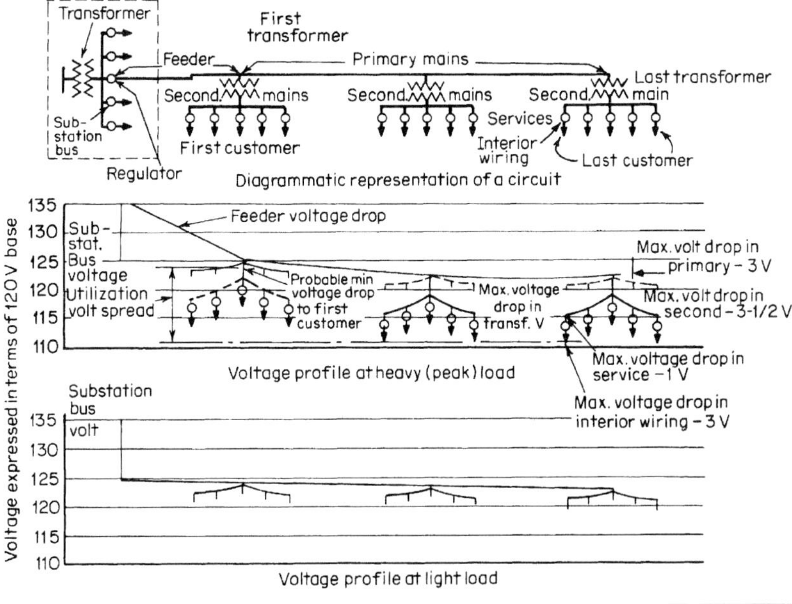


**FIGURE 7-15** Distribution circuit with voltage profiles at heavy and light loads.

***Voltage Control.*** As implied in Fig. 7-15, usually there is voltage control equipment in the substation consisting of load-tap changers on the power transformers or bus or feeder voltage regulators. This regulating equipment can control only the voltage *level* of the primary system. It can have no effect on the voltage *spread* between the first and last customers on the feeder.

There are several procedures which can be taken to correct for increasing voltage drops as the load on the feeders grows; among them are capacitors and supplementary feeder-voltage-regulator installations.

The effect of capacitor application is illustrated in Fig. 7-16, where the load is assumed to be uniformly distributed along the feeder, and a capacitor bank is installed as indicated. The capacitor produces a voltage rise because of its leading current flowing through the inductive reactance of the feeder. As is seen in the figure, this voltage rise increases linearly from zero at the substation to its maximum value at the capacitor location. Between the capacitor location and the remote end of the feeder, the rise due to the capacitor is at its maximum value.

When the capacitor voltage-rise profile is combined with the original feeder profile, the resulting net profile is obtained. The capacitor has increased the voltage *level* all along the feeder, resulting also in a reduced voltage *spread.*

In practical applications, the capacitor bank can be a permanently connected or "fixed" bank as shown or an automatically switched bank. The fixed bank is limited in size by the allowable voltage rise during light-load conditions, and therefore may not produce sufficient voltage rise during heavy-load conditions. It can be supplemented by additional switched capacitors which automatically switch on at heavy-load conditions and off again as the load decreases.

The effect of applying a supplementary feeder-voltage regulator is shown in Fig. 7-17. Note that the regulator produces no voltage effect between the source and the regulator location and its entire boost effect is between the regulator location and the remote end of the feeder.

A typical primary feeder serves distributed loads, as well as concentrated loads, and may also have shunt capacitors and supplementary voltage regulation, such that all these previous concepts must be employed in studying voltage conditions.

***Voltage Regulation.*** Voltage regulation in distribution substations usually is accomplished by individual feeder-voltage regulators or by automatic load-tap-changing equipment in the substation transformers. Individual feeder-voltage regulators are advantageous where feeders of differing lengths and diverse load characteristics are supplied from the same substation bus. Automatic load-tap-changing equipment in the power transformer provides voltage control on the substation bus, or group regulation, when feeder lengths and load characteristics are reasonably homogeneous.

Voltage control is needed to compensate not only for the voltage regulation in the subtransmission system and substation transformer, which is measurable at the substation, but also for the voltage regulation which occurs in the distribution transformers and in the primary and secondary systems beyond the substation. The latter portion of the overall system voltage regulation is a function of the load flow and system impedances and cannot be measured directly at the substation. Therefore, the

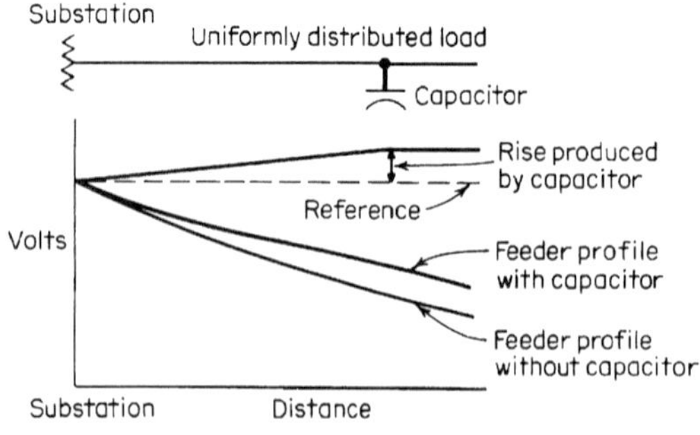


**FIGURE 7-16** Effect of shunt-capacitor application.

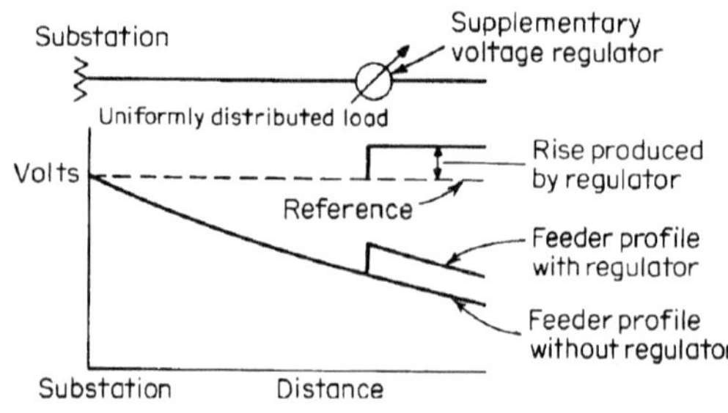


**FIGURE 7-17** Effect of supplementary voltage regulator.

control systems of the voltage regulators or tap-changing equipment not only sense the voltage at the substation but also usually contain a "line-drop compensator" which simulates the voltage drop between the station and some point in the distribution system and controls the regulating equipment accordingly. Switched shunt capacitor banks sometimes are installed at the distribution substation as part of the overall system voltage control.

***Feeder-Voltage Regulator.*** In the typical radial primary system, it is often necessary to regulate the voltage of each feeder separately by means of feeder-voltage regulators. These regulators may be of single-phase or 3-phase construction. The former are available in sizes from 25 to more than 400 kVA, the latter from 500 to 2000 kVA. For distribution-system application they are commonly available for voltages from 2.5 to 34.5 kV grounded wye. Regulators commonly are capable of raising or lowering the voltage delivered to the feeder by 10% and normally are rated on this basis.

Modern voltage regulators all are of the step-voltage type, which has completely supplanted the earlier induction-voltage regulators. The step-voltage regulator basically is an autotransformer which has numerous taps in the series winding. Taps are charged automatically under load by a switching mechanism which responds to a voltage-sensing control in order to maintain voltage as close as practicable to a predetermined level. The voltage-sensing control receives its inputs from potential and current transformers and provides control of system voltage level and bandwidth. In addition, it permits selection of line-drop compensation and provides features such as operation counter, time-delay selection, test terminals, and control switch.

Most feeder-voltage regulators are of the 32-step design. Since they usually operate over a range of voltage of 20%, the voltage change per step is ⅝%. If the full range of regulation of ±10% is not required, the regulators can carry more than rated current. For example, operating with a range of ±5%, 160% of rated current can be carried.

***Line-Drop Compensator.*** In simplified terms, the regulator voltage (local voltage) is stepped down by means of a potential transformer and fed to the control system, where it is compared with the desired and preset voltage level. If the actual voltage deviates from the preset level by more than ±½ of the bandwidth, which also is preset by the operator, the tap-changing mechanism operates, after a preset time delay, to return the voltage within the preset band. From a practical point of view, the minimum bandwidth is twice the size of the voltage step, or 2 × ⅝% = 1.25%. Maintaining a small bandwidth is important in reducing voltage variations and in making full use of the allowable system voltage drop.

The line-drop compensator consists of adjustable resistance and reactance components and is preset to simulate system impedance. By means of a current transformer, current proportional to load current is circulated through the resistance and reactance, producing a voltage signal which is combined with the signal from the local voltage. The net result is that the line-drop compensator causes a higher voltage to be held at the voltage regulator during periods of heavy load. In this way, a constant voltage is held at some point in the system, as determined by the compensator setting. This helps to achieve the goal of minimizing the voltage change with varying loads at any location.

***Supplementary Voltage Regulation.*** In some long primary circuits, such as rural feeders, it is often necessary to provide voltage regulation in addition to that incorporated in substation equipment because of large voltage drops in the system. This supplementary voltage regulation usually is improved by single-phase automatic step regulators in the smaller ratings. These regulators are suitable for pole mounting.

***Bus Regulation.*** Bus regulation at the distribution substation usually is provided by automatic load-tap-changing equipment built into the substation transformer or by large step-voltage regulators.

***Switched Shunt Capacitors.*** Switched shunt capacitors are often applied at distribution substations or out on the primary feeders to accomplish a portion of the overall voltage-regulation job. Most utilities apply shunt capacitors primarily as a tool in economic system design. Usually fixed (unswitched) shunt capacitors are applied to bring the light-load power factor to more or less 100%. Then, additional automatically switched shunt capacitor banks are added to achieve an economic full-load power factor, which is usually in the order of 95% to 100%.

These capacitors, in addition to their economic functions, such as reducing losses and releasing system capacity, improve system conditions substantially. Usually additional voltage control is needed, however, and this is most economically accomplished with voltage-regulating equipment.

***Voltage Regulation on Secondary Wire.*** The growing penetration levels of DER and fast-charging electric vehicle loads are creating dynamic voltage conditions in the distribution grids. These voltage dynamics increase the switching operations of the electromechanical voltage regulation devices installed on the primary, effectively reducing their service life. Additionally, the primary side voltage regulation equipment is typically slow in response due to their time delay settings for operation. For example, the voltage regulators are set to wait for a time delay that range from 30 to 60 s when the measured voltage crosses the preset voltage band. While such time delay settings may improve the service life of these assets by reducing the number of tap operations, they degrade the voltage control performance especially in the presence of DER with power output variabilities. Furthermore, the sparse locations of the voltage regulation equipment on the primary side and their discretized tap or switching operations do not offer direct and precise control over the secondary side load voltages. The problems associated with the primary side voltage control equipment have led to the emergence of several advanced voltage control technologies that provide voltage regulation on the secondary wire. These technologies include universal power flow controller (UPFC), static var compensator (SVC), and the smart inverters of DER such as PV, wind, and energy storage.

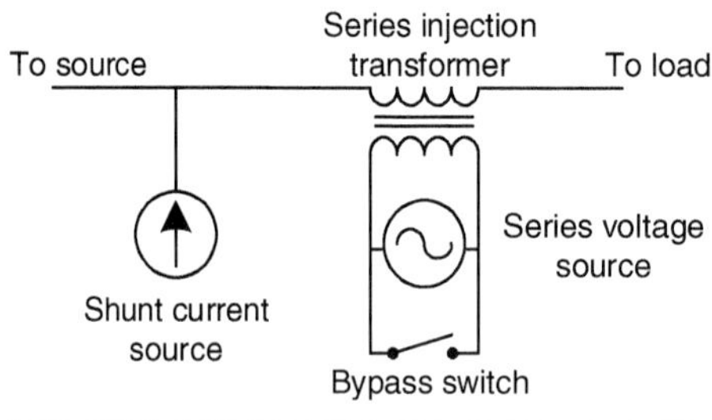


**FIGURE 7-18** UPFC architecture.

The UPFC is a single-phase power electronics based series-connected device that can be installed on a service transformer secondary (on a 240-V bus). Figure 7-18 shows the UPFC architecture. The device specifications of a commercially available 50-kVA UPFC unit are given in Table 7-4. This device can be configured to operate in any one of the three functional modes: voltage regulation mode, power factor correction mode, and complementary mode. In voltage regulation mode, the UPFC regulates the voltage at the output side of the device to a set point value by injecting a series voltage. The injected voltage can be either positive or negative depending upon the requirement. In power factor correction mode, the device injects capacitive or inductive reactive power to regulate the power factor at the input side of the device to a preset value. The device tries to accomplish both voltage regulation and power factor correction in the complementary mode.

**TABLE 7-4** Specifications of UPFC

| Specification | Details |
|---|---|
| Rating | Single-phase, 60 Hz, 50 kVA |
| Load voltage regulation range | ±10% with 0.5% accuracy |
| Var compensation range | 10% of rating (leading of lagging), i.e., up to 5 kvar |
| Response time | Less than 1 cycle |
| Operation modes | Mode 1: Voltage regulation |
| | Mode 2: Power factor correction |
| | Mode 3: Both voltage regulation and power factor correction |

The SVC is a single-phase shunt-connected device designed to be installed on a service transformer secondary. This device has several capacitor stages, as shown in Fig. 7-19, to inject required amount of capacitive reactive power into the circuit. The specifications of a commercially available 20-kvar unit are shown in Table 7-5. The SVC offers an option to configure a voltage set point. Using proprietary control algorithms, the device then switches in required number of capacitor stages to maintain its terminal voltage at the configured voltage set point value. The capacitor stages are switched out as necessary when the SVC terminal voltage exceeds beyond 3 V above the voltage set point.

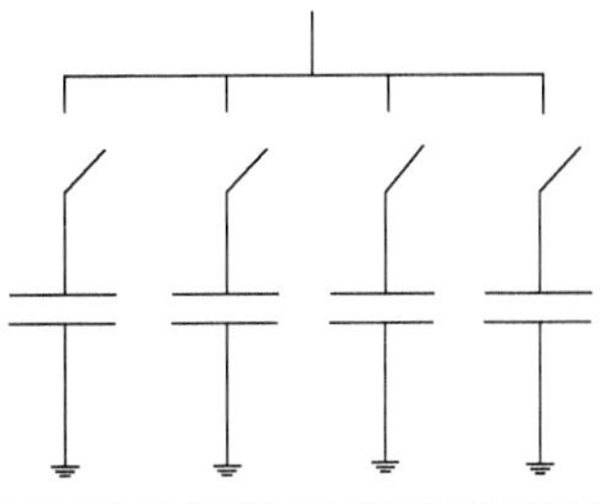

**FIGURE 7-19** SVC architecture.

**TABLE 7-5** Specifications of SVC

| Specification | Details |
|---|---|
| Rating | Single-phase, 60 Hz, 50 kvar |
| Capacitor stages | 4 capacitor stages each rated for 4.375 kvar at 240 V |
| Voltage boost characteristic | Low threshold to switch in the capacitor stages |
| | Start switching out the capacitor stages when terminal voltage is above 3 V of target voltage |
| Extreme thresholds | >252 V switch all stages out |
| | <216 V switch in all stages |

The smart inverters of DER such as PV, wind, and battery energy storage offer several functional modes. They can be configured to regulate the voltage at their interconnection points by controlled reactive power injection. Such active voltage regulation by the DER was not permitted by the past standards. Accordingly, the DER was recommended to be set to inject active power only at unity power factor. However, updated standards as on date permit active voltage regulation by the DER. Consequently, research efforts have begun to select appropriate smart inverter settings to obtain effective voltage regulation by DER.

## 7.9 OVERCURRENT PROTECTION

***General Principles.*** Coordination of overcurrent protection devices means their proper arrangement in series along a distribution circuit so that they function to clear faults from the lines and equipment in accordance with a prearranged sequence of operation. Fuse cutouts, automatic circuit reclosers, sectionalizers, and relayed circuit breakers are the overcurrent protective devices most commonly used. Ratings and characteristics can be obtained from appropriate product bulletins of the manufacturers.

When the protective devices are properly applied and coordinated,

They can eliminate service outages resulting from temporary faults.

They reduce the extent of outages, that is, the number of users affected.

They are helpful in locating the fault, thereby reducing the duration of interruptions.

***Main-Line Sectionalizing.*** Usually, the first protective device on a primary feeder is a circuit breaker or a power-class recloser located in the substation. If the circuit is overhead, the circuit breaker often is provided with reclosing relays so that it operates in much the same manner as a recloser. If the circuit is primarily underground, reclosing is not generally used.

If portions of the main feeder and long branches extend beyond the zone of protection of the relayed breaker or recloser at the substation, additional overcurrent protective equipment usually will be installed out on the main feeder. Manually operated sectionalizing equipment such as pole-top disconnecting switches or solid blade cutouts also are installed at strategic locations along the main feeder to

Provide a convenient means of isolating faults so that repairs can be made after other parts of the feeder are restored to service

Provide means of connecting the feeder to adjacent feeders so that service can be maintained to most customers while repair or maintenance operations are taking place

On underground feeders, this sectionalizing equipment is often in the form of 3-phase, manually operated, load-break switches.

***Branch-Circuit Protection.*** It is exceedingly important to isolate faults on branch and subbranch lines, even short ones, in order to maintain service on the rest of the feeder. Not only does the branch-circuit protection protect the rest of the feeder, but it helps to pinpoint the location of the fault.

Also, there is usually much more mileage and much more exposure in the branch circuit or laterals than in the feeder main. The simple expulsion-fuse cutout is almost universally used for branch and subbranch overcurrent protection. It may be used in combination with reclosers.

On underground feeders, the lateral circuits usually are fused at the point where the main feeder is tapped to establish the lateral. Often, the fuses for several lateral circuits are grouped into a sectionalizing equipment which may also incorporate main-feeder and load-break sectionalizing switches.

***Temporary Fault Protection.*** On overhead distribution circuits, a large portion of the faults are of a temporary nature or are potentially of a temporary nature. For example, some types of transitory faults include momentary contacts with tree limbs and lightning flashover of insulators or crossarms where no sustained 60-Hz short-circuit current is established and no protective devices operate. Other types of faults which result in 60-Hz follow current can be of a transient nature if the circuit voltage can be removed quickly for a short period of time and then restored after the fault path has recovered adequate dielectric strength. Such faults can result from lightning flashovers, bird or animal contacts, conductors swinging together, etc. Reclosers and reclosing breakers provide the function of fault deenergization, pause for deionization of the arc path, and reestablishment of voltage.

If the fault has disappeared during the "dead time," the reclosure is successful. If not, one or more additional reclosing cycles may be attempted. If the fault persists after the prescribed number of reclosing operations, the breaker or recloser will lock open, or the fault will be removed by operation of a fuse or sectionalizer.

It should be recognized that the reclosing function is provided to eliminate the effects of *temporary* faults only. If all faults were of a permanent nature, reclosing would be pointless. Also, temporary faults on branch circuits result in a momentary outage to all customers on the feeder when reclosing is used. Some utilities, in an effort to reduce the number of momentaries, are allowing the branch fuse to blow for temporary faults. (This is done by eliminating the instantaneous trip.) While this procedure reduces the number of momentaries seen by customers, it has the negative effect of creating a substantial interruption out of a temporary fault condition for the customers on the affected branch.

To provide effective protection against temporary faults, all parts of the feeder should be within the zone of a reclosing device. That is, if the station recloser or relayed circuit-breaker sensing does not reach to the remote ends of the circuit, it should be supplemented with reclosers out on the line. (The term *reach* here is used with the meaning of "sense" faults or "sense and operate" for faults.)

***Permanent Fault Protection.*** Permanent faults are those which require repairs, maintenance, or replacement of equipment by the utility operating department before voltage can be restored at the point of fault. System overcurrent protection is provided to disconnect the faulted portion of the system automatically so that an outage is experienced by a minimum number of consumers. Isolation of permanent faults is usually accomplished by the operation of fuse cutouts. It is also achieved in some cases by operation (to lock out) of reclosers, circuit breakers, or sectionalizers.

***Combination of Permanent and Temporary Fault Protection.*** If all faults were of a permanent nature, low-cost fuse cutouts would be the best solution for primary line protection. If all faults were temporary, automatic reclosing devices capable of covering the entire circuit would be the best solution. In actual practice, both kinds of faults occur, and the problem becomes one of selecting the type of device or combination of devices to provide best overall results. For selection of a system of overcurrent protection, it is necessary to give proper consideration to many factors such as importance of service, total number of faults per year, ratio of temporary to permanent faults, cost to utility of service interruptions, and annual charge on investment.

***Selection of Overcurrent Protective Equipment—General.*** The one-line diagram of a distribution circuit, as shown in Fig. 7-20, will show how a well-coordinated installation of overcurrent protective equipment can be made.

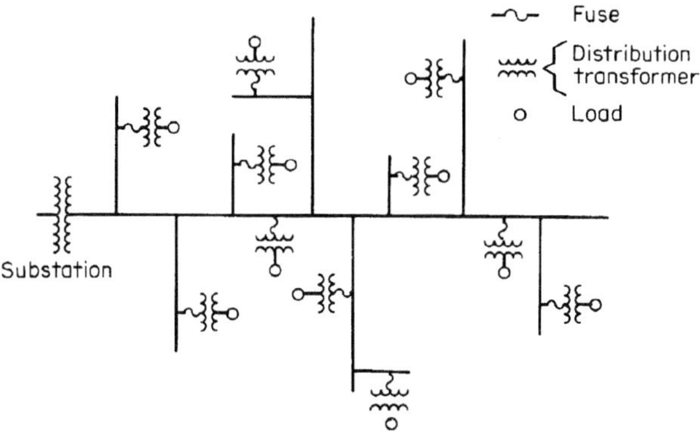


**FIGURE 7-20** Distribution feeder.

At the left is the substation, which steps down the voltage from high-voltage subtransmission level to primary-distribution voltage level. It is at this point that the distribution system starts. A distribution substation usually has a number of radial 3-phase feeders radiating from it. However, for the purposes of illustration, only a single feeder will be considered, and it is shown extending to the right from the substation. At various points along the feeder, branch lines or laterals are tapped off and in some cases subbranches are tapped from these branches. There are, of course, loads (residences, stores, garages, etc.) all along the feeder, branches, and subbranches. Only a few of these loads are shown, for the sake of clarity of the diagram.

It is general practice to install a fuse on the primary (incoming) line side of each distribution transformer, as shown in Fig. 7-20. This may be a transformer internal fuse or an external fuse installed in a cutout. Transformer fusing will be discussed later. Figure 7-20 shows the basic system to which additional overcurrent protective equipment must be added to ensure good service continuity.

To properly apply overcurrent protective equipment to this system, it will be necessary to know the highest and lowest (maximum 3-phase and minimum line-to-ground or line-to-line) values of short-circuit currents which can flow if a fault should occur where the feeder leaves the substation, at each branch junction point, and at each subbranch junction point, as well as the minimum line-to-ground short-circuit current which could flow if a fault should occur at the end of any of the branches or subbranches. These short-circuit currents may be calculated easily by conventional methods.

***Clearing Nonpersistent or Temporary Faults.*** Operating records, as well as numerous studies, indicate that a reduction of 75% to 90% in the number of total outages on an overhead system can be attained by the installation of automatic reclosing devices (automatic circuit recloser or reclosing circuit breaker). The recloser or breaker will open the circuit "instantaneously" when a fault occurs, and reclose it after a short period of time.

Referring to Fig. 7-20, automatic circuit reclosers will be applied to protect the entire system against temporary faults. To achieve this sort of protection, the first recloser should be installed on the main feeder at the substation or the power circuit breaker at the substation should be equipped with overcurrent and reclosing relays.

In applying reclosers to do this job, certain factors must be considered: (1) The voltage rating of the recloser must be high enough to meet the requirements of the system. (2) Load current, or the amount of current which flows at the point of installation of the recloser under full-load conditions, should not exceed the amount of current which the manufacturer has rated the recloser to carry continuously (continuous-current rating). Recloser ratings are usually selected to be 140% of the peak load current of the circuit. This allows for normal load growth. (3) The highest value of short-circuit current which will flow through the recloser should not be greater than the interrupting current rating of the recloser. Typically, a recloser will have a continuous rating of 560 A or less and an interrrupting rating of 16,000 A or less. A breaker, on the other hand, will usually handle at least 1200 A continuously and up to about 40 kA under short-circuit conditions.

Referring to Fig. 7-21, a recloser or breaker with reclosing relays will be located at *A* to meet the three application principles mentioned above. This device will be depended on to clear nonpersistent faults which occur in the feeder, branches, or subbranches, anywhere within its protective orbit zone *A* (shown by dotted line in Fig. 7-21). This protective zone extends to the point where the minimum available short-circuit current, as determined by calculation, is equal to the smallest value of current which will cause the device to operate. This value of current required to operate the recloser or breaker is called minimum pickup current. For a recloser it is usually equal to *twice the continuous current rating of the recloser.* A fault beyond this zone may not cause the recloser or breaker *A* to operate, and therefore, another recloser, *B*, with a lower minimum pickup current rating, should be installed just inside of zone *A*, thus resulting in so-called overlapping protection.

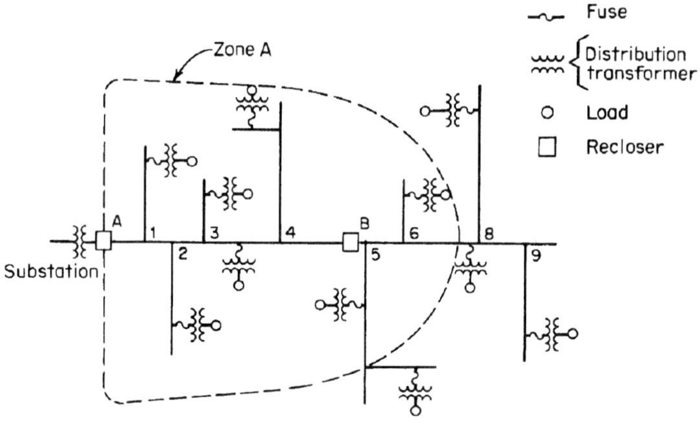


**FIGURE 7-21** Distribution feeder with automatic reclosers.

This second recloser, *B* in Fig. 7-21, is placed on the source side (side nearest source of power) of branch 5 so that it can protect the end of this branch from nonpersistent faults which may not cause recloser *A* to operate. It is applied according to the same considerations as was the recloser at *A*. It will be assumed that a fault on the feeder or any branch or subbranch beyond (to the right of) *B* will cause enough current to flow to operate the recloser at *B*. Every point on the entire circuit is now protected against nonpersistent faults because every point is within the protective zone of some reclosing device. Obviously, if every point were not within the protective orbit of some reclosing device, another recloser would have to be installed still farther out on the line.

***Clearing Persistent Faults.*** The first requirement of protecting the circuit against nonpersistent or transient faults has been taken care of by recloser application. It is necessary now to concentrate on the second and third requirements, that is, confining persistent faults to the shortest practical section of line and making persistent faults easy to locate.

If a permanent fault occurs anywhere on the system beyond a recloser, the recloser will operate once, twice, or three times instantaneously, depending on adjustment, in an attempt to clear the fault. However, since a persistent fault will still be on the line at the end of these operations, it must be cleared by some means other than the instantaneous recloser operations. For this reason, the recloser is provided with one, two, or three time-delay operations, depending on adjustment. These additional operations are purposely slower (time-delay operations) to provide coordination with fuses or to allow the fault to "self-clear." If the fault is still on the line after the last opening, the recloser will not close in but lock open.

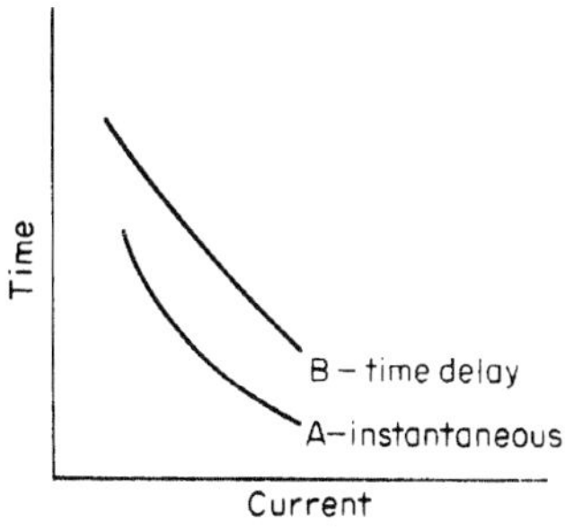


**FIGURE 7-22** Recloser tripping characteristics.

Referring to Fig. 7-22, curve *A* represents the instantaneous tripping characteristic with respect to time for the first and second opening of a conventional automatic circuit recloser. Curve *B* represents the tripping characteristics for the third and fourth openings. Following the fourth trip on time delay, the recloser will lock out and must be manually reclosed after the cause of the fault has been remedied.

A persistent fault on a branch or subbranch line *should not* cause a recloser to lock open, since a fault on a relatively unimportant subbranch could shut down the entire circuit, in addition to being extremely difficult to locate. Therefore, some means should be employed to confine outages due to persistent faults to the branch or subbranch on which they occur. This may be done in either of two ways.

One method by which persistent faults can effectively be dealt with is illustrated in Fig. 7-23. A fuse cutout is installed at each branch or subbranch junction to confine outages due to persistent faults to the branch or subbranch on which they occur, that is, fuses 1, 2, 3, 4, etc.

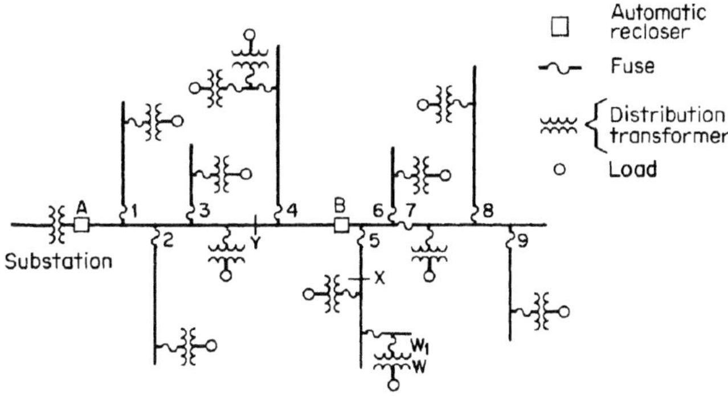


**FIGURE 7-23** Distribution feeder with automatic reclosers and fuse cutouts.

The fuse cutout to be installed at a particular location must be of sufficiently high voltage rating to meet the voltage requirements of the circuit. Its continuous current rating must be equal to or greater than the full-load current at the point of installation. Its interrupting rating must be high enough so that it will successfully open the circuit for any persistent fault occurring beyond it. This may be checked by comparing the interrupting rating of the cutout with the maximum available short-circuit current calculated for the point on the system where the cutout is to be installed.

For an ideal system, when the correct ratings of fuse links are used throughout the system, *no fuse will be blown or even damaged by a temporary fault beyond it; that is, the recloser will open the circuit one, two, or three times on instantaneous operations without the fuse link being damaged.* In many systems, however, where short-circuit levels are very high, it is sometimes impossible to prevent even

**TABLE 7-6** Automatic Recloser and Fuse Range of Coordination*

| Recloser rating, rms A (continuous) | | Fuse link ratings, rms A | | | | | | | |
|---|---|---|---|---|---|---|---|---|---|
| | | 25T | 30T | 40T | 50T | 65T | 80T | 100T | 140T |
| | | Range of coordination, rms A | | | | | | | |
| 50 | Min | 190 | 480 | 830 | 1200 | 1730 | 2380 | | |
| | Max | 620 | 860 | 1145 | 1510 | 2000 | 2525 | | |
| 70 | Min | 140 | 180 | 365 | 910 | 1400 | 2000 | 2750 | |
| | Max | 550 | 775 | 1055 | 1400 | 1850 | 2400 | 3200 | |
| 100 | Min | 200 | 200 | 200 | 415 | 940 | 1550 | 2280 | |
| | Max | 445 | 675 | 950 | 1300 | 1700 | 2225 | 3050 | |
| 140 | Min | | 280 | 280 | 280 | 720 | 710 | 1750 | |
| | Max | | 485 | 810 | 1150 | 1565 | 2075 | 2875 | |
| 200 | Min | | | | 400 | 400 | 400 | 880 | 3200 |
| | Max | | | | 960 | 1380 | 1850 | 2600 | 4000 |
| 280 | Min | | | | | | 620 | 620 | 1350 |
| | Max | | | | | | 1500 | 2200 | 4000 |

*Recloser sequence: two instantaneous plus two standard time-delay operations.

the largest fuse from operating during a temporary fault. On a permanent fault, the first fuse link on the source side of the fault will be blown, and the circuit thus will be opened by the blowing of the fuse during the third or fourth (time-delay) operation of the recloser, before the recloser will lock open. Hence, the fault will be isolated by the fuse, and the recloser will reset automatically, restoring service everywhere except beyond the blown fuse. The recloser should never lock open on a permanent fault beyond the fuse if it has been properly coordinated with the recloser. Extensive coordination tables are available, as illustrated in Table 7-6, to simplify and facilitate the job of coordinating reclosers with fuse links.

***Recloser-Fuse Coordination.*** Figure 7-24 shows the time-current characteristic curves of the automatic circuit recloser similar to those shown in Fig. 7-22. On these curves, the time-current (TC) characteristics of a fuse *C* are superimposed. It will be noted that fuse curve *C* is made up of two parts; that is, the upper portion of the curve (low current range) represents the total clearing-time TC curve, and the lower portion (high current range) represents the melting TC curve for the fuse. The intersection points of the fuse curves *C* with the recloser curves *A* and *B* illustrate the limits between which coordination will be expected. Basically, this is correct within the interest of simplicity. However, to establish intersection points *a* and *b* accurately and to prepare coordination charts, it is necessary

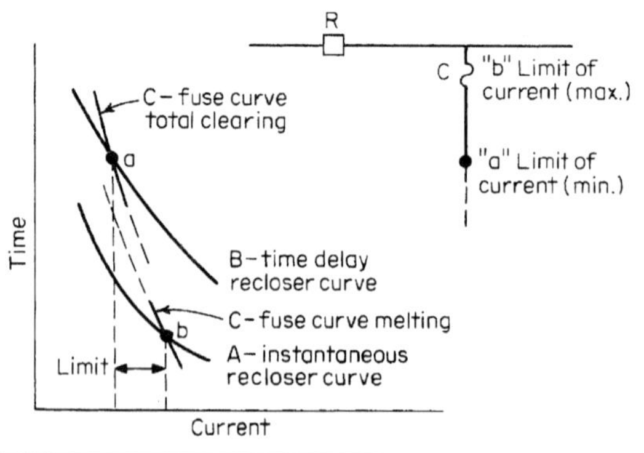


**FIGURE 7-24** Recloser and fuse time-current characteristics.

that the characteristic curves of both recloser and fuse be shifted, or modified, to take into account alternate heating and cooling of the fusible element as the recloser goes through its sequence of operations. For example, if the fuse is to be protected for two instantaneous openings, it is necessary to compute the heat input to the fuse during these two instantaneous recloser operations.

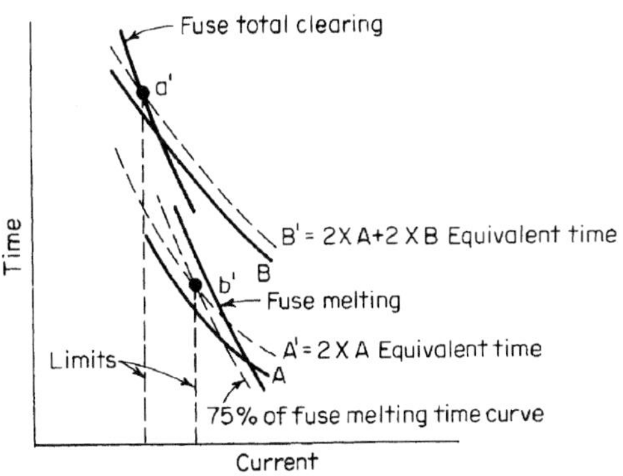


**FIGURE 7-25** Recloser and fuse time-current characteristics.

Curve $A'$ in Fig. 7-25 is the equivalent TC characteristic of two instantaneous openings ($A$) and is compared with the fuse-damage curve, which is 75% of the melting-time curve of the fuse. This will establish the high current limit of satisfactory coordination indicated by intersection point $b'$. To establish the low current limit of successful coordination, compare the total heat input to the fuse represented by curve $B'$, which is equal to the sum of two instantaneous ($A$) plus two time-delay ($B$) operations, with the total clearing-time curve of the fuse. The point of intersection is indicated by $a'$.

On the basis of all corrections added, the fuse will coordinate successfully with recloser between the current limits of $a'$ and $b'$.

To further clarify what is meant by coordination within prescribed limits, refer to Fig. 7-23—branch 5 and recloser $B$—and also Fig. 7-25 to establish how coordination is achieved between the limits of $a'$ and $b'$. Assume that fuse 5 beyond recloser $B$ is to be protected against blowing or being damaged during two instantaneous operations of the recloser *in the event of a transient fault at X*. If the maximum calculated short-circuit current at the fuse location does not exceed the magnitude of current indicated by $b'$, the fuse will be protected against blowing during all transient faults. By observation of the characteristics in Fig. 7-25, for any magnitude of short-circuit current less than $b'$ but greater than $a'$, the recloser will trip on its instantaneous characteristic once or twice to clear the fault before the fuse-melting characteristic is approached. On the other hand, *if the fault at X is persistent*, the fuse at 5 should blow before the recloser $B$ locks out. If the minimum (line-to-ground) calculated short-circuit current available *at the end of branch* 5 is substantially greater than the current indicated by $a'$, the fuse will blow (Fig. 7-25) in accordance with the total clearing characteristic, probably before the first time-delay characteristic of the recloser is approached.

The correct fuse link for any application may be selected by comparing its TC characteristics curve with those of the recloser and making certain allowances and corrections as shown. However, tables have been prepared similar to Table 7-6 to simplify greatly the job of coordinating reclosers with fuse links. This table shows the maximum and minimum currents at which certain ratings of fuse links will coordinate with certain ratings of reclosers. The only requirement in their use is a knowledge of the available short-circuit currents and load currents on the system.

Other sequences of recloser operation can be employed, but one instantaneous and two time-delay operations is the combination most widely used. In some cases, it is necessary to coordinate recloser operation with a relayed breaker at the substation. The principles of coordination are similar to the previous discussions, but a detailed study is beyond the scope of this handbook. This is also true of the application requirements for power-class reclosers for substation and line protection.

***Fuse-to-Fuse Coordination.*** It may be desirable to use more than two fuses in series beyond a recloser in order to reduce the number of consumers affected by an outage. An example of this would be the fuses at points 7, 8 and at transformers on branch 8 in Fig. 7-23. The coordination of these fuses in series beyond the recloser $B$ may be accomplished by coordinating adjacent fuses first with each other and then with the recloser in the manner just outlined.

Figure 7-26 illustrates the general principle of coordinating fuses in series. Fuse 7 is called the *protected fuse*, and fuse 8 is called the *protecting fuse*. For perfect coordination, fuse 8 must clear the circuit

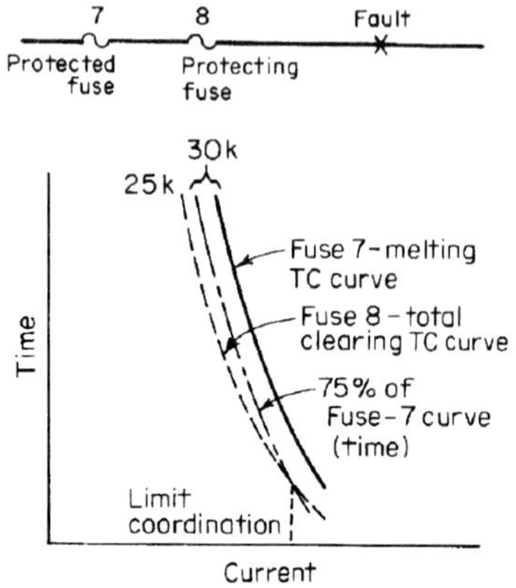


**FIGURE 7-26** Fuse time-current characteristics.

during a fault anywhere beyond it, such as at X, before fuse 7 is damaged or partially melted. From this can be seen the requirement for melting-time–current curves plotted to minimum values and total-clearing-time–current curves plotted to maximum values for each fuse-link rating. *Total-clearing-time* curves represent the total time, including melting time and arcing time, plus manufacturing tolerance, that it takes the fusible elements to clear the circuit. *Melting-time* curves represent the minimum time, based on factory test, at which the fusible element melts for various currents. From the melting-time curves, *damaging-time* curves can be determined by applying a factor of safety. It usually is suggested that the damaging-time curve be made by taking 75% of the melting time (in seconds) of a particular size at various current values.

To establish coordination of two fuses in series, it is necessary to compare the total-clearing-time–current curve of the protecting fuse with the damage-time–current curve of the protected fuse. If there is no intersection of these two curves throughout their entire current range, coordination or selectivity can be expected. Where there is an intersection of the curves, the current value indicated by the point of intersection will establish the limit of selectivity.

Because of the inherent characteristics of fuses, the maximum available short-circuit current in that section (determined by calculation) controlled by the protecting link (8 in Fig. 7-26) is the determining current which establishes coordination possibilities.

Most fuse-link manufacturers publish tables which make coordination very simple. These tables eliminate the necessity of comparing actual fuse-characteristic curves. Table 7-7 is illustrative of tables used for fuse-to-fuse coordination. The values in the left-hand column are the protecting fuse ratings and the values across the top are the protected fuse ratings. The numerical values in the table show the magnitude of current or curve intersection points at which, or below which, fuse 7 will be protected by fuse 8. These current magnitudes are maximum values; in other words, for any short-circuit current greater than that shown, fuse 7 will be damaged. Hence, a larger-rated fuse will have to be selected for location 7 or else its position must be changed.

***Isolation by Sectionalizer.*** Another method of isolating persistent faults is to install a device, known as a *sectionalizer*, at locations where a fuse might otherwise be used. A sectionalizer is a device

**TABLE 7-7** Fuse Ratings

| Type K EEI-NEMA ratings, A, of the protecting fuse links (8 in diagram) | Type K EEI-NEMA ratings, A, of the protected fuse links (7 in diagram) | | | | | | | | |
|---|---|---|---|---|---|---|---|---|---|
| | 6K | 8K | 10K | 12K | 15K | 20K | 25K | 30K | 40K |
| | Max short-circuit rms A to which fuse links will be protected | | | | | | | | |
| 1K | 135 | 215 | 300 | 395 | 530 | 660 | 820 | 1100 | 1370 |
| 2K | 110 | 195 | 300 | 395 | 530 | 660 | 820 | 1100 | 1370 |
| 3K | 80 | 165 | 290 | 395 | 530 | 660 | 820 | 1100 | 1370 |
| 5-A series Hi-surge | 14 | 133 | 270 | 395 | 530 | 660 | 820 | 1100 | 1370 |
| 6K | | 37 | 145 | 270 | 460 | 620 | 820 | 1100 | 1370 |
| 8K | | | 133 | 170 | 390 | 560 | 820 | 1100 | 1370 |
| 10-A series Hi-surge | | 16 | 24 | 260 | 530 | 660 | 820 | 1100 | 1370 |
| 10K | | | | 38 | 285 | 470 | 720 | 1100 | 1370 |
| 12K | | | | | 140 | 360 | 660 | 1100 | 1370 |
| 15K | | | | | | 95 | 410 | 960 | 1370 |
| 20K | | | | | | | 70 | 700 | 1200 |
| 25K | | | | | | | | 140 | 580 |

***Source:*** General Electric Company.

which counts the operations of a backup automatic-interrupting device such as a recloser. It has no interrupting capacity of its own but operates in a predetermined coordination scheme to open a faulted lateral before the backup device locks out.

The sectionalizer opens the circuit after a predetermined number (usually two or three) of operations of a reclosing device. Its opening operation occurs during a period when the reclosing device is open. It can be used to replace a lateral sectionalizing fuse or to replace a lateral recloser where interrupting requirements have grown beyond the capability of the recloser. Among its operating advantages are

It allows coordination with breakers or reclosers where fault current is above 5000 A. Such coordination usually is impossible with expulsion fuses.

It can provide a new sectionalizing point on an existing circuit without upsetting existing overcurrent coordination, since the device operates as a counter and does not introduce another level of time-current coordination.

### *Equipment Protection*

*General.* It is necessary to provide overcurrent protection for distribution equipment such as capacitors and distribution transformers

To protect the system from the effects of equipment failures

To reduce the probability of violent failures

To indicate the location of the fault

A detailed discussion of all aspects of overcurrent protection of equipment is beyond the scope of this handbook. However, because of its importance, a few comments will be included regarding the overcurrent protection of distribution transformers.

*Self-Protected Transformers.* The term *self-protected distribution transformer* is applied to units which incorporate an internal primary expulsion fuse, a direct-mounted arrester, and an internal secondary circuit breaker. The low-voltage circuit breaker protects the transformer from excessive overload and from some of the faults originating on the secondary system. The expulsion fuse has the sole function of removing a failed transformer from the system.

The rating of the internal expulsion fuse usually is quite large compared with the continuous current rating of the transformer, perhaps 10 to 14 times. This is done

1. To ensure that the fuse is not damaged by the maximum tripping current of the circuit breaker
2. To minimize the possibility of extraneous fuse blowing because of lightning current effects

Another reason is that fuse removal and replacement may require that the transformer be taken to a shop facility.

Transformer internal expulsion fuses are installed at the factory and are given a designating number rather than an ampere rating for coordination purposes. For a 7200-V transformer, the internal expulsion fuse, often called *weak link*, has an interrupting capacity of about 3000 A. Weak links for higher-voltage transformers have somewhat lower interrupting capacity.

Despite the fact that self-protected transformers often are installed at locations on the system where the interrupting capacity of the weak link may be exceeded for a solid fault, experience over the years has been excellent, probably because most transformer failures begin as relatively low fault-current turn-to-turn failures. As the fault current progressively becomes larger, the fuse will operate well before its interrupting capacity is exceeded. Thus, while high-current transformer faults can occur, their frequency of occurrence is very small.

However, there is growing concern among utility companies regarding the occasional violent failures of transformers, and many users are using, or are considering the use of, current-limiting fuses as one method to minimize the energy input into a failed transformer.

The secondary circuit breaker is depended on to provide protection against excessive transformer loads and secondary system faults that occur within its zone of protection, or *reach*. Its TC

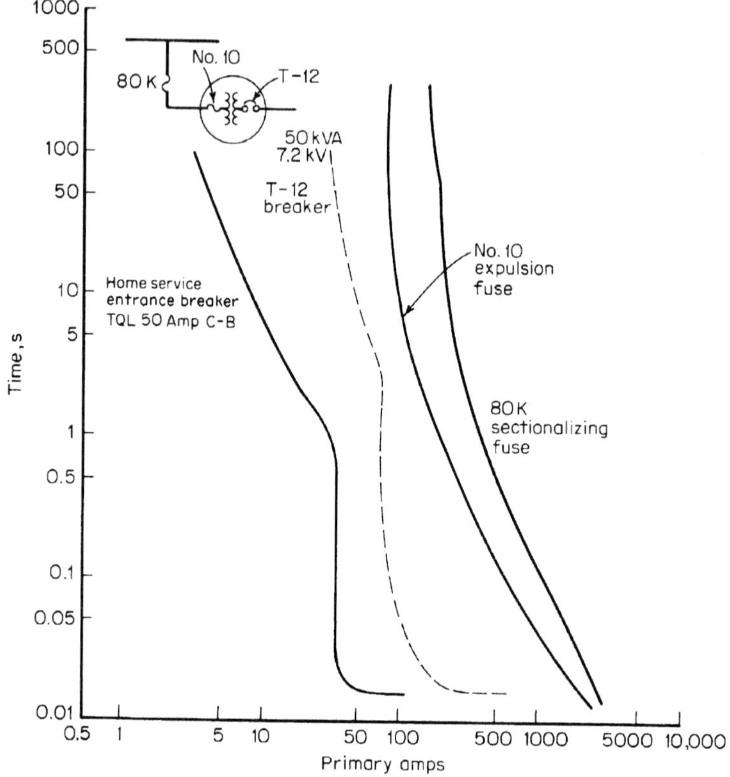


**FIGURE 7-27** Overcurrent coordination for self-protected distribution transformer.

characteristic should be such that it will always operate before the primary fuse suffers any damage, as illustrated in Fig. 7-27. On the other hand, the breaker should not operate for faults beyond the customer's service-entrance-protective equipment. Likewise, the internal primary fuse should operate to clear transformer faults before damage occurs to the line sectionalizing fuses back toward the source.

*Conventional Transformers.* Conventional distribution transformers usually are protected by separately mounted expulsion fuse cutouts in series with the primary winding. No secondary overcurrent protection is provided, so protection against extreme overloads or secondary faults, if any, must come from the primary fuse. Therefore, the size of the primary fuse is relatively much smaller than for the self-protected transformer, usually being chosen in the range of 2 to 3 times the full-load current of the transformer.

It is desirable to keep the fuse rating as low as possible consistent with certain application limitations.

1. When a transformer is energized by the closing of its cutout or operation of a recloser or other switch, a large "magnetizing inrush" current can occur. Initially, this current can be as much as 20 or more times normal, rapidly decaying to normal in a short time—perhaps ½ to 1 s or more. The primary fuse link must be large enough to avoid damage by the magnetizing inrush current, so it usually is selected at least large enough to carry 12 times rated transformer current for 0.1 s without damage.
2. The primary fuse should not be damaged by lightning currents or arrester discharge currents (depending on connection used) or large magnetizing currents which can result from saturation

of the core due to lightning currents. Many utilities assign an arbitrary minimum fuse size which they will employ. With expulsion fuses, 10- or 15-A rating is often designated as the minimum size.

With a fuse *rating* of 2 to 3 times rated transformer current, the minimum melting current under long-time conditions will be in the range of 4 to 6 times transformer rating. Consequently, little overload protection is obtained.

In the absence of overload protection, many utilities count on a transformer load-management program or seasonal load-survey techniques to keep their "burnouts" at an acceptable level. Also, the primary fuse has a limited reach as far as secondary faults are concerned; therefore, secondary faults can occur which cannot be "seen" by the fuse. Often these faults—especially on underground systems—will burn clear.

***Expulsion Cutouts.*** Distribution expulsion cutouts are by far the most common type of protective device used on overhead primary-distribution systems. The open-type cutout has generally supplanted the porcelain-enclosed style. The cutout consists of an insulating structure and a hinged fuse tube of hollow cylindrical construction which contains the fuse link. When the fuse link melts, the ensuing arc impinges on the wall of the fibrous tube holder (and usually a small auxiliary tube), generating gas which provides the expulsion action needed to extinguish the fault current. Separation of the fuse link also releases the cutout-latching mechanism so that the fuse holder falls to the open position and can readily be located by operating personnel. The fuse holder also can be switched manually with a switch stick, much like a disconnect switch. In some cases, a solid blade is used in place of the fuse holder to provide a disconnecting function. The cutout also can be provided with load-breaking accessories so that it can be used as a load-break switch.

Generally cutouts are available in 100- and 200-A continuous-current ratings for fuses and 300- or even 400-A with solid blades. Cutouts are available with voltage ratings for all the common primary system voltages and interrupting capacities generally from 1200 up through 16,000 A symmetrical and more.

Fuse links for cutouts are available with a variety of TC characteristics. However, the two most widely used types are the Type K (fast) fuse links and the Type T (slow) links with characteristics as defined in ANSI C37.43. Both types have certain application advantages and disadvantages which must be evaluated by the utility. Ordinarily, a given utility uses one type or the other, not both. Use of the Type K links is believed to be somewhat greater than use of Type T.

Other common types of primary fuses employ fusible elements immersed in oil.

Fuses are not widely used in electric utility secondary systems, with the notable exception of secondary network systems, where *limiters* are frequently used in the secondary cable circuits. Limiters are fusible elements whose TC characteristics are coordinated with the cable size and insulation characteristics to prevent damage to the cable when faults do not burn clear or self-extinguish.

***Current-Limiting Fuses.*** The use of current-limiting fuses in distribution systems has been growing. The fuse generally is constructed of silver wire or ribbon fusible elements—often several in parallel—spirally wound on a core or spider and packed in a quartz-sand filler in a sealed cylindrical glass or epoxy-glass container. Provisions for suitable electrical connections are made at the ends. When operation takes place under high-fault-current conditions, the fusible element melts almost instantaneously at a series of reduced sections all along its length. The resulting arc dissipates its heat rapidly into the surrounding sand, melting the sand around the arc into a glass-like structure called a *fulgorite*. This action builds up the apparent resistance of the fuse extremely rapidly, resulting in a "back voltage" greater than system voltage. Thus, the fault current is limited to a value much less than the available system fault currrent.

Current-limiting fuses are characterized by

1. High-current interrupting ability. Interrupting ratings of 50,000 A symmetrical or greater are commonly available.
2. Operation is noiseless, and there is no expulsion of the arc or arc products. Thus, the fuse can be "packaged" into relatively confined space in transformers and protective equipment, making it extremely attractive for use on underground systems.
3. In the current-limiting mode of operation, the interrupting time is very fast, one-half cycle or less.

**4.** Current-limiting action and fast operation reduce the amount of $I^2t$ (or fault energy) let through into failed equipment, thereby reducing resultant damage. In the case of distribution transformers applied on systems of high available fault current, protection by current-limiting fuses can virtually eliminate violent failures due to high fault current.

*General-purpose current-limiting fuses* are designed to clear fault currents over a broad range. They are defined by ANSI Standard C37.40-3.2.2.2 as fuses capable of interrupting all currents from the maximum interrupting current down to the current causing melting of the fusible element in 1 h. Current-limiting fuses inherently are excellent fault-current interrupters in the high current range. Typical general-purpose fuses operate in the current-limiting mode at fault currents equal to *approximately* 25 times rated current or larger.

Special design and construction techniques are required to obtain clearing of low-fault-current values. For operating times greater than about 0.01 s, the fuses have TC characteristics which are plotted on log-log coordination paper in the same manner as expulsion fuse characteristics.

*Backup current-limiting fuses* are defined by ANSI Standard 37.40-3.2.2.1 as fuses capable of interrupting all currents from the rated maximum interrupting current down to the rated minimum interrupting current as given by the manufacturer. The low current clearing must be accomplished by an auxiliary device, most commonly an expulsion fuse. In this case, the TC characteristics are a composite of the two fuses as shown in Fig. 7-28. The backup current-limiting fuse can be retrofitted into existing pole-type distribution transformer installations which have expulsion fuse protection only.

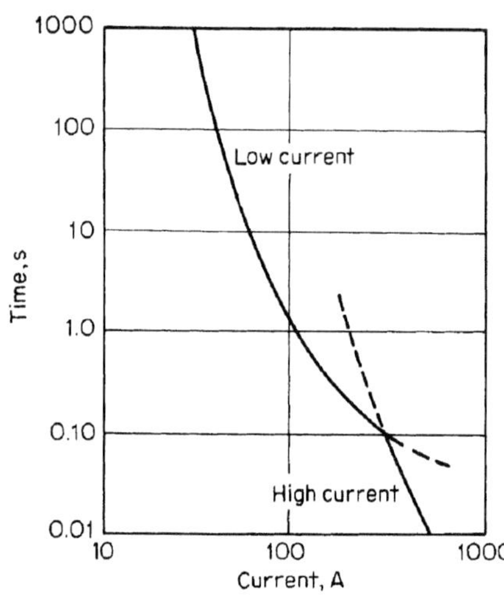


**FIGURE 7-28** Time-current characteristic for backup current-limiting fuse in series with expulsion fuse.

## 7.10 OVERVOLTAGE PROTECTION

***Lightning.*** Lightning is the most frequent cause of overvoltages on distribution systems. Basically, lightning is a gigantic spark resulting from the development of millions of volts between clouds or between a cloud and the earth. It is akin to the dielectric breakdown of a huge capacitor.

The voltage of a lightning stroke may start at hundreds of millions of volts between the cloud and earth. Although these values do not reach the earth, millions of volts can be delivered to the building, tree, or distribution line struck. In the case of overhead distribution lines, it is not necessary that a stroke contact the line to produce overvoltages dangerous to equipment. This is so because "induced voltages" caused by the collapse of the electrostatic field with a nearby stroke may reach values as high as 300 kV.

The amount of current in a stroke is a statistical quantity, depending on the energy in the cloud and the voltage difference between the cloud and the earth at the start of the stroke. A few stroke currents in excess of 200,000 A have been measured; however, 50% of all stroke currents are less than 15,000 A.

The time duration of the current flow in the majority of the high-current strokes is only tens or hundreds of microseconds. Typically, the current rises to its maximum in 0.5 to 10 μs, decreases to half value in 20 to 50 μs, and falls to zero within 100 to 200 μs. On a 60-Hz basis, these are extremely short times if one considers that one-half cycle is equivalent to $1/120$ s or 1,000,000/120 = 8333 μs. Numerous field investigations have established the numerical statistics which apply to lightning.

In summary, lightning can produce voltages dangerous to the distribution system and all its component equipment. It poses a major threat to service continuity and must be coped with by means of distribution surge arresters.

***Arrester Selection.*** Choosing an arrester rating for a distribution system is based on the system's line-to-ground voltage and the way it is grounded. The limiting condition for an arrester does not usually have anything to do with the magnitude of the surges (switching or lightning) that it might see. This is in contrast to the selection of arresters for transmission. In distribution, rating of the arrester is based on the maximum steady-state line-to-ground voltage the arrester might see. This limiting condition is normally caused when there is a line-to-ground fault on one of the other phases.

According to ANSI Standard C62.22, "Guide for the Application of Metal-Oxide Surge Arresters for Alternating-Current Systems," proper application of arresters on distribution systems requires knowledge of "(1) the maximum normal operating voltage of the power system, and (2) the magnitude and duration of temporary overvoltages (TOV) during abnormal operating conditions. This information must be compared to the arrester MCOV rating and to the arrester TOV capability."

The MCOV of the arrester is, however, somewhat easier to define because it is approximately 84% of the arrester duty cycle rating. What this means is that a 10-kV duty cycle rated arrester, typically used for a 13.2-kV system, could be operated continuously with a maximum continuous line-to-ground voltage of 8.4 kV or less.

Table 7-8, from ANSI C62.22, shows the commonly applied voltage ratings of metal-oxide arresters for distribution systems. All these duty cycle ratings are the same as the rating for the older gapped silicon carbide arresters except at the 13.8-kV level. Typically, a 13.8-kV, 4-wire, multigrounded system has used 10-kV gapped arresters. Today, most of these same utilities are still using 10-kV MOVs. Some utilities, however, have recognized that the 10-kV arrester is very marginal and possibly should be replaced by a 12-kV rating to be on the more conservative side.

**TABLE 7-8** Commonly Applied Voltage Ratings of Metal-Oxide Arresters on Distribution Systems[a]

| System voltage, kV rms | | Commonly applied arrester voltage ratings, kV rms duty cycle voltage ratings (MCOV)[b] | | |
|---|---|---|---|---|
| Nominal voltage | Maximum voltage range B[c] | 4-Wire multigrounded neutral wye | 3-Wire low-impedance[d] grounded[e] | 3-Wire high impedance[d] grounded |
| 2400 | 2540 | | | 3 (2.55) |
| 4160Y/2400 | 4400Y/2540 | 3 (2.55) | 6 (5.1) | 6 (5.1) |
| 4260 | 4400 | | | 6 (5.1) |
| 4800 | 5080 | | | 6 (5.1) |
| 6900 | 7260 | | | 9 (7.65) |
| 8320Y/4800 | 8800Y/5080 | 6 (5.1) | 9 (7.65) | |
| 12000Y/6930 | 12700Y/7330 | 9 (7.65) | 12 (10.2)[f] | |
| 12470Y/7200 | 13200Y/7620 | 9 (7.65) or 10 (8.4) | 15 (12.7)[f] | |
| 13200Y/7620 | 13970Y/8070 | 10 (8.4) | 15 (12.7)[f] | |
| 13800Y/7970 | 14605Y/8430 | 12 (10.1) | 15 (12.7)[f] | |
| 13800 | 14520 | | | 18 (15.3) |
| 20780Y/12000 | 22000Y/12700 | 15 (12.7) | 21 (17.0)[f] | |
| 22860Y/13200 | 24200Y/13870 | 18 (15.3) | 24 (19.5)[f] | |
| 23000 | 24340 | | | 30 (24.4) |
| 24940Y/14400 | 26400Y/15240 | 18 (15.3) | 27 (22.0)[f] | |
| 27600Y/15930 | 29255Y/16890 | 21 (17.0) | 30 (24.4)[f] | |
| 34500Y/19920 | 36510Y/21080 | 27 (22.0) | 36 (29.0)[f] | |

[a]Spacer cable circuits have not been included—there has been insufficient experience with the application of metal-oxide arresters on spacer cable circuits to include them in this table.

[b]For each duty cycle rating, the maximum continuous operating voltage (MCOV) is also listed.

[c]See ANSI C84.1-2016.

[d]Low impedance circuits are typically 3-wire, unigrounded at the source. High impedance circuits are generally ungrounded (i.e., delta). Additional information regarding system grounding is contained in ANSI C62.92, Part 1.

[e]Line-to-ground fault duration not to exceed 30 min. For longer durations consult manufacturers' temporary overvoltage capability.

[f]Individual case studies may show lower voltage ratings may be used.

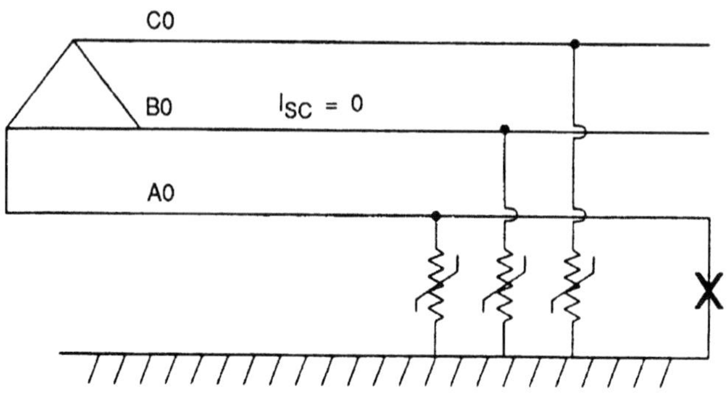


**FIGURE 7-29** Line-to-ground fault on a delta system.

*TOV.* How much voltage shift which will occur is a function of the type of system grounding. For example, on a delta system, a line-to-ground fault will cause a full offset; that is, the line-to-ground voltage will become the line-to-line voltage. Figure 7-29 illustrates this condition. As can be seen, when a phase has a fault there is no current because the transformer is delta-connected. For a 4-wire multigrounded system, there is less voltage rise (Fig. 7-30). The arresters connected from nominal line-to-neutral voltage multiplied by the product of the regulation factor 1.05 and the voltage rise factor 1.2 results in a factor of 1.25 times nominal line-to-neutral system voltage. For an MOV type arrester, this voltage is compared with the TOV rating of the MOV. Because the MOV arrester is more sensitive to poor grounding, poor regulation, and the reduced saturation sometimes found in new transformers, it is generally recommended that a 1.35 factor be considered for MOVs.

A summary of this and other recommendations is as follows:

Open-wire multigrounded system:

Rating = nominal line-to-ground (L-G) voltage × 1.25 (gapped)

Rating = nominal L-G voltage × 1.35 (MOV)

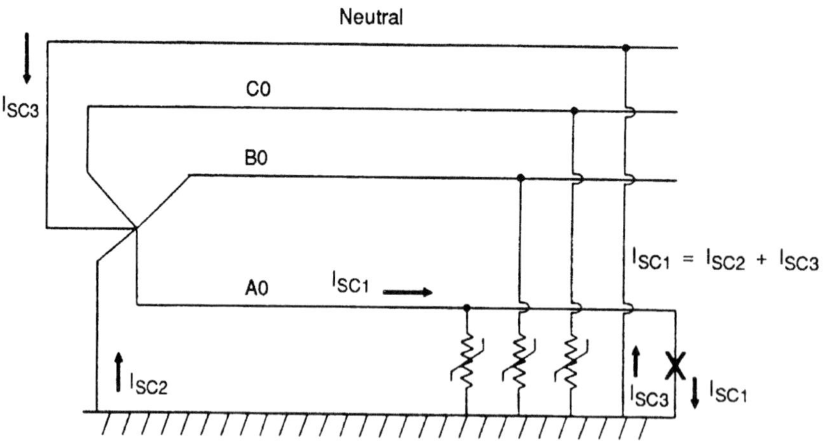


**FIGURE 7-30** Line-to-ground fault on a grounded-wye system.

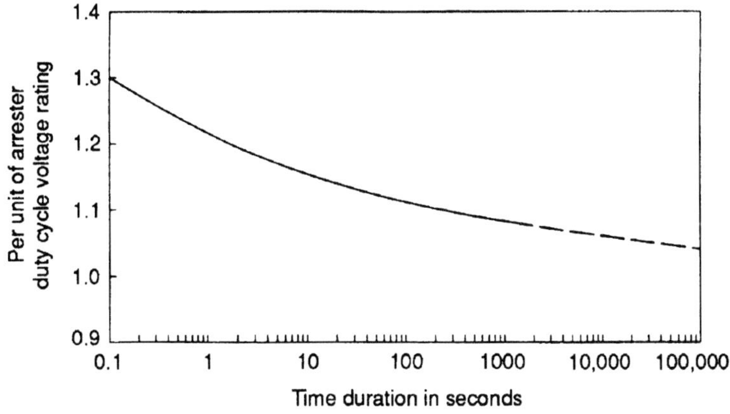


**FIGURE 7-31** Example TOV curve (consult arrester manufacturer for exact curve).

Spacer cable systems:

$$\text{Rating} = \text{nominal L-G voltage} \times 1.5$$
$$\text{Unigrounded rating} = \text{nominal L-G voltage} \times 1.4$$

The temporary overvoltage capability of the arrester as a function of time is shown in Fig. 7-31.

### *Insulation Coordination*

*Margins for Overhead Equipment.* It is important to note that application of arresters for transmission and distribution is different. In transmission, lightning is of secondary concern in surge arrester application. The primary concern is switching surges. On a distribution circuit, however, the relatively low voltage and short lines tend to make switching surges minimal, and consequently, lightning is of primary importance.

Reflection of this fact can be seen in typical characteristics published for distribution-class arresters, as shown in Tables 7-9 and 7-10. As can be seen, protective characteristics are shown for front-of-wave sparkover and *IR* discharges but not for switching surge waves (as shown for higher rated transmission arresters).

**TABLE 7-9** Distribution Arrester Characteristics from Handbook—Silicon Carbide

| Arrester rating, kV RMS | Maximum ANSI front-of-wave sparkover, kV crest | | Maximum discharge voltage, kV crest at indicated 8 × 20-μs impulse current | | |
|---|---|---|---|---|---|
| | With disconnector | Externally gapped | 5000 A | 10,000 A | 20,000 A |
| 3 | 14.5 | 31 | 11 | 12 | 13.5 |
| 6 | 28 | 51 | 22 | 24 | 27 |
| 9 | 39 | 64 | 33 | 36 | 40 |
| 10 | 43 | 64 | 33 | 36 | 40 |
| 12 | 54 | 77 | 44 | 48 | 54 |
| 15 | 63 | 91 | 50 | 54 | 61 |
| 18 | 75 | 105 | 61 | 66 | 74 |
| 21 | 89 | — | 72 | 78 | 88 |
| 27 | 98 | — | 87 | 96 | 107 |

**TABLE 7-10** Distribution Arrester Characteristics from Handbook—MOV (Heavy Duty)

| Arrester rating, kV rms | MCOV, kV rms | Front-of-wave protective level,* kV crest | Maximum discharge voltage, 8 × 20-μs current wave | | |
|---|---|---|---|---|---|
| | | | 5 kA | 10 kA | 20 kA |
| 3 | 2.55 | 10.7 | 9.2 | 10.0 | 11.3 |
| 6 | 5.10 | 21.4 | 18.4 | 20.0 | 22.5 |
| 9 | 7.65 | 32.1 | 27.5 | 30.0 | 33.8 |
| 10 | 8.40 | 35.3 | 30.3 | 33.0 | 37.2 |
| 12 | 10.2 | 42.8 | 36.7 | 40.0 | 45.0 |
| 15 | 12.7 | 53.5 | 45.9 | 50.0 | 56.3 |
| 18 | 15.3 | 64.2 | 55.1 | 60.0 | 67.6 |
| 21 | 17.0 | 74.9 | 64.3 | 70.0 | 78.8 |
| 24 | 19.5 | 84.3 | 72.3 | 78.8 | 88.7 |
| 27 | 22.0 | 95.2 | 81.7 | 89.0 | 100.2 |
| 30 | 24.4 | 105.9 | 90.9 | 99.0 | 111.5 |
| 36 | 30.4 | 124.8 | 107.0 | 116.6 | 131.3 |

*Based on a 10-kA current impulse that results in a discharge voltage cresting in 0.5 μs.

The two protective characteristics normally used for insulation coordination are

*Front-of-wave sparkover.* This is the first thing that happens to the gapped arrester—it sparks over. It is compared with the fast front equipment insulation characteristics such as the chopped wave insulation level of the transformer. An MOV has no gap but does have an equivalent sparkover, as shown in Table 7-10.

*IR discharge at 10 kA.* After the arrester sparks over the gap, the lightning current discharges through the block material. Standards recommend that a 10-kA discharge level be used for coordination purposes. (Discharge characteristics across a MOV are very similar to gapped silicon carbide arresters, so the margin calculation is virtually identical.)

Distribution equipment is normally defined as being in a voltage class such as 15 or 25 kV. Most utility equipment is operated in the 15-kV class. A distribution transformer in the 15-kV class is defined by the following insulation characteristics:

60-Hz, 1-min withstand = 34 kV

Chopped wave (short-time) = 110 kV at 1.8 μs

Basic insulation level (BIL) = 95 kV

Assuming a 12,470-V, 4-wire system (7200 V L-G), we would select the arrester rating based on the rules developed in the preceding section, i.e., a 9-kV arrester (gapped).

We can see from Tables 7-9 and 7-10 that a 9-kV gapped arrester has a sparkover of 39 kV and an *IR* discharge at 10 kA of 36 kV. This could be plotted with the transformer characteristics as in Fig. 7-32.

Standards recommend 20% margins calculated by the formula

$$\text{Margin} = \frac{\text{insulation withstand} - \text{protective level}}{\text{protective level}} \times 100$$

Two margins are calculated, one for the chopped wave and one for the full wave (BIL) of the transformer. These calculations are performed as follows:

$$\text{Margin} = \frac{110-39}{39} \times 100\% = 182\% \text{ (chopped wave)}$$

$$\%\ \text{Margin} = \frac{95-36}{36} \times 100\% = 164\% \text{ (BIL)}$$

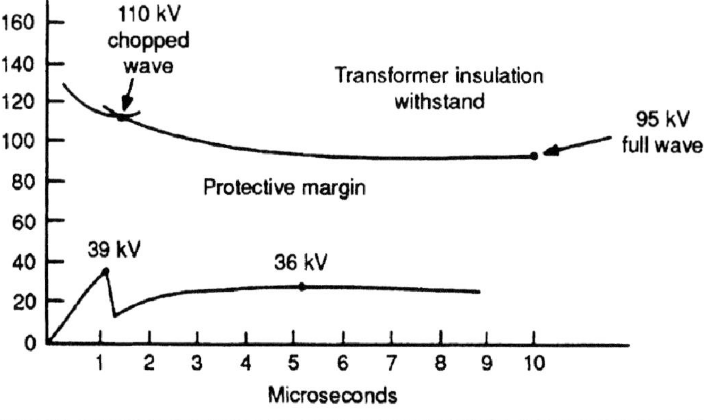


**FIGURE 7-32** Insulation coordination.

As can be seen, these margins (182% and 164%) are greatly in excess of the recommended 20% and consequently show good protection practice. If we were using an MOV, we would simply use the equivalent sparkover or compare only the *IR* discharge and the BIL, since this is the lesser of the two margins. The margins would be similar.

*Margins for Underground Equipment.* If the system is underground, we must be more concerned with the phenomena of traveling waves and the consequent doubling of voltage surges at an open point. For example, a typical underground residential design is shown in Fig. 7-33. A surge entering

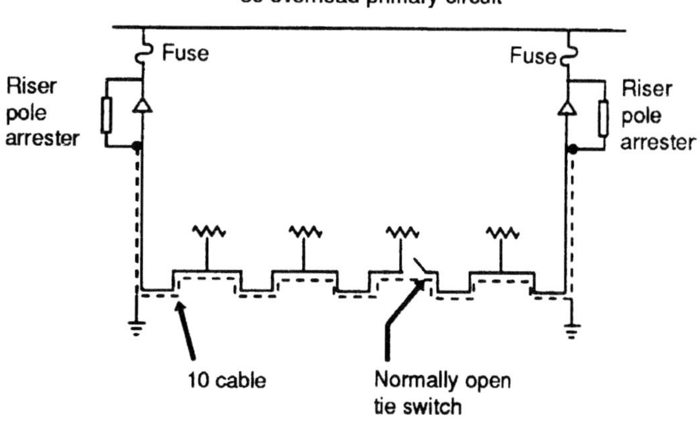


**FIGURE 7-33** Underground lateral.

the cable will travel to the open point where its voltage will double, as shown in the figure, and start on its way back.

This reflected wave plus the incoming waves impose approximately twice the normal voltage on the entire cable and all the equipment connected to it (Fig. 7-34). For example, if we had an arrester with a 36-kV *IR* discharge level (we are only considering

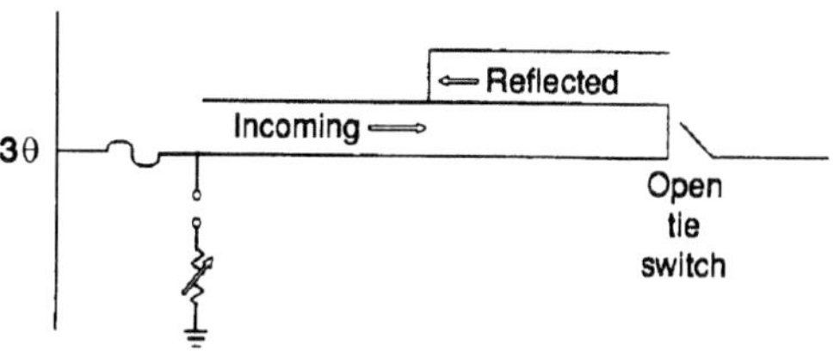


**FIGURE 7-34** Reflected surge voltage at open point.

BIL margin), we would now expect to see 72 kV imposed across the insulation of this equipment. The new margin would then be calculated as follows:

$$\%\text{ Margin} = \frac{95-72}{72} \times 100 = 32\%$$

For higher voltage levels, where the margins become considerably less, it may be necessary to use an arrester at the open point or riser-pole arresters having lower discharge levels.

## 7.11 DISTRIBUTION TRANSFORMERS

Distribution transformers convert electrical energy from primary voltages (2.4 to 34.5 kV) to utilization voltages (120 to 600 V). Momentary drops in lighting voltage caused by the starting current of motors often necessitate use of separate transformers where 3-phase motors 20 hp and larger and 1-phase motors 6.5 hp and larger must be served from radial circuits.

***Standard Ratings of Single-Phase Distribution Transformers.*** By agreement between users and manufacturers, certain features of line-transformer design have been standardized for sizes up to 500 kVA and for voltages up to 67,000 V. Capacities are 10, 15, 25, 37½, 50, 75, 100, 167, 250, 333, and 500 kVA.

Voltage ratings on primary windings are 2400/4160Y, 4800/8320Y, 7200/12,470Y, 12,470GrdY/7200, 7620/13,200Y, 13,200GrdY/7620, 12,000, 13,200/22,860GrdY, 13,200, 13,800GrdY/7970, 13,800/23,900GrdY, 13,800, 14,400/24,940GrdY, 16,340, 24,940GrdY/14,400, 19,920/34,500GrdY, 34,500GrdY/19,920, 22,900, and 34,400. On the secondary side, windings are built for 3-wire operation at voltages of 120/240 or for 240/480. For some of the larger kVA sizes, secondary side windings are available at voltages from 2400 to 7970 V. Bushings for secondary terminals are located on the side of the case, except that primary bushings for 7200 V and higher are cover-mounted. Supporting lugs are arranged to permit mounting either by bolting to the pole or by hanging on crossarms. Where necessary, provision is made for a grounding connection to the case or from the secondary neutral terminal to the case. Similar standards have been promulgated by ANSI for 3-phase pole-type transformers up to 500 kVA.

Electrical characteristics typical of single- and 3-phase transformers of the 12470Y/7200V class are given in Table 7-11. Distribution transformers with different primary voltages will have values only slightly different from those shown in Table 7-11. Transformer regulation for a kVA load of power factor cos $\theta$, at rated voltage, can be calculated from the formula

$$\text{Percent regulation} = \frac{\text{kVA load}}{\text{kVA transformer}}\left[\%IR\cos\theta + \%IX\sin\theta + \frac{(\%IX\cos\theta - \%IR\sin\theta)^2}{200}\right]$$

Transformers are installed on poles in the following ways: transformers 100 kVA and smaller are bolted directly to the pole, and sizes 167 to 500 kVA have support lugs attached to the transformer and are intended for bolting to adapter plates for direct pole mounting or hung on crossarms by means of steel hangers attached securely to the transformer.

Banks of three single-phase transformers are hung side by side on heavy double arms, usually located low on the pole, or on a "cluster" bracket which spaces them around the pole. Three or more transformers 167 kVA and larger are installed on a platform supported by two poles set 10 to 15 ft apart. The transformer-platform structure is often placed on the customer's premises to reduce the distance that secondaries must be run and to avoid pole congestion on public thorough fares.

Transformers are installed in street vaults, in manholes, on pads at ground level, subsurface, or within buildings. When installed within buildings where the possibility of submersion is remote, the overhead or inside types of transformer and cutout are used. Transformer vaults within a building are of fireproof construction, except when transformers are dry type or filled with nonflammable liquid.

**TABLE 7-11** Typical Electrical Characteristics of Single-Phase and 3-Phase 60-Hz Distribution Transformers Loss factors can vary according to evaluation factors.

| Size, kVA | Percent *IR* | Percent *IX* | Percent *IZ* | Percent no-load loss | Percent load loss |
|---|---|---|---|---|---|
| Pole-type single-phase transformers—voltage rating 7200/12,470Y to 120/240 V | | | | | |
| 10 | 1.6 | 1.4 | 2.1 | 0.59 | 1.65 |
| 15 | 1.3 | 1.0 | 1.6 | 0.51 | 1.28 |
| 25 | 1.2 | 1.7 | 2.1 | 0.38 | 1.26 |
| 37½ | 1.3 | 1.9 | 2.3 | 0.37 | 1.31 |
| 50 | 1.1 | 1.8 | 2.1 | 0.36 | 1.10 |
| 75 | 1.0 | 2.1 | 2.3 | 0.34 | 1.03 |
| 100 | 1.0 | 2.1 | 2.3 | 0.32 | 1.02 |
| 167 | 1.0 | 2.0 | 2.2 | 0.29 | 0.96 |
| 250 | 1.0 | 2.3 | 2.5 | 0.23 | 0.99 |
| 333 | 0.9 | 2.4 | 2.6 | 0.21 | 0.90 |
| 500 | 0.8 | 2.5 | 2.6 | 0.20 | 0.82 |
| Pad-mounted 3-phase transformers—voltage ratings 12,470Y/7200 to 208Y/120 V | | | | | |
| 75 | 1.0 | 3.0 | 3.2 | 0.52 | 0.95 |
| 112.5 | 1.1 | 3.2 | 3.4 | 0.40 | 1.15 |
| 150 | 1.0 | 3.4 | 3.5 | 0.39 | 0.96 |
| 225 | 1.0 | 3.4 | 3.5 | 0.36 | 0.98 |
| 300 | 1.0 | 3.8 | 3.9 | 0.33 | 0.97 |
| 500 | 1.0 | 3.9 | 4.0 | 0.27 | 0.97 |

***Pole-Mounted Regulators.*** Today 16- or 32-step regulators built for pole mounting cover the customary ±10% voltage range. Voltage-level control and line-drop compensators give them essentially the same characteristics as the larger station regulators.

Open-Δ connection enables small power customers to receive 3-phase service from two transformers connected to a 3-phase circuit, thus reducing the investment in transformers. Open-Δ from a 3-wire system is the usual Δ connection with one transformer omitted. The connection from a 4-wire system is shown in Fig. 7-35, 2-phase wires and neutral being used on the primary side of the transformers. Current in each of two single-phase transformers connected in open Δ is 73% greater than in each of three transformers connected in closed Δ.

The *Scott connection* shown in Fig. 7-36 gives an accurate transformation but requires one of the transformers to have an 86.6% tap and the other to have a 50% tap.

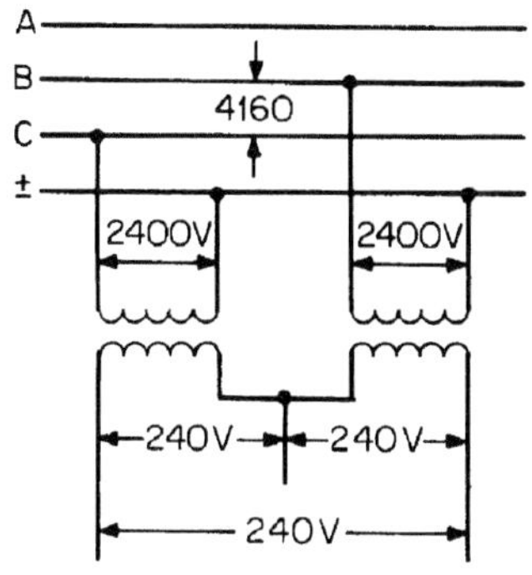


**FIGURE 7-35** Open Y connection from 4-wire, 3-phase system.

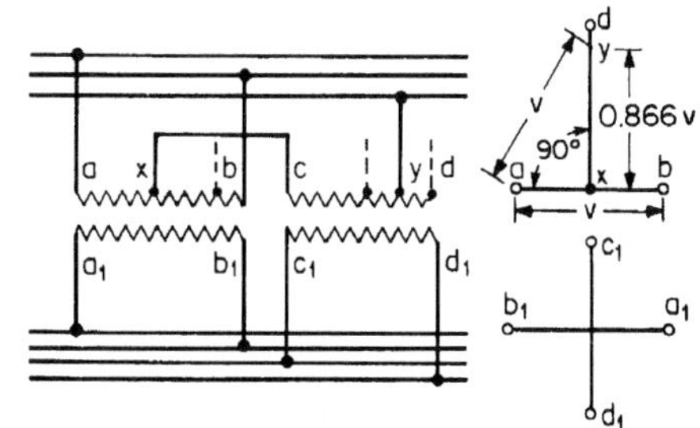


**FIGURE 7-36** Balanced T- or Scott-connected transformer connection for 3- to 2-phase transformation.

## 7.12 SECONDARY RADIAL DISTRIBUTION

*Secondary mains* operate at utilization voltage and serve as the local distributing main. In early commercial radial systems, secondary mains that supply general lighting and small power are usually separate from mains that supply 3-phase power because of the dip in voltage caused by starting motors. This dip in voltage, if sufficiently large, causes an objectionable lamp flicker.

Single-phase secondary mains supplying general lighting and small power are usually 3-wire mains operating at 120 V line-to-neutral and 240 V line-to-line. Incandescent lamps, fans, heating devices, small fractional-horsepower motors, and other appliances rated 115 or 120 V are supplied from the 120-V line and neutral. Electric ranges, larger single-phase motors up to 6.5 hp, and large appliances rated 230 or 240 V are supplied 240 V. Some utilities supply these loads at 120/208 V.

Three-phase secondary mains are commonly operated 3-wire 240 V. Some utilities offer 208Y/120-V 3-phase 4-wire service. The 3-phase mains are on the same poles or in the same duct line (but in separate ducts) with single-phase lighting mains. Separate single-phase and 3-phase services are extended to customers who require both types of service. In large commercial and industrial installations, power is often delivered at 480 V to effect an economy in conductor investment.

European practice is to supply 230 V to lighting and appliances from a 230/440-V system. This effects a savings in distribution and interior wiring but results in less efficient incandescent lamps and other small appliances.

In America, large commercial buildings and factories are served at 480Y/277 V because most permanent lighting is fluorescent, which operates efficiently at 277 V, and 480 V is well suited for the numerous 3-phase motors. Such installations have small dry-type transformers to supply 120 V for portable lights, convenience outlets, and tools and for business machines; these transformers are located near the 120-V loads and supplied from the 480-V system.

Fractional-horsepower motors up to about ¾ hp are regularly supplied by single-phase 120-V mains. Industry committees, sparked by sudden acceptance of home air conditioning, several years ago agreed to permit starting currents not to exceed 50 A for 115-V motors. Special design enabled motors up to ¾ hp to meet this limitation. Larger motors up to 6.5 hp are usually served at 240 V, although 3- and 6.5-hp motors may require extra care in distribution design to avoid troublesome flicker. Motors larger than 6.5 hp are usually connected 3-phase. Three-phase service is not usually supplied in residential areas.

***Light and Power from One Secondary Main.*** In a *radial* system, 3-phase service is sometimes supplied from a separate secondary main if voltage is affected by elevator motors or other intermittently used load. If separation of light and power service is not necessary, the nature of the connection may depend upon the relative size of light and power loads. When power load is predominant, lighting load may be served by providing additional capacity in one of the transformers and bringing in a neutral from it for the lighting service. The neutral for lighting service is sometimes derived from a transformer connected to one phase of 240- or 480-V power circuits giving 120/240 V for lighting. This is the usual procedure where power is served at 480 V. When the lighting load is predominant, service is often provided at 208Y/120V, 4-wire.

***Transformer and Secondary-Main Economy, Overhead Distribution.*** Several independent studies have been made to determine the proper combination of transformer and radial secondary main that provides satisfactory voltage regulation and costs a minimum per kVA of load served. All these studies indicate that for 120/240-V single-phase distribution, overhead secondary mains should be three No. 1/0 to three No. 4/0 aluminum, the latter being preferred when air conditioning or heating is to be served.

Permissible length of the three No. 1/0 aluminum secondary mains depends on the load density. On the assumption of evenly distributed loads and 3% drop in the mains, for 15 kW/1000 ft, the permissible length is 600 ft, and for 30 kW/1000 ft, 400 ft. Widespread use of ranges and motor-driven appliances establishes an additional limit for flicker at 200 to 300 ft.

Transformer size should be such that the initial peak load is between 75% and 100% of rated capacity. In medium-load densities, 25- and 50-kVA transformers will fulfill this requirement. Transformers should be allowed to remain in service until their winter peak load reaches at least

150% to 180% of rated capacity. When this occurs, the "hot spot" winding temperature is approaching 110°C—the maximum safe temperature.

Load growth should be taken care of by *installing additional transformers* and cutting radial secondaries or by *increasing the size* of the existing transformers where secondary-main regulation permits. The three No. 1/0 to 4/0 aluminum single-phase secondary mains should not be replaced by larger conductors to improve secondary-main regulation. Additional transformers should be installed and parts of the existing mains transferred to the new transformers.

Underground systems should also be designed initially with capacity for growth. In order to accomplish this, many utilities in underground residential distribution (URD) work do not use secondary mains. Rather, one transformer is used to supply four to six homes by installing service drops large enough for future loads from the transformer to each home. With this system design it is relatively easy to change out the transformer to a larger size when the load grows.

Pad-mounted transformers can be sized and operated the same as overhead-type transformers. Advantage can be taken of the short-time overload capability given by ANSI C57.91, "Guide for Loading Mineral Oil Immersed Overhead-Type Distribution Transformers with 55C or 65C Average Winding Rise." Subsurface transformers in close-fitting cylindrical vaults require special baffles and chimney specified by the manufacturer in order that they might be loaded the same as an overhead-type transformer.

Subway-type transformers should not be replaced or relieved of load until the calculated hot-spot winding temperature exceeds 110°C, provided, of course, that voltage at the ends of the secondary is satisfactory. To calculate hot-spot winding temperature, the maximum load and top-oil (or case) temperatures must be measured. Maximum case temperature has been found to be within 3°C of top-oil temperature. It is assumed in making the calculation that the difference between hot-spot-winding and top-oil temperature is 20°C at full load and that this difference varies as the square of load. This is a conservative assumption. For example, assume maximum case temperature 67°C when 130% load is on the transformer. Then the calculated winding hot-spot temperature is given by

$$67°\text{C} + 3°\text{C} + 20°\text{C}(1.30)^2 = 114°\text{C}$$

Fans to supplement natural air movement have been used to boost safe capability of vault transformers.

Table 7-12 gives the voltage drop per 10,000 A · ft for single-phase and 3-phase secondaries for a variety of load power factors. The underground portion of the table can be used for underground systems and also overhead systems where triplex cable construction is employed. The overhead part of the table gives the voltage-drop information for overhead aluminum conductors on racks. The table can be used to determine voltage drop quickly on any secondary circuit if load, circuit length, and conductor size are known.

All values in the table are for aluminum conductors at 50°C temperature. Values for copper can be determined with satisfactory accuracy by using the table for a conductor of equivalent resistance; that is, use an aluminum conductor two sizes larger than the copper conductor.

In using the table, the first thing required is the number of ampere-feet involved in the problem. This is obtained by multiplying the amperes per phase by length of circuit in feet. (For single-phase, use number of feet between source and load; impedance of return circuit is included in table.) Divide this ampere-feet by 10,000 to determine the multiplier to be used with values in the table. For the proper voltage, conductor size, and power factor, find the voltage-drop factor in the table and multiply by the multiplier determined previously. This will be the absolute line-to-neutral volts difference (drop) between the sending and receiving ends of the circuit. Dividing by the line-to-neutral voltage of sending end or receiving end and multiplying by 100 will express this as a percentage of sending- or receiving-end voltage, respectively.

*Example.* Given a 3-phase 60-Hz secondary 500 ft in length, which consists of No. 4/0 aluminum conductor cable; conductor temperature 50°C; receiving-end load 100 kVA at 0.8 power factor lagging; receiving-end line-to-line voltage 480.

$$\text{A} \cdot \text{ft} = \frac{100}{\sqrt{3} \times 0.48} \times 500 \text{ ft} = 60{,}142, \text{ or } 6.014 \text{ times tabular value}$$

**TABLE 7-12** Voltage Drops per 10,000 A · ft* for Single-Phase and 3-Phase Secondaries, 60 Hz

| Conductor size | Voltage: 120/240-V single-phase, lagging power factor 0.7 | 0.8 | 0.9 | 0.95 | 1.00 | 208Y/120 V, 240 V, 480Y/277 V, and 480-V 3-phase, lagging power factor 0.7 | 0.8 | 0.9 | 0.95 | 1.00 |
|---|---|---|---|---|---|---|---|---|---|---|
| Underground or triplex secondary† | | | | | | | | | | |
| Aluminum: | | | | | | | | | | |
| No. 2 | 4.524 | 5.042 | 5.530 | 5.752 | 5.858 | 2.262 | 2.521 | 2.765 | 2.876 | 2.929 |
| No. 1 | 3.690 | 4.084 | 4.450 | 4.606 | 4.646 | 1.845 | 2.042 | 2.225 | 2.303 | 2.323 |
| No. 1/0 | 3.002 | 3.304 | 3.574 | 3.686 | 3.684 | 1.501 | 1.652 | 1.787 | 1.843 | 1.842 |
| No. 2/0 | 2.458 | 2.684 | 2.880 | 2.954 | 2.920 | 1.229 | 1.342 | 1.440 | 1.477 | 1.460 |
| No. 3/0 | 2.028 | 2.194 | 2.334 | 2.380 | 2.318 | 1.014 | 1.097 | 1.167 | 1.190 | 1.159 |
| No. 4/0 | 1.684 | 1.804 | 1.898 | 1.920 | 1.840 | 0.842 | 0.902 | 0.949 | 0.960 | 0.920 |
| 350 kcmil | 1.166 | 1.218 | 1.238 | 1.228 | 1.114 | 0.583 | 0.609 | 0.619 | 0.614 | 0.557 |
| Overhead secondary‡ | | | | | | | | | | |
| Aluminum: | | | | | | | | | | |
| No. 2 | 5.530 | 5.888 | 6.146 | 6.192 | 5.860 | 2.801 | 2.974 | 3.095 | 3.112 | 2.930 |
| No. 1/0 | 3.932 | 4.088 | 4.150 | 4.102 | 3.700 | 2.002 | 2.074 | 2.097 | 2.067 | 1.850 |
| No. 2/0 | 3.372 | 3.456 | 3.448 | 3.368 | 2.940 | 1.722 | 1.758 | 1.746 | 1.700 | 1.470 |
| No. 4/0 | 2.516 | 2.504 | 2.406 | 2.284 | 1.840 | 1.294 | 1.282 | 1.225 | 1.158 | 0.920 |
| 336.4 kcmil | 1.940 | 1.876 | 1.792 | 1.594 | 1.160 | 1.006 | 0.968 | 0.918 | 0.813 | 0.580 |
| 477 kcmil | 1.646 | 1.556 | 1.392 | 1.248 | 0.820 | 0.858 | 0.808 | 0.718 | 0.640 | 0.410 |
| 795 kcmil | 1.336 | 1.224 | 1.042 | 0.892 | 0.480 | 0.704 | 0.642 | 0.543 | 0.462 | 0.240 |

***Note:*** 1 in = 25.4 mm; 1 ft = 0.3048 m. Regulation of copper conductors can be estimated with reasonable accuracy the same as that of aluminum conductors two sizes larger.

*Values in the table give the difference in absolute value between sending-end and receiving-end line-to-neutral voltages of balanced 3-phase circuit and phase-to-phase or phase-to-neutral voltages of single-phase circuit.

†Underground cable impedances are based on 50°C conductor temperature with close triangular spacing of cable using typical solid-dielectric insulation, 100% insulation level, single conductor.

‡Overhead conductor impedances are based on 50°C conductor temperature with 8-in equivalent spacing for single-phase and 10-in spacing for 3-phase.

From Table 7-12 for No. 4/0 cable, 0.8 of the value is 0.902. Line-to-neutral voltage drop is $0.902 \times 6.014 = 5.425$. This is $5.425/277 \times 100 = 1.96\%$ voltage drop on basis of receiving end.

## 7.13 BANKING OF DISTRIBUTION TRANSFORMERS

***Banking.*** Tying together the secondary mains of adjacent transformers supplied by the same primary feeder is known as *banking*. The practice of banking, when used, is usually applied to the secondaries of single-phase transformers, and all transformers in a bank must be supplied from the same phase of the primary circuit. The use of banking is not as prevalent as it was formerly. Banked distribution transformers differ from the low-voltage ac network in that one circuit supplies all transformers where secondaries are banked together, whereas different circuits supply adjacent transformers in an ac low-voltage network. Only a few companies operate their transformers banked.

Advantages claimed for banking compared with secondary radial distribution are (1) reduction in lamp flicker caused by starting motors; (2) less transformer capacity required because of greater load diversity among a larger group of customers; (3) better average voltage along the secondary; and (4) greater flexibility for load growth. There are two general types of secondary banking: the grid type and the straight-line type, as shown in Figs. 7-37 and 7-38.

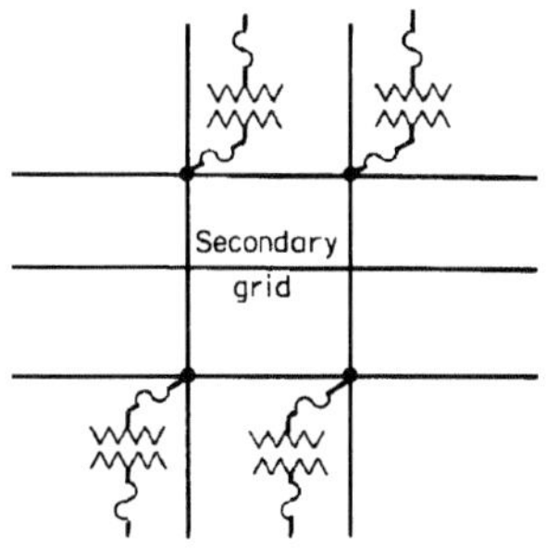


**FIGURE 7-37** Fuse application in grid systems.

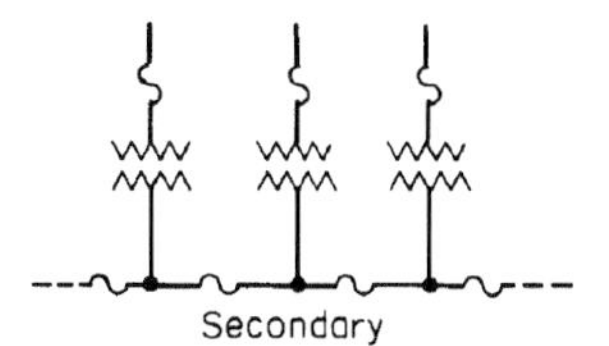


**FIGURE 7-38** Fuse application in straight-line systems.

## 7.14 APPLICATION OF CAPACITORS

***Power Factor Correction.*** It is desirable to add shunt capacitors in the load area to supply the lagging component of current. The cost is frequently justified by the value of circuit and substation capacity released and/or reduction in losses. Installed cost of shunt capacitors is usually least on primary distribution systems and in distribution substations.

The application of a shunt capacitor to a distribution feeder produces a uniform voltage boost per unit of length of line, out to its point of application. Therefore, it should be located as far out on the distribution system as practical, close to the loads requiring the kilovars. There are some cases, particularly in underground distribution, where secondary capacitors are economically justified despite their higher cost per kilovar.

Development of low-cost switching equipment for capacitors has made it possible to correct the power factor to a high value during peak-load conditions without overcorrection during light-load periods. This makes it possible for switched capacitors to be used for supplementary voltage control. Time clocks, temperature, voltage, current, and kilovar controls are common actuators for capacitor switching.

***Capacitor Installations.*** Capacitors for primary systems are available in 50- to 300-kvar single-phase units suitable for pole mounting in banks of 3 to 12 units. Capacitors should be connected to the system through fuses so that a capacitor failure will not jeopardize system reliability or result in violent case rupture. To ensure that the proper fuse protection is provided, the installed capacitor fuse ratings are listed in Tables 7-13 and 7-14 and the probability of rupture is shown in Table 7-15.

***Effect of Shunt Capacitors on Voltage.*** Proposed permanently connected capacitor applications should be checked to make sure that the voltage to some customers will not rise too high during light-load periods. Switched capacitor applications should be checked to determine that switching the capacitor bank on or off will not cause objectionable flicker. The curves in Fig. 7-39 can be used to compute voltage rise.

***Effect of Shunt Capacitors on Losses.*** The maximum loss reduction on a feeder with distributed load is obtained by locating capacitor banks on the feeder where the capacitor kilovars is equal to twice the load kilovars beyond the point of installation. This principle holds whether one or more than one capacitor bank is applied to a feeder.

Capacitor kilovars up to 70% of the total kilovar load on the feeder can be applied as one bank with little sacrifice in the maximum feeder-loss reduction possible with several capacitor banks.

**TABLE 7-13** Recommended Group Fusing, K- or T-Rated Links (Floating-Y Banks)

| | 3-Phase kilovar | | | | | | | | | | |
|---|---|---|---|---|---|---|---|---|---|---|---|
| Volts | 150 | 300 | 450 | 600 | 900 | 1,200 | 1,350 | 1,800 | 2,400 | 2,700 | 3,600 |
| 2,400 | 40K | — | — | — | — | — | — | — | — | — | — |
| 4,160 | 25 | 40 | 65[a,b] | 80K[a,d] | — | — | — | — | — | — | — |
| 4,800 | 20 | 40 | 50[c] | — | — | — | — | — | — | — | — |
| 7,200 | 12 | 25 | 40 | 50K[c] | 80[a,e] | — | — | — | — | — | — |
| 8,320 | 12 | 25 | 30 | 40 | 65[f] | 80K[h] | — | — | — | — | — |
| 12,470 | 8 | 15 | 25 | 30 | 50[g] | 65[i] | 65[a] | 80K[j] | — | — | — |
| 13,200 | 8 | 15 | 20 | 25 | 40 | 50 | 65[a] | 80K[j] | 100K[a,k] | — | — |
| 13,800 | 6 | 12 | 20 | 25 | 40 | 50 | 65[a] | 80[k] | 100K[a,k] | — | — |
| 14,400 | 6 | 12 | 20 | 25 | 40 | 50K | 65[a] | 80[k] | — | — | — |
| 20,800 | — | 8 | 12 | 20 | 25 | 40 | 40 | 50 | 65 | 80[a] | 100K[a] |
| 21,600 | — | 8 | 12 | 15 | 25 | 30 | 40 | 50 | 65 | 80[a] | — |
| 23,000 | — | 8 | 12 | 15 | 25 | 30 | 40 | 50 | 65 | 80T[a] | — |
| 23,900 | — | 8 | 12 | 15 | 25 | 30 | 30 | 50 | 65 | 65 | 80K |
| 24,900 | — | 8 | 12 | 15 | 25 | 30 | 30 | 50 | 65[j] | 65 | 80K |
| 34,500 | — | — | 8 | 10 | 15 | 20 | 25 | 30 | 40 | 50 | 65 |

***Notes:*** Fusing is in safe zone unless otherwise shown. Max bank size for 50 kvar units is 600 kvar. Max bank size for 100 kvar units is 1200 kvar. Max bank size for 150 kvar units is 1800 kvar. Max bank size for 200 kvar units is 2400 kvar.

[a]Zone 1.
[b]150-kvar units only.
[c]Zone 1 for 50-kvar units.
[d]200-kvar units only.
[e]300-kvar units only.
[f]Zone 1 for 100- or 150-kvar units.
[g]Zone 1 for 100-kvar units.
[h]Zone 1 for 200-kvar units. Not suitable for 100-kvar units.
[i]Zone 1 for 100- and 200-kvar units.
[j]For 200-kvar and larger only, zone 1 for 200 kvar units.
[k]For 300-kvar and larger only.

**TABLE 7-14** Recommended Group Fusing, K- or T-Rated Links

Grounded-Y- and Δ- Connected Banks

| | 3-Phase kilovar | | | | | | | | | | |
|---|---|---|---|---|---|---|---|---|---|---|---|
| Volts | 150 | 300 | 450 | 600 | 900 | 1,200 | 1,350 | 1,800 | 2,400 | 2,700 | 3,600 |
| 2,400 | 40 | 80 | — | — | — | — | — | — | — | — | — |
| 4,160 | 25 | 50 | 80 | 100 | — | — | — | — | — | — | — |
| 4,800 | 20 | 40 | 65 | 80 | 140 | — | — | — | — | — | — |
| 7,200 | 15 | 30 | 40 | 65 | 80 | — | — | — | — | — | — |
| 8,320 | 12 | 25 | 40 | 50 | 80 | 100 | — | — | — | — | — |
| 12,470 | 8 | 15 | 25 | 40 | 50 | 65 | 80 | 100 | 140 | — | — |
| 13,200 | 8 | 15 | 25 | 30 | 50 | 65 | 80 | 100 | 140 | — | — |
| 13,800 | 8 | 15 | 25 | 30 | 50 | 65 | 65 | 100 | 140 | 140 | — |
| 14,400 | 8 | 15 | 20 | 30 | 40 | 65 | 65 | 80 | 140 | 140 | — |
| 20,800 | 6 | 10 | 15 | 20 | 30 | 40 | 50 | 65 | 80 | 100 | 140 |
| 21,600 | 6 | 10 | 15 | 20 | 30 | 40 | 40 | 65 | 80 | 80 | 140 |
| 23,000 | 6 | 10 | 15 | 20 | 25 | 40 | 40 | 50 | 80 | 80 | 100 |
| 23,900 | 6 | 8 | 12 | 20 | 25 | 40 | 40 | 50 | 80 | 80 | 100 |
| 24,900 | 6 | 8 | 12 | 15 | 25 | 40 | 40 | 50 | 65 | 80 | 100 |
| 34,500 | 6 | 6 | 10 | 12 | 20 | 25 | 25 | 40 | 50 | 50 | 80 |

***Notes:***

1. Refer to Table 7-15 for fuse sizes within fault current limits.
2. Maximum link size for each unit—check Table 7-11 for all:

| | | |
|---|---|---|
| 50 kvar | 65K, 30T | Check Table 7-3 |
| 100 kvar | 80K, 50T | Check Table 7-3 |
| 150 kvar | 100K, 50T | Check Table 7-3 |
| 200 kvar | 100K, 65T | Check Table 7-3 |
| 300 kvar and up | 140K, 80T | Check Table 7-3 |

3. Ratio of fuse continuous current rating to nominal capacitor current is 1.65 minimum.

**TABLE 7-15** Coordination Table: Grounded-Y- and Δ-Connected Banks

Maximum fault current for zone indicated.

| | 50 kvar unit | | 100 kvar unit | | 150 kvar unit | | 200 kvar unit | | 300 and 400 kvar unit | |
|---|---|---|---|---|---|---|---|---|---|---|
| Fuse link | Safe zone | Zone 1 | Safe zone | Zone 1 | Safe zone | Zone 1 | Safe zone | Zone 1 | Safe zone | Zone 1 |
| 30 K and lower | 2900 | 3900 | 4000 | 5300 | 4600 | 6300 | 5400 | 7000 | 5800 | 7000 |
| 40 K | 2700 | 3900 | 4000 | 5300 | 4600 | 6300 | 5400 | 7000 | 5800 | 7000 |
| 50 K | 2000 | 3700 | 3900 | 5300 | 4600 | 6300 | 5400 | 7000 | 5800 | 7000 |
| 65 K | — | 2400 | 2800 | 5300 | 4000 | 6300 | 5400 | 7000 | 5800 | 7000 |
| 80 K | — | — | 700 | 3500 | 2200 | 5500 | 4100 | 7000 | 5000 | 7000 |
| 100 K | — | — | — | — | — | 2800 | 1700 | 6300 | 2800 | 7000 |
| 140 K | — | — | — | — | — | — | — | 1800 | — | 3500 |
| 20 T and lower | 2900 | 3900 | 4000 | 5300 | 4600 | 6300 | 5400 | 7000 | 5800 | 7000 |
| 25 T | 2200 | 3900 | 4000 | 5300 | 4600 | 6300 | 5400 | 7000 | 5800 | 7000 |
| 30 T | 800 | 2800 | 3200 | 5300 | 4200 | 6300 | 5400 | 7000 | 5800 | 7000 |
| 40 T | 220 | 1000 | 1700 | 4300 | 3000 | 6300 | 4500 | 7000 | 5600 | 7000 |
| 50 T | — | 200 | 400 | 2500 | 1100 | 4000 | 2800 | 7000 | 4200 | 7000 |
| 65 T | — | — | — | 500 | — | 2100 | 1600 | 5500 | 2500 | 6800 |
| 80 T | — | — | — | — | — | — | — | 3500 | 1000 | 5000 |
| 100 T | — | — | — | — | — | — | — | — | — | 2200 |

***Note:*** Safe zone—rupture probability less than 10%. Zone 1—rupture probability 10% to 50%.

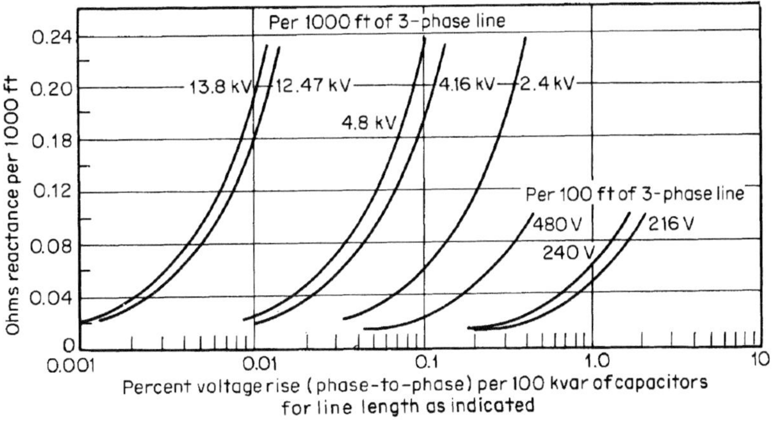


**FIGURE 7-39** Curves of voltage rise caused by capacitor application.

A rule of thumb for locating a single capacitor bank on a feeder with uniformly distributed loads is that the maximum loss reduction can be obtained when the capacitor kilovars of the bank is equal to two-thirds of the kilovar load on the feeder. This bank should be located two-thirds of the distance out on the distributed feeder portion. Deviation of the capacitor bank location from the point of maximum loss reduction by as much as 10% of the total feeder length does not appreciably affect the loss benefit. Therefore, in practice, in order to make the most out of the capacitor's loss reduction and voltage benefits, it is best to apply the capacitor bank just beyond the optimum loss-reduction location.

## 7.15 LINE CONDUCTORS

***Conductor Factors.*** Copper and aluminum are the metals most used as conductors in distribution systems. Proportions are fixed by the combined effect of conductivity, weight, strength, and cost. Recent years have seen such a shift in availability and cost that aluminum has gained almost universal use in distribution, supplanting copper, which was preferred for many years.

***Conductor Materials.*** Aluminum has the advantage of about 70% less weight for a given size, but its conductivity is only about 61% that of annealed copper. For distribution, it is commonly rated as equivalent to a copper conductor two AWG sizes smaller, which has almost identical resistance. Its tensile strength is less than copper, and it is commonly used, particularly in the smaller sizes, by stranding aluminum around a steel core of proper size to give the desired tensile strength. In larger sizes, the tensile-strength requirements of distribution are satisfied by stranded aluminum without the reinforcing steel. Another way of obtaining high tensile strength is to combine steel with copper or aluminum wires. Steel is combined with copper in a high-strength strand known as Copperweld, which has 30% to 40% of the conductivity of a copper conductor of equal size. In a similar manner aluminum and steel conductor can be combined into what is known as Alumoweld.

Both copper and aluminum are suitable for use as substation buses, being available in flat bars, tubes, and rods. For very heavy currents, channel shapes are used to make up box-type buses, which are the most economical for such applications.

***Use of Copper.*** Where copper is used for overhead circuits with span lengths of 200 ft or more, it is commonly used in the hard-drawn form because of its greater tensile strength. For common types of local distribution circuit where spans are shorter and flexibility is desirable, medium-hard-drawn, or annealed, copper is used. Mechanical connectors are extensively used for joints and taps on overhead copper.

*Underground copper cables* are usually made of standard soft copper because of its greater flexibility. The smaller size of copper conductors helps to offset unfavorable price levels because of savings in insulating and sheathing material as well as the ability to put maximum carrying capability in a given size of duct.

***Use of Aluminum.*** In rural line work, where long spans and conductors of high tensile strength are an economic necessity, the combined requirements of conductivity and strength have been met with aluminum stranded around a steel core sized to give the required strength. Such a cable is known as *aluminum cable steel-reinforced* and is commonly designated as ACSR. Development of high-strength aluminum alloys has led to such alternative cables as aluminum conductor alloy-reinforced (ACAR) and all-aluminum-alloy conductor (AAAC), which also combine conductivity with tensile strength. Urban distribution uses ACSR and all-aluminum conductors. Stranded aluminum is common where large conductors are required.

*Underground Aluminum Cables.* The development of such synthetic insulations as polyethylene has made aluminum almost universally used for underground distribution. In the smaller sizes for URD, a solid conductor is often applied rather than stranded construction. *Jointing of aluminum* requires special care to secure good contact and to guard against corrosion. Jointing is often done with compression devices, although mechanical connectors packed with corrosion-inhibiting compound can be used.

***Use of Steel.*** Steel conductors are rarely used for distribution circuits because of their high resistance. But steel with a heavy covering of copper, known as Copperweld,[b] or with a heavy covering of aluminum, known as Alumoweld,* has conductivity approaching 40% that of copper and can be

[b]Copperweld and Alumoweld are registered trademarks of the Copperweld Bimetallic Group.

used in some applications. Such coated conductors are also very attractive as high-strength strands or reinforcements for composite cables, which get improved conductivity from strands of hard-drawn copper over the Copperweld or hard-drawn aluminum over the Alumoweld.

Conductors reinforced with steel have impedances which increase somewhat as current density increases. Voltage drops are correspondingly higher than those of copper or aluminum conductors of equal conductivity.

Copperweld and Alumoweld are generally more durable than galvanized-steel cables. They have therefore been used to some extent for guy cables. They are also used widely for shield wires.

## 7.16 RESISTANCE AND REACTANCE OF OVERHEAD LINES

Resistance depends primarily on the conductor size and type of conductor used. Reactance depends not only on the conductor size but also on the equivalent delta spacing between the conductors. Table 7-16 provides the physical and electrical characteristics for commonly used overhead conductors.

The conductor reactance may be separated into two parts—the internal reactance of the conductor including the area around the conductor of 1-ft radius and the external reactance of the conductor beyond the 1-ft radius. The total reactance ($X$) per conductor is equal to

$$X = X_1 + X_2$$

where $X_1$ = reactance of the conductor at 1 ft
$X_2$ = reactance of the conductor beyond 1 ft
$= 0.0528 \log_{10} S - 0.057$ (in ohms per 1000 ft)

Table 7-16 provides the reactance values in ohms/1000 ft for various conductor types and sizes. Figure 7-40 gives the $X_2$ values in ohms/1000 ft for various equivalent spacings between conductors.

For single-phase circuits, the equivalent spacing is the distance between conductors. For three-phase circuits, the equivalent spacing, $S$, is given by the following formula:

$$S = \sqrt[3]{s_{ab} \times s_{ac} \times s_{bc}}$$

where $s_{ab}$, $s_{ac}$, and $s_{bc}$ represent the phase-to-phase spacings of the conductors.

For three-phase circuits, $X$ is commonly referred to as the positive sequence impedance.

Conductor resistances in Table 7-16 are given at 25°C. To correct for other temperatures ($T$ in °C), use the following:

For aluminum:

$$R_T = R_{25} \times \frac{228 + T}{253}$$

For copper:

$$R_T = R_{25} \times \frac{234.5 + T}{259.5}$$

**TABLE 7-16** Characteristics of Overhead Distribution Line Conductors

| ACSR | | | Diameter, | Weight, | Ampacity (80°C/40°C) | | $R$(25°C) | $X_1$(at 1 ft) |
|---|---|---|---|---|---|---|---|---|
| AWG | Strands | kcmil | inches | #/1000 ft | Still air | 2 ft/s | Ω/1000 ft | Ω/1000 ft |
| 6 | 6/1 | 26.25 | 0.198 | 36.2 | 55 | 85 | 0.675 | 0.128 |
| 4 | 6/1 | 41.74 | 0.250 | 57.6 | 75 | 120 | 0.425 | 0.126 |
| 2 | 6/1 | 66.37 | 0.316 | 91.6 | 110 | 165 | 0.267 | 0.125 |
| 1/0 | 6/1 | 105.54 | 0.398 | 145.6 | 150 | 225 | 0.168 | 0.124 |
| 2/0 | 6/1 | 133.1 | 0.447 | 183.7 | 175 | 260 | 0.134 | 0.122 |
| 3/0 | 6/1 | 167.8 | 0.502 | 231.6 | 210 | 305 | 0.106 | 0.118 |
| 4/0 | 6/1 | 211.6 | 0.563 | 292.1 | 245 | 355 | 0.084 | 0.110 |
| | 26/7 | 266.8 | 0.642 | 366.8 | 290 | 410 | 0.066 | 0.088 |
| | 26/7 | 336.4 | 0.721 | 462.4 | 340 | 480 | 0.053 | 0.086 |
| | 26/7 | 397.5 | 0.783 | 546.4 | 380 | 535 | 0.045 | 0.084 |
| | 26/7 | 477.0 | 0.858 | 655.7 | 430 | 605 | 0.037 | 0.082 |
| | 26/7 | 556.5 | 0.927 | 765.0 | 480 | 670 | 0.032 | 0.080 |
| | 26/7 | 795.0 | 1.108 | 1093.0 | 620 | 850 | 0.022 | 0.076 |
| **All aluminum—hard drawn** | | | **Diameter,** | **Weight** | **Ampacity (80°C/40°C)** | | **$R$(25°C)** | **$X_1$(at 1 ft)** |
| **AWG** | **Strands** | **kcmil** | **inches** | **#/1000 ft** | **Still air** | **2 ft/s** | **Ω/1000 ft** | **Ω/1000 ft** |
| 4 | 7 | | 0.232 | 39.0 | 75 | 115 | 0.424 | 0.114 |
| 2 | 7 | | 0.292 | 62.0 | 105 | 160 | 0.267 | 0.109 |
| 1/0 | 7 | | 0.368 | 98.5 | 145 | 215 | 0.168 | 0.103 |
| 2/0 | 7 | | 0.414 | 124.3 | 170 | 250 | 0.134 | 0.101 |
| 3/0 | 7 | | 0.464 | 156.7 | 200 | 290 | 0.106 | 0.098 |
| 4/0 | 7 | | 0.522 | 197.6 | 240 | 340 | 0.084 | 0.095 |
| | 7 | 266.8 | 0.586 | 249.1 | 280 | 400 | 0.066 | 0.092 |
| | 19 | 336.4 | 0.666 | 315.7 | 330 | 465 | 0.053 | 0.088 |
| | 19 | 397.5 | 0.724 | 373.0 | 370 | 520 | 0.045 | 0.086 |
| | 19 | 477.0 | 0.793 | 447.6 | 425 | 590 | 0.037 | 0.084 |
| | 19 | 556.5 | 0.856 | 522.0 | 465 | 645 | 0.032 | 0.082 |
| | 37 | 795.0 | 1.026 | 746.0 | 605 | 820 | 0.022 | 0.079 |
| **Copper—hard drawn** | | | **Diameter,** | **Weight** | **Ampacity (80°C/40°C)** | | **$R$(25°C)** | **$X_1$(at 1 ft)** |
| **AWG** | **Strands** | **kcmil** | **inches** | **#/1000 ft** | **Still air** | **2 ft/s** | **Ω/1000 ft** | **Ω/1000 ft** |
| 8 | 1 | 16.51 | 0.1285 | 50 | 50 | 80 | 0.656 | 0.126 |
| 6 | 1 | 26.25 | 0.162 | 80 | 70 | 110 | 0.413 | 0.121 |
| 4 | 3 | 41.74 | 0.254 | 128 | 110 | 161 | 0.263 | 0.114 |
| 2 | 7 | 66.37 | 0.292 | 205 | 145 | 210 | 0.167 | 0.109 |
| 1/0 | 7 | 105.5 | 0.368 | 326 | 200 | 285 | 0.105 | 0.1035 |
| 2/0 | 7 | 133.1 | 0.414 | 411 | 240 | 335 | 0.083 | 0.101 |
| 3/0 | 7 | 167.8 | 0.464 | 518 | 280 | 390 | 0.066 | 0.098 |
| 4/0 | 7 | 211.6 | 0.522 | 653 | 330 | 450 | 0.053 | 0.095 |
| | 19 | 250 | 0.574 | 772 | 375 | 510 | 0.045 | 0.092 |
| | 19 | 300 | 0.629 | 926 | 425 | 575 | 0.037 | 0.090 |
| | 19 | 350 | 0.679 | 1081 | 475 | 635 | 0.032 | 0.088 |

***Note:*** Ampacity based on 80°C conductor, 40°C ambient; emissivity, $\varepsilon = 0.2$ for aluminum, 0.5 for copper.
Total reactance per phase $= X_1 + X_2$.
$X_2$ = external reactance of conductor beyond 1 ft is obtained from Fig. 7-41.

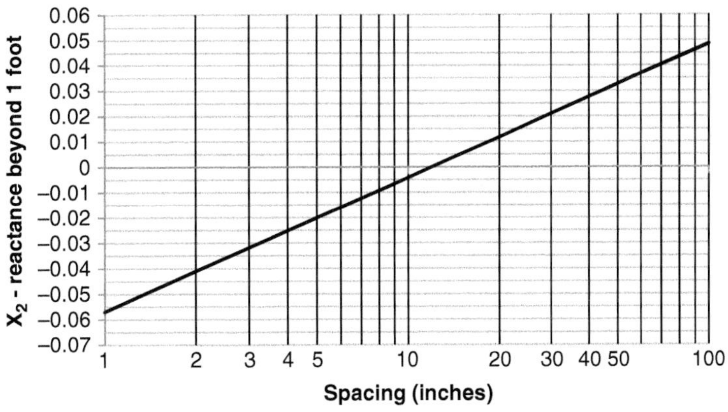


**FIGURE 7-40** External conductor reactance beyond one foot in ohms per 1000 ft.

## 7.17 UNDERGROUND RESIDENTIAL DISTRIBUTION

During the past 40 years, the evolution of underground distribution systems, particularly single-phase systems to serve residential areas (URD), has proceeded at a rapid rate. For a so-called mature industry, the rate of change has been phenomenal. Costs have been steadily reduced through the introduction of new system concepts, improved installation practices, and the development of specialized equipment.

Nearly every utility in the United States now has a policy covering the installation of URD in new residential tracts. Conditions vary all the way from a substantial differential payment by the developer to a no-charge basis by the utility, although the developer usually is requested to assist with excavation. In addition, a number of states have established legal requirements mandating that all new residential developments in excess of a given number of homes be served by an underground distribution system. As a result, perhaps as many as two-thirds of new residential dwelling units are being served underground.

***Cost.*** Underground distribution systems often cost more than comparable overhead systems. What are the principal factors contributing to the rapid growth of URD? These include

- Greater public interest in the aesthetic appearance of residential communities.
- Reduced cost of underground equipment and installations brought about by
  - Solid dielectric insulated cables—lower-cost—suitable for direct burial without duct systems
  - Factory-built cable terminations and splices of low cost easily prepared in the field by ordinary line crews without the aid of highly trained cable splicers
  - Mass production of specialized equipment such as pad-mounted transformers and accessories
  - Improved installation technique and equipment

***Performance.*** Most observers are of the opinion that the frequency of faults is lower on underground systems than on overhead systems and that the faults are not so likely to "bunch up" because of storm conditions. However, faults are much more difficult and time-consuming to find, to isolate, and to repair on underground systems. This, coupled with the fact that many operating procedures cannot be performed on an underground system while it is energized and that it is impossible to make many of the temporary improvisations on underground circuits that can be accomplished on overhead systems, has led to the development of protective and sectionalizing equipment such as switches and separable cable connectors which often are physically integrated as accessory devices in

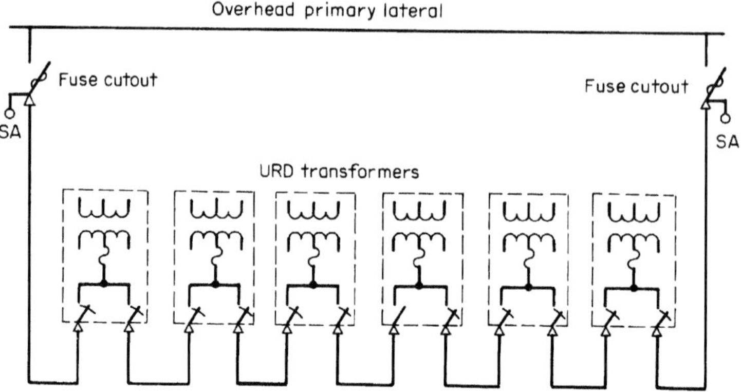


**FIGURE 7-41** URD system derived from existing overhead circuits.

the underground distribution transformers. In addition, several utilities with significant amounts of older cable have found that they are having more unexpected reliability problems as cables have aged.

Service-restoration requirements also have resulted in primary-system designs which operate as a normally open loop as shown in Fig. 7-41. In the case of a cable fault, this facilitates location and isolation of the failure and more rapid service restoration to all customers on the unfaulted portion of the primary loop. It is estimated that about 85% of primary URD systems are being operated as loops, the remainder being radial. Where radial laterals are used, many utilities provide portable, aboveground cables so that faulted cables can be bypassed temporarily and service restored while repairs are being made.

***How Much Does Undergrounding Improve Electric Reliability?*** Comparative reliability data indicate that the frequency of outages on underground systems can be substantially less than for overhead systems. However, when the duration of outages is compared, underground systems lose much of their advantage. The data show that the frequency of power outages on underground systems is only about one-third of that of overhead systems. A 2000 report issued by the Maryland Public Service Commission concluded that the impact of undergrounding on reliability was "unclear." In a 2003 study, the North Carolina Commission summarized 5 years of underground and overhead reliability comparisons for North Carolina's investor-owned electric utilities—Dominion North Carolina Power, Duke Energy, and Progress Energy Carolinas. The data indicate that the frequency of outages on underground systems was 50% less than for overhead systems, but the average duration of an underground outage was 58% longer than for an overhead outage. In other words, for the North Carolina utilities, an underground system suffers only about half the number of outages of an overhead system, but those outages take 1.6 times longer to repair. Based on this data, Duke Power concluded, "Underground distribution lines will improve the potential for reduced outage interruption during normal weather, and limit the extent of damage to the electrical distribution system from severe weather-related storms." However, once an interruption has occurred, underground outages normally take significantly longer to repair than a similar overhead outage.

***Reliability Characteristics of Overhead and Underground Power Lines***

- Overhead lines tend to have more power outages primarily due to trees coming in contact with overhead lines.
- It is relatively easy to locate a fault on an overhead line and repair it. A single line worker, for example, can locate and replace a blown fuse. This results in shorter duration outages.
- Underground lines require specialized equipment and crews to locate a fault, a separate crew with heavy equipment to dig up a line, and a specialized crew to repair the fault. This greatly increases the cost and the time to repair a fault on an underground system.

- In urban areas, underground lines are four times more costly to maintain than overhead facilities.
- Underground lines have a higher failure rate initially due to dig-ins and installation problems. After 3 or 4 years, however, events that affect failures become virtually nonexistent.
- As underground cables approach their end of life, failure rates increase significantly and these failures are extremely difficult to locate and repair. Maryland utilities report that their underground cables are becoming unreliable after 15 to 20 years and reaching their end of life after 25 to 35 years.
- Pepco found that customers served by 40-year-old overhead lines had better reliability than customers served by 20-year-old underground lines.
- Two Maryland utilities have replaced underground distribution systems with overhead systems to improve reliability.
- Water and moisture infiltration can cause significant failures in underground systems when they are flooded, as often happens in hurricanes.
- Due to cost or technical considerations, it is unlikely that 100% of the circuit from the substation to the customer can be placed entirely underground. This leaves the circuit vulnerable to the same types of events that impact other overhead lines, for example, high winds and ice storms.

***Transformers.*** The heart of the URD system is the single-phase distribution transformer because the primary cable terminations, switching and sectionalizing equipment, and overcurrent protective equipment usually are housed in the transformer enclosure. Thus most operating procedures require access to one or more distribution transformers. Two general types of single-phase transformers are in use.

*Pad-Mounted.* Figure 7-42 shows the predominant type of transformer being used for URD. The transformer shown is called the *mini-pad.* The term *pad* derives from the fact that transformers in this category usually are installed on concrete slabs, or pads.

The electrical functions of URD transformers cover essentially the same range as pole-type units. For reasons of safety, of course, they must be built in tamper-resistant configurations with no exposed electrically energized parts because of the proximity of such transformers to the general public.

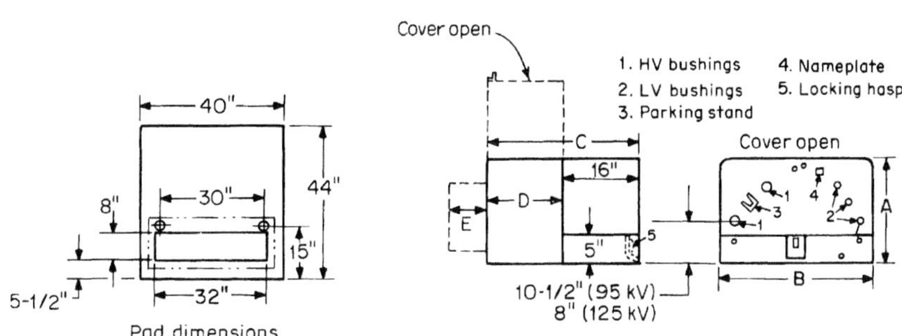


95- and 125-kV BIL (35 kV GrdY and below)

| kVA | Dimensions, in A | B | C | D | E |
|---|---|---|---|---|---|
| 15 | 24 | 36 | 33.9 | 17.9 | |
| 25 | 24 | 36 | 33.9 | 17.9 | |
| 37.5 | 24 | 36 | 33.9 | 17.9 | |
| 50 | 24 | 36 | 33.9 | 17.9 | 8.0 |
| 75 | 24 | 36 | 40.0 | 24.0 | 8.0 |
| 100 | 26 | 36 | 40.0 | 24.0 | 8.0 |
| 167 | 32.5 | 36 | 40.0 | 24.0 | 8.0 |

**FIGURE 7-42** Mini-pad distribution transformer. (*General Electric Company.*)

The mini-pad in Fig. 7-42 has its cover open. The two primary bushings at the upper left are for use with load-break, separable insulated connectors, or elbows. This results in a "dead-front" configuration which is required to achieve the low-height mini-pad construction. The three 120/240-V bushings are at the right-hand side.

Many other combinations of pad-mounted construction and accessory equipment are available, including "live-front" primary connections with stress cones for the cables, internal or external primary fuses and switches, secondary circuit breakers, etc. Refer to appropriate product bulletins of the manufacturers or handbooks for further equipment details. Generally, the loadability of pad-mounted transformers is comparable with that of pole types.

*Residential Subsurface Transformers (RST).* Although usage of pad-mounted transformers predominates, a number of residential subsurface transformers are used. The RSTs are installed in relatively tight-fitting vaults with the cover grating of the vault at ground level.

Cooling is accomplished by natural convection of the air, although some users increase the efficiency of circulation by means of special chimneys to direct and control the circulation. With properly designed and installed chimneys, the loadability of RSTs is equal to that of pole types.

The RSTs must be submersible and therefore utilize dead-front primary cable terminations, usually the separable insulated connectors or "elbows." Provisions for operation of accessories such as switches, fuses, and circuit breakers are located on the cover of the transformer so that they can be operated by a member of the line crew standing on the surface of the ground. Usually the vault is too small for a person to enter.

***Primary Cables.*** Primary URD cables are almost universally of the single-conductor concentric-neutral type employing polyethylene or cross-linked polyethylene insulation. Specifically, the use of TRXLPE and EPR are increasing in usage. Ordinary polyethylene is a thermoplastic which melts at temperatures in the order of 110°C. The process of "cross-linking" polyethylene converts it into a thermosetting material which does not have a melting point, per se.

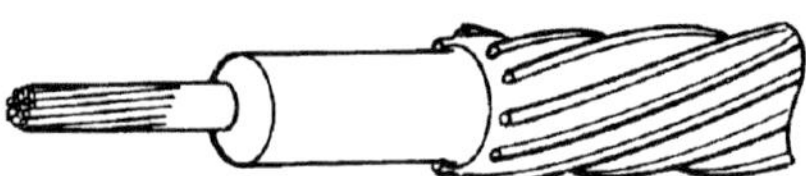

**FIGURE 7-43** Concentric-neutral type of primary URD cable.

Figure 7-43 shows a section of primary URD cable. The central conductor is the energized phase conductor, and the external concentric wires serve as the neutral.

Corrosion of the copper concentric-neutral wires of primary URD cables results in reduced cross-sectional area of the wires, increasing their resistance. In some instances, the continuity of the wires is destroyed. Neutral corrosion may cause safety and operating problems on the URD circuit. Corrosion occurs when the neutral wires become anodic, which results in loss of metal. The wires may become anodic due to nearby dissimilar metals or to variations in soil characteristics along the cable route. Determining the location and extent of corrosion damage is a complex procedure which may involve surveys, testing, and in some cases, excavation. Corrective actions for existing cables include replacing portions of the cable, reestablishing the neutral circuit, and installing sacrificial anodes for cathodic protection. Corrosion in new installations can be controlled by the proper selection of materials, cable construction, type of installation, and cathodic protection. The use of jacketed concentric neutral cable to reduce the problem has been increasing over the years and was used by over 80% of respondents in *Transmission and Distribution*'s 1990 survey on underground distribution practices. Most utilities directly bury the primary cables, although the trend to conduit is increasing. Often the URD cables are placed in the same trench as the telephone cables.

The precise calculation of voltage drop in direct-buried, concentric-neutral primary cables is quite complex because a portion of the single-phase load current flows in the concentric-neutral conductors and a portion in the earth surrounding the cable. Also, there may be an induced circulating current in the neutral conductors. Typical values of voltage drop per 100,000 A · ft are shown in Table 7-17. To use the table, calculate the ampere-feet as the product of the current in the phase conductor and the distance in feet between the source and the load. The effects of direct burial on impedance of the return current path are included in the tabulated voltage drops.

**TABLE 7-17** Single-Phase Voltage Drops per 100,000 A · ft* for 15- and 35-kV Direct-Buried Concentric-Neutral Cables (Loop Values)

| | Voltage class | | | | | | | | | |
|---|---|---|---|---|---|---|---|---|---|---|
| | 15 kV | | | | | 35 kV | | | | |
| | Lagging power factor | | | | | | | | | |
| Conductor size | 0.7 | 0.8 | 0.9 | 0.95 | 1.00 | 0.7 | 0.8 | 0.9 | 0.95 | 1.00 |
| Underground primary | | | | | | | | | | |
| Aluminum: | | | | | | | | | | |
| Concentric-neutral—direct-buried, cross-linked polyethylene, conductor 70°C, neutral 60°C, earth resistivity 90 Ω-cm³, full insulation | | | | | | | | | | |
| No. 2 | 44.1 | 46.6 | 48.1 | 48.1 | 44.8 | | | | | |
| 1/0 | 30.9 | 32.8 | 34.0 | 34.1 | 32.0 | 31.3 | 33.0 | 34.1 | 34.1 | 31.7 |
| 2/0 | 25.3 | 27.0 | 28.1 | 28.3 | 26.8 | 25.8 | 27.3 | 28.3 | 28.4 | 26.6 |
| 4/0 | 17.0 | 18.2 | 19.1 | 19.3 | 18.5 | 17.5 | 18.6 | 19.3 | 19.5 | 18.4 |

*Values in the table give the difference in absolute value between sending-end and receiving-end line-to-neutral voltages, in volts.

***Secondary Cables.*** Usually three polyethylene-insulated, single-conductor cables are used for the 120/240-V secondaries and services. These may be separate cables or of triplex construction. In some cases a bare copper neutral conductor is used. The secondary and service cables are usually directly buried.

***Homes Served per Transformer.*** There is an optimal number of homes to serve from each transformer depending on the load per home, size of lots, and type of system to be used. For a given load per home and lot size, the cost per kVA of transformer decreases as the number of homes increases. This is so because increasingly larger transformers would be used.

However, as the number of homes per transformer increases, the cost of the secondary and service system increases because of the larger secondary cable required. Since the total cost is the sum of those costs, an optimum number of homes per transformer will exist.

In making such an economic study, it is necessary to examine the secondary-service-system voltage drop. A detailed study also should evaluate transformer and cable losses for the various arrangements.

Four to eight homes served per transformer seems to be reasonably typical of present practice. Larger loads per home and larger lot sizes favor a smaller number of homes per transformer. Conversely, smaller loads per home and smaller lot sizes favor serving more homes from each transformer.

## 7.18 UNDERGROUND SERVICE TO LARGE COMMERCIAL LOADS

Large commercial loads constitute one of the major segments of utility distribution systems, especially in built-up areas where underground supply systems are a requirement. Demands range from a few hundred to many thousands of kVA per customer, and the engineering and design time to provide adequate service facilities is substantial. Each job is special, requiring selection of appropriate and correctly sized equipment, negotiation of space and layout with building owners or their consultants, and frequently a detailed discussion of facilities, charges, rates, and contracts. The best tool the distribution engineer has is an adequate knowledge of the systems and components which are available, together with guidelines on their cost and reliability. Beyond this, engineering common sense and reasonable operating practices must be combined with the other factors in order to arrive at a decision on the method of service.

***Characteristics of Large Commercial Loads.*** All large commercial loads generally involve the following factors:

1. *Loads* are in the range of 300 to 4000 kVA or more. The larger loads (even up to values of 50 or 75 MVA) are normally supplied by multiples of lower-capacity services.
2. *Utilization voltage* is 480Y/277, although smaller loads may be 208Y/120 and some of the larger institutional loads may be 4160Y/2400 (with the customer providing further step-down).
3. Individual *service size* is limited to about 4000 A by availability of service entrance switching, maximum fuse or breaker sizes, largest commercial wiring systems, and a growing "gut feeling" that this represents enough eggs in any one basket. Providing adequate interrupting capacity is also a definite factor, and single transformers above 4000 A may be priced as specials.
4. *Installation space* is limited and has a high value to the owner. Utility equipment must be as compact as possible and should not require exceptional customer requirements for auxiliaries.
5. Each job is one-of-a-kind and requires much *custom engineering* as well as detailed coordination with the owner of the building facilities. Complex commercial considerations are also involved, covering rates, ownership of facilities, contracts, and future maintenance responsibilities.
6. Service quality must be high, as to both voltage regulation and continuity. Frequent interruptions are not tolerable, and long planned interruptions are not feasible. Service complaints when expressed are long and loud.

***Service Arrangements.*** Several basic service arrangements can be considered for these loads:

1. Radial
2. Primary loop
3. Primary selective
4. Secondary selective
5. Spot network

If radial service were adequate, there would be no need for the succeeding systems because the radial system is the least complex and the least expensive. Unfortunately, when the supply system is underground, it also is the least reliable and generally is unsatisfactory except in special cases. The principal drawback of the radial system is its exposure to long interruptions due to component failure and the necessity for repeated planned interruptions for routine maintenance or new construction.

These five basic service systems are illustrated in Fig. 7-44, which also shows a basic main feeder system of two similar underground feeders.

***Radial System.*** The radial system is exposed to many interruption possibilities, the most important of which are those due to primary cable failure or transformer failure. Either event will be accompanied by a long interruption, reported nominally by utilities as 10 to 12 h. Both components have finite failure rates, and such interruptions are expected and statistically predictable. The system will be satisfactory *only* if the interruption frequency is very low and if there are ways to operate the system without planned outages.

***Primary Loop.*** A great improvement is obtained by arranging a primary loop, which provides two-way feed at each transformer. In this manner, any section of the primary can be isolated, without interruption, and primary faults are reduced in duration to the time required to locate a fault and do the necessary switching to restore service. The cable in each half of the loop must have capacity enough to carry all the load. The additional cable exposure will tend to increase the frequency of faults, but not necessarily the faults per customer. The addition of a loop tie switch at the open point also introduces the possibility of a single equipment fault causing an interruption to both halves of the loop. Murphy's law generally applies to these situations. Automatic loop switching to reduce interruption duration further is very difficult to arrange and is not normally applied to these systems.

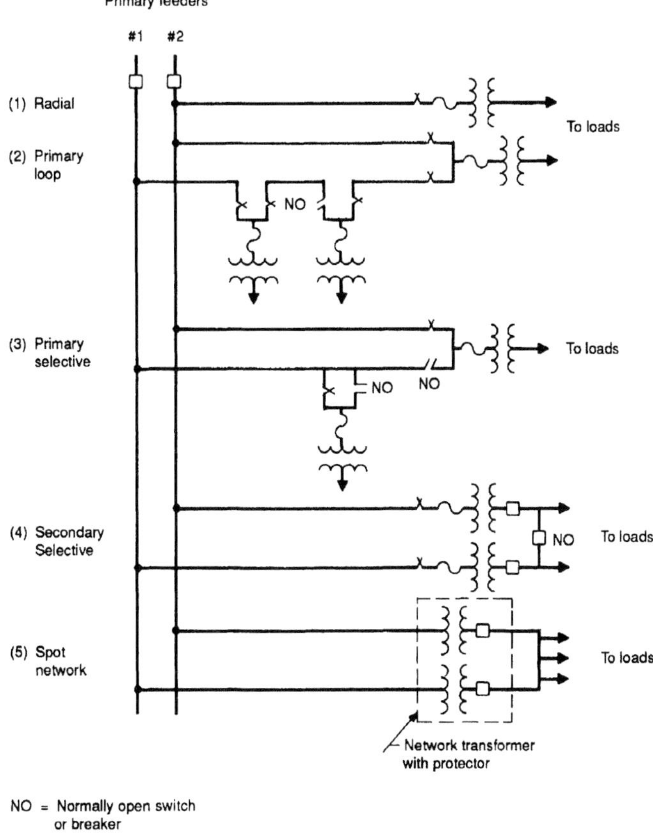


**FIGURE 7-44** Five basic service systems.

***Primary Selective.*** This system uses the same basic components as in the primary loop but arranged in a dual or main/alternate scheme. Each transformer can "select" its source, and automatic switching is frequently used. When automatic, the interruption duration can be limited to 2 to 3 s. Each service represents a potential two-feeder outage (if the open switch fails), but under normal contingencies, service restoration is rapid and there is no need to locate the fault (as with the loop) prior to doing the switching. This scheme is in popular use on many underground systems. Switching times can be improved to less than ½ cycle with the use of a static transfer switch (STS).

***Secondary Selective.*** This service system uses two transformers and low-voltage switching. It is not in popular use by utilities for 480-V service but is common in industrial plants and on institutional properties. Primary operational switching is eliminated and with it some causes of difficulty. Duplicate transformers virtually eliminate the possibility of a long interruption due to failure. Load is divided between the two units, and automatic transfer is employed on loss of voltage to either load. There must be close coordination of utility and customer during planned transfers, and the split responsibility is probably the principal reason for its limited use as a service system.

***Secondary Spot Network.*** Maximum service reliability and operating flexibility are gained by a spot network using two or more transformer/protector units in parallel. The low-voltage bus is continuously energized by all units, and automatic disconnection of any unit is obtained by sensitive reverse power relays in the protector. Maintenance switching of primary feeders can be done without customer interruption or involvement. Spot networks are common in downtown, high-density areas and have been applied frequently in outlying areas for large commercial services where the supply feeders can be made available. This system also represents the most compact and reliable arrangement of components for service in underground systems.

## 7.19 LOW-VOLTAGE SECONDARY-NETWORK SYSTEMS

Distributed or grid-type secondary network systems have been used for many years by electric utility companies to serve high-density load areas in the downtown section of cities. Secondary networks are used in about 90% of the cities in this country having a population of 100,000 or more and in more than one-third of all cities with population between 25,000 and 100,000.

The service voltage is 208Y/120 V supplying light and power loads in stores, hotels, restaurants, office buildings, apartment houses, and in some cases individual residences. The systems and equipment are entirely underground, and the 208Y/120-V portion consists of grids of interconnected cables supplied at numerous points by network transformers which feed the grid through network protectors.

A given secondary network is supplied by several primary feeders suitably interlaced through the area in order to achieve acceptable loading of the transformers under emergency conditions and to provide a system of extremely high service reliability. Primary voltages are found in the range of 5 to 34.5 kV, with the 15-kV class predominating. See Fig. 7-45.

The number and routing of the primary feeders are usually based on the assumption that the loss of one or two feeders will not cause a service interruption. For example, the design bogey may be such that the network can operate satisfactorily during the forced outage of one feeder when another feeder is out of service for repairs or maintenance (single contingency).

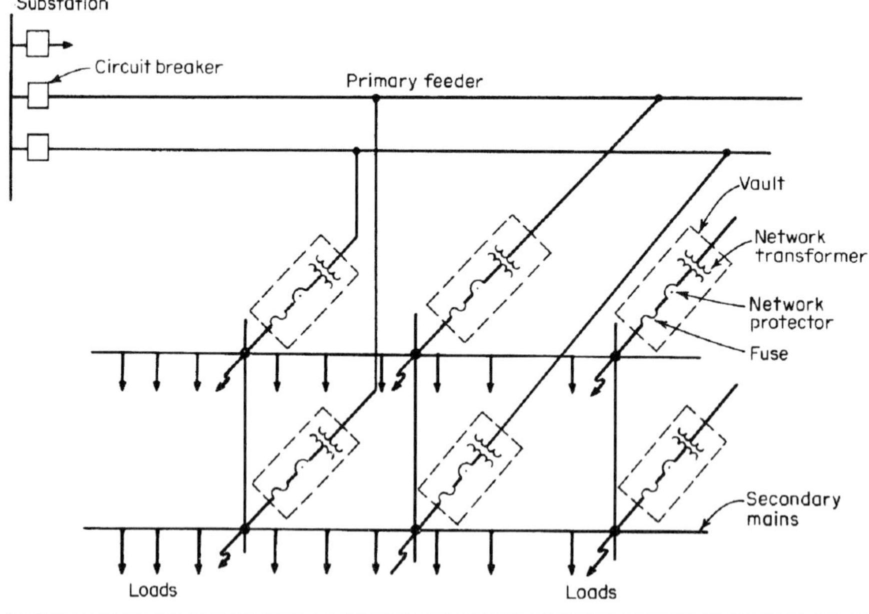


**FIGURE 7-45** Schematic diagram of small segment of a secondary network.

The secondary cable system is designed so that the loss of one transformer will not cause low voltage or a service interruption. Secondary cable faults in 208Y/120-V networks are allowed to burn clear or are cleared by means of limiters, which essentially are fuses having characteristics proportioned to protect the cable and to coordinate with other protective devices. Secondary faults usually will *not* burn clear on 480Y/277-V networks, so limiters are used extensively in these systems. Usually the secondary mains consist of two or more cables in parallel so that failure of one cable does not result in a service interruption.

***Network Transformers.*** The network protector is mounted on one end of a network transformer, and the primary disconnecting and grounding switch is usually on the other end. In some installations the network protector is isolated from the network transformer, with the connection between the low-voltage terminals of the transformer and the protector made with insulated low-voltage cables. The typical network transformer is 3-phase, 216Y/125 V, oil-cooled, in a heavy corrosion-resistant tank suitable for installation in subsurface vaults under streets or sidewalks. Occasionally, there are installations in dry locations where submersible construction is not required. Five hundred kVA is a very common rating, although 750- and 1000-kVA units are available and in use. (For *spot* networks of 480 Grd Y/277 V, transformer ratings are available through 2500 kVA.) There are only two or three distributed street networks at 480Y/277 V in the United States.

***Network Protectors.*** The network transformer is connected to the secondary network through a network protector (NWP) as shown in Fig. 7-45. The NWP is an air circuit breaker with relays and auxiliary devices and backup fuses, all enclosed in a metal case, which usually is physically mounted on the secondary side of the transformer. The functions of the relays are

1. To open the NWP on power-flow reversal, or in case of a fault in the transformer or in the primary feeder
2. To reclose the NWP when the voltage of the primary feeder is of the correct magnitude and phase relation with respect to the network voltage so that when the NWP closes, power (watts) and vars will flow *from* the feeder *into* the network

Thus, if there is a fault on a primary feeder, it is cleared by operation of the feeder breaker at the substation and the opening of all network protectors on transformers supplied by that feeder. Also, if a feeder breaker is opened manually in preparation for maintenance work on the feeder, all NWPs on that feeder should open automatically because of the reverse power flow caused by excitation of the transformers from the low-voltage side.

***Cables.*** All primary and secondary cables are routed along the streets in duct lines as indicated in Fig. 7-45. Loads are served along the streets and at intersections as shown. Primary cables traditionally have been paper-insulated, lead-covered (PILC), but the solid-dielectric insulated cables have gained rapid acceptance. Secondary cables have commonly used rubber insulating materials, but polyethylene and ethylene propylene rubber insulations now are used extensively. Manholes at street intersections are large enough to hold numerous cable connections and limiters and to allow workers to pull and splice cables.

***Continuity of Service.*** Continuity of service is the outstanding advantage of a network system. When a failure occurs in a primary feeder or in a transformer, the faulty feeder is automatically disconnected, and service continues without interruption. Secondary cable faults are allowed to burn clear or are cleared by means of limiters without loss of service. Substations supplying networks are so designed that typical substation faults will not shut down the network; this is further enhanced by careful interlacing of the primary feeders through the load area and their connection to different bus sections in the substation. It is strongly recommended that a given secondary network be supplied from only one substation. If a network is supplied, for instance, from two different substations, it is possible under some system conditions for power to flow from one substation to the other through the secondary grid and network transformers. Should this occur, some network protectors could "see" reverse power flow and open, thus resulting in the undesirable disconnection of these transformers from the network.

***Network Size.*** A secondary network supplied by five or more feeders will keep transformer loadings at 125% or less during the outage of one primary feeder. If the feeders are in the 15-kV class, each feeder could easily supply six 1000-kVA or twelve 500-kVA network units. Thus five feeders interlaced could supply a 30,000-kVA network under idealized conditions. With 500-ft-square blocks, 1000 kVA per block corresponds to 112 MVA of load per square mile. Some utilities plan for the emergency outage of one feeder while a second is out of service for maintenance.

In general, 208Y/120-V networks are in the order of 30,000 to 40,000 kVA in size. There are many networks smaller than this range and some larger. One important limitation to the size of a secondary network is the ability to restore service after the network has been shut down.

***Spot Networks.*** New commercial buildings in existing 208Y/120-V network areas usually have very large electric loads. These loads frequently are supplied by 480Y/277-V *spot* networks, since it is impractical to handle individual loads much larger than about 200 kVA from the 208Y/120-V street networks, and 480Y/277 V is an excellent voltage for supplying large commercial buildings. In some cases the spot networks are supplied from primary feeders which also serve a distributed network. Early 480Y/277-V spot networks used the same overcurrent protective devices employed successfully for clearing faults in the 208Y/120-V networks. Included are the network relays in the protector for detecting faults on the primary feeders, and the network protector fuses, cable limiters, and service fuses for detecting and isolating faults in the secondary systems. Some 208Y/120-V systems do not use cable limiters, as faults in the 208Y/120-V network systems are self-clearing under some circumstances.

Operating experience with the 480Y/277-V spot network systems showed that some faults were arcing in nature, drawing significantly less current than that available for a bolted fault. Many of these faults would not self-clear or would self-clear only after extensive damage was done at and around the original point of fault. Also, some of these faults did not draw sufficient current to blow fuses, or else blew fuses only after significant damage was done at the point of fault.

As a result, some utilities have installed devices for detecting and clearing low-current arcing faults in the 480-V portions of the spot-network system. Heat sensors and ground-fault relays are the most commonly installed devices for the detection function. Clearing has been accomplished by tripping of the network protector, which is effective only for faults downstream from the protector terminals. In a few instances, vacuum circuit breakers or interrupters have been installed on the high-voltage side of the network transformer to clear faults in the associated network transformer, network protector, and other portions of the 480-V system.

***Network Monitoring.*** Remote monitoring capability has been added to a few in-service secondary network systems to automatically gather data needed for the operation and planning of the system. Heretofore, such data were obtained from manual measurements at vaults and manholes. With remote monitoring, the data are continuously collected and transmitted to an operations center or other location for manual and computer analysis. Telephone lines, power-line carrier, two-way radio, and fiberoptic cable have been used as the communications links from the network vaults and other monitored sites of the substation or elsewhere. Virtually any quantity which can be digitized can be monitored. Examples of monitored quantities are the load on the protectors, protector position (open or closed), network protector fuses status (okay or blown), network transformer temperature, vault temperature, bus voltage, and vault water level. In several monitoring systems with two-way communications, it is possible to remotely trip or close the network protectors.

***High-Rise Buildings.*** Primary-voltage feeders are being used as the riser feeder in high-rise buildings. A rule of thumb is that if an apartment building is 10 floors or more, it is more economical from an overall point of view to use the primary voltage rather than utilization voltage for the riser feeder. Similarly, for a commercial building with 480Y/277-V utilization, a building of 50 floors or more usually justifies the use of primary voltage feeders as risers in the building.

The primary system pattern within a high-rise building is usually a loop as shown in Fig. 7-46 or multiple as shown in Fig. 7-47. This will allow cable faults to be manually isolated by proper switching so that cable faults will cause only short interruptions to customers. Customers fed radially through a transformer will be without service if the transformer fails until a temporary connection can be made to an adjacent transformer. For more important loads like elevators, hall lighting, and fire pumps, better reliability is often obtained by using a spot network or low-voltage selective system.

Dry-type transformers using air as the insulating medium of the transformers are the most desired for high-rise buildings. This is so because no special provisions have to be made in the transformer room, such as fireproofing for oil-filled or venting the transformer to the outside of the building for nonflammable liquid-filled transformers. Many of the transformer rooms for apartment buildings and some commercial buildings are in the core of the building, which makes it difficult to use the liquid-filled transformer. However, transformers for supplying heavy loads, such as air conditioning in commercial buildings, usually can be located against an outside wall on machinery floors of the building. This makes it relatively easy to vent a nonflammable liquid-filled transformer to the outside of the building. Hence network transformers are often used for this application.

The primary load-break switch or load-break connector with a fuse can be arranged for either the multiple or loop type of feed where the short-circuit current available is within their rating. Usually the primary short-circuit current available is in the 8000- to 10,000-A range. The current-limiting type of fuse is often used for this "inside-the-building" application because it does not discharge ionized gases or noise during interruption.

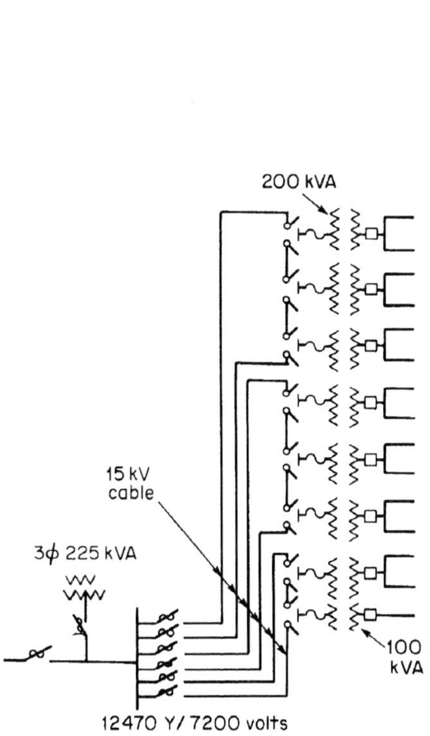


**FIGURE 7-46** Schematic diagram of a looped primary system.

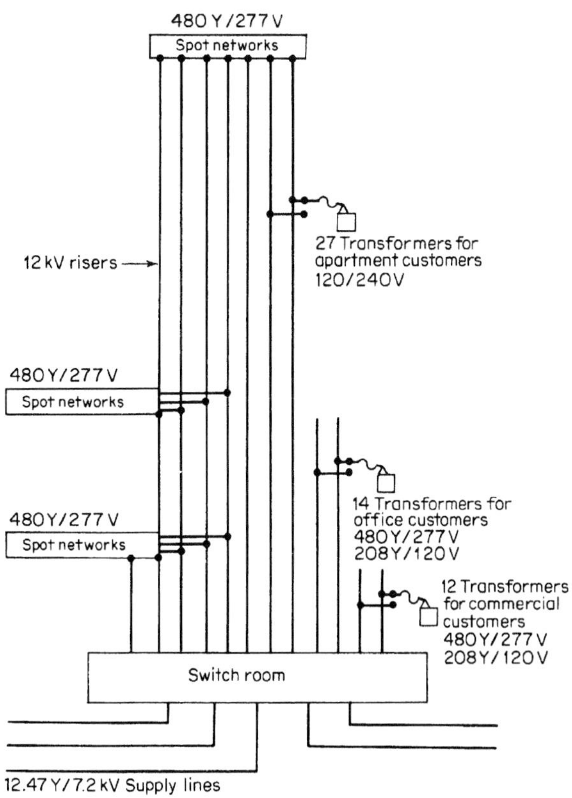


**FIGURE 7-47** One-line diagram of the multiple primary system for John Hancock Center.

## 7.20 UNDERGROUND CABLES

***Types of Cables.*** Underground distribution systems have been in use for many years in the downtown built-up areas of American cities. In most instances these are secondary network systems with facilities installed beneath streets and sidewalks, and the cables are usually installed in conduit or duct systems. For primary voltage circuits from 5 to 35 kV, paper-insulated, lead-covered (PILC), 3-conductor cable has been used extensively. Single-conductor secondary cables with rubber insulation and neoprene jacket are common. More recently, single-conductor polyethylene-insulated cables are being used for both primary and secondary. Copper conductor predominated in the past, but aluminum has nearly displaced copper in new installations, except where existing duct space is limiting.

In residential and suburban areas, new underground distribution systems to serve commercial loads, such as shopping centers and commercial and industrial parks, often employ direct-buried cables; conduits may be provided in locations where subsequent excavation would be excessively expensive or inconvenient. Aluminum conductors are almost universal. For primary cables, solid-dielectric insulation is used almost exclusively, with cross-linked polyethylene and EPR insulations predominating. Concentric-neutral wires are common. Secondary cables in these systems generally have aluminum conductors and solid-dielectric insulation, with cross-linked polyethylene being the most common. The secondary neutral is usually an insulated conductor, although there is some use of bare copper neutrals.

For most distribution circuits in the 5-kV class or higher, the cables employ a shielded construction. *Shielding* is the use of a conducting or semiconducting material on the surface of insulating material to confine the electric field to the insulation proper and to avoid undesired concentrations of electric stress. Shielding is used on the outer surface of the cable insulation or directly over the main conductor, or both. Outside shielding, often in the form of metallic tapes, metallic sheaths, or concentric wires, must be effectively grounded. This shielding also provides a return path for short-circuit current in the event of cable failure and protects workers from the shock of charging current.

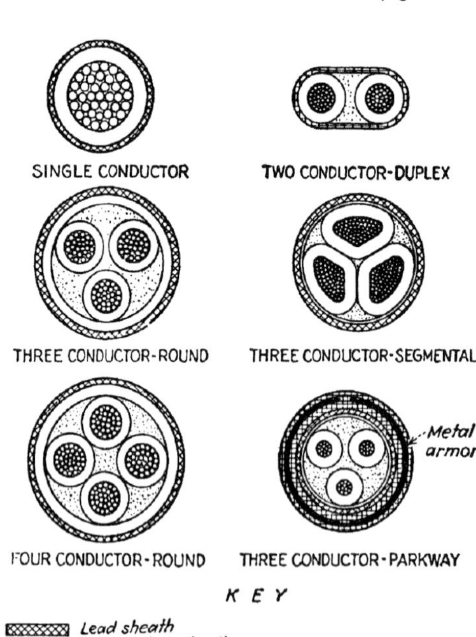


**FIGURE 7-48** Cross sections of typical cables.

***Number of Conductors.*** Cables can be classified as single-conductor, 2-conductor, 3-conductor, etc., according to the number of separately insulated conductors enclosed by a single sheath or jacket (Fig. 7-48).

***Cable Insulation.*** Electric supply cables are insulated with a wide variety of insulating materials depending on voltage ratings, type of service, installation conditions, etc. In the past, the following have been commonly used:

1. Rubber and rubberlike for 0 to 35 kV
2. Varnished cambric for 0 to 28 kV
3. Impregnated paper of the solid type for voltages up to 69 kV and with pressurized gas or oil up to 345 kV or higher

These insulation systems usually require a sheath or suitable jacket to prevent infiltration of moisture, loss of oil, gas, or impregnant, and to provide protection against corrosion and electrolysis. In some cases, an armor overlay is used to provide mechanical protection. With impregnated-paper insulation of the solid type, a lead sheath is usually provided.

A wide variety of joints, splices, and terminations is used, depending on the voltage, cable insulation, number of conductors, jacketing or sheathing material, and method of shielding. Joints, splices, and terminations are discussed in more detail later in this section.

Single-conductor cables are used, of course, in single-phase primary system for residential service and normally are used in single-phase or 3-phase secondary systems where many taps and connections are involved. Single-conductor cables also are frequently used in direct-buried, 3-phase primary systems. Three-conductor primary cables are often used in duct systems where they have the advantage of occupying only one duct. Several typical cables are shown in Fig. 7-48.

At the present time, solid-dielectric insulating materials such as tree retardant, cross-linked polyethylene, and EPR are receiving the widest application in underground distribution systems, both direct-buried and duct systems. Principal reasons for the wide usage of these insulations are

1. Low cost.
2. Suitability for direct burial or for use in duct systems.
3. Sheath or jacket not generally required.
4. Much easier to tap, splice, and terminate than systems such as solid impregnated paper. Factory-made splices, connectors, and terminations are available and widely used.
5. Excellent mechanical and electrical properties.

Insulation thickness for typical cross-linked, polyethylene-insulated, nonjacketed distribution cables are given in Table 7-18. Thickness for most ratings of non-cross-linked polyethylene cables are essentially the same.

**TABLE 7-18** Insulation Thickness for Cross-Linked, Thermosetting, Polyethylene-Insulated Cable

| Rated circuit voltage, phase-to-phase volts | Conductor size, AWG or kcmil | Insulation thickness for 100% and 133% insulation levels | |
|---|---|---|---|
| | | mils | mm |
| 0–600 | 14–9 | 45 | 1.14 |
| | 18–2 | 60 | 1.52 |
| | 1–4/0 | 80 | 2.03 |
| | 225–500 | 95 | 2.41 |
| | 525–1,000 | 110 | 2.79 |
| 601–2,000 | 4–9 | 60 | 1.52 |
| | 8–2 | 70 | 1.78 |
| | 1–4/0 | 90 | 2.29 |
| | 225–500 | 105 | 2.67 |
| | 525–1,000 | 120 | 3.05 |
| 2,001–5,000 | 8–1,000 | 90 | 2.29 |
| 5,001–8,000 | 6–1,000 | 115 | 2.92 |
| 8,001–15,000 | 2–1,000 | 175 | 4.45 |
| 15,001–25,000 | 1–1,000 | 260 | 6.60 |
| 25,001–28,000 | 1–1,000 | 280 | 7.11 |
| 28,001–35,000 | 1/0–1,000 | 345 | 8.76 |

***Note:*** *100% level* applied where system overcurrent protection is such that ground faults are cleared within 1 min. Applies to the great majority of distribution systems. *133% level* applied where clearing time of 100% level cannot be met, but there is assurance of fault clearing within 1 h. Minimum-size conductors should be in accordance with above values to limit maximum voltage stress on the insulation at the conductor to a safe value.

***Source:*** Adapted from IPCEA Pub. S-66-524, NEMA Pub. WC-7-1471. Revision No. 3, September 1974.

***Cable Diameters.*** Overall diameter $D$ of a cable may be computed from the diameter of its conductors $d$, the thickness of its conductor insulation $T$, its belt insulation $t$, and its lead sheath $S$, as follows:

Single-conductor: $$D = d + 2T + 2S \tag{7-13}$$

2-conductor: $$D = 2(d + 2T + t + S) \tag{7-14}$$

3-conductor: $$D = 2.155(d + 2T) + 2(t + S) \tag{7-15}$$

4-conductor: $$D = 2.414(d + 2T) + 2(t + S) \tag{7-16}$$

These formulas apply to conductors of circular cross section. For sector-type 3-conductor cables, the overall diameter

$$D_3 = D - 0.35d \quad \text{approx.} \tag{7-17}$$

***Electrical Characteristics of Cable.*** Skin effect is an ac phenomenon whereby alternating current tends to flow more densely near the outer surface of a conductor than near the center. That is, the magnetic-flux linkages of current near the center of the conductor are relatively greater than the linkages of current flowing near the surface of the conductor. The net effect is that the effective resistance of the cable is greater for alternating current than for direct current. This effect increases as the conductor size increases and as the frequency increases. It is also a function of the relative resistance of the conductor material, being less for materials of higher resistance; for example, the skin effect for a given diameter of cable is great if the material is copper rather than aluminum. Because of skin effect, large cables are sometimes built up over a central core of nonconducting material.

The nonuniform distribution of alternating current across the cross section of the cable also has the effect of reducing the effective internal inductance of the cable. Usually, this effect is extremely small in distribution circuits and is neglected.

It should be noted that magnetic flux linking the cable because of nearby current also can affect the cross-sectional distribution of current and can significantly change the effective ac resistance of the cable for multiconductor cables or cables in the same duct. This is known as the *proximity effect.* Most tables of conductor characteristics list factors which combine the results of the skin effect and proximity effect.

If an insulated cable has an outer metallic wrapping such as sheaths, metal pipes, or concentric-neutral conductors installed in such a manner that induced circulating currents can flow normally in these external conductors, losses will occur in these circuits, reducing the ampacity of the cable.

***Skin-Effect Coefficients.*** Skin-effect and proximity-effect coefficients are given in Table 7-19 for copper and aluminum conductors at 25°C. To determine the skin effect on the effective resistance of a single-conductor 1000-kcmil copper cable operating at 25°C, refer to Table 7-19, where the dc resistance is 0.01079 Ω/1000 ft and the skin-effect coefficient is 1.067. The effective resistance at 60 Hz is 1.067 × 0.01079 = 0.0115 Ω/1000 ft, 6.7% greater than for direct current. For a similar 2000-kcmil, the increase in resistance for alternating current is 23.3%; the ampacity of the cable is reduced to 100/1.233 = 81.1%.

The last two columns of Table 7-19 give the coefficients for multiconductor cables or cables in the same duct. They are used in the same manner as in the previous examples. For the larger conductors, the derating factor is substantial.

***Electrostatic Capacitance.*** The capacitance of a shielded or concentric-neutral single-conductor cable is

$$C = \frac{0.00736K}{10^6 \log_1 (D/d)} \tag{7-18}$$

where $C$ = capacitance, farads/1000 ft
$K$ = dielectric constant of insulation
$D$ = diameter over the insulation
$d$ = diameter over the conductor shield

**TABLE 7-19** DC Resistance and Correction Factors for AC Resistance

| Conductor size, AWG or kcmil | DC resistance, Ω/1000 ft at 25°C* | | AC resistance multiplier | | | |
|---|---|---|---|---|---|---|
| | | | Single-conductor cables† | | Multiconductor cables‡ | |
| | Copper | Aluminum | Copper | Aluminum | Copper | Aluminum |
| 8 | 0.6532 | 1.071 | 1.000 | 1.000 | 1.00 | 1.00 |
| 6 | 0.4110 | 0.6741 | 1.000 | 1.000 | 1.00 | 1.00 |
| 4 | 0.2584 | 0.4239 | 1.000 | 1.000 | 1.00 | 1.00 |
| 2 | 0.1626 | 0.2666 | 1.000 | 1.000 | 1.01 | 1.00 |
| 1 | 0.1289 | 0.2114 | 1.000 | 1.000 | 1.01 | 1.00 |
| 1/0 | 0.1022 | 0.1676 | 1.000 | 1.000 | 1.02 | 1.00 |
| 2/0 | 0.08105 | 0.1329 | 1.000 | 1.001 | 1.03 | 1.00 |
| 3/0 | 0.06429 | 0.1054 | 1.000 | 1.001 | 1.04 | 1.01 |
| 4/0 | 0.05098 | 0.08361 | 1.000 | 1.001 | 1.05 | 1.01 |
| 250 | 0.04315 | 0.07077 | 1.005 | 1.002 | 1.06 | 1.02 |
| 300 | 0.03595 | 0.05897 | 1.006 | 1.003 | 1.07 | 1.02 |
| 350 | 0.03082 | 0.05055 | 1.009 | 1.004 | 1.08 | 1.03 |
| 500 | 0.02157 | 0.03538 | 1.018 | 1.007 | 1.13 | 1.06 |
| 750 | 0.01438 | 0.02359 | 1.039 | 1.015 | 1.21 | 1.12 |
| 1000 | 0.01079 | 0.01796 | 1.067 | 1.026 | 1.30 | 1.19 |
| 1500 | 0.00719 | 0.01179 | 1.142 | 1.058 | 1.53 | 1.36 |
| 2000 | 0.00539 | 0.00885 | 1.233 | 1.100 | 1.82 | 1.56 |

***Note:*** 1 ft = 0.3048 m.

*To correct to other temperatures, use the following:

For copper: $R_T = R_{25}[(234.5 + T)/259.5]$

For aluminum: $R_T = R_{25}[(228 + T)/253]$, where $R_T$ is the new resistance at temperature $T$(°F) and $R_{25}$ is the tabulated resistance.

†Includes only skin effect (use for cables in separate ducts).

‡Includes skin effect and proximity effect (use for triplex, multiconductor, or cables in the same duct).

***Charging Current.*** The charging current of a single-conductor cable is

$$I = \frac{0.0463 EfK}{1000 \log_1 (D/d)} \tag{7-19}$$

where $E$ = voltage to neutral, kV

$f$ = frequency, Hz

$I$ = amperes per 1000 ft, charging current

For overhead circuits at distribution voltages and power frequencies, the charging current usually is negligible. It may become significant in high-voltage transmission circuits, as discussed in Sec. 5. For insulated cables, the charging current is relatively greater than in overhead circuits because of close spacing and the higher dielectric constant of the cable insulation; $K = 1$ for air and 3.3 for impregnated paper. For unfilled polyethylene $K = 2.3$, and it may run as high as 2.9 for filled, cross-linked polyethylene.

***Geometric Factors.*** Charging current of 3-phase three-core cable is affected by arrangement of conductors (round or sector) and by relative thicknesses of conductor insulation $T$ and belt insulation $t$. These relations have been put into usable form by working out logarithmic denominators of the equation for various ratios of thickness of insulation to diameter of conductor. This has been termed the *geometric factor*. Charging current of a three-core 3-phase cable is

$$I = \frac{3 \times 0.106 EfK}{1000 G_2} \text{ A/1000 ft} \tag{7-20}$$

**TABLE 7-20** Table of Geometric Factors of Cables

| Ratio $\frac{T+t}{d}$ | G Single conductor | Sector factor | Three-conductor cables | | | | | |
|---|---|---|---|---|---|---|---|---|
| | | | $G_1$ at ratio $t/T$ | | | $G_2$ at ratio $t/T$ | | |
| | | | 0 | 0.5 | 1.0 | 0 | 0.5 | 1.0 |
| 0.2 | 0.34 | ... | 0.85 | 0.85 | 0.85 | 1.2 | 1.28 | 1.4 |
| 0.3 | 0.47 | 0.690 | 1.07 | 1.075 | 1.08 | 1.5 | 1.65 | 1.85 |
| 0.4 | 0.59 | 0.770 | 1.24 | 1.27 | 1.29 | 1.85 | 2.00 | 2.25 |
| 0.5 | 0.69 | 0.815 | 1.39 | 1.43 | 1.46 | 2.10 | 2.30 | 2.60 |
| 0.6 | 0.79 | 0.845 | 1.51 | 1.57 | 1.61 | 2.32 | 2.55 | 2.95 |
| 0.7 | 0.88 | 0.865 | 1.62 | 1.69 | 1.74 | 2.55 | 2.80 | 3.20 |
| 0.8 | 0.96 | 0.880 | 1.72 | 1.80 | 1.86 | 2.75 | 3.05 | 3.45 |
| 0.9 | 1.03 | 0.895 | 1.80 | 1.89 | 1.97 | 2.96 | 3.25 | 3.70 |
| 1.0 | 1.10 | 0.905 | 1.88 | 1.98 | 2.07 | 3.13 | 3.44 | 3.87 |
| 1.1 | 1.16 | 0.915 | 1.95 | 2.06 | 2.15 | 3.30 | 3.60 | 4.05 |
| 1.2 | 1.22 | 0.921 | 2.02 | 2.13 | 2.23 | 3.45 | 3.80 | 4.25 |
| 1.3 | 1.28 | 0.928 | 2.08 | 2.19 | 2.29 | 3.60 | 3.95 | 4.40 |
| 1.4 | 1.33 | 0.935 | 2.14 | 2.26 | 2.36 | 3.75 | 4.10 | 4.60 |
| 1.5 | 1.39 | 0.938 | 2.20 | 2.32 | 2.43 | 3.90 | 4.25 | 4.75 |
| 1.6 | 1.44 | 0.941 | 2.26 | 2.38 | 2.49 | 4.05 | 4.40 | 4.90 |
| 1.7 | 1.48 | 0.944 | 2.30 | 2.43 | 2.55 | 4.17 | 4.52 | 5.05 |
| 1.8 | 1.52 | 0.946 | 2.35 | 2.49 | 2.61 | 4.29 | 4.65 | 5.17 |
| 1.9 | 1.57 | 0.949 | 2.40 | 2.54 | 2.67 | 4.40 | 4.76 | 5.30 |
| 2.0 | 1.61 | 0.952 | 2.45 | 2.59 | 2.72 | 4.53 | 4.88 | 5.42 |

For impregnated-paper cable, $K$ is 3.3, and the equation for 60-Hz circuits becomes

$$I = \frac{3\times 3.3\times 0.106\times 60E}{1000G_2} = 0.063\frac{E}{G_2} \quad \text{amperes} \tag{7-21}$$

Values of $G$ for single-conductor and $G_2$ for 3-conductor cable may be taken from Table 7-20.

***Geometric Factors of Cables.*** See Table 7-20. Intermediate values may be found by interpolation.

*Example.* Find 60-Hz charging kVA for 33-kV cable having three 350,000-cmil sector-type conductors each with 10/32 in of paper and a 5/32-in belt.

$$T = 0.313 \text{ in} \qquad t = 0.156 \text{ in} \qquad d = 0.681 \text{ in} \qquad t/T = 0.5$$

$$(T+t)/d = (0.313 + 0.156)/0.681 = 0.69; \; E = 33/1.73 = 19 \text{ kV}$$

Interpolating in Table 7-20, we find $G_2 = 2.78$.

For sector-type cable, $G_2$ must be multiplied by the sector factor for 0.69, which is seen to be 0.86 in the sector-factor column in Table 7-20. For such a cable,

$$G_2 = 0.86 \times 2.78 = 2.39 \quad \text{and} \quad I = (0.063 \times 19)/2.39 = 0.5\text{A}/1000 \text{ ft}$$

Charging kVA = $3IE = 3 \times 0.5 \times 19 = 28.5$ kVA/1000 ft, and for a cable having a length of 20 mi it would be $20 \times 5.28 \times 28.5 = 3010$ kVA. For single-conductor cables, $t = 0$ and $(T + t)/d = T/d$, which is used to get the value of $G$ from the values for single-conductor cable in Table 7-20.

***Cable Terminations.*** A cable termination must perform several functions:

1. Provide means for electrical connection of the cable to an equipment or circuit.
2. Control the electrostatic stresses so that there is no electrical-discharge activity in the termination at design voltage levels. One important consideration is to control the voltage stresses where the change is from a uniform radial field within the (shielded) cable to a new configuration beyond the termination of the shield. Other considerations are to provide adequate flashover and creepage strength to nearby grounds.
3. Prevent loss of gas or liquid insulation impregnant from the cable, where needed, or from the termination.
4. Provide suitable mechanical and/or hermetic termination of the sheath, where used.
5. Serve as a load-break switch or separable connection, where needed.

Many types of terminations are in use, ranging from those made by hand in the field to factory-made types requiring very little work in the field. Two broad classifications are live-front and dead-front. The former involves exposed, bare electrical connections and possibly lengths of unshielded cable. The latter type of termination is completely enclosed in a semiconducting or metallic structure essentially at ground potential, such that it can be touched without hazardous shock while the equipment is energized.

***Ambient Earth Temperatures.*** Ambient earth temperatures vary with geographic location, season, and depth. Average daily air temperature at a given location follows a more or less sinusoidal curve over the seasons of the year. As a result, the earth ambient temperatures also exhibit an annual sinusoidal variation. At depths greater than 1 to 2 ft, there is essentially no daily variation in earth ambient, but there is a seasonal variation which is greater at shallow depths, decreasing with depth. In addition, as the depth increases, the variation in ambient temperature increasingly lags behind the daily ambient air temperature curve; at a depth of 6 ft, this lag may be as great as 6 to 8 weeks. At depths of 20 to 30 ft, the earth temperature remains practically constant at about the mean annual air temperature. As a result, the earth temperature tends to increase with depth in the winter and decrease with depth in the summer. At a depth of 3½ ft, typical temperatures are as follows:

| | Temperature, °C | |
|---|---|---|
| | Summer | Winter |
| Northern U.S. | 20–25 | 2–15 |
| Southern U.S. | 25–30 | 10–20 |

***Maximum Allowable Conductor Temperature.*** The ICEA temperature ratings for polyethylene (thermoplastic) and cross-linked-polyethylene-insulated power cables are

| | Max. conductor temperature, °C | |
|---|---|---|
| Insulation | Normal operation | Emergency overload |
| Polyethylene | 75 | 90 |
| Cross-linked polyethylene | 90 | 130 |

Maximum conductor temperatures for impregnated paper-insulated cables are given in Table 7-21, as adapted from Publication P-46-426 of the IPCEA.

***Ampacity of Cables.*** There is a growing use of single-conductor, solid-dielectric power cables for important 3-phase distribution circuits in the 15- to 35-kV class. Various shielding systems are in use including concentric wires, ribbons, and tapes. In many cases, on 4-wire primary-distribution circuits, the concentric wires are used as neutral conductors. When the shields are bonded together

**TABLE 7-21** Maximum Conductor Temperatures for Impregnated-Paper-Insulated Cable

| | Conductor temperature, °C | | |
|---|---|---|---|
| Rated voltage, kV | Normal operation | Emergency operation | |
| | Solid-type multiple conductor belted | | |
| 1 | 85 | 105 | |
| 2–9 | 80 | 100 | |
| 10–15 | 75 | 95 | |
| | Solid-type multiple conductor shielded and single conductor | | |
| 1–9 | 85 | 105 | |
| 10–17 | 80 | 100 | |
| 18–29 | 75 | 95 | |
| 30–39 | 70 | 90 | |
| 40–49 | 65 | 85 | |
| 50–59 | 60 | 75 | |
| 60–69 | 55 | 70 | |
| | Low-pressure gas-filled | | |
| 8–17 | 80 | 100 | |
| 18–29 | 75 | 95 | |
| 30–39 | 70 | 90 | |
| 40–46 | 65 | 85 | |
| | Low-pressure oil-filled and high-pressure pipe type | | |
| | | 100 h | 300 h |
| 15–17 | 85 | 105 | 100 |
| 18–39 | 80 | 100 | 95 |
| 40–162 | 75 | 95 | 90 |
| 163–230 | 70 | 90 | 85 |

*Source:* Copyright 1962 by Insulated Power Cable Engineers Association. Used by permission.

and grounded at multiple locations, circulating currents can flow in the shields, resulting in $I^2R$ losses and appreciable heating effect. Such losses may be significant when the cables are spaced.

The AIEE-IPCEA ampacity tables do not include the effects of circulating-current losses, but these effects are included in the ampacity tables of IPCEA Publication P-53-426, NEMA Publication WC50-1976, "Ampacities Including Effect of Shield Losses for Single-Conductor Solid-Dielectric Power Cable 15 kV through 35 kV (Copper and Aluminum Conductors)." The ampacity data in Table 7-22 have been taken from this publication and apply to directly buried, solid-dielectric power cable.

Table 7-23 has been adapted from the IPCEA Publication P-53-426 (and NEMA Publication WC50-1976) to illustrate typical ampacities of single-conductor, solid-dielectric power cables installed in underground ducts. The type of installation is assumed to be directly buried fiber or plastic duct of inside diameter nominal pipe size to provide a minimum diametral clearance of 0.75 in between cable outside diameter and inside diameter of the duct. The assumed arrangement of the ducts is shown in Fig. 7-49.

## 7.21 FEEDERS FOR RURAL SERVICE

***Basic Conditions.*** Rural distribution differs from urban in that consumers are farther apart and load units are generally small. Since distances are great, the primary system voltage should be the 15-kV class or higher, and the load per mile being low requires the cost of feeder construction to be as low as

**TABLE 7-22** Ampacity of Single-Conductor Solid Dielectric Power Cable Installed Direct Buried

| Cond. size | Neutral size | Ampacity | °C* | W/ft$^2$ |
|---|---|---|---|---|
| 1/0 | Full | 241 | 66 | 46.2 |
| 1/0 | ½ | 245 | 66 | 45.2 |
| 1/0 | 1/3 | 246 | 66 | 44.8 |
| 1/0 | 1/6 | 247 | 66 | 44.2 |
| 4/0 | Full | 339 | 71 | 47.4 |
| 4/0 | ½ | 349 | 70 | 45.7 |
| 4/0 | 1/3 | 355 | 69 | 44.8 |
| 4/0 | 1/6 | 361 | 69 | 43.6 |
| 500 | 1/3 | 513 | 75 | 45.5 |
| 500 | 1/6 | 544 | 74 | 43.1 |
| 500 | 1/12 | 566 | 73 | 40.9 |
| 500 | 1/18 | 575 | 72 | 40.2 |
| 750 | 1/3 | 575 | 75 | 42.3 |
| 750 | 1/6 | 624 | 75 | 40.8 |
| 750 | 1/12 | 671 | 74 | 38.5 |
| 750 | 1/18 | 690 | 73 | 37.2 |
| 1000 | 1/6 | 675 | 76 | 39.8 |
| 1000 | 1/12 | 748 | 76 | 37.9 |
| 1000 | 1/24 | 799 | 75 | 35.6 |
| 1000 | 1/36 | 819 | 74 | 34.8 |

***Note:*** 90°C aluminum conductor.
Single circuit—three cables spaced.
25°C earth ambient.
90 rho soil resistivity.
*Corresponding earth interface temperature in °C.
***Source:*** From IEEE Std 835-1994 (© 1994 IEEE).

**TABLE 7-23** Ampacity of Single-Conductor Solid Dielectric Power Cable Installed in Underground Ducts

| Cond. size | Neutral size | Ampacity 75% LF | Ampacity 100% LF |
|---|---|---|---|
| 250 | 1/3 | 272 | 239 |
| 250 | 1/6 | 279 | 245 |
| 250 | 1/12 | 282 | 249 |
| 250 | 1/18 | 284 | 250 |
| 350 | 1/3 | 316 | 277 |
| 350 | 1/6 | 329 | 288 |
| 350 | 1/12 | 337 | 295 |
| 350 | 1/18 | 339 | 298 |
| 500 | 1/3 | 364 | 317 |
| 500 | 1/6 | 386 | 337 |
| 500 | 1/12 | 402 | 351 |
| 500 | 1/18 | 407 | 356 |
| 750 | 1/3 | 414 | 359 |
| 750 | 1/6 | 449 | 389 |
| 750 | 1/12 | 480 | 417 |
| 750 | 1/18 | 493 | 428 |
| 1000 | 1/6 | 490 | 424 |
| 1000 | 1/12 | 536 | 464 |
| 1000 | 1/24 | 571 | 494 |
| 1000 | 1/36 | 584 | 506 |

***Note:*** 90°C aluminum conductor.
Two circuits—three cables spaced 7.5 inches.
25°C earth ambient.
90 rho soil resistivity.
***Source:*** From IEEE Std 835-1994 (© 1994 IEEE).

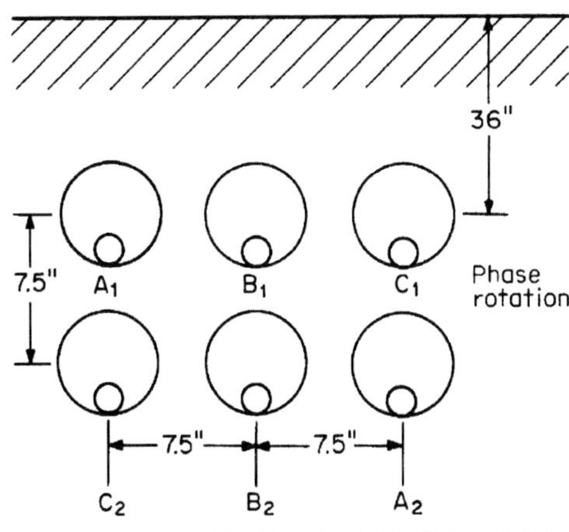


**FIGURE 7-49** Arrangement of ducts.

is consistent with a reasonable degree of permanence and reliability. One transformer per customer is required in many cases. Rural construction since the late 1930s has made electric service available to practically every farm. Efforts are directed now to bolstering capacity to serve the growing loads.

***Poles and Spans.*** Design of overhead lines for rural service differs from that of urban lines in several respects. Costs are reduced by using longer spans and as few accessories as possible. Longer spans mean greater sag and higher poles to get proper clearance at the low point of the span. The increase in sag may, however, be reduced by use of higher tensile stresses in conductors. This is possible when steel is employed in conjunction with copper or aluminum wires. Steel is combined with copper in a high-strength strand known as Copperweld, which has 30% or 40% of the conductivity of a copper conductor of equal size, or in a similar aluminum and steel conductor known as Alumoweld. When greater conductivity is needed, one or more strands of hard copper are stranded with or around the Copperweld or hard aluminum strands around Alumoweld. Steel is also stranded with aluminum wires into ACSR conductor. Such types of conductor have ample conductivity for rural lines, and they have been used widely. In level country, spans of 400 to 600 ft are practical, while in hilly country, spans of 800 to 900 ft are occasionally possible.

***Cable.*** Design of underground circuits for rural service is similar to that of urban underground circuits. Concentric-neutral cables are likely to be used in both types of systems. Because the rural circuits have longer uninterrupted runs of cable, there is a better opportunity to plow in the cable rather than digging trenches. Plowing results in lower installed costs per unit length of cable. In fact, some electric suppliers report that the total cost of a rural underground system is less than the cost of an overhead system to serve the same load. However, this is often NOT the case, due to the NESC requirement that the concentric neutral be grounded four times per mile. Such grounding does not require splicing the cable, but it does require skinning back the outer jacket, applying a grounding kit to the exposed neutral conductors, resealing the jacket, and bonding the neutral to an electrode with the grounding kit jumper.

***Location of Circuits.*** Rural-service circuits are run along main highways, where the largest number of users may be reached. Branches along intersecting roads are extended as may be warranted by service requirements. In some cases, private rights of way, maintained for transmission lines, may be utilized.

***Voltage.*** Rural circuits may be extended 5 to 50 mi from the point of supply, and voltage used for primary distribution must be chosen accordingly. Loadings are often so small that the minimum size of conductor required for dependable strength for overhead or cable insulation for underground is sufficient to meet requirements of voltage drop and line loss. This is particularly true when the higher voltages are used.

The most common primary voltage used in rural areas is 12,470Y/7200 V, 4-wire for normal load densities and 24,940Y/14,400 V for very light load densities. There is a trend toward using both 24,940Y/14,400 V and 34,500Y/19,900 V for all types of rural areas.

Single-phase circuits are most economical for the usual light loads found in rural areas and where power units do not exceed 10 hp. Vee-phase circuits consisting of 2-phase conductors and the neutral are an economical method of supplying 3-phase loads using open-wye-open-delta transformer banks. Full 3-phase, 4-wire construction will be desirable for many areas. In some cases there may be relatively small 3-phase loads in a single-phase area. Often these loads can be supplied economically from a single-phase system by means of a phase converter, the output of which is 3-phase voltage.

Limitations of voltage and distance are illustrated by the following table showing kilowatt-miles corresponding to a 5% line drop at 80% power factor for a circuit of 1/0 ACSR, or its equivalent in other metals.

Kilowatts × miles for 5% voltage drop, power factor 80%

| System | 4.16 kV | 12.47 kV | 24.94 kV | 34.5 kV |
|---|---|---|---|---|
| Single-phase | 82 | 737 | 2,949 | 5,646 |
| 3-phase | 488 | 4,375 | 17,919 | 33,815 |

Values for other sizes are approximately in proportion to relative cross section. In order to determine specific voltage-drop values for 3-phase overhead and underground circuits, refer to Table 7-3.

***Conductors and Spans.*** Because of the economy of using long spans, the choice of span lengths and conductor strength is of much importance in planning rural lines. Single-phase lines are commonly taken from a 3-phase system with neutral grounded. The grounded conductor is carried on a bracket about 2 ft below the phase wire, which rests on an insulator carried on the top of the pole. No crossarm is required, except on a main line of more than one phase.

While the conductivity of No. 4 ACSR or Copperweld may be thermally adequate for the greater part of a rural system, system economics can dictate the use of larger conductors. The strength of No. 4 ACSR or Copperweld is usually ample for spans of 350 to 600 ft, depending upon design-loading conditions. Conductors should be sagged in accordance with the conductor manufacturer's recommendations.

***Transformer Installations.*** Transformers usually supply not more than one or two customers, and sizes, therefore, are small compared with the average used in urban work, 10 to 15 kVA being average for single-phase installations. Where points of use are more than about 500 ft apart, it is usually most economical to provide separate transformers. When two users are within this distance, they can be served by placing a transformer between them and constructing a secondary. Rural loads on some systems have grown to the point where 15- and 25-kVA transformers are required.

Transformer capacity usually may be selected on the basis of loading the transformer to 150% of nameplate rating for peak loads lasting for 1 to 2 h. Pumping for drainage or irrigation is likely to require rated capacity more nearly equal to load.

***Stray Voltages.*** Stray voltages on dairy farms may cause lowered milk production and increased mastitis in dairy cattle. Dairy cattle are particularly sensitive to low magnitudes of voltage. Voltages on the order of 0.5 V occurring between metal stanchions or metal drinking cups and the concrete floor may be troublesome. It should be noted that the same symptoms may be due to other causes and that stray voltages are not always to blame. In particular, a large proportion of such problems are due to noncompliance with NEC (ANSI/NFPA 70) grounding and bonding requirements. Other problems are often caused by contaminated water, feed, or other hygiene issues.

One characteristic of the common-neutral distribution system, in which the neutral conductor is common to both the primary and secondary systems, is the multiplicity of ground connections between the neutral conductor and earth. Unbalanced load conditions on either the farm secondary system or the utility primary system result in current flow in the neutral conductor. Due to the multiplicity of ground connections, some portion of the neutral current flows in earth. Quite often, violations of the National Electrical Code (NFPA70) on the farm may create situations conducive to stray voltage.

One way to minimize stray voltages in the dairy barn is to cast wire mesh into the concrete floor and to bond the mesh and all metal structures together to establish equal potentials. In cases where current flow in the utility primary neutral is identified as a cause of stray voltages, it may be necessary to isolate the primary and secondary neutrals of the distribution transformer serving the farm. The NESC addresses this situation in Section 97D.

## 7.22 DEMAND AND DIVERSITY FACTORS

***Demand Factor.*** The ratio of maximum demand to total load connected, expressed as a percentage, is termed the *demand factor* of an installation. For example, if a residence having equipment connected with a total rating of 6000 W has a maximum demand of 3300 W, it has a demand factor of 55%. Demand factors of various types of large loads are helpful in designing systems, particularly those in buildings. As an example, a single household electric clothes dryer, of course, has a demand factor of 100%, but 25 dryers in a group have a demand factor of 33%. Similarly, three to five all-electric apartments in a multifamily dwelling have a demand factor of 45%. The lower the demand factor, the less system capacity required to serve the connected load. However, summer air conditioning and winter electric heating are loads that make for high demand factors.

***Coincidence or Diversity Factor.*** The *coincidence factor* is defined as the ratio of the maximum demand of the load as a whole, measured at its supply point, to the sum of the maximum demands of the component parts of a load. The *diversity factor* is the reciprocal of the coincidence factor. Coincidence factors can be applied to known consumer demands for estimating the loading of distribution transformers, lines, and other facilities. Coincidence factors for residential consumers can vary over a wide range for different types of consumers. The coincidence factor for a large group of consumers with no major appliance might be as low as 30%, whereas a group of electric-heating consumers might be as high as 90%.

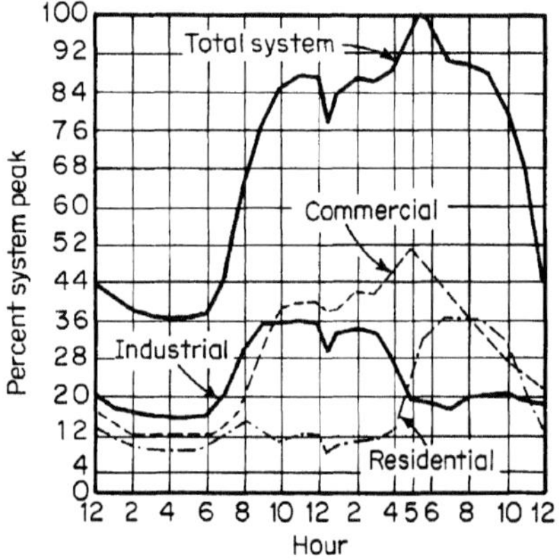


**FIGURE 7-50** Characteristic metropolitan load pattern.

***Diversity Between Classes of Users.*** The *daily-load curve* of a utility is a composite of demands made by various classes of users. The load curve on the day of maximum total system peak occurs when class loads gang up to create this maximum demand for the year. This is not necessarily the day, and usually is not the day, of any particular class peak. Class load curves on the day of system peak are illustrated in Fig. 7-50.

Air-conditioning loads have shifted these curves for many systems to cause daytime peaks during hot weather in the summer. Electric house heating builds heavy morning and evening loads during cold weather in the winter.

***Diversity in the Feeder System.*** The diversity of demands of transformers on a radial feeder makes the maximum load on the feeder less than the sum of the transformer loads. The diversity factors of a feeder vary greatly depending on load conditions. Some typical diversity factors are given in Table 7-24. The diversity factor of lighting feeders ranges from 1.1 to 1.5, while that of mixed light-and-power feeders is likely to be 1.5 to 2 or more. At the substation there is also a diversity factor of 1.05 to 1.25 between the sum of feeder maxima and the substation maximum. A large system has a further diversity factor between substations of 1.05 to 1.25. Total diversity factors in a large system are somewhat as in Table 7-24.

**TABLE 7-24** Diversity Factors

| Elements of system between which diversity factors are stated | Diversity factors for | | | |
|---|---|---|---|---|
| | Residence lighting | Commercial lighting | General power | Large users |
| Between individual users | 2.0 | 1.46 | 1.45 | |
| Between transformers | 1.3 | 1.3 | 1.35 | 1.05 |
| Between feeders | 1.15 | 1.15 | 1.15 | 1.05 |
| Between substations | 1.1 | 1.10 | 1.1 | 1.1 |
| From users to transformer | 2.0 | 1.46 | 1.44 | |
| From users to feeder | 2.6 | 1.90 | 1.95 | 1.15 |
| From users to substation | 3.0 | 2.18 | 2.24 | 1.32 |
| From users to generating station | 3.29 | 2.40 | 2.46 | 1.45 |

## 7.23 DISTRIBUTION SYSTEM LOSSES

About 8% of the total output of a large power system is lost or unaccounted for. Much of this loss is in the distribution system. Since investment must be made in facilities to supply these losses, they should be an important consideration in the engineering design of the system. A knowledge of their magnitude is essential and they should not be omitted from overall comparisons of alternative facilities without a study of each specific situation.

***Line Losses.*** The line losses, which are the sum of the $I^2R$, or resistance losses, can be easily found when the currents at peak load are known. Simplifying assumptions often can be made in making these calculations. For instance, if the load can be considered as being uniformly distributed, the losses are the same as if the total load were concentrated at a point one-third of the way out on the feeder.

***Transformer Losses.*** Transformers have a no-load loss as well as a load loss. The transformer no-load loss is independent of load, whereas the load loss will vary as the square of the current. These losses for distribution transformers are usually published as no-load and total loss when the transformer is operating at rated voltage and rated kVA. The load loss at full-load current is the difference between total and no-load losses.

***Working Principles.*** The problem of converting kWh of lost energy to dollars and cents has resulted in considerable controversy among system operators because of the difficulty of determining the value of the energy. It is not the purpose of this handbook to take sides in the controversy but rather to show the principles involved so that engineers will be able to evaluate losses using appropriate system costs.

The cost of supplying losses can be broken down into two major parts:

1. Energy component, or production cost to generate kWh losses
2. Demand component, or annual costs associated with system investment required to supply the peak kW of loss

The two components of cost usually are combined into a single figure either in terms of cents per kilowatthour of total energy loss or as dollars per kilowatt of peak loss. Expressing losses in terms of dollars per kilowatt is usually called *capitalized* cost of losses, and it has some advantage in that it shows directly the amount of money that could be economically spent to save 1 kW of loss. However, the expression of cost of losses in cents per kilowatthour is usually a more convenient form to use in most engineering studies.

The cost of losses depends on the point in the system at which they occur. The farther out on the system, the greater value losses have. One kilowatt of loss saved on the secondary system is worth more than 1 kW loss at generation because of the cumulative effect of increments of losses as they pass through various elements of the system.

In calculating loss, present-day or future cost of system investment should be used. The primary interest is to find the incremental investment, in dollars, required to supply an incremental load in kilowatts.

Opinions differ widely as to the degree to which the demand component of losses shall be evaluated. This ranges all the way from the dollar cost per kilowatt for future system expansion to no value at all for this component. The great majority of utility engineers prefer to assign full value to the demand component of losses.

***Responsibility Factor.*** Owing to diversity between classes of loads (i.e., residential, industrial, etc.) on a distribution system, peak loads on distribution, transmission, and generation usually do not occur at the same time. Therefore, a loss which contributes 1 kW to the distribution system peak might contribute less than this to transmission and production plant peak because its maximum does not occur at the same time as the transmission or generation peak. This introduces *peak responsibility factors* used for evaluating cost of losses in various parts of the system.

***Loss Factor.*** If the peak conductor losses of line or transformer have been calculated, it will still be necessary to know the loss factor or percent equivalent hours before it is possible to calculate the actual losses over a period of time. *Loss factor* is usually defined as the ratio of the average power loss, over a designated period of time, to the maximum loss occurring in that period. The term can refer to any part or all of the electric system. It is sometimes referred to as the *load factor of the losses.*

A corollary to loss factor is the term *equivalent hours.* This is defined as the number of hours per day, week, month, or year of peak load necessary to give the same total kilowatthours of loss as that produced by the actual variable load over the selected period of time. The period of time for distribution studies is usually 1 year, and it is obvious that *percent equivalent hours* has the same meaning as the term *percent loss factor.*

***Relation Between Loss Factor and Load Factor.*** Definitions of *loss factor* and *load factor* are quite similar. (*Load factor* is defined as the ratio of average power demand over a stipulated period of time to the peak or maximum demand for that same interval.) Care should be taken that the latter is not used in place of loss factor when considering system losses. There is a relationship between the two factors which depends on the shape of the load curve. Because resistance losses vary as the square of the load, it can be shown that the value of loss factor can vary between the extreme limits of load factor and load factor squared. A number of typical load curves have been studied to determine this relationship for distribution feeders and distribution transformers. The relation is shown in Fig. 7-51. Note that loss factor is always less than load factor except where they are both unity, as would be the case for transformer core losses. The relationship between load factor and loss factor at the distribution transformer can be expressed by the empirical formula

$$\text{Loss factor} = 0.15 \text{ load factor} + 0.85 \text{ (load factor)}^2$$

It should be noted that when the shape of the load curve is known or can be reasonably estimated, the loss factor should be calculated directly and not determined by the empirical formula.

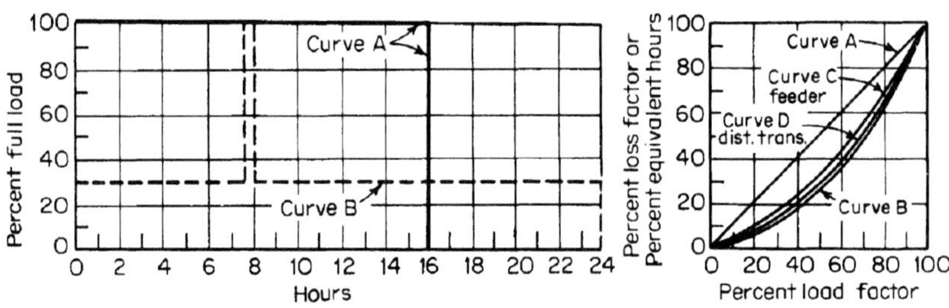


**FIGURE 7-51** Relationship between load factor and loss factor or equivalent hours.

***Cost of Losses.*** The two parts of the cost to supply losses are as follows:

$$\text{Energy component} = 8760 F_L E$$
$$\text{Demand component} = F_S P$$

where $F_L$ = loss factor of load
$E$ = cost of energy, dollars/kWh
$F_S$ = responsibility factor
$P$ = annual cost of system capacity, dollars/kW · year

Annual cost of losses can be combined into one value, in terms of either dollars per kilowatthour or dollars per kilowatt-year of peak loss, with the following formulas:

$$\text{Cost of losses, \$/kWh} = \frac{F_S P}{8760 F_L} + E$$

$$\text{Capitalized cost of losses, \$/kW-year} = F_S P + 8760\ F_L E$$

## 7.24 RELIABILITY

Reliability has always been a major consideration for utilities. In recent years there has been even more interest because loads are becoming more sensitive to even small system disturbances and concern has been expressed that deregulation, with associated cuts in personnel and budgets, will negatively impact system reliability. Distribution reliability is in a state of change. Some of these changes are

- Standardization of indices.
- Mandatory indices in some states.
- Performance-based rates.
- Sags and momentaries equaling outages for some loads.
- Contract penalties for interruptions, sags, etc.
- Maintenance budgets being reduced.

There are many ways to measure reliability. Some of the more common indices used by utilities are CAIDI, SAIDI, and SAIFI, as illustrated in Fig. 7-52.

These three indices are defined for sustained interruptions of 5 minutes or longer:

*SAIFI [system average interruption frequency index (sustained interruptions)]*. The system average interruption frequency index is designed to give information about the average frequency of sustained interruptions per customer over a predefined area. In words, the definition is

$$\text{SAIFI} = \frac{\text{total number of customer interruptions}}{\text{total number of customers served}}$$

To calculate the index, use the following equation:

$$\text{SAIFI} = \frac{\Sigma N_i}{N_T}$$

*SAIDI (system average interruption duration index)*. This index is commonly referred to as Customer Minutes of Interruption or Customer Hours, and is designed to provide information about the average time the customers are interrupted. In words, the definition is

$$\text{SAIDI} = \frac{\Sigma\text{customer interruption durations}}{\text{total number of customers served}}$$

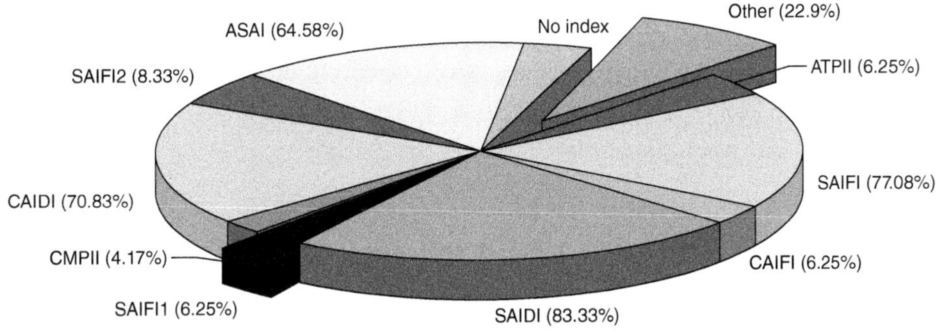


**FIGURE 7-52** Percentage of companies using reliability indices (48 companies responded to survey).

To calculate the index, use the following equation:

$$\text{SAIDI} = \frac{\Sigma r_i N_i}{N_T}$$

*CAIDI* (*customer average interruption duration index*). CAIDI represents the average time required to restore service to the average customer per sustained interruption. In words, the definition is

$$\text{CAIDI} = \frac{\Sigma \text{customer interruption durations}}{\text{total number of customers interruptions}}$$

To calculate the index, use the following equation:

$$\text{CAIDI} = \frac{\Sigma r_i N_i}{\Sigma N_i} = \frac{\text{SAIDI}}{\text{SAIFI}}$$

Values of these indices vary widely depending on many factors, including climate (snow, wind, lightning, etc.), system design (radial, looped, primary selective, secondary network, etc.), and load density (urban, suburban, and rural). Typical values seen by utilities in the United States are

| SAIDI | SAIFI | CAIDI |
|---|---|---|
| 96 min/yr | 1.2 int/yr | 80 min/yr |

Some utilities are already measuring indices to reflect system disturbances, other than interruptions, that cause sensitive loads to misoperate. One of these, the momentary average interruption event frequency index ($\text{MAIFI}_E$), is an index to record momentary outages caused by successful reclosing operations of the feeder breaker or line recloser. This index is very similar to SAIFI, but it tracks the average frequency of momentary interruption events. In words, the definition is

$$\text{MAIFI}_E = \frac{\text{total number of customer momentary interruption events}}{\text{total number of customers served}}$$

To calculate the index, use the following equation:

$$\text{MAIFI}_E = \frac{\Sigma\, \text{ID}_E N_i}{N_T}$$

*Note.* Here, $N_i$ is the number of customers experiencing momentary interruptions events and $\text{ID}_E$ equals interrupting device events during reporting period. This index does not include the events immediately preceding a lockout.

Another proposed index to reflect voltage sags caused by faults on other parts of the system is the system average rms (variation) frequency index$_{\text{threshold}}$ ($\text{SARFI}_{\%V}$). This index records the number of specified short-duration rms variation per system customer. Voltage threshold allows assessment of compatibility for voltage-sensitive devices. To calculate the index, use the following equation:

$$\text{SARFI}_{\%V} = \frac{\Sigma N_i}{N_T}$$

where $\%V$ = rms voltage threshold 140, 120, 110, 90, 80, 70, 50, 10
$N_i$ = number of customers experiencing rms $< \%V$ for variation $i$ (rms $> \%V$ for $\%V > 100$)
$N_T$ = total number of system customers

It is inevitable that more and more utilities will adopt some of these so-called power quality indices as their customers demand even better power for their sensitive loads.

In these days of reduced budgets, when utilities are being required to increase reliability, some of the techniques which cost very little or even nothing to achieve the goal of greater power quality are as follows:

- Purchase better-quality equipment.
- Shorten lead lengths on arresters.
- Use open-tie protection on underground systems.
- Use higher fuse ratings for transformers and laterals.
- Increase the number of homes per transformer.
- Pay attention to proper grounding.
- Use predictive reliability computer analysis to optimize designs.

## 7.25 EUROPEAN PRACTICES

In a time of deregulation and privatization, it has become common practice for a utility in one part of the world to own a utility in another country. While generation and transmission have relatively similar practices in all parts of the globe, distribution practices are considerably different depending on whether the system is based on American or European practices and standards. The following is a brief comparison of the two systems to familiarize engineers with the fact that in many ways distribution system operation and philosophy are so varied that direct comparison becomes extremely difficult.

***Voltage Levels.*** In the United States, primary voltage levels can be just about anything. Figure 7-53 shows some of the more common voltage levels in the United States, with 13.8 kV probably being the most popular for the distribution primary. European voltage levels are much more standardized. Thus 30, 20, and 10 kV are used throughout the world where European standards are practiced.

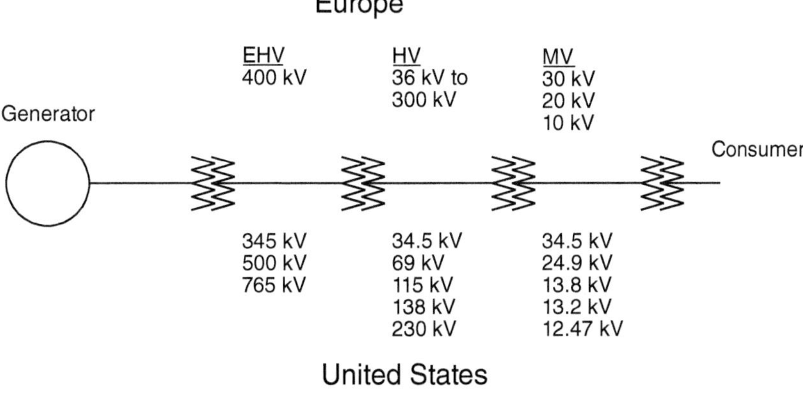


**FIGURE 7-53** European and U.S. primary voltage levels.

***30-kV/10-kV Distribution.*** Figure 7-54 is a typical European system design, showing a unigrounded system at the 10-kV level where most of the distribution loads are supplied. This voltage level tends to be radial but can be networked. The 30-kV system, on the other hand, tends to be looped and is a 3-wire delta system sometimes using a grounding transformer to facilitate overcurrent protection.

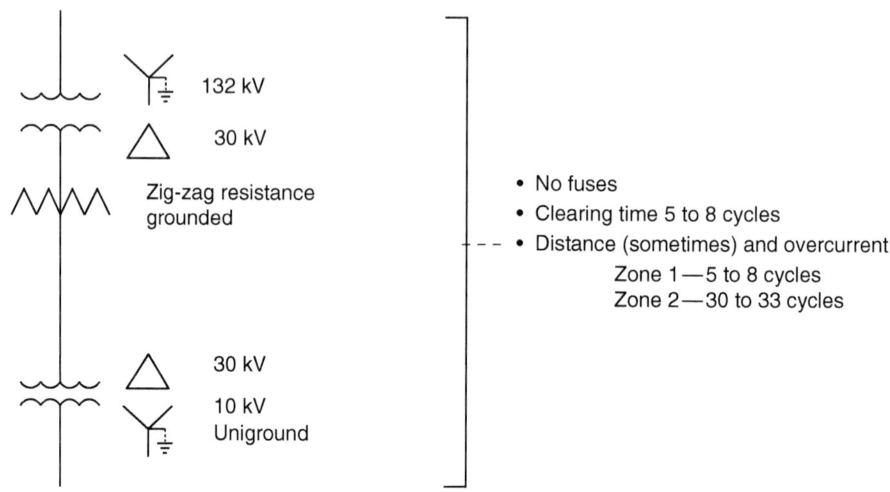


**FIGURE 7-54** European distribution system grounding practices.

***Residential Distribution.*** Table 7-25 illustrates some of the major differences in philosophy between the two designs which makes comparison difficult. The major difference is that European practice tends to use 3-phase transformers with a much larger kVA rating to supply many more homes. Higher-density loads and higher secondary voltages are part of the reason for this difference.

***Distributed Resources (DR) on the Distribution System.*** Distributed resources is a term which includes a variety of small generation technologies, including fuel cells, photovoltaics, microturbines, reciprocating engines, wind turbines, etc., with and without battery storage, which could be installed on the distribution system. In addition, many of these resources are becoming more modular in design, so that manufacturing economies of scale are driving their costs down. Generally, distributed resources are small in size, ranging from less than 1 kW to a few hundred kWs. The practical size limit for generators on the distribution system is in the area of around 35 MW.

The justification for introducing these newer technologies include the renewable resource aspect and lower environmental impact along with niche opportunities in particular regions when one or more of these technologies might excel. It is claimed that more DR means less T&D investment. Without storage, many DR technologies provide an additional energy source but no demand or investment savings. Even then improved reliability is often touted as a benefit, but this needs to be properly evaluated in the context of what technology is being discussed. Without careful engineering, a particular MW level of DR penetration can adversely affect the distribution system.

**TABLE 7-25** 1-Phase vs. 3-Phase Residential Distribution

| Difference | United States | Europe |
|---|---|---|
| Distribution transformers | Single phase, 25–100 kVA typical | Three phase, 300–500 kVA |
| Secondary voltage | 120/240 V, 3-wire | 400 wye/230 V, 4-wire |
| Service voltage | 120/240 V | 230 V |
| Secondaries/transformer | 1 | 5–10 |
| Homes/transformer | 4–6 | 100–200 |

Distribution designs are based on a number of principles that can be upset by DR including:

- Most distribution circuits are radial in nature with power flow from the substation to the loads.
- Voltage control with line drop compensation through LTC transformers or voltage regulators presumes a dropoff of load and voltage as one proceeds from the substation and, further, that power flow is from the source side to the load side.
- Most overcurrent protective device selection and coordination is based on higher fault current magnitudes at the substation and declining fault current magnitudes toward the end of the circuit.
- Reclosing intervals of circuit breakers and reclosers and reclosing practice in general is based on reasonable fault clearing times and the expectation that the reclosing device will close into a deenergized line.
- Utility restoration practices are based on having a limited number of known sources controlled by switches and other protective devices with known states (energized or deenergized).
- Harmonics are limited to larger known sources and distributed smaller sources.
- Most 3-phase transformer connections on distribution are grounded-wye.

With the arrival of distribution resources, many of these *normal conditions* now change, including

- Bidirectional power flow and fault current flow along portions of the circuit.
- Load magnitude dropping off along the feeder and then reaching a step change at the DR location(s) and dropping off beyond that point.
- Fault current profile showing the increase in total fault current due to added generation as well as the bidirectional fault current flows from the DRs depending on fault location.
- Automatic reclosing for utility circuit breakers and reclosers will have to be supervised with some form of voltage check and possible time delay. An alternative is to use transfer-tripping schemes to assure that the DR is off-line before reclosing takes place.
- Many DR advocates seek a delta-wye connection. Under backfeed conditions, a single-line-to-ground fault on the primary will overstress the surge arresters on the unfaulted phases.
- Photovoltaics and many small wind generators are dc machines that rely on invertors to produce an ac waveform and also produce high harmonic content. Induction-type machines draw reactive power from the power system. The combination introduces both voltage drop and power quality concerns.

IEEE Std 1547-2003, Standard for Interconnecting Distributed Resources with Electric Power Systems, was developed to provide a uniform standard for the interconnection of distributed resources with electric power systems. It provides requirements relevant to the performance, operation, testing, safety considerations, and maintenance of the interconnection. Follow-on standards projects to Std 1547 are intended to address the different DR technologies and engineering concerns about the interconnection issues.

A logical concern is—at what level of distributed resources does one have to be concerned with some of these potential issues becoming genuine problems? Unfortunately, we do not yet have an answer to that question.

## 7.26 BIBLIOGRAPHY

### 7.26.1 Books for General Reference

Blume, L. F., Boyajian, A., Camilli, G., Lennox, T. C., Minneci, S., and Montsinger, V. M.: *Transformer Engineering*, Wiley, 1951.

Burke, James J.: *Power Distribution Engineering: Fundamentals and Applications*, Marcel Dekker, 1994.

Edison Electric Institute: *Underground Systems Reference Book*, Edison Electric Institute, 1957.

Grainger, J. J. and Stevenson, W. D.: *Power System Analysis*, McGraw-Hill, 1994.

Greenwood, A.: *Electrical Transients in Power Systems*, Wiley-Interscience, 1971.

Padullaparti, H. V., Nguyen, Q., and Santoso, S.: Advances in volt-var control approaches in utility distribution systems, IEEE Power and Energy Society General Meeting, Boston, MA, 2016.

Lewis, W. W.: *Protection of Transmission Systems against Lightning*, Wiley, 1950.

Pansini, A. I.: *Electrical Distribution Engineering*, McGraw-Hill, 1983.

Peterson, H. A.: *Transients in Power Systems*, Dover Publications, 1951.

Short, T. A.: *Electrical Power Distribution Handbook*, CRC Press, 2004.

Thue, W. A.: *Electrical Power Cable Engineering*, Marcel Dekker, 1999.

## 7.26.2 Manufacturers' Publications

*Distribution Data Book*, GET-1008M, General Electric Company, 1980.

*Distribution System Feeder Overcurrent Protection*, GET-6450, General Electric Company, 1977.

*Distribution Transformer Guide*, ABB, October 1991.

*Electric Utility Engineering Reference Book—Distribution Systems*, Westinghouse Electric Corporation, vol. 3, 1965.

*Electrical Distribution—System Protection*, Cooper Power Systems, 3rd ed., 1990.

*Electrical Transmission and Distribution Reference Book*, Westinghouse Electric Corporation, 1964.

*Getting Down to Earth*, Biddle Instruments, 1982.

*Overhead Conductor Manual*, Southwire Company, 1994.

*Power Cable Manual*, Southwire Company, 1997.

## 7.26.3 IEEE Transactions (Formerly AIEE)

### *System*

Bankus, H. M. and Gerngross, J. E.: Unbalanced Loading and Voltage Unbalance on Three-Phase Distribution Transformer Banks, 1954, vol. 73, pt. III, p. 367.

Bankus, H. M. and Gerngross, J. E.: Combined Single-Phase and Three-Phase Loading of Open-Delta Transformer Banks, *Power Apparatus and Systems*, February 1958, pp. 1337–1343.

Easley, J. H. and Shula, W. E.: Cost and Reliability Evaluation of Four Underground Primary Distribution Feeder Plans, Transactions Paper, Conference Record—1974 Underground Transmission and Distribution Conference, 74-CH0832-6-PWR, pp. 436–443.

Mitchell, C. F., Sweeney, J. O., and Cantwell, J. L.: An Economic Analysis of Distribution Transformer Application, *Power Apparatus and Systems*, December 1959, pp. 1196–1202.

Nickel, D. L.: Distribution Transformer Loss Evaluation. I—Proposed Techniques, *Power Apparatus and Systems*, vol. PAS-100, no. 2, February 1981, pp. 788–797.

Ward, D. J., Griffith, D. C., and Burke, J. J.: Power Quality—Two Different Perspectives, *Trans. on Power Delivery*, vol. 5, no. 3, July 1990, pp. 1501–1513.

### *System Planning*

Anderson, A. S. and Thiemann, V. A.: Distribution Secondary Conductor Economics, *Power Apparatus and Systems*, February 1960, pp. 1839–1843.

Blake, D. K.: Some Observations on the Economic Benefits in Going from One System Voltage Level to a Higher System Voltage Level, *Power Apparatus and Systems*, vol. 71, pt. III, pp. 585–592.

Campbell, H. E., Ender, R. C., Gangel, M. W., and Talley, V. C.: Economic Analysis of Distribution Systems, *Power Apparatus and Systems*, August 1960, pp. 423–443.

Jones, A. I., Smith, B. E., and Ward, D. J.: Considerations for Higher Voltage Distribution, *Trans. on Power Delivery*, April 1992, pp. 782–788.

Rudasill, C. L. and Ward, D. J.: Distribution Underground Cable Evaluation, *Trans. on Power Delivery*, July 1997, vol. 12, no. 3, pp. 1398–1403.

Sarkas, R. H. and Thacker, H. B.: Distribution System Load Characteristics and Their Use in Planning and Design, *Power Apparatus and Systems*, August 1957, pp. 564–573.

Schultz, N. R.: Distribution Primary Feeder $I^2R$ Losses, *Power Apparatus and Systems*, March/April 1978, vol. PAS-97, no. 2, pp. 603–609.

Smith, J. A.: Determination of Economical Distribution Substation Size, *Power Apparatus and Systems*, October 1961, pp. 663–670.

Smith, J. A.: Economics of Primary Distribution Voltages of 4.16 through 34.5 kV, *Power Apparatus and Systems*, October 1961, pp. 670–683.

Van Wormer, F. C.: Some Aspects of Distribution Load Area Geometry, *Power Apparatus and Systems*, December 1954, pp. 1343–1349.

Webler, R. M., Gangel, M. W., Carter, G. K., Zeman, A. L., and Ender, R. C.: Secondary Distribution System Planning for Load Growth, *Power Apparatus and Systems*, December 1963, pp. 908–927.

### Overvoltage and Overvoltage Protection

Burke, J. J., Sakshaug, E. C., and Smith, S. L.: The Application of Gapless Arresters on Underground Distribution Systems, *Trans. on Power Apparatus and Systems*, March 1981, vol. 100, pp. 1234–1243.

Clayton, J. M. and Hileman, A. R.: A Method of Estimating Lightning Performance of Distribution Lines, *Power Apparatus and Systems*, 1954, vol. 73, pt. III, p. 953.

Headrickson, P. E., Johnson, I. B., and Schultz, N. R.: Abnormal Voltage Conditions Produced by Open Conductors on Three-Phase Circuits Using Shunt Capacitors, *Power Apparatus and Systems*, 1953, vol. 72, pt. III, p. 1183.

Hopkinson, R. H.: Better Surge Protection Extends URD Cable Life, *Trans. on Power Apparatus and Systems*, October 1984, vol. 103, pp. 2827–2836.

Hopkinson, R. H.: Ferroresonance during Single-Phase Switching of Three-Phase Distribution Transformer Banks, *Power Apparatus and Systems*, April 1965, vol. PAS-4, pp. 289–293. *Discussion*, June 1965, pp. 514– 517.

Kershaw, S. S., Gaibrois, G. L., and Stump, K. B.: Applying Metal Oxide Surge Arresters on Distribution Systems, *Trans. on Power Delivery*, January 1989, vol. 4, no.1, pp. 301–307.

Mancao, R. T., Short, T. A., and Burke, J. J.: Application of MOVs in the Distribution Environment, *Trans. on Power Delivery*, January 1994, vol. 9, no. 1, pp. 293–305.

Sakshaug, E. C., Kresge, J. S., and Miske, S. A., Jr.: A New Concept in Station Arrester Design, *Trans. on Power Apparatus and Systems*, March/April 1977, vol. 96, p. 647.

Short, T. A.: Distribution Lightning Performance Calculations, *IEEE Computer Applications in Power*, November 1991.

Task Force Report-Investigation and Evaluation of Lightning Protective Methods for Distribution Circuits—Part I, Model Study and Analysis; Part II, Applications and Evaluation, *Power Apparatus and Systems*, August 1969, vol. PAS-88, no.8, pp. 1232–1247.

Working Group of Surge Protective Devices Committee: Voltage Rating Investigation for Application of Lightning Arresters on Distribution Systems, *Trans. on Power Apparatus and Systems*, May/June 1972, vol. 91, no. 3, pp. 1067–1074.

### Overcurrent and Overcurrent Protection

Arndt, R. H., Koch, R. E., and Schultz, N. R.: Concept Alternatives and Application Considerations in the Use of Current-Limiting Fuses for Transformer Protection, Transactions Paper, Conference Record—1974 Underground Transmission and Distribution Conference, 74-CH0832-6-PWR, pp. 259–267.

Auer, G. G., Ender, R. C., and Wylie, R. A.: Digital Calculation of Sequence Impedances and Fault Currents for Radial Primary Distribution Circuits, *Power Apparatus and Systems*, February 1961, pp. 1264–1277.

Burke, J. J. and Lawrence, D. J.: Characteristics of Fault Currents on Distribution Systems, January 1984, PAS vol. 103, no. 1, pp. 1–6.

Harner, R. H.: Secondary-Fault Recovery Voltage Investigation, *Power Apparatus and Systems*, February 1968, vol. PAS-87, no.2, pp. 463–487.

IEEE Tutorial Course on Application and Coordination of Reclosers, Sectionalizers, and Fuses, Publication 80 EHO157-8-PWR, 1980.

#### *Voltage Regulation and Kilovar Supply*

Barger, J. V. and Smith, D. R.: Impedance and Circulating Current Calculations for UD Multi-Wire Neutral Circuits, *Power Apparatus and Systems*, May–June 1972, vol. PAS-91, no.3, pp. 992–1006.

Grainger, J. J. and Lee, S. H.: Optimum Size and Location of Shunt Capacitors for Reduction of Losses on Distribution Feeders, *Power Apparatus and Systems*, March 1981, vol. PAS-100, no. 3, pp. 1105–1118.

Johnson, I. B., Schultz, A. J., Schultz, N. R., and Shores, R. B.: Some Fundamentals on Capacitance Switching, *Power Apparatus and Systems*, August 1955, pp. 727–736.

Neagle, N. M. and Samson, D. R.: Loss Reduction from Capacitors Installed on Primary Feeders, *Power Apparatus and Systems*, October 1956, pp. 950–959.

#### *Grounding*

Application Guide and Methods of Substation Grounding, AIEE Group on Substation Grounding Practices, 1954, vol. 73, pt. m, p. 271.

Mancao, R. T., Myers, A., and Burke, J. J.: The Effect of Distribution System Grounding on MOV Selection, *Trans. on Power Delivery*, January 1993, vol. 8, p. 1.


## 7.26.4 Standards and Standards Publications


American National Standards Institute (ANSI) C2-2012, National Electrical Safety Code.

ANSI C84.1-2016, Voltage Ratings for Electric Power Systems and Equipment (60-Hz).

ANSI/ICEA S-94-649 Concentric Neutral Cables Rated 5-46 kV, 2004.

ANSI/ICEA S-97-682 Utility Shielded Power Cable Rated 5-46 kV, 2006.

IEEE Std 835-1994 (R2006), Power Cable Ampacity Tables.

NEC 2011, NFPA 70, National Electric 01 Code.


## 7.26.5 Periodicals


#### *System*

Burke, J. J.: Utility Characteristics Affecting Sensitive Industrial Loads, *Power Quality Assurance Magazine*, November–December 1996.

Gangel, M. W. and Propst, R. F.: Investigating Distribution Transformer Load Characteristics, *Distribution Magazine*, July 1961, p. 6.

#### *System Planning*

Brown, P. G., Propst, H. R., and Tice, J B.: Unity Power Factor Is Essential to Emergency Kilowatt Transportation, *Electric Forum Magazine*, Fall 1975, p. 10.

Campbell, H. E.: Serving Critical Loads, *Distribution Magazine*, 1966, 4th quarter, p. 9.

Hayes, R. H. and Hill, O. L.: Progress in Remote Line Switch Control, *Transmission and Distribution*, June 1975, p. 52.

Van Wormer, F. C.: Design and Operation of Spot Networks, *Distribution Magazine*, 1966, 2d/3d quarter p. 5; 1966, 4th quarter, p. 19.

#### *Overvoltage and Overvoltage Protection*

Auer, G. G.: Basic Considerations in Lighting Protection of URD Systems, *Distribution Magazine*, April 1968, p. 16.

Barker, P. P. and Burke, J. J.: Protecting Underground Distribution Systems, *Electric Light and Power*, April 1991.

#### *Overcurrent and Overcurrent Protection*

Howard, S.B. and Stroebed, R. W.: Can Single-Phase Cutouts Be Applied to Three-Phase Circuits, *Distribution Magazine*, 1964, 2d quarter, p. 4.

Lasseter, J. A.: Burndown Tests on Bare Conductors, *Electric Light and Power*, December 15, 1956, p. 94.

### *Voltage Regulation and Kilovar Supply*

Gangel, M. W.: Compensator Settings Made Easier, *Distribution Magazine*, pt. 1, April 1960, p. 22; pt. 2, July 1960, p. 18.

Schultz, N. R.: Calculating Voltage Drop and Power Loss, *Distribution Magazine*, January 1969, p. 11.

### *Underground Systems*

Van Wormer, F. C.: Underground Distribution Systems for Residential Areas, *Distribution Magazine*, pt. 1, January 1959, p. 3; pt. 2, April 1959, p. 12; pt. 3, April 1960, p. 16; pt. 4, April 1962, p. 3; pt. 5, April 1963, p. 22.


## 7.26.6 Miscellaneous Publications


### *System*

Beaty, H. W.: 10th Annual T&D Construction Survey, *Electrical World*, September 1, 1975, pp. 35–42.

Dudas, J. and Fletcher, C.: Underground Cable Specification Advances and Installation Practices of the Largest Investor Owned Utilities, Fall Insulated Conductors Committee Meeting, 2004.

Gangel, M. W. and Propst, R. F.: Transformer Characteristics Correlated to Loading: Power Distribution Conference, University of Texas, October 1963.

RUS Specifications and Drawings for 12.5/7.2 kV Line Construction 5/83, Bulletin 50-3, 2005.

RUS Specification and Drawings for 34.5/19.9 kV Distribution Line Construction (11–86), Bulletin 50-4.

RUS Specifications and Drawings for Underground Electric Distribution (3–90), Bulletin 50-6, 2000.

RUS Specifications and Drawings for Underground Cable Installation, Doc. 345-152, Form 515d, 1989.

### *System Planning*

Campbell, H. E.: Today and Tomorrow, Underground Distribution to High Rise Buildings, IEEE Conference Record-Special Technical Conference on Underground Distribution, 31C35, September 1966, pp. 223–239.

Crawford, J. W. and Hamner, F. G.: Demand and Diversity Characteristics of Residential Loads, Southeastern Electric Exchange, Engineering and Operating Conference, April 1963.

Load Growth Forces Higher Voltages, *Electrical World*, June 1, 1974, pp. 154–163.

### *Overcurrent Protection*

Beaty, H. W.: Special Report-Switching and Overcurrent Protection for Distribution Systems, *Electrical World*, April 1, 1974, pp. 41–56.

Campbell, H. E.: Implication of Increased Short-Circuit Duty on Residential Distribution Systems, American Power Conference, vol. 35, 1973, pp. 1098–1104. Underground Systems.

IEEE Conference Record-1974 Underground Transmission and Distribution, 74CHO832-6-PWR and 74CHO832-6-PWR (SUP.), April 1–5, 1974.

IEEE Conference Record-1991 Transmission and Distribution Conference, 911CH3070-0, September 1991.

Lewis, S. M.: URD Survey Report, *Transmission and Distribution*, July 1973, pp. 88–95.

Specifications and Drawings for Underground Electric Distribution, RUS Bulletin 50-6, Rural Utilities Service, U.S. Department of Agriculture, March 1990.

Underground Corrosion Control Guide, NRECA Research Project, August 1982.

# 8 SMART GRIDS AND MICROGRIDS


## Anurag K. Srivastava

*Associate Professor, School of Electrical Engineering and Computer Science, Washington State University, Pullman, Washington*

## Sayonsom Chanda

*Research Engineer, Idaho National Lab, Idaho Falls, Idaho*

## Nikos Hatziargyriou

*Professor of Power Systems, School of Electrical and Computer Engineering, National Technical University of Athens, Athens, Greece, and CEO, Hellenic Electricity Distribution Network Operator SA (HEDNO)*

## Jianhui Wang

*Department of Electrical Engineering, Southern Methodist University, Dallas, Texas, and Energy Systems Division, Argonne National Laboratory, Argonne, Illinois*

Advancements in computational and communication technologies enabled the power grid to be "smart" with enhanced situational awareness, better decision support, distributed intelligence, and automated response. The large penetration of distributed energy resources (DERs) and push for energy security require a shift from traditional centralized control architecture and passive distribution system to a hierarchical and distributed control with active distribution systems and microgrids [1].

Based on needs of today's economy and "prosumer" expectations, a modern power grid needs to have the following characteristics :

- Managing the impact of variability in power generation from renewable resources and facilitating power transfers across regions
- Informed and engaged consumers capable of participating in demand response and having ownership over small-scale distributed energy resources
- An increased emphasis on power quality, while giving consumers a wide variety of price options to choose their services from.
- Improved and more secure data acquisition of system states and focus on preventive control to minimize impact to consumers
- Automatic event detection and response to problems
- Resilience to unpredictable events and uncertainties, natural disasters, and rapid restoration capabilities
- Support higher penetration of renewable energy and energy security
- Support high reliability and resiliency using microgrid infrastructure

The traditional vertical power system infrastructure needs to be collectively upgraded in order to make it suitable for existing and future needs. The next-generation power grid needs to be more robust, more reliable, and more resilient with active distribution systems and microgrids.

In the following subsections, we analyze smart grids and microgrids in depth, and elaborate on key aspects pertaining to their development and deployment across the world.

# 8.1 INTRODUCTION TO SMART GRIDS AND MICROGRIDS

## 8.1.1 Smart Grid System

*Smart grid* is the name given to an evolved power system, engineered and managed to keep up with ongoing changes in the electric grid given renewable integration, environmental impact, active distribution system, and enhanced uncertainty. A series of renovations and paradigm shifts in technologies related to power generation, transmission, and distribution control will contribute to the eventual emergence of smart grids across all nations. Since smart grids are essentially subjected to constant evolution and developments, a working definition of smart grid should be attributed by:

- Secure communication among connected components of power system, as needed for system operation and control supported by adoption of advanced networking and computational solutions
- Adaptability to new operating conditions
- Automation, self healing capabilities, and resilience of the power grid
- Environment friendliness and ability to support distributed renewable resources
- Curtail peak demands and defer investments in large-scale power generation units
- Support demand response, distributed intelligence, and customer choice

According to the European Technology Platform on Smart Grids [2], a smart grid is "an electricity network that can intelligently integrate the actions of all users connected to it generators, consumers and those that do both in order to efficiently deliver sustainable, economic and secure electricity supplies." Smart grid leverages innovative products, modern computing algorithms, data analytics, and real-time services together with intelligent measurement, monitoring, control, communication, and self-healing technologies to (1) make information available to customers about their choice of services; (2) enable customers to participate in optimizing demand and supply in the energy grid; (3) enhance reliability of operation of generators across a broad spectrum of sizes and types; (4) minimize impact on the environment caused due to operation of the power grid; and (5) improve system security, resiliency, and reliability.

The U.S. Department of Energy (DOE) has envisioned smart grids to have several components, as shown in Fig. 8-1 [3]. In order to enable all the aforementioned attributes, real-time measurements, and situation assessment will play a critical role in their operation.

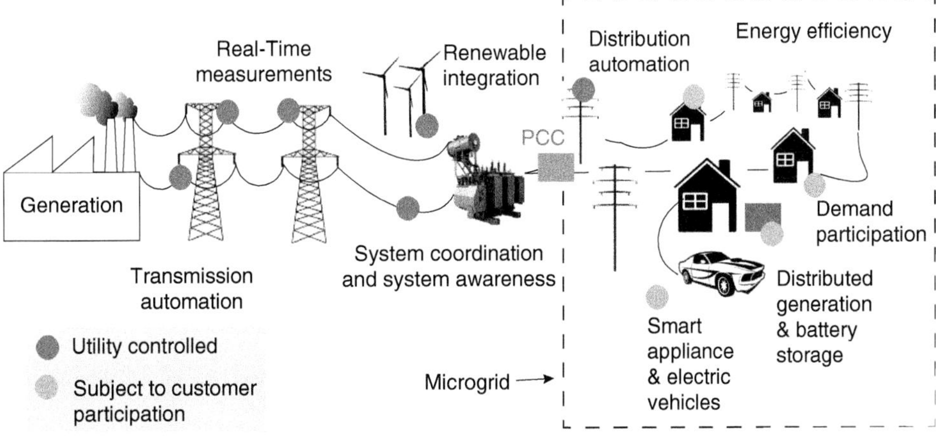


**FIGURE 8-1** Smart grid architecture.

According to [1], the smart grid architecture can be grouped into distinct sub-systems, which mutually interact with each other by means of communication links and other information flow mechanisms (such as optimization signals). Smart devices for monitoring and control form part of the generation components' real-time information processes. In a smooth functioning smart grid, the sub-systems must be integrated in the operation of both primary energy delivery resources (such as sub-stations), as well as DERs closer to points of consumption. Along with the interface, a smart grid will comprise:

1. *Transmission Subsystem Component.* It is important to include real-time measurement and computation tools to perform dynamic optimal power flow analysis, state estimation, and stability assessment. Real-time monitoring based on synchrophasors from phasor measurement units (PMUs), and communication technologies are the most important and widely used tools for developing smart transmission functionality.
2. *Distribution Subsystem Component.* At the distribution level, intelligent support schemes are required in the smart grid that can actively monitor automation using smart meters, communication between consumers and utility control, energy management components, and advanced metering infrastructure (AMI). Distribution system automation refers to the ability to the grid to expedite and automate fault detection, self-heal from a large number of faults and disturbances, enable voltage optimization and intelligent load transfer, automate billing, and keeping the consumers informed about real-time pricing, and facilitating demand response [4]. The distribution system should also be able to isolate the consumers in case it observes large disturbances in the power grid, and continue to supply its consumers using local on-site power generation or storage tools. Such distribution systems that is coupled to the main transmission grid at normal times (using at special connection point called "point of common coupling" or PCC) and can strategically isolate itself is broadly referred to as a microgrid. Other features of a microgrid will be discussed in detail in later subsections. In Fig. 8-1, a microgrid has been shown within the dashed lines.
3. *Demand Subsystem Component.* Tools are needed to facilitate demand-side management to disseminate service information to the consumers for demand response, aggregate customer-side power generation, and aid in deferring the use of expensive generators.
4. *Storage Subsystem Component.* Power system consumption and renewable generation do not coincide under most operating scenarios. Thus, it is important to integrate a controllable energy absorption and injection resource into the system to absorb system uncertainties.

Self-monitoring and self-healing attributes will be crucial to ensure smooth operation of the smart grid. The control technology component should be capable of adapting to variations in generation and demand, and be robust enough to mitigate grid congestion and override events that can cause significant instability across the grid. Bringing transmission and distribution systems at same level of sophistication in control will open new possibilities in terms of energy markets, and is perceived as one of the most significant contributions of the smart grid.

Modernization of the legacy power grid into smart grid will be the result of improvements across the grid, as well as reflect modern lifestyle of the customers. The introduction of modern variable speed drive systems have replaced many traditional industrial and consumer induction motor loads across the grid. These induction motors (used in pumps, fans, washers, etc.) offer better control capabilities, and are significantly more efficient than their respective predecessors. Use of digital computers at homes, commercial places, and industries have surged astronomically. Purely resistive light bulbs have been replaced by energy-efficient compact fluorescent lamps (CFLs) and light emitting diodes (LED). Increase in more homes opting for installing affordable rooftop photovoltaic (PV) arrays,[a] using more energy-efficient appliances, constructing more thermally efficient buildings, and industries migrating to better equipment are reducing the amount of electricity needed from power companies. This is changing the traditional business models of the regulated utility industry. Market surveys show that electric vehicles (EVs) are growing at the rate of 18.6% in the United States compounded annually, according to [5]. If

[a]The number of homes in the United States with solar PV installations has grown from 15,500 in 2004 to more than 600,000 by the end of 2014.

this trend continues to hold or increase, EV charging will amount to a large fraction of total electricity demand from the power grid. Thus, changes in customer preferences also impact the future growth of smart grid technologies. In view of such dynamically shifting landscape of power systems, microgrids provide more reliable power supply to the consumers.

### 8.1.2 Microgrid System

Microgrids are distribution systems with a certain level of automation and self-sufficiency, demand-side integration of distributed generation (DGs), and renewable energy sources (RES). The transition of active distribution networks into microgrids that can island themselves with identified physical boundary from the larger grid in an automated manner and connect back to the larger grid as needed offers great flexibility in distribution network operation. Microgrids are identified as "building blocks of smart grids" [6–9].

There can be several perspectives for defining a microgrid, which are summarized below:

- *Customer Point of View.* Microgrids are resource for resilience in and enhanced reliability of energy supply, with co-generation of thermal and electricity capabilities.
- *Grid Operator Point of View.* Microgrids are considered as a singular controllable aggregation of loads or micro-generation resources, which, depending upon financial incentives can provide ancillary grid services, or aid in meeting energy demands to customers at a local level.

Several definitions of microgrids have been provided across regulatory bodies [10]. A commonly accepted definition of the microgrids was put forwarded by a CIGRE Working Group (CIGRE WG 6.22 Definition of Microgrid) [11]:

*Microgrids are electricity distribution systems containing loads and distributed energy resources, (such as distributed generators, storage devices, or controllable loads) that can be operated in a controlled, coordinated way either while connected to the main power network or while islanded.*

Though one standard definition does not completely describe to the philosophy of microgrids, all definitions echo the following points:

(a) Microgrid is seen as an entity that aggregates local micro-generation sources, energy storage infrastructure and demand resources located in a low voltage distribution grid.

(b) A microgrid can operate in either grid-connected mode or operate when islanded from the main grid.

(c) The most important distinction between a legacy grid penetrated by micro-generation resources and a microgrid is the control, management, and coordination of resources.

(d) A microgrid operator can independently carry out tasks including aggregation of multiple generators, prioritize loads, and have control over its own emissions to serve multiple economic, technical, and environmental goals of the network, without explicitly involving the utility.

According to some working definitions, microgrid must have the ability to voluntarily island and reconnect itself to the grid. A microgrid must be characterized by coordinated control and supervision of the generation resources and interaction with energy providers and consumers to ensure continuity of the power supply during contingencies. A distinction must also be made between microgrids and virtual power plants (VPP): these two terms cannot be used interchangeably. A VPP, also known as captive power plants in certain countries, is an aggregation of distributed microgeneration resources which can be operated as a single controllable entity. A VPP can be used instead of conventional power plants to provide higher system efficiency and increase flexibility of the overall network. In a microgrid, DERs are usually co-located with the local distribution network, downstream from the PCC. In a VPP, DERs may not be co-located with the local network; thus, they are coordinated over wide geographical areas. As of writing this, VPP aggregated energy production are more suited to participate in existing energy markets. The installed capacity of microgrids is relatively small (from few kW to several MW), while VPPs can be larger in capacity.

**TABLE 8-1** Differences between Existing Power Grid and Smart Grid

| | Existing system | Smart grid |
|---|---|---|
| Architecture | Monolithic | Modular and agile |
| Generation profile | Most power production using fossil fuel based, nuclear, hydropower plants | Mix of renewable energy resources and existing resources, with the intent of completely shifting toward renewable generation |
| Decision driver | Generation companies, utilties | Customers have opportunity to ensure that power companies prioritize their prerogative |
| Data-driven | Operational data and asset management are not strongly coupled to form control decisions | Leverage advances in big data and Internet of things capability of modern appliances to boost energy efficiency |
| Vulnerability | Highly vulnerable to cyber attacks and cascading power outages | Self-healing networks that make the entire power network robust and resilient with minimal power outage-related downtimes |

Microgrids play a key role in integration of a large variety of micro-generation resources and DERs for utilities. Different micro-generation sources, such as roof-top solar panels and photovoltaics, fuel cells, micro-turbines, and wind turbines with a rated power ranging up to 100 kW can be integrated in the sub-transmission networks. Such units are usually cited closest to their owners, and other users who can benefit directly from these resources if needed. Micro-generation resources have proven to be a very viable option to meet rapidly increasing customer demand for reliable, and economical electric power.

Microgrids are poised to play a crucial role in energy markets, and become a revenue resource for certain communities [12]. Energy markets vary widely in structure, regulation policies, and organization [13–15]. On the other hand, microgrid's roles in the energy market are distinguishable by the level of their DER aggregation. In some microgrids, a simple collection of collaborative independent market players encompass all micro-generation resources and consumers. As in the case of transmission networks, the operational structure of a microgrid mainly depends on the ownership of micro-generation resources. Table 8-1 summarizes some key differences between contemporary microgrid operation and design, and legacy power delivery infrastructures.

## 8.2 SMART GRID FUNDAMENTALS

In this subsection, fundamental architecture of the smart grid is discussed. A brief description of the communication technologies, and standards associated toward smart grid development is presented. Cyber-security risks and standards associated with smart grids are also discussed.

### 8.2.1 Smart Grid Domains

Smart grid is characterized by active interactions of grid components enabled by embedded processors, sensors, data sharing, communications and distributed control at the transmission level, and at low voltage (LV) distribution system consumers. Real-time measurements obtained at distribution systems using smart meters and mechanisms such as demand response adopted by customers have also shaped the vision of smart grid.

The smart grid is composed of seven high-level groups, or domains (1) Customer, (2) Market, (3) Service Provider, (4) Operations, (5) Bulk Generation, (6) Transmission, and (7) Distribution [3]. Transmission and distribution systems, with advanced automation, has been at the core of smart grid development, while the latest smart grid technologies focus on helping commercial, industrial, and even residential consumers to participate more actively in producing small amounts of energy for emergency use and even storing them using batteries. Smart grid domain comprises modern, communication-capable devices, systems, and programs. Technologies have been developed to enable, and continually evolve to improve the integration and cooperation of these distinct domains into a singular smart grid, such that it holds a much higher value proposition than a legacy grid.

The smart grid framework is shown in Fig. 8-2. Each domain has its own management and technological boundaries and control areas, with distinguishing input and output sets.

Figure 8-3 [3] shows a broad overview of a few major relationships that are being deployed and demonstrated across the smart grid domains. The overview helps engineers and system planners to identify smart grid components as might be necessary, possible communications paths in the smart grid, and for identifying potential intra and inter-domain interactions. In Fig. 8-3, an actor is either a human or a hardware or software-based computer system that can engage in communication with the smart grid. Domain gateway actor refers to actors that interfaces with other actors in other domains or in other networks, and "comms path" refers to communication channels between smart grid domains.

## 8.2.2 Enabling Technologies for Smart Grid

Real-time situational awareness and communication between connected components play a crucial role in upgrading a legacy power grid to a smart grid. Smart grid requires instruments for real-time, accurate, and secure measurements at high, medium, and low voltage levels of the power grid. Cohesively, the network of these components provide the data necessary for operation of the power grid and provide informed control decisions on trends in the power market. The smart grid infrastructure is also extremely efficient in detecting outages and interruption of power lines. Between the late 1990s and 2010, there have been several new inventions and innovations in the power industry and communication technology enabling the smart grid to perform the aforementioned tasks with high efficiency.

Advances in communication technology triggered the smart grid era, as scientists and engineers began to realize the potential of two-way communication between power system components to improve multiple aspects of the infrastructure, such as stability and reliability. Industrial acceptance of the benefits assured by two-way communication led to the development of high-speed, interactive infrastructure for gathering and analyzing power grid data for real-time control and operations. Using pricing signals, and through the implementation of time-of-use tariff in regions with typically high peak demands, congestion in transmission lines could be reduced, along with superior power systems protection capabilities, and improved power quality and voltage profile at consumer ends.

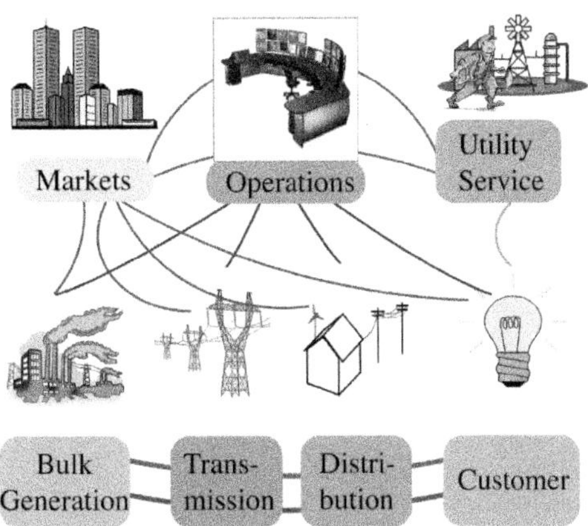


**FIGURE 8-2** Smart grid architecture and framework [3].

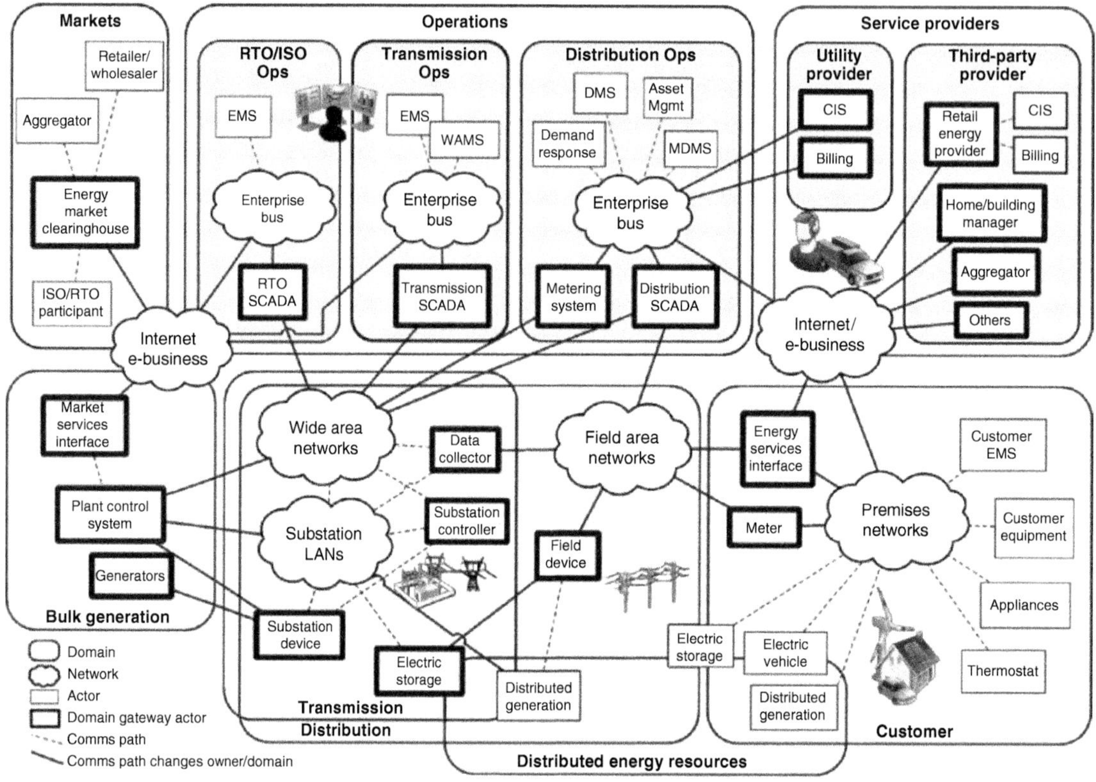


**FIGURE 8-3** Interaction between multiple domains of the smart grid [3].

From initial deployment projects and cost-benefit analyses, it was understood that smart grid can be enabled only if the communication network is characterized by:

(a) High bandwidth to transfer large amounts of telemetry data.
(b) Standard IPs, such as IPv4 and IPv6 (preferred).
(c) Encrypted communication for cyber-security.

The communications technologies that can be engineered to meet the above mentioned requisites can be used to establish a nation-wide, end-to-end, fully integrated smart grid. Examples of communication technologies are multiprotocol label switching (MPLS), worldwide interoperability for microwave access (WiMax), broadband over power lines (BPL), and wireless fidelity (WiFi). MLPS is commonly used for data transmission between nodes using high-performance telecommunications capabilities. WiMax is used over the Internet for point to multi-point data transmission. BPL facilitated powerline communication over the Internet, while WiFi is mostly used for setting up wireless local area networks for smart devices. Additional technologies include optical fiber, mesh, and multi-point spread spectrum.

### 8.2.3 Implementation of Smart Grid

Overhaul of legacy systems, standards and protocols forms a significant part of the plan for grid modernization using smart grid concepts. New technologies of energy generation are being integrated with the power grid, rapid growth in the use of power electronic and electronic based interfaces for connecting and controlling, and reduced inertia in the grid due to diminishing fraction of power generated using rotating machines—raises concerns for safe and stable implementation of the smart grid. In addition, improving the physical network with robust material and new equipment, the vulnerable cyber-security of the communication channels that overlays the physical infrastructure needs to be protected using diverse, ever-evolving means to keep it ahead of malicious hackers.

Thus, in order to migrate a critical infrastructure such as the power grid to a new cyber-physical architectural framework, there are needs of appropriate standards and protocols to piece it all together, so that components can operate at peak performance without conflict.

***Interoperability in Smart Grids.*** Interoperability among several components of the smart grid depends upon the open architecture of technologies and the underlying software systems. This facilitates interaction among various devices with other systems and technologies [25]. In 2004, recognizing the importance of interoperability of mutiple components in a smart grid, the GridWise Architecture Council (GWAC) published interoperability concepts to facilitate the integration of diverse smart grid technologies. In 2007, Energy Independence and Security Act tasked the National Institute of Standards and Technology (NIST) to design a robust framework that will include various protocols, models and standards to bring greater levels of interoperability of smart grid devices and systems. In 2009, NIST launched the Smart Grid Interoperability Panel (SGIP) to involve experts and encourage community-wide discussions leading to contributions to NIST's interoperability objectives. As a deliverable of that task, "NIST Framework and Roadmap for Smart Grid Interoperability Standards" was made available to the public in 2010. Since that publication, it has undergone several revisions, and more revisions are likely. The document presents a well-rounded, solid foundation for a secure, interoperable smart grid that can be adopted by many utilities worldwide to bootstrap their grid modernization efforts.

Interoperability in smart grid will allow a network to seamlessly and autonomously integrate diverse components—such as power generation resources, monitoring devices, protection equipment, transmission, distribution, and substation equipment, management and communication equipment, renewable, and emergency power resources to work together—with minimal human intervention and downtime due to incompatibility.

According to [1], simply adhering to or "merely having complete compliance to applicable smart grid standards" is not enough to enable interoperability among a large number of smart grid devices. In other words, interoperability ensures usage and interpretation of common semantics across the entire network.

The IEEE (Institute of Electrical and Electronic Engineers) American National Standards project (P2030) is a key resource for benchmarking smart grid interoperability, in conjunction with the IEEE 1547 series of standards addressing distributed resources interconnection with the grid. These standards have an emphasis on the "system of systems" approach toward developing and deploying end-to-end smart grids. The International Electrotechnical Commission (IEC) projects have internal project-level standards to meet particular objectives of specific smart grid projects. The smart grids task force was set up by the European Commission in 2009 to advise on issues related to smart grid deployment and development. It consists of five expert groups who focus on specific areas.

Expert Group 1 aims to explore smart grid services and operation, and how best to deliver smart grids for the benefit of the energy system and its users. The focus of attention for this expert group is the technical standards and provisions designed to allow the interoperability of systems and technologies within a smart grid environment. Based on work of this expert group, the European Commission has issued mandates to European Standardisation Organisations (ESOs)—CEN, CENELEC, and ETSI—to develop and update technical standards:

1. Mandate M/490 for smart grids (March 2011)
2. Mandate M/468 for electric vehicles (June 2010)
3. Mandate M/441 for smart meters (March 2009)

**TABLE 8-2** Summary of Commonly Used Standards Used to Implement Smart Grids

| Standard body | Standard | Applications to smart grid |
|---|---|---|
| International Electrotechnical Commission (IEC) | IEC 61850 | Substation automation, distributed generation, SCADA communications, distribution automation<br>Distribution management and AMI back office interface |
| | IEC 61968 | Advanced Metering Infrastructure communications |
| | IEC TC 13, 57 | |
| Institute of Electrical and Electronic Engineers (IEEE) | IEEE 802.3 | Ethernet |
| | IEEE 802.11 | WiFi WiMax |
| | IEEE 802.16 | Zigbee |
| | IEEE 802.15.4 | |
| Internet Engineering Task Force (IETF) | RFC 791 | Internet Protocol (IP) |
| | RFC 793 | Transport Control Protocol (TCP) |
| | RFC 1945 | HyperText Transfer Protocol (HTTP) |
| | RFC 2571 | Simple Network Management Protocol (SNMP) |
| | RFC 3820 | Internet X509 Public Key Infrastructure (PKI) Security |
| American National Standards Institute (ANSI) | ANSI C12.19 | Metering tables internal to the meter |
| | ANSI C12.22 | Communications for metering table |
| National Institute of Standards and Technology (NIST) | SP-800.53 | Recommended security controls for federal information systems (USA) |
| | SP-800.82 | Guide to SCADA security |
| North American Reliability Corporation (NERC) | CIP 002-09 | Bulk power standards with regards to critical cyber-assets Identification, security management controls, incident reporting, response planning and recovery plans for critical assets |

***Smart Grid Standards.*** Standards are the specifications that have been established after many years of research and operational experience to ensure smooth functioning of a smart grid. It ensures interoperability among the components of the smart grid, safety, stability, and reliability of the overall power systems infrastructure. Many standards development committees and bodies, including the NIST, IEC, IEEE, IETF, ANSI, NERC, and the World Wide Web Consortium (W3C) are addressing interoperability issues for a broad range of problems concerning the power grid. Table 8-2 gives a summary of the globally recognized standards of the smart grid that is shaping the modernization of the power system.

Since the smart grid is under rapid development, standards need to be developed and globalized at an unprecedented fast pace than the traditional time taken to develop, accept, and implement standards worldwide. NIST has adopted a robust framework for the longer term evolution of the standards and establishment of testing and certification procedures.

***Smart Grid Cyber Security.*** Communication-driven grid modernization has ushered in a lot of positive changes in the power grid, but not without vulnerabilities. Table 8-3 shows the differences between vulnerabilities in a legacy power systems and the smart grid.

**TABLE 8-3** Comparison of Threats Faced by Legacy Power Systems and Smart Grids

| | Threats faced by legacy system | Threats faced by smart grid |
|---|---|---|
| Impact | Direct physical damage | Indirect damage to resources, physical assets, by intrusion through software systems. |
| Location of attacks | Local | Local, remote, coordinated across multiple geographically distributed sites. |
| Occurrene | Infrequent (usually single episodes) | Attacks can happen multiple times if the system vulnerabilities are not completely removed, or the hackers continue to discover new methods to intrude into the system using software systems. |
| Duration of attack | Immediate damage | Attacks may be designed to be surreptitious with long incubation periods, being dormant and undetectable for extended periods before damaging the system. |
| Restoration | Restoration after attack was easier to implement | Attacker may continue to prevent restoration after gaining access to the system. |

The importance of two-way communication between interconnected power systems components in a smart grid cannot be overstated. As a result, smart grid cyber security gains equal, if not greater importance in modernizing the power grid. Since the modern end-to-end power grid is dependent heavily on internet-based communication, the entire network is exposed to a larger number of vulnerabilities from malicious hackers. The intent of a cyber-attack may range from being a playful ruse of a skilled cyber rogue to targeted terrorism intended to cripple national security or trigger financial loss.

Today the electric power system does not have adequate measures to guarantee protection against malicious cyber-attacks, which makes it highly vulnerable. Various incidents and security concerns in the past have indicated the extent to which the nations power grid is vulnerable and the urgent need to protect it against cyber-attacks. Advanced threats, such as "Stuxnet," present highly sophisticated, stealthy, and evolving attacks that encompass both traditional IT and control system environments to target physical systems [16].

Recent North American Electric Reliability Corporation (NERC) directives make it mandatory for utilities to perform cyber-security risk assessment to take preventive and corrective measures. In addition, the adoption of smart grid technologies will significantly increases the attack surface, which further underscores the importance of cyber security and its demands for more substantial information and communication dependencies. The U.S. DOE has documented attack resiliency as a primary requirement for the next-generation grid [17]. Intelligent cyber-attacks can significantly affect a power systems security and adequacy by negating the effect of system redundancy and other existing defense mechanisms. The National Institute for Standards and Technology (NIST) report NISTIR 7628 has thoroughly enumerated cyber security requirements for the smart grid.

The NERC report [18] identifies State Estimators, Contingency Analysis, Wide-Area Monitoring and Control Systems, Special Protection Systems, and PMUs as some of the critical components which impact reliability of the bulk power system. NIST's Guidelines for smart grid cyber security report identifies a need for intrusion detection methods tailored to the specialized smart grid devices and systems and the contextualized understanding of their usage. In addition, DOEs 2011 report, Roadmap to Achieve Energy Delivery Systems Cybersecurity identifies nearterm requirements for intrusion detection and response capabilities [19]. The NERC cyber-attack task force (CATF) has introduced strategies to provide attack tolerance within the grid [20].

For effective prevention, detection, and response of cyber-attacks, the cyber-physical characteristics of the electric grid should be taken into consideration from both the network infrastructure and power applications perspective. Thus a multilayered defense architecture (defense-in-depth), as shown in Fig. 8-4, with representative applications in wide-area monitoring, protection, and control will provide a holistic solution to enhance cyber-security of smart grids.

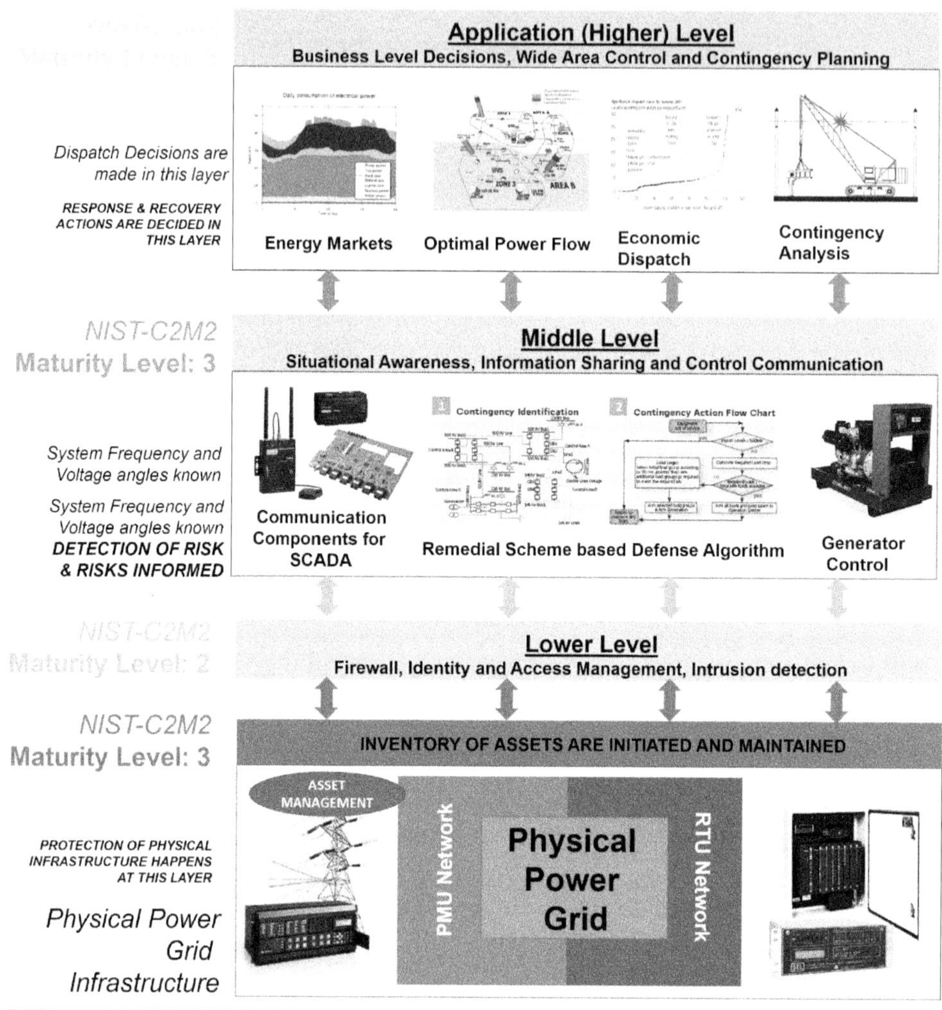


**FIGURE 8-4** Multi-layer defense architecture considering cyber-physical characteristics for WAMPAC.

Cyber-attacks can be used to steal customer data, override control decisions implemented by a utility or an operator, perpetrate a denial of service attack (DoS) to components that a community would depend on during contingent situations. Due to the cyber-attack, the power grid is subject to operational failures and loss of synchronization. Cyber-attacks will also lead to system-wide operational failures, resulting into damage of critical power system components. Such events will invariably interrupt the power supply and cause cascading disruptions in other critical infrastructures that depend on the power grid, as well as severe customer inconveniences and financial losses.

***How to Implement Cyber-Security.*** Cyber-security implementation is still evolving and proactively implemented security protocols to defend the power system against unauthorized software activity, injected into the communication channel by computer or computer terminals and the protection of other physical assets from modification or damage from accidental or malicious misuse of computer-based control facilities. A cyber-security coordination task force has been established [21] to oversee the

**TABLE 8-4** Time Latency for Getting Different Classes of Data

| Time requirements | Data availability for the specific applications |
|---|---|
| Less than 4 ms | Protective relaying |
| Sub seconds | Transmission wide-area situational awareness monitoring |
| Seconds | Substation and feeder SCADA data |
| Minutes | Monitoring noncritical equipment and some market pricing information |
| Hours | Consumer energy meter reading and longer-term market pricing information |
| Days/Weeks/Months | Collecting long-term data such as power quality information |

standardization of processes for designing grid security at the architectural level. As apparent from research and experiences worldwide, the security requirements of a Smart Grid is different from other critical infrastructures.

In order to secure communications in the smart grid, it is important to objectively classify the problems into one of the following three types (1) data security, (2) data integrity, and (3) data privacy.

One of the most popular ways to monitor system attack is by observing the time taken for data communication, and bandwidth used for the communication. If sustained anomalies in data latency and bandwidth are observed, a compromised power system is a possibility and raises a "data security" concern. The usual time latency of the availability of data from different domains of the smart grid is shown in Table 8-4.

Other approaches to ensure cyber security in smart grids are to ensure data integrity is maintained. It refers to the ability of an operator to authenticate or validate the source and quality of data. The modification or destruction of original data leads to loss of data integration. Intrusion by a cyber attacker or due to some undetected component malfunction will compromise data integrity. Smart grid cybersecurity must also ensure operators, utilities, and users have fair rights to privacy of their data, and whenever data is exchanged, the transaction must happen only through secured channels.

Commonly adopted approaches deployed by utilities to ensure cyber-security in smart grid are:

- Continuous monitoring of the system's functionality and data flows to proactively detect malicious behavior and system anomalies.
- Data integrity must be upheld through encryption, hardening of existing firewalls and restricting access of personnel to mission-critical computers, computing networks or facilities.
- Use of backups to restore power grid operating channels to last known secure state prior to an attack, and then upgrading the security.
- Intrusion detection incident response.
- Staying up-to-date on security standards and upgrading security patches and firmware on products as soon as they are are released into the public domain.
- Hire ethical hackers (known as white-hat hackers in certain countries) to perform penetration testing ("pen-testers") to analyze the cyber-security of the power grid.

### 8.2.4 Cyber-Physical Interdependencies

With increasing integration of Information and Communication Technologies (ICT), the smart grids could achieve efficient and reliable grid operation and control. The physical power system and the communication networks are more dependent on each other, and forms the so-called "cyber-physical system (CPS)." While achieving the benefits, the inter-dependencies might intensify the risk of failure and have adverse effects on the resilience of the power system [22, 23]. The failures in the communication network contribute factors to degrade the stability of power systems that eventually causes blackouts. As a result of the interdependency, failures occurring to nodes/links in communication networks can cause failures to nodes or functionality in the power grid. Longer outage can cause failure of servers and components of the communication network. Thus, small fraction of failed nodes in one network can lead to a cascading process of failures between other networks, resulting into widespread damage to the entire system. Power grid black-outs are examples of consequences of cascading failures of power system nodes. Blackout events, such as the Italy 2003 blackout, can be traced back to a singular point of failure in a power station, which triggered a series of failures between power stations and communication servers that depended on the power grid [24].

Quantitative evaluation of cyber-physical inter-dependencies is of significance for understanding the reliability and resilience of power systems as CPS, and could provide guidance on the systems operation, control, and planning. The existing methodologies to study cyber-physical inter-dependencies in the literature can be divided into two categories: analytical-based methods and co-simulation-based methods.

1. The analytical-based methods apply graph-based models to represent both systems and their interactions. Reference [26] is the first work applying this method to analyze the interdependencies of CPS, where both the power system and communication system are modeled as a graph, and the interdependency is modeled as the links between the related nodes in each network. Although the work derived some analytical conclusion on the interdependency features, the drawback is that the model is oversimplified and lacks capturing characteristics of each individual system. The following work [25] add DC power flow model in the power system network model, but the communication network model is still simple graph without including dynamic features. The work in [26] abstracted the communication network using information flow to capture the unique features of the control-system communications. The existing approaches cannot provide a general analytical framework to quantify the interdependency features of CPS without losing key features of each individual system.
2. Co-simulation-based methods aim to develop integrated models of both power systems and communication networks and study the inter-dependencies using different scenarios. There are several existing simulation software, either open-source or commercial, to simulate power systems and communication systems individually. However, the co-simulation development to connect both systems is a non-trivial task. The most significant part of the challenge lies in setting up the connection, transmitting, receiving, and synchronizing the data between the connected simulators using their respective software interfaces. Time synchronization of the discrete and disparate events are crucial and challenging, mainly because each simulator have different time settings, which are not synchronized by any clock. Continuous time modeling in power system simulations where the power system state variables are time-dependent functions by nature of the system. The communication networks are modeled by using software-defined packet switching networks (i.e., IP-based technologies). Cyber events in the power grid are modeled as discrete event systems characterized by events such as sending and receiving of packets, latency, etc. The advantage of the co-simulation-based methods is that it can capture the complex characteristics in each individual system more accurately than the analytical-based method; thus could provide a high-fidelity results. However, the disadvantage comes from the need of a large amount of time taken to analyze number of scenarios to study the inter-dependency properties. For different systems the scenarios and the corresponding results may differ; thus the studies should be done case-by-case and are not easy to apply in a general case.

## 8.3 SMART GRID INFRASTRUCTURE

### 8.3.1 Smart Meter Infrastructure

A smart meter is an electronic device that records consumption of electric energy typically every 15 minutes and sends the information to the utility for monitoring and billing. The U.S. DOE

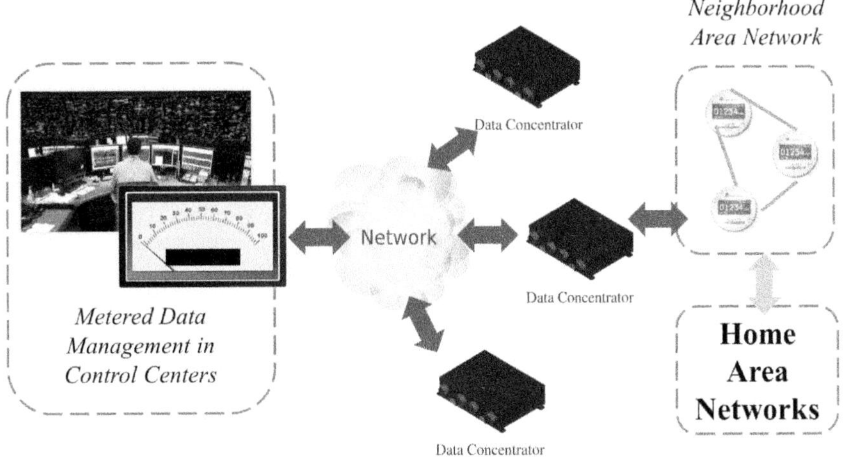


**FIGURE 8-5** A typical AMI system architecture.

defines AMI as "an integrated system of smart meters, communications networks, and data management systems that enables two-way communication between utilities and customers". AMI is also known as Smart Meter Infrastructures. Customer systems that are considered part of the AMI are in-home displays, home energy management systems (HEMS), and other customer-facing technologies used to communicate information from utility to the consumer. Combination and coordination of these resources enable smart grid technologies in residences, commercial buildings, and industrial facilities.

The most important role AMI plays is to retrieve customer usage data for billing automation. Billing applications of AMI do not pose stringent timing requirements on the AMI system, as the data is typically collected at 15 minutes intervals, though sometimes it is more or less frequent, and the bill is sent out once a month or more.

Figure 8-5 illustrates a typical AMI architecture in smart grid. Within the residential home, the smart meter is installed to monitor and record the energy usage of appliances. Some smart appliances may provide operational flexibility (e.g., clothes washers/dyers, dish washers, electric vehicles, etc.) so that the demand-side management can be achieved via the home area network (HAN) and HEMS, which is connected to the smart meter. The data of the smart meters of several homes (i.e., in a community) will be transmitted to data concentrator via neighborhood area network (NAN), which can be implemented using either power line communications or wireless mesh network. The collected data will then be sent to meter data management system (MAMS) in control center via wide area network (WAN). This hierarchical structure of the AMI provides a scalable solution for implementing large amount of smart meters.

Besides the billing application, smart meters also provide wide range of applications with the potential to benefit the utilities. For example, through two-way communication capabilities, the customers can receive the real-time electricity price signals, and accordingly they can alter/shift the power usage to save the energy bills while the utilities can reduce peak load. Smart meters also have the potential to facilitate the outage management and restoration. The smart meters are able to report the outage information to the utilities' outage management system (OMS), so the outage areas can be identified in a faster way than the traditional customers' trouble call systems. Smart meters can also push real-time notifications to operators when power is restored, or even allow utilities to make "ping" requests to meters in the regions affected by the event to estimate the extent to which the grid is affected. It can also be used to verify the restoration progress, enabling field crews to be deployed more efficiently, thus reducing the restoration time. Other functionalities of smart meter infrastructure include improved energy diagnostics from more detailed load profiles, losses and theft detection, and ability for a retail energy service provider to manage its revenues through more effective financial management. All of these applications can serve to improve the efficiency of the grid or reduce costs and improve the customer experience.

## 8.3.2 Synchrophasor Infrastructure

Synchrophasors are time-synchronized measurements of both the magnitude and phase angle of the sine waves representing the voltage or current in the electric power grid. Synchrophasor technologies and systems use monitoring devices called PMUs. PMUs can measure the instantaneous voltage, current, and frequency at specific locations in the power grid. The voltage and current measurements are used to characterize the delivery of electric power from generation plants to end-user loads, while frequency helps to monitor the demand-supply ratio in the power grid. The measurement is then time-stamped, based on the Global Positioning System (GPS) clock. This ensures that all measurements taken by PMUs in different locations and collected by other transmission operators are collated accurately. In this way, PMUs can help to reconstruct a very complete picture of transmission system operations across the power grid.

Synchrophasor infrastructure includes PMUs, phasor data concentrators (PDCs), and the underlying communication networks. PMUs are used to gather and transmit time-stamped measurements of the power system, typically about 200 times faster than traditional SCADA. The PMUs comprise bus voltage phasors and branch current phasors, along with geo-spatial information and other user-defined bits. They are equipped with sensitive sensing devices that can observe the state of the system with high accuracy at several locations at the same time, and then synchronize the obtained information to one instant of time. The high accuracy is ensured by taking voltage and current analog sampling up to 6 to 12 times in a cycle, in conjunction with GPS clock data for reporting, and compute phasors and frequency about 15 to 60 times per second. This helps in advanced monitoring of the power system, including dynamics. If anomalies are observed, appropriate actions can be taken while there is still room to recuperate without interrupting the continuity of power supply to any load. Usage of PMUs, invented in 1988, has proliferated to about 1700 installations in North America.[b] Several modern, microprocessor based relays and disturbance fault recorders are bundled together with PMUs, which help improve wide area monitoring. PMUs greatest role is to aid in power system automation and form an integral part of grid modernization efforts. PMUs and their various applications for enabling smart grid are discussed in greater depth in Sec. 8.3.

PMUs transmits the data gathered to the PDCs, which are specially designed servers with appropriate software parts to receive data streams from many PMUs and other PDCs, time-align synchrophasor data from a large number of sources, and forward the data to control centers or next-level PDCs (e.g., superPDCs). PDCs also perform data-quality checks, monitor the performance of the PMUs, and feed a data archive. Increasingly, PDC functionality can be located within the grid at transmission substations, aggregating local PMU data and feeding it to local applications and actions, as well as passing the data upstream to multiple applications and operations centers. The underlying communication systems for PMU network include a WAN and gateways that provide access to the network [27]. Figure 8-6 shows a typical synchrophasor infrastructure, which has a hierarchical architecture. Wide area monitoring systems (WAMS) are designed and deployed by the utilities to operate the transmission grid at its optimal capacity, by using synchrophasor measurements from multiple locations and leveraging that information to create a demand-supply balance in the grid.

Synchrophasor data specifications are defined in IEEE Standard C37.118-2011 [28], including data reporting rate, data quality, data frame format, etc. Synchrophasor technology is used for online (near real-time) operations. Data obtained from synchrophasors are also used for deeper off-line analytics to gain insights about operation of the grid, and then use them to improve grid reliability, stability and efficiency, and lowering operational and maintenance costs. Online applications include wide-area monitoring and visualization, oscillation detection, frequency stability monitoring, voltage stability monitoring, disturbance detection, state estimation, islanding, restoration, etc. Offline applications include post-event analysis and power system model validation.

WAMS technologies are under active developments and can prevent the spread of disturbances across the power grid, as it is capable of providing real-time information on stability and operating safety margins for the power grid. Using its network of distributed sensors, time-synchronized data and GPS, WAMS can detect disturbances and generate early warnings to operators, so that they have

[b]*Source*: NASPI Synchrophasor Fact Sheet 2014.

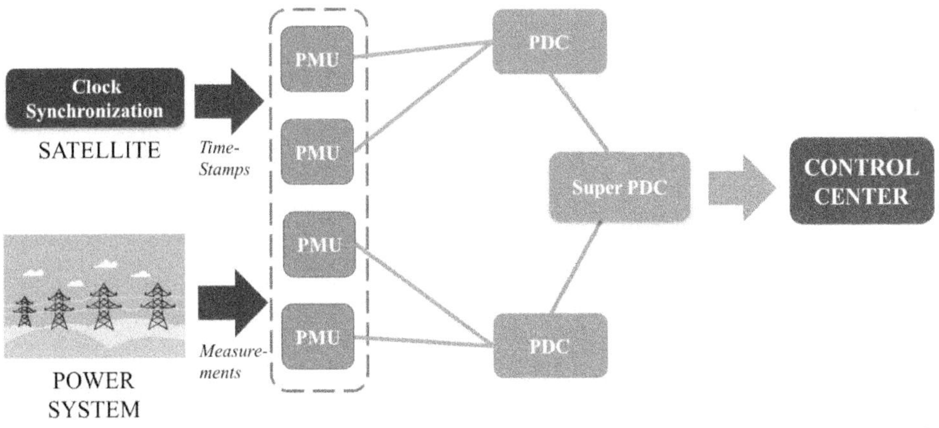


**FIGURE 8-6** Illustration of a typical synchrophasor infrastructure.

ample time to act in a manner to mitigate the possibility of system-wide blackouts. The other benefits of the expanding synchrophasor infrastructure are as follows:

1. PMU technology will help refine the engineering models that simulate individual power plants and operational behavior of large power system interconnections. Better models will translate into more accurate identification of system operating points and thresholds.
2. PMU technology can prevent potential power plant damages, and even help in stopping cascading failures by enabling the operators to run the power system closer to operational limits by analysing real-time data and comparing them to historical outage scenario data. As an example, let us consider the use of synchrophasor data by Bonneville Power Administration (BPA) in the western United States. BPA own and maintains a 4.8 GW inter-tie to facilitate power exchange between the Pacific Northwest and California. Without synchrophasor data, BPA frequently operates below capacity to be conservative about the safety and security of the tie-line. PMU-technology deployments and simulation helped in predicting that energy flows in the inter-tie can be increased by 100 MW or more using synchrophasors to take real-time control actions. This automation will translate into approximately monetary gains worth \$35 million to \$75 million over 40 years without any new high-voltage capital investments.

In another example, BPA used archived synchrophasor data on the actual performance of the 1100 MW Columbia Nuclear Generating Station to validate and calibrate the plants dynamic model. This helped in avoiding to take the plant offline for manual tests every 5 years for reliability inspection. Thus, the online data analysis method resulted into savings worth \$700,000 for Energy Northwest, the owners of the plant. Apart from the monetary savings, data analytics helped in improving the plants model used for behavior prediction.

### 8.3.3 Distribution Automation Infrastructure

Unlike electric transmission systems, automation of distribution systems are recent trends in the power system industry. Infrequent manual set-point changes were deemed to be sufficient for management and operation of the distribution system. However, due to improvements in power distribution system automation, capacitor bank switches can be controlled to turn on or off based on local signals, such as load demand, reactive power support requirement, time of day or current. Automatic reclosers deployed as a part of distribution automation (DA) attempt reclosing a set number of times before locking out after a local fault has been detected. Lateral fuses would blow if the current became too high.

DA aims to extend the intelligent control over electrical power grid functions to the distribution level and beyond. The implementation of DA will facilitates utilities to improve the operation of distribution grid effectively and efficiently via distribution management system (DMS), and better handle the challenges of increasing penetration of DERs and demand response.

To implement DA functionalities properly, it is important to interconnect control systems, communication networks, and devices installed in the field. For example, to include fault location, isolation, and service restoration (FLISR) capabilities within the DMS, an underlying communication network must exist between field devices such as smart relays and automated feeder switches. The network should be coordinated such that the system reliability and resilience can be improved. Another example of voltage/volt-ampere reactive (VAR) control includes field devices such as automated capacitors, voltage regulators, and voltage sensors which are managed by DMS via communication networks.

Evidently, power system communication networks play a vital role in the DA development. Communication networks help in the process of acquiring large amount of data from sensors, filter the data, and generate and propagate appropriate control signals to operate equipment. To better manage these geographically dispersed field devices, multi-layered communication systems are usually applied between information and control systems and field devices. For two-layer communications system case, the first layer of the network connects substations and the DMS at control center, and consists of high-speed, fiberoptic or microwave communications systems. As of 2017, many utilities prefer to use existing supervisory control and data acquisition (SCADA) communications systems for this layer due to legacy and reliability reasons. The second layer of the network is used to connect substations with multiple field devices and employs commonly used protocols for wireless networks or power line carrier communications [29].

Information and control systems collect the measurements from field sensor devices and provide strategies for the field devices to actuate to achieve DA functionalities. Generally, there are two types of approaches: centralized and decentralized. Centralized schemes involve coordination among field devices and centralized control systems. Decentralized approaches use local control mechanism to operate automated equipment. In most cases, a combination of centralized and decentralized approaches, that is, hierarchical control mechanism, are employed to achieve better trade-off between response time requirements and controllability.

## 8.4 CONTROL AND OPERATION OF SMART GRIDS

This subsection presents the common concepts associated with ensuring smooth control and operations of the smart grid.

### 8.4.1 Demand Response

According to the Federal Energy Regulatory Commission (FERC), the definition of demand response (DR) is "Changes in electric usage by demand-side resources from their normal consumption patterns in response to changes in the price of electricity over time, or to incentive payments designed to induce lower electricity use at times of high wholesale market prices or when system reliability is jeopardized".

"Changes in electric usage" mentioned in the definition above refer to customer participation in DR. DR schemes will work most effectively with loads that can have significant operational flexibility. Thus appliances that are suitable for DR can be used to "shift" electricity usage in time and bring benefits for both the grid and the consumers connected to the grid. Customers, if adequately educated and informed by utility and provided with financial incentive, will opt to use smart dish washer or smart laundry systems whose start time can be modified to serve the goals of the utility. Proliferation in the use of electric vehicles is being hailed by experts as the most important emerging load type that has the potential to offer flexibility in the energy demand. A large battery-load can be considered a valuable asset for demand response because individual battery charging can be controlled,

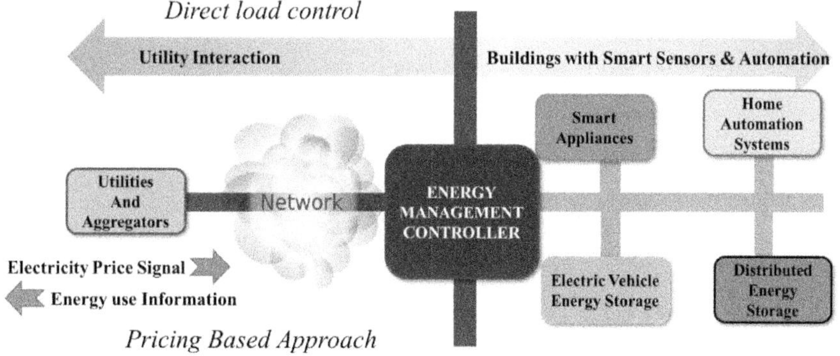


**FIGURE 8-7** Two types of demand response techniques [13].

coordinated, and effectively optimized, by aggregating multiple batteries across the whole network. From the point of view of utilities, peak shaving will bring in the following benefits:

- Reduction in the cost of drawing power from efficient generation during periods of peak demand.
- Reduce congestion in the transmission grid.
- Improve grid stability.

From the point of view of the customer, DR has the most important advantage of helping them reduce their spending on energy consumption. Despite these advantages, utilities still need to design definitive, scalable, and robust means of implementing DR schemes and properly informing customers about their incentives to participate in DR programs. Transactive signals, discussed later in the section, provide a framework for implementing DR schemes at the utility level.

Current techniques on DR can be divided into two categories: direct load control approach and pricing-based approach [30], as shown in Fig. 8-7.

1. *Direct Load Control (DLC).* Existing DLC programs offered by most utilities in the United States are contract-based. Customers who want to participate in the DR program must formally agree to giving the utilities the option to remotely shut down appliances during peak periods or during unforeseen power supply emergencies. In exchange, the customers receive credit on energy consumption bills for their participation. "Contracted Direct Load Control" by Wisconsin Public Service and "Distribution Load Relief Program" by ComEdison are examples of DLC implemented in the United States.
2. *Pricing-Based Approaches.* In this approach, the electric utility controls customers' appliances indirectly by sending pricing signals. Specifically, variations in wholesale electricity prices are integrated to retail electricity prices and thus eventually affect customers' cost of energy consumption. Due to variations in time of use energy tariff, customers may be motivated to shift usage of some high-energy appliances from high-price periods to low-price periods. This type of pricing-based approach have been implemented by utilities like Ameren Illinois and ComEdison in the United States.

Based on the level of aggregation, DR implementation and participation in the layered structure of the electricity market is divided into three categories [31] as shown in Fig. 8-8.

1. *Consumer Premise Level.* The DR implemented in this level minimizes the cost of energy consumption by scheduling operating times of several large-load appliances within a home. This is coordinated by the energy management controller (EMC) that can communicate with appliances as well as the utility.

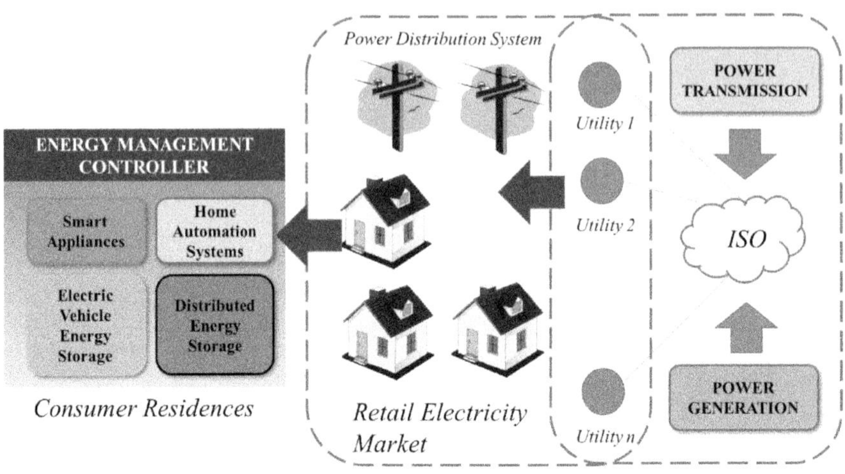


**FIGURE 8-8** Three levels of DR aggregation in electricity market [31].

2. *Retail Market Level.* The EMC's interaction with appliance-level and the utilities can influence the utilities to design incentives to encourage the customers modify their power usage profile, either directly or indirectly (as discussed above). This will help to decrease peak demand and improve stability and reliability of the grid during periods of high demand.
3. *Wholesale Market Level.* Large-scale aggregation of DR resources can participate in and impact the wholesale electricity market. Depending on the Independent System Operators (ISOs) market design, DR may provide energy, reserve ancillary service (A/S), or capacity in the wholesale electricity market [32]. Currently, several trials of DR programs have been implemented by some ISOs in the United States, e.g., New York ISO (NYISO), PJM [33] and ISO New England (ISO-NE). At the same time, stimulus of DR in the wholesale electricity market is also provided by regulatory institutions like FERC. For example, FERC Order No. 719 [34] and No. 745 [35] specify how ISOs can permit Demand Response Providers (DRPs) to bid DR on behalf of retail customers directly into the ISOs organized markets, and get compensated for the service at the locational marginal price (LMP).

## 8.4.2 Distributed Control

Smart grids proactively employ state-of-the-art technologies in communications, computing, and control to improve the efficiency, reliability, sustainability, and stability of the electric power grid. The trend of high penetration of DERs integration and the increasing deployment of large numbers of sensors and controllable devices as a result of DA pose great challenges on the corresponding control mechanisms. Since centralized control schemes usually suffer from computation scalability, communication, and robustness issues, distributed control may provide a viable solution to address these challenges. For example, distributed control usually involves local communications and computation so that the scheme is of low communication infrastructure costs and low computation burden compared to the centralized method. In addition, distributed control does not rely on a central controller so the single point of failure can be avoided. Examples of distributed control techniques include:

1. *Consensus-Based Techniques.* In networks of agents, consensus means to reach an agreement regarding a certain quantity of interest that depends on the state of all agents. A consensus-based algorithm is an interaction rule that specifies the information exchange between an agent and all of its neighbors on the network [36]. Graph Laplacian and their spectral properties are important

graph-related metrics that play a crucial role in convergence analysis of consensus algorithm, as the network topology impacts the convergence and the design of optimal consensus algorithm. By carefully designing the agent states, the global distributed control objective can be achieved by just using local interaction among neighboring agents. Examples of using consensus-based algorithms for power system applications in a distributed manner include economic dispatch problem [30], demand response [30], and distribution system restoration [37].

2. *Decomposition-Based Techniques.* Decomposition-based techniques aim to decompose the original optimization problem into a number of sub-problems that were solved iteratively until convergence. This type of approach usually exploits the separable structure of the optimization problem. Several decomposition methods can be applied to optimization problems, e.g., dual decomposition, augmented Lagrangians, and alternating direction method of multipliers (ADMM). Decomposition-based techniques have been applied in several power system applications in a distributed manner, e.g., reactive power control [38], optimal power flow [39], state estimation [40].

### 8.4.3 Distributed Energy Resources

DERs are energy sources connected to a distribution grid in a dispersed manner, such as PV, wind turbines, small hydro, combined heat and power (CHP), and fuel cell, as well as distributed energy storage units, e.g., batteries, flywheels, and ultra-capacitors. While DERs can be defined based on their energy sources and technologies, from an electrical perspective, DERs can be categorized into two classes: (1) inverter-based generation, e.g., PV, and (2) rotational machine-based generation, e.g., CHP. The energy storage technologies, regardless of whether they are inverter-based, like battery storage, or rational machine-based, like pump storage, have to be modelled differently because they can operate in either the generation mode or load mode.

High penetration of DERs poses a series of challenges to distribution grids. The traditional distribution grid is passive, and radially configured with one-way power flow. Due to the connections of DERs along the distribution feeder circuits, distribution networks will become active and involve two-way power flow. The network topology may or may not be radial, depending on how a utility manages and configures its feeders. Therefore, the voltage profiles and power flow directions in the individual feeder circuits will not be obviously observed from the topology and need to be dynamically determined from real-time measurements and/or load-flow or state-estimation calculations from time to time. The static and dynamic operational characteristics of individual DERs, as well as their connection/disconnection to/from the grid, will have direct impacts on operational reliability and power delivery quality. As a result, most of applications in DMS need adjustment to accommodate the changes from the integration of DERs.

With high-level penetration of DER in a distribution grid, it may be necessary to have a Distributed Energy Resources Management System (DERMS) to aggregate, control, and manage the operation of the DER dispersed widely in the network. Its key objectives and functionalities mainly consist of effectively organizing, managing, optimizing, and controlling DERs for maximum economic benefits, enhanced grid operation reliability and service quality, including how the DERs are aggregated or grouped, e.g., at the substation level, feeder level, or even at the feeder section level, as well as in other ways, such as by generation types, capacities, response rates or other characteristics. The optimal schedules may be allocated to the individual DER through disaggregation for actual execution. A DERMS may directly monitor and control the individual DER or may coordinate through shared SCADA, AMI, field area network (FAN), or other communication-capable applications. A DERMS solution illustration (by OATI [41]) is shown in Fig. 8-9.

### 8.4.4 Information Technology and Data Management

With the development of the smart grid, a large number of sensing and measurement devices are being deployed in the grids, which generate large volume and different types of data. While these data have the potential to improve the grid operation efficiency and reliability, utilities are facing significant

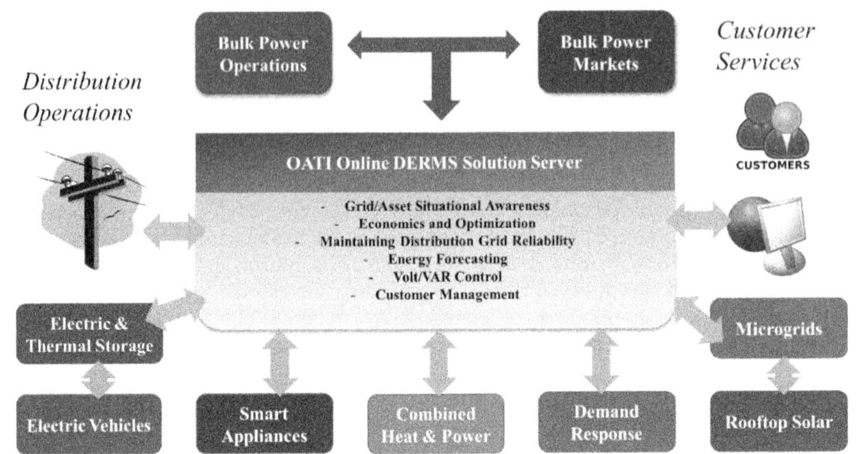


**FIGURE 8-9** Illustration of a DERMS solution [35].

challenges on how to manage these data effectively and efficiently such that the information behind the data can be better retrieved and utilized. These challenges include:

- Data storage costs can explode due to increased data volumes from large numbers of smart grid devices, e.g., smart meters, PMUs, field measurement devices in DA, asset monitoring sensors, etc. The traditional, relational database technologies are not suitable for ever increasing data volume challenge [42].
- Time requirements for the data are diverse due to different types of power system applications. For example, for the power grid monitoring and disturbance detection, it involves large volumes of high-velocity data, and the analysis of the data should be completed in a strict deadline; while for post-event analysis, or billing applications, the time requirements are not stringent. In this sense, the data management solution needs to handle diverse time requirements.
- System integration is also a difficult task. Many smart grid applications are composite applications that draw on data and functions from multiple systems, thus it is challenging to design and implement architectures that allow easy data access, sharing, and collaboration between systems. In addition, most of the data generated are unstructured, so a unified data structure is necessary to facilitate the system integration.

Unlike the IT systems, the big data issues in the power systems have their own characteristics which are highly related to the features of power system operation and control. In this sense, the off-the-shelf data management solutions from IT vendors may not be tailored well to power system requirements, e.g., they may include unnecessary features and there is lack of alignment with operational processes. However, the data management solutions from operational technology (OT) vendors may not have as good analytics and IT integration capabilities as for IT vendors. Thus, it will be beneficial for collaboration of both IT and OT to develop a better data management solution to enable smart grid development.

## 8.5 MICROGRID FUNDAMENTALS

### 8.5.1 Islanding in a Microgrid

Islanding of a microgrid is a strategically implemented control action designed to prevent propagation of failures, which might have had far-reaching consequences. Islanding is achieved by controlling the circuit breakers located in the utility point of common coupling, also known as PCC.

***Planned Islanding.*** There are three primary scenarios to transition to island a microgrid voluntarily, also known as intentional islanding. These situations are summarized as:

- Emergency islanding when a utility grid outage occurs and significant power resources (by means of renewable or diesel generators) are available within the microgrid to supply its own critical loads (or all loads).
- Planned islanding due to forecasts of severe weather or other factors
- Black start when a utility grid outage occurs and sufficient DG is not online

Intentional islanding of a microgrid is initiated by a microgrid operator, or by an MGMS, when notified about an impending emergency situation. In this scenario, the generators are brought online to bring the power exchange at the PCC to zero. The MGMS is programmed to calculate the load shed plan for fast balancing of load and generation within the microgrid. If critical loads are identified, critical loads are re-prioritized, and the system is brought to stable operating condition.

***Unplanned Islanding.*** Unplanned islanding is the disconnection of a microgrid from the main grid without any prior warning provided to the operator or the MGMS. For example, utility outages occur and PCC breaker opens, leading to unintentional islanding of the microgrid. In such scenarios, the MGMS is programmed to assign a frequency to the largest generator in the microgrid, so that it becomes "grid forming". Critical loads are picked up and the generators in the microgrid are resynchronized using volt/var management of isochronous motors if there is one available.

***Communication Infrastructure.*** Some of the common means of setting up communications within a microgrid are listed below:

1. *Dedicated Copper Wiring.* Large generators (if connected as distributed energy resource within the microgrid) use dedicated copper wiring for control. Though they tend to be very reliable connections, connecting large numbers of individual DG systems with copper systems will drive the installations costs very high.
2. *Continuous-Carrier Power Line Communications Carriers.* Though no longer common, some utilities still use power-line communication for automating their meter reading process. These connections have been replaced by modern technologies due to reliability reasons. The lack of reliability is attributed to the fact that the power line communications carrier signal is lost if the connection to the utility is lost. However, this kind of communication can be used very effectively for inverter anti-islanding control in microgrids. Other reasons why the industry has moved away from power line communications are high cost of installation, low bandwidth of the communication channel, and high energy requirement for data transmission.
3. *Ad hoc Mesh Networks.* These networks comprise small-scale, smart network topologies that can discover the presence of other devices and automatically set-up a "cooperative mesh network" [43]. They are capable of establishing a giant infrastructure with end-to-end routing links, consisting of small resilient sub-networks.
4. *Dedicated Fiber-Optic Link.* Control centers and substations under these control centers commonly use fiber-optic links for dedicated, secure communications. Such connections ensure high communication speed and reliability that facilitates real-time data exchange and communication with synchrophasors.
5. *Ethernet.* Ethernet is used to set-up communications building-level deployment of power system automation. However, they must also be connected to a wide-area-network technology for proper control. It must be noted that wide-area networks lack reliability required to support protection functions.
6. *Wireless Local Area Network (IEEE802.11).* Like the Ethernet, wireless local area network are most commonly used inside and outside of buildings to provide short distance transmission of wireless data.

7. *Wireless Metropolitan Area Networks (IEEE802.16d).* This technology is often used in cities or crowded spaces to transmit data wirelessly to another node spaced about 2 km apart, with no need for line-of-sight antenna configurations. It can also be used for up to 10 km wireless data transmission with unobstructed path.
8. *Personal area networks.* Provides short-distance (few meters) wireless communications.
9. *BACnet.* This communication protocol was developed by the American Society of Heating, Refrigerating, and Air Conditioning Engineers (ASHRAE), BACnet is internationally accepted standard for control networking. It provides one of the data link/physical layers of BACnet. BACnet is used extensively in deploying building automation technologies, including smart homes, homes with roof-top solar generation units, and homes and commercial buildings that are technically equipped for utility controlled demand response.

### 8.5.2 Black Start in Islanded Microgrids

Energizing disconnected loads when a microgrid is in an islanded state is called *black start.* Black start in microgrids are designed to energize the loads by reconnecting the loads to DGs, either seamlessly or with minimum downtime. During the phase of black start, the operating states of the loads being reconnected must be carefully considered as the capability of the microgrid must be able to withstand start-up voltage and current surges and maintain acceptable voltage profile. Thus, local, low-level load controllers (LCs) must be developed in close coordination with the MGMS. Operation of the MGMS must also be tested under dynamic operating conditions. The blackstart functionality in microgrids help:

1. Assure distribution system operation during power outages caused due to failures in the transmission system.
2. Improve power supply reliability.
3. Reliability of power to critical loads.

In order to improve the resilience of the power system and ensure that black start capabilities of the microgrid can be leveraged during all types of contingencies, several sources within the advanced microgrid must have black-start capabilities. It is a common practice to install stand-by power supply and a monitoring and control scheme in the microgrid control center.

A consensus on common standards for black start microgrid restoration is still evolving. In order to facilitate black-start with limited resources, loads are prioritized by utilities and only the critical loads are picked up first. Black start functionalities within microgrids will require continuous amendments to available standards (such as IEEE 519 for harmonics and voltage with specifications for microgrids).

### 8.5.3 Anti-Islanding and Islanding Control

With an increase in power electronics interface based renewable energy resources, such as PV systems, connected to utility systems, the risk of formation of unintentional islands also increases. Utilities have to strike the balance while trying to keep their systems secure, and not limiting customers with roof-top PV units. However, many utilities lack adequate experience and expertise with these systems, disabling operators from taking informed control decisions. In the absence of such certainty, utilities must operate carefully, often leading operators to operate with more caution than necessary, and far below safe and acceptable operation limits. Anti-islanding protection is both critical and costly. In order to observe whether a distribution system is at risk of islanding—(1) passive approaches, such as monitoring rate of change of voltage and frequency, (2) active approaches such as impedance monitoring, or (3) communication-based approaches, such as using direct transfer trip (DTT) communications with breakers, power line communications, remote terminal units, SCADA integration of inverters, etc. can be used.

Due to the expense in unintentional islanding detection mechanisms, the objective is to reduce the number of cases of unnecessary application of additional protection while giving utilities a basis on which to request additional study in cases where it is indispensable.

There are several cases in which the literature, accumulated experience, and physical reasoning suggest that unintentional islanding is so unlikely as to be considered impossible for all practical purposes. According to suggested guidelines by industry experts on microgrid islanding [44], those cases include the following:

- *The Two-Thirds Rule.* If the aggregated nameplate AC rating of all DG systems within the potential island is less than the minimum real power load within the potential island—a sustained islanded operation of a microgrid is not possible, eliminating the need of anti-islanding protection. If PV is the only type of DG in the potential island, then the value that should be used is the minimum load during daylight hours. Considering that load and PV output both rise during the morning hours, the time at which the fraction of PV output to load may realistically become meaningful is not at dawn, but rather closer to 10 AM, at which point feeder load is likely to be above the required minimums. In the case in which the net DG rating is below the specified loading fraction, after the switch opens, the loads voltages decrease rapidly. "Some fraction" refers to 77% (88% squared), because below this level, the voltage decreases to less than 0.88 p.u. and the inverter enters an operational state in which IEEE 1547 requires a 2-second trip, but this is strictly true only for impedance loads. Thus, a practical observation says that a sustained island is not possible if the sum of the AC nameplate ratings of all the DG in a potential island is less than two-thirds of the minimum feeder load within the potential island. The two-thirds fraction is somewhat conservative (since the actual fraction is 77%) and easy to remember. The caveat to this rule is that reliable data on minimum load must be available. Also, if IEEE 1547 is changed from existing 2016 standards to allow low-voltage ride through (LVRT) capability, the two-thirds rule will need modifications.
- *Reactive Supply Imbalance.* If the reactive power supply and demand within the potential unintended island cannot be balanced, a sustained island cannot be formed.

  Since most loads and power system components connected to distribution systems absorb VArs, there must be a source of VAr in the potential island in order for islanding to be sustained. The most obvious VAr source is capacitance, which is often installed for power factor correction or occur as a consequence of underground cabling. PV inverters operate at unity power factor, but, increasingly, larger inverters are being designed to operate at a fixed power factor according to a schedule or command, thus behaving like a VAr sink. If the load VAr demand is larger than the VAr sources within the island, then the chances of a sustained unintended islanding are minimized. This is due to the rise in system frequency beyond the IEEE 1547 specified limit of 60.5 Hz, triggering frequency relays to disrupt power supply in the distribution system. The mechanism of this frequency change is the phase locked loop (PLL) used by the inverters to synchronize to the grid frequency. When the grid source is lost, the PLL will change the frequency of the inverters output current to bring the inverters voltage and current into a phase relationship the PLL is programmed to maintain (usually, zero). Thus, in such cases, anti-islanding protection is not required.
- If DTT is properly implemented, only a failure of the DTT communications system would result in a failure to detect an unintentional island.

The utility operators may choose to exercise advanced diagnostics in which the potential island contains large capacitors, and the power factor within that potential island is very close to 1.0. Similarly, systems with a very large numbers of inverters, or with inverters from several different manufacturers, systems that have both inverters and rotating generators are known to have unintended islanding frequently. As operational practice, the approach suggested by Sandia National Laboratory in a assessment guideline document for minimizing DG unintentional islanding risk can be adopted by microgrid operators to prevent unintentional islanding [44]:

**Steps:**

(a) Determine whether the aggregate power output rating of all DG installed in the microgrid exceed two-thirds of the minimum demand of the downstream feeders.

(b) Determine if the total reactive power demand in the microgrid is within 1% of the total aggregate capacitor rating within the island.

(c) Determine if the potential unintended island is made up of a combination of inverter-based DG and traditional rotating machinery, and that the total AC ratings of the rotating DG exceed 25% of the total AC rating are all DGs in the potential island. If all of the rotating machine AC ratings are greater than 25% of the total DG, then further study may be prudent. If the sum of all rotating machine AC ratings is less than 25% of the total DG, then proceed to Step 4.

(d) Sort the inverters by manufacturer, compute and sum up the total AC rating of each manufacturers product within the potential island, and determine each manufacturers percentage of the entire DG-power production. Based on the two-thirds rule discussed in the last page, the risk of unintentional islanding can be assumed to be minimal if more than two-thirds of the total DG is from a single manufacturer.

## 8.6 CONTROL AND OPERATION OF MICROGRIDS

A control and management architecture is required in order to facilitate full integration of local generation and active load management into modern, specialized LV distribution systems, such as microgrids. The coordinated control of a large number of DERs can be achieved in a variety of manners: ranging from decentralized hierarchical controls to vertical tightly coupled control of demand, supply, and operation. The control strategy adopted by a particular microgrid operator depends on the share of responsibilities assumed by a central controller and the local device-level controllers of the DGs and controllable loads. The control of a microgrid is coordinated by the use of Microgrid Central Controller (MGCC), which is also known as microgrid management systems (MGMS). A basic framework for microgrid control is shown in Fig. 8-10, and distinctions between centralized and decentralized control of microgrids are presented in Table 8-5.

The LC are low-level or device controllers, responsible for controlling and monitoring DERs, storage devices, and customer loads. A typical LC could either be an integrable piece of hardware or a software installed in either the digital interface of the power electronic components, electronic meters, or any other device deployed in the field with requisite computing and communicating capabilities.

The MGMS provides the primary point of interaction between the microgrid and players in distribution system management and control, such as the distribution system operator (DSO), or the energy services company (ESCO). The MGMS in distribution microgrids can be assigned different roles: ranging from the main responsibility for the maximization of the microgrid value to optimal operation of the local LCs, or meeting certain environmental criteria. The MGMS can be configured to generate reference set-points for the MCs, for data acquisition, or for passive observation and supervision of the operation of the microgrid's load and controllers. The MGMS is typically located inside the substation and the MGMS human-machine interface includes a specialized

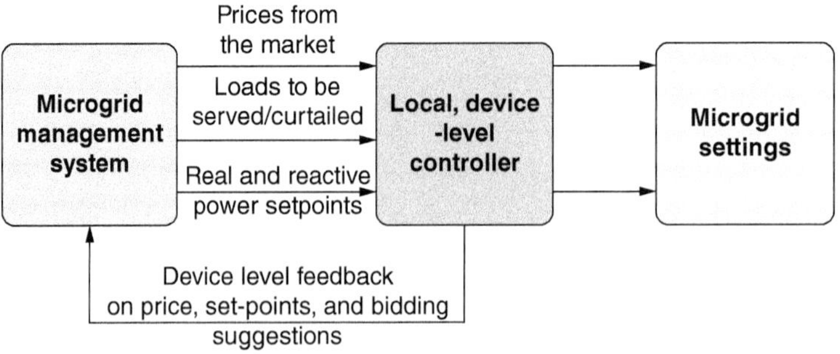


**FIGURE 8-10** Microgrid management system operation schematic.

**TABLE 8-5** General Differences between Centralized and Decentralized Control of Microgrids

| | Centralized control | Decentralized control |
|---|---|---|
| DG ownership | Single owner of assets | Multiple owners of assets |
| Network operators | Usually have lesser automation and employs personnel to manage network operations like traditional distribution systems | Higher dependence on software-based control of multiple assets across different voltage levels and geographical locations |
| Market participation | Feasible | Some units may not participate in energy trading |
| Optimization of control | Complex algorithms may be deployed by single owners | Open nature of the control system requires less complex algorithms for active participation of assets of multiple owners |
| Communication requirements | High | High |
| Collaboration among assets | Feasible | Some units may compete against each other in the same microgrid |

software routines of functionalities depending on the role of the MGMS. Depending upon the vendor, sometimes the MGMS software can be integrated with the DMS.

## 8.6.1 Centralized Control

A commonly used architecture for centralized microgrid control is shown in Fig. 8-12. The power distribution system (i.e. the microgrid) is interfaced with the prime-mover based PQ and VSI controllers through a converter mechanism, as shown in the illustration of a typical microgrid control architecture. Microgrids can be centrally managed by extending and properly adapting the functionalities of existing energy management system (EMS) functions. The basic feature of centralized control is that decisions about the operation of the DER are taken by the microgrid operator or ESCO at the MGCC level. The MGCC is equipped, among other things, with scheduling routines that provide optimal setpoints to the MCs, based on the overall optimization objectives.

## 8.6.2 Hierarchical Control Levels

There are typically three levels of controls in hierarchical control of power systems. Such control topologies are not exclusive to microgrids. The Union for the Coordination of Transmission of Electricity (UCTE, Continental Europe) has defined a hierarchical control for large power systems [45]. However, the same hierarchical control theory for large transmission systems cannot be ported into microgrids without significant adaptation.

Original hierarchical control systems for power grids were designed to operate over large high inertia synchronous machines and inductive networks. However, in microgrids that have large power electronic interfaces to integrate renewable energy resources, there are very limited inertia, and the networks have high *R*/*X* ratios.

However, like hierarchical control in transmission systems, the primary control in microgrids deals with the inner control of the DG units by adding virtual inertia and controlling their output impedance. The secondary control serves to restore the amplitude and frequency deviations resulting from the virtual inertia and output virtual impedance. The tertiary control is used to monitor and control the power flows between the main grid and the microgrid at the PCC.

### 8.6.3 Inner Control Loops

The use of intelligent power interfaces between the electrical generation sources and the microgrid is a critical component. These interfaces consists of DC-AC inverters, that operate as current-source inverters (CSIs) and as voltage-source inverters (VSIs) depending upon specific operating scenarios. Such inverters consist of an inner current loop and a PLL to continuously stay synchronized with the grid, and the VSIs, which consist of an inner current loop and an external voltage loop. In order to inject current to the grid, CSIs are commonly used, while in island or autonomous operation, VSIs are needed to keep the voltage stable, and within constraints. It is not required for VSIs to have external reference points to maintain synchronism. VSIs can provide voltage ride through capability and power quality enhancement to the DG. VSIs are responsible for controlling the power import from, or the power export to the main grid. Thus, VSIs facilitate the ability of the microgrid to operate in both grid-connected and islanded modes (Fig. 8-11).

VSIs and CSIs have closely associated roles in modern microgrids. The VSIs are integrated with energy storage devices, and maintain a fixed frequency and voltage level within the microgrid. The CSIs are integrated with low capacity DERs, such as wind turbines or solar panel cells. It must be noted that these DG inverters be used to perform the duties of the VSIs if required. Thus, it is feasible to have multiple CSIs and VSIs, or VSIs exclusively connected in parallel to form a microgrid. When these inverters are inter-connected to operate in grid-connected mode, they change their core functionality from being VSI to CSI.

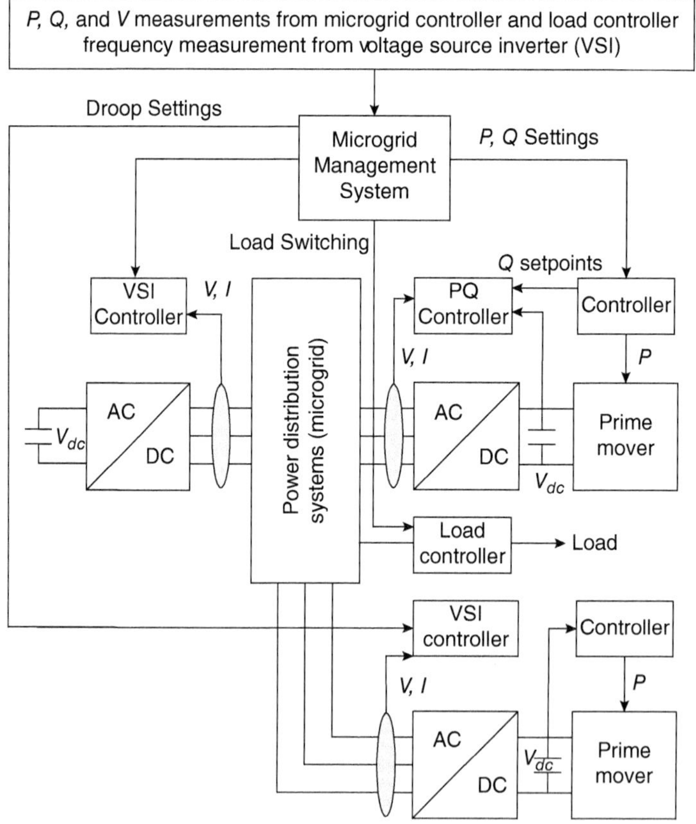


**FIGURE 8-11** Typical microgrid control architecture.

### 8.6.4 Primary Control

The key concept of primary control is to implement the inertial behavior of a synchronous generator, which proportionally decreases the frequency due to increase in active power in the microgrid. Primary control level alters the amplitude of voltage and the frequency reference used by internal voltage and current control loops. When connecting multiple VSIs are connected in parallel, it is possible to detect circulating active and reactive powers. Thus, the primary control principle can be used by VSIs to implement the P/Q droop method in microgrids, which is decribed by:

$$f = f_* - G_P(s).(P - P_*) \tag{8-1}$$

$$V = V_* - G_Q(s).(P - P_*) \tag{8-2}$$

where $P$ and $Q$ are the active and reactive powers, $P_*$ and $Q_*$ are active and reactive power references, $V$ and $f$ are the amplitude and frequency of the output voltage reference, $V_*$ and $f_*$ are the amplitude and frequency references, and $G_P(s)$ and $G_Q(s)$ are active and reactive power transfer functions. The DC gains of $G_P(s)$ and $G_Q(s)$ compensators contribute to static $\Delta f/\Delta V$ and $\Delta Q/\Delta V$ deviations, which are important to maintain the system in sync with other rotating machinery, and for maintaining the voltage stability of the microgrid. Proportional droop terms of the compensators $G_P(s)$ and $G_Q(s)$ parameters are $m$ and $n$, respectively. They are described mathematically as follows:

$$m = \frac{\Delta f}{P_{\max}} \tag{8-3}$$

$$n = \frac{\Delta V}{2Q_{\max}} \tag{8-4}$$

where $\Delta V$ and $\Delta f$ are the maximum tolerated voltage and frequency deviations; and, $P_{\max}$ and $Q_{\max}$ are the maximum active and reactive powers delivered by the inverter, respectively.

Figure 8-12 shows the correlation between the $Pf$ and $QV$ droops and the $PQ$ circle of a DG or DER unit. The DG or DER unit has the capability to inject active power ($P > 0$) and store energy ($V < 0$)

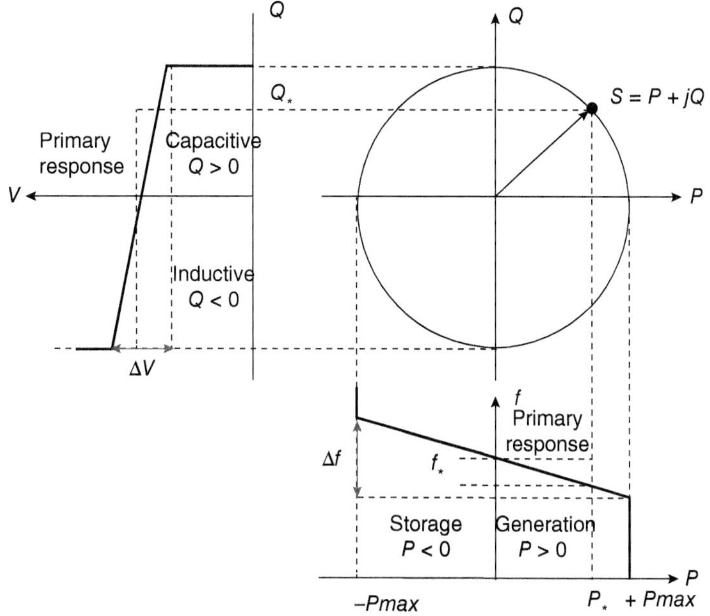


**FIGURE 8-12** *PQ* circle and *Pf* and *QV* droop primary control relationship.

while supplying reactive power ($Q > 0$ for capacitor-behavior) or absorbing reactive power ($Q$ for inductor-behavior). In the conventional droop method used by large power systems, the output impedance of high inertia synchronous generators and the line impedance is mostly inductive. In systems with a significant amount of power electronics integration, the output impedance depends on the control techniques adopted by the inner control loops. Also, the line impedance in low-voltage applications has approximately pure resistive values. Thus, the control droops in Eqs. (8-1) and (8-2) can be subjected to Park Transformation, based on the impedance angle $\theta$ of the system.

$$f = f* - G_P(s)[(P - P_*) \sin\theta - (Q - Q_*) \cos\theta] \tag{8-5}$$
$$V = V* - G_Q(s)[(P - P_*) \cos\theta + (Q - Q_*) \sin\theta] \tag{8-6}$$

The primary control technique depends upon a virtual output impedance loop. The output voltage of the loop is expressed as:

$$v_0^* = V_{ref} - Z_D(s) \tag{8-7}$$

where $V_{ref}$ is the reference voltage reference obtained from Eqs. (8-5) and (8-6) with $V_{ref} = V \sin(2\pi ft)$ and $Z_D(s)$ represents the transfer function of the virtual output impedance. Under usual operating scenarios, the $V_{ref}$ is used to enforce inductive behavior at the frequency of the grid. The virtual impedance $Z_D(s)$ must be selected such that it exceeds the sum of impedance of the inverter and the line impedance, such that the total equivalent output impedance is close to the values of $Z_D(s)$. The virtual output impedance $Z_D(s)$ is approximately equal to the total series impedance of a synchronous generator.

Virtual output impedance has features that distinguish it from a physical impedance, such as: (1) virtual output impedance has zero power loss, and (2) it is possible to implement resistance without reducing operating efficiency. Also, virtual impedance control loop helps to achieve two additional things:

1. Control Over Inverter Output Impedance. This can be achieved by adjusting the phase angle of Eqs. (8-6) and (8-7) according to the known $R/X$ ratio of the line impedance $\theta = \arctan(X/R)$ and the angle of the output impedance at the frequency of the grid.
2. Harmonic current sharing and hot-swap operation are feasible by means of having virtual output impedance [46–48].

## 8.6.5 Secondary Control

A secondary control can be used to compensate for the excursions in frequency and amplitude and frequency due to variations in load or generation inside the microgrid. It operates by sensing the frequency and amplitude levels in the microgrid $f_{MG}$ and $V_{MG}$ are sensed and comparing them with the reference frequency $f_{*MG}$ and reference voltage $V_{*MG}$, respectively. The errors are filtered by means of compensators ($\delta f$ and $\delta V$); and then the filtered frequency voltage and amplitude values are propagated to all connected units to restore the output voltage frequency and amplitude.

The secondary control is used in power systems to limit grid frequency deviation within allowable limit. It comprises a proportional-integral (PI)-type controller, to allow load-frequency control commonly used in several European microgrids; and automatic gain controller generally used in the United States. In the case of AC microgrids, the frequency and amplitude restoration controller parameters $G_f$ and $G_V$ can be designed as follows:

$$\delta f = k_p f(f_{*MG} - f_{MG}) + k_i f(f_{*MG} - f_{MG})dt + \Delta fs \tag{8-8}$$
$$\delta V = k_p V(V_{*MG} - V_{MG}) + k_i V(V_{*MG} - V_{MG})dt \tag{8-9}$$

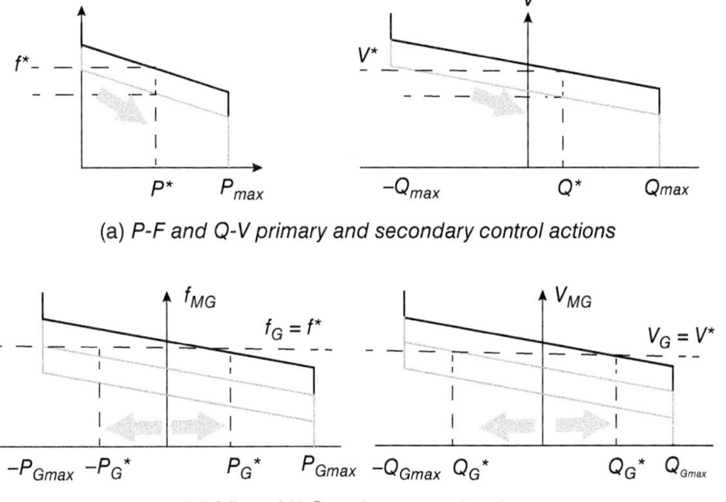


**FIGURE 8-13** Microgrid control mechanism.

where $k_p f$, $k_p V$, $k_i f$ and $k_i V$ are the control parameters of the secondary control compensator, and $\Delta f_s$ is a synchronization flag variable which remains equal to zero when the microgrid is operating in islanded mode. It must be noted that $\delta f$ and $\delta V$ must not surpass the maximum tolerated frequency and amplitude deviations, specific to the microgrid.

Figure 8-13(a) shows the primary and secondary control actions over the $P-f$ and $Q-E$ characteristics. Secondary control in a droop-controlled microgrid enables the frequency and amplitude restoration process. The restoration process also implies corresponding increase in real and reactive demands. If secondary control is not correctly implemented, both frequency and amplitude of the microgrid will be dependent on the incumbent load of the microgrid.

### 8.6.6 Tertiary Control

While operating the microgrid when it is connected to the main grid, the power flow in the microgrid is controllable by means of modifying the frequency or by varying the steady state phase angle. These actions also correspond to variation of the voltage amplitude inside the microgrid [48]. By measuring the $P/Q$ at the PCC, $P_G$ and $Q_G$, they can be compared with the desired $P_{*G}$ and $Q_{*G}$ and controlled as follows:

$$f_{*MG} = k_p P\,(P_{*G} - P_G) + k_{iP} \int (P_{*G} - P_G)dt \tag{8.10}$$

$$V_{*MG} = k_p Q\,(Q_{*G} - Q_G) + k_{iQ} \int (Q_{*G} - Q_G)dt \tag{8.11}$$

where $k_{pP}$, $k_{iP}$, $k_{pQ}$, and $k_{iQ}$ are the control parameters of the tertiary control compensator. Note that $f_{*MG}$ and $V_{*MG}$ will become saturated if they operate outside of their respective maximum tolerated limits. These variables $f_{*MG} = f_G$ and $V_{*MG} = V_G$ are depenedent on the secondary control. When the microgrid is in grid-connected mode, the synchronization process commences, and $f_{*MG}$ and $V_{*MG}$ must match with their corresponding values in the power grid to enable successful synchronization. After synchronization is complete, these signals can be described by the tertiary control.

Depending on whether $P_{*G}$ and $Q_{*G}$ is positive or negative, the active and reactive-power flows can be independently imported or exported. Figure 8-13(b) shows the tertiary control action, which facilitates the interchange of $P$ and $Q$ at the PCC. It also helps maintain the observability of the bi-directionality of the power flow in the microgrid. The grid has constant amplitudes and frequency ($E_G = E_*$ and $f_G = f_*$), represented by horizontal lines. The amount of $P$ and $Q$ exchanged between the microgrid and the grid ($P_G$ and $Q_G$) can be determined geometrically from the intersection region of the droop characteristics of the horizontal lines of the grid and the operating lines of the microgrid. Thus, the generated power $P_G$ is controllable by variation of the microgrid reference frequency $f_{*MG}$. In this formulation, it is easy to remember that the microgrid injects $P$ to the grid if $f_{*MG} > f_G$, and the microgrid absorbs active power P from the grid if $f_{*MG} < f_G$. It also leads to variation of the power angle. For reactive-power, similar equations can be framed.

It can be seen from Eqs. (8-10) and (8-11), that, if $k_{iP} = 0$ and $k_{iQ} = 0$, the tertiary control will behave as the primary control of the microgrid. It will thus be able to allow the interconnection of multiple microgrids, as discussed in Sec. 8.5.8. Thus, this control loop can be leveraged to improve the power quality at the PCC. To enable voltage dip ride-through capabilities in the microgrid, it must be capable of injecting reactive power to the grid, thus achieving inner voltage stability. Observe that, if $k_{iQ} = 0$, the microgrid will automatically inject $Q$ as soon as a voltage sag is identified, or will automatically absorb reactive power when there is a voltage surge in the grid. This provides low voltage ride through capability to the MicroGrid.

Islanding detection is used to disconnect both the tertiary control references and the integral terms of the reactive-power PI controllers to avoid voltage instabilities, particularly during scenarios of unplanned islanding. In events of unplanned islanding, the tertiary control will endeavor to absorb active power $P$ from the grid. If it fails to do so, the frequency will start to decrease. If the thresholds of tolerance are crossed, the microgrid islands itself from the grid for its own safety.

## 8.6.7 Operation of Multiple Microgrids

The microgrid concept can be extended to the development of a new concept of the multiple microgrids, or "multi-microgrids" (MMG). In addition to developments in microgrid implementation, MMG concept will require a complete re-design of a distribution system control architecture as well as the development of new management tools or the adaptation of existing DMS tools. MMG architecture corresponds to a high-level structure created at the medium voltage (MV) level, consisting of several LV microgrids and DG units connected to the adjacent network of MV feeders. A large number of LV networks with micro-sources and loads, that are no longer passive elements of the distribution grid, then need to be operated together in a coordinated way. Therefore, the system to be managed grows in complexity and dimension, requiring a completely new control and management architecture.

An effective management of this MMG can be achieved using hierarchical control architecture, where control will be coordinated by an intermediate controller, the m-MGMS controller (i.e. Multiple Microgrid Management System), to be installed at the MV bus level of a substation, supervised by a DSO. An m-MGMS controller will also have to deal with technical and commercial constraints and contracts, in order to manage the MMG both in grid-connected operating mode and in grid disconnected modes, during emergencies. Key functionalities of m-MGMS will be:

- Volt-VAR support.
- Frequency support through active coordination between DG and loads across the entire MMG network.
- Control scheduling in energy markets, if available.
- State estimation.
- Restoration based on load prioritization, and other emergency functions as deemed by the operation.

## 8.7 SMART GRID DEPLOYMENT PROJECTS

Here are some examples of smart grid deployment projects:

### 8.7.1 Duke Energy

This project involved implementation of AMI and DA systems in five states in the United States (there were major deployments in North Carolina and Ohio, while smaller deployments of DA were carried out in Indiana, Kentucky, and South Carolina). Duke Energy also took the initiative to launch pilot-scale programs for electricity pricing. Their efforts included deployment of peak-time discounts, time-of-use rates, and critical peak pricing. Customers (who participated in these pilot programs) were actively using Web portals, controlling their home temperature using smart thermostats, and had installed direct load control devices (such as smart dishwasher, washer and dryer, etc.) to reduce their electricity consumption and peak demand. The project utilized multiple distributed resources at both the customer and distribution level. Energy storage using Li-ion batteries were installed to be used by multiple applications. Several photo-voltaic units were also commissioned for installation for residential generation.

### 8.7.2 Pacific Gas and Electric Company

In 2014, Pacific Gas and Electric Company (PG&E) introduced its "Grid of Things" vision to mobilize its grid modernization efforts to aid in the optimization of DERs. The "Grid of Things" integrated smart energy devices and novel technologies with the grid. PG&E encouraged their customers (who owned small scale DERs and smart appliances) to participate in their large-scale effort to "greater value from PG&E's energy technology investments." PG&E encouraged purchase and installation of rooftop solar, EVs, energy storage, demand response technologies, etc., through a large number of customer-facing monetary initiatives. Through their efforts, PG&E is aiming to establish four critical capabilities for the smart grid: (1) integrate environment-friendly, DERs (2) aid decision-making for both operators and customers (3) oversee grid automation and self-healing capabilities of networks, and (4) enable customer participation.

### 8.7.3 State Grid Corporation of China (SGCC)

SGCC proposed to develop the "Strong & Smart Grid" in 2009. The project laid its foundations by developing an ultra-high-voltage (UHV) to support the development of automated and smart subsystems of the power grid at all levels. SGCC approach toward smart grid deployment was essentially a top-down approach toward restructuring the grid, that included meticulous planning before construction was commissioned.

## 8.8 SMART GRID DEMONSTRATION PROJECTS

Main research and development challenges in smart grid include the ability to include more RES connected to distribution networks, improving the efficiency of customer participation in electricity markets, secure energy supply, improve network reliability, deploy network automation in medium and low voltage networks, smarter data-driven demand management, optimal use of the controllable loads (like EVs), storage technologies, and microgrids. However, in order to move from theory to practice, several pilot projects were deemed to be necessary by industry and regulatory bodies by several nations across the globe.

The Smart Grid Demonstration Program (SGDP) in United States was launched to demonstrate the potential of existing and emerging smart grid concepts to be integrated to prove technical, operational, and business-model feasibility for modernizing the power grid. FERC aims to "demonstrate new and more cost-effective smart grid technologies, tools, techniques, and system configurations that significantly improve on the ones commonly used today."

The program comprised 32 projects in the two broad areas: Smart Grid Regional Demonstrations (16 projects) and Energy Storage Demonstrations (16 projects). Smart Grid Regional Demonstrations focused on regional smart grid demonstrations to validate the feasibility and determine the costs-to-benefits ratio for expensive smart grid investments. The project also aimed to validate new smart grid business models that could easily facilitate scale-up of the developed smart grid technologies. Energy Storage Demonstrations emphasised research and development of energy storage technologies—flywheels, pump-hydro storage, compressed air energy storage systems for load shifting, batteries, frequency regulation design, ramp control design. The research and development projects also focused on the grid integration of renewable resources.

## 8.8.1 Pacific Northwest Smart Grid Demonstration

The U.S. DOE partnered with industries in late 2009 to partially sponsor Pacific Northwest Smart Grid Demonstration (PNWSGD). The PNWSGD project can be counted as one of the largest and most comprehensive demonstrations of electricity grid modernization. PNWSGD project successfully accomplished demonstrations across several states in the United States, and received cooperation from multiple electric utilities, including rural electric coops, public utilities, municipal utilities, and other investor-owned utilities. The local objectives for these systems were (i) improved reliability, (ii) energy conservation, (iii) improved efficiency, and (iv) demand responsiveness.

The demonstration pioneered the development of a "transactive system" that enabled easy coordination of many of the project's DERs and demand-responsive components. The transactive energy framework helped in countering of challenges associated with unpredictability in load demands and renewable energy production. The transactive energy framework could also be used by multiple utilities to meet the regional goals. The transactive system showed the feasibility and advantages of being able to coordinate electricity supply, and distribution, and engaging end-users by making all participants in the energy supply chain involved.

A utility in the Pacific Northwest region involved in the project, Avista Utilities, demonstrated a high level of integration among the demonstrated asset systems across transmission and distribution system, down to end-user appliances, as shown in Fig. 8-14. "ZigBee" smartmeters were deployed for

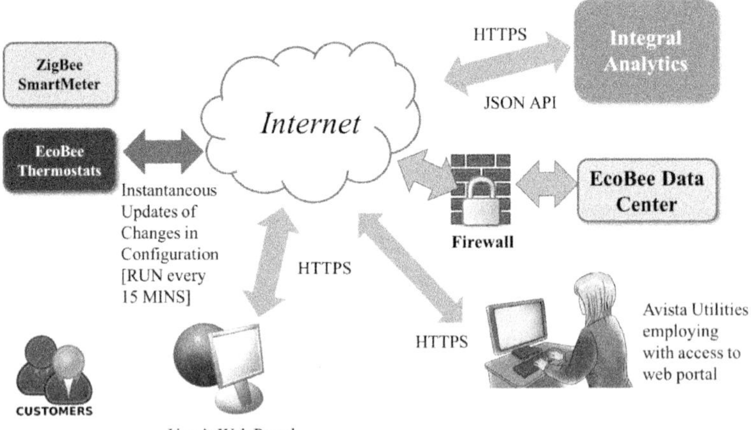


**FIGURE 8-14** Technologies installed by Avista Utilities during smart grid demonstration in the Pacific Northwest power system [49].

each customer in Pullman, Washington, along with "ecobee" smart thermostats in certain homes. Data was sampled every 5 minutes and updated to the system every 15 minutes over standard, secure IP, HTTPS. This data was visible to customers and utility alike through Web-based interfaces, as well as used for analytics for operation and control of the modern grid. A key lesson learned during this project was that despite the well-structured integration, benefit rendered by each individual component could not be quantified in the demonstration. While the integration of systems is one of the important aspects of a smart grid, a type of unit testing of the individual subsystems is recommended for cost-benefit analysis before commissioning other projects.

### 8.8.2 Southern California Edison Company

Southern California Edison Company (SCEC) demonstrated smart grid benefits in Irvine, California, with a total investment of $80 million. Like PNWSGC, SCEC also demonstrated an end-to-end, easy-to-scale approach to develop and deploy a smart grid system from transmission to customer end-use appliances, such as smart washer/dryer and electric vehicles. From the lessons learned from this project, several gaps have been identified in the interoperability of several smart devices in the grid. Results and insights from the SCEC smart grid project have been incorporated in many smart grid related standards.

### 8.8.3 NSTAR Electric and Gas Corporation

NSTAR demonstrated smart grid benefits in a densely populated urban area of United States, which faces extreme seasonal demands. The project was based in and around Boston, Massachusetts. This project showcased the successful use of smart, sensor-based data observation and measurement instrumentation in distribution systems in Boston, Massachusetts. Using the results of the demonstration project, NSTAR leveraged several state-of-the-art big data analytics tools and techniques to gain further understanding of the power grid status and behavior. They upgraded their system to perform proactive maintenance activities that can lead to significant reliability of their operations. The results from the project were also used to increase the system's capability to accommodate renewable DERs in the grid. Analysts are able to extrapolate the experiences gained from the project to secondary area network grids in large urban areas with similar climates, such as New York City, Chicago, and Philadelphia.

### 8.8.4 The Boeing Company

Smart grid demonstration projects are not limited to utilities and federal organizations like U.S. Department of Energy, but several private corporate-driven investigations have been commissioned to evaluate the move toward smart grid. World's leading aeroplane and space equipment manufacturer, The Boeing Company, launched smart grid projects in several of its corporate and manufacturing campuses in the United States, viz., St Louis, Missouri, Sunnyvale, California, and Huntington Beach, California. The investment was about $17 million. The Boeing Company endeavored to demonstrate an advanced smart grid technology with military-grade cyber security for optimizing regional transmission system planning and operation by enabling wide-area situational awareness, coordination, and collaboration. The project team included regional transmission operators and utilities, impacting more than 90 million customers across 21 states in the United States.

### 8.8.5 National Rural Electric Cooperative Association

National Rural Electric Cooperative Association (NRECA) is the organization that represents the interests of over 900 electric cooperatives in the United States, to various legislatures. Independent electric utilities are not-for-profit and are owned by their members. NRECA invested about $67 million in grants and projects to demonstrate smart grid viability in transmission and distribution system in sparsely populated cities and counties in the United States. Their efforts led to the installation and operation of a suite of diverse smart grid technologies such as smart meters and PMUs. The projects also assisted in aggregation of data from 17 rural electric cooperatives across 11 states. Deployed grid modernization infrastructure included more than 130,000 meters and over 18,000 demand response switches. The project also installed hundreds of voltage sensors and fault detectors in the networks it covered. The demonstration data from all sites and locations were collected and organized into a single database. Several analytical results were obtained from the subsequent studies, such as self-healing feeder design for higher system reliability, advanced Volt/VAr for total demand, peak pricing, DER location optimization, research on customer's appliance control, and time-of-use tariffs for providing better incentives for rural customers.

### 8.8.6 GRID4EU

GRID4EU, an European smart grid demonstration project, was launched in 2011 with an investment worth 25.5 million euros by the European Union (EU). The project amassed overall expenditure in the tune 54 million euros prior to successful completion, making it one of the biggest smart grid investments by EU.

GRID4EU led to the formation of a consortium of six major European energy distributors (alphabetically CEZ, Enel Distribuzione, ERDF, Iberdrola, CEZ Distribuce, RWE, and Vattenfall Eldistribution). The project investigated the feasibility and benefits of deploying smart grids in certain regions of Europe as well as researched on increasing renewable energy integration, proliferation of use of EVs, power system automation, loss minimization, utility as well as customer-side demand management, energy storage, and energy efficiency. Six demonstrators joined resources to investigate innovative solutions that can directly lead to improvement of MV and LV network automation technologies to face the constraints introduced by the increased amount of DER and new usages (e.g., electric vehicles, heat pumps) and reduce energy losses and maintain or harden the system to cyber-physical attacks. Demand response and investigation of optimal integration of renewable resources were also key research topics in the smart grid demonstration projects carried out by GRID4EU.

### 8.8.7 Internationalization of Smart Grid Goals

Several countries in North America and Europe have signed bilateral assistance commitments to among neighboring countries to modernize the electric grid. Many countries in Asia, South America, and Africa have launched intensified efforts to encourage the development of indigenous smart grids. International coordination will provide multiple benefits, such as:

- Development of world-wide industry standards will ensure the expansion of the market for smart grid vendors and suppliers across the world. An internationally coordinated approach will lead to increase in exchange exports of their smart grid solutions and technologies, and deployment services overseas, which in turn will increase further innovation and job growth in the power sector worldwide.
- The use of international standards results in manufacturing efficiency and encourages an "open market" environment, which will drive competition and improvement of products and services in the smart grid industry.

## 8.9 TRENDS IN SMART GRID DEPLOYMENT AND FUTURE OUTLOOK

The definition of smart grid is often contextual, and specific to national or regional interests. Thus, the valuation of technologies that are critical to smart grids vary from utility to utility. A few more years of experimental operation and learning from different smart grid strategies will be required to efficiently estimate the total costs and benefits of smart grid technologies, especially as several utilities across the world begin to mine new and large data-sets from deployed smart grid systems. Utilities, federal and regional regulators have a diverse experiences with smart grid technologies. Thus it is common to come across differing opinions and views on costs and benefits. As a result, investment decisions and deployment rates are determined at very region-specific projects motivated by individual local energy goals, availability of funds and investments, and the level of smart grid expertise and experience at individual utilities. In the United States, DOE and Electric Power Research Institute (EPRI) invested time and resources to design an organized framework for utilities to proactively compute costs and benefits, based on observations from prior smart grid demonstrations projects. Analysis of previous smart grid projects leads to development of new smart grid paradigms. The advances are further augmented by progress in technology, algorithms, metering devices, forecasting tools, and computing technologies. The evolution of smart grid depends strongly on the insights from historical data, and additional benefits are generated by including a larger number of enabling technologies to existing smart grid systems [50].

### 8.9.1 Distribution Automation Projects

DA technologies significantly improves resilience and reliability of the power grid. The benefits also extend to marked improvements in the system's operational efficiencies across the network. Grid modernization projects in the United States that installed automated feeder switches have reported that outage times have been reduced to 56% shorter durations, and the frequency of such outages dropped by 11% to 49%. Following a severe windstorm on July 5, 2012, utilities in Chattanooga, Tennessee were able to quickly restore power to more than 40,000 customers (i.e. more than half the impacted customers) within 2 seconds using automated feeder switching. Apart from being able to avoid outage damages to residents and businesses, the utility also did not lose revenue worth $1.4 million as it was able to restore power more quickly.

Several utilities across the world are inclined to adopt smart grid technologies to optimize voltage and reactive power levels in certain distribution circuits dynamically. Smart devices are being able to save 2.2% energy reductions and 1.8% peak load reductions per distribution circuit on average over an extended period of time [51–53].

Multiple ARRA projects have deployed CVR within their distribution systems [54]. By automating capacitor banks on their lines (the concept was first reported in the 1970s), a few utilities (especially in California) are making an active endeavor to reduce peak energy demands [55, 56]. Several subsequent studies have successfully replicated the energy saving algorithms. From prior CVR-based projects, it is possible to estimate that substantial energy efficiency improvements (approximately 6500 MW) are possible by using CVR technologies alone [57].

### 8.9.2 Synchrophasor Applications Projects

Modernization and upgrading the transmission system refers to more widespread installation of digital, high precision measuring equipment to monitor and control local operations of the grid. Synchrophasor technology, which uses time-synchronized devices called PMUs to measure the instantaneous voltage, current, and frequency at substations, is installed in a large number of locations to improve wide-area monitoring and control of the transmission system. The data obtained from the PMUs are propagated across the network in real time to advanced software applications that allow engineers and operators to identify points of system instability, monitor, detector, or even predict frequency and voltage excursions. Synchrophasor technology has given power system operators the ability to have real-time (and sometimes even proactive) insights about the system behavior and its likelihood to exceed acceptable operating limits. Thus, such insights are very useful to take corrective and preventative measures for disturbances before grid stability is very severely impacted. It is important to know that data obtained from the PMUs also facilitate coordination and control of generators of all scales and sizes, including renewable resources (e.g., solar farms, wind power plants, etc.).

The 2003 Northeast blackout in the United States was largely attributed to the lack of situational awareness for grid operators. In order to take measures to prevent such large blackouts in the future, widespread PMU installations can be used. PMUs installed in the substations can provide time-stamped data 30 times per second or faster, which is approximately a hundred times faster than conventional SCADA technology. Deployment of PMU-based synchrophasor technology includes phasor data concentrators (commonly known as PDCs) in the power grid.

## 8.9.3 Microgrid Deployment

DOE launched Advanced Microgrid Program in 2013 to encourage development of microgrids across the United States [58]. The Advanced Microgrid Program (AMP) intended improve the resilience and reliability of microgrids, initially limited to less than 10 MW capacity. AMP laid great emphasis on improving existing communications technology by emphasising on data security and consumer privacy. AMP also leverage "adaptive logic" to ensure maintaining a balance between the system's energy resources and the energy storage. By focusing on designing innovative interfaces for advanced autonomous operation (islanded operation) with rest of the existing legacy power grid, AMP intended to make it easier from control and operations point of view. Deployment of smart inverters and controllers helped in increasing interactivity of low voltage power equipment with the transmission level EMS and facilitate demand response in the low voltage network.

## 8.9.4 Smart Grid Vision

Developed and developing nations around the world are investing significant resources in modernizing their power grids to increase reliability and sustainability of the energy critical infrastructure. These efforts include developing and installing new equipment to facilitate the transition to renewable sources of energy, and adding additional communication capabilities in increasingly larger number of nodes in the power grid. A communication-driven redesign of power system operation and control must require modifications in security protocols and business models followed by utilities today, worldwide. The customer evolution into "prosumers" and their increased participation in the distribution grid will have impact on energy markets as well. With greater levels of customer-side generation and higher efficiency in transmission and distribution, the traditional utility business model is likely to see significant revisions [59]. A "transactive energy framework" is being proposed as the common ground in which utilities, consumers, and other participants in the energy market will be able to determine the most appropriate technologies, configurations, and system designs that will make the power grid more reliable and economical. The transactive energy framework also insists on maintaining optimality in power flow and meeting financial goals of the participants [60].

Modernization of the power grid is a cost-intensive venture, and it will be prudent to have long-term investment strategies in place. The tasks and responsibilities need to be shared with organizations and utilities at all levels of the power grid in order to minimize duplication of efforts, or having two entities compete for resources to meet the same overall objective. The future power grid will see benefit from advances in material science for improving storage and insulation requirements of the grid, reducing losses, and increasing ability for more power transfer at higher reliability. Advancements in the areas of parallel and distributed high power computing will be necessary to solve complicated optimization problems in real-time. Improved information technology resources will also be required to ensure security and scalability of computational complexity caused by additional uncertainty factors introduced in the smart grid. It cannot be stressed enough that cyber-physical security need to be embedded in smart grid synchrophasor applications, DA, and microgrid controllers as well as in other digital devices being adopted with the smart power grid. Thus, in other words, the deployment of resilient and efficient smart grid calls for interdisciplinary contributions from many domains of science, society, and technology.

## 8.10 REFERENCES

[1] Momoh, J., *Smart Grid: Fundamentals of Design and Analysis*. John Wiley & Sons, vol. 63, 2012.

[2] "European technology platform for the electricity networks of the future," *European Technology Platform on Smartgrids*, 2010.

[3] Locke G. and P. D. Gallagher, "NIST framework and roadmap for smart grid interoperability standards, release 1.0," *National Institute of Standards and Technology*, vol. 33, 2010.

[4] Wiginton, L., Nguyen, H., and Pearce, J. M., "Quantifying rooftop solar photovoltaic potential for regional renewable energy policy," *Computers, Environment and Urban Systems*, vol. 34, no. 4, pp. 345–357, 2010.

[5] Pelletier, S., Jabali, O., and Laporte, G., "Goods distribution with electric vehicles: Review and research perspectives," *Technical Report CIRRELT-2014-44*. CIRRELT, Montréal, Canada, 2014.

[6] Lasseter, R., Akhil, A., Marnay, C., Stephens, J., Dagle, J., Guttromson, R., Meliopoulous, A., Yinger, R., and Eto, J., "The certs microgrid concept, white paper on integration of distributed energy resources," *California Energy Commission, Office of Power Technologies-US Department of Energy, LBNL-50829, http:// certs. lbl. gov*, 2002.

[7] Hatziargyriou, N., and Strbac, G., "Microgrids: A possible future energy configuration?" *IEA Seminar Distributed Generation: Key Issues, Challenges and Roles*, 2004.

[8] Hatziargyriou, N., Asano, H., Iravani, R., and Marnay, C., "Microgrids," *IEEE Power and Energy Magazine*, vol. 5, no. 4, pp. 78–94, 2007.

[9] Katiraei, F., Iravani, R., Hatziargyriou, N., and Dimeas, A. "Microgrids management," *IEEE Power and Energy Magazine*, vol. 6, no. 3, pp. 54–65, 2008.

[10] Agrawal, P., "Overview of DOE microgrid activities," *Symposium on Microgrid, Montreal*, June, vol. 23, 2006.

[11] "Microgrids evolution roadmap, microgrids 1: Engineering, economics, & experience," *CIGRE. Working Group C6.22*, 2015.

[12] Venkataramanan, G. and Marnay, C., "A larger role for microgrids," *IEEE power and energy magazine*, vol. 6, no. 3, pp. 78–82, 2008.

[13] Tsikalakis, A. G. and Hatziargyriou, N. D., "Centralized control for optimizing microgrids operation," *IEEE power and energy society general meeting*, 2011, pp. 1–8.

[14] Jiayi, H., Chuanwen, J., and Rong, X., "A review on distributed energy resources and microgrid," *Renewable and Sustainable Energy Reviews*, vol. 12, no. 9, pp. 2472–2483, 2008.

[15] Hatziargyriou, N. D., Dimeas, A., Tsikalakis, A. G., Lopes, J. P., Kariniotakis, G., and Oyarzabal, J., "Management of microgrids in market environment," *International Conference on Future Power Systems*, vol. 18, Amsterdam, 2005.

[16] Falliere, N., Murchu, L., and Chien, E., "W32. Stuxnet Dossier: Version 1.3," *Symantec Security Response*, 2010. [Online]. Available: http://www.symantec.com/content/en/us/enterprise/media/security\response/whitepapers/w32\ stuxnet\ dossier.pdf.

[17] Batchelor, B., Han, D., and Kim, E., "Annual energy review (FY2010)," National Energy Technology Laboratory (NETL), Tech. Rep., 2010.

[18] McCurley, P., Whitaker, V., Bacik, S., Kotting, C., Myrda, P., Siegfried, T., and Ilic, M., "Reliability considerations from the integration of smart grid," *North American Electric Reliability Corporation*, 2010.

[19] "Roadmap to achieve energy delivery systems cybersecurity," *Energy Sector Control Systems Working Group, Energetics, Inc, URL https://www. control-systemsroadmap.net/ieRoadmap\%20Documents/roadmap.pdf*, 2011.

[20] Johnson, S., "Critical infrastructure protection committee update," *North American Electric Reliability Corporation (NERC)*, 2011.

[21] Creery, A., and Byres, E., "Industrial cybersecurity for power system and scada networks," *Record of Conference Papers Industry Applications Society 52nd Annual Petroleum and Chemical Industry Conference*. IEEE, 2005, pp. 303–309.

[22] Bajpai, P., Chanda, S., and Srivastava, A. K., "A novel metric to quantify and enable resilient distribution system using graph theory and choquet integral," *IEEE Transactions on Smart Grid*, 2016.

[23] Chanda, S., and Srivastava, A. K., "Defining and enabling resiliency of electric distribution systems with multiple microgrids," *IEEE Transactions on Smart Grid*, vol. 7, no. 6, pp. 2859–2868, 2016.

[24] Buldyrev, S. V., Parshani, R., Paul, G., Stanley, H. E., and Havlin, S., "Catastrophic cascade of failures in interdependent networks," *Nature*, vol. 464, no. 7291, pp. 1025–1028, 2010.

[25] Parandehgheibi, M., Modiano, E., and Hay, D., "Mitigating cascading failures in interdependent power grids and communication networks," *International Conference on Smart Grid Communications (SmartGridComm)*. IEEE, 2014, pp. 242–247.

[26] Xin, S., Guo, Q., Sun, H., Zhang, B., Wang, J., and Chen, C., "Cyber-physical modeling and cyber-contingency assessment of hierarchical control systems," *IEEE Transactions on Smart Grid*, vol. 6, no. 5, pp. 2375–2385, 2015.

[27] N. A. S. Initiative et al., "Phasor gateway technical specifications for north american synchro-phasor initiative network," *Pacific Northwest National Laboratory*, Richland, Washington, 2009.

[28] P. S. R. Committee et al., "IEEE standards for synchrophasor measurements for power systems-ieee std c37. 118.1-2011," New York, USA, 2011.

[29] Fang, X., Misra, S., Xue, G., and Yang, D., "Smart grid the new and improved power grid: A survey," *IEEE Communications Surveys & Tutorials*, vol. 14, no. 4, pp. 944–980, 2012.

[30] Chen, C., Wang, J., and Kishore, S., "A distributed direct load control approach for large-scale residential demand response," *IEEE Transactions on Power Systems*, vol. 29, no. 5, pp. 2219–2228, 2014.

[31] Chen, C., "Topics in demand response for energy management in smart grid," *Ph.D. Dissertation*, Lehigh University, 2013.

[32] Rahimi, F., and Ipakchi, A., "Demand response as a market resource under the smart grid paradigm," *IEEE Transactions on Smart Grid*, vol. 1, no. 1, pp. 82–88, 2010.

[33] Manual, P., "11: Energy & ancillary services market operations," *Prepared by Forward Market Operations group at PJM*, 2013.

[34] "719, wholesale competition in regions with organized electric markets," *Federal Energy Regulatory Commission*, 2008.

[35] F. E. R. Commission et al., "Demand response compensation in organized wholesale energy markets," Washington DC, March, vol. 15, p. 2, 2011.

[36] Olfati-Saber, R., Fax, J. A., and Murray, R. M., "Consensus and cooperation in networked multi-agent systems," *Proceedings of the IEEE*, vol. 95, no. 1, pp. 215–233, 2007.

[37] Chen, C., Wang, J., Qiu, F., and Zhao, D., "Resilient distribution system by microgrids formation after natural disasters," *IEEE Transactions on Smart Grid*, vol. 7, no. 2, pp. 958–966, 2016.

[38] Šulc, P., Backhaus, S., and Chertkov, M., "Optimal distributed control of reactive power via the alternating direction method of multipliers," *IEEE Transactions on Energy Conversion*, vol. 29, no. 4, pp. 968–977, 2014.

[39] Erseghe, T., "Distributed optimal power flow using admm," *IEEE Transactions on Power Systems*, vol. 29, no. 5, pp. 2370–2380, 2014.

[40] Kekatos, V., and Giannakis, G. B., "Distributed robust power system state estimation," *IEEE Transactions on Power Systems*, vol. 28, no. 2, pp. 1617–1626, 2013.

[41] "Tame the impacts of distributed energy resources-the electric grid of the future," *OATI: http://www.oati.com/Solution/Smart-Energy/distributed-energy-resource-management, September 2016.*

[42] "Managing big data for smart grids and smart meters," *Internet Security Group, IBM Corporation Whitepaper*, 2012.

[43] "Smart grid research and development multi-year program plan," *US Department of Energy*, 2011.

[44] Ropp, M. and Ellis, A., "Suggested guidelines for assessment of dg unintentional islanding risk," *Sandia National Laboratories, Nat. Renewable Energy Lab., Albuquerque, New Mexico, Rep.* SAND2012-1365, 2012.

[45] UCTE "Union for the co-ordination of transmission of electricity," Brussels, Belgium, 2004.

[46] Guerrero, J. M., De Vicuna, L. G., Matas, J., Castilla, M., and Miret, J., "Output impedance design of parallel-connected ups inverters with wireless load-sharing control," *IEEE Transactions on Industrial Electronics*, vol. 52, no. 4, pp. 1126–1135, 2005.

[47] Guerrero, J. M., Matas, J., De Vicuna, L. G. D. V., Castilla, M., and Miret, J., "Wireless-control strategy for parallel operation of distributed-generation inverters," *IEEE Transactions on Industrial Electronics*, vol. 53, no. 5, pp. 1461–1470, 2006.

[48] Guerrero, J. M., Vasquez, J. C., Matas, J., Castilla, M., and de Vicuña, L. G., "Control strategy for flexible microgrid based on parallel line-interactive ups systems," *IEEE Transactions on Industrial Electronics*, vol. 56, no. 3, pp. 726–736, 2009.

[49] Hammerstrom, D., Johnson, D., Kirkeby, C., Agalgaonkar, Y., Elbert, S., Kuchar, O., et al., "Pacific northwest smart grid demonstration project technology performance report volume 1: Technology performance," *PNWD-4445 volume*, vol. 1, 2015.

[50] Gellings, C., "Estimating the costs and benefits of the smart grid: a preliminary estimate of the investment requirements and the resultant benefits of a fully functioning smart grid," *Electric Power Research Institute (EPRI), Technical Report (1022519)*, 2011.

[51] US Department of Energy, "Drivers and barriers for the application of conservation voltage reduction practices by electric utilities," Report on Smart Grid Projects, 2014.

[52] El-Hawary, M. E., "The smart gridstate-of-the-art and future trends," *Electric Power Components and Systems*, vol. 42, no. 3-4, pp. 239–250, 2014.

[53] Chanda, S., Shariatzadeh, F., Srivastava, A., Lee, E., Stone, W., and Ham, J., "Implementation of non-intrusive energy saving estimation for volt/var control of smart distribution system," *Electric Power Systems Research*, vol. 120, pp. 39–46, 2015.

[54] US Department of Energy, "Application of automated controls for voltage and reactive power management—initial results," Report on Smart Grid Investment Grant Program, December 2012.

[55] Lauria, D., "Conservation voltage reduction (cvr) at northeast utilities," *IEEE Transactions on Power Delivery*, vol. 2, no. 4, pp. 1186–1191, 1987.

[56] Barker, P. P., and De Mello, R. W., "Determining the impact of distributed generation on power systems. I. radial distribution systems," in *Power Engineering Society Summer Meeting, 2000. IEEE*, vol. 3. IEEE, 2000, pp. 1645–1656.

[57] Schneider, K. P., Tuffner, F., Fuller, J., Singh, R., et al. "Evaluation of conservation voltage reduction (cvr) on a national level," *Pacific Northwest National Laboratory report*, 2010.

[58] Smith, M., and Ton, D., "Key connections," *IEEE Power & Energy Magazine*, vol. 11, no. 4, pp. 22–7, 2013.

[59] Kind, P., "Disruptive Challenges: Financial Implications and Strategic Responses to a Changing Retail Electric Business," 2013.

[60] Melton, R. B., "Gridwise transactive energy framework (draft version)," Pacific Northwest National Laboratory (PNNL), Richland, WA (US), Tech. Rep., 2013.

# 9 WIND POWER GENERATION

Zhe Chen

*Professor of Electrical Engineering, Department of Energy Technology, Aalborg University, Aalborg, Denmark*

David Infield

*Professor of Renewable Energy Technologies, Department of Electronic and Electrical Engineering, University of Strathclyde, Glasgow, United Kingdom*

Nikos Hatziargyriou

*Professor of Power Systems, School of Electrical and Computer Engineering, National Technical University of Athens, Athens, Greece, and CEO, Hellenic Electricity Distribution Network Operator SA (HEDNO)*

# 9.1 INTRODUCTION

## 9.1.1 Wind Power Introduction

Wind is air flow caused by differences in the atmospheric pressure within the planetary boundary layer. Early civilizations made use of the power in the wind, in particular wind energy was used to propel boats along the Nile River as early as 5000 B.C. Windmills were used to pump water in China, while vertical-axis windmills were used to grind grain in Persia and the Middle East by 200 B.C. Windmills appeared in Europe during the Middle Ages and were refined by the Dutch for draining lakes and marshes of the Rhine River Delta. In the late 19th century, windmills were commonly used to pump water for farms [1].

The first documented electricity-generating wind turbine was for battery charging; it was designed and installed by an academic, Professor James Blyth, to light his holiday house in Marykirk, Scotland in July 1887 [2]. With the development of electric power generating technologies, small wind turbines were used to supply farms and homesteads remote from power networks. By the 1930s, wind generators for electricity generation were common on farms without grid connection in the United States [3, 4]. More recently larger wind generators have been developed for connection to electricity grids for commercial electricity power generation.

Denmark has played a central role in modern wind power development. By 1900, there were about 2500 windmills for driving mechanical loads such as pumps and mills, producing an estimated combined peak power of about 30 MW. The largest machines were mounted on 24-m towers with four-bladed, approximately 23-m-diameter rotors. In 1957, Johannes Juul installed a 24-m-diameter wind turbine at Gedser. It was a three-bladed, horizontal-axis, upwind, and stall-regulated turbine and ran from 1957 until 1967. This became the basis for standard Danish turbines, many of which were manufactured and installed in the remainder of the 20th century, some of which are still in operation.

However, large-scale modern wind turbines came to the fore only as a result of the oil crisis of the 1970s when Denmark turned seriously to renewable energy technologies, where incentives have been provided for larger wind turbines since the 1980s. Following this lead, more countries such as the United Kingdom, Germany, Spain, and the United States started to promote wind turbine developments from the early 1990s. Later other countries, notably China and India, joined in the promotion of wind power development. The Danish company Vestas led the field, having the largest wind turbine market share worldwide for a number of years in early 2000s. As a result of a dramatic growth in installation in recent years, China now leads the world in installed capacity.

***Power in the Wind.*** Wind carries *kinetic energy* (E), which can be expressed for an air mass ($M$ in kg) moving with a speed $v$ (m/s), as

$$E = \frac{1}{2} M v^2 \quad \text{(J)}$$

The air mass intercepted by a wind turbine rotor with swept area ($A$ in m$^2$) over a time interval $\Delta t$ is the product of the air density $\rho$ and the air volume which is $A \times v \times \Delta t$. Therefore, the corresponding kinetic energy in the air is

$$\Delta E = \frac{1}{2} (\rho A v \Delta t) v^2 \quad \text{(J)}$$

and the power in the wind $P_W$ is

$$P_W = \frac{\Delta E}{\Delta t} = \frac{1}{2} \rho A v^3 \quad \text{(W)} \tag{9-1}$$

It can be seen that the wind power has a cubic relationship with the wind speed. The Beaufort scale shown in Table 9-1, taken from [5], is an empirical measure for describing wind conditions based on observed sea situations. Modern wind turbines are normally designed with a cut in wind speed between 3 and 5 m/s and a cutout wind speed about 25 m/s.

**TABLE 9-1** Specifications and Equivalent Speeds [5]

| Beaufort wind scale | Mean wind speed Knots | Mean wind speed $ms^{-1}$ | Limits of wind speed Knots | Limits of wind speed $ms^{-1}$ | Wind descriptive terms |
|---|---|---|---|---|---|
| 0 | 0 | 0 | <1 | <1 | Calm |
| 1 | 2 | 3 | 1–3 | 1–2 | Light air |
| 2 | 5 | 3 | 4–6 | 2–3 | Light breeze |
| 3 | 9 | 5 | 7–10 | 4–5 | Gentle breeze |
| 4 | 13 | 7 | 11–16 | 6–8 | Moderate breeze |
| 5 | 19 | 10 | 17–21 | 9–11 | Fresh breeze |
| 6 | 24 | 12 | 22–27 | 11–14 | Strong breeze |
| 7 | 30 | 15 | 28–33 | 14–17 | Near gale |
| 8 | 37 | 19 | 34–40 | 17–21 | Gale |
| 9 | 44 | 23 | 41–47 | 21–24 | Strong gale |
| 10 | 52 | 27 | 48–55 | 25–28 | Storm |
| 11 | 60 | 31 | 56–63 | 29–32 | Violent storm |
| 12 | — | — | 64+ | 33+ | Hurricane |

*Notes:*
1. These values refer to well-developed wind waves of the open sea.
2. The lag effect between the wind getting up and the sea increasing should be borne in mind.
3. The official term is strong gale; however, the U.K. Met Office uses the descriptive term severe gale.

## 9.1.2 Wind Resource Characteristics

The design and operation of wind turbines reflects an understanding of the wind resource and its temporal and spatial characteristics. More detailed accounts can be found in text books such as [6] and [7]. The purpose here is to provide a very brief introduction to essential characteristics so as to inform the later discussion of wind turbine technology and control.

***Temporal Variations in Wind Speed.*** Wind speed varies in time across a wide range of temporal scales. Due to external and man-made impacts, climate is forever changing and with it the wind. European historical records show periods in the past with more storms and higher wind speeds, and the annual mean wind speed can vary significantly from year to year. Within the year, systematic changes in weather result in a pattern of changing winds. In Europe, winds are higher in winter than summer. Wind speed varies with passing weather systems; these take typically 3 to 5 days to pass and are known as synoptic variations. Then there are systematic day night variations, known as diurnal. And finally there is the short-term variation of wind speed known as turbulence which is caused by a combination of mechanical mixing as wind flows over terrain features and buoyancy effects due to solar heat input to the atmosphere.

Inter-annual variations in wind speed are not normally modelled explicitly but account is taken of them in wind farm site assessment by looking backward to the mean wind speed over the past 10 to 20 years and assuming that this is the best guide to the future over a similar timescale. Variations of wind speed within the year are dealt with statistically by fitting the 2-parameter Weibull distribution function to 10 min (or sometimes hourly) mean values of wind speed recorded over a year. The Weibull probability density function is given by

$$P(U)\left(\frac{k}{C}\right)\left(\frac{U}{C}\right)^{k-1}\exp\left[-\left(\frac{U}{C}\right)^{k}\right] \tag{9-2}$$

where $k$ is known as the shape parameter as it has a significant impact on the shape of the distribution, and $C$ is referred to as the scale parameter because it is approximately proportional to the annual mean wind speed. Figure 9-1 shows an example of Weibull distribution for $k = 2.2$ and $C = 10$.

There are a number of ways identifying the parameters that give a good fit with measured data, including approximations based on simple statistics, linear fitting to ln(ln) scaled data, and maximum

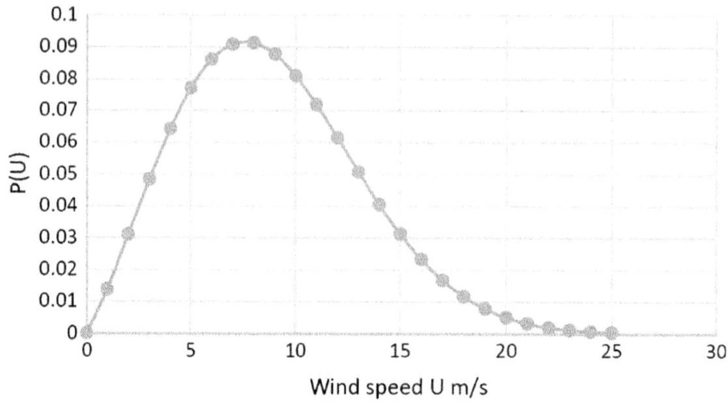


**FIGURE 9-1** Weibull probability distribution for $k = 2.2$ and $C = 10$.

likelihood. The $k$ parameter for most sites is about 2. For the case when it is exactly 2, the distribution is known as the Raleigh distribution. Integrating the Weibull density function gives the cumulative probability distribution:

$$Q(U < V) = 1 - \exp[-(V/C)^k] \tag{9-3}$$

These probability distributions are used together with wind turbine power curves to estimate the energy yield at a given site.

Turbulence intensity is a standard measure of how much the wind varies within a 10-min period [7]. It is defined as $\sigma/U$, where $\sigma$ is the standard deviation of wind speed measured typically at 1 Hz over the 10-min period, and $U$ is the 10-min mean wind speed. Above about 5 m/s this ratio is relatively constant and independent of wind speed for a specific site with a given fetch. It is important as it impacts directly on the fatigue loading of the wind turbine, and also affects power quality.

A particular aspect of wind speed variation that has direct design consequences for wind turbines is the extreme wind that can be expected at a site. This will determine the ultimate loads that the wind turbine must be designed to survive. Although the Weibull distribution has a high wind speed tail indicative of the highest expected 10-min (or hourly) average values, the extreme wind speeds that provide the peak structural loads are gusts, commonly taken to be 3-s gusts. It turns out that estimation of the likelihood of such gusts at a given site requires an appropriate extreme value probability distribution to be fitted to historical data, typically these make use of double exponential functions. A common distribution used to model gust occurrence is

$$P(\le \hat{U}) = e^{-e^{-\alpha(\hat{U} - U_m)}} \tag{9-4}$$

where $\hat{U}$ is the gust wind speed, $U_m$ is the mode of the distribution, and $\alpha$ is known as the dispersion factor. Engineers often talk about a return period (in years) for a particular extreme wind value. This is defined as

$$R = \frac{1}{1 - P(\le \hat{U})} \tag{9-5}$$

Figure 9-2 shows an example with $U_m = 20$ m/s and a dispersion $(1/\alpha) = 5$ m/s.

The risk associated with exceeding a design extreme wind speed for a turbine at a given site over a design life of $N$ years is given by

$$1 - [P(\le \hat{U})]^N \tag{9-6}$$

For the same example parameters considered above the risk as a function of extreme/design wind speed is shown in Fig. 9-3.

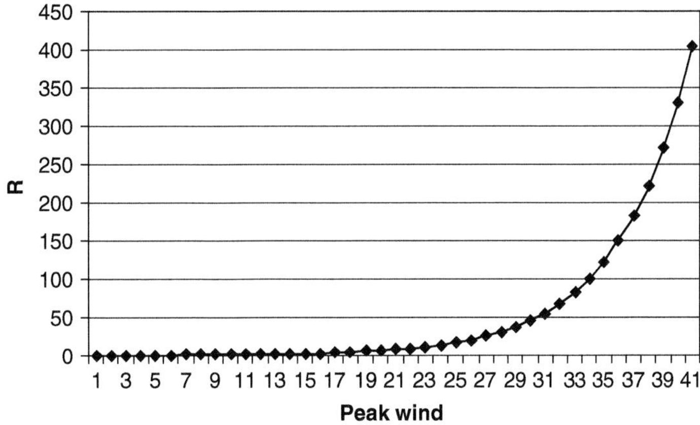


**FIGURE 9-2** Relationship between return period and extreme wind speed for a site with $U_m$ = 20 m/s and $(1/\alpha)$ = 5 m/s.

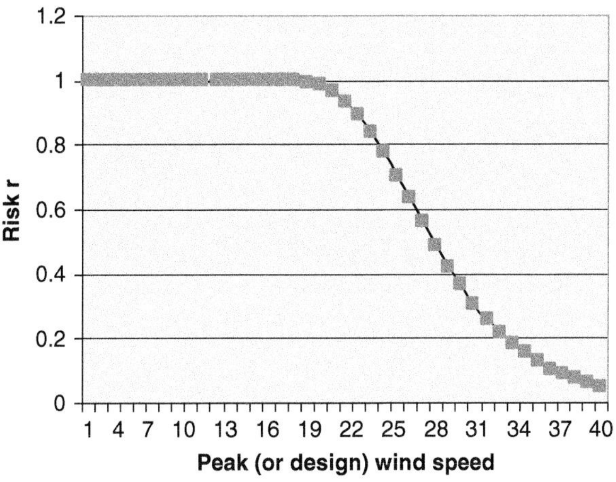


**FIGURE 9-3** Risk as a function of the design wind speed for $U_m$ = 20 m/s and $(1/\alpha)$ = 5 m/s.

***Spatial Variation in Wind Speed.*** The most obvious form of spatial wind speed variation is with height above the ground which is called wind shear caused by the roughness of the ground retarding wind speed near the surface, in principle to zero at the surface itself (the so-called zero slip condition of fluid dynamics). Wind speed above the surface can be regarded as forming a planetary boundary layer. This layer ranges from hundreds of meters up to 1 or 2 km, depending on atmospheric conditions. At the top of the boundary layer wind flow is smoothly driven by pressure gradients and unaffected by the surface. Wind turbines with heights typically around 100 m are well within the boundary layer and thus subject to wind shear. Wind speeds at the top of the rotor can be considerably greater than those at the bottom. The higher the hub height of the turbine the greater the wind speed in general, and the lower the turbulence intensity.

The atmosphere is categorized in terms of stability. Unstable conditions involve potentially large convective currents, these expand the depth of the boundary layer, and through increased mixing, tend to have lower wind shear. At the other extreme is the stable boundary layer with lower turbulence and higher wind shear and a low boundary layer height. Particular cases of such boundary layers can involve temperature inversion and also low-level jets. In between these two stability cases is neutral stability; this is the simplest in terms of the physics and is where mechanical mixing with the ground surface dominates over thermal gradients. It applies onshore when wind speeds are high and thus has tended to be the main case considered by wind engineers. Care must be taken though as non-neutral conditions are now understood to be more widespread than previously thought, especially offshore.

For neutral atmospheric conditions the relationship between wind speed and height is given by

$$U(z) = \left(\frac{U^*}{k}\right) \cdot \ln\left(\frac{z}{z_0}\right) \tag{9-7}$$

where $U(z)$ is the wind speed at height $z$ meters above the ground surface, $k$ is the Von Karman constant (0.4), $U^*$ is the friction velocity, and $z_0$ the surface roughness length (a length in meters that characterizes the roughness of the surface). Because $U^*$ is very difficult to measure it is more common to recast this so called log law in terms of wind speed measured at a reference height:

$$U(z) = U(z_r)\left[\frac{\ln(z/z_0)}{\ln(z_r/z_0)}\right] \tag{9-8}$$

This equation is easy to use to estimate the wind speed at an unknown height provided its value (typically an average value) is known at a reference height, often 10 m in the case of met office data, together with the surface roughness length (that can be estimated from knowledge of the ground cover).

As already indicated, turbulence decreases with height above the ground. An approximate relationship for neutral conditions gives $\sigma = 2.5U^*$. This can be combined with Eq. (9-7) to give

$$I(z) = \frac{1}{\ln(z/z_0)} \tag{9-9}$$

The approximate variation of turbulence intensity with height.

The discussion so far has concentrated on the wind speed component in the downwind (or longitudinal) direction. Sometimes this is denoted by a lower case $u$. Wind however is more complex and includes cross wind (lateral) and vertical components, these are labelled with subscripts $v$ and $w$, respectively. Turbulence is short-term wind speed variation in time (say as measured by an anemometer on a mast). For limited distances, these time variations translate into different locations downstream as if the wind speed variations were convected as a solid block downstream; this effect has the name of Taylor's frozen wake hypothesis. Just as there are spatial variations in wind speed downstream, there will also be variations laterally and vertically, and these can apply to all three wind energy components. It is conventional to calculate length scales for each of these cases that give an idea of the scale in the different dimensions of typical turbulent eddies. This characterization of three-dimensional wind turbulence is important due to the impact on wind turbine rotor loads and is covered in the IEC design standards.

One last aspect of turbulence relates to the frequency content as measured by the spectral density function which specifies how much turbulent variation is present at each frequency. Standard spectra due to Von Karman and Kaimal are widely used in wind analysis. Related to three-dimensional turbulence is the connection between turbulent wind variations at different points in space and at different frequencies. Here the coherence function is used as a measure; it is, in effect, a normalized form of the cross-spectral density function. There are simplified models for coherence that are used in rotor aerodynamic analysis. Commonly the coherence of wind variations across the rotor are assumed to decay exponentially with distance.

### 9.1.3 Wind Turbines

A wind turbine is a device that converts the kinetic energy in wind into mechanical power and then to electrical power. Wind turbines are designed and manufactured in a wide range of capacities from the smallest (measured in tens of watts) for low-power applications such as street lights, auxiliary power sources of boats or caravans or domestic power supplies, to large MW wind turbines, usually connected together to form wind farms for commercial scale electrical power generation into national power grids. Two generic types of wind turbine can be distinguished: horizontal axis wind turbines (the most common by far) and vertical axis wind turbines. Figure 9-4 shows the two types of wind turbines [8].

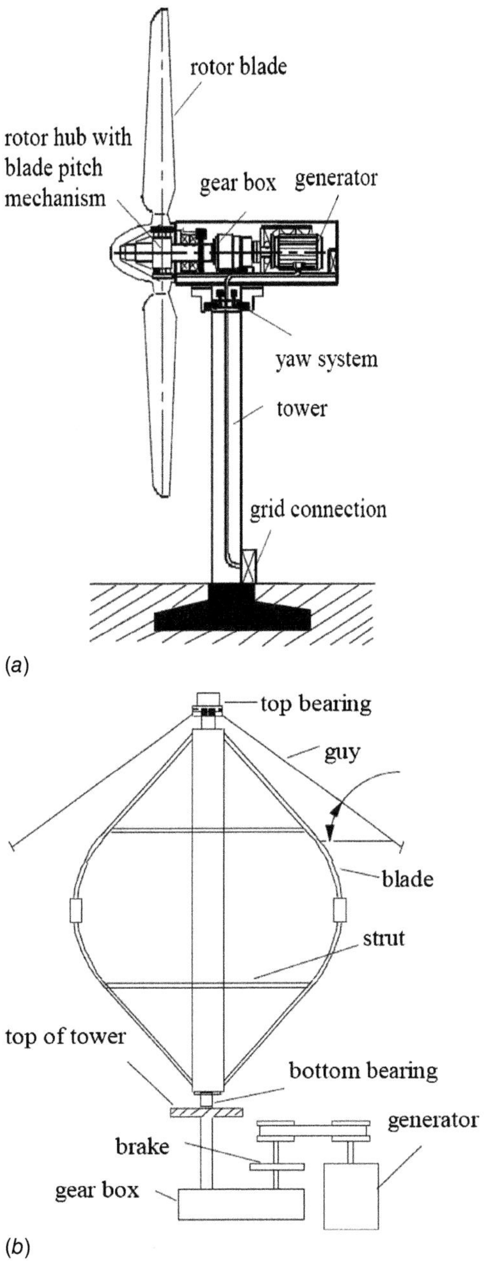

**FIGURE 9-4** Two types of wind turbines [8]. (*a*) Schematic horizontal wind turbine system; (*b*) schematic vertical axis (Darrieus) wind turbine system.

***Horizontal Axis Wind Turbines.*** Horizontal-axis wind turbines (HAWTs) have a horizontal main rotor shaft and an electrical generator located in a nacelle at the top of the tower. The nacelle must be turned (yawed) so as to keep the turbine facing into the wind at all times. Small turbines may be yawed by a simple tail vane (known as passive yawing), while larger turbines have an active yaw control system comprising generally a servo motor(s) and yaw error sensor (usually a wind vain mounted on the nacelle away from the rotor).

A HAWT is usually operated upwind, that is, with the turbine blades rotating in front of the supporting tower with regard to the wind direction. Turbine blades are sufficiently stiff to prevent the blades from being bent into contact with the tower, even in the highest operational winds. Downwind turbines have also been built which can make use of passive yawing, but they are subject to increased fatigue loading due to the reduced wind experienced behind the tower (tower shadow). This can also generate undesirable noise. Most HAWTs used for commercial electricity production these days have three blades and operate upwind.

The tower height and the blade length of the wind turbines tend to increase with the rated generation capacity of the wind turbine. In low-wind-speed areas, higher towers and longer blades may be used than for high-wind-speed areas for a given rated power output. Offshore wind turbines up to 8 MW are now available commercially, and have a blade length up to 80 m. Generally, the larger a wind turbine, the slower is its rotational speed. This follows from the aerodynamics of wind turbine rotors as will be discussed later. Turbines larger than a few tens of kilowatts will commonly use a gear box to increase the rotational speed to match that required for the electrical generator (generally 1500 rpm). Gearboxes can be avoided by using a multipole generator and this arrangement is known as direct drive. All turbines are equipped with control and protective arrangements to adjust the wind turbine in normal operation and to stop the wind turbine under an emergency. Most turbines these days employ blade pitch regulation, and this too is subject to normal and emergency control.

***Vertical-Axis Wind Turbines.*** Vertical-axis wind turbines (VAWTs) have the vertical main shaft. One advantage of VAWT is that the turbine does not need to yawed into the wind. A small VAWT may be easily integrated into a building structure. The generator and gearbox may be placed near the ground for ease of maintenance.

However, a VAWT normally has a lower aerodynamic efficiency than a HAWT, and thus lower energy production for a given rotor swept area. The VAWT blades in each rotational cycle are subject widely changing effective wind direction (angle of attack) and thus loads, which results in power fluctuation and high fatigue loading of the blades, shaft, and tower. Although efforts have been made to improve the performance of the VAWT, most present wind turbines for commercial electricity production are of HAWT type, which will be the main focus in this section.

***Wind Turbine Classification.*** Wind turbines may be classified based on the wind conditions at the site in question. IEC 61400-1 classifies wind turbines from class I to class IV, with A or B referring to the different levels of turbulence (Table 9-2).

**TABLE 9-2** IEC Classification of Wind Turbines [9]

| Wind turbine class | I | II | III | S |
|---|---|---|---|---|
| $V_{ave}$ (m/s) | 10 | 8.5 | 7.5 | User defined |
| $V_{ref}$ (m/s) | 50 | 42.5 | 37.5 | |
| $V_{50,gust}$ (m/s) | 70 | 59.5 | 52.5 | |
| $I_{ref}$ | A | 0.16 | | |
| | B | 0.14 | | |
| | C | 0.12 | | |

In the table (IEC61400-1: 2005):

1. Rayleigh distribution is assumed, i.e., a Weibull distribution with $k = 2$.
2. $V_{ave}$ is the annual mean wind speed at hub height; $V_{ref}$ is the 50-year extreme wind speed over 10 min; $V_{50,gust}$ is the 50-year extreme gust over 3 s; $I_{ref}$ is the mean turbulence intensity at 15 m/s.
3. A, B, and C are the categories of higher, medium and lower turbulence intensity characteristics, respectively.

A wind turbine rotor extracts kinetic energy from the wind and converts it to mechanical power for driving a generator. The power captured by a wind turbine is given in Eq. (9-10)

$$P_M = c_p(\theta,\lambda)P_W = c_p(\theta,\lambda)\frac{1}{2}\rho A v^3 = c_p(\theta,\lambda)\frac{1}{2}\rho\pi R^2 v^3 \tag{9-10}$$

where $P_M$ is the mechanical power extracted from the wind; $\rho$ is the air density; $R$ is the turbine rotor radius; $v$ is the wind speed; $c_p$ is the aerodynamic power coefficient which depends for a given rotor design on the pitch angle $\theta$ and the tip speed ratio $\lambda$.

The tip speed ratio is the ratio between the rational speed of the blade tip and the wind speed. It can be expressed in terms of $R$, $v$, and the rotational speed $\omega_t$, as

$$\lambda = \frac{R\omega_t}{v} \tag{9-11}$$

Clearly, in order to achieve the maximum power output possible at a given wind speed, the wind turbine should be operated at its maximum aerodynamic efficiency, known as $c_{p,\max}$. A group of $c_p - \lambda$ curves for different blade pitch settings, $\theta$, is illustrated in Fig. 9-5 [10]. It can be seen that the highest

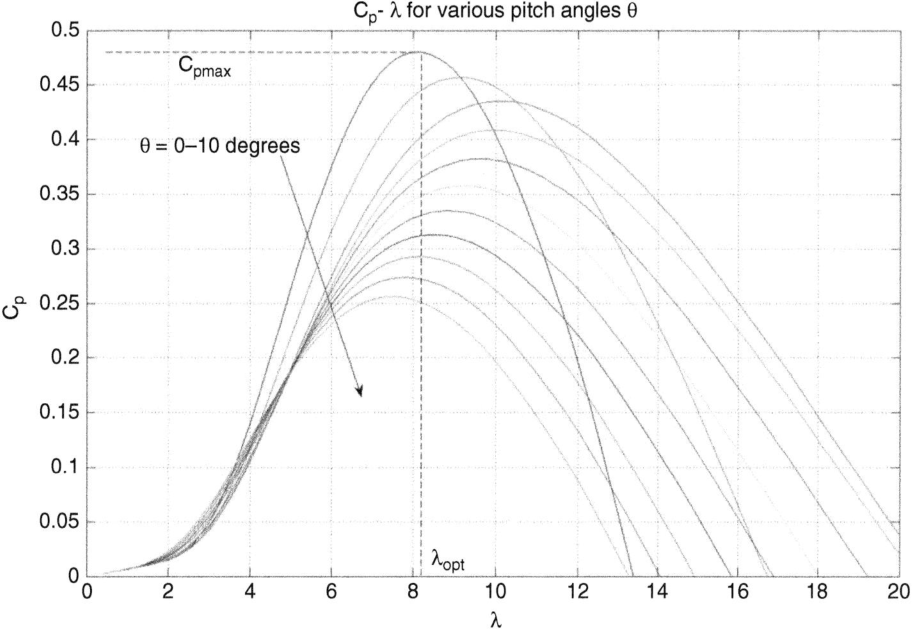


**FIGURE 9-5** $c_p$-$\lambda$ curves for various pitch angles.

power coefficient is achieved at a pitch angle near to zero. Also, for a fixed $\theta$, there is a $\lambda$ corresponding to the power coefficient $c_p$, which gives the maximum power extraction from the wind. This tip speed ratio associated with $c_{p,\max}$ is referred to as $\lambda_{\max}$. Therefore, for maximum power point operation, the wind turbine should be operated with a pitch angle near zero and a tip speed ratio of $\lambda_{\max}$. Considering both the real-world fluctuations in wind speed and Eq. (9-3), it can be seen that optimal operation is achieved by varying the operational speed of the wind turbine $\omega_t$, in proportion to the wind speed so as to stay at $\lambda_{\max}$. Furthermore, variable speed operation of a wind turbine allows the turbine rotor to store or release its kinetic energy, thus reducing power fluctuations and structural loads. In general, for a variable speed wind turbine, the turbine rotational speed is so controlled to keep the tip speed ratio constant at $\lambda_{\max}$ over a range of wind speeds. Generally, if the wind is below the rated speed, a variable speed wind turbine often operates at its maximum efficiency to extract as much power as possible from the wind. When the wind is above the rated speed, the pitch angle is controlled to make the wind turbine operate at a lower efficiency to keep the wind turbine power at its rated power and to limit the mechanical loads as well.

***Wind Turbine Power Curves.*** A wind turbine power curve describes the net output power of a wind turbine as a function of the 10-min averaged wind speed. Data is collected, ideally on a flat and open site, across a wide range of wind conditions. Wind speed is measured at the turbine hub height on a met mast positioned from 2 to 2.5 diameters upstream of the wind turbine. Net power output is measured at the point where the turbine is connected to the power network. Ten minute average values of wind speed and power are recorded. Air temperature and pressure is also measured so that account can be taken of air density variations. The test air density is calculated from Eq. (9-12), where $T$ is the measured air temperature in kelvins, and $B$ is the barometric pressure in mille bars.

$$\rho = 1.225\left[\frac{288.15}{T}\right]\left[\frac{B}{1013.3}\right] \tag{9-12}$$

Following the IEC Standard, the wind speed, rather than the power, is corrected:

$$V_C = V_M \left[\frac{\rho}{1.225}\right]^{1/3} \tag{9-13}$$

where $V_M$ is the measured wind speed and $V_C$ is the density corrected wind speed that will be used to calculate the power curve. The power curve is conventionally calculated using the data reduction technique known as binning. An example power curves is discussed later in the section and shown in Fig. 9-34.

***Wind Turbine Gear Boxes.*** A wind turbine is subject to range of structural loads caused by the manner in which the rotor blades extract energy from the wind. Only the torque component driving the generator is useful for producing electricity. The other loads, such as thrust, are transferred to the ground through the tower. The drivetrain of a wind turbine comprises in general a low speed shaft, a gearbox, a high-speed shaft and a generator. In addition there are additional necessary components, such as power electronics which are used for interfacing a variable speed generator to the fixed frequency ac grid. These play a very important role in determining wind turbine performance, and can also be considered as a part of the wind turbine drivetrain.

The wind turbine gearbox connects with the low-speed shaft attached with the turbine blades on one side, to the high speed generator on the other side. A gearbox may consist of a number of gear conversion stages to transfer the power with lower speed and higher torque into the power with higher speed and lower torque for driving a relative high speed generator, for example, converting the rotor speed in the range 30–60 to 1,000–1,500 rpm for a six- or four-pole high-speed generator or from about 10 to 60–100 rpm for a multipole medium speed generator. A combination of parallel and planetary gears may be used [11]. Further, the torque loading on the gearbox increases significantly with the increase of wind turbine power rating and the associated reduction of wind turbine shaft rotational speed. The increase of gearbox torque seriously challenges the reliability of the wind turbine system because gearbox failure is a major contributor to the wind turbine downtime, and in addition is expensive to repair. Consequently, there has been increasingly interest in direct-drive arrangements using generators with high numbers of poles to eliminate the gearbox from the drivetrain. However, direct-drive designs have their own mechanical design challenges in part associated with very large diameter generators and associated bearing support systems. More operational data is required to determine whether these direct-drive wind turbines are in fact more reliable.

***Simple Lumped Mass Model of the Wind Turbine Drivetrain.*** A number of factors may affect the drivetrain loading, including gearbox and generator designs, power electronic and control strategies. A simplified mechanical model with two lumped masses, shown in Fig. 9-6 and represented by Eqs. (9-14) to (9-17), is normally used to represent the mechanical system for torque transmission. This two-mass model, comprises two lumped masses connected by shafts with stiffness and damping, where the larger

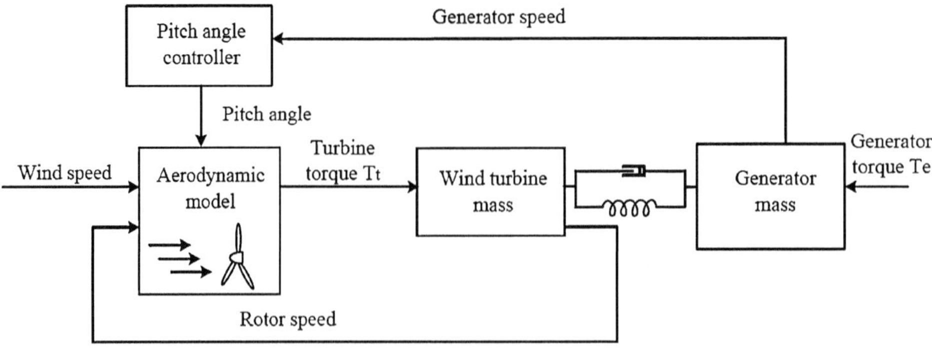


**FIGURE 9-6** A wind power system model with two mass blocks.

mass corresponds to the turbine rotor inertia and a smaller mass represents the generator inertia. The gearbox inertia is often ignored in these models (or it can be split between the two lumped masses), although its ratio must be accounted for. Such a two-mass model has been used for investigating wind turbine dynamics during a power grid fault [12, 13].

$$J_t \frac{d\omega_t}{dt} = T_t - T_{sh} \tag{9-14}$$

$$J_g \frac{d\omega_g}{dt} = T_{sh} - T_e \tag{9-15}$$

$$T_{sh} = D_{sh} \frac{d\varphi}{dt} + K_{sh}\varphi \tag{9-16}$$

$$\frac{d\varphi}{dt} = \omega_t - \omega_g \tag{9-17}$$

where $J_t$ is the wind turbine rotor inertia; $J_g$ is the generator inertia; $T_t$ is the wind turbine rotor torque; $T_{sh}$ is the shaft torque; $T_e$ is the generator torque; $\omega_t$ is the wind turbine rotor speed; $\omega_g$ is the generator rotor speed; $\varphi$ is the shaft twist angle; $K_{sh}$ is the shaft stiffness; and $D_{sh}$ is the damping coefficient of the shaft. The variables here are presented as per unit values.

***Generators.*** The generator is an important component in wind power conversion systems. The input of a wind generator is mechanical power transmitted from the high-speed shaft, the output of the generator is electrical power sent into the electrical grid, often these days through power electronic interfaces.

A conventional generator consists of two main components, a stator and a rotor. The space between the rotor and stator is "air gap," where an electromagnetic field rotates. The electromagnetic field is produced by the currents flowing in generator windings (normally stator windings) and, if the generator is a synchronous machine, the field will be contributed by the excitation current in a field winding or by permanent magnets. The speed of the electromagnetic field is the synchronous speed $n_s$, which is related to the frequency of stator ac current and the number of pole pairs in the generator by

$$n_s = \frac{60 f_s}{pp} \tag{9-18}$$

where $n_s$ = synchronous rotational speed (rpm)
$f_s$ = ac system frequency (Hz)
$pp$ = number of pole pairs

There are two types of generators in common use: the synchronous generator and the asynchronous generator, the latter is also called induction generator. A synchronous generator always rotates at the synchronous speed while an induction generator may run at a speed different from the synchronous speed.

A synchronous generator can be built as a direct-drive generator with a large number of pole pairs to rotate at a low speed, for example, at the wind turbine's rotor speed, then the generator can be directly connected to the wind turbine rotor allowing the gearbox to be eliminated.

For an induction machine, the slip $s$ is defined as

$$s = \frac{n_s - n_r}{n_s} \tag{9-19}$$

$n_r$ is speed of the generator (rpm).

Ignoring the power losses in the induction generator, the power balance may be expressed as

$$P_m = P_s + P_r \tag{9-20}$$

$$P_r = -sP_s \tag{9-21}$$

$P_m$ is the mechanical power, $P_s$ is the stator power, and $P_r$ is the rotor power. Conventionally, if an induction machine rotates at a super synchronous speed, it is in generation mode and the slip is negative. The rotor power of an asynchronous generator is proportional to the slip *s*, and a larger slip (absolute value) means a larger amount of electrical energy going through the rotor circuit.

Both induction generators and synchronous generators are used in wind turbines in various formats [14], the main types of the wind power generators are described below.

***Squirrel Cage Induction Generator.*** Squirrel cage induction generators (SCIGs) were the most widely used for wind turbines during the 1980s and 1990s. A three-stage gearbox was usually used in the drivetrain to connect the low-speed turbine shaft to a four- or six-pole induction generator with the corresponding synchronous speeds of 1500 or 1000 rpm, respectively. A higher pole pair number would decrease the efficiency of the induction generator, thus is rarely used. Consequently, a gearbox is necessary for an induction generator based wind turbines.

The induction generator may be directly connected to the power grid. The upwind, stall regulated, three bladed wind turbine with a grid direct connection of SCIG were widely installed in Denmark in 1980s, which is also referred as "Danish concept."

These directly grid connected squirrel cage rotor induction generator based wind turbines have a very narrow range of speed variation, (reflecting the SCIG slip of typically 1%), which are thus also called fix speed wind turbines. Within this limited speed range, a linear characteristic between the torque and the slip of the generator may be kept, which presents a damping effect. The torque and speed characteristic of an induction machine is shown in Fig. 9-7.

***Wound Rotor Induction Generators.*** A wound rotor induction generator (WRIG) can have a same type of stator as that of the SCIG; however, the rotor is equipped with three-phase windings. In wind power applications, the rotor of a WRIG may be either connected to a resistor, which absorbs the power into the rotor, or connected to the ac grid through a power electronic circuit which can control the power flow in the rotor circuit and through this the generator speed. The latter is also called doubly fed induction generator (DFIG).

*Wound Rotor Induction Generator with Rotor Resistance Control.* Figure 9-8 shows the torque characteristics of an induction generator with different rotor resistance values. It can be seen, for the same level of torque, the larger resistance of the rotor windings, the larger the slip is. Therefore, by changing the effective resistance value using a power electronic control circuit, the slip and the generator speed can be controlled.

The generator has to run at a super-synchronous speed mode for power generation (power into the rotor circuit), and the rotor power has to be dissipated as heat in the rotor circuit, therefore, a higher slip means more power lost in the rotor, a typical variable speed range of WRIG with rotor resistance control is limited within 10% above the synchronous speed. The connection between the rotating rotor and the external stationary circuit is usually made using brushes and slip rings, which introduces additional costs and increases maintenance requirements, and thus is a drawback in comparison with the simple cage rotor induction machine based wind turbines, but a larger speed variation range is an advantage, which can reduce the stress and power fluctuations. Vestas' "Opti-slip" scheme puts the controllable resistor on the rotor so that the brushes and slip rings are removed, but the power dissipation in the rotor circuit remains a problem.

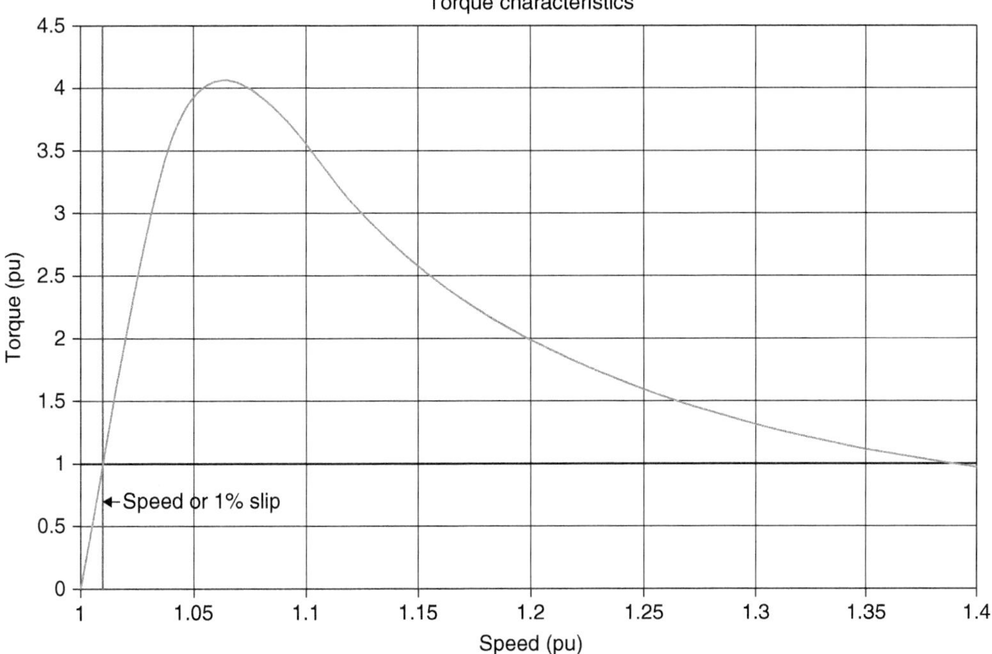


**FIGURE 9-7** Torque and speed characteristics of an induction machine.

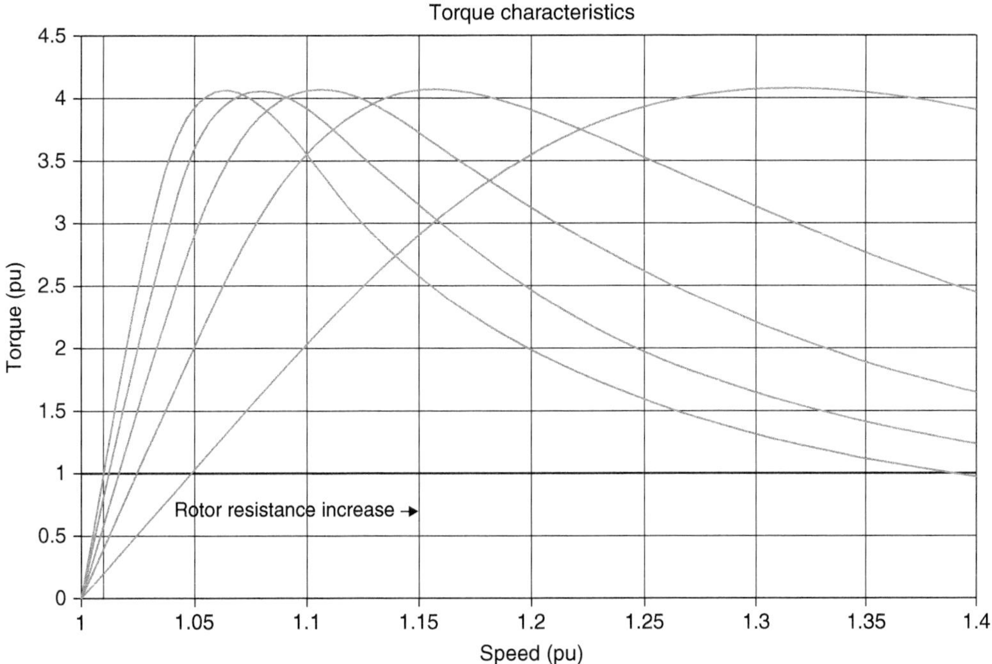


**FIGURE 9-8** Torque and speed characteristics of a wound rotor induction machine.

*Wound Rotor Induction Generator with Power Electronic Converters.* In a doubly fed induction generator (DFIG) based wind turbine, the stator winding of a DFIG is directly connected to the grid, and the rotor winding of the generator is connected to a power electronic system through slip rings, and then to the grid. The power electronic system can control the rotor power flow and thus the rotor speed by adjusting the frequency of the rotor circuit. This concept supports a wide speed range of operation, depending on the rotor power, thus the rating of the power electronic systems. The generator can operate in both super synchronous speed ($s$ <0) and subsynchronous speed ($s$ >0) depending on the power flow direction in the rotor circuit. The rating of the power electronic system is a trade-off between the cost and the desired speed range. Normally a 30% speed variation adopted and the rating of power electronic converters is also 30% of the generator rated power.

***Synchronous Generators.*** A synchronous generator in a wind turbine may not be directly connected to the grid due to lack of damping, unless a continuous gear ratio changing gearbox system is in place. Most commercial synchronous generators for wind turbines use a full-scale power electronic interface. Compared with the variable speed concept of a DFIG, the full-scale power converter can perform smooth grid connection over a wider speed range, and furthermore the fully rated power electronic interface can provide good fault ride-through capability by decoupling the generator from the grid, so that a grid fault may not directly affect the generator. However, it has a higher cost and a slightly higher power loss since all the generated power has to go through the power electronic converters.

Similar to the stator of a SCIG or a WRIG, a synchronous generator generally has a stator made of laminations with uniform slots for stator windings. There are two main approaches to rotor excitation, as outlined earlier. In the first, a field winding is used and supplied by a dc current, this is called an electrically excited synchronous generator (EESG); and in the second, the field is supplied directly by permanent magnets (PMs) attached to the rotor, this is called a permanent magnet synchronous generator (PMSG).

*Multipole Synchronous Machines.* A synchronous generator can be made with multipoles, and then can operate at a low synchronous speed as a direct-drive wind generator without a gearbox, or alternatively to provide medium speed generator compatible with a reduced ratio gearbox than would be used conventionally. The multipole synchronous generator–based wind turbine has a simplified drivetrain, and expected to deliver higher overall efficiency, reliability, and availability in comparison with a normal gearbox drivetrain. The direct-drive generator rotates at a low speed as wind turbine rotor speed, and is subject to higher torque than the generator with gear driven. A higher torque requires a larger size of the generator, therefore, the direct-drive generators (either field excited or PM generator) have a larger diameter for accommodating large number of poles with a reasonable pole pitch.

*Electrically Excited Synchronous Generator.* An EESG is usually built with a rotor carrying a field winding. The stator may carry a three-phase winding quite similar to that of an induction machine while a multiphase arrangement is also possible. The rotor may have salient or cylindrical poles. Salient poles may be used in low-speed machines. The pole pitch of the field excited machine must be large enough for the excitation windings and pole shoes, consequently, a direct-drive EESG generator is a large, heavy, and expensive solution. The large diameter generator may also create difficulties for transportation and installation.

The field current is controllable, for example, through slip rings and brushes by an adjustable dc power supply, or brushless exciter with a rectifier on the rotating shaft. Thus, the EESG has the advantage of a controllable field, for instance, controlling the flux to minimize the losses, although noting that field current losses are inevitable.

*Permanent Magnet Synchronous Generator.* The stator winding of a permanent magnet excited generator may be the same as that of a wound rotor field synchronous machine, but the rotor is fitted with permanent magnets which provide excitation instead of a field winding with a dc field current supply. The use of rare-earth permanent magnets with high-energy density can reduce the generator weight, eliminate the rotor windings and associated insulation degradation and power losses; however, a PMSG might suffer from decreased performance in harsh atmospheric conditions.

Permanent magnet machines require less cooling since the rotor winding and the associated losses are eliminated, which allows permanent magnet generators to be built with a smaller frame than an equivalent wound rotor synchronous machine. These advantages are achieved at the expense of compromising field voltage/current controllability, but the problem is alleviated with the power electronics interface which can provide certain control functionalities.

The PMSG is becoming more attractive for direct-drive wind turbines. Compared with electrically excited machines, PMSGs have a number of economic and technical advantages, such as

- Higher efficiency, energy yield, and better thermal characteristics due to absence of the field winding
- No need for power supply to provide field excitation
- Higher reliability due to the absence of slip rings and brushes

However, PM machines have some disadvantages:

- High cost of PM materials
- Difficulties to assemble
- Potential demagnetization of PM at harsh conditions

There are various structures for permanent magnet machines; radial-flux PM (RFPM) machines, constructed with high number of permanent magnet poles being glued onto the rotor surface are often used for commercial wind turbines.

Many other types of wind generators have also been investigated, such as the stator interior permanent magnet machine (SIPM) [15, 16], switched reluctance generators [17, 18], claw pole generators [19, 20]; brushless doubly fed induction generators (BDFIG) [21, 22], hybrid excitation doubly salient machines [23], etc. Some recent development work includes superconducting generators [24–26] and magnetic gear generators [27, 28].

*Superconducting generators* allow significant increases in power density compared to normal wound copper or permanent magnet machines. Different types of rotating machine using superconducting conductors are possible. For example, using superconducting conductors in the field winding of a synchronous machine, current densities can be significantly increased in comparison with the conventional wound field synchronous generators. The high-current density may make it possible to build more compact machines. Other expected advantages of a superconducting machine include improved efficiency, reduced noise, and vibration.

*Magnetic gear integrated generators* can replace a mechanical gearbox and a gear driven generator. A magnetic gear integrated generator may consist of a magnetic gear with a standard wound stator. The magnetic gear employs rare-earth magnets, and the problems associated with the mechanical gearbox, such as reliability and noise, etc., can be removed altogether. Magnetic gears also have several advantages, including increased efficiency, lubricant-free operation, physical isolation between input and output shafts, inherent overload protection and reduced maintenance costs. A cross sectional of a proposed generator is shown in Fig. 9-9 [28]. The torque produced by this rotor is then geared up in the magnetic gear and carried out via the steel pole pieces, which results in a compact, high-torque density machine.

### *Power Electronics for Wind Generator Interface*

*Power Electronic Semiconductors and Converters.* Power electronic systems are used to convert electrical power from one form to another so as to match the required characteristics in various applications. Power electronics is an enabling technology for many renewable energy based power

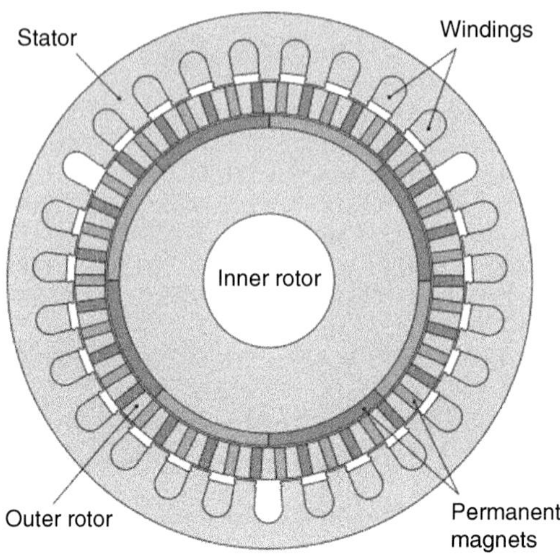


**FIGURE 9-9** Configuration of a magnetic gear generator [28].

generation technologies, such as wind turbines and PV panels. Most renewable energy based generators may only operate at the maximal efficiency point under a set of specific conditions (rotational speed, voltage, etc.) which may not directly match the specifications of the connected grid or load. For example, a wind turbine can have the maximum energy conversion efficiency if it changes operational speed to follow the variation in wind speed (keeping a $\lambda$ corresponding the maximum $c_p$). This results in a varying frequency of the generated power. In order to send the generated power into a grid which has a fixed frequency (either 50 or 60 Hz), power electronic technologies are used to convert the variable frequency power into the fixed grid frequency.

Power electronic technology, including semiconductor devices, circuit topologies, modulation, and control methods [29], will be introduced here with reference to the applications in wind power technology.

The main power circuit of a power electronic converter is constructed by semiconductor devices. Semiconductors may be classified as noncontrollable (diodes), ac voltage commutated (thyristors) and self-commutated (GTOs, IGBTs, IGCTs, etc.) devices. In the power range of wind turbines, IGBTs are most commonly used semiconductors. IGBTs' advantages include simple gate drivers and protection (both overcurrent and overvoltage), possible snubberless operation and a sufficiently high switching speed. IGBTs are available at various voltage and current ratings. New materials, such as silicon carbide (SiC) are being actively researched for power electronic semiconductors; such devices could significantly reduce power losses and dramatically increase the power density of the converters.

The most common converters are dc link type of converters, which may be further divided into voltage source converters (VSCs) and current source converters (CSCs). The most common power electronic converters used in wind power industry are voltage source converters which use a dc capacitor to smooth the dc voltage. Such a VSC is able to perform four-quadrant operation, that is, transferring active power in both directions (from ac to dc and vice versa) and independently absorbing or producing reactive power at both sides. The ac output of a VSC may be a pulse width modulation (PWM) voltage waveform for reducing harmonics. The PWM high-frequency switching techniques are used to move the harmonics and interharmonics into the range of some kHz, which are relatively easy to remove using small filters.

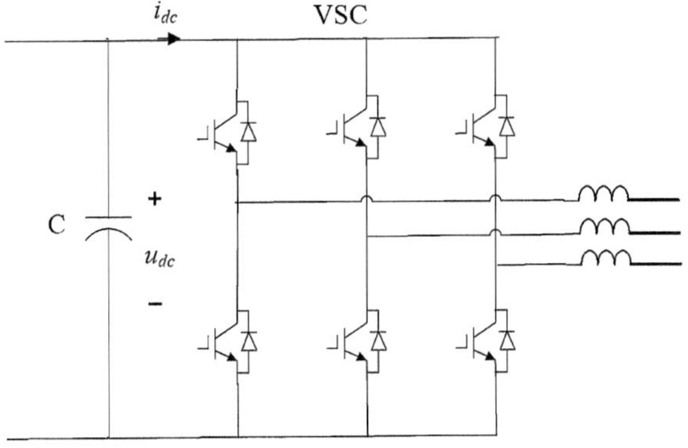


**FIGURE 9-10** Schematic of six-pulse three-phase two-level VSC power electronic converters.

The VSC may have various configurations; two level converters, as shown in Fig. 9-10, are the most simple and basic configuration.

VSCs have other configurations for high-power and lower-harmonics operation, for example, multilevel configurations, including the neutral-point clamped (NPC) converter, flying capacitor converter (FLC), modular multilevel converters ($M^2LC$ or $M^2C$ or MMC), and cascade H-bridge (CHB) converter, etc.

The one-leg circuit topologies of NPC, FLC, and MMC converters are shown in Fig. 9-11*a*, *b*, and *c*. Multilevel converters can vary their output voltage between several levels by switching on and

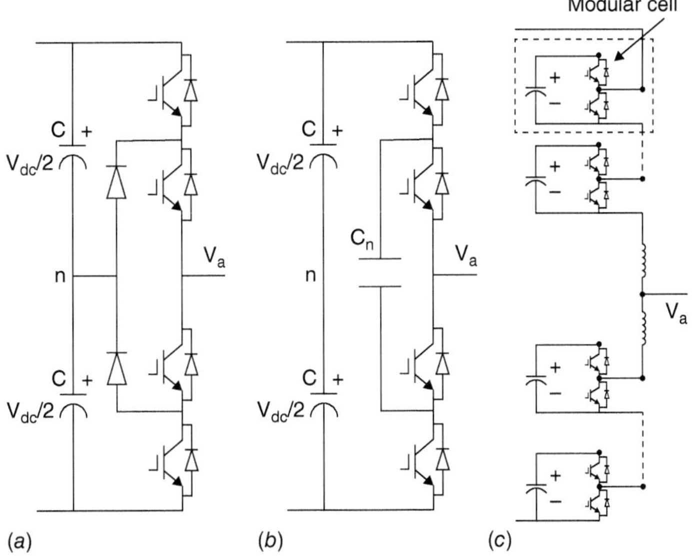


**FIGURE 9-11** One leg/single-phase multilevel voltage source converters. (*a*) One leg of three-level neutral point clamped (NPC)VSC; (*b*) one leg of three-level flying capacitor (FLC) VSC; (*c*) one leg of a modular multilevel converter (MMC) VSC.

off the semiconductors, which can provide improved voltage waveforms, reduced filtering requirements, and reduced common mode voltage and EMI impacts. In a multilevel converter, dynamic voltage sharing and balancing may be achieved by combing the structure design with appropriate control techniques.

A modular multilevel converter (MMC) leg is constructed by connecting a number of converter cells and two inductors as shown in Fig. 9-11*c*. Each cell has its own capacitor which could be charged or discharged by controlling the relevant semiconductors of the cell. The capacitors in the cells can function as distributed dc capacitors. The MMC has the feature of easy construction and assembly, and is suitable for high- or medium-voltage power conversion. MMC applications associated with wind power industry include STATCOMs and VSC-HVDC, the latter can be used for connecting offshore wind power plants to onshore grids and in the future may be used to create a so called super-grid, where a number of offshore wind farms and the power networks are connected with VSC-HVDCs.

*Modulation Techniques of Voltage Source Converters.* Modulation is the process of switching the semiconductors in a power electronic converter in order to produce a desired output by following a reference.

Pulse width modulation (PWM) techniques are very important in modern power electronics and affect power electronic converter performance in many aspects, such as the power losses, harmonics, electromagnetic interference (EMI), etc. Many different PWM techniques have been developed. The PWM methods for VSC applications may be classified as voltage control PWM and current control PWM schemes. The voltage control methods normally produce a controllable ac voltage fundamental component while the current control PWM scheme usually switches the semiconductors to drive the output current by following a current reference waveform, there are different current control PWM schemes, including hysteresis controller, ramp comparison controller, predictive controller, etc. A ramp comparison controller is shown in Fig. 9-12 [8].

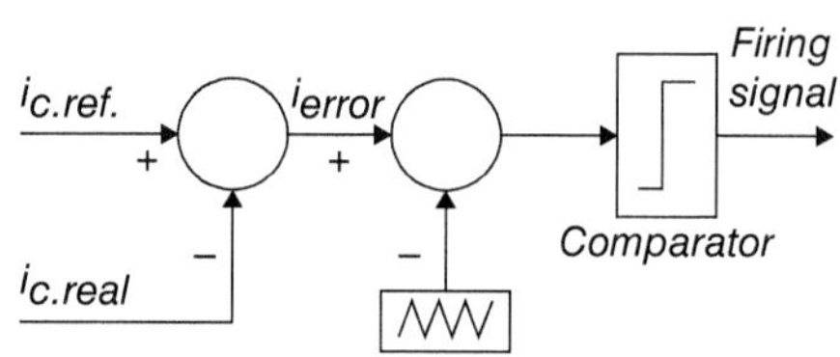


**FIGURE 9-12** Ramp comparison controller of current control PWM schemes [8].

*Voltage Control PWM Techniques for VSCs.* Among many voltage control PWM techniques, the space vector PWM (SVPWM) approach has the advantage of higher efficiency over other methods and thus is commonly used. The principle of the SVPWM is illustrated in Fig. 9-13. The desired voltage $V_{des}$ can be synthesized by three-phase output voltages $v_a$, $v_b$, and $v_c$ in space vector form (harmonics are ignored):

$$\bar{V}_{des} = \sqrt{\frac{2}{3}}(v_a + av_b + a^2 v_c) \tag{9-22}$$

where $a = e^{j120^\circ}$

$$v_a = V_m \sin\omega t$$

$$v_b = V_m \sin(\omega t - 120^\circ)$$

$$v_c = V_m \sin(\omega t + 120^\circ)$$

$V_m$ is the amplitude of the fundamental component.

For a three-phase two-level VSC, there are eight possible switching positions, including two zero output voltage vectors and six basic voltage vectors as shown in Fig. 9-13*a*. The corresponding voltage vectors are depicted in Fig. 9-13*b*.

A desired voltage vector may be synthesized by selecting two of the six basic voltage vectors plus zero voltage vectors as shown in Fig. 9-13*c*, together with their corresponding dwelling periods. For

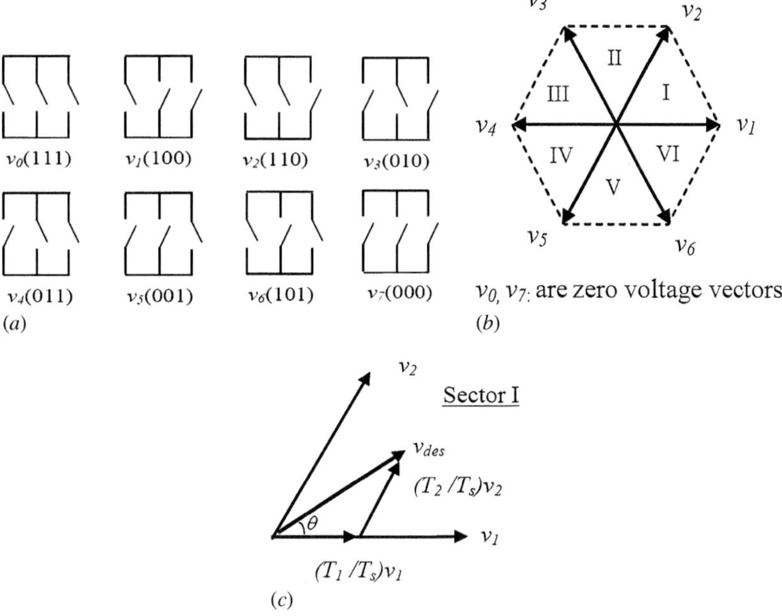


**FIGURE 9-13** The space vector PWM (SVPWM) scheme for two-level VSC. (*a*) Eight switch positions; (*b*) basic voltage vectors; (*c*) voltage vector composition.

example, if the reference voltage vector is located in sector I as shown in Fig. 9-13*c*, then it may be composed by voltage vectors of $v_1$, $v_2$, and zero voltage vectors $v_0$ and $v_7$. The relationships between the desired voltage vector and its two corresponding basic voltage vectors can be expressed as

$$\bar{V}_{\text{des}} = v_1 \frac{T_1}{T_s} + v_2 \frac{T_2}{T_s} \tag{9-23}$$

where $T_s$, $T_1$, and $T_2$ are the dwelling time for $\bar{V}_{\text{des}}$, $v_1$, and $v_2$, respectively. The voltage space vector may be described as

$$T_1 \cdot \sqrt{\frac{2}{3}} V_d \cdot \begin{bmatrix} 1 \\ 0 \end{bmatrix} + T_2 \cdot \sqrt{\frac{2}{3}} V_d \cdot \begin{bmatrix} \cos 60^\circ \\ \sin 60^\circ \end{bmatrix} = T_s \cdot \sqrt{\frac{2}{3}} V_d \cdot k \cdot \begin{bmatrix} \cos\theta \\ \sin\theta \end{bmatrix} \tag{9-24}$$

where $k = |V_{des}| / \sqrt{2/3}\, V_d, 0 \le \theta \le 60^\circ$, and $V_d$ is the dc-link voltage. $T_1$, $T_2$, and $T_0$ can be solved from Eq. (9-24) as

$$T_1 = T_s \cdot k \cdot \frac{\sin(60-\theta)}{\sin 60^\circ}$$

$$T_2 = T_s \cdot k \cdot \frac{\sin\theta}{\sin 60^\circ}$$

$$T_7 = T_0 = \frac{T_s}{2} - T_2 - T_1$$

The shorter of the cycle time $T_s$ is, the smaller the possible deviation between the reference voltage vector and the voltage synthesized by the available voltage vectors. However, a shorter cycle time means higher switching frequency, and higher switching losses. Some work has been done to minimize the switching losses, for example, by arranging the switching sequence so that the transition from one state to another is performed by only one switch.

***Power Control of Voltage Source Converters.*** In a wind power system, a power electronic system is normally required to deliver the desired active and reactive power to meet required operational specifications (voltage, frequency etc.). The controller of a power electronic converter should present the desired references to the PWM generation function block which then drives the converter. A common control approach is vector control, based on a rotational reference frame.

*Reference Frames for Vector Control.* Figure 9-14 shows three sets of reference frames: *a-b-c* three-phase stationary reference frame, $\alpha$-$\beta$ two-axis stationary reference frame, and *d-q* two-axis rotational reference frame rotating at speed $\omega$.

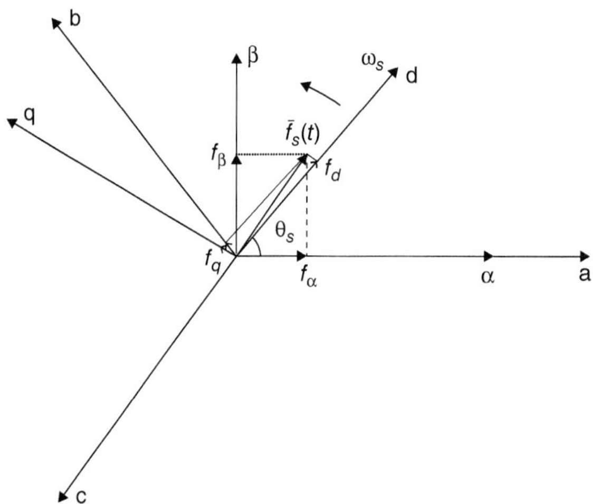


**FIGURE 9-14** Coordination reference frames (*a-b-c* three-phase stationary reference frame, $\alpha$-$\beta$ two-axis stationary reference frame and *d-q* two-axis rotational reference frame).

A set of balanced three-phase variables $(f_a, f_b, f_c)$ in *a-b-c* reference frame may be, respectively, represented by the variables $(f_\alpha, f_\beta)$ in a $\alpha$-$\beta$ two-axis stationary reference frame with Eq. (9-25) and the variables $(f_d, f_q)$ in a *d-q* two-axis rotational reference frame with Eqs. (9-26) and (9-27).

$$\begin{bmatrix} f_\alpha \\ f_\beta \end{bmatrix} = \frac{2}{3}\begin{bmatrix} 1 & -1/2 & -1/2 \\ 2 & -\sqrt{3}/2 & \sqrt{3}/2 \end{bmatrix}\begin{bmatrix} f_a \\ f_b \\ f_c \end{bmatrix} \tag{9-25}$$

$$\begin{bmatrix} f_d \\ f_q \end{bmatrix} = \frac{2}{3} \begin{bmatrix} \cos(\theta) & \cos(\theta - 2\pi/3) & \cos(\theta + 2\pi/3) \\ \sin(\theta) & \sin(\theta - 2\pi/3) & \sin(\theta + 2\pi/3) \end{bmatrix} \begin{bmatrix} f_a \\ f_b \\ f_c \end{bmatrix} \tag{9-26}$$

where

$$\theta = \int \omega dt = \tan^{-1} \frac{f_\beta}{f_\alpha} \tag{9-27}$$

The control of the active power and reactive power can be decoupled by vector control techniques in the rotating $d$-$q$ reference frame. The $d$-$q$ current components of an ac current with a frequency of w will become dc components, $i_d$ and $i_q$, in a $d$-$q$ reference frame rotating at the same speed of $\omega$. If one of the $d$-$q$ axes is aligned with the ac voltage at the converter terminal, the current component on voltage axis will correspond to active power, while the current component on the other axis would represent reactive power. For example, if the voltage is aligned to the $d$ axis of the reference frame, $i_d$ is related to the active power and $i_q$ is associated with the reactive power, and vice versa.

*AC Voltage Oriented Vector Control.* The circuit diagram shown in Fig. 9-15, is used to illustrate the derivation of equations, where $u_{sabc}$ are the three-phase ac source voltages, which may be represented by a space vector voltage $u_s$, as shown in Fig. 9-16, $u_{Cabc}$ are the three-phase converter voltages, $i_{sa}, i_{sb}, i_{sc}$ are the three-phase currents (represented by a space vector voltage $i_s$) from the converter to the ac source, $R_g$ and $L_g$ are the inductor resistance and inductance $C$ is the dc-link capacitor, and $i_{dcC}, i_{dcR}$ are the dc currents.

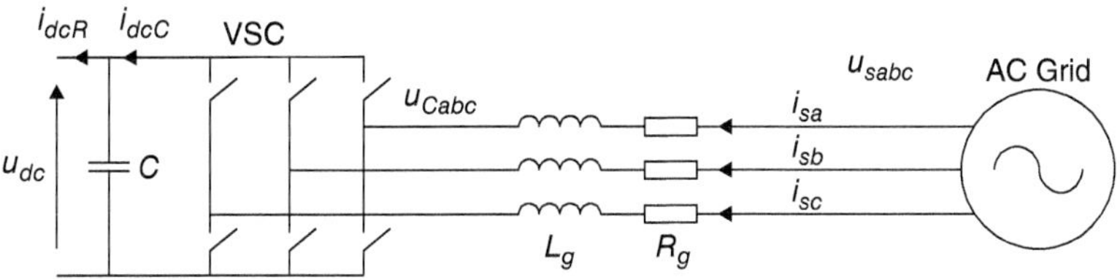


**FIGURE 9-15** A voltage source converter connected to an ac source.

The voltage equations for the circuit shown in Fig. 9-15 may be represented as

$$\begin{bmatrix} u_{sa} \\ u_{sb} \\ u_{sc} \end{bmatrix} = R_g \begin{bmatrix} i_{sa} \\ i_{sb} \\ i_{sc} \end{bmatrix} + L_g \frac{d}{dt} \begin{bmatrix} i_{sa} \\ i_{sb} \\ i_{sc} \end{bmatrix} + \begin{bmatrix} u_{Ca} \\ u_{Cb} \\ u_{Cc} \end{bmatrix} \tag{9-28}$$

Using the *abc* to *dq* transformation [Eq. (9-26)], the corresponding equations in the *dq*-reference frame rotating at $\omega_s$ [rad/s] ($\omega_s = 2\pi f_s$, $f_s$ is the electrical frequency of the ac system) are

$$\begin{aligned} u_{sd} &= R_g i_{sd} + L_g \frac{di_{sd}}{dt} - \omega_s L_g i_{sq} + u_{Cd} \\ u_{sq} &= R_g i_{sq} + L_g \frac{di_{sq}}{dt} + \omega_s L_g i_{sd} + u_{Cq} \end{aligned} \tag{9-29}$$

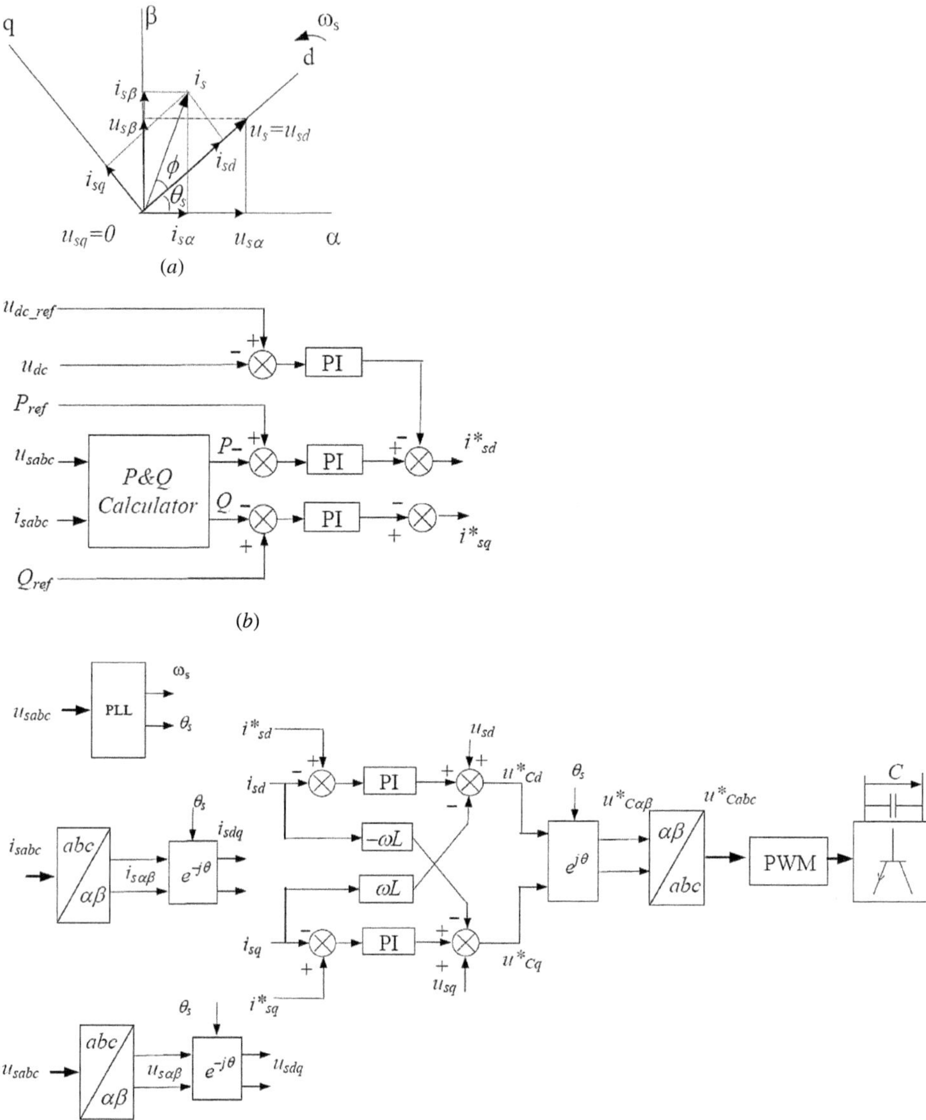


**FIGURE 9-16** VOC on *d-q* reference frame with ac voltage on *d* axis. (*a*) Vector diagram; (*b*) outer loop power controller for reference currents; (*c*) inner loop current controller.

where $u_{sd}, u_{sq}$ are the grid voltages, $u_{Cd}, u_{Cq}$ are the converter voltages, $i_{sd}, i_{sq}$ are the currents, all in *d-q* rotating axis frame. Active power $P_g$ and reactive power $Q_g$ are given by

$$
\begin{aligned}
P_g &= \frac{3}{2}(u_{sd}i_{sd} + u_{sq}i_{sq}) \\
Q_g &= \frac{3}{2}(u_{sq}i_{sd} - u_{sd}i_{sq})
\end{aligned}
\tag{9-30}
$$

The angular position $\theta_s$ of the $dq$ reference frame may be calculated as

$$\theta_s = \int \omega_s dt = \tan^{-1}\frac{u_{s\beta}}{u_{s\alpha}} \tag{9-31}$$

where $u_{s\alpha}$ and $u_{s\beta}$ are the ac source voltage components in the stationary $\alpha\beta$ axes.

Aligning the $d$ axis of the reference frame along the ac source voltage position given by Eq. (9-31), the vector components of voltages $u_{sq}$ is zero, and $u_{sd}$ is equal to the amplitude of the ac source voltage $u_s$. Then the active and reactive power flow between the ac source and the converter will be proportional to $i_{sd}$ and $i_{sq}$, respectively, as given in Eq. (9-32). Therefore, the $P_g$ and $Q_g$ can be, respectively, controlled by controlling $i_{sd}$ and $i_{sq}$.

$$\begin{aligned} P_g &= \frac{3}{2}u_{sd}i_{sd} \\ Q_g &= -\frac{3}{2}u_{sd}i_{sq} \end{aligned} \tag{9-32}$$

The currents $i_{sd}$ and $i_{sq}$ can be driven by converter voltages, $u_{Cd}$ and $u_{Cq}$, respectively. Considering Eq. (9-29), the required voltage $u^*_{Cd}$ and $u^*_{Cq}$ can be expressed as

$$\begin{aligned} u^*_{Cd} &= -u'_{Cd} + (u_{sd} + \omega_s L_g i_{sq}) \\ u^*_{Cq} &= -u'_{Cq} + (u_{sd} - \omega_s L_g i_{sd}) \end{aligned} \tag{9-33}$$

where

$$\begin{aligned} u'_{Cd} &= R_g i_{sd} + L_g \frac{di_{sd}}{dt} \\ u'_{Cq} &= R_g i_{sq} + L_g \frac{di_{sq}}{dt} \end{aligned} \tag{9-34}$$

A vector control scheme with the ac source voltage vector being aligned with the $d$ axis is also called voltage oriented control (VOC). Block diagrams of VOC are shown in Fig. 9-16, where $u^*_{Cabc}$ are the three-phase converter voltage references, $u^*_{Cd}$ and $u^*_{Cq}$ are the corresponding components in $d$-$q$ axis frame, $i^*_{sd}$, $i^*_{sq}$ are the reference of the ac source currents in $d$-$q$ axis frame. The $i_{sq}$ is related to the reactive power flow between the grid and the converter. If a zero reactive power exchange between the grid and the converter is preferred, the reference value, $i^*_{sq}$, may be set as zero. In Fig. 9-16*b*, the dc link voltage reference may also be used, instead of active power reference, in the calculation of active current reference component to keep a desired dc voltage.

The vector controller normally has a nested loop structure where an inner control loop is designed with a significantly shorter time constant than the outer loop. Consequently, it is possible to design the outer control loop while neglecting the dynamics of the inner control loop, which will ease the controller design. The PI-controllers, usually used due to their simplicity and robustness, may be designed using standard methods. As shown in Fig. 9-16, the fast inner current loops controls the $dq$-axis currents (Fig. 9-16*c*), and the outer slower loops are used for torque/power and reactive power control (Fig. 9-16*b*). The $d$-component current reference is generated from an active power reference, while the $q$-current component reference is set to follow a reactive power reference.

*Back to Back Voltage Source Converters (B2B-VSC).* A back to back (B2B) voltage source converter system shown in Fig. 9-17 is the most commonly used power electronic interface system in wind power conversion systems, in either DFIG system or fully rated power electronic systems. The B2B system consists of two bidirectional VSCs connected back to back. One is on the generator side and the other one is on the grid side; these two converters share a common dc link supported by a dc capacitor.

A VSC has the ability to control both the active and reactive power at its ac terminals. The reactive power of the two converters can be controlled separately. For example, the reactive power at

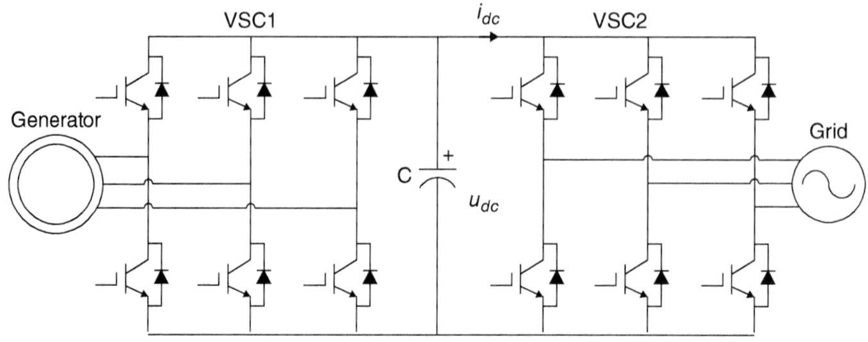


**FIGURE 9-17** A two-level back-to-back voltage source converter (B2B-VSC) system.

the generator side can be controlled to minimize the generator power loss or to keep a constant generator terminal voltage; while the reactive power of the grid side can be controlled as zero (unity power factor) or at a value desired for system operation. However, the active power of the two converters is related. Ignoring the power loss in the system, the active power entering from the generator side should be equal to the active power delivered into the grid, vice versa, so that a desired dc-link voltage can be kept. Consequently, a coordination of active power control of the two converters is needed to keep the dc-link voltage within the specified range. Therefore, the dc-link voltage can be used as a control variable for one of the VSCs. Normally, one converter performs the power control by following a specified power reference (such as shown in Fig. 9-18, and

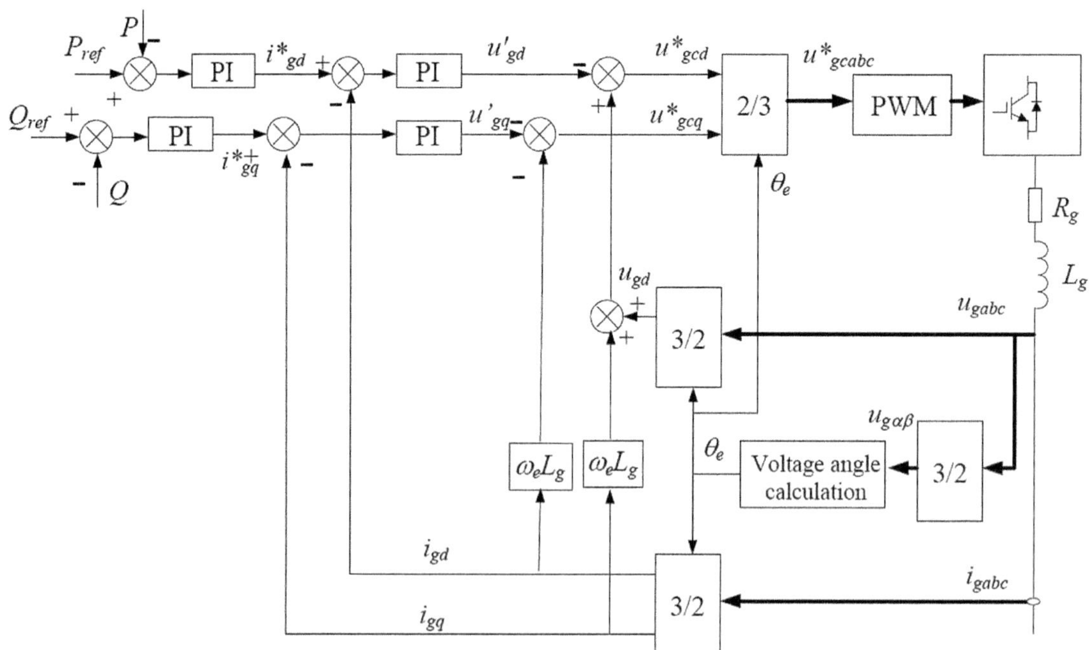


**FIGURE 9-18** P, Q control of a VOC PWM VSC (ac voltage aligned on $d$ axis).

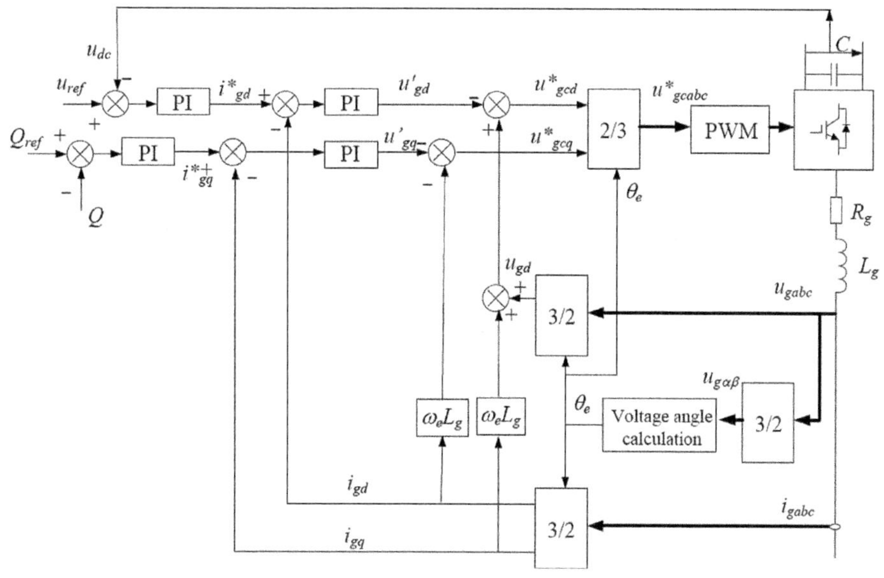


**FIGURE 9-19** $u_{dc}$, Q control of a VOC PWM VSC (ac voltage aligned on $d$ axis).

the other converter regulates the dc link voltage as shown in Fig. 9-19 which uses the $i_{sd}$ reference derived from the dc-link voltage error.

The tasks of controlling active power and dc link voltage may be taken by either one of the converters or shared by the two converters. The roles played by the converters is not critical under normal operation. However, in case of a grid fault, the grid side converter may not be able to function appropriately [29]. The basic control principles and schemes described above can be used for various types of B2B VSC configurations.

*Three-Level Neutral Point Diode Clamped Back-to-Back Topology (3L NPC-B2B).* A three-level neutral point diode clamped back-to-back (*3L NPC-B2B*) converter is shown in Fig. 9-20. A 3L-B2B can produce more output voltage levels thus less *dv*/*dt* stresses, which results in smaller filter size. The mid-point voltage fluctuation may be dealt with by controlling the status of redundant switches. However, the power loss may be unequally distributed between the outer and inner devices in a leg which could lead to derated converter power capacity.

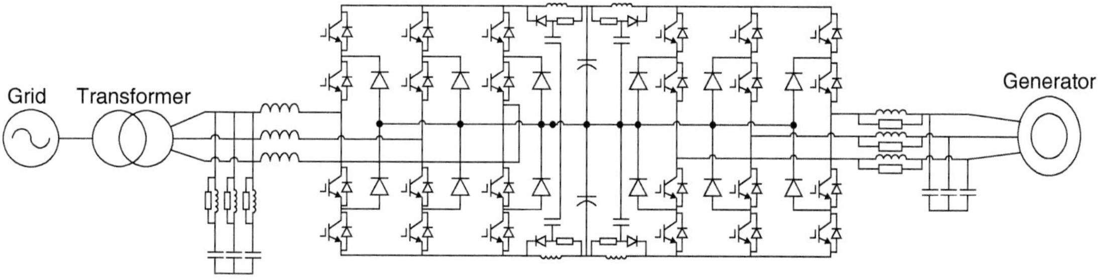


**FIGURE 9-20** Three-level neutral point clamped back-to-back converter (*3L-NPC B2B*).

*Power Electronic Conversion System with Diode Rectifier, DC/DC Converter-VSC.* For a synchronous generator, an electromagnetic excited or a permanent magnet (PM) excited generator, the converter at the generator side does not have to provide excitation for the generator; therefore, a diode rectifier may be used for simplicity and low cost [8, 30–32].

A variable speed PM generator will output a voltage with variable frequency and magnitude, which can be supplied through a diode rectifier into a variable-voltage dc link, then a dc/dc converter may be placed between the generator-rectifier and the grid side VSC as shown in Fig. 9-21 [8, 33]. The dc/dc converter separates the dc link into two voltage levels, one at the rectifier terminal for active power control of the generator-rectifier and the other at the inverter terminal to stable a dc voltage so that the grid side converter can have a high utilization. For the dc/dc converter, an interleave configuration such as the system illustrated in Fig. 9-22 may also be used for providing higher power capacity and better power quality [34].

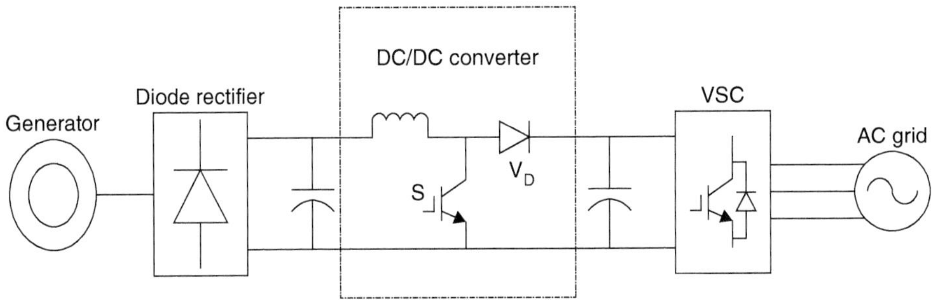


**FIGURE 9-21** Power electronic conversion system with diode rectifier, DC/DC converter-VSC [8].

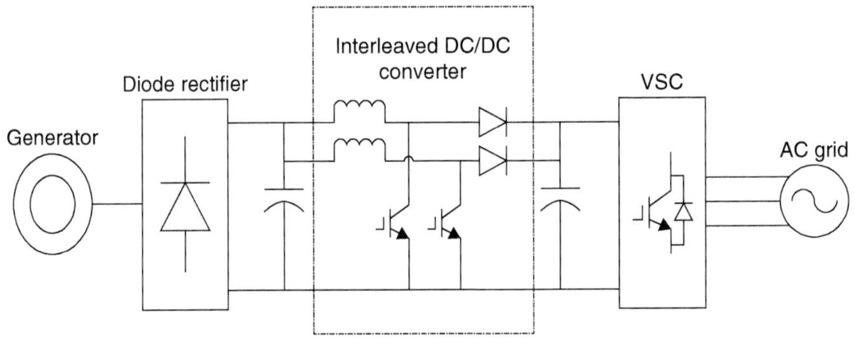


**FIGURE 9-22** Power electronic system with an interleaved DC/DC converter [33].

Similar to the case of back to back VSC converters, the control tasks described for B2B VSC could also be performed except that there is no reactive power control ability with the power electronics on the generator side. The function of active power and dc voltage control may be shared by the dc/dc converter and the grid side VSC.

***Multimodular Diode Rectifier Systems.*** Depending on the generator configuration, the power electronic rectifier system can have different configurations. For a modular PM generator [30], the

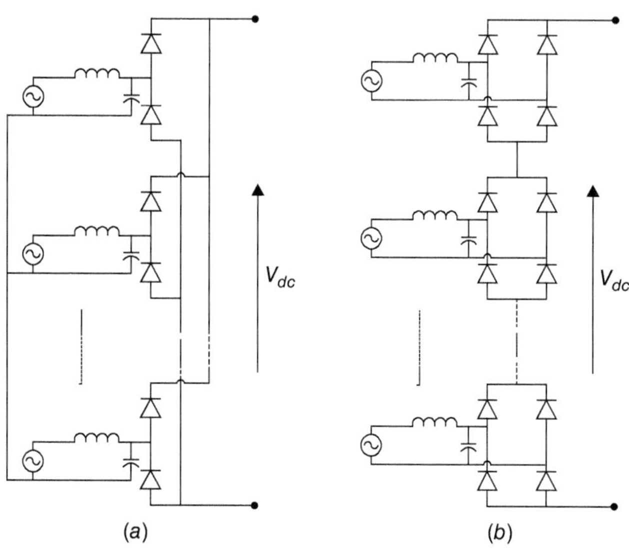


**FIGURE 9-23** Diode rectifier configurations for modular PM generators [8, 31]. (*a*) Star stator coil connection; (*b*) H-bridge stator coil connection.

corresponding modular diode rectifier configuration shown in Fig. 9-23 may be used [8, 31]. The outputs of each coil and rectifier may be connected in parallel to the dc link. The dc-link voltage may have very little ripple because of the large number of diode H-bridges.

***Multivoltage Source Converter Topologies.*** Various voltage source converters may be further connected into multiconverter topologies for higher power delivery and enhanced power quality. Figure 9-24 shows a 24-pulse voltage source converter system with a multiwinding transformer [8]; the output voltage waveform can be synthesized as a 24-step voltage waveform. Another possible multiconverter system is shown in Fig. 9-25.

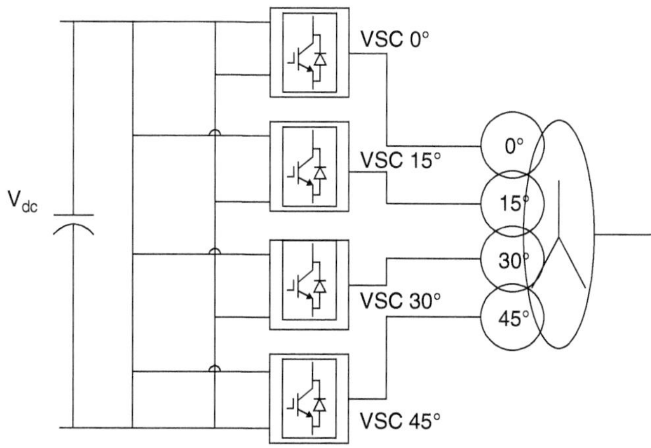


**FIGURE 9-24** A 24-pulse VSC with a multiwinding transformer [8].

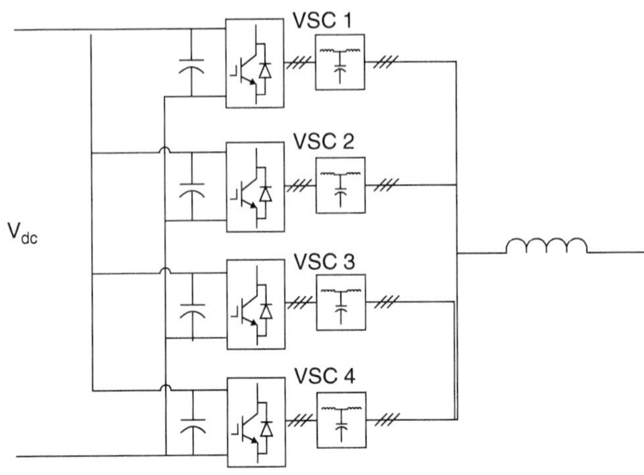


**FIGURE 9-25** Multi-interleaved VSC grid interface.

In addition to the VSCs discussed above, other types of power electronic converters may also be used, such as thyristor or silicon controlled rectifiers (SCR) and current source converters (SCR-CSC) as shown in Fig. 9-26 [8, 35, 36] which can be further configured as high-pulse converters, a 12-pulse SCR-CSCs is shown in Fig. 9-27.

*An Example of Comparative Study.* A comparative power circuit design study [37] has been conducted for a wind power generation system with a full-scale power electronic B2B VSC configuration. The studied converter systems include two-level, three-level neutral point clamped and three-level flying capacitor B2B converters. The system cost, power losses, and power quality were evaluated, including the converter rating, semiconductor rating/dc link voltage level, filter design, modulation methods, and control strategies. The general considerations for designing a power electronic system for wind application have been illustrated, with relevant factors discussed in some detail.

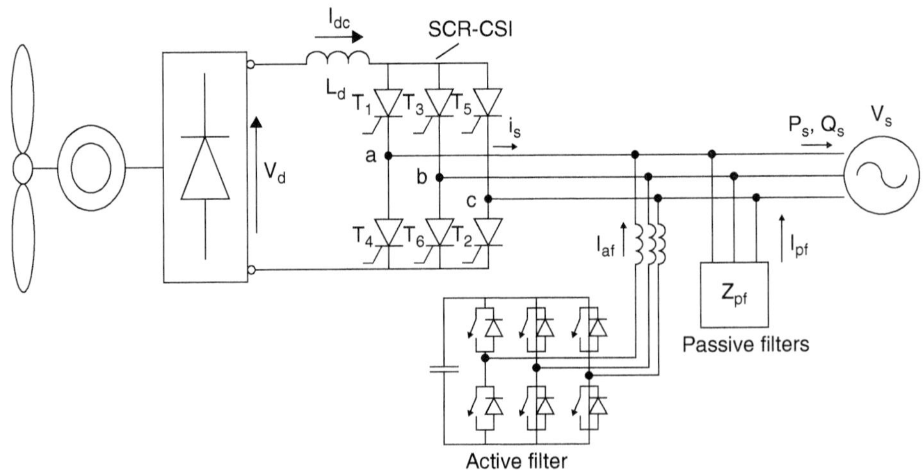


**FIGURE 9-26** SCR-CSC grid converter with parallel connected hybrid compensators [8, 35, 36].

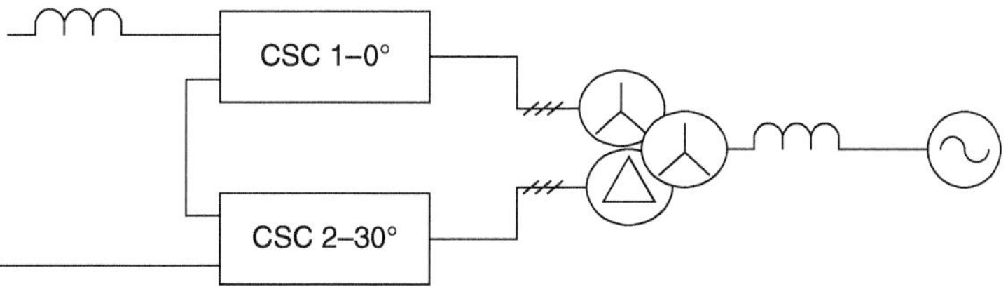


**FIGURE 9-27** A 12-pulse SCR-CSCs [8, 35, 36].

## 9.2 WIND ENERGY CONVERSION SYSTEMS

### 9.2.1 Wind Turbine Concepts

As already mentioned, most commercial wind turbines are horizontal axis, have three blades, and operate upwind of the tower. But the drivetrain, including gearbox, generator, and power electronic interface, may have various different configurations as described above, which consequently determine the characteristics and performance of the wind energy conversion systems. For example, depending on whether a gearbox is used (and also what type of a gearbox), the generator could be at a high speed about 1000 to 1500 rpm, a medium speed of around 100 rpm or a low speed about 10 rpm. Induction generators, including squirrel cage rotor, wound rotor and DFIG normally use multistage gearbox drivetrains and operate at a high speed. A synchronous generator system (either electrically excited or with permanent magnet excitation) may operate at a high speed with a three-stage gearbox or at a medium speed with a single-stage gearbox or at a low speed of direct drive. In this subsection, main types of wind turbines widely used on the commercial market are presented with reference to types 1, 2, 3, and 4 (or A, B, C, and D) wind turbines [38, 39] as summarized in Table 9-3.

**TABLE 9-3** Types of Wind Turbines

| Wind turbine types | Description |
|---|---|
| 1 | Fixed speed wind turbine (FSWT) with squirrel cage induction generator (SCIG) |
| 2 | Wound rotor induction generator (WRIG) with rotor resistance control variable slip |
| 3 | Wound rotor induction generator with B2B VSCs in rotor circuit (DFIG) |
| 4 | Full-scale power electronic (FSPE) interfaced generators (EESG, PMG, SCIG, etc.) |

***Multistage Geared Squirrel Cage Rotor Induction Generator (Type 1: Fixed speed wind turbine).*** A fixed speed wind turbine (FSWT), sometimes also called the *Danish concept*, was widely used in early wind turbine systems. A FSWT is based on a squirrel cage induction generator (SCIG), the high-speed generator is coupled with the low-speed turbine rotor via a gearbox and its stator is directly connected to the power grid through a step-up transformer. The wind power entering the system can be limited aerodynamically by stall or active stall or pitch control. The basic configuration of the fixed speed system is shown in Fig. 9-28. To limit the possible voltage drop caused by high inrush currents when connecting an induction generator to a power system, a thyristor soft-starter is commonly used to limit the inrush current to a level below two times of the generator rated current. After the connection to the grid has been completed, the soft starter is short circuited by a contactor which carries the full load current with a reduced power loss. The soft-starter can also effectively dampen the torque peaks associated with the peak currents and hence reduce the loads on the gearbox. A SCIG needs reactive power for excitation to operate; capacitor banks are connected in parallel to the generator and can be switched in or out according to the active power production to compensate for the reactive power consumption.

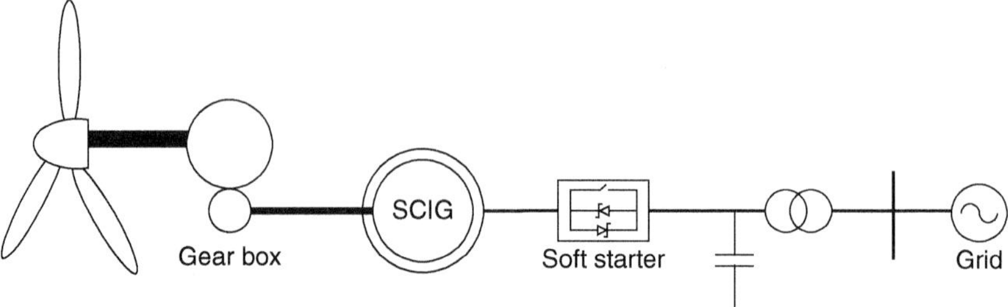


**FIGURE 9-28** Fixed speed wind turbine with multi-stage geared squirrel cage rotor induction generator.

Because the speed variation range of the induction generator is very small, it is referred to as a "fixed speed wind turbine." The advantages of SCIG wind turbines are their simplicity, robustness, and cheap construction, and the fact that no synchronization device is required.

There are some drawbacks of SCIG-based wind turbines. Rotating at an almost fixed speed under all wind speeds is a disadvantage for the aerodynamic efficiency; in addition, it cannot perform maximum power point operation, the noise from the rotor blades may be a problem at low wind speeds, and under a fixed speed operation the fluctuating wind power may cause torque pulsations and create higher loads on the drivetrain. A more expensive mechanical construction may be required to absorb these higher mechanical loads, and the power fluctuations may cause flicker problems on the network. Lastly, such turbines cannot ride through a grid fault and it requires a stiff power grid to enable stable operation.

SCIG wind turbines have made use of pole number changing during the operation; in practice, this is done by changing the connection of the stator windings. The generator operates with a higher number of pole pairs (e.g., $pp = 3$) at low wind speeds; while it is switched into a lower number of pole pairs (e.g., $pp = 2$) at higher wind speeds. A lower rotational speed at low wind speeds yields a higher aerodynamic efficiency and a lower noise emission.

***Multistage Geared Wound Rotor Induction Generator (Type 2: Limited variable speed wind turbine).*** This type of wind turbine is equipped with a wound rotor induction generator (WRIG). The stator winding of the WRIG is directly connected to the power grid and the rotor winding is connected through slip rings and brushes to an external variable resistor illustrated in Fig. 9-29. The equivalent resistance of the resistor can be adjusted by a power electronic system, thus the slip and the rotor speed of the generator can be controlled. The generator may have a range of speed variation up to 10% above synchronous speed. The rotor resistance may also be such controlled in power system

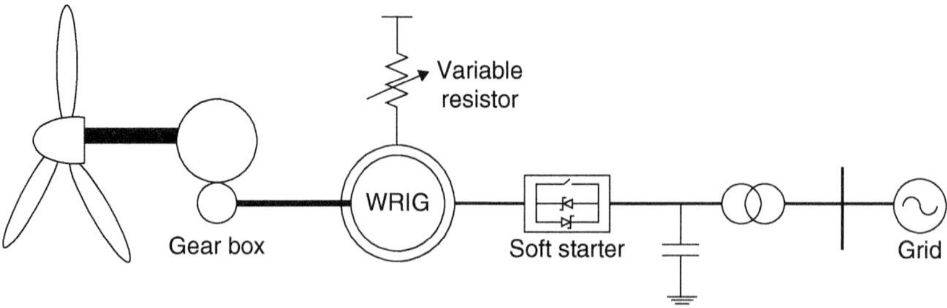


**FIGURE 9-29** Limited variable speed wind turbine with multi-stage geared wound rotor induction generator.

transients to help fault-ride through [13]. Compared to type 1 wind turbine, a type 2 wind turbine is slightly aerodynamically more efficient and has better power quality and lower mechanical stress on the drivetrain. The gearbox, soft starter, and shunt capacitor banks are all required as in a fixed speed wind turbine.

Both SCIG and WRIG need to operate at a super-synchronous speed for power production, also draw reactive power which might be supplied from the installed compensation equipment, such as capacitor banks. For a wind farm with those types of generators, a static var compensator (SVC) with thyristor-switched capacitors or STATCOM may be used to provide normal operational reactive power compensation as well as dynamic compensation to improve the dynamic responses of the wind farm, such as fault-ride through capability.

***Multistage Geared DFIG System (Type 3: Variable speed wind turbine with DFIG).*** The doubly fed induction generator (DFIG), as shown in Fig. 9-30, is a popular modern wind turbine. A DFIG is also a wound rotor induction machine, with the stator directly connected to the electrical grid, and the rotor circuit connected to a B2B-VSC system through slip rings and brushes. The two VSCs share a common dc link, one VSC connected to the rotor of the DFIG and the other connected to the grid. The slip is varied with the power flowing through the power electronic circuit. By controlling the active power of the converter, it is possible to vary the rotational speed of the generator and, thus, the speed of the wind turbine rotor. The converter system is usually so rated to enable a wide range of variable speed operation (normally ±30% of the synchronous speed). This concept is aerodynamically more efficient and has lower drivetrain mechanical stress and power fluctuations. The smaller converter also makes it more cost-efficient.

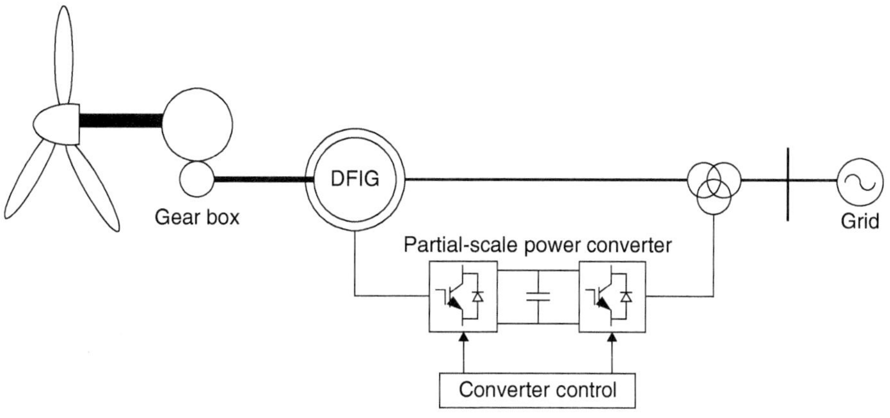


**FIGURE 9-30** DFIG-based wind turbine.

The generator can deliver energy to the grid at both super-synchronous and subsynchronous speeds. The advantages of a DFIG are that the speed can be changed within a sufficient range and for a limited converter cost. Compared with the Optslip concept, the rotor energy, instead of being dissipated, is fed into the grid through the B2B-VSCs.

The power electronic converters have the ability to control the active and reactive power delivered to the grid. The reactive power to the grid from the generation unit can be controlled as zero or to a desired value specified by the system operator within the converter rating limit.

The doubly fed induction generator system also enables the application of special operation strategies and provides the high-quality power to the grid, for example, the acoustical noise from the wind turbines can effectively be reduced by operating the wind turbine system at a lower speed when the wind becomes quiet. The doubly fed induction generator solution needs neither a soft-starter nor a reactive power compensator.

The drawbacks of a DFIG system are as follows:

- A multistage gearbox is necessary in the drivetrain. A gearbox is inevitably associated with some drawbacks, such as heat dissipation from friction, regular maintenance, and audible noise.
- The connection of the rotating rotor circuit with a stationary power electronic converter requires brushes and slip rings, which demand regular maintenance, increase the power losses and are a potential cause of machine failures.
- A grid fault may cause a significant increase of the stator and rotor currents for a short time (~100 ms), which would result in high stress on the insulation of the windings and high torque loading on the drivetrain, consequently affecting the lifetime of the turbine system.
- The system may have difficulties in dealing with fault-ride through and post-fault restoration.

***Full Rated Power Electronic Systems (Type 4: Variable speed wind turbine with full power converter).*** In this concept (Fig. 9-31), the wind power generator is connected to the power grid through a full-scale back-to-back power electronic converter system, which decouples the generator from the power grid and thus enables the wind turbine variable speed operation over a wide range. Compared to the type 3 wind turbine, full-scale converter power electronics (FCPE) provide better controllability and grid support capability.

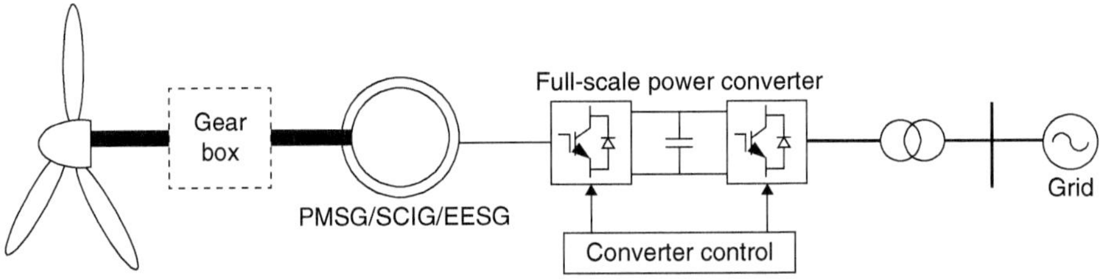


**FIGURE 9-31** Full-scale power electronic converter wind turbine configuration.

In this configuration, the grid performance of the wind turbine system is largely determined by the power electronic interface and control. The power electronic converter enables the system to control active and reactive power very fast. However, the losses in the converter are higher because all power has to be processed by the power electronic converters.

Various types of generators at different rotational speeds could all be used in this system, including the following generators:

- Multistage geared high-speed generators
- Direct-drive low-speed synchronous generators
- One-stage geared medium-speed synchronous generators

These types of wind turbines are discussed below:

***Multistage Geared High-Speed Generators and Full-Scale Power Electronic System.*** Compared with the SCIG-based fixed-speed system, the FCPE SCIG has the feature of flexible control with a variable-speed operation at all wind speeds, better performances of reactive power control and smooth grid connection. Compared to the multistage geared DFIG system, a multistage geared SCIG is a brushless generator and has the better efficiency and grid-fault ride through capability due to the full-scale power electronic system, though the converter is larger and more expensive.

***Direct-Drive Low-Speed Synchronous Generator and Full Rated Power Electronic System.*** Gearboxes are problematic for wind turbines. The fluctuating aerodynamic loads may require a large safety factor for designing purposes. The gearboxes for large wind turbines are more expensive per kW than for smaller turbines because the nonlinearly increased torque, that is, the torque increases more quickly than the power in wind turbines because larger wind turbines operate at lower rotational speeds. The direct-drive concept can simplify the drivetrain, avoid the expensive gearbox and associated problems of reliability, efficiency, maintenance requirements of the gearbox. However, direct-drive generators have large diameter, heavy weight, and tend to be more expensive.

The multipole synchronous generators can be PM based or field excited. Enercon produces direct-drive synchronous generators with electrically exited rotor windings, the generator size is large, for example, the ENERCON E112 wind turbine has a power capacity of 4.5 MW, a generator mass about 220 ton and diameter of 12 m, the weight of the tower head mass (nacelle plus rotor) is 500 ton [40–42] while most other direct-drive designs are based on permanent magnet (PM) generators.

***Medium-Speed Synchronous Generator with Reduced Gear and Full Rated Power Electronic System.*** With the increase of wind turbine capacity and the decrease of wind turbine rotor speeds, either gearboxes become more heavy and expensive as with a multistage geared system, or generators become larger and more expensive as with a direct-drive system. Between the above two cases, a hybrid system with a single stage gearbox and a medium-speed generator is a further option, which may result in a simple and more reliable gearbox and a not excessively large generator.

This concept has a lower generator cost than the direct-drive concept, and a lower gearbox cost, higher availability, and operating reliability than the multistage geared drive concept. An overall optimal system design may be obtained [43], which may be able to minimize capital costs, or maximize reliability or minimize the cost/kWh of the wind energy.

The Multibrid M5000 5-MW wind turbine has some distinctive features, including a highly compact integrated slowly rotating drive system, comprising a single main bearing, a single-stage gearbox and a medium speed PM generator (58 to 147 rpm). With a tower head mass of 310 tons, the M5000 is apparently a light wind turbine for its size, and is rated at 5 MW [42].

***Summary.*** This subsection has described various types of wind turbines drivetrains, wind power generators, power electronic interfaces, and common commercial wind turbine types.

The important distinctions between different drivetrains are related to the generators used and power electronics [44, 45]. A type 1 wind turbine is the simplest wind turbine consisting of a multistage gearbox and a SCIG. A type 2 wind turbine has a wound rotor induction generator with a controllable rotor resistance, the generator slip can be continuously adjusted, and so is the rotational speed accordingly. The allowable speed variations are typically about 10% above synchronous speed. A type 3 wind turbine consists of a wound rotor induction generator with the rotor connected to the grid through a partial rated power electronic converter, called DFIG. The speed variation range is proportional to the power going through the rotor circuit, normally ±30% around the synchronous speed. The type 3 wind turbine is quite popular in modern multi-MW wind turbines. The variable-speed operation of the wind turbine can lead to a higher aerodynamic efficiency, reduced noise, and mechanical stress. All induction generators, including squirrel cage rotor, wound rotor with rotor resistance control and DFIG, normally require a multistage gearbox and have a synchronous speed 1500 or 1000 rpm (2 or 3 pairs of poles).

Type 4 wind turbine use a full-scale power electronic interface; the generators can be induction generators or synchronous generators. An induction generator based type 4 wind turbine will need a multistage gearbox while a synchronous generator in type 4 wind turbine system can use a multistage gearbox like an induction generator or a single-stage gearbox or without a gearbox.

A gearbox is a heavy and expensive component and a direct-drive system without a gearbox is attractive for a range of reasons. However, a direct-drive multi-MW wind turbine leads to a large and heavy generator. An alternative hybrid drive approach, using a medium-speed generator with one- or two-stage gearbox, can reasonably reduce the dimension and improve the reliability. The bearing, gearbox, and permanent magnet generator may be fitted directly into a cast-iron frame with the generator bearings being part of the gearbox assembly.

Power electronic systems are vulnerable components within a wind turbine system, and their loading and thus the thermal performance is a very important factor affecting their reliability. The development of new power electronic devices, circuits, and control, such as silicon carbide devices, could deliver promising improvements in performance.

## 9.2.2 Control of Wind Turbine Systems

A wind turbine should be highly efficient and reliable and capable of delivering the captured power with high quality in normal operation, and maintaining safety under emergencies. In this aspect, the functions of control and protection are very important and can involve a number of devices, in particular the pitch angle controller, the power electronic converter, yaw control, reactive power compensators, and protection/braking systems. In this subsection, the main methods of controlling active power, pitch control, and power electronic control are introduced. Then reactive power control approaches, such as capacitor banks, static VAR compensators (SVCs), STATCOMs, etc., are presented.

***Active Power Control.*** The maximum design power capacity (the rated power) of a typical wind turbine is normally reached at a rated wind speed (around 10 to 15 m/s) while the normal wind speed range for operating a wind turbine is between 4 and 25 m/s. The main active power controls are the pitch control (to control the input power) and the power electronic control (to control the output power or electrical loading). When the wind speed is below the rated wind speed, the power electronic control may be applied on the generator side, by adjusting the electromagnetic torque so as to move the system toward the desired operational speed for maximum power production. If the wind speed exceeds the rated wind speed, the wind turbine input power is limited at the rated power by power limitation methods, such as passive aerodynamic stall induced on the blades or active pitch of the blades or moving the system to a less efficient operation point (speed) by electrical loading control. The wind turbine power may even be controlled by yawing it. However, such yaw control may bring additional loading to the structure.

***Control of Input Power with Aerodynamic Approaches.*** Power control methods used in modern wind turbines may be classified into three categories: passive stall (the blade is fixed and operated at fixed speed, stall occurs along the blade at high wind speed due to an increase in the angle of attack), the active stall control (the blade angle is adjusted to create controllable stall along the blades), and the pitch control (the blades are turned out of the wind at high wind speed). The first two alternatives have often been used in earlier fixed-speed wind turbines, while the pitch control turbine is commonly used in variable speed wind turbines. The passive stall turbine has basically no active control action in normal operation condition. The active stall turbine may pitch the blade to optimize the power production when the wind speed is below the rated wind speed and keep the rated power reference when wind speed is higher than the rated wind speed.

*Fixed Blade Passive Stall Wind Turbines.* When the angle of attack of air flow onto an airfoil exceeds a certain critical angle, the air flow separates from the airfoil surface resulting in a collapse of the lift force and a dramatic increase in drag. The blades contribution to torque and also power thus falls quickly. The stall process is illustrated in Fig. 9-32. This feature is used for passive stall wind turbines to limit the power input.

In a fixed blade wind turbine, the blades are mounted on the rotor at a fixed angle. The attack angle of the air flow incident to the blades is determined by the blade speed and the wind speed. The fixed-pitch blades are designed slightly twisted to optimize the angle of attack along the blade and to ensure stall occurs gradually starting from the blade root, when the wind speed reaches the rated wind speed. The higher the wind speed, the greater the area of the blade that is in stall so that the captured power is limited.

The advantages of stall regulated wind turbines are that they are cheap and simple and have no moving parts. However, the power limitation is not very precise and the output power may not be

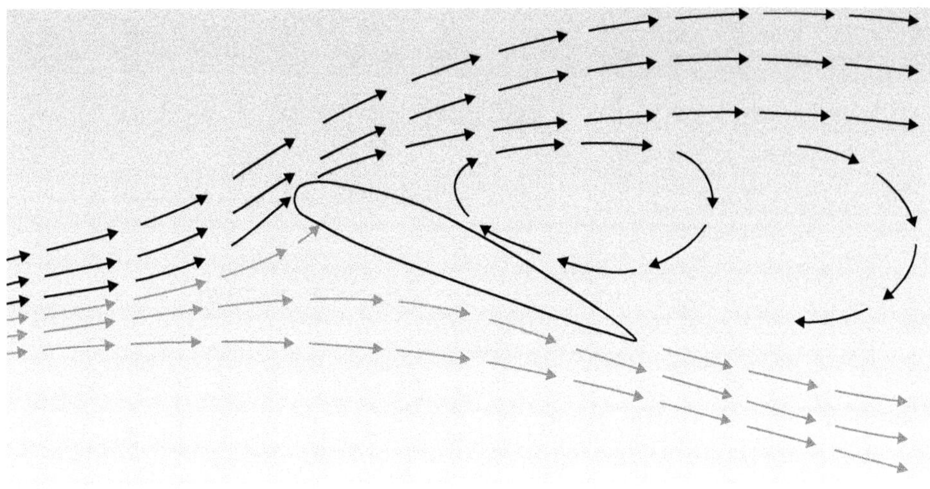

**FIGURE 9-32** Turbulence and wind turbine stall [46].

held at the constant rated power. Stall may lead to an output power drop in the high wind speeds. The output power of a stall regulated turbine is also affected by air density and ambient turbulence levels, and by aerodynamic influences such as dirt on the blades. Furthermore, the blades are heavier than for a pitch controlled blade as a consequence of higher thrust loads. With the increase of wind turbine size, stall regulated rotors tend to be heavier and more expensive. This is one of the reasons that large modern wind turbines tend to use pitch rather than stall regulation.

*Blade Pitched Wind Turbines (Pitch Regulated Wind Turbines).* In a blade pitched turbine, the blades are mounted on the rotor hub with turntable bearings and the blade can be controlled by rotating the blade around its long axis during operation. The aerodynamic torque can be reduced by pitching the blade either toward feather (reducing the angle of attack) or toward stall (increasing the angle of attack). In high winds, a wind turbine with adjustable pitch can provide smooth power, because the pitch angle can be continuously adjusted to change the lift force, for example, to reduce lift force and thereby to limit the wind power capture. The pitch mechanism is usually driven by hydraulics or electric motors, and the rotation speed of changing the pitch angle is limited by the strength of the pitching mechanism and blade inertia. More importantly, the blade pitch system can be used to stop the wind turbine by acting as an aerodynamic brake, preventing excessive over speed when a turbine loses its electrical loading.

*Blade Pitched Wind Turbines (Active Stall Regulated Wind Turbines).* Active stall turbines use the pitching blades and stall phenomena for power limitation. The blades are pitchable, like the blades in a pitch-controlled system, but, instead of pitching the blades out of the wind, it pitches to produce stall. In low and medium wind speeds the pitch may be controlled to yield maximum power output for a fixed speed turbine. However, in order to keep the rated power, the blades are adjusted toward the opposite direction comparing with a normal controlled pitch by increasing the angle of attack, inducing stall and limiting the power at the desired level.

Comparing with a passive stall system, the active stall system can present a smooth power curve without output power reduction in high wind speeds and without overshoot beyond the rated power. However, stall with associated turbulence and high thrust produces loads on the blades which would require them to be strong and thus somewhat heavy.

***Pitch Angle Controller.*** A generic pitch angle controller is illustrated in Fig. 9-33, where the control parameter X can be the power or the torque or the speed or even a combination of the above variables.

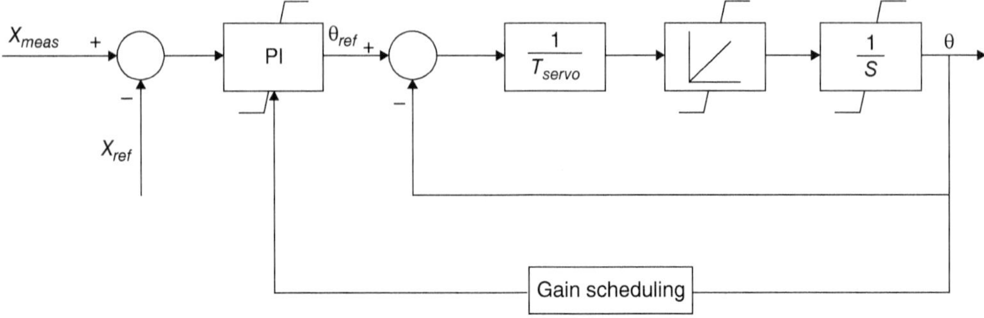


**FIGURE 9-33** Generic pitch angle controller for variable speed wind turbines.

The pitch angle control can be implemented with a PI controller. The time constant $T_{servo}$ corresponds to the delay caused by the servo system, which can be a hydraulic system or an electrical drive system. Gain scheduling control may be used to compensate for the nonlinear aerodynamic characteristics of the rotor. The limiters represent the limits to the pitch control dynamics and pitching angle.

In general, the pitch angle can be controlled to: (1) optimize the power generation of wind turbines at a wind speed below the rated wind speed; (2) prevent the mechanical power from exceeding the rated power when the wind is above the rated wind speed; (3) protect the turbine mechanical construction from overloading and risk of damages; and (4) stop or shut down the wind turbine in normal or emergency situation. Most pitch control system work in a collective manner, that is, all blades are turned at the same time with a same angle. The blades, however, experience different wind speed (at different positions) and thus different loads, consequently individual pitch control methods have been investigated to reduce these loads and improve the power quality [47, 48].

The typical power-wind speed curves for stall and pitch regulated turbines (active stall and pitch control) with a cutout wind speed of 25 m/s are shown in Fig. 9-34; note that cutout wind speeds may vary depending on the wind turbines manufacturers. It can be seen that the output power increases with the increase of wind speed if the wind speed is below the rated wind speed. For a stall controlled wind turbine, there is overshoot of output power around the rated wind speed and then the output power reduces at higher wind speeds.

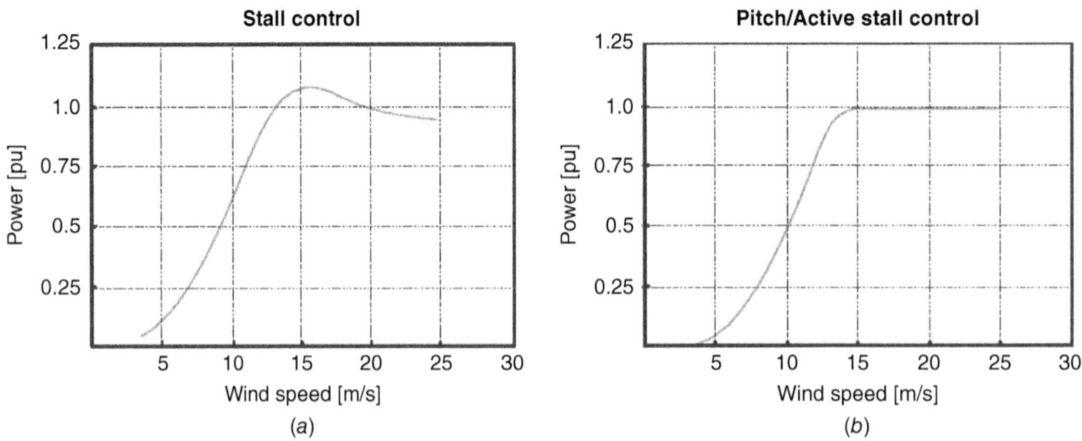


**FIGURE 9-34** Characteristics of wind turbine input power (*a*) stall control (*b*) pitch/active stall control.

For a pitch controlled variable speed wind turbine, the simplest control mechanism using only electromagnetic torque control, with little or no pitch action, is applied if wind speed is below the rated wind speed so that the wind turbine speed is adjusted to optimize power capture or reduce mechanical stress, while the pitch control is used to limit the wind power input for the conditions of above-rated wind speeds. Under such a scheme, the torque/power controller and the pitch controller may be designed separately, thus the design of controllers becomes easier. Furthermore, wind turbines have to be stopped around the cut off wind speed to avoid damage from the excessive mechanical stresses, the cut off wind speed is normally about 25 to 30 m/s.

Research and development work has been conducted on improving the wind turbine aerodynamic performance, for example, vortex generators (VGs) have been developed to improve the aerodynamic properties of the blades by energizing the flow around the blade surface and reducing flow separation. The performance of the entire turbine in terms of power, loads, and service life can be therefore improved [49].

***Control of Output Active Power of a Wind Power Generator.*** Wind turbine operational speed and efficiency is related to the variations in and balance between the input power and output power; the output power and electrical (air gap) torque loading of a wind turbine maybe adjusted using different approaches.

The traditional wind turbines with induction generators (type 1 wind turbine) are connected directly to grid, the rotor speed is almost fixed by the grid frequency and the output active power may not be actively controlled.

For type 2 wind turbine, the wound rotor induction generator with the rotor circuit connected to a controllable resistor, the operation point may be changed by adjusting the value of the rotor resistor, however, that rotor resistor control may only be used for over synchronous speed situation, the active power propositional to the rotor speed is dissipated in the rotor circuit.

Only variable speed wind turbines, the power electronic converter–based DFIG (type 3) and the full-scale power electronic converter–based wind turbine (type 4) have the possibility to effectively control the active output power of a wind power generator and the operation points.

*Control of Output Active Power of Variable Speed Wind Turbines.* The most commonly used power electronic system in variable speed wind turbines are back-to-back-connected (B2B) power electronic voltage source converters (VSCs). The VSC control described previously in this section is used. Type 3 variable speed wind turbine uses the partially rated B2B VSCs in the rotor circuit of DFIG, while type 4 variable speed wind turbine normally uses a full-scale B2B VSCs. In both cases, the B2B VSCs control the electrical power output of the generator, thus, the loading electrical torque, to affecting the wind turbine efficiency and dynamic performance. An overall control system of a DFIG is illustrated in Fig. 9-35.

***DFIG Wind Turbine.*** Figure 9-35 sketches an overall control system of a variable speed DFIG wind turbine. Two control loops with different bandwidths are

- Wind turbine controller
- DFIG B2B VSCs controller

The DFIG B2B VSC controller further consists of two controllers, which can control the two VSCs, respectively, with a coordinative way, for example,

- *Rotor side converter controller* controls the active and reactive power from the stator to the grid.
- *Grid side converter controller* controls the dc link voltage $U_{dc}$ and reactive power from the rotor circuit to the grid.

The reactive power from the DFIG to the grid may be controlled by the rotor side VSC through the stator or the grid side VSC or by shared control of both VSCs.

The wind turbine controller, as an upper level controller, provides reference signals to the pitch controller and the power controller as illustrated in Fig. 9-35. The set point $P_{ref}$ for power control may be obtained from a lookup table as a function of the measured rotor speed $\omega_{mean}$. The lookup table

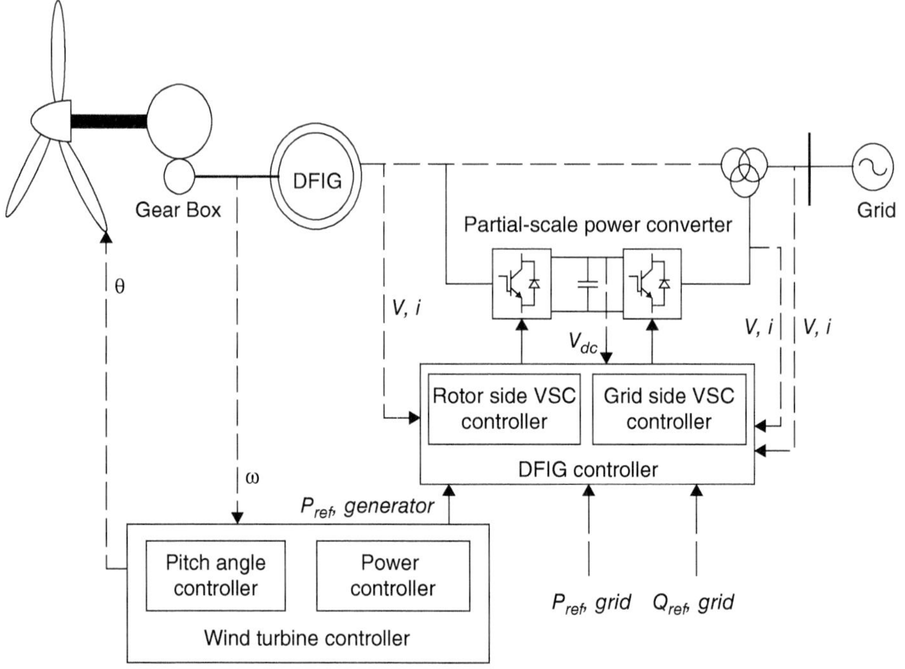


**FIGURE 9-35** Overall control system of variable speed wind turbine with doubly fed induction generator.

is created to represent the relationship between the operational speed and power in the speed range below the rated wind speed; for higher wind speeds, the rotor speed is limited around a nominal rated value by control of the blade pitch angle $\theta_{\text{pitch}}$.

***Full-Scale Power Electronic Interfaced Wind Turbine.*** For a full-scale power electronic interfaced variable speed wind turbine, a similar control structure to that of DFIG may be used, however, the power electronic converter system has to handle all the power from the generator. Similar to the DFIG's control scheme presented previously, the controller of both converters in a full-scale power electronic interfaced wind turbine may also use the structure of two cascaded control loops. The faster inner current controller, regulating the currents to the reference values, is specified by the slower outer loop power controller. The current controller provides reference signals in the $d$ and $q$ axes for the PWM-controlled VSCs in a similar way to that described previously.

Active power control needs the coordination of the generator side VSC and the grid side VSC, the two converters can control active power and the dc link voltage, respectively, or exchange the two control variables or share the two control variables, as previously discussed. For example, one converter can follow a preferred active power reference, the other converter balances the dc link voltage, which ensures the power transfer from the generator terminals to the grid side.

A grid fault would cause the voltage drop at the terminal of a wind turbine and cause the wind turbine to accelerate due to the unbalance between the aerodynamic torque and electrical load torque which drops due to the voltage drop. The voltage drop may even disable the function of the grid side converter. One scheme is to use the generator side VSC to control the generator terminal voltage and dc link voltage. A potential advantage of the control strategy is to reduce the risk of dc link overvoltage [50]. During a grid fault, the generator-side converter is less affected by the fault since it is not directly coupled to the grid. It may therefore, in addition, employ protection methods such as a dc link chopper to absorb the energy and protect the system.

***Summary of Active Power Control of Modern Wind Turbines.*** Wind power technology development is undertaken to produce the maximum energy at the lowest possible cost, that is, at the minimum price per kWh. This also means that the control strategy should maximize energy production over the turbine lifetime. Thus, the active power reference may be so set to maximize available wind power, while the control strategy aims to reduce loads on the structure and the mechanical components, in order to reduce the maintenance cost and increase the reliability and lifetime of the turbine. Furthermore, the performance of wind turbines or wind farms has to meet the requirements of the power grid especially when large-scale wind power integration is concerned.

The active power control schemes for a typical modern wind turbine can be as follows:

1. *Below the Rated Wind Speed.* Under this situation, the loading on turbine structure is generally small, a common control target is to realize the maximal power capture. In this operation region, the generator torque is controlled while the pitch angle is kept constant because the power capture is maximized for a particular value of the pitch angle (near zero degree). A control method may be to set the generator torque reference $T_{gen}$ as $T_{gen} = k\omega^2$, or generator power reference $P_{gen}$ as $P_{gen} = k\omega^3$, where $\omega$ is the generator speed and the constant $k$ is chosen so that the generator reference corresponds to the condition that $\lambda = \lambda_{opt}$.
2. *Above the Rated Wind Speed.* In this condition, the control objective is to keep the power output of the turbine at the rated value and the associated loads on the turbine structure within the designed limits. The strategy for regulating power is to use the pitch control to keep generator power at the rated value or keep generator torque constant and regulate generator speed at the rated value.

In addition to maximize energy capture and limit structure loads, the drivetrain damping may also be built into the control function, because drivetrain vibrations may seriously impact the gearbox life and the replacement of a failed gearbox is expensive and time consuming, which results in the loss of wind energy.

***Reactive Power Control.*** For variable speed wind turbines, either DFIG or full-scale power electronic interfaced wind turbines, the generator side VSC and the grid side VSC could control their related reactive power separately as previously discussed.

For a DFIG system, the grid reactive power is the sum of the reactive power from grid side VSC and the reactive power from the generator stator, the latter is controlled by the generator side VSC. Therefore, the exchanged reactive power between the wind turbine system and the grid can be controlled coordinately by the two VSCs.

For the full-scale power electronic interfaced wind turbines, the reactive power control at the generator side is completely decoupled from that at the grid side, therefore, the reactive power to the grid will depend only on the grid side power electronic converter. The grid side reactive power may be set by the grid operator, for example, following a unit power factor or a specified reactive power or an ac voltage reference, while the generator side converter can optimize the reactive power for the generator operation, for example, keeping a constant generator terminal voltage or minimizing the generator power loss as described below.

*Keep the Generator Terminal Voltage to the Rated Value.* The control strategy keeps the generator at the rated voltage and prevents the possible overvoltage and saturation in case of overspeeds. Under this control strategy, the generator presents a varying reactive power demand with the changes of active power production. Such reactive power demand needs to be met by the generator side power electronic converter, which should consequently be sized appropriately.

*Keep Zero Generator Reactive Power.* The control strategy keeps the generator at zero reactive power, that is, a unit power factor, to minimize the generator current and thus the generator power losses, then the stator voltage would be varied with the active power production.

The traditional wind turbines with induction generators without power electronic interfaces are normally equipped with capacitor banks using mechanical contactors; the capacitor banks are typically controlled in intervals of 1 to 10 min, which may not be fast enough to support the required grid voltage control. In order to provide a faster reactive power control, static VAR compensator (SVC), such as the thyristor controlled reactors and thyristor switched capacitors, or STATCOM, can be used to provide fast reactive power control and significantly reduce the switching transients of mechanical contactors, such reactive power compensation devices may be installed at the wind farm level to be more cost-effective.

***Wind Turbine Braking Systems.*** Several types of braking methods may be used for wind turbines, including aerodynamic brakes, mechanical brakes and electrical brakes. Aerodynamic brakes do not require large forces to overcome the rotor torque and can be easily made fail-safe type system. For a pitch adjustable turbine, the aerodynamic brakes are directly applied on the turbine rotor blades by pitching the blade to stop the wind turbine; for a stall wind turbine, a small tip section of the blade is normally made turnable as an aerodynamic brake to stop the turbine. Mechanical disc brakes are commonly located on the high-speed shaft to act as the parking brake and/or to be used in conjunction with aerodynamic brakes. Electrical dynamic brake may be another option by applying electrical torque/loading on the generator for stopping the turbine.

### 9.2.3 Wind Farms

A wind farm consists of a group of wind turbines located closely together at a site and connected together electrically. It is a cost-effective way for commercial electricity production in terms of infrastructure, maintenance, management, and control coordination. A large wind farm may consist of several hundred wind turbines and cover a large area. Many large wind farms have been built around world, including some very large wind farms, for example, the Gansu Wind Farm in China (6000 MW) and the Alta Wind Energy Center in California, United States (1320 MW).

***Site Selection and Assessment.*** Wind farm sites are selected based on a number of factors, one of the most important factors is the wind resource, which for flatter terrains, may be obtained from a wind atlas, and validated with wind measurements.

In more complex terrain, the technique of measuring correlate predict is widely applied. In this approach 10-min averages of wind speed are measured over a year or more, usually at a range of heights, together with wind direction, from a met mast installed at the site. Concurrent wind data is obtained from one or more nearby meteorological stations. These data are grouped by direction sector of a chosen size, and correlations between the site data and the met office data calculated. Least squares linear fitting is most commonly used for this, and the scaling ratio (from the fit gradient) is then used to recalibrate the long term (say 20-year record) from the met site(s) to the candidate wind farm site. The long-term resource, often represented by a fitted Weibull distribution, is used to estimate the expected energy yield from the site for wind turbines with a known power curve. This may be adjusted to account for any height differences between the measured data and the intended wind turbine hub height. The probability calculation to give the mean power output is

$$\overline{P} = \int_{u=0}^{\infty} p(U)\,P(U)\,dU \tag{9-35}$$

where the function $P(U)$ is the wind turbine power curve giving power $P$ in terms of wind speed $U$. $p(U)$ is the fitted Weibull probability distribution. In practice, this integral is computed numerically. For a wind farm on $N$ similar turbines the mean power is adjusted to reflect the expected wind turbine availability (typically 98% onshore, but usually lower offshore, sometimes significantly) and for wake losses through a wind farm wake efficiency factor of typically 0.9.

***Wind Farm Layout and Wake Effects.*** Wind turbine interaction in a wind farm is dominated by wake effects, which reduce the wind speed seen by turbines operating in the wake of downstream turbines. The overall effect is to reduce the output of the wind farm. Operation in the wake of one or more wind turbines also results in higher fatigue damage due to the higher levels of turbulence created in the wake region. In earlier times, some simple rules were used for the placements of wind turbines within a wind farm. For example, a separation distance of 8D (D: rotor diameter of the wind turbine) to 12D was often used in the main wind direction and 3D to 5D in the cross-wind direction. As a result, some offshore wind farms, where wakes persist for greater distances due to lower ambient wind turbulence, exhibited disappointingly low wind farm wake efficiency factors. The process of wake development is now acknowledged to be more complex than previously thought, and widely use simplified wake models are presently treated with care. Atmospheric stability is known to impact on wake development but most models do not properly take this into account. There are also so-called *large array effects*, where very large wind farms actually impact on the boundary layer itself and some models attempt to compensate for this. The complexity and importance of accurate wake modelling has prompted a considerable interest in using CFD, both by researchers but also by commercial wind farm designers. Finally, in principle, the placement of wind turbines can be optimized by using an objective function (e.g., annual wind energy production or cost of wind energy) linked to the individual wind turbine positions. This can work well onshore where differences in elevation can impact significantly in individual turbine output due to wind speed-up effects that can be estimated by simplified flow codes, or also by CFD.

***Wind Farm Electrical System.*** The voltage level of electrical equipment is often designed with reference to the amount of power associated with the equipment. The wind power generators are normally designed for operation at low voltage (i.e., below 1 kV) due to the relatively small capacity of the wind turbine (below 10 MW). The collection system of a wind farm is an electrical network for collecting the power from individual wind turbines together for supply to the power grid. The wind farm collection network may deal with a few hundred or even thousands of MW, and the desired voltage level is at normal distribution system voltages, for example, 10, 20, 35, and 60 kV. Normally, an individual wind turbine is equipped with a transformer stepping up from the generator terminal voltage to the voltage of the collection system within a wind farm. If the wind farm is small and close to the grid connection point, the wind farm may be connected to the main grid at the collection system voltage level. If the wind farm is large and distant from the grid connection point, a wind farm transformer may be needed to further increase the voltage from the collection system voltage to the network voltage, which may be at transmission level. In such a situation, the wind farm needs a substation, which is similar to the substation used with conventional large-scale generation plant. In the substation, the wind farm main transformers are located with other equipment, such as busbars, circuit breakers, switchgear, as well as wind farm level reactive power compensation devices, such as capacitors, reactors, SVCs, and STATCOMs if required. Figure 9-36 illustrates a typical wind farm electrical system configuration.

***Wind Farm Control.*** Active and reactive power control of each individual wind turbine has been previously discussed. A wind farm monitoring and control system communicates with all the wind turbines in the wind farm, collecting local information from each wind turbine and issuing the control commands to the wind turbines in the wind farm. With the recently increase in wind power penetration experienced by many power system operators, wind farms, especially large wind farms, are treated to an extent as conventional generators which should be able to follow demands for given levels of active and reactive power. Consequently, an advanced wind farm control system needs to coordinate with the individual wind turbine controllers to optimize the operation of the wind farms and to support the grid, and to make a wind farm resemble conventional generators, providing active power (frequency) and reactive power (voltage) regulation in normal operation, and emergency control during faults.

A wind farm control system may have a hierarchical structure with a central controller to control the active and reactive power following a specified operational mode in normal operation; during a transient response, such as a fault-ride through, the control action makes use of the individual wind turbine controller for fast response, while the reactive compensation equipment at wind farm level would also be used to effectively support the system voltage during a power system transient.

According to the power reference command from a power system operator, a wind farm controller must determine the power reference set points for each individual wind turbines in the wind farm, so that each wind turbine controller can contribute to meeting the overall contributions from the wind farm required by the system operator.

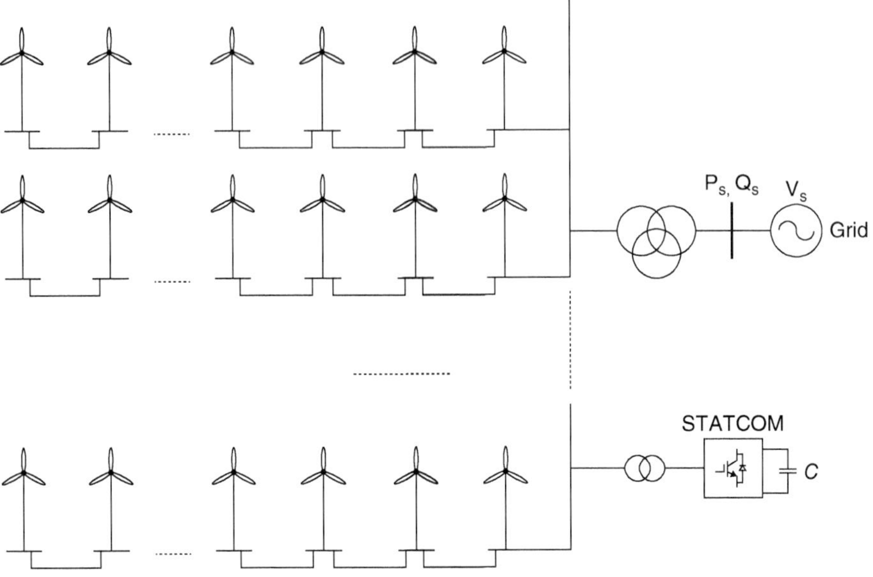


**FIGURE 9-36** Layout for a wind farm with ac collection system.

A generic wind farm control system may have a structure as illustrated in Fig. 9-37. The wind farm controller receives the active and reactive power set points from the grid operator. The system voltage, current and frequency measured at the point of common coupling (PCC) are used to calculate the delivered active and reactive power, together with data from individual wind turbine in the wind farm. The data from the wind turbines may include wind speed and/or available power at the individual wind turbines in a short time frame (i.e., from seconds to minutes).

The system operator may specify the operational modes, discussed later in this section, such as balance control, delta control, power ramp rate limiter, frequency control curve, reactive power, or voltage control to make the wind farms to operate like conventional generation plant. Such selectable options are illustrated in Fig. 9-37. The operator may also specify the required amount of power reserve capacity. The possible upper limits of the active and reactive power outputs of the wind farm can be decided based on the current conditions of the wind farm. The upper limit of active power may be the rated power of the whole wind farm or the sum of the available power of each individual wind turbines in the wind farm. The reactive power limit is related to the active power being delivered and reactive power compensation equipment at wind farm level.

Active power control will be performed for active power and/or frequency control, while the reactive power control would be conducted for the reactive power and/or voltage control. Both the frequency and voltage control functions can be implemented at the wind farm level. Frequency control (droop response) is limited by the active power reserve set with associated derating of the turbine, while the voltage control is limited by the maximum available reactive power. In case that only wind farm's active or reactive power production needs to be controlled, the voltage or frequency control modes may not be selected. The power control loops can consist of a PI controller with anti-wind-up to ensure a correct output for the whole wind farm.

The dispatch units convert the control signals for the whole wind farm into power reference signals for each individual wind turbines and send these out to the wind turbines within the wind farm. The conventional dispatch scheme is often based on a proportional distribution of the available active and reactive power, respectively. However, new control methods are being investigated for improving power quality [51–53] and realizing optimization [54–56].

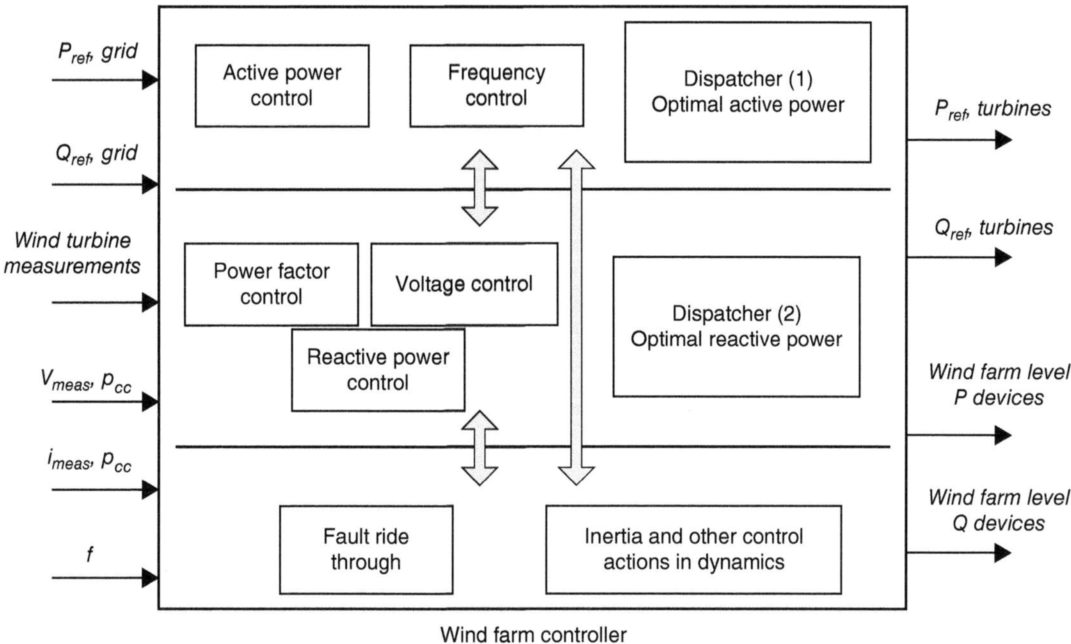


**FIGURE 9-37** Illustrative structure of a wind farm controller.

## 9.3 OFFSHORE WIND POWER

*Offshore wind power* refers to the wind turbines or wind farms installed at sea. Europe is the global leader in offshore wind power. Most of the world's offshore wind power is currently installed in the North, Baltic, Irish Seas, and the English Channel. The world's first offshore wind farm (Vindeby) was constructed in Denmark in 1991. Offshore wind plays an important role in achieving Europe's renewable energy targets. Many other countries and regions outside of Europe have also shown interests in offshore wind power, including China, India, Japan, South Korea, Taiwan, and the United States.

Onshore wind energy is one of the most cost-effective renewable technologies, but its development has been limited by available sites and also notably by public hostility reflecting concerns about noise and visual impacts. Offshore wind farms are normally far from shore and thus have much reduced impacts so that objections from local communities are much reduced than for onshore developments, although not completely absent.

While the offshore wind industry has grown dramatically over the last several decades, and numerous investigations related to various aspects of offshore wind power have been conducted, there are still some unresolved environmental concerns, for example, issues associated with the marine environment, in particular the risk of seabirds being struck by wind turbine blades, the impacts of the noise associated with the turbine foundation installation, and the physical presence of offshore wind farms on the behavior of fish and seabirds.

Because the sea has lower surface roughness (smaller $z_0$), good wind conditions pertain with low turbulence levels and as a result, offshore wind turbines may provide higher energy production than the land-based wind farms per MW of installed capacity. The offshore wind turbines also generally have higher capacity factors than wind turbines at onshore and near shore locations.

However, the equipment installation and service/maintenance of offshore wind farms are more expensive, which is a significant challenge for the development of offshore wind farms. Few service vessels can lift the large and heavy components required for offshore wind turbines, which are now rated up to 8 MW. Special service vehicles for installation as well as for maintenance are in high demand. Figure 9-38 shows the cost distribution of offshore wind power.

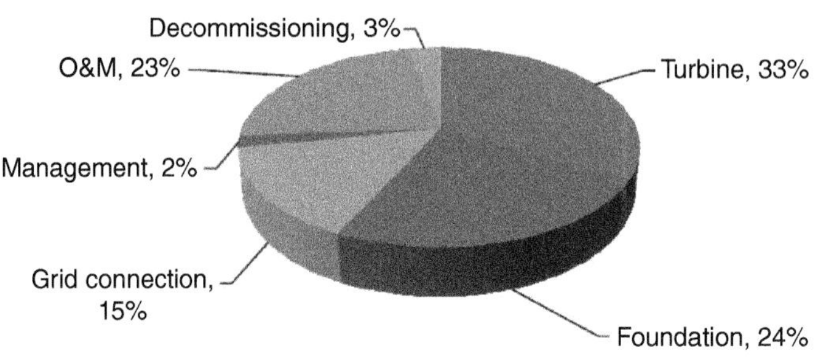


**FIGURE 9-38** Costs distribution for offshore wind facilities [57].

Currently, offshore wind technology is still under development. Research and development activities are being conducted to advance the technology and make offshore wind more efficient and cost competitive. Key approaches to cost reduction include

- Reduction of the turbine structure/weight
- New drivetrain to eliminate or reduce the size of gearboxes
- Turbine load analysis and mitigation, considering the dynamic coupling between translational (surge, sway, and heave), rotational (roll, pitch, and yaw) loading and turbine motions, as well as the dynamic characterization of mooring lines for floating systems
- Turbine and rotor designs to increase the efficiency and reliability and reduce weight
- New generators and power electronics to increase the efficiency and reliability
- Improving wind farm performance, considering interactions between wind turbines and also wind farms (e.g., improved wake models)
- Advanced maintenance strategies, remote monitoring, diagnostics, prognosis and health-monitoring systems, for improving reliability, reducing turbine down time, and operation and maintenance costs.
- Economic modeling and optimization of the overall wind farm system.

Although offshore wind remains expensive compared to onshore, the huge potential of the offshore wind resource and the reduced environment concerns provide strong motivation, so that many countries and stakeholders are actively involved in the investigation and development of offshore wind technology.

### 9.3.1 Offshore Wind Turbines

One of the significant differences between offshore and onshore wind turbines is the foundation. Offshore wind turbine design is a challenge because the natural frequencies of offshore wind turbine structures are close to the forcing frequencies from the wind, wave, and 1p (rotor frequency) and 2p/3p (blade shadowing frequency) loading effects. Thus, the dynamic responses, such as deflection/rotation, are likely to be amplified and may cause excessive fatigue damage, and consequently, a reduced lifetime. The design should ensure that the structure's natural frequency is well away from the forcing frequencies.

The design of the support structure is highly dependent on the local project conditions (water depth and soil characteristics). Various types of foundations have been proposed for offshore wind turbines (Fig. 9-39), which may be classified into two major types: bottom fixed wind

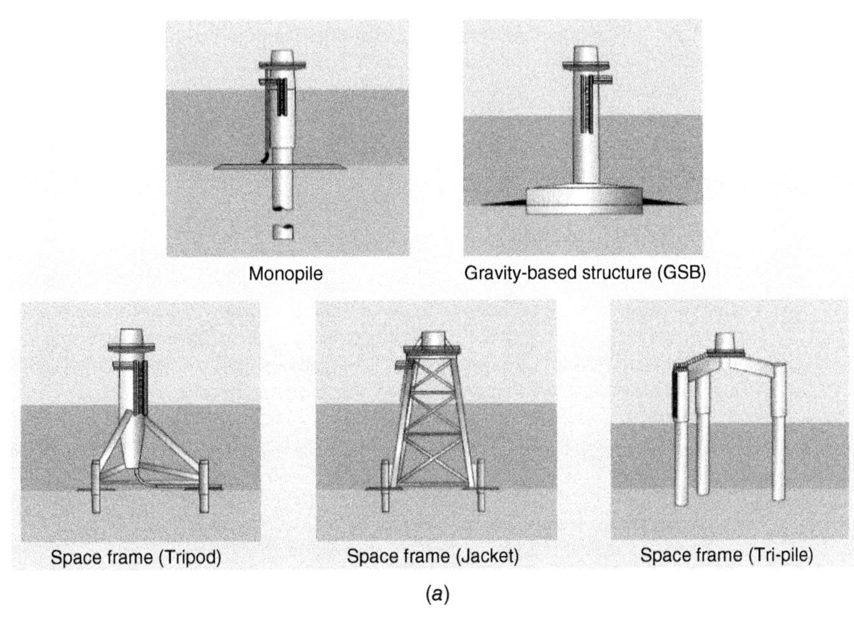


(*a*)

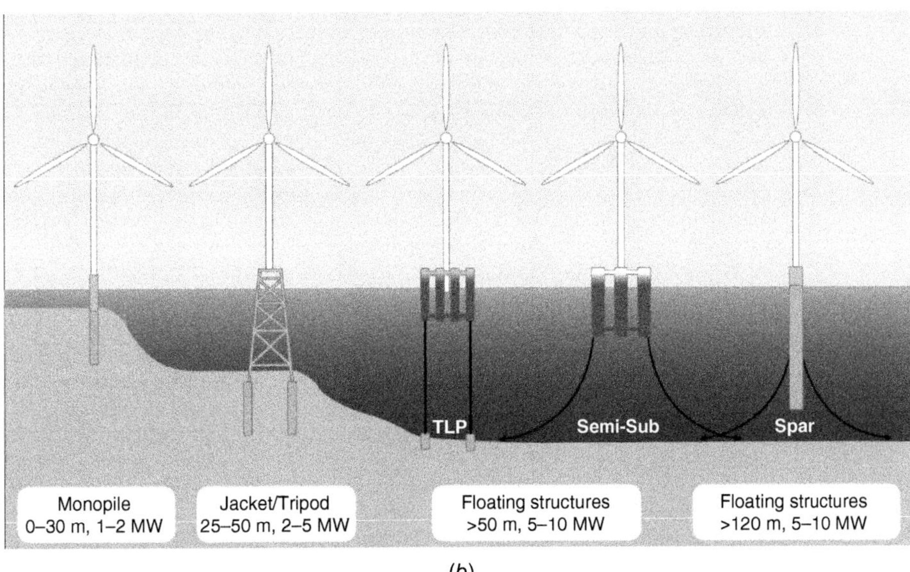


(*b*)

**FIGURE 9-39** Various types of offshore wind turbine foundations [58]. (*a*) Seabed fixed wind turbine foundations (*EWEA*); (*b*) wind turbine foundations with water depth (*Principle Power*).

turbine foundations for use in shallow and moderate depth waters; and the floating structures for deep waters (over 60 m). The bottom fixed type of foundation includes monopile, gravity base, tripod pile, tripod caisson, and jacket caisson. A number of options exist for floating supports, including spar buoy (ballast stabilizing type), tension leg platform (mooring line stabilizing type), and semi-submersible [58].

The high offshore installation costs and their scaling are driving the trend to larger wind turbines. As discussed previously, larger wind turbines normally have lower speeds and thus must deal with higher torques. As a result direct drive generators without a gearbox, or hybrid drives with a reduced ratio gearbox, are often considered. For example, the 8-MW wind turbines of MHI Vestas V164, Gamesa and Areva the Adwen AD-180 use medium-speed geared drivetrains; while Siemens SWT-8.0-154 8MW and Enercon E-126 7.5MW wind turbines use a direct-drive drivetrain.

## 9.3.2 Offshore Wind Farms

Similar to on land wind farms, the wind turbines in an offshore wind farm are normally connected via a medium voltage network. Each of the wind turbines has a step-up transformer to increase the voltage from the wind turbine's low voltage to a higher voltage, for example 36 kV, for connecting to the wind farm power collection system. The transformer and related medium voltage switch gear will be usually installed inside the wind turbine's tower or nacelle.

The electrical connection of an offshore wind farm may have a similar layout to that of an onshore wind farm, except that all the electrical connection are made with sea cables, and all installations have to be installed using special equipment, which are normally quite expensive. Three Danish offshore wind farm configurations are shown in Fig. 9-40. Due to the geographical constraints and other issues, a wind farm may take an irregular layout as shown in Fig. 9-40*c*.

The offshore wind farm configuration must be assessed to identify the optimal design, that is, the one which minimizes cost of wind energy considering the costs of investment, operation and maintenance, cabling losses, and wake losses within the wind farm [60–63].

A medium voltage dc network connecting the wind turbines in a wind farm may be a possible alternative for a wind farm collection system. In this situation, some dc electrical equipment becomes necessary, such as dc circuit breakers for clearing faults in a dc circuit, dc/dc transformers for transforming medium dc voltage of the collection system to high dc voltage for a dc transmission system. Relevant work has been reported [64–66].

***Offshore Wind Power Transmission.*** If an offshore wind farm is close to shore and the wind farm capacity is not too large, the wind farm may be directly connected to the onshore grid with an ac submarine cable at the same voltage level of the power collection system. This has the attraction of using simple, low cost, and mature technology.

However, the larger offshore wind farms now tend to be located further from the shore. Such wind farms may have to transfer power via a high-voltage transmission system, therefore, an offshore substation at the wind farm site is needed to step up the medium voltage of the wind farm power collection system to a high voltage for transmission and connection to the onshore grid.

An offshore substation is normally installed on a platform with a structure similar to that used in the oil and gas industry, and thus more expensive than an onshore substation. The onshore substation can be of conventional design, and may make use of an existing substation if sufficient spare capacity exists.

Either high voltage alternating current (HVAC) or high voltage direct current (HVDC) cable transmission systems may be used for transferring power from an offshore wind farm to an onshore grid. HVAC is often used for the situation where the offshore wind farm is not too far from the onshore grid. However, the capacitive nature of the ac submarine cables limits HVAC application in high-power and long-distance transmission because the ac cable produces reactive current, which limits the capacity available for transferring active power and results in significant

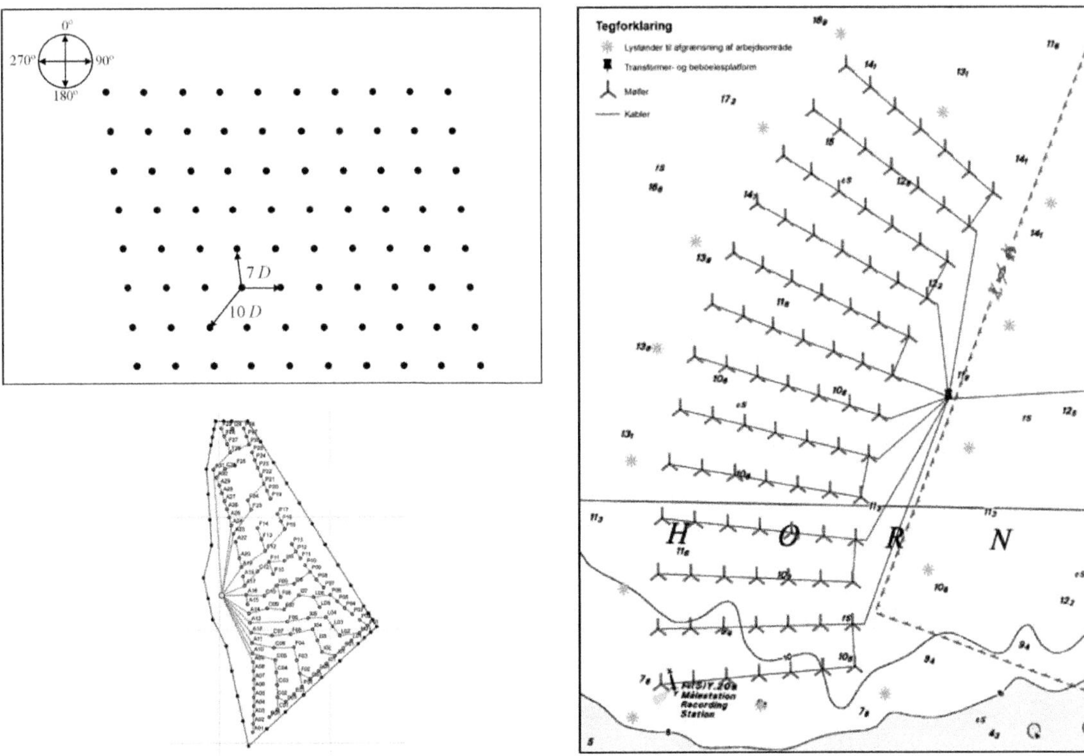


**FIGURE 9-40** Danish offshore configurations [59]. (*a*) Horns Rev I offshore wind farm; (*b*) Horns Rev II offshore wind farm; (*c*) Anholt offshore wind farm.

power losses. The amount of generated reactive current is propositional to the cable length and the cable voltage. It is possible to increase the transmission distance by installing the reactive power compensation along the transmission system, but this would be very expensive for undersea cable transmission. The installation of compensation reactors at the terminals can extend the transmission distance, whereas the effect is limited as shown in Fig. 9-41 [67]. Some investigations on the methods for increasing the transmission distance have been conducted, for example, to use a lower frequency, such as 16⅔ Hz, where capacitive reactive power effect of the cables is reduced. However, this requires larger transformers and reactors, and extra frequency converters for onshore grid connection.

If a large-size offshore wind farm is remotely located, the excessive charging current and power losses make the HVAC cable option unsuitable, and HVDC is the obvious choice. HVDC has some advantages, for example, good controllability of power flow, interconnection of asynchronous grids, no cable reactive charging current and lower cable power loss and higher transmission capacity. Furthermore, ac transmission lines need to be designed for the peak voltage of the ac sine waveform. However, the effective power that can be transmitted through an ac line is related to the root mean squared (RMS) value of the voltage, a factor of $1/\sqrt{2}$ of the peak value. This means that for the design with the same maximum voltage, a dc line may carry more power than an ac line. HVDC links are normally point to point connected with one converter station at each end which suits the case of normal offshore wind farms. A three-phase ac system needs more cables than a dc transmission system,

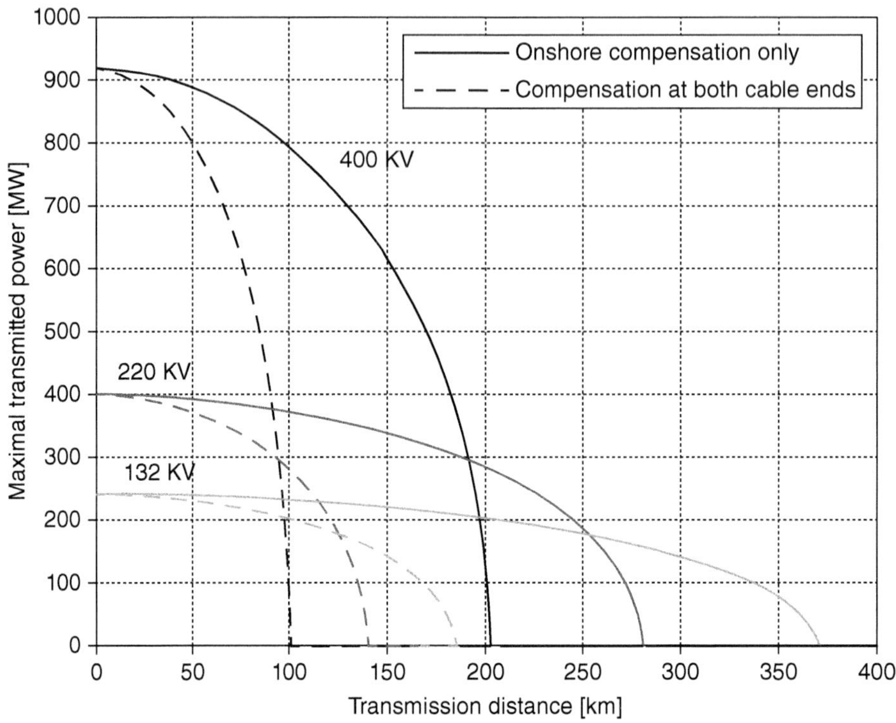


**FIGURE 9-41** Limits of ac cables transmission capacity for three voltage levels, 132, 220, and 400 kV [67].

but a dc transmission system needs more expensive equipment at the converter stations, including power electronic converters. Consequently, for shorter distances, the cost of the equipment outweighs the savings in the cost of the transmission cables. Over longer distances, the cost and capacity limitation on the cable becomes more significant, which makes HVDC economically advantageous in comparison with HVAC systems. For underwater transmission systems, the cable losses due to capacitance are much greater than overhead lines that makes HVDC economically advantageous at a much shorter distance than on land overhead lines.

Figure 9-42 illustrates a general relationship between the cost of HVAC and HVDC systems and the transmission distance. The substation cost for a dc system is higher than an ac system but the lower line/cable cost mean HVDC is more cost-effective for long-distance transmission. HVAC cable systems are favorable for transmission distance up to about 50 to 70 km [68].

The HVDC technology can be further categorized into conventional line-commuted converter based HVDC (LCC-HVDC) using thyristor technology and voltage source converter–based (VSC) HVDC (VSC-HVDC) using IGBTs. The LCC-HVDC has been in commercial application for more than half century with good reliability and availability. It offers higher voltage level and transmission capacity compared to VSC-HVDC. VSC-HVDC converters work at a higher frequency than the LCC-HVDC converters, and produces considerably less harmonics at the cost of the increased switching losses. However, LCC-HVDC requires a strong ac network for commutation as well as a comparatively large filters and reactive power compensation equipment at a converter station, thus a large offshore substation converter station, which limits LCC-HVDC's application in offshore wind industry.

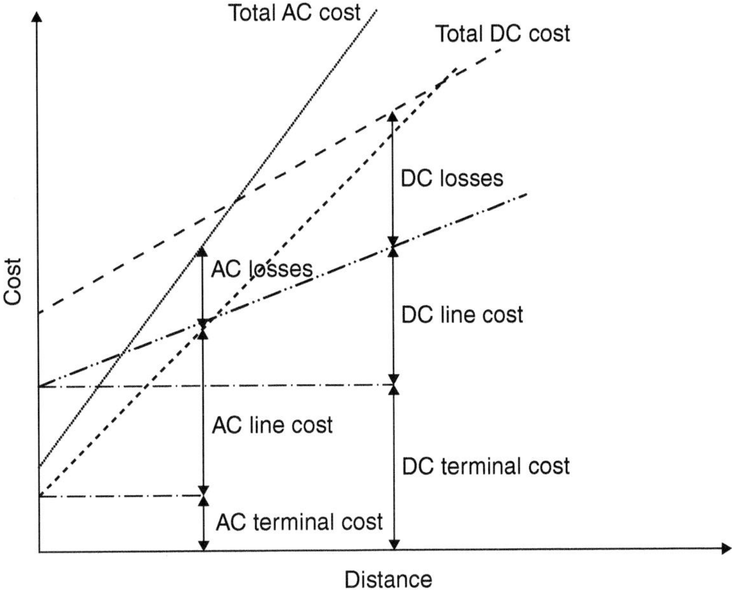


**FIGURE 9-42** Break-even point of HVAC and VSC-HVDC systems.

On the other hand, VSC-HVDC does not require a strong external ac grid. VSC-HVDC converter stations are relatively compact, which includes circuit breakers, the interface transformer, ac side harmonic filters, and converter side harmonic filters, while the dc link consists of dc subsea cable, dc capacitor, and dc harmonic filters. The converter station is much more compact than a LCC-HVDC converter station, and may be positioned in location where it can cost-effectively collect the power from the offshore wind turbines and transfer the power into the onshore grid.

The VSC has the ability of performing independent reactive and active power control, black start, plus fast, and reversible power flow control. The above beneficial characteristics make VSC-HVDC a more attractive transmission system for large-scale offshore wind farms. On the other hand, VSCs have a limited current capability. Overcurrents, even for very short durations, may result in thermal stresses that degrade or cause permanent damage to the semiconductors. Under a grid fault, the VSC will not be able to provide the short circuit current as a conventional ac power system. A protection strategy for a dc overvoltage may be to use a dc chopper. When the dc link voltage exceeds a threshold value, the chopper is put into operation to dissipate the excess energy and prevent the possible overvoltage of the dc circuit and converters, and over speed of the generators. A possible configuration of an offshore wind farm dc collection system and HVDC transmission system is illustrated in Fig. 9-43.

***Submarine Power Cables.*** *Submarine power cables* are an important component in offshore wind power transmission systems. A submarine power cable normally consists of a concentric assembly of conductor, insulation, and protective layers. The conductor is made from copper or aluminum wires. Different types of electric insulation around the conductor are used, including

- Self-contained fluid-filled cables, including oil filled or gas filled
- Paper insulated (lapped insulated) cables, including mass-impregnated, etc.
- Extruded cables including EPR (ethylene propylene rubber), PE (polyethylene), XLPE (cross-linked polyethylene)

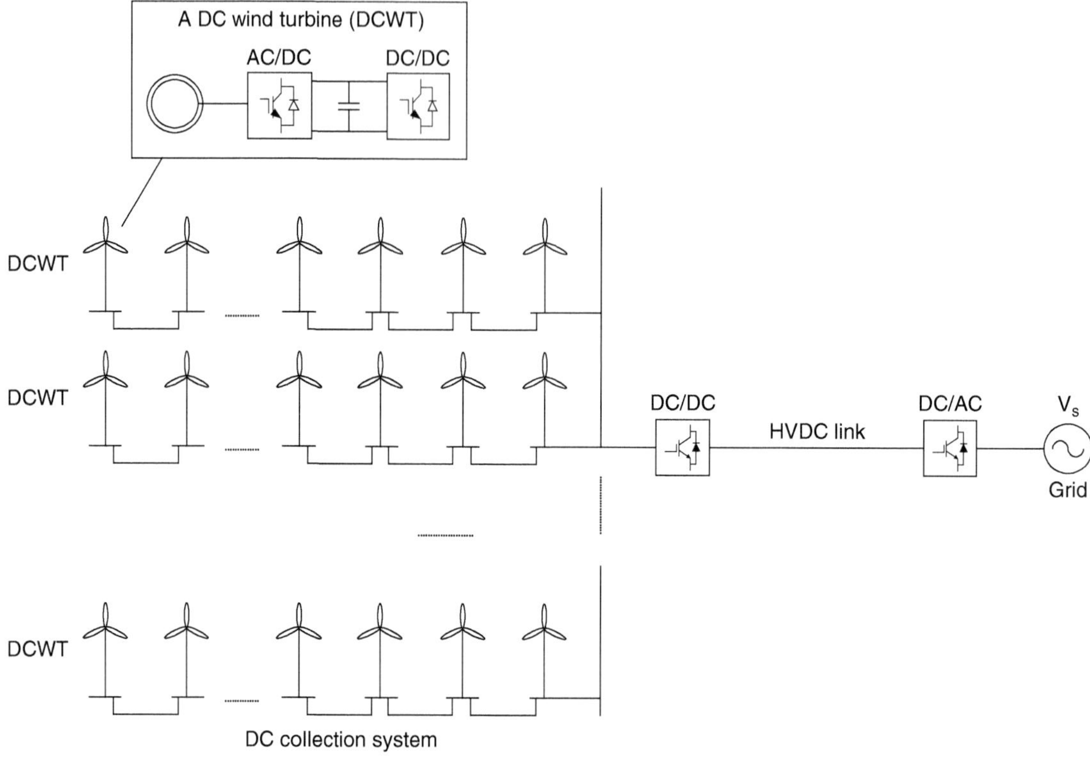


**FIGURE 9-43** A possible configuration of an offshore wind farm DC collection system and HVDC transmission system.

The conductor and the around insulation layers form the cable core. In single-core cables the core is surrounded by a concentric armoring. In three-core cables, three cable cores are laid up in a spiral configuration and then the armoring is applied. The armoring consists often of steel wires, soaked in bitumen for corrosion protection. The armoring is sometimes equipped with nonmagnetic metallic materials (stainless steel, copper, brass) to reduce the power losses caused by the alternating magnetic field in ac cables. Modern power cables, for example, cables for the interconnection of offshore wind turbines, may carry optical fibers for data transmission or measurement. Figure 9-44 shows a number of types of subsea cables [69], and high voltage cable construction is illustrated in Fig. 9-45.

Extruded cables are an important part of a VSC-HVDC link. The environmental benefits are: the cable insulation is PE based and not dangerous, the risk of oil spill is eliminated in XLPE cables. Furthermore, they are lighter and more compact, therefore, very competitive in comparison with the other types of cables. For VSC-HVDC system the maximum voltage level with XLPE submarine cables presently available on the market is ±320 kV [70]. However, technology improves quickly and higher voltage and power levels are expected.

*Examples of Offshore Wind Power Transmission Systems* [71]. Various types of offshore wind farm transmission systems for connecting to the onshore grid have been adopted. A few examples are introduced here.

As mentioned, some offshore wind farms do not have an offshore substation and their wind turbines are directly connected to the onshore power grid via medium ac voltage cables. Examples are: the United Kingdom's first offshore wind farm, North Hoyle (60 MW, 2003, 12 km offshore) connected to the shore via 33 kV cables; and the U.K. offshore wind farms, Kentish Flats (90 MW, 2005, 8.9 km offshore), and Scroby Sands (60 MW, 2004, 2.5 km offshore).

| |  |  | 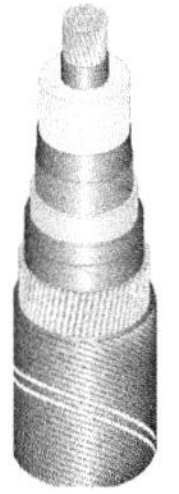 | 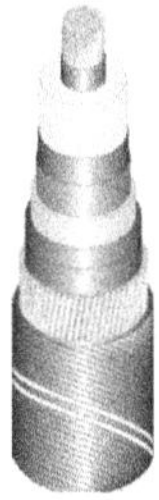 | 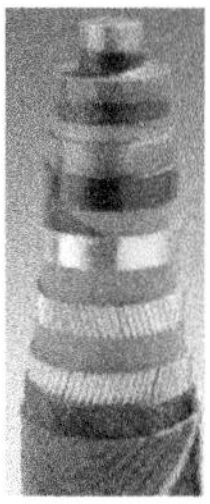 |
|---|---|---|---|---|---|
| Rated Voltage | 33 kV ac | 150 kV ac | 420 kV ac | 320 kV dc | 450 kV dc |
| Insulation | XLPE, EPR | XLPE | Oil/paper or XLPE | Extruded | Mass Impregnated |
| Typical application | Supply of small islands, connection of offshore wind turbines | Supply to large islands, offshore platform export cables | Crossings of rivers, straits with large transmission capacity | Long distance connections of offshore platforms or wind parks | Interconnection of power grids over long distances |
| Max. length | 20-30 km | 70-150 km | <50 km | >500 km | >500 km |
| Typical rating | 30 MW | 180 MW | 700 MW/3 cables | 1000 MW/ cable pair | 600 MW/ cable |

**FIGURE 9-44** Submarine power cable types [69].

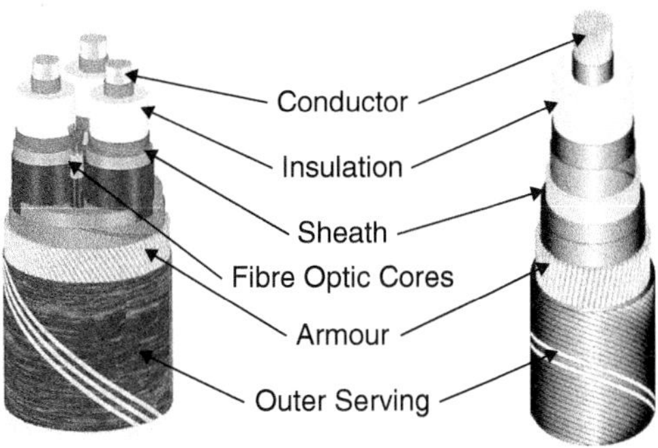


**FIGURE 9-45** HV cable construction [69].

Denmark's Horns Rev wind farm constructed in 2002, a main European offshore pilot project with a capacity of 160 MW, is the first offshore wind farm using an offshore transformer substation, which connected to the onshore grid through a 15-km three-core ac cable with a rated voltage of 150 kV, while Denmark's 165.6 MW Nysted wind farm, constructed in 2003, also has an offshore transformer station, connected to the onshore grid through a 10-km 132-kV ac submarine cable. The United Kingdom's 90 MW Barrow wind farm, which was completed in 2006, also uses an offshore substation, which is connected to the onshore grid with a 7-km 132-kV ac cable.

HVDC BorWin1 is the first VSC-HVDC in the world for connecting an offshore wind farm to an onshore grid, the system was built by ABB_on the basis of experiences of Gotland HVDC Light, the world's first commercial HVDC transmission using VSC converters. HVDC BorWin1 connecting BorWin Alpha, the first VSC-HVDC station in the world installed on an offshore platform, to Diele substation, consists of a 125-km of submarine cable and 75 km of underground cable, and can transfer 400 MW power at a bipolar voltage of ±150 kV. BorWin1 uses two-level converters in which several hundred IGBTs are connected directly in series in each valve.

HVDC BorWin2 also uses a VSC-HVDC, built by Siemens with the modular multilevel converter (MMC) technology, 4.5 kV IGBTs are used as the switching elements. The VSC-HVDC link has a rated voltage of ±300 kV and a transmission capacity of 800 MW. The offshore converter is located on the BorWin Beta platform, the onshore converter station is also at Diele. Each converter station is equipped with two, three-phase transformers, each rated at 590 MVA (offshore) and 585 MVA (onshore). Each converter valve is equipped with a dry-type valve reactor of 50 mH. The cables are the extruded cross-linked polyethylene insulation supplied by Prysmian. The total cable length from BorWin Beta to the onshore substation is 200 km, of which 125 km is sea cable and 75 km is underground land cable.

## 9.4 WIND POWER IN POWER SYSTEMS

### 9.4.1 Challenges of Large-Scale Wind Power Integration and Power System Requirements [72]

Wind turbines present significantly different features from conventional power plant using synchronous generators. Some of main issues are as follows:

- Wind power is variable and challenging to accurately predict, making it difficult to participate in power dispatch and electricity market.
- Wind power lacks reliability and wind power fluctuations may cause problems to power system balancing; large penetrations of wind power increase the demand for reserve capacity and ancillary services.
- Wind power plant is less flexible and less controllable in comparison to conventional power generation.
- Wind power conversion system may behave differently from conventional generators in terms of inertia, frequency control, reactive power regulation, and power system dynamic response.
- Modern wind power plants are mostly power electronic interfaced, the response speed and short circuit power are different from conventional generators. They may present different impedance characteristics over a range of frequencies which may create a risk of harmonic resonance.
- Wind power plants may be located in geographically unfavorable areas, far away from consumers/load centers, with large amounts of fluctuating power needing to be transmitted over long distance transmission lines, which may cause transmission line instability.

In order to realize stable, reliable, and economical operation of power grids with high wind power penetration, changes need to be made both to the power network (including conventional generators) and the wind plant. Wind turbine manufacturers and wind farm operators have been improving the performance of the wind turbines, while the grid operators have taken actions to secure the reliable operation of a grid with large-scale wind power penetration. Grid codes are part of the process.

***Power System Requirements: GRID CODES [73].*** Grid codes relating to wind power specify the requirements wind turbines must meet before connecting to a power grid. A grid code covers many technical aspects, in general, it includes steady-state performance (frequency, voltage, active and reactive power, and power quality); dynamic performance (frequency gradients, start-stop, active power ramp rates, reactive power and voltage dynamic control, fault-ride through, inertia, overvoltage and protection); communication and control; simulation models, certification/verification; commissioning and performance verification.

When wind power penetration level was low, the wind turbines or wind farms were simply treated as industrial load in some grid codes which may be specified by a distribution system operator (DSO), for example, may only require power factor, harmonic, inrush current, etc. However, the rapid increase of wind power penetration level forces many transmission system operators (TSOs) to set up and continuously update the grid codes for wind turbine connection and operation. In this subsection, the grid codes are briefly introduced, some contents of the Danish grid code are used as illustrative examples, including the regulation of frequency and voltage, control of active and reactive power, fault ride-through capability, etc.

***Operation Range of Frequency and Voltage.*** Electrical equipment connected to the grid is designed for a set of specified conditions; frequency and voltage are two very important parameters. In normal operation of a conventional power system, the frequency and voltage are respectively used to judge the balance of active and reactive power production and consumption. For example, a frequency rise indicates more active power production than consumption, and a voltage increase means more reactive power generation than consumption. A power system operator needs to maintain the power system frequency and voltage within specified ranges by respectively adjusting the active power and reactive power production to meet the active power and reactive power consumption. The nature of the wind energy resource may prevent a wind farm from adjusting its active power production freely. At the power grid level, more reserve may be needed for balancing. Traditionally the system operators would arrange for reserve to cover variations, but increasingly they expect wind farm operators to take some responsibility for balancing. For larger wind turbines, operation outside the normal conditions may be required for some specified time periods so that the system operator can have time to restore normal operation. Figure 9-46 shows the specified frequency/voltage operation range for wind farm between 25 kW and 1.5 MW [73].

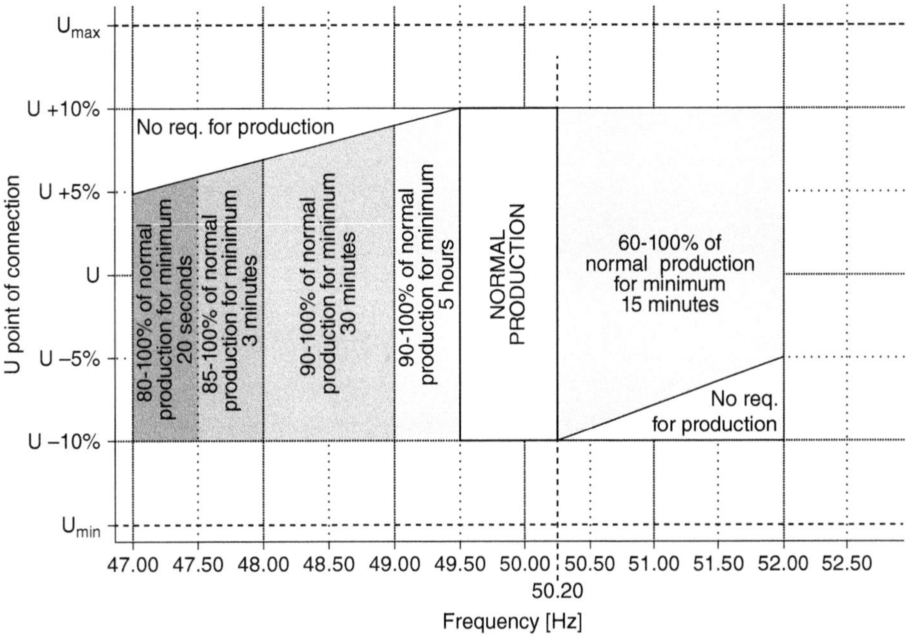


**FIGURE 9-46** The required frequency/voltage operation range for wind power plants at 25 kW to 1.5 MW [73].

To contribute to frequency regulation, the wind turbines are required to increase or decrease power output in line with power system frequency variation. The wind power output should be reduced if the frequency is higher than a specified range, on the other hand, the wind turbine needs to increase power production if the frequency is lower than a specified range. Figure 9-47 illustrates a required frequency power characteristic for a wind farm, where $f_1, f_2, \ldots$ can be specified by the system operator, while the active power production may be kept unchanged within a dead band between $f_2$ and $f_3$, and the linearly regulated with possible different slops in other frequency ranges.

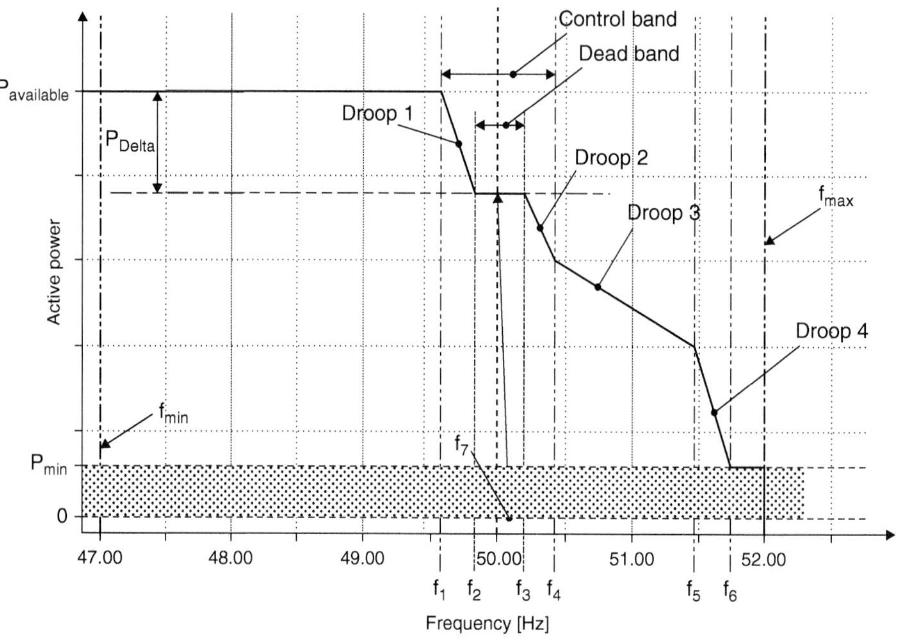


**FIGURE 9-47** Frequency control for wind power plants with a power output higher than 25 MW [73].

A number of operation modes should be made available at a wind farm and be ready to be put into operation, those operation modes may include

- *Balance Control.* Wind farm production can be adjusted up or down to specified levels.
- *Delta Control.* Wind farm is operated with a certain constant reserve capacity in relation to its available power production capacity, such reserve power may be used in a frequency/power control.
- *Power Ramp Rate Limiter.* This specifies the speed at which the wind farm power production can be adjusted.
- *Automatic Frequency Control.* A wind farm can automatically adjust its active power production in response to frequency variation at the measuring point.
- *Reactive Power Control.* A wind farm may be required to produce or absorb a specified amount of reactive power.
- *Automatic Voltage Control.* A wind farm can automatically adjust its reactive power production in response to voltage variation at the measuring point.

Also the power production of a wind turbine should be able to be adjusted to any value between the maximum available power and the minimum power of the wind turbine within a specified period.

Wind farms are required to have the capability of providing a certain amount lagging/leading reactive power to contribute to reactive power balance and voltage support in both normal operation and fault situations. Grid codes often require wind turbines to be able to operate within a range of power factor and reactive power with reference to active power production. Figure 9-48 illustrates reactive power requirements for wind power plants.

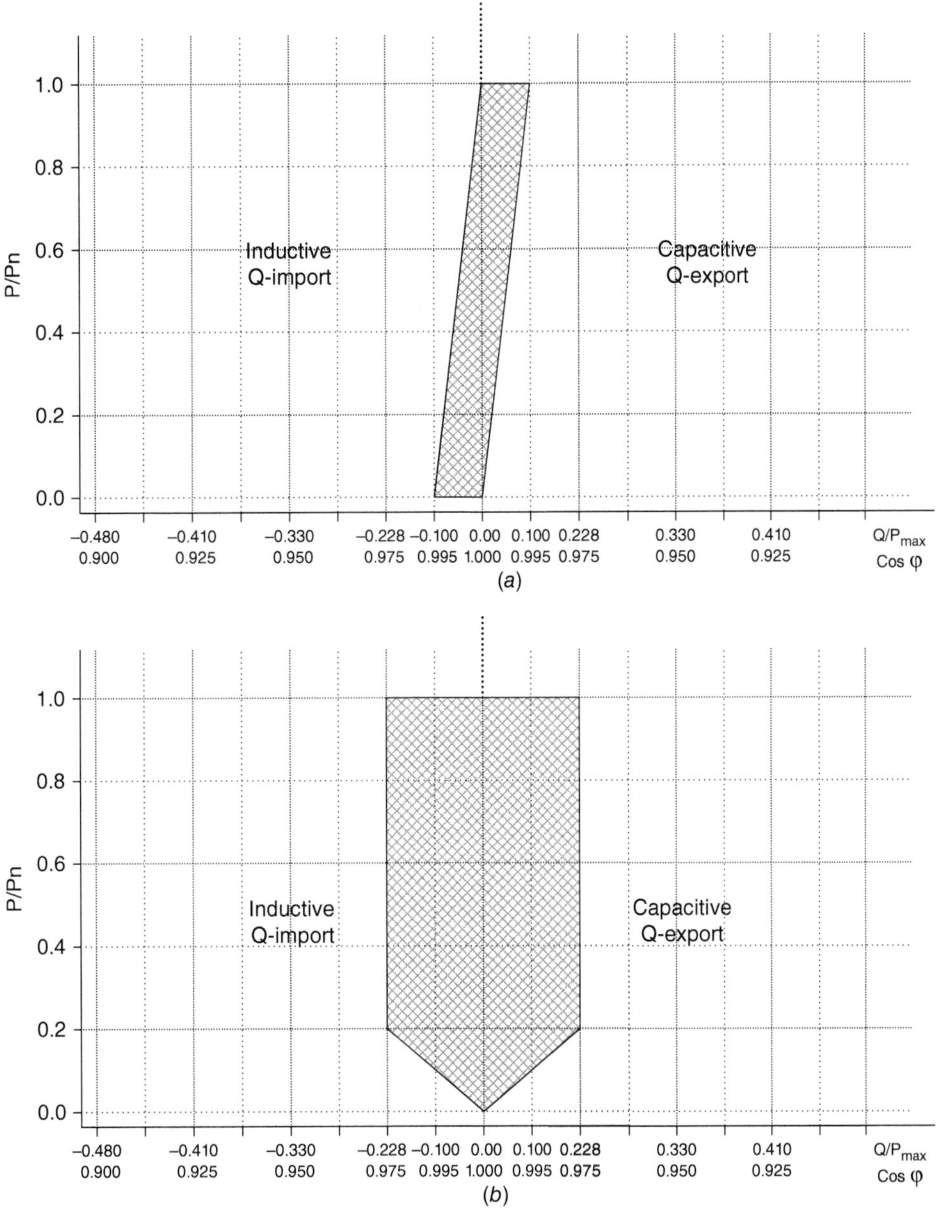


**FIGURE 9-48** Requirements for reactive power [73]. (*a*) Power range of wind turbine: 1.5 kW to 1.5 MW; (*b*) power range of wind turbine: 1.5 to 25 MW; (*c*) power range of wind turbine: >25 MW.

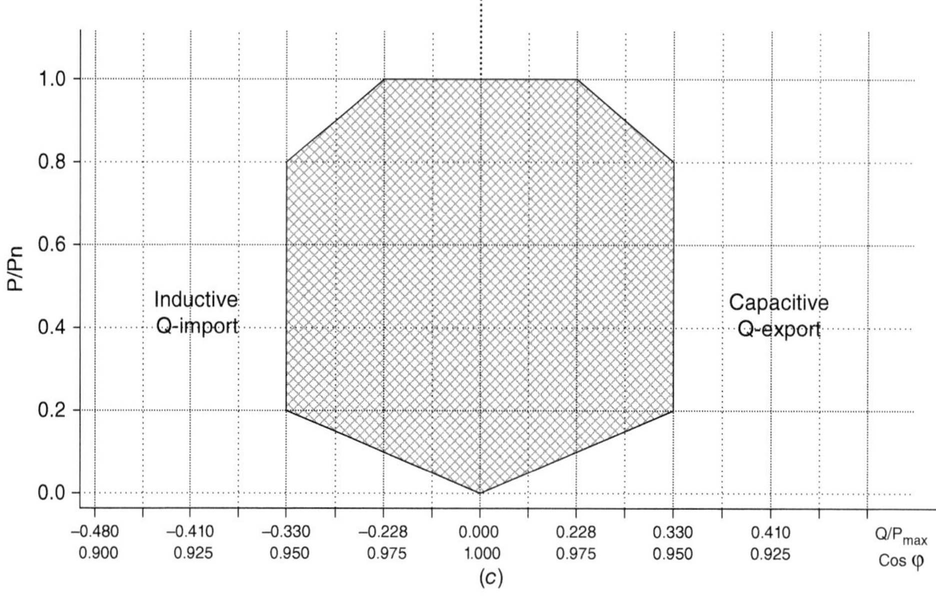


(c)

**FIGURE 9-48** *(Continued)*

From Fig. 9-48, it can be clearly seen that the demand on the reactive power capacity becomes greater if the size of the wind power plant increases. Sometimes, a wind farm may also be required to help keeping the voltage at the point of common coupling (PCC) within the required range.

Early wind turbines based on conventional induction generators are normally only required to keep a good power factor, which is achieved by using switching capacitor banks or other reactive power compensation devices, such as SVCs. DFIG and full-scale power electronic interfaced wind power generators can contribute to reactive power regulation with power electronic interface systems. If more and flexible reactive power capacity is required, additional reactive power compensation systems, such as SVCs and STATCOMs (SVGs), may be used at the wind farm level to provide support in both normal and transient conditions.

## 9.4.2 Fault-Ride through and Reactive Power Support in Grid Faults

To help power system recovery after a grid fault, grid codes require that wind turbines have the capability of fault-ride through, which means that the wind turbine systems should remain connected to electricity system during a grid fault so that the wind turbines can quickly support the grid recovery after the grid fault is cleared. The disconnection of wind turbines from the grid may cause significant loss of generation capacity if wind penetration is high, adding to the challenge of maintaining the power system and reducing system security.

Under a power system fault, a wind turbine will experience a rapid voltage variation. Since most power system faults are temporary, that is, faults are usually cleared quickly by auto-reclosure. The severity and the time period of such a voltage variation will determine whether the wind turbine must not be disconnected. The requirements are normally specified as a characteristic, as shown in Fig. 9-49, where area A is the normal operation zone and area B is the fault-ride through zone, that is, wind turbines should not be disconnected from the grid if the voltage is in area B. To help system voltage recovery, reactive power support is very important and thus is also specified by grid codes, such an example is illustrated in Fig. 9-50. The grid code may also require that wind turbines be able to withstand more than one independent fault within an interval of a few minutes.

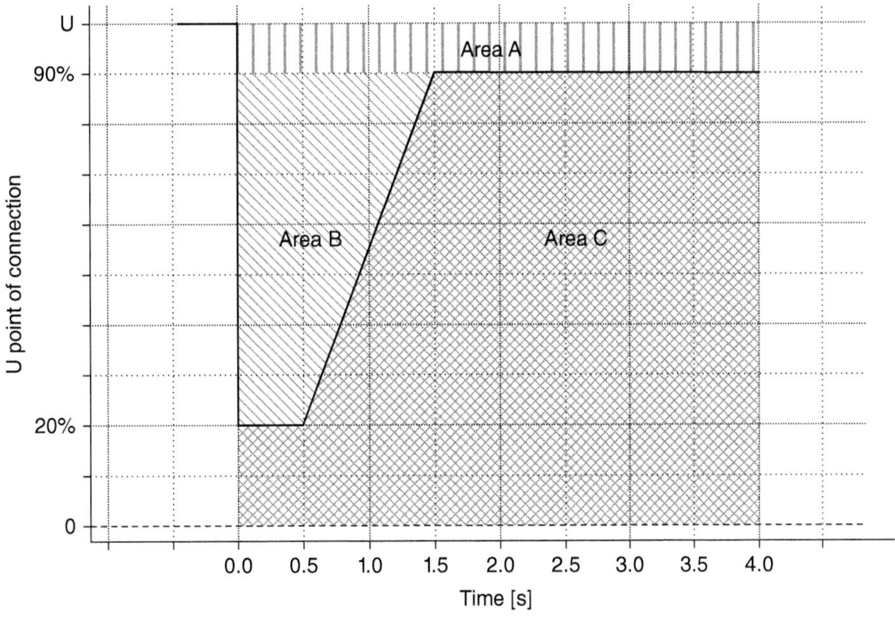


**FIGURE 9-49** Requirement for tolerance of voltage drops (power output >1.5 MW) [73].

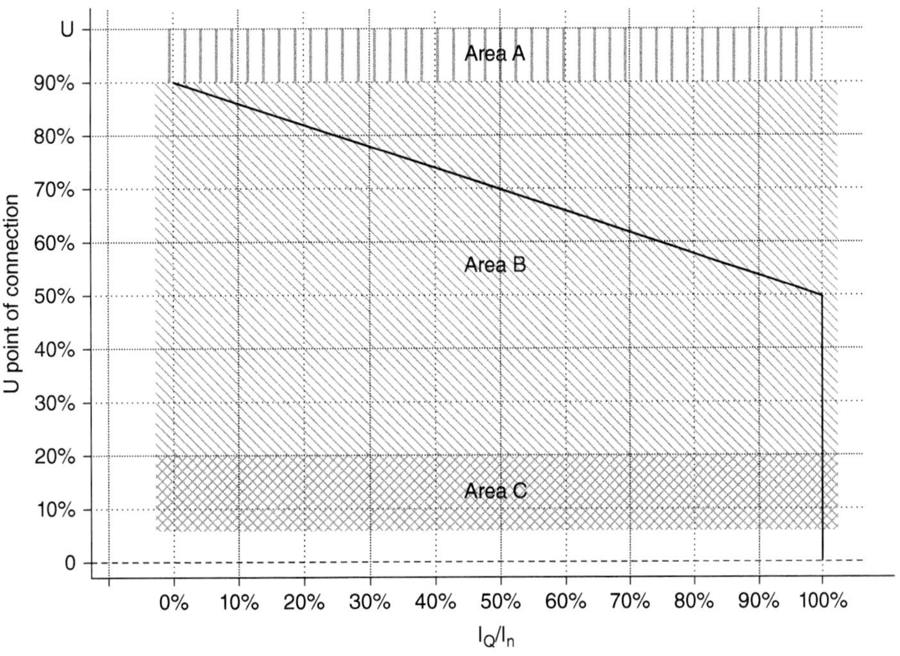


**FIGURE 9-50** Requirement for reactive power supply during voltage drops (power output >1.5 MW) [73].

Grid codes also specify requirements for power quality, for example, rapid voltage variations, flicker, harmonics, and inter-harmonics. Monitoring and communication systems are recommended, such as a supervisory control and data acquisition (SCADA) system for wind power plants so that the information about operation and control can be communicated. In most situations, the wind farm operator is required to provide key information to the power system operator, while, more detailed signals are used by the wind farm owner to monitor and control the wind farm. Information generally includes voltage, current, frequency, active power, reactive power, wind speed, wind direction, regulation capability, temperature, and rotating speed.

Different power system operators in different region/countries have different requirements in their grid codes for wind power. The requirements may vary significantly from one country to another depending on the grid conditions. For instance, a weak or isolated power network would require more restrictive fault-ride-through requirements (lower voltage and longer duration) than a strongly inter-connected power system.

### 9.4.3 Enhancing the Controllability of Wind Power for Large-Scale Penetration in Power Systems

***Wind Power Plant in Power Systems.*** With the increase of wind power penetration, wind farms have had to an extent take over the role of conventional generators in some networks. In the future it will be expected that wind turbines and wind farms work as wind power plants (WPPs), that is, like a conventional generator to contribute to voltage and frequency control. Many efforts have been made to enhance WPP performance; some approaches are briefly discussed here.

***WPP Controller.*** A WPP may consist of a large number of wind turbines, which are distributed in a relative large area. These wind turbines may be subject to very different operational conditions, such as wind speed, wind direction, and turbulence intensity. The wind turbines in a WPP need to be monitored and controlled in a coordinated manner, which requires significant computational effort and modern control and communication hardware. The two-level hierarchical structure, consisting of a wind power plant controller and individual wind turbine controllers discussed previously, may be used. The WPP controller communicates with the higher level control center and determines the operating points for wind turbines to meet grid demands. The turbine controllers respond to the power set point from WPP controller. Fast and accurate set point tracking is essential for efficiency so any communication delay between the wind turbine controllers and the wind power plant controller must be assessed in the context of possible unstable operation. The control strategy design should consider the wind power characteristics and grid code requirements, such as active power control, reactive power control, fault-ride through, power quality [74, 75], as well as other issues, including inertia, resonance, and oscillations.

***Improvement of WPP Power Forecast Accuracy.*** The limited predictability of wind power presents a challenge for a WPP to perform well in grid operation and in the electricity market. As the percentage of wind energy in power generation capacity grows, it becomes increasingly important to improve the accuracy of wind power forecasts so as to help in operational planning, scheduling, and contractual agreements in power market. If the installed wind power capacity is large, a small error in the wind speed forecast can result in a significant error in the active power prediction. Optimization of regulation and reserve power would enable cost effective utilization of system resources. Such optimization will depend on the accuracy of wind forecasts. Improved forecasts can reduce the need for reserve capacity and ancillary services provision thus increasing profits and minimizing risks.

***Active Power-Frequency Control of WPP.*** Conventional synchronous generators, using frequency and voltage control systems, can automatically adjust the production of active and reactive power in line with variations of power system frequency and voltage. The primary and secondary power and frequency controls are set at different response timescales to reduce respectively any power imbalance and to bring the frequency back into the specified range. Conventional generation units are normally equipped with governor control, which works as primary frequency control to reduce power imbalance; secondary control is used to increase or decrease of power production of the specified generators in order to restore the frequency to its nominal value. Also, a conventional synchronous generator supports the system frequency by storing or releasing its kinetic energy in response to the speed/frequency variation.

In order to enable a WPP contribute to the system inertia and frequency control, controllers may directly adjust the WPP output power reference commands in response to frequency variations [76]. For example, the WPP can increase its power output in response to a frequency drop through droop control. However, in order to deliver the required energy, the WPP needs either to be equipped with an energy storage system or to keep some spinning reserve, that is, be operated at a power level below the maximum available wind power level, which means a reduced utilization of the wind energy, and hence reduced revenues.

While the wind farms are considered like other generating facilities to participate in the system frequency control, an overall optimization may be needed to decide how to distribute the primary and secondary control capacities among conventional power plants, WPPs, other generation units and possible energy storage systems.

***Energy Storage System for WPP.*** Variable speed wind turbines can store or release energy by speeding up or slowing down the turbine rotor in a similar way as a flywheel to smooth out short-term power spikes. However, dedicated energy storage systems may be needed to deal with long time and large-scale power fluctuations, especially in an islanded system or a system with weak interconnections. One way of fully exploiting the fluctuating wind power and providing a stable electricity supply from a WPP would be to combine the wind turbines with an energy storage system [77], such as hydro storage, compressed air energy storage systems, batteries, flywheels, or other types of energy storage. Energy storage systems could work well technically; however, further improvements on the energy storage technologies are required to make them more competitive from an economic perspective.

From an overall power system perspective, large-scale system storage systems, such as hydropower or pump storage systems, are very useful. Fast-controllable hydro can be used for real-time balancing of power in areas where a large part of electricity power is provided by noncontrollable primary source like wind power.

***WPP Reactive Power and Voltage Control.*** As indicated in grid codes, large wind farms are required to have the ability to control both active and reactive power. The simplest way of controlling reactive power is to use switched capacitor banks. For the wind farm with the fixed speed wind turbines equipped with conventional induction generators, the active power production and the reactive power absorption are strongly coupled. Thus, the active power fluctuations can result in similar fluctuations of the reactive power absorption, and consequently, the appropriate dynamic reactive power compensation equipment is needed to smooth the possible voltage fluctuations. Furthermore, reactive power can also be controlled to damp power system oscillations [78].

In a modern WPP, power electronics systems are normally used to interface wind turbines. These power electronic converters can contribute to reactive power control in order to compensate for the voltage variation caused by active power fluctuations, so that the fluctuations of the grid voltage can be limited within the specified range.

Despite this, a WPP may need a centralized reactive power compensation system, for example, capacitor banks, reactors, SVCs or STATCOMs for providing voltage regulation.

***WPP in Power System Transients: Wind Turbines Fault-Ride Through.*** During a short-circuit fault in a power system where a WPP is connected, the short circuit current may result in a voltage drop at the WPP terminal. Due to the voltage dip, the output electrical power and the electromagnetic torque of the wind generators are significantly reduced, while the mechanical torque may be still applied to the wind turbines from the wind. Consequently, the turbines and generators will be accelerated due to the resulting torque unbalance.

After the fault clearance, the voltage tends to recover. If the voltage is not able to return to its normal range or the generator speed is too high, there may be insufficient electromagnetic torque to balance the mechanical torque. Hence, the machine would continue to accelerate. If this happens, the wind turbine may have to be disconnected, and thus fails to ride-through the fault. Such wind turbine disconnection should be avoided because losing a significant part of the power generation capacity could threaten the security of the power system.

To ensure the security of a power system with high wind power penetration level, a WPP should be able to ride through the fault. Some modelling and control methods [79–83] to improve the fault-ride through capability are presented as follows:

- Wind turbine protection and control
- Fast power control (at the levels of wind turbines and WPP)
- Dynamic reactive power compensation (at the levels of wind turbines and WPP)
- Fast response energy storage/consumption system

Some countries require fault-ride through systems to be experimentally tested before being accepted for connection to the power system.

***Power Oscillation Damping of WPP.*** Fluctuating wind power, including the 3p component, may excite or amplify power system oscillations. Several studies have investigated the impacts of wind power on power system oscillations and small signal stability, for example [84].

The active and reactive power of a WPP can be controlled to damp power system oscillations. Several power oscillation damping controllers have been studied, including active power modulation, reactive power modulation, and a combination of both active power and reactive power modulations. Such a damping controller may be implemented within the WPP controller. Wind turbine mechanical resonance frequencies may be within the typical range for power system oscillations, and so attention should be paid to avoid exciting oscillations at mechanical resonance frequencies of the wind turbine.

### 9.4.4 Enhancing the Power System Capability of Accepting Large-Scale Wind Power

***Interconnected Networks and Virtual Power Plants.*** Wind power fluctuates with wind speed, while the electrical grid must maintain a balance between the supply and the demand, normally by adjusting the power generation to follow the load variation. Large concentrated wind farms, such as offshore wind farms, may inject more significant power fluctuations into power systems than those wind turbines or wind farms distributed across a large area. Such power fluctuations may challenge the power balance and frequency control of power systems. For example, the offshore wind farm Horns Rev I (160 MW) in Western Denmark produces more significant active power fluctuations than the aggregated onshore wind power (over 3000 MW) in western Danish power system. A large geographical spread of wind power generation will reduce variability, increase predictability and decrease the occurrences of near zero or peak outputs. Geographical averaging is very useful to smooth wind power variations in all timescales.

In order to deal with the power balance, the power system will need a sufficient amount of regulating power and appropriate arrangements for power exchange with neighboring power systems could also ease the task. It is clear that interconnection of power grids and distribution of wind power farms will help to smooth power fluctuations so as to reduce the challenge to system regulation and stability.

Wind turbines distributed on land may be connected to distribution voltage level, such installations are relatively small in capacity individually, but can play an important role if a large number of such units exist in a power system, like the Danish power system. These distributed wind turbines and wind farms together with some other local generation units within the distribution systems may be organized as a virtual power plant (VPP). The VPP may make a combined contribution to the power balance of the power system and can participate in the power market in a similar way to a conventional power plant, a VPP could also include the smart use of electricity household level, adjusting the use of dishwasher, heaters, refrigeration, and electric vehicles in response to a power market [85, 86].

A power dispatch center can perform the function of controlling a VPP by sending control signals, such as reference operating points, to a wind farm controller which then can perform the control of the wind turbines in the wind farm, for example, responding to frequency control, reactive power control and wind power curtailment.

***Wind Power and Electricity Price, Demand Side Management, and Flexible Generation Units.*** Similar to other power generators, wind power can participate in power markets. Because the wind resource is free, high wind availability tends to drive down the electricity price as shown in Fig. 9-51 [87]. There may also be congestion in power transmission, especially during periods with high wind power generation. If the available transmission capacity cannot allow the power transmission freely, the constrained line may separate the area into two independent pricing areas. In the area with excess supply of power, the electricity price could be lower.

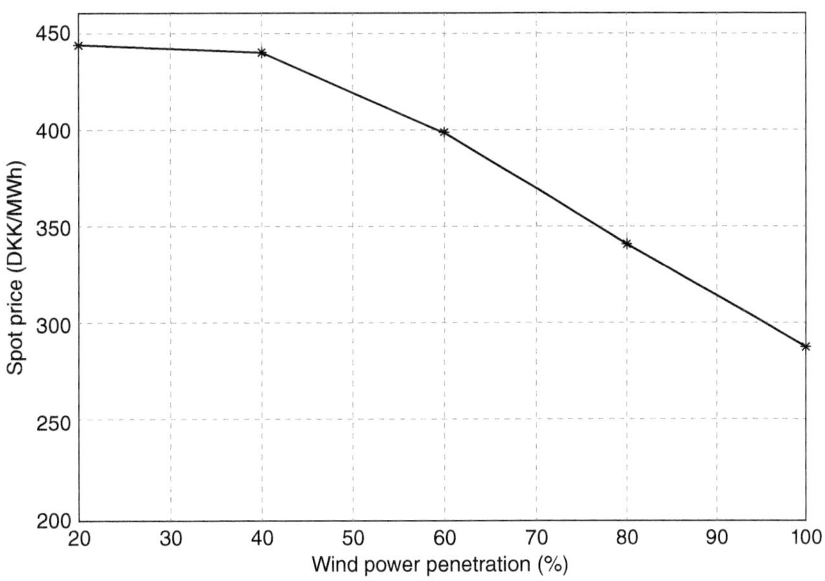


**FIGURE 9-51** Electricity price versus wind power production [87].

The electricity price may play a role of load regulation in a smart grid. The variation of electricity price could motivate the intelligent operation of electrical load and energy storage, including both electrical and thermal energy storage, such as electric vehicles, electrical boilers, heat pumps, household batteries, hydrogen fuel cells, and other types of storages, the integration and interaction of these units with the grid can contribute to power smooth and efficient use of wind power, and will affect the economics and security of the whole energy system, including thermal systems [88].

Nonelectrical energy systems, like heat storage, can also contribute, since a heat storage could decouple the heat production and electrical power production, giving more freedom for high efficiency CHPs (combine heat and power units) to perform power balance control and economic operation.

***Wind Power Transmission and Weak Grid Operation.*** In some areas, the power system may not be designed to transfer the increased scale of wind power generation; consequently, the wind power transmission may have to be limited to reflect the possible power system constraints and bottlenecks. Actions, such as grid reinforcement and expansion, wind power curtailment and energy storage, may have to be taken to handle grid congestion and constraints violation. The cost of the transmission may have to be considered for selecting a wind farm location.

Grid reinforcement may be necessary to maintain adequate transmission when wind power penetration increases. The construction of new lines may also be a prerequisite for reaching regions with a high wind resource. However, in some areas, the required network upgrades, especially the construction of new lines, may be a very lengthy process. The utilization of existing power lines may be increased by operating them at a higher capacity, assisted by temperature monitoring and dynamic line rating. Improving the cross-border electricity exchange between different networks is also a method for alleviating congestion. If controllable power plants are available within the congested area, coordinated automatic generation control (AGC) is also an option.

There are also other ways of increasing the transmission capacity of the network, including

1. Increasing system capacity by increasing voltage level, adding transformers or new lines,
2. Improving the power flow distribution in the network to fit better with the line capacities, for example, by installing new facilities, like phase shifting transformers, or devices to increase voltage support (such as static VAR compensators) [89].
3. Performing dynamic security assessment and preventive control to ensure the power system stability [90].

A wind farm may be connected to a weak grid which has a low short circuit capacity. A low short circuit ratio tends to give power quality and stability problems [91]. The problem may be mitigated by tuning the wind turbine/farm controllers, using a wind turbine power electronic interface, installing reactive power compensation devices and network enhancement.

***Wind Power Curtailment.*** Reinforcing a transmission network may completely remove a bottleneck, however it may not be economically justified. Because the equivalent full load hours of a wind farm are normally only 2000 to 4000 h/year, this means that the wind turbines operate below their power rating for most of the year.

Wind power curtailment to relieve congestion is considered as another solution for large-scale wind power integration in situations with limited or no grid reinforcement. Wind power curtailment may be achieved through either manual curtailment, AGC or frequency control.

***Wind Power in Island Systems.*** Island systems are typically weak grids. Their electrification conventionally relies on conventional generating units, running on light fuel or heavy fuel oil, which leads to high operational costs. Because of lack of interconnections with neighboring systems, frequency stability is of primary concern [92]. Primary frequency control is performed by fast generation units, such as internal combustion diesel engines, combined-cycle units and peaking gas turbines, while secondary control is made either via automatic generation control systems, where available, or manually by the system operators. Due to the high operational costs, wind power exploitation becomes particularly attractive. With increasing wind power penetration levels, frequency response of wind turbines becomes a key factor to ensure secure integration into isolated electrical grids.

Frequency controllers for variable speed wind turbines can incorporate any of the two fundamental frequency response methods of inertia and droop control or their combination. In principle, inertia control is incorporated in the individual wind turbine controllers of a wind farm, while droop control can be applied in the central wind farm controller.

According to the inertia control principle, the wind turbine responds to the rate of change of the frequency. This feature is often referred to as "virtual inertia" effect, as it introduces an output power term proportional to $2H_{WT}\frac{df}{dt}$ [93, 94].

The droop control consists of providing an output power term proportional to the deviations of frequency:

$$\Delta P_{\text{droop}}^{\text{WT}} = K_{\text{droop}}\left(f - f_o\right) \tag{9-36}$$

where $f_o$ is the reference frequency. The gain $K_{\text{droop}}$ is the inverse of the droop parameter R:

$$R = \frac{\Delta f}{\Delta P} = \frac{1}{K_{\text{droop}}} \tag{9-37}$$

Values of 3% to 5% are common for $R$ in conventional generators. In droop control a dead band may also be included, as shown in Fig. 9-47.

The effects of the frequency response methods in an autonomous island grid have been studied by simulating the model system of Rhodes island shown in Fig. 9-52 and the results are presented in Fig. 9-53 [95]. This model includes two thermal power plants comprising steam turbines two 15 MW, gas turbines 28 MW and diesel two 15.6 MW, and five wind farms, of different technologies, namely type 3 (WFA1 and WFA2, DFIG of 17 MW), type 4 (WFB1 and WFB2, PMSG of 21 MW), and type 1 (WFC, SCIG of 12 MW). The operating scenario comprises load demand of 74.8 MW, thermal power generation of 35 MW, and wind power of 40.34 MW, that is, an instantaneous wind power penetration in excess of 50%. In such conditions, the system inertia decreases significantly and frequency control becomes challenging. A severe disturbance provoked by the sudden loss of the largest conventional generator (a 14-MW steam unit, operating at rated power) is considered. All wind farms operate at constant wind speed and frequency control is implemented to all variable speed wind turbines, assuming a 10% balance power level.

The responses of the system frequency under the different frequency control methods are shown in Fig. 9-50. Without frequency control a frequency excursion of approximately 1.4 Hz is noted. Droop control alone achieves only a marginal improvement of the minimum frequency, while the maximum rate of change of the frequency (ROCOF) immediately after the event remains the same. When inertia control is applied, either alone or in combination with the droop controller (combined control), the frequency excursion is reduced by about 50% and the rate of frequency change is also lower. Notably, an increased delay is observed in the restoration of frequency to its final value when inertia control is applied, either alone or in a combined control.

In island systems, the maximum frequency excursion and the ROCOF are the most crucial parameters, as they can trigger the load shedding protection systems. For this reason, it can be stated that the combined control achieves the best overall performance, in spite of the delay in frequency restoration, as the frequency remains within acceptable limits throughout the whole transient event.

***Offshore Super Grid and Wind Power.*** A super grid is a large transmission network connecting a number of nations across a wide geographic area and can transfer volumes of electricity over long distances. It is also referred as a "mega grid." Such super grids not only have intelligent features in the wide area transmission level but can also integrate local smart grids where the distributed renewable energy integration makes the energy flow become bidirectional, and turns the local smart grid into a virtual power plant.

A super grid consists of high voltage transmission network. On land either HVAC systems or HVDC systems may be used, while HVAC system is easy for interconnection into a mesh network, and most existing HVDC systems are point to point transmission, with only a few multiterminal configurations. VSC-HVDC would be able to support traditional power systems, and thyristor technology–based HVDC systems [96] could play an important role in super grids.

Offshore wind power will play an important role in meeting European future energy demand. It is possible that the European Union could be supplied only with the wind power. However, the variability of wind cannot secure a reliable power source. An interconnected European offshore super grid would connect the wind farms dispersed across a wide geographic area, ranging from the Baltic

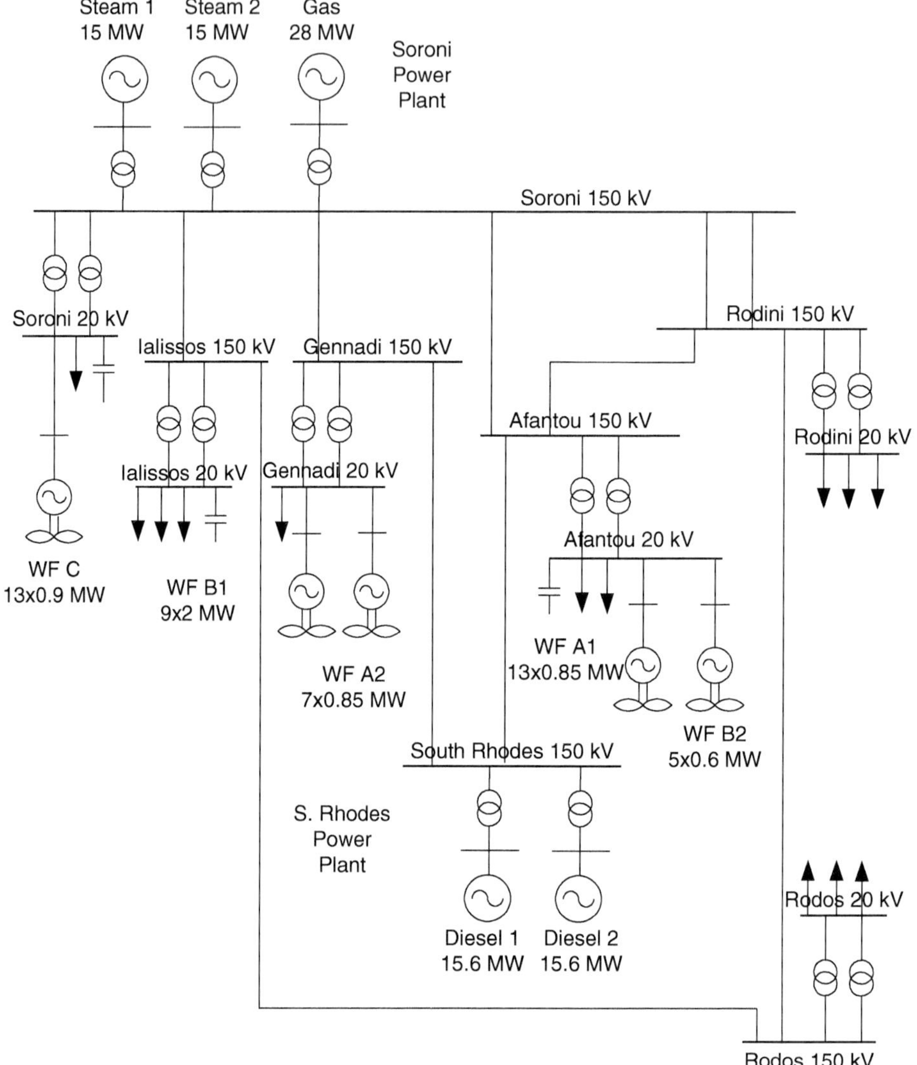


**FIGURE 9-52** The power system of Rhodes island.

Sea to the Mediterranean and Atlantic. Such an offshore wind farm grid could provide the EU with a clean, sustainable, and secure power supply. The offshore super grid can help to solve the problem of the variability of wind by exploiting the meteorological characteristics to balance the fluctuations of wind power as the wind is likely always to be blowing somewhere. A lower power output from a wind farm may be balanced by a simultaneous high power output from another wind farm several hundred kilometers away. An ocean super grid will create a network between a number of countries to balance the fluctuations of wind energy and to solve the problem of the variability of wind, and may turn this fluctuating energy source into a quite stable power source. Furthermore, it is possible to combine the production of the fluctuating wind power with that from some dispatchable power

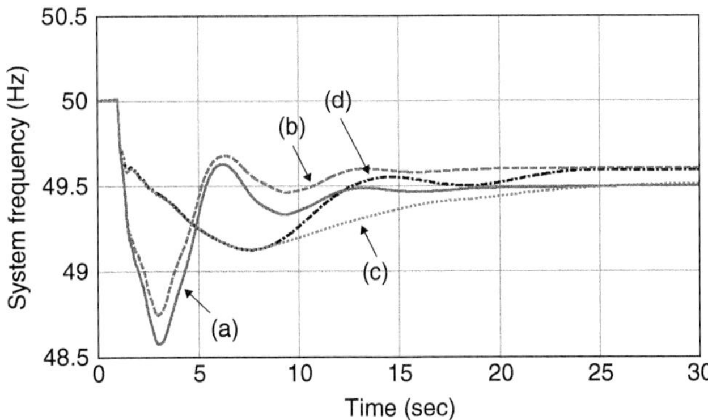


**FIGURE 9-53** System frequency following the outage of the largest thermal unit under different frequency control methods. (*a*) No frequency control; (*b*) droop control; (c) inertia control; (*d*) combined control.

sources located at a distance, such as hydropower generation units, in order to keep an overall stable power supply. Hydropower plants may be used as energy storage to export or import power when the produced wind power is not enough or is excessive.

Furthermore, the electricity markets in Europe require an adequate transport capacity among different power market zones to enable effective competition and the trade of electricity. Therefore, enhancing the suitability of the grid for the increased inter-market electricity transport is important for both the wind industry and electricity market development. The super grid could have an important function in distribution and transmission of renewable energy between the relevant countries and it can introduce additional flexibility to the power system operation and control.

A super grid will facilitate an efficient international electricity market for Europe. For example, when the wind is very strong, the surplus wind energy produced in Denmark can be exported to Norway which has many hydropower plants, the Norwegian hydropower plants can stop production and hydro energy is stored; when the wind does not blow in Denmark, the Norwegian hydropower plants can produce the electricity and send it to Denmark.

The important benefits of an offshore super grid include

- Increase the interconnection and transmission capacity within the relevant countries/areas.
- Reduce overall wind power fluctuations, improve the accuracy of short-term wind power forecasting.
- Increase the utilization of renewable energy, particular wind energy and reduce the curtailment of wind energy.
- Optimally use energy storage units and fast generating units, such as hydropower resources, to meet the rapid changes in demand and/or renewable energy production.
- Optimally deploy power reserve.
- Lower the cost of power in all connected countries by sharing the most efficient power plants within these countries.
- Significantly increase the use of renewables, decrease dependence of imported fuels and reduce greenhouse gas emission.

A study has optimized a vast grid covering North Africa, Eastern Europe, Norway, and Iceland, with a number of scenarios, wind, concentrated solar power (CSP), nuclear, etc., and the results showed that all European power could largely come from wind energy, with relatively low amounts of combustion plants needed during low wind periods. Furthermore, the study showed that no new storage would be required; existing hydro is sufficient. The total cost of energy would be at the same level as or lower than that at present according to this study [97].

A number of European offshore super grids have been proposed, one example is shown in Fig. 9-54, from [98]. Other proposed schemes include *Baltic Energy Market Interconnection Plan, Europagrid, North Sea Offshore Grid, EU PowerNet, ISLES, Low Grid,* and *High Grid.*

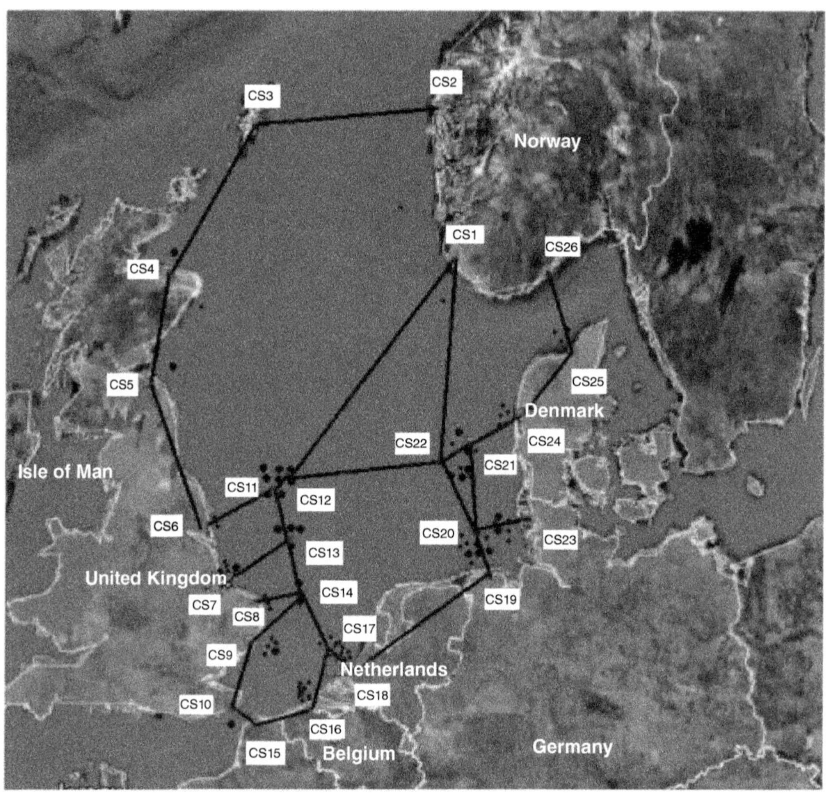


**FIGURE 9-54** A proposed offshore super grid in North Sea [98].

A VSC-HVDC multiterminal system can be constructed if three or more HVDC wind farm converter stations are connected to a network with cables. In this case, only one converter may be used to regulate the dc voltage of the system and the other converters control the import or export power of the HVDC system. Figure 9-55 illustrates a multiterminal HVDC system.

***HVDC Link Example.*** *HVDC link NorNed* was manufactured by ABB and Nexans Norway AS and is jointly owned (50/50) by two TSOs, Statnett in Norway and TenneT in The Netherlands. The construction work started in 2006 and the link was commissioned in May 2008. LCC-HVDC technology is used. The cable has a length of 580 km with a capacity of 700 MW at ±450 kV dc voltage and connects the Netherlands network, a part of European network UCTE synchronous area, to the Norwegian network, a part of the Nordic synchronous area.

The cable has the capacity of transmitting power in both directions for balancing generation and consumption in both networks. In a typical situation, the consumption in Netherlands is higher

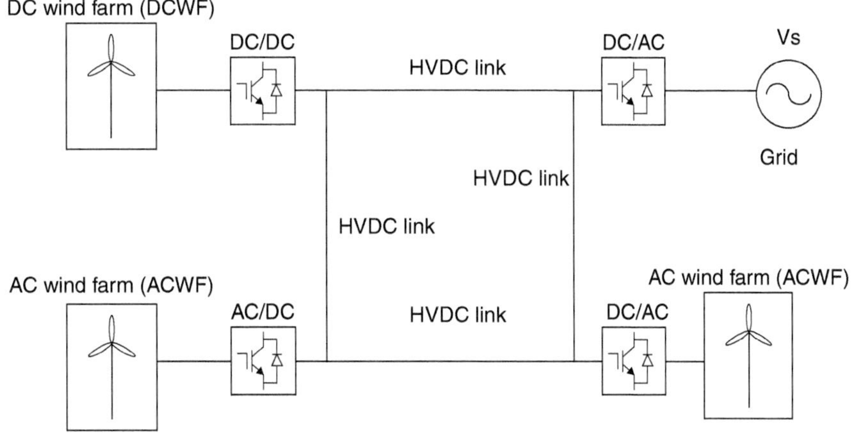


**FIGURE 9-55** Illustration of a multiterminal HVDC system.

during the day, the cheaper hydro power is sent to the Netherlands via the cable from Norway. At night the power flow reverses its direction because Norway consumes more electricity during the night than during the day. The Netherlands export at night gas-produced electricity via the cable allowing the Norwegian reservoirs to fill up for the day time.

## 9.5 ACKNOWLEDGMENT

The authors of this section would like to thank Dr. Eduard Muljadi, of the National Renewable Energy Laboratory in Golden, Colorado, for his support.

## 9.6 REFERENCES

1. Wind Energy Foundation, "History of Wind Energy." 2016. http://windenergyfoundation.org/about-wind-energy/history/.
2. Hardy, C., "Renewable Energy and Role of Marykirk's James Blyth." *The Courier.* Dundee, D.C. Thomson & Co. Archived from *the original* on 10 July 2010.
3. https://en.wikipedia.org/wiki/History_of_wind_power.
4. History of Wind Energy in Encyclopedia of Energy Vol. 6.
5. Beaufort wind force scale, http://www.metoffice.gov.uk/guide/weather/marine/beaufort-scale.
6. Lalas, D. P. and Ratto, C. F., (eds), *Modelling of Atmospheric Flow Fields*, World Scientific, Singapore, 1996.
7. Burton, T., Jenkins, N., Sharpe, D., and Bossanyi, E., *Wind Energy Handbook*, 2nd ed., John Wiley & Sons, Ltd, 2011.
8. Chen, Z., "Advanced Wind Energy Converters Using Electronic Power Conversion," Ph.D., University of Durham, UK, 1997.
9. http://onlinelibrary.wiley.com/doi/10.1002/9781118900116.app2/pdf, Accessed April 2017.
10. Siegfried Heier, "Grid Integration of Wind Energy Conversion Systems," John Wiley & Sons, Ltd., United Kingdom, 1998. ISBN 0-471-97143-X.
11. Chen, Z., "Wind Turbine Drive Train Systems," in *Wind Energy Systems, Optimising Design and Construction for Safe and Reliable Operation*, edited by John D. Sørensen and Jens N. Sørensen, Woodhead Publishing, Ltd., United Kingdom, 2011. ISBN: 978-1-84569-580-4.

12. Bronzeado, H. S., Feitosa, E. A. N., Rosas, P. A. C., Miranda, M. S., de Barros, M. E. M., and Rohatgi, J., "Investigation of the Behaviour of Wind Turbines under Low Turbulence Wind Conditions and Their Interaction with the Distribution Grid," *Wind Eng.*, vol. 24, no. 2, pp. 101–109, 2000.

13. Chen, Z., Hu, Y., and Blaabjerg, F., "Stability Improvement of Induction Generator-Based Wind Turbine Systems," *IET Proc.—Renewable Power Generation*, vol. 1, no. 1, pp. 81–93, Mar. 2007.

14. Li, H. and Chen, Z., "Overview of Different Wind Generator Systems and Their Comparisons," *IET Proc.—Renewable Power Generation*, vol. 2, no. 2, pp. 123–138, Jun. 2008.

15. Zhang, J., Chen, Z., and Cheng, M., "Design and Comparison of a Novel Stator Interior Permanent Magnet Generator for Direct-Drive Wind Turbines," *IET Proc.—Renewable Power Generation*, vol. 1, no. 4, pp. 203–210, Dec. 2007. ISSN 1752-1416.

16. Jianzhong, Z., Cheng, M., and Chen, Z., "Investigation of a New Stator Interior Permanent Magnet Machine," *IET Proc.—Electric Power Applications*, vol. 2, no. 2, pp. 77–87, Mar. 2008.

17. Liu, X., Park, K., and Chen, Z., "A Novel Excitation Assistance Switched Reluctance Wind Power Generator," *IEEE Transactions on Magnetics*, vol. 50, no. 11, pp. 1–4, 2014.

18. Lobato, P., Dente, J. A., Martins, J. F., and Pires, A. J., "Scale Models Formulation of Switched Reluctance Generators for Low Speed Energy Converters," *IET Electric Power Applications*, vol. 9, no. 9, pp. 652659, 2015.

19. Melcescu, L., Popescu, M., Covrig, M., Cistelecan, M., and Craiu, O., "Numerical and Experimental Analysis of Two Permanent Magnet Claw Poles Wind Generators," 18th International Conference on Electrical Machines, 2008.

20. Keysan, O., Radyjowski, P., Burchell, J., and Mueller, M. A., "Towards More Reliable and Cost Effective Superconducting Generators for Wind Turbines," *7th IET International Conference on Power Electronics,* Machines and Drives (PEMD 2014), Apr. 8–10, 2014.

21. Gorginpour, H., Oraee, H., and McMahon, R. A., "A Novel Modeling Approach for Design Studies of Brushless Doubly Fed Induction Generator Based on Magnetic Equivalent Circuit," *IEEE Transactions on Energy Conversion*, vol. 28, no. 4, pp. 902–912, 2013.

22. van der Blij, N. H., Strous, T. D., Wang, X., and Polinder, H., "A Novel Analytical Approach and Finite Element Modelling of a BDFIM," *2014 International Conference on Electrical Machines (ICEM)*, pp. 346–352, 2014.

23. Chen, Z., Wang, B., Chen, Z., and Yan, Y., "Comparison of Flux Regulation Ability of the Hybrid Excitation Doubly Salient Machines," *IEEE Transactions on Industrial Electronics*, vol. 61, no. 7, pp. 3155–3166, Jul. 2014.

24. He, J., Tang, Y., Li, J., Ren, L., Shi, J., Wang, J., Qu, R., Su, L., Li, X., Xu, Y., and Zhu, Z., "Conceptual Design of the Cryogenic System for a 12 MW Superconducting Wind Turbine Generator," *IEEE Transactions on Applied Superconductivity*, vol. 24, no. 3, pp. 1–5, 2014.

25. Liu, D., Polinder, H., Abrahamsen, A. B., and Ferreira, J. A., "Topology Comparison of Superconducting Generators for 10-MW Direct-Drive Wind Turbines: Cost of Energy Based," *IEEE Transactions on Applied Superconductivity*, vol. 27, no. 4, pp. 1–7, 2017.

26. Shafaie, R., Amirkhanloo, F., and Kalantar, M., "Toward an Optimum Design of Large-Scale HTS Synchronous Generator for Wind Turbine Applications," *IEEE Transactions on Applied Superconductivity*, vol. 26, no. 2, pp. 1–8, 2016.

27. Wang, R.-J., Brönn, L., Gerber, S., and Tlali, P. M., "Design and Evaluation of a Disc-Type Magnetically Geared PM Wind Generator," *4th International Conference on Power Engineering, Energy and Electrical Drives*, pp. 1259–1264, 2013.

28. Zhang, X., Liu, X., and Chen, Z., "A Novel Coaxial Magnetic Gear and Its Integration with Permanent-Magnet Brushless Motor," *IEEE Tarnation on Magnetics*, vol. 52, no. 7, pp. 1–4, Jul. 2016.

29. Chen, Z., "An Overview of Power Electronic Converter Technology for Renewable Energy Systems," in *Direct-Drive Wind and Marine Energy Systems*, edited by Markus Mueller, Woodhead Publishing, Ltd., United Kingdom, 2013.

30. Chen, Z. and Spooner, E., "A Modular, Permanent-Magnet Generator for Variable speed Wind Turbines," *IEE International Conference EMD'95*, Conference Publication No. 412, pp. 453–457, 1995.

31. Chen, Z., Spooner, E., Norris, W. T., and A. C.Williamson, "Capacitor-Assisted Excitation of Permanent-Magnet Generators," *IEE Proc.—Electr. Power Applications*, vol. 145, no. 6, pp. 497–508, Nov. 1998.

32. Chen, Z. and Spooner, E., "Grid Interface Options for Variable-Speed, Permanent-Magnet Generators," *IEE Proc.—Electr. Power Applications*, vol. 145, no. 4, pp. 273–283, Jul. 1998.

33. Chen, Z. and Spooner, E., "Voltage Source Inverters for High-Power, Variable-Voltage DC Power Sources," *IEE Proc.—Generation, Transmission and Distributions*, vol. 148, no. 5, pp. 439–447, Sep. 2001.

34. Li, J., Wu, X., Hu, S., Zhao, B., and Xu, H., "Direct-Drive Wind Power Generator System Based Interleaved Boost Converter," *Proceedings of ISES World Congress 2007* (Vol. I–V), vol. 6, pp. 2301–2304, 2007.

35. Chen Z. and Spooner, E., "Current Source Thyristor Inverter and Its Active Compensation System," *IEE Proc.—Generation, Transmission and Distributions*, vol. 150, no. 4, pp. 447–454, Jul. 2003.

36. Chen, Z., "Compensation Schemes for a SCR Converter in Variable Speed Wind Power Systems," *IEEE Transactions on Power Delivery*, vol. 19, no. 2, pp. 813–821, Apr. 2004.

37. Zeng, X., Chen, Z., and Blaabjerg, F., "Efficiency Comparisons of Full-Size Converters for Large Variable-Speed Wind Turbine Systems," *13th European Conference on Power Electronics and Applications*, EPE2007, Sep. 2007.

38. Muljadi, E., Singh, M., and Gevorgian, V., "Fixed-Speed and Variable-Slip Wind Turbines Providing Spinning Reserves to the Grid," National Renewable Energy Laboratory, IEEE Power & Energy Society. General Meeting, Vancouver, BC, 2013.

39. Singh, M., Muljadi, E., Jonkman, J., Gevorgian, V., Girsang, I., and Dhupia, J., "Simulation for Wind Turbine Generators—with FAST and MATLAB-Simulink Modules," Technical Report NREL/TP-5D00-59195, Apr. 2014.

40. McDonald, A. S., Mueller, M. A., and Polinder, H., "Comparison of Generator Topologies for Direct-Drive Wind Turbines Including Structural Mass," in *Proc. of the International Conference on Electrical Machines (ICEM)*, pp. 360.1–360.7, Sep. 2006.

41. Bang, D., Polinder, H., Shrestha, G., and Ferreira, J. A., "Review of Generator Systems for Direct-Drive Wind Turbines," European Wind Energy Conference , 2011.

42. Offshore Class Prototypes http://www.wwindea.org/technology/ch01/en/1_3_4.html.

43. Li, H., Chen, Z., and Polinder, H., "Optimization of Multibrid Permanent Magnet Wind Generator Systems," *IEEE Transactions on Energy Conversion*, vol. 24, no. 1, pp. 82–92, Mar. 2009. ISSN: 0885-8969

44. Blaabjerg F. and Chen, Z., "Power Electronic for Modern Wind Turbines, in *Power Electronics Series*, Morgan & Claypool Publishers, 2006. ISBN: 1598290320

45. Chen, Z., Guerrero, J. M., and Blaabjerg, F., "A Review of the State-of-the-Art of Power Electronics for Wind Turbines," *IEEE Transactions on Power Electronics*, vol. 24, no. 8, pp. 1859–1875, Aug. 2009.

46. Wind turbine blade tip analysis using CFD, https://sites.google.com/site/ee535test/andrew-baker.

47. Zhang, Y., Chen, Z., and Cheng, M., "Proportional Resonant Individual Pitch Control for Mitigation of Wind Turbines Loads," *IET Renewable Power Generation*, vol. 7, no. 3, pp. 191–200, May 2013.

48. Zhang, Y., Chen, Z., Hu, W., and Cheng, M., "Flicker Mitigation by Individual Pitch Control of Variable Speed Wind Turbines with DFIG," *IEEE Transactions on Energy Conversion*, vol. 29, no. 1, pp. 20–28, Mar. 2014.

49. Dvorak, P., "How Vortex Generators Improve Wind Turbine Performance," http://www.windpowerengineering.com/design/mechanical/blades/vortex-generators-improve-wind-turbine-performance/.

50. Chen, Z., "Power Electronic Converter Systems for Direct Drive Renewable Energy Applications," in *Direct-Drive Wind and Marine Energy Systems*, edited by Markus Mueller, Woodhead Publishing, Ltd., United Kingdom, 2013.

51. Sun, T., Chen, Z., and Blaabjerg, F., "Flicker Study on Variable Speed Wind Turbines with Doubly Fed Induction Generators," *IEEE Transactions on Energy Conversion*, vol. 20, no. 4, pp. 896–905, Dec. 2005.

52. Hu, W., Chen, Z., Wang, Y., and Wang, Z., "Flicker Mitigation by Active Power Control of Variable Speed Wind Turbines with Full-Scale Back-to-Back Power Converters," *IEEE Transactions on Energy Conversion*, vol. 24, no. 3, pp. 640–649, Sep. 2009.

53. Zhang, Y., Hu, W., Chen, Z., Cheng, M., and Hu, Y., "Flicker Mitigation Strategy for a Doubly Fed Induction Generator by Torque Control," *IET Renewable Power Generation*, vol. 8, no. 2, pp. 91–99, Mar. 2014.

54. Zhao, B., Li, H., Wang, M., Chen, Y., Liu, S., Yang, D., Yang, C., Hu, Y., and Chen, Z., "An Optimal Reactive Power Control Strategy for a DFIG-Based Wind Farm to Damp the Sub-Synchronous Oscillation of a Power System," *Energies*, vol. 7, no. 5, pp. 3086–3103, May 2014.

55. Zhang, B., Hou, P., Hu, W., Soltani, M., Chen, C., and Chen, Z., "A Reactive Power Dispatch Strategy with Loss Minimization for a DFIG-Based Wind Farm," *IEEE Transactions on Sustainable Energy*, vol. PP, no. 99, pp. 1–10, Jan. 2016.

56. Tian, J., Zhou, D., Su, C., Chen, Z., and Blaabjerg, F., "Reactive Power Dispatch Method in Wind Farms to Improve the Lifetime of Power Converter Considering Wake Effect," *IEEE Trans. Sustainable Energy*, 08, Sep. 2016.

57. http://www.offshorewindenergy.org/CA-OWEE/Resource_and_Economics_3.html, Accessed April 2017.

58. Deep water—the next step for offshore wind energy, www.ewea.org/fileadmin/files/library/publications/reports/Deep_Water.pdf, Accessed April 2017.

59. https://www.google.co.uk/search?q=danish+offshore+wind+farm+layouts&tbm=isch&imgil=7-ieJCLDMIZiiM%253A%253BEazaDd6DF9ED8M%253Bhttp%25253A%25252F%25252Frsta.royalsocietypublishing.org%25252Fcontent%25252F373%25252F2035%25252F20140100&source=iu&pf=m&fir=7-ieJCLDMIZiiM%253A%252CEazaDd6DF9ED8M%252C_&usg=__8Gvutj6fColUNA37TWMyeZzHZyI%3D&biw=863&bih=687&ved=0ahUKEwigvd-4xtnSAhUGDCwKHemcCg4QyjcIJQ&ei=Br_JWKCiOoaYsAHpuapw#tbm=isch&q=layouts+of+Horns+Rev+offshore+wind+farms+&, Accessed April 2017.

60. Zhao, M., Chen, Z., and Blaabjerg, F., "Probabilistic Capacity of a Grid Connected Wind Farm Based on Optimization Method," *Renewable Energy*, vol. 31, pp. 2171–2178, 2006.

61. Zhao, M., Chen, Z., and Blaabjerg, F., "Optimization of Electrical System for Offshore Wind Farms via Genetic Algorithm," *IET Proc.—Renewable Power Generation*, vol. 3, no. 2, pp. 205–216, Jun. 2009. ISSN 1752-1416.

62. Hou, P., Hu, W., Chen, C., and Chen, Z., "Optimization of Offshore Wind Farm Cable Connection Layout Considering Levelised Production Cost Using Dynamic Minimum Spanning Tree Algorithm," *IET Renewable Power Generation*, vol. 10, no. 2, pp. 175–183, 2016.

63. Hou, P., Hu, W., Soltani, M., Chen, C., and Chen, Z., "Combined Optimization for Offshore Wind Turbine Micro Siting," *Applied Energy*, vol. 189, pp. 271–282, Mar. 2017.

64. Deng F. and Chen, Z., "Control of Improved Full-Bridge Three-Level DC/DC Converter for Wind Turbines in a DC Grid," *IEEE Transaction on Power Electronics*, vol. 28, no. 1, pp. 135–143, 2013.

65. Deng F. and Chen, Z., "Design of Protective Inductors for HVDC Transmission Line within DC Grid Offshore Wind Farms," *IEEE Transactions on Power Delivery*, vol. 28, no. 1, pp. 75–83, 2013.

66. Deng F. and Chen, Z., "Operation and Control of a DC Grid Offshore Wind Farms under DC Transmission System Faults," *IEEE Transactions on Power Delivery*, vol. 28, no. 3, pp. 1356–1363, 2013.

67. Negra, N. B., Todorovic, J., and Ackermann, T., "Loss Evaluation of HVAC and HVDC Transmission Solutions for Large Offshore Wind Farms," *Electric Power Systems Research*, vol. 76, no. 11, pp. 916–927, Jul. 2006.

68. HVDC vs. HVAC cables for offshore wind, http://analysis.windenergyupdate.com/offshore-wind/hvdc-vs-hvac-cables-offshore-wind.

69. Donaghy, R., "HV Submarine Cable Systems Design, Testing and Installation" CIGRE, Ireland Technical Seminar, Oct. 6, 2010.

70. Vrana, T. K., "Review of HVDC Component Ratings: XLPE Cables and VSC Converters," *IEEE International Energy Conference (ENERGYCON)*, 2016.

71. http://www.4coffshore.com/windfarms/, Accessed April 2017.

72. Chen Z. and Blaabjerg, F., "Wind Farm—A Power Source in Future Power Systems," *Renewable and Sustainable Energy Reviews*, vol. 13, nos. 6, 7, pp. 1288–1300, Aug./Sep. 2009. ISSN: 1364-0321.

73. Technical regulation 3.2.5 for wind power plants with a power output greater than 11 kW, Energynet.dk.

74. Chen, P., Siano, P., B. Bak-Jensen, and Chen, Z., "Stochastic Optimization of Wind Turbine Power Factor Using Stochastic Model of Wind power," *IEEE Transactions on Sustainable Energy*, vol. 1, no. 1, pp. 19–29, Apr. 2010.

75. Chen Z. and Spooner, E., "Grid Power Quality with Variable-Speed Wind Turbines," *IEEE Transactions on Energy Conversion*, vol. 16, no. 2, pp. 148–154, Jun. 2001.

76. Wang, H., Chen, Z., and Jiang, Q., "Optimal Control Method for Wind Farm to Support Temporary Primary Frequency Control with Minimised Wind Energy Cost," *IET Renewable Power Generation*, vol. 9, no. 4, pp. 350–359, 2015.

77. Wang, L., S.-Chen, S., Lee, W.-J., and Chen, Z., "Dynamic Stability Enhancement and Power Flow Control of a Hybrid Wind and Marine-Current Farm Using SMES," *IEEE Transactions on Energy Conversion*, vol. 24, no. 3, pp. 626–639, Sep. 2009.

78. Zhao, B., Li, H., Wang, M., Chen, Y., Liu, S., Yang, D., Yang, C., Hu, Y., and Chen, Z., "An Optimal Reactive Power Control Strategy for a DFIG-Based Wind Farm to Damp the Sub-Synchronous Oscillation of a Power System," *Energies*, vol. 7, no. 5, pp. 3086–3103, May 2014.

79. Muljadi, E., Zhang, Y. C., Gevorgian, V., and Kosterev, D., "Understanding Dynamic Model Validation of a Wind Turbine Generator and a Wind Power Plant," *Energy Conversion Congress and Exposition (ECCE)*, 2016 IEEE, pp. 18–22, Sep. 2016.

80. Li, H., Zhao, B., Yang, C., Chen, H. W., and Chen, Z., "Analysis and Estimation of Transient Stability for a Grid-Connected Wind Turbine with Induction Generator," *Renewable Energy*, vol. 36, no. 5, pp. 1469–1476, May 2011.

81. Wei M. and Chen, Z., "Fast Control Strategy for Stabilizing Fixed-Speed Induction Generator Based Wind Turbines in an Islanded Distributed System," *IET Proc.—Renewable Power Generation*, vol. 7, no. 2, pp. 144–162, Mar. 2013.

82. Jun, Y., Hui, L., Zhe, C., Xianfeng, X., Xiyin, C., Qing, L., and Yong, L., "Enhanced Control of a DFIG-Based Wind-Power Generation System with Series Grid-Side Converter under Unbalanced Grid Voltage Conditions," *IEEE Transaction on Power Electronics*, vol. 28, no. 7, pp. 3167–3181, 2013.

83. Zhu, R., Chen, Z., Wu, X., and Deng, F., "Virtual Damping Flux Based LVRT Control for DFIG-Based Wind Turbine," *IEEE Transaction Energy Conversion*, vol. 30, no. 2, pp. 714–725, Jun. 2015.

84. Su, C., Hu, W., Chen, Z., and Hu, Y., "Mitigation of Power System Oscillation Caused by Wind Power Fluctuation," *IET Renewable Power Generation*, vol. 7, no. 6, pp. 639–651, Dec. 2013.

85. Wen, L., W.Hu, H.Lund, and Chen, Z., "Electric Vehicles and Large-Scale Integration of Wind Power—the Case of Inner Mongolia in China," *Applied Energy*, vol. 104, pp. 445–456, 2013.

86. Hu, W., Su, C., Chen, Z., and B. Bak-Jensen, "Optimal Operation of Plug-In Electric Vehicles in Power Systems with High Wind Power Penetrations," *IEEE Transaction on Sustainable Energy*, vol. 4, no. 3, pp. 577–585, 2013.

87. Hu, W., Chen, Z., and B. Bak-Jensen, "Analysis of Electricity Price in Danish Competitive Electricity Market," *IEEE Power and Energy Society General Meeting*, pp. 22–26, Jul. 2012.

88. Li, J., Fang, J., Zeng, Q., and Chen, Z., "Optimal Operation of the Integrated Electrical and Heating Systems to Accommodate the Intermittent Renewable Sources," *Applied Energy*, vol. 167, pp. 244–254, Apr. 1, 2016.

89. Rather, Z. H., Chen, Z., Thøgersen, P., and Lund, P., "Dynamic Reactive Power Compensation of Large-Scale Wind Integrated Power System," *IEEE Transactions on Power Systems*, vol. 30, no. 5, pp. 2516–2526, Sep. 2015.

90. Liu, C., Sun, K., Hussain, Z., Chen, Z., Bak, C. L., Thøgersen, P., and P. Lund "A Systematic Approach for Dynamic Security Assessment and the Corresponding Preventive Control Scheme Based on Decision Trees," *IEEE Transaction on Power Systems*, vol. 29, no. 2, pp. 717–730, Mar. 2014.

91. Abulanwar, E., Hu, W., Chen, Z., and Iov, F., "Adaptive Voltage Control Strategy for Variable-Speed Wind Turbine Connected to a Weak Network," *IET Renewable Power Generation*, vol. 10, no. 2, pp. 238–249, Feb. 2016.

92. Hatziargyriou, N., Margaris, I., Stavropoulou, I., Papathanassiou, S., and Dimeas, A., "Non-Interconnected Island Systems: The Greek Case," *IEEE Electrification Magazine*, 2017.

93. Lalor, G., Mullane, A., and M. J. O'Malley, "Frequency Control and Wind Turbine Technologies," *IEEE Transactions on Power Systems*, vol. 20, no. 4, Nov. 2005.

94. Ramtharan, G., Ekanayake, J. B., and Jenkins, N., "Frequency Support from Doubly Fed Induction Generator Wind Turbines," *IET Renew. Power Gener.*, vol. 1, no. 1, pp. 3–9, 2007.

95. Margaris, I. D., Papathanassiou, S. A., Hatziargyriou, N. D., Hansen, A. D., and Sorensen, P., "Frequency Control in Autonomous Power Systems with High Wind Power Penetration," *IEEE Transactions on Sustainable Energy*, vol. 3, no. 2, pp. 189–199, 2012.

96. Liu Y. and Chen, Z., "A Flexible Power Control Method of VSC-HVDC Link for the Enhancement of Effective Short-Circuit Ratio in a Hybrid Multi-Infeed HVDC System," *IEEE Transaction on Power System*, vol. 28, no. 2, May 2013, pp. 1568–1581.

97. Czish, G., "Low Cost but Totally Renewable Electricity Supply for a Huge Supply Area, a European/Trans-European Example," http://transnational-renewables.org/Gregor_Czisch/projekte/LowCostEuropElSup_revised_for_AKE_2006.pdf.

98. Grimaldi, A., Chen, Z., Chen, P., Siano, P., Piccolo, A., "Designing Offshore Super Grid for the Combined Operation of Offshore Wind Farms and Hydro Storage," *International Journal On Power System Optimization*, vol. 2, no. 1, pp. 149–158, 2011. ISSN: 0975-458X.

# 10

# SOLAR POWER GENERATION AND ENERGY STORAGE

Benjamin Kroposki
*Director, Power Systems Engineering Center, National Renewable Energy Laboratory, Golden, Colorado*

Robert Margolis
*Principal Energy Analyst, National Renewable Energy Laboratory, Golden, Colorado*

Mark Mehos
*Program Manager, Concentrating Solar Power, National Renewable Energy Laboratory, Golden, Colorado*

Jim Eyer
*Principal and Senior Analyst, E&I Consulting, Oakland, California*

Rahul Walawalkar
*President and Managing Director, Customized Energy Solutions India Pvt. Ltd., and Executive Director, India Energy Storage Alliance, Pune, India*

Haresh Kamath
*Senior Program Manager, Energy Storage and Distributed Generation, Electric Power Research Institute, Palo Alto, California*

## 10.1 SOLAR ENERGY BASICS

*BY BENJAMIN KROPOSKI, ROBERT MARGOLIS, AND MARK MEHOS*

Solar power generation may be classified as either direct or indirect. Direct solar power involves only one transformation into a usable form. The two examples discussed in this section include photovoltaic (PV) conversion of sunlight directly into electricity and concentrating solar power (CSP) applications wherein sunlight is concentrated to heat a surface that, in turn, transfers the energy to a fluid. Indirect solar power generation involves more than one transformation to reach a usable form and it is not discussed in detail in this section. Examples of indirectly solar generation include (1) vegetation that uses photosynthesis to convert solar energy to chemical energy, which can later be burned as fuel to generate electricity; (2) energy obtained from oil, coal, and peat that originated as solar energy, was captured by vegetation in the remote geological past, and became fossilized; (3) hydroelectric dams and wind turbines that are indirectly powered by solar energy through their interactions with the Earth's atmosphere and the resulting weather phenomena; (4) that energy obtained from methane (natural gas) may be derived from solar energy as either a biofuel or fossil fuel; and (5) ocean thermal energy production, which uses the thermal and gradients that are present across ocean depths to generate power.

PV and CSP technologies directly use sunlight to generate electricity. However, they do it in different ways and use different forms of the sun's radiation. PV—or solar electric—systems use semiconductor solar cells to convert sunlight directly into electricity. In contrast, CSP—or solar thermal electric—systems use mirrors to concentrate sunlight and exploit the sun's thermal energy. This energy heats a fluid that can be used to drive a turbine or piston, thus producing electricity. The Earth's surface receives sunlight in either a direct or diffuse form as shown in Fig. 10-1.

Direct sunlight is solar radiation whose path comes directly from the sun's disk and is known as direct normal irradiance (DNI). This is the form used by CSP systems and concentrating PV systems, wherein the reflection or focusing of the sun is essential to the electricity-generating process. Flat-plate, or non-concentrating, PV systems can also use this type of sunlight, but it can also produce electricity from diffuse sunlight. The map in Fig. 10-2 shows the annual direct-normal solar radiation in the United States; the highest concentration of direct sunlight is in the southwest United States.

The other component to solar radiation is diffuse, which refers to sunlight that reaches the earth's surface after passing through a thin cloud cover or having been first reflected off of particles or surfaces. Global radiation is the sum of the direct and diffuse components of sunlight. Global radiation, as well as direct or diffuse radiation, can be used by flat-plate PV systems to generate electricity. Figure 10-3

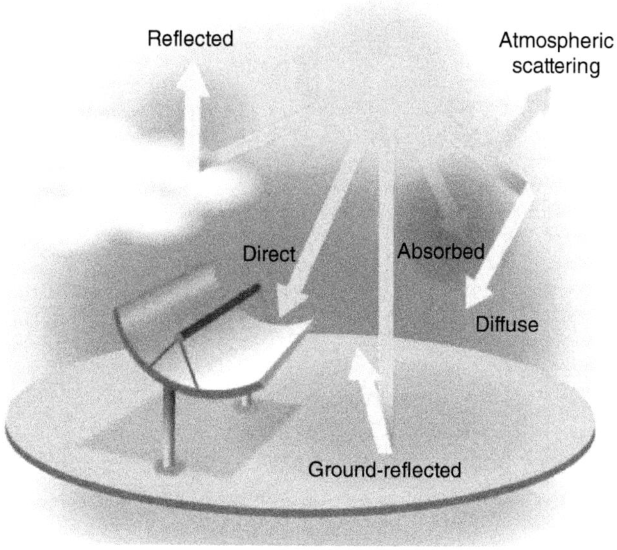


**FIGURE 10-1** Components of solar radiation. (*NREL.*)

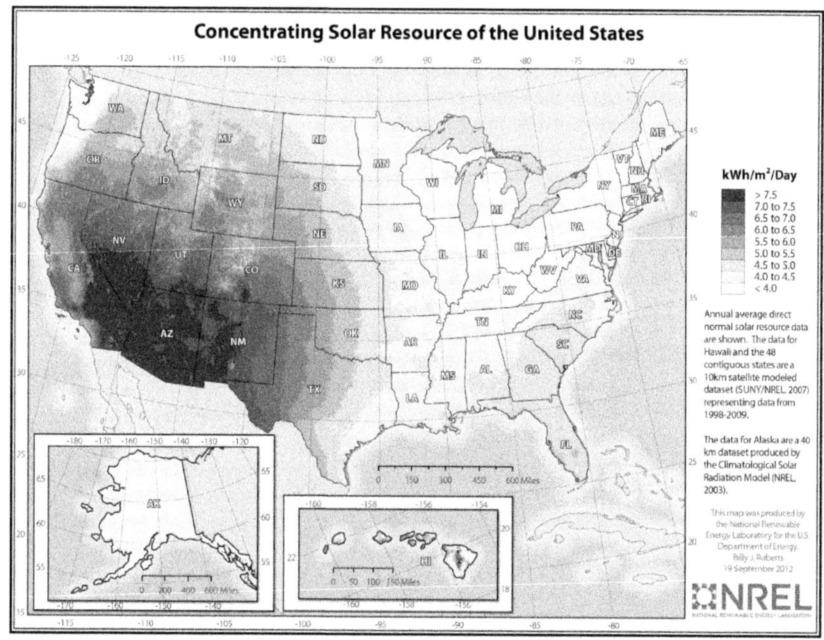


**FIGURE 10-2** Solar radiation map of the United States, showing the intensity of direct-normal sunlight averaged during a year. (*NREL.*)

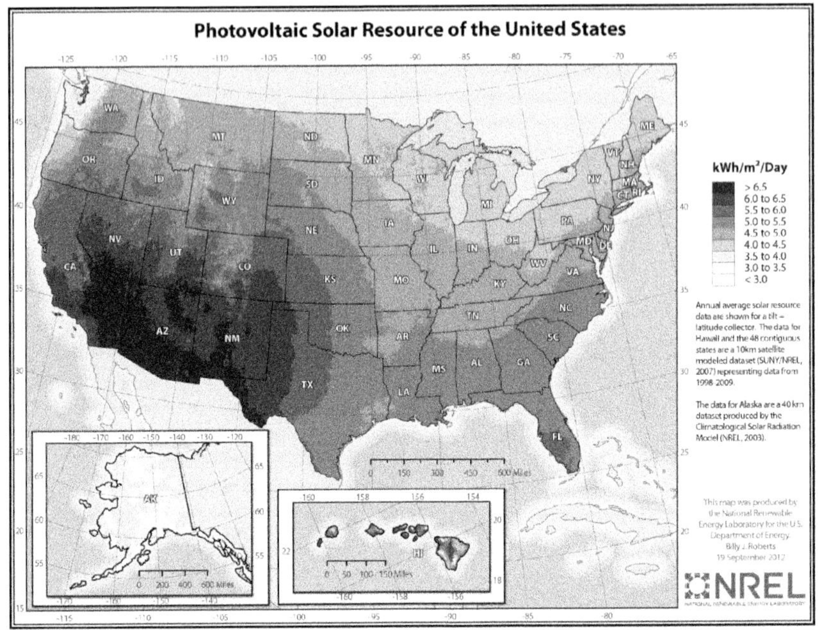


**FIGURE 10-3** Annual solar radiation on a latitude-tilt surface in the United States. (*NREL.*)

shows the annual solar radiation on a latitude tilt surface for the United States. For fixed tilt systems, latitude tilt typically produces the maximum power output on an annual basis. This figure shows very high solar resources for the southwestern United States, however, relatively good production exists throughout the entire United States as well as throughout most areas around the world.

The sun emits radiation over a wide range of wavelengths. Most of the solar radiation is in the wavelength range from 290 to 3000 nm and is referred to as broadband solar radiation. Figure 10-4

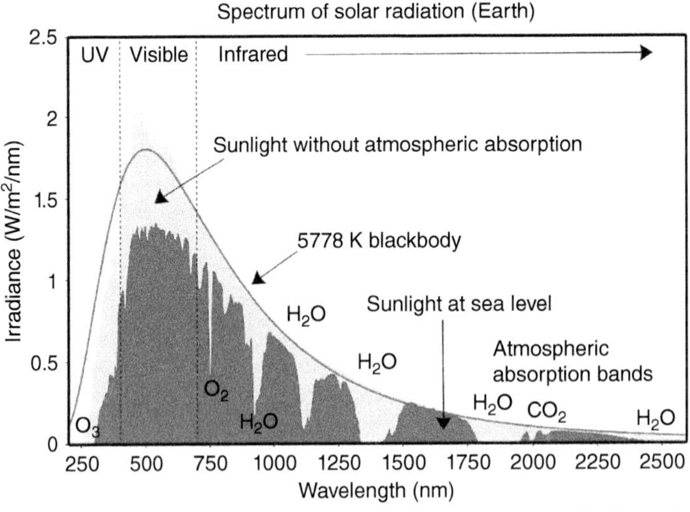


**FIGURE 10-4** Solar irradiance spectrum from 250 to 2500 nm. (*NREL.*)

shows terrestrial (dark shade) and extraterrestrial spectrum (light shade) of DNI. Atmospheric absorption bands are due to water vapor and other gases in the atmosphere.

The total radiant power from the sun is remarkably constant. In fact, the solar output has commonly been called the solar constant, but it does vary slightly over time due to cycles in the number of sunspots (cooler, dark areas on the sun) and changes in Earth's elliptical orbit. The measured variation resulting from the sunspot cycle is ±0.2%, only twice the precision of the most accurate radiometers measuring the irradiance in space. There is, however, some large variability in a few spectral regions, especially the ultraviolet (wavelengths less than 400 nm), caused by solar activity. The annual variations in solar irradiance due to Earth's orbit result in changes of ±3%.

### 10.1.1 Bibliography


- M. Sengupta, A. Habte, S. Kurtz, A. Dobos, S. Wilbert, E. Lorenz, T. Stoffel, D. Renné, C. Gueymard, D. Myers, S. Wilcox, P. Blanc, and R. Perez, "*Best Practices Handbook for the Collection and Use of Solar Resource Data for Solar Energy Applications*," NREL/TP-5D00-63112, February 2015.
- T. Stoffel, D. Renné, D. Myers, S. Wilcox, M. Sengupta, R. George, C. Turchi, "*CONCENTRATING SOLAR POWER Best Practices Handbook for the Collection and Use of Solar Resource Data*," NREL/TP-550-47465, September 2010.


## 10.2 PHOTOVOLTAIC TECHNOLOGIES

*BY BENJAMIN KROPOSKI, ROBERT MARGOLIS, AND MARK MEHOS*

A variety of materials can directly convert sunlight into electricity via the "photovoltaic" effect. These PV materials include silicon, thin films, and multi-junctions technologies from the III-V elements. These materials are manufactured into cells and integrated into PV modules. Typically silicon materials and thin film technologies are made into flat-plate modules. Advanced III-V materials typically use concentrated sunlight to reduce the cost of the module. These materials and designs are discussed in the following subsections.

### 10.2.1 Crystalline Silicon

Silicon was one of the first materials to be used in early PV devices, and it continues to dominate the commercial solar cell market at more than 90% of the market share. Cells using silicon have been labeled as first-generation PV. Pure silicon is "doped" with minute amounts of other elements such as boron and phosphorus, which produces positive-type and negative-type semiconductor materials, respectively. Putting these two materials into contact with one another creates a built-in potential field. When this semiconductor device is illuminated, the energy of the sunlight frees electrons that then move out of the cell—due to the potential field—into wires that form an electrical circuit. This "photovoltaic" effect requires no moving parts and does not use up any of the material in the process of generating electricity. As shown in Fig. 10-5, a typical solar cell consists of a glass or plastic cover or other encapsulant, an antireflective surface layer, a front contact to allow electrons to enter a circuit, a back contact to allow them to complete the circuit, and the semiconductor layers wherein the electrons begin and complete their journey.

### 10.2.2 Thin-Film Photovoltaic

Second-generation PV devices are a more recent development and rely on layers of semiconductor materials that are much thinner than those in silicon cells. The thickness of a crystalline silicon

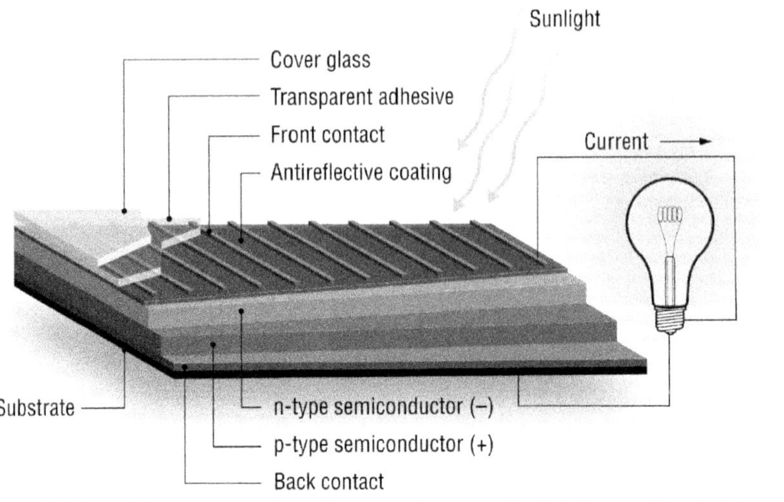


**FIGURE 10-5** Basic components of a silicon PV cell. (*NREL*.)

(c-Si) cell may be 170 to 200 μm ($10^{-6}$ m) thick, whereas the active region in a thin-film cell is on the order of only 2 to 3 μm thick. (For comparison, a human strand of hair has a thickness of approximately 80 μm.) If silicon is used, it is typically in the form of amorphous silicon (a-Si), which has no discernible crystal structure; in addition, microcrystalline silicon thin-film devices are also under development. But other thin-film materials have also been developed and commercialized, including cadmium telluride (CdTe) and copper indium gallium diselenide (CIGS). These PV devices require much less material than traditional c-Si devices.

Thin films, compared to crystalline silicon cells, generally have a lower solar conversion efficiency—which is the percentage of the sun's power shining on the cell that is converted into electric power by the cell. For example, if 1000 W of solar power illuminate a cell, and 200 W of electricity are generated, then the cell has a solar conversion efficiency of 20%. A commercial silicon cell may have an efficiency of around 20%, whereas a commercial CdTe cell is approximately of 11%. The thin-film cell uses less material and can be deposited with a method that is much less energy-intensive than that for silicon. Less material also equates to lighter weight. And some thin-film technologies do not rely on rigid cells, rather, they can be deposited on flexible substrates of stainless steel or plastic. Flexibility may be a desirable aspect depending on the application. Thus, in theory, thin-film PV should be less expensive to manufacture and easier to integrate into a wide range of applications; however, in practice, crystalline silicon costs have remained at a competitive level by reducing manufacturing costs.

### 10.2.3 Concentrating Photovoltaic

Another type of second-generation PV device is the high-efficiency multijunction cell that uses compounds from the group III and group V elements of the Periodic Table of Elements. An example of this type of multijunction cell (see Fig. 10-6) is a top layer of gallium indium phosphide, a middle layer of gallium arsenide, and a bottom layer of germanium. Very high efficiencies—over 40%—can be generated by this scheme. This is because each layer in this multijunction cell is designed to absorb and use a different portion of the solar spectrum. Figure 10-6 shows how the top layer would absorb shorter wavelengths (blue light), while the bottom cell would absorb longer wavelengths (red light).

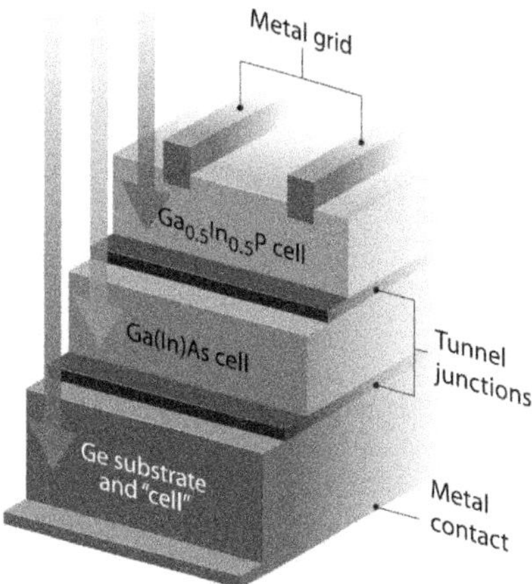


**FIGURE 10-6** Typical multijunction solar cell design showing three layers, each of which absorbs a different portion of the solar spectrum to use in generating electricity. (*NREL.*)

PV cells from these III-V materials are expensive to produce and use concentrating systems wherein a relatively inexpensive lens or mirror can be used to focus sunlight on only a small area of cells. For example, if a 10-in by 10-in lens focuses that amount of incident sun onto a 0.5-in by 0.5-in cell, the concentration factor is 400 times (100 in$^2$/0.25 in$^2$). This cell with the lens can produce as much power as a 10-in by 10-in cell without a lens, but at approximately 1/400th the cell cost. Figure 10-7 shows a typical basic concentrator unit that consists of a lens to focus the light, cell assembly, housing element, and secondary concentrator to reflect off-center light rays onto the cell, a mechanism to dissipate excess heat produced by concentrated sunlight, and various contacts and adhesives. The module depicted uses 12 cell units in a 2 × 6 matrix. These basic units may be combined in any configuration to produce a module with the desired power output.

Concentrating systems currently under development range in concentration levels from tens (10×) to hundreds (100×). Although they are not suitable for small projects, concentrating systems could be very effective in large-scale power generation. One downside to concentrating PV systems is that they require a tracking mechanism that keeps the modules always pointed at the sun. This can increase the overall cost of the system as well as be an additional point of maintenance. The challenge is in developing concentrating systems that balance the overall system-level costs so that they can be competitive in the marketplace.

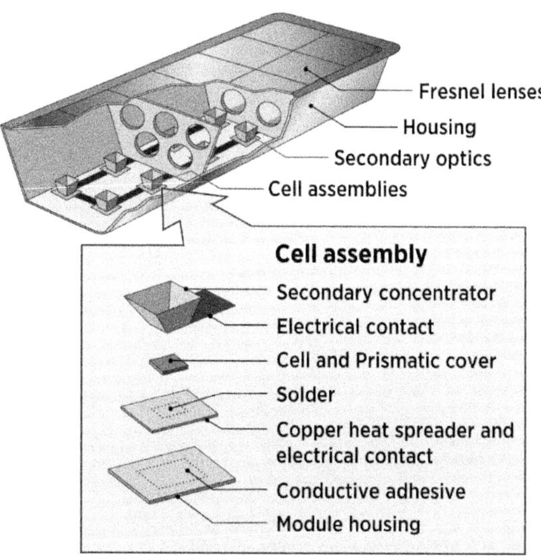


**FIGURE 10-7** Concentrating PV system that uses a lens to concentrate sunlight onto a high-efficiency solar cell. (*NREL.*)

## 10.2.4 Future Technologies

Scientists are exploring approaches for third-generation solar cells. One pathway is that of very high-efficiency cells, with their attendant high costs. The push is toward reaching the theoretical limits of various material systems and device configurations. The highest-efficiency device to date is a GaInP/GaAs/GaInAs multijunction, which has a conversion efficiency of 40.8% under a concentration of 326 times. There is a push to lower the costs of these by one or two orders of magnitude. The world record efficiency for a two-junction cell at one sun is 31.1%.

The other pathway is that of very low-cost cells, but they also have lower efficiencies. An example is the dye-sensitized cell, which operates under a completely different physical paradigm that uses dye molecules adsorbed onto very small spheres of titanium dioxide. To date, this photoelectrochemical device has been able to generate electricity on small areas at efficiencies exceeding 10% on small areas. Again, this efficiency is relatively low, but the simplicity of the materials and structure make it very inexpensive to manufacture.

Other third-generation approaches, some of which are still only in the conceptual phase, include advanced cells based on quantum dots, organic PV, intermediate-band cells, and multiple-exciton generation. Verified efficiencies are typically in the single digits; however, there is considerable opportunity for technological innovation and improvement, and the potential for very low manufacturing costs. One new technology that holds promise is PV cells based on Perovskites. Perovskites are new polycrystalline thin film technologies that have demonstrated efficiencies higher than 20%.

## 10.2.5 Photovoltaics Balance of Systems

Balance of systems (BOS) includes all of the components of a PV system beyond the actual PV module that produces the power. A frame structure may be needed to hold the module, keep it oriented toward the sun, and stabilize it in the outdoor elements such as wind and snow. The mounting structure can be relatively simple and installed on a flat roof or complex with PV modules mounted onto structural elements that track the sun's position throughout time.

PV systems produce direct-current (DC) electricity. If alternating current (AC) is required, as in most grid-tied applications, the BOS must include an inverter. This component usually decreases the overall system efficiency by approximately 5% to 10%, and typically has the greatest reliability problems of any component in the system. Figure 10-8 shows the components of a typical grid-connected PV system.

System efficiency can be boosted by attaching a tracker to the solar modules. Single-axis trackers (Fig. 10-9) aligned with the axis in a north-south direction allow the module to follow the sun's progress across the sky from east to west during the day. Dual-axis trackers further refine the module's orientation, allowing the sun to always illuminate the cells that are perpendicular to the plane of the

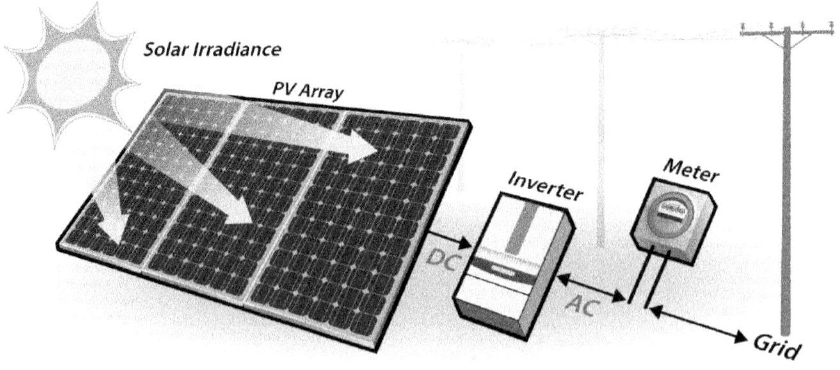


**FIGURE 10-8** Typical components of a grid-connected PV system. (*NREL*.)

**FIGURE 10-9** Single-axis tracking PV system. (*NREL.*)

module. This geometry facilitates the maximum energy output from the system, but it increases costs and also typically requires additional spacing to prevent one module from shading an adjacent one.

Flat-plate PV systems do not need trackers, but they will produce approximately 25% more energy if one is used. Concentrating PV systems require direct-normal radiation, making a tracking system essential. These concentrating systems also require a lens or mirror to focus the sunlight onto the solar cells. Fresnel lenses extruded from acrylic are one option for concentration, but mirrors can also provide concentration. These requires a sturdy frame and robust tracking mechanism to maintain optical accuracy and structural integrity against wind.

In most cases, residential and commercial PV systems are directly connected to the grid without energy storage. Users draw power from the utility during hours of darkness or on cloudy days. As we move toward a smart grid in the future, more systems may incorporate battery energy storage to increase reliability if battery costs decrease or grid reliability is an issue. One potential way to incorporate energy storage with PV systems could be to integrate them with the use of plug-in electric hybrid vehicles if these vehicles are used on a large scale.

If users have stand-alone systems—with no grid connection—then the BOS will include batteries and charge controllers to produce electricity at night or during cloudy conditions. The batteries store excess power that is generated from the PV array to be used later, and the charge controllers regulate the current to the batteries to prevent overcharging.

### 10.2.6 Examples of Photovoltaic Systems

PV systems are unique power plants in that they can come in a range of sizes from 300 W to 500 MW. Residential systems are typically from 1 to 10 kW and connected to the grid via single-phase inverters. Figure 10-10 shows a typical residential PV system configuration with the addition of local batteries for backup power. The PV system produces DC electricity and connected through a charge controller to the battery bank. The charge controller regulates battery charging and will turn down the power from the PV system if the batteries have reached full charge. The dual-purpose inverter converts DC power from the batteries to AC power to either sell back to the grid or provide local power to the backup AC loads.

Commercial PV systems range from 10 kW to 5 MW and are typically installed on the rooftops of stores. These systems use three-phase inverters to connect to the grid.

Utility scale system can range from 1 MW to more than 500 MW. Currently, the largest installed PV systems in the United States are over 500 MW. These systems use three-phase inverters to connect to the grid and connect to the grid at both distribution and transmission systems. Figure 10-11 shows a typical layout of a utility scale PV system. Large utility scale systems typically

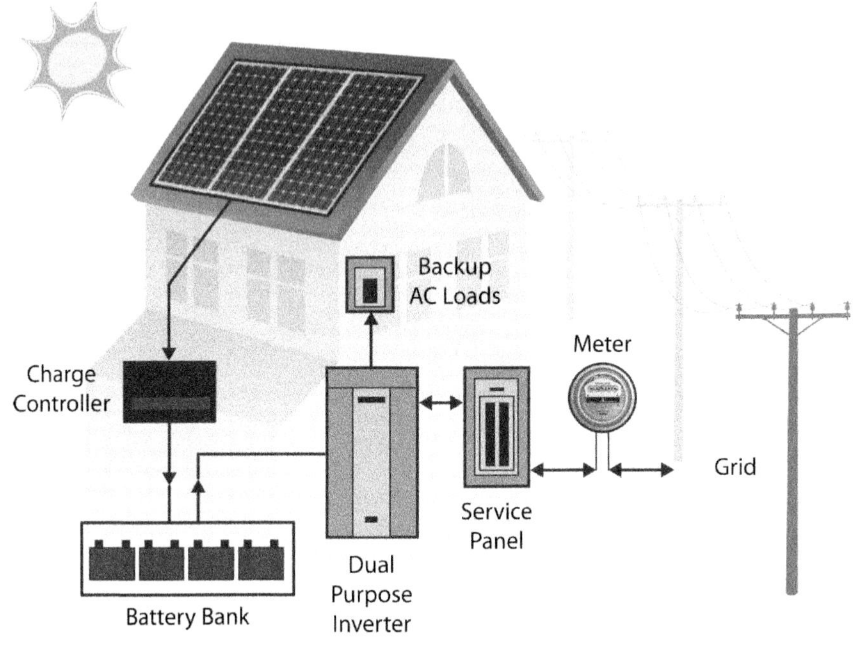


**FIGURE 10-10** Typical residential PV system with battery backup. (*NREL.*)

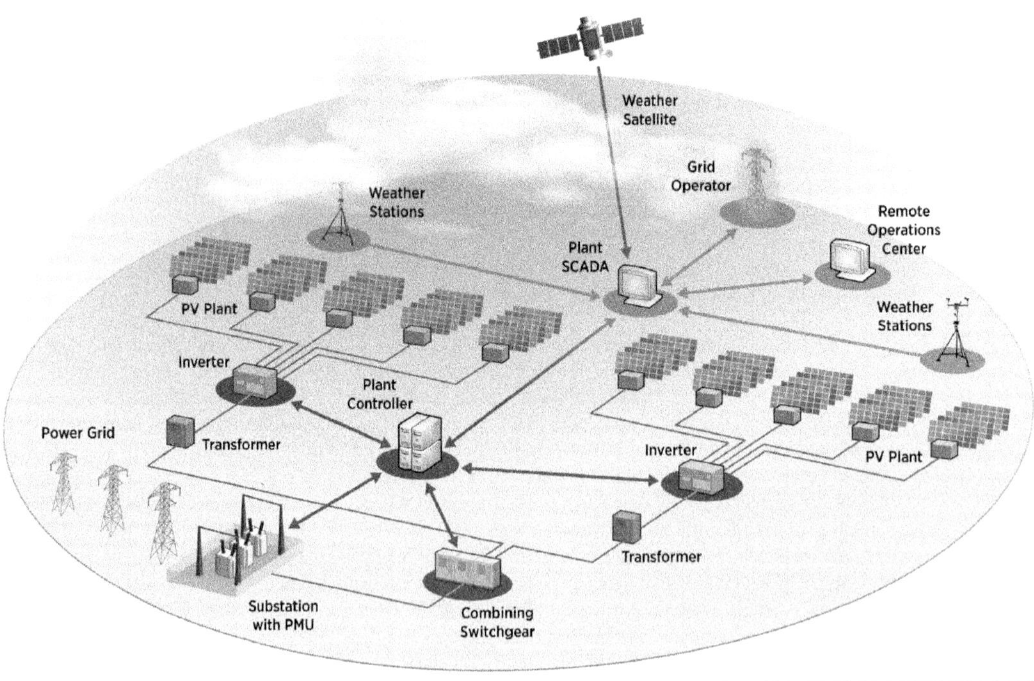


**FIGURE 10-11** Utility scale PV system. (*NREL.*)

have a plant controller that will coordinate power output from all the PV inverters within the plant. The size of a typical inverter for these systems ranges from 250 kW to 4 MW, and there can be several hundred inverters to coordinate. For very large systems, a plant supervisory control and data acquisition (SCADA) system is used to communicate the operational state back to the grid operator. The grid operator may also request that the plant curtail power or provide other grid services to the grid.

### 10.2.7 Bibliography


- V. Gevorgian and B. O'Neill, *"Advanced Grid-Friendly Controls Demonstration Project for Utility-Scale PV Power Plants,"* NREL/TP-5D00-65368, January 2016.
- B. Kroposki, R. Margolis, and D. Ton, *"Harnessing the Sun—An Overview of Solar Technologies,"* IEEE Power and Energy Magazine, May/June 2009.
- M. Morjaria, D. Anichkov, V.Chadliev, and S. Soni, *"A Grid-Friendly Plant—The Role of Utility-Scale Photovoltaic Plants in Grid Stability and Reliability,"* IEEE Power and Energy Magazine, May/June 2014.


## 10.3 CONCENTRATING SOLAR POWER TECHNOLOGIES

*BY BENJAMIN KROPOSKI, ROBERT MARGOLIS, AND MARK MEHOS*

CSP technologies use mirrors to reflect and concentrate sunlight onto receivers that collect the solar energy and convert it to heat. This thermal energy can then be used to produce electricity via a turbine (e.g., steam, air, supercritical carbon dioxide) or heat engine driving a generator. CSP systems are typically classified by how the various systems collect solar energy. The three main systems described below are linear, power tower, and dish/engine systems. BOS and thermal energy storage (TES) will also be discussed.

### 10.3.1 Linear Concentrating Solar Power Systems

Linear CSP collectors capture the sun's energy with large mirrors that reflect and focus the sunlight onto a linear receiver tube. The receiver contains a fluid that is heated by the sunlight and then used to create superheated steam that spins a turbine to drive a generator to produce electricity. Alternatively, steam can be generated directly in the solar field. In this method, no heat exchanger is used, but the system uses more costly pressure-rated piping throughout the entire solar field and typically has a lower operating temperature although recent years have seen as shift toward more efficient tower technology (see next subsection). Linear concentrating collector fields consist of a large number of collectors in parallel rows that are typically aligned in a north-south orientation to maximize both annual and summertime energy collection. With a single-axis sun-tracking system, this configuration enables the mirrors to track the sun from east to west during the day, ensuring that the sun reflects continuously onto the receiver tubes.

The predominant CSP systems currently in operation in the United States are linear concentrating units that use parabolic trough collectors (see Fig. 10-12). In typical systems, the receiver tube is positioned along the focal line of each parabola-shaped reflector. The tube is fixed to the mirror, and the heated fluid—commonly a high-temperature oil—flows through and out of the field of solar mirrors to where it is used to create steam, and then it is sent directly to the turbine.

The Solar Energy Generating Systems (SEGS) plants in the Mojave Desert of California currently have an aggregated output capacity of 359 MW. There are three additional operating plants in the southwestern United States each with capacities of at least 250 MW.

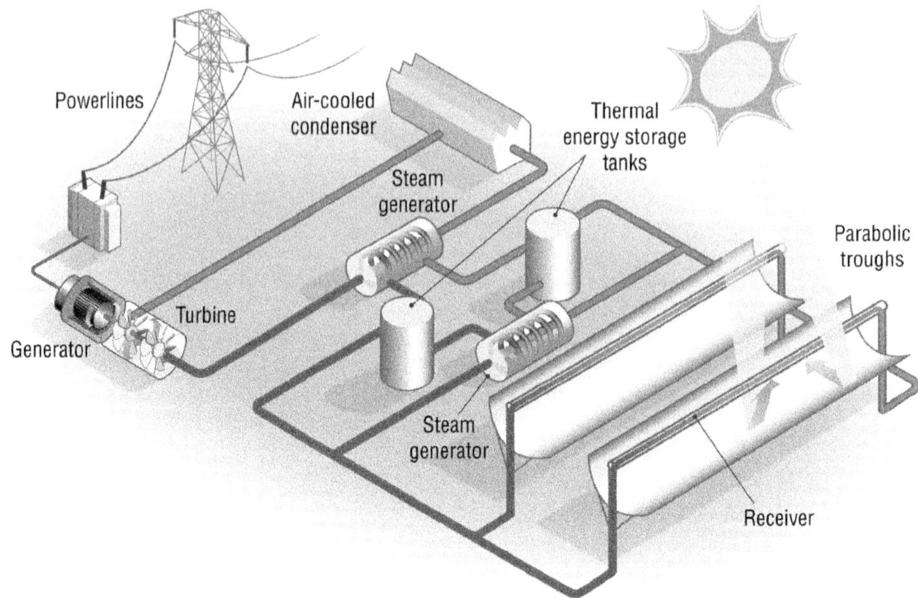


**FIGURE 10-12** Schematic showing the basic operation of a parabolic trough CSP system. (*NREL.*)

In addition, individual systems can be collocated with power plants such as the Martin Next Generation Solar Energy Center in Florida. This plant is an integrated solar combined cycle that is used as an auxiliary heat supply to the steam cycle of the natural gas plant. The capacity of this type of plant would be constrained only by the transmission capacity and the availability of contiguous land area.

Trough designs can incorporate thermal storage. In such systems, the collector field is oversized to heat a storage system during the day that can be used in the evening to generate additional steam to produce electricity. Parabolic trough plants can also be designed as hybrid systems that use fossil fuel to supplement the solar output during periods of low solar radiation. In such a design, a natural-gas-fired heater or gas-steam boiler/reheater is used. Troughs may also be integrated with combined-cycle natural-gas and coal-fired plants to improve the plant heat rate or provide a peaking boost to the steam turbine in a combined-cycle plant, much like a duct burner does.

A second linear concentrating technology is the linear Fresnel reflective system (see Fig. 10-13). Flat or slightly curved mirrors mounted on trackers on the ground are configured to reflect sunlight onto a receiver tube fixed in space above these mirrors. A small parabolic mirror is sometimes added atop the receiver to further focus the sunlight.

## 10.3.2 Power Tower Concentrating Solar Power Systems

CSP systems that use a power tower design, numerous, flat, sun-tracking mirrors, known as heliostats, focus sunlight onto a receiver at the top of a tower (see Fig. 10-14). There is a movement toward smaller heliostats (<20 $m^2$) versus the "large" 100 $m^2$ heliostats. Some are as small as a few $m^2$. A heat-transfer fluid heated in the receiver is used to generate steam, which, in turn, is used in a conventional turbine generator to produce electricity. Some power towers use water/steam as the heat-transfer fluid but, given the high cost of storing large amounts of steam over an extended period, the trend world-wide is to use molten salt as the HTF because of its superior

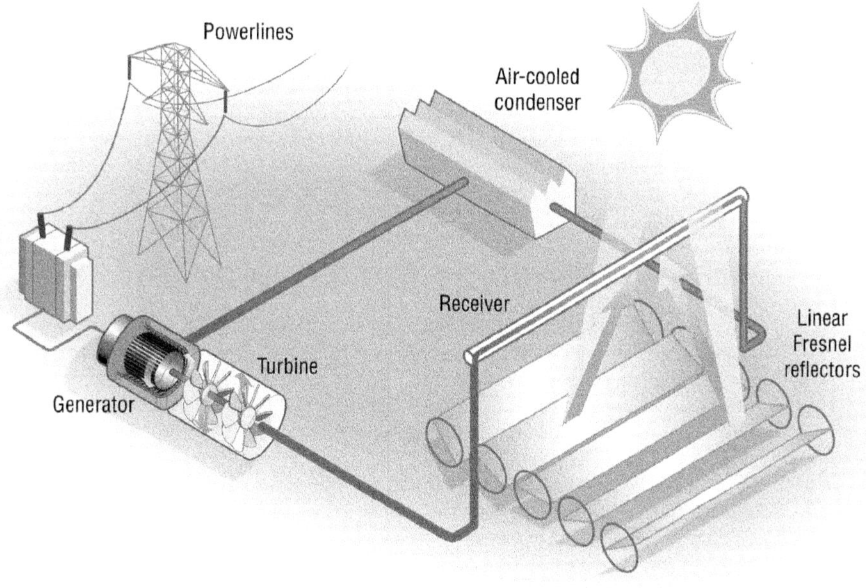


**FIGURE 10-13** Schematic showing the basic operation of a linear Fresnel reflective system. (*NREL*.)

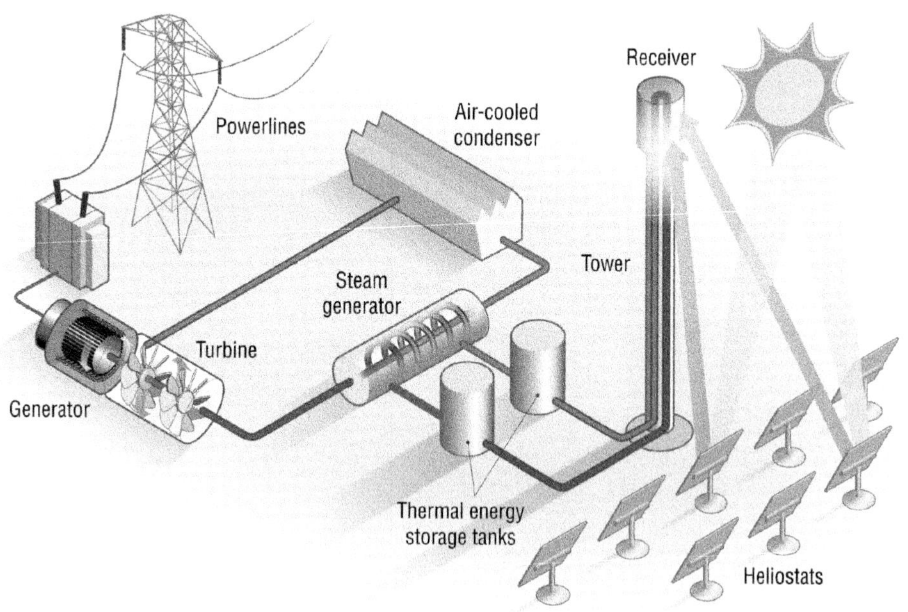


**FIGURE 10-14** Schematic showing the basic operation of a power tower or central receiver CSP system with TES. (*NREL*.)

heat-transfer and energy-storage capabilities. Individual commercial plants can be sized to produce up to 200 MW of electricity.

Currently, the largest operational power tower design in the world is the 392 MW Ivanpah Solar Power Facility in San Bernadino County, California. The Crescent Dunes Solar Energy Project is a 110 MW power tower system located near Tonopah, Nevada and it includes a molten salt energy storage component. Large power towers look to be cost-competitive and offer high solar-to-electric conversion efficiency. Molten salt based tower projects bid outside of the United States point to a trend toward significant cost reductions in this technology.

## 10.3.3 Dish/Engine Concentrating Solar Power Systems

Dish/engine systems produce relatively small amounts of electricity compared to other CSP technologies—typically in the range from 3 to 25 kW. A solar concentrator, or dish, gathers the solar energy coming directly from the sun. The resulting beam of concentrated sunlight is reflected onto a thermal receiver that collects the solar heat (see Fig. 10-15). The dish is mounted onto a structure that tracks the sun continuously throughout the day to reflect the highest percentage of sunlight possible onto the thermal receiver.

The power conversion unit includes the thermal receiver and the engine/generator. The thermal receiver is the interface between the dish and the engine/generator. It absorbs the concentrated beams of solar energy, converts them to heat, and transfers the heat to the engine/generator. A thermal receiver can be a bank of tubes with a cooling fluid—usually hydrogen or helium—that typically is the heat-transfer medium and also the working fluid for an engine. Alternate thermal receivers are heat pipes, wherein the boiling and condensing of an intermediate fluid transfers the heat to the engine.

The engine/generator system is the subsystem that takes the heat from the thermal receiver and uses it to produce electricity. Currently, the most common type of heat engine used in dish/engine systems is the Stirling engine. A Stirling engine uses the heated fluid to move pistons and create mechanical power. The mechanical work, in the form of the rotation of the engine's crankshaft, drives a generator and produces electric power.

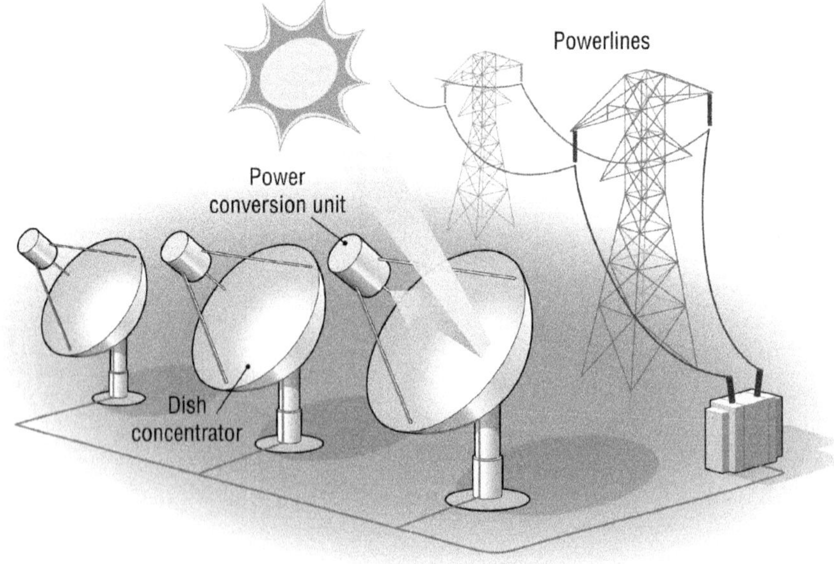


**FIGURE 10-15** Schematic showing the basic operation of a dish/engine CSP system. (*NREL.*)

### 10.3.4 Thermal Energy Storage

In a CSP system, the solar field includes the mirrors, trackers, and receivers needed to collect the sun's thermal energy. A second component is the power plant, which may include a turbine and generator or a Stirling heat engine. A third component is TES, which addresses one of the challenges facing the widespread use of solar energy—the problem of reduced or curtailed energy production when the sun sets or is blocked by clouds.

TES provides a workable solution to this challenge. In a CSP system, if the receiver contains oil or molten salt as the heat-transfer medium, then the thermal energy can be stored for later use. This allows CSP systems to be a cost-competitive option for providing clean, renewable energy. Several TES technologies have been tested and implemented since 1985. Two-tank systems have one tank containing fluid at a high temperature and another containing fluid at a lower temperature. In the direct two-tank system, the heat-transfer fluid from the collectors and in the tanks is the same; in an indirect two-tank system, the heat-transfer fluid transfers its heat to the fluid in the tanks through a heat exchanger.

Research on advanced heat-transfer fluids and novel thermal-storage concepts seeks to increase efficiency and reduce costs for TES. Scientists are working to identify and characterize novel fluids and materials that possess physical and chemical properties needed to improve thermal storage. For example, phase-change materials (PCMs), as described above, allow large amounts of energy to be stored in relatively small volumes, potentially lowering the costs relative to other storage concepts. Initially, PCMs were considered for use with parabolic trough plants that use a synthetic heat-transfer fluid designed to withstand high temperatures in the solar field. PCM thermal storage is still being considered for applications with direct steam, air (or other gas), or S-$CO_2$-based working fluids. Additional options under investigation include single-tank thermocline systems, which sets up a temperature gradient in a tank that contains a solid material such as sand, and phase change systems.

Incorporating TES into CSP power plants allows utilities to enhance dispatchability. As TES technologies improve and allow for longer storage periods and lower costs, more utilities may consider CSP as a viable alternative to or supplement for power plants that depend solely on fossil fuels.

### 10.3.5 Bibliography


- Mehos, M., Turchi, C., Jorgenson, J., Denholm, P., Ho, C., and Armijo, K., National Renewable Energy Laboratory, "*On the Path to SunShot: Advancing Concentrating Solar Power Technology, Performance, and Dispatchability,*" NREL/TP-5500-65688, 2016.


## 10.4 SOLAR ENERGY MARKET

*BY BENJAMIN KROPOSKI, ROBERT MARGOLIS, AND MARK MEHOS*

Renewable energy technologies such as PV and CSP are expected to become a larger part of our energy portfolio during the next couple of decades. The reasons include the growing concerns about climate change, the adoption of renewable portfolio standards (RPS) and incentives, and the accelerated reduction of system costs. As PV and CSP technologies mature, they have the potential to supply a significant share of electricity demand.

### 10.4.1 Photovoltaic Costs

PV costs have decreased dramatically during the past 5 years, whereas deployments have significantly increased. Figure 10-16 shows the medium system price for utility scale PV systems as well as the number of GW of installed systems.

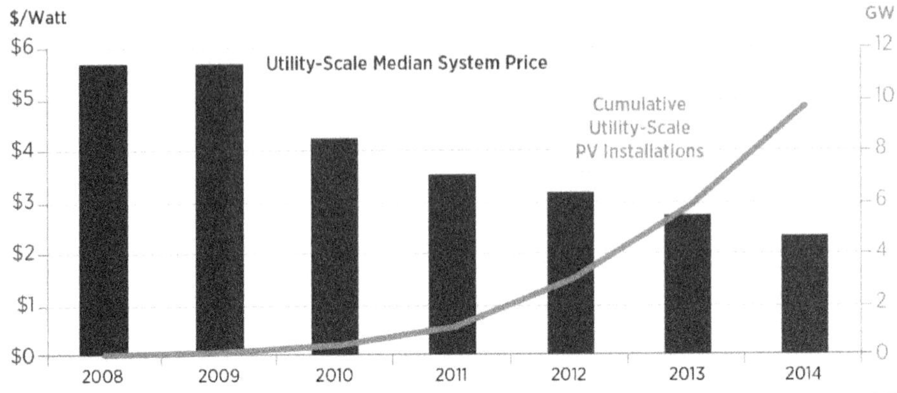


**FIGURE 10-16** Price and cumulative installations of utility-scale PV in the United States. (*DOE.*)

From 2008 to 2014 there was a 59% decline in PV prices. This shows that in the United States, solar is starting to reach cost parity with other forms of electricity generation. Prices and installations are on a similar pace for distributed PV systems. Figure 10-17 shows information on residential PV systems.

Trends suggest that PV systems will become less expensive due to research on lower cost technologies, increased efficiencies, and reduced manufacturing costs. The U.S. Department of Energy has a goal of reaching installed system costs of $1/W for utility scale systems by 2020 as part of its SunShot program. Figure 10-18 shows the historic, current, and future costs for PV systems in the three different categories (residential, commercial, and utility scale). It also shows the cost components of a PV system (BOS costs, BOS hardware, inverter, and module).

Figure 10-18 shows that it is expected that PV modules will see the greatest decrease in cost during the next 5 years, however, BOS-Soft costs will also need to drop significant to meet long-term targets.

## 10.4.2 Photovoltaic Market Share

PV makes up most of the deployed solar systems world-wide. Crystalline PV (both mono-crystalline and multi-crystalline) still dominates the market with 80% to 95% of all modules shipped for decades. Figure 10-19 shows the annual shipments of different PV technologies throughout time.

As of 2014, 177GW of cumulative PV was installed around the world, with the largest shares in Europe, China, and the United States. Figure 10-20 shows the level of deployment in various regions around the world.

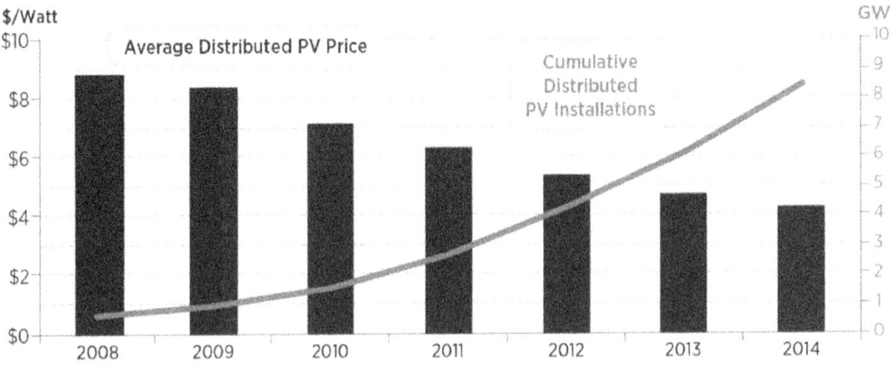


**FIGURE 10-17** Price and cumulative installations of residential PV in the United States. (*DOE.*)

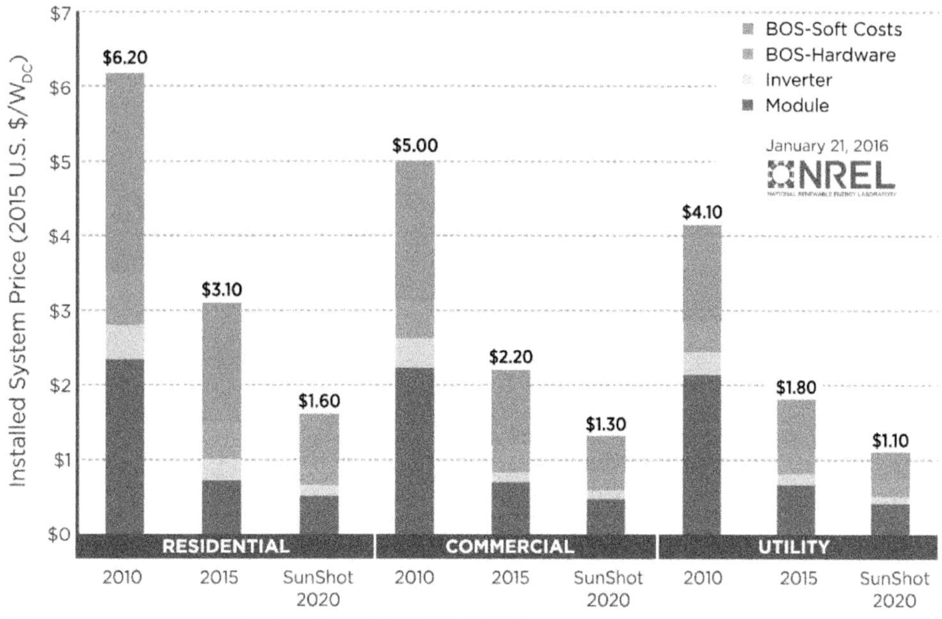


**FIGURE 10-18** Historic, current, and future prices of PV systems. (*NREL.*)

### 10.4.3 Concentrating Solar Power Costs

CSP costs have also declined during the past 5 years, but not as drastically as PV. Figure 10-21 shows the levelized cost of energy (LCOE) for historical, current, and future projections for CSP. In 2010, the cost was approximately 20.5 cents/kWh. In 2015, costs ranged from 19 cents/kWh for trough systems to 12 cents/kWh for power towers. The primary reason for this nearly 30% drop in LCOE has

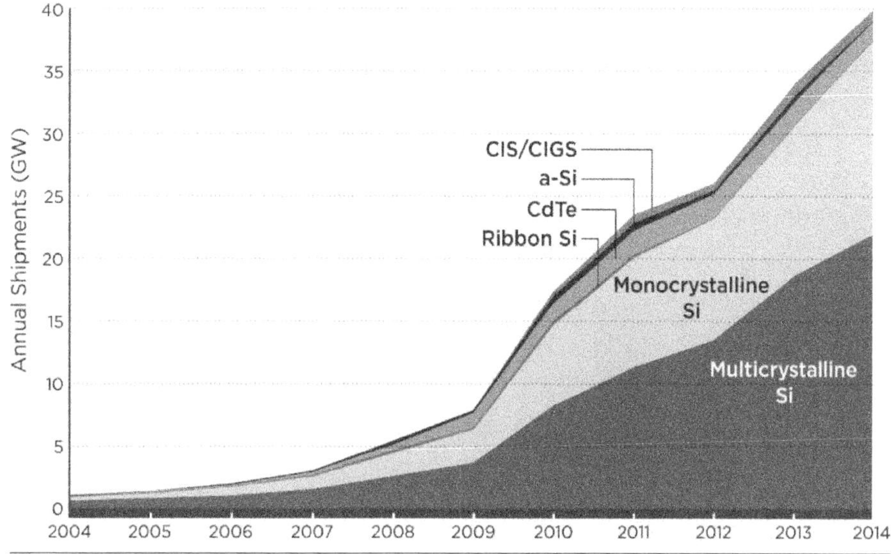


**FIGURE 10-19** Annual shipments of different PV technologies. (*Mints.*)

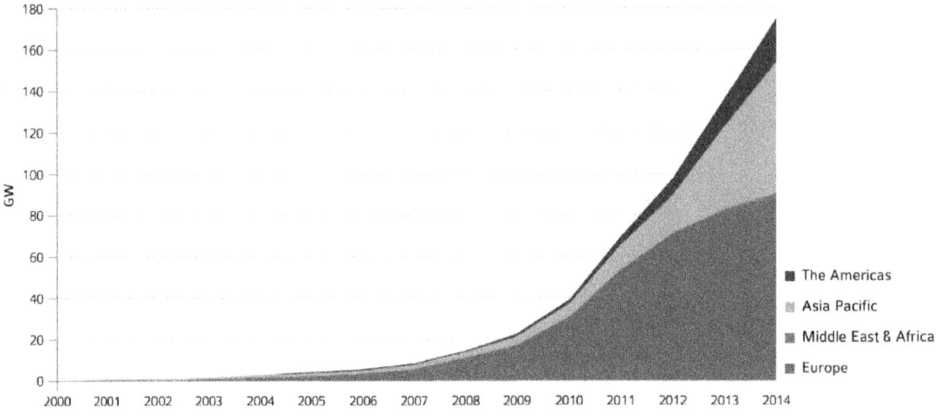


**FIGURE 10-20** Regional deployments of PV systems. (*International Energy Agency.*)

been the introduction of commercially deployed molten-salt power-tower technology. As discussed earlier, recent power tower bids outside of the United States in the range of 6.5 to 8 cents/kwh demonstrate the potential for further rapid reductions in costs for CSP systems. The goal is to have power tower costs at 6 cents/kWh by 2020.

Although the costs for troughs and towers have declined, CSP acceptance and deployment has been negatively impacted by the declining cost of PV technology. This situation can be mitigated when considering the flexibility offered by CSP with TES. Figure 10-22 shows the results from a study that compared the combined operational and capacity benefits of CSP with TES relative to

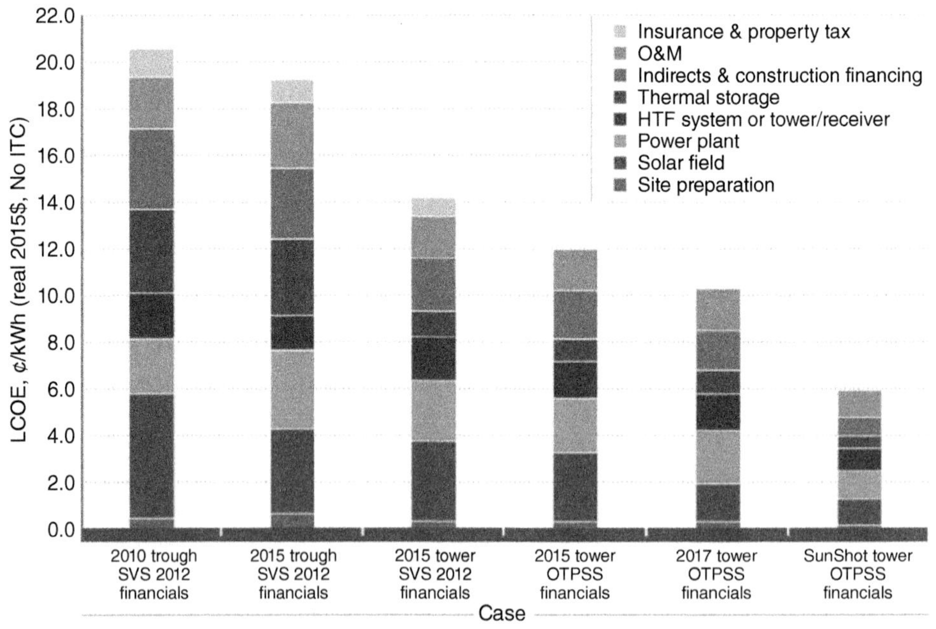


**FIGURE 10-21** LCOE from CSP systems. (*NREL.*)

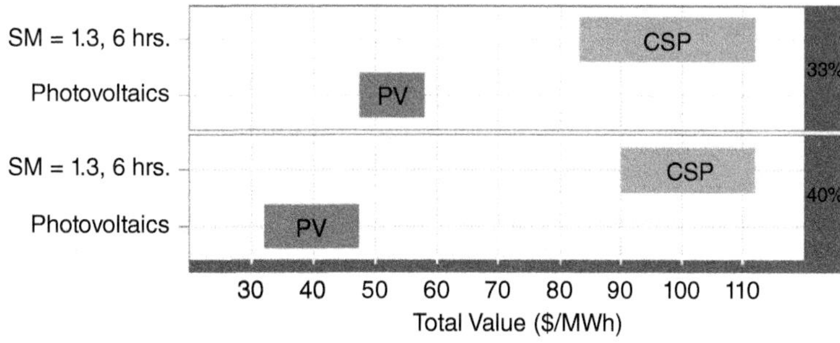


**FIGURE 10-22** Total value, which includes operational and capacity value, of CSP with TES and PV under 33% and 40% RPS scenarios. (*NREL.*)

PV under varying levels of renewable penetration in California. In the analysis a variety of Solar multiples (SM) and TES sizes were examined. The analysis found that the value of CSP, compared to variable-generation PV, demonstrated an increase in value of up to 6 cents/kWh under a 40% RPS. This analysis and other similar analyses show that the high grid value of CSP-TES, not just the LCOE, must be considered when evaluating the portfolio of renewable energy technology options.

Reducing the cost of CSP systems is vital to successful deployment in future, non-subsidized markets. However, the value of the dispatchability offered by CSP integrated with TES plays an important role in differentiating CSP from variable-generation renewable technologies such as PV and wind. This value will become increasingly apparent as PV and wind gain a greater penetration into the regional power markets.

### 10.4.4 Concentrating Solar Power Market Share

Figure 10-23 shows that CSP capacity has grown significantly since 2009. This growth has been particularly concentrated in Spain and the United States, although other countries began increasing CSP capacity at a greater rate starting in 2013. In the United States, the growth in CSP capacity has been driven primarily by policy support at the state and federal levels. State-driven RPS, combined with a 30% federal investment tax credit and federal loan guarantees, provided the opportunity for CSP developers to the kick-start construction of CSP plants throughout the Southwest.

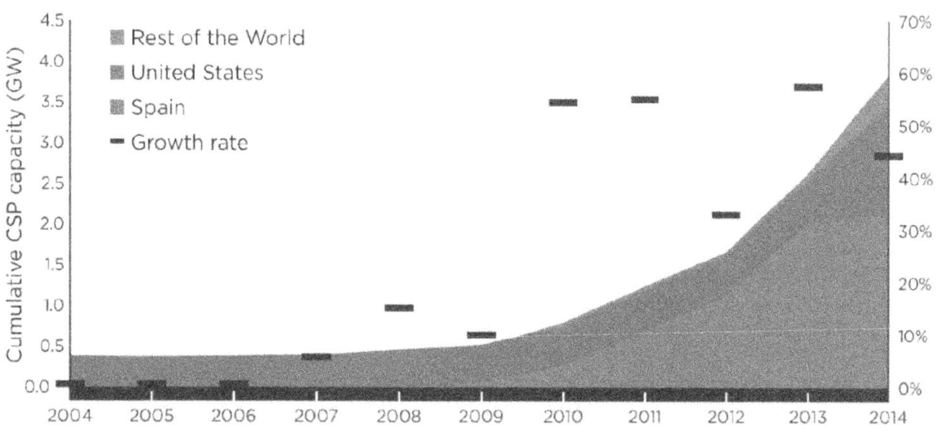


**FIGURE 10-23** Cumulative capacity of CSP systems. (*International Energy Agency.*)

### 10.4.5 Bibliography


- Denholm, P., Wan, Y., Hummon, M. and Mehos, M., "*An Analysis of Concentrating Solar Power with Thermal Energy Storage in a California 33% Renewable Scenario*," NREL Report No. TP-6A20-58186 (2013).
- "*IEA Technology Roadmap: Solar Thermal Electricity*," OECD/IEA, Paris, 2014.
- Mehos, M., Turchi, C., Jorgenson, J., Denholm, P., Ho, C., and Armijo, K., "*On the "Path to SunShot: Advancing Concentrating Solar Power Technology, Performance, and Dispatchability,"* National Renewable Energy Laboratory, NREL/TP-5500-65688, 2016.
- Mints, P., "*Photovoltaic Manufacturer Shipments: Capacity, Price, and Revenues 2015*," *SPV Market Research*, April 2015.
- "*Trends 2015 in Photovoltaic Applications—20th Edition,*" International Energy Agency Report IEA-PVPS T1-27:2015.
- U.S. Department of Energy, "*Revolution...Now—The Future Arrives for Five Clean energy Technologies—2015 Update*," November 2015. *http://energy.gov/sites/prod/files/2015/11/f27/Revolution-Now-11132015.pdf.*
- Woodhouse, M., Jones-Albertus, R., Feldman, D., Fu, R., Horowitz, K., Chung, D., Jordan, d., and Kurtz, S., "*On the Path to SunShot: The Role of Advancements in Solar Photovoltaic Efficiency, Reliability, and Cost,*" National Renewable Energy Laboratory, NREL/TP-6A20-65872, 2016.


## 10.5 GRID INTEGRATION OF SOLAR ENERGY

*BY BENJAMIN KROPOSKI, ROBERT MARGOLIS, AND MARK MEHOS*

For the past 100 years, the electric power system has evolved to one that is based on large, central station generation that delivers electricity from large generators through a transmission and distribution (T&D) system to the customer loads. The main principals of this system are to deliver reliable electricity at the least cost. As the power system has evolved, customers have found additional attributes are also important including sustainability and resilience of the grid. To achieve sustainability goals while maintaining reliable, resilient, and affordable operation, variable renewables such as wind and solar energy need to be integrated into electric power system operations and provide similar grid services as conventional generators. Solar energy is unique, it can be integrated into either the transmission or distribution system. The rest of this subsection discusses grid integration of solar technologies.

### 10.5.1 Transmission Integration

The transmission system was designed to move electric power from large generators to distribution systems at load sites. Transmission systems are typically classified by voltage levels that range from 115 to 1000 kV and can be either AC or DC. Bulk power is moved from generators that are often located in remote locations to cities or other loads. Large-scale PV and CSP systems are often located in remote areas that have good solar access and available land. The output from these systems is then converted to grid compatible AC power via a three-phase inverter for PV or a synchronous generator for CSP plants. Because CSP plants use a synchronous generator to connect to the grid, their output characteristics are well known to utilities. Often they incorporate some amount of local energy storage, which reduces the variability of the system output.

Large PV plants on the other hand typically do not include any energy storage. Their output is variable and uncertain due to the weather conditions at the site. These systems present challenges to large-scale, cost-effective grid integration by reducing the value of PV energy (if the energy cannot be used at a particular time) and capacity (if the plant capacity cannot be counted on). One challenge to integrating PV is the fact that PV output is coincident with the solar irradiance within the same longitude. This means that PV power will be produced time coincident with other plants and is the highest at midday. This can potentially cause overgeneration of PV power that cannot be absorbed by the net load on the system. If this occurs, the PV system may need to be curtailed to reduce power. This situation can also cause other conventional generation plants to be turned off, and thus they are not able to respond to dynamic conditions on the grid.

To accommodate higher levels of variable generation on the grid, the electric power system must increase its flexibility, that is, its ability to balance supply and demand throughout multiple timescales. A variety of approaches that can increase grid flexibility have been used to increase the amount of variable generation. These include

- System operation—Changing the way the grid is scheduled and dispatched. Because solar can change rapidly, shorter time-steps for grid operations can help accommodate variable generation. This can include changes to market rules where markets exist.
- Flexible conventional generation—Allowing conventional forms of generation to be flexible when they are run or to adding controls that increase the ramp rates of generators.
- Reserves and stability services from PV—Allowing inverter-based PV plants to provide voltage and frequency control and provide services similar to those of synchronous generators.
- Transmission and coordination—Balancing supply and demand throughout larger areas to reduce the net variability of both load and renewable resources by coordinating assets over a larger geographic footprint.
- Demand-side management—Increasing or decreasing load to match available generation to lessen curtailment of PV system output.
- Energy storage—Providing a variety of grid services and can store excess solar output during times of low load.

### 10.5.2 Distribution Integration

PV is one of the few generation technologies that can be integrated at the distribution system level. Distribution voltages range from 4 to 35 kV. This voltage level is then stepped down further to a voltage (less than 1 kV) provided to customer loads. PV can provide benefits such as deferring system upgrades, managing voltage using advanced inverters (with or without batteries), and reducing distribution losses, however the integration of PV can present challenges to distribution system operations because the distribution systems are typically designed to provide one-way power flow from the substation to the customer loads. PV output is variable during short timescales (seconds to minutes) because of passing clouds and longer timescales because of the diurnal nature of the sun. This variability produces important and potentially challenging interactions with the distribution system—particularly with larger distributed PV systems and higher penetrations of PV—including managing the distribution voltage, potentially increasing wear and tear on electromechanical utility equipment, and complicating the configuration of circuit breakers and other protection systems. Some of these challenges are explained below.

- Voltage regulation—Distribution circuits are designed to maintain voltage with a nominal ±5% window so that loads will not be damaged. Capacitor banks, voltage regulators, and tap changers on transformers are all used to regulate voltage on a distribution circuit. Adding PV systems that inject only real power will cause voltages to rise and may impact the operations of the equipment on the grid that regulates voltage. If not properly planned for, PV systems may cause the existing voltage-regulating equipment to misoperate or increase the number of operations due to changes in voltage and current on the distribution circuit from the PV. This concern can be alleviated with the use of smart inverters that can locally regulate voltage or run at a non-unity power factor.

- Protection coordination—Protection of the distribution system is done by protective relays, circuit breakers, reclosers, and fuses. These devices are intended to operate under fault conditions and are designed to protect the grid components (e.g., conductors and transformers). Coordination of the protective device operation is set in minimize customer disruptions when a fault occurs. When PV is added to a system, the coordination of the protection system should be reexamined to determine if the addition of the PV system fault current causes any protection devices to not be coordinated with each other correctly.
- Unintentional islanding—Islanding occurs when parts of the electric grid are isolated from the main grid and maintain enough generation to supply the local load. If the electrical island is not planned, then it is unintentional. This can cause the distribution system to operate outside of normal voltage and frequency parameters as well as cause a safety risk to utility workers. Unintentional islanding should be avoided and modern PV inverters utilize anti-islanding techniques that can quickly identify these situations and shut down the units.
- Secondary networks—Most distribution systems are radial in nature from the substation to the loads. Secondary networks are used in dense urban areas and provide increased reliability by having multiple parallel connection points to a customer power network. Integrating distributed generation with these types of systems requires special planning and coordination with the network protection system.

## 10.5.3 Bibliography


- Denholm, Paul, Kara Clark, and Matt O'Connell, "*On the Path to SunShot: Emerging Issues and Challenges in Integrating High Levels of Solar into the Electrical Generation and Transmission System*," Golden, CO: National Renewable Energy Laboratory, NREL/TP-6A20-65800, 2016.
- J. Bebic, R. Walling, K. O'Brien, and B. Kroposki, "*The Sun Also Rises—Planning for Large-Scale Solar Power,*" *IEEE Power and Energy Magazine*, May/June 2009.
- Coddington, Michael, and Jeff Smith, "*Current Utility Screening Practices, Technical Tools, Impact Studies, and Mitigation Strategies for Interconnecting PV on the Electric Distribution Systems,*" EPRI Report #3002003277, Palo Alto, CA: EPRI, 2014.
- R. Butler Hoke, J. Hambrick, and B. Kroposki, "Steady-State Analysis of Maximum Photovoltaic Penetration Levels on Typical Distribution Feeders," *IEEE Transactions on Sustainable Energy*, vol. 4, no. 2, pp. 350–57, 2013. doi:10.1109/TSTE.2012.2225115.
- J. Keller and B. Kroposki, "*Understanding Fault Characteristics of Inverter-Based Distributed Energy Resources*," NREL/TP-5500-46698, NREL, 2010.
- Y. Liu, J. Bebic, B. Kroposki, J. de Bedout, and W. Ren. "*Distribution System Voltage Performance Analysis for High-Penetration PV*," In IEEE Energy 2030 Conference, ENERGY 2008, 1–8, 2008. doi:10.1109/ENERGY.2008.4781069.
- M. McGranaghan, T. Ortmeyer, D. Crudele, T. Key, J. Smith, and P. Barker, "*Renewable Systems Interconnection Study: Advanced Grid Planning and Operations*," SAND2008-0944, Sandia National Laboratories, 2008.
- A. Mills, M. Ahlstrom, M. Brower, A. Ellis, R. George, T. Hoff, B. Kroposki, C. Lenox, N. Miller, M. Milligan, J. Stein, and Y. Wan, "Understanding Variability and Uncertainty of Photovoltaics for Integration with the Electric Power System," *IEEE Power & Energy Magazine*, vol. 9, no. 3, May/June 2011.
- Palmintier, Bryan, Robert Broderick, Barry Mather, Michael Coddington, Kyri Baker, Fei Ding, Matthew Reno, Matthew Lave, and Ashwini Bharatkumar, "*On the Path to SunShot: Emerging Issues and Challenges in Integrating Solar with the Distribution System*," Golden, CO: National Renewable Energy Laboratory, NREL/TP-5D00-65331, 2016.
- Palmintier, E. Hale, T. Hansen, W. Jones, D. Biagioni, K. Baker, H. Wu, J. Giraldez, H. Sorensen, M. Lunacek, N. Merket, J. Jorgenson, and B. Hodge, "*Integrated Distribution-Transmission Analysis for Very High Penetration Solar PV*," National Renewable Energy Laboratory, NREL/TP-5D00-65550, 2016.

## 10.6 ENERGY STORAGE

*BY JIM EYER, RAHUL WALAWALKAR, AND HARESH KAMATH*

### 10.6.1 Introduction

Energy storage has a long history. Thousands of years ago, people learned to use the potter's wheel to smooth the pulsed application of kinetic power, simplifying the production of symmetrical pottery. For centuries, dams have been used to hold back water so the potential energy can be used later. Practical electrochemical battery storage has existed for over 150 years.

Energy storage technology development is accelerating in the modern era. New applications and markets such as mobile electronics, electric transportation, and renewable energy-fueled generation have encouraged increased investment in the field; new technologies in materials science, nanotechnology, and power electronics are being harnessed to improve the cost, life, and performance of storage products.

As the technology advances, there is a growing recognition of the important roles that storage can and will play in the future. Newer application drivers of storage technology development include (1) a generally more diverse and possibly less flexible electric resource base that includes increasing deployment of renewable energy with variable output and high-efficiency generation with constant output; (2) electric vehicles (EVs); (3) increasing interest in utility-interactive distributed resources and microgrids; (4) smart grid; (5) electricity users' electricity "bill management" such as participation in demand response (DR) programs; and (6) increasing price transparency in electric energy, power, and ancillary services markets.

This subsection includes an overview of the various types of energy storage devices and systems used to store energy in a variety of forms but whose input and output are usable electric or thermal energy. It includes a brief description of the theory underlying the operation. Also covered is the range of energy storage benefits and applications.

### 10.6.2 Storage Characteristics

This subsection describes some of the general performance characteristics of all energy storage technologies. Specific storage technologies are addressed in Sec. 10.6.3.

***Power, Energy, and Discharge Duration.*** All energy storage devices provide power at a specified level or levels and contain a fixed quantity of usable energy. The amount of energy stored determines the discharge duration—the amount of time that the storage can provide power at its rated power output.

Storage systems can usually discharge at rates other than their rated power output. At lower rates of discharge, storage systems typically (but not always) exhibit longer run times and better efficiency. Systems with large parasitic loads or inherent losses may actually exhibit lower efficiency at lower rates of discharge.

Higher rates of discharge result in shorter run times, lower efficiency and, typically, more wear and tear on the equipment, than a discharge at the rated capacity.

It is important to recognize that these effects are not linear. For instance, a system that discharges for 4 hours at its rated output of 10 kW (delivering 40 kWh) may discharge for substantially less than 2 hours at a 20-kW load (delivering less than 40 kWh).

Operation at greater than rated output—even for a few seconds or minutes—can be a valuable capability. For instance, a storage system could be used to provide the nominal amount of power required to serve a regularly occurring need (e.g., peak demand reduction) while the same storage could provide power at a higher output level for urgent needs that occur infrequently and that last for a few to several minutes at a time. (Note that manufacturers often limit the number of instances in which such operation with output exceeding nominal power is permissible.)

***Energy and Power Density, Footprint and Volume.*** The amount of power that a storage system can deliver, per unit volume occupied by the system, is called the *power density*. The amount of energy that can be stored per unit volume is called the *energy density*.

The footprint and space (volume) requirements for energy storage are directly related to the power and energy density of the underlying storage technology. Depending on the storage technology, floor area and/or space constraints may be a challenge, especially in heavily urbanized areas.

***Specific Energy and Specific Power.*** The amount of power that a storage system can deliver per unit mass of the system is called the *specific power*. Similarly, the amount of energy that can be stored in a storage device per unit mass is called the *specific energy*.

The weight of an energy storage system is directly related to the specific power and specific energy of the underlying storage technology. This is a consideration in installations, as the foundation must be strong enough to support the weight of the system, as well as in determining the shipping costs for the system.

Confusingly, the term *power density* is sometimes used to describe *specific power*, and the term *energy density* is sometimes used to describe *specific energy*. It is useful to clarify which quantity is being discussed, either by including the units of measurement, or by using the descriptors *gravimetric* (for quantity per unit mass) and *volumetric* (for quantity per unit volume). Table 10-1 summarizes these four quantities.

**TABLE 10-1** Specific Power and Energy, Power and Energy Density

| Quantity name | Definition | Unit of measurement |
|---|---|---|
| Specific power (gravimetric power density) | Power output per unit mass | Watts/kilogram |
| Power density (volumetric power density) | Power output per unit volume | Watts/liter |
| Specific energy (gravimetric energy density) | Energy content per unit mass | Watt-hours/kilogram or joules/kilogram |
| Energy density (volumetric energy density) | Energy content per unit volume | Watt-hours/liter or joules/liter |

***Storage System Round-Trip Efficiency.*** All energy transfer and conversion processes have losses. Energy storage is no different. The term "storage system round-trip efficiency" (often simplified as "efficiency") describes the amount of energy that comes out of a storage system during discharge relative to the amount put into the storage system on the immediately preceding charge.

It is important that this efficiency calculation should account for any auxiliary loads that are part of the typical energy storage configuration. Manufacturers sometimes provide DC-DC round trip efficiency numbers, but when sizing the system and estimation of operational costs, one must consider the AC-AC round trip efficiency.

Typical values for AC-AC round trip efficiency are 60% to 75% for most aqueous electrochemical batteries; 75% to 90% for advanced electrochemical batteries (including lithium-ion); 70% to 80% for CAES (not including fuel inputs); 75% to 80% for pumped hydroelectric; and 90+% for state-of-the flywheel storage, capacitors, and superconducting magnetic energy storage (SMES).

***Storage Lifetime.*** All energy storage media degrade with time, as well as with use (i.e., with each charge-discharge cycle). This leads to two measures of the storage technology's life: calendar life (the number of years a storage system is expected to last) and cycle life (the number of charge-discharge cycles a system can be expected to sustain before failure). Both types of degradation depend on the type of storage technology, operating parameters, environmental conditions, and other variables. This is especially important for electrochemical batteries.

For some storage technologies—especially batteries—the extent to which the system is emptied (discharged) on each discharge cycle also affects the storage media's useful life. Discharging a small portion of stored energy is a "shallow" discharge and discharging most or all of the stored energy is a "deep" discharge. For these technologies, a shallow discharge is usually less damaging to the storage medium than a deep discharge.

It is very important that before selecting energy storage system, one should consider the type of duty cycle that is expected from the energy storage system, as well as any extreme duty cycle conditions that may be experienced by the system in operational life. This could have significant impact on the selection of the energy storage system, as well as sometimes can be handled by configuration changes to selected energy storage system.

Note that many battery vendors can produce storage media with extra service life (relative to the baseline product) to accommodate additional charge-discharge cycles and/or deeper discharges. Of course, there is usually a corresponding incremental cost for the superior performance. To the extent that the storage medium degrades and must be replaced during the expected useful life of the storage system, the cost for that replacement must be added to the variable operating cost (VOC)of the storage system.

***Response Time.*** Storage response time is the amount of time required to go from no discharge to full discharge. At one extreme, under almost all conditions, storage has to respond quite rapidly if used to provide capacity on the margin in lieu of T&D capacity. That is because the output from T&D equipment (i.e., wires and transformers) changes nearly instantaneously in response to demand.

In contrast, consider storage used in lieu of generation capacity. In this application, storage does not need to respond as quickly because generation tends to respond relatively slowly to demand changes. Specifically, some types of generation—such as engines and combustion turbines—take several seconds to many minutes before generating at full output. For other generation types, such as those fueled by coal and nuclear energy, the response time may be hours.

Most types of storage have a response time of several seconds or less. Compressed air energy storage (CAES) and pumped hydroelectric storage (PHS) tend to have a response time of several minutes, though they still respond quickly enough to serve several important applications.

***Ramp Rate.*** An important storage system characteristic for some applications is the ramp rate—the rate at which power output can be changed. Ramp rate is measured in MW/Min for most of the power system components. Generally, storage ramp rates are rapid (i.e., output can be increased or decreased quite rapidly). Depending on the application requirements, some storage developers can design storage to have the necessary ramp rate and system life.

***Charge Rate.*** Charge rate—the rate at which storage can be charged—is an important criterion because energy storage must be recharged before it can again serve load. If storage cannot recharge quickly enough, then it will not have enough energy to provide the necessary service. In most cases, storage charges at a rate that is similar to the rate at which it discharges. In some cases, storage may charge more rapidly or more slowly, depending on the availability of charging power, capacity of the power conditioning equipment, and the condition and/or physics of the energy storage medium.

***Self-Discharge, Parasitic Losses, and Standby Losses.*** Energy storage, when charged, tends to lose stored energy over time even if it is not being discharged. This phenomenon of "energy leakage" is referred to as *self-discharge.*

The mechanisms for self-discharge vary substantially with the type of storage. In batteries, self-discharge is typically caused by side chemical reactions that occur in the electrochemical cell. These reactions can be reversible (i.e., the energy can be returned to the battery through recharge) or irreversible (meaning the reaction causes a permanent reduction of capacity in the battery).

Other types of storage also exhibit some variety of self-discharge. For example, PHS may suffer from evaporation of water in the upper reservoir, and CAES systems may lose energy through air leakage. Thermal storage may also lose the stored thermal energy due to transfer of heat between storage area and the surroundings.

In addition to these inherent losses, most storage systems must continuously supply a small amount of power to auxiliary equipment, such as controls, heating and cooling systems. The power drawn for such equipment is called a *parasitic load* because this power is unavailable for the principal applications of the storage system.

Taken together, the self-discharge and parasitic loads are called the *standby losses* for the energy storage system. If the storage system is isolated from the grid, these standby losses can accumulate over time. The length of time it takes for a fully charged storage system to self-discharge to some minimum level of operability is sometimes called the *energy retention time.*

When the storage system is connected to a primary power source (such as the grid or a generator), it is often charged at a low rate—just high enough to replace the energy lost through standby losses and remain at a constant state of charge. The power a storage system requires to remain at full charge is often referred to as the *standby power.*

Standby loss is important primarily in calculation of efficiency. This is especially true for storage in applications with long discharge periods or long times between discharges, where the standby losses can accumulate over time.

***Power Quality.*** The power quality of the output from storage systems is an important consideration, especially for grid-connected systems. Ideally, the waveform output should be as close as possible to a sine wave. While some low-cost power electronics systems can produce output with substantial deviation from this ideal (called *distortion*), most modern grid-connected storage systems include filters that produce a very good approximation of a sine wave.

It is also important for storage output voltage to remain reasonably constant. Depending on the circumstances, voltage should remain within about 5% to 8% of the rated value. Power factor required should exceed 0.85 for smaller applications and higher for larger systems.

Harmonic currents in distribution equipment can pose a significant challenge. Harmonic currents are components of a periodic wave whose frequency is an integral multiple of the fundamental frequency. For instance, if the fundamental frequency of the utility power line frequency is 60 Hz, harmonic currents might exist with frequencies of $3 \times 60$ Hz (180 Hz) or $7 \times 60$ Hz (420 Hz). Total harmonic distortion (THD) is the contribution of all the individual harmonic currents to the distortion of the current wave form.

***Storage System Reactive Power Capability.*** One application (voltage support) and one incidental benefit (power factor correction) described herein involve storage whose capabilities include absorbing and injecting reactive power (expressed in units of volt-amperes reactive or VARs). This feature is commonly referred as "VAR support" and is actually a function of the power conditioning system, not the storage technology itself.

***Communications and Control.*** Storage used for most applications addressed in this report must receive and respond to appropriate control signals. In some cases, storage may have to respond to a dispatch control signal. In other cases, the signal may be driven by a price or prices. Storage response to a control signal may be a simple ramp up or ramp down of power output in proportion to the control signal. A more sophisticated response may be needed to ensure that the physical limitations of the storage medium are also considered during the response.

***Modularity.*** Energy storage is often built by combining relatively small, modular blocks rather than as a single large system. One attractive feature of modular energy storage is the flexibility that system "building blocks" provide. Modularity allows for more optimal levels and types of capacity and/or discharge duration because modular resources allow utilities to increase or decrease storage capacity, when and where needed, in response to changing conditions. Modular capacity provides attractive means for utilities to address uncertainty and to manage risk associated with "lumpy" utility T&D investments by substituting small, relatively inexpensive and temporary storage installations for large, expensive, and permanent T&D infrastructure.

***Transportability.*** Transportability can be a valuable feature of storage systems for at least two reasons. First, transportable storage can be (re)located where it is needed most and/or where benefits are most significant. Second, some locational benefits only last for 1 or 2 years. Perhaps the most compelling example of the latter is T&D upgrade deferral (described in the subsection Storage in Lieu of Transmission Capacity).

With transportable systems, utilities can deploy the energy storage system at an optimal location for few years and then move the system to another location when the load growth or other upgrades to the T&D systems take place. This offers increased value for the same asset over its life. Given these considerations, transportability may significantly enhance the prospects that lifecycle benefits will exceed lifecycle cost.

### 10.6.3 Storage Technologies

***Introduction.*** Energy can be stored in chemical, mechanical, and thermal forms using electric or magnetic fields. Energy can be converted from those forms to another form by one or more intermediate processes. The most common electrical energy storage approaches and forms are summarized in Fig. 10-24.

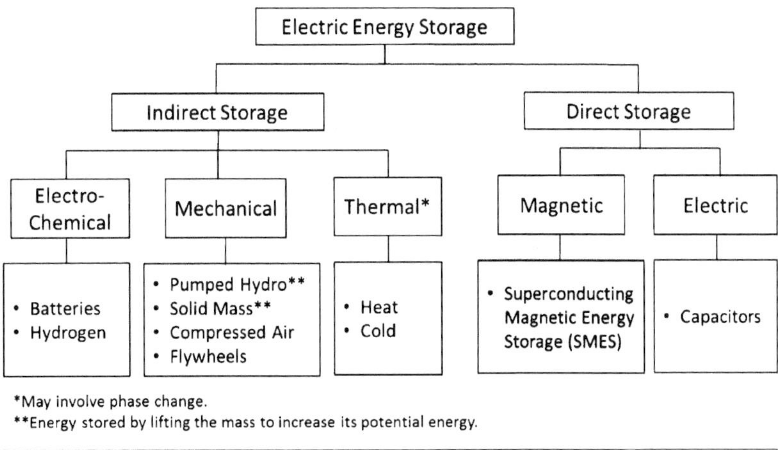


**FIGURE 10-24** Electric energy storage types. (*E&I Consulting.*)

Note that, in some cases, the energy input is electric energy and the energy output is not electricity. Consider electric energy used to create heat or cold that is stored and used later.

***Storage Systems.*** Typically, energy storage systems involve three key functions: (1) convert energy from the type/form available (input energy) to the type to be stored, (2) contain and store the energy, and (3) convert the stored energy to the type/form needed (output energy). A fourth function—commonly referred to as *interconnection*—is needed if the storage interacts with an external power system. Additional equipment needed is referred to as *balance-of-plant*, or BOP. Common BOP components include controls, communications and monitoring sub systems, buildings, support structures, conductors, and air conditioning equipment.

***Power Conditioning Units.*** Electric storage technologies are typically restricted to a narrow voltage and current range during both input and output. For this reason, most electric storage systems require a power electronics device, usually called a *charger*, which converts AC power from the grid to the voltage and current range accepted by the storage technology. Another power electronics device is often used to convert the output power of the storage system into the form needed by the end application. If the end application requires AC power, the latter device is called an *inverter*. Most grid storage applications require inverters of some sort. In many instances, the charger and inverter are combined into a single device called the *power conditioning system (PCS)* or *power conditioning unit (PCU)*. The PCU, in concert with an included control system, provides charge control, discharge control, reactive power control, necessary power quality, and synchronization of the storage system's output with AC power from the grid.

If storage will be charged with energy from the grid or will inject energy into the grid, it must meet applicable interconnection requirements. At the distribution level, an important reference is the Institute of Electronics and Electrical Engineers (IEEE) Standard 1547. States and utilities may have more specific interconnection rules and requirements.

The PCU must be specified in terms of: (1) DC voltage and current, (2) AC voltage and current, and (3) reactive power control capability. These parameters also affect the cost of the PCU. In general, PCU cost is determined by the desired current capability for the system. Higher voltage systems tend to be more cost-effective, as they reduce the necessary current capability for the PCU at a given power level. However, higher operating voltages can also complicate the design of the storage system.

***Balance of Plant.*** In addition to the storage technology and the power conditioning unit, most energy storage systems also require a number of auxiliary systems and constructions to properly operate. These items are collectively referred to as the BOP. The BOP typically includes:

- Structural components such as the foundation or concrete pad on which systems are installed
- Enclosures or buildings that protect the system against the environment
- Heating, ventilation, and cooling subsystems
- Safety and maintenance-related subsystems

In costing the system, installation and commissioning costs are often also included in the BOP. The installed cost of the storage system is thus divided into three components: (1) storage technology, (2) PCU, and (3) BOP.

***Electrochemical Energy Storage.*** In electrochemical storage systems, energy is stored in chemically active materials that, when connected electrically, react to release energy through a DC electric current. In some systems, the reaction can be caused to run in the other direction through the application of a DC voltage, effectively storing electrical energy in chemical form.

Electrochemical batteries, in which the chemical reactants are contained in closed cells, are the most commonly used form of electrochemical storage. However, other forms of electrochemical storage are possible. For instance, hydrogen can be created (through the electrolysis of water), stored, and used later to generate electricity via fuel cells, engines, and combustion turbines.

*Electrochemical Batteries.* Batteries are composed of one or more electrochemical cells. Each cell contains three components: two chemically reactive electrodes through which current flows into and out of the battery and the battery electrolyte. The electrodes are connected to an external load or source through the battery's negative or positive terminal as shown in Fig. 10-25, which depicts a basic lead-acid battery cell.

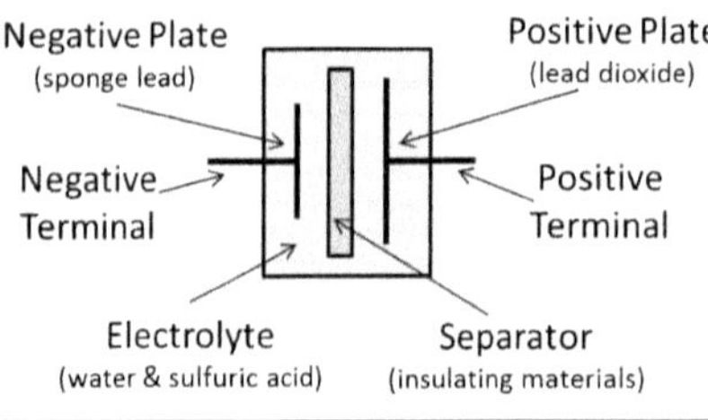


**FIGURE 10-25** Lead-acid battery cell elements. (*E&I Consulting.*)

The electrode connected to the negative terminal (i.e., the negative plate in Fig. 10-25) is made of material that, in the charged state, can electrochemically react so as to release an electron while releasing energy. (During battery discharge this electrode is the anode and when the battery is being charged it is the cathode.)

The electrode connected to the positive terminal (i.e., the positive plate in Fig. 10-25) is made of material that, in the charged state, can electrochemically react so as to absorb an electron while releasing energy. These two materials are bridged by an electrolyte, which serves as a path for ionic conduction. (During battery discharge this electrode is the cathode and when the battery is being charged it is the anode.)

The two different materials comprising the electrodes are bridged by an electrolyte, which serves as a path for ionic conduction.

If the electrodes are not connected by an electrically conductive path, the two electrode materials remain electrochemically inactive. When a conductive path is established, the electrochemical reactions begin, and the cell begins to generate a DC electric current through the path. An electrical load along the path can be powered by this current. The reaction will proceed until the electrodes' materials are fully reacted.

If current is pushed in the other direction through a secondary battery cell, the electrodes' materials gradually return to their charged state. The charge and discharge process can be repeated a number of times, but mechanical fatigue and other degradation mechanisms will gradually reduce the available capacity on discharge. Eventually, the cell capacity will degrade to the point that the cell is no longer useful as a storage system.

Many applications require voltages much higher than those produced by an individual cell, so cells are often connected in series to generate a higher voltage. These series strings are sometimes connected in parallel to allow a higher current level. These series-parallel combinations of cells are called batteries.

Cells are characterized fundamentally by their chemistry, which describes the composition of the two electrodes and the electrolyte. There are many possible battery chemistries, ranging from the relatively familiar zinc-carbon (the primary batteries used in flashlights) and lead-acid (the starting battery in vehicles) to more exotic chemistries such as sodium sulfur and zinc-bromine. Each battery chemistry has its unique characteristics, such as the nominal voltage, life, energy density, and efficiency.

*Lead-Acid Batteries.* The lead-acid battery is one of the oldest and best-known rechargeable battery types. Lead-acid battery technology relies on reduction of lead dioxide to lead sulfate at the positive electrode and the simultaneous oxidation of the lead to lead sulfate at negative plate. The electrolyte—sulfuric acid—is consumed and energy is discharged during this process. Energy is stored by reversing these reactions.

The principal advantages of lead-acid batteries are their extremely low cost and widespread availability. Most other commercial battery chemistries are superior to lead-acid in technical performance, but very few are as inexpensive. For this reason, lead-acid batteries are still a leading option for many applications whose energy requirements range from a few tenths of a kWh to many kWh. Common applications include submarines, forklifts, uninterruptible power supplies (UPSs), vehicle engine starting, and short-term emergency power systems for several applications including telephones, computers, and nuclear power stations. Lead-acid batteries have also been used occasionally for electric utility applications such as energy time-shift, maintaining frequency stability, and for spinning reserves.

The amount of energy stored is proportional to the voltage (2.08 V per cell) and to the amount of lead. Lead, which has a high molecular weight and is an inherently inefficient chemical for battery energy storage, is used for carrying current in a cell. As a result, the amount of lead required per kWh of storage is 50 to 100 lb. So, a potential drawback of lead-acid batteries is its low energy density—about 10 to 25 Wh/lb (watt-hour per pound) depending also on design and operation.

As with all batteries, the life of lead-acid batteries is highly dependent on both the design and the operating conditions. Lead-acid batteries built for cycling can last up to 2000 cycles if properly maintained. Flooded lead-acid batteries designed for standby operation can last for 20 years with proper maintenance. Sealed lead-acid batteries, also known as valve-regulated lead-acid (VRLA) batteries, are somewhat less durable and may require more frequent replacement.

Recent research into lead-acid batteries has included the development of lead-acid battery electrodes with substantially increased plate surface area. This can increase power density and reduce the key failure mechanisms that limit lead-acid battery life. This design can enable use of lead-acid batteries for both energy and power applications.

*Nickel-Cadmium Batteries.* Nickel-cadmium (Ni-Cd) batteries, originally developed for military applications, became the first rechargeable battery widely used in consumer electronics. The chemical reaction involves reduction of nickel oxide to nickel hydroxide and oxidation of cadmium to cadmium hydroxide in an alkaline electrolyte. The voltage of this couple (1.3 V) is less than that of lead-acid, and the energy density (about 20 to 25 Wh/lb) is slightly higher. In addition, nickel-cadmium batteries have substantially better cycle life than lead-acid. However, nickel and cadmium are significantly more expensive than lead, leading to a higher price. Additionally, cadmium is a highly toxic material, making manufacture and disposal of these batteries problematic.

For many years, nickel-cadmium was used in applications for which the higher energy density and longer life outweighed the substantial cost advantage of lead-acid batteries, such as satellites, military applications, consumer electronics, and power tools. In recent years, however, nickel-cadmium batteries have been largely replaced by other chemistries, such as nickel-metal hydride and lithium-ion.

Some nickel-cadmium batteries exhibit a condition known as *voltage depression*, commonly (but mistakenly) referred to as the *memory effect*. This condition occurs when the battery is repeatedly discharged to a partial state of charge and then recharged. The undischarged active material gradually decays to a lower energy state, causing a drop in battery voltage that makes the battery appear to "remember" the partial state of charge. This "memory" can be removed by simply discharging the battery fully and then fully charging it.

Voltage depression is a real phenomenon that affects some types of nickel and cadmium electrodes, but is actually rare in most modern nickel-cadmium batteries and non-existent in any other type of rechargeable battery. The "memory effect" is notable today only because some users, having experienced it with early nickel-cadmium batteries, erroneously associate it with reduced capacity in any rechargeable battery, even those that do not contain nickel or cadmium electrodes. In such cases, the reduced capacity is caused by some other factor, such as temperature or normal degradation over life.

*Nickel-Metal Hydride Batteries.* Nickel-metal hydride (NiMH) batteries are secondary batteries that are somewhat similar to nickel-cadmium batteries, although they employ a hydrogen-absorbing alloy for the anode instead of cadmium. As with nickel-cadmium batteries, the active cathode material is nickel. Nickel-metal hydride batteries have somewhat better energy density and specific energy than nickel-cadmium batteries and are much easier to dispose of because of the absence of cadmium. They also have somewhat better rate capacity, though they are more sensitive to the effects of temperature.

*High-Temperature Sodium Batteries.* Certain beta-alumina ceramics exhibit unusually high ionic conduction of sodium ions at temperatures above 250°C. There are a number of high-temperature sodium battery chemistries that take advantage of this phenomenon. They are unusual in that the electrolyte is solid, while the electrodes are liquid (a reverse from most batteries, which have solid electrodes and liquid electrolyte). Sodium-sulfur (Na-S) and sodium-nickel chloride (NaNiCl) are the two most prominent members of this family. Both chemistries have been used in electricity grid storage applications.

Sodium-sulfur batteries offer both a relatively high energy density relatively high power density given the high-conductivity of the ceramic electrolyte. The high operating temperature and the presence of liquid sodium has made the use of these batteries in mobile applications a challenge, but sodium sulfur batteries have been deployed extensively in grid applications, with over 350 MW of installed capacity as of this writing.

Sodium-nickel chloride batteries are similar to sodium-sulfur batteries: they operate at high temperatures and the molten sodium is separated from the cathode by a beta-alumina electrolyte. The cathode is comprised of molten nickel-chloride salt.

Sodium-nickel chloride batteries are less widely used to date but are believed to also have great potential in grid applications.

High-temperature sodium batteries can be used for applications requiring both high energy density and high power density making them viable for a somewhat broad range of uses.

*Lithium-Based Batteries.* An increasingly important battery category is lithium-ion (Li-ion) based chemistries. The term "lithium-ion" describes a family of related battery chemistries that all operate based on the lithium-ion as a charge carrier, in place of the hydrogen ion or hydroxide ion used in conventional aqueous batteries. The various lithium-ion chemistries share some general performance characteristics and are manufactured in very similar ways. Lithium-ion batteries have become the dominant technology for consumer electronics and for many battery-driven tools. They are now finding adoption in vehicle applications as well as grid applications.

Key features of lithium-ion batteries include: high cell voltage (between 3.0 and 4.0 V per cell), excellent performance at high temperatures, high energy density, low self-discharge, and high charge efficiency. With improvement in manufacturing techniques and massive investment in production facilities, lithium-ion has become the most cost-effective energy storage technology for consumer electronics, in terms of cost per unit energy stored.

There are some notable disadvantages to lithium-ion batteries. First, they have a relatively short track record vis-à-vis other more mature battery types, with only about two decades of commercial experience. Second, lithium-ion batteries tend to be relatively sensitive to overcharge and overdischarge, therefore require sophisticated control circuitry to charge and operate optimally.

The useful life of lithium-ion batteries is affected by battery age (calendar life) as well as how much it is cycled (cycle life). Historically, lithium-ion batteries have been limited by calendar life, but this has not impacted consumer electronics since most small gadgets are not expected to live beyond 2 or 3 years. Naturally, this is a greater concern for vehicle and stationary applications. Recent effort has been focused on extending both calendar life and cycle life of lithium-ion batteries for these applications.

Lithium-ion batteries are sometimes considered more dangerous than conventional batteries, but the truth is somewhat more complex. Despite their name, lithium-ion batteries are not at all similar to the lithium primary batteries often used in low-power applications such as watches or calculators. Lithium-ion batteries contain no metallic lithium, and the chemical reactions are fundamentally different from those in lithium primary batteries.

Nonetheless, lithium-ion batteries are generally safer than lithium primary batteries, although they must be properly treated and operated to ensure safety and reliability. All batteries concentrate energy in a relatively small volume, so they are potentially dangerous in a fire. Pound for pound, lithium-ion batteries hold more energy than other batteries, so they naturally release more energy in such an incident. Lithium-ion batteries contain inflammable organic electrolytes which increases the danger in a conflagration.

At the same time, because this danger is more readily acknowledged for lithium-ion batteries, system designers have typically designed lithium-ion-powered products with more safety features, improving the overall safety record for battery-powered products since their introduction.

While lithium-ion batteries contain some toxic materials such as cobalt, they are less dangerous in both amount and degree of toxicity than the active materials in lead-acid batteries or nickel-cadmium batteries.

*Flow Batteries.* Several advanced batteries employ flowing electrolytes, hence the term "flow batteries." For those battery types, the chemical reactants are stored in separate vessels, outside of the cells where the chemical reactions occur. An example is shown in Fig. 10-26.

Flow batteries offer some important advantages, primarily related to decoupling the chemicals from the reaction vessel. For example, storage power and energy can be scaled separately. Managing electrolytes is relatively straightforward given that they can be removed from the containment vessel and replaced with new or refurbished electrolyte. Similarly, if cells need to be replaced or serviced, no electrolyte is involved. Flow batteries require additional subsystems, especially those to store and enable flow to/from batteries, which may add to system complexity and increase possible points of failure.

*Vanadium Redox Batteries.* One promising flow battery technology is the vanadium-redox battery. The basic chemical process involves reduction and oxidation of vanadium ions in acidic aqueous solution. Vanadium can exist in four different oxidation states that form stable, water soluble species so only one element is needed. All four of those valence states are used in vanadium redox batteries.

Vanadium redox battery electrodes are made of stable porous carbon materials that are not altered—physically or chemically—as dissolved salts contact with them. Each cell provides 1.26 V at 25°C.

The electrolyte for vanadium redox batteries is durable and maintenance costs are low. These batteries operate at room temperature and have relatively low self-discharge rates. Many charge-discharge discharge cycles are possible without significant decrease in power and energy storage

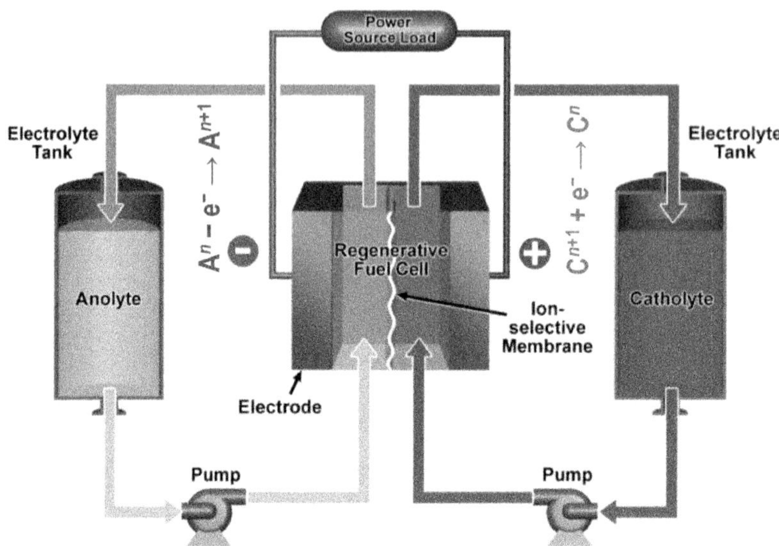


**FIGURE 10-26** Redox flow battery. (U.S. Department of Energy, Pacific Northwest National Laboratories.)

capacity. Finally, vanadium redox flow batteries can be left in fully discharged state for extended periods and can be quickly recharged with limited effect on performance and life.

*Zinc-Bromine Batteries.* Zinc-bromine flow batteries use zinc and bromine ions as the active species. During discharge, both ion types are dissolved in their respective electrolytes. When charged; zinc is a solid that precipitates onto the negative electrode. That characteristic makes zinc-bromine battery sizing somewhat complicated relative to sizing for other flow battery types. Zinc-bromine batteries have evaluated for several utility applications, particularly energy time-shifting and T&D deferral.

*Other/Advanced Batteries.* A number of less familiar battery types (chemistries and formats) have been and are being investigated and developed. Leading advanced battery systems include zinc-chloride, zinc-halogen, iron-chromium, zinc-halogen, zinc-air, advanced lead-acid, and lead-carbon. Just a few of the other less familiar and more exotic chemistries include silver-zinc (AgZn), nickel-iron (NiFe), and nickel-hydrogen ($NiH_2$).

Some of these advanced batteries employ low-cost, readily available active materials, and/or relatively simple electromechanical reactions, which should lead to good durability. Most advanced battery types are comparatively expensive—compared to the more established battery types. A key reason for this is that low manufacturing volumes do not yet allow for economies-of-scale. Integration and system design requirements for these technologies are also not yet well-established.

*Electrolysis of Water for Hydrogen Fuel.* Electrolysis of water involves use of electric energy to disassociate the hydrogen and oxygen within water. The hydrogen (and possibly the oxygen) is stored and used later for energy-related needs: as fuel for combustion processes, to produce heat or as a feedstock for fuel cells. In most circumstances, hydrogen must be stored at high pressure or the energy density may be unacceptable. Greater energy densities could be achieved by liquefaction, but the mechanical and cryogenic technologies required are expensive and energy intensive. One of the major attractions of electrolysis for energy storage is that the feedstock is water. Water is regenerated when the hydrogen is recombined with oxygen during discharge.

*Fuel Cells.* A fuel cell is an electrochemical device wherein the chemical energy of a fuel is converted directly into electric power. Though, technically, fuel cells generate electricity, they are mentioned herein because they can be used like a battery to convert stored energy into electricity.

Fuel cells have an even longer history than the lead-acid battery. The first hydrogen-oxygen fuel cell—similar to the modern "phosphoric-acid" fuel cell—was demonstrated in principle in the mid-1800s. More modern versions were developed primarily for power generation in space and later for terrestrial stationary and transportation applications.

Unlike batteries—which involve sealed/closed systems wherein the battery is fully discharged once reactants are depleted of electrical charge—fuel cells are open systems that will continue to generate as long as reactants are supplied from an external source. Like flow batteries, reactive chemicals (or fuel) for fuel cells are stored in and supplied from external vessels. A generic fuel cell-based power processing system showing the major plant processes is shown in Fig. 10-27.

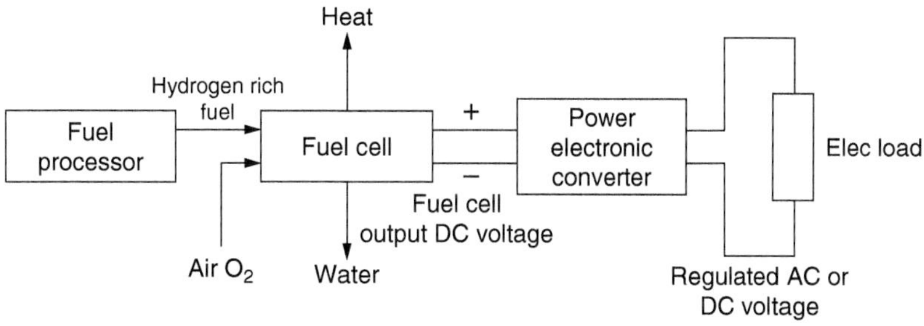


**FIGURE 10-27** Typical fuel cell-based power processing system.

As shown in Fig. 10-27, there are three major elements of a fuel cell system. The first step that occurs is the fuel processor which generates high purity hydrogen gas. This is accomplished in two primary ways. First, fuel cells used for generation-only can use hydrocarbon fuel (especially natural gas) that is "reformed" into a hydrogen-rich gas. Second, fuel cells used for electricity storage rely on hydrogen fuel produced via electrolysis (disassociation of hydrogen and oxygen in water) when the price is low for electricity input. The hydrogen or the hydrogen and oxygen are stored for later use (although they could be used real-time, if needed for generation).

The second step is conversion of the fuel to electric energy. The hydrogen or hydrogen rich fuel reacts with the anode and oxygen (stored or in air) reacts with the cathode. The hydrogen atoms at the anode are split into positive protons and negative electrons. The electrons follow an external path, through the load and then to the cathode.

The third step is the power conditioning using power electronics converters.

Fuel cells are often characterized by the type of electrolyte employed. Leading fuel cell types are phosphoric acid (PA), proton exchange membrane (PEM), molten carbonate (MC), solid oxide (SO), direct methanol (DM) and alkaline fuel cell. Each type of fuel cell has advantages and disadvantages.

Performance and characteristics are very similar to those for batteries. Output is DC. Each operates at a characteristic voltage. Efficiency ranges from 60% to as high as 90% if waste heat is utilized for combined heat and power (CHP) or combined cooling, heat and power (CCHP).

### *Mechanical Energy Storage*

*Flywheel Energy Storage.* Flywheel energy storage systems (FESS) employ kinetic energy stored in a rotating mass with very low frictional losses. Electric energy input accelerates the mass to speed via an integrated motor-generator. The energy is discharged by drawing down the kinetic energy using the same motor-generator.

The amount of energy that can be stored energy is proportional to the object's moment of inertia times the square of its angular velocity. To optimize the energy-to-mass ratio, the flywheel must spin at the maximum possible speed. Rapidly rotating objects are subject to significant centrifugal forces, however, while dense materials can store more energy, they are also subject to higher centrifugal force and thus may be more prone to failure at lower rotational speeds than low-density materials. Therefore, tensile strength is more important than the density of the material.

Low-speed flywheels are built with steel and rotate at rates up to 10,000 rpm. More advanced FESSs achieve attractive energy density, high efficiency, and low standby losses (over periods of many minutes to several hours) by employing four key features: (1) rotating mass made of fiberglass resins or polymer materials with a high strength-to-weight ratio, (2) a mass that operates in a vacuum to minimize aerodynamic drag, (3) mass that rotates at high frequency, and (4) air or magnetic suspension bearing technology to accommodate high rotational speed. (Advanced FESSs operate at a rotational frequency in excess of 100,000 rpm with tip speeds in excess of 1000 m/s.) A schematic of a modern FESS is shown in Fig. 10-28.

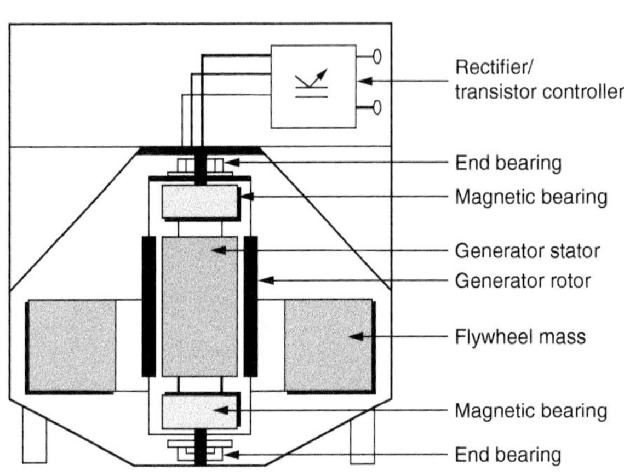


**FIGURE 10-28** Flywheel storage system.

FESSs are relatively expensive means to store significant amounts of energy, with costs that are at least twice that of the lead-acid battery (per kWh of storage capacity). Nonetheless, they have several advantages over chemical energy storage. They have higher energy density and better capability to exchange energy frequently. They also have very fast response and ramp rates. In fact, they can go from full discharge to full charge and from full charge to full discharge within a few seconds or less.

One disadvantage of flywheel storage is that power and energy storage are contained in the same vessel. So, scaling energy and power separately is challenging. To achieve more energy storage, more devices—with both power and energy storage subsystems—are added.

FESS are increasingly important as high power, relatively low energy applications. They are especially attractive for applications requiring frequent cycling given that they incur limited life reduction if used extensively (i.e., they can undergo many partial and full charge-discharge cycles with trivial wear per cycle).

FESSs are especially well-suited to several applications including electric service power quality and reliability, ride-through while *gen-sets* start-up for longer term backup, area regulation, fast area regulation, and frequency response. FESSs may also be valuable as a subsystem in hybrid vehicles that stop and start a lot and as a component of track-side or on-board regenerative braking systems.

*Pumped Hydroelectric Storage.* Pumped hydroelectric storage (PHS) is a mature and well-known way to use electric energy input to generate mechanical energy used to increase the potential energy of water. Energy is stored by pumping water to an upper reservoir using a turbine connected to an integrated motor-generator. The energy is discharged as the water flows to a lower reservoir through the turbine/motor-generator which converts the kinetic energy of flowing water to electrical energy—much like turbine generators in conventional hydroelectric generation plants.

Key elements of PHS systems include an upper reservoir, waterways, a pump, a turbine, a motor, a generator, and lower reservoir as shown schematically in Fig. 10-29.

Typically, PHS storage systems have involved large scale (hundreds of MW) and use of existing or natural waterways and lakes. However, growing environmental concerns and concerns about water

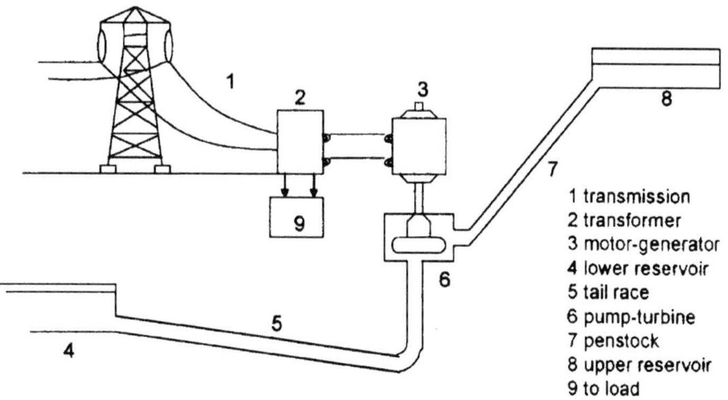


**FIGURE 10-29** Pumped hydroelectric energy storage.

quality have lead developers to locate PHS plants away from existing/natural water bodies that are otherwise suitable (i.e., those with "good head"). In fact, environmental concerns may be the single most important barrier to increased deployment.

Some systems involve reservoirs and/or penstocks that are enclosed, to reduce evaporative losses. Two notable variations on the PHS concept are (1) modular PHS and (2) underground PHS. Fully contained modular PHS systems rated at hundreds of kW to a few MW have been developed; though, deployment has been limited. Underground PHS entails a lower reservoir that is located underground, in a suitable geologic formation, possibly including abandoned mines. A key advantage of underground PHS is that it can be sited on flat terrain and have a modest effect on the surrounding (above ground) environment. However, underground PHS plants may have to have very large scale (perhaps GW) to be economically attractive.

The amount of energy that can be stored depends on the volume of available water, the elevation of the upper reservoir relative to the lower reservoir (head), and pumping efficiency. The amount of energy that can be recaptured depends on evaporative water losses and on the efficiency of the turbine and motor generator.

Power—from a given turbine/motor-generator—is driven by the head (height of the upper reservoir) and the flow rate of the water.

Consider two examples as an indication of the relationship between head and volume: for PHS with head of 1200 ft, 36 $ft^2$ of water is required for generation of 1 kWh whereas PHS with head of 120 ft requires 360 $ft^2$ of water to generate 1 kWh.

Hydraulic turbines have a maximum efficiency of around 95% and pumps are less efficient operating around 90%. Therefore, PHS plants will have maximum efficiency of around 85%. With state-of-the-art technology, typically, about 65% to 80% of the energy stored is recouped.

PHS is also comparatively responsive compared to most types of thermal generation. And, newer PHS employing variable speed turbines can be very flexible electric supply resources. For example, typically, PHS can achieve full output within about 10 seconds.

PHS seems poised to be an important energy storage resource for the electricity marketplace of the future given advantages, including: (1) it is familiar and proven, (2) it is mature and the technology is improving, (3) competitive cost, (4) low incremental cost for energy storage capacity, and (5) good or even fast response.

Given its many attractive features and characteristics, PHS storage can serve a broad array of electric supply applications including electric energy time-shift and capacity, most ancillary services, and renewables integration.

*Compressed Air Energy Storage.* Compressed air energy storage (CAES) converts electric energy to mechanical energy which is used to compress air to charge a robust geologic formation or above ground vessel. In most basic terms, during charging the compressor is powered with electricity to pressurize the storage cavern/vessel. During discharge, the pressurized air is used to operate a generator.

For grid-scale CAES plants constructed to date, the prime mover is a modified configuration of a combustion turbine (CT). Several configurations have been proposed. Other variations include use of piston-driven engines.

Above ground vessels may include high-pressure tanks or pipes. Geologic formations that may be suitable for CAES include salt domes, excavated rock, evacuated/depleted natural gas reservoirs, and aquifers. Most geologic formations are naturally occurring, though salt domes may be humanmade. Technically viable geologies exist in many areas.

Depending on the specific geologic conditions involved, it may be necessary to remove most of the heat of compression—to be used for other processes or stored, if possible—before the air is injected into the reservoir.

Basic CAES systems have relatively low overall efficiency due mostly to thermal losses: a significant amount of heat is generated during air compression and air expansion/decompression generates a significant amount of cold.

Three notable CAES variations are *diabatic*, *adiabatic*, and *isothermal*. Diabatic CAES dissipates heat of compression to the atmosphere. Though thermal losses can be significant, diabatic CAES systems are relatively simple. In most cases, the air must be reheated before it can be used for generation. Adiabatic CAES includes retention (storage) of the heat of compression, to heat the air when it is expanded to generate. The heat can be stored in a variety of media such as oil, molten salt solutions, or even rock or concrete. Isothermal CAES involves compression and expansion at operating temperatures that result in constant heat exchange to the environment. Given the elaborate plant required, isothermal CAES may only be practical for low power levels.

Most commercial designs involve combustion turbine-based, diabatic systems. They are comprised of combustion turbines whose compressor and expander sections are physically separated by clutches from the motor generator. A basic diagram for the system is shown in Fig. 10-30.

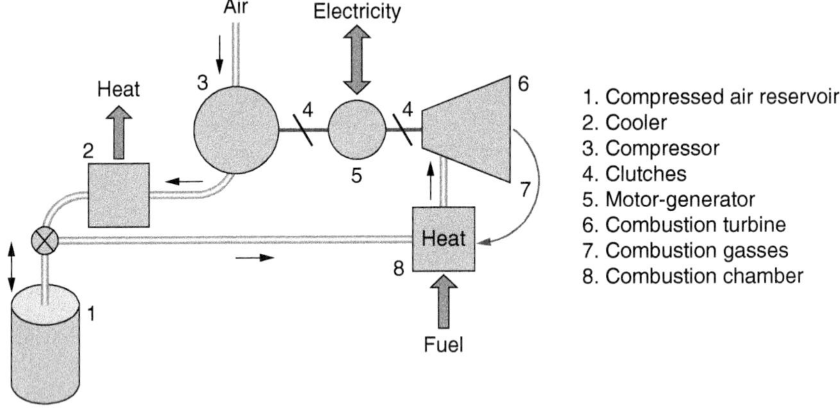


**FIGURE 10-30** Combustion turbine-based CAES plant. (*E&I Consulting.*)

The system's compressor uses electricity to compress and inject the air into the reservoir. Ideally some of the heat of compression is recovered and utilized. During discharge (i.e., generation), the air is expanded through the turbine to drive a generator. Before the air passes into turbine, it is heated to offset the cold generated during air decompression to raise the air temperature to that needed for combustion turbine operation. Heat can be from a thermal store (including stored heat of compression—as in adiabatic CAES) or from combustion of fuel, especially natural gas or from waste heat from the turbine.

For an advanced CAES system, incorporating a heat exchanger to recoup heat from the turbine exhaust (to heat the pressurized air the energy inputs) energy input—per kWh out—would be approximately 0.78 kWh (2662 Btus) of electricity for compression and 4000 Btus of fuel during the generation phase. Though direct storage efficiency tends to exceed 75%, net efficiency is about 51.2% (3413 Btus/kWh/(2662 Btus + 4000 Btus).

CAES system power ratings can range from several tens of kW to hundreds of MW. At the low end are systems using microturbines and piston-based engines. At the high end are systems employing bulk-generation-scale turbines, typically ranging from 50 to 300 MW. Long discharge durations—many hours to a few days—are possible given that reservoirs can be quite large and losses tend to be small.

Most key challenges for CAES involve reservoir geology: identification, verification, and uncertainty about reservoir dynamics both during operation at any given time, and more importantly changes over time. Some of the early CAES projects have realized that finding or developing appropriate reservoir could take number of years and significant investment.

Larger scale CAES can deliver a broad range of benefits including electric supply energy (time-shift) and capacity, ancillary services, and dynamic operating benefits. If well located, they could also be part of coordinated/regional transmission congestion management efforts.

***Thermal Energy Storage.*** At the most basic level, TES involves storage of heat or cold for use at a later time. Heat could come from a variety of sources, especially waste heat from processes, solar thermal arrays, and direct gain from insolation. Heat can be stored in an array of simple and advanced materials, ranging from rocks and oil to molten salts. Cold is typically stored in the form of chilled water or ice.

In general, energy density and efficiency for thermal storage are low. For heat storage, the available heat usually has relatively low temperature. TES can have non-trivial standby energy losses (e.g., if insulation is not adequate). Cold storage can be efficient enough to be financially attractive.

Heat TES may be best suited to applications requiring relatively low temperatures, especially space heating and some processes such as crop drying.

Within the context of electricity storage, thermal storage is relevant to the extent that when the thermal energy is used it offsets the need for electric energy and/or power during times when the power system/grid is serving their peak demand and when price is high. In that regard cold storage is most compelling. Using stored cold during electric utilities' peak demand periods in lieu of real-time operation of air conditioning equipment can yield significant benefits.

***Electrical and Magnetic Energy Storage***

*Superconducting Magnetic Energy Storage.* SMES stores energy in the magnetic field generated by the DC current flowing through a superconducting wire in the form of a coil. The coil is made of superconducting material, meaning that electric current can flow within the material without electrical resistance (i.e., without direct energy losses).

Because energy is stored as circulating current, SMES can be charged and discharged quite rapidly. Although parasitic losses associated with refrigeration can add up, SMES is quite efficient (95% or better) when energy is not stored for extended periods.

A common distinction used to characterize a superconducting material is whether the *critical temperature* is "high" or "low." (Critical temperature is the temperature below which superconduction occurs.) High-temperature superconductors can operate using liquid nitrogen whose temperature is 77K (−321°F). Low-temperature superconductors (nitrogen at 25K or -415°F) require more exotic and more expensive refrigeration.

In addition to the coil and associated infrastructure, SMES systems also include power conditioning systems. Two common types used are (1) current source converter (CSC) and (2) voltage source converter (VSC)—used in conjunction with a DC-DC chopper for charging and for discharging.

SMES power ($P$) and energy ($E$) are expressed as follows:

$$E = \frac{1}{2}LI^2 \quad P = \frac{dE}{dt} = LI\frac{dI}{dt} = VI$$

where $L$ is the inductance of the superconducting coil, $I$ is DC current in the coil and $V$ is voltage across the coil.

The three SMES operation modes—charge, discharge, and standby—are controlled by way of the voltage applied across the coil as follows. The coil is charged or discharged by applying a positive or negative voltage, respectively. The SMES system enters a standby mode operation when the average voltage is zero and the current in the coil is constant.

Though still relatively expensive, SMES has the potential to serve several important applications. They are especially well-suited to T&D applications requiring high power output for short durations such as power oscillation damping, dynamic voltage stability, tie line control, under-frequency load shedding reduction, circuit breaker reclosing, and subsynchronous resonance damping. They could also be used to improve local power quality [1].

*Capacitors.* Capacitors store electric energy as an electrostatic charge. An increasing array of larger capacity capacitors—sometimes called "supercapacitors" —store significantly more electric energy than conventional capacitors.

Capacitors are especially well-suited to being discharged quite rapidly, to deliver a significant amount of energy over a short period of time (i.e., they are attractive for high-power applications that require short or very short discharge durations).

The specific energy of a normal capacitor is relatively low: on the order of 0.5 Wh/kg. More advanced supercapacitors can store several times more energy per kg. To increase energy density, some advanced capacitors employ carbon nanotubes and polymers. Polymers—long chemical chains that are comprised of many molecules—used can undergo oxidation and reduction which is how energy is stored. The carbon nanotubes are quite porous, so the polymers can reside in many tiny spaces for a significant increase in active surface area compared to conventional capacitors. More recent capacitor research involves ceramics, which have superconducting properties that could be used in lieu of carbon nanotubes.

Commercial supercapacitors rated at several hundred to a few thousand farads are available. Commercially available supercapacitors reportedly can provide 500,000 to 2.5 million charge/recharge cycles.

A capacitor storage system with grid interface is shown in Fig. 10-31.

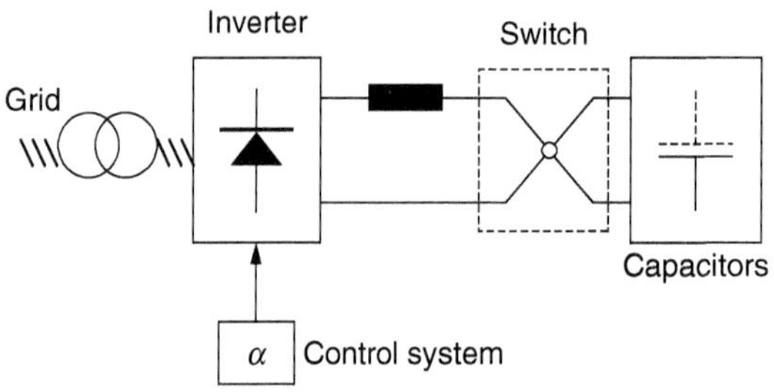

**FIGURE 10-31** Capacitor storage system with grid connectivity.

Supercapacitors are well-suited to serve power applications as they can provide a large portion of stored energy very quickly. Advanced capacitors may even be suitable for some of the same short-duration applications served by batteries, given their increasing energy density and given that they are less susceptible to some of the phenomenon that limit batteries' attractiveness, such as: (1) fewer lifetime charge-discharge cycles, (2) much slower recharge time, (3) more limited discharge rates (per kW), and (4) lower efficiency due in part to higher thermal losses.

Despite capacitors' lower energy density, they are used instead of batteries for applications requiring high power for short durations. For example, supercapacitors have been used by the United States military to start engines in submarines and large land vehicles. In fact, capacitors are used for storage in a broad array of circumstances such as handheld electronic devices, as a key element of hybrid vehicles' power systems and for conventional engine based vehicles (instead of batteries).

Grid-connected supercapacitors could be used for power quality and reliability related applications, especially voltage sags. They could also be used in hybrid systems with other storage that is better suited to significant energy storage, for a storage system that can provide high short-duration power and constant long duration power. Consider a storage system serving motors. In that example, the capacitor would provide surge current for motor startup and possibly to serve rapid increases in power draw while batteries could serve the load during normal operations.

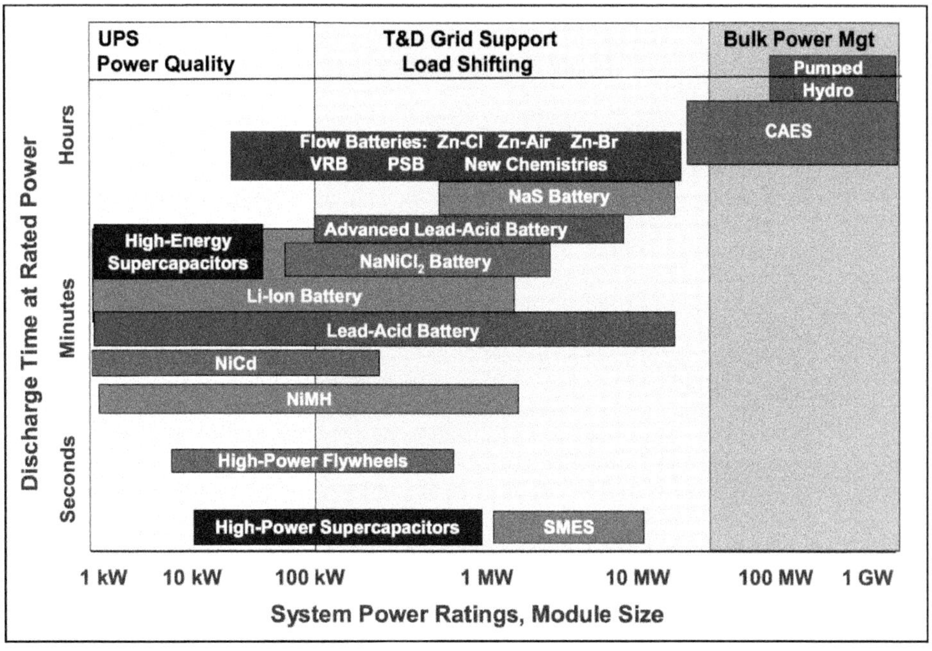


**FIGURE 10-32** Comparison of storage technologies based on power and discharge duration [2].

***Energy Storage Comparison Based on Power and Energy.*** Figure 10-32 provides a high-level comparison of key energy storage technologies based on two key characteristics: power and discharge duration.

## 10.6.4 Economics

***Energy Storage Economics Can Be Complicated.*** Given the unique characteristics and multifaceted nature of the storage opportunity, assessing the economics of energy storage is more challenging than assessing the economics of either generation or demand-side alternatives such as direct load control, for several reasons. Most notable are the following.

First, storage requires good assessment of and management of electricity input-related costs *and* output-related costs. Second, storage can often provide more benefits than generation or demand-side alternatives, though assessing many benefits in specific circumstances is an art. Third, assessing incremental storage equipment wear and related cost can be challenging for some technologies and thus some benefit/cost assessments are also challenging. Fourth, charge control can be complicated by the need to accommodate the unique characteristics of a wide array of storage types, especially for batteries.

One of the major drivers for interest in storage has been the price transparency provided by electricity markets. Most electricity markets allow monetization of various attributes of energy storage through energy, capacity, and ancillary services (including frequency regulation, synchronous reserves, non-synchronous reserves, voltage support, black start, etc.). In addition, most electricity markets also have locational pricing for these value streams, thus helping in identification of appropriate location to build energy storage. At the same time, since these markets are dynamic and require in-depth understanding of market scenarios for long-term forecasting, project developers still do face some uncertainty.

***Power and Energy Related Facility Cost.*** In basic terms, the two primary storage cost elements tend to be (1) power-related facility cost and (2) energy-related facility cost. Of course, cost for some portions of the facility may not fit neatly into those categories (e.g., grid interconnection, fuel handling, controls, and communication).

Power-related cost—expressed in units of \$/kW—is for plant and equipment needed to convert available energy to electric energy and/or to manage electricity flows and/or to manage the characteristics of electricity input and output. For storage whose output is DC, the primary power equipment is the power conditioning system including the inverter. For PHS and CAES, the primary power-related costs are for turbines, motors and generators.

Energy-related cost—expressed in units of \$/kWh of storage—is for plant and equipment needed to contain the stored energy. Batteries may contain reactants (and cells) in the same enclosure, or reactants may be stored outside of the battery (cells) in tanks for use as needed. CAES energy is stored in caverns, tanks, or pipes. Pumped hydroelectric energy is stored in the upper reservoir. Flywheels store energy in a spinning mass. In SMES systems, energy is stored within the superconducting coil.

The ratio of power-related to energy-related facility cost varies, depending mostly on the technology and the circumstances. For large CAES and hydroelectric storage, the energy-related cost—mostly for large reservoirs—tends to be relatively low compared to power-related costs. For SMES and flywheels, the energy-related equipment costs are relatively high in relation to those for power equipment. For capacitors, the cost for energy storage is high compared to power-related cost. The relationship varies for batteries, depending on the discharge duration.

Readers should note that storage plant cost is expressed using two primary unit cost metrics: (1) \$/kWh of storage capacity and (2) \$/kW for a system with a specified discharge duration. To understand the difference, consider the example illustrated in Table 10-2. A storage system with an all-in installed cost of \$85,000 has a nominal power rating of 100 kW and discharge duration of 5 hours.

In this example, the unit cost for the storage system as \$85,000 ÷ 500 kWh = \$170 per kWh of storage. The alternative is to express unit cost for the storage system—with 5 hours of discharge duration—as \$850/kW. Recent costs for a variety of energy storage technologies can be found in the *DOE/EPRI Electricity Storage Handbook in Collaboration with NRECA* [2].

***Variable Operating Cost.*** Storage VOC (as distinct from plant capital cost or plant financial carrying charges) consists of two key components: (1) energy-related costs and (2) operating costs not related to energy. Further, non-energy operating costs include at least four elements: (1) labor associated with plant operation, (2) plant maintenance, (3) equipment wear leading to loss of life, and (4) decommissioning and disposal cost. VOC is a key factor affecting the cost-effectiveness of storage in many circumstances. It is especially important for high-use applications involving many charge-discharge cycles.

The energy-related component of VOC cost for storage consists of all costs incurred to purchase energy used to charge the storage, including the cost to purchase energy needed to make up for (round trip) energy losses. An example: For a storage system with 75% efficiency, if the unit price for energy used for charging is $\$0.05/\text{kWh}_{\text{in}}$, then the plant energy cost is $\$0.05/\text{kWh}_{\text{in}}/0.75 = \$0.067/\text{kWh}_{\text{out}}$.

**TABLE 10-2** Electricity Storage Plant Unit Cost Metrics Comparison

| Plant characteristic | Plant total cost | Plant unit cost |
|---|---|---|
| 100 kW, 5 hours (500 kWh) | Power related costs:<br>\$100/kW × 100 kW = \$10,000 | \$85,000/500 kWh = **\$170/kWh** |
| \$100/kW for power conditioning (and balance of plant) | Energy related costs:<br>\$150/kWh × 5 hours × 100 kW = \$75,000 | \$85,000/100 kW = |
| \$150/kWh for storage "vessel" | **Total plant cost = \$85,000** | **\$850/kW with a 5 hour discharge duration** |

So, ideally, storage for high-use applications should have relatively high or very high efficiency and relatively low VOC. Otherwise, the total cost to charge then discharge the storage is somewhat-to-very likely to be higher than the benefit. That can be a significant challenge for some storage types and applications.

Plant maintenance costs are incurred to undertake normal, scheduled, and unplanned repairs and replacements for equipment, buildings, grounds, and infrastructure. Fixed maintenance costs are the same magnitude irrespective of how much the storage is used. Variable maintenance costs are proportional to the frequency and duration of storage use. Plant maintenance costs are highly circumstance-specific.

If specific equipment or subsystems within a storage system are expected to wear out during the expected life of the system, then a "replacement cost" will be incurred. In such circumstances, a "sinking fund" is needed to accumulate funds to pay for replacements when needed. That replacement cost is treated as a variable cost (i.e., the total cost is spread out over each unit of energy output from the storage plant). Replacement cost is highly technology-, circumstance- and application-specific.

In some cases, labor may be required for storage plant operation. Fixed labor costs are the same magnitude irrespective of how much the storage is used. Variable labor costs are proportional to the frequency and duration of storage use. More specifically, in some situations labor is required to operate larger storage facilities and/or "blocks" of aggregated storage capacity whereas little or no labor may be needed for smaller/distributed systems that tend to be designed for autonomous operation.

Energy storage project owners can, in such cases, utilize third party market operations centers that can offer remote monitoring, scheduling, and dispatch of the energy storage facilities using SCADA systems. Such remote and automated systems also become mandatory for energy storage systems participating in markets for frequency response or synchronous reserve, where system operators cannot rely on human intervention due to restrictions on response time (typically a few seconds to less than 5 minutes).

Completing the example above involving a 75% efficient storage system that: (1) has a non-energy-related VOC of $\$0.04/\text{kWh}_{out}$ and (2) is charged with energy costing $\$0.05/\text{kWh}_{in}$; the total VOC is $\$0.067/\text{kWh}_{out}$ ($\$0.05/\text{kWh}_{in}/0.75$) plus $\$0.04/\text{kWh}_{out} = \$0.107/\text{kWh}_{out}$.

***Energy Storage Benefits Introduction.*** This subsection provides brief descriptions of the numerous ways that grid-interactive storage can be beneficial. The emphasis is on individual benefits that could be associated with storage use, including reduced or avoided costs and increased profit. As described in Sec. 10.6.5, combinations of those individual benefits comprise the elements of value propositions that underpin specific use cases (applications).

In many cases, storage will be specified for specific needs and applications. In those cases, the benefit for storage is either known or presumed to exceed the cost for the storage. For example, batteries used for UPS applications are used to reduce or avoid electric service outage-related costs.

In a growing number of cases, additional benefits may also accrue and in some situations more than one benefit may be needed so that benefits exceed cost. So, it can be helpful to be familiar with the spectrum of storage benefits. Six broad storage use categories and individual benefits within each category are shown in Table 10-3 and described briefly below.

***Electric Supply Benefits.*** Electric supply benefits are related to reduced need for and use of generation equipment. The *energy time-shift* benefit involves purchase of energy when demand and price are low, for later use when demand and price are high. The supply capacity benefit implies generation equipment: storage use during peak demand periods reduces the need for generation capacity.

***Grid Operations Benefits.*** Grid operations benefits involve avoided or reduced cost for or revenues for providing what are often called *ancillary services* in the electric utility realm. Ancillary services complement the primary electric supply resources in an electric supply system, to enable controlled and stable grid operations.

*Load following* is generation whose output varies throughout the day as system load picks up (i.e., in the morning) and winds down (i.e., in the evening).

*Area regulation* (a.k.a. frequency regulation or just regulation) is needed to reconcile moment-to-moment discrepancies between supply and demand. Without area regulation, the AC frequency of

**TABLE 10-3** Storage Use Categories and Benefits (*E and I Consulting*)

| Category | Benefits |
|---|---|
| Electric supply | 1. Electric energy time-shift |
| | 2. Electric supply capacity |
| Grid operations (Ancillary services) | 3. Load following (hours) |
| | 4. Area regulation (seconds) |
| | 5. Fast area regulation (seconds) |
| | 6. Frequency response (sub-second) |
| | 7. Ramping (minutes) |
| | 8. Electric supply reserve capacity |
| | 9. Voltage support |
| | 10. Black start |
| Grid infrastructure | 11. Transmission support |
| | 12. Transmission congestion relief |
| | 13. Transmission and distribution upgrade deferral |
| | 14. Substation onsite power |
| End-user | 15. Time-of-use energy cost management |
| | 16. Demand charge management |
| | 17. Electric service reliability |
| | 18. Electric service power quality |
| Renewables integration | 19. Renewables energy time-shift |
| | 20. Renewables generation capacity firming |
| | 21. Variable renewables operational integration |
| Incidental and other benefits | 22. Generation dynamic operating benefits |
| | 23. Reduced generation fossil fuel use |
| | 24. Reduced generation air emissions |
| | 25. Increased (gt and d) asset utilization |
| | 26. Reduced T and D "$I^2R$" energy losses (net) |
| | 27. Avoided transmission access charges |
| | 28. Reduced T and D investment risk |
| | 29. Power factor correction |
| | 30. Flexibility (especially modular and transportable) |

the grid would vary unacceptably. Traditionally, area regulation was provided with thermal generation, whose output cannot be varied rapidly. By contrast, output from most types of storage can be varied somewhat or even very rapidly. So, use of storage for regulation frees up generation resources to be used, more effectively, as supply capacity.

Storage provides "up" regulation—needed to offset a momentary deficit of power for the grid relative to demand—by increasing output and/or reducing charging. Conversely, storage provides "down" regulation—needed to offset a momentary excess of power for the grid relative to demand—by increasing charging and/or reducing discharging.

Notably, storage can provide up regulation *and* down regulation equal to its entire rated capacity given that output can be varied between full charge and full discharge. By contrast, generation can provide 50% of its capacity for up regulation and 50% for down regulation or it could provide all of its output for either up or down regulation. That is because generation can only provide *output* and thus output can be varied from none to full power. Typically, though, generation used for regulation is operated at part load (e.g., 90% of full load) for power generation with the balance (10% in the example) used for regulation. See Fig. 10-33 for a graphical depiction of those relationships.

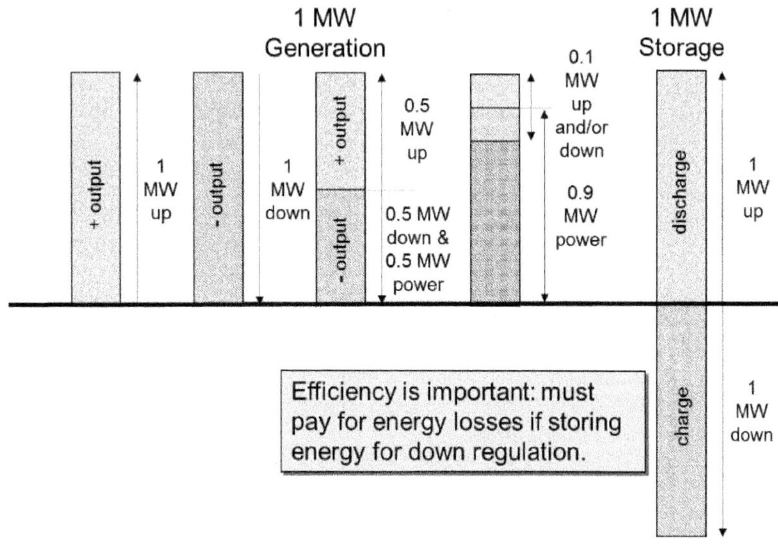


**FIGURE 10-33** Generation and storage operation for area regulation. (*E&I Consulting.*)

For traditional regulation, generators need to respond to a 4- to 6-second control signal, but can receive credit for regulation capacity based on amount of movement it can achieve during 5-minute period (i.e., 5 × Ramp Rate MW/Min). Given that some of the storage technologies can (1) respond almost instantaneously and (2) withstand frequent charge discharge cycles; they can be used for an emerging version of area regulation called *fast area regulation* (or just fast regulation).

Fast area regulation requires storage output to vary between full charge and full discharge within a few seconds. A concrete example is storage that responds to regulation signal that includes what the utility grid operators call the area control error (ACE) signal, which changes every few seconds.

Among other advantages of using energy storage for regulation: Doing so can result in reduced generation equipment wear, fuel use, and air emissions relative to regulation provided by "slow" generation (operated at part load). And, given the relative effectiveness of providing regulation with a rapid response resource—compared to generation-based regulation—less regulation capacity is needed to maintain the grid frequency.

Another emerging ancillary service is *frequency response*. That involves storage that responds to actual frequency excursions (i.e., from the fundamental 50 or 60 Hz.), quite rapidly (i.e., in just a few cycles, less than 1 second). Unfortunately, currently there is no market mechanism established to recognize and pay for a frequency response service.

*Ramping* service is needed for grid systems with a significant penetration of variable resources like solar and wind generation. Ramping occurs when the variable generation's output changes rapidly, such as times when wind speed varies significantly or when numerous clouds pass over solar collectors. Most conventional/thermal generation types are not well-suited to provide this emerging ancillary service because generation output cannot be varied quickly enough and such variations increase wear, fuel use, and air emissions (per kWh generated). Most types of storage could provide ramping service more efficaciously than generation.

Reserves are, essentially, backup generation capacity that is used if there is an unexpected electric supply shortfall or system emergency such as tripping of a generator or transmission line. *Spinning reserves* (a.k.a synchronous reserve) involve generation that is synchronized to the grid and being operated—at lower than maximum economic output—so it can respond quickly to pick up load rapidly if/when another supply resource becomes unavailable. Additional non-synchronous reserves include generation that can be started quickly or relatively quickly (e.g., within 10 minutes) to pick up load. Storage can provide reserves quite effectively. First, most storage types can pick up load quickly even when they are not being operated. And, stand-by losses and air emissions are low or non-existent, depending on the type of storage. Finally, when storage is being charged it can provide two times its capacity as reserves, if charging is stopped and discharging commences. But, typically, electric systems do not require such reserves during off peak times, when sufficient generation is available for providing reserves.

*Voltage support* is needed to offset reactive power within the grid system, at the supply level. It is a locational service as reactive power cannot be transmitted for long distances. It is provided to the grid by injecting or absorbing reactive power (to/from the grid) to adjust voltage or power factor. The service is provided by generators equipped with a synchronous condenser (sometimes referred to as a synchronous compensator). In most cases voltage support is needed to offset effects from inductive loads. Storage that is capable of providing reactive power can be used in lieu of generation-based voltage support resources for this service. And, because reactive power does not transmit well over long distances, storage located at or near offending loads can be more responsive and effective. (Note that this ancillary service provided at the electric supply level is distinct from power factor correction which is accomplished, locally, using capacitors which cannot inject reactive power, rather they compensate for low power factor).

*Black start* resources are used to begin the re-energizing process after a grid-wide outage. Black start resources are specially equipped to start up when the grid is not energized. A common approach to black start is to equip specific generation facilities with diesel gen-sets that provide power for generation facility start-up without the need for energy from the grid. Typically this requires capability to run in stand-alone mode (without additional support from grid for 10 to 16 hours) while rest of the electric grid is being restored as per the grid restoration plans developed by each system operator. As a result, only a few storage types are well-suited to provide black start functionality.

***Grid Infrastructure Benefits.*** Storage can play a significant role in support of or even in lieu of T&D capacity, in a variety of ways.

Transmission support involves storage used within the transmission system to manage or reduce electrical effects that decrease the dynamic load carrying capacity of specific portions of the system. Fast-responding energy storage with just a few seconds of discharge duration could enhance transmission throughput by (1) improving dynamic stability, (2) providing active real and/or reactive power modulation at sub-synchronous resonance modal frequencies, (3) by reducing the voltage dip that follows a system disturbance, (4) improving voltage stability, and (5) reducing load shedding needed to manage under-frequency conditions which occur during large system disturbances [3–5].

An increasing challenge for grid planners and operators is transmission congestion that occurs electrically upstream from load centers, during times when demand for electricity is high. To address the congestion, storage would be installed downstream from the bottlenecks, charged at night when there is no congestion and when $I^2R$ energy losses and ambient temperature are low, so the energy can be discharged to serve load when transmission congestion would have occurred. One critical advantage of energy storage is that it can be installed in most of the load centers where traditional generators cannot be sited due to direct emission restrictions and to some extent due to noise-related constraints.

T&D upgrade deferral and life extension are accomplished using relatively small amounts of storage—located electrically downstream from the targeted T&D equipment—to serve peak load. For deferral, the amount of load served is the amount that would exceed the load carrying capacity of the existing T&D equipment. For life extension, storage would reduce peak load on the T&D equipment to a level where degradation of the equipment is reduced to an acceptable level, to reduce operating temperature and in some cases to reduce/avoid ground faults.

Deferral and life extension can be quite attractive because a small or relatively small amount of storage—whose power rating is a few percent of the affected T&D equipment's load carrying capacity rating—can defer the need for expensive and/or large T&D upgrades. And, depending on the circumstances, storage deployed for deferral/life extension may not have to be operated much, if at all. For example, if the affected T&D equipment serves load with a summer peak and if weather—in the respective year—is mild (the highest temperatures encountered are below normal) then load may not exceed the T&D load carrying capacity. Finally, deferral may be especially valuable in situations with uncertainty about: (1) the magnitude and timing of load additions or (2) availability of resources needed to make a given upgrade before load exceeds the capacity.

In most cases, the benefit (per kW of storage needed) diminishes for deferral and/or life extension over a few years as load grows. As a result, increasing amounts of storage power and energy are needed to serve load that exceeds the T&D equipment's rating. So, eventually, it is not practical or cost-effective to continue to defer upgrades or to extend equipment life.

Electric utilities use a significant number of UPS-like systems at their substations so there is electricity during grid outages. That is needed so on-site operations including communications and controls can operate without energy from the grid.

***Electricity End-User Benefits.*** Energy storage can be used to reduce cost for electricity in two ways: (1) time-shift *energy* from times when energy prices are low to times when energy prices are high and (2) reduce peak electric demand (maximum *power* draw) during times when *demand charges* and the marginal cost of electric energy generation are highest (demand charges are assessed by the utility based on the end-user's maximum power draw). The energy prices and demand charges are specified in the applicable utility tariff.

Electricity end-users may also use storage—alone or as part of a system that includes generation—to reduce cost associated with (1) unacceptable electric service reliability (i.e., outage frequency and/or duration are unacceptable) and (2) unacceptable power quality (e.g., voltage sags and overvoltage, surges and spikes, harmonics, etc.).

The benefit is very location and application specific. In most cases it is based primarily on: (1) the need to maintain service for critical loads such as emergency lighting and medical equipment and more generally, (2) the avoided cost for lost productivity and/or damaged materials or product and/or electricity-using equipment damage.

Recent changes in market rules under the direction of Federal Electricity Regulatory Commission (FERC) have opened up competitive electricity markets for DR resources, where end users can participate in these organized electricity markets using energy storage units installed behind the meter.

***Renewables Integration.*** Energy storage can be an important element of an overall approach to integration of increasing amounts of renewable energy (RE) generation, especially RE generation with variable output.

First, storage can be used to time-shift electric energy from RE generation that occurs when demand is low. That is especially compelling for wind generation—much of which occurs at night—and for solar generation whose maximum output is not coincident with peak demand. Similarly, storage could be used to "firm" variable RE generation so the generation can be relied upon for service like a dispatchable generator whose output can be constant (e.g., combustion turbines or diesel gen-sets.)

From the system operators point of view, the energy storage can be utilized to optimize transmission capacity required to integrate renewable sources that have relatively lower capacity factors as shown in wind output curve for a wind farm in Texas in Fig. 10-34.

Aside from those electric supply related benefits, storage can be used to manage operational challenges posed by RE generation variability. Specifically, storage can compensate for unexpected RE generation availability shortfalls, such as those that can occur due to incorrect wind or cloud forecasts. Storage can also be used to offset rapid RE generation output variations—ramping. And, storage can be used to manage RE generation's negative effect(s) on power quality.

***Other Benefits.*** Energy storage can provide benefits that do not fall into any obvious category. What follows are brief descriptions of several of them. In some cases, one or more of these benefits

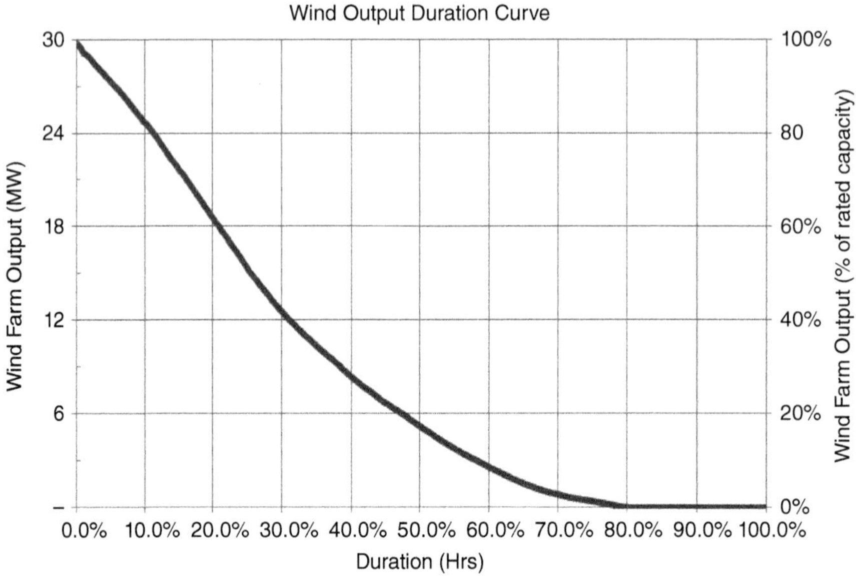


**FIGURE 10-34** Wind generation output duration curve.

will be significant and for others some of them may just be incidental. Nonetheless, they may be important considerations for storage system benefit/cost assessments and during system design.

A potentially important group of interrelated benefits that storage may provide is sometimes referred to as *dynamic operating benefits*. They are possible if storage use enables more optimized operation of electric generation, including (1) fewer generation start-ups, (2) reduced part load operation of the generation, and (3) reduced output variability and ramping. The benefit accrues due to (1) reduced fuel use and cost (per kWh delivered), (2) reduced air emissions (per kWh delivered), and (3) reduced wear on equipment which reduces maintenance cost, increases equipment life and may increase generation reliability [6].

For storage connected to a large electric grid, the dynamic operating benefits may be somewhat marginal or incidental, though that may change if storage penetration is high and as air emissions become a bigger issue. However, dynamic operating benefits can be a primary reason to use storage for small, isolated, or island power systems.

Use of storage can reduce generation fuel use and air emissions in several ways. One way is via the dynamic operating benefits described just above. Also, storage deployment can enable more use of baseload generation whose fuel efficiency is relatively high and whose air emissions are relatively low or even nil (e.g., natural gas fueled combined cycle generation, hydroelectric generation, nuclear generation, and geothermal generation). Finally, *distributed storage* charged at night may lead to reduced $I^2R$ losses (per kWh delivered).

In many situations, storage will lead to increased use (utilization) of utility assets. That is, the same generation assets can be used to generate more energy and the same T&D can be used to transmit and deliver more energy. Consider an example: A utility that serves peak demand which is growing such that more generation and T&D capacity will be needed. Furthermore, the utility has unused generation and T&D capacity during off-peak periods. For that situation, installing distributed storage would enable the utility to use existing generation and T&D capacity to charge the distributed storage when demand is low. In addition to deferring the need for additional generation and T&D capacity, the utility delivers more energy using the existing capacity—increasing asset utilization. The benefit is especially compelling for utilities or locations for which peak load occurs infrequently and/or for just a few hours per year.

Storage used to serve electric utility loads can reduce T&D $I^2R$ energy losses because, in most cases, storage is charged when the T&D equipment is lightly loaded and because ambient temperatures are lowest, usually at night. Furthermore, presumably, the stored energy is discharged when at times when losses are highest. Importantly, given that losses are a function of the square of the current, somewhat modest load reductions can have a somewhat significant effect.

The benefit for reduced $I^2R$ losses is twofold. First, the net loss (on-peak minus off-peak) represents a direct generation fuel use reduction (and, often, generation operating at night is the cleanest and most efficient generation). Second, generation and transmission infrastructure is needed to generate and transmit the extra energy needed to make up for the losses. So, reducing losses during high demand periods could reduce the amount of electricity generation and a T&D infrastructure capacity needed.

To one extent or another, electricity is transmitted across the high voltage transmission system. Of course, there is a charge for that "service," sometimes referred to as *transmission access charges*. If storage is charged at night, transmission access charges might be reduced in two ways. First, in an increasing number of areas, electricity prices reflect the location and time of energy delivery/use. So, storage charging at night may occur when transmission access charges are low. Second, in some areas transmission cost is incurred over specific periods of time, such as 1 month or 1 year. In that situation, there is no incremental transmission access charge incurred (for charging) and the capacity is freed up during peak demand times for the owner to use or sell.

If the storage system has the necessary features, it could be used for local power factor correction, in lieu of capacitors that utilities almost always use for power factor correction. The benefit is likely to be modest because capacitors are quite inexpensive compared to the cost for a storage system with the necessary characteristics. Nonetheless, there may still be non-trivial avoided cost that could make a given value proposition more attractive.

Given all the ways that storage can provide benefits, it can be a flexible alternative to generation and T&D. If the storage is modular and/or transportable, then it provides even more flexibility. Although quantifying the benefit for flexibility may be challenging, it is real and may be significant—especially when there is significant uncertainty and/or attractive market opportunities (e.g., to reduce cost and/or to increase revenue/profit). One way to think of the flexibility is to characterize the *real options* that they can provide where real options can be defined as "actual business endeavors that are possible because an asset is owned or can be used." For example, storage could be (1) used to purchase especially low priced off-peak energy that becomes available unexpectedly, (2) sold, rented, or leased to another user, or (3) used in ways that enable utilities to delay or avoid other investments.

### 10.6.5 Energy Storage Value Propositions and Applications

***Introduction.*** Exhaustive coverage of the myriad *storage value propositions* and applications is not possible given all of the benefit combinations that are possible. So, what follows are descriptions of benefit combinations—value propositions—that are intended to represent the spectrum of important storage applications.

The concept of value propositions is emphasized because merely specifying individual storage "uses" does not capture the richness of the storage alternative. Another reason is that it may be necessary to "stack" more than one benefit to justify the cost. Or, said another way, storage may be too expensive for a specific/individual primary application unless additional benefits can also be realized.

***Power Applications versus Energy Applications.*** A common designation assigned to storage applications is whether they are *power* applications or *energy* applications. Power applications involve high power, sometime frequent output for relatively short periods of time (e.g., a few seconds to a few minutes). Storage technologies used for power applications usually have capacity to store fairly modest amounts of energy per kW of rated power output. Important storage technologies for power applications include capacitors, SMES, Li-ion batteries, and flywheels. Conversely, energy applications require discharge of many minutes to several hours at-or-near the storage system's nominal power rating. So, energy applications require storage that is well-suited to storing a significant amount of energy per kW of nominal. Storage technologies that are well-suited to energy applications include CAES, pumped hydroelectric, thermal energy storage, and most battery types.

***Benefits Stacking.*** A value proposition is characterized by one or more combined (stacked) benefits. An *attractive* value proposition is one whose benefits exceed costs including return on investment. (Depending on the stakeholders involved, other considerations that could increase the attractiveness of a given value proposition could include the amount of up-front capital needed, opportunity cost and payback period.)

In some (perhaps many), situations, storage used for just one benefit may justify the cost. But, in many circumstances, it is necessary to combine two or more benefits to justify the cost. So, it is important for storage specifiers and designers to consider the important concept of combining (stacking) benefits when assessing the need and value for storage.

Of course, benefits must be compatible if they are to be combined. A combination of benefits is *technically* compatible if the same storage system has the characteristics and performance needed for all of the applications. Benefits are *operationally* compatibility if there are no operational conflicts that may arise during operation for any individual benefit.

***Storage Peaker.*** Storage is a potentially compelling source of energy *and* power needed to serve peak electric demand served by a power system. For most situations, the standard approach used by utilities to serve peak demand is use of simple cycle combustion turbine (CT) "peaker" generation. CTs are mature, familiar, and responsive. However, CT peaker plants tend to have relatively low fuel efficiency and they tend to have relatively high air emissions (per kWh of energy generated), especially when operated at part load. Other common generation-related means to address peak demand include (1) reciprocating engines (especially diesel gen-sets), (2) less responsive natural gas fired steam generation plants and, in some cases, (3) hydroelectric generation, and (4) photovoltaics and solar thermal generation. An increasingly important non-generation peaking resource is *load management* (i.e., controlled reduction of electricity users' demand based on reliability based or price responsive control signal), often referred to as DR.

Storage tends to be much more responsive than generation-based peaking resources because, for most types of storage: (1) start-up is somewhat to very quick, (2) output can be varied rapidly (i.e., can be used for fast ramping), and (3) operation at part load poses a limited efficiency penalty.

There are two versions of storage peakers: (1) bulk/central facilities comprised of one large plant at sites that tend to be somewhat or very remote from load centers and (2) modular/distributed storage located near or within load centers. Bulk/central peakers are most likely to involve pumped hydroelectric and CAES, though many modular storage units could be co-located at a central site and operated as one resource. Modular/distributed storage peakers could be installed at or near loads if locational benefits or other advantages related to distributed deployment are significant.

The primary benefit is related to the reduced need for generation equipment. An important incremental benefit is the value of the time-shifted energy. Several other circumstance-specific benefits could also accrue. Storage used as a peaker during peak demand periods could also provide ancillary services during times when demand is lower. Depending on circumstances (primarily location), the storage peaker could provide potentially significant benefits related to (1) transmission congestion relief, (2) T&D upgrade deferral or life extension, (3) local voltage support and power factor correction, and (4) improved electric service reliability or power quality. If the peaker is modular and distributed, then the value may be somewhat higher.

Depending on the type of fuel and generation used for charging of storage used as a peaker, net fuel use and/or emissions associated with storage peakers may be comparable or perhaps even lower than those for CT peakers [7]. So, depending on the circumstances, a modular distributed storage peaker could provide several benefits such that the total exceeds that of a CT peaker significantly [8].

***Conventional and Fast Area Regulation and Frequency Response.*** Three individual storage benefits can be significant enough to exceed storage cost: (1) conventional area regulation, (2) fast area regulation, and (3) frequency response. These benefits may also be elements of other value propositions. There are at least three key reasons for this. First, in a growing number of regional grid control areas market-based prices for area regulation exist. Second, storage used for area regulation can effectively provide two services simultaneously: up regulation and down regulation. Third, given most storage types' fast or even very fast ramp rate—compared to thermal generation that is used for most area regulation—fast storage used for area regulation is more effective than generation-based area regulation and it can provide frequency response.

***Merchant Storage.*** Perhaps the most attractive value proposition for large-scale storage is what might be called the "merchant value proposition." It involves continuous optimization among cost and possible benefits (real options, described in the subsection Energy Storage Benefits Introduction).

Key benefits include (1) renewable and/or wholesale electric energy time-shift, (b) electric supply capacity, and (c) most or all ancillary services. Depending on the location, other potentially significant benefits could include transmission support and/or transmission congestion relief. The optimal combination of benefits varies depending on several criteria, including: time-of-day, season, location, the available mix of electric supply resources, and longer term electricity demand trends.

Two variations of the merchant value proposition are

1. The plant owner participates in the wholesale market for energy, electric supply capacity, ancillary services and in the future merchant providers might be enabled to provide-related transmission service(s).
2. The plant owner establishes a contract with a utility or other third party to provide specific services at specific times (generically referred to as a power purchase agreement or PPA).

One attractive merchant storage value proposition involves aggregation of distributed storage—and possibly distributed generation and DR—into a block of electric resources that could be operated like a large-scale merchant plant. Benefits include those for the large-scale merchant plant plus other locational benefits.

Aggregation—of DR resources—is already taking place. Further, the smart grid is expected to include many of the protocols and the infrastructure needed to coordinate and control many diverse and dispersed resources.

***Baseload Renewables "Capacity Shift."*** Storage could be used in conjunction with baseload renewable energy resources—those fueled by geothermal or biomass or other biofuel—to increase the generators' value. In that case, storage is charged at night (using energy from the baseload RE generation) so that it can be discharged later when the capacity and energy are much more valuable. The primary benefit is reduced need for conventional peaking generation capacity. That benefit is coupled with the enhanced value of the energy that is time-shifted. If the storage is distributed, then other locational benefits could also apply.

***Variable Renewables Generation Operational Integration.*** Another application may involve wind "smoothing"—using a shorter duration and fast response storage unit to smooth the minute to minute variability in the wind and optimizing the transmission and ramp rate requirement for conventional generators.

***Bulk Wind Generation to Distributed Storage.*** Integration of wind generation into the electric supply and transmission systems poses some well-characterized challenges. A significant portion of wind generation occurs at night, when the value is low. Also, electric energy produced by wind generation at night contributes to an increasing number of hours during which "negative prices" prevail (i.e., electricity supply exceeds demand and output from the generation that is on-line cannot be reduced without significant penalty).

One attractive response is to use distributed storage—located electrically downstream from constrained or congested transmission corridors/equipment—in conjunction with large central wind farms. The storage is charged at night when (1) wind generation is operating, (2) the value and price of electric energy is low, (3) transmission capacity loading is light, and (4) T&D $I^2R$ energy losses are lowest. This value proposition could involve a bilateral contract between (1) the wind generation owner and (2) the storage owner or aggregated distributed merchant storage.

***Storage in Lieu of Transmission Capacity.*** Similar to storage for T&D upgrade deferral and life extension, storage could be deployed, long term, in lieu of additional transmission capacity. Consider transmission enhancements needed to (1 ) meet new load on the margin and/or (2) to address NERC reliability violations. A helpful example of the latter is a load pocket that is served by two subtransmission lines, neither of which can serve more than 90% of the peak demand in the load pocket. So, if either of the lines is out of service there may be a reliability violation. The normal response would be to upgrade both subtransmission circuits. Another alternative is to install enough storage to reliably serve about 10% of the peak demand. Several other important benefits may also accrue, especially for electric supply (energy time-shift and capacity) and increased asset utilization.

***T&D Upgrade Deferral or Life Extension.*** This application involves use of storage to either (1) defer or avoid the need for a T&D equipment upgrade or (2) to extend the life of existing T&D equipment.

The benefit for this application is especially compelling because it can be quite significant. Adding a small amount of storage on the margin can (1) delay the need for a significant replacement and/or "lump" addition to T&D capacity or (2) reduce loading on existing equipment such that the equipment's life is extended.

In most situations, storage for this application is needed for relatively few hours per year because load only exceeds T&D equipment's load carrying capacity during the most extreme peaks. Storage for this application may not have to be discharged at all if demand does not grow as expected or if load does not exceed the equipment's rating (usually due to mild weather leading to modest air conditioning use).

In most cases, it eventually becomes less expensive to make upgrades because as peak demand grows so does the amount of storage needed. So the benefit—per kW of storage needed—often diminishes rapidly after just a few years. However, the storage can continue to deliver other benefits after T&D upgrades are made. If the storage is transportable, it can be moved to another location for similar and/or other benefits. If the storage cannot be moved, it can be used for many other benefits as characterized just below.

T&D deferrals and life extensions can provide an important primary benefit for a wide spectrum of possible value propositions. First, consider that the value of the benefit can be hundreds of dollars per kW of storage installed, for one, two or even several years. The same storage system can also be used to provide most of the benefits characterized herein, circumstances permitting. Of special note are on-peak electric supply capacity and energy because peak loading on T&D equipment is likely to be coincident with system-wide peak demand periods. Other possibly important benefits include (1) local voltage support, (2) improved power quality and reliability, (3) reduced $I^2R$ losses, and (4) dynamic operating benefits [9].

***Distributed Grid-Connected Photovoltaics Integration.*** Distributed grid-connected PV will play an increasingly important role as an electricity generation resource. On-site and/or nearby storage used in conjunction with that PV could (1) address challenges related to extensive deployment of rooftop PV and (2) provide some important benefits that are not possible with PV alone.

Consider three challenges that storage can address. First, an especially important PV-related operational challenge is output variability caused by (1) the daily variation of solar energy and (2) rapid, dramatic, short-duration output variations due to passing clouds—ramping. Storage can "fill-in" when PV output falls off, so output is nearly constant. Second, when distributed PV's output exceeds local demand and/or exceeds the distribution equipment's load carrying capacity, storage can absorb the excess and/or enable electric vehicle charging using locally generated electric energy. Third, a non-trivial portion of PV energy output occurs when the value of the energy is relatively low (e.g., mornings and weekends). In fact, many PV systems' output is not especially well-correlated with the peak loading. Electricity generated during those times could be stored for use when it is more valuable.

Beyond addressing those specific challenges, by virtue of its distributed deployment, numerous storage benefit combinations (i.e., value propositions) are possible. The benefit possibilities are even greater if the PV-storage system is part of an aggregated distributed resource.

One of the most notable benefits is related to the aforementioned preference for constant output from the PV-storage system: Renewables generation capacity firming or demand charge management. Also, energy discharged from the storage can provide an energy time-shift benefit or time-of-use (TOU) energy cost management benefit. Depending on location, the PV-storage system may reduce transmission congestion and/or allow for T&D upgrade deferral. The PV-storage system could also improve local electric service reliability and/or power quality related impacts (primarily, negative effects on voltage and introduction of current harmonics).

Note also some technical synergies between PV and modular storage, which means that some costs and some equipment and hardware can be shared (e.g., design and construction-related costs). Especially notable are synergies that allow sharing of power conditioning equipment (with modest additional cost). The PV-storage system could also including sharing of equipment and costs related to wiring, communication and controls and utility interconnection.

***Community Energy Storage.*** Community Energy Storage (CES) is a concept that was developed by American Electric Power Corporation—a large U.S. electric utility—as a unifying theme for small battery systems. CES installations are located near residential customers; they are rated at a few tens of kW and have 2 to 3 hours of discharge duration. CES also includes state-of-the-art communications and controls. Figure 10-35 shows an example of CES.

**FIGURE 10-35** Community energy storage as utility infrastructure. (*American Electric Power.*)

CES provides several benefits, especially those for electric supply (energy and capacity), T&D deferral, increased reliability, and distributed RE generation integration.

A novel facet of the CES concept is that special attention was given to making the CES facilities' aesthetics appear like that of conventional utility infrastructure, as shown in the photo above. And CES could serve as an important element of smart grid.

***Distributed Storage for On-Peak EV Charging.*** EVs will be an important load for utilities in the future. Clearly, most charging of EVs is expected to be done during off-peak hours (e.g., at night and on weekends) when electric demand and energy prices are low. However, at least some EV charging will occur during peak demand times. If charging occurs in parts of the grid that do not have sufficient capacity then two basic alternatives are (1) upgrade the conventional infrastructure to add needed capacity (possibly including generation and/or transmission and/or distribution) or (2) install distributed storage, electrically downstream from congestion points, to serve the added demand on-peak locally.

***Microgrids and Islanded Electrical Systems.*** *Microgrids* and islanded electrical systems involve either (1) remote/isolated power systems or (2) a portion of a utility's distribution system. In either case, they must be able to operate autonomously.

Some refer to microgrids as "peer-to-peer power networks" because they enable use of a variety of resources (including distributed resources) that can be co-optimized and shared within a network of loads ranging from residences to high-use buildings, such as hospitals, offices, industrial complexes, and data centers. Loads within a microgrid remain connected to the utility "macro" grid when practical. However, generation and storage located within the microgrid can (1) generate power to serve loads locally which reduces reliance on the macro grid, (2) provide power to the macro grid, or (3) allow for continuous operation within the microgrid during macro grid outages.

Some microgrid examples include

- Isolated/remote communities that are not connected to a central grid.
- Actual islands with a small electrical system with small resources.
- Military "forward bases."
- Parts of a utility's grid that becomes electrically isolated from the greater electric system.

Storage is likely to be a key element of many microgrids and islanded electrical systems. In general, storage serves as the system hub to (1) reconcile minute-to-minute and diurnal differences between supply and demand and (2) to coordinate and optimize performance of individual elements of the islanded system, including dispatchable generation and load and variable power sources such as solar and wind generation. That optimizes the benefits from variable power sources while reducing part load and variable operation of dispatchable generation (leading to more efficient and less polluting operations). Like storage in larger electric systems, storage in a microgrid or islanded system offsets the need for peaking generation and load reduction.

***Uninterruptible Power Supply.*** The uninterruptible power supply (UPS) value proposition is well established and well understood. And, there is significant ongoing demand for and use of UPSs. The value proposition involves reduced/avoided costs related to poor power quality and/or unacceptable electric service reliability. Of particular note are (1) need to maintain electric service to critical safety-related loads such as emergency lighting and medical equipment and (2) cost avoided for lost productivity and/or damaged materials or product and/or electricity-using equipment damage.

So, in general, UPSs provide power to loads for allow enough time for an orderly shutdown or for a transition to another longer duration auxiliary power source such as diesel fueled back-up generation. UPSs can also compensate for poor power quality that can shut down operations and damage electricity using equipment.

The discharge duration of most smaller (e.g., under desk) UPSs is a relatively short 0.5 to 15 minutes needed to enable orderly shutdown while systems designed to provide facility-wide power may have discharge durations of many seconds—needed for smooth transfer to generation-related backup and/or for load shedding—to many minutes to a few hours if the UPS will serve load for extended periods.

An especially attractive facet of this value proposition is the prospect of using the same storage for UPS service *plus* end-user bill management.

***End-User Bill Management.*** Electricity end-users can use storage to manage (i.e., reduce) the cost associated with electric service. The benefit—reduced cost for electric service—can be significant. Electricity end-users who qualify for a tariff involving TOU pricing for energy can purchase and store energy at night when the TOU energy price is low, for use when the TOU price is high (normally during weekday afternoons). Furthermore, electricity end-users who qualify for a tariff that includes both TOU energy pricing and demand charges can purchase and store energy at night when the TOU price and demand charge are low, and the energy can be discharged when the TOU price and demand charge are high (typically during week-day afternoons, especially during summer).

The same storage system could be used to time-shift energy from on-site RE generation, especially PV, and it could provide the same function as an uninterruptible power supply. As utility market-mechanisms and pricing evolve and become richer, the same storage could be aggregated and operated in concert with other distributed resources.

***Reduce On-Peak Air Conditioning Loads.*** Air conditioning (A/C) is one of the most expensive loads for utilities to serve. Perhaps most importantly, the generation and/or transmission and/or distribution capacity needed to serve small A/C is used relatively little—reflecting poor asset utilization. That is important because the cost for the capacity on the margin (needed to serve A/C) is spread across relatively few units of energy (delivered), which can have a significant marginal effect on the cost-of-service.

There are other reasons that A/C is such an expensive load to serve. It tends to occur when demand is already high, therefore requiring use of the most expensive, least efficient, and most polluting electricity generated (e.g., by simple cycle combustion turbines). Also, T&D $I^2R$ losses are most significant when grid equipment is most heavily loaded and when ambient temperatures are high. It is also notable that small A/C compressor motors can be the worst offender during grid-wide voltage emergencies [10].

Distributed storage used to serve or to offset A/C load, locally, could yield significant benefits, whether that storage is owned by the utility or their customers. Most savings would be for reduced GT&D capacity-related needs, especially if electric supply shortages or transmission congestion exist. Reduced on-peak energy use is also notable. Other potentially important benefits include increased asset utilization (of baseload or intermediate duty generation and existing T&D equipment) and reduced T&D $I^2R$ energy losses. The same storage could also offset the need for some ancillary services, especially voltage support, and it could improve local electric service reliability and power quality.

There are two notable versions of this use of storage. One is use of *electric* storage—charged when demand is low—to provide power directly to A/C systems in lieu of real-time use of power from the grid, during peak demand periods. The other approach makes use of *cold* storage (TES) whereby electricity is used to create cold (i.e., as chilled water or ice) at night. That cold is used when A/C is needed, which reduces or eliminates the need to generate that cold real-time, during peak demand periods, when electric energy and power are expensive. Furthermore, if operating A/C equipment at night—when ambient temperatures are relatively low—cold production is relatively more efficient than production that occurs when/as the cold is needed.

## 10.6.6 References

1. Ribeiro, Paulo F., Johnson, Brian K., Crow, Mariesa L., Arsoy, A., Liu Yilu. *Energy Storage Systems for Advanced Power Applications.* Proceedings of the IEEE, vol. 89, No. 12, December 2001.
2. Akhil, Abbas, Georgianne Huff, Aileen Currier, Benjamin Jaun, Dan Rastler, Stella Chen, Andrew Cotter, Dale Bradshaw, William Gauntlett, *DOE/EPRI Electricity Storage Handbook in Collaboration with NRECA*, Sandia National Laboratories Report SAND2015-1002, February 2015.
3. Electric Power Research Institute. *Reassessment of Superconducting Magnetic Energy Storage (SMES) Transmission System Benefits.* Electric Power Research Institute Report #1006795. March 2002.
4. Torre, William V., DeSteese, J. G., Dagle, J. E., *Evaluation of Superconducting Magnetic Energy Storage for San Diego Gas and Electric.* Electric Power Research Institute Report #106286 2572-14. August 1997.
5. Mears, D., Gotschall, H., *EPRI-DOE Handbook of Energy Storage for Transmission and Distribution Applications.* Electric Power Research Institute Report #1001834. December 2003.
6. Schainker, Dr. Robert, et al., "Emerging Strategies for Energy Storage." *EPRI Journal,* vol. 14, No. 5 (July/August 1989), pp. 12.

7. Eyer, J., Corey, G., *Energy Storage for the Electricity Grid: Benefits and Market Potential Assessment Guide.* Sandia National Laboratories, Energy Storage Program, Office of Electric Transmission and Distribution, U.S. Department of Energy. Sandia National Laboratories Report # SAND2010-0815. February 2010.
8. Lin, J., Damato, G., *Energy Storage—A Cheaper and Cleaner Alternative to Natural Gas-Fired Peaker Plants.* California Energy Storage Alliance Whitepaper. February 8, 2011.
9. Eyer, J., *Electric Utility Transmission and Distribution Upgrade Deferral Benefits from Modular Electricity Storage.* Sandia National Laboratories, Energy Storage Program, Office of Electric Transmission and Distribution, U.S. Department of Energy. Sandia National Laboratories Report # SAND2009-4070. June 2009.
10. Kirby, B., Hirst, E., *Ancillary Service Details: Voltage Control.* Oak Ridge National Laboratory, Energy Division. Sponsored by The National Regulatory Research Institute. Oak Ridge National Laboratory Report #ORNL/CON-453. December 1997.

# 11 SUBSTATIONS

Diane Watkins

*Manager, Substation Field Engineering, Xcel Energy, Denver, Colorado; Senior Member, IEEE*

George W. Becker

*Senior Substation Engineer, POWER Engineers, Inc., Fort Mill, South Carolina; Senior Member, IEEE*



## 11.1 AIR-INSULATED SUBSTATIONS

*BY DIANE WATKINS*

### 11.1.1 Function of Substations

***Transmission and Distribution Systems.*** In large, modern ac power systems, the transmission and distribution systems function to deliver bulk power from generating sources to users at the load centers. Transmission systems generally include generation switchyards, interconnecting transmission lines, autotransformers, switching stations, and step-down transformers. Distribution systems include primary distribution lines or networks, transformer banks, and secondary lines or networks, all of which serve the load area, and an increasing amount of distributed generation (DG). A switching station is a type of substation where no transformers are utilized. A switchyard is a type of substation located adjacent to a generator step-up unit (GSU) transformer. Transition stations, whose sole purpose is to change from overhead to underground design, are not generally considered substations. Some substations contain dc facilities that are also air-insulated and converted to ac via high-power electronics.

### 11.1.2 Design Objectives

As an integral part of the transmission or distribution systems, the substation functions as a connection and switching point for generation sources, transmission or subtransmission lines, voltage or reactive power support, distribution feeders, and step-up and step-down transformers. The design objective for the substation is to provide maximum reliability and flexibility for system and environmental requirements, while minimizing total lifecycle costs.

***Voltage Levels.*** The selection of optimal system voltage levels depends on the load to be served and the distance between the generation source and the load. Many large power plants are located great distances from the load centers to address energy sources or fuel supplies, cooling methods, site costs and availability, and environmental concerns. For these reasons, the use of transmission voltages in North America range from 69 to 765 kV (some 1100 to 1200 kV have been installed in Russia, India, and China). Common voltage classes used in the United States for major substations include 69, 115, 138, 161, and 230 kV (considered *high voltage* or *HV class*) and 345, 500, and 765 kV (considered *extra-high voltage* or *EHV class*). Voltages of 1100 through 1500 kV are referred to as *ultra-high voltage* or *UHV class*. Distribution system substations operate at secondary voltage levels from 4 to 46 kV.

***Design Considerations.*** Many factors influence the selection of the proper type of substation for a given application, such as voltage level, load capacity, environmental considerations, reliability, site space limitations, and transmission-line right-of-way requirements. While also considering the installation cost of equipment, labor, and land, every effort must be made to select a substation type that will satisfy all requirements at minimum costs. The major substation costs are reflected in the number of power transformers, circuit breakers, and disconnecting switches, buswork and their associated structures and foundations. Therefore, the bus layout and switching arrangement selected will determine the number of devices that are required and in turn the overall cost. The choice of insulation levels and coordination practices also affects cost, especially at EHV. A drop of one level in basic insulation level (BIL) can significantly reduce the cost of major electrical equipment. A careful analysis of alternative switching schemes is essential and can result in considerable savings by choosing the minimum equipment necessary to satisfy system requirements.

A number of factors must be considered in the selection of bus layouts and switching arrangements for a substation to meet system and station requirements. A substation must be safe, reliable, economical, and as simple in design as possible. The design also should provide for further expansion, flexibility of operation and clearances, maintenance access, equipment removal, and low maintenance costs.

The physical orientation of the transmission-line routes often dictates the substation's location, orientation, and bus arrangement. This requires that the selected site allow for a convenient arrangement of the lines to be accomplished, to minimize line crossings or rearrangement.

For reliability, the substation design should reduce the probability of a total substation outage caused by faults or equipment failure and should permit rapid restoration of service after a fault or failure occurs. The layout also should consider how future additions and extensions can be accomplished without interrupting service.

Other factors should be considered as well, as they can impact reliability and overall cost of the facility; factors include fire protection, fault duty, arc flash, animal deterrents, oil containment, and community acceptance. Severe damage can result from a fire, so it is important to recognize the hazards from oil-filled equipment or other combustible material, especially in conjunction with surface grade and adjacent equipment or controls. Fire protection measures can mitigate or lessen the hazards regarding many areas of concern, including life safety, detection, signaling, spacing, separation, material selection, emergency access, and active suppression.

As the amount of load and generation increases on the interconnected grid, the level of fault duty likely also increases, especially as substation busses become larger and more interconnected.

It is important therefore to validate that existing installations are within fault level design parameters, and to provide significant design margin for new installations which allow future fault duty increases. Fault duty also has a substantial impact on arc flash levels, in addition to fault clearing time. To improve safety, increasing attention is being directed toward calculating and mitigating arc flash levels, particularly for utility workers. It is important therefore to consider work proximity during operations, maintenance, commissioning, and even safe distances for general visitors. Arc flash hazards can be mitigated by assessing hazard risk categories (HRC), arc flash boundaries (AFB), minimum working distance (MWD), and flame resistant (FR)-rated clothing and personal protective equipment (PPE).

Open-air designs are susceptible to adverse impacts of animal contacts at distribution voltage levels (separation distances), and depending on the size of the animal, can affect higher voltage levels. Therefore, it is important to consider the local animal types and migrations, clearance distance requirements, and applicable mitigation methods in the substation design. Common animal types include squirrels, birds, cats, snakes, and raccoons. Barriers are one form of mitigation and can involve small-mesh fencing, climbing guards, solid walls, or even electrified fencing. Certain mitigations are effective with some animal types, yet possibly not others. It is important to consider bird nesting habits as well, and utilize equipment and structural designs that discourage nesting locations. Deterrents are another mitigation approach, and can include ultrasonic or other noise devices, chemical repellents, plastic predators, spined perches, or poisons. And finally, there are numerous insulating or isolating products available at distribution voltages, such as insulating tape, pre-formed insulating covers, rigid insulating discs, spray-on coatings, and more recently, insulated structure members.

For sites in close proximity to navigable waters or designated wetlands, the containment of large oil spills is integral for environmental regulations and minimizing cleanup costs. Containment criteria can include discharge volume and controls, waterway proximity, surface grade and treatment, maintenance access, and underlying soil.

In addition to oil containment, other aspects of community acceptance include design considerations for noise, aesthetics, lighting, electromagnetic fields, adjacent uses, and hazardous materials.

IEEE has numerous guidelines addressing these topics in more detail, which are listed in the Reference section.

***Bus Schemes.*** The substation design or scheme selected determines the electrical and physical arrangement of the switching equipment. Different bus schemes can be selected as emphasis is shifted between the factors of safety, reliability, economy, and simplicity dictated by the function and importance of the substation.

The substation bus schemes used most often are

1. Single bus
2. Main and transfer bus
3. Double bus, single breaker
4. Double bus, double breaker
5. Ring bus
6. Breaker and a half

Some of these schemes may be modified by the addition of bus-tie breakers, bus sectionalizing devices, breaker bypass facilities, and extra transfer buses. Figures 11-1 to 11-6 show one-line diagrams for the typical schemes listed above.

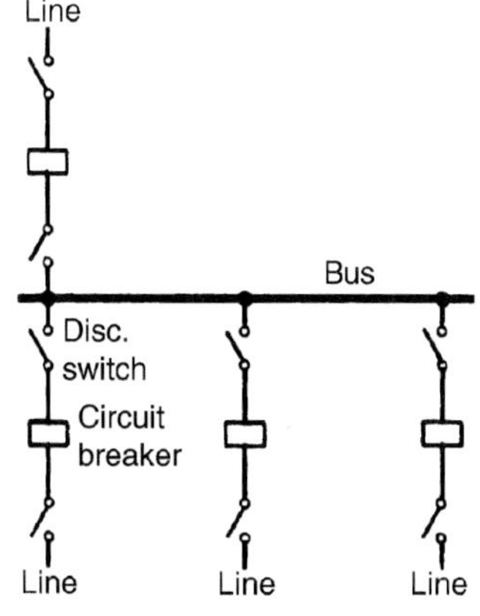


**FIGURE 11-1** Single bus.

***Single Bus.*** The single-bus scheme (Fig. 11-1) is not normally used for major substations. Dependence on one bus can result in a serious outage in the event of breaker or bus failure without the use of mobile equipment. The station must be deenergized in order to carry out bus maintenance or add bus extensions. Although the protective relaying is relatively simple for this scheme, the single-bus scheme is considered inflexible and subject to complete outages of extended duration.

***Main and Transfer Bus.*** The main- and transfer-bus scheme (Fig. 11-2) adds a transfer bus to the single-bus scheme. An extra bus-tie circuit breaker is provided to tie the main and transfer buses together.

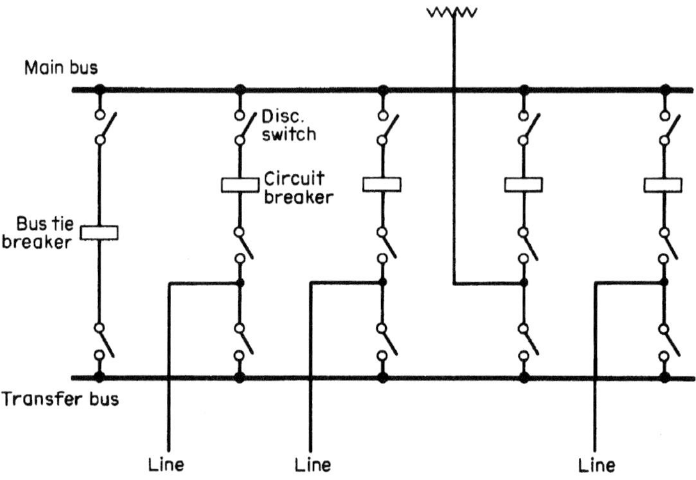


**FIGURE 11-2** Main and transfer bus.

When a circuit breaker is removed from service for maintenance, the bus-tie circuit breaker is used to keep that circuit energized. Unless the protective relays are also transferred, the bus-tie relaying must be capable of protecting transmission lines or generation sources. This is considered rather unsatisfactory because relaying selectivity is poor.

A satisfactory alternative consists of connecting the line and bus relaying to current transformers located on the lines rather than on the breakers. For this arrangement, line and bus relaying need not be transferred when a circuit breaker is taken out of service for maintenance, with the bus-tie breaker used to keep the circuit energized.

If the main bus is ever taken out of service for maintenance, no circuit breakers remain to protect any of the feeder circuits. Failure of any breaker or failure of the main bus can cause complete loss of service of the station.

Due to its relative complexity, disconnect-switch operation with the main- and transfer-bus scheme can lead to operator error and a possible outage. Although this scheme is lower in installation cost and enjoys some popularity, it may not provide as high a degree of reliability and flexibility as required.

***Double Bus, Single Breaker.*** This scheme uses two main buses, and each circuit includes two bus selector disconnect switches. A bus-tie circuit (Fig. 11-3) connects to the two main buses and, when closed, allows transfer of a feeder from one bus to the other bus without deenergizing the feeder circuit by operating the bus selector disconnect switches. The circuits may all operate from either the no. 1 or no. 2 main bus, or half the circuits may be operated off either bus. In the first case, the station will be out of service for bus or breaker failure. In the second case, half the circuits will be lost for bus or breaker failure.

In some cases circuits operate from both the no. 1 and no. 2 bus, and the bus-tie breaker is normally operated closed. For this type of operation, a very selective bus-protective relaying scheme is required to prevent complete loss of the station for a fault on either bus. Disconnect-switch operation becomes quite involved, with increased possibility of operator error, or outage. The double-bus, single-breaker scheme is relatively restricted in reliability and is not normally used for larger substations.

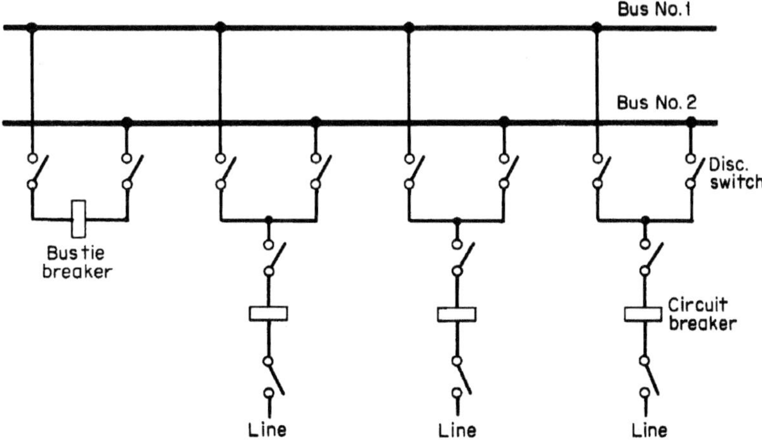


**FIGURE 11-3** Double bus, single breaker.

***Double Bus, Double Breaker.*** The double-bus, double-breaker scheme (Fig. 11-4) requires two circuit breakers for each feeder circuit. Normally, each circuit is connected to both buses. In some cases, half the circuits operate on each bus. For these cases, a bus or breaker failure would cause loss of only half the circuits, which could be rapidly corrected through switching. The physical location of the two main buses must be selected in relation to each other to minimize the possibility of faults spreading to both buses. The use of two breakers per circuit makes this scheme more expensive; however, it does represent a high degree of reliability.

***Ring Bus.*** In the ring-bus scheme (Fig. 11-5), the breakers are arranged in a ring with circuits connected between breakers. There are the same number of circuits as there are breakers. During normal

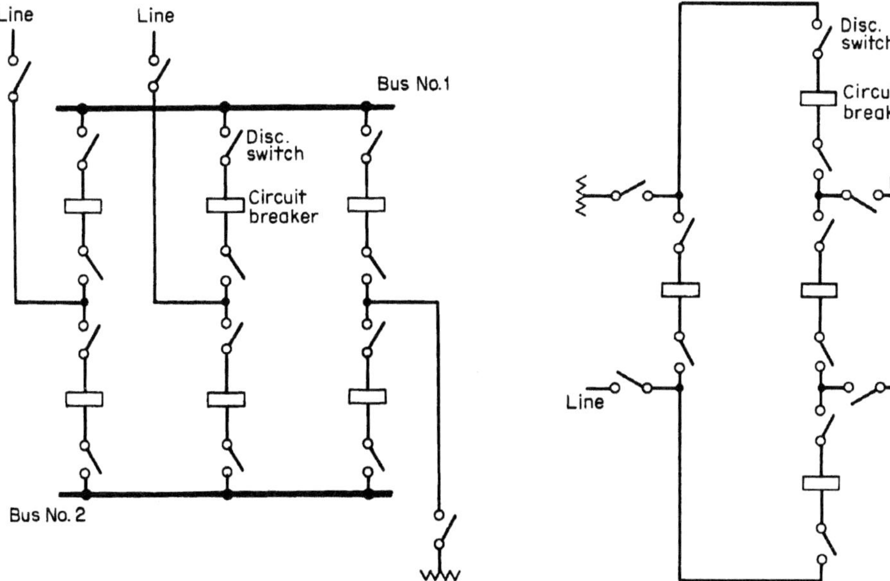


**FIGURE 11-4** Double bus, double breaker.

**FIGURE 11-5** Ring bus.

operation, all breakers are closed. For a circuit fault, two breakers are tripped, and in the event that one of the breakers fails to operate to clear the fault, an additional circuit will be tripped by operation of breaker-failure backup relays. During breaker maintenance, the ring is broken, but all lines remain in service.

Ideally the circuits connected to the ring are arranged so that sources are alternated with loads. For an extended circuit outage, the line-disconnect switch may be opened, and the ring can be closed. No changes to protective relays are required for any of the various operating conditions or during maintenance.

The ring-bus scheme is relatively economical in cost, has good reliability, is flexible, and is normally considered suitable for substations up to a limit of six circuits. Protective relaying and automatic reclosing are more complex than for previously described schemes. It is common practice to build major substations initially as a ring bus; for more than five outgoing circuits, the ring bus can then be converted to the breaker-and-a-half scheme, based on effective initial ring bus arrangement and relay schemes.

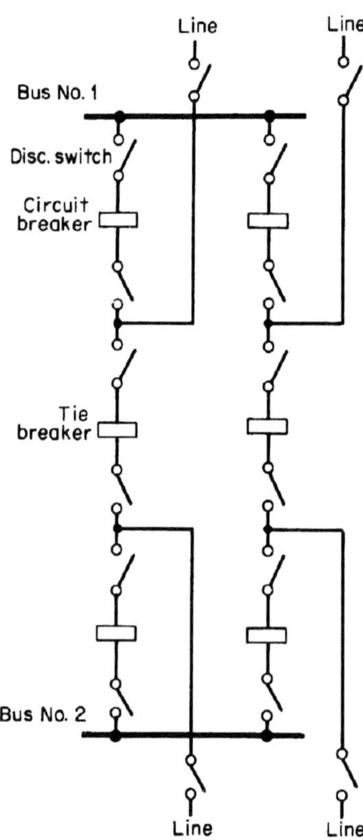


**FIGURE 11-6** Breaker-and-a-half scheme.

***Breaker and a Half.*** The breaker-and-a half scheme (Fig. 11-6), sometimes called the *three-switch scheme*, has three breakers in series between two main buses. Two circuits are connected between the three breakers, hence the term *breaker and a half.* This pattern is repeated along the main buses so that one and a half breakers are used for each circuit.

Under normal operating conditions, all breakers are closed, and both buses are energized. A circuit is tripped by opening the two associated circuit breakers. Tie-breaker failure will trip one additional circuit, but no additional circuit is lost if a line trip involves failure of a bus breaker. Either bus may be taken out of service at any time with no loss of service. With sources connected opposite to loads, it is possible to operate with both buses out of service. Breaker maintenance can be done with no loss of service, no relay changes, and simple operation of the breaker disconnects.

The breaker-and-a-half arrangement is more expensive than the other schemes, with the exception of the double-breaker, double-bus scheme, and protective relaying, and automatic reclosing schemes are more complex than for other schemes. However, the breaker-and-a-half scheme is superior in flexibility, reliability, and safety. Also it can be space efficient if aligned with incoming and outgoing lines or transformers.

## 11.1.3 Reliability Comparisons

The various schemes have been compared to emphasize their advantages and disadvantages, based upon the economic justification of a particular degree of reliability. Determining the degree of reliability involves an appraisal of current and anticipated operating conditions and the continuity of service required by the load to be served. Table 11-1 contains a summary of the comparison of switching schemes to show advantages and disadvantages.

**TABLE 11-1** Summary of Comparison of Switching Schemes

| Switching scheme | Advantages | Disadvantages |
|---|---|---|
| 1. Single bus | 1. Lowest cost and amount of equipment. | 1. Failure of bus or any circuit breaker results in shutdown of entire substation.<br>2. Difficult to do any maintenance.<br>3. Bus cannot be extended without completely deenergizing substation.<br>4. Can be used only where loads can be interrupted or have other supply arrangements. |
| 2. Main and transfer | 1. Low initial and ultimate cost.<br>2. Any breaker can be taken out of service for maintenance without impacting circuit.<br>3. Potential devices may be used on the main bus for relaying. | 1. Requires one extra breaker for the bus tie.<br>2. Switching is somewhat complicated when maintaining a breaker.<br>3. Failure of bus or any circuit breaker results in shutdown of entire substation. |
| 3. Double bus, single breaker | 1. Permits some flexibility with two operating buses.<br>2. Either main bus may be isolated for maintenance.<br>3. Circuit can be transferred readily from one bus to the other by use of bus-tie breaker and bus selector disconnect switches. | 1. One extra breaker is required for the bus tie.<br>2. Four switches are required per circuit.<br>3. Bus protection scheme may cause loss of substation if all circuits are connected to that bus.<br>4. High exposure to bus faults or breaker failure.<br>5. Line breaker failure takes all circuits connected to that bus out of service.<br>6. Bus-tie breaker failure takes entire substation out of service. |
| 4. Double bus, double breaker | 1. Each circuit has two dedicated breakers.<br>2. Has flexibility in permitting circuits to be connected to either bus.<br>3. Any breaker can be taken out of service for maintenance.<br>4. High reliability. | 1. Most expensive.<br>2. Would lose half of the circuits for breaker failure if circuits are not connected to both buses. |
| 5. Ring bus | 1. Low initial and ultimate cost.<br>2. Flexible operation for breaker maintenance.<br>3. Requires only one breaker per circuit.<br>4. Does not use main bus.<br>5. Each circuit is fed by two breakers.<br>6. All switching is done with breakers.<br>7. Each circuit has two dedicated breakers. | 1. If a fault occurs during a breaker maintenance period, the ring can be separated into two sections provided operating conditions can be met.<br>2. Automatic reclosing and protective relaying circuitry rather complex, yet no bus differential is required.<br>3. If a single set of relays is used, the circuit must be taken out of service to maintain the relays (common on all schemes).<br>4. Requires potential devices on all circuits since there is no definite potential reference point. These devices may be required in all cases for synchronizing, live line, or voltage indication.<br>5. Breaker failure during a fault on one of the circuits causes loss of one additional circuit owing to operation of breaker-failure relaying. |
| 6. Breaker and a half | 1. Most flexible operation.<br>2. High reliability.<br>3. Breaker failure of bus side breakers removes only one circuit from service.<br>4. All switching is done with breakers.<br>5. Simple operation; no disconnect switching required for normal operation; fewer switches.<br>6. Either main bus can be taken out of service at any time for maintenance.<br>7. Bus failure does not remove any circuits from service.<br>8. Each circuit has two dedicated breakers. | 1. 1½ breakers per circuit.<br>2. Relaying and automatic reclosing are somewhat involved since the middle breaker must be responsive to either of its associated circuits. |

***Note:*** Any breaker can be removed for maintenance without interrupting load.

### 11.1.4 Arrangements and Equipment

Once a determination of the switching scheme best suited for a particular substation application is made, it is necessary to consider the station arrangement and equipment that will satisfy the many physical requirements of the design. Available to the design engineer are the following:

1. Conventional outdoor air-insulated open-type bus-and-switch arrangement substations (using either a strain bus or rigid bus design)
2. Metal-clad or metal-enclosed substations
3. Gas (sulfur hexafluoride)–insulated substations
4. Hybrid approach

Outdoor open-type bus-and-switch arrangements generally are used because of their lower cost, but they are larger in overall physical size. Metal-clad substations generally are limited to 38 kV. Gas-insulated substations are generally the highest in cost but smallest in size.

***Substation Components.*** The electrical equipment in a typical substation can include the following:

Circuit breakers
Disconnecting switches
Grounding switches
Current transformers
Voltage transformers or capacitor voltage transformers
Coupling capacitors
Line traps
Surge arresters
Power transformers
Shunt reactors
Current-limiting reactors
Station buses and insulators
Grounding systems
Series capacitors
Shunt capacitors
High-power electronics, control buildings, batteries and station service, revenue metering, telecommunications

Modern circuit breakers are most commonly insulated with an inorganic, nonflammable gas called sulfur hexafluoride ($SF_6$). $SF_6$ is considered a greenhouse gas, and its emission may be regulated by certain agencies.

***Support Structures.*** In order to properly support, mount, and install the electrical equipment, structures made of steel, aluminum, wood, or concrete and associate foundations are required. Structures need to withstand operating loads, environmental requirements, and local soil types. The typical open-type substation requires strain structures to support the transmission-line conductors; support structures for disconnecting switches, current transformers, potential transformers, lightning arresters, and line traps, capacitor voltage transformers; and structures and supports for the strain and rigid buses in the station.

The design of supporting structures is affected by the phase spacings and ground clearances required, by the types of insulators, by the length and weight of buses and other equipment, and by wind and ice loading. For data on wind and ice loadings, see National Electric Safety Code®, IEEE Standard C2-2012, or latest edition. For required clearances and phase spacings, see IEEE Std 1427-2006 Table 3.

Other structural and concrete work required in the substation includes site selection and preparation, roads, control houses, manholes, conduits, ducts, drainage facilities, catch basins, oil containment, and fences.

### 11.1.5 Site Selection

Civil engineering work associated with the substation should be initiated as early as possible in order to ensure that the best available site is selected. This work includes a study of the topography and drainage patterns of the area together with a subsurface soil investigation. The information obtained from the subsurface soil investigation also will be used to determine the design of the substation foundations. For large substations or substations located in areas with poor soils, it may be necessary to obtain additional subsurface soil tests after final selection of the substation site has been made. The additional information should fully describe the quality of the soil at the site, since the data will be used to design equipment foundations. Other design considerations should be factored in, including space for construction laydown, access, physical security, noise, aesthetics, water supply, and adjacent land uses or utilities.

***Open-Bus Arrangement.*** An air-insulated, open-bus substation arrangement consists essentially of open-bus construction using either rigid- or strain-bus design such as the breaker-and-a-half arrangement shown in Fig. 11-7; the buses are arranged to run the length of the station and are located toward the outside of the station. The transmission-line exits cross over the main bus and

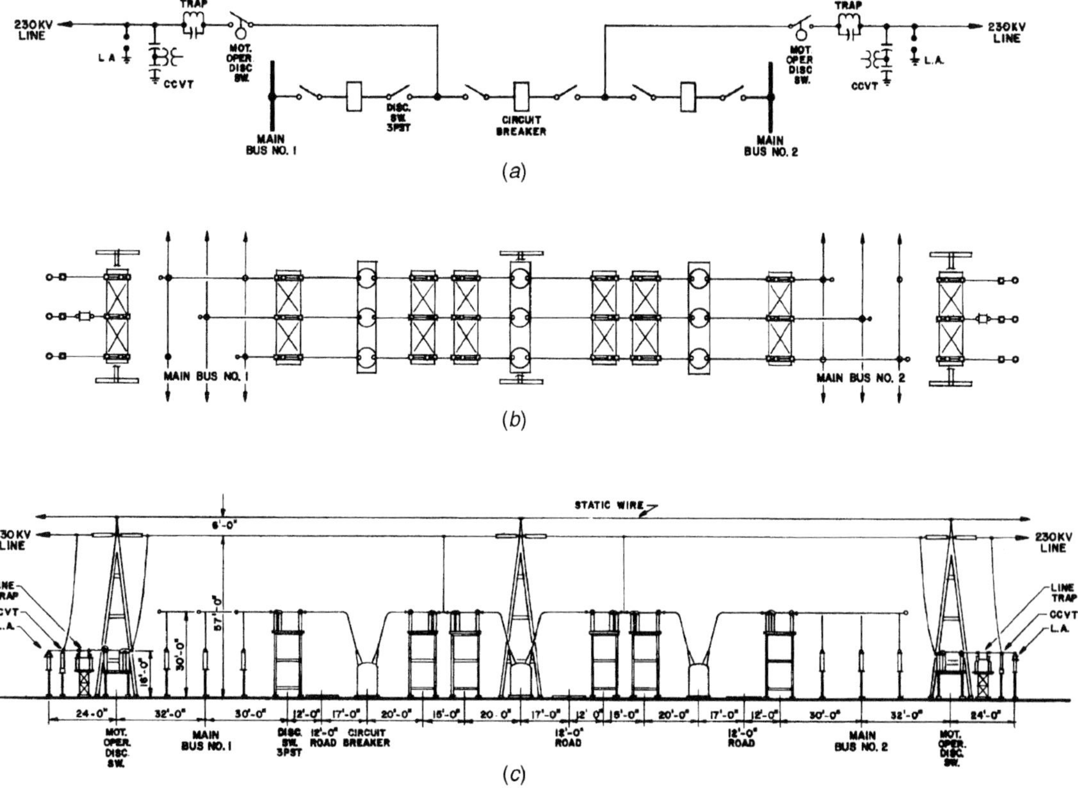


**FIGURE 11-7** Typical conventional substation layout, breaker-and-a-half scheme. (*a*) Main one-line diagram; (*b*) plan; (*c*) elevation.

are dead-ended on takeoff tower structures. The line drops into the bay (a common term for the segment of main bus where one circuit terminates) and connects to the disconnecting switches and circuit breakers.

Use of this arrangement requires three distinct levels of bus to make the necessary crossovers and connections to each substation bay. Typical dimensions of these levels at 230 kV are 16 ft for the main bus location above ground, 30 ft high for the bus spanning the bay, and 57 ft for the highest level of bus (see Fig. 11-7). An alternate consideration is to position the main bus at the 30 ft level, and the bus spanning the bay at 16 ft high.

This arrangement, in use since the mid-1920s and widely used by many electric utilities, has the advantage of requiring a minimum of land area per bay and relative ease of maintenance, and it is ideally suited to a transmission-line through-connection, where a substation must be inserted into a transmission line.

***Inverted Bus.*** An alternate arrangement is the inverted-bus, breaker-and-a-half scheme for EHV substations. A typical layout is outlined in Fig. 11-8. A one-line diagram of a station showing many variations of the inverted-bus scheme is presented in Fig. 11-9. With this arrangement, all outgoing circuit takeoff towers are located in the outer perimeter of the substation, eliminating the crossover of line or exit facilities. Main buses are located in the middle of the substation, with all disconnecting switches, circuit breakers, and bay equipment located outboard of the main buses. The end result of the inverted-bus arrangement presents a very low profile station with many advantages in areas where beauty and aesthetic qualities are a necessity for good public relations. The overall height of the highest bus in the 230-kV station just indicated reduces from a height of 57 ft above ground in the conventional arrangement to a height of only 30 ft above ground for the inverted-bus low-profile scheme.

***Proximity to Other Installations.*** Appropriate room should be provided to allow the perimeter fence or wall to be inset from the property line enough to limit the opportunity for (1) personnel to simultaneously touch the station fence and an adjacent fence or (2) equipment or stored items on adjacent property to provide elevated access for personnel to climb over the station fence. Where it is necessary in constrained locations for the station fence to be so close to another fence or other item that can be simultaneously contacted, a nonconducting fence section, barrier, or other mitigating system should be used. Where the station fence must be close enough to other facilities or areas where materials may be placed that would provide assistance in climbing the station fence, a taller fence section will be required.

NOTE: Part 1 of the National Electrical Safety Code (NESC) ANSI C2-2012 specifies clearances from station fences to energized parts within the station. Also see Sec. 11.1.7 in this handbook.

### 11.1.6 Substation Buses

Substation buses are an important part of the substation because they carry electric currents in a confined space. They must be carefully designed to have sufficient structural strength to withstand the maximum stresses that may be imposed on the conductors, and in turn on the supporting structures, due to short-circuit currents, high winds, ice, and seismic loadings.

During their early development, HV class substations were usually of the strain-bus design. The strain bus is similar to a transmission line and consists of a conductor such as ACSR (aluminum cable steel reinforced), copper, or high-strength aluminum alloy strung between substation structures. EHV substations normally use the rigid-bus approach and enjoy the advantage of low station profile and ease of maintenance and operation (see Fig. 11-8). The mixing of rigid- and strain-bus construction is normally employed in the conventional arrangement shown in Fig. 11-7. Here, the main buses use rigid-bus design, and the upper buses between transmission towers are of strain-bus design. A typical design at 765 kV uses a combination of both rigid and strain buses (Fig. 11-10).

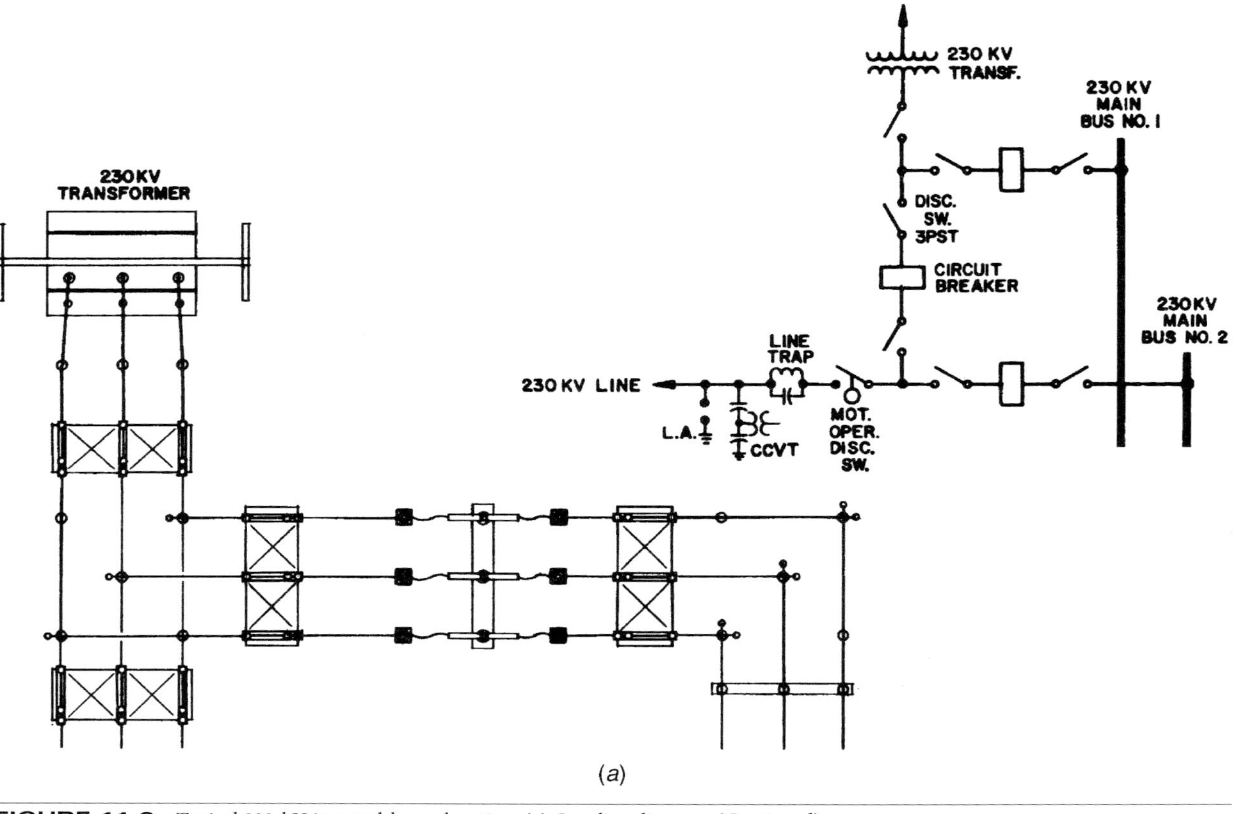


**FIGURE 11-8** Typical 230-kV inverted-bus substation. (*a*) One-line diagram. (*Continued*)

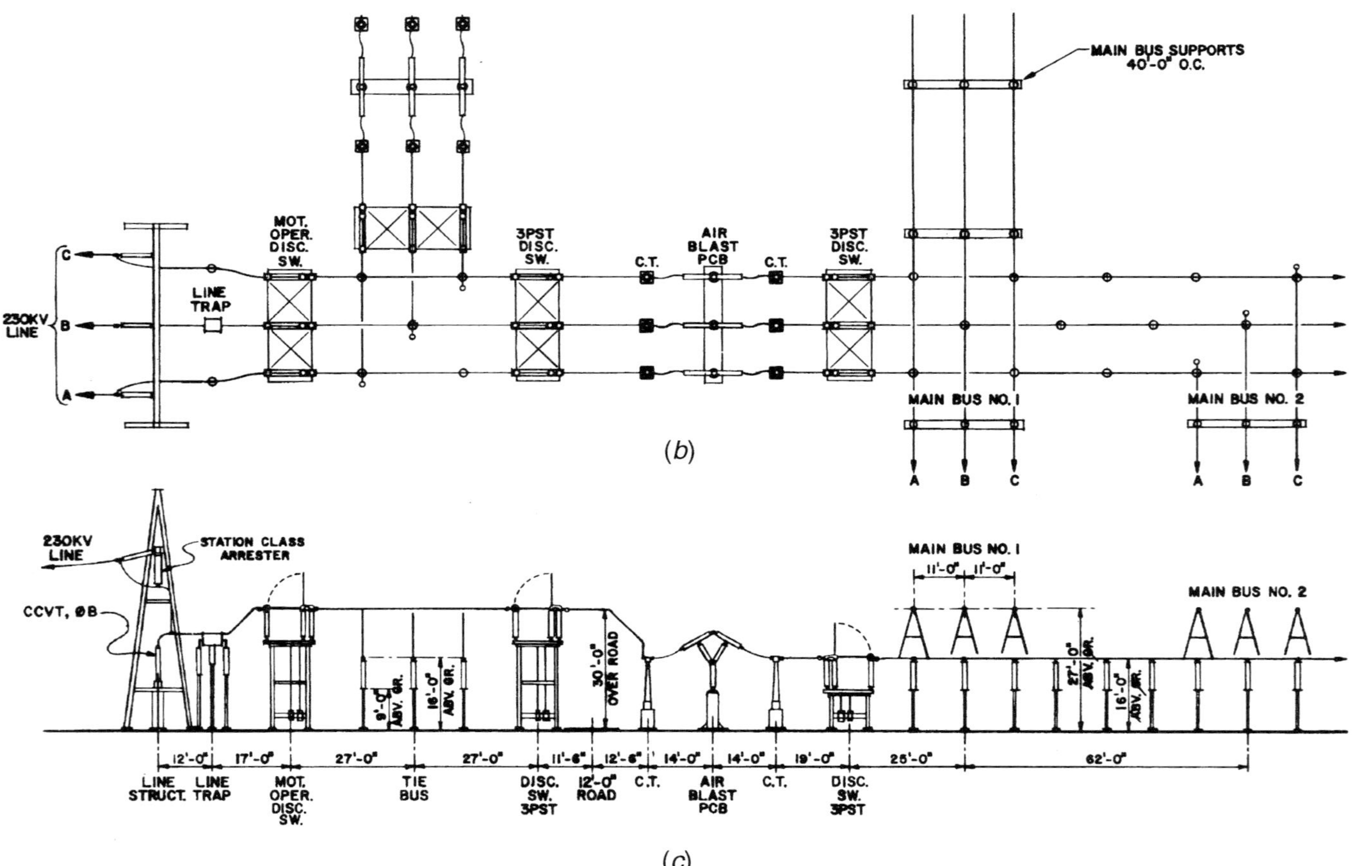


**FIGURE 11-8** Typical 230-kV inverted-bus substation. (*b*) plan; (*c*) elevation. (*Continued*)

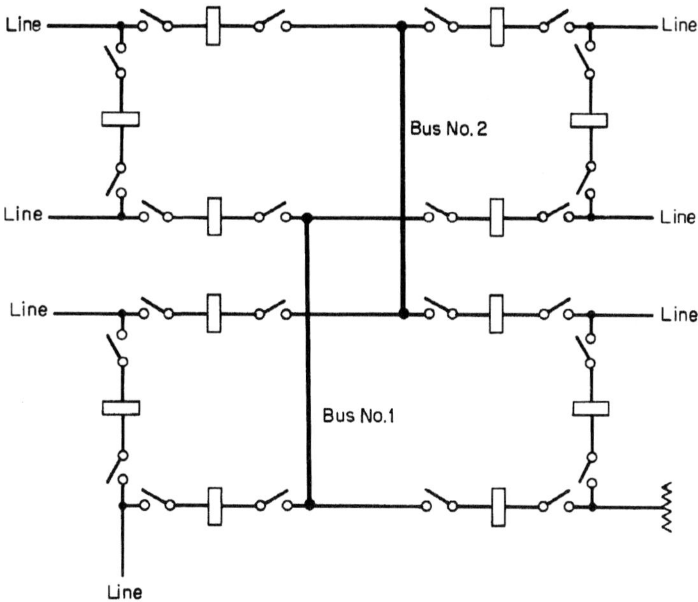


**FIGURE 11-9** EHV substation, low-profile, inverted breaker-and-a-half scheme.

A comparison of rigid and strain buses in Table 11-2 indicates that careful consideration should be given to selection of the proper type of bus to use.

The design of station buses depends on a number of elements, which include the following:

1. Current-carrying capacity
2. Short-circuit stresses
3. Minimum electrical clearances
4. Interconnection flexibility to provide for thermal expansion and seismic interaction

The current-carrying capacity of a bus is limited by the heating effects produced by the current. Buses generally are rated on the basis of the temperature rise, which can be permitted without danger of overheating equipment terminals, bus connections, and joints.

The permissible temperature rise for plain copper and aluminum buses is usually limited to 30°C above an ambient temperature of 40°C. This value is the accepted standard of IEEE, NEMA, and ANSI. This is an average temperature rise; a maximum or hot-spot temperature rise of 35°C is permissible. Many factors enter into the heating of a bus, such as the type of material used, the size and shape of the conductor, the surface area of the conductor and its condition, skin effect, proximity effect, conductor reactance, ventilation, and inductive heating caused by the proximity of magnetic materials.

***Rigid-Bus Material.*** Rigid-bus materials in general use are aluminum and copper. Hard-drawn aluminum, especially in the tubular shape, is the most widely used material in HV and EHV open-type outdoor stations. Aluminum has the advantage of being about one-third the weight of copper and requires little maintenance. The proper use of alloys of aluminum will provide the rigidity needed to serve as a bus material. For a given current rating and for equal limiting temperatures, the required area of aluminum bus is about 133% of the area of the copper bus. Copper and aluminum

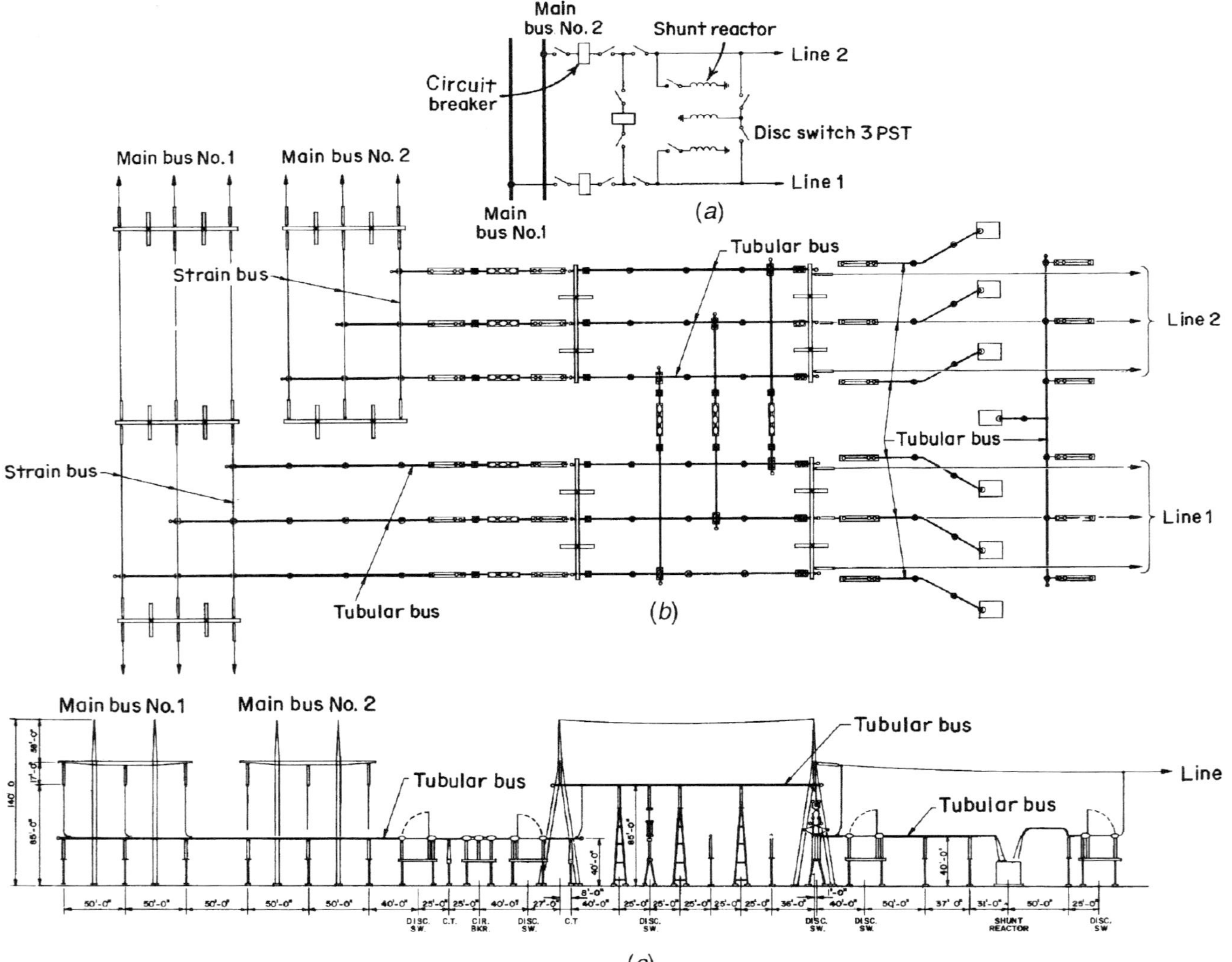


**FIGURE 11-10** A 765-kV substation using both rigid- and strain-bus design. (*a*) Main one-line diagram; (*b*) plan; (*c*) elevation.

**TABLE 11-2** Comparison of Substation Bus Designs

| Bus type | Advantages | Disadvantages |
|---|---|---|
| Rigid | • Less steel is used, and structures are of a simpler design<br>• Conductors are not under constant strain<br>• Individual pedestal-mounted insulators are more accessible for cleaning<br>• Bus is lower in height, has a distinct layout, and can be segregated for maintenance<br>• Low profile provides good visibility of the conductors and apparatus and gives a good appearance to the substation<br>• Clearances are fixed between rigid parts<br>• Higher ampacity in a smaller space | • More insulators and supports usually needed, thus requiring more insulators to clean<br>• More sensitive to structural deflections, causing misalignment problems and possible damage to the bus<br>• Usually requires more land area<br>• Comparatively more expensive<br>• Larger diameter busses require special details for temporary personal ground attachments |
| Strain | • Comparatively lower cost<br>• Typically requires less land area<br>• Fewer structures required | • Requires larger structures and foundations<br>• Insulators more difficult to clean due to height<br>• Painting of high structures can be costly and hazardous<br>• Emergency conductor repairs more difficult |

tubing, as well as other special shapes, are also used for low-voltage distribution substation buses. Type of connection needs consideration as well, for example, welded versus bolted or swaged style.

***Skin Effect.*** *Skin effect* in a conductor carrying an alternating current is the tendency toward crowding of the current into the outer layer, or "skin," of the conductor due to the self-inductance of the conductor. This results in an increase in the effective resistance of the conductor and in a lower current rating for a given temperature rise. Skin effect is very important in heavy-current buses where a number of conductors are used in parallel, because it affects not only each conductor but also each group of conductors as a unit.

Tubing has less skin-effect resistance than rod or flat conductors of the same cross section, and tubing with a thin wall is affected the least by skin effect. Aluminum conductors are affected less by skin effect than copper conductors of similar cross section because of the greater resistance of aluminum.

***Proximity Effect.*** *Proximity effect* in a bus is distortion of the current distribution caused by induction between the leaving and returning conductors. This distortion causes a concentration of current in the parts of the buses nearest together, thus increasing their effective resistance. The proximity effect must be taken into account for buses carrying alternating current. The effect is less on three-phase buses than on single-phase buses.

***Tubular Bus.*** Tubular conductors used on alternating current have a better current distribution than any other shape of conductor of similar cross-sectional area, but they also have a relatively small surface area for dissipating heat losses. These two factors must be balanced properly in the design of a tubular bus. Furthermore, tubing can be bent to provide desired angles or offsets.

Tubing provides a relatively large cross-sectional area in minimum space and has the maximum structural strength for equivalent cross-sectional area, permitting longer distances between supports. In outdoor substations, spans of up to 40 and 50 ft with 6-in-diameter copper or aluminum tubes are considered practicable. The use of long spans reduces the number of insulator posts to a minimum. Current-carrying capacities of copper and aluminum tubular buses of different dimensions are shown in Tables 11-3 and 11-4.

***Thermal Expansion.*** Thermal expansion and contraction of bus conductors is an important factor in bus design, particularly where high-current buses or buses of long lengths are involved. An aluminum bus will expand 0.0105 in/ft of length for a temperature rise of 38°C (100°F). In order to protect insulator supports, disconnecting switches, and equipment terminals from the stresses caused by this

**TABLE 11-3** Current Ratings for Bare Copper Tubular Bus, Outdoors
(40°C ambient temperature, 98% conductivity copper, frequency 60 Hz, wind velocity 2 ft/s at 90° angle)

| Nominal size | Outside diameter, in | Inside diameter, in | Current ratings, A | | |
|---|---|---|---|---|---|
| | | | 30°C rise | 40°C rise | 50°C rise |
| | | Standard pipe sizes | | | |
| ½ | 0.840 | 0.625 | 545 | 615 | 675 |
| ¾ | 1.050 | 0.822 | 675 | 765 | 850 |
| 1 | 1.315 | 1.062 | 850 | 975 | 1080 |
| 1¼ | 1.660 | 1.368 | 1120 | 1275 | 1415 |
| 1½ | 1.900 | 1.600 | 1270 | 1445 | 1600 |
| 2 | 2.375 | 2.062 | 1570 | 1780 | 1980 |
| 2½ | 2.875 | 2.500 | 1990 | 2275 | 2525 |
| 3 | 3.500 | 3.062 | 2540 | 2870 | 3225 |
| 3½ | 4.000 | 3.500 | 3020 | 3465 | 3860 |
| 4 | 4.500 | 4.000 | 3365 | 3810 | 4305 |
| | | Extra-heavy pipe sizes | | | |
| ½ | 0.840 | 0.542 | 615 | 705 | 775 |
| ¾ | 1.050 | 0.736 | 760 | 875 | 970 |
| 1 | 1.315 | 0.951 | 1000 | 1140 | 1255 |
| 1¼ | 1.660 | 1.272 | 1255 | 1445 | 1600 |
| 1½ | 1.900 | 1.494 | 1445 | 1650 | 1830 |
| 2 | 2.375 | 1.933 | 1830 | 2080 | 2325 |
| 2½ | 2.875 | 2.315 | 2365 | 2720 | 3020 |
| 3 | 3.500 | 2.892 | 2970 | 3365 | 3710 |
| 3½ | 4.000 | 3.358 | 3380 | 3860 | 4255 |
| 4 | 4.500 | 3.818 | 3840 | 4350 | 4850 |

***Note:*** 1 in = 25.4 mm; 1 ft/s = 0.3048 m/s.
***Source:*** From Anderson Electric Technical Data, Table 13.

expansion, provisions should be made by means of expansion joints and bus-support clamps, which permit the tubing to slide.

***Flexible Interconnections.*** To provide for physical displacement during seismic motion, flexibility is desired where the bus connects to equipment. A helpful resource for information on this design consideration is IEEE Std 1527-2006 "Recommended Practice for Design of Flexible Buswork Located in Seismically Active Areas."

***Bus Vibration.*** Long tubular-bus spans have experienced vibration caused by wind blowing across the bus. Over time, this vibration can damage the bus and the equipment connected to the bus. The vibration can be eliminated or reduced by inserting a length of cable inside the tubular bus.

***Bus Spacing.*** The spacing of buses in substations is largely a matter of design experience. However, in an attempt to arrive at some standardization of practices, minimum electrical clearances for standard basic insulation levels were established and published by the AIEE Committee on Substations. The data are summarized in AIEE Paper 54-80, which appeared in *Transactions* (June 1954, p. 636). This guide, shown in Table 11-5, provides minimum clearance recommendations for electric transmission systems designed for impulse-withstand levels up to and including 1175 kV BIL.

Ongoing studies attempt to extend the clearance recommendations to include the EHV range. The data published in 1954 are satisfactory to withstand anticipated switching-surge requirements of electric systems rated 161 kV and below. For systems rated 230 kV and above, more accurate determination of the switching-surge characteristics of insulation systems was required before final clearance recommendations could be made.

**TABLE 11-4** Current Ratings for Bare Aluminum Tubular Bus, Outdoors
(Ratings based on 30°C over 40°C ambient, frequency 60 Hz, wind velocity 2 ft/s crosswind)

| Nominal size | Outside diameter, in | Inside diameter, in | Current ratings, A | |
|---|---|---|---|---|
| | | | 6063-T6* | 6061-T6† |
| ASA Schedule 40 (standard pipe size) | | | | |
| ½ | 0.840 | 0.622 | 405 | 355 |
| ¾ | 1.050 | 0.824 | 495 | 440 |
| 1 | 1.315 | 1.049 | 650 | 575 |
| 1¼ | 1.660 | 1.380 | 810 | 720 |
| 1½ | 1.900 | 1.610 | 925 | 820 |
| 2 | 2.375 | 2.067 | 1150 | 1020 |
| 2½ | 2.875 | 2.469 | 1550 | 1370 |
| 3 | 3.500 | 3.068 | 1890 | 1670 |
| 3½ | 4.000 | 3.548 | 2170 | 1920 |
| 4 | 4.500 | 4.026 | 2460 | 2180 |
| 5 | 5.563 | 5.047 | 3080 | 2730 |
| ASA Schedule 80 (extra-heavy pipe size) | | | | |
| ½ | 0.840 | 0.546 | 455 | 400 |
| ¾ | 1.050 | 0.742 | 565 | 500 |
| 1 | 1.315 | 0.957 | 740 | 655 |
| 1¼ | 1.660 | 1.278 | 930 | 825 |
| 1½ | 1.900 | 1.500 | 1070 | 945 |
| 2 | 2.375 | 1.939 | 1350 | 1200 |
| 2½ | 2.875 | 2.323 | 1780 | 1580 |
| 3 | 3.500 | 2.900 | 2190 | 1940 |
| 3½ | 4.000 | 3.364 | 2530 | 2240 |
| 4 | 4.500 | 3.826 | 2880 | 2560 |
| 5 | 5.563 | 4.813 | 3640 | 3230 |

*6063-T6 = 53% IACS typical.
†6061-T6 = 40% IACS typical.
***Note:*** 1 in = 25.4 mm.
***Source:*** Data from Aluminum Company of America.

**TABLE 11-5** Minimum Electrical Clearances for Standard BIL Outdoor Alternating Current

| kV class* | BIL level, kV withstand† | Minimum clearance to ground for rigid parts, in‡ | Minimum clearance between phases (or live parts) for rigid parts, in, metal to metal§ | Minimum clearance between overhead conductors and grade for personnel safety inside substation, ft¶ | Minimum clearance between wires and roadways, inside substation enclosure, ft |
|---|---|---|---|---|---|
| 7.5 | 95 | 6 | 7 | 8 | 20 |
| 15 | 110 | 7 | 12 | 9 | 20 |
| 23 | 150 | 10 | 15 | 10 | 22 |
| 34.5 | 200 | 13 | 18 | 10 | 22 |
| 46 | 250 | 17 | 21 | 10 | 22 |
| 69 | 350 | 25 | 31 | 11 | 23 |
| 115 | 550 | 42 | 53 | 12 | 25 |
| 138 | 650 | 50 | 62 | 13 | 25 |
| 161 | 750 | 58 | 72 | 14 | 26 |
| 230 | 825 | 65 | 80 | 15 | 27 |
| 230 | 900 | 71 | 89 | 15 | 27 |
| | 1050 | 83 | 105 | 16 | 28 |
| | 1175 | 94 | 113 | 17 | 29 |

*Coordinate kV class and BIL when choosing minimum clearances.
†The values above are recommended minimums but may be decreased in line with good practice, depending on local conditions, procedures, etc.
‡The values above apply to 3300 ft above sea level. Above this elevation, the values should be increased according to IEEE Standard C37.30-1992.
§These recommended minimum clearances are for rigid conductors. Any structural tolerances, or allowances for conductor movement, or possible reduction in spacing by foreign objects should be added to the minimum values.
¶These minimum clearances are intended as a guide for the installation of equipment in the field only, and not for the design of electric devices or apparatus, such as circuit breakers and transformers.
1 in = 25.4 mm; 1 ft = 0.3048 m.

## 11.1.7 Clearance Requirements

In 1972, the Substations Committee of the IEEE published Trans. Paper T72 131-6, which established recommendations for minimum line-to-ground electrical clearances for EHV substations based on switching-surge requirements. The recommendations are based on a study of actual test data of the switching-surge strength characteristics of air gaps with various electrode configurations as reported by many investigators. The results are shown in Table 11-6 and include minimum line-to-ground clearances for EHV system voltage ratings of 345, 500, and 765 kV. The clearances given

**TABLE 11-6** Minimum Electrical Clearances for EHV Substations Based on Switching Surge and Lightning Impulse Requirements
(Line to ground)

| System voltage, kV | | Transient voltage | | SS clearances, in | | BIL clearances, in | |
|---|---|---|---|---|---|---|---|
| Nom. | Max. | PU SS | Withstand SS crest, kV | Equivalent SS CFO, kV | Line to ground | Withstand BIL, kV | Line to ground |
| 345 | 362 | 2.2 | 650 | 785 | 84 | 1050 | 84 |
| | | 2.3 | 680 | 821 | 90 | | |
| | | 2.4 | 709 | 857 | 96 | | |
| | | 2.5 | 739 | 893 | 104 | 1300 | 104 |
| | | 2.6 | 768 | 928 | 111 | | |
| | | 2.7 | 798 | 964 | 118 | | |
| | | 2.8 | 828 | 1000 | 125 | | |
| | | 2.9 | 857 | 1035 | 133 | | |
| | | 3.0 | 887 | 1071 | 140 | | |
| 500 | 550 | 1.8 | 808 | 976 | 124 | 1550 | 124 |
| | | 1.9 | 853 | 1031 | 132 | | |
| | | 2.0 | 898 | 1085 | 144 | 1800 | 144 |
| | | 2.1 | 943 | 1139 | 156 | | |
| | | 2.2 | 988 | 1193 | 168 | | |
| | | 2.3 | 1033 | 1248 | 181 | | |
| | | 2.4 | 1078 | 1302 | 194 | | |
| | | 2.5 | 1123 | 1356 | 208 | | |
| | | 2.6 | 1167 | 1410 | 222 | | |
| | | 2.7 | 1212 | 1464 | 238 | | |
| | | 2.8 | 1257 | 1519 | 251 | | |
| 765 | 800 | 1.5 | 982 | 1186 | 166 | 2050 | 167 |
| | | 1.6 | 1047 | 1265 | 185 | | |
| | | 1.7 | 1113 | 1344 | 205 | | |
| | | 1.8 | 1178 | 1423 | 225 | | |
| | | 1.9 | 1244 | 1502 | 246 | | |
| | | 2.0 | 1309 | 1581 | 268 | | |
| | | 2.1 | 1375 | 1660 | 291 | | |
| | | 2.2 | 1440 | 1739 | 314 | | |
| | | 2.3 | 1505 | 1818 | 339 | | |
| | | 2.4 | 1571 | 1897 | 363 | | |
| | | 2.5 | 1636 | 1976 | 389 | | |
| | | 2.6 | 1702 | 2055 | 415 | | |

*Notes:*

**1.** Minimum clearances should satisfy either maximum switching-surge or BIL duty requirement, whichever dictates the larger dimension.

**2.** For installations at altitudes in excess of 3300 ft elevation, it is suggested that correction factors, as provided in IEEE C37.30-1992, be applied to withstand voltages as given above.

SS: switching surge
CFO: critical flashover
1 in = 25.4 mm.

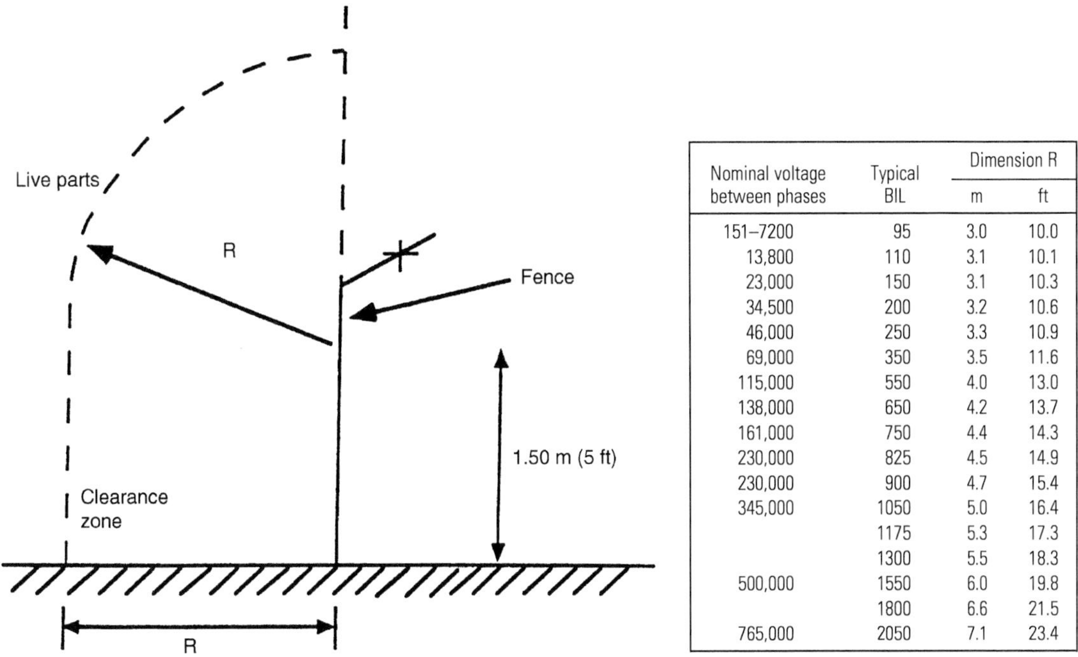


| Nominal voltage between phases | Typical BIL | Dimension R | |
|---|---|---|---|
| | | m | ft |
| 151–7200 | 95 | 3.0 | 10.0 |
| 13,800 | 110 | 3.1 | 10.1 |
| 23,000 | 150 | 3.1 | 10.3 |
| 34,500 | 200 | 3.2 | 10.6 |
| 46,000 | 250 | 3.3 | 10.9 |
| 69,000 | 350 | 3.5 | 11.6 |
| 115,000 | 550 | 4.0 | 13.0 |
| 138,000 | 650 | 4.2 | 13.7 |
| 161,000 | 750 | 4.4 | 14.3 |
| 230,000 | 825 | 4.5 | 14.9 |
| 230,000 | 900 | 4.7 | 15.4 |
| 345,000 | 1050 | 5.0 | 16.4 |
| | 1175 | 5.3 | 17.3 |
| | 1300 | 5.5 | 18.3 |
| 500,000 | 1550 | 6.0 | 19.8 |
| | 1800 | 6.6 | 21.5 |
| 765,000 | 2050 | 7.1 | 23.4 |

**FIGURE 11-11** Substation fence clearance requirements. (*National Electrical Safety Code, IEEE C2-2012.*)

in Table 11-5 are considered adequate for both line-to-ground and phase-to-phase values for the voltage classes up through 230 kV nominal system voltage where air-gap distances are dictated by impulse (BIL) withstand characteristics. The National Electric Safety Code, IEEE Standard C2-2012, also includes clearance requirements to the substation fence (Fig. 11-11). The 2012 edition of the NESC allows the fence to have lesser clearances to energized parts if either (1) a solid fence section is placed adjacent to the energized parts [Rule 110A2] or (2) the energized part is guarded from contact by personnel outside the station fence [Rule 124A1].

The Substations Committee of the IEEE successfully issued IEEE Standard P1427 in 2006, Guide for Recommended Electrical Clearances and Insulation Levels in Air-Insulated Electrical Power Substations.

Considerable information has been published by CIGRE relative to establishing phase-to-phase air clearances in EHV substations as required by switching surges. The CIGRE method is based on nearly simultaneous and equal opposite-polarity surge overvoltages in adjacent phases. The phase-to-ground surge overvoltage is multiplied by a factor of up to 1.8 (the theoretical maximum phase-to-phase voltage would be twice the phase-to-ground surge overvoltage). The estimated value of phase-to-phase overvoltage is then compared with obtained clearances. Refer to an article in CIGRE, *Electra*, no. 29, 1973, "Phase-to-Ground and Phase-to-Phase Air Clearances in Substations," by L. Paris and A. Taschini.

Suggested values of phase-to-phase clearances for EHV substations based on the CIGRE method are shown in Table 11-7. The table was formulated by choosing various phase-to-ground transient voltage values such as are used in Table 11-6. These values of phase-to-ground overvoltage were multiplied by a factor of 1.8 to arrive at a value of estimated phase-to-phase transient overvoltages. An equivalent phase-to-phase critical flashover value of voltage is next assumed by multiplying the switching-surge phase-to-phase voltage by 1.21. Finally, this value is compared with data in the CIGRE article prepared by Paris and Taschini to arrive at air-clearance values based on switching-surge impulse voltages.

**TABLE 11-7** Suggested Electrical Clearances for EHV Substations Based on Switching Surge Requirements and Including U.S. Utility Practice
(Phase to phase)

| System voltage, kV | | Transient voltage | | SS clearances, in* | | Present practice |
|---|---|---|---|---|---|---|
| Nom. | Max. | SS L-G PU | Withstand SS crest, kV | Equivalent L-L SS CFO, kV | Rod to rod withstand* | U.S. utility phase spacing, ft |
| 345 | 362 | 2.2 | 650 | 1405 | 103 | 15–18 |
| | | 2.6 | 768 | 1660 | 128 | |
| | | 3.0 | 887 | 1915 | 159 | |
| 500 | 550 | 1.8 | 808 | 1745 | 138 | 20–35 |
| | | 2.2 | 988 | 2135 | 190 | |
| | | 2.5 | 1123 | 2425 | 239 | |
| | | 2.8 | 1257 | 2715 | 294 | |
| 765 | 800 | 1.5 | 982 | 2120 | 189 | 45–50 |
| | | 1.8 | 1178 | 2545 | 261 | |
| | | 2.1 | 1375 | 2970 | 356 | |
| | | 2.4 | 1571 | 3395 | 480 | |

*Note:* 1 in = 25.4 mm; 1 ft = 0.3048 m.
*The values of L-L switching-surge clearances are based on the use of SS L-G crest voltages multiplied by 1.8. This value of L-L SS voltage is then multiplied by 1.21 to indicate an SS CFO value of voltage used to determine the clearances. For a description of method used, refer to CIGRE report by L. Paris and A. Taschini, Phase-to-Ground and Phase-to-Phase Air Clearances in Substations, CIGRE, *Electra*, no. 29, 1973, pp. 29–44.
L-G: line-to-ground; L-L: line-to-line; SS: switching surge; CFO: critical flashover.

EHV substation bus phase spacing is normally based on the clearance required for switching-surge impulse values plus an allowance for energized equipment projections and corona rings. This total distance may be further increased to facilitate substation maintenance.

## 11.1.8 Mechanical and Electrical Forces

A station bus must have sufficient mechanical strength to withstand short-circuit stresses. Two factors are involved: (1) the strength of the insulators and their supporting structure and (2) the strength of the bus conductor.

A simple guide for the calculation of electromagnetic forces exerted on buses during short-circuit conditions is stated in ANSI Standard C37.32-2002, High-Voltage Switches, Bus Supports, and Accessories Schedules of Preferred Ratings, Construction Guidelines and Specifications. Also, the 2011 update of IEEE Std 605, Guide for Bus Design in Air-Insulated Substations, contains equations and examples for calculating short-circuit forces.

The electromagnetic force exerted between two current-carrying conductors is a function of the current, its decrement rate, the shape and arrangement of conductors, and the natural frequencies of the complete assembly, including mounting structure, insulators, and conductors. Obviously, it is not feasible to cover each and every case with one simple equation, even if some approximations are made, because of the large number of variables involved, including the wide range of constants for support structures.

The force calculated by the following equation is that produced by the maximum peak current. In most cases, the calculated force is higher than that which actually occurs, due to inertia and flexibility of the systems, and this fact tends to compensate for the neglect of resonant forces. The equation, therefore, is sufficiently accurate for practical conditions.

$$F = M\frac{5.5\times I^2}{S\times 10^7} \qquad (11\text{-}1)$$

**TABLE 11-8** Multiplying Factor ($M$) for Calculation of Electromagnetic Forces

| Circuit | Amperes ($I$) expressed as | Multiplying factor ($M$) |
|---|---|---|
| dc | Max. peak | 1.0 |
| ac, 3-phase | Max. peak | 0.866 |
| ac, 3-phase | rms asymmetrical | $(0.866 \times 1.63^2) = 2.3$ |
| ac, 3-phase | rms symmetrical | $(0.866 \times 2.82^2) = 6.9$ |
| 1 phase of 3 phase or 1 phase | Max. peak | 1.0 |
| 1 phase of 3 phase or 1 phase | rms asymmetrical | $(1.63^2) = 2.66$ |
| 1 phase of 3 phase or 1 phase | rms symmetrical | $(2.82^2) = 8.0$ |

where $F$ = pounds per foot of conductor
$M$ = multiplying factor
$I$ = short-circuit current, A (defined in Table 11-8)
$S$ = spacing between centerlines of conductors, in

After determining the value of $I$, select the corresponding $M$ factor from Table 11-8.

Structures with long spans held in tension by strain insulators cannot be calculated for stresses by the preceding procedure, but approximate estimates can be made by following the procedure generally used for calculating mechanical stresses in transmission-line conductors.

The total stress in an outdoor bus is the resultant of the stresses due to the short-circuit load together with the dead, ice, wind, and seismic loads.

1. *Buses up to 161 kV.* The distance between phases and the character of the bus supports and their spacing and horizontal orientation are such that wind loading usually may be neglected. Ice load of ½ in is usually considered.
2. *Buses for 230 kV and higher voltages.* The spacing between phases is usually so large that the mechanical effects of short-circuit currents may not be the determining factor, and such buses, when designed properly for the mechanical loads only, may be found to also satisfy the electrical short-circuit current requirements. However, short-circuit duties on modern systems continue to rise, and the electrical forces should be checked by Eq. (11-1).

Deflections and stresses on aluminum buses can be determined by referring to Tables 11-8 and 11-9. All loads are assumed to be uniformly distributed. Loading includes the dead load of the bus and, in addition, includes ice loadings of ½- and 1-in coating on the bus. Wind loads are assumed to be 8 lb/ft$^2$ of the projected area of tubing including ½ in of ice. Large deflections should be avoided even if the maximum bending stress is found to be within safe limits. It is generally satisfactory, in approximation of bus diameter, to allow 1 in of bus outside diameter for every 10 ft of bus span. Refer to the foot notes below Tables 11-9 and 11-10 for the method of support and number of spans.

Stresses on disconnecting switches under short-circuit conditions may be sufficient to open them, with disastrous results; therefore, modern switch designs embody locks, or overtoggle mechanisms, to prevent this from occurring. The force on the switchblade varies as the square of the current. This force will be increased if the return circuit passes behind the switch and will vary inversely with the distance from the center of the switchblade to the center of the return conductor.

Bus supports are designed for definite cantilever strength, expressed in inch pounds and measured at the cap supporting the conductor clamp. Ample margin of safety with regard to insulation and structural strength should be provided, manufacturers' data should be checked carefully, and units should be so selected that allowable values for the particular units are not exceeded. Good practice recommends that the working load must not exceed 40% of the published rating, and short-circuit loads must not exceed the insulator published rating. These loads should include forces for ultimate short-circuit growth and worst mechanical loading.

**TABLE 11-9** Aluminum Round Tubular Bus Bar Deflections and Stresses
(Standard iron pipe sizes)

| IPS size, in | Loading | Span, ft | | | | | | | | | | | | | |
|---|---|---|---|---|---|---|---|---|---|---|---|---|---|---|---|
| | | 20 | | 25 | | 30 | | 35 | | 40 | | 45 | | 50 | |
| | | Deflection, in | Stress, lb/in² | Deflection, in | Stress, lb/in² | Deflection, in | Stress, lb/in² | Deflection, in | Stress, lb/in² | Deflection, in | Stress, lb/in² | Deflection, in | Stress, lb/in² | Deflection, in | Stress, lb/in² |
| 1¼ | Bare | 1.45 | 2010 | 3.54 | 3135 | | | | | | | | | | |
| | ½" ice | 3.94 | 5445 | 9.61 | 8510 | | | | | | | | | | |
| | ½" ice + 8 lb wind | 5.12 | 7090 | 12.51 | 11075 | | | | | | | | | | |
| | 1" ice | 7.57 | 10470 | 18.48 | 16360 | | | | | | | | | | |
| 1½ | Bare | 1.09 | 1725 | 2.66 | 2700 | | | | | | | | | | |
| | ½" ice | 2.83 | 4475 | 6.90 | 6990 | | | | | | | | | | |
| | ½" ice + 8 lb wind | 3.61 | 5715 | 8.81 | 8930 | | | | | | | | | | |
| | 1" ice | 5.28 | 8365 | 12.90 | 13070 | | | | | | | | | | |
| 2 | Bare | 0.68 | 1350 | 1.67 | 2110 | 3.45 | 3040 | | | | | | | | |
| | ½" ice | 1.65 | 3265 | 4.03 | 5100 | 8.35 | 7345 | | | | | | | | |
| | ½" ice + 8 lb wind | 2.05 | 4055 | 5.00 | 6340 | 10.38 | 9125 | | | | | | | | |
| | 1" ice | 2.95 | 5845 | 7.21 | 9135 | 14.95 | 13150 | | | | | | | | |
| 2½ | Bare | 0.47 | 1130 | 1.15 | 1765 | 2.38 | 2540 | 4.42 | 3455 | | | | | | |
| | ½" ice | 0.96 | 2310 | 2.36 | 3610 | 4.89 | 5200 | 9.05 | 7080 | | | | | | |
| | ½" ice + 8 lb wind | 1.14 | 2730 | 2.78 | 4270 | 5.77 | 6150 | 10.70 | 8370 | | | | | | |
| | 1" ice | 1.61 | 3845 | 3.92 | 6010 | 8.13 | 8655 | 15.06 | 11780 | | | | | | |
| 3 | Bare | 0.31 | 910 | 0.76 | 1425 | 1.58 | 2050 | 2.93 | 2790 | 5.00 | 3640 | | | | |
| | ½" ice | 0.61 | 1775 | 1.49 | 2775 | 3.08 | 3995 | 5.71 | 5440 | 9.74 | 7105 | | | | |
| | ½" ice + 8 lb wind | 0.71 | 2060 | 1.72 | 3220 | 3.58 | 4635 | 6.62 | 6310 | 11.30 | 8240 | | | | |
| | 1" ice | 0.98 | 2860 | 2.39 | 4465 | 4.96 | 6430 | 9.19 | 8755 | 15.68 | 11435 | | | | |
| 3½ | Bare | 0.24 | 790 | 0.58 | 1230 | 1.20 | 1775 | 2.22 | 2415 | 3.79 | 3155 | 6.06 | 3995 | | |
| | ½" ice | 0.45 | 1490 | 1.09 | 2330 | 2.26 | 3355 | 4.19 | 4565 | 7.15 | 5960 | 11.46 | 7545 | | |
| | ½" ice + 8 lb wind | 0.51 | 1710 | 1.25 | 2670 | 2.59 | 3845 | 4.81 | 5230 | 8.20 | 6835 | 13.14 | 8650 | | |
| | 1" ice | 0.70 | 2350 | 1.72 | 3670 | 3.57 | 5280 | 6.61 | 7190 | 11.27 | 9390 | 18.05 | 11885 | | |

| | | | | | | | | | | | | | | | |
|---|---|---|---|---|---|---|---|---|---|---|---|---|---|---|---|
| 4 | Bare | 0.19 | 695 | 0.45 | 1090 | 0.94 | 1565 | 1.74 | 2130 | 2.97 | 2785 | 4.76 | 3525 | 7.25 | 4350 |
| | ½" ice | 0.34 | 1275 | 0.83 | 1995 | 1.72 | 2870 | 3.19 | 3910 | 5.45 | 5105 | 8.72 | 6465 | 13.30 | 7980 |
| | ½" ice + 8 lb wind | 0.39 | 1450 | 0.94 | 2265 | 1.96 | 3260 | 3.62 | 4435 | 6.18 | 5795 | 9.90 | 7335 | 15.09 | 9055 |
| | 1" ice | 0.53 | 1975 | 1.28 | 3085 | 2.66 | 4440 | 4.93 | 6045 | 8.42 | 7895 | 13.49 | 9990 | 20.55 | 12330 |
| 4½ | Bare | 0.15 | 620 | 0.36 | 970 | 0.76 | 1400 | 1.40 | 1905 | 2.39 | 2490 | 3.83 | 3150 | 5.83 | 3890 |
| | ½" ice | 0.27 | 1115 | 0.65 | 1740 | 1.35 | 2505 | 2.51 | 3410 | 4.28 | 4455 | 6.85 | 5640 | 10.44 | 6960 |
| | ½" ice + 8 lb wind | 0.30 | 1255 | 0.73 | 1960 | 1.52 | 2820 | 2.82 | 3840 | 4.81 | 5015 | 7.71 | 6345 | 11.75 | 7835 |
| | 1" ice | 0.41 | 1695 | 0.99 | 2650 | 2.06 | 3810 | 3.81 | 5190 | 6.51 | 6780 | 10.42 | 8580 | 15.89 | 10590 |
| 5 | Bare | 0.12 | 555 | 0.29 | 870 | 0.61 | 1250 | 1.12 | 1705 | 1.92 | 2225 | 3.07 | 2815 | 4.69 | 3475 |
| | ½" ice | 0.21 | 970 | 0.51 | 1520 | 1.06 | 2185 | 1.96 | 2975 | 3.35 | 3885 | 5.37 | 4920 | 8.18 | 6070 |
| | ½" ice + 8 lb wind | 0.23 | 1085 | 0.57 | 1695 | 1.18 | 2440 | 2.19 | 3320 | 3.74 | 4335 | 5.99 | 5490 | 9.13 | 6775 |
| | 1" ice | 0.31 | 1455 | 0.77 | 2275 | 1.59 | 3275 | 2.94 | 4455 | 5.02 | 5820 | 8.04 | 7365 | 12.26 | 9095 |
| 6 | Bare | 0.08 | 465 | 0.20 | 725 | 0.42 | 1040 | 0.79 | 1420 | 1.34 | 1850 | 2.15 | 2345 | 3.28 | 2895 |
| | ½" ice | 0.14 | 775 | 0.34 | 1210 | 0.71 | 1745 | 1.32 | 2375 | 2.25 | 3105 | 3.60 | 3930 | 5.49 | 4850 |
| | ½" ice + 8 lb wind | 0.15 | 855 | 0.38 | 1335 | 0.78 | 1925 | 1.45 | 2615 | 2.48 | 3420 | 3.97 | 4325 | 6.05 | 5340 |
| | 1" ice | 0.21 | 1135 | 0.50 | 1770 | 1.04 | 2550 | 1.92 | 3470 | 3.28 | 4530 | 5.26 | 5735 | 8.02 | 7080 |

***Note:*** The tabulated deflections are for single-span, simply supported buses. Deflections for fixed-end buses are one-fifth of the values given above, and the deflections for continuous buses for the center spans are also one-fifth of the values above. The deflections for the end spans are two-fifths of the values given. The stresses given in the above table are the stresses in the outer fibers as calculated for simply supported beams with a uniformly distributed load. 1 in = 25.4 mm; 1 ft = 0.3048 m; 1 lb = 0.4536 kg; 1 lb/in$^2$ = 6.895 kPa.

***Source:*** From *Kaiser Aluminum Electrical Conductor Technical Manual.*

**TABLE 11-10** Aluminum Round Tubular Bus Bar Deflections and Stresses
(Extra-heavy pipe sizes)

| | | Span, ft | | | | | | | | | | | | | |
|---|---|---|---|---|---|---|---|---|---|---|---|---|---|---|---|
| | | 20 | | 25 | | 30 | | 35 | | 40 | | 45 | | 50 | |
| IPS size, in | Loading | Deflection, in | Stress, lb/in² | Deflection, in | Stress, lb/in² | Deflection, in | Stress, lb/in² | Deflection, in | Stress, lb/in² | Deflection, in | Stress, lb/in² | Deflection, in | Stress, lb/in² | Deflection, in | Stress, lb/in² |
| 1¼ | Bare | 1.54 | 2130 | | | | | | | | | | | | |
| | ½" ice | 3.54 | 4900 | | | | | | | | | | | | |
| | ½" ice + 8 lb wind | 4.42 | 6110 | | | | | | | | | | | | |
| | 1" ice | 6.47 | 8945 | | | | | | | | | | | | |
| 1½ | Bare | 1.15 | 1825 | 2.82 | 2855 | | | | | | | | | | |
| | ½" ice | 2.53 | 4005 | 6.17 | 6255 | | | | | | | | | | |
| | ½" ice + 8 lb wind | 3.09 | 4895 | 7.55 | 7645 | | | | | | | | | | |
| | 1" ice | 4.47 | 7085 | 10.92 | 11070 | | | | | | | | | | |
| 2 | Bare | 0.72 | 1425 | 1.76 | 2225 | | | | | | | | | | |
| | ½" ice | 1.46 | 2890 | 3.57 | 4520 | | | | | | | | | | |
| | ½" ice + 8 lb wind | 1.73 | 3430 | 4.23 | 5360 | | | | | | | | | | |
| | 1" ice | 2.46 | 4870 | 6.01 | 7610 | | | | | | | | | | |
| 2½ | Bare | 0.49 | 1185 | 1.21 | 1850 | 2.50 | 2665 | | | | | | | | |
| | ½" ice | 0.89 | 2125 | 2.17 | 3320 | 4.49 | 4780 | | | | | | | | |
| | ½" ice + 8 lb wind | 1.01 | 2420 | 2.46 | 3780 | 5.11 | 5440 | | | | | | | | |
| | 1" ice | 1.40 | 3345 | 3.41 | 5225 | 7.07 | 7525 | | | | | | | | |
| 3 | Bare | 0.33 | 955 | 0.80 | 1495 | 1.66 | 2150 | 3.07 | 2925 | | | | | | |
| | ½" ice | 0.56 | 1625 | 1.36 | 2540 | 2.82 | 3660 | 5.23 | 4980 | | | | | | |
| | ½" ice + 8 lb wind | 0.62 | 1815 | 1.52 | 2840 | 3.15 | 4085 | 5.84 | 5560 | | | | | | |
| | ½" ice | 0.85 | 2465 | 2.06 | 3850 | 4.28 | 5545 | 7.93 | 7550 | | | | | | |
| 3½ | Bare | 0.25 | 825 | 0.60 | 1290 | 1.25 | 1860 | 2.32 | 2530 | 3.96 | 3305 | | | | |
| | ½" ice | 0.41 | 1360 | 1.00 | 2125 | 2.07 | 3060 | 3.83 | 4165 | 6.53 | 5440 | | | | |
| | ½" ice + 8 lb wind | 0.45 | 1500 | 1.10 | 2345 | 2.28 | 3380 | 4.23 | 4600 | 7.21 | 6010 | | | | |
| | 1" ice | 0.60 | 2015 | 1.48 | 3145 | 3.06 | 4530 | 5.67 | 6170 | 9.67 | 8055 | | | | |

| | | | | | | | | | | | | | | | |
|---|---|---|---|---|---|---|---|---|---|---|---|---|---|---|---|
| 4 | Bare | 0.19 | 725 | 0.47 | 1135 | 0.98 | 1635 | 1.82 | 2230 | 3.10 | 2910 | 4.97 | 3680 | 7.58 | 4545 |
| | ½" ice | 0.31 | 1165 | 0.76 | 1820 | 1.57 | 2620 | 2.91 | 3565 | 4.97 | 4655 | 7.96 | 5895 | 12.13 | 7275 |
| | ½" ice + 8 lb wind | 0.34 | 1270 | 0.83 | 1990 | 1.72 | 2865 | 3.18 | 3900 | 5.43 | 5095 | 8.70 | 6445 | 13.26 | 7955 |
| | 1" ice | 0.45 | 1690 | 1.10 | 2640 | 2.28 | 3800 | 4.22 | 5170 | 7.20 | 6755 | 11.54 | 8550 | 17.59 | 10555 |
| 4½ | Bare | 0.16 | 650 | 0.38 | 1015 | 0.79 | 1460 | 1.46 | 1990 | 2.49 | 2600 | 4.00 | 3290 | 6.09 | 4060 |
| | ½" ice | 0.24 | 1015 | 0.59 | 1585 | 1.23 | 2285 | 2.28 | 3110 | 3.90 | 4060 | 6.24 | 5135 | 9.51 | 6340 |
| | ½" ice + 8 lb wind | 0.26 | 1100 | 0.65 | 1720 | 1.34 | 2475 | 2.48 | 3370 | 4.23 | 4405 | 6.77 | 5575 | 10.32 | 6880 |
| | 1" ice | 0.35 | 1445 | 0.85 | 2260 | 1.76 | 3255 | 3.26 | 4430 | 5.55 | 5785 | 8.90 | 7320 | 13.56 | 9040 |
| 5 | Bare | 0.13 | 580 | 0.31 | 905 | 0.63 | 1305 | 1.17 | 1775 | 2.00 | 2320 | 3.20 | 2935 | 4.88 | 3625 |
| | ½" ice | 0.19 | 885 | 0.47 | 1380 | 0.97 | 1990 | 1.79 | 2710 | 3.05 | 3535 | 4.89 | 4475 | 7.45 | 5525 |
| | ½" ice + 8 lb wind | 0.21 | 950 | 0.50 | 1490 | 1.04 | 2140 | 1.93 | 2915 | 3.29 | 3810 | 5.26 | 4820 | 8.02 | 5950 |
| | 1" ice | 0.27 | 1240 | 0.65 | 1935 | 1.35 | 2785 | 2.51 | 3795 | 4.28 | 4955 | 6.85 | 6270 | 10.44 | 7745 |
| 6 | Bare | 0.09 | 485 | 0.21 | 755 | 0.44 | 1090 | 0.82 | 1485 | 1.40 | 1940 | 2.25 | 2455 | 3.43 | 3030 |
| | ½" ice | 0.13 | 700 | 0.31 | 1095 | 0.64 | 1580 | 1.19 | 2150 | 2.04 | 2810 | 3.26 | 3555 | 4.97 | 4390 |
| | ½" ice + 8 lb wind | 0.13 | 745 | 0.33 | 1165 | 0.68 | 1675 | 1.27 | 2280 | 2.16 | 2980 | 3.46 | 3775 | 5.27 | 4660 |
| | 1" ice | 0.17 | 950 | 0.42 | 1485 | 0.87 | 2140 | 1.61 | 2910 | 2.75 | 3800 | 4.41 | 4810 | 6.72 | 5940 |

***Note:*** The tabulated deflections are for single-span, simply supported buses. Deflections for fixed-end buses are one-fifth of the values given above, and the deflections for continuous buses for the center spans are also one-fifth of the values above. The deflections for the end spans are two-fifths of the values given. The stresses given in the above table are the stresses in the outer fibers as calculated for simply supported beams with a uniformly distributed load. 1 in = 25.4 mm; 1 ft = 0.3048 m; 1 lb = 0.4536 kg; 1 lb/in$^2$ = 6.895 kPa.

***Source:*** From *Kaiser Aluminum Electrical Conductor Technical Manual.*

### 11.1.9 Protective Relaying

***Protective Relaying.*** A substation can employ many relaying systems to protect the equipment associated with the station and the system; the fundamental components of a substation requiring protective relays are as follows:

1. Transmission and distribution lines emanating from the station
2. Step-up and step-down transformers
3. Station buses
4. Breakers
5. Shunt and series reactors
6. Shunt and series capacitors

Protective relays are shown on the substation one-line diagrams most frequently as a circle with a number inside, the number identifying the type of relay and its function. Some common numbers are

- 87—differential (line, transformer, bus)
- 50/51—instantaneous/time overcurrent (transformer, bus, feeder, cap bank)
- 21—impedance/distance relay (lines, transformers; 21-1: ~zone 1, 21-2: ~zone 2)
- 27/59—under-/overvoltage

In addition, the lines on a diagram emanating from a relay symbolize the current and potential transformers connected to the relay. The current transformers are an indication of the element the relay is protecting—for example, a bus or transformer. Substations serving bulk transmission system circuits must provide a high order of reliability and security in order to provide continuity of service to the system. More and more emphasis is being placed on very sophisticated relaying systems which must function reliably and at high speeds to clear line and station faults while minimizing false tripping.

Since the early 1970s, protective relays have evolved from the electromechanical relays that used magnets and induction disks, to electrostatic relays that used traditional circuit board components, to the current modern microprocessor-based relay. As this evolution took place, many of the control and communication functions within a substation were integrated with these relays. The modern microprocessor-based relay digitizes the substation current and voltage quantities through analog to digital convertors, and the protection functions are numerical calculations based on these quantities. Oftentimes a single microprocessor-based relay has the protective capability of several traditional electromechanical relays.

Most EHV and UHV systems now use two or more sets of protective relays for lines, buses, and transformers. The use of two separate sets of relays, operating from separate potential and current transformers and from separate station batteries, allows for the testing of relays without the necessity of removing the protected line or bus from service. Test switches or otherwise should be provided for the relay scheme to allow efficient function testing. For more difficult relaying applications, such as EHV lines using series capacitors in the line, some utilities use three sets of relays to provide the protection systems. We will discuss protective relays based on the elements of the substation they protect.

***Step Distance.*** The fundamental transmission line protection scheme involves distance relays that derive its protection functions by using the secondary current and voltage to determine an effective impedance. That impedance is used to determine whether a fault condition is located within the protected line segment. A standard step distance scheme (see Figure 11-12) uses several impedance elements to provide protection on transmission lines. Each distance element is referred to as a zone of protection. A step distance scheme uses the secondary impedance of the line and compares it to what the relay is measuring. If the relay sees a fault that is less than the secondary impedance of the line segment, it will determine that the fault is on the line section and trip. The first zone or zone 1 is usually set to trip with no time delay. In order to prevent over tripping zone 2 and 3 are time delayed

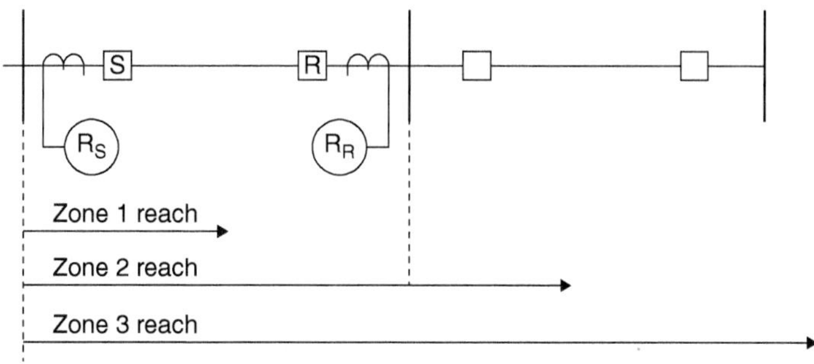


**FIGURE 11-12** Step distance relaying.

to allow breakers at the remote substation to isolate a fault condition first. Distance relays are often set as follows:

- Zone 1 is usually set to 70% to 80% of the line impedance.
- Zone 2 is usually set to 100% of the protected line plus some margin of the next shortest line out of the remote substation (usually time delayed 20 to 30 cycles).
- Zone 3 is usually set to 100% of the protected line plus 100% of the longest line out of the remote substation (usually time delayed 60 to 90 cycles).

***Communication-Based Relays.*** Transmission-line relay terminals are located at the substation and can employ many different types of communication-based relaying schemes. There are two types of communication-based relay schemes employed on transmission lines. Schemes 1 to 4 shown below use the distance elements described in the step distance section to provide instantaneous clearing of faults along the entire length of the transmission line. This is implemented by providing the ability of the relays to communicate to each other from one end of the line to the other end. If both ends of a line detect an internal fault through its distance elements those relays can trip instantaneously rather than waiting for a zone 2 or 3 element to time out. Schemes 5 to 7 implement variations of current-based differential relaying in that they do not use voltage but use the current at each end of the line to detect internal fault conditions. The basic communication-based protection schemes can be described as follows:

1. Direct underreaching (DUTT)
2. Permissive underreaching (PUTT)
3. Permissive overreaching (POTT)
4. Directional comparison (DCB or DCUB)
5. Phase comparison
6. Differential
7. Pilot wire

***Direct Underreaching Fault Relays.*** These relays at each terminal of the protected line sense fault power flow into the line. Their zones of operation must overlap but not overreach any remote terminals. The operation of the relays at any terminal initiates both the opening of the local breaker and the transmission of a continuous remote tripping signal to effect instantaneous operation of all remote breakers. For example, in Fig. 11-13, for a line fault near bus *A*, the fault relays at *A* open (trip) breaker *A* directly and send a transfer trip signal to *B*. The reception of this trip signal at *B* trips breaker *B*.

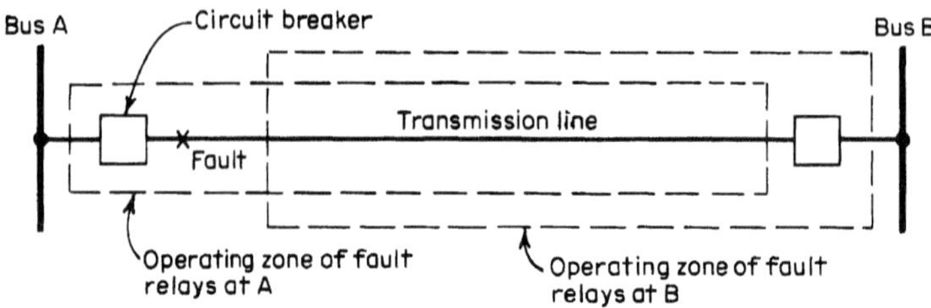


**FIGURE 11-13** Fault-relay operating zones for the underreaching transfer trip transmission-line pilot relaying system.

***Permissive Underreaching Relays.*** The operation and equipment for this system are the same as those of the direct underreaching system, with the addition of fault-detector units at each terminal. The fault detectors must overreach all remote terminals. They are used to provide added security by supervising remote tripping. Thus, the fault relays operate as shown in Fig. 11-13 and the fault detectors operate as shown in Fig. 11-14. As an example, for a fault near *A* in Fig. 11-13, the fault relays at *A* trip breaker *A* directly and send a transfer trip signal to *B*. The reception of the trip signal plus the operation of the fault detector relays at *B* (Fig. 11-14) trip breaker *B*.

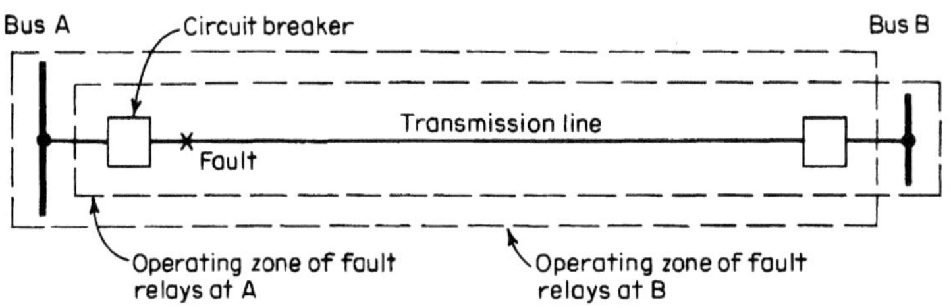


**FIGURE 11-14** Fault-relay operating zones for the overreaching transmission-line pilot relaying system.

***Permissive Overreaching Relays.*** Fault relays at each terminal of the protected line sense fault power flow into the line, with their zones of operation overreaching all remote terminals. Both the operation of the local fault relays and a transfer trip signal from all the remote terminals are required to trip any breaker. Thus, in the example of Fig. 11-14 for the line fault near *A*, fault relays at *A* operate and transmit a trip signal to *B*. Similarly, the relays at *B* operate and transmit a trip signal to *A*. Breaker *A* is tripped by the operation of the fault relay *A* plus the remote trip signal from *B*. Likewise, breaker *B* is tripped by the operation of fault relay *B* plus the remote trip signal from *A*.

***Directional-Comparison Relays.*** The channel signal in these systems is used to block tripping in contrast to its use to initiate tripping in the preceding three systems. Fault relays at each terminal of the protected line section sense fault power flow into the line. Their zones of operation must overreach all remote terminals. Additional fault-detecting units are required at each terminal to initiate the channel-blocking signal. Their operating zones must extend further or be set more sensitively than the fault relays at the far terminals. For example, in Fig. 11-15 the blocking zone at *B* must extend further behind breaker *B* (to the right) than the operating zone of the fault relays at *A*. Correspondingly, the blocking zone at *A* must extend further out into the system (to the left) than the operating zone of the fault relays at *B*.

For an internal fault on line *AB*, no channel signal is transmitted (or if transmitted, it is cut off by the fault relays) from any terminal. In this absence of any channel signal, fault relays at *A* instantly

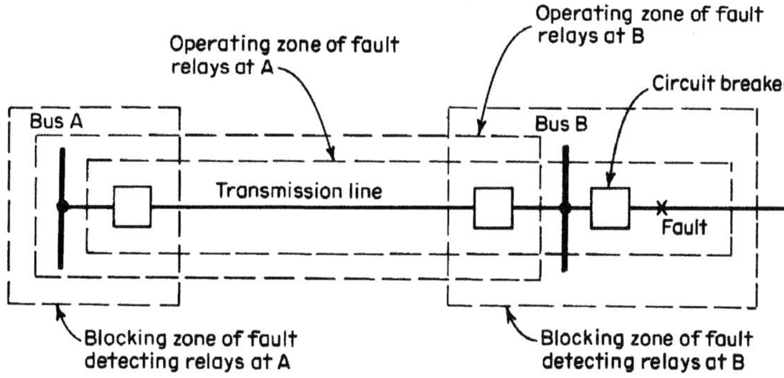


**FIGURE 11-15** Fault and blocking relay operating zones for the directional-comparison transmission-line pilot relay system.

trip breaker *A*, and fault relays at *B* instantly trip breaker *B*. For the external fault to the right of *B* as shown in Fig. 11-15, the blocking zone relays at *B* transmit a blocking channel signal to prevent the fault relays at *A* from tripping breaker *A*. Breaker *B* is not tripped because the *B* operating zone does not see this fault.

***Phase-Comparison Relays.*** The three line currents at each end of the protected line are converted into a proportional single-phase voltage. The phase angles of the voltages are compared by permitting the positive half-cycle of the voltage to transmit a half-wave signal block over the pilot channel. For external faults, these blocks are out of phase so that alternately the local and then the remote signal provide essentially a continuous signal to block or prevent tripping. For internal faults, the local and remote signals are essentially in phase so that approximately a half-cycle of no channel signal exists. This is used to permit the fault relays at each terminal to trip their respective breakers.

***Differential Protection.*** Modern line protection relays implement differential protection often using high speed communication channels that are either multiplexed on fiber optic cables or use direct fiber optic connections from one end of the line to the other. Implementation of the differential element is often done on A, B, and C phase current elements as well as having ground and negative sequence current differential protection.

***Pilot-Wire Relaying.*** Pilot-wire relaying is an adaptation of the principle of differential relaying to line protection and functions to provide high-speed clearing of the line for faults anywhere on the line. Pilots include wire pilot (using a two-wire pair between the ends of the line), carrier-current pilots, microwave pilots, fiber-optics pilots, and the use of audio-tone equipment over wire, carrier, fiber-optics, or microwave. The transmission lines may have two or more terminals each with circuit breakers for disconnecting the line from the rest of the power system. All the relaying systems described can be used on two-terminal or multiterminal lines. The relaying systems program the automatic operation of the circuit breakers during power-system faults.

***Station Bus Protection.*** Station bus protection deserves very careful attention because bus failures are, as a rule, the most serious that can occur to an electrical system. Unless properly isolated, a bus fault could result in complete shutdown of a station. Many methods are employed to protect the station buses.

Because of the high magnitude of currents encountered during bus faults, current transformers may saturate and thus cause false tripping during external faults. The possibility of ac and dc saturation during faults makes it mandatory that current transformers used for bus differential protection be accurate and of the best quality possible. Also, in high impedance differential relay applications, current transformers should be matched to provide identical ratios and similar saturation characteristics. Bus protection can be generally characterized as a full bus differential that monitors all current flowing into a bus or as a partial bus differential that is monitoring associated lines.

***Full Bus Differential Relay.*** There are basically two types of full bus differential relays. The first type is referred to as a high-impedance bus differential relay. This type of relay is designed and implemented so that it drives the current transformers into a saturated state during an internal bus fault condition. This type of relay requires that CT ratios are matched and saturation characteristics are similar. The relay monitors the voltage across the saturated CT terminals to determine if the CT is saturating; therefore, the pickup of the differential relay is in secondary voltage. Many of the early bus differential relays operated off this principle. With the evolution of relays into microprocessors, the modern bus differential relays are actual differential relays in that the relay sums the current vectors for each phase and terminal to determine a true differential current across the bus differential zone of protection. These relays would be characterized as a low impedance differential relay. This type of relay requires that the current transformers do not saturate and therefore calculate an accurate differential current. These relays also have the advantage of working for different CT ratios. In some stations with extremely high fault current that may cause saturation issues a low impedance bus differential relay may not be feasible.

***Partial Bus Differential Relay.*** On substations where CT saturation is not a concern or all elements of the substation do not have CTs available, a partial bus differential relay may be applied. A partial bus differential usually consists of a time overcurrent relay wired to the summed CT circuits off a bus for operating as a bus differential relay. A variation on this theme is a time overcurrent relay that monitors just the summed CTs associated with a bus but not the transformers that may be tapped off the bus section. This style of bus protection is sometimes referred to as a partial bus differential. In a partial bus differential application, it becomes necessary to coordinate the over current with overcurrent relays on any tapped transformers connected to the bus section. It is also important to verify that the magnitude of the differential current due to a fault on the bus is significantly higher than the differential current due to low side fault on any unmonitored transformers tapped off the bus section.

***Transformer Protection.*** Transformers may be subjected to short circuits between phase and ground, open circuits, turn-to-turn short circuits, and overheating. Interphase short circuits are rare and seldom develop as such initially, since the phase windings are usually well separated in a three-phase transformer. Faults usually begin as turn-to-turn failures and frequently develop into faults involving ground.

It is highly desirable to isolate transformers with faulty windings as quickly as possible to reduce the possibility of oil fires, with the attendant resulting cost for replacement. Differential protection is the preferred type of transformer protection due to its simplicity, sensitivity, selectivity, and speed of operation. If the current-transformer ratios are not perfectly matched, taking into account the voltage ratios of the transformer, autotransformers, or auxiliary current transformers are sometimes required in the current-transformer secondary circuits to match the units properly so that no appreciable current will flow in the relay operating coil, except for internal fault conditions.

In applying differential protection to transformers, somewhat less sensitivity in the relays is usually required, as compared with generator relays, since they must remain nonoperative for the maximum transformer tap changes that might be used. It is also necessary to take into account the transformer exciting inrush current that may flow in only one circuit when the transformer is energized by closing one of its circuit breakers. As a rule, incorrect relay operation can be avoided by imposing a slight time delay for this condition.

*Voltage-load tap-changing (LTC) transformers* may be protected by differential relays. The same principles of applying differential protection to other transformers hold here as well. It is important that the differential relay be selected carefully so that the unbalance in the current-transformer secondary circuits will not in any case be sufficient to operate the relay under normal conditions. It is suggested that the current transformers be matched at the midpoint of the tap-changing range. The current-transformer error will then be a minimum for the maximum tap position in either direction. *Current-transformer and relay connections* for various types of differential protection are indicated (1) in Fig. 11-16 for a Y-delta transformer and (2) in Fig. 11-17 for a three-winding Y-delta-Y transformer. Modern microprocessor-based relays usually have the ability to compensate for delta-Y connected

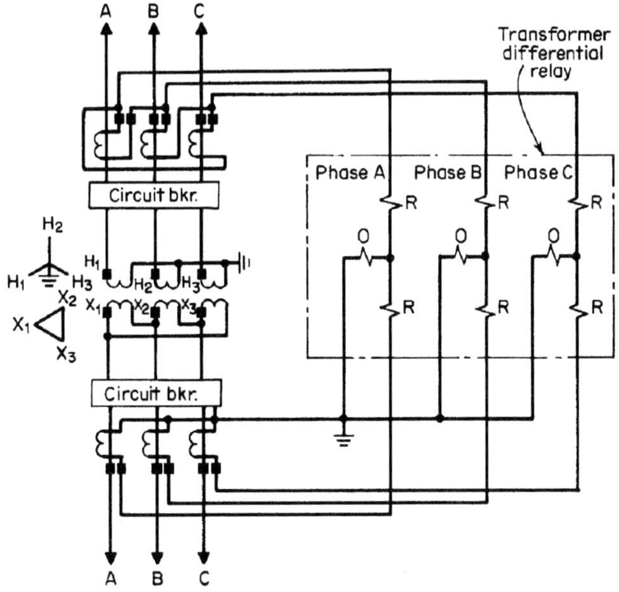


**FIGURE 11-16** Transformer differential protection for a Y-Δ transformer.

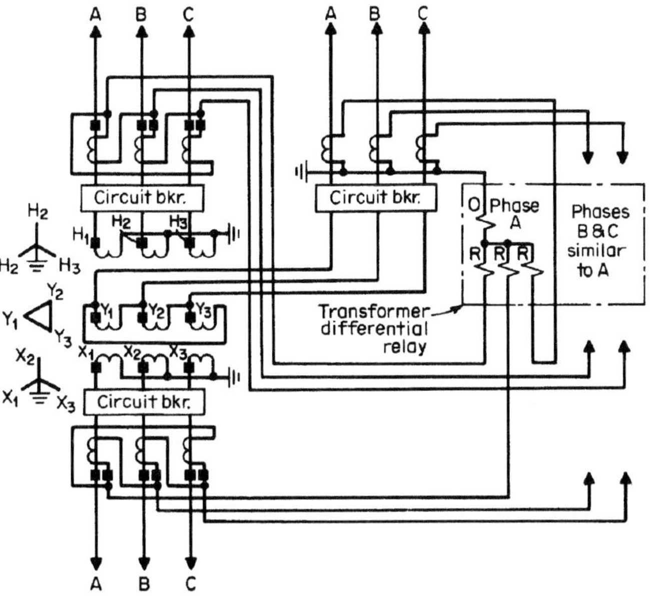


**FIGURE 11-17** Transformer differential protection for a Y-Δ-Y transformer.

transformers through internal compensation settings. This ability allows the use of Wye-connected CT circuits regardless of the transformer windings. However, if utilizing an older style transformer differential relay that cannot be set to compensate for the delta/wye transformation, there are two rules frequently used in laying out the wiring for differential protection of transformers whose main windings are connected in Y and delta. They are

1. The current transformers in the leads to the Y-connected winding should be connected in delta; current transformers in the leads to a delta-connected winding should be connected in Y.
2. The delta connection of the current transformers should be a replica of the delta connection of the power transformers; the Y connection of the current transformers should be a replica of the Y connection of the power transformers.

*Current transformers* that will give approximately 5-A secondary current at full load on the transformer should be chosen. This will not be possible in all cases, particularly for transformers having three or more windings, since the kVA ratings may vary widely and may not be proportional to the voltage ratings.

*Overcurrent protection* should be applied to transformers as the primary protection where a differential scheme cannot be justified or as "backup" protection if differential is used. Frequently, faster relaying may be obtained for power flow from one direction by the use of power-directional relays.

*Transformer overheating protection* is sometimes provided to give an indication of overtemperature, rarely to trip automatically. Overload relays of the replica type may be connected in the current-transformer circuits to detect overloading of the unit. Others operate on top-oil temperature, and still others operate on top-oil temperature supplemented with heat from an adjacent resistor connected to a current transformer in the circuit. A recently developed sensor using a glass chip sensitive to temperature changes employs fiber-optics techniques to measure winding hot-spot temperatures.

Gas- or oil-pressure relays are available for attachment to the top or side of transformer tanks to indicate winding faults, which produce gas or sudden pressure waves in the oil. Rapid collection of gas or pressure waves in the oil, due to short circuits in the winding, will produce fast operation. New, more sophisticated methods to detect incipient failures by frequent monitoring of gas samples are being developed.

***Circuit-Breaker Protection.*** In recent years, great emphasis has been placed on the need to provide backup protection in the event of failure of a circuit breaker to clear a fault following receipt of a trip command from protective relays. For any fault, the protective relays operate to trip the necessary circuit breakers. In addition, these same protective relays, together with breaker-failure fault-detector relays, will energize a timer to start the breaker-failure backup scheme. If any breaker fails to clear the fault, the protective relays will remain picked up, permitting the timer to time-out and trip the necessary other breakers to clear the fault.

Circuit-breaker failure can be caused by loss of dc trip supply, blown trip fuses, trip-coil failure, failure of breaker trip linkages, or failure of the breaker current-interrupting mechanism. The two basic types of failures are (1) mechanical failure and (2) electrical failure of the breaker to clear the fault. Mechanical failure occurs when the breaker does not move following receipt of a trip command because of loss of dc trip supply, trip coil failure, or trip linkage failure. Electrical failure occurs when the breaker moves in an attempt to clear a fault on receipt of the trip command but fails to break the fault current because of misoperation of the current interrupter itself.

In order to clear faults for these two types of breaker failures, two different schemes of protection can be employed. Conventional breaker-failure schemes consist of using instantaneous current-operated fault detectors, which pick up to start a timer when fault relays operate. If the breaker fails to operate to clear the fault, the timer times out and trips necessary breakers to clear the fault. However, if the breaker operates correctly to clear the fault, enough time must be allowed in the timer setting to ensure reset of the fault-detector relay. Total clearing times at EHV using this scheme are quite fast and usually take 10 to 12 cycles from the time of fault until the fault is cleared. For those faults where mechanical failure of the breakers occurs, an even faster scheme is used.

This scheme depends on a breaker auxiliary switch (normally open type 52-A contact) to initiate a fast timer. The auxiliary switch is specially located to operate from breaker trip linkages to sense actual movement of the breaker mechanism. If the breaker failure is mechanical, the breaker-failure timer is actuated through the auxiliary switch when the protective relays operate. The advantage of using the auxiliary switch is the extremely fast reset time of the breaker-failure timer that can be realized when the breaker operates correctly. Schemes in use with the fast breaker-failure circuit can attain total clearing times of 7.5 cycles when a breaker failure occurs.

***Shielding and Grounding Practices for Control Cables.*** For several years, the increased application of solid-state devices for protective relaying and control and for electronic equipment, such as audio tones, carrier and microwave equipment, event recorders, and supervisory control equipment, in EHV substations resulted in many equipment failures. Many of these failures have been attributed to transients or surges in the control circuits connected to the solid-state devices. Failures due to transients or surges have been experienced even with conventional electromechanical devices.

The failures being experienced are attributed to the use of EHV (345 kV and higher voltage levels) as well as the presence of unusually high short-circuit currents. One of the major sources of transient voltages is the switching of capacitances, for example, the operation of a disconnect switch, which generates high-magnitude, high-frequency oscillatory surge currents. The transient magnetic fields associated with these high-frequency surge currents are both electrostatically and magnetically coupled to cables in the area. Induced voltages have been reported to be as high as 10 kV in cables without shielding, and the frequencies of these induced voltages have been reported to be as high as 3 MHz.

In order to avoid insulation breakdown at 10-kV crest and possible false operation of relays, it is important that station design includes necessary precautions to limit the undesirable surges and control circuit transients to an acceptable minimum.

In any station design there are several precautions that can be taken. All cable circuits that are used in a substation should be run radially, with each circuit separated from any other circuit and with both supply and return conductors contained within the same cable. If a conductor is routed from the control house to a point in the switchyard with the return circuits following different paths, loops may be formed that are inductive and are subject to magnetically induced voltages. However, when the two conductors involved are both affected by the same field, the voltage appearing between them at the open end should be essentially zero. More information on substation cable shielding practices can be found in IEEE Std 525-2007. In addition, consider utilizing fiber-optic cable for digital signals and information.

Because of ground-mat potential differences and longitudinally induced voltages in the radial circuits, proper cable shielding is necessary to maintain lowest possible voltages on the cable leads. The cables that require shielding include control, current, and potential transformer circuits. The shield should be of as low resistance as possible, and it should be connected to the ground grid at least at both ends. To reduce penetration of a magnetic flux through the nonferrous shield (lead, copper, bronze, etc.), a current must flow in the shield to produce a counterflux, which opposes the applied flux. Ground-grid conductors should be placed in parallel to and in close proximity to the shield to maintain as low a resistance between the ends as possible and also to form a small loop to reduce the reactance between ground and the shield. Without close coupling of the conductor and ground shield, the propagation time of the two paths could differ so that a voltage impulse could arrive at the receiving end with a time difference, hence causing an unwanted voltage difference.

All control, potential-transformer, and current-transformer cables should be shielded, with the shield grounded at the switchyard end and at the control-house end. In addition, each group or run of conduits and cables should be installed with a separate No. 4/0 bare stranded copper cable buried directly in the ground and grounded and bonded to the control-cable shield at each end of each cable. The bare copper cable should run as closely as possible to the cable run. The heavy cable functions to provide a low-resistance path in an attempt to prevent heavy fault currents from flowing in the shield and to reduce reactance between ground and shield.

In order to limit induced voltages, the control-cable runs should be installed, where possible, at right angles to high-voltage buses. Where it is necessary to run parallel to a high-voltage bus for any appreciable distance, the spacing between cables and high-voltage buses should be made as great as possible. Distances of at least 50 ft should be maintained.

It is further considered good practice to have both current-transformer and potential-transformer leads installed with the ground for the secondary wye neutral made at the control-house end rather than at the switchyard end. Any rise due to induced voltages will be concentrated at the switchyard and will ensure operator safety at the control switchboard in the control house.

The shield can be grounded by using a flexible tinned copper braid of from ½ to 1 in wide. The shielded-cable outer insulation is peeled back, exposing the sheath. The 1-in braid is wrapped around the sheath and soldered carefully to it. The other end of the braid is connected to a lug, and solder should be run over the lug to the braid connection. The lug is then bolted securely to the ground bus bar. The flexible copper braid circuits should be kept as short as possible and should be run directly to the ground bus without any bends, if possible.

It should be pointed out that the shields should be grounded at multiple points rather than at a single point, because of the tendency to lose any advantage from single-point grounding at 50 kHz and above. As an example, assume that one input and one output terminal of a system are grounded, each at different points on a common ground plane. A small noise voltage will usually exist across these ground points because of currents flowing in the finitely conductive ground plane. If either the load or source ground is lifted, a ground loop is no longer formed, and coupling of unwanted signals is minimized. This is the advantage of having one physical ground.

Removal of one of the ground connections achieves a single-point ground only for dc and low-frequency signals. At higher frequencies, ground loops will be created by capacitance coupling. Frequencies below 50 kHz are considered the arbitrary crossover point for single-point grounding.

At EHV, the transient voltages above 50 kHz represent the more serious problem; for this reason, all cable shields should be grounded at least at two points. It should be noted that shielding of control cables is normally provided for substations operating at voltage levels of 138 kV and above.

### 11.1.10 Substation Grounding

Grounding at substations is highly important. The functions of a grounding system are listed below:

1. Provide the ground connection for the grounded neutral for transformers, reactors, and capacitors.
2. Provide the discharge path for lightning rods, arresters, gaps, and similar devices.
3. Ensure safety to operating personnel by limiting potential differences, which can exist in a substation.
4. Provide a means of discharging and deenergizing equipment in order to proceed with maintenance on the equipment.
5. Provide a sufficiently low-resistance path to ground to minimize rise in ground potential with respect to remote ground.

Substation safety requirements call for the grounding of all exposed metal parts of switches, structures, transformer tanks, metal walkways, fences, steelwork of buildings, switchboards, instrument-transformer secondaries, etc. so that a person touching or near any of this equipment cannot receive a dangerous shock if a high-tension conductor flashes to or comes in contact with any of the equipment listed. This function in general is satisfied if all metalwork between which a person can complete contact or which a person can touch when standing on the ground is so bonded and grounded that dangerous potentials cannot exist. This means that each individual piece of equipment, each structural column, etc., must have its own connection to the station grounding mat. It is also helpful to utilize a high-resistance surface layer such as crushed rock to provide increased resistance and safety for personnel standing on the native soil, particularly when the soil is wet.

A most useful source of information concerning substation grounding is contained in the comprehensive guide IEEE Standard 80-2013, IEEE Guide for Safety in AC Substation Grounding Period. Much of the following information is based on recommendations stated in the IEEE Standard 80.

NOTE: If a station fence is located so close to items not connected to the station ground grid (such as a conductive fence on neighboring property) that they can be simultaneously contacted, a nonconducting station fence section should be used.

*The basic substation ground system* used by most utilities takes the form of a grid of horizontally buried conductors. The reason that the grid or mat is so effective is attributed to the following:

1. In systems where the maximum ground current may be very high, it is seldom possible to obtain a ground resistance so low as to ensure that the total rise of the grounding system potential will not reach values unsafe for human contact. This being the case, the hazard can be corrected only by control of local potentials. A grid is usually the most practical way to do this.
2. In HV and EHV substations, no ordinary single electrode is adequate to provide needed conductivity and current-carrying capacity. However, when several are connected to each other and to structures, equipment frames, and circuit neutrals which are to be grounded, the result is necessarily a grid, regardless of original objectives. If this grounding network is buried in soil of reasonably good conductivity, this network provides an excellent grounding system.

*The first step in the practical design of a grid or mat* consists of inspecting the layout plan of equipment and structures, and noting all operating locations for personnel. A continuous cable should surround the grid perimeter to enclose as much ground as practical and to avoid current concentration and hence high gradients at projecting ground cable ends. Within the grid, cables should be laid in parallel lines and at reasonably uniform spacing. They should be located, where practical, along rows of structures or equipment to facilitate the making of ground connections. The preliminary design should be adjusted so that the total length of buried conductor, including cross connections and rods, is at least equal to that required to keep local potential differences within acceptable limits.

*A typical grid system* for a substation might comprise 4/0 bare stranded copper cable buried 12 to 18 in below grade and spaced in a grid pattern of about 10 by 20 ft. (Other conductor sizes, burial depths, and grid conductor spacings, however, are frequently used.) At each junction of 4/0 cable, the cables would be securely bonded together, and there might also be connected a driven copper-covered steel rod approximately ⅝ in. in diameter and approximately 8 ft long. In very high-resistance soils it might be desirable to drive the rods deeper (lengths approaching 100 ft are recorded). In cases where rod depth is restricted, consider utilizing ground wells. A typical grid system usually extends over the entire substation yard and sometimes a few feet beyond the fence which surrounds the building and equipment. Figure 11-18 shows a grounding plan for a typical EHV substation operating at 345 kV.

In order to ensure that all ground potentials around the station are equalized, the various ground cables or buses in the yard and in the substation building should be bonded together by heavy multiple connections and tied into the main station ground. This is necessary in order that appreciable voltage differences to ground may not exist between the ends of cables, which may run from the switchyard to the substation building.

Heavy ground currents, such as those that may flow in a transformer neutral during ground faults, should not be localized in ground connections (mats or groups of rods) of small area in order to minimize potential gradients in the area around the ground connections. Such areas should have reinforced wire sizes where necessary to handle adequately the most severe condition of fault-current magnitude and duration.

*Copper cables or straps* are usually employed for equipment-frame ground connections. However, transformer tanks are sometimes used as part of the ground path for lightning arresters mounted thereon. Similarly, steel structures may be used as part of the path to ground if it can be established that the conductivity, including that of any joints, is and can be maintained as equivalent to the copper conductor that would otherwise be required. Studies by some utilities have led to their successful use of steel structures as part of the path to the ground mat from overhead ground wires, lightning arresters, etc. Where this practice is followed, any paint films, which might otherwise introduce a high-resistance joint should be removed and a suitable joint compound applied or other effective means taken to prevent subsequent deterioration of the joint from oxidation.

Connections between the various ground leads and the cable grid and connections within the cable grid are usually clamped, welded, or brazed. Ordinary soldered connections are to be avoided because of possible failure under high fault currents or because of galvanic corrosion.

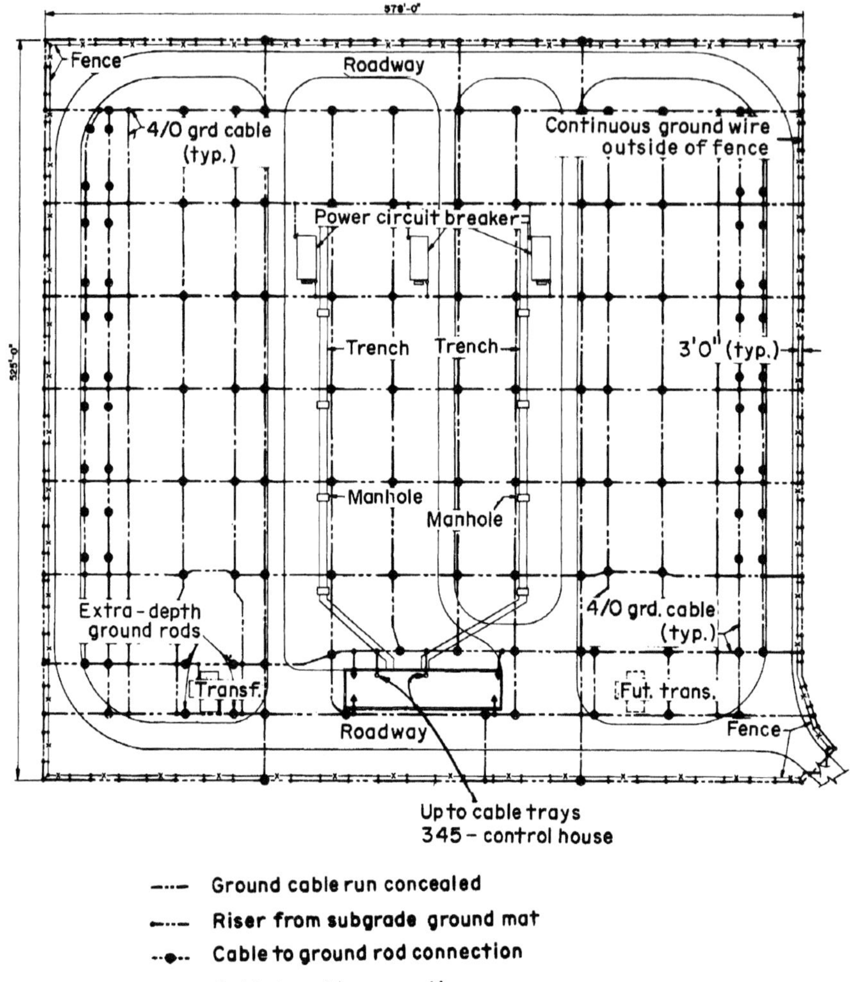


**FIGURE 11-18** Grounding plan for a 345-kV substation.

Each element of the ground system (including grid proper, connecting ground leads, and electrodes) should be so designed that it will

1. Resist fusing and deterioration of electric joints under the most adverse combination of fault-current magnitude and fault duration to which it might be subjected.
2. Be mechanically rugged to a high degree, especially in locations exposed to physical damage.
3. Have sufficient conductivity so that it will not contribute substantially to dangerous local potential differences.

Adequacy of a copper conductor and its joints against fusing can be determined from Table 11-11 and by referring to Fig. 11-19.

**TABLE 11-11** Minimum Copper Conductor Sizes to Avoid Fusing

| Time duration of fault, s | Circular mils per ampere | | |
|---|---|---|---|
| | Cable only | With brazed joints | With bolted joints |
| 30 | 40 | 50 | 65 |
| 4 | 14 | 20 | 24 |
| 1 | 7 | 10 | 12 |
| 0.5 | 5 | 6.5 | 8.5 |

If the substation is on soil of high resistivity so that it is impossible to obtain suitably low resistance from rods driven within the station, it is possible to reduce the resistance by extending the main ground grid outside the enclosed substation area to a secondary ground mat located adjacent to the substation. The effective resistance of the complete grounding system can be lowered appreciably by the use of a more extensive grid area and of additional grid conductor length. An important reason for trying to lower grid resistance is to minimize ground-potential rise with respect to remote ground during ground faults.

*Ground-potential rise* depends on fault-current magnitude, system voltage, and ground-system resistance. The current through the ground system multiplied by its resistance measured from a point remote from the substation determines the ground-potential rise with respect to remote ground. The current through the grid is usually considered to be the maximum available line-to-ground fault current. For example, a ground fault of 15,000 A flowing into a ground grid with a value of 0.5 Ω resistance to absolute earth would cause an *IR* drop of 7500 V. The 7500-V *IR* drop due to the fault current could cause serious trouble to communications lines entering the station if the communications facilities are not properly insulated or neutralized.

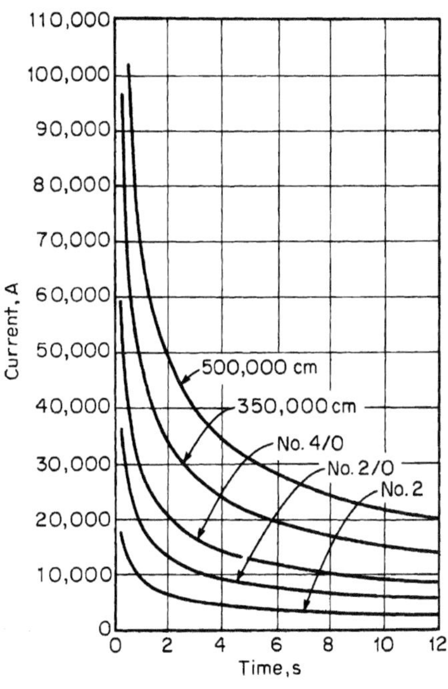


**FIGURE 11-19** Short-time fusing curves for copper cable.

*Low-resistance station grounds* are frequently difficult to obtain. In such cases, the use of driven grounds will provide the most convenient means of obtaining a suitable ground connection. The arrangement and number of driven grounds will depend on the station size and the nature of the soil. The ground mat of Fig. 11-18 has a value measured to be on the order of 0.5 Ω. The best soils for ground mats are wet and marshy, with clay or clay loam as the next best. Sand and sandy soils are of higher resistance, making it difficult to obtain low-resistance ground connections.

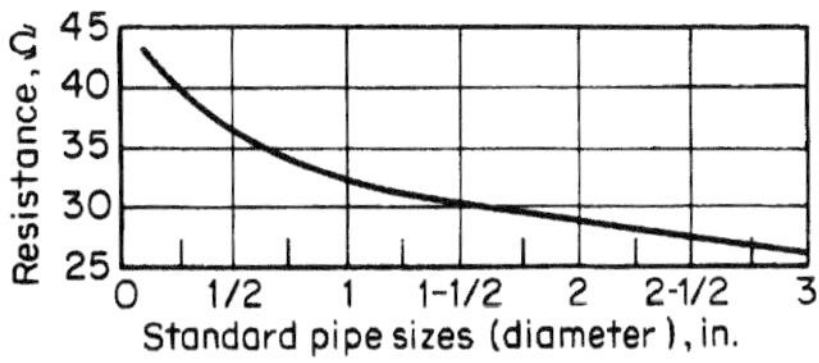


**FIGURE 11-20** Relation between pipe diameter and ground resistance. (*NBS Technologic Paper No. 108, June 1918.*)

The size of the rods used is determined mainly by the depth to which they must be driven, although small rods can be driven to considerable depths by the use of driving collars. Figure 11-20 shows the relationship between rod size and resistance obtained.

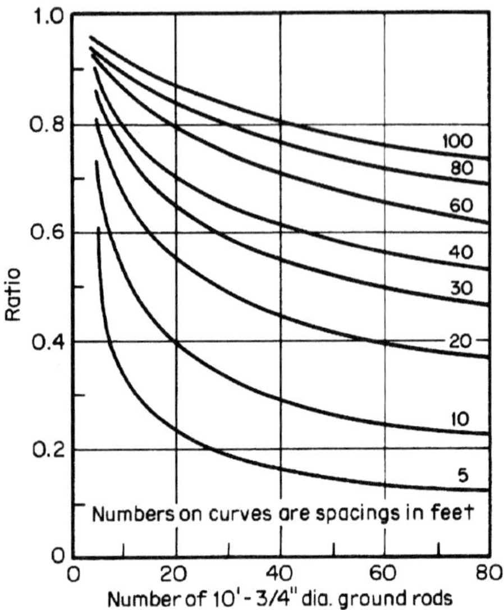


**FIGURE 11-21** Ratio of conductivity of ground rods in parallel on an area to that of isolated rods. (*H. B. Dwight, Trans. AIEE, vol. 55, p. 1936.*)

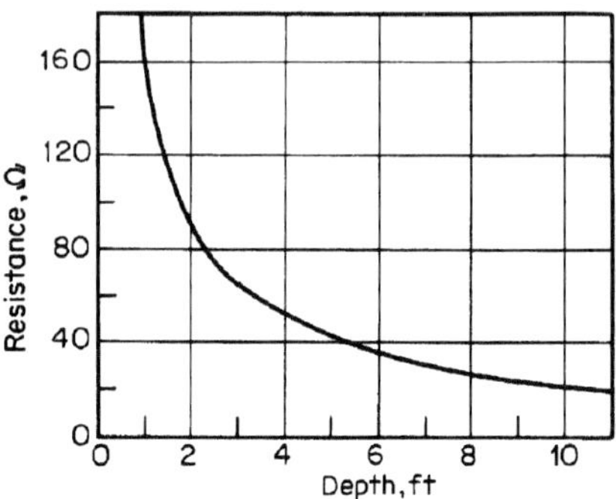


**FIGURE 11-22** Variation of resistance of driven pipes with depth. Soil fairly wet. External diameter of pipe is 24.9 mm (1.02 in). (*NBS Technologic Paper No. 108, June 1918.*)

Driving more rods in a given space will help reduce resistance, but the reduced resistance is not a function of the number of rods. Figure 11-21 shows the effect on resistance of spacing and number of rods in square areas. These curves apply to ¾-in by 10-ft rods. The rods or pipes for permanent stations should be of noncorroding materials. Figure 11-22 shows the effect of increased length of rods in uniform soil. Usually the improvement is much greater than indicated because the rods penetrate into better-conducting earth as they are driven deeper. In addition, where the ground can become frozen, rods must be driven below the frost line to obtain low resistance.

In general, it is advisable to obtain reduced ground resistance by the use of a more extensive mat and more ground rods rather than by treating the earth around the rods with salt because of the impermanence of the treatment. However, treatment of the soil is sometimes the only means whereby suitable resistance can be obtained.

It is not possible to describe all methods of obtaining ground connections of suitably low resistance. The problem sometimes presents great difficulties and calls for considerable extra expense. Locating substations on solid rock with little or no topsoil should be avoided, since the cost of obtaining a low-resistance ground would be excessive. Such a ground would require the use of an extensive counterpoise system with many drilled "wells," in which electrodes would be inserted in treated filling, with provision made for renewing the treatment.

***Measuring Ground Resistance.*** The measurement of ground resistance is necessary both at the time of initial energization of a substation and at periodic intervals thereafter to ensure that the value of ground resistance does not increase appreciably. The measurement of the resistance of a ground connection with respect to absolute earth is somewhat difficult. All results are approximations and require careful application of the test equipment and selection of reference ground points. A useful source of information for measuring ground resistance is IEEE Std 81-1983 "Guide for Measuring Earth Resistivity, Ground Impedance, and Earth Surface Potentials of a Ground System." There are several methods of testing ground resistance, but all of them are similar in that two reference ground connections are used and a suitable source of current is required for the test. Some form of alternating current is circulated through the ground under test in amounts from a few milliamperes, as in bridge methods and with some of the patented ground testers, up to 100 A or more.

The amount of current used depends on the method, and methods using very small currents will give results as accurate as methods using heavy currents if the ground under test is one for which the test method is suitable.

Methods of testing ground resistance fall into three general groups:

1. *Triangulation or three-point methods,* in which two auxiliary test grounds and the point to be measured are arranged in a triangular configuration. The series resistance of each pair of ground points in the triangle is determined by measuring the voltage across and the current through the ground resistance being measured. Resistance measurements are made by the voltmeter-ammeter method or by means of a suitable bridge. For accurate results, the resistance of the auxiliary grounds and the ground under test should be of the same order of magnitude, and results may be meaningless if the test grounds have more than 10 times the resistance of the ground under test. This method is suitable for measuring the resistance of tower footings, isolated ground rods, or small grounding installations. It is not suitable for measurement of low-resistance grounds such as the ground grid at large substations.
2. *Ratio methods,* in which the series resistance $R$ of the ground under test and a test probe is measured by means of a bridge which operates on the null-balance principle. A calibrated slide-wire potentiometer is connected to the two ground connections, with the slider of the potentiometer connected to a second test probe. The potential of the slider to ground is adjusted to zero or null. If $D$ is the total slide-wire resistance and $d_1$ is the resistance from the slider to the ground under test, the resistance $R$ of the ground under test is $(d_1/D) \times R$. The vibrometer and the groundometer, self-contained test instruments, make use of this principle. This method is much more satisfactory than triangulation methods, since ratios of test-probe resistance to the resistance of the ground under test run as high as 300:1 with test instruments such as the groundometer. Although this method has its limitations in testing low-resistance grounds of large areas, suitable readings can be obtained by locating the test probes in a straight line, in a direction 90° from the substation fence, and with the distance of the farthest probe twice the width of the substation. Best accuracy can be attained by taking measurements at the greatest possible distance from the ground grid being measured.
3. *Fall-of-potential methods,* in which close-in reference grounds are included, usually less than 1000 ft from the ground under test. The principle of the fall-of-potential method using close-in reference grounds is illustrated by Fig. 11-23. A fixed probe is driven in the ground at point $C_2$ with a movable probe $P_2$ set at various points in a straight line between $C_2$ and the ground mat $G$ under test. Either alternating current or direct current is circulated through ground $G$ and fixed

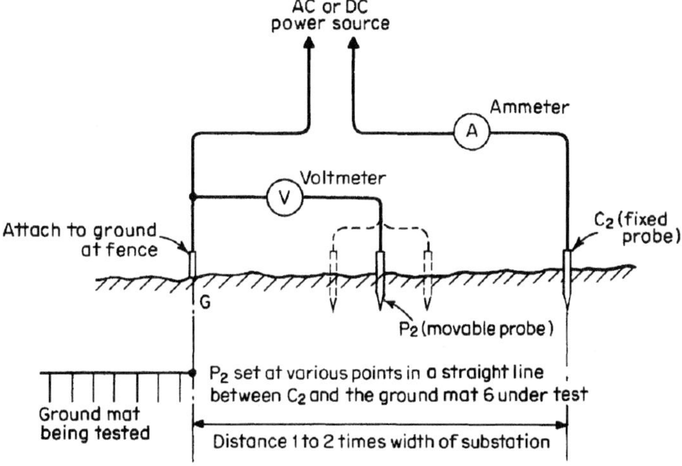


**FIGURE 11-23** Field setup for making ground-resistance tests by means of the fall-of-potential method.

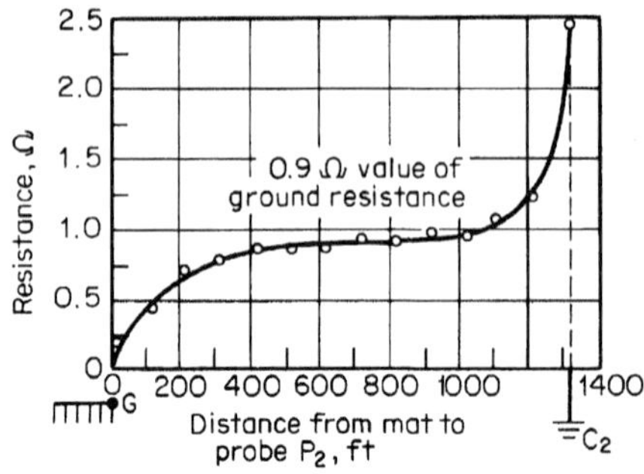


**FIGURE 11-24** Ground-resistance curve for a substation ground mat.

test probe $C_2$. A voltmeter is connected between point $G$ and probe $P_2$, and an ammeter is connected to observe current flow through probe $C_2$. Voltmeter readings $E$ are taken simultaneously with ammeter readings $I$. The reading $E/I$, which equals the resistance in ohms, is plotted in Fig. 11-24. The resistance shown on the flat part of the curve or at the point of inflection is taken as the resistance of the ground. This method may be subject to considerable error if stray currents are present. It is normally applied by using several test-probe readings at 10% intervals of the distance from $G$ to $C_2$, with the test probe located midway between $G$ and $C_2$. Self-contained test instruments which make use of this method are available; among them are the ground ohmer and Megger ground tester. These instruments give better results than the voltmeter-ammeter method, since they are designed to eliminate the effects of stray currents.

Considerable emphasis has been placed on the use of computer programs to calculate the design parameters of substation ground systems. These programs normally employ methods detailed in IEEE Guide 80, with the following typical input data required:

1. System voltage, symmetric rms single-phase-to-ground fault currents, and the clearing time of faults
2. Length and width of substation area
3. Estimated value of ground resistivity in ohm-meters; sometimes in multiple layers
4. Assumed value of ground conductor length
5. Cross section of conductors available

The following typical information is derived from the program:

1. Size, total length, and number of strands of copper ground conductor
2. Spacing of main grid configuration along width and length
3. Expected ground-mat resistance
4. Depth of grid below ground level
5. Tolerable limits and maximum values of step and touch potentials

It should be noted that the step and touch potentials are defined as follows:

$E_{step}$ is the tolerable potential difference between any two points on the ground surface, which can be touched simultaneously by the feet.

$E_{touch}$ is the tolerable potential difference between any point on the ground where a person may stand and any point which can be touched simultaneously with either hand.

## 11.1.11 Transformers

***Transformer Connections.*** *Delta-delta–connected* transformers are used mainly on the lower transmission voltages. This is due to the fact that the complete winding must be insulated for full line-to-line voltage; for voltages above 73 kV, the cost increase is appreciable over Y-connected transformers with graded insulation. The delta-delta–connected transformers have one advantage in that the bank can be operated open delta at 86.6% of the capacity of the two remaining transformers.

*The delta-star connection* is in common use for both step-up and step-down purposes. When used as a step-up transformer, the high-tension winding is Y-connected; when used for step-down purposes, the low-tension winding is usually Y-connected in order to provide a grounded neutral for secondary transmission or for primary distribution.

*Delta-connected* high-tension windings, however, are seldom used for transmission voltages of 138 kV and above. The delta-Y connection almost completely suppresses the triple harmonics with the neutral solidly grounded. Triple harmonics, which can appear on power systems are the third and its odd multiples. Y-connected windings on the higher voltages are usually provided with graded insulation, the neutral-end turns of which may have very little insulation if the neutral is solidly grounded. If neutral impedance (reactor or resistor) is used, neutral insulation must be equal to or greater than the maximum *IZ* drop of the neutral impedance. If the neutral is to be left ungrounded on either grounded neutral systems or ungrounded neutral systems, the neutral insulation should be the same as it is on the line side to avoid traveling-wave troubles.

*Star-star–connected (Y-Y) transformers* are used infrequently on high-voltage transmission systems. When used with both neutrals grounded, if single-phase or three-phase shell type, they must be used with Y-connected generators, and a solid neutral connection must be provided between the generator, or generators, and the low-tension transformer neutral in order to minimize triple-harmonic troubles. The various types of Y-Y–connected transformers can be used with both neutrals ungrounded with satisfactory results or with neutrals grounded if of the three-phase core type. The triple harmonics are nearly suppressed in three-phase core-type transformers.

Star-star–connected transformers with a delta-connected third winding (tertiary) overcome the difficulties of the simple Y-Y connection. The tertiary winding may be for the suppression of harmonics only, in which case no connections are brought out with three-phase transformers. Y-delta-Y transformers are frequently used to supply two distribution voltages or a distribution voltage and a secondary-transmission voltage. If the service supplied from the delta-connected winding is four-wire three-phase, the neutral must be obtained from a separate grounding transformer. A common use for the tertiary winding is to provide substation station-service power to operate station auxiliary equipment. Three-winding transformers, all windings of which are used, are frequently rated with two outputs: (1) the individual output of each secondary winding alone with the other secondary winding carrying no load and (2) a simultaneous loading rating in which each secondary winding is given a rated loading with the primary-winding loading the resultant of the two secondary loadings.

*Autotransformers* are generally used for transforming from one transmission voltage to another when the ratio is 3:1 or less. Such transformers are normally connected in Y with the neutral solidly grounded and when so connected should be provided with a closed delta tertiary winding of adequate capacity for the suppression of harmonics, for ground-fault duty, and to provide station-service power. The tertiary is frequently used to provide a supply of distribution voltage. Autotransformers are superior to separate-winding transformers owing to their lower cost, greater efficiency, smaller size and weight, and better regulation. Autotransformers also may be obtained with zigzag-connected windings or with delta-connected windings. Both these types are free from triple-harmonic troubles but in general are more expensive.

*Delta-connected autotransformers* have a possible disadvantage in that they insert a phase shift into the transformation, which means that the system being served must be radial or else it must be served by similar transformations at other points. Operating, reliability, and economic considerations should dictate whether the design should be single-phase, or three-phase, and whether voltage regulation is required. Oftentimes a spare single-phase unit is utilized on-site to address loss of another phase. Larger autotransformers are typically available as single-phase only.

***Transformer Loading Practice.*** Because of the varying load cycle of most transformers, it is customary to permit loading considerably in excess of the transformer nameplate rating. There may be limitations on the transformer imposed by bushings, leads, tap changers, cables, disconnecting switches, circuit breakers, etc. Good engineering design, however, will permit operation without these limitations.

The increase in transformer loading is limited by the effect of temperature on insulation life. High temperature decreases the mechanical strength and increases the brittleness of fibrous insulation and makes transformer failure increasingly likely even though the dielectric strength of the insulation may not be seriously decreased. Overloading should be limited then by giving consideration to the effect on insulation life and transformer life. For recurring loads, such as the daily load cycles, the transformer would be operated for normal life expectancy. For emergencies, either planned or accidental, loading would be based on some percentage loss of life.

In a typical case for a failure of part of the electrical system, a 2.5% loss of life per day for a transformer may be acceptable. Loading recommendations based on the evaluation of the loss of insulation life as affected by temperature are contained in IEEE Standard C57.91-1995, Guide for Loading Mineral-Oil-Immersed Transformers. NEMA Publ. TR98-1964 contains corresponding recommendations for loading power transformers with 65°C average winding rise insulation systems. IEEE Standard C57.91-1995 states that an average loss of life of 1% per year or 5% in any one emergency operation is considered reasonable.

Daily overload cycles consistent with normal life expectancy for air-cooled power transformers at 30°C ambient temperature are given in Table 11-12, which is a condensation of data taken from IEEE Standard C57.91-1995. For a listing of transformer loading above normal with some sacrifice of life expectancy, data given in NEMA Publ. TR98-1964, Part 3, are condensed in Table 11-13.

**TABLE 11-12** Percent Daily Peak Load for Normal Life Expectancy with 30°C Cooling Air

| Duration of peak load, h | Self-cooled with % load before peak of | | | Forced-air-cooled up to 133% of self-cooled rating, with % load before peak of | | | Forced-air-cooled over 133% of self-cooled rating, or forced-oil-cooled, with % load before peak of | | |
|---|---|---|---|---|---|---|---|---|---|
| | 50% | 70% | 90% | 50% | 70% | 90% | 50% | 70% | 90% |
| 0.5 | 189 | 178 | 164 | 182 | 174 | 161 | 165 | 158 | 150 |
| 1 | 158 | 149 | 139 | 150 | 143 | 135 | 138 | 133 | 128 |
| 2 | 137 | 132 | 124 | 129 | 126 | 121 | 122 | 119 | 117 |
| 4 | 119 | 117 | 113 | 115 | 113 | 111 | 111 | 110 | 109 |
| 8 | 108 | 107 | 106 | 107 | 107 | 106 | 106 | 106 | 105 |

Ambient temperature affects load capacity by an amount depending on the type of cooling as shown in Tables 11-12 and 11-13. For changes from this average ambient temperature, transformer ratings may be adjusted as shown in Table 11-14. The table applies to both the 55°C and the 65°C average winding-temperature-rise transformers. For the ambient temperature of air-cooled transformers, use the average value over a 24-h period or 10°C under the maximum during the 24-h period, whichever is higher.

The following temperatures and load limitations are generally applied to transformers. The temperature of the top oil should never exceed 100°C. The maximum hot-spot winding temperature should not exceed 150°C for 55°C rise transformers or 180°C for 65°C rise transformers. Short-time peak loading for ½ h or more should not exceed 200% rating. At abnormally high temperatures it may be necessary to remove some oil in order to avoid overflow or excessive pressure.

## 11.1.12 Surge Protection

A substation should be designed to include safeguards against the hazards of abnormally high voltage surges that can appear across the insulation of electrical equipment in the station. The most severe

**TABLE 11-13** Allowable Peak Loads (in Multiples of Maximum Nameplate Rating) for Moderate Sacrifice of Life Expectancy with 30°C Cooling Air

| Duration of peak load, h | Hottest-spot temperature reached, °C | Life loss in percent not more than | Self-cooled (OA) with % load before peak of | | | | Forced-air-cooled (OA/FA) up to 133% of self-cooled rating with % load before peak of | | | | Forced-air-cooled (OA/FA/FA) over 133% of self-cooled rating or forced-oil-cooled (FOA or OA/FOA/FOA) with % load before peak of | | | |
|---|---|---|---|---|---|---|---|---|---|---|---|---|---|---|
| | | | 50% | 70% | 90% | 100% | 50% | 70% | 90% | 100% | 50% | 70% | 90% | 100% |
| ½ | 171 | 0.25 | 2.00 | 2.00 | 2.00 | 1.96 | 2.00 | 1.95 | 1.85 | 1.80 | 1.64 | 1.60 | 1.54 | 1.51 |
| | 180 | 0.50 | 2.00 | 2.00 | 2.00 | 2.00 | 2.00 | 2.00 | 1.95 | 1.90 | 1.69 | 1.66 | 1.60 | 1.57 |
| 1 | 163 | 0.25 | 1.96 | 1.89 | 1.80 | 1.74 | 1.77 | 1.72 | 1.65 | 1.61 | 1.47 | 1.45 | 1.49 | 1.39 |
| | 180 | 1.00 | 2.00 | 2.00 | 1.99 | 1.94 | 1.93 | 1.88 | 1.81 | 1.78 | 1.57 | 1.55 | 1.52 | 1.50 |
| 2 | 155 | 0.25 | 1.68 | 1.63 | 1.57 | 1.53 | 1.53 | 1.50 | 1.47 | 1.44 | 1.33 | 1.32 | 1.31 | 1.30 |
| | 171 | 1.00 | 1.83 | 1.79 | 1.71 | 1.64 | 1.66 | 1.64 | 1.60 | 1.58 | 1.42 | 1.41 | 1.39 | 1.39 |
| | 180 | 2.00 | 1.91 | 1.83 | 1.71 | 1.64 | 1.74 | 1.71 | 1.65 | 1.61 | 1.47 | 1.46 | 1.44 | 1.43 |
| 4 | 147 | 0.25 | 1.44 | 1.41 | 1.39 | 1.37 | 1.35 | 1.34 | 1.33 | 1.32 | 1.24 | 1.23 | 1.23 | 1.23 |
| | 163 | 1.00 | 1.55 | 1.52 | 1.47 | 1.44 | 1.47 | 1.46 | 1.45 | 1.45 | 1.32 | 1.32 | 1.32 | 1.32 |
| | 180 | 4.00 | 1.55 | 1.52 | 1.47 | 1.44 | 1.51 | 1.50 | 1.47 | 1.46 | 1.40 | 1.40 | 1.39 | 1.39 |
| 8 | 139 | 0.25 | 1.28 | 1.27 | 1.27 | 1.26 | 1.24 | 1.24 | 1.24 | 1.24 | 1.18 | 1.18 | 1.18 | 1.18 |
| | 155 | 1.00 | 1.38 | 1.37 | 1.36 | 1.36 | 1.36 | 1.36 | 1.36 | 1.36 | 1.27 | 1.27 | 1.27 | 1.27 |
| | 171 | 4.00 | 1.38 | 1.37 | 1.36 | 1.36 | 1.42 | 1.42 | 1.41 | 1.41 | 1.35 | 1.35 | 1.35 | 1.35 |

***Note:*** For forced-air-cooled transformers, the peak loads are calculated on the basis of all cooling being in use during the period preceding the peak load. When operating without fans, use the tables for OA transformers. Differences in cooling methods used with forced-oil-cooled transformers result in differences in peak-load-carrying ability. Consult the manufacturer before applying loads above the values given in the table.

***Source:*** Based on capability tables in *NEMA Publ.* TR98, Part 3.0

**TABLE 11-14** Effect of Ambient Temperature on kVA Capacity

| Type of cooling | % of rated kVA decrease in capacity for each °C increase over 30°C air | % of rated kVA increase in capacity for each °C decrease under 30°C |
|---|---|---|
| Self-cooled—OA | 1.5 | 1.0 |
| Forced-air-cooled—OA/FA, OA/FA/FA | 1.0 | 0.75 |
| Forced-air-cooled—FOA, OA/FOA/FOA | 1.0 | 0.75 |

overvoltages are caused by lightning strokes and by switching surges. The main methods to prevent these overvoltages from causing insulation failures include

1. Use of surge arresters
2. Equipment neutral grounding
3. Proper selection of equipment impulse insulation level
4. Proper selection and coordination of equipment basic insulation levels
5. Careful study of switching-surge levels that can appear in the substation

The main device used to prevent dangerous overvoltages, flashovers, and serious damage to equipment is the surge arrester. The surge arrester conducts high surge currents, such as can be caused by a lightning stroke, harmlessly to ground and thus prevents excessive overvoltages from appearing across equipment insulation. For a detailed description of the characteristics and application of arresters, refer to Sec. 27.

The important consideration in applying surge arresters and in selecting equipment insulation levels depends greatly on the method of grounding used. Systems are considered to be effectively grounded when the coefficient of grounding does not exceed 80%. Similarly, systems are noneffectively grounded or ungrounded when the coefficient of grounding exceeds 80%.

A value not exceeding 80% is obtained approximately when, for all system conditions, the ratio of zero sequence reactance to positive sequence reactance ($X_0/X_1$) is positive and less than 3 and the ratio of zero sequence resistance to positive sequence reactance ($R_0/X_1$) is positive and less than 1.

What this says in effect is that if neutrals are grounded solidly everywhere and if a ground occurs on one of the conductors, then the voltage that can appear on the healthy phases cannot exceed 80% of normal phase-to-phase voltage.

Thus, the *coefficient of grounding* is defined as the ratio of maximum sustained line-to-ground voltage during faults to the maximum operating line-to-line voltage. On many HV and EHV systems, the coefficient of grounding may be as low as 70%.

Surge-arrester ratings are normally selected on the basis of the coefficient of grounding; thus, for effectively grounded systems, the 80% arrester is selected when using the conventional gap-type arrester. When using the gapless metal oxide arrester, a lower-value arrester may be selected based on the maximum continuous operating voltage (MCOV) equal to the maximum normal line-to-neutral voltage. For example, a 115-kV system (maximum operating voltage equals 121 kV) can use a 97-kV conventional arrester, that is, 80% of 121 kV, when operating on a solidly grounded system, and can use a gapless-type metal oxide arrester rated 70 kV. It should be noted that other factors, such as resonant conditions and system switching, could increase the value of the coefficient of grounding and thus should be studied in each individual system.

*The impulse insulation level* of a piece of equipment is a measure of its ability to withstand impulse voltage. It is the crest value, in kilovolts, of the wave of impulse voltage that the equipment must withstand. However, at EHV, the switching-surge insulation level may be lower than the corresponding impulse level, and thus the switching-surge level becomes the dominant factor in establishing insulation levels.

Basically, the coordination of insulation in a substation means the use of no higher-rated arrester than required to withstand the 60-Hz voltage and the choice of equipment insulation levels that can be protected by the arrester. Careful study of switching-surge levels that can occur at the substation as determined, for example, by transient network analyzer studies also can be used to determine and coordinate proper impulse insulation and switching-surge strength required in substation electrical equipment.

## 11.1.13 References for Air-Insulated Substations


### *Books*

*Alcoa Aluminum Bus Conductor Handbook.* Pittsburgh, PA, Aluminum Company of America, 1957.

*Applied Protective Relaying.* Newark, NJ, Westinghouse Electric Corporation.

Blume, L. F., Boyajian, A., Camilli, G., Lennox, T. S., Minneci, S., and Montsinger, V. M.: *Transformer Engineering.* New York, John Wiley & Sons; 1951.

*Electrical Transmission and Distribution Reference Book.* East Pittsburgh, PA, Westinghouse Electric Corporation, 1950. Mason, C. R.: *The Art and Science of Protective Relaying.* New York, John Wiley & Sons, 1967.

Van C. Warrington, A. R.: *Protective Relays, Their Theory and Practice,* vol. 1. London, Chapman & Hall, 1971.

Van C. Warrington, A. R.: *Protective Relays, Their Theory and Practice,* vol. 2. London, Chapman & Hall, 1974.

### *Station Design*

AIEE Committee Report: A Guide for Minimum Electrical Clearances for Standard Basic Insulation Levels, *Trans. AIEE, Power Appar. Syst.,* June 1954, vol. 73, pp. 636–641.

AIEE Committee Report: Basic Structural Design for Transmission Substations Including Light Metals, *Electr. Eng.,* April 1952, vol. 71, pp. 344–350.

Colombo, A., Sartorio, G., and Taschini, A.: Phase to Phase Air Clearances in EHV Substations as Required by Switching Surges, *CIGRE Paper* 33-11, 1972.

Committee Report: Design Standardization Methods and Techniques for Substation Facilities (Bibliogr.), *Trans. AIEE, Power Appar. Syst.,* October 1964, vol. 83, pp. 1029–1034.

Committee Report: 500 kV AC Substation Design Criteria, Summary of Industry Practices, *IEEE Trans., Power Appar. Syst.,* 1969, vol. 88, pp. 854–861.

Committee Report: Minimum Line-to-Ground Electrical Clearances for EHV Substations Based on Switching Surge Requirement, *IEEE Trans., Power Appar. Syst.,* 1972, vol. 91, pp. 1924–1930.

Committee Report: 700/765 kV AC Substation Design Criteria, A Summary of Industry Practices, *IEEE Trans., Power Appar. Syst.,* 1970, vol. 89, pp. 1521–1524.

Dolan, P. R., and Peat, A. J.: Design of the First 500 kV Substations on the Southern California Edison Company System, *IEEE Trans., Power Appar. Syst.,* 1967, vol. 86, pp. 531–539.

Hertig, G. E.: High- and Extra-High-Voltage Substation Design and Economic Comparisons, *Trans. AIEE,* 1963, vol. 81, pp. 832–840.

IEEE Standard C2-2012, National Electrical Safety Code.

IEEE Standard 693-2005 Recommended Practice for Seismic Design of Substations.

IEEE Standard 979-2012 Guide for Substation Fire Protection.

IEEE Standard 980-2013 Guide for Containment and Control of Oil Spills in Substations.

IEEE Standard 1127-2013 Guide for the Design, Construction, and Operation of Electric Power Substations for Community Acceptance and Environmental Compatibility.

IEEE Standard 1264-2015 Guide for Animal Deterrents for Electric Power Supply Substations.

IEEE Standard 1427-2006 Guide for Recommended Electrical Clearances and Insulation levels in Air-Insulated Electrical Power Substations.

Paris, L., and Taschini, A.: Phase-to-Ground and Phase-to-Phase Air Clearances in Substations, *CIGRE, Electra,* 1973, no. 29. (Recommended by CIGRE S.C. 23 and CIGRE S.C. 33.)

Scherer, H. N.: 765 kV Station Design, *IEEE Trans. Power Appar. Syst.,* 1969, vol. 88, pp. 1372–1376.

### *Bus Construction*

Attri, N. S., and Edgar, J. N.: Response of Bus Bars on Elastic Supports Subjected to a Suddenly Applied Force, *IEEE Trans., Power Appar. Syst.,* 1967, vol. 86, pp. 636–650.

Committee Report: Use of Aluminum for Substation Busses, *IEEE Trans., Power Appar. Syst.,* 1963, vol. 82, pp. 72–102.

Dwight, H. B.: Skin Effect and Proximity Effect in Tubular Conductors, *Trans. AIEE,* February 1922, pp. 189–198.

Fischer, E. G.: Seismic Design of Bus Runs and Supports, *IEEE Trans., Power Appar. Syst.*, 1973, vol. 92, pp. 1493–1500.

Foti, A.: Design and Application of EHV Disconnecting Switches, *Trans. AIEE, Power Appar. Syst.*, October 1965, vol. 84, pp. 868–876.

Higgins, T. J.: Formulas for Calculating Short Circuit Forces between Conductors of Structural Shape, *Trans. AIEE,* October 1943, vol. 62, pp. 659–663.

Higgins, T. J.: Formulas for Calculating Short Circuit Stresses for Bus Supports for Rectangular Tubular Conductors, *Trans. AIEE,* August 1942, vol. 61, pp. 578–580.

IEEE Standard C37.30-1997, Requirements for High-Voltage Switches.

IEEE Standard C37.30.1-2011, Requirements for AC High-Voltage Air Switches Rated Above 1000 V.

IEEE Standard C37.32-2002 American National Standard for High-Voltage Switches, Bus Supports, and Accessories, Schedules of Preferred Ratings, Construction Guidelines, and Specifications.

IEEE Standard 605-2011 Guide for Bus Design in Air-Insulated Substations.

IEEE Standard 1527-2006 Recommended Practice for the Design of Flexible Buswork Located in Seismically Active Areas.

Milton, R. M., and Chambers, F.: Behavior of High-Voltage Busses and Insulators during Short Circuits, *Trans. AIEE,* August 1955, vol. 74, pp. 742–749.

NEMA Standard SG6-1995, Power Switching Equipment.

Rayleigh, J. W. S.: *Aeolian Tones.* New York, Cambridge University Press, 1920.

Schurig, O. R., and Sayre, M. F.: Mechanical Stresses in Bus Bar Supports During Short Circuits, *J. AIEE,* April 1925, vol. 44, pp. 365–372.

Taylor, D. W., and Stuehler, C. M.: Short Circuit Tests on 138 kV Busses, *Trans. AIEE,* August 1956, vol. 75, pp. 739–747.

Temple, G., and Brickley, W. G.: *Rayleigh's Principle.* New York, Oxford University Press, 1933.

Wagner, C. F.: Current Distribution in Multi Conductor Single Phase Buses, *Electr. World,* March 18, 1922, vol. 79, no. 11.

### *System Protection*

Blackburn, J. L.: Future Automatic Switching of EHV Transmission Lines—Development and Application of Solid-State Relays, *Proc. Am. Power Conf.,* 1965, vol. 27, pp. 998–1008.

Boyaris, E., and Guyot, W. S.: Experience with Fault Pressure Relaying and Combustible Gas Detection in Power Transformers, *Proc. Am. Power Conf.,* 1971, vol. 33, pp. 1116–1126.

Chadwick, J. W., and Goff, L. E.: Development of a Static Single-pole Relaying Scheme for the TVA 500-kV System, *Proc. Am. Power Conf.,* 1971, vol. 33, pp. 1127–1133.

Committee Report: Relaying the Keystone 500 kV System, *IEEE Trans.,* June 1968, vol. 87, no. 5, no. 6, pp. 1434–1439.

Elmore, W. A.: Some Guidelines for Selecting a Solid-State Transmission Line Relaying System, *Westinghouse Eng.*, March 1972, vol. 32, no. 2, pp. 50–59.

Emanuel, A. E., and Vora, J. P.: Sensor Coil for Internal Fault Protection of Shunt Reactors, *IEEE Trans.*, November-December 1974, vol. 93, no. 6, pp. 1917–1926.

Forford, T., and Linders, J. R.: A Half Cycle Bus Differential Relay and Its Applications, *IEEE Trans.*, July-August 1974, vol. 93, no. 4, pp. 1110–1120.

Horowitz, S. H., and Seeley, H. T.: Relaying the AEP 765 kV System, *IEEE Trans.,* September 1969, vol. PAS-88, no. 9, pp. 1382–1389.

IEEE Committee Report: Bibliography of Relay Literature 1995.

IEEE Committee Report: Ground Relaying Practices and Problems: A Power System Relaying Committee Survey, *IEEE Trans., Power Appar. Syst.,* May 1966, vol. PAS-85, no. 5, pp. 524–532.

IEEE Standard C37.90.1-2012 Surge Withstand Capability (SWC) Tests for Relays and Relay Systems Associated with Electric Power Apparatus.

IEEE Standard C37.90-2005, Relays and Relay Systems Associated with Electric Power Apparatus.

IEEE Standard C37.91-2008, Guide for Protecting Power Transformers.

Korponay, N., and Ungrad, H.: The Requirements Made of Current Transformers by High-Speed Protective Relays, *Brown Boveri Rev.,* June 1968, vol. 55, no. 6, pp. 289–297.

Narayan, V.: Distance Protection of HV and EHV. Transmission Lines, *Brown Boveri Rev.,* July 1971, vol. 58, no. 7, pp. 276–286.

Schumm, G. P.: The Philosophy of Protective Relaying in the United States and Europe, *Proc. Am. Power Conf.,* 1971, vol. 33, pp. 1105–1115.

Sutton, H. J.: The Application of Relaying on an EHV System, *IEEE Trans.*, April 1967, vol. 86, no. 4, pp. 408–415.

Sykes, J. A., and Morrison, I. F.: A Proposed Method of Harmonic Restraint Differential Protection of Transformers by Digital Computer, *IEEE Trans.*, May-June 1972, vol. 91, no. 3, pp. 1266–1272.

Ungrad, H.: Back-up Protection, *Brown Boveri Rev.*, June 1968, vol. 55, no. 6, pp. 297–305.

Ungrad, H.: Distance Relays with Signal Transmission for Main and Backup Protection, *Brown Boveri Rev.*, July 1971, vol. 58, no. 7, pp. 293–304.

Vanderleck, J. M.: Measurement of Composite Error of Relay-type Current Transformers, *Ontario Hydro Res. Q.*, 1967, vol. 19, no. 3, pp. 15–18.

### *Shielding of Control Cables*

Dietrick, R. E., Ramberg, H. C., and Barber, J. C.: BPA Experience with EMI Measurements and Shielding in EHV Substations, *Proc. Am. Power Conf.*, 1970, vol. 32, pp. 1054–1061.

Kotheimer, W. C.: Control Circuit Transients, Pt. 1, *Power Eng.*, January 1969, vol. 73, no. 1, pp. 42–46.

Kotheimer, W. C.: Control Circuit Transients, Pt. 2, *Power Eng.*, February 1969, vol. 73, no. 2, pp. 54–56.

IEEE Standard 525-2007, Guide for the Design and Installation of Cable Systems in Substations.

Sutton, H. J.: Transients Induced in Control Cables Located in EHV Substation, *IEEE Trans.*, July-August 1970, vol. 89, no. 6, p. 1069.

### *Grounding*

AIEE Committee Report: Application Guide on Methods of Substation Grounding, *Trans. AIEE, Power Appar. Syst.*, April 1954, vol. 73, pp. 271–275.

Bellasi, P. L.: Impulse and 60-Cycle Characteristics of Driven Grounds, *Trans. AIEE,* March 1941, vol. 60, pp. 123–128.

Committee Report: Principles and Practices in Grounding, *Edison Electr. Inst. Ser. Rep.* D9, October 1936.

Eaton, J. R.: Grounding Electric Circuits Effectively, I, II, III, *Gen. Electr. Rev.*, June, July, and August 1941.

EPRI Final Report EL-2682: Analysis Techniques for Power Substation Grounding Systems, vol. 1, Design Methodology and Tests.

IEEE Standard 80-2013, Guide for Safety in AC Substation Grounding.

IEEE Standard 81-2012, Guide for Measuring Earth Resistivity, Ground Impedance, and Earth Surface Potentials of a Grounding System.

IEEE Standard 142-2007, Recommended Practice for Grounding of Industrial and Commercial Power Systems.

Kinyon, A. L.: Earth Resistivity Measurements for Grounding Grids, Trans. *AIEE, Power Appar. Syst.*, December 1961, vol. 80, pp. 795–800.

## 11.2 GAS-INSULATED SUBSTATIONS

*BY GEORGE W. BECKER*

### 11.2.1 Introduction and Overview

Gas-insulated substations (GIS) provide many benefits related to the design of the high-voltage transmission system. Their compact size, modular construction, immunity from environmental hazards (e.g., pollution, salt spray, chemical agents), and inherent high security in areas where substation equipment needs to be protected from sabotage and attack make them ideal design alternatives for system expansion, reliability upgrades, and aging infrastructure replacements.

To construct, they require 10% to 25% of the space compared to a conventional air-insulated substation (AIS) described in Sec. 11.1. GIS provide minimal fire/explosion risks, and no pollution risk from leaked oil. They produce no noise or radio interference and are considered arc proof for electrical fault occurrences.

Section 11.2 provides an overview of the design of gas-insulated substations. This subsection summarizes the general characteristics of gas-insulated equipment; $SF_6$ properties in gas-insulated substations; the GIS gas sealing system; enclosure and conductor design features, voltage withstand, and ampacity; and GIS typical layout and enclosure designs. These subject areas include each of the major design components in GIS equipment [1].

## 11.2.2 General Characteristics of Gas-Insulated Equipment

The foundation of the electrical capabilities of gas-insulated substations is the dielectric capability of the insulating fluid used in the equipment and its enclosures that hold the insulating fluid. The definition of a GIS is the following, a compact, multicomponent assembly, enclosed in a grounded metallic housing in which the primary insulating medium is $SF_6$ and which normally includes buses, switches, circuit breakers, and other associated equipment [2].

The two key elements of the overall GIS are the insulating medium ($SF_6$ gas), and the gas-insulated switchgear enclosure. The enclosure contains the insulating medium under pressure and is defined as a grounded part of the gas-insulated metal-enclosed switchgear assembly. Retaining the insulating gas under the prescribed conditions is necessary to maintain the required insulation level, protect the equipment against external influences, and provide a high degree of insulation from live energized parts [2].

It is important for engineers to have a basic understanding of the physics of gas-insulated switchgear systems, gaseous insulation, and the properties of the insulating medium.

The dielectric design of GIS is based on a simple formula. The electrical stress $E$ at any given radial point $r$ in the system is determined by the relationship:

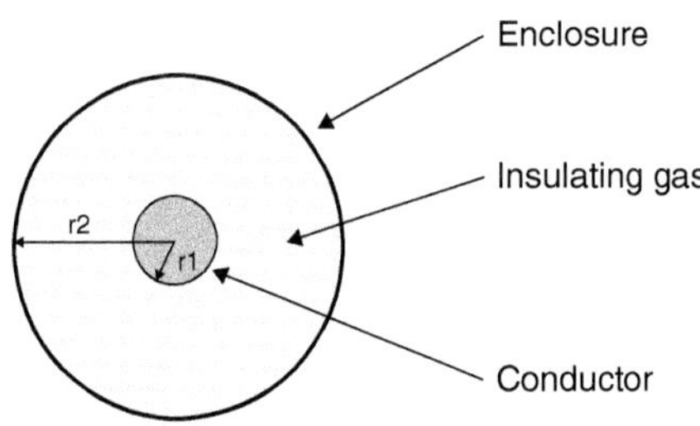


**FIGURE 11-25** Dimensions of gas-conductor enclosure [3].

$$E = \frac{U}{r \ln(r_2/r_1)}$$

where $U$ is the applied voltage, and $r_2$ and $r_1$ are the enclosure and conductor radii, respectively (Fig. 11-25).

The most economical use of materials occurs when $\ln(r_2/r_1) = 1$.

The functionality of gas-insulated equipment is based on a coaxial design that has a circular conductor centered within a grounded metallic enclosure supported by insulators (see Fig. 11-25). The volume between the conductor and the enclosure contains the dielectric fluid—in this case, the insulating gas. The goal is to minimize and distribute the electric field in the system. The maximum electric field inside a GIS remains nearly constant over a wide range of ratios of enclosure to conductor radii (Fig. 11-26) [3, 4].

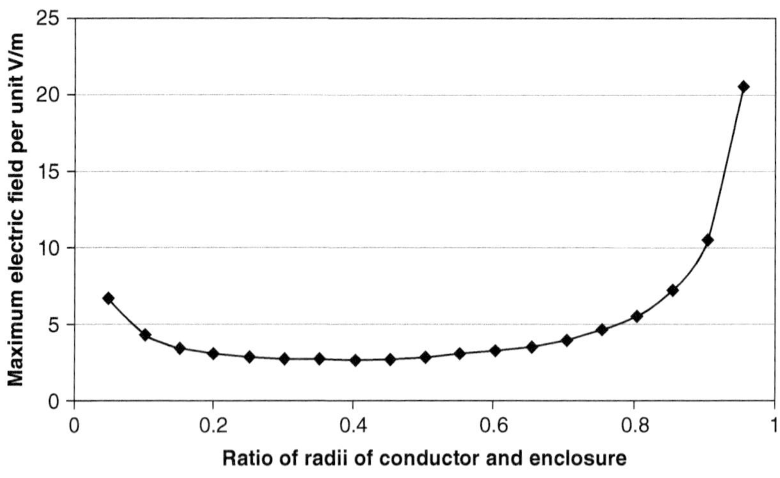


**FIGURE 11-26** Electric field and enclosure dimensions [3].

The integrity of the dielectric insulating gas is essential to the electrical performance of the gas-insulated substation equipment from both the basic insulation level (BIL) perspective and the arc interruption perspective.

Sulphur hexafluoride ($SF_6$) or mixtures of $SF_6$ with other gases are the primary insulating medium used in GIS. The first ionization potential of $SF_6$ is very close to that of nitrogen ($N_2$), which is another gas that has been widely used as an insulating medium for electrical equipment. The $SF_6$ molecule is much larger than the nitrogen molecule. The resultant mean free path for an electron-to-molecule collision in $SF_6$ is about one-third the distance in nitrogen. This makes $SF_6$ an excellent insulating medium, along with its other positive qualities related to electrical insulation in contrast to nitrogen or air.

The dimensional comparison between air insulated equipment and gas-insulated equipment provides an understanding of the general design features of GIS and GIL and the compact nature of the equipment. Figure 11-27 depicts the dimensional differences for required phase spacing for an air insulated system versus a gas-insulated system for the same application [1].

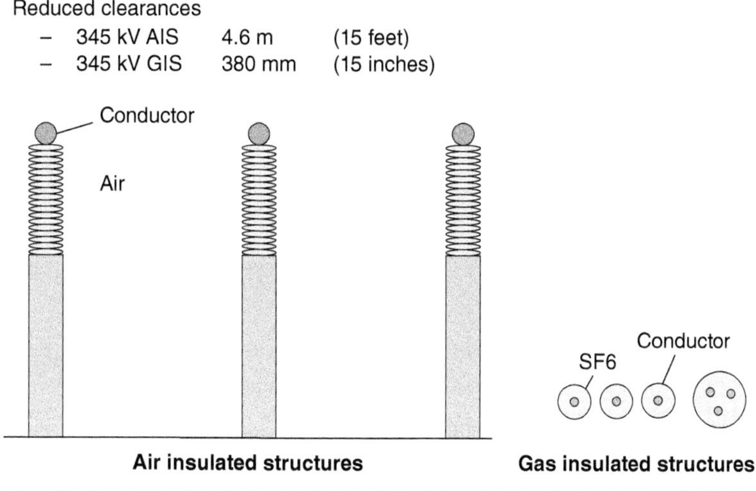


**FIGURE 11-27** Air-insulated and gas-insulated equipment clearances [3].

## 11.2.3 $SF_6$ Properties in Gas-Insulated Substations

Sulfur hexafluoride gas is a man-made gas. It is colorless, odorless, and extremely stable over varied temperature and pressure ranges. It is a large octahedral nonpolar molecule with a molecular weight of 146.05, which is about five times the weight of air. The vapor pressure of $SF_6$ is 320 psig at 70°F (21°C), and its freezing point is −83.5°F (−64°C). Heat and arcs break down $SF_6$. However, in its pure state, it largely reforms into $SF_6$. Figure 11-28 compares an $SF_6$ and $N_2$ molecule.

A comparison of $SF_6$ with $N_2$ shows why $SF_6$ has remarkable properties in terms of voltage withstand and current interruption. The fact that $SF_6$ in its pure state has the property of reversibility of arc dissociation (meaning it tends to recombine to reform $SF_6$ after arc exposure), makes it valuable from the standpoint of switching device chambers in that it does not require maintenance after arc interruption.

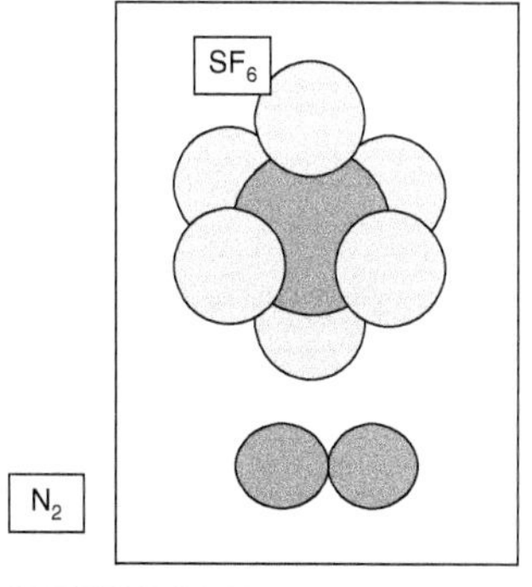


**FIGURE 11-28** $SF_6$ molecule versus $N_2$ molecule.

A comparison of $SF_6$ and $N_2$ shows how the physical and electrical properties of $SF_6$ are ideal for high-voltage switchgear. $SF_6$ is 125,000 times the mass of an electron and it is five times the mass of $N_2$ with the same high ionization potential. The ionization potential is the amount of energy required to remove an electron from an atom or molecule in the gaseous state. A higher ionization potential results in better insulating properties of a molecule.

The size of the $SF_6$ molecule, coupled with increased density of the gas, provides for an excellent dielectric. Free electrons such as those emanating from an energized conductor require a certain distance to accelerate to a sufficient velocity to have enough energy to cause ionization. The average distance between molecules is a function of the collision diameter and density. The collision diameter of $SF_6$ versus $N_2$ is 36.5 angstroms versus 16.1 angstroms (Å). The $SF_6$ molecule is sufficiently large so as to minimize the mean free path of electrons in the dielectric space.

When the $SF_6$ gas density is increased using pressurization, the mean free path of electrons is all but eliminated.

$SF_6$ gas is electronegative, which means that it has a strong affinity for electron attachment with a very low energy requirement (0.1 eV or less). This simply means that free electrons are often captured by the $SF_6$ molecule. The mechanisms for capture are the following:

$$\text{Resonance capture: } e + SF_6 \rightarrow SF_6^-$$

$$\text{Dissociative attachment: } e + SF_6 \rightarrow SF_5^- + F$$

The impact of the electronegativity of the insulating gas on dielectric breakdown is such that it attracts free electrons and holds them, thereby increasing the dielectric capability of the gas-insulated system.

To summarize why $SF_6$ is a good insulating gas:

- The $SF_6$ molecule is large, which prevents the migration of free electrons causing collisions.
- $SF_6$ has a high electron affinity which helps to capture free electrons.
- The ionization potential of $SF_6$ is relatively high; it does not give up electrons freely.

$SF_6$ also has good properties during interruption of high currents. $SF_6$ has excellent heat transfer properties, and it significantly lowers the temperatures and heat in arcs compared to $N_2$ or air. It has better thermal conductivity in the temperature ranges of electrical arcs than $N_2$ or air. The thermal time constant of $SF_6$ is 1/100 the time of air, which means it dissipates heat very quickly. The electronegativity properties of $SF_6$ help to stop conduction in arcing. But most importantly, $SF_6$ has rapid dielectric recovery, and it returns to its insulating state faster than other insulating gases.

$SF_6$ is stable but does decompose at temperatures over 350°F (117°C). Some gaseous decomposition and arc by-products are $SO_2$, $SOF_2$, $SF_4$, HF, $S_2F_{10}$ ($SF_5$), $SO_2F_2$, $SOF_4$ ($SF_4$), and $CF_4$. Some solid by-products (typically as a powder known as switching dust) are $WF_6$, $WO_3$, $AlF_3$, and $CuF_2$.

Many of these substances are toxic. $S_2F_{10}$ is very toxic even at low levels, and HF requires $H_2$, from a source of $H_2O$. Many by-products react with moisture that can form HF on skin, lungs, eyes, etc., and can be a health hazard to humans and animals [1].

### 11.2.4 GIS Sealing System

The modules of GIS and GIL at flange joining points are joined using a bolted flange arrangement with an "O" ring seal system for the enclosure and a sliding plug-in contact for the conductor. Internal parts of the GIS are supported by cast epoxy insulators. These insulators may provide a gas barrier between parts of the GIS, or they may be cast with holes in the epoxy to allow $SF_6$ gas to pass from one side to the other. The barrier insulators are used to create separate gas compartments in order to limit the gas volume in any one enclosure (gas zone).

There are generally two types of flange arrangements in GIS:

1. The double O-ring sealing design, where barrier insulators are installed and the opposing flange surfaces mate up with the surface of the barrier insulator. This design requires the installation of two O-rings, one on each face of the opposing flanges (Fig. 11-29).
2. The single O-ring sealing design, where there is no barrier insulator and the enclosure flanges mate up directly. This design requires the installation of one O-ring between the faces of the opposing flanges (Fig. 11-30) [1].

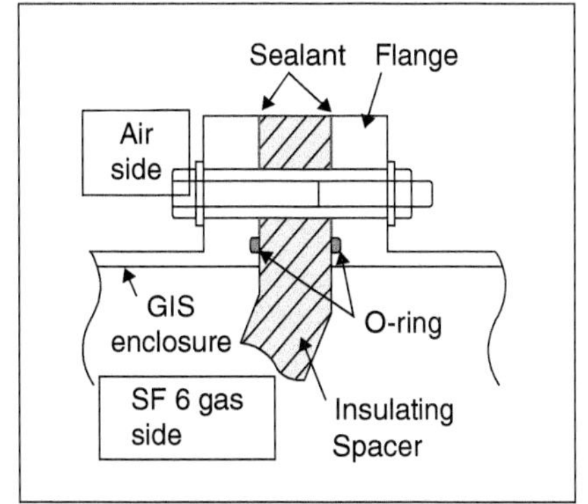


**FIGURE 11-29** Double O-ring sealing design.

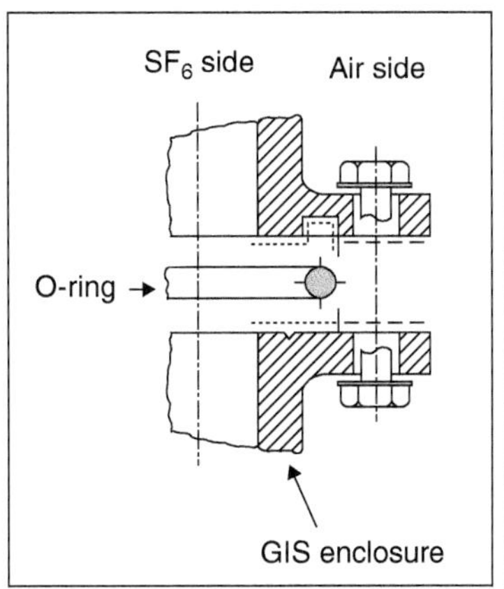


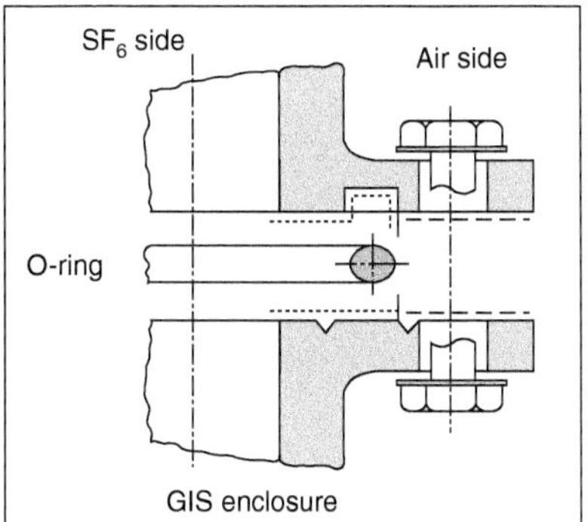


**FIGURE 11-30** Single O-ring sealing design.

### 11.2.5 Enclosure and Conductor Design Features, Voltage Withstand, and Ampacity

As in Sec. 11.2.2, the dielectric design of GIS is based on a coaxial arrangement of the current carrying conductor centered within a grounded metallic enclosure with insulating gas. The conductor (bus) is supported by conical insulators that maintain the electrical space clearance necessary for the specific operation of the equipment at maximum voltage ratings with the pressurized insulating gas as the dielectric fluid. The most economical use of materials and electric field distribution occurs when $\ln(r_2/r_1) = 1$. This leads to a multitude of possible enclosure/conductor diameters, depending on the insulating gas space required for specific operating voltages and whether pure $SF_6$ or a gas mixture such as $N_2 + SF_6$ is used.

Figure 11-31 shows a typical straight unit of GIL or GIS bus utilizing an $N_2 + SF_6$ mixture for the insulating gas. This example provides an understanding of the design parameters associated with the sizing of the conductor/enclosure, the BIL performance, and the ampacity of the conductor.

Conductor voltage stress is higher than enclosure voltage stress for a range of diameters. The optimal design ratio is shown in Fig. 11-32. It is the GIS designer's goal to design the conductor and enclosure diameters and geometric concentricity to be as close as possible to $\ln(r_2/r_1) = 1$.

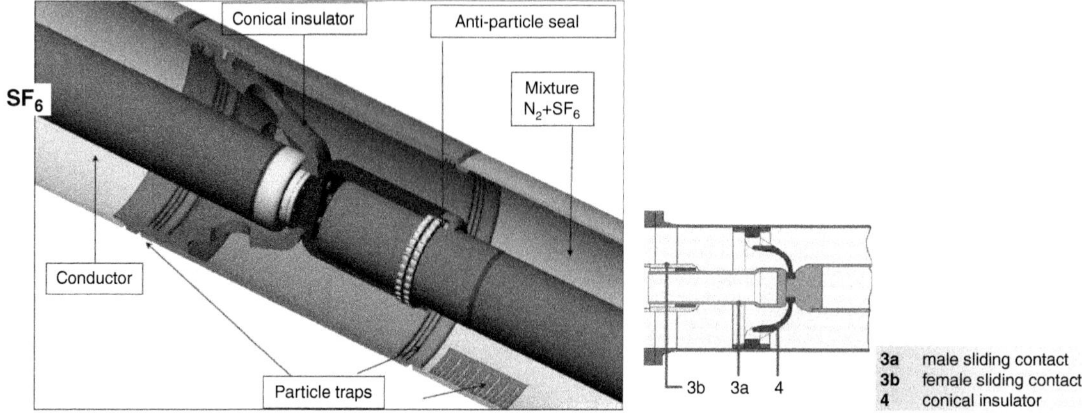


**FIGURE 11-31** GIS bus and enclosure cross section [3].

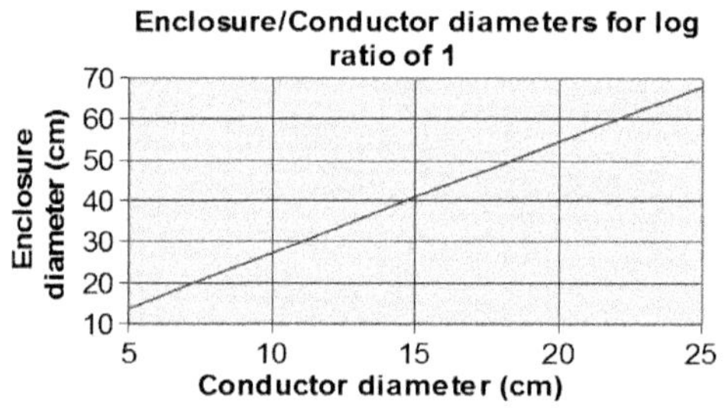


**FIGURE 11-32** Enclosure-to-conductor design ratio [3].

Because the conductor voltage stress is higher than the enclosure voltage stress, a negative impulse voltage (BIL) determines the overall system size. Figure 11-33 shows the relationship between BIL performance and enclosure design diameter for a range of enclosure diameters with an insulating gas ($SF_6$) pressure of 4 bar (58 psi).

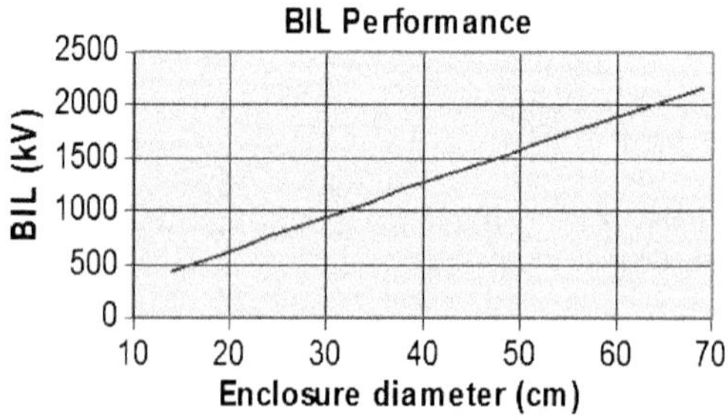


**FIGURE 11-33** BIL-to-conductor diameter design ratio [3].

In conjunction with the design requirements for BIL, the properties of the materials used for the conductor and the enclosure determine the ampacity of the GIS or GIL. The enclosure diameter has a large influence on the mutual inductance of the enclosure, as well as the heat transfer capability of the thermal interface between the insulating gas and the enclosure. Figure 11-34 shows the relationship between the continuous current carrying capability of the conductor and the design diameter for a range of enclosure diameters with an insulating gas ($SF_6$) pressure of 4 bar (58 psi).

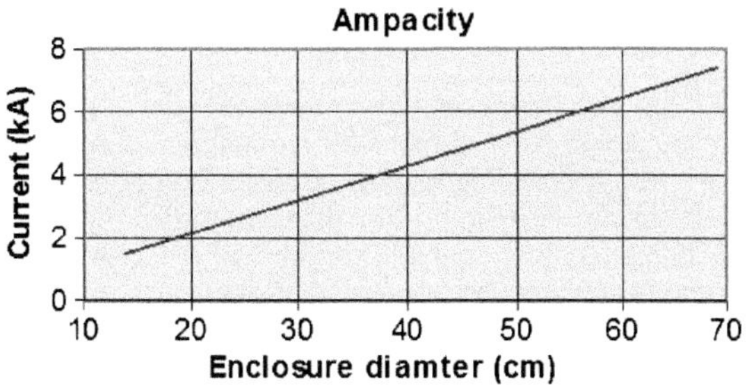


**FIGURE 11-34** Ampacity-to-enclosure diameter design ratio [3].

Short circuit withstand capability in GIS is strictly a function of the strength and spacing of the insulating spacers and the thickness and diameter of the conductor. The rated short-time withstand current, the peak withstand current, and the duration of the short circuit are the basic dimensioning parameters for GIS design. These values have a significant impact on the electromechanical forces experienced by the insulators and conductors, and on the thermal rise of the contact system. Table 11-15 is a list of the typical short-circuit current ratings of GIS related to their voltage classes.

**TABLE 11-15** Short Circuit Rating by Voltage Class

| Maximum voltage rating class, kV | Short circuit current rating range, kA |
|---|---|
| 52–72.5 | 16–25 |
| 100–170 | 25–50 |
| 245–300 | 25–63 |
| 362–550 | 31.5–100 |
| 800 | 50–100 |
| 1100 | 50–100 |

Table 11-16 shows the typical rated duration of short circuit used in the design of the GIS. The duration of the short circuit is dependent on the electrical system protection schemes and is a symmetrical function. The duration of the short circuit current directly impacts the design of the GIS; this duration should be minimized. The location and strength of the barrier insulators are dependent on the available mechanical forces due to the short circuit [1].

**TABLE 11-16** Relative Rated Short Circuit Duration Design Criteria

| Duration | Rated short circuit duration, s |
|---|---|
| Short | 0.5 |
| Standard | 1.0 |
| Long | 2.0 |
| Very long | 3.0 |

## 11.2.6 GIS Typical Layout and Enclosure Designs

Modern GIS employ two basic enclosure types: three-phase and classic single-phase. The maximum voltage rating of three-phase enclosures ranges from 1 to 245 kV. At present, the gas-insulated technologies of three-phase enclosures available are not economical at maximum operating voltages above 245 kV because of the diameter of the enclosure cross sections. Equipment that operates at system voltages above 245 kV is generally designed with individual phases in single-phase enclosures. There are no major established performance differences between the three-phase enclosure and the single-phase enclosure GIS, except for the applications of the equipment related to transient recovery voltage (TRV) requirements.

The three-phase enclosed GIS systems offer the advantage of compactness and modularity. The three-phase design requires that all three conductors are housed within a large-diameter aluminum enclosure. At higher-voltage withstand levels, the enclosure cross-sectional diameter increases based on the electrical stress relationship discussed earlier.

The three-phase enclosed GIS contains all of the switching devices, bus bars, and sensing transformers necessary for complete circuit breaker bay positions.

A three-phase enclosed GIS system has fewer parts, fewer enclosure materials, and less insulating gas than a comparable single-phase enclosed design. The three-phase enclosed GIS also combines the functionality of components in single-module enclosures, such as a combination disconnect switch and grounding switch.

Figure 11-35 shows a typical three-phase enclosed GIS design, which includes the following components:

- Circuit breaker (CB)
- Bay control cabinet with protective relays (control cabinet)
- Circuit breaker operating mechanism (CB operating mech)
- Main bus bars with provisions for disconnect and grounding switches
- Current transformers (CT)
- Voltage (potential) transformers (PT)

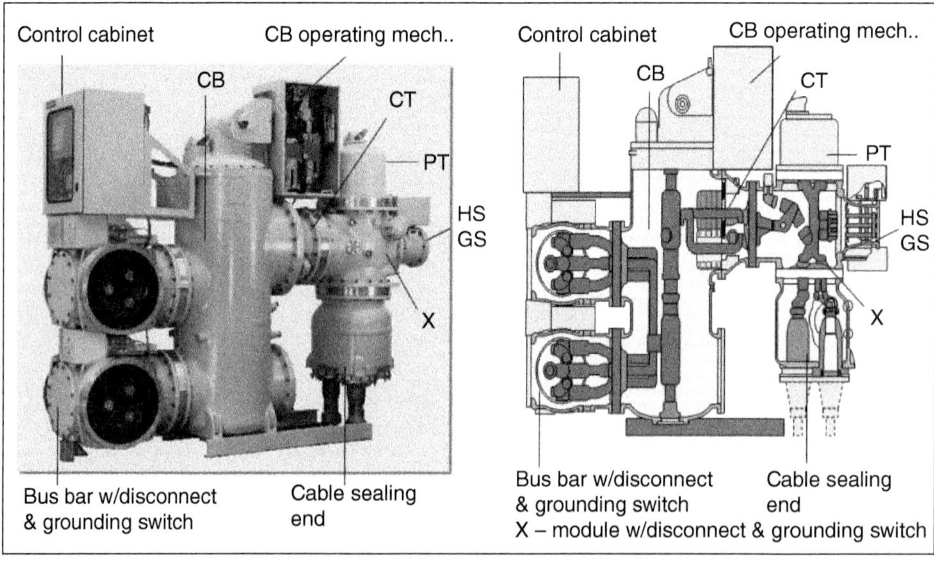


**FIGURE 11-35** Design of a three-phase enclosure GIS (145 kV) [3].

- Disconnect switch and grounding switch module (X)
- High-speed grounding switch for cable termination (HS-GS)
- Cable sealing end with cable terminators

The single-phase enclosed GIS design is a modular assortment of components connected together to form a circuit breaker bay position. Figure 11-36 represents the individual components of the modular structure in schematic form as well as sample photographs of the component modules.

The single-phase enclosed GIS is a highly standardized arrangement, with each module having only one function. Unlike the three-phase enclosed design where there are component combinations

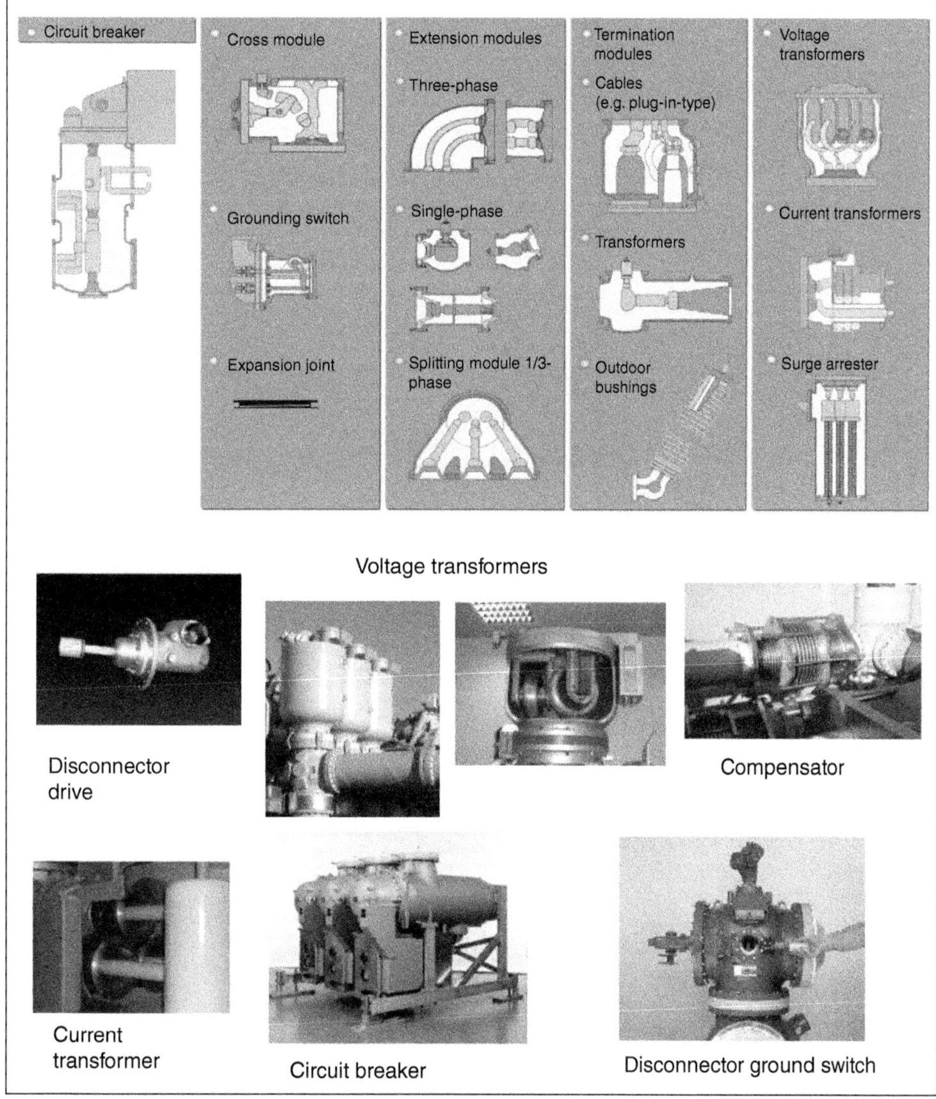


**FIGURE 11-36** Modular structure of three-phase enclosed GIS [3].

in a single enclosure, the single-phase design specifically segregates individual components based on their specific function. The components are arranged in pressure-resistant, gas-tight enclosures according to their functions. That is why the single-phase design is more well suited to extra high-voltage applications where the voltage withstand levels are greater, such as maximum operating voltages above 245 kV.

A wide range of design variations can be constructed using 15 to 20 different component modules. These arrangements can vary from a simple single-bus design to a complex breaker and one-half or double-bus, double-breaker design.

Figure 11-37 shows a typical single-phase enclosed GIS design, which includes the following components:

- Circuit breaker
- Bay control cabinet with protective relays
- Circuit breaker operating mechanism
- Main bus bars
- Disconnect switches
- Grounding switches

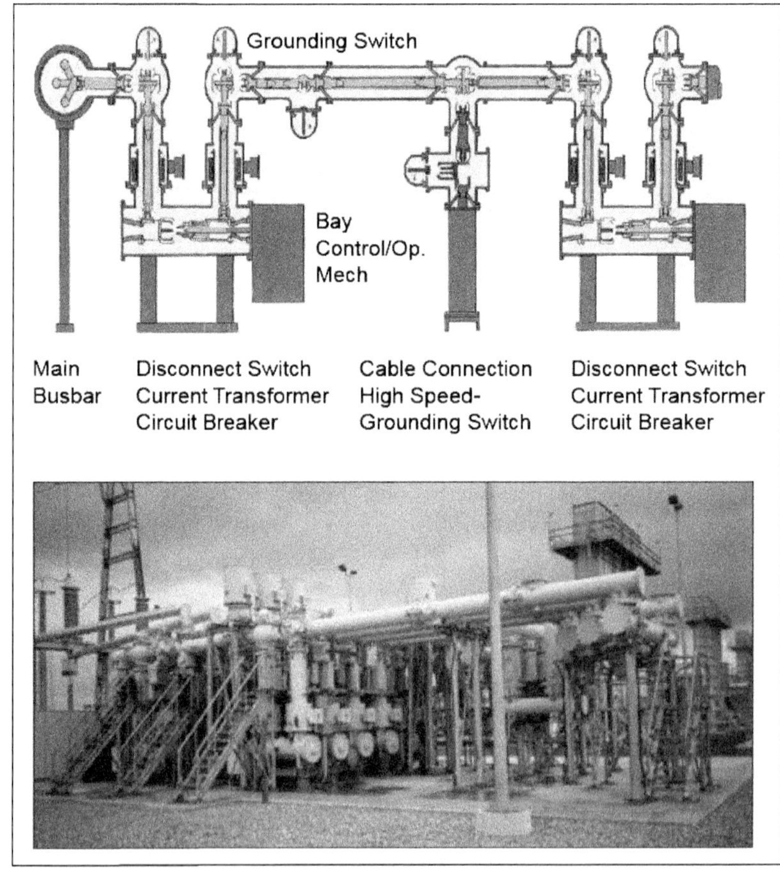


**FIGURE 11-37** Classic single-phase enclosed GIS (242 kV) [3].

- High-speed grounding switches
- Cable connections

The single-phase enclosed GIS design, like the three-phase enclosed design, is a modular assortment of components connected together to form a circuit breaker bay position. Figure 11-38 represents the individual components of the modular structure assembled in schematic form as well as sample photographs of the component modules and a one-line diagram [1].

Figure 11-39 represents a complete GIS installation and isometric view.

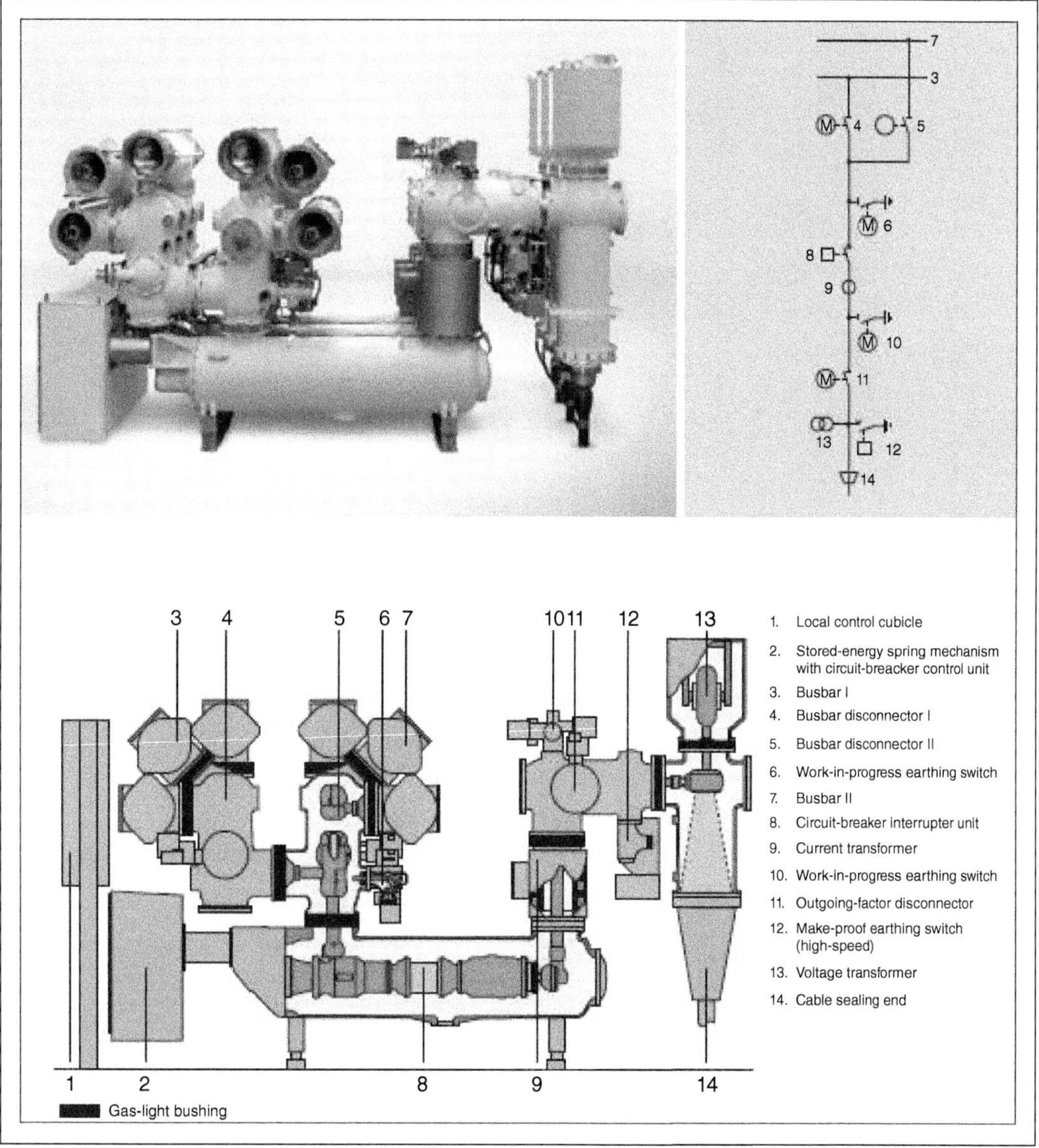


**FIGURE 11-38** Modular structure of single-phase enclosed GIS. (*Siemens AG.*)

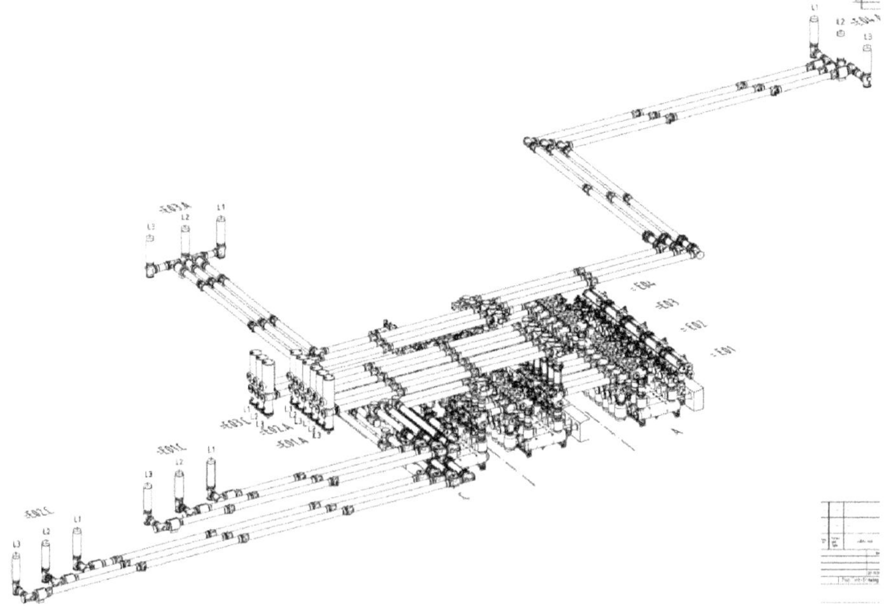

**FIGURE 11-39** GIS installation and isometric view. (*Siemens AG.*)

### 11.2.7 References for Gas-Insulated Substations

1. G. Becker, EPRI, "Gas-Insulated Substations and Lines Guidebook: 2015 Update," EPRI 3002005938, December 2015.
2. IEEE Standard C37.122-2010, "IEEE Standard for High Voltage Gas-Insulated Substations Rated Above 52 kV," 2010.
3. G. Becker, H. Koch, J. Brunke, P. Grossmann, M. Etter, A. Arora, S. Parsi, A. Ficheux, D. Mitchell, and S. Pravakar, "IEEE Gas-Insulated Substations and Line Tutorial," 2014.
4. Jean-Louis Habert, "GIS Substation Design and Execution—HV and EHV GIS Application and Design Considerations," IEEE, 2014.

# 12

# SWITCHGEAR AND POWER COMPONENTS

David S. Johnson

*Consultant, Pittsburgh, Pennsylvania*

Jeffrey H. Nelson

*Manager, Project Development, Transmission, Tennessee Valley Authority, Chattanooga, Tennessee*

T. W. Olsen

*Retired Former Manager, Technology, Energy Management Division, Medium Voltage and Systems, Siemens Industry, Inc., Wendell, North Carolina*

Michael W. Wactor

*Technical Director, Corporate Product Development, Powell Industries, Inc., Houston, Texas*

Kenneth Long

*Engineering Manager, Transmission & Distribution, Stantec Consulting Ltd., Portland, Oregon*

Hamid R. Sharifnia

*Manager of Engineering, Stantec Consulting Ltd., Portland, Oregon*

Mark McVey

*Principal Engineer, Dominion, Richmond, Virginia*

## 12.1 CIRCUIT BREAKERS

*BY DAVID S. JOHNSON, JEFFREY H. NELSON, AND T. W. OLSEN*

Definitions of terms used in this subsection can be found in the IEEE standards and application guides referenced in this subsection and/or in the *IEEE Standards Dictionary* (available at http://ieeexplore.ieee.org/xpls/dictionary.jsp?tag=1).

### 12.1.1 History of Development

In the early days of electrification (around 1890), switches were of the hand-operated, knife-blade type.

***Air Switches.*** With increasing current and voltage, spring-action driving mechanisms were developed to reduce contact burning by increasing the speed of the opening operation. Later, main contacts were supplemented with arcing contacts of special material and shape, which opened after and closed before the main contacts. This improvement eliminated arcing damage on the main contacts.

Further improvements of the air switch were the brush-type contact with a wiping and cleaning function, insulating barriers between arc chutes, and magnetic blowout coils with excellent arc-extinguishing properties. These features, as well as the horn gap contact, are still in use in low-voltage ac and dc air circuit breakers.

***Bulk-Oil Circuit Breaker.*** Around 1900, in order to cope with the new requirement for "interrupting capacity," ac switches were immersed in a tank of oil. Oil is very effective in quenching the arc and establishing the dielectric strength of an open break after current zero. Deion grids, oil-blast features, pressure-tight joints and vents, new operating mechanisms, and multiple-break interrupters were introduced over several decades to make the bulk-oil circuit breaker a reliable apparatus for system voltages up to 362 kV.

***Minimum-Oil Circuit Breaker.*** This mainly live-tank variant of bulk-oil circuit breakers was developed after 1930 to reduce overall size and oil volume, and used mainly in Europe. They employed special low-oil-volume interrupting chambers of extra-light weight. The arc is cooled and extinguished effectively by means of current-dependent oil streams in different directions and supported by oil injection. The interrupters are mounted on porcelain or molded-resin supports, thus avoiding oil as an insulating medium to ground.

***Air-Blast Circuit Breaker.*** Further increase of system voltages and generating capacities triggered the search for faster and stronger circuit breakers utilizing oilless arc interruption. After 1940, the air-blast circuit breaker was developed, making use of the good insulating and arc-quenching properties of dry and clean compressed air, at pressures up to 25 bars.

Figure 12-1 shows a typical air-blast circuit breaker of modular design, installed in 1950.

Further development of the air-blast circuit breaker led to a two-cycle interrupting time, extra-heavy interrupters, and the constant-pressure control system.

***Air-Magnetic Circuit Breaker.*** The air-magnetic circuit breaker uses a combination of a strong magnetic field (coil or soft iron plates) with a special arc chute to lengthen and cool the arc until the system voltage cannot maintain the arc any longer. This interrupting principle was applied mainly in the distribution voltage range in metal-clad switchgear from the late 1930s to the 1980s.

***$SF_6$ Gas Circuit Breakers.*** $SF_6$ gas circuit breakers were first developed in the early 1950s by Westinghouse, following the discovery of the excellent arc quenching and insulating properties of $SF_6$ gas

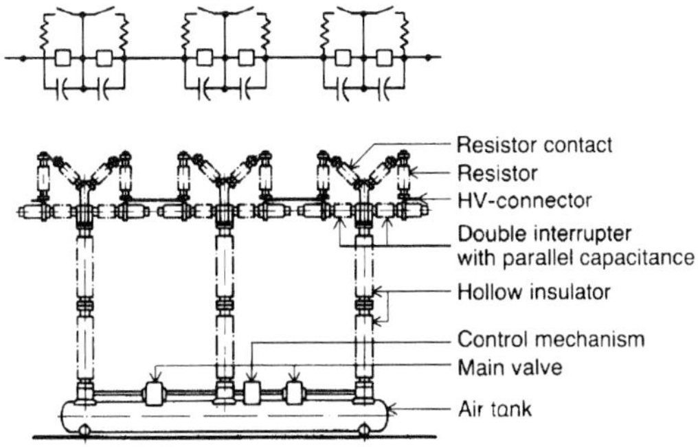


**FIGURE 12-1** Outdoor air-blast circuit breaker 230 kV, 1000 A, 16 kA, 3-cycle interrupting time.

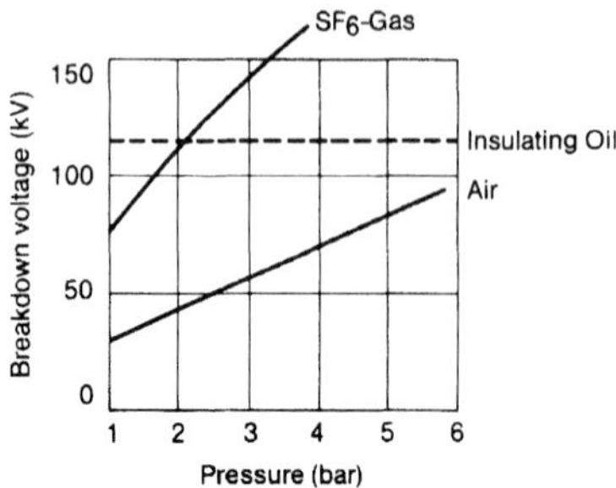


**FIGURE 12-2** Breakdown voltage of oil, air, and $SF_6$ gas as a function of pressure at 38 mm (1½ in) electrode distance.

(see Fig. 12-2). Both live-tank and dead-tank designs were introduced from the late 1950s into the 1960s. $SF_6$ remains the dominant insulating and arc quenching medium at higher voltages (72.5 kV and above) even today, though concerns about greenhouse warming potential of $SF_6$ are forcing a search for alternatives.

Dead-tank $SF_6$ gas circuit breakers were incorporated into gas-insulated substations (GIS) up to 245 kV in the mid-1960s, and to 800 kV in the 1980 timeframe. GIS offers space savings and environmental advantages over conventional outdoor substations, using the reduced insulation clearances possible with $SF_6$ gas.

***Vacuum Circuit Breakers.*** The first application of vacuum interrupter technology was in the 1950s for capacitor switching. Vacuum circuit breakers employ a vacuum interrupter, which includes the contact system. The contacts within the vacuum interrupter are driven, through a bellows, by a low-energy mechanism. The contact stroke of vacuum circuit breakers is short, which leads to very reliable, low-mechanical-energy designs. Vacuum interrupters have now improved in their range of voltages, short-circuit currents, and continuous current capabilities, such that most medium voltage application requirements can be met. Vacuum circuit breaker designs have become the dominant technology for outdoor circuit breakers and metal-clad switchgear applications at 38 kV and below. The voltage range of vacuum circuit breakers is increasing, with applications to 72.5 kV and even to 145 kV in recent years, particularly in Asia.

## 12.1.2 Fundamentals

***Design Fundamentals.*** Circuit breakers are mechanical switching devices capable of making and breaking currents under either normal or specified abnormal (short-circuit) conditions on the power system. Though circuit breakers are primarily defined by their protective capabilities, continuous current ratings, and ratings under abnormal short-circuit conditions, they also perform switching duties under a myriad of other system conditions, each of which has its own set of switching stresses.

Circuit breakers are rated primarily by power frequency voltage, insulation levels [BIL, switching impulse, power-frequency withstand (hi-pot) voltage], continuous current, short-circuit current, and interrupting time. Reference is made to IEEE C37.04[1] and IEEE C37.06[2] for values of ratings typically applied to circuit breakers, and to IEEE C37.100[3] for definitions of ratings subjects. (The IEEE PES Switchgear Committee is currently working on the development of a new Standard for Common Definitions for Switchgear which will be designated C37.100.5, and after it is completed C37.100 will be withdrawn.)

Circuit breakers employ a variety of media for high-voltage insulation and/or current interruption. The type of media employed in a specific design is often designated as a prefix in the naming of the circuit breaker, for example, vacuum circuit breaker, or sulfur hexafluoride ($SF_6$) gas circuit breaker.

Circuit breaker designs typically consist of the following construction elements: (1) contact system or interrupter, at high-voltage, (2) insulation between the contact system and ground potential ($SF_6$ gas, porcelain, molded resin), (3) operating mechanism and related control systems, and (4) an insulated link between the operating mechanism and the high-voltage contact system.

Circuit breakers are often categorized as being of either "dead tank" or "live tank" design. In the dead-tank design, the interrupting contact system is enclosed in a grounded tank, typically surrounded by an insulating fluid (oil) or gas ($SF_6$) (see Figs. 12-3 and 12-4 for examples of dead-tank circuit breakers). The electrical current enters the tank through high-voltage entrance bushings (Fig. 12-5), passes through the contact system, and then exits through another high-voltage entrance bushing, which allows current transformers, necessary for protective relaying or metering, to be installed around the conductor in the entrance bushings on both sides of the contact system. In the live-tank design, the interrupting contact system is supported by insulators above ground potential contained

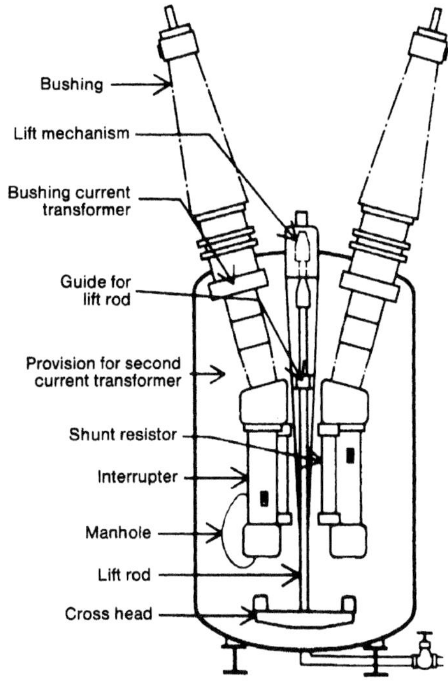


**FIGURE 12-3** Outline of a dead-tank 161-kV outdoor oil circuit breaker.

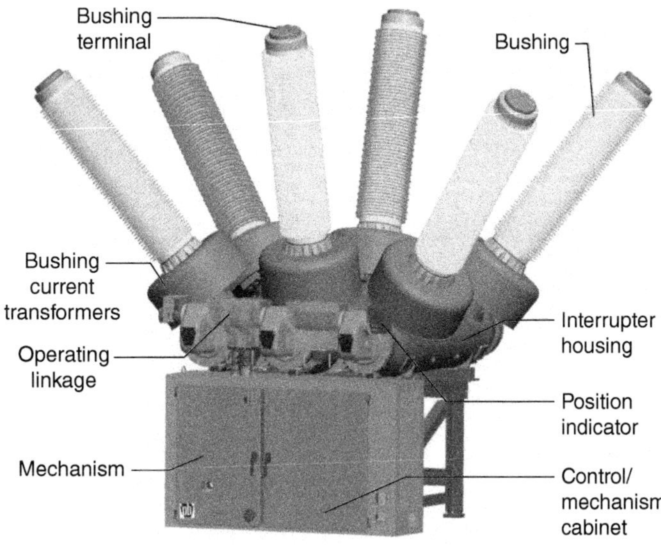


**FIGURE 12-4** Dead-tank $SF_6$ circuit breaker rated 245 kV, 63 kA.

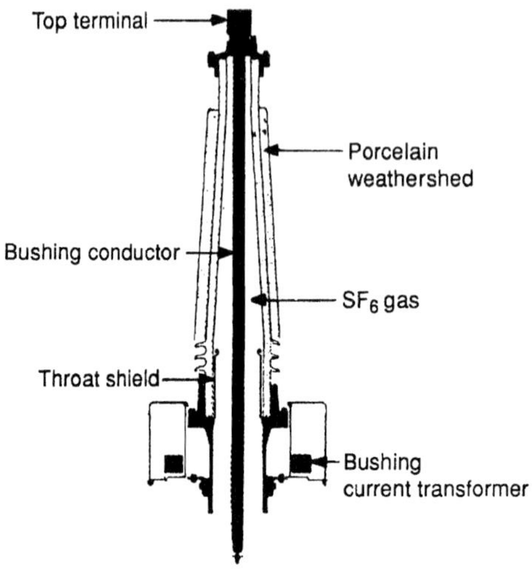


**FIGURE 12-5** Gas-filled bushing.

in a housing energized at the system voltage (see Figs. 12-6 and 12-7 for examples of live-tank circuit breakers). Live-tank designs offer no location to place current transformers directly on the circuit breaker, and therefore external current transformers must be independently placed adjacent to the circuit breaker, typically on one side of the contact system.

***Tripping Facilities.*** Tripping facilities, including circuit breaker controls, and the mechanical tripping mechanism of the circuit breaker operating mechanism, are vital to ensure proper operation during all conditions.

***Short-Circuit Duty.*** The *short-circuit duty* is determined by the maximum short-circuit current that the power system, connected to the circuit breaker at the time of short circuit, can pass through the circuit breaker to a point just beyond the circuit breaker, at the instant the circuit breaker contacts open. The short-circuit current is determined by the characteristics of synchronous and induction machines connected to the system at the time of the short circuit, the impedance between them and the point of short circuit, and the elapsed time between the initiation of the short circuit and the parting of the circuit breaker contacts.

In *calculating* short-circuit currents of high-voltage circuits, it is ordinarily sufficiently accurate to take into account only the reactance of the machines and circuits, whereas in low-voltage circuits, resistance as well as reactance may enter into the calculation. In dc circuits, resistance only is ordinarily sufficient.

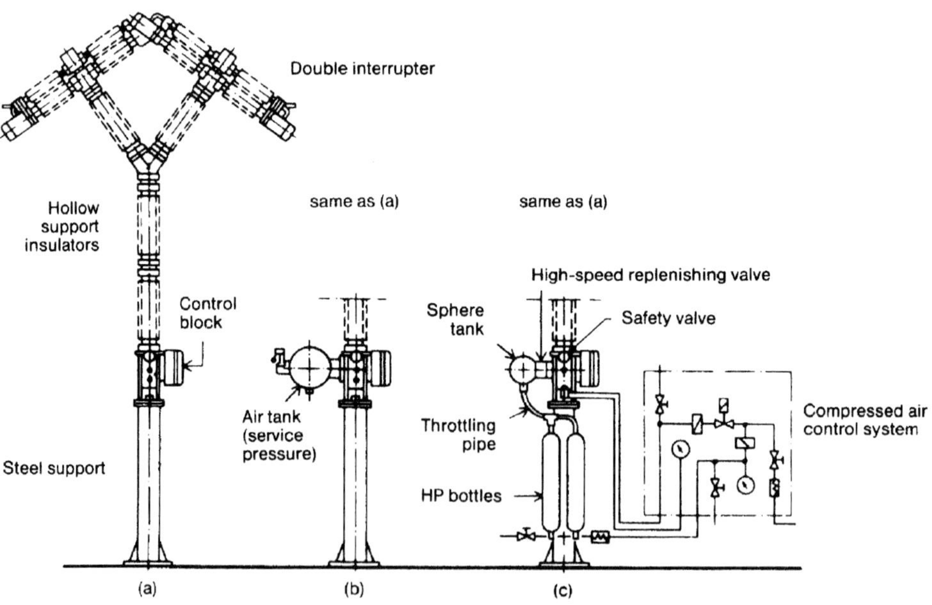


**FIGURE 12-6** Modular setup of a 362-kV outdoor air-blast circuit breaker with two uprating steps: (*a*) without air tank, low braking capability; (*b*) with air tank, standard breaking capability; (*c*) with constant-pressure air supply and high breaking capability.

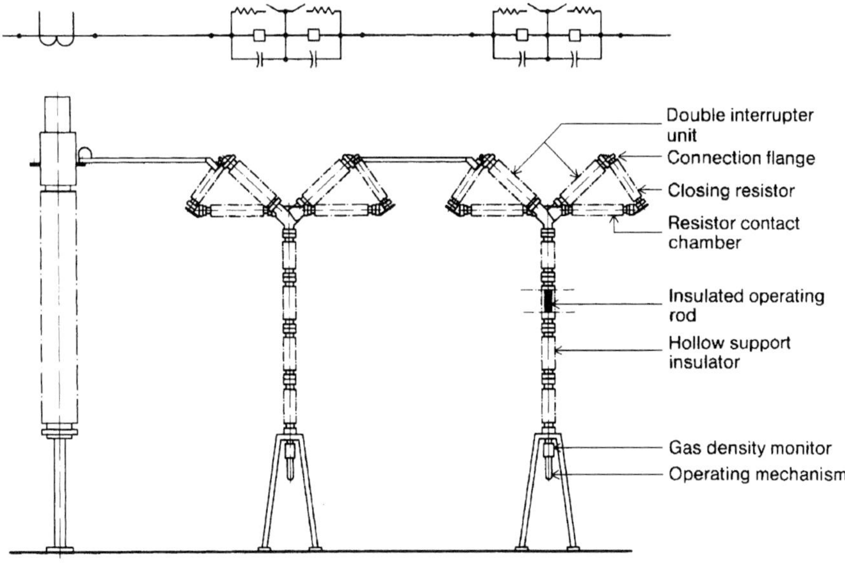


**FIGURE 12-7** Live-tank $SF_6$ outdoor circuit breaker and current transformer arrangement, 800 kV, 3000 A, 40 kA.

For first *approximations*, the reactance and typical time-decrement curves of the synchronous machines may be used. For close calculations, the actual reactance and time characteristics of the equipment should be used, and calculation made for single- as well as three-phase faults. The "per unit" impedance system and the "internal-voltage" method, using "symmetrical components," are often used in more exact calculations. Programs are available for digital computer studies of system short-circuit currents, both balanced three-phase and phase-to-ground.

***Total Current Basis.*** Historically (prior to 1964), high-voltage circuit breaker interrupting ratings were defined in terms of kilovolt amperes (kVA).

If a fault initiation occurs at a time close to voltage zero, the first few cycles of the short-circuit current is made up of a sinusoidal ac component superimposed on a decaying dc component [Fig. 12-8 shows single-phase short-circuit currents: (a) symmetrical and (b) asymmetrical]. The circuit breaker must interrupt the energy in the arc caused by the current at the time when the circuit breaker contacts part. The current at contact separation could be asymmetrical or symmetrical depending on when the contacts part, and on initial asymmetry. At any point, the total rms, or total current, can be expressed as the rms value of the ac and dc components of current:

$$I_T = \sqrt{(ac)^2 + (dc)^2} \tag{12-1}$$

where $I_T$ is total rms current at any instant.

Since the standards (up to 1964) were based on the interrupting rating on the total current, including the dc component, at the instant of contact separation, it became known as the "total current basis of rating."

***Symmetrical Current Basis.*** In the 1964 revision to IEEE C37.04,[1] Rated Short-Circuit Current was defined as, "...the highest value of the symmetrical component of the short-circuit current in rms amperes, measured from the envelope of the current wave at contact separation, which the circuit breaker is required to interrupt at rated maximum voltage..." for a specified duty cycle. The rating methodology later became known as the "symmetrical basis of rating," to distinguish it from the previous "total current" method of ratings. Modern ratings are expressed in symmetrical kiloamperes (kA),

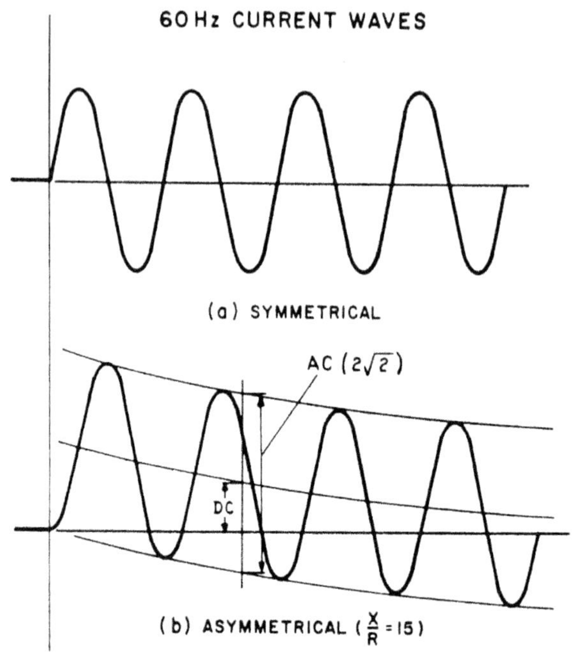


**FIGURE 12-8** Single-phase short-circuit current waves: (*a*) symmetrical (*b*) asymmetrical.

the symmetrical basis of rating. Most short-circuit current calculation software programs express the system short-circuit current at a fault in symmetrical kA. The rated short-circuit current in rms kiloamperes is referred to as the rated maximum voltage in kilovolts.

The ratings structure and tables of ratings for ac high-voltage circuit breakers are found in IEEE C37.04[1] and IEEE C37.06.[2]

***AC Interruption.*** Interruption in an ac circuit occurs at current zero. During the following half-cycle the *recovery voltage* will build up across the circuit breaker main contacts. The typical appearance of recovery voltage will differ in inductive, resistive, and capacitive circuits (see Fig. 12-9). When opening an inductive circuit, the recovery voltage will rise suddenly at a high rate because current interruption occurs at the moment of system voltage peak. This case requires a rapid increase of dielectric strength of the open contact gap.

When interrupting resistive load, current and voltage pass through zero at about the same moment. The recovery voltage will therefore rise at a moderate rate and no particular problems are imposed on the circuit breaker.

At the moment of interruption of capacitive current, the capacitance is fully charged. The recovery voltage rises slowly during the first half-cycle, but continues to rise to a value twice the system voltage. This may lead to restrikes, undesired network oscillations, and overvoltages.

The waveforms of short-circuit current and transient recovery voltage in a simplified network system are shown in Fig. 12-10. At the moment of fault-current interruption the two sections, source side (S) and load side (L), of the network are decoupled and oscillated independently about their driving voltages. The difference of these two transients appears across the open contacts of the circuit breaker pole. The behavior of this transient recovery voltage is determined by the circuit parameters. The still-moving, or already fully-open, circuit breaker contacts must be able to withstand the recovery voltage. The most severe stress for the open contact gap is the *initial peak* and the *rate of rise* (kV/μs) of the recovery voltage.

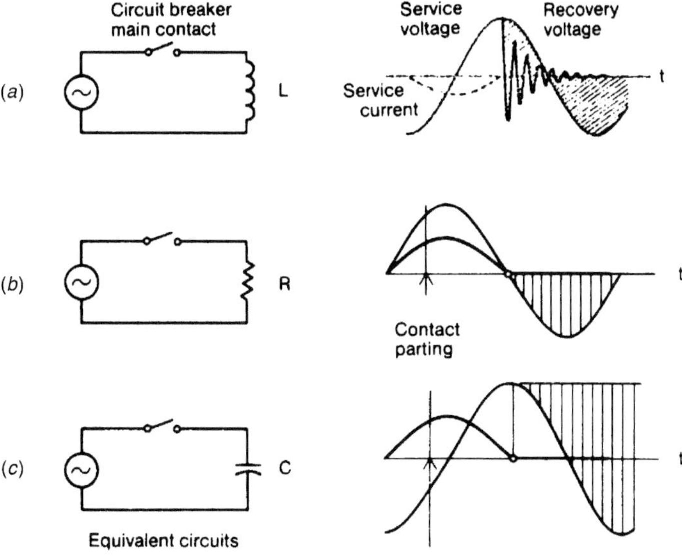


**FIGURE 12-9** Typical shape of recovery voltage on interruption: (*a*) induction current; (*b*) resistive current; (*c*) capacitive current.

If the recovery voltage exceeds the gap dielectric strength, the arc will restrike and current will continue until the next current zero, when interruption will again be attempted. The rate of rise of recovery voltage is a function of the constants of the circuit which supplies power through the circuit breaker. The larger the adjacent capacitance to ground before the major inductance limiting the fault current, the slower will be the rise of the recovery voltage. Some circuit breakers modify the recovery-voltage characteristics by limiting the current, modifying its power factor, and so on.

***DC Interruption.*** Interruption in a dc circuit (see Fig. 12-11) is basically different from ac interruption. After contact parting, the arc is lengthened and cooled and consequently the arc

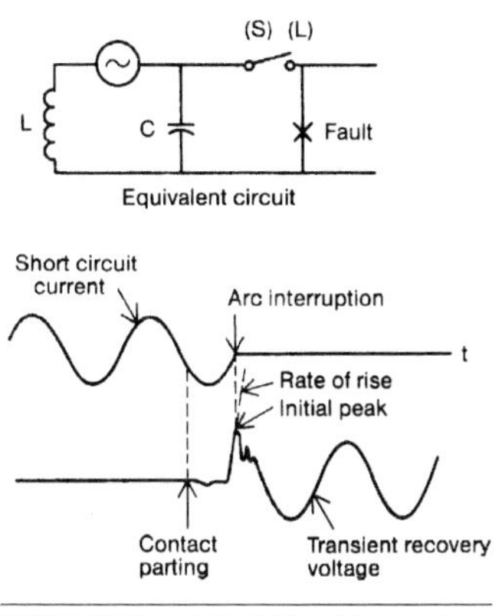


**FIGURE 12-10** Alternating-current interruption; typical shape of short-circuit current and transient recovery-voltage.

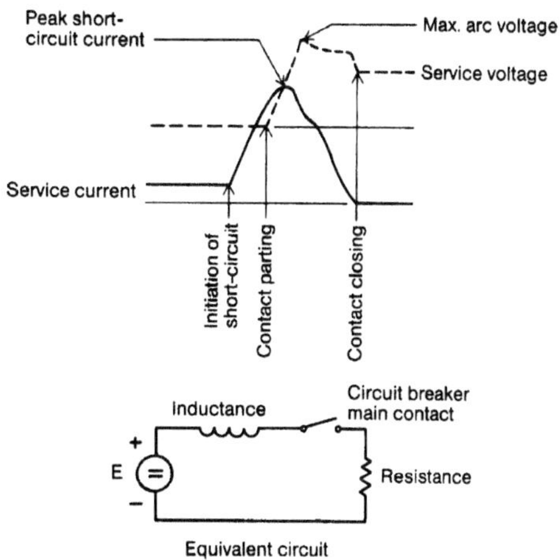


**FIGURE 12-11** Direct-current interruption; typical shape of short-circuit current and recovery voltage.

voltage is rising. The current will extinguish after the change in current *di/dt* becomes negative and the arc voltage rises above service voltage. DC circuit breakers must therefore operate rapidly, in order to allow the arc voltage to build up in a few milliseconds. The energy originating from the source and inductance will have to be absorbed by the arc. Magnitude and duration of short-circuit current depend on the magnitude of network inductance. With rising inductance, the short-circuit current will decrease, whereas the duration of the arc will increase.

### 12.1.3 Severe Interrupting Conditions

The following severe cases of circuit-breaker switching conditions have to be considered carefully: terminal fault, short-line fault, out-of-phase switching, switching of small inductive currents, switching of capacitive currents, and closing on a fault. The chosen examples are typical only; they are uniformly based on similar and simplified network configurations and are restricted to single-phase fault conditions. Other switching conditions may also be important.

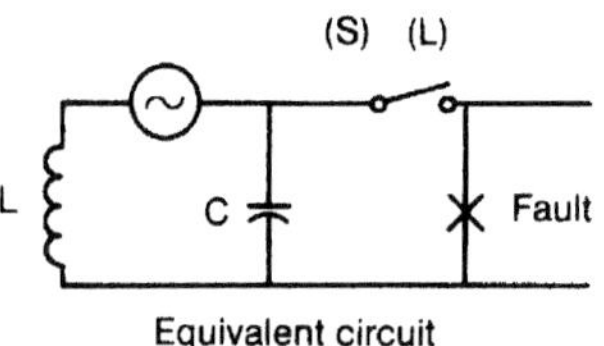


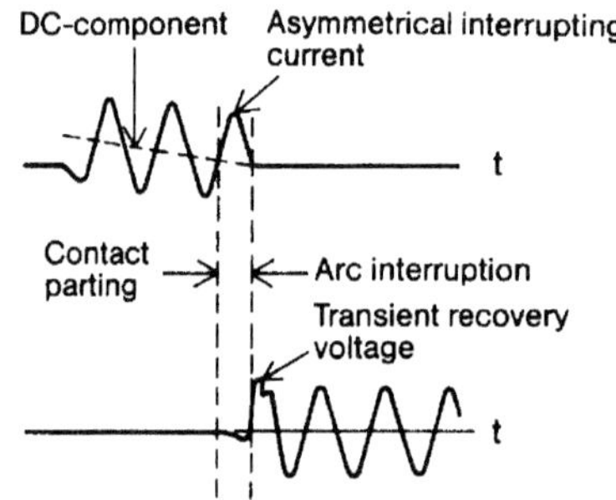


**FIGURE 12-12** Principle of terminal-fault interruption, equivalent circuit; typical shape of short-circuit current and transient recovery voltage.

***Terminal Fault.*** After interruption of short-circuit current, the recovery voltage oscillates toward the service frequency driving voltage via an initial peak. The natural frequency is determined by the inductance and capacitance of the driving system (Fig. 12-12).

The dc component of the short-circuit current depends on the instant of fault initiation, the time constants of the network components like generators, transformers, cables, and high-voltage lines and their reactances of the zero-sequence and the positive-sequence networks. The recovery voltage will accordingly vary depending on the location of the circuit breaker within the network.

***Short-Line Fault.*** In the case of a short-line fault, a section of line lies between the circuit breaker and the fault location (Fig. 12-13). After the short-circuit current has been interrupted, the oscillation at the line side (L) of the circuit breaker assumes a superimposed "saw-tooth" shape. The rate of rise of this line oscillation is directly proportional to the effective surge impedance and the time rate of change of current (*di/dt*) at current zero. The component on the supply side (S) basically exhibits the same waveform as a terminal fault. The circuit breaker is stressed by the difference between these two voltages. Because of the high frequency of the line oscillation, the transient recovery voltage has a very steep initial rate of rise. Since the initial rate of rise increases with increasing rate of current change, the limiting interrupting capability of many circuit breaker designs is determined by the short-line fault.

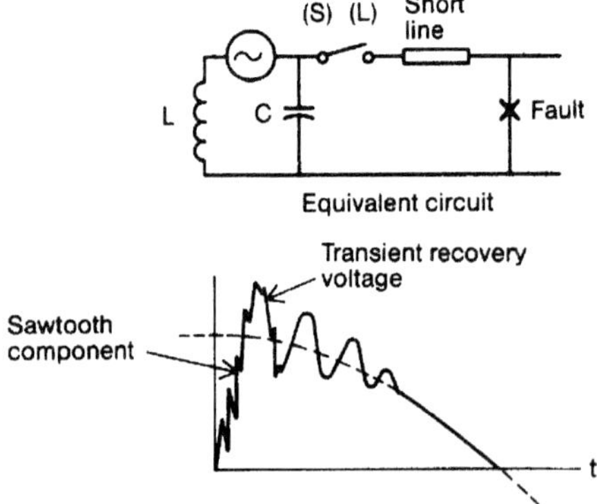


**FIGURE 12-13** Principle of short-line fault interruption, equivalent circuit; typical shape of recovery voltage.

***Out-of-Phase Switching.*** Two network systems with driving voltages E1 and E2 are connected via a transmission line (Fig. 12-14). Since the circuit is closed via the closed circuit breaker, the resulting driving voltage is equal to the sum of the two system voltages. Driving voltage

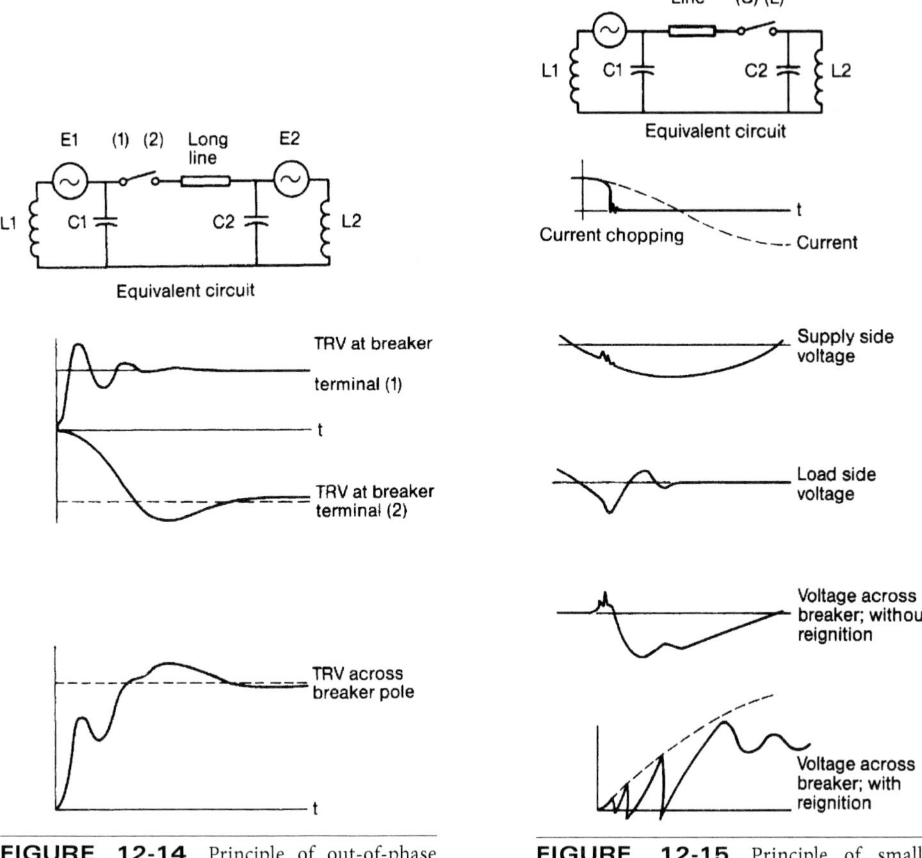


**FIGURE 12-14** Principle of out-of-phase switching, equivalent circuit; typical shape of recovery voltage.

**FIGURE 12-15** Principle of small-inductive-current interruption; equivalent circuit; typical shape of current and voltage.

E2 may, for example, exceed voltage E1 by the voltage drop across the transmission line. After opening the circuit breaker, the transient recovery voltages of the disconnected networks oscillate independently.

The circuit breaker is stressed by the difference of these two voltages. In the case of disconnection of long lines, the recovery voltage across the circuit breaker could be increased because of the Ferranti effect, where the voltage of the receiving end can be up to 15% higher than that at the sending end if the line is lightly loaded.

***Interruption of Small Inductive Currents.*** This occurs (see Fig. 12-15) when disconnecting unloaded transformers, reactors, or compensating coils. An arc is produced between the contacts when the circuit breaker is opened. The arc voltage is approximately constant at higher currents, since the arc energy is removed only by convection. With small currents, the arc voltage increases as a result of arc looping and a change in the cooling mechanism.

When approaching current zero, the arc current begins to oscillate as a result of interaction with the system; that is, it becomes unstable.

As a result of the high-oscillation frequency the current interruption may occur prior to the natural zero passage, which can be regarded as an instantaneous change in current, and is called *current chopping*. The chopping current is affected not only by the properties of the circuit breaker but also to a great extent by the system parameters.

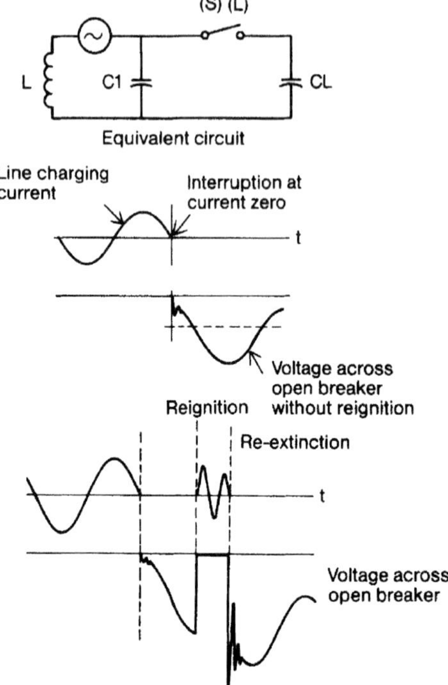


**FIGURE 12-16** Principle of capacitive current interruption, equivalent circuit; typical shape of current and voltage.

Energy at the disconnected load side (L) oscillates with the natural frequency of the capacitances local to the circuit breaker. The maximum voltage is attained at the moment when all the energy is converted into capacitive energy.

As a result of the resistive losses, the voltage on the disconnected load side decays to zero. During current chopping, the circuit breaker is stressed by the supply-side voltage on one side and by the load voltage on the other side. The supply-side voltage is at a maximum, since the load is highly inductive. The load side voltage is the oscillating voltage as the energy exchanges from inductive energy to capacitive energy. This load side voltage will have a high frequency of up to several thousand cycles per second. During this increasing stress, reignition across the circuit breaker may occur.

However, the arc is immediately extinguished again because of the low current and the process begins anew. Hence, the reignition also helps reduce the energy stored in the disconnected circuit.

***Interruption of Capacitive Currents.*** Capacitive currents occur while disconnecting lines, unloaded cables, or capacitor banks (Fig. 12-16).

Current chopping may occur at a low instantaneous current value during interruption of capacitive currents, but this does not lead to overvoltages. After interruption of current, the voltage at the load capacitance (L) remains at the peak value of the power frequency voltage, whereas the voltage on the source side (S) oscillates with the power frequency voltage. The difference between the two voltages appears across the circuit breaker with an amplitude of more than double the rated voltage peak. If the circuit breaker cannot withstand this higher voltage, *restriking* may occur. Restriking is similar to energizing transmission lines with a trapped charge. After restriking, transient current flows through the circuit breaker, which is of higher frequency than that of the system and which can again be interrupted during the reignition process. After *reextinction*, the load capacitance (L) is charged to the potential of the peak value of the equalizing process, whereas the circuit-breaker terminal on the source side (S) recovers to the system voltage. A very high differential voltage appears across the circuit breaker, which may lead to renewed restriking and even switching failures. Basically the same phenomenon occurs during disconnection of capacitor banks. Restrike-free interruption of capacitive currents is thus highly desirable, but a circuit breaker cannot be guaranteed to be restrike-free. Present-day standards recognize that restrike performance must be defined by probabilities and define restrike performance of circuit breakers in classes (C0, C1, and C2), each with different levels of restrike probability (see Sec. 12.1.4).

To determine the voltage stresses on the circuit breaker, the grounding condition of the supply system and capacitor bank, and the arrangement of the capacitor bank, have to be taken into account, as these affect the differential voltage across the breaker during capacitive switching.

***Closing on a Fault.*** This (see Fig. 12-17) directs the stress onto the circuit breaker contact system, particularly as regards the electrodynamic and thermal forces. The current and voltage stress is different during closing on (1) symmetrical or (2) asymmetrical short-circuit current.

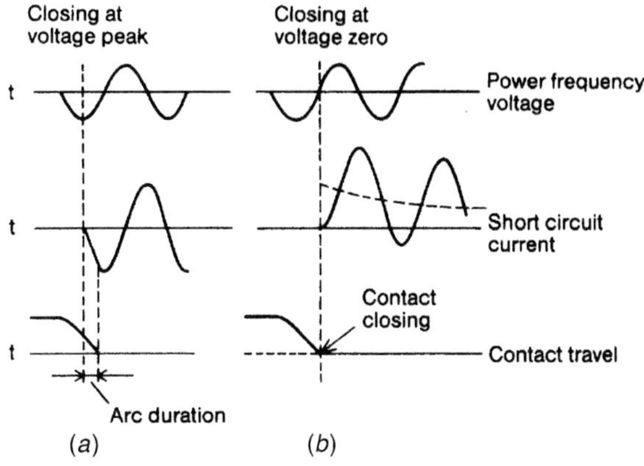


**FIGURE 12-17** Stress on contact when closing on a fault, contact travel related to (*a*) symmetric, and (*b*) asymmetrical short-circuit currents.

The deciding factor is the moment of pre-strike (prior to contact touch) relative to the phase angle of system voltage. In case pre-strike (and consequently ignition of the pre-strike arc) occurs at the voltage maximum, the short-circuit current will appear symmetrical. The other extreme case takes place with the moment of closing at voltage zero. Here the asymmetrical short-circuit current contains the maximum dc component. A contact system designed for high-velocity closing operation will be subjected to a shorter pre-arcing time and consequently to reduced contact burning when closing on short-circuit currents. High contact velocity operation is therefore not only important for opening but also for circuit-breaker closing.

### 12.1.4 High-Voltage Circuit Breaker Ratings and Capabilities

Particular reference is made to the rating structures and preferred ratings for ac high-voltage circuit breakers per the latest standard revisions of IEEE C37.04[1] and IEEE C37.06.[2] (NOTE: A revision of IEEE C37.04 is currently underway within the IEEE Switchgear Committee that will merge IEEE C37.06 into IEEE C37.04, resulting in one document that will contain the ratings structure and preferred ratings.)

Historically, high-voltage circuit breakers are defined as indoor or outdoor. Typically, indoor circuit breakers are connected to shielded cable systems. Outdoor circuit breakers may be connected to either shielded cable, or to unshielded conductors in overhead line systems. The revised IEEE C37.04[1] now defines new terms *Class S1* and *Class S2*.

A *Class S1* circuit breaker is intended to be used in a shielded cable system. The definition of cable systems is restricted to systems of rated voltages higher than 1 kV and less than 100 kV.

A *Class S2* circuit breaker is intended to be used in a line system or in a cable system with a direct connection, without shielded cable, to overhead lines. Typically in line systems, no cable is connected on the supply side of the circuit breaker.

***Rated Maximum Voltage.*** High-voltage circuit breakers are designed for voltage ratings as defined in IEEE C37.04[1] and IEEE C37.06.[2] The rated maximum voltage is the highest continuous rms phase-to-phase voltage for which the circuit breaker is designed, and is the upper limit for operation. Circuit breakers have to be dimensioned to withstand the maximum voltages as specified.

For circuit breakers rated in accordance with ANSI C37.06-1987[4] (or earlier), the range between upper and lower limit is defined by voltage range factor K. Current-interrupting capabilities vary within this range in inverse proportion to the operating voltage.

For circuit breakers rated in accordance with ANSI C37.06-1997[4] (or later), the current-interrupting capability is a constant kA value at any voltage equal to or lower than the rated maximum voltage.

***Rated Power Frequency.*** The rated power frequency of a circuit breaker is the frequency at which it is designed to operate. Standard frequencies, per IEEE C37.04[1] are 50 and 60 Hz. Applications at other frequencies require special consideration and the user should consult with the circuit breaker manufacturer.

***Rated Continuous Current.*** The rated continuous current is the current, in rms amperes, at rated power frequency that a circuit breaker has to be able to carry continuously without exceeding specified temperature rise limits in a specified ambient temperature range. Attention must be paid to reduction factors arising from the altitude at the site of installation and class of insulation material.

For guidance on continuous load current capability based on actual ambient temperature see IEEE C37.010.[5]

***Rated Dielectric Withstand Capability.*** Rated dielectric withstand capability of a circuit breaker is its voltage withstand capability with specified magnitudes and waveshapes of voltage applied under specified conditions. Preferred ratings for the dielectric withstand and external creepage insulation to ground (for Class S2 circuit breakers) are specified in IEEE C37.06.[2]

The dielectric withstand levels are determined by the rated withstand test voltages including: dry power frequency withstand voltage; wet power frequency withstand voltage (where applicable); full wave lightning impulse withstand voltage; chopped wave lightning impulse withstand voltage (where applicable); dry and wet switching impulse withstand voltage (where applicable). Reduction factors apply for sites at an altitude in excess of 3300 ft (1000 m).

For multiple-break circuit breakers, equal voltage distribution over the series breaks is achieved by grading capacitors paralleled to the interrupting chambers. (The need for multiple breaks has reduced considerably in recent years due to increased voltage capability per interrupter break, leading to single break designs even at 550 kV.)

Coordination between internal and external insulation, as well as insulation coordination between interrupters and ground insulation, has to be properly designed to reduce the probability of flashover inside the circuit breaker or over the open break. Note that dielectric coordination differs between dead-tank circuit breakers and live-tank circuit breakers.

***Rated Standard Operating Duty.*** The current standard operating duty cycle of a circuit breaker is

Open – T – close-open – 3 min – close-open (usually shown as O – T – CO – 3 min – CO)

T is defined as either 15 or 0.3 s, depending on whether the circuit breaker is rated for high-speed reclosing; this distinction is important in application. Even circuit breakers rated for high-speed reclosing must still be allowed a 0.3-s delay to allow for proper recovery of the dielectric strength of the air at the fault location following the initial fault interruption.

For historic oil and air-magnetic circuit breakers, the standard operating duty cycle was

Open – 15 s – close-open (usually shown as O – 15 s – CO)

***Rated Interrupting Time.*** The rated interrupting time is the maximum permissible interval between the energizing of the trip circuit at rated control voltage and operating pressure, and the interruption of the current in all poles.

The interrupting time during close-open operations may exceed the rated interrupting time by either ½ cycle (for 2 and 3 cycle circuit breakers) or by 1 cycle (for 5 cycle or greater circuit breakers).

***Rated Short-Circuit Current.*** High-voltage ac circuit breakers must be able to interrupt the rated short-circuit current (kA, symmetrical). The rated short-circuit current is the highest value of the symmetrical component of the three-phase, short-circuit current in rms amperes at the instant of primary arcing contact separation that the circuit breaker shall be required to interrupt at rated maximum voltage on the standard operating duty.

The following required capabilities are related to the rated short-circuit current be fixed ratios, as defined in IEEE C37.04.[1]

- Symmetrical interrupting capability for three-phase faults
- Asymmetrical interrupting capability for three-phase faults
- Rated closing, latching, and short-time current carrying capability
- Required reclosing capability

Circuit breakers shall be capable of performing the required closing-latching-carrying and interrupting duties in immediate succession. For circuit breakers rated in accordance with ANSI C37.06-1997[4] (or later), the closing and latching capability (peak kA) at 60 Hz is 2.6 times rated short-circuit current (kA, symmetrical), 2.5 times for 50 Hz. For circuit breakers rated in accordance with ANSI C37.06-1987[4] (or earlier), the closing and latching capability (rms asymmetrical kA) is 1.6K times rated short circuit (kA, symmetrical) at 60 Hz. The short-circuit current-carrying capability is determined by the rated short-time current (kA, rms symmetrical).

For historic oil and air-magnetic circuit breakers the standard operating duty cycle was open – 15 s – close-open (usually shown as O – 15 s – CO). For additional operations, and/or any close operation in the duty cycle with a time delay of less than 15 s after an opening operation, the interrupting rating and related required capabilities of the oil or air-magnetic circuit breaker have to be derated. All operations within a 15-min period are considered part of the same duty cycle and a duty cycle shall have no more than five opening operations. For guidance on interrupting capability for reclosing service for oil and air-magnetic circuit breakers manufactured after 1960 refer to IEEE C37.010.[5] Circuit breakers manufactured prior to IEEE C37.7-1960[6] have different methodologies for calculation of a reduced interrupting capability for reclosing.

***Rated Transient Recovery Voltage.*** Each transient recovery voltage (TRV) rating is defined for a three-phase circuit breaker in IEEE C37.04[1] and IEEE C37.06.[2] The TRV rating for a three-phase circuit breaker is defined by an envelope of required withstand capability. At rated maximum voltage, a circuit breaker must interrupt three-phase terminal faults at its rated short-circuit current where the TRV does not exceed the rated TRV envelope. (See IEEE C37.011[8] for guidance in applying TRV ratings for high-voltage circuit breakers.)

Two types of TRV waveforms at rated short-circuit current are used based on the voltage class of the equipment:

- A four-parameter TRV for system voltages of 100 kV and above
- A two-parameter TRV for system voltages up to 100 kV

The two-parameter TRV is also used at all rated voltages for the TRVs for fault currents that are 30% or less of the rated short-circuit current.

Prior to the IEEE C37.04b-2008[7] amendment to IEEE C37.04-1999, the TRV envelopes were defined by a 1-cosine wave for circuit breakers rated below 100 kV, and by the higher of an exponential waveform and a 1-cosine waveform for circuit breakers rated 100 kV and above.

***Rated Capacitance Current Switching.*** Capacitance current switching ratings and test requirements were completely revised with the publication of IEEE C37.04a-2003[9] and IEEE C37.09a-2005.[21] The ratings of a circuit breaker for capacitance current switching shall include the following, where applicable:

- Rated line-charging breaking current applicable to all outdoor (or Class S2) circuit breakers
- Rated cable-charging breaking current applicable to all indoor (or Class S1) circuit breakers
- Rated single capacitor bank breaking current
- Rated back-to-back capacitor bank breaking current
- Rated back-to-back capacitor bank inrush making current and frequency

Preferred capacitance current switching ratings are given in IEEE C37.06.[2] Each capacitance current switching rating has a probability of restrike classification. Three classes of restrike performance are specified in IEEE C37.04a-2003[9] and IEEE C37.04b-2008.[7]

- Class C0: Probability of restrike up to one restrike per operation (similar to old general purpose circuit breaker)
- Class C1: Low probability of restrike (similar to old definite purpose circuit breaker)
- Class C2: Very low probability of restrike (intended to have about 1/10 or less probability of restrike than a Class C1 circuit breaker)

***Out-of-Phase Switching Current Capability.*** This rating is for circuit breakers expected to be used for switching the connection between two parts of a power system during out-of-phase conditions, by as much as full-phase opposition. Currently this is an optional rating. The preferred rating for circuit breakers that have an out-of-phase switching current rating is 25% of the rated short-circuit current.

## 12.1.5 High-Voltage Circuit Breaker Construction

***Operating Mechanism.*** Opening and closing of power circuit breakers under service conditions is seldom performed manually, since most circuit breakers are installed in systems designed for remote control providing specific redundancy. Various types of circuit breaker drive mechanisms are therefore used, such as (1) dc solenoids, (2) solenoids operated from an ac source through a dry-type rectifier, (3) compressed air, (4) high pressure oil, (5) charged spring, (6) electric motor.

*Automatic reclosing* of circuit breakers in overhead line feeders is frequently used to restore service quickly after a line trips out because of lightning or other transitory fault. Instantaneous or time-delay reclosing may be provided with a lockout to prevent more than one to several successive recloses, as desired. If the fault is cleared before the lockout feature operates, the reclosing device resets itself, permitting a complete cycle of reclosing at a subsequent fault.

The circuit-breaker-operating mechanism has to cope with the increasing requirements in interrupting and current-carrying capability as well as with shorter operating times. Simplicity of design, robustness, and reliability has to ensure safe operation of this vital link between the electrical system controls and the interrupter. The principle of a pneumatic drive is sketched for an extra-high-voltage circuit breaker which functions according to the differential piston principle in Fig. 12-18. A pneumatic interlocking device in connection with the $SF_6$ gas system ensures that the circuit breaker always remains in the defined open or closed position even on loss of air pressure. Besides opening and closing functions, effective damping of the highly accelerated moving parts is incorporated.

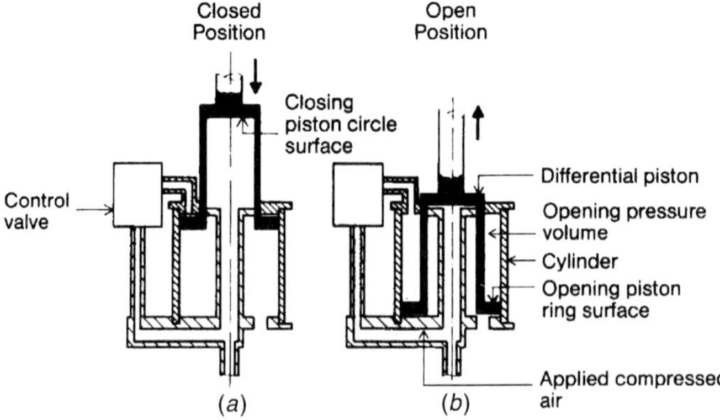


**FIGURE 12-18** Principle of the drive system for an $SF_6$ outdoor circuit breaker: (*a*) closed position; (*b*) open position.

***Accessories.*** Circuit breakers may be equipped with a wide range of accessories, either required, like pressure controls, gas-density monitors, safety valves, and position indicator, or optional, such as a choice of different release, alarms, or auxiliary contacts. To illustrate the importance of accessories for safe and reliable circuit breaker operation, Fig. 12-19 shows the $SF_6$ gas monitoring system of a high-voltage $SF_6$ outdoor circuit breaker.

The insulation and breaking capacity of an $SF_6$ circuit breaker depends on the gas density. It is assumed that the volume remains constant during temperature variations, whereas the pressure of the $SF_6$ is highly dependent on temperature change. Hence, to monitor the state of the gas, it is logical to supervise the density of the gas, instead of the gas pressure.

The density monitor operates according to the principle of a temperature-compensated pressure gage, the characteristics of which correspond to the constant-density line. The $SF_6$ gas pressure acts on a metal bellows, the movement of which is transmitted by a transfer mechanism with a bimetal disk to the microswitch.

The density monitor is set for the operating pressure based on the rated density requirements of the circuit breaker. Figure 12-19 shows a typical pressure-temperature diagram for this type of circuit breaker, with a minimum pressure of 5 bars, measured at 20°C, based on an $SF_6$ density of 31 kg/m$^3$. The density monitor emits a signal at 5.2 bars, indicating that refilling is necessary. If the pressure drops below 5 bars, operation of the circuit breaker is blocked.

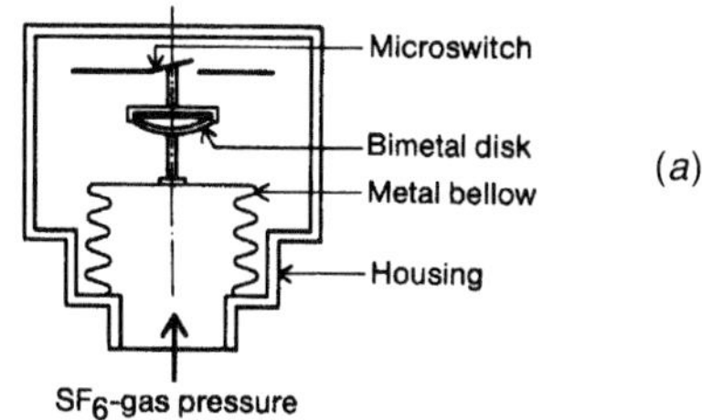


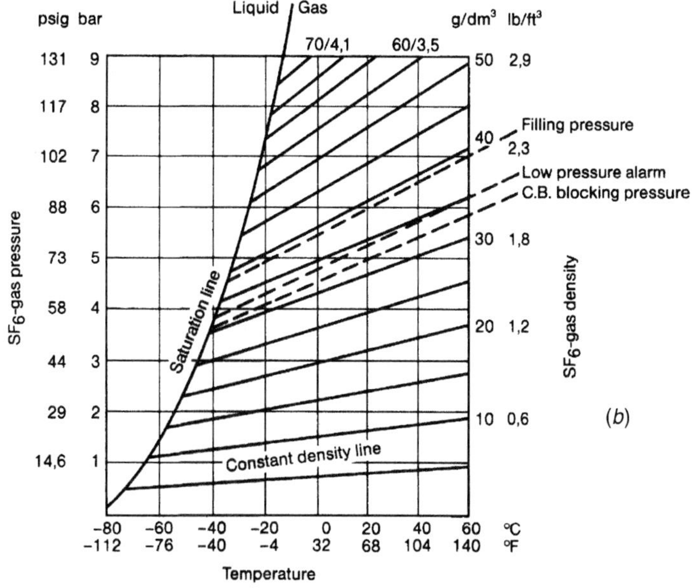


**FIGURE 12-19** (*a*) Arrangement and (*b*) pressure-temperature diagram for $SF_6$ gas-density monitor system for an outdoor circuit breaker. (*Brown Boveri.*)

## 12.1.6 Testing and Installation of High-Voltage Circuit Breakers

***Development and Design Testing of Circuit Breakers.*** Developing high-voltage circuit breakers requires attention to several different technical problems, including withstanding high voltages, short-circuit interruption, continuous current capability, mechanical endurance, and environmental conditions. Testing programs are carried out to ensure that the circuit breaker is capable of withstanding these stresses. Test procedures for ac high-voltage circuit breakers are specified in IEEE C37.09.[10]

***High Voltage.*** The circuit breaker must carry current at high voltage, and be able to withstand transient voltage surges at much higher levels (lightning strikes, for example). Typical test voltage levels are shown in Table 12-1 for IEEE rated breakers.

Designing for these voltages requires specialized engineering software and knowledge. Dimensions are set for insulating gaps between contacts, and between live parts and the grounded surrounding structures (tanks, shields, etc.). Electromagnetic finite element analysis software is the main tool for this task. $SF_6$ circuit breakers use pressurized sulfur hexafluoride gas for insulation and arc quenching. This is because pressurized $SF_6$ has substantially higher dielectric strength than air at atmospheric pressure, meaning gaps can be much smaller for the same voltage, and $SF_6$ is vastly better for arc quenching (see Fig. 12-2).

High-voltage testing requires "high-voltage laboratory" capability with both power frequency withstand voltage "hi-pot" test capability, and also voltage surge ("lightning impulse" and "switching impulse") withstand test capability.

***Short-Circuit Interruption.*** The most important function of circuit breakers is to interrupt short-circuit currents. This is to protect generators, transmission lines, transformers, and other components of the transmission system. Typical short-circuit requirements of high-voltage systems are 25 to 63 kA, though there is an increasing need for 80+ kA. During a short circuit (fault) the circuit breaker is subjected to high current immediately before interruption, followed by high voltage across the contact gap immediately after interruption. Designing for this capability involves engineering simulations and computational fluid dynamic analysis. Testing is difficult, as the momentary test power requirements can greatly exceed even that available on the transmission grid. Therefore, so-called "synthetic" techniques are used for this testing. Synthetic testing involves separate sources for the high current and high voltage, and only combining them during a very brief window of time. This greatly reduces the power requirements of simulating short-circuit interruption.

The high current is typically supplied from dedicated "short-circuit generators," which are specialized machines in the 1000 to 3000 short-circuit MVA class. High voltage is supplied from a

**TABLE 12-1** Typical Test Voltages for Outdoor High-Voltage AC Circuit Breakers

| Rated maximum voltage, kV | Power frequency withstand voltage (1 min dry) kV, rms | Full wave withstand (BIL) kV, peak |
|---|---|---|
| 15.5 | 50 | 110 |
| 38 | 80 | 200 |
| 72.5 | 160 | 350 |
| 145 | 310 | 650 |
| 170 | 365 | 750 |
| 245 | 425 | 900 |
| 362 | 555 | 1300 |
| 550 | 860 | 1800 |
| 800 | 960 | 2050 |

"high-voltage synthetic circuit," which is a combination of capacitor banks, reactors, triggering circuits, and computer controls. This synthetic circuit can produce a high-voltage waveshape approximating real system recovery voltage.

In a real world short-circuit interruption, a transient voltage occurs following interruption which tries to reestablish the arc. This transient voltage must be synthesized by the test circuit and the required wave shapes differ widely between different circuit breaker switching duties. A typical test involves initiating a short circuit with the high current circuit, and then at a "target" current zero simultaneously firing the high-voltage synthetic circuit and switching out the high current circuit (see Fig. 12-20). The firing of the high-voltage synthetic circuit provides correct short-circuit current *di/dt* at the target current zero, and also provides the correct recovery voltage waveshape after successful interruption. Variations of the basic synthetic test circuit are used to simulate different switching and interrupting duties on the power system. Unit testing, testing only one interrupting break of a multi-break circuit breaker, is accepted by all current standards, with some qualification.

***Continuous Current Capability.*** Circuit breakers must carry a rated continuous current without exceeding allowable temperature limits. This is demonstrated by testing with a suitable test transformer and current regulator.

***Mechanical Endurance.*** Circuit breakers are intended to operate for many years in service without significant maintenance. They also must be capable of withstanding many switching operations over that life. Testing to at least 2000 switching operations is required by standards for Class S2 circuit breakers, though circuit breakers may be required to meet higher values depending on the application and required ratings.

***Environmental.*** Circuit breakers are applied in all temperature zones and in many severe seismic zones. Climatic testing is carried out to verify operation at temperature extremes and to verify the performance of heating systems, when required. For example, $SF_6$ circuit breakers may require that the insulating $SF_6$ gas remain heated to a temperature of at least –30°C (depending on specific design pressure for the breaker) so that liquefaction does not reduce the gas density. Reduction of

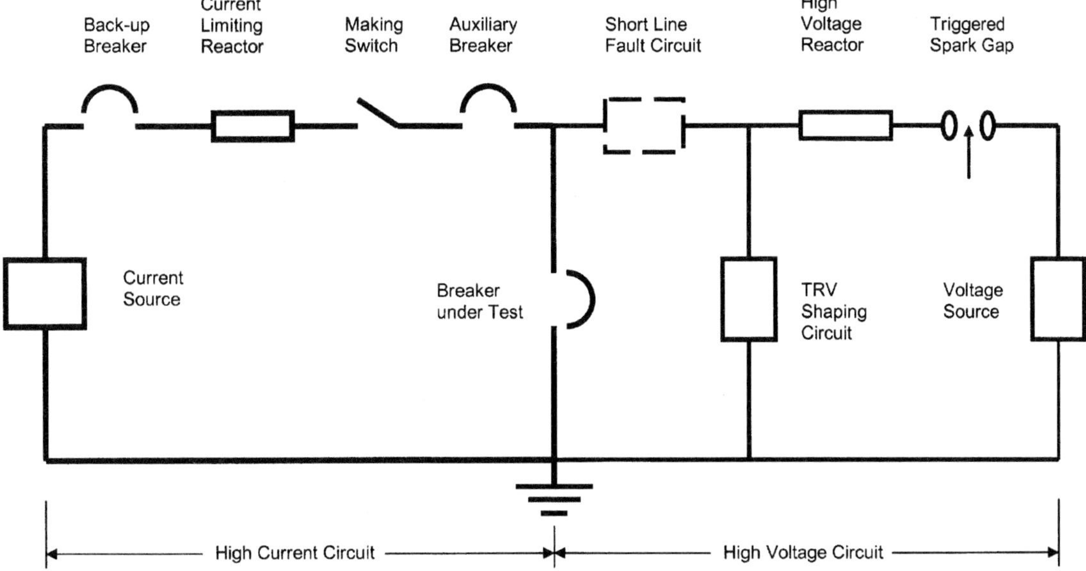


**FIGURE 12-20** Parallel current injection test circuit.

the insulating gas density below allowable levels may reduce insulating and interrupting capability such that the circuit breaker can no longer meet its rated performance. Alternatively, mixing $SF_6$ with another gas, such as $CF_4$, can allow operation at lower temperatures without heaters or derating.

Seismic withstand capability is demonstrated either by shake-table testing, dynamic analysis, or static coefficient method, depending on the voltage class. Typically, shake-table testing is required at 169 kV and above, with dynamic analysis being acceptable from 121 kV to less than 169 kV, and static coefficient method from 35 kV to less than 121 kV. Standard requirements for seismic specification, testing, and application are detailed in IEEE 693.[11]

***Production Tests.*** Every circuit breaker is subjected to a series of routine production tests primarily intended to prove design conformance and quality. These tests typically include high-voltage power frequency withstand voltage tests ("hi-pot"), mechanical operation and timing tests, fluid/gas leakage tests (when applicable), and control circuit operation verification.

***Installation and Field Tests.*** These tests are carried out on-site according to specific user's or manufacturer's instructions. Modern $SF_6$ circuit breakers up to 245 kV are often shipped fully assembled with a slight overpressure of $SF_6$, thus eliminating evacuation and gas-filling procedures on-site.

***Service and Maintenance.*** With rising system voltages, currents, interrupting ratings, and the requirement for uninterrupted power supply, circuit breaker reliability becomes more and more important. Besides influencing factors of (1) design, (2) quality assurance, and (3) testing, which are mainly a responsibility of the circuit breaker manufacturer, maximum attention must be paid to the maintenance during service. Maintenance instructions for different makes and types of circuit breakers may differ considerably in details and extent, but all strive to obtain maximum circuit breaker reliability despite longer maintenance intervals, smaller inventories of exchange parts, and shorter maintenance hours. Efforts are made to find the easiest way of handling service without influencing neighboring equipment and consequently obtaining the lowest service costs. Utility maintenance staffs, standardizing groups, and circuit breaker developers have taken into account these requirements. The progression of designs from oil circuit breaker to air-blast and finally the $SF_6$ and vacuum circuit breakers indicate a considerable minimizing of maintenance combined with maximum reliability.

## 12.1.7 Application and Selection of High-Voltage Circuit Breakers

The proper selection and application of circuit breakers is an extremely important element in the design of an electrical system. Circuit breakers are relied on to separate a defective portion of the system from the remainder to prevent the spread of damage and to permit the good portion to continue in service.

Application conditions and considerations for ac high-voltage circuit breakers are outlined in the latest revision of IEEE C37.010,[5] C37.011,[8] C37.012,[12] and C37.015.[13]

Among others, the following criteria have to be considered when selecting a circuit breaker:

*System data*, such as maximum system voltage, insulation level, short-circuit requirements, line or cable parameters.

*Switching conditions*, such as service currents; switching of unloaded transformers, unloaded lines and cables, choke coils, capacitors, generators, and motors; interrupting short-circuit currents and performing special duties such as out-of-phase switching, evolving fault, closing on a fault, closing on long lines; duty cycle, reclosing, and operating times.

*Service requirements*, such as special application for industrial plants, hazardous plants, arc-furnace transformer duty, railway duty, marine duty, maintenance, and operation.

*Site of installation*, altitude above 3300 ft, climatic conditions, humidity, wind load, ice, air contamination, space requirements, environmental requirements, earthquake, connection to and function with other switchyard and network components, open installation, or metal-clad switchgear.

High-voltage circuit breakers have been applied from 1 kV up to 1100 kV. They are available in either live-tank or dead-tank design, and may incorporate various insulating and interrupting media as discussed in Sec. 12.1.1.

In today's market, $SF_6$ circuit breakers dominate above 38 kV, with vacuum circuit breakers being dominant in the "medium" voltage range between 2.4 and 38 kV. Air magnetic circuit breakers were historically common up to 15 kV in metal-clad switchgear. Before the advent of air magnetic circuit breakers, oil circuit breakers were prevalent in these voltage ranges, and in outdoor circuit breakers. Even though oil circuit breakers are out of production today, many remain in service.

Indoor circuit breakers in the United States have been historically magnetic air or the less common air-blast types. Today vacuum circuit breakers dominate for indoor application.

For guidance in circuit breaker selection and application, refer to the latest revisions of IEEE C37.12,[14] C37.010,[5] C37.011,[8] and C37.012.[12]

***Ratings.*** Standard short-circuit currents range from 12.5 to 63 kA, with some applications requiring 80 kA. Commercially available continuous current ratings range from 600 to 5000 A. Preferred ratings are defined in IEEE C37.04[1] and C37.06.[2]

Ratings differentiate between indoor (Class S1) and outdoor (Class S2) service. Preferred ratings are further established for capacitive current switching, dielectric withstand and external insulation, transient recovery voltage capabilities, switching surge factors for line closing, control voltages, reclosing times, and operation endurance capabilities.

Refer to IEEE C37.04[1] and C37.06[2] for the rating structure and preferred ratings.

***Switching Considerations.*** Circuit breakers are designed to perform the closing and reclosing operations as per standard requirements.

When operated to close on long lines, extra-high-voltage circuit breakers require special measures to keep switching overvoltages within specified limits. Such measures may be single or multiple step *closing resistors*, synchronous closing at the moment of voltage zero, or *polarity-controlled-closing*, which means closing during the period of equal polarity at the line and source side of the circuit breaker.

When operated to close on capacitor banks, special measures may be taken to limit transient currents and voltages. Such measures may be closing resistors; controlled closing at the moment of voltage zero for grounded wye capacitor banks; or controlled closing on ungrounded wye capacitor banks where the first phase is closed at the moment of voltage zero and the other two phases are closed at a point where the voltage difference between the two phases is zero.

When operated to close on power transformers or shunt reactors, special measures may be taken to limit inrush transient currents and transient voltages. Such measures may be single or multiple step closing resistors, or controlled closing at the moment of voltage peak.

The magnitude of overvoltages on energizing and reenergizing is influenced by the nature and variables of the power system. Parameters of supply side and line must be taken into account in order to compute the overvoltages or to determine them using transient network analyzers or transient analysis software, such as EMTP, ATP, or PSCAD.

For a summary of the magnitude of overvoltages occurring when energizing high-voltage lines, based on numerous studies and measurements in high-voltage networks, see Table 12-2. Surge arresters may also be used to limit switching overvoltages.

***Environmental Considerations.*** Outdoor circuit breakers are generally available with special bushings that provide increased creepage distance for installation sites with highly contaminated air. For heavily polluted atmospheres, spray washing of live or deenergized circuit breakers may be an additional measure. Because of the method of design with enclosed ground insulation, the GIS circuit breaker is not influenced by atmospheric pollution.

**TABLE 12-2** Overvoltages Occurring When Energizing High-Voltage Lines

| Prevailing condition | Overvoltage factor (per unit) |
|---|---|
| 1. Line with trapped charge, no compensation, no means of reduction employed | >3 |
| 2. Line without trapped charge, no compensation, no closing resistors, or with trapped charge, no closing resistor, but polarity-dependent closing | 2.0–2.8 |
| 3. Same as 2, but with compensation | 2.0–2.5 |
| 4. Single-stage closing resistors, compensated line | ≤2.0 |
| 5. Two-stage closing resistors, optimum compensation | ≤1.7 |
| 6. Two-stage closing resistors, combined with polarity-dependent closing, or compensation with optimized multistage closing resistors | 1.5 |

***Altitude Correction.*** For installation at altitudes above 3300 ft (1000 m), altitude correction factors have to be applied. Altitude correction factors are covered in IEEE C37.100.1.[15] The values of rated maximum voltages and insulation levels are multiplied by the factors to obtain the values for the application.

***Oil Circuit Breakers.*** Oil circuit breakers are out of production today, but many remain in service. These circuit breakers were classified as either dead-tank "bulk oil" circuit breakers, or as live-tank "minimum oil" circuit breakers. Oil circuit breakers use oil as both an arc quenching and insulating medium, with dead-tank "bulk oil" designs using oil as the primary insulation to ground, within a grounded tank.

Dead-tank "bulk oil" circuit breakers consist of a steel tank partly filled with oil, through the cover of which are high-voltage entrance bushings. Contacts at the bottom of the bushings are bridged by a conducting crosshead carried by a wood or composite lift rod. The circuit breaker typically opens by spring action, separating the interrupting contacts, and also further separating an isolation break below the contact system. Accelerating springs are used to increase the speed of opening. In some designs the crosshead is opened with a rotary motion by springs. Circuit breakers with three poles in one tank were made up to 69 kV. Higher voltages had separate tanks for each pole. Figure 12-3 shows a typical dead-tank "bulk oil" circuit breaker, and Fig. 12-21 shows a typical interrupter.

Minimum oil circuit breakers were developed mainly in Europe to reduce space and the quantity of oil in circuit breakers. They were manufactured for indoor applications up through 38 kV and outdoor applications up through 800 kV. The layout and interrupter details of a medium voltage minimum-oil circuit breaker are shown in Fig. 12-22.

***Vacuum Circuit Breakers.*** Progress in high-vacuum technology and circuit breaker development, combined with improved manufacturing and testing methods, has opened a growing area for vacuum circuit breaker application, concentrating, but not limited to voltages up to 38 kV, continuous current ratings up to 4000 A, and covering all standard interrupting ranges. Vacuum interrupters are available today up to 145 kV, though acceptance is a work in progress. The principal design of a vacuum interrupter is shown in Fig. 12-23.

Two contacts are mounted on an insulating envelope from which virtually all air has been evacuated. One contact is stationary, the other movable. Vacuum interruption has the inherent advantage of moving a lightweight contact for only a very small distance in an almost perfect dielectric medium. This results in safe, quiet, and fast switching and interruption of load or fault currents.

The moving contact is opened up to full gap distance by means of a driving mechanism. A metal-vapor arc discharge thus occurs in the contact gap through which the current flows until the next current zero. The arc is quenched at current zero.

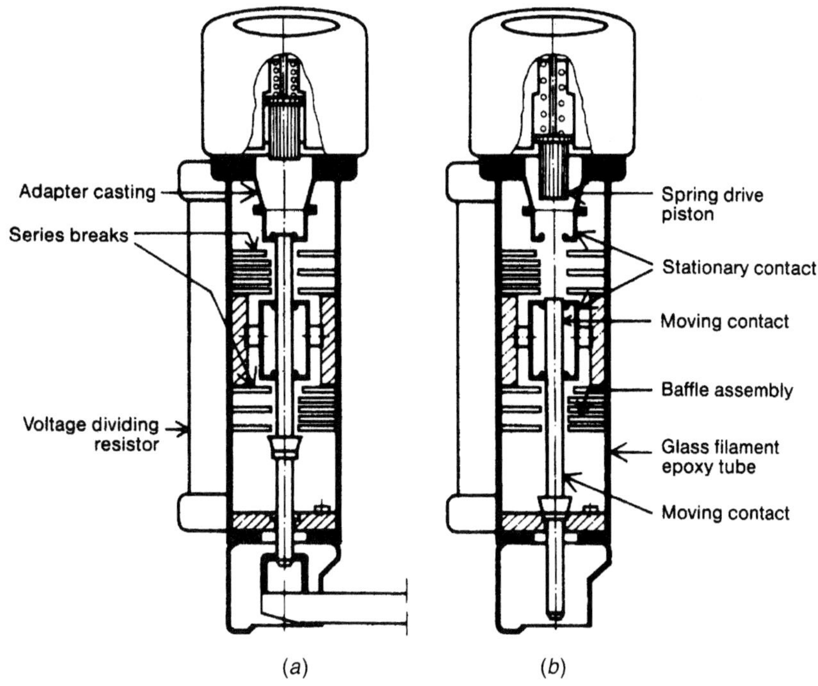


**FIGURE 12-21** Details of a 161-kV outdoor oil circuit breaker interrupter: (*a*) closed position; (*b*) open position.

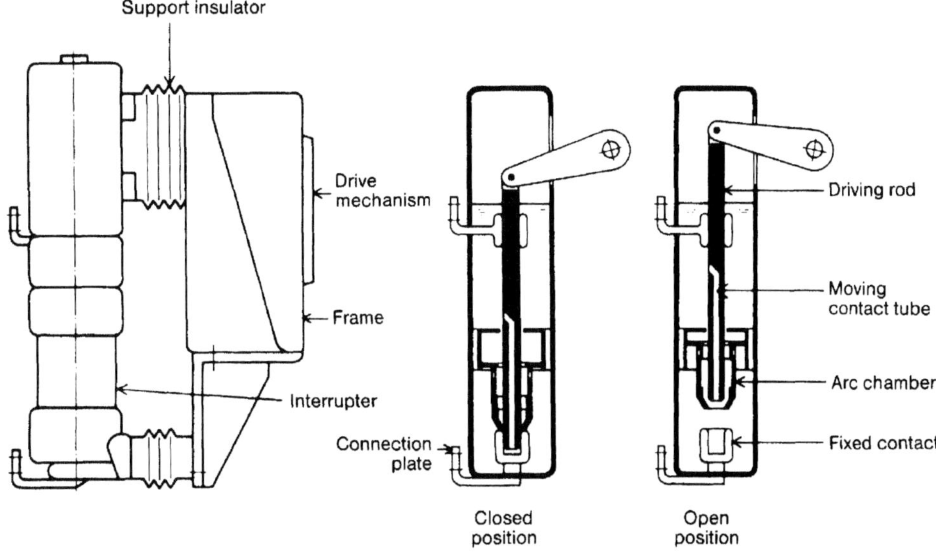


**FIGURE 12-22** Outline and interrupter details of a 15-kV, 3-pole minimum-oil circuit breaker. (*Courtesy of ABB T&D Company, Inc.*)

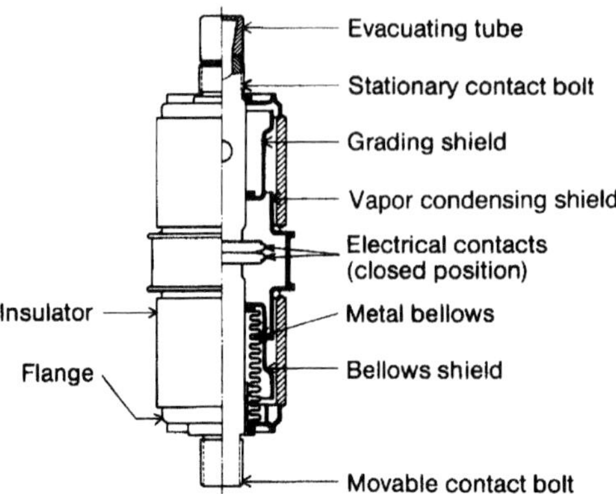


**FIGURE 12-23** Partial section of vacuum interrupter 23 kV, 2000 A, 21 kA. (*ABB T&D Company, Inc.*)

The metal-vapor plasma is fully deionized within a few microseconds by diffusion and recombination so that the conduction path very quickly recovers its dielectric strength. Figure 12-24 shows details of a horizontal-drawout vacuum circuit breaker. One or more interrupters may be utilized in series per pole. Vacuum interrupters may additionally be protected against outside influences by an insulating casing. They may also be fitted with hand- or motor-charged stored-energy-operated mechanisms or magnetic actuators.

Because of their fast closing and opening times, vacuum circuit breakers are particularly suitable for automatic reclosing and synchronizing duty. Breaking of the short-circuit currents with very steep initial rise of transient recovery voltage is possible due to restoration of the dielectric strength of the contact gap within a few microseconds. The steep rise of dielectric strength over the whole current range offers a high capacitive-current-switching capability. Switching of unloaded transmission lines and cables can therefore reliably be performed.

***Air Magnetic Circuit Breakers.*** Medium voltage air magnetic circuit breakers are no longer manufactured. This type of circuit breaker is usually stored-energy mechanism-operated and interrupts the main circuit in the normal atmosphere under the influence of a strong magnetic field which acts to force the arc deep into a specially designed arc chute (see Fig. 12-25). The arc chute cools and lengthens the arc to a point where the arc cannot be maintained by the voltage of the system, and interruption is accomplished. The zone between the main contacts is clear of ionized air by the time interruption is obtained in the arc chute, and so restriking at this point is avoided. Since the magnetic effect is not great at low currents such as small load, transformer magnetizing, and cable-charging current, all designs used an air-pump "puffer" actuated by the operating mechanism that blows a blast of air across the arc and thereby ensures its entering the arc chute and giving rapid interruption

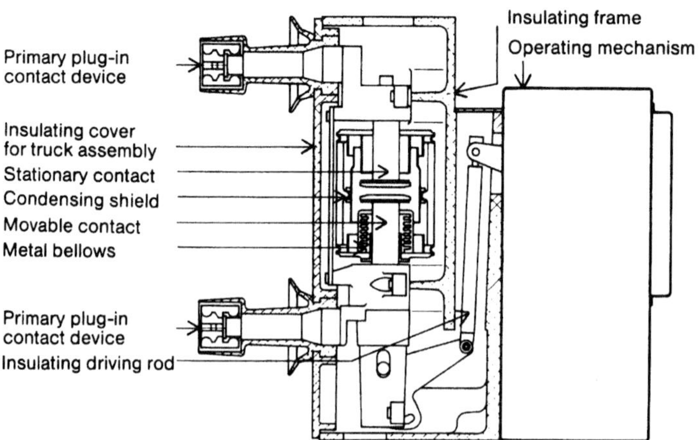


**FIGURE 12-24** Outline and interrupter details of a 15-kV horizontal drawout vacuum circuit breaker 2000 A, 28 kA. (*ABB T&D Company, Inc.*)

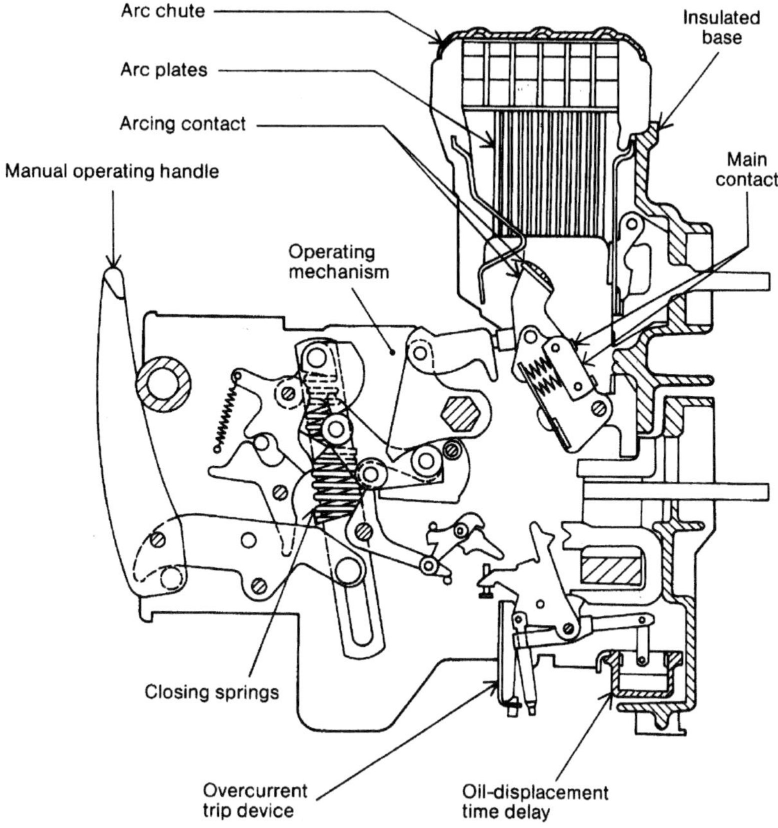


**FIGURE 12-25** Typical low-voltage air circuit breaker with magnetic air chutes; breaker in the open position.

at the low-current values. When the circuit breaker is opened, the arc transfers from the main arcing contacts to fixed arc runners which are within the arc chute. The magnetic field is produced by coils in the main-current circuit, in some cases wound around a magnetic core which magnetizes soft-iron plates in the sides of each arc chute. Some designs did not require an iron core.

Magnetic air circuit breakers were available in any of the ratings of Table 2 of ANSI Standard C37.06-1987[4] (or earlier) through 15 kV. All were designed for use in metal-clad enclosures. Figure 12-26 shows the horizontal-drawout type of circuit breaker in a metal-clad enclosure. Although the design shown is for indoor use only, the same circuit breakers are placed in weatherproof housings for outdoor service. When they are so used, suitable heaters are put in the housings to avoid internal moisture condensation.

***Air-Blast Circuit Breakers.*** Air blast circuit breakers are no longer manufactured. These circuit breakers fulfilled the heavy-duty requirements of circuit breakers in high-voltage systems. They have been used to provide the indoor ratings up to 38 kV. They were, however, mainly used in outdoor applications up to 800 kV. Today they have been replaced by $SF_6$ technology at most ratings above 38 kV, and by vacuum technology up to 38 kV.

Air-blast circuit breakers have been used for special applications as (1) generator circuit breakers with continuous current ratings of up to 42 kA and higher, (2) arc-furnace transformer circuit

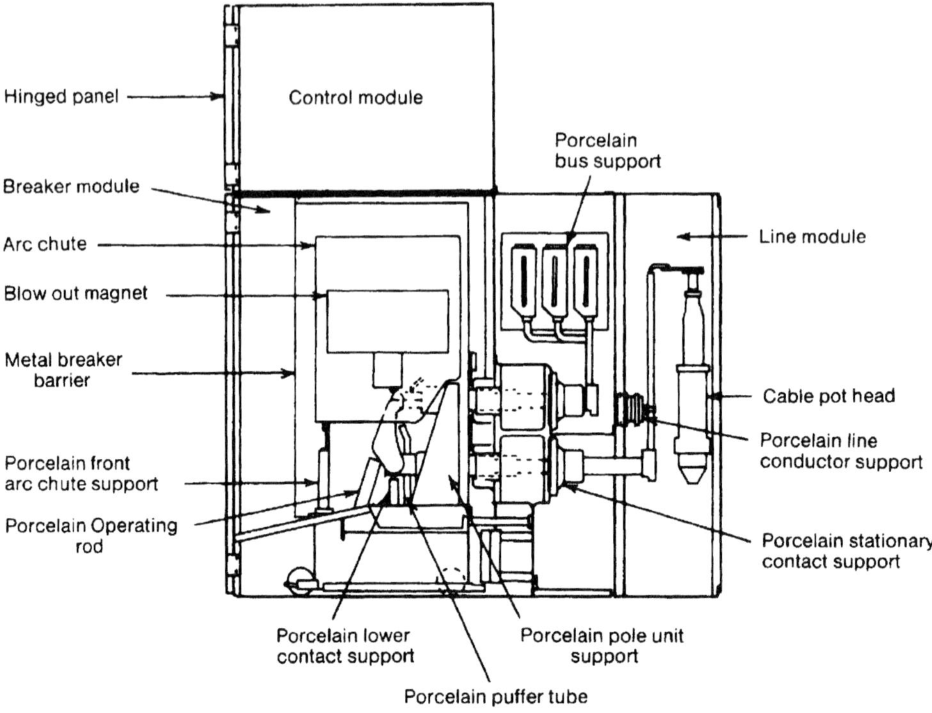


**FIGURE 12-26** Horizontal-drawout, metal-clad magnetic circuit breaker in service position.

breakers with an extra-high number of switching operations (20 to 50 operations per day), and (3) extra high interrupting currents. Air-blast circuit breakers were usually fixed-mounted, but a variety of circuit breaker types were available truck-mounted for application in drawout metal-clad switchgear. All air blast circuit breakers make use of dry and clean air compressed to 15 to 25 bars, depending on the make and types of circuit breaker. The compressed air is used to operate the circuit breaker as well as to serve as the medium for arc quenching and insulation.

Continuous current ratings up to 5000 A were possible. Total breaking time of 2 cycles (from energizing of trip coil until arc extinguishing) was first achieved by air-blast designs in the 1960s. Special designs allowed for even shorter breaking time. Some 69-kV circuit breakers are equipped with sequential isolators, but the bulk of designs did not integrate the isolator to form part of the circuit breaker. Some older designs employed separate chambers for opening and closing operation, but later air-blast circuit breakers perform opening and closing with the same contact system. Closing resistors and/or opening resistors, with some designs, were often used. Limited voltage capability per break led to many interrupter breaks at higher voltages, up to 12 to 15 per phase. Voltage distribution over the multiple breaks of one pole was usually achieved by parallel grading capacitors.

***Generator Circuit Breakers.*** Generator circuit breakers represent another class rated for very high continuous currents and short-circuit currents, typically at generator voltages. Generator circuit breakers are often incorporated into iso-phase buses and can include other switchgear components for measuring current, detecting faults, and grounding.

Generator circuit breakers are available up to 50 kA continuous current and up to 220 kA interrupting current. Three technologies are employed (1) air blast at the higher ratings (see Fig. 12-27), (2) $SF_6$ self-blast at medium power levels (up to 120 kA), and (3) vacuum for lower power levels

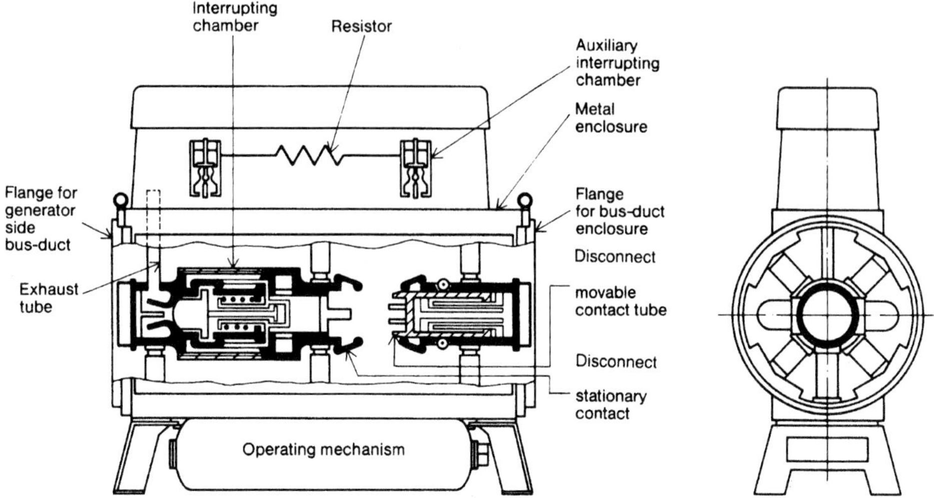


**FIGURE 12-27** Outline and interrupter details of a generator air-blast circuit breaker-type DR, 36 kV, up to 50 kA with forced cooling, 200 kA.

(up to 80 kA). For continuous currents above 20 kA the generator circuit breaker is usually equipped with a forced cooling system, using water, for example. Generator circuit breakers have been available since the 1960s.

Advantages of using generator circuit breakers include the following:

- Simplification of operation, especially during commissioning and recommissioning; because the generator can be handled as a separate unit, isolated from the main and unit transformers.
- Fault protection between the generator and transformer. Two zones of protection are created and generator faults are cleared by the opening of the generator circuit breaker alone.
- Unbalanced load protection of the generator.
- Protection of the generator from transformer faults.
- Reliability/availability increase.

Historically generator circuit breakers have been of air-blast design with pneumatic operators. This is the technology still used today for large nuclear and fossil fuel power plants (up to 1500 MW), and large pumped storage installations. The design has a tubular housing and is horizontal.

Newer designs utilize $SF_6$ self-blast technology and hydraulic operators. These are rated for application to smaller power plants (gas turbine/cogeneration, for example) from 60 to 400 MW and smaller pumped storage installations. Vacuum generator circuit breakers are increasingly used for generators from 10 to 80 MW and higher.

***$SF_6$* Circuit Breakers.** Sulfur hexafluoride ($SF_6$) gas has proven to be an excellent arc quenching and insulating medium for circuit breakers. $SF_6$ is a very stable compound, inert up to about 500°C, nonflammable, nontoxic, odorless, and colorless. At a temperature of about 2000 K, $SF_6$ has a very high specific heat and high thermal conductivity, which promotes cooling of the arc plasma just before and at current zero, and thus facilitates quenching of the arc. The electronegative behavior of the $SF_6$, that is, the property of capturing free electrons and forming negative ions, results in high dielectric strength and also promotes rapid dielectric recovery of the arc channel after arc quenching.

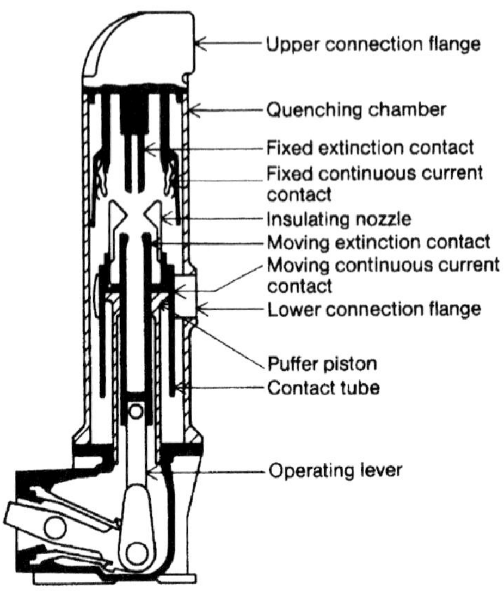


**FIGURE 12-28** Section of a $SF_6$ puffer-piston indoor circuit breaker, 23 kV.

$SF_6$ circuit breakers are available for all voltages up to 1200 kV, continuous currents up to 5000 A for conventional circuit breakers (higher for generator circuit breakers), and short-circuit interruption up to 80 kA.

$SF_6$ circuit breakers of the indoor type have been incorporated into some designs of metal-clad switchgear (see Fig. 12-28). Outdoor designs include both dead-tank (see Fig. 12-4) and live-tank circuit breakers (see Fig. 12-7).

Over the years, $SF_6$ circuit breakers have reached a high degree of reliability; thus they can cope with all known switching phenomena. Their closed-gas system eliminates external exhaust during switching operations and thus adapts to environmental requirements. Their compact design considerably reduces space requirements and building and installation costs. In addition, $SF_6$ circuit breakers require very little maintenance. Each pole is equipped with one or more interrupters; stored energy, spring, hydraulic, or pneumatic driving mechanisms are provided for each pole or three-pole unit. Gas-density monitors are standard.

$SF_6$ gas circuit breakers were initially of the two-pressure type, in which high pressure gas for interruption is compressed to 15 to 20 bars and then stored for later interrupting duty. Later designs employed the puffer principle, in which interrupting pressure is developed during the contact motion itself, and no high pressure gas is stored. Puffer interrupters require high mechanical forces and energies to be supplied from the operating mechanism to achieve the $SF_6$ pressures needed for interruption. The interrupting principle of an $SF_6$ puffer-type interrupter is shown in Fig. 12-29.

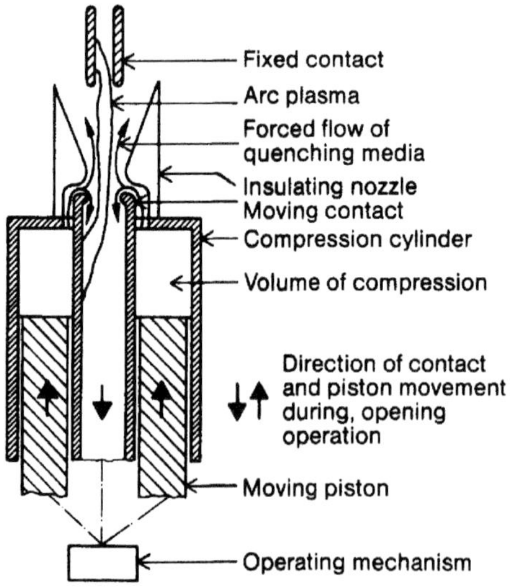

The latest designs of $SF_6$ gas circuit breakers reduce mechanical force requirements by using the arc energy itself to develop the interrupting pressure; these designs are referred to as *self-blast* or *thermal-assist* circuit breakers. Further force/energy reductions are achieved using double motion designs, in which both contacts move to achieve a higher relative speed, saving kinetic energy.

Figure 12-30 illustrates the opening sequence of a typical puffer circuit breaker. In the closed position the current flows over the continuous current contacts and the complete volume of the circuit breaker pole is under the same pressure of $SF_6$ gas.

The precompression of the $SF_6$ gas commences with the opening operation. The continuous current contacts separate and the current is transferred to the arcing contacts.

At the instant of separation of the arcing contacts, the pressure required to extinguish the arc is reached. The arc produced is drawn and at the same time exposed to the gas, which escapes through the ring-shaped space between

Fixed contact
Arc plasma
Forced flow of quenching media
Insulating nozzle
Moving contact
Compression cylinder
Volume of compression
Direction of contact and piston movement during, opening operation
Moving piston
Operating mechanism

**FIGURE 12-29** Principle of puffer-type arc interrupters.

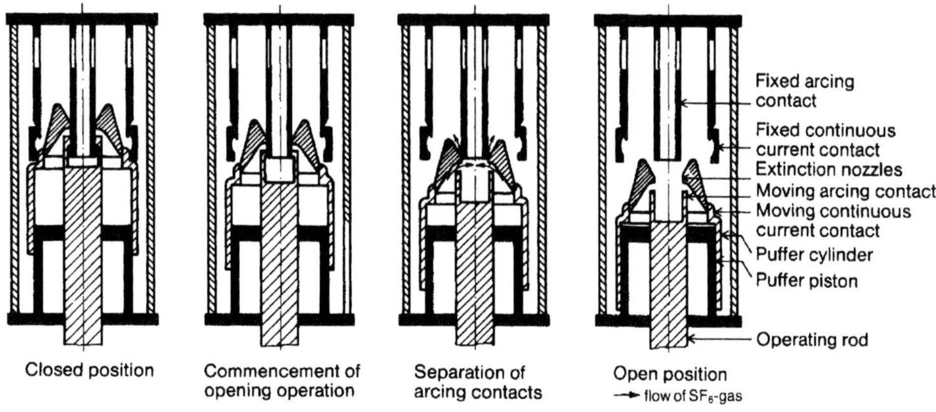


**FIGURE 12-30** Principle of $SF_6$ puffer-type interrupter showing four positions during opening operation.

the extinction nozzle and the moving arcing contact. The escaping gas has the effect of a double blast in both axial directions.

Until the open position is reached, $SF_6$ gas flows out of the puffer cylinder. The existing overpressure maintains stability of the dielectric strength until the full open position of the contacts at the rated service pressure is reached.

The *self-blast principle* of interruption is illustrated in Fig. 12-31. In the case of high-current interruption, arc energy heats the gas, resulting in a pressure rise in the static volume, $V_1$. This pressure then quenches the arc at an ensuing current zero. In the low-current case an auxiliary puffer (volume $V_2$) generates sufficient pressure for interruption. Necessary force requirements for the mechanical system are therefore drastically reduced.

All ancillary equipment, including the oil pump and accumulator associated with the drive, form a modular assembly that is mounted directly on the circuit breaker, thus eliminating installation of piping on the site. The *metal-enclosed GIS circuit breaker* is provided with the necessary items to fit into the substation arrangement (see Fig. 12-32). The main equipment flanges of the circuit breaker are fitted with contact assemblies to accept the isolator moving contacts. Other equipment modules can be coupled to the same flanges. On the fixed-contact end of the circuit breaker, provision is made for coupling two modules, facilitating the mounting of an extension module to connect the second busbar isolator.

Dead-tank $SF_6$ circuit breakers typically employ gas-filled *bushings*, illustrated in Fig 12-5. Such bushings are usually integral to the circuit breaker itself and are not interchangeable with other apparatus bushings. Electrical grading is provided by a lower throat shield. Ring-type bushing current transformers are located at the base of the bushing. Potential taps are not generally available in $SF_6$ bushings because of the lack of a capacitive grading structure.

Porcelain alternatives, such as composites, have been used to provide greater safety (explosion resistance), easier handling (lighter and nonbrittle), seismic performance (lighter and stronger), and improved pollution performance.

*Current transformers* (CTs) for dead-tank circuit breakers are of the ring-type bushing design. Outdoor circuit breakers of the live-tank layout are generally provided with free-standing CTs of the paper-oil-insulated or $SF_6$ design.

For oil-filled CTs hermetical seal of oil is either of the fixed design with gas cushion or of the pressure-free bellows type. Up to six magnetic cores can be provided per CT unit, generally in multiple ratios for 5 A or 1 A secondary by means of secondary taps. Primary-current ratings up to 2000 A normally employ the wound-type design with two or more turns. Higher primary currents up to 6000 A require the inverted or head design, with a straight tube as single-turn primary winding and the core and secondary-winding assembly arranged at the CT top to limit temperature rise and to increase the mechanical withstand capability of the CT. The latter design has its full main insulation on the secondary winding.

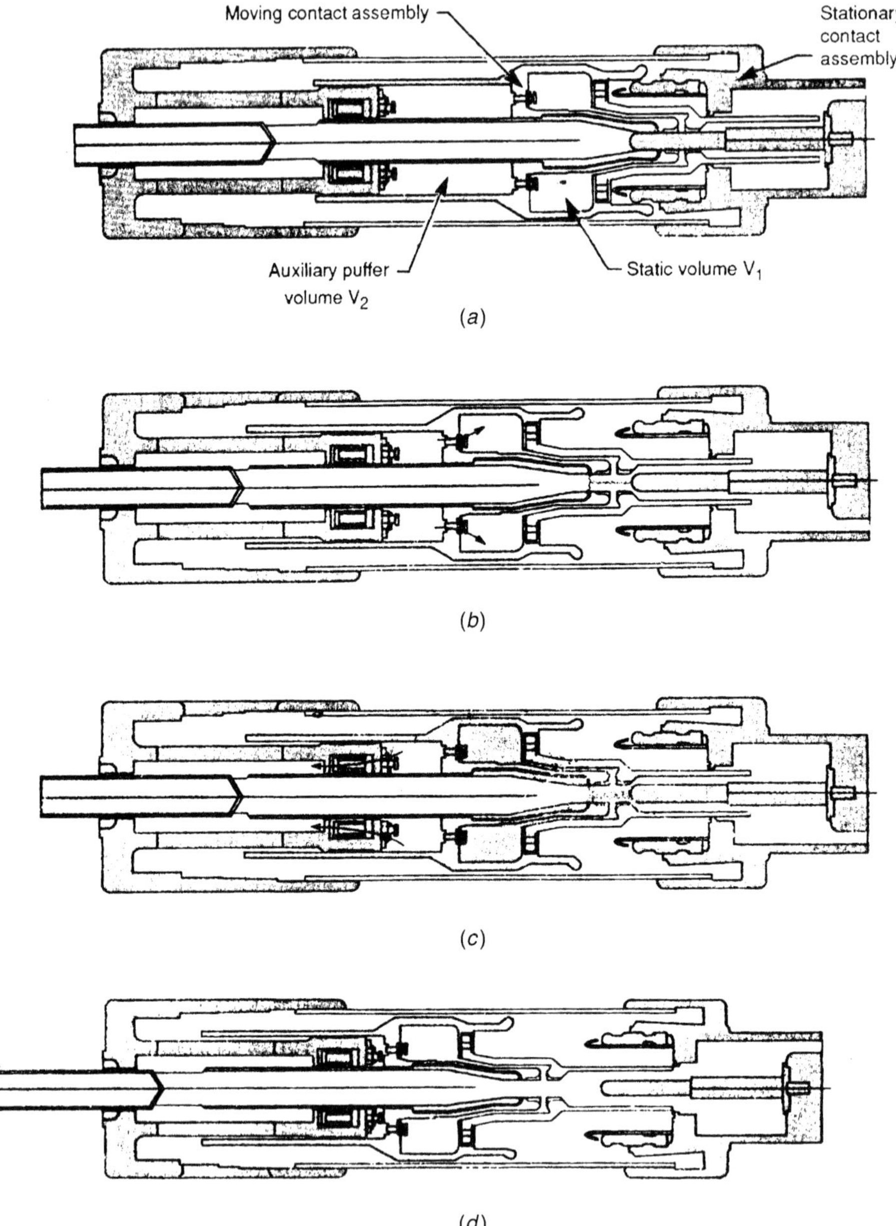


**FIGURE 12-31** Self-blast principle of interruption: (*a*) full-closed position; (*b*) low-current interruption; (*c*) high-current interruption; (*d*) full-open position.

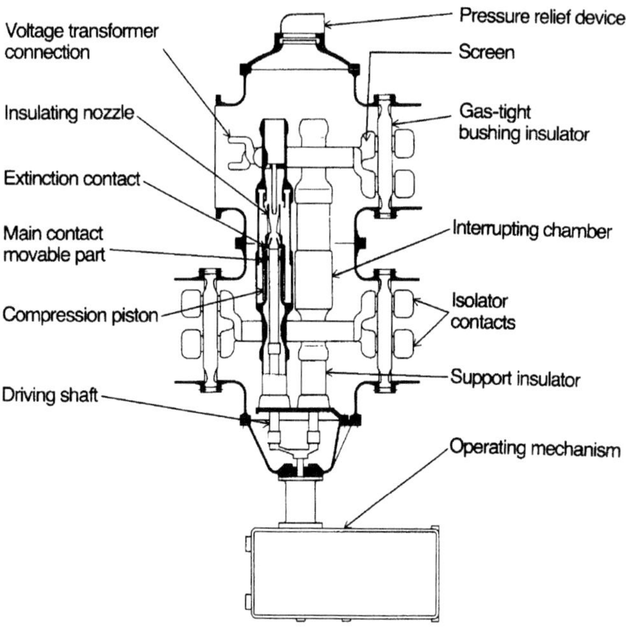


**FIGURE 12-32** Section of a 145 kV $SF_6$ circuit breaker for gas-insulated substation (GIS) type ELK.

Free-standing CTs are available for all output and accuracy requirements for modern system relaying and measuring for voltages up to 1100 kV. For the upper voltage ranges, free-standing CTs are normally provided with separate potential layers. The CTs are generally dimensioned for the same dielectric and mechanical characteristics chosen for the related circuit breaker.

## 12.1.8 Low-Voltage Circuit Breaker Ratings

The following ratings should be specified by the user in the application of low-voltage circuit breakers in an enclosure.

- Rated maximum voltage
- Rated continuous current
- Rated power frequency (for ac circuit breakers, 50 or 60 Hz)
- Rated short-time current
- Rated short-circuit current
- Rated peak current (for dc circuit breakers)

Standard electrically and manually operated circuit breakers are listed in ratings up to and including 6000 A ac and 12,000 A dc. Electrically operated circuit breakers are available in higher current ratings for special applications. Standard circuit breakers are rated on the basis of a temperature rise on the contacts not to exceed 85°C above an ambient outside the enclosure of 40°C. Voltage ratings are 254 to 1000 V ac and 250 to 3200 V dc.

The short-time current ratings are based on three-phase symmetric short-circuit currents; the single-phase short-circuit current ratings are 87% of these values. For details refer to the latest revision of IEEE C37.13.[17]

## 12.1.9 Low-Voltage Circuit Breaker Construction

***Assembly Variations.*** The circuit breakers are usually installed in a metal-enclosed cubicle for dead-front or drawout type of construction. Metal barriers between circuit breakers and busbars provide increased safety in service.

Hand operation by means of a lever is common, even on large circuit breakers. Electric operation by means of a solenoid or motor mechanisms for 48, 125, or 250 V dc, or 120 or 240 V ac is available on all air circuit breakers and on higher-rated molded-case circuit breakers.

Circuit breakers are supplied with an overcurrent trip mechanism which may be of the instantaneous or the time-delay type, or a combination of both. Trip devices are adjustable over a wide range of ratings. Other trip devices and arrangements may be used, for example undervoltage trips, shunt trips, reverse current, or overcurrent relays.

Multiple-pole circuit breakers are commonly used in practically all capacities, one pole being used for each ungrounded line of a circuit, that is, a two-pole circuit breaker for a three-wire grounded circuit or a single-pole circuit breaker for a two-wire grounded circuit.

Circuit breakers can usually be equipped with auxiliary contacts, alarm contacts, pushbutton control, position indicator, and key interlock. The widely used drawout type of circuit breaker may be moved into and locked in the connected, test, and disconnected positions and/or completely withdrawn.

Refer to the latest revisions of IEEE C37.13,[17] C37.14,[18] C37.16,[16] and C37.17.[19]

***Air Circuit Breakers.*** The usual construction of an air circuit breaker (Fig. 12-25) makes use of two fixed terminals mounted one above the other in a vertical plane, which, when the circuit breaker is closed, are bridged under heavy pressure by a bridging member operated by a system of linkages. Arcing contacts close before and open after the main contacts. The arcing contacts are easily renewable. The circuit breaker is held closed by a latch which may be tripped electrically or mechanically. Modern circuit breakers are trip-free.

Many circuit breakers use a solid bridging member with spring-mounted self-aligning contacts. The contact surfaces are made of silver so that oxidation will not cause excessive resistance and overheating.

Arcing contacts of modern circuit breakers use a silver-tungsten or copper-tungsten alloy which is arc-resisting.

Barriers between poles are generally furnished with circuit breakers on ac and dc circuits 250 V and above, and special arc chutes, quenchers, or deionizing chambers are also used throughout the available lines of air circuit breakers. These devices are made in different forms by different manufacturers and serve to improve the interrupting performance of the circuit breaker and to decrease the arcing time.

***Molded-Case and Insulated-Case Circuit Breakers.*** This circuit breaker is completely enclosed within a ruggedly constructed molded case of insulating material. It has received wide acceptance in industry and is particularly adaptable in large buildings and industrial plants. The molded-case circuit breaker, in smaller sizes, is adaptable in home lighting circuits where convenience of automatic protection with manual reset of the circuit breaker is desired.

Continuous current ratings range from 15 to 6000 A; the interrupting ratings are from 5 to 45 kA within the standard range. High interrupting ratings up to 200 kA are available.

For details of technical data, application, and accessories refer to manufacturers catalogs.

***Current-Limiting Circuit Breakers.*** Low-voltage switchgear is often connected to systems with high or extra-high short-circuit currents. The standard-range circuit breaker cannot satisfy these requirements. Figure 12-33 outlines different methods to solve the problem. The current-limiting circuit breaker with high interrupting capacity offers a technically sound and economical solution.

Current-limiting circuit breakers operate extremely rapidly. Interruption takes place within the first half-cycle of short-circuit current, and act to limit the peak instantaneous current.

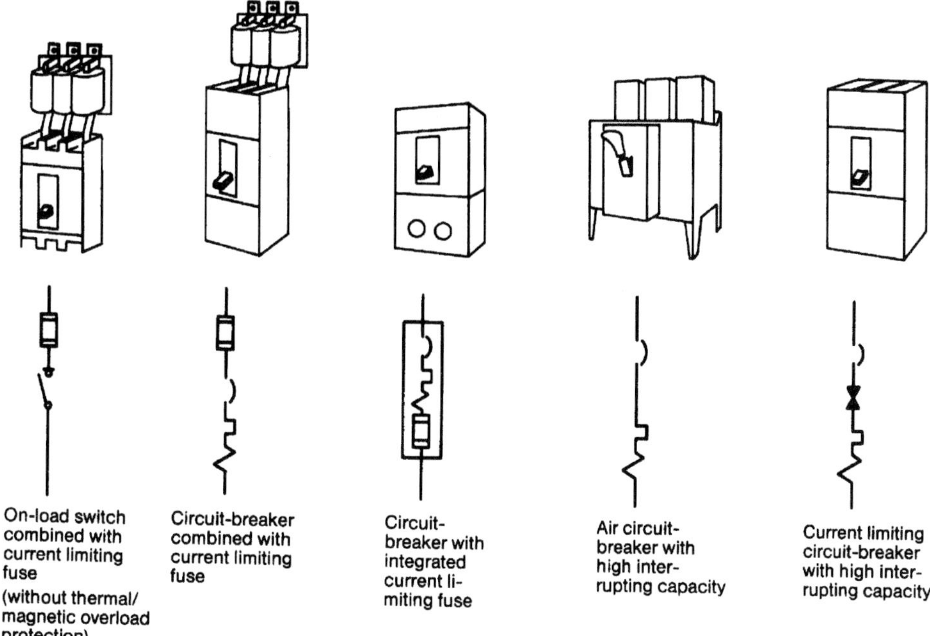


**FIGURE 12-33** Methods of current-limiting in low-voltage circuits.

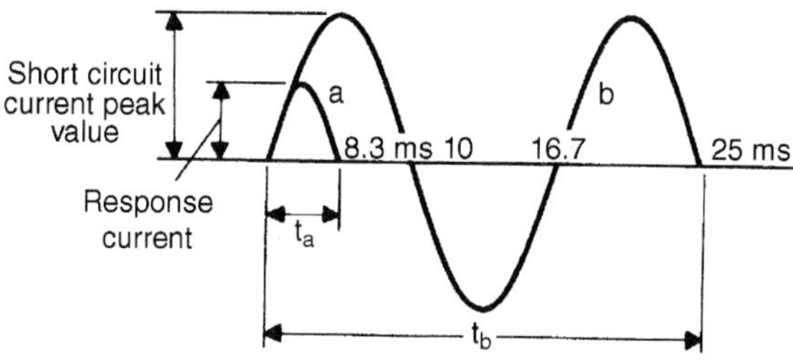


**FIGURE 12-34** Current wave (*a*) with limitation, and (*b*) without limitation; $t_a$ = total break time *a*; $t_b$ = total break time *b*.

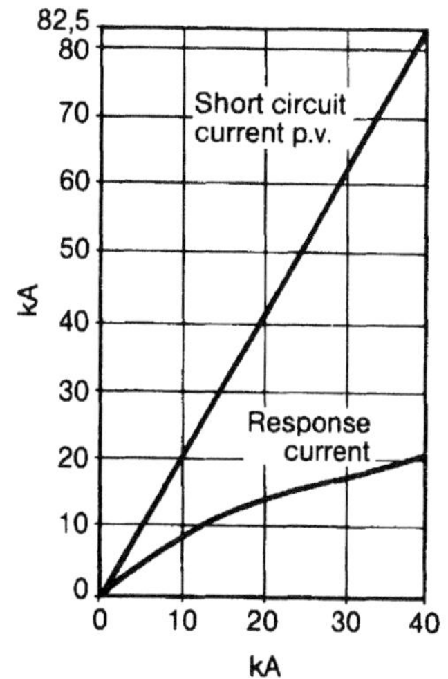


**FIGURE 12-35** Current-limiting capability of a motor-protection circuit breaker, 100-A continuous current rating.

Figure 12-34 illustrates the current curve, and Fig. 12-35 shows the current-limiting characteristic of a 100-A circuit breaker. With an initial symmetric short-circuit current of 40 kA, the prospective peak value would be 82.5 kA, considering a dc component of 50% and power factor of 0.25. By using a current-limiting circuit breaker, the peak value is limited to about 20 kA. The mechanical stress on the conductors is thus reduced considerably. The contacts in current-limiting circuit breakers are so arranged that the interruption is assisted by the electrodynamic action of the short-circuit current. The higher the short-circuit current, the faster the interruption takes place.

Because of the short opening time, the current-limiting circuit breaker, with suitable accessories, can be used to protect power electronic components. Rectifier circuits omitting fuses, for example can be built in this way.

## 12.1.10 Application of Low-Voltage Circuit Breakers

***Application.*** Air circuit breakers are used on dc and ac circuits for reactive current switching, and for the protection of general lighting, power, motor circuits, and traction power systems.

Distinction is made between various protection classes and different service and ambient conditions. For selection of a circuit breaker, type and rating, operating speed, selectivity with fuses, and rated voltage must be taken into account.

Further consideration has to be given to severe or hazardous service conditions like tropical climate or marine- or explosion-proof installations.

Reference is made to IEEE C37.13,[17] C37.14,[18] C37.16,[16] and C37.17,[19] UL-489.[20]

## 12.1.11 References

1. IEEE C37.04, *Standard Rating Structure for AC High-Voltage Circuit Breakers.*
2. IEEE C37.06, *Standard for AC High-Voltage Circuit Breakers Rated on a Symmetrical Basis—Preferred Ratings and Related Required Capabilities for Voltages above 1000 V.*
3. IEEE C37.100, *Standard Definitions for Power Switchgear.*
4. ANSI C37.06, *Standard for AC High-Voltage Circuit Breakers Rated on a Symmetrical Basis—Preferred Ratings and Related Required Capabilities* (up to the year 2000, C37.06 was an ANSI document under the ASC C37. The 2009 version is an ANSI/IEEE document. IEEE C37.06 will be made obsolete when revisions to IEEE C37.04 incorporate the preferred ratings values into C37.04).
5. IEEE C37.010, *Application Guide for AC High-Voltage Circuit Breakers Rated on a Symmetrical Current Basis.*
6. ASA C37.7-1960, *Interrupting Rating Factors for Reclosing Service.*
7. IEEE C37.04b-2008, *Standard Rating Structure for AC High-Voltage Circuit Breakers Rated on a Symmetrical Basis—Amendment 2: To Change the Description of Transient Recovery Voltage for Harmonization with IEC 62271-100.*
8. IEEE C37.011, *Application Guide for Transient Recovery Voltage for AC High-Voltage Circuit Breakers.*
9. *IEEE C37.04a-2003, Standard Rating Structure for AC High-Voltage Circuit Breakers Rated on a Symmetrical Current Basis—Amendment 1: Capacitance Current Switching.*
10. *IEEE C37.09, Standard Test Procedure for AC High-Voltage Circuit Breakers Rated on a Symmetrical Current Basis.*
11. IEEE 693, *Recommended Practice for Seismic Design of Substations.*
12. IEEE C37.012, *Application Guide for Capacitance Current Switching for AC High-Voltage Circuit Breakers.*
13. IEEE C37.015, *Guide for the Application of Shunt Reactor Switching.*
14. IEEE C37.12, *Guide for Specification of High-Voltage Circuit Breakers (over 1000 Volts).*
15. IEEE C37.100.1, *Standard of Common Requirements for High-Voltage Power Switchgear Rated Above 1000 V.*
16. IEEE C37.16, *Standard for Preferred Ratings, Related Requirements, and Application Recommendations for Low-Voltage (635 V and below) and DC (3200 V and above) Power Circuit Breakers* (IEEE C37.16 is now obsolete, as preferred ratings have been incorporated into IEEE C37.13 and IEEE C37.14).
17. IEEE C37.13, *Standard for Low-Voltage AC Power Circuit Breakers Used in Enclosures.*
18. IEEE C37.14, *Standard for Low-Voltage DC Power Circuit Breakers Used in Enclosures.*
19. IEEE C37.17, *Standard for Trip Devices for AC and General Purpose DC Low-Voltage Power Circuit Breakers.*

20. UL-489, *Standard for Safety Molded-Case Circuit Breakers, Molded-Case Switches, and Circuit-Breaker Enclosures.*

21. IEEE C37.09a-2005, *Standard Test Procedure for AC High-Voltage Circuit Breakers Rated on a Symmetrical Current Basis—Amendment 1: Capacitance Current Switching.*

## 12.1.12 Bibliography on Circuit Breakers

Balestrero, A., Ghezzi, L., Popov, M., Tribulato, G., and van der Sluis, L., *Black Box Modeling of Low-Voltage Circuit Breakers*, IEEE Transactions on Power Delivery, vol. 25, Issue 4, Oct. 2010, Page(s): 2481–2488.

Balestrero, A., Ghezzi, L., Popov, M., Tribulato, G., and van der Sluis, L., *Current Interruption in Low-Voltage Circuit Breakers*, IEEE Transactions on Power Delivery, vol. 25, Issue 1, Jan. 2010, Page(s):206–211.

Bergman, W. J. B., *Selecting Circuit Breaker Monitoring*, 2001 IEEE/PES Transmission and Distribution Conference and Exposition, vol. 2, 28 Oct.-2 Nov. 2001, Page(s):1071–1076.

*Bibliography of Switchgear Literature: IEEE Committee Report*, IEEE Transactions on Power Delivery, vol. 5, Issue 1, Jan. 1990, Page(s):177–188.

*Bibliography of Switchgear Literature: IEEE Committee Report*, Veverka, E. F.; Schmunk, E. W., McCall, L. V., IEEE Transactions on Power Delivery, vol. 10, Issue 2, April 1995, Page(s):824–844.

*Bibliography of Switchgear Literature: IEEE Committee Report*, Glinkowski, M. T.; Schmidt, L., Veverka, E. F., IEEE Transactions on Power Delivery, vol. 13, Issue 1, Jan. 1998, Page(s):135–156.

Browne, T. E. Jr., ed., *Circuit Interruption—Theory and Techniques*, Marcel Dekker Inc., 1984.

Brunke, J. H., *Application of Metal Oxide Surge Arresters for the Control of Line Switching Transients*, Paper Presented at *Insulation Coordination Seminar—CRA Centennial Meeting*, Toronto, Ontario, May 1991.

Brunke, J. H., *Application of Power Circuit Breakers for Switching Capacitive and Small Inductive Currents*, Presented at the IEEE/PES General Meeting, 24 July 2008, available at: http://www.ewh.ieee.org/soc/pes/switchgear/presentations/tech_pres.html.

Bruettner, D. E., Colclaser, R. G., and Wagner, C. L., *Thermal Requirements of Resistors Used in Circuit Breakers for Voltage Control*, IEEE Transactions, vol. PAS-80, 1970.

Canay, I. M., *Comparison of Generator Circuit-Breaker Stresses in Test Laboratory and Real Service Condition*, IEEE Transactions on Power Delivery, vol. 16, Issue 3, Jul. 2001, Page(s):415–421.

Colcaser, R. G., Berkebile, L. E., and Buettner, D. E., *The Effect of Capacitors on the Short-Line Fault Component of Transient Recovery Voltage*, IEEE Transactions, vol. PAS-90, 1971.

Dufournet, D., *Recent Evolution of High-Voltage $SF_6$ Circuit-Breakers*, IEE Colloquium on Physics of Power Interruption, 31 Oct. 1995, Page(s):3/1–3/3.

Dufournet, D. and Montillet, G. F., *Transient Recovery Voltages Requirements for System Source Fault Interrupting by Small Generator Circuit Breakers*, IEEE Transactions on Power Delivery, vol. 17, Issue 2, Apr. 2002, Page(s):474–478.

Dufournet, D. and Smith, R. K., *Transient Recovery Voltages for High-Voltage Circuit Breakers*, Presented at the IEEE/PES General Meeting, 24 Jul. 2008, available at: http://www.ewh.ieee.org/soc/pes/switchgear/presentations/tech_pres.html.

Dufournet, D. and Hu, J., *Revision of IEEE C37.011 Guide for the Application of Transient Recovery Voltages for AC High-Voltage Circuit Breakers*, IEEE Transactions of Power Delivery, vol. 27, Issue 2, Apr. 2012, Page(s):1018–1022.

Franck, C. M., *HVCT Circuit Breakers: A Review Identifying Future Research Needs*, IEEE Transactions on Power Delivery, vol. 26, Issue 2, Apr. 2011, Page(s):998–1007.

Freeman, W., *Seismic Considerations of Circuit Breakers*, Presented at the IEEE/PES General Meeting, 24 Jul. 2008, available at: http://www.ewh.ieee.org/soc/pes/switchgear/presentations/tech_pres.html.

Garzon, R. D., *High-voltage Circuit Breakers: Design and Applications*, Marcel Dekker Inc., 1997.

Glinkowski, M. T.; Gutierrez, M. R.; Braun, D., *Voltage Escalation and Reignition Behavior of Vacuum Generator Circuit Breakers During Load Shedding*, IEEE Transactions on Power Delivery, vol. 12, Issue 1, Jan. 1997, Page(s):219–226.

Greenwood, Allan, *Electrical Transients in Power Systems*, Wiley-Interscience, 1971.

Hall, W. M. and Gregory, G. D., *Short-Circuit Ratings and Application Guidelines for Molded-Case Circuit Breakers*, IEEE Transactions on Industry Applications, vol. 35, Issue 1, Jan.-Feb. 1999, Page(s):135–143.

Hedman, D. E., Johnson, I. B., Titus, C. H., and Wilson, D. O., *Switching of Extra-High-Voltage Circuits II—Surge Reduction with Circuit-Breaker Resistors, IEEE Transactions*, vol. PAS-83, 1964.

Heiermeier, H., *Testing of Reactor Switching for UHV Circuit Breakers*, IEEE Transactions on Power Delivery, vol. 30, Issue 3, Jun. 2015, Page(s):1172–1178.

Janssen, A. L. J., Brunke, J. H., Heising, C. R., and Lanz, W., *CIGRE WG 13.06 Studies on The Reliability of Single Pressure $SF_6$-Gas High-Voltage Circuit-Breakers*, IEEE Transactions on Power Delivery, vol. 11, Issue 1, Jan. 1996, Page(s):274–282.

Kimblin, C. W. and Long, R. W., *Low-Voltage Power Circuit Breakers and Molded Case Circuit Breakers—A Comparison of Test Requirements*, 1999 IEEE Industrial and Commercial Power Systems Technical Conference, 2–6 May 1999, Page(s):7.

Koch, Herman J., *Gas Insulated Substations*, Wiley-IEEE Press, Aug. 2014.

Lambert, S. R., *Insulation Coordination for AC High-voltage Circuit Breakers*, Presented at the IEEE/PES General Meeting, 24 Jul. 2008, available at: http://www.ewh.ieee.org/soc/pes/switchgear/presentations/tech_pres.html.

Landry, M., Turcotte, O., and Brikci, F., *A Complete Strategy for Conducting Dynamic Contact Resistance Measurements on HV Circuit Breakers*, IEEE Transactions on Power Delivery, vol. 23, Issue 2, Apr. 2008, Page(s):710–716.

Legate, A. C., Brunke, J. H., Ray, J. J., and Yasuda, E. J., *Elimination of Closing Resistors on EHV Circuit Breakers*, IEEE Transactions on Power Delivery, vol. 3, Issue 1, Jan. 1988, Page(s):223–231.

Lester, G. N. and Nelson, J. H., *History of Circuit Breaker Standards*, Presented at the IEEE/PES General Meeting, 24 July 2008, available at: http://www.ewh.ieee.org/soc/pes/switchgear/presentations/tech_pres.html.

McCabe, A. K., Seyrling, G., Mandeville, J. D., and Willieme, J. M., *Design and Testing of a Three-Break 800 kV $SF_6$ Circuit Breaker with Zno Varistors for Shunt Reactor Switching*, IEEE Transactions on Power Delivery, vol. 7, Issue 2, Apr. 1992, Page(s): 853–861.

Meyer, J. M. and Rufer, A., *A DC Hybrid Circuit Breaker with Ultra-Fast Contact Opening and Integrated Gate-Commutated Thyristors*, IEEE Transactions of Power Delivery, vol. 21, Issue 2, Apr. 2006, Page(s):646–651.

Musa, Y. I., Keri, A. J. F., Halladay, J. A., Jagtiani, A. S., Mandeville, J. D., Johnnerfelt, B., Stenstrom, L., Khan, A. H., and Freeman, W. B., *Application of 800-kV $SF_6$ Dead Tank Circuit Breaker with Transmission Line Surge Arrester to Control Switching Transient Overvoltages*, IEEE Transactions on Power Delivery, vol. 17, Issue 4, Oct. 2002, Page(s): 957–962.

Nelson, J. H., *Electric Utility Considerations for Circuit Breaker Monitoring*, 2001 IEEE/PES Transmission and Distribution Conference and Exposition, vol. 2, 28 Oct.-2 Nov. 2001, Page(s):1094–1097.

Peelo, David F., *Current Interruption Transients Calculation*, Wiley Press, Apr. 2014.

Peelo, D. F., Polovick, G. S., Sawada, J. H., Diamanti, P., Presta, R., Sarshar, A., and Beauchemin, R., *Mitigation of Circuit Breaker Transient Recovery Voltages Associated with Current Limiting Reactors*, IEEE Transactions on Power Delivery, vol. 11, Issue 2, April 1996, Page(s):865–871.

Popov, M., Smeets, R. P. P, van der Sluiss, L., DeHerdt, H., and Decercq, F., *Experimental and Theoretical Analysis of Vacuum Circuit Breaker Prestrike Effect on a Transformer*, IEEE Transactions on Power Delivery, vol. 23, Issue 3, Jul. 2009, Page(s):1266–1274.

Ribeiro, J. R., and McCallum, M. E., *An Application of Metal Oxide Surge Arresters in the Elimination of Need for Closing Resistors in EHV Breakers*, IEEE Transactions on Power Delivery, vol. PD-4, 1989.

Roybal, D. D., *Standards and Ratings for the Application of Molded-Case, Insulated-Case, and Power Circuit Breakers*, IEEE Transactions on Industry Applications, vol. 37, Issue 2, Mar.-Apr. 2001, Page(s):442–451.

Runde, M., *Failure Frequencies for High-Voltage Circuit Breakers, Disconnectors, Earthing Switches, Instrument Transformers, and Gas-Insulated Switchgear*, IEEE Transactions on Power Delivery, vol. 28, Issue 1, Jan 2013, Page(s):529–530.

Smeets, R. P. P., *High-Power Testing of Circuit Breakers Needs a Proper Choice of Test-Circuits*, Presented at the IEEE/PES General Meeting, 24 Jul. 2008, available at: http://www.ewh.ieee.org/soc/pes/switchgear/presentations/tech_pres.html.

Smith, R. K., *Tests Show Ability of Vacuum Circuit Breaker to Interrupt Fast Transient Recovery Voltage Rates of Rise of Transformer Secondary Faults*, IEEE Transactions on Power Delivery, vol. 10, Issue 1, Jan. 1995, Page(s):266–273.

Smith, R. K. and Dufournet, D., *The Harmonization of IEEE and IEC Transient Recovery Voltage Waveforms*, Presented at the IEEE/PES General Meeting, 24 July 2008, available at: http://www.ewh.ieee.org/soc/pes/switchgear/presentations/tech_pres.html.

Steurer, M., Frohlich, K., Holaus, W., and Kaltenegger, K., *A Novel Hybrid Current-Limiting Circuit Breaker for Medium Voltage: Principle and Test Results*, IEEE Transactions on Power Delivery, vol. 18, Issue 2, Apr. 2003, Page(s):460–467.

Sweetser, C., Bergman, W. J., Montillet, G., Mannarino, A., O'Donnell, E. J., Long, R.W., Nelson, J., Gavazza, R., and Jackson, R., *Strategies for Selecting Monitoring of Circuit Breakers*, IEEE Transactions on Power Delivery, vol. 17, Issue 3, July 2002, Page(s):742–746.

Swindler, D. L., Schwartz, P., Hamer, P. S., and Lambert, S. R., *Transient Recovery Voltage Considerations in the Application of Medium-Voltage Circuit Breakers*, IEEE Transactions on Industry Applications, vol. 33, Issue 2, Mar.-Apr. 1997, Page(s):383–388.

Valentine, R. D., *A Perspective of Low-Voltage Circuit Breaker Interrupting Rating*, IEEE Transactions on Industry Applications, vol. 36, Issue 3, May–Jun. 2000, Page(s):916–919.

Valdes, M. E., Cline, C., Hansen, S., and Papallo, T., *Selectivity Analysis in Low-Voltage Power Distribution Systems with Fuses and Circuit Breakers*, IEEE Transactions on Industry Applications, vol. 46, Issue 3, Mar.-Apr. 2010, Page(s):593–602.

Wagner, C. L. and Bankoske, J. W., *Evaluation of Surge Suppression Resistors in High-Voltage Circuit Breakers*, IEEE Transactions, vol. PAS-86, 1967.

Wagner, C. L., *Circuit Breaker Application*, Westinghouse Printing, 1983.

Wagner, C. L., Dufournet, D., and Montillet, G. F., *Revision of the Application Guide for Transient Recovery Voltage for AC High-Voltage Circuit Breakers of IEEE C37.011: A Working Group Paper of the High-Voltage Circuit Breaker Subcommittee*, IEEE Transactions on Power Delivery, vol. 22, Issue 1, Jan, 2007, Page(s):161–166.

Yanabu, S., Zaima, E., and Hasegawa, T., *Historical Review of High-Voltage Switchgear Developments in the 20th Century for Power Transmission and Distribution System in Japan*, IEEE Transactions on Power Delivery, vol. 21, Issue 2, Apr. 2006, Page(s):659–664.

York, R. A., *Interrupting Mediums used in High-Voltage Circuit Breakers*, Presented at the IEEE/PES General Meeting, 24 Jul. 2008, available at: http://www.ewh.ieee.org/soc/pes/switchgear/presentations/tech_pres.html.

## 12.2 SWITCHGEAR ASSEMBLIES

*BY JEFFREY H. NELSON, MICHAEL W. WACTOR, AND T. W. OLSEN*

Definitions of terms used in this subsection can be found in the IEEE standards and application guides referenced in this subsection and/or in the *IEEE Standards Dictionary* (available at http://ieeexplore.ieee.org/xpls/dictionary.jsp?tag=1). The term "low-voltage" as used in this subsection refers to rated voltages up to 1000 V ac and 3200 V dc. The term "medium-voltage" as used in this subsection refers to rated voltages above 1000 V ac up to 38 kV ac (up to 52 kV ac for metal-enclosed gas-insulated switchgear).

Switchgear assemblies cover a wide range of low-voltage and medium-voltage structures that are generally factory-assembled and are divided into the following main groups: (1) metal-enclosed low-voltage power circuit breaker switchgear, (2) medium-voltage metal-clad switchgear, (3) metal-enclosed interrupter switchgear, (4) metal-enclosed bus, (5) metal-enclosed gas-insulated switchgear, and (6) switchboards. IEEE C37.20.1,[1] C37.20.2,[2] C37.20.3,[3] C37.20.9,[10] C37.23,[4] and NEMA PB 2[5] apply. Any of these equipment types may be rated as *"arc resistant metal-enclosed switchgear"* by successfully meeting the requirements of IEEE C37.20.7.[6]

### 12.2.1 Metal-Enclosed Low-Voltage Power Circuit Breaker Switchgear

Metal-enclosed low-voltage power circuit breaker switchgear indicates a design, which contains low-voltage ac or dc power circuit breakers in individual grounded metal compartments. The circuit breakers can be either stationary or drawout; manually or electrically operated; fused or unfused; and either three-pole, two-pole or single-pole construction. The switchgear may also contain associated control, instruments, metering, protective and regulating equipment as necessary. Definitions, ratings, design and production tests, construction requirements, and guidelines for application, handling, storage, and installation are covered in IEEE C37.20.1.[1]

Low-voltage metal-enclosed switchgear is typically installed in industrial plants, utility and distributed-generation facilities, and commercial buildings for the protection and distribution of power for loads such as lighting, machinery, motor control centers, elevators, air conditioning, blowers, compressors, fans, pumps, and motors. Low-voltage switchgear is available in ac ratings up to 1000 V and 6000 A continuous and in dc ratings up to 3200 V and 12000 A continuous. Short-circuit current ratings are available up to 200 kA.

## 12.2.2 Metal-Clad Switchgear

Metal-clad switchgear is used for low- and medium-capacity circuits, for indoor and outdoor installations with nominal voltages of 2.4 to 34.5 kV and continuous current ratings typically up to 4000 A. Short-circuit withstand current ratings of switchgear assemblies are at least equal the ratings of the lowest rated circuit breaker used. Metal-clad switchgear is commonly used for the control and protection of apparatus used for power generation, conversion, and transmission and distribution.

The term "metal-clad switchgear" indicates a design of equipment specifically enhanced with features intended to isolate primary circuit components and restrict the communication of faults between sections.

Metal-clad switchgear is characterized by the following required features. The main switching and interrupting device is of the removable (drawout) type arranged with a mechanism for moving it physically between connected and disconnected positions and equipped with self-aligning and self-coupling primary disconnecting devices and disconnectable control wiring connections. Mechanical interlocks are provided for proper operating sequence under normal operating conditions. Ground connections are provided for all removable elements to ensure that the frame and mechanism are grounded until the primary circuit is disconnected and the removable element is moved a safe distance.

Primary bus conductors and connections (live parts) are covered with insulating material throughout.

All live parts are enclosed within grounded metal compartments. Major parts of the primary circuit, including the circuit switching or interrupting devices, buses, voltage transformers, and control power transformers are completely enclosed by grounded metal barriers that have no intentional openings between compartments. Additionally, instruments, meters, relays, secondary control devices, and their wiring are isolated by grounded metal barriers from all the primary circuit elements with the exception of short lengths of wire such as at instrument transformer terminals.

Circuit breakers are generally the vacuum type, although air-magnetic circuit breakers were used for many years and $SF_6$ circuit breakers are also available. Circuit breaker disconnection is typically accomplished by horizontal-drawout design, illustrated in Fig. 12-36; however, earlier designs used a vertical lift design as the connection means. Interlocks are provided in metal-clad assemblies to prevent disconnecting or connecting the circuit breaker while in the closed condition and to prevent circuit breaker operation while moving between disconnected and connected position or vice versa. The metal-clad assembly is equipped with shutters to protect personnel from coming in contact with the high-voltage circuits when the circuit breaker is removed from the compartment. A circuit breaker test position is standard to allow circuit breaker control with the main contacts (primary disconnecting devices) removed from the primary circuit, but maintaining auxiliary and ground contacts between compartment and circuit breaker truck.

Definitions, ratings, design and production tests, construction requirements, and guidelines for application, handling, storage, and installation are covered in IEEE C37.20.2.[2]

Ground and test (G&T) devices are accessories available for metal-clad switchgear that may be temporarily inserted in place of a drawout circuit breaker for the purpose of grounding the main bus and/or external circuits connected to the switchgear assembly and/or primary circuit testing. Electrical and manual G&T device types are generally supplied for temporary circuit maintenance procedures for insertion in place of the circuit breaker. Electrical and manual G&T devices are required to have dielectric withstand, momentary current withstand, and short-time current withstand ratings at least equal to the circuit breaker they are intended to temporarily replace. Complete definitions,

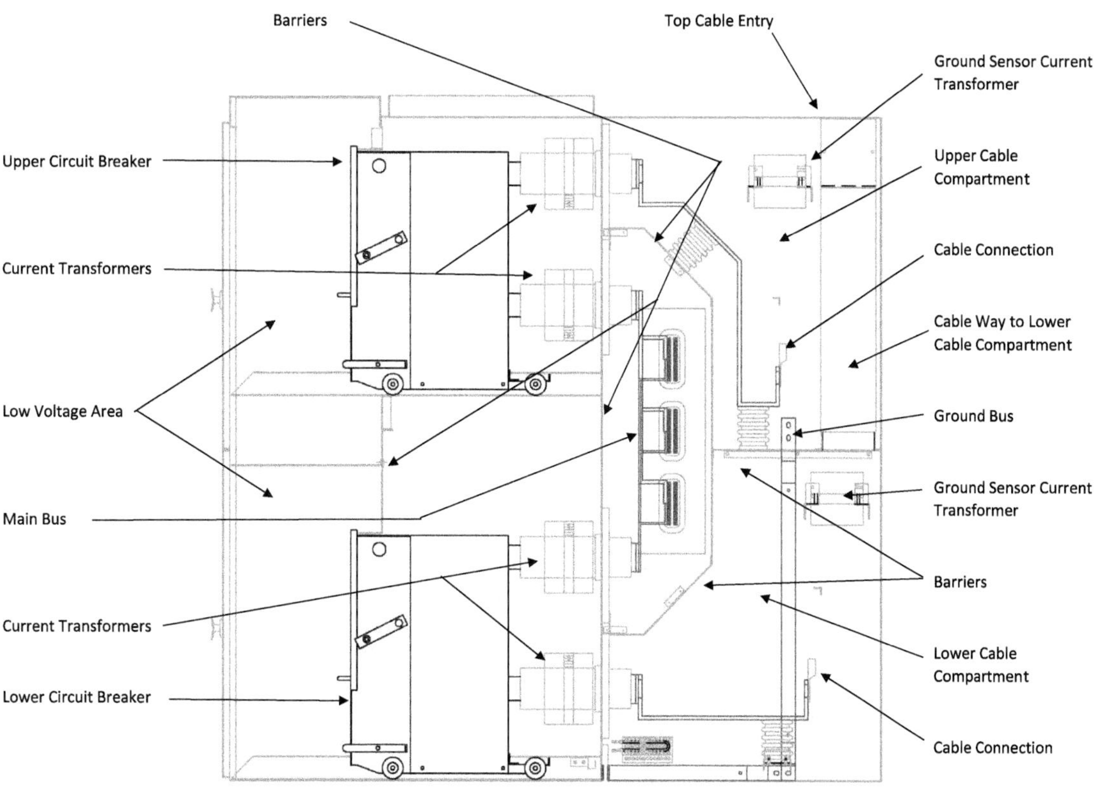


**FIGURE 12-36** Side view of a typical 15 kV metal-clad switchgear unit using horizontal-drawout circuit breaker design.

ratings, functional component requirements, design and production tests, construction and application of G&T devices are covered in IEEE C37.20.6.[11]

Electrical G&T devices are equipped with power-operated ground-making switches. An electrical G&T device is required to have a closing and latching current capability equal to that of the circuit breaker it is intended to temporarily replace. Test ports and test probes may also be provided to enable high-potential testing of primary circuits.

Manual G&T devices ground the primary circuit by means of either manually connected grounding conductors or a manually operated switch. Manual G&T devices are not required to have a close and latch rating.

## 12.2.3 Metal-Enclosed Interrupter Switchgear

Metal-enclosed interrupter switchgear assemblies include the following equipment as required: interrupter (interrupter switches and/or stationary-mounted circuit breakers), bare bus and connections, selector switches, power fuses [current-limiting or non-current-limiting (expulsion)], control and protective equipment, instrumentation, meters, and instrument transformers. The interrupter switches and power fuses may be stationary or withdrawable (drawout). When switches and fuses are withdrawable, mechanical interlocks are provided for proper operating sequence. Also, automatic shutters are provided which cover primary circuit elements when the withdrawable device is in the disconnected, test, or removed position.

Definitions, ratings, design and production tests, construction requirements, and guidelines for application, handling, storage, and installation for metal-enclosed interrupter switchgear are covered in IEEE C37.20.3.[3]

Metal-enclosed interrupter switchgear is typically used in industrial or commercial environments where continuous load currents are low and frequent switching is not required. Interrupter switches will interrupt load currents up to their rated continuous current capability. Fuses can be installed to provide short-circuit protection. For example, if the interrupter switchgear is connected to other switching equipment, fuses can be installed in the connection between the two to prevent an interruption of one assembly from a fault in the other assembly.

Typical applications for interrupter switchgear include main service disconnect, transformer primary and secondary switching, medium-voltage switchgear, primary and feeder circuit switching. The switching device may be manually operated or motor operated. Motor operated designs are often applied in an automatic transfer scheme.

Metal-enclosed interrupter switchgear is typically available in ac ratings above 1 kV up to 38 kV and up to 2000 A continuous current. Short circuit withstand ratings have to be equivalent to the ratings of the switching and protective equipment used or to the rating of the current transformers used.

## 12.2.4 Metal-Enclosed Gas-Insulated Switchgear

Medium-voltage metal-enclosed gas-insulated switchgear (MEGIS) is a type of equipment available internationally since the early 1980s, and since the 1990s in the United States, see Fig. 12-37. IEEE is in the process of developing a standard for this type of switchgear, identified as IEEE C37.20.9.[10]

MEGIS usually employs fixed-mounted circuit breakers with associated disconnect switches, but may be available with drawout type circuit breakers. When supplied with disconnect switches, the arrangement often includes a grounding switch to allow use of the circuit breaker to ground the outgoing circuit for maintenance purposes.

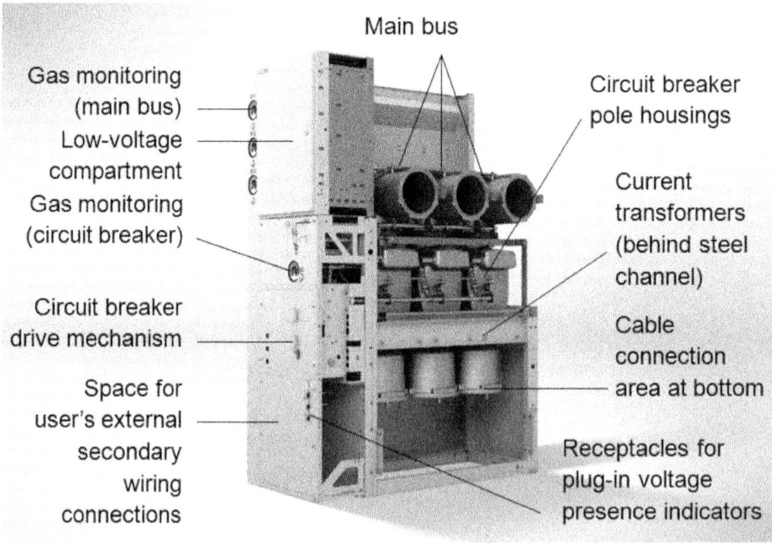


**FIGURE 12-37** Typical medium-voltage metal-enclosed gas-insulated switchgear (MEGIS).

MEGIS switchgear is available for up to 52 kV ac, and up to 3000 A continuous current, and up to 40 kA rms symmetrical short-circuit current.

Mechanical interlocks between the circuit breakers and their associated disconnect switches are provided for proper operating sequence. Primary buses and connections are normally bare (uninsulated) as the insulation is provided by the gas used. All primary conductors are enclosed in grounded metal compartments.

The enclosure is gas-tight, with maximum permissible leakage rate of 0.5% per compartment per year if the gas in the enclosure can be replenished, and 0.1% per compartment per year if the design is not intended to require replenishment of the gas once installed.

A variety of designs are available, including isolated-phase and nonsegregated phase designs. In the isolated-phase design, each phase is independently housed in a metal enclosure, providing phase-ground isolation and eliminating the possibility of phase-to-phase faults inside the switchgear. In nonsegregated phase designs, all three phases are housed in a common enclosure.

A gas, other than atmospheric air, is used in some or all portions of the MEGIS. Often, the gas (typically, sulfur-hexafluoride or $SF_6$, although other types of gases are under investigation) is used only for insulation, with interruption accomplished using a vacuum interrupter, but some designs employ the gas for interruption as well as insulation.

MEGIS originated in Europe in the early 1980s and its usage has grown significantly due at least in part to extremely compact size. A typical installation for 38 kV application can be as small as 25% of the size of a comparable air-insulated metal-clad switchgear lineup for the same system.

## 12.2.5 Metal-Enclosed Bus

Metal-enclosed bus is an assembly of conductors with associated connections, joints, and insulating supports with a grounded metal enclosure. Metal enclosed buses have four basic types of construction: (1) nonsegregated-phase, (2) segregated-phase, (3) isolated-phase, and (4) cable bus.

Rated voltages of ac metal-enclosed bus assemblies range from 635 V through 38 kV, and dc metal-enclosed bus assemblies range from 300 V through 3200 V. Definitions, service conditions, ratings, testing, construction requirements, and application guidelines for metal-enclosed bus are covered in IEEE C37.23.[4] An informative guide for calculating losses in isolated-phase bus is also included.

***Nonsegregated-Phase Metal-Enclosed Bus.*** Nonsegregated-phase metal-enclosed bus is a type of design in which all phase conductors, with their associated connections, joints, and insulating supports, are enclosed in a common metal housing without barriers between phases, see Fig. 12-38. When associated with metal-clad switchgear, the phase conductors of a non-insulated bus assembly entering the switchgear assembly and connecting to the switchgear bus are covered with insulating material equivalent to the switchgear insulation system. Enclosures that are totally enclosed are preferred, but ventilated enclosures can be provided in indoor applications.

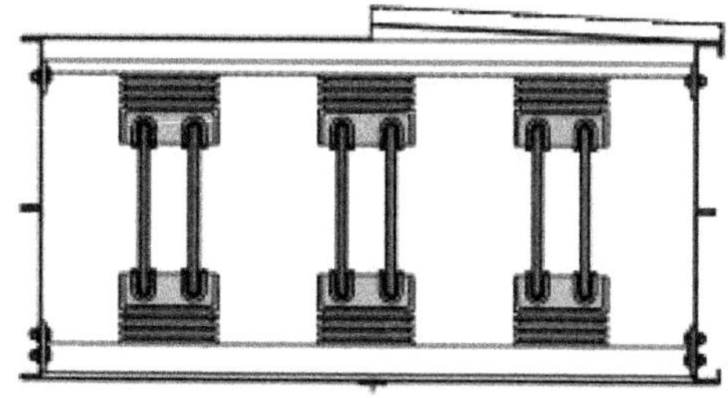

**FIGURE 12-38** Typical nonsegregated-phase metal enclosed bus. (*Courtesy of Powell Industries, Inc.*)

Nonsegregated-phase metal-enclosed bus is utilized on circuits, which require higher reliability than can be obtained with the application of power cables. Typical applications are the connections between transformers and switchgear assemblies, connections from switchgear assemblies to rotating apparatus, tie connections between switchgear assemblies, connections between motor control centers and large motors, and as main generator leads for small generators.

Preferred continuous self-cooled current ratings for nonsegregated-phase are available up to 12,000 A for 1000 V ac and all dc voltage ratings, 6000 A for 4.76 kV through 15.5 kV, and 3000 A above 15.5 kV through 38 kV. Short-time withstand current ratings (rms symmetrical) up to 85 kA for up to 1000 V ac, 63 kA for 4.76 kV up to 15 kV, and 40 kA for above 15 kV up to 38 kV are available, and up to 120 kA for dc ratings.

***Segregated-Phase Metal-Enclosed Bus.*** Barriers may be installed between the phase conductors to segregate the conductors and the assembly is then referred to as segregated-phase metal-enclosed bus, see Fig. 12-39. This design is also used on circuits, which require a higher degree of reliability.

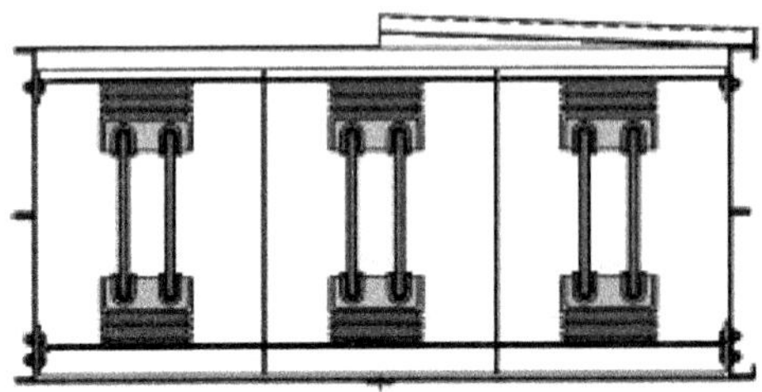

**FIGURE 12-39** Typical segregated-phase metal enclosed bus. (*Courtesy of Powell Industries, Inc.*)

Segregated-phase bus is primarily used as generator leads in power plants, but it is also applied in heavy industrial environments and as tie connections in metal-enclosed substations.

Preferred continuous self-cooled current ratings for segregated-phase are available up to 12,000 A for 635 V ac and all dc voltage ratings, 6000 A for 4.76 kV through 15.5 kV, and 3000 A above 15.5 kV through 38 kV. Short-time withstand current ratings (rms symmetrical) up to 85 kA for up to 1000 V ac, 63 kA for 4.76 kV up to 15 kV, and 40 kA for above 15 kV up to 38 kV are available, and up to 120 kA for dc ratings.

***Isolated-Phase Metal-Enclosed Bus.*** Isolated-phase metal-enclosed bus (iso-phase bus) is a type of design in which each phase is enclosed in an individual metal housing, and an air space is provided between the housings. It is considered to be the safest, most practical, and most economical way of preventing phase-to-phase short circuits by means of construction methods. The bus may be self-cooled or forced-cooled by circulating air or liquid. Definitions, ratings, design and production tests, construction requirements, and application guidelines for metal-enclosed bus are covered in IEEE C37.23.[4]

Briefly, the iso-phase bus has the following features:

1. Proof against contact; locked electrical premises not necessary
2. Faults only in the form of ground faults; protection against fault spreading to more than one phase
3. Field forces, static and dynamic, only between enclosures and conductor, not between phases
4. Protection against contamination and moisture
5. No losses in surrounding conductive material (grilles, railings, concrete reinforcements, lines, etc.)

Isolated-phase buses are available up through 38 kV and include continuous current ratings from about 1.2 kA up to 24 kA self-cooled, or 40 kA with forced cooling. The momentary current ratings have to match the rating of attached equipment. With high current ratings, more attention must be paid to the following:

1. Progressive rise of conductor temperature due to skin effects
2. Heating of surrounding conducting material by the magnetic field of conductors
3. High forces on main or component conductors in the event of a short circuit

In an enclosure with sections of tube insulation (sectional enclosure), eddy currents exist with values as large as the conductor current. These give rise to heat losses, and so the magnetic field of the main conductor is not always compensated for sufficiently. An important technical feature of the iso-phase bus, therefore, is the electrically continuous enclosure. The tubes enclosing each phase have electric conducting joints throughout their length and are short-circuited across the three phases at both ends. The enclosure thus constitutes a secondary circuit to the conductors (Fig. 12-40). The currents in the enclosures reach almost the corresponding conductor currents, depending on the resistance of the enclosure, but are of the opposite direction. The magnetic field outside the enclosure is

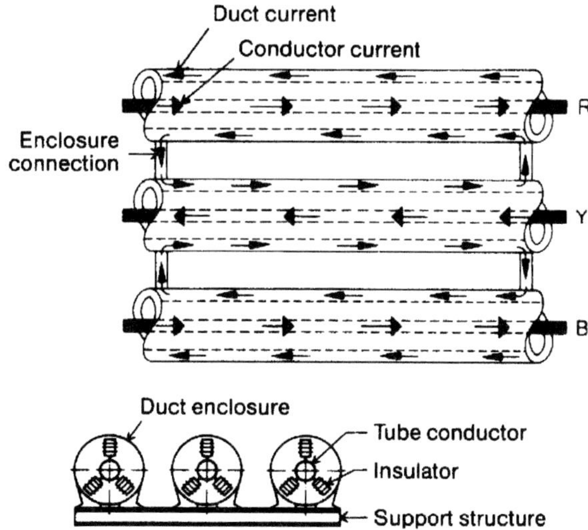


**FIGURE 12-40** Three-phase arrangement of an isolated-phase bus and principle of enclosure connection; according to Kirchhoff's law sum of conductor currents (+) and sum of enclosure currents (−) is zero.

almost completely eliminated, and thus there are no external losses or field forces between the phases. Connections to machines and switchgear must be adaptable and removable.

Current transformers for measurement and protection are of the bushing type or are integrated into the bus enclosure at a suitable place. Voltage transformers can be contained in the bus enclosure or mounted in separate instrument boards. The same applies to protective capacitors. Care must be taken that branch lines are adequately dimensioned with regard to thermal short-circuit strength.

The reliability of iso-phase bus can be enhanced by employing means to maintain the air pressure in the enclosure. Although iso-phase bus enclosures are generally leak proof, the large number of dismantleable joints may cause a slight leakage and might lead to moisture condensation during a plant shutdown. Supplying the bus enclosure with filtered, precompressed air at slight positive pressure ensures that the airflow is only outward; contamination of the conductors is not possible. Drying the air by precompressing prevents condensation.

Short-circuiting and grounding facilities are usually required in the bus system design to protect the generator and also for maintenance grounding purposes. Manually positioned links and straps are sufficient for small unit ratings; motor-operated grounding switches are recommended for higher capacities. A typical isolated-phase bus arrangement of a power station including generator circuit breaker is shown in Fig. 12-41.

## 12.2.6 Switchboards

Floor-mounted deadfront switchboards typically consist of an enclosure, molded case or low-voltage power circuit breakers, fusible or non-fusible switches, instruments, metering equipment, monitoring equipment and/or control equipment, and are fitted with associated interconnections and supporting structures. Switchboards can consist of one or more sections which are electrically and mechanically interconnected. Main disconnect devices can be mounted individually or be an integral part of a panel assembly. Definitions, ratings, design and production tests, construction requirements, and guidelines for application, handling, storage, and installation are covered in NEMA PB 2[5] and UL 891.[9]

Switchboards are typically installed in industrial plants, utility and co-generation facilities, and commercial and residential buildings for the distribution of electricity for light, heat, and power.

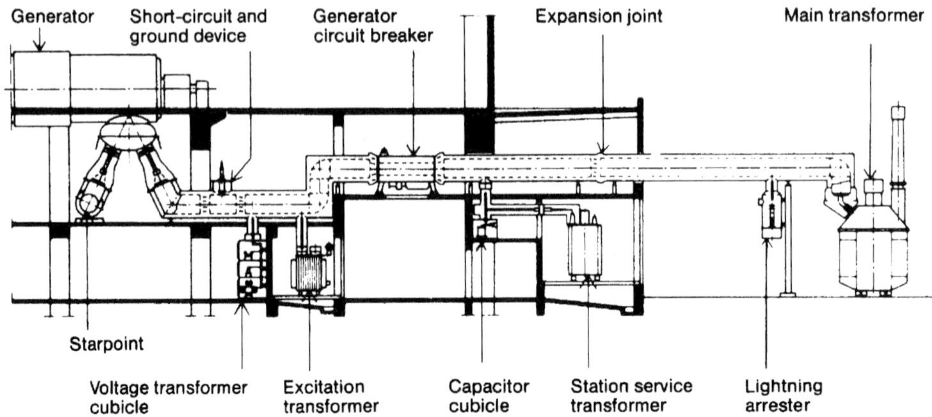


**FIGURE 12-41** Generating-plant isolated-phase bus arrangement with generator circuit breaker type DR.

They are typically available in voltage ratings of 600 V or less, continuous current ratings of 6000 A or less, and short-circuit current ratings up to 200 kA.

## 12.2.7 Arc Resistant Metal-Enclosed Switchgear

The term "arc resistant switchgear" indicates a design in which the equipment has met the requirements of IEEE C37.20.7.[6] During internal arcing tests, the switchgear assembly is subjected to an internal arcing fault in key locations throughout the assembly for a specified current level and duration and the equipment performance is evaluated against five basic criteria. The arcing fault is initiated by a small wire placed across the primary conductors which vaporizes when current flows, providing an ionized air path for the arc. The preferred current level for this test is the short-time current rating of the equipment and the preferred duration for current flow is 0.5 s for all equipment unless that value exceeds the equipment short-time rating. Other values are permissible when stated by the manufacturer and indicated on the nameplate. The equipment is evaluated for its ability to mitigate conditions which could be hazardous to personnel working nearby. Until recently, the arc resistant rating was only applicable to low-voltage switchgear qualified to IEEE C37.20.1[1] and medium-voltage switchgear qualified to IEEE C37.20.2[2] or C37.20.3.[3] In the 2017 edition of IEEE C37.20.7,[6] the equipment types included were extended to cover more types of equipment, including metal-enclosed bus (IEEE Std C37.23[3]), medium-voltage controllers (UL 347[7]), low-voltage motor control centers (UL 845[8]), switchboards (UL 891[9] and NEMA PB 2), and MEGIS (IEEE Std C37.20.9[10]). Figure 12-42 illustrates the typical modifications made to a medium-voltage switchgear design for vacuum-type circuit breakers when converting to arc resistant performance. Definitions, ratings, test requirements, and guidelines for application and installation are covered in IEEE C37.20.7.[6]

## 12.2.8 Station-Type Cubicle Switchgear

Another type of switchgear assembly that was previously used is station-type cubicle switchgear. Station-type cubicle switchgear is no longer manufactured, but is briefly discussed here for historical purposes. Requirements for station-type cubicle switchgear were included in IEEE Std C37.20.2[2] in the 1987 edition, but were eliminated from later editions as this type was no longer manufactured.

The term "station-type cubicle switchgear" indicates a design in which the major component parts of a circuit, such as buses, circuit breakers, disconnecting switches, and current and voltage transformers, are in separate metal housings, and the circuit breakers are of the stationary

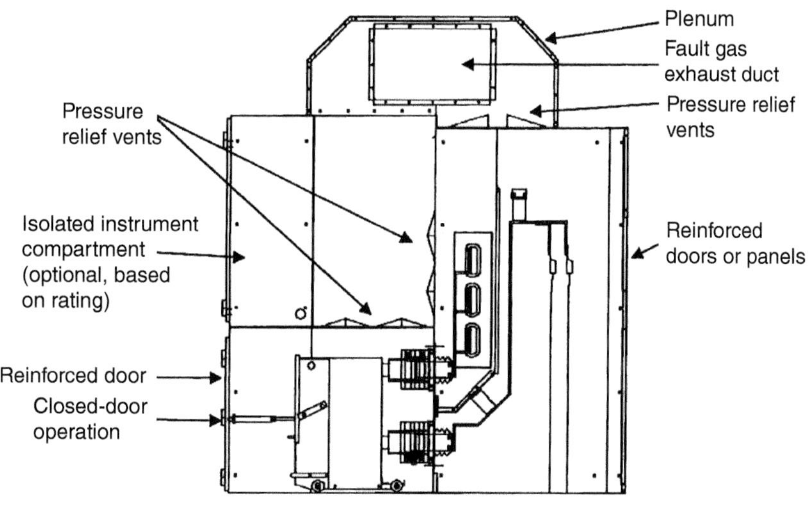


**FIGURE 12-42** Side view of a typical 15-kV metal-clad switchgear unit using horizontal-drawout vacuum circuit breaker design with modifications for arc resistant performance.

type (Fig. 12-43). Phase segregation in station-type cubicle switchgear was required, in which a three-phase metal housing is divided into three single-phase compartments by means of metal barriers.

Metal-enclosed station-type switchgear was used in industrial, commercial, and utility installations, generally for voltages of 14.4 to 69 kV, and continuous current ratings up to 5000 A.

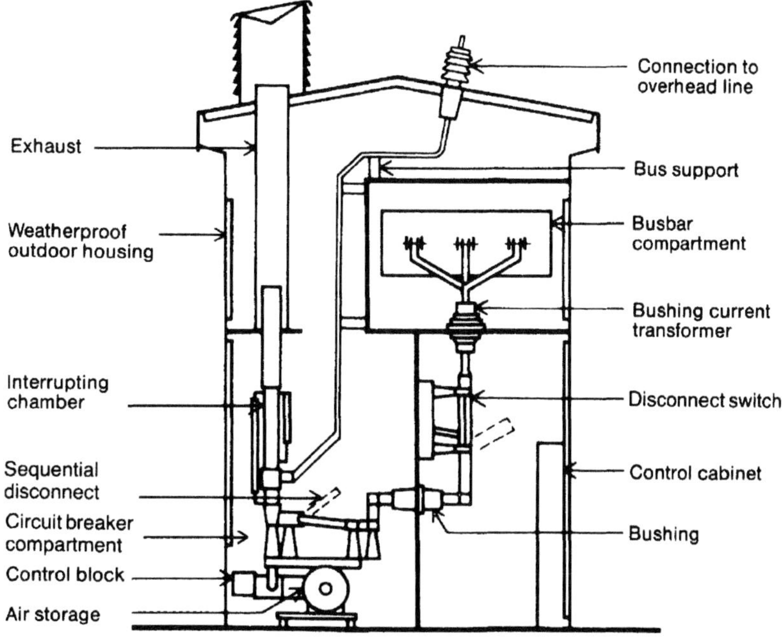


**FIGURE 12-43** Metal-enclosed station-type switchgear cubicle for outdoor installation, equipped with a heavy-duty, air-blast circuit breaker.

## 12.2.9 References

1. IEEE C37.20.1, *Standard for Metal-Enclosed Low-Voltage Power Circuit Breaker Switchgear.*
2. IEEE C37.20.2, *Standard for Metal-Clad Switchgear.*
3. IEEE C37.20.3, *Standards for Metal-Enclosed Interrupter Switchgear.*
4. IEEE C37.23, *Standard for Metal-Enclosed Bus.*
5. NEMA PB 2, *Standard for Deadfront Distribution Switchboards.*
6. IEEE C37.20.7, *Guide for Testing Switchgear Rated up to 52kV for Internal Arcing Faults.*
7. UL 347, *Medium-Voltage AC Contactors, Controllers, and Control Centers.*
8. UL 845, *Motor Control Centers.*
9. UL 891, *Switchboards.*
10. IEEE C37.20.9, *Metal Enclosed Switchgear Rated 1 kV to 52 kV Incorporating Gas Insulation Systems.*
11. IEEE C37.20.6, *Standard for 4.76 to 38 kV Rated Ground and Test Devices Used in Enclosures.*

## 12.2.10 Bibliography on Switchgear Assemblies

*Bibliography of Switchgear Literature: IEEE Committee Report*, IEEE Transactions on Power Delivery, vol. 5, Issue 1, Jan. 1990, Page(s):177–188.

*Bibliography of Switchgear Literature: IEEE Committee Report*, Veverka, E. F., Schmunk, E. W., McCall, L.V., IEEE Transactions on Power Delivery, vol. 10, Issue 2, Apr. 1995, Page(s):824–844.

*Bibliography of Switchgear Literature: IEEE Committee Report*, Glinkowski, M. T., Schmidt, L., Veverka, E.F., IEEE Transactions on Power Delivery, vol. 13, Issue 1, Jan. 1998, Page(s):135–156.

*Bibliography of Switchgear Literature, 1992–1996*, IEEE/PES Switchgear Committee webpage, http://www.ewh.ieee.org/soc/pes/switchgear/index.htm.

Bowen, J. and Burse, T.A., *Medium-Voltage Replacement Breaker Projects*, IEEE Transactions on Industry Applications, vol. 38, Issue 2, Mar.-Apr. 2002, Page(s):pp. 584–595.

Bridger, B., Jr., *Comparison of ANSI/IEEE and IEC requirements for Metal-Clad Switchgear*, IEEE Transactions on Industry Applications, vol. 33, Issue 1, Jan.-Feb. 1997, Page(s):216–225.

Bridger, B., Jr., Burse, T. A., and Wactor, M. W., *Design Considerations for 38 kV Metal-Clad Switchgear Using Vacuum Interrupting Technology*, Proceedings of the 1994 IEEE Power Engineering Society Transmission and Distribution Conference, 10–15 Apr. 1994, Page(s):15–20.

Conangla, A. and White, H. F., *Isolated-Phase Bus Enclosure Loss Factors*, IEEE Transactions on Power Apparatus and Systems, vol. PAS-87, Jul. 1968, Page(s):1622–1628.

Dwight, H. B., *Electrical Coils and Conductors*, New York; McGraw Hill, 1945.

Dwight, H. B., *Some Proximity Effect Formulas for Bus Enclosures*, IEEE Transactions on Power Apparatus and Systems, vol. PAS-83, Dec 1964, Page(s):1167–1172.

Eblen, M. L., Short, T. A., and Lee, W., *Medium-Voltage Arc Flash in Switchgear and Live-Front Transformers*, IEEE Transactions on Industry Applications, vol. 52, Issue 6, Nov.-Dec. 2016, Page(s):5280–5288.

Elgar, E. C., Rehder, R. H. and Swerdlow, N., *Measured Losses in Isolated-Phase Bus and Comparison with Calculated Values*, IEEE Transactions on Power Apparatus and Systems, vol. PAS-87, Aug. 1968, Page(s):1724–1730.

Garzon, R., *The Arc Terminator*, IEEE Industry Applications Magazine, vol. 9, Issue 3, May–Jun. 2003, Page(s):51–55.

Heberlein, G. E., Jr., Malkowski, C., Jr., and Cibulka, M. J., *The Effect of Altitude on the Operation Performance of Low-Voltage Switchgear and Controlgear Components*, IEEE Transactions on Industry Applications, vol. 38, Issue 1, Jan.-Feb. 2002, Page(s):189–194.

IEEE C37.13, *Standard for Low-Voltage AC Power Circuit Breakers Used in Enclosures.*

IEEE C37.14, *Standard for DC (3200 V and below) Power Circuit Breakers Used in Enclosures.*

IEEE C37.20.4, *Standard for Indoor AC Switches (1 kV–38 kV) for use in Metal-Enclosed Switchgear.*

IEEE C37.21, *Standard for Control Switchboards.*

IEEE C37.24, *Guide for Evaluating the Effect of Solar Radiation on Outdoor Metal-Enclosed Switchgear.*

IEEE C37.81, *Guide for Seismic Qualification of Class 1E Metal-Enclosed Power Switchgear Assemblies.*

Kalkstein, E.W., Doughty, R. L., Paullin, A. E., Jackson, J. M., Ryner, J. L., *Safety Benefits of Arc-Resistant Metalclad Medium-Voltage Switchgear*, IEEE Transactions on Industry Applications, vol. 31, Issue 6, Nov.-Dec. 1995, Page(s):1402–1411.

Koul, S., *Comparitive Requirements of IEC and IEEE Standards for Medium-Voltage Switchgear*, IEEE Transactions on Power Delivery, vol. 24, Issue 4, Oct. 2009, Page(s):1912–1923.

Land, H. B., *Determination of the Cause of Arcing Faults in Low-Voltage Switchboards*, IEEE Transactions on Industry Applications, vol. 44, Issue 2, Mar.-Apr. 2008, Page(s):430–436.

Land, H. B., *The Behavior of Arcing Faults in Low-Voltage Switchboards*, IEEE Transactions on Industry Applications, vol. 44, Issue 2, Mar.-Apr. 2008, Page(s):437–444.

Lav, C. T., Staley, D. B., and Olsen, T. W., *Practical Design Considerations for Application of GIS MV Switchgear*, IEEE Transactions on Industry Applications, vol. 40, Issue 5, Sept.-Oct 2004, Page(s):1427–1434.

Lee, W., Sahni, M., Methaprayoon, K., Chiman, K., Zhubing, R., and Sheeley, J. M., A *Novel Approach for Arcing Fault Detection for Medium-/Low-Voltage Switchgear*, IEEE Transactions on Industry Applications, vol. 45, Issue 4, Jul.-Aug. 2009, Page(s):1475–1483.

Nemoller, A. B., *Isolated-Phase Bus Enclosure Currents*, IEEE Transactions on Power Apparatus and Systems, vol. PAS-87, Aug. 1968, Page(s):1714–1718.

Pihler, J., Ticar, I., and Vorsic, J., *Design and Development of Medium Voltage Metal-Clad Switchgear with Metal Partition Walls*, IEEE Transactions on Power Delivery, vol. 18, Issue 2, Apr. 2003, Page(s):475–479.

Rochette, D., Clain, S., and Gentils, F., *Numerical Investigations on the Pressure Wave Absorption and the Gas Cooling Interacting in a Porous Filter, During an Internal Arc Fault in a Medium-Voltage Cell*, IEEE Transactions on Power Delivery, vol. 23, Issue 1, Jan. 2008, Page(s):203–212.

Shah, K. R., Cinsavich, A. L., and De Silva, P., *Impact of Arc Flash Hazards on Medium-Voltage Switchgear*, IEEE Transactions on Industry Applications, vol. 44, Issue 6, Nov.-Dec. 2008, Page(s):1859–1863.

Valdes, M. E., Purkayastha, I., Papallo, T., *The Single-Processor Concept for Protection and Control of Circuit Breakers in Low-Voltage Switchgear*, IEEE Transactions on Industry Applications, vol. 40, Issue 4, Jul.-Aug. 2004, Page(s):932–940.

Wactor, M., Olsen, T. W., Ball, C. J., Lemmerman, D. J., Puckett, R. J., and Zawadzki, J., *Strategies for Mitigating the Effects of Internal Arcing Faults in Medium-Voltage Metal-Enclosed Switchgear*, 2001 IEEE/PES Transmission and Distribution Conference and Exposition, vol. 1, 28 Oct.-2 Nov. 2001, Page(s):323–328.

Wilkie, E., *Comparison of ANSI/IEEE and IEC requirements for low-voltage switchgear*, IEEE Transactions on Industry Applications, vol. 40, Issue 6, Nov.-Dec. 2004, Page(s):1656–1664.

## 12.3 FUSES AND SWITCHES

*BY KENNETH LONG AND HAMID R. SHARIFNIA*

### 12.3.1 Fuses

In electronics and electrical engineering, a *fuse* is a type of low-resistance component that acts as a sacrificial device to provide overcurrent protection, interrupting either the load or source circuit. Its essential component is a metal wire or strip that melts when excessive current flows through it, thereby interrupting the circuit that it connects. Short circuits, overloading, mismatched loads, or device failure are the prime reasons for excessive current. Fuses may be used as alternatives to circuit breakers in many cases.

The fuses to be considered are current sensitive devices, designed to serve as the intentional weak link in the electrical circuit. Their function is to provide protection of discrete components, or of complete circuits, by reliably melting under current overload conditions. This subsection will cover some important facts about fuses, selection considerations, and standards.

A fuse interrupts an excessive current so that further damage by overheating or fire is prevented. Wiring regulations often define a maximum fuse current rating for circuits. Overcurrent protection devices are essential in electrical systems to limit threats to human life and property damage. The time and current operating characteristics of fuses are chosen to provide adequate protection without needless interruption. Slow blow fuses are designed to allow harmless short-term currents over their rating while still interrupting a sustained overload. Fuses are manufactured in a wide range of current and voltage ratings to protect wiring systems and electrical equipment. Self-resetting fuses automatically restore the circuit after the overload has cleared, and are useful in environments where a human replacing a blown fuse would be difficult or impossible.

***Resistance.*** The resistance of a fuse is usually an insignificant part of the total circuit resistance. Since the resistance of fractional amperage fuses can be several ohms, this fact should be considered when using them in low-voltage circuits.

The fuse parameters and application concepts presented herein should be well understood to properly select a fuse for a given application. Since these are only a few of the contributing parameters, application testing is strongly recommended and should be used to verify performance in the circuit/application.

***Ambient Temperature.*** Refers to the temperature of the air immediately surrounding the fuse and is not to be confused with "room temperature." The fuse ambient temperature is appreciably higher in many low-voltage cases, because it is enclosed (as in a panel mount fuse holder) or mounted near other heat producing components, such as resistors, transformers, etc.

***Rating.*** For 25°C ambient temperatures, it is recommended that fuses be operated at no more than 75% of the nominal current rating established using the controlled test conditions. These test conditions are part of UL/CSA/ANCE (Mexico) 248-14 "Fuses for Supplementary Overcurrent Protection," whose primary objective is to specify common test standards necessary for the continued control of manufactured items intended for protection against fire, etc. Some common variations of these standards include: fully enclosed fuse holders, high contact resistances, air movement, transient spikes, and changes in connecting cable size (diameter and length). Fuses are essentially temperature-sensitive devices. Even small variations from the controlled test conditions can greatly affect the predicted life of a fuse when it is loaded to its nominal value, usually expressed as 100% of rating.

***Breaking Capacity.*** Also known as interrupting rating or short-circuit rating, this is the maximum approved current which the fuse can safely break at rated voltage.

***Current Rating.*** The nominal amperage value of the fuse. Manufacturer established value of current which the fuse can carry, based on a controlled set of test conditions. Catalog Fuse part numbers include series identification and amperage ratings.

***Voltage Rating.*** The voltage rating, as marked on a fuse, indicates that the fuse can be relied upon to safely interrupt its rated short-circuit current in a circuit where the voltage is equal to, or less than, its rated voltage.

***Classification.*** Fuses can be classified into three categories. The first is low-voltage fuses operating up through 600 V ac. Most of these devises are tested and approved by Underwriters' Laboratories, Inc., and are marketed in a wide variety of characteristics and physical configurations. The second classification is medium-voltage through 44 kV (250 kV BIL), and the third is high-voltage fuses through 169 kV. Electric utilities use these fuses to protect transmission-distribution-class equipment and by large industrial complexes which have their own electrical distribution systems.

*Low-Voltage Fuses.* ANSI establishes low-voltage fuse standards, NEMA, or Underwriters' Laboratories. Their characteristics include a voltage class, an ampere rating, and an interrupting rating and, for some classes of fuses, a current-limiting rating. Cartridge fuses are classified in the following voltage classes: not over 250 V ac, not over 300 V ac, and not over 600 V ac. Fuses should not be used for dc applications unless recommended by the manufacturer. Most low-voltage fuses must be used in accordance with the National Electrical Code. Exceptions include those used in ships, railways, aircraft, and automotive vehicles other than mobile homes and recreational vehicles.

The standard lines of low-voltage fuses are available in several steps of ampere capacity, each of which is a different physical size (see Table 12-3).

A minimum of 10,000-A interrupting capacity is typical in low-voltage fuses, but some sizes and types can interrupt up to 200,000 A ac or 100,000 A dc. The interrupting rating is the highest RMS symmetrical alternating current which the fuse can interrupt at rated voltage.

Low-voltage current-limiting fuses are designed so that non-current-limiting fuses cannot be inserted into the fuse holder as a direct replacement. Thus, Class K fuses which are interchangeable with Class H fuses are not permitted in the "current-limiting" label.

A low-voltage current-limiting fuse successfully and safely interrupts all available currents within its specified interrupting rating and within its current-limiting range, and limits the clearing time at rated voltage to an interval equal to or less than the first major current loop. These fuses also limit peak let-through current to a value less than the normal peak current that would be otherwise possible without current-limiting availability. Two electrical measurements express the current-limiting characteristics of current-limiting fuses: (1) maximum peak let-through current, which is the maximum instantaneous value of current passed by the fuse during time of operation; and (2) maximum clearing $I^2t$ (amperes-squared-seconds), an expression of the energy available because of current flow during the clearing time of operation.

**TABLE 12-3** Typical Dimension Grouping of Low-Voltage Fuses

| Class | Volts | Amperes |
|---|---|---|
| G | 300 | 0–15 |
| G | 300 | 16–20 |
| G | 300 | 21–30 |
| G | 300 | 31–60 |
| H, K | 250, 600 | 0–30 |
| H, K | 250, 600 | 31–60 |
| H, K | 250, 600 | 61–100 |
| H, K | 250, 600 | 101–200 |
| H, K | 250, 600 | 201–400 |
| H, K | 250, 600 | 401–600 |
| L | 600 | 601–800 |
| L | 600 | 801–1200 |
| L | 600 | 1201–1600 |
| L | 600 | 1601–2000 |
| L | 600 | 2001–2500 |
| L | 600 | 2501–3000 |
| L | 600 | 3001–4000 |
| L | 600 | 4001–5000 |
| L | 600 | 5001–6000 |

*Medium- and High-Voltage Fuses.* A medium-voltage fuse is defined as any fuse (above 600 V and less than 48.3 kV) or fuse device used to isolate an electric short circuit from an electrical distribution system. A high-voltage fuse is defined as any fuse rated above 48.3 kV and used for this same purpose (see Table 12-4). Classes of fuses, or fused devices, are enclosed cutouts and fuses, open cutouts and fuses, open-link cutouts and fuses, current-limiting fuses, power fuses, and oil-immersed protective links.

**TABLE 12-4** Classes of High-Voltage Fuses or Fused Devices and Applicable Standards

| Class device | Standard |
|---|---|
| Distribution cutouts and fuse links | ANSI C37.42 |
| Distribution oil cutouts and fuse links | ANSI C37.44 |
| Power fuses | ANSI C37.46 |
| Current-limiting fuses | ANSI C37.47 |

Medium- and high-voltage fuses are used to protect potential transformers, distribution or power transformers, and lateral taps from main distribution feeder circuits. They are often used as sectionalizing devices on main feeder circuits. The ampacity and interrupting rating of these devices range up to 400 A, 20,000 A RMS symmetrical at 7200 V for medium-voltage fuses, and 400 A, 40,000 A RMS symmetrical at 169,000 V for high-voltage fuses. Fuses are generally used in electrical series with other fuses or circuit-protective devices. Care must be taken in coordinating the time-current characteristics for proper isolation of the electric circuit during fault and overload conditions. The goal with fuse selectivity is to isolate just the failed section and minimize the extent of any outage to maintain power to as much of the load as possible.

Distribution fuse links for use with expulsion cutouts are available with many different time-current characteristics. Figure 12-44 shows the minimum melting time-current characteristics for NEMA Type K fuse links.

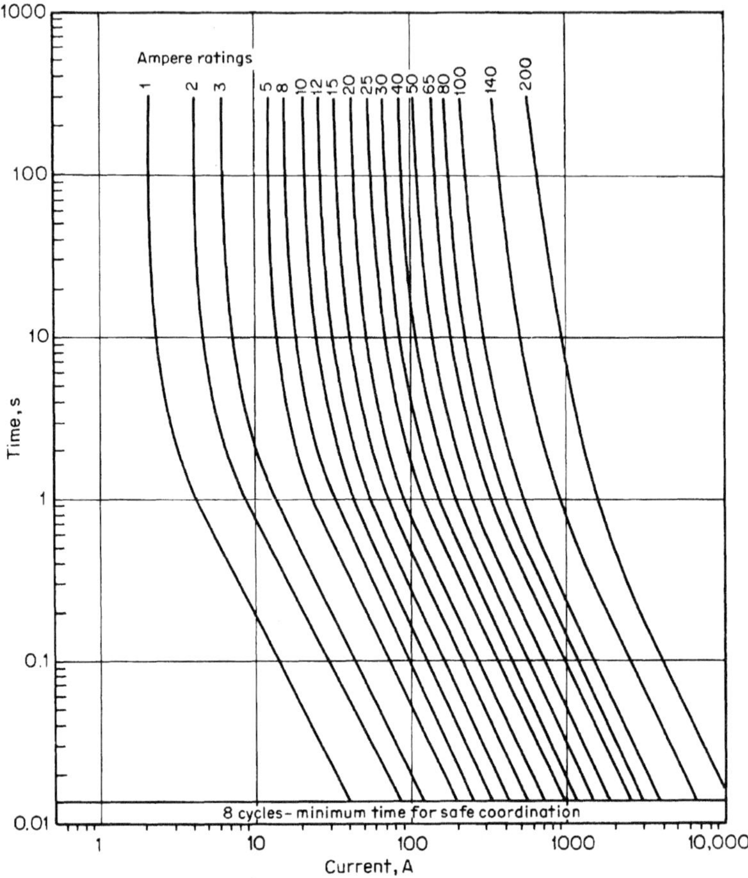


**FIGURE 12-44** Minimum melting time-current characteristics of a NEMA Type K fuse. (*A.B. Chance Company.*)

To standardize medium-voltage fuse-link characteristics, NEMA has adopted time-current characteristics for three basic fuse-link types: Type H (high surge), Type K (fast), and Type T (slow). These fuse links are designed to have the same time-current characteristics regardless of manufacturer. A wide variety of non-standardized fuse-link characteristics are also available. Fuse-link characteristics are usually based on tests at an ambient temperature of 25°C and no initial load. For characteristics at other ambient or for preloading variations, consult the individual manufacturer.

*Time-Current Curve.* The graphical presentation of the fusing characteristic, time-current curves are generally average curves which are presented as a design aid but are not generally considered part of the fuse specification. Time-current curves are extremely useful in defining a fuse, since fuses with the same current rating can be represented by considerably different time-current curves. The fuse specification typically will include a life requirement at 100% of rating and maximum opening times at overload points (usually 135% and 200% of rating depending on fuse standard characteristics). A time-current curve represents average data for the design; however, there may be some differences in the values for any one given production lot. Samples should be tested to verify performance, once the fuse has been selected (Fig. 12-45).

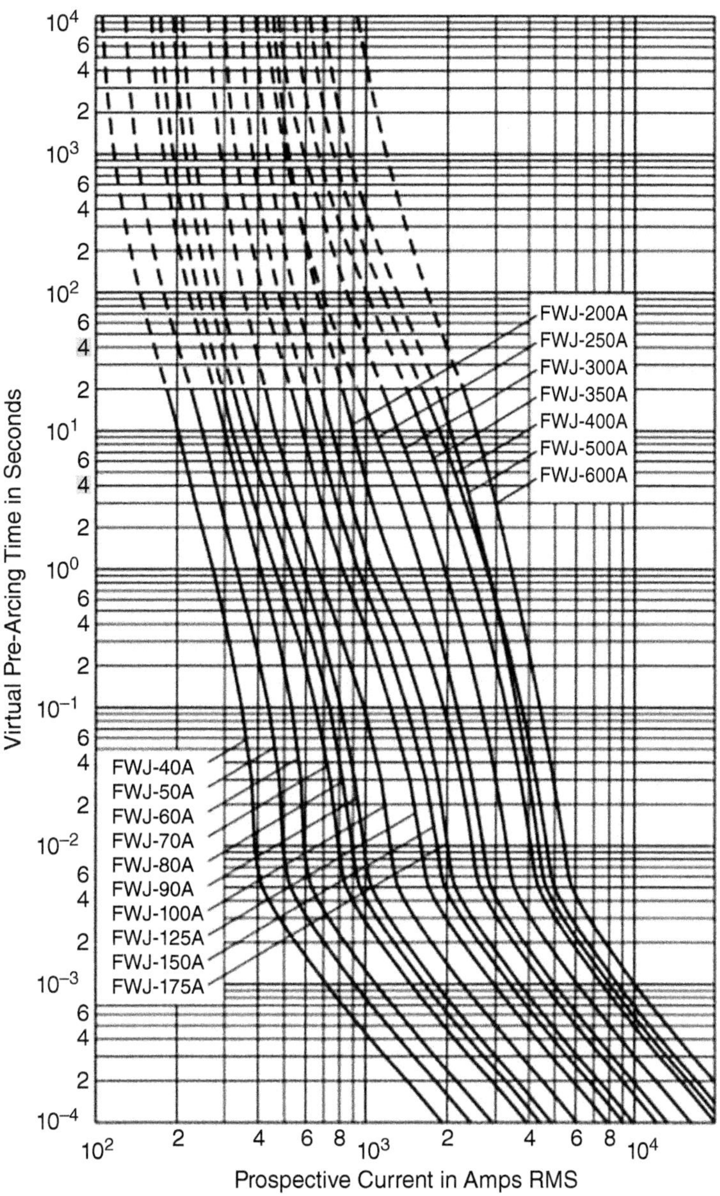


**FIGURE 12-45** Time-current curve. (*Edison Fuse Full Line Catalog, # 1005.*)

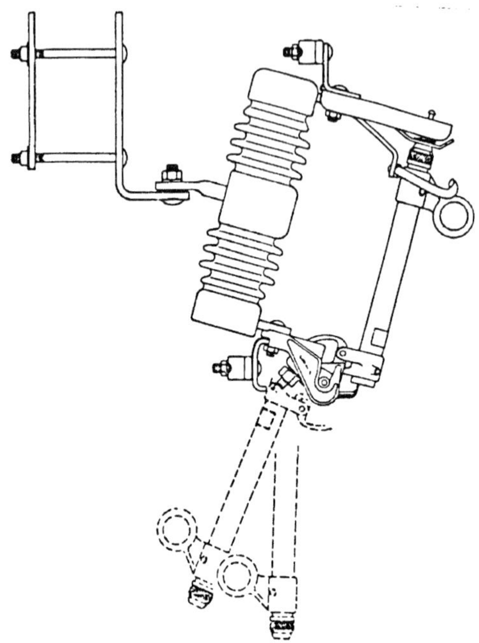

**FIGURE 12-46** Open-type distribution fuses cutout. (*Hubbell Power System/Chance.*)

The characteristics of various medium- and high-voltage protective devices used on distribution power systems are

1. Expulsion cutouts
   a. Open type: 20 kA asymmetrical maximum interrupting current (IC), 5 to 38 kV, violent in operation at high faults, low cost, both initial and refusing. Maximum continuous-current rating of 200 A can be fitted with a solid blade for conversion to a disconnect switch with a rating of 300 A. Figure 12-46 shows a typical open-type cutout.
   b. Enclosed type: 8- to 10-kA asymmetric maximum interrupting current, used primarily where safety codes dictate use. Enclosed cutouts are no longer manufactured and are being replaced with open-type cutouts.
   c. Open link: Generally, less than 3000 A IC, 200 A maximum continuous current, applied on rural lines and/or small transformers generally at 27 kV and below.
2. Oil cutouts: Considerable application in the past, especially in underground vaults; however, low interrupting current now poses serious underrating problems.
3. Liquid fuses: Nonviolent, low interrupting current (8 to 10 kA maximum), now obsolete and can be replaced with fuse limiters.
4. Power fuses: Reduced arc energy, somewhat less violent than cutouts on high faults, rated to 20,000 A IC; both initial purchase and replacement expensive. Figure 12-47*a* shows an indoor-type power fuse.
5. Under-oil protective link: 3500 A asymmetric maximum IC, violent in operation, low cost, contaminates insulating oil.
6. General-purpose current-limiting fuses: Nonviolent, current-limiting, high interrupting current (50 kA), requires coordination study, generated peak arc voltage, not affected by system transient recovery voltage, both initial purchase and replacement expensive.
7. High-range backup current-limiting fuses: Current-limiting, high interrupting current (50 kA), requires a low-current interrupting device in series, operates only at high currents, does not affect existing system coordination, low refusing cost on majority of outages because of only blowing expulsion link, not affected by system transient recovery voltage.
8. Vacuum fuses: Function with no external arcing or violence. They are nonrenewable and the associated cost is high. They have not found widespread application.
9. $SF_6$ fuses: Function with no external arcing or violence. They utilize rotating-arc technology and have a 12.5-kA interrupting current rating.

All expulsion-principle fuses depend on arc-quenching material—bone fiber, liquid solutions, or boric acid powder—to develop water vapor and/or other gases to cool the arc from the melted fuse link. These fuses have no energy-limiting ability and require a natural current-zero crossing to successfully interrupt a short-circuit current. Figure 12-47*b* shows a cross section of a power type fuse.

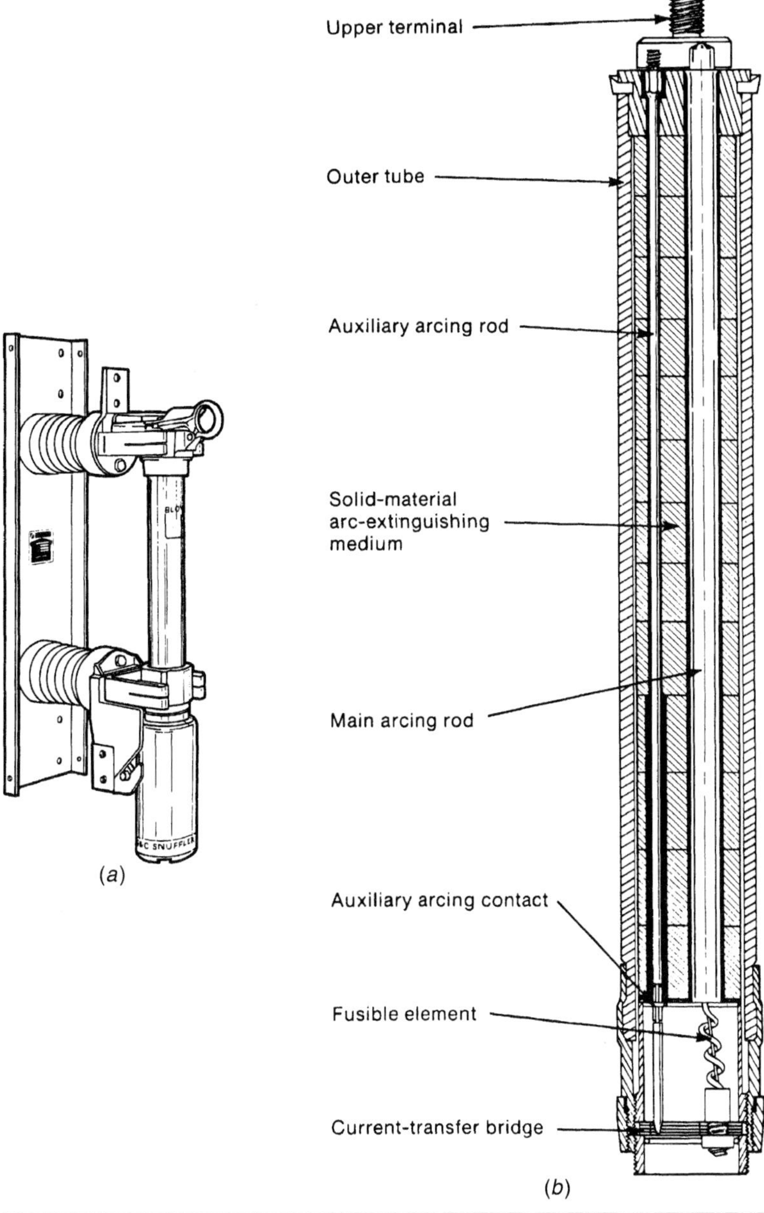


**FIGURE 12-47** (*a*) Indoor-type power fuse; (*b*) cross-sectional view of power-fuses refill unit. (*S&C Electric Company.*)

*Self-Resetting Fuses.* So-called self-resetting fuses use a thermoplastic conductive element known as a polymeric positive temperature coefficient (or PPTC) thermistor that impedes the circuit during an overcurrent condition (by increasing device resistance). The PPTC thermistor is self-resetting in that when current is removed, the device will cool and revert to low resistance. These devices are often used in aerospace/nuclear applications where replacement is difficult, or on a computer motherboard so that a shorted mouse or keyboard does not cause motherboard damage.

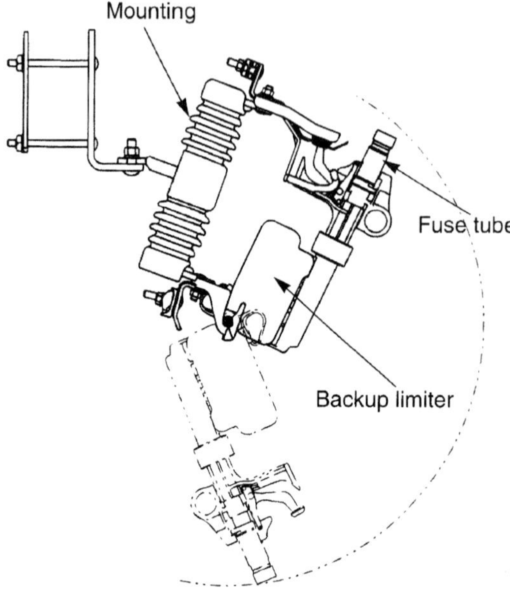


**FIGURE 12-48** Fuse limiters. (*S&C Electric Company.*)

*Current-Limiting Fuses.* High-voltage current-limiting fuses for use on distribution systems have two distinct classes: (1) general-purpose current-limiting fuses, devices which will successfully interrupt currents which will melt the fusible element in 1 h or more on through the maximum interrupting current that it is rated to clear; and (2) backup current-limiting fuses, devices which have a definite minimum interrupting rating as specified by the manufacturer. These devices require other protective devices in electrical series to interrupt fault currents below its minimum interrupting rating. There are also backup current-limiting fuses referred to as fuse limiters. These devices combine the series fuse and the current-limiting fuse into one package that handles just like a cutout and fits into the cutout mounting. The fuse limiter is available at 15 through 38 kV, and in current ratings through 20 kA. It is specifically designed for protecting overhead distribution transformers (see Fig. 12-48).

Three important parameters should be known about high-voltage current-limiting fuses:

1. Continuous current rating: The maximum current that the fuse is designed to carry continuously.
2. Peak arc voltage: Maximum voltage generated by the current-limiting fuse. If wire-wound, the voltage value is a function of fault current. If it is a ribbon-element fuse, the voltage is a function of applied voltage across the fuse.
3. $I^2t$ clearing: Maximum allowed by the current-limiting fuse. This measures the energy-limiting effect of the fuse.

Care in application of current-limiting fuses per voltage rating must be maintained. In general, these fuses should not be applied to circuits with a voltage less than 50% of the fuse-voltage rating to avoid excessive peak arc voltages.

It is equally important that fuses not be exposed to system recovery voltages more than their rating.

*Fuse Holders.* In many applications, fuses are installed in fuse holders. These fuses and their associated fuse holders are not intended for operation as a "switch" for turning power "on" and "off."

*Fuses in Enclosures.* High-voltage fuses may be mounted in enclosures for several applications: industrial service entrance switchgear, pad-mounted switchgear, or transformers for underground circuits, or in enclosures for subsurface applications. Most fuses will require special adaptation. Power fuses are fitted with a muffler to reduce the intensity of the exhaust gases when used in enclosures. Current-limiting fuses are supplied with special seals to prevent the ingress of fluid when applied under oil such as in transformers. Fuse cutouts are not recommended for use in enclosures or vaults.

Derating of fuses in enclosures may have to be considered because of restricted heat transfer. Consult the manufacturer.

*Electronically Controlled Protective Devices.* Electronically controlled protective devices offer greater flexibility and accuracy using state-of-the-art electronics. An electronic power fuse and an electronic sectionalizer for use at distribution voltage levels are two examples of this technology.

The electronic power fuse utilizes an electronic control module to provide current sensing, time-current characteristics, and control power for the fuse. High-speed interruption of fault currents to 40,000 A is provided by an interrupting module. The devices are completely self-contained (non-venting) and require no external power source. These power fuses are used in metal-enclosed switchgear, pad-mounted switchgear, and metal-enclosed fuse gear. They are available to 600 A continuous current and 4.16 through 25 kV. Three families of time-current characteristics are available. Figure 12-49 shows an electronic power fuse.

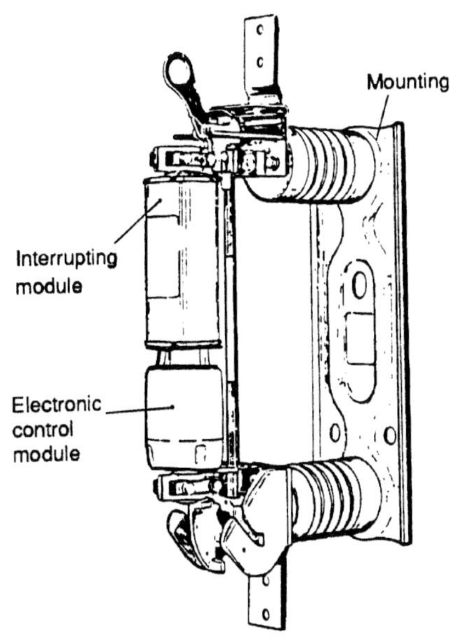


**FIGURE 12-49** Electronic power fuse. (*S&C Electric Company.*)

The electronic sectionalizer consists of an electronic module which fits into the mounting of a standard open-type cutout. The sectionalizer counts the number of fault current pulses allowed by an upstream recloser and operates to an open position during the recloser's open time. The sectionalizer has no time-current characteristics. It is generally used to replace a fuse at the distribution lateral where fuse coordination is difficult or impossible. Sectionalizer have no fault-interrupting capability and must be used in conjunction with an upstream recloser. Sectionalizer are available in continuous current ratings to 200 A; count settings of 1, 2, 3 or 4; and distribution voltage ratings from 15 to 38 kV. The electronic modules require no external power source. An electronic sectionalizer is shown in Fig. 12-50*a*.

Other devices that can be used to replace a distribution lateral fuse include single-phase recloser and single-phase dropout recloser. These devices provide improved protection performance on laterals with frequent occurrences of momentary faults. They include fault interrupting capability so there is

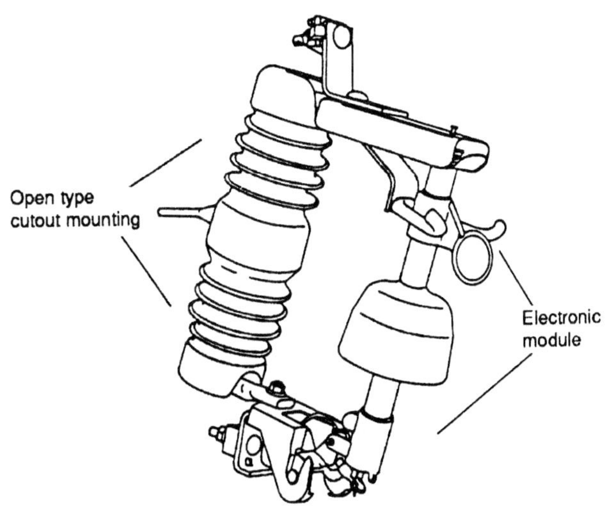


**FIGURE 12-50*a*** Electronic sectionalizer. (*Hubbell Power Systems/Chance.*)

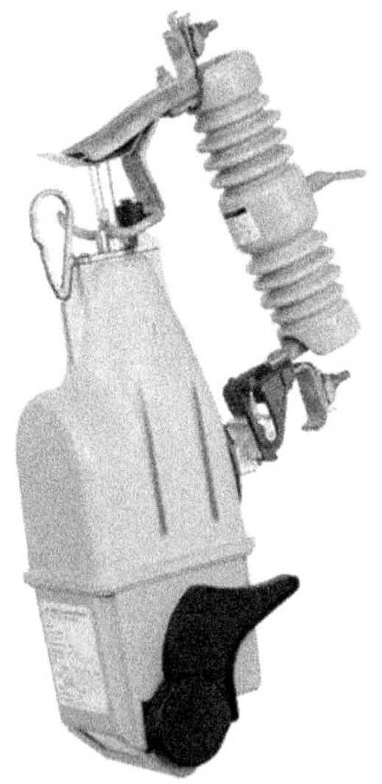

**FIGURE 12-50*b*** Single-phase dropout recloser. (*S&C Electric Company.*)

no momentary outage upstream from the device for lateral faults. The devices will eliminate permanent outages that would result from lateral fuses responding to temporary faults. The dropout style includes a self-powered, electronically controlled vacuum interrupter that fits into a standard cutout mounting (see Fig. 12-50*b*).

*Thermal Fuses.* A low-voltage thermal fuse is often found in consumer equipment such as coffee makers or hair dryers or transformers powering small consumer electronics devices. They contain a fusible, temperature-sensitive alloy which holds a spring contact mechanism normally closed. When the surrounding temperature gets too high, the alloy melts and allows the spring contact mechanism to break the circuit. The device can be used to prevent a fire in a hair dryer for example, by cutting off the power supply to the heater elements when the air flow is interrupted (e.g., the blower motor stops or the air intake becomes accidentally blocked). Thermal fuses are a "one shot," non-resettable device which must be replaced once they have been activated (blown).

### 12.3.2 Switches

Disconnecting switches are used primarily for isolation of equipment such as buses or other live apparatus. They are used for sectionalizing electric circuits such as buses or lateral circuits or even portions of main feeders for special purposes such as testing and maintenance. Standards pertaining to disconnect switches are listed in Table 12-5. Generally, these devices are not rated to break load current except when equipped with auxiliary devices. However, these switches must be designed to carry rated load currents and remain closed for momentary current flow such as fault currents. Fault currents more than a rating may cause the switch to be blown open by the magnetic forces due to the short-circuit current.

**TABLE 12-5** Standards Related to Disconnect Switches

| | |
|---|---|
| Ratings and Application Guide | ANSI C37.32 |
| Rated Control Voltages | ANSI C37.33 |
| Test Code | ANSI C37.34 |
| Operation and Maintenance | ANSI C37.35 |
| Loading Guide | ANSI C37.37 |

There are three classes of disconnect switches:

1. Station
2. Transmission
3. Distribution

Switches can be further categorized as group-operated or hookstick-operated, and loadbreak or non-loadbreak types. Overhead group-operated switches have a vertical drive linkage down the support structure or pole to a handle that operates all three phases of the switch. Hookstick-operated switches replace the vertical drive linkage with a latched switch blade that can be accessed and operated with the pulling action of a hookstick. This mechanism eliminates the requirement of any equipment on the pole and shortens the installation time.

The basic insulation level (BIL) of station class equipment is normally higher than for transmission or distribution class equipment. Station equipment ranges from 2.4 to 800 kV at present. Disconnect switches rated up through 5000 A continuous current are available. Manual or motor-operated switching can be provided.

The design of disconnecting switches demands considerable attention to the contact surfaces. Consideration must be given to the rigors of extreme environments.

High-pressure contacts are generally the form used to provide the current transfer. Current densities of 100,000 A/in$^2$ are common when using silver for contact points. Contact pressures as high as 500,000 lb/in$^2$ ensure that good cleaning action is achieved and keeps the current transfer points free from contamination. Reverse current lop jaw type contacts are also employed to apply addition force to the contacts during through fault conditions. Vertical-break and double-end-break style switches are commonly used in areas where ice build-up is prevalent, due to the rotational rolling movement of the blade and contact during the opening operation cycle.

Transmission disconnecting switches equipped with load break capability are generally used as load-management tools. Increasing needs for transmission lines and decreasing availability of right-of-way makes automatic switching of transmission load desirable.

Load management is often achieved during "dead time" by switching the proper disconnect automatically through sensing loss of voltage. There are also loadbreak switches available for systems up through 230 kV at 2000 A and 3000 A interrupting ratings. These switches use in-line or shunt $SF_6$ interrupters to isolate the circuits.

The objective of load management is to minimize outage time and allow for more efficient utilization of substation capacity at the distribution level. There is a growing interest in automating distribution class switches to achieve load-management objectives.

Distribution disconnecting switches are the method of providing for both single- and three-phase sectionalizing. As reliability demands increase, the utility must provide more sectionalizing or switching capability or suffer large and longer outages during faults. Hence, at 25 kV one may find some switching capability every 2000 to 3000 ft of overhead conductor. Figure 12-51*a* shows a typical group-operated distribution loadbreak switch. Figure 12-51*b* shows an upright, horizontal, hookstick-operated distribution load-break switch.

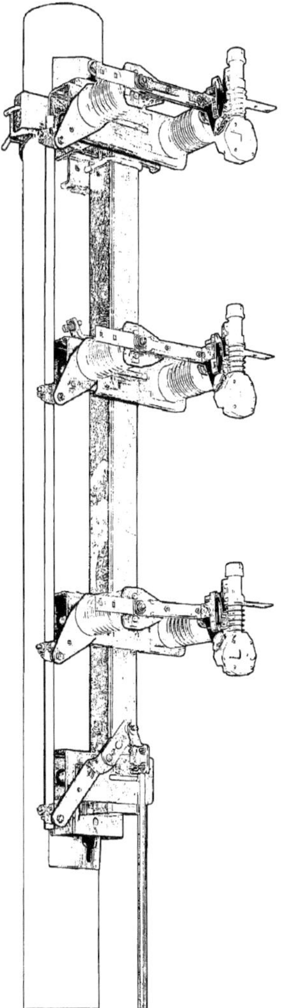

**FIGURE 12-51*a*** Gang-operated, phase-over-phase, distribution loadbreak switch. (*S&C Electric Company.*)

Single-phase disconnect switching with load-interrupter capability can be applied where ferroresonance is not a problem. This type of switching is found on single-phase circuits. Single-phase switching of a heavily loaded three-phase circuit is not desirable.

Group-operated switches with loadbreak capability interrupt these loads without concern for ferroresonance problems. In application, these three-phase switches can be mounted in either horizontal, phase-over-phase (tiered), or vertical (riser poles) configurations. To ensure proper operation, the mounting should be as rigid as possible. Care must be exercised in proper alignment of blades and interrupter shunt contacts. Attention to these matters allows proper operation for repeatable switching duty without any need for adjustment in the field.

***Load-Interrupter Devices.*** Load-interrupter devices, when combined with disconnecting switches, provide the economical capability of switching load currents. Generally, these interrupters are auxiliary devices and are not continuous-duty in terms of carrying load current. They are often referred to as shunt interrupters. This load interruption can be achieved by:

1. Use of an interrupter paralleling the main contacts just prior to opening and interrupting in this auxiliary chamber after the main contacts open. This is typically accomplished with an expulsion type device or vacuum switch.

**FIGURE 12-51*b*** Hookstick-operated three-phase distribution loadbreak switch. (*S&C Electric Company.*)

2. Use of a blast of $SF_6$ gas to effectively lengthen and cool the arc resulting from the main contacts opening.

Figure 12-52 shows an expulsion-type load interrupter used on distribution disconnecting switches to assist in interrupting load current.

***Switches for Underground Circuits.*** The continuing trend toward underground distribution circuits increases the need for pad-mounted switchgear. These are available in both live- and dead-front configurations, with the latter growing in popularity.

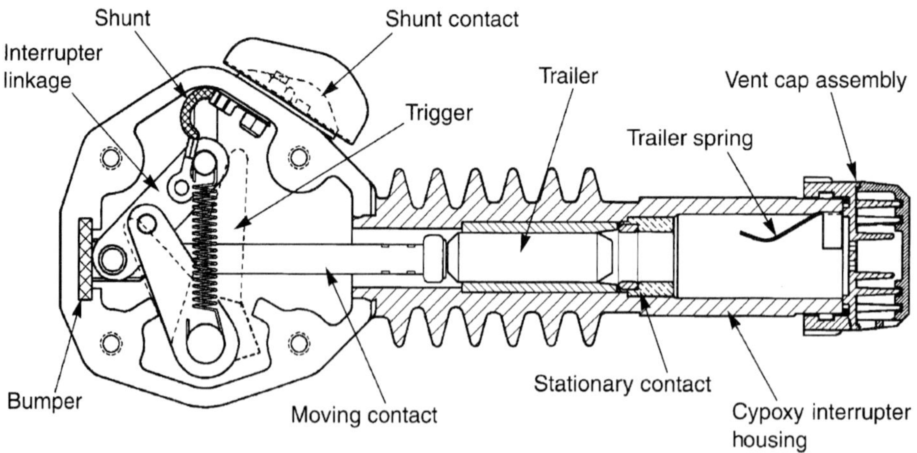


**FIGURE 12-52** Load interrupter, expulsion type, used in distribution systems. (*S&C Electric Company.*)

Live-front switchgear is typically air-insulated and utilizes in-air switches for loadbreak operation and power fuses for fault interruption. All components are directly accessible and operable.

Dead-front switchgear is typically air-, oil-, or gas-insulated. Air-insulated dead-front switchgear is like live-front switchgear except that components are isolated within grounded compartments and are not accessible when energized.

Oil-insulated dead-front switchgear typically uses under-oil switches or vacuum interrupters for loadbreak operation and current-limiting fuses or vacuum interrupters for fault interruption. These have the advantage of being more compact than air-insulated units. The disadvantage is that the oil insulation can become contaminated following arc interruption or from external contaminants.

$SF_6$-insulated switchgear uses components like those used in oil-insulated switchgear. Contamination of $SF_6$ is less of a concern, although $SF_6$-insulated enclosures must be more carefully designed and constructed to ensure gas integrity.

Dead-front designs generally provide increased isolation from energized components, but at the expense of operating simplicity and visual confirmation. External connections are made by means of separable insulated connectors. These connectors generally must be removed to provide a visible break when working on cable or equipment. Some more recent dead-front designs allow visual confirmation of internal visible breaks on both switches and fault interrupters, and some provide integral grounding of cables. This eliminates the need to move elbows and provides isolation from energized components, while also offering the visual confirmation provided in live-front switchgear.

In addition, recent designs have become available either with provisions for motor operation or fully integrated with motor operation and controls for use with SCADA systems.

Oil-insulated and $SF_6$-insulated switchgear can also be submersible and thus used in subsurface applications where space is at a premium or aesthetics are critical, such as in metropolitan areas. Refer to ANSI C37.72 for standards governing dead-front pad-mounted switchgear, and ANSI/IEEE C37.71 for subsurface load-interrupting switches.

## 12.4 CIRCUIT SWITCHERS

*BY KENNETH LONG AND HAMID R. SHARIFNIA*

*Circuit switchers* are mechanical switching devices suitable for frequent switching operations; capable of making, carrying, and breaking currents under normal circuit conditions; capable of making, and carrying for a specified time, currents under specified abnormal conditions; and capable of breaking currents under certain other specified abnormal circuit conditions. They are not necessarily capable of high speed reclosing. Some configurations include an integral open-gap disconnecting device. Circuit switchers available today use $SF_6$ as an interrupting medium and insulation for the interrupting contacts, and may be equipped with a trip device activated by a protective relaying device to open the circuit switcher's interrupting contacts automatically under specified abnormal conditions, such as overcurrent or faults.

A circuit switcher, like a circuit breaker, must carry normal load currents within a specified temperature range to prevent damage to key components such as contacts, linkage, terminals, and isolating device parts. Principal designating parameters of a circuit switcher are maximum operating voltage, BIL, rated load current, interrupting current, momentary and short time currents, reactive power switching requirements, whether a disconnecting or isolation device is required, whether a trip device is required, whether manual or motorized operation is required, and if motorized the station service and control voltages.

A circuit switcher essentially combines the functions of a circuit breaker (without high-speed reclosing capability) and a disconnecting switch (by providing visible isolation). Several models are also available which match the interrupting speed of a circuit breaker. A circuit switcher provides a cost-effective alternative means of transformer protection and switching, line and loop switching, capacitor or reactor switching, and load management.

Evolution of the circuit switcher concept provides a more in-depth understanding of its application versatility and its limitations.

### 12.4.1 History of Circuit Switcher Development

After World War II, the drive to electrify the remaining rural and sparsely populated areas of the United States was renewed. Providing fully rated circuit breakers for switching loaded circuits was frequently beyond budget limitations. This created a need for new transmission and subtransmission voltage circuit-switching devices. One such device could be described as a load interrupter. It appeared in a wide variety of forms. Most were attachments to disconnect switches.

Initially, most of these devices used low-volume oil as an interrupting medium. Ablative gas generating devices and later vacuum displaced oil. With rare exceptions, these devices had deficiencies.

In the mid-1950s, $SF_6$ gas was first employed as an interrupting medium. The application was an interrupter attachment for disconnect switches.

Whereas ablative devices and vacuum bottles were limited to approximately 30-kV recovery voltage per gap, this single-gap $SF_6$ device was readily applied on 138-kV systems for up to 600 A load switching. Most of these vacuum, ablative, and $SF_6$ devices were shunted into the circuit during the disconnect switch opening process. As the 1960s approached, the circuit switcher was born. It appeared as an in-line device. While the first version employed several ablative devices in series, it soon evolved into the use of $SF_6$ as a medium. Because of the unfavorable experience with the earlier devices, the general acceptance of the circuit switcher took much effort and considerable time. A typical installation is shown in Fig. 12-53.

Applications for circuit switchers have been primarily for high-side power transformer switching under load and fault protection. The circuit switcher provides load-switching capability and protection for faults that originate on the secondary or low-voltage side of the substation transformer. The zone of protection for circuit switchers in this application is typically from the current transformers inside the transformer on the high-voltage bushings to the secondary feeder breakers, or in some

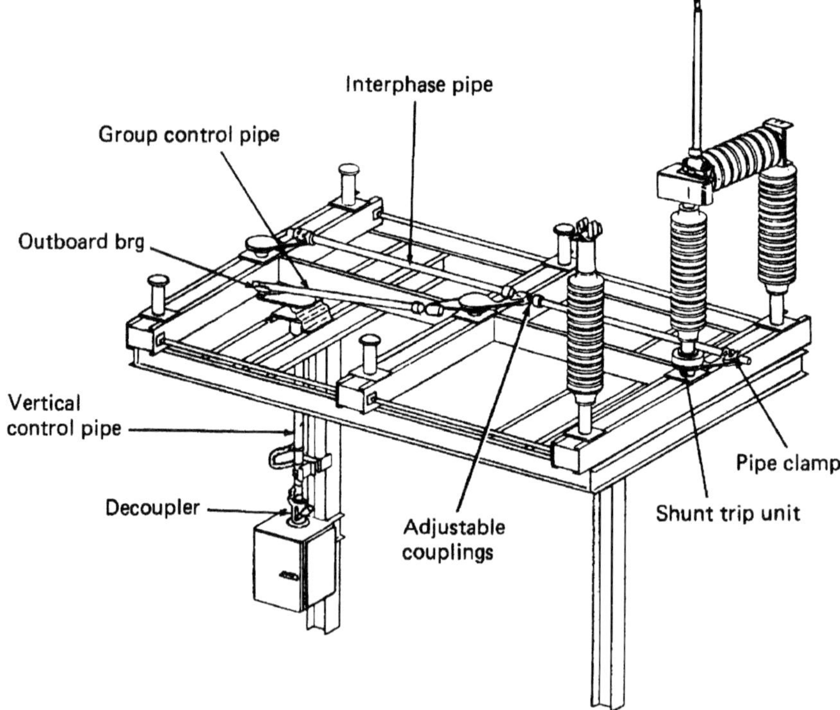


**FIGURE 12-53** Schematic of typical three-pole arrangement. Two poles have been deleted to clarify mechanical drive-train arrangement.

cases the low-side main circuit breaker. There is generally shorter strike distance on the secondary bus and more exposure to flashover from wildlife and other causes. Therefore, circuit switchers are specifically tested to interrupt current and withstand the higher transient recovery voltages (TRVs) associated with faults initiated on the secondary of the transformer and cleared by the high-side interrupting device. For applications where the available high-side short-circuit current exceeds the device's capability, blocking relays can be used. However, in most distribution substation applications blocking or time delay is not necessary.

New devices have come on the market which can also be used in the following applications:

***Capacitor Switching.*** The special purpose $SF_6$ capacitors switcher is used for routine switching of single step or back-to-back steps, either grounded or ungrounded. Its pre-insertion closing resistors or inductors provide transient suppression to minimize the detrimental effects of voltage transients on sensitive equipment and to minimize the detrimental effects of current transients on utility equipment's.

The key features of this device include circuit making in $SF_6$ rather than in air, single mechanism spring operator for reliable long-life operation, and single gap per phase puffer interrupters having long contact life and multi-time fault closing capability.

In addition, the $SF_6$ switching device equipped with a pre-insertion resistor reduces the inrush current thus eliminating the need for inrush current reactors that are commonly used with vacuum switching devices. Note, outrush current limiting reactors may still be required in some back-to-back switching applications.

Circuit switchers used in capacitor switching applications have the following features:

- Closing resistors provide the most reliable and consistently repeatable voltage and current transient suppression available in the market.
- Interrupting contacts which use $SF_6$ as an insulation medium and direct it to extinguish the arc are designed and tested for restrike-free performance
- Makes and breaks circuit in $SF_6$ gas rather than in air or vacuum
- Simple, cost effective, mechanical design provides repeatability
- Eliminates the need for inrush reactors

***Reactor Switching.*** This application can impose a severe duty on the connected system, switching device, and the shunt reactor itself. Due to the relatively small inductive current, the interrupting device attempts to clear at a forced current zero; deviation from which results in current chopping. If the interrupter's contacts have not separated enough to sustain the system voltage, a re-ignition of the arc will occur. These high magnitude and high frequency re-ignitions can shorten the life of the reactor and the switching device.

Circuit switchers used in shunt reactor switching applications have the following features:

- Special nozzle/contact design resulting in very low probability of re-ignitions
- Increased reactor life due to less stress on its insulation system
- Patented interrupter minimizes probability and magnitude of re-ignitions
- Reduced turn-to-turn voltage stress on reactor windings
- Simplified design compared to that of a circuit breaker improves reliability
- Local visual indication of gas pressure provided by color coded temperature compensated gas gauge
- Common gas system with gas density switch with low-pressure alarm and low-pressure lockout for remote status monitoring
- Compact candlestick type design can fit in tight spaces

### 12.4.2 General Construction

Today, most circuit switchers are designed as live-tank type interrupting chambers, using $SF_6$ gas puffer-type interrupting contacts. In the closed position, the contacts are surrounded by a flow guide and piston assembly which is ready to mechanically generate a "puff" of $SF_6$ to cool and deionize the arc that is established prior to circuit interruption. The moving cylinder attached to the contact assembly, is driven by the main opening spring-charged mechanism, causing the gas to be pressurized by the stationary piston. The stationary contact "follows" the moving contact as the piston assembly achieves the prepressurized gas condition. When the contacts (which are hollow tubes) part, an arc is established and the gas flow divides into two parts and flows down the stationary and moving contact tubes. The alternating nature of the arc current waveform results in two current zeros every cycle. If the arc is sufficiently "hot" or conductive through the $SF_6$ dielectric medium, the current will reestablish itself. Modern circuit switchers have sufficient $SF_6$ density such that at the first current zero the interrupting contacts can stop the arc from reestablishing itself, providing the necessary dielectric strength to sustain arc interruption. This entire process from trip signal initiation to current interruption requires from 3 to 8 cycles or 50 to 133 ms in modern circuit switchers.

Figure 12-54 illustrates a typical "blade-disconnect model" circuit switcher with the interrupter and blade connected in series. For opening, the operator receives a trip signal when the relay system detects an abnormal condition within the specified range. By discharging its opening spring, the operator actuates the interrupters to interrupt the circuit. Once the interrupter is open, the blade opens to achieve visible isolation. The blade-hinge mechanism is actuated directly by rotating a support insulator through the driver mechanism. As the blades are opened, both the closing and opening springs are recharged.

For closing, the reverse rotation of the insulators allows the blades to begin closing. The blades close slowly since they are not used to pick up the circuit for this design. Once the blades are in the closed position, the operator receives a signal to release the closing spring and close the contacts in the interrupters to complete the circuit. If the unit has closed into a faulted circuit condition that provides a trip signal, the opening process may proceed immediately since the opening spring is charged and ready.

In some circuit switcher designs, the movement of the interrupting contacts is initiated with a shunt trip designed to achieve the proper opening speed. For models of this type without shunt trip, opening is accomplished by rotating the insulator to the point where the driver opening spring would normally be tripped by the shunt trip's rotation. This configuration is used where protection duty is not a function of the circuit switcher. For this design, the closing operation may be achieved by first closing the interrupter during the opening stroke of the blade. Then, when a close operation is called for, all that is necessary is to close the blade because the interrupter is already closed. Since contact is established in air for this type of closing, high-speed operation of

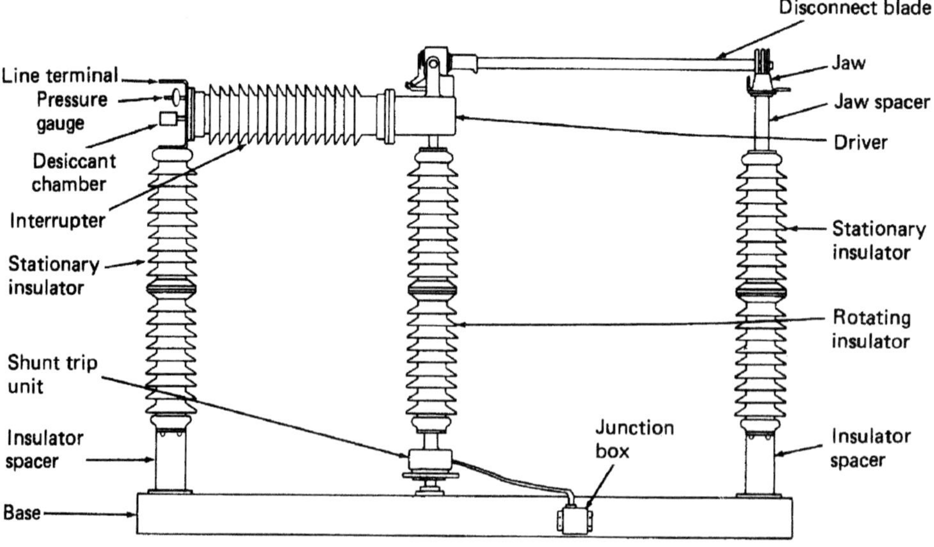


**FIGURE 12-54** Single pole of blade-type circuit switcher. (*S&C Electric Company.*)

the blade is necessary to minimize the impact of picking up the circuit with the blades. Both methods of closing in the interrupter or with the blade have been proven over many years of field use.

Circuit switchers are available with integral blades and in bladeless configurations. The units with the integral disconnect can help to provide a visible gap on the same structure with the interrupters. This can be a significant benefit in locations where there is limited real estate. For bladeless circuit switchers, a separate disconnect switch is used to isolate the circuit after the interrupters are open. In these installations the circuit switcher is used to open and close the circuit, and the disconnect switch provides the isolation and visible gap.

***Transformer Protective Devices.*** There is another type of device available for transformer protection that provides the same type of protection as a circuit switcher. It is designed for application on the high-side of substation transformers from 69 to 138 kV. These devices offer a three-cycle 31,500 A interrupting rating with electronically linked pole units. They have been tested for interrupting the high TRVs from secondary-side faults and provide an economical alternative to other protective devices. They are supplied in conjunction with a separate or integrated disconnecting device on the high-voltage bus side of the protective device.

The transformer protective device is different from circuit switchers in the operating sequence. The standard version includes electronically tripped interrupters with manual reset of each interrupter. A disconnect is used to isolate the circuit and provide a visible gap after the interrupters are open. With the disconnect open, the interrupters are closed and charged manually or with charging motors. Since the individual interrupters are closed and charged slowly, they are not used to pick up the circuit. Once the interrupters are closed, the disconnect is used to pick up the transformer. Figure 12-55 shows a typical substation transformer protective device.

### 12.4.3 Ratings

***Short-Circuit Current.*** Circuit switchers are "specific-duty" or "medium-fault" interrupting devices and generally fulfill a requirement between high-power fuses and circuit breakers in a

**FIGURE 12-55** Substation transformer protective device. (*S&C Electric Company.*)

transmission or distribution system. Circuit switcher designs to date do not encompass instantaneous reclosing capability since their most common application of transformer protection precludes this requirement. Typical short-circuit current interrupting ratings are from 4 to 40 kA rms symmetrical, depending on type of interrupting duty and voltage rating.

When interruption occurs, a TRV is imposed in microseconds across the interrupting device because of system adjustments to the new state before a steady-state condition is achieved known as the normal-frequency recovery voltage. TRV attempts to reestablish the arc by either thermal reignitions or dielectric breakdown of the interrupting gap. Refer to ANSI definitions and ratings. The TRV seen by the high-side interrupting device after clearing a fault that initiated on the secondary of the transformer is high with high frequency because of the characteristics of transformers.

*Continuous Current.* A continuous current rating is the designated limit of current in rms amperes that can be carried continuously under usual service conditions and in an ambient temperature not more than 40°C without exceeding temperature limits assigned to the various materials comprising the current-carrying parts or that are in contact with these parts. For further information, refer to ANSI C37.04, C37.30, and IEC 694.

Circuit-switcher continuous current ratings are 1200, 1600, and 2000 A. These ratings are typically well above the requirements for the transformer protection application based on size of the transformer being protected and the maximum continuous current of these transformers.

*Short-Time Current.* This is a dual rating to verify the capability of the circuit switcher to carry abnormally high currents in the closed position for short periods of time. Short-time current includes the following:

1. Rated momentary current is the total current in asymmetric rms amperes (ac and dc components), which the device can carry for 10 cycles on a 60-Hz basis. These ratings can approach 50 times the continuous current ratings, and provide an assurance that the device will withstand the high electromagnetic forces from initial transient conditions of a short circuit tending to bring about mechanical damage or contact separation.
2. Rated 1-second or 3-second current is 62.5% of the rated momentary current and verifies that the current carrying parts will withstand the heating effect without excessive annealing or contact welding. Common circuit switcher momentary ratings are 60, 70, and 80 kA.

*Close-and-Latch Current.* For those circuit switchers that make the circuit in the interrupting device, a close-and-latch rating is required as is a close rating for the type that closes the circuit on a disconnect switch blade. This rating verifies the capability of the contact structure and operating mechanism to withstand the forces developed by closing in on a fault. Close-and-latch ratings are 30 and 40 kA. Limited-duty close ratings for the blade type are of the same magnitudes.

*Rated Voltage.* Rated voltages for circuit switchers are generally stated in terms of maximum system design voltages that indicate the upper limit at which the device is designed to operate. Circuit switchers are most commonly applied at 72.5 through 242 kV but are also available for special applications at 362 kV.

To provide safe phase-to-ground insulation levels and across the device when in the open position, the circuit switcher must withstand certain specified magnitudes and wave shapes of test voltages without flashover or puncture of any of its insulation systems. This is the rated dielectric strength and consists of:

1. 1-minute dry and 10-second wet low-frequency (60-Hz) withstand voltage.
2. Dry lightning-impulse ($1.2 \times 50$ μs wave shape) withstand voltage. It is the positive-polarity withstand level that is the basic-insulation-level rating (BIL in kilovolts).
3. Dry and wet switching-impulse ($250 \times 2500$ μs) withstand voltage. This rating applies only at 362 kV and above.
4. Dry chopped-wave impulse withstand voltage. The requirement for this rating has yet to be determined for a circuit switcher but is a standard rating for a circuit breaker. If applicable, this rating would apply to a bladeless model circuit switcher having graded gaps in series per pole.

*Interrupting Time.* The rated interrupting time is the maximum permissible interval between energizing the shunt trip coils at rated control voltage and the interruption of the main circuit in all poles when interrupting a current within the required interrupting capabilities. This interrupting time may be modified under certain asymmetrical current conditions as detailed in ANSI C37.04. Circuit switchers are available from 3- to 8-cycle interrupting times.

*Duty Cycle.* Operating duty is the short-circuit current to be interrupted, closed upon, etc. The duty cycle is a stated sequence of closing and opening operations. Various types of circuit switchers will have different duty cycles. An example is O-17s-CO cycle is where the first operation is an open, followed by 17 seconds to permit the operating mechanism and blade to recycle, and then to repeat the close-open operation cycle.

*Corona-RIV.* A corona-free rating is commonly established to prevent radio and television interference. This is referred to as the radio influence voltage (RIV) and should be less than 500 μV as determined by a test circuit per NEMA Publication No. 107. As a rule, a circuit switcher that produces no visible corona under dark conditions at 105% of rated voltage will have a negligible RIV level.

### 12.4.4 Selection and Application

The versatility of circuit switchers and related circuit-interrupting devices requires careful selection of the ratings, components, and accessories to be specified for a given protection and/or switching duty. The following criteria must be considered:

1. System data
   a. Nominal service voltage
   b. Maximum continuous current
   c. Through-fault current withstand
   d. Basic impulse (insulation) level
2. Circuit-protection duty
   a. Present and future available fault currents
   b. Length of overhead lines or underground cables
   c. Transformer ratings, impedance, and connections
   d. TRV
   e. Interrupting time, maximum
3. Switching duty
   a. Inductive currents (unloaded transformers)
   b. Small capacitive currents (unloaded lines and cables)
   c. Inductive/capacitive currents (choke coils, capacitors, grounded or ungrounded)
   d. Closing on fault
   e. Closing of long lines

Approximately 80% of all circuit switchers are applied on the primary or high-voltage side of a substation transformer for switching and fault-protection duty. By means of protective relaying, faults which occur within the protection zone of the transformer can be detected to bring about tripping of the primary-side circuit switcher. Should the fault occur on the secondary side, which is most frequently the case, the transformer impedance will limit the magnitude of fault current at the circuit switcher's location, and is generally less than the available primary-fault current. Hence, the use of the term "inherent secondary-fault current," and for some circuit switchers, a dual short-circuit current rating for primary or secondary faults.

Associated with these reflected secondary faults are the inherent capacitance and inductance of the transformer windings, which generally produce higher TRVs than is the case with a primary fault. These higher TRVs can increase the difficulty of the interruption process, while conversely, the impedance-limited fault current is easier to interrupt. Circuit switchers are tested for this condition as defined in the nearly completed ANSI C37.016 which is the new standard for circuit switchers.

This test standard which applies to circuit switchers defines a secondary fault test with a TRV that is more severe than the circuit breaker test standard. The circuit switcher demonstrates its interrupting capability for this duty by passing these tests. In applications where the primary fault currents are greater than the switcher's interrupting capability, the fault can often be cleared by source-side circuit breakers having interrupting times shorter than those of a circuit switcher. It is also common to block tripping of the circuit switcher when the fault current is greater than its rating and pass this duty on to the source-side breaker.

Circuit switchers and interrupters are generally available in mounting arrangements like isolating air-disconnect switches. Since they are not freestanding, their application and that of the supporting structure design and/or optional freestanding pedestal is influenced by phase spacing's desired, terminal height, altitude, atmospheric elements, seismic conditions, and wind, ice, and terminal pad loadings.

The phase spacing recommended in the ANSI C37.32 differentiates between the minimums for vertical-break disconnect switches (also applying to bus supports) and the increased spacing's required for horn gap switches (switching devices utilizing arcing in air as an operating mode). From a safety viewpoint, it is important to consider the method of operation in addition to the duty applications of the device as a determination in selecting the phase spacing.

Elevation of the installation site, in addition to the temperature, humidity, and air-contamination characteristics, significantly affects the dielectric strength and coordination of these circuit-interrupting devices. The guidelines established in ANSI C37.30 for site elevations coupled with a further analysis of these other factors may dictate the use of higher-voltage-rated units or extra-leakage-distance porcelain standoff insulators.

Seismic, wind, ice, and terminal pad loadings must also be considered in the design of the supporting structure.

***Testing.*** Industry standards specifically for circuit switchers are included in ANSI C37.016. Prior to this standard, applicable sections of existing standards for circuit breakers and disconnect switches had been used as a guide in establishing definitions, ratings, capabilities, and test procedures for circuit switchers (see ANSI C37.09 and C37.34). In addition, testing for secondary fault currents and the associated higher TRVs has been completed by some circuit switcher manufacturers.

Tests can be considered to consist of four different groupings:

1. Type or design tests are for proving the capability of the switcher to meet the specified ratings.
2. Reliability tests are part of a quality assurance program to demonstrate that the circuit-switcher design has achieved a specified mechanical reliability. They consist of accelerated aging or prototype tests to improve on the established design reliability, demonstration, or pilot run tests to assure the reliability requirements have been met, and acceptance or production tests to verify that production units comply with the demonstrated design reliability.
3. Routine or production tests are done to ensure that the production is in accordance with established procedures. This includes testing of individual components and subassemblies.
4. Installation testing to assure that the circuit switcher will perform its intended function in the system.

***Accessories and Maintenance.*** As discussed in a preceding section, circuit switcher manufacturers offer optional freestanding support pedestals at various heights in addition to many other accessories, some of which include bypass and grounding switches. Other options include the capability of monitoring the circuit switcher status through SCADA.

The maintenance requirements are specified by each manufacturer. Typical inspection recommendations will include a general visual check of the insulator condition along with checking the $SF_6$ gas density levels and the connections. If the circuit switcher includes an integral disconnect, the contacts should be inspected and wiped clean of the old grease, and a light coating of new grease applied. In general, an overall inspection of the circuit switcher should be used to identify any condition that is not normal.

## 12.5 AUTOMATED FEEDER SWITCHING SYSTEMS

*BY KENNETH LONG AND HAMID R. SHARIFNIA*

The method known as SCADA has long been used to monitor and control transmission systems, providing the operational flexibility and speed required for efficient and reliable performance. The use of SCADA in the distribution system is becoming increasingly important as utilities seek to improve service to their customers. The acronym *SCADA* is being replaced by the generic term *distribution automation* (DA), which incorporates the principle of computers operating switching and other control devices automatically in response to events in the system. Automated switching of distribution feeder circuits provides significant improvements in reliability, enhances operational flexibility, and increases productivity of both utility personnel and distribution lines.

Automated feeder switching systems utilize controls, sophisticated algorithms, and data communications to go beyond the reliability benefits of more traditional reclosers and automatic sectionalizers. An automated switching system can be programmed to not only reduce outage times, but also do so without subjecting unfaulted circuits to "bumps" caused by intentional closing into faults required by fault-detection or troubleshooting schemes. Further, an integrated fault-isolation and restoration system can incorporate real-time evaluation of pre-fault loads into the algorithm to prevent overloading of a backup circuit when transferring load from a faulted circuit. This allows the utility planner more flexibility in circuit design by increasing the amount of load that can normally be carried by a given circuit.

Such systems, with current and voltage monitoring, also provide the utility with a convenient way to monitor distribution feeder circuits, thus enabling the utility to take immediate action in the event of current or voltage excursions exceeding normal operating limits. Modern systems also monitor the communications network for abnormal conditions, and will implement path switching or if necessary block automated responses to avoid miss operation. These features when implemented make it possible to improve power quality and continuity of service for the customer.

Automated feeder switching systems can also provide the means to optimize feeder and substation loading by enabling the shifting of load from one feeder to another in a very short period. This same capability can yield hard-dollar cost savings associated with deferment of capital projects when coupled with planning practices that take advantage of the new technologies.

### 12.5.1 Examples of Automated Feeder Switching Systems

A large investor-owned electric utility in New York needed to dramatically improve reliability in response to a challenge from the New York public utility commission. After analyzing its system as well as the nature and locations of faults, it was decided that an automated feeder switching system would provide the quickest route to needed improvements in electric service reliability.

A typical feeder "loop" is shown in Fig. 12-56. It consists of two circuits joined by a normally open tie switch. In turn, each feeder circuit will have a normally closed automated switch installed at the electrical half-way point on the circuit. All switches are equipped with voltage and current monitoring as well as communication and control units (CCUs) housing the switch controls, power supplies, remote-terminal units (RTUs), and radios for establishing communication with the primary station.

Faults occurring in section 2 will result in a trip of substation breaker BKR A. The control at switch SW1 will sense the passage of fault current and the subsequent loss of voltage precipitated by the opening of BKR A. A sectionalizing algorithm will initiate and open the switch after a predetermined time delay set to coordinate with the recloser interval of BKR A. The station breaker is

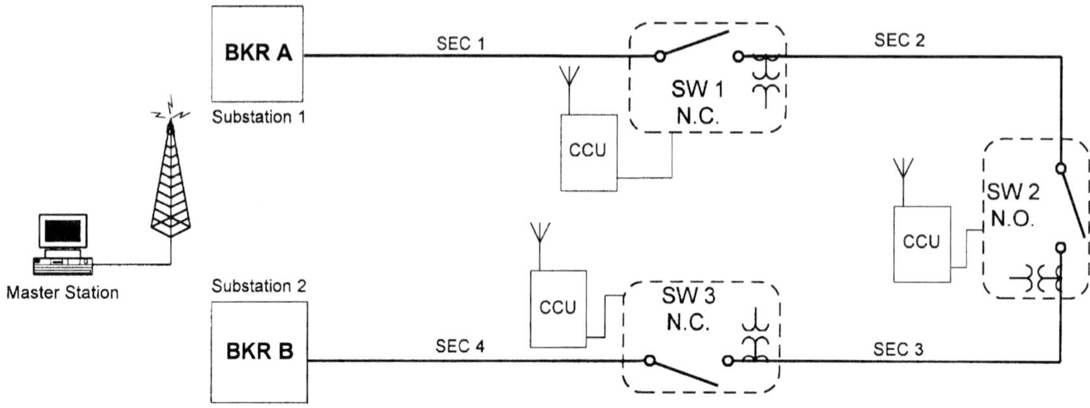


**FIGURE 12-56** Typical automated feeder circuit loop.

reclosed to determine if the fault condition is temporary or permanent. A typical reclosing breaker may reclose up to three times on some systems. Once SW1 is open and the fault is isolated, BKR A will reclose, restoring service to the customers in section 1. At this point, the crew can proceed directly to SW1 and commence the search for the fault from there, further reducing the amount of time for fault location.

Faults occurring in section 1 will result in a lockout of substation BKR A. At this point, SW1 will see a loss of voltage and transmit an alarm to the primary station. The primary will then interrogate SW1 and determine that it experienced loss of voltage, but no overcurrent. This condition indicates a section 1 fault. The primary station will then initiate the section 1 fault algorithm and interrogate SW1 to determine the load in section 2 just prior to the fault. The primary then interrogates SW3 for load information.

If the addition of section 2 load to the existing load on section 3 will not result in an overload condition, then the master station will instruct SW2 to close, restoring service to section 2.

The automation system can be operated in three modes. The first is traditional SCADA, where the dispatcher must identify and analyze circuit conditions before deciding on restoration procedures, which must then be sent to the field devices individually. In the second mode, the primary station computer will identify and analyze the circuit condition and make an operational recommendation to the dispatcher through the human-machine interface (HMI), who can then choose to accept the recommendation and execute a simple acknowledge command which will automatically issue appropriate commands to field devices. The third mode identifies and analyzes circuit conditions and automatically issues commands to field devices to isolate faulted sections and restore service to unfaulted sections. These systems can also be used to automatically return the circuit to its normal operating configuration after receiving confirmation from the operator through the HMI. It should be noted that the system can utilize many switching devices on a given circuit loop like that shown in Fig. 12-56. The figure was simplified for clarity in this example.

A large metropolitan utility wanted to improve electric service to its customers throughout its service area. However, it did not have a SCADA primary station at the beginning of the project and needed a way to improve service without installing one. In this case, the utility installed a system which utilizes distributed intelligence in the switch controllers to do all analysis as well as fault isolation and circuit restoration. Such a system operates in much the same way as described in the example above. The key difference is that this system does not utilize the first two modes of operation: traditional SCADA and suggesting operational recommendations. The advantage is the ability to install the system without the need for a communications network or SCADA primary station. This system does its work automatically without human intervention and the entire fault location, isolation, and service restoration occurs in less than 1 minute. It should be noted that the system can be easily disabled in the field for line work if desired and can be upgraded to work with a primary control station.

### 12.5.2 Automated Switches

A key part of an automated feeder switching system is the automated switch. The term "automated" in this context means the switch is designed for use on a fully automated system, a SCADA system, or in a system using distributed intelligence. To be automated, existing switches may be retrofitted with motor operators, current and voltage sensors, RTUs, and communication devices to allow the remote operation necessary to realize the benefits available with automated feeder switching systems. However, switches designed for occasional, manual operation may not be entirely suitable for operation on an automated distribution circuit feeder. Manual switches are typically not designed to be operated the hundreds of times required by a fully automated system over the life of a typical switch. Nor are they ordinarily designed for duty-cycle fault-closing to allow the system operator to inadvertently close into a fault from the SCADA primary station and still leave the switch in an operable condition.

There are several switches designed specifically for automation systems (Fig. 12-57). Such switches incorporate design features such as those described below that make them particularly suitable for use in an automated feeder switching system:

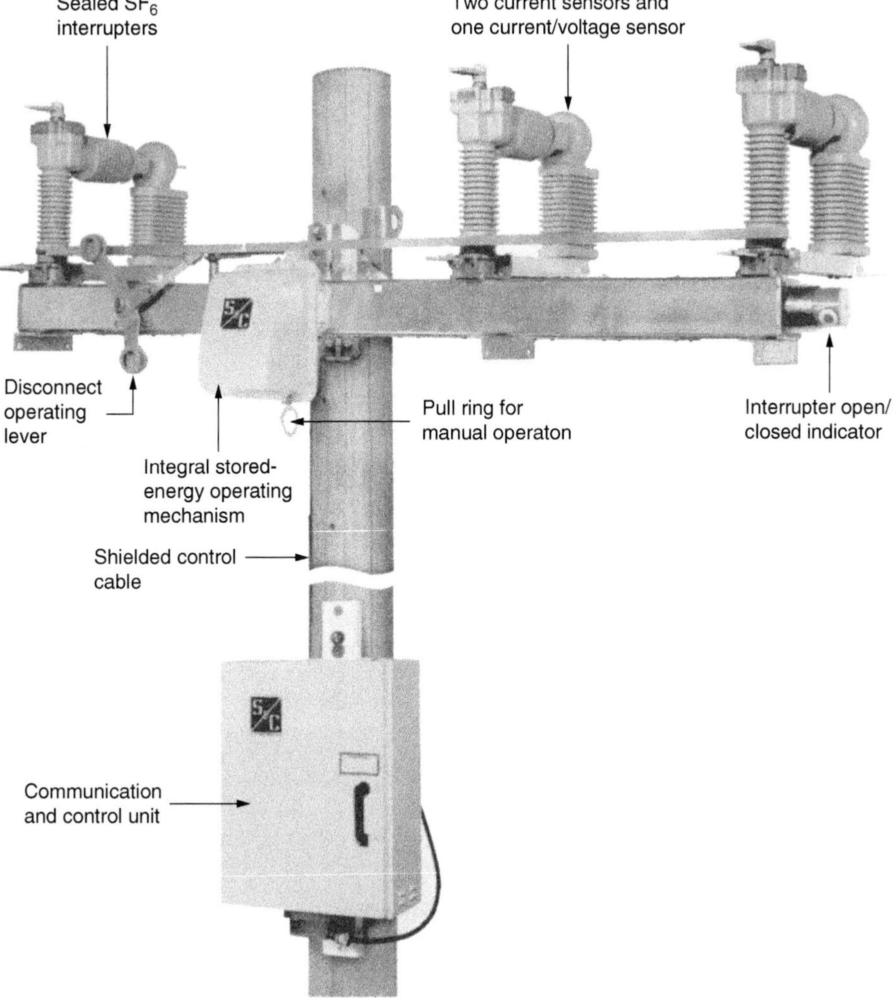


**FIGURE 12-57** Automated switches with integrated sensing. (*S&C Electric Company.*)

1. *Duty-cycle fault-closing* allows the switch to be closed into a typical fault several times before experiencing damage severe enough to render the switch inoperable.
2. *Integrated voltage and current sensors* provide the ability to monitor voltages, currents, and loads that are in turn used as inputs to algorithms to effect automated switching for fault isolation and restoration, and for shifting loads for circuit optimization.
3. *Integrated operating mechanisms* enable the switches to be operated remotely via computer commands. Integration with the switch ensures optimum operation without the need for cumbersome ground-to-switch linkages.
4. *Integrated load interrupters* are typically designed to allow operation under most weather conditions since it will not be possible to visibly inspect the switch for ice or other problems prior to operation.
5. *Integrated control power sources* eliminate the need to rely on locally available control power sources, or to install such power sources.
6. *Integrated visible air-gap isolation* provides a visible air gap when required as a safety clearance point for certain types of overhead line work.

An associated control package should include switch-operating controls, a local/remote switch, backup power for dead-line SCADA operation, a remote-terminal unit, and data-communication devices. The entire package should be assembled and tested for proper operation by a single supplier to eliminate the need for the utility to perform the integration.

Similar automated switching devices are also available for underground distribution systems. In underground switchgear applications, the control interface should be isolated from the high-voltage compartments of the switchgear. Underground distribution systems commonly employ multifunction protective relaying devices with control logic to perform localized automated decision making.

## 12.6 POWER CAPACITORS

*BY JEFFREY H. NELSON AND MARK McVEY*

Definitions of terms used in this subsection can be found in the IEEE standards and application guides referenced in this subsection and/or in the *IEEE Standards Dictionary* (available at http://ieeexplore.ieee.org/xpls/dictionary.jsp?tag=1).

### 12.6.1 System Benefits of Power Capacitors

Power capacitors provide several benefits to power systems. Among these include power factor correction, system voltage support, increased system capacity, reduction of power system losses, reactive power support, and power oscillation damping.

***Power Factor Correction.*** In general, the efficiency of power generation, transmission, and distribution equipment is improved when it is operated near unity power factor. The most cost-effective way to achieve near unity power factor is with the application of capacitors. Capacitors provide a static source of leading reactive current and can be installed in close proximity to the load. Thus, the maximum efficiency may be realized by reducing the magnetizing (lagging) current requirements throughout the system. Table 12-6 is a simple tool that can be used to determine the kilovars (kvar) required for correcting the system power factor. Find the row that corresponds to the existing power factor and the column that corresponds to the desired power factor and select the kilowatt multiplier where these intersect. Then simply multiply this factor by the power system load in kilowatts to determine the kilovars required to be installed to achieve the desired power factor.

**TABLE 12-6** Power Factor Correction Kilowatt Multipliers

| Present power-factor percentage | Corrected power-factor percentage | | | | | | | | | | | | | | | | | | | | |
|---|---|---|---|---|---|---|---|---|---|---|---|---|---|---|---|---|---|---|---|---|---|
| | 80 | 81 | 82 | 83 | 84 | 85 | 86 | 87 | 88 | 89 | 90 | 91 | 92 | 93 | 94 | 95 | 96 | 97 | 98 | 99 | 100 |
| 50 | 0.982 | 1.008 | 1.034 | 1.060 | 1.086 | 1.112 | 1.139 | 1.165 | 1.192 | 1.220 | 1.248 | 1.276 | 1.306 | 1.337 | 1.369 | 1.403 | 1.442 | 1.481 | 1.529 | 1.590 | 1.732 |
| 51 | .937 | .962 | .989 | 1.015 | 1.041 | 1.067 | 1.094 | 1.120 | 1.147 | 1.175 | 1.203 | 1.231 | 1.261 | 1.292 | 1.324 | 1.358 | 1.395 | 1.436 | 1.484 | 1.544 | 1.687 |
| 52 | .893 | .919 | .945 | .971 | .997 | 1.023 | 1.050 | 1.076 | 1.103 | 1.131 | 1.159 | 1.187 | 1.217 | 1.248 | 1.280 | 1.314 | 1.351 | 1.392 | 1.440 | 1.500 | 1.643 |
| 53 | .850 | .876 | .902 | .928 | .954 | .980 | 1.007 | 1.033 | 1.060 | 1.088 | 1.116 | 1.144 | 1.174 | 1.205 | 1.237 | 1.271 | 1.308 | 1.349 | 1.397 | 1.457 | 1.600 |
| 54 | .809 | .835 | .861 | .887 | .913 | .939 | .966 | .992 | 1.019 | 1.047 | 1.075 | 1.103 | 1.133 | 1.164 | 1.196 | 1.230 | 1.267 | 1.308 | 1.356 | 1.416 | 1.559 |
| 55 | .769 | .795 | .821 | .847 | .873 | .899 | .926 | .952 | .979 | 1.007 | 1.035 | 1.063 | 1.090 | 1.124 | 1.156 | 1.190 | 1.228 | 1.268 | 1.316 | 1.377 | 1.519 |
| 56 | .730 | .756 | .782 | .808 | .845 | .860 | .887 | .913 | .940 | .968 | .996 | 1.024 | 1.051 | 1.085 | 1.117 | 1.151 | 1.189 | 1.229 | 1.277 | 1.338 | 1.480 |
| 57 | .692 | .718 | .744 | .770 | .796 | .822 | .849 | .875 | .902 | .930 | .958 | .986 | 1.013 | 1.047 | 1.079 | 1.113 | 1.151 | 1.191 | 1.239 | 1.300 | 1.442 |
| 58 | .655 | .681 | .707 | .733 | .759 | .785 | .812 | .838 | .865 | .893 | .921 | .949 | .976 | 1.010 | 1.042 | 1.076 | 1.114 | 1.154 | 1.202 | 1.263 | 1.405 |
| 59 | .618 | .644 | .670 | .696 | .722 | .748 | .775 | .801 | .828 | .856 | .884 | .912 | .939 | .973 | 1.005 | 1.039 | 1.077 | 1.117 | 1.165 | 1.226 | 1.368 |
| 60 | .584 | .610 | .636 | .662 | .688 | .714 | .741 | .767 | .794 | .822 | .850 | .878 | .905 | .939 | .971 | 1.005 | 1.043 | 1.083 | 1.131 | 1.192 | 1.334 |
| 61 | .549 | .575 | .601 | .627 | .653 | .679 | .706 | .732 | .759 | .787 | .815 | .843 | .870 | .904 | .936 | .970 | 1.008 | 1.048 | 1.096 | 1.157 | 1.299 |
| 62 | .515 | .541 | .567 | .593 | .619 | .645 | .672 | .698 | .725 | .753 | .781 | .809 | .836 | .870 | .902 | .936 | .974 | 1.014 | 1.062 | 1.123 | 1.265 |
| 63 | .483 | .509 | .535 | .561 | .587 | .613 | .640 | .666 | .693 | .721 | .749 | .777 | .804 | .838 | .870 | .904 | .942 | .982 | 1.030 | 1.091 | 1.233 |
| 64 | .450 | .476 | .502 | .528 | .554 | .580 | .607 | .633 | .660 | .688 | .716 | .744 | .771 | .805 | .837 | .871 | .909 | .949 | .997 | 1.058 | 1.200 |
| 65 | .419 | .445 | .471 | .497 | .523 | .549 | .576 | .602 | .629 | .657 | .685 | .713 | .740 | .774 | .806 | .840 | .878 | .918 | .966 | 1.027 | 1.169 |
| 66 | .388 | .414 | .440 | .466 | .492 | .518 | .545 | .571 | .598 | .626 | .554 | .682 | .709 | .743 | .775 | .809 | .847 | .887 | .935 | .996 | 1.138 |
| 67 | .358 | .384 | .410 | .436 | .462 | .488 | .515 | .541 | .568 | .596 | .624 | .652 | .679 | .713 | .745 | .779 | .817 | .857 | .905 | .966 | 1.108 |
| 68 | .329 | .355 | .381 | .407 | .433 | .459 | .486 | .512 | .539 | .567 | .595 | .623 | .650 | .684 | .716 | .750 | .788 | .828 | .876 | .937 | 1.079 |
| 69 | .299 | .325 | .351 | .377 | .403 | .429 | .456 | .482 | .509 | .537 | .565 | .593 | .620 | .654 | .866 | .720 | .758 | .798 | .840 | .907 | 1.049 |
| 70 | .270 | .296 | .322 | .348 | .374 | .400 | .427 | .453 | .480 | .508 | .536 | .564 | .591 | .625 | .657 | .691 | .729 | .769 | .811 | .878 | 1.020 |
| 71 | .242 | .268 | .294 | .320 | .346 | .372 | .399 | .425 | .452 | .480 | .508 | .536 | .563 | .597 | .629 | .663 | .701 | .741 | .783 | .850 | .992 |
| 72 | .213 | .239 | .265 | .291 | .317 | .343 | .370 | .396 | .423 | .451 | .479 | .507 | .534 | .568 | .600 | .634 | .672 | .712 | .754 | .821 | .963 |
| 73 | .186 | .212 | .238 | .264 | .290 | .316 | .343 | .369 | .396 | .424 | .452 | .480 | .507 | .541 | .573 | .607 | .645 | .685 | .727 | .794 | .936 |
| 74 | .159 | .185 | .211 | .237 | .263 | .289 | .316 | .342 | .369 | .397 | .425 | .453 | .480 | .514 | .546 | .580 | .618 | .658 | .700 | .767 | .909 |
| 75 | .132 | .158 | .184 | .210 | .236 | .262 | .289 | .315 | .342 | .370 | .398 | .426 | .453 | .487 | .519 | .553 | .591 | .631 | .673 | .740 | .882 |

*(Continued)*

**TABLE 12-6** Power Factor Correction Kilowatt Multipliers (*Continued*)

| Present power-factor percentage | Corrected power-factor percentage | | | | | | | | | | | | | | | | | | | | |
|---|---|---|---|---|---|---|---|---|---|---|---|---|---|---|---|---|---|---|---|---|---|
| | **80** | **81** | **82** | **83** | **84** | **85** | **86** | **87** | **88** | **89** | **90** | **91** | **92** | **93** | **94** | **95** | **96** | **97** | **98** | **99** | **100** |
| **76** | .105 | .131 | .157 | .183 | .209 | .235 | .262 | .288 | .315 | .343 | .371 | .399 | .426 | .460 | .492 | .526 | .564 | .604 | .652 | .713 | .855 |
| **77** | .079 | .105 | .131 | .157 | .183 | .209 | .236 | .262 | .289 | .317 | .345 | .373 | .400 | .434 | .466 | .500 | .538 | .578 | .620 | .687 | .829 |
| **78** | .053 | .079 | .105 | .131 | .157 | .183 | .210 | .236 | .263 | .291 | .319 | .347 | .374 | .408 | .440 | .474 | .512 | .552 | .594 | .661 | .803 |
| **79** | .026 | .052 | .078 | .104 | .130 | .156 | .183 | .209 | .236 | .264 | .292 | .320 | .347 | .381 | .413 | .447 | .485 | .525 | .567 | .634 | .776 |
| **80** | .000 | .026 | .052 | .078 | .104 | .130 | .157 | .183 | .210 | .238 | .266 | .294 | .321 | .355 | .387 | .421 | .459 | .499 | .541 | .608 | .750 |
| **81** | | .000 | .026 | .052 | .078 | .104 | .131 | .157 | .184 | .212 | .240 | .268 | .295 | .329 | .361 | .395 | .433 | .473 | .515 | .582 | .724 |
| **82** | | | .000 | .026 | .062 | .078 | .105 | .131 | .158 | .186 | .214 | .242 | .269 | .303 | .335 | .369 | .407 | .447 | .489 | .556 | .698 |
| **83** | | | | .000 | .026 | .052 | .079 | .105 | .132 | .160 | .188 | .216 | .243 | .277 | .309 | .343 | .381 | .421 | .463 | .530 | .672 |
| **84** | | | | | .000 | .026 | .053 | .079 | .106 | .134 | .162 | .190 | .217 | .251 | .283 | .317 | .355 | .395 | .437 | .504 | .645 |
| **85** | | | | | | .000 | .027 | .053 | .080 | .108 | .136 | .164 | .191 | .225 | .257 | .291 | .329 | .369 | .417 | .478 | .620 |
| **86** | | | | | | | | .026 | .053 | .081 | .109 | .137 | .167 | .198 | .230 | .265 | .301 | .343 | .390 | .451 | .593 |
| **87** | | | | | | | | | .027 | .055 | .082 | .111 | .141 | .172 | .204 | .238 | .275 | .317 | .364 | .425 | .567 |
| **88** | | | | | | | | | | .028 | .056 | .084 | .114 | .145 | .177 | .211 | .248 | .290 | .337 | .398 | .540 |
| **89** | | | | | | | | | | | .028 | .056 | .086 | .117 | .149 | .183 | .220 | .262 | .309 | .370 | .512 |
| **90** | | | | | | | | | | | | .028 | .058 | .089 | .121 | .155 | .192 | .234 | .281 | .342 | .484 |
| **91** | | | | | | | | | | | | | .030 | .061 | .093 | .127 | .164 | .206 | .253 | .314 | .456 |
| **92** | | | | | | | | | | | | | | .031 | .063 | .097 | .134 | .176 | .223 | .284 | .426 |
| **93** | | | | | | | | | | | | | | | .032 | .066 | .103 | .145 | .192 | .253 | .395 |
| **94** | | | | | | | | | | | | | | | | .034 | .071 | .113 | .160 | .221 | .363 |
| **95** | | | | | | | | | | | | | | | | | .037 | .079 | .126 | .187 | .328 |
| **96** | | | | | | | | | | | | | | | | | | .042 | .089 | .150 | .292 |
| **97** | | | | | | | | | | | | | | | | | | | .047 | .108 | .251 |
| **98** | | | | | | | | | | | | | | | | | | | | .061 | .203 |
| **99** | | | | | | | | | | | | | | | | | | | | | .142 |

For example, with a load of 200 kW at 77% power factor, how many capacitive kilovars are needed to correct to a power factor of 95%? At the point where the row for the 77% existing power factor and the column for the desired power factor of 95% intersect, we find a kilowatt multiplier of 0.5. Therefore, the following calculation can be made to determine the kvar required to achieve 95% power factor.

$$\begin{aligned} 3\emptyset \text{ kvar} &= 3\emptyset \text{ load (kW)} \times \text{kilowatt multiplier} \\ &= 200 \text{ kW} \times 0.5 \\ &= 100 \text{ kvar} \end{aligned} \tag{12-2}$$

***System Voltage Support.*** Power systems are predominately inductive in nature and during peak load conditions or during system contingencies there can be a significant voltage drop between the voltage source and the load. Application of capacitors to a power system results in a voltage increase back to the voltage source, and also past the application point of the capacitors in a radial system. The actual percentage increase of the system voltage is dependent upon the inductive reactance of the system at the point of application of the capacitors. The short-circuit impedance at that point is approximately the same as the inductive reactance; therefore, the three-phase short-circuit current at that location can be used to determine the approximate voltage rise. The following rule-of-thumb equation is commonly used.

$$\Delta V \approx \frac{kvar_C}{kVA_{SC}} \times 100\% \tag{12-3}$$

where $\Delta V$ is percent voltage rise at the point of the capacitor installation, $kvar_C$ is capacitor three-phase kvar, and $kVA_{SC}$ is system three-phase short-circuit kVA at the point of the capacitor installation.

$$kVA_{SC} = \sqrt{3} \times V_{LL} \times I_{SC} \tag{12-4}$$

where $V_{LL}$ is system line-to-line (phase-to-phase) voltage, $I_{SC}$ is three-phase short-circuit current at the point of the capacitor installation.

***Increased System Capacity.*** The application of shunt or series capacitors can affect the power system capacity. Application of shunt capacitors reduces the inductive reactive current on the power system, and thus reduces the system kVA loading. This can have the effect of increasing system capacity to serve additional load.

Series capacitors are typically used to increase the power carrying capability of transmission lines. Series capacitors insert a voltage in series with the transmission line that is opposite in polarity to the voltage drop across the line, which decreases the apparent reactance and increases the power transfer capability of the line. The reduction in reactance often improves system stability. Many systems are not limited by thermal capacity but system stability.

***Power System Loss Reduction.*** The installation of capacitors can reduce the current flow in a power system. Since losses are proportional to the square of the current, a reduction in current will lead to reduced system losses.

***Reactive Power Support.*** Capacitors can help support steady-state stability limits and reactive power requirements at generators.

***Power Oscillation Damping.*** Controlled series capacitors can provide an active damping for power oscillations that many large power systems experience. They can also provide support after significant disturbances to the power system and allow the system to remain in synchronous operation.

## 12.6.2 Capacitor Units

The terms capacitor unit, power capacitor or capacitor may be used interchangeably.

The base standard which covers the ratings and requirements for capacitor units is IEEE 18.[1] This standard applies to capacitors rated 216 V or higher, 2.5 kvar or more, and designed for shunt connection on alternating current (ac) power systems operating at a nominal power frequency of 50 or 60 Hz. IEEE 18[1] covers service conditions, ratings and capabilities, manufacturing, and testing of shunt capacitor units.

IEEE 824[2] specifies ratings, capabilities and testing requirements in addition to IEEE 18[1] for capacitor units to be utilized in series capacitor banks.

IEEE 1531[3] covers considerations to take into account when specifying units to be used in a harmonic filter bank.

There are typically two voltage classes of capacitor units applied and they are divided into the categories of low-voltage capacitors (below 1000 V) and high-voltage capacitors (1000 V and above).

***Low-Voltage Capacitor Units.*** Low-voltage capacitor units are typically applied in industrial power systems. They are generally available in voltage ratings from 240 to 600 V over the range of 2.5 to 100 kvar three-phase. Low-voltage capacitors are typically dry-type, metalized polypropylene film (see Fig. 12-58). Typical voltage and kvar ratings are listed in IEEE 18.[1]

***High-Voltage Capacitor Units.*** High-voltage capacitor units are typically connected to industrial power systems or to utility transmission and distribution power systems. The units are generally available in voltage ratings from 2.4 to 25 kV. Modern manufacturing techniques continue to decrease the size of high-voltage capacitor units, which continues to increase the available kvar ratings for units. Typical voltage and kvar ratings are listed in IEEE 18.[1]

The capacitor elements in present day high-voltage capacitor units are manufactured using an *all-film dielectric*, typically two layers of polypropylene film between two layers of thin aluminum foil. The capacitor elements are connected in a series/parallel combination to achieve the desired voltage and kvar rating. The elements are placed in a metal enclosure with bushings for external connections and impregnated with a dielectric fluid. The unit is then hermetically sealed (see Fig. 12-59).

***Capacitor Unit Ratings.*** IEEE 18[1] establishes the following standard ratings for capacitor units, as applied under the normal service conditions and ambient temperatures defined in the standard:

1. Voltage, rms (terminal-to-terminal)
2. Terminal to case (or ground) insulation class

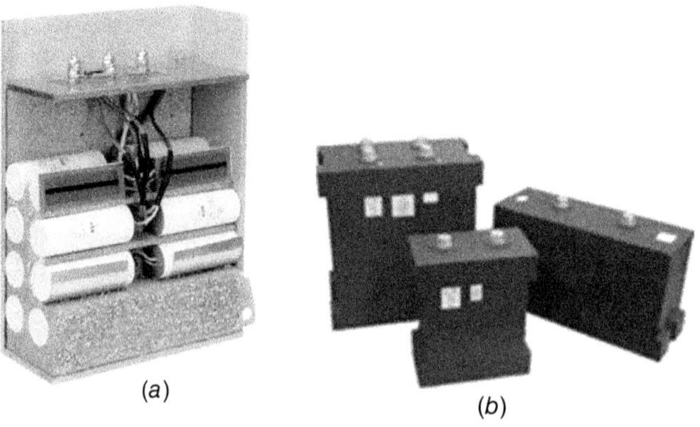


**FIGURE 12-58** Typical low-voltage capacitor units with self-healing metalized polypropylene capacitor elements: (*a*) liquid-free steel enclosure with three-phase connection terminals, and (*b*) single-phase epoxy encapsulated.

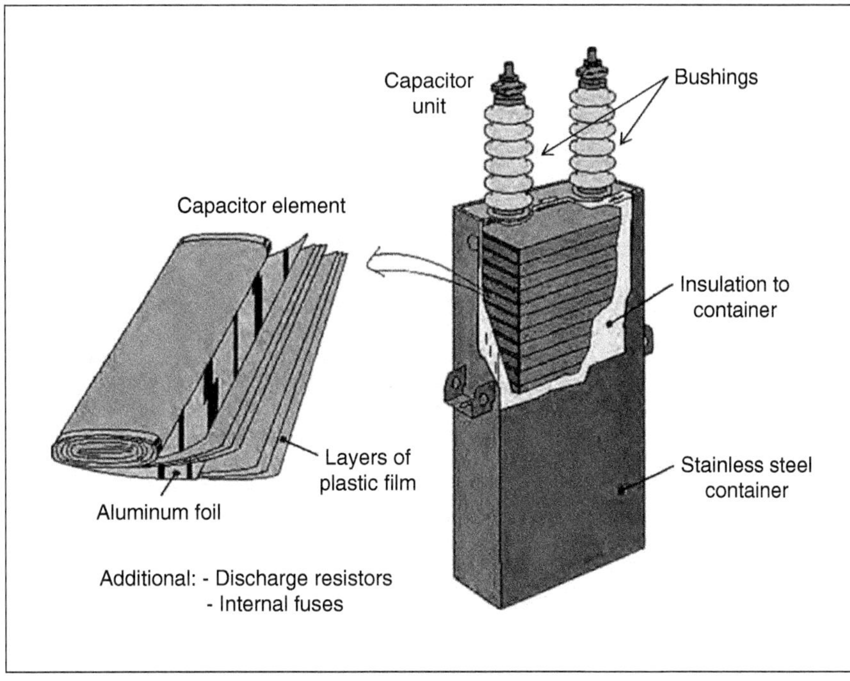


**FIGURE 12-59** Typical high-voltage capacitor unit.

3. Reactive power
4. Number of phases
5. Frequency

Capacitors are intended to be operated at or below their rated voltage. They are designed to operate at overvoltages of 10% for bank contingencies, such as the failure of an individual unit or element, or system contingencies, such as high voltage. Note that the overvoltage capability of the individual capacitor elements inside the unit is 110% of its portion of the unit voltage rating. The capacitor unit is capable of operating continuously under contingency conditions provided that the peak voltage does not exceed 120% of rated peak voltage, including harmonics. The maximum peak voltage rating of the capacitor including harmonics, but excluding transients, is equal to

$$V_{pk} = 1.2\sqrt{2}(V_{\text{rms}}\text{rated}) \tag{12-5}$$

Also, the unit is designed to operate continuously if the current does not exceed 135% of nominal rms current based on rated kvar and rated voltage, and the kvar output does not exceed 135% of rated kvar. The total rms current is equal to

$$I_{\text{rms}} = \sqrt{(I_1)^2 + (I_2)^2 + \cdots + (I_n)^2} \tag{12-6}$$

where $n$ is harmonic number.

The kvar output of the capacitor increases as the square of the applied voltage:

$$kvar_{E_2} = kvar_{E_1} \frac{(E_2)^2}{(E_1)^2} \tag{12-7}$$

For example, a 200-kvar, 7200-V capacitor unit will supply 242 kvar at 7920 V:

$$kvar_{E_2} = 200 \times \frac{(7920)^2}{(7200)^2}$$
$$kvar_{E_2} = 242 \text{ kvar}$$

***Transient Overvoltage and Overcurrent Withstand Capabilities.*** When the installation of a switched rack or bank is contemplated, other nearby capacitor equipment must be evaluated.

Capacitor banks switched in a back-to-back configuration can create very high peak magnitude and high frequency current transients. Also, when switching a capacitor bank without any type of transient suppression a voltage transient is generated, typically in the range of 300 to 800 Hz. This undamped voltage transient can couple through a power transformer to a lower voltage system and be transmitted several miles away. If a capacitor bank is installed on the lower voltage system a resonant condition may exist. If the resonant frequency of the LC circuit on the high-voltage bus where the capacitor bank is being switched is approximately equal to the LC circuit at the lower voltage capacitor bank, the voltage transient will be magnified. This phenomenon is known as voltage magnification.

These switching conditions could result in high overvoltage and overcurrent transients. The use of series reactors, pre-insertion inductors or resistors, and/or special switching devices is sometimes required to reduce these transients to lower levels.

Transient overvoltage and overcurrent capabilities for shunt capacitors are covered in IEEE 1036.[4] Prior to 2002, they were covered in IEEE 18.[1]

***Capacitance Tolerance.*** In the 2002 revision to IEEE 18, the allowable capacitance tolerance range for shunt capacitor units was changed. In previous versions of IEEE 18 the allowable capacitance tolerance range of each unit was "–0% to +15%" of the nominal value based on rated kvar, voltage and frequency, measured at 25°C uniform case and internal temperature. In the 2002 revision, it was changed to " –0% to +10%." However, with today's modern manufacturing practices, capacitance tolerances typically do not exceed +5% for standard units.

A tighter tolerance range may be specified for units to be utilized in a harmonic filter capacitor bank, for example: " –2% to +2%" of the nominal capacitance. This tolerance may be applied to the capacitance of the entire harmonic filter bank instead of the individual units.

Capacitance tolerances for series capacitors are specified on a per phase basis for the entire bank (see IEEE 824[2]).

***Discharge Resistors.*** Under certain conditions, when line voltage is removed from a power capacitor the possibility exists that the unit will retain an extremely high charge even days later. This characteristic of retaining such a charge is demonstrated by the high-efficiency and low-loss operation of a power capacitor. To eliminate this hazard, all power capacitors contain internal-discharge resistors. This resistor assembly will reduce the terminal voltage from line voltage to 50 V within 5 min of deenergization for a capacitor rated higher than 600 V ac, and within 1 min for capacitors rated 600 V ac or less.

***Capacitor Unit Losses.*** The cost of operating high-voltage capacitors is lower per kvar than low-voltage capacitors because of the basic difference in dielectric materials, which allows high-voltage capacitors to be operated more efficiently. Present day high-voltage capacitors operate at a lower watts loss per kvar than do low-voltage capacitors. For example, capacitors that utilize an *all-film dielectric* may operate with losses of less than 0.1 W per kvar. Low-voltage capacitors using *metalized polypropylene dielectric* may experience losses of near 0.5 W per kvar.

***Capacitor Tank Rupture.*** Capacitor tank rupture of high-voltage capacitors will occur if the total energy applied to the capacitor unit under failure conditions is greater than the ability of the capacitor tank to withstand such energy. Tank-rupture curves are essential to the correct selection of fuse links for overcurrent protection of externally fused capacitors. A typical tank rupture curve is shown in Fig. 12-60.

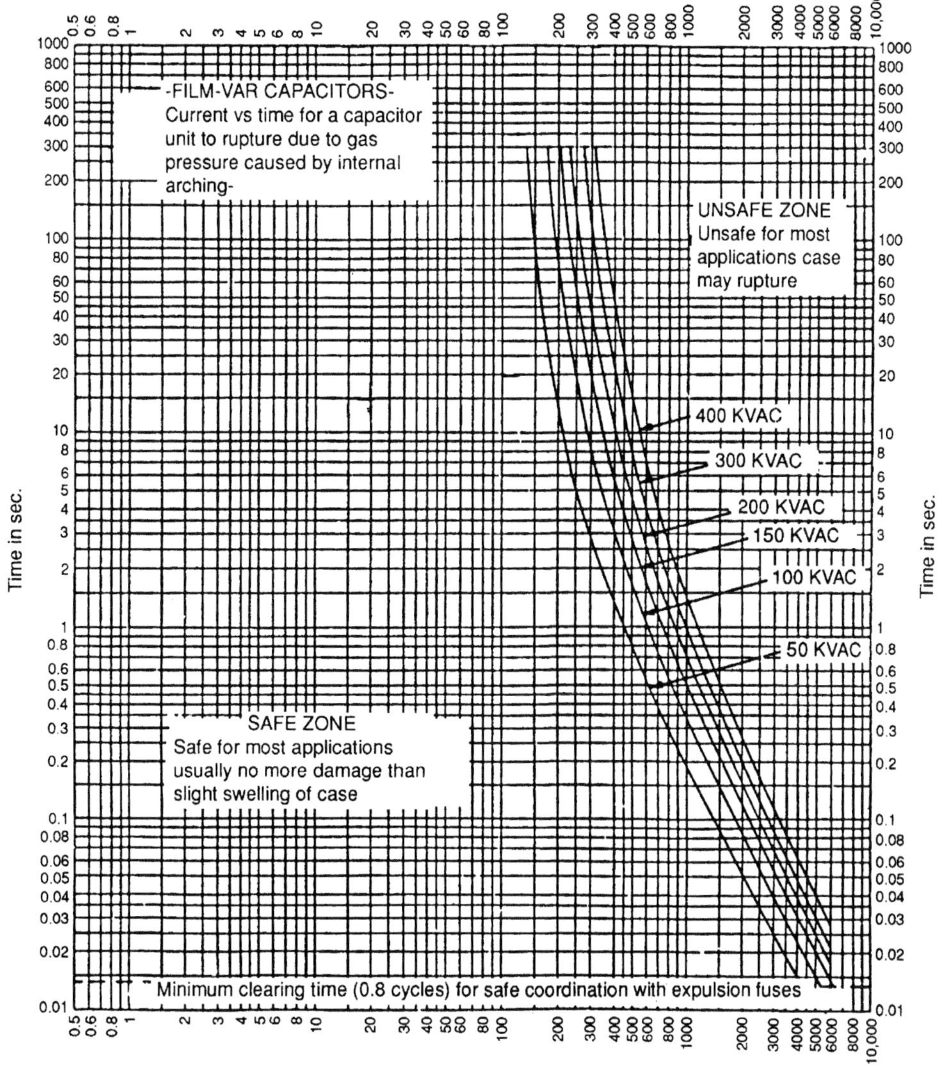


**FIGURE 12-60** Typical high-voltage tank-rupture curve.

***Temperature.*** Although very efficient, power capacitors do consume some power and generate heat. This heat must be adequately ventilated when enclosed or exposed to higher than normal ambient temperatures.

Maximum and minimum operating temperatures for power capacitors are specified in IEEE 18. Capacitors are designed for continuous operation at a maximum ambient temperature of 46°C over a 24-hour period, with a peak of 55°C. The minimum ambient temperature for continuous operation is –40°C. Different requirements for maximum or minimum ambient temperatures can be specified by the user. Lower required minimum ambient temperatures may affect the maximum ambient at which the unit can operate, and conversely, higher required maximum ambient temperatures may affect the minimum ambient at which the unit can operate.

Capacitor applications must be designed for adequate overvoltage and "corona" capabilities, since the partial discharge (corona) characteristics vary with temperature. It is necessary to consider the full range of temperatures to which the capacitors will be exposed, in both the energized and deenergized modes.

## 12.6.3 Shunt Capacitors

The most common application of power capacitors is shunt connected capacitors and they are either energized continuously or switched on and off during load cycles.

***Common Shunt Capacitor Connections.*** Figure 12-61 shows four of the most common capacitor connections: three-phase grounded wye, three-phase ungrounded wye, three-phase delta, and single phase. Grounded or ungrounded wye connections are usually made on medium- and high-voltage systems, whereas delta and single-phase connections are usually made on low-voltage systems.

Grounded-wye capacitors can bypass some line surges to ground and therefore exhibit a certain degree of self-protection from transient voltages and lightning surges. The grounded-wye connection also provides a low impedance path for harmonics.

If the capacitors are electrically connected ungrounded-wye, the maximum fault current would be limited to three times line current. If the available fault current exceeds the tank rupture curve the use of current limiting fuses must be considered.

***Low-Voltage Shunt Capacitors.*** Low-voltage shunt capacitors are typically applied in industrial power systems, in voltage ratings from 240 to 600 V over the range of 2.5 to 100 kvar three-phase. They can be used for power factor improvement and voltage support on low-voltage systems where adjustable speed drives, power electronics, and other industrial equipment are applied.

Low-voltage capacitors made with metalized polypropylene are self-healing. That is, the conductors consist of very thin layers of metal deposited on the film dielectric. In the event of short circuit, the conductor vaporizes to eliminate the fault with negligible loss of capacitance and continued operation. External fuses are not required on these units although some users add them for additional protection of their connections and for ground faults.

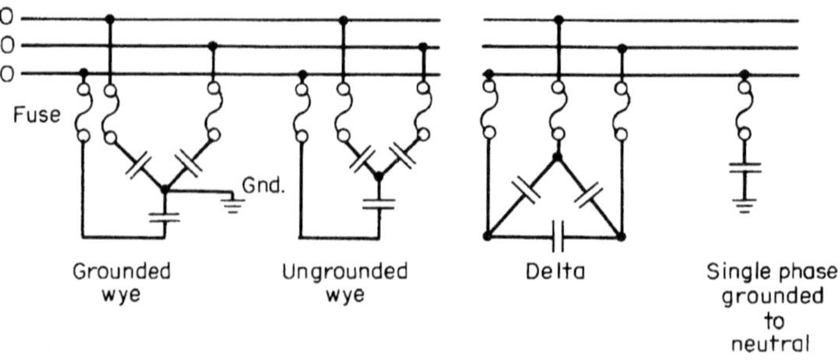


**FIGURE 12-61** Common methods for connecting capacitors.

**FIGURE 12-62** Typical pole-mounted distribution capacitor bank.

***Distribution Shunt Capacitors.*** Capacitors for overhead distribution systems can be pole-mounted, typically in banks of 300 to 3600 kvar, at nearly any system voltage up to 34.5 kV phase-to-phase. Pad-mounted capacitor equipments are used for underground distribution systems in the same range of sizes and voltage ratings. See Fig. 12-62 for a typical pole-mounted capacitor bank.

The majority of the power capacitor equipment installed on medium-voltage distribution systems is connected grounded-wye. Grounded-wye capacitors are desirable on distribution lines where single-phase reclosers are applied. For example, if a recloser opens a single phase of a three-phase distribution line with an ungrounded-wye capacitor bank applied down circuit of the recloser, the neutral will shift on the capacitor bank and apply a voltage on the opened phase resulting in an unsafe system condition.

With the grounded-wye connection, tanks and frames of switching equipment are at ground potential. This also provides increased personnel safety. In addition, grounded-wye connections provide for faster fuse operation in case of a capacitor failure.

Distribution shunt capacitors can be protected with individual unit fuses or group fused. Tank rupture curves are essential to the correct selection of fuse links for overcurrent protection of any capacitor installation. Fuse selections should be based upon the coordination of the fuse-link maximum clearing curve (Fig. 12-63) and the high-voltage capacitor tank-rupture curve (Fig. 12-60).

Several fundamental principles must be observed in the selection of fuses for capacitor applications:

1. The fuse link must be capable of continuously carrying 135% of the nominal capacitor current as a minimum. Higher values may be required when high harmonic currents are present.
2. The fuse cutout must have sufficient interrupting capacity to successfully handle the available fault current, clearing voltage, and available energy before the capacitor tank ruptures. Also, the fuse cutout voltage rating should be properly rated for the application. (If the capacitor bank is not grounded then phase-to-ground rated cutouts should not be used.)
3. The fuse link must withstand, without damage, the normal transient current during bank energization or de-energization. Similarly, it must withstand the capacitor unit's discharge current during a terminal-to-terminal short.
4. For ungrounded-wye banks, maximum fault current is usually limited to three times normal line current. The fuse link must clear within 5 min at 95% of available fault current.

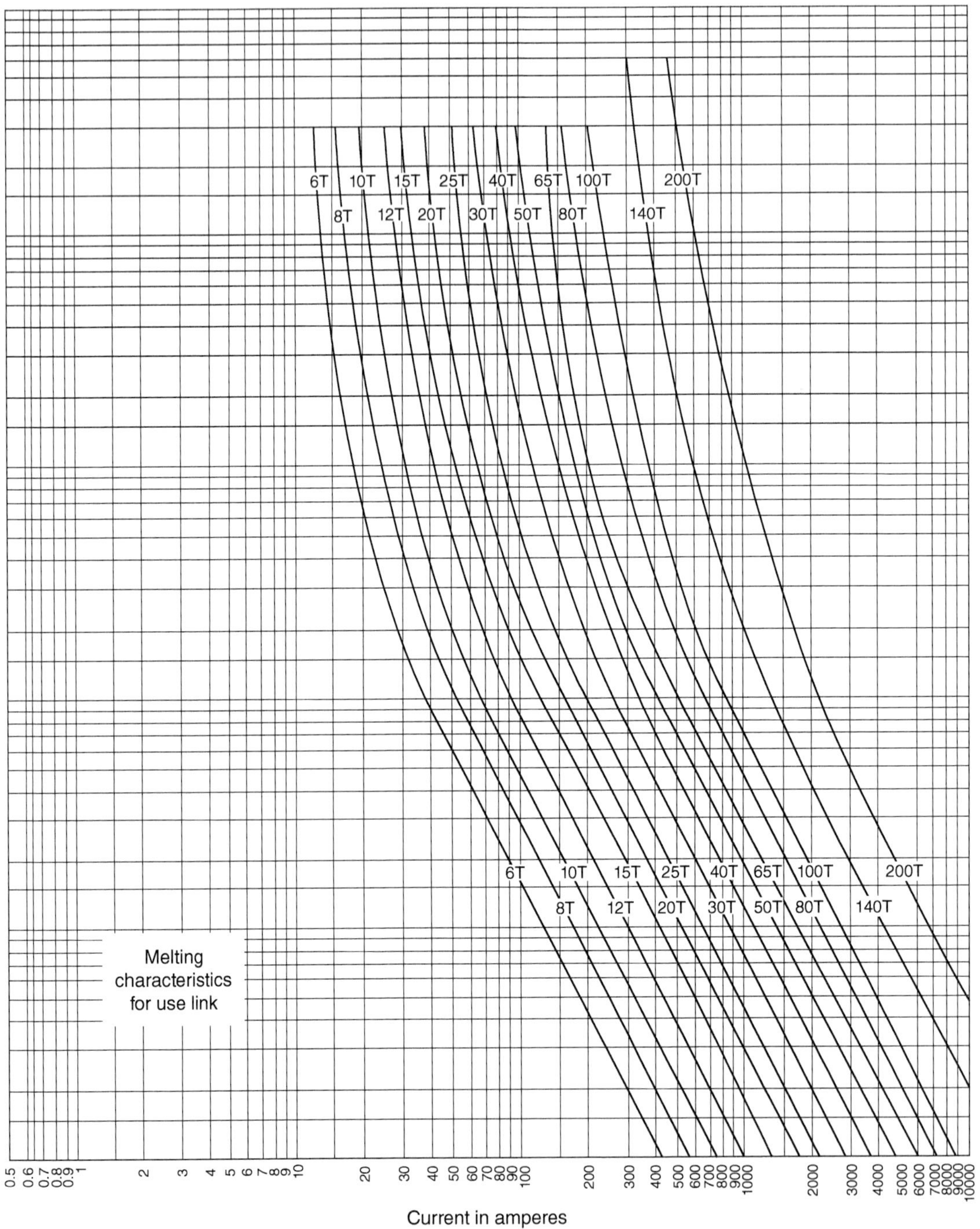


**FIGURE 12-63** Typical time-current curve for an expulsion fuse.

5. For effective capacitor protection, maximum asymmetric rms fault current should not exceed the current value at the intercept point of the capacitor tank-rupture time-current characteristic (TCC) curve and the minimum time shown on the fuse maximum-clearing TCC curve.
6. The maximum-clearing TCC curve of the fuse link must coordinate with the tank-rupture TCC curve of the capacitor unit.

For more detailed information on the application of shunt capacitors in distributions systems refer to IEEE 1036.[4] For application guidance on external fuses for shunt capacitors refer to IEEE C37.48.[5]

***Substation Shunt Capacitor Banks.*** Substation shunt capacitor banks are usually installed grounded-wye, ungrounded-wye, or delta. Delta connected shunt banks are usually at lower voltages, for example, 2400 V. Grounded-wye shunt capacitor bank installations are typically more economical than ungrounded-wye banks because the neutral does not have to be insulated for the system voltage. This advantage increases with system voltage. Also, the transient recovery voltages (TRV) for grounded-wye banks are less than ungrounded-wye, thus reducing the TRV requirements for switching equipment.

A disadvantage of grounded-wye shunt capacitor banks is that there is the possibility of high frequency switching transients in the ground grid that could induce transients in relay and control cables resulting in erroneous relay operations. The effect of these transients can be mitigated by specific neutral grounding methods of the capacitor bank installation. Two common methods of neutral grounding utilized are single point grounding and peninsula type grounding. For single point grounding, the neutrals of all capacitor banks at a specific voltage within the station are connected together and tied to the substation ground grid at only one point, effectively keeping the transients out of the ground grid. For peninsula grounding, a ground grid is established in the capacitor area and the neutral of each capacitor bank is connected to the peninsula ground grid conductors, then the peninsula ground grid is connected to the main station ground grid at only one point. This isolates the transients from the main ground grid. These grounding methods are discussed in more detail in IEEE 1036.[4]

Ungrounded-wye shunt capacitor banks may be applied at any system voltage as long as the insulation and TRV requirements are considered.

For shunt banks that have only one series group, it is more desirable to install an ungrounded-wye bank. In a grounded-wye bank with one series group, if one unit completely shorts and the phase-to-ground fault current is too high it will rupture the unit. With an ungrounded-wye bank, the fault current is limited to three times the bank current.

There are three common substation shunt capacitor bank types utilized today: externally fused, internally fused, and fuseless. These three types are briefly described in this subsection. For more detailed information on the application and protection of substation shunt capacitor banks, refer to IEEE 1036,[4] IEEE C37.99,[6] and IEEE C37.48.[5]

***Externally Fused Shunt Capacitor Banks.*** An externally fused shunt capacitor bank is made up of series groups of parallel connected capacitor units. Each unit is individually fused with an expulsion or current-limiting fuse mounted external to the unit. A typical schematic for an externally fused shunt capacitor bank is shown in Fig. 12-64.

A capacitor unit voltage rating is usually chosen to limit the number of series groups to as few as possible, while achieving the desired voltage rating of the bank. This usually results in the simplest design and highest sensitivity for unbalance protection.

Fewer series groups may mean more units in parallel in each series group. This could require the use of current limiting unit fuses instead of expulsion type fuses. When a unit fails in an externally fused bank the other units in parallel with it will discharge into the failed unit through its individual fuse. When the energy from this discharge exceeds the energy capability of an expulsion fuse, the user will have to specify current limiting fuses which have a higher energy capability. To avoid this, the user can specify a lower-voltage rating for the units and design the bank with more series groups and fewer units in parallel. Another option is to split the bank in two sections, called a double-wye configuration.

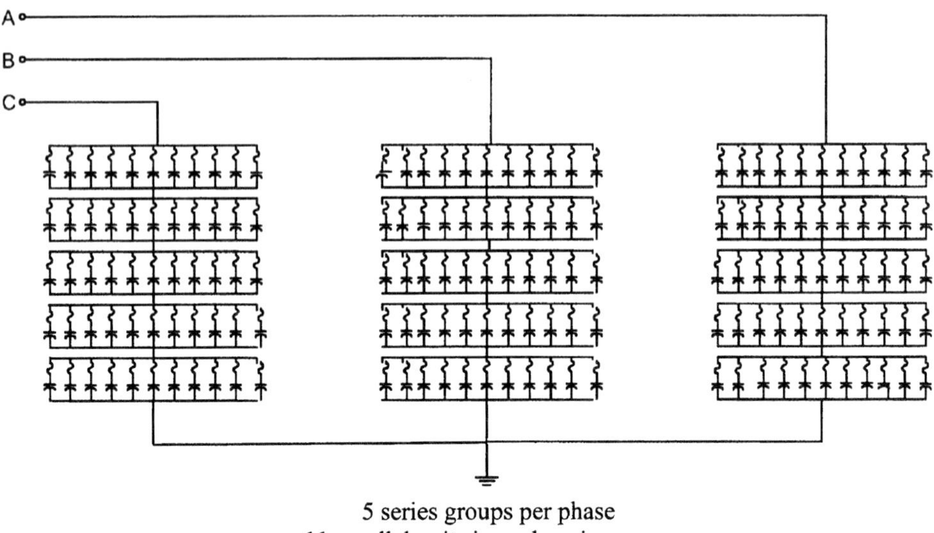


**FIGURE 12-64** Schematic of a typical externally fused shunt capacitor bank.

For principles of fuse selection for the individual units refer to the earlier section on distribution shunt capacitors. For more detailed guidance on fusing refer to IEEE C37.48[5] and IEEE C37.48.1.[7]

It is also important to keep a minimum number of units in parallel in each series group. When a fuse operates and removes a unit from a series group the voltage will increase on the remaining units in parallel with it. It is desirable to have enough units in parallel so that the 110% overvoltage capability will not be exceeded for the loss of one unit. There are curves available to help determine the voltage on remaining capacitor units in a series group based on the percentage of capacitor units removed from the series group. These curves are located in IEEE 1036,[4] which covers the application of shunt capacitors. The curves were previously located in an appendix of IEEE 18,[1] but were removed from the 2002 revision. A typical externally-fused shunt capacitor bank is shown in Fig. 12-65.

**FIGURE 12-65** Typical externally fused shunt capacitor bank.

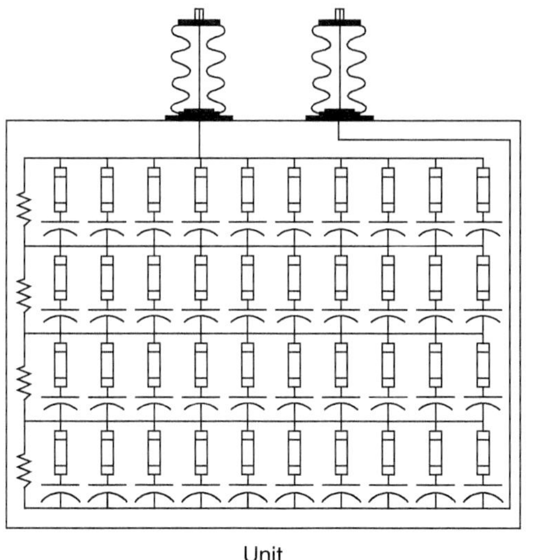


**FIGURE 12-66** Internal connection diagram of a typical internally fused capacitor unit.

***Internally Fused Shunt Capacitor Banks.*** An internally fused capacitor bank is made up of series and/or parallel connected capacitor units. The units are made up of capacitor elements connected in series/parallel combinations to achieve the desired rated voltage and kvar rating. A fuse is put in series with each element inside the unit case. An internally fused unit is shown schematically in Fig. 12-66. When a failure of a capacitor element occurs, its individual fuse will operate to isolate the element. The unit typically has a large number of elements in parallel so that the loss of one element does not have a significant increase in voltage on the remaining elements in parallel with the failed element. The kvar rating of internally fused units is typically much larger than externally fused units.

Internally fused capacitor banks typically have fewer units in parallel and more series groups than externally fused capacitor banks. A typical schematic for an internally fused shunt capacitor bank is shown in Fig. 12-67. It is usually desirable to have at least two units in parallel. If one unit has a large number of failed elements, the unit in parallel will help keep down

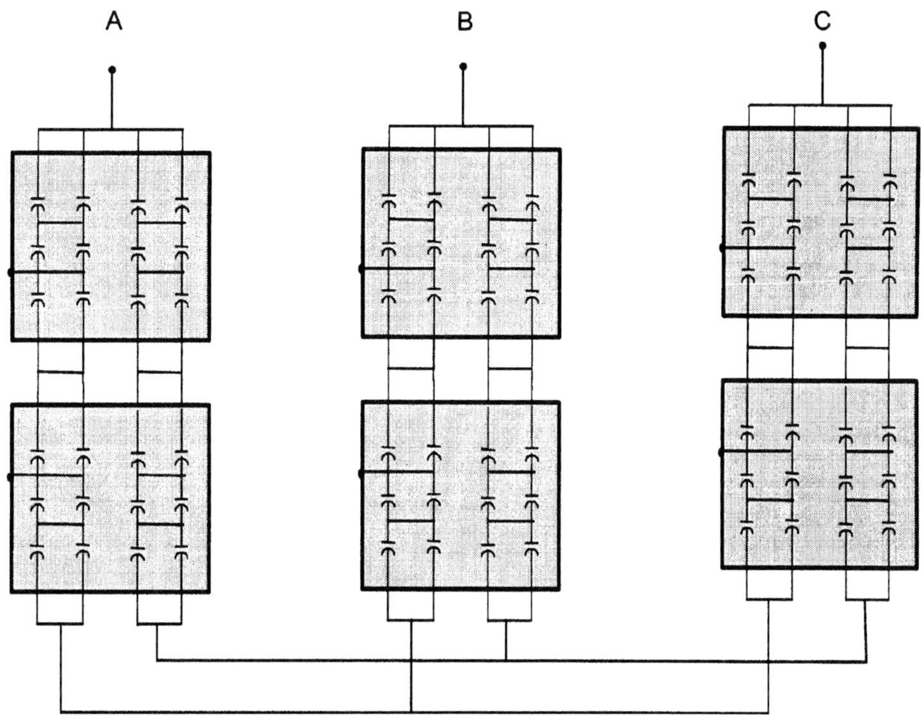


**FIGURE 12-67** Schematic of a typical internally fused shunt capacitor bank.

**FIGURE 12-68** Example of an internally fused shunt capacitor bank.

the terminal-to-terminal voltage. The maximum number of capacitor units that can be put in parallel is determined by the energy capability of the internally fused unit. A typical internally fused capacitor bank is shown in Fig. 12-68.

***Fuseless Shunt Capacitor Banks.*** A fuseless shunt capacitor bank is not simply an externally fused capacitor bank with the fuses removed. The capacitor bank has a different arrangement of capacitor units designed to operate without fuses. A fuseless shunt capacitor bank is made up of units connected in series phase-to-ground or phase-to-neutral. This arrangement of units is called a series string. The unit voltage is selected to achieve the desired voltage rating of the bank. The kvar rating of the unit is chosen to achieve a desired reactive output per each string. Then the bank kvar rating can be increased in these increments by putting multiple series strings of capacitors in parallel. A schematic for a typical fuseless shunt capacitor bank is shown in Fig. 12-69. The fuseless capacitor bank is made up of parallel groups of series units, while the externally fused bank is made up of series groups of parallel units.

Fuseless capacitor bank installations were made possible with the design of the *all-film dielectric* capacitor unit, described earlier in this section. The common failure mode of an element in an *all-film dielectric* unit is for it to fail shorted. When it shorts, the aluminum foil melts and welds together creating a low-resistance point that will continue to carry the current through the capacitor indefinitely.

Typically, kvar ratings of units in fuseless capacitor banks are larger than units used in externally fused capacitor banks. This is because fuseless banks do not have the same design consideration of

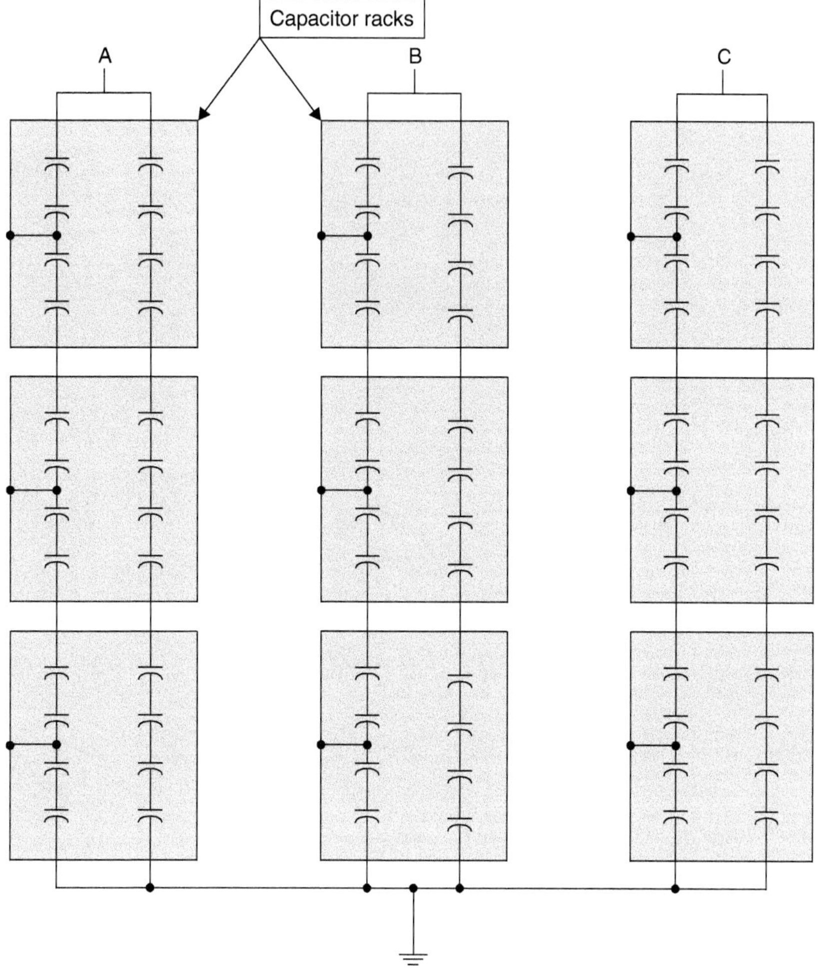


**FIGURE 12-69** Schematic of a typical fuseless shunt capacitor bank. (*Two parallel series strings of 12 units in series.*)

keeping a minimum number of units in parallel for unbalance protection, as discussed previously for externally fused banks. With the units connected in series, if one element in a unit fails the voltage will increase proportionately across the remaining elements in series with it. For example: if a unit has 10 series groups of elements inside each unit and there are five units in a series string, then there will be 50 elements in series phase-to-neutral. If one element fails then the voltage on the remaining elements in series will increase by 1/50th of the total phase-to-neutral voltage. The unbalance protection would usually be set to trip when the overvoltage has exceeded the 110% capability of the elements.

When designing a fuseless bank and choosing the rated voltage and kvar of the units, keep in mind the continuous current capability and the transient current capability of the units. With the series string design of a fuseless bank, there are typically fewer parallel paths for these currents. Also, keep in mind maintainability of the bank. Larger kvar units weigh more and may require special rigging or considerations. A typical fuseless shunt capacitor bank is shown in Fig. 12-70.

**FIGURE 12-70** Example of a fuseless shunt capacitor bank.

***Harmonic Filter Shunt Capacitor Banks.*** Non-linear devices or loads, such as rectifiers and arc furnaces, can create harmonic distortion on the power system. Excessive harmonic voltages or currents can create problems for equipment and the electrical system in general. One common way to eliminate these harmonics is to install a passive harmonic filter near the device or load to shunt some of the harmonic current and thereby reduce the harmonic current flowing into the system. A schematic for a typical harmonic filter capacitor bank is shown in Fig. 12-71. The need for harmonic filters can be on low-, medium-, or high-voltage electrical systems.

For low-voltage systems the dry-type, metalized film capacitors are typically used and an inductor is used to create an LC circuit approximately tuned to the frequency of the harmonic being generated.

For medium- and high-voltage systems the all-film, oil-filled dielectric capacitors are commonly used and, depending on the system voltage, a dry-type or oil-filled reactor is used to create an LC circuit approximately tuned to the frequency of the harmonic being generated. The capacitor portion for medium- and high-voltage harmonic filter banks can be externally fused, internally fused, or fuseless.

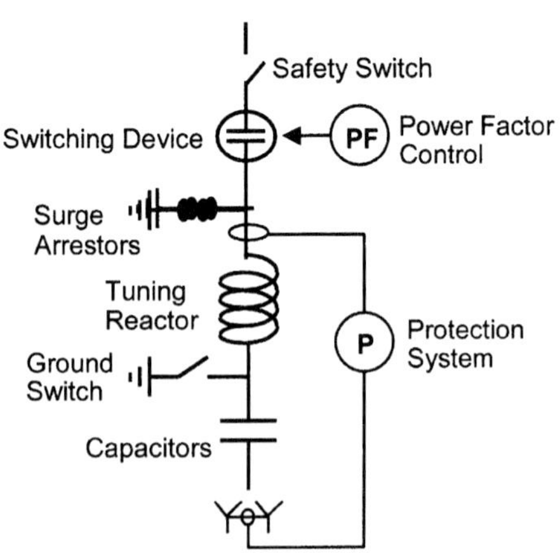


**FIGURE 12-71** Typical schematic of a harmonic filter shunt capacitor bank.

Harmonic filter capacitor banks may look similar to conventional shunt capacitor banks, but there are different criteria to consider when designing a harmonic filter bank. In general, the following should be considered when designing a harmonic filter bank:

1. Reactive power (kvar) requirements
2. Harmonic limitations
3. Normal system conditions, including ambient harmonics
4. Normal harmonic filter conditions
5. Contingency system conditions, including ambient harmonics
6. Contingency harmonic filter conditions

For detailed guidance in designing low-, medium-, and high-voltage harmonic filter capacitor banks refer to IEEE 1531.[3] A typical harmonic filter shunt capacitor bank is shown in Fig. 12-72.

**FIGURE 12-72** Typical high-voltage harmonic filter shunt capacitor bank.

***Protective Relaying for Shunt Capacitor Banks.*** Protection of shunt capacitor banks includes bank and system protection schemes.

Bank protection schemes usually involve some type of unbalance protection to trip the bank for overvoltage due to unit (element) failures, overcurrent protection for faults in the bank other than in units, and phase overcurrent or negative sequence protection for rack-to-rack flashovers in a phase-over-phase bank design.

System protection schemes can involve phase voltage relays for system overvoltage conditions, bus overcurrent, ground-overcurrent, or differential protection for faults in the capacitor installation or major capacitor bank failure, harmonic overload protection for excessive harmonic currents in the system, undervoltage relays for system outages, and breaker failure relays.

The following is a list of basic considerations for protection of shunt capacitor banks:

1. Type of shunt capacitor bank (externally fused, internally fused, or fuseless)
2. Bank connection (grounded-wye, ungrounded-wye, double-wye, etc.)
3. Capacitor bank design and arrangement of units
4. Normal and contingency system conditions
5. Point of connection to the power system
6. Capacitor unit capabilities
7. Impact of geomagnetic induced currents on protection and harmonic loading
8. Overvoltage on remaining capacitor units
9. Installation grounding for grounded banks (single-point or peninsula grounding)

For detailed guidance on the protection of conventional and harmonic filter shunt capacitor banks refer to IEEE C37.99.[6]

## 12.6.4 Series Capacitor Banks

***Fixed (Conventional) Series Capacitors.*** A fixed series capacitor bank is an assembly of different components inserted in series with a power line to reduce the apparent reactance, increase the power transfer capability, and improve system stability. The major components of a fixed series capacitor bank include capacitors, varistors, bypass gaps, bypass switches, discharge current limiting reactors, insulated structures, and protection and control systems. The inserted reactance is primarily established by the reactance of the capacitors. A schematic for a typical fixed series capacitor bank is shown in Fig. 12-73. The capacitors can be of the externally fused, internally fused, or fuseless design. Series capacitor banks are more commonly applied on higher-voltage transmission systems,

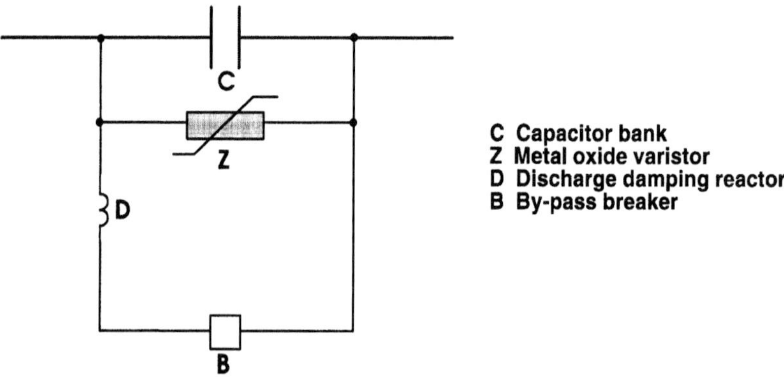


**FIGURE 12-73** Schematic of fixed conventional series capacitor bank.

**FIGURE 12-74** Typical fixed series transmission series capacitor bank.

but they have been applied on distribution systems. See Fig. 12-74 for a typical high-voltage series capacitor bank.

For detailed information on the design, application and protection of fixed transmission series capacitor banks refer to IEEE 824[2] and "*Series Compensation of Power Systems*."[8] For guidance in the functional specification of transmission fixed-series capacitor banks refer to IEEE 1726.[9] For information on distribution series capacitors, the user can refer to an IEEE Transactions paper written by the IEEE Series Capacitor Working Group.[10]

For information on protection of transmission series capacitor banks refer to IEEE C37.116.[11] General information about the protection of series capacitor banks can also be found in the IEEE Special Publication TP-126-0.[12]

***Thyristor-Controlled Series Capacitors.*** A thyristor-controlled series capacitor (TCSC) is a series capacitor bank paralleled with a thyristor-controlled reactor. In the controlled operation mode, periodic gating results in thyristor currents at a frequency higher than the power frequency, which add with the line current. This boosts the capacitor voltage beyond the level that would be obtained by the line current alone. Also, controlled operation provides dynamic var compensation to control line current or through power. A schematic for a TCSC bank is shown in Fig. 12-75. The ability to control TCSC reactance rapidly provides a degree of transient stability and subsynchronous resonance mitigation, which is an advantage a TCSC has over a fixed series capacitor bank. A typical TCSC is shown in Fig. 12-76. For detailed information on thyristor controlled series capacitors refer to IEEE 1534.[13]

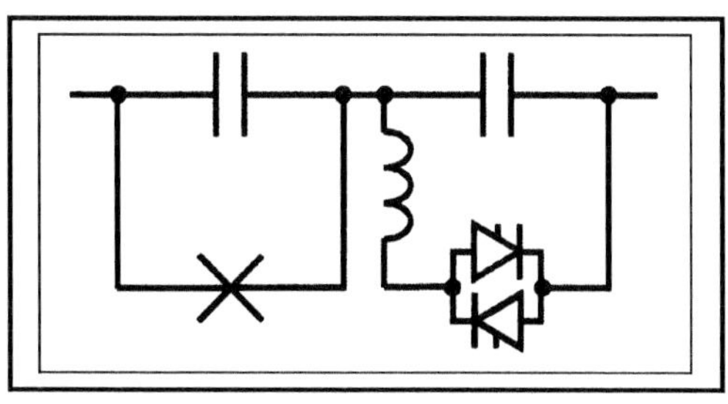

**FIGURE 12-75** Typical schematic of a thyristor controlled series capacitor bank.

**FIGURE 12-76** Typical thyristor controlled series capacitor bank.

### 12.6.5 Capacitor Switching

This section will not go into the details of specifying equipment used to switch and protect capacitor banks. A bibliography of documents providing rating, application, and specification information on switching equipment for shunt and series capacitors is provided at the end of this section for the user's reference.

Many users apply fixed inductors, referred to as transient limiting inductors (TLI), to limit transient switching currents. The IEEE PES Capacitor Subcommittee developed a Technical Report, PES-TR16,[14] to provide guidance on uses and sizing of TLIs, where to locate them, and the drawbacks of installing them. In addition, it provides appendices with information on capacitor bank switching transients and switching transient mitigation methods.

### 12.6.6 References

1. IEEE 18, *Standard for Shunt Power Capacitors.*
2. IEEE 824, *Standard for Series Capacitor Banks in Power Systems.*
3. IEEE 1531, *Guide for Application and Specification of Harmonic Filters.*
4. IEEE 1036, *Guide for the Application of Shunt Power Capacitors.*
5. IEEE C37.48, *Guide for the Application, Operation, and Maintenance of High-Voltage Fuses, Distribution Enclosed Single-Pole Air Switches, Fuse Disconnecting Switches, and Accessories.*
6. IEEE C37.99, *Guide for the Protection of Shunt Capacitor Banks.*
7. IEEE C37.48.1, *Guide for the Operation, Classification, Application, and Coordination of Current-Limiting Fuses with Rated Voltages 1-38 kV.*
8. P. M. Anderson and R. G. Farmer, *Series Compensation of Power Systems*, PBLSH! Inc., 1996, Internet - http://www.pblsh.com/CDsBooks.

9. IEEE 1726, *Guide for the Functional Specification of Fixed-Series Capacitor Banks for Transmission System Applications.*
10. "Considerations for the application of series capacitors to radial power distribution circuits," IEEE Transactions on Power Delivery, vol. 16, Issue 2, April 2001, Page(s):306–318.
11. IEEE C37.116, *Guide for Protective Relay Application to Transmission-Line Series Capacitor Banks.*
12. IEEE Special Publication TP-126-0, *Series Capacitor Bank Protection.*
13. IEEE 1534, *Recommended Practice for Specifying Thyristor Controlled Series Capacitors.*
14. PES-TR16, *Transient Limiting Inductor Applications in Shunt Capacitor Banks*, Nov. 2014.

## 12.6.7 Bibliography

### *Capacitor Literature*

Benitez, J., *Application of Capacitors for Power Factor Correction of Industrial Electrical Distribution Systems*, Industry Applications Society 39th Annual Petroleum and Chemical Industry Conference, 1992, Record of Conference Papers., 28–30 Sept. 1992, Page(s):77–86.

Blooming, T. M. and Carnovale, D. J., *Capacitor Application Issues*, IEEE Transactions on Industry Applications, vol. 44, No 4, July-Aug 2008, Page(s):1013.

Bonner, J. A., Hurst, W. M., Rocamora, R. G., Dudley, R. F., Sharp, M. R., and Twiss, J. A., *Selecting Ratings for Capacitors and Reactors in Applications Involving Multiple Single-Tuned Filters*, IEEE Transactions on Power Delivery, vol. 10, Issue 1, Jan. 1995, Page(s):547–555.

Capacitive Current Switching—State of the Art, Electra No. 155, Aug. 1994, Page(s):33.

Chang, G. W., Chao, J. P., Huang, H. M., Chen, C. I., and Chu, S. Y., *On Tracking the Source Location of Voltage Sag and Utility Shunt Capacitor Switching;* IEEE Transactions on Power Delivery, vol. 23, No. 4, Oct. 2008, Page(s):2124.

Dugan, R. C. and Kennedy, B. W., *Predicting Harmonic Problems Resulting from Customer Capacitor Additions for Demand-Side Management*, IEEE Transactions on Power Systems, vol. 10, Issue 4, Nov. 1995, Page(s):1765–1771.

Guillermin, C., Dujeu, O., and Lupin, J., *Metallized Film Power Capacitors End-of-Life Study Through Monitored Destruction Tests*, IEEE Transactions on Power Delivery, vol. 28, Issue 1, 2013, Pages:368–375.

Hammond, P. W., *A Harmonic Filter Installation to Reduce Voltage Distortion from Static Power Converters*, IEEE Transactions on Industry Applications, vol. 24, Issue 1, Part 1, Jan.-Feb. 1988, Page(s):53–58.

Jowder, F. A. R. A and Ooi, B., *Series Compensation of Radial Power System by a Combination of SSSC and Dielectric Capacitors*, IEEE Transactions on Power Delivery, vol. 20, No. 1, Jan. 2005, Page(s):458–465.

Kalyuzhny, A., *Switching Capacitor Bank Back-to-Back to Underground Cables*, IEEE Transactions on Power Delivery, vol. 28, Issue 2, Apr. 2013, Page(s):1128–1137.

Li, C., *Initial Voltage Change on Capacitor Switching*, IEEE Transactions on Power Delivery, vol. 27, Issue 1, Jan. 2012, Page(s):452–454.

McGranaghan, M. F. and Mueller, D. R., *Designing Harmonic Filters for Adjustable-Speed Drives to Comply with IEEE-519 Harmonic Limits*, IEEE Transactions on Industry Applications, vol. 35, Issue 2, Mar.-Apr. 1999, Page(s):312–318.

Mendis, S. R., Bishop, M. T., McCall, J. C., Hurst, W. M., *Overcurrent Protection of Capacitors Applied on Industrial Distribution Systems*, IEEE Transactions on Industry Applications, vol. 29, Issue 3, May–Jun. 1993, Page(s):541–547.

Miske, S. A.(Working Group paper of the IEEE Capacitor Subcommittee), *Considerations for the Application of Series Capacitors to Radial Power Distribution Circuits*, IEEE Transactions on Power Delivery, vol. 16, Issue 2, Apr. 2001, Page(s):306–318.

Nelson, J. H., Rostron, J. R., and Luke, D. J., *Two-Stage, 161-kV, Fuseless Shunt Capacitor Bank*, in Proc. 2014 IEEE/PES Transmission and Distribution Conference and Exposition, vol. 12, 14-17 Apr. 2014.

Nepveux, F. J., *Protection of Tuned Capacitor Banks*, IEEE Transactions on Industry Applications, vol. 44, No. 4, Jul.-Aug. 2008, Page(s):973.

Peggs, J. F., Powell, P. W., Grebe, T. E., *Innovations for Protection and Control of High Voltage Capacitor Banks on the Virginia Power System*, Proceedings of the 1994 IEEE Power Engineering Society Transmission and Distribution Conference, 10-15 Apr. 1994, Page(s):284–290.

Peeran, S. M. and Cascadden, C. W. P., *Application, Design, and Specification of Harmonic Filters for Variable Frequency Drives*, IEEE Transactions on Industry Applications, vol. 31, Issue 4, Jul.-Aug. 1995, Page(s):841–847.

*Power Capacitor Bibliography*, IEEE/PES Capacitor Subcommittee webpage, http://grouper.ieee.org/groups/td/cap/.

*Power Capacitor for Harmonic Filter Bibliography*, IEEE/PES Capacitor Subcommittee webpage, http://grouper.ieee.org/groups/td/cap/.

*Series Power Capacitor Bibliography*, IEEE/PES Capacitor Subcommittee webpage, http://grouper.ieee.org/groups/td/cap/.

Sevigny, R., Menard, S., Rajotte, C., and McVey, M., *Capacitor Measurement in the Substation Environment: A New Approach*, 2000 IEEE ESMO-2000 IEEE 9th International Conference on Transmission and Distribution Construction, Operation and Live-Line Maintenance Proceedings, 8–12 Oct. 2000, Page(s):299–305.

Thiel, P. H., Harder, J. E., and Taylor, G. E., *Fuseless Capacitor Banks*, IEEE Transactions on Power Delivery, vol. 7, Issue 2, Apr. 1992, Page(s):1009–1015.

**_Standards and Application Guides for Equipment Used to Switch Power Capacitors_**

IEEE 824, *Standard for Series Capacitor Banks in Power Systems.*

IEEE 1247, *Standard for Interrupter Switches for Alternating Current, Rated above 1000 Volts.*

IEEE C37.04, *Standard Rating Structure for AC High-Voltage Circuit Breakers.*

IEEE C37.012, *Application Guide for Capacitance Current Switching for AC High-Voltage Circuit Breakers.*

IEEE C37.016, *Standard for Circuit Switchers.*

IEEE C37.66, *Standard Requirements for Capacitor Switches for AC Systems, 1-38 kV.*

IEC 62271-109, *Alternating Current Series Capacitor By-Pass Switches.*

# 13

# POWER TRANSFORMERS

Pavlos S. Georgilakis

*National Technical University of Athens (NTUA), Athens, Greece; Senior Member, IEEE*

## 13.1 TRANSFORMER THEORY AND PRINCIPLES

### 13.1.1 Introduction

A power transformer is a static device that, by electromagnetic induction, transmits electrical power from one alternating voltage level to another without changing the frequency. It has two or more windings of wire wrapped around a ferromagnetic core. These windings are not electrically connected, but they are magnetically coupled, that is, the only connection between the windings is the magnetic flux present within the core.

One of the transformer windings, the primary winding, is connected to an alternating current (ac) electric power source. The second transformer winding, the secondary winding, supplies electric power to loads. If the transformer has three windings, then the third winding, the tertiary winding, also supplies electric power to loads.

The electrical energy received by the primary winding is first converted into magnetic energy that is reconverted back into a useful electrical energy in the secondary winding (and tertiary winding, if it exists).

A transformer is called a step-up transformer if its secondary winding voltage is higher than its primary winding voltage. In a step-up transformer, the primary winding is also called the low-voltage winding and the secondary winding is also called the high-voltage winding. On the other hand, if the transformer secondary winding voltage is lower than its primary winding voltage, the transformer is called a step-down transformer. In a step-down transformer, the primary winding is also called the high-voltage winding and the secondary winding is also called the low-voltage winding.

The transformer is an electrical machine that allows the transmission and distribution of electrical energy simply and inexpensively, since its efficiency is from 95% to 99%, that is, the transformer operates more efficiently than most electrical devices. This means that the transformer changes one ac voltage level to another while keeping the input power, that is, the power at the first voltage level is practically equal to the output power, that is, the power supplied to the loads. In a step-up transformer, the secondary voltage is higher than the primary voltage, which means that the secondary current has to be lower than the primary current to keep the input power equal to the output power. Because the power losses in the transmission lines are proportional to the square of the current in the transmission lines, raising the transmission voltage and reducing the resulting transmission currents by a factor of 10 with step-up transformers reduces transmission line losses by a factor of 100. That is why step-up transformers are used in power generating stations so that more power can be transmitted efficiently long distances. Step-down transformers are used in power distribution networks, factories, commercial buildings, and residences to reduce the voltage to a level at which the equipment and appliances can operate. Transformers play also a key role in the interconnection of power systems at different voltage levels. Without the transformer, it would simply not be possible to use electric power in many of the ways it is used today. Consequently, transformers occupy prominent positions in the electric power system, being the vital links between power generating stations and points of electric power utilization.

### 13.1.2 Magnetic Materials

In the context of transformer manufacturing, the importance of magnetic materials is twofold. First, through their use it is possible to obtain large magnetic flux densities with relatively low levels of magnetizing force, which plays an important role in the performance of a transformer. Second, since magnetic materials can be used to constrain and direct magnetic fields in well-defined paths, in transformers the magnetic materials are used to maximize the coupling between the windings as well as to lower the excitation current required for transformer operation.

The relationship between the magnetic field density $B$ (Wb/m$^2$ or $T$) and the magnetic field intensity $H$ (A · t/m) for a magnetic material is both nonlinear and multivalued. In general, the characteristics of the material cannot be described analytically. They are commonly presented in graphical form as a set of empirically determined curves based on test samples of the material. The most common curve used to describe a magnetic material is the $B-H$ curve or hysteresis loop. For many engineering applications it is sufficient to describe the material using the dc or normal magnetization curve, which is a curve drawn through the maximum values of $B$ and $H$ at the tips of the hysteresis loops. Such a curve is shown in Fig. 13-1.

### 13.1.3 Induced Voltage

In magnetic circuits with windings, such as Fig. 13-2, when the magnetic field in the core varies with time, an induced voltage $e$ is produced at the terminals, which is calculated by Faraday's law:

$$e = N \cdot \frac{d\phi}{dt} = \frac{d\lambda}{dt} \tag{13-1}$$

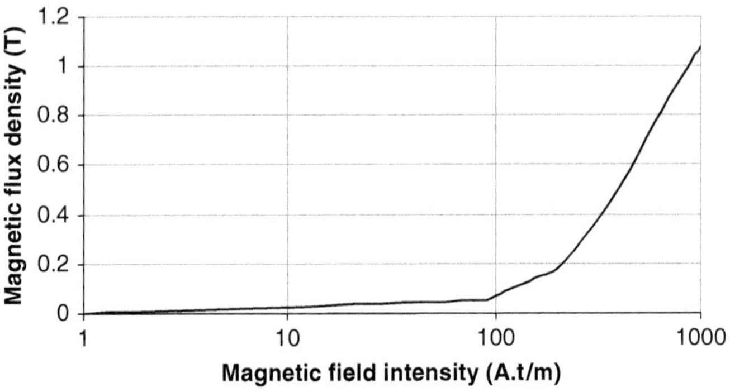


**FIGURE 13-1** Magnetization curve.

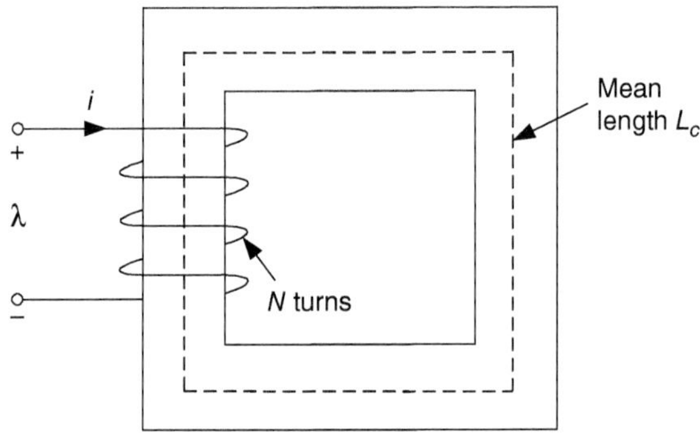


**FIGURE 13-2** Magnetic circuit.

where $N$ is the number of turns, $\phi$ is the time-varying magnetic flux, and $\lambda$ is the flux linkage of the winding (coil) in weber-turns (Wb·t).

Let us suppose that the magnetic flux $\phi$ is sinusoidal:

$$\phi = \Phi_{\text{max}} \cdot \sin(\omega \cdot t) \tag{13-2}$$

where $\Phi_{\text{max}}$ is the maximum flux and $\omega$ (rad/s) is the angular frequency.

By combining Eqs. (13-1) and (13-2), the induced voltage $e$ is computed from the following equation:

$$e = \omega \cdot N \cdot \Phi_{\text{max}} \cdot \cos(\omega \cdot t) \tag{13-3}$$

The effective value $E$ of the induced voltage $e$ is

$$E = \frac{\omega \cdot N \cdot \Phi_{\text{max}}}{\sqrt{2}} = \frac{2 \cdot \pi \cdot f \cdot N \cdot \Phi_{\text{max}}}{\sqrt{2}} \Rightarrow$$

$$E = 4.44 \cdot f \cdot N \cdot \Phi_{\text{max}} \tag{13-4}$$

where $f$ (Hz) is the frequency.

The maximum flux $\Phi_{max}$ is also computed as follows:

$$\Phi_{max} = B_{max} \cdot A_{eff} \tag{13-5}$$

where $B_{max}$ is the maximum magnetic induction (peak magnetic flux density) and $A_{eff}$ is the effective core cross-section area of the magnetic flux.

By combining Eqs. (13-4) and (13-5), the effective value $E$ of the induced voltage $e$ is computed from the following equation:

$$E = 4.44 \cdot f \cdot N \cdot B_{max} \cdot A_{eff} \tag{13-6}$$

### 13.1.4 Equivalent Circuit

The elementary transformer magnetic circuit is shown in Fig. 13-3, where for simplicity the primary and secondary windings are shown on opposite legs of the core. The primary winding has $N_1$ turns and the secondary winding has $N_2$ turns.

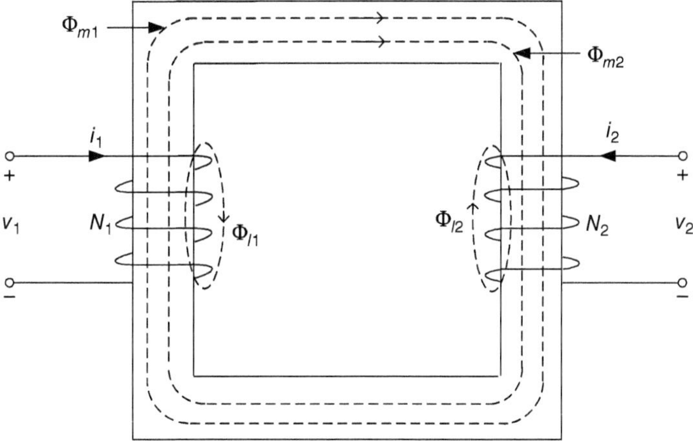


**FIGURE 13-3** Transformer magnetic circuit.

The leakage flux $\Phi_{l1}$ is generated by current $i_1$ flowing in winding 1 (primary) and it links only the turns of winding 1. The leakage flux $\Phi_{l2}$ is produced by current $i_2$ flowing in winding 2 (secondary) and it links only the turns of winding 2.

The magnetizing flux $\Phi_{m1}$ is generated by current $i_1$ flowing in winding 1 and it links all the turns of windings 1 and 2. The magnetizing flux $\Phi_{m2}$ is generated by current $i_2$ flowing in winding 2 and it links all the turns of windings 1 and 2.

The transformer $T$ equivalent circuit is as shown in Fig. 13-4, where $\mathbf{i}_\varphi$ is the phasor of the excitation current (also called no-load current), $\mathbf{i}_C$ is the core-loss current, and $\mathbf{i}_M$ is the magnetizing current. The parameter $\alpha$ denotes the transformer turns ratio or voltage ratio, that is,

$$\alpha = \frac{N_1}{N_2} \tag{13-7}$$

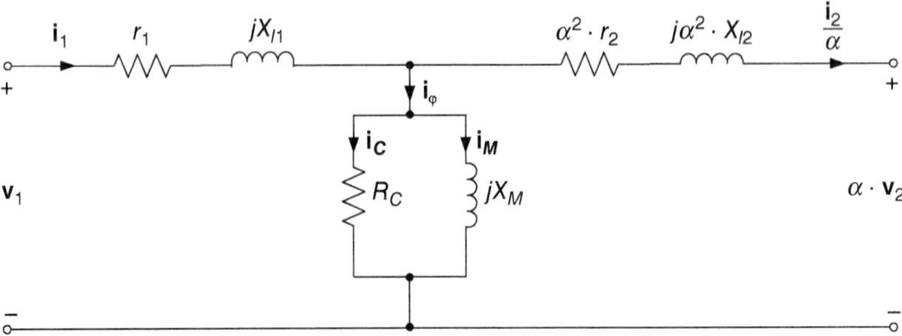


**FIGURE 13-4** Transformer T equivalent circuit with winding 1 being the reference winding.

In Fig. 13-4, $r_1$ and $r_2$ denote the resistance of the primary and secondary winding, respectively, while $X_{l1}$ and $X_{l2}$ denote the reactance of the primary and secondary winding, respectively. The resistance $R_C$ and the reactance $X_M$, shown in Fig. 13-4, model the core excitation effects. The resistances $r_1$ and $r_2$, shown in Fig. 13-4, model the transformer copper losses of the primary and secondary winding, respectively.

Moving the excitation branch (shunt branch) representing the excitation current out from the middle of the $T$ circuit of Fig. 13-4 to either the primary or the secondary windings, as in Figs. 13-5 and 13-6, respectively, often can appreciably reduce the computational effort involved. Error is introduced by neglecting the voltage drop in the primary or the secondary leakage impedance caused by the excitation current, but this error is insignificant in most problems involving power transformers.

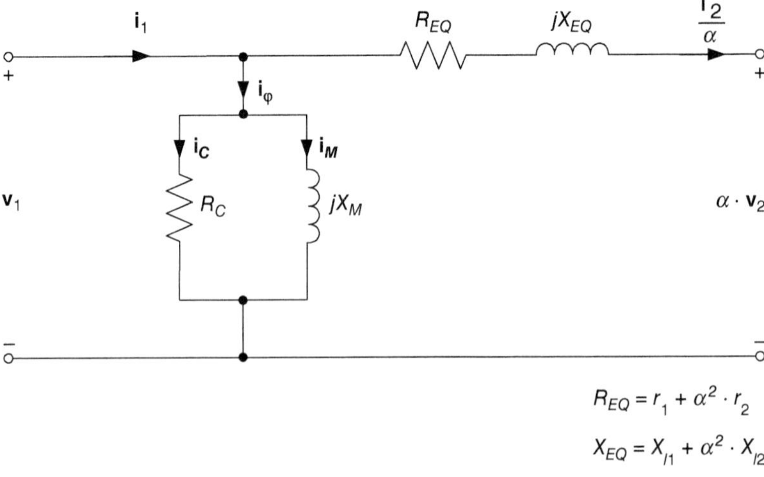


**FIGURE 13-5** Transformer approximate equivalent circuit referred to the primary winding of the transformer.

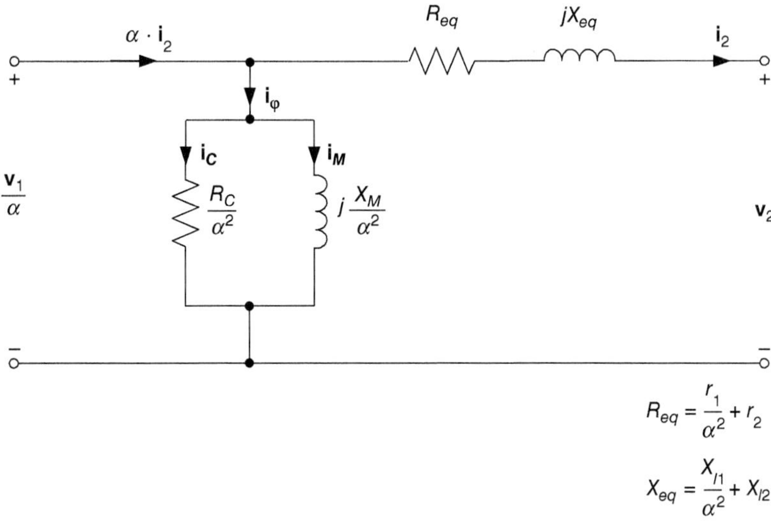


**FIGURE 13-6** Transformer approximate equivalent circuit referred to the secondary winding of the transformer.

### 13.1.5 Derivation of Equivalent Circuit Parameters

In order to determine the parameters of the transformer approximate equivalent circuit of Figs. 13-5 and 13-6, the following two tests are used:

1. *Open-circuit test* measured from secondary side. During this test, the transformer primary side is open-circuited and rated voltage $V_{OC}$ is applied on the secondary winding, while the current $I_{OC}$ and the power $P_{OC}$ on the secondary side are measured.
2. *Short-circuit test* measured from primary side. During this test, the transformer secondary side is short-circuited and appropriate voltage $V_{SC}$ is applied on the primary winding so as to obtain full-load primary current $I_{SC}$, while the input power $P_{SC}$ is measured.

The open-circuit test yields the values for the excitation branch $R_C$ and $X_M$ (referred to the secondary side).

The magnitude of the excitation admittance (referred to the secondary side) is

$$Y_{EX} = \frac{I_{OC}}{V_{OC}} \tag{13-8}$$

and the angle of the excitation admittance is

$$\theta_{EX} = -\cos^{-1}\left(\frac{P_{OC}}{V_{OC} \cdot I_{OC}}\right) \tag{13-9}$$

so the excitation admittance is calculated as follows:

$$\mathbf{Y}_{EX} = Y_{EX}\angle\theta_{EX} = G_C - jB_M \tag{13-10}$$

where

$$R_C = \frac{1}{G_C} \quad \text{and} \quad X_M = \frac{1}{B_M} \tag{13-11}$$

The short-circuit test yields the values for the equivalent series impedance $\mathbf{Z}_{EQ} = R_{EQ} + jX_{EQ}$ (referred to the primary side).

The magnitude of the equivalent series impedance (referred to the primary side) is calculated as follows:

$$Z_{EQ} = \frac{V_{SC}}{I_{SC}} \tag{13-12}$$

and the angle of the equivalent series impedance is computed as follows:

$$\theta_{EQ} = \cos^{-1}\left(\frac{P_{SC}}{V_{SC} \cdot I_{SC}}\right) \tag{13-13}$$

and the equivalent series impedance is calculated as follows:

$$\mathbf{Z}_{EQ} = Z_{EQ}\angle\theta_{EQ} = R_{EQ} + jX_{EQ} \tag{13-14}$$

## 13.2 TRANSFORMER ELECTRICAL CHARACTERISTICS

### 13.2.1 Rated Power

The rated power (MVA) of a transformer is the output that can be delivered at rated secondary voltage and rated frequency without exceeding the specified temperature rise limitations.

The rated power $S_n$ of a three-phase transformer is calculated using the following formula:

$$S_n = \sqrt{3} \cdot U_n \cdot I_n \tag{13-15}$$

where $U_n$ is the rated voltage and $I_n$ is the rated current of the transformer.

Similarly, the rated power $S_n$ of a single-phase transformer is calculated by the following formula:

$$S_n = U_n \cdot I_n \tag{13-16}$$

### 13.2.2 Temperature Rise

The temperature rise is the difference between the temperature of the part under consideration (usually the average winding rise or the hottest-spot winding rise) and the ambient temperature.

The average winding temperature rise of a transformer is the arithmetic difference between the average winding temperature of the hottest winding and the ambient temperature. The top-oil temperature rise is the arithmetic difference between the top-oil temperature (the temperature of the top layer of the insulating liquid in a transformer) and the ambient temperature.

Typical characteristics for oil-immersed transformers:

- The average temperature rise of the winding is 65 K, that is, 65°C above the ambient temperature.
- The top-oil temperature rise is 60 K, that is, 60°C above the ambient temperature.

### 13.2.3 Ambient Temperature

The ambient temperature is the temperature of the air into which the heat of the transformer is dissipated.

The rated power of the transformer is typically calculated for the following conditions:

- Maximum ambient temperature of 40°C
- Average daily ambient temperature of 30°C
- Average annual ambient temperature of 20°C

### 13.2.4 Altitude of Installation

The rated power of a transformer is valid for installation altitudes up to 1000 m. If the transformer is going to be installed at an altitude higher than 1000 m, this should be mentioned in the transformer specification.

### 13.2.5 Impedance Voltage

The impedance voltage or short-circuit impedance or short-circuit voltage is the percentage of the rated primary voltage that has to be applied at the transformer primary winding, when the secondary winding is short-circuited, in order to have the rated current at the primary winding.

The impedance voltage is very important because it represents the transformer's impedance. The higher the short-circuit impedance, the higher the voltage regulation. The lower the short-circuit impedance, the higher the short-circuit current, in case of a short circuit. Based on short-circuit impedance, the following are determined: the voltage regulation due to transformer loading, the distribution of loads in case of parallel operation of transformers, and the short-circuit current.

### 13.2.6 No-Load Losses

Core loss is the power dissipated in the magnetic core subjected to a time-varying magnetizing force. Core loss includes hysteresis and eddy current losses of the core.

No-load losses or excitation losses are incident to the excitation of the transformer. No-load losses include core loss, dielectric loss, conductor loss in the winding due to excitation current, and conductor loss due to circulating current in parallel windings.

Table 13-1 presents the five lists ($E_0$, $D_0$, $C_0$, $B_0$, and $A_0$) of no-load losses for transformers from 50 to 2500 kVA with 24 kV maximum voltage according to EN 50464-1: 2007. It can be seen that for the same rated power, $A_0$ has the lowest and $E_0$ has the highest no-load losses.

### 13.2.7 Load Losses

Load losses are incident to the carrying of a specified load. Load losses include $I^2 \cdot R$ loss in the current carrying parts (windings, leads, busbars, bushings), eddy losses in conductors due to eddy currents, and stray loss induced by leakage flux in the tank, core clamps, or other parts.

**TABLE 13-1** Lists of No-Load Losses According to EN 50464-1: 2007

| | No-load losses, W | | | | | |
|---|---|---|---|---|---|---|
| Rated power, kVA | $E_0$ | $D_0$ | $C_0$ | $B_0$ | $A_0$ | Short-circuit impedance, % |
| 50 | 190 | 145 | 125 | 110 | 90 | 4 |
| 100 | 320 | 260 | 210 | 180 | 145 | 4 |
| 160 | 460 | 375 | 300 | 260 | 210 | 4 |
| 250 | 650 | 530 | 425 | 360 | 300 | 4 |
| 315 | 770 | 630 | 520 | 440 | 360 | 4 |
| 400 | 930 | 750 | 610 | 520 | 430 | 4 |
| 500 | 1100 | 880 | 720 | 610 | 510 | 4 |
| 630 | 1300 | 1030 | 860 | 730 | 600 | 4 |
| 630 | 1200 | 940 | 800 | 680 | 560 | 6 |
| 800 | 1400 | 1150 | 930 | 800 | 650 | 6 |
| 1000 | 1700 | 1400 | 1100 | 940 | 770 | 6 |
| 1250 | 2100 | 1750 | 1350 | 1150 | 950 | 6 |
| 1600 | 2600 | 2200 | 1700 | 1450 | 1200 | 6 |
| 2000 | 3100 | 2700 | 2100 | 1800 | 1450 | 6 |
| 2500 | 3500 | 3200 | 2500 | 2150 | 1750 | 6 |

**TABLE 13-2** Lists of Load Losses According to EN 50464-1: 2007

| Rated power, kVA | Load losses, W | | | | Short-circuit impedance, % |
|---|---|---|---|---|---|
| | $D_k$ | $C_k$ | $B_k$ | $A_k$ | |
| 50 | 1,350 | 1,100 | 875 | 750 | 4 |
| 100 | 2,150 | 1,750 | 1,475 | 1,250 | 4 |
| 160 | 3,100 | 2,350 | 2,000 | 1,700 | 4 |
| 250 | 4,200 | 3,250 | 2,750 | 2,350 | 4 |
| 315 | 5,000 | 3,900 | 3,250 | 2,800 | 4 |
| 400 | 6,000 | 4,600 | 3,850 | 3,250 | 4 |
| 500 | 7,200 | 5,500 | 4,600 | 3,900 | 4 |
| 630 | 8,400 | 6,500 | 5,400 | 4,600 | 4 |
| 630 | 8,700 | 6,750 | 5,600 | 4,800 | 6 |
| 800 | 10,500 | 8,400 | 7,000 | 6,000 | 6 |
| 1000 | 13,000 | 10,500 | 9,000 | 7,600 | 6 |
| 1250 | 16,000 | 13,500 | 11,000 | 9,500 | 6 |
| 1600 | 20,000 | 17,000 | 14,000 | 12,000 | 6 |
| 2000 | 26,000 | 21,000 | 18,000 | 15,000 | 6 |
| 2500 | 32,000 | 26,500 | 22,000 | 18,500 | 6 |

Table 13-2 presents the four lists ($D_k$, $C_k$, $B_k$, and $A_k$) of load losses for transformers from 50 to 2500 kVA with 24 kV maximum voltage according to EN 50464-1: 2007. It can be seen that for the same rated power, $A_k$ has the lowest and $D_k$ has the highest load losses.

For example, it is said that a transformer has a combination of losses of $C_kC_0$, if its load losses belong to list $C_k$ and its no-load losses belong to list $C_0$. More specifically, one transformer with rated power of 1000 kVA and combination of losses $C_kC_0$, has load losses equal to 10,500 W (see Table 13-2) and no-load losses equal to 1100 W (see Table 13-1).

## 13.2.8 Rated Voltages

The rated primary voltage (input voltage) is the voltage at which the transformer is designed to operate. The rated primary voltage determines the basic insulation level (BIL) of the transformer, according to international standards (IEC 60076). The BIL is a basic transformer characteristic, since it indicates the ability of the transformer to withstand the overvoltages that can appear in the network. The calculation of the winding insulation is based on the BIL.

The rated secondary voltage (output voltage) is the voltage at the terminals of the secondary winding at no-load, under rated primary voltage and rated frequency.

## 13.2.9 Vector Group

The vector group determines the phase displacement between the primary and the secondary winding.

The primary or secondary windings can be connected in different ways in order to have a three-phase transformer. These connections are the following:

- D(d): delta connection for primary (secondary) winding
- Y(y): star connection for primary (secondary) winding
- Z(z): zigzag connection for primary (secondary) winding
- N(n): the neutral exists in primary (secondary) winding for connection outside the transformer

### 13.2.10 Frequency

The frequency at which the transformer is designed to operate is 50 or 60 Hz in accordance with the network frequency.

### 13.2.11 Short-Circuit Current

The short-circuit current is composed of the asymmetrical and the symmetrical short-circuit current. The asymmetrical short-circuit current stresses the transformer mechanically, while the symmetrical short-circuit current stresses the transformer thermally.

In case of sudden short-circuits, mechanical forces increase many times and can be dangerous for the transformer. In some cases, steady state short-circuit current reaches as high as 10 to 15 times the transformer rated current. Since the mechanical forces are proportional to the square of the current, they increase to as much as 100 to 225 times the mechanical forces at rated current. Such large mechanical forces can cause appreciable damage to the transformer. Hence, the transformer windings must be designed and constructed to withstand the mechanical forces during short circuits.

### 13.2.12 No-Load Current

The no-load current or excitation current represents the current that the transformer absorbs, when rated voltage is applied to the primary winding and the secondary winding is open-circuited. The no-load current is expressed as a percentage of the value of the rated primary current.

### 13.2.13 Voltage Regulation

The voltage regulation $\Delta V$ of a transformer is defined as the difference in the magnitude of the secondary voltage at no-load $v_{2,nl}$ and its value when loaded $v_2$ divided by $v_2$ with the primary voltage is held constant:

$$\Delta V = \frac{v_{2,nl} - v_2}{v_2} \tag{13-17}$$

Let us consider the simplified transformer equivalent circuit of Fig. 13-7, where the effects of the excitation branch on voltage regulation are ignored, so the equivalent transformer impedance is

$$\mathbf{z}_{eq} = R_{eq} + jX_{eq} \tag{13-18}$$

In Fig. 13-7, it is assumed that $\mathbf{v}_2$ is a reference phasor (zero phase angle), that is, $\mathbf{v}_2 = v_2\angle 0^0$, and the load $\mathbf{z}_L$ has power factor $\cos\theta_L$ lagging, thus $\mathbf{i}_L = i_L\angle -\theta_L$.

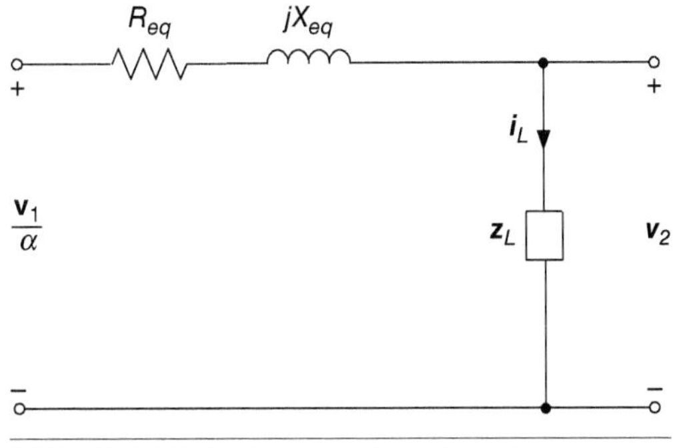


**FIGURE 13-7** Determination of transformer voltage regulation.

Using basic circuit analysis, it can be concluded from Fig. 13-7 that

$$\Delta V = \frac{v_{2,nl} - v_2}{v_2} = \frac{\frac{v_1}{\alpha} - v_2}{v_2} = \frac{v_2 \cdot \left|1 + \frac{\mathbf{z}_{eq}}{\mathbf{z}_L}\right| - v_2}{v_2} \Rightarrow$$

$$\Delta V = \left|1 + \frac{\mathbf{z}_{eq}}{\mathbf{z}_L}\right| - 1 \tag{13-19}$$

where in Eq. (13-19), the symbol $|\mathbf{w}|$ denotes the magnitude of the complex number **w**.

Equation (13-19) can be further simplified and the following expression for the calculation of voltage regulation is obtained:

$$\Delta V = \frac{i_L}{v_2} \cdot (R_{eq} \cdot \cos\theta_L + X_{eq} \cdot \sin\theta_L) + \frac{1}{2} \cdot \left[\frac{i_L}{v_2}\right]^2 \cdot (X_{eq} \cdot \cos\theta_L - R_{eq} \cdot \sin\theta_L)^2 \tag{13-20}$$

Equivalently, the voltage regulation can also be calculated by the following equation:

$$\Delta V = \frac{S}{S_n} \cdot (e_r \cdot \cos\theta_L + e_x \cdot \sin\theta_L) + \frac{1}{2} \cdot \left[\frac{S}{S_n}\right]^2 \cdot (e_r \cdot \sin\theta_L - e_x \cdot \cos\theta_L)^2 \tag{13-21}$$

where

$$e_r = \frac{LL}{S_n} \tag{13-22}$$

and

$$e_x = \sqrt{U_k^2 - e_r^2} \tag{13-23}$$

where $S$ (VA) is the transformer load, $S_n$ (VA) is the transformer rated power, $\cos\theta_L$ is the power factor ($\theta_L$ is positive for lagging load and negative for leading load), $LL$ (W) are the transformer load losses, and $U_k$ (%) is the short-circuit impedance of the transformer.

### 13.2.14 Efficiency

The power efficiency of any electrical machine is defined as the ratio of the useful power output, $P_{\text{out}}$ (W), to the total power input, $P_{\text{in}}$ (W). The efficiency can be defined by simultaneously measuring the output and the input power. However, this measurement is expensive and difficult, especially for large machines. Moreover, in the case of high-efficiency machines (e.g., transformer), higher precision can be achieved, if the efficiency is expressed through the losses. Consequently, the transformer efficiency $n$ is calculated using the following formula:

$$n = \frac{P_{\text{out}}}{P_{\text{in}}} = \frac{P_{\text{out}}}{P_{\text{out}} + \text{losses}} = \frac{S \cdot \cos\theta_L}{S \cdot \cos\theta_L + \text{losses}} \tag{13-24}$$

where $S$ is the transformer load (VA), *losses* are the transformer losses (W) and $\cos\theta_L$ is the power factor.

The transformer efficiency is increased with a decrease of transformer losses.

The transformer losses are divided into no-load losses and load losses. The no-load losses are constant, while the load losses are proportional to transformer load. Consequently, the efficiency of transformer is calculated using the following formula:

$$n = \frac{S \cdot \cos\theta_L}{S \cdot \cos\theta_L + NLL + LL \cdot \left(\frac{S}{S_n}\right)^2} \tag{13-25}$$

where $NLL$ are the no-load losses (W), $LL$ are the load losses (W), and $S_n$ is the rated power of the transformer (VA).

If $L$ is the per-unit load:

$$L = \frac{S}{S_n} \tag{13-26}$$

then by substituting Eq. (13-26) into Eq. (13-25), we obtain the following expression for transformer efficiency:

$$n = \frac{L \cdot S_n \cdot \cos\theta_L}{L \cdot S_n \cdot \cos\theta_L + NLL + LL \cdot L^2} \tag{13-27}$$

Taking $L$ as an independent variable, the value of $L$ that maximizes efficiency is calculated as follows:

$$\frac{dn}{dL} = 0 \Rightarrow \frac{d}{dL}\left[\frac{L \cdot S_n \cdot \cos\theta_L}{L \cdot S_n \cdot \cos\theta_L + NLL + LL \cdot L^2}\right] = 0 \Rightarrow$$

$$\frac{(L \cdot S_n \cdot \cos\theta_L + NLL + LL \cdot L^2) \cdot S_n \cdot \cos\theta_L - L \cdot S_n \cdot \cos\theta_L \cdot (S_n \cdot \cos\theta_L + 2 \cdot L \cdot LL)}{(L \cdot S_n \cdot \cos\theta_L + NLL + LL \cdot L^2)^2} = 0 \Rightarrow$$

$$L \cdot S_n^2 \cdot \cos^2\theta_L + NLL \cdot S_n \cdot \cos\theta_L + L^2 \cdot LL \cdot S_n \cdot \cos\theta_L - L \cdot S_n^2 \cdot \cos^2\theta_L$$
$$-2 \cdot L^2 \cdot LL \cdot S_n \cdot \cos\theta_L = 0 \Rightarrow$$

$$NLL \cdot S_n \cdot \cos\theta_L - L^2 \cdot LL \cdot S_n \cdot \cos\theta_L = 0 \Rightarrow NLL - L^2 \cdot LL = 0 \Rightarrow$$

$$L_{\text{opt}} = \sqrt{\frac{NLL}{LL}} \tag{13-28}$$

As can be seen from Eq. (13-28), the optimum per-unit load, $L_{\text{opt}}$, that is, the per-unit load that maximizes transformer efficiency is independent of the power factor of the load.

Substituting Eq. (13-28) into Eq. (13-27), we obtain the following expression for the maximum efficiency:

$$n_{\max} = \frac{\sqrt{\frac{NLL}{LL}} \cdot S_n \cdot \cos\theta_L}{\sqrt{\frac{NLL}{LL}} \cdot S_n \cdot \cos\theta_L + NLL + LL \cdot \frac{NLL}{LL}} \Rightarrow$$

$$n_{\max} = \frac{\sqrt{NLL} \cdot S_n \cdot \cos\theta_L}{\sqrt{NLL} \cdot S_n \cdot \cos\theta_L + 2 \cdot NLL \cdot \sqrt{LL}} \tag{13-29}$$

*Example 13-1.* Consider a three-phase transformer with rated power 630 kVA, rated primary voltage 20 kV, rated secondary voltage 0.4 kV, no-load losses 1200 W, load losses 9300 W, and short-circuit impedance 6%. Draw the transformer efficiency curves versus per unit load for power factor 1.0 and 0.8.

*Solution.* The maximum efficiency corresponds to the following per-unit load:

$$L_{\text{opt}} = \sqrt{\frac{NLL}{LL}} = \sqrt{\frac{1\,200}{9\,300}} = 0.36$$

At maximum efficiency, the transformer load is

$$S_{\text{opt}} = L_{\text{opt}} \cdot S_n = 0.36 \cdot 630\,000 \Rightarrow S_{\text{opt}} = 226{,}800 \text{ VA}$$

For power factor 1.0, the maximum efficiency is

$$n_{\max} = \frac{S_{\text{opt}} \cdot \cos\theta_L}{S_{\text{opt}} \cdot \cos\theta_L + NLL + LL \cdot L_{\text{opt}}^2} \Rightarrow$$

$$n_{\max} = \frac{226\,800 \cdot 1.0}{226\,800 \cdot 1.0 + 1\,200 + 9\,300 \cdot 0.36^2} \Rightarrow n_{\max} = 98.95\%$$

For power factor 0.8, the maximum efficiency is

$$n_{\max} = \frac{226\,800 \cdot 0.8}{226\,800 \cdot 0.8 + 1\,200 + 9\,300 \cdot 0.36^2} \Rightarrow n_{\max} = 98.69\%$$

Similarly, the efficiency is computed for per-unit load varied from 0.1 to 1.2 and the obtained efficiency curves versus per-unit load are shown in Fig. 13-8.

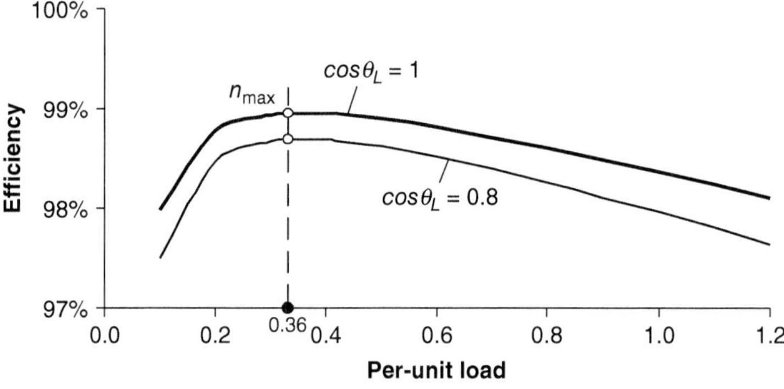


**FIGURE 13-8** Efficiency curves versus per-unit load for Example 13-1.

## 13.3 TRANSFORMER TYPES

The transformers are classified into various categories, according to their

- Use
- Cooling method
- Insulating medium
- Core construction

### 13.3.1 Classification According to Transformer Use

Transformers are classified according to their use into the following categories:

1. *Distribution Transformers.* They are used in distribution networks in order to transmit energy from the medium voltage (MV) network to the low voltage (LV) network of the consumers. Their rated power usually ranges from 50 to 1600 kVA.

2. *Power Transformers.* They are used in high-power generating stations for voltage step-up and in transmission substations for voltage step-up or step-down. Usually they are of power greater than 2 MVA.
3. *Autotransformers.* They are used for voltage transformation within relatively small limits, for connection of electric energy systems of various voltages, and for starting alternating current motors.
4. *HVDC Transformers.* They are key components in HVDC transmission systems. The HVDC transformers react as coupling elements between the connected ac grids and high-power rectifiers, and are necessary for adapting the voltage. They insulate the rectifier itself from the ac grid and generate a phase shift.
5. *Test Transformers.* They are used for the execution of performance tests with high or ultrahigh voltage.
6. *Special Power Transformers.* They are used for special applications, for example, traction systems, furnaces, and welding.
7. *Instrument Transformers.* They are used for the accurate measurement of voltage or current.
8. *Telecommunication Transformers.* They are used in telecommunication applications aiming at the reliable reproduction of a signal over a wide range of frequency and voltage.

### 13.3.2 Classification According to Transformer Cooling Method

The identification of oil-immersed transformers according to the cooling method is expressed by a four-letter code. The first letter expresses the internal cooling medium in contact with the windings. The second letter identifies the circulation mechanism for the internal cooling medium. The third letter identifies the external cooling medium. The fourth letter identifies the circulation mechanism for external cooling medium. For example, if the internal cooling medium is mineral oil, which is circulated by natural flow, and the external cooling medium is air, which is circulated by natural convection, then this cooling method is coded as ONAN (oil natural air natural).

In power transformers, various cooling methods are used, including oil circulation by pumps, or forced air circulation by fans, or both of the above. As a result, the following cooling methods exist:

1. ONAF: oil natural air forced
2. OFAN: oil forced air natural
3. OFAF: oil forced air forced
4. OFWF: oil forced water forced

### 13.3.3 Classification According to Transformer Insulating Medium

Transformers are classified according to their insulating medium into the following categories:

1. *Oil-Immersed Transformers.* The insulating medium is mineral oil or synthetic (silicon) oil.
2. *Dry Type Transformers.* The cooling is implemented with natural air circulation and the windings are usually insulated with materials of H or F class. The materials of H class are designed to operate, in normal conditions, at temperatures up to 180°C and the materials of F class at temperatures up to 155°C.
3. *Resin Type Transformers.* The resin type transformer is a dry type transformer insulated with epoxy resin cast under vacuum.

### 13.3.4 Classification According to Transformer Core Construction

Construction of the magnetic circuit of three-phase transformers can be implemented, alternatively, as follows:

1. *With Three Legs (Vertical Limbs).* The magnetic flux of one leg must flow through the other two legs and the flux also flows through the windings of the other phases, that is, the transformer has no free return of the flux.
2. *With Five Legs (Vertical Limbs).* Free return of the flux through the external legs.

There are two different technologies for stacking the electrical steel sheets of the magnetic material of the core:

1. *Stack Core.* The layers of the sheets of the magnetic material are placed one over the other and vertical and horizontal layers are overlapped.
2. *Wound Core.* The magnetic circuit is of shell type and the sheets are wound.

Two different materials are used for core construction:

1. *Silicon Steel Sheet.* The silicon steel sheet that is used for core construction is an alloy consisting of 97% iron and 3% silicon. This material is crystalline. The silicon steel sheets have thickness from 0.18 to 0.5 mm. There are also silicon steel sheets for operation at high magnetic induction (Hi-B).
2. *Amorphous Metal Sheet.* The amorphous metal sheet that is used for core construction is an alloy consisting of 92% iron, 5% silicon, and 3% boron. This material is not crystalline. It has 70% lower no-load loss than silicon steel. The thickness of the amorphous metal sheet is 0.025 mm, that is, it is about 10 times thinner than the typical thickness of silicon steel sheet.

## 13.4 TRANSFORMER CONNECTIONS

Three-phase transformer connections can be made either by using a three-phase transformer or three single-phase transformers. The primary and secondary windings of any three-phase transformer can be independently connected in either a wye (Y) or a delta (Δ). This gives a total of four possible connections for a three-phase transformer:

1. Y–Y: The primary and secondary windings are connected in a Y.
2. Y–Δ: The primary is connected in a Y and the secondary is connected in a Δ.
3. Δ–Y: The primary is connected in a Δ and the secondary is connected in a Y.
4. Δ–Δ: The primary and secondary windings are connected in a Δ.

In a Y-connected winding the current flowing through each phase winding is equal to the line current. As a result, the Y connection, also called star connection, is the typical choice of connection for the highest voltages (lowest currents) and when the neutral is intended for loading.

In a Δ-connected winding the current flowing through each phase winding is equal to the line current divided by $\sqrt{3}$. As a result, the Δ-connected winding is advantageous in large power transformers when the current is high and the voltage is relatively low, like the low voltage winding of generator step-up transformers. Moreover, the Δ-connected winding requires $\sqrt{3}$ times as many turns as a Y-connected winding for the same voltage.

When Y connection is used in the one winding of the transformer, the other winding should preferably be Δ-connected, especially when the neutral of the Y-connected winding is intended to be loaded. The Δ-connected winding provides ampere-turn balance for the zero sequence current flowing through the neutral and each phase of the Y-winding, which results in a reasonable zero sequence impedance.

Another connection is the zigzag connection, also called interconnected Y connection and Z connection. The sole purpose of zigzag connection is to establish a neutral point for grounding purposes.

### 13.4.1 Y-Y Connection

The Y-Y connection of a three-phase step-down transformer is shown in Fig. 13-9. The high-voltage (primary winding) terminals are designated with $H_1$, $H_2$, and $H_3$. The low-voltage (secondary winding) terminals are designated with $X_1$, $X_2$, and $X_3$. The Y-Y connection (or Y-connected autotransformer) may be used to interconnect two delta systems and provide suitable neutrals for grounding both of them. A Y-connected autotransformer may be used to interconnect two Y systems that already have neutral grounds, for reasons of economy.

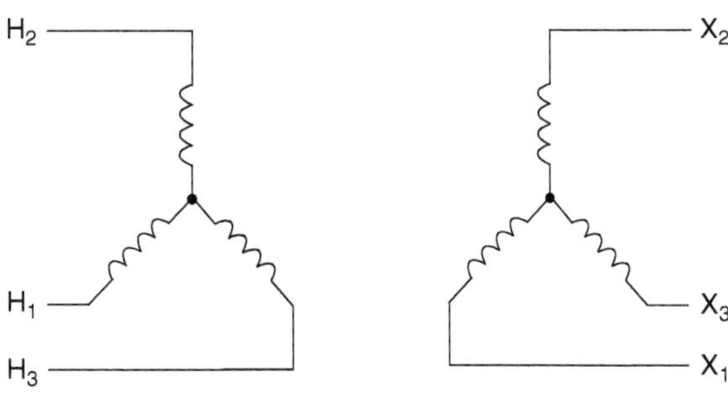


**FIGURE 13-9** Y-Y connection.

The Y-Y connection has two very serious disadvantages:

1. The third-harmonic voltages can be large; they can be even larger than the fundamental voltage itself.
2. If the loads of the transformer are unbalanced, then the phase voltages of the transformer can become very unbalanced.

The above two problems can be solved using one of the following two methods:

1. Solidly ground the neutrals of the transformer, especially the neutral of the primary winding. The grounding of the neutral permits the additive third-harmonic components to cause a current flow in the neutral (instead of building up large third-harmonic voltages). Moreover, the neutral provides a return path for the current imbalances of the load.
2. Add a tertiary (third) winding connected in Δ to the transformer bank. In this case, the third-harmonic components of voltage in the Δ will add up, creating a circulating current flow in the winding, which suppresses the third-harmonic components of voltage.

### 13.4.2 Y-Δ Connection

The Y-Δ connection of a three-phase transformer is shown in Fig. 13-10. The Y-Δ connection has no problem with third-harmonic components in its voltages, because they are consumed in a circulating current on the Δ side. The Y-Δ connection is also more stable to unbalanced loads, because the Δ partially redistributes any imbalance that may occur.

### 13.4.3 Δ-Y Connection

The Δ-Y connection of a three-phase transformer is shown in Fig. 13-11. The Δ-Y connection has the same advantages as the Y-Δ connection.

The Δ-Y step-up and the Y-Δ step-down connections are without question the best for high-voltage transmission systems. They are economical in cost, and provide a stable neutral whereby the high-voltage system may be directly grounded or grounded through resistance of such value as to damp the system critically and prevent the possibility of oscillation.

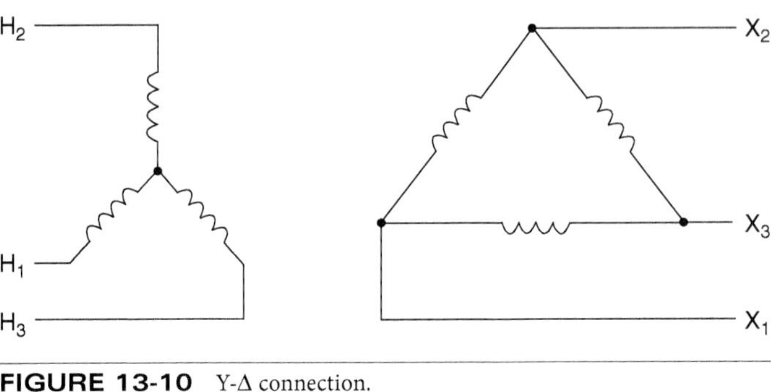


**FIGURE 13-10** Y-Δ connection.

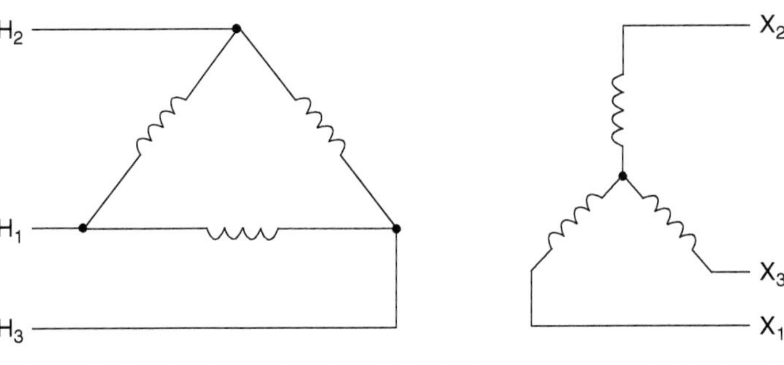


**FIGURE 13-11** Δ-Y connection.

## 13.4.4 Δ-Δ Connection

The Δ-Δ connection of a three-phase transformer is shown in Fig. 13-12. The Δ-Δ connection may be used as step-up transformer for moderate voltages.

The Δ-Δ connection has no problems with unbalanced loads or harmonics. Another advantage of the Δ-Δ connection, if composed of three single-phase transformers, is that one single-phase transformer can be removed (in case of damage), and the resulting connection that is called open Δ connection operates at $1/\sqrt{3}$ or 57.7% of the rating of the three-phase transformer.

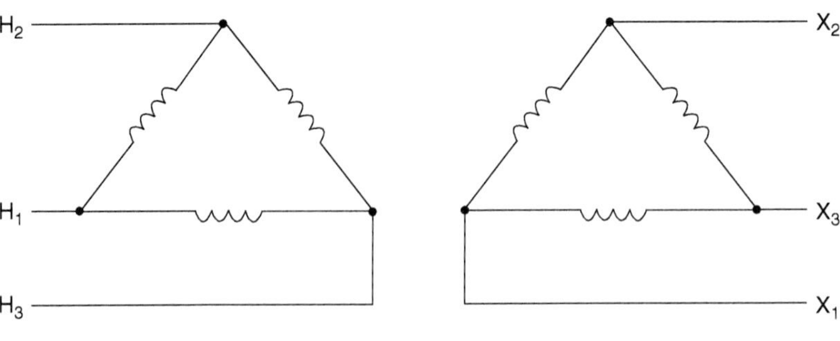


**FIGURE 13-12** Δ-Δ connection.

The main disadvantage of Δ-Δ connection is that the neutral is not available. The insulation is more costly because the Δ-Δ connection has higher ground voltages during transient voltages or system fault. The Δ-connection insulation costs increase with increasing voltage. The Δ-Δ connection is typically limited to a maximum system voltage of 345 kV.

### 13.4.5 Z-Connection

The zigzag connection is also called Z connection as well as interconnected Y connection. The Z connection of the high-voltage (primary) winding of a three-phase transformer is shown in Fig. 13-13, where $H_1$, $H_2$, and $H_3$ are the high voltage terminals, while $H_0$ is the neutral. The sole purpose of Z connection is to establish a neutral point for grounding purposes.

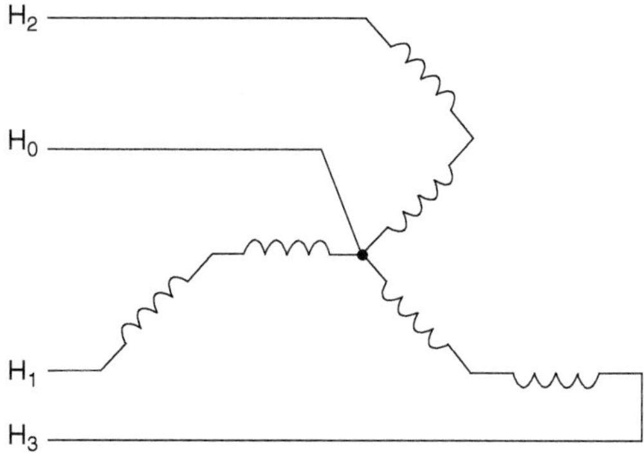

**FIGURE 13-13** Z connection.

The Z connection may be used with either a Δ-connected winding or a Y-connected winding for step-up or step-down operation. For example, if a Δ winding is used as a primary winding and a Z winding is used as a secondary winding, the Δ-Z connection is created. The Δ-Z connection provides the same advantages as the Δ-Y connection, like third harmonic suppression and ground current isolation. An additional advantage with the Δ-Z connection is that there is no phase angle displacement between the primary and the secondary windings. Therefore, the Δ-Z connection can be used in the same manner as Y-Y and Δ-Δ connections without introducing any phase shifts in the windings.

## 13.5 STEP-VOLTAGE REGULATORS

An important objective of any electrical system is to provide consumers with a supply voltage compatible with their utilization equipment. Each electrical device is designed to operate at a certain rated voltage for optimum efficiency and maximum lifetime. In case of power distribution feeders without distributed energy resources, the voltage magnitudes decrease along the feeder, so the consumers at the end of the feeder have the lowest voltage.

Power distribution systems must be designed in such a way that the voltage magnitudes always remain within a specified range as required by standards. This is achieved through the use of transformer on-load tap changers, switched and fixed capacitors, and single-phase and three-phase voltage regulators. Section 13.6 is devoted to tap changers.

The step-voltage regulators are designed, manufactured, and tested according to the IEEE Standard C57.15. A step-voltage regulator is defined as a regulating autotransformer in which the voltage of the regulated circuit is controlled in steps by means of taps and without interrupting the load.

The choice of using a three-phase or three single-phase voltage regulators depends on the cost and the need for unbalanced operation. A three-phase voltage regulator has no capability to correct the voltage imbalances caused by unequal loading. If feeders have similar load characteristics and voltage regulation requirements and if the incoming supply voltage is balanced, then the distribution side of the substation bus can be controlled by a three-phase voltage regulator. Consequently, if the load is mainly three-phase or consistently balanced, a three-phase voltage regulator may be the better choice. However, most rural power distribution systems contain only a small percentage of balanced three-phase loads. Therefore, single-phase step-voltage regulators dominate the power distribution market. They are used in substations having up to 30 MVA as well as in power distribution feeders and laterals.

## 13.5.1 Regulation Method

Currently, the most common single-phase step-voltage regulator has 32 steps, which provide ±10% voltage regulation with 0.625% voltage regulation per step. This means that the following 33 voltage regulation values can be obtained:

$$-10.000\% \quad -9.375\% \quad \ldots \quad -0.625\% \quad 0.000\% \quad 0.625\% \quad \ldots \quad 9.375\% \quad 10.000\%$$

A typical single-phase step-voltage regulator involves a shunt winding, a series winding, and a bridging reactor or preventive autotransformer. The series winding has eight tapped sections of 2×0.625% or 1.25% voltage each, giving a total of 10% voltage regulation.

A preventive autotransformer (or bridging reactor) is required to maintain continuity during a tap change and to provide impedance for limiting the amount of current to be interrupted by the tap changer. The preventive autotransformer is used as a voltage divider while bridging across a section of the tapped series winding to assist in providing the 0.625% voltage regulation steps.

Step-voltage regulators being tapped autotransformers have part of one winding common to both the primary and the secondary circuits associated with that winding. In other words, the shunt (primary) winding is both electrically and magnetically connected to the series (secondary) winding. The shunt (exciting) winding is common to both primary and secondary; the series winding is connected in series with load (output) current.

The process of moving from one voltage-regulator tap to the adjacent voltage-regulator tap consists of closing the circuit at one tap before opening the circuit at the other tap. The movable tap-changer contacts move through stationary taps alternating in eight bridging and eight nonbridging (symmetrical) positions. Figure 13-14*a* shows the two movable tap-changer contacts on a symmetrical position, with the center tap of the reactor at the same voltage. This is the case at tap position neutral and all evenly numbered tap positions.

Figure 13-14*b* shows an asymmetrical position that is realized when one tap connection is open before transferring the load to the adjacent tap. At this position, all of the load current flows through one-half of the reactor, magnetizing the reactor, and the reactance voltage is introduced into the circuit for about 25–35 ms during the tap change.

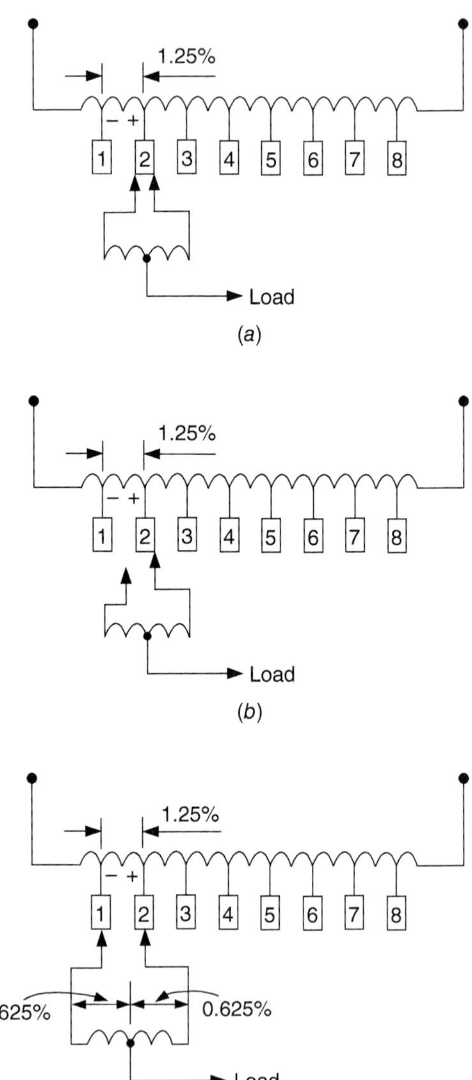


**FIGURE 13-14** (*a*) Two movable contacts on the same stationary contact on a symmetrical position. (*b*) One movable contact on stationary contact on an asymmetrical position. (*c*) Two movable contacts on adjacent stationary contacts.

Figure 13-14*c* shows the movable contacts in a bridging position; voltage change is one-half of the 1.25% tap voltage of the series winding because of its center tap and movable contacts located on adjacent stationary contacts. This is the case at all oddly numbered tap positions. At the bridging position, a circulating current is established through the preventive autotransformer or a center tapped bridging reactor, and load voltage is seen to be the average voltage of the taps being bridged. To minimize the arc duration, a quick-break mechanism accelerates the moving contacts. The circulating current ($I_C$), resulting from the two contacts being at different positions (reactor energized with 1.25% tap voltage), is limited by the reactive impedance of this circuit.

## 13.5.2 Regulator Technical Characteristics

The kVA rating of a step-voltage regulator is computed as follows:

$$S_r = V_r \cdot I_r \cdot VR \tag{13-30}$$

where $V_r$ is its rated voltage, $I_r$ is its rated load current, and $VR$ is its per-unit range of voltage regulation. For example, the kVA rating of a step-voltage regulator with 7.62 kV rated voltage, 100 A rated current, and with ±10% voltage regulation ($VR = 0.1$), is computed as follows:

$$S_r = 7.62 \times 100 \times 0.1 \Rightarrow S_r = 76.2 \text{ kVA}$$

The purpose of the step-voltage regulator is the activation of the monitored tap changer to change taps to maintain the correct regulator output voltage. The regulator control settings are the values the regulator user has selected as control parameters for the regulated voltage. Tables are provided with the regulators to use in computing the appropriate settings. The basic regulator control settings are the following:

1. *Set Voltage.* The set voltage depends on the regulator rating and the system voltage on which it is installed.
2. *Band Width.* To avoid a hunting condition of the regulator, a band width is set to define the limits (total range) of acceptable voltage about the set voltage. For example, if the set voltage is 120 V and the band width is 2 V, then the regulator will be in-band if the output is in the range from 119 to 121 V.
3. *Time Delay.* The time delay is the time duration outside of the prescribed band required before tap-changer actuation. A general recommendation is a minimum time delay of 15 s. This time length covers the vast majority of temporary voltage swings due to equipment starting and cold-load pickup.
4. *Line-Drop Resistive and Reactive Compensation.* The use of line-drop compensation will cause the regulator to hold the voltage-level setting at a point remote from the regulator, rather than at the regulator location. The classic illustration of the application of line-drop compensation involves a load center some miles from the substation. It is required to hold a given voltage at the load center. Given that the line is inductive in nature, this implies holding a higher voltage at the substation, the incremental voltage increase being a function of the line impedance (resistive and reactive) and the line current. Thus, the line-drop resistive and reactive compensation settings reflect the resistive and reactive line-drop to be anticipated between the regulator and the load when the system is carrying peak load current.

## 13.5.3 Regulator Control Functions

Highly sophisticated, microprocessor-based digital control circuitry is included with the voltage regulator. All regulator controls comprise three major parts:

1. A voltage-sensing device, which monitors the output voltage of the regulator and sends a signal to the motor drive circuitry.
2. An amplifying or switching section, which delays and transmits the signal to the tap changer motor.
3. A motor drive circuit which responds to the signal by closing relay contacts or actuating electronic switches that cause the motor to operate the tap changer drive mechanism.

### 13.5.4 Bypassing Voltage Regulators

One of the advantages of the application of step-voltage regulators versus on-load tap-changing transformers is the ability of voltage regulators to be bypassed in service such that load interruption is not necessary during installation procedures. Taking step-voltage regulators in or out of service involves opening or closing a bypass switch between the source and the load bushings. It is critical that proper switching instructions are followed to avoid the extremely high circulating current that will appear in the series winding if by mistake bypassing occurs at other than the neutral tap position. Consequently, it is imperative the step-voltage regulator is placed in the neutral position before it is installed in or removed from the line. More specifically, to place the step-voltage regulator in service, first the source switch is closed. The regulator is checked out by running the tap changer in the raise and lower directions. The step-voltage regulator is returned to the neutral position, and the load switch is closed. After the load switch is closed, the bypass switch may be opened. To take the regulator out of service, the procedure is reversed. First the step-voltage regulator is run to the neutral position, the bypass switch is closed, and then finally the source switch is opened.

### 13.5.5 Three-Phase Voltage Regulators

In three-phase installations, the choice of using a three-phase voltage regulator or three single-phase voltage regulators depends on the cost and the need for unbalanced operation. If the load is mainly three-phase or consistently balanced, a three-phase step-voltage regulator may be the better choice. However, most rural power distribution systems contain only a small percentage of balanced three-phase loads. Consequently, three-phase step-voltage regulators are less common in the power distribution industry than the single-phase step-voltage regulators.

The design of three-phase regulators is similar to single-phase regulators. More specifically, three single-phase regulators are located in the same tank with their tap changers ganged together and being monitored and operated by one control. The three-phase regulator has only one of its phases monitored to provide a voltage and current supply to the control. Consequently, all three phases are regulated based on the monitoring of only one phase.

In three-phase installations, it is a very common application to use banks of three single-phase step-voltage regulators. Configuring the regulators in Y, closed Δ, and open Δ can be acceptable alternatives, depending on system conditions. In deciding on the proper and safe connection for a given application, three basic phenomena should be considered: (1) third harmonics, (2) system line surges, and (3) line faults.

### 13.5.6 Voltage Regulator Developments

Recently, innovation associated with step-voltage regulators has concentrated on the electronic control. The very nature of digital controls now offered routinely with new step-voltage regulators facilitates the mathematical manipulation of the measured line voltage and current into system parameters of interest to the user. Consequently, controls are available that display voltage, current, power factor, voltage and current harmonics, active and reactive power, various time-integrated demand quantities, and other conditions of interest.

The control typically operates with one set of configuration settings that are programmed or changed through the human-machine interface or one of the available communications channels using interface software. Alternative configurations may be available allowing for the control to be programmed with an additional set of configuration settings.

An additional feature available on many controls is the ability to communicate this information to a remote location so the regulator control becomes the sensory apparatus of the supervisory control and data acquisition system. Common communication interfaces available with the control are Fiber Optic, Ethernet, RS232, and RS485.

Most step-voltage regulators are installed in power distribution networks with a well-defined power flow from source to load. However, some power distribution networks have interconnections or loops in which the direction of power flow through the regulator may change. For optimum power distribution system performance, step-voltage regulators, installed on such networks, have the capability of detecting this reverse power flow and then sensing and controlling the load-side voltage of the regulator, regardless of the direction of power flow.

In distribution systems, the increasing levels of distributed generation pose new challenges to power distribution utilities in their use of step-voltage regulators. Traditionally, power distribution networks have been used purely to transmit energy from the transmission network down to lower voltage levels. A distributed generator delivering power directly to the distribution network can reverse the normal direction of power flow in a regulator. In the electronic control of the step-voltage regulator, options are available for handling different types of scenarios that give rise to reverse power flow conditions.

## 13.6 TAP CHANGERS

Power systems must be designed in such a way that the voltage magnitudes always remain within a specified range as required by standards. As already discussed in Sec. 13.5, one method of voltage control is through the use of tap changers in transformers. More specifically, the transformer is equipped with taps in the winding. The voltage control is achieved by changing the turn ratio. The tap changer changes the voltage ratio of a transformer by adding turns to or subtracting turns from either the primary or the secondary winding. Consequently, the transformer is equipped with a regulating or tap winding that is connected to the tap changer. The execution of tap-change operation can be done with the transformer either deenergized or energized and as a result, two types of tap changers are available:

1. Deenergized tap changer (DETC). The DETC, which is also called off-circuit tap changer, is designed to be operated only when the transformer is deenergized, that is, only when the transformer is disconnected on both the primary and the secondary winding. The DETCs are available as hand-driven (they are operated with a hand wheel or a hand crank) or motor-driven devices. The DETC is mainly used when the variation in the nominal operating voltage (which requires a tap changer operation) is not expected frequently. Figure 13-15 shows the basic connections of a DETC.

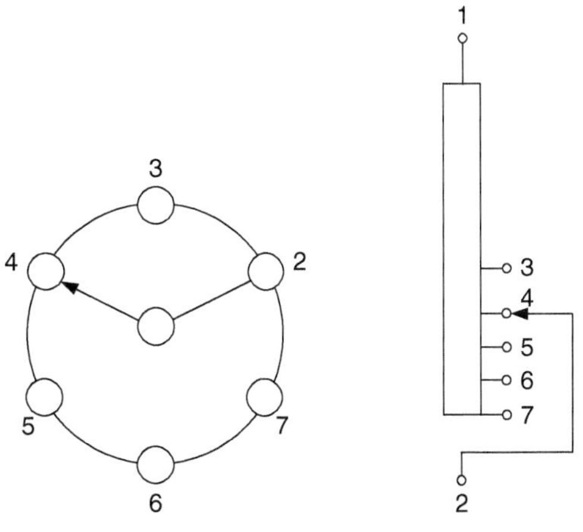

**FIGURE 13-15** Basic connections of a de-energized tap changer.

2. Load tap changer (LTC). The LTC, which is also called on-load tap changer, is a device that connects different taps of the tapped windings of transformers without

interrupting the load. The LTC must be capable of switching from one tap position to another without interrupting the flow of the current to the load as well as without creating a short circuit between any two taps of the transformer winding. The LTC is mainly used when the variation in the nominal operating voltage (which requires a tap changer operation) is expected frequently.

The selection between the DETC and the LTC depends on the frequency with which it is required to change taps as well as the size and importance of transformer. The rest of the subsection is devoted to LTC.

### 13.6.1 Types of Load Tap Changers

The LTC must be capable of switching from one tap position to another without interrupting the flow of the current to the load as well as without creating a short circuit between any two taps of the transformer winding. According to the switching principle, two types of LTC have been used:

1. *Resistor-Type LTC.* In this case, resistors are used to handle the transition from one tap to another. This type of transition requires the resistors to be capable of withstanding the full load current plus the circulating current during the transition. To reduce the energy absorption requirement for the resistors, the duration of the complete transition has to be minimized. That is why resistor-type LTCs normally have very fast transients.
2. *Reactor-Type LTC.* In this case, reactors are used to handle the transition from one tap to another. These reactors do not have to dissipate as much energy as the resistors. The reactors mainly use reactive energy, which does not produce any heat. Consequently, the reactor-type LTCs can be designed to withstand the full load plus the circulating current for long periods, even continuously.

According to the location of the tap changer, there are two types of LTC:

1. *In-Tank.* The majority of resistor-type LTCs and the DETCs are installed inside the transformer tank.
2. *Separate Compartment.* The reactor-type LTCs are installed in a separate compartment, which is normally welded to the transformer tank.

According to the switching technology, two types of LTC have been used:

1. *Oil-Type LTC.* In this case, the transformer oil is used as an insulating, cooling, lubricating, and arc-quenching medium. The LTC is immersed in transformer oil, and switching contacts make and break current under oil. This conventional LTC technology has reached a very high level and is capable of meeting most requirements of the transformer manufacturer. This applies to the complete voltage and power fields of today, which will probably remain unchanged in the foreseeable future.
2. *Vacuum-Type LTC.* The switching principles of vacuum-type LTCs differ from those of conventional oil-type LTCs. A simple duplication of the switching contacts of a conventional LTC with vacuum interrupters would lead to a solution that is unnecessarily more expensive and greater in volume. Therefore, special designs with special switching principles were created for the vacuum-type LTCs. The vacuum switching technology entirely eliminates the need for an online oil filtration system. Moreover, the vacuum switching technology offers reduced downtimes with increased availability of the transformer, simplified maintenance logistics, and substantial savings for the end user. Today's resistor-type and reactor-type LTCs are based more and more on vacuum interrupters.

### 13.6.2 Applications of Load Tap Changers

The control of transformer ratio under load is a desirable means of regulating the voltage of high-voltage feeders and of primary networks. It may be used for the control of the bus voltage in large distribution substations. It finds a wide field of application in controlling the ratio on step-up transformers operating from power stations whose bus voltage must be varied to suit local distribution.

In industrial applications, it is used for the control of current in a variety of furnace operations and electrolytic processes. It also furnishes a convenient means for voltage regulation of concentrated industrial loads.

A lot of LTC equipment is installed at points of interconnection between systems or between power stations, in order to control the interchange of reactive current, or, in other words, to control the power factor in the tie line. This reactive current may be highly undesirable, especially as it may add to the burden on a fully loaded generating system. It can be increased, eliminated, or reversed by inserting a suitable small ratio of transformation between the systems. It can be varied in amount and in direction of flow to suit varying system conditions, if this ratio is variable and under the control of a station operator. Inserting such a ratio of transformation in a tie line by means of tap-changing equipment is equivalent in its effect on the flow of reactive current to raising or lowering the voltage on one of the systems. Current can be exchanged at any power factor from zero lag to zero lead, without interfering with the voltage maintained on either system.

### 13.6.3 Phase-Shifting Transformers

A phase-shifting transformer (PST) has the ability to control the phase angle of the voltage. Recently, there is a steady increase in PSTs that are used to control the power flow on transmission lines in meshed power systems.

Transformers used to derive phase-angle control do not differ materially, either mechanically or electrically, from those used for in-phase control. In general, phase-angle control is obtained by interconnecting the phases, that is, by deriving a voltage from one phase and inserting it in another.

The simple arrangement shown in Fig. 13-16 illustrates a single-core Δ-connected autotransformer in which the series windings are so interconnected as to introduce into the line a quadrature voltage. One phase only is printed in solid lines so as to show more clearly how the quadrature voltage is obtained. The terminals of the common winding are connected to the midpoints of the series winding in order that the in-phase voltage ratio between the primary lines *ABC* and secondary lines *XYZ* is unity for all values of phase angle introduced between them.

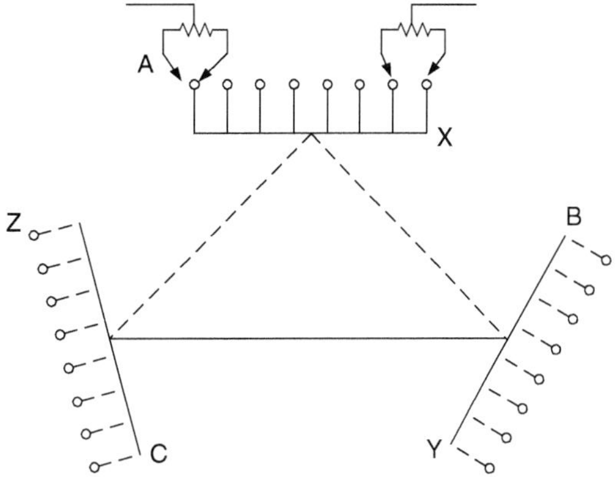

**FIGURE 13-16** Phase-shifting regulating transformer: single-core Δ-connected common winding for low-voltage systems.

As large high-voltage systems have become extensively interconnected, a need has developed to control the transfer of real power between systems by means of phase-angle-regulating transformers. The most commonly used circuit for this purpose is the two-core, four-winding arrangement shown in Fig. 13-17. The high-voltage common winding is Y connected, with reduced insulation at the neutral for economy of design, and a series transformer is employed so that low-voltage-switching equipment may be used.

The use of phase-shifting transformers has the following benefits:

- Reduction of overall system losses by elimination of circulating currents
- Improvement of power system capability by appropriate load management
- Improvement of power factor of the power system
- Control of power flow to meet contractual requirements

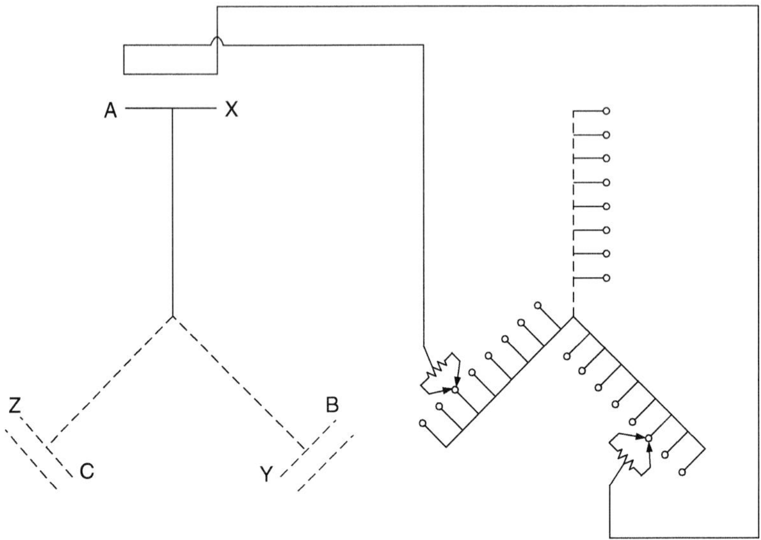


**FIGURE 13-17** Phase-shifting regulating transformer: two-core Y-connected common winding for high-voltage systems.

## 13.7 TRANSFORMER DESIGN

The objective of transformer design optimization (TDO) is to design the transformer so as to minimize the transformer manufacturing cost, that is, the sum of materials cost plus labor cost, subject to constraints imposed by international standards and transformer user specification.

The aim of transformer design is to obtain the dimensions of all parts of the transformer in order to supply these data to the manufacturer. The transformer design should be carried out based on the specification given, using available materials economically in order to achieve low cost, low weight, small size and good operating performance.

The transformer design is worked out using various methods based on accumulated experience realized in different formulas, equations, tables, and charts. Transformer design methods vary among transformer manufacturers.

While designing a transformer, much emphasis should be placed on lowering its cost by saving materials and reducing to a minimum labor-consuming operations in its manufacture. The design should be satisfactory with respect to dielectric strength and mechanical endurance, and windings must withstand dynamic and thermal stresses in the event of short circuit.

In order to meet the above requirements, the transformer designer should be familiar with the prices of basic materials used in the transformer. He should also be familiar with the amount of labor consumed in the production of transformer parts and assemblies.

### 13.7.1 Problem Formulation

The TDO problem is formulated as follows: minimize an objective function subject to several constraints.

Among the various objective functions of the TDO problem, the most commonly used objective functions are

1. The minimization of transformer total owning cost. This is mainly used when designing transformers for electric utilities, since these users usually evaluate the cost of losses when purchasing transformers.
2. The minimization of transformer manufacturing cost. The transformer manufacturing cost includes the materials cost and the labor cost. This objective function is mainly used when designing transformers for industrial and commercial users, since these users usually do not evaluate the cost of losses when purchasing transformers.

The constraints of the TDO problem are related to transformer operation, manufacturing capabilities, and transformer user special needs.

### 13.7.2 Objective Function

In the bibliography of transformer design, several objective functions are optimized: (1) minimization of total owning cost, (2) minimization of manufacturing cost, (3) minimization of main materials cost, (4) minimization of active part cost, (5) minimization of active part mass, and (6) maximization of transformer rated power.

### 13.7.3 Constraints

The most commonly used constraints of the TDO problem are the following: (1) induced voltage constraint, (2) turns ratio constraint, (3) no-load loss constraint, (4) load loss constraint, (5) total loss constraint, (6) impedance voltage constraint, (7) magnetic induction constraint, (8) heat transfer constraint, (9) temperature rise constraint, (10) efficiency constraint, (11) no-load current constraint, (12) voltage regulation constraint, (13) impulse voltage constraints, and (14) tank dimensions constraints.

### 13.7.4 Solution Methods

The general TDO problem is a complex constrained mixed-integer nonlinear programming problem. The TDO problem is further complicated by the fact that the objective function is discontinuous.

Several mathematical programming and heuristic optimization methods have been proposed for the solution of the TDO problem. Mathematical programming methods for the solution of TDO include geometric programming and mixed-integer nonlinear programming. Heuristic methods for the solution of TDO include multiple design method, genetic algorithm, and particle swarm optimization.

### 13.7.5 Multiple Design Method

Transformer manufacturers usually use the multiple design method to solve the TDO problem. The multiple design method is a heuristic technique that assigns many alternative values to the design variables (unknowns of the TDO problem) so as to generate a large number of alternative designs and finally to select the design that satisfies all the problem constraints with the optimum value of the objective function, for example, with the minimum transformer manufacturing cost.

The optimum design is implemented through the following steps:

1. Initially the input variables are entered in a computer program. Many different values of the design variables are given, so many candidate solutions are considered. For example, the magnetic circuit is typically designed using the specific no-load loss curve of the transformer, which has the form of Fig. 13-18.

2. A computer program determines which candidate solutions are acceptable and which are rejected (they violate one or more of the constraints).
3. The acceptable solutions are sorted according to their manufacturing cost. The optimum transformer corresponds to the least-cost solution.

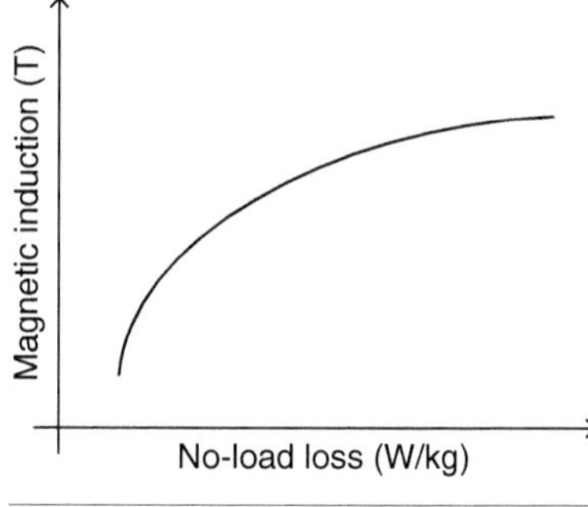


**FIGURE 13-18** Transformer specific no-load loss curve.

If all the candidate solutions are rejected, then the computer file of nonacceptable solutions should be studied and the reasons for rejection understood.

Generally, the following cases may appear: (1) necessity to decrease or increase no-load loss, (2) necessity to decrease or increase load loss, and (3) necessity to decrease or increase impedance voltage.

In fact, the multiple design method is a repetitive transformer design process. For example, Table 13-3 shows how changing core and conductor design can reduce no-load and load losses but also affects the cost of the transformer, when we try to further improve the optimum design.

**TABLE 13-3** Loss Reduction Alternatives

| | Impact of loss reduction technique on | | |
|---|---|---|---|
| How to decrease no-load loss and load loss | No-load loss | Load loss | Cost |
| *To decrease no-load loss:* | | | |
| A. Use lower-loss core material | Lower | No change | Higher |
| B. Decrease flux density by | | | |
| 1. Increasing core cross-section area (CSA) | Lower | Higher | Higher |
| 2. Decreasing volts per turn | Lower | Higher | Higher |
| C. Decrease flux path length by decreasing conductor CSA | Lower | Higher | Lower |
| *To decrease load loss:* | | | |
| A. Decrease current density by increasing conductor CSA | Higher | Lower | Higher |
| B. Decrease current path length by | | | |
| 1. Decreasing core CSA | Higher | Lower | Lower |
| 2. Increasing volts per turn | Higher | Lower | Lower |

## 13.8 TRANSFORMER INSULATION

Transformer windings and leads have to operate at high voltages relative to the core, tank, and structural elements. Moreover, different windings and even parts of the same winding operate at different voltages. Consequently, some type of insulation must be provided between these various parts to avoid voltage breakdown or corona discharges. Although normal operating voltages are quite high (10-500 kV), the transformer must be designed to withstand even higher voltages that can occur when lightning strikes the power system or when power is suddenly switched on or off in some part of the electrical system.

Insulation systems in power transformers consist of a fluid (either liquid or gas) together with solid materials. Currently, the standard insulating media for power transformers is mineral oil and cellulose.

Within the core and coil assembly, insulation can be divided into two fundamental groups: major insulation and minor insulation. Major insulation separates the high-voltage and low-voltage windings, and the windings to core. Minor insulation may be used between the parts of individual coils or windings depending on construction. Finally, turn insulation is applied to each strand of conductor and/or groups of strands forming a single turn.

The life of a power transformer is limited by the life of its insulation system. When the insulation ultimately reaches its end of life, the transformer will fail due to a dielectric or mechanical failure. Fortunately, the mineral oil can be drained from the transformer and reclaimed or replaced. However, the solid insulation cannot be replaced and consequently is the limiting factor in the life of a transformer. From the moment a transformer is placed in service, both the liquid and solid insulation begin a slow but irreversible degradation process.

With the steady increase in power transmission voltages, the voltage ratings of power transformers have also increased making insulation cost a significant part of transformer cost. Moreover, insulation space impacts the cost of core and winding, as well as the quantity of mineral oil in the transformer. Verification of insulation design is essential for enhancing reliability and optimizing transformer cost. Robust insulation design methods, use of appropriate insulating materials, and controlled manufacturing processes ensure quality and reliability of power transformers.

## 13.8.1 Oil-Insulated Transformers

Low-cost, high-dielectric strength, excellent heat transfer characteristics, and ability to recover after dielectric overstress make mineral oil the most widely used transformer insulating material. The oil is reinforced with solid insulation in various ways. The major insulation usually includes barriers of wood-based paperboard (pressboard), the barriers usually alternating with oil spaces. Because the dielectric constant of the mineral oil is 2.2 and that of the solid is approximately 4.0, the dielectric stress in the oil ends up being higher than that of the pressboard, and the design of the structure is usually limited by the stress in the oil.

Voltage-time curves represent the relationship between voltage and time to breakdown. The higher the voltage the lower is the time that causes a breakdown. Figure 13-19 shows a typical impulse voltage-time curve of oil-gap in a transformer.

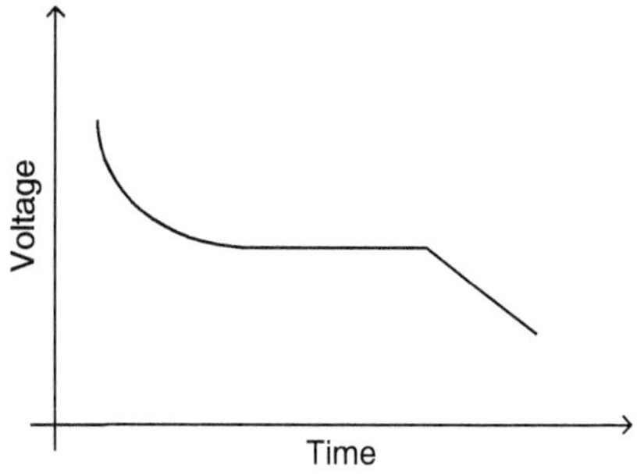

**FIGURE 13-19** Typical impulse voltage-time curve of oil-gap in a transformer.

The insulation on the conductors of the windings may be enamel or wrapped paper that is either wood-based or nylon-based. The use of insulation directly on the conductor actually inhibits the formation of potentially harmful streamers in the oil, thereby increasing the strength of the structure. Again, the limit of dielectric strength is usually that of mineral oil.

Heavy paper wrapping is also usually used on the leads coming from the winding. In this case, the insulation serves to reduce the stress in the oil, by moving the interface from the surface of the conductor (where the stress is high) to a distance away from the conductor (where the stress is considerably lower). Again, the stress in the oil determines the amount of paper required, and the thermal considerations establish the minimum size of the conductor for the necessary insulation.

## 13.8.2 Alternative Fluids of Liquid-Insulated Transformers

In addition to mineral oil, there are currently other alternative dielectric fluids used in liquid-insulated transformers. These alternative fluids are mainly used because of their enhanced thermal and flammability properties. Mineral oil has a fire point of 165°C. This value has proved to be too low for transformers installed inside or near buildings. Large transformers for indoor use must either be of the dry type (containing no liquid) or use a less flammable liquid, for example, silicon fluid (fire point of 350°C) or natural ester (fire point of 360°C).

Currently, silicon-based or fluorinated hydrocarbons are used, where the additional expense of a fire-resistant liquid offsets additional building cost for a transformer vault. Combustion-resistant vegetable-oil-based dielectric coolants and natural esters are also becoming increasingly common as alternatives to naphthenic mineral oil. Esters are nontoxic to aquatic life, are readily biodegradable, and have a lower volatility and a higher flash point than mineral oil.

### 13.8.3 Gas-Insulated Transformers

Gas insulation systems include fluorogases, nitrogen, and air. Fluorogases have better dielectric strength than the nitrogen or air. Although their heat transfer characteristics are poorer than oil, they are better than nitrogen or air because of their higher density. Both dielectric strength and heat transfer capability increase with pressure. In fact, the dielectric strength at 3 atm gage pressure (where some fluorocarbon-insulated transformers operate) can approach that of oil. The gas insulation is reinforced with solid insulation used in the form of barriers, layer or disk insulation, turn insulation, and lead insulation similar to oil-immersed transformers. It is usually economical to operate fluorogas-insulated transformers at higher temperatures than oil-insulated transformers. Suitable solid insulating materials include glass, asbestos, mica, high-temperature resins, and ceramics.

Nitrogen and air insulated transformers are generally limited to 34.5 kV and lower operating voltages. Air-insulated transformers in clean locations are frequently ventilated to the atmosphere. In contaminated atmospheres a sealed construction is required, and nitrogen is generally used at approximately 1 atm and some elevated operating temperatures.

### 13.8.4 Design of Operating Structures

Three factors must be considered in the evaluation of the dielectric capability of an insulation structure: (1) the voltage distribution must be calculated between different parts of the winding, (2) the dielectric stresses are then calculated knowing the voltages and the geometry, and (3) finally the actual stresses can be compared with breakdown or design stresses to determine the design margin.

Voltage distributions are linear when the flux in the core is established. This occurs during all power frequency test and operating conditions and to a great extent under switching impulse conditions. Switching impulse waves have front times in the order of tens to hundreds of microseconds and tails in excess of 1000 μs. These conditions tend to stress the major insulation and not inside of the winding. For shorter-duration impulses, such as full wave, chopped wave, or front wave, the voltage does not divide linearly within the winding and must be determined by calculation or low-voltage measurement. The initial distribution is determined by the capacitive network of the winding. For disk and helical windings, the capacitance to ground is usually much greater than the series capacitance through the winding. Under impulse conditions, most of the capacitive current flows through the capacitance to ground near the end of the winding, creating a large voltage drop across the line-end portion of the coil.

The capacitance network for shell form and layer-wound core form results in a more uniform initial distribution because they use electrostatic shields on both terminals of the coil to increase the ratio between the series and to ground capacitances. Static shields are commonly used in disk windings to prevent excessive concentrations of voltages on the line-end turns by increasing the effective series capacitance within the coil, especially in the line-end sections. Interleaving turns and introducing floating metal shields are two other techniques that are commonly used to increase the series capacitance of the coil.

Following the initial period, electrical oscillations occur within the windings. These oscillations impose greater stresses from the middle parts of the windings to ground for long-duration waves than for short-duration waves. Very fast impulses, such as steep chopped waves, impose the greatest stresses between turns and coil portions. Note that switching impulse transient voltages are of two types: asperiodic and oscillatory. Unlike the asperiodic waves discussed earlier, the oscillatory waves can excite winding natural frequencies and produce stresses of concern in the internal winding insulation.

Transformer windings that have low natural frequencies are the most vulnerable because internal damping is more effective at high frequencies.

Dielectric stresses existing within the insulation structure are determined using direct calculation (for basic geometries), analog modeling, or most recently, sophisticated finite-element computer programs.

Allowable stresses are determined from experience, model tests, or published data. For liquid-insulated transformers, insulation strength is greatly affected by contamination and moisture. The relatively porous and hygroscopic paper–based insulation must be carefully dried and vacuum impregnated with oil to remove moisture and gas to obtain the required high dielectric strength and to resist deterioration at operating temperatures. Gas pockets or bubbles in the insulation are particularly destructive to the insulation because the gas (usually air) not only has a low dielectric constant (about 1.0), which means that it will be stressed more highly than the other insulation, but also air has a low dielectric strength.

High-voltage dc stresses may be imposed on certain transformers used in terminal equipment for dc transmission lines. Direct-current voltage applied to a composite insulation structure divides between individual components in proportion to the resistivities of the material. In general, the resistivity of an insulating material is not a constant but varies over a range of 100:1 or more, depending on temperature, dryness, contamination, and stress. Insulation design of high-voltage dc transformers in particular requires extreme care.

## 13.9 TRANSFORMER COOLING

Core loss, copper loss in windings, stray loss in windings, and stray loss due to high-current field are mainly responsible for heat generation inside the transformer. The heat generated due to all these losses must be dissipated without allowing the core, winding, and structural parts to reach a temperature that will deteriorate the insulation. If the insulation is subjected to temperatures higher than the permitted value for a long time, the insulation gets aged and it loses its insulation properties, severely affecting transformer life.

In oil-immersed transformers, the mineral oil provides a medium for both insulation and cooling. Heat from core, windings, and structural parts is dissipated by means of oil circulation. Finally, the heat is transmitted either to atmospheric air or water. Figure 13-20 shows the oil natural air natural (ONAN) method of cooling in a power transformer. More specifically, the heat developed in the active part (core and windings) is passed onto the surrounding oil through convection (heat transfer mechanism). The oil temperature increases and the oil specific gravity reduces, so the oil flows upward and then into the coolers. The oil heat is dissipated along the colder surfaces of the coolers, the specific gravity of the oil increases, and the oil flows downward and it enters the tank from the inlet at the bottom. In ONAN cooling, the heat dissipation from the oil to the atmospheric air is done by natural means.

In small rating transformers, the tank surface area may be able to dissipate heat directly to the atmosphere. Bigger rating transformers usually require much larger dissipating surface in the forms of radiators/tubes. If the number of radiators is small, they are preferably mounted directly on the tank in order to reduce the overall dimensions. When the number of radiators is large, they are mounted on a separate structure called radiator bank.

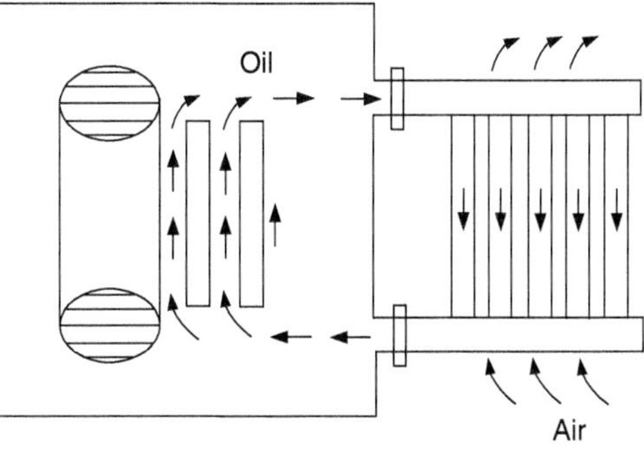


**FIGURE 13-20** Oil natural air natural (ONAN) method of cooling.

Fan-cooled transformers use external fans to improve heat dissipation from the radiators, and sometimes internal pumps to circulate the oil through the radiators (and sometimes also through cooling ducts in the core and coils). Forced-oil-cooled transformers use external oil-to-air heat exchangers requiring both air fans and oil pumps for all operating conditions. Water-cooled transformers usually have the oil withdrawn from the transformers at the top of the tank, pumped through an external cooler, and returned to the bottom of the tank.

Accurate estimation of temperatures on all surfaces is very important in the design of transformers to determine the operating magnetic induction in the core and the current densities in the windings. Moreover, it helps check the adequacy of cooling arrangements provided for the core and windings.

## 13.9.1 Heat Transfer Mechanisms

Three different heat transfer mechanisms take place in a transformer: conduction, convection, and radiation. In oil-immersed transformers, convection is the most important and conduction is the least important. Rigorous mathematical treatment of these heat transfer mechanisms is very difficult and consequently transformer designers mostly rely on empirical formulas.

The temperature drop ($\Delta\theta$ in °C) across the insulation, which is due to the conduction heat transfer mechanism, can be computed by the basic thermal law:

$$\Delta\theta = \frac{Q \cdot t_i}{k \cdot A} \tag{13-31}$$

where $Q$ is the heat flow (power loss) in W, $t_i$ is the insulation thickness in m, $k$ is the thermal conductivity in W/(m · °C), and $A$ is the cross-section area in $m^2$.

The heat transfer ($P_R$ in W), which is due to the radiation heat transfer mechanism, is computed by the Stephan-Boltzmann law:

$$P_R = \eta \cdot E \cdot A_R \cdot \left(T_s^4 - T_a^4\right) \tag{13-32}$$

where $\eta = 5.67 \times 10^{-8}$ W/($m^2 \cdot K^4$) is the Stephan-Boltzmann constant, $E$ is the surface emissivity, $A_R$ is the surface area for radiation in $m^2$, $T_s$ is the average temperature of radiating surface in K, and $T_a$ is the ambient air temperature in K. The surface emissivity depends on the surface finish and the type of paint applied on the surface.

The heat dissipation from the core and windings occurs mainly due to convection. When a heated surface (core, windings) is immersed in a fluid (oil), heat flows from the surface to the cooling medium (oil). The heat flow ($Q_c$ in W), which is due to the convection heat transfer mechanism, is computed by the formula:

$$Q_c = h \cdot A_h \cdot (T_h - T_f) \tag{13-33}$$

where $h$ is the heat transfer coefficient in W/($m^2 \cdot$ °C), $A_h$ is the area of the heated surface for convection in $m^2$, $T_h$ is the temperature of the heated surface in °C, and $T_f$ is the temperature of the fluid (oil) in °C.

## 13.9.2 Calculation of Temperatures

The IEEE C57.91 standard provides the methodology to calculate the oil temperatures and the hottest-spot temperatures and estimate the expenditure of insulation life expected during a specified load and ambient cycle for a specified transformer design.

The ultimate top-oil rise for several time constants, $TOR_u$, is computed as follows:

$$TOR_u = \left(\frac{TOR_r \cdot K^2 + R}{R+1}\right)^n \tag{13-34}$$

where $TOR_r$ is the rated top-oil rise at the nameplate rating load, $K$ is the per-unit load, $R$ is the ratio of the transformer no-load and load loss, and $n$ is the oil exponent that is between 0.7 and 1.0 (typically 0.8).

Equation (13-34) provides the ultimate top-oil rise above an ambient temperature that a given transformer with a rated top-oil rise above ambient will reach if loaded to a load of $K$ for period of time exceeding several time constants. For most liquid-immersed transformers, the top-oil time constant is several hours.

The ultimate hottest-spot temperature gradient above the top-oil temperature $HSG_u$ is computed as follows:

$$HSG_u = HSG_r \cdot K^{2m} \tag{13-35}$$

where $HSG_r$ is the hottest-spot gradient above the $TOR_r$ (top-oil rise at rated load), $K$ is the per-unit load, and $m$ is the winding exponent that is between 0.8 and 1.0 (typically 0.8).

Given a daily load cycle and ambient temperature profile, and the top-oil and winding time constants, the temperatures of the top-oil and the hottest-spot can be calculated for any time during a cycle. If the load $K$ is rising, the instantaneous top-oil rise, $TOR_t$, for the time interval $t$ is computed by

$$TOR_t = T_i + (T_u - T_i) \cdot (1 + e^{-t/t_c}) \tag{13-36}$$

If the load $K$ is falling:

$$TOR_t = T_u + (T_i - T_u) \cdot e^{-t/t_c} \tag{13-37}$$

where $T_u$ is the ultimate top-oil temperature rise at load $K$, $T_i$ is the initial top-oil temperature rise at the end of the previous moment in time, $t$ is the time interval (hours), and $t_c$ is the oil time constant (hours).

If the load $K$ is rising, the instantaneous hot-spot gradient $HSG_t$ for the time interval $t$ is computed by

$$HSG_t = g_i + (g_u - g_i) \cdot (1 + e^{-t/w_c}) \tag{13-38}$$

The instantaneous hottest-spot temperature $HS_t$ for the time interval $t$ is computed by

$$HS_t = T_t + TOR_t + HSG_t \tag{13-39}$$

where $T_t$ is the ambient temperature during the time interval $t$.

## 13.9.3 Design Requirements

International standards (IEEE, IEC) set the requirements that define the transformer thermal characteristics. The average ambient temperature of the air in contact with the cooling equipment for a 24 h period shall not exceed 30°C, and the maximum ambient temperature shall not exceed 40°C.

The average winding temperature rise above ambient shall not exceed 65°C for a specified continuous-load current. The industry uses this criterion instead of the more relevant hot-spot temperature rise because manufacturers can check it by simply measuring the resistance of the windings during temperature-rise tests. The manufacturer must guarantee that the transformer meets this criterion.

The hot spot rise (hottest-conductor temperature rise) in the winding shall not exceed 80°C under the same ambient conditions. This 80°C limit does not ensure that the hot-spot rise is 15°C higher than the average winding temperature rise.

International standards specify requirements for operation of the transformer above rated voltage. For example, at no load, the voltage shall not exceed 110%. At full load, the voltage shall not exceed 105%.

The nameplate rating (kVA or MVA) of a transformer is the continuous load at rated voltage that produces an average winding temperature rise within the 65°C limit and hot-spot temperature rise within the 80°C limit.

### 13.9.4 Insulation Aging

The average, practical life of a transformer is probably 30 to 50 years. The aging of insulation impacts transformer life. The Arrhenius reaction equation is the basic principle for thermal-aging calculations of transformer insulation:

$$L = A \cdot e^{B/T} \tag{13-40}$$

where $L$ is the insulation life in hours, $T$ is the absolute temperature in K, while $A$ and $B$ are constants that depend on the aging rate, the end-of-life definition, and the condition of the insulation system.

Normal life of cellulose insulation is the time in years for a transformer operated with a constant 110°C hot-spot winding temperature to reach its defined end-of-life criteria. As the hottest-spot conductor temperature moves below 110°C, the life expectancy increases rapidly, and vice versa. A rule of thumb is that the life doubles for every 8°C decrease in operating temperature. The insulation life halves for every 8°C increase in operating temperature.

Users seldom operate transformers such that the hot-spot winding temperature is above 110°C. Even when peak loads cause 140°C hot-spot winding temperatures, the total time that the transformer operates above 110°C is relatively short, maybe less than 200 to 400 h/year.

## 13.10 TRANSFORMER SOUND LEVELS

The transformer noise is mainly due to the magnetostriction of the electrical steel sheets of the magnetic circuit. In general, a transformer operating at low magnetic induction has low noise level. Other sources of transformer noise are the windings and the cooling equipment. Transformers located in residential areas should have sound level as low as possible.

Table 13-4 presents the sound power levels of the five lists ($E_0$, $D_0$, $C_0$, $B_0$, and $A_0$) of Table 13-1 for transformers from 50 to 2500 kVA with 24 kV maximum voltage according to EN 50464-1: 2007. It can be seen from Tables 13-1 and 13-4 that for the same rated power, $A_0$ has the lowest, and $E_0$ has the highest no-load losses and sound power levels.

**TABLE 13-4** Lists of Sound Power Levels According to EN 50464-1: 2007

| | Sound power level, dB | | | | | |
|---|---|---|---|---|---|---|
| Rated power, kVA | $E_0$ | $D_0$ | $C_0$ | $B_0$ | $A_0$ | Short-circuit impedance, % |
| 50 | 55 | 50 | 47 | 42 | 39 | 4 |
| 100 | 59 | 54 | 49 | 44 | 41 | 4 |
| 160 | 62 | 57 | 52 | 47 | 44 | 4 |
| 250 | 65 | 60 | 55 | 50 | 47 | 4 |
| 315 | 67 | 61 | 57 | 52 | 49 | 4 |
| 400 | 68 | 63 | 58 | 53 | 50 | 4 |
| 500 | 69 | 64 | 59 | 54 | 51 | 4 |
| 630 | 70 | 65 | 60 | 55 | 52 | 4 |
| 630 | 70 | 65 | 60 | 55 | 52 | 6 |
| 800 | 71 | 66 | 61 | 56 | 53 | 6 |
| 1,000 | 73 | 68 | 63 | 58 | 55 | 6 |
| 1,250 | 74 | 69 | 64 | 59 | 56 | 6 |
| 1,600 | 76 | 71 | 66 | 61 | 58 | 6 |
| 2,000 | 78 | 73 | 68 | 63 | 60 | 6 |
| 2,500 | 81 | 76 | 71 | 66 | 63 | 6 |

### 13.10.1 Source of Sound

There are three different sources of transformer noise:

1. Core vibration due to magnetic forces and mainly due to magnetostrictive forces. The core is the predominant source of transformer noise. Core steel laminations undergo elongations and contractions (magnetostriction) as flux through them varies. This magnetostriction is nonlinear and independent of flux direction. Hence, noise is emitted in even multiples of the power frequency, that is, 120, 240, 360 Hz, etc., for a system with 60-Hz power frequency. The harmonic components decrease in magnitude as the mode of vibration goes up. However, an overexcited transformer or core-resonance may produce abnormally high third or higher harmonic frequencies.
2. Load noise mainly due to electromagnetic forces in the windings. Load noise is caused by vibrations in transformer windings, tank walls, and magnetic shields due to the electromagnetic forces resulting from leakage fields generated by load currents. These electromagnetic forces are proportional to the square of the load currents. The frequency of load noise is usually twice the power frequency.
3. Noises from cooling equipment. All rotary machinery on a transformer, including fans and pumps, produce noise with a broadband frequency spectrum. This "white noise" can have various magnitude and directionality, depending on the design of the fans and pumps and on their arrangement.

### 13.10.2 Sound Measurement

The sound level, expressed in decibels (dB), is measured using the so-called A-weighted scale that closely follows the sensitivity of the human ear. There are two main methods of sound measurement: sound pressure measurement and sound intensity measurement. The details of these methods are given in IEEE C57.12.90 and C57.12.91 and in IEC 60076-10 standards.

The A-weighted sound pressure is the most commonly used method for determining sound levels in transformers. Sound pressure is a scalar quantity that requires simple instrumentation. Sound pressure measurements are quite sensitive to the ambient sound levels on the test location. Consequently, appropriate corrections for the ambient sound level and reflected sound from the surrounding surfaces must be also quantified to calculate the correct sound level of the transformer. The A-weighted ambient sound pressure levels must be measured immediately before and after the measurements on a transformer. According to the sound pressure measurement, the A-weighted sound power level, $L_{WA}$ in dB, is computed by

$$L_{WA} = L_{PA} + 10 \cdot \log(S) \tag{13-41}$$

where $L_{PA}$ is the A-weighted average sound pressure level in dB, and $S$ is the area of the radiating surface in $m^2$.

The A-weighted sound intensity method measures the sound power radiated through a unit area per unit time. Sound intensity is a vector quantity and the method measures directional sound. Consequently, it is less affected by ambient sound. This means that the sound intensity method can provide more accurate measurements in the presence of ambient sound. The sound intensity method is especially suitable for transformers with very low sound levels. However, sound intensity measurements require higher skill personnel and more sophisticated instrumentation. According to the sound intensity measurement, the A-weighted sound power level, $L_{WA}$ in dB, is computed by

$$L_{WA} = L_{IA} + 10 \cdot \log(S) \tag{13-42}$$

where $L_{IA}$ is the A-weighted average sound intensity level in dB, and $S$ is the area of the radiating surface in $m^2$.

### 13.10.3 Sound Level Reduction

Manufacturers have at their disposal a variety of means of obtaining sound reduction. Most measures aim at reducing noise generation:

1. *Reducing Core Vibration.* Because magnetostriction is a function of flux intensity, a manufacturer's first option is to reduce induction levels of transformers. This has an additional advantage of reducing no-load losses. Alternatively, grades of steel having a different magnetostrictive characteristics can be substituted for the same induction-level design. A step-lap design can also reduce noise emission from joints. Finally, the designer has to anticipate the natural frequencies of the core mechanical structure and avoid their coincidence with harmonics of 120 Hz (for 60 Hz power frequency).
2. *Reducing Cooling Equipment Noises.* The most significant noise reduction measure for cooling equipment is to reduce fan rotational speed or adjust the fan blade incidence angle. There is ample supply of low-noise designs, ranging from low speed to encased fans, from which manufacturers can choose.

When all possibilities of noise emission are exhausted and still further noise reduction is required, some sort of a mass-damper or absorption system has to be incorporated on or outside the tank structure. Moderate reductions can be realized by the use of barriers within the tank. Some of these are "soft" barriers, which operate on the principle of absorbing vibrational energy from the core and reducing its transmission to the tank. Others are "mass" barriers, which operate on the principle of loading the tank to decrease its magnitude of vibration for given energy transmission from the core. To achieve large sound reductions (as much as 25 to 30 dB), some manufacturers employ complete external enclosures of steel. For smaller substation units, these enclosures can be preassembled and shipped in place over the transformer tank.

To reduce the sound level of an existing transformer, the most satisfactory method has been found to be the erection of barrier walls on one or more sides of the transformer. The attenuation that can be achieved depends on the transmission loss through the barrier, the diffraction over and around the barrier, and the pressure buildup between the tank and the barrier. Transmission loss through a barrier wall is a function of the mass of the wall. Structural requirements of most practical masonry barriers ensure sufficient mass to produce 25 to 40 dB attenuation through the wall. The effectiveness is usually limited by diffraction around the edges of the barrier.

## 13.11 TRANSFORMER OPERATION

### 13.11.1 Loading Practice

The temperature limitation of loading must be considered. Ordinarily, the kVA that a transformer should carry is limited by the effect of reactance on regulation or by the effect of load loss on system economy. At times, it is desirable to ignore these factors and increase the kVA load until the effect of temperature on insulation life is the limiting factor. High temperature decreases the mechanical strength and increases the brittleness of fibrous insulation, making transformer failure increasingly likely, even though the dielectric strength of the insulation material may not be seriously decreased. Overloading of transformers should be limited by reasonable consideration of the effect on insulation life and the probable effect on transformer life. The insulation life of a transformer is defined as the time required for the mechanical strength of the insulation material to lose a specified fraction of its initial value. Loss of 50% of the tensile strength is the usual basis for evaluating conductor insulation for power transformers.

The temperature of the top oil should never exceed 110°C for power transformers with a 55°C average winding-rise insulation system or 110°C for those with a 65°C average winding-rise insulation system. The consequence of exceeding these limits could be oil overflow or excessive pressure. The winding hot spot should not exceed 150°C for the 55°C average winding-rise insulation system or 180°C for the 65°C average winding-rise insulation system. These limitations are based principally on a concern for rates of insulation aging, but it should be noted that free bubbles may be evolved at hot-spot temperatures above 140°C, with consequent weakening of dielectric strength. The peak short-duration loading should never exceed 200% of rating, except for transformers rated over 100 MVA a limit of 150% of rating is recommended. This reflects a concern for stray flux heating in large units.

### 13.11.2 Parallel Operation of Transformers

To supply a load above the rating of an existing transformer, one or more transformers are connected in parallel with the existing transformer. It is usually economical to install another transformer in parallel instead of replacing the existing transformer by a single larger transformer. Moreover, it is better to have a parallel transformer for reliability reasons.

The parallel operation of two or more transformers is feasible, when the following requirements are met:

1. Their voltage ratio (turns ratio) should be the same (the permitted tolerance is according to ANSI/IEEE and IEC standards).
2. Their vector groups should be the same. In other words, the transformers must have the same inherent phase angle difference between primary and secondary terminals, the same polarity and the same phase sequence.
3. Their series impedance should be the same (the permitted tolerance is according to ANSI/IEEE and IEC standards).

The first two requirements ensure that the open-circuit secondary voltages of the transformers are closely matched in order to avoid excessive circulating currents when the parallel connections are made.

The third requirement ensures that when transformers having the same series impedance are connected in parallel, the load currents will split in proportion to the kVA ratings of the transformers. Consequently, transformers with different kVA ratings can be successfully operated in parallel as long as their series impedance values are all approximately the same.

Let us consider that two transformers with equal voltage ratios are operated in parallel. Figure 13-21 shows the simplified equivalent single-phase circuit referred to the secondary winding of these two transformers, which feed a load **I** at voltage $\mathbf{V}_2$. As can be seen in Fig. 13-21, the excitation branch is ignored in both transformers; the first transformer is represented by its series impedance $\mathbf{Z}_A = R_A + jX_A$ and the second transformer is represented by its series impedance $\mathbf{Z}_B = R_B + jX_B$. Since the two transformers are operated in parallel, the voltage drop across the two transformers is the same:

$$\mathbf{I}_A \cdot \mathbf{Z}_A = \mathbf{I}_B \cdot \mathbf{Z}_B = \mathbf{I} \cdot \mathbf{Z} \tag{13-43}$$

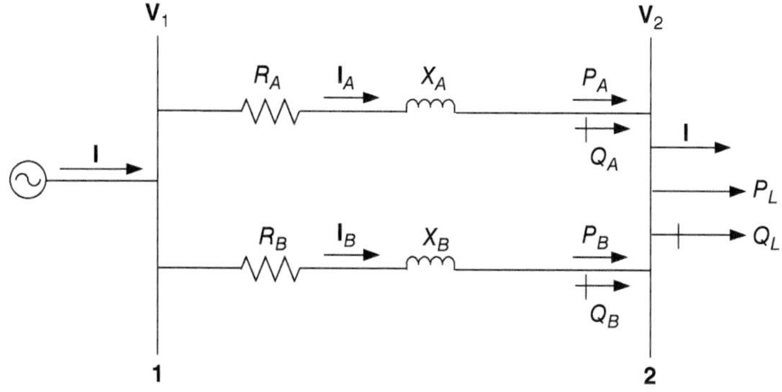


**FIGURE 13-21** Simplified equivalent single-phase circuit referred to the secondary winding of two transformers operated in parallel.

where $\mathbf{Z}$ is the equivalent series impedance of the two transformers, which is given by

$$\mathbf{Z} = \frac{\mathbf{Z}_A \cdot \mathbf{Z}_B}{\mathbf{Z}_A + \mathbf{Z}_B} \tag{13-44}$$

By combining Eqs. (13-43) and (13-44), the following equations are obtained:

$$\mathbf{I}_A = \frac{\mathbf{Z}_B}{\mathbf{Z}_A + \mathbf{Z}_B} \cdot \mathbf{I} \tag{13-45}$$

$$\mathbf{I}_B = \frac{\mathbf{Z}_A}{\mathbf{Z}_A + \mathbf{Z}_B} \cdot \mathbf{I} \tag{13-46}$$

Since the voltage is $\mathbf{V}_2$ at the load terminals, the first transformer supplies complex power $\mathbf{S}_A$ (active power $P_A$ and reactive power $Q_A$) that is given by

$$\mathbf{S}_A = P_A + jQ_A = \mathbf{V}_2 \cdot \mathbf{I}_A^* \tag{13-47}$$

and the second transformer supplies complex power $\mathbf{S}_B$ (active power $P_B$ and reactive power $Q_B$) that is given by

$$\mathbf{S}_B = P_B + jQ_B = \mathbf{V}_2 \cdot \mathbf{I}_B^* \tag{13-48}$$

Equation (13-43) can also be written as follows:

$$\frac{\mathbf{I}_A}{\mathbf{I}_B} = \frac{I_A \cdot e^{j\theta_A}}{I_B \cdot e^{j\theta_B}} = \frac{I_A}{I_B} \cdot e^{j(\theta_A - \theta_B)} = \frac{\mathbf{Z}_B}{\mathbf{Z}_A} = \frac{Z_B \cdot e^{j\varphi_B}}{Z_A \cdot e^{j\varphi_A}} = \frac{Z_B}{Z_A} \cdot e^{j(\varphi_B - \varphi_A)} \tag{13-49}$$

When $\varphi_B - \varphi_A = 0$, then $\theta_A - \theta_B = 0$, and Eq. (13-49) can be written as follows:

$$\frac{I_A}{I_B} = \frac{Z_B}{Z_A} = \frac{S_{nA} \cdot u_{kB}}{S_{nB} \cdot u_{kA}} \tag{13-50}$$

where $S_{nA}$ and $S_{nB}$ is the rated power of the first and second transformer, respectively, while $u_{kA}$ and $u_{kB}$ is the short-circuit voltage of the first and second transformer, respectively. Equation (13-50) shows that when $u_{kA} = u_{kB}$, then there is proportionality between load currents $I_A$ and $I_B$, impedances $Z_A$ and $Z_B$, and rated power outputs $S_{nA}$ and $S_{nB}$. Moreover, when $u_{kA} = u_{kB}$, the load current $I_A$ is proportional to the rated power $S_{nA}$, and the load current $I_B$ is proportional to the rated power output $S_{nB}$. When $u_{kA} \neq u_{kB}$, the transformer with the smaller impedance takes the larger load current, which is no longer proportional to the rated power output of that transformer.

*Example 13-2.* Two three-phase transformers are connected in parallel and feed a three-phase load of 1700 kVA with power factor 0.8 lagging. Both transformers have rated primary voltage 20 kV, rated secondary voltage 0.4 kV, rated power 1000 kVA, per unit resistance 2%, and per unit impedance 6%. The voltage at the sending bus (20 kV) is $\mathbf{V}_1 = 1.05\angle 0^0$ pu. Compute the loading of the two transformers.

*Solution.* The base power of the power system is $S_{\text{Base}} = 1000$ kVA. The reactance of each transformer is

$$X = \sqrt{Z^2 - R^2} = \sqrt{0.06^2 - 0.02^2} \Rightarrow X = X_A = X_B = 0.057 \text{ pu}$$

The impedance of each transformer is

$$\mathbf{Z} = R + jX \Rightarrow \mathbf{Z} = \mathbf{Z}_A = \mathbf{Z}_B = (0.02 + j0.057) \text{ pu}$$

The per unit load is

$$\mathbf{S}_L = \frac{1700 \cdot (0.8 + j0.6)}{1000} \Rightarrow \mathbf{S}_L = P_L + jQ_L = (1.36 + j1.02)\ \text{pu}$$

The single-line diagram of the power system is shown in Fig. 13-22. Using a power flow software, the solution of the power flow problem of Example 13-2 is $\mathbf{V}_2 = 1.0075\angle -1.53°$ pu.

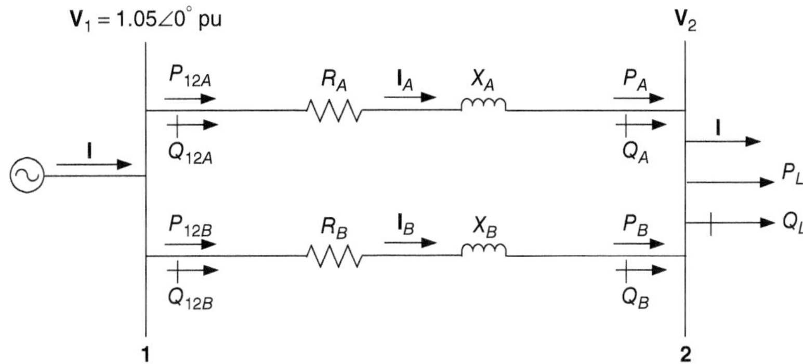


**FIGURE 13-22** Single-line diagram of the power system of Example 13-2.

The total current is

$$\mathbf{I} = \frac{\mathbf{S}_L^*}{\mathbf{V}_2^*} = \frac{1.36 - j1.02}{1.0075\ \angle 1.53°} \Rightarrow \mathbf{I} = 1.6874\angle -38.40°\ \text{pu}$$

The current in the first transformer is

$$\mathbf{I}_A = \frac{\mathbf{Z}_B \cdot \mathbf{I}}{\mathbf{Z}_A + \mathbf{Z}_B} = \frac{\mathbf{I}}{2} \Rightarrow \mathbf{I}_A = 0.8437\angle -38.40°\ \text{pu}$$

The current in the second transformer is

$$\mathbf{I}_B = \frac{\mathbf{Z}_A \cdot \mathbf{I}}{\mathbf{Z}_A + \mathbf{Z}_B} = \frac{\mathbf{I}}{2} \Rightarrow \mathbf{I}_B = 0.8437\angle -38.40°\ \text{pu}$$

The complex power that the first transformer delivers is

$$\mathbf{S}_A = \mathbf{V}_2 \cdot \mathbf{I}_A^* \cdot S_{\text{Base}} = (1.0075\angle -1.53°) \cdot (0.8437\angle 38.40°) \cdot 1000 \Rightarrow$$

$$\mathbf{S}_A = P_A + jQ_A = 680\ \text{kW} + j510\ \text{kVAR} \Rightarrow \mathbf{S}_A = 850\angle 36.87°\ \text{kVA}$$

The complex power that the second transformer delivers is

$$\mathbf{S}_B = \mathbf{V}_2 \cdot \mathbf{I}_B^* \cdot S_{\text{Base}} = (1.0075\angle -1.53°) \cdot (0.8437\angle 38.40°) \cdot 1000 \Rightarrow$$

$$\mathbf{S}_B = P_B + jQ_B = 680\ \text{kW} + j510\ \text{kVAR} \Rightarrow \mathbf{S}_B = 850\angle 36.87°\ \text{kVA}$$

Since the two transformers have the same rated power, the same voltage ratio, and the same impedance, the load (1700 kVA) is equally distributed between them. The complete power flow results are shown in Fig. 13-23.

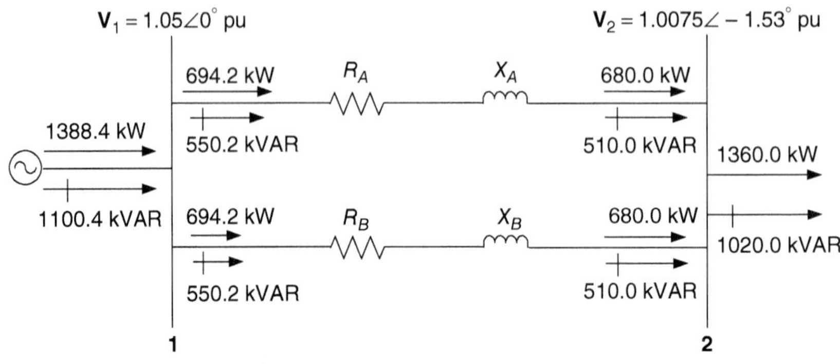


**FIGURE 13-23** Power flow results of Example 13-2.

An alternative approach to compute the loading (apparent power) $S_A^a$ and $S_B^a$ of the two transformers is as follows:

$$S_A^a = \frac{\frac{S_{rA}}{Z_A}}{\frac{S_{rA}}{Z_A}+\frac{S_{rB}}{Z_B}} \cdot S_L = \frac{\frac{1000}{0.06}}{\frac{1000}{0.06}+\frac{1000}{0.06}} \cdot 1700 \Rightarrow S_A^a = 850 \text{ kVA}$$

$$S_B^a = \frac{\frac{S_{rB}}{Z_B}}{\frac{S_{rA}}{Z_A}+\frac{S_{rB}}{Z_B}} \cdot S_L = \frac{\frac{1000}{0.06}}{\frac{1000}{0.06}+\frac{1000}{0.06}} \cdot 1700 \Rightarrow S_B^a = 850 \text{ kVA}$$

It can be seen that the alternative approach provides the same results as the accurate power flow approach because the two transformers have the same rated power, the same voltage ratio, and the same impedance.

*Example 13-3.* Two three-phase transformers are connected in parallel and feed a three-phase load of 1700 kVA with power factor 0.8 lagging. Both transformers have rated primary voltage 20 kV, rated secondary voltage 0.4 kV, rated power 1000 kVA, and per unit resistance 2%. The first transformer has per unit impedance 6%. The second transformer has per unit impedance 8%. The voltage at the sending bus (20 kV) is $\mathbf{V}_1 = 1.05\angle 0°$ pu. Compute the loading of the two transformers.

*Solution.* The base power of the power system is $S_{\text{Base}} = 1000$ kVA. The reactance of the first transformer is

$$X_A = \sqrt{Z_A^2 - R_A^2} = \sqrt{0.06^2 - 0.02^2} \Rightarrow X_A = 0.057 \text{ pu}$$

The impedance of the first transformer is

$$\mathbf{Z}_A = R_A + jX_A \Rightarrow \mathbf{Z}_A = (0.02 + j0.057) \text{ pu}$$

The reactance of the second transformer is

$$X_B = \sqrt{Z_B^2 - R_B^2} = \sqrt{0.08^2 - 0.02^2} \Rightarrow X_B = 0.077 \text{ pu}$$

The impedance of the second transformer is

$$\mathbf{Z}_B = R_B + jX_B \Rightarrow \mathbf{Z}_B = (0.02 + j0.077)\ \text{pu}$$

Using a power flow software, the solution of the power flow problem of Example 13-3 is $\mathbf{V}_2 = 1.0022\angle -1.86^\circ$ pu.

The total current is

$$\mathbf{I} = \frac{\mathbf{S}_L^*}{\mathbf{V}_2^*} = \frac{1.36 - j1.02}{1.0022\angle 1.86^\circ} \Rightarrow \mathbf{I} = 1.696\angle -38.73^\circ\ \text{pu}$$

The current in the first transformer is

$$\mathbf{I}_A = \frac{\mathbf{Z}_B \cdot \mathbf{I}}{\mathbf{Z}_A + \mathbf{Z}_B} \Rightarrow \mathbf{I}_A = 0.970\angle -36.59^\circ\ \text{pu}$$

The current in the second transformer is

$$\mathbf{I}_B = \frac{\mathbf{Z}_A \cdot \mathbf{I}}{\mathbf{Z}_A + \mathbf{Z}_B} \Rightarrow \mathbf{I}_B = 0.728\angle -41.58^\circ\ \text{pu}$$

The complex power that the first transformer delivers is

$$\mathbf{S}_A = \mathbf{V}_2 \cdot \mathbf{I}_A^* \cdot S_{\text{Base}} = (1.0022\angle -1.86^\circ)\cdot(0.970\angle 36.59^\circ)\cdot 1000 \Rightarrow$$
$$\mathbf{S}_A = P_A + jQ_A = 799.1\ \text{kW} + j553.9\ \text{kVAR} \Rightarrow \mathbf{S}_A = 972.3\angle 34.73^\circ\ \text{kVA}$$

The complex power that the second transformer delivers is

$$\mathbf{S}_B = \mathbf{V}_2 \cdot \mathbf{I}_B^* \cdot S_{\text{Base}} = (1.0022\angle -1.86^\circ)\cdot(0.728\angle 41.58^\circ)\cdot 1000 \Rightarrow$$
$$\mathbf{S}_B = P_B + jQ_B = 560.9\ \text{kW} + j466.1\ \text{kVAR} \Rightarrow \mathbf{S}_B = 729.2\angle 39.72^\circ\ \text{kVA}$$

The complete power flow results are shown in Fig. 13-24.

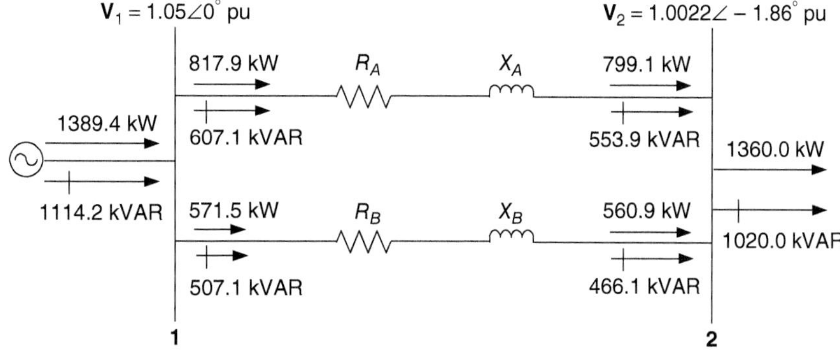


**FIGURE 13-24** Power flow results of Example 13-3.

An alternative approximate approach to compute the loading (apparent power) $S_A^a$ and $S_B^a$ of the two transformers is as follows:

$$S_A^a = \frac{\dfrac{S_{rA}}{Z_A}}{\dfrac{S_{rA}}{Z_A}+\dfrac{S_{rB}}{Z_B}} \cdot S_L = \frac{\dfrac{1000}{0.06}}{\dfrac{1000}{0.06}+\dfrac{1000}{0.08}} \cdot 1700 \Rightarrow S_A^a = 971.4 \text{ kVA}$$

$$S_B^a = \frac{\dfrac{S_{rB}}{Z_B}}{\dfrac{S_{rA}}{Z_A}+\dfrac{S_{rB}}{Z_B}} \cdot S_L = \frac{\dfrac{1000}{0.08}}{\dfrac{1000}{0.06}+\dfrac{1000}{0.08}} \cdot 1700 \Rightarrow S_B^a = 728.6 \text{ kVA}$$

In comparison to the accurate power flow approach, the error of the alternative approximate approach is

$$e_{S_A} = \frac{S_A^a - S_A}{S_A} \cdot 100\% = \frac{971.4 - 972.3}{972.3} \cdot 100\% \Rightarrow e_{S_A} = -0.09\%$$

$$e_{S_B} = \frac{S_B^a - S_B}{S_B} \cdot 100\% = \frac{728.6 - 729.2}{729.2} \cdot 100\% \Rightarrow e_{S_B} = -0.09\%$$

*Example 13-4.* Two three-phase transformers are connected in parallel and feed a three-phase load of 2700 kVA with power factor 0.8 lagging. Both transformers have rated primary voltage 20 kV, rated secondary voltage 0.4 kV, and per unit resistance 2%. The first transformer has rated power 1000 kVA and per unit impedance 6%. The second transformer has rated power 2500 kVA and per unit impedance 8%. The voltage at the sending bus (20 kV) is $\mathbf{V}_1 = 1.05\angle 0°$ pu. Compute the loading of the two transformers.

*Solution.* The base power of the power system is $S_{\text{Base}} = 1000$ kVA. The reactance of the first transformer is

$$X_A = \sqrt{Z_A^2 - R_A^2} = \sqrt{0.06^2 - 0.02^2} \Rightarrow X_A = 0.057 \text{ pu}$$

The impedance of the first transformer is

$$\mathbf{Z}_A = R_A + jX_A \Rightarrow \mathbf{Z}_A = (0.02 + j0.057) \text{ pu}$$

At the base power $S_{\text{Base}}$, the resistance of the second transformer is

$$R_B = 0.02 \cdot \frac{1000}{2500} \Rightarrow R_B = 0.008 \text{ pu}$$

At the base power $S_{\text{Base}}$, the magnitude of the impedance of the second transformer is

$$Z_B = 0.08 \cdot \frac{1000}{2500} \Rightarrow Z_B = 0.032 \text{ pu}$$

The reactance of the second transformer is

$$X_B = \sqrt{Z_B^2 - R_B^2} = \sqrt{0.032^2 - 0.008^2} \Rightarrow X_B = 0.031 \text{ pu}$$

The impedance of the second transformer is

$$\mathbf{Z}_B = R_B + jX_B \Rightarrow \mathbf{Z}_B = (0.008 + j0.031)\text{ pu}$$

Using a power flow software, the solution of the power flow problem of Example 13-4 is $\mathbf{V}_2 = 1.0046\,\angle -1.84°$ pu.

The total current is

$$\mathbf{I} = \frac{\mathbf{S}_L^*}{\mathbf{V}_2^*} = \frac{1.36 - j1.02}{1.0046\angle 1.84°} \Rightarrow \mathbf{I} = 2.688\angle -38.71°\text{ pu}$$

The current in the first transformer is

$$\mathbf{I}_A = \frac{\mathbf{Z}_B \cdot \mathbf{I}}{\mathbf{Z}_A + \mathbf{Z}_B} \Rightarrow \mathbf{I}_A = 0.936\angle -35.45°\text{ pu}$$

The current in the second transformer is

$$\mathbf{I}_B = \frac{\mathbf{Z}_A \cdot \mathbf{I}}{\mathbf{Z}_A + \mathbf{Z}_B} \Rightarrow \mathbf{I}_B = 1.754\,\angle -40.45°\text{ pu}$$

The complex power that the first transformer delivers is

$$\mathbf{S}_A = \mathbf{V}_2 \cdot \mathbf{I}_A^* \cdot S_{\text{Base}} = (1.0046\angle -1.84°) \cdot (0.936\angle 35.45°) \cdot 1000 \Rightarrow$$
$$\mathbf{S}_A = P_A + jQ_A = 782.8\text{ kW} + j520.3\text{ kVAR} \Rightarrow \mathbf{S}_A = 939.9\angle 33.61°\text{ kVA}$$

The complex power that the second transformer delivers is

$$\mathbf{S}_B = \mathbf{V}_2 \cdot \mathbf{I}_B^* \cdot S_{\text{Base}} = (1.0046\angle -1.84°) \cdot (1.754\angle 40.45°) \cdot 1000 \Rightarrow$$
$$\mathbf{S}_B = P_B + jQ_B = 1377.2\text{ kW} + j1099.7\text{ kVAR} \Rightarrow \mathbf{S}_B = 1762.4\angle 38.61°\text{ kVA}$$

The complete power flow results are shown in Fig. 13-25.

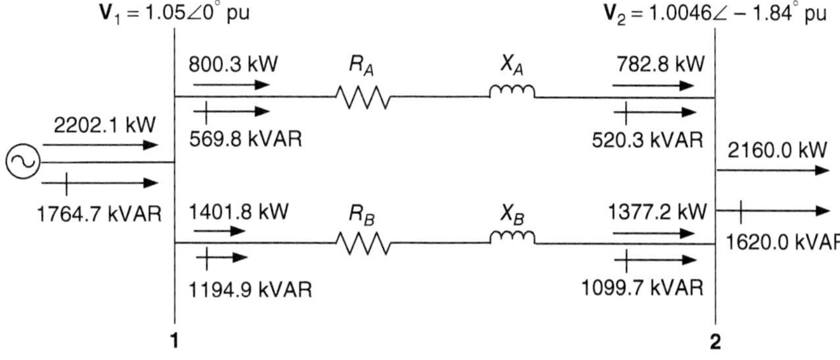


**FIGURE 13-25** Power flow results of Example 13-4.

An alternative approximate approach to compute the loading (apparent power) $S_A^a$ and $S_B^a$ of the two transformers is as follows:

$$S_A^a = \frac{\frac{S_{rA}}{Z_A}}{\frac{S_{rA}}{Z_A}+\frac{S_{rB}}{Z_B}} \cdot S_L = \frac{\frac{1000}{0.06}}{\frac{1000}{0.06}+\frac{2500}{0.08}} \cdot 2700 \Rightarrow S_A^a = 939.1 \text{ kVA}$$

$$S_B^a = \frac{\frac{S_{rB}}{Z_B}}{\frac{S_{rA}}{Z_A}+\frac{S_{rB}}{Z_B}} \cdot S_L = \frac{\frac{2500}{0.08}}{\frac{1000}{0.06}+\frac{2500}{0.08}} \cdot 2700 \Rightarrow S_B^a = 1760.9 \text{ kVA}$$

In comparison to the accurate power flow approach, the error of the alternative approximate approach is

$$e_{S_A} = \frac{S_A^a - S_A}{S_A} \cdot 100\% = \frac{939.1 - 939.9}{939.9} \cdot 100\% \Rightarrow e_{S_A} = -0.09\%$$

$$e_{S_B} = \frac{S_B^a - S_B}{S_B} \cdot 100\% = \frac{1760.9 - 1762.4}{1762.4} \cdot 100\% \Rightarrow e_{S_B} = -0.09\%$$

## 13.12 TRANSFORMER TESTING

Power transformer reliability in service is considerably enhanced by the selection of appropriate tests and the specification of correct test levels. ANSI/IEEE standards C57.12.00, C57.12.90, C57.98, and C57.113 contain detailed, authoritative information on testing power and distribution transformers. These standards are very important because they facilitate precise communication and understanding between transformer manufacturers and users. These standards identify critical features, and provide minimum requirements for safe and reliable operation of power and distribution transformers.

Transformer tests are classified, in accordance with IEC 60076-1 standard, as follows:

- Type tests
- Routine tests
- Special tests

### 13.12.1 Type Tests

Type tests, which are made on one transformer from every transformer type, are the following:

1. *Temperature Rise Test.* The temperature rise test procedure is typically performed according to IEC 60076-2. The objective of this test is to verify guaranteed temperature rises for oil and windings.
2. *Lightning Impulse Test.* The lightning impulse test procedure is typically performed according to IEC 60076-3. A standard full-wave lightning impulse curve is shown in Fig. 13-26. The lightning impulse test checks if the transformer can withstand overvoltages. These overvoltages are caused by (1) traveling waves (caused by lightning) in transmission lines, (2) sudden on/off switching of breakers, and (3) short circuits. It should be noted that the lightning impulse test is a routine test for transformers with higher voltage for equipment $U_m$ greater than 72.5 kV and a type test for $U_m \leq 72.5$ kV.

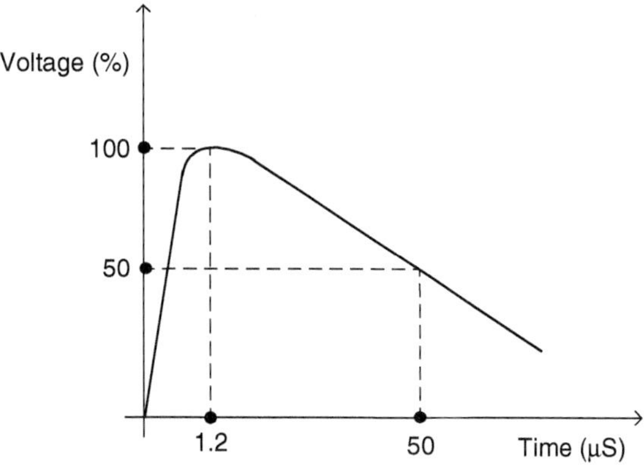


**FIGURE 13-26** A standard full-wave lightning impulse curve.

## 13.12.2 Routine Tests

Routine tests are performed on every transformer separately, and include

1. *Winding Resistance.* The winding resistance is defined as the direct current (dc) resistance of a winding. The procedure for the measurement of windings resistance is typically performed according to IEC 60076-1. During this test the resistance of each winding is measured and the temperature is recorded. The test is performed with direct current.
2. *Voltage Ratio and Check of Phase Displacement.* Measurement of the voltage ratio is typically performed according to IEC 60076-1. The objective of the test is to compare the measured values of the transformer ratio with the respective guaranteed values. For the transformer, the voltage ratio is equal to the turns ratio, namely, $V_1/V_2 = N_1/N_2$, where $V_1$ and $V_2$ is the phase voltage of the primary and secondary winding, respectively, and $N_1$ and $N_2$ is the number of turns of the primary and secondary winding, respectively.
3. *Impedance Voltage.* Measurement of impedance voltage is typically performed according to IEC 60076-1. The impedance voltage, which is expressed as a percentage of the rated voltage, represents the transformer's impedance. The IEC standard requires the impedance voltage to be calculated at the reference temperature of 75°C. The transformer impedance voltage is guaranteed by the manufacturer and is verified for the customer during the impedance voltage routine test.
4. *Load Loss.* The transformer load loss is guaranteed by the manufacturer and is verified for the customer during the load loss routine test. The measurement of load loss is implemented with the secondary winding short-circuit and by increasing the voltage of the primary winding until the current of the primary winding reaches its nominal value. The load losses are calculated at the reference temperature of 75°C according to the IEC standard.
5. *No-Load Current and No-Load Losses.* The measurement is typically performed according to IEC 60076-1. The no-load current represents the real value of current that is required to magnetize the magnetic core. The no-load losses represent the power that is absorbed by the transformer core when rated voltage and rated frequency are applied to one winding (e.g., secondary) and the other winding (e.g., primary) is open circuited.
6. *Dielectric Routine Tests.* The dielectric routine tests are the following:
   - *Applied Voltage Dielectric Test.* The duration of the test, according to IEC 60076-3, is 1 min. With this specific test, the following are checked: (1) the insulation between primary and secondary windings, (2) the insulation between the tested winding and the tank, and (3) the insulation between the tested winding and the magnetic circuit.

- *Induced Voltage Dielectric Test.* A three-phase voltage, twice the rated voltage, is applied to the transformer for 1 min. However, the doubling of the voltage will double the magnetic induction resulting in transformer saturation and, consequently, there is a danger of the transformer being destroyed. In order to avoid saturation, the frequency is also doubled, so the magnetic induction remains constant. Consequently, during this test, the volts per turn and therefore the volts per layer are doubled. This test verifies the dielectric strength between turns and layers.

### 13.12.3 Special Tests

Special tests are not included in the category of type or routine tests and are executed after agreement between customer and manufacturer. The special tests are the following:

1. Dielectric special tests.
2. Determination of capacitances of windings-to-earth and between windings.
3. Short-circuit withstand test. According to this test, the transformer is subjected to successive short circuits of 0.5 s duration and the transformer must withstand these short circuits. Since this test requires high power, it is executed in special test centers.
4. *Determination of Sound Levels.* The transformer is energized at no-load and at rated voltage and rated frequency, so the noise peripheral to the transformer can be measured.
5. Measurement of the harmonics of the no-load current.
6. Measurement of insulation resistance and/or measurement of dissipation factor of the insulation system capacitances.
7. Radio interference voltage.
8. Measurement of zero-sequence impedance.

## 13.13 TRANSFORMER PROTECTION

The power transformer has to be protected against faults occurring on any part of the power system as well as against faults occurring within the transformer.

The transformer has to be protected against the effects of power system faults, that is, overloads, short circuits, lightning, and earth faults.

The transformer has also to be protected against the effects of faults arising in the transformer, that is,

- Insulation breakdown, which results in short circuits or ground faults. Insulation breakdown between windings or between windings and the core can be caused by insulation aging due to long time overtemperature, oil contamination and oil leakage, corona discharges in the insulation, transient overvoltages due to thunderstorms or network switching, and current forces on the windings due to high current external faults.
- Overheating due to overexcitation, which can damage the metal parts of the transformer.

When a fault occurs in a transformer, the damage is usually severe. The transformer has to be transported to a workshop to be repaired, which needs considerable time. It is always difficult to operate a power transmission system with a power transformer out of service.

In order to prevent faults and to minimize the damage in case of a fault, power transformers are equipped with protective relays and monitors. The choice of protective equipment varies depending on transformer size, importance of transformer within the power system, voltage level, winding connection and design.

### 13.13.1 Overcurrent Protection

A transformer may be subjected to overcurrents ranging from just in excess of nameplate rating to as much as 10 or 20 times rating. Currents up to about twice rating normally result from overload conditions on the system, while higher currents are a consequence of system faults. When such overcurrents are of extended duration, they may produce either mechanical or thermal damage in a transformer, or possibly both. At current levels near the maximum design capability (worst-case through-fault), mechanical effects from electromagnetically generated forces are of primary concern. The pulsating forces tend to loosen the coils, conductors may be deformed or displaced, and insulation may be damaged. Lower levels of current principally produce thermal heating. For all current levels, the extent of the damage is increased with time duration.

Whatever the cause, magnitude, or duration of the overcurrent, it is desirable that some component of the system recognizes the abnormal condition and initiates action to protect the transformer. Fuses and protective relays are two forms of protective devices in common use. A fuse consists of a fusible conducting link that will be destroyed after it is subjected to an overcurrent for some period of time, thus opening the circuit. Typically, fuses are employed to protect distribution transformers and small power transformers up to 5000 to 10,000 kVA. Traditional relays are electromagnetic devices that operate on a reduced current derived from a current transformer in the main transformer line to close or open control contacts, which can initiate the operation of a circuit breaker in the transformer line circuit. Relays are used to protect all medium and large power transformers.

All protective devices, such as fuses and relays, have a defined operating characteristic in the current-time domain. This characteristic should be properly coordinated with the current-carrying capability of the transformer to avoid damage from prolonged overloads or through faults. Transformer capability is defined in general terms in a guide document, ANSI/IEEE C57.109, IEEE guide for liquid-immersed transformers through-fault-current duration. The format of the transformer capability curves is shown in Fig. 13-27. The solid curve, A, defines the thermal capability for all ratings, while the dashed curves, B (appropriate to the specific transformer impedance), define mechanical capability. For proper coordination on any power transformer, the protective device characteristic should fall below both the mechanical and thermal portions of the transformer capability curve.

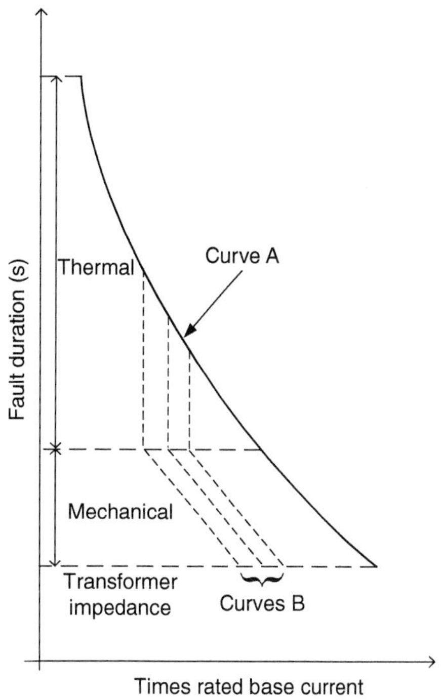


**FIGURE 13-27** Transformer through-fault protection curves.

### 13.13.2 Protection against Lightning

Impulse insulation level may be demonstrated by factory impulse-voltage tests using $1.5 \times 50$-μs full waves and chopped waves. The full wave demonstrates the BIL for traveling waves coming into the station over the transmission line. The chopped wave demonstrates strength against a wave traveling along the transmission line after flashing over an insulator some distance away from the transformer. These waves do not simulate direct lightning strokes on or near the transformer terminals, which would result in the application of a steep-front wave to the transformer winding. Such strokes are usually avoided by ground wires or protecting grounded structures.

A transformer may be subjected to severe lightning voltages as a result of a direct stroke to the transformer terminal, adjacent bus, or transmission line. Less severe voltages may result from strokes on a distant part of the system or from strokes to ground near the system. Since lightning voltage may exceed the insulation strength of the transformer, protection is necessary.

Voltage-time curves are used in evaluating protection, because for short times the insulation strength changes significantly with duration of voltage. Protection is effective if the voltage-time curve of the transformer is above the voltage-time curve of the protective equipment, so that for any time duration the kilovolts insulation strength of the transformer exceeds the protective level at the same duration. The voltage-time curves of transformer insulation have considerable "turn-up," that is, for durations under 10 μs the kilovolts insulation strength is much greater. Rod gaps in air are unsuitable for protecting transformers, because they have even more turn-up than do transformers.

The modern surge arrester has very little turn-up and is an essential adjunct to the transformer whenever there is lightning exposure.

The required surge arrester rating depends on the effectiveness of the neutral grounding. The rating is expressed in percent of rated line-to-line power-frequency voltage that the arrester will withstand. Effectiveness of system grounding is described by the ratios of the zero-sequence resistance and impedance to the positive-sequence resistance and impedance. An 80% arrester is commonly used when the ratio of zero sequence to positive sequence is between 0.5 and 1.5 for resistance and between 1 and 3 for impedance. Lower ratios may permit 75% or 70% arresters. Higher ratios may require 85% or 90% arresters. Use of the 100% protective level is not economical at high voltages.

## 13.14 TRANSFORMER NAMEPLATE INFORMATION

Every power and distribution transformer has a metal nameplate attached to transformer tank that provides very important information on how the transformer has to be connected and operated.

According to IEEE C57.12.00 standard, for transformers rated above 500 kVA, the minimum information to be provided on transformer nameplate is the following:

- Manufacturer name
- Transformer serial number
- Month/year of transformer manufacturing
- Cooling class
- Number of phases
- Operating frequency
- Power rating (kVA or MVA)
- Voltage ratings
- Tap voltages
- Rated temperature rise
- Polarity for single-phase transformers
- Phasor or vector diagram for polyphase transformers
- Percent impedance
- Basic lightning impulse insulation level (BIL) of each winding and each bushing
- Approximate mass of core and windings, tank and fittings, insulating oil, total weight, and heaviest piece
- Connection diagram
- References for installation and operating instructions
- The word "transformer" or "autotransformer"
- Suitability for step-up operation

- Maximum positive and negative operating pressures of the oil preservation system
- Maximum negative pressure of the tank for vacuum filling
- Liquid level below the top surface of the highest point of the highest manhole flange at 25°C
- Change in liquid level per 10°C change in liquid level
- Oil volume of each transformer compartment
- Type of insulating liquid
- Conductor material of each winding

There is some flexibility in the standards for the layout of transformer nameplate. Each manufacturer arranges the data somehow differently. Some customers specify a particular layout that suits their needs. Figure 13-28 shows one typical layout of power transformer nameplate.

<table>
<tr><td colspan="2">Manufacturer data</td></tr>
<tr><td>Serial number</td><td>Impedances</td></tr>
<tr><td colspan="2">Rated power and rated voltages</td></tr>
<tr><td>Physical layout</td><td>Vector diagram</td></tr>
<tr><td colspan="2">Connection diagram<br>Winding taps and ratings</td></tr>
<tr><td>Weights</td><td>BIL ratings</td></tr>
<tr><td colspan="2">Oil levels and tank operating pressures</td></tr>
</table>

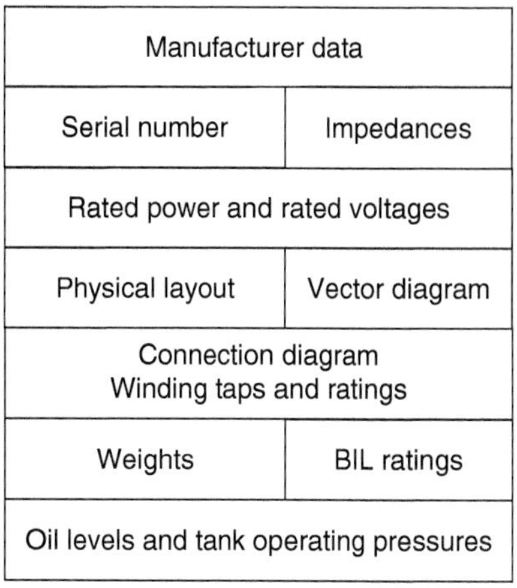

**FIGURE 13-28** Layout of power transformer nameplate.

## 13.15 TRANSFORMER INSTALLATION AND MAINTENANCE

Proper installation and maintenance of power transformers is needed to ensure long life in service. General requirements from IEEE C57.93 and from manufacturer's instructions are summarized below. Most power transformers have additional manufacturer's installation and maintenance instructions that should be carefully observed. These instructions, including safety practices, should be followed for the protection of workers and the transformer.

Transformers are tested and inspected before shipment. Smaller transformers are shipped filled with oil and fully assembled when shipping clearances and weights permit. For most large-power transformers, the external components, such as coolers and bushings, must be removed to meet shipping dimensions. Also, the oil is removed to reduce weight, and the unit is secured in dry gas.

### 13.15.1 Inspection on Arrival

Before removal from the car, inspect for shipping damage. If damage is found, a claim should be filed with the carrier and the manufacturer should be notified.

Transformers shipped oil-filled should be inspected for evidence or leakage or entrance of moisture during shipment. If the transformer is received in damaged condition, tests should be made to check the transformer for dryness. Oil samples taken from the bottom valve should be tested for moisture limits. Where bushings are shipped in place, insulation power-factor measurements can be used. Power factors of new transformers range from 0.2% to 0.5% at 25°C. In any case, a power factor measurement may be helpful for comparison with test values recorded by the manufacturer prior to shipment.

Transformers shipped gas-filled are fitted with a connection to which a compound pressure/vacuum gage can be connected. The gage should have a range of +10 lb/in$^2$. A positive or negative pressure indicates that the tank is tight; a continuous zero reading indicates a probable leak. If there is a reason to question dryness, the moisture can be estimated by a dew-point measurement of the gas in the transformer. Add dry gas immediately to about 3-lb/in$^2$ (gage) pressure and check for leaks.

Dew point is sensitive to temperature changes; therefore it is essential that the insulation temperature be accurately determined. Some instruments will not provide accurate dew-point readings below 32°F. The final degree of insulation dryness should be determined by measuring the insulation power factor or insulation resistance. If the measured dew point exceeds an acceptable value, contact the manufacturer for recommended course of action.

A preliminary internal inspection is not normally required unless shipping damage is apparent.

### 13.15.2 Oil Sampling

Samples of oil should be taken from the bottom. An oil-sampling valve is provided at the bottom of the transformer tank for this purpose. A metal or glass thief tube can be conveniently used to obtain a bottom sample from an oil barrel. Test samples should be taken only after the oil has settled for some time, varying from 8 h for a barrel to several days for a large transformer. Cold oil is much slower in settling. In drawing samples of oil from a sampling valve, some oil should first be discarded so that the sample will come from the bottom of the container and not from the sampling pipe. Examine a sample in a clear glass container for free water, which in any quantity is readily observable. The sample container should be a large-mouthed glass bottle, 1 qt or larger, with cork or glass stopper. The bottle should be carefully cleaned and dried before being used. Bottles should be of amber color if samples are to be stored to be tested later for color or sludge-forming characteristics. Refer to ASTM D 923 standard practice for important details of oil-sampling technique.

### 13.15.3 Testing for Oil Dielectric Strength

The testing fixture should be cleaned thoroughly to remove any particles or fibers and rinsed out with a portion of the oil to be tested. The testing fixture should be filled with oil, and both oil and fixture should be at room temperature. Allow 3 min for air bubbles to escape before applying voltage. Tests are made by two methods. ASTM D 877 uses 1-in-diameter square-edge electrodes spaced 0.10 in apart and a rate of voltage rise of 3000 V/s. ASTM D 1816 uses special radius-surface electrodes spaced 0.04 in apart, with continuous oil circulation, and a rate of voltage rise of 500 V/s. The latter test is more sensitive to slight moisture or particulate contamination. In either case, the average voltage for five breakdowns is taken as the dielectric strength of the oil. Strength of new oil should exceed the minimum value for good oil as shown in Table 13-5.

**TABLE 13-5** Dielectric Strength of Oil

| kV average dielectric strength by ASTM D 877-1982 | kV average dielectric strength by ASTM D 1816-1982 | Condition of oil |
|---|---|---|
| 30 or over | 29 or over | Good |
| 26–29 | 23–28 | Usable |
| Under 26 | Under 23 | Poor |

### 13.15.4 Filtering to Increase Dielectric Strength

If the oil tests below "good," it should be filtered to remove impurities and moisture. It is best to discharge filtered oil into a clean, dry tank and avoid mixing with unfiltered oil. If the filtered oil must be discharged back into the transformer tank, the oil should be withdrawn from the bottom filter-press valve and, after filtering, returned through the top filter-press valve. Oil should not be filtered while the transformer is energized, because the dielectric strength may be temporarily reduced by aeration. If no facilities are available for making dielectric tests, send a sample to the manufacturer marked with the serial number of the transformer.

### 13.15.5 Drying the Core and Coils

This should be necessary only if an accident occurs during shipping, storage, or service. If possible, determine the extent of the moisture and manner in which it entered the tank. The manufacturer should be contacted for recommendations concerning additional checks and strips for drying out the transformer. If drying is necessary, one or more of the following methods may be used, depending on the facilities available:

- Heat and vacuum (method 1)
- Vacuum only (method 2)
- Heat in oil (method 3)
- Heat in air (method 4)

A low value of residual moisture can be attained most rapidly by method 1. Method 2 is effective but requires longer time and better vacuum equipment. Method 3 is very slow and not as effective as the vacuum method. Method 4 is recommended only for smaller, low-voltage transformers.

### 13.15.6 Time Required for Drying

This can range from 72 h to 3 weeks depending on the size and condition of the transformer and the method of drying. In general, the use of a high vacuum and a cold trap is faster and more efficient than heat alone.

### 13.15.7 Insulation Resistance

This will indicate the degree of dryness only when the transformer is dried without oil. If the initial insulation resistance is measured at room temperature, it may be high, although the insulation is not dry, but as the transformer is heated up, it will drop rapidly.

As the drying proceeds at a constant temperature, the insulation resistance will generally increase gradually until toward the end of the drying period, when it increases quite rapidly and then levels off at a high value. The drying should continue until the resistance is constant for a period of 12 h.

### 13.15.8 Insulation Power-Factor Reading

These readings (at 60 Hz) will indicate the degree of dryness. The power factor will first increase as the temperature increases and then will gradually decrease as drying progresses. Drying should continue until the power factor is constant for a period of 12 h. If power factor is measured on transformers dried in oil by the short-circuit method, the power factor should be used to supplement oil tests as a measure of dryness.

### 13.15.9 Filling without Vacuum

This method should be practiced only on low-voltage transformers. Check manufacturer's recommendations. Use extreme care to keep moisture out of the core and coils. The tank should not be opened to the atmosphere until the core and coils are under oil, unless vacuum filling is available. The oil shipping tank or oil drums should not be opened until their temperature is the same as or higher than that of the surrounding air and the transformer is in place and ready to receive the oil. Metal or synthetic rubber hose should be used for filling, because transformer oil is contaminated by natural rubber. Oil should never be added to a transformer without passing through a filter press.

Static charges can be developed when transformer oil flows in pipes, hoses, or tanks. Oil leaving a filter press may be charged to over 50,000 V. To accelerate dissipation of the charge in the oil, ground the filter press, the tank, and all bushings or winding leads during oil flow into any tank. Conduction through oil is slow; therefore it is desirable to maintain these grounds for at least 1 h after the oil flow has ended.

Avoid explosive gas mixtures in any container into which oil is flowing. Arcs can occur along the surface of the charged oil even though all metal is grounded.

## 13.15.10 Filling with Vacuum

The vacuum line should be connected to a tapped opening on a cover-mounted shipping plate or to a valve near the top of the tank. An opening of 5.08 cm minimum is recommended. The oil line can be connected to a suitable opening on a cover-mounted shipping plate or the top filter-press valve. The oil line should always be connected at the top of the tank so that the oil can be deaerated as it enters.

Transformers with operating voltage less than 161 kV and with core and coils not exposed to the atmosphere should be filled under vacuum better than 25 mmHg absolute pressure. The vacuum should be held 4 h before filling and continued during filling until the core and coils are covered. The vacuum can then be removed for installation of bushings and the remaining oil added without vacuum. Transformers with an operating voltage of 161 kV and above or transformers with core and coils exposed to the atmosphere should be completely filled under a vacuum better than 2 mmHg absolute pressure. A 2-mm vacuum should be held until the tank is filled to the 25°C level.

The filling rate should be under 1500 gal/h to facilitate evacuation and complete oil filling of all air pockets and voids. It is also recommended that the core and coil temperature be above 10°C to prevent "frost" formation.

Especially for extra high voltage transformers, once transformer is full, oil circulation is usually required to ensure full dehumidification and degassing of the oil together with full oil impregnation of the insulation structures. Larger transformers may have thick insulation structures that should be fully impregnated for appropriate dielectric performance. It can be seen from Fig. 13-29 that the depth of oil impregnation is strongly affected by time and temperature of the oil.

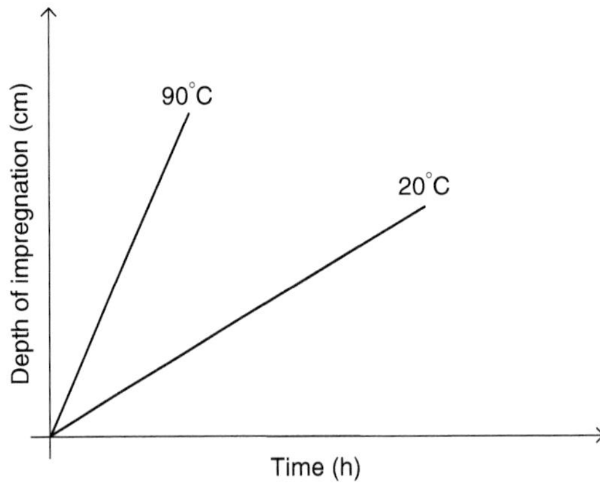


**FIGURE 13-29** Insulation impregnation chart.

## 13.15.11 Energization

When the voltage is first applied, it should, if possible, be brought up slowly to its full value so that any wrong connection or other trouble will be discovered before damage results. After full voltage has been applied successfully, the transformer should preferably be operated for a short period without load.

When the transformer is first energized it should be kept under close observation for the first 8 h. Check and record the oil temperature, the winding temperature, the tank pressure, and the ambient temperature. Watch particularly for any sudden changes. After 7 days of operation, check for oil leaks and for abnormally high usage of nitrogen if the transformer is equipped with Inertaire. Stop all oil and gas leaks. The observation should continue on a daily schedule for 7 days and then weekly for the first month of operation. An oil sample should be taken during the first month of operation for gas-in-oil analysis. This analysis should be repeated annually. Heat exchangers, tap changer, pumps, fans, etc. should be serviced per manufacturer's instructions.

### 13.15.12 Internal Inspection of In-Service Transformers

This is not necessary unless there is a specific indication of a problem. Oil analysis is a good method to discover potential problems. Sludging of the oil, low dielectric strength, moisture in the oil, or the presence of combustible gases are conditions that may merit an internal inspection of the transformer.

Generation of combustible gas usually indicates internal trouble (not necessarily serious). Analysis of the gas sometimes helps to identify the source. If collection of combustible gas continues without discoverable cause, partial discharge voltage measurement may establish whether or not there is an internal fault.

Severe system disturbances, incidence of through-fault, or a circuit-breaker operation would also be reason for an internal inspection of a transformer.

### 13.15.13 Operating without Cooling

A liquid-cooled transformer should not be run continuously, even at no load, without the cooling liquid. In an emergency, forced-oil air-cooled transformers may be operated without fans and pumps (1) at rated load for approximately 1 h, starting at full-load temperature rise, (2) at rated load for approximately 2 h, starting cold (at ambient temperature), (3) at rated voltage and no load for approximately 6 h, starting at full-load temperature rise, and (4) at rated voltage and no load for approximately 12 h, starting cold.

When only a portion of the cooling equipment is operating, the transformer may be operated at reduced load approximately as indicated in Table 13-6.

**TABLE 13-6** Operation with Limited Cooling Equipment

| Percent of cooling equipment in operation | Percent of rated load that may be carried |
|---|---|
| 33 | 50 |
| 40 | 60 |
| 50 | 70 |
| 80 | 90 |

## 13.16 TRANSFORMER CONDITION MONITORING AND ASSESSMENT

Transformer failures are due to any combination of electrical, thermal, or mechanical stresses. Figure 13-30 shows a typical failure distribution for power transformers with on-load tap changers. It can be seen from Fig. 13-30 that the majority of failures are related with the on-load tap changers and the windings. Figure 13-31 shows a typical failure distribution for transformers with off-load tap changers. It can be seen from Fig. 13-31 that the majority of failures are related with the terminals and the windings.

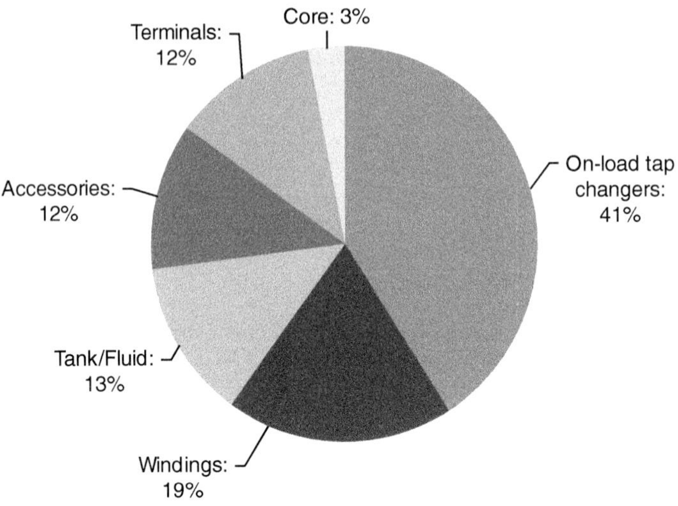


**FIGURE 13-30** Failure distribution for power transformers with on-load tap changers.

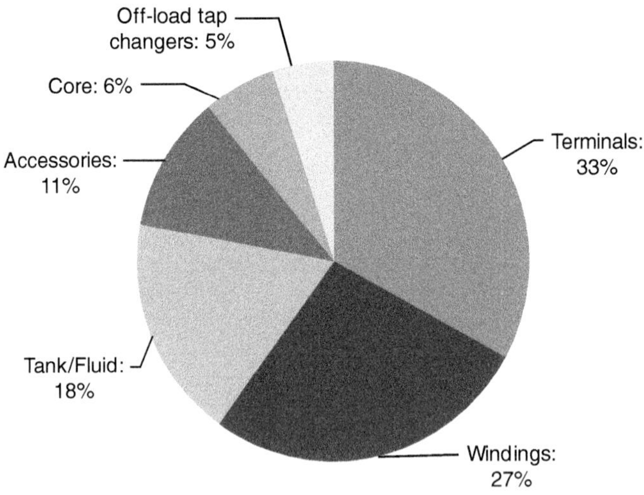


**FIGURE 13-31** Failure distribution for transformers with off-load tap changers.

Transformer failures are sometimes catastrophic and almost always include irreversible internal damage of the power transformer. Some of such failures may lead to high cost for replacement or repair. Moreover, an unplanned outage of a power transformer is very uneconomical. Since some faults may occur, it is important to closely monitor online and offline the power transformers.

There are three monitoring strategies for power transformers:

1. *Reliability-Centered Monitoring Strategy.* Various relays have been developed that respond to severe power failures requiring immediate removal of the faulty transformer from service, which means that outages are inevitable. This monitoring strategy cannot detect incipient faults.
2. *Time-Based Monitoring Strategy.* At regular time intervals, various offline tests are applied to detect possible incipient faults. This monitoring strategy is labor intensive and not cost-effective.
3. *Condition-Based Monitoring Strategy.* It employs advanced fault diagnosis techniques in order to detect online and offline incipient faults. It supplies information for transformer condition in real time, it processes this information and then it can determine when maintenance is needed. This monitoring strategy helps obtain the maximum practicable operating efficiency and the optimum life of power transformers. Moreover, this strategy minimizes the risks of premature failures and optimally schedules transformer maintenance.

In recent years, there is a trend to move from time-based monitoring to condition-based monitoring, also called condition monitoring and assessment (CMA) of power transformers. More specifically, nowadays, in CMA, monitoring is online and monitoring equipment is permanently mounted on the transformer. Since in developed countries failure rates of power transformers are low, typically 0.2% to 2.0% per year and per transformer, the monitoring equipment for failure prevention must be cost-effective.

There are four categories of diagnostic methods:

1. *Methods for Thermal-Related Faults.* These methods include dissolved gas analysis, degree of polymerization, furanic compounds analysis, and thermography.
2. *Methods for Dielectric-Related Faults.* These methods include localization of partial discharges and characterization of the type of partial discharges, also known as partial discharge analysis methods.
3. *Methods for Mechanical-Related Faults.* These methods include frequency response analysis and leakage inductance.
4. *Methods for General Degradation.* These methods include dielectric response, oil analysis, and furanic compounds analysis.

Dissolved gas analysis (DGA) is currently the most widely used preventive maintenance technique for online monitoring of oil-immersed transformers. During normal operations, there is a slow degradation of the mineral oil and certain gases are dissolved in the oil. However, during an electrical fault inside a transformer, gases are produced much more rapidly. Methods for interpretation of DGA are given in the IEEE C57.104 international standard.

Nowadays, the frequency response analysis (FRA) has received worldwide attention for detecting mechanical problems, for example, deformations or displacements in the windings and the core sheets, because these types of mechanical faults are difficult to locate using other techniques. Two FRA measurement techniques exist: impulse FRA (IFRA) and sweep FRA (SFRA). Details for FRA can be found in the international standards, for example, IEEE C57.149 and IEC 60076-18.

In CMA, the term "condition monitoring" is defined as online collection of data and includes sensor development, measurement techniques, and data acquisition. The term "condition assessment" is defined as the process that contains interpretation of data acquired by condition monitoring. Condition assessment may also include offline measurements on transformers. Figure 13-32 shows the processes that are involved in CMA of power transformers.

Nowadays, several manufacturers offer specialized cost-effective products for online CMA of power transformers. The products are generally easy to install and use. Besides the standard functions, the products are usually modular and expandable for additional requirements that may be needed in the future. The products usually include expert systems, that is, algorithms that analyze the online acquired data and provide recommendations and information regarding transformer operation and potential need for maintenance. The benefits of using the online CMA products include reduced probability of unexpected outages, condition-based maintenance cutting maintenance costs, instant diagnosis and data for condition assessment, cost-effective solution, and low overall investment cost due to product customization with parameterization.

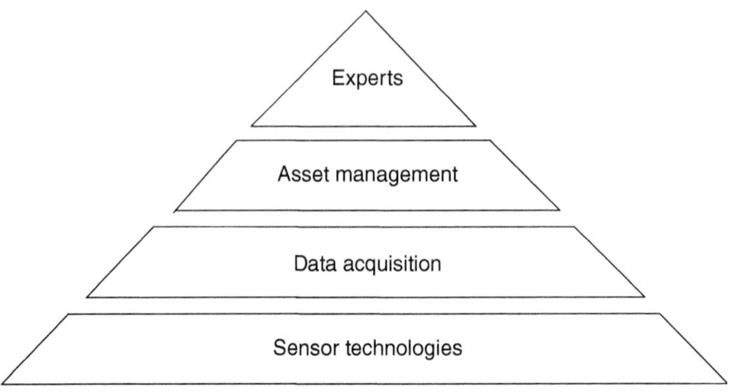


**FIGURE 13-32** Processes involved in condition monitoring and assessment of power transformers.

## 13.17 TRANSFORMER LOSS EVALUATION AND SELECTION

Energy efficient transformers cost more but use less energy than low-efficiency transformers. The decision as to whether to purchase a low-cost, inefficient transformer or a more expensive, energy-efficient transformer, is primarily an economic one. The common practice used by electric utilities for determining the cost-effectiveness of distribution transformers is based on the total owning cost (TOC) method, where TOC is equal to the sum of transformer-purchasing price plus the cost of transformer losses throughout the transformer lifetime.

There are various methods to compute the total owning cost of transformers. Some methods compute TOC for industrial and commercial users of transformers, such as the method of Sec. 13.17.1. Other methods compute TOC for electric utilities, such as the method of IEEE C57.120 standard. Recently, TOC method has been developed that additionally takes into account the environmental cost due to transformer losses.

### 13.17.1 Evaluation Methodology

The method of this subsection can be applied by the industrial and commercial users of transformers in order to evaluate and select their transformers. Among the various transformer offers, the most cost-effective transformer is the one that minimizes the total owning cost, *TOC*, which is computed as follows:

$$TOC = BP + A \cdot NLL + B \cdot LL \tag{13-51}$$

where *BP* is the transformer bid price ($) or purchasing price, *NLL* is the transformer no-load loss (W), *LL* is the transformer load loss (W), *A* is the no-load loss factor ($/W), and *B* is the load loss factor ($/W).

The no-load loss factor is computed by

$$A = PV_m \cdot EP \cdot HPY \cdot 10^{-3} \tag{13-52}$$

where *EP* is the electricity price ($/kWh), *HPY* is the number of hours per year that the transformer operates, and $PV_m$ is the present value multiplier that is computed by

$$PV_m = \frac{(1+d)^N - 1}{d \cdot (1+d)^{N-1}} \tag{13-53}$$

where *d* is the annual discount rate, and *N* is the number of years of transformer lifetime.

The load loss factor is computed by

$$B = A \cdot L^2 \tag{13-54}$$

where $L$ is the per-unit load of transformer.

The transformer annual energy loss (kWh/year) is

$$EL = (NLL + LL \cdot L^2) \cdot HPY \cdot 10^{-3} \tag{13-55}$$

The annual energy savings ($/year) by selecting and using transformer $j$ instead of transformer $i$ are given by the formula:

$$ES = (EL_i - EL_j) \cdot EP \tag{13-56}$$

where $EL_i$ is the annual energy loss of transformer $i$.

The simple payback (years) by selecting and using transformer $j$ instead of transformer $i$ is calculated as follows:

$$SP = \frac{BP_j - BP_i}{(EL_i - EL_j) \cdot EP} \tag{13-57}$$

where $BP_i$ is the bid price of transformer $i$.

*Example 13-5.* Table 13-7 shows two competing offers for three-phase, oil-immersed, distribution transformers with 250-kVA rated power. The electricity price is 0.12 $/kWh. The discount rate is 7%. The transformer operates 8760 h/year. The transformer lifetime is 30 years. Compute the total owning cost of the two offers if the per-unit load is varied from 0.0 to 1.0 with step 0.1.

**TABLE 13-7** Data for Two Competing Transformer Offers

| Parameter | Offer 1 | Offer 2 |
|---|---|---|
| Rated power (kVA) | 250 | 250 |
| No-load loss (W) | 650 | 425 |
| Load loss (W) | 4,200 | 2,750 |
| Bid price ($) | 11,750 | 14,800 |

*Solution.* The computation of the total owning cost of offer 1 will be shown for the case the per-unit load $L$ is 0.4.

The present value multiplier is

$$PV_m = \frac{(1+d)^N - 1}{d \cdot (1+d)^{N-1}} = \frac{(1+0.07)^{30} - 1}{0.07 \cdot (1+0.07)^{30-1}} \Rightarrow PV_m = 13.2777$$

The no-load loss factor is

$$A = PV_m \cdot EP \cdot HPY \cdot 10^{-3} = 13.2777 \cdot 0.12 \cdot 8760 \cdot 10^{-3} \Rightarrow A = 13.9575 \text{ \$/W}$$

The load loss factor is

$$B = A \cdot L^2 = 13.9575 \cdot 0.4^2 \Rightarrow B = 2.2332 \text{ \$/W}$$

The total owning cost of offer 1 for per-unit load 0.4 is

$$TOC = BP + A \cdot NLL + B \cdot LL = 11750 + 13.9575 \cdot 650 + 2.2332 \cdot 4200 \Rightarrow TOC = \$\ 30202$$

The total owning cost of the offers 1 and 2 for the different values of per-unit load is shown in Table 13-8 and Fig. 13-33.

**TABLE 13-8** Total Owning Cost for the Two Competing Transformer Offers of Example 13-5

| Per-unit load | Total owning cost of offer 1, $ | Total owning cost of offer 2, $ |
|---|---|---|
| 0.0 | 20,822 | 20,732 |
| 0.1 | 21,409 | 21,116 |
| 0.2 | 23,167 | 22,267 |
| 0.3 | 26,098 | 24,186 |
| 0.4 | 30,202 | 27,873 |
| 0.5 | 35,478 | 30,328 |
| 0.6 | 41,926 | 34,550 |
| 0.7 | 49,547 | 39,540 |
| 0.8 | 58,340 | 45,297 |
| 0.9 | 68,306 | 51,822 |
| 1.0 | 79,444 | 59,115 |

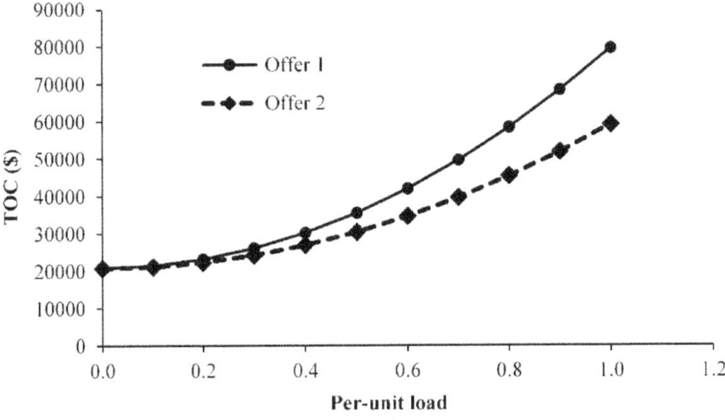


**FIGURE 13-33** Total owning cost for the two offers of Table 13-7 as a function of per-unit load.

*Example 13-6.* Table 13-9 shows nine competing offers for three-phase, oil-immersed, distribution transformers with 1000-kVA rated power. The electricity price is 0.12 $/kWh. The discount rate is 7%. The transformer operates 8760 h/year. The transformer lifetime is 30 years. The per-unit load is 0.5.

Compute the simple payback and the savings due to the selection of the offer with the lowest total owning cost instead of the offer with the lowest bid price. Moreover, if the electricity price is varied from 0.08 to 0.16 $/kWh with step 0.01 $/kWh, compute the simple payback and the savings due to the selection of the offer with the lowest total owning cost instead of the offer with the lowest bid price.

*Solution.* Figure 13-34 ranks the offers of Table 13-9 from the lowest to the highest bid price. If the purchasing criterion is simply the lowest bid price, then offer O4 is the best choice, as can be seen from Fig. 13-34.

The computation of the annual energy losses and the total owning cost of Offer O1 will be shown for the case the electricity price *EP* is 0.12 $/kWh.

**TABLE 13-9** Data for Nine Competing Transformer Offers

| Offer | No-load loss, W | Load loss, W | Bid price, $ |
|---|---|---|---|
| O1 | 1700 | 10,500 | 28,630 |
| O2 | 1400 | 10,500 | 29,600 |
| O3 | 1100 | 10,500 | 32,820 |
| O4 | 1700 | 13,000 | 27,150 |
| O5 | 1400 | 13,000 | 27,420 |
| O6 | 1100 | 13,000 | 30,650 |
| O7 | 1700 | 9,500 | 30,500 |
| O8 | 1400 | 9,500 | 31,550 |
| O9 | 1100 | 9,500 | 35,300 |

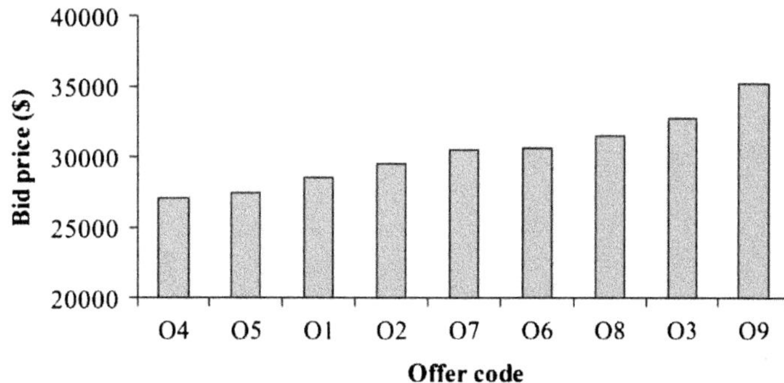


**FIGURE 13-34** Ranking of offers of Table 13-9 according to the bid price.

The annual energy losses of Offer O1 are

$$EL = (NLL + LL \cdot L^2) \cdot HPY \cdot 10^{-3} = (1700 + 10500 \cdot 0.5^2) \cdot 8760 \cdot 10^{-3} \Rightarrow$$

$$EL = 37887 \text{ kWh/year}$$

The present value multiplier is

$$PV_m = \frac{(1+d)^N - 1}{d \cdot (1+d)^{N-1}} = \frac{(1+0.07)^{30} - 1}{0.07 \cdot (1+0.07)^{30-1}} \Rightarrow PV_m = 13.2777$$

The no-load loss factor is

$$A = PV_m \cdot EP \cdot HPY \cdot 10^{-3} = 13.2777 \cdot 0.12 \cdot 8760 \cdot 10^{-3} \Rightarrow A = 13.9575 \text{ \$/W}$$

The load loss factor is

$$B = A \cdot L^2 = 13.9575 \cdot 0.5^2 \Rightarrow B = 3.4894 \text{ \$/W}$$

The total owning cost of Offer O1 is

$$TOC = BP + A \cdot NLL + B \cdot LL = 28630 + 13.9575 \cdot 1700 + 3.4894 \cdot 10500 \Rightarrow$$

$$TOC = \$88,996$$

The annual energy losses and the total owning cost of the nine offers of Table 13-9 are shown in Table 13-10. Figure 13-35 ranks the offers of Table 13-10 from the lowest to the highest total owning cost. If the purchasing criterion is the lowest total owning cost, then Offer O9 is the best choice, as can be seen from Fig. 13-35.

**TABLE 13-10** Annual Energy Losses and Total Owning Cost for the offers of Table 13-9 When the Electricity Price is 0.12 $/kWh

| Offer | Annual energy losses, kWh/year | Total owning cost, $ |
|---|---|---|
| O1 | 37,887 | 88,996 |
| O2 | 35,259 | 85,779 |
| O3 | 32,631 | 84,812 |
| O4 | 43,362 | 96,240 |
| O5 | 40,734 | 92,322 |
| O6 | 38,106 | 91,365 |
| O7 | 35,697 | 87,377 |
| O8 | 33,069 | 84,240 |
| O9 | 30,441 | 83,802 |

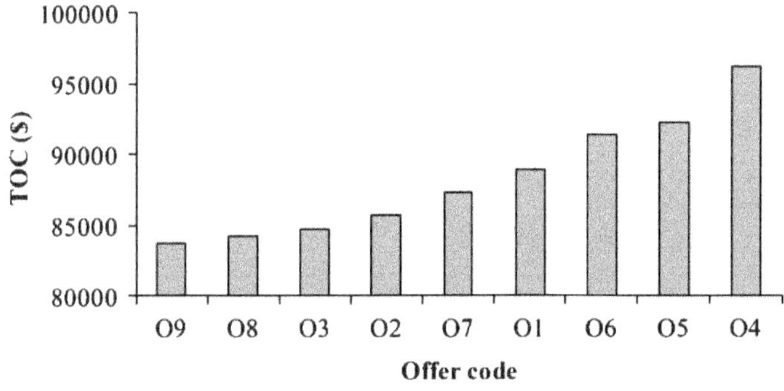


**FIGURE 13-35** Ranking of offers of Table 13-9 according to the total owning cost.

Figure 13-35 shows that offers O9 and O4 have the lowest and highest total owning cost, respectively.

Due to the selection of the offer with the lowest total owning cost (i.e., Offer O9, as Fig. 13-35 shows) instead of the offer with the lowest bid price (i.e., offer O4, as Fig. 13-34 shows), the savings in total owning cost are

$$S_{TOC} = TOC_4 - TOC_9 = 96{,}240 - 83{,}802 \Rightarrow S_{TOC} = \$12{,}438$$

The annual energy savings by selecting Offer O9 (with the lowest total owning cost) instead of Offer O4 (with the lowest bid price) are

$$ES = (EL_4 - EL_9)\cdot EP = (43{,}362 - 30{,}441)\cdot 0.12 \Rightarrow ES = 1550.52\ \$/\text{year}$$

The simple payback (years) by selecting Offer O9 (with the lowest total owning cost) instead of Offer O4 (with the lowest bid price) is

$$SP = \frac{BP_9 - BP_4}{(EL_4 - EL_9)\cdot EP} = \frac{35{,}300 - 27{,}150}{(43{,}362 - 30{,}441)\cdot 0.12} \Rightarrow SP = 5.26\ \text{years}$$

All the previous calculations have been done for the case the electricity price is 0.12 $/kWh. The same calculations are done again for all the considered electricity prices and the results are presented in Table 13-11. Moreover, the results of Table 13-11 are plotted in Figs. 13-36 through 13-38.

**TABLE 13-11** Savings due to the Selection of the Offer Corresponding to the Minimum TOC instead of the Offer O4 corresponding to the minimum bid price

| | Offer with minimum *TOC* | | | Offer O4 | | Selection of offer with minimum *TOC* instead of Offer O4 | | |
|---|---|---|---|---|---|---|---|---|
| *EP* ($/kWh) | Offer | *EL*, kWh/year | *TOC*, $ | *EL*, kWh/year | *TOC*, $ | *ES*, $/year | *SP*, years | $S_{TOC}$, $ |
| 0.08 | O8 | 33,069 | 66,676 | 43,362 | 73,210 | 823 | 5.34 | 6,533 |
| 0.09 | O8 | 33,069 | 71,067 | 43,362 | 78,967 | 926 | 4.75 | 7,900 |
| 0.10 | O8 | 33,069 | 75,458 | 43,362 | 84,725 | 1,029 | 4.27 | 9,267 |
| 0.11 | O9 | 30,441 | 79,760 | 43,362 | 90,482 | 1,421 | 5.73 | 10,722 |
| 0.12 | O9 | 30,441 | 83,802 | 43,362 | 96,240 | 1,551 | 5.26 | 12,437 |
| 0.13 | O9 | 30,441 | 87,844 | 43,362 | 101,997 | 1,680 | 4.85 | 14,153 |
| 0.14 | O9 | 30,441 | 91,886 | 43,362 | 107,755 | 1,809 | 4.51 | 15,869 |
| 0.15 | O9 | 30,441 | 95,928 | 43,362 | 113,512 | 1,938 | 4.21 | 17,584 |
| 0.16 | O9 | 30,441 | 99,970 | 43,362 | 119,269 | 2,067 | 3.94 | 19,300 |

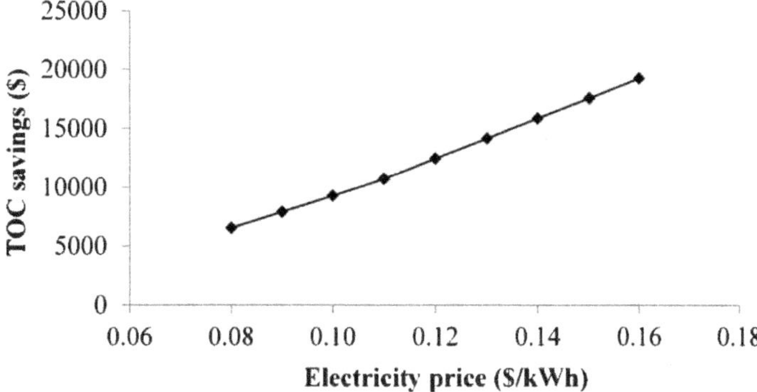


**FIGURE 13-36** Total owning cost savings due to selection of the offer corresponding to the minimum TOC instead of the offer corresponding to the minimum bid price. These savings are plotted as a function of electricity price.

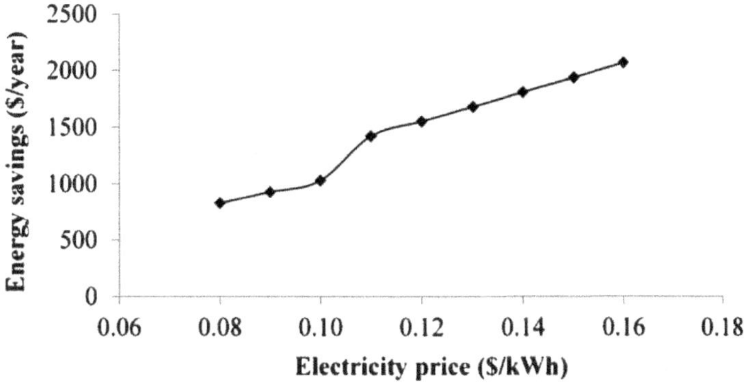


**FIGURE 13-37** Annual energy savings due to selection of the offer corresponding to the minimum TOC instead of the offer corresponding to the minimum bid price.

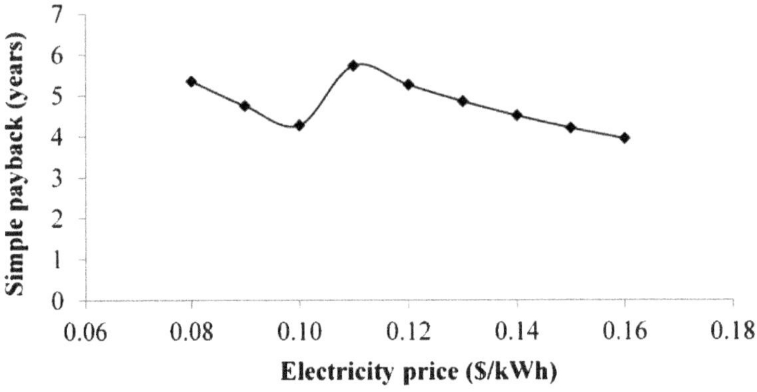


**FIGURE 13-38** Simple payback due to selection of the offer corresponding to the minimum TOC instead of the offer corresponding to the minimum bid price.

It can be seen from Table 13-11 that if the electricity price is 0.08, or 0.09, or 0.10 $/kWh, then the most cost-effective offer is Offer O4, since it has the lowest total owning cost, while for all the other considered electricity prices, Offer O9 is the most cost-effective offer.

## 13.18 TRANSFORMER STANDARDS

Transformer manufacturing is based on international standards as well as on specific customer needs. From time to time, some of the standards may be modified and in that case they are republished. Tables 13-12 presents IEEE/ANSI standards for transformers. Tables 13-13 presents IEC standards for transformers.

**TABLE 13-12** IEEE/ANSI Standards for Transformers

| Standard | Year | Description |
|---|---|---|
| C57.12.00 | 2015 | General requirements for liquid-immersed distribution, power, and regulating transformers |
| C57.12.01 | 2015 | General requirements for dry-type distribution and power transformers |
| C57.12.10 | 2010 | Requirements for liquid-immersed power transformers |
| C57.12.20 | 2011 | Requirements for overhead type distribution transformers, 500 kVA and smaller; high voltage 34,500 V and below; low voltage 7,970/13,800 Y V and below |
| C57.12.21 | 1992 | Requirements for pad-mounted, compartmental-type self-cooled, single-phase distribution transformers with high voltage bushings; high voltage 34,500 GRYD/19,920 V and below; low voltage 240/120 V; 167 kVA and smaller |
| C57.12.23 | 2009 | Submersible single-phase transformers: 167 kVA and smaller; high voltage 25,000 V and below; low voltage 600 V and below |
| C57.12.24 | 2009 | Submersible, three-phase transformers, 3750 kVA and smaller: high voltage, 34,500 GrdY/19,920 V and below; low voltage, 600 V and below |
| C57.12.35 | 2013 | Bar coding for distribution transformers and step-voltage regulators |
| C57.12.37 | 2015 | Electronic reporting of distribution transformer test data |
| C57.12.40 | 2011 | Network, three-phase transformers, 2500 kVA and smaller; high voltage, 34,500 GrdY/19,920 and below; low voltage, 600 V and below; subway and vault types (liquid immersed) |
| C57.12.44 | 2014 | Requirements for secondary network protectors |
| C57.12.50 | 1981 | Requirements for ventilated dry-type distribution transformers, 1–500 kVA, single-phase, and 15–500 kVA, three-phase, with high voltage 601–34500 V, low voltage 120–600 V |

**TABLE 13-12** IEEE/ANSI Standards for Transformers (*Continued*)

| Standard | Year | Description |
|---|---|---|
| C57.12.51 | 2008 | General requirements for ventilated dry-type power transformers, 501 kVA and larger, three-phase, with high voltage 601–34500 V, low voltage 208Y/120 V to 4160 V |
| C57.12.52 | 2012 | General requirements for sealed dry-type power transformers, 501 kVA and higher, three-phase, with high voltage 601–34,500 V, low voltage 208Y/120–4,160 V |
| C57.12.55 | 1987 | Transformers used in unit installations, including unit substations |
| C57.12.56 | 1986 | Test procedure for thermal evaluation of insulation systems for ventilated dry-type power and distribution transformers |
| C57.12.58 | 1991 | Guide for conducting a transient voltage analysis of a dry-type transformer coil |
| C57.12.59 | 2015 | Guide for dry-type transformer through-fault current duration |
| C57.12.60 | 2009 | Test procedure for thermal evaluation of insulation systems for dry-type power and distribution transformers, including open-wound, solid-cast, and resin-encapsulated transformers |
| C57.12.70 | 2011 | Terminal markings and connections for distribution and power transformers |
| C57.12.80 | 2010 | Terminology for power and distribution transformers |
| C57.12.90 | 2015 | Test code for liquid-immersed distribution, power, and regulating transformers |
| C57.12.91 | 2011 | Test code for dry-type distribution and power transformers |
| C57.15 | 2009 | Requirements, terminology, and test code for step-voltage regulators |
| C57.18.10 | 1998 | Practices and requirements for semiconductor power rectifier transformers |
| C57.91 | 2011 | Guide for loading mineral-oil-immersed transformers and step-voltage regulators |
| C57.93 | 2007 | Guide for installation and maintenance of liquid-immersed power transformers |
| C57.94 | 2015 | Recommended practice for installation, application, operation, and maintenance of dry-type distribution and power transformers |
| C57.96 | 2013 | Guide for loading dry-type distribution and power transformers |
| C57.98 | 2011 | Guide for transformer impulse tests |
| C57.100 | 2011 | Test procedure for thermal evaluation of insulation systems for liquid-immersed distribution and power transformers |
| C57.104 | 2008 | Guide for the interpretation of gases generated in oil-immersed transformers |
| C57.105 | 1978 | Guide for application of transformer connections in three-phase distribution systems |
| C57.110 | 2008 | Recommended practice for establishing liquid-filled and dry-type power and distribution transformer capability when supplying nonsinusoidal load currents |
| C57.111 | 1989 | Guide for acceptance of silicone insulating fluid and its maintenance in transformers |
| C57.113 | 2010 | Recommended practice for partial discharge measurement in liquid-filled power transformers and shunt reactors |
| C57.116 | 2014 | Guide for transformers directly connected to generators |
| C57.120 | 1991 | Loss evaluation guide for power transformers and reactors |
| C57.121 | 1998 | Guide for acceptance and maintenance of less-flammable hydrocarbon fluid in transformers |
| C57.123 | 2010 | Guide for transformer loss measurement |
| C57.124 | 1991 | Recommended practice for the detection of partial discharge and the measurement of apparent charge in dry-type transformers |
| C57.125 | 2015 | Guide for failure investigation, documentation, analysis, and reporting for power transformers and shunt reactors |
| C57.127 | 2007 | Guide for the detection and location of acoustic emissions from partial discharges in oil-immersed power transformers and reactors |
| C57.129 | 2007 | General requirements and test code for oil-immersed HVDC converter transformers |
| C57.130 | 2015 | Guide for the use of dissolved gas analysis applied to factory temperature rise tests for the evaluation of mineral oil-immersed transformers and reactors |
| C57.131 | 2012 | Requirements for load tap changers |
| C57.134 | 2013 | Guide for determination of hottest-spot temperature in dry-type transformers |
| C57.135 | 2011 | Guide for the application, specification, and testing of phase-shifting transformers |
| C57.136 | 2000 | Guide for sound level abatement and determination for liquid-immersed power transformers and shunt reactors rated over 500 kVA |
| C57.138 | 1998 | Recommended practice for routine impulse test for distribution transformers |
| C57.140 | 2006 | Guide for evaluation and reconditioning of liquid-immersed power transformers |
| C57.142 | 2010 | Guide to describe the occurrence and mitigation of switching transients induced by transformers, switching device, and system interaction |
| C57.143 | 2012 | Guide for application for monitoring equipment to liquid-immersed transformers and components |
| C57.149 | 2012 | Guide for the application and interpretation of frequency response analysis for oil-immersed transformers |
| C57.150 | 2012 | Guide for the transportation of transformers and reactors rated 10,000 kVA or higher |

**TABLE 13-13** IEC standards for transformers

| Standard | Year | Description |
|---|---|---|
| 60076-1 | 2011 | Power transformers—Part 1: General requirements |
| 60076-2 | 2011 | Power transformers—Part 2: Temperature rise for liquid-immersed transformers |
| 60076-3 | 2013 | Power transformers—Part 3: Insulation levels, dielectric tests and external clearances in air |
| 60076-4 | 2002 | Power transformers—Part 4: Guide to the lightning impulse and switching impulse testing—power transformers and reactors |
| 60076-5 | 2006 | Power transformers—Part 5: Ability to withstand short circuit |
| 60076-6 | 2007 | Power transformers—Part 6: Reactors |
| 60076-7 | 2005 | Power transformers—Part 7: Loading guide for oil-immersed power transformers |
| 60076-8 | 1997 | Power transformers—Part 8: Application guide |
| 60076-10 | 2016 | Power transformers—Part 10: Determination of sound levels |
| 60076-11 | 2004 | Power transformers—Part 11: Dry-type power transformers |
| 60076-12 | 2008 | Power transformers—Part 12: Loading guide for dry-type power transformers |
| 60076-13 | 2006 | Power transformers—Part 13: Self-protected liquid-filled transformers |
| 60076-14 | 2013 | Power transformers—Part 14: Liquid-immersed power transformers using high-temperature insulation materials |
| 60076-15 | 2015 | Power transformers—Part 15: Gas-filled power transformers |
| 60076-16 | 2011 | Power transformers—Part 16: Transformers for wind turbine applications |
| 60076-18 | 2012 | Power transformers—Part 18: Measurement of frequency response |
| 60076-19 | 2013 | Power transformers—Part 19: Rules for the determination of uncertainties in the measurement of the losses on power transformers and reactors |
| 60137 | 2008 | Insulated bushings for alternating voltages above 1000 V |
| 60156 | 1995 | Insulating liquids—determination of the breakdown voltage at power frequency—test method |
| 60214-1 | 2014 | Tap changers—Part 1: Performance requirements and test methods |
| 60214-2 | 2004 | Tap changers—Part 2: Application guide |
| 61378-1 | 2011 | Convertor transformers—Part 1: Transformers for industrial applications |
| 61378-2 | 2001 | Convertor transformers—Part 2: Transformers for HVDC applications |
| 61378-3 | 2015 | Convertor transformers—Part 3: Application guide |
| 61558-1 | 2005 | Safety of power transformers, power supplies, reactors and similar products—Part 1: General requirements and tests |

## 13.19 FURTHER READING


Amoiralis, E. I., Georgilakis, P. S., Tsili, M. A., and Kladas, A. G.: Global transformer optimization method using evolutionary design and numerical field computation, *IEEE Transactions on Magnetics*, Vol. 45, pp. 1720–1723, Mar. 2009.

Beaty, H. W. and Fink, D. G.: *Standard Handbook for Electrical Engineers, 16th ed.*, McGraw-Hill, New York, 2013.

Bengtsson, C.: Status and trends in transformer monitoring, *IEEE Transactions on Power Delivery*, Vol. 11, No. 3, pp. 1379–1384, Jul. 1996.

Chapman, S. J.: *Electric Machinery Fundamentals, 4th ed.*, McGraw-Hill, New York, 2005.

Del Vecchio, R. M., Poulin, B., Feghali, P. T., Shah, D. M., and Ahuja R.: *Transformer Design Principles: With Applications to Core-Form Power Transformers, 2nd ed.*, CRC Press, New York, 2010.

Georgilakis, P. S.: Decision support system for evaluating transformer investments in the industrial sector, *Journal of Materials Processing Technology*, Vol. 181, pp. 307–312, Jan. 2007.

Georgilakis, P. S.: Recursive genetic algorithm-finite element method technique for the solution of transformer manufacturing cost minimisation problem, *IET Electric Power Applications*, Vol. 3, pp. 514–519, Nov. 2009.

Georgilakis, P. S.: *Spotlight on Modern Transformer Design*, Springer, London, 2009.

Georgilakis, P. S. and Amoiralis, E. I.: Distribution transformer cost evaluation methodology incorporation environmental cost, *IET Generation, Transmission, and Distribution*, Vol. 4, No. 7, pp. 861–872, Jul. 2010.

Georgilakis, P. S., Doulamis, N. D., Doulamis, A. D., Hatziargyriou, N. D, Kollias, S. D.: A novel iron loss reduction technique for distribution transformers based on a combined genetic algorithm-neural network approach, *IEEE Transactions on Systems, Man and Cybernetics, Part C*, Vol. 31, pp. 16–34, Feb. 2001.

Georgilakis, P., Hatziargyriou, N., Paparigas, D.: AI helps reduce transformer iron losses, *IEEE Computer Applications in Power*, Vol. 12, pp. 41–46, Oct. 1999.

Georgilakis, P. S. and Kagiannas, A. G.: A novel validated solution for lightning and surge protection of distribution transformers, *International Journal of Electrical Power and Energy Systems*, Vol. 63, pp. 373–381, Dec. 2014.

Georgilakis, P. S., Katsigiannis, J. A., Valavanis, K. P., Souflaris, A. T.: A systematic stochastic petri net based methodology for transformer fault diagnosis and repair actions, *Journal of Intelligent and Robotic Systems*, Vol. 45, pp. 181–201, Feb. 2006.

Georgilakis, P. S., Tsili, M. A., Souflaris, A. T.: A heuristic solution to the transformer manufacturing cost optimization problem, *Journal of Materials Processing Technology*, Vol. 181, pp. 260–266, Jan. 2007.

Harlow, J. H.: *Electric Power Transformer Engineering, 3rd ed.*, CRC Press, New York, 2012.

Kulkarni, S. V. and Khaparde, S. A.: *Transformer Engineering Design and Practice*, Marcel Dekker, New York, 2004.

Olivares-Galván, G. C., Georgilakis, P. S., Ocon-Valdez, R.: A review of transformer losses, *Electric Power Components and Systems*, Vol. 37, pp. 1046–1062, Sep. 2009.

Tang, W. H. and Wu, Q. H.: *Condition Monitoring and Assessment of Power Transformers Using Computational Intelligence*, Springer, London, 2011.

Wang, M., Vandermaar, A. J., Srivastava, K. D.: Review of condition assessment of power transformers in service, *IEEE Electrical Insulation Magazine*, Vol. 18, pp. 12–25, Nov./Dec. 2002.

Winders, J. J., Jr.: *Power Transformers Principles and Applications*, Marcel Dekker, New York, 2002.

14

# ELECTRIC MACHINES: GENERATORS

Dan M. Ionel

*Professor and L. Stanley Pigman Chair in Power, Department of Electrical and Computer Engineering, University of Kentucky*

Erik Abromitis

*Generator Engineering, Siemens Energy, Inc.*

Samuel A. Drinkut

*Generator Engineering, Siemens Energy, Inc.*

Franklin T. Emery

*Generator Engineering, Siemens Energy, Inc.*

Om P. Malik

*Professor Emeritus, Department of Electrical and Computer Engineering, University of Calgary*

Osama A. Mohammed

*Professor, Department of Electrical and Computer Engineering, Florida International University*

Vandana Rallabandi

*Research Engineer, Department of Electrical and Computer Engineering, University of Kentucky*

Narges Taran

*Research Engineer, Department of Electrical and Computer Engineering, University of Kentucky*

## 14.1 PRIME MOVERS

### 14.1.1 Steam Prime Movers

Steam prime movers are either reciprocating engines or turbines. Reciprocating engines offer low speed (100 to 400 r/min), high efficiency in small sizes (less than 500 hp), and high starting torque. Steam turbines are a product of the twentieth century and completely dominate the field of power generation and variable-speed applications in ship propulsion (through gears), centrifugal pumps, compressors, and blowers. Modern steam turbine types include condensing turbines most commonly in power plants, noncondensing or back pressure turbines widely used for process steam applications, reheat turbines also almost exclusively used in power plants, and induction turbines.

Steam engine types include (1) simple D side engines rated less than 0.01 hp and are used for auxiliary drive and (2) single cylinder counterflow and uniflow engines rated less than 1000 hp and are used for generator or equipment drives.

The basic thermodynamic cycle is shown in Fig. 14-1. The network of the cycle is represented by the area enclosed within the diagram and is represented by the mean effective pressure (mep), that is, the net work (area) divided by the length of the diagram. The power output is computed by the "plan" equation:

$$\text{hp} = \frac{p_m Lan}{33{,}000} \tag{14-1}$$

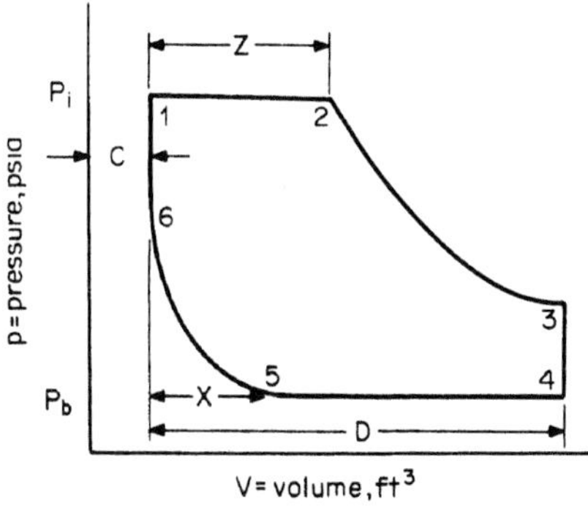


| Where | Usual value |
|---|---|
| $D$ = displacement | 0.05–20 ft² |
| $C$ = clearance | 0.03–0.2 |
| $Z$ = cutoff, fraction of $D$ | 0.1–0.6 |
| $X$ = compression, fraction of $D$ | 0.1–0.8 |
| $p_i$ = initial pressure | 100–300 lb/in² (abs.) |
| $p_b$ = back pressure | 2–30 lb/in² (abs.) |
| $p_m$ = mean effective pressure | 50–125 lb/in² (abs.) |

**FIGURE 14-1** Pressure-volume diagram for a steam-engine cycle. Phase 1-2, constant-pressure admission at $P_i$; phase 2-3, expansion, $pv = C$; phase 3-4, release; phase 4-5, constant-pressure exhaust pipe at $P_b$; phase 5-6, compression, $pv = C$; phase 6-1, constant-volume admission.

where hp is horsepower; $p_m$ is mep, pounds per square inch; $L$ is length of stroke, feet; $a$ is net piston area, square inches; and $n$ is number of cycles completed per minute.

Basically, steam turbines are a series of calibrated nozzles through which heat energy is converted into kinetic energy which, in turn, is transferred to wheels or drums and delivered at the end of a rotating shaft as usable power.

Turbines are built in two distinct types: impulse and reaction. *Impulse turbines* have stationary nozzles, and the total stage pressure drop is taken across them. The kinetic energy generated is absorbed by the rotating buckets at essentially constant static pressure. Increased pressure drop can be efficiently utilized in a single stage (at constant wheel speed) by adding a row of turning vanes or "intermediates" which are followed by a second row of buckets. This is commonly called a *Curtis* or 2-row stage.

In the *reaction* design, both the stationary and rotating parts contain nozzles, and an approximately equal pressure drop is taken across each. The pressure drop across the rotating parts of reaction-design turbines requires full circumferential admission and much closer leakage control. The first stage of the turbine must be designed to pass the maximum flow through the unit at rated inlet steam conditions.

The turbine stage efficiency is defined as the actual energy delivered to the rotating blades divided by the ideal energy released to the stage in an isentropic expansion from $P_1$ to $P_2$ of the stage. The most important factors determining the stage efficiency are the relationship of the mean blade speed to the theoretical steam velocity, the aspect ratio (blade length/ passage width), and the aerodynamic shape of the passages. Figure 14-2 describes the typical variation in nozzle and bucket efficiencies with velocity ratio and nozzle height.

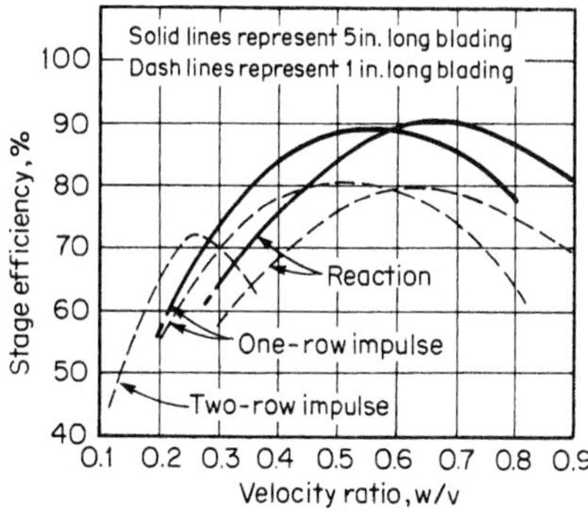


**FIGURE 14-2** Approximate relative efficiencies of turbine stage types.

A 100% efficiency cannot be obtained because of friction in the blading and clearance between the stationary and rotating parts, and because the nozzle angle cannot be zero degrees. Axial clearance increases in the stages further from the thrust bearing to satisfy the need to maintain a minimum clearance at extreme operating conditions when the differential expansion between the light rotor and heavy casing is at its worst. To reduce this leakage, radial spill-bands are used which are thin, metal-strip seals. There is also leakage around the nozzles between the bore of the blade ring or nozzle diaphragm and the drum or rotor that must be kept to a minimum.

There are some other losses associated with the reduced efficiency. The rotation losses consist of losses due to the rotation of the disks, the blades, and shrouds. A carryover loss occurs on certain stages when the kinetic energy of the steam leaving the rotating blades cannot be recovered by the following stage because of a difference in stage diameters or a large axial space between adjacent stages. Exhaust loss, nozzle end loss, and supersaturation and moisture loss are other types of losses that need to be considered and minimized.

The internal used energy of the stage is obtained by multiplying the isentropic energy available to the stage by the stage efficiency. The sum of the used energies of all stages in the turbine represents the total used energy of the turbine. The internal efficiency of the turbine can be obtained by dividing the total used energy by the overall isentropic available energy from throttle pressure and temperature conditions to the exhaust pressure.

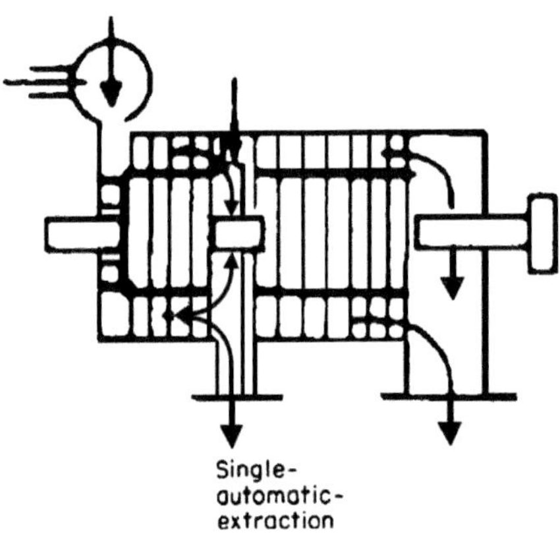


**FIGURE 14-3** Basic construction of an example noncondensing single automatic extraction turbine. (*General Electric.*)

Other losses which must be accounted for to arrive at the turbine overall efficiency include valve-stem and shaft-end packing leakages and bearing and oil-pump losses. Determination of the overall efficiency of a turbine and its driven equipment must take into account the losses of gears or generators and their bearings as well.

Since the early 1900s, horizontal-shaft units have been universally used. Horizontal units may be single-shaft or double-shaft, with single, double, or triple steam cylinders on one shaft. These modern units may be throttle or multiple-nozzle governed, have one or more steam extraction points, and exhibit innumerable variations in construction. Figure 14-3 shows an example modern noncondensing single automatic extraction turbine with construction details illustrated in cross section in Fig. 14-4.

Steam turbines require a number of systems and components to provide control and protective capability. These may be divided into two functional categories: (1) primary control systems and (2) secondary and/or protective control systems. *Primary control systems* may be further subdivided into control valves and associated operating gear, speed/load control, and pressure control. *Secondary or protective systems* consist of overspeed limiting devices, emergency valves, trip devices, and associated alarm devices.

## 14.1.2 Steam-Turbine Applications

Central station turbines usually are either a 60 MW nonreheat steam turbine which is typically installed in small utility plants, or a 600 to 800 MW tandem composed single reheat steam turbine typically used in large fossil fired central stations, or a nuclear steam turbine with a capacity of 1000 to 1300 MW.

In many industrial plants, large amounts of electric power and process steam at various pressure levels are needed. Industrial users can generate their own power and charge a large part of the cost to the process, because the steam is also needed. Plants have historically been built and expanded with various condensing and noncondensing extraction turbine types, as the process requires, and in sizes ranging from 1 to 200 MW.

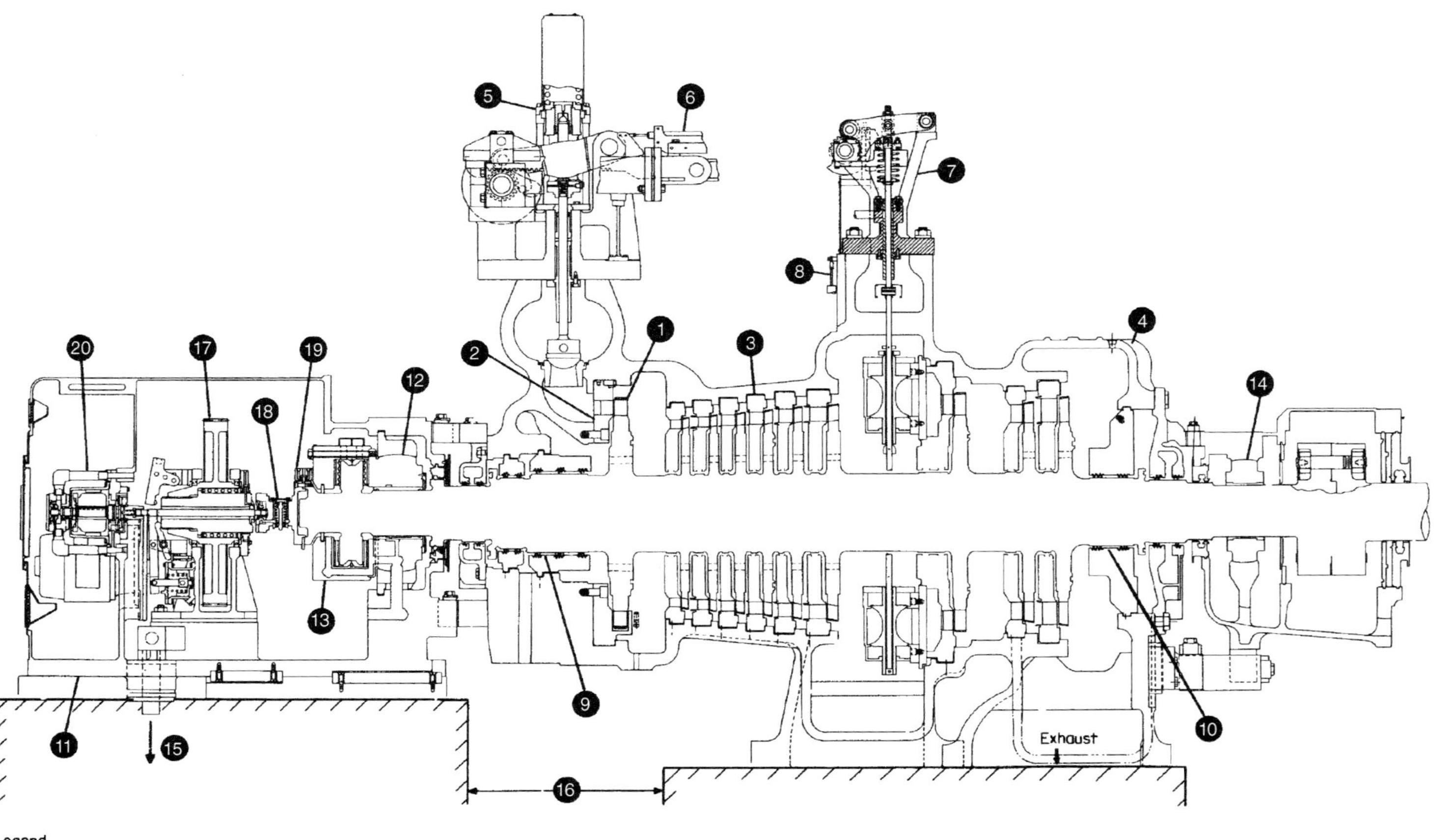


Legend
1. Bucket wheel
2. Nozzle plate
3. Nozzle diaphragm
4. High pressure head casing
5. Main steam control valves
6. Power actuator
7. Extraction control valves
8. Power actuator
9. HP Labyrinth packing rings
10. LP Labyrinth packing rings
11. Front standard
12. Bearing No.1
13. Thrust bear
14. Bearing No. 2
15. To lubrication system
16. Foundation
17. Turning gear
18. Overspeed governor
19. Thrust wear and failure device
20. PMG drive assembly

**FIGURE 14-4** Cross section of a modern single-automatic-extraction noncondensing steam turbine showing construction details. (*General Electric.*)

The steam turbine is used extensively as the prime mover for ship propulsion at ratings above 10,000 shp. They are used almost exclusively in naval combat ships (aircraft carriers and nuclear submarines) as well as for large auxiliary supply ships. Applications are predominantly nonreheat, but because of the steadily rising cost of fuel, reheat applications are gaining popularity.

### 14.1.3 Steam-Turbine Performance

The steam turbine constitutes the expansion portion of a vapor cycle, which requires separate devices, including a boiler, turbine, condenser, and feedwater pump, to complete the cycle. This vapor cycle for steam power plants is commonly called the Rankine cycle (Figs. 14-5 and 14-6) and is less efficient than the Carnot cycle because the exhaust vapor is completely liquefied to facilitate pumping, and because superheat is added at increasing temperature. The work of the cycle is equal to $h_1 - h_2$ minus the small pump work $h_4 - h_3 = v_3(P_4 - P_3)/J$ required, and the heat added to the cycle is equal to $h_1 - h_4$. Therefore

$$\text{Rankine-cycle efficiency} = \frac{(h_1 - h_2) - v_3(P_4 - P_3)/J}{h_1 - h_4} \tag{14-2}$$

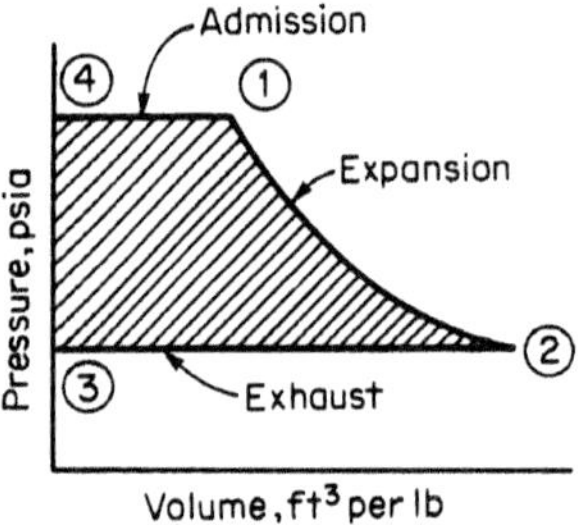


FIGURE 14-5 Pressure-volume diagram for the Rankine cycle: phase 4-1, constant-pressure admission; phase 1-2, complete isentropic expansion; phase 2-3, constant-pressure exhaust. Crosshatched area represents the work of the cycle.

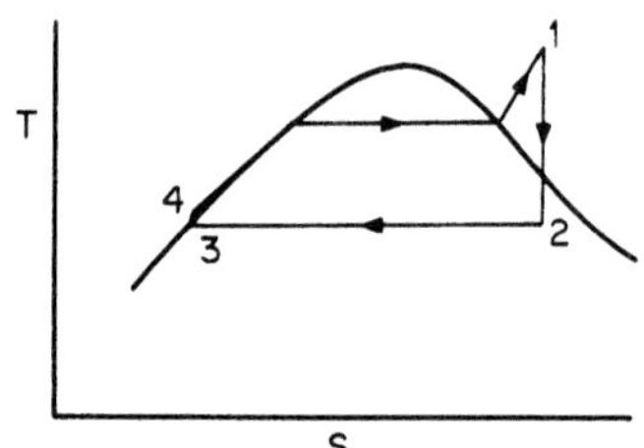


FIGURE 14-6 T-S diagram; nonreheat, nonextracting turbine cycle.

Cycle efficiency is not commonly used when comparing plant efficiencies because it is only indirectly determined, compared with heat rate, which can be quickly measured.

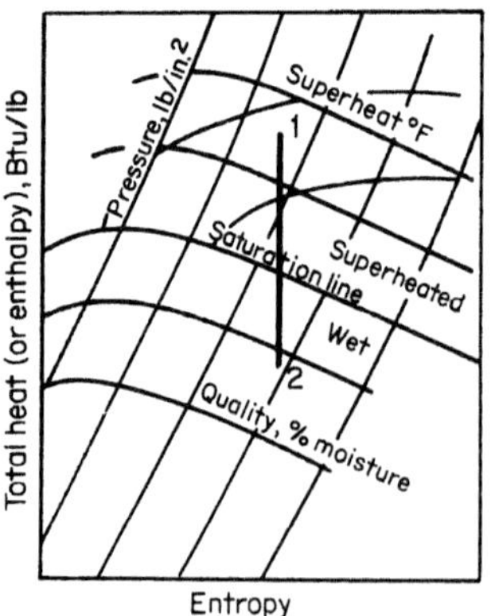


FIGURE 14-7 Steam chart (Mollier diagram).

In the process industries, such as paper and petrochemical, large amounts of steam are used. Considerable by-product power can be generated by raising the boiler pressure above the process pressure and expanding the steam through a noncondensing turbine before exhausting it to the process. In this cycle, no heat is rejected because the exhaust steam is required for process and the thermodynamic cycle efficiency of this power is affected only by the boiler efficiency, the auxiliary losses chargeable to the power generation, and the mechanical and electrical losses of the turbine and generator.

The station heat rate is used to measure power plant performance, but it is of little use in evaluating the specific pieces of equipment in the cycle. The engine efficiency of the steam turbine defines its actual performance to the ideal performance. The Rankine-cycle work of the turbine is most conveniently obtained by use of the Mollier diagram (Fig. 14-7), where

$$\Delta W = h_1 - h_2 \tag{14-3}$$

and $\Delta W$ = Rankine-cycle work in Btu/lb, $h_1$ = steam enthalpy at throttle in Btu/lb, $h_2$ = steam enthalpy at exhaust in Btu/lb, and $h_1$ and $h_2$ are at the same entropy (vertical line). The actual work of a real turbine is less than the ideal Rankine-cycle work, with engine efficiency defined as

$$\text{Engine efficiency} = \frac{\text{actual work, Btu/lb}}{\text{Rankine-cycle work, Btu/lb}} \tag{14-4}$$

Steam can be extracted at several stages in the turbine to heat feedwater being returned to the boiler. In the Rankine cycle (Figs. 14-5 and 14-6), it was shown that the feedwater was heated from $h_4$ to $h_1$ in the boiler. By raising the temperature $h_4$ entering the boiler close to the saturation temperature in the drum, less fuel will be consumed in evaporating each pound of steam to $h_1$ conditions. The heat in the extracted steam is added to the feedwater without loss, and the heat rejected to the condenser decreases as extraction flow increases. The kilowatts do not decrease inversely with extraction flow, however, as partial expansion is made down to the extraction stages. The result is an improvement in heat rate.

The actual heat rate of a regenerative cycle must be determined from a heat balance prepared by using the extraction conditions available from the turbine and the heater characteristics as specified. Improvement in the steam cycle has been rapid. The advancement of steam conditions, regenerative feed heating, reheating, and size of unit has brought the overall station heat rate down from 16,000 Btu/kWh to less than 8800 Btu/kWh on the best station.

The gas turbine developed very rapidly as a prime mover because of the improved steam cycle. During the 1950s, several exhaust-fired combined-cycle power plants were built, utilizing the exhaust gas from the gas turbine as the air supply for a fired main steam generator. After the Northeast Blackout of 1965, a large number of gas turbines were installed in the United States to serve as black start and peaking capacity units. As a result, they have become well established and accepted as a prime mover for peaking capacity.

Combined-cycle interest was renewed with the development of non-radiant-heat recovery steam generators, and the electric generation in combined-cycle plants changed from 80% to 90% steam-cycle power to 70% gas-cycle power. Since 1970, utilities have installed increasing numbers of this breed of combined-cycle plant, as they offer low initial cost, consume about one-third of the water used by straight steam plants, and provide a station heat rate 5% to 10% better than the most efficient steam plants. The major obstacle to universal acceptance of the steam-and-gas combined cycle is its fuel dependency on clean gaseous or liquid fuels.

The automatic-extraction turbine provides the capability of delivering extraction steam at more than one process pressure simultaneously. When a condensing element is used, the kilowatt output of the unit can be maintained if the process flow varies, and will permit generation in excess of by-product power capability. The base efficiency for an automatic-extraction turbine is less than that of a straight condensing or straight noncondensing turbine because of (1) the introduction of a second control stage and accompanying parasitic losses, (2) the partial-load loss resulting when the high-pressure section of the automatic-extraction unit is passing only the nonextraction flow, and (3) the decreased pressure ratio of each section of the unit.

## 14.1.4 Gas Turbines

Internal combustion engines, such as conventional automotive engines, operate on the Otto cycle; injection engines operate on the diesel cycle; and the gas or combustion turbine operates on the Brayton cycle (Fig. 14-8), also called the *gas-turbine simple cycle*. Referring to Fig. 14-9*a*, an axial or centrifugal compressor delivers the compressed air to the combustion system, and fuel is burned to increase the fluid temperature.

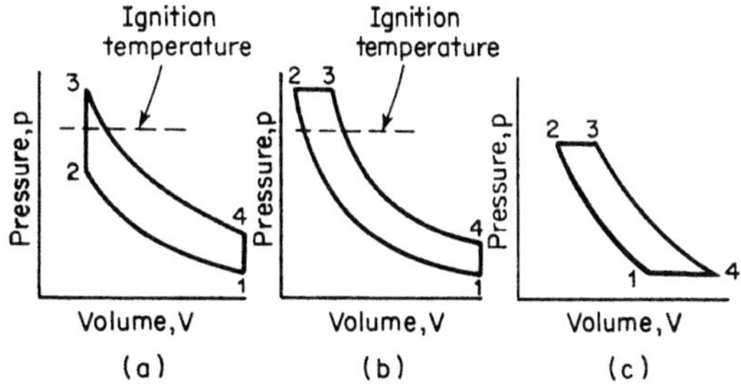


**FIGURE 14-8** Ideal indicator cards (pressure-volume diagrams) for internal combustion cycles; (*a*) Otto cycle; (*b*) diesel cycle; (*c*) Brayton cycle. In general, phase 1-2 represents isentropic compression; phase 2-3, heat addition at constant pressure or volume; phase 3-4, isentropic expansion; and phase 4-1, heat rejection at constant pressure or volume.

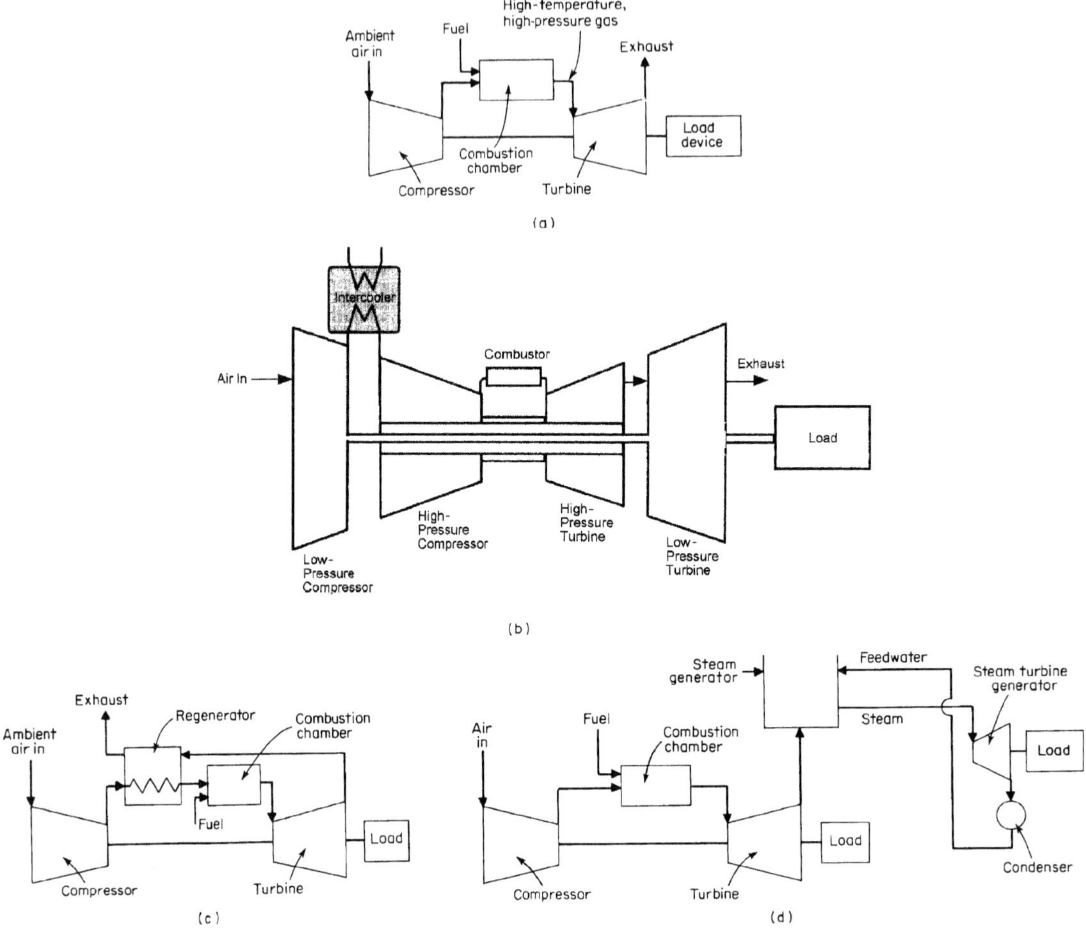


**FIGURE 14-9** Typical gas-turbine cycles: (*a*) open, (*b*) intercooled, (*c*) regenerative, and (*d*) combined.

The products of combustion expand through the turbine, producing sufficient power to drive the compressor and the load. Some compressed air typically bypasses the combustor and is used to cool the turbine parts. The highest-pressure turbine airfoils contain internal cooling passages in order to maintain the metal temperature at acceptable levels for durability, while the gas path temperature is considerably higher than the metal temperature, to achieve high power and efficiency.

An improvement in power output and efficiency can be obtained through the use of an intercooler (Fig. 14-9*b*), in which air is cooled after part of the compression process. The intercooler reduces the work of compression of the high-pressure compressor and allows higher airflow and overall pressure ratio to be attained, while reducing the temperature of the cooling air for the turbine section.

Another efficiency improvement can be obtained from a regenerator or recuperator (Fig. 14-9*c*), which exchanges heat from the exhaust to the combustor inlet to reduce the fuel required to heat the gas. The highest efficiencies are available from combined cycles, where the gas turbine exhaust heat produces steam to drive a steam-turbine generator (Fig. 14-9*d*). The steam turbine output is obtained with no additional fuel input.

The most effective utilization of the fuel input is available through *cogeneration*, or combined heat and power. A typical cycle has the gas-turbine generating power and producing steam from its exhaust heat. The steam is sent to an industrial process, in some cases after generating some power in a noncondensing steam turbine. When credit is taken for the heat sent to process plus the power generated, efficiencies exceeding 80% are commonly achieved.

Simple cycles and combined cycles are by far the most commonly employed gas-turbine cycles. Some regenerative cycles were developed in the 1960s and early 1970s, but durability problems with the regenerators prevented further use. Newer regenerative cycle development started in the mid-1990s. Intercooled cycles are being studied principally as possible derivatives of commercial aircraft engines.

Component efficiencies, airflow, pressure ratio, and turbine inlet (or firing) temperature are the major factors affecting gas-turbine output and efficiency. Typical multistage axial component efficiencies are in the 86% to 93% range. Material developments and turbine cooling techniques permit turbine rotor inlet temperatures to exceed 2500°F (1370°C) for the most advanced units.

The most familiar application of gas turbines has been for aircraft propulsion, where the turbine drives only the compressor and the remaining energy is used for thrust. The industrial gas-turbine industry started in the late 1950s, following the development of aircraft gas turbines. Generation of electric power, mechanical drive (principally gas or oil pipeline compression), and marine propulsion are the three applications of industrial gas turbines. Over 90% of the gas-turbine applications, as measured in megawatts of power produced, are in electric power generation. The majority of the navies of the world use gas turbines to propel most of their surface ships.

In electric power generation service, the combination of lower capital cost, shorter installation time, high efficiency, and environmental advantages compared to steam-turbine-based power plants has resulted in gas-turbine-based power plants having a major market for new power generation equipment.

Much work has been done to adapt gas turbines to coal fuel. Coal gasification, utilized to remove contaminants, is particularly attractive since the gas-turbine combined cycle can be integrated into the gasification process for improved efficiency. Air extracted from the gas turbine can be used as the source of oxidant for the gasification process, and the steam produced in the process can be expanded through the steam turbine.

## 14.2 DIRECT-CURRENT GENERATORS

### 14.2.1 The DC Machine

***Applications.*** The most important role played by the dc generator is the power supply for the important dc motor. It supplies essentially ripple-free power and precisely held voltage at any desired value from zero to rated levels. This is truly dc power, and it permits the best possible commutation on the motor because it is free of severe waveshapes of dc power from rectifiers. It has excellent response and is particularly suitable for precise output control by feedback control regulators. It is also well suited for supplying accurately controlled and responsive excitation power for both ac and dc machines.

***General Construction.*** Figure 14-10 shows the parts of a medium or large dc generator. They differ from ac machines in having a commutator and the armature on the rotor. They also have salient poles on the stator, and, except for a few small cases, they have commutating poles between the main poles.

***Construction and Size.*** Small dc machines have large surface-to-volume ratios and short paths for heat to reach dissipating surfaces. The cooling requires little more than the means to blow air over the rotor and between the poles. The rotor punchings are mounted solidly on the shaft, with no air passages through them.

Larger units, with longer, deeper cores, use the same construction, but with longitudinal holes through the core punchings for cooling air to circulate. Medium and large machines must have large heat-dissipation surfaces and effectively placed cooling air, or "hot spots" will develop. Their

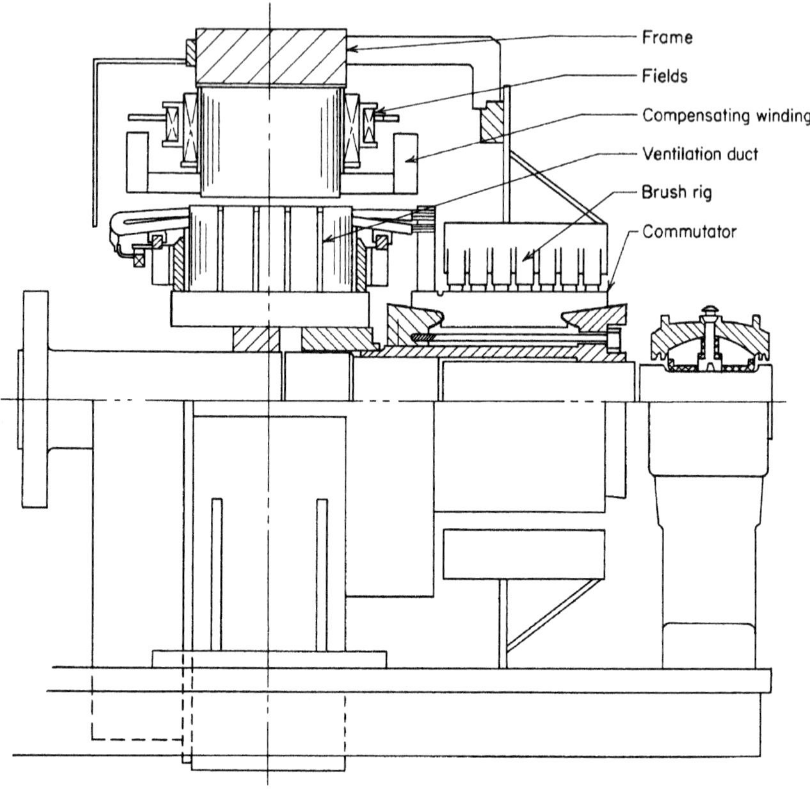


**FIGURE 14-10** The dc machine.

core punchings are mounted on arms to permit large volumes of cool air to reach the many core ventilation ducts and also the ventilation spaces between the coil end extensions.

***Design Components.*** The armature-core punchings are usually of high-permeability electrical sheet steel, 0.017 to 0.025 inch thick, and have an insulating film between them. Small and medium units use "doughnut" circular punchings, but large units, above about 45 inch in diameter, use segmental punchings shaped as shown in Fig. 14-11, which also shows the fingers used to form the ventilating ducts.

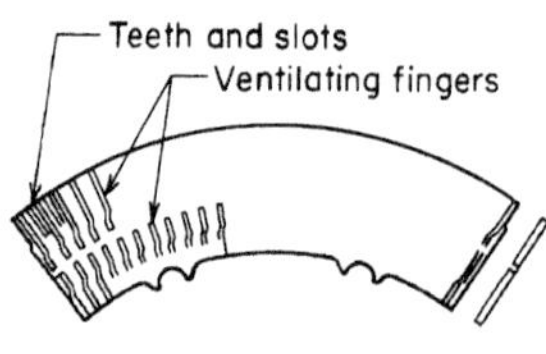


**FIGURE 14-11** Armature segment for a dc generator showing vent fingers applied.

The main- and commutating-pole punchings are usually thicker than rotor punchings because only the pole faces are subjected to high-frequency flux changes. These range from 0.062 to 0.125 inch thick, and they are normally riveted.

The frame yoke is usually made from rolled mild steel plate, but, on high-demand large generators for rapidly changing loads, laminations may be used. The solid frame has a high magnetic time constant of about half a second due to currents induced by changes in load, thereby making laminated frames with smaller time constants of about 0.05 to 0.005 seconds, more viable for applications with rapidly changing loads.

The commutator is truly the heart of the dc machine. It is made up of hard copper bars drawn accurately in a wedge shape. These are separated from each other by mica plate segments, whose thicknesses must be held accurately for nearly perfect indexing of the bars and for no skew. This thickness is 0.020 to 0.050 inch, depending on the size of the generator and on the maximum voltage that can be expected between the bars during operation. The mica segments and bars are clamped between two metal V-rings and insulated from them by cones of mica. On very high speed commutators of about 10,000 ft/min, shrink rings of steel are used to hold the bars. Mica is used under the rings.

Carbon brushes ride on the commutator bars and carry the load current from the rotor coils to the external circuit. The brush holders hold the brushes against the commutator surface by springs to maintain a fairly constant pressure and smooth riding.

## 14.2.2 General Principles

***Electromagnetic Induction.*** A magnetic field is represented by continuous lines of flux considered to emerge from a north pole and to enter a south pole. When the number of such lines linked by a coil is changed (Fig. 14-12), a voltage is induced in the coil equal to 1 V for a change of $10^8$ linkages/s (Mx/s) for each turn of the coil, or $E = (\Delta\phi T \times 10^{-8})/t$ V.

If the flux lines are deformed by the motion of the coil conductor before they are broken, the direction of the induced voltage is considered to be into the conductor if the arrows for the distorted flux are shown to be pointing clockwise and outward if counterclockwise. This is the generator action (Fig. 14-13).

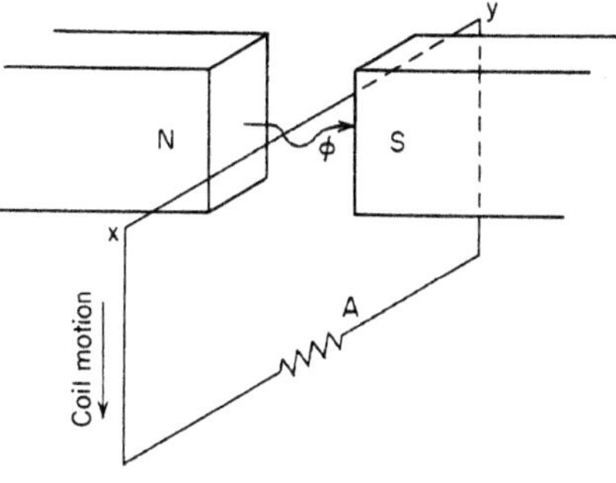


**FIGURE 14-12** Generated emf by coil movement in a magnetic field.

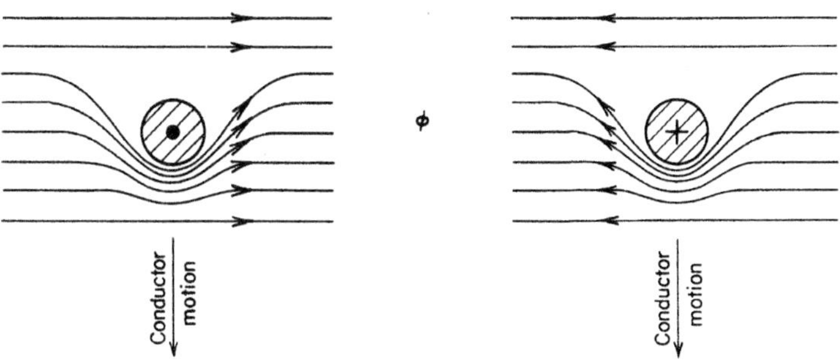


**FIGURE 14-13** Direction of induced emf by conductor movement in a magnetic field.

***Force on Current-Carrying Conductors in a Magnetic Field.*** If a conductor carries current, loops of flux are produced around it (Fig. 14-14). The direction of the flux is clockwise if the current flows away from the viewer into the conductor, and counterclockwise if the current in the conductor flows toward the viewer.

If this conductor is in a magnetic field, the combination of the flux of the field and the flux produced by the conductor may be considered to cause a flux concentration on the side of the conductor, where the two fluxes are additive and a reduction on the side where they oppose. A force on the conductor results that tends to move it toward the side with reduced flux (Fig. 14-15). This is motor action.

***Generator and Motor Reactions.*** It is evident that a dc generator will have its useful voltage induced by the reactions described above. An external driving means must be supplied to rotate the armature so that the conductor loops will move through the flux lines from the stationary poles. However,

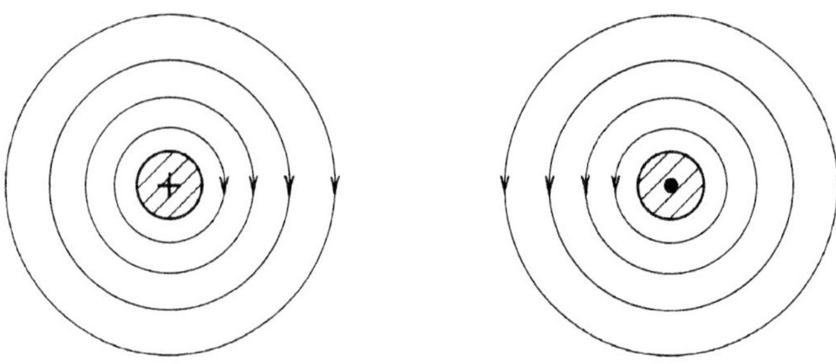

**FIGURE 14-14** Magnetic fields caused by current-carrying conductors.

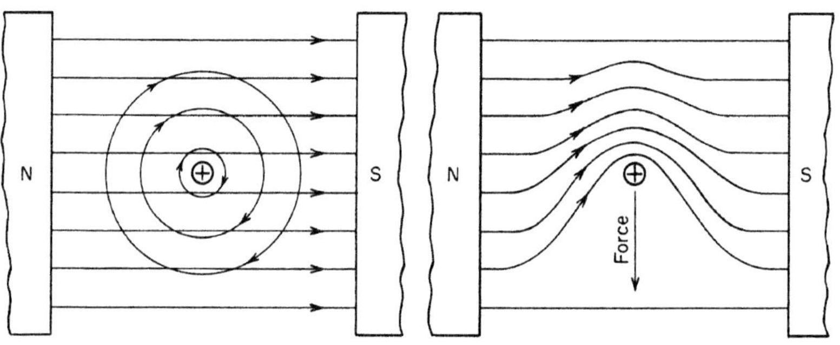


**FIGURE 14-15** Force on a current-carrying conductor in a magnetic field.

these conductors must carry current for the generator to be useful, and this will cause retarding forces on them. The prime mover must overcome these forces.

In the case of the dc motor, the conductor loops will move through the flux, and voltages will be induced in them. These induced voltages are called the "counter emf," and they oppose the flow of currents which produce the forces that rotate the armature. Therefore, this emf must be overcome by an excess voltage applied to the coils by the external voltage source.

***Direct-Current Features.*** Direct-current machines require many conductors and two or more stationary flux-producing poles to provide the needed generated voltage or the necessary torque. The direction of the current flow in the armature conductors under each particular pole must always be correct for the desired results (Fig. 14-16). Therefore, the current in the conductors must reverse at some time while the conductors pass through the space between adjacent north and south poles.

This is accomplished by carbon brushes connected to the external circuit. The brushes make contact with the conductors by means of the commutator.

To describe commutation, the Gramme-ring armature winding (which is not used in actual machines) is shown in Fig. 14-17. All the conductors are connected in series and are wound around a steel ring. The ring provides a path for the flux from the north

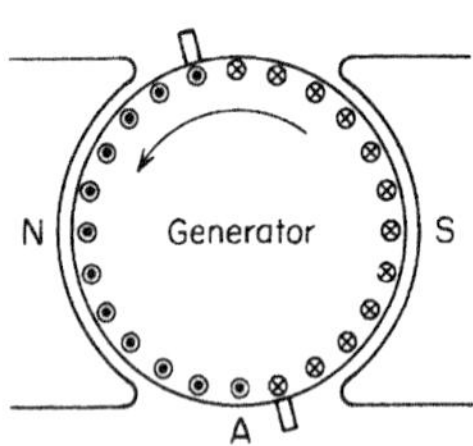


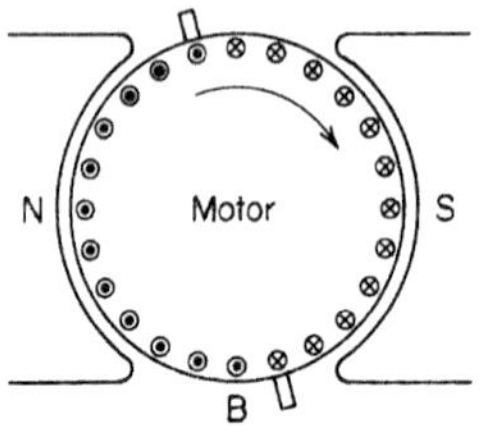


**FIGURE 14-16** Direction of current in generator and motor.

to the south pole. Note that only the outer portions of the conductors cut the flux as the ring rotates. Voltages are induced as shown. With no external circuit, no current flow, because the voltages induced in the two halves are in opposition. However, if the coils are connected at a commutator $C$ made up of copper blocks insulated from each other, brushes $B-$ and $B+$ may be used to connect the two halves in parallel with respect to an external circuit and the currents will flow in the proper direction in the conductors beneath the poles.

As the armature rotates, the coil $M$ passes from one side of the neutral line to the other and the direction of the current in it is shown at three successive instants at $A$, $B$, and $C$ in Fig. 14-17. As the armature moves from $A$ to $C$ and the brush changes contact from segment 2 to segment 1, the current in $M$ is automatically reversed. For a short period, the brush contacts both segments and short circuits the coil. It is important that no voltage be induced in $M$ during that time, or the resulting circulating currents could be damaging. This accounts for the location of the brushes so that $M$ will be at the neutral flux point between the poles.

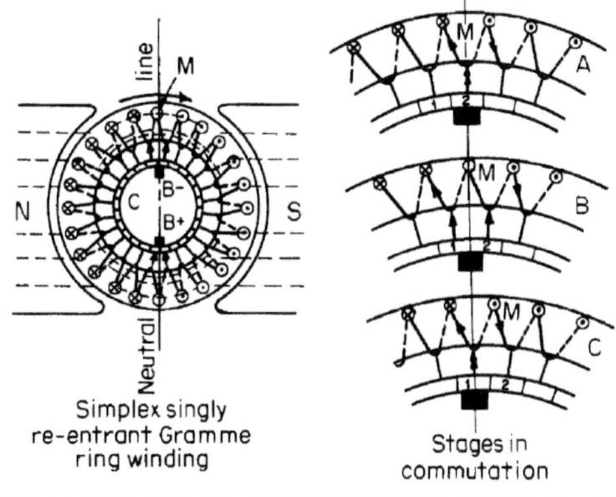


**FIGURE 14-17** Principle of commutation.

***Field Excitation.*** Flux from the main poles is obtained by winding conductors around the pole bodies and passing current through them. This current may be supplied in different ways. When a generator supplies its own exciting current, it is "self-excited." When current is supplied from an external source, it is "separately excited." When excited by the load current of the machine, it is "series excited."

## 14.2.3 Armature Reactions

***Cross-Magnetizing Effect.*** Figure 14-18*a* represents the magnetic field produced in the air gap of a 2-pole machine by the magnetomotive force (mmf) of the main exciting coils, and part *b* represents the magnetic field produced by the mmf of the armature winding alone when it carries a load current. If each of the $Z$ armature conductors carries $I_c$ A, then the mmf between $a$ and $b$ is equal to $ZI_c/p$ At. That between $c$ and $d$ (across the pole tips) is $\psi ZI_c/p$ At, where $\psi$ = ratio of pole arc to pole pitch. On the assumption that all the reluctance is in the air gap, half the mmf acts at $ce$ and half at $fd$, and so the cross-magnetizing effect at each pole tip is

$$\frac{\psi ZI_c}{2p} \quad \text{ampere-turns} \tag{14-5}$$

for any number of poles.

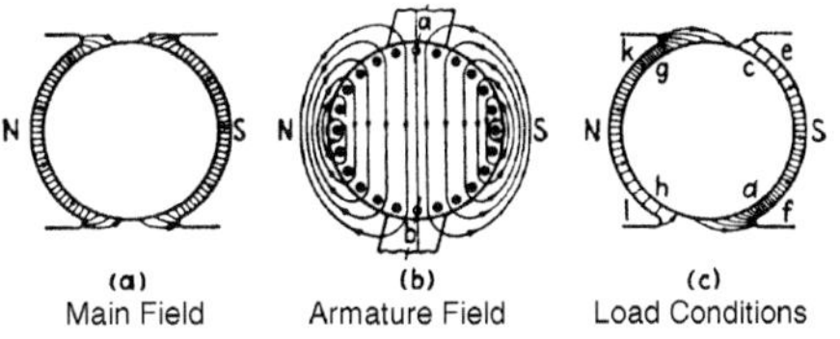


**FIGURE 14-18** Flux distribution in (*a*) main field, (*b*) armature field, and (*c*) load conditions.

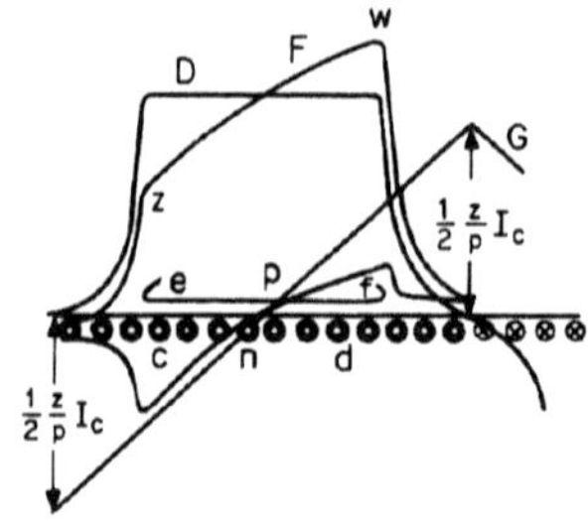


**FIGURE 14-19** Flux distribution in a large machine with $p$ poles.

***Field Distortion.*** Figure 14-18$c$ shows the resultant magnetic field when both armature and main exciting mmfs exist together; the flux density is increased at pole tips $d$ and $g$ and is decreased at tips $c$ and $h$.

***Flux Reduction Due to Cross-Magnetization.*** Figure 14-19 shows part of a large machine with $p$ poles. Curve $D$ shows the flux distribution in the air gap due to the main exciting mmf acting alone, with flux density plotted vertically. Curve $G$ shows the distribution of the armature mmf, and curve $F$ shows the resultant flux distribution with both acting. Since the armature teeth are saturated at normal flux densities, the increase in density at $f$ is less than the decrease at $e$, so that the total flux per pole is diminished by the cross-magnetizing effect of the armature.

***Demagnetizing Effect of Brush Shift.*** Figure 14-20 shows the magnetic field produced by the armature mmf with the brushes shifted through an angle $\theta$ to improve commutation. The armature field is no longer at right angles to the main field but may be considered the resultant of two components, one in the direction $OY$, called the "cross-magnetizing component," and the other in the direction $OX$, which is called the "demagnetizing component" because it directly opposes the main field. Figure 14-21 gives the armature divided to show the two components. As can be seen, the demagnetizing ampere-turns per pair of poles are

$$\frac{ZI_c}{p} \times \frac{2\theta}{180} \quad \text{ampere-turns} \tag{14-6}$$

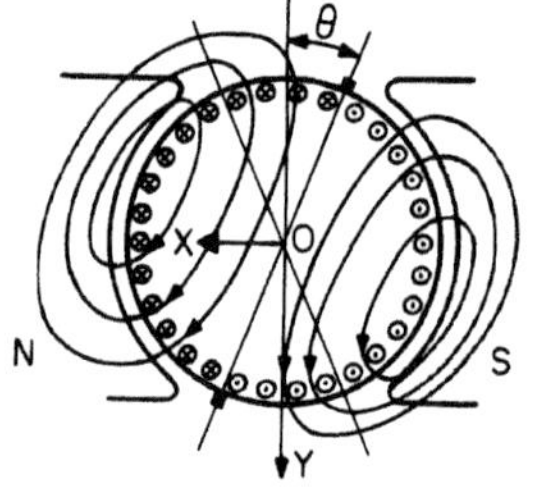


**FIGURE 14-20** Demagnetizing effect.

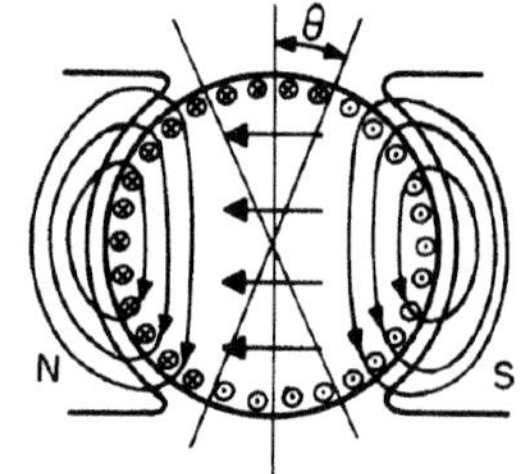


**FIGURE 14-21** Cross-magnetizing effect.

where $2\theta/180$ is about 0.2 for small noncommutating pole machines where brush shift is used. The demagnetizing ampere-turns per pole would be

$$0.1ZI_c/p \quad \text{ampere-turns} \tag{14-7}$$

***No-Load and Full-Load Saturation Curves.*** Curve 1 of Fig. 14-22 is the no-load saturation curve of a dc generator. When full-load current is applied, there is a decrease in useful flux, and therefore a drop in voltage $ab$ due to the armature cross-magnetizing effect (see paragraph on flux reduction, above). A further voltage drop from brush shift is counterbalanced by an increase in excitation $bc = 0.1\ ZI_c/p$; also a portion $cd$ of the generated emf is required in overcoming the voltage drop from the current in the internal resistance of the machine. The no-load voltage of 240 V requires 8000 At. At full load at that excitation, the terminal voltage drops to 220 V. To have both no-load and full-load voltages equal to 240 V, a series field of 10,700 – 8000 = 2700 At would be required.

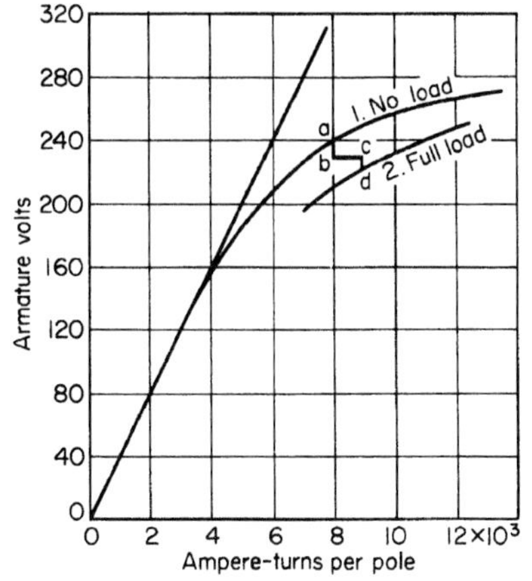


**FIGURE 14-22** Saturation curves—dc generator.

## 14.2.4 Commutation

***Commutation Defined.*** The voltages generated in all conductors under a north pole of a dc generator are in the same direction, and those generated in the conductors under a south pole are all in the opposite direction (Fig. 14-23). The currents will flow in the same direction as induced voltages in generators and in the opposite direction in motors. Thus, as a conductor of the armature passes under a brush, its current must reverse from a given value in one direction to the same value in the opposite direction. This is called "commutation."

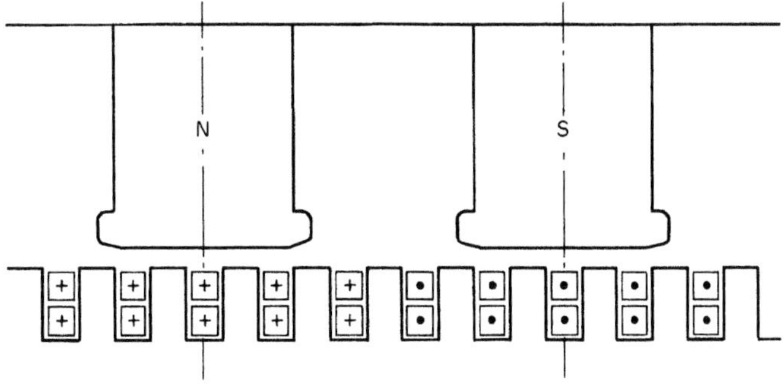


**FIGURE 14-23** Conductor currents.

***Conductor Current Reversal.*** If commutation is "perfect," the change of the current in a coil will be linear, as shown by the solid line in Fig. 14-24. Unfortunately, the conductors lie in steel slots, and self- and mutual inductances cause voltages to be induced in the coils short-circuited by the brushes. These result in circulating currents that tend to prevent the initial current change, delaying the reversal. In extreme cases, the delay may be as severe as indicated by the dotted line of Fig. 14-24. Because the current must be reversed by the time the coil leaves the brush (when there is no longer any path for circulating currents), the current remaining to be reversed at *F* must discharge its energy in an electric arc from the commutator bar to the heel of the brush. This is commutation sparking. It can burn the edges of the commutator bars and the brushes. However, most large and heavy-duty dc machines have some nondamaging sparking, and "sparkless" commutation is not required by accepted standards. However, commutation must not require undue maintenance.

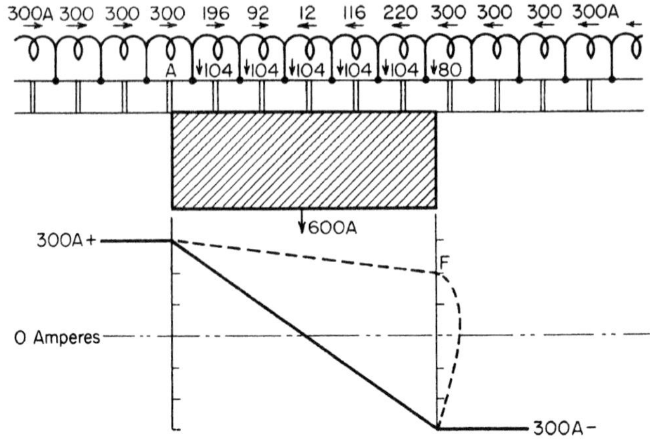

**FIGURE 14-24** Commutation.

***Commutating Poles.*** The beneficial factors that limit the circulating currents in coils being commutated are not adequate to prevent serious delays in current reversal. Other means must be taken to prevent sparking. Stationary poles may be provided midway between the main poles. (Fig. 14-25). The windings on these *commutating poles* carry the load current. The commutating pole (C in the figure) flux is proportional to the rotor conductor currents and, theoretically, can cancel the voltages induced in the coils being commutated by the slot leakage flux. In the case of the dc motor, the current reverses in both the armature and the commutating field, and proper canceling is maintained.

Almost all modern dc machines use commutating poles, although some small machines have only half as many as main poles.

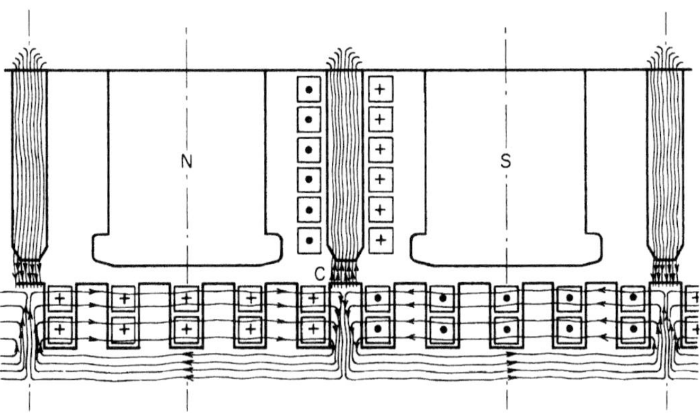

**FIGURE 14-25** Slot-leakage flux and commutating-pole flux.

***Compensating Windings.*** Although the commutating pole is a good solution for commutation, it does not prevent distortion of the main-pole flux by armature reaction. If the pole face is provided with another winding and connected in series with the load, it can set up an mmf equal and opposite to that of the armature. This would tend to prevent distortion of the air-gap field by armature reaction. Such windings are called *compensating windings* and are usually provided on medium-sized and large dc machines to obtain the best possible characteristics. They are also often needed to make machines less susceptible to flashovers.

Compensation of armature reaction effectively reduces the armature circuit inductance. This makes the machine to be less susceptible to the bad effects of $L(di/dt)$ voltages caused by very fast load current changes.

### 14.2.5 Cooling and Ventilation

***Cause of Temperature Rise.*** The losses in a dc machine cause the temperature of the parts to rise until the difference in temperature between their surfaces and the cooling air is great enough to dissipate the heat generated.

Permissible measured temperature rises of the parts are limited by the maximum "hot-spot" temperature that the insulation can withstand and still have reasonable life. The maximum surface temperatures are fixed by the temperature gradient through the insulation from the hot spot to the surface.

The IEEE Insulation Standards have established the limiting hot-spot temperatures for systems of insulation. The *American National Standards Institute* Standard C50.4 for dc machines gives typical gradients for those systems, listing acceptable surface and average copper temperature rises above specified ambient-air temperatures for various machine enclosures and duty cycles. Typical values are 40°C for Class A systems, 60°C for Class B, and 80°C rise for Class F systems on armature coils. Class H systems usually contain silicones and are seldom used on medium and large dc machines. Silicone vapors can cause greatly accelerated brush wear at the commutator and severe sparking, particularly on enclosed machines.

***Temperature Gradients in Rotor Coils.*** In case of the coils, the hot spot is probably at the core centerline and near the center of the conductor. Heat will probably travel along the conductor to the end turn and also through the insulation to the iron. The amount of heat flowing in each direction is difficult to calculate. Also, variations in the coils, such as resin fill and tightness in the slots, make heat conductivity factors difficult to predict.

***Heating of End Connections of Armature Windings.*** Small machines often have "solid" end windings banded down on insulated "shelf"-type coil supports. Larger machines are more heavily loaded per unit volume and usually have narrow coil supports, air spaces between the end turns, and ventilating air scouring both the top and bottom surfaces of the coil extensions.

***Commutator Heating.*** In a modern DC motor, the commutator diameter ranges from 55% to 85% of the rotor core, and the commutator necks joining the bars with the rotor winding extensions are usually separated from one another by air spaces, so that, when the armature revolves, air circulation is set up as shown by the arrows. The heat to be dissipated is that due to brush friction and the brush contact $I^2R$ losses. There may be other losses due to poor commutation, brush chattering, and commutator surface. If commutation is very good and the brush riding is excellent, the temperature will be lower.

### 14.2.6 Losses and Efficiency

***Armature Copper $I^2R$ Loss.*** At 75°C the resistivity of copper is $8.25 \times 10^{-7}$ Ω/in³. Thus, for an armature winding of $Z$ conductors, each with a length of $\overline{L}_t/2$ (half the mean length turn of the coil), each with a cross-sectional area of $A$ and arranged in several parallel circuits, the resistance is

$$R_a = Z\frac{\overline{L}_t}{2A}\frac{8.25\times10^{-7}}{(\text{circuits})^2} \quad \text{ohms} \tag{14-8}$$

The $\overline{L}_t$ is best found by layout, but an approximate value is

$$\overline{L}_t = 2[(1.35)(\text{pole pitch})+(\text{rotor length})\times 3] \tag{14-9}$$

There are also eddy current losses in the rotor coils, but these may be held to a minimum by conductor stranding. Some allowance for these is included in the load loss.

***Compensating, Commutating, and Series Field $I^2R$ Losses.*** These fields also carry the line current, and the $I^2R$ losses are easily found when the resistance of the coils is known. Their $\overline{L}_t$ is found from sketch layouts. At 75°C

$$R = T\frac{\overline{L}_t}{A}\frac{8.25\times10^{-7}}{(\text{circuits})^2}p \qquad \text{ohms} \tag{14-10}$$

where $R$ is the field resistance in ohms, $T$ the number of turns per coil, $p$ the number of poles, $\overline{L}_t$ the mean length of turn, and $A$ the area of the conductor. The total of these losses ranges from 60% to 100% of the armature $I^2R$ for compensated machines and is less than 50% for noncompensated machines.

The brush $I^2$ loss is caused by the load current passing through the contact voltage drop between the brushes and the commutator. The contact drop is assumed to be 1 V.

$$\text{Brush } I^2R \text{ loss} = 2(\text{line amperes}) \qquad \text{watts} \tag{14-11}$$

***Load Loss.*** The presence of load current in the armature conductors results in flux distortions around the slots, in the air gap, and at the pole faces. These cause losses in the conductors and iron that are difficult to calculate and measure. A standard value has been set at 1% of the machine output.

$$\text{Load loss} = 0.01(\text{machine output}) \tag{14-12}$$

***Shunt Field Loss.*** Heating calculations are concerned only with the field copper $I^2R$ loss. It is customary, however, to charge the machine with any rheostat losses in determining efficiency. Thus

$$\text{Shunt field and rheostat loss} = I_f V_{ex} \qquad \text{watts} \tag{14-13}$$

where $I_f$ is the total field current and $V_{ex}$ is the excitation voltage.

***Core Loss.*** The flux in any portion of the armature passes through $p/2$ c/r (cycles per revolution) or through $(p/2)[(\text{r/min})/60]$ Hz.

The *iron losses* consist of the *hysteresis loss,* which equals $K\beta^{1.6}fw$ watts, and the *eddy current loss,* which equals $K_e(\beta ft)^2 w$ watts. $K$ is the hysteresis constant of the iron used, $K_e$ is a constant inversely proportional to the electrical resistance of the iron, $\beta$ is the maximum flux density in lines per square inch, $f$ is the frequency in hertz, $w$ is the weight in pounds, and $t$ is the thickness of laminations in inches.

The *eddy loss* is reduced by using iron with as high an electrical resistance as is feasible. Very high resistance iron has a tendency to have low flux permeability and to be mechanically brittle and expensive. It is seldom justified in dc machines. The loss is kept to an acceptable value by the use of thin core laminations, 0.017 to 0.025 inch thickness.

Another significant loss is the *pole-face loss.* As the armature rotates and the teeth move past the pole face, emfs are induced which tend to cause currents to flow across the pole face. These losses are included in the core loss.

Unfortunately, there are other losses in the core that may differ widely even on duplicate machines and that do not lend themselves to calculation. These include:

1. Loss due to filing of slots.
2. Losses in the solid spider, core end plates, and coil supports from leakage fluxes may be appreciable.
3. Losses due to nonuniform distribution of flux in the rotor core are difficult to anticipate.

***Brush Friction Loss.*** This loss varies with the condition of the commutator surface and the grade of carbon brush used. A typical machine has about 8-W loss/(in$^2$ of brush contact surface)(1000 ft/min) of peripheral speed when normal brush pressure of 2½ lb/in$^2$ is used.

$$\text{Brush friction} = (8)\,(\text{contact area})\,\frac{\text{peripheral velocity}}{1000} \tag{14-14}$$

***Friction and Windage.*** Most large dc machines use babbitt bearings and many small machines use ball or roller bearings, although both types of bearing may be used in machines of any size. The bearing friction losses depend on the speed, the bearing load, and the lubrication. The windage losses depend on the construction of the rotor, its peripheral velocity, and the machine restrictions to air movement. The two losses are lumped in most estimates because it is not practical to separate them during machine testing.

Figure 14-26 shows typical values of friction and windage losses for various rotor diameters referred to rotor velocities.

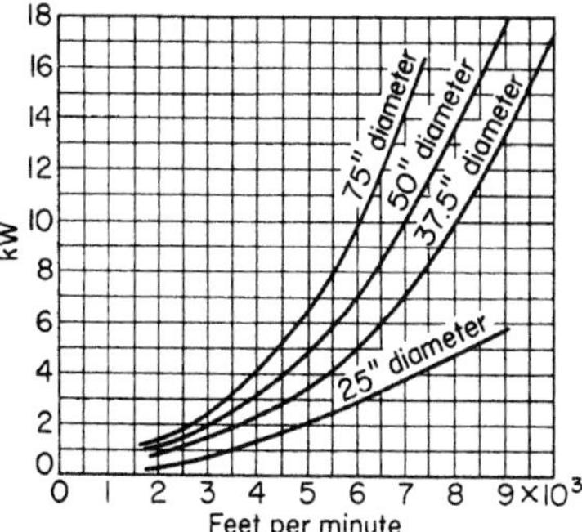


**FIGURE 14-26** Friction and windage versus rotor velocity.

***Efficiency.*** The efficiency of a generator is the ratio of the output to its input. The prime mover must supply the output and, in addition, the sum of all the losses. This is the input.

$$\text{Efficiency} = \frac{\text{output}}{\text{input}} = \frac{\text{output}}{\text{output} + \text{losses}} \tag{14-15}$$

## 14.2.7 Generator Characteristics

The voltage regulation of a dc generator is the ratio of the difference between the voltage at no load and that at full load to the rated-load voltage. The characteristic is normally drooping as the load is increased. However, it can rise because of series field effects or the action of circulating currents of commutation at very low voltage operation.

For a dc generator, the terminal-voltage equation is

$$TV = E - IR = [(K\phi_t)(\text{r/min}) - IR] \tag{14-16}$$

where $E$ is the induced emf, $IR$ is the armature circuit drop, $K$ is a constant depending on the machine design, and $\phi_t$ is the total main-pole flux of the generator.

The regulation curves are easily calculated by using the no-load and full-load saturation curves. The effect of the excitation method is found by the use of the field and rheostat $IR$ line for self-excited machines and by the constant-ampere-turn line for separate excitation.

A separately excited compensated generator which is shunt-wound will have a voltage-load characteristic which will approach a straight line; it droops to full load an amount equal to the percent $IR$ drop. There is little or no flux loss due to armature reaction or brush shift.

At voltages 10% or less of rated, the main-field strength is so weak that currents circulating in the coils short-circuited by the brushes at commutation may cause an increase in main-pole flux with load that causes a rising characteristic. These armature coils loop the main poles and their ampere-turns produce direct axis flux. A rising voltage characteristic can be undesirable, particularly if the generator supplies a dc motor whose speed is caused to rise with load, since this causes instability.

A separately excited noncompensated dc generator which is shunt-wound has a nonlinear loss of flux due to armature reaction as the load current is increased. The characteristics of such dc generators drop at an ever increasing rate with load increase, giving a curve that is concave downward.

A self-excited noncompensated dc generator which is shunt-wound has its shunt-field excitation decreased as the terminal voltage drops. This results in a reduction of main-field ampere-turns and a loss of still more flux. This gives a severe droop which may be so great that, above a certain peak-load

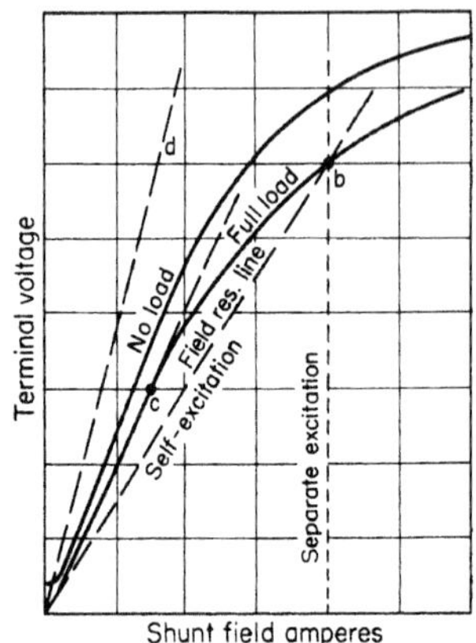


FIGURE 14-27 No-load and field-load saturation curves.

current, the terminal voltage will not be high enough to provide enough field current to maintain the voltage and load current and the voltage will collapse, as shown in *d* of Fig. 14-27.

***Instability of Self-Excited Generators.*** A self-excited dc generator is unstable if the rheostat line does not make a definite intersection with the load-saturation curve. The shunt-field current is fixed by the terminal voltage, and the resistance is in the shunt-field circuit. Instability will exist if the slope of the rheostat line is nearly equal to or greater than the slope of a line tangent to the operating point on the saturation curve. In Fig. 14-27, point *b* is a stable operating condition, but point *c* is not, because a decrease in voltage decreases the shunt field ampere turns, and this produces a further decrease in voltage.

If the field circuit resistance were set at *d*, the self-excited generator would never build up beyond residual voltage. Another cause of failure to build up may be the connection of the shunt field. If the current flow due to residual voltage is such that it tends to kill the flux producing the residual voltage, no buildup occurs.

***Series Generators.*** Curve 1 of Fig. 14-28 shows the relation between voltage and current if there is no armature resistance or armature reaction. This is actually the no-load curve of the machine obtained by separately exciting the series field. Curve 2 shows the actual relation between load current and terminal voltage. The total voltage drop is made up of a part caused by the decrease in flux by armature reaction and a part caused by the *IR* drop of the armature, brushes, and series fields.

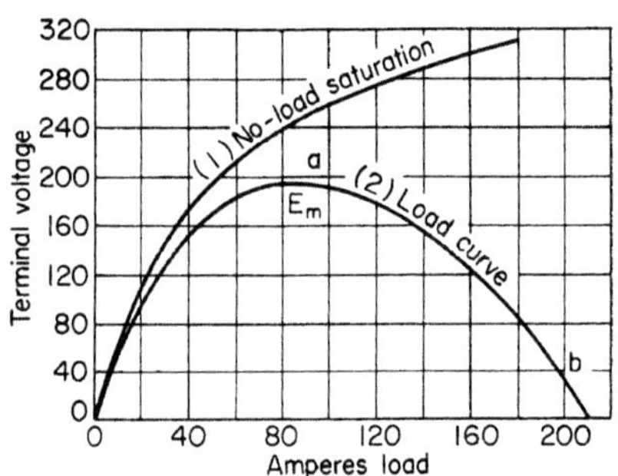


FIGURE 14-28 Characteristic curves of a series generator.

***Field Time Constants.*** The major delay in change of output voltage by an excitation change is caused by the inductance of the main fields. The time constant of the shunt field is the ratio of its inductance in henries to its resistance in ohms, and this ratio represents the time in seconds required for 63% of a field current change to occur when the excitation voltage is suddenly changed. In the case of an example 2500-kW generator, a mean main-field inductance over the voltage range from zero to rated is 6.20 H. The main-field resistance is 2.21 Ω. The field time constant is therefore 2.8 s.

***Armature-Circuit Time Constants.*** Compensating windings effectively lower the inductances of the armature circuit. An example 2500 kW DC machine has an armature-circuit inductance of 0.0001929 H and a circuit resistance of 0.00398 Ω for a time constant of 0.048 s. These are typical values for large dc machines. Smaller noncompensated units have longer time constants.

### 14.2.8 Testing

***Factory Tests.*** These depend on the size, application, and design of the dc generator. The American National Standards Institute (ANSI) C50.4 for dc machines includes lists of recommended tests for dc generators and motors. The IEEE Test Code for dc machines covers recommended methods to be used for these tests.

## 14.2.9 Special Generators

***General.*** The adaptability of the dc generator for specific uses has led to the development of many special generators. These machines over the years made a significant contribution to industrial progress. However, most of these special applications have disappeared or are now being met with other devices such as silicon controlled rectifiers or programmed control of field currents to the main dc generator.

***Synchronous Converters.*** Of all the special generators, this was one of the earlier and most widely used. It was the principal dc power source for streetcars and interurban lines. It was a most ingenious device, combining in a single armature and winding, an ac motor taking its current from the lines through slip rings at the rear and a dc generator providing dc power from a commutator on the front end. Because the flow of the currents was in opposition, the resulting rotor winding could be small in cross section. A single stator provided flux for both functions. With the decline of street railway systems, the synchronous converter disappeared.

***Rotating Regulators.*** These dc machines had trade names like Rototrol, Regulex, and Amplidyne. They, too, have been replaced by solid-state devices. In addition to having fields for feedback intelligence, response was enhanced using self-excited shunt fields tuned to the air-gap line or by means of cross-magnetization from armature reaction.

***Three-Wire Devices.*** Because three-wire dc circuits are no longer in use, balancer sets and three-wire generators are relics in school labs or museums.

***Homopolar or Acyclic DC Generators.*** The single-pole machine principle still fascinates electrical engineers and several research and development labs continue to study new arrangements of its basic parts. Fundamentally, it consists of a single conductor moving through a uniform single-direction flux with a collector at each end of the conductor. The output is a steady ripple-free pure dc current and no commutation. Currents reaching 270,000 A at 8 V were provided by one commercial unit shown in Fig. 14-29. Recent efforts have been mainly to use liquid metals to take the large currents from the rotating collectors and to obtain higher voltages by connecting units in series. Some success has been possible, but restricting the sodium potassium to the collector area has proved difficult.

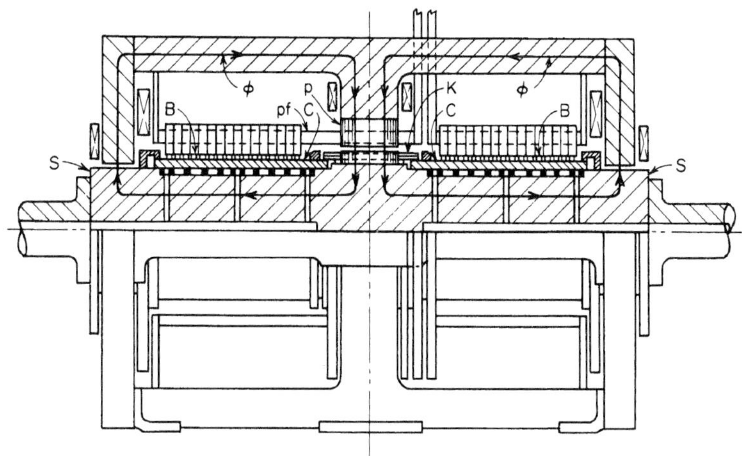


**FIGURE 14-29** Brush-type homopolar generator.

# 14.3 ALTERNATING-CURRENT GENERATORS

## 14.3.1 Introduction

This subsection deals with ac electric machines that convert mechanical power into electrical power. Such generators can be either synchronous generators or induction generators. Rotational speed of a synchronous generator is exactly at a speed that is synchronized with the ac power frequency, and this rotational speed is kept constant with varying loading conditions. Rotational speed of an induction generator is slightly above synchronous speed, and this rotational speed varies slightly with varying loading conditions. Induction generators find their major power generation application in wind turbine power generation. Synchronous ac generators dominate present-day commercial power generation by fossil fuels, nuclear reactors, and hydraulic turbines. All discussions of ac generators in this subsection are focused upon synchronous generators.

AC synchronous generators range in size and capability from very modest machines that are rated at a few hundred watts to the largest machines that are rated at 2000 MW.

## 14.3.2 Topology

All synchronous generators convert mechanical power into electrical power using an intermediate reservoir of energy in which the energy is stored in the form of magnetic fields.

The magnetic fields are typically established by electrical currents circulated in stationary ac windings, called the armature, and rotating dc windings, called the field, and these magnetic fields are circulated within the generator through high permeability steel structures. In such a generator, the ac winding is electrically connected to an electrical power system and physically mounted on the stationary member of the generator (the stator), and the dc winding is electrically connected to a dc power source and physically mounted on the rotating member of the generator (the rotor). The ac winding usually has three phases. Small generators can be built to have a single phase. The dc field winding could be replaced with permanent magnets. The electromagnetic interaction between the ac armature winding and the dc field winding provides the basis for ac power generation.

In a generator, like that illustrated in Fig. 14-30, the magnetic circuit consists of a steel stator core mounted inside the steel stator case and a steel rotor that is supported on bearings that are

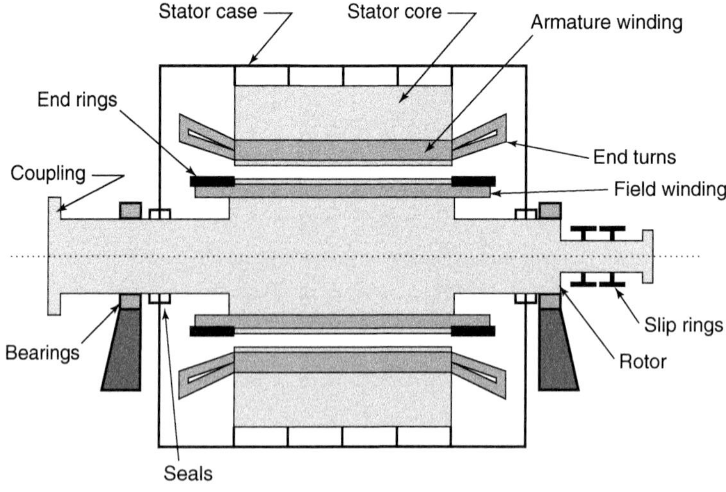


**FIGURE 14-30** Elements of an ac generator.

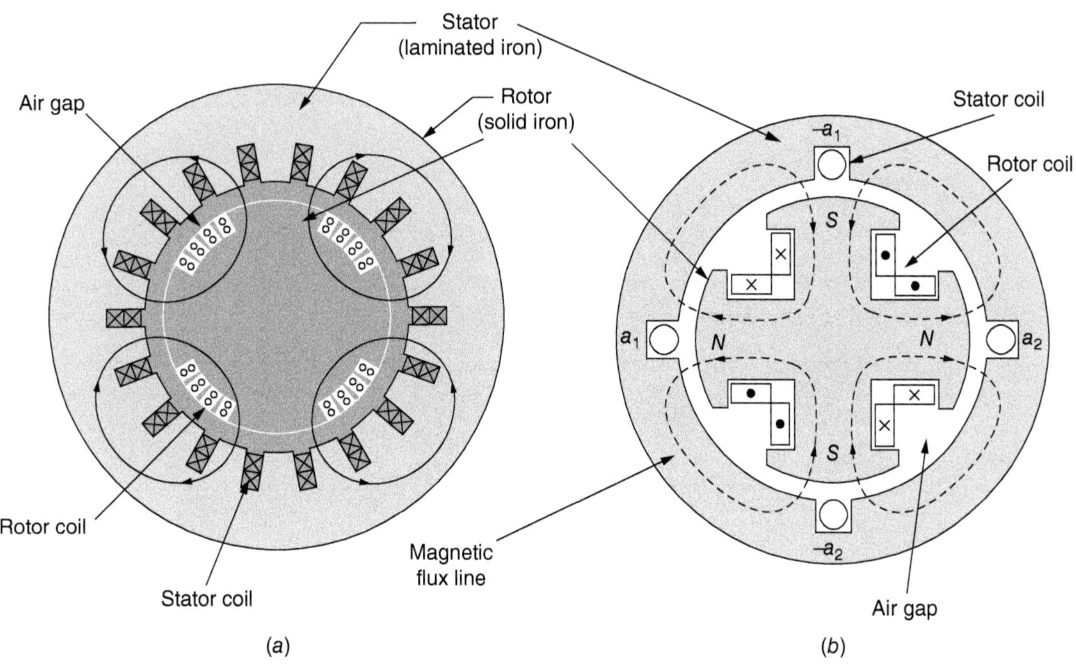


**FIGURE 14-31** 4-Pole generator (a is round rotor, b is salient pole).

either set into the case or separately mounted to the foundation. The coils of the armature winding are mounted in the stator core, and the coils of the field winding are mounted on the rotor. Armature winding electrical coils for generators of the type shown in Fig. 14-30 are typically deployed in radial slots formed in the inner diameter of the stator, and field winding electrical coils are typically deployed in radial slots formed in the outer diameter of the rotor, as illustrated in Fig. 14-31.

In a synchronous machine the armature (stator) winding must be designed for the same number of poles as the field (rotor) winding. The torque is produced by the magnetic interaction between a succession of rotating north-south magnetic-field pole pairs produced by the polyphase currents in the polyphase armature winding and the magnetic field poles on the rotor rotating at the same speed as the armature magnetic field. For a machine with the number of poles as $P$, the relationship between the electrical frequency $f_e$ (Hz) and mechanical speed $N$ (rpm)is

$$N \cdot P = 120 \cdot f_e \tag{14-17}$$

Synchronous generators can be a round-rotor machine or a salient-pole machine. Round-rotor machines are constructed of a cylindrical rotor. Therefore, the air gap between rotor and stator is uniform. On the other hand, the rotor of a salient-pole machine is constructed of a number of pole pieces mounted on a central rotor shaft. Figure 14-31 represents a four pole generator with round-rotor and a salient pole generator.

> As illustrated in Fig. 14-31, the magnetic flux passes through the magnetic circuit of the machine which is made up of the airgap, the stator teeth and backiron, the rotor poles, and the shaft section. The function of the magnetic circuit is to carry flux that links the armature conductors to produce voltage.

In both round-rotor and salient-pole generators, the magnetic flux passing through the rotors does not vary in time, and the magnetic flux passing through the stator core does vary periodically in time at the electrical line frequency. Consequently, the rotors can be made of solid steel, but the stator cores must be made of thousands of thin layers (laminations) of highly permeable electrical steel. Each layer of stator core steel is coated with a thin layer of electrical insulation. Electrical insulation is also used to isolate conductors and strands from each other and ground. The thermal classes of insulation material, most frequently used in ac generators, are Classes 130, 155, and 180.

### 14.3.3 Operation

The synchronous operation may be described in two elementary ways, referred to as the mmf *method* and the *flux method*. These are described here, assuming a simple, linear, round-rotor model for the machine. It should be noted that this model will, of necessity, be modified later to fully understand operation of the machine.

***MMF Method.*** A principal feature of a synchronous generator is the mutual inductance between the field and the phase windings. Assuming a three-phase machine, the mutual inductances between the field winding and the three-phase windings are

$$M_{af} = M \cos p\phi \tag{14-18}$$

$$M_{bf} = M \cos\left(p\phi - \frac{2\pi}{3}\right) \tag{14-19}$$

$$M_{cf} = M \cos\left(p\phi + \frac{2\pi}{3}\right) \tag{14-20}$$

where $M$ is the peak value of mutual inductance and $\phi$ is the angle, in mechanical radian, between the axes of the field winding and the armature (stator) phase winding designated *a*. If it is further assumed that phase-phase inductances, both self- and mutual, are not a function of rotor position, the use of energy methods gives a simple expression for machine torque:

$$T = -pMi_a i_f \sin p\phi - pMi_b i_f \sin\left(p\phi - \frac{2\pi}{3}\right) - pMi_c i_f \sin\left(p\phi + \frac{2\pi}{3}\right) \tag{14-21}$$

Noting that the sum of phase currents is, under balanced conditions, zero and that the mutual phase-phase inductances are equal, this is

$$\lambda_a = (L_a - L_{ab})i_a + MI_f \cos p\theta = L_d i_a + MI_f \cos p\theta \tag{14-22}$$

where $L_d$ denotes the *d*-axis synchronous inductance taking the field axis the direct or *d*-axis.

This flux is described by the equivalent circuit of Fig. 14-32, where,

$$E_{af} = j\omega MI_f e^{j\delta} \tag{14-23}$$

and $\delta$ is the phase angle between *internal* voltage $E_{af}$ and *terminal* voltage $V$, and $X_d = \omega L_d$.

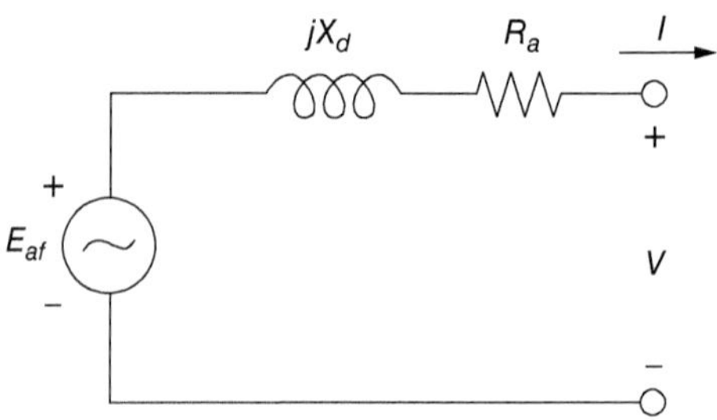


**FIGURE 14-32** Steady-state equivalent circuit ($R_a$ is neglected for analysis).

Assume $R_a << X_d$. Then $R_a$ the armature resistance, is negligible.
If the generator is connected to a voltage source (i.e., if $V$ is fixed), terminal current is

$$I = \frac{V - E_{af}e^{j\delta}}{jX_d} \tag{14-24}$$

If the rotor turns at a constant angular velocity $\omega/p = 2\pi f_e/p$, where $\omega = 2\pi f_e$ electrical rad/s, the field current is held constant at a value of $I_f$ and the three stator currents are sinusoids in time, with the same peak amplitude $I$, and with phases that differ by 120°

$$p\phi = \omega t + \delta_i \tag{14-25}$$

$$i_a = I \cos \omega t \tag{14-26}$$

$$i_b = I\cos\left(\omega t - \frac{2\pi}{3}\right) \tag{14-27}$$

$$i_c = I\cos\left(\omega t + \frac{2\pi}{3}\right) \tag{14-28}$$

torque is

$$T = -\frac{3}{2} pMII_f \sin\delta_i \tag{14-29}$$

Note that torque is proportional to the product of the two current amplitudes and to the sine of the phase angle between the field and armature current distributions. Further, the torque is acting in a direction so as to align the two current distributions.

***Flux Method.*** The flux method for estimating machine torque focuses on voltage (and hence flux) induced in the machine stator. If $L_a$ is the phase self-inductance and $L_{ab}$ is the phase-phase mutual inductance, flux linked by armature phase $a$ is

$$\lambda_a = L_a i_a + L_{ab} I_b + L_{ab} I_c + MI_f \cos p\theta \tag{14-30}$$

Real and reactive powers into the terminals of phase $a$ are

$$P_a = -\frac{1}{2}\frac{VE_{af}}{X_d}\sin\delta \tag{14-31}$$

$$Q_a = \frac{1}{2}\frac{V^2}{X_d} - \frac{1}{2}\frac{VE_{af}}{X_d}\cos\delta \tag{14-32}$$

Considering all three phases, total generated power is

$$P = -3P_a = \frac{3}{2}\frac{VE_{af}}{X_d}\sin\delta \tag{14-33}$$

Phasor diagrams illustrating the operation of a round-rotor synchronous generator are shown in Fig. 14-33. When the machine is overexcited, terminal current lags, I, terminal voltage. When the generator is underexcited, terminal current leads terminal voltage.

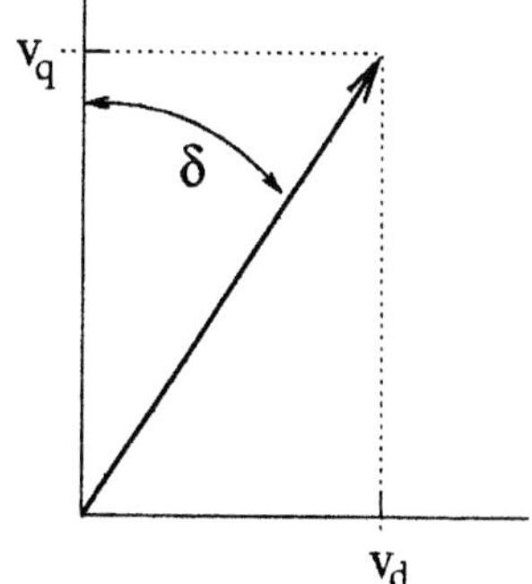


**FIGURE 14-33** Direct- and quadrature-axis voltages.

### 14.3.4 Two-Reaction Theory or d-q Axis Transformation

Salient-pole generators, such as hydroelectric generators, have armature inductances that are a function of rotor position, making analysis one step more complicated. The key to analysis of such machines is to separate mmf and flux into two orthogonal components. The two components are aligned with the direct axis and the quadrature axis of the machine (Fig. 14-34). The direct axis is aligned with the field winding, while the quadrature axis leads the direct axis by 90°. Then, if $\phi$ is the angle between the direct axis and the axis of phase $a$, flux linking phase $a$ is

$$\lambda_a = \lambda_d \cos\phi - \lambda_q \sin\phi \tag{14-34}$$

Then, in steady-state operation, if $V_a = d\lambda_a/dt$ and $\phi = \omega t + \delta$, we obtain

$$V_a = -\omega\lambda_d \sin\phi - \omega\lambda_q \cos\phi \tag{14-35}$$

or

$$V_d = -\omega\lambda_q = V\sin\delta \tag{14-36}$$

$$V_q = \omega\lambda_d = V\cos\delta \tag{14-37}$$

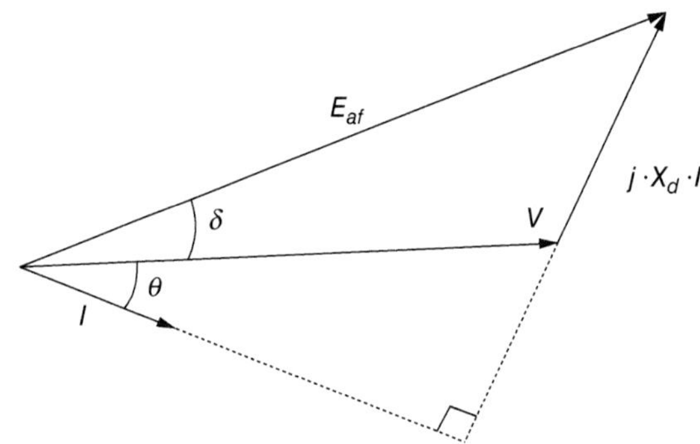


(*a*) Overexcited (lagging power factor).

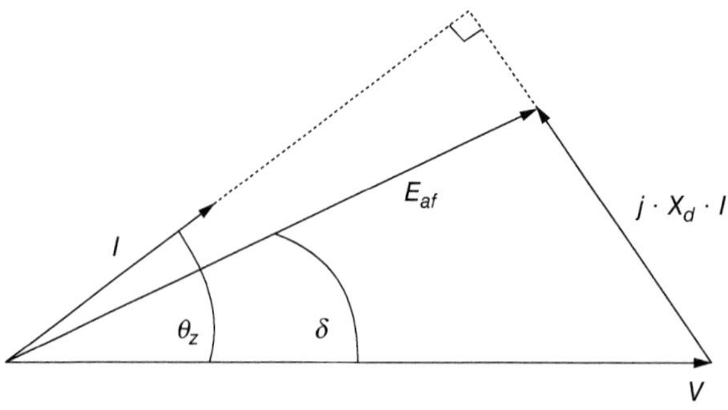


(*b*) Underexcited (leading power factor).

**FIGURE 14-34** Round-rotor synchronous generator.

One might think of the *voltage* vector as leading the *flux* vector by 90°. If the machine is linear, fluxes are given by

$$\lambda_d = L_d I_d + M I_f \tag{14-38}$$

$$\lambda_q = L_q I_q \tag{14-39}$$

Note that, in general, $L_d \neq L_q$, and for wound-field machines, $L_d > L_q$. Terminal voltage now has these components

$$V_d = -\omega\lambda_q = -\omega L_q I_q = V \sin\delta \tag{14-40}$$

$$V_q = \omega\lambda_d = \omega L_d I_d + \omega M I_f = V \cos\delta \tag{14-41}$$

which is easily inverted to produce

$$I_d = \frac{V\cos\delta - E_{af}}{X_d} \tag{14-42}$$

$$I_q = -\frac{V\sin\delta}{X_q} \tag{14-43}$$

where $X_d = \omega L_d$, $X_q = \omega L_q$, and $E_{af} = \omega M I_f$.

In the complex frame of reference

$$V = V_d + jV_q \tag{14-44}$$

$$I = I_d + jI_q \tag{14-45}$$

complex power is, in the sense of a generator

$$P + jQ = -\frac{3}{2}VI^* = -\frac{3}{2}[(V_d I_d + V_q I_q) + j(V_q I_d - V_d I_q)] \tag{14-46}$$

or

$$P = \frac{3}{2}\left[\frac{VE_{af}}{X_d}\sin\delta + \frac{V^2}{2}\left(\frac{1}{X_q} - \frac{1}{X_d}\right)\sin 2\delta\right] \tag{14-47}$$

$$Q = -\frac{3}{2}\left[\frac{V^2}{2}\left(\frac{1}{X_q} + \frac{1}{X_d}\right) - \frac{V^2}{2}\left(\frac{1}{X_q} - \frac{1}{X_d}\right)\cos 2\delta - \frac{VE_{af}}{X_d}\cos\delta\right] \tag{14-48}$$

### 14.3.5 Machine Size and Utilization

Generators produce torque through interaction between magnetic flux density and current over the surface of the stator, and reaction torque through the same type of interaction over the surface of the rotor. The stator and rotor face each other across the air gap. Power produced is

$$P = \omega_{\text{mech}} T = 2\pi\frac{f}{p}T = 2\pi\frac{N}{60}T \tag{14-49}$$

where torque produced is

$$T = 2\pi R^2 l\sigma \tag{14-50}$$

where $f$ = electrical frequency, Hz
$N$ = mechanical speed, r/min
$R$ = stator inner radius
$l$ = active length
$\sigma$ = average value of air gap shear stress, given approximately by

$$\sigma \approx \frac{1}{\sqrt{2}} B_1 K_z \tag{14-51}$$

where $B_1$ is the peak value of fundamental magnetic flux density at the stator surface and $K_z$ is the effective surface current density root mean square (rms) of the armature. The effective surface current density is ampere-turns per unit of periphery, modified by pitch and distribution factors, and by power factor.

Shear stress normally increases with pole pitch for a particular voltage and number of poles because the deeper armature slots and greater field coil space allow more ampere-conductors per unit of periphery. Typical shear stress levels for indirectly cooled, salient-pole generators are shown in Fig. 14-35. Typically shear stress is higher for directly cooled machines and a consequence of increased current density.

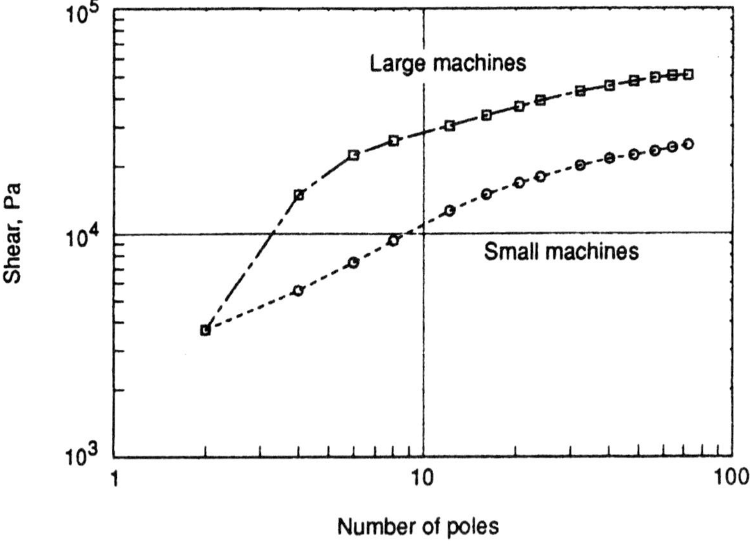


**FIGURE 14-35** Typical shear stress, salient-pole, air-cooled generators.

Another rough estimate of machine size employs the *utilization factor*:

$$c = \frac{S}{D^2 l N} \tag{14-52}$$

where $S$ is machine rating, $D$ and $l$ are outside diameter and length, respectively, and $N$ is rotational speed. The utilization factor depends on unit rating (high-speed machines) or unit rating per pole (low-speed machines) and on cooling.

### 14.3.6 Electromagnetics

Sketches of the major elements of synchronous generators, together with field "flux forms," are shown in Figs. 14-36 and 14-37. Here the generator is "unrolled" and shown as if the air gap were flat. Note that the main flux distribution in the air gap is not exactly sinusoidal. Thus the voltage produced will contain time harmonics. Assign $C_n$ to be the amplitude of the $n$th-space harmonic of the flux density, relative to the peak amplitude $B_0$.

The fundamental flux per pole is

$$\Phi_1 = \frac{2RlB_0C_1}{p} \tag{14-53}$$

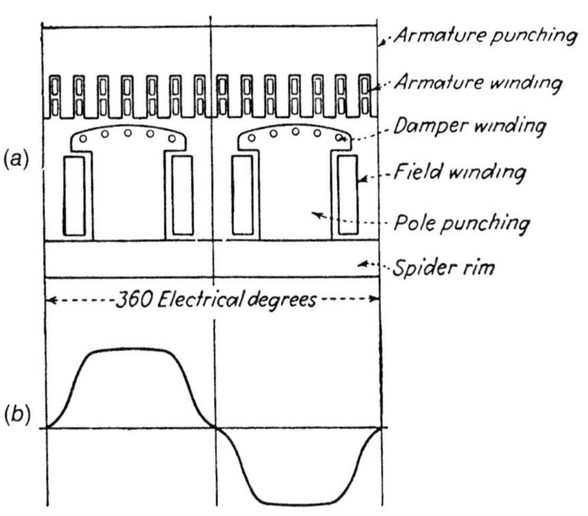


**FIGURE 14-36** (*a*) Generalized sketch of one pair of poles for a salient-pole machine; (*b*) flux form for typical pair of poles (current in field winding only).

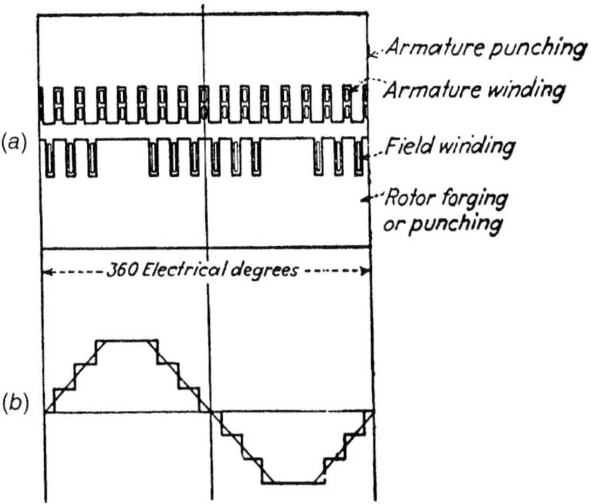


**FIGURE 14-37** (*a*) Generalized sketch of one pair of poles for cylindrical rotor machine; (*b*) flux form for a typical pair of poles (current in the field winding only).

The rms amplitude of generated voltage is, for the time, fundamental

$$V_1 = \frac{1}{\sqrt{2}} \omega N_a k_w \Phi_1 \tag{14-54}$$

where $V_1$ is the rms value of the time fundamental voltage induced in a phase and $N_a$ is the number of turns in the phase. In a three-phase machine connected in wye, line-line voltage is $\sqrt{3}$ times the phase voltage. The winding factor $k_{wn}$ is explained in Eqs. (14-55) through (14-57).

As the voltage waveform is not perfectly sinusoidal, time harmonic voltage will be induced that can be estimated by

$$V_n = V_1 \frac{C_n}{C_1} \frac{k_{wn}}{k_w} \tag{14-55}$$

It should be noted that certain time harmonics, those referred to as the *triplen* harmonics (odd factors of 3), are induced in all three phases of a three-phase machine in phase. Thus, if line-line voltage is measured, these harmonics (orders 3, 9, 15, . . .) will turn out to have zero amplitude.

The winding factor is $k_{wn} = k_{pn} k_{bn} k_{sn}$. Its component parts are called the *pitch factor*

$$k_{pn} = \cos \frac{n\alpha}{2} \tag{14-56}$$

the *breadth factor*

$$k_{bn} = \frac{\sin \frac{mn\gamma}{2}}{m \sin \frac{n\gamma}{2}} \tag{14-57}$$

and, when applicable, the *skew factor*

$$k_{bn} = \frac{\sin \frac{n\theta_s}{2}}{\frac{n\theta_s}{2}} \tag{14-58}$$

where $\alpha$ = *pitch angle*: the electrical angle between coil halves of the armature winding (this is generally a bit less than $\pi$ to reduce the impact of higher harmonics and to make armature end windings shorter)

$m$ = number of slots per pole per phase

$\gamma$ = electrical angle between slots

$\theta_s$ = electrical angle of skew

Here, the relevant angles are stated in *electrical* terms. The electrical angle is $p$ times the physical angle. Thus, in a 4-pole machine with 72 slots, the electrical angle between slots is $2 \times 360/72 = 10°$.

In many ac generators, the skew angle is zero for which the skew factor is unity. In some cases, the stator is skewed with respect to the rotor (or vice versa) to reduce the effects of slot openings in inducing currents in rotor parts with a consequent effect on rotor surface losses and noise.

## 14.3.7 Armature Reaction

Current in the armature conductors produces an mmf which has the same number of poles as the field structure. The fundamental harmonic of this mmf rotates at synchronous speed, and adds to the field mmf in a vector sense as shown in Fig. 14-38. For generators supplying reactive current to an inductive load, the net effect of the armature reaction is to oppose the field mmf, requiring additional field current to sustain flux.

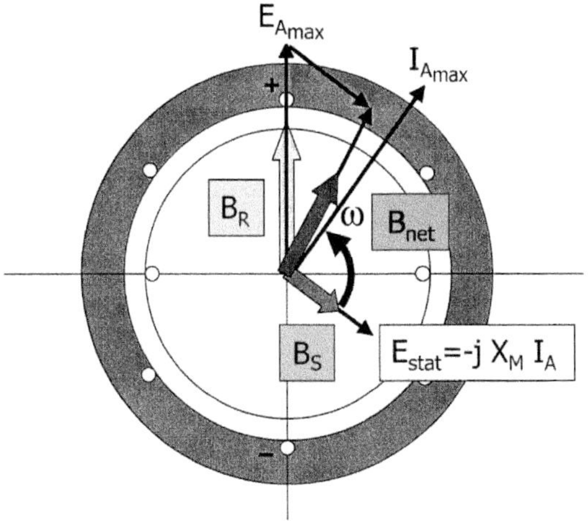


**FIGURE 14-38** Armature reaction (vector analysis of magnetic fields).

The peak value of the space fundamental of armature reaction current for a three-phase machine is

$$F_a = \frac{3}{2}\frac{4}{\pi}\sqrt{2}N_a I k_p k_b k_s \tag{14-58}$$

where $I$ is rms terminal current, $N_a$ is the number of series turns per phase, and the winding factors are as defined above.

Space harmonic components of armature reaction current can produce losses in the rotor as they turn asynchronously.

### 14.3.8 Capability Diagram

In normal operation, real power is dictated by the prime mover, and reactive power is determined by the real power and by field current.

Equations (14-47) and (14-48) are approximate ways of estimating the real and reactive power output of a generator as a function of field current and torque angle. (Eq. 14-59 is provided for additional clarification.) If these are cross-plotted a way of representing the capability of an ac generator emerges as shown in Fig. 14-39. This is called a *capability diagram* for obvious reasons and four limits are shown.

$$\begin{aligned} P &= 3\cdot V_\phi \cdot I_A \cdot \cos(\theta_z) = 3\cdot V_\phi \left[\frac{E_A \cdot \sin(\delta)}{X_s}\right] \\ Q &= \left(\frac{3\cdot V_\phi}{X_s}\right)\cdot X_S \cdot I_A \sin(\theta_z) = 3\cdot V_\phi \cdot I_A \sin(\theta_Z) \end{aligned} \tag{14-59}$$

1. The field winding limit is generally related to the thermal capability of the field winding and limits the operation of the generator at high reactive power conditions.
2. The armature winding limit is generally related to the thermal capability of the armature winding and is typically a limit on total kVA (kilovoltampere) output of the machine.

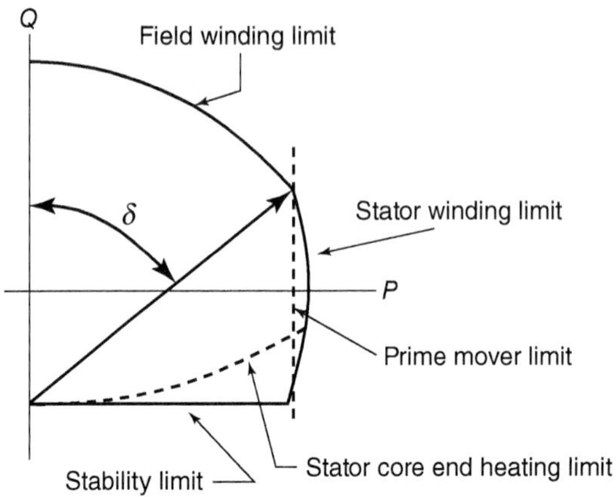


**FIGURE 14-39** Capability diagram, round-rotor machine.

3. The stability limit is related to torque angles that are nearly at the peak of the torque-angle curve (90° for round-rotor machines).
4. Often, the configuration of magnetic flux is such that, at high reactive power absorption (negative Q), there is axial flux in the core ends, leading to excessive heating and limiting the reactive power that can be absorbed by the machine.

Very often the prime-mover power rating is plotted on the capability chart. It is, of course, a line of constant real power. The heating related limits (armature, field, and core end) may be functions of the state of the cooling system of the machine, such as hydrogen pressure. (See Sec. 14.3.12 on cooling.)

## 14.3.9 Saturation Curves and Excitation

Alternating-current generators are usually operated with at least part of the magnetic circuit partially saturated, so that the linear model of machine operation implied by some of the foregoing discussion does not give exactly the right answers. What follows is an approximate way of estimating excitation requirements for an ac generator.

It should be noted that numerical methods, employing finite elements, are now available and capable of making even more accurate estimates of machine performance including excitation requirements.

Figure 14-40 shows open- and short-circuit characteristic curves for an ac generator. These curves are taken with the generator operating at rated speed. Important curves are

1. The air gap line is the extrapolation of open-circuit voltage versus field current at low levels of field current.
2. The no-load saturation curve is the actual open-circuit voltage versus field current characteristic.
3. The short-circuit saturation curve typically does not show saturation. It represents a measurement of current in the terminals of the generator, with the terminals short-circuited, versus field current.
4. The rated current, zero power factor saturation curve shows voltage at the terminals of the machine versus field current if the machine is operated with rated armature current at zero power factor.

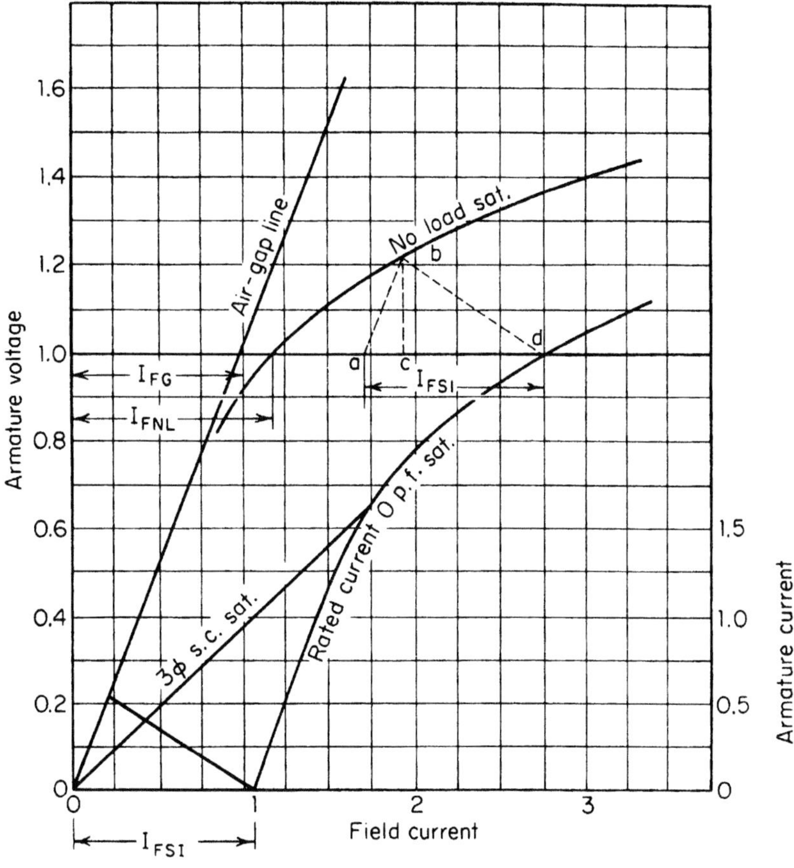


**FIGURE 14-40** Typical saturation curves of an ac generator showing graphic determination of Potier reactance (quantities are in per unit values).

where $I_{FG}$ is the field current required to excite the air gap, $I_{FSI}$ is the field current required to compensate for direct-axis armature current, and $I_{FS}$ is the field current required to compensate for saturation. Note that this method does not account for saturation of the quadrature axis. Armature resistance is generally neglected.

Addition of the three components of current is shown in Fig. 14-41. In this figure, the field current that compensates for armature reaction is added at the power factor angle $\theta$. The current $I_{FS}$ is the current required to compensate for direct axis saturation. $I_{FS}$ is the distance from the air-gap line to the no-load saturation curve at the voltage corresponding to the flux level in the magnetic circuit. This voltage is referred to as the voltage behind Potier reactance. It is estimated as shown in Fig. 14-42.

To find the (fictitious) Potier reactance, refer back to Fig. 14-40. Note that for the zero power factor test, all the fluxes are on the direct axis, so they add directly. Two of the current components, $I_{FG}$ and $I_{FSI}$, are easily determined. The difference between field current for the zero power factor test and the sum of these two currents is $I_{FS}$. Since this corresponds to the distance between the air-gap line and saturation curve at the voltage behind Potier reactance, it determines that voltage. For the zero power factor test, $I_a x_p$ is a vertical line of length $x_p$ since $I_a = 1$. The distance between the saturation curve and air-gap line is the same as $I_{FS}$ at a voltage found by casting a line parallel to the air-gap line from $I_F - I_{FSI}$. This is *ab* on Fig. 14-40. Then $x_p$ corresponds to *bc* on the same figure.

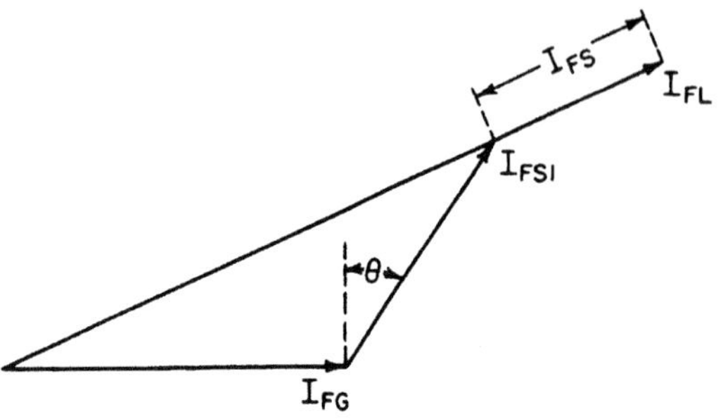


**FIGURE 14-41** ANSI method of calculating load excitation.

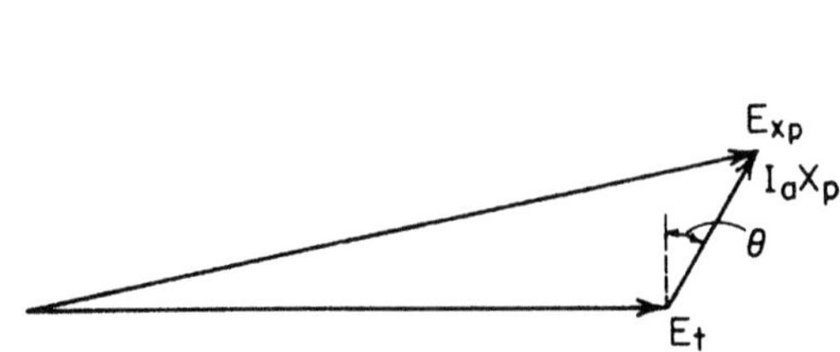


**FIGURE 14-42** Calculation of voltage behind Potier reactance.

### 14.3.10 Armature Windings

A wide variety of winding types may be used to produce a desired voltage with the desired number of phases and a suitable waveshape.

Fractional slot windings, in which the number of slots per pole per phase is not an integer, have coil groups that differ from one another. These can be arranged to produce balanced voltages under circumstances that are beyond the scope of this discussion.

At power frequencies (50 or 60 Hz), the skin depth in copper is on the order of 1 cm so that it is usually necessary to subdivide armature conductors into a number of parallel strands. In form-wound coils, these strands are usually rectangular to allow for good space factor. To prevent circulating currents between parallel strands, it is necessary to employ transposition to ensure that voltages induced in each strand are approximately the same.

Figure 14-43 shows an example winding diagram. For the purposes of this figure, the machine is shown "rolled out flat," with the dotted lines on either side representing the same azimuthal location.

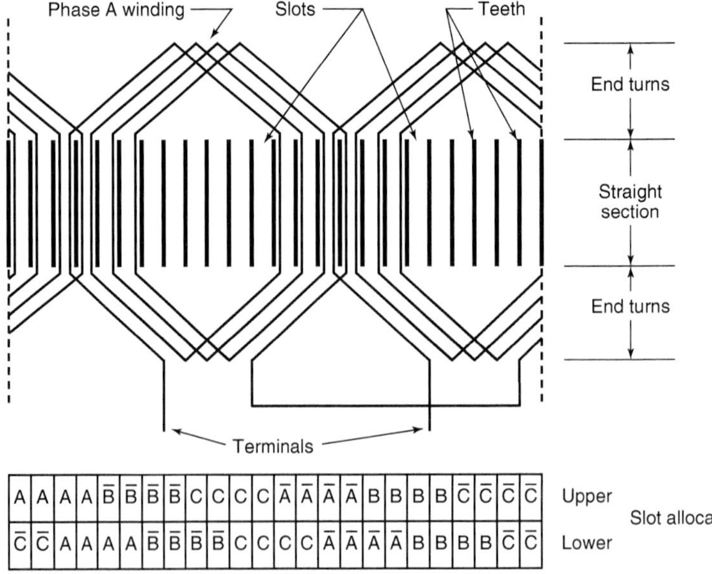


| | | | | | | | | | | | | | | | | | | | | | | | | |
|---|---|---|---|---|---|---|---|---|---|---|---|---|---|---|---|---|---|---|---|---|---|---|---|---|
| Upper | A | A | A | A | $\bar{B}$ | $\bar{B}$ | $\bar{B}$ | $\bar{B}$ | C | C | C | C | $\bar{A}$ | $\bar{A}$ | $\bar{A}$ | $\bar{A}$ | B | B | B | B | $\bar{C}$ | $\bar{C}$ | $\bar{C}$ | $\bar{C}$ |
| Lower | $\bar{C}$ | $\bar{C}$ | A | A | A | A | $\bar{B}$ | $\bar{B}$ | $\bar{B}$ | $\bar{B}$ | C | C | C | C | $\bar{A}$ | $\bar{A}$ | $\bar{A}$ | $\bar{A}$ | B | B | B | B | $\bar{C}$ | $\bar{C}$ |

**FIGURE 14-43** Armature in 24 slots, 5/6 pitch.

In this case, the machine has 24 slots, each with two half-coils, as shown in the slot allocation section of the drawing, at the bottom of the figure. The upper part of the figure shows how one phase of the winding would be laid out. This drawing shows a lap type winding (the most commonly used) with a $5/6$ pitch. In a 24-slot, two-pole winding a full-pitch coil would span 12 slots, while in the $5/6$ pitch winding the coils span 10 slots.

Armature cores are built up of thin laminations with thin electrical insulation in between, to reduce the eddy currents. Successive layers or groups of layers are staggered to minimize the effect of the joints in the magnetic circuit. The core is clamped between pressure plates and fingers to support it with sufficient pressure to prevent undue vibration of the laminations. The armature windings are fitted tightly in the slots and secured radially by wedges driven into suitable notches at the air-gap end of the slots. It is necessary that the stator coil ends be able to resist the abnormal forces associated with short circuits. A supporting structure may be employed for this purpose. Coil supports ordinarily are designed to suit the need of a particular machine.

The pole pieces of salient-pole alternators may be built up of steel laminations, both as manufacturing convenience and a means of limiting the loss in their air-gap surfaces due to pulsations in air-gap flux. The field coils, wound directly on the poles or preformed and then mounted on the poles, are suitably insulated from the poles for the voltages associated with normal and transient operation. The shaft may be integral with the body, as in the case of a forging, or may be bolted to or inserted into the body. The rotor of a round-rotor machine is cylindrical in shape with axial slots provided in its body for the field coils. The body is usually a steel forging with integral shaft ends. The field coils are wound in axial slots in the rotor body, held in place by heavy slot wedges and by retaining rings over the coil ends. Wound filed excitation may be replaced with permanent magnet field excitation. These rotors can have surface mounted PMs (SPM) or interior PMs (IPM) which correspond to the round-rotor and salient-pole machines, respectively. The elimination of the filed windings results in a simpler construction in case of the PM synchronous generators.

Although, antifriction bearings are occasionally used on alternators of smaller ratings, the great majority are furnished with oil-lubricated babbitted bearings. For horizontal shafts at small ratings ring-oiled bearings are used, but at higher ratings recirculation of externally cooled oil is used.

Two principal types of thrust bearings are used on vertical alternators: the pivoted-shoe type and the spring type. The adjustable *pivoted-shoe* type, introduced in the United States by Albert Kingsbury, consists of a flat rotating collar or runner of steel or fine-grained cast iron resting on a stationary member consisting of several babbitted segmental shoes pivoted near their center on adjusting screws, which, by changing the elevation of the shoes, can provide equal loading on each. The screws also permit small adjustments in rotor elevation to correct generator and turbine clearances.

The bearings are immersed in oil. In operation, a thin, wedge-shaped film of oil is formed between the runner and the shoe. The oil is continuously circulated by the rotation of the runner and is cooled by either radiation or water cooling, usually within the oil bath but occasionally by an external system. Some of the larger bearings are cooled by means of water circulated through tubes embedded below the babbitt surface.

The *spring*-type bearing is inherently self-equalizing; that is, each shoe carries very nearly the same amount of load. A variation of the pivoted-shoe bearing, in which the shoes are supported on a system of interconnected levers, provides the same self-equalizing feature.

## 14.3.11 Mechanical Construction

Generators consist of a rotor, a stator, and those mechanical parts required to hold the machine together and assure proper alignment, cooling, and other aspects of operation. Design criteria for the mechanical aspects of machine construction are based not only on anticipated steady state operation but also on abnormal aspects of operation such as overspeed and short circuits.

Essential variations in machine construction are salient pole versus round rotor and horizontal versus vertical shaft.

The prime mover determines if the shaft is horizontal or vertical. Most generators driven by steam turbines, gas turbines, or reciprocating engines have horizontal shafts. Larger, low-speed hydraulic turbines have vertical shafts, while smaller, higher-speed turbines can have shafts that are vertical, horizontal, or even inclined.

### 14.3.12 Losses and Efficiency

Efficiency of a generator is defined as

$$\eta = \frac{P_{\text{out}}}{P_{\text{in}}} \cong 1 - \frac{P_{\text{dissipation}}}{P_{\text{in}}} \tag{14-60}$$

Dissipation or losses include:

*Windage and friction* includes loss from the bearings, shearing of the air or hydrogen at the rotor surface, mechanical work done on the air or hydrogen that flows through the rotor, and power expended by the cooling fans. This loss is generally not a function of operating point, but it does vary with the pressure of the cooling gas inside the generator. Provided that the cooling gas pressure is kept constant, this loss is usually considered to be independent of load.

*Core loss* is caused by hysteresis and eddy currents in the core laminations. It also includes losses induced in structural parts of the machine that are exposed to stray alternating magnetic flux at no-load conditions. Core loss is a function only of terminal voltage, and it is usually considered to be independent of load.

*Field loss* is ohmic loss caused by field current flowing in the field winding. This loss depends on both the field current and the temperature of the field winding. It thus varies with load.

*Armature loss* is ohmic loss caused by armature current flowing in the armature winding. This loss is defined by the square of the armature current multiplied by the dc resistance of the armature winding, corrected for temperature. This loss varies with load.

*Stray load loss* explains sources of loss not adequately covered by the other categories. It includes eddy current loss in the armature and losses in structural elements exposed to magnetic fields arising from armature current and to those parts of the rotor surface affected by armature leakage fields. Stray loss is generally proportional to the square of armature current and may, with only small error, be expressed as a fraction of armature loss.

Dissipation in generators appears as heat which must be removed. This heat appears in the armature conductors, field-winding conductors, stator core, rotor surface, and other structural elements of the machine. Cooling of armature and field conductors may be direct or indirect; the difference is direct contact of the cooling medium with the conductor or contact through electrical insulation.

Alternating-current generators may be cooled by air, hydrogen, water, or (very infrequently) oil. In large machines, no matter what the cooling medium, heat is transferred to water in heat exchangers that are located within the machine case. Fans used in electric machines may be of either radial flow or axial flow, and a wide variety of cooling paths are used.

Figure 14-44 represents an example of ventilation paths in a cooling system.

### 14.3.13 Testing of AC Generators

Tests are performed on generators to determine the performance characteristics and machine parameters.

Field and armature resistances are typically small, so measurements should be made using a 4-wire technique. It is important that resistance be measured at a known temperature so that correction can be made to actual operating temperature.

The generator is driven by a motor to rated speed and excitation varied to produce terminal voltage over a range, typically from perhaps 30% to 120% that of rated. Some caution is required

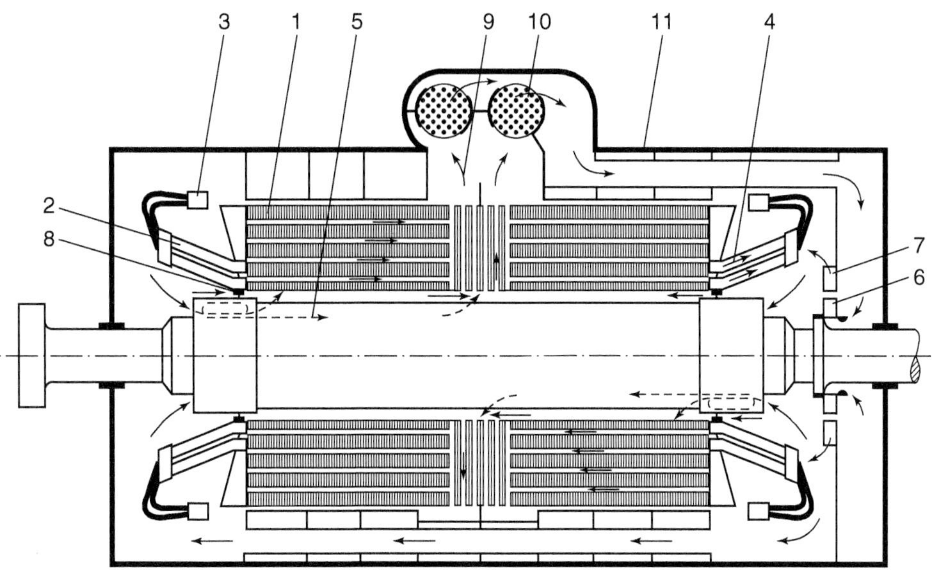


**FIGURE 14-44** Cooling system, combination of gas and water. (1) stator core; (2) water-cooled stator winding; (3) water manifold; (4) water flow; (5) rotor body; (6) radial-flow fan; (7) diffuser; (8) air-gap baffle; (9) gas flow; (10) hydrogen cooler; (11) stator housing. (*ABB.*)

here, particularly for large machines in which excessive flux can damage the core. Open-circuit losses may be established by open circuit saturation curve if the drive motor is well characterized and input power is measured.

Short-circuit saturation curve test is similar to the open-circuit test, except the armature terminals are short-circuited and excitation varied to produce armature current over some convenient range. Windage and friction losses may be inferred from power input at zero excitation. Stray load loss may be estimated as the difference between input power at rated armature current and the sum of friction and windage and armature $I^2R$.

For a relatively small generator, the zero power factor saturation curve can be determined by running the machine with its shaft unloaded, driven by a second generator.

By adjusting the excitation on the ac generator under test and excitation on the second generator, it is possible to measure the zero power factor saturation curve.

For large generators for which this "back-to-back" method is not practical, the zero power factor curve is usually determined by numerical methods.

Deceleration may be used for determining losses if the inertia of the machine is known. Since, if the shaft of a machine is unloaded, power dissipated is

$$P_w = \omega_m J \frac{d\omega_m}{dt} = \frac{d}{dt}\left\{\frac{1}{2} J \omega_m^2\right\} \tag{14-61}$$

where $\omega_m$ is mechanical speed, deceleration through synchronous speed can give a good measure of dissipation. The test may be run with the machine operating either at open-circuit or short-circuit conditions, or at zero excitation. It is usually run from a slight overspeed. This test can be used to determine an unknown inertia from known losses and observed deceleration.

Heat runs are tests performed by operating the generator at some condition until the temperature stabilizes. Heat runs at open-circuit, short-circuit, and zero power factor may be combined to estimate temperature rise in actual operation. In large machines, good estimates of dissipation may be made by measuring the temperature rise of coolant (e.g., water). This is an alternative or supplement to measuring input power to the drive motor or machine deceleration $\omega_m$.

### 14.3.14 Dynamic Models

In applying an ac generator to power systems, it is important to understand the dynamic performance of the machine. This subsection describes one way of estimating the dynamic performance of a generator.

Voltages, currents, and impedances are often expressed as *per unit* values, that is, a base for each quantity is assumed and every ordinary variable is compared with that base. Thus, when a machine is operating at its rated condition, both voltage and current magnitude could be said to be "one per unit." Multiplying voltage and current yields power, while dividing voltage by current yields impedance magnitude, so these can be expressed in per unit terms as well. The general form of the per unit (*pu*) is shown in Eq. (14-62).

$$pu = \frac{\text{actual}}{\text{base}} \tag{14-62}$$

Generally, the base apparent power or total power ($S_{\text{base}}$) and the base voltage ($V_{\text{base}}$) are specified; all other base values are determined from these two.

Typical electrical models of electric machines represent the relationship between voltage and current in the various windings of the machine. Thus, the three-phase windings and the field winding are represented. Important dynamics, however, arise from currents in elements of the rotor surface, the rotor body or amortisseur or damper bars, or some combination of all of these factors. These are typically represented by equivalent windings, too. For the purpose of this discussion, only one additional winding (referred to here as the "damper") will be included for each axis of the rotor.

The analysis begins with normalized variables referred to as the *direct* and *quadrature axes* of the machine. The per unit fluxes are

$$\begin{bmatrix} \psi_d \\ \psi_{kd} \\ \psi_f \end{bmatrix} = \begin{bmatrix} x_d & x_{ad} & x_{ad} \\ x_{ad} & x_{kd} & x_{fd} \\ x_{ad} & x_{fd} & x_f \end{bmatrix} \begin{bmatrix} i_d \\ i_{kd} \\ i_f \end{bmatrix} \tag{14-63}$$

$$\begin{bmatrix} \psi_q \\ \psi_{kq} \end{bmatrix} = \begin{bmatrix} x_q & x_{aq} \\ x_{aq} & x_{kq} \end{bmatrix} \begin{bmatrix} i_q \\ i_{kq} \end{bmatrix} \tag{14-64}$$

where the variables $\psi$ represent per unit fluxes and $i$ represents per unit currents. Subscripts $d$ and $q$ represent the direct and quadrature axes, respectively, while the subscripts $a$, $k$, and $f$ represent the armature, damper, and field. There are only two armature variables here, where, strictly speaking, a third would be required. Ordinarily, the "zero" axis variable can be ignored as under the usually balanced operation of generators current in that winding is zero.

The equivalent circuits shown in Fig. 14-45 represent the same flux-current relationship as do Eqs. (14-63) and (14-64), with winding resistances added if

$$x_d = x_{al} + x_{ad} \tag{14-65}$$

$$x_{fd} = x_{ad} + x_{akd} \tag{14-66}$$

$$x_{kd} = x_{fd} + x_{kdl} \tag{14-67}$$

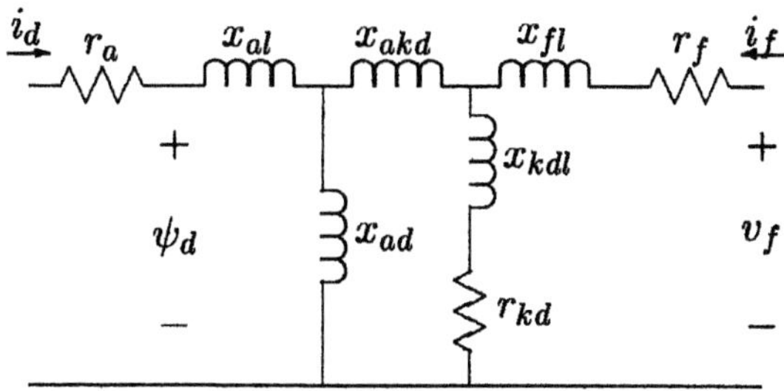


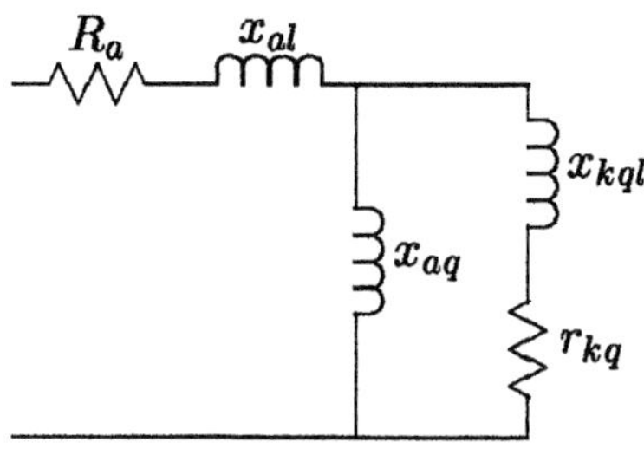


**FIGURE 14-45** Direct and quadrature-axis equivalent circuits.

$$x_f = x_{fd} + x_{fl} \tag{14-68}$$

$$x_q = x_{al} + x_{aq} \tag{14-69}$$

$$x_{kq} = x_{kql} + x_{aq} \tag{14-70}$$

The interesting dynamics of the rotor can be described by these two equivalent circuits.

The various reactance and resistance parameters of the equivalent circuits of Fig. 14-45 may be calculated from first principles or may be measured.

The equivalent circuits of Fig. 14-45 are only approximations to the actual performance of ac generators. In some cases it will be necessary to use higher-order models. Generally these are represented by multiples of the damper winding with different magnitudes and time constants. These are beyond the scope of this discussion.

Voltages are produced by time variations of fluxes and by rotation of the machine. Translated into per unit measurement the voltages are

$$\upsilon_d = \frac{1}{\omega_0}\frac{d\psi_d}{dt} - \frac{\omega}{\omega_0}\psi_q + r_a i_d \tag{14-71}$$

$$\upsilon_q = \frac{\omega}{\omega_0}\psi_d + \frac{1}{\omega_0}\frac{d\psi_q}{dt} + r_a i_q \tag{14-72}$$

$$\upsilon_f = \frac{1}{\omega_0}\frac{d\omega_f}{dt} + r_f i_f \tag{14-73}$$

$$\upsilon_{kd} = \frac{1}{\omega_0}\frac{d\psi_{kd}}{dt} + r_{kd} i_{kd} \tag{14-74}$$

$$\upsilon_{kq} = \frac{1}{\omega_0}\frac{d\psi_{kq}}{dt} + r_{kq} i_{kq} \tag{14-75}$$

$$\upsilon_0 = \frac{1}{\omega_0}\frac{d\psi_0}{dt} + r_a i_0 \tag{14-76}$$

These expressions may be turned into a concise simulation model, suitable for use in modern computing apparatus for estimating the performance of an ac generator. This is simply done by isolating the first-order time derivatives.

The state variables are the two stator fluxes $\psi_d$, $\psi_q$, two "damper" fluxes $\psi_{kd}$, $\psi_{kq}$, field $\psi_f$, and rotor speed $\omega$ and torque angle $\delta$. The most straightforward way of stating the model employs currents as auxiliary variables

$$\begin{bmatrix} i_d \\ i_{kd} \\ i_f \end{bmatrix} = \begin{bmatrix} x_d & x_{ad} & x_{ad} \\ x_{ad} & x_{kd} & x_{fd} \\ x_{ad} & x_{fd} & x_f \end{bmatrix}^{-1} \begin{bmatrix} \psi_d \\ \psi_{kd} \\ \psi_f \end{bmatrix} \tag{14-77}$$

$$\begin{bmatrix} i_q \\ i_{kq} \end{bmatrix} = \begin{bmatrix} x_q & x_{aq} \\ x_{aq} & x_{kq} \end{bmatrix}^{-1} \begin{bmatrix} \psi_q \\ \psi_{kq} \end{bmatrix} \tag{14-78}$$

Then the state equations are

$$\frac{d\psi_d}{dt} = \omega_0 \upsilon_d + \omega\psi_q - \omega_0 r_a i_d \tag{14-79}$$

$$\frac{d\psi_q}{dt} = \omega_0 \upsilon_q - \omega\psi_d - \omega_0 r_a i_q \tag{14-80}$$

$$\frac{d\psi_{kd}}{dt} = -\omega_0 r_{kd} i_{kd} \tag{14-81}$$

$$\frac{d\psi_{kq}}{dt} = -\omega_0 r_{kq} i_{kq} \tag{14-82}$$

$$\frac{d\psi_f}{dt} = -\omega_0 r_f i_f \tag{14-83}$$

$$\frac{d\omega}{dt} = \frac{\omega_0}{2H}(T_e + T_m) \tag{14-84}$$

$$\frac{d\delta}{dt} = \omega - \omega_0 \tag{14-85}$$

and

$$T_e = \psi_d i_q - \psi_q i_d \tag{14-86}$$

For a practical simulation, $v_d$, $v_q$, and mechanical torque $T_m$ must be specified.

It should be clear from examining the equivalent circuits of Fig. 14-45 that the behavior of the machine, meaning its flux/current relationship, is a function of the speed of any given disturbance. This leads to approximate analyses of ac generators, which recognize the frequency dependence of the rotor elements.

*Synchronous* reactances $x_d$ and $x_q$ are applicable when the rotor is moving in synchronism with the mmf produced by the armature currents, or when the deviation in speed is very small.

*Transient* reactance $x_d'$ is applicable when armature mmf is changing with time, as in electromechanical transients or "swings." This effective reactance is reduced by the tendency of (nearly) short-circuited field winding to trap or hold flux nearly constant for periods comparable to or shorter than its time constant. This reactance is

$$x_d' \approx x_{al} + x_{ad} \,||\, (x_{fd} + x_{fl}) \tag{14-87}$$

In some types of generators there may be an identifiable transient reactance on the quadrature axis, less than the quadrature-axis synchronous reactance, but in others the applicable reactance to use for the quadrature axis in transient events is just $x_q$.

For transient events that occur very rapidly, such as switching events and terminal short circuits, the effective reactances are the *subtransient* reactances. These result from the tendency of the amortisseurs (or even rotor iron) to support or trap flux. They are

$$x_d'' = x_{al} + x_{ad} \,||\, (x_{akd} + x_{kdl} \,||\, x_{fl}) \qquad x_d'' = x_{al} + x_{aq} \,|x_{kql} \tag{14-88}$$

The foregoing reactances are applicable when the armature mmf and rotor are rotating in synchronism or nearly so. The *negative-sequence* reactance $x_2$ is applicable with an armature mmf rotating backward at synchronous speed while the rotor is rotating forward. Negative-sequence currents arise from certain types of unbalanced operation. The negative-sequence reactance is approximately the average of subtransient reactances

$$x_2 = ½(x_d'' + x_q') \tag{14-89}$$

*Zero-sequence* reactance is applicable to situations in which all three phases of the armature have identical currents such as would arise from a ground fault. Such currents do not produce substantial air-gap flux, and even some of the components of armature leakage are reduced, so the zero-sequence inductance is quite small.

Torque-angle curves for operation of a synchronous generator may be written for both the synchronous and transient mode of operation. In per unit, these torque-angle curves are for the steady state case

$$t_e = \frac{e_{af}}{x_d} \sin\delta + \left(\frac{1}{x_q} - \frac{1}{x_d}\right) \sin 2\delta \tag{14-90}$$

and for the transient case

$$t_e = \frac{e_q'}{x_d'} \sin\delta + \left(\frac{1}{x_q} - \frac{1}{x_d'}\right) \sin 2\delta \tag{14-91}$$

where $\delta$ is the torque angle, $e_{af}$ is the internal voltage, and $e'_q$ is the *voltage behind transient reactance*, and the terminal voltage is one per unit.

The relationship between power, which is directly proportional to torque, and displacement (power) angle is shown in Fig. 14-46. Note that the torque-angle curve for transient operation has a higher peak torque, although the two curves coincide at the steady-state operating point.

Disturbances from steady-state operation due to changes in prime mover torque, load changes, or faults cause load angle to change, usually with dynamic swings as described in this subsection. Generally, the most serious transient swing results from complete loss of load as from a nearby short-circuit. Under such a circumstance the generator, under the influence of prime mover torque,

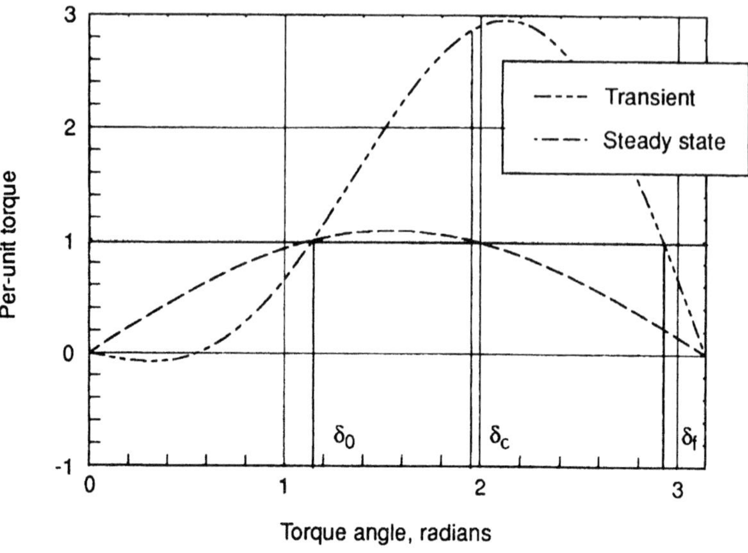


**FIGURE 14-46** Torque-angle curves, steady state, and transient.

accelerates. The torque angle increases quadratically with time. Even after the power output is restored, the machine will swing forward until load torque stops its advance. A simple approximate way of estimating the maximum angle achieved is to use the *equal-area* criterion. The area of a disturbance is related to energy contributed to rotation, and for every (stable) situation the positive and negative areas must balance.

Pictured in Fig. 14-46 is a "critical" swing. The area underneath the prime mover torque curve from $\delta_0$ to $\delta_c$ is the area that would be contributed if all load torque were zero during a period of acceleration between those two angles. If the area *above* the prime-mover torque between $\delta_c$ and $\delta_f$ is equal to or greater than this first area, the machine will regain synchronous operation. This establishes a *critical angle.* In turn, the *critical clearing time* $t_c$ is established by

$$\delta_c - \delta_0 = \frac{1}{2}\frac{\omega_0}{2H}t_c^2 \tag{14-92}$$

If an ac generator is operated open-circuited at rated terminal voltage and then its terminals are suddenly short-circuited, the following currents flow in the terminal leads:

$$i_a = \left[\frac{e_{af}}{x_d} + \left(\frac{e_{af}}{x'_d} - \frac{e_{af}}{x_d}\right)e^{-(t/T'_d)} + \left(\frac{e_{af}}{x''_d} - \frac{e_{af}}{x'_d}\right)e^{-(t/T''_d)}\right]\cos(\omega t - \theta_0) - \frac{e_{af}}{x''_d}\cos\theta_0 e^{-(t/T_a)}$$

$$i_b = \left[\frac{e_{af}}{x_d} + \left(\frac{e_{af}}{x'_d} - \frac{e_{af}}{x_d}\right)e^{-(t/T'_d)} + \left(\frac{e_{af}}{x''_d} - \frac{e_{af}}{x'_d}\right)e^{-(t/T''_d)}\right]\cos\left(\omega t - \theta_0 - \frac{2\pi}{3}\right) - \frac{e_{af}}{x''_d}\cos\left[\theta_0 - \frac{2\pi}{3}e^{-(t/T_a)}\right]$$

$$i_c = \left[\frac{e_{af}}{x_d} + \left(\frac{e_{af}}{x'_d} - \frac{e_{af}}{x_d}\right)e^{-(t/T'_d)} + \left(\frac{e_{af}}{x''_d} - \frac{e_{af}}{x'_d}\right)e^{-(t/T''_d)}\right]\cos\left(\omega t - \theta_0 + \frac{2\pi}{3}\right) - \frac{e_{af}}{x''_d}\cos\left[\theta_0 + \frac{2\pi}{3}e^{-(t/T_a)}\right]$$

These currents are shown in Fig. 14-47.

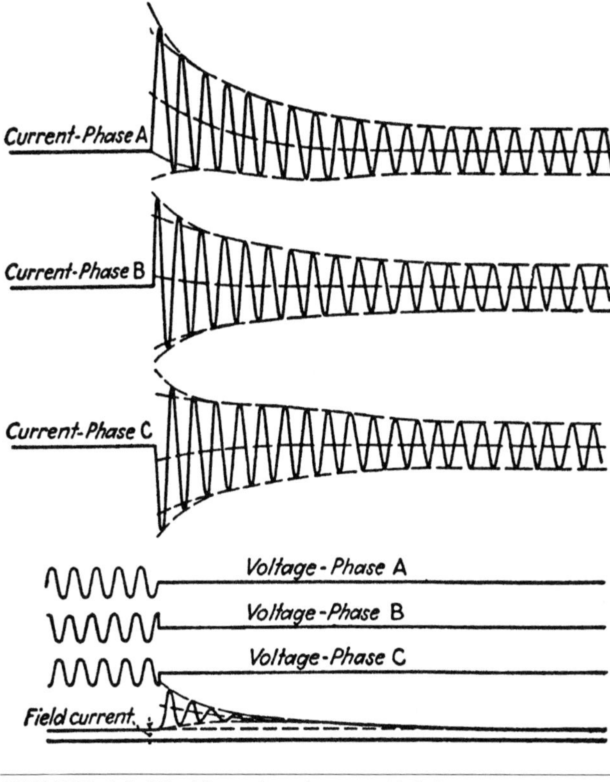


**FIGURE 14-47** Three-phase (symmetric) fault.

## 14.3.15 Special AC Generators

***Permanent Magnet Synchronous Generators.*** In recent years, permanent magnet synchronous (PMSG) generators have gained popularity, especially due to developments in magnetic materials and power electronics, which made possible new designs with high efficiency and compact dimensions. The construction and operation of PMSG is similar to that of ac synchronous machines, which have been previously discussed, with a main difference in the rotor field excitation that is provided by permanent magnets (PM), as illustrated in Fig. 14-48.

A variety of designs is possible for PMSG rotors, including SPM, interior permanent magnet (IPM), and other topologies, as discussed in Sec. 15. Relatively low per-unit inductance and high back emf may be preferable for generator applications in order to ensure a high power factor. Such a requirement typically favors the SPM designs of the type shown in Fig. 14-48, in which the equivalent electromagnetic air gap is substantially larger as it includes the mechanical air gap and the magnet thickness.

Large size and power PMSG typically employ a stator with multiple slots per pole and phase and distributed windings. In this case, short pitching the winding may result in improved performance. For example, a winding with a short pitch of 5/6, which can be implemented with a core design having two slots per pole and phase by employing a two layer arrangement with some of the slots shared by coil sides belonging to different phases, reduces the 5th and 7th order harmonics, resulting in a more sinusoidal back emf waveform, reduced 6th order harmonic torque pulsations (ripple), and lower rotor losses.

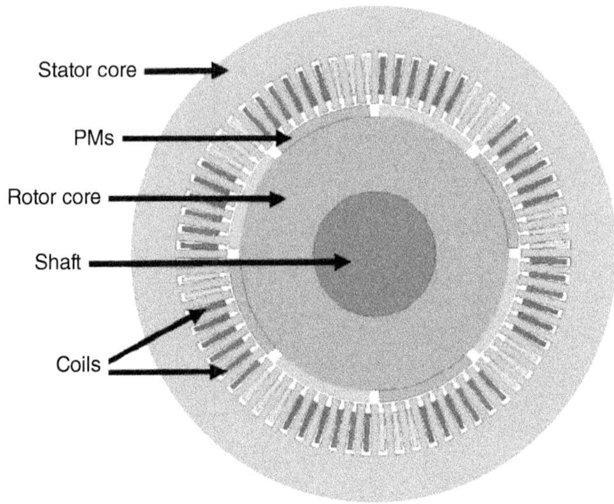


**FIGURE 14-48** Cross-sectional view of a permanent magnet synchronous generator (PMSG). In this example, the rotor includes six magnets (PM) of alternating polarity attached to the surface of the rotor core facing the air gap. The stator has a core with slots in which a three-phase six-pole winding with distributed coils is placed.

Smaller PMSG may employ fractional slot stator designs, for example for three-phase machines, with three slots for each pole pair, such as 9 slots 6 pole or 12 slots 8 pole combinations, or with the total number of slots equal to plus or minus 2, or plus or minus 1, the number of (rotor) poles. Such example combinations include 9 slots and 6 poles, 9 slots and 8 poles, and 12 slots and 10 poles. The concentrated coils only span one tooth pitch and are not overlapping with other phase coils resulting in compact end turns. Fractional slot concentrated windings may be of the single layer type, in which only alternate teeth are wound, or of the double layer type, with all teeth wound. The latter present lower fundamental winding factors, but have smaller end turns. As compared with distributed winding designs, concentrated winding machines, may result in copper weight saving and possibly lower copper losses, but they have larger mmf harmonics potentially leading to increased core losses and torque pulsations. Furthermore, the increased leakage reactance may be disadvantageous for generators.

Applications for relatively low power PMSG include, among others, portable generation systems and small hydro. The use of damper bars in the rotors of PMSG employed in such distributed energy generation applications may improve stability.

Latest generations of wind turbines, electric traction systems, and distributed generation applications benefit of the PMSG advantages in terms of increased efficiency and specific power density. Because the PM magnetization is basically fixed, specific controls are required in order to ensure functionality, as discussed in the following.

***Wind Turbine Generators Including PMSG, Induction, and Doubly Fed Induction Types.*** In variable-speed wind turbines, typically for multi-MW utility applications, PMSG or electromagnetically excited synchronous generators employ a power electronics system with two back to back ac/dc and dc/ac converters connected via a dc link (Fig. 14-49). Power electronics is required in order to interface the variable voltage at the generator terminals, which is caused by the changes in wind speed, with the fixed grid frequency and voltage. A similar approach can be employed in conjunction with squirrel cage rotor induction generators. The drive system includes multiple control loops.

Rotor field oriented control maybe used for the generator and the active current reference is derived from the speed loop. The reactive component of current is maintained at zero for PMSG, and derived from the rotor flux reference for an induction machine. The grid side and the generator side controllers cooperate to maintain the dc bus voltage at the required level, to ensure synchronization with the grid and deliver active and reactive power.

Wind generators may include a gearbox or be directly coupled to the turbine rotor. Both configurations have their advantages and limitations. In the configuration that has been proven over the years, a gearbox is included in order to increase the rated speed of the generator and hence reduce the torque and size of the electrical machine, resulting in a drive system with components of comparable dimensions. In direct drive arrangements, the gearbox is eliminated, the number of components is reduced leading to potential reliability improvements and reduced maintenance costs. As the rotor and generator speed are very low in this case, the torque is extremely high, resulting in electric machines of very large dimensions with outer diameters in the order of meters for multi-MW power output. In order to maintain the generator frequency to reasonable levels the machine design employs very high number of poles. Both electromagnetically excited and PMSG have been implemented in direct drive wind turbines.

Apart from the conventional radial flux configuration, the axial flux type has been proposed for wind turbine generators. In such a configuration, the rotor and the stator are substantially disk shaped, are placed to face each other axially. One advantage is that the area available for torque production is proportional to the square of the diameter, unlike in radial flux machines, where it is proportional only to the diameter. Axial flux generators may be realized in a single stator single rotor configuration, or may use a dual stator single rotor, or dual rotor single stator design.

In variable wind turbine systems, such as the one illustrated in Fig. 14-49, the power electronics ensures full energy conversion and has the same rating as the generator. Reduced rating for the electronics can be achieved by employing a doubly fed induction generator (DFIG), in which case an ac-ac bidirectional converter is only connected to the rotor electric circuit, as shown in Fig. 14.50, and typically rated to only a quarter to a third of the generator power, depending on the maximum speed range. The ac/ac converter maybe implemented through two back-back ac-dc and dc-ac converters, respectively. Other solutions, such a direct ac-ac matrix converter without a stiff dc link are also possible, in principle. The stator is directly connected to the grid.

Historically, this solution has been advantageous in terms of reduced cost for the power electronics, but more recently its applicability was limited due to new grid code requirements for reactive

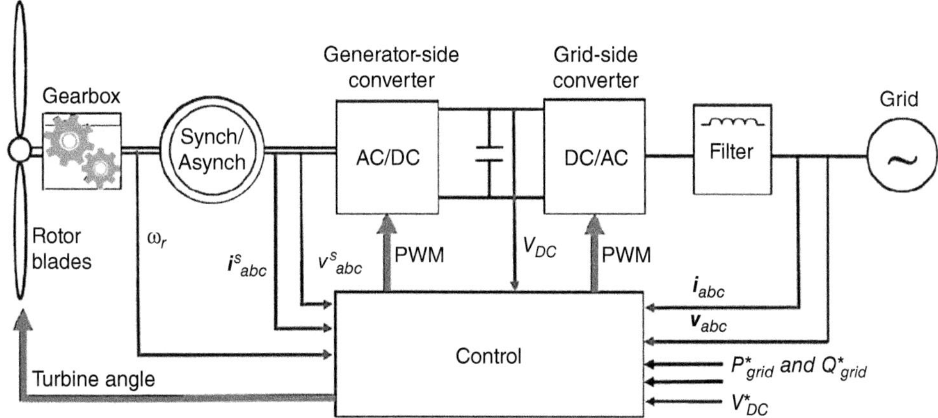


**FIGURE 14-49** Schematic of a variable speed wind turbine drive and power conversion system, including back-to-back power electronic converters and associated controls. PMSG or squirrel-cage rotor induction generators maybe employed. Direct drive implementations do not use a gearbox.

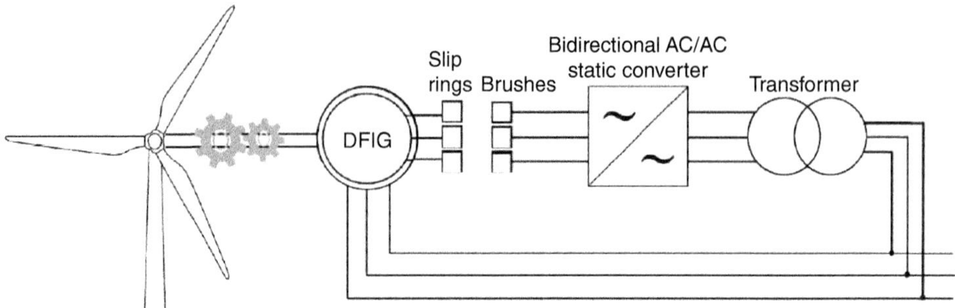


**FIGURE 14-50** Schematic of a wind turbine employing a DFIG. The power electronic converter from the rotor side is rated for partial power, smaller than that of the generator. The bi-directional ac/ac converter may consist of two back-back ac-dc and dc-ac converters, respectively.

power capability and reactive power, which are better met with full conversion power electronic systems. A DFIG generator is generally controlled by orienting along the stator field. The grid side converter is controlled by orienting along the grid voltage, in order to maintain the dc bus voltage and provide the required reactive power. The DFIG may operate at sub-synchronous speeds, such that mechanical speed, $\omega_m$, is lower than the synchronous speed, $\omega_1$, and the stator produces and the rotor absorbs electrical power. During super-synchronous operation, both the stator and the rotor deliver electrical power output (Fig. 14-51).

In order to address the maintenance and reliability issues associated with the use of slip rings, brushless doubly fed machines have been proposed. Typical designs of this kind include two independent sets of windings in the stator, each with a different number of poles, and with one winding connected to the grid and the other, "control" winding, connected to a partially rated power electronics inverter. The rotor has a special construction that may incorporate a nested variable reluctance structure with flux barriers or slots and rotor bars with specific end connections. As an interesting operational feature, if the "control" winding is dc supplied, the machine operates similarly to a wound rotor synchronous machine, and, if this winding if short-circuited, the machine operates similar to a cage rotor induction machine. Such machines typically have a low power factor, a disadvantage that offsets their benefits and limits widespread application.

For small wind turbines, that do not require full variable speed operation, simple induction generators with shunt and series capacitors for self-excitation, and without power electronics converters may also be used. In terms of recent research for multi-MW wind turbines, multiphase generators, for example with nine phases, have been proposed as they may have advantages, such as smaller ratings of the power electronic switches equipment due to lower current per phase, and better fault tolerance and high redundancy. Superconducting generators with stationary dc field coils have been considered as a longer-term solution, especially for the extremely large wind turbines for offshore installations.

***Other Special Electric Generators.*** In critical standby, remote and distributed generation applications with power ratings up to the MW range, the electric generator is a subsystem of a generator set (gen set), which includes a prime mover, such as a diesel engine, and associated switched gear and electronic controls. The synchronous electric generator system shown in Fig. 14-52 is a low voltage 2.5 MW unit that incorporates three separate electric machines on the same shaft, that is the main generator (on the right side of the figure) providing the output power, an exciter to power the rotor field winding of the main generator, and a small auxiliary PMSG used to power an electronic regulator for the control the output voltage and other system characteristics.

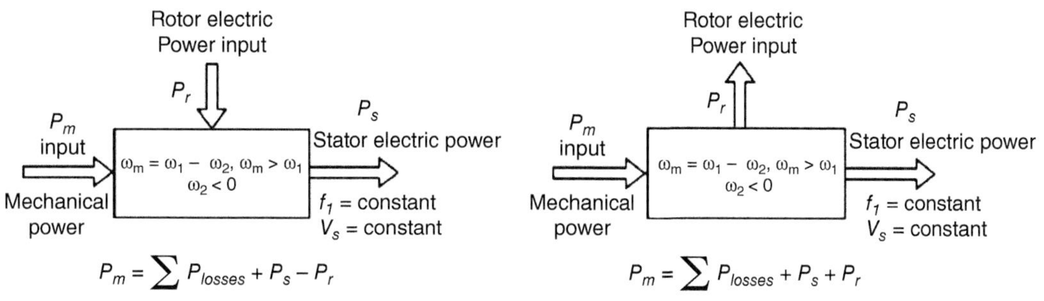


**FIGURE 14-51** Power flow in the sub-synchronous (left) and super-synchronous modes of operation for a DFIG.

**FIGURE 14-52** Rendering of an example low voltage 2.5 MW electromagnetically excited field wound salient-pole synchronous generator, incorporating on the same shaft the main generator, an exciter, and a small auxiliary PMSG. (*Courtesy of Regal Beloit Corp.*)

Claw pole generators are widely used in automotive applications in order to produce electrical energy on-board. Such a machine includes a stator with a multi-phase winding and a rotor with a core that is typically made of cast steel and specially shaped to include claws surrounding a circular coil (Fig. 14-53). When supplied with dc current, through slip rings and brushes, the rotor coil produces a magnetic field, which is directed by the claws into a multi-pole radial air-gap field. Typically, these generators have stators with one or two slots per pole per phase and 12, or up to 18, poles. The three-phase ac stator output is converted to dc for on-board battery connection using a full bridge diode rectifier. The robust and rather inexpensive construction have maintained the claw pole generator as a popular solution in the automotive industry. Variations of this topology include stationary dc windings and additional magnets in the rotor.

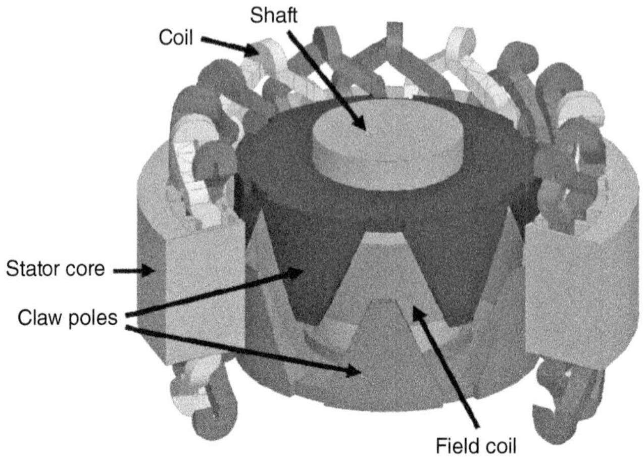


**FIGURE 14-53** Example claw pole synchronous generator with three-phase stator winding and 12 poles. Each of the rotor "claws" attached to one rotor disk has the same polarity.

Drive trains for electric traction as, for example, those for hybrid electric vehicles require the electric machines to operate in motoring and generating modes, or employ separate electric motors and generators. During regenerative braking, energy is generated and stored in the battery leading to an increase in system efficiency, especially in urban driving cycles. These machines are typically of the radial flux type and are coupled to gearboxes. In-wheel direct drive machines employing, for example, axial flux PM topologies, such as the one depicted in Fig. 14-54 are also possible.

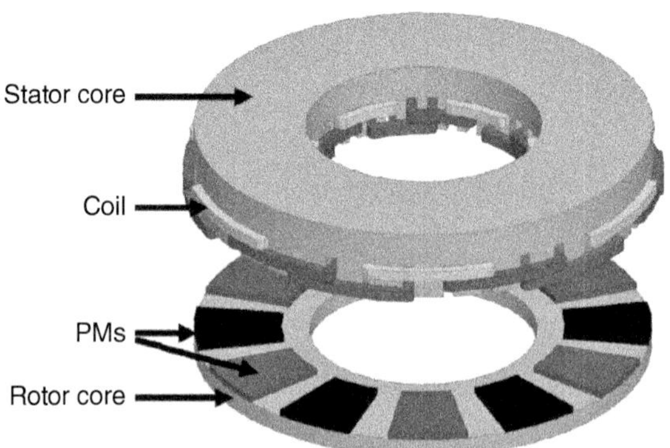


**FIGURE 14-54** Example of an axial flux PM motor/generator used for in-wheel electric traction applications, such as racing solar cars. During regenerative braking, the machine operates like a generator and the energy generated is stored in a battery.

## 14.4 BIBLIOGRAPHY

Alerich, W. N. and Keljik, J., *Electricity 4: DC Motors & Generators*, Albany, NY: Delmar, 2001.

Alerich, W. N., *Electricity 3: DC Motors & Generators, Controls, Transformers*, Albany, NY: Delmar, 1981.

Avallone, E. A., Baumeister III, T., and Sedegh, A., *Marks' Electronic Standard Handbook for Mechanical Engineers*, New York: McGraw-Hill, 2006.

Bapat, Circulating Current Losses in Electric Machines and Measures to Reduce Them (title translated); Techn. Mitt. AEG-TELEFUNKEN 63, 1973.

Bartheld, R., Organizational Structure of IEC TC2, *Conference Proceedings, IEEE Winter Power Meeting*, New York, 1999.

Berrong, D. B., McCown, W. R., Winnie, P. D., and Montgomery, L. W., Designing Central Station Turbine Generators for the Year 2000 and Beyond, *CIGRE Session*, 1998.

Blalock, G. C., *Direct-Current Machinery*, New York: McGraw-Hill, 1947.

Bloch, H. P., *A Practical Guide to Steam Turbine Technology*, New York: McGraw-Hill, 1996.

Block, Heinz P. and Murari Singh, 2008. www.amazon.com.

Boldea, I., *Synchronous Generators*. CRC Press, 2015.

Boldea, I., *Variable Speed Generators*, CRC Press, 2nd edition, 2015.

Boyse, M. P., *Gas Turbine Engineering Handbook*, 4th ed., Burlington, MA: Elsevier, Inc, 2011.

Carlin, P. W., Laxson, A. S., and Muljadi, E. B., "The History and State of the Art of Variable-Speed Wind Turbine Technology," NREL Technical Report, Feb., 2001.

Chalmers, B. J. and Spooner, E., "An Axial-Flux Permanent-Magnet Generator for a Gearless Wind Energy System," in *IEEE Transactions on Energy Conversion*, vol. 14, no. 2, pp. 251–257, Jun. 1999.

Chapman, S. J., *Electric Machinery Fundamentals*, 4th ed., New York: McGraw-Hill, 2005.

Chapman, S. J., *Electric Machinery Fundamentals*, New York: McGraw-Hill, 1998.

Clayton, A. E., *The Performance and Design of Direct Current Machines, A Textbook for Students at Universities and Technical Schools*, London: Pitman, 1947.

Clayton, A. E., *The Performance and Design of Direct Current Machines*, London: Pitman, 1959.

Detinko, F. M., Cooper, G. D., and Montgomery, L. W., Mechanical Design of a New Hydrogen Inner-Cooled Modular Generator Line, *International Joint Power Generation Conference and Exposition*, October 1992.

Drinkut, S. A. and Hurley, J. D., AC Generators and Generator Protection, Section 4.1, *Standard Handbook of Power Plant Engineering*, 2nd ed., New York: McGraw-Hill, 1997.

Elliott, T. C., Chen, K., and Swanskamp, R., *Standard Handbook of Powerplant Engineering*, New York: McGraw-Hill, 1998.

Emery, F. T. and Weddleton, R. F., Latest Advances Associated with Insulation Systems of High Voltage Stator Coils, *IEEE Symposium on Electrical Insulation*, June 1996.

Emery, F. T., The Application of Conductive and Stress Grading Tapes to Vacuum Pressure Impregnated, High Voltage Stator Coils, *IEEE Electrical Insulation Magazine*, July/August, 1996.

Fitzgerald, A. E., Kingsley, C., Jr., and Umans, S. D., *Electric Machinery*, New York: McGraw-Hill, 1990.

Gott, B. E. B., Advances in Turbogenerator Technology, *IEEE Electrical Insulation Magazine*, July/August, 1996.

Gott, B. E. B., McCown, W. R., Montgomery, L. W., and Michalec, J. R., Implications of Differences between the ANSI C50 Series and the IEC 60034 Series Standards for Large Cylindrical Rotor Synchronous Machines, *Panel Discussion IEEE-PES Summer Meeting, Berlin, Germany*, July 1997.

Gott, B. E. B., McCown, W. R., Montgomery, L. W., and Michalec, J. R., Implications of Differences between the ANSI C50 Series and the IEC 34 Series Standards for Large Cylindrical Rotor Synchronous Machines, *IEEE PES Panel Session on Harmonizing, Berlin, Germany*, 1997.

Gott, B. E. B., McCown, W. R., Montgomery, L. W., and Michalec, J. R., Progress in Revision of IEEE/ANSI C50 Series of Standards for Large Steam and Combustion Turbine Generators and Harmonization with the IEC 34 Series, *International Conference on Electric Machines and Drives*, Boston, MA, June 2001.

Gott, B. E. B., McCown, W. R., Montgomery, L. W., and Michalec, J. R., Update of Revision of ANSI C50 Series of Standards for Large Synchronous Machinery and Harmonization with IEC 34 Series, *Conference Proceedings, IEEE Winter Power Meeting*, New York, 1999.

Granicher, W., Air Cooled Turbogenerators up to 240 MVA, *Proceedings of the American Power Conference*, vol. 52, 1990.

Haase, H., Largadier, H., and Suter, J., Air Cooled Turbogenerators in the 200 MVA Class, *Brown Boveri Review*, vol. 73, March 1986.

Haldemann, J., Transpositions in Stator Bars of Large Turbogenerators, *IEEE Transactions on Energy Conversion*, vol. 19, no. 3, Sept. 2004.

Heller, S., *Direct Current Motors and Generators: Repairing, Rewinding, and Redesigning*, New Canaan, CT, Datarule, 1982.

Hubert, C. I., *Electric Machines: Theory, Operating Applications, and Controls*, 2nd ed., Upper Saddle River, NJ: Prentice Hall, 2002.

Intichar, L. and Kulig, T. S., Development of Turbogenerators in Recent Years and in the Future, pp. 136–141, vol. 1, *Proceedings of International Conference on the Evolution and Modern Aspects of Synchronous Machines*, August 1991.

Jovanovic, M. G., Betz, R. E., and Jian, Yu, "The use of doubly fed reluctance machines for large pumps and wind turbines," in *IEEE Transactions on Industry Applications*, vol. 38, no. 6, pp. 1508–1516, Nov/Dec 2002.

Kaminski, C. A., Panel Discussion of Issues Related to Harmonization of Standards for Electrical Machines, *Panel Discussion, IEEE-PES Winter Meeting*, New York, 1999.

Kiameh, P., *Power Generation Handbook*, New York: McGraw-Hill, 2003.

Kloeffer, R. G., Brenneman, J. L., and Kerchner, R. M., *Direct Current Machinery*, New York: Macmillan, 1950.

Kurihara, K., "Effects of Damper Bars on Steady-State and Transient Performance of Interior Permanent-Magnet Synchronous Generators," in *IEEE Transactions on Industry Applications*, vol. 49, no. 1, pp. 42–49, Jan.-Feb. 2013.

Kurtz, M. and Stone, G. C., In-service Partial Discharge Testing of Generator Insulation, *IEEE Transactions on Electrical Insulation*, vol. EI-14, no. 2, Apr. 1979.

Liu, X., Lin, H., Zhu, Z. Q., Yang, C., Fang, S., and Guo, J., "A Novel Dual-Stator Hybrid Excited Synchronous Wind Generator," in *IEEE Transactions on Industry Applications*, vol. 45, no. 3, pp. 947–953, May-Jun., 2009.

Liu, Y., Noe, M., and Doppelbauer, M., "Feasibility Study of a Superconducting DC Direct-Drive Wind Generator," in *IEEE Transactions on Applied Superconductivity*, vol. 26, no. 4, pp. 1–6, June 2016.

Langsdorf, A. S., *Principles of Direct-Current Machines*, New York, London: McGraw-Hill, 1940.

Lister, E. C., and Rusch, R. J., *Electric Circuits and Machines*, New York: McGraw-Hill, 1993.

Liwschitz-Garik, M., *Direct-Current Machines*, Princeton, NJ: Van Nostrand, 1956.

McCown, W. R., Winnie, P. D., and Montgomery, L. W., Trends in Electric Generator Development, *American Power Conference*, vol. 59, 1997.

Muller, S., Deicke, M., and De Doncker, R. W., "Doubly Fed Induction Generator Systems for Wind Turbines," in *IEEE Industry Applications Magazine*, vol. 8, no. 3, pp. 26–33, May/Jun 2002.

Nelson, R. J. and Montgomery, L. W., Electrical Design of a Modular Line of Two Pole Hydrogen Inner-Cooled Generators, *American Power Conference*, vol. 54, 1992.

Nelson, R. J., Drinkut, S. A., and Gregory, M. D., ANSI vs. IEC Standards for Turbine Generators, *Proceedings of the American Power Conference*, vol. 54, 1992.

Nilsson, N. E., Report on the Working Group to Revise ANSI C50.41, *Conference Proceedings, IEEE, Winter Power Meeting*, New York, 1999.

Nippes, P. I., and Nilsson, N. E., International Harmonization of Standards Detailed Report, *IEEE Transactions on Energy Conversion*, vol. 14, no. 4, pp. 1318–1322, December 1999.

Nippes, P. I., and Nilsson, N. E., International Harmonization of Standards, *IEEE IEMDC'97*, 1997.

Nippes, P. I., IEC-US Standards Comparison by the IEEE-PES-EMC Task Force on Standards Harmonization, February 27, 1997.

Polinder, H., Ferreira, J. A., Jensen, B. B., Abrahamsen, A. B., Atallah, K., and McMahon, R. A., "Trends in Wind Turbine Generator Systems," in *IEEE Journal of Emerging and Selected Topics in Power Electronics*, vol. 1, no. 3, pp. 174–185, Sept. 2013.

Potgieter, J. H. J. and Kamper, M. J., "Double PM-Rotor, Toothed, Toroidal-Winding Wind Generator: A Comparison With Conventional Winding Direct-Drive PM Wind Generators Over a Wide Power Range," in *IEEE Transactions on Industry Applications*, vol. 52, no. 4, pp. 2881–2891, Jul.-Aug. 2016.

Prieto-Araujo, E., Junyent-Ferré, A., Lavèrnia-Ferrer, D., and Gomis-Bellmunt, O., "Decentralized Control of a Nine-Phase Permanent Magnet Generator for Offshore Wind Turbines," in *IEEE Transactions on Energy Conversion*, vol. 30, no. 3, pp. 1103–1112, Sept. 2015.

Protsenko, K. and Xu, D., "Modeling and Control of Brushless Doubly-Fed Induction Generators in Wind Energy Applications," in *IEEE Transactions on Power Electronics*, vol. 23, no. 3, pp. 1191–1197, May 2008.

Rebhi, R., Ibala, A. and Masmoudi, A., "MEC-Based Sizing of a Hybrid-Excited Claw Pole Alternator," in *IEEE Transactions on Industry Applications*, vol. 51, no. 1, pp. 211–223, Jan.-Feb. 2015.

Rieger, K., *D-C Generators and Motors.* Scranton, PA, International Correspondence Schools, 1968.

Ringland, W. L. and Rosenberg, L. T., A New Stator Coil Transposition for Large Machines, *AIEE Transactions*, 1959.

Ruelle, G., Guillard, J. M., Bennett, R., and Jackson, R., Development of Large Air Cooled Generators for Gas Turbines and Combined Cycles, Paper 11-201, *CIGRE Session*, 1992.

Sedlazeck, K., Adelmann, W., et al., Influence of Customer's Specifications Upon Design Features of the EPR Turbogenerator, *CIGRE Session*, 2002.

Shlyakhin, P., *Steam Turbines: Theory and Design*, Honolulu, HI: University Press of the Pacific, 2005.

Siskind, C. S., *Direct-Current Machinery*, New York: McGraw-Hill, 1952.

Stephan, C., Baer, J., Zimmerman, H., Neidhofer, G., and Egli, R., New Air-cooled Turbogenerator in the 300 MVA Class, *ABB Review*, Jan. 1996.

Soong, W. L., Kahourzade, S., Liaw, C. Z., and Lillington, P., "Interior PM Generator for Portable AC Generator Sets," in *IEEE Transactions on Industry Applications*, vol. 52, no. 2, pp. 1340–1349, Mar.-Apr. 2016.

Williamson, S. and Ferreira, A. C., "Generalised theory of the brushless doubly-fed machine. 2. Model verification and performance," in *IEE Proceedings—Electric Power Applications*, vol. 144, no. 2, pp. 123–129, Mar 1997.

Williamson, S., Ferreira, A. C. and Wallace, A. K., "Generalised theory of the brushless doubly-fed machine. I. Analysis," in *IEE Proceedings—Electric Power Applications*, vol. 144, no. 2, pp. 111–122, Mar. 1997.

Woodruff, E. B., Lammars, H. B., and Lammars, T. F., *Steam Plant Operation*, New York: McGraw-Hill, 2005.

Woods, E. J., Standards Harmonization: What Next, *Conference Proceedings of the IEEE Winter Power Meeting*, New York, 1999.

Young, E. L., *D-C Machines*, Scranton, PA, International Correspondence Schools, 1975. (Based on material provided by Scott Hancock; rev. by E. L. Young.)

ANSI C42.10, *Definitions of Electrical Terms.*

ANSI C50.22, *Recommended Guide for Testing Insulation Resistance of Rotating Machinery.*

IEC 60034-1, *Rotating Electrical Machines, Ratings and Performance.*

IEC 60034-2, *Rotating Electrical Machines—Methods for Determining Losses and Efficiency from Tests (excluding machines for traction vehicles).*

IEC 60034-3, *Rotating Electrical Machines—Specific Requirements for Cylindrical Rotor Synchronous Machines.*

IEC 60034-4, *Rotating Electrical Machines—Methods for Determining Synchronous Machine Quantities from Test.*

IEC 60034-18, *Functional Evaluation of Insulating Systems (various parts).*

IEC 62114, *Electrical Insulation Systems (EIS)—Thermal Classification.*

IEEE C50.12, *Standard for Salient-Pole 50 and 60 Hz Synchronous Generators and Generator/Motors for Hydraulic Turbine Applications Rated 5 MVA and Above.*

IEEE C50.13, *Standard for Cylindrical-Rotor 50 and 60 Hz Synchronous Generators Rated 10 MVA and Above.*

IEEE 115, *Test Procedures for Synchronous Machines.*

IEEE Std. 286-2000, *Recommended Practice for Measurement of Power Factor Tip-Up of Electric Machinery Stator Coil Insulation.*

IEEE 1110, *Guide for Synchronous Generator Modelling Practices and Applications in Power System Stability Analyses.*

15

# ELECTRIC MACHINES: MOTORS AND DRIVES


Dan M. Ionel

*Professor and L. Stanley Pigman Chair in Power, Department of Electrical and Computer Engineering, University of Kentucky, Lexington, Kentucky*

Om P. Malik

*Professor Emeritus, Department of Electrical and Computer Engineering, University of Calgary, Calgary, Alberta, Canada*

Vandana Rallabandi

*Research Engineer, Department of Electrical and Computer Engineering, University of Kentucky, Lexington, Kentucky*

Narges Taran

*Research Engineer, Department of Electrical and Computer Engineering, University of Kentucky, Lexington, Kentucky*

## 15.1 GENERAL

***Types of Electric Motors.*** Electric motors provide motive power to a wide variety of domestic and industrial machinery. Successful motor application depends on selecting a type of motor which satisfies the kinetic starting, running, and stopping requirements of the driven machinery. There are several methods of classifying electric motors. Based on the electric power supply, motors are classified as direct current (DC) and alternating current (AC) motors. Figure 15-1 shows further classification of ac and dc motors based upon the stator and rotor construction.

NEMA classification according to the variability of speed includes constant-speed motors such as ac synchronous motors; induction motors with low, medium, or high slip; dc shunt-wound motors; varying-speed motors such as dc series motors or repulsion motors; and variable-speed motors such as dc shunt-, series-, and compound-wound motors.

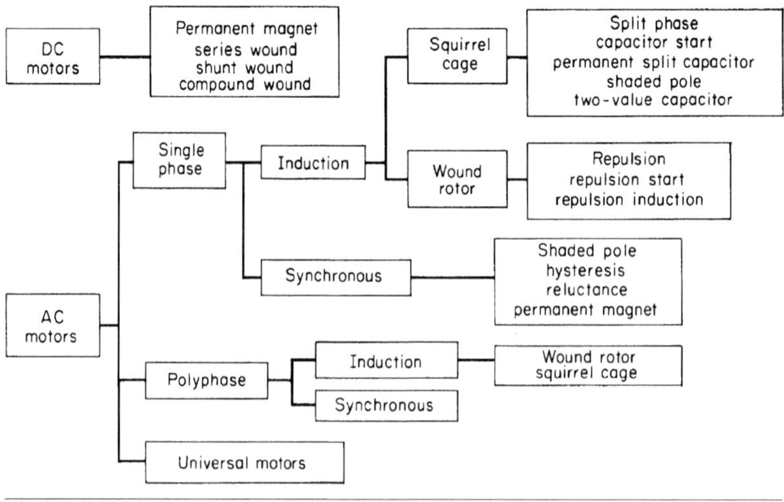


**FIGURE 15-1** Classification of ac and dc motors.

## 15.2 DC MOTORS

DC motors are used in a wide variety of industrial applications because of the ease with which the speed can be controlled. The speed-torque characteristic may be varied to almost any useful form. Continuous operation over a speed range of 8:1 is possible. While ac motors tend to stall, dc motors can deliver over 5 times the rated torque (power supply permitting).

### 15.2.1 Classification

Permanent magnet motors are available in fractional horsepower ratings, while wound-field dc motors are classified as (1) shunt motor, in which the field winding is connected in parallel with the armature; (2) series motor, in which the field winding is connected in series with the armature; and (3) compound motor, which has a series-field and shunt-field winding. The construction of dc motors with a wound field is practically identical to that of dc generators; with minor adjustment, the same dc machine may be operated either as a dc generator or as a motor. Permanent magnet dc motors have fields supplied by permanent magnets that create two or more poles in the armature by passing magnetic flux through it. The magnetic flux causes the current-carrying armature conductors to create a torque. This flux remains basically constant at all motor speeds—the speed-torque and current-torque curves are linear.

***Shunt Motors.*** DC shunt motors are suitable for application where almost constant speed is needed at any control setting or where appreciable speed range (by field control) is needed. The field circuit connection is shown in Fig. 15-2*a*.

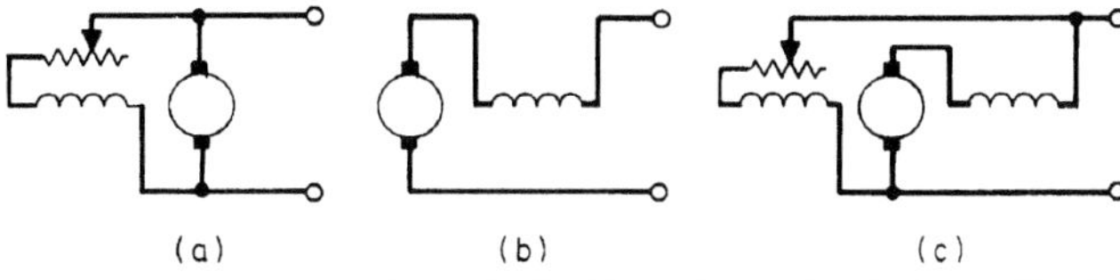


**FIGURE 15-2** Field circuit connections of dc motor.

Since a motor armature revolves in a magnetic field, an emf is generated in the conductors which is opposed to the direction of the current and is called the counter emf. The applied emf must be large enough to overcome the counter emf and also to send the armature current $I_a$ through $R_m$, the resistance of the armature winding, commutator, and the brushes; or

$$E_a = E_b + I_a R_m \quad \text{volts} \tag{15-1}$$

where $E_a$ is applied emf and $E_b$ is counter emf. Since the counter emf at zero speed, that is, at starting, is identically zero and since normally the armature resistance is small, it is obvious in view of Eq. (15-1) that, unless measures are taken to reduce the applied voltage, excessive current will circulate in the motor during starting. Normally, starting devices consisting of variable series resistors are used to limit the starting current of motors.

The torque of a motor is proportional to the number of conductors on the armature, the current per conductor, and the total flux in the machine. The formula for torque is

$$\text{Torque} = 0.1592 Z \phi I_a \frac{\text{poles}}{\text{paths}} \text{Nm} \tag{15-2}$$

where $Z$ is the total number of armature conductors, $\phi$ is the total flux per pole in webers, and $I_a$ is the armature current in amperes taken from the line.

$$E_b = E_a - I_a R_m = Z\phi \frac{\text{r/min}}{60} \frac{\text{poles}}{\text{paths}} \quad \text{volts} \tag{15-3}$$

or

$$\text{Speed} = 60 \frac{E_a - I_a R_m}{Z\phi} \frac{\text{paths}}{\text{poles}} \tag{15-4}$$

For a given motor, the number of armature conductors $Z$, the number of poles, and the number of armature paths are constant. The torque can therefore be expressed as

$$\text{Torque} = \text{constant} \times \phi I_a \tag{15-5}$$

and the speed, likewise, is expressed as

$$\text{Speed} = \text{constant} \times (E_a - I_a R_m)/\phi \tag{15-6}$$

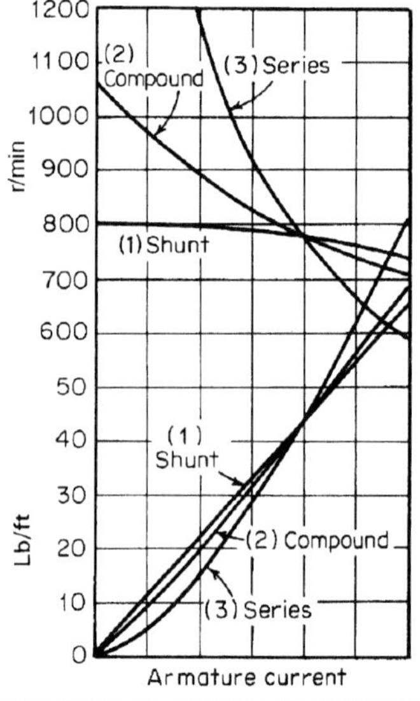


**FIGURE 15-3** Motor characteristics.

In the case of the shunt motor, $E_a$, $R_m$, and $\phi$ are constant, and the speed and torque curves are shown as curves 1 (Fig. 15-3); the effective torque is less than that generated by the torque required for the windage and the bearing and brush friction. The drop in speed from no load to full load seldom exceeds 5%; indeed, since $\phi$, the flux per pole, decreases with increase of load, owing to armature reaction, the speed may remain approximately constant up to full load.

***Speed and Torque of Series Motors.*** Equations (15-5) and (15-6) apply to motors of all continuous-current types. In the case of series motors, the flux $\phi$ increases with the armature current $I_a$; the torque would be proportional to $I_a^2$ were it not that the magnetic circuit becomes saturated with increase of current. Since $\phi$ increases with load, the speed drops as the load increases. The speed and torque characteristics are shown in curves 3 (Fig. 15-3).

If the load on a series motor becomes small, the speed becomes very high, so that a series motor should always be geared or direct-connected to the load. If it were belted and the belt were to break, the motor would run away and would probably burst.

For a given load, and therefore for a given current, the speed of a series motor can be increased by shunting the series winding or by short-circuiting some of the series turns so as to reduce the flux. The speed can be decreased by inserting resistance in series with the armature.

***Compound Motors.*** Compound motor connections are shown in Fig. 15-2*c*. As it has both the shunt and series connected field windings, characteristics of the *compound motor* are a compromise between the shunt and the series motors. By adjusting the direction of flow of current in the series field winding, the flux produced by the series field can assist (cumulative compounding) or oppose (differential compounding) the flux produced by the shunt field winding. In cumulative compounding, the series winding, which assists the shunt winding, the flux per pole increases with the load, so that the torque increases more rapidly and the speed decreases more rapidly than if the series winding were not connected. However, the motor cannot run away under light loads, because of the shunt excitation. The speed and torque characteristics for such a machine are shown in curves 2 (Fig. 15-3).

The speed of a compound motor can be adjusted by armature and field rheostats, just as in the shunt machine.

***Power Supplies.*** Power supplies to dc motors may be batteries, a dc generator, or rectifiers. The permanent-magnet and miniature motors use battery power supplies. Large integral-horsepower dc motors such as rolling-mill motors use dc generators as the power supply. Most fractional-horsepower and integral-horsepower dc motors operate with rectifier power supplies.

***Permanent-Magnet DC Motors.*** Permanent-magnet (PM) motors are available in fractional and low integral-horsepower sizes. They have several advantages over field-wound types. Excitation power supplies and associated wiring are not needed. Reliability is improved, since there are no exciting field coils to fail, and there is no likelihood of overspeed due to loss of field. Efficiency and cooling are improved by elimination of power loss in an exciting field, and the torque-versus-current characteristic is more nearly linear. Finally a PM motor may be used where a totally enclosed motor is required for a continuous-excitation duty cycle.

Temperature effects depend on the kind of magnet material used. Integral-horsepower motors with Alnico-type magnets are affected less by temperature than those with ceramic magnets because flux is constant. Ceramic magnets ordinarily used in fractional-horsepower motors have characteristics that vary about as much with temperature as do the shunt fields of excited machines.

Disadvantages are the absence of field control and special speed-torque characteristics. Overloads may cause partial demagnetization that changes motor speed and torque characteristics until magnetization is fully restored. Generally, an integral-horsepower PM motor is somewhat larger and more expensive than an equivalent shunt-wound motor, but total system cost may be less.

A PM motor is a compromise between compound-wound and series-wound motors. It has better starting torque, but approximately half the no-load speed of a series motor. In applications where compound motors are traditionally used, the PM motor could be considered where slightly higher efficiency and greater overload capacity are needed. In series-motor applications, cost consideration may influence the decision to switch.

***Power Supplies.*** Power supplies to dc motors may be batteries, a dc generator, or rectifiers. The PM and miniature motors use battery power supplies. Large integral-horsepower dc motors such as rolling-mill motors use dc generators as the power supply. Most fractional-horsepower and integral-horsepower dc motors operate with rectifier power supplies.

***Losses and Efficiency.*** Power losses in dc motors are due to bearing friction, brush friction, windage, eddy currents and hysteresis in the armature core and pole faces, brush contact-drop, $I^2R$ losses in the armature, commutator, and field windings, and stray load losses. Typical values of total losses in industrial motors are 4% to 10% of the output. The bearing friction and brush friction losses are proportional to the speed of the motor, while the windage loss is proportional to the square of the speed. Eddy current loss in the armature teeth and in the armature core is proportional to the square of the speed and to the square of the air-gap flux density. Hysteresis loss in the armature teeth and core is proportional to the speed and the square of the flux density in the air gap. Brush contact drop is typically 1 V per brush arm for carbon-graphite brushes and 0.25 V for metal-graphite. Stray load losses are due to eddy currents in armature conductors, brush short-circuit losses in the commutator, and additional core loss arising from distortion of the magnetic field due to armature reaction. The efficiency of the dc motor is defined as

$$\eta = (\text{input electric power} - \text{losses})/\text{input power} \times 100\% \qquad (15\text{-}7)$$

Typically efficiency of dc machines are provided in Fig. 15-4.

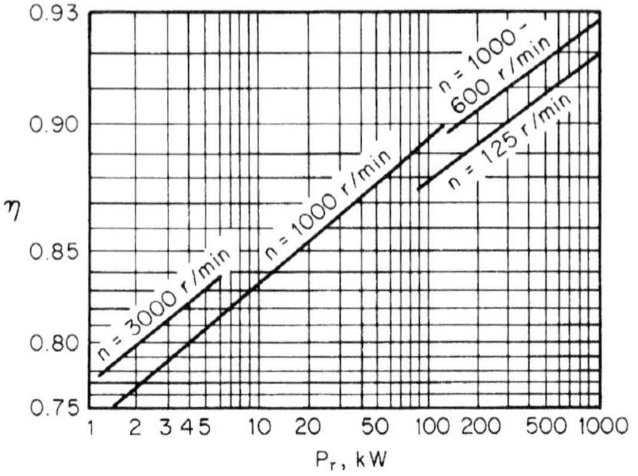


**FIGURE 15-4** Typical efficiency curves of dc machines.

## 15.2.2 DC Motor Starting

*Direct-current motors* of small capacity may be started by connecting the motor directly to line voltage. Motors rated 2 hp or more generally require a reduced-voltage starter. The reduced voltage for starting is obtained by using resistance in series with the motor armature or by varying the armature supply voltage. Manual or magnetic control may be used.

DC motors in adjustable-voltage, adjustable-speed drives are started by turning the speed control up from zero to the desired speed, or by internal circuits that ramp the armature voltage to the desired value. Starting equipment, other than the armature-voltage rectifier or generator, is not required.

*Direct-current manual starters* are satisfactory for applications that do not require frequent starting and stopping and where the starter can be mounted near the operator without requiring long motor leads. Across-the-line starters provide the simplest means of starting small dc motors. Manually operated switches for this service are available in sizes up to 1.5 hp at 115 V and 2 hp at 230 V. For larger motors resistance is connected in series with the motor armature to limit the current inrush on starting. A manually operated means is then provided for removing the resistor from the circuit in a series of steps. Starters are available in the faceplate type, the multiple-switch type, and the drum type. The faceplate type is built for motors up to 35 hp, 115 V, and 50 hp, 230 V. It consists of a movable lever and a series of stationary contact segments to which sections of resistor are connected. The resistor sections are short-circuited one at a time by moving the lever across the segments.

Manual starters have generally been replaced by push-button-operated magnetic control that incorporates overload protection and other safety features.

*Direct-current magnetic starters* are used for applications where case and convenience of operation are important; where the starter is operated frequently; where the motor is located at a distance from the operator; where automatic control by means of a pressure switch, limit switch, or similar device is desired; and for large motors which require the switching of heavy currents. Resistance is connected in series with the motor armature to limit the initial current and is then short-circuited in one or more steps.

For larger motors a series of magnetic contactors is used, each of which cuts out a step of armature resistance. The magnetic contactors are operated as the motor starts by one of two methods called *current-line acceleration* and *time-limit acceleration;* the starting time is always matched to the burden of the load. Time-limit acceleration is advantageous where the starting time of the motor must be integrated into a timing sequence for an overall machine or process. Examples of each will be given.

Figure 15-5 shows a type of *time-limit acceleration* where the operation of contactors, and therefore the rate of acceleration, is governed by a *magnetically operated definite time relay*. This time relay operates on the principle of discharging a capacitor, thus obtaining a definite time period, which is unaffected by changes in temperature and load or by dust and dirt. With the motor at rest, a circuit is obtained through a normally closed contact on *M* to energize the *CT* timing-delay coil and to charge capacitor *C*1. Contacts *CT*1 and *CT*2 on relay *CT* are open with the relay energized. Capacitor *C*2 is charged through the normally closed contact on the 2*A* contactor. Pressing the "start" button energizes the main contactor *M*, which maintains itself through a normally open interlock finger. Relay *FA* is energized, and its contact *FA*1 short-circuits the field rheostat. The motor accelerates from rest to a speed determined by the value of the *R*1-*R*3 resistor. The circuit to timing relay *CT* is opened by the interlock on *M*, and capacitor *C*1 discharges through the *CT* coil and the *AB* resistor. Contacts *CT*1 and *CT*2 can be individually adjusted to close at any time during the capacitor discharge period. Closing *CT*1 energizes the 1*A* contactor, which short-circuits the *R*1-*R*2 resistor. The motor then accelerates to a speed determined by the value of the *R*2-*R*3 resistor step. Closing *CT*2 energizes 2*A* and connects the motor across the line, permitting it to accelerate to normal speed. Relay *FA* is deenergized when 2*A* closes. Contacts *FA*1 and *FA*2 remain closed for a definite time because of the discharge of capacitor *C*2 through the *FA* relay coil. When contact *FA*1 opens, a resistance is inserted in the motor shunt field, equivalent to the field-rheostat resistance and *XY* resistance in parallel. Opening *FA*2 disconnects *XY*, and the motor runs at a speed determined by the setting of the field rheostat.

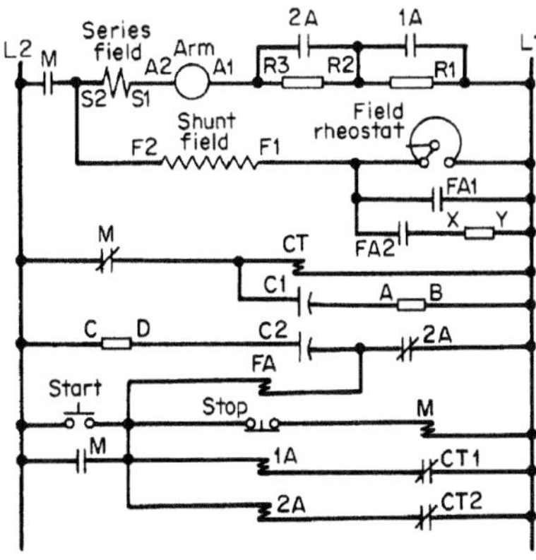


**FIGURE 15-5** Time-limit acceleration with definite-time relay.

*Direct-current starters with current-limit acceleration* are designed to half the starting operation whenever the required starting current exceeds an adjustable predetermined value, the starting operation being resumed when the current falls below this limit. With current-limit acceleration, the time required to accelerate will depend entirely upon the load. When the load is light, the motor will accelerate rapidly, and when it is heavy, the motor will require a longer time to accelerate. For this reason a current-limit starter is not so satisfactory as a time-limit starter for drives having varying loads. Time-limit starters are simpler in construction, accelerate a motor with lower current peaks, use less power during acceleration, and always accelerate the motor in the same time regardless of variations in load. Current-limit starters are desirable for motors driving high-inertia loads. A typical current-limit starter is shown in Fig. 15-6. The relays *SR*1-*SR*2 have normally closed contacts connected in series with the coils of the accelerating contactors. The coils of these relays are connected in the main motor circuit. The relays are provided with an adjustment so they can be set to close on a selected value of current.

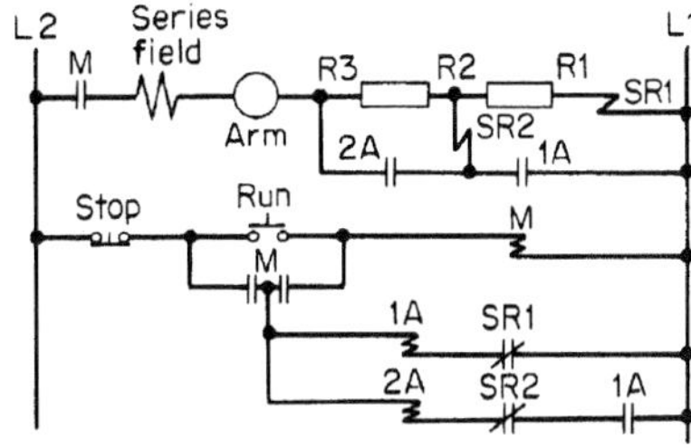


**FIGURE 15-6** Direct-current series-relay accelerations.

Pressing the "run" button energizes the main contactor *M*, which closes and connects the motor to the line in series with the *R*1-*R*3 resistor. Motor current will flow through the *SR*1 coil, and its contacts will open rapidly and prevent 1*A* from closing. When the motor has accelerated enough to bring the line current down to the value for which *SR*1 is set, the relay contacts will close; a circuit is then provided for 1*A*, which closes, cutting out the first step of resistance and short-circuiting the *SR*1 oil. Current now flows through the *SR*2 coil and the 1*A* contacts. *SR*2 relay contacts open and prevent 2*A* from being energized. The motor accelerates again, and when the current falls to the value for which *SR*2 is set, its contacts close, energizing 2*A* and connecting the motor across the line.

*Magnetic controllers for large dc motors* are manufactured in forms to suit the application. The controllers are available in the following forms: (1) nonreversing, without and with dynamic braking; (2) nonreversing with speed regulation by field control, without and with dynamic braking; and (3) reversing with dynamic braking, without and with speed regulation by field control.

### 15.2.3 Basic Speed Control Methods

Speed of a dc motor is controlled either by varying the voltage across the armature, the field winding, or both. Series-parallel combinations are an effective means of reducing armature voltage and motor speed. This method is applied in cam-controlled traction motors. Two identical motors are connected in parallel or in series. When in parallel, full voltage is applied across each motor, causing it to run at base speed. When in series, the motor speeds are essentially one-half of base speed. Field-series resistance in shunt motors weakens the field, which causes the motors to run above the base speed. Speed range as high as 8:1 may be obtained in special motors. Armature-series resistance used with shunt or series motors produces motor speed below the base speed. In the series motor the field winding is also affected by the armature-series resistance, producing greater effect on the speed-torque characteristic than for the shunt motor where the field is constant. Speed control by this method is usually limited to approximately 50% of the base speed. The above-speed control method results in power losses in the external resistors.

## 15.3 INDUCTION MACHINES

### 15.3.1 Theory of the Polyphase Induction Motor

***Principle of Operation.*** An induction motor is simply an electric transformer whose magnetic circuit is separated by an air gap into two relatively movable portions, one carrying the primary and the other the secondary winding. Alternating current flowing in the primary winding, when an alternating voltage is applied from an electric power system, induces a voltage in the secondary and an opposing current flows in the secondary winding when the latter is short-circuited or closed through an external impedance. Relative motion between the primary and secondary structures is produced by the electromagnetic forces corresponding to the power thus transferred across the air gap by induction. The essential feature that distinguishes the induction machine from other types of electric motors is that the secondary currents are created solely by induction, as in a transformer, instead of being supplied by a dc exciter or other external power source, as in synchronous and dc machines.

Induction motors are classified as squirrel-cage motors and wound-rotor motors. The secondary windings on the rotors of squirrel-cage motors are assembled from conductor bars short-circuited by end rings or are cast in place from a conductive alloy. The secondary windings of wound-rotor motors are wound with discrete conductors with the same number of poles as the primary winding on the stator. The rotor windings are terminated on slip rings on the motor shaft. The windings can be short-circuited externally by the brushes bearing on the slip rings, or they can be connected to resistors or solid-state converters for starting and speed control.

***Construction Features.*** The normal structure of an induction motor consists of a cylindrical rotor carrying the secondary winding in slots on its outer periphery and an encircling annular core of laminated steel carrying the primary winding in slots on its inner periphery. The primary winding is commonly arranged for 3-phase power supply, with three sets of exactly similar multipolar coil groups spaced one-third of a pole pitch apart. The superposition of the three stationary, but alternating, magnetic fields produced by the 3-phase windings produces a sinusoidally distributed magnetic field revolving in synchronism with the power-supply frequency, the time of travel of the field crest from 1-phase winding to the next being fixed by the time interval between the reaching of their crest values by the corresponding phase currents. The direction of rotation is fixed by the time sequence of the currents in successive phase belts and so may be reversed by reversing the connections of one phase of a 2- or 3-phase motor.

Figure 15-7 shows the cross section of a typical polyphase induction motor, having in this case a 3-phase 4-pole primary winding with 36 stator and 28 rotor slots. The primary winding is composed of 36 identical coils, each spanning 8 teeth, one less than the 9 teeth in one pole pitch. The winding is, therefore, said to have 8/9 pitch. As there are three primary slots per pole per phase, phase A comprises four equally spaced "phase belts," each consisting of three consecutive coils connected in series. Owing to the short pitch, the top and bottom coil sides of each phase overlap the next phase on either side. The rotor, or secondary, winding consists merely of 28 identical copper or cast-aluminum bars solidly connected to conducting end rings on each end, thus forming a "squirrel-cage" structure. Both rotor and stator cores are usually built on silicon-steel laminations, with partly closed slots, to obtain the greatest possible peripheral area for carrying magnetic flux across the air gap.

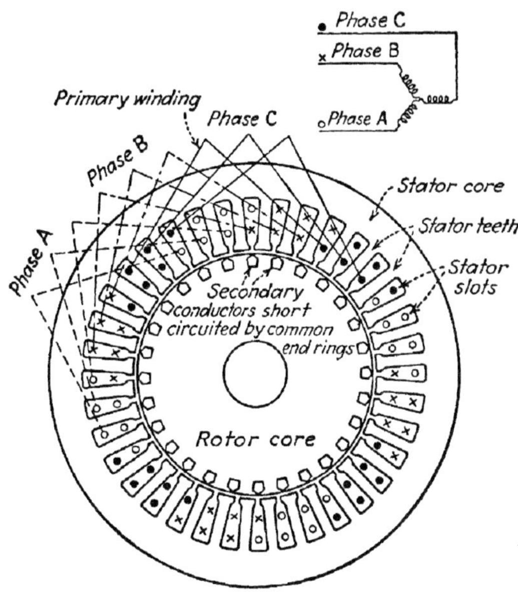


**FIGURE 15-7** Section of squirrel-cage induction motor, 3-phase, 4-pole, 8/9-pitch stator winding.

***The Revolving Field.*** The key to understanding the induction motor is a thorough comprehension of the revolving magnetic field.

The rectangular wave in Fig. 15-8 represents the mmf, or field distribution, produced by a single full-pitch coil, carrying $H$ At. The air gap between stator and rotor is assumed to be uniform, and the effects of slot openings are neglected. To calculate the resultant field produced by the entire winding, it is most convenient to analyze the field of each single coil into its space-harmonic components, as indicated in Fig. 15-8 or expressed by the following equation:

$$H(x)=\frac{4H}{\pi}\left(\sin x+\frac{1}{3}\sin 3x+\frac{1}{5}\sin 5x+\frac{1}{7}\sin 7x+\cdots\right) \tag{15-8}$$

When two such fields produced by coils in adjacent slots are superposed, the two fundamental sine-wave components will be displaced by the slot angle $\theta$, the third-harmonic components by the angle $3\theta$, the fifth harmonics by the angle $5\theta$, etc. Thus, the higher space-harmonic components in the resultant field are relatively much reduced as compared with the fundamental. By this effect of distributing the winding in several slots for each phase belt, and because of the further reductions due to fractional pitch and to phase connections, the space-harmonic fields in a normal motor are reduced to negligible values, leaving only the fundamental sine wave components to be considered in determining the operating characteristics.

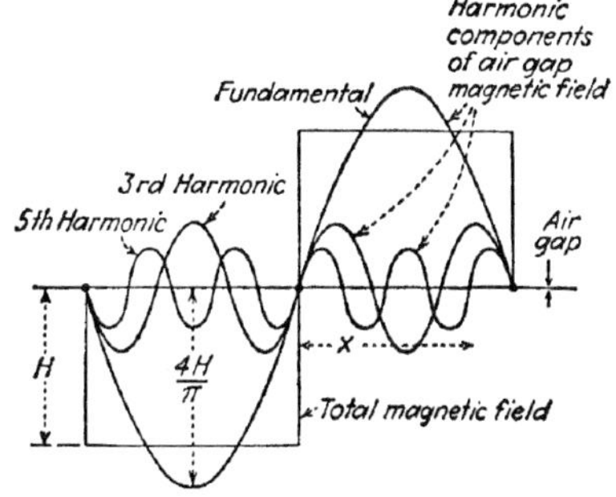


**FIGURE 15-8** Magnetic field produced by a single coil.

The alternating current flowing in the winding of each phase, therefore, produces a sine-wave distribution of magnetic flux around the periphery, stationary in space but varying sinusoidally in time in synchronism with the supply frequencies. Referring to Fig. 15-9, the field of phase A at an angular

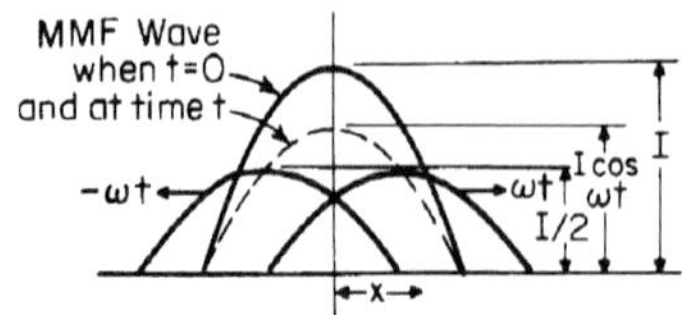


FIGURE 15-9 Resolution of alternating wave into two constant-magnitude waves revolving in opposite directions.

distance $x$ from the phase axis may be represented as an alternating phasor $I \cos x \cos \omega t$ but may equally well be considered as the resultant of two phasors constant in magnitude but revolving in opposite directions at synchronous speed:

$$I \cos x \cos \omega t = \frac{I}{2}[\cos(x - \omega t) + \cos(x + \omega t)] \tag{15-9}$$

Each of the right-hand terms in this equation represents a sine-wave field revolving at the uniform rate of one pole pitch, or 180 electrical degrees, in the time of each half cycle of the supply frequency. The synchronous speed $N_s$ of a motor is therefore given by

$$N_s = \frac{120f}{P} \quad \text{r/min} \tag{15-10}$$

where $f$ is the line frequency in hertz and $P$ is the number of poles of the winding.

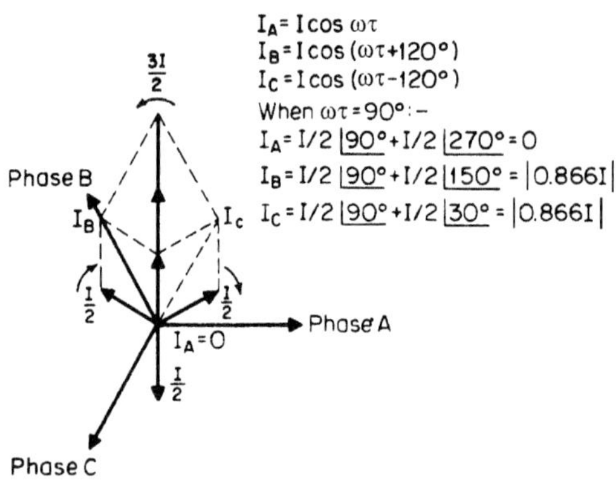


FIGURE 15-10 Resolution of alternating emf of each phase into oppositely revolving constant-magnitude components shown at instant when phase A current is zero ($\omega\tau = 90°$).

Considering phase A alone (Fig. 15-10), two revolving fields will coincide along the phase center line at the instant its current is a maximum. One-third of a cycle later, each will have traveled 120 electrical degrees, one forward and the other backward, the former lining up with the axis of phase B and the latter with the axis of phase C. But at this moment, the current in phase B is a maximum, so that the forward-revolving B field coincides with the forward A field, and these two continue to revolve together. The backward B field is 240° behind the backward A field, and these two remain at this angle, as they continue to revolve. After another third of a cycle, the forward A and B fields will reach the phase C axis, at the same moment that phase C current becomes a maximum. Hence, the forward fields of all three phases are directly additive, and together they create a constant-magnitude sine-wave-shaped synchronously revolving field with a crest value 1.5 times the maximum instantaneous value of the alternating field due to one phase alone. The backward-revolving fields of the three phases are separated by 120°, and their resultant is therefore zero so long as the 3-phase currents are balanced in both magnitude and phase.

If a 2-phase motor is considered, it will have two 90° phase belts per pole instead of three 60° phase belts, and a similar analysis shows that it will have a forward-revolving constant-magnitude field with a crest value equal to the peak value of one phase alone and will have zero backward-revolving fundamental field. A single-phase motor will have equal forward and backward fields and so will have no tendency to start unless one of the fields is suppressed or modified in some way.

While the space-harmonic-field components are usually negligible in standard motors, it is important to the designer to recognize that there will always be residual harmonic-field values which may cause torque irregularities and extra losses if they are not minimized by an adequate number of slots and correct winding distribution. An analysis similar to that given for the fundamental field shows that in all cases the harmonic fields corresponding to the number of primary slots (seventh and nineteenth in a nine-slot-per-pole motor) are important and that the fifth and seventh harmonics on 3-phase, or third and fifth on 2-phase, may also be important.

The third-harmonic fields and all multiples of the third are zero in a 3-phase motor, since the mmf's of the three phases are 120° apart for both backward and forward components of all of them. Finally, therefore, a 3-phase motor has the following distinct fields:

1. The fundamental field with $P$ poles revolving forward at speed $N_s$.
2. A fifth-harmonic field with $5P$ poles revolving backward at speed $N_s/5$.
3. A seventh-harmonic field with $7P$ poles revolving forward at speed $N_s/7$.
4. Similar thirteenth, nineteenth, twenty-fifth, etc., forward-revolving and eleventh, seventeenth, twenty-third, etc., backward-revolving harmonic fields.

Figure 15-11 shows a test speed-torque curve obtained on a 2-phase squirrel-cage induction motor with straight (unspiraled) slots. The torque dips due to three of the forward-revolving fields are clearly indicated.

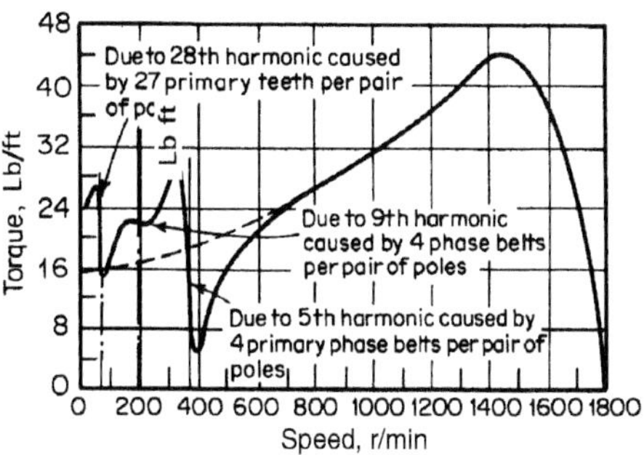


**FIGURE 15-11** Speed-torque curve of 2-phase motor showing harmonic torque.

***Torque, Slip, and Rotor Impedance.*** When the rotor is stationary, the revolving magnetic field cuts the short-circuited secondary conductors at synchronous speed and induces in them line-frequency voltage and currents flow. To supply the secondary $IR$ voltage drop, there must be a component of voltage in time phase with the secondary current, and the secondary current, therefore, must lag in space position behind the revolving air-gap field. A torque is then produced corresponding to the product of the air-gap field by the secondary current times the sine of the angle of their space-phase displacement.

At standstill, the secondary current is equal to the air-gap voltage divided by the secondary impedance at line frequency, or

$$I_2 = \frac{E_2}{Z_2} = \frac{E_2}{R_2 + jX_2} \tag{15-11}$$

where $R_2$ is the effective secondary resistance and $X_2$ is the secondary leakage reactance at primary frequency.

The speed at which the magnetic field cuts the secondary conductors is equal to the difference between the synchronous speed and the actual rotor speed. The ratio of the speed of the field relative to the rotor to synchronous speed is called the slip $s$

$$s = \frac{N_s - N}{N_s}$$

or

$$N = (1 - s)N_s \tag{15-12}$$

where $N$ is the actual and $N_s$ is the synchronous rotor speed.

As the rotor speeds up, with a given air-gap field, the secondary induced voltage and frequency both decrease in proportion to $s$. Thus, the secondary voltage becomes $sE_2$, and the secondary impedance $R_2 + jsX_2$, or

$$I_2 = \frac{sE_2}{R_2 + jsX_2} = \frac{E_2}{(R_2/s) + jX_2} \tag{15-13}$$

The only way that the primary is affected by a change in the rotor speed, therefore, is that the secondary resistance as viewed from the primary varies inversely with the slip.

***Analysis of Induction Motors.*** Induction motors can be analyzed by equivalent circuit. Other methods such as circle diagram can also be used with less accuracy for detailed calculations and design. That is why this subsection focuses on analysis based on equivalent circuit.

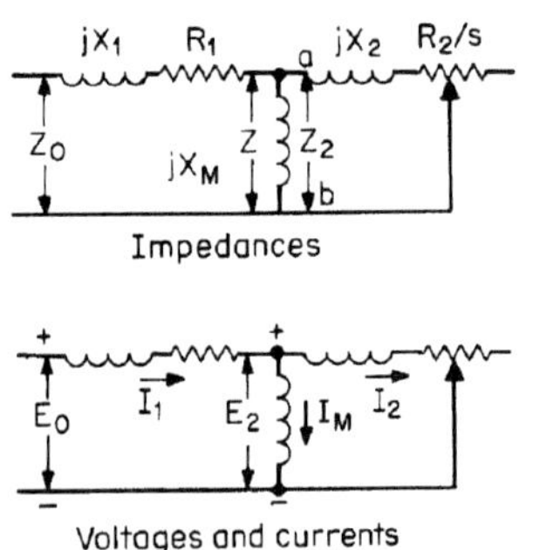


**FIGURE 15-12** Equivalent circuit of polyphase induction motor.

Figure 15-12 shows the polyphase motor circuit usually employed for accurate work. Some advantages of using this circuit for analysis are that it facilitates the derivation of simple formulas, charts, or computer programs for calculating torque, power factor, and other motor characteristics and that it enables impedance changes due to saturation or multiple squirrel cages to be readily taken into account.

Formulas for calculating the constants from test data are given in Table 15-1, and their definitions are given in Table 15-2.

Inspection of the circuit reveals several simple relationships which are useful for estimating purposes. The maximum current occurs at standstill and is somewhat less than $E_0/X$. Maximum torque occurs when $s = R_2/X$, approximately, at which point the current is roughly 70% of the standstill current.

Hence, the maximum torque is approximately equal to $E^2/2X$. This gives the basic rule that the percent maximum torque of a low-slip polyphase motor at a constant impressed voltage is about half the percent starting current.

By choosing the value of $R_2$, the slip at which maximum torque occurs can be fixed at any desired value. The maximum-torque value itself is affected, not by changes in $R_2$, but only by changes in $X$ and to a slight degree by changes in $X_M$.

The magnetizing reactance $X_M$ is usually 8 or more times as great as $X$, while $R_1$ and $R_2$ are usually much smaller than $X$, except in the case of special motors designed for frequent-starting service.

The equivalent circuit of Fig. 15-12 shows that the total power $P_{g1}$ transferred across the air gap from the stator is

$$P_{g1} = mI_2^2 \frac{R_2}{s} \tag{15-14}$$

**TABLE 15-1** Formulas for Calculating Circuit Constants from Test Data for 3-Phase Motors

$X = \frac{f}{f_t}\sqrt{\frac{V^2}{3I^2} - \left(\frac{W}{3I^2}\right)}$ (see text)

$X_1 = X_2 = 0.5X$ for single squirrel-cage or wound-rotor motors

$X_1 = 0.4X$ and $X_2 = 0.6X$ for low-starting-current motors

$W_H + W_F = W_{RL} - 3I_M^2 R_1$ (see text)

$W_s$ (see text)

$X_M = \frac{E_0}{I_M} - X_1$

**TABLE 15-2** Definitions of Equivalent-Circuit Constants

Unless otherwise noted, all quantities except watts, torque, and power output are per phase for 2-phase motors and per phase Y for 3-phase motors:

$E_0$ = impressed voltage (volts) = line voltage ÷ √3 for 3-phase motors
$I_1$ = primary current (amperes)
$I_2$ = secondary current in primary terms (amperes)
$I_M$ = magnetizing current (amperes)
$R_1$ = primary resistance (ohms)
$R_2$ = secondary resistance in primary terms (ohms)
$R_0$ = resistance at primary terminals (ohms)
$X_1$ = primary leakage reactance (ohms)
$X_2$ = secondary leakage reactance referred to primary (ohms)
$X = X_1 + X_2$
$X_0$ = reactance at primary terminals (ohms)
$X_M$ = magnetizing reactance (ohms)
$Z_1$ = primary impedance (ohms)
$Z_2$ = secondary impedance in primary terms (ohms)
$Z_0$ = impedance at primary terminals (ohms)
$Z$ = combined with secondary and magnetizing impedance in parallel (ohms)
$s$ = slip (expressed as a fraction of synchronous speed)
$N$ = synchronous speed (revolutions per minute)
$m$ = number of phases
$f$ = rated frequency (hertz)
$f_t$ = frequency used in locked-rotor test
$T$ = torque (foot-pounds)
$W_0$ = watts input
$W_H$ = hysteresis and eddy current core loss (watts)
$W_F$ = friction and windage (watts)
$W_{RL}$ = running light watts input
$W_s$ = stray-load loss (watts)

The total rotor copper loss is evidently

$$\text{Rotor copper loss} = mI_2^2R_2 \tag{15-15}$$

The internal mechanical power $P$ developed by the motor is therefore

$$\begin{aligned} P = P_{g1} - \text{rotor copper loss} &= mI_2^2\frac{R_2}{s} - mI_2^2R_2 \\ &= mI_2^2R_2\frac{1-s}{s} \\ &= (1-s)P_{g1} \end{aligned} \tag{15-16}$$

We see, then, that of the total power delivered to the rotor, the fraction $1 - s$ is converted to mechanical power and the fraction $s$ is dissipated as rotor-circuit copper loss. The internal mechanical power per stator phase is equal to the power absorbed by the resistance $R_2\,(1 - s)/s$. The internal electromagnetic torque $T$ corresponding to the internal power $P$ can be obtained by recalling that mechanical power equals torque times angular velocity. Thus, when $\omega_s$ is the synchronous angular velocity in mechanical radians per second

$$P = (1-s)\omega_s T \tag{15-17}$$

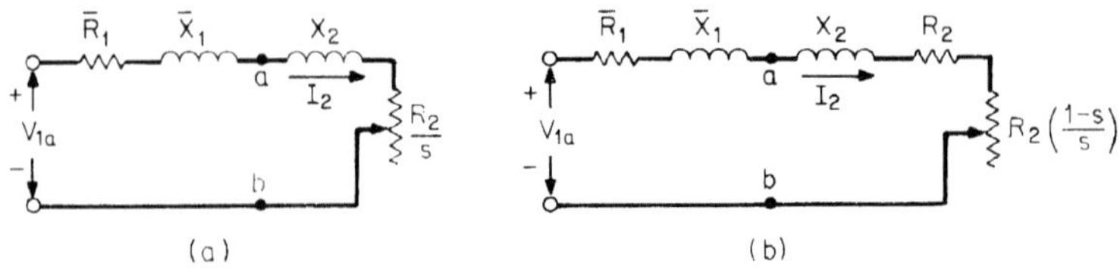


**FIGURE 15-13** Induction-motor equivalent circuit simplified by Thévenin's theorem.

with $T$ in newton-meters. By use of Eq. (15-16)

$$T = \frac{1}{\omega_s} m I_2^2 \frac{R_2}{s} \qquad (15\text{-}18)$$

***Torque and Power.*** Considerable simplification results from application of Thévenin's network theorem to the induction-motor equivalent circuit. Thévenin's theorem permits the replacement of any network of linear circuit elements and constant phasor voltage sources, as viewed from two terminals by a single phasor voltage source $E$ in series with a single impedance $Z$. The voltage $E$ is that appearing across terminals $a$ and $b$ of the original network when these terminals are open-circuited; the impedance $Z$ is that viewed from the same terminals when all voltage sources within the network are short-circuited. For application to the induction-motor equivalent circuit, points $a$ and $b$ are taken as those so designated in Fig. 15-12. The equivalent circuit then assumes the forms given in Fig. 15-13. So far as phenomena to the right of points $a$ and $b$ are concerned, the circuits of Figs. 15-12 and 15-13 are identical when the voltage $V_{1a}$ and the impedance $R_1 + jX_1$ have the proper values. According to Thévenin's theorem, the equivalent source voltage $V_{1a}$ is the voltage that would appear across terminals $a$ and $b$ of Fig. 15-12 with the rotor circuits open and is

$$V_{1a} = E_0 - I_0(R_1 + jX_1) = E_0 \frac{jX_M}{R_1 + jX_{11}} \qquad (15\text{-}19)$$

where $I_M$ is the zero-load exciting current and

$$X_{11} = X_1 + X_M$$

is the self-reactance of the stator per phase and very nearly equals the reactive component of the zero-load motor impedance. For most induction motors, negligible error results from neglecting the stator resistance in Eq. (15-19). The Thévenin equivalent stator impedance $R_1 + jX_1$ is the impedance between terminals $a$ and $b$ of Fig. 15-12, viewed toward the source with the source voltage short-circuited, and therefore is

$$\bar{R}_1 + j\bar{X}_1 = R_1 + jX_1 \qquad \text{in parallel with } jX_M$$

From the Thévenin equivalent circuit (Fig. 15-13) and the torque expression [Eq. (15-18)], it can be seen that

$$T = \frac{1}{\omega_s} \frac{m V_{1a}^2 (R_2/s)}{(\bar{R}_1 + R_2/s)^2 + (\bar{X}_1 + X_2)^2} \qquad (15\text{-}20)$$

The slip at maximum torque, $s_{\max T}$, is obtained by differentiating Eq. (15-20) with respect to $s$ and equating to zero:

$$s_{\max T} = \frac{R_2}{\sqrt{\bar{R}_1^2 + (\bar{X}_1 + X_2)^2}}$$

The corresponding maximum torque is

$$T_{\max} = \frac{1}{\omega_s} \frac{0.5 m V_{1a}^2}{\bar{R}_1 + \sqrt{\bar{R}_1^2 + (\bar{X}_1 + X_2)^2}}$$

## 15.3.2 Construction and Manufacturing of Induction Motors

A typical construction of NEMA frame squirrel motors is illustrated in an exploded view in Fig. 15-14, showing the main components: stator frame and wound core, rotor core with squirrel cage, incorporating end rings, shaft, bearings, end-frames, terminal box, fan and fan cover. The stator and rotor cores are manufactured from thin laminations of silicon steel or cold rolled motor laminated steel. The rotor and stator laminations are stamped one by one with a progressive die and a press. The high-productivity process punches the rotor laminations inside the stator laminations and separates them through a single cut. After assembling the rotor core and die casting an aluminum cage through the rotor slots, machining is performed on the outer rotor surface in order to ensure the small air-gap required for induction motors. Stator coils are form of random wound from conductive, copper or aluminum, wire with electric insulation and inserted in the stator core. Electric insulation is also provided in between the coils and the core, to the top of the slot, and in between the phases. End coils are compacted electrical connections in between the coils and the terminals and the winding is finally impregnated according to the insulation scheme. Construction and manufacturing are similar to those employed for other types of ac electric machines, such as those incorporating permanent magnets, which are inserted or attached to the rotor and magnetized. The construction is assembled within a frame and end-frames are attached in order to close the structure and support the shaft and bearings that enable rotation and the transmission of motoring torque.

**FIGURE 15-14** Exploded view of a NEMA frame squirrel-cage induction motor illustrating the main components: stator frame and wound core, rotor core with squirrel cage, incorporating end rings, shaft, bearings, end-frames, terminal box, fan and fan cover. (*Image courtesy of Regal Beloit Corporation.*)

## 15.3.3 Testing of Polyphase Induction Machines

Proof of guaranteed performance, the determination of torque or efficiency of driven machines, and the evaluation of design changes are some of the purposes that require accurate tests of induction machines. Normally, running-light, locked-rotor, resistance, and dielectric tests only are made on standard motors. Input-output tests or segregated-loss tests are made when accurate efficiency determination is required. The inconvenience of making input-output tests and the inaccuracies inherent in any method which determines the losses as a small difference between two large quantities make the segregated-loss methods of test preferable in many cases. Such tests are especially necessary when actual performance under the varying conditions of service is to be determined from a limited number of factory or laboratory test runs. Experience has shown that the equivalent-circuit method of calculation enables accurate predictions of efficiency and other performance data to be made, provided the circuit "constants" are determined in advance by careful tests.

***Running Light Test.*** The motor is run at no load with normal frequency and voltage applied, until the watts input becomes constant. On slip-ring motors, the brushes are short-circuited. Readings of amperes and watts are taken at one or more values of impressed voltage, with rated frequency maintained. Accurately balanced phase voltages and a sine-wave form of voltage are necessary for good results, requiring operation of the test alternator and transformers well below magnetic saturation. The watts input at rated voltage will be the sum of the friction and windage, core loss, and no-load primary $I_2R$ loss.

The motor impedance per phase is determined from the volts, amperes, and watts readings. The total resistance component for a 3-phase motor is

$$R = \frac{W}{3I^2} \quad \Omega/\text{phase Y} \tag{15-21}$$

and the reactance component is

$$X = \sqrt{\frac{V_2}{3I^2} - R^2} \quad \Omega/\text{phase Y} \tag{15-22}$$

where $W$ is the watts input, $I$ is the line current, and $V$ is the voltage between lines.

Normally, the primary and secondary leakage-reactance values $X_1$ and $X_2$ are assumed equal, each having the value $X/2$.

***Locked-Rotor Test.*** The rotor is locked so it cannot rotate; a reduced voltage of rated frequency is applied to the terminals; and readings of volts, watts, and amperes are taken. Readings should be taken quickly, and the temperature of the windings should be observed before and after the test to minimize errors due to changing resistance values.

***Slip Test.*** Whenever feasible, a current-slip curve should be taken under actual load conditions, with rated voltage and frequency maintained at the motor terminals. Measurements at a few points in the neighborhood of full-load current are usually sufficient; but for slip-ring motors a wider range should be covered, owing to the variable resistance and should, therefore, be measured with a slip meter or stroboscopically. The slip-meter method makes use of a revolution counter differentially geared to the motor under test and to a small synchronous motor driven from the same power supply at the same synchronous speed. Care must be taken to correct the observed values of slip for the difference between the test temperature and the standard value of 75°C or the temperature attained in a full-load heat run with an ambient temperature of 25°C.

In practice, the value of current corresponding to an assumed value of $R_s2/s$ is calculated exactly by the equivalent circuit; the corresponding value of $s$ is read off the slip-current curve; and the true value of $R_2$ is obtained by multiplying $R_2/s$ by this value of $s$. However, $R_2$ may be approximately determined as follows:

Very roughly, the secondary resistance is equal to

$$R_2 = K\frac{E \cdot s}{I_1} \quad \text{approx} \cdot \Omega/\text{phase} \tag{15-23}$$

where $E$ is the terminal voltage per phase, $s$ is the ratio of revolutions per minute of slip to synchronous speed, and $I_1$ is the observed phase current.

The coefficient $K$ varies over a range of about 1 to 1.2, depending on the motor characteristics and the value of the test load.

In case direct slip measurements are not practicable, the value of $R_2$ determined by Eq. (15-21) in a low-frequency locked-rotor test may be used. Or, in the case of a wound rotor, the actual resistance between slip rings may be measured and multiplied by the square of the ratio of primary to secondary volts to obtain the resistance referred to primary. The voltage ratio is obtained by measurement of primary and secondary voltages at standstill with the slip rings open-circuited. Averages of several rotor positions are taken to avoid errors due to possible unbalance.

***Stray-Load Loss Tests.*** Stray-load losses, $W_s$, are defined as the excess of the total measured losses above the sum of the friction and windage, core, and copper losses calculated for the conditions of load from the no-load tests described above. These extra losses are made up chiefly of high-frequency core losses and rotor $I^2R$ losses caused by the pulsations of the leakage-reactance fluxes produced by load currents. While the stray-load losses may be determined by direct input-output tests with a dynamometer or calibrated driving motor, the result is a small difference between two large quantities and so accuracy is very difficult to obtain. Whenever such tests are made, it is desirable to repeat them with the direction of power flow reversed, so the measurement errors may be substantially canceled out.

***Performance Calculations.*** From the foregoing tests, all the circuit constants may be determined, enabling the equivalent-circuit calculations to be carried out. To facilitate this, the formulas for calculating the constants as defined in Table 15-2 are collected in Table 15-1.

***Temperature Tests.*** Temperature tests are made to determine the temperature rise of insulated windings under load conditions. ANSI Standards specify a limiting temperature for continuous-rated machines of 50°C by thermometer or 60°C by either the resistance- or the embedded-detector method for Class A insulating materials and corresponding values of 70°C by thermometer and 80°C by resistance or embedded detector for Class B insulation.

The preferred method of making a full-load temperature test is to maintain nameplate voltage, current, and frequency until the temperature becomes constant, readings being taken every half hour. When constant temperature is reached, the motor is stopped as quickly as possible and additional thermometers are applied to the rotating parts as soon as these have come to rest.

### 15.3.4 Characteristics of Polyphase Induction Motors

***Types.*** All polyphase induction motors may be classified as squirrel-cage or wound-rotor, and may be of the single-speed or multispeed type.

*Squirrel-Cage Motors.* All integral-horsepower induction-motor design categories can mechanically withstand the magnetic stresses and locked-rotor torques of full-voltage line starting. The torque- and current-speed curves for Design A, B, C, and D squirrel-cage motors are shown in Fig. 15-15. Design B motors are most widely used; they have starting-torque and line-starting current characteristics suitable for most power systems. Design C and D motors have higher torque than Class B motors. For motors of all designs, the percentage torques tend to decline with increased hp rating cost.

***Wound-Rotor Motors.*** An insulated winding, usually 3 phase, is provided on the rotor; the terminal of each phase is connected to a slip ring on the shaft. The stationary brushes, which bear on the slip rings, are connected to external adjustable resistances or solid-state converters by which power can be removed from, or injected into, the rotor to adjust the speed. Speed-torque and speed-current curves for a typical wound-rotor motor for various amounts of external resistance are shown in Fig. 15-16. The numbers on the curves refer to the percent external resistance; 100% resistance gives rated torque at standstill.

Wound-rotor motors are normally started with relatively high external resistance and this resistance is short-circuited in steps as the motor comes up to speed. This procedure allows the motor to deliver high-starting and accelerating torques, yet draw relatively light line current.

The curves of Fig. 15-16 indicate that the external resistance reduces the speed at which the motor will operate with a given load torque.

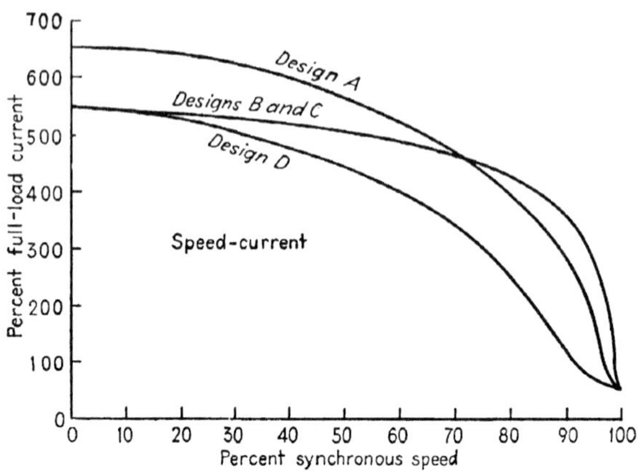


**FIGURE 15-15** Typical speed-current curves for squirrel-cage induction motors.

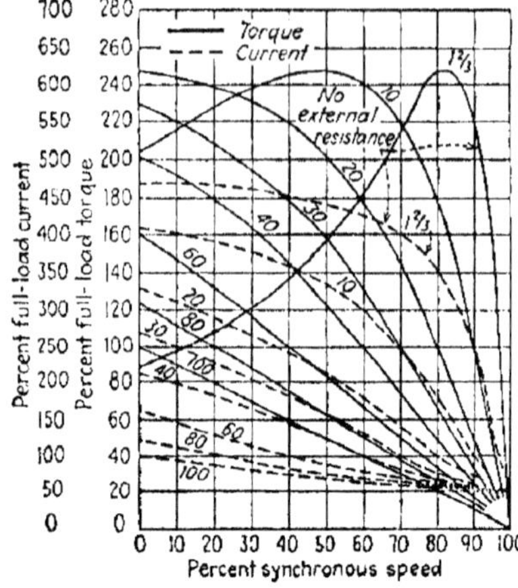


**FIGURE 15-16** Speed-torque and speed-current curves of typical wound-rotor induction motor.

***Efficiency and Power Factor.*** Typical full-load efficiencies and power factors of standard Design B squirrel cage induction motors are given in Figs. 15-17, and 15-18, respectively. The efficiencies of Design A motors are generally slightly lower, and those of Design D motors considerably lower. The power factors of Design A squirrel-cage induction motors are slightly higher, and those of Design C are slightly lower. Energy-efficient motors are those whose design is optimized to reduce losses. Comparative efficiencies of standard and energy-efficient motors of NEMA Design B are shown in Fig. 15-19.

***Full-Load Current.*** With the efficiency and power factor of a 3-phase motor known, its full-load current may be calculated from the formula

$$\text{Full-load current} = \frac{746 \times \text{hp rating}}{1.73 \times \text{efficiency} \times \text{pf} \times \text{voltage}} \tag{15-24}$$

where the efficiency and power factor are expressed as decimals.

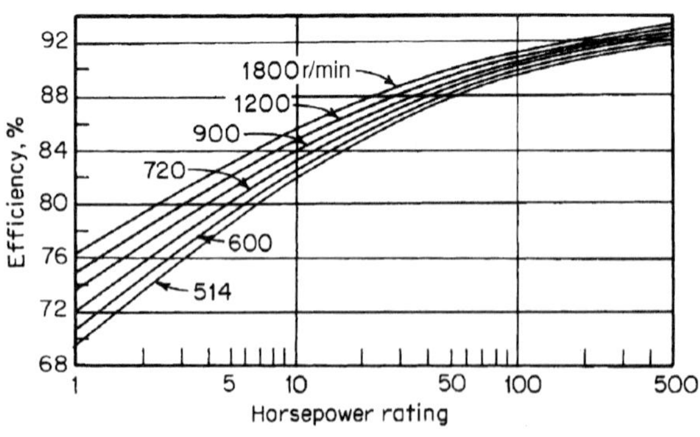


**FIGURE 15-17** Typical full-load efficiencies of Design B squirrel-cage motors.

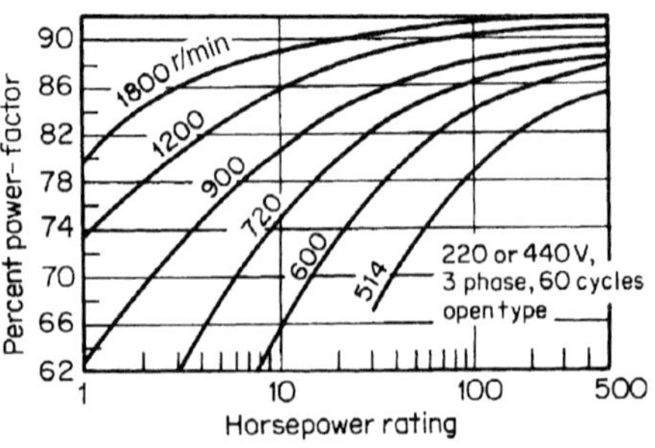


**FIGURE 15-18** Typical full-load power factors of Design B squirrel-cage motors.

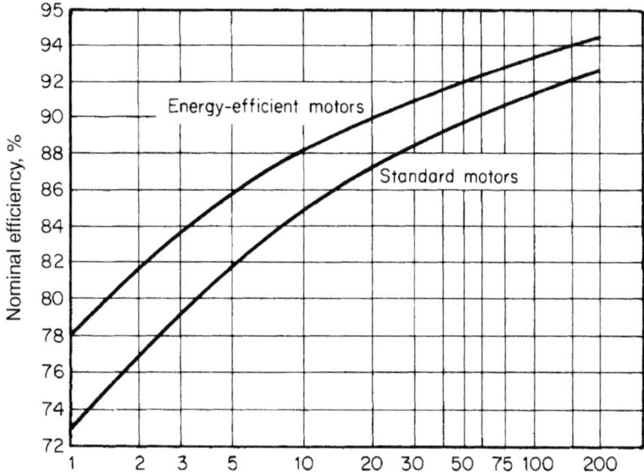


**FIGURE 15-19** Nominal efficiencies for NEMA Design B, 4-pole motors, 1800 r/min; standard vs. energy-efficient motors.

***Starting Methods.*** Wound-rotor motors are invariably started on full voltage but with external resistance in the secondary circuit. Ordinarily sufficient resistance is provided to give 100% torque at standstill, which means that 100% current will be drawn from the line. If a higher torque is required to start the load, less external resistance must be used, and the current drawn is proportionately higher. As the motor accelerates, the external secondary resistance is short-circuited in one or more steps.

### 15.3.5 Single-Phase Induction Motors

***General Theory.*** If one supply line to a polyphase induction motor is opened, the motor will not develop any starting torque, although if it is already operating, it will continue to run at a slightly reduced speed, with a somewhat lower breakdown torque. The crux of the single-phase motor problem, therefore, is in providing auxiliary means for starting.

The magnetic field of a single-phase winding carrying alternating current may be represented as a phasor stationary in space but alternating in time, or as the sum of two equal and oppositely revolving field phasors, which are constant in magnitude. In a polyphase motor, the backward-revolving field phasors of the several phases cancel each other, and the forward-revolving ones add directly, giving a uniform revolving field. In the single-phase motor, means are provided to reduce the backward field, but this field has always some remaining magnitude (except at one particular load in the case of certain capacitor-run motors), and consequently a single-phase induction motor always has extra losses and a double-frequency pulsating torque not possessed by a polyphase motor.

A simple way to visualize the effects of this backward field is to consider that the forward- and backward-revolving fields are separately produced by the same stator current; that is, they are connected in series. Each field may then be treated as a separate polyphase induction motor, the forward field having a slip $s$ with respect to the rotor, and the other a slip $2 - s$. At standstill, both values of slip are unity, and the two circuits are identical. At all times, the net torque developed is equal to the difference of the separate torques produced by the two fields. On this basis, the single-phase induction motor equivalent circuit is given by Fig. 15-20.

The values of $R_1$, $X_1$, $R_2$, $X_2$, and $X_M$ are the impedance constants derived by measurements across the single-phase terminals. Since half the total air-gap impedance at standstill is due to each field, the magnetizing and secondary impedance values are divided by 2 to obtain the values corresponding to the separate fields.

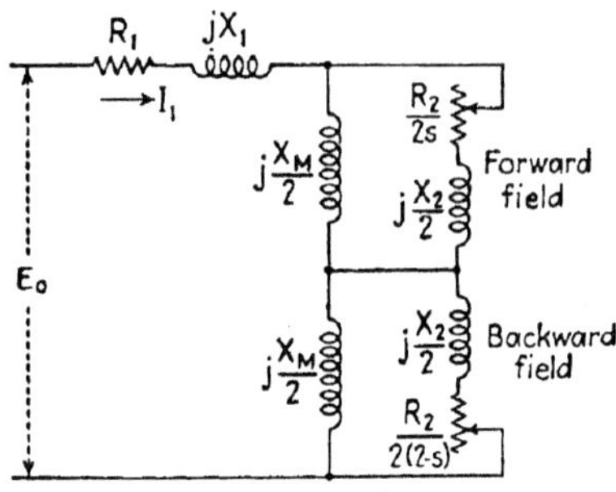


**FIGURE 15-20** Equivalent circuit of single-phase induction motor.

Inspection of this circuit reveals several interesting properties of the motor. At full speed, $s$ is very small, and the backward field appears as an external series impedance of $R_2/4 + j(X_2/2)$. The corresponding loss $I^2R_2/4$ represents the power delivered to the rotor by the backward field. However, there is an equal loss due to the rotor's being driven forward at speed $1 - s$ against the backward-field torque; so the total loss caused by the backward field is $I^2R_2/2$, approximately. Since the backward-field rotor currents occur at double-line frequency, any double squirrel-cage or deep-bar rotor design which had an increased resistance at high frequency would greatly increase the power losses, and such designs, therefore, are seldom used for single-phase motors. The breakdown torque of a single-phase motor may be approximately calculated for a polyphase induction motor, if the impedance of the backward-revolving field is considered as a series impedance added in the primary circuit of the polyphase motor. Hence, any increase in the secondary resistance of a single-phase motor actually reduces the breakdown torque and lowers the speed at which breakdown occurs.

Another interesting characteristic is the double-frequency torque pulsation. The double-frequency current in the rotor reacting on the slip-frequency forward magnetic field evidently produces a torque pulsation, even at no load. Physically, the no-load part of the pulsating torque provides the means for supplying and removing the magnetic field twice each cycle in the axis at right angles to the stator winding, and the additional part under load corresponds to the double-frequency pulsation of the single-phase power input to the rotor. To prevent objectionable transmitted vibration and noise from this cause, it is usual to mount single-phase machines on supports with torsional elasticity of some type, often rubber rings encircling the bearing housings in the case of fractional-horsepower motors.

One of the issues of single-phase induction motors is the starting torque. The simplest way of providing a single-phase induction motor with starting torque is to place a permanently short-circuited winding of relatively high resistance in the stator at an electrical angle of 30 to 60 from the main winding. Usually this auxiliary winding, called a "shading coil," consists of an uninsulated copper strip encircling approximately one-third of a pole pitch. The current induced in the shading coil, by the portion of the main field linking it, reduces the magnitude of this flux and also causes it to lag in time phase. In consequence, the air-gap field has two components, an undamped alternating flux and a damped flux displaced both in space and in time. Shaded-pole motors are used only in very small sizes normally below 50 W output. Principal applications are for desk fans and air circulators, where their simplicity, low torque, and low cost are well suited to the requirements.

A considerably greater starting torque can be obtained by providing a separate starting winding, or auxiliary phase, 90° displaced in space from the main winding of a single-phase induction motor. This extra winding is normally wound with fewer turns of a much smaller size of wire, so that it has a considerably greater resistance to reactance ratio than the main winding, and it is connected directly across the power supply, in parallel with the main winding. Just as in the case of the shaded-pole motor, the field of the auxiliary winding is displaced in time and in space, so that its vectorial combination with the main field gives a much larger forward than backward field component. The motor can be reversed by reversing either the main or the auxiliary winding. This design of single phase induction motors is called split-phase motor.

Another way of obtaining single-phase-motor starting torque is to provide a dc winding and commutator on the rotor, with a single pair of short-circuited brushes for starting and a centrifugal mechanism, which short-circuits the entire commutator as the motor approaches full speed. This gives a pure repulsion-motor starting characteristic with very high torque per ampere and pure single-phase induction-motor operating characteristics. These motors are known as repulsion-start induction-run motor and are widely used in sizes up to about 5 hp. Low-cost low-voltage capacitors have proved extremely useful in improving the performance of split-phase motors. By inserting an external series capacitor in the auxiliary winding circuit and making this winding with many more turns of much lower resistance, the angle of phase split can be increased to 90, or even more, and the coincident increase in the turn ratio a permits a further decrease in the auxiliary winding current. Thus, the capacitor-start motor gives an adequate starting torque for a reasonable starting current and at the same time has so much greater thermal capacity than a resistance split-phase motor, by virtue of the reduced winding-current density, that it is satisfactory for nearly all industrial single-phase motor applications.

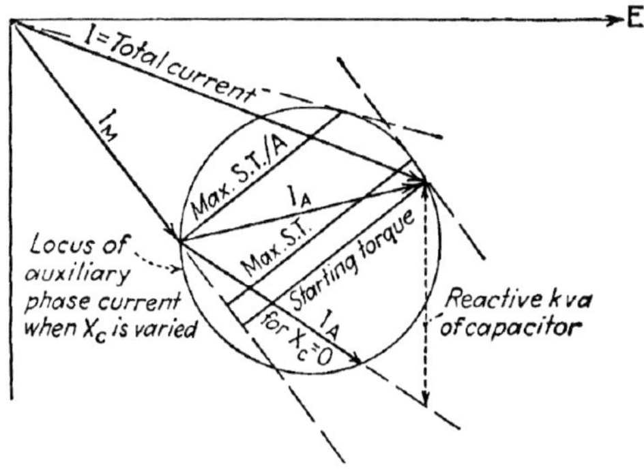


**FIGURE 15-21** Capacitor-motor, starting-torque diagram.

Figure 15-21 illustrates a convenient method of determining the best size of capacitor to use with a given motor. $I_M$ represents the locked-rotor current in the main winding and $I_A$ the current in the auxiliary winding. With no external capacitor, $X_C = 0$, and the motor becomes a plain resistance split type. As $X_C$ is increased, $I_A$ moves ahead in time phase, following a circular locus, increasing the torque and reducing the total current drawn from the line. Points of maximum starting torque and maximum starting torque per ampere are indicated on the diagram.

***Horsepower, Speed, and Voltage Ratings.*** Standard horsepower and speed ratings of single-phase motors are given in Table 15-3. Motors built in frames having a continuous rating of less than 1 hp, open type, at 1700 to 1800 r/min are designated *fractional-horsepower* motors, and those built in larger frames are called *integral-horsepower* motors.

Both capacitor and split-phase motors are available in the multispeed as well as the single-speed type. They are used principally for belt and direct drive of centrifugal and propeller fans and are of the variable-torque class. The multispeed motors for fan drive allow a change in fan speed without changing pulleys, which is essential where remote or automatic control of the rate of air delivery is required.

**TABLE 15-3** Standard Horsepower and Speed Ratings—Single-Phase Constant-Speed Motors

| Standard horsepower ratings | | | | | |
|---|---|---|---|---|---|
| 1/20 | 1/6 | 1/2 | 1½ | 5 | 15 |
| 1/12 | 1/4 | 3/4 | 2 | 7½ | 20 |
| 1/8 | 1/3 | 1 | 3 | 10 | 25 |

| Standard speed ratings | | |
|---|---|---|
| Rpm 60 cycles | Fractional hp | Integral hp |
| 3600 | 1/20—1 | 1½—25 |
| 1800 | 1/20—3/4 | 1—25 |
| 1200 | 1/20—1/2 | 3/4—25 |
| 900 | 1/20—1/3 | 1/2—25 |
| Rpm 50 cycles | | |
| 3000 | 1/20—1 | 1½—20 |
| 1500 | 1/20—3/4 | 1—20 |
| 1000 | 1/20—1/2 | 3/4—20 |
| 750 | — | 1/2—20 |

***Temperature Rise.*** The standard temperature rises and service factors for single-phase motors are the same as for polyphase motors.

***Efficiencies and Power Factors.*** Typical efficiencies and power factors of the various types of induction motor that might be used to fill the requirements of the different ratings are shown in Fig. 15-22. Repulsion-start induction-run motors have about the same efficiencies and power factors except in the 1½- to 3-hp range, where they are lower. Repulsion-induction motors have roughly the same efficiencies but higher power factors.

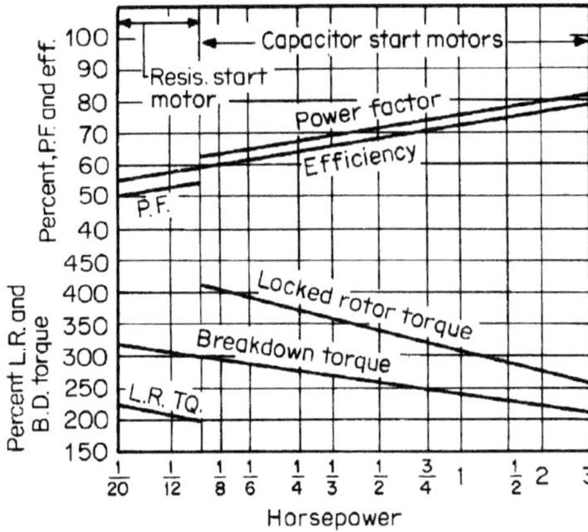


**FIGURE 15-22** Typical operating characteristics of 1800-r/min single-phase motors.

***Single-Phase Motor Characteristics.*** The full-load current of a single-phase motor is equal to

$$\frac{746 \times \text{hp}}{\text{Efficiency} \times \text{voltage} \times \text{pf}} \tag{15-25}$$

where the efficiency and power factor are expressed as decimals.

Characteristics of a 60-Hz, 4-pole, 1800-r/min, single-phase motor are shown in Fig. 15-22.

The horsepower rating of a single-phase motor is defined by its breakdown torque. Thus, any 1800-r/min motor with a breakdown torque between 31.5 and 40.5 oz · ft is, by definition, a ⅓-hp motor. The value used for definition is the minimum of the range of manufacturing variation for that particular design.

### 15.3.6 Induction Motor Online Starting

Selection of an induction motor ac starter is a compromise between requirements and cost. The primary requirements of the starter, obviously, are that the motor starting torque shall be adequate to start the load under worst-case line voltage and load conditions; also, that the line current shall not exceed limits set by the utility or plant voltage dip.

The secondary requirements in starter selection include smoothness of acceleration, maintenance, power factor, reliability, and efficiency. The selection of a closed-transition starter depends upon whether the motor and the supply line can withstand the peak current at the time the starter transfers the motor to full voltage.

Alternating-current across-the-line starters are simple in construction, easy to install and maintain, and inexpensive. A typical starter consists of a 3-pole contactor with a thermal overload relay for protecting the motor. The starter connects the motor directly to the line, impressing full voltage on the motor terminals. It is particularly suitable for squirrel-cage motors. Since these starters connect the motor directly to the supply lines, the motor will draw an inrush current of 6 to 10 times running current. In the majority of installations this is not objectionable and will not damage the motor or the driven machinery. When the starting inrush must be lower, some form of reduced-voltage starting must be used. The common types of starters are autotransformer, primary-resistance, part winding Y-Δ, and solid state.

## 15.4 SYNCHRONOUS MOTORS

### 15.4.1 Basics of Synchronous Motor Operation

***Definition.*** A synchronous motor is a machine that transforms electric power into mechanical power. The average speed of normal operation is exactly proportional to the frequency of the system to which it is connected. Unless otherwise stated, it is generally understood that a synchronous motor has field poles excited with direct current.

***Types.*** The synchronous motor is built with one set of ac polyphase distributed windings, designated the *armature*, which is usually on the stator and is connected to the ac supply system. The configuration of the other member, usually the rotor, determines the type of synchronous motor. Motors with dc excited field windings on salient-pole or round rotors, rated 200 to 100,000 hp and larger, are the dominant industrial type. In the brushless synchronous motor, the excitation (field current) is supplied through shaft-mounted rectifiers from an ac exciter. In the slip-ring synchronous motor, the excitation is supplied from a shaft-mounted exciter or a separate dc power supply. Synchronous-reluctance motors rated below 5 hp, usually supplied from adjustable-speed drive inverters, are designed with a different reluctance across the air gap in the direct and quadrature axis to develop reluctance torque. The motors have no excitation source for synchronous operation. Synchronous motors below 1 hp usually employ a permanent-magnet type of rotor. These motors are usually driven by a transistor inverter from a dc source; they are termed *brushless dc motors.*

***Theory of Operation.*** The operation of the dc separately excited synchronous motor can be explained in terms of the air-gap magnetic-field model, the circuit model, or the phasor diagram model of Fig. 15-23.

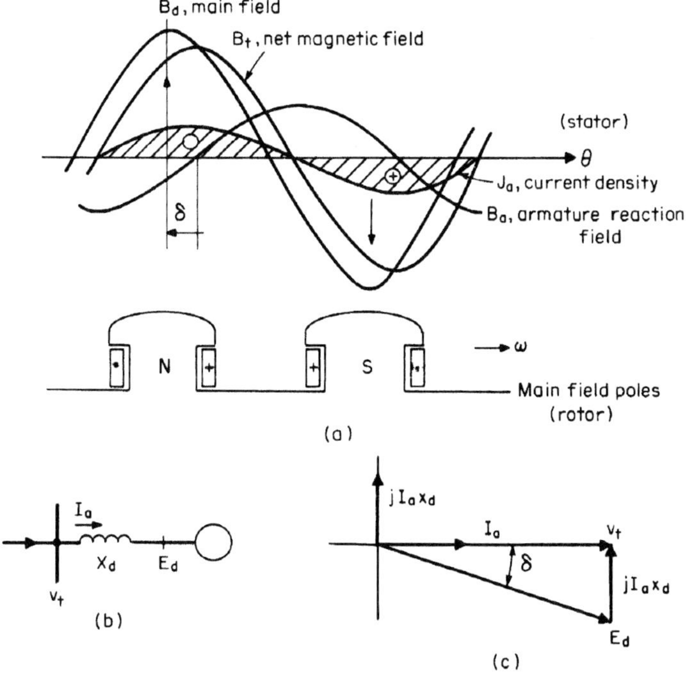


**FIGURE 15-23** Operation of synchronous motor: (*a*) air-gap magnetic-field model; (*b*) circuit model; (*c*) phasor-diagram model.

In the magnetic-field model of Fig. 15-23*a*, the stator windings are assumed to be connected to a polyphase source, so that the winding currents produce a rotating wave of current density $J_a$ and radial armature reaction field $B_a$ as explained below. The rotor carrying the main field poles is rotating in synchronism with these waves. The excited field poles produce a rotating wave of field $B_d$. The net magnetic field $B_t$ is the spatial sum of $B_a$ and $B_d$; it induces an air-gap voltage $V_{ag}$ in the stator windings, nearly equal to the source voltage $V_t$. The current-density distribution $J_a$ is shown for the current $I_a$ in phase with the voltage $V_t$, and pf = 1. The electromagnetic torque acting between the rotor and the stator is produced by the interaction of the main field $B_d$ and the stator current density $J_a$, as a $J \times B$ force on each unit volume of stator conductor. The force on the conductors is to the left ($-\phi$); the reaction force on the rotor is to the right ($+\phi$), and in the direction of rotation.

The operation of the synchronous motor can be represented by the circuit model of Fig. 15-23*b*. The motor is characterized by its synchronous reactance $x_d$ and the excitation voltage $E_d$ behind $x_d$.

The model neglects saliency (poles), saturation, and armature resistance, and is suitable for first-order analysis, but not for calculation of specific operating points, losses, field current, and starting.

The phasor diagram of Fig. 15-23*c* is drawn for the field model and circuit model described previously with unity power factor at the motor terminals. The phasor diagram neglects saliency and armature resistance. The phasors correspond to the waves in the field model. The terminal voltage $V_t$ is generated by the field $B_t$; the excitation voltage $E_d$ is generated by the main field $B_d$; the voltage drop $jI_a x_d$ is generated by the armature reaction field $B_a$; and the current $I_a$ is the aggregate of the current-density wave $J_a$. The power angle $\delta$ is that between $V_t$ and $E_d$, or between $B_t$ and $B_d$. The excitation voltage $E_d$, in per-unit (pu),[a] is equal to the field current $I_{fd}$, in pu, on the air-gap line of the no-load (open-circuit) saturation curve of the machine.

***Power-Factor Correction.*** Synchronous motors were first used because they were capable of raising the power factor of systems having large induction-motor loads. Now they are also used because they can maintain the terminal voltage on a weak system (high source impedance), they have lower cost, and they are more efficient than corresponding induction motors, particularly the low-speed motors. Synchronous motors are built for operation at pf = 1.0, or pf = 0.8 lead, the latter being higher in cost and slightly less efficient at full load.

The selection of a synchronous motor to correct an existing power factor is merely a matter of bookkeeping of active and reactive power. The synchronous motor can be selected to correct the overall power factor to a given value, in which case it must also be large enough to accomplish its motoring functions; or it can be selected for its motoring function and required to provide the maximum correction that it can when operating at pf = 0.8 lead. In Fig. 15-24, a power diagram shows how the active and reactive power components $P_s$ and $Q_s$ of the synchronous motor are added to the components $P_i$ and $Q_i$ of an induction motor to obtain the total $P_t$ and $Q_t$ components, the kVA$_t$, and the power factor. The $Q_s$ of the synchronous motor is based on the rated kVA and pf = 0.8 lead, rather than the actual operating kVA.

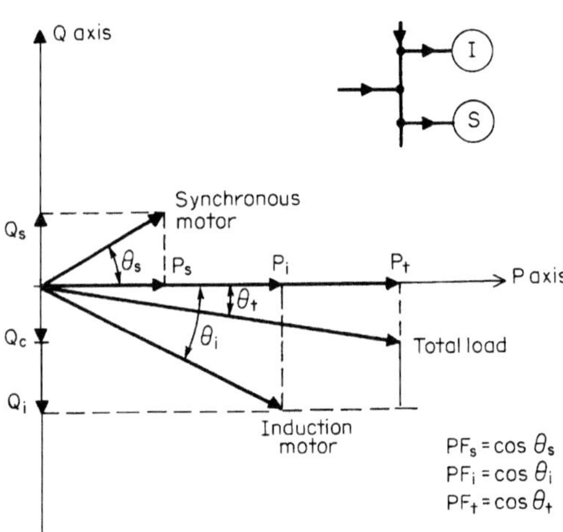


**FIGURE 15-24** Power diagram of induction motor and synchronous motor operating in parallel, showing component and net values of *P* and *Q*.

The synchronous motor can support the voltage of a weak system, so that a larger-rating synchronous motor can be installed than an induction motor for the same source impedance. With an induction motor, both the *P* and *Q* components produce voltage drops in the source impedance. With a synchronous motor operating at leading power factor, the *P* component produces a voltage drop in the source resistance, but the *Q* component produces a voltage rise in the source reactance that can offset the drop and allow the terminal voltage to be normal. If necessary, the field current of the synchronous motor can be controlled by a voltage regulator connected to the motor bus. The leading current of a synchronous motor is able to develop a sufficient voltage rise through the source reactance to overcome the voltage drop and maintain the motor voltage equal to the source voltage.

***Starting.*** The interaction of the main field produced by the rotor and the armature current of the stator will produce a net average torque to drive the synchronous motor only when the rotor is revolving at speed *n* in synchronism with the line frequency *f*; $n = 120\,f/p$, $p$ = poles. The motor must

[a]A reference unit for expressing all parameters on a common reference base. One pu is 100% of the chosen base.

be started by developing other than synchronous torques. Practically, the motor is equipped with an induction-motor-type squirrel-cage winding on the rotor, in the form of a damper winding, in order to start the motor.

The motor is started on the damper windings with the field winding short-circuited, or terminated in a resistor, to attenuate the high "transformer"-induced voltages. When the motor reaches the lowest slip speed, practically synchronous speed, the field current is applied to the field winding, and the rotor poles accelerate and pull into step with the synchronously rotating air-gap magnetic field. The damper windings see zero slip and carry no further current, unless the rotor oscillates with respect to the synchronous speed.

***Torque Definitions.*** *Locked-rotor torque* is the minimum torque, which the synchronous motor will develop at rest for all angular positions of the rotor, with rated voltage at rated frequency applied.

*Pull-in torque* is the maximum constant-load torque under which the motor will pull into synchronism, at rated voltage and frequency, when its rated field current is applied.

In addition, the *pull-up torque* is defined as the minimum torque developed between standstill and the pull-in point. This torque must exceed the load torque by a sufficient margin to assure satisfactory acceleration of the load during starting.

The *reluctance torque* is a component of the total torque when the motor is operating synchronously. It results from the saliency of the poles and is a manifestation of the poles attempting to align themselves with the air-gap magnetic field. It can account for up to 30% of the pull-out torque.

The *synchronous torque* is the total steady-state torque available, with field excitation applied, to drive the motor and the load at synchronous speed. The maximum value as the motor is loaded is developed at a power angle $\delta = 90°$ for the round rotor motor. In a salient-pole motor, the maximum torque is developed at an angle less than 90 due to the presence of reluctance torque.

***Synchronization.*** Synchronization is the process by which the synchronous motor "pulls into step" during the starting process, when the field current is applied to the field winding. Initially, the rotor is revolving at a slip with respect to the synchronous speed of the air-gap magnetic-field waves. The rotor torque, produced by the damper windings, is in equilibrium with the load torque at that slip. The ability of the rotor to accelerate and synchronize depends upon the total inertia ($Wk^2$), the initial slip, and the closing angle of the poles with respect to the field wave at the instant field current is applied.

***Damper Windings.*** Damper windings are placed on the rotors of synchronous motors for two purposes: for starting and for reducing the amplitude of power-angle oscillation. The damper windings consist of copper or brass bars inserted through holes in the pole shoes and connected at the ends to rings to form the equivalent of a squirrel cage. The rings can extend between the poles to form a complete damper. Synchronous motors with solid pole shoes, or solid rotors, perform like motors with damper windings.

***Exciters.*** DC separately excited synchronous motors are provided with a shaft-driven exciter to supply the field power. Exciters are classified into slip-ring types and brushless types. The slip-ring type consists of a dc generator, whose output is fed into the motor field winding through slip rings and stationary brushes. The brushless type consists of an ac generator, with rotating armature and stationary field; the output is rectified by solid-state rectifier elements mounted on the rotating structure and fed directly to the motor field winding. In each type, the motor field current is controlled by the exciter field current. Typical kilowatt ratings for exciters for 60-Hz synchronous motors are given in MG1-21.16 as a function of hp rating, speed, and power factor. For a given hp rating, the excitor kW increases as the speed is reduced, and as the power factor is shifted from pf = 1.0 to pf = 0.8 lead.

During starting, the motor field winding must be disconnected from the exciter and loaded with a resistor, to limit the high induced voltage, to prevent damage to the rectifier elements of the brushless type, and to prevent the circulation of ac current through a slip-ring-type dc exciter. The switching is done with a contactor for the slip-ring type, and with thyristors on the rotating rectifier assembly for the brushless type. Except for the disconnection for starting, the synchronous-motor excitation system is practically the same as for an ac generator of the same rating.

***Standard Ratings.*** Standard ratings for dc separately excited synchronous motors are given in NEMA MG1-1978, Part 21. Standard horsepowers range from 20 to 100,000 hp. Speed ratings extend from 3600 r/min (2-pole) to 80 r/min (90-pole) for 60-Hz machines, and five-sixths of the values for 50-Hz machines. The power factor shall be unity or 0.8 leading. The voltage ratings for 60-Hz motors are 200, 230, 460, 575, 2300, 4000, 4600, 6600, and 13,200 V. It is not practical to build motors of all horsepower ratings at these speeds and voltages.

***Efficiency.*** Efficiency shall be determined at rated output, voltage, frequency, and power factor. The following losses shall be included in determining the efficiency: (1) $I^2R$ loss of armature and field; (2) core loss; (3) stray-load loss; (4) friction and windage loss; and (5) exciter loss for shaft-driven exciter. The resistances should be corrected for temperature. Typical synchronous motor efficiencies are shown in Fig. 15-25.

The 0.8 pf synchronous motor, because of the increased copper loss, is lower in efficiency; its efficiency is closer to that of the induction motor at high speed, but better at low speed.

***Standard Tests.*** Tests on synchronous motors shall be made in accordance with the latest issues of *IEEE Test Procedure for Synchronous Machines*, IEEE Standard 115, and ANSI C50.10-1965. The following tests shall be made on motors completely assembled in the factory and furnished with shaft and complete set of bearings: resistance test of armature and field windings; dielectric test of armature and field windings; mechanical balance; current balance at no load; and direction of rotation. The following tests may be specified on the same or duplicate motors: locked-rotor current; temperature rise; locked-rotor torque; overspeed; harmonic analysis and TIF; segregated losses; short-circuit tests at reduced voltage to determine reactances and time constants; field-winding impedance; and speed-torque curve.

The following tests shall be made on all motors not completely assembled in the factory: resistance and dielectric tests of armature and field windings. The following field tests are recommended after installation: resistance and dielectric tests of armature and field windings not completely assembled in the factory; mechanical balance; bearing insulation; current balance at no load; direction of rotation. The following field tests may be specified on the same or duplicate motors: temperature rise; short-circuit tests at reduced voltage to determine reactances and time constants; field-winding impedance.

The dielectric test for the armature winding shall be conducted for 1 min, with an ac rms voltage of 1000 V plus twice the rated voltage. For machines rated 6 kV and above, the test may be conducted with a dc voltage of 1.7 times the ac rms test value. The dielectric test for the field winding depends upon the connection for starting. For a short-circuited field winding, the ac rms test voltage is 10 times the rated excitation voltage, but no less than 2500 V, nor more than 5000 V. For a field winding closed through a resistor, the ac rms test voltage is twice the rms value of the *IR* drop, but not less than 2500 V, where the current is the value that would circulate with a short-circuited winding. When a test is made on an assembled group of several pieces of new apparatus, each of which has passed a high-potential test, the test voltage shall not exceed 85% of the lowest test voltage for any part of the group. When a test is made after installation of a new machine, which has passed its high-potential test at the factory and whose windings have not since been disturbed, the test voltages should be 75% of the original values.

## 15.4.2 Synchronous Motor Online Starting

***Methods of Starting.*** The method used to start a synchronous motor depends on two factors: the required torque to start the load and the maximum starting current permitted from the line. Basically, the motor is started by using the damper windings to develop asynchronous (induction) torque or by using an auxiliary motor to bring the unloaded motor up to synchronous speed. Solid-state converters have also been used to bring up to speed large several-hundred-MVA synchronous motor/generators for pumped storage plants.

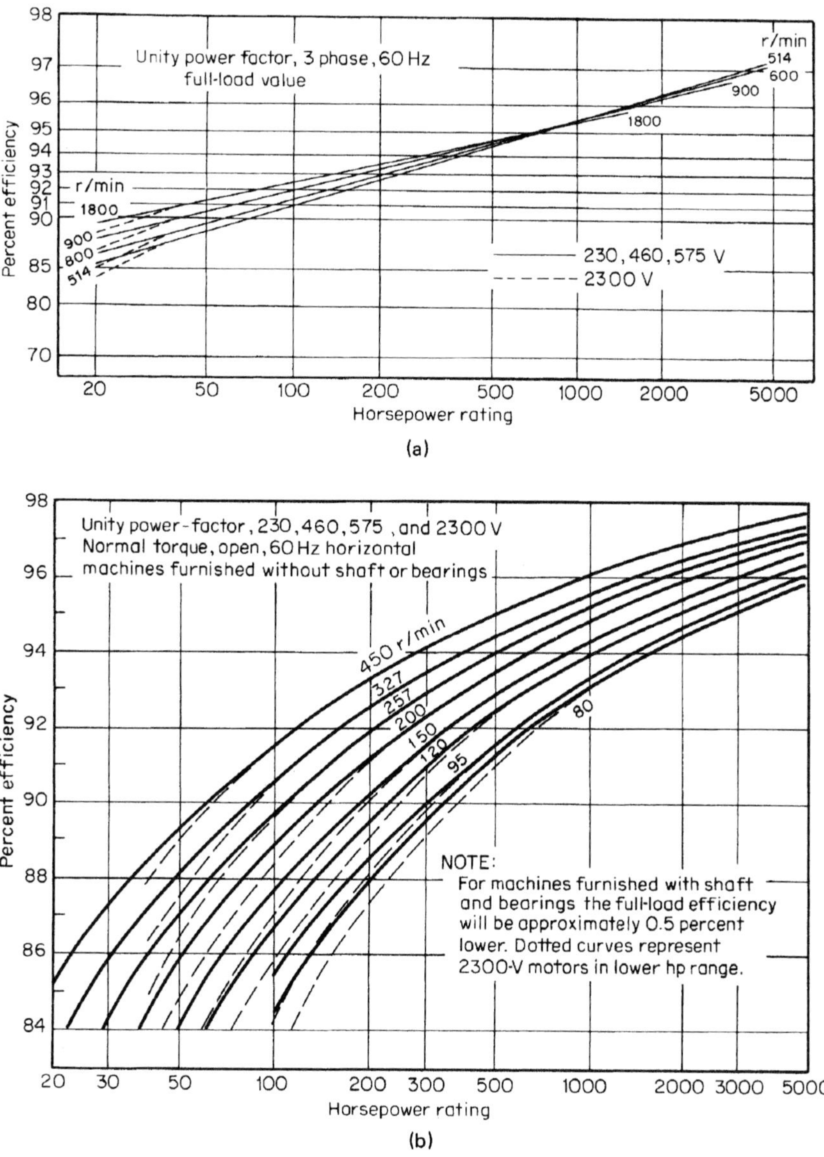


**FIGURE 15-25** Full-load efficiencies of (*a*) high-speed general-purpose synchronous motors and (*b*) low-speed synchronous motors.

Synchronous-motor starters of the full-voltage type connect the motor directly to the supply lines. The field winding is short-circuited through a discharge resistor during the starting period. The field is connected to the dc lines when the motor is at a speed near synchronism. Reduced-voltage starters connect the motor to a reduced voltage for starting and transfer to full voltage at a speed just below synchronism. This transfer may be controlled by a time relay or a frequency relay. The field is energized either immediately before or immediately after the full-voltage switch closes. Most modern synchronous motors obtain their field voltage from a brushless exciter on the shaft.

## 15.5 BASIC METHODS OF SPEED CONTROL

Speed control of electric motors may be obtained by various means. The design of a speed-regulating controller is determined by the type of motor with which it will be used.

Multispeed squirrel-cage motors are suitable for applications that require up to four operating speeds but that do not require speed control between these fixed speeds. However, a solid-state inverter plus a single-speed motor might be less costly than a multispeed motor, and provide much more cost-efficient operation.

Controllers for multispeed ac squirrel-cage motors may be either the drum type or the magnetic type. Drum controllers are widely used, because the many changes in connections required to obtain different speeds can be readily accomplished. Drum controllers can be used with reconnected winding or separate-winding-type motors and with constant-torque, variable-torque, or constant-horsepower motors. Lowvoltage and overload protection can be obtained by using a magnetic contactor and overload relay. When complete control by push buttons or other pilot devices is required, magnetic contactors are used to change the motor connections. Controllers of this type can be arranged to permit starting at any speed or to permit starting only at the slowest speed and changing to each higher speed in sequence. With the developments in power electronics, the mechanical/electromagnetic type motor speed control schemes are being replaced by solid-state based motor drives that provide smoother, more flexible, efficient, and economical speed control of electric motors.

### 15.5.1 Primary Voltage Control

AC squirrel-cage motors are inherently constant-speed motors when supplied directly from utility lines. Narrow-speed-range control is obtained by adjusting the primary voltage on Design D motors using saturable reactors or solid-state phase-controlled thyristors in the stator circuits. Wide-speed-range control is obtained by adjusting the primary frequency and voltage on Design B motors using motor-alternator sets or solid-state frequency converters. The frequency of 60-Hz motors is typically adjusted from 3 to 120 Hz. From 3 to 60 Hz, the voltage is raised proportional to frequency so that the motor can deliver its full rated and breakdown torque. From 60 to 120 Hz, the voltage is kept constant so that the motor can deliver its rated horsepower.

Speed is controlled with thyristors in each of the lines to the stator of the induction motor as shown in Fig. 15-26*a*. Retarding the firing angles of the thyristors reduces the stator voltage of the motor.

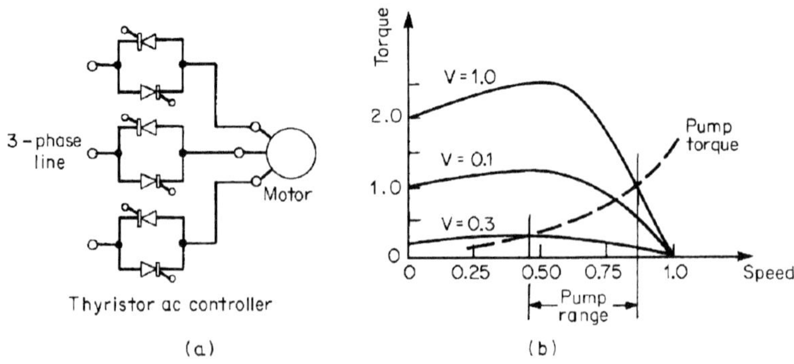


**FIGURE 15-26** Primary-voltage control: (*a*) circuit of controller; (*b*) torque-speed characteristic at three-stator voltages; pump characteristic and range for 10% to 100% power.

The torque at each speed is reduced as $V_2$, as shown in Fig. 15-26*b*. NEMA Design D motors ensure a sufficient range of descending torque in which the motor can stably drive its load. The power loss in the rotor is proportional to the torque × slip. With pump and fan loads, as shown in Fig. 15-26*b*, the torque is reduced as speed, so that the rotor power loss is acceptable at reduced speed. Typical ranges of pump and fan operation are 50% to 100% speed, 10% to 100% power.

Single-phase ac versions of these regulators are still used in the speed control of single-phase ac motors in applications such as ceiling fans. A variable ac voltage is applied across the motor terminals by varying the turn on angle of the thyristors, thereby leading to variable speed operation. The motor currents are not sinusoidal, particularly at lower speeds, leading to torque pulsation and acoustic noise.

### 15.5.2 Speed Control of Slip Ring Induction Motors

*AC slip-ring motor control* requires that power be extracted from the rotor windings via the slip rings to reduce the motor speed, that is, increase the slip. Three methods are used: (1) secondary resistors, (2) rotor-power recovery by auxiliary rotating machines, and (3) rotor-power recovery by auxiliary solid-state rectifier and converter. The auxiliary systems recover the electric energy that would be dissipated in the secondary resistors.

*AC slip-ring motor secondary-resistor speed regulators* consist of a contactor to connect the primary of the motor to the supply lines and some form of resistance-switching device for the secondary circuit. The switching device may be a three-arm faceplate controller, a drum, or magnetic contactors. Regulating devices differ from starting devices in that the switching means can remain continuously on any one of the resistor steps. The motor will therefore operate continuously at a reduced speed, as determined by the amount of resistance remaining in the motor circuit. The use of secondary resistance for speed control is not an efficient method because of the power loss in the resistor. The amount of speed reduction obtained will vary directly with the load on the motor. Speed controllers of this type are usually designed for 50% speed reduction. Under favorable conditions, however, motors can be operated at 75% speed reduction.

The resistors are of the same type as the resistors used for armature regulation of dc motors.

Rotor-power recovery drives are often classified as either constant horsepower or constant torque; the designation refers to the inherent limitation in power based on full current and flux in the main machine. In the first scheme (constant-horsepower drive), the slip energy is converted into mechanical power and then returned to the main motor shaft. Since horsepower is a function of the product of torque and speed, such motors have high torque at low speeds and lower torque at higher speeds. In drives using this arrangement, the auxiliary machine is mounted on or mechanically geared to the main motor shaft (Fig. 15-27*a*).

In the second scheme (constant-torque drive), the slip energy is converted into electric power of the frequency and voltage of the supply circuit and is returned or fed back into the line. Since this power is not delivered to the main motor shaft, the auxiliary machine is not mechanically attached to the shaft but is separately driven. As the limiting torque of the main motor is constant, the maximum horsepower output is proportional to the operating speed (Fig. 15-27*b*).

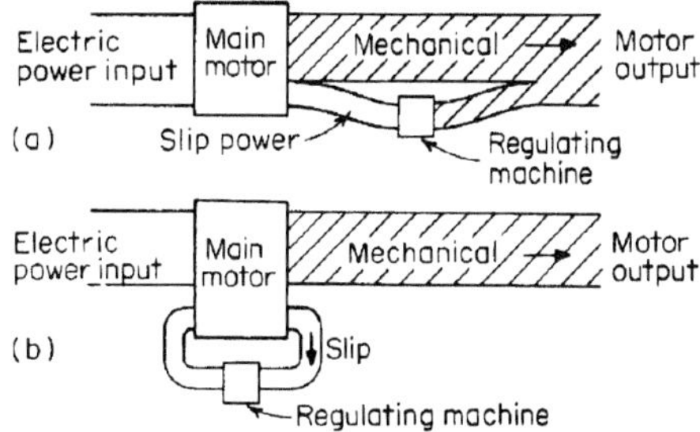


**FIGURE 15-27** (*a*) Constant-horsepower drive; regulating machine, coupled to main motor, returns power mechanically; (*b*) constant-torque drive; regulating machine, mechanically separate from main motor, returns slip power electrically.

The classical recovery systems using auxiliary machines are termed the *Scherbius drive* and the *Krämer drive*. The former employs ac commutator machines; the latter relies on a dc link and rotary converters. A variation of the Krämer drive uses a synchronous motor and a dc generator in place of the rotary converter and a constant-speed set feeding the slip power back into the line. This drive has been used for a number of large wind-tunnel drives. It is particularly adapted to a wide range of speed control and to minimum disturbance on starting.

## 15.6 VARIABLE SPEED DRIVES WITH POWER ELECTRONICS

### 15.6.1 DC Drives

DC machine drives were historically very popular due to the relative straightforward implementation of controls for power electronics. This is because the speed of a dc machine can be controlled through the armature voltage and the torque through the field current.

An ac source and phase-controlled rectifiers, such as half-wave and semiconverters, provide single quadrant, that is, positive voltage and current operation, while fully controlled converters can be used to operate the machine in two quadrants, that is, positive and negative voltage and positive current. For four-quadrant operation, dual converters, that is, back-back thyristor bridges, may be employed. It should be noted that rectifiers draw harmonic currents from the ac source, leading therefore to a poor power factor and input filters are typically required for performance improvement.

Alternatively, dc-dc converters or choppers fed from a constant dc voltage, such as a battery or diode bridge rectifier, can be used to feed a variable dc voltage to the armature. These converters may employ IGBTs or MOSFETs. Configurations allowing only motoring or both motoring and regeneration are available.

Both the phase controlled rectifiers and the dc-dc converters lead to discontinuous armature voltage across the armature, and, for some operating conditions, particularly low speed and low load, this might lead to discontinuous currents deteriorating the dynamic performance of the machine as well as the input power factor. For choppers, the use of a high-switching frequency may mitigate these limitations.

A widely acknowledged drawback of dc drives is represented by the substantial maintenance required by the use of commutators and brushes in the electric motor. Advancements over the last decades in electric controls made possible the increased use of ac motor technologies and reduced the need for dc drives.

### 15.6.2 AC Drives—Three-Phase Inverters

***Two-Level Inverters.*** An ac electric machine, such as an induction motor, requires variable voltage and frequency in order to achieve variable speed operation. Supply can be provided from a dc source via a dc-ac power electronics inverter or from an ac source via an ac-dc-ac combined converter. The ac-dc input stage of the ac-dc-ac converter may be an uncontrolled rectifier, such as a diode bridge rectifier. Uncontrolled rectifiers, which tend to have a low poor power factor, owing to the rather large value of the dc bus capacitance required in order to maintain a stiff dc link voltage, are typically used in conjunction with passive or active power factor correction circuits. The addition of a boost converter at the output of the uncontrolled rectifier and the appropriate control of its duty cycle also improves the power factor. Implementations of the ac-dc conversion stage with controllable power electronic switches, such as IGBTs and MOSFETs, have the advantage of ensuring bidirectional power flow.

The dc-ac converter/inverter may be a voltage source or current source converter. Typically, a voltage source inverter (VSI), such as the one schematically illustrated in Fig. 15-28, requires a stiff dc link and employs IGBTs or MOSFETs switching at high frequencies. The output ac voltage at the VSI terminals includes a fundamental frequency sine wave and high-frequency harmonic components, the frequency of which depends on the switching frequency employed. The high-order harmonics may be filtered out, to a certain extent, by the inductance of the electric machine, as exemplified in Fig. 15-29.

Different modulation techniques are available for the VSI inverters. These include sine-triangle pulse width modulation (PWM), hysteresis current regulation, and space vector PWM. According to the sine-triangle PWM algorithms, the gating pulses for the power switches, for example, IGBTs, are generated through the comparison of a sinusoidal modulating function with a high-frequency triangular carrier wave. The frequency of the sinusoidal modulating wave is equal to the fundamental frequency

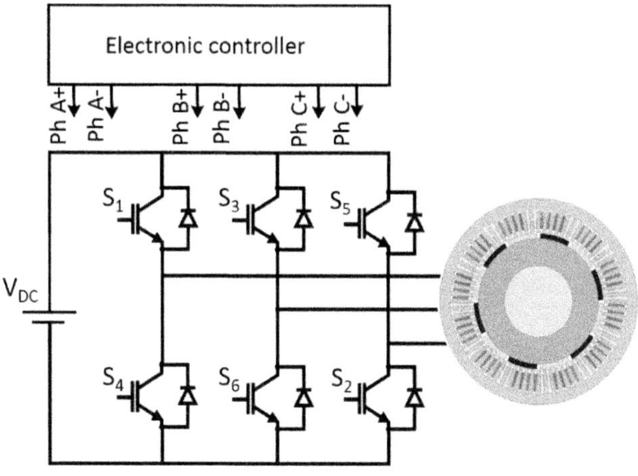


**FIGURE 15-28** A two-level 3-phase voltage source inverter (VSI), which can be employed to drive ac machines, such as of the induction (asynchronous), wound rotor synchronous or PM synchronous type. The inverter output has three wires and the motor windings are typically wye (star) connected with an isolated neutral point.

of the output voltage. The magnitude of the output voltage is controlled by changing the ratio of the amplitude of the modulating signal to the amplitude of the carrier wave, a parameter known as the modulation index. The output ac voltage is directly proportional to the modulation index ($m$), as long as $0 < m < 1$. For $m > 1$, low-frequency harmonics begin to appear in the output voltage of the inverter. The maximum output ac voltage obtained for $m = 1$ is smaller than the dc link voltage. A higher ac voltage may be obtained for the same dc bus voltage if other types of modulating signals are used. Examples of this type include the use of a sine wave with a third harmonic component or the use of space vector modulation that improves the dc bus utilization.

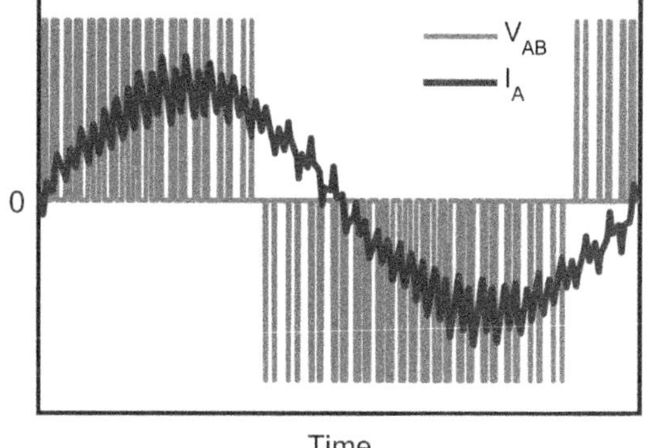


**FIGURE 15-29** Phase current and line voltage waveforms in a two-level VSI. The high-frequency components of the inverter output voltage are filtered by the machine inductance, rendering the currents quasi-sinusoidal.

***Special Inverters.*** In the case of back-to-back controllable ac-dc-ac converters, specific algorithms allow the improvement of the power factor. Such a configuration has also the advantage of allowing power flow from the machine to the input source, which provides regenerative braking. Alternative means of ensuring bidirectional power flow include the use of thyristors for the input rectifier; a solution that may result in a poor power factor. If the electric machine has a leading power factor, another option is to use a current source inverter (CSI). In a CSI, thyristors may be employed in place of IGBTs or MOSFETs, and the configuration maybe advantageous, in principle, for large power ratings. Nevertheless, thyristors cannot be switched at high frequencies and, as a result, the current harmonic content is high leading to torque ripple and additional losses. Thyristor converters or line-commutated inverters may be used for large synchronous machine drives, in which case the machine is overexcited in order to ensure operation at a leading power factor.

Direct ac converters, also known as matrix converters, have also been developed. These converters are built with a large number of switches and do not require a stiff dc link.

For large power drives, the higher rated voltages impose an upper limit on the switching frequency thereby increasing torque ripple and loss. In this case, multilevel inverters, which yield a voltage output with reduced harmonic content, even at relatively low-switching frequencies, may be employed.

## 15.7 SPEED CONTROL OF AC MOTORS

### 15.7.1 Stability and Dynamics

The first-order differential mechanical equation for a motor, under the assumption of rigid coupling, is given by

$$T_{em} - T_L = J\frac{d\omega}{dt} + B\omega$$

where $T_{em}$ is the electromagnetic torque, $T_L$ is the load torque, $J$ is the moment of inertia, $\omega$ is the rotor speed and $B$ is the damping coefficient due to friction. In general, for stable operation, the following condition must be met:

$$\frac{dT_{em}}{d\omega} < \frac{dT_L}{d\omega}$$

For line-fed machines, only some zones of the torque-speed curve lead to stable operation. With the use of power electronic controls, the torque-speed characteristic of the machine can be shaped as desired, subject to the availability of dc bus voltage, enabling thus, operation over a wide range of conditions.

### 15.7.2 Scalar Control

***Open Loop Scalar Control.*** The control of induction motors according to this algorithm, which is schematically depicted in Fig. 15-30, requires maintaining the ratio of terminal voltage to frequency constant over a range of speeds. If the stator resistance and leakage reactance are neglected, this results in a substantially constant air-gap flux, providing constant maximum torque. The stator frequency and voltage can be varied continuously in order to maintain the mechanical speed of the motor constant for any load. This requires the knowledge of the slip, which in traditional schemes may be obtained under the assumption of a linear relationship between torque and slip. This approach may lead to errors at low speeds and slip compensation, as well as voltage boosting, may be employed in order to overcome this limitation.

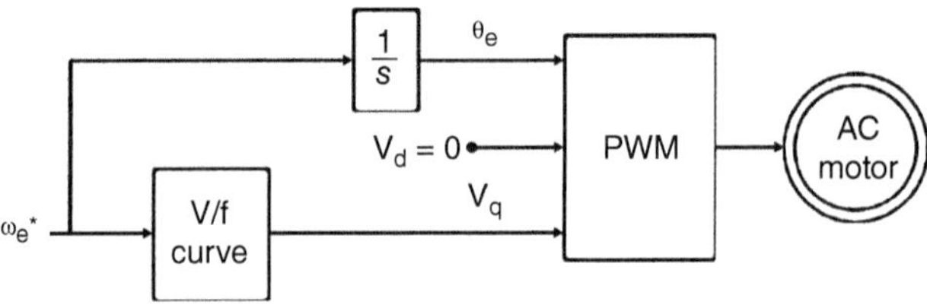


**FIGURE 15-30** Diagram for open loop scalar control of an ac motor, which is typically employed for induction machines.

***Closed Loop Scalar Control.*** In this scheme, the motor speed is measured, fed back, and compared with the set speed. The output of the speed controller is represented by the required slip, which along with rotor speed, is then used to determine the operating supply frequency for the motor.

### 15.7.3 Vector Control

Vector control requires the independent control of torque and flux. While servicing a step change in load torque, a vector controlled machine shifts very rapidly to the new operating point with a better dynamic response, as compared with scalar control. The vector control ensures, in principle, that the stator currents are derived from set values of flux and torque.

In case of an induction motor, this is achieved by regulation of the rotor, stator, or air-gap flux according to a scheme as shown in Fig 15-31. Transformations orienting the reference frame along the rotor, stator or air gap flux are employed in the process. In direct field orientation, the information required for the transformations is obtained using flux sensors or flux observers. In contrast, in case of indirect field orientation, the slip is calculated as a function of motor parameters and an encoder provides information about the rotor position. The slip depends on the desired reference frame of orientation and is a function of the parameters of the motor. The $d$-axis current is derived from an outer flux loop and the value of the set flux depends on the operating speed of the motor. The currents are maintained at their desired references with the aid of the inverter.

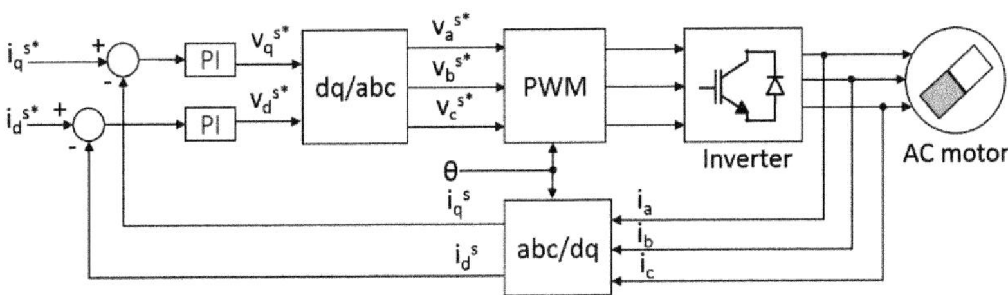


**FIGURE 15-31** Diagram for vector control of ac motors. The determination of the transformation angle ($\theta$) depends upon the type of machine, i.e., induction or synchronous.

In a surface mounted permanent magnet synchronous motor, the stator currents are transformed to a reference frame oriented along the $d$ axis of the rotor. The current along the $d$ axis, or the flux producing current, is maintained at zero, while the $q$-axis component of the current or the torque producing current is derived from an outer speed loop. The rotor position information required for the transformation is provided by a position encoder or maybe estimated in a sensorless algorithm. For salient permanent magnet synchronous motors, the optimum value of $d$-axis current may not be zero, in order to take advantage of the reluctance torque component. In such a case, the desired values of $d$ and $q$ currents may be obtained from the maximum torque per ampere characteristics.

### 15.7.4 Direct Torque Control

The advantage of this method for decoupled control of torque and flux is that it does not require reference frame transformations nor position encoders. The stator flux and torque are maintained at their respective references by the application of appropriate PWM pulses to the inverter. The method can be applied to both synchronous and asynchronous machines and requires flux and torque estimators.

# 15.8 ELECTRONICALLY OPERATED MOTORS OF THE PERMANENT MAGNET AND RELUCTANCE TYPE

## 15.8.1 Permanent Magnet Synchronous Motors

The rotor of permanent magnet synchronous motors (PMSM) includes PMs placed in different configurations, such as surface mounted, interior, and spoke (Fig. 15-32). The rotor topology influences the motor parameters, for example, inductances, and characteristics, such as saliency ratio and reluctance torque. Both high-energy rare earth, NdFeB or SmCo, magnets and less expensive ferrites can be used for PMSMs. Stators may employ distributed windings, similar to those of an induction motor, or nonoverlapping concentrated windings placed around the teeth.

Each motor topology has its own advantages recommending it for specific applications. For example, PMSMs with concentrated windings may have shorter end turns leading to reduced copper weight and losses, but have a magnetic field rich in high-order harmonics, yielding high stator and rotor core losses, such that the motor may be better suited for low-speed high-torque drives. PMSMs are employed in a variety of high-efficiency applications, including electric vehicles, industrial and automotive drive systems, and home appliances.

The operation of a PMSM is similar to a wound rotor synchronous motor with fixed excitation determined by the PM field. A back electromotive force (emf) with a substantially sinusoidal waveform, the magnitude and frequency of which are linearly dependent on the rotational speed, is induced across the stator windings. The stator coils, when excited with quasi-sine-wave regulated three-phase currents, of the type illustrated in Fig. 15-29, produce a sinusoidally distributed rotating mmf in the air gap, which interacts with the PM rotor field and produce electromagnetic torque. In PMSMs the torque is produced due to the interaction between the stator and rotor mmf's, and in designs with interior permanent magnets an additional reluctance torque is generated because of the rotor saliency. These machines, particularly of the spoke PM type, with a large number of rotor poles also present flux concentration, which serves to increase the magnetic loading. This allows the use of weaker and less expensive magnets, such as ferrites.

Electronic controls and power electronic inverters are employed for PMSMs. Rotor position information is provided by encoders or is estimated by sensorless algorithms. Drives with a PMSM are able to operate with suitable electric controls and motor parameters on a variety of characteristics, including the typical electric traction curve that includes a constant torque region up to the base speed and constant power at higher speeds.

A typical construction of PMSMs is of the radial air-gap type having a cylindrical rotor placed inside and coaxially with a stator, as shown in Fig. 15-32, or have the rotor placed outside a cylindrical stator in what is commonly referred to as an "inside-out" arrangement. PMSMs may also be designed with substantially disk shaped stators and rotors separated by one or multiple axial air-gaps in what

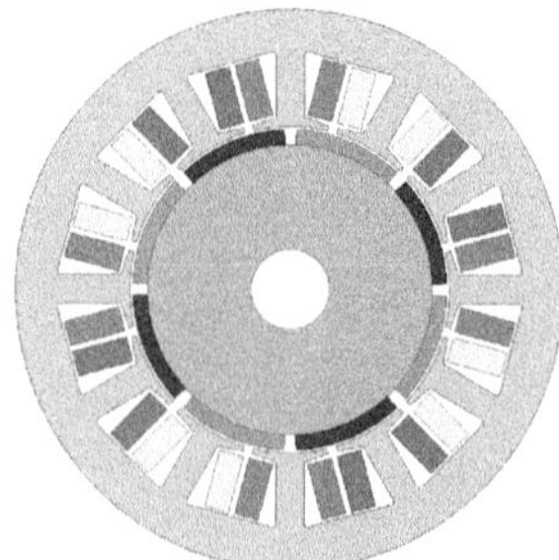
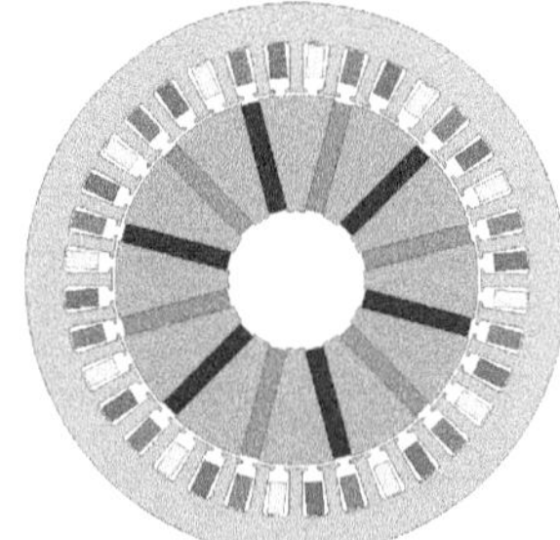

**FIGURE 15-32** Example configurations of 3-phase permanent magnet synchronous motors (PMSM). From left to right: surface permanent magnet rotor (SPM) with 8-poles and a 12-slot stator with concentrated coils around the teeth; interior permanent magnet rotor (IPM) with 4-poles and a stator with a distributed winding with two slots per pole and phase; spoke permanent magnet rotor with radially magnetized PMs, 12-poles, and a stator with a distributed winding with one slot per pole and phase.

is commonly referred to as a "pancake" type construction. Due to the substantial magnetic attraction forces between the stator and the rotor, single-sided axial flux machines exhibit an unbalanced magnetic pull. Configurations with multiple rotors and stators are also possible, as discussed for example in the handbook section on generators.

## 15.8.2 Brushless DC Motors

Motors with PMs in the rotor that are designed such that the back emf has a substantially trapezoidal waveform shape are best suited to be operated with quasi-square wave currents, in which case, they are referred to as of the brushless dc (BLDC) type. The power electronics inverter configuration for BLDC motor drives is the same as for PM synchronous motors (PMSM) with quasi-sine wave currents, which sometimes are referred to as of the brushless ac (BLAC) type, but the electronic control is different, so that the phase currents are regulated as illustrated, in principle, in Fig. 15-33.

In this case, each phase conducts for 120 electrical degrees during the positive half cycle and for another 120 electrical degrees during the negative half cycle, such that for most of the time only two phase are conducting and contribute to the torque production. Commutation occurs every 60 electrical degrees, and the combination between the associated transients and the real waveforms for back emf and currents may cause torque ripple, which is not shown in the schematic illustration of principles from Fig. 15-33 and may be significant in real applications. Hall sensors are typically employed for rotor pole detection and for enabling phase commutation. The diodes, which are placed antiparallel to the power electronic switches, conduct in the commutation period. The rate of current rise and fall, following a phase energization and deenergization, respectively, is determined by the speed and hence the back emf and the BLDC motor time constant, including the inductance. At high speeds, in order to allow the current to raise fast enough, phases are turned on in advance of the of the pole commutation.

BLDC motor drives may be cost-effective as they employ rather inexpensive Hall sensors and control electronics and, at least in principle, under ideal conditions, the motor can deliver for the same stator winding losses 15% higher torque output than in the PMSM/BLAC mode. Nevertheless, sinusoidal drives are typically reported to be advantageous in real applications in terms of lower torque ripple, lower acoustic noise, and higher efficiency.

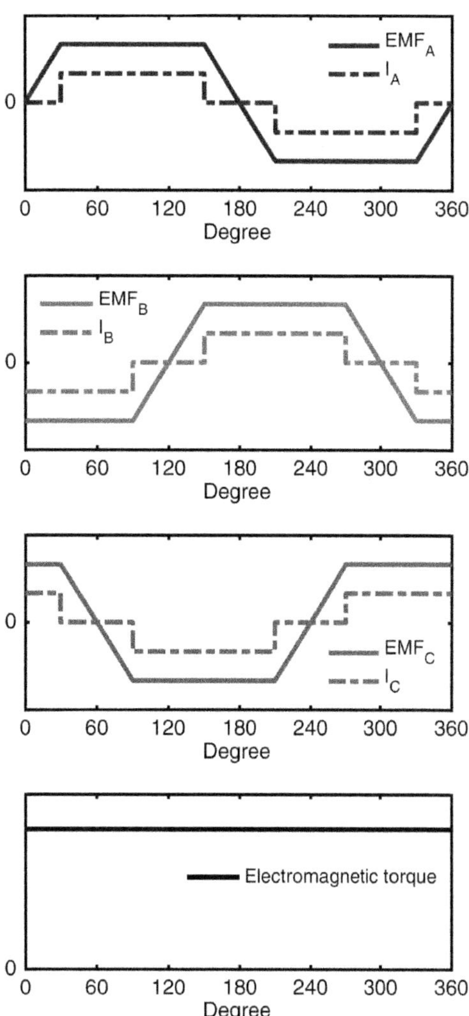


**FIGURE 15-33** Phase currents and back emfs, and the electromagnetic torque illustrating the ideal operation of a brushless dc motor (BLDC). Two phases produce electromagnetic torque at any given time. Under real operation, the rate of raise and fall of the current is limited by the operating speed, supply voltage and motor parameters, and the combination of transients related to phase switching and the actual waveforms for back emf and currents may cause significant current ripple, which is not included in the figure.

### 15.8.3 Switched Reluctance Motors

SRMs have magnetic saliency both on the laminated steel stator and rotor cores resulting in a doubly salient configuration, as shown in Fig. 15-34. Excitation is provided through multiphase stator windings that typically employ compact coils concentrated around the teeth. Torque is produced by energizing in appreciate sequence the phase windings, which tends to move the rotor from an unaligned toward an aligned position. The inductance of a phase winding increases as the rotor moves from the unaligned to the aligned position. Motoring torque is produced for as long as the phase winding remains energized when the inductance is increasing. The phase current must be switched off by the time the rotor reaches the aligned position for which the torque is zero. If the phase would continue to carry current beyond this position, a braking torque would be produced. The direction of the torque, being the outcome of attractive force corresponding to an equivalent electromagnet, represented by the stator active phase and the ferromagnetic rotor, is independent of the polarity of current. Neglecting saturation, it can be proved that the torque is given by

$$T = \frac{i^2}{2}\frac{dL}{d\theta}$$

where $i$ is the phase current, $L$ is the phase inductance, and $\theta$ is the rotor position.

A typical power electronic inverter employed for SRM consist of an asymmetric H-bridge with the schematic illustrated in Fig. 15-35. The use of other inverter topologies with reduced number of switches per phase has been subject of research.

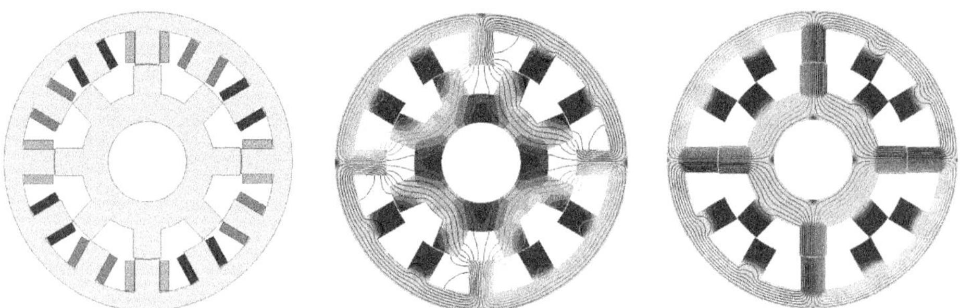

**FIGURE 15-34** Cross-section of a 3-phase, 12-slot, 8-pole SRM (left). Flux lines and magnetic flux density illustrating substantially higher magnetic loading in the aligned position (right) than in the unaligned position (center).

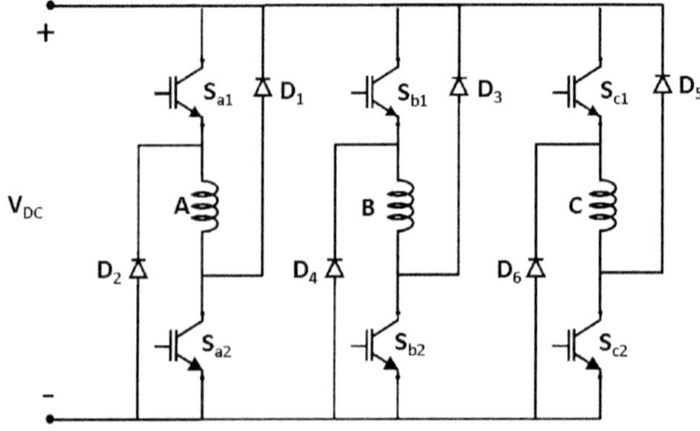


**FIGURE 15-35** Circuit diagram for an asymmetric bridge inverter typically employed for SRMs.

In an SRM, positive motoring torque can be produced by each phase at the most for only 180 electrical degrees, as in a symmetrical machine, the inductance profile is rising for half the electrical cycle and falling for the other half. Thus, in single-phase and two-phase machines there are positions of zero torque, which are eliminated in designs with three or more phases. Due to the double salient construction with concentrated coils, the mutual coupling between the phases is negligible and the SR has inherently improved fault tolerance.

Three-phase SRMs can be designed with several different stator slot/rotor pole combinations, such as 6/4, 6/8, 6/10, and their multiples. During typical ideal operation, only one phase conducts at a time, and thus each phase conducts for 120 electrical degrees (Fig. 15-36).

The rotational speed of the SRM is a function of the number of rotor poles, $P$, and stator supply fundamental frequency $f$,

$$N = \frac{60f}{P}$$

The motor phases are energized at "firing" angles based on the information provided by a position sensor or estimated by a sensorless control algorithm. The unaligned inductance being typically small, the rate of rise of current is high. Phase turn off is initiated as the rotor approaches the aligned position when the inductance is significantly higher, resulting in a slower rate of fall of current as compared to its rate of rise. Phase commutation leads to torque ripple and acoustic noise, which are cited as major limitations of this machine. The main losses in SRMs include stator copper losses and core losses in both the stator and rotor. As the flux density in different parts of the magnetic circuit has high peak values and high-frequency components the core loss can be significant. SRM drives can have very wide constant power operating regions. Operation in this case may require advance turning on the phases and continuous phase conduction.

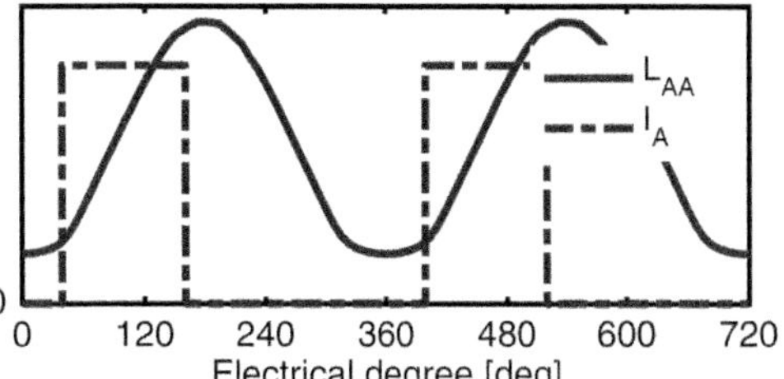


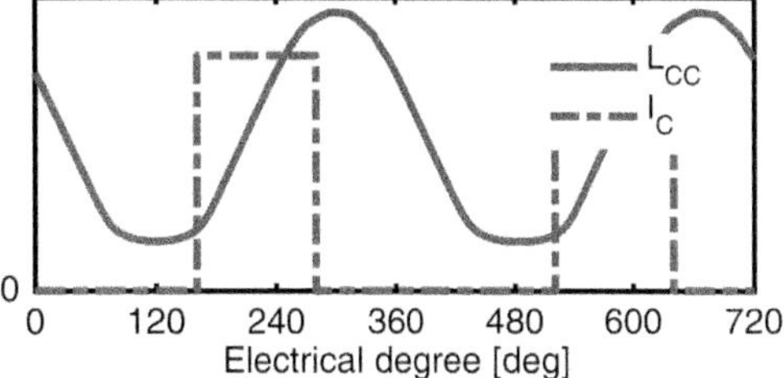


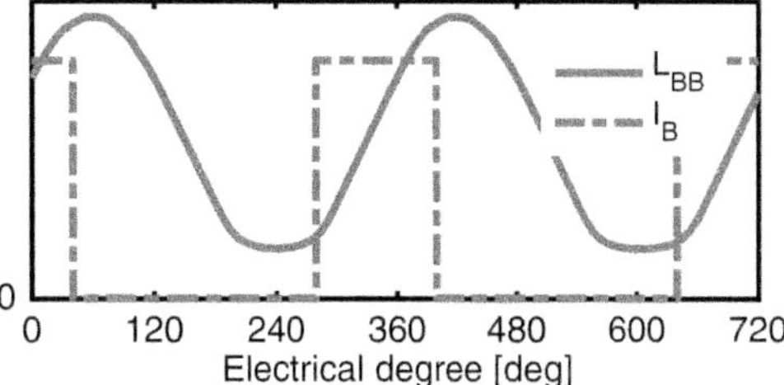


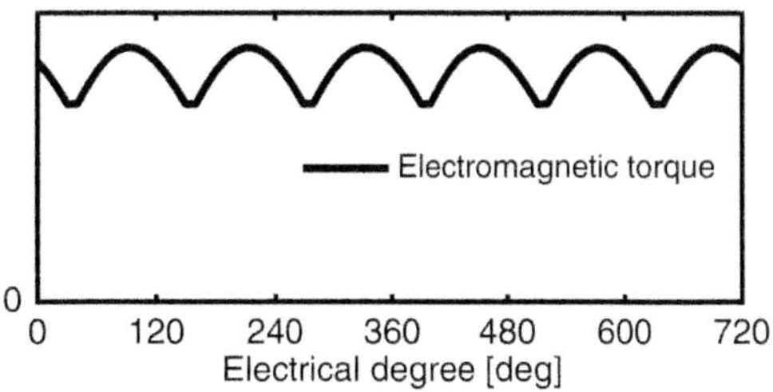


**FIGURE 15-36** Phase currents and inductance profile, and the electromagnetic torque illustrating the ideal operation of an SRM. Only one phase produces torque at a given time. In real operation, the rates of rise and fall of current are limited by the phase inductance and applied voltage, the waveforms deviate from the ideal profiles shown in the figure and the torque ripple maybe substantially higher.

Owing to its simple construction, lack of permanent magnets, wide constant power region with potential high efficiency, SRMs attracted substantial industrial and research interest over the last decades. For example, variations of this machine with full pitched windings, or with segmented rotor constructions, or with high number of rotor poles, as well as specific converters with reduced number of switches per phase have been proposed and studied.

### 15.8.4 Synchronous Reluctance Motors

Synchronous reluctance motors (SynRM) have a stator similar to that of an induction machine with distributed windings and a laminated rotor designed in order to achieve highly different reluctance along a $d$ and $q$ axis, respectively, resulting a high saliency ratio. The torque in an SynRM is produced because of the tendency of the rotor to align its highest magnetic permeance axis along the axis of the rotating field. Maximizing the saliency ratio is key to the design in order to improve the torque output. The typical implementation is with rotor laminations incorporating slots and bridges in order to create preferential magnetic flux guides or channels. Other rotor topologies that were subject of research included axially laminated versions.

A power electronics inverter is typically employed and a position encoder or a sensor-less means of rotor position estimation is required in order to determine the phase of the applied stator currents. An SynRM can be vector controlled, similarly to a PMSM and IM. On line start-up, without an inverter maybe possible by incorporating in the rotor a squirrel cage, similar to that of an induction motor. The major advantage of SynRMs over PMSMs is represented by the lack of permanent magnets. Benefits of SynRMs above induction machines include significantly reduced, virtually zero, rotor losses and hence the potential for higher efficiency. Ranges of SynRMs of the very high IE4 efficiency class have become recently commercially available.

The power factor of SynRMs is relatively low and an improvement in this respect can be achieved by the addition of permanent magnets optimally placed in the high-reluctance rotor structure. Such designs, typically referred to as permanent magnet (PM) assisted SynRM, have typically higher specific torque, efficiency, and constant power speed range.

## 15.9 OTHER SPECIAL MOTORS

***Multiphase Motors.*** AC motors with more than three phases may provide the benefits of increased redundancy, higher torque density, lower rating of power electronics, reduction of the dc bus voltage ripple, lower mmf harmonics and torque ripple. For example, in the case of 3-phase IMs and PMSMs, the fifth and seventh harmonics from the the stator mmf produce sixth harmonic torque pulsations and cause additional losses in the rotor. The effect of these can be mitigated by using a multiphase stator configuration with two sets of three-phase windings shifted by 30° electrical from each other, wherein the torque ripple has a 12th harmonic of lower magnitude. Furthermore, in a PM motor, particularly of the BLDC type, there may be a significant third harmonic component in the back emf and, for example, in a 5-phase third harmonic currents can be used to produce additional torque.

***Stator PM Motors.*** Such motors incorporate in the stator windings and PMs. The rotor contraction is similar to that from an SRM. In one topology, the stator has a salient structure, also similar to that of an SRM, and the PMs are placed in the stator back iron such that flux linked with the concentrated stator coils is a function of rotor position. In such a motor, the torque is produced both in the positive and negative half cycles and the flux linking the phase winding does not reverse.

In another topology, typically referred to as a flux-switching machine, the PMs are placed in the center of the stator teeth, around which concentrated coils are placed. As the salient pole rotor moves through one rotor pole pitch, the flux linked with the coil varies from a positive maximum to zero, to negative maximum. The topology may have advantages in terms of specific torque output, but additional space is required by the PMs and manufacturing is complex. In contrast, in a motor of the flux reversal type, the PMs are placed on the face of the stator teeth facing the air gap. This approach simplifies the manufacturing. A typical drawback is a low power factor, as not all the PMs are simultaneously active.

Stator PM machines have the advantages that most of the losses occur in the stator, which simplifies the cooling. The reliability is also higher because the permanent magnets are stationary. Similar to other PM synchronous machines, these machines can be driven with a vector controlled voltage source inverter.

***Pseudo-Direct Drive PM Motors.*** These motors include, apart from a stator and a rotor, another active component in the form of a modulator resulting in very high torque output that may benefit especially low-speed applications of the direct-drive type. The fundamental component of the stator armature mmf is different from the number of PM rotor poles and the interaction is made possible through the modulator, which may be rotating or stationary. Different variations of this topology exists, including constructions in which the modulator is stationary and represented by notches that profile the stator tooth tips facing the air gap or is constructed as a separate body with alternating ferromagnetic and nonmagnetic sections. Low-power factor may be an issue, which may not be significant if the speed is substantially low and torque is high, such that the size of the electric machine is large and its cost is substantially higher than that of power electronics—or may require special design measures.

## 15.10 AC COMMUTATOR MOTORS

***Classification.*** As compared with the induction motor, the ac commutator motor possesses two of the advantages of the dc motor: a wide speed range without sacrifice of efficiency and superior starting ability. In the induction motor, the starting torque is limited by the small space-phase displacement between the air-gap flux and the induced secondary current and by magnetic saturation of the flux paths. In the ac commutator motor, on the other hand, the air-gap flux and current are held at the optimum space-phase displacement by proper location of the brush axis, and the secondary current is not limited by magnetic saturation, giving high torque per ampere at starting. Furthermore, the series commutator motor may be operated far above the induction-motor synchronous speed, giving high power output per unit of weight.

***Single-Phase Straight Series Motor.*** An ordinary dc series motor, if constructed with a well-laminated field circuit, will operate (although unsatisfactorily) if connected to a suitable source of single-phase alternating current. Since the armature is in series with the field, the periodic reversals of current in the armature will correspond with simultaneous reversals in the direction of the flux, and consequently the torque will always be in the same direction. But the inductance of the motor will be so great that the current will lag far behind the voltage, and the motor will have a very low power factor. The entire amount of armature flux produced along the brush axis generates a reactive voltage in the armature, which must be overcome by the applied voltage, without performing any useful function.

The simple single-phase series motor has therefore two major faults, low power factor and poor commutation at low speeds, confining its use to fractional horsepower and very high-speed applications.

***Single-Phase Compensated Series Motor.*** In all, except the smallest sizes, it is usual to employ a compensating winding on the stator, in series with the armature and so arranged that its mmf as nearly as possible counteracts the armature mmf. A commutating winding is also frequently used, which somewhat overcompensates the armature reaction along the interpolar, or commutating-zone, axis and so provides a voltage to aid the current reversal, just as in a dc motor. By these means, the flux along the brush axis is reduced to a small fraction of its uncompensated value, and the power factor of the motor is greatly improved. Further improvement of the power factor is secured by using a smaller air gap and correspondingly fewer field ampere-turns than in an uncompensated motor, thus reducing the reactive voltage in the series field to a minimum.

***Universal Motors.*** Small series motors up to about ½-hp rating are commonly designed to operate on either direct current or alternating current and so are called *universal motors*. Universal motors may be either compensated or uncompensated; the latter type is used for the higher speeds and smaller ratings only. Owing to the reactance voltage drop, which is present on alternating current but absent on direct current, the motor speed is somewhat lower for the same load ac operation, especially at high loads. On alternating current, however, the increased saturation of the field magnetic circuit at the crest of the sine wave of current may materially reduce the flux below the dc value, and this tends to raise the ac speed. It is possible, therefore, to design small universal motors to have approximately the same speed-torque performance over the operating range, for all frequencies from 0 to 60 Hz. On a typical compensated-type ¼-hp motor, rated at 3400 r/min, the 60-Hz speed may be within 2% of the dc speed at full-load torque but 15% or more lower at twice normal torque, while on an uncompensated motor the speed drop will be materially greater.

## 15.11 MOTOR-PROTECTING DEVICES

*Fuses* should be provided for motor circuits, in accordance with the NEC. The current rating of the fuse must be considerably higher than the current rating of the motor, or the fuse will blow when the motor is started. For that reason fuses do not provide adequate overload protection. They furnish protection for the motor only in case of a short circuit or a very heavy overload. Their primary purpose is to protect the circuit rather than the motor.

*Magnetic-type overload relays* are operated by direct magnetic action of the motor current on a plunger. The relay consists of a series coil connected in the motor circuit and a plunger which is pulled up into the center of the coil when a certain value of current has been reached. When the plunger is lifted, a contact is tripped, opening the motor contactor-coil circuit and disconnecting the motor from the line. The tripping current can be varied by adjusting the initial position of the plunger with respect to the coil. Time delay in tripping is obtained by attaching a small oil dashpot to the plunger. The time delay can be adjusted so that the overload will not trip on the starting-current inrush but will trip on small sustained overloads.

*Thermal overload relays* are available in the bimetallic type and the fusible-alloy type. The bimetallic type has two heaters in series with the circuit to be protected, and above these heaters are two strips of bimetallic material, which act as latches for the contact members. Bending of the bimetallic strips under heating of overload current will release the latches and allow the contacts to open. The fusible-alloy type has two heaters, each surrounding a thermal element consisting of a small tube, inside which is a loose-fitting shaft. The tube and shaft are rigidly joined by a special low-melting eutectic alloy. On overload, the increased current drawn melts the alloy, allowing the shaft to turn and the contacts to open.

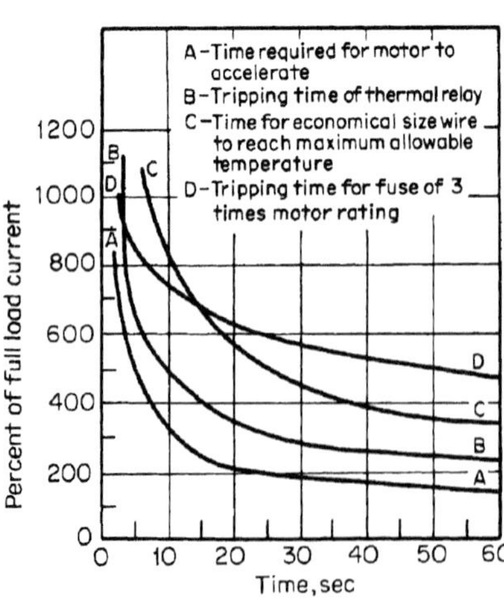


**FIGURE 15-37** Characteristics of thermal overload relays.

Characteristics of a typical thermal overload are shown in Fig. 15-37. An inspection of these curves shows that the thermal overload adequately protects the wiring, that the fuse blows first on short-circuit current, and that the thermal relay allows the motor ample time to accelerate. A thermal overload has a tripping characteristic which corresponds closely to the heating characteristics of a motor and, therefore, provides an ideal protecting means. An overload coil should be selected so that the maximum permissible output can be obtained from the motor. A motor rated 40°C rise on the basis of 40°C ambient temperature will have a final safe temperature of approximately 95°C and will operate at 15% overload continuously without overheating. An overload coil should therefore be selected having an ultimate tripping current equivalent to 15% overload on the motor. A continuous overload of 15% would therefore ultimately trip the thermal relay. For overloads in excess of 15%, the tripping time would be shorter than the time required for the motor to reach a dangerous temperature.

*Low-voltage protection* is the effect of a device, operative on the reduction or failure of voltage, to cause and maintain the interruption of power to the main circuit.

With magnetic controllers, this protection is obtained by using some form of 3-wire master switch. Should the line voltage drop to a low value or fail altogether, the main-line contactor will open and remain open, stopping the motor. To restart, it is necessary to push the "start" button. This type of control should always be used where the unexpected restarting of a motor after voltage failure may be dangerous to workers or equipment.

*Low-voltage release* is the effect of a device, operative on the reduction or failure of voltage, to cause the interruption of power to the main circuit but not to prevent the reestablishment of the main circuit on return of voltage. Such protection is obtained when a 2-wire pilot device, for example, a snap switch, float switch, or pressure switch, is used.

*Phase-failure protection* is the effect of a device, operative upon the failure of power in one wire of a polyphase circuit, to cause and maintain the interruption of power in all the wires of the circuit.

*Phase-reversal protection* is the effect of a device, operative on the reversal of the phase rotation in a polyphase circuit, to cause and maintain the interruption of power in all wires of the circuit. Protection of this type is necessary on elevators, where reversing of the phases would cause the car to start in a direction opposite to that in which the operator expects it to move.

*Field-failure protection* is usually provided in controllers for dc shunt- and compound-wound motors. The coil of a relay is connected in series with the motor shunt field, and a normally open contact of the relay is connected in the stop circuit. If the field circuit is opened, the relay will be deenergized and the motor will be disconnected from the line. This prevents overspeeding the motor owing to an open circuit in the field. A *field protective relay* is used to insert resistance in series with the shunt field whenever the motor is not running. The coil of the relay is connected in parallel with the main switch coil, and a normally open relay contact is used to short-circuit a step of resistor in the field circuit. The resistor should be designed to reduce the voltage across the field to one-half line voltage. This reduces the field wattage to one-fourth the normal value and prevents overheating the field with the motor at standstill.

A *field-discharge resistor* should be provided for 230-V motors rated 7½ hp or more and for 550-V motors rated 5 hp or more whenever the shunt-field circuit must be opened. The ohmic value of a discharge resistor should be 1 to 3 times the ohms in the field. If a resistance of three times the field ohms is used, the induced voltage, when the circuit is opened, will be four times normal line voltage. This voltage, caused by the inductance of the field, must be limited to prevent damage to the insulation of the field windings. On nonreversing controllers without dynamic braking, the shunt field can be connected behind the main contactor and the field allowed to discharge through the motor armature.

## 15.12 BIBLIOGRAPHY

### 15.12.1 Books and Publications


Alerich, W. N., and Herman, S. L., *Electric Motor Control*, Delmar Publishing, Albany, New York, 1998.

Andreas, J. C., *Energy-Efficient Electric Motors, Selection and Application*, Marcel Dekker, New York, 1992.

Beaty, H. W., and Kirtley, J. L., Jr., *Electric Motor Handbook*, McGraw-Hill, New York, 1998.

Boldea, S. Nasar, *Electric Drives*, 2nd ed., CRC Press, 2006.

Bose, B. K., *Adjustable Speed AC Drive Systems*, IEEE Press, New York, 1980.

Chapman, S. J., *Electric Machinery Fundamentals*, McGraw-Hill, New York, 1998.

Chau, K. T., "Vernier Permanent Magnet Motor Drives," in *Electric Vehicle Machines and Drives: Design, Analysis and Application*, 1, Wiley-IEEE Press, 2015.

Cochran, P. L., *Polyphase Induction Motors, Analysis, Design, Application*, Marcel Dekker, New York, 1989.

Deodhar, R. P., Andersson, S., Boldea, I., and Miller, T. J. E., "The Flux-Reversal Machine: A New Brushless Doubly Salient Permanent-Magnet Machine," in *IEEE Transactions on Industry Applications*, vol. 33, no. 4, pp. 925–934, Jul./Aug. 1997.

Dorf, R. C. (ed.), *The Electrical Engineering Handbook*, 2nd ed., CRC Press, Boca Raton, Florida, 1997.

EL-Refaie, M., "Fractional-Slot Concentrated-Windings Synchronous Permanent Magnet Machines: Opportunities and Challenges," in *IEEE Transactions on Industrial Electronics*, vol. 57, no. 1, pp. 107–121, Jan. 2010.

Emadi, Ali, *Energy-Efficient Electric Motors*, Marcel Dekker, New York, 2005.

Fitzgerald, A. E., Kingsly, C., Jr., and Umans, S. D., *Electric Machinery*, 6th ed., McGraw-Hill, New York, 2005.

Gieras, J. F., Wang, R. J., and Kamper, M. J., *Axial Flux Permanent Magnet Brushless Machines*, Springer, Netherland, 2008.

Hamdi, E. S., and Hamdi, H. S., *Design of Small Electrical Machines: Design and Measurement in Electronic Engineering*, John Wiley & Sons, New York, 1994.

Hendershot, J. R., and Miller, T. J. E., *Design of Brushless Permanent Magnet Machines*, 2nd ed., Motor Design Books LLC, 2010.

Hughes, A., *Electric Motors and Drives: Fundamentals, Types and Applications*, Butterworth-Heinemann, Oxfordshire, England, 1993.

Ishizaki, T. T., Takasaki, K., and Nishikata, S., "Theory and Optimum Design of PM Vernier Motor," *1995 Seventh International Conference on Electrical Machines and Drives (Conf. Publ. No. 412)*, Durham, 1995, pp. 208–212.

Kaiser, J., *Electrical Power: Motors, Controls, Generators, Transformers*, Goodheart-Willcox, Tinley Park, Illinois, 1998.

Keljik, J., *Electrical Motors and Motor Controls*, Delmar Publishing, New York, 1995.

Mecrow, B. C., and Jack, A. G., "A New High Torque Density Permanent Magnet Machine Configuration," *International Conference on Electric Machines,* 1990.

Miller, T. J. E., *Switched Reluctance Machines and Their Control*, Oxford University Press, 1993.

Munoz-Garcia, A., Lipo, T. A., and Novotny, D. W., "A New Induction Motor V/f Control Method Capable of High-Performance Regulation at Low Speeds," in *IEEE Transactions on Industry Applications*, vol. 34, no. 4, pp. 813–821, Jul./Aug. 1998.

Murphy, J. M. D., *Thyristor Control of AC Motors*, Pergamon, New York, 1973.

Nasar, S. A., and Boldea, I., *Linear Motion Electric Machines*, Wiley, New York, 1976.

Novotny, D. W., and Lipo, T. A., *Vector Control and Dynamics of AC Drives*, Clarendon

Press, Oxford, England, 1996 (Imprint of Oxford University Press).

Parsa, L., "On Advantages of Multi-Phase Machines," *31st Annual Conference of IEEE Industrial Electronics Society, IECON 2005*, 2005.

Rodriguez, J., Jih-Sheng Lai, and Fang Zheng Peng, "Multilevel Inverters: A Survey of Topologies, Controls, and Applications," in *IEEE Transactions on Industrial Electronics*, vol. 49, no. 4, pp. 724–738, Aug. 2002.

Say, M. G., *Alternating Current Machines*, 5th ed., Wiley, New York, 1983.

Sen, P. C., *Principles of Electric Machines and Power Electronics,* 2nd ed., Wiley, New York, 1996.

Shoults, D. R., Rife, C. J., and Johnson, T. C., *Electric Motors in Industry*, Wiley, New York, 1942.

Smeaton, R. W., *Motor Application and Maintenance Handbook*, McGraw-Hill, New York, 1969.

Subrahmanyam, V., *Electric Drives: Concepts and Applications*, McGraw-Hill, New York, 1996.

Wheeler, P. W., Rodriguez, J., Clare, J. C., Empringham, L., and Weinstein, A., "Matrix Converters: A Technology Review," in *IEEE Transactions on Industrial Electronics*, vol. 49, no. 2, pp. 276–288, Apr. 2002.

Zhu, Z. Q., "Switched Flux Permanent Magnet Machines—Innovation Continues," *2011 International Conference on Electrical Machines and Systems*, Beijing, 2011, pp. 1–10.

## 15.12.2 Websites

American National Standards Institute: http://ansi.org/IEEE Press: http://shop.ieee.org/

Library of Congress: http://catalog.loc.gov/McGraw-Hill.Professional Books: http://books.mcgraw-hill.com/ Wiley

Publications: http://wiley.com/

http://www.ti.com/lit/an/sprabq8/sprabq8.pdf

http://www.cypress.com/file/253451/download

# 16

# POWER ELECTRONICS


## Alex Q. Huang

*Dula D. Cockrell Centennial Chair in Engineering, Univesity of Texas at Austin*

## Xu She

*Lead Electrical Engineer, GE Global Research*





Grateful acknowledgment is given to past contributors to this section: Amit Kumar Jain and Raja Ayyanar.

# 16.1 INTRODUCTION

## 16.1.1 Role of Power Electronic Converters

Power electronics is an enabling technology that achieves conversion of electric power from one form to another utilizing a combination of high-power semiconductor devices and passive components—chiefly transformers, inductors, and capacitors. The input and output may be ac or dc and may differ in magnitude and frequency. The conversion sometimes involves multiple stages with two or more converters connected in a cascade. The end goals of a power electronic converter are to achieve high efficiency of conversion, minimize size and weight, and achieve desired regulation of the output. Figure 16-1 shows power electronic converters in a generic application.

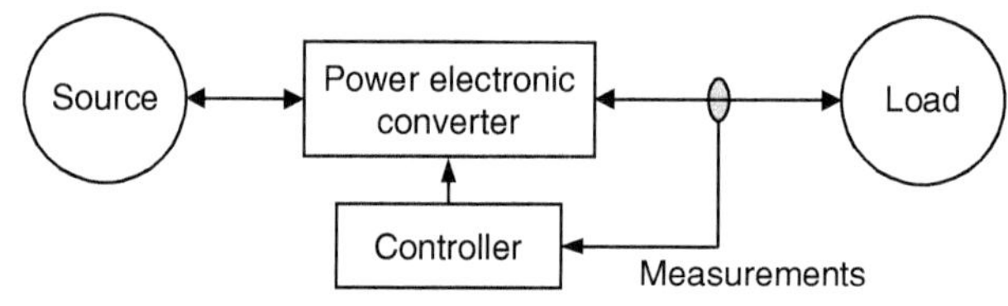


**FIGURE 16-1** Application of power electronic converters.

## 16.1.2 Application Examples

Power electronic converters can be classified into four different types on the basis of input and output—*dc-dc*, *dc-ac*, *ac-dc*, and *ac-ac*, named with the first part referring to the input and the second to the output. The diode bridge rectifier is the front end for most low-power converters. It converts line frequency ac (e.g., from a wall outlet) to an unregulated dc voltage, and the process is commonly called rectification. In a *dc-dc* converter both the input and the output are dc, and in the simplest case the output voltage needs to be regulated in presence of variation in load current and changes in the input voltage. A computer power supply has a diode bridge front end followed by a dc-dc converter, the combination of which converts line frequency ac voltage to several regulated dc

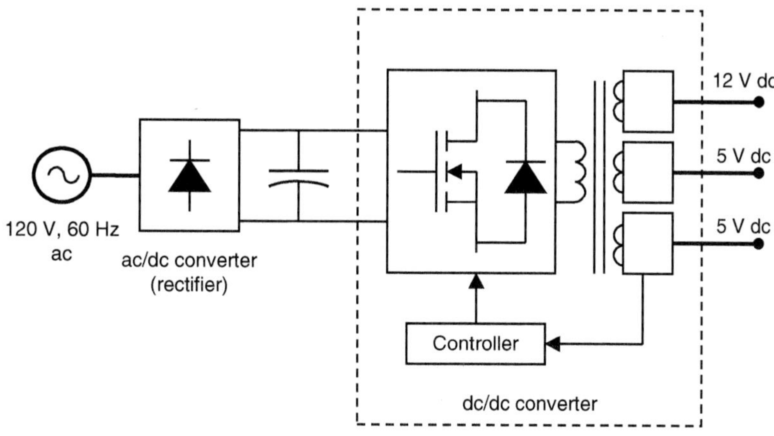


**FIGURE 16-2** Computer power supply.

voltages (Fig. 16-2). Electronic ballasts for compact fluorescent lamps consist of a line frequency rectifier followed by a dc to high-frequency ac converter (frequency range of 20 to 100 kHz) whose output is connected to a resonant tank circuit that includes the load. In an adjustable speed motor drive application (Fig. 16-3), the input is a three-phase ac supply, and the output is a three-phase ac whose magnitude and frequency are varied for optimum steady state operation and dynamic requirements of the drive.

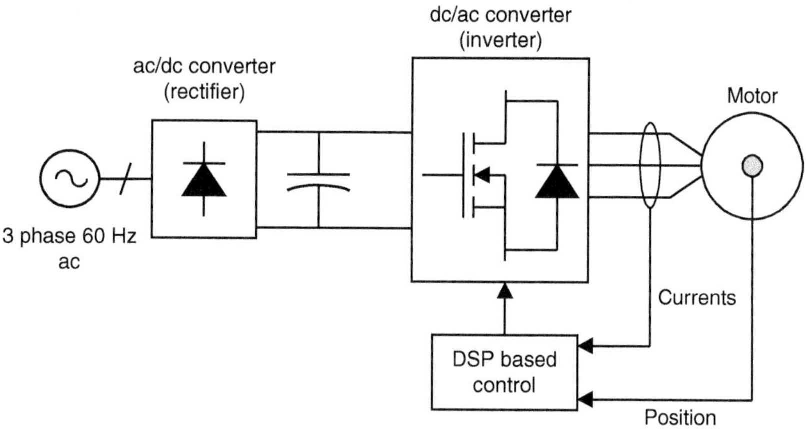


**FIGURE 16-3** Adjustable speed motor drive.

Development of power semiconductors with very high voltage and current ratings has enabled the use of power electronic converters for utility applications. In transmission systems, power electronic converters are being utilized to control power flow, damp power oscillations, and enhance system stability. At the distribution level, power electronic converters are used for enhancing power quality by means of dynamic voltage restorers, static var compensators, and active filters. Power electronic converters also play a significant role in grid connection of distributed generation and especially renewable energy sources; their functions include compensation for steady state and dynamic source characteristics leading to optimal energy transfer from the source, and protective action during contingencies.

Future automotives are expected to have a large number of power electronic converters performing various functions, e.g., electric power steering, active suspension, control over various loads, and transferring power between the conventional 14-V bus and the *42-V Power Net*,[1] which has already been adopted in some high-end vehicles. Hybrid electric and all-electric vehicles also utilize controlled power electronic converters for interfacing the battery and motor/generator.

The proliferation of power electronics connected to the utility grid has also led to power quality concerns due to injection of harmonic currents by grid-connected inverters, and highly distorted currents drawn by diode bridge rectifiers. Due to fast transients of voltages and currents within power converters, they can be a source of electromagnetic emissions leading to electromagnetic interference. Several solutions to limit and correct for these effects have therefore been developed.

## 16.2 PRINCIPLES OF SWITCHED MODE POWER CONVERSION

This subsection presents some basic principles that are common to the analysis of most switch mode power converters.

### 16.2.1 Bi-Positional Switch

The most basic component of a switch mode power converter is the bi-positional switch shown in Fig. 16-4*a*. Nodes 1 and 2 of the switch are invariably connected across a dc voltage source (or across a big capacitor whose voltage is close to a constant dc), and pole *A* of the switch is in series with a dc current source (or a big inductor whose current is close to a constant dc). This bi-positional switch, which is also referred to as a switching power pole, switches at very high frequencies, and is controlled by the signal $q_A(t)$. The switched pole *A* voltage and the input current based on the control signal $q_A(t)$ are listed in Table 16-1, and the corresponding waveforms shown in Fig. 16-4*b*.

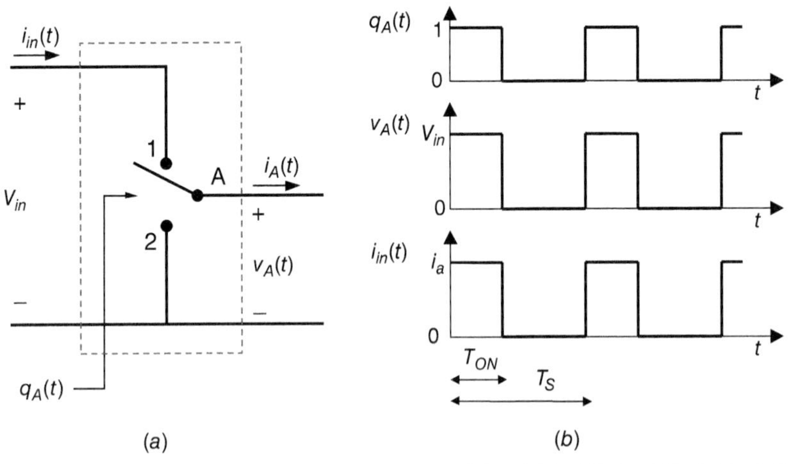


**FIGURE 16-4** (*a*) Bi-positional switch, (*b*) switching waveforms.

Figure 16-5*a* shows the electronic implementation of a complete bi-positional switch using MOSFETs. This implementation can support pole current in either direction and is useful for applications where current direction can reverse. In most dc to dc converter applications, the current through the pole *A* is unidirectional, and hence, the implementation shown in Fig. 16-5*b* is sufficient to realize the bi-positional switch.

**TABLE 16-1** States of a Bi-Positional Switch

| $q_A(t)$ | Switch position | MOSFET and diode state | Pole voltage and input current |
|---|---|---|---|
| 1 | 1 | S1 and D1 ON, S2 and D2 OFF | $v_A = V_{in}$, $i_{in} = i_A$ |
| 0 | 2 | S1 and D1 OFF, S2 and D2 ON | $v_A = 0$, $i_{in} = 0$ |

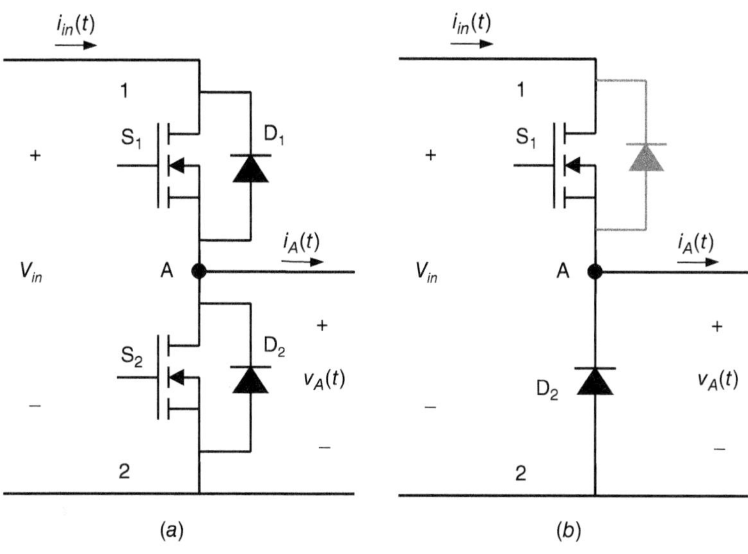


**FIGURE 16-5** Electronic implementation of bi-positional switch: (*a*) for bidirectional pole current, (*b*) for unidirectional pole current.

## 16.2.2 Pulse Width Modulation

The concept of pulse width modulation (PWM) is central to all switch mode power converters. PWM refers to the control of the average value of a switching variable, for example, $v_A(t)$ in Fig. 16-4*b*, by controlling or modulating its pulse width. Some basic concepts and definitions necessary to understanding PWM are presented here.

***Duty Ratio.*** The frequency at which the bi-positional switch is switched on and off is denoted by $f_s$, and the corresponding time period by $T_s$ $(= 1/f_s)$. The transition between the two states of the switch occurs in a very small duration compared to $T_s$. The time for which the switch remains in position 1 during a switching period is denoted by $T_{on}$. The duty ratio $d$ of the bi-positional switch is then defined as the ratio of on-time to total time period

$$d = \frac{T_{on}}{T_s} \tag{16-1}$$

***Averaging.*** Currents and voltages in power electronic converters have (a) high-frequency components corresponding to the switching frequency of the bi-positional switch elements, and (b) low-frequency components due to slower variations caused by change in load demand, source magnitude, and changes in reference value of the desired outputs. For dynamic control and steady-state analysis, the low-frequency components are of primary interest. To study these components, it is sufficient to study their averages over one switching time period. It should be noted that the averaging presented here[2] is a very basic form of the general averaging method[3] and has limitations in terms of validity w.r.t. the switching frequency. However, this simplification is good enough for most practical purposes, and can be confidently used for steady state and dynamics up to one-fifth the switching frequency. Throughout this section the averaged variables, that is, averaged over one switching period, are denoted by a '-' on top of the variables. Thus, the averaged value of $x(t)$ is given by

$$\bar{x}(t) = \frac{1}{T_s}\int_{t-T_s}^{t} x(\tau)\, d\tau \tag{16-2}$$

In steady state, the average values of $q_A(t)$ and $\upsilon_A(t)$ are given by

$$\bar{q}_A = \frac{1}{T_s}\int_0^{T_s} q_A(\tau)\, d\tau = \frac{1}{T_s}\int_0^{T_{on}} 1\, d\tau = \frac{T_{on}}{T_s} = d \tag{16-3}$$

$$\bar{\upsilon}_A = \frac{1}{T_s}\int_0^{T_s} \upsilon_A(\tau)\, d\tau = \frac{1}{T_s}\int_0^{T_{on}} V_{in}\, d\tau = d\cdot V_{in} \tag{16-4}$$

In general, the averaged quantities can be time varying, since the pulse widths of the switching waveform can vary with time. Thus,

$$\bar{q}_A(t) = d(t) \tag{16-5}$$

$$\bar{\upsilon}_A(t) = d(t)\cdot V_{in} \tag{16-6}$$

As an example of PWM, we can regulate the average value of $\upsilon_A(t)$ in Fig. 16-4*b* by varying the duty ratio $d$. If $V_{in} = 10$ V, $f_s = 100$ kHz $\Rightarrow T_s = 10\ \mu s$, then $T_{on} = 5\ \mu s \Rightarrow d = 0.5$, and $\upsilon_A = 5$ V, etc. By varying the duty ratio sinusoidally a low-frequency ac voltage can be synthesized from a dc voltage as illustrated in Fig. 16-6.

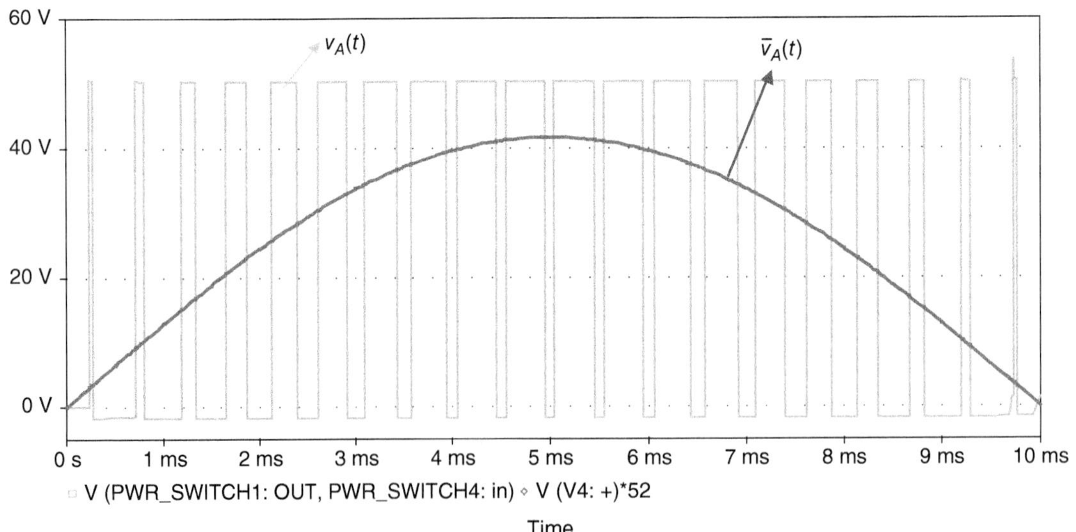


**FIGURE 16-6** AC synthesis using PWM.

### 16.2.3 Concept of Steady State

A converter is said to be in dc steady state when all its waveforms exactly repeat in each switching period, that is, $x(t) = x(t - T_s)\forall t$, where $x$ is any of the converter variables. With reference to Eq. (16-2), it is clear that in steady state the average value of any variable is constant. Analysis of steady-state operation is essential to determine ratings and design of the power stage components in the converter, viz., power semiconductor devices, inductor, capacitors, and transformers. Important concepts that enable steady-state analysis from a circuit viewpoint are discussed below. It should be remembered that these are only valid during steady-state operation.

***Steady-State Averages of Inductor Voltage and Capacitor Current.*** The instantaneous *v-i* relationship for an inductor is

$$v_L(t) = L\frac{di_L(t)}{dt} \quad \text{or} \quad i_L(t) = i_L(0) + \frac{1}{L}\int_0^t v_L(\tau)d\tau \tag{16-7}$$

where $v_L(t)$ is the voltage across the inductor and $i_L(t)$ is the current flowing through the inductor. Since $i_L(T_s) = i_L(0)$ in steady state, from the integral form of Eq. (16-7) it follows that

$$\bar{v}_L = \frac{1}{T_s}\int_0^{T_s} v_L(\tau)d\tau = 0 \tag{16-8}$$

The above relationship can also be derived directly in terms of the averaged quantities as below

$$\bar{v}_L(t) = L\frac{d\bar{i}_L(t)}{dt} = 0 \quad [\text{since } \bar{i}_L(t) \text{ is constant in steady state}] \tag{16-9}$$

This is referred to as *volt-second balance* in an inductor. Figure 16-7 shows a typical steady-state waveform of an inductor voltage for many power converters. As seen, the positive area is exactly cancelled by the negative area, making the average value zero. It may be mentioned that during the start-up transient, $\bar{v}_L$ remains positive for several switching cycles, allowing the inductor current to rise from zero to its final steady-state value.

In a similar fashion it can be shown that in steady state the average current through a capacitor is zero. This is referred to as *ampere-second balance* in a capacitor. Note that though the average value of the capacitor current is zero, its root mean square (RMS) value, which is one of the main selection criteria for a capacitor, can be substantial depending on the converter topology.

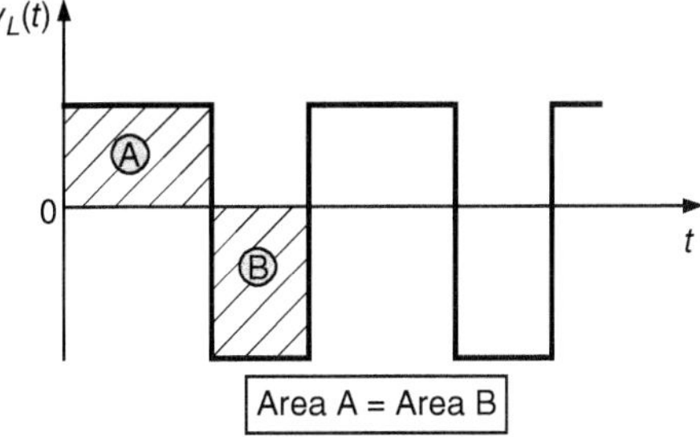


**FIGURE 16-7** Volt-second balance for an inductor.

***Power Balance.*** For analytical purposes, it is often useful to neglect all losses in the converter and consider input power to be equal to the output power, again in an average sense.

$$P_{\text{in}} = \bar{v}_{\text{in}}\bar{i}_{\text{in}} = P_o = \bar{v}_o\bar{i}_o \tag{16-10}$$

This implies that there is no increase or decrease in the energy stored in inductors and capacitors over one switching time period. This is valid for the input-output of the entire converter as well as any intermediate stage.

***Kirchoff's Laws for Averages.*** Just like the instantaneous quantities, the averaged quantities also obey Kirchoff's current and voltage laws. The sum of average currents entering a node is zero. The proof follows from interchanging the order of summation (for individual currents) and integration (over a switching time period):

$$\sum_k \bar{i}_k = \frac{1}{T_s}\sum_k \left[\int_0^{T_s} i_k\right] = \frac{1}{T_s}\int_0^{T_s}\left[\sum_k i_k\right] = 0 \qquad (\text{since} \sum_k i_k \equiv 0) \tag{16-11}$$

Similarly, the sum of average voltages in a circuit loop is zero.

$$\sum_k \bar{v}_k = 0 \tag{16-12}$$

### 16.2.4 Power Loss in the Bi-Positional Switch

Electronic implementations of the bi-positional switch shown in Fig. 16-5*a* and *b* use semiconductor switches and diodes, and they have significant power loss. The power loss can be divided into two kinds—conduction loss and switching loss.

With reference to Fig. 16-5*b*, when the MOSFET is ON there is a nonzero voltage across it. Similarly the diode has a forward voltage drop while it is conducting. Both of these lead to power loss whose sum averaged over one switching time period is called *conduction loss*.

A finite time interval is required to transition from one state to the other: (MOSFET ON and diode OFF) to (MOSFET OFF and diode ON), and vice versa. While the MOSFET is turning OFF, the diode cannot conduct until it is forward biased. As the voltage across the MOSFET increases from near zero to the full input voltage $V_{\text{in}}$, it conducts the full output current. Once the diode is forward biased the current starts transferring from the MOSFET to the diode. During the reverse transition, first current is transferred from the diode to the MOSFET, and then the voltage across the MOSFET reduces from $V_{\text{in}}$ to the conduction voltage drop. Thus, the MOSFET incurs significant power loss during both transitions. The above description is simplified and there are other phenomena which contribute to loss during the transitions. The diode also has power loss during the transitions. The sum of losses in the MOSFET and diode during the transitions averaged over one switching time period is called the *switching loss*. Switching power loss increases with increase in switching frequency and increase in transition times. Sum of the conduction and switching loss, computed as averages over one complete switching periods, gives the total power loss shown in Fig. 16-8.

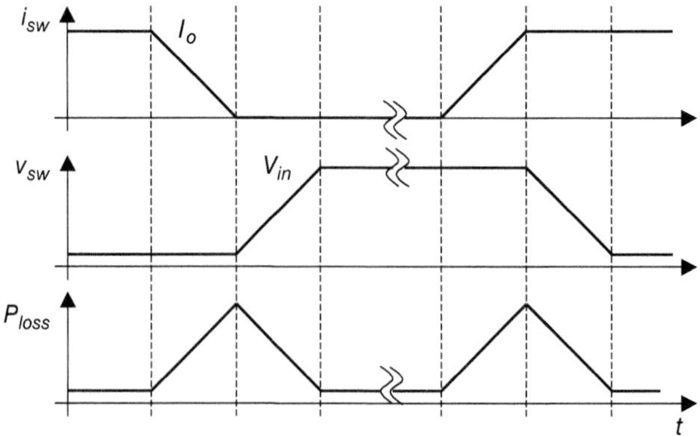


**FIGURE 16-8** Switching transients in bi-positional switch implementation.

Similar losses occur in the realization of Fig. 16-5*a*. When S1 is turned off by its control signal, current $i_A(t)$ transfers to D2, the antiparallel diode of S2. After this transition S2 is turned on and the current transfers from the diode to the MOSFET channel (which can conduct in either direction). A short-time delay, called *dead time*, is required between the ON signals for S1 and S2. The dead time prevents potential shorting of the input voltage, also known as *shoot-through fault*.

The nonidealities of nonzero voltage drop and switching times will be neglected for analysis of power electronic converters presented throughout this section. However, these are extremely important in design and selection of components for a real power converter.

## 16.3 DC TO DC CONVERTERS

DC-DC converters represent one major area in power electronics. In a dc-dc converter, the input and output may differ in magnitude, the output may be electrically isolated from the input, and the output voltage may have to be regulated in the presence of variation in input voltage and load current. In a typical power distribution system for digital systems, several lower magnitude dc voltages are derived from a common input using one or more converters. Battery-powered portable devices use converters that boost the input 1.5 V cell voltage to 3.3, 5, or 9 V. Most of these converters have unidirectional power flow from input to output. The presentation here is limited to the basic converter types. The interested reader is referred to text books which deal with details of these converters.[4–8]

### 16.3.1 Buck Converter

The buck converter is used to step down an input voltage to a lower magnitude output voltage. Figure 16-9*a* shows the schematic of a buck converter. A power MOSFET and diode combination is shown for implementation of the bi-positional switch with unidirectional output current. The bi-positional switch is followed by an L-C low-pass filter which attenuates the high-frequency switching component of the pole A voltage and provides a filtered dc voltage at the output. A high switching frequency is desirable to reduce the size of the filter, the higher limit depending on the power level of the converter and the semiconductor devices used. The final choice of switching frequency depends on several factors: size, weight, efficiency, and cost. It is usually above the audible range and frequencies above 100 kHz are very common.

***Operation.*** The input voltage $V_{\text{in}}$ is assumed to remain constant within a switching cycle. The inductor $L$ and capacitor $C$ are sufficiently large so that the inductor current $i_L$ and output voltage $v_o$ do not change significantly within one switching cycle. The load is represented by the resistor $R_L$. Under steady-state operation it is assumed that the inductor current is always greater than zero. The MOSFET is turned ON in response to signal $q_A(t)$ for $T_{\text{on}} = DT_s$, where $D$ represents the steady-state duty ratio. During this time $v_A = V_{\text{in}}$ and $i_{\text{in}} = i_L$. When the MOSFET is turned OFF, the inductor current flows through diode $D_1$ leading to $v_a = 0$ and $i_{\text{in}} = 0$. Since the average voltage across the inductor is zero, $\bar{v}_L = 0$, the average output voltage is given by

$$\bar{v}_o = \bar{v}_a = DV_{\text{in}} \tag{16-13}$$

The average current through the capacitor C, $\bar{i}_c$, is zero. Thus, $\bar{i}_L = I_o$ and the input current is given by

$$\bar{i}_{\text{in}} = DI_o \tag{16-14}$$

From the above equations it is clear that the output voltage is lower than the input voltage and output current is higher than the input current. Also, power balance for averaged quantities can be verified

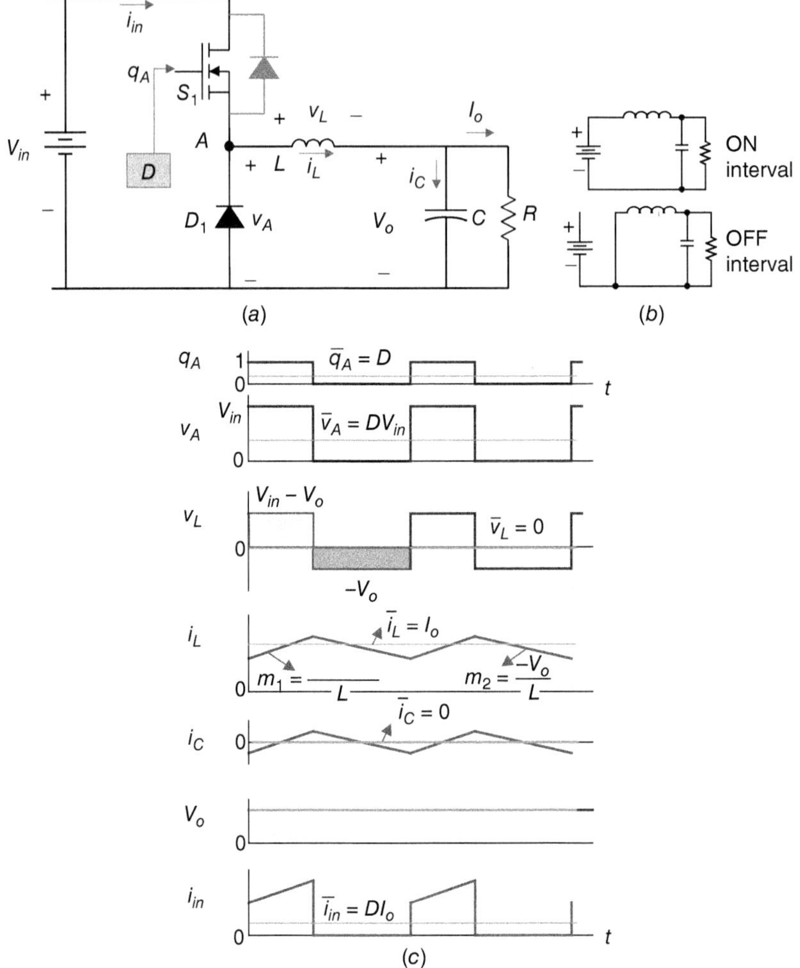


**FIGURE 16-9** Buck converter: (*a*) circuit, (*b*) equivalent circuits during ON and OFF intervals, (*c*) steady-state waveforms.

from Eqs. (16-13) and (16-14). Within a switching cycle, instantaneous values of the inductor current and capacitor voltage vary as follows:

$$\begin{array}{lll} \text{MOSFET ON:} & L\dot{i}_L = V_{\text{in}} - v_o & C\dot{v}_o = i_L - v_o/R_L \\ \text{MOSFET OFF:} & L\dot{i}_L = -v_o & C\dot{v}_o = i_L - v_o/R_L \end{array} \tag{16-15}$$

Equivalent circuits for the two intervals and instantaneous waveforms are shown in Fig. 16-9*b* and *c*.

***Component Selection.*** Usually the inductor and capacitor are significantly large so that within a switching period $v_o$ can be assumed constant in computation of $i_L$. This leads to the linear variation of $i_L$ shown in Fig. 16-9*c* with a *peak-to-peak ripple*, $\Delta I_L$, given by

$$\Delta I_L = \frac{V_o(1-D)T_s}{L} \tag{16-16}$$

In most designs the inductance value is chosen to limit $\Delta I_L$ between 10% and 30% of the full load current. Since the average capacitor current is zero, the instantaneous capacitor current is approximately equal to the ripple component of the inductor current.

$$i_C(t) = i_L(t) - I_o \tag{16-17}$$

The peak-to-peak capacitor voltage ripple resulting from the capacitor current can then be derived as

$$\Delta V_o = \frac{\Delta I_L \cdot T_s}{8C} \tag{16-18}$$

Capacitors used for filtering in most dc-dc converters are electrolytic capacitors which are characterized by significant effective series resistance (ESR) and effective series inductance (ESL). These parasitics also contribute to the output voltage ripple and should supplement Eq. (16-18) in the choice of capacitors. Film or ceramic capacitors, which have significantly lower ESR and ESL, are often used in addition with the electrolytic capacitors.

The MOSFET has to be rated to block a voltage greater than $V_{in}$, and conduct an average current greater than $I_{in}$. Power dissipation and temperature considerations usually require MOSFETs to be rated for two to three times the maximum input average current expected. In addition, the peak MOSFET current, equal to the maximum peak of the inductor current, should not exceed its maximum current rating. The diode has to be rated to block $V_{in}$, and conduct an average current greater than the maximum output current. Diodes are usually chosen with ratings approximately two times the expected maximum current. Final selection of power semiconductors is based on a combination of acceptable power loss, cooling effort required to dissipate the power loss, maximum junction temperature with the available cooling mechanism, and reliability for expected operating conditions.

***PWM Control Implementation.*** As evident from Eq. (16-13) the duty ratio of the switch controls the output voltage. In response to variation in input voltage and load current, the duty ratio has to be changed by a feedback-controlled system as shown in Fig. 16-10*a*. The error between the reference

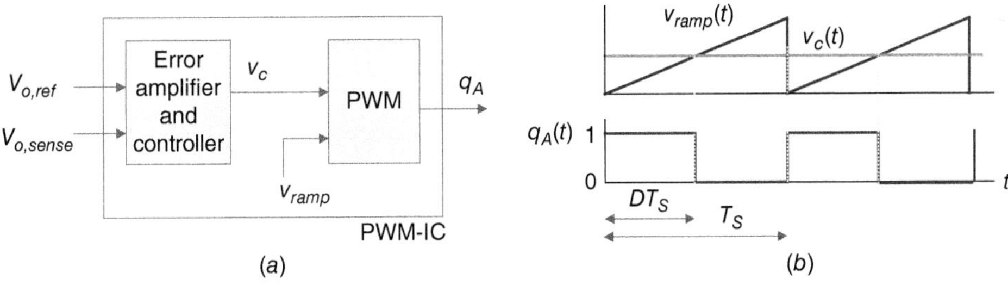


**FIGURE 16-10** PWM generation: (*a*) ramp comparison, (*b*) control block diagram.

and actual output voltage is given to an appropriately designed error compensating amplifier, the output of which is a control voltage $v_c$. This control voltage is compared with a constant frequency sawtooth waveform. The output of the comparator is the switching signal $q_A(t)$ which determines the ON or OFF state of the MOSFET. When the output voltage is lower than the reference value, the control voltage increases, leading to an increase in the duty ratio, which in turn increases the output voltage. The error amplifier and comparator, and several other features, are available in a single integrated circuit (e.g., UC3825A) available from several manufacturers.[9–12] This integration of components leads to reduction in overall size and cost.

### 16.3.2 Boost Converter

As evident from the name the boost converter is used to step up an input voltage to a higher magnitude output voltage. Figure 16-11*a* shows the schematic. In this case the MOSFET is in the lower position while the diode is in the upper position. The inductor is on the input side and the output has a purely capacitive filter.

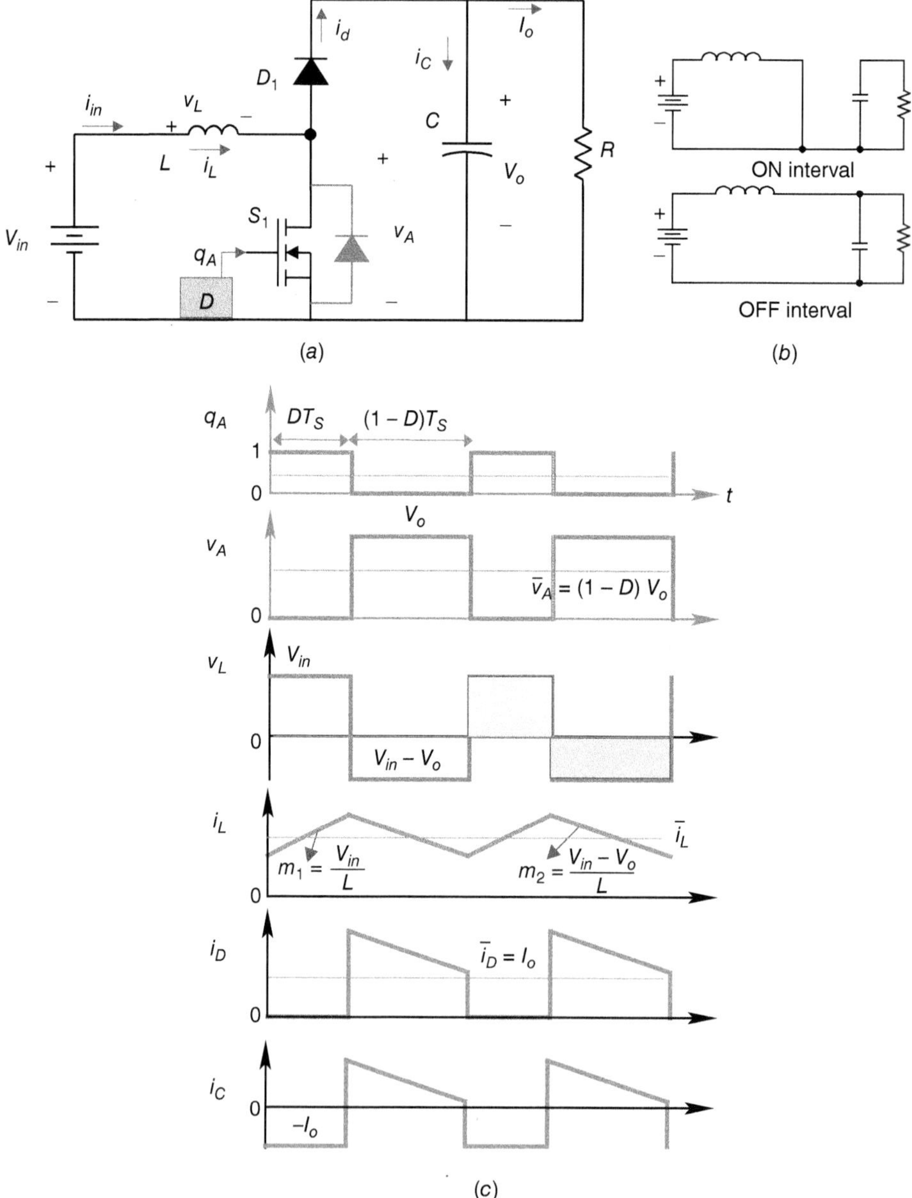


**FIGURE 16-11** Boost converter: (*a*) circuit, (*b*) operating states, and (*c*) steady-state waveforms.

Assumptions made for analysis of buck converter are made here as well. When the MOSFET is ON in response to $q_A(t) = 1$, diode $D_1$ is OFF, and the inductor current increases due to a positive voltage across it (Fig. 16-11*b*). When the MOSFET is switched OFF, the inductor current flows through diode $D_1$ and its magnitude decreases as energy is transferred from the inductor to the output capacitor and load. Instantaneous values of the variables during ON and OFF intervals are

$$\begin{array}{lllll} \text{MOSFET ON} & \upsilon_A = 0; & i_d = 0 & L\dot{i}_L = V_{\text{in}}; & C\dot{\upsilon}_o = -\upsilon_o/R_L \\ \text{MOSFET OFF} & \upsilon_A = \upsilon_o; & i_d = i_L & L\dot{i}_L = V_{\text{in}} - \upsilon_o; & C\dot{\upsilon}_o = i_L - \upsilon_o/R_L \end{array} \tag{16-19}$$

Noting that $\bar{\upsilon}_L = 0$, the averaged pole $A$ voltage is $\bar{\upsilon}_A = V_{\text{in}} = (1 - D)V_o$. In addition, utilizing $\bar{i}_C = 0$ the steady state conversion ratios for the boost converter can be obtained as follows:

$$\frac{V_o}{V_{in}} = \frac{1}{1-D} \quad \text{and} \quad \frac{I_{in}}{I_o} = \frac{1}{1-D} \tag{16-20}$$

From the above equation it is evident that the output voltage is always higher than the input voltage, and conversely, the output current is always lower than the input current by the same ratio. Waveforms of the boost converter variables are shown in Fig. 16-11*c*. The PWM implementation is the same as in the buck converter (Fig. 16-10), with the control objective being regulation of output voltage to a desired value.

The buck and boost converters are capable of either decreasing or increasing the input voltage magnitude but not both. The buck-boost converter is the third basic converter which can be used obtain an output voltage both lower and higher than the input voltage; since the output voltage is usually maintained constant, this implies that the input voltage may be higher or lower than the output voltage. A drawback of the buck-boost converter is that the output voltage polarity is inverted with respect to the input voltage return. The SEPIC converter (single-ended primary inductor converter) provides buck and boost gain without polarity inversion but at the expense of additional components. The Ćuk converter, derived from the buck-boost converter utilizing the duality of current and voltage, is another basic dc-dc converter topology.[4,5] None of the above converters have electrical isolation between the input and the output, however, isolated versions for all of these can be derived.

### 16.3.3 Flyback Converter

Figure 16-12*a* shows the buck-boost converter circuit. Discussion of this converter in its original configuration is omitted here. Instead, its electrically isolated version known as the flyback converter is described. The flyback converter is very common for low-power applications. It has the advantage of providing electrical isolation with low component count. Derivation of the flyback converter from the buck-boost converter is shown in Fig. 16-12*a*. The flyback converter has a coupled inductor instead of an inductor with just one winding. The primary winding is connected to the input while the secondary is connected to the output. The circuit diagram is shown in Fig. 16-12*b*. The coupled inductor is represented by an inductor on the primary and an ideal transformer between the primary and secondary. The coupling coefficient is assumed to be 1. Assuming the primary to secondary turns ratio is 1 : *n*, we get

$$\bar{\upsilon}_L = V_{\text{in}} DT_s - \frac{V_o}{n}(1-D)T_s = 0 \tag{16-21}$$

$$\Rightarrow \frac{V_o}{V_{\text{in}}} = n\frac{D}{1-D} \tag{16-22}$$

$$\Rightarrow \frac{I_{\text{in}}}{I_o} = n\frac{D}{1-D} \tag{16-23}$$

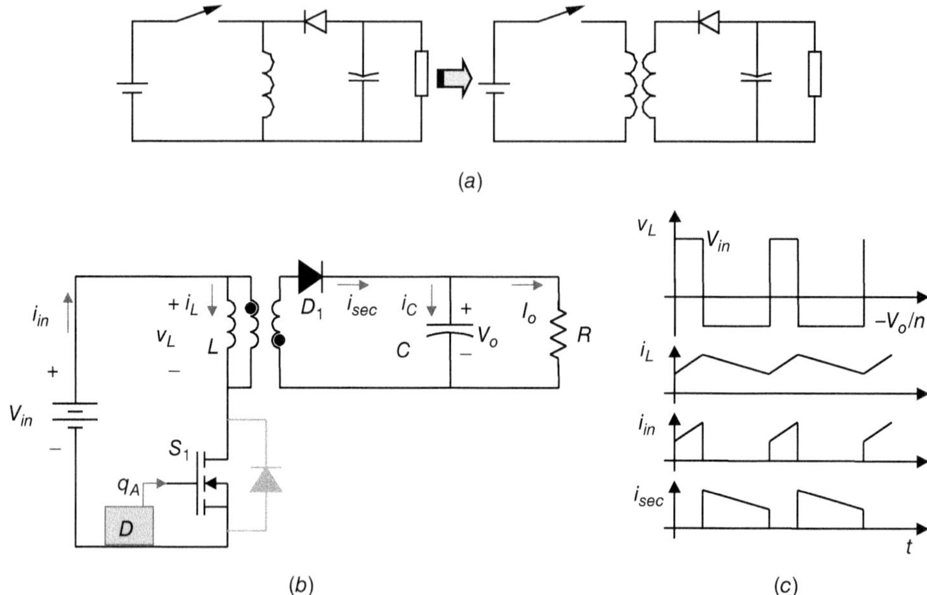


**FIGURE 16-12** Flyback converter: (*a*) derivation from buck-boost, (*b*) circuit, (*c*) steady-state waveforms.

Although the analysis presented here assumes that the inductor current never goes to zero (called *continuous conduction mode* or CCM), it is very common to design flyback converters so that the inductor current does go to zero within each switching cycle. This operation, known as *discontinuous conduction mode* (DCM), leads to simplification of control design for flyback converters.[5] It should be noted that requirement of electrical isolation is not the only reason that a transformer or (coupled inductor) is used. Another important reason is that transformer turns ratio leads to better utilization of power semiconductor devices.

## 16.3.4 Full Bridge DC-DC Converter

Figure 16-13*a* shows the full bridge dc-dc converter which is derived from the buck converter. The bridge circuit formed by switches $S_1$, $S_2$, $S_3$, and $S_4$, converts the input dc voltage to a high-frequency ac (≥ 100 kHz) which is applied to the primary of transformer T1. The high frequency results in a small size for the transformer. After isolation, the high-frequency ac at the secondary of the transformer is rectified by the center tapped diode bridge rectifier formed by $D_1$ and $D_2$, and subsequently filtered by the L and C as in a buck converter. The topology is very popular for power levels greater than 500 W, when isolation is required.

Steady-state operating waveforms for the converter are shown in Fig. 16-13*b*. With switches $S_3$ and $S_4$ OFF, $S_1$ and $S_2$ are turned ON simultaneously for $T_{on} = DT_s/2$, thereby applying a positive voltage across the transformer primary $T_{1,\text{prim}}$ and secondary $T_{1,\text{sec1}}$. During this time diode $D_1$ conducts and a positive voltage appears across the inductor. With all the switches OFF, the transformer primary and secondary voltages are zero and the inductor current splits equally between diodes $D_1$ and $D_2$. In the second half of the switching cycle $S_1$ and $S_2$ are OFF, while $S_3$ and $S_4$ are simultaneously turned on $DT_s/2$. The rectified voltage waveform is similar to that in the buck converter and is at double the switching frequency of each switch. The magnetizing flux in the transformer is bi-directional

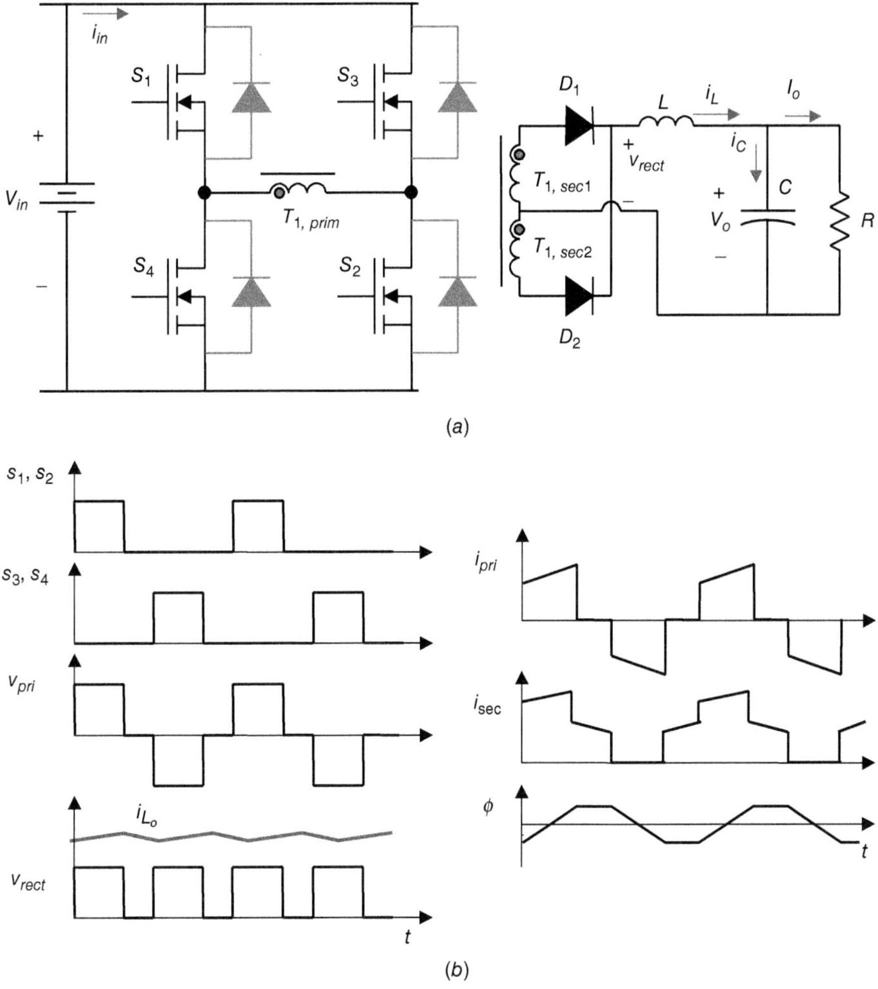


**FIGURE 16-13** Full bridge dc to dc converter: (*a*) circuit, (*b*) steady-state waveforms.

(Fig. 16-13*b*), resulting in better utilization of the core as discussed in the subsection "Transformer Design." The conversion ratio is similar to the buck converter, but scaled by the secondary to primary transformer turns ratio.

## 16.3.5 Other Isolated DC-DC Converters

Several other isolated converters are based on the buck converter. Figure 16-14*a* shows the *forward converter*. The operation and conversion ratio is similar to the buck converter. However, the output voltage is scaled by the transformer turns ratio, an additional winding and diode ($D_R$) are needed to reduce the core flux to zero in each switching cycle, and an additional diode ($D_2$) is required at the output. The forward and flyback converters have unidirectional core flux and are limited to low-power applications. The *push-pull* converter (Fig. 16-14*b*), also derived from the buck converter, is better suited for higher power levels, limited by the voltage rating of the switches required. Details of these converters can be found in Refs. 4 and 5.

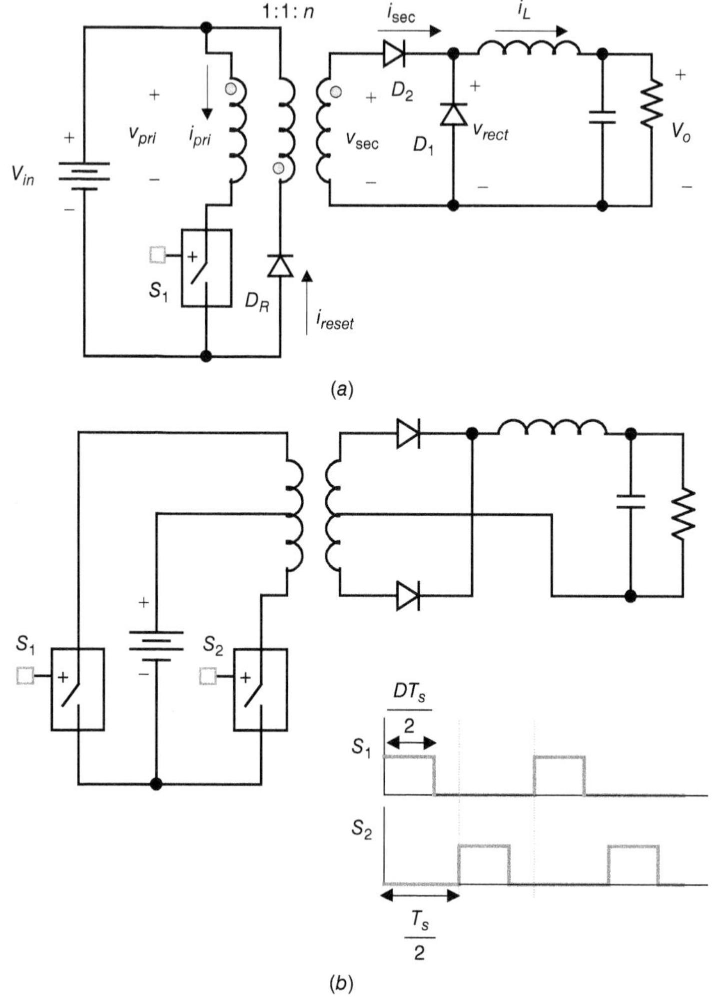


**FIGURE 16-14** Other isolated converters: (*a*) forward converter, (*b*) push-pull converter.

## 16.4 FEEDBACK CONTROL OF POWER ELECTRONIC CONVERTERS

In the last subsection, we saw that the steady-state output of a dc-dc converter, usually the output voltage, is controlled by the duty ratio. To account for changes in load current, input voltage, losses, and nonidealities in the converter, feedback based automatic control is required. Figure 16-15 shows a block diagram of output voltage control for a buck converter. The laplace domain control block diagram is also shown. The sensed output voltage is multiplied by a feedback gain $G_{FB}(s)$ before being compared with a reference value. The error is fed to an appropriate error compensator which generates a control voltage $v_c$ that is converted to duty ratio $d$ by the PWM block.

Toward designing a suitable controller, we will first describe a dynamic model of the power converter and then a simple loop shaping control design method based on input to output bode plots. It is possible to design more complex controllers in order to meet specific requirements, and the

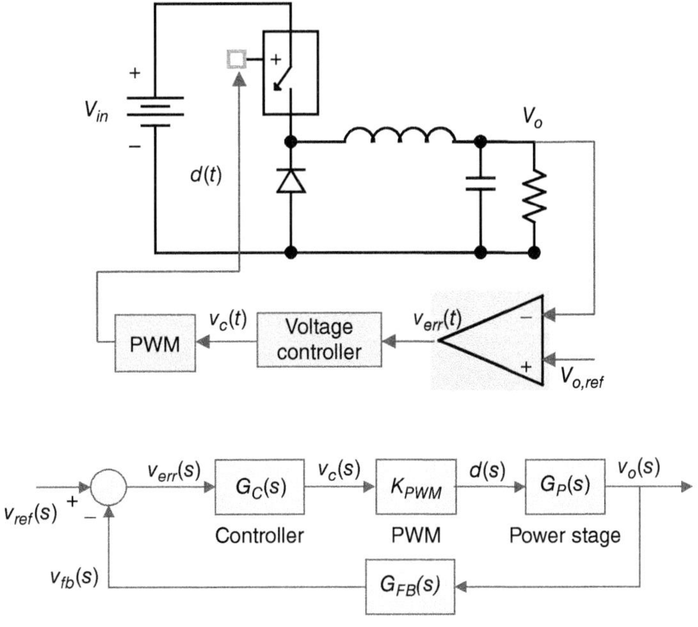


**FIGURE 16-15** Block diagram of output voltage control for a buck converter.

converter topology or operating method may also be modified to make the control design easier (e.g., DCM operation of flyback converters). To keep the explanation simple it is assumed that the converter operates in CCM.

## 16.4.1 Dynamic Modeling

The power converter essentially consists of the PWM block and the power stage itself. The feedback gain $G_{FB}$ is usually a constant. The PWM block shown in Fig. 16-10 converts the input control voltage $\bar{v}_c(t)$ to a duty ratio $d(t)$. From geometrical considerations

$$d(t)/\bar{v}_c(t) = 1/\hat{V}_{\text{ramp}} = K_R \tag{16-24}$$

where $\hat{V}_{\text{ramp}}$ is the peak value of the ramp $v_{\text{ramp}}(t)$. The power stage transfer function from $d(s)$ to $\bar{v}_o(s)$ can be derived using one of the methods stated below.

***Dynamics of Averaged Quantities.*** As stated earlier, the bi-positional switch approach and averaging are valid for analyzing low-frequency dynamics ($< f_s/5$) of the power converter. Unlike steady-state analysis, under dynamic conditions $\bar{v}_L \neq 0$ and $\bar{i}_C \neq 0$. Averaging the instantaneous state Eq. (16-25) over one switching cycle, dynamics of the averaged inductor current and capacitor voltage in a buck converter are

$$L\dot{\bar{i}}_L = d(t) \cdot V_{\text{in}} - \bar{v}_o \tag{16-25}$$

$$C\dot{\bar{v}}_o = \bar{i}_L - \bar{v}_o/R_L \tag{16-26}$$

Here the time varying duty cycle $d(t)$ is the control input and the averaged output voltage $\bar{v}_o$ is the output that has to be regulated. The situation for the buck converter is simple because the model described by Eqs. (16-25) and (16-26) is linear if $V_{\text{in}}$ and $R_L$ are assumed constant, for which case exact transfer functions describing large signal behavior can be derived. For boost and buck-boost converters, the averaged state equations involve terms with multiplications of $d(t)$ with a state variable. Thus, the model has to be linearized, and small signal dynamics obtained at different operating points are utilized for linear control design. It is of course possible to design large signal control based on the nonlinear model at the expense of mathematical complexity.[22] However, ease of design, and simple cost-effective implementation has made linear design the preferred method in most power electronic converters in the low to medium power range.

***Averaged Circuit Representation.*** Instead of writing averaged state equations explicitly as in Eqs. (16-25) and (16-26), an averaged circuit representation of the bi-positional switch can be derived and substituted in different converter circuits.[23–25] As shown in Fig. 16-16*a* the bi-positional switch

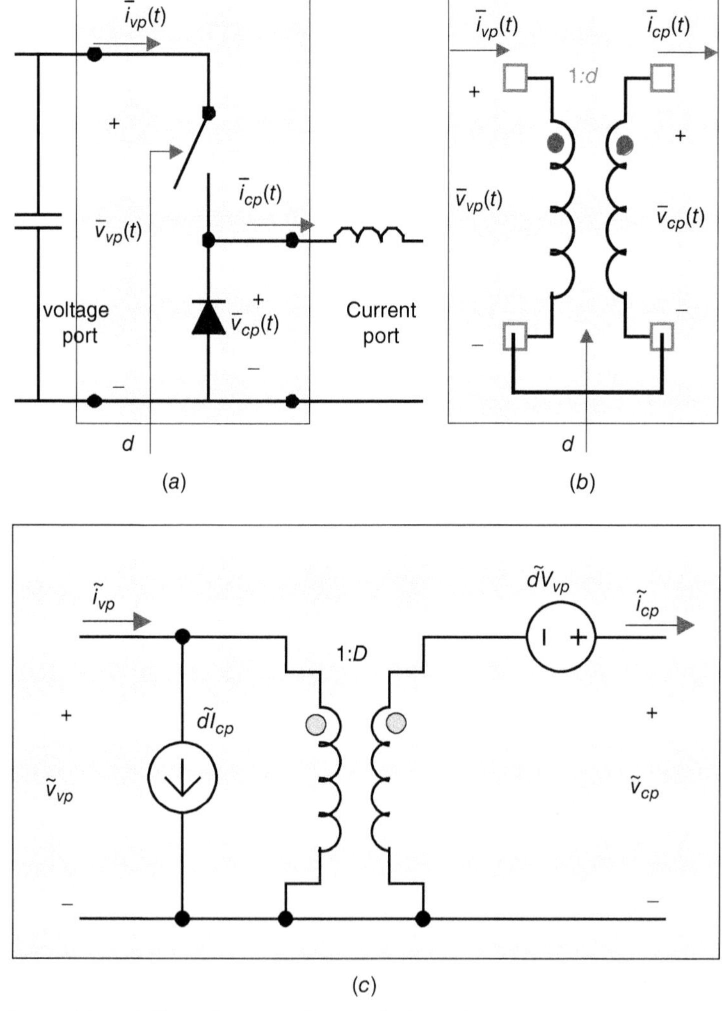


**FIGURE 16-16** Bi-positional switch: (*a*) two-port network, (*b*) average representation, (*c*) small signal model.

can be considered as a two-port network with a voltage port ('$_{vp}$') at the input and a current port ('$_{cp}$') at the output. The average voltage and currents of the two ports are related as

$$\bar{v}_{cp}(t) = d(t) \cdot \bar{v}_{vp}(t) \tag{16-27}$$

$$\bar{i}_{vp}(t) = d(t) \cdot \bar{i}_{cp}(t) \tag{16-28}$$

The relations in the above equations correspond to those of an ideal transformer with turns ratio of $1 : d(t)$. Thus, for analysis purposes, the bi-positional switch can be modeled as an ideal transformer whose turns ratio $d(t)$ can be controlled as shown in Fig. 16-16*b*. This representation is extremely useful in conjunction with circuit simulators which can perform operating point (dc bias) calculations, linearization, and ac analysis. Parasitic effects, like series resistances of inductors and capacitors, can be easily incorporated in the averaged circuit model. Circuit simulators like SPICE,[26] Saber, and Simplorer are commonly used for this purpose. The small signal circuit representation of the averaged circuit obtained via linearization is shown in Fig. 16-16*c*. Quantities in upper case indicate operating point values, while the quantities with a '~' indicate small signal perturbations about the operating point. This representation can be utilized to derive small signal transfer functions using circuit analysis techniques.

## 16.4.2 Control Design

For a dc-dc converter the main control objectives are stability, zero steady state error, specified transient response to step change in reference and step change in disturbance inputs (load and input voltage), and robustness to parametric changes. Transfer functions of different components – $K_R$, $G_{PS}(s)$, and $G_{FB}(s)$—are obtained as described above. The error compensator is then designed so that the open-loop transfer function $G_{OL}(s)$ has a specified gain cross-over frequency and phase margin. The gain cross-over frequency determines the response time of the controlled converter to changes in reference voltage and load current. Phase margin is usually in the range of 45° to 60° depending on the overshoot tolerable. Details on relation between gain cross-over frequency, phase margin, and transient response can be found in any textbook on linear control.[27]

An integrator (pole at origin) is added in $G_c(s)$ to obtain zero steady-state error. Zeroes are added at appropriate locations to obtain required phase margin. The dc gain of $G_c(s)$ is adjusted to achieve the required cross-over frequency. Finally, to improve noise immunity, poles may be added for fast roll-off of the gain after the cross-over frequency. A systematic loop shaping procedure along with implementation suited to dc-dc converters is described in Ref. 28.

***Example: Voltage Control of a Buck Converter.*** A controller has to be designed to regulate the output voltage of a buck converter to a constant value. The specifications, parameters, and control requirements are listed in Table 16-2. Using the methods described above, the duty ratio to output voltage transfer function can be derived to be

$$\frac{\tilde{v}_0(s)}{\tilde{d}(s)} = \frac{V_{in}(1+sCR_{\mathrm{ESR}})}{1+s[CR_{\mathrm{ESR}}+L/R_L]+s^2LC[1+R_{\mathrm{ESR}}/R_L]} \tag{16-29}$$

**TABLE 16-2** Control Design Example

| Specifications | | Parameters | | Requirements | |
|---|---|---|---|---|---|
| Input voltage | 20 to 30 V dc | $L$ | 75 $\mu$H | Cross-over frequency | 8 kHz |
| Output voltage | 15 V dc | $C$ | 47 $\mu$F | Phase margin | >60° |
| Max. load current | 5 A | $R_{ESR}$ | 0.3 Ω | | |
| Switching frequency | 200 kHz | $K_R$ | 1 | | |

The transfer function has a complex pole pair due to the L-C filter, and a left half zero due to the ESR of the output capacitor. The compensator designed in accordance with the aforementioned considerations is

$$G_c(s) = \frac{K_c(1+s/\omega_z)^2}{s(1+s/\omega_{p1})(1+s/\omega_{p2})} \tag{16-30}$$

where $K_c = 2011$, $\omega_z = 2\pi \times 2556$ rad/s, $\omega_{p1} = 2\pi \times 11.3e3$ rad/s, and $\omega_{p1} = 2\pi \times 80e3$ rad/s. Representation of the controlled converter using ORCAD PSpice[29] is shown in Fig. 16-17*a*. The bi-positional switch has been replaced by a two-port network which models a transformer with controllable turns ratio. The compensated and uncompensated transfer functions obtained using ac analysis are shown in Fig. 16-17*b*. The gain cross-over frequency and the corresponding phase of the compensated transfer function are indicated. Figure 16-17*c* shows dynamic response of the controlled converter when a step change in load is applied at 0.2 ms.

## 16.4.3 Current Mode Control

In most converters, the inductor current is an *internal* state of the power converter. Changes in the input voltage and duty ratio are first reflected in the inductor current and subsequently in the output voltage. Thus, controlling the inductor current can lead to better performance. Figure 16-18 shows a cascaded control structure where the internal current controller is about an order of magnitude faster than the outer voltage loop. The average value of the inductor current is controlled to a reference that is generated by the error compensator for the voltage control loop. The current controller is designed using the transfer function from the duty ratio to the inductor current. For voltage controller, the current control loop is assumed to be ideal, that is, $i_L = i_{L,\text{ref}}$; this is justified since the current controller is much faster than the voltage controller. The voltage compensator is then designed utilizing the inductor current to the output voltage transfer function. In a buck or buck-derived topology, the average inductor current is equal to the load current. Thus, fast control over the inductor current effectively mitigates steady state and transient variations in the input voltage without their affecting the output voltage. This current control method is called *average current control.*

Another popular method is *peak current mode control.* In this method the peak value of the inductor current is controlled to the reference value (generated by the voltage control loop) in each switching cycle. Peak current mode control has the additional advantage of balancing the positive and negative flux excursions in transformer isolated topologies like full-bridge and push-pull. However, peak current mode control requires extra precautions to avoid subharmonic and chaotic operation.[4,5,30]

## 16.4.4 Other Control Techniques

The basic modeling method described above is applicable to other types of converters (like dc-ac) as long as low-frequency behavior is being studied. Dynamics of the filter elements may of course be different. In dc-ac converters the control objective is usually to track a moving reference (e.g., sinusoidal control voltage or current). Using a stationary to rotating frame transformation, commonly called the *abc to dq transformation*, the tracking problem can often be reduced to a regulation problem. If current control is implemented in the stationary frame, where the control objective is to track a sinusoidally varying reference, then either average current control or *hysteretic current control* is used. In hysteretic current control, the current is maintained in a band about the reference value. If the current error is below the lower limit of the band a positive voltage is applied across the inductor (switch ON in a buck converter) to increase the current. To reduce the current a negative or zero

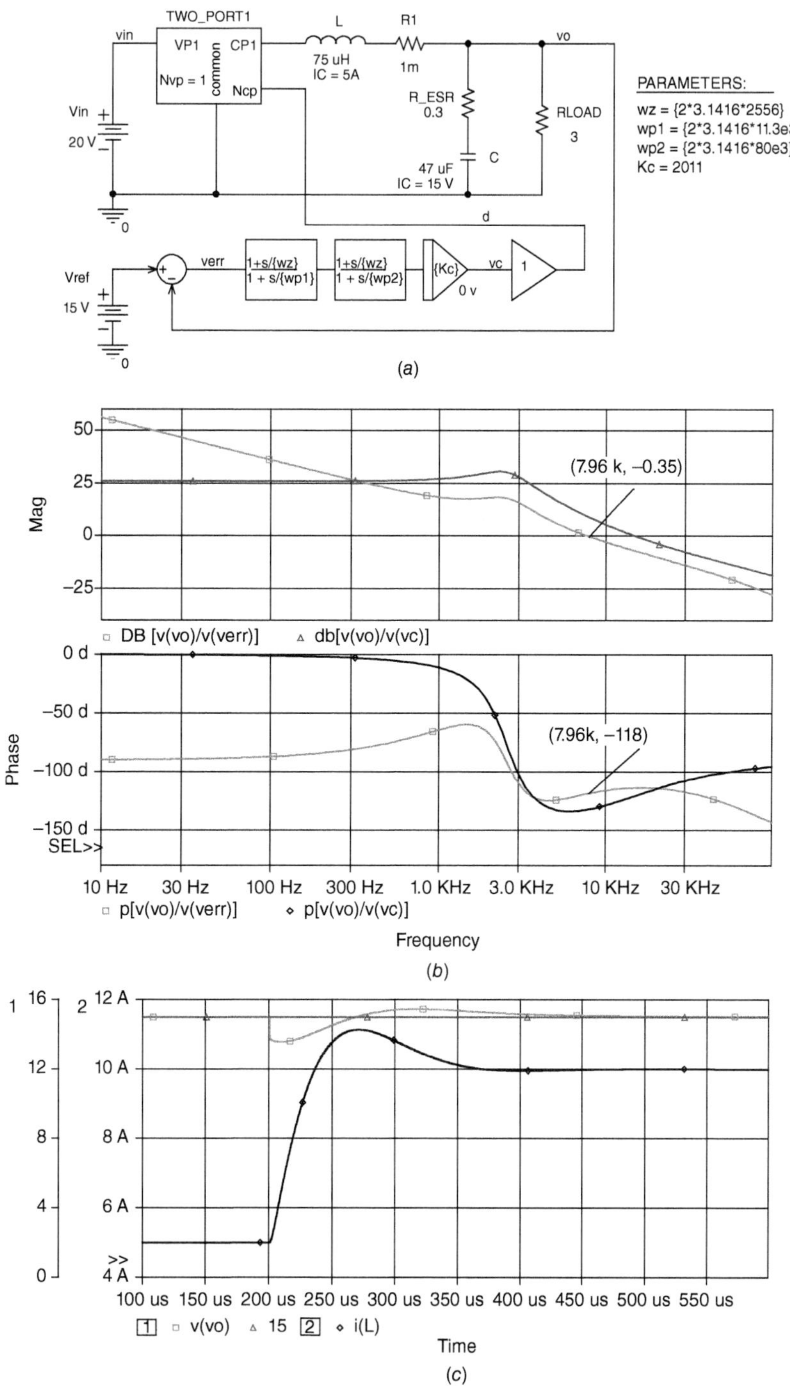


**FIGURE 16-17** Controlled buck converter: (*a*) averaged representation using PSpice, (*b*) open-loop bode plots, (*c*) transient response.

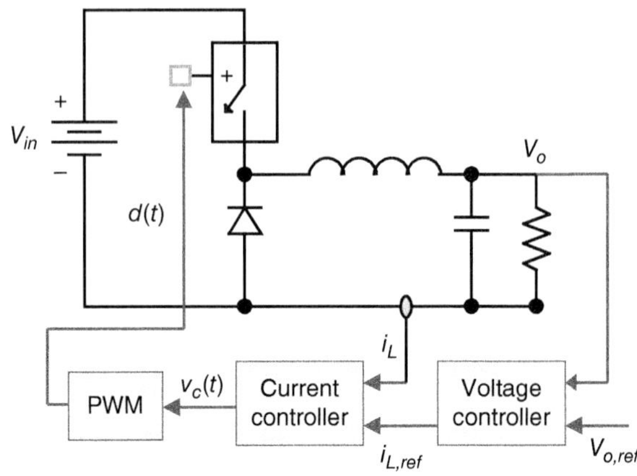


**FIGURE 16-18** Average current control of a buck converter.

voltage is applied across the inductor. With hysteretic control the rise and fall times are only limited by the power components of the converter. However, it has the disadvantage of variable switching frequency, whose instantaneous value depends on a combination of several factors. Several other techniques have been proposed for control of power electronic converters: sigma-delta, sliding mode, dead beat, etc. Details of these control techniques can be found in Ref. 31.

Control design described above is relevant for analog implementation. Due to advances in digital system implementation in terms of computation capability, power consumption, and cost, there has been significant effort in a complete digital implementation of the power converter control and PWM generation mechanisms. So far, digital control is only used in high-power converters, where the overall cost justifies the cost of digital and interface components. However, there is significant effort in extending the benefits of digital control to lower-power converters.[32–34] The benefits include reduced sensitivity to compensator component variation, possibilities to adapt the controller based on operating point or variation in power converter component values over lifetime or production spreads, and inclusion of sophisticated methods that rely on more direct computation. In digital implementations, predictive current control is commonly used to reduce the effect of sensing and computational delays.

## 16.5 DC TO AC CONVERSION: INVERSION

Dc to ac converters constitute a significant portion of power electronic converters. These converters, also called *inverters*, are used in applications such as electric motor drives, uninterruptible power supplies (UPS), and utility applications such as grid connection of renewable energy sources. Inverters for single-phase ac and three-phase three-wire ac systems are described in this subsection.

### 16.5.1 Single-Phase AC Synthesis

In an ac system both the voltage and the current should be able to reverse in polarity. Further, the voltage and current polarities may or may not be the same at a given time. Thus, a dc to ac converter implementation should be able to output a voltage independent of current polarity. In the full bridge dc to dc converter shown in Fig. 16-19*a* the primary circuit consisting of four controlled switches, also called *H-bridge*, has two bi-positional switch implementations. Each bi-positional switch has bidirectional current capability but only positive output voltage ($v_{AN}$ >, $v_{BN} > 0$). However, based

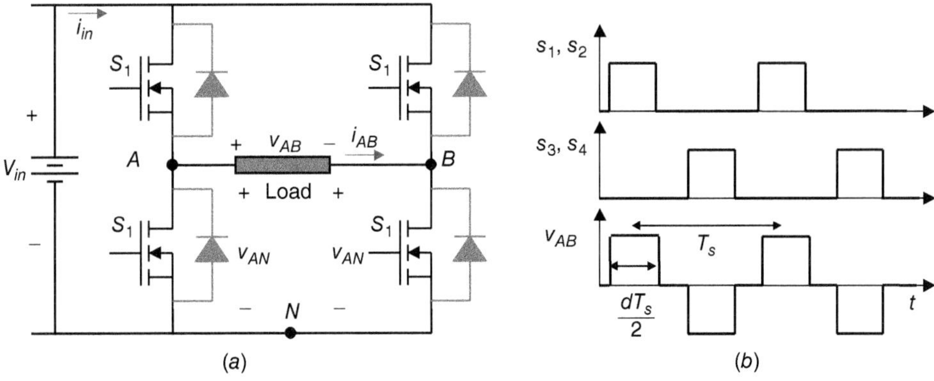


**FIGURE 16-19** Single-phase inverter: (*a*) circuit, (*b*) quasi-square wave synthesis.

on the duty cycles, the difference of the outputs, $V_{AB} = v_{AN} - v_{BN}$, can reverse in polarity. Thus the H-bridge is used for synthesizing single-phase ac voltage from a dc voltage.

***Quasi-Square Wave Inverter.*** The simplest form of dc to ac conversion, albeit with poor quality, is synthesis of quasi-square wave ac instead of a pure sine wave. Diagonally opposite switches in the H-bridge are turned on simultaneously. The pulse width of each pair is controlled to adjust the magnitude of the fundamental component, while the switching frequency is equal to the required output frequency. The synthesized voltage waveform is shown in Fig. 16-19*b*. The peak value of fundamental and harmonic components are

$$V_{AB,n} = \frac{4V_{in}}{n\pi}\sin(n\pi d/2) \qquad n \text{ odd} \tag{16-31}$$

where $d$ is the duty ratio and $n$ the harmonic number. This converter is widely used for low cost low power UPS applications where the voltage waveform quality is not important. Incandescent lighting, universal input motors, and loads with a diode bridge or power factor corrected front end (discussed in Sec. 16.8) are not affected by the voltage waveform quality. The load current, $i_{AB}$, has harmonics based on the load characteristics. Sometimes an LC filter is added at the output to reduce the harmonic content. Low power low cost inverters such as those used to generate ac from 14 V dc in automobiles usually have quasi-square wave voltage output.

***Single-Phase Sinusoidal Voltage Synthesis.*** For applications requiring low voltage and current distortion high-frequency PWM is utilized to generate a sinusoidally varying average voltage. The power converter used is the H-bridge shown in Fig. 16-19*a*. The duty ratio for each bipositional-switch, also called one leg of the inverter, is varied sinusoidally. The switching signals are generated by comparison of a sinusoidally varying control voltage with a *triangle* wave as shown in Fig. 16-20. Equations relating the control voltages, duty ratios, and the averaged output voltages are as follows:

$$v_c = \hat{V}_c \cdot \sin(\omega_m t) \tag{16-32}$$

$$v_{cA}(t) = v_c = \hat{V}_c \cdot \sin(\omega_m t) \tag{16-33}$$

$$v_{cB}(t) = -v_c = -\hat{V}_c \cdot \sin(\omega_m t) \tag{16-34}$$

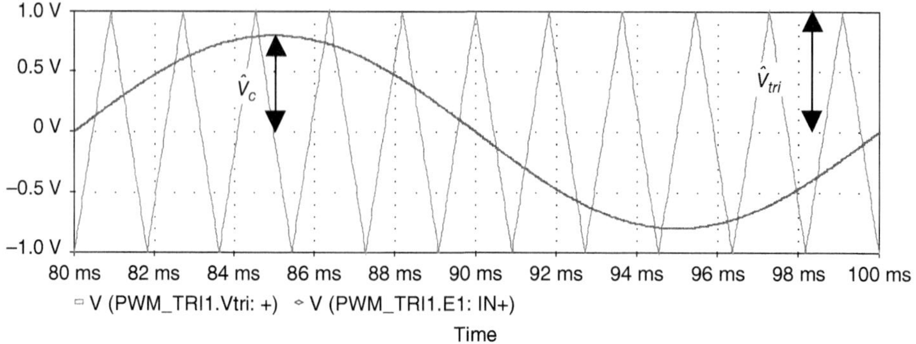


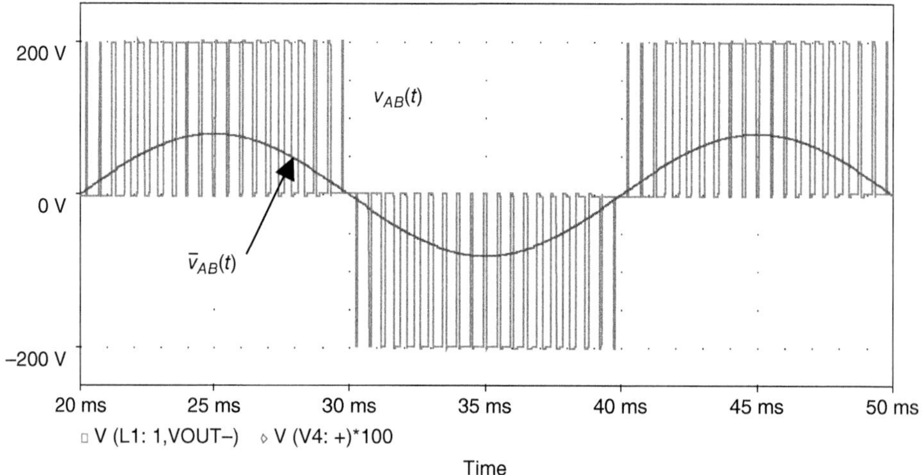


**FIGURE 16-20** Single-phase sinusoidal ac synthesis waveforms.

$$d_A(t) = \frac{1}{2}\left(1 + \frac{v_{cA}}{\hat{V}_{\text{tri}}}\right) = \frac{1}{2}[1 + m \cdot \sin(\omega_m t)] \tag{16-35}$$

$$d_B(t) = \frac{1}{2}\left(1 + \frac{v_{cB}}{\hat{V}_{\text{tri}}}\right) = \frac{1}{2}[1 - m \cdot \sin(\omega_m t)] \tag{16-36}$$

$$\bar{v}_{AN}(t) = d_A(t) \cdot V_{\text{in}} = \frac{1}{2}[1 + m \cdot \sin(\omega_m t)] \cdot V_{\text{in}} \tag{16-37}$$

$$\bar{v}_{BN}(t) = d_B(t) \cdot V_{\text{in}} = \frac{1}{2}[1 - m \cdot \sin(\omega_m t)] \cdot V_{\text{in}} \tag{16-38}$$

$$\bar{v}_{AB}(t) = [d_A(t) - d_B(t)] \cdot V_{\text{in}} = m \cdot V_{\text{in}} \cdot \sin(\omega_m t) = \underbrace{(V_{\text{in}}/\hat{V}_{\text{tri}})}_{k_{\text{PWM}}} \cdot v_c(t) \tag{16-39}$$

Here $\hat{V}_c$ and $\hat{V}_{\text{tri}}$ are peak values of control voltage and the triangle wave respectively, $m = \hat{V}_c/\hat{V}_{\text{tri}} \in [0, 1]$ is the modulation index, $\omega_m = 2\pi f_m$ is the angular frequency of the sinusoid to be synthesized, while

$d_A(t)$ and $d_B(t)$ are duty ratios of switches $S_1$ and $S_3$, respectively. In Eq. (16-39) $k_{PWM}$ may be regarded as the gain of the power converter that amplifies the control signal $v_c(t)$ to the average output voltage $\bar{v}_{AB}(t)$. The maximum peak value of the output voltage, obtained for $m = 1$, is $V_{in}$. This is significantly lower than that obtainable with the quasi-square wave inverter ($4V_{in}/\pi$). However, harmonics in the output voltage are significantly reduced and are at much higher frequencies: $k \cdot f_s \pm l \cdot f_m$, where $k$ and $l$ are integers such that $k + l$ is odd.[4] The switching frequency is much higher than the output frequency $f_m$, which has a maximum value of about 50/60 Hz for standard applications or 400 Hz for aerospace applications.

If the load is inductive, such as a motor, the current harmonics are much lower than the voltage harmonics. For several applications maximum harmonic content for the voltage and current output from the inverter is specified. In these cases an L-C filter similar to the buck converter is used. Depending upon the application a two-stage L-C filter or a two-stage notch filter (to suppress the dominant switching frequency harmonics) may be used. Further, it has to be ensured that when connected to the load, the filter is adequately damped by a combination of passive selection and the control loop. This aspect is particularly important for line connected applications where the inverter is supplying power to the utility grid.

Equation (16-37) can be rewritten as

$$\bar{v}_{AN}(t) = d_A(t) \cdot V_{in} = \frac{V_{in}}{2} + \frac{k_{PWM}}{2} \cdot v_c(t) \tag{16-40}$$

This clearly shows that on an average basis the "neutral point" for the output of one inverter leg is $V_{in}/2$ above "N," that is, at the mid-point of the input dc bus. Thus using the same H-bridge a split-phase ac (two ac voltages 180° out of phase with a common return) can be generated if the center point of the dc bus is available as the neutral connection for the output. This type of configuration is commonly used in generating ±120/240 from the same inverter. Furthermore, using three legs instead of two the converter can generate three-phase voltages with a neutral connection, with the flexibility that the three phases may be loaded independently. Common applications are inverters for interfacing photovoltaic systems to the utility grid and exporting power from vehicles.

## 16.5.2 Three-Phase AC Synthesis

The last observation in the previous subsection leads us to three-phase inverters without a neutral connection. The circuit consists of three legs, one for each output with a common dc link as shown in Fig. 16-21*a*. Using sine triangle PWM with control voltages offset by 120° (instead of 180° as in the single-phase case) we obtain:

$$v_{c,k}(t) = \hat{V}_c \cdot \sin(\omega_m t) \tag{16-41}$$

$$\bar{v}_{kN}(t) = \frac{V_{in}}{2} + \frac{k_{PWM}}{2} \cdot v_{c,k}(t) \tag{16-42}$$

$$k = A, B, C \tag{16-43}$$

The zero sequence component of the output voltages, $v_z = (v_{AN} + v_{BN} + v_{CN})/3 = V_{in}/2$, does not appear in the line-to-line voltages, and since there is no neutral connection to the inverter, zero sequence currents do not flow.

The maximum peak value of the output line-to-line voltages is $\hat{V}_{LL} = (\sqrt{3}/2)V_{in}$. Using square wave inversion, similar to that for the single-phase case, we can obtain higher magnitude for the fundamental component of the output voltages at the cost of adding harmonics. However, if, instead of all the harmonics, only the fundamental and those harmonics of the square wave that contribute zero sequence component (triplen harmonics) are retained, the output voltage amplitude increases without adding harmonics to the line-to-line voltages and the line currents. Usually, addition of the third

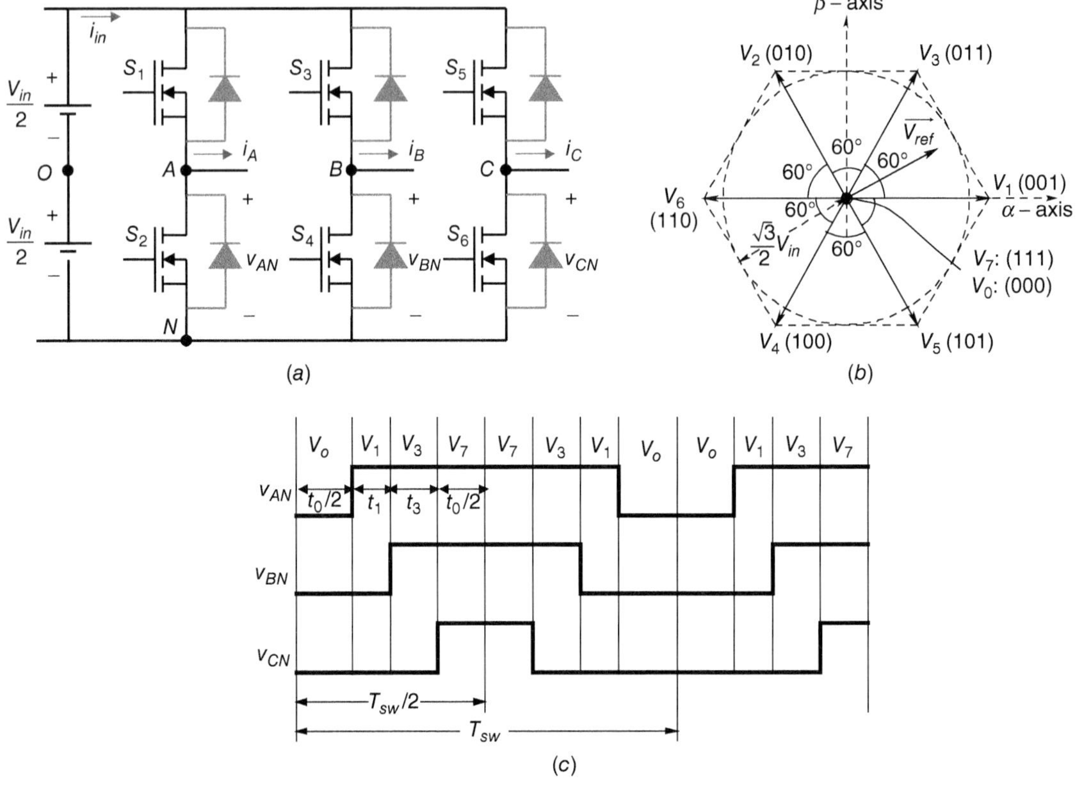


**FIGURE 16-21** Three-phase ac synthesis: (*a*) converter, (*b*) output voltage vectors, (*c*) instantaneous waveforms.

harmonic component is sufficient.[35, 36] As described in Refs. 37 and 38 the most effective method is to add the following zero sequence component to the control voltages for each phase:

$$v_{cz}(t) = -\frac{1}{2}\left\{\max[v_{cA}(t), v_{cB}(t), v_{cC}(t)] + \min[v_{cA}(t), v_{cB}(t), v_{cC}(t)]\right\} \tag{16-44}$$

In terms of output voltage generation, this is equivalent to space vector modulation (SVM).

### 16.5.3 Space Vector Modulation

This method has become extremely popular for three-phase inverters in the low to medium power range. A very brief description will be presented here and details can be found in Refs. 31, 35, and 37.

For three-phase systems with no zero sequence component, that is, $v_z = (v_{AN} + v_{BN} + v_{CN})/3 = 0$, the three-phase quantities are linearly dependent and can be transformed to a two-phase orthogonal system commonly called the $\alpha\beta$ system. Quantities in the $\alpha\beta$ system can be represented by complex numbers and as two-dimensional vectors in a plane, called *space vectors*. The transformation from the *abc* to $\alpha\beta$ quantities is given by

$$\vec{v}_{\alpha\beta}(t) = v_\alpha(t) + j\cdot v_\beta(t) = e^{j0}\cdot v_a(t) + e^{j2\pi/3}\cdot v_b(t) + e^{j4\pi/3}\cdot v_c(t) \tag{16-45}$$

With negative sequence components absent, $\alpha$ and $\beta$ components of steady state sinusoidal *abc* quantities are also sinusoids with constant amplitude and a 90° phase difference between them. Under transient conditions they are arbitrary time varying quantities. Thus, for balanced sinusoidal conditions, the space vector $\vec{v}_{\alpha\beta}(t)$ rotates in counter clockwise direction with angular frequency equal to frequency of the *abc* voltages, and describes a circle of radius$(3/2)\,\hat{V}_{ph}$, $\hat{V}_{ph}$ being the peak of the phase voltage.

The instantaneous output voltages of the three-phase inverter shown in Fig. 16-21*a* can assume eight different combinations based on which of the six MOSFETs are on. The space vectors for these eight combinations are shown in Fig. 16-21*b*. For example, vector $V_4$ denoted by (100) corresponds to switch states $v_{AN} = V_{in}$, $v_{BN} = 0$, and $v_{CN} = 0$. The vectors $V_0(000)$ and $V_7(111)$ have zero magnitude and are called *zero vectors.*

Synthesis utilizing the idea of space vectors is done by dividing one switching time period into several time intervals, for each of which a particular voltage vector is output by the inverter. These time intervals and the vectors applied are chosen so that the average over one switching time period is equal to the desired output voltage vector. For the reference voltage vector $\vec{v}_{ref}$, shown in Fig. 16-21*b*, the nonzero vectors adjacent to it ($V_1$ and $V_3$), and the zero vectors ($V_0$ and $V_7$) are utilized as shown in Fig. 16-21*c*. Relative values of time intervals $t_1$ and $t_3$ determine the direction, while ratio of $t_0$ to the switching time period determines the magnitude of the output vector synthesized. The formulae for time intervals are as follows:

$$t_1 = \frac{T_{sw}}{2}\left(\frac{|\vec{v}_{ref}|}{V_{in}}\frac{\cos(30°+\theta)}{\cos(30°)}\right) \tag{16-46}$$

$$t_3 = \frac{T_{sw}}{2}\left(\frac{|\vec{v}_{ref}|}{V_{in}}\frac{\cos(90°-\theta)}{\cos(30°)}\right) \tag{16-47}$$

$$t_0/2 = T_{sw}/2 - (t_1 + t_3) \tag{16-48}$$

where $\theta$ is the angle of the vector $\vec{v}_{ref}$ measured from the $\alpha$–axis. The maximum obtainable average vector lies along the hexagon connecting the six nonzero vectors. As stated earlier, balanced three-phase sinusoidal quantities describe a circle in the $\alpha\beta$ plane. Thus, to synthesize distortion free and balanced three-phase sinusoidal voltages, the circle must be contained within the hexagon, that is, with a maximum radius of $\sqrt{3}/2 \cdot V_{in}$. This gives the maximum peak value of line-to-line voltage obtained with SVM as $\hat{V}_{LL} = V_{in}$. This is significantly higher than that obtained using sine triangle PWM: $\sqrt{3}/2 \cdot V_{in}$. Further, the sequence and choice of vectors applied can be optimized to minimize number of switchings and ripple in the resulting currents.[39] There are several variations of SVM, each suited to a different application. SVM can be easily implemented digitally using microcontrollers or digital signal processors (DSPs), and is advantageous in control of three-phase ac machines using *vector control* and *direct torque control* (DTC).[40–44] Experimental waveforms for an SVM inverter are shown in Fig. 16-22.

## 16.5.4 Multilevel Converters

The converter topologies described so far are based on a two-level converter leg (bi-positional switch), where the output voltage of each leg ($v_{AN}$) can be either zero or $V_{in}$. The converters are therefore called two-level converters. In two-level converters, all the switches have to block the full dc bus voltage ($V_{in}$). For high-power applications IGBTs and GTOs are used as the semiconductor switches. These have higher voltage and current ratings and lower on-state voltage drop compared to power MOSFETs, but cannot switch as fast. In some applications like some motor drives and utility applications, even the voltage ratings of available IGBTs and GTOs is not sufficiently high. Simple series connection, to achieve a higher blocking voltage, has problems of steady-state and dynamic voltage sharing. Moreover, due to the low switching frequency of high-power switches, the output voltage and current quality deteriorates. These issues are addressed by multilevel converters. In a multilevel converter,[45,46] the output of each

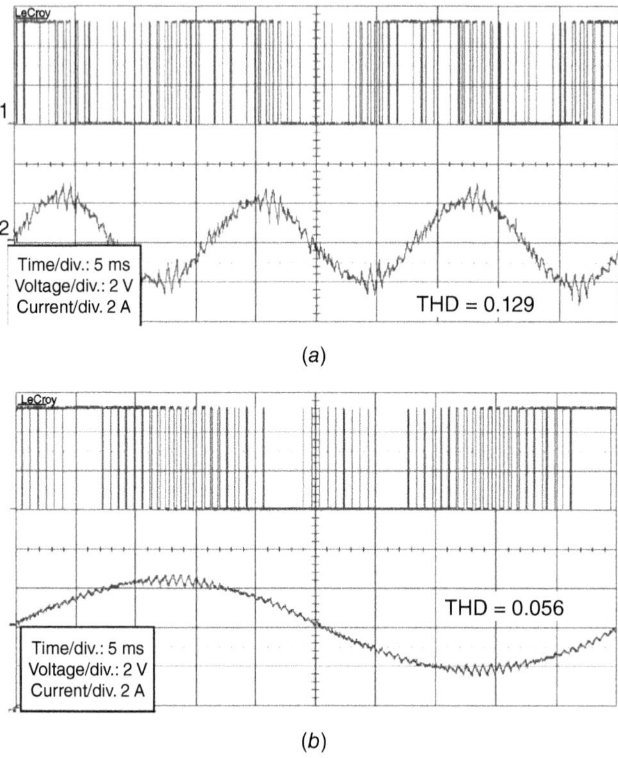


**FIGURE 16-22** Experimentally measured PWM signal and line current for one phase of a three-phase SVM inverter: (*a*) 60 Hz synthesis; (*b*) 20 Hz synthesis.

phase leg can attain more than two levels leading to improved quality of the output voltage and current. The circuit comprising each leg and its proper operation ensure that voltage blocked by the switches reduces as the number of levels is increased. In addition, multilevel converters are modular to some extent, thereby making it easy to scale voltage ratings by increasing the number of "cells."

***Multilevel PWM.*** For two-level PWM, comparison of the control voltage with a triangle wave generates the switching signal for the top switch, while the bottom switch is controlled in complement to the top switch. Each of these two states corresponds to the two levels of the output voltage. For multilevel converters, there are more than two effective switch states, each of which corresponds to an output voltage level. For example, in a three-level converter there are three effective states $q(t) = 0, 1, 2$, corresponding to output voltage levels $v_{AN}(t) = 0, V_{in}/2, V_{in}$. The control voltage $v_c(t)$ is compared with two triangle waves to obtain two switching signals $q_1(t)$ and $q_2(t)$, and the effective switching signal can be obtained as $q(t) = q_1(t) + q_2(t)$ as shown in Fig. 16-23. The output voltage is then given by $v_{AN} = q(t) \cdot (V_{in}/2)$. Switching signals for the individual switches are derived using $q(t)$ and the circuit topology. For the waveforms in Fig. 16-23, $f_s = 60$ Hz and $V_{in} = 2$ kV. Since the $v_{AN}$ waveform is closer to desired sinusoid in the three level case, the output voltage has lower THD even if the switching frequency is low. For three-phase converters, space-vector-based PWM can be used for generating the switching signals,[47] the advantage in the multilevel case compared to the two-level case being the significantly higher number of output voltage vectors.

***Multilevel Converter Topologies.*** The chief multilevel converter topologies are diode clamped, flying capacitor, and cascaded full bridge.

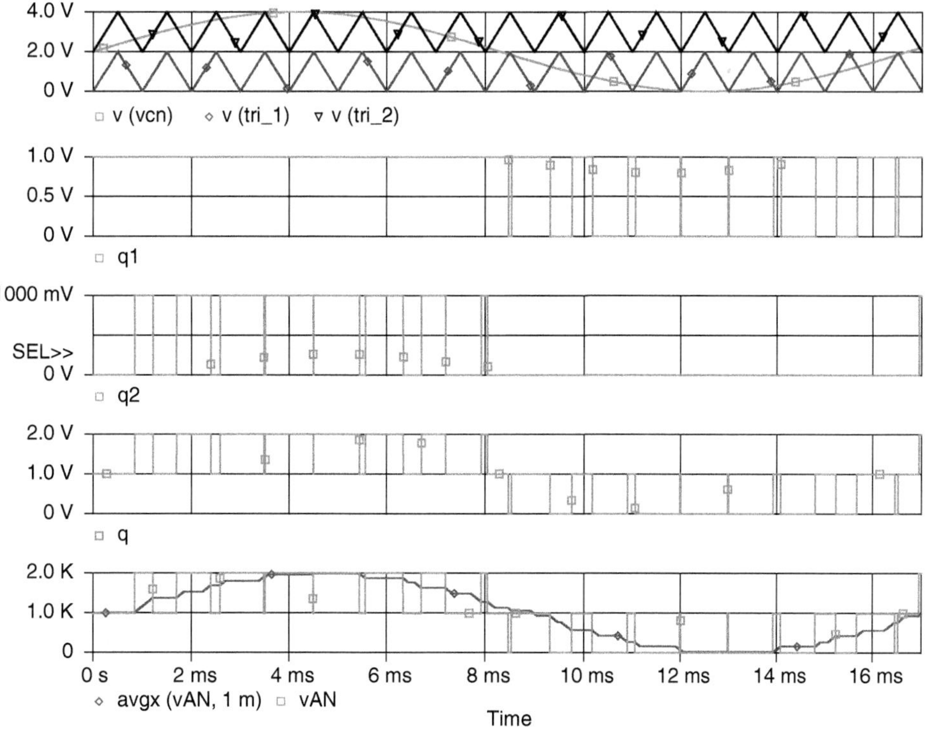


**FIGURE 16-23** Multilevel triangle comparison.

*Diode Clamped Converter.* Figure 16-24*a* shows one phase leg of a three-level diode-clamped converter.[48] The input dc bus is split by means of capacitors. Pairs of switches are turned on to obtain three different voltage levels for the output voltage $v_{AN} = 0$, $V_{in}/2$, $V_{in}$ as shown in Fig. 16-24*c*. It is evident that this circuit acts like a tri-positional switch connecting the output to one of three positions of the input dc bus. The minimum voltage at point "*b*1," and the maximum voltage at point "*b*2," is clamped to $V_{in}/2$ by the blocking diodes $D_{b1}$ and $D_{b2}$, respectively. Thus, all the switches have to block $V_{in}/2$ during their off state. This topology can be extended to more number of levels. However, it is eventually limited by the voltage rating of blocking diodes, which have to block increasing voltages as the number of levels is increased. One-phase leg of a five-level version is shown in Fig. 16-24*b*.

*Flying Capacitor Converter.* Figure 16-25 shows the topology of a three-level flying capacitor converter. The basic idea here is that the capacitor *C* is charged to half the input dc voltage by appropriate control of the switches. The capacitor can then be inserted in series with the output voltage, either adding or subtracting $V_{in}/2$, and thereby giving three output voltage levels.

*Cascaded Full Bridge Converters.* In this scheme,[49] single-phase H-bridges shown in Fig. 16-19*a* are connected in series at the output to form one single-phase circuit as shown in Fig. 16-26*a*. Three separate circuits are required for a three-phase implementation. A delta connection of cascaded converters is shown in Fig. 16-26*b*. Since all the H-bridges are same, the circuit is modular and can be scaled by adding more H-bridges. However, dc sources at the input of all H-bridges have to be isolated from each other. It is also possible to combine different types of H-bridges—IGBT-based fast switching type and GTO-based slower switching type—or have different dc bus voltage magnitudes in different bridges to optimize losses or increase effective number of levels.[50] One example of the

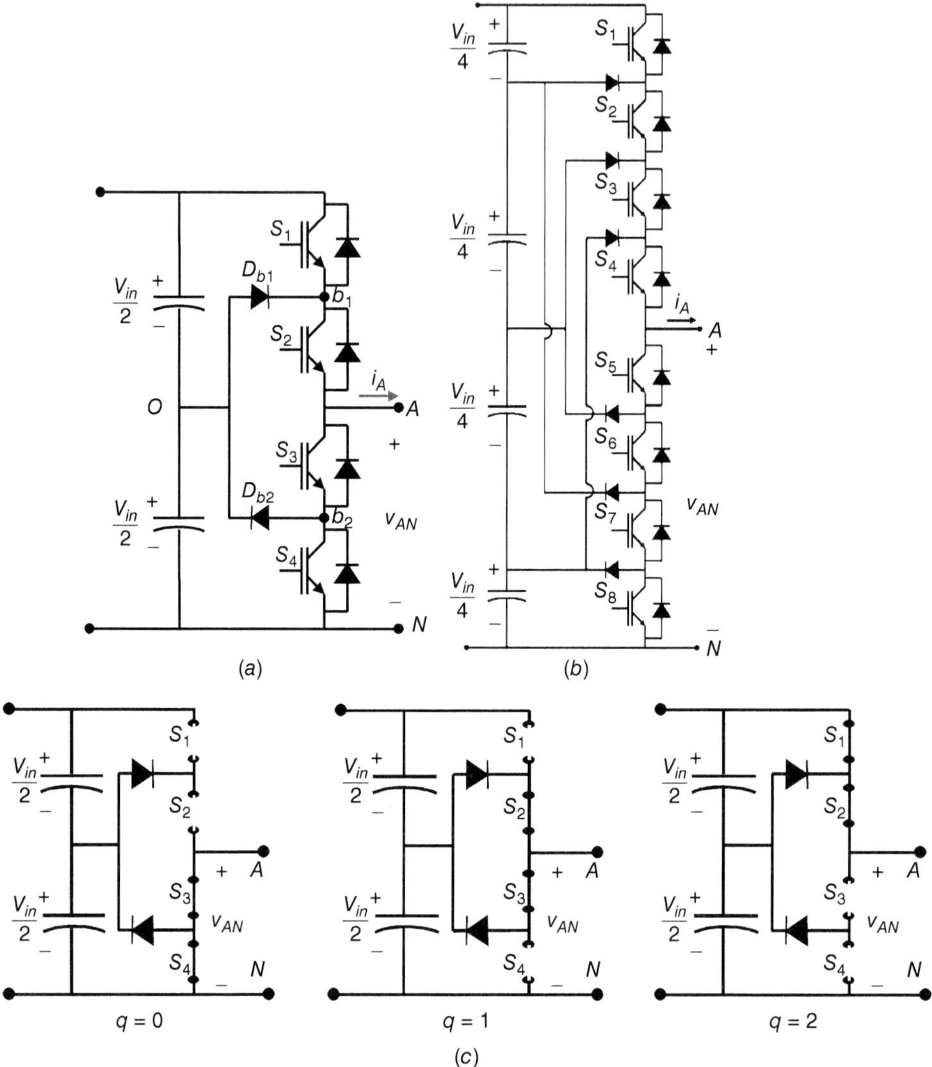


**FIGURE 16-24** Diode clamped converters: (*a*) one phase of a three-level converter, (*b*) one phase of five-level converter, (*c*) switching states in a three-level converter.

cascaded approach is the multilevel drive offering from Robicon, now a part of Siemens.[51] In some solar inverters the dc input (PV panel) is common and the isolation is carried out by transformers at the output of the H-bridges; the transformer secondaries are then connected in series to obtain the stepped waveform construction of the ac voltage.

*Other Multilevel Converters.* The recently proposed modular multilevel converter (MMC)[52] uses series connected cells that together generate the required voltage for each phase. The major advantage of this approach is that a common high voltage dc exist in the MMC which is very useful for forming a common dc link between an inverter and a rectifier. It also maintains the modular

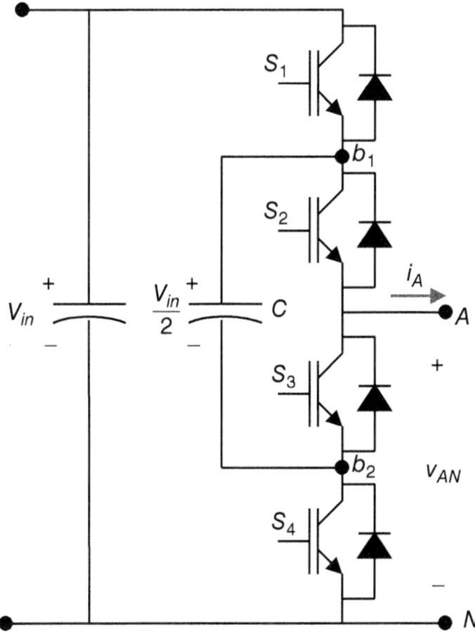


**FIGURE 16-25** Three-level flying capacitor converter.

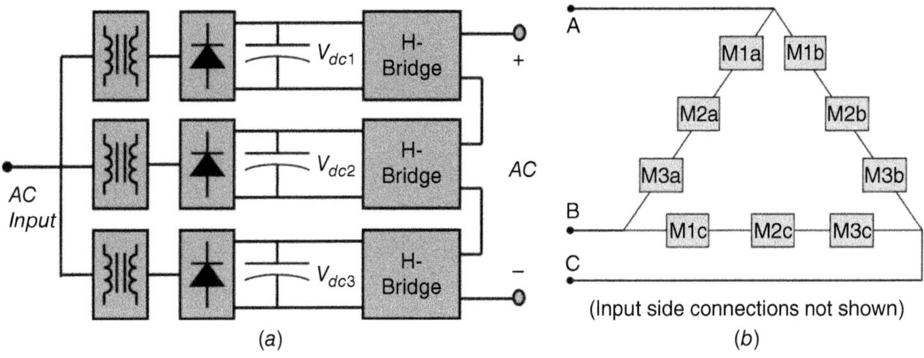


**FIGURE 16-26** Cascaded converters: (*a*) one phase; (*b*) three-phase connection in delta.

scalability and redundancy. A more detailed description of the MMC can be found in Sec. 16.9.8. Other types of multilevel converters proposed recently are the inter-connected multilevel converter[53] and the Hexagram converter.[54]

## 16.6 AC TO DC CONVERSION: RECTIFICATION

Ac to dc converters, or rectifiers, are used at the input of almost all line connected electronic equipment. Electronic devices that are powered directly from line and do not have regulation requirements use single- and three-phase diode bridge rectifiers for converting line frequency ac to an uncontrolled dc voltage. For control over the output dc voltage, thyristor-based rectifiers are used. Power factor corrected front end converters provide output voltage regulation as well as near unity power factor.

### 16.6.1 Single-Phase Diode Bridge Rectifier

Figure 16-27*a* shows the circuit of a single-phase diode bridge rectifier with a purely capacitive output filter. Due to its simplicity and low cost this circuit is preferred for low-power applications such as input stages of ac to dc adapters and computer power supplies. Diodes conduct in pairs to transfer energy from the input to the output when the input line voltage exceeds the output dc voltage in magnitude. Diodes $D_1$ and $D_4$ conduct when $v_s > v_o$, while $D_2$ and $D_4$ conduct when $-v_s > v_o$. The capacitor $C_d$ gets charged by high-current pulses during these small intervals near the peak of $v_s$, and discharges with the almost constant load current during the rest of the line cycle as shown in Fig. 16-27*b*. The output dc voltage is approximately equal to the peak of the line voltage minus the forward voltage drop of two diodes. The capacitor value is chosen on the basis of the maximum load current and allowable output voltage ripple. The line current has significant harmonic content as shown in Fig. 16-27*c*. Source inductance of the line, common for regular utility supply, leads to lower peak input current, larger conduction times for the diodes, and reduced magnitude of the output voltage.

To quantify the line current distortion the following definitions are commonly used.

*Total harmonic distortion* (THD) is the ratio of rms values of the distortion component to the fundamental component, expressed as a percentage:

$$\mathrm{THD} = \frac{I_{\mathrm{dist}}}{I_1} \times 100 = \frac{\sqrt{(I^2 - I_1^2)}}{I_1} \times 100 \tag{16-49}$$

*Real power* is the actual value of power consumed computed as an average over one line cycle:

$$P_{\mathrm{real}} = \frac{2\pi}{\omega} \int_0^{2\pi/\omega} v_s(t) i_s(t) dt = VI_1 \cos(\phi_1) \tag{16-50}$$

*Apparent power* is the product of the rms values of the input voltage and current:

$$P_{\mathrm{app}} = V \cdot I \tag{16-51}$$

*Power factor* (PF) is defined as the ratio of real power to apparent power:

$$\mathrm{PF} = \frac{P_{\mathrm{real}}}{P_{\mathrm{app}}} = \frac{VI_1 \cos(\phi_1)}{VI} = \frac{I_1}{I} \cdot \cos(\phi_1) \tag{16-52}$$

where $V$, $I$, and $I_1$ denote the rms value of the voltage, current, and fundamental component of the current, respectively, $\phi_1$ is the phase angle of the fundamental component of the current with respect to the input voltage (assumed sinusoidal), and $I_{\mathrm{dist}}$ is the rms value of the distortion component of the input current. The term $\cos(\phi_1)$ is called the *displacement power factor*, while the term $I_1/I$ is called the *distortion power factor*.

For the circuit values of Fig. 16-27*a*, the load current is 0.84 A, the peak line current $\hat{I}_s = 19.4$ A, rms line current $I_s = 3.8$ A, rms of the fundamental component $I_{s1} = 1.18$ A, $THD = 280\%$, and the power factor $= (1.18/3.8) \cdot \cos(2.0°) = 0.31$. The quality of the input current can be improved significantly if an inductive filter is used at the output of the rectifier. With a high enough inductance, the output current can be maintained nearly constant. This leads to a square waveshape for the input current as shown in Fig. 16-27*d*, which has a THD of 48% and a power factor of 0.9. With the inductive filter, the output voltage has an average value equal to the average value of a rectified sinusoid, that is, $2\hat{V}_s/\pi$, where $V_s$ is the peak value of the input phase voltage. Inductive output filter is preferable for medium power applications so that the input current has lower harmonic content.

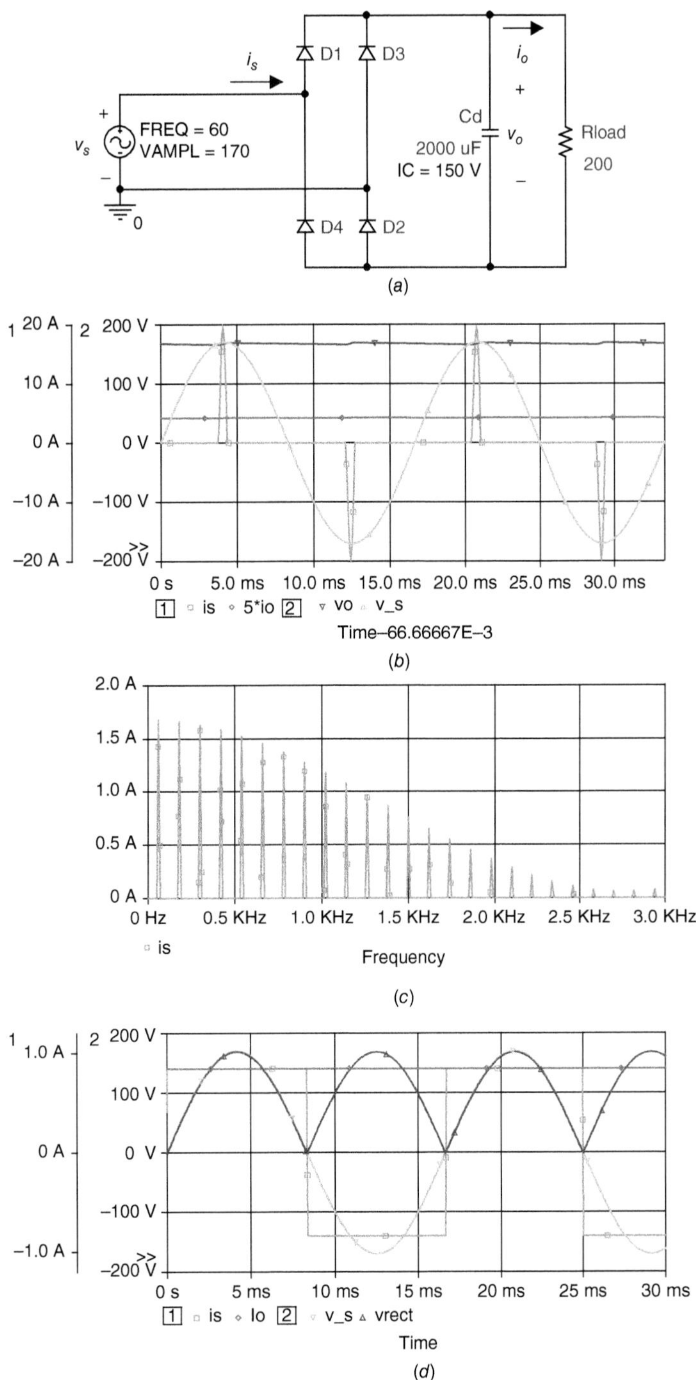


**FIGURE 16-27** Single-phase diode bridge rectifier: (*a*) circuit, (*b*) waveforms, (*c*) line current harmonics, (*d*) waveforms with inductive filter.

### 16.6.2 Three-Phase Diode Bridge Rectifier

Figure 16-28*a* shows a three-phase diode bridge rectifier with an inductive output filter. The operation is similar to the single-phase case. Diodes conduct in pairs—one from the upper three and one from the lower three. Cathodes of diodes *D*1, *D*3, and *D*5 are connected together, so the diode with the highest voltage at its anode conducts. The converse holds for diodes *D*2, *D*4, and *D*6. The rectified voltage follows the envelope of the line voltages and their negatives: $v_{\text{rect}} = \max(|v_{ab}|, |v_{bc}|, |v_{ca}|)$. This rectifier is also called the six-pulse rectifier because the voltage at the output of the diode bridge, $v_{\text{rect}}$, has six pulses in every line cycle. The average output voltage across the load is $V_o = \bar{v}_{\text{rect}} = (3/\pi)\,\hat{V}_{LL}$, $\hat{V}_{LL}$ being the peak value of the line-to-line voltage. The input line currents can be derived considering which diodes are conducting at a given time. They have quasi-square waveshapes as shown in Fig. 16-28*b*. The harmonic distortion is lower than in the single-phase case with inductive filter: THD = 31%, and PF = 0.955. If the output filter is purely capacitive, the output voltage is equal to $\hat{V}_{LL}$, while the input currents are significantly distorted (Fig. 16-28*c*) and have harmonics at $(6m \pm 1)f$, where $f$ is the line frequency and $m$ is an integer. As in the single-phase rectifier with capacitive output filter, THD of the input current depends significantly on the source impedance.

The quality of input current and power factor generally improve when going from single phase to three phase, and can be further improved with higher number of phases if voltages with appropriate phase difference are generated from the utility supplied three phases. The output filter requirements also reduce as the number of phases is increased. With six voltage sources phase shifted by 30°, 12 diodes can be utilized to generate a 12-pulse rectifier. Isolated voltage sources phase shifted by 30° can be obtained using a wye-delta connected three-phase transformer. Other phase shifts are generated by vectorial combination of appropriately scaled, isolated, voltages that are obtained from the input three-phase voltages using line frequency transformers. Rectifiers with pulse numbers 12, 18, and 24 are common for medium and high-power applications that require good power factor and low THD but do not have stringent constraints on size and weight.

### 16.6.3 Controlled Thyristor Rectifiers

Diode bridge rectifiers do not have any regulation capability and the output dc voltage varies with changes in line and load. This drawback is overcome by controlled thyristor rectifiers. Thyristor rectifiers are primarily used in medium to high-power applications where regulation of the output dc voltage is required but line current quality and power factor are not important (or can be corrected externally). Increasing concerns for power quality have resulted in reduced applications for these converters. High-power dc motor drives, especially those used in traction, battery chargers, and high voltage dc (HVDC) transmission are the most common uses for these converters.

To understand the operation of thyristor rectifiers it is first necessary to know the basic terminal characteristics of thyristors. Thyristors, also called silicon-controlled rectifiers (SCRs), are high-power semiconductor devices which can block voltage of either polarity and conduct current in one direction (from anode to cathode). They can be switched on by applying a current pulse to their gate terminal when forward biased (positive voltage from anode to cathode), and can be switched off only by reducing the device current to zero.

***Single-Phase Thyristor Rectifier.*** Figure 16-29*a* shows a single-phase *fully controlled* thyristor rectifier. The output has to be inductive for proper operation. For analysis presented here, it is assumed that $I_o$ is constant and that there is no source impedance. During the positive half of the line cycle ($v_s > 0$), $T_1$ and $T_4$ are switched on after a delay angle $\alpha$ from the zero crossing of $v_s$. The angle $\alpha$ is commonly called the *firing angle*. With $T_1$ and $T_4$ on, $i_s = I_o$ and $v_{\text{rect}} = v_s$. When $v_s$ reverses in polarity, thyristors $T_1$ and $T_4$ keep conducting since the current through them has not been reduced to zero. During the negative half cycle, $T_2$ and $T_3$ are switched on after angle $\alpha$ from the zero crossing of $v_s$. At this point current is transferred from ($T_1$, $T_4$) to ($T_2$, $T_3$). In reality there is some finite source inductance, due to which the current transfer takes some time.[4] Once $T_2$ and $T_3$ start conducting, $i_s = -I_o$ and there is a reverse voltage across $T_1$ and $T_4$ that keeps them in the

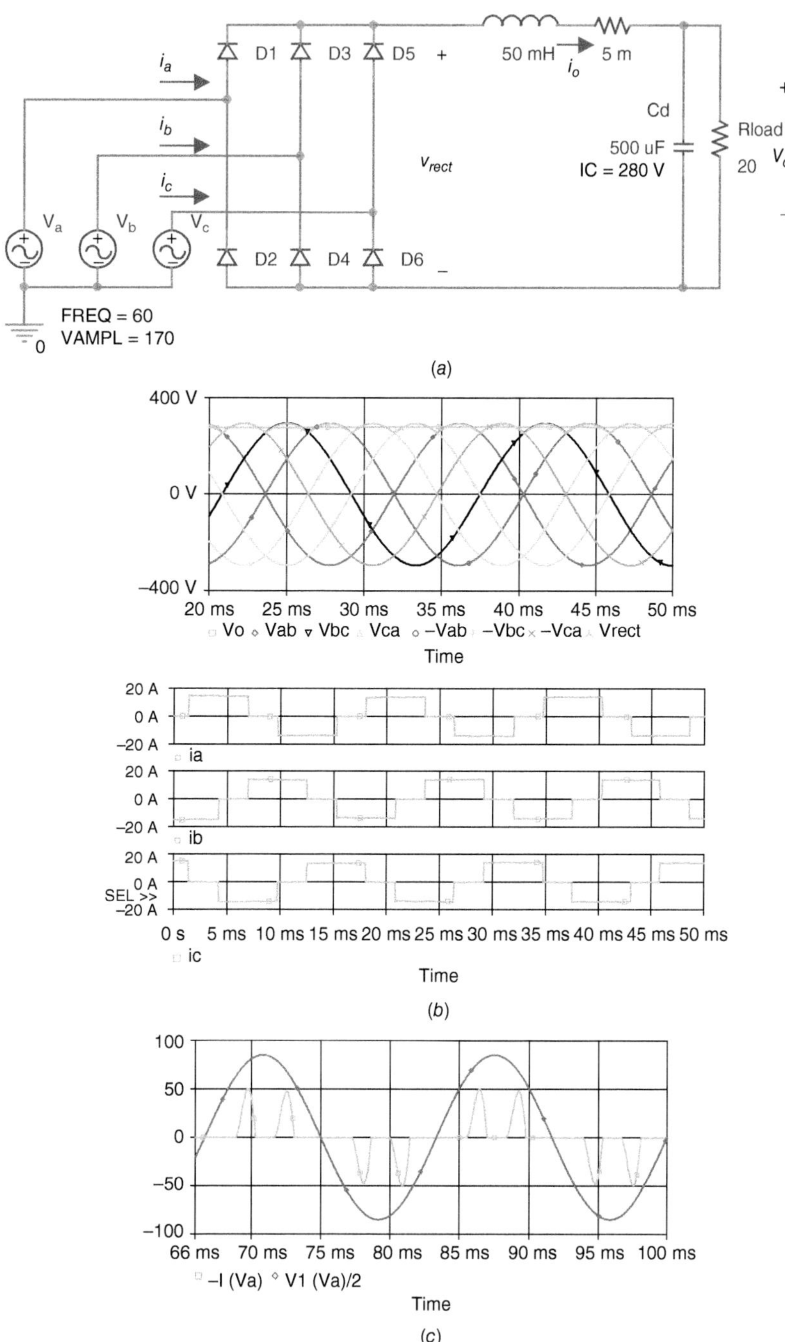


**FIGURE 16-28** Three-phase diode bridge rectifier: (*a*) circuit, (*b*) waveforms with inductive filter, (*c*) line current waveform with capacitive filter.

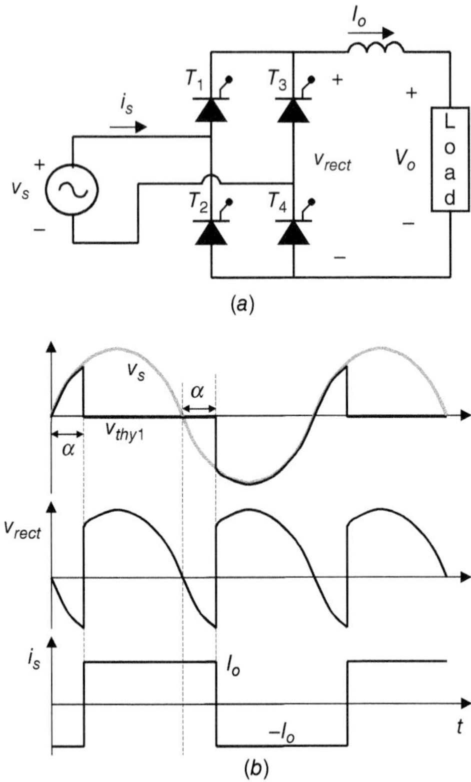


**FIGURE 16-29** Single-phase thyristor rectifier: (*a*) circuit, (*b*) waveforms.

off state. Steady-state operating waveforms are shown in Fig. 16-29*b*. The average dc voltage across the load is given by

$$V_o = \bar{v}_{\text{rect}} = \frac{1}{\pi}\int_{\alpha}^{\alpha+\pi} \hat{V}_{ph}\sin(\omega t) \qquad d(\omega t) = \frac{2\cos(\alpha)}{\pi}\cdot\hat{V}_{ph} \tag{16-53}$$

$V_o$ can be controlled by varying $\alpha \in [0°, 180°]$. It is maximum for $\alpha = 0°$ where the thyristor rectifier behaves exactly like a diode bridge rectifier, and zero for $\alpha = 90°$. For $\alpha > 90°$, $V_o < 0$, and power is transferred from the dc side to the ac side. Where bidirectional power flow is not required, thyristors $T_2$ and $T_4$ are replaced by diodes resulting in a *half controlled* rectifier.[55] THD of the input current is same as that in the diode bridge rectifier, but the fundamental component of the input current lags the input voltage by angle $\alpha$, leading to a displacement power factor of cos($\alpha$). Thus, regulation of output voltage is achieved at the expense of lower power factor.

***Three-Phase Thyristor Rectifier.*** The three-phase thyristor rectifier is shown in Fig. 16-30*a*. Similar to the single-phase case, each thyristor is switched with a delay of angle $\alpha$ after the anode to cathode voltage across it becomes positive. Each thyristor conducts for 120°, so the input line currents are quasi-square waves with magnitude equal to $I_o$ as shown in Fig. 16-30*b*. The average output voltage is $V_o = [3\cos(\alpha)/\pi]\,\hat{V}_{LL}$. For $\alpha > 90°$ power flows from the dc side to the ac side, and the converter acts like an inverter. For unidirectional power flow, the three lower thyristors can be replaced with

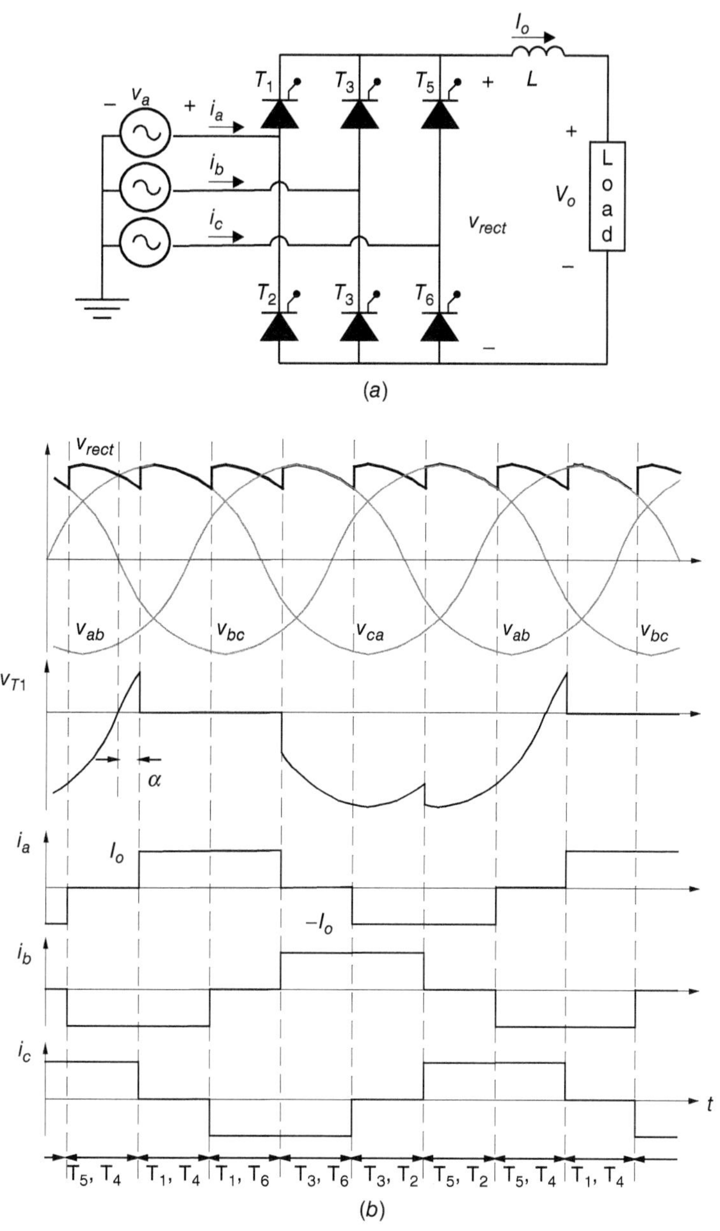


**FIGURE 16-30** Three-phase thyristor rectifier: (*a*) circuit, (*b*) waveforms.

diodes to give a half-controlled version. As with diode bridge rectifiers, thyristor bridges can also be used to obtain 12 (or higher) pulse rectification, resulting in lower THD for the input current and reduced size for the output filter. Due to the bidirectional power flow capability of this converter and very high voltage and current rating of thyristors, series connected thyristor rectifiers are utilized in HVDC transmission systems.[56]

## 16.7 AC TO AC CONVERSION

In applications where a controllable three-phase ac voltage has to be synthesized, the most common strategy is to first rectify line frequency ac to obtain a dc voltage, and then use a three-phase inverter. The dc link requires a substantial electrolytic capacitor, which filters the dc voltage and also provides energy storage for short duration line voltage sags and interruptions. Capacitors add significant size and cost, and electrolytic capacitors also have the problem of lower reliability. To reduce the number of stages from two to one, and to eliminate the electrolytic capacitor, there has been a significant research effort in direct ac to ac conversion.

### 16.7.1 Thyristor Cyclo-Converters

These converters have been used extensively for direct ac-ac conversion in high power variable frequency ac drives.[55] In these converters a low-frequency ac waveform is synthesized by a piecewise combination of the available input ac voltages. However, they have limited control over the magnitude, frequency, and quality of the output voltage, and quality of the input line current.

### 16.7.2 Matrix Converters

Recently, matrix converters utilizing controllable bidirectional switches and PWM have been developed for direct ac-ac conversion. As the name suggests, the converter consists of a matrix of switches connecting each input phase to each output phase as shown in Fig. 16-31*a*. All the switches, denoted by square boxes in the figure, need to have bidirectional voltage blocking and current conduction capabilities. So far, a single semiconductor switch with these capabilities has not been invented. Thus, the switch has to be realized using a combination of existing power devices. Three different implementations are shown in Fig. 16-31*b*.

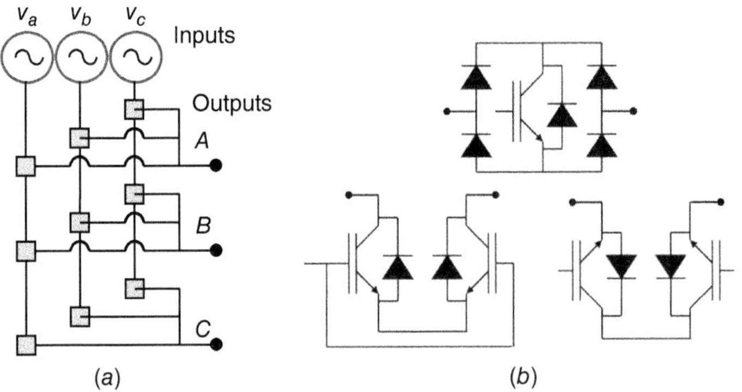


**FIGURE 16-31** Matrix converters: (*a*) three-phase matrix converter, (*b*) bidirectional switch implementations.

At any time instant each of the output phases is connected to one of the input phases, and more than one output phase may be connected to the same input phase. Selecting appropriate switches and utilizing PWM, output voltages with continuously variable amplitude and frequency can be synthesized. The synthesis is most easily understood by means of space vectors.[31,57] The total number of meaningful switching combinations are 27. Out of these, 6 lead to output voltage space vectors rotating at the input line frequency, 18 lead to stationary output voltage space vectors, while 3 lead to zero vectors. As in the dc to three-phase ac case, on an average basis, a desired output voltage vector can be synthesized by utilizing a combination of the stationary nonzero and zero space vectors. There is sufficient flexibility to ensure that the input power factor is unity. Considerable research has also been done to ensure proper operation of matrix converters under unbalanced line conditions. However, as shown in Ref. 58 the output voltage of a matrix converter has a theoretical limitation of $V_{LL,op} = (\sqrt{3}/2)V_{LL,\text{in}}$, which is considerably lower than that obtainable with an ac-dc-ac configuration ($V_{LL,\text{in}}$). In addition, clamping circuits and small input and output filters are required for proper operation.

So far matrix converters have not been commercially successful. This is chiefly due to the number and cost of bidirectional switches, limitation on the maximum amplitude of the output voltages, and lack of energy storage, required for *ride-through* capability during short duration line failures. Some of these challenges have been countered by introduction of bidirectional switch modules,[59,60] and *sparse matrix topologies*[61] that require fewer number of switches.

### 16.7.3 AC Regulator

These converters consist of two thyristors connected in antiparallel configuration in each of three phases between the source and load. They are used for starting of line connected three-phase ac machines. During startup, the firing angle of the thyristors is ramped up from 0° in order to limit the inrush current. Once steady state is reached either the firing angle is maintained at close to 180° (continuous conduction) or the thyristors are shorted out by a three-phase relay.

## 16.8 APPLICATIONS OF POWER ELECTRONIC CONVERTERS

The foregoing subsections described the principles and control of different power converters. This subsection describes the requirements of specific applications, and how power electronic converters are utilized for these. The next subsection is devoted to utility applications of power electronics.

### 16.8.1 DC Power Supplies

DC power supplies are required for powering different components in electronic equipment. Their specific features and level of sophistication depend on the application. For example, in a *bias supply*, required for various analog components in a power converter, the main requirements are isolation and a (relatively) large tolerance in the output voltage level. The input to the bias supply may be derived from a regulated dc bus so that line regulation is not required. A simple push-pull converter without an output inductor can be used for this purpose. For use inside a sensitive instrument the power supply requirements would include: strignent regulation with respect to line and load, good dynamic response, compliance with EMI and input power quality standards, and adequate energy storage for normal operation during short-duration line failures. Features commonly required in dc power supplies are

*Output voltage regulation* with respect to line and load. For high-current low-voltage power supplies there may be an additional requirement of regulating the voltage at the load terminals. This requirement, called remote sensing, accounts for voltage drops in the connecting wires. Dynamic requirements of response time and overshoot/undershoot in the output to step change in load current are also specified.

*Output current limit.* This may be a fixed value, or have a *foldback* characteristic.[6] With the latter, once the output current exceeds a certain value the current limit is varied as a function of output voltage—decreasing as the output voltage decreases.

*Isolation.* Electrical isolation between output and input.

*Soft start.* This is required to limit the inrush current for initial charging of capacitors.

*Holdup time.* For line powered applications, there is usually a requirement for holdup time, time for which the power supply should operate normally in the absence of the input voltage. Usually energy storage is provided by adding capacitors; sometimes auxiliary capacitors charged to a higher voltage to store more energy are used.

*Sleep mode.* Typically power supplies have a low efficiency at light load. Thus, in battery-powered and portable applications light load condition is detected and the power supply operation is changed to reduce its losses, e.g., by reducing the switching frequency.

*Power factor corrected front end.* For single-phase applications, it also helps in operation with universal input voltage range (100–240 V, 50/60 Hz).

*EMI compliance* with a specified standard.

*Environmental compliance* for temperature, humidity, and altitude of operation.

Some features can be implemented with standard PWM control ICs leading to reduced number of components and design simplification. Commercial availability of low-power dc/dc converter modules with standard input and output voltages has increased significantly in recent years. Thus, a custom power converter design may not be required, unless there are special requirements as in space, defense, and some instrumentation applications.

A specific application of dc power supplies is in digital systems. All digital systems need a power electronic converter to provide the requisite supply voltages. With the digital supply voltages going down and the clock frequencies going up, the requirements are increasingly toward low voltages and high currents. For a high-speed microprocessor the required values are around 2 V and a few hundred amperes. The dynamic requirements are also very demanding since the microprocessor load can go from almost zero current to full load current in a few microseconds. Furthermore, digital components operating at different supply voltage levels may be used in a complete digital system. For these applications, the intermediate bus architecture is used. First a 12-V (or a lower voltage) bus is derived from the input supply using an intermediate bus converter (IBC); the IBC may also provide isolation from the input. This low-voltage bus is then input to several point-of-load (POL) converters, which are located close to the loads and provide the required steady state and dynamic voltage regulation. POL converters do not provide isolation between input and output. IBCs and POLs with standard input and output voltages are commercially available as modules from several manufacturers.[17,72]

## 16.8.2 Electric Drives

***Introduction.*** Traditionally electric motors have been powered by direct connection to ac line, or to dc voltage obtained from a rectifier. However, this usually results in inefficient operation due to lack of control. Electric motors powered by appropriately controlled power electronic converters lead to significant increase in the overall system efficiency due to the advantages of variable speed operation.[73] Electric motor load has more than half the share of electric power consumption in the United States, of which about half is in industrial applications. Thus, any increase in system efficiency due to electronic-controlled motor drives can lead to large savings for the company and overall electricity consumption. Moreover, performance advantages of fast dynamic response and very accurate control over speed and position are obtained.

Squirrel cage induction machines (IMs) are the most widely used electric motors. This is due to the advantages of simple, robust, and low-cost construction, ease of powering, and lower maintenance requirement due to absence of brushes. Thus, control of induction machines for better performance and improved efficiency has been researched extensively. The invention of vector control,[40] and direct torque control[41,42] have made the dynamic performance of induction machines similar to that of a dc machine. The popularity of vector control was also aided by development of PWM inverters and SVM described in Sec. 16.5. Developments in permanent magnets and power electronics, and increasing concerns for size, weight, and efficiency have led to significant interest in permanent magnet synchronous machines (PMSM) and brushless dc (BLDC) machines. PMSM and BLDC motors have the highest operating efficiency and the highest power density of all the motors.[74] Further, due to development of power electronics, the switched reluctance machine (SRM), the first electromechanical machine invented in the early 1800s, has also been commercialized.[75,76] This machine has a simple, robust, and inexpensive construction, and is powered from an inherently robust power converter topology. SRMs are most popular in the low to medium power range and high-speed applications requiring medium performance. Squirrel cage induction, PMSM, synchronous reluctance, and BLDC drives require a three-phase inverter for powering the stator windings and no electrical excitation for the rotor. SRMs have a special power converter that has independent sections for individual phases. For illustration, a vector-controlled ac drive and an SRM drive are discussed here. Details of these drives may be found in Refs. 35, 37, 43, 44, and 73–76.

***Vector-Controlled AC Drive.*** Figure 16–32 shows the basic block diagram of a vector-controlled PMSM drive. The structure is identical for the squirrel cage induction motor drive. The input stage is a single- or three-phase diode bridge rectifier, the output of which is filtered to obtain a dc. The input current during initial charging of the dc bus capacitor is limited using either (a) parallel combination of a controllable switch and a current limiting resistor connected in series with the main power path, or (b) a half-controlled thyristor rectifier. The dc voltage is converted to variable frequency variable magnitude three-phase ac by means of a three-phase PWM inverter. During *dynamic braking*, energy stored in the inertia of the machine and load is transfered back to the electrical circuit. If the rectifier has bidirectional power transfer capability, this energy can be transfered to the source. Otherwise the energy charges the dc bus capacitance. To limit the dc bus voltage a brake resistor is connected in parallel with the dc bus by means of a controlled switch.

The switching frequency of the inverter in ac drives is usually between 20 and 30 kHz, since the machine inductance is usually sufficient to filter the output current. Thus, IGBTs are commonly used for this application. MOSFETs are used if the dc voltage is below 600 V and a higher switching frequency is required in order to reduce the output current ripple. Intelligent power modules (IPMs) containing all the power semiconductor components along with gate drive and protection circuits in a single package are now available from several manufacturers.[77–81] Use of IPMs results in considerable design simplification, size reduction, and smaller design cycles.

The drive control is carried out using a microcontroller or DSP, with the use of DSPs becoming more popular due to increased computational requirements. In an industrial application, the DSP

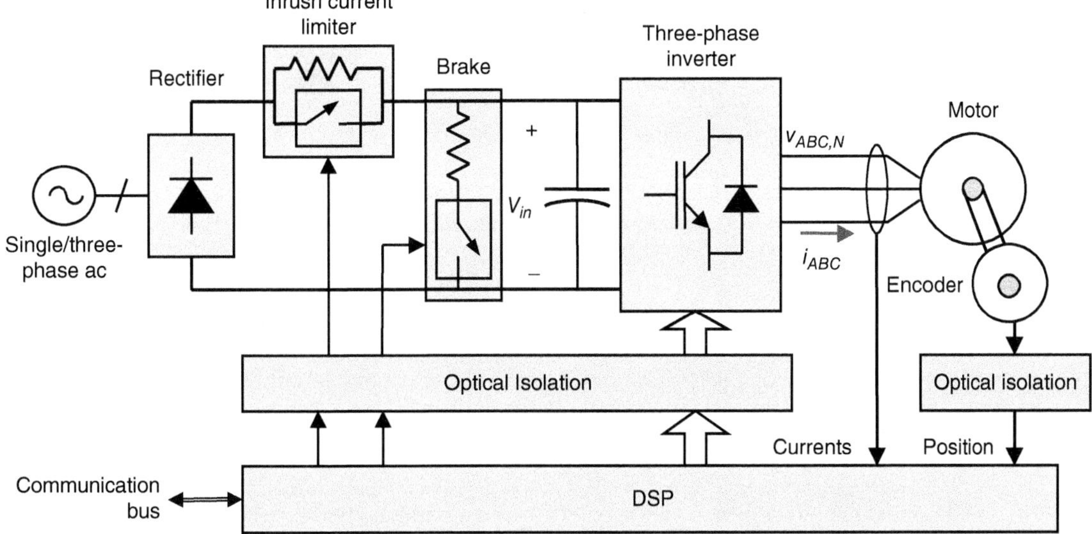


**FIGURE 16-32** Vector-controlled three-phase ac drive.

controls the machine in response to command signals (for speed or position) obtained over a communication interface bus. AC machines are usually controlled using either vector control or direct torque control. From the power electronics point of view, this requires control of the output currents with a high-control bandwidth (of the order of 1 kHz). This is done in a cascade manner: the current control algorithm generates reference voltage vectors which are then synthesized using space vector PWM. Both current control and PWM are implemented on the DSP. The final output of the DSP are switching signals for each of the three inverter legs and the brake switch. These signals are optically isolated and input to the gate drive circuitry for each of the controlled switches. Output currents of the inverter are sensed using either hall effect sensors or sense resistors. The analog current signal is converted to digital form using the internal ADCs of the DSP. The dc bus voltage may also be sensed for feedforward of input line variations and for dynamic braking. An optical encoder is used if direct rotor position sensing is required. Since the encoder outputs are already in digital form they can be easily interfaced. However, due to close proximity of the encoder and the machine windings, which are subject to pulsed voltages generated by the inverter, these signals have high common mode noise. To solve this problem, differential mode line drivers and receivers, and optical isolation of encoder signals are commonly employed.

***SRM Drive.*** The basic structure of an SRM drive is similar to an ac motor drive. However, the power converter is quite different owing to the machine characteristics. Fig. 16-33*a* shows the cross-section

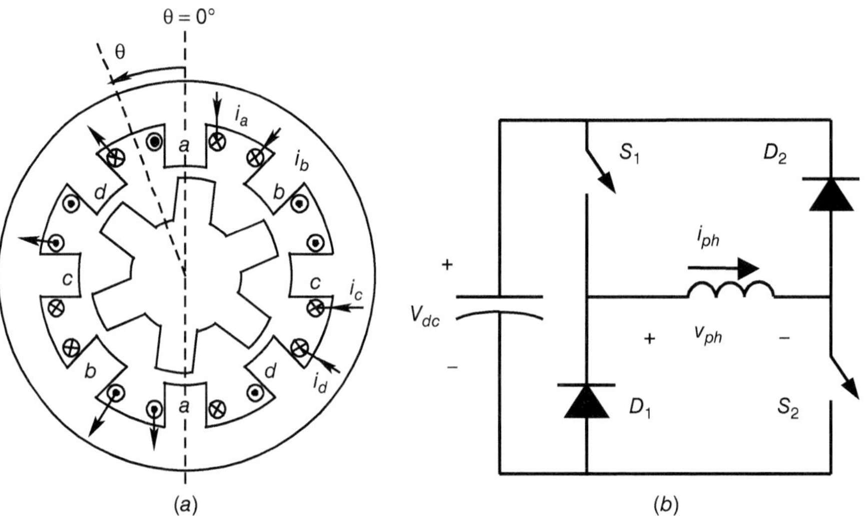


**FIGURE 16-33** SRM drives: (*a*) cross-section of a four-phase 8/6 SRM, (*b*) asymmetric bridge converter.

of a four-phase SRM. The SRM has saliency on the stator as well as the rotor. Each phase consists of concentrated coils wound on diametrically opposite stator poles. If phase *a* carries current $i_a$, then the closest set of rotor poles are attracted to the stator phase *a* poles. Once the rotor poles are aligned with the stator poles there is no torque on the rotor and $i_a$ has to be reduced to zero. By energizing and de-energizing phases in sequence, a continuous rotation can be achieved.

To build up and reduce winding current, the converter needs to have bidirectional voltage capability. However, since the direction of torque is independent of the current polarity, only unidirectional current capability is required. At low speeds the back emf of the machine is very low so the winding current has to be controlled usually by hysteretic current control. This is done by letting the current freewheel with a zero voltage across the winding terminals, commonly called application of a zero voltage loop (ZVL). To determine which phase needs to be energized or de-energized, rotor position information is required; this may be sensed or estimated, and discrete position information is sometimes sufficient.[76] The total torque produced by the machine has significant ripple, and unless instantaneous torque control is a requirement, average torque control is implemented. The average torque depends on the turn-on and turn-off angles (rotor positions w.r.t. the phase winding where that phase is energized and de-energized, respectively), and the current magnitude if current control is used.[75] The control is implemented based on results of numerical simulation and then tuned experimentally.

Ray et al. published one of the earliest papers on power converters for SRM,[82] and suggested the asymmetric bridge converter, which is also known as the classical converter. The circuit topology, shown in Fig. 16-33*b*, is similar to a two-switch forward converter[4] and fullfils all the basic requirements. Only one phase is shown, the circuit being identically for the other phases with a common connection to the input dc bus. The basic operating modes of the converter are

| | | |
|---|---|---|
| Energization | $(S_1, S_2)$ on, $(D_1, D_2)$ off | $v_{ph} = V_{dc}$ |
| De-energization | $(D_1, D_2)$ on, $(S_1, S_2)$ off | $v_{ph} = -V_{dc}$ |
| ZVL | $(S_1, D_2)$ on, $(S_2, D_1)$ off, or | $v_{ph} = 0$ |
| | $(S_2, D_1)$ on, $(S_1, D_2)$ off | |

where all switches and diodes have been assumed ideal and the winding resistance has been neglected. This converter is extremely reliable and fault tolerant. The problem of "shoot-through" due to spurious turn-on of the two controlled switches in the same leg is avoided, since the phase winding is in series with the controlled switches. Shoot-through can be a major problem in full-bridge dc-dc converters and three phase inverters. Although this converter topology has become very popular and a packaged module similar to three-phase ac inverters is now available,[83] there are several other converter topologies which may be suitable for specific applications. A detailed comparison of SRM converter topologies can be found in Refs. 75 and 84–86.

DC motor drives use a dc-dc converter—buck, buck-boost, or full bridge depending on the requirements—and armature current control for controlling the instantaneous torque. The field current may also be controlled to reduce back emf at high speeds. Their popularity has been decreasing due to maintenance requirements and speed limitations of brushes. As the cost of power electronics reduces, better and low-cost permanent magnets are developed, better control and estimation techniques are implemented on lower-cost DSPs, and energy efficiency gains priority, it is expected that PMSM and BLDC drives will increase their share in the motor drives market. SRMs are also expected to gain ground in high-speed and harsh environment applications where energy efficiency is not the only driving factor.

### 16.8.3 Battery Charging

Battery charging is a very large part of the ac-dc and dc-dc converter applications. The end use for batteries is in telecommunications, electric and hybrid electric vehicles, portable electronics, and energy storage for improving power system stability. Recently, storage for energy arbitrage and firming generation output of renewable sources is also gaining more attention.

For a lead acid battery, there are three standard operating modes for the battery charger: (a) constant current (or bulk) charging during low state of charge, (b) constant voltage charging after about 80% state of charge, and (c) "float" or trickle charge after the battery reaches its open circuit voltage. Thus, the requirement on the power converter is to operate at its maximum current rating over a significant output voltage range with high efficiency. In addition, isolation from input line and power factor correction are usually required.

In recent years lithium-ion (Li-ion) batteries have gained widespread usage for portable applications like laptop computers, PDAs, and cell phones. They have also been proposed for automotive applications. So far nickel metal hydride (NiMH) batteries are used in commercially available hybrid electric vehicles (HEV) while high-end electric vehicles use lithium-based battery chemistries.[87] The advantages of Li-ion batteries are high-energy density, higher cell voltage of about 4.2 V, and low self-discharge rate.[88] The disadvantages are high sensitivity to electrical stress and limited temperature range. The cell voltage has to be monitored for overvoltage (which is very close to the nominal open circuit voltage) to avoid catastrophic failure, and undervoltage to maintain battery life. Linear chargers are used for single-cell batteries requiring low-charging current, while switch mode converters are used for high-voltage and high-current charging. The charging technique is similar to lead acid—constant current followed by constant voltage. However, pulse charging (in combination with periodic discharge and periodic relaxation pulses) has also been proposed for improving battery life. A further complication with Li-ion is that the dependence of state of charge on open circuit cell voltage is only on a small voltage window (3.2–3.8 V). Thus, in the absence of a detailed cell model, a coulomb counting technique, which measures the charging and discharging current and estimates the self-discharge rate, is normally used to determine the battery state.

## 16.8.4 Fluorescent Lamps and Solid-State Lighting

About 30% of the total electricity generated in the United States is consumed for lighting.[89] Thus, there is significant interest in increasing energy efficiency of lighting mechanisms. Incandescent lighting although the most inefficient is still the cheapest in terms of $/lumens. Compact fluorescent lamps (CFLs) have gained ground due to their energy efficiency and are now commercially available as replacements for incandescent bulbs. In fact, in the United States, incandescent bulbs will not be available starting January 2012. Besides lighting, fluorescent lamps that emit ultraviolet light are used in several industrial applications like sterilization and curing (drying of coatings).

CFLs and industrial lamps are usually powered using a switched mode converter called an electronic ballast. In one implementation, shown in Fig. 16-34*a*, the converter consists of a full- or half-bridge inverter with its output connected to an L-C-C (all in series) tank circuit. The dc voltage input to the inverter may be obtained using either a diode bridge rectifier or a power factor corrected front-end. The lamp filaments are connected in series with the L-C-C tank, with the two filaments being in parallel with one of the capacitors ($C_p$). The other capacitor ($C_s$) provides dc blocking and also tunes the tank operation to a desirable characteristic. When the lamp is off (gas inside not ionized), it acts as an open circuit, so the inverter is effectively connected to a series resonant L-C circuit. To ionize the gas inside the lamp, the filaments have to be heated sufficiently followed by application of a high voltage [~ (kV)] across the lamp for a short time. Once the lamp is on (gas ionized) it acts as a resistive load, and the inverter load is then a series parallel tank circuit—series L-C-C with the equivalent resistance of the lamp in parallel with one capacitor. The inverter outputs a high-frequency square wave, typically in the range of 20 to 100 kHz; lamp operation at these frequencies results in increased light output. The switching frequency is varied to control the ballast operation. Initially, the tank circuit and the lamp are excited at a frequency significantly higher than the resonance frequency. During this time, called the preheat time, the lamp filaments increase in temperature. After a suitable delay the switching frequency is reduced toward resonance so that the voltage across the lamp increases until the gas inside the lamp ionizes and the lamp "ignites." After ignition, the lamp acts like a resistive load, and the lamp output power is controlled indirectly by regulating the phase difference between the inverter current and voltage. The tank circuit also ensures resonant transitions of the converter switches[90] leading to reduced losses. Several ICs that implement full control of a fluorescent lamp including preheat, ignition, and dimming control are commercially available.[81] Operating waveforms from an electronic ballast are shown in Figs. 16-34*b* and 16-34*c*.

High-intensity discharge (HID) lamps have a large share of commercial lighting such as street lighting, sports facilities, etc. They are of three types—mercury vapor, sodium vapor, and metal halide. The ballast requirements of these lamps differ from those of fluorescent lamps. Operation of these lamps at higher frequencies does not improve light output. Further, they exhibit acoustic resonance in the 10- to 100-kHz range. Thus electronic ballasts for these lamps operate at around 400 Hz. So far electronic ballasts for these lamps have not become very popular.

Power supplies for plasma cutting have requirements similar to fluorescent lamp ballasts. Initially, a high voltage (~10 kV) is applied to produce an arc between two high-voltage electrodes.

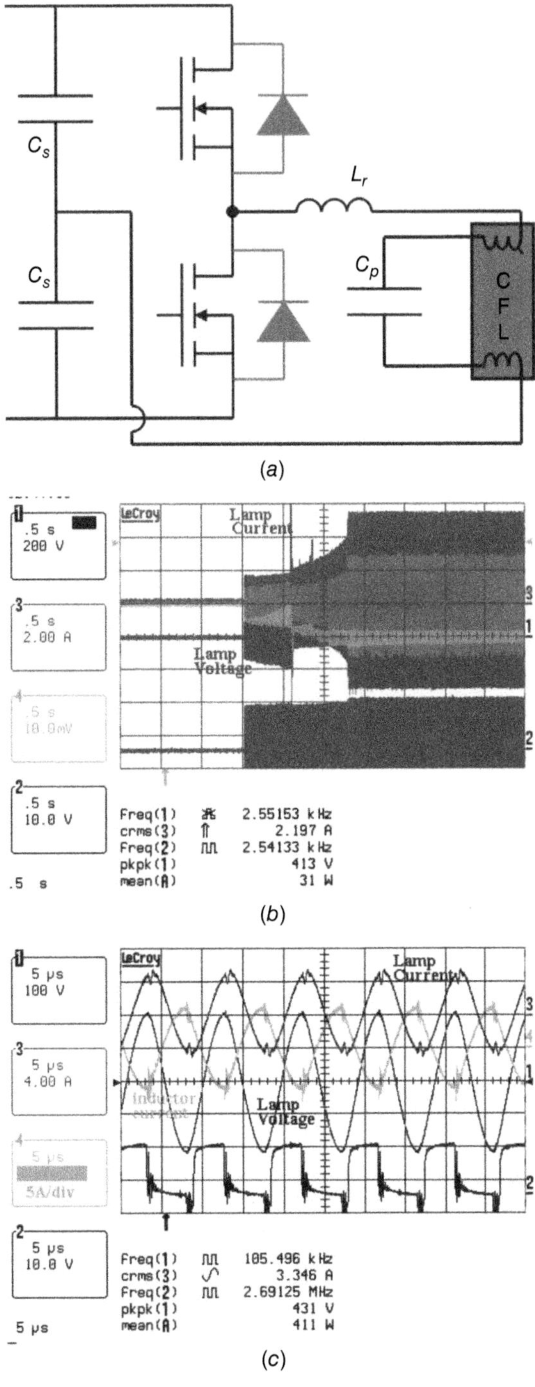


**FIGURE 16-34** (*a*) compact fluorescent ballast circuit; (*b*) start-up waveforms for a 1-kW ballast (ignition occurs 0.6 s after turn-on); (*c*) steady-state waveforms of a 1-kW ballast.

This is followed by *arc transfer*, transfer of the conducting ionized gas between the electrodes to the space between the negative electrode and the metal being cut (workpiece). Once the arc is transfered, the cutting procedure requires a current-controlled supply (at lower voltage) connected across the negative electrode and the workpiece. Since there are three terminals, two separate power supplies can be used—one for generating the initial high voltage, and the other for the actual cutting operation.

The U.S. Department of Energy has identified solid-state lighting as a means to increase lighting efficiencies in the near future.[89] Recent and expected breakthroughs in semiconductor and organic LEDs are the main impetus behind this. The major reasons to promote the LED are longer lifetime, mercury-free structure, energy saving property, and flexibility of color mixing and control. The LED can be connected in series, parallel, or combination of both, depending on the applications. A switched mode converter with a dc current regulated output is used to power the LED string. The output current is regulated to provide dimming control and to avoid failure due to overcurrent.[91] At present LED lighting is not economically competitive with the alternatives in main stream applications although several innovations are being made to reduce the system level cost of adopting LEDs. When LEDs start replacing HID lamps, incandescent bulbs, and CFCs for commercial/residential lighting, they will open up a huge market for efficient and low-cost power electronic converters.

## 16.8.5 Automotive Applications

Automotive applications of power electronics are in two domains: electric and hybrid electric drive trains, and auxiliary systems controlled by power converters.

***HEVs, PHEVs, and EVs.*** Hybrid electric vehicles (HEVs) are now gaining commercial success due to their higher mileage, lower emissions, tax breaks, and rising oil prices. The basic ideas[92] in an HEV are

- Operate engine at optimal efficiency, achieved at high speeds.
- During initial acceleration and at low speeds, power is supplied by a battery-powered motor leading to reduced idling losses and emissions.
- High acceleration power is derived from both the engine and the battery.
- During high speeds, the battery is charged from a generator connected to the engine. The motor used for generating mechanical power can also be used as a generator.
- During braking, the generator feeds energy back to the battery.

Another variation is the so-called plug-in hybrid electric vehicle (PHEV). A PHEV operates from the energy stored in the battery for small distances (25 to 100 mi), e.g., commuting from home to work. The battery is then charged from a standard wall outlet or dedicated charging station. For longer distances the PHEV would switch to generating energy from gasoline. For the long range, besides the standard gasoline-engine approach, an alternate approach is to generate electricity from a generator set mounted on the vehicle.

Fully electric vehicles (EV) have also been introduced notably by Tesla,[93] Nissan,[94] and Chevrolet.[95] These operate from batteries (typically Li-ion) alone and are expected to have a range of 100 mi for a single charge. Means to increase their range via fast charging stations and quick battery swaps are being developed. On-board power electronic requirements for these are similar to HEVs. Charging stations with different levels of charging are currently under standardization by the SAE and IEEE.

Hybrid and completely electric vehicles with fuel cells as the power source and some means of energy storage are also being developed by several auto manufacturers. These require power converters to interface fuel cell, energy storage element, motor/generator, and other electronically controlled loads.

***Auxiliary Systems.*** In recent years electronically controlled load in automobiles has increased significantly. Further increase is expected due to more comfort features, and potential replacement of some mechanical systems by all electrical systems like active suspension and power steering. The present bus voltage in automotive systems (14 V), decided by the alternator charging voltage, implies very high currents for the expected high-power consumption. To keep the current levels manageable, the *42-V Power Net* was proposed, and adopted by most auto companies and suppliers for future generation automobiles.[1] The choice of 42 V has been made on the basis of safety considerations, load dump overvoltage transients, and optimal utilization of silicon in power semiconductor devices. It is expected that there will be a dual voltage system, with both 42 V and 14 V loads. Thus, a dc to dc converter, possibly bidirectional, will be required for interconnection between the two system voltages. Furthermore, converters with sophisticated controls will be required for features like active suspension and power steering.[96]

## 16.9 UTILITY APPLICATIONS OF POWER ELECTRONICS

### 16.9.1 Introduction

The electric utility grid can be improved significantly leading to reduction in energy loss and operating costs, increased adoption of renewable generation sources, and more reliable operation, all with the use of mature technologies. This is the basic premise of the smart grid efforts for grid modernization in the United States. The smart grids umbrella covers advances from different fields like renewable generation, battery chemistry, grid monitoring and control, information technology, electric-sourced transportation, and conservation at point of use, to name a few. Power electronics is a key enabler for smart grids and has the potential to change the landscape of power generation, transmission, distribution, and end use.

Renewable generation is a fast growing component of the generation mix as a result of several advances in generation and grid interface technology, and government incentives. Possibility of large-scale renewable adoption has posed challenges of variability in generated power, capacity underutilization, and effects on system dynamics. Power electronics technology in the form of grid interface converters for the generation sources and energy storage devices, and grid control devices is therefore an important component of sustainable generation.

Currently in their infancy, electric vehicles represent a big shift in transportation from a dependence on oil to electric power that may eventually be generated from renewable sources. The effort ambitious in scope and daunting in infrastructure requirements is being realized in small steps trying out a variety of technologies addressing different aspects. Power electronic converters are a critical component in electric sourced transportation both for energy conversion within the vehicle and grid interface of the vehicles.

Growing energy demand, coupled with economic, environmental, and political restrictions on newer generation and transmission infrastructure, means that the existing resources operate near their stability limits. As a result dynamic instability, interarea oscillations, voltage instability cascading to major blackouts have become issues of real concerns today. Power electronic technology in the form of FACTS devices is the key component that increases stability limits of the grid infrastructure.

Several modern loads such as the processing plants in semiconductor industry or data centers require clean and uninterrupted power. Power electronic-based power quality solutions are essential for these loads to mitigate problems such as voltage sags, harmonics, and flicker in line voltage.

Widespread use of power electronics in power systems is further fueled by dramatic advances in power semiconductor materials and devices, especially those based on silicon carbide (SiC),[97,98] and advances in the fields of wide area power system monitoring and communication. This subsection briefly describes the major power electronics applications in power systems, namely, interfacing renewable generation and storage with electric grid, flexible ac transmission systems (FACTS), and custom power. Figure 16-35 shows the interconnected power system network with some of the major power electronics highlighted.

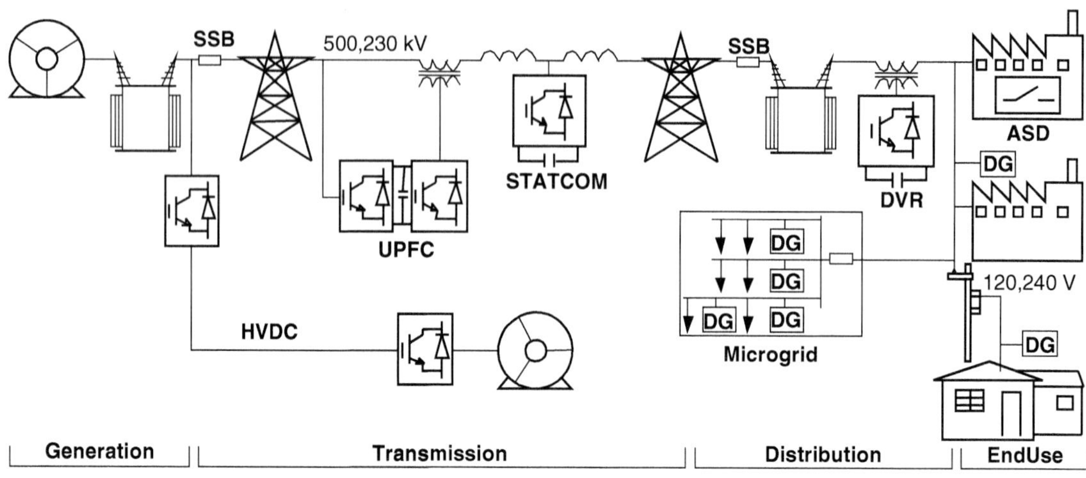


**FIGURE 16-35** Power electronics applications in power systems.

## 16.9.2 Renewable and Distributed Generation Interface

Renewable generation is an important energy option for the 21st century and a key element of the restructuring of electric grid. Renewable sources such as solar, wind, and ocean energy and hydrogen (with photovoltaics for hydrogen generation) represent one of the most promising paths to sustainable energy.

Photovoltaics (PV), wind energy, fuel cells, and micro turbines are among the most promising technologies at present. Most of these resources require a power electronic converter to interface with the power system network. Fuel cells and PV require dc to ac conversion, micro turbines require high frequency to line frequency conversion, and generators used to capture wind energy are controlled through a rotor side power converter. Further, these may also require power electronics-controlled energy storage.

***Grid Interface for Photovoltaics.*** Figure 16-36 shows a schematic diagram of the converter that can be used to interconnect photovoltaics with the utility. The input voltage from a PV array may vary from 50 to 500 V. This is converted to a well regulated and isolated dc voltage through a high frequency dc-dc converter. The dc link voltage is then converted to the required 60 Hz ac voltage by using a PWM voltage source converter. In a simple grid connection of PV, there is no need for the intermediate dc bus other than enabling the use of mature converter topologies for dc-dc and dc-ac conversion.

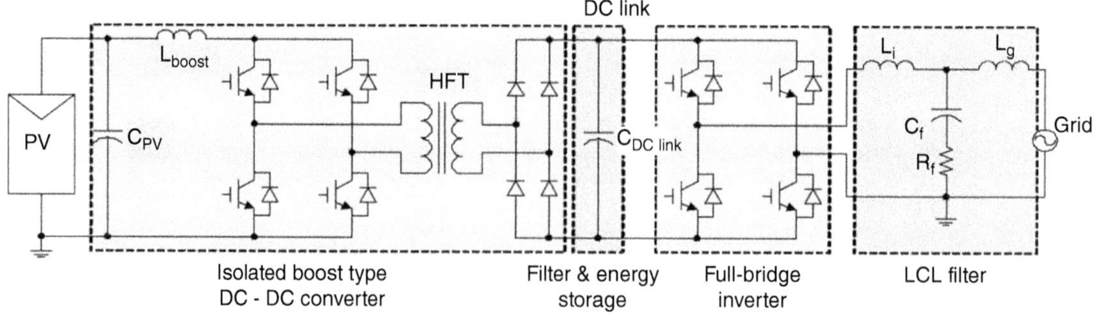


**FIGURE 16-36** Single-phase grid interface for photovoltaics.

Thus some PV inverters now use direct dc-ac conversion without an intermediate dc link: the dc output of PV panels is converted to high frequency (order of 10 kHz) ac, which after the required scaling by a high-frequency transformer is then converted to low-frequency ac using a cyclo-converter. This approach, although it requires bidirectional switches is now gaining popularity due to reducing cost of power semiconductor devices and elimination of high-voltage electrolytic capacitors.

An important feature of a PV interface is maximum power point tracking (MPPT) circuitry designed to derive the maximum possible energy from solar radiation by suitably adjusting the current drawn from the solar panel. The cost of the power converter is quite small and its energy conversion efficiency very high relative to the PV panel. Thus, system level options that increase total energy capture via fine-grain MPPT using multiple power converters have been proposed (see Fig. 16-37). In the basic *micro-inverter* approach each PV panel has a grid interface

**FIGURE 16-37** PV collection architectures.

micro-inverter and panel level MPPT is performed.[99] The scheme improves performance with respect to different solar conditions for different panels (e.g., shading on one panel will not affect performance of other panels) and manufacturing differences between panels. A variation of this approach is to convert the output of each PV panel with MPPT to a regulated dc and then use a central inverter to interface with the grid. This provides an opportunity to integrate energy storage at the intermediate dc bus. Going further, subpanel level MPPT can be performed by using several converters for each panel. This strategy accounts for partial shading of panels and differences between cells within the panel. It has been shown that the scheme increases total energy output even though an extra conversion stage is used.[100,101]

An emerging trend in PV inverters especially at high power ratings (MW) is the so-called smart inverters which provide grid interactive functionalities including reactive power support, voltage regulation, storage management, controllable ramp rates, and fault ride through.

Fuel cells can also be interfaced to the grid using a similar configuration as shown in Fig. 16-38. Since fuel cells have very little over current or short-circuit rating, an energy storage capacitor is installed at the input side of the converter.

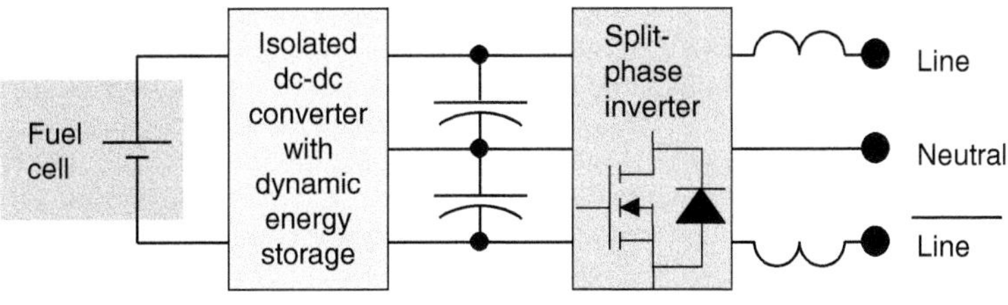


**FIGURE 16-38** Fuel cell connection to the grid.

***Grid Interface for Wind Turbine.*** Wind energy is widely considered to be the fastest growing alternate energy source. The cost of wind energy in large wind farms located in good wind sites, at about 5 cents/kWh, is now competitive with the conventional utility generation and is expected to reduce further. The total installed capacity of wind energy worldwide was 215,000 MW for the world and 42,432 MW for the United States as of June 2011.

Large wind farms consist of several multimegawatt wind turbines that are interconnected with the utility grid through a medium voltage collector network. Doubly fed induction generators (DFIG) with a wound rotor and an ac/dc/ac PWM converter, as shown in Fig. 16-39, is the most widely used technology for wind generation for high power. The main advantage of the DFIG-based generation is that it allows extraction of maximum energy from the wind at varying wind speeds, with the power converter rated only for about 20% to 25% of the total power. The stator winding is connected directly to the utility grid while the rotor is supplied with controlled, variable frequency currents by the PWM converter. By appropriate control of the rotor currents the machine can generate power from subsynchronous to super synchronous speeds. Another advantage of DFIG is that the grid side converter can generate or absorb reactive power.

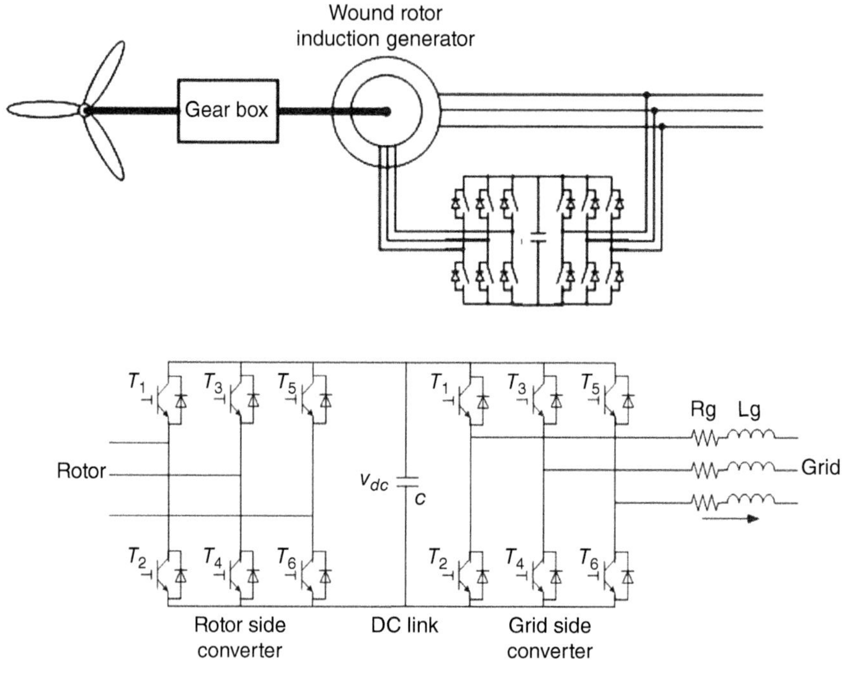


**FIGURE 16-39** Grid connection of a doubly fed induction generator.

Permanent magnet synchronous machines (PMSM) are widely used in wind turbines at low-power levels (up to a few kW). These low-power systems usually have a diode bridge rectifier at the machine terminals and the generated dc power is then used to charge batteries or converted to ac via an inverter. PMSMs with fully rated converters are also beginning to be widely used in high-power wind energy generators. A typical configuration for the converter of PMSM wind generators consists of back-to-back PWM voltage source converters rated for the peak power and sharing a common dc link as shown in Fig. 16-40. In this configuration, the permanent magnet machine is fully decoupled from the grid, making it easier to provide fault and zero voltage ride through, and also provide enhanced grid support features. In addition the PMSM machines offer higher efficiency, wider range of speed control, smaller size, and lower requirements on maintenance due to the absence of slip rings.

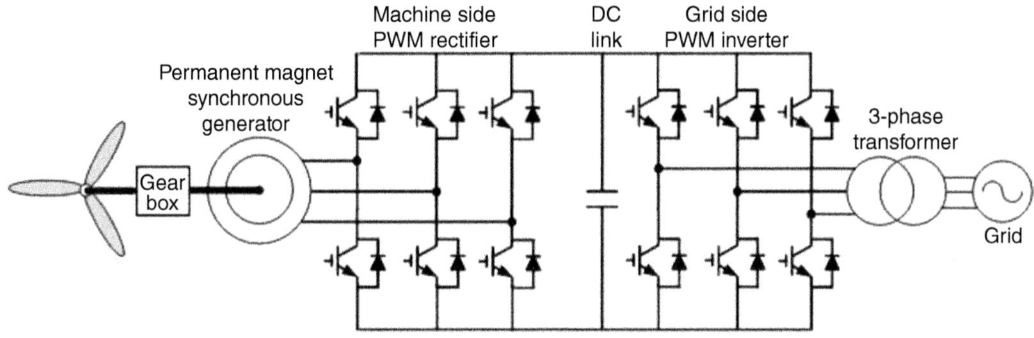


**FIGURE 16-40** Permanent magnet generator system for a wind turbine.

### 16.9.3 Distributed Generation and Microgrids

Apart from serving as an environment-friendly energy source, the distributed generation (DG) systems are expected to provide various other benefits. For example, the concept of microgrids is gaining prominence.[102] Microgrid is a cluster of distributed energy resources (DER) with power converters, energy storage, and loads, which can be controlled together and present to the grid as a single entity. With suitably designed power converters, and with coordinated control, the microgrids can enhance stability, provide relief to transmission congestion, and provide reactive power support. Another promising approach is to use distributed micro sources for combined heat and power (CHP).[103] Due to the various ancillary functions expected, inverter-based DGs are becoming more widespread compared to traditional reciprocating engines.

The interconnect standards for DG are still evolving, like the IEEE 1547 standard for interconnecting distributed resources with electric power systems.[104] IEEE 1547 specifies the requirements for the DG to disconnect from the grid under deviations in voltage magnitude and frequency or under grid outages. Much work is still needed to understand the effect of large penetration of DG on fault currents, protection, and dynamic interactions with the power system. Standards work currently underway is looking at enhancing the role of DG. The IEEE standard on smart grid technology currently being formulated is expected to include interconnection requirements and added functionalities for DER and electric-sourced transportation.

### 16.9.4 Electric-Sourced Transportation

Large-scale adoption of electric and plug-in electric vehicles will dramatically change the grid load profile and usage. Besides charging at residences, the limited range of these vehicles on one charge requires charging infrastructure similar to gas stations. Pilot installations of charging infrastructure are now being carried out in different parts of the United States. Three levels of charging with increasing power and decreasing charge times have been proposed. Level 1 charging is from a standard power outlet in the house. Level 2 charging is from ±120 V with an 80-A service. Level 3 charging uses a higher voltage dc interface. Power converters are required for charging stations external to vehicles and chargers within vehicles. Needless to say this is a potentially big application of power electronics. Integration of vehicle charging with a smart grid will include time of day-based charging in order to flatten the daily load profile seen by the utility, and thereby shift more generation to base-loaded plants.

Besides acting as a load, several functions for these vehicles and the charging infrastructure have been proposed due to the presence of the battery and power converters. Among these are energy arbitrage, integration with demand response or load control, frequency regulation, reactive power control, and renewable firming. These functionalities are expected to offset the high cost of batteries. However, their ramifications on battery lifetime and capacity reduction are still being evaluated.

## 16.9.5 Flexible AC Transmission Systems

FACTS is a collective term for different types of power electronic devices/converters-based systems that are capable of controlling power flow in high-voltage ac transmission systems.[105,106] With advances in power semiconductor devices, PWM methods, and control theory, the use of FACTS devices has seen a significant increase. FACTS devices are capable of the following major functions:

- Control power flow along desired transmission corridors, which is critical for a deregulated utility; they can also minimize loop flows.
- Increase transmission capacity without requiring new transmission infrastructure.
- Improve transient, dynamic, and voltage stability, and provide damping for interarea oscillations.

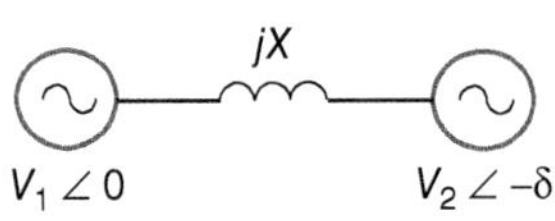


**FIGURE 16-41** Single-line diagram of a two-bus system.

Different types of FACTS devices control different parameters of the transmission system like the effective line impedance, bus voltage magnitudes, or phase angles to control power flow and to increase stability margins. Consider the single line diagram of a two bus system shown in Fig. 16-41. The real power flow in the connecting transmission line is given by

$$P = \frac{V_1 V_2}{X} \sin(\delta) \tag{16-54}$$

where $V_1$, $V_2$ are the magnitudes of sending and receiving end voltages, respectively, $\delta$ is the phase angle between the two voltages, and $X$ is the series line impedance.[107] FACTS devices control one or more of these three parameters to control power flow and improve stability.

***Transient Stability.*** FACTS devices have the ability to enhance both transient and dynamic stability of power system networks, thereby enabling increased power flow through existing transmission lines. Transient instability occurs when a major disturbance like fault, line outage, or loss of generation results in large rotor angle deviations leading to loss of synchronism. The rotor angle deviation is governed by the swing equation given in Eq. (16-55).

$$\frac{2H}{\omega_o} \frac{d^2\delta}{dt^2} = P_m - P_e \tag{16-55}$$

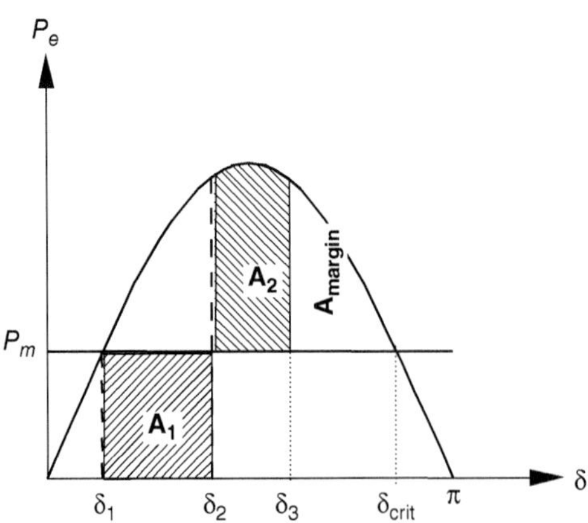


**FIGURE 16-42** Equal area criteria for transient stability.

where $H$ is the inertia constant in MJ/MVA, $\omega_o$ is the synchronous speed, while $P_m$ and $P_e$ are the mechanical power input and electrical power output, respectively. During a fault, the electrical energy drawn from the generator reduces significantly, while mechanical power input remains roughly constant, leading to increasing rotor angle. If the fault is not cleared before a critical time, transient instability occurs. The critical clearing time depends on the electrical power output during the fault and immediately after fault clearance. Since power flow can be controlled continuously using FACTS devices, power during and after a fault can be controlled to improve the stability margin of the system.

Transient instability is often studied using the equal area criterion as shown in Fig. 16-42. Initially the mechanical input power input is equal to the electrical power transmitted, at an angle $\delta_1$. A fault at the generator makes the electrical power zero while the mechanical input power remains the same, leading to increase in rotor angle from $\delta_1$ to $\delta_2$, at which point the fault is cleared. During this interval, the stored kinetic energy in the machine increases, and increase in kinetic energy is equal to the area $A_1$ in Fig. 16-42. After the fault is cleared the electrical power transmitted is higher (due to increased phase angle) than the mechanical power input. Hence, the machine begins to decelerate. However, the phase angle increases further due to the stored kinetic energy. The maximum angle is reached at $\delta_3$, when the decelerating energy represented by the area $A_2$ becomes equal to the accelerating area $A_1$. If the phase angle extends beyond $\delta_{crit}$, then the system is unstable since decelerating energy cannot balance the accelerating energy. The area $A_{margin}$ between $\delta_3$ and $\delta_{crit}$ represents the transient stability margin of the system. FACTS devices can improve the margin by dynamically changing the $P - \delta$ characteristics of the system.

***Thyristor-Controlled Series Capacitor.*** The earlier FACTS devices were predominantly thyristor based, like the thyristor-controlled series capacitor (TCSC) and static VAR compensator (SVC).[106] TCSC is a series-connected FACTS device that controls the effective impedance of the transmission line. Figure 16-43*a* shows the basic schematic of a TCSC, which consists of a capacitor in parallel with a thyristor-controlled reactor (TCR). TCR is a series combination of an inductor and a pair of phase-controlled thyristors. By suitably controlling the firing angle of the thyristors, the reactance (inductive) of TCR, and therefore, the effective fundamental impedance of TCSC, can be controlled continuously. The relationship between the firing angle and the effective TCSC impedance is highly nonlinear. The firing angle, $\alpha$, is measured from the zero crossing of the capacitor voltage. As an example, corresponding to an installed capacitive impedance of 0.5 pu (per unit) and inductive reactance of 0.1667 pu, the effective impedance of TCSC can be controlled from about 4 pu capacitive to 2 pu inductive. The effect of TCSC control on the transient stability margin is illustrated in Fig. 16-43*b*. In the figure *s* indicates the degree of compensation. One of the major advantages of TCSC, when compared with uncontrolled series compensation, is the ability to mitigate subsynchronous resonance (SSR).[106]

***Voltage Source Converter-Based FACTS.*** The newer FACTS devices are based on voltage source converters (VSC) implemented using fully controllable devices such as GTO, MCT, IGCT, and IGBT.[108] Within their voltage and current ratings, the VSC-based FACTS devices are capable of injecting any suitable, controlled voltages and/or currents at the line frequency. The main advantages of these FACTS devices, compared to the thyristor-based devices, are the speed of response and the extended control range which is mostly independent of the line operating conditions. The main VSC based FACTS devices are the static compensator (STATCOM), the static synchronous series compensator (SSSC), and the unified power flow controller (UPFC).

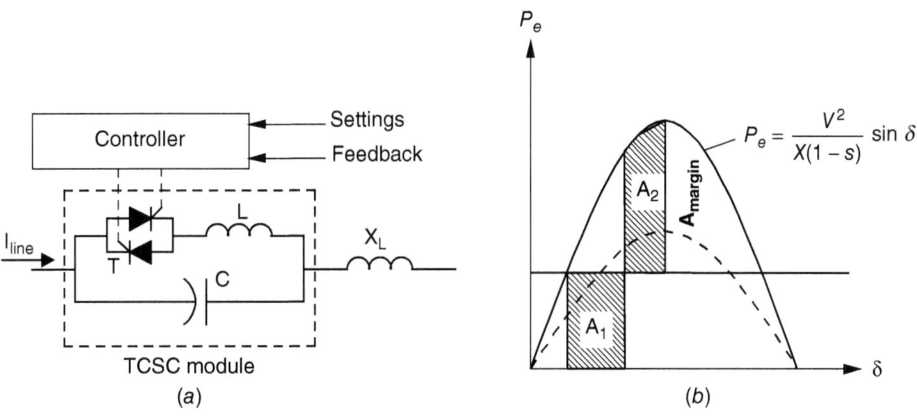


**FIGURE 16-43** TCSC: (*a*) schematic, (*b*) enhancement of transient stability.

*Static Synchronous Compensator.* Static Synchronous Compensator (STATCOM) is a shunt FACTS device capable of injecting controlled currents at the point of connection with the transmission system.[109,106] The injected current is usually in phase quadrature (leading or lagging) with the line voltage, so that only reactive power is supplied or consumed by the STATCOM. If real power capability is present, through the use of active energy sources or large energy storage systems, then the injected current can have different phase relationships with the line voltage, thereby extending its control range.

Figure 16-44*a* shows the schematic of a STATCOM connected at the midpoint of a two bus transmission system model. The voltage source converter is capable of generating the required fundamental voltage such that the current injected into the system has the desired phase and magnitude to control power flow. The voltage at the dc link is kept constant by large capacitor banks. Losses in the system

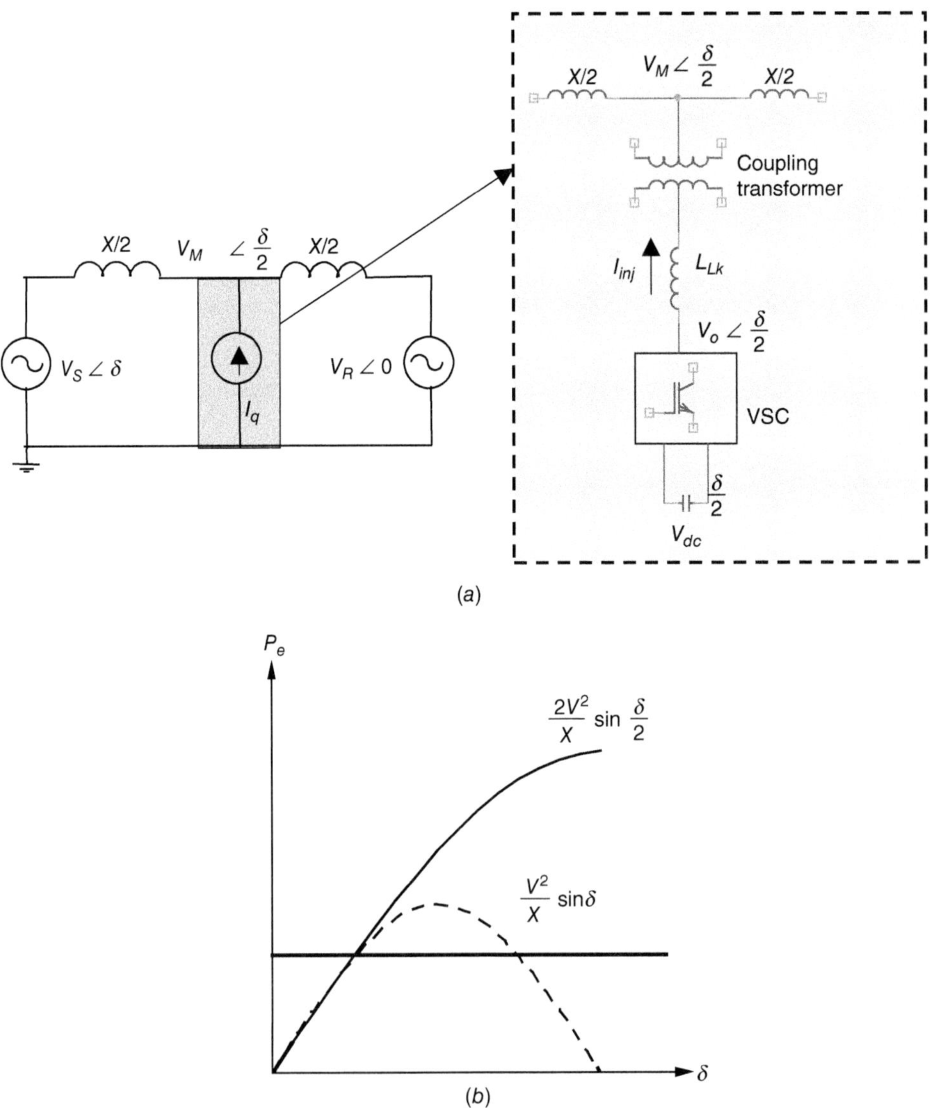


**FIGURE 16-44** STATCOM: (*a*) mid-point connection, (*b*) variation of power with phase angle $\delta$.

are compensated, and the capacitor voltage maintained, by drawing a small real power from the transmission system. The voltage source converter is connected to the transmission line through a line frequency coupling transformer, which enables the STATCOM to work with lower voltage switches.

The output voltage of STATCOM (neglecting losses) is controlled to be in phase with the line voltage. Hence, the system can be modeled as two in-phase, line frequency voltage sources, connected by a reactor (usually the leakage inductance of the coupling transformer), which results in the current $I_{\text{inj}}$ being purely reactive. Referring to Fig. 16-44*a*, if the magnitude of $V_o$ is larger than $V_m$, then the STATCOM feeds reactive power into the system, and if $V_o$ is smaller, it absorbs reactive power.

Referring to Fig. 16-44*a*, the amplitudes of the sending end, midpoint, and receiving end voltages are assumed to be equal for simplicity ($V_s = V_m = V_r = V$). The STATCOM compensation at the midpoint effectively segments the transmission line into two independent parts, each with an effective line reactance of $X/2$. Neglecting losses, the real power flow is the same in both parts, and can be derived as given in Eq. (16-56).

$$P = \frac{V^2}{(X/2)} \sin\left(\frac{\delta}{2}\right) \tag{16-56}$$

where $\delta$ is the angle between the sending and receiving end voltages. Figure 16-44*b* shows the variation of real power flow with phase angle, as determined by Eq. (16-56). The curve corresponding to no compensation is also shown for comparison. It can be clearly seen that the STATCOM significantly improves the transient stability margin, that is, for a given fault clearing time, STATCOM allows a much higher real power to be transmitted. In the case of power system oscillations, such as inter-area oscillations, the shunt compensation is varied dynamically to provide damping.

*Static Synchronous Series Compensator.* The SSSC is a series connected device that injects a synchronous line frequency voltage, normally in quadrature with the line current. SSSC controls power flow by controlling the line voltage amplitude, phase angle, and effective line impedance.

*Unified Power Flow Controller.* The UPFC is a versatile FACTS device that combines the functions of a STATCOM and an SSSC, and extends their capability to inject shunt current or series voltage that involves real power flow as well.[110] With UPFC, the real and reactive power can be controlled independently. UPFC is capable of controlling all the power system parameters such as voltage magnitudes, phase angles, and effective line impedance simultaneously, and therefore, meet multiple control objectives. Figure 16-45*a* shows the schematic diagram of the UPFC. It consists of two voltage source converters with separate controllers but sharing a common dc link with dc storage capacitors.

In present installations of UPFC, most of the control functions are performed by the series converter by injecting a voltage $V_{\text{inj}}$ whose phase is independent of the line current and can vary practically from 0° to 360°. The magnitude of the injected voltage can also be varied continuously within the rating of the series converter. The main function of the shunt converter is to provide the real power exchanged by the series converter with the system. It may be noted that the real power exchanged by the series converter is ultimately derived from the transmission line, but, the reactive power is absorbed or supplied locally by the series converter and does not need to come from the transmission system. The shunt converter can be operated at unity power factor or can be controlled to provide additional functions beyond supporting real power needs of the series converter.

Figure 16-45*b* highlights the capabilities of the series converter of the UPFC, namely control of voltage magnitudes, phase angles, and impedance.[106] With the injected voltage in phase or antiphase with the line voltage the UPFC provides voltage regulation or magnitude control. For a given voltage rating of the series converter, the in-phase addition provides maximum voltage magnitude control. For impedance control, the magnitude of the injected voltage is proportional to the line current and the phase is in quadrature (leading or lagging). In phase angle control, the magnitude and angle of the injected voltage are controlled such that the sending end voltage has the required phase angle without any change in the magnitude. These three features can be combined to achieve multifunction power flow control.

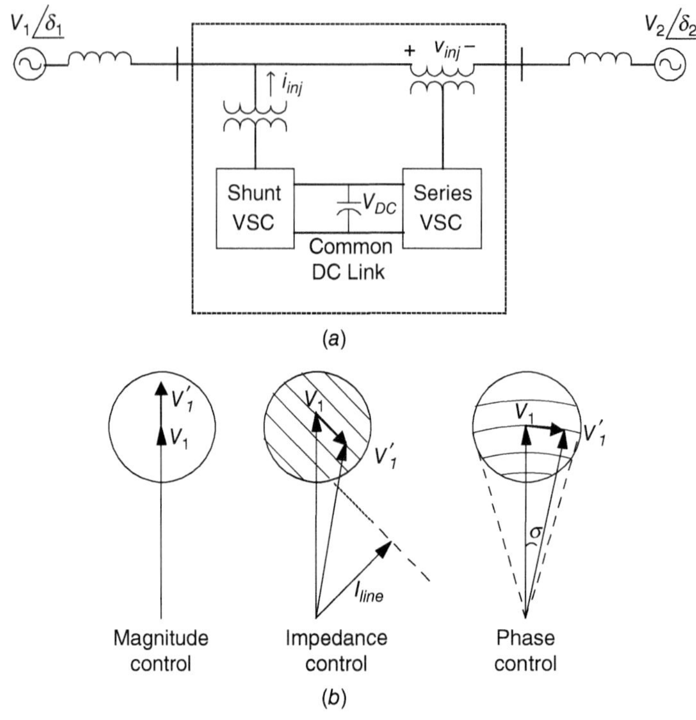


**FIGURE 16-45** Unified power flow controller: (*a*) schematic, (*b*) control capabilities.

## 16.9.6 Custom Power

The digital age loads, e.g., processing plants in semiconductor industry and data centers, require clean and uninterrupted power. These loads are highly intolerant to (even momentary) power quality problems such as voltage sags or interruptions, harmonics in line voltage, phase unbalance, and flicker in supply voltage. Power electronic systems that mitigate power quality problems in utility distribution systems (1 to 38 kV) are defined as custom power devices.[111–113]

Similar to FACTS, the custom power devices can be connected in shunt or series with the distribution line or a combination of both. The major custom power devices are the dynamic voltage restorer (DVR), distribution static compensator (DSTATCOM) and the unified power quality controller (UPQC). The DVR is a series connected device that injects a controlled voltage to compensate for voltage sags and other momentary disturbances. The DSTATCOM is a shunt connected device injecting controlled currents at the point of common coupling to compensate for power quality problems in the load current. UPQC combines the features of DVR and DSTATCOM.

***Dynamic Voltage Restorer.*** Short duration voltage sags are the predominant power quality events, with estimated revenue lost per event of more than $1M for pharmaceutical industries. Power acceptability curves that quantify voltage disturbances in terms of magnitude of these sags (and swells) and duration of the disturbance have been developed.[114] The most popular of the power acceptability curves is the CBEMA curve shown in Fig. 16-46. This was developed by the Computer Business Equipment Manufacturers Association (CBEMA), now the Information Technology Industry Council (ITIC). The semiconductor industry has its own standard called SEMI F47, developed by the Semiconductor Equipment and Materials Institute (SEMI).

Dynamic voltage restorers, which are among the most installed custom power devices, protect sensitive equipment against short-term voltage disturbances.[115] Figure 16-47 shows the schematic

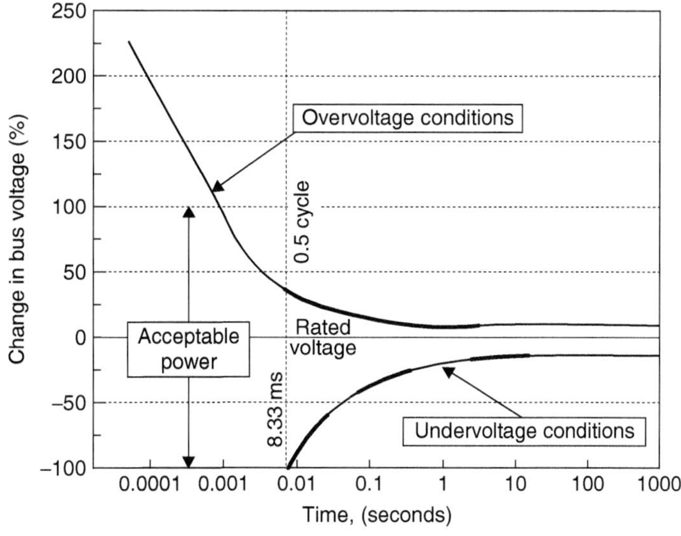


**FIGURE 16-46** The CBEMA curve.

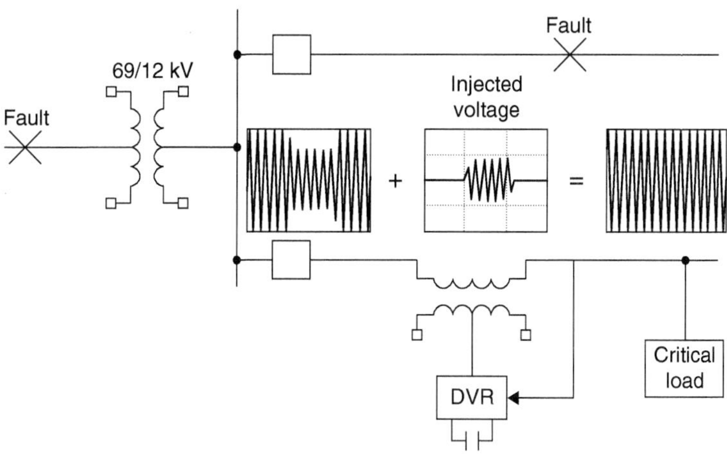


**FIGURE 16-47** Application of a DVR.

of a DVR. As seen, DVR is a voltage source converter-based series connected device that injects a line frequency voltage of appropriate magnitude and phase such that the voltage across the sensitive loads is always well regulated, and any disturbances in the input voltage is not propagated to the load. The voltage source converter is implemented using IGBT switches, which operate at frequencies in the range of tens of kHz. They have fairly high control bandwidth and can respond to voltage disturbances in a small fraction of the line frequency cycle.

When the DVR is not connected to an active dc source and cannot handle real power in steady state, the injected voltage is constrained to be in phase quadrature with the load current. With this mode of control, the magnitude of sags that a DVR can correct becomes a function of the load power factor, and at higher power factor (close to unity) only a smaller voltage disturbance can be corrected.

Since most of the sag events are of short duration, many of the installed DVRs rely on the large energy storage capacitor to supply real power for a short duration, and not constrain the injected voltage to be in-phase quadrature. Several other installations use a separate rectifier to supply real power to the dc link from the distribution system.[116]

***DSTATCOM.*** The distribution STATCOM has similar structure as that of the STATCOM used in transmission systems, and injects controlled currents. However, the main objectives of DSTATCOM are quite different. The load currents in distribution system can be unbalanced and contain reactive and harmonic components. Standards such as IEEE 519 and IEC 61000 place limits on maximum permissible harmonic currents for various types of equipment and voltage levels.[62,63] The DSTATCOM with closed loop control, injects correction currents such that the compensated load draws balanced, fundamental, unity power factor current.

*Unified Power Quality Conditioner.* The unified power quality conditioner (UPQC) combines the features of a DVR and DSTATCOM and can inject current in shunt and voltage in series simultaneously. Figure 16-48 shows the schematic of a UPQC. It has the same structure as that of UPFC used in transmission systems, consisting of two voltage source converters sharing a common dc link. One of the converters is connected in series with the distribution line injecting controlled voltages and the other converter is connected in shunt and injects controlled currents. Therefore, the UPQC can simultaneously correct for unbalances and distortion in line voltage as well as load currents.

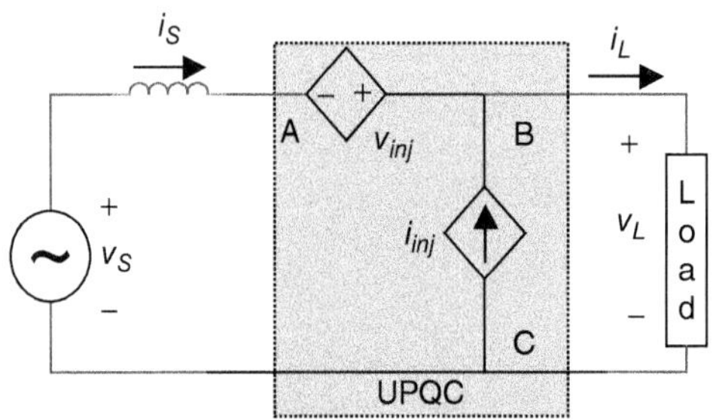

**FIGURE 16-48** Unified power quality conditioner.

Solid state switches used to connect critical loads to multiple feeders or to break short-circuit currents, hence improving power quality, are also considered as custom power devices. These are referred to as network reconfiguring devices and include solid-state current limiter (SSCL), solid-state breaker (SSB), and solid-state transfer switch (SSTS).[111] These are much faster than the conventional mechanical switches and hence, significantly enhance the reliability of the distribution system.

## 16.9.7 Solid State Transformers

Power generation, transmission, and distribution are the three main parts of the modern power system, in which the power transformers play a critical role. The power transformers enable a high efficiency and long distance power transmission by boosting the voltage to a higher one on the generation side. On the distribution system side, this high voltage is stepped down to lower voltages for the industrial, commercial, and residential use. The traditional line frequency (60 Hz to 50 Hz) transformer (LFT) is totally a passive device. Its advantages are high efficiency and high reliability since it has been in use for more than a century. It could also be said that the invention and development of the LFT is a key reason that today's electric grid is dominantly an ac grid. The main disadvantage of the LFT is its lack of any control functionality. These functionalities are increasing required by today's grid, especially on the destitution side of the grid, also known as the edge of the grid. These functionalities, include, but not limited to reactive power support and voltage support, power flow control, harmonic mitigation, voltage ride through, and fault isolation. Also the size and weight of LFT is general very high and hence there is a desire to reduce them if possible.

Some of the above mentioned functions are currently provided by solutions invented to address these issues on a case by case basis. Tap changer and capacitor banks, for example, are used to address the reactive power support and voltage control issue. Power electronics based solutions such as SVC and STATCOM are also used to provide reactive power and voltage control. Active power filter (APF) is used to mitigate the harmonic current in the grid. DVR is used to provide low voltage ride through capability. However, a much better and elegant solution is to replace the LFT with a solid-state transformer (SST) or power electronics transformer (PET). The motivation is that the SST will embed almost all of the above mention smart functions into a single equipment. Equipment with communication capability, it also allows a much better and faster control from a utility operator. The size and weight of the SST is also much smaller than the LFT because the galvanic isolation transformer is operating at a frequency significantly higher than 60 Hz.

Shown in Fig. 16-49, the SST is a power electronic device that replaces the traditional LFT by means of high frequency transformer isolated ac-ac conversion technique. The basic operation of

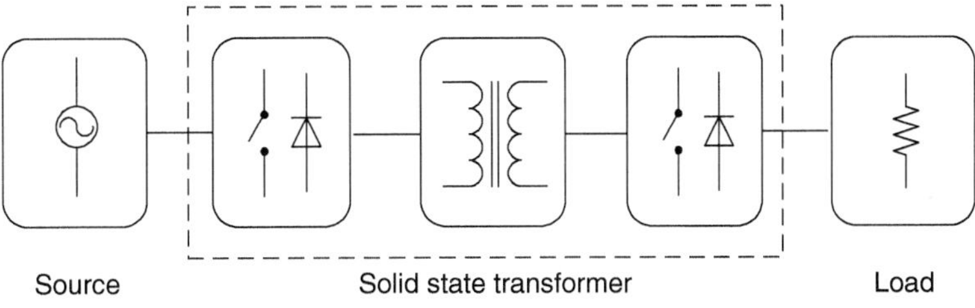


**FIGURE 16-49** Solid State transformer structure.

the SST is firstly to change the 50/60 Hz ac voltage to a high frequency one (normally in the range of several kHz to tens of kHz), then this high-frequency voltage is stepped up/down by a high-frequency transformer with dramatically decreased volume and weight, and finally shaped back into the desired 50/60 Hz one to feed the load. Voltage step down is also achieved by properly selecting the turns ratio of the high-frequency transformer. Depends on the application, at least one side of the SST is connected to very high voltages, making SST a very challenging application for power electronics. As an example, a single phase SST for residential customer in the United States may require the input side to operate with a 7.2 kV ac distribution grid voltage with a peak value of 10 kV. This will require semiconductor devices with breakdown voltages of more than 15 kV if simple two level topologies are used. Needless to say, the progress in power electronics technology, such as the multilevel converter topology and ultra-high voltage power semiconductor devices, is making the SST concept more practical today than many years ago. For example, 15 kV MOSFET transistor based on wide bandgap semiconductor material SiC has been proposed and researched as a key technology for SST application[142,146] while 6.5 kV Si IGBT device and multilevel topology are used in a SST prototypes.[145] The demand for smart grid functionalities and better renewable energy integration is another main driver for the world wide effort in investigating SST applications in power system.[143]

Numerous power electronics topology could be applied to the SST, and they can be classified as shown in Fig. 16-50. Each SST architecture has its own pros and cons in terms of efficiency, weight, and functionality. Type A SST represents a possible low cost, high reliability, and light weight solution due to the simple one-stage conversion configuration. Four-quadrant power devices will be needed in case of bidirectional power flow operation condition while unidirectional switches can be used if the power flow is in one direction. However, the lack of the dc link makes them unsuitable for applications where reactive power compensation is required. In addition, disturbances on one side may also affect the other side, which is one of the drawbacks of the LFT. Type B and Type C SSTs adopt a two-stage configuration, with an isolation stage on either the high- or low-voltage side. Four-quadrant power devices may also be needed on the ac side of the isolation stage for bidirectional power flow. Compared with the Type A configuration, reactive power compensation is possible for Type B and C SSTs if suitable topologies are chosen. Type B also offers a low-voltage dc link that can be used to power dc loads and dc resources such as PV and energy storage devices. However, Type B configuration is quite challenging from power electronics point of view since zero-voltage-switching (ZVS) is hard to be guaranteed in such a wide input range circuit, and also mature multilevel topologies cannot be easily applied on the high-voltage side. In this condition, high switching losses may not be avoided without sacrificing switching frequency, leading to lower efficiency, and

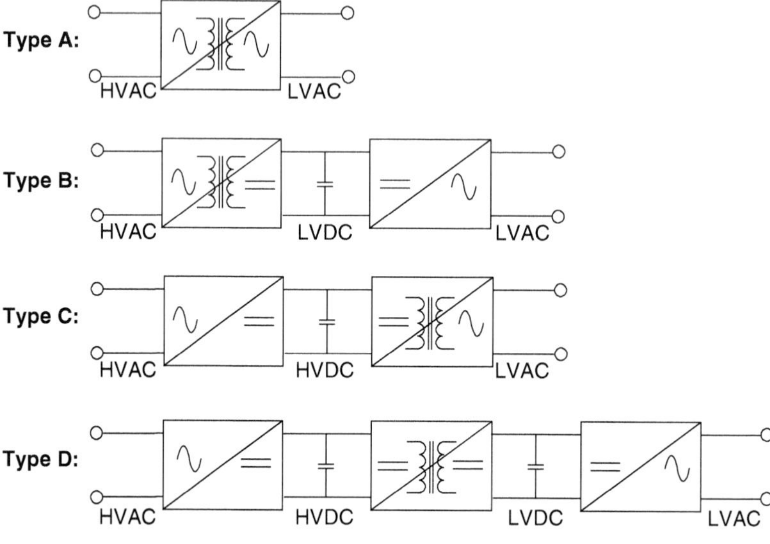


**FIGURE 16-50** Classification of SST topologies.

difficult thermal management of the power devices. Type C topologies may not face such a problem since lots of works have been done for the high-voltage ac/dc conversion. Nonetheless, the lack of the low-voltage dc bus makes the integration of renewable resources on the low-voltage (potentially residential) side unfeasible. Most of the SST topologies developed so far use the structure shown in Type D, in which three stage structures are adopted. This structure can leverage many of the existing topologies, and optimized design could be achieved.

If the semiconductor breakdown voltage is not high enough for the intended ac grid voltage, modular multilevel topologies can be used to increase the voltage handing capability. A typical implementation is shown in Fig. 16-51[144] which also has three conversion stages, so it is of Type D SST but with a modular multilevel implementation instead of a simple two level implementation. The cascaded multilevel converter is utilized on the high-voltage side with identical (low voltage) H-bridges, thus low-voltage power device can be adopted. Several dc-dc converters with relatively high switching frequency (in the range of a few kHz to tens of kHz decided by the power devices) are

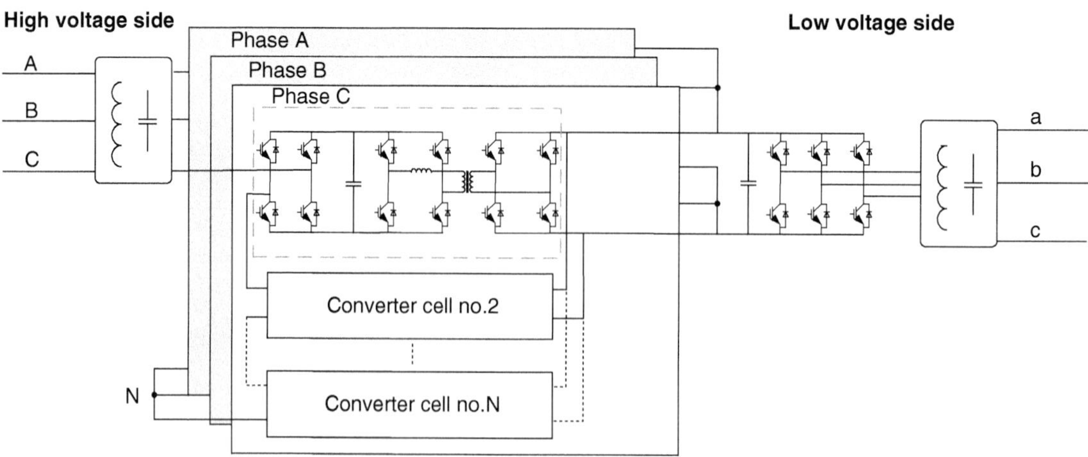


**FIGURE 16-51** A modular type SST for high-voltage and high-power application.

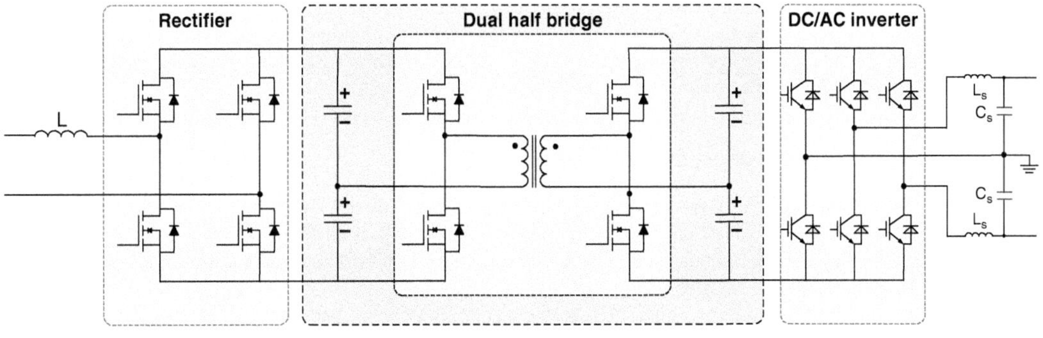


**FIGURE 16-52** A Type D SST with a two level converter using 15 kV SiC MOSFET.[146]

then parallel connected to a common low-voltage dc bus. In the last stage, conventional low-voltage source inverter technologies can be used. It is worth to point out that there is no limitation for the number of modules since this is the key feature of the cascaded type multilevel converter. The number of the modules is depended on the operating voltage and power device adopted. Single phase or three-phase SST can both be implemented with this architecture as shown in Fig. 16-51.

If the semiconductor device breakdown voltage is high enough, SST can be implemented with a simple two level topology, such as the one shown in Fig. 16-52.[146] In this case, six 15 kV SiC MOSFETs are used to implement a 20 kVA Type D SST. The isolated dc/dc converter operates at 20 kHz due low switching loss under the ZVS condition. The current rating required for the high-voltage MOSFETs is only 10 A, which is very low due to the low KVA rating needed for residential applications.

As already mentioned, the SST is not a simple replacement of the LFT, it provides additional functions in addition to isolation and voltage conversion, such as voltage regulation, load regulation, harmonic compensation, etc. The future distribution system architecture is proposed as shown in Fig. 16-53. The existing power distribution network is depicted on the left part, where traditional

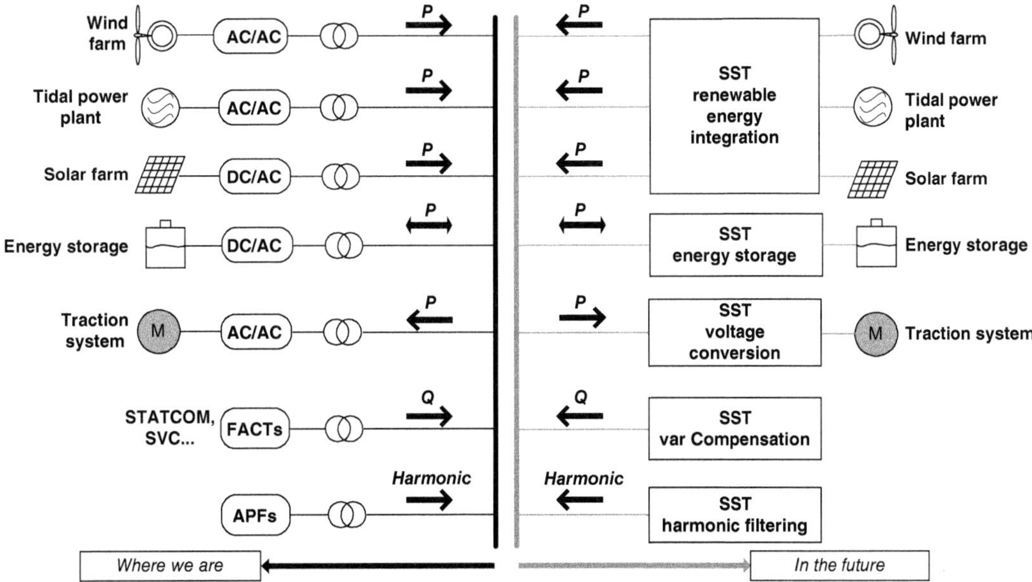


**FIGURE 16-53** Vision of SST enabled future distribution power grid.

transformers are adopted to interface renewable energy resources, energy storage devices, FACTs devices, and loads to the distribution line. The right hand side demonstrates conceptually how the future distribution system looks like with the SSTs as the main interface. It is clear that the SST could significantly change the architecture of future power distribution grid.

### 16.9.8 Modular Multilevel Converters

Multilevel converters are attractive for two major reasons, the ability to increase the operating voltages and an improved ac waveform quality. Among many different types of multilevel converter topologies, MMC is a relatively new family member.[147,148] It has gained its popularity quickly because of its advantages such as distributed location of capacitive energy storages, modular and scalable design, low switching frequency, transformer/inductor-less grid connection, and the availability of a common dc link which is very useful for HVDC transmission system application.[149] Figure 16-54*a*

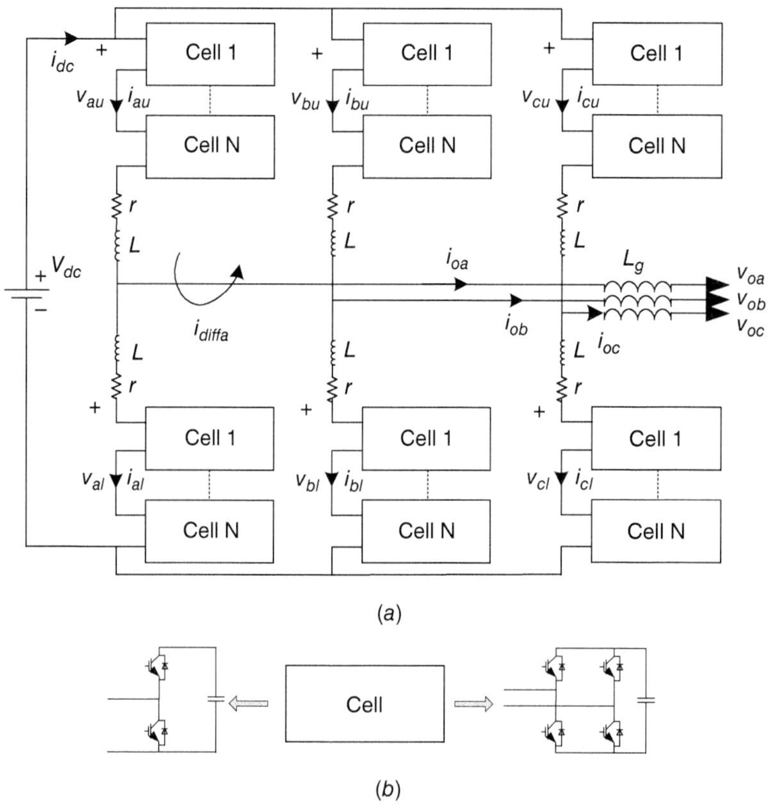


**FIGURE 16-54** Configuration of a three-phase MMC converter: (*a*) power circuit, and (*b*) cell configuration.

shows the basic topology of the MMC. It consists of two arms per phase leg, in which several switching cells are connected in series together with an arm inductor. The switching cells are controlled to synthesize the desired ac phase voltage, and at the same time balance the distributed capacitor voltage. The arm inductor is necessary to limit the circulating current that flows within the converter.[150]

Different switching cells could be adopted in MMC, while the most common ones are half-bridge circuit and full-bridge circuit as shown in Fig. 16-54*b*.

The control architecture of MMC could be formed by the following modeling process by taking phase A as an example. Define the direction of circulating current as shown in Fig. 16-54, the arm current can be expressed as

$$i_{au} = \frac{i_{oa}}{2} + i_{\text{diffa}} \quad i_{al} = -\frac{i_{oa}}{2} + i_{\text{diffa}} \tag{16-57}$$

Thus the circulating current is

$$i_{\text{diffa}} = \frac{1}{2}(i_{au} + i_{al}) \tag{16-58}$$

According to Kirchoff's voltage law (KVL)

$$-\frac{V_{dc}}{2} + v_{au} + L\frac{di_{au}}{dt} + ri_{au} + L_g\frac{di_{oa}}{dt} + v_{oa} = 0 \tag{16-59}$$

$$-\frac{V_{dc}}{2} + v_{al} + L\frac{di_{al}}{dt} + ri_{al} - L_g\frac{di_{oa}}{dt} - v_{oa} = 0 \tag{16-60}$$

Substituting (16-57) into (16-59) and (16-60)

$$-\frac{V_{dc}}{2} + v_{au} + \frac{L}{2}\frac{di_{oa}}{dt} + L\frac{di_{\text{diffa}}}{dt} + r\frac{i_{oa}}{2} + ri_{\text{diffa}} + L_g\frac{di_{oa}}{dt} + v_{oa} = 0 \tag{16-61}$$

$$\frac{V_{dc}}{2} - v_{al} + \frac{L}{2}\frac{di_{oa}}{dt} - L\frac{di_{\text{diffa}}}{dt} + r\frac{i_{oa}}{2} - ri_{\text{diffa}} + L_g\frac{di_{oa}}{dt} + v_{oa} = 0 \tag{16-62}$$

From (16-61) and (16-62)

$$v_{au} - v_{al} + (L + 2L_g)\frac{di_{oa}}{dt} + ri_{oa} + 2v_{oa} = 0 \tag{16-63}$$

$$V_{dc} - (v_{al} + v_{au}) - 2L\frac{di_{\text{diffa}}}{dt} - 2ri_{\text{diffa}} = 0 \tag{16-64}$$

Rewriting the equations

$$v_{oa} = \frac{v_{al} - v_{au}}{2} - \left(\frac{L}{2} + L_g\right)\frac{di_{oa}}{dt} - \frac{r}{2}i_{oa} = e_a - \left(\frac{L}{2} + L_g\right)\frac{di_{oa}}{dt} - \frac{r}{2}i_{oa} \tag{16-65}$$

$$\frac{V_{dc}}{2} - \frac{(v_{al} + v_{au})}{2} = L\frac{di_{\text{diffa}}}{dt} + ri_{\text{diffa}} = v_{\text{diffa}} \tag{16-66}$$

Considering (16-65) and (16-66), the arm voltage references are given as

$$v_{al\_\text{ref}} = \frac{V_{dc}}{2} + e_a - v_{\text{diffa}} \tag{16-67}$$

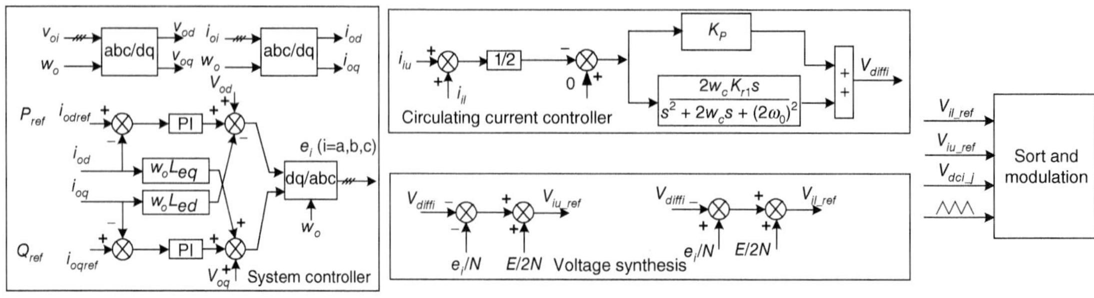


**FIGURE 16-55** Example control architecture of the MMC.

$$v_{au\_ref} = \frac{V_{dc}}{2} - e_a - v_{\text{diffa}} \tag{16-68}$$

It is shown in (16-67) and (16-68) that the control objectives for MMC are to generate the $e_i$ $(i = a, b, c)$ by system controller and the $v_{\text{diffi}}$ $(i = a, b, c)$ by the circulating current controller. Figure 16-55 shows one of control architecture of MMC.[150] The system controller under $d$-$q$ coordinate is implemented to regulate the active and reactive power flow. The circulating current is suppressed by an additional proportional resonant (PR) controller, and the capacitor voltage balanced is achieved with proper sorting and modulation algorithm.

MMC has been investigated in various applications, such as HVDC, medium-voltage drive, etc.[148] Figure 16-56 shows the HVDC transmission system based on MMC, transferring the off shore wind energy to onshore load centers. One MMC operates like an ac-dc rectifier while the other one acts as a dc-ac inverter. Compared with the traditional two-level voltage source converters, it significantly reduces the switching loss as well as increases the modularity. Because both MMCs are voltage source converter, they can also generate reactive power on both sides of the ac terminals. Several commercial installations of MMC based HVDC system have been reported. The dc link voltage can be as high as ±350 kV dc. In this case, many MMC cells based on high-voltage IGBT, IGCT, and ETO can be used.[151] Figure 16-57 illustrates the operation of MMC for medium-voltage drive application. However due to the low frequency voltage fluctuation issue, it may only fit for applications with limited speed ranges, such as fan/blower applications.

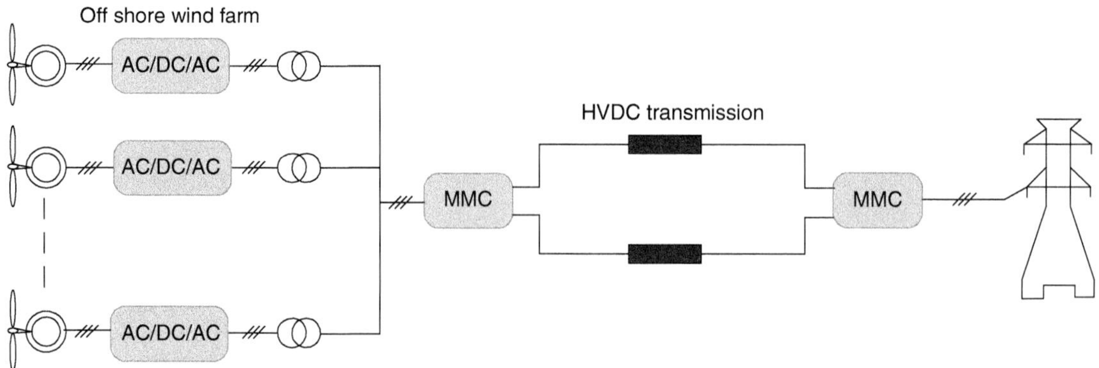


**FIGURE 16-56** Off-shore wind farm based on MMC-HVDC transmission.

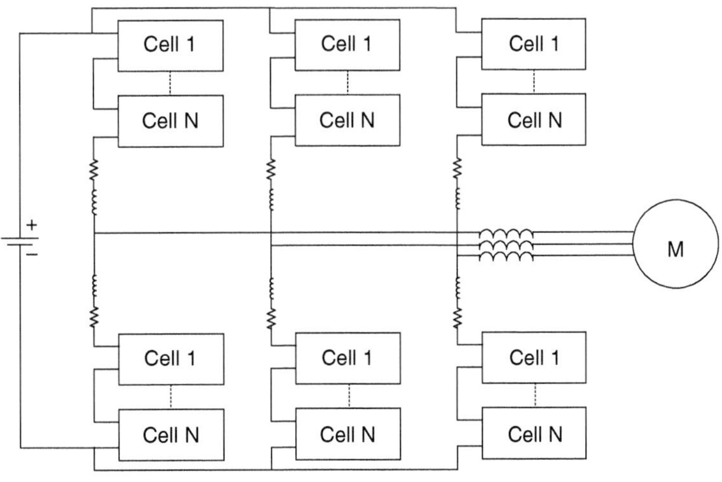


**FIGURE 16-57** Medium-voltage drive based on MMC.

## 16.10 COMPONENTS OF POWER ELECTRONIC CONVERTERS

This subsection describes the individual components that constitute a power converter. These include power semiconductor devices and passive components (inductors, transformers, capacitors). A detailed description of the structure and physics of power semiconductor devices is beyond the the scope of the discussion here. The interested reader is referred to standard books on this subject.[4,117,118] Details of magnetics, material properties, and capacitors are also likewise not covered. Unlike semiconductor devices and capacitors, very few magnetics are available as standard products from manufacturers, and therefore require custom design. Magnetics design is covered in significant detail in Ref. 119. Properties of magnetic materials, even though they follow some generic trends, have to be obtained from magnetic core manufacturers. Similarly, for details of capacitors, application notes and datasheets supplied by manufacturers have to be relied on.

### 16.10.1 Silicon Power Semiconductor Devices

Power electronic circuits require high-power semiconductor switches and diodes. An ideal switch should have the following characteristics: full control over switch state (on/off), zero-voltage drop during on state, infinite impedance during off-state, and instantaneous transition between states. Diodes should have very low-voltage drop during conduction, infinite impedance in off-state, and instantaneous transition. Practical devices have nonideal characteristics, and different devices capitalizing on one advantage while sacrificing some other have been developed. Figure 16-58 shows the circuit symbols of common power semiconductor devices. These are listed below with their voltage, current, and switching limitations for silicon power devices. A discussion of power devices based on wide bandgap (WBG) materials such as SiC and GaN can be found in 16.10.2.

**Diodes**: line frequency, fast recovery, ultra-fast recovery, and Schottky—in increasing order of switching speed, and decreasing order of reverse voltage rating

**MOSFETs** (metal oxide semiconductor field effect transistor): good for low voltage ~100 s of volts, high switching frequency (> 100 kHz)

**IGBTs** (insulated gate bipolar transistors): good from a few hundred volts to about 6.5 kV, currents up to 1.2 kA, and switching frequency up to 30 kHz

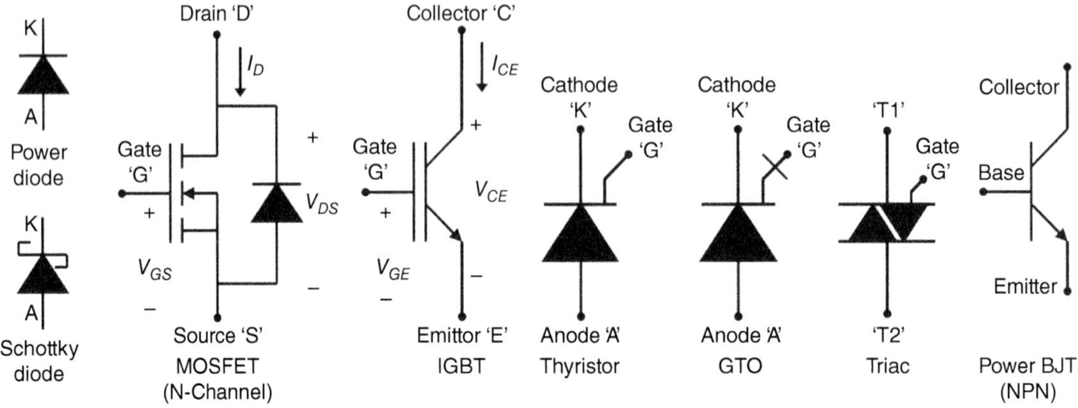


**FIGURE 16-58** Circuit symbols of common power semiconductor devices.

**Thysristors or SCRs** (silicon-controlled rectifiers): good for very high voltage and current [~ (kV) and (kA)], and low-power moderate performance applications

**GTOs** (gate turn-off thyristors): good for very high voltage [~ (kV)], high-current applications [~ (kA)], with switching frequency up to a few kHz

**Miscellaneous**: IGCT (integrated gate commutated thyristor), ETO (emitter turn-off thyristor), MCT (MOS-controlled thyristor), BJTs (bipolar junction transistors), triacs, diacs

The ratings and device types listed above are those made with doped silicon. Recently introduced devices that are made with silicon carbide and gallium nitride semiconductors are discussed later in 16.10.2.

***Diodes.*** There are essentially two types of power semiconductor diodes: PN junction and metal-semiconductor junction (Schottky).

PN junction power diodes have an additional $N^-$ (lightly doped with N type impurities) drift region, so the overall structure is $PN^-N$. The depletion layer of the PN junction extends in the $N^-$ region when the diode is reverse biased, and its length determines the maximum reverse voltage the diode can block. Intrinsically, the $N^-$ region has a high resistance. However, when the diode is forward biased there is injection of excess carriers in this region resulting in a low effective resistance. This phenomena is commonly called *conductivity modulation*. The on-state voltage drop across the diode consists of the PN junction drop and the resistive drop in the drift region; typically it ranges from 0.6 to 1 V. There is a small delay in going from off-state to on-state due to the time required for the carriers to build up. During turn-off, the excess carriers in the drift region have to be removed. Thus, for a short time, the diode conducts in the reverse direction with a high voltage across it. This phenomena, known as reverse recovery, leads to significant power loss and becomes one of the limiting factors in high-frequency circuits. PN junction-type power diodes are classified as

- Line frequency rectifiers: for rectification of 50/60 Hz utility input.
- Fast and ultra-fast recovery diodes: for high-frequency rectification. These have recovery time ranging from a couple of microseconds to 10 s of nanoseconds.

Schottky diodes are based on metal-semiconductor junctions. These junctions have a lower junction potential leading to a lower forward voltage drop. Silicon-based Schottky power diodes have forward voltage drop ranging from 0.3 to 0.6 V and can withstand reverse voltages up to 200 V. As opposed to PN junction diodes, Schottkys are majority carrier devices, so they do not have any reverse recovery. Compared with PN junction diodes, Schottky diodes have higher reverse leakage current therefore Si Schottky diode is typically only available at lower voltages. Silicon-based Schottky diodes are suited for very high-frequency, low-voltage, and high-current rectification. Recently, silicon carbide (SiC)-based Schottky diodes have been developed and are now commercially available with rating up to 1700 V.[120,121] The absence of reverse recovery makes them suitable for high-voltage high-frequency rectification, albeit at a higher cost. Because the resistance of the SiC Schottky is much lower than Si, a much higher Schottky barrier can be used (typically 0.7 to 0.9 V). This significantly reduces the reverse leakage current and much high voltage rating SiC Schottky diode and its variants can be developed. Research prototypes of 15,000 V has been demonstrated.

Another development is the introduction of higher-voltage silicon diodes (e.g., Qspeed from power integrations[122]) that rely on innovative structure, combining benefits of Schottky and PN junction characteristics, and achieve higher blocking voltage capability with insignificant reverse recovery. Single dies rated up to 300 V and series combination of two devices in a single package rated up to 600 V are available at this time. In terms of performance and cost, these diodes lie in between ultrafast silicoh PN diodes and silicon carbide Schottky diodes.

***MOSFETs.*** Unlike signal level MOS devices which are fabricated laterally, power MOSFETs have a vertically diffused structure. For an N-channel MOSFET the doping is of the form $N^+PN^-N^+$. The drain is the $N^+$ terminal next to the $N^-$ region, while the source is the $N^+$ region next to the P region. A positive voltage on the isolated gate terminal produces an electron channel in the P-region allowing current to flow from drain to source (or source to drain). Further, the P-type body is shorted to the source terminal resulting in an intrinsic diode with anode at the source and cathode at the drain. Although this diode is not very good in performance, it is useful for most power electronic circuits. Both P and N channel power MOSFETs are available, but N-channel MOSFETs are more prevelant due to their lower on-state resistance. The discussion here is therefore restricted to N-channel MOSFETs.

The steady-state *V-I* characteristics for an N-channel MOSFET are shown in Fig. 16-59*a*. For $V_{GS} < V_{th}$, the MOSFET acts as an open circuit from drain to source; $V_{th}$, the threshold voltage, is in the range of 2 to 4 V. For $V_{GS} > V_{th}$, the MOSFET follows the characteristics shown in Fig. 16-59*a*. In amplifier circuits, MOSFETs are operated in their active region, where the drain current $I_D$ is almost independent of the drain to source voltage $V_{DS}$. Power MOSFETs are operated in the ohmic region where $I_D$ is proportional to $V_{DS}$, and the MOSFET behaves like a resistance. The effective on-resistance, designated $R_{DS}$, depends on $V_{GS}$, $I_D$, and the junction temperature $T_j$. For MOSFETs, $R_{DS}$ increases with increasing temperature, a property useful in paralleling of devices to obtain higher current carrying capacity. The maximum value of $V_{GS}$ is usually ±20 V; for logic level power MOSFETs, it is limited to ±10 V. For most MOSFETs, increasing $V_{GS}$ beyond 10 V does not have significant effect on $R_{DS}$. MOSFETs are rated for (a) maximum drain to source breakdown voltage ($BV_{DSS}$) (b) maximum continuous average current for a specified temperature (e.g., $I_{D25}$ at 25°C), and (c) a safe operating area (SOA) in terms of $V_{DS}$, $I_D$, and time duration for which $I_D$ flows. In high current applications, current capacity of leads and bond wires, and thermal performance of the device package also impose an additional constraint.

MOSFET gate drive circuits usually run on a nominal $V_{cc}$ = 15 or 12 V, and switch $V_{GS}$ between $V_{cc}$ and 0 V. For very high current applications and improved noise sensitivity, a negative $V_{GS}$ may be applied during the off-time. Since the gate is insulated from the source, there is no dc current flow from gate to source. However, depending on their ratings power, MOSFETs have a significant input capacitance $C_{iss}$. Thus, to increase $V_{GS}$ from 0 to 15 V in a very short time (20 to 100 ns) a significant current (order of 1 A) is required. Figure 16-59*b* shows a basic MOSFET gate drive circuit. Another circuit shown in Fig. 16-59*c* has faster turn-off due to the additional path through $R_{g,off}$ and a negative voltage for the off-state. Several gate drive ICs that can source and sink current up to a few amperes, while providing several other auxiliary functions, are commercially available.[9,12] For *high-side switches*, switches whose source voltage changes with the switch state, as is the case for the MOSFET in a buck converter, an isolated gate drive is required. The isolation is provided by using either high-frequency transformers or high-speed optocouplers. There are several standard gate drive configurations, each with its own pros and cons. The best source of these circuits are application notes from device manufacturers.[81]

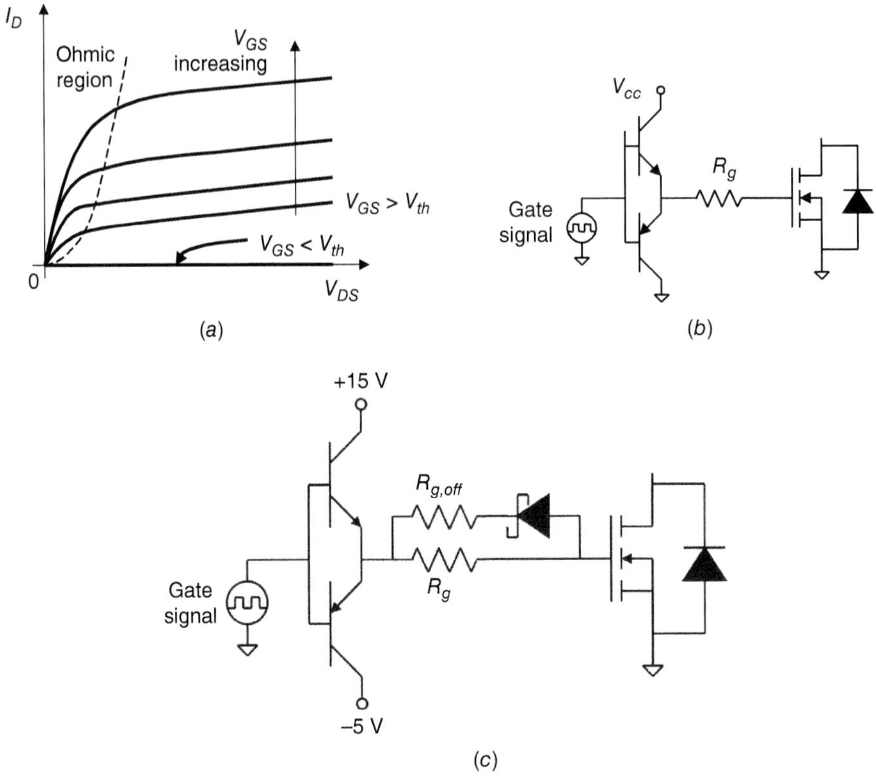


**FIGURE 16-59** MOSFETs: (*a*) *V-I* characteristics; (*b*) simple gate drive circuit; (*c*) fast turn-off gate drive.

Power MOSFETs can be used up to a few 100 kHz and in some applications in the MHz range. To increase operating frequencies further, *RF Power MOSFETs* have been introduced.[123] These can be operated in the 10-MHz range and are expected to reduce converter size and weight considerably. MOSFET based on SiC and GaN are now being introdcued and they offer even faster switching speeds.

***Super-Junction MOSFETs.*** Transistors and diodes that operate based on the majority carrier electrons (such as MOSFET, JFET, and Schottky diode) all have a resistive voltage-current relationship when conducting current. The resistance is theoretically related to the square of the breakdown voltage $BV$ as $R = 4BV^2/A_{\text{chip}}\, \mu_n E_c^3$ where $A_{\text{chip}}$ is the semiconductor chip area, $\mu_n$ is the electron mobility and $E_c$ is the critical electric field causing avalanche breakdown. Therefore, the resistance increases quickly as the voltage rating of these devices go up. Remember the conduction loss is proportional to $I^2R$, therefore the current handling capability of a power device decreases quickly when the breakdown voltage goes higher. Reducing the resistance can only be achieved by increasing chip area $A_{\text{chip}}$. Practical chip size is limited to below 1 cm$^2$ because of an increased cost due to a decreasing manufacturing yield. Power modules where parallel devices are used are equivalent to increasing $A_{\text{chip}}$. The resistance can also be reduced by using different materials that have higher $\mu_n$ and $E_c$. This approach is the motivation behind the interest in WBG power semiconductor materials because they have much higher $E_c$ than silicon.

The resistance's relationship to the breakdown voltage can be significantly improved to a much better linear relationship $R = 2 \times d \times BV/A_{\text{chip}}\, \mu_n E_c^2$ if the super-junction structure is used. This structure changed the way the PN junction is formed in a typical transistor, resulting in above significantly improved R-BV relationship. For this reason, power MOSFETs at higher voltages, especially for voltages beyond 500 V, super-junction MOSFET should be used. Several manufacturers, including Infineon Technologies and ST Microelectronics have introduced the super junction MOSFETs under various name, such as the CoolMOS products from Infineon. The breakdown voltage of these commercial products range from 500 to 1200 V. For power MOSFET constructed using WBG materials, the need to use super-junction concept is much less urgent since the resistance is significantly reduced by the much larger $E_c$.

In Si super-junction MOSFETs, there is a significantly increased PN junction area. This causes the device to have a very bad reverse recover behavior when the MOSFET's body diode is used as the freewheeling diode. For this reason, Si super-junction MOSFET should not be used where the current flow is bidirectional such as a motor drive.

***IGBTs.*** The structure of IGBTs is similar to MOSFETs. Some simulation models for MOSFETs use a MOS gated bipolar junction transistor although the operation is more complicated than this simple model implies. IGBTs have significantly higher voltage and current ratings compared to MOSFETs, and their on-state voltage drop is also lower. In addition, several IGBT chips are paralleled inside one package to form an IGBT module with a significantly higher current rating. IGBTs modules are commercially available in voltage rating up to 6.5 kV, and currents up to 3600 A. The gate drive requirements of an IGBT are similar to that of MOSFETs. It is turned on by applying a positive gate source voltage (typically 15 V), and turned-off by applying a smaller negative voltage (about −5 V). Compared to MOSFETs, IGBTs have much longer switching times. Unlike MOSFETs, IGBTs are minority carrier devices, so their turn-off is characterized by a "tail current." Their switching frequencies are generally limited to 30 kHz, and the maximum switching frequency reduces with power level. The IGBT only conducts current in one direction, therefore typically a fast recovery diode will have to be packaged together to allow reverse current flow needed in applications such as motor drive inverters and PV inverters.

***Thyristor and Similar Devices.*** Thyristors, also called silicon-controlled rectifiers (SCRs), are high-power semiconductor devices which can block voltage of either polarity and conduct current in one direction only (from anode to cathode). They can be switched on by applying a current pulse to their gate terminal (with return path through cathode) when there is a positive voltage from anode to cathode. They are characterized by a latching action due to positive feedback of conductivity modulation in the PNPN structure: once the anode current reaches a threshold, the device continues to conduct until the current is reduced to zero by the external circuit. Their turn-off cannot be effected via the gate. Thyristors are available in very high current and voltage ratings and have a very low conduction voltage drop. Thyristors are mostly used for ac to dc (or vice versa) power conversion, where the device current is expected to reduce to zero. Normally, thyristors are switched at or close to the ac system frequency. They are ideally suited for high-power dc motor drives, and utility applications such as HVDC and controlled reactors.

*Gate turn-off (GTO)* thyristors are high-power devices, with their low-end ratings overlapping with the high-end rating of IGBTs. Unlike a thyristor, they can be turned on and turned off using the gate terminal, although the gate drive is more complex compared to a MOSFET or an IGBT. The switching times for GTOs are of the order of 10 $\mu$s so their maximum switching frequency is in the kHz range. They are used exclusively in very high power applications like motor drives, FACTs devices, and active filters. GTOs are commercially available with ratings up to 6 kV and 6 kA. Enhancements to GTOs have led to the development of IGCT (integrated gate commutated thyristor)[124] or GCT (gate commutated turn-off thyristor),[125] ETO (emitter turn-off thyristor), and MTO (MOS turn-off thyristor).[118] These enhancements lead to much faster switching speed in the order or 1~2 $\mu$s, and the elimination of the turn-off snubber. Overall circuit efficiency improves over conventional GTO based converters. At present there is still a strong research effort in development of high-power devices with high switching speeds. Ultra-high-voltage devices based on SiC are also being developed.

Two thyristors connected in antiparallel (anode of one to the cathode of other) can conduct current in either direction and block voltages of either polarity. Triacs realize this functionality in a single semiconductor device. However, voltage and current ratings of triacs are very low compared to that of thyristors and their performance is inferior to that of thyristors. Solid-state switches utilize triacs and antiparallel connected thyristors.

Power *bipolar junction transistors* (BJTs) have been almost completely replaced by MOSFETs and IGBTs, due to ease of control and higher switching frequency. However, they are still used in some applications like linear power supplies.

## 16.10.2 Wide Bandgap Power Semicoductors Devices

Si power devices have been developed and improved over the last 60 years. As silicon power devices are approaching their limitations in terms of additional conduction and switching loss reduction, researchers turned their attentions to power devices developed on WBG materials that have higher bandgap ($E_g$) than silicon material. The direct benefit is a significantly increased breakdown critical electric field ($E_c$) which can be utilized to design power devices with much lower conduction resistance $R_{on}$ for a given chip area. For semiconductor switches, such as MOSFET/BJT/JFET, the conduction resistance or on-resistance is theoretically related to the square of the breakdown voltage $BV$ as $R = 4BV^2/A_{chip}\mu_n Ec^3$, where $A_{chip}$ is the semiconductor chip area, $\mu_n$ is the electron mobility and $E_c$ is the critical electric field causing avalanche breakdown. WBG materials typically have a much higher $E_c$ therefore the resistance can be significantly reduced. For example, in SiC material, the critical electric field is about 10× time higher than silicon, therefore the resistance could theoretically be reduced by about 1000 times for the same chip size $A_{chip}$. Commercial devices have not reached this level of predicted reductions yet. But it has so far achieved about 100× reduction at a breakdown voltage of 1200 V. This comparison is with Si 1200 V power MOSFET. The improvements over Si SJ MOSFET or Si IGBT is approximately 10× and 3×, respectively.

Because of this large reduction in conduction resistance, power devices based on unipolar current conduction (MOSFET, JFET, etc.) are sufficient to cover a large spectrum of power electronics applications. The needs for bipolar devices are much less than the silicon case where silicon power MOSFET could not be scaled economically to higher than 600 V, even if the super junction (SJ) MOSFET concept is used. Therefore in high-voltage applications where the current state-of-the-art device is the silicon bipolar IGBT, WBG devices also have an upper hand in switching speed due to the unipolar current conduction mechanism. The switching speed is fundamentally determined by the parasitic capacitance in the device. Higher switching speed can be utilized to increase the system switching frequency without a major penalty in conversion efficiency. This can result in higher power density. The large bandgap also results in much lower leakage currents in the WBG power devices hence the intrinsic capability of these devices to operate at higher junction temperatures $T_j$ is excellent. On the other hand, silicon power devices have higher leakage currents, which limits the device's ability to operate above 125°C. However, due to the unipolar current conduction mechanism, the WBG power devices conduction resistance $R_{on}$ increases with the temperature which will limit the extent of the high $T_j$ benefit. Increasing $T_j$ also have an impact on other supporting components in a WBG converter such as the gate driver circuit and packaging materials. Therefore, the high temperature roadmap for WBG power device is not currently a high priority but is actively studied.

There are many materials that can be classified as WBG materials. Currently only SiC and Gallium Nitride (GaN) power devices are commercially introduced after several decades of development. There are research developments underway for power devices based on other WBG materials such as diamond, AlN, and $Ga_2O_3$.

***SiC Power Devices.*** SiC power devices that have been commercially introduced are Schottky diodes, SiC MOSFET, SiC JFET, and SiC BJT. All of these devices conduct currents based on only one charge carrier, the electron, and all have extremely high switching speed. For users who are familiar with Si Schottky diodes and MOSFETs, one can view the development as a significant expansion of the power ratings served by these devices.

***SiC Schottky Diodes.*** SiC diodes can be constructed as a Schottky diode or a PN junction diode. This is shown in Fig. 16-60. Currently only Schottky diodes are commercialized. Among the Schottky

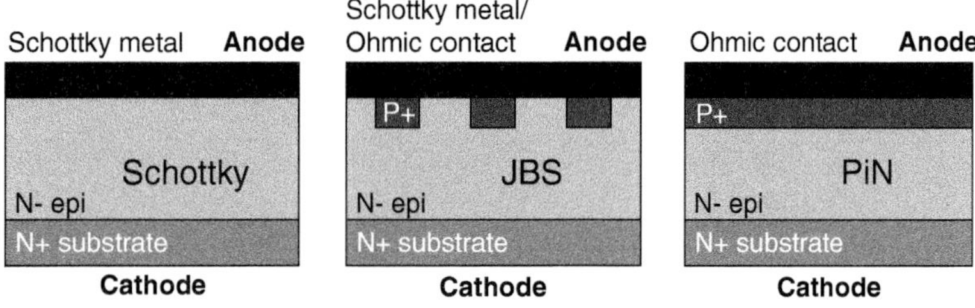


**FIGURE 16-60** Three basic SiC diode structures.

diode family, there are two major variations, the standard Schottky diode and the JBS diode. JBS is used to reduce the leakage current of the Schottky as well as to improve the surge current capability. Fig. 16-61 shows the ratings of the discrete SiC Schottky diodes available from the major market players. The typical voltage ratings are 600 V, 650 V, 1.2 kV, and 1.7 kV. There are also some 3.3 kV and 8 kV products available as shown in Fig. 16-61; however, the current rating is currently limited. For instance, the current rating for the 8 kV SiC diode is only 50 mA.

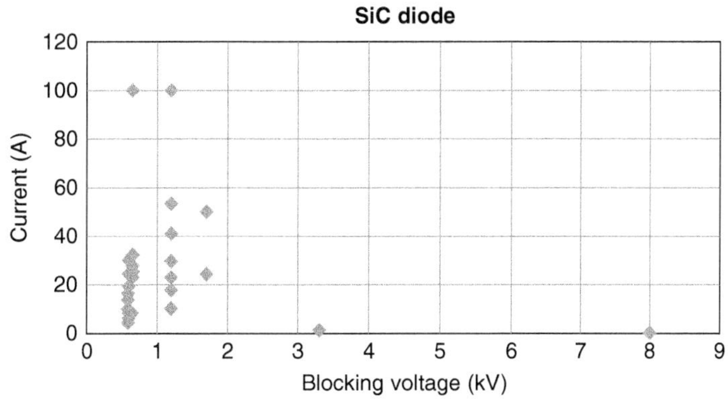


**FIGURE 16-61** Commercially available SiC Schottky diodes ratings @25°C.

The key characteristics of a SiC Schottky diode are shown in Fig. 16-62. In the forward state the device can be modeled as a constant knee voltage $V_T$ plus a resistance $R_{on}$. Both are temperature dependent, with $V_T$ decreasing as temperature increases while $R_{on}$ increasing as the temperature rises. The device has no reverse recover current except that contributed by the charging of the non-linear junction capacitance as shown in Fig. 16-62*b*. Due to this zero reverse recovery loss behavior, SiC Schottky diode are replacing Si PN junction diodes in high frequency converter applications such as flyback converters. SiC Schottky diode is also co-packaged with Si IGBT to form a high performance hybrid IGBT module in application such as PV inverters and motor drives.

Research prototype devices of SiC PIN and JBS diode of more than 10 kV have been also been reported. Figure 16-63*a* shows a comparison of two 10 kV prototype devices forward conductions. PIN diode shows a better conduction at higher current levels due to the bipolar conduction mechanism. However, the PIN diode will have a noticeable reverse recovery current, as shown in Fig. 16-63*b*.

***SiC Three Terminal Switches.*** Commercially introduced devices are SiC MOSFET, JFET, and BJT. Figure 16-64 shows the ratings of available devices on the market in discrete packages. Power modules

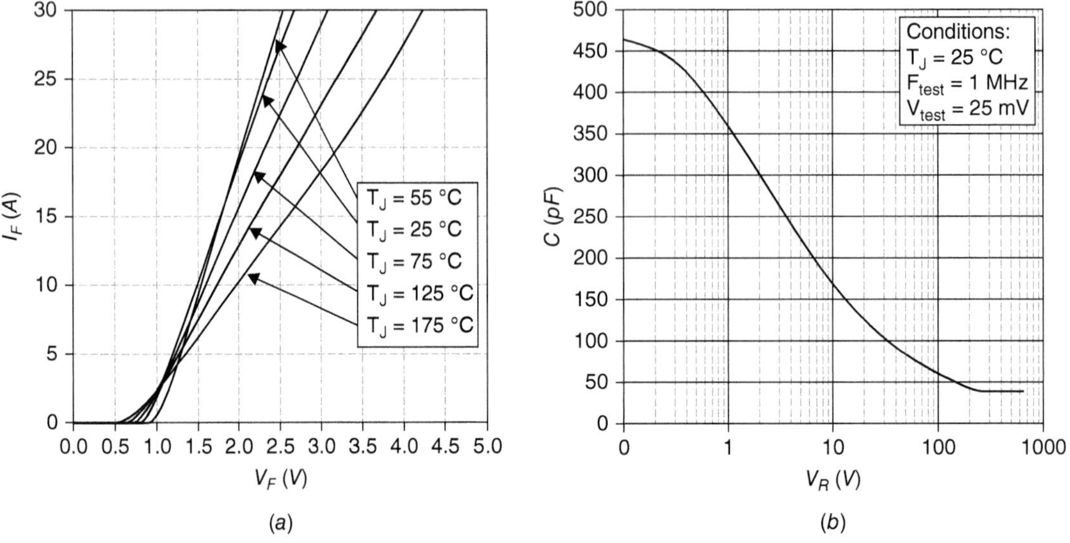


**FIGURE 16-62** Key characteristics of SiC Schottky diode: (*a*) forward I-V, (*b*) reverse capacitance.[152]

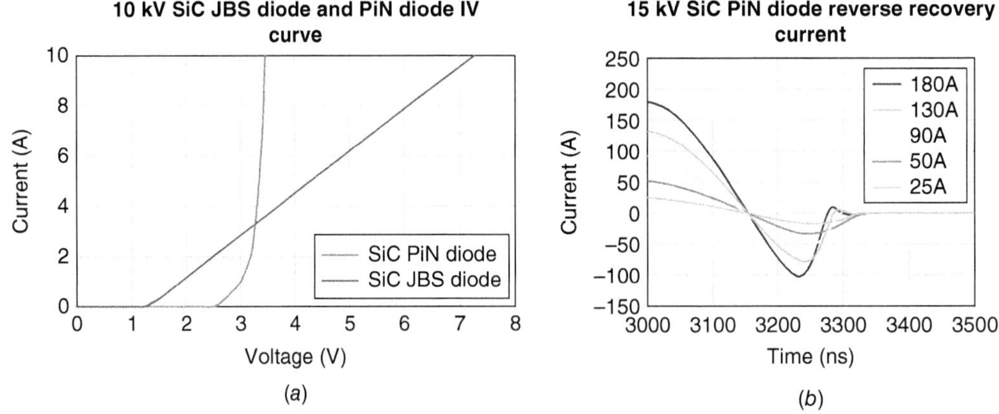


**FIGURE 16-63** (*a*) 10 kV SiC JBS diode and PiN diode *I-V* curves at room temperature, normalized to 0.32 cm² chip area. (*b*) Reverse recover current tested at $V_R = 7$ kV for a 15 kV SiC PiN diode ($T = 25°C$).

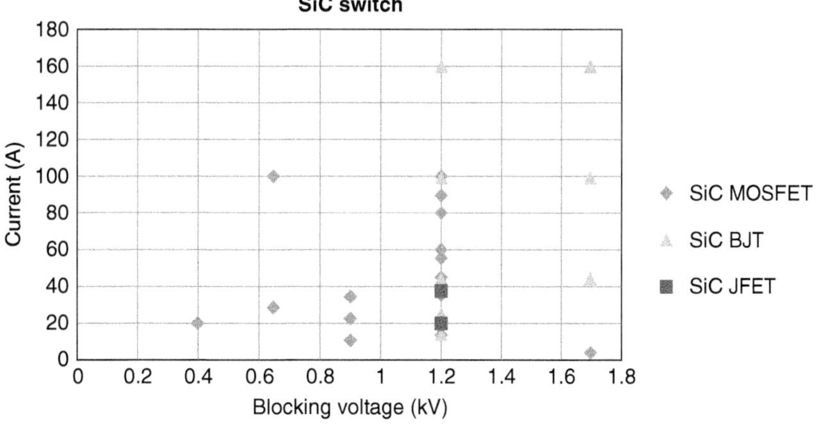


**FIGURE 16-64** Commercially available SiC discrete switch ratings @25°C.

with higher ratings are also available such as the recently introduced 900 V/1.25 mohm SiC power MOSFET module in which several MOSFETs are placed in parallel to increase the current rating and reduce the on-resistance.[153] MOSFETs on the market are all N-channel normally off device with a threshold voltage of several volts. On the other hand, most JFETs are N channel normally on devices with a threshold voltage of around −13 V.[154] These JFETs therefore require a negative gate voltage of more than −15 V to keep it in the off state. This is not a desirable characteristic in power converters because it presents a short circuit during the converter start up. The JFET with a series connected Si MOSFET (so called JFET cascode structure as shown in Fig. 16-65) could be used to eliminate this issue.[154] Since the BJT does not have a channel region, its on-resistance, is lower than SiC MOSFET. However, the current driving characteristic makes it less attractive for users who are familiar with the MOSFET drive interface. On the hand, the SiC BJT's switching speed is similar to that of the MOSFET due to the absence of any sizable minority carrier storage in the drift region.

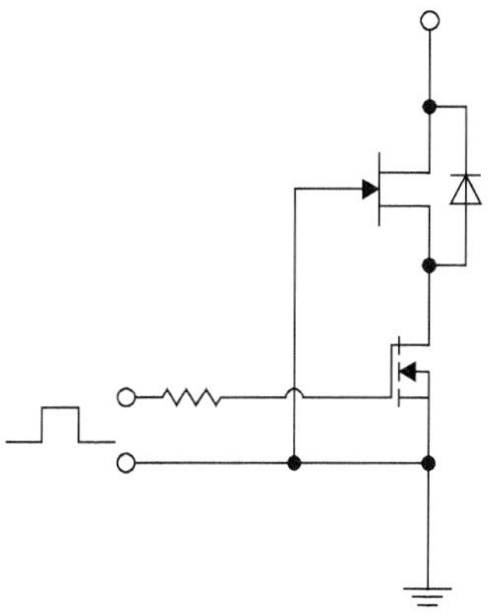

**FIGURE 16-65** JFET Cascode configuration which combines a normally on SiC JFET with a Si MOSFET to form a three terminal device.

Due to the familiar MOS gate-driving interface, it is expected that in the future the SiC MOSFET will dominate the market from 600 to 6500 V, competing or replacing Si IGBTs. Currently, 1200 V SiC MOSFETs have been introduced by several vendors. 600 V SiC MOSFETs are also introduced.[155] The SiC MOSFETs have two basic structures, a planar SiC MOSFET and a trench SiC MOSFET as shown in Fig. 16-66, both have a vertical current flow path inside a SiC wafer. These structure are very similar to those used in Si MOSFETs. From a user point of view, one can consider that SiC MOSFETs expand the Si MOSFETs power ratings significantly and will compete and replace Si IGBTs from 1200 to 6500 V due to SiC MOSFET's significant advantage in lowering the switching losses. The switching speed of the SiC MOSFET is fast because the device's speed is only limited by the charging and discharging of the parasitic capacitance, which has been reduced due to significantly reduced chip size for a given current rating. 3.38 MHz operation has recently been demonstrated using a 1200 V SiC MOSFET.[156] Another important characteristic of the SiC MOSFET is its 3rd quadrant operation such as the one shown in Fig. 16-67*a*.[157] By turning on the device in the 3rd quadrant, the device exhibits the same low resistive *I-V* characteristic as in the 1st quadrant, hence allowing the converter to reduce the reverse conduction losses in applications such as PV inverters and electric motor drives. The SiC

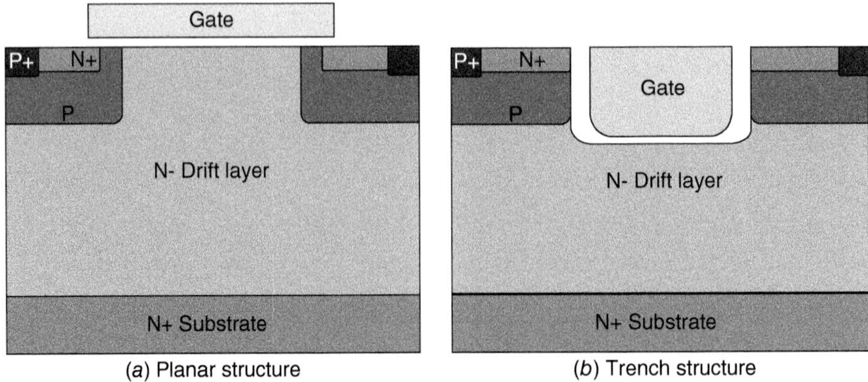


**FIGURE 16-66** Two typical SiC MOSFET structures.

MOSFET's body diode only conducts during the short dead time hence the conduction loss is minimized. The reverse recovery current of the body diode is also very small, as shown in Fig. 16-67*b*, due to very poor bipolar operations in SiC PN junction diodes.

There are various research activities underway to study SiC power switches for even higher voltages. For example, 27 kV SiC IGBT[158] and 22 kV SiC GTO/ETO have been reported.[159,160] Figure 16-68 shows a comparison of three devices forward conduction capabilities. The SiC MOSFET, IGBT, and GTO are all designed for a 15 kV voltage rating. It shows the advantage of bipolar devices at higher currents. High temperature characteristics of SiC IGBT and GTO are also significantly better than the MOSFET. On the other hand, the MOSFET is significantly faster during switching.

Generally speaking, SiC power devices enable a more efficient power conversion system. It can also be used to improve the power density by operating at higher frequencies. Figure 16-69 shows the SiC device application roadmap predicted by Yole.[161] SiC has already found its commercial applications in power factor correction (PFC), lighting, solar, railway traction, uninterrupted power supply (UPS), and will enter into drive, wind, electrical vehicle (EV) soon. Except for the technical design challenges,

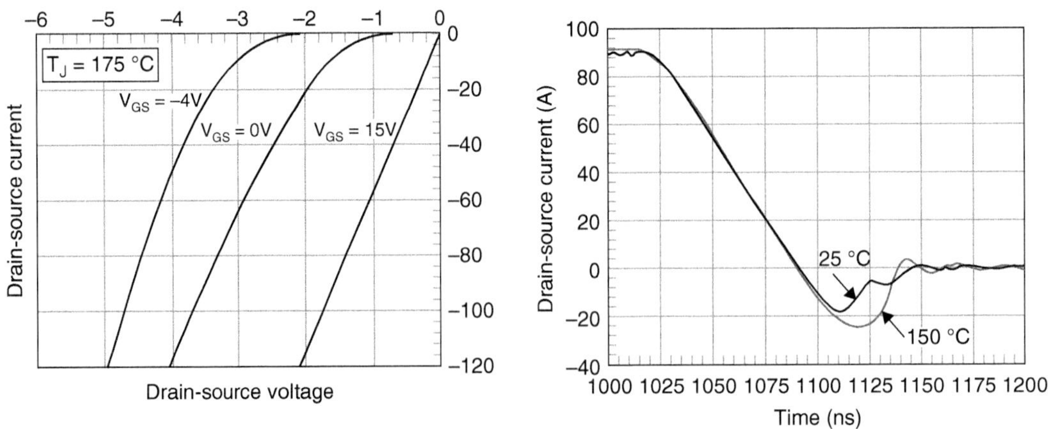


**FIGURE 16-67** (*a*) 3rd Quadrant characteristics of the 900 V, 10 mΩ MOSFET at 175°C.[6] (*b*) Reverse recovery characteristics of the 900 V/10 mohm SiC MOSFET, $V_R = 600$ V.[6]

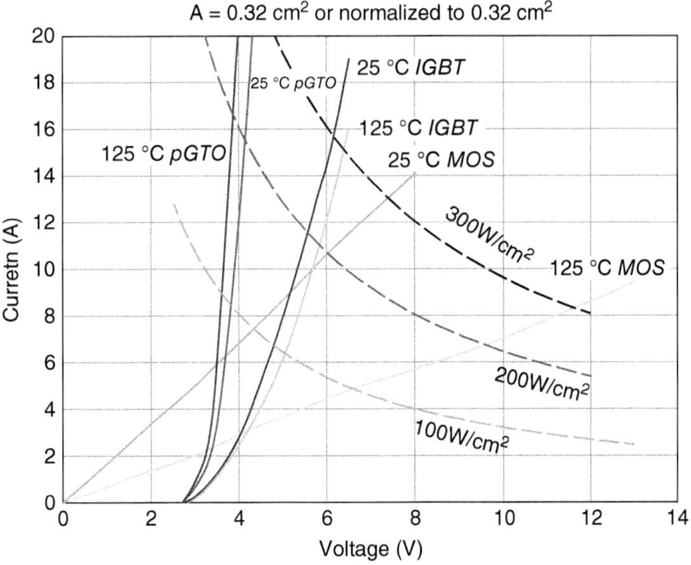


**FIGURE 16-68** *I-V* curve comparison of 15 kV SiC p-GTO, IGBT, and MOSFET, at 25°C and 125°C.

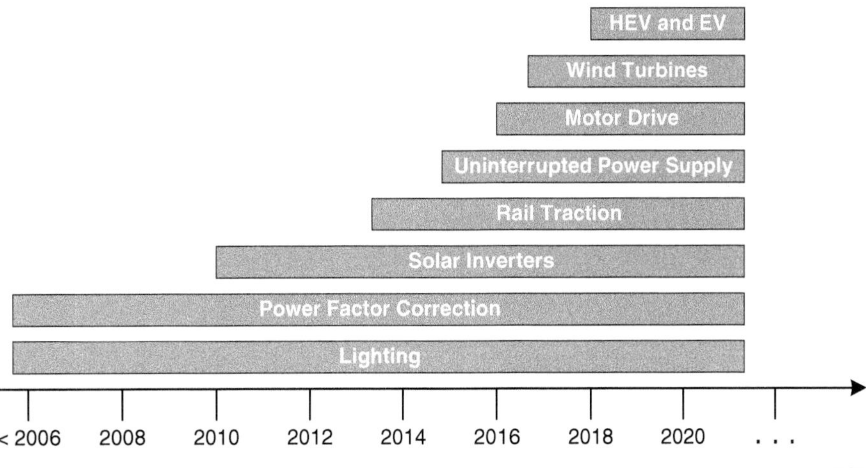


**FIGURE 16-69** SiC device application roadmap predicted by Yole.[10]

balancing higher cost of SiC device with improved system cost/performance is critical. The cost/W is one of the most concerned figure of merit from user perspective. Depending on the applications, the higher device cost does not necessary lead to higher system cost. In fact, the cost issue of SiC devices should be looked at from two aspects. First of all, the cost of SiC MOSFET decreases with the volume production. With larger volumes, better manufacturing process with larger wafers, and improved device performance, the cost/Amp of SiC devices will be reduced. Secondly, it is admitted that the absolute cost of SiC device is higher than Si due to intrinsic higher material cost. However, the cent/W cost of the

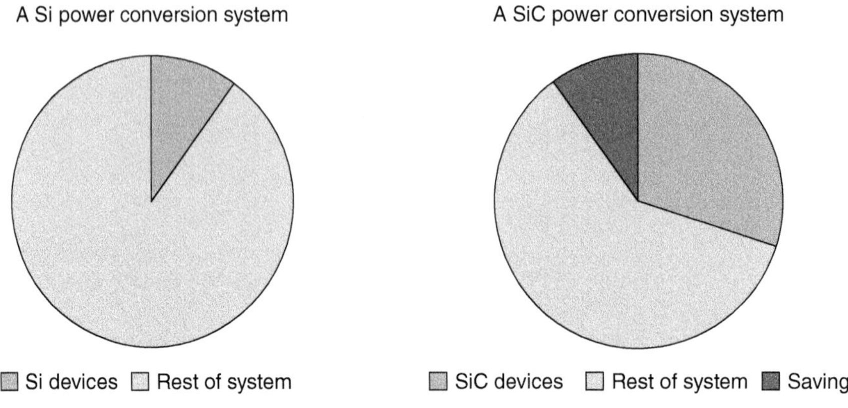


**FIGURE 16-70** Potential system cost benefit from SiC power devices.

overall system could potentially be reduced with a better balance of plant system design. Figure 16-70 shows the potential system cost saving by using a SiC power conversion systems. The saving may come from a smaller passive component, less cooling requirement, higher absolutely power rating, etc. It is demonstrated in Ref. 162 that the cost of a 17 kW solar inverter could be reduced by 20% with SiC JFET and SiC diode. Additionally, operational cost reduction gained from efficiency improvement could also justify the higher capital cost, such as in the UPS application.[163]

***GaN Heterojunction Field Effect Transistor.*** Gallium nitride (GaN) is another WBG material that has 10× increased critical electric field than Si, therefore it has also been intensively studied for power device applications. However, the material growth technique for GaN as well as the manufacturing process capability is significantly lag behind SiC, therefore there is currently no GaN power MOSFET developed or commercialized. Instead, GaN power transistor are developed based on the GaN Heterojunction FET (HFET) structure and commercial HFETs from 30 to 600 V are currently available on the market by an increased number of vendors. GaN HFET is developed on GaN-on-Si wafer and the current conduction path is lateral through the 2DEG formed by the GaN/AlGaN heterojunction as shown in Fig. 16-71. This structure results in lower cost since only a thin layer needs to be grown on a cheap Si substrate. The manufacturing techniques for the GaN HFET are also very similar to the Si CMOS process. These attributes result in the rapid commercialization of GaN HFETs in the last 5 years. Although the lateral structure is considered a disadvantage when compared to a

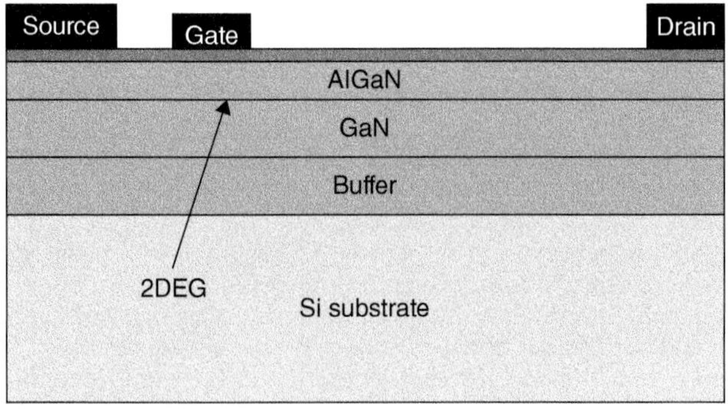


**FIGURE 16-71** GaN HFET.

**TABLE 16-3** Comparison of 600 V Si SJ MOSFET, SiC MOSFET, and GaN HFET

| 600 V FETs | $R_{on}$ (mohm) | Ciss (nF) | FOM1 ($R_{on}$ × Ciss) | Coss(nF) @400 V | FOM2 ($R_{on}$ × Coss) | Qrr(μC) | FOM3 ($R_{on}$ × Qrr) |
|---|---|---|---|---|---|---|---|
| Si SJ | 37 | 7.24 | 267 | 0.38 | 14 | 36 | 1332 |
| SiC MOS | 120 | 1.2 | 144 | 0.09 | 10.8 | 0.053 | 6.3 |
| GaN HFET | 25 | .52 | 13 | 0.13 | 3.25 | 0.113 | 2.8 |

Si SJ: Infineon IPW65R037C6. SiC MOSFET: Rohm SCT2120AF GaN HFET: GaNSystem GS66516T.

vertical power device due to the ineffective utilization of the chip area, GaN HFET's $R_{on}$ reduction is still very impressive when compared with Si power MOSFET, thanks to the high channel mobility (~2000 cm$^2$/V·s) and the elimination of the substrate conduction resistance $R_{sub}$. Furthermore, the lateral HFET has even lower junction capacitance (Ciss, Coss) due to the lateral geometry, making them even more attractive in high frequency applications such as flyback converters.

Table 16-3 compares device performance currently achieved by 600 V SiC MOSFET and GaN HFET and the best silicon SJ MOSFET. The important performance-dependent parameters are the three normalized figure of merits (FOMs) which can be obtained from the datasheet. Smaller FOM1 represents the advantage of fast gate driving capability. Both SiC and GaN are superior to SJ MOSFET but the advantage of the GaN over SiC is also clear. FOM2 represents the reduction in switching losses in hard switched or soft switched converters. Again, both SiC and GaN are superior to Si while there is still a clear advantage for GaN. Finally, FOM3 represents the reverse recovery loss reduction in SiC and GaN. Dramatic improvement is possible in both SiC and GaN when operate as a rectifier in the 3rd quadrant. As a matter of fact, the reverse recovery current and the associated losses, which have been a major headache for silicon power diodes, can be considered eliminated in future WBG converters using WBG Schottky diodes or 3rd quadrant operation of the FETs. The GaN HFET can also operate as a synchronous rectifier, similar to the case of SiC MOSFET, reducing further the reverse conduction loss. Due to the lateral device construction and its limitations, GaN HFET is expected to be primarily manufactured at breakdown voltages below 1200 V. SiC MOSFET will dominate the markets for higher-voltage devices. Future devices based on vertical GaN structure may change this conclusion.

The low $R_{on}$ and fast switching capability of the GaN HFET could have a dramatic impact on future power supplies by moving toward higher efficiency and higher power density. A record power density of 143 W/inch$^3$ has been achieved in the Google Little Box competition in which 600 V GaN FETs are used.[164] It can also simplify the power delivery architecture with substantial energy savings. A possible GaN enabled power supply roadmap is shown in Fig. 16-73 when compared to today's architecture shown in Fig. 16-72. At the front end, the Totem-Pole bridgeless PFC (Fig. 16-74*a*)[166] could be utilized to replace the currently used CCM PFC, achieving 99% efficiency while reducing the footprint due to >1 MHz operating frequency. Figure 16-74*b* shows a design of a 3 kW GaN prototype using two phase Totem-pole PFC topology. The GaN devices operate with a variable frequency under ZVS condition, and >99% efficiency have been measured. In Fig. 16-73 power delivery system, a single isolated

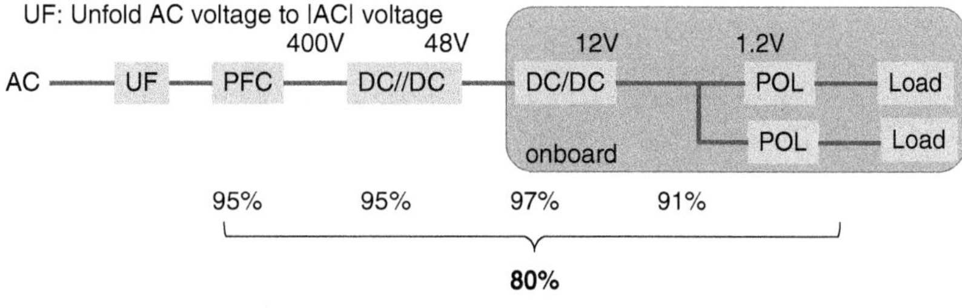


**FIGURE 16-72** Today's low-voltage power delivery system.

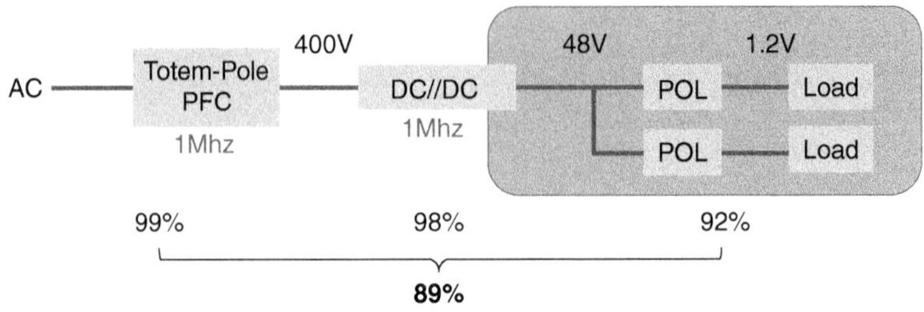


**FIGURE 16-73** Simplified power deliver system with GaN.

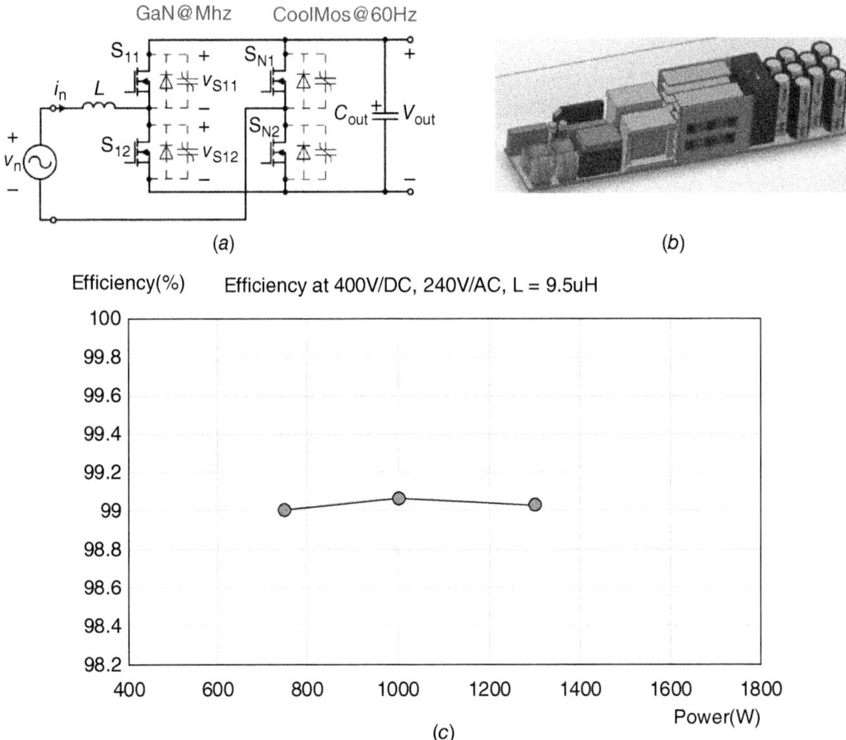


**FIGURE 16-74** (*a*) Totem-pole bridgeless PFC circuit where 600 V GaN switches at high frequency with ZVS condition. (*b*) A prototype design of a 2.3 kW GaN Totem-Pole PFC. (*c*) Measured efficiency exceeds 99%.

dc/dc stage based on soft switched converter topologies such as LLC[167] or phase-shift half-bridge[168] topology can be used to replace the current two-stage solution by eliminating the intermediate bus. The high switching frequency can be used to provide fast regulation of the 48 V (or 12 V) bus. 98.3% efficiency was achieved in Ref. 167 while more than 98% efficiency was also achieved by the phase-shift half-bridge as reported in Ref. 168. The circuit topology and results from Ref. 168 are shown in Fig. 16-75. Finally, the 48 V-to-1 V POL converter can be achieved by using 80 to 100 V GaN HFET in a non-isolated Buck converter.[169] Thanks to the extremely fast switching speed, the ultra-low duty ratio and small dead time can be achieved. 92% peak efficiency has been reported in Ref. 169. With these innovations, the efficiency of the low-voltage power delivery system (universal ac input to POL) could be improved by 9%. This will be a substantial achievement if realized on a large scale.

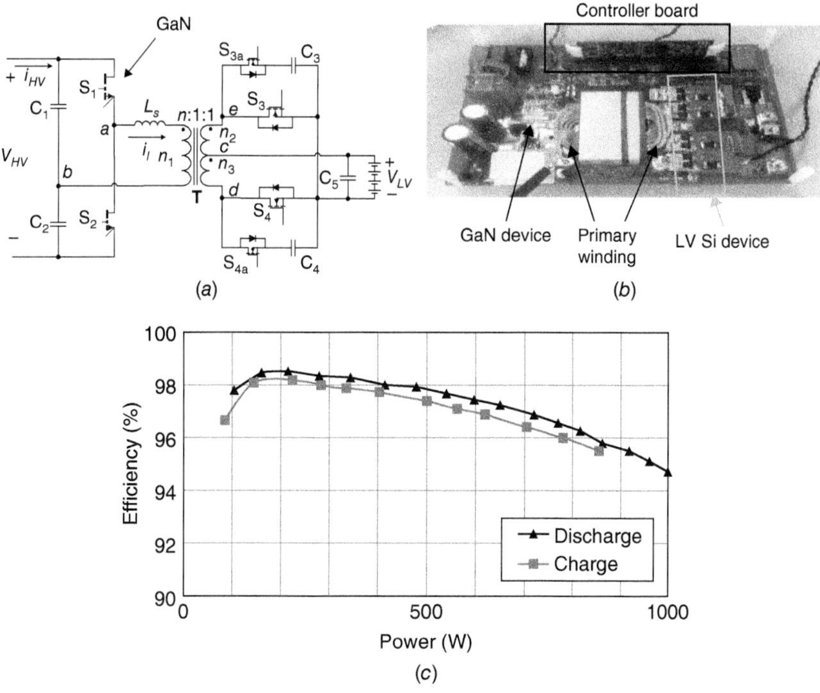


**FIGURE 16-75** (*a*) Phase shift half-bridge push-pull topology with active clamp. (*b*) A 1 kW prototype. (*c*) Measured efficiency exceeds 98%.

### 16.10.3 Magnetic Components

In power electronic converters three types of magnetic components are used: single winding inductors (for filtering current and aiding in resonant transitions in some circuits), multi-winding coupled inductors (to provide filtering and isolation), and transformers (for isolation and stepping up/down voltage). Unlike semiconductor devices, these have to be custom designed using available cores, wires, etc. The primary consideration for their design are size/weight and power loss. Design procedures using the common *area product method* is presented here. Another approach is the *core geometry method*, which can be found in Refs. 5 and 119. It is assumed that the reader is familiar with basics of electromagnetism and magnetic circuits. An E-E type core will be used for illustration, but the method is applicable to any core shape. The E-E type core construction, along with relevant definitions of its geometry, is shown in Fig. 16-76. The complete core is formed with two E cores with possibly an air gap between them. The coils are wound over a plastic bobbin placed on the outside of the center leg of the E sections.

Due to high-frequency magnetic fields, significant eddy currents are induced in the magnetic core and windings, and due to high-frequency electric field there can be significant capacitive currents between windings. Eddy currents lead to significant losses and necessitate the use of high-resistivity magnetic core materials and thinner wires (which may be paralleled for high current capacity) for the coils. Loss due to magnetic hysterisis also increases with increase in frequency.

***Magnetic Core Materials.*** Inductors use one of three different kinds of core materials, depending on the currents they are supposed to carry. The three core material types are silicon steel for low-frequency filtering, powdered iron for high-frequency filtering, and ferrites for carrying high-frequency currents. Transformers in power converters always carry high-frequency currents and therefore use ferrite cores.

For low-frequency filtering, where the inductor primarily carries a dc component with a 120-Hz ripple (as in-line frequency rectifiers), standard silicon steel laminations can be used to form the core. Silicon

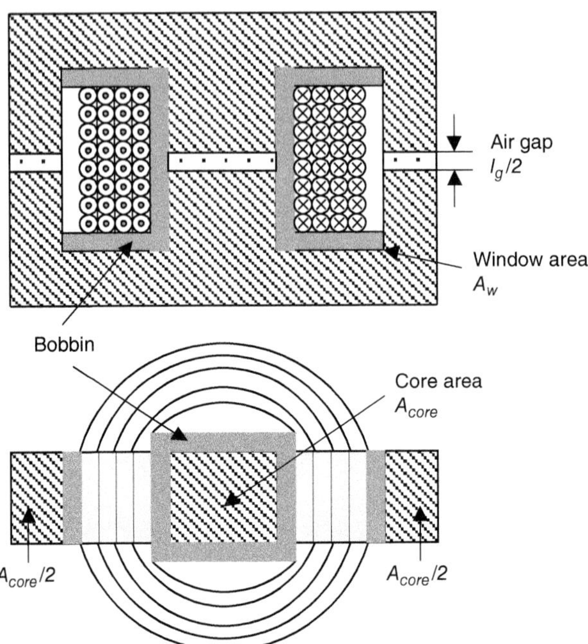


**FIGURE 16-76** Vertical and horizontal sections of an E-E core inductor.

steel has high-saturation flux density (~1.8 Tesla) and high-relative permeability ($\mu_r$ ~40,000). This results in a small core size without incurring significant core loss. However, the relative permeability reduces with increase in flux density, so the effective inductance value changes with the dc component of the current.

For inductors that are supposed to carry currents consisting of a large dc component and a small high-frequency ac component (e.g., the inductor used at the output of a dc/dc converter), powdered iron or MolyPermalloy powder (MPP) cores are used. These cores have high resistivity leading to lower eddy current losses, high-saturation flux density (up to 1.4 Tesla) leading to lower core size, and distributed air gap leading to a low relative permeability. For most of these materials the relative permeability changes with flux density. The cores are available in toroidal and E shapes. Toroids are difficult to wind but can be used to make very good quality inductors. There are a variety of powdered iron and MPP materials with different relative permeability, loss characteristics, maximum operating temperature, and aging characteristics.[136,137]

For transformers and inductors that have to carry significant high-frequency current, the powdered iron material has unacceptable loss due to eddy currents and hysterisis. Ferrite materials, with very high resistivity, lower weight, but low-saturation flux density (~0.4 Tesla) are suitable for these. Common ferrite materials are 3F3 from Ferroxcube and PC44 from TDK. Cores made from these materials are available in a wide variety of shapes and sizes: toroids, pot core, E, PQ, RM, etc. Custom cores are commonly used for high-power applications.

Nano-crystalline materials are a recent innovation in high-power, high-frequency (HF) magnetics.[138–140] These have several advantages over iron- and cobalt-based amorphous materials, and Ni-Zn and MN-Zn ferrites: higher-saturation flux density (Bsat) in the range of 1.2 to 1.4 Tesla, compared to 0.4 to 0.5 Tesla for ferrites and 0.5 to 0.6 Tesla for cobalt-based amorphous metals; lower core losses; high operating temperature with stable characteristics up to 200°C; low magneto-striction; and high initial permeability.

For power converters, these materials have the following applications (1) inductors and transformers due to high Bsat, low losses, high operating temperature, and low magneto-striction; (2) common mode chokes due to high initial permeability and high-saturation flux density. These materials are characterized by high eddy current loss. To minimize the loss, the cores are manufactured by depositing thin films on insulating plastic material which is wound as toroids or U-U cores. For transformers and inductors, the cores are particularly useful for high-efficiency and small-size in high-power medium frequency (>1 kW, 10–200 kHz) range.

***Inductor Design.*** For inductors, the design requires specified values of the inductance ($L$), and the peak and rms current ($\hat{I}$ and $I_{rms}$) it has to carry. The relations between peak current and core area ($A_{core}$), rms current and window area ($A_w$), and the expression for the *area product*, $A_p = A_{core} \cdot A_w$, are as follows:

$$\lambda = L\hat{I} = N\Phi = NB_{max} \cdot A_{core}$$

$$\Rightarrow A_{core} = \frac{L\hat{I}}{NB_{max}} \tag{16-69}$$

$$I_{rms} = J \cdot A_{cond} = J \cdot \frac{A_w \cdot k_w}{N}$$

$$\Rightarrow A_w = \frac{NI_{rms}}{Jk_w} \tag{16-70}$$

$$A_p = A_w \cdot A_{core} = \frac{L\hat{I}I_{rms}}{JB_{max}k_w} \tag{16-71}$$

Here $\lambda$ is the flux linkage, $N$ is the number of turns of the winding, $\Phi$ is the magnetic flux corresponding to peak current $\hat{I}$, $B_{max}$ is the maximum flux density (may be different from the saturation flux density), $J$ is the current density in the winding, $A_{cond}$ is the cross-sectional area of one conductor in the winding, and $k_w$ is called the winding factor is an emperical factor used to indicate the fraction of window area utilized by the copper of the windings. In the expression for $A_p$, the numerator consists of specified quantities, while the denominator has quantities which are chosen for design. $B_{max}$ may be chosen close to the saturation flux density depending on allowable core loss; $J$ is chosen in the range of 2.5 to 8 A/mm$^2$, depending on allowable conduction loss, mechanism for heat removal, and the allowable temperature rise; $k_w$ is in the range of 0.3 of 0.8 depending on space taken by bobbin and insulation, and the method of winding. Once the area product has been calculated, a core with the desired value of $A_p$ can be chosen from manufacturers' catalogs. The number of turns and the air gap required to achieve the desired inductance value are then calculated as

$$N = \frac{L\hat{I}}{B_{max}A_{core}} \tag{16-72}$$

$$L = N^2/\Re = N^2\mu_o A_{core}/l_g \tag{16-73}$$

$$\Rightarrow l_g = N^2\pi_o A_{core}/L \tag{16-74}$$

where $\Re$ is the core reluctance. The above equation assumes that $\mu_r/l_m \gg 1/l_g$, where $l_m$ is the magnetic path length of the core. The conductor size is calculated as $A_{cond} = k_w \cdot A_w/N$. If the current is expected to have a significant high-frequency component, the radius of the conductors should not exceed the *skin depth* of copper at the expected frequency. Several conductors may then be paralleled to achieve the required value of $A_{cond}$. Finally, $P_{cond}$, conduction loss in the windings, and $P_{core}$, loss in the core, should be calculated. $P_{cond}$ requires an estimate of the winding resistance, while $P_{core}$ is computed using manufacturer-supplied empirical loss curves. Core loss in an inductor depends on the operating flux density $B$ (related to the average value of the current), and change in flux density during each switching cycle $\Delta B$ (related to the expected current ripple). It is generally accepted that $P_{cond} = P_{core}$ indicates a good design. However, power loss in the core can be dissipated more easily compared to windings (especially if there are several layers). Thus, it may be desirable to have $P_{cond} < P_{core}$. This affects the choice of $B_{max}$ and $J$ values used above.

Powdered-iron cores have distributed air gaps and the above procedure is not directly applicable to them. For selection of these cores, the manufacturers suggest a number of cores based on a metric similar to the product $L\hat{I}I_{rms}$. The final core selection can then be made based on allowable losses ($P_{cond}$ relating to $J$, and $P_{core}$ relating to $B_{max}$), and temperature rise. The cores have a specified $A_L$ *value*, the inductance obtained from the core in nH/turns$^2$. The number of turns required to obtain the inductance value then can be computed using the specified $A_L$ value. Finally, the inductance variation with dc current should be checked to make sure that the required value is obtained at the maximum operating current. Software packages are available from manufacturers for initial core selection and then detailed analysis of chosen design. Pictures of some inductors are shown in Fig. 16-77.

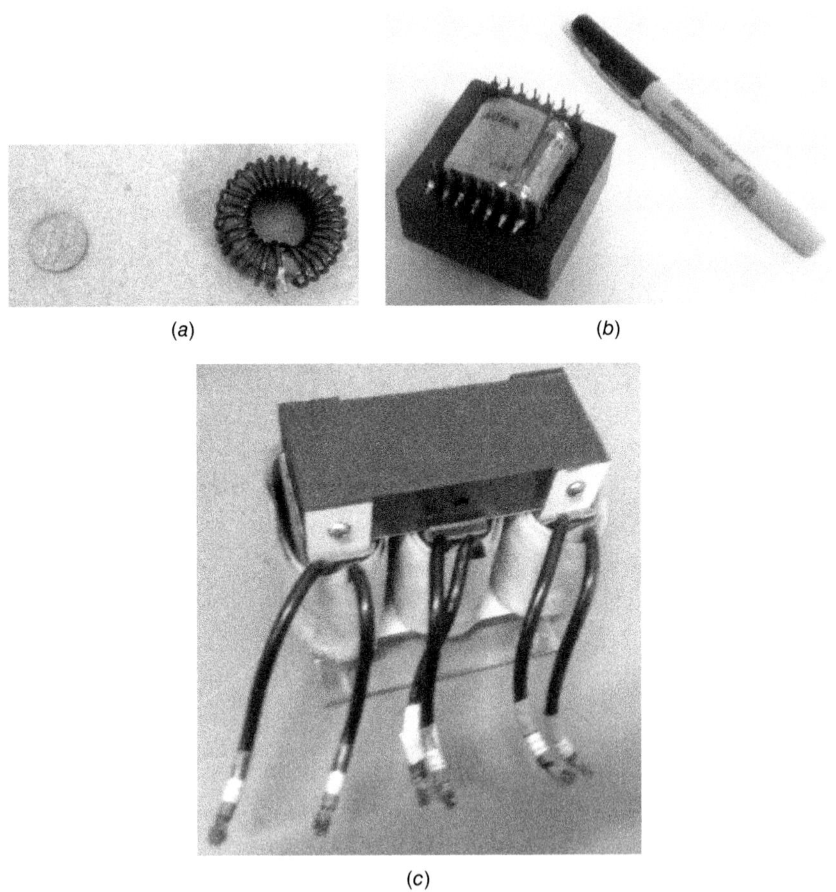

(*a*) (*b*) (*c*)

**FIGURE 16-77** Pictures of some inductors: (*a*) single-layer toroid; (*b*) 5-A inductor for 1.25-kW, 100-kHz buck converter made with E-E powdered iron core and six parallel strands of solid copper wire; (*c*) 60-Hz three-phase line inductor made with silicon steel laminations.

***Transformer Design.*** For a transformer, the design specifications are the turns ratio, rms current in all the windings, and the maximum volt-second product (Vsec, the maximum product of voltage and time for which the voltage is applied) expected across the primary winding. The window area can be related to the rms current, the core area to the maximum volt-second product, and the resulting area product calculated as follows:

$$k_w \cdot A_w = N_p \cdot A_{\text{cond},p} + \sum_m (N_{sm} \cdot A_{\text{cond},sm})$$

$$\Rightarrow A_w = \frac{N_p}{Jk_w} \cdot \left\{ I_{p,\text{rms}} + \sum\nolimits_m [(N_{sm}/N_p) \cdot I_{sm,\text{rms}}] \right\} \tag{16-75}$$

$$\Delta\lambda = [V\,sec] = N_p \Delta\Phi = N_p \Delta B_{\max} \cdot A_{\text{core}}$$

$$\Rightarrow A_{\text{core}} = \frac{[V\,sec]}{N_p \Delta B_{\max}} \tag{16-76}$$

$$A_p = A_w \cdot A_{\text{core}} = \frac{[V\,sec] \cdot \left\{ I_{p,\text{rms}} + \sum_m [(N_{sm}/N_p) \cdot I_{sm,\text{rms}}] \right\}}{J \Delta B_{\max} k_w} \tag{16-77}$$

The subscripts "$_p$" and "$_s$" indicate primary and secondary respectively, while the index $m$ refers to the secondary number. In the above equations $\Delta B_{\max}$ is the maximum *change* in flux density; this equals the maximum value of $B_{\max}$ for single quadrant operation (as in a forward converter), and twice $B_{\max}$ when flux direction reverses (as in full-bridge dc/dc converters). Core loss has more than quadratic dependence on $\Delta B_{\max}$. For transformers the value of $B_{\max}$ is therefore chosen much lower than the saturation flux density (about 0.1 to 0.25 Tesla). After a core is chosen by utilizing the computed $A_p$, the number of primary turns are calculated as

$$N_p = \frac{[V\,sec]}{B_{\max} A_{\text{core}}} \tag{16-78}$$

In high-frequency transformers the magnetizing current can be a significant part of the primary current and should be accounted for in the above calculations. Conductor selection for transformers is similar to that for inductors. For very high-frequency and high-current applications, use of insulated copper foils and *Litz wire* are alternatives to use of multiple parallel conductors. A 1.25-kW, 100-kHz transformer is shown in Fig. 16-78. It has interleaved primary and secondary windings. The primaries are realized using copper foils, while the secondaries have solid copper wire with several strands in parallel.

It should be noted that the procedures described above are rudimentary and several other factors have to be accounted for in a real design. These include: ambient temperature, temperature rise of the core (which affects the saturation flux density and the core loss significantly), *hot spot* temperature, winding loss due to *fringing fields* near the air gap, and winding loss due to *proximity effect*. In transformers, interleaving primary and secondary windings helps in reducing the proximity loss and leakage inductance at the expense of increased interwinding capacitance. To eliminate the winding process, planar magnetics that use tracks on multilayer PCBs to form the windings have been developed. Planar magnetics also have the benefit of a low profile. To reduce the number of discrete magnetic components, there has been significant research in integrated magnetics, realizing the functionality of several magnetic components using a single magnetic component.

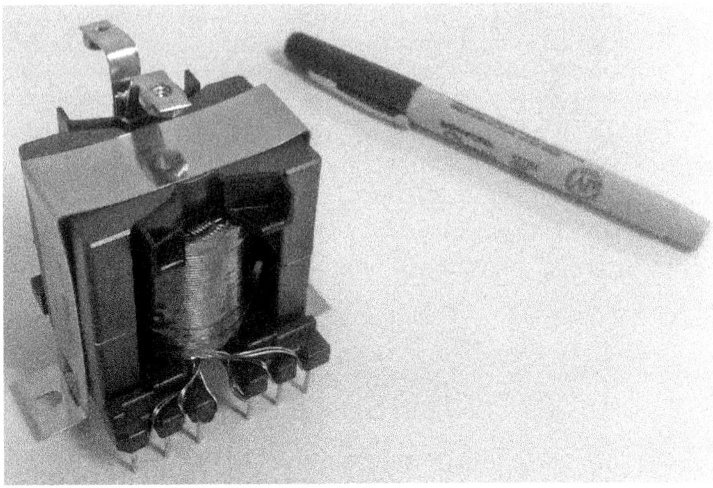

**FIGURE 16-78** A 1.25-kW, 100-kHz transformer using PQ type ferrite cores; high current primary windings built with copper foil; secondaries made with paralleled solid copper wires; primaries and secondaries interleaved to minimize leakage inductance.

### 16.10.4 Capacitors

There are several types of capacitors each with different advantages and limitations. For power electronic circuits the choice criteria are: capacitance value, voltage rating, ESR, and ESL, and the ripple and ac current capacity at different frequencies. All these can be found as specifications in manufacturers datasheets. ESR is either specified as a number at different frequencies (typically 120 Hz and 20/100 kHz), or in terms of the *dissipation factor*, $tan\delta = ESR \times (2 \times f \times C)$. If ESL is specified, it may be either as a number or in terms of the series resonance frequency, $f_r = 1/(2\pi\sqrt{ESL \times C})$. The main capacitor types used in power electronic circuits are listed in Table 16-4. The list is by no means

**TABLE 16-4** Capacitors Types Commonly Used in Power Electronic Circuits

| Capacitor type | Characteristics | Application |
|---|---|---|
| Electrolytic | Polarized (unipolar voltage only)<br>High density, high ESR and ESL<br>Low reliability<br>Voltage ratings up to 500 V<br>Capacitance up to 100 s of mF<br>Rated for ripple current capacity | DC input and output filters<br>Very short time energy storage |
| Ceramic and multilayer ceramic | Very low ESL and ESR<br>Low capacitance values (up to ~100 $\mu$F)<br>Maximum capacitance value reduces with increasing voltage rating | Paralleled for low-voltage output filtering, bypass for gate drives<br>High-voltage capacitors (up to 1 kV) used in snubber circuits |
| Metallized Polyester film | Low ESR and ESL<br>Low capacitance values (up to a few $\mu$F)<br>High voltage ratings (few kV)<br>Bigger than ceramic and electrolytic | In input and output dc filters: to suppress switching transients, ac filters<br>Snubber circuits |
| Metallized Polypropylene film | Very low ESR and ESL<br>Low capacitance values (up to a few $\mu$F)<br>Very high rms current, and high voltage ratings | In resonant converters: for carrying high frequency ac current |

exhaustive and there are several other capacitor types suited to different applications, e.g., tantalum, mica, and high-power film capacitors.

### 16.10.5 Heat Sinks

Power loss in power electronic circuits occurs in power semiconductor devices, windings and cores of the magnetic components, capacitors subjected to high-ripple current, and auxiliary circuits like gate drives. The resulting heat has to be removed from the components and eventually transferred to the atmosphere/ambient surroundings due to the following reasons. Power semiconductors have a maximum junction temperature rating beyond which they fail, and on-state resistance of devices like MOSFETs increases with junction temperature leading to reduced efficiency. Electrolytic capacitors can fail when heated beyond their ratings and their expected lifetime reduces with increase in temperature. Increased temperatures in magnetics can lead to higher power loss, poor magnetic core characteristics, insulation breakdown, and shorter lifetimes.

Most semiconductor devices are cooled by mounting their package on a heat sink. A heat sink is essentially a piece of metal designed to dissipate heat by convection and radiation (by maximizing the heat sink surface area). The heat transfer from the device to the heat sink is via conduction. For high-power applications, the power semiconductor die is mounted on direct bonded copper (DBC) consisting of two sheets of copper with an insulating ceramic in between. One side of the DBC holds the die while the other side is soldered to a metal-base plate. This assembly of the die, DBC, base plate, and electrical terminals forms a power device module. The module is then mounted on a heat sink. For lower power/voltage applications the package of the device is not electrically isolated from its electrical terminals. In such cases an interface material like Silpads[141] with good thermal conductivity and high dielectric breakdown voltage has to be used. For magnetics, resins may be used to conduct heat from the windings and the core to the metal heat sink. The heat sink itself may be cooled naturally, by forced convection using fans to blow air on the heat sink surface, or liquid cooled by circulating a liquid through the mass of the heat sink and cooling the liquid by a radiator. Simple steady analysis and design of heat sinks is carried out using specified *thermal resistances*, ratio of temperature difference to heat transferred across the material, of all the materials in the heat transfer path. For pulsed power applications, a transient thermal model, usually consisting of first order lags between two consecutive interfaces, is used.

## 16.11 REFERENCES


1. J. G. Kassakian, J. M. Miller, and N. Traub, "Automotive electronics power up," *IEEE Spectrum*, vol. 37, no. 5, May 2000, pp. 34–36.
2. R. Middlebrook and S. Ćuk, *Advances in Switched-Mode Power Conversion*. CA, USA: TESLAco, 1981, vol. 1.
3. H. Khalil, *Nonlinear Systems*, 3rd ed. Upper Saddle River, NJ: Prentice Hall, 2002.
4. N. T. M. Undeland and W. P. Robbins, *Power Electronics: Converters, Applications, and Design*, 3rd ed. Hoboken, NJ: John Wiley & Sons, 2003.
5. R. W. Erickson and D. Maksimovic, *Fundamentals of Power Electronics*, 2nd ed. Modify. New York: Kluwer Academic Publishers, 2001.
6. A. I. Pressman, *Switching Power Supply Design*, 2nd ed. New York: McGraw-Hill, 1997.
7. K. Billings, *Switchmode Power Supply Handbook*, 2nd ed. New York: McGraw-Hill, 1999.
8. R. P. Severns and G. E. Bloom, *Modern DC-to-DC Switchmode Power Converter Circuits*, 1st ed. New York: Van Nostrand Reinhold Company Inc., 1985.
9. Texas Instruments company website. [Online]. Available: http://www.ti.com/
10. National Semiconductors company website. [Online]. Available: http://www.national.com/
11. Maxim company website. [Online]. Available: http://www.maxim-ic.com/
12. Linear Technologies company website. [Online]. Available: http://www.analog.com/

13. R. Ayyanar and N. Mohan, "Novel soft-switching dc-dc converter with full zvs-range and reduced filter requirement part i: Regulated-output applications," *IEEE Transactions on Power Electronics*, vol. 16, no. 2, Mar. 2001, pp. 184–192.
14. R. Ayyanar and N. Mohan, "Novel soft-switching dc-dc converter with full ZVS-range and reduced filter requirement part ii: Constant-input, variable-output applications," *IEEE Transactions on Power Electronics*, vol. 16, no. 2, Mar. 2001, pp. 193–2001.
15. On Semiconductor company website. [Online]. Available: http://www.onsemi.com/
16. Power Integrations company website. [Online]. Available: http://www.powerint.com/
17. Texas Instruments power management webpage. [Online]. Available: http://www.ti.com
18. G. Schrom, P. Hazucha, F. Paillet, D. Rennie, S. Moon, D. G. T. Kamik, P. Sun, T. Nguyen, M. Hill, K. Radhakrishnan, and T. Memioglu, "A 100 MHz eight-phase buck converter delivering 12a in 25 $mm^2$ using air-core inductors," in *Proc. of the Annual IEEE Applied Power Electronics Conference and Exposition (APEC2007)*, Anaheim, CA, Feb. 2007, pp. 1048–2334.
19. Proc. of international workshop on power supply on chip (2008 and 2010). [Online]. Available: http://pwrsocevents.com
20. R. Jain and S. Sanders, "A 200 MA switched capacitor voltage regulator on 32 nm CMOS and regulation schemes to enable DVFS," in *European Power Electronics Conference (EPE '11)*, Birmingham, UK, Aug. 2011.
21. J. Li, A. Stratakos, A. Schultz, and C. Sullivan, "Using coupled inductors to enhance transient performance of multi-phase buck converters," in *Proc. of the Annual IEEE Applied Power Electronics Conference and Exposition (APEC 2004)*, vol. 2, Washington, DC, Feb. 2004, pp. 1289–1293.
22. S. R. Sanders and G. C. Verghese, "Lyapunov-based control for switched power converters," *IEEE Transactions on Power Electronics*, vol. 7, no. 1, Jan. 1992, pp. 17–24.
23. V. Vorpérian, "Simplified analysis of PWM converters using model of a PWM switch part i: continuous conduction mode," *IEEE Transactions on Aerospace and Electronic Systems*, vol. 26, no. 3, May 1990, pp. 490–496.
24. V. Vorpérian, "Simplified analysis of PWM converters using model of a PWM switch part ii: discontinuous conduction mode," *IEEE Transactions on Aerospace and Electronic Systems*, vol. 26, no. 3, May 1990, pp. 497–505.
25. N. Mohan, *First Course on Power Electronics and Drives*. Minneapolis, MN: MNPERE, 2003.
26. SPICE. University of Berkeley. [Online]. Available: http://www-cad.eecs.berkeley.edu/Software/software.html
27. K. Ogata, *Modern Control Engineering*, 4th ed. Upper Saddle River, NJ: Prentice Hall, 2001.
28. H. D. Veneble, "The k facor: a new mathematical tool for stability analysis and synthesis," in *Proc. of the International Solid-State Power Electronics Conference (POWERCON 10)*, San Deigo, CA, Mar. 1983, pp. 1–12.
29. *ORCAD PSpice A/D User's Guide*, Cadence Design Systems, 2003.
30. G. C. Verghese, *Nonlinear Phenomena in Power Electronics: Bifurcations, Chaos, Control, and Applications*. Hoboken, NJ: John Wiley & Sons, 2001.
31. M. P. Kazmierkowski, R. Krishnan, and F. Blaabjerg, *Control in Power Electronics: Selected Problems*. Maryland Heights, MO: Elsevier Science & Technology Books, 2002.
32. Digital power controllers. Silicon Labs. [Online]. Available: https://www.silabs.com/products/power-over-ethernet/powered-devices
33. Digital power. Texas Instruments. [Online]. Available: http://www.ti.com/ww/en/analog/digital-power/index.html
34. B. Patella, A. Prodic, A. Zirger, and D. Maksimovic, "High-frequency digital controller IC for dc/dc converters," in *Proc. of the Annual IEEE Applied Power Electronics Conference and Exposition (APEC 2002)*, vol. 1, Dallas, TX, Mar. 2002, pp. 374–380.
35. B. K. Bose, *Power Electronics and Variable Frequency Drives: Technology and Applications*. NJ: IEEE Press, 1996.
36. B. K. Bose, *Modern Power Electronics and AC Drives*. Upper Saddle River, NJ: Prentice Hall, 2001.
37. N. Mohan, *Advanced Electric Drives*. Minneapolis, MN: MNPERE, 2001.
38. R. C. Panaitescu and N. Mohan, "A simple space-vector PWM algorithm for VSI-fed ac motor drives," in *Proc. IEEE Applied Power Electronics Conference and Exposition (APEC'02)*, vol. 1, Dallas, TX, Mar. 2002, pp. 72–75.

39. G. Narayanan, V. Ranganathan, D. Z. H. Krishnamurthy, and R. Ayyanar, "Space vector based hybrid PWM techniques for reduced current ripple," *IEEE Transactions on Industrial Electronics*, vol. 55, no. 4, Apr. 2008, pp. 101–112.
40. F. Blashke, "The principle of field orientation as applied to the new transvector closed loop control system for rotating field machines," *Siemens Review*, vol. 34, no. 5, 1972, pp. 217–220.
41. M. Depenbrock, "Direct self-control (DSC) of inverter-fed induction machine," *IEEE Transactions on Power Electronics*, vol. 3, no. 4, 1988, pp. 420–429.
42. I. Takahashi and T. Noguchi, "A new quick-response and high efficiency control strategy of an induction motor," *IEEE Transactions on Industry Applications*, vol. 22, no. 5, 1986, pp. 820–827.
43. P. Vas, *Sensorless Vector and Direct Torque Control.* New York: Oxford University Press, 1998.
44. D. W. Novotony and T. A. Lipo, *Vector Control and Dynamics of AC Drives.* Oxford, UK: Claredon Press, 1996.
45. K. Corzine, *The Power Electronics Handbook,* T. L. Skvarenina, ed. New York: CRC Press, 2002.
46. J. Rodriguez, J. Lai, and F. Peng, "Multilevel inverters: a survey of topologies, controls, and applications," *IEEE Transaction on Industrial Electronics*, vol. 40, no. 4, Aug. 2002, pp. 724–738.
47. B. P. McGrath, D. G. Holmes, and T. Lipo, "Optimized space vector switching sequences for multilevel inverters," *IEEE Transactions on Power Electronics*, vol. 18, no. 6, Nov. 2003, pp. 1293–1301.
48. A. Nabae, I. Tkahashi, and H. Akagi, "A new neutral-point clamped PWM inverter," *IEEE Transactions on Industry Applications*, vol. IA-17, no. 5, Sep.-Oct. 1981, pp. 518–523.
49. L. M. Tolbert, F. Z. Peng, and T. G. Habetler, "Multilevel converters for large electric drives," *IEEE Transactions on Industry Applications*, vol. 35, no. 1, Jan.-Feb. 1999, pp. 36–44.
50. G. R. Kamath and N. Mohan, "Series-connected, all-hybrid converters for utility interactive applications," in *Proc. of the 23rd International Conference on Industrial Electronics, Control and Instrumentation (IECON '97)*, vol. 2, New Orleans, LA, Nov. 1997, pp. 726–731.
51. Siemens automation website. [Online]. Available: http://www.automation.siemens.com
52. A. Lesnicarand R. Marquardt, "An innovative modular multilevel converter topology suitable for a wide power range," in *Proc. of the 2003 IEEE Power Tech Conference*, vol. 3, Bologna, Italy, Jun. 2003, pp. 1–6.
53. E. Cengelci, S. U. Sulistijo, B. O. Woo, P. Enjeti, R. Teoderescu, and F. Blaabjerg, "A new medium-voltage PWM inverter topology for adjustable-speed drives," *IEEE Transactions on Industry Applications*, vol. 35, no. 3, May-Jun. 1999, pp. 628–637.
54. J. Wen and K. M. Smedley, "Synthesis of multilevel converters based on single and/or three-phase converter building blocks," *IEEE Transactions on Power Electronics*, vol. 23, no. 3, May 2008, pp. 1247–1256.
55. G. K. Dubey, S. R. Doradla, A. Joshi, and R. M. K. Sinha, *Thyristorised Power Controllers.* New York: John Wiley & Sons, 1986.
56. K. R. Padiyar, *HVDC Power Transmission Systems.* New Delhi, India: Wiley Eastern Ltd., 1993.
57. P. W. Wheeler, J. Rodrigueza, J. C. Clare, L. Empringham, and A. Weinstein, "Matrix converters: A technology review," *IEEE Transactions on Industry Electronics*, vol. 49, no. 2, Apr. 2002, pp. 276–278.
58. A. Alesina and M. G. B. Venturini, "Analysis and design of optimum-amplitude nine-switch direct ac-ac converters," *IEEE Transactions on Power Electronics*, vol. 4, no. 1, Jan. 1989, pp. 101–112.
59. J. Chang, T. Sum, and A. Wang, "Highly compact ac-ac converter achieving high voltage transfer ratio," *IEEE Transactions on Industrial Electronics*, vol. 49, no. 2, pp. 345–352, 2002.
60. Economac: the first all-in-one IGBT module for matrix converters. Eupec. [Online]. Available: http://www.eupec.com
61. J. Kolar, M. Baumann, F. Schafmeister, and H. Ertl, "Novel three-phase ac-dc-ac sparse matrix converter," in *Proc. of Annual IEEE Applied Power Electronics Conference and Exposition (APEC'02)*, vol. 2, 2002, pp. 777–791.
62. IEEE Recommended Practices and Requirements for Harmonic Control in Electrical Power Systems, IEEE Std. 519-1992, 1992.
63. Electromagnetic Compatibility Part 3: Limits, International Electrotechnical Commission (IEC) Std. 1000-3, 1995.
64. R. Ayyanar and A. K. Jain, *The Power Electronics Handbook,* T. L. Skvarenina, ed. New York: CRC Press, 2002.

65. V. Rao, A. K. Jain, K. K. Reddy, and A. Behal, "Experimental comparison of digital control techniques for single phase power factor correction," *IEEE Transactions on Industrial Electronics*, vol. 55, no. 1, Jan. 2008, pp. 67–78.
66. PFC smart power module. Fairchild Semiconductor. [Online]. Available: http://www.fairchildsemi.com/products/discrete/spm/pfc_spm.html
67. Accusine power correction system. Scheider Electric. [Online]. Available: http: //products.schneider-electric.us/products-services/products/power-management-products/reactivar-power-factor-correction-capacitors/accusine-power-correction-system/
68. M. J. Nave, *Power Line Filter Design for Switched-Mode Power Supplies*. Van Nostrand Reinhold, 1991.
69. EMC/EMI products. Schaffner. [Online]. Available: http://www.schaffner.com
70. Federal Communications Commission website. [Online]. Available: http://www.fcc.gov
71. Verband deutscher elektrotechniker (association of German electrical engineers). [Online]. Available: https://www.vde.com/en
72. Artesyn Technologies company website. [Online]. Available: http://www.artesyn.com/powergroup
73. N. Mohan, *Electric Drives: An Integrative Approach*. Minneapolis, MN: MNPERE, 2003.
74. T. J. Miller, ed., *Brushless Permanent-Magnet and Reluctance Motor Drives*. Oxford, UK: Oxford University Press, 1989.
75. T. J. E. Miller, *Switched Reluctance Motors and their Control*. Oxford, UK: Magna Physics Publishing and Clarendon Press, 1993.
76. T. J. E. Miller, ed., *Electronic Control of Switched Reluctance Machines*. Oxford, UK: Newnes, 2001.
77. Powerex, Inc. company website. [Online]. Available: http://www.pwrx.com
78. Semikron company website. [Online]. Available: http://www.semikron.com
79. Mitshubishi electric semiconductor website. [Online]. Available: http://www.mitsubishielectric.com/semiconductors/
80. Eupec company website. [Online]. Available: http://www.eupec.com
81. International Recitifier company website. [Online]. Available: http://www.irf.com
82. W. F. Ray and R. M. Davis, "Inverter drive for doubly salient reluctance motor: its fundamental behaviour, linear analysis, and cost implications," *IEE Journal on Electric Power Applications*, vol. 2, no. 6, Dec. 1979, pp. 185–193.
83. Fcas50sn6: Smart power module for SRM. Fairchild Semiconductor. [Online]. Available: www.fairchildsemi.com/
84. A. K. Jain, "Two phase modeling, experimental characterization, and power converter with fast demagnetization for switched reluctance motor drives," Ph. D. dissertation, University of Minnesota, Minneapolis, MN, 2003.
85. M. Barnes and C. Pollock, "Power electronic converters for switched reluctance drives," *IEEE Transactions on Power Electronics*, vol. 13, no. 6, Nov. 1998, pp. 1100–1111.
86. S. Vukosavic and V. R. Stefanovic, "SRM inverter topologies: a comparative evaluation," *IEEE Transactions on Industry Applications*, vol. 27, no. 6, Nov.-Dec. 1991, pp. 1034–1047.
87. Roadster. Tesla Motors. [Online]. Available: http://www.teslamotors.com/roadster
88. B. Masoud, "Charging ahead with Li-ion and Li-pol batteries," *Power Electronics Technology*, vol. 28, no. 7, Jul. 2002, pp. 26–29.
89. Building technologies program: Lighting research and development website. U.S. Department of Energy. [Online]. Available: http://www.eere.energy.gov/buildings/tech/lighting/
90. R. L. Steigerwald, "A comparison of half-bridge resonant converter topologies," *IEEE Trans. on Power Electronics*, vol. 3, no. 2, Apr. 1988, pp. 174–182.
91. P. Greenland and W. Burns, "Powering next-generation solid-state lighting," *Power Electronics Technology*, vol. 30, no. 5, May 2004, pp. 34–39.
92. M. Ehsani, Y. Gao, and A. Emadi, *Modern Electric, Hybrid Electric, and Fuel Cell Vehicles: Fundamentals, Theory, and Design*. New York: CRC Press, 2009.
93. Tesla Motors company website. [Online]. Available: http://www.teslamotors.com/
94. Nissan Leaf. Nissan. [Online]. Available: http://www.nissanusa.com/leaf-electric-car/index

95. Chevy Volt. Chevrolet. [Online]. Available: http://www.chevrolet.com/volt-electric-car/

96. J. G. Kassakian, "Automotive electrical systems—the power electronics market of the future," in *Proc. of Annual IEEE Applied Power Electronics Conference and Exposition (APEC'00)*, vol. 1, New Orleans, LA, Feb. 2000, pp. 3–9.

97. J. Cooper and A. Agarwal, "SIC power-switching devices—the second electronics revolution?" *Proceedings of the IEEE*, vol. 90, no. 6, Jun. 2002, pp. 956–968.

98. A. Elasser and T. P. Chow, "Silicon carbide benefits and advantages for power electronic circuits and systems," *Proceedings of IEEE*, vol. 90, no. 6, Jun. 2002, pp. 969–986.

99. Enphase Energy company website. [Online]. Available: http://www.enphase.com

100. L. Linares, R. W. Erickson, S. MacAlpine, and M. Brandemuehl, "Improved energy capture in series string photovoltaics via smart distributed power electronics," in *Proc. of the Annual IEEE Applied Power Electronics Conference and Exposition (APEC 2009)*, vol. 2, Washington, DC, Feb. 1996, pp. 904–910.

101. Solarmagic IC products. Texas Instruments.

102. R. Lasseter, "Microgrids," in *Proc. IEEE Power Engineering Society Winter Meeting*, vol. 1, New York, Jan. 2002, pp. 305–308.

103. T. M. Kaarsberg and J. M. Roop, "Combined heat and power: how much carbon and energy can manufacturers save?" *Aerospace and Electronic Systems Magazine*, vol. 14, no. 1, Jan. 1999, pp. 7–12.

104. IEEE Standard for Interconnecting Distributed Resources with Electric Power Systems, IEEE Std. 1547, 2003.

105. N. G. Hingorani, "Flexible ac transmission," *IEEE Spectrum*, vol. 30, no. 4, Apr. 1993, pp. 40–45.

106. N. G. Hingorani and L. Gyugyi, *Understanding Facts: Concepts and Technology of Flexible AC Transmission Systems*. New York: John Wiley & Sons, 1999.

107. P. Kundur, *Power System Stability and Control*. New York: McGraw-Hill, 1994.

108. L. Gyugyi, "Application characteristics of converter-based facts controllers," in *Proc. International Conference on Power System Technology (PowerCon'00)*, vol. 1, Perth, WA, Australia, Dec. 2000, pp. 391–396.

109. K. Sen, "Statcom-static synchronous compensator: theory, modeling, and applications," in *Proc. IEEE Power Engineering Society Winter Meeting*, vol. 2, New York, Jan.-Feb. 1999, pp. 1177–1183.

110. L. Gyugyi, C. Schauder, S. Williams, T. Rietman, D. Torgerson, and A. Edris, "The unified power flow controller: a new approach to power transmission control," *IEEE Transactions on Power Delivery*, vol. 10, no. 2, Apr. 1995, pp. 1085–1097.

111. A. Ghosh and G. Ledwich, eds., *Power Quality Enhancement Using Custom Power*. Boston: Kluwer Academic Publishers, 2002.

112. N. G. Hingorani, "Introducing custom power," *IEEE Spectrum*, vol. 32, no. 6, Jun. 1995, pp. 41–48.

113. IEEE Standard on Custom Power, in Draft Stage, IEEE Std. P1409.

114. J. Kyei, R. Ayyanar, G. Heydt, R. Thallam, and J. Blevins, "The design of power acceptability curves," *IEEE Transactions on Power Delivery*, vol. 17, no. 3, Jul. 2002, pp. 828–833.

115. J. Nielsen, M. Newman, H. Nielsen, and F. Blaabjerg, "Control and testing of a dynamic voltage restorer (DVR) at medium voltage level," *IEEE Transactions on Power Electronics*, vol. 19, no. 3, May 2004, pp. 806–813.

116. N. Woodley, "Field experience with dynamic voltage restorer (ddvrtmmv) systems," in *Proc. IEEE Power Engineering Society Winter Meeting*, vol. 4, New York, Jan.-Feb. 2000, pp. 2864–2871.

117. J. B. Baliga, *Power Semiconductor Devices*. New York: PWS Publishing Company, 1995.

118. T. L. Skvarenina, ed., *The Power Electronics Handbook, Part I: Power Electronic Devices*. New York: CRC Press, 2002.

119. W. T. McLyman, *Transformer and Inductor Design Handbook*, 3rd ed. New York: Marcel Dekker, 2002.

120. Infineon Technologies company website. [Online]. Available: http://www.infineon.com

121. Cree Semiconductors company website. [Online]. Available: http://www.cree.com

122. Qspeed family of diodes. Power Integrations. [Online]. Available: http://www.powerint.com/qspeed

123. Ixys Rf company website. [Online]. Available: http://www.ixysrf.com

124. Abb power semiconductors website. ABB. [Online]. Available: http://www.abb.com

125. Mitsubishi electric power semiconductors website. Mitsubishi Electric. [Online]. Available: http://www.mitsubishielectric.com/semiconductors/

126. R. J. Callanan, A. Agarwal, A. Burk, M. Das, B. Hull, F. Husna, A. Powell, J. Richmond, S. Ryu, and Q. Zhang, "Recent progress in SIC DMOSFETS and JBS diodes at CREE," in *34th IEEE Industrial Electronics Annual Conference (IECON 2008)*, Orlando, FL, Nov. 2008, pp. 2885–2890.

127. A. Hefner, S. H. Ryu, B. Hull, D. Berning, C. Hood, J. M. Ortiz-Rodriguez, A. Rivera-Lopez, T. Duong, A. Akuffo, and M. Hernandez-Mora, "Recent advances in high-voltage, high-frequency silicon-carbide power devices," in *IEEE Industry Applications Conference (IAS 2006)*, vol. 1, Oct. 2006, pp. 330–337.

128. A. Jain, D. McIntosh, M. Jones, and B. Ratliff, "Performance of a 25-kW 700-V galvanically isolated bidirectional dc-dc converter using 1.2 kV silicon carbide MOSFETS and Schottky diodes," in *2011 International Conference on Silicon Carbide and Related Materials (ICSCRM '11)*, Cleveland, OH, Sep. 2011, pp. 1–4.

129. A. Jain, D. McIntosh, M. Jones, and B. Ratliff, "A 2.5 kV to 22 V, 1 kW radar decoy power supply using silicon carbide semiconductors," in *European Power Electronics Conference (EPE '11)*, Birmingham, UK, Aug. 2011, pp. 1–10.

130. Semisouth Laboratories, Inc. company website. [Online]. Available: http://www.semisouth.com

131. GeneSiC Semiconductor company website. [Online]. Available: http://www.genesicsemi.com

132. Rohm Semiconductor company website. [Online]. Available: http://www.rohm.com

133. Transic company website. Fairchild Semiconductor. [Online]. Available: http://www.transic.com/

134. J. Wang, G. Wang, S. Bhattacharya, and A. Q. Huang, "Comparison of 10-kV SiC power devices in solid-state transformer," in *IEEE Energy Conversion Congress and Exposition (ECCE 2010)*, Atlanta, Sep. 2010, pp. 3284–3289.

135. L. Stevanovic. (2009) Advanced components for high speed, high-MW drives. [Online]. Available: http://www.nist.gov/pml/high megawatt/upload/7 2-approved-Stevanovic.pdf

136. Micrometals, Inc. company website. [Online]. Available: http://www.micrometals.com/

137. Magnetics, Inc. company website. [Online]. Available: http://www.mag-inc.com/

138. M. Willard, "Nanocrystalline inductor materials for power electronic applications," *Passive Compoent Industry*, Jul.-Aug. 2005.

139. Vacuumschmelze company website. [Online]. Available: www.vacuumschmelze.de

140. Finemet products. Metglas, Inc. [Online]. Available: http://www.metglas.com

141. The Bergquist Company website. [Online]. Available: www.bergquistcompany.com/

142. Alex Q. Huang, Mariesa L. Crow, Gerald Thomas Heydt, Jim P. Zheng, and Steiner J. Dale, "The future renewable electric energy delivery and management (FREEDM) system: The energy internet," *Proceedings of the IEEE*, vol. 99, no. 1, Jan. 2011, pp. 133–148.

143. Xu She, Alex Huang, and Rolando Burgos, "Review of solid state transformers and their application in power distribution system," *IEEE Journal of Emerging and Selected Topics in Power Electronics*, vol. 1, no. 3, Sep. 2013.

144. Xu She, Xunwei Yu, Fei Wang, and Alex Q. Huang, "Design and demonstration of a 3.6 kV-120 V/10 kVA solid state transformer for smart grid application," *IEEE Transactions on Power Electronics*, vol. 29, no. 8, pp. 3982–3996.

145. Chunhong Zhao, Drazen Dujic, Akos Mester, Juergen K. Steinke, Michael Weiss, Silvia Lewdeni-Schmid, Toufann Chaudhuri, and Philippe Stefanutti, "Power electronic traction transformer—medium voltage prototype," *IEEE Transactions on Industrial Electronics*, vol. 61, no. 7, Jul. 2014, pp. 3257–3268.

146. Fei Wang, Gangyao Wang, Alex Q. Huang, Wensong Yu, and Xijun Ni, "Design and operation of a 3.6 kV high performance solid state transformer based on 13 kV SiC MOSFET and JBS diode," in Proc. *IEEE Energy Conversion Congress and Exposition (ECCE)*, 2014, pp. 4553–4560.

147. A. Lesnicar and R. Marquardt, "An innovative modular multilevel converter topology suitable for a wide power range," in Proc. *IEEE Bologna Power Tech Conference Proceedings*, 2003, vol. 3.

148. Maryam Saeedifard and Reza Iravani, "Dynamic performance of a modular multilevel back-to-back HVDC system," *IEEE Transactions on Power Delivery*, vol. 25, no. 4, Oct. 2010, pp. 2903–2912.

149. Makoto Hagiwara and Hirofumi Akagi, "Control and experiment of pulse width-modulated modular multilevel converters," *IEEE Transactions on Power Electronics*, vol. 24, no. 7, Jul. 2009, pp. 1737–1746.

150. Xu She and Alex Q. Huang, "Circulating current control of double-star chopper-cell modular multilevel converter for HVDC system," in Proc. *IEEE IECON*, 2012, pp. 1234–1239.

151. Gahzal Falahi and Alex Q. Huang, "Design consideration of an MMC-HVDC system based on 4500 V/4000 A emitter turn-off (ETO) thyristor," in Proc. *IEEE ECCE*, 2015, pp. 3462–3467.

152. Wolfspeed SiC Schottky Diode Datasheet, Part No. C3D20060D, www.wolfspeed.com.

153. J. Casady, et al., "Ultra-low (1.25 mΩ) on-resistance 900 V SiC 62 mm half-bridge power modules using new 10 mΩ SiC MOSFETs," PCIM Europe 2016, pp. 1–8.

154. Infineon 1200 V CoolSiC™ Power Transistor, Part No. IJW120R070T1, www.infineon.com.

155. Rohm 650 V SiC MOSFET datasheet, Part No. SCT2120AF, www.rohm.com.

156. Guo, Suxuan, Zhang, Liqi, Lei, Yang, Li, Xuan, Xue, Fei, Yu, Wensong, Huang, and Alex Q., "3.38 Mhz operation of 1.2 kV SiC MOSFET with integrated ultra-fast gate drive," in Wide Bandgap Power Devices and Applications (WiPDA), 2015 IEEE 3rd Workshop on, vol., no., 2–4 Nov. 2015, pp. 390–395.

157. V. Pala, G. Wang, B. Hull, S. Allen, J. Casady, and J. Palmour, "Record-low 10 mohm SiC MOSFETs in TO-247, rated at 900V," *2016 IEEE Applied Power Electronics Conference and Exposition (APEC)*, Long Beach, CA, 2016, pp. 979–982.

158. E. Van Brunt, et al., "27 kV, 20 A 4H-SiC n-IGBTs," Mat. Sci. Forum, vol. 821–823, pp. 847–850, 2015.

159. L. Cheng et al., "20 kV, 2 cm2, 4H-SiC gate turn-off thyristors for advanced pulsed power applications," *2013 19th IEEE Pulsed Power Conference (PPC)*, San Francisco, CA, 2013, pp. 1–4.

160. Song, Xiaoqing, Huang, Alex Q., Lee, Mengchia, Peng, Chang, Cheng, Lin, O'Brien, Heather, Ogunniyi, Aderin, Scozzie, Charles, Palmour, and John, "22 kV SiC emitter turn-off (ETO) thyristor and its dynamic performance including SOA," Power Semiconductor Devices & IC's (ISPSD), 2015 IEEE 27th International Symposium on, vol., no., 10–14 May 2015, pp. 277–280.

161. Kamel Madjour, "Silicon carbide market update: From discrete devices to modules…" presented at PCIM Europe, Germany, 2014.

162. Ulrich Schwarzer, Stefan Buschhorn, and Klaus Vogel, "System benefits for solar inverters using SiC semiconductor modules," in Proc. PCIM Europe, 2014, pp.787–794.

163. Stefan Buschhorn and Klaus Vogel, "Saving money: SiC in UPS applications," in Proc. PCIM Europe, 2014, pp. 765–771.

164. https://www.littleboxchallenge.com/.

165. Y. Hayashi, H. Iso, D. Hara, and A. Matsumoto, "Current-fed GaN front-end converter for ISOP-IPOS converter-based high power density dc distribution system," *(EPE'15 ECCE-Europe), 2015 17th European Conference on*, Geneva, 2015, pp. 1–10.

166. Z. Liu, X. Huang, M. Mu, Y. Yang, F. C. Lee, and Q. Li, "Design and evaluation of GaN-based dual-phase interleaved MHz critical mode PFC converter," *2014 IEEE Energy Conversion Congress and Exposition (ECCE)*, Pittsburgh, PA, 2014, pp. 611–616.

167. M. Domb, "High power density, high efficiency 380 V to 52 V LLC converter utilizing E-Mode GaN switches," *PCIM Europe 2016*, pp. 1–7.

168. F. Xue, R. Yu, W. Yu, and A. Q. Huang, "Distributed energy storage device based on a novel bidirectional Dc-Dc converter with 650 V GaN transistors," *2015 IEEE PEDG*, Aachen, 2015, pp. 1–6.

169. Texas Instrument, SNVU520–May 2016 "Using the LMG5200POLEVM-10 48 V to Point of Load EVM."

17

# POWER SYSTEM ANALYSIS

Francisco de León

*Associate Professor, Department of Electrical and Computer Engineering, NYU Tandon School of Engineering, New York University, Brooklyn, New York*

Tianqi Hong

*Research Fellow, Department of Electrical and Computer Engineering, NYU Tandon School of Engineering, New York University, Brooklyn, New York*

Ashhar Raza

*Research Fellow, Department of Electrical and Computer Engineering, NYU Tandon School of Engineering, New York University, Brooklyn, New York*

## 17.1 INTRODUCTION

Since as far back as the 1970s, when mainframe computers became popular, the analysis of electric power systems is performed (almost exclusively) using digital computer simulations. Because of its high cost, experimentation in an actual power system is carried out very seldom. Additionally, the results from computer simulations are generally very accurate; therefore, power engineers today trust many of the analytical tools commercially (and royalty-free) available. Today (2017), and for many years already, most power system simulations of large power systems can be performed in personal computers and laptops.

This section describes the concepts of what is called in the electric power industry *power system steady-state analysis* (or "power system analysis" for short). Power system steady-state analysis includes two very large areas of study: power flow and faults analyses. The former is, without any doubt, the most important study used today to design and operate the power system. It consists of the solution (in steady state) of the electrical circuit with the particularities of a power system (i.e., loads do not behave as constant impedances as in an electric circuit, but behave closer to constant power). In other words, with power flow analysis one computes the voltage at every location (bus or node) and the flow of power in every link (transmission line, cable, transformer, etc.). Those two quantities (bus voltage and power flow) are of paramount importance since all power equipment requires a narrow voltage range to operate correctly and power flow determines the size (and therefore the cost) of the equipment. The latter, power system fault analysis actually refers to the transient state when a fault occurs. Faults are very common in power systems because large portions are exposed to the elements, ageing, animal intrusion, and other internal and external situations. The calculation of fault currents is performed very efficiently in steady state even when faults last only for a few cycles.

Traditionally power system analysis is performed independently for transmission and distribution systems. There are several reasons for this even when the entire system is interconnected: (1) transmission and distribution companies frequently have different owners; (2) for a transmission system, the distribution system (of a city for example) can be represented as lumped load (or several lump loads). For a distribution system, the transmission system can be seen as an impedance in series with an infinite bus, say its Thevenin equivalent; (3) transmission systems mostly operate balanced and thus a single phase equivalent (the positive sequence) can be used for power flow studies. Distribution systems are commonly unbalanced (or even contain fewer than three phases) and therefore the positive sequence equivalent does not apply. Both transmission and distribution systems, use the technique called symmetrical components transformation to perform unbalanced fault (or short-circuit) analysis, such as single-line-to-ground fault.

This section starts with the presentation of the basic concepts of complex numbers and phasors as they apply to power system analysis. Next, the circuit models of the major pieces of equipment are described. This is followed by the derivation of the per unit system used in power engineering, the method of symmetrical components, and sequence impedances. The section ends with the main two subsections: one dedicated to power flow analysis and the other to fault analysis.

## 17.2 PHASOR ANALYSIS AND COMPLEX POWER

### 17.2.1 Complex Numbers

A complex number $F$ can be represented as

$$F = a + jb$$

where $j = \sqrt{-1}$ stands for a unity imaginary number (in power systems, we use $j$ rather than $i$ for $\sqrt{-1}$ because $i$ is reserved for electrical current). $a$ is the real part of the complex number $F$ and $b$ is a real number representing the imaginary part of the complex number $F$. The complex number $F$ can be represented by a vector in the complex plane as illustrated in Fig. 17-1.

According to Fig. 17-1, the complex number $F$ can be represented in trigonometric form as

$$F = |F|(\cos\theta + j\sin\theta)$$

Using Euler's formula $F$ can be further rewritten as follows:

$$F = |F|e^{j\theta} = |F|\angle\theta$$

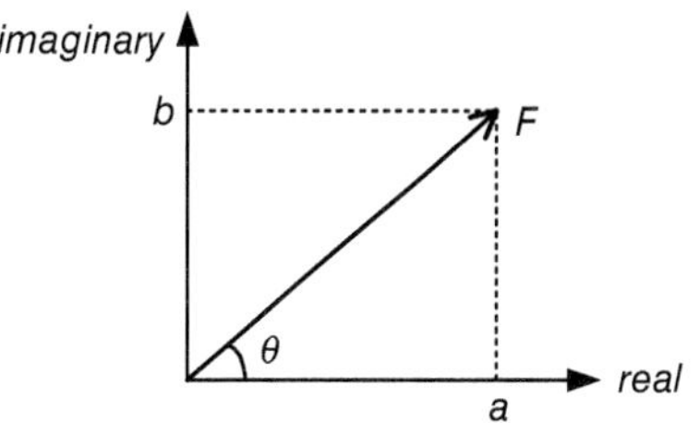


**FIGURE 17-1** Graphic representation of a complex number $F$.

where $|F|$ is the modulus (or magnitude) of $F$ and $\theta$ is the angle of $F$ ($\theta = \arg(F)$). The representation of a complex number as $|F|\angle\theta$ is a very convenient way used by electrical engineers to denote a complex number by its magnitude and angle. The relationships between $|F|$, $\theta$, $a$, and $b$ are given below (all obtained from simple trigonometry):

$$|F| = \sqrt{a^2 + b^2}$$

$$\theta = \arctan\left(\frac{b}{a}\right)$$

$$a = |F|\cos\theta \text{ and } b = |F|\sin\theta$$

Some useful functions and operations for complex numbers are listed in Table 17-1 (let $F_1 = a_1 + jb_1$ and $F_2 = a_2 + jb_2$).

**TABLE 17-1** Useful Functions and Operations for Complex Numbers

| |
|---|
| $\mathrm{Re}[F_1] = a_1$ and $\mathrm{Im}[F_1] = b_1$ |
| $F_1^* = a_1 - jb_1$ |
| $F_1 \pm F_2 = (a_1 \pm a_2) + j(b_1 \pm b_2)$ |
| $F_1F_2 = \|F_1\|\|F_2\|e^{j(\theta_1+\theta_2)} = \|F_1\|\|F_2\|\angle(\theta_1+\theta_2) = (a_1a_2 - b_1b_2) + j(a_1b_2 + a_2b_1)$ |
| $\frac{F_1}{F_2} = \frac{\|F_1\|}{\|F_2\|}e^{j(\theta_1-\theta_2)} = \frac{\|F_1\|}{\|F_2\|}\angle(\theta_1-\theta_2) = \frac{a_1a_2 + b_1b_2}{\|F_2\|^2} + j\frac{a_2b_1 - a_1b_2}{\|F_2\|^2}$ |

## 17.2.2 Phasors and Phasor Diagrams

AC current is represented in time domain by the following expression:

$$i(t) = \sqrt{2}I_{\mathrm{rms}}\cos(\omega t + \varphi_i) \tag{17-1}$$

where $I_{\mathrm{rms}}$ is the root mean square value (rms) of the ac instantaneous current $i(t)$; $\omega$ is the angular frequency of $i(t)$ and $\varphi_i$ is the phase angle. The phasor notation of ac current $i(t)$ is

$$I = I_{\mathrm{rms}}\angle\varphi_i \tag{17-2}$$

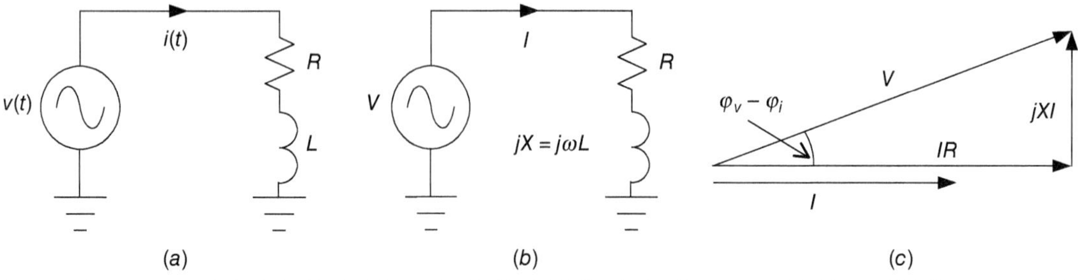


**FIGURE 17-2** (*a*) Electric circuit in time domain; (*b*) electric circuit in phasor notation; (*c*) phasor diagram of the electric circuit.

Similarly, the phasor notation of the instantaneous ac voltage $v(t)$ is

$$V = V_{\text{rms}}\angle\varphi_v \tag{17-3}$$

where $V_{\text{rms}}$ is the rms value of the ac voltage $v(t)$; $\varphi_v$ is its phase angle. For a linear electrical circuit, shown in Fig. 17-2*a*, we have from Kirchhoff voltage law (KVL):

$$v(t) = Ri(t) + L\frac{d}{dt}i(t) \tag{17-4}$$

Assume $v(t)$ is a cosine function with its angular frequency equals to $\omega$, $i(t)$ is also a cosine function in steady state with the same frequency. Hence, the above equation can be rewritten as

$$V_{\text{rms}}\cos(\omega t + \varphi_v) = RI_{\text{rms}}\cos(\omega t + \varphi_i) - \omega LI_{\text{rms}}\sin(\omega t + \varphi_i) \tag{17-5}$$

According to this equation, a phasor electrical circuit can be obtained shown in Fig. 17-2*b*. Based on the properties of complex numbers, a phasor diagram can be plotted to represent the electrical relationships of the electrical circuit; see Fig. 17-2*c*.

### 17.2.3 Complex Power

The active power $P$ of the circuit shown in Fig. 17-2*a* is defined as the average of the instantaneous power:

$$P = \frac{1}{T}\int_0^T v(t)i(t)dt = V_{\text{rms}}I_{\text{rms}}\cos(\varphi_v - \varphi_i) = V_{\text{rms}}I_{\text{rms}}\cos\theta \tag{17-6}$$

where $\theta$ is the angle of the load impedance; cos $\theta$ is the power factor of the circuit and it is a measure of how much the system is utilized to carry active power. The reactive power $Q$ of the circuit is

$$Q = V_{\text{rms}}I_{\text{rms}}\sin\theta \tag{17-7}$$

The apparent power $|S|$ of the circuit is

$$|S| = V_{\text{rms}}I_{\text{rms}} \tag{17-8}$$

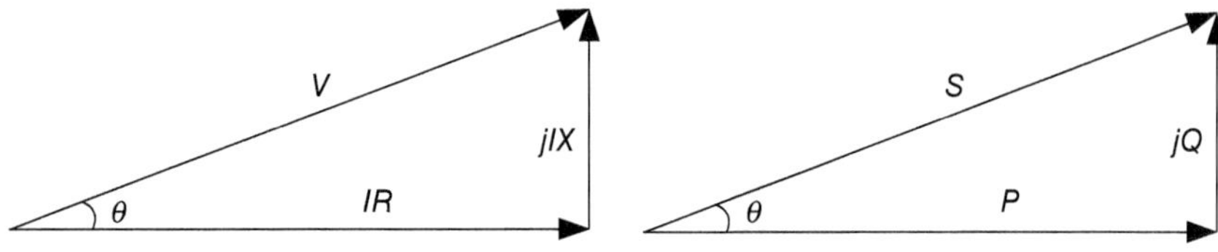


**FIGURE 17-3** Comparison between circuit and power phasor diagrams.

According to the above equations, we have the following relationships between $P$, $Q$, and $|S|$:

$$P = |S|\cos\theta, Q = |S|\sin\theta, \text{and} |S| = \sqrt{P^2 + Q^2}.$$

Hence, a complex power can be defined as

$$S = P + jQ = |S|\angle\theta \tag{17-9}$$

The phasor diagram of the relationship between each power quantities is drawn in Fig. 17-3. The power triangle and the circuit triangle shown in Fig. 17-2*c* are similar triangles.

## 17.3 COMPONENT MODELING

### 17.3.1 Generator

Synchronous generators are the most commonly used source of energy in power systems. There are two types of synchronous generators: cylindrical (nonsalient pole) synchronous and salient-pole synchronous generators. The simplified structures of those two types of synchronous generators are shown in Fig. 17-4. Cylindrical generators are commonly driven by high-speed steam turbines and contain a few poles, typically 2 or 4. Salient-pole generators are driven by low-speed hydraulic turbines and contain several dozen poles. On a salient pole rotor two axes can be defined. Regardless of their construction, both types of generators have the same equivalent circuit and the parameters are given by the manufacturer.

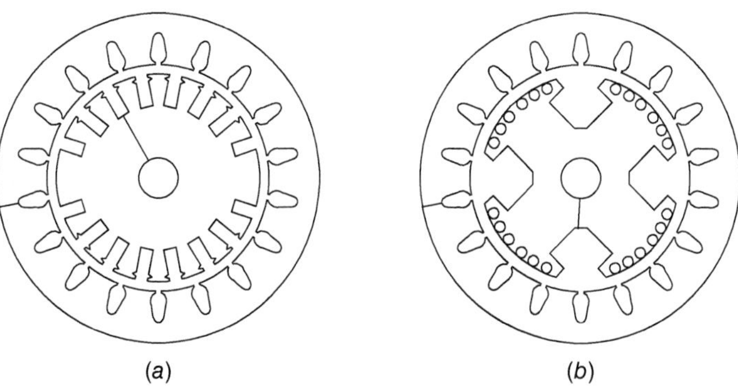


**FIGURE 17-4** Structures of synchronous generators. (*a*) Cylindrical synchronous generator; (*b*) salient-pole synchronous generator.

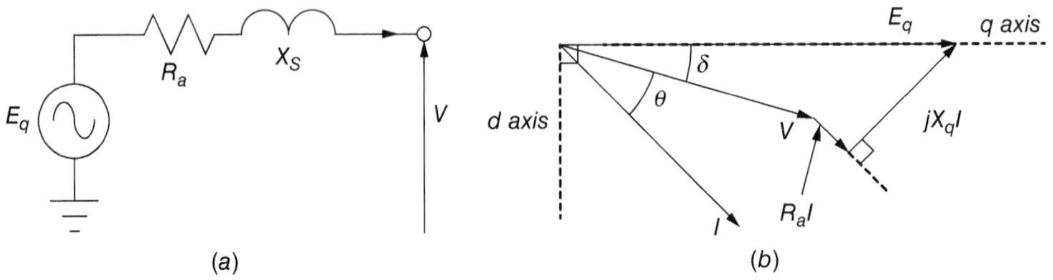


**FIGURE 17-5** Equivalent circuit and phasor diagram of nonsalient pole synchronous generator. (*a*) Equivalent circuit; (*b*) phasor diagram.

Assume the common case when the power factor is lagging. Let the terminal voltage and current phasors be $V$ and $I$, respectively. Taking the terminal voltage as reference we have

$$V = |V| \angle 0 \text{ and } I = |I| \angle \theta \tag{17-10}$$

where $|V|$ and $|I|$ is the magnitudes of phasors $V$ and $I$; $\theta$ is the power factor angle. Let $X_S$ be the leakage reactance between the rotor and stator windings. The steady-state equivalent circuit of a generator is illustrated in Fig. 17-5*a*. According to the equivalent circuit, we can write

$$E_q = |E_q| \angle \delta = V + (R_a + jX_S)I \tag{17-11}$$

where $E_q$ is the effective internal voltage, $R_a$ is the armature resistance per phase. The corresponding phasor diagram is plotted in Fig. 17-5*b*.

### 17.3.2 Transmission Lines

This subsection discusses the different methods of modeling transmission lines. The best model for a line depends on its length. Hence, there are three different models for the transmission line: short line model, medium line model, and the long line model. Calculations of line parameters are not discussed in this subsection.

***Short Line Model.*** Lines shorter than 50 miles long are modeled as short lines. In the short-line model the capacitance of the line is ignored. The complex impedance of the line $Z$ is written as

$$Z = (r + j\omega L)l \tag{17-12}$$

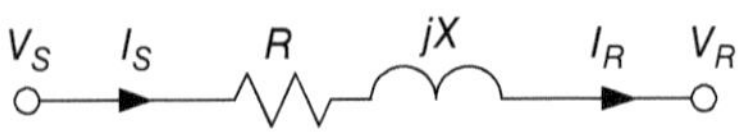


**FIGURE 17-6** Short-line model.

where $r$, $L$ are the per-phase resistance and inductance of the line per unit length and $l$ is the length of the line. $\omega$ is the operating frequency. This model is drawn as shown in Fig. 17-6.

$V_S$, $V_R$, $I_S$, $I_R$ are the phasors of the sending and receiving end voltages, and sending and receiving end currents, respectively. The application of Kirchhoff's laws (KVL and KCL) reveals that

$$\begin{aligned} V_S &= I_R(R + jX) + V_R \\ I_S &= I_R \end{aligned} \tag{17-13}$$

In matrix form, the above equation can be written as

$$\begin{bmatrix} V_S \\ I_S \end{bmatrix} = \begin{bmatrix} 1 & R+jX \\ 0 & 1 \end{bmatrix} \begin{bmatrix} V_R \\ I_R \end{bmatrix} \tag{17-14}$$

To make it more general, the above equation is written as a two-port circuit as

$$\begin{bmatrix} V_S \\ I_S \end{bmatrix} = \begin{bmatrix} A & B \\ C & D \end{bmatrix} \begin{bmatrix} V_R \\ I_R \end{bmatrix} \tag{17-15}$$

where $A = 1$, $B = R + jX$, $C = 0$, and $D = 1$.

An important measure of the operation of a line is voltage regulation. Voltage regulation is defined as percent change in voltage between no-load operation and full load operation as follows:

$$V_{\text{reg}}(\%) = \frac{|V_{R,\,\text{no load}}| - |V_{R,\,\text{full load}}|}{|V_{R,\,\text{full load}}|} \times 100\% \tag{17-16}$$

Efficiency of the line is given by

$$\eta = \frac{S_R}{S_S} \tag{17-17}$$

***Medium Line Model.*** As the length of the line increases, ignoring the capacitance yields erroneous results. For lines greater than 50 miles and less than 150 miles, the medium line model is used. In this model, the shunt admittance of the line $Y$ is divided in two equals parts and added to each end of the line. This is called the nominal $\pi$ model of the line and it is illustrated in Fig. 17-7.

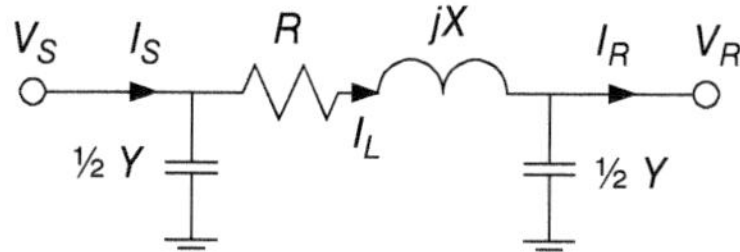


**FIGURE 17-7** Medium-length line model.

Voltage and current equations for the nominal $\pi$ medium line model are written as (from KVL and KCL):

$$\begin{aligned} V_S &= I_L(R+jX) + V_R \\ I_L &= V_R\left(\frac{Y}{2}\right) + I_R \\ I_S &= I_L + V_S\left(\frac{Y}{2}\right) \end{aligned} \tag{17-18}$$

Substituting $I_L$ into the other two equations, we get

$$\begin{aligned} V_S &= \left((R+jX)\left(\frac{Y}{2}\right)+1\right)V_R + I_R(R+jX) \\ I_S &= Y\left(1+\frac{(R+jX)Y}{4}\right)V_R + \left(1+\frac{(R+jX)Y}{2}\right)I_R \end{aligned} \tag{17-19}$$

In matrix form, the equations are written as

$$\begin{bmatrix} V_S \\ I_S \end{bmatrix} = \begin{bmatrix} A & B \\ C & D \end{bmatrix} \begin{bmatrix} V_R \\ I_R \end{bmatrix} \tag{17-20}$$

where

$$A=\left((R+jX)\left(\frac{Y}{2}\right)+1\right), B=(R+jX), C=Y\left((R+jX)\left(\frac{Y}{4}\right)+1\right), D=\left((R+jX)\left(\frac{Y}{2}\right)+1\right)$$

***Long Line Model.*** For transmission lines longer than 150 miles, the medium line model does not yield accurate results. This is because the medium line model considers lumped parameters, whereas the long line model considers distributed parameters. A long transmission line model is shown in Fig. 17-8. The line is divided into several smaller sections of length $\Delta x$, where the parameters can be considered lumped.

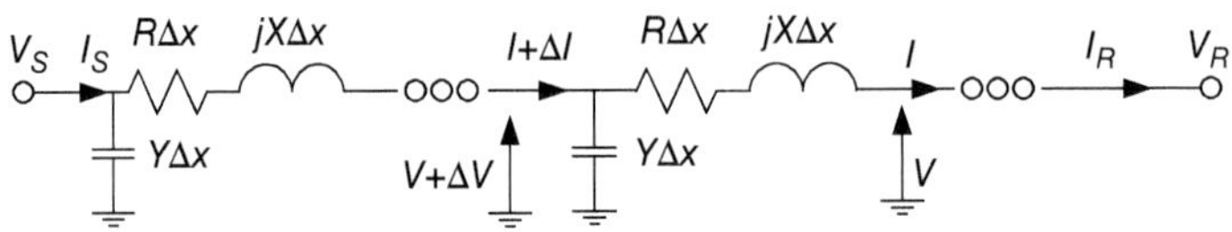


**FIGURE 17-8** Long line model.

To derive the two port circuit parameters of the long transmission line, let us consider the circuit shown in Fig. 17-8.

From the circuit it can be seen that

$$\Delta V = I(R+jX)\Delta x \tag{17-21}$$

$$\frac{\Delta V}{\Delta x} = I(R+jX) \tag{17-22}$$

As $\Delta x \to 0$,

$$\frac{\partial V}{\partial x} = I(R+jX) \tag{17-23}$$

Similarly from the circuit, we can see that

$$\Delta I = (V+\Delta V)Y\Delta x \tag{17-24}$$

$$\Delta I = VY\Delta x + Y\Delta V\Delta x \tag{17-25}$$

Neglecting the very small term $Y\Delta V\Delta x$, we get

$$\frac{\Delta I}{\Delta x} = VY \tag{17-26}$$

As $\Delta x \to 0$

$$\frac{\partial I}{\partial x} = VY \tag{17-27}$$

Taking the derivative of Eq. (17-23) w.r.t. $x$,

$$\frac{\partial^2 V}{\partial x^2} = (R+jX)\frac{\partial I}{dx} \tag{17-28}$$

Substituting $\partial I/\partial x$ from Eq. (17-27)

$$\frac{\partial^2 V}{\partial x^2} = (R + jX)VY \tag{17-29}$$

Solving the differential equation with the following boundary conditions: $V = V_R$, $I = I_R$ at $x = 0$ and $V = V_S$, $I = I_S$ at $x = l$, we can write

$$\begin{bmatrix} V_S \\ I_S \end{bmatrix} = \begin{bmatrix} A & B \\ C & D \end{bmatrix} \begin{bmatrix} V_R \\ I_R \end{bmatrix} \tag{17-30}$$

$$A = \cosh \delta l; \qquad B = Z_C \sinh \delta l; \qquad C = \frac{\sinh \delta l}{Z_C}; \qquad D = \cosh \delta l$$

where

$$Z_C = \sqrt{\frac{R + jX}{Y}}\ \Omega; \qquad \delta = \sqrt{Y(R + jX)}$$

### 17.3.3 Transformers

Transformers make possible the long-distance transmission of electric power by changing the current and voltage to the appropriate level for each application. Additionally, transformers are used to regulate voltage, perform reactive power control, and limit short-circuit currents. In an electric power system, transformers can be either three-phase units or three single-phase units forming a three-phase bank. Popular transformer connections include Y – Y, Δ – Δ, and Y – Δ. Y connections are frequently grounded.

***Model of Two-Winding Transformers.*** The accepted equivalent circuit for a two-winding transformer operating in steady state is given in Fig. 17-9*a*. $R_T$ and $X_T$ are the equivalent resistance and leakage reactance of the transformer. For most power system steady-state studies the small magnetizing current that transformers draw when energized can be neglected. According to the equivalent circuit of the two-winding transformer, the relationship between primary and secondary quantities can be obtained as

$$I_p = \frac{V_p - NV_s}{N^2(R_T + jX_T)} = \frac{V_p - NV_s}{N^2 Z_T} = \left(V_p - \frac{V_s}{c}\right)c^2 Y_T \tag{17-31}$$

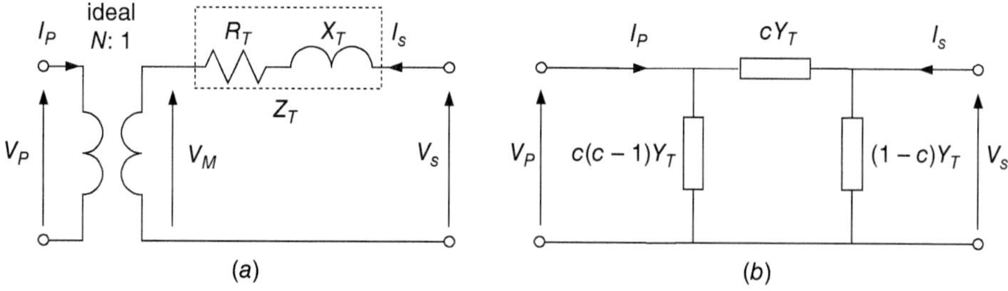


**FIGURE 17-9** Equivalent circuits for two-winding transformers; (*a*) standard equivalent circuit with ideal transformer; (*b*) $\pi$ equivalent circuit. Both consider that the tap can be different than nominal.

$$I_s = \frac{NV_s - V_p}{N(R_T + jX_T)} = (NV_s - V_p)\frac{Y_T}{N} = \left(\frac{V_s}{c} - V_p\right)cY_T \tag{17-32}$$

where $V_p$ and $I_p$ are the primary voltage and current; $V_s$ and $I_s$ are the secondary voltage and current, respectively; $N$ is the turns ratio, which is a complex number with magnitude and angle. The magnitude gives the transformation ratio (given by the number of turns) and the angle gives the phase shift due to the connection.

To integrate the transformer model into power flow calculations, we revise the equivalent circuit shown in Fig. 17-9*a* into a $\pi$ model; see Fig. 17-9*b*. One can easily see that their output is identical.

The analysis above can be directly applied to three-phase transformers. The factor $\sqrt{3}$ is caused by Y – Δ connection that can be considered into the parameter $N$. The phase shift caused by Y – Δ connection is normally not considered for balanced conditions. For unbalanced operation it becomes essential to consider the phase shift of the different voltage levels explicitly.

***Model of Three-Winding Transformers.*** The equivalent circuit of a three-winding transformer is shown in Fig. 17-10*a*. To obtain the equivalent impedances of $Z_p$, $Z_s$, and $Z_t$, we need to first calculate $Z_{ps}$, $Z_{pt}$, and $Z_{st}$ from standard short-circuit tests, $Z_{ps}$ is the leakage impedance measured in primary with secondary shorted and tertiary open. $Z_{pt}$ is the leakage impedance measured in primary with tertiary shorted and secondary open. $Z_{st}$ is the leakage impedance measured in secondary with tertiary shorted and primary open. Then the equivalent impedances $Z_p$, $Z_s$, and $Z_t$ can be obtained as follows:

$$\begin{aligned} Z_p &= \frac{1}{2}(Z_{ps} + Z_{pt} - Z_{st}) \\ Z_s &= \frac{1}{2}(Z_{ps} + Z_{st} - Z_{pt}) \\ Z_t &= \frac{1}{2}(Z_{pt} + Z_{st} - Z_{ps}) \end{aligned} \tag{17-33}$$

Note that frequently one of the impedances is negative. This does not affect steady-state calculations in any. The model is mathematically correct. Through wye-delta transformation, a delta equivalent circuit of three-winding transformers can be obtained; see Fig. 17-10*b*. The corresponding $\pi$ model of each terminal can be obtained as same as the two-winding transformer described in the previous subsection.

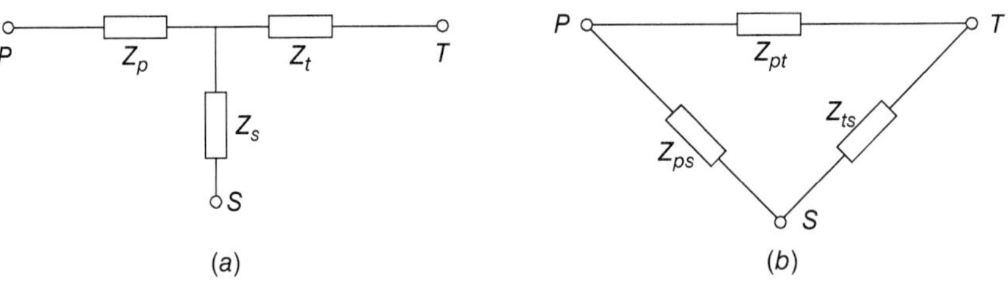


**FIGURE 17-10** One-line diagram of three-winding transformer; (*a*) three single-phase transformers in Y connection; (*b*) equivalent three single-phase transformers in Δ connection.

***Model of Phase-Shifting Transformers.*** The equivalent circuit of the phase-shifting transformer is illustrated in Fig. 17-11. Note that the turn's ratio now needs to be a complex number. According to the equivalent circuit, we have

$$V_p(I_p)^* = -V_M(I_s)^* \tag{17-34}$$

$$V_p = |N|\angle\alpha \cdot V_M = |N|(\cos\alpha + j\sin\alpha)V_M = (a_T + jb_T)V_M \tag{17-35}$$

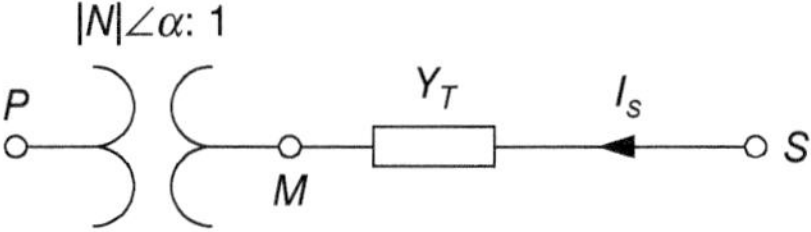


**FIGURE 17-11** One-line diagram of phase-shifting transformer.

Solving for $I_p$, $I_s$ from the two equations above, yielding

$$I_P = -\left(\frac{1}{a_T + jb_T}\right)^* I_s = \frac{Y_T}{a_T - jb_T}(V_M - V_s) = \frac{Y_T}{a_T^2 + b_T^2}[V_p - (a_T + jb_T)V_s] \tag{17-36}$$

$$I_s = \frac{Y_T}{a_T + jb_T}[(a_T + jb_T)V_s - V_p] \tag{17-37}$$

Because the admittance from primary to secondary is not equal to the admittance from secondary to primary we cannot create a $\pi$ equivalent circuit for the phase-shifting transformer.

### 17.3.4 Loads

There are several ways to represent a load for power system studies. In this subsection, we will look at some of the common models used by various pieces of software.

***Constant Load Model.*** The most common model used in load flow calculations is constant load model. Loads are represented by constant *P* and constant *Q*. This makes the problem nonlinear because Ohm's law does not apply to constant power loads (in contrast with constant impedance loads). Since the problem is nonlinear sometimes a solution does not exist. To avoid such conditions in a constant PQ model, a cut-off voltage is commonly used. Below this cut-off voltage, the model is converted to constant impedance model to guarantee a solution. Most power electronics loads are constant power loads. Induction motors driving pumps, which account for a vast majority of modern day loads, are also very close to constant power in the working region. Therefore, most power flow software model loads as constant PQ.

***Constant Impedance Model.*** In this method, the load is modeled as constant impedance. It means load is modeled as constant resistance, constant inductance, and constant capacitance. This kind of modeling reduces the circuit to a linear model and can be solved very easily in phasor domain using traditional circuit solvers. Today, very few of the devices are constant impedance devices. Common examples of constant impedance devices are incandescent bulbs and resistive heaters. These devices do not account for a very large proportion of the load demanded today. When performing power flow calculations below rated voltage, constant impedance is not a very wise choice because constant impedance models show a reduction in power.

***ZIP Coefficients Model.*** As mentioned in the above two subsections, most loads are neither constant impedance nor constant power. Rather they are a combination of constant power, constant impedance, and constant current. A load modeled as ZIP coefficients load is written as follows:

$$P_i = P_o\left(Z_p\left(\frac{V_i}{V_o}\right)^2 + I_p\left(\frac{V_i}{V_o}\right) + P_p\right) \tag{17-38}$$

$$Q_i = Q_o\left(Z_q\left(\frac{V_i}{V_o}\right)^2 + I_q\left(\frac{V_i}{V_o}\right) + P_q\right) \tag{17-39}$$

where $P_o$, $Q_o$ and $V_o$ are rated active power, reactive power, and voltage magnitude (not phasor), respectively. $P_i$, $Q_i$ are active and reactive power consumed at operating voltage $V_i$. $Z_p$, $I_p$, $P_p$ are ZIP parameters for the active power, similarly $Z_q$, $I_q$, $P_q$ are ZIP parameters for reactive power. As can be seen in the ZIP equation there are three parts: the first part can be seen of as the constant impedance component, the second part can be thought of as constant current component, and the last part is the constant power component. At $V_i = V_o$ the equations becomes

$$P_o = P_o(Z_p + I_p + P_p) \tag{17-40}$$

$$Q_o = Q_o(Z_q + I_q + P_q) \tag{17-41}$$

which implies that

$$Z_p + I_p + P_p = 1 \tag{17-42}$$

$$Z_q + I_q + P_q = 1 \tag{17-43}$$

In reference [1], ZIP coefficient for most commonly found loads are tabulated. Since the ZIP model is most appropriate to describe loads, most modern day power flow pieces of software include an option to model loads as ZIP coefficients.

## 17.4 PER-UNIT SYSTEM

In power system studies, voltage, current, and power are often written as dimensionless fraction with respect to a base rather than actual values in volts, amperes, and watts. This is called per-unit system. There are several advantages to such a system, the most obvious advantage of this system is that it immediately provides a reference to the engineer regarding the "health" of the quantity. For example, a transmission level voltage of 100 kV on a 138-kV base system will be written as 0.72 pu (100/138). An engineer uninformed of the working transmission level voltage cannot make a decision whether 100 kV is low, high, or appropriate voltage. On the other hand, if one sees a voltage of 0.72 pu, the conclusion is that this voltage is 28% less than the "normal" value. The per-unit system is very important for the description of voltages because most loads in a power system will not work well if the voltage is below 90% or above 110% of rated. Thus one can very easily see if the operating is not correct when presented in per unit. The same logic is applied to current and power. As one can deduce, the selection of the base quantity with respect to which the per-unit values are written is very important to have a meaningful per-unit value. We shall go over an example later in the section that shows how bases for various quantities are selected.

Some of the other advantages of per-unit system are

- Per-unit voltages and currents are the same in line-to-ground and line-to-line calculations. Therefore, the $\sqrt{3}$ factor is not seen in per-unit system.
- Voltages and currents on either side of transformers are the same in per-unit system, which makes for much easier calculations.
- Calculations are vastly simplified when all values are converted to per-unit system with values around 1 pu.
- Comparison of per-unit impedance across the transformers gives an idea about voltage drop in the transformer. It is difficult to gauge this information from impedance values in ohm. For two identical power rating transformer, the one with a lower per-unit impedance produces lesser voltage drop.
- Per-unit system gives an idea about the relative magnitude of the quantity. In a list where quantities are written in per unit, it is very easy to search for low, high, and appropriate magnitude quantities, regardless of the base.

### 17.4.1 Calculation Formulas

The basic formula for all per-unit calculations is this:

$$\text{Quantity}_{\text{pu}} = \frac{\text{Quantity}_{\text{actual}}}{\text{Quantity}_{\text{base}}} \tag{17-44}$$

From this we derive the following set of equations:

$$S_{\text{pu},\,3\phi} = \frac{S_{\text{actual},\,3\phi}}{S_{\text{base},\,3\phi}} \tag{17-45}$$

$$S_{\text{pu},\,1\phi} = \frac{S_{\text{actual},1\phi}}{S_{\text{base},1\phi}} = \frac{3\,S_{\text{actual},1\phi}}{3\,S_{\text{base},1\phi}} = \frac{S_{\text{actual},3\phi}}{S_{\text{base},3\phi}} = S_{pu,3\phi} \tag{17-46}$$

$$V_{\text{pu},\,L-G} = \frac{V_{\text{actual},\,L-G}}{V_{\text{base},\,L-G}} \tag{17-47}$$

$$V_{\text{pu},\,L-L} = \frac{V_{\text{actual},\,L-L}}{V_{\text{base},\,L-L}} = \frac{\sqrt{3}\,V_{\text{actual},\,L-G}}{\sqrt{3}\,V_{\text{base},\,L-G}} = \frac{V_{\text{actual},\,L-G}}{V_{\text{base},\,L-G}} = V_{\text{pu},\,L-G} \tag{17-48}$$

Similarly, it can be shown that $I_{\text{pu,phase}} = I_{\text{pu,line to line}}$. This simplifies the calculations because the factor of $\sqrt{3}$ does not exist in a per-unit system.

To calculate the base quantities, we start by selecting $V_{\text{base}}$ and $S_{\text{base}}$. As is the norm in power system studies, any voltage quantity is always line-to-line rms voltage unless otherwise mentioned and power is three-phase power. After selecting the appropriate voltage and power bases, we calculate the bases for impedance, admittance, and current in the following manner:

$$S_{\text{base}} = \sqrt{3}\,V_{\text{base}}\,I_{\text{base}} \tag{17-49}$$

Thus,

$$I_{\text{base}} = \frac{S_{\text{base}}}{\sqrt{3}\,V_{\text{base}}} \tag{17-50}$$

$$Z_{\text{base}} = \frac{V_{\text{base},L-G}}{I_{\text{base}}} = \frac{V_{\text{base},L-G}}{\dfrac{S_{\text{base}}}{\sqrt{3}V_{\text{base}}}} = \frac{V_{\text{base}}^2}{S_{\text{base}}} \tag{17-51}$$

$$Y_{\text{base}} = \frac{1}{Z_{\text{base}}} = \frac{S_{\text{base}}}{V_{\text{base}}^2} \tag{17-52}$$

Once we have calculated the appropriate bases we use the following formulas to calculate the per-unit values:

$$S_{\text{pu}} = \frac{S_{\text{actual}}}{S_{\text{base}}}, \quad V_{\text{pu}} = \frac{V_{\text{actual}}}{V_{\text{base}}}, \quad I_{\text{pu}} = \frac{I_{\text{actual}}}{I_{\text{base}}}, \quad Z_{\text{pu}} = \frac{Z_{\text{actual}}}{Z_{\text{base}}}, \quad Y_{\text{pu}} = \frac{Y_{\text{actual}}}{Y_{\text{base}}} \tag{17-53}$$

Reactance and resistance per-unit values are computed from

$$X_{\text{pu}} = \frac{X_{\text{actual}}}{Z_{\text{base}}}, \quad R_{\text{pu}} = \frac{R_{\text{actual}}}{Z_{\text{base}}} \tag{17-54}$$

### 17.4.2 Converting pu Values from One Base to Another

Let us consider that we have voltage, current, and power in old $S_{base}$ and $V_{base}$. To convert these per-unit values from old base to per-unit values according to a new base we use the following formulas:

$$V_{pu,new} = V_{pu,old} \frac{V_{base,old}}{V_{base,new}} \tag{17-55}$$

$$S_{pu,new} = S_{pu,old} \frac{S_{base,old}}{S_{base,new}} \tag{17-56}$$

Thus,

$$I_{pu,new} = I_{pu,\,old} \frac{I_{base,old}}{I_{base,new}} = I_{pu,old} \frac{S_{base,old}}{\sqrt{3} * V_{base,old}} \frac{\sqrt{3} * V_{base,new}}{S_{base,new}} = I_{pu,old} \left(\frac{S_{base,old}}{S_{base,new}}\right)\left(\frac{V_{base,new}}{V_{base,old}}\right) \tag{17-57}$$

$$Z_{pu,new} = Z_{pu,old} \frac{Z_{base,old}}{Z_{base,new}} = Z_{pu,old} \frac{V^2_{base,old}}{S_{base,old}} \frac{S_{base,new}}{V^2_{base,new}} = Z_{pu,old} \left(\frac{S_{base,new}}{S_{base,old}}\right)\left(\frac{V_{base,old}}{V_{base,new}}\right)^2 \tag{17-58}$$

*Example 17-1.* Convert all the voltages and impedances to per-unit in the following one-line diagram circuit.

We start by selecting voltages in each zone. There are three zones as shown in Fig. 17-12. Zones are demarcated by transformers. $S_{base}$ is same in all the zones. We select $S_{base}$ and $V_{base1}$ arbitrarily. Most power system studies use $S_{base}$ as 100 MVA.

$$S_{base} = 100 \text{ MVA} = S_{base1} = S_{base2} = S_{base3}$$

$$V_{base1} = 13.8 \text{ kV}$$

$$V_{base2} = V_{base1} \frac{138}{13.2} = 144.27 \text{ kV}$$

$$V_{base3} = V_{base2} \frac{13.8}{138} = 14.43 \text{ kV}$$

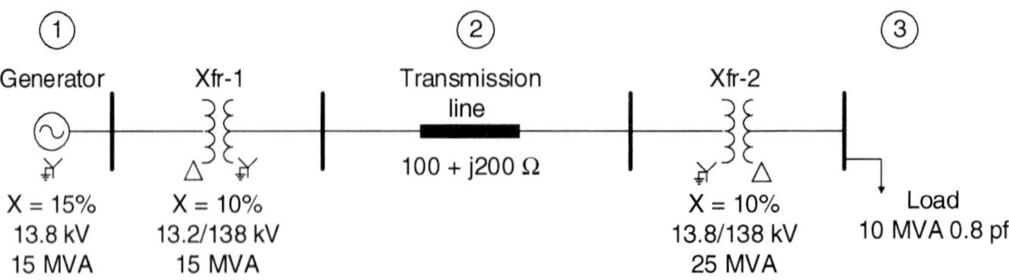


**FIGURE 17-12** One-line diagram in physical quantities to be converted to per unit.

Using formula for $Z_{\text{base}}$, we get

$$Z_{\text{base1}} = \frac{V_{\text{base1}}^2}{S_{\text{base1}}} = \frac{(13.8\times10^3)^2}{100*10^6} = 1.90\ \Omega$$

$$Z_{\text{base2}} = \frac{V_{\text{base2}}^2}{S_{\text{base2}}} = \frac{(144.27\times10^3)^2}{100*10^6} = 208.13\ \Omega$$

$$Z_{\text{base3}} = \frac{V_{\text{base3}}^2}{S_{\text{base3}}} = \frac{(14.43\times10^3)^2}{100\times10^6} = 2.08\ \Omega$$

$$Z_{\text{line,pu}} = \frac{Z_{\text{line}}}{Z_{\text{base2}}} = \frac{100+j200}{208.13} = 0.48+j0.96$$

Per-unit values are already given for transformers and generators by the manufacturers at their power base. We need to convert them to the system base that we have selected in this problem. For this we use the above equations to get

$$X_{xfr1,\text{pu, new}} = X_{xfr1,\text{pu,old}} * \left(\frac{S_{\text{base,new}}}{S_{\text{base,old}}}\right)\left(\frac{V_{\text{base,old}}}{V_{\text{base,new}}}\right)^2 = 0.1\left(\frac{100}{15}\right)\left(\frac{13.2}{13.8}\right)^2 = 0.609955\ \text{pu}$$

$$Z_{xfr2,\text{pu, new}} = Z_{xfr2,\text{pu,old}} * \left(\frac{S_{\text{base,new}}}{S_{\text{base,old}}}\right)\left(\frac{V_{\text{base,old}}}{V_{\text{base,new}}}\right)^2 = 0.1\left(\frac{100}{25}\right)\left(\frac{13.8}{14.43}\right)^2 = 0.365835\ \text{pu}$$

$$Z_{G,\text{pu, new}} = Z_{G,\text{pu,old}} * \left(\frac{S_{\text{base,new}}}{S_{\text{base,old}}}\right)\left(\frac{V_{\text{base,old}}}{V_{\text{base,new}}}\right)^2 = 0.15\left(\frac{100}{15}\right)\left(\frac{13.8}{13.8}\right)^2 = 1\ \text{pu}$$

$$S_{\text{load,pu}} = \frac{S_{\text{load}}}{S_{\text{base}}} = \frac{10}{100} = 0.1\ \text{pu}$$

$$V_{g,\text{pu}} = \frac{13.8}{13.8} = 1\ \text{pu}$$

The new circuit in per-unit system looks like the one depicted in Fig. 17-13.

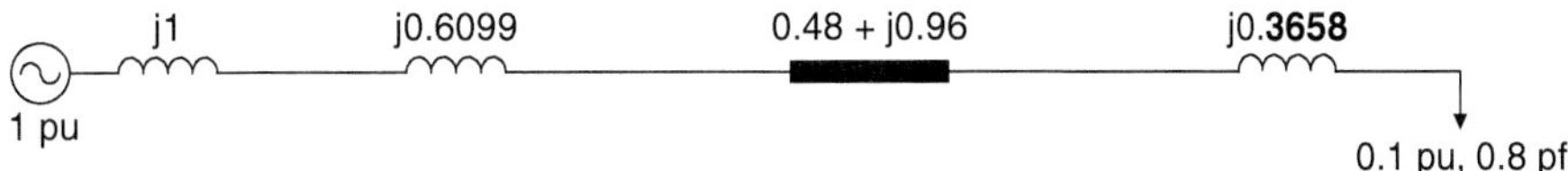


**FIGURE 17-13** One-line diagram in per-unit system.

## 17.5 SYMMETRICAL COMPONENTS

Symmetrical components method is a way of representing $N$ unbalanced vectors as a sum of $N$ set of balanced vectors. This concept was first introduced by C. L. Fortescue. Symmetrical components are particularly useful in fault analysis, where a balanced system becomes unbalanced due to a single-line-to-ground or a line-to-line fault. The use of symmetrical components will be demonstrated in this subsection using three-phase systems since existing power systems are three phase.

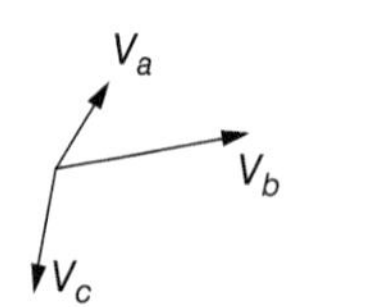

**FIGURE 17-14** Unbalanced phasors.

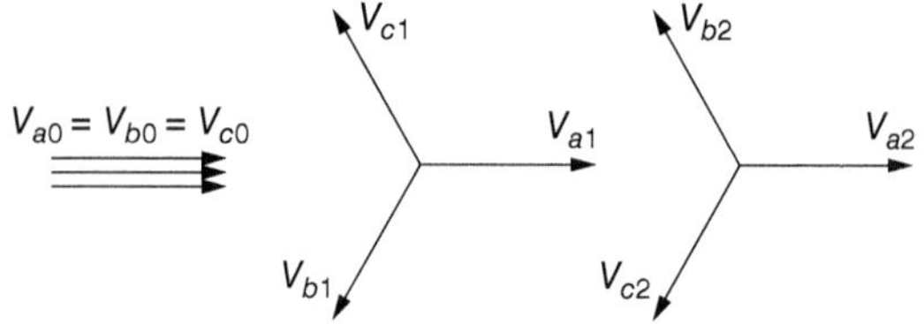


**FIGURE 17-15** Zero, positive, negative, and sequence component.

In a three-phase system, the three sets of balanced vectors are called positive, negative, and zero sequences. Let us consider a set of unbalanced vectors as shown in Fig. 17-14. There exist three sets of balanced vectors such that adding them up yields the above unbalance system as shown in Fig. 17-15.

The three balanced set of vectors are

$$V_{a1} = V_1\angle 0 \qquad V_{a2} = V_2\angle 0 \qquad V_{a0} = V_0\angle 0$$
$$V_{b1} = V_1\angle 240 \qquad V_{b2} = V_2\angle 120 \qquad V_{b0} = V_0\angle 0$$
$$V_{c1} = V_1\angle 120 \qquad V_{c2} = V_2\angle 240 \qquad V_{c0} = V_0\angle 0$$

Adding the balanced set of vectors, we get

$$\begin{aligned} V_a &= V_1\angle 0 + V_2\angle 0 + V_0\angle 0 \\ V_b &= V_1\angle 240 + V_2\angle 120 + V_0\angle 0 \\ V_c &= V_1\angle 120 + V_2\angle 240 + V_0\angle 0 \end{aligned} \tag{17-59}$$

In matrix form, we can write it as

$$\begin{bmatrix} V_a \\ V_b \\ V_c \end{bmatrix} = \begin{bmatrix} 1 & 1 & 1 \\ 1 & a^2 & a \\ 1 & a & a^2 \end{bmatrix} \begin{bmatrix} V_{a0} \\ V_{a1} \\ V_{a2} \end{bmatrix} \tag{17-60}$$

where $a = 1\angle 120$ is a phasor that when multiplied by another phasor rotates the second by 120°. The above expression can be written in matrix form as

$$V_{abc} = T\ V_a^{012} \tag{17-61}$$

$T$ is called the symmetrical components transformation matrix. Inversely, we can also say $V_a^{012} = T^{-1}V_{abc}$, where $T^{-1}$ is the inverse symmetrical components transformation matrix given by

$$T^{-1} = \frac{1}{3}\begin{bmatrix} 1 & 1 & 1 \\ 1 & a & a^2 \\ 1 & a^2 & a \end{bmatrix} \tag{17-62}$$

The following relations on "$a$" are fulfilled:

$$a = 1\angle 120°$$
$$a^2 = 1\angle -120° = 1\angle 240°$$

$$
\begin{aligned}
a^* &= a^2 \\
(a^2)^* &= a \\
a^3 &= 1 \\
a^4 &= a \\
a+a^2+1 &= 0 \\
a+a^2 &= -1 \\
a-a^2 &= j\sqrt{3} \\
a^2-a &= -j\sqrt{3}
\end{aligned}
\tag{17-63}
$$

*Example 17-2.* Consider the following set of unbalanced vectors:

$$
\begin{aligned}
V_a &= 100\angle 10 \\
V_b &= 100\angle -80 \\
V_c &= 80\angle 60
\end{aligned}
$$

Calculate the sequence components and draw them.

*Solution.* We know that $V_a^{012} = T^{-1}V_{abc}$. Using this, we write

$$
\begin{bmatrix} V_{a0} \\ V_{a1} \\ V_{a2} \end{bmatrix} = \frac{1}{3}\begin{bmatrix} 1 & 1 & 1 \\ 1 & a & a^2 \\ 1 & a^2 & a \end{bmatrix}\begin{bmatrix} V_a \\ V_b \\ V_c \end{bmatrix} = \begin{bmatrix} 52.10\angle -4.34^\circ \\ 71.81\angle 3.29^\circ \\ 30.47\angle 145^\circ \end{bmatrix}
$$

To help visualize, let us draw the original unbalanced vectors and the set of balanced vectors.

Adding the symmetrical components in Fig. 17-17 yields the set of unbalanced vectors shown in Fig. 17-16. Even though in power system studies we only use three sets, there is no limit on number of vectors for this method. Useful Matlab code for generating the transformation matrix of a system of any order is given below.

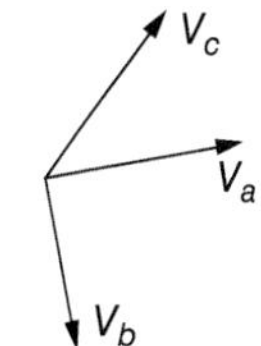


**FIGURE 17-16** Original unbalanced vectors (not to scale).

```
function T=TransM(n)
% TransM outputs the tranformation matrix for
% a set of n vectors.
T=ones(n);
a=cos(2*pi/n)+1i*sin(2*pi/n);
for j=2:n
    T(2,j)=a^(n+1-j);
end
for k=2:n
    for l=3:n
    T(l,k)=(a^-(k-1)).*T(l-1,k);
    end
end
end
```

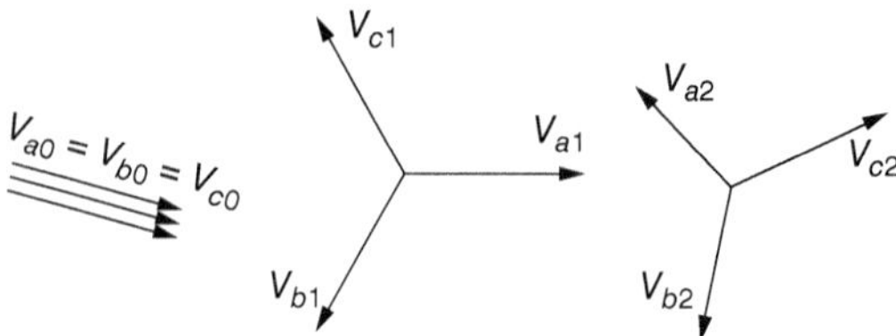


**FIGURE 17-17** Set of symmetrical balanced vectors (not to scale).

## 17.6 SEQUENCE IMPEDANCES

Sequence components are mainly used in fault studies. Even a balanced system becomes unbalanced when a fault other than a three-phase fault occurs, for example, a single-line-to-ground or double-line with or without ground. It is easier to divide the unbalanced circuit into three different balanced sequence circuits. Converting voltage into sequence components has been discussed above. In this subsection, sequence impedances are discussed. Sequence impedance is discussed for balanced transmission lines and transformers.

### 17.6.1 Transmission Lines

In this subsection, we discuss the case of sequence components for a balanced transmission line. This happens only when the line is continuously transposed. The self-impedance is $Z_s$ and mutual impedance is $Z_m$ as shown in Fig. 17-18. The voltage current relationship is given by

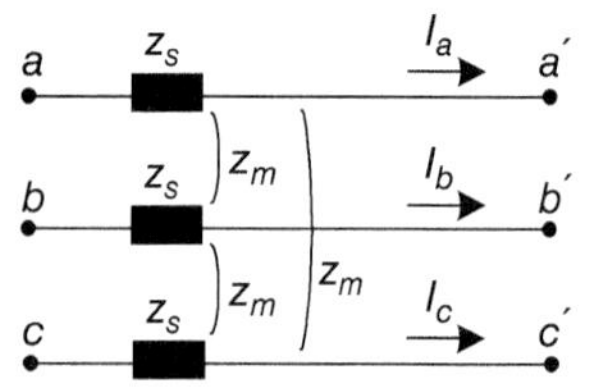

$$\begin{bmatrix} V_{aa'} \\ V_{bb'} \\ V_{cc'} \end{bmatrix} = \begin{bmatrix} Z_s & Z_m & Z_m \\ Z_m & Z_s & Z_m \\ Z_m & Z_m & Z_s \end{bmatrix} \begin{bmatrix} I_a \\ I_b \\ I_c \end{bmatrix} \tag{17-64}$$

**FIGURE 17-18** Transmission line showing the self and mutual inductances.

Converting voltage and currents to the sequence components, the equation becomes

$$T\begin{bmatrix} V_{aa'0} \\ V_{aa'1} \\ V_{aa'2} \end{bmatrix} = \begin{bmatrix} Z_s & Z_m & Z_m \\ Z_m & Z_s & Z_m \\ Z_m & Z_m & Z_s \end{bmatrix} T \begin{bmatrix} I_{a0} \\ I_{a1} \\ I_{a2} \end{bmatrix} \tag{17-65}$$

where $T$ is the symmetrical components transformation matrix. Multiplying both sides by $T^{-1}$ on the left gives

$$\begin{bmatrix} V_{aa'0} \\ V_{aa'1} \\ V_{aa'2} \end{bmatrix} = T^{-1}\begin{bmatrix} Z_s & Z_m & Z_m \\ Z_m & Z_s & Z_m \\ Z_m & Z_m & Z_s \end{bmatrix} T \begin{bmatrix} I_{a0} \\ I_{a1} \\ I_{a2} \end{bmatrix} \tag{17-66}$$

Simplifying yields

$$\begin{bmatrix} V_{aa'0} \\ V_{aa'1} \\ V_{aa'2} \end{bmatrix} = \begin{bmatrix} Z_s + 2Z_m & 0 & 0 \\ 0 & Z_s - Z_m & 0 \\ 0 & 0 & Z_s - Z_m \end{bmatrix} \begin{bmatrix} I_{a0} \\ I_{a1} \\ I_{a2} \end{bmatrix} \tag{17-67}$$

The diagonal terms of the matrix are the sequence impedances for a balanced transmission line.

$$\begin{aligned} Z_0 &= Z_s + 2Z_m \\ Z_1 &= Z_s - Z_m \\ Z_2 &= Z_s - Z_m \end{aligned} \tag{17-68}$$

### 17.6.2 Transformers

In the case of three-phase transformers, the sequence circuits depend on the type of connections made at the primary and secondary. We look at five cases: (1) Yg-Yg, both neutrals grounded; (2) Yg-Y only one neutral grounded; (3) Yg-Δ, Y grounded; (4) Y-Y, ungrounded neutrals; and (4) $\Delta - \Delta$.

In all these cases, the positive and negative sequence impedances are the same as described above. Each connection has a different zero sequence circuit.

***Y-Y, Both Neutrals Grounded (Yg-Yg).*** The voltage equation on the primary side is

$$V_A = V_{AN} + V_N \tag{17-69}$$

$V_{AN}$ is the primary side voltage across the winding of the ideal transformer. $V_N$ is the voltage across the neutral. Converting to sequence components, we have

$$V_A^0 + V_A^1 + V_A^2 = V_{AN}^0 + V_{AN}^1 + V_{AN}^2 + 3Z_1 I_A^0 \tag{17-70}$$

where $I_A^0$ is zero sequence current on the primary side $Z_1$ is grounding resistance on the primary side.

Similarly, on the secondary side

$$V_a^0 + V_a^1 + V_a^2 = V_{an}^0 + V_{an}^1 + V_{an}^2 - 3Z_2 I_a^0 \tag{17-71}$$

where $I_a^0$ = zero sequence current on the secondary side. $Z_2$ is grounding resistance on the secondary side.

The negative sign in Eq. (17-70) is to account for the direction of the current. Using the transformation ratio, we can write the following relations between primary and secondary voltages and currents:

$$V_{an} = nV_{AN}; \qquad I_a = \frac{I_A}{n} \tag{17-72}$$

$n$ being the turns ratio. Substituting for it in Eq. (17-71), we have

$$V_a^0 + V_a^1 + V_a^2 = nV_{AN}^0 + nV_{AN}^1 + nV_{AN}^2 - \frac{3Z_2 I_A^0}{n} \tag{17-73}$$

Dividing both sides by $n$, we get

$$\frac{V_a^0 + V_a^1 + V_a^2}{n} = V_{AN}^0 + V_{AN}^1 + V_{AN}^2 - \frac{3Z_2 I_A^0}{n^2} \tag{17-74}$$

Substituting Eq. (17-70), we can write

$$\frac{V_a^0 + V_a^1 + V_a^2}{n} = V_A^0 + V_A^1 + V_A^2 - 3Z_1 I_A^0 - \frac{3Z_2 I_A^0}{n^2} \tag{17-75}$$

By equating terms of same sequence one can obtain the following expressions:

$$V_a^1 = nV_A^1; \qquad V_a^2 = nV_A^2$$

$$\frac{V_a^0}{n} = V_A^0 - \left[3Z_1 + \frac{3Z_2}{n^2}\right] I_A^0 \tag{17-76}$$

Therefore,

$$Z_0 = Z + 3Z_1 + 3Z_2' \tag{17-77}$$

where $Z$ = leakage impedance
$Z_1$ = primary neutral impedance
$Z_2'$ = secondary neutral impedance referred to primary

***Yg-Y, Only One Neutral Grounded.*** In this case, no zero-sequence current can flow as one of the neutral is ungrounded. Therefore, the zero sequence circuit in the case of a Y-Y transformer with only neutral grounded is shown as an open circuit between primary and the secondary.

***Y-Δ, Y Grounded.*** In this case, the zero sequence current flows only on the Y side. The zeros sequence component in this case is $Z_0 = 3Z_N$, where $Z_N$ is the grounding impedance. No zero-sequence current can flow in the Δ side.

***Y-Δ, Ungrounded Neutral.*** If the Y side is not grounded, there can be no zero-sequence current on the Y side. As discussed before, there can be no zero-sequence current in the Δ side.

***Δ − Δ Transformer.*** In this case, there can be no zero-sequence current either on the primary or the secondary sides. Though there can be circulating zero sequence current in the delta winding.

A summary of zero-sequence circuits for all different transformer connections is shown in Fig. 17-19.

## 17.7 POWER FLOW

As mentioned in the Introduction, power flow analysis is the most important study necessary to plan and operate the system. The basic idea is to obtain the solution of the electrical circuit (i.e., compute voltage at every node and power flowing in every component, line and transformer) in steady state using the models described in the previous subsections of this section. Different from a typical electrical circuit, the power flow problem is nonlinear because the loads are nonlinear.

### 17.7.1 Power Flow Problem

There exist several formulations of the power flow problem each one with advantages and disadvantages. They all start from the application of Kirchhoff current law (KCL) to all buses (nodes) in the system. Consider the connections to bus $k$ as depicted in Fig. 17-20, where $y_{km} = 1/z_{km} = g_{km} + j\,b_{km}$ are

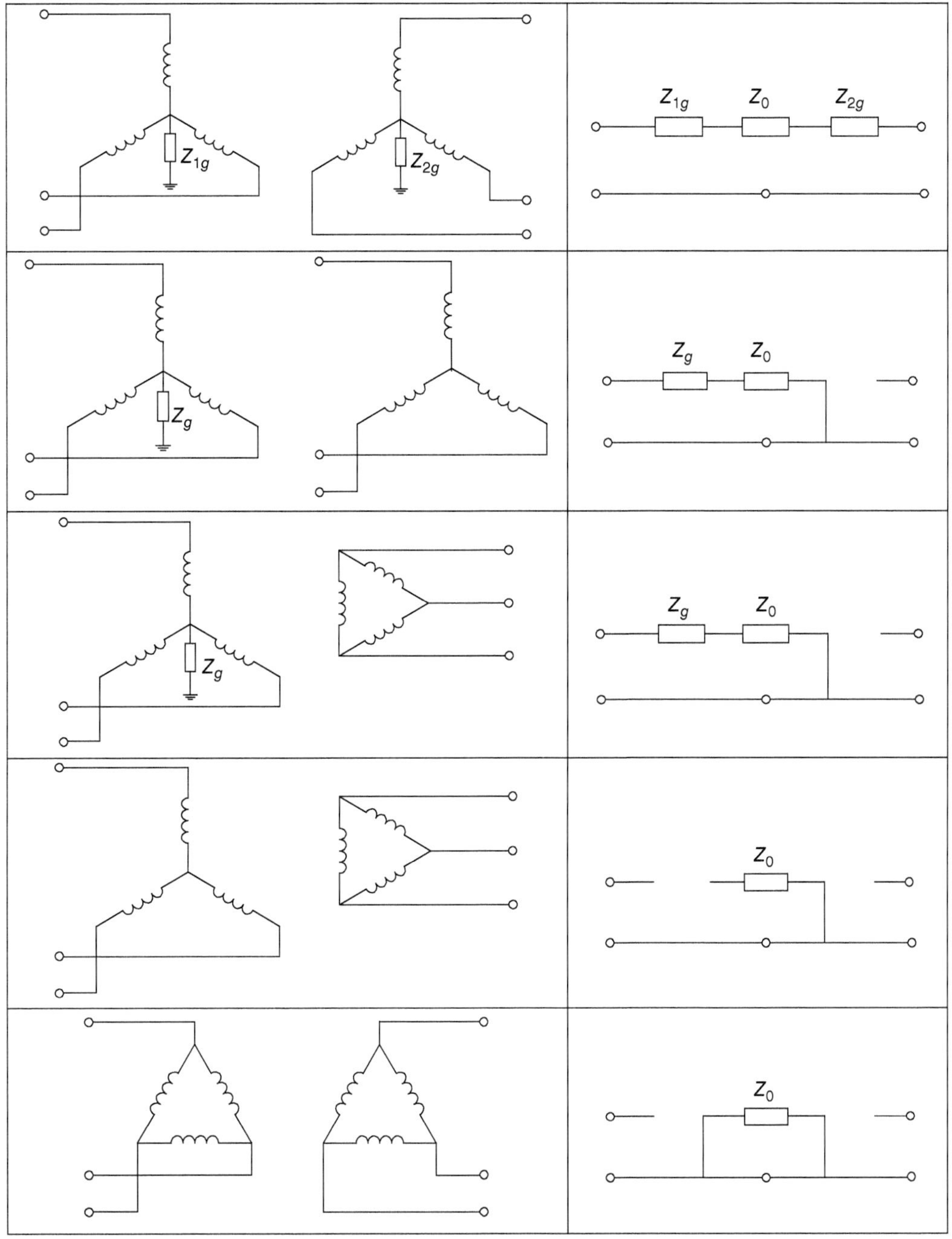


**FIGURE 17-19** Connection of the zero sequence impedance for the most common transformer configurations.

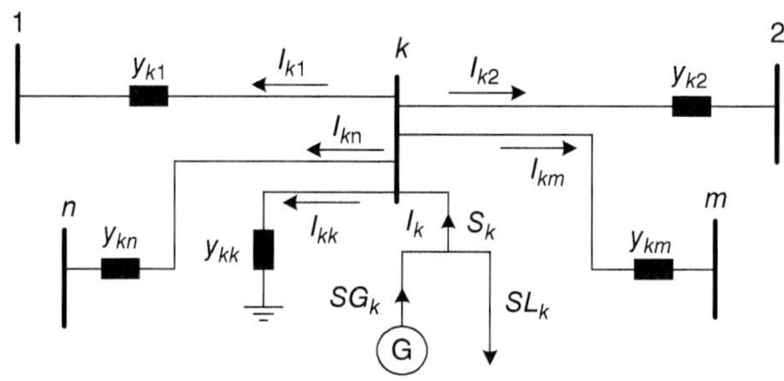


**FIGURE 17-20** Connections to bus $k$.

the positive sequence (series) admittances of the lines connecting to the other buses in the system. The shunt admittances at bus $k$ of all lines have been lumped into admittance $y_{kk}$. $SG_k$ $(= PG_k + j\,QG_k)$ is the power generated (or injected) at bus $k$. $SL_k$ $(= PL_k + j\,QL_k)$ is the power consumed (or extracted) at bus $k$; in other words the load. Let $S_k = SG_k - SL_k$ be the total power injected at bus $k$, which could be a positive or negative number. $n$ is the total number of buses in the system. Because the solution is obtained by node (rather than by mesh) it is better to use the positive sequence admittances of the lines rather than their impedance $z_{km}$. Note that, due to the size of the system, the total number of buses is very large (tenths of thousands), while the number of direct connections to a given bus is very small (a hand full at most). Let us assume that all currents $I_{km}$ are in the direction departing from node $k$ (power extractions). Then, the current $I_k$ injected at bus $k$ can be computed using KCL applied to node $k$ as

$$I_k = \sum_{m=1}^{n} I_{km} = \sum_{\substack{m=1 \\ m \neq k}}^{n} y_{km}(V_k - V_m) + y_{kk}V_k \tag{17-78}$$

where $V_k$ and $V_m$ are the voltages of buses $k$ and $m$. The first term was obtained from Ohm's law as $I_{km} = y_{km}(V_k - V_m)$ while the second is the current in the shunt admittances of lines and transformers. By combining elements, Eq. (17-78) can be written as

$$I_k = \sum_{m=1}^{n} y_{km}V_k - \sum_{\substack{m=1 \\ m \neq k}}^{n} y_{km}V_m \tag{17-79}$$

The above equation, when applied to all buses in the system, can be written in matrix for using the so-called $Y_{\text{bus}}$ matrix as follows:

$$\begin{bmatrix} I_1 \\ I_2 \\ \vdots \\ I_k \\ \vdots \\ I_n \end{bmatrix} = \begin{bmatrix} Y_{11} & Y_{12} & Y_{1k} & Y_{1n} \\ Y_{21} & Y_{22} & Y_{2k} & Y_{2n} \\ & & & \\ Y_{k1} & Y_{k2} & Y_{kk} & Y_{kn} \\ & & & \\ Y_{n1} & Y_{n2} & Y_{nk} & Y_{nn} \end{bmatrix} \begin{bmatrix} V_1 \\ V_2 \\ \vdots \\ V_k \\ \vdots \\ V_n \end{bmatrix} \tag{17-80}$$

This matrix can be built by inspection. The elements of the main diagonal are the sum of all the admittances directly connected to a given node and the off-diagonal elements are the negative of the admittances between nodes as follows:

$$Y_{kk} = \sum_{m=1}^{n} y_{km}; \qquad Y_{km} = -y_{km} \tag{17-81}$$

Please note that lower case $y$ is used to represent line/transformer admittances while upper case $Y$ is used to represent the elements of the $Y_{\text{bus}}$ matrix. In matrix form, we have

$$[I] = [Y_{\text{bus}}][V] \tag{17-82}$$

The $Y_{\text{bus}}$ matrix as presented above is ill-conditioned because the connections to ground are weak since only the admittances to ground of lines are considered. To solve the problem a reference (or slack) bus is selected for the solution of the power flow problem.

Let us now develop row $k$ of Eq. (17-80):

$$I_k = \sum_{m=1}^{n} Y_{km} V_m \tag{17-83}$$

Using the definition of complex power in terms of voltage and current, we have

$$S_k^* = P_k - jQ_k = V_k^* I_k = V_k^* \sum_{m=1}^{n} Y_{km} V_m \tag{17-84}$$

This equation needs to be solved for all $n$ nodes giving the following set of nonlinear algebraic equations:

$$\begin{aligned} S_1^* &= P_1 - jQ_1 = V_1^* \sum_{m=1}^{n} Y_{1m} V_m \\ S_2^* &= P_2 - jQ_2 = V_2^* \sum_{m=1}^{n} Y_{2m} V_m \\ S_n^* &= P_n - jQ_n = V_n^* \sum_{m=1}^{n} Y_{nm} V_m \end{aligned} \tag{17-85}$$

The unknowns are the bus voltages ($V_i$). Once the bus voltages are known, the currents in all lines are computed from Ohm's law $I_{km} = y_{km}(V_k - V_m)$. The power flow is computed from its definition as $S = VI^*$. In contrast with the current, the receiving end and sending end powers are different; the difference is the power consumed (or loss) in the line. Thus for line connecting buses $k$ and $m$, we have

$$S_{km} = V_k I_{km}^* \neq S_{mk} = V_m I_{mk}^* \tag{17-86}$$

Note that $I_{km} = -I_{mk}$, but $V_k \neq V_m$, thus $S_{km} \neq -S_{mk}$. The recommended numerical solution method depends on the type of system (transmission or distribution) and the reactance over resistance ratio of the lines involved.

### 17.7.2 Bus Types

Because of its operation characteristics, buses in a power system are classified into three types: load buses (PQ), generator buses (PV), and slack bus. Most buses in a power system are load buses and contain customer loads. They are called PQ buses because their consumed power $SL_k = PL_k + j\,QL_k$

is known. Some may even contain generation, but this generation is either not sufficient to supply all the load or it is operated in a way that the generators behave as negative loads (positive injected $S_k$). Regardless of whether $SG_k$ is larger, equal or smaller than $SL_k$ they are mathematically treated the same. For these buses the unknowns are the voltage magnitudes and angles.

Buses containing generators are called PV buses because in addition to an active power ($P$) injection, the generator is operated in a way that the voltage magnitude ($V$) is controlled to a specified value. The unknown variables are the voltage angle ($\varphi$) and the reactive power ($Q$). The voltage is controlled by varying the reactive power injection. Frequently, during the process of controlling voltage, the limits of reactive power are exceeded and the PV bus is operated, from that point on, as a PQ bus with fixed reactive power generation. Under those conditions, voltage control is not possible any longer.

Typically one generator bus (the largest), but sometimes more are needed, is designated as a slack bus. The slack bus specifies the reference angle and voltage magnitude. The slack generator, in addition to providing a reference to the system of equations, supplies the losses in the system and does not participate in the solution of the set of equations. The system order is reduced to $n - 1$ for a solution containing only one slack bus (or $n - n_{sb}$ for $n_{sb}$ slack buses). Voltage magnitude and angle are considered known and the power injected (active and reactive) is computed after the solution is obtained. For multislack bus solutions a rule needs to be established on how to share the system losses; sharing by generator rating is recommended.

### 17.7.3 Transmission Systems

There is a gamut of solution methods for the power flow equations for transmission systems. Here only the most commonly used methods will be described.

***Gauss Seidel.*** Historically, Gauss-Seidel was the first method used for the solution of the power flow equations. For this method, it is best to use the complex form given in Eq. (17-85) for $n - 1$ buses. Let us develop the sum corresponding to the $k$ row as

$$S_k^* = V_k^* \sum_{m=1}^{k-1} Y_{km} V_m + V_k^* Y_{kk} V_k + V_k^* \sum_{m=k+1}^{n-1} Y_{km} V_m \tag{17-87}$$

Solving for $V_k$ we have:

$$V_k = -\frac{1}{Y_{kk}} \sum_{m=1}^{k-1} Y_{km} V_m + \frac{S_k^*}{Y_{kk} V_k^*} - \frac{1}{Y_{kk}} \sum_{m=k+1}^{n-1} Y_{km} V_m \tag{17-88}$$

Starting from a suitable initial guess $V_k^{(0)}\ \forall\ k$, Eq. (17-88) can be solved iteratively as follows:

$$V_k^{(j+1)} = -\frac{1}{Y_{kk}} \sum_{m=1}^{k-1} Y_{km} V_m^{(j+1)} + \frac{S_k^*}{Y_{kk} V_k^{*(j)}} - \frac{1}{Y_{kk}} \sum_{m=k+1}^{n-1} Y_{km} V_m^{(j)}; \qquad k = 1, 2, \ldots, n-1 \tag{17-89}$$

If a better initial condition is not known, a flat start is used (all voltage magnitudes equal to 1.0 pu and zero angle): $V_k^{(0)} = 1 + j0\ \forall\ k$. Frequently, power flow solutions are obtained in sequence, thus the last converged values are a good start for the next case when the conditions are similar. The stopping criterion for all power flow solution methods is the nodal power mismatch, that is, the difference between the power injected at every bus and the power flowing in the lines. Power mismatches are computed as

$$\Delta S_k = S_k - V_k \sum_{m=1}^{n} Y_{km}^* V_m^* \tag{17-90}$$

In this way, even when the numerical solution is approximate, the solution is accurate because balance of power is used as the converge criterion. Relative or absolute power mismatches can be used. The former gives equal weight for each node, while the latter gives more weight to the larger loads and generators.

The Gauss-Seidel method has a robust convergence when the diagonal elements ($Y_{kk}$) of $Y_{\text{bus}}$ are dominant. This happens for strongly meshed systems with solid ground connections. The Gauss-Seidel method is very easy to program. Additionally, in each equation there are only a few nonzero elements, which makes it very efficient. On the negative side it has only a linear convergence characteristic and the number of iterations (and simulation time) increase with the size of the system. Divergence is not very frequent, but it happens when the system is weakly meshed. Acceleration factors are successfully used as follows:

$$\Delta V_k = \left( -\frac{1}{Y_{kk}} \sum_{m=1}^{k-1} Y_{km} V_m^{(j+1)} + \frac{S_k^*}{Y_{kk} V_k^{*(j)}} - \frac{1}{Y_{kk}} \sum_{m=k+1}^{n-1} Y_{km} V_m^{(j)} \right) - V_k^{(j)} \tag{17-91}$$

$$V_k^{(j+1)} = V_k^{(j)} + \alpha \Delta V_k \tag{17-92}$$

where the acceleration factor $\alpha$ is a number between 1 and 2. Frequently $\alpha = 1.6$ is selected for the solution of the power flow.

***Newton Raphson.*** The Gauss-Seidel method has problems converging when the system is relatively weak (radial systems). It also takes a large number of iterations when the system is large. For these systems, the Newton-Raphson method can be used to solve the set of Eq. (17-85). It is better to separate the real and imaginary parts. For this we use Euler's formula $V = |V| e^{j\theta} = |V|(\cos\theta + j\sin\theta)$

$$P_k = \text{Re}[V_k I_k^*] = \text{Re}\left[ V_k \sum_{m=1}^{n-1} Y_{km}^* V_m^* \right] = |V_k| \sum_{m=1}^{n-1} |Y_{km}||V_m| \cos(\theta_k - \theta_m - \phi_{km}) \tag{17-93}$$

and

$$Q_k = \text{Im}[V_k I_k^*] = \text{Im}\left[ V_k \sum_{m=1}^{n-1} Y_{km}^* V_m^* \right] = |V_k| \sum_{m=1}^{n-1} |Y_{km}||V_m| \sin(\theta_k - \theta_m - \phi_{km}) \tag{17-94}$$

where $\theta_k$ and $\theta_m$ are the angles of the bus voltages and $\phi_{km}$ is the angle of the line impedance given by

$$\phi_{km} = \arctan\left( -\frac{X_{km}}{R_{km}} \right) \tag{17-95}$$

The Newton-Rapson (NR) method is derived from the expansion of a nonlinear equation $f(x) = 0$ in Taylor series around an initial point and chopping the series at the second term as follows:

$$f(x^{(0)} + \Delta x^{(0)}) \approx f(x^{(0)}) + \left( \frac{df(x^{(0)})}{dx} \right) \Delta x^{(0)} \approx 0 \tag{17-96}$$

From where we get

$$\Delta x^{(0)} \approx -\left( \frac{df(x^{(0)})}{dx} \right)^{-1} f(x^{(0)}) \tag{17-97}$$

A value closer to the solution is computed as follows:

$$x^{(1)} = x^{(0)} + \Delta x^{(0)} = x^{(0)} - \left( \frac{df(x^{(0)})}{dx} \right)^{-1} f(x^{(0)}) \tag{17-98}$$

For $N$ equations the NR method becomes

$$\begin{aligned} f_1(\mathbf{x}^{(0)}) + \left(\frac{df_1}{dx_1}\right)^{(0)} \Delta x_1 + \left(\frac{df_1}{dx_2}\right)^{(0)} \Delta x_2 + \cdots + \left(\frac{df_1}{dx_N}\right)^{(0)} \Delta x_N \approx 0 \\ f_2(\mathbf{x}^{(0)}) + \left(\frac{df_2}{dx_1}\right)^{(0)} \Delta x_1 + \left(\frac{df_2}{dx_2}\right)^{(0)} \Delta x_2 + \cdots + \left(\frac{df_2}{dx_N}\right)^{(0)} \Delta x_N \approx 0 \\ \vdots \\ f_N(\mathbf{x}^{(0)}) + \left(\frac{df_N}{dx_1}\right)^{(0)} \Delta x_1 + \left(\frac{df_N}{dx_2}\right)^{(0)} \Delta x_2 + \cdots + \left(\frac{df_N}{dx_N}\right)^{(0)} \Delta x_N \approx 0 \end{aligned} \tag{17-99}$$

which can be written in matrix form as

$$\begin{pmatrix} \left(\frac{df_1}{dx_1}\right)^{(0)} & \left(\frac{df_1}{dx_2}\right)^{(0)} & \cdots & \left(\frac{df_1}{dx_N}\right)^{(0)} \\ \vdots & \vdots & \ddots & \vdots \\ \left(\frac{df_N}{dx_1}\right)^{(0)} & \left(\frac{df_N}{dx_2}\right)^{(0)} & \cdots & \left(\frac{df_N}{dx_N}\right)^{(0)} \end{pmatrix} \begin{pmatrix} \Delta x_1 \\ \Delta x_2 \\ \vdots \\ \Delta x_N \end{pmatrix} = -\begin{pmatrix} f_1(\mathbf{x}^{(0)}) \\ f_2(\mathbf{x}^{(0)}) \\ \vdots \\ f_N(\mathbf{x}^{(0)}) \end{pmatrix} \tag{17-100}$$

or

$$[J]\Delta\mathbf{x} = -\mathbf{f}(\mathbf{x}^{(0)}) \tag{17-101}$$

When applied to the following power flow equations for $P_k$ and $Q_k$

$$\begin{aligned} fp_k &= \sum_{m=1}^{n-1} |V_k||Y_{ik}||V_m| \cos(\theta_k - \theta_m - \phi_{km}) - P_k = 0 \qquad \text{for } k = 1, 2, \ldots, n-1 \\ fq_k &= \sum_{m=1}^{n-1} |V_k||Y_{ik}||V_m| \sin(\theta_k - \theta_m - \phi_{km}) - Q_k = 0 \qquad \text{for } k = 1, 2, \ldots, n - n_{pv} - 1 \end{aligned} \tag{17-102}$$

we get

$$\begin{pmatrix} \left(\frac{dfp_1}{d\theta_1}\right)^{(0)} & \left(\frac{dfp_1}{d\theta_2}\right)^{(0)} & \cdots & \left(\frac{dfp_1}{d|V_{n-1}|}\right)^{(0)} \\ \vdots & \vdots & \ddots & \vdots \\ \left(\frac{dfq_{n-1}}{d\theta_1}\right)^{(0)} & \left(\frac{dfq_{n-1}}{d\theta_2}\right)^{(0)} & \cdots & \left(\frac{dfq_{n-1}}{d|V_{n-1}|}\right)^{(0)} \end{pmatrix} \begin{pmatrix} \Delta\theta_1^{(0)} \\ \Delta\theta_2^{(0)} \\ \vdots \\ \Delta|V_{n-n_{pv}-1}^{(0)}| \end{pmatrix} - \begin{pmatrix} P_1 \\ P_2 \\ \vdots \\ Q_{n-1} \end{pmatrix} = -\begin{pmatrix} fp_1(\mathbf{x}^{(0)}) \\ fp_2(\mathbf{x}^{(0)}) \\ \vdots \\ fq_{n-1}(\mathbf{x}^{(0)}) \end{pmatrix} \tag{17-103}$$

where

$$\mathbf{x} = [\theta_1 \quad \theta_2 \quad \cdots \quad \theta_{n-1} \quad |V_1| \quad |V_2| \quad \cdots \quad |V_{n-n_{pv}-1}|]^T \tag{17-104}$$

The above equation can be written in matrix notation as follows:

$$\begin{bmatrix} \dfrac{\partial fp}{\partial \theta} & \dfrac{\partial fq}{\partial |V|} \\ \dfrac{\partial fp}{\partial \theta} & \dfrac{\partial fq}{\partial |V|} \end{bmatrix} \begin{bmatrix} \Delta\theta \\ \Delta|V| \end{bmatrix} = \begin{bmatrix} P - fp \\ Q - fq \end{bmatrix} = \begin{bmatrix} \Delta P \\ \Delta Q \end{bmatrix} \tag{17-105}$$

which is frequently written as

$$\begin{bmatrix} H & N \\ J & L \end{bmatrix} \begin{bmatrix} \Delta\theta \\ \Delta|V| \end{bmatrix} = \begin{bmatrix} \Delta P \\ \Delta Q \end{bmatrix} \tag{17-106}$$

Submatrix $[H]$ is of rank $n - 1 \times n - 1$. Submatrix $[L]$ is of rank $n - n_{pv} - 1 \times n - n_{pv} - 1$, where $n$ is the total number of buses, and $n_{pv}$ is the number of generator (PV) buses. Remember that we need one bus to be the slack and thus the base rank is $n - 1$. There exist analytical expressions for the terms in Eq. (17-105). Perhaps the easier to program are those published in the book by El-Hawary [2]. Here is the Matlab code to efficiently build the Jacobian matrix.

```
function [Jaco] = JACO(G,B,P,Q,N,NG,NL,vm,an,e,f)
% JACOBIAN according to El-Hawary pp. 328 - 334
% formed and calculated sparse

N1 = N - 1 ;
a  = sparse(N1,N1) ;
b  = sparse(N1,N1) ;
H = sparse(N1,N1) ;
N = sparse(N1,NL) ;
J = sparse(NL,N1) ;
L = sparse(NL,NL) ;

for i = 2 : N
    i1 = i - 1 ;
   inx = find(B(i,:)) ;
   ns  = size(inx,1) ;
   for k = 1 : ns
      j  = inx(k) ;
      if j > 1
        j1 = j - 1  ;
        j2 = j - NG ;
        if i == j
          H(i1,j1) = - Q(i) - B(i,i) * vm(i)^2 ;
          if j > NG
           N(i1,j2) =  P(i) + G(i,i) * vm(i)^2 ;
          end
        else
          a(i,j)      =  G(i,j) * e(j) - B(i,j) * f(j) ;
          b(i,j)      =  G(i,j) * f(j) + B(i,j) * e(j) ;
          H(i1,j1)  =   a(i,j) * f(i) -  b(i,j) * e(i) ;
```

```
                if j > NG
                 N(i1,j2) =   a(i,j) * e(i) +  b(i,j) * f(i) ;
                end
              end
              if i > NG
                 j3 = j - NG ;
                 i2 = i - NG ;
                 if i == j
                   J(i2,j1)   =  P(i) - G(i,i) * vm(i)^2 ;
                   if j > NG
                     L(i2,j3) = Q(i) - B(i,i) * vm(i)^2 ;
                   end
                 else
                   J(i2,j1)  = - ( a(i,j) * e(i) + b(i,j) * f(i) );
                   if j > NG
                    L(i2,j3) =    a(i,j) * f(i) -  b(i,j) * e(i) ;
                   end
                 end
              end
            end
         end
end

Jaco = [ H  N
            J  L ] ;
```

In the above Matlab code and the rest of the subsection we have separated $Y_{km}$ in its real (conductance) and imaginary (susceptance) components as follows:

$$Y_{km} = G_{km} + jB_{km} \tag{17-107}$$

The Newton-Raphson power flow algorithm is as follows:

Step 1. Establish an initial guess

$$\begin{bmatrix} \theta^{(0)} \\ |V|^{(0)} \end{bmatrix} = \begin{bmatrix} \theta_k^{(0)} \\ |V_k^{(0)}| \end{bmatrix} \tag{17-108}$$

If there is no better information, then use the flat start $\theta_k = 0$ & $|V_k| = 1.0\ \forall k$.

Step 2. Compute the power mismatch

$$\begin{aligned} \Delta P_k &= P_k - \sum_{m=1}^{n-1} |V_k||Y_{km}||V_m|\cos(\theta_k - \theta_m + \phi_{ik}) \qquad \text{for } k = 1, 2, \ldots, n-1 \\ \Delta Q_k &= Q_k - \sum_{m=1}^{n-1} |V_k||Y_{km}||V_m|\sin(\theta_k - \theta_m + \phi_{ik}) \qquad \text{for } k = 1, 2, \ldots, n - n_{pv} - 1 \end{aligned} \tag{17-109}$$

Step 3. Test convergence: Is power mismatch smaller than a tolerance?

Step 4. Compute (or update) the Jacobian matrix.

Step 5. Solve the following equation:

$$\begin{bmatrix} H & N \\ J & L \end{bmatrix} \begin{bmatrix} \Delta\theta \\ \Delta|V| \end{bmatrix} = \begin{bmatrix} \Delta P \\ \Delta Q \end{bmatrix} \tag{17-110}$$

Step 6. Add the previous value and the error

$$\begin{bmatrix} \theta^{(j+1)} \\ |V|^{(j+1)} \end{bmatrix} = \begin{bmatrix} \theta^{(j)} \\ |V|^{(j)} \end{bmatrix} + \begin{bmatrix} \Delta\theta^{(j)} \\ \Delta|V|^{(j)} \end{bmatrix} \tag{17-111}$$

Repeat steps 2 to 6 until convergence is obtained.

***Fast Decoupled.*** It is well known to electric power engineers that the flow of active power ($P$) is strongly coupled to bus angle differences ($\theta_{km}$) and that reactive power ($Q$) flow is strongly coupled to bus voltage magnitude differences $|V_{km}|$. The same physical phenomenon happens to the Newton power flow equations and the effect of submatrices [$N$] and [$J$] of the Jacobian matrix is much smaller than [$H$] and [$L$]. Therefore, a decoupled form, neglecting [$N$] and [$J$] is possible. An even better method was developed by Stott and Alsaç under the name of Fast Decoupled Load Flow (FDLF) [3]. A series of physically founded simplifications and assumptions are made to produce a very efficient algorithm. None of the simplifications affects the accuracy of the method because the convergence criterion is still the balance of power at every bus.

The assumptions of the FDLF include the angle difference between buses is small and line conductance is much smaller than its susceptance. This is mathematically expressed as follows:

$$\begin{aligned} &\cos(\theta_{km}) \approx 1 \\ &G_{km}\sin(\theta_{km}) \ll B_{km} \\ &Q_k \ll B_{kk}V_k^2 \end{aligned} \tag{17-112}$$

The above assumptions are substituted in [$H$] and [$L$] to obtain the following FDLF equations:

$$\begin{aligned} \left[\frac{\Delta P}{V}\right] &= [B'][\Delta\theta] \\ \left[\frac{\Delta Q}{V}\right] &= [B''][\Delta V] \end{aligned} \tag{17-113}$$

Matrices [$B'$] and [$B''$] have the same structure as [$H$] and [$L$] of the Newton power flow, but contain only network admittances. Therefore, they are constant and can be factorized in LU factors only once at the beginning of the calculation.

***DC Power Flow.*** The dc power flow is a method used to solve ac power flow with real number calculations only as we do with a dc circuit, form where the name derived. For robust power systems one can assume that the voltage magnitudes are all equal to 1.0 per unit. Therefore, reactive power does not circulate. Also one can assume that in all transmission lines and transformers $R_{km} << X_{km}$. This is true for very large components. Therefore the power flow equations are simplified as follows:

$$P_{km} = \frac{\theta_{km}}{X_{km}} = \frac{\theta_k - \theta_m}{X_{km}} \tag{17-114}$$

The above equation is analogous to Ohm's law for the solution of dc circuits. $P_{km}$ is akin to dc current, $\theta_{km}$ is analogous to voltage and the reactance $X_{km}$ of the lines takes the place of the resistance of the dc circuit. When applied to a system Eq. (17-114) becomes a matrix equation as

$$[B][\theta] = [P] \tag{17-115}$$

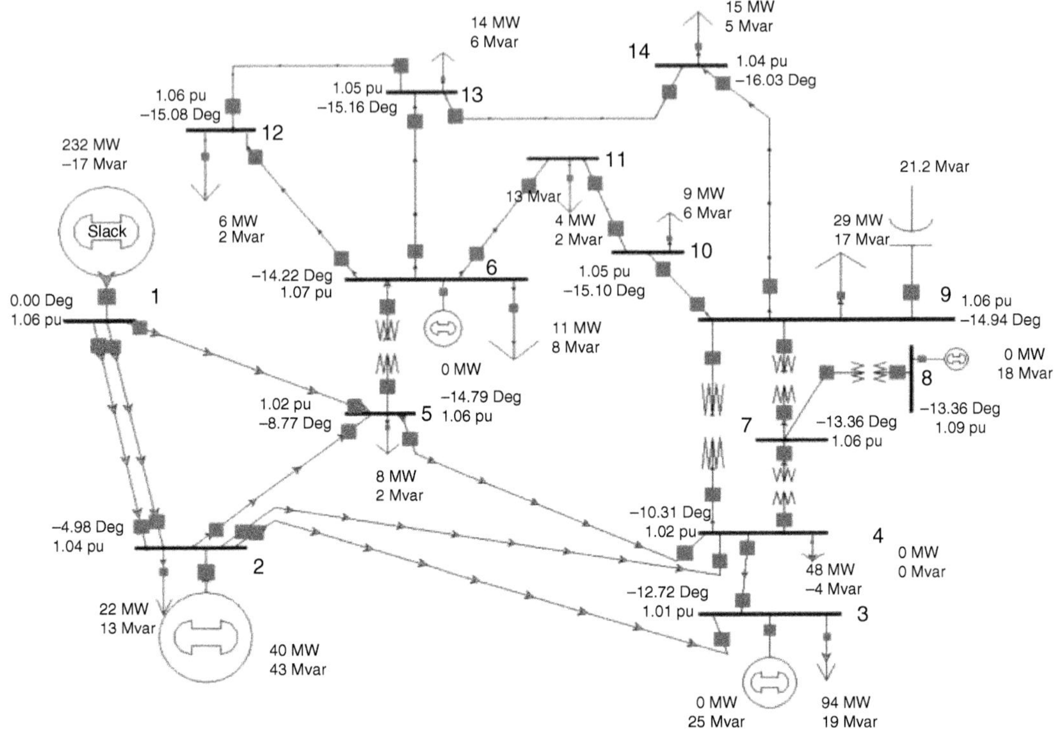


**FIGURE 17-21** IEEE 14 bus power flow test case.

*Example 17-3.* The IEEE has published a series of test systems ranging from 14 buses to 300 buses. The information, data (in common data format), and results can be downloaded from: http://www2.ee.washington.edu/research/pstca/. The converged results of the 14 bus test system are presented in Fig. 17-21 (extracted from the academic version of the PowerWorld simulator: https://www.powerworld.com/). Table 17-2 presents the Gauss-Seidel iteration results. Convergence is obtained after 180 very fast iterations.

**TABLE 17-2** Bus Voltages Obtained from the Gauss-Seidel Iterations

| | Iter0 | Iter1 | Iter2 | Iter3 | Iter4 | ... | Iter 180 |
|---|---|---|---|---|---|---|---|
| Bus 1 | 1.06000 | 1.06000 | 1.06000 | 1.06000 | 1.06000 | | 1.06000 |
| Bus 2 | 1.04500 | 1.04500 | 1.04500 | 1.04500 | 1.04500 | | 1.04500 |
| Bus 3 | 1.01000 | 1.01000 | 1.01000 | 1.01000 | 1.01000 | | 1.01000 |
| Bus 4 | 1.00000 | 1.00242 | 1.02178 | 1.02440 | 1.02661 | | 1.01770 |
| Bus 5 | 1.00000 | 1.01563 | 1.02726 | 1.02860 | 1.02871 | | 1.01956 |
| Bus 6 | 1.07000 | 1.07000 | 1.07000 | 1.07000 | 1.07000 | | 1.07000 |
| Bus 7 | 1.00000 | 1.03220 | 1.04558 | 1.05079 | 1.06134 | | 1.06153 |
| Bus 8 | 1.09000 | 1.09000 | 1.09000 | 1.09000 | 1.09000 | | 1.09000 |
| Bus 9 | 1.00000 | 1.01169 | 1.02819 | 1.03551 | 1.05120 | | 1.05594 |
| Bus 10 | 1.00000 | 1.00249 | 1.02611 | 1.03270 | 1.04676 | | 1.05099 |
| Bus 11 | 1.00000 | 1.03215 | 1.04430 | 1.04769 | 1.05488 | | 1.05691 |
| Bus 12 | 1.00000 | 1.02695 | 1.04801 | 1.04919 | 1.05266 | | 1.05519 |
| Bus 13 | 1.00000 | 1.03608 | 1.04306 | 1.04473 | 1.04871 | | 1.05038 |
| Bus 14 | 1.00000 | 1.00374 | 1.01633 | 1.02129 | 1.03208 | | 1.03554 |

### 17.7.4 Distribution Systems

Different from transmission systems, distribution systems are in general radial. In particular the medium voltage feeders are almost universally operated radially (no loops and no meshes). This is done to simplify the protection schemes and reduce cost at the expense of reduced reliability.

As for transmission systems, a large number of techniques have been proposed for the power flow solution. Here we only describe the method known as the backward/forward sweeps, which is considered the most efficient in existence so far.

Consider the circuit depicted in Fig. 17-22. The substation is represented by its Thevenin equivalent circuit (known voltage $V_0$ and known impedance $Z_{01}$). Each line is represented by its series impedance $Z_{ij}$. The shunt capacitors can be included in the loads, which are represented by their complex power $S_k$. In distribution system analysis it is common to represent the loads as ZIP coefficients (part constant impedance "Z," part constant current "I," and part constant power "P" as mentioned above in Sec. 17.4.2). The method described here applies in the same fashion regardless of load model being used.

The solution begins by assuming initial bus voltages (when lacking of a better guess, use the flat start ($V_k^{(0)} = 1 + j0 \; \forall \; k$). Starting from the far end (downstream) at node $n$ the load current (which is the same of the line between buses $n - 1$ and $n$) is computed from

$$I_n^{(0)} = \frac{S_n^*}{V_n^{*(0)}} \tag{17-116}$$

The next current upstream (backward sweep) is computed by adding the current downstream plus the load current at the $n - 1$ bus as follows:

$$I_{n-1}^{(0)} = I_n^{(0)} + \frac{S_{n-1}^*}{V_{n-1}^{*\;(0)}} \tag{17-117}$$

The same rationale is applied iteratively as:

$$I_k^{(0)} = I_{k+1}^{(0)} + \frac{S_k^*}{V_k^{*(0)}} \qquad k = n-1, n-2, \ldots, 1 \tag{17-118}$$

Once the source (substation) current $I_1^{(0)}$ is computed, the forward sweep is applied to compute a new set of voltages (using the voltage drop across the lines) as follows:

$$V_k^{(1)} = V_{k-1}^{(0)} + \Delta V_k^{(0)} = V_{k-1}^{(0)} + z_{k,k-1} I_k^{(0)} \qquad k = 1, 2, \ldots, n \tag{17-119}$$

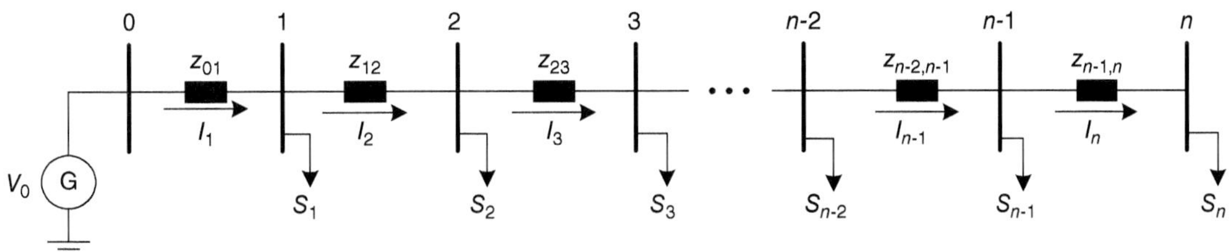


**FIGURE 17-22** Radial distribution system.

The iterative process is as follows:

$$\begin{cases} I_n^{(j)} = \dfrac{S_n^*}{V_n^{*(j)}} \\ I_k^{(j)} = I_{k+1}^{(j)} + \dfrac{S_k^*}{V_k^{*(j)}}; & k = n-1, n-2, \ldots, 1 \\ V_k^{(j+1)} = V_{k-1}^{(j)} + z_{k,k-1} I_k^{(j)}; & k = 1, 2, \ldots, n \end{cases} \tag{17-120}$$

The above equations are iterated, $j$ = 0, 1, 2, … until the bus voltages change less than a certain tolerance.

Actual distribution systems have several laterals branching from the main feeders. Branches also split from branches, but a given load is always supplied from only one path to the substation. Take for example the system depicted in Fig. 17-23. The backward/forward method can be used in the same manner as described above taking care of adding all the currents of the downstream line sections. For this it is convenient to align branches as per their depth in the system.

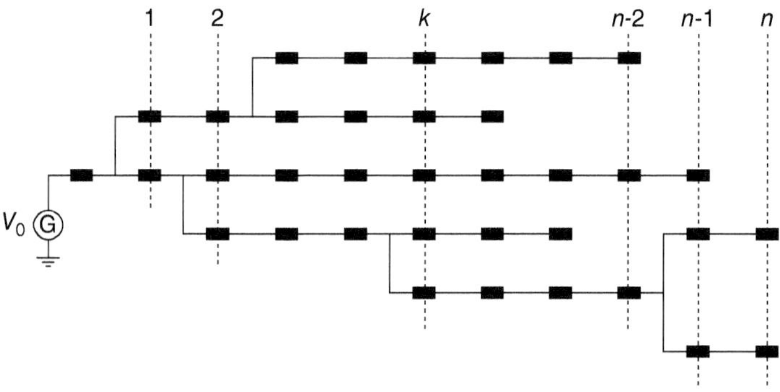


**FIGURE 17-23** Distribution system with laterals branching.

In contrast with transmission systems, a characteristic of distribution systems is their unbalanced nature. Not only lines are not transposed, but the loads are not balanced between phases. In fact, commonly there are single- (or two-) phase sections that even sometimes merge back to three-phase circuits after long unbalanced runs. Therefore, the power-flow calculations are not performed using the positive sequence impedances, but three-phase line models are used. Lines are represented by their phase impedance as follows:

$$\left[Z_{km}^{abc}\right] = \begin{bmatrix} z_{aa} & z_{ab} & z_{ac} \\ z_{ba} & z_{bb} & z_{bc} \\ z_{ca} & z_{cb} & z_{cc} \end{bmatrix} \tag{17-121}$$

Equation (17-82) is modified as

$$\begin{cases} \begin{bmatrix} I_{\text{load}_a} \\ I_{\text{load}_b} \\ I_{\text{load}_c} \end{bmatrix}_k^{(j)} = [Y_{\text{load}}]_k \begin{bmatrix} V_a \\ V_b \\ V_c \end{bmatrix}^{(j)}; & k=1,2,\ldots,n \\ \begin{bmatrix} I_a \\ I_b \\ I_c \end{bmatrix}_k^{(j)} = \begin{bmatrix} I_a \\ I_b \\ I_c \end{bmatrix}_{k+1}^{(j)} + \begin{bmatrix} I_{\text{load}_a} \\ I_{\text{load}_b} \\ I_{\text{load}_c} \end{bmatrix}_k^{(j)}; & k=n-1,n-2,\ldots,1 \\ \begin{bmatrix} V_a \\ V_b \\ V_c \end{bmatrix}_k^{(j+1)} = \begin{bmatrix} V_a \\ V_b \\ V_c \end{bmatrix}_{k-1}^{(j)} + \begin{bmatrix} z_{aa} & z_{zb} & z_{ac} \\ z_{ba} & z_{bb} & z_{bc} \\ z_{ca} & z_{cb} & z_{cc} \end{bmatrix}_{k,k-1} \begin{bmatrix} I_a \\ I_b \\ I_c \end{bmatrix}_k^{(j)}; & k=1,2,\ldots,n \end{cases} \tag{17-122}$$

Matrices $[Y_{\text{load}}]_k$ are determined from the connection of the load (wye, delta, single- or two-phase, etc.). The backward/forward sweeps method, as described above, strictly works only for radial systems. The compensation method has been successfully used to solve looped systems and even heavily meshed secondary networks. Consider the single-loop system depicted in Fig. 17-24*a*. To apply the backward/forward sweeps method, the system is "radialized" as shown in Fig. 17-24*b*. The system is now solved, but because the loop has been opened, a voltage difference appears between the broken terminals. To compensate for the current that should have circulated from the lower section of the circuit to the load ($S_2$), one can connect a current source ($I$) between the open points as shown in Fig. 17-24*c*. The compensation current source is computed (from Ohm's law) as

$$I = \frac{\Delta V}{z_1 + z_2} \tag{17-123}$$

Note that $z_1 + z_2$ is the impedance around the loop and therefore $I$ is the current that would fully compensate for the voltage difference caused by breaking the loop. Note that the method is still iterative because the solution of the radial system is computed iteratively starting from an initial guess. The current that is injected or subtracted is simply treated as a load for the next iteration of the radial system.

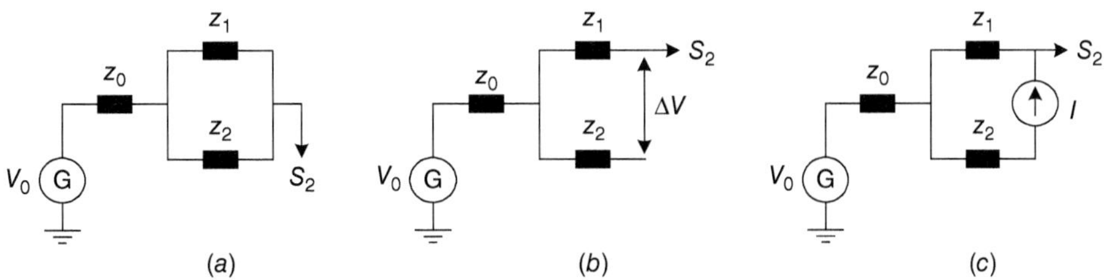


**FIGURE 17-24** Process of the compensation method.

The method has been generalized to any number of meshes, but Eq. (17-85) becomes a matrix equation that needs to be solved at each iteration. The order of the matrix equation is equal to the number of loops that need to be broken to make the system radial.

### 17.7.5 Commercial and Royalty-Free Software

There exists a large variety of commercial and royalty-free software to perform power system analysis. The standard piece of software to perform power system analysis and used by most utilities is Siemens PSS/E. It is impossible to mention them all, but among the commercial ones commonly used in the industry we can list: Eaton/CYME, ETAP, Digsilent, PowerWorld, SKM, etc. A complete list of them with contact information can be found in the print version of the IEEE Power and Energy Magazine. Reference [4] has reviewed many of them in 2005.

Among the many royalty-free available pieces of software for power system steady-state simulation we can mention two: OpenDSS and the Matlab-based PSAT. The authors of this section have used OpenDSS (developed by EPRI) successfully to perform a large number of utility projects.

An updated (2017) list and description of commercial and royalty-free software can be found here: http://www.openelectrical.org/wiki/index.php?title=Power_Systems_Analysis_Software.

*Example 17-4.* The IEEE/PES (Power and Energy Society) has published a comprehensive series of distribution test feeders ranging from 4 buses to 8500 buses, which can be downloaded from here: https://ewh.ieee.org/soc/pes/dsacom/testfeeders/.

The systems are actual utility feeders distilled for public use. The purpose is for the systems to serve as benchmarks for software developers. The systems encompass more than power flow, they include distributed generation, short-circuit tests, small, medium and large systems, etc. The configuration of the 8500 test feeder is shown in Fig. 17-25. The results are not copied here, but can be downloaded from the link provided.

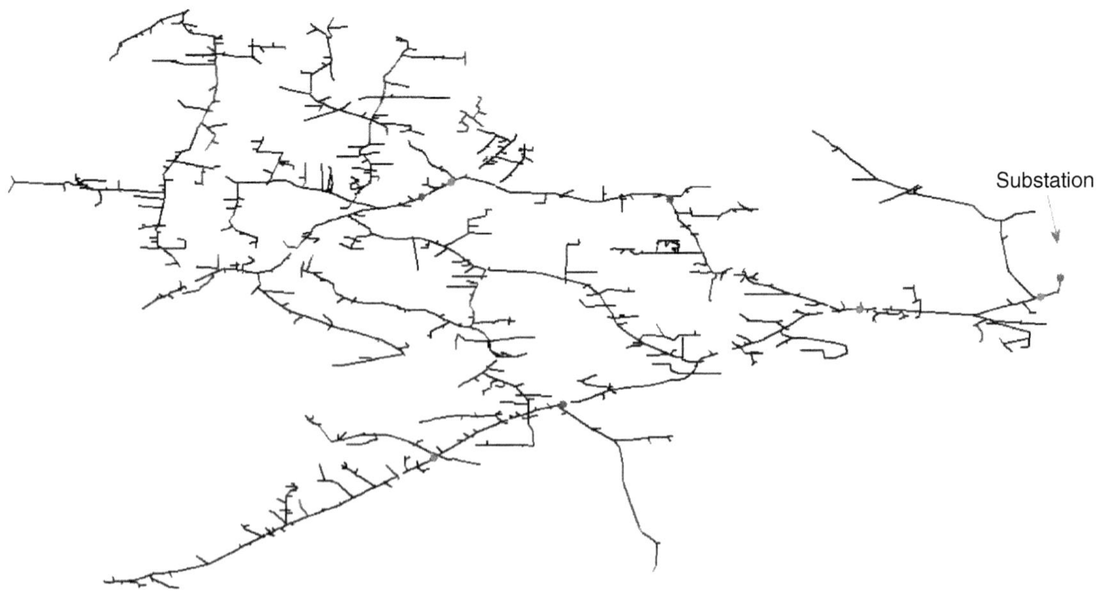


**FIGURE 17-25** IEEE 8500 bus test feeder.

## 17.8 SHORT CIRCUIT

Faults are a fact of the everyday operation of a power system. Because large sections of the system are exposed to the elements (lightning, wind, ice, etc.) and animal intrusion also occurs, the system needs to be designed to withstand such events. Short-circuit current calculations in power systems are of paramount importance for the design of the protection system. The rating of breakers, bus bars, bus ways, etc., is determined by the thermal and mechanical effects of short-circuit currents. The proper settings of relays is established by the short-circuit currents. The purpose is to stop the short-circuit currents by isolating the smallest possible portion of the system and disconnecting the smallest number of customers.

The study of faulted power systems is most commonly performed for symmetrical (three-phase) faults (LLL) and the nonsymmetrical single-line-to-ground fault (SLG). Three-phase faults produce the largest short-circuit power, but single-line-to-ground faults sometimes produce larger short-circuit currents. The symmetrical components transformation described above is used for nonsymmetrical faults. Other nonsymmetrical faults, which are not studied as much, include: line-to-line (LL), and line-to-line-to-ground fault (LLG).

### 17.8.1 Modeling Assumptions

There are two recognized methods for computing the short-circuit currents: the ANSI method and the IEC method. The numerical methods to obtain the solution are the same, but there are differences in the assumptions and therefore the two methods compute slightly different short-circuit current values.

ANSI recommends using a pre-fault voltage of 1.0 pu at the faulted bus and neglecting the load current. IEC recommends accounting for the position of the transformer taps and adjusting the voltage accordingly. IEC also recommends considering the pre-fault loading to determine the initial voltage. Both standards recommend neglecting static loads and positive sequence line charging. IEC recommends considering the negative sequence impedance matrix for unbalanced faults, while in ANSI the positive sequence matrix is used instead. Both standards recommend neglecting the zero sequence line charging for solidly grounded systems. For isolated grounding, the IEC requires the modeling of the zero sequence line charging. In conclusion the circuit is linearized for both standards and all the tools of circuit theory are applied. The short circuit currents are determined by inverting the admittance matrix by column or using the bus impedance matrix.

### 17.8.2 Symmetrical Faults

The easiest way to understand the calculation of the short-circuit currents for three-phase faults (LLL) is using the bus impedance matrix $[Z_{\text{bus}}]$, which is the inverse of the $[Y_{\text{bus}}]$ matrix used above for power flow, but using ground (rather than the slack bus) as reference. Consider the system of equations given in Eq. (17-80), repeated here for convenience, which was obtained from the application of Kirchhoff current law in all buses:

$$\begin{bmatrix} I_1 \\ I_2 \\ \vdots \\ I_k \\ \vdots \\ I_n \end{bmatrix} = \begin{bmatrix} Y_{11} & Y_{12} & Y_{1k} & Y_{1n} \\ Y_{21} & Y_{22} & Y_{2k} & Y_{2n} \\ & & & \\ Y_{k1} & Y_{k2} & Y_{kk} & Y_{kn} \\ & & & \\ Y_{n1} & Y_{n2} & Y_{nk} & Y_{nn} \end{bmatrix} \begin{bmatrix} V_1 \\ V_2 \\ \vdots \\ V_k \\ \vdots \\ V_n \end{bmatrix} \tag{17-124}$$

or

$$[I]=[Y_{\text{bus}}]\,[V]$$

Then

$$[V]=[Y_{\text{bus}}]^{-1}[I]=[Z_{\text{bus}}][I] \tag{17-125}$$

which can be written as

$$\begin{pmatrix} Z_{11} & Z_{12} & \cdots & Z_{1n} \\ Z_{21} & Z_{22} & \cdots & Z_{2n} \\ \vdots & \vdots & \ddots & \vdots \\ Z_{n1} & Z_{n2} & \cdots & Z_{nn} \end{pmatrix}\begin{pmatrix} I_1 \\ I_2 \\ \vdots \\ I_n \end{pmatrix}=\begin{pmatrix} V_1 \\ V_2 \\ \vdots \\ V_n \end{pmatrix} \tag{17-126}$$

where the diagonal elements ($Z_{ii}$) are the driving point impedances of every bus with respect to the reference bus and the off-diagonal elements ($Z_{ij}$) are the transfer impedances between buses. The diagonal elements ($Z_{ii}$) are the Thevenin equivalent impedance, therefore the short circuit current at node $i$ ($I_{SCi}$) is computed as:

$$I_{SCi}=\frac{V_i}{Z_{ii}} \tag{17-127}$$

In general, $V_i = 1.0$ pu and the three-phase short-circuit current (in pu) simply becomes the inverse of the driving point impedance:

$$I_{SCi}=\frac{1}{Z_{ii}} \tag{17-128}$$

Plugging Eq. (17-27) back into Eq. (17-26), we can compute the voltages at buses during the fault as

$$\begin{pmatrix} V_{F1} \\ V_{F2} \\ \vdots \\ V_{Fn} \end{pmatrix}=\begin{pmatrix} Z_{11} & Z_{12} & \cdots & Z_{1n} \\ Z_{21} & Z_{22} & \cdots & Z_{2n} \\ \vdots & \vdots & \ddots & \vdots \\ Z_{n1} & Z_{n2} & \cdots & Z_{nn} \end{pmatrix}\begin{pmatrix} 0 \\ 0 \\ I_{SCi} \\ I_n \end{pmatrix} \tag{17-129}$$

From the knowledge of all faulted voltages, we can compute the short-circuit currents flowing in every branch of the system as

$$I_{km} = y_{km}(V_{Fk}-V_{Fm}) = (V_{Fk}-V_{Fm})/z_{km} \tag{17-130}$$

Remember that $y_{km}$ (or $z_{km}$) is the admittance (or impedance) connecting buses $k$ and $m$.

### 17.8.3 Nonsymmetrical Faults

For the study of nonsymmetrical faults, SLG, LL, and LLG the use of the symmetrical components method is recommended. Let us assume that the bus impedance matrices have been calculated for the positive, negative, and zero sequences, named $[Z_{\text{bus}}^{(1)}]$, $[Z_{\text{bus}}^{(2)}]$, and $[Z_{\text{bus}}^{(0)}]$, respectively. The diagonal elements ($Z_{ii}$) of each matrix ($Z_{1ii}$ $Z_{2ii}$ $Z_{0ii}$) are used to construct the equivalent sequence circuits for bus $i$ shown in Fig. 17-26.

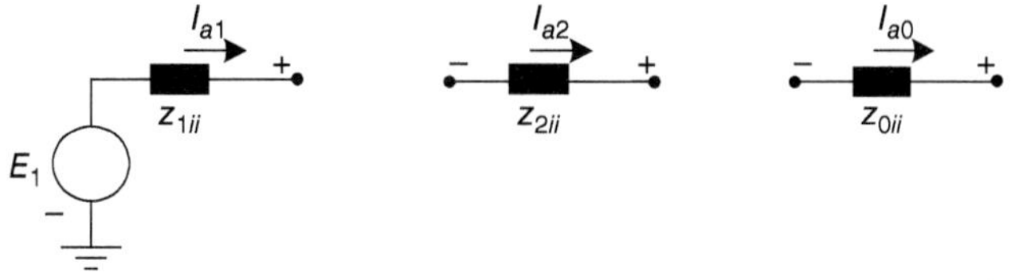


**FIGURE 17-26** Positive, negative, and zero sequence circuits.

The connection of the circuits depends of the type of fault that one wants to compute as described below. Only SLG and LL faults are described.

***Single-Line-to-Ground Fault.*** For a single-line-to-ground (SLG) fault the conditions can be obtained from the circuit shown in Fig. 17-27. One can see that

$$V_a = Z_F I_a \tag{17-131}$$

$$[I] = \begin{bmatrix} I_a \\ I_b \\ I_c \end{bmatrix} = \begin{bmatrix} I_a = I_F \\ 0 \\ 0 \end{bmatrix} \tag{17-132}$$

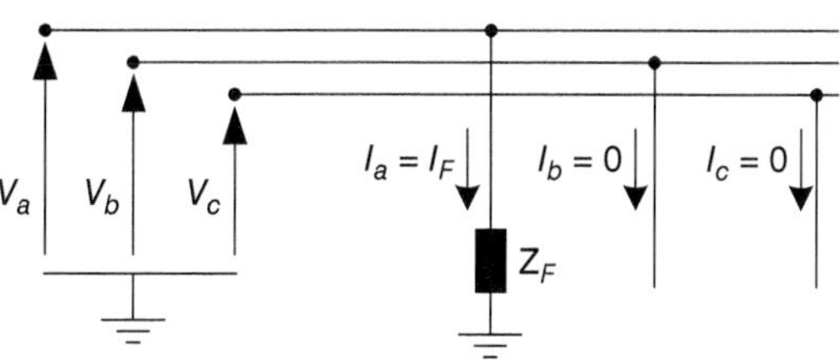


**FIGURE 17-27** Voltages and currents for a single-line-to-ground fault.

From the definition of the symmetrical components transformation we have

$$\begin{bmatrix} V_a \\ V_b \\ V_c \end{bmatrix} = \begin{bmatrix} 1 & 1 & 1 \\ 1 & a^2 & a \\ 1 & a & a^2 \end{bmatrix} \begin{bmatrix} V_{a0} \\ V_{a1} \\ V_{a2} \end{bmatrix} \tag{17-133}$$

and

$$\begin{bmatrix} I_a \\ 0 \\ 0 \end{bmatrix} = \begin{bmatrix} 1 & 1 & 1 \\ 1 & a^2 & a \\ 1 & a & a^2 \end{bmatrix} \begin{bmatrix} I_{a0} \\ I_{a1} \\ I_{a2} \end{bmatrix} \tag{17-134}$$

then

$$\begin{aligned} V_a &= V_{a0} + V_{a1} + V_{a2} \\ I_a &= I_{a0} + I_{a1} + I_{a2} \end{aligned} \tag{17-135}$$

Substituting Eq. (17-132) into the inverse symmetrical components transformation

$$\begin{bmatrix} I_{a0} \\ I_{a1} \\ I_{a2} \end{bmatrix} = \begin{bmatrix} 1 & 1 & 1 \\ 1 & a^2 & a \\ 1 & a & a^2 \end{bmatrix}^{-1} \begin{bmatrix} I_a \\ 0 \\ 0 \end{bmatrix} = \frac{1}{3} \begin{bmatrix} 1 & 1 & 1 \\ 1 & a & a^2 \\ 1 & a^2 & a \end{bmatrix} \begin{bmatrix} I_a \\ 0 \\ 0 \end{bmatrix} \tag{17-136}$$

we get $I_{a0} = I_{a1} = I_{a2} = I_a/3$ or $I_a = 3(I_{a0} + I_{a1} + I_{a2}) = 3I_{a1}$. Therefore, the fault current is computed by connecting the three circuits in series since $I_{a0} = I_{a1} = I_{a2}$. Note that the series connection is also consistent with $V_a = V_{a0} + V_{a1} + V_{a2}$. The fault current ($I_F = I_a = 3I_{a1}$) is computed by substituting $I_a$ and $V_a$ into Eq. (17-131) and we get

$$V_{a0} + V_{a1} + V_{a2} = Z_F(3I_{a1}) = (3Z_F)I_{a1} \tag{17-137}$$

Thus the corresponding circuit becomes as shown in the Fig. 17-28. Note that the phase fault impedance needs to be multiplied by 3 in the sequence diagram because of condition [Eq. (17-137)]. Solving for $I_{a1}$, we get

$$I_{a1} = \frac{E_1}{Z_{1ii} + Z_{2ii} + Z_{0ii} + 3Z_F} \tag{17-138}$$

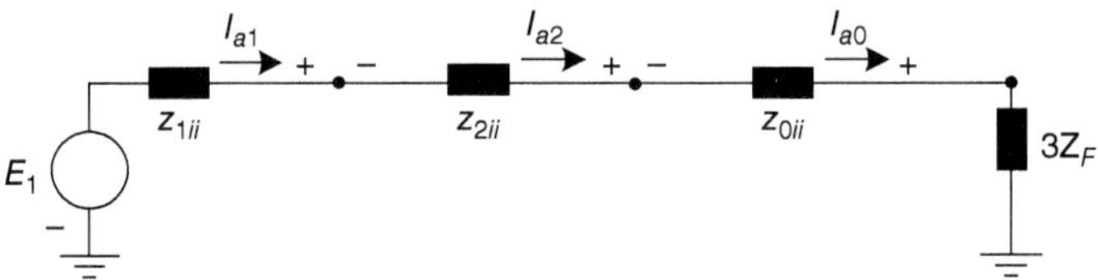


**FIGURE 17-28** Series connection of the sequence equivalent circuits for a single-line-to-ground fault.

which yields a phase fault current given by

$$I_a = I_F = \frac{3E_1}{Z_{1ii} + Z_{2ii} + Z_{0ii} + 3Z_F} \tag{17-139}$$

***Line-to-Line Fault.*** The case of a line-to-line (LL) fault, with no ground involved, is illustrated in Fig. 17-29. As one can see the following conditions are satisfied for phase current and voltages:

$$\begin{aligned} I_a &= 0 \\ I_b &= -I_c = I_F \\ V_b - V_c &= Z_F I_b = -Z_F I_c \end{aligned} \tag{17-140}$$

As before we apply the inverse symmetrical components transformation as follows:

$$\begin{bmatrix} I_{a0} \\ I_{a1} \\ I_{a2} \end{bmatrix} = \frac{1}{3} \begin{bmatrix} 1 & 1 & 1 \\ 1 & a & a^2 \\ 1 & a^2 & a \end{bmatrix} \begin{bmatrix} 0 \\ I_b \\ -I_b \end{bmatrix} = \frac{1}{3} \begin{bmatrix} 0 \\ -(a+a^2)I_b \\ (a^2+a)I_b \end{bmatrix} \tag{17-141}$$

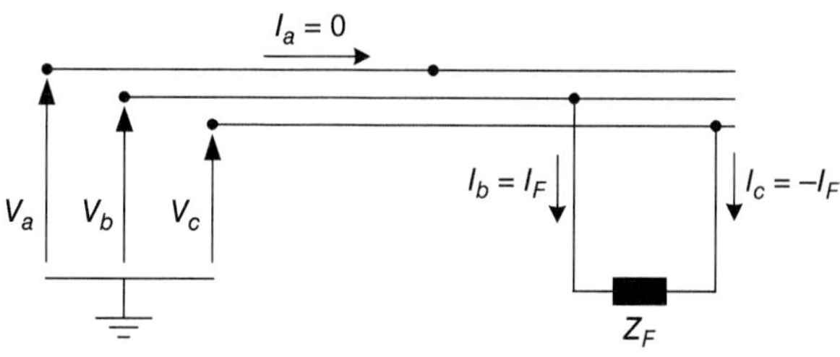


**FIGURE 17-29** Voltages and currents for a line-line fault.

The sum $a + a^2 = -1$, yielding

$$I_{a0} = 0$$
$$I_{a1} = -I_{a2} = -\frac{1}{3} I_b \tag{17-142}$$

The voltage relationship is obtained from the symmetrical components transformation matrix as follows:

$$V_b = V_{a0} + a^2 V_{a1} + a V_{a2}$$
$$V_c = V_{a0} + a V_{a1} + a^2 V_{a2} \tag{17-143}$$

and for the current we get

$$I_b = I_{a0} + a^2 I_{a1} + a I_{a2} \tag{17-144}$$

Then

$$V_b - V_c = (a^2 - a) V_{a1} + (a - a^2) V_{a2} \tag{17-145}$$

which substituted in (17-140) yields

$$V_b - V_c = (a^2 - a)(V_{a1} - V_{a2}) = Z_F I_b \tag{17-146}$$

Using the current $I_b$, we get

$$V_b - V_c = (a^2 - a)(V_{a1} - V_{a2}) = Z_F (I_{a0} + a^2 I_{a1} + a I_{a2}) \tag{17-147}$$

Substituting the values, we have

$$V_b - V_c = (a^2 - a)(V_{a1} - V_{a2}) = Z_F (a^2 - a) I_{a1} \tag{17-148}$$

Thus we finally arrive at the relation for the symmetrical components as

$$V_{a1} - V_{a2} = Z_F I_{a1} \tag{17-149}$$

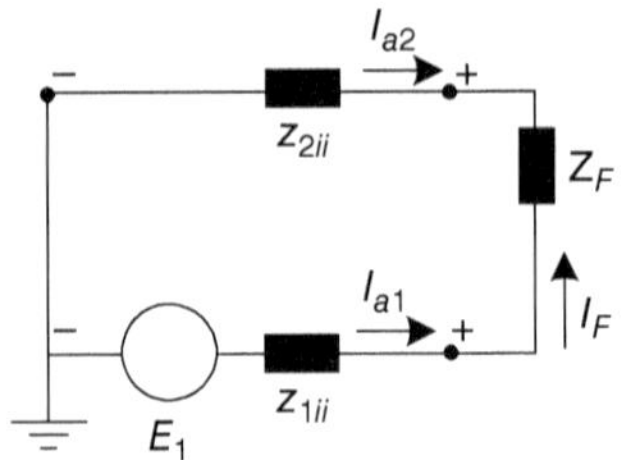


**FIGURE 17-30** Parallel connection of the positive and negative sequence equivalent circuits for a line-to-line fault.

From Eqs. (17-142) and (17-149), we can conclude that the positive and negative sequence circuits should be connected in parallel and that the zero sequence circuit does not participate. This is shown in Fig. 17-30.

The positive (and negative) sequence fault current is computed as

$$I_{a1} = -I_{a2} = \frac{E_1}{Z_{1ii} + Z_{2ii} + Z_F} \tag{17-150}$$

Finally, the phase fault current is computed from the symmetrical components transformation from

$$I_F = I_b = a^2 I_{a1} + a I_{a2} = (a^2 - a) I_{a1} = -j\sqrt{3} I_{a1} \tag{17-151}$$

$$I_F = I_b = -I_c = -j\frac{\sqrt{3}E_1}{Z_{1ii} + Z_{2ii} + Z_F} \tag{17-152}$$

### 17.8.4 Inversion of the Admittance Matrix by Column

Let us assume that voltage sources (generators and motors) are connected to buses 1 to $k-1$ and that a bolted three-phase fault occurs at node $k$; thus $V_k = 0$. Nodes $k+1$ to $n$ have only passive loads that do not contribute to the short circuit current. To compute the positive sequence short circuit current ($I_{sc}$), we assume that we know all the current injected at the healthy nodes (Norton equivalents given as $E_{\text{source}}/Z_{\text{source}}$). The node voltages (except $V_k = 0$) are unknown. Therefore, we need to remove the $k$th row and column of the $Y_{\text{bus}}$ matrix. The reduced admittance matrix is conveniently decomposed into triangular factors consisting of a lower $[L]$ matrix and an upper $[U]$ matrix as follows ($[L][U] = [Y]$):

$$\begin{pmatrix} l_{11} & 0 & 0 & \cdots & 0 \\ l_{21} & l_{22} & 0 & \cdots & 0 \\ \vdots & & & & \\ l_{k-1,1} & l_{k-1,2} & l_{k-1,3} & \cdots & 0 \\ l_{k+1,1} & l_{k+1,2} & l_{k+1,3} & \cdots & 0 \\ \vdots & & & & \\ l_{n1} & l_{n2} & l_{n3} & \cdots & l_{nn} \end{pmatrix} \begin{pmatrix} 1 & u_{12} & u_{13} & \cdots & u_{1n} \\ 0 & 1 & u_{23} & \cdots & u_{2n} \\ \vdots & & & & \\ 0 & 0 & 0 & \cdots & u_{k-1,n} \\ 0 & 0 & 0 & \cdots & u_{k+1,n} \\ \vdots & & & & \\ 0 & 0 & 0 & \cdots & 1 \end{pmatrix} \begin{pmatrix} V_1 \\ V_2 \\ \vdots \\ V_{k-1} \\ V_{k+1} \\ \vdots \\ V_n \end{pmatrix} = \begin{pmatrix} I_1 \\ I_2 \\ \vdots \\ I_{k-1} \\ 0 \\ \vdots \\ 0 \end{pmatrix} = \begin{pmatrix} 1/Zsc_1 \\ 1/Zsc_2 \\ \vdots \\ 1/Zsc_{k-1} \\ 0 \\ \vdots \\ 0 \end{pmatrix} \tag{17-153}$$

Since ANSI standards assume that all voltages are 1.0 pu, the right-hand-side vector is simply the inverse of the short-circuit impedance ($1/Zsc$) of the generator/motor buses. The elements of the $[L]$ and $[U]$ matrices are computed alternating a column of $[L]$ and a row of $[U]$ as follows:

$$\begin{aligned} & l_{i1} = Y_{i1} && ;\ i = 1, \ldots, n \\ & u_{1i} = Y_{1i}/Y_{11} && ;\ i = 1, \ldots, n \\ & l_{i2} = Y_{i2} - l_{i1}u_{12} && ;\ i = 2, \ldots, n \\ & u_{2i} = (Y_{2i} - l_{21}u_{1i})/l_{22} && ;\ i = 2, \ldots, n \\ & l_{i3} = Y_{i3} - l_{i1}u_{13} - l_{i2}u_{23} && ;\ i = 3, \ldots, n \\ & \ldots \end{aligned} \tag{17-154}$$

The above equations are obtained by the simple multiplication of matrices [$L$] and [$U$] and making the results equal to [$Y$]. The Matlab code to replace the [$Y_{bus}$] matrix by its triangular factors is given below.

```
% Factorize the [Y] matrix in place (no extra memory)
for i = 2 : n
    y(i,1) = y(i,1)/ y(1,1) ;
    for j = 2 : n
        sum = 0 ;
        if ( j >= i )
            % compute elements of [U] by column
            for k = 1 : i-1
                sum = sum + y(i,k) * y(k,j) ;
            end
            y(i,j) = y(i,j) - sum ;
        else
            % compute elements of [L] by row
            for k = 1 : j-1
                sum = sum + y(i,k) * y(k,j) ;
            end
            y(i,j) = y(i,j) - sum ;
            y(i,j) = y(i,j) / y(j,j) ;
        end
    end
end
```

The voltages of all buses during the short circuit are computed by forward and backward substitutions. Let us substitute $[X] = [U][V]$ in $[L][U][V] = [I]$ to get

$$[L][X]=[I] \tag{17-155}$$

Expanding the last equation, we get

$$\begin{pmatrix} l_{11} & 0 & 0 & \cdots & 0 \\ l_{21} & l_{22} & 0 & \cdots & 0 \\ \vdots & & & & \\ l_{k-1,1} & l_{k-1,2} & l_{k-1,3} & \cdots & 0 \\ l_{k+1,1} & l_{k+1,2} & l_{k+1,3} & \cdots & 0 \\ \vdots & & & & \\ l_{n1} & l_{n2} & l_{n3} & \cdots & l_{nn} \end{pmatrix} \begin{pmatrix} X_1 \\ X_2 \\ \vdots \\ X_{k-1} \\ X_{k+1} \\ \vdots \\ X_n \end{pmatrix} = \begin{pmatrix} 1/Zsc_1 \\ 1/Zsc_2 \\ \vdots \\ 1/Zsc_{k-1} \\ 0 \\ \vdots \\ 0 \end{pmatrix} \tag{17-156}$$

which can be easily solved with the following simple forward substitutions:

$$\begin{aligned} X_1 &= (1/Zsc_1)/l_{11} \\ X_2 &= (1/Zsc_2 - l_{21}X_1)/l_{22} \\ &\cdots \\ X_{k-1} &= \left(\frac{1}{Zsc_{k-1}} - \sum_{j=1}^{k-1} l_{kj}X_j\right)/l_{k-1,k-1} \end{aligned} \tag{17-157}$$

$$X_{k+1} = \left(-\sum_{j=1}^{k+1} l_{kj} X_j\right) / l_{k+1,k+1}$$
$$\ldots$$
$$X_n = \left(-\sum_{j=1}^{n} l_{kj} X_j\right) / l_{nn}$$

Finally, $[X] = [U][V]$ is expanded as

$$\begin{pmatrix} 1 & u_{12} & u_{13} & \cdots & u_{1n} \\ 0 & 1 & u_{23} & \cdots & u_{2n} \\ \vdots & & & & \\ 0 & 0 & 0 & \cdots & u_{k-1,n} \\ 0 & 0 & 0 & \cdots & u_{k+1,n} \\ \vdots & & & & \\ 0 & 0 & 0 & \cdots & 1 \end{pmatrix} \begin{pmatrix} V_1 \\ V_2 \\ \vdots \\ V_{k-1} \\ V_{k+1} \\ \vdots \\ V_n \end{pmatrix} = \begin{pmatrix} X_1 \\ X_2 \\ \vdots \\ X_{k-1} \\ X_{k+1} \\ \vdots \\ X_n \end{pmatrix} \tag{17-158}$$

The solution for $[V]$ can be obtained with backward substitutions as

$$\begin{aligned} V_n &= X_n \\ V_{n-1} &= X_{n-1} - u_{n-1} X_n \\ V_{n-2} &= X_{n-2} - u_{n-2,n-1} X_{n-1} - u_{n-2,n} X_n \\ &\ldots \\ V_{k-1} &= X_{k-1} - \sum_{j=k+1}^{n} u_{kj} X_j \\ &\ldots \\ V_1 &= X_1 - \sum_{j=2}^{n} u_{1j} X_j \end{aligned} \tag{17-159}$$

Now that when all bus voltages are known, we can compute the short-circuit currents in all branches from Ohm's law. In particular, for the branches connected to bus $k$, we write

$$I_{km} = y_{km}(V_k - V_m) \tag{17-160}$$

The total short-circuit current at bus $k$ is computed as

$$Isc_k = \sum_{m=1}^{n} I_{km} \tag{17-161}$$

The above solution works very well when the short circuit currents for a given faulted bus are needed. If all (or many) buses need to be studied for short-circuit currents, as is frequency the case, the procedure needs to be repeated for each new calculation. Different rows and columns would be eliminated from $Y_{\text{bus}}$ and therefore the computation time would be large. Repeated refactorization can be avoided by recognizing that the circuit is linear and therefore the principle of superposition applies.

When unbalanced faults are to be analyzed, the process of inverting the admittance matrices by column is repeated only for the zero sequence bus impedance matrix. The negative sequence bus impedance matrix is the same as the positive sequence bus matrix. Note that there are two particularities for the zero sequence bus impedance matrix that need to be considered:

1. The mutual coupling of lines sharing the same right of way. This is not necessary for the positive sequence, but it is important for the zero sequence impedance since the zero sequence currents in all three phases are the same. This complicates the calculation of the $Y_{bus}$ matrix that no longer can be computed by inspection. The impedance matrix of the mutually coupled lines needs to be obtained first. Then this (small) matrix is inverted and incorporated into the $Y_{bus}$ matrix.
2. The grounding connection of transformers, Y – Y, Y – Δ, etc., as described above, changes the connection of the zero sequence impedances. Sections of the circuit become disconnected from each other every time that we have an ungrounded transformer. Therefore the zero sequence bus admittance matrix $Y_{bus0}$ is diagonal by blocks, which makes the calculation of it inverse very convenient and efficient. This frequently results in a smaller zero sequence fault impedance in comparison with the positive (or negative) sequence impedance and the fault currents are larger than for a three-phase fault.

*Example 17-5.* The IEEE/PES (Power and Energy Society), in addition to publishing power flow case studies, has also published systems to test faults. The link is the same: https://ewh.ieee.org/soc/pes/dsacom/testfeeders/. Here we present a case study with the academic version of ASPEN http://www.aspeninc.com. ASPEN is perhaps the most advanced commercially available program for the study of short circuits in power systems, including breaker rating, a protection relay settings.

Figure 17-31 shows a circuit as drawn in ASPEN for the study of short circuit. Two faults are presented: three-phase (LLL) and single-line-to-ground (SLG). The positive, negative and zero sequence impedances of all components are given in Tables 17-3 and 17-4. The resulting current contributions

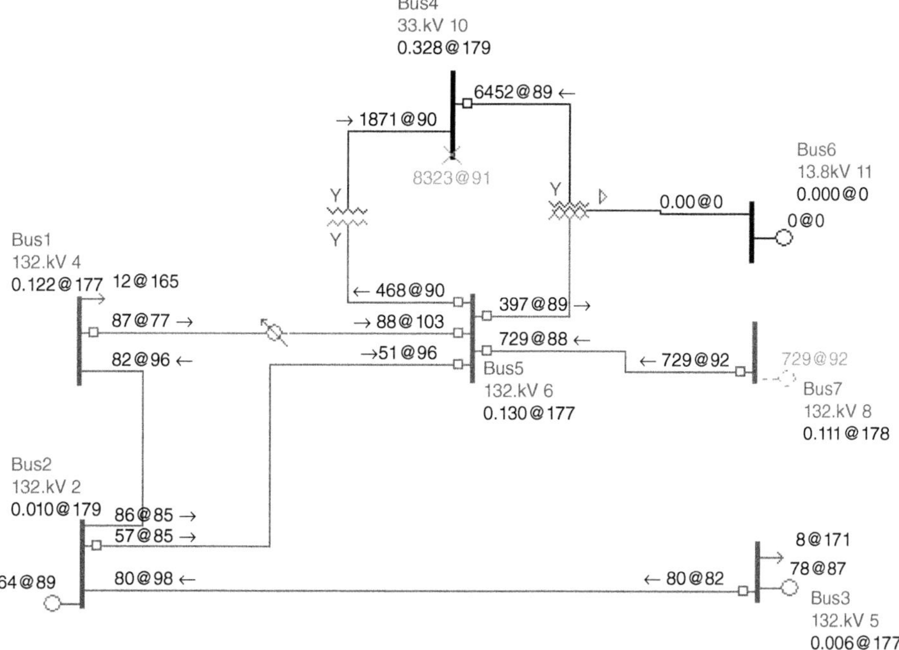


**FIGURE 17-31** Circuit for the study of short circuits.

**TABLE 17-3** Impedance Values

| From bus | To bus | Element | $R_0$ | $X_0$ | $R_1$ | $X_1$ | $B_0$ | $B_1$ |
|---|---|---|---|---|---|---|---|---|
| 1 | 2 | Line | 0.256 | 1.693 | 0.095 | 0.346 | 0.023 | 0.062 |
| 2 | 3 | Line | 0.010 | 0.061 | 0.004 | 0.014 | 0.001 | 0.002 |
| 2 | 5 | Line | 0.409 | 2.747 | 0.153 | 0.534 | 0.037 | 0.103 |
| 5 | 7 | Line | 0.005 | 0.034 | 0.002 | 0.007 | 0.000 | 0.001 |
| 1 | 2 | Line | 0.256 | 1.693 | 0.095 | 0.346 | 0.023 | 0.062 |
| 5 | 4 | Transformer | 0 | 0.556 | 0 | 0.556 | — | — |

**TABLE 17-4** Three-Winding Transformer Impedance Values

| Bus 1 | Bus 2 | Bus 3 | $X_{ps1}$ | $X_{pt1}$ | $X_{st1}$ | $X_{ps0}$ | $X_{pt0}$ | $X_{st0}$ |
|---|---|---|---|---|---|---|---|---|
| 5 | 4 | 6 | 0.318 | 0.416 | 0.318 | 0.318 | 0.416 | 0.318 |

**TABLE 17-5** Short Circuit Results

| Fault type | Phase A (A) | Phase B (A) | Phase C (A) | Positive seq. (A) | Negative seq. (A) | Zero seq. (A) |
|---|---|---|---|---|---|---|
| LLL | 8210 $\angle -90.7$ | 8210 $\angle -149.7$ | 8210 $\angle -29.3$ | 8210 $\angle -90.7$ | 0 | 0 |
| SLG | 8322 $\angle -90.9$ | 0 | 0 | 2773.4 $\angle -90.9$ | 2773.4 $\angle -90.9$ | 2773.4 $\angle -90.9$ |

are given in the diagram for the SLG fault. Table 17-5 shows the phase and sequence fault currents for both cases. Note that because the faulted bus has ungrounded transformers in the path from the sources, the short-circuit current for the SLG fault is slightly larger than that for the LLL fault.

## 17.9 REFERENCES


1. Bokhari, A., Alkan, A., Dogan, R., Diaz Aguiló, M., de León, F., Czarkowski, D., Zabar, Z., Birenbaum, L., Noel, A., and Uosef, R. E., "Experimental Determination of the ZIP Coefficients for Modern Residential, Commercial and Industrial Loads," *IEEE Transactions on Power Delivery*, vol. 29, no. 3, pp. 1372–1381, Jun. 2014.
2. El-Hawary, Mohamed E., *Electrical Power Systems: Design and Analysis*, Wiley-IEEE Press, 1995.
3. Stott, B., and Alsaç, O., "Fast Decoupled Load Flow," *IEEE Transactions on Power Apparatus and Systems*, vol. PAS-93, no. 3, pp. 859–869, May 1974.
4. Bam, L., and Jewell, W., "Review: Power System Analysis Software Tools," IEEE 2005 Power Engineering Society General Meeting.

# 18 POWER SYSTEM OPERATIONS

Hong Chen
*Senior Lead Engineer, PJM Interconnection, United States*

Jianwei Liu
*Senior Lead Engineer, PJM Interconnection, United States*

Jay Giri
*Director, GE Grid Software Solutions, United States*

Simon Tam
*Manager of Transmission Operations, PJM Interconnection, United States*

Mike Bryson
*Vice President of Operations, PJM Interconnection, United States*

Patrick Panciatici
*Scientific Advisor, RTE France, France*

Federico Milano
*Professor, University College Dublin, Ireland*

Jian Zhou
*Director, East China Grid, China*

Simon Bartlett
*Professor, University of Queensland, Australia*

S. K. Soonee
*Adviser, POSOCO, India*

## S. R. Narasimhan

*Additional General Manager, POSOCO, India*

## S. C. Saxena

*Deputy General Manager, POSOCO, India*

## K. V. N. Pawan Kumar

*Senior Engineer, POSOCO, India*

## 18.1 OVERVIEW

*BY HONG CHEN*

The major components of power systems are generation resources, demand resources or load, connected by transmission and distribution facilities. A power system is considered as the largest machine (or control system) in the world [1].

Power systems are normally interconnected to reduce total generation requirements and production cost, and enhance reliability. Frequency and voltage are the two most important parameters of an interconnected power system. They are maintained at prescribed nominal ranges for stable system operations and equipment safety. Section 18.2 reviews the fundamentals of power balance and frequency control.

The ultimate objective of power system operation is to keep the lights on. Power system operation is guided by the basic circuit theory: Ohm's and Kirchhoff's Laws. All facilities have physical limitations. As a control system, power system also has its dynamic characteristics and constraints. Section 18.3 discusses transmission operation and security.

Energy management system (EMS) is an important tool to assist power system operation, in terms of monitoring, control, and analysis. The details are covered in Sec. 18.4.

Because of the significant amount of equipment in operation simultaneously, regular maintenance is required. Outage scheduling is critical to system operation planning. The general practices in Independent System Operator (ISO) systems are discussed in Sec. 18.5.

Regulatory issues are another leading forces on market and system operation. Their impact on system operation is examined in Sec. 18.6.

Different regions have different operation practices. Operation experiences in Europe, China, Australia, and India are covered in Secs. 18.7.1, 18.7.2, 18.7.3, and 18.7.4, respectively.

## 18.2 POWER BALANCE AND FREQUENCY CONTROL

*BY HONG CHEN AND JIANWEI LIU*

The common practices of ensuring power balance are presented in the following sequence: power balance fundamental, forecasting, reserves, generation scheduling and dispatch, and frequency control.

### 18.2.1 Power Balance Fundamental

Electricity demand or load is constantly changing in the system, every hour, every minute, and every second. It is significantly impacted by weather conditions and pattern of human activities. At current stage, it is still challenging to store electricity on a large scale. Due to limited energy storage devices, generation has to be balanced with demand at all times, which is a moving target.

Since the generation is largely based on synchronous machines, if the total generation in the system is not balanced with the total system demand, system frequency changes. If generation is higher than demand, frequency increases; if generation is less than demand, frequency decreases, as shown in Fig. 18-1. The nominal frequency in North America is 60 Hz. In other parts of the world, frequency is maintained at 50 Hz.

For interconnected power systems, the interchanges with neighboring systems are also important components in keeping power balance. Some of the transactions can be scheduled ahead of time based on the specified rules. Therefore, power balance equation can be expressed by Eq. (18-1):

$$\text{Total Generation} = \text{Total Demand} + \text{Total Loss} + \text{Net Interchange} \tag{18-1}$$

where Total Loss is the energy lost in the system equipment; and Net Interchange is the net flow out of the interconnected system, and equals the difference between total Exports and total Imports.

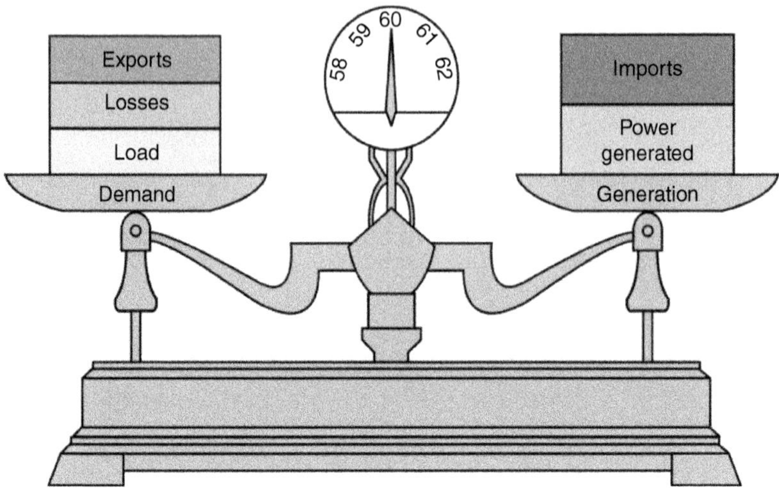


**FIGURE 18-1** Power balance and frequency impact.

### 18.2.2 Forecasting

***Load Forecast.*** Electricity load varies during the day and with seasons. The main load types are industry load, commercial load, lighting, residential load, etc. Each load type has unique characteristics. Typical load patterns exist in most systems. For example, in the Northeast of United States, during winter time, heating and lighting load determines the morning and evening peaks; during summer time, air conditioning determines the summer peak load.

Short-term load forecast (STLF) is critical to schedule and dispatch generation to maintain power balance. Many factors can impact load, therefore, they are accounted for in STLF. Main impacting factors are time, temperature, humidity, cloud covering, special social events, such as holidays or weekends. With electricity market in place and demand responses, electricity price also becomes an impacting factor. When scheduling generation 1 day to 1 week ahead, load is normally forecasted hourly for 24 hours or 168 hours ahead of time. When dispatching generation in real time, very short-term load forecast (VSTLF) is used to forecast load every 5 minutes for 1 to 3 hours ahead.

The main forecasting methods used in industry are similar days methods according to season, day of work, and weather effects, time series methods such as autoregressive moving average (ARMA) and autoregressive integrated moving average (ARIMA), artificial neural network (ANN), fuzzy logic, and hybrid method. The correlation of load to temperature, humidity, light intensity, wind speed, and actual weather conditions is analyzed based on historical data.

***Interchange Forecast.*** Unscheduled interchange (or loop flows) could occur from time to time, due to contract path scheduling practices. Interchanges are volatile and are significantly impacted by market dynamics. The uncertainty of interchanges poses challenge to balancing operation, and becomes one of the primary uncertainties faced by system operators today.

Efforts have been started to forecast them, using ARIMA, ANN, random forecast, support vector regression (SVR), Bayesian model aggregation (BMA), with the potential to include weather and price information of neighboring control areas as impacting factors.

***Wind Power Forecast and Solar Power Forecast.*** Unlike conventional power plants, wind power and solar power are very volatile and intermittent, therefore, not often dispatchable. Their outputs are often forecasted by system operators for generation scheduling and dispatch, as well as by plant owners for power trading. With increased penetration of wind and solar, their output forecast becomes very important to the overall system operation.

Wind power output is a function of wind speed, and solar power output depends on the incoming radiation and on the solar panel characteristics, with seasonal variations and daily cycles. Both wind power and solar power are highly volatile and uncertain, therefore, hard to forecast.

For short-term forecast up to 4 to 6 hours ahead, naïve methods based on the last measured values are often used as the reference methods. Advanced approaches use the predictions of meteorological variables as inputs: physical approach based on plant model characteristics and statistical approach capturing the relationship between meteorological predictions and power output, relying primarily on past data to "train" models. In practice, the physical approach and the statistical approach can be blended. More comprehensive stochastic learning techniques through master optimization process (MOP) have also gained attention.

### 18.2.3 Reserves

There exist volatility and uncertainties on both generation and load sides. Generator tripping and sudden load increases cause the frequency to spike low while sudden large load decreases cause the frequency to spike high. To mitigate the associated power imbalance, reserves are needed in the system to control normal frequency deviation and to survive large disturbances.

Reserves are the flexible unused available real power response capacity hold to ensure continuous match between generation and load during normal conditions and effectively respond to sudden system changes, such as loss of generation, sudden load changes, etc. Reserves are critical to maintain system reliability.

Reserves are secured across multiple timescales to respond to different events. Terminologies and rules vary in different systems, but they all share some fundamental characteristics. In general, some reserve types are for nonevent continuous needs; and others are for contingency events (e.g., loss of generator or facility tripping) or longer timescale events (e.g., load ramps and forecast errors). They are further categorized based on response time, online/offline status and physical capabilities.

In North America, according to North America Electric Reliability Corporation (NERC), operating reserves are defined as "that capability above firm system demand required to provide for regulation, load forecast error, equipment forced and scheduled outages, and local area protection. It consists of spinning and non-spinning reserve [2]." Reserves are often categorized as 30 minutes supplemental reserve, 10 minutes non-spinning reserve, 10 minutes spinning reserve and regulating reserve, etc. Operating reserves often refer to the generation available within 30 minutes from online or offline reserve resources. Figure 18-2 shows the general operating reserve categories.

***Regulating Reserves.*** Regulating reserves are often procured in both upward and downward directions to respond to normal load changes, that is, short-term load deviations that may cause the

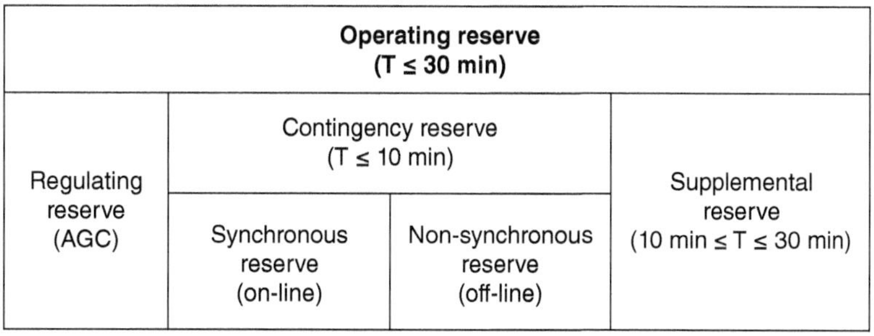

<table>
<tr><td colspan="4">Operating reserve<br>(T ≤ 30 min)</td></tr>
<tr><td rowspan="2">Regulating reserve (AGC)</td><td colspan="2">Contingency reserve<br>(T ≤ 10 min)</td><td rowspan="2">Supplemental reserve<br>(10 min ≤ T ≤ 30 min)</td></tr>
<tr><td>Synchronous reserve<br>(on-line)</td><td>Non-synchronous reserve<br>(off-line)</td></tr>
</table>

**FIGURE 18-2** Operating reserve categories.

system to operate above or below normal frequency, e.g., 60 Hz in North America. With regulating reserves, the performance of power systems can be better controlled to provide minute-to-minute system power balance by adjusting power output of generating units connected to automatic generation control (AGC) system. They are the reserves responsive to AGC command and only carried in regulating resources. The details of AGC are covered in Sec. 18.4.2.2.

Regulating resources are fast responding resources, having specific telecommunication, control and response capabilities to increase or decrease their output in response to a regulating control signal to control for frequency deviation. In many systems, regulating resources are required to respond to regulating signals within 4 seconds, and must increase or decrease their outputs at their ramp capability rates.

In some systems, regulating resources are further divided into two types: one for traditional regulating resources, such as steam units and hydro units, which move relatively slow and can sustain for a long time; one for fast response resources, such as battery resources, which can move very fast but cannot sustain due to limited energy storage for short-term energy storage resources. The fast and slow types of regulating resources often receive different regulation signals, and are paid differently as well based on mileage and performance [3].

***Contingency Reserves.*** Contingency reserves are used for the loss of supply, e.g., generation losses. NERC refers to contingency reserves as on/off-line reserves available within 15 minutes. Some systems require response within 10 minutes. The severity of the event determines how quickly the reserves have to be picked up.

Spinning or synchronous reserves are unused synchronized capacity and interruptible load, which is automatically controlled and made available within a given period of time, e.g., 10 minutes in most of the systems in North America. Synchronous reserves can be provided by on-line generators, synchronous condensers, load of pumped hydro resources, and interruptible load.

Non-spinning or non-synchronous reserves are real power capability not currently connected to the system but can be available within a specified time period, which may vary in different systems. In some systems, quick start reserves refer to the generation available within 10 minutes. The resources that qualify for this type of reserves are mainly run-of-river hydro, pumped hydro, combustion turbines, combined cycle units, diesels, and interruptible demand resources.

***Supplemental Reserves.*** Supplemental reserves are reserve capability that can be fully converted into energy or load that can be removed from the system within a 10- to 30-minute interval following a request from the system operator. Resources providing such a reserve do not need to be electrically synchronized to the system.

***Reserve Requirements.*** Reserve requirements are lower-limit reliability requirements. Synchronous reserves, non-synchronous reserves, and supplemental reserves have a priority sequence based on the level of reliability they provide. Higher priority reserves can also qualify for the requirements of lower priority reserves, that is, synchronous reserves can also qualify for a requirement which requires non-synchronous or supplemental reserves; and non-synchronous reserves can also qualify for a requirement which requires supplemental reserves.

Reserve requirements are set to protect system against instantaneous load variations, load forecasting error and loss of generation capabilities, for frequency control and area protection.

Reserve requirements are often set differently in different systems. Common practices are based on the largest contingency of the system. NERC BAL-002 standard requirement is to maintain at least enough contingency reserve to cover the most severe single contingency [2]. Each region/system has different operation practices. For example, in New York system, 10-minute synchronous reserve requirement is set as one-half of the largest single contingency [4]; while PJM's synchronous reserve requirement is set as the largest single contingency [5].

For regulating reserves, NERC does not impose explicit requirements, just to maintain sufficient regulating reserves to meet NERC control performance standards (CPS1, CPS2, or BAAL) [2].

The location of the reserves is also important due to transmission congestion. If there is not enough available transmission capacity, the affected zones need to have certain reserve requirements met by the reserve resources within the zones. Reserve deliverability is often evaluated in system operation. Zonal reserve requirements are set in most of the North America regional transmission operator (RTO) systems.

With increasing penetration level of intermittent renewable resources, the reserve requirements are being re-evaluated and adjusted to account for increased variability. For example, in Electric Reliability Council of Texas (ERCOT), forecasted wind output is factored in setting the regulating and contingency reserve requirements [6].

### 18.2.4 Generation Scheduling and Dispatch

All generation resources have their physical limitations, such as time to start, minimum run time, minimum down time, minimum and maximum output, ramp rates, turn-around time, mill points, etc.

Different types of generation have different characteristics. In general, nuclear units do not start up and shut down often, generate at relatively constant level, and their incremental costs are almost zero. Therefore, they serve base load. Steam units often start up and shut down slowly, and have relatively low incremental costs. They serve base load as well. Gas units can start up and shutdown quickly and often called "fast start units." Traditionally, they have relatively high incremental costs, and often run during peak hours. For this reason, they are also called "peaking units." With the increase of gas supply, the incremental costs of fast start units have decreased significantly. As a result, these units run much longer now than just the peak hours. Finally, hydro units often have fixed amount of energy for a day or a week. Therefore, hydro plants are normally scheduled for peak hours.

As described in Sec. 18.2.2, intermittent renewable resources, such as wind and solar, are often considered neither schedulable nor dispatchable. Their outputs are often forecasted for system operation.

To balance generation with demand and maintain system frequency, some generation (normally slow start generation) has to be scheduled way ahead of time based on forecasted load. As the time is closer to real-time, more generation (normally fast start generation) is committed to balance demand. Every 5 minutes, dispatchable generation is moved up or down to follow the load, and the output of regulating resources is adjusted based on AGC signal, often referred as secondary frequency control. The details of AGC will be covered in Sec. 18.4.2.2. The governor control of generators is often called as primary frequency control. In a summary, generation is staged to balance with load and maintain system frequency [7].

***Unit Commitment.*** Unit commitment (UC) is the key mathematical optimization problem for centralized generation scheduling both before the operating day and within the operating day. Before the operating day, UC provides the hourly generation schedules for the entire day, that is, 24 hours; within the operating day, UC provides the generation schedules for the look ahead hours.

- Inputs:
  - Forecasted load
  - Interchange schedules
  - Reserve requirements
  - Generation production cost model: such as startup cost, shutdown cost, no load cost, and incremental cost curve
  - Generation parameters: time to start, minimum run time, minimum down time, minimum and maximum output limits, reserve limits, ramp rates
  - Prescheduled status or output
  - Derations
  - Network loss
- Outputs:
  - Hourly on/offline status of each generator
  - Hourly megawatt output for each online generator

*Mathematical Model.* UC is modeled as mathematical optimization problem. The objective is to minimize the total production cost, or maximize the total social welfare under the market environment when offers and bids are submitted by market participants.

- System constraints:
  - Power balance constraint
  - Reserve requirement constraints
- Unit constraints:
  - Ramp up/down constraints
  - Start-up/shutdown ramping constraints
  - Minimum on/off time constraints
  - Power generation limit constraints: minimum and maximum
  - Reserve constraints

Lagrange relaxation and mixed integer programming (MIP) have been used extensively in the industry to solve for large scale UC problems.

***Security Constrained Unit Commitment.*** Security constrained unit commitment (SCUC) provides an hourly UC solution that is physically feasible and economically viable. It refers to the economic scheduling of generating units for serving the hourly demand and meeting the reserve requirements, while satisfying temporal and operation limits of generation and transmission facilities in contingency-based power systems. SCUC is the key decision-making component in today's power system operation, especially for large-scale systems.

Comparing to UC, transmission topology, parameters and flow limits are also included as inputs, so that operation constraints of transmission network are also considered, for both base and contingency cases.

SCUC is a large-scale mixed integer optimization problem with a large number of binary variables, continuous and discrete control variables, and a series of prevailing equality and inequality constraints. Due to its computational complexity, it is a class of nondeterministic polynomial-time hard (NP-hard) problem.

Direct current (DC) network constraints are often used for simplicity, with linear sensitivity factors. Alternate current (AC) network constraints are used for accurate network representation. Benders decomposition methods are generally used to solve the SCUC problem [8]. To account for the volatility brought by increased penetration of renewable resources, stochastic unit commitment is being considered and evaluated for practical operation.

***Security Constrained Economic Dispatch.*** Security constrained economic dispatch (SCED) is the key algorithm for generation dispatch. It determines the most economic dispatch for all online generators, to serve forecasted load, meet reserve requirements, while satisfying all generation and transmission limitations.

- Inputs:
  - Forecasted load
  - Interchange schedule
  - Reserve requirements
  - Generation incremental cost
  - Generation parameters: minimum and maximum output, reserve limits, ramp rates
  - Prescheduled output
  - Derations
  - Transmission topology

  - Transmission parameters
  - Transmission flow limitations
- Outputs:
  - Megawatt output for each online generator
  - Reserve assignment for each generator

*Mathematical Model.* SCED is modeled as mathematical optimization problem as well. The objective is also to minimize the total production cost, or maximize the total social welfare under market environment.

- System constraints:
  - Power balance constraints
  - Reserve requirement constraints
  - Transmission security constraints
- Unit constraints:
  - Ramp up/down constraints
  - Power generation limit constraints: minimum and maximum
  - Reserve constraints

Comparing to SCUC, SCED is a much easier problem to solve, commonly solved by Linear Programming in practical systems.

### 18.2.5 Area Control Error and Control Performance

The system-wide mismatch between generation and load results in frequency deviation from scheduled frequency.

***Area Control Error.*** In North America, area control error (ACE) is used to identify the imbalance between generation and load (including interchange). Imbalance is measured indirectly based on the frequency deviation of the system. ACE is measured based on Eq. (18-2)

$$ACE = [NI_A - NI_S] - [(10 \times B) \times (F_A - F_S)] - I_{ME} \tag{18-2}$$

where $NI_A$ represents actual net interchange, $NI_S$ represents scheduled net interchange, $B$ represents frequency bias constant, which is an estimate of system frequency response. $F_A$ represents the actual frequency, and $F_S$ represents scheduled frequency. $I_{ME}$ represents Interchange metering error [9]. The ACE is measured in megawatts.

System operators in North America operate in accordance with NERC resource and demand balancing (i.e., BAL) standards, to ensure their capability to utilize reserves to balance resources and demand in real-time, and to return interconnection frequency within defined limits following a reportable disturbance. NERC control performance standards (CPS) define a standard of minimum control performance for each Balancing area: CPS1 and CPS2. An alternative metric, balancing authority ACE limit (BAAL), has been used as a substitute for CPS2.

***CPS1.*** The CPS1 indicates the relationship between the ACE and the system frequency on a 1-minute average over a period (usually 12 months). It is the measure of short-term error between load and generation, and is defined as

$$AVG_{\text{period}}\left(\left[\frac{ACE}{-10B}\right]_{\text{clock-minute}} [\Delta f]_{\text{clock-minute}}\right) \le \varepsilon_1^{\,2} \tag{18-3}$$

where $\Delta f$ is the clock-minute average of interconnection frequency error. $B$ is the control area frequency bias and $[\cdot]_{\text{clock-minute}}$ represents the function of clock-minute average. $\varepsilon_1$ is a constant derived from a targeted frequency bound set by NERC.

For compliance with NERC, the CPS1 should not be less than 100%. If a control area closely matches generation with load, or if the mismatch causes system frequency to be driven closer to nominal frequency, the CPS1's performance will be good.

***CPS2.*** CPS2 would place boundaries on CPS1 to limit net unscheduled power flows that are unacceptably large. It sets limits on maximum average ACE for every 10-minute period. It imposes a limit on the magnitude of short-term ACE value. CPS2 prevents excessive generation/load mismatch even if a mismatch is in the proper direction.

The average ACE for each of the six 10-minute periods during the hour, that is, for the 10-minute periods ending at 10, 20, 30, 40, 50, and 60 minutes past the hour, must be within specific limit, referred to as L10.

***Balancing Authority ACE Limit.*** To meet the BAAL standard requirement, the ACE signal should satisfy the constraint

$$\text{BAAL}_{\text{Low}} \ll \text{ACE} \ll \text{BAAL}_{\text{High}} \tag{18-4}$$

where $\text{BAAL}_{\text{Low}}$ and $\text{BAAL}_{\text{High}}$ are set based on NERC BAL-001-2 standard. $\text{BAAL}_{\text{Low}}$ is calculated based on the low-frequency trigger limit, and $\text{BAAL}_{\text{High}}$ is calculated based on the high-frequency trigger limit.

When actual frequency is equal to scheduled frequency, $\text{BAAL}_{\text{Low}}$ and $\text{BAAL}_{\text{High}}$ do not apply. When actual frequency is less than scheduled frequency, $\text{BAAL}_{\text{High}}$ does not apply, and when actual frequency is greater than scheduled frequency, $\text{BAAL}_{\text{Low}}$ does not apply.

NERC disturbance control standard (DCS) (BAL-002-0) is used by each balancing area to monitor control performance during recovery from disturbance conditions, which states that ACE must return either to zero or to its pre-disturbance level within 15 minutes following the start of the disturbance.

### 18.2.6 Frequency Control

The governor control of generators is referred to as primary frequency control. Under normal operation, the small frequency deviation can be attenuated by the primary frequency control. For larger frequency deviation, AGC is responsible to restore system frequency. For serious load generation imbalance associated with rapid frequency change following a significant fault, contingency reserves are often deployed and economic dispatch is adjusted as well.

Emergency frequency control and related protection schemes must be the last option to decrease the risk of cascade faults, additional generation events, load/network and separation events. Load shedding is the most popular emergency control scheme, by curtailing some parts of system load. If the frequency falls below a specified frequency threshold following a serious disturbance, underfrequency load shedding is then triggered to protect against excessive frequency decrease.

### 18.2.7 Impact of Intermittent Renewable Resources on Frequency Control

Intermittent renewable resources, such as wind and solar, are highly volatile, not quite dispatchable and are hard to forecast as well. Increased penetration of these resources increases system operation uncertainty and causes significant challenges to power balancing, impacting on system reliability and control performance. In some systems, cloud covering could become the largest contingency.

Stochastic methods provide natural solutions to account for increased system uncertainties. However, they are computationally challenging, especially for real-world problems.

Existing tools, e.g., EMS, SCUC, and SCED, are all based on deterministic methods, therefore, they cannot reflect uncertainties well. Planning adequate reserves becomes an important issue in today's system operation. For example, regulation reserve requirements may need to be increased to account for a larger amount of fluctuating wind power or solar power.

Designing intermittent generators with the full range of performance capability, which is comparable to conventional synchronous generators, is also becoming possible. For example, wind power facilities can be equipped to provide governing control and participate in AGC.

With the advancement of storage technologies and attractive market design, energy storages have become excellent frequency control resources, especially in the markets where they are compensated correspondingly for their fast responsive performances [3]. Also, the combination of energy storages with renewable resources has been proved to provide better system performance.

## 18.3 TRANSMISSION OPERATION AND SECURITY

*BY HONG CHEN AND JIANWEI LIU*

Power system operation follows basic physical laws: Ohms law and Kirchhoff's laws. Operating reliability, or security, plays the key role in system operation.

Network (transmission and distribution) has limited capability to transfer power from generation to load due to facility thermal, stability and/or voltage limits. Power transfer can be restricted to any of these limits, or a combination of them. Network security constraints are non-linear, especially stability limits and voltage limits. Security constrained optimal power flow (OPF) is a fundamental tool to ensure a secure operation.

### 18.3.1 Security Criteria

Uncertainty exists in all aspects of power system operation. A facility can trip or malfunction at any time. Facility tripping can overload other facilities. Therefore, the system has to be operated in a manner that it will stay within its technical limits under normal system conditions and also under the conditions that another facility trips.

"Contingency" refers to the sudden loss of a generating unit, transmission line, transformer, or breaker. A single contingency may disconnect multiple generating facilities (e.g., a plant with single connection to bulk power system), or multiple transmission facilities (e.g., radial lines with tapping substation).

The historical practice is the N-1 contingency criteria, which means that when a facility trips, no system limit violation occurs. Under certain conditions, for example, hurricane onset, a possible event resulting in the failure or malfunction of one or more facilities, that is, N-k, is considered.

### 18.3.2 Facility Thermal Limitation

Network facilities, such as transmission lines and transformers, have thermal ratings limiting the amount of current or apparent power that can be carried. Exceeding the thermal limits of transmission lines can cause the conductors to sag and stretch due to overheating, which could further result in faults or fires. Most equipment can be safely overloaded by a certain degree. The key is how big the overload is and how long it lasts. Typically, thermal ratings are set to allow a specified overload for a specified period of time.

Due to thermal capabilities, the flow on any facility has to be within its thermal limit, under normal system conditions and contingency conditions as well.

In North America, normal continuous ratings and emergency ratings (long term and short term) are specified for each facility [10]. Some systems have load dump ratings as well [11, 12]. Ambient temperature can affect facility thermal ratings significantly. Some systems have their thermal ratings corresponding to different temperature sets, such as PJM [11]. Dynamic line ratings are being implemented or investigated in many systems [6, 13]. The severity of thermal limit exceeding often determines corrective actions and time to correct with load shedding [11].

Power flow analysis and contingency analysis are utilized to determine the actual flow and contingency flow on the facilities. The actual flow on the facilities often come from measurements and/ or state estimation, and is continuously monitored in EMS.

### 18.3.3 System Stability Limitation

As a control system, power systems are also subject to stability limitations, that is, system should be able to return to the stable states after a disturbance. There are two main stability categories experienced in a power system, namely, angular stability and voltage stability. Each category can be further divided based on how big the disturbance: small perturbation and large disturbance. According to [14], there is also a mid-term/long-term stability which involves large voltage and frequency shift.

Voltage is the key to the overall stability of a power system. Angular stability is related to the angular separation between points in the power system; and voltage stability is related to the magnitude of the system voltages and reactive power reserves. Often, angular and voltage instability go together.

***Angular Stability.*** A power system is composed of many synchronous machines. Angular stability has to be maintained for the synchronization of the grid, to ensure that system torque and power angle remain controllable. The angles change as system conditions change. An interconnected power system loses synchronization when the power transfer rises to extreme large magnitudes that power angles reach excessive values. Following a large disturbance, transient stability becomes the concern. Power system may become instable for a period of time: angles may reach high magnitudes and rapidly change over a wide range. Synchronous generators are critical to the transient stability analysis. When torque/ power angles are too large, and disturbances occur, magnetic bonds of generators may be lost. The system becomes angle unstable when system operators lose their ability to control angles and power flows.

Stability analysis is often used to determine stability limits. Many power systems restrict their real power transfers due to transient stability concerns. In general, those are the power systems with long transmission lines and remote generation. Detailed stability analysis is covered in Section 20.

***Voltage Stability.*** Voltage stability is the ability of a power system to maintain adequate voltage magnitudes. In a voltage stable system, both power and voltage are controllable. Voltage stability is mainly a function of power system load. The reactive component of the load has a greater impact than the real component. Excessive loading in the power system leads to deficiencies in reactive power and the system is no longer able to support voltage. A voltage collapse could then occur. The shortage of reactive power drives to voltage instability.

When a power system experiences a voltage collapse, system voltages decay to a level from which they are unable to recover. Voltage collapse is a process during which voltage instability leads to loss of load in a part of the power system. The effects of a voltage collapse are more serious than those of a typical low-voltage scenario from system perspective. As a consequence of voltage collapse, the entire systems may experience a blackout. Restoration procedures are then required to restore the power system.

As power systems are pushed to transfer more and more power, the likelihood of the occurrence of a voltage collapse becomes greater. Voltage stability is mainly a concern in heavily loaded systems, and has been responsible for major network collapses in recent years [15]. Often, system transfer capabilities are limited by steady state voltage stability limits.

### 18.3.4 Voltage Limits

All equipment is designed to operate at certain rated supply voltages. Large deviations could cause damage to system equipment. High voltages can lead to the breakdown of equipment insulation, cause transformer over-excitation, and adversely affect customer equipment. Low voltages can impact power system equipment and operations in numerous ways.

Voltage control is closely related to the availability of reactive power. Amount of available reactive support often determines the power transfer limit. Heavy power transfers are the main cause of low voltages due to the reactive power losses. Lightly loaded transmission lines are the main cause of high voltages. Capacitors, reactors, load tap changers (LTCs), and static VAR compensators (SVCs) are the equipment aimed at controlling system voltages. For example, reactive support from capacitor is often needed to help prevent low-voltage problem. Synchronous machines can also provide voltage control. In system operation, reactive reserves need to be maintained and voltage deviations need to be controlled.

### 18.3.5 Maintain Network Security

EMS is the primary tool to monitor and evaluate current system state and simulated post-contingency states, for pre-/post-contingency thermal and voltage limits, stability limits, as well as assess voltage collapse for reactive interfaces.

Reactive transfer interfaces are often defined across the transmission paths to prevent voltage criteria violation and voltage collapse. The interface limits are used to limit the total flow over the interfaces. The reactive interface limits are either pre-contingency active power limits, or post-contingency active power limits. PV curves are often used to determine reactive interface limits.

When a facility overloads, the general practice is to apply corrective actions with a little or no cost first, such as transformer tap adjustments, phase shift regulators (PAR) adjustments, capacitor/reactor switching, line switching, and curtailment of non-firm transactions. Adjusting generation real power output via re-dispatch can be used pre-contingency to control post-contingency operation. Cost-effective generation redispatch is mostly achieved through SCED application for congestion management. If the above actions do not relieve an actual or simulated post-contingency violation, then emergency procedures can be directed, including dropping or reducing load as required.

The following control actions are often used to control low-voltage or high-voltage conditions:

- Switch capacitors in-service for low-voltage conditions and out-of-service for high-voltage conditions
- Switch reactors out-of-service for low-voltage conditions and in-service for high-voltage conditions
- Adjust variable reactor tap positions
- Adjust voltage set points of SVC
- Operate synchronous condensers
- Change transformer tap positions
- Change generation excitation
- Adjust generation MW output
- Adjust transactions
- Adjust PARs
- Switch facilities in/out of service

Thermal and voltage constraints are often controlled cost effectively on a pre-contingency basis. Due to the high cost associated with pre-contingency control and low probability of contingency events, some systems have started considering post-contingency congestion management under certain conditions.

### 18.3.6 Impact of Intermittent Renewable Resources on Transmission Security

Intermittent renewable resources, such as wind and solar, can often cause local congestions with flow over thermal limits or low-voltage limits. Their volatile nature poses challenge to the congestion management, resulting in volatile dispatch signals. Due to the low or zero operating cost of these resources, often, more dispatchable resources, such as steam units, are dispatched up or down to relief congestion. Sometimes, the output of these intermittent renewable resources has to be curtailed for congestion control. More advanced dispatch algorithms, such as risk-based dispatch, are being developed to accommodate high intermittent renewable impact.

### 18.3.7 References


1. Gómez-Expósito, A., Conejo, A. J., and Cañizares, C., "*Electric Energy Systems: Analysis and Operation*", CRC Press, Boca Raton, FL, USA, 2009.
2. www.nerc.com.
3. www.pjm.com
4. "Ancillary Services Manual," New York Independent System Operator, Dec. 2015.
5. "PJM Manual 13: Emergency Operations," PJM, Jan. 1, 2016.
6. www.ercot.com.
7. Chen, H., "*Power Grid Operation in a Market Environment: Economic Efficiency and Risk Mitigation,*" Wiley-IEEE Press, USA, 2016.
8. Fu, Y., Shahidehpour, M., and Li, Z., "Security-Constrained Unit Commitment with AC Constraints," *IEEE Transaction On Power Systems*, vol. 20, no. 3, Aug. 2005.
9. "Balancing and Frequency Control," NERC, Jan. 26, 2011.
10. "System Operating Limit Definition and Exceedance Clarification," NERC, Aug. 4, 2014.
11. "PJM Manual 3: Transmission Operations," PJM, Dec. 1, 2015.
12. www.iso-ne.com.
13. www.entsoe.eu.
14. Kundur, P., "*Power System Stability and Control,*" McGraw-Hill, Inc, USA, 1994.
15. Hossain, J., and Pota, H. R., "*Robust Control for Grid Voltage Stability: High Penetration of Renewable Energy,*" Springer, New York, USA, 2014.


## 18.4 ENERGY MANAGEMENT SYSTEM

*BY JAY GIRI*

### 18.4.1 Introduction

The big Northeastern U.S. blackout of 1965 was a watershed event for the electric power industry. Millions of dollars of lost business revenue, as well as intense consumer dissatisfaction, have been attributed to this blackout. This unprecedented event alerted the United States public to the importance of grid reliability and security, to ensure the lights stay on at all times.

The EMS at a utility control center collects real-time measurements to monitor current grid conditions. The EMS is a suite of analytics that synthesize these measurements to provide the grid operator with information to identify current problems and potential future threats to grid security.

Managing the grid is becoming more challenging because of evolving grid influences, such as growth of variable renewable generation resources, distributed generation, microgrids, demand response, and customer engagement programs. Concurrently, however, there are nascent technologies and other advances that improve our ability to manage future grid operations. These technologies include new sub-second synchrophasor measurements and analytics; advances in high performance computing, visualization platforms, digital relays, cloud computing, etc. Other advances include adding more intelligence at substations and distribution systems, as well as self-managed microgrids and wide area monitoring systems.

One key initiative is to create a "predict and mitigate" proactive grid management paradigm to enable better anticipation, so that timely decisions can be made to mitigate problems before they spread across the grid.

This subsection describes the historical development of EMS control centers and their evolution to manage uncertainties in the future grid.

***The Energy Management System.*** The EMS is a centralized facility of software and hardware whose primary purpose is to ensure that the transmission system operating conditions stay safe, reliable, and secure around the clock. The EMS is deemed the nerve center or brain of utility grid management.

Since there is a tremendous volume of constantly changing conditions across the grid, the challenge is to sift through this data to identify conditions that are potential imminent problems that need operator attention. Conditions need to be monitored in a timely, periodic manner in order to immediately detect any adverse conditions, as soon as they arise, so that corrective actions could be implemented to mitigate potentially harmful conditions that could lead to a wide spread grid collapse.

A major challenge is to convert vast amounts of data into useful information. Another challenge is to show the operator this information in an intelligent, concise manner, so as to facilitate prompt decision making. Timely visualization of real-time grid conditions is essential for successful grid operations.

The EMS has evolved based on a centralized command and control grid management paradigm. The goal very simply is "Keep the lights on."

Figure 18-3 is an overview of the control room at a typical large modern EMS. There are multiple operator positions along with their respective monitors and communication systems. These include operators who manage generation, transmission, switching, interchanges with neighbors and a supervisor. The EMS also has large wallboard screen displays which can be monitored by all the control room operators.

**FIGURE 18-3** A modern EMS control center.

## 18.4.2 EMS Subsystems

Figure 18-4 shows the main EMS functional subsystems. There are three operator-related subsystems and a system-related infrastructure subsystem. The operator-related subsystems are:

1. SCADA—supervisory control and data acquisition
2. Generation—monitoring and control
3. Network—analysis and optimization

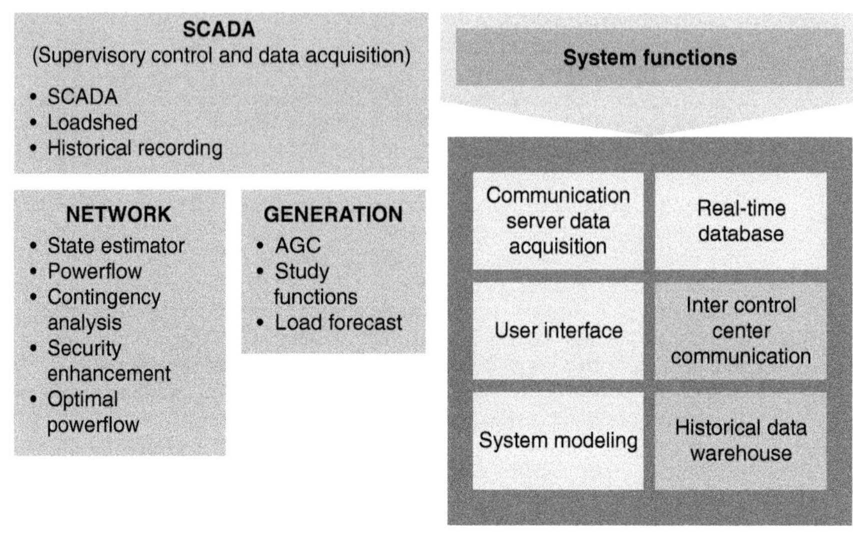


**FIGURE 18-4** EMS functions.

The system-related subsystem is primarily for support engineers, software engineers, and IT staff. Their function is to maintain the software and hardware infrastructure, database models, displays, interfaces to other entities and other ancillary functions, to ensure that the EMS environment is up to date and available around the clock.

Figure 18-5 shows a typical hardware configuration of the EMS.

Figure 18-6 shows the multiple subsystem application layers.

Figure 18-7 shows the various periodicities and cycles at which the EMS analytics run. The various EMS subsystems are described next.

***SCADA.*** SCADA receives real-time power system measurements typically every 2 to 4 seconds. It consists of a master station that communicates with remote terminal units (RTUs) for monitoring the major components of the transmission system: generating plants, transmission lines and transformers, substations. RTUs transmit device status and measurements to, and receive control commands from the master station. Communication is generally via dedicated circuits.

The main SCADA functions include:

- *Data Acquisition.* Provides telemetered measurements and status information.
- *Supervisory Control.* Allows operator to remotely control devices, e.g., open and close circuit breakers. A “select before operate” procedure is used for greater safety.
- *Tagging.* Identifies a device subject to operating restrictions and prevents unauthorized operation.
- *Alarms.* Informs operators of unplanned events and undesirable operating conditions.
  - Alarms are sorted by criticality, area of responsibility, and chronology. Acknowledgment may be required.
- *Logging.* Logs all operator entries, alarms, and other key data.
- *Load Shedding.* Provides automatic and operator-initiated tripping of load in response to system emergencies.
- *Trending.* Plots measurements on selected time scales.
- Communication circuit configuration management.

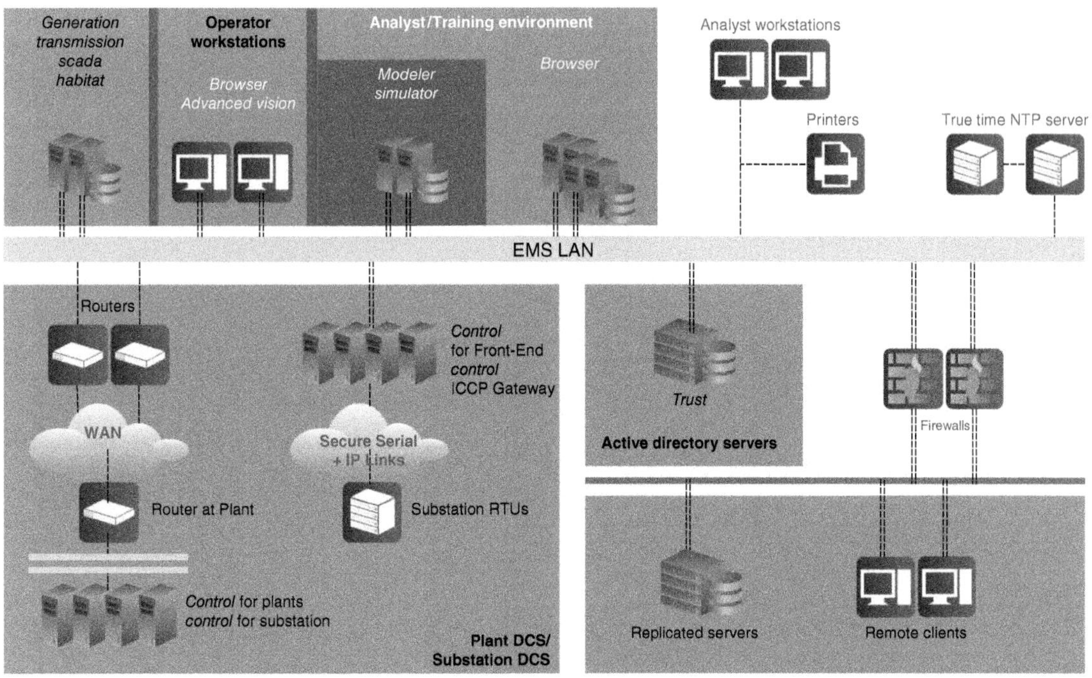


**FIGURE 18-5** Typical EMS hardware.

***Generation Monitoring and Control.*** The generation subsystem is responsible for monitoring the generating plants and for sending them controls to increase or lower generation based on system conditions.

AGC is the primary generation function that typically runs every 4 to 10 seconds. The objective is to adjust the generation to continually meet load while minimizing production costs. The major

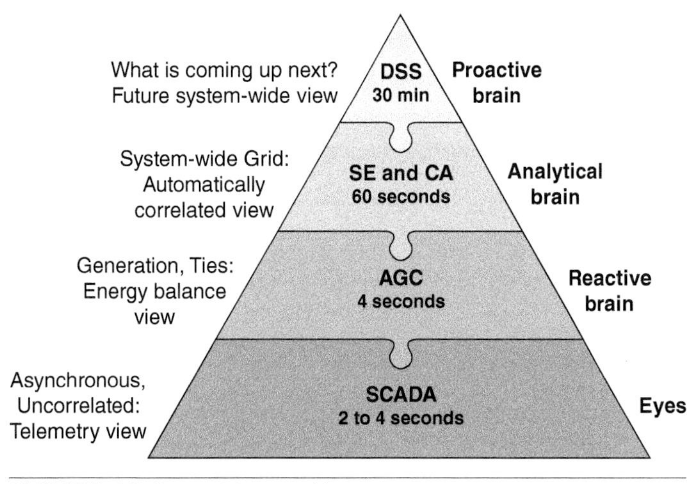


**FIGURE 18-6** EMS functional layers.

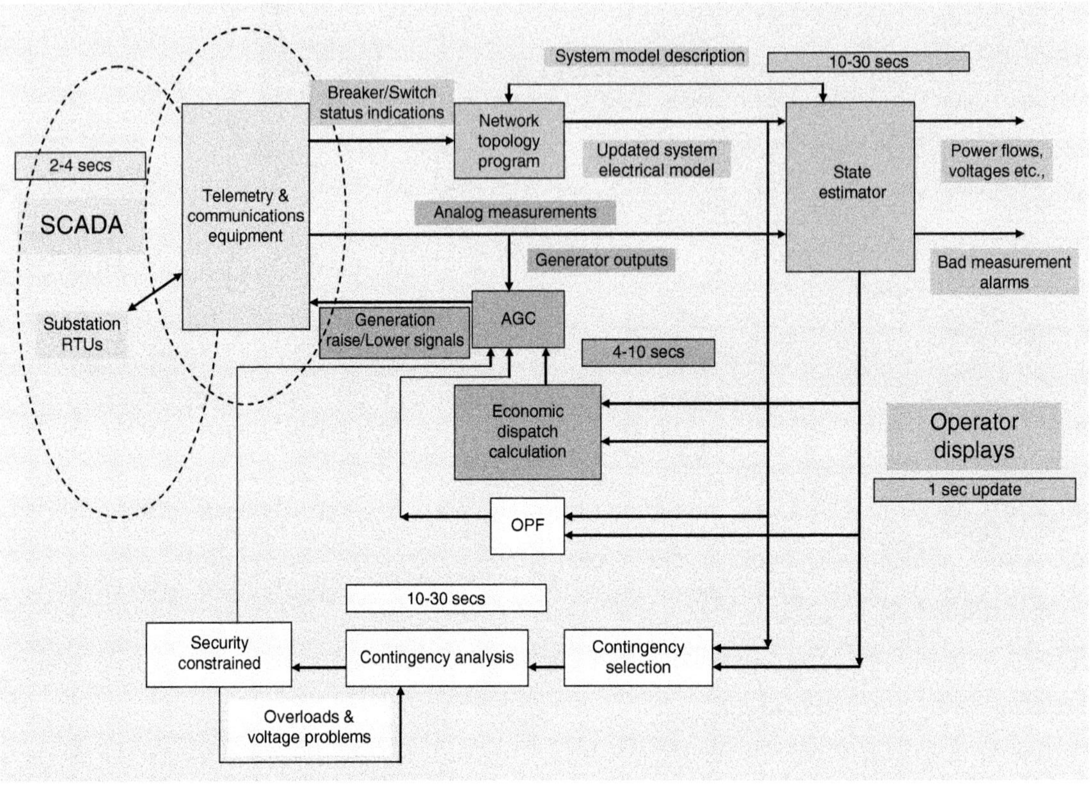


**FIGURE 18-7** EMS application cycles.

functions are load-frequency control (LFC) and economic dispatch (ED). Other functions are reserve monitoring, interchange scheduling, and related monitoring and recording functions.

*Load-Frequency Control.* The primary objectives of LFC are:

1. Maintain frequency at its nominal value (50 Hz or 60 Hz).
2. Maintain net power interchanges with neighboring control areas at scheduled values.
3. Allocate generation across plants to minimize overall system production cost.

The LFC philosophy is that each utility of the interconnection should closely follow its own load during normal operation, and during emergencies, should provide support to the interconnection according to its relative size. In order to provide this support the utility needs to maintain adequate regulation margin as well as adequate response capability.

The objective of the AGC control logic is to ensure "good prompt" control without "excessive unnecessary" movement of units. AGC controller tuning is system-specific and depends on the unit type (thermal, hydro, gas turbine, etc.) and the characteristics of the communication system.

*Economic Dispatch.* Since all generating units have different costs of generation, it is necessary to optimally allocate these units to meet load at minimum cost. The minimum dispatch cost is when the incremental cost of all the generators is equal. This allocation logic is performed by the economic dispatch function within AGC and is run typically every 60 seconds.

*Reserve Monitoring.* Reserve capacity is required in case generation is lost or load suddenly increases. Typical reserves include spinning (already online) and operating (available within 10 minute or 30 minute).

*Interchange Transaction Scheduling.* AGC needs to account for contractual exchanges of power between utilities. The net interchange (sum of all the buy and sale agreements) is used to augment the AGC generation target.

*Forecasting and Accounting.* Grid operation also needs to monitor and consider

- *System Load Forecast.* Forecasts system energy demand for a specified forecast period.
  - Generally a mix of forecast intervals (5, 15, 30, 60 minutes) with time horizons from 15 minutes, to hours, to multiple days.
  - With the growth of unpredictable renewable energy resources, it is important to have forecast updates more frequently—typically every few hours.
- *Unit Commitment.* Determines start-up and shut-down times for most economical operation of thermal generating units for each hour of a specified period of 1 to 7 days.
- *Fuel Scheduling.* Determines the most economical choice of fuel consistent with plant requirements, fuel purchase contracts, and stockpiled fuel.
- *Hydro-Thermal Scheduling.* Determines the optimum schedule of thermal and hydro energy production for each hour of a study period up to 7 days while ensuring that hydro and thermal constraints are not violated.
- *Transaction Evaluation.* Determines the optimal incremental and production costs for exchange (purchase and sale) of additional blocks of energy with neighboring companies.
- *Transmission Loss Minimization.* Recommends controller actions to be taken in order to minimize overall power system network losses.
- *Security Constrained Dispatch.* Determines optimal outputs of generating units to minimize production cost while ensuring that a network security constraint is not violated.
- *Production Cost Calculation.* Calculates actual and economical production costs for each generating unit on an hourly basis.

***Network Analysis and Optimization.*** The goal of grid security is to be able to survive probable potential contingencies. A contingency is defined as the unplanned loss of a major component such as: transmission lines, transformers, generators, substations, etc. Survival means that the system stabilizes and continues to operate at acceptable voltage and frequency levels without loss of load. The following steps are sequentially performed:

1. Determine the state of the system based on either current measurements.
2. Process a pre-defined list of contingencies to determine the consequences of each contingency.
3. Determine preventive or corrective actions for contingencies which represent unacceptable risk.

The following is a typical real-time sequence of network analytics performed at the EMS:

- *Topology Processor.* Processes real-time status measurements to determine an electrical connectivity (bus) model of the power system network.
- *State Estimator.* Uses real-time status and analog measurements to determine the "best" estimate of the state of the power system.
  - It uses a redundant set of measurements; calculates voltages, phase angles, and power flows for all components in the system; and reports overload conditions.
- *Contingency Analysis.* Assesses the impact of contingencies on the state of the power system and identifies potentially harmful contingencies that could cause operating limit violations.

- *Optimal Power Flow.* Recommends controller actions to optimize a specified objective function (such as system operating cost or losses) subject to a set of power system operating constraints.
- *Security Enhancement.* Recommends corrective control actions to be taken to alleviate an existing or potential overload in the system while ensuring minimal operational cost.
- *Preventive Action.* Recommends control actions to be taken in a "preventive" mode before a contingency occurs to preclude an overload situation if the contingency were to occur.
- *Bus Load Forecasting.* Uses real-time measurements to adaptively forecast loads for the electrical connectivity (bus) model of the power system network.
- *Transmission Loss Factors.* Determines incremental loss sensitivities for generating units; calculates the impact on losses if the output of a unit were to be increased by 1 MW.

Other offline study analytics include:

- *Power Flow.* Determines the steady-state conditions of the power system network for a postulated generation and load pattern.
- *Short-Circuit Analysis.* Determines fault currents for single-phase and three-phase faults at different fault locations.

## 18.4.3 Dispatcher Training Simulator

A dispatcher training simulators (DTS) is an offline simulation environment which models the EMS with a power system simulation model in order to provide a realistic environment for operators to practice normal, every-day operating tasks and procedures, as well as experience emergency operating situations. Training activities can be safely practiced with the simulator responding in a manner similar to the actual power system.

The DTS can be used to recreate past actual operational scenarios and to formulate and develop system restoration procedures. Scenarios can be created, saved, and reused. NERC has been emphasizing the need for black start training—the DTS provides an invaluable environment for this.

The DTS can also be used to evaluate the functionality and performance of new real-time EMS functions and for tuning AGC in an off-line, secure environment.

Figure 18-8 shows the three main subsystems of the DTS:

1. The energy control system
2. Power system dynamic simulation
3. Instructional system

***Energy Control System.*** The energy control system (ECS) emulates normal EMS functions and is the only part of the DTS with which the trainee interacts. It consists of the supervisory control and data acquisition (SCADA) system, generation control system, and all other EMS functions.

***Power System Simulation.*** This subsystem simulates the dynamic behavior of the power system. System frequency is simulated using the "long-term dynamics" system model, where frequency of all units is assumed to be the same. The prime-mover dynamics are represented by models of the units, turbines, governors, boilers, and boiler auxiliaries. The network flows and states (bus voltages and angles, topology, transformer taps, etc.) are calculated at periodic intervals. Relays are modeled, and they emulate the behavior of the actual devices in the field.

***Instructional System.*** This subsystem includes the capabilities to start, stop, restart, and control the simulation. It also includes making save cases, retrieving save cases, reinitializing to a new time, and initializing to a specific real-time situation.

It is also used to define event schedules. Events are associated with both the power system simulation and the ECS functions. Events may be deterministic (occur at a predefined time), conditional (based on a predefined set of power system conditions being met), or probabilistic (occur randomly).

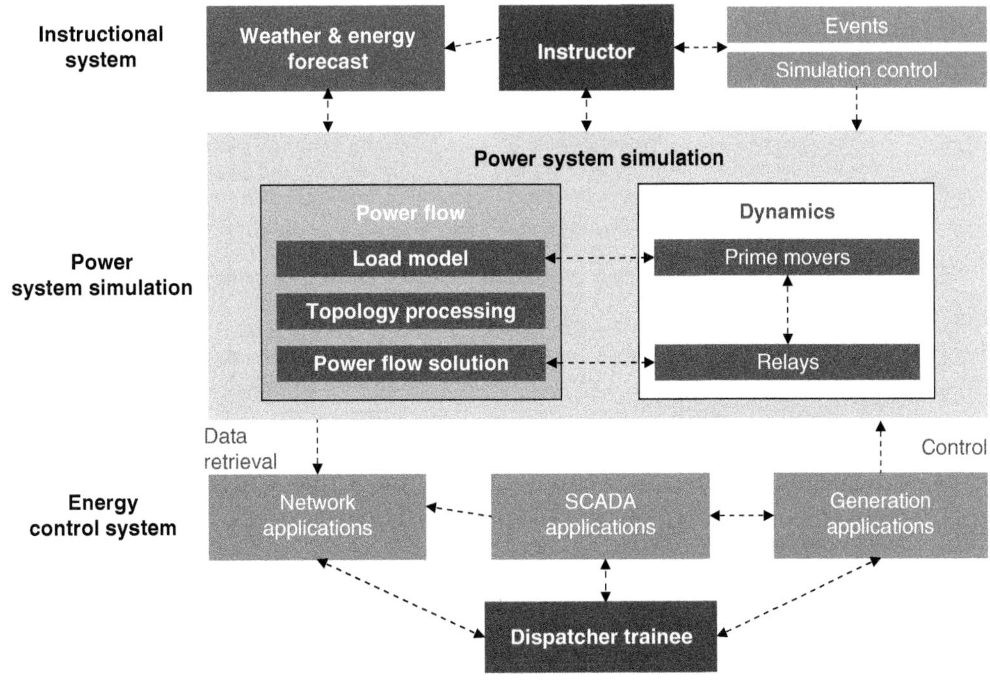


**FIGURE 18-8** DTS.

### 18.4.4 Recent EMS Trends

A major recent trend related to the EMS is the growth of synchrophasor measurement units. Cyber-security issues are also an emerging requirement for EMS. These are described in the following subsections.

Other trends that impact the EMS include:

- Electricity markets—market dispatch replacing economic dispatch in some utilities
- Interfaces and integration with decentralized management systems such as:
  - Distribution management systems
  - Substation automation systems
  - Microgrids
  - Renewable and distributed generation resources
  - Demand response programs

### 18.4.5 Next Generation EMS

***Synchrophasor Measurements in EMS.*** In the past few years, phasor measurement units (PMUs) have been deployed in greater numbers at utilities globally. PMUs provide synchronized, sub-second measurements, which provide prompt and higher resolution visibility of grid conditions. Advances in visualization capabilities coupled with the availability of PMU data dramatically improves the ability to manage grid operations more effectively, which in turn helps grid operators make faster and better decisions to maintain grid integrity.

New control center applications are continually being developed to use this new type of PMU synchronized measurement technology to further improve the ability to maintain the integrity of the power system.

The objective is to provide the operator "eyes" to always be aware of current system conditions and potential problems that might be lurking ahead. As the volume and frequency of measurement data grows—especially with the growth of sub-second synchronous measurement, such as PMUs—it is of utmost importance, to convert this data tsunami into relevant useful information that can be concisely shown on an operator display screen and to allow prompt decisions to be made with confidence.

Figure 18-9 shows the next generation modern EMS with synchrophasor measurements and analytics.

Figures 18-10, 18-11, and 18-12 show typical EMS operator displays in a modern EMS.

Figure 18-13 shows how the modern EMS is evolving to meet the changing grid environment. It is becoming a utility's central hub of information for the increasing plethora of various grid management processes and solutions.

***Cyber Security in EMS.*** In the United States, the NERC Critical Infrastructure Protection (CIP) standards 002 through 009 require utilities and other responsible entities to place critical cyber assets within an electronic security perimeter. The electronic security perimeters must be subjected to vulnerability analyses, use access control technologies, and include systems to monitor and log the electronic security perimeter access. The Federal Energy Regulatory Commission (FERC) requires responsible entities involved in bulk electricity transmission to adhere to the NERC CIP standards. So as the EMS evolves with new measurements and analytics, the next generation EMS must carefully consider these emerging cyber-security requirements.

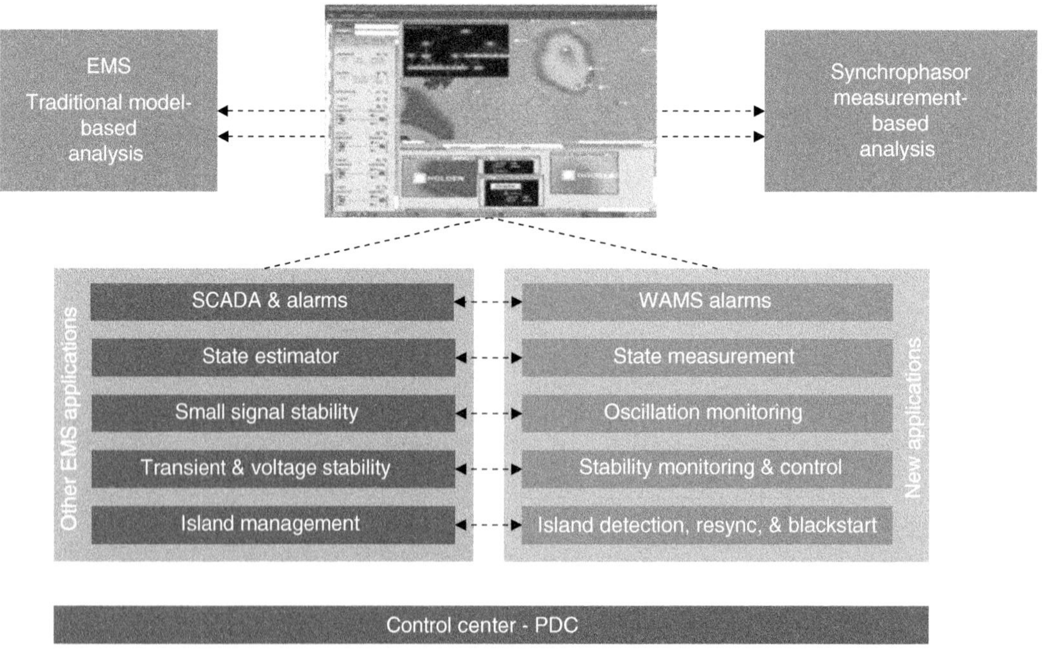


**FIGURE 18-9** Next generation EMS.

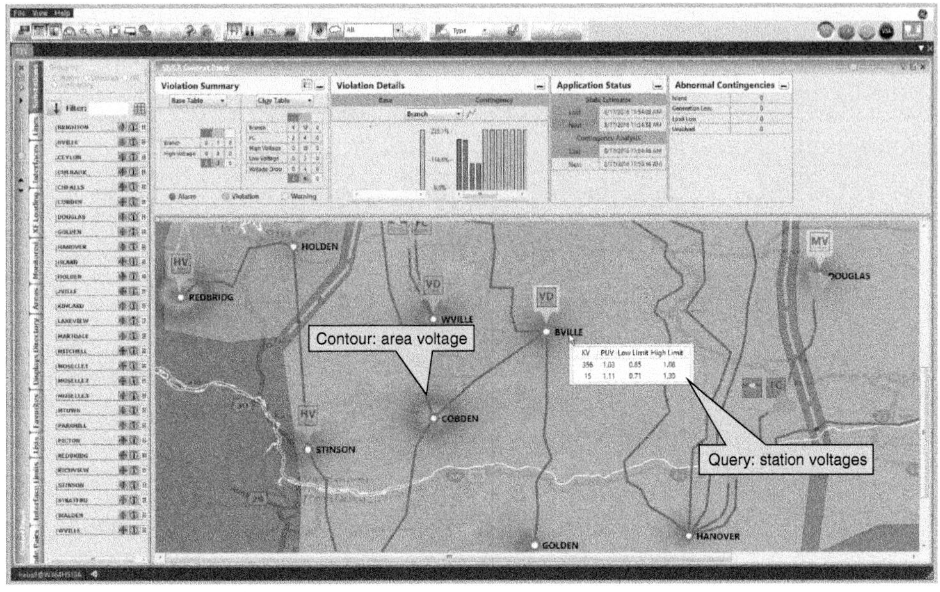


**FIGURE 18-10** Integrated SCADA and GIS displays.

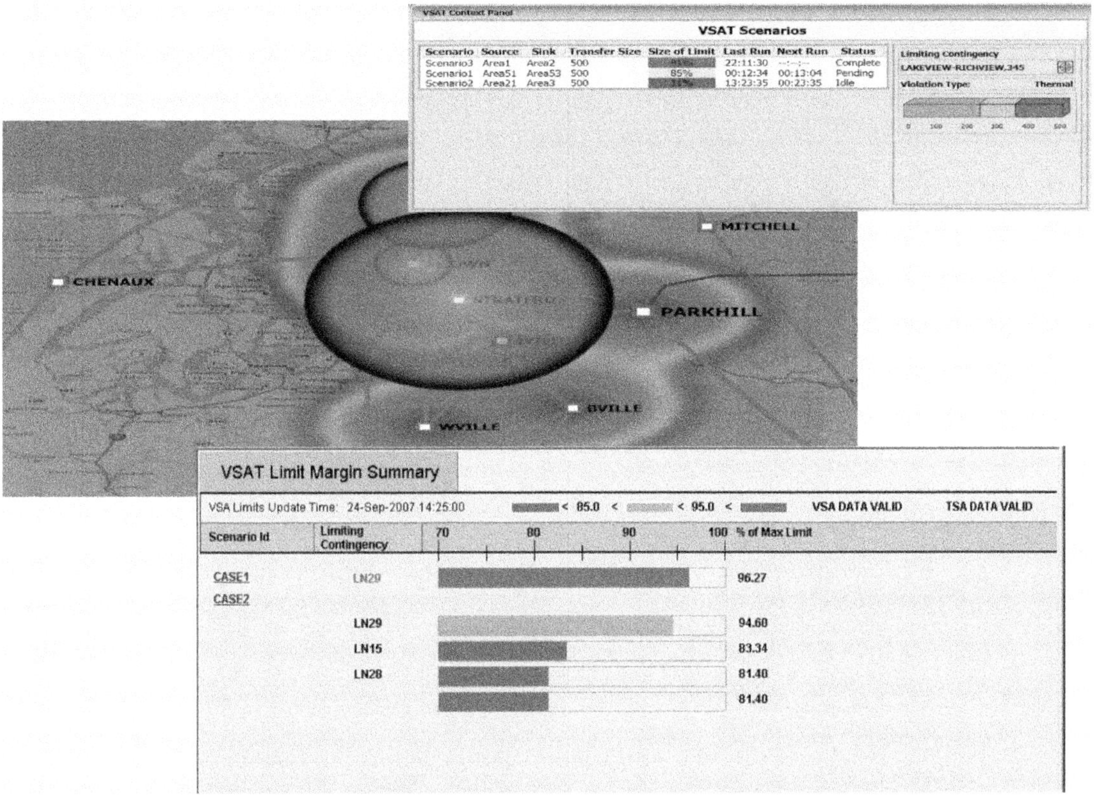

VSAT Context Panel

**VSAT Scenarios**

| Scenario | Source | Sink | Transfer Size | Size of Limit | Last Run | Next Run | Status |
|---|---|---|---|---|---|---|---|
| Scenario3 | Area1 | Area2 | 500 | [illegible] | 22:11:30 | --:--:-- | Complete |
| Scenario1 | Area51 | Area53 | 500 | 85% | 00:12:34 | 00:13:04 | Pending |
| Scenario2 | Area21 | Area3 | 500 | [illegible] | 13:23:35 | 00:23:35 | Idle |

Limiting Contingency
LAKEVIEW-RICHVIEW,345
Violation Type: Thermal

CHENAUX
MITCHELL
PARKHILL
BVILLE
WVILLE

**VSAT Limit Margin Summary**

VSA Limits Update Time: 24-Sep-2007 14:25:00 < 85.0 < < 95.0 < VSA DATA VALID TSA DATA VALID

| Scenario Id | Limiting Contingency | % of Max Limit |
|---|---|---|
| CASE1 | LN29 | 96.27 |
| CASE2 | | |
| | LN29 | 94.60 |
| | LN15 | 83.34 |
| | LN28 | 81.40 |
| | | 81.40 |

**FIGURE 18-11** Voltage stability—locations and controls.

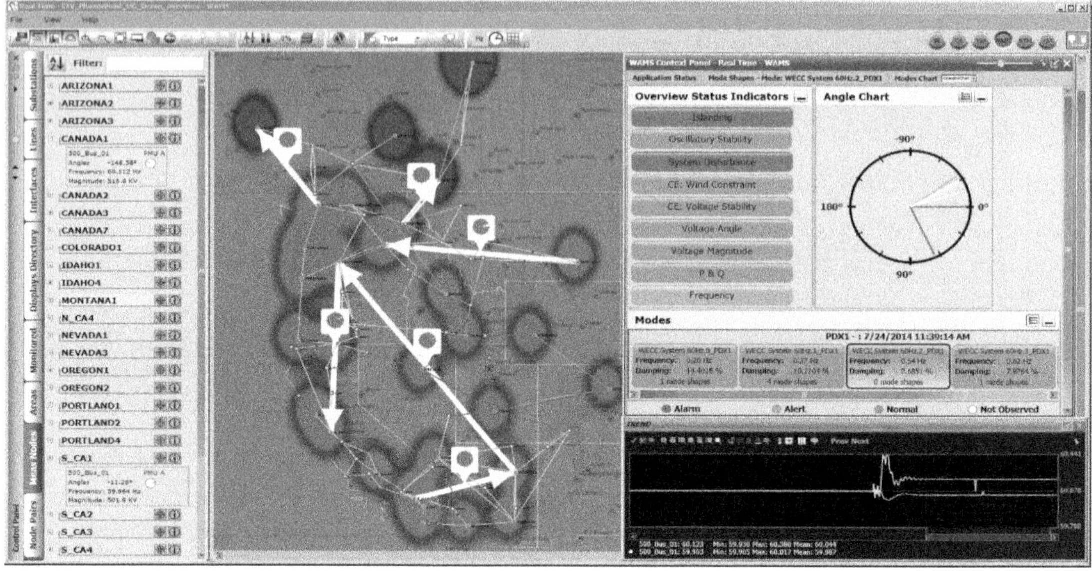


FIGURE 18-12 Visualization of dynamic clusters and frequency.

Future EMS development also needs to be designed to protect against potential network attacks. The primary threats to process control systems include: data injection, command injection, and denial of service.

### 18.4.6 Proactive Grid Management

Most control center operator decisions today are still quite *reactive*. Current information is used to reactively make an assessment of the current state. Operators then mentally extrapolate to make decisions to avoid imminent potential problems. The next step is to help operators make decisions that are *proactive*.

Advances in hardware, software, and visualization technologies are helping us solve larger analytical problems and to visualize them more quickly. This provides an opportunity to perform look ahead predictive analysis in real time in order to "see what is coming up" and to facilitate proactive decision making to avoid potential imminent problems.

### 18.4.7 Summary

When introduced in the 1970s, the EMS was designed to achieve the following grid management objectives: reliability, security, and efficiency. Today the objectives also include resiliency and flexibility—resiliency to withstand unforeseen events such as super storms, etc.—and flexibility to accommodate new types of distributed energy resources and other new stakeholders.

The EMS was originally designed as a single, centralized command and control system, which relied on imperfect real-time measurements that had variable latencies. Today, with the growth of synchrophasor PMU measurements, this premise is no longer valid—we know precisely when they were measured—and we can create a set of perfectly synchronized grid measurements. This facilitates development of a new genre of EMS analytics that utilize perfectly synchronized measurements.

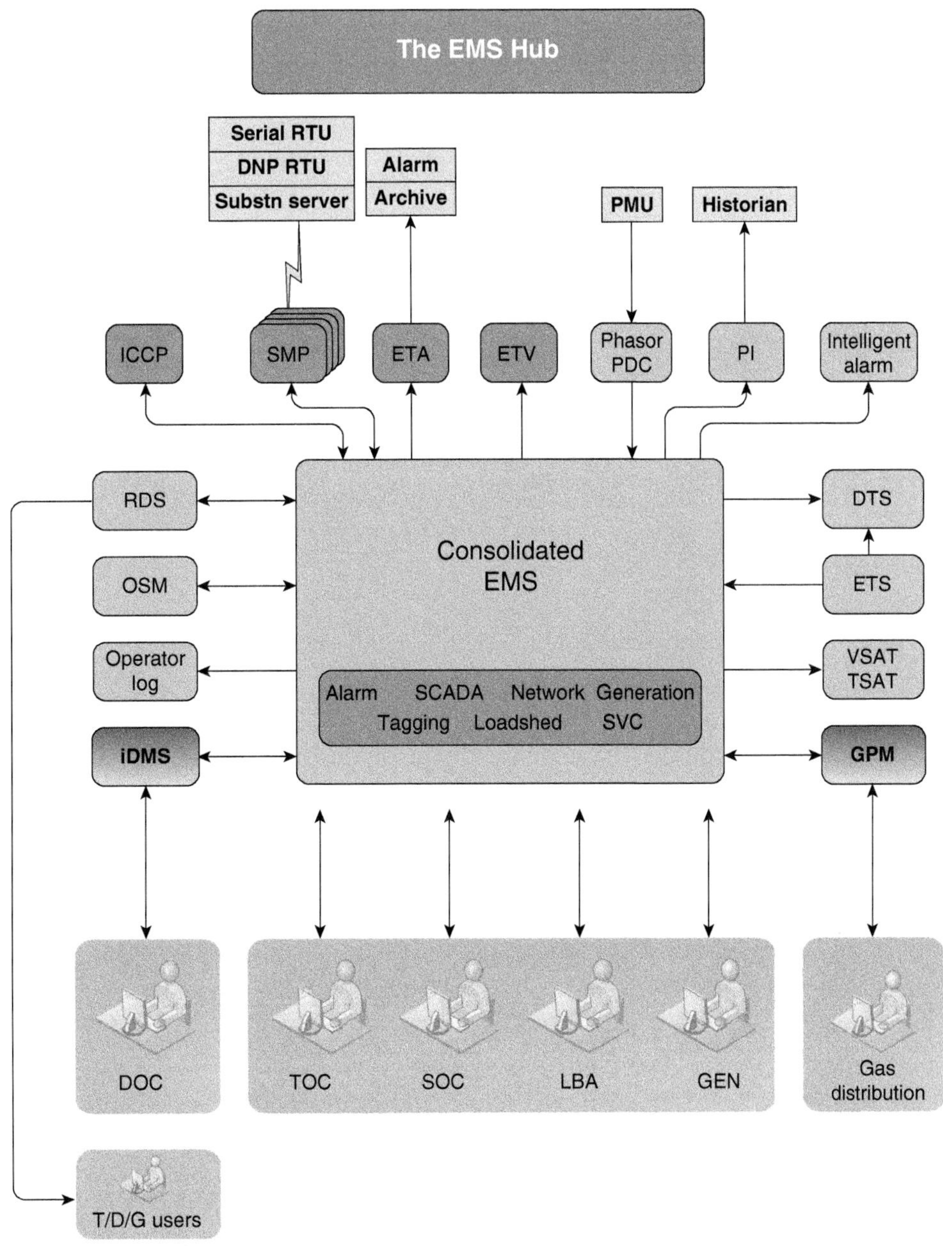


**FIGURE 18-13** The EMS hub.

Many large utilities have been moving toward more decentralized management. For example, independent system operators (ISOs) are comprised of many self-managed transmission owners (TOs). Some large grids have implemented a 3- to 4-level hierarchy of EMS functions—one such is the 4-level Indian national grid.

Substation automation solutions are also growing which ensure better local grid management. Distribution management systems (DMS) are also being deployed to provide active management of the lower-voltage distribution network. This benefits the high-voltage transmission network, since many problems which originate in the low-voltage grid are being managed locally.

With the proliferation of distributed energy resources, microgrids, and decentralized intelligence, the utility business model continues to evolve—and to keep pace, the EMS needs to evolve as well. Regions of the interconnected grid are becoming more autonomous and actively self-managed—hence more predictable—and in turn, grid management will undoubtedly benefit.

Customer engagement and demand response programs are helping flatten the load profile. This is also leading the way to evolve from a traditional generation following grid management paradigm to a load following paradigm.

Advances in hardware, software, and visualization technologies are facilitating faster than real time look ahead predictive analysis, in order to facilitate proactive decision-making to ensure grid reliability.

Today, uncertainties in the grid are growing at a faster pace than ever before. Nevertheless, nascent creative solutions show significant promise to ensure accomplishing our enduring mission "to always keep the lights on!"

## 18.4.8 Acknowledgments

The effort of hundreds of GE Grid Software Solutions staff is gratefully acknowledged—their dedication, passion, and creativity helped develop EMS technology over the past many decades. Utility EMS customers are also acknowledged for providing opportunities to develop the modern-day EMS.

## 18.4.9 Further Reading


1. Giri, J., Podmore, R., et al., "An Advanced Dispatcher Training Simulator," *IEEE PAS Transactions*, PAS-101(1), 17–24, 1982.
2. Phadke, A. G. and Thorp, J. S., "*Synchronized Phasor Measurements and Their Applications*", Springer: New York, 2008.
3. Giri, J., Sun, D., Avila-Rosales, R., "Wanted: A More Intelligent Grid," *IEEE Power & Energy Magazine*, Digital Object Identifier 10.1109/MPE.2008.000000, Jan./Feb. 2009.
4. Giri, J., "System-Wide Computerization of India's Grid Operations," *Homi Bhabha and the Computer Revolution* (editors R. K. Shyamasundar and M. A. Pai), Oxford University Press: Mumbai, 2011.
5. Giri, J., Parashar, M., Trehern, J., Madani, V., "The Situation Room," *IEEE Power & Energy Magazine*, Digital Object Identifier 10-1109/MPE_2012.2205316, Sep./Oct, 2012.
6. Giri, J., "Benefits of Synchrophasors in Operation of the Future Grid," Presentation made at the Cigre US National Committee, 2012 Grid of the Future Symposium, Kansas City, Oct. 2012.
7. Giri, J., Morris, T., Huang, Z., et al., *Transmission Systems, Smart Grids—Infrastructure, Technologies and Solutions* (editor S. Borlase), CRC Press: Boca Raton, FL, pp. 156–162 and pp. 188–210, 2013.
8. Madani, V., Jampala, A., Parashar, M., Giri J., "*Advanced EMS Applications Using Synchrophasor Systems for Grid Operations*," IEEE T&D, Chicago, Apr. 2014.
9. Avila-Rosales, R., Fairchild, R., Giri, J., et al., "*An Advanced Training Simulator for Synchrophasor Applications*," Cigre US, National Committee, 2014 Grid of the Future Symposium, Houston, Oct. 2014.
10. Giri, J., "Proactive Management of the Future Grid", *IEEE Power & Energy Technology Systems Journal* 2, http://ieeexplore.ieee.org/stamp/stamp.jsp?tp=&arnumber=7080837, 2015.

## 18.5 OUTAGE SCHEDULING IN ISO SYSTEMS

*BY SIMON TAM*

### 18.5.1 General Principles

Transmission outage management is very important in maintaining power system reliability. TOs have the right and obligation to maintain and repair their portion of the transmission system. The ISO/RTO has the responsibility to study and to approve transmission facility outage requests prior to removal of the equipment from service. The ISO/RTO has to coordinate scheduled outages of transmission facilities with planned generation outages that may affect power system operation.

The ISO/RTO maintains a list of reportable transmission facilities which TOs are required to obtain outage approval from the ISO/RTO. TOs should submit tentative dates of all transmission outages of reportable transmission facilities to the ISO/RTO as far in advance as possible. The ISO/RTO establishes procedures and timelines for the scheduling, coordinating, studying, approving, and notifying of the transmission outage submitted by TOs. These procedures and timelines are kept up-to-date by periodic reviews and revisions.

Under certain conditions such as extreme weather, peak load, heightened homeland security, etc., the ISO/RTO will evaluate the need to operate the power grid in a more conservative manner. Actions that may be taken under these special circumstances include, but are not limited to, canceling or rescheduling transmission outages and returning out-of-service equipment back to service.

### 18.5.2 Scheduling Transmission Outage Requests

Each transmission owner submits the tentative dates of all planned transmission outages of reportable transmission facilities to the ISO/RTO as far in advance as possible and provide updates on outage dates as soon as practical. The ISO/RTO maintains a planned transmission outage schedule for a period of at least the next 13 months. The latest planned transmission outage schedule is posted online on a public domain, such as the open access same-time information system (OASIS). Planned transmission outages are given priority based on the time of submission. The ISO/RTO periodically reviews all submissions of planned transmission outages and assesses the effect of proposed transmission outages upon the integrated operation of the transmission system using established operating reliability criteria.

When a transmission owner submits an outage request to the ISO/RTO, the request includes the following information:

- Date
- Facility and associated elements
- All line and transformers that will be taken out of service or open-ended as a result of the scheduled outage must be included in the outage request. For example, an outage request for circuit breaker work that open-end a line must include the line as being out of service in the request. This will ensure proper posting of all outages on the ISO/RTO website.
- Planned switching times
- Job description
- Availability/emergency return time

***Transmission Outage Submission Requirements.*** The ISO/RTO must clearly define the transmission outage submission rules which have to be explicitly communicated to all TOs. The rules determine whether an outage request can be approved or not.

Here are some examples of ISO/RTO outage submission rules:

- To obtain "On Time" status for outage duration of 5 days or less, the transmission owner must submit the outage request before the 1st of the month prior to the month of the requested start date.
- To obtain "On Time" status for outage duration of more than 5 days, the transmission owner must submit the outage request before the 1st of the month 6 months in advance of the requested start date.
- "On Time" outage will be approved, provided it does not jeopardize system reliability.
- "Late" outage may be cancelled if it jeopardizes system reliability or causes congestion requiring out-of-merit dispatch operations in some markets.

***Coordinating Outage Requests with Other Transmission Owners.*** In the event that a scheduled outage of one transmission owner affects the availability of another transmission owner's facility, it is the responsibility of the transmission owner initiating the request to notify the affected transmission owner for their consideration. If agreeable to the directly connected transmission owner, the initiating transmission owner may submit the outage request to the ISO/RTO

***Coordinating Transmission Outage Requests with Planned Generation Outages.*** The ISO/RTO and TOs are to coordinate transmission outages with planned outages for generators. The ISO/RTO resolves potential transmission and generation outage conflicts based on system reliability by performing the following actions:

- Review the transmission and generator maintenance schedules to coordinate major transmission and generator outages and communicate with the outage submitting entities in attempting to minimize the anticipated constrained operations
- Recommend adjustments to transmission outage schedules throughout the year to coincide with planned generator outages within the ISO/RTO and surrounding balancing authorities

### 18.5.3 Processing Transmission Outage Requests

The ISO/RTO is required to maintain system reliability under all planned outage conditions. Transmission outage requests are studied to ensure:

- No actual flows exceed normal thermal ratings
- No actual voltages exceed normal voltage limits
- No post-contingency overloads exceed emergency thermal ratings
- No post-contingency voltages exceed emergency voltage limits
- No non-converged contingencies
- All defined transfer limits are observed [A transfer limit is the MW flow limitation across an interface to protect the system from large voltage drops or collapse caused by any viable contingency—in some regions this may be referred to an interconnection reliability operating limit (IROL)]
- No defined stability limits are violated

If it appears that the expected outage will adversely impact system reliability, the ISO/RTO will determine if a better window of opportunity exists for this work to be scheduled. The ISO/RTO will deny the outage request if there are no satisfactory mitigation measures to address all the identified reliability concerns.

If conflicting outages that are submitted for the same timeframe from different TOs, the outage submitted first for that timeframe will have priority.

The ISO/RTO is to communicate to the TOs the final status of all outage requests (either approved or denied) at least a few days before the requested start of the outage.

## 18.6 IMPACT OF REGULATORY ISSUES ON POWER SYSTEM OPERATIONS

*BY MIKE BRYSON*

By now you have read an overview of a few of the aspects of power system operations to include power balance, frequency control, EMS systems, and outage scheduling. This subsection focuses on some of the impacts of regulatory issues on system operations. It explores the more prominent U.S. regulatory agencies but also touches on state regulatory bodies as well as international regulatory bodies. That discussion also includes some of the more conspicuous regulations introduced by those bodies and addresses the impact of some of these regulations on system operations.

One of the more active and relevant U.S. Federal Regulatory Agencies is the FERC which is an independent agency that regulates the interstate transmission of natural gas, oil, and electricity. FERC also regulates natural gas and hydropower projects.[a] Congress established FERC with the Department of Energy Organization Act of 1977[b] in response to a national concern of the United States dependency on foreign energy sources and a general lack of national energy policy.[c] The Department of Energy (DOE) whose mission is to ensure America's security and prosperity by addressing its energy, environmental, and nuclear challenges through transformative science and technology solutions were also established as part of this act.[d]

While FERC has initiated numerous regulations since the years of its establishment, there are a couple of noteworthy orders in the last few decades. In April 1996, FERC issued order 888.[e] In the introduction to the final rule, FERC spelled out its purpose:

> "The legal and policy cornerstone of these rules is to remedy undue discrimination in access to the monopoly owned transmission wires that control whether and to whom electricity can be transported in interstate commerce. A second critical aspect of the rules is to address recovery of the transition costs of moving from a monopoly-regulated regime to one in which all sellers can compete on a fair basis and in which electricity is more competitively priced."

The more significant of the two aspects was the requirement for utilities to file an open access transmission tariff. This fundamentally allowed energy producers and purchasers to conduct energy transactions over the bulk transmission system regardless of who had built the infrastructure or who owned it. While the rule pushed traditional utilities into putting in place processes to make available for sale unused capacity on the transmission system, it also established a significant set of business rules and financial transactions which has impacted system operations for years. The rule required a public website to allow customers to view available transmission capacity and request "transmission service" for purchase. These websites became known as "OASIS" sites. Once transmission service was purchased, it quickly became necessary for system operators to manage the actual day-to-day transactions and their impact on system operations.

This may be a good segue to discuss NERC, which has been mentioned earlier in this section. In June 1968, the National Electric Reliability Council (NERC) is established by the electricity industry in response to the 1965 blackout and the recommendation of the Federal Power Commission, which is the predecessor to FERC. Nine regional reliability organizations are formalized under NERC.[f] Eventually this organization is formalized as the electric reliability organization (ERO) envisioned in the Energy Policy Act of 2005.[g] The new North American Electric Reliability Corporation (NERC) is a not-for-profit international regulatory authority whose mission is to assure the reliability of the bulk power system in North America. NERC develops and enforces reliability standards; annually assesses

[a]http://www.ferc.gov/about/about.asp.
[b]https://www.gpo.gov/fdsys/pkg/STATUTE-91/pdf/STATUTE-91-Pg565.pdf.
[c]http://energy.gov/management/office-management/operational-management/history/brief-history-department-energy.
[d]http://www.energy.gov/about-us.
[e]http://www.ferc.gov/legal/maj-ord-reg/land-docs/rm95-8-00v.txt.
[f]http://www.nerc.com/AboutNERC/Documents/History%20AUG13.pdf.
[g]http://energy.gov/sites/prod/files/2013/10/f3/epact_2005.pdf.

seasonal and long-term reliability; monitors the bulk power system through system awareness; and educates, trains, and certifies industry personnel. NERC is the ERO for North America, subject to oversight by the FERC and governmental authorities in Canada. NERC's jurisdiction includes users, owners, and operators of the bulk power system, which serves more than 334 million people.[h]

Back in 1997, NERC was still a voluntary organization of industry participants, but it developed the initial "etag"[i] concept to allow system operators to manage energy transfers between operating regions. As described above, NERC eventually became the ERO reporting to FERC under Energy Policy Act of 2005. As the ERO, NERC is responsible for mandatory reliability standards, so the voluntary "etag" spreadsheet developed in 1997[j] is now part of the mandatory interchange scheduling and coordination (INT) standards.[k] System operators have been using some form of the NERC tagging system for the past 19 years to plan the system, model the impact of transactions on reliability of the system, monitor transactions in real time, and curtail and reload transactions that have a negative effect on system operations.

In February 2007, FERC issued an additional order—Order Nos. 890 and 890-A and 890-B—as *Improvements in Open Access Transmission, e.g., Transmission Planning (Part 1)* to add additional clarity and standardization to the open access standard it had established in 1996.[l] This order also addressed consistency in transmission planning standards across regions.

Another regulatory agency that impacts system operations is the United States Environmental Protection Agency (EPA).

> "The mission of EPA is to protect human health and the environment. EPA's purpose is to ensure that all Americans are protected from significant risks to human health and the environment based on the best available scientific information; federal laws protecting human health and the environment are enforced fairly and effectively; environmental protection is an integral consideration in U.S. policies concerning natural resources, human health, economic growth, energy, transportation, agriculture, industry, and international trade, and these factors are similarly considered in establishing environmental policy."[m]

The EPA was established on December 2, 1970.[n] From the time of the initiating act, The Clean Air Act of 1970 that established the EPA, the agency has been establishing escalating standards for air and water emissions which govern the way power generation operates on the system. These rules affect the way units operate, particularly in hot weather conditions. For example, the rules limit air emissions, water temperature, and water quality conditions which, depending on the circumstances, impact how and whether generators operate.

In December 2011, the EPA announced standards to limit mercury, acid gases and other toxic pollution from power plants.[o] This rule defined pollution limits that individual facilities must meet by a set date. These rules set technology-based emissions limitation standards for mercury and other toxic air pollutants, reflecting levels achieved by the best-performing sources currently—coal- and oil-fired electric generating units (EGUs) with a capacity of 25 MW or greater. The EPA gave existing generators up to 4 years to comply with Mercury Air Toxic Standards (MATS).

> "From data reports provided to the Energy Information Administration (EIA), about 16 gigawatts of generating capacity will be retired in 2015, of which nearly 13 gigawatts is coal-fired. The coal-fired capacity will be retired primarily because of EPA's Mercury and Air Toxics Standards (MATS), which requires coal- and oil-fired electric generators to meet stricter emissions standards by incorporating emissions control technologies or retire the generators."[p]

[h]http://www.nerc.com/Pages/default.aspx.
[i]http://www.nerc.com/docs/oc/is/Interchange_Reference_Guidlines_V2_2012_02_17_Final.pdf.
[j]http://www.nerc.com/comm/OC/Operating%20Manual%20DL/opman_20140825.pdf.
[k]http://www.nerc.net/standardsreports/standardssummary.aspx.
[l]http://www.ferc.gov/industries/electric/indus-act/oatt-reform.asp.
[m]https://www.epa.gov/aboutepa/our-mission-and-what-we-do.
[n]https://www.epa.gov/aboutepa/origins-epa.
[o]https://www3.epa.gov/mats/actions.html.
[p]http://instituteforenergyresearch.org/analysis/eia-13-gigawatts-of-coal-capacity-to-retire-in-2015-due-to-epa-regulation/.

As a result of this rule and other factors—such as innovative extraction methods that allow for the cheap recovery of Shale natural gas—an unprecedented quantity of generation has retired. The impact of the retirement of this magnitude had a significant effect on system operations in the regions where these units retired. While ultimately the regions were able to be operated reliably, new rules to compensate for the loss of some of the reliability parameters of the retiring units needed to be added. Much of the replacement generators were renewable or gas-fired generation.[q] Renewable generation is an intermittent energy source—only providing needed electricity during certain hours of the day, and additional gas generation added pressure to the gas pipeline infrastructure during peak energy demand periods such as the winter months. System operators needed to make adjustments in the way units are scheduled and cycled throughout the days given the changing dynamics of the generation mix.

In the coming decades, the generation mix may also be affected by the EPA's Clean Power Plan (CPP).

> "On August 3, 2015, President Obama and EPA announced the Clean Power Plan – a historic and important step in reducing carbon pollution from power plants that takes real action on climate change. Shaped by years of unprecedented outreach and public engagement, the final Clean Power Plan is fair, flexible and designed to strengthen the fast-growing trend toward cleaner and lower-polluting American energy. With strong but achievable standards for power plants, and customized goals for states to cut the carbon pollution that is driving climate change, the Clean Power Plan provides national consistency, accountability and a level playing field while reflecting each state's energy mix. It also shows the world that the United States is committed to leading global efforts to address climate change."[r]

While the CPP rule is still under review by the Supreme Court, many states have already initiated the process to develop plans to address the targets by the year 2030. This rule is likely to also drive a number of both retirements and new entry in power generation which will in turn require system operators to adapt to changes in future resource mixes.

State regulatory agencies also affect system operations. One example of this is the Public Utilities Commission of Ohio (PUCO) which

> "affects every household in Ohio. That's because the PUCO regulates providers of all kinds of utility services, including electric and natural gas companies, local and long distance telephone companies, water and wastewater companies, rail and trucking companies. The PUCO was created to assure Ohioans adequate, safe and reliable public utility services at a fair price. More recently, the PUCO gained responsibility for facilitating competitive utility choices for Ohio consumers."[s]

Most states rely on utilities commission. These commissions are responsible for a number of areas including electricity. While FERC governs electricity at the wholesale level, the states make the rules at the residential, retail, and distribution levels. This could include areas such as competitive retail electric service providers, interconnection services, electric reliability, safety and customer service standards enforcement, and certification to operate as a provider of competitive retail electric services, etc.[t]

Most of this discussion has dealt with regulations from federal and state agencies in the United States. Similar regulations promulgated by parallel state and federal utility commissions and agency exist around the world, all effecting the way operators have to run their systems on a daily basis. One example of this is the Australian Energy Market Commission (AEMC). The AEMC is

> "The rule maker for Australian electricity and gas markets. We make and amend the National Electricity Rules, National Gas Rules and National Energy Retail Rules. We also provide market development advice to governments... The Reliability Panel forms part of the AEMC's institutional arrangements that

[q]http://www.eia.gov/todayinenergy/detail.cfm?id=20292.
[r]https://www.epa.gov/cleanpowerplan/fact-sheet-overview-clean-power-plan.
[s]http://www.puco.ohio.gov/puco/index.cfm/how-the-puco-works-for-you/#sthash.neAZ54E3.dpbs.
[t]http://www.puco.ohio.gov/puco/index.cfm/rules/#sthash.Cp9QRE7g.dpbs.

support the national electricity system. The Panel's core functions relate to the safety, security and reliability of the national electricity system."[u]

This has been a brief overview of just a few of the regulations and regulatory agencies which have affected system operators at the wholesale or bulk level over the years. Additional details for the agencies and regulations are available at some of the referenced websites.

Regulations are adopted by government agencies for a myriad of reasons—safety, jobs, industry growth, defense, and many others. Regulations that are addressed at the energy industry are bound to have a direct impact on some aspects of system operations whether it be the resource mix to serve energy demand, or the energy market rules which provide incentives for resources to respond to changes in system conditions.

## 18.7 INTERNATIONAL EXPERIENCES

### 18.7.1 Power System Operation Practices in European Countries

*BY PATRICK PANCIATICI AND FEDERICO MILANO*

***European Power System.*** The European power system is one of the largest power systems in the world. It has been built and improved during the last 100 years, even though the period of maximum expansion began after WWII. The system is characterized by a vast diversity in the generation mix, the voltage levels, and the structure of the grid, which is densely meshed in some areas and mostly radial in some peninsulas.

In recent years, the transition toward a low-carbon economy associated with the shutdown of nuclear power plants has led to rapid changes in the generation mix all over Europe. This trend is clearly shown by some statistics related to the European Union section of the grid (EU-28,[v] Source: Eurostat), as reported in the subsections below.

*Generation Mix in Europe (EU-28).* EU-28 total electricity generation in 2013 was 3262 TWh, which is a 1.1% decrease compared with the same figure in 2012. For the first time, renewable energy resources, with a total production of 890 TWh, have the highest share in electricity production (27.2%), followed closely by nuclear power plants (26.9%) and coal fired plants (26.7%) then gas (16.6%), oil (1.9%) and non-renewable waste (0.8%).

Over the last two decades, there have been significant changes in the structure of renewable energy resources used for electricity production. In 1990, 94.2% of electricity obtained from renewable sources was produced from hydro power plants, while in 2013 the share of hydro energy was less than half of that. In 2013, the figures of renewable energy sources used for electricity production were 45.4% hydro, 26.5% wind, 9.2% solid biofuel, 9.1% solar PV, 6.0% biogas, 2.1% municipal renewable waste, 0.7% geothermal energy, and 1% other sources.

The trend in electricity production of nuclear power plants shows a moderate increase from 1999 to 2004. From 2004 to 2013, however, the production of nuclear power fell by 13.0%. Germany accounted for the sharpest decrease during that period (−41.8%), as several nuclear power plants were shut down. In the same period, electric energy produced from coal, which had been decreasing since the 90s, rose by 9.7% from 2009 to 2012 and experienced a decrease of 3.3% in 2013.

The installed electrical capacity increased by 70% in the period from 1990 to 2013. The figures of the installed capacity changed significantly over this period. Comparing the situation in 2013 with previous decades, the share of installed capacity of combustible fuels decreased to 50%, the

[u]http://www.aemc.gov.au/.

[v]EU-28 consists of the following countries: Austria, Belgium, Bulgaria, Croatia, Cyprus, Czech Republic, Denmark, Estonia, Finland, France, Germany, Greece, Hungary, Ireland, Italy, Latvia, Lithuania, Luxembourg, Malta, Netherlands, Poland, Portugal, Romania, Slovakia, Slovenia, Spain, Sweden, United Kingdom.

share of hydro to 16% and the share of nuclear to 13%. On the other hand, the share of wind increased to 12% and the share of solar to 9%, while geothermal and tide, wave and ocean have remained negligible.

*Electricity Consumption (EU-28).* During the last decades, the European electricity consumption in the EU-28 has significantly increased (32.5%) in the period from 1990 to 2008. In 2009, due to the worldwide financial and economic crisis, consumption decreased by 5.2% but recovered immediately in 2010 almost back to the 2008 level. Overall, in the period from 1990 to 2013, electricity consumption increased by 28.1%. Residential and services sectors are the main responsible for this growth. It is expected, however, that two aspects will affect this trend in the next decade.

On one hand, strong incentives and policies to improve the efficiency of consumption processes, e.g., building insulation, and optimization of industrial process. On the other hand, the gradual substitution of fossil-fuel with electricity, e.g., electrical transportation and building heating. The resulting trend is not easy to anticipate, thus further complicating generation and grid planning.

*Exchanges of Electricity.* The creation of the internal energy market for electricity and gas in EU-28 has fostered the exchanges of electricity between European countries. As a consequence, cross border capacities have been potentiated and better utilized through shorter-term grid capacity allocation mechanisms. In the past, the interconnections were utilized mainly to ensure the reliability of the European power system and energy was exchanged based on long-term constant contracts. In 2013, Italy was the largest net importer; France and Germany were the main net exporters (see Fig. 18-14).

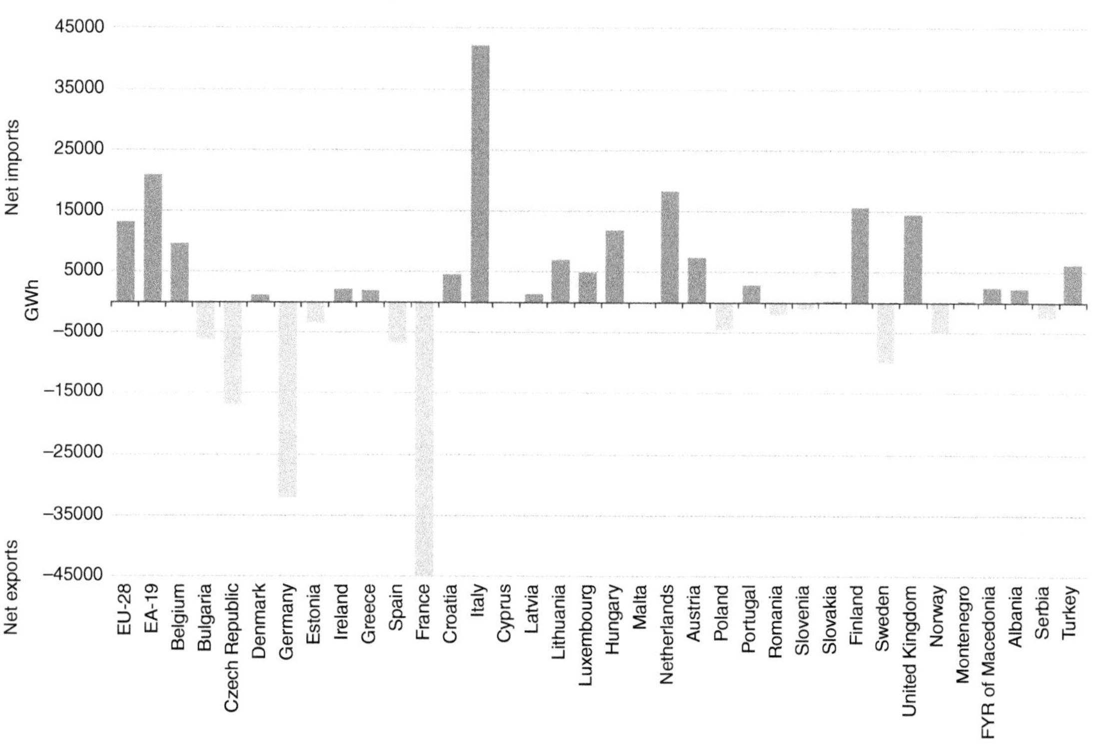


**FIGURE 18-14** Electricity net imports by country in 2013. (*Eurostat.*)

The yearly generation exported and consumption imported by each EU-18 countries in 2014 are shown in Figs. 18-15 and 18-16, respectively. These quantities are defined as follows.

*Yearly Generation Exported.* The share of generation of a country which is physically exported by a country to its neighbors is the ratio between the country's net exports and generation.

*Yearly Consumption Imported.* The ratio between its imports and its consumption represents the share of its yearly consumption which is covered by physical imports from its neighbors.

As it can be observed in Figs. 18-15 and 18-16, in 2014, 10 countries of ENTSO-E's perimeter export more than 10% of their annual national generation to neighboring countries; whereas 11 countries of ENTSO-E import more than 10% of their annual internal consumption from other ENTSO-E countries.

The situation is evolving rapidly and the volatility of exchanges increased with the integration of more and more renewable energy sources (wind, solar) in the system. Figure 18-17 shows the exchanges between France and Germany. During sunny days in the south of Germany; Germany exports around noon and imports during the night.

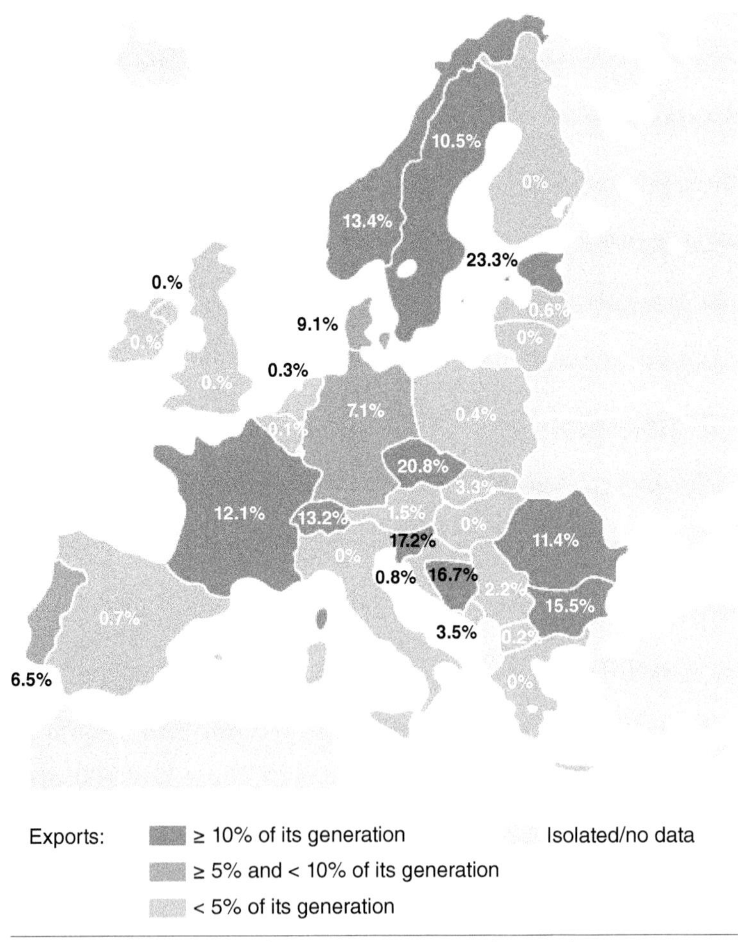


**FIGURE 18-15** Share of yearly generation exported in 2014 by EU-28 countries. (*ENTSO-E.*)

**Share of yearly consumption imported**

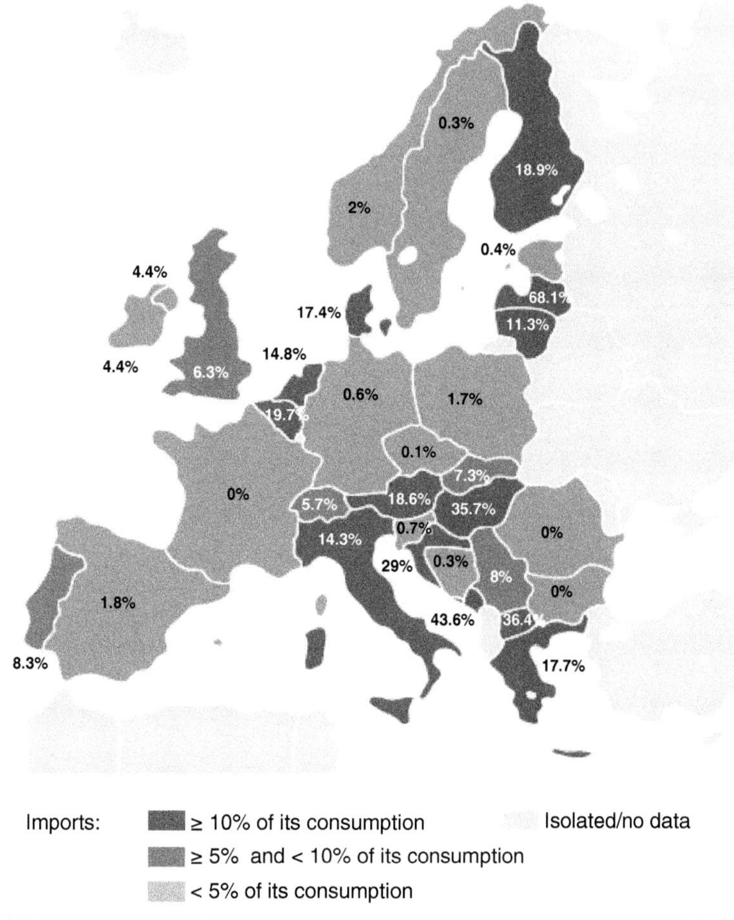


**FIGURE 18-16** Share of yearly consumption imported in 2014 by EU-28 countries. (*ENTSO-E.*)

*Synchronous Areas.* There are four synchronous areas in Europe. The nominal frequency is 50 Hz in all areas.

1. The continental Europe area (CE) includes part or all of Austria, Belgium, Bosnia and Herzegovina, Bulgaria, Croatia, Czech Republic, Denmark (western part), France, Germany, Greece, Hungary, Italy, Luxembourg, Macedonia, Montenegro, the Netherlands, Poland, Portugal, Romania, Serbia, Slovakia, Slovenia, Spain, and Switzerland.
2. Great Britain (GB) alone constitutes a synchronous area.
3. The all-island Irish system (IRE) composed of Republic of Ireland and Northern Ireland is another synchronous area.
4. The synchronous inter-Nordic system (NE) includes the transmission grids of Sweden, Norway, eastern Denmark, and Finland.

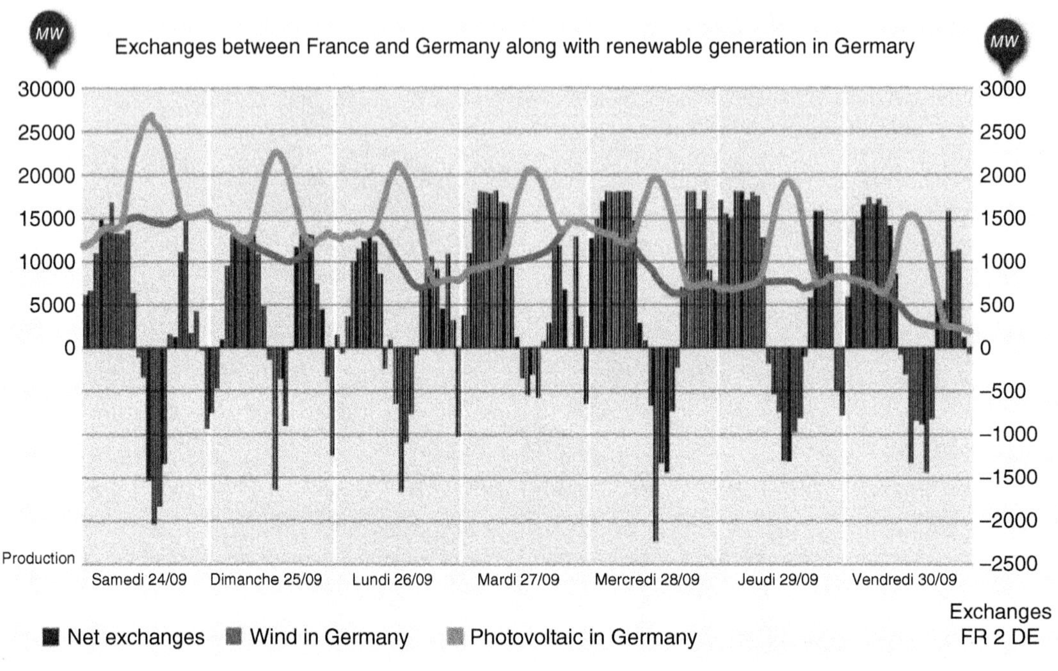


**FIGURE 18-17** Power exchanges between France and Germany. (*RTE.*)

A small west electricity island of Ukraine is synchronized with the grid of CE. Albania is operating the national grid synchronously with the synchronous grid of CE. Note that also the grids of Morocco, Algeria, and Tunisia are synchronized with the European grid through the Gibraltar ac link. Finally, the grid of Turkey was synchronized to CE in April 2015.

Despite being EU Member States for more than a decade, Lithuania, Latvia, and Estonia, have remained a part of the huge synchronous Russian Power System (IPS/UPS). In December 2015, two HVDC interconnections connecting Lithuania to Poland and Sweden have been officially inaugurated.

*ENTSO-E.* ENTSO-E is the association of European TSOs for electricity (see Fig. 18-18). The TSOs are entities that operate independently from other electricity market players and are responsible for the bulk transmission of electric power on the main high-voltage transmission networks. TSOs provide grid access to the electricity market players (i.e., generating companies, traders, suppliers, distributors, and directly connected customers) according to non-discriminatory and transparent rules. In order to ensure the security of supply, they also guarantee the safe operation and maintenance of the system. In Europe, TSOs are generally owners of grid's assets and in charge of the development of the grid infrastructure.

*Diversity of European Power Systems.* There are significant differences between the grids of each country in Europe. These differences are not limited to the geographical diversity (e.g., coast lines, plains, and mountains) and meteorological conditions (e.g., Arctic, Continental, Atlantic, and Mediterranean), which are clearly not identical all over the continent and power systems have been designed taking into account these constraints. The urbanization and the density of population are also quite different, leading again to different choices for the grid design. Finally, the variety of

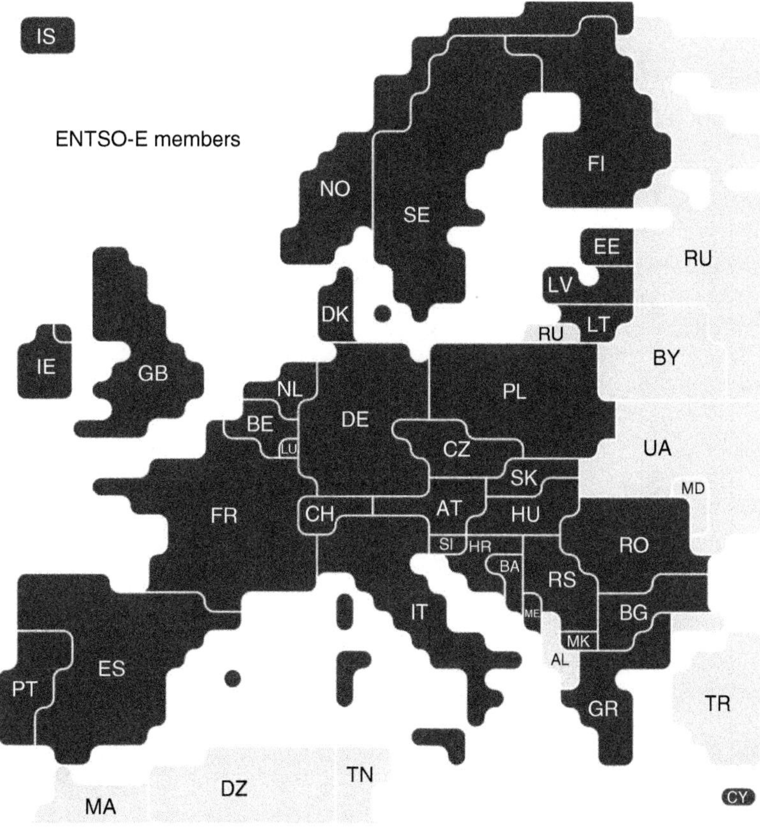


**FIGURE 18-18** The 42 TSOs from 35 countries that are members of ENTSO-E. (*ENTSO-E.*)

national energy policies amplifies even more these differences. Table 18-1 shows statistics illustrating this large diversity. Note that small generating units (<0.5 MW) are not included in the statistics of Table 18-1.

There is an obvious correlation between the total generating capacity, GDP and population per country as shown in Fig. 18-19. However, the total length of electrical circuits over 200 kV per country is not well correlated with generation capacity per country (see Fig. 18-20). Table 18-2 provides some figures for most of the European countries. Note that there is a bias in this analysis because in some countries there is a large portion of 150 kV grids which is in fact a part of the transmission grid but it is difficult to define a clear limit between transmission and distribution grid, as discussed in the following subsection.

*Distribution Grid vs. Transmission Grid.* In Europe, there are around 2400 electricity distribution companies with 260 million customers covering 99% of residential customers and small businesses. However, the definitions of distribution and transmission grids are not consistent all over the continent. While it is commonly accepted as a discriminating parameter, the voltage level is actually not the unique feature that defines the limit between distribution and transmission grid. Moreover, the organization of the electrical sector of each European country had largely impacted the national definition of this limit. Another commonly used criterion to distinguish between transmission and distribution system is the fact that, historically, distribution grids are generally passive and operated in

**TABLE 18-1** Diversity of the European Power Systems (*EUROSTAT, ENTSO-E, and Wikipedia*)

| | | | | | | Net generating capacity on December 31st 2015 (MW) | | | | | | | |
|---|---|---|---|---|---|---|---|---|---|---|---|---|---|
| ISO country code | EU28 | Country name | Population | Area (km²) | GDP (M. US$) | Hydro | Nuclear | Wind | Solar | Biomass | Fossil fuels | Others | Total (G) |
| AL | N | Albania | 2 886 026 | 28 748 | 34 282 | 1 428 | | 42 | | | 186 | | 1 656 |
| AT | Y | Austria | 8 581 500 | 83 855 | 437 582 | 13 569 | | 1 981 | 404 | 417 | 7 540 | 315 | 24 226 |
| BA | N | Bosnia and Herzegovina | 3 871 643 | 51 197 | 41 127 | 2 060 | | | | | 1 578 | | 3 638 |
| BE | Y | Belgium | 11 258 434 | 30 528 | 534 230 | 1 424 | 5 926 | 2 172 | 2 958 | 1 210 | 7 599 | | 21 289 |
| BG | Y | Bulgaria | 7 202 198 | 110 994 | 55 824 | 3 198 | 2 000 | 701 | 1 041 | 64 | 5 706 | | 12 710 |
| CH | N | Switzerland | 8 211 700 | 41 285 | 493 126 | 3 350 | 3 220 | 110 | 1 360 | | 88 | | 8 128 |
| CY | Y | Cyprus | 847 008 | 9 251 | 23 263 | 0 | | 155 | | | 1 478 | | 1 633 |
| CZ | Y | Czech Republic | 10 419 743 | 78 866 | 205 270 | 2 253 | 4 040 | 277 | 2 067 | 850 | 11 140 | | 20 627 |
| DE | Y | Germany | 81 089 331 | 357 021 | 3 874 437 | 9 610 | 10 793 | 43 429 | 38 411 | 6 915 | 78 746 | | 187 904 |
| DK | Y | Denmark | 653 357 | 43 075 | 342 362 | 7 | | 5 082 | 781 | 510 | 7 540 | 2 | 13 922 |
| EE | Y | Estonia | 1 313 271 | 45 227 | 26 506 | 8 | | 301 | | 101 | 2 575 | | 2 984 |
| ES | Y | Spain | 46 439 864 | 504 030 | 1 406 538 | 20 336 | 7 573 | 23 003 | 6 967 | 747 | 46 974 | 588 | 106 188 |
| FI | Y | Finland | 5 471 753 | 338 424 | 272 649 | 3 263 | 2 752 | 1 082 | | 1 919 | 8 420 | 245 | 17 681 |
| FR | Y | France | 66 352 469 | 643 801 | 2 833 687 | 25 421 | 63 130 | 10 312 | 6 191 | 1 703 | 22 553 | | 129 310 |
| GB | Y | Great Britain | 62 661 449 | 222 058 | 2 906 607 | 3 866 | 8 981 | 8 503 | | 1 377 | 47 914 | | 70 641 |
| GR | Y | Greece | 10 846 979 | 131 990 | 237 970 | 3 242 | | 1 775 | 2 444 | 52 | 9 922 | 135 | 17 570 |
| HR | Y | Croatia | 4 225 316 | 56 594 | 57 073 | 2 112 | | 384 | 44 | 26 | 1 770 | | 4 336 |
| HU | Y | Hungary | 9 855 571 | 93 030 | 136 989 | 57 | 1 887 | 328 | 29 | 246 | 5 541 | 88 | 8 176 |
| IE+NI | Y | Ireland | 6 731 551 | 91 825 | 294 246 | 534 | | 2 400 | | 7 | 6 409 | 251 | 9 601 |
| IS | N | Iceland | 332 529 | 102 775 | 14 488 | 1 972 | | 2 | | | 660 | | 2 633 |
| IT | Y | Italy | 61 438 480 | 301 338 | 2 147 744 | 22 150 | | 8 750 | 19 100 | 3 700 | 66 020 | | 119 720 |
| LT | Y | Lithuania | 2 921 262 | 65 200 | 48 288 | 126 | | 290 | 69 | 77 | 2 332 | 900 | 3 794 |
| LU | Y | Luxembourg | 562 958 | 2 586 | 65 683 | 1 334 | | 60 | 116 | 11 | 495 | 21 | 2 037 |
| LV | Y | Latvia | 1 986 096 | 64 589 | 31 972 | 1 556 | | 70 | 1 | 113 | 1 144 | | 2 884 |
| MK | N | Republic of Macedonia | 2 069 162 | 25 713 | 22 147 | 634 | | 37 | | | 800 | | 1 471 |
| NL | Y | Netherlands | 17 155 169 | 41 543 | 880 716 | 38 | 486 | 3 641 | 1 429 | 398 | 25 572 | 674 | 32 238 |
| NO | N | Norway | 9 875 378 | 385 178 | 363 290 | 31 200 | | 860 | | 32 | 1 600 | | 33 692 |
| PL | Y | Poland | 38 005 614 | 312 685 | 547 894 | 2 355 | | 5 186 | 87 | 612 | 29 232 | | 37 472 |
| PT | Y | Portugal | 10 374 822 | 92 390 | 229 948 | 6 146 | | 4 826 | 429 | 613 | 6 519 | | 18 533 |
| RO | Y | Romania | 19 861 408 | 238 391 | 199 093 | 6 339 | 1 300 | 2 923 | 1 249 | 112 | 8 496 | | 20 419 |
| RS | N | Serbia | 7 041 599 | 88 361 | 100 180 | 3 009 | | | | | 5 549 | | 8 558 |
| SE | Y | Sweden | 9 790 000 | 449 964 | 570 591 | 5 886 | 10 000 | 3 745 | 79 | 822 | 2 280 | | 22 812 |
| SI | Y | Slovenia | 2 062 874 | 20 273 | 49 570 | 1 233 | 696 | 3 | 263 | 17 | 1 470 | 40 | 3 722 |
| SK | Y | Slovakia | 5 403 134 | 49 035 | 99 869 | 2 533 | 1 940 | 3 | 532 | 259 | 2 718 | | 7 985 |

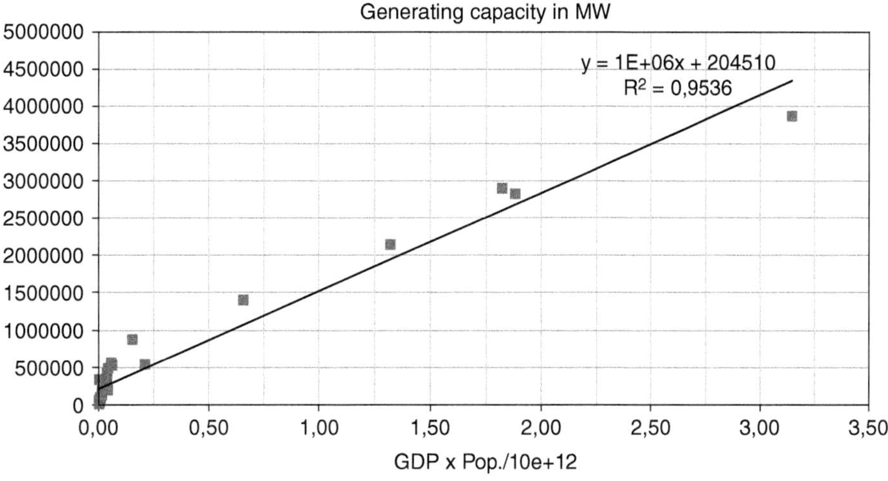


**FIGURE 18-19** Total generating capacity vs. GDP per population.

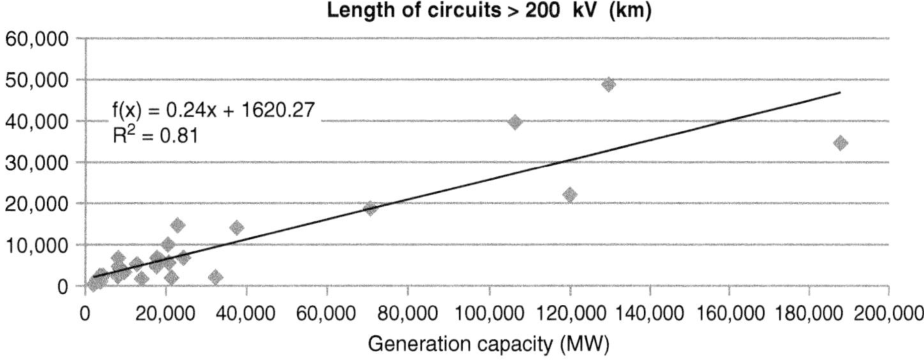


**FIGURE 18-20** Correlation between generation capacity and length of circuits.

radial mode, while transmission systems are active and meshed. However, due to the high penetration of renewable sources, this distinction is not valid anymore.

This has led to the current situation where most European DSOs partly operate high voltage networks as well. The DSOs of only six EU member states—namely, Cyprus, Estonia, France, Italy, Lithuania, and Latvia—operate exclusively low and medium voltage lines. Figure 18-21 illustrates the voltage levels operated by DSOs and/or TSOs in the European power system.

***Power System Operation in European Countries.*** The objective of power system operation (PSO) is to ensure a secure, sustainable, and efficient electrical supply for all the European citizens. That means the best possible utilization of existing grid assets allowing the usage of the most efficient and sustainable electrical generations. Since the grids have technical limitations and finite capacities, as it would be economically inefficient to design grids with no limitations, the PSO effectively consists in finding the trade-off between efficiency, security and economic dispatch. PSO is also essential to

**TABLE 18-2** Lengths of Circuits per EU-28 Countries (*ENTSO-E*)

| Lengths of circuits > 200 kV on December 31st 2015 (km) | | |
|---|---|---|
| ISO country code | L: Total length (km) | G: Total generation capacity (MW) |
| AT | 6 729 | 24 226 |
| BA | 2 390 | 3 638 |
| BE | 1 929 | 21 289 |
| BG | 5 256 | 12 710 |
| CH | 6 700 | 8 128 |
| CZ | 5 526 | 20 627 |
| DE | 34 615 | 187 904 |
| DK | 1 641 | 13 922 |
| EE | 1 692 | 2 984 |
| ES | 39 613 | 106 188 |
| FI | 6 690 | 17 681 |
| FR | 48 723 | 129 310 |
| GB | 18 630 | 70 641 |
| GR | 4 698 | 17 570 |
| HR | 2 458 | 4 336 |
| HU | 4 640 | 8 176 |
| IE+NI | 3 275 | 9 601 |
| IS | 857 | 2 633 |
| IT | 21 931 | 119 720 |
| LT | 1 812 | 3 794 |
| LU | 259 | 2 037 |
| LV | 1 360 | 2 884 |
| NL | 2 000 | 32 238 |
| PL | 13 990 | 37 472 |
| PT | 6 147 | 18 533 |
| RO | 9 966 | 20 419 |
| RS | 3 975 | 8 558 |
| SE | 14 567 | 22 812 |
| SI | 997 | 3 722 |
| SK | 2 332 | 7 985 |

cope with low frequency stressed conditions in particular during planned outages and to maintain grid assets. PSO is based on the definition of reliability criteria, as defined below.

*Reliability Criteria.* Power systems are affected by external factors (threats) including:

- Natural hazards
- Human/technical errors
- Terrorist attacks, sabotages

These threats are not very likely but their impact is unacceptable (i.e., large and long lasting blackouts) for our modern societies, which is more and more dependent on electrical supply. Clearly, it is impossible to design and to operate a power system in order to cope with all the possible threats so that they do not have any impact on the electrical supply. The well-accepted N-1 contingency criterion states that the system should be able to withstand at all times a credible contingency, that is, unexpected failure or outage of a system component (such as a line, transformer, or generator—in such a way, that the system is capable of accommodating the new operational situation without violating operational security limits. Credible contingencies are, in this case, the outage of a single component. This criterion is based on the outage of a component, regardless the threat causing it. For example, in case of a permanent short circuit (the threat is generally lightning), local protections open breakers to isolate the short circuit causing the outage of a power line.

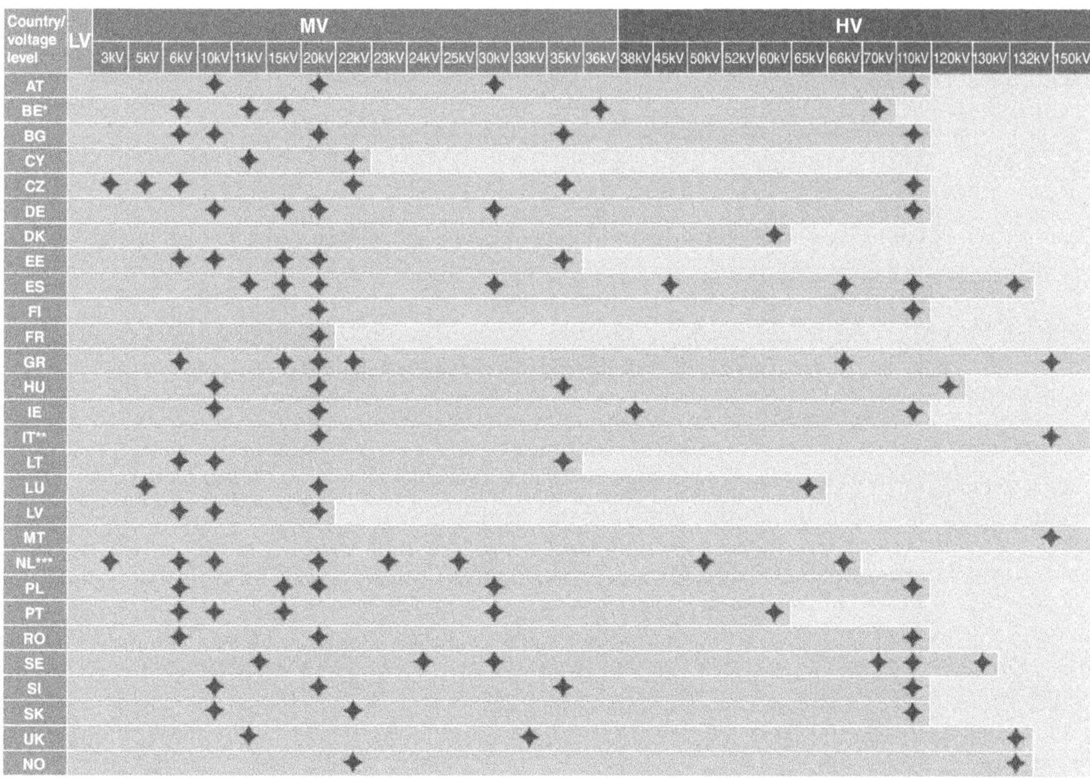


**FIGURE 18-21** Voltage levels operated by DSOs and TSOs in the European power system. Asterisks in the figure indicate the actual utilized voltage levels. (*Eurelectric*.)

In the past, the N-1 criterion was implemented in a purely preventive way. The operating conditions of the power system were selected in order to cope with all credible contingencies without taking any action immediately after the occurrence of a failure or an outage. Such an implementation is not optimal and indeed one needs to make costly decisions (increase of the grid losses, adjustment of generation, postponing of maintenance works, etc.) even if the failure/outage never occurs. The substation automation and advanced ICT system (SCADA/EMS) in control centers allowed moving towards a smarter implementation using corrective actions (remedial actions) through local automated controls or simple rules in control centers.

The N-1 criterion is a very basic risk based approach since it does not consider explicitly the probability of failure/outage (fuzzy concept of credible contingency) and assuming that the state of the system is perfectly known even for day-ahead assessment. With the massive integration of renewables, in the European System, the N-1 criterion must be reviewed and this can be done by taking into account the joint probability of a failure/outage and uncertainties on the system state.

The concept of joint probability is better explained through an example. Let us consider a wind farm with 100 MW of installed capacity and assume that this power plant will never produce more than 30 MW in average over a year. If the N-1 contingency criterion were strictly enforced, the associated upgrade of the grid should imply to build two power lines of 100 MW capacity each to properly accommodate the maximum generation of the wind power plant. In case of line outage, in fact, the other line could stand the total power flow. However, this solution is clearly non-economical as the transmission line would be underutilized for the majority of the time.

A much more reasonable solution is to build two power lines of 50 MW each. The probability to have an outage of power line simultaneously with a wind speed larger than the average annual value, in fact, is very low. This is only a simple example and complex systems require sophisticated analysis.[w] Reliability risk analysis is so important that ENTSO-E as well as most European TSOs have defined their own criteria. Some of these are described in the following subsections.

*N-1 Criterion: ENTSO-E Rules and Some TSO Practices.* ENTSO-E defines general principles regarding "reliability criteria" focusing on coordination. For Central Continental Europe, the document UCTE[x] OH—Policy 3: Operational Security—provides a set of rules, some of which are summarized below. The rules reported in the remainder of this subsection are "mandatory" for all the TSOs of Central Continental Europe.

*N-1 Contingency Criterion—ENTSO-E Policies.* *A1-S1.* Any event of the contingency list (normal and exceptional types of contingencies considered in the contingency list) must not endanger the security of interconnected operation. After any of these contingencies the operational condition within the TSO's responsibility area must not lead to the triggering of an uncontrollable cascading outage propagating across the borders or having an impact outside the borders of the TSO itself—which can be summarized as the "no cascading with impact outside my borders" principle.
Following conditions must be fulfilled after the implementation of *remedial actions*[y]:

- *A1-S1.1. Power Flow Pattern Within Security Limits.* All current values of individual network elements of a responsibility area must remain under control avoiding the impact of cascading effects outside the area.
- *A1-S1.2. Voltage Deviation.* Any contingency of the contingency list should not cause a voltage drop outside acceptable operating limits within the TSO's responsibility area, which can initiate voltage collapse and cascading outages with impact outside the area.
- *A1-S1.3. Locally Limited Consequences.* As long as the secure operation of the interconnected system is ensured, locally limited and predictable consumption outages can be tolerated by the TSO within its responsibility area.
- *A1-S1.4. Limiting Cascading Effects.* If the TSO anticipates the risk of a cascading effect with impact outside its borders, it must inform the affected neighbor(s). The neighbors complement their respective security computation in order to check the cascading risk at home and to prepare remedial actions by mutual agreement.
- *A1-S2. Coordination for Exceptional Type of Contingency.* It is the responsibility of the operator of the concerned network elements to establish the list of exceptional contingencies utilized for security calculations and to communicate this list to the neighboring TSOs. Such list shall be based on the likelihood of occurrence of the event.
- *A1-S3. N-1 Security Calculations.* Each TSO has to perform N-1 security calculations to assess the effects of contingencies on their responsibility area concerning power flow and voltage patterns.
- *A1-S3.2. Calculations in the Operational Planning Phase.* The N situation has to be determined by *load flow calculations*[z] on the basis of *adequate forecast.*[aa] Each TSO has to perform N-1 simulations for all the contingencies of the contingency list.
- *A1-S3.3. Calculations in Real-Time Operation.* The N situation has to be determined by state estimation on the basis of measurements and topology. Each TSO must perform an automatic N-1 simulation for all the contingencies of the contingency list in real time.

[w] R. Billington, R. Karki, A. K. Verma, eds. "Reliability and Risk Evaluation of Wind Integrated Power Systems," Springer, 2013.
[x] UCTE is the former name of Continental Europe section of ENTSO-E.
[y] ENTSO-E defines remedial actions as normal means to cope with "credible" contingencies.
[z] Here, the expression "load flow" is not precise. Remedial actions must be taken into account. Time domain simulations or, at least, quasi-steady-state simulations (sequence of power flow computations) should be solved.
[aa] In this context, "adequate forecast" means the most likely system status. This appears to be a deterministic approach even for day-ahead assessment despite the increasing amount of uncertainties.

- *A1-S3.3.1. Frequency of Calculation.* The automatic N-1 simulation must run periodically, at least every 15 minutes in real time.

*N-1 Contingency Criterion—TSO Practices.* Some TSOs go beyond the minimal rules defined by ENTSO-E. In any case, the method to define the contingency list (credible contingencies) is not imposed by ENTSO-E and some TSO have some specific rules to define this list. Moreover, N-2 contingencies can be added to the list. For example, if two power lines are installed on the same towers at least during a significant length (typically more than 10 km). In some countries, power lines with two different voltage levels can be found on the same tower. In this case, the N-2 contingency consists in the simultaneous outage of both lines. The addition of N-2 contingencies can be also based on forecast weather conditions, e.g., if a thunderstorm is expected.

More and more remedial actions are used to ensure the system reliability and some TSOs must simulate these remedial actions in their contingencies analysis using a sequence of power flows or time domain simulations. In particular, phase shifting transformers are installed in order to control the power flows and some of them are equipped with local automated control which changes the tap position automatically when the current is higher than a given threshold.

The period with which real-time contingency analysis is repeated is generally 5 minute.

In order to improve the coordination, groups of TSOs had created coordination centers: CORESO (www.coreso.eu) and TSC (www.tscnet.eu). In March 2016, ENTSO-E approved a Regional Security Coordination Initiative and each TSO must be now part of at least one initiative.[ab]

Coordination centers perform security analysis using a full model of the transmission grid of their area in day ahead, intraday and real time based on data provided by all European TSOs. These coordination centers focus on cross border issues: a failure/outage in country A could induce a constraint violation in country B and an efficient remedial action could involve another country C. In case of possible cross border impacts, a coordination center alerts all involved TSO main control centers and propose coordinated actions. The coordination centers are service providers and coordination facilitators. The decisions remain to the TSO, which has the legal responsibility to manage the grid.

*Balancing Principles: Load-Frequency Control and Active Power Reserves.* This subsection describes relevant balancing principles defined by ENTSO-E and rules in the network code on load-frequency control and reserves. This network code defines three levels of controls, as follows:

1. Frequency Containment (FC): FC shall aim at containing the system frequency deviation after an incident within a pre-defined range which is similar to the traditional primary load-frequency control.
2. Frequency Restoration (FR): FR shall aim at restoring the system frequency to its nominal frequency of 50 Hz which is similar to the traditional secondary load-frequency control.
3. Replacement Reserves (RR): RR replace the activated reserves to restore the available reserves in the system or for economic optimization which is similar to the tertiary load-frequency control.

The network code defines frequency quality parameters for each synchronous area. Some of these parameters are shown in Table 18-3.

The network code defines also the principles to select the required reserves associated for the three levels of controls. For the frequency containment reserves (FCR), the capacity required for the synchronous area shall at least cover the Reference Incident of the synchronous area, based on a deterministic analysis and respecting the frequency quality defining parameters. The size of the reference incident must respect the two following conditions:

1. *For the Synchronous Area CE.* The reference incident is the absolute value of the largest imbalance that may result from an instantaneous change of active power of one or two power generating modules or one or two HVDC interconnectors connected to the same electrical node or the

[ab]More information is available at: https://www.entsoe.eu/news-events/announcements/announcements-archive/Pages/News/historical-agreement-on-regional-operational-coordination.aspx.

**TABLE 18-3** ENTSO-E Network Code: Frequency Quality Parameters per Synchronous Area

| | CE | GB | IRE | NE |
|---|---|---|---|---|
| Standard frequency range | ±50 mHz | ±200 mHz | ±200 mHz | ±100 mHz |
| Maximum instantaneous frequency deviation | 800 mHz | 800 mHz | 1000 mHz | 1000 mHz |
| Maximum steady-state frequency deviation | 200 mHz | 500 mHz | 500 mHz | 500 mHz |
| Time to recover frequency | *not used* | 1 min | 1 min | not used |
| Frequency recovery range | *not used* | ±500 mHz | ±500 mHz | not used |
| Time to restore frequency | 15 min | 10 min | 20 min | 15 min |
| Frequency restoration range | *not used* | ±200 mHz | ±200 mHz | ±100 mHz |
| Alert state trigger time | 5 min | 10 min | 10 min | 5 min |
| Maximum number of minutes outside the standard frequency range per year | 15 000 (3%) | 15 000 (3%) | 10 500 (2%) | 15 000 (3%) |

maximum instantaneous loss of active power consumption due to the tripping of one or two connection points.

2. *For the Synchronous Areas GB, IRE, and NE.* The reference incident is the largest imbalance that may result from an instantaneous change of active power of a single power generating module, single demand facility, single HVDC interconnector or from a tripping of an ac-line or the maximum instantaneous loss of active power consumption due to the tripping of one or two connections points, separate for positive and negative direction.

The typical value for the FCR is 3000 MW in CE area. The sharing between the countries based on the previous year statistics gives for example around 700 MW for France, 600 MW for Germany, and 100 MW for Belgium. The interested reader can find in "The Supporting Document for the Network Code on Load-Frequency Control and Reserves" more details on all the processes discussed above.[ac]

*Thermal Limits Management.* The thermal limits of transmission assets are the main constraints which are manageable in control centers by operators. Indeed, these are not fast processes (generally last few minutes) and the operators have time to act in order to relieve these constraints. Overhead power lines, underground/undersea cables, and transformers have different characteristics but all these branches have some thermal limits which impose a maximum capacity.

Generally, each TSO defines maximum loading capacity versus duration of the loading by branch. The following are relevant quantities:

- Permanently admissible transmission loading (PATL): this is the loading in Amps, MVA or MW that can be accepted by a branch for an unlimited duration.
- Temporarily admissible transmission loading (TATL) that is an overload corresponding to a fixed percentage of the PATL for a given duration is allowed (for example, 115% of the PATL can be accepted during 15 minutes).
- Several specific couples (TATL, admissible duration) are calculated for each branch taking into accounts its particular configuration (e.g., age, health index, and local climate) and condition of functioning (e.g., for a given line, it can be defined one TATL acceptable during 20 minutes and another one acceptable during 10 minutes).
- Tripping current without delay (TC) is the maximum admissible instantaneous current. This value can also be applied considering potential bottlenecks due to substations. Note that, effectively, PATL is a temporarily admissible transmission loading which could last indefinitely and TC is a TATL which could not last more than one second. We have PATL < TATL(d1) < ...< TATRL(dn) < TC; d1 > d2> .. > dn; with di the duration.

[ac]The document is available at http://networkcodes.entsoe.eu/wp-content/uploads/2013/08/130628-NC_LFCR-Supporting_Document-Issue1.pdf.

*Management of the PATL and the Overload Conditions.* The fulfillment of the PATL and other overloading conditions (TATL and admissible duration) can be undertaken in different ways. For example:

- By the dispatcher in control centers who can be supported for that by his SCADA/EMS; in real time, in case of violation of the PATL or TATL respectively, no local device will order the tripping of the branch. In that case, the dispatcher decides whether to open the network element or not.
- By local overload protections in which several conditions (TATL; admissible duration) can be implemented. If the loading has not come back under a given TATL after its allowed duration, the local protection will order immediately and automatically the tripping of the network element without any possible action of the dispatcher as to stop this process. In this case, the dispatcher must act before the local protection to relieve the violation. This type of protection is imposed by laws in some countries (for example in France) to ensure the safety around the electrical assets.
- The tripping current without delay (TC) can be monitored by a distance protection or by an overload protection, but in any case, the tripping of the line will be ordered immediately without any possible remedial actions performed by the dispatcher, and is effective at the end of the cycle of the protection device (less than 1 second in case of a distance protection, less than 1 minute in case of an overload protection).

*Definition of the Temporarily Admissible Transmission Loading.* Thermal limits obviously depend on the local temperature, wind, moisture, etc., around the branch counterbalancing or affecting the heating effect of the current in the branch (i.e., Joule effect). Hence, the TATL are defined as per branch depending on local weather conditions. TATL can be assigned a single value for the whole duration of a year but, more often, seasonal values are utilized, e.g., four values, corresponding to the four seasons are used in France. The definition of the seasons can be different for each branch. In some cases, also day/night values are used.

This approach is a first step toward dynamic rating. With the integration of wind generation in power system, TSOs try to exploit the relevant fact that when the wind blows, a high power must be transmitted but, in some favorable conditions, the wind cools down overhead power lines, thus increasing the overall grid capacity. Different types of local devices are installed to monitor the overhead power lines (and, indirectly, their current sag). Some meteorological based approaches are also proposed. The main challenge is that the benefits are mainly brought by the avoidance of costly preventive actions and accurate forecast of the dynamic capacities is a critical issue. Moreover, the complexity to operate such dynamic system is not negligible and dedicated software tools must be implemented to help the operators. The main current operational application of dynamic rating is the utilization in local controls which curtail the power generated by wind farm.[ad]

*Actions to Manage Overloading.* In general, TSOs gives priority to inexpensive actions based on the exploitation of the flexibility of the grid. Example of such non-costly actions are breaker switching; changes of PST tap positions and adjustments flows of HVDC links in parallel with ac power lines. These actions are inexpensive as they have no direct impact on the economic dispatch.

TSOs can also enforce adjustments of generations, loads or flows of HVDC connected to another synchronous area but, in this case, these actions have a measurable direct impact on the market and the TSOs have to pay for these services. Unfortunately, in Europe, there is still no agreement on the price of these services, which are currently defined by each national regulatory framework. This lack of harmonization is a barrier to improve the coordination among TSOs.

*Voltage/Reactive Power Management.* Voltage and reactive power are mostly local phenomena. These dynamics are generally fast (i.e., less than a minute) and possible issues are mainly managed by local closed loop controls. The mission of the operators in the control centers is to define the relevant

[ad]Potential of Improved Wind Integration by Dynamic Thermal Rating of Overhead Lines Tilman Ringelband, Matthias Lange, Martin Dietrich, and Hans-Jürgen Haubrich. http://ieeexplore.ieee.org/stamp/stamp.jsp?arnumber=5281897.

set points for the local controls and to select the reactive power reserves (procurement of ancillary services).

The main objectives are:

- to maintain voltage magnitudes inside the contractual range for the transmission grid users: generators, industrial loads and distribution grids;
- to minimize transmission grid losses; and
- to avoid voltage collapse that can trigger uncontrollable cascading failures.

Some TSOs can change remotely the voltage set points of the automatic voltage regulators (AVR) of the generating units and large wind/photo-voltaic solar farms generally on the high voltage side of the step up transformers. Other TSOs (e.g., in France and Italy) have installed decades ago, centralized closed loop secondary voltage control which controls the voltage at some pilot nodes sending automatically, typically each 10 seconds, a new set point to the local AVR of all the generating units of an area. With this regard, TSOs must take into account the reactive power capabilities of the generating units which are generally part of the performances defined in the connection contract.

In order to manage the dynamic reactive reserve of the generating units, the TSOs can use capacitor/reactor banks which can be switched remotely and by local automated controls based on voltage thresholds. Some TSOs installed SVCs in area without generating units or with costly generating units which are not started very often based on market prices. Some other TSOs negotiate special contacts to be able to run some generating units in synchronous condenser mode to avoid voltage problems. Moreover, the TSO in Germany transformed a stopped nuclear power plant (Biblis) in synchronous condenser to support the voltage in central Germany far away from the wind power in the north and photo-voltaic solar power in the south.[ae]

During summer, the power lines are not loaded and they produce reactive power boosting the voltage. The TSOs must open power lines in order to manage the possible overvoltage. Sometimes, they have to install reactors because opening line becomes incompatible the N-1 criterion avoiding overloading.

*Emergency Controls (Defense Plans).* As previously mentioned, it is impossible and totally inefficient to design and operate a power system in order to avoid any possible impact on the electrical supply against every conceivable threat. Emergency controls and defense plans are designed to mitigate the impact of low probability events, in particular, to speed up the restoration.

*Under Frequency Load Shedding.* The most common emergency control installed all over European power system is the under frequency load shedding. The objective is to keep the system frequency beyond a minimum threshold because generating units are equipped with a local protection that disconnects the unit in case of low frequency, hence creating uncontrollable cascading failures and possibly very large blackouts. This minimum frequency is defined in the connection contract of each generating unit.

The ENTSO-E network code for requirements for grid connection applicable to all generators defines the expected behavior of generating units in case of under frequency. For the continental Europe area, the behavior is defined in Table 18-4.

**TABLE 18-4** Expected Behavior of Generating Units in Case of Under Frequency as Defined by the ENTSO-E Network Code

| Frequency range | Time period for operation |
|---|---|
| 47.5—48.5 Hz | To be defined by each TSO but not less than 30 minutes |
| 48.5—49.0 Hz | To be defined by each TSO but not less than the period for 47.5—48.5 Hz |
| 49.0—51.0 Hz | Unlimited |

[ae]Further details are available at: http://www.energy.siemens.com/mx/pool/hq/automation/automation-control-pg/sppa-e3000/Electrical_Solutions/BiblisA_RWE-Power-AG_electrical-solutions_generator_synchronous-condenser_sppa-e3000.pdf.

The generating units are not allowed to trip instantaneously if the frequency is larger than 47.5 Hz but they can trip after certain duration; they are allowed to reduce their active power if the frequency is below 49 Hz. Moreover, at this time, there are no mandatory rules for the generating units connected to the distribution grids.

Under frequency load shedding (UFLS) schemes are generally triggered if the frequency is lower than 49 Hz. Different implementations are used with different thresholds associated with different percentages of amount of load shedding; some utilize the Rate of Change of Frequency (ROCF).

The system disturbance on the 4th of November 2006 is a good illustration of the UFLS efficiency. On the evening of the 4th of November 2006, there were significant east-west power flows as a result of international power trade and the obligatory exchange of wind feed-in inside Germany. The tripping of several high-voltage lines, which originated in Northern Germany, split the Continental European grid into three separate areas (west, north-east, and south-east) with significant power imbalances in each area (see Fig. 18-22).

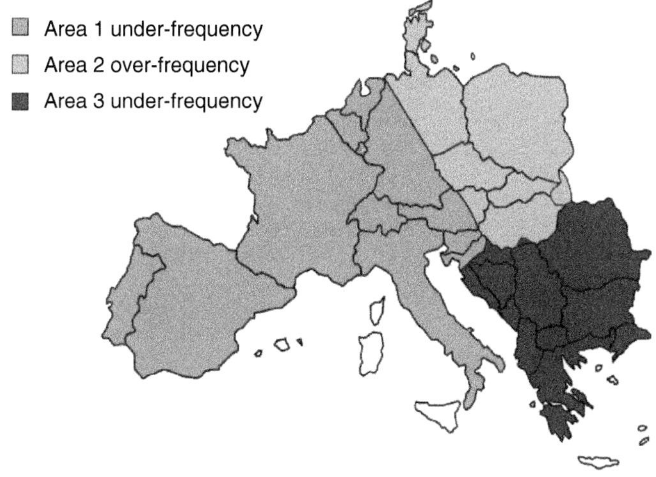


**FIGURE 18-22** Areas in which the ENTSO-E system split during the system disturbance on the 4th of November 2006. (*ENTSO-E.*)

In both under-frequency areas (west and south-east), there were sufficient generation reserves and load shedding allowed to restore the normal frequency within about 20 minutes. The imbalance between supply and demand as a result of the splitting was further increased in the first instants due to a significant amount of tripped generation connected to the distribution grid.

In the over-frequency area (north-east), the lack of sufficient control over generation units contributed to deteriorate system conditions in this area (long-lasting over-frequency with severe overloading on high-voltage transmission lines). Moreover, the uncontrolled operation of dispersed generation (mainly wind and combined-heat-and-power) during the disturbance complicated the process of re-establishing normal system conditions.

Full resynchronization of the UCTE system was completed 38 minutes after the separation. The involved TSOs were able to restore a normal situation in all European countries in less than 2 hours.

Due to the adequate performance of automatic countermeasures (mainly UFLS) in each individual TSO control area and additional manual actions by TSOs a few minutes after the separation, a further deterioration of the system conditions and a Europe-wide black-out could be avoided but the severe frequency drop caused an interruption of supply for more than 15 million European households during less than one hour.

During the incident, in the Western area, load and pump[af] shedding comply with the values declared by TSOs in defense plans. A total of about 17,000 MW of consumption was shed and 1600 MW of pumps was shed. Whereas the load shedding related to the imbalance caused by the splitting of the grid amounted to about 9000 MW, additional load shedding was necessary due to tripping of generation.[ag]

*Other Emergency Control Schemes.* Emergency control schemes other than the UFLS exist in the European power system. For example, to prevent the loss of synchronism of large generators caused by the possible failure of local protection, there exists a control scheme that allows clearing fast enough three-phase short circuits. Another emergency control scheme deals with voltage collapse of importing areas with a lack of local voltage/reactive support. The voltage collapse can cause the tripping of the nearest generators leading to uncontrollable cascading failures. This happened, for example, in January 1987, in Brittany, a western region of France. On load tap changers (OLTC) play a key role in this mechanism. As a consequence of the blackout in Brittany, a dedicated emergency control was designed to stop (or, at least, to postpone) this phenomenon. Three actions are undertaken, sequentially, if the voltage magnitude does not recover: (1) the reduction of voltage set point of OLTC (−5%); (2) the blocking of OLTC; and (3) under-voltage load shedding (UVLS).

*Interaction With Grid Maintenance and Outage Management.* Grid assets must be maintained and maintenance must be performed without any impact on the electrical supply. Planned maintenance activities are generally performed when the system is not in stressed conditions. In the north part of Europe, this is the case, in spring and summer and fortunately weather conditions are also favorable to perform outdoor works. But some unexpected breakdowns must be fixed as soon as possible whatever the season. A strong coordination between the maintenance activity and the system operation is necessary. Indeed, in order to perform the maintenance work, some equipment must be de-energized. A specific and unusual topology in the substation must be chosen. Beforehand, this configuration must be analyzed carefully because it can impact the stability of the system in case of short circuit due to too large currents and/or too long fault clearing times. Another important issue is that the N-1 contingency criterion must be enforced also in this case. Maintenance can be postponed if existing conditions are consistently different than the expected ones (e.g., more wind power or consumption). This is a very challenging and costly decision which must be made almost in real time.

*Interaction With Electricity Markets.* In Europe, day ahead electricity markets are operated by market operators which are never the TSOs. At the beginning of the deregulation, electricity markets were national but, nowadays, the European Commission is pushing toward a single internal electricity market and national markets are merging in a common initiative based on multilateral market coupling.

The target model is the "flow based market coupling."[ah] This mechanism is already in place in the Continental West Europe (CWE). The old explicit allocation of grid capacities is not in-place anymore in most part of Europe. The allocation is achieved implicitly at the coupled clearing of different market places. In order to perform the balance between demand and supply, a grid model is required, which is provided by the TSOs. For the sake of simplicity, highly sought by market players and market operators, it has been decided to use a dc approximation of the grid, the so called PTDF (power transfer distribution factors), that accounts for the impact of the balance of each market place on the flows of critical branches. This model is clearly approximated but is able to ensure that the power flows in critical branches are lower than branch maximum capacities.

In CWE, the TSOs have developed a common process to compute the PTDF and coordination centers (e.g., CORESO and TSC) perform the associated tasks. All data must be provided before the market clearing, generally the morning before day-ahead market clearing. This is a complex process, which requires a forecast of the most likely base case and then the derivation of all N-1 cases as well

[af]Pumped-storage hydroelectric plants.

[ag]Full details on the are available at https://www.entsoe.eu/fileadmin/user_upload/_library/publications/ce/otherreports/Final-Report-20070130.pdf.

[ah]http://www.cigre.org/content/download/17044/680596/version/1/file/C5_204_2012.pdf.

as associated remedial actions. Uncertainty is included by reducing the maximum capacity of each critical branch based on a statistical analysis. This procedure leads to a large set of linear inequalities which is compressed using a presolving method.[ai]

Once the market is cleared, each market participant submits to the TSOs its unit commitment that matches his portfolio bid. Then, the TSOs solve a security analysis taking into account remedial actions to check that there are all constraints are satisfied. If some unit commitment has to be modified, The TSOs must pay for this service. This process pushes TSOs to give priority to inexpensive grid flexible actions in order to have a minimal impact on the unit commitment defined by the electricity market. This procedure can be seen, in fact, as a pragmatic relaxation of a large complex optimization problem, which would be otherwise intractable.

## 18.7.2 Power System Operation Practices in China

*BY JIAN ZHOU AND JIANWEI LIU*

This subsection focuses on power system operation practices in China—the balancing of electricity demand and supply subject to the physical constraints of the power system. This subsection is divided into four main subsections: Overview, Power System Operations in China, Ultra-High Voltage (UHV) and Smart Grid, and Smart Grid Dispatching and Control System.

***Overview.*** With the rapid development of the power grid, the generation capacity in China is growing very fast. At the end of the 2013, in terms of the total installed generation capacity, China overtakes the United States for the first time. The total capacity is 1250 GW and the renewable energy is 390 GW at that time. In 2015, according to the National Bureau of Statistics of China, the total installed capacity reached 1506 GW; the total power demand of society reached about 5550 TWH, and the anticipated power demand of 2020 will reach 7670 TWH. In addition, the power resources are far away from load centers. 76% of coal and 80% of wind resources are located in the northern areas. 80% of hydropower resources are located in the Southwest. Over 2/3 of the power demand concentrates in East and Central China. Long-distance and high-voltage transmission systems are necessary choice for massive development of wind power. Meanwhile the fluctuation of the renewable energy greatly influences the security of the grid. All these factors put forward a huge challenge to the dispatching system.

After the electricity "Plant-Grid Separation" reform in early 2002, the assets of State Electric Power Corporation were divided into the five "power generation groups" that retained the power plants and two grid companies. The two grid companies created were the State Grid Corporation of China (SGCC) and a smaller China Southern Power Grid Company (CSGC). Six regional branch belonging to the SGCC in Beijing, they are SGCC North China Branch, SGCC Central China Branch, SGCC Northwest China Branch, SGCC East China Branch, SGCC Northeast China Branch, and SGCC Southwest China Branch.

Under the regulations, the National Energy Administration (NEA), have the authority to determine the responsibilities of dispatch organizations (DOs). DOs are currently power dispatching and control centers (PDCC) within the SGCC (or SPGC), the regional branch companies, the provincial grid companies, the prefecture-level and county-level electricity supply companies.

The organizational hierarchy laid out in the Regulations is based on a principle of "unified dispatch and multi-level management." Multilevel management is based on a five-level hierarchy of DOs, each with a separate jurisdiction and function. They are National Power Dispatching and Control Center (NPDCC), the Regional Dispatching and Control Centers (RDCC), the Provincial Dispatching and Control Centers (PDCC), Prefecture-level Dispatching and Control Centers (MDCC), and county-level Dispatching and Control Centers (CDCC). The three principal actors within this five-level hierarchy are NPDCC, RDCC, and PDCC, which are responsible for scheduling and balancing most of

[ai]Presolving is a preprocessing used in linear programming in order to reduce very efficiently the number of redundant variables or constraints without any approximation. http://lpsolve.sourceforge.net/5.5/Presolve.htm.

the system. As a general principle, scheduling and balancing responsibilities among DOs are separated according to geography and voltage levels, with PDCC responsible for managing the 220-kV provincial grids and generators that are dispatched to meet within-province demand, and RDCC responsible for higher voltage (330 to 500 kV) provincial interconnections and generators that are dispatched across provinces. The NPDCC, SGCC's dispatching and control center, has jurisdiction over regional grid interconnections and generators that are dispatched across regions. MDCC and CDCC are responsible for implementing dispatch instructions from PDCC, monitoring frequency and voltage conditions in local grids, and managing local generators and load. MDCC control any generating units in their geographic area that are not under the control of a more senior DO, as well as lower-voltage (<110 kV) sub-transmission and distribution substations and lines in their jurisdiction. CDCC typically control any remaining generating units that are in their jurisdiction, as well as substations less than 110 kV and distribution lines less than 35 kV. MDCC and CDCC are responsible for demand planning within their jurisdictions, a process that is coordinated across the province by the PDCC.

Each level in the five-level hierarchy of dispatch organizations develops detailed dispatch rules and procedures, which are contained in operating procedures for dispatch. Topics covered in these procedures include management responsibilities, procedures for frequency and voltage control and contingency management, and rules for equipment repair schedules and information exchange between dispatch organizations.

*Power System Operations in China.* The rules and regulations currently governing electricity dispatch in China are stipulated in a 1993 State Council regulatory directive, Grid Dispatch Regulations, which was revised in 2011. This document allocates authority and responsibility for dispatch, sets an organizational hierarchy, and specifies a basic process for and rules governing dispatch. Implementation instructions for the regulations were provided in the then-Ministry of Electric Power's Implementation Measures for Grid Dispatch Regulations, which was released in 1994. According to these rules, the DOs are the power system operators who arrange for the secure operation of the power system in China and ensures the security and quality of supply of electricity to consumers. They are the nerve center of the electricity generation and transmission system. Usually, four phases of work can be distinguished in system operation:

- Operational planning, extending between 1 or 2 days to several years (say 2 to 3) ahead
- Day-ahead or short-term, between 1 or 2 hours and 1 or 2 days ahead
- Real-time, between 1 or 2 hours ahead to an hour or two after the event
- Post-event, covering collection and analysis of data from the previous phases, this may extend from a few hours to a few months after the event.

Each DO will have its own organizational structures for accomplishing this work. For instance, the system operation department will cover the operational planning, and responsible for the overall assessment of medium to long-term power system security, stability, and reliability. It carries out power system studies, and assesses impact of new generating plants on the power system as well as the transmission expansion plans of electricity transmission licensees. The dispatching scheduling department will look after the day-ahead and short-time period, the dispatching control department response for real-time and post-event timescale, who monitors and controls the power generation, transmission and dispatch system to maintain power system security. Operators in the department work round the clock to monitor, control and direct the generation and transmission of electricity throughout China. They are responsible for the reliable and secure bulk transfer of electricity from the generating stations to the load centers. The protective relay department responsible for detect defective lines or apparatus or other power system conditions of an abnormal or dangerous nature and to initiate appropriate control circuit action. In addition, the automation department is responsible for the operation of the EMS used to monitor and control the electricity generation and transmission network, and also provides support for the information technology needs of the DO.

In 2007, the National Development and Reform Commission (NDRC), State Electricity Regulatory Commission (SERC), and the Ministry of Environmental Protection (MEP) announced a pilot "energy efficient" dispatch system, in Guangdong, Guizhou, Henan, Jiangsu, and Sichuan provinces.

This system specifies a dispatch order, with renewable, large hydropower, nuclear, and cogeneration units given priority over conventional thermal units, and conventional thermal units within each category (e.g., coal-fired units) dispatched according to efficiency (heat rates) and emissions rate. This system sought to achieve two goals (1) to reduce wind and hydropower curtailment, by prioritizing it in dispatch order; and (2) to capture additional fuel cost and coal resource savings above and beyond what the generation rights trading system induced voluntarily by administratively forcing less energy efficient generators to cede more operating hours to more efficient ones. In 2010, energy efficient dispatch was extended to the five provinces in the southern grid.

Provincial planning agencies, typically provincial Economy and Information Commissions (EICs) are responsible for planning annual generator output. For provinces that do not use energy efficient dispatch, each year provincial agencies develop an annual generator output plan, which is based on a recommended plan drafted by the PDCC and approved by the provincial grid company. The plan is typically drawn up in October and finalized and distributed in December. Annual and monthly output totals from this plan are included in annual contracts for generators.

Annual output plans are intended to guarantee operating hours for generators, subject to system constraints, and are not intended to be granted targets. When these changes occur, adjustments are made on the basis of target completion rates. Nationally, as a general principle, SERC's 2003 Temporary Measures for Transparent, Fair, and Just Electricity Dispatch requires that adjustments to generators' annual contracts be made on the basis of fairness.

For provinces that use energy efficient dispatch, provincial planning agencies develop annual and quarterly "dispatch order tables," which create an ordered ranking of generating units for dispatch, on the basis of which PDCC create an annual unit commitment plan. Within technology categories (e.g., coal, natural gas), units are ranked in order of ascending heat rates, and, if ties exist, in order of ascending emissions rates.

Under current law, grid companies are required to give renewable energy priority in dispatch and, by extension, in output planning. Priority dispatch in China takes two forms. First, SERC's 2007 Regulatory Measures for Grid Companies' Full purchase of renewable energy requires grid companies to purchase all renewable energy, regardless of dispatch system, subject to grid security constraints. Second, in provinces that use energy efficient dispatch, non-fossil fuel resources are prioritized in dispatch order.

SERC established the first, and still current, formal definition of ancillary services in China in its 2006 Temporary Measures for Ancillary Services Management for Interconnected Generators. These definitions classified ancillary services into "basic" and "compensated" services. The operating reserve is part of the compensated services. The 1993 Grid Dispatch Regulations explicitly required grid companies to hold some capacity in reserve. The 1994 Implementation Measures specified three categories of reserves and, for each category, reasonable reserve levels: regulation reserves address short-term fluctuations in load and load forecast error and should be 2% to 5% of "peak generator load;" contingency reserves, which respond to equipment failure and should be approximately 10% of peak generator load, but not less than the largest unit in the regional grid; and maintenance reserves, which are held to cover units undergoing routine maintenance, and should be 8% to 15% of peak generator load. The sum of these three reserves, according to the Implementation Measures, should not be less than 20% of peak generator load.

***Ultra-High Voltage and Smart Grid.*** Since the 21st century, China's rapidly-developing economy and increasing energy demands have provided a strong impetus to the construction of large-scale power bases of coal-fired, hydro, nuclear and renewable energy, in particular wind and solar generation. Meanwhile, the geographically reverse distribution of China's energy resources and energy consumption creates the urgent demand for optimized allocations of resources over extensive areas and on a large scale. This calls for leveraging the strengths of robust smart grids in high-capacity and long-distance transmission to realize the intensive development of power bases across different energy types and the export of energy for consumption. In recent years, China has vigorously promoted the development of UHV and smart grid technologies.

Ultra-high voltage (UHV) technologies cover UHV ac and dc transmission technologies, representing the state-of-art level of grid technologies in the world. The UHV technologies were developed by Chinese engineers through practical projects since 2004, not only demonstrated leading technologies, but also enabled domestic brand of UHV equipment. As at the end of 2014, SGCC attained a total of 705 patents in UHV transmission technology, including 318 invention patents and 387 utility model and appearance design patents.

In recent years, a collection of UHV ac and dc transmission projects have been developed in China, fully confirming the safety, economics and environment-friendliness of UHV transmission. The successful construction and operation of these projects clearly demonstrate the feasibility of building a global energy network on UHV technology. As at the end of 2014, three UHV ac lines and six UHV dc lines were completed and commissioned in China. In June 2014, the National Energy Administration issued the notice on accelerating the construction of 12 major transmission channels under the action plan for atmospheric pollution control, including four UHV ac and five UHV dc projects. SGCC is responsible for constructing eight of these projects. In November 2014, work commenced on two UHV ac projects and one UHV dc project. As at the end of 2014, SGCC's UHV transmission lines in operation and under development measured a total length of nearly 16,000 km, with a transformer (converter) capacity of approximately 0.16 billion kVA (kW).

UHV ac transmission projects lay the foundation for building main robust grids, realizing power transmission through regional interconnections and forming a strong grid at the receiving end to provide structural support for UHV dc, high-capacity and multi-circuit transmission. The three UHV ac projects are Southeastern Shanxi-Nanyang-Jingmen 1000 kV UHV ac transmission experimental and demonstration project, Huainan-Northern Zhejiang-Shanghai 1000 kV same-tower double-circuit UHV ac transmission project, Northern Zhejiang-Fuzhou 1000 kV same-tower double-circuit UHV ac transmission project.

Through a UHV dc transmission project, long-distance, high-capacity and point-to-point transmission of electric power from large energy bases to load centers can be achieved. SGCC has completed and commissioned four UHV dc transmission lines, effectively transmitting electric energy from coal, hydro, wind, and solar energy bases in the western, northern, and southwestern regions to load centers in support of eastern and central China's socio-economic development. The six UHV dc projects are Xiangjiaba-Shanghai ±800 kV UHV dc transmission demonstration project, Jinping-Southern Suzhou ±800 kV UHV DC transmission project, Southern Hami-Zhengzhou ±800 kV UHV dc transmission project, Xiluodu-Western Zhejiang ±800 kV UHV DC transmission project, ±800 kV Chusui UHV DC transmission project, and ±800 kV Puqiao UHV DC transmission project.

Based on China's practical condition, SGCC put forward the strategy to build a strong and smart grid with UHV grid as the backbone and coordinated development of subordinate grids at all levels, featured as being IT-based, automated and interactive. SGCC will follow the principle of overall planning, unified standards, pilot project first, comprehensive progress. It has compiled the outline of smart grid planning, and laid down the plan for special subjects, including SGCC master plan, the plan for grid smartness, distribution grid, communication network, and information network. Being IT-based is the fundamental approach to achieve high integration, sharing, and utilization of real-time and non-real-time information for smart grid; being automated is an important means of strong and smart grid, which comprehensively improves automation level of grid operation and control by highly efficient user data acquisition and integrated applications; being interactive is the inherent requirement of strong and smart grid, which achieves friendly interaction and effective coordination among different parts of the grid by real-time information communication and analysis so as to improve user-experience, promote safe, efficient, and environment-friendly power consumption. The major features of a smart grid are:

1. "Strong" and "smart" are two basic requirements for modern grid development. "Strong" is the foundation while "Smart" is the key. The two are correlated and indispensable from each other.
2. Strong and smart grid is an intelligent power system, encompassing power generation, transmission, transformation, distribution, consumption, and dispatching. Each of these parts needs to be well coordinated to bring the advantage of the overall system to full play.
3. Strong and smart grid marks a remarkable shift in the function of power grid. The role of power grid will expand in an all-round way. It is no longer simply a carrier and platform for the transmission and distribution of electricity, but rather a highly integrated intelligent platform for

internet of things, internet network, communication network, radio, and TV networks. At the same time, the boundary between generation-side and demand-side will blur.

Therefore, strong and smart grids will not only constitute an important material foundation for socio-economic development, but become the nerve centrals of modern society. SGCC will implement the construction of strong and smart grid in three stages, following the principle of "overall planning, unified standards, pilot project first, comprehensive progress."

Technological innovation in smart grids has contributed to the improvement of safety, adaptability, economics, and interactivity in electric system operations. China has had all-round breakthroughs in smart-grid equipment monitoring, system operation, smart interactivity, information and communication, and other technologies. At the end of 2014, SGCC owned 11,312 patents related to smart grids, including 1,622 invention patents and 9,690 utility model and apparent design patents.

In recent years, CSGC has identified smart grid construction as one of its priorities and has been developing the smart grid with the idea of "3C Green Grid." The idea refers to be more smart, efficient, reliable, and environmentally-friendly in the development of the power grid by building and upgrading the original power grid with advanced technologies of computer, communication, and control. 3C Green Grid is a specific form in which CSG develops smart grid. The overall goal of CSG smart grid development can be summarized as "Four Improvements and One Promotion," that is, improve the level of safe and stable operation of the power system, improve the utilization efficiency of the system and the assets, improve the energy efficiency management and the service for consumers, improve the capabilities for optimal allocation and efficient utilization of resources, and promote the construction and development of a resource-efficient and environmentally-friendly society.

***Smart Grid Dispatching and Control System.*** The power grid dispatching system is the "nerve center" of the entire power grid system, which ensures the secure, reliable, and economical operation of the power grid. As technologies have matured, the power grid dispatching automation system has grown from a mono-functional SCADA system into a multi-functional system with SCADA as the basis and EMS as the core.

In 2009, a smart grid dispatching and control system, called as D5000, was put into operation in SGCC as the dispatching automation system. D5000 provides key functions such as SCADA, EMS, and wide area measurement system (WAMS).

The D5000 system consists of one platform and four application subsystems: real-time monitoring and pre-alarm system (RTS), operation planning and scheduling (OPS), security checking system (SCS) and operation management system (OMS). It has the following features:

1. *Integrity.* The characteristics of the five-level dispatching structure in China and their correlations are well considered to modulate the function and standardize the interfaces. The supporting platform of the system is open, flexible and user friendly. Sharing of the real-time data, graphics, preventive analysis results and application functions are implemented. It meets the requirements of the dispatching system for smart grid and the backup dispatching system, and fits to the integrated operation of the NDCC, the RDCCs, and all the PDCCs.
2. *Standardization.* The system fully supports the advanced international and national technical standards, and supports the latest IEC standard series, such as IEC 61970, IEC 61850, IEC 61968, and IEC 62351, etc. SGCC has actively participated in developing new international standards, and plans to prepare the IEC standards on model description, simple service, dynamic message, graph description, and naming rules within the IEC Technical Committee 57 (IEC TC57).
3. *Advancement.* International cutting-edge technologies including SOA, zone-based security architecture, equipment oriented standard model and unified visual MMI are introduced in the system.
4. *Security.* Based on the principles of the security protection for the secondary part of power system and according to corresponding requirements of China's regulations on classified protection of information security, the system adopts the advanced security technologies based on the security certificates and security tags for dispatching and strengthens the security protection of

the supporting platform and all application functions. All the computers, communication equipment, operating systems, and databases conform to the security requirements. The base platform and four types of application subsystems are independently developed by Chinese companies. A security protection system is built to integrate the operation and control functions in a dispatching center, and integrates the systems among the five-level dispatching centers, covering all directly dispatched substations and power plants.

The main characteristic of the D5000 system is the Horizontal Application Integration and Vertical Control Center Interaction. This unique platform integrates multiple independent systems into one, while vertical dispatching centers can cooperate to monitor and control the whole grid.

In the D5000 system, a multi-level shared, intelligent root cause analysis and online fault diagnosis technology were implemented. It can quickly locate the fault, provide suggestions, and push the alarm information automatically to the dispatcher on duty. The awareness of adjacent fault is shortened from tens of minutes to a few seconds. All these improve the collaborative capabilities to clear the fault.

With the online Dynamic Security Analysis (DSA) and cascading event tracking function, the D5000 system can help the operator to handle the cascading fault. The dynamic alert analysis is implemented every 15 minutes in normal operation state. As soon as a large disturbance event occurs, the dynamic alert analysis is triggered. The event-triggered analysis and the decision support ensure security of grid operation.

Online low frequency oscillation detection, which combines dynamic monitoring and small disturbance function, was also implemented in the system. Unit identification algorithm is used to quickly identify large low frequency oscillation mode.

With multiple time scale coordinated dispatch, the generation scheduling function deals with monthly, weekly, day-ahead, intra-day, real time power scheduling, accommodating wind power, hydropower, and other clean energy resources.

Two standards, CIM/E power system model description and CIM/G graphic description were applied in D5000 system.

Power system data markup language—E language specification (hereafter referred to as "the E language") is based on object-oriented abstraction of IEC 61970-301 power system public data model Common Information Model (CIM) and aimed to resolve the low efficiency problem of CIM described under XML. The instance data formed by E language is a kind of tagged plain text data. It can describe various kinds of simple or complex data models concisely and efficiently through a few mark symbols and syntax. The more data it processes, the more efficient it is. In addition, E language is more suitable for people's habit rather than XML and is simpler in computer processing. E language combines traditional relationship-oriented data descriptive pattern in power system with the object-oriented CIM. This combination not only retains high efficiency and inherits long-term achievements, but also absorbs the strength of object-oriented method (e.g., class inheritance). Therefore, the E language is concise, efficient, and fit for the power system. Both E language and XML follow CIM fundamental object class. The power system models described by E language and XML language can be used for bi-directional conversion.

The power system graphic description language—G language specification, based on IEC 61970-453 CIM graphics interchange, is a new efficient graphic descriptive language for long SVG data and slow network transmission. It uses XML to describe power flow graphics, and line diagram for graphics storage and exchange.

### 18.7.3 Power System Operation Practices in Australia

*BY SIMON BARTLETT*

***Introduction.*** Australia has a similar land area to China and United States yet its population of only 24 million, similar to Shanghai, is spread across the entire continent. Most of the population live within several hundred kilometers of the coast in the eastern and south eastern states, including the island of Tasmania. Other significant population centers are in south-western Australia and in

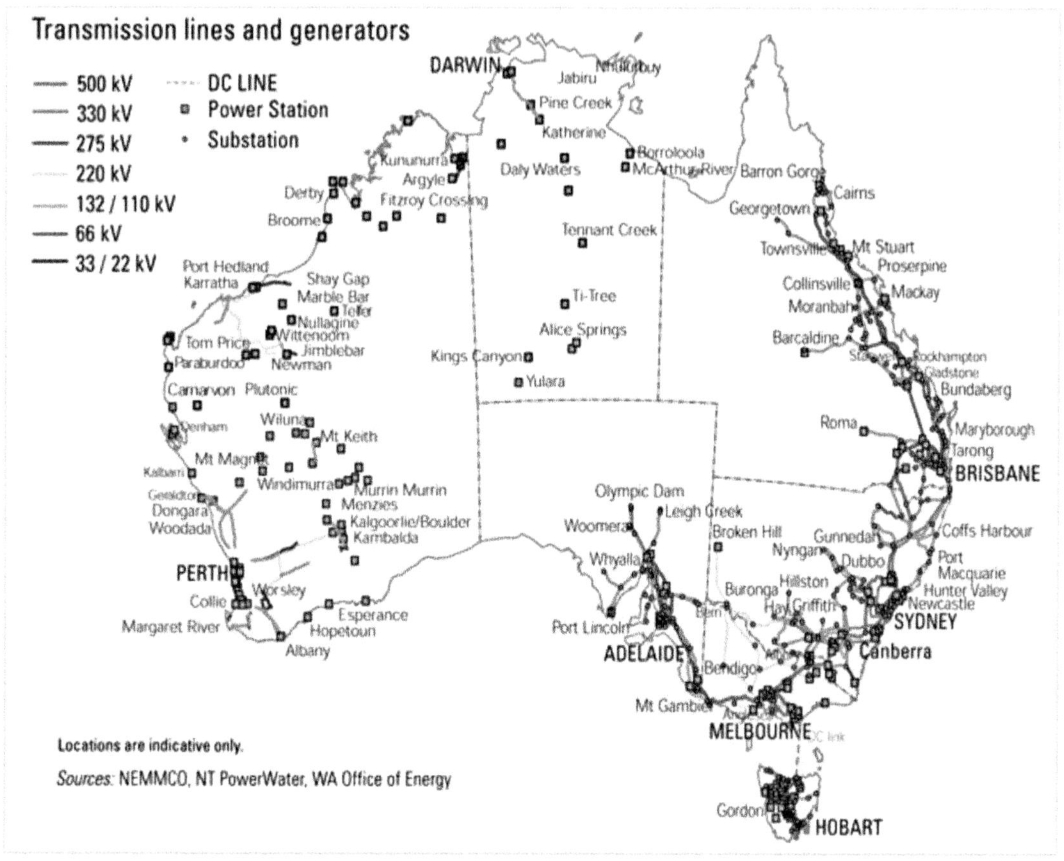


**FIGURE 18-23** Australian electricity supply systems.

the north at Darwin. Australia's secondary and tertiary industries including mineral processing are located in the same regions, however there are numerous remote mine-sites and small communities scattered across the continent.

As illustrated in Fig. 18-23, "Australia's Electricity Supply Systems" matches its population pattern, with an interconnected grid servicing the coastal regions of the five eastern states, Queensland, NSW, Victoria, South Australia and Tasmania where approximately 90% of the electricity is generated and consumed. There are also smaller power systems in south-western Western Australia; the iron-ore region in north-west Western Australia; and around Darwin in the Northern Territory. The remote mines and communities elsewhere in the country have their own isolated power supplies supplied mostly by diesel generators.

The eastern interconnected power system is shown in Fig. 18-24 and is known as the NEM. It covers a large area with one of the world's longest ac transmission networks stretching over 5000 km from Port Douglas in Far North Queensland to Port Lincoln in South Australia. Tasmania is connected to the mainland via the single undersea HVDC Basslink cable. The NEM total maximum demand is around 35 000 MW with an installed generating capacity of around 50 000 MW. The operation of the NEM interconnected power system and electricity market is overseen by the Australian Electricity Market Operator (AEMO) from control centers in Brisbane and Sydney.

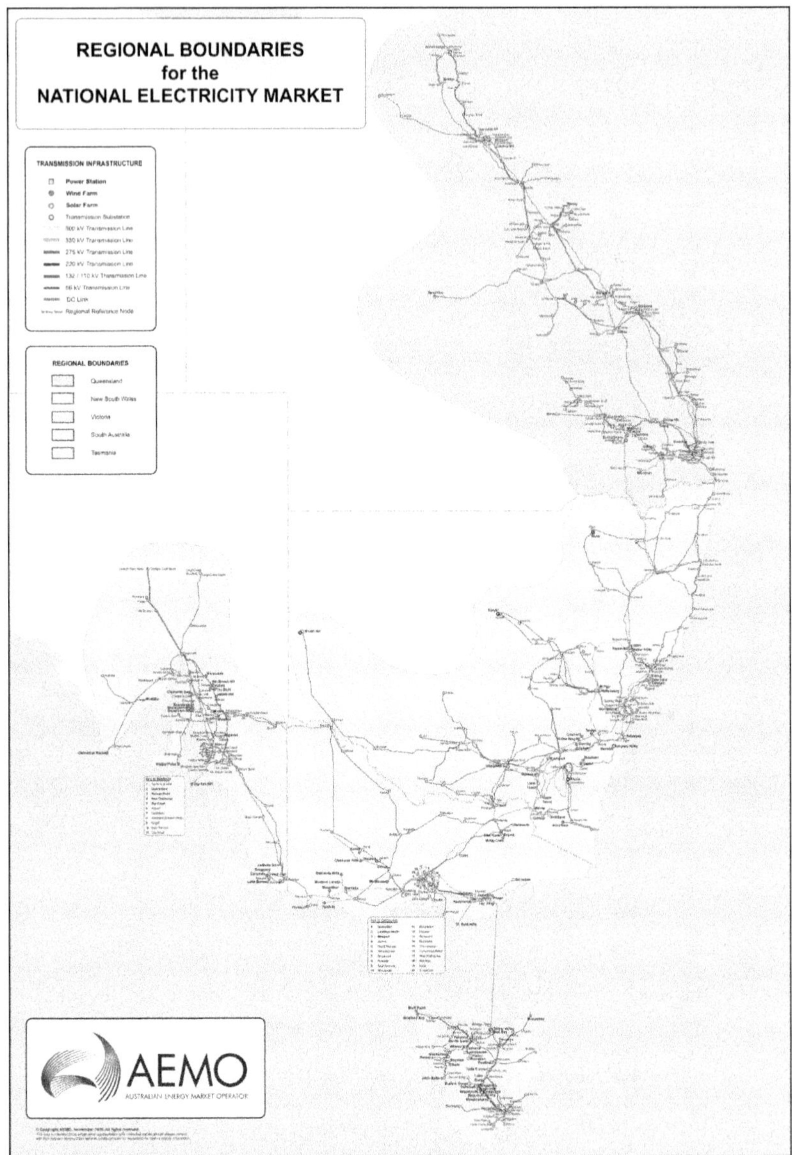


**FIGURE 18-24** Australian national electricity market.

This subsection highlights NEM power system operating practices that are fundamentally different to those found in large international power systems in North America, Europe, and China. These are driven by the vastness of the country, combined with its low population and scarce funds to invest in electricity infrastructure over such long distances. Other differentiating factors include a high occurrence of natural disasters (cyclones, droughts, floods, bushfires), an electricity market heavily constrained by transmission limitations and losses, and vast undeveloped renewable resources that are located far away from the existing transmission grid.

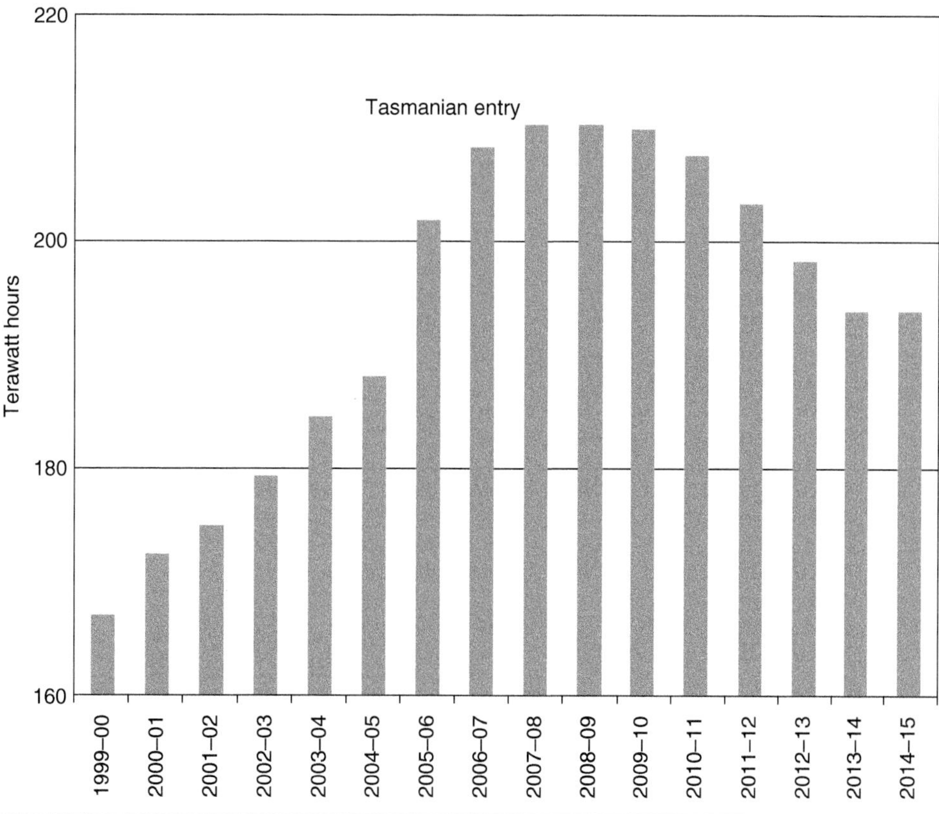


**FIGURE 18-25** Growth in NEM total electrical energy consumption (Twh).

***Electricity Demand Growth.*** As illustrated in Fig. 18-25, the NEM's electrical energy consumption was consistently growing at about 2% until the Global Financial Crisis in 2008/09. Electricity consumption has since been reducing by about 2% due to the lower economic growth energy efficiency measures, the adoption of residential PV and energy conservation incentivized by the increasing electricity prices. Forecasts of electricity growth is flat or continue to decline, except for the new liquefied natural gas (LNG) export industry in Queensland which is expected to need some 800 MW continuously for upstream gas processing and compression.

***Generation Mix and Characteristics.*** Australia's power generation has traditionally been coal-fired, based on the low priced open-cut coal resources in the eastern states, being black coal in Queensland and New South wales (NSW) and brown coal in Victoria. There has been increased gas fueled generation in recent years; however, that is now declining as LNG exports to Asia from Queensland approach full capacity. Coal currently accounts for some 70% of Australian electricity generation, gas 15%, renewable energy 14%, and other 1%. Despite Australia having almost 25% of the world's uranium, there is no nuclear power generation in Australia.

There is some 9000 MW of existing hydro-electric and pumped storage plant in the NEM, however it is mostly water storage peaking plant operating at low annual capacity factors of only some 15%.

***Transmission Network.*** The NEM transmission networks originated as isolated state based systems that became progressively interconnected, some quite recently (e.g., Queensland in 2001 and Tasmania in 2006). Consequently, the main interconnected grid is a range of different voltages (e.g.,

275 kV, 330 kV, and 500 kV), with most interstate interconnections being long double circuit, single tower construction. Given that the interstate interconnectors are typically more than 600 km long, the interconnected network is characterized by relatively low capacity, high transmission losses, large angular differences, and continuous angular oscillations. It is not unusual for there to be up to 90 degrees angular difference across the NEM with continuous oscillations occurring at 2-second intervals which are almost always adequately damped. Compared to European and US networks, the NEM has been likened to "riding a racehorse."

Power system stability is a continuous consideration within the NEM transmission network, with the limiting factor sometimes changing among voltage stability, transient stability and oscillatory stability for some interconnections as system conditions vary during the day.

Coordinating transmission circuit outages can be critically important for maintaining power system security and for minimizing expensive constraints on the dispatch of generators and subsequent electricity market outcomes. For these reasons, AEMO has oversight of all scheduled transmission outages, and transmission companies are required to publish planned outages of critical transmission lines up to 2 years in advance.

***Distribution Networks.*** There are 13 distribution companies in the NEM, who own and operate the networks 66 kV and below as well as some radial 132/110 kV networks. They typically have greater freedom to operate these networks without AEMO oversight, although they coordinate outages of those parts of their networks interfacing with the transmission networks. An increasing area of operational challenge is within the low-voltage distribution network due to the very high penetration of solar PV which is causing voltage regulation issues.

***Electricity Market.*** The NEM has a mature fully competitive electricity market that has been operating since 1998 and is regarded internationally as being successful. Over the last 18 years, it has successfully stimulated significant investments in new generating capacity, renewable generation, plant retirements, while maintaining power system security and avoiding interventions.

The Australian electricity market is a compulsory, energy only market, which incorporates transmission security constraints and transmission losses on the main interconnections to optimize market outcomes and dispatch. It simultaneously incorporates frequency control ancillary services which are also bid into market along with generator energy bids. Bid prices can vary from negative $1,000/MWh to $13 800/MWh (megawatt-hour). The market is optimized and generation is dispatched every 5 minutes, with gate closing just before each dispatch interval.

The very fast market framework, with ancillary services re-optimized every 5 minutes, means that there is no need for a balancing market and that secondary and tertiary frequency response is effectively managed by the market systems re-optimizing the next 5-minute dispatch interval.

There is a voluntary over-the-counter trading for financial contracts as well as electricity futures trading by the Sydney Electricity Futures Exchange which is quite fluid, and have market volumes many times the spot market volumes. These are financial markets only, hence have no involvement or implications for the system operations.

The risk and cost of transmission network congestion constraining the economic dispatch and the market is borne by generators and retailers. There is no requirement for the transmission companies or the market operator to be responsible for these costs through an uplift payment. There are, however, some regulatory incentive mechanisms for transmission companies to minimize their network outages, particularly those that cause significant market impacts. None-the-less, spot market prices can often vary significantly from state to state as illustrated in Fig. 18-26.

The NEM has competitive, non-compulsory ancillary service markets that also operate on a 5-minute dispatch interval basis and are co-optimized with the energy markets. There are eight frequency response markets; regulation raise and lower, and 6-second, 60-second and 5-minute raise and lower frequency response services for contingencies. Black start services are tendered via competitive tendering process which requires primary or secondary black start services for each state.

The optimal generation dispatch is determined for each 5-minute dispatch interval and controlled via AGC run by AEMO. As the market systems factor in interconnection limits and losses, there is no need for separate tie-line control systems to manage interconnection flows. The frequency regulating

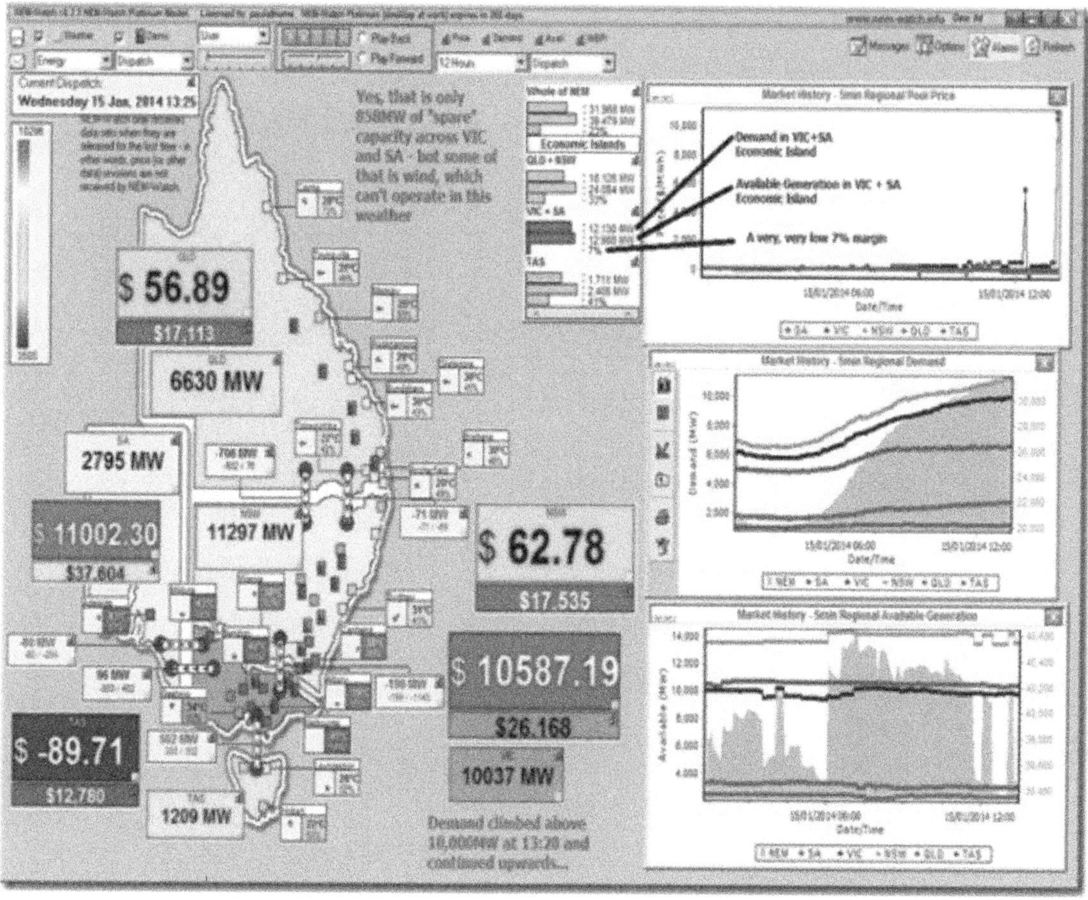


**FIGURE 18-26** High variability in market prices between states.

services are also controlled by AEMO's AGC system. All contingency frequency response services are controlled by local governor controls or by under-frequency relays.

There is a compulsory NEM wide under-frequency load shedding (UFLS) scheme with settings coordinated by AEMO and security coordinators from each state. UFLS relay settings for the mainland NEM states start at 49 Hz (hertz) and extend down to 47.5 Hz, for a nominal system frequency of 50 Hz. The system frequency is managed to within 0.5 Hz of 50 Hz for credible single contingencies.

***Growth in Renewables and Operational Challenges.*** Figure 18-27 illustrates the growth in renewable generation in the NEM over the last 10 years, which reached approximately 14% in 2014/15. The variation in hydro energy reflects rainfall variations from year to year. There has been significant growth in wind-power. Australia has a renewable energy target of generating 20% of electricity from new renewables by 2020, that is, excluding traditional hydro-electric generation. So far, a total of 4000 MW of wind-power and 5000 MW of Photovoltaic (PV) generation has been installed across the country, taking Australia to about half-way toward its 2020 target.

Figure 18-28 shows that the best NEM wind resources are located in the southern states. In South Australia, wind-power and solar PV developments are already exceeding the state's total demand and

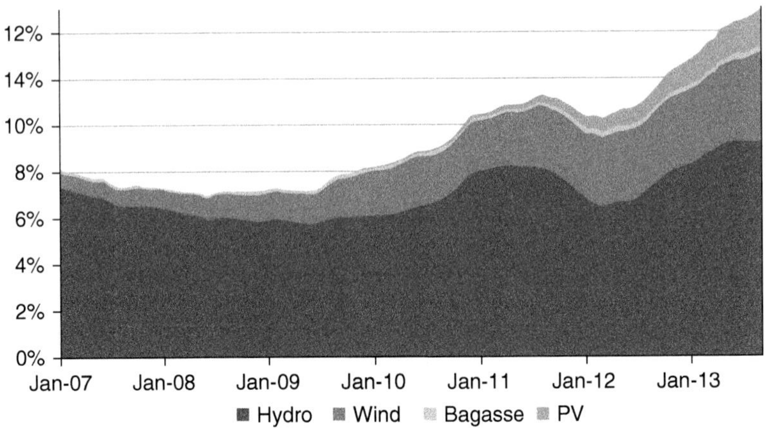


**FIGURE 18-27** Growth in renewable generation.

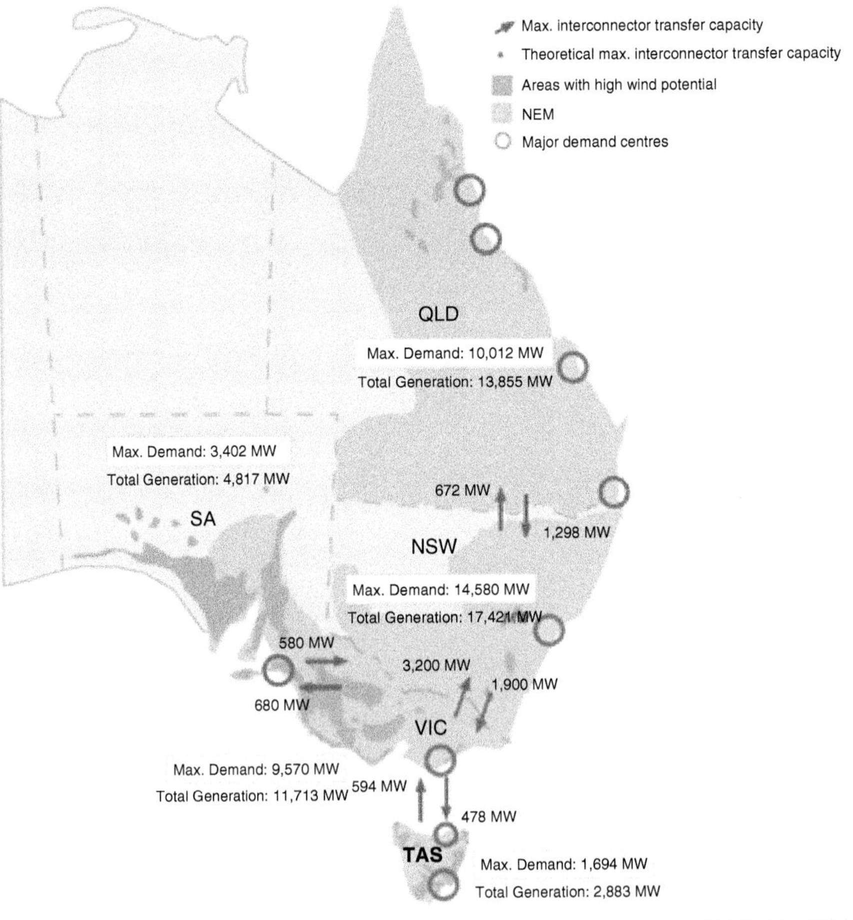


**FIGURE 18-28** Best wind power in southern states.

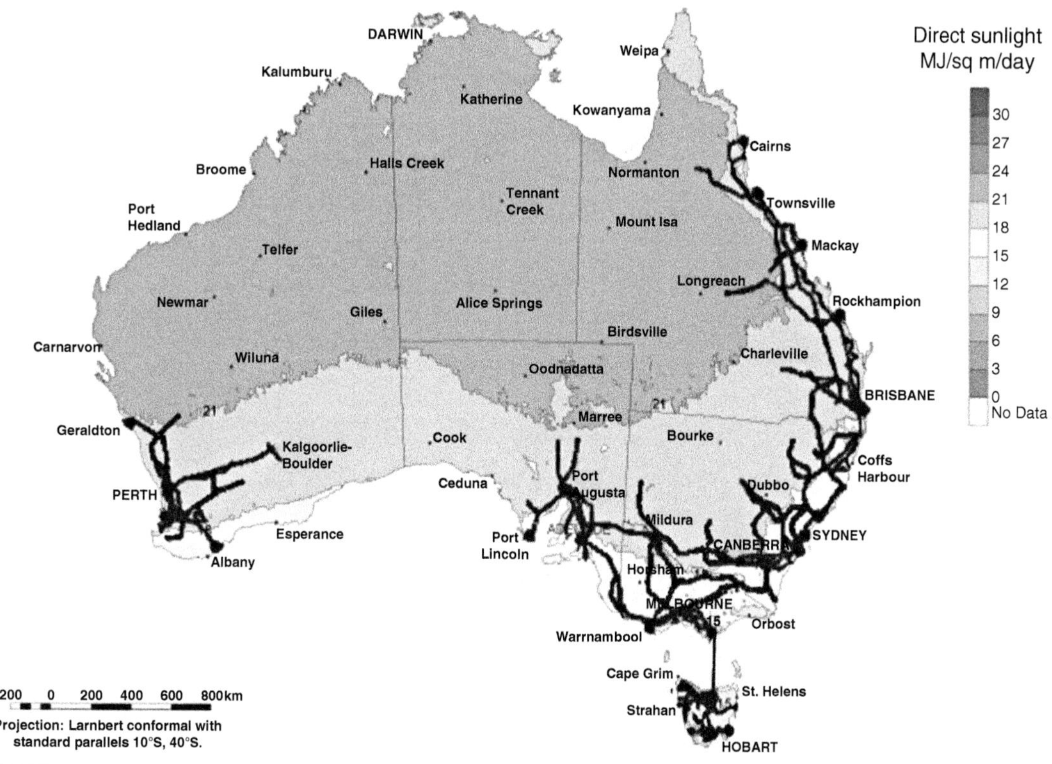


**FIGURE 18-29** Best solar resources in the northern states.

congesting the single ac interconnection to the eastern states. At time of high wind generation, the South Australian pool price is often depressed and this is driving the closure of conventional fossil-fueled plant. Synchronous inertia levels can therefore become so low that frequency regulation and rate-of-change-of-frequency concerns can constrain the operation of the South Australian power system.

As illustrated in Fig. 18-29, the best solar energy resources are located inland in the northern parts of the country, however the existing transmission grid only reaches inland at a few locations. Solar PV developments already exceed 4000 MW and are primarily low capacity, roof-top, PV systems installed by householders. Queensland and South Australia already average more than 25% penetration of rooftop PV with some suburbs exceeding 75% of households with PV. The voltage regulation issues are driving a need to provide better visibility of the MV and LV distribution networks to distribution operators so they can monitor and manage high voltages on their low-voltage networks.

***Natural Disasters.*** Australia has a reputation for extremes of weather and natural disasters. Long droughts can exhaust water reservoirs supplying hydro-electric power stations. On the mainland, this can be managed operationally by electricity market and dispatch processes, however, Tasmania is totally dependent on hydro-electric and wind resources, and a single HVDC cable inter-connection to the mainland. The recent severe drought, which is coincided with an extended forced outage of the HVDC cable necessitated energy conservation measures and the emergency installation of 200 MW of diesel generators in Tasmania. Severe drought can also limit cooling water supplies to inland thermal power stations which may necessitate intervention by the relevant state governments and AEMO to develop and implement long-term strategies for optimizing available cooling water and the associated power supply risks.

Severe bush-fires can create risks of ionized gases causing flashovers of multiple transmission lines on a single transmission line easement and have been known to cause the simultaneous tripping of up to three single-phase transmission circuits on the main interconnections, leading to severe under-frequency load shedding of several states. This risk is managed operationally by AEMO, acting on advice from transmission companies, to determine whether multiple transmission line outages could be a credible risk due to the proximity and intensity of bushfires, which are tracked by satellite.

Severe tropical cyclones are risks to Queensland and can result in wind speeds up to 200 mph in the coastal regions where transmission and distribution lines can be severely damaged. As shown in Fig. 18-30, it is a satellite image of category 5 cyclone Yasi that collapsed transmission towers and distribution networks between Townsville and Cairns in 2011. Operational management of power system cyclone damage and restoration rests with the transmission and distribution operators in collaboration with state and regional emergency service organizations, who can provide access to assistance from the Australian armed services.

**FIGURE 18-30** Cyclones—200 mph winds impacting Queensland network.

Severe floods are another natural disaster that severely impact the power system through the inundation of transmission substations or towers and low lying parts of the distribution networks. Whilst these are more localized and predictable occurrences, the areas impacted and numbers of customers affected can be large, particularly if major urban substations are flooded, as shown in Fig. 18-31. Operational management is undertaken by the relevant transmission and distribution companies in collaboration with state and regional emergency services.

***Power System Security, Stability, Outages.*** The NEM is generally planned and operated to maintain a secure power system which is defined as continuing to operate in a Satisfactory Operating State, even during and after the occurrence of a single credible contingency: being the tripping of the largest generator or transmission line. A Satisfactory Operating States requires all equipment to be operating safely, frequency and voltage kept within the limits in the National Electricity Rules, and no cascading tripping of the network or wide-scale power supply interruptions. The power system is required to be restored to a Secure Operating State within 30 minutes following a contingency.

AEMO is responsible for maintaining power system security through the procurement and dispatch of sufficient ancillary services, and the oversight of the interconnected transmission network.

**FIGURE 18-31** Flooded 275/110 kV substation—Brisbane 2011.

Transmission companies determine and advise AEMO of the transmission limit equations for their transmission network. These enable transient, voltage, oscillatory stability limits, and thermal limits to be calculated in real time from linear equations that are used by AEMO to constrain generation dispatch depending on the real-time values of system loads, generation commitment and dispatch, transmission flows and synchronous inertia. AEMO operators may also use on-line measurements of power system security as well as real-time phasor measurements.

Planned outages of transmission lines must be submitted to AEMO for approval up to 2 years in advance and published for the benefit of market participants.

***Examples of Significant Power System Incidents.*** There has not been a complete power system blackouts of the NEM since its inception in 1998. There have been a number of significant incidents including:

**(a)** Simultaneous tripping of approximately 3000 MW of generation in NSW, over 7 to 8 generating units, that caused under-frequency load shedding of around 1500 MW across the NEM on two occasions. Neither resulted in the separation of the interconnected network.

**(b)** Bush-fire tripping of interconnection between NSW and Victoria, resulting in significant under frequency load shedding in Victoria and South Australia. The interconnection between Victoria and South Australia also tripped.

**(c)** Separation of South Australian to Victoria interconnection on a number of occasions due to simultaneous tripping of several generators in South Australia, or protection mal-operations on the interconnection. Under-frequency operation in South Australia but no further transmission outages occurred.

**(d)** Tripping of the only two transmission circuits supplying Northern Queensland due to various causes on several occasions, resulting in the collapse of the North Queensland power system (as expected), followed by restoration of the transmission corridor and customer supply.

In conclusion, Australia's power system operating practices have adapted to countries vast distances and low population density together with the geography, topography and extreme climatic conditions. Innovative electricity market and grid security operational practices have been incorporated in the power system operational framework to manage the technical limitations of a long, sparse transmission network with remote renewable energy resources.

### 18.7.4 Power System Operation Practices in India

*BY S. K. SOONEE, S. R. NARASIMHAN, S. C. SAXENA, AND K. V. N. PAWAN KUMAR*

***Introduction.*** Electricity is the Item 38 in List III of the Seventh Schedule of the Constitution of India in concurrent list, and both the central and state governments have jurisdiction. In practice, this has meant that the Central government is responsible for all inter-state and international generation and transmission. The state government is responsible for generation, transmission, and distribution within the state. The main body responsible for the development of electricity sector in India is the Ministry of Power, which has the overarching responsibility for policy of power sector.

***Institutional Arrangement.*** Regulatory regime has been implemented in India with State Electricity Regulatory Commission (SERC) at the state level and Central Electricity Regulatory Commission (CERC) at the inter-state level. An Appellate Tribunal for Electricity (ATE) has also been established for the appellate process. In order to harmonize practices across different regulatory commissions, there is a "Forum of Regulators (FOR)" headed by the Chairperson of the CERC (see Fig. 18-32).

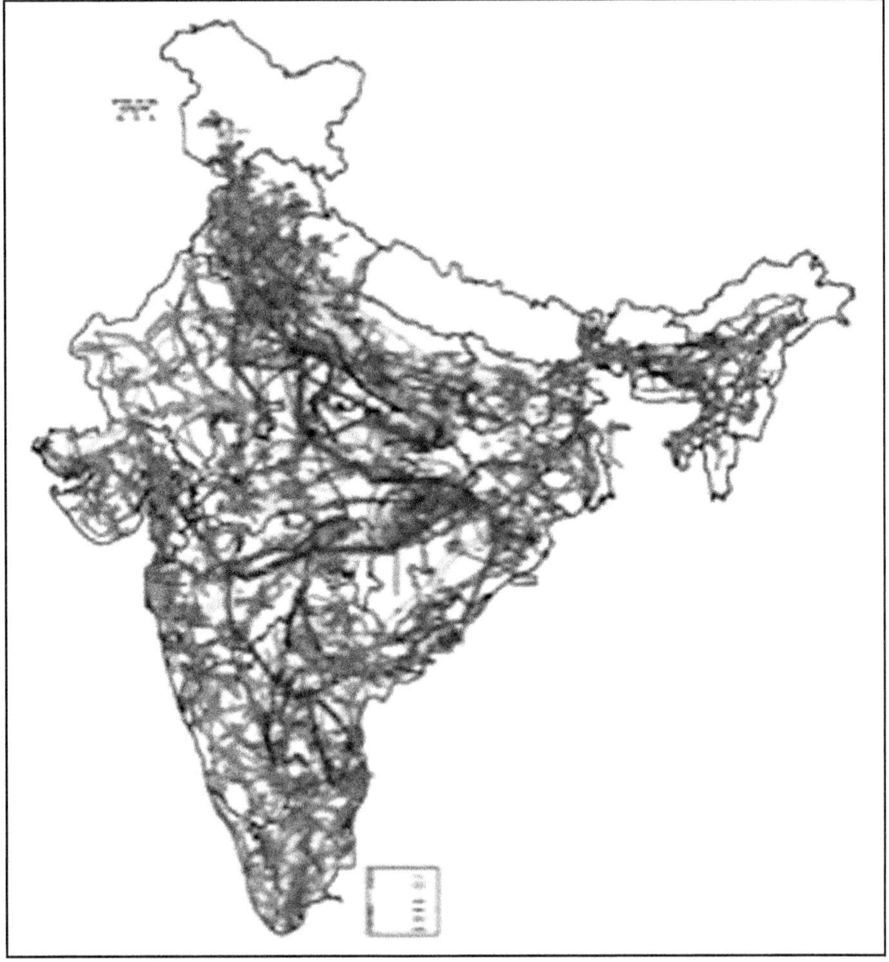

**FIGURE 18-32** Highly meshed Indian grid.

Central Electricity Authority (CEA) is the main technical body at the central level which is responsible for formulation of technical standards in the sector. CEA is also responsible for power system planning and integrated resource planning in coordination with the central transmission utility (CTU) and the state transmission utilities (STUs). There are multiple transmission licensees, both private sector and public sector, operating in the country. Grid management in India is hierarchical in nature through the national, regional and state load dispatch centers. Power system operation is being carried out on a regional basis in coordination with constituent states and union territories (Fig. 18-33).

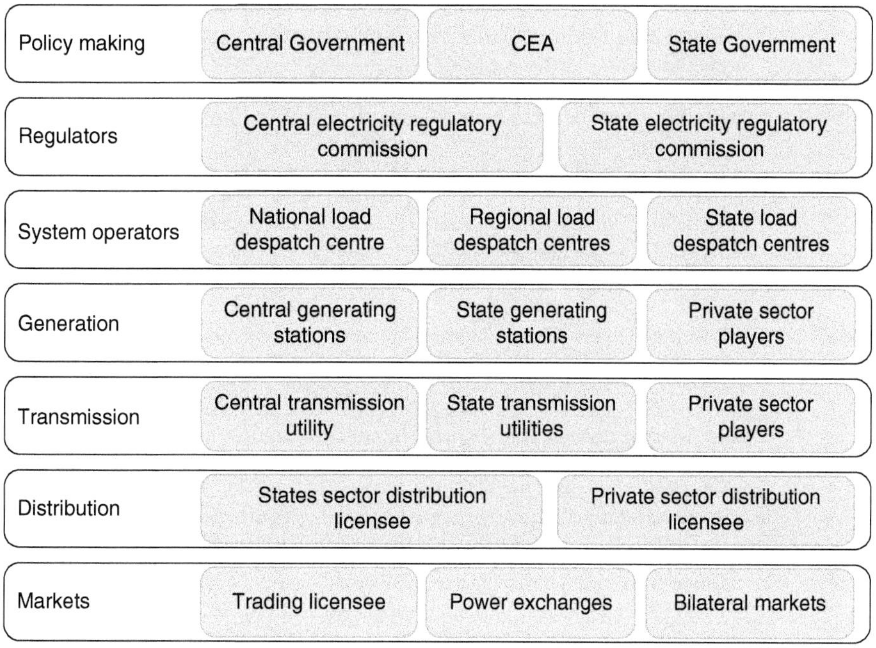

| | | | |
|---|---|---|---|
| Policy making | Central Government | CEA | State Government |
| Regulators | Central electricity regulatory commission | | State electricity regulatory commission |
| System operators | National load despatch centre | Regional load despatch centres | State load despatch centres |
| Generation | Central generating stations | State generating stations | Private sector players |
| Transmission | Central transmission utility | State transmission utilities | Private sector players |
| Distribution | States sector distribution licensee | | Private sector distribution licensee |
| Markets | Trading licensee | Power exchanges | Bilateral markets |

**FIGURE 18-33** Structure of Indian electricity.

The country is electrically demarcated into five regions namely, northern, western, southern, eastern, and north eastern region. The 29 states and 7 union territories of India are the constituents of the electrically demarcated five regions. The regional grids are operated as loose power pools with decentralized scheduling and dispatch, in which the states shall have operational autonomy.

A vibrant electricity market is functional in the country with two power exchanges operating in the single physical delivery market. Renewable Energy Certificates (RECs) are also traded through these power exchanges. A variety of products in different time frames are also available in over-the-counter (OTC) market segment. Financial products in the electricity market may be introduced in the future.

***Power System Operation.*** As per the Indian Electricity Act, 2003 (herein after, "the Act"), State Load Dispatch Centers (SLDC) is the apex statutory body to ensure integrated operation of the power system in a state. The SLDC is responsible for optimum scheduling and dispatch of electricity within a state, in accordance with the contracts entered into with the licensees or the generating companies operating in that state. SLDC exercises supervision and control over the intra-state transmission system. SLDC is responsible for carrying out real-time operations for grid control and dispatch of electricity within the state through secure and economic operation of the state grid in accordance with the grid standards and the state grid code.

As per the Act, each of the five regions has a Regional Load Dispatch Center (RLDC), which is the apex body to ensure integrated operation of the power system in the concerned region. The RLDCs are responsible for carrying out real time operations for grid control and dispatch of electricity within the region through secure and economic operation of the regional grid in accordance with the grid standards and the grid code notified by the Central Electricity Regulatory Commission (CERC). As per the Act, at national level, National Load Dispatch Centre (NLDC) has been established for optimum scheduling and dispatch of electricity among the regions. NLDC has been designated as apex body to ensure integrated operation of national power system. NLDC also coordinates all international transactions with the neighboring countries of Bhutan, Nepal, Bangladesh, and Myanmar.

System operation includes close monitoring of the system parameters, such as frequency, voltages, line loadings, while facilitating inter-state/inter-regional power exchanges in accordance with Indian Electricity Grid code (IEGC).

***Frequency Control.*** The nominal frequency of operation in the Indian grid is 50 Hz (see Fig. 18-34). In compliance to the IEGC section 5.2 (m), all Users, RLDC, SLDC STUs, CTU, and NLDC take all possible measures to ensure that the grid frequency always remains within the operational frequency band, that is, 49.90 to 50.05 Hz. All regional entities shall ensure that the generating units synchronized with the grid provide primary response in line with sections 5.2 (f), 5.2 (g), and 5.2 (h) of IEGC. SLDC/distribution licensee and bulk consumer initiate action to restrict drawal of its control area, from the grid, within the net drawal schedule in line with IEGC 5.4.2(a). A roadmap for provision of reserves and secondary frequency control has been laid out by the CERC which when implemented should further stabilize the frequency.

***Switching Coordination.*** In line with Regulation 6(1) of the Central Electricity Authority (Grid Standards) regulations 2010, no entity shall introduce an element in the inter-state transmission system (ISTS) of regional grid without the concurrence of respective RLDC in the form of an operation code. In case a new power system element is likely to be connected with the ISTS or is to be energized for the first time, from the ISTS, the applicant User/STU/CTU/licensee/generator shall send a separate request in advance along (at least 1 week) with the relevant details.

***Voltage Management.*** All generating units keep their AVRs in operation and power system stabilizers (PSS) in AVRs tuned in line with clause 5.2 (k) of IEGC. The drawee utilities/constituent states take action in regard to VAR exchange with the grid looking at the topology and voltage profile of the exchange point. In compliance to the IEGC section 5.2 (s), all Users, RLDC, SLDC STUs, CTU, and NLDC take all possible measures to ensure that the grid voltage always remains within the following operating range, as shown in Fig. 18-35.

***Power System Element Outage Management.*** The list of important elements of the regional grid is prepared by respective RLDCs. Outage planning would be done centrally only in respect of these lines, ICTs and generating units, allowing sufficient discretion to the SLDCs/STUs in respect of other lines

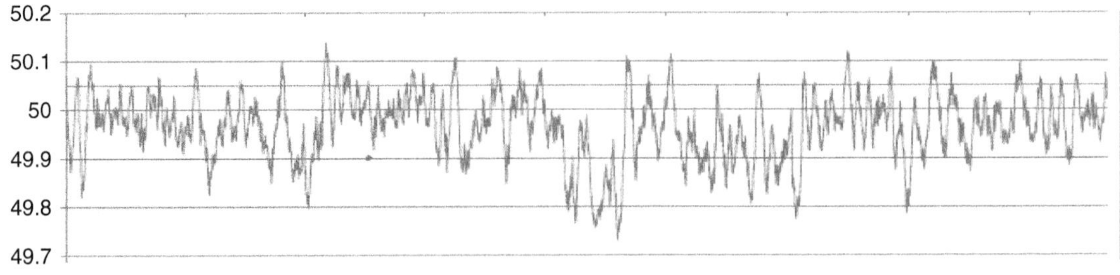


**FIGURE 18-34** Typical frequency profile of Indian grid.

and units. Power system studies are done, at regional level for intra-regional outages and at national level for inter-regional outages, to assess the effect of outages on the grid security. The annual outage plans are reviewed on quarterly and monthly basis.

***Defence Mechanism.*** The respective Regional Power Committee (RPC) forum formulates a suitable under frequency relay load shedding scheme for the respective Region. Under defence mechanisms for the system, the Automatic Under Frequency Relay with set points and df/dt at different conditions are made available as a part of operating procedure. Schemes are formulated considering the largest single credible contingency occurring in the system and based on both flat frequency as well as ROCF. Generally, such scheme is meant to take care of unforeseen contingencies.

| | Normal rating | |
|---|---|---|
| Nominal | Maximum | Minimum |
| 765 | 800 | 728 |
| 400 | 420 | 380 |
| 220 | 245 | 200 |
| 132 | 145 | 120 |
| 110 | 121 | 99 |
| 66 | 72 | 60 |
| 33 | 36 | 30 |

**FIGURE 18-35** Voltage operating range (in kV).

### *Power Market Operation*

*Demand Estimation.* Demand estimation is essentially the responsibility of SLDCs as per the grid code. RLDCs give instructions to SLDCs mainly pertaining to the security of the regional grid and such instructions have to be complied by all SLDCs/other stakeholders. The need for demand control by the state arises on account of the following conditions.

1. Variations in demand from estimated/forecasted values
2. Unforeseen generation/transmission outages resulting in reduced power availability
3. Heavy reactive power demand causing low voltages
4. Sudden Variations in Renewable Generation
5. Commercial reasons

*Scheduling and Dispatch.* A multi-lateral coordinated model has been implemented in India for scheduling. At the regional level, the RLDCs are responsible for scheduling generation from Inter State Generating Stations (ISGS), drawal schedules of states and any market related trades/transactions. The power system under each SLDC constitutes a notional control area in which the states have full operational autonomy and their SLDCs shall have the total responsibility for scheduling and dispatch of their own generation (including generation of their captive licensees), regulating the demand of their customers, scheduling of their drawal from ISGS, arranging any bilateral exchanges and power exchange transactions scheduling and regulating their net drawal from the regional grid.

For the purpose of scheduling each day is divided into 96 blocks of 15 minutes duration each and for each block respective RLDC would intimate each SLDC the drawal schedule and to each ISGS the generation schedule in advance. A weekly settlement cycle is followed at the inter-state level. Transmission losses are estimated in advance on a weekly basis by the RLDCs and are payable in kind.

The net drawal schedule of any regional entity would be the sum of the ex-power plant schedules from different ISGS and any bilateral exchange agreed with other "Regional Entities" in any region and the net of collective (PX) transactions of the concerned control area minus the estimated transmission losses. The "Regional Entity" would be required to maintain their actual drawal from the grid close to such "net drawal schedule" by regulating their own generation and/or consumer's load.

The dispatch schedule to each ISGS/Ultra Mega Power Plants (UMPP)/ISTS connected generating stations/other generating stations that come under Regional Control Area shall be the sum of the requisitions made by each of the beneficiaries from the respective generating station, and subject to the maximum and minimum value criteria or any other technical constraint as indicated by RLDC.

Deviations are handled in accordance with the CERC Regulations on Deviation Settlement Mechanism, which specifies the allowable deviation volume limits, the deviation price vector and penalties/surcharge for violations. Any deviation beyond the prescribed limit of CERC Deviation Settlement Regulations attracts penalties/surcharges and persistent violations are reported to the CERC by the concerned RLDC.

***Open Access—Introduction of Competition in the Power Sector in India.*** Promotion of competition in the electricity industry in India is one of the key objectives of the Electricity Act, 2003 (the Act). In line with the mandate of the Electricity Act and the National Electricity Policy, competitive procurement of generation and transmission has been implemented in the country. Transmission Section 62 and 63 of the Electricity Act, 2003 provide for tariff regulation and determination of tariff of generation, transmission, wheeling and retail sale of electricity by the Appropriate Commission. In order to facilitate competitive procurement of power by the distribution licensees, MoP has provided Standard Bidding Guidelines and Standard Bidding Documents. The bidding process is carried out by the designated Bid Process Coordinators (BPC) namely, Power Finance Corporation (PFC) Ltd. and Rural Electrification Corporation Ltd. (REC).

Section 66 of the Electricity Act 2003 mandates Appropriate Commission to promote the development of a market (including trading) in power in such manner as may be specified and shall be guided by the National Electricity Policy. The Electricity Act, 2003 recognizes trading as a separate licensed activity. CERC issues license for inter-state trading under CERC (Procedure, Terms and Conditions for grant of trading license and other related matters) Regulations, 2009. The trading licensees undertake electricity transactions through bilateral and power exchanges. As on 31.03.2015, there are 47 trading licensees.

There are two power exchanges in India namely Indian Energy Exchange (IEX) and Power Exchange of India Ltd. (PXIL). CERC (Power Market) Regulations, 2010 prescribes various aspects such as registration process, eligibility criteria, procedures, prudential norms, risk management, IT infrastructure, shareholding pattern, ownership, and governance structure of Power Exchange, as shown in Fig. 18-36.

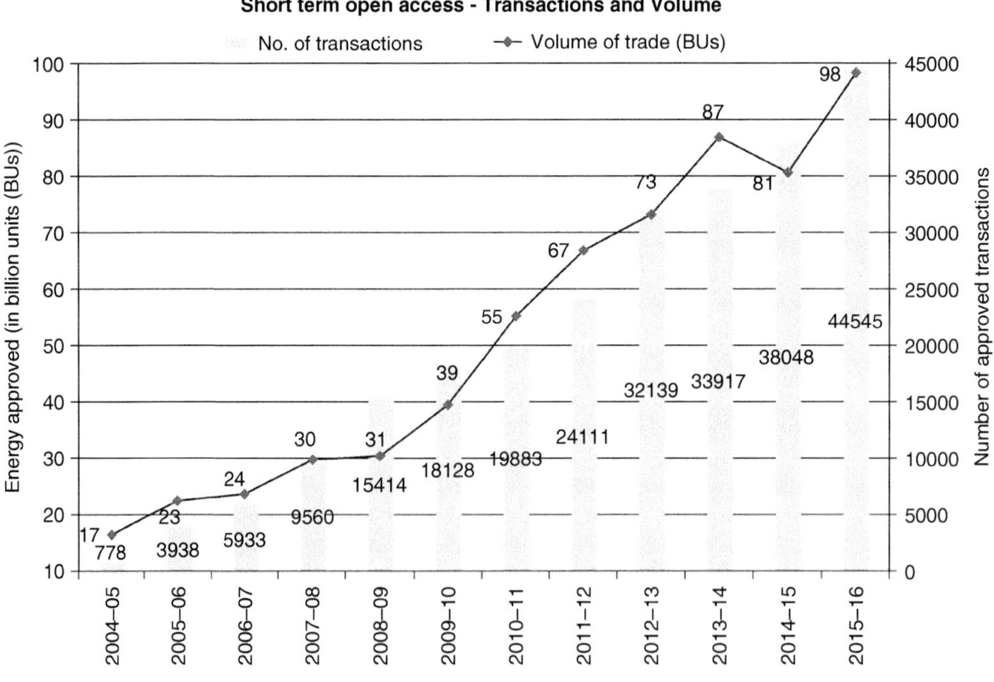


**FIGURE 18-36** Short-term open access transactions.

Private sector participation has been encouraged by the government in power sector. With the advent of open access and de-licensing of generation, many private sector players established generation and participate in the open electricity market through bilateral contracts and power exchanges. As of April 2016, there is 125 GW of installed capacity in the private sector out of total all India installed generation capacity of 302 GW.

***Transfer Capability Assessment and Congestion Management.*** Assessment of total transfer capability (TTC), transmission reliability margin (TRM) and available transfer capability (ATC) for import and export of power for reliable system operation and for facilitating non-discriminatory open access in transmission. The transfer capability assessment within the region is carried out by respective RLDC and across regions/bid-areas by NLDC 3 months in advance. However, depending on the system condition like major variation in generation, loads, transmission element availability etc., the TTC/ATC will be revised on day-ahead basis for power exchange and same day basis for bilateral transactions to facilitate maximum power availability with reliable grid operation. To ensure line loadings are within safe permissible limits, critical corridors forming flow gates are identified based on the studies carried out by RLDC/NLDC, for continuous monitoring of power flow on real time. The flow gates are modified based on the system studies from time to time depending on the system network changes. The transactions already scheduled/deviations/infirm injections may be curtailed by respective RLDC/NLDC in the event of transmission constraints, congestion in the grid, or in the interest of grid security.

***Metering, Accounting, and Settlement System.*** As per Clause 6.4.21 of IEGC, Central Transmission Utility (CTU) installs special energy meters on all interconnections between the regional and other identified points for recording of actual net MWh energy interchanges and average frequency on 15-minute time block basis and MVARh drawls under low-/high-voltage conditions. The installation, operation, and maintenance of special energy meters is in accordance with Central Electricity Authority (installation and operation of meters) Regulations 2006 and Amendment Regulations, 2010.

All the energy meter data reaches respective RLDC preferably by every Monday but not later than noon of Tuesday. The computation of block wise net injection of each ISGS/generating station whose dispatch scheduling comes under the responsibility of RLDC and actual net drawal of each beneficiary regional entities is carried out based on the above meter readings received at SRLDC. The preparation and issue of the weekly deviation/reactive/congestion charge accounts is done by the RPC Secretariat on the basis of the meter data and implemented schedules and congestion messages forwarded by respective RLDC.

***Power System Visualization and PMU/WAMS.*** EMS modules implemented in the different control centers (NLDC/RLDCs/SLDCs) consist of various applications covering wide spectrum of functions like pre-dispatch, real-time dispatch, and post-dispatch functions. The EMS tools such as state estimation, contingent analysis, optimum power flow, load forecasting, resources scheduling has provided system operators greater flexibility in taking appropriate and judicious decisions during the course of operation.

A strong communication backbone network is the key to the smooth functioning of the grid. Keeping this in view; a dedicated communication backbone system has also been established under Unified Load Dispatch Centre (ULDC) scheme consisting of fiberoptic, microwave, and power line carrier communication (for the end link data flow). Also in order to provide reliability and redundancy in data communication channel, an alternate physical route between RLDCs and NLDC have been implemented which has greatly improved the data availability/reliability.

PMUs have been implemented in India and at present, over 100 PMUs are installed, as shown in Fig. 18-37. PMU data provides wide-area visibility across the bulk power system in ways that let grid operators understand real-time conditions, see early evidence of emerging grid problems, and better diagnose, implement and evaluate remedial actions to protect system reliability and carry out signature analysis. Phasor data concentrator (PDC) at NLDC has been installed and all regional pilot projects have been integrated at national level.

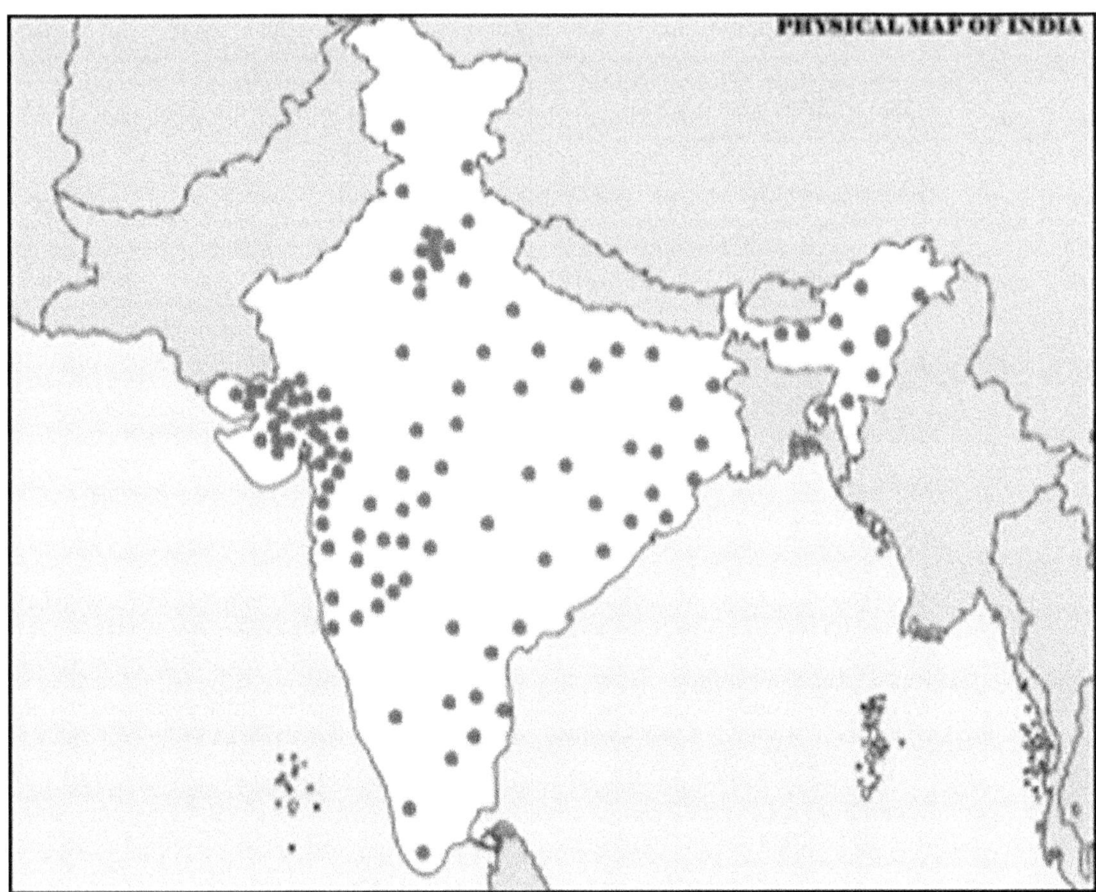


**FIGURE 18-37** Spread of PMUs across India.

A full-fledged unified real time dynamic state measurement (URTDSM) scheme is being implemented for pan-India installation of 1740 PMUs. The regulatory approval by the CERC has been given for URTDSM scheme. It is being implemented in phases which involves installation of PMUs on substations at 400 kV level and above in the state and central grids, all generating stations at 220 kV level and above HVDC terminals, important inter-regional connection points, inter-national connection points etc., provision of phasor data concentrators (PDCs) at all SLDCs, RLDCs, NLDC, and strategic locations in states along with development of visualization aids. Optical fiber communication network is being strengthened along with associated terminal equipment. Analytics application development is in progress for line parameter estimation, Vulnerability analysis on distance relay characteristics, linear state estimator, supervised zone-3 blocking, CT/CVT parameter validation, and control schemes for improving system security.

Various reports on synchrophasors initiative [1–4] have been prepared which give deep insights on utilization of PMUs for monitoring and analysis of oscillations, computation of frequency response characteristic, validation of correct operation of protection system, island detection and re-synchronization of grid, dynamic model validation, visualization of special events such as HVDC frequency controller testing, power system stabilizer testing, turbine valving, and monitoring of cyclones.

### *References*

1. http://www.nldc.in/WebsiteData/Reports/OtherReports/Synchrophasors%20Initiative%20in%20India%5FJune%202012.pdf.
2. http://www.nldc.in/WebsiteData/Reports/OtherReports/Synchrophasors%20Initiatives%20in%20India%20Decmber%202013%20%2D%20Web.pdf.
3. http://www.nldc.in/WebsiteData/Reports/OtherReports/Report%5FPower%20System%20oscillations%5FIndia%5FPOSOCO%5FFinal.pdf.
4. http://www.nldc.in/WebsiteData/Reports/OtherReports/Report%20on%20Low%20Frequency%20Oscillation%20in%20Indian%20Power%20System%20%5FMarch%202016.pdf.

# 19

# POWER SYSTEM PROTECTION

Héctor J. Altuve Ferrer

*Distinguished Engineer and Dean, Schweitzer Engineering Laboratories, Inc. (SEL) University, Pullman, Washington*

## 19.1 PROTECTION FUNDAMENTALS

Power systems are exposed to faults and other abnormal operation conditions that may cause human safety risks, damage equipment, affect power quality, and trigger major disturbances leading to widespread outages. Protection and control systems must take preventive or corrective actions during power system disturbances.

The first line of defense is the protection of power system components. Protection at this level detects faults and abnormal conditions and disconnects protected equipment. The power system also requires system level protection functions (system integrity protection systems). These functions include underfrequency and undervoltage load shedding, power system out-of-step protection, and automatic generator shedding.

The main function of power system protection is to disconnect protected equipment for faults and abnormal conditions. Another important function is to facilitate restoration by providing an indication of the fault. Modern microprocessor-based relays send this information over a communications channel. Power system operators then use this information in real time to make more efficient restoration decisions.

Protective devices include fuses, automatic circuit reclosers, sectionalizers, low-voltage breakers, and protective relays. This section covers relay protection systems.

Modern power systems have reduced redundancy and operate close to the security limits. These systems may experience angular and voltage stability problems that can lead to major system collapses. To prevent stability problems in relatively weak power systems, we need fast and reliable protection and control systems.

Renewable energy sources are proliferating in today's power systems. As a result, generation is becoming distributed, variable, and more difficult to dispatch. These small sources, typically connected through power electronics systems, have weak contribution to faults, determined not by the laws of physics but by control algorithms. Protection systems need to be sensitive enough to detect these small fault currents.

In addition, deregulation of the electric power industry requires better power system protection and control. Open energy markets create the need for power transfer from certain sources to specific load regions. These power flows can overload parts of the transmission system and create further reliability problems.

### 19.1.1 Relay Protection Systems

In a relay protection system, or relaying system, used in medium- and high-voltage systems (Fig. 19-1), the protection scheme provides fault detection, and the breaker provides current interruption. The protection system may include protective relays, current transformers (CTs), voltage transformers (VTs), breakers, dc supply systems, communications channels, and control cables. Protective relays detect faults and abnormal conditions and close output contacts to initiate breaker tripping. VTs and CTs isolate the relays from the primary system and reduce the primary voltage and current values. Breakers receive the tripping signal from the relays and open the primary circuit to interrupt the fault current. The dc supply system provides the breaker tripping current when relay output contacts close. When protecting power lines, a communications channel may be necessary for the relays at both line terminals to exchange information.

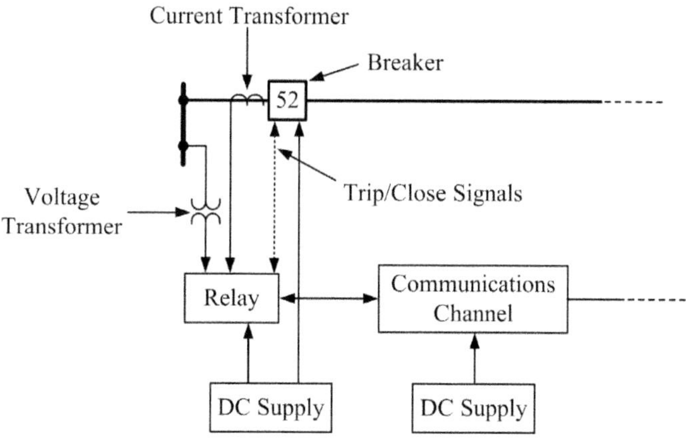


**FIGURE 19-1** Relay protection system.

### 19.1.2 Primary and Backup Protection

Since protection systems may fail to operate, we typically apply primary and backup protection. This is particularly important for short-circuit protection because the short circuit is the most frequent and damaging type of system failure.

Primary protection operation should be as fast as possible, preferably instantaneous (no intentional time delay). Figure 19-2 shows the one-line diagram of a power system section that illustrates the basic concepts of primary short-circuit protection. Breakers connect adjacent power system components, allowing the protection system to completely isolate faulted equipment. The relays create the primary protection zones, represented by dotted contours in the figure. In many cases, CT location defines the limits of the primary protection zones. For a fault inside a zone, the relays trip all the breakers within that zone.

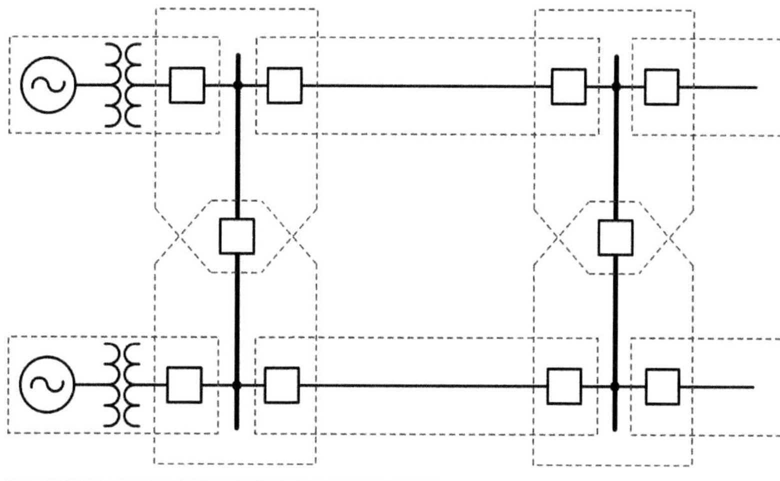

**FIGURE 19-2** Relays create primary protection zones.

Adjacent protection zones overlap to provide full primary protection coverage in the power system. A fault in the overlapping areas causes the tripping of more breakers than the breakers needed to isolate the fault. Therefore, the overlapping areas need to be as small as possible.

When the relays receive information from CTs located at all zone boundaries, they create a closed protection zone, also called a unit protection zone. When the relays measure only part of the zone boundary currents, the zone limits are not well defined in the direction of the interconnections without CTs. These limits vary with the power system operation conditions and the type of protection system applied. In this case, we have an open, or nonunit, protection zone.

A backup protection system operates when the primary protection system fails or is out of service. Failure of any primary protection system component could cause a failure to operate. We design primary and backup protection systems to avoid common-mode failures.

Figure 19-3 illustrates the concept of backup protection. For the fault shown in the figure, Breaker 5 and Breaker 6 should operate as the primary protection. If Breaker 5 does not trip, there are two possibilities for removing the fault current contribution from A, B, E, and F: open Breaker 1, Breaker 3, and Breaker 8; or open Breaker 2 and Breaker T. In either case, backup protection should be time-delayed to give the primary protection an opportunity to operate first.

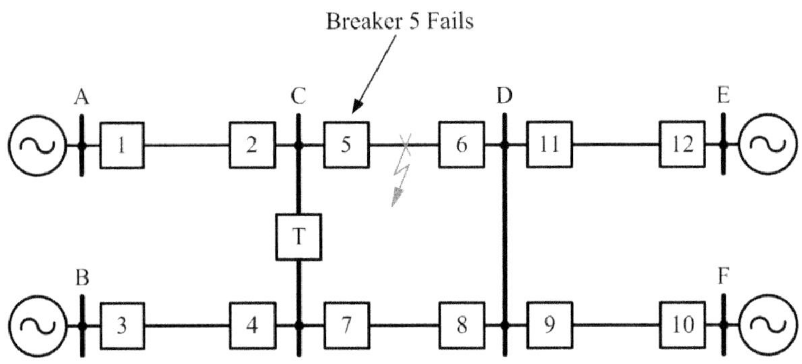


**FIGURE 19-3** Backup protection operates when primary protection fails or is out of service.

Breaker 1, Breaker 3, and Breaker 8, which are located in remote substations, provide remote backup protection. A protection system often provides remote backup for several power system components. For example, the protection system at Substation A is the remote backup for Line BC, both CD lines, and Bus C. Remote backup protection is economical (provided by protection systems existing in remote substations), which makes it widely used in distribution and subtransmission systems. However, operational drawbacks of remote backup protection limit its application in transmission systems. For example, tripping Breaker 1 for the fault shown in Fig. 19-3 disconnects the healthy Line AC. Furthermore, the protection system at Substation A may not be sensitive enough to detect low-current faults on Line BC and both the CD lines. Finally, relays at Substation A need to time-coordinate with the slowest primary protection systems they back up, which may result in very slow fault clearance.

Breaker 2 and Breaker T, which are located in the same substation as the primary protection, provide local backup protection. This protection is more expensive than remote backup because additional equipment is needed to duplicate some protection system elements. Operational advantages of local backup over remote backup make it widely used in transmission systems. Multifunction microprocessor-based relays reduce the cost of local backup protection and facilitate its application in subtransmission and distribution systems.

### 19.1.3 Protection System Functional Characteristics

Correct protection system performance includes tripping when required (for faults or abnormal conditions in protected equipment) and not tripping for other system conditions, such as normal load or external faults. During most of its life, the protection system is in a correct, no-operation condition. In a broader sense, the protection system performs correctly when primary protection operates correctly to clear the fault, backup protection does not operate, and as a result, the faulted equipment is isolated in the expected time.

Incorrect protection system performance includes failures to trip and false tripping or misoperations. A failure to trip is not tripping when required. A false trip or misoperation is a trip that occurs when protected equipment is healthy. Misoperations can occur during normal operation conditions or during external faults or other system disturbances, such as power swings.

The most important protection system functional characteristics are reliability, selectivity, speed of operation, and sensitivity.

Reliability expresses the ability of the protection system to consistently perform correctly. Reliability has two facets, dependability and security. Dependability expresses the ability of the protection system to trip when required (for in-zone faults). Security expresses the ability of the protection system to not trip when not required (when there is not an in-zone fault). They are usually inversely related, but good protection schemes can increase both to improve overall reliability. Security is challenged every time faults occur in adjacent protection zones, but dependability is only challenged for in-zone faults. Protection system false trips indicate a bias toward dependability in the system design, whereas failures to trip indicate a bias toward security.

There is a bias among protection engineers toward dependability in protection system design. For example, [1] reports approximately 94% of protection performance problems resulting in false trips. This bias reflects the fact that power systems are redundant to a certain extent; they generally tolerate the loss of additional components better than uncleared faults. In modern power systems, however, this concept is changing. For example, some wide-area protection systems, where security is very important, use two-out-of-three voting schemes.

Selectivity is the ability of a protection system to eliminate a fault with the least disconnection of system components. Selectivity is a measure of how well a protection element can differentiate between an in-zone and an out-of-zone fault. Selective protection operation implies that primary protection eliminates faults and that backup protection operates only when primary protection fails. Selectivity is inherent in the type of protection element.

Speed of operation is the ability of the protection system to provide fast fault clearing. Fast operation is important for preserving system stability, reducing equipment damage, improving power quality, and enhancing human safety. Protection system operating time includes protection scheme operating time and breaker current interrupting time. Speed can be impacted by inherent and intentional delays. An intentional delay may be required to coordinate with adjacent zone protection elements or to ride through transient conditions. Speed is directly related to selectivity. More selective protection elements can be set with low or no intentional delay, while less selective elements must rely on a delay to coordinate with adjacent zone elements.

Sensitivity is the ability of the protection system to detect even the lowest current magnitude faults within the protected zone. Protection sensitivity is important to ensure the detection of high-impedance faults or the reduced contribution to faults from small, distributed generators. Sensitivity is related to selectivity as well. Protection elements that are set with more sensitivity are less selective and vice versa. Selectivity, speed, and sensitivity need to be balanced to produce a reliable protection system.

## 19.2 TIME-SYNCHRONIZED MEASUREMENTS

Time-synchronized, accurate measurements improve power system performance and reliability. Many protection, control, and monitoring applications today require high-accuracy time (less than 1 μs), including time-synchronized phasors (synchrophasors), line current differential protection, fault locating, and IEC 61850 Sampled Values messaging. Additionally, regulations now require the use of accurate time distribution across power transmission networks. Time-synchronized oscillographic reports and synchrophasor measurements from multiple intelligent electronic devices (IEDs) are valuable tools for analyzing system disturbances.

For example, synchrophasors enable power system engineers to directly measure system state, display this information to system operators, and create better solutions for local and wide-area protection, control, and monitoring applications [2]. Engineers can build systems where relays and meters measure and communicate synchrophasors to one or more processing devices every fraction of a second. The processing devices, in turn, assess the overall system state and respond accordingly.

### 19.2.1 Time Synchronization

The Global Positioning System (GPS), an accurate and globally defined time source, provides a reference for time-synchronized measurements. The U.S. military developed the GPS satellite network as a private navigation system, but now this useful time source is available to anyone. Sensitive receivers can detect GPS signals. Russia's Global Navigation Satellite System (GLONASS) is another fully functional system. Other systems under development include China's BeiDou-2 Navigation Satellite System, the European Union's Galileo Positioning System, Japan's Quasi-Zenith Satellite System, and India's Indian Regional Navigation Satellite System.

Accurate time based on satellite systems is now widely available in substations and control facilities. Advancements in satellite receiver technology have made these devices economical for wide-scale deployment.

The conventional method of high-accuracy time distribution is provided by individual or multiple satellite clocks collocated at each substation at which this service is required. Typical time-distribution systems used in power system applications consist of a satellite receiver with one or more IRIG-B outputs [3], which are directly connected to IEDs or are daisy-chained from one IED to another [4].

New methods of accurate time distribution over wide areas are being implemented via terrestrial private fiber-optic systems such as a synchronous optical network (SONET) or Ethernet. Advantages to these systems are the following:

- Fewer time sources are required.
- Time signals distributed over a fiber-optic communications system are immune to external radio interference.

These systems, also known as terrestrial time-distribution systems (TTDSs), use multiple time sources (e.g., GPS and accurate time references), combine them, and then produce a best-estimate time signal. This method addresses the issues of single satellite clock systems. Further, even if all external time sources are lost, the TTDSs have the ability to provide a relative time throughout the system (i.e., the time being distributed may not be correlated with absolute time, but all nodes within the system are time-synchronized).

Figure 19-4*a* shows an IEEE Standard 1588 [5] implementation that uses the Ethernet network to distribute time [6]. Specialized Ethernet switches allow distributing time from redundant master clocks. Figure 19-4*b* depicts a SONET-based system that uses time-division multiplexing (TDM) techniques [6]. In this network, Stratum 1 level clocks normally provide the transport frequency source [7]. These clocks are required for the network timing and can be extended to provide high-accuracy absolute time to the network as well.

### 19.2.2 Synchrophasors

A synchrophasor is calculated from data samples using a common time signal as a reference [8]. Using the received IRIG-B signal to time-stamp the acquired voltage and current values, IEDs estimate phasors that are synchronized with the time reference (Fig. 19-5).

At nominal frequency, an integer number of waveform periods fit within each one-second interval. Therefore, the instantaneous angle measurement $\delta_x$ remains constant at nominal frequency. If the signal is at off-nominal frequency, the instantaneous angle varies according to the frequency deviation from nominal.

In the power system steady state, total generation equals total load, plus losses, and machine speeds are almost equal to synchronous speed at system nominal frequency. In a two-source system, active and reactive power transfer depend on the voltage magnitudes and the angle $\delta$ between source voltages. Previously, state estimators used voltage magnitude and power measurements to estimate $\delta$ because it could not be measured directly. Now, time-synchronized measurements allow direct measurement of $\delta$ and therefore the system state.

Synchronous generators accelerate or decelerate to adapt to changing power transfer requirements that occur during system disturbances. The angular displacements of the generators in a system provide information about the system dynamics. Synchrophasors allow measuring these angular displacements.

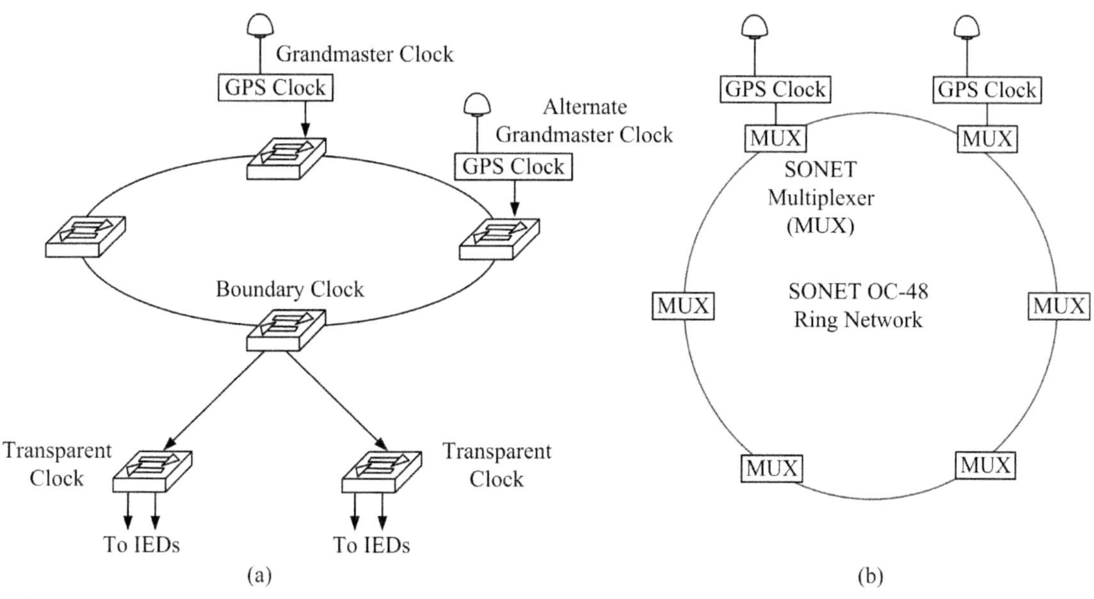


**FIGURE 19-4** Time-distribution systems. (*a*) Ethernet-based system using IEEE Standard 1588; (*b*) SONET/TDM-based system.

Frequency has special significance in real-time phasor measurement. Phasors are a steady-state concept applied at a single frequency. In an actual network, system frequency changes constantly, based on system dynamics. Synchrophasors provide an accurate method of measuring system frequency.

## 19.2.3 Time-Synchronized Systems

A time-synchronized system may include the following components [2]:

- *Time sources.* Provide synchronized time-code signals at low cost.
- *Phasor measurement devices.* Provide synchrophasor values. Today, relays, meters, and other IEDs perform this function without the expense of a separate device.
- *Synchrophasor processors.* Align, concentrate, and mathematically process synchrophasor data for real-time protection, control, and monitoring.
- *Communications networks.* Allow devices to exchange synchrophasor values via serial and Ethernet communications.
- *Application software.* Runs on a synchrophasor processor or a generic computer platform and provides human-machine interface capability, plus protection, control, and monitoring.

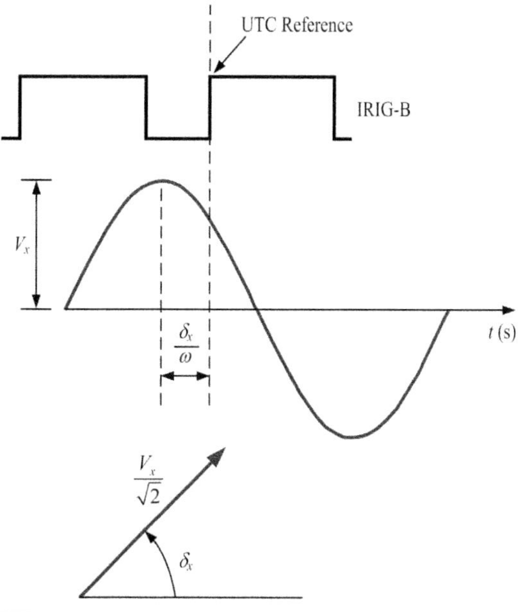


**FIGURE 19-5** Synchrophasor representation using common time reference.

The composition, configuration, and types of devices of a time-synchronized system depend on the specific application. A simple system can be a device measuring synchrophasor data for display purposes. More complex systems may include tiered synchrophasor processors and multiple phasor measurement devices at different stations for wide-area protection, control, and monitoring.

## 19.3 LINE PROTECTION

In radial systems, fault current flows in only one direction. Overcurrent protection using relays, automatic reclosers, and fuses is the typical protection for radial distribution lines.

In looped systems, fault current can flow in either direction at the relay location, depending on the fault location. Looped lines are common in transmission systems, and are increasingly used in subtransmission and distribution systems. Protection of looped lines can use any of these principles:

- Directional overcurrent protection.
- Distance protection.
- Directional comparison protection.
- Phase comparison protection.
- Current differential protection.

### 19.3.1 Overcurrent Protection

Overcurrent protection uses current magnitude information to detect faults in the protected equipment, based on the fact that short-circuit currents are generally larger than load currents. In most cases, a separation between the load current and the fault current values makes it possible to set the overcurrent device pickup current between these two current values.

In heavily loaded, long distribution feeders, the fault current could be less than the load current for short circuits at the end of the feeder. For such cases, the phase overcurrent elements cannot be set to detect faults at the end of the line. Using a negative-sequence overcurrent element can eliminate this sensitivity limitation of the phase overcurrent protection [2, 9]. Balanced load current is positive-sequence current; therefore, a negative-sequence element can be set for a very low pickup current. This element responds to unbalanced faults.

The fault impedance may limit the ground fault current to values close to or even below the load current. A zero-sequence overcurrent element, which can be set below the load current, improves overcurrent protection sensitivity for ground faults.

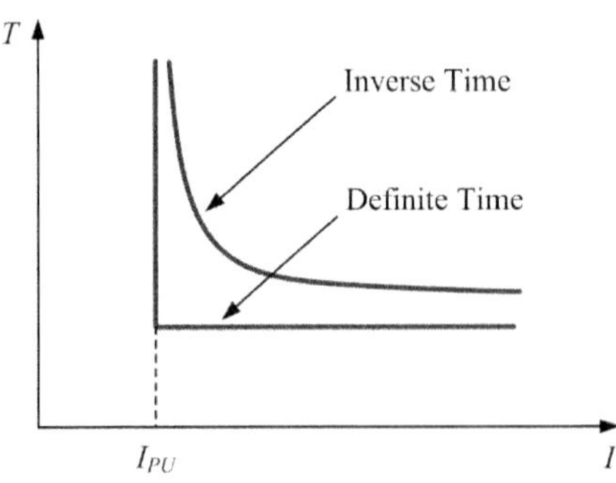


**FIGURE 19-6** Time-current characteristics of time-delayed overcurrent elements.

Overcurrent elements sense the input current and issue a trip signal to the breaker when the current is above the pickup value. Overcurrent elements are classified according to their operating time characteristics as follows:

- Instantaneous overcurrent elements.
- Time-delayed overcurrent elements (Fig. 19-6):
  - Definite-time overcurrent elements.
  - Inverse-time overcurrent elements.

Figure 19-7 depicts a typical overcurrent protection scheme. The CTs are wye-connected. The phase overcurrent elements (50, 51) measure the phase currents. The ground overcurrent elements (50N, 51N) measure the wye neutral (residual)

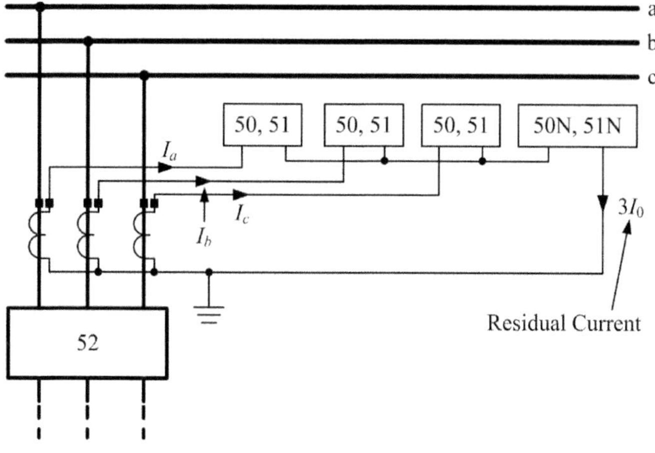


**FIGURE 19-7** Overcurrent protection scheme.

current, equal to the sum of the phase currents (or three times the zero-sequence current). In the past, this scheme required four electromechanical relays. Digital technology permits implementation of the scheme in one multifunction relay.

We set the 51 element pickup value above the maximum load current. We set the 51N element pickup value above the zero-sequence unbalance current, which is typically between 10% and 30% of the phase current, depending on the type of feeder. The higher sensitivity of the ground element, as compared to the phase elements, is important for detecting low-current ground faults. However, when conductors fall to ground in overhead distribution lines, the fault impedance can be so high that conventional ground elements often fail to detect such faults. Reference [2] describes methods to improve ground overcurrent protection sensitivity.

Figure 19-8 depicts a radial system with two inverse-time overcurrent devices to illustrate overcurrent protection coordination. To ensure selectivity, the upstream (backup) device must wait for

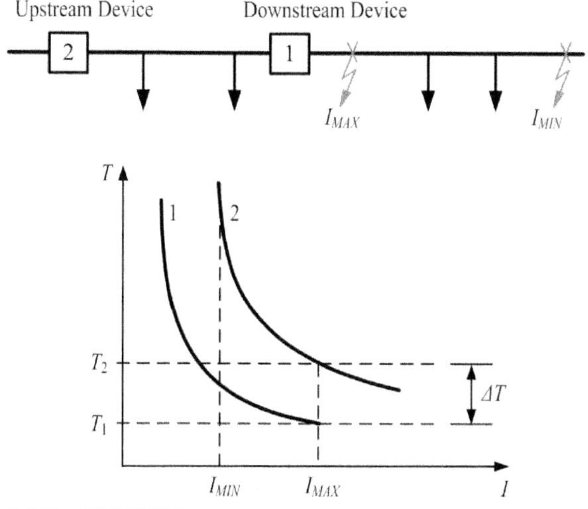


**FIGURE 19-8** Overcurrent protection is time-delayed to provide coordination.

the downstream (primary) device to clear the fault. Otherwise, customers between the upstream and downstream devices will lose service unnecessarily. The time-current curve of the upstream device must be above the downstream device curve, with some time margin. Unfortunately, this level of selectivity requires a sacrifice in speed. Fault clearing time increases as the fault moves to the line sections located closer to the source, where the fault current is higher. The undesirable result is longer clearing times for more severe faults.

A solution to this problem is to add instantaneous overcurrent elements, which provide faster tripping. These elements can be coordinated only when the feeder impedance creates a substantial difference in fault current between close-in and remote faults. In tightly knit distribution systems, such as those in urban areas, in manufacturing plants, and aboard ships, these elements become difficult or impossible to coordinate, because the fault current may not change significantly from the source to the farthest load.

### 19.3.2 Directional Overcurrent Protection

Directional overcurrent protection uses directional elements to supervise the operation of overcurrent elements. Although directional overcurrent protection is relatively simple and economical, its high-speed protection zone reach varies with changes in the source impedance, and it requires setting revisions for power system changes.

Phase directional overcurrent protection (67) must be set above load current, which limits protection sensitivity and speed. Overcurrent elements that respond to negative- or zero-sequence current can be set below load current. Adding a negative-sequence overcurrent element improves sensitivity and speed for phase-to-phase faults. Ground directional overcurrent protection (67N), which responds to the zero-sequence current, is more sensitive than phase directional overcurrent protection and ground distance protection.

Figure 19-9*a* illustrates the application of directional overcurrent protection to a system with several sources. The arrows represent the protection tripping direction. The relays are oriented toward the protected lines. This orientation divides the system protection into two independent groups: the relays "looking" to the right and those "looking" to the left. The directionality divides the coordination process into two independent processes. A relay needs to be coordinated only with the other relays in its group. Figure 19-9*b* depicts a looped system with only one source. In this case, all the relays must be directional except the relays adjacent to the source bus. For line faults close to the source bus, the system is inherently directional. That is, fault current can flow only out of the bus and into the lines. Thus, there is no need for directional relays in these locations.

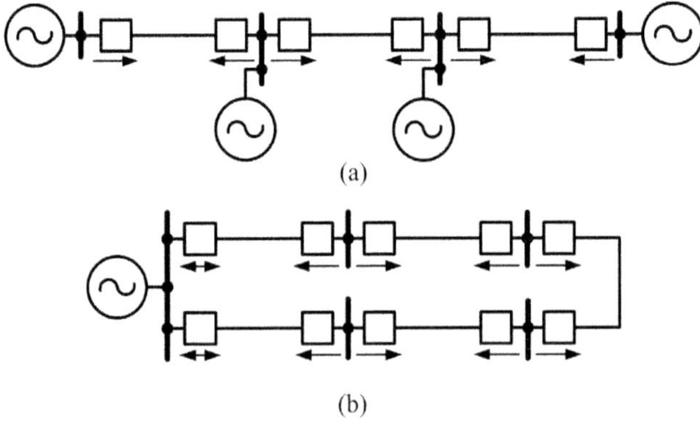


**FIGURE 19-9** Application of directional overcurrent protection. (*a*) Multisource system; (*b*) looped system.

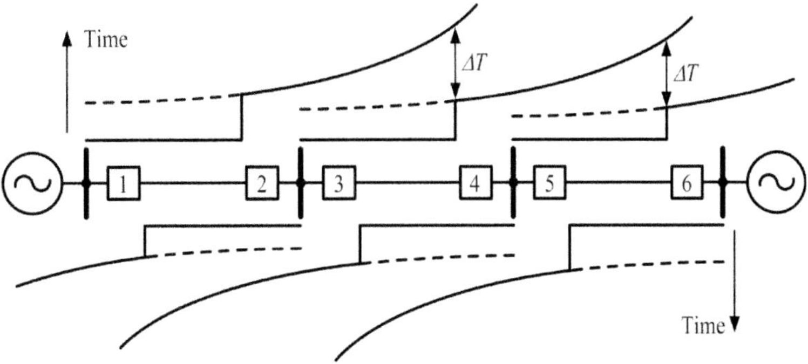


**FIGURE 19-10** Time-distance characteristics of directional overcurrent relays protecting lines of a two-source power system.

Figure 19-10 shows the time-distance characteristics of six directional overcurrent relays protecting three lines that interconnect two sources. Instantaneous overcurrent elements underreach the remote line terminals to prevent misoperation for faults on the adjacent lines or the remote bus. Inverse-time overcurrent elements overreach the adjacent lines to provide remote backup protection. A coordination time interval $\Delta T$ that includes the breaker fault clearing time provides coordination between adjacent inverse-time overcurrent elements. Figure 19-10 shows that faults located in midline sections cause instantaneous operation at both line terminals. Faults close to a line terminal cause instantaneous operation of the relay that is closer to the fault and time-delayed operation of the remote-terminal relay.

A phase directional overcurrent protection scheme (67) uses three phase overcurrent elements supervised by directional elements. A ground directional overcurrent protection scheme (67N) uses a ground overcurrent element that responds to the zero-sequence current and is supervised by a ground directional element.

Traditional 67 schemes use three directional elements; each element typically responds to a phase-to-phase voltage (polarizing quantity) and the current of the opposite phase (operating quantity) [10]. Traditional 67N schemes use a ground directional element that responds to the zero-sequence voltage or current as the polarizing quantity and to the zero-sequence current as the operating quantity [10].

Modern relays provide better directional discrimination solutions. A family of relays includes the following elements [2]:

- Voltage-polarized directional element (32P) measuring positive-sequence impedance that responds to three-phase faults.
- Voltage-polarized directional element (32Q) measuring negative-sequence impedance that responds to phase-to-phase, phase-to-phase-to-ground, and single-phase-to-ground faults.
- Voltage-polarized directional element (32V) measuring zero-sequence impedance that responds to phase-to-phase-to-ground and single-phase-to-ground faults.
- Current-polarized directional element (32I) that responds to phase-to-phase-to-ground and single-phase-to-ground faults.

These relays use the 32P and 32Q elements for phase fault protection. The 32Q element has priority over the 32P element. For ground fault protection, these relays use the 32Q, 32V, and 32I elements. The relay selects the best polarizing and operating quantities for each ground fault, based on system conditions [11]. Applying this solution eliminates the need to make choices and compromises.

The 32P element compares the angle of the measured positive-sequence impedance $\bar{Z}_1$ with two thresholds. Equation (19-1) shows that this element declares a forward fault when the angle of $\overline{Z}_1$ is within the threshold angle range and declares a reverse fault when the angle of $\overline{Z}_1$ is outside the angle range.

$$(-90° + Z1ANG) < Arg\left[\overline{Z}_1\right] < (90° + Z1ANG) \text{ (Forward fault)} \tag{19-1}$$

where *Arg* $\overline{Z}_1$ is the angle of the measured positive-sequence impedance $\overline{Z}_1$; *Z1ANG* is a relay setting, equal to the line positive-sequence impedance angle.

Figure 19-11 shows the operating characteristic of the 32P element.

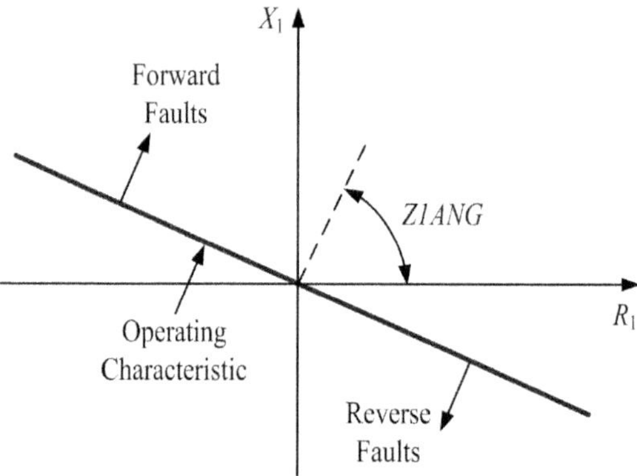


**FIGURE 19-11** The positive-sequence, voltage-polarized directional element (32P) provides directional discrimination for three-phase faults.

The 32Q element calculates the scalar quantity $z_2$ using Eq. (19-2) [2, 12].

$$z_2 = \frac{\text{Re}\,[\overline{V}_2\ (\overline{I}_2 \bullet 1\angle Z1ANG)^*]}{I_2^2} \tag{19-2}$$

where $\overline{V}_2$ is the negative-sequence voltage; $\overline{I}_2$ is the negative-sequence current; $I_2$ is the magnitude of $\overline{I}_2$; *Z1ANG* is a relay setting, equal to the line positive-sequence impedance angle.

For forward faults, the directional element measures $z_2 = -\left|\overline{Z}_{2S}\right|$, corresponding to the negative-sequence impedance of the equivalent system behind the relay. For reverse faults, the scalar quantity is $z_2 = \left|\overline{Z}_{2L} + \overline{Z}_{2R}\right|$, corresponding to the negative-sequence impedance of the equivalent system in front of the relay, equal to the sum of the line negative-sequence impedance and the negative-sequence impedance of the equivalent system beyond the remote line terminal [2].

The 32Q element compares $z_2$ with two thresholds. If $z_2$ is below a forward-fault threshold, the 32Q element declares a forward fault. If $z_2$ is above a reverse-fault threshold, the 32Q element declares a reverse fault.

The 32V element uses Eq. (19-3) to calculate the scalar quantity $z_0$ [2, 12].

$$z_0 = \frac{\text{Re}\left[3\overline{V}_0\ (3\overline{I}_0 \bullet 1\angle Z0MTA)^*\right]}{(3I_0)^2} \tag{19-3}$$

where $\overline{V}_0$ is the zero-sequence voltage; $\overline{I}_0$ is the zero-sequence current; $I_0$ is the magnitude of $\overline{I}_0$; *Z0MTA* is the maximum torque angle of the 32V element (a relay setting).

For forward faults, the directional element measures $z_0 = -\left|\overline{Z}_{0S}\right|$. For reverse faults, the measured scalar quantity is $z_0 = \left|\overline{Z}_{0L} + \overline{Z}_{0R}\right|$ [2].

The 32V element makes directional decisions in the same way as the 32Q element. The element compares $z_0$ against the forward-fault and reverse-fault thresholds to determine the direction of the ground fault.

Figure 19-12 shows the operating characteristics of the 32Q element and the calculated $z_2$ values for forward and reverse faults. For simplicity, we consider the power system to be purely inductive and assume the maximum torque angle (*Z1ANG*) is set to 90°. We use $\overline{Z}_{1L}$ instead of $\overline{Z}_{2L}$ in Fig. 19-12. Recall that $\overline{Z}_{1L} = \overline{Z}_{2L}$ for power lines. The 32V element characteristics are similar to those of the 32Q element.

Because the $z_2$ values for forward and reverse faults are of opposite signs, there is ample margin for setting the forward-fault and reverse-fault thresholds. The characteristic shown in Fig. 19-12*a* provides adequate dependability, security, and sensitivity for most applications. This characteristic has threshold settings around zero, such as *Z2F* = – 0.3 and *Z2R* = 0.3 ohms secondary for relays with 5 A nominal current. The settings are *Z2F* = – 1.5 and *Z2R* = 1.5 ohms secondary for relays with 1 A nominal current. You should conduct fault studies to validate these settings [13].

As mentioned earlier, the characteristic shown in Fig. 19-12*a* fits most applications. Two exceptions are the following:

- Series-compensated line applications.
- Applications with extremely strong sources behind the relay. For example, when the equivalent source impedance is less than 0.5 ohms secondary for a 5 A rated relay or less than 2.5 ohms secondary for a 1 A rated relay.

In these applications, you should use the operating characteristic shown in Fig. 19-12*b*. You can set the forward-fault (*Z2F*) and reverse-fault (*Z2R*) thresholds at approximately one-half of the

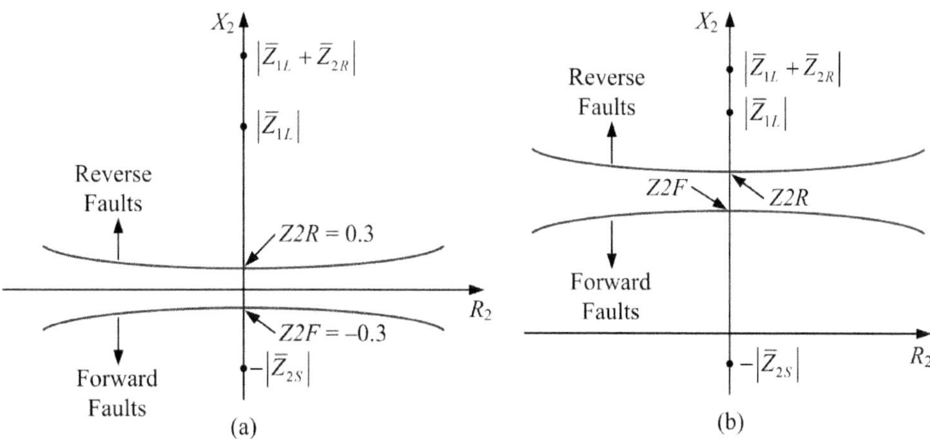


**FIGURE 19-12** Operating characteristics of the negative-sequence, voltage-polarized directional element. (*a*) Characteristic that fits most applications; (*b*) characteristic applicable to series-compensated lines and to cases when the source behind the relay is very strong.

positive-sequence line impedance. Separate these thresholds to create a security band between them. Possible settings are $Z2F = Z_{1L}/2$ ohms secondary and $Z2R = Z_{1L}/2 + 0.1$ ohms secondary. You should conduct fault studies to validate these settings.

The 32I element uses Eq. (19-4) to calculate the scalar quantity $T$ [2, 12].

$$T = \mathrm{Re}\left[3\bar{I}_0 \bullet \bar{I}^*_{POL}\right] \tag{19-4}$$

where $3\bar{I}_0$ is the operating current; $\bar{I}_0$ is the zero-sequence current; $\bar{I}^*_{POL}$ is the complex conjugate of the polarizing current $\bar{I}_{POL}$.

The 32I element compares $T$ with two thresholds. When $T$ is positive and above the positive threshold, the element declares a forward ground fault. When $T$ is negative and below the negative threshold, the element declares a reverse ground fault.

### 19.3.3 Distance Protection

Distance protection uses voltage and current information to determine the distance to the fault. A distance protection scheme may include phase distance elements (21) and ground distance elements (21N), which provide three or more protection zones. Source-impedance changes have almost no effect on the high-speed zone reach of distance protection.

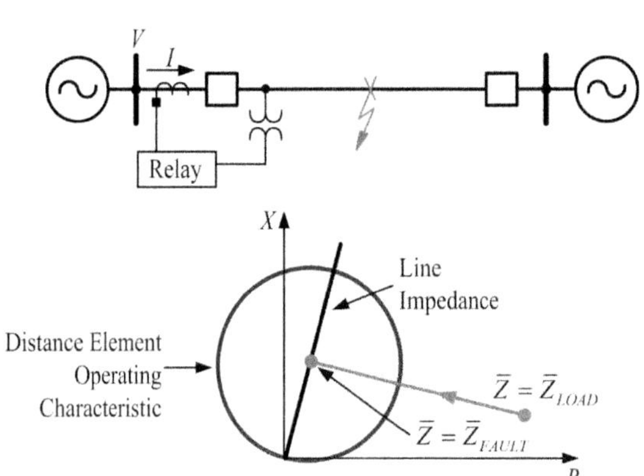


**FIGURE 19-13** Impedance-plane representation of distance element operation.

Directional overcurrent and distance schemes provide primary line protection and also provide backup protection to adjacent lines and buses. These schemes do not require communications channels, but their high-speed protection zones do not cover 100% of the protected line. Thus, clearing of line-end faults may be time delayed.

We use the complex impedance plane, or $R$-$X$ plane, to analyze distance element operation (Fig. 19-13). We plot the element operating characteristic, shown as a mho circle in Fig. 19-13; we also plot the apparent or measured impedance $\bar{Z}$, given by Eq. (19-5). The distance element operates when the measured impedance enters the element operating characteristic.

$$\bar{Z} = \frac{\bar{V}}{\bar{I}} \tag{19-5}$$

where $\bar{V}$ is the element voltage input signal; $\bar{I}$ is the element current input signal.

For normal load, the measured impedance ($\bar{Z} = \bar{Z}_{LOAD}$ in Fig. 19-13) is the result of the power flow through the relay location. Typically, $\bar{Z}_{LOAD}$ plots close to the real axis of the impedance plane.

When a fault occurs on the protected line, the measured impedance rapidly changes from $\bar{Z} = \bar{Z}_{LOAD}$ to $\bar{Z} = \bar{Z}_{FAULT}$. With the proper input signals to the distance element, for bolted (no fault resistance) faults, $\bar{Z}_{FAULT}$ equals the positive-sequence impedance of the line section between the relay location and the fault location. This impedance is proportional to the electrical distance to the fault.

A distance protection scheme typically includes one instantaneous zone (Zone 1) and two definite-time delayed zones (Zone 2 and Zone 3). Zone 1 and Zone 2 provide primary line protection, Zone 2 backs up the remote bus and partially backs up the adjacent lines, and Zone 3 provides remote backup protection for the whole length of the adjacent lines. Figure 19-14 shows the time-distance characteristics of six distance relays protecting three lines that interconnect two sources. The coordination time interval $\Delta T$ provides coordination between Zone 1 and Zone 2 and between Zone 2 and Zone 3 of adjacent distance relays.

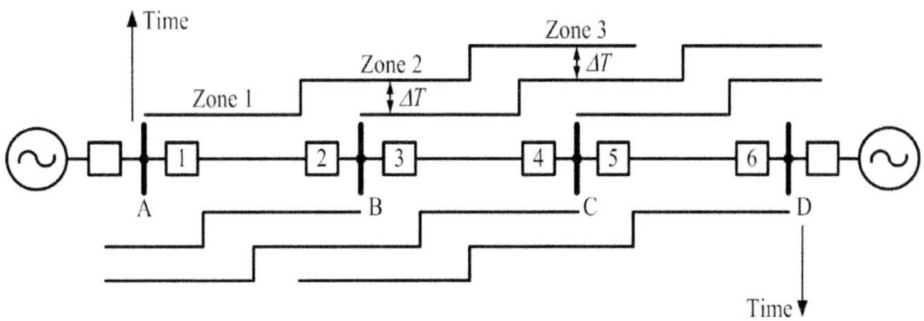


**FIGURE 19-14** Time-distance characteristics of distance relays protecting lines of a two-source power system.

Figure 19-14 shows that midline faults cause instantaneous Zone 1 operation at both line terminals. Should Zone 1 fail to operate, Zone 2 or Zone 3 of the remote backup relay operates. Faults close to a line terminal cause Zone 1 operation of the relay closer to the fault and Zone 2 operation of the remote-terminal relay. When this sequential fault clearing is unacceptable, you can apply either a directional comparison scheme (Sec. 19.3.4) or a line current differential scheme (Sec. 19.3.5).

Instantaneous Zone 1 must underreach the remote line terminal to prevent misoperation for faults on the adjacent lines or the remote bus. Typical reach settings are 70% (for short lines) to 90% (for long lines) of the line positive-sequence impedance. Time-delayed Zone 2 must overreach the protected line to ensure coverage of the remote bus and the line-end section not covered by Zone 1. A typical reach setting is 120% of the line positive-sequence impedance. Time-delayed Zone 3 should overreach the longest adjacent line under all system operation conditions.

When Zone 2 does not reach beyond Zone 1 of the adjacent line relays, it is possible to time-coordinate the Zone 2 elements with the Zone 1 elements of the adjacent line relays and with the remote bus relays, as shown in Fig. 19-14. When an adjacent line is short, it may not be possible to prevent Zone 2 from overreaching the Zone 1 of the adjacent line relays. In this case, the Zone 2 elements must time-coordinate with the Zone 2 elements of the adjacent line relays.

When Zone 3 does not reach beyond Zone 2 of the adjacent line relays, the Zone 3 elements must time-coordinate with the Zone 2 elements of the adjacent line relays, as shown in Fig. 19-14. Long Zone 3 reach settings make it difficult to avoid adjacent Zone 3 overlap and require long time delays. Long-reaching distance zones are also prone to misoperation for load conditions. For these reasons, local backup protection, including breaker-failure protection, is the preferred backup protection for high-voltage and extra-high-voltage transmission lines [2].

Protecting a three-phase transmission line for all possible fault types generally requires three phase distance elements and three ground distance elements. Phase elements respond to three-phase, phase-to-phase, and phase-to-phase-to-ground faults. Ground elements respond to single-phase-to-ground and, in some implementations, phase-to-phase-to-ground faults.

Table 19-1 summarizes the input signals to traditional phase and ground distance elements. Reference [14] and other books provide the mathematical derivation of these input signals. For bolted faults, phase and ground elements that receive only faulted-phase information (referred to as the fault loop

**TABLE 19-1** Voltage and Current Input Signals to Traditional Phase and Ground Distance Elements

| Distance elements | | Voltage ($\bar{V}$) | Current ($\bar{I}$) |
|---|---|---|---|
| Phase elements | AB | $\bar{V}_a - \bar{V}_b$ | $\bar{I}_a - \bar{I}_b$ |
| | BC | $\bar{V}_b - \bar{V}_c$ | $\bar{I}_b - \bar{I}_c$ |
| | CA | $\bar{V}_c - \bar{V}_a$ | $\bar{I}_c - \bar{I}_a$ |
| Ground elements | AG | $\bar{V}_a$ | $\bar{I}_a + \bar{k}_0 \bullet \bar{I}_r$ |
| | BG | $\bar{V}_b$ | $\bar{I}_b + \bar{k}_0 \bullet \bar{I}_r$ |
| | CG | $\bar{V}_c$ | $\bar{I}_c + \bar{k}_0 \bullet \bar{I}_r$ |

elements) measure the positive-sequence impedance of the faulted line section, as Eq. (19-6) shows. For example, for a BC fault, $\bar{V} = \bar{V}_b - \bar{V}_c$ and $\bar{I} = \bar{I}_b - \bar{I}_c$ in Eq. (19-6).

$$\bar{Z} = \frac{\bar{V}}{\bar{I}} = m\bar{Z}_{1L} \tag{19-6}$$

where $\bar{Z}_{1L}$ is the line positive-sequence impedance; $m$ is the distance to the fault in per unit of $Z_{1L}$.

Ground distance elements require the phase currents to be compensated by the residual current $\bar{I}_r$ [Eq. (19-7)] times a multiplying factor $\bar{k}_0$ [Eq. (19-8)].

$$\bar{I}_r = \bar{I}_a + \bar{I}_b + \bar{I}_c \tag{19-7}$$

where $\bar{I}_a$ is the measured A-phase current; $\bar{I}_b$ is the measured B-phase current; $\bar{I}_c$ is the measured C-phase current.

$$\bar{k}_0 = \frac{\bar{Z}_{0L} - \bar{Z}_{1L}}{3\bar{Z}_{1L}} \tag{19-8}$$

where $\bar{Z}_{0L}$ is the line zero-sequence impedance.

Another approach is to use the zero-sequence current $\bar{I}_0$ instead of the residual current $\bar{I}_r$ in the ground distance element input signals. In this approach, the value of $\bar{k}_0$ is three times the value given by Eq. (19-8).

Modern distance elements use other voltages or even currents as polarizing quantities instead of the voltage input signals listed in Table 19-1 [2].

Distance elements make phase or amplitude comparison of signals derived from the measured voltages and currents to create operating characteristics. Electromechanical relays perform torque comparison. Most solid-state analog relays use time-coincidence phase-comparison techniques. Microprocessor-based relays use torque-like products and other methods to create their operating characteristics [15].

Figure 19-15 shows two typical distance element operating characteristics. The operating characteristic of the mho element is a circle that crosses at the origin of the coordinates, so the relay responds to forward faults and not to reverse faults. Mho distance elements are often used for line phase and ground fault protection.

Quadrilateral distance elements are often used for line ground fault protection and for phase fault protection of short lines. Quadrilateral distance elements are preferred over mho distance elements

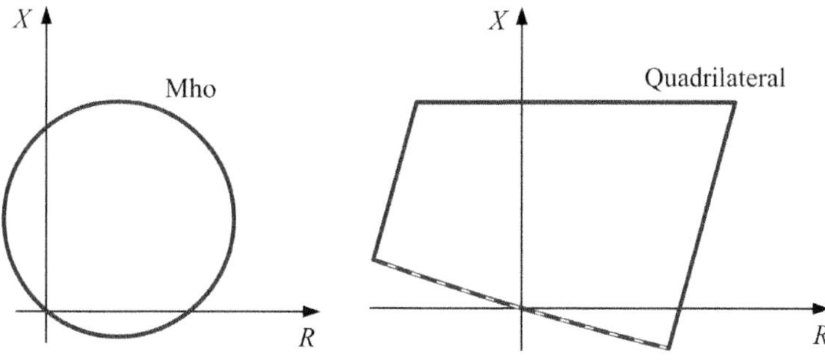


**FIGURE 19-15** Typical distance element operating characteristics.

for ground fault protection of impedance-grounded systems. Shaping a quadrilateral distance characteristic requires three types of elements:

- Reactance element (top line).
- Resistance elements (right-side and left-side lines).
- Directional element (supervises the reactance and resistance elements).

The quadrilateral distance element operates when all elements operate.

The mho element described in [15] maps all points of a mho circle onto a single point on a straight line by using Eq. (19-9) to calculate the per unit distance to the fault *m*. The element compares *m* with a reach setting. This approach is computationally efficient, because each fault loop needs only a single calculation for all protection zones.

$$m = \frac{\mathrm{Re}\left[\bar{V} \bullet \bar{V}_P^*\right]}{\mathrm{Re}\left[\bar{Z}_{1L} \bullet \bar{I} \bullet \bar{V}_P^*\right]} \tag{19-9}$$

where *m* is the distance to the fault in per unit of $Z_{1L}$; $\bar{Z}_{1L}$ is the line positive-sequence impedance; $\bar{I}$ is the element current input signal; $\bar{V}$ is the element voltage input signal; $\bar{V}_P$ is the polarizing voltage.

Figure 19-16 shows the operating characteristics on the impedance plane of self-polarized and polarized mho elements. In this figure, $\bar{Z} = \bar{V}/\bar{I}$ and $\bar{Z}_P = \bar{V}_P/\bar{I}$. The self-polarized mho element uses $\bar{V}$ as the polarizing voltage (any of the voltages of Table 19-1). Its characteristic passes through the origin of coordinates, as shown in Fig. 19-16*a*. However, $\bar{V}$ is very small for close-in, bolted faults, and the element may not perform well because the polarizing signal is not a reliable reference. The polarized mho element adds unfaulted-phase information to the polarizing quantity $\bar{V}_P$ to solve this problem. Typical polarizing quantities are the positive-sequence voltage and unfaulted-phase voltages. These polarizing quantities expand the circle to the third quadrant for forward faults (Fig. 19-16*b*) and reduce it on the first quadrant for reverse faults (Fig. 19-16*c*). The mho characteristic has a fixed reach point, defined by the replica impedance $\bar{Z}_R$ (a relay setting). However, even positive-sequence voltages and unfaulted-phase voltages are very small for close-in, bolted, three-phase faults. Memory polarization solves this problem and also the voltage inversion problems in series-compensated line applications [2].

Under ideal conditions, distance elements should measure the positive-sequence impedance of the line section between the relay and the fault. Problems such as infeed effect, fault resistance, magnetic mutual coupling, load encroachment, effect of unfaulted phases, coupling-capacitor voltage

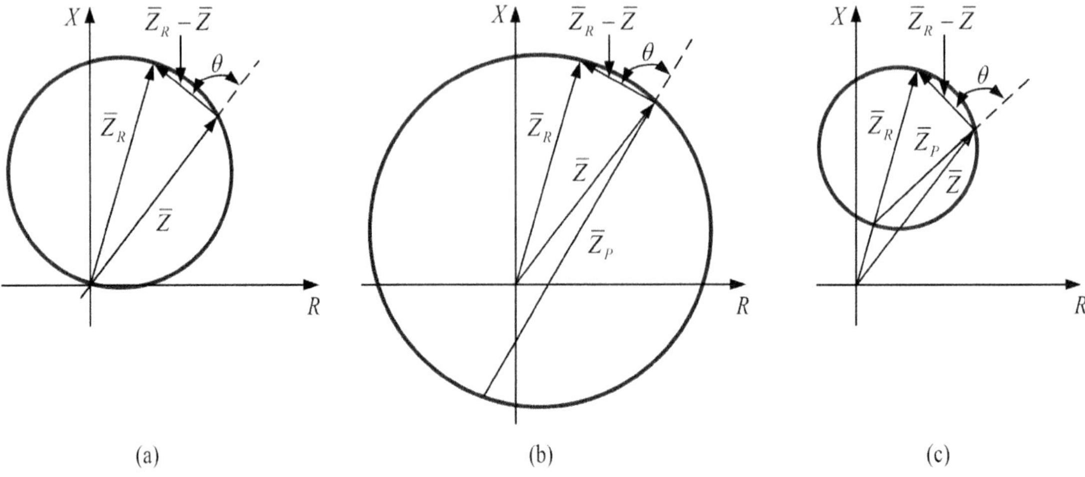


**FIGURE 19-16** Mho element characteristics. (*a*) Self-polarized mho element; (*b*) polarized mho element responding to a forward fault; (*c*) polarized mho element responding to a reverse fault.

transformer transients, and loss of potential, cause distance element measuring errors. Reference [2] examines these problems and provides solutions.

### 19.3.4 Directional Comparison Protection

Pilot protection, or teleprotection, uses a communications channel to compare information from the line terminals and provide high-speed fault clearing for 100% of the protected line. Pilot protection includes:

- Directional comparison schemes.
- Current-based schemes:
  - Phase comparison schemes.
  - Current differential schemes.

In a directional comparison scheme, forward- and reverse-looking instantaneous directional overcurrent or distance elements provide information for the scheme logic at each line terminal. The forward-looking elements are set to overreach the remote terminal with enough margin to detect all in-section faults. For an internal fault, both forward-looking elements operate. For an external fault, one forward-looking and one reverse-looking element operate. The scheme uses this information at each line terminal to provide fast tripping for internal faults. Underreaching elements at each terminal provide instantaneous protection, which is independent of the communications-assisted tripping logic.

Directional comparison schemes do not require a high-bandwidth channel because the relays exchange information on the status of their directional or distance elements. Typical bandwidth requirements are 0.5 to 1.5 kHz for analog channels and 9.6 kbps for digital channels. The most common communications channels for directional comparison schemes are power line carrier, microwave, optical fiber, and point-to-point radio.

The most widely used directional comparison schemes are the following:

- Permissive underreaching transfer trip (PUTT).
- Permissive overreaching transfer trip (POTT).
- Directional comparison blocking (DCB).
- Directional comparison unblocking (DCUB).

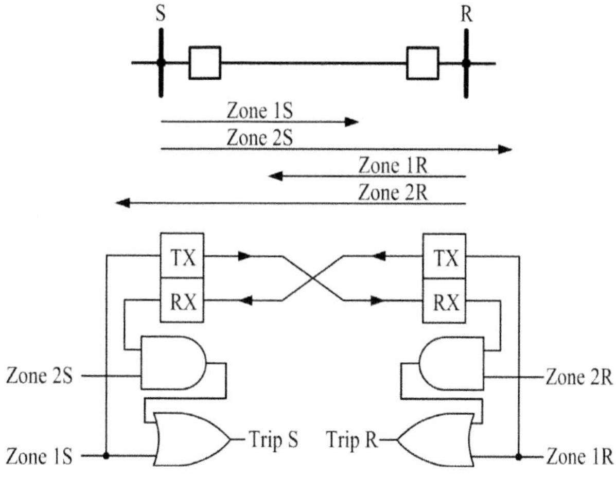


**FIGURE 19-17** Permissive underreaching transfer trip logic.

The PUTT scheme (Fig. 19-17) uses an instantaneous, forward-looking, underreaching element (a directional overcurrent or distance element, referred to as Zone 1 element in the figure) to trip the local breaker and send a permissive trip signal to the remote terminal. An instantaneous, forward-looking, overreaching element (referred to as Zone 2 element in the figure) supervises tripping at each line terminal. The Zone 2 element can be an overreaching directional overcurrent or distance element. If the remote Zone 2 element detects a fault, the remote relay trips the breaker when it receives the permissive signal. Using the overreaching element to supervise tripping makes this scheme secure under noisy channel conditions. Because the PUTT scheme uses an underreaching element to send permission, it does not send a permissive signal for out-of-section faults. For this reason, PUTT schemes do not require additional supervisory logic to maintain security under current-reversal conditions on parallel lines [2]. However, the underreaching element limits the fault-resistance coverage of PUTT schemes.

The POTT scheme (Fig. 19-18) uses an overreaching Zone 2 element to trip the local breaker and send a permissive trip signal to the remote terminal. The Zone 2 element can be an overreaching directional overcurrent or distance element. If the remote Zone 2 element detects a fault, the remote relay trips the breaker when it receives the permissive signal. The overreaching elements of POTT schemes provide better fault-resistance coverage than those of the PUTT schemes because of their

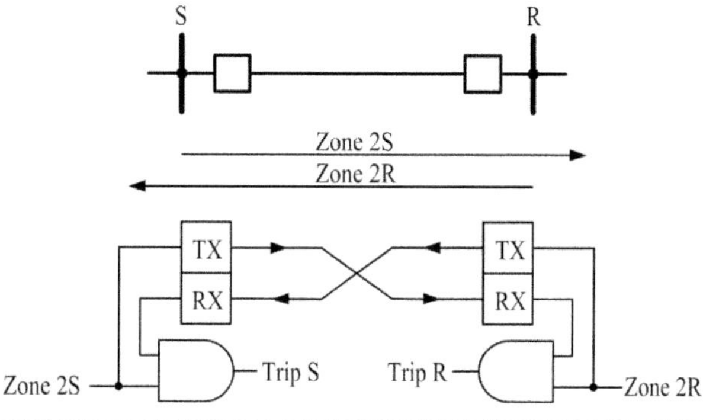


**FIGURE 19-18** Permissive overreaching transfer trip logic.

higher sensitivity. Because the scheme uses an overreaching element to send permission, in parallel line applications, it may misoperate for a current-reversal condition in the healthy line. In addition, if one of the relays fails to detect a forward fault due to a weak source supplying the fault (weak infeed) or if the breaker is open, the POTT scheme will not clear the fault at high speed. Current-reversal, echo, and weak-infeed logic solve these problems. Some relays provide a hybrid scheme that includes the POTT logic and the additional supervisory logic [16]. In other relays, the user adds the supervisory logic when required.

Unlike PUTT and POTT schemes, which send a tripping signal when they detect a fault in the forward direction, the DCB scheme (Fig. 19-19) sends a trip-blocking signal when it detects a fault in the reverse direction. The DCB scheme uses an instantaneous, reverse-looking element (referred to as Zone 3 element in the figure) to send the blocking signal. The Zone 3 element can be a directional overcurrent element or a reversed Zone 3 distance element. If the local Zone 3 element detects a reverse fault, it sends a blocking signal to the remote terminal. If the remote Zone 2 element detects a fault, the remote relay trips the breaker after a short coordinating time delay *CTD*, if it does not receive the blocking signal. In many applications, a fast, nondirectional element sends the blocking signal. In these cases, the blocking signal is quickly shut off if the fault is in the forward direction.

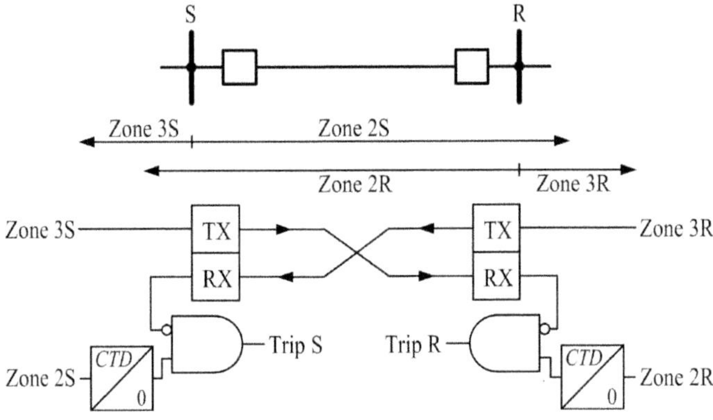


**FIGURE 19-19** Directional comparison blocking logic.

The DCUB scheme continuously transmits a guard signal under normal conditions. When the channel is healthy, the DCUB scheme operates similarly to POTT schemes. If the local Zone 2 element detects a fault, the guard signal is shut off, and a trip signal is sent. The remote relay detects the change in signals from guard to trip. If the remote Zone 2 element also detects a fault, the remote relay trips the breaker. The DCUB scheme also uses logic that ensures tripping for internal faults that result in a loss of channel. This logic permits a trip for a predefined time (100 to 150 ms) if the DCUB scheme detects a loss of channel simultaneously with the Zone 2 element detecting a fault, even if the DCUB scheme does not receive a trip signal. After the predefined time, the DCUB scheme is locked out to prevent misoperations on long-term loss-of-channel conditions.

An important consideration in selecting directional comparison schemes is the effect of a communications channel failure on the scheme. Blocking schemes are biased toward dependability: they can trip for internal faults even when the channel is out of service. However, these schemes may overtrip for out-of-section faults that occur while the channel is not available. Hence, blocking schemes are suitable for channels that share the same path with the protected line and that can be affected by the line fault, such as power line carrier and fiber-optic cable embedded in the overhead line shield wire.

Permissive tripping schemes are faster and have higher security than blocking schemes, but lower dependability: if the communications channel is not operational, the scheme fails to operate for in-section faults. For channels not affected by line faults, such as wireless or separately routed fiber-optic links, permissive tripping schemes can provide faster tripping and take advantage of continuous channel monitoring. The DCUB scheme is as fast as tripping schemes and can be applied with channels that share the same path with the protected line.

### 19.3.5 Current Differential Protection

Line current differential protection (87L) compares current information from all the line terminals over a communications channel (Fig. 19-20). Phase comparison protection, a particular type of current differential protection, only compares current phase angle information, which reduces the channel bandwidth requirements. However, line current differential protection is more sensitive than phase comparison protection [17]. Line current differential protection is more secure and dependable than other types of protection in response to the effects of unbalances, power swings, mutual coupling, and voltage inversion [2]. It performs well for evolving, intercircuit, and cross-country faults [2].

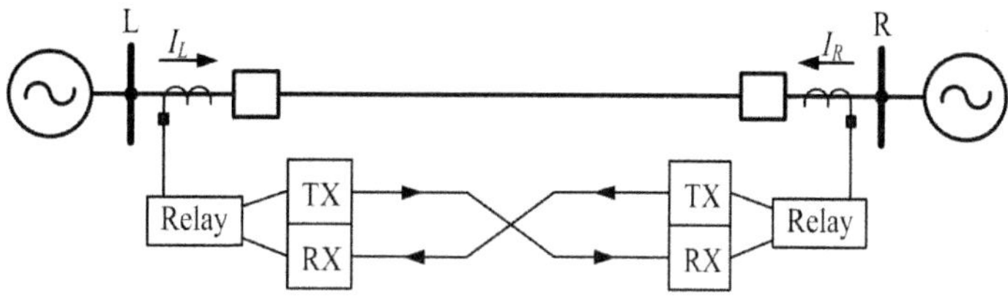


**FIGURE 19-20** Line current differential schemes compare current information from all the line terminals.

The first line current differential schemes used a pilot wire channel (a telephone line), which limited its application to short lines. The other analog communications channels were only suited to directional comparison or phase comparison protection. Now, modern digital channels also support line current differential schemes. A microprocessor-based line current differential scheme requires a fiber-optic or digital microwave channel with a bandwidth of 56 kbps or higher. References [2] and [18] provide comprehensive coverage of line current differential protection.

Microprocessor-based line current differential schemes compare digitized current samples or current phasor values. Each relay that receives current data from all line terminals operates in a master mode: it time-aligns data, executes a differential protection algorithm, and trips its breaker when required. Each relay that serves data, but does not receive all the remote data, operates in a slave mode: a direct transfer trip received from a remote master relay allows a slave relay to trip its breaker.

The channel time delay causes a fictitious phase shift between the local current and the received remote current(s). Line current differential schemes align (synchronize) current samples or phasors to prevent the errors caused by this phase shift. Data synchronization methods include the following:

- *Channel-based synchronization.* This method measures the two-way channel time delay by time-stamping the exact instants that test signals are sent and received. It assumes the clock offset between the two relays is equal to half the measured two-way channel time delay. Channel-based synchronization works well for symmetrical channels. Channel asymmetry causes a fictitious phase shift between the aligned local and received currents.
- *External-time-reference-based synchronization.* This method uses an external time reference for synchronization as a complement or replacement for channel-based synchronization. Reliance on a global satellite system and additional devices is not an ideal differential protection solution. A wide-area terrestrial time-distribution system using a SONET or Ethernet network is an excellent choice for modern line current differential schemes [19].

Traditional line current differential schemes use percentage differential elements, which compare operating current $I_{OP}$ with restraining current $I_{RT}$. The element generates a tripping signal if $I_{OP}$ is greater than a percentage of $I_{RT}$ and is also greater than a minimum pickup current. The element operating characteristic is typically a plot of $I_{OP}$ as a function of $I_{RT}$ (Sec. 19.4.1).

The current-ratio complex plane, or alpha plane, provides a way to analyze differential element operation [20]. You can superimpose on the alpha plane the element characteristic and the current-ratio locus resulting from the fault or abnormal condition in the power system. This method is equivalent to analyzing distance element operation on the impedance plane. For a two-terminal line (Fig. 19-20), the alpha-plane complex variable is the ratio $\overline{I}_R/\overline{I}_L$ of the remote current to the local current.

Figure 19-21 depicts the alpha-plane representation of power system load and fault conditions [2, 17, 21]. For ideal through-current conditions (power flow or external faults with no CT errors), the magnitudes of $\overline{I}_L$ and $\overline{I}_R$ are equal, and their phases are 180° apart. Therefore, $\overline{I}_R/\overline{I}_L = 1\angle 180°$. Hence, through-current conditions ideally plot at point $1\angle 180°$ on the alpha plane. Channel asymmetry causes the current ratio to rotate around the origin on the alpha plane. The ratio magnitude is unchanged. The rotation angle equals the angle error caused by channel asymmetry. For example, a 1 ms error rotates the current ratio 21.6° when the system frequency is 60 Hz. Figure 19-21 shows how channel asymmetry expands the through-current dot into a circular arc.

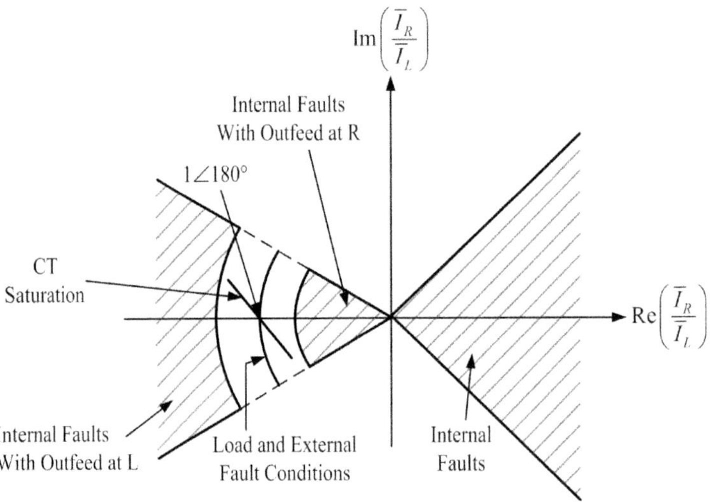


**FIGURE 19-21** Alpha-plane representation of power system load and fault conditions.

For internal faults, currents $\overline{I}_L$ and $\overline{I}_R$ are in phase under ideal conditions. These faults plot on the positive real axis of the alpha plane. In general, the angles of $\overline{I}_L$ and $\overline{I}_R$ for internal faults depend on the source voltage angles (prefault power flow condition) and the angles of the system impedances at both sides of the fault point. Figure 19-21 shows the internal fault region on the alpha plane as an angular sector that accommodates source voltage and impedance angle differences. Channel asymmetry further expands the internal fault region. Negative- and zero-sequence differential elements do not respond to balanced power flow. Hence, the alpha-plane internal fault region is narrower for these elements, because it only considers system impedance angle differences.

Line-charging current and tapped load cause current to flow into the line at all terminals. They traverse in the internal fault region of the alpha plane. The traditional method for avoiding misoperation is setting the differential element pickup above these currents, but this method desensitizes the differential scheme. To accommodate line-charging current, one modern relay uses negative- and zero-sequence differential elements [17]. Another relay uses a line-charging current compensation algorithm that estimates the line-charging current and subtracts it from the relay measured currents [19]. These relays include overcurrent elements that respond to the total fault current and can time-coordinate with tapped load overcurrent protection. The line current differential elements supervise the overcurrent elements. These solutions allow the relay to accommodate line-charging current and tapped load without sacrificing sensitivity.

For some internal faults, the current flows out of the line at one terminal [2, 17]. High-resistance internal faults with fault current less than load current cause outfeed conditions. In series-compensated lines, outfeed occurs when the reactance from one of the sources to the fault point is capacitive. A three-terminal line with a strong external parallel tie between two terminals may experience outfeed at one terminal for some internal faults. For internal faults with outfeed at one line terminal, the angle between $\overline{I}_L$ and $\overline{I}_R$ is close to 180°, and their magnitudes are not equal. These faults plot close to the negative real axis of the alpha plane but do not coincide with the $1\angle 180°$ point, as shown in Fig. 19-21. Channel asymmetry further expands these fault regions.

When a CT saturates, the fundamental component of the secondary current decreases in magnitude and advances in angle. When the local CT saturates and the CT at the remote terminal of the protected line does not saturate, the current-ratio magnitude of the phase currents increases and its phase angle decreases, moving the operating point upward and to the left from the $1\angle 180°$ point (Fig. 19-21). When the remote CT saturates and the local CT does not saturate, the current-ratio magnitude decreases and its phase angle increases, moving the operating point downward and to the right. Because of the effect of the current dc offset on CT saturation and the relay filtering transients, the current ratio actually describes a time-dependent irregular trajectory. For simplicity, Fig. 19-21 depicts this trajectory as a straight line.

A differential element must restrain for through-load conditions and external faults. Therefore, its restraining region on the alpha plane must include the point $1\angle 180°$. The element must trip for internal faults, so the restraining region must exclude the shaded areas shown in Fig. 19-21.

Figure 19-22 shows the alpha-plane differential element characteristic that is available in some modern relays [17]. The restraining region includes the point $1\angle 180°$ and perfectly fits the shape of the through-load and external fault region depicted in Fig. 19-21. Setting *87LANG* determines the angular extent of the restraining region. Setting *87LR* determines the restraining region outer radius. The inner radius is the reciprocal of *87LR*. The differential element operates when the current ratio $\overline{I}_R/\overline{I}_L$ leaves the restraining region and the differential current magnitude $|\overline{I}_L + \overline{I}_R|$ is above a minimum pickup value (a relay setting). The alpha-plane differential characteristic is the best choice for line protection because it is more tolerant of channel asymmetry than traditional percentage differential characteristics [2, 17, 21].

Combining phase and sequence-component differential elements in the same relay [2] provides very fast operation for phase faults and very high fault-resistance coverage for ground faults. Some modern relays contain five differential elements: three phase elements, one negative-sequence element, and one zero-sequence element. The phase elements provide high-speed operation for high-current faults. Negative- and zero-sequence elements provide sensitive protection for unbalanced faults without compromising security.

Reference [19] describes an advanced differential protection scheme for multiterminal lines. Some characteristics of this scheme are the following:

- Uses generalized phase, negative-sequence, and zero-sequence alpha-plane differential elements for multiterminal line applications.
- Consolidates all the currents in each line terminal into one partial differential current and one partial restraining current to optimize the use of the channel bandwidth. The partial currents feed into the generalized alpha-plane calculations.
- Provides external fault detection logic at each line terminal for higher security.
- Provides line-charging current compensation for higher sensitivity.
- Accommodates in-line transformers without sacrificing sensitivity.
- Can use direct or multiplexed fiber-optic channels and also wide-area SONET or Ethernet networks.
- Provides channel-based and/or external-time-reference-based data synchronization.

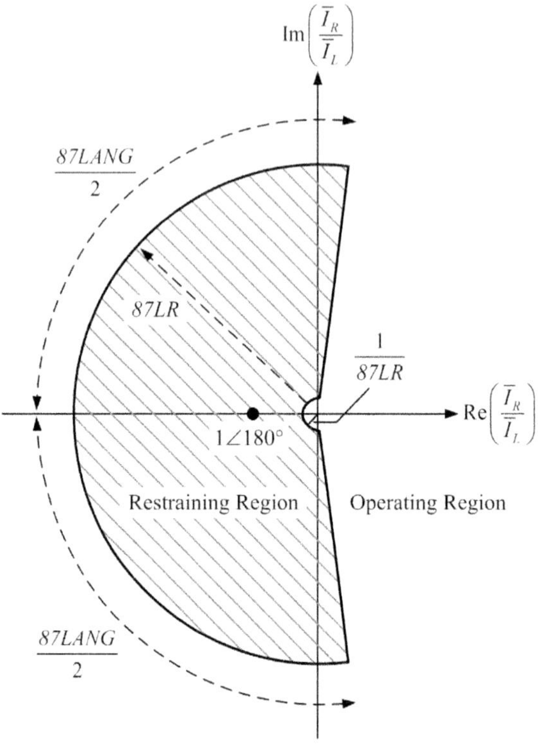


**FIGURE 19-22** Alpha-plane differential element characteristic.

### 19.3.6 Time-Domain Protection

Faults launch traveling waves (TWs) that travel close to the speed of light and get reflected and transmitted at buses and other discontinuities according to their corresponding characteristic impedances. In the initial stage of the fault, which lasts a few microseconds, the power system behaves as a distributed parameter network. After a few roundtrip reflections, TWs recombine into stationary waves and the power system behaves as a lumped-parameter resistive-inductive-capacitive network. This second stage of the fault lasts a few milliseconds. Finally, when the signals settle down, the faulted power system reaches a steady state and can be represented as a fundamental-frequency network.

Relay input signals include the different frequency components created by this transient process. The TW component frequency is on the order of hundreds of kilohertz. The transient lumped-parameter network components have frequencies on the order of hundreds of hertz. The steady-state fault component has the power system frequency.

Most electromechanical, solid-state, and microprocessor-based relays operate on phasor-based principles. These relays apply filtering so that they respond to the fundamental-frequency signal components. Extracting these components implies a filter delay on the order of one fundamental-frequency cycle. We can understand this simply from the following observation: in order to accurately measure a frequency component, we need to observe this component for the duration of its cycle [22].

The filtering necessary for phasor measurement results in relay operating times of about one fundamental-frequency cycle, with the best-case times approaching half a cycle for close-in high-current faults, as shown in Fig. 19-23. This particular relay has a dual-filter scheme that uses a full-cycle mho element in parallel with a high-speed, half-cycle element [2, 23].

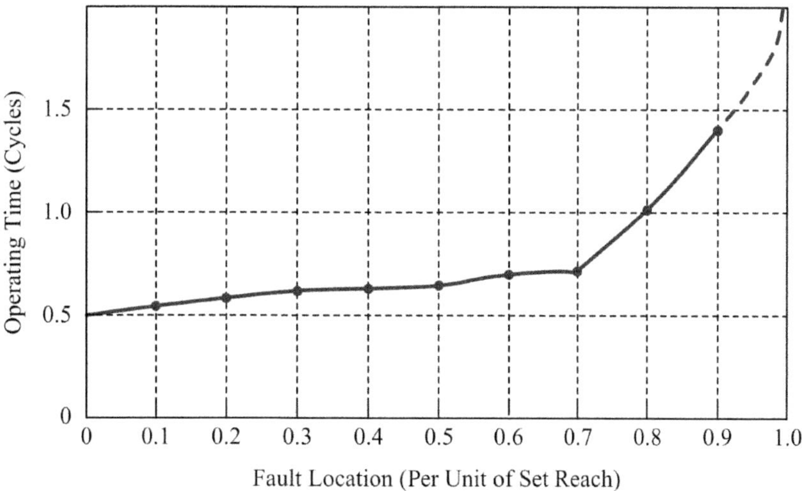


**FIGURE 19-23** Typical operating times of a phasor-based distance relay with a dual-filter scheme.

As a result, primary protection schemes typically operate in one to one-and-a-half cycles, and the breakers interrupt current in one-and-a-half to three cycles, so faults are typically cleared in three to four cycles. Modern power systems require even faster fault clearing to improve power system stability, enhance public and utility personnel safety, limit equipment wear, improve power quality, and reduce property damage. Every millisecond saved in fault clearing time means more power can be transferred. Reference [24] shows that on a particular line, a one-cycle reduction in fault clearing time increased the power transfer by 250 MW, amounting to about 15 MW/ms.

Today's electronics and communications technology allows meeting the need for fast fault clearing of modern power systems. High sampling rates, data storage, processing power, and communications capabilities of new relay hardware platforms allow us to further improve line protection operating times.

Today, time-domain relays are becoming available for ultra-high-speed line protection. These relays use TW principles as well as incremental-quantity principles to provide fast and secure line protection. A new time-domain relay includes the following elements:

- Elements that use TW information to provide ultra-high-speed protection, with possible operating times that are below 1 ms. The speed of light is the limiting factor. TWs from a fault anywhere on a 100 mi line reach both terminals within 600 μs.
- Elements that use transient lumped-parameter network components and reliably operate in milliseconds. These elements use incremental quantities (signals that appear due to a fault and do not contain load quantities).

References [22] and [25] describe the theory, implementation, and performance evaluation of time-domain line protection elements. These elements are an incremental-quantity directional element (TD32) and a TW directional element (TW32) suitable for a POTT scheme, an incremental-quantity distance element (TD21), and a TW line current differential element (TW87). Let us briefly consider the TW32 and TW87 operating principles, which require the relay to sample line currents and voltages at a rate suitable for measuring TWs (e.g., 1 MHz).

Figure 19-24 illustrates the TW32 element operating principle. The relative polarity of the voltage ($v_{TW}$) and current ($i_{TW}$) TWs convey information of the fault direction. Because of the CT connection, the two signals have opposite polarities for forward faults and the same polarity for reverse faults. Hence, a simple directional element checks the relative polarities of the current and voltage traveling waves.

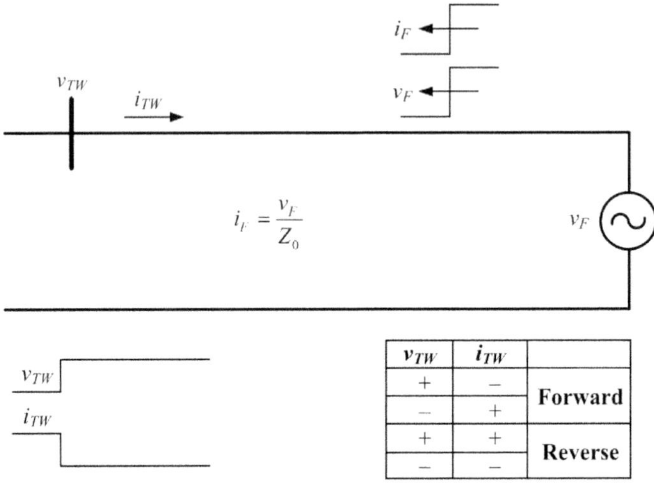


| $v_{TW}$ | $i_{TW}$ | |
|---|---|---|
| + | – | Forward |
| – | + | |
| + | + | Reverse |
| – | – | |

**FIGURE 19-24** TW32 element operating principle.

A practical TW32 implementation uses the logic shown in Fig. 19-25 instead of just checking the TW polarities [25]. The logic calculates a TW torque as a product of the TW current and the sign-inverted TW voltage (so the torque is positive for forward faults) and integrates the torque over time. For security, the integrator is released only if the voltage and current TWs are above minimum levels. The logic checks the output of the integrator ($E_{FWD}$) after a set time period (on the order of tens to hundreds of microseconds) from the beginning of the disturbance. The *TW32* output asserts when $E_{FWD}$ exceeds a security margin. Each TW coming from the line direction (forward faults) will integrate up, and each TW coming from behind the relay (reverse faults) will integrate down. The first TW is higher in magnitude than the subsequent reflections. As a result, the integrated value is a reliable indication of the fault direction even as multiple reflected TWs are integrated over the set time period.

The TW87 element uses only current information and is based on the concept that the fault is the source of the current TWs. Hence, the following is true:

- For external faults and other surges, the current TWs enter the line at one terminal and leave the line at the other terminal after the line propagation time.

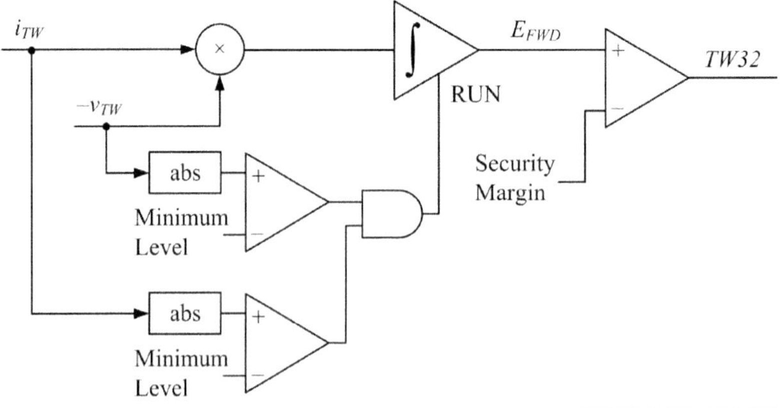


**FIGURE 19-25** Simplified logic of the TW32 element.

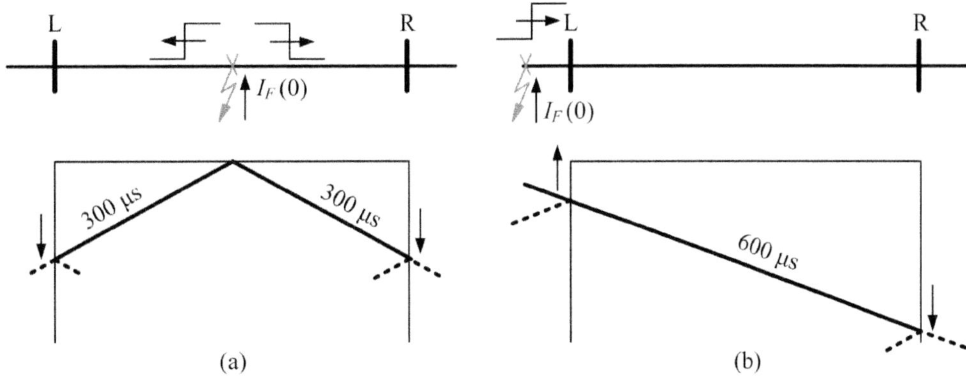


**FIGURE 19-26** Application of the current-based TW differential principle to a line with a propagation time equal to 600 μs. (*a*) Internal midline fault; (*b*) external fault.

- For internal faults, the current TWs leave the line at both terminals. The difference between their arrival times to both line terminals is less than the line propagation time.
- Because of the CT connection, the measured current TWs have the same polarity at both line terminals for internal faults. For external faults, the measured current TWs have opposite polarities at both line terminals.

In this new TW differential principle, each differential element measures the local current TW and receives the time-stamped current TW measurement from the remote line terminal. Each element adds the time-aligned current TWs to create an operating quantity. Each element subtracts from the local TW the remote TW measured with a time difference equal to the line propagation time to create a restraining quantity. For internal faults (same polarity TWs), the operating quantity is greater than the restraining quantity (a tripping condition). For external faults (opposite polarity TWs), the operating quantity is smaller than the restraining quantity (a nontripping condition).

Figure 19-26 shows the application of the TW differential principle to a line with a propagation time equal to 600 μs. We use the Bewley lattice diagram to represent TWs on a distance-time plane. For a mid-line fault (Fig. 19-26*a*), the current TWs arrive at both line terminals at the same time, and they are measured with the same polarity. The operating quantity (sum of the time-aligned TWs) is large, and the restraining quantity (difference of the TWs 600 μs apart) is small. For internal faults at other locations we get the same result. For an external fault (Fig. 19-26*b*), the first TW travels the entire line. This TW is measured with opposite polarities at both line terminals. The operating quantity is small, and the restraining quantity is large.

## 19.4 TRANSFORMER PROTECTION

Transformer failures are expensive. Consider the cost of repairing the transformer, the cost of energy not delivered because of transformer unavailability, and the possible additional costs of spreading damage to adjacent equipment or of a major system blackout. Transformer failures are also dangerous.

Internal and external faults and other abnormal operation conditions, such as overload, overexcitation, and overvoltage, stress transformers. The transformer protection scheme must disconnect the transformer before extensive damage occurs in the transformer and/or the power system.

### 19.4.1 Differential Protection

The differential protection (87) principle follows Kirchhoff's current law (KCL) in applications where a metallic per-phase connection exists between the terminals of protected equipment. These applications include buses, lines, generator and motor stators, and reactors. The sum of the terminal currents in each phase (also called the differential current) ideally equals zero for through-current conditions (i.e., no internal faults) and equals the fault current for internal faults that violate KCL for the protected apparatus. The relay trips the protected apparatus when it detects a differential current. The differential protection zone is precisely defined by the location of the CTs, which provides high selectivity. This high selectivity allows a differential relay to trip quickly without needing any time delay to coordinate with other relays.

Transformer differential protection follows a different principle. The individual windings of a power transformer are galvanically isolated (except in the case of autotransformers), and KCL cannot be directly applied. In transformer applications, the differential principle follows the ampere-turn (AT) balance between the core legs [26]. The sum of the ATs around any closed magnetic circuit loop ideally equals zero for transformer through-current conditions (i.e., no internal faults). An internal fault (including a turn-to-turn fault) upsets the AT balance and can be detected by the transformer differential elements.

In each phase, you can parallel the CTs connected at all protected equipment terminals to obtain the differential current. Differential overcurrent protection and high-impedance differential protection use this connection, which requires dedicated CTs [2]. Modern differential relays calculate the differential current mathematically and do not require dedicated CTs.

The differential current resulting from CT errors, including unequal CT saturation, can cause undesirable differential element operation for external faults. In a differential scheme, the worst-case scenario for an external fault occurs when the CT on the faulted branch circuit saturates completely, while the other CTs remain unsaturated.

The differential current caused by CT errors increases with the through-fault current (the current that flows through the differential protection zone for an external fault). Based on this fact, researchers developed the percentage differential element, which compares the operating current $I_{OP}$ with the restraining current $I_{RT}$ that is a measure of the through current. Figure 19-27 shows a typical differential element connection diagram.

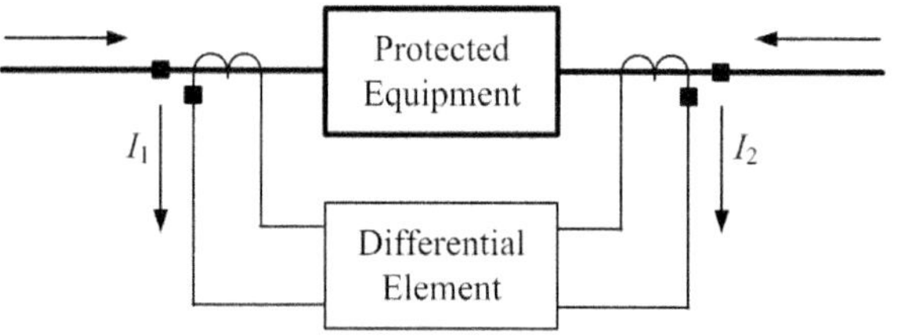


**FIGURE 19-27** Typical differential element connection diagram.

Equation (19-10) defines the operating current $I_{OP}$.

$$I_{OP} = |\overline{I}_1 + \overline{I}_2| \tag{19-10}$$

where $\overline{I}_1$ and $\overline{I}_2$ are the differential element input currents.

The restraining current $I_{RT}$ is a designer's choice. Equations (19-11) to (19-13) are the most common expressions for $I_{RT}$. Equations (19-12) and (19-13) also apply to differential elements with more than two restraint elements.

$$I_{RT} = k|\overline{I}_1 - \overline{I}_2| \tag{19-11}$$

$$I_{RT} = k\left(\left|\overline{I}_1\right| + \left|\overline{I}_2\right|\right) \tag{19-12}$$

$$I_{RT} = \text{Max}\left(\left|\overline{I}_1\right|, \left|\overline{I}_2\right|\right) \tag{19-13}$$

where $k$ is a scaling factor, usually equal to 1 or 0.5.

The differential element generates a tripping signal if the operating current $I_{OP}$ is greater than the minimum pickup current $I_{PU}$ and is also greater than a percentage of the restraining current $I_{RT}$.

$$I_{OP} > I_{PU} \tag{19-14}$$

$$I_{OP} > SLP\ I_{RT} \tag{19-15}$$

where $SLP$ is the slope, a relay setting.

Figure 19-28 shows the differential element operating characteristic as a scalar plot of $I_{OP}$ as a function of $I_{RT}$. The characteristic is a straight line with a slope equal to $SLP$ and a horizontal straight line defining the element minimum pickup current $I_{PU}$. The operating region is located above the characteristic, and the restraining region is below the characteristic. The figure also represents the fictitious operating current resulting from CT errors for external faults. For low fault currents, the CTs behave linearly and the error current is a linear function of the restraining current. For higher fault currents, the CTs saturate and cause a nonlinear growth of the operating current. The slope characteristic of the percentage differential element provides security for external faults that cause CT saturation. Two solutions that further increase the element security for high-current external faults are a dual-slope differential characteristic and an adaptive characteristic with a slope that increases upon external fault detection [2].

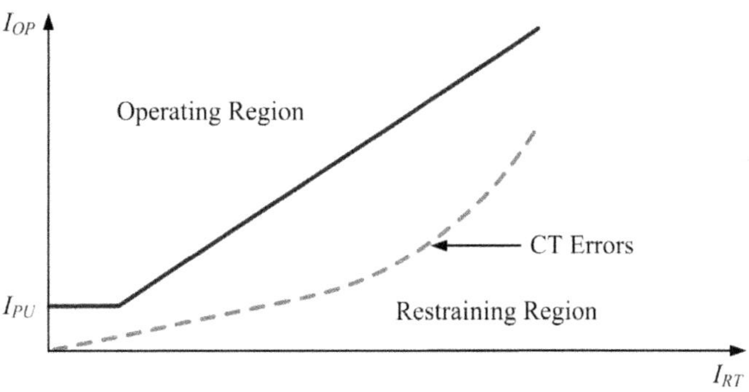


**FIGURE 19-28** Percentage differential element operating characteristic.

CT saturation is only one cause of undesirable differential current. In transformer applications, there are other possible causes [2]:

- *Mismatch between the CT ratios and the transformer turns ratio.* This mismatch creates a fictitious differential current that can cause a differential element misoperation. The solution is to perform current scaling. Electromechanical transformer differential relays have physical transformer taps for scaling the currents. Compensation is rarely perfect because the number of available taps is limited. In contrast, microprocessor-based transformer relays can fully compensate for the current amplitude differences resulting from the mismatch. The relay calculates $TAP$ values based on the transformer MVA rating, transformer winding voltage ratings, and CT ratios and connections (wye or delta). The relay uses the calculated $TAP$ values to scale the secondary currents to a common base.

- *Phase shift between the transformer primary and secondary currents for delta-wye connections.* This phase shift also creates a fictitious differential current. The solution is to perform phase-shift compensation. Electromechanical relays require using wye- and delta-connected CTs, as appropriate, to mirror the transformer connections. Modern transformer relays perform the phase-shift compensation internally, which allows using wye-connected CTs on both sides of the transformer.
- *Presence of a grounded-wye transformer winding inside the differential protection zone.* For external ground faults, the positive- and negative-sequence currents enter and exit the protection zone, resulting in a nonoperation condition for the differential element. However, the zero-sequence current only passes through the differential element CTs on the grounded-wye side of the transformer, creating a differential current that could cause a relay misoperation. By applying delta compensation to the currents of the wye winding, microprocessor-based relays remove the zero-sequence current, preventing it from getting to the differential element. The delta connection of the CTs removes the zero-sequence current in transformer differential schemes with electromechanical relays.
- *Magnetizing inrush currents during transformer energization, voltage recovery after the clearance of an external fault, or energization of a parallel transformer.* Transformer magnetizing current only passes through the source-side differential element CTs, creating a differential current that could cause a differential element misoperation. The harmonic content of the differential current serves to differentiate faults from inrush conditions. Harmonics can be used to either restrain or block the transformer differential element. Harmonic restraint methods use harmonic components of the differential current to provide additional differential element restraint. The presence of harmonics desensitizes the differential element. Harmonic blocking methods block the differential element when the ratio of the harmonic content to the fundamental component of the differential current is above a preset threshold. Other methods for discriminating internal faults from inrush conditions directly recognize the wave-shape distortion of the differential current.
- *High transformer excitation currents caused by overexcitation.* The high excitation current produced by transformer overexcitation could cause a differential element misoperation. The harmonic content of the differential current serves to differentiate faults from overexcitation conditions. Transformer differential elements use fifth-harmonic blocking to prevent misoperation on overexcitation.

## 19.4.2 Restricted Earth-Fault Protection

Transformer differential protection provides excellent sensitivity for phase-to-phase and most phase-to-ground winding faults. However, the phase current is low for ground faults close to the transformer neutral. Differential elements that respond to phase currents have low sensitivity for these faults. On the other hand, the neutral current is very high for these faults. Restricted earth fault (REF) protection, which responds to neutral current, can detect ground faults close to the transformer neutral quickly and reliably.

Microprocessor-based relays with REF elements calculate the REF operating current using the currents measured by all the CTs of the REF protection zone. These enhanced elements only require connection of the transformer neutral CT to the relay to complete the REF protection zone. The scheme does not require a dedicated set of CTs and can compensate for different CT ratios.

In the past, REF protection used current-polarized directional relays or high-impedance differential relays. Modern microprocessor-based relays generally use directional elements, which are stable for external faults with heavy CT saturation. For example, the relay can use a zero-sequence, current-polarized directional element (32I) that measures the phase angle between the transformer neutral current and the residual current at the transformer terminals.

The 32I element uses Eq. (19-16) to calculate the scalar quantity $T$.

$$T = \text{Re}\left[\overline{I}_X \bullet \overline{I}_Y^*\right] \tag{19-16}$$

where $\overline{I}_X$ is derived from the residual current at the transformer terminals; $\overline{I}_Y^*$ is the complex conjugate of $\overline{I}_Y$, which is the transformer neutral current.

The 32I element compares $T$ with two thresholds. When $T$ is positive and above the positive threshold, the element declares an internal ground fault. When $T$ is negative and below the negative threshold, the element declares an external ground fault. To improve security, the 32I element calculates $T$ only when $I_X$ and $I_Y$ are above threshold values. This element can be set with high sensitivity because of its inherent security for external faults with CT saturation. This security comes from the fact that the neutral current $\overline{I}_Y$ is measured using a single CT. Uneven CT saturation for external phase faults may cause a fictitious zero-sequence value of $\overline{I}_X$, but $\overline{I}_Y$ will equal zero, resulting in $T = 0$.

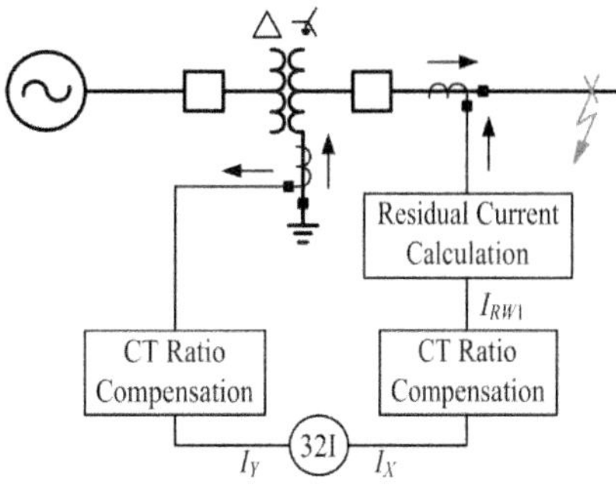


**FIGURE 19-29** REF protection for two-winding transformers using a current-polarized directional element.

Figure 19-29 shows an REF element connected to protect a two-winding transformer. The relay uses the phase currents to calculate the Winding 1 residual current $\overline{I}_{RW1}$. Then, the relay multiplies this current and the secondary neutral current by the corresponding CT ratios to calculate currents $\overline{I}_X$ and $\overline{I}_Y$ in amperes primary. Finally, the relay calculates $T$ using Eq. (19-16). For the external ground fault shown in this figure, $\overline{I}_X$ and $\overline{I}_Y$ are 180° out of phase, and $T$ is negative. For internal ground faults, $\overline{I}_X$ and $\overline{I}_Y$ are in phase, and $T$ is positive.The REF element includes logic (not shown in Fig. 19-29) to ensure element operation for internal faults when the wye-side breaker is open ($\overline{I}_X = 0$) or the external system has no ground sources.

## 19.4.3 Overexcitation Protection

Transformer overexcitation can occur when the ratio of the per unit voltage to per unit frequency at the secondary terminals exceeds 1.05 at full load, 0.8 power factor, or 1.1 at no load [27]. Transformers may temporarily exceed their continuous volts-per-hertz capability. Transformer manufacturers provide information on the short-term volts-per-hertz capability as a function of time. The overexcitation limit is either a curve or a set point with a time delay.

During overexcitation, the increase in exciting current can cause the transformer differential element to operate. However, the differential element may not operate for some overexcitation conditions that could damage the transformer or may operate too quickly. Transformers typically withstand overexcitation conditions longer than differential element operating times. Premature tripping of a transformer during a system disturbance can make the situation worse. In microprocessor-based transformer relays, fifth-harmonic blocking prevents differential element tripping during transformer overexcitation.

For transformer overexcitation protection, use a volts-per-hertz (24) element, especially on large network transformers and generator step-up transformers. This element is available in multifunction transformer relays that also include voltage inputs.

The volts-per-hertz element uses Eq. (19-17) for calculating the ratio of the measured voltage to frequency in per unit of the rated quantities, which is proportional to the transformer magnetic flux.

$$\varphi = \frac{V}{f} \bullet \frac{f_{NOM}}{V_{NOM}} \tag{19-17}$$

where $\varphi$ is the estimated magnetic flux value in per unit; $V$ is the measured voltage; $f$ is the measured frequency; $V_{NOM}$ is the transformer rated voltage; $f_{NOM}$ is the rated frequency.

Microprocessor-based relays provide several types of time curves for the volts-per-hertz element. Some relays also allow user-programmable curves. Set the volts-per-hertz element to coordinate with the volts-per-hertz capability curve of the protected transformer.

### 19.4.4 Overcurrent Protection

Traditionally, overcurrent protection was often the main protection for small transformers. This approach precludes fast fault clearing because overcurrent elements must coordinate with protective devices in adjacent zones. Thus, internal faults could cause severe damage to the transformer. The low cost of modern relays makes differential protection feasible even for small transformers.

Transformer differential elements, sudden-pressure relays, and Buchholz relays do not protect the transformer from through-fault damage. Overcurrent protection provides this primary protection function. Coordinate it with the transformer through-fault capability curves to take the transformer out of service before damage from an uncleared external fault occurs. IEEE Standard C57.109 [28] provides transformer through-fault capability curves.

Overcurrent protection can also provide backup protection that is independent of the transformer primary protection but at significantly reduced sensitivity and speed.

Overcurrent devices provide some transformer overload protection by detecting heavy transformer overloads. For better overload protection and monitoring, we recommend using elements based on transformer thermal models [2].

Figure 19-30 depicts a typical overcurrent relay protection scheme for a delta-grounded wye transformer. In the past, this scheme required several relays. Today, one multifunction transformer relay can provide all overcurrent protection functions, in addition to transformer differential and other functions. Two multifunction transformer relays provide fully redundant transformer protection with higher reliability and less cost than a scheme that uses individual relays.

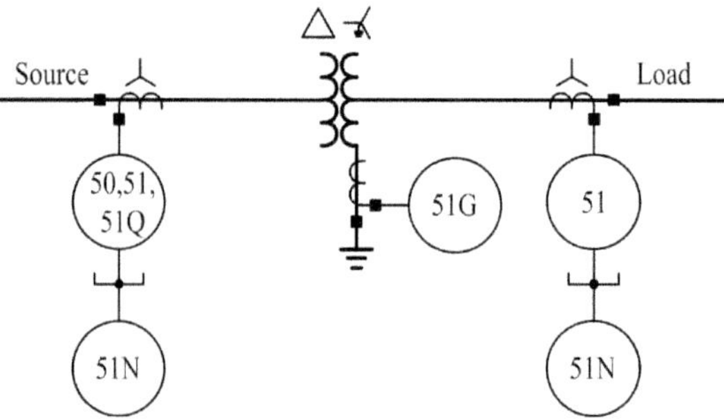


**FIGURE 19-30** Typical transformer overcurrent relay protection scheme.

Primary-side instantaneous phase overcurrent elements (50) provide high-speed primary protection for internal phase faults. These elements should trip the transformer high-side breaker and/or the transformer lockout relay.

Primary-side inverse-time phase overcurrent elements (51) provide backup protection for internal phase faults and also provide primary protection for phase faults between the transformer and the secondary-side main breaker. These elements should trip the high-side breaker and/or the lockout relay.

The primary-side residual overcurrent element (51N) provides sensitive primary protection for ground faults in the delta winding if the source is a grounded system. This element should trip the high-side breaker and/or the lockout relay.

The phase overcurrent elements on the delta side of the transformer are relatively insensitive to ground faults on the wye side. The inverse-time ground overcurrent element (51G) provides primary protection for these faults. Transformers do not remove negative-sequence currents caused by ground faults. Therefore, if a neutral CT is not available, a primary-side negative-sequence overcurrent element (51Q) can provide primary protection for ground faults on the wye side and backup protection for all other unbalanced faults. The 51G and 51Q elements should trip the high-side breaker and/or the lockout relay.

Secondary-side inverse-time phase (51) and ground (51N) overcurrent elements provide primary protection for secondary-side bus faults and backup protection for feeder faults. These elements should trip the low-side breaker.

### 19.4.5 Sudden-Pressure and Gas-Accumulation Protection

Sudden-pressure and gas-accumulation relays provide partial redundancy for transformer differential protection, because they respond only to faults inside the transformer tank.

Sudden-pressure relays (63) monitor the pressure inside the transformer tank and trip when a fault inside the tank causes a sudden pressure change. Sudden-pressure relays provide sensitive detection of low-grade faults, such as turn-to-turn faults; however, sudden-pressure relays can misoperate for the following reasons:

- Winding movement on severe through-fault or seismic events.
- Arcing over of the microswitch contact.
- Operator error when maintaining the oil-preservation system.

Supervising sudden-pressure relays with microprocessor-based relays improves their security, increasing sensitivity and reliability in overall transformer protection [2].

Incipient transformer faults under oil typically generate combustible gases. Gas-accumulation relays detect these gases and provide an early warning. Gas-accumulation relays allow drawing off the gas for analysis to identify the problem. The Buchholz relay is a variation of this type of relay. These relays may misoperate for the same reasons as sudden-pressure relays.

## 19.5 BUS PROTECTION

Buses are the nodes in electric power networks. When a bus fault occurs, all branches supplying current to that node must be opened to clear the fault.

Bus faults are rare but can be very disruptive. The high fault current can cause severe and costly damage to power system equipment, power system transient instability, and service disruption to customers. A fault at a critical bus that is left uncleared for a long enough time can result in a wide-area blackout. The consequences of a bus cleared in error are basically the same. For these reasons, bus protection must provide fast operation for all bus faults and high security for external faults.

The selection of a bus protection method depends on the bus configuration (if it is fixed or switchable), the availability and characteristics of CTs, and the availability of disconnect switch auxiliary contacts (for switchable bus arrangements) [2].

Microprocessor technology improves bus protection performance and economics. Microprocessor-based bus protection schemes are faster, more reliable, more selective, more sensitive, and more economical than older bus protection schemes.

The major types of bus protection schemes are as follows [2, 29]:

- Differential protection:
  - Differential overcurrent protection.
  - High-impedance differential protection.
  - Low-impedance percentage differential protection.
- Partial differential protection.

- Zone-interlocked protection.
- Fault-bus protection.
- Protection using remote time-coordinated relays that overlap the bus zone.

This subsection covers the high-impedance and low-impedance differential, partial differential, and zone-interlocked bus protection schemes.

## 19.5.1 High-Impedance Differential Protection

The high-impedance differential scheme obtains the differential current from parallel connected CTs. The scheme introduces a high-value resistor (called a stabilizing resistor) in the differential branch of the circuit. The stabilizing resistor reduces the differential current that results from heavy CT saturation during external faults.

Figure 19-31 shows the relay connection for a typical application. A low-impedance overcurrent element (87) senses the current flowing through the stabilizing resistor $R_S$. The current through the 87 element is proportional to the voltage $V_S$ across the relay. If CT saturation occurs for an external fault, the high impedance of the relay will force the current through the lower impedance path of the saturated CT. Very little current will then flow through the 87 element. For this small current, the relay will not operate.

For internal faults, the voltage produced by the CTs drives a significant differential current through the stabilizing resistor that causes the 87 element to operate. The resulting high voltage across the relay, which can reach dangerous levels, typically drives all CTs into saturation. The relay circuit includes a voltage limiter, such as a metal-oxide varistor (MOV), to limit this voltage to a safe level (Fig. 19-31). Shorting the stabilizing resistor and MOV with an auxiliary lockout relay contact (86a) prevents damage by limiting the energy absorbed by these components when the relay trips.

The scheme shown in Fig. 19-31 includes an optional instantaneous overcurrent element (50). Use this element when a sustained breaker-failure initiation (BFI) signal is required [2]. When the 87 element operates and trips the lockout relay, the high-impedance path is short-circuited, and the 87 element drops out. The differential current will continue to flow until all breakers successfully

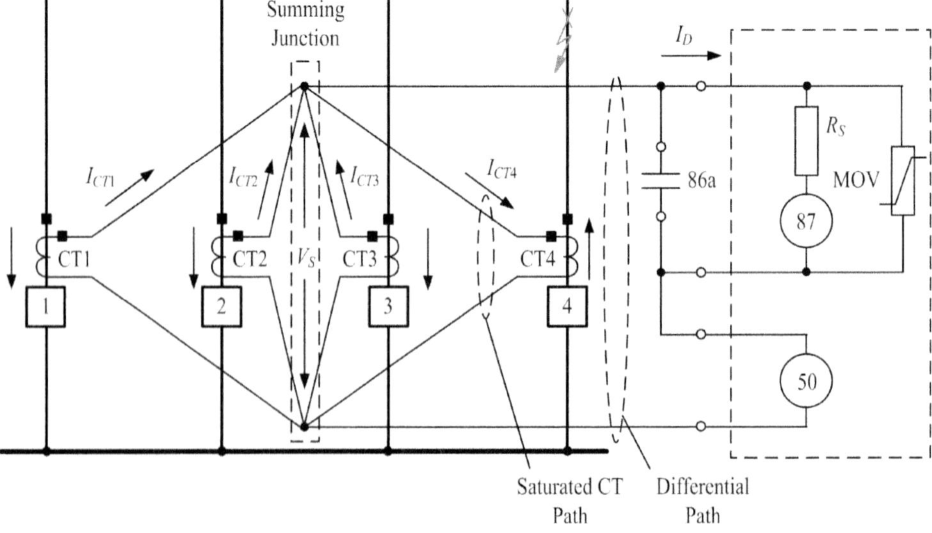


**FIGURE 19-31** Bus high-impedance differential protection scheme.

interrupt the fault, so the 50 element remains asserted and sustains the BFI signal. The 50 element can also improve scheme dependability by sustaining the trip condition if an MOV failure short-circuits the high-impedance path when high voltage is present.

The high-impedance differential scheme is very sensitive because it only takes a small differential current flowing through the stabilizing resistor to produce voltage high enough to trip. This high sensitivity makes it generally suitable for application in high-impedance grounded systems.

This scheme requires dedicated CTs on all bus terminals. When available CTs comply with the scheme requirements, the high-impedance bus differential scheme is a very effective and economical solution, especially for buses with many branch circuits and simple configurations.

## 19.5.2 Low-Impedance Percentage Differential Protection

Although widely used, the high-impedance bus differential scheme is less effective when the CTs do not meet the scheme requirements or when the substation bus arrangement is complex. A percentage differential scheme with multiple restraint inputs is the most versatile solution for bus protection. Section 19.4.1 describes the operation principle of percentage differential elements.

Figure 19-32 shows the application of a percentage differential relay to a bus with $N$ terminals. The relay measures all the bus terminal currents (no CT paralleling required). A percentage differential element uses Eqs. (19-18) and (19-19) to calculate the operating ($I_{OP}$) and restraining ($I_{RT}$) currents.

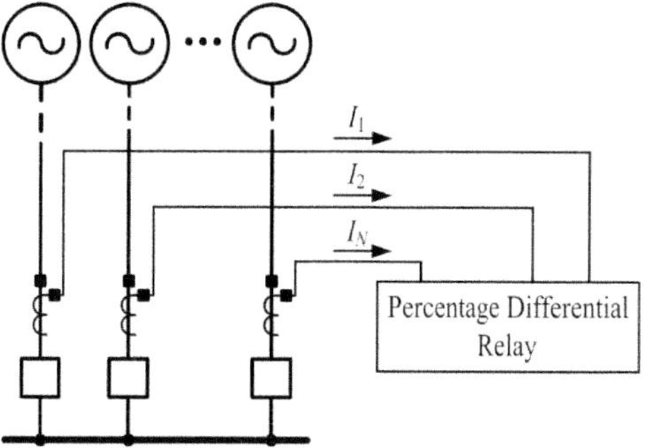


**FIGURE 19-32** Bus percentage differential protection scheme.

$$I_{OP} = \left|\bar{I}_1 + \bar{I}_2 + \ldots + \bar{I}_N\right| \tag{19-18}$$

$$I_{RT} = k\left(\left|\bar{I}_1\right| + \left|\bar{I}_2\right| + \ldots + \left|\bar{I}_N\right|\right) \tag{19-19}$$

where $k$ is a scaling factor.

The percentage differential element generates a tripping signal if the operating current $I_{OP}$ is greater than the minimum pickup current $I_{PU}$ [Eq. (19-14)] and is also greater than a percentage of the restraining current $I_{RT}$ [Eq. (19-15)]. Figure 19-28 shows the operating characteristic of a percentage differential element.

The number of available restraint inputs is usually the main limitation of a multiple restraint differential relay. In the past, engineers paralleled CTs when the bus had more CT circuits than available restraint inputs. This practice sacrificed security, because if one of the paralleled CTs were to saturate, it would not contribute adequate restraint. Paralleling CTs is no longer necessary with the availability of modern bus differential relays.

Traditional percentage differential relays lack security for external faults in substations with heavy CT saturation problems. High-impedance differential relays are difficult to apply with CTs that have low voltage ratings. Advanced microprocessor-based bus protection schemes solve these problems. These systems are a good choice for the following applications:

- Where the CTs present saturation problems.
- Where CTs of different ratios must be shared with other protection schemes.
- Where the substation has a complex bus arrangement.

Advanced microprocessor-based bus protection schemes perform these requirements [30, 31]:

- Fast operating times for all bus faults.
- Security for external faults with heavy CT saturation.
- Security during normal switching conditions.
- Security with subsidence current present after clearing an external fault.
- Security when a CT is open or short-circuited.
- Minimum operating time for external-to-internal evolving faults.

### 19.5.3 Partial Differential Protection

Partial differential protection schemes (Fig. 19-33) use inverse-time overcurrent relays connected to paralleled CTs that only measure the bus source currents. Each relay provides primary protection to its corresponding bus section and backup protection to the feeders connected to this bus section. Therefore, each overcurrent relay must coordinate with the overcurrent relays of these feeders. The result is time-delayed bus fault clearing.

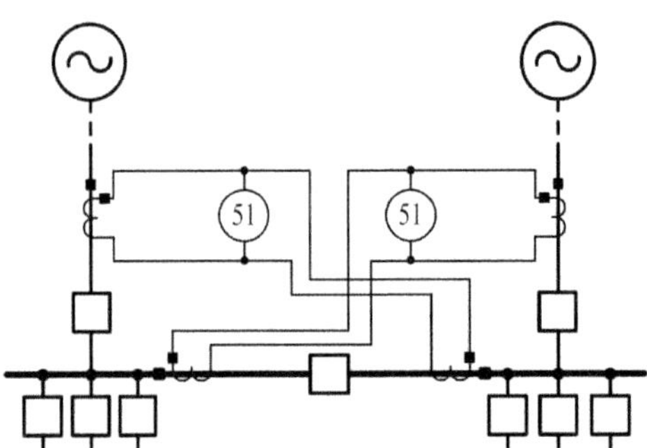


**FIGURE 19-33** Bus partial differential protection scheme.

Partial differential protection schemes are used when not all bus branch circuits have CTs available to provide a complete differential zone. Previously, these schemes were also used when feeder CTs were unsuitable for differential protection because they had different ratios and/or characteristics than the CTs of the source circuits and the bus-tie breaker. Some modern microprocessor-based relays can tolerate high CT ratio mismatch values. The zone-interlocked schemes discussed in Sec. 19.5.4 are also suitable for applications with dissimilar CTs. Whenever possible, use differential or zone-interlocked schemes for bus primary protection and use partial differential schemes only for backup protection.

### 19.5.4 Zone-Interlocked Protection

An alternative to bus differential protection is a zone-interlocked scheme, which uses information from the relays on each of the bus branch circuits to determine whether a fault is internal or external to the bus. Each relay sends status information on its branch circuit. For buses with multiple sources, the scheme requires directional relays on the source circuits. Reference [2] describes this application. In radial systems, the scheme only requires overcurrent relays. To implement this scheme with existing devices, use the directional and overcurrent elements available in the branch circuit multifunction relays.

The performance of this scheme is almost independent of the ratio, characteristics, and performance of the CTs. This substantial independence makes the scheme suitable for substations that have CTs of different types and/or ratios, especially when high operating speed is not required.

Zone-interlocked schemes are only applicable for single-breaker configurations where the branch circuits connect to only one bus through one breaker. Another disadvantage is that the scheme requires a small time delay for coordination with the blocking elements and is more complex than a dedicated bus differential scheme. Also, the scheme does not provide the desired zone overlap when the branch circuits have bus-side CTs.

Bus faults are more common at distribution levels because small animals can easily bridge the small clearances between the grounded steel structures and energized buswork. The most common practice is to clear bus faults with transformer backup protection, which has a time delay to coordinate with feeder protection. Fault clearing times are typically between 0.5 and 1.0 s. However, the high fault current and the number and duration of bus faults can reduce the operating life of the power transformer, the most important and expensive equipment in the substation.

In radial substations, a simple and economical zone-interlocked blocking scheme, sometimes called a fast bus-tripping scheme, provides relatively high-speed fault clearing for buses that do not have differential protection. Instead of relying on a traditional coordination interval in the bus main relay, this scheme only requires a short delay to allow the feeder relays to block the bus main relay for an external fault. This scheme can operate for bus faults in approximately two to three cycles.

You can implement the scheme as shown in Fig. 19-34 (Ref. [2] describes the logic for this application):

1. Program an instantaneous overcurrent element in each of the feeder relays to close an output contact when a fault occurs on the feeder.
2. Wire the blocking contacts from each of the feeder relays in parallel to an input on the bus main relay (input IN104 in this example).
3. Delay overcurrent elements in the bus main relay only long enough to allow sensing of the blocking contacts; a typical coordination delay is approximately one to two cycles.

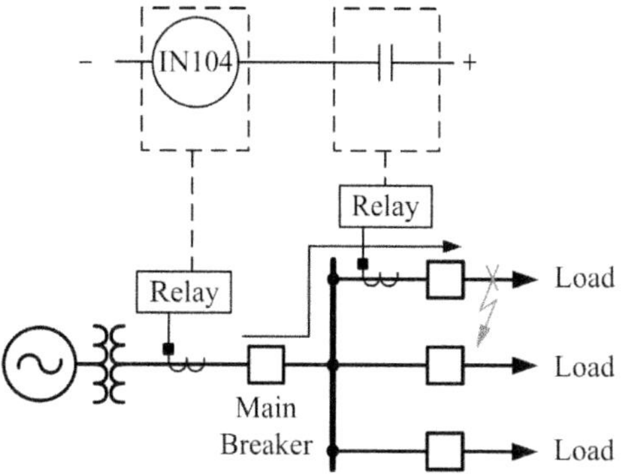


**FIGURE 19-34** Fast bus-tripping scheme application.

## 19.6 GENERATOR PROTECTION

Power generating stations represent approximately half of the capital investment in an electric power system. Synchronous generators are complex electromechanical systems that may experience many harmful operation conditions. Generator outages caused by faults, abnormal operation conditions, or generator protection misoperations are costly. A complete generator protection system must include a variety of protection functions to respond to faults and also to abnormal operation conditions to prevent future failures.

Modern microprocessor-based generator protection relays include all the protection functions required in most installations. Multifunction relays can provide even small-capacity generators with complete protection at low cost. Selecting the protection functions that a particular generator needs and determining appropriate setting values require a thorough knowledge of the protected machine. References [2, 32, 33] provide comprehensive coverage of generator protection.

### 19.6.1 Generator Connections and Grounding

In large utility generating stations, generators are typically connected directly to the step-up transformer as shown in Fig. 19-35. This arrangement is referred to as a generator-transformer unit. One or two unit auxiliary transformers may also be connected to the generator terminals.

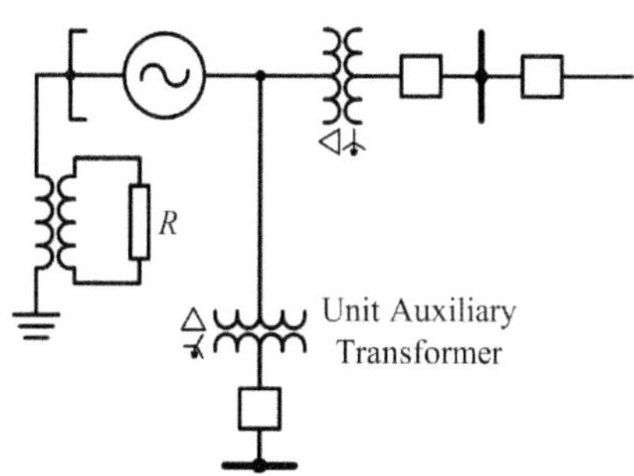


**FIGURE 19-35** Generator-transformer unit arrangement.

High-resistance generator grounding is common in generator-transformer units. This method uses a transformer connected between the generator neutral

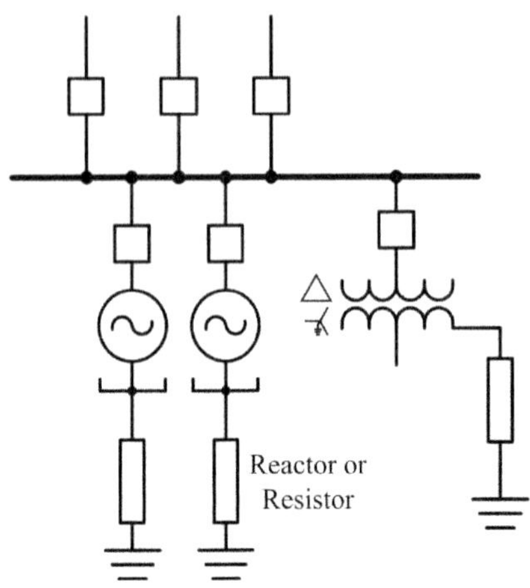


**FIGURE 19-36** Direct connection of generators to a distribution system.

and ground, and a resistor connected to the transformer secondary, as shown on Fig. 19-35. Typically, high resistance limits ground fault currents to 15 A or less to prevent stator core lamination damage for generator stator ground faults.

Figure 19-36 depicts the direct connection of small generators to a medium-voltage distribution system, which is typically used in industrial power systems. In this arrangement, generator neutral grounding is generally determined by the distribution system grounding needs, and is typically a low-resistance grounding. This grounding method uses a resistor connected between the generator neutral and ground. The ground fault current is usually no higher than 150% of generator rated current.

The main disadvantage of generator low-resistance grounding is that the high ground fault current can cause significant stator core lamination damage, which increases the cost of the generator repair.

## 19.6.2 Stator Phase Fault Protection

Stator phase faults (three-phase, phase-to-phase, and phase-to-phase-to-ground faults) require very fast clearing times because of the high fault currents and their potentially destructive effect. Fault current continues to flow for seconds after the generator main and field breakers trip and the prime mover valve closes, because of the magnetic energy trapped within the machine. Generator phase fault protection must trip and shut down the generator and prime mover and transfer unit auxiliaries to a standby source with minimal delay.

Stator phase protection schemes for large generators often use differential elements (87) connected as shown in Fig. 19-37. High fault currents and a high generator impedance *X*/*R* ratio may cause heavy CT saturation during external faults. Typically, percentage differential elements with variable or dual-slope characteristics provide adequate security.

The stator windings of certain types of generators have coils with multiple turns. An interturn insulation failure can cause a turn-to-turn fault. Stator phase differential protection does not detect turn-to-turn faults. For these generators, dedicated turn-to-turn fault protection must trip and shut down the generator and prime mover and transfer unit auxiliaries to a standby source with minimal delay.

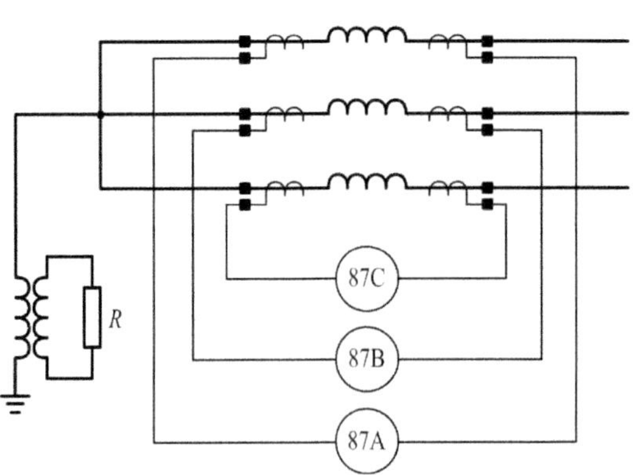


**FIGURE 19-37** Generator differential protection scheme.

For generators with two or more parallel circuits per phase, users can provide turn-to-turn fault protection by splitting the circuits into two groups and comparing the currents of these groups. During normal operation, the current difference between two groups of the same phase is very small. Turn-to-turn faults cause an unbalance current that increases the current difference.

Traditional split-phase turn-to-turn fault protection schemes use an overcurrent relay with instantaneous and time-delayed elements that is connected to measure the current difference. In large capacity generators, split-phase protection requires connecting a CT to each circuit group and connecting the CTs in parallel to obtain the current difference. Uneven CT saturation for external faults causes a fictitious current difference. The instantaneous element, set above the maximum current difference caused by CT saturation, provides high-speed operation. The time-delayed element has a lower pickup setting to improve the scheme sensitivity.

A split-phase protection scheme using a self-balanced CT connection (Fig. 19-38) minimizes the chance of CT saturation. This scheme uses flux-summing, low-ratio CTs. The conductors carrying the circuit group currents pass in opposite directions through the CT window. The overcurrent elements only receive the difference current. The small size of the CT window limits the size of the conductor and, therefore, the size of generators that can be protected.

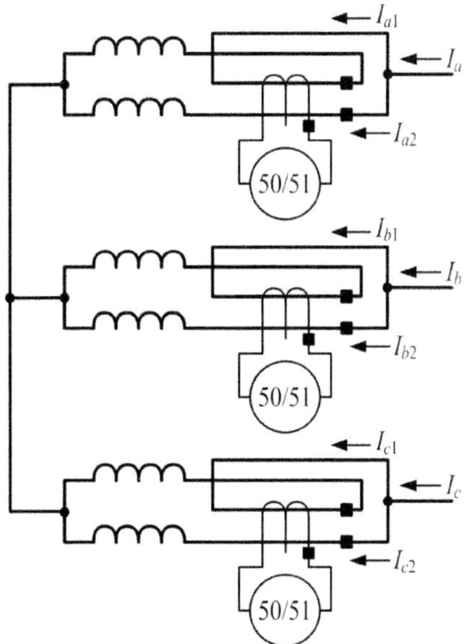


**FIGURE 19-38** Generator split-phase protection using flux-summing CTs.

### 19.6.3 Stator Ground Fault Protection

Stator phase fault protection has low sensitivity for ground faults because grounding impedance and fault resistance limit ground fault current magnitude. In addition, ground faults close to the generator winding neutral point cause very low fault currents. Generators need separate sensitive stator ground fault protection that generally trips and shuts down the generator and prime mover and transfers unit auxiliaries to a standby source.

For high-resistance-grounded generators, you can use the voltage at the neutral point of the stator winding to detect ground faults. This voltage is close to zero when the system is balanced; it rises significantly for the system unbalance caused by stator ground faults. The neutral voltage is at maximum magnitude for faults at the generator terminals and decreases as the fault location moves closer to the neutral point. As shown in Fig. 19-39, an overvoltage element (59GN) connected to the grounding transformer secondary detects stator ground faults. The 59GN element responds to the fundamental-frequency voltage; its pickup must be above the neutral voltage that results from normal system unbalance.

The magnitude of the voltage across the 59GN element depends directly on the ground fault location inside the generator winding. The 59GN will measure the full generator phase-to-ground voltage if the fault occurs at the generator terminals, and it will measure zero voltage if the ground fault occurs at the generator neutral point. Therefore, a 59GN element does not protect a region of approximately 5% of the total winding length, starting from the neutral point.

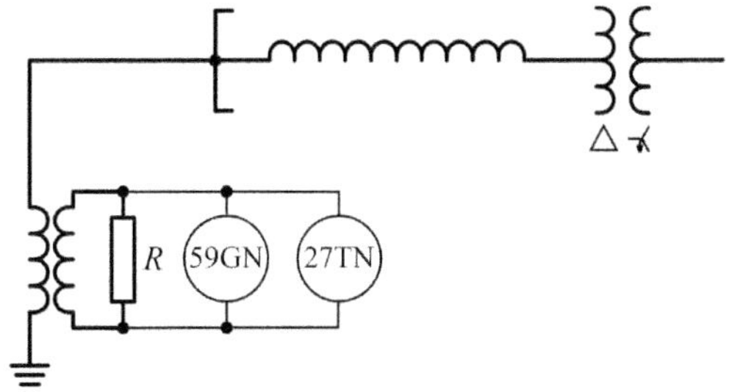

**FIGURE 19-39** Ground fault protection for high-resistance grounded generators using fundamental-frequency and third-harmonic voltage.

Generator windings normally produce varying amounts of third-harmonic voltage, depending on machine construction and loading. Under normal operation conditions, the neutral voltage and the terminal voltage have a certain third-harmonic content. Ground faults close to the neutral point cause the neutral third-harmonic voltage to decrease and the terminal third-harmonic voltage to increase. Ground faults close to the generator terminals increase the neutral third-harmonic voltage and decrease the terminal third-harmonic voltage.

A third-harmonic undervoltage element (27TN) measuring the neutral voltage can detect ground faults in the proximity of the neutral point (Fig. 19-39). Other solutions include a third-harmonic overvoltage element measuring the terminal voltage and a third-harmonic voltage differential element that responds to the difference between the neutral and terminal third-harmonic voltages. Setting these elements requires knowing the normal values of third-harmonic voltages over the full range of generator output conditions. Modern generator protection relays can provide these measurements.

By combining the fundamental-frequency neutral overvoltage element and the third-harmonic undervoltage element or the voltage differential element, the relay provides 100% protection of the stator windings for ground faults.

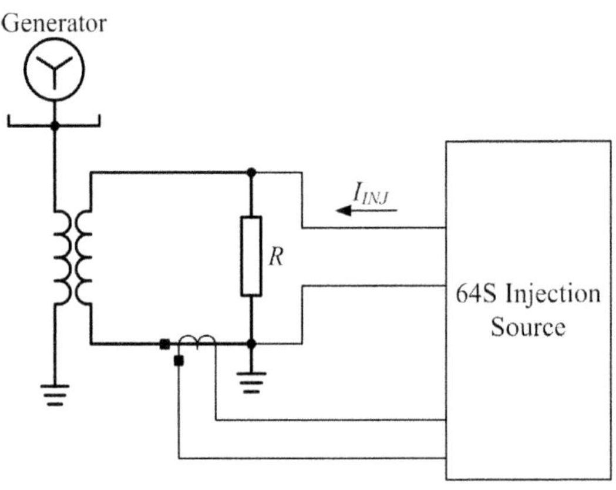


**FIGURE 19-40** Ground fault protection for high-resistance grounded generators using the low-frequency injection method.

Some generators do not produce sufficient third-harmonic voltage to apply third-harmonic-based ground fault protection. If a generator does not produce adequate third-harmonic voltage, a signal injection method is an alternative for 100% stator ground fault protection. Figure 19-40 shows a protection module (64S) that injects a multifrequency current signal ($I_{INJ}$ in the figure) into the generator neutral [34]. The 64S module estimates the leakage resistance and capacitance of the stator circuit, and compares the insulation leakage resistance to a setting value. This scheme provides protection even when the generator is in turning gear and during startup, which allows for continuous supervision and reliable detection of stator winding insulation failure before the generator is put online.

For low-resistance grounded generators, you can use a time-delayed overcurrent element 51N to provide stator ground fault protection. Connect the 51N element to a low-ratio CT placed on the generator neutral. Since this element also responds to ground faults beyond the generator terminals, it needs to time coordinate with the distribution circuit overcurrent elements. Coordination requirements limit the sensitivity and speed of the 51N element for stator ground faults.

Restricted-earth fault (REF) elements provide fast, selective, and sensitive ground fault protection for low-resistance grounded generators. In this application, REF protection compares the current flowing in the generator neutral with the zero-sequence current measured at the generator terminals. REF protection only responds to ground faults located between the neutral CT and the CTs at the generator terminals and does not require any coordination time delay.

### 19.6.4 Rotor Fault Protection

The field winding of a generator is wound on the rotor and ungrounded. A dc voltage source or exciter connected to the field winding produces the field current. In many generators, a stationary exciter injects current into the field winding through brushes and collector rings. Rotating exciters include dc generators and ac generators with rectifiers. Modern static excitation systems receive power from the machine terminals or the power system. A step-down excitation transformer feeds a three-phase controlled rectifier bridge that converts ac voltage into dc voltage. Another alternative is a brushless excitation system, in which the ac exciter armature and the rectifier bridge are mounted on the same rotating shaft system as the field. In this subsection, we discuss rotor fault protection for generators that have brushes and collector rings.

Insulation deterioration or breakdown can cause the field winding to contact the rotor core. The first ground fault does not affect generator operation. However, the chance of a second fault increases, because the first fault establishes a ground reference. A second ground fault shorts part of the field winding and causes unbalanced air gap fluxes in the generator. This unbalanced field can cause excessive rotor vibration that damages the machine. The localized excessive heat generated by the fault current causes uneven rotor temperatures, which also leads to excessive vibration. The level of magnetic field unbalance depends on the location of the ground fault points.

Rotor field ground fault protection (64F) detects the first ground fault and provides an alarm or initiates generator tripping. Users that apply alarming normally keep the generator in service. If the ground fault persists for some time, the operator removes the generating unit from service in an orderly manner.

Traditional rotor field ground fault detection methods include the following:

- *Voltage divider method.* This method uses a voltage divider circuit and a voltage relay, which is connected across the divider midpoint and ground.
- *DC injection method.* In this method, a dc power supply, in series with a voltage relay, is connected across the excitation negative terminal and ground.
- *AC injection method.* This method is similar to the dc injection method, but it uses an ac power supply.

A new field ground module that uses a dc-switching injection method can detect field ground faults in generators that have brushes and collector rings [2]. The module periodically measures the rotor winding insulation resistance and transmits this value through a fiber-optic link to the generator protection relay, which then compares this value with threshold values for alarm or trip.

The module connects to the generator field circuit as shown in Fig. 19-41 to measure the field resistance to ground. This device applies square-wave voltage through two resistors $R$ at the points where the exciter connects to the field winding. In this figure, $R_f$ is the value of the fault resistance between the rotor winding and ground, and $C_{fg}$ is the capacitance between the field circuit and ground. By measuring the voltage across a sensing resistor $R_s$, the dc-switching injection method calculates $R_f$. Under normal conditions, $R_f$ equals the insulation resistance, which has a very high value. For ground faults, $R_f$ takes a smaller value.

The dc-switching injection method provides a fault resistance calculation that is independent from the fault location and the field voltage.

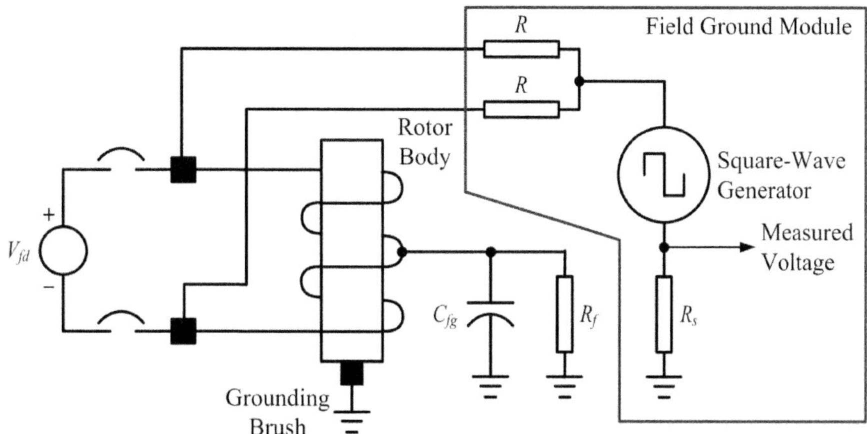


**FIGURE 19-41** The dc-switching injection method measures the field resistance to ground. The measured resistance is very high under normal conditions and lower for ground faults.

### 19.6.5 Current Unbalance Protection

Synchronous generators may experience many abnormal operation conditions. This and the following subsections describe some of the functions required to protect the generator and the prime mover (and in some cases, to also protect the power system) against generator abnormal operation conditions.

When a generator operates with unbalanced three-phase stator currents, the resulting negative-sequence current generates a magnetic flux that rotates in opposition to the rotor rotation. This reverse-rotating flux induces double-frequency currents in the rotor, which may cause high temperatures on some parts of the rotor and eventually cause machine damage.

System asymmetries and unbalanced loads may cause low-magnitude, long-duration negative-sequence currents. IEEE Standard C50.12 [35] and IEEE Standard C50.13 [36] specify the permissible continuous negative-sequence current for generators, which is between 5% and 10% of the machine rated current.

Unbalanced short circuits and open phases generally cause high-magnitude negative-sequence currents. Generators can withstand these currents for a short time. Equation (19-20) expresses the short-time generator capability as a function of the integrated negative-sequence current that the generator can withstand during a time interval *T*. The *K* factor depends on the generator type and capacity. References [35] and [36] specify the *K* factors for various types of generators.

$$I_2^2 T = K \tag{19-20}$$

where $I_2$ is the stator negative-sequence current.

An inverse-time negative-sequence overcurrent element provides generator current unbalance protection (46). Equation (19-21) describes the time-current characteristic as implemented in a modern relay [2].

$$T = \frac{46Q2K}{\left(\dfrac{I_2}{I_{NOM}}\right)^2} \tag{19-21}$$

where *T* is the element operating time in seconds; *46Q2K* is the time-dial setting; $I_{NOM}$ is the generator rated current.

Figure 19-42 shows that the element time-current characteristics are a family of straight lines with slope equal to −2 when plotted using a log-log scale. The element picks up when the per unit negative-sequence current is greater than a preset *46Q2P* threshold. Set *46Q2P* to a value slightly greater than the permissible continuous negative-sequence current. Set *46Q2K* at a value slightly smaller than the generator *K* factor.

### 19.6.6 Loss-of-Field Protection

A generator may totally or partially lose its excitation as a result of accidental field breaker tripping, field open circuit, field short circuit (slip-ring flashover, for example), voltage regulator failure, or loss of the excitation system.

When a generator loses excitation, the rotor field gradually extinguishes, and the magnetic coupling between rotor and stator magnetic fields eventually diminishes to a point where the machine loses synchronism. The rotor speed increases to a value for which the machine, operating as an induction generator, produces the active power demanded by the power system in this new condition. Operating as an induction machine, the generator draws large amounts of reactive power from the system, which causes high stator-current values (approximately two to four times rated current) and depresses the voltage. Additionally, slip-frequency eddy currents circulate in the rotor. The magnitude of these currents is proportional to the generated power.

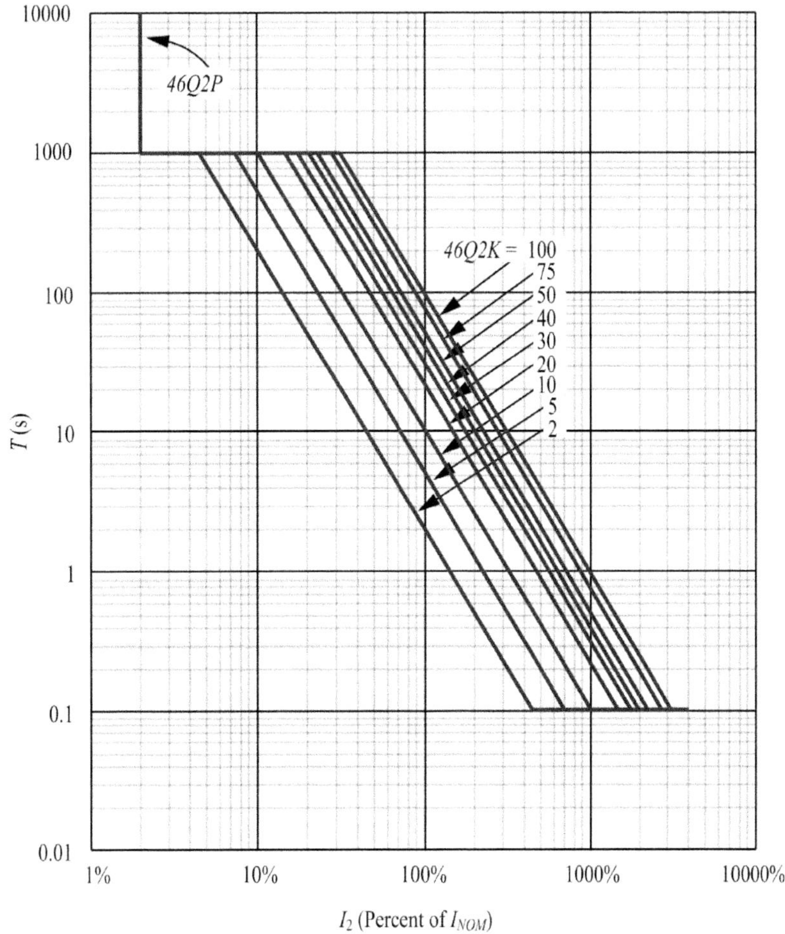


**FIGURE 19-42** Time-current characteristics of a modern negative-sequence overcurrent element for generator current unbalance protection.

The machine and the power system are at risk when a generator loses excitation. The generator may suffer rotor or stator overheating and experience large pulsating torques. No general guidelines exist for the length of time a generator may operate without excitation. Some generator manufacturers provide this information. The reactive-power deficit may cause a voltage collapse, especially if a large generator connected to a weak system loses excitation. Another possible problem is the loss of steady-state stability. When these problems arise, the system may lose voltage or synchronous stability in a few seconds.

Loss-of-field (LOF) protection (40) should provide an early alarm to permit the operator to restore the field for an accidentally tripped field breaker. After a time delay, the protection must trip the main generator breaker and the field breaker and transfer unit auxiliaries. Tripping steam turbine stop valves may also be necessary [32].

Use a distance mho element to detect LOF; the generator terminal voltage and current are input signals to this element. Figure 19-43 shows the offset mho element characteristic with the originally recommended settings [37]. In this figure, $X_d$ and $X_d'$ are the direct-axis synchronous and transient generator reactances, respectively. The impedance measured by the mho element when the generator

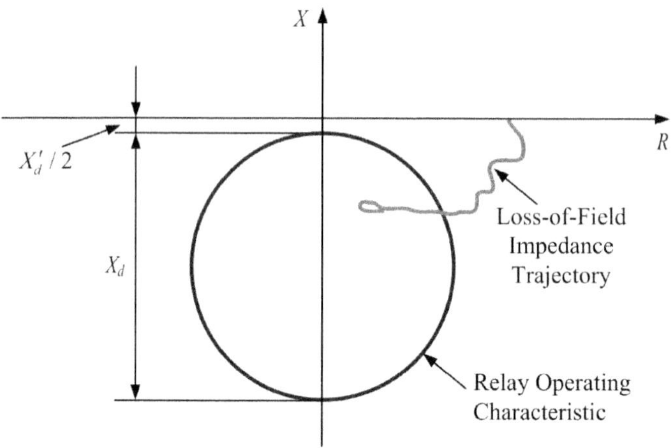


**FIGURE 19-43** LOF protection using a negative-offset mho element.

loses its field describes a trajectory on the impedance plane that starts at the impedance value corresponding to the generator initial load and ends oscillating in a region of the fourth quadrant. The location of this region depends on the generator load [38].

During stable and unstable power swings, the impedance measured by the LOF element also describes a trajectory on the impedance plane (not shown in Fig. 19-43). The element may misoperate if the impedance penetrates the operating characteristic. Delaying operation of the LOF element enhances security. For old generators with typical $X_d$ values of 1.1 to 1.2 p.u., a time-delay setting of 0.1 s normally ensures security for power swings. The larger $X_d$ values (1.5 to 2 p.u.) of modern generators may require longer time-delay settings. Perform transient stability studies to determine element time-delay settings.

Modern LOF protection elements have two-zone characteristics suitable for generators with large direct-axis reactance. References [2] and [32] describe these elements.

### 19.6.7 Motoring Protection

Generator motoring occurs when the energy supply to the prime mover is cut off while the generator is connected to the power system. The generator operates as a synchronous motor driving the prime mover. This operation condition will not harm the generator, but may damage the prime mover. In addition, the mechanical load that the prime mover presents to the generator, when the generator is operating as a synchronous motor, may be high. This load represents an active-power loss for the power system.

For steam turbines, motoring causes overheating and may damage turbine blades and other turbine parts. Steam turbines may overheat even when the generator is operating at no load or with low power output. Turbine manufacturers provide information on the permissible time that steam turbines may operate in a motoring condition. Other types of prime movers may have different problems during motoring. Hydraulic turbines may suffer cavitation of the blades on low water flow during motoring. Gas turbines may have gear problems when rotating as a mechanical load. Diesel-engine generating units are in danger of explosion and fire from unburned fuel.

Motoring protection is typically used to trip and shut down all types of generating units for inadvertent motoring conditions except hydroelectric units designed to operate as synchronous condensers [32]. A time-delayed directional power element (32) detects the active power reversal caused by the motoring condition. The element setting depends on the type of prime mover. Time delay prevents misoperation for power swings caused by system disturbances or when synchronizing the machine to the system. Typical delay values are tens of seconds. A motoring protection element generally trips the main generator breaker and the field breaker, transfers the auxiliaries, and provides a trip signal to the prime mover [32].

Another application for 32 elements is sequential tripping of generators. To prevent overspeed, sequential tripping requires detecting a motoring condition following the steam turbine valve closing, to ensure steam flow is completely shut off before opening the generator breaker. You can use a 32 element to monitor the reverse power condition and perform tripping interlock in sequential tripping.

### 19.6.8 Other Generator Protection Functions

Other generator protection functions include the following [2, 32]:

- Stator thermal protection.
- Field thermal protection.
- Overexcitation protection.
- Overvoltage protection.
- Abnormal frequency protection.
- Out-of-step protection.
- Inadvertent energization protection.
- External fault backup protection.

## 19.7 MOTOR PROTECTION

Electric motors include induction (asynchronous) motors and synchronous motors. Induction motors may have squirrel cage or wound rotors. Synchronous motors have a rotor dc winding fed by an excitation system. These motors typically have a squirrel cage winding for starting.

In this subsection, we focus on medium-voltage squirrel cage induction motor protection, which is covered in more detail in [39–41].

### 19.7.1 Induction Motor Basic Concepts

The stator of an induction motor has a distributed three-phase winding that creates a rotating magnetic field when a three-phase voltage is applied to it. Equation (19-22) defines the field rotating speed or synchronous speed $n_s$.

$$n_s = \frac{120f}{p} \tag{19-22}$$

where $n_s$ is the synchronous speed in revolutions per minute; $f$ is the power system frequency in hertz; $p$ is the number of stator poles.

As noted before, induction motor rotors can have either a squirrel cage or a wound rotor winding. The squirrel cage winding consists of aluminum or copper bars placed in the slots of the rotor and running almost parallel to the shaft. Rings connect the bar ends to form a short-circuited assembly. The rotor bar assembly resembles a squirrel cage. The wound rotor winding is a three-phase winding, similar to the stator winding. Winding terminals are connected to three slip rings. Stationary brushes that press against the slip rings allow connecting the rotor winding to an external circuit, such as a three-phase resistor for high-staring torque applications or speed control purposes. The squirrel cage induction motor is widely used in the industry because it is simpler, more rugged, and more economical than the wound rotor motor.

The stator rotating magnetic field cuts the rotor bars and induces voltage in the conductors. The voltage drives a large current through the short-circuited bars. The field of the rotor current opposes the stator field and produces a torque to accelerate the rotor. At synchronous speed, there would be no relative motion between the rotor and the stator field and no induced voltage. Consequently, the rotor attains a speed slightly short of the synchronous speed at a point where the induced current supplies the load torque and losses. The slip $s$ is the difference between the synchronous speed $n_s$ and the rotor speed $n_r$, in per unit of $n_s$, as Eq. (19-23) shows.

$$s = \frac{n_s - n_r}{n_s} \tag{19-23}$$

Motor manufacturers provide information in the motor nameplate. For large motors, manufacturers generally also provide test data sheets and the characteristic current and torque graphs of the motor. These characteristic graphs include the torque versus speed and current versus speed graphs.

Figure 19-44 depicts the typical variation of induction motor current and torque as functions of speed during the starting process. It shows the distinctive characteristic of the induction motor to drive very high current until the peak (or breakdown) torque develops near full speed. The locked-rotor current (labeled *LRA* in the figure) may be three to seven times or more of the motor rated full-load current (labeled *FLA*).

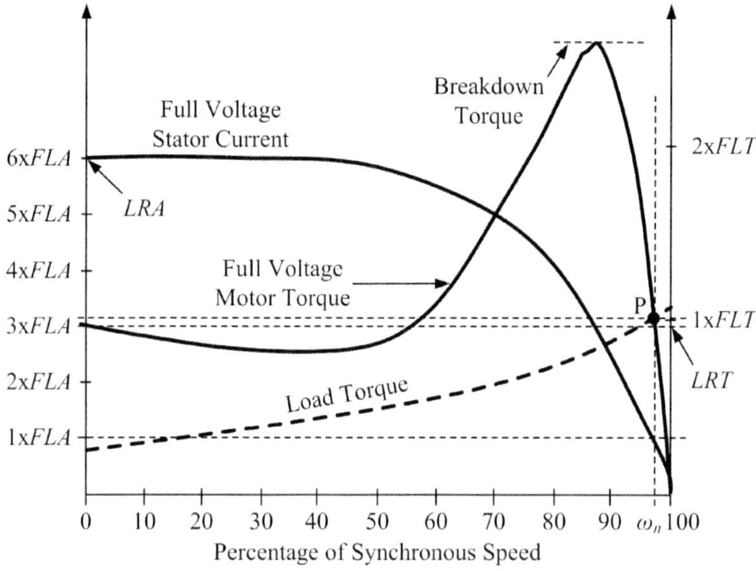


**FIGURE 19-44** Induction motor characteristic graphs.

In some cases, the manufacturer provides partial information about the torque graph, such as the following values:

- Rated torque (*FLT*).
- Locked-rotor torque (*LRT*) in percent of the rated torque.
- Breakdown torque in percent of the rated torque.

Manufacturers also provide the motor thermal limit curves, which specify the motor thermal capability. A thermal limit curve is a plot of the maximum possible safe time as a function of the motor winding current for conditions other than normal operation. They represent three conditions:

1. Locked rotor.
2. Starting and accelerating.
3. Running overload.

IEEE Standard 620 [42] provides the guidelines for presenting thermal limit curves of squirrel cage motors. The document states that the curves shall represent two initial conditions: the machine initially at ambient temperature, and the machine initially at rated operating temperature. Plots of the motor starting current as a function of time at 100% and 80% of rated voltage should also be included. The thermal limit curves show only two of the possible conditions of a first-order thermal process, where a balance of heat storage and heat loss determines temperature.

Figure 19-45 shows the thermal limit curves for a 2250 hp, two-pole motor and the plot of the motor starting current at rated voltage. The motor starting time is typically shorter than the locked-rotor time, as shown in the figure. This allows the operating time of an inverse-time overcurrent relay to be set long enough to let the motor start, yet short enough to prevent the motor from exceeding the locked-rotor time. However, for high-inertia loads, the starting current may encroach on the locked-rotor limit curve, resulting in an insufficient coordination time margin. The motor appears to overheat and the inverse-time overcurrent relay cannot be set to avoid a trip. A microprocessor-based relay with an accurate motor thermal model can properly protect the motor.

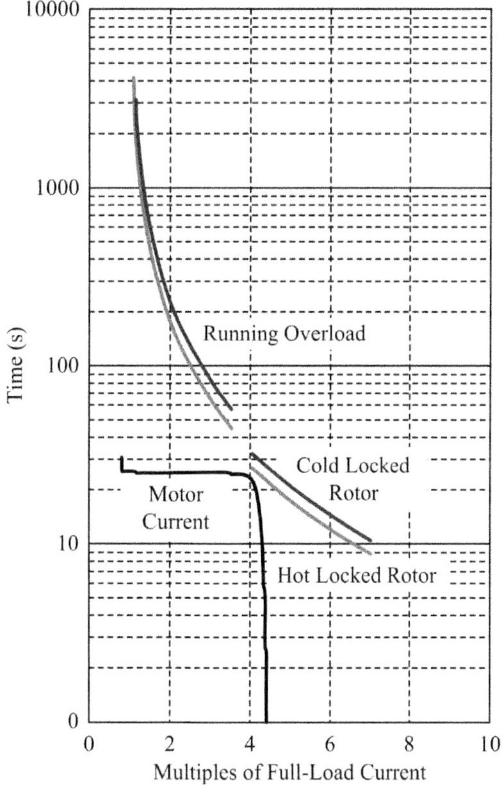


**FIGURE 19-45** Thermal limit curves of a 2250 hp, two-pole induction motor.

Motor insulation systems lose their physical and dielectric integrity over time from mechanical stress, contaminants in the insulation, and heat. Excessive temperatures substantially accelerate the decay process. The effect of elevated temperature is to reduce the ability of the insulation to withstand electrical or mechanical stress. The temperature level at which the insulation should be protected is a matter of judgment with some guidance in standards (such as NEMA Standard MG-1 [43]). Insulation failures cause motor internal faults.

Motor overheating may result from continuous or intermittent overload, unbalanced or low-voltage operation, locked-rotor conditions, higher than design ambient temperature or clogged cooling ducts. Induction motors have a large heat storage capacity, and slight balanced overloads for short periods of time do not produce damaging temperature excursions. Motor operation under voltage unbalance conditions causes negative-sequence current. The rotor negative-sequence resistance is higher than the positive-sequence resistance. As a result, unbalanced motor operation may cause excessive rotor heating even for normal load currents. Locked-rotor current causes rotor temperature to rise so quickly that there is little time for heat lost before reaching the limiting temperature.

Motor thermal limit curves only specify the motor thermal capability for locked-rotor and balanced, slow-varying overload conditions. These curves do not show the effect of unbalanced, fast-varying, or intermittent overload conditions.

In summary, typical induction motor problems include the following [41]:

- Electrical faults (phase and ground faults).
- Overheating resulting from:
  - Overload (continuous or intermittent).
  - Unbalanced operation.
  - Locked rotor (failure to start or load jams).
  - High ambient temperature.
  - Clogged cooling ducts.
- Other abnormal conditions:
  - Abnormal voltage.
  - Abnormal frequency.
  - Reversed phases.
  - Frequent starts.
  - Load loss.

### 19.7.2 Thermal Protection

The purpose of motor thermal protection is to allow the motor to start and run within the manufacturer's published guidelines, but trip if the motor thermal energy exceeds allowable values because of overloads, unbalanced operation, locked-rotor conditions, high ambient temperature, clogged ventilation ducts, or too frequent or prolonged starting.

There are two methods of motor thermal protection. One method uses resistive temperature detectors (RTDs) to directly measure the motor temperature. The other method measures the motor stator currents (and in some cases voltages) and estimates the rotor and stator thermal energy levels.

RTD-based protection provides good stator thermal protection for balanced, slow-varying overloads. It may also detect motor overheating because of high ambient temperature or clogged ventilation ducts. Additional RTDs on motor and drive gearings provide temperature information for the relay to detect impending mechanical problems. RTDs may also provide ambient temperature information.

Unfortunately, the slow response of RTDs reduces their value to detect fast motor heating during the starting process. Additionally, the heating problems during starting conditions are mostly manifested in the rotor, making RTD transmission difficult and expensive. Finally, RTDs fail to provide reliable motor thermal information for unbalanced conditions and for fast-varying or intermittent overload conditions. RTD-based thermal protection is a good solution for large motors, but it must be complemented with current-based thermal protection.

In the past, users relied on inverse-time phase overcurrent elements and a separate negative-sequence overcurrent element to detect currents that could lead to overheating. However, inverse-time overcurrent elements do not respond properly to the motor thermal dynamics during variable or intermittent load conditions. These elements neither accurately track the excursions of conductor temperatures nor account for the motor thermal history.

A better approach for current-based motor thermal protection is to use mathematical thermal models that account for the $I^2R$ heating caused by positive- and negative-sequence currents. These mathematical models use current information to calculate the motor thermal energy in real time. The thermal energy is then compared to thermal limit trip and alarm thresholds to prevent overheating from overload, unbalanced operation, locked rotor, or too frequent or prolonged startups [39, 44, 45]. The thermal models are defined by motor nameplate and thermal limit data.

An electric motor behaves like a heat transfer system. The motor body is heated up by a heat source, which is given by the $I^2R$ losses. For a constant current condition, all the heat source energy is transferred to ambient by convection, conduction, and radiation, and the motor temperature is constant. When the motor current increases, the additional thermal energy is initially absorbed by the motor body, its temperature increases, and part of that energy is transferred to the cooler ambient. During some time, defined by the motor thermal time constant, the temperature rises to a new equilibrium value.

The first-order equation of the heat transfer system has the same form as the equation expressing the voltage rise in an electrical resistive-capacitive circuit fed by a current source. Therefore, we can use the electrical analog circuit shown in Fig. 19-46 to visualize the motor thermal model. This electric analog of the heat transfer system includes a heat source, modeled as a current source; a thermal capacitance $C_{TH}$, modeled as a capacitor that represents the motor body capacity to absorb thermal energy; a thermal resistance $R_{TH}$, modeled as a resistor that represents the motor capacity to transfer heat to ambient; and a comparator, to compare the present thermal energy estimate $U$ with a thermal trip threshold $U_{TRIP}$. Figure 19-46 also shows the time variation of $U$ when the motor experiences a sudden overload condition. The thermal model trips when $U = U_{TRIP}$. The operating time is $T_{TRIP}$ in this example.

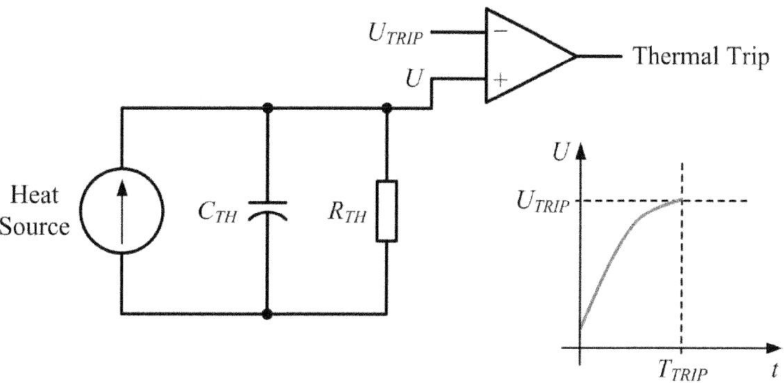


**FIGURE 19-46** Electric analog circuit of the motor first-order thermal model.

Three thermal models describe the induction motor. The rotor thermal model includes an adiabatic model for the motor starting process and a first-order model for the motor normal running condition. The stator thermal model is a first-order model that represents the motor normal running condition. To define the rotor and stator thermal models, we need to determine the characteristics of each component of the electric analog circuit depicted in Fig. 19-46.

For the rotor thermal model, we need to consider that, because of the skin effect, the current distribution on the rotor bar cross-sectional area depends on the rotor current frequency. The rotor current frequency $f_r$ equals $s$ times the power system nominal frequency $f$.

When the motor is in a normal running condition, the very low rotor current frequency (0.6 Hz for $s = 0.01$ at 60 Hz, for example) causes the current to be uniformly distributed on the rotor bar area. The rotor resistance $R_N$ at the rated running condition is low.

At the beginning of the starting process, with the rotor at rest, $s = 1$, and $f_r = f$. At this higher frequency, the rotor current flows in a fraction (approximately one-third) of the rotor bar cross-sectional area. With the current flowing in approximately one-third of the area, the locked-rotor resistance $R_L$ is roughly three times the running resistance $R_N$. The combination of high rotor starting current and increased rotor resistance produces very high $I^2R$ heating during the starting process.

For example, a typical starting current of six times the rated current with $R_L = 3\ R_N$ causes the $I^2R$ heating to be approximately $6^2(3) = 108$ times the heating at rated current [39]. Consequently, the extreme temperature caused by the high starting current must be tolerated for a limited time to allow the motor to start. This 60 Hz skin effect is called the deep bar effect. It is an important phenomenon used by motor designers to increase the starting torque by increasing the initial rotor resistance. As the motor accelerates, $s$ decays, and the current occupies more and more of the rotor bar cross-sectional area. The rotor positive-sequence resistance $R_1$ varies linearly from the $R_L$ value to the $R_N$ value during the starting process.

Equation (19-24) gives the positive-sequence rotor resistance $R_1$ [39, 44, 45].

$$R_1 = (R_L - R_N)s + R_N \tag{19-24}$$

Motor operation under three-phase unbalanced supply voltages causes unbalanced stator currents. The resulting negative-sequence current causes a magnetic field that rotates in opposition to the rotor rotation. The field rotates with respect to the rotor at twice synchronous speed, less the slip value, and the induced rotor current has a frequency equal to $(2 - s)f$. Because of the skin effect, this high-frequency current concentrates close to the rotor bar surface and occupies a very small portion (approximately one-sixth) of the rotor bar cross-sectional area. With the current flowing in approximately one-sixth of the area, the rotor negative-sequence resistance $R_2$ is approximately six times the running resistance $R_N$. This concentration of current causes excess heating that can quickly overheat and damage the rotor.

We obtain the expression for the negative-sequence resistance $R_2$ by replacing $s$ with the negative-sequence slip $2 - s$ in Eq. (19-24):

$$R_2 = (R_L - R_N)(2 - s) + R_N \tag{19-25}$$

In an induction motor, the rotor heat source is the $I^2R$ losses caused by the positive- and negative-sequence currents. We can obtain factors expressing the relative heating effect of positive- and negative-sequence currents by dividing Eqs. (19-24) and (19-25) by the running resistance $R_N$. Equation (19-26) provides the heat source expression, valid for the starting and running conditions.

$$\textit{Heat Source} = \frac{R_1}{R_N} I_1^2 + \frac{R_2}{R_N} I_2^2 \tag{19-26}$$

Figure 19-47 depicts the adiabatic rotor thermal model for motor starting, assuming the rotor positive- and negative-sequence resistances to be constant during the starting process. During this process, the motor heats up very fast without losing thermal energy to the surrounding environment. Heat transfer is limited by the fact that forced cooling is absent at the initial locked-rotor condition and increases slowly as the motor gains speed. In this adiabatic process, the motor temperature grows continuously until the starting current decreases to a normal value or the motor is disconnected. For this reason, the model does not include a thermal resistance. After successfully starting, the motor cools down and eventually reaches the normal running temperature.

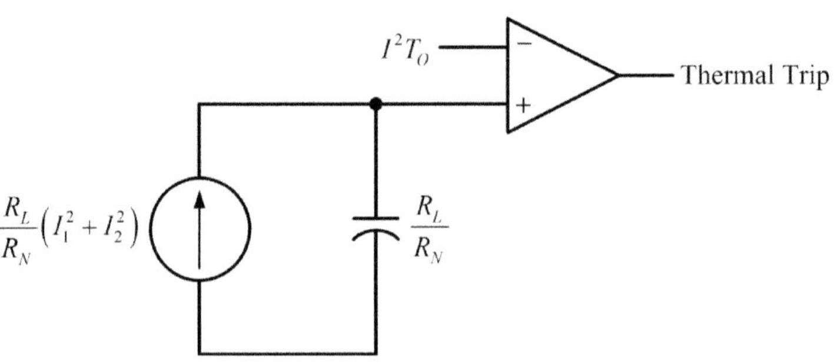


**FIGURE 19-47** Rotor thermal model for the motor starting condition.

For the locked-rotor condition, $s = 1$. Hence, both the positive-sequence slip $s$ and the negative-sequence slip $2 - s$ equal unity. From Eqs. (19-24) and (19-25), we get

$$R_1 = R_2 = R_L \tag{19-27}$$

Substituting Eq. (19-27) into Eq. (19-26), we get the heat source expression for the locked-rotor condition:

$$\text{Heat Source} = \frac{R_L}{R_N}(I_1^2 + I_2^2) \tag{19-28}$$

In order to determine the thermal trip threshold of the model we use the motor safe starting time, which is indicated by the thermal limit curves shown in Fig. 19-45. The cold locked-rotor characteristic specifies the time for the starting current to heat the rotor to the limit temperature with the motor initially at ambient temperature. The hot locked-rotor characteristic specifies the time for the starting current to heat the rotor to the limit temperature with the motor initially at rated operating temperature.

The rotor model trip threshold (Fig. 19-47) should equal the allowable temperature rise for starting from rated operating temperature:

$$\text{Thermal Trip Threshold} = I_L^2 T_O \tag{19-29}$$

where $I_L$ is the locked-rotor current in per unit of the rated current $I_N$; $T_O$ is the safe locked-rotor time from rated operating temperature.

The thermal capacitance $C_{TH}$ of the model is selected equal to $R_L/R_N$ (Fig. 19-47). With $C_{TH} = R_L/R_N$, the thermal energy estimate $U$ reaches the trip value in the locked-rotor time when the motor positive-sequence current $I_1$ equals the locked-rotor current $I_L$.

Figure 19-48 depicts the rotor thermal model for the motor running condition. When the motor is running, it transfers thermal energy to the surroundings through radiation, conduction, and convection. The rotor running thermal model simulates this heat transfer through the thermal resistor $R_{TH}$.

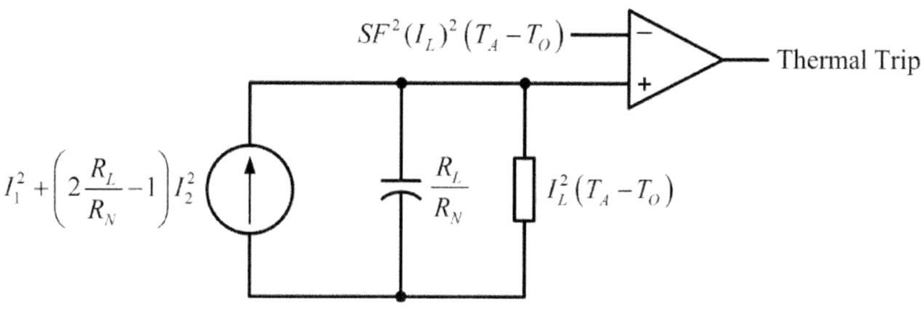


**FIGURE 19-48** Rotor thermal model for the motor running condition.

For the motor running condition, $s \approx 0$. Hence, the positive-sequence slip $s$ is close to zero, and the negative-sequence slip $2 - s$ is close to two. From Eqs. (19-24) and (19-25), we get

$$R_1 = R_N \tag{19-30}$$

$$R_2 = 2R_L - R_N \tag{19-31}$$

Substituting Eqs. (19-30) and (19-31) into Eq. (19-26), we get the rotor heat source expression for the motor running condition:

$$\text{Heat Source} = I_1^2 + \left(2\frac{R_L}{R_N} - 1\right)I_2^2 \tag{19-32}$$

The thermal trip threshold in this case equals the thermal energy level representing the motor rated operating temperature times the motor service factor squared:

$$\mathit{Thermal\ Trip\ Threshold} = SF^2U_O = SF^2(I_L)^2(T_A - T_O) \tag{19-33}$$

where $U_O$ is the motor thermal energy level at rated operating temperature; $SF$ is the motor service factor; $T_A$ is the safe locked-rotor time from ambient temperature.

The value of the thermal resistance $R_{TH}$ should reflect the fact that the motor will reach a thermal energy level representing the rated operating temperature when 1 p.u. of positive-sequence current flows in the motor for a long time. Because the positive-sequence heating factor equals one in the running model, and 1 p.u. of $I_1$ squared equals one, the value of the thermal resistance $R_{TH}$ equals the thermal energy level at the motor rated operating temperature $U_O = I_L^2(T_A - T_O)$.

As in the rotor model for the starting condition, the thermal capacitance $C_{TH}$ is selected equal to $R_L/R_N$ (Fig. 19-48).

Relays using the described thermal models automatically select the rotor thermal model state to use based on the magnitude of the sum of positive- and negative-sequence currents [39, 44, 45]. The relay declares a starting state when this current exceeds 2.5 p.u. of the motor rated current. In this starting condition, the model assumes the rotor resistance remains constant at the locked-rotor value ($R_r = R_L$). The relay declares a running state when the current becomes less than 2.5 p.u. In this state, the trip threshold decreases exponentially from the locked-rotor value to the running condition value using the motor thermal time constant. This method emulates the motor temperature that decreases to the steady-state running condition after the motor starts. In the running condition, the model assumes the rotor resistance has the rated running value ($R_r = R_N$).

Assuming the $R_L/R_N$ ratio to be equal to three, the model thermal capacitance $C_{TH}$ equals three. From Eqs. (19-28) and (19-32), the heat sources for the rotor thermal models are the following:

$$\text{Starting model: } \mathit{Heat\ Source} = 3(I_1^2 + I_2^2) \tag{19-34}$$

$$\text{Running model: } \mathit{Heat\ Source} = I_1^2 + 5I_2^2 \tag{19-35}$$

The running overload curves (Fig. 19-45) show the stator thermal limit. These curves fit the time-current Eq. (19-36), which has the form of the thermal characteristic equation defined in Clause 3.1.2 of IEC Standard 255-8 [46].

$$t = \tau \bullet \ln\left(\frac{I^2 - I_0^2}{I^2 - SF^2}\right) \tag{19-36}$$

where $\tau$ is the stator thermal time constant; $I$ is the stator current in per unit of rated current; $I_0$ is the initial current in per unit of rated current; $SF$ is the motor service factor.

The overload curves shown in Fig. 19-45 are asymptotic to a current equal to the service factor, which heats the stator to its temperature limit. These curves are taken as the trip threshold. The stator thermal time constant can be determined by a heat run, where a load current is applied and the temperature rise is measured at regular time intervals. The temperature will rise exponentially, and the thermal time constant will be the time it takes the temperature to reach 63.2% of its final value. In the case of the 2250 hp motor, the time constant $\tau$ is 4940 s.

Figure 19-49 depicts the equivalent circuit that corresponds to the stator first-order thermal model. In this case, the thermal capacitance $C_{TH}$ equals the motor thermal time constant $\tau$. By assigning a value of 1 to the thermal resistor $R_{TH}$, we get a value equal to $\tau$ for the time constant $R_{TH}\, C_{TH}$ of the equivalent circuit. Since the positive- and negative-sequence currents have the same heating effect on the stator, the heat source equals $I_1^2 + I_2^2$. With $R_{TH} = 1$, the trip threshold should equal the service factor squared. Another alternative for the stator thermal model equivalent circuit results when we assign a value of

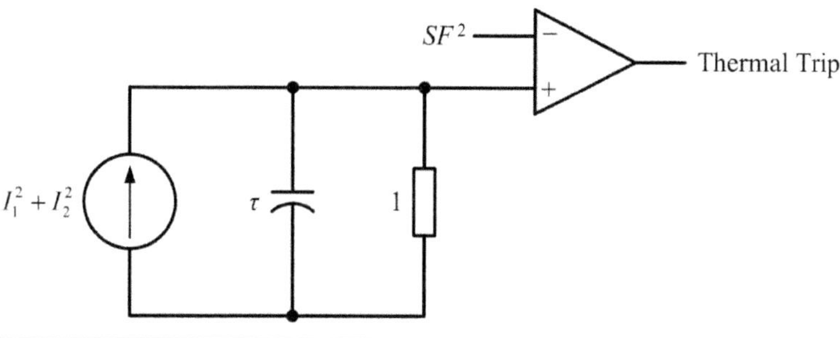


**FIGURE 19-49** Stator thermal model for the motor running condition.

$I_L^2\,(T_A - T_O)$ to the thermal resistance $R_{TH}$ and a value of $SF^2\,(I_L)^2\,(T_A - T_O)$ to the thermal trip setting. The heat source and the thermal capacitance $C_{TH}$ do not change in this case.

### 19.7.3 Short-Circuit Protection

Phase and ground overcurrent protection limits the damage caused by faults in the motor stator windings and leads. Phase overcurrent elements also provide a certain level of ground fault protection. However, ground overcurrent elements provide higher sensitivity for ground fault detection.

Figure 19-50 shows two motor overcurrent protection schemes. With modern relays, the scheme in Fig. 19-50*a* can include the following elements: phase overcurrent instantaneous elements (50);

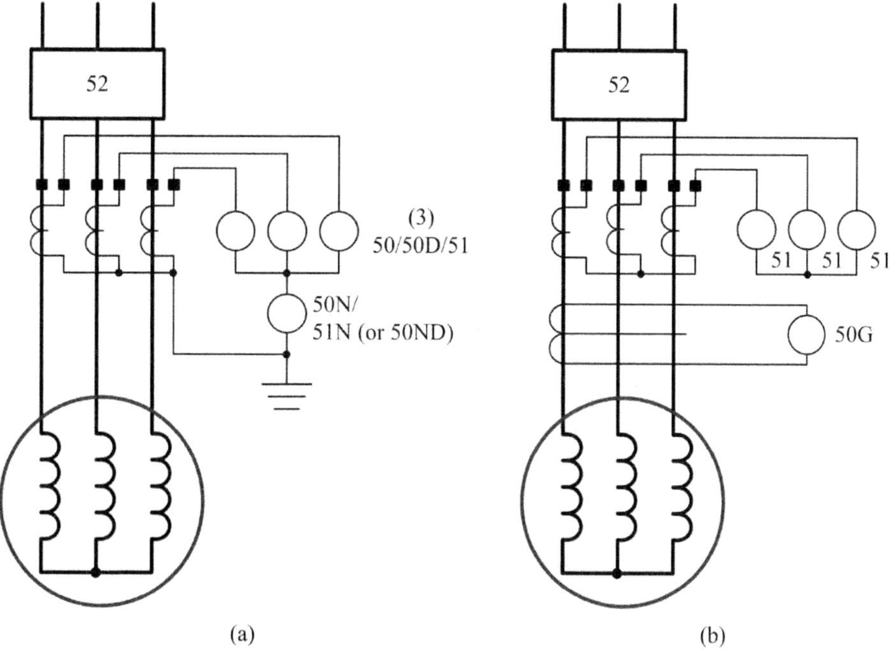


**FIGURE 19-50** Motor overcurrent protection schemes. (*a*) The ground elements measure the CT residual current; (*b*) a flux-summing CT provides the zero-sequence current.

phase short-time, definite-time elements (50D); phase overcurrent inverse-time elements (51); a ground overcurrent instantaneous element (50N); and a ground overcurrent inverse-time element (51N) or a ground short-time, definite-time element (50ND). In this scheme, ground fault detection sensitivity is limited by the high phase CT ratios and by the fictitious residual current caused by uneven CT saturation during the motor starting process. Figure 19-51 depicts a typical coordination chart for the overcurrent elements included in the Fig. 19-50*a* scheme.

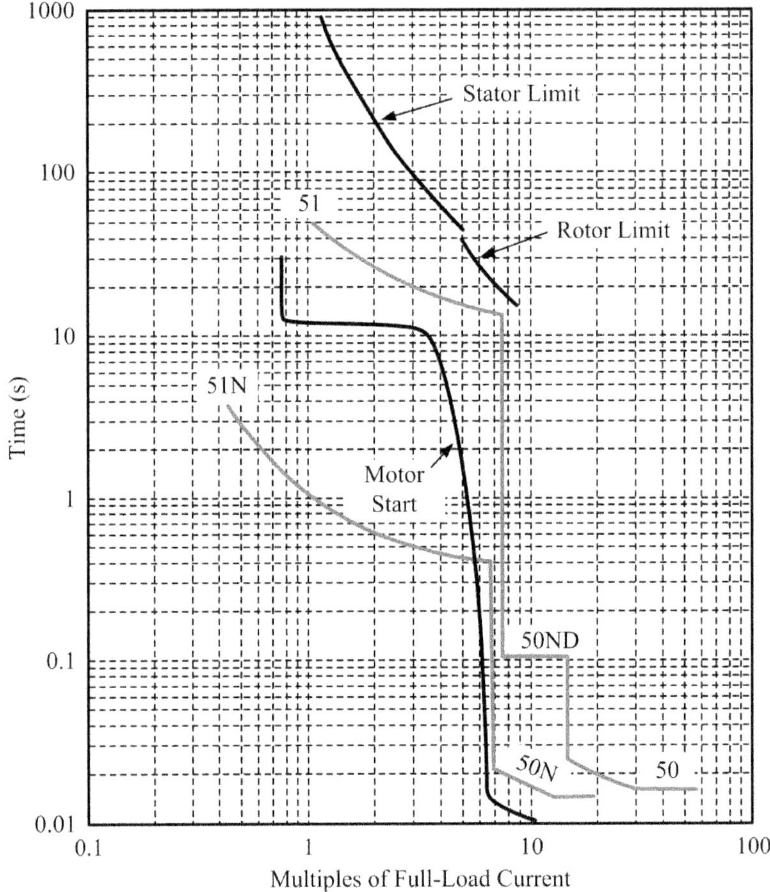


**FIGURE 19-51** Typical coordination chart for the overcurrent protection scheme shown in Fig. 19-50*a*.

To improve sensitivity, we can use a ground element (50G) connected to the secondary of a flux-summing CT, as shown in Fig. 19-50*b*. Since this CT measures three times the zero-sequence current, it can have a considerably lower ratio than phase CTs (a typical ratio is 50/5). In addition, this scheme avoids the false residual current problem of the Fig. 19-50*a* scheme.

In high-capacity medium voltage motors, we can also apply differential protection. The protection scheme is similar to the scheme used in generators (Fig. 19-37). Three CTs are usually located in the switchgear in order to include the motor cables in the protection zone. The remaining three CTs are located in the neutral connection of the motor. Six leads must be brought out from the motor as specified at the time of purchase.

### 19.7.4 Other Motor Protection Functions

Motor protection must cover other abnormal operation conditions, such as the following:

- Load jam and load loss are abnormal operation conditions that require detection and protection. We can provide load-jam protection with a definite-time overcurrent element. This function is armed once the motor is running to ride through high starting currents. Load-jam conditions can also be detected by a well-designed motor thermal element; however, the specific load-jam protection trips faster than the thermal element. A load-loss condition is caused by an issue with the driven load, such as a broken shaft. For load-loss detection, we can use time-delayed undercurrent or underpower elements, which are enabled once the motor is running.
- During each startup, the components of the induction motor (stator coil, rotor bars, and rotor end rings) are subjected to very high mechanical and thermal stresses due to the high value of the starting current and inadequate cooling before the motor comes up to speed and forced cooling becomes effective. The cumulative effect of frequent startups can cause critical component temperatures to rise and exceed the design limits. This condition can result in damage to the rotor or insulation system. We must limit the number of startups per hour to a safe value, recommended either by the manufacturer or by NEMA Standard MG-1 [43]. We can use a startup counter to protect the motor from excessive starting.
- The rotor running thermal model presented earlier accounts for the negative-sequence heating and provides current unbalance protection under all operation conditions. However, some motor protection relays lack a thermal model. In other relays, the thermal model does not account for the negative-sequence rotor heating. In these cases, a separate negative-sequence overcurrent element should provide current unbalance protection. A pickup current of $3I_2 = 1.5$ A with a definite time delay of 4 s are typical settings [39]. You can also use this negative-sequence overcurrent element to complement a thermal element that accounts for the negative-sequence heating.
- In some applications, such as pumps, the head causes the motor pump to reverse rotation direction after the motor is shut down. It is not desirable to restart the motor during this condition because the excessive torques can damage the motor and driven equipment. The solution is to block a restart for some time until after the motor and pump have stopped turning.
- Induction motors are capable of operating at rated balanced load with a ±10% variation from normal voltage. Because motors are constant kVA loads, the current increases with decreasing voltage. In addition, torque decreases as a function of the voltage squared. Consequently, low voltage causes increased temperature and the possibility of stalling, in the worst case. Undervoltage elements with an appropriate time delay provide motor undervoltage protection. The time delay allows overriding transient voltage sags caused by motor starting and short circuits on adjacent sections of the power system. However, the application of undervoltage protection depends on the service provided by the motor. A general approach is to trip nonessential motors for undervoltage conditions to allow the voltage to recover and to keep the essential motors in operation.

## 19.8 WIDE-AREA PROTECTION AND CONTROL

Modern power systems typically have reduced redundancy and operate close to their security limits. They are prone to angle and voltage stability problems that could lead to major system collapse.

Local systems protecting transmission lines, generators, transformers, and buses are the first line of defense against power system disturbances. Reliable, fast, and selective fault clearance normally prevents cascading events that could cause wide-area disturbances. However, modern power systems also need wide-area protection and control systems (system integrity protection systems) that detect abnormal system conditions and take preplanned, corrective actions to minimize the risk of wide-area disruptions and to increase power system transfer capability [2]. This subsection reviews power system out-of-step protection, underfrequency and undervoltage load shedding, and automatic generator shedding. Reference [2] provides a comprehensive coverage of modern wide-area protection, control, and monitoring systems.

### 19.8.1 Power Swing Blocking and Out-of-Step Tripping

Power system disturbances cause oscillations in machine rotor angles that result in power flow swings. Depending on the disturbance severity and the control actions, the system may return to a new equilibrium state (a stable power swing) or may have an out-of-step condition, a loss of synchronism between groups of generators.

Fast fault clearing prevents power system transient stability problems that could cause major blackouts. Power swings may cause undesirable power system protection operation that could have a compounding effect on the power system disturbance. Hence, protection systems must trip fast for faults and remain secure during power swings.

Figure 19-52 shows the angular behavior of a simple two-machine system during stable and unstable power swings. In a stable system, the angle $\delta$ between the machine voltages oscillates in a damped mode around a final value $\delta_1$. The system reaches a new steady-state operation point. No protection operation is required during stable power swings. In an unstable system, $\delta$ grows monotonically, and the machines lose synchronism. Unstable system operation is undesirable, as it creates high currents and power flows, as well as high and low voltages. Unstable system operation also causes severe generator torque oscillations. Voltage fluctuations may also affect the power station auxiliaries, which may need to be tripped. The protection systems must detect unstable system oscillations and make the appropriate tripping decisions in order to divide the system into electrical islands.

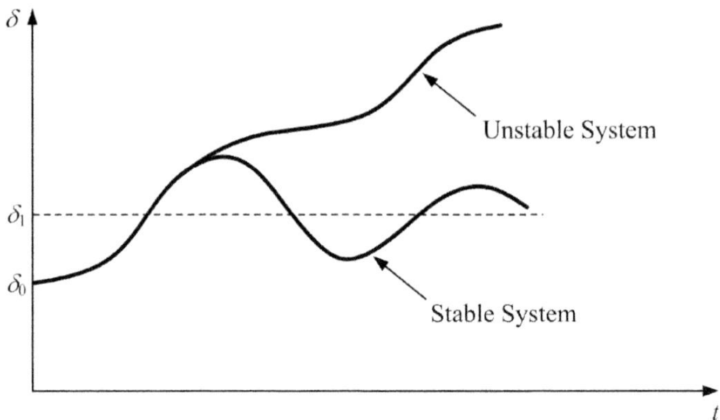


**FIGURE 19-52** Power system angular behavior during stable and unstable power swings.

Current and voltage fluctuations during the power swing may cause undesirable operation of distance, phase directional overcurrent, phase overcurrent, and phase undervoltage elements. Current differential protection schemes are immune to power swings because these swings only cause through current in the differential scheme.

The impedance measured by distance elements tends to oscillate during power swings as a result of voltage and current oscillations. The voltage near the electrical center of the system decreases while the current increases, which appears as an underimpedance condition. The measured impedance may penetrate the element operating characteristic and cause a misoperation.

Figure 19-53 depicts typical impedance trajectories for stable and unstable power swings. In a stable power swing, $\delta$ grows to a given maximum value (typically between 90° and 120°) during the first oscillation and then recovers. The impedance trajectory typically does not leave the first (and eventually fourth) quadrant or the second (and eventually third) quadrant, depending on the initial system operation condition. During an unstable power swing, $\delta$ continues to grow (or to decrease)

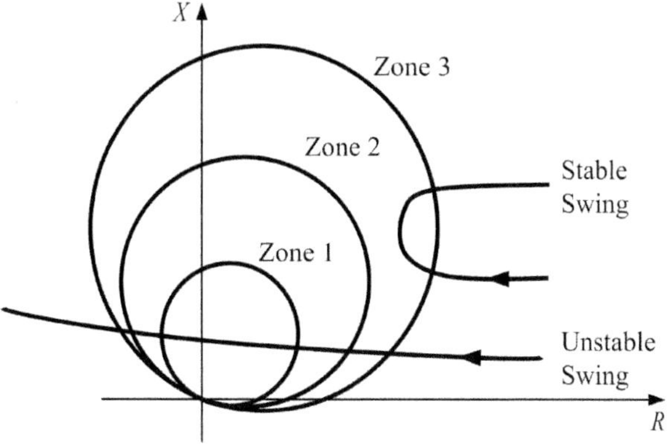


**FIGURE 19-53** Power swings may cause undesirable distance element operation.

monotonically and never recovers. The resulting impedance trajectory moves from the first or fourth quadrant to the second or third quadrant, or vice versa, depending on the initial system operation condition. Figure 19-53 shows that the measured impedance can penetrate distance element operating characteristics. The element will trip if the measured impedance stays inside the operating characteristic for a time greater than the zone time delay. Fast-operating Zone 1 elements and directional comparison tripping schemes are particularly sensitive to power swings. However, Zone 2 and Zone 3 elements may also misoperate for slow power swings.

The protection philosophy for power swings is to avoid tripping for stable power swings and to separate the power system into islands for unstable power swings, to prevent wide-area blackouts and equipment damage. The power swing blocking (PSB) function (68) discriminates between faults and power swings and blocks relay elements prone to operate during power swings. The out-of-step tripping (OOST) function (78) discriminates between stable and unstable power swings and initiates system islanding during loss of synchronism.

Most power swing detection methods use measured impedance information. Other methods estimate the swing-center voltage and its rate of change or use synchrophasor information [2].

Power swing detection is often based on the rate of change of the measured impedance [47]. The traditional approach is to compare the measured phase or positive-sequence impedance with an element characteristic on the impedance plane. This characteristic may consist of circles, blinders, or polygons.

Figure 19-54 shows a characteristic composed of two concentric polygons, defined on the positive-sequence impedance plane. You can use this characteristic to provide PSB and OOST functions. Set the outer power swing detection characteristic (OPSD) so that it does not operate for the maximum load conditions. For PSB applications, set the inner characteristic (IPSD) to be outside the largest distance element characteristic you want to block (shown as a mho circle in Fig. 19-54). For OOST applications, set the IPSD characteristic so that it does not operate for the worst stable power swing. An alternative to the use of two concentric polygons is a characteristic with four blinders.

The PSB logic using the characteristic shown in Fig. 19-54 measures the impedance rate of change to discriminate power swings from faults. The measured impedance changes slowly during power swings because of large generator rotor inertias. The impedance rate of change is very fast during a system fault. The PSB logic issues a blocking signal when the measured positive-sequence impedance stays between the OPSD characteristic and the IPSD characteristic longer than a blocking delay (a relay setting).

The impedance rate of change also helps discriminate between stable (slower) and unstable (faster) power swings. The OOST logic based on the Fig. 19-54 characteristic issues a tripping signal when the positive-sequence impedance remains between the OPSD characteristic and the IPSD characteristic longer than a tripping delay (a relay setting) and then enters the IPSD characteristic. The tripping

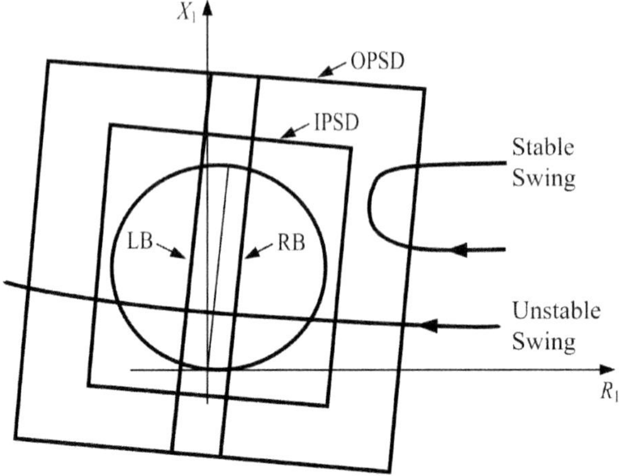


**FIGURE 19-54** Power swing detection characteristic composed of two concentric polygons, applicable to PSB and OOST functions.

delay is shorter than the blocking delay. This trip-on-the-way-in (TOWI) occurs before the first pole slip, when the angle between the equivalent generators at both sides of the relay is less than 180°.

You can also program the OOST logic to trip after the timer expires and the impedance enters and then exits the IPSD characteristic. This trip-on-the-way-out (TOWO) occurs after the first pole slip. To select between TOWI and TOWO, conduct transient stability studies and evaluate the ability of the breakers to interrupt current with a significant voltage across the contacts.

Another impedance-based power swing detection method uses the characteristic shown in Fig. 19-55, which is applicable to the OOST function in generators. Two blinders divide the impedance plane into three areas. The blinders are located at both sides of the system impedance. During faults or stable swings, the apparent impedance moves from Area 1 to Area 2 or from Area 3 to Area 2, depending on the initial system operation condition. On the other hand, an

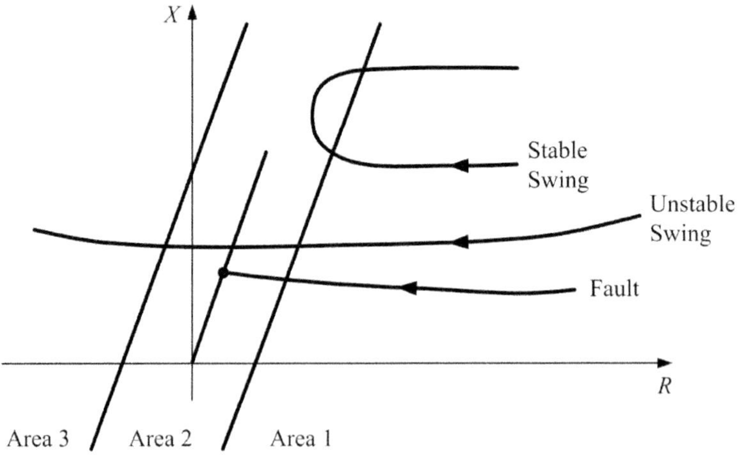


**FIGURE 19-55** Power swing detection characteristic composed of two blinders, applicable to the OOST function in generators.

unstable swing causes the apparent impedance to cross all three areas of the impedance plane. This information can be used as the basis for the OOST logic. This logic initiates generator breaker tripping after the first pole slip (TOWO).

The swing center of a two-source equivalent power system is the location where the voltage magnitude equals zero when the angle between the source voltages is 180°. The swing-center voltage provides information for power swing detection. References [2] and [48] describe a PSB method that uses the rate of change of the positive-sequence swing-center voltage. This method does not require user settings or stability studies for proper application.

Reference [2] also describes synchrophasor-based power swing detection methods.

### 19.8.2 Underfrequency Load Shedding

When a power system experiences a sudden increase of load (tripping of a tie line, for example) or a sudden generation loss, the resulting overload condition causes the frequency to decrease. If the available generators cannot increase output fast enough, the magnitude of the frequency drop and its rate of change depend on the magnitude of the overload, the effect of frequency-sensitive loads, which demand less active power at lower frequencies, and the power system inertia constant, which represents the total rotational inertia of all generating units and loads.

Generator operation at abnormal frequency subjects steam turbine blades to mechanical resonance and fatigue. Manufacturers typically provide data on permissible turbine operating time at specified frequency bands. The manufacturer could require that blade fatigue damage be considered cumulative during the turbine life. Generator abnormal frequency protection trips the generator breaker when the generating unit operates out of the specified frequency bands. The abnormal frequency operation limitations of combustion gas turbines are similar to those of steam turbines. Hydraulic turbines have broader frequency operating ranges than steam and combustion gas turbines.

At low frequencies, the power station auxiliaries driven by induction motors have reduced output. These auxiliaries include boiler feed pumps, fans supplying combustion air, and in nuclear stations, coolant fluid pumps. The result is reduced power station capability, which limits the generation reserve available for frequency control.

During power system underfrequency disturbances, the resulting reduction of power plant output compounds the problem. If the frequency reaches the setting values of generator abnormal frequency protection, generation unit tripping further increases the active power deficit and can cause a major blackout.

The solution to this problem is a wide-area underfrequency (81) load-shedding scheme that disconnects load to restore the generation-load balance before generation units begin to trip on underfrequency. In most disturbances, load-shedding schemes restore the system frequency close to its rated value.

Traditional load-shedding schemes consist of underfrequency relays located at selected feeders, which are grouped in several load-shedding stages. When the frequency drops below the first-stage setting value, the corresponding feeders are disconnected from the system. If the frequency continues to drop, other load-shedding stages are activated. Some underfrequency load-shedding schemes have as many as five underfrequency stages set at, for example, 59.2, 59, 58.8, 58.6, and 58.4 Hz in a 60 Hz system. Ideally, all underfrequency load-shedding schemes should have the same operating time so that the schemes act simultaneously across the power system when the frequency drops. A typical operating time is six cycles. Some traditional schemes also shed load groups at the same frequency, but with different time delays. This time-delayed operation allows for temporary frequency drops.

Schemes based on frequency drop typically work well for small overloads. For large overloads, frequency drops very fast, and a scheme that takes into account frequency drop and frequency rate of change provides better performance.

Designing underfrequency load-shedding schemes requires determining the total amount of load to shed, the number of load groups, and the frequency and delay settings for each load group. Schemes that also respond to frequency rate of change require determining this setting for each load group. Scheme design involves extensive dynamic simulation studies. These studies should consider system topology (and possible formation of islands), types of generation and loads, operation conditions, allowable frequency bands, and types of frequency relays. Once the scheme is in operation, systematic analysis of its operation using disturbance records provides data for improving the scheme.

Underfrequency load-shedding scheme design considerations include the following:

- The total amount of load to shed should be no less than the maximum expected system overload in order to restore frequency to a normal level. The analysis to determine this value should consider system operation as a whole and the possibility of system islanding. Typical load values to shed are 25% to 50% of maximum demand.
- System underfrequency relays should be coordinated with generator underfrequency protection. This setting ensures that load shedding prevents the frequency from falling to a value where generators start tripping, thus aggravating the disturbance. In some cases, you have to set some system underfrequency relays below the generator underfrequency protection setting values. In these cases, the system underfrequency relays should operate faster than the generator relays to ensure coordination.
- Frequency settings for the load groups depend on the types of generating units and frequency elements. Typical values of the minimum separation between load group frequency settings are 0.3 Hz for hydraulic generation and 0.2 Hz for thermal generation.

Traditional underfrequency load-shedding schemes trip predetermined feeders when underfrequency conditions exist for a certain amount of time. The system has no information on the amount of power that has been shed. Sometimes, more load is shed than necessary. Ideally, we want to shed just the amount of load necessary for the system to recover and to optimize power delivery. We can develop an underfrequency load-shedding scheme that adapts to system operation conditions. Such a scheme can be designed with existing microprocessor-based relays, taking advantage of relay programmability, mathematical calculation ability, and the ability to communicate with other IEDs.

The adaptive scheme shown in Fig. 19-56 reads the amount of power $P_{SD}$ that the system must shed according to system operation conditions in real time [49]. At the same time, the scheme has information about the power demand $P_1$ and $P_2$ of the feeders. Additionally, load priority can be programmed into the scheme to trip less critical loads first. Load shedding and system performance can be optimized with this information.

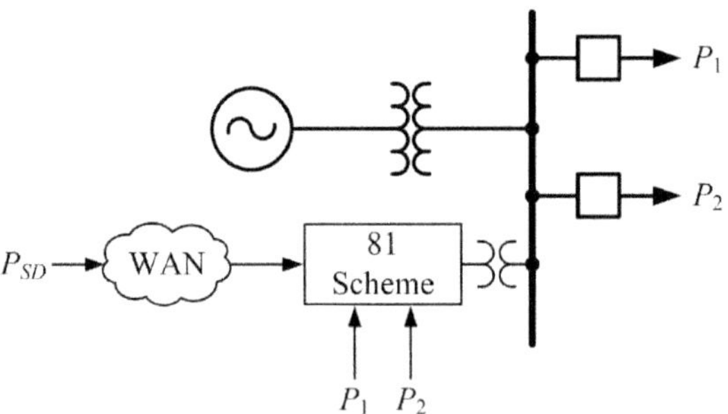


**FIGURE 19-56** Adaptive underfrequency load-shedding scheme.

The scheme reads the system active-power load-shedding requirements $P_{SD}$. When an underfrequency condition exists, the scheme trips Feeder 1 if the Feeder 1 active power $P_1$ is greater than $P_{SD}$. Otherwise, the scheme checks if the load $P_2$ in Feeder 2 is greater than $P_{SD}$. If $P_2$ is greater than $P_{SD}$, the scheme trips Feeder 2. If neither of the above loading conditions occurs, the scheme trips both feeders. With this scheme, one feeder, Feeder 2, could remain in service during an underfrequency disturbance when Feeder 1 load exceeds the system load-shedding requirements for a specific substation. Similarly, Feeder 1 could remain in service during the underfrequency disturbance when the Feeder 2 load exceeds the system load-shedding requirements for a specific substation.

### 19.8.3 Undervoltage Load Shedding

When a power system experiences a sudden deficit of reactive power (losing a large reactive power source, for example), the resulting condition causes the voltage to decrease. If the reactive power sources available cannot increase output fast enough, the system may experience a voltage collapse.

Load shedding is the last resort in preventing a system voltage collapse; definite-time undervoltage elements (27) are traditionally used for load shedding [50]. One disadvantage of this load-shedding method is that the tripping time is fixed independently of the type of load. However, voltage collapse depends on the type of load. For example, induction motors demand more reactive power from the system than other loads. A load-shedding scheme should identify loads demanding the most reactive power and shed these loads as soon as possible without affecting other loads in the system [49].

The system shown in Fig. 19-57 consists of two transmission lines that feed a load with constant power. Figure 19-58 shows the PV curves of this system for normal operation conditions and for operation with one transmission line out of service [2]. Point A is the operation point for a normal operation condition; $P_A$ is the active power that the load consumes for this condition. When one of the lines opens, the system PV curve changes, and Point C is the new critical operating point. The amount of load to be shed $P_{SD}$ must be greater than $P_A - P_C$ to restore stable operation conditions in the system. If $P_{SD} = P_A - P_B$, which is greater than $P_A - P_C$, the new operating point is B.

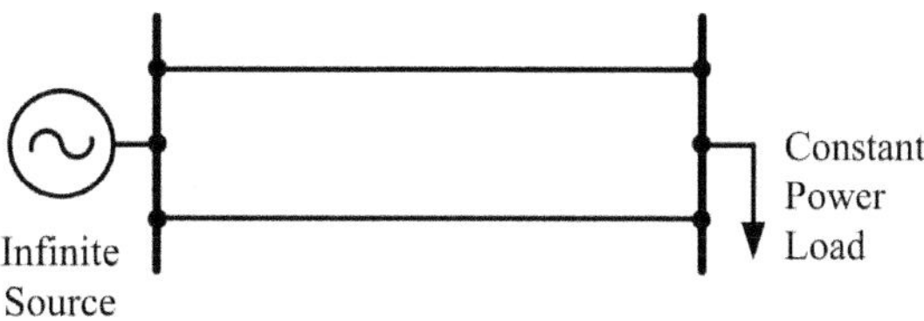


**FIGURE 19-57** Two-transmission line power system feeding a constant power load.

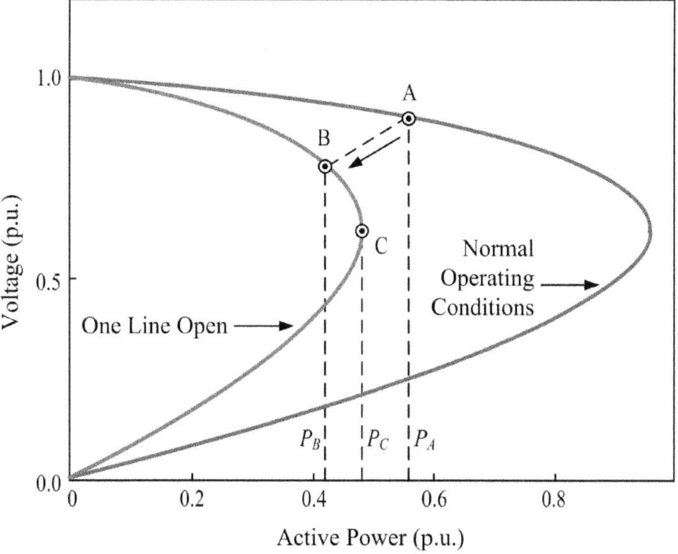


**FIGURE 19-58** PV curves of Fig. 19-57 system for normal operation conditions and for operation with one of the transmission lines out of service.

Undervoltage load-shedding schemes can have distributed or centralized topologies:

- Distributed schemes use relays associated with load groups. Using feeder multifunction relays for undervoltage load shedding reduces scheme cost. These schemes respond to local voltage only.
- Centralized schemes use a processor that receives voltage information from several locations, makes load-shedding decisions, and sends tripping commands to remote loads. These schemes measure voltage at critical system nodes and can also monitor load flow over important tie lines, which provides better information about impending voltage collapse conditions. These schemes require communications channels.

Figure 19-59 shows a distributed undervoltage load-shedding scheme using a multifunction relay. The positive-sequence undervoltage element (27V1) operates with some time delay to shed the feeder load when the voltage drops below a threshold value. Old schemes used three phase undervoltage elements (27P) instead of a 27V1 element. The positive-sequence voltage is a better measure of the system voltage level than individual phase voltage measurements.

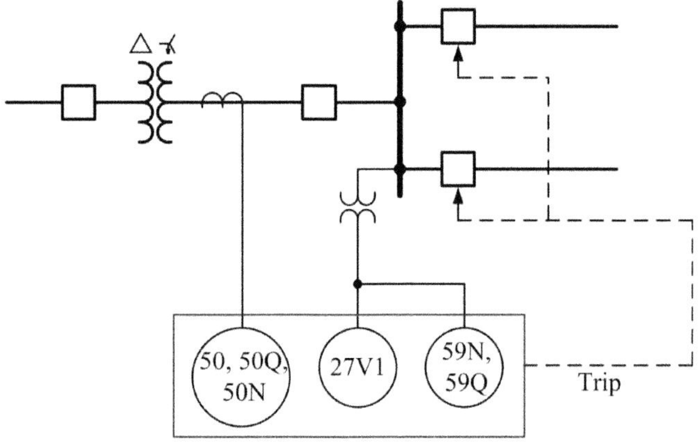


**FIGURE 19-59** Distributed undervoltage load-shedding scheme.

The scheme requires supervision to avoid misoperation for voltage sags caused by faults:

- Phase instantaneous overcurrent elements (50) provide supervision to prevent undervoltage scheme operation for feeder phase faults. We can add instantaneous negative-sequence (50Q) and/or ground (50N) overcurrent elements to detect unbalanced faults, including ground faults.
- A ground overvoltage element (59N) detects the zero-sequence voltage caused by ground faults. The load-shedding scheme operates only when there is no zero-sequence voltage present. A negative-sequence overvoltage element (59Q) is an alternative to the 59N element.

Figure 19-60 depicts the distributed load-shedding scheme logic. If an undervoltage condition exists, and there is no overcurrent and no zero-sequence or negative-sequence voltage, the timer starts. After a certain time *T* (1 s, for example), the scheme sheds the feeder load. Schemes with different delays provide staggered load shedding to minimize voltage overshooting conditions after the load is shed.

Figure 19-61 shows a more elaborate power system model to study voltage stability [49, 51]. The system has constant impedance load at Bus 8 and a combination of constant impedance and constant current loads at Bus 11. Figure 19-62 shows the voltage magnitudes at Bus 8 and Bus 9 after two of the five transmission lines that connect Bus 6 and Bus 7 open. The voltages at both buses drop immediately after the lines open. The voltage magnitude at Bus 9 is about 4% lower than the voltage

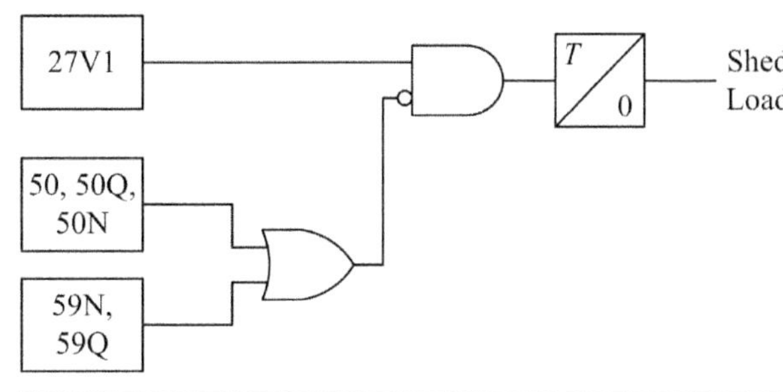


**FIGURE 19-60** Distributed undervoltage load-shedding scheme logic.

magnitude at Bus 8. The load at Bus 11 is the most critical one for voltage stability because of the under-load tap changer (ULTC) of the transformer feeding this load. The ULTC operates to raise the voltage at Bus 11. This action increases the reactive power consumption at Bus 11 and causes the voltages at Bus 8 and Bus 9 to continue dropping. Definite-time undervoltage elements at Bus 8 and Bus 9 (set to 95% of rated voltage as shown in Fig. 19-62) trip after a preset time delay to drop both loads. This undervoltage element operation does not consider the fact that the voltage at Bus 9 is lower than the voltage at Bus 8.

We can use an inverse-time undervoltage element to provide faster undervoltage element operation at the buses where the voltage drops deeper and faster in order to minimize system disturbance. Another advantage of this approach is that the inverse-time element operating time increases during system voltage recovery because the voltage magnitude increases over time. That is, inverse-time undervoltage elements take advantage of load dynamics for optimizing load shedding to prevent system voltage collapse while minimizing overshedding.

Figure 19-63 shows an inverse-time undervoltage element characteristic. The element starts timing when the voltage drops below a threshold value and trips in a time $T_U$ that is a function of the per unit voltage magnitude $V/V_{NOM}$.

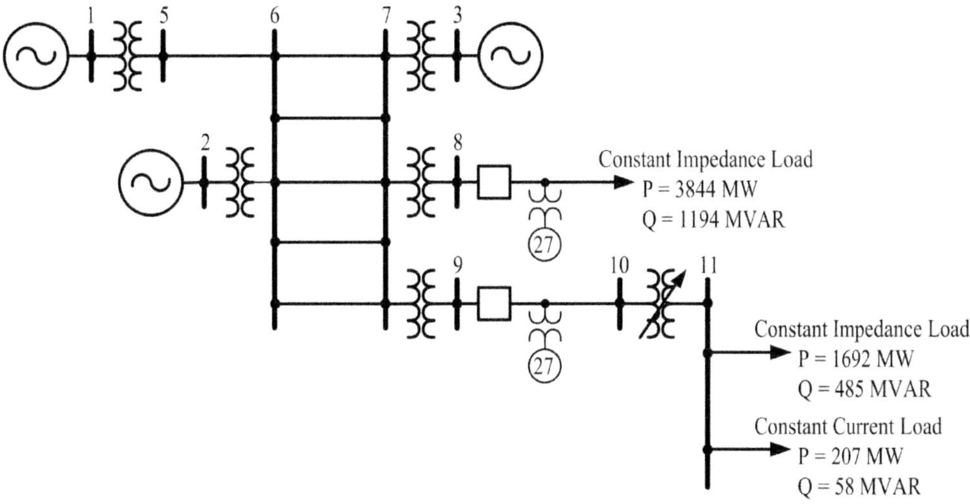


**FIGURE 19-61** Power system model to study voltage stability.

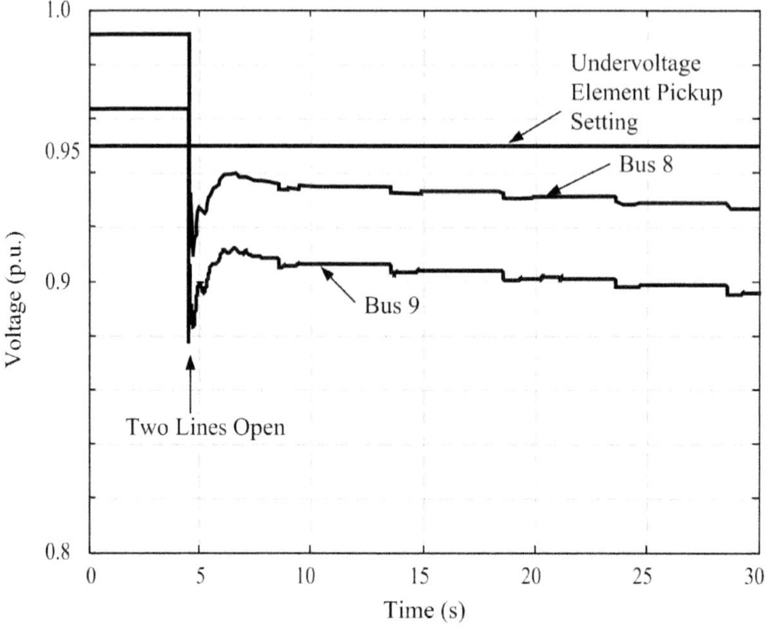


**FIGURE 19-62** Voltage magnitudes at Bus 8 and Bus 9 of Fig. 19-61 system. When two lines open, both voltages drop below 95% (the setting of definite-time undervoltage elements).

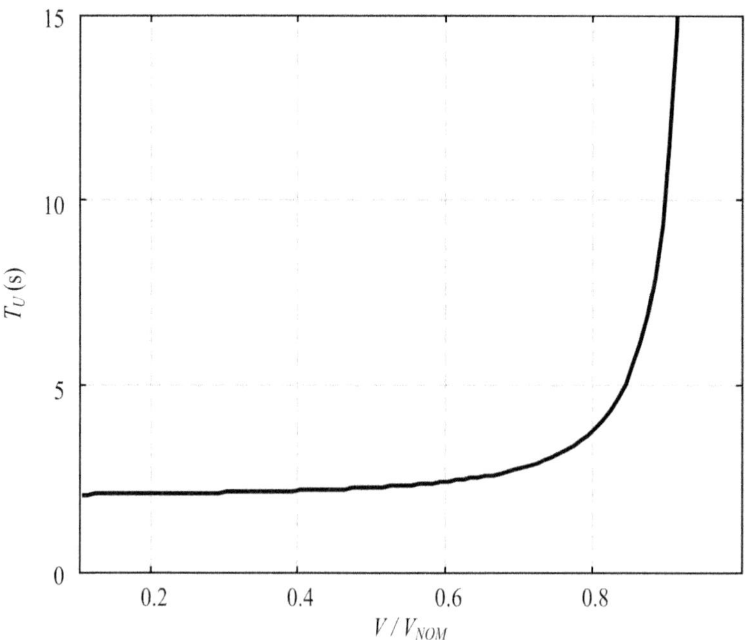


**FIGURE 19-63** Inverse-time undervoltage element characteristic.

Figure 19-64 depicts the effect of applying inverse-time undervoltage elements at Bus 8 and Bus 9 of the system shown in Fig. 19-61. The undervoltage element at Bus 9, the bus with the lowest voltage, operates faster than the element at Bus 8 because of the inverse-time characteristic. This operation drops Bus 11 load. After the load at Bus 11 drops, the system returns to stable conditions, without dropping the load at Bus 8.

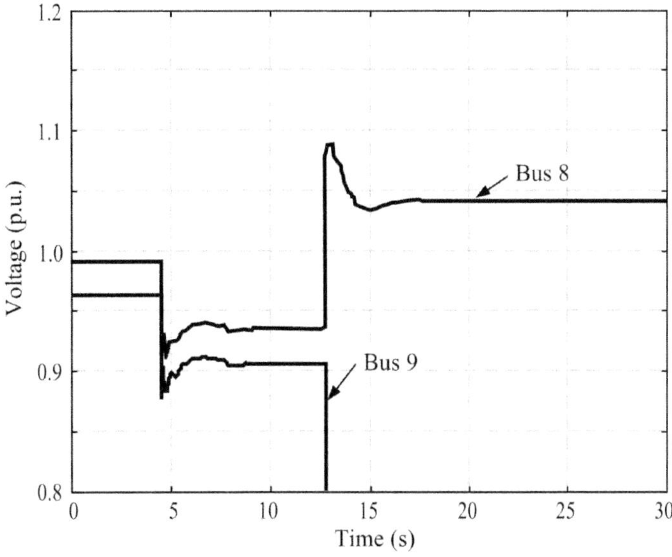


**FIGURE 19-64** Voltage magnitudes at Bus 8 and Bus 9 of Fig. 19-61 system. The voltage at Bus 8 recovers immediately after the inverse-time undervoltage element at Bus 9 trips to drop the Bus 11 load.

Figure 19-65 depicts a centralized undervoltage load-shedding scheme that monitors undervoltage conditions in three areas of a power system with inadequate reactive power reserve [52]. The scheme also monitors the reactive power output of four synchronous condensers. The synchronous condensers are located in Area 3. The voltage measurements are taken from 230 kV buses.

Figure 19-66 depicts the scheme logic. The total reactive power output of the four synchronous condensers enables and disables the undervoltage load-shedding scheme. The scheme is enabled when the synchronous condensers operate close to their rated output. In this condition, the synchronous condensers may not be able to provide system reactive power requirements during some disturbances. If the voltage drops below 97% of the minimum normal voltage in Area 1 and Area 3 or in Area 2 and Area 3 for longer than a $T_1$ delay, the scheme sheds Load Block 1. Load Block 2 and Load Block 3 are shed after $T_2$ and $T_3$ delays to provide staggered load shedding.

## 19.8.4 Automatic Generator Shedding

Line tripping during severe disturbances reduces power transfer over critical transmission links and may cause the system to lose synchronism. As a consequence, the OOST schemes will divide the system in islands and many generators will be tripped. Automatic, selective tripping of generating units upon detection of line tripping is a preventive action that may preserve power system stability. Utilities typically design automatic generation-shedding schemes (AGSSs) to operate for double contingencies. For most AGSSs, the double contingencies of interest occur when two parallel lines are lost simultaneously. Loss of two major transmission lines will severely stress most power systems.

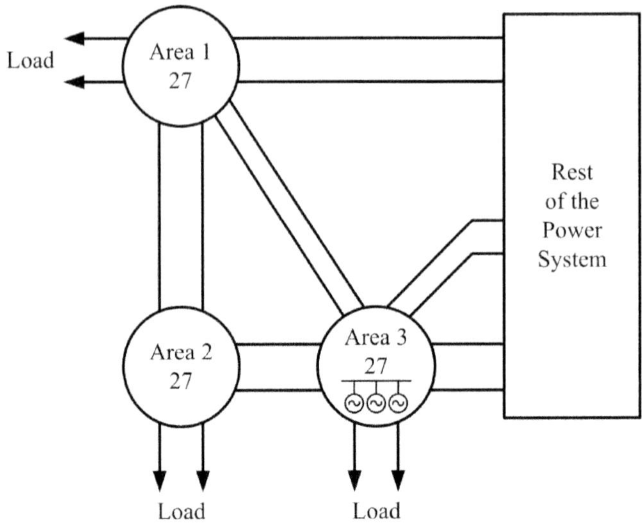


**FIGURE 19-65** Centralized undervoltage load-shedding scheme.

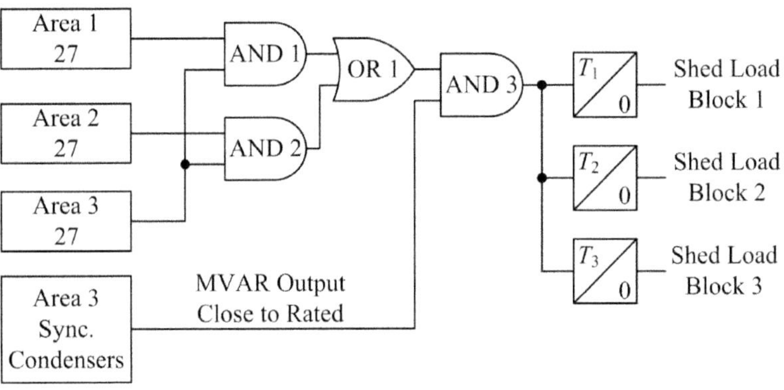


**FIGURE 19-66** Centralized undervoltage load-shedding scheme logic.

For transmission links with several lines and intermediate substations, existing AGSSs monitor network topology and power transfer capability using open-line detectors to enable generation shedding, select generators to trip, or activate tripping commands. Open-line detectors are based on the following:

- Breaker auxiliary contact signals (52a or 52b).
- Undercurrent elements and/or underpower elements.

Usually, these AGSSs use information from both terminals of each transmission line to determine whether the line is open. This method requires twice as many open-line detectors as transmission lines.

This subsection describes a synchrophasor-based AGSS as an example of the many possible applications of synchrophasors for wide-area protection, control, and monitoring. This system uses information on the angle difference $\delta$ between two buses to detect open-line conditions. The two-machine system shown in Fig. 19-67 illustrates this concept. During steady-state operation conditions, the voltage magnitudes of the power system buses are close to 1 p.u. The active power transfer capability depends mainly on voltage angle difference $\delta$ and transmission link reactance. This reactance depends on the number of lines and transformers in service between the two buses. When transmission lines are lost during a system disturbance, the link reactance and angle difference increase to maintain the same active power transfer between the two buses.

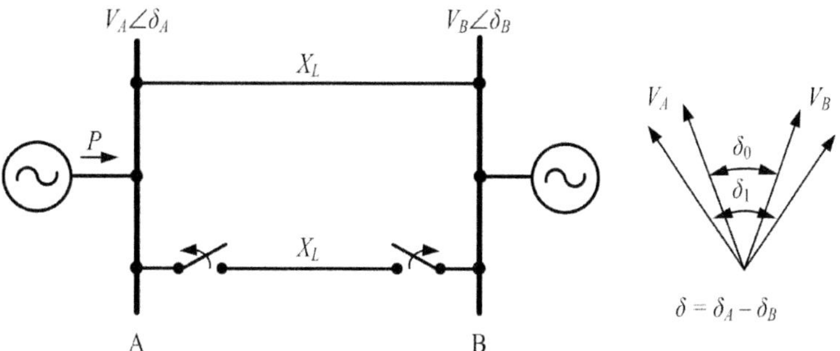


**FIGURE 19-67** Angle difference serves to detect an open-line condition.

Figure 19-68 illustrates the active-power transfer capability as a function of voltage angle difference $\delta$ and the operating point of the power system, both during normal operation conditions and after transmission lines are lost because of a system disturbance. The increase in transmission link reactance reduces the system power transfer capability and causes the angle difference to increase from $\delta_0$ to $\delta_1$.

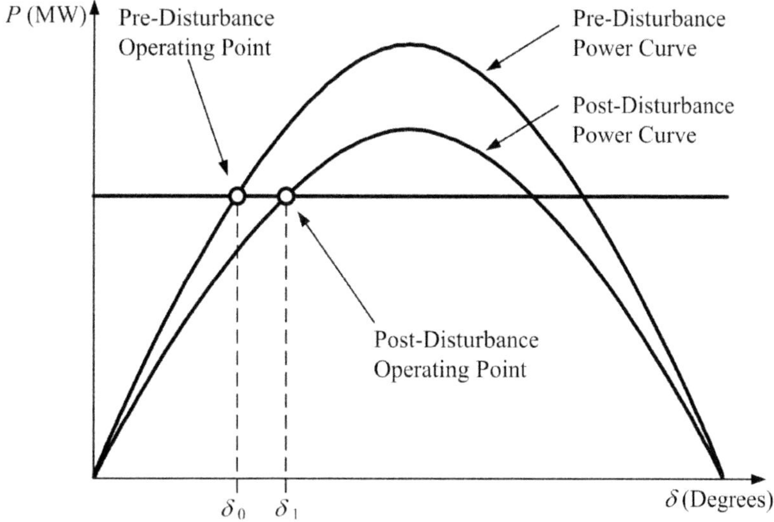


**FIGURE 19-68** Angle difference increases when the parallel line opens.

Figure 19-69 shows the section of the utility power system where the synchrophasor-based AGSS was installed [53]. In this power system, two 400 kV transmission links and one 115 kV subtransmission network interconnect two hydroelectric power plants, called Power Plant 1 and Power Plant 2

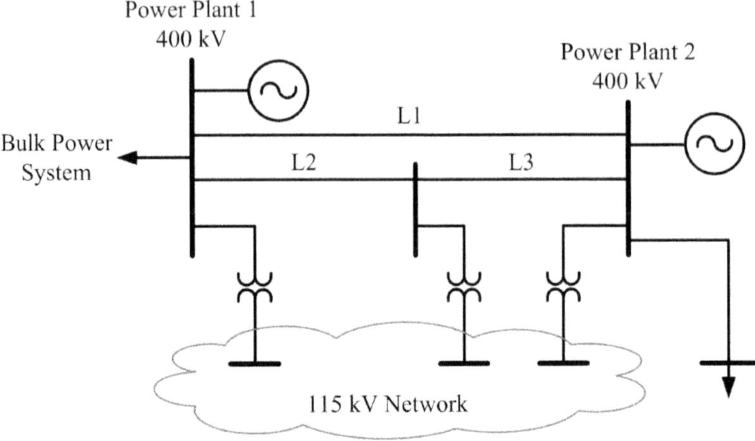


**FIGURE 19-69** Simplified diagram of the utility power system.

in the figure. When two of the 400 kV parallel lines between Power Plant 1 and Power Plant 2 trip, the shedding of all generators except one at Power Plant 2 avoids uncontrolled power flow over the subtransmission network. The generator that remains in service feeds the load in the isolated area.

The AGSS installed at Power Plant 2 includes relays that measure the 400 kV bus positive-sequence voltages at the Power Plant 1 and Power Plant 2 buses, respectively. The relay at Power Plant 1 sends the voltage information to the relay at Power Plant 2. The AGSS monitors the voltage angle difference between Power Plant 1 and Power Plant 2 and detects the loss of the 400 kV links. The AGSS needs only two synchrophasor voltage measurements and one communications channel to monitor the system. Therefore, the AGSS has fewer points of failure, which makes it more reliable than a system using traditional open-line detectors.

## 19.9 REFERENCES

1. North American Electric Reliability Corporation, "Misoperations Report," prepared by the Protection System Misoperation Task Force, Apr. 2013. Available: http://www.nerc.com/docs/pc/psmtf/PSMTF_Report.pdf.
2. H. J. Altuve Ferrer, and E. O. Schweitzer, III (eds.), *Modern Solutions for Protection, Control, and Monitoring of Electric Power Systems.* Schweitzer Engineering Laboratories, Inc., Pullman, WA, 2010.
3. *IRIG Serial Time Code Formats*, IRIG Standard 200-04, 2004.
4. K. Behrendt and K. Fodero, "The Perfect Time: An Examination of Time-Synchronization Techniques," in *60th Annual Georgia Tech Protective Relaying Conference*, Atlanta, GA, May 2006.
5. *IEEE Standard for a Precision Clock Synchronization Protocol for Networked Measurement and Control Systems*, IEEE Standard 1588, 2008.
6. K. Fodero, C. Huntley, and D. Whitehead, "Secure, Wide-Area Time Synchronization," in *12th Annual Western Power Delivery Automation Conference*, Spokane, WA, Apr. 2010.
7. E. O. Schweitzer, III, D. Whitehead, S. Achanta, and V. Skendzic, "Implementing Robust Time Solutions for Modern Power Systems," in *14th Annual Western Power Delivery Automation Conference*, Spokane, WA, Mar. 2012.
8. *IEEE Standard for Synchrophasors for Power Systems*, IEEE Standard C37.118, 2005.
9. A. F. Elneweihi, E. O. Schweitzer, III, and M. W. Feltis, "Improved Sensitivity and Security for Distribution Bus and Feeder Relays," in *18th Annual Western Protective Relay Conference*, Spokane, WA, Oct. 1991.

10. C. R. Mason, *The Art and Science of Protective Relaying.* John Wiley and Sons, New York, 1956.
11. A. Guzmán, J. B. Roberts, and D. Hou, "New Ground Directional Elements Operate Reliably for Changing System Conditions," in *23rd Annual Western Protective Relay Conference*, Spokane, WA, Oct. 1996.
12. J. B. Roberts and A. Guzmán, "Directional Element Design and Evaluation," in *21st Annual Western Protective Relay Conference*, Spokane, WA, Oct. 1994.
13. D. Costello, M. Moon, and G. Bow, "Use of Directional Elements at the Utility-Industrial Interface," in *31st Annual Western Protective Relay Conference,* Spokane, WA, Oct. 2004.
14. V. Cook, *Analysis of Distance Protection.* Letchworth, Hertfordshire: Research Studies Press Ltd., 1985.
15. E. O. Schweitzer, III and J. B. Roberts, "Distance Relay Element Design," in *19th Annual Western Protective Relay Conference*, Spokane, WA, Oct. 1992.
16. W. Tucker, A. Burich, M. Thompson, R. Anne, and S. Vasudevan, "Coordinating Dissimilar Line Relays in a Communications-Assisted Scheme," in *67th Annual Conference for Protective Relay Engineers,* College Station, TX, Apr. 2014.
17. J. B. Roberts, D. A. Tziouvaras, G. Benmouyal, and H. J. Altuve, "The Effect of Multiprinciple Line Protection on Dependability and Security," in *55th Annual Georgia Tech Protective Relaying Conference*, Atlanta, GA, May 2001.
18. H. J. Altuve Ferrer, B. Kasztenny, and N. Fischer (eds.), *Line Current Differential Protection: A Collection of Papers Representing Modern Solutions.* Schweitzer Engineering Laboratories, Inc., Pullman, WA, 2014.
19. H. Miller, J. Burger, N. Fischer, and B. Kasztenny, "Modern Line Current Differential Protection Solutions," in *36th Annual Western Protective Relay Conference*, Spokane, WA, Oct. 2009.
20. A. R. van C. Warrington, *Protective Relays: Their Theory and Practice*, Volume One. Chapman and Hall Ltd., London, 1962.
21. B. Kasztenny, G. Benmouyal, H. J. Altuve, and N. Fischer, "Tutorial on Operating Characteristics of Microprocessor-Based Multiterminal Line Current Differential Relays," in *38th Annual Western Protective Relay Conference*, Spokane, WA, Oct. 2011.
22. E. O. Schweitzer, III, B. Kasztenny, A. Guzmán, V. Skendzic, and M. V. Mynam, "Speed of Line Protection—Can We Break Free of Phasor Limitations?" in *41st Annual Western Protective Relay Conference*, Spokane, WA, Oct. 2014.
23. D. Hou, A. Guzmán, and J. B. Roberts, "Innovative Solutions Improve Transmission Line Protection," in *24th Annual Western Protective Relay Conference*, Spokane, WA, Oct. 1997.
24. R. B. Eastvedt, "The Need for Ultra-Fast Fault Clearing," in *3rd Annual Western Protective Relay Conference*, Spokane, WA, Oct. 1976.
25. E. O. Schweitzer, III, B. Kasztenny, and M. V. Mynam, "Performance of Time-Domain Line Protection Elements on Real-World Faults," in *42nd Annual Western Protective Relay Conference*, Spokane, WA, Oct. 2015.
26. B. Kasztenny, M. Thompson, and N. Fischer, "Fundamentals of Short-Circuit Protection for Transformers," in *63rd Annual Conference for Protective Relay Engineers*, College Station, TX, Mar. 2010.
27. *IEEE Guide for Protecting Power Transformers*, IEEE Standard C37.91, 2008.
28. *IEEE Guide for Liquid Immersed Transformers Through-Fault-Current Duration*, IEEE Standard C57.109, 1993.
29. *IEEE Guide for Protective Relay Applications to Power System Buses*, IEEE Standard C37.234, 2009.
30. A. Guzmán, C. Labuschagne, and B. L. Qin, "Reliable Busbar and Breaker Failure Protection with Advanced Zone Selection," in *31st Annual Western Protective Relay Conference*, Spokane, WA, Oct. 2004.
31. A. Guzmán, B. L. Qin, and C. Labuschagne, "Reliable Bus Protection with Advanced Zone Selection," *IEEE Transactions on Power Delivery,* vol. 20, no. 2, pp. 625–629, Apr. 2005.
32. *IEEE Guide for AC Generator Protection*, IEEE Standard C37.102, 2006.
33. *IEEE Guide for Generator Ground Protection*, IEEE Standard C37.101, 2006.
34. P. Soñez, F. Vicentini, V. Skendzic, M. Donolo, S. Patel, Y. Xia, and R. Scharlach, "Injection-Based Generator Stator Ground Protection Advancements," in *41st Annual Western Protective Relay Conference*, Spokane, WA, Oct. 2014.
35. *Salient-Pole 50 Hz and 60 Hz Synchronous Generators and Generator/Motors for Hydraulic Turbine Applications Rated 5 MVA and Above*, IEEE Standard C50.12, 2005.

36. *IEEE Standard for Cylindrical-Rotor 50 Hz and 60 Hz Synchronous Generators Rated 10 MVA and Above*, IEEE Standard C50.13, 2014.

37. C. R. Mason, "A New Loss of Excitation Relay for Synchronous Generators," *AIEE Transactions*, vol. 68, pt. II, pp. 1240–1245, 1949.

38. R. Sandoval, A. Guzmán, and H. J. Altuve, "Dynamic Simulations Help Improve Generator Protection," in *33rd Annual Western Protective Relay Conference*, Spokane, WA, Oct. 2006.

39. S. E. Zocholl, *AC Motor Protection*. Schweitzer Engineering Laboratories, Inc., Pullman, WA, 2003.

40. *IEEE Guide for AC Motor Protection*, IEEE Standard C37.96, 2012.

41. J. L. Blackburn, *Protective Relaying, Principles and Applications*, 2nd ed. Marcel Dekker, Inc., New York-Basel, 1998.

42. *IEEE Guide for the Presentation of Thermal Limit Curves for Squirrel Cage Induction Machines*, IEEE Standard 620, 1996.

43. *Motors and Generators*, NEMA Standard MG-1, 2014.

44. S. E. Zocholl, E. O. Schweitzer, and A. Aliaga-Zegarra, "Thermal Protection of Induction Motors Enhanced by Interactive Electrical and Thermal Models," *IEEE Transactions on Power Apparatus and Systems*, vol. PAS-103, no. 7, pp. 1749–1755, Jul. 1984.

45. E. O. Schweitzer and S. E. Zocholl, "Aspects of Overcurrent Protection of Feeders and Motors," in *PEA Relay Committee Spring Meeting*, Matamoras, PA, May 1995.

46. *Electrical Relays—Part 8: Thermal Electric Relays*, IEC Standard 255-8, 1990.

47. D. A. Tziouvaras and D. Hou, "Out-of-Step Protection Fundamentals and Advancements," in *30th Annual Western Protective Relay Conference*, Spokane, WA, Oct. 2003.

48. G. Benmouyal, D. Hou, and D. A. Tziouvaras, "Zero-Setting Power-Swing Blocking Protection," in *31st Annual Western Protective Relay Conference*, Spokane, WA, Oct. 2004.

49. A. Guzmán, D. Tziouvaras, E. O. Schweitzer, III, and K. E. Martin, "Local and Wide Area Network Protection Systems Improve Power System Reliability," in *31st Annual Western Protective Relay Conference*, Spokane, WA, Oct. 2004.

50. C. W. Taylor, "Concepts of Undervoltage Load Shedding for Voltage Stability," *IEEE Transactions on Power Delivery*, vol. 7, no. 2, pp. 480–488, Apr. 1992.

51. P. Kundur, *Power System Stability and Control*. McGraw-Hill, New York, 1994.

52. Protection Aids to Voltage Stability Working Group, "Summary of System Protection and Voltage Stability," *IEEE Transactions on Power Delivery*, vol. 10, no. 2, pp. 631–637, Apr. 1995.

53. E. Martínez, N. Juárez, A. Guzmán, G. Zweigle, and J. León, "Using Synchronized Phasor Angle Difference for Wide-Area Protection and Control," in *33rd Annual Western Protective Relay Conference*, Spokane, WA, Oct. 2006.

20

# POWER SYSTEM STABILITY AND CONTROL


Arturo R. Messina

*Professor, Graduate Studies Program in Electrical Engineering, Center for Research and Advanced Studies, Guadalajara, Mexico*

Emilio Barocio

*Professor, Graduate Studies Program in Electrical Engineering, Universidad de Guadalajara, Guadalajara, Mexico*

Kai Sun

*Associate Professor, University of Tennessee, Knoxville, Tennessee*

Daniel Ruiz-Vega

*Professor, Graduate Program in Electrical Engineering, Instituto Politécnico Nacional, Mexico*

Nilanjan Senroy

*Associate Professor, Indian Institute of Technology, Delhi, India*

Sukumar Mishra

*Professor, Indian Institute of Technology, Delhi, India*

## 20.1 SMALL-SIGNAL DYNAMIC PERFORMANCE AND STABILITY

*BY ARTURO R. MESSINA AND EMILIO BAROCIO*

### 20.1.1 System Response to Small Disturbances

Power system dynamic response to small disturbances encompasses a wide range of temporal and spatial scales from fast electromagnetic to slow electromechanical phenomena. The temporal scales are to some degree, dynamically coupled with each other, but it is often convenient to analyze transient phenomena separately using various analytical approaches and simulation tools [1–4].

The description and understanding of oscillatory processes is a central question in power system small-signal stability analysis and control. From a mathematical point of view, any physical system disturbed from an equilibrium condition will oscillate under some circumstances or physical conditions [5]. In this sense, small-signal stability can be defined as the ability of a physical system to regain a state of operating equilibrium after being subjected to a small disturbance or change in the system [5, 6]. The instability that results occurs in the form of increasing or sustained oscillations associated with the interaction of system generators, wind farms, control devices, and other dynamic elements. Oscillations may manifest themselves in many ways depending on the nature of the disturbance, system structure, operating mode, and the level of stress and control characteristics.

Experience shows that the oscillations can be triggered by events or disturbances, or result from changes in the initial operating condition and interactions between dynamic devices and have spatial scales ranging from local to global, and temporal scales ranging from 0.1 Hz to hundreds of hertz. Knowledge of the fundamental characteristics of the system oscillation modes provides valuable information about the stability of the oscillatory phenomena and may be helpful in identifying machines and other dynamic devices and their control systems involved in the exchange of oscillating energy and the design or modifications of controllers [2]. Small signal stability analysis techniques are also of interest to design appropriate countermeasures for the phenomenon.

The class of oscillatory processes is exceptionally large. Unstable system oscillations can be caused by electromechanical and electromagnetic interactions, small random disturbances or be the result of cyclic or random forcing of the system and manifest in different ways. In some cases, oscillations can exhibit amplitude and phase modulation, be nonperiodic or exhibit nonlinear and time-varying characteristics.

Power system oscillatory processes can be categorized and classified in a number of different ways, including their interaction characteristics, operating characteristics, and the nature of system oscillations. Three main types of oscillatory processes are usually recognized:

1. *Electromechanical oscillations* involving the interaction between the rotor masses of system generators. Such dynamic phenomenon commonly arises from minor perturbations or even without apparent cause, and is mostly exhibited at interconnections between electric machines. Oscillations may emerge from the interaction of large groups of closely coupled machines connected by weak links. In some cases, system oscillations may result from a combination of system events involving operational changes, highly stressed transmission grids, weak transmission system, load levels, and inadequate control. Protection and control actions are required to stop the propagation of the system disturbance and restore the system to a normal state. References [1, 2] provide a historical perspective of the development of oscillatory phenomena in North America.
2. *Subsynchronous resonance and subsynchronous torsional interactions* between turbine-generator units and series capacitive compensated transmission lines occur when the negative damping of the series compensated system exceeds the inherent mechanical damping of the shaft at a specific subsynchronous torsional frequency [7]. Torsional oscillations may result in fatigue and loss of life of the turbine-governor shaft segments. Sources of potential interaction are the fast-acting control systems of high-voltage direct current (HVDC) converters, power system stabilizers, FACTS controllers, and other power electronic based equipment [7, 8]. Other related phenomena concern the potential interaction between a series compensated line and fast acting controls used on HVDC systems and FACTS devices. Subsynchronous control interactions (SSCI) between modern power electronics based wind turbine generators and a nearby series compensated line have also been observed in field tests and actual system operation.
3. *Control mode oscillations* associated with inadequate tuning of control systems. These include interactions between power system controllers such as power system stabilizers (PSSs), FACTS devices, AVRs, static VAR compensators (SVCs) and torsional modes, and interactions between the control systems of modern wind farms and other major system controllers [9].

The use of emerging technologies such as inverter-based wind and photovoltaic resources and the increasing size and complexity of modern power systems is expected to introduce new interaction and phenomena. These interactions, in turn, may influence the nature of existing oscillatory processes. Examples of typical oscillatory processes involving low- and high-frequency oscillations are shown in Fig. 20-1.

***Power System Models for Small-Signal Stability Studies.*** To study the system response to small perturbations, the nonlinear power system model is *linearized* around the predisturbance operating condition to extract modal properties. As a general statement, the governing equations of motion of the system are assumed to be nonlinear,

$$\begin{aligned} \dot{\mathbf{x}} &= \mathbf{f}(\mathbf{x},\mathbf{y},\mathbf{u}), \quad \mathbf{x}(\mathbf{0}) = \mathbf{x}_o \\ \mathbf{0} &= \mathbf{g}(\mathbf{x},\mathbf{y}), \quad \mathbf{y}(\mathbf{0}) = \mathbf{y}_o \end{aligned} \tag{20-1}$$

where $\mathbf{x}$ is the $n$-dimensional vector of system states (the trajectory of the system in the space of $n$ dimensions), $\mathbf{y}$ is the $r$-dimensional vector of algebraic states, and $\mathbf{u}$ is the $p$-dimensional vector of system inputs; $\mathbf{x}_o$ is the vector of initial conditions or an initial perturbation state vector at $t = 0$.

Alternative models can be obtained for the study of dynamic phenomena involving the electrical dynamics of the transmission network when each of the power system components is represented by a linear differential equation [10].

*Equilibrium Points.* Equilibrium solutions of Eq. (20-1) satisfy $\mathbf{f}(\mathbf{x}_o, \mathbf{y}_o, \mathbf{u}_o) = \mathbf{0}$, $\mathbf{g}(\mathbf{x}_o, \mathbf{y}_o) = \mathbf{0}$. Physically, this means that the sum of all forces acting on the system is zero. In the case of an unstable equilibrium condition, the system will be displaced from the equilibrium position.

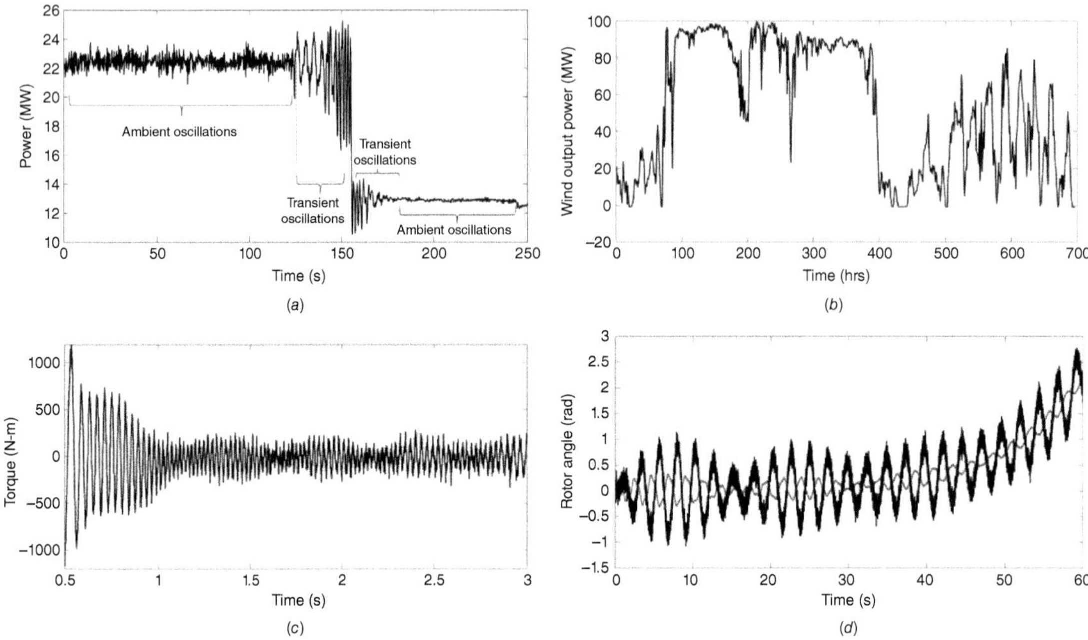


**FIGURE 20-1** Examples of oscillatory system response. (*a*) Combined ambient and transient oscillations; (*b*) wind power fluctuations; (*c*) torsional oscillations in wind farms; (*d*) combined ambient and forced oscillations. (Note the time scales.)

*Stability of the Solutions.* The nominal stability of an equilibrium point of the system is established using Lyapunov analysis. An equilibrium point of Eq. (20-1) is stable if all the eigenvalues of the Jacobian **f** evaluated at the equilibrium have negative real parts.

***Methods of Analysis of Small Signal Stability.*** Quantitative analysis of the equation of motion for small perturbations involves numerical solution of a simplified (linear) version of Eq. (20-1). A typical and practical approach is to assume that system motion is limited to small variations from a reference condition or equilibrium point.

Modeling practices to obtain the state representation Eq. (20-1) have traditionally focused on two main approaches which have different advantages and disadvantages: (1) Linearization of system equations about an initial operating condition, and (2) numerical approximation of the state representation.

*Analytical Methods.* In many practical cases, the system oscillatory modes can be analyzed by eigenvalue decomposition techniques applied to linearized models. As a special case, the power series expansion of Eq. (20-1) about a stable equilibrium point (SEP), $\mathbf{x}_o$, is [11]

$$\begin{cases} \dot{\mathbf{x}} = \mathbf{f}(\mathbf{x},\mathbf{y})\Big|_{\substack{\mathbf{x}=\mathbf{x}_o\\ \mathbf{y}=\mathbf{y}_o}} + \dfrac{\partial \mathbf{f}(\mathbf{x},\mathbf{y})}{\partial \mathbf{x}}\Bigg|_{\substack{\mathbf{x}=\mathbf{x}_o\\ \mathbf{y}=\mathbf{y}_o}} \Delta\mathbf{x} + \dfrac{\partial \mathbf{f}(\mathbf{x},\mathbf{y})}{\partial \mathbf{y}}\Bigg|_{\substack{\mathbf{x}=\mathbf{x}_o\\ \mathbf{y}=\mathbf{y}_o}} \Delta\mathbf{x} + \dfrac{1}{2!}\dfrac{\partial^2 \mathbf{f}(\mathbf{x},\mathbf{y})}{\partial^2 \mathbf{x}}\Bigg|_{\substack{\mathbf{x}=\mathbf{x}_o\\ \mathbf{y}=\mathbf{y}_o}} \Delta\mathbf{x}^2 + \dfrac{1}{2!}\dfrac{\partial^2 \mathbf{f}(\mathbf{x},\mathbf{y})}{\partial^2 \mathbf{y}}\Bigg|_{\substack{\mathbf{x}=\mathbf{x}_o\\ \mathbf{y}=\mathbf{y}_o}} \Delta\mathbf{x}^2 + \cdots \\ \mathbf{0} = \mathbf{g}(\mathbf{x},\mathbf{y})\Big|_{\substack{\mathbf{x}=\mathbf{x}_o\\ \mathbf{y}=\mathbf{y}_o}} + \dfrac{\partial \mathbf{g}(\mathbf{x},\mathbf{y})}{\partial \mathbf{x}}\Bigg|_{\substack{\mathbf{x}=\mathbf{x}_o\\ \mathbf{y}=\mathbf{y}_o}} \Delta\mathbf{x} + \dfrac{\partial \mathbf{g}(\mathbf{x},\mathbf{y})}{\partial \mathbf{y}}\Bigg|_{\substack{\mathbf{x}=\mathbf{x}_o\\ \mathbf{y}=\mathbf{y}_o}} \Delta\mathbf{y} + \cdots \end{cases} \tag{20-2}$$

Conceptually, the linearization procedure is equivalent to assuming that the state variables are of the form $\mathbf{x}_o + \Delta\mathbf{x}$, $\mathbf{y}_o + \Delta\mathbf{y}$, where $\mathbf{x}_o$, $\mathbf{y}_o$ characterizes the reference condition (the equilibrium condition), and $\Delta\mathbf{x}$, $\Delta\mathbf{y}$ represents a perturbation from the reference condition.

*Differential-Algebraic Equation (DAE) Power System Models.* Linearization of the nonlinear DAE model Eq. (20-2) about a given system operating condition neglecting second- and higher-order terms yields the sparse model

$$\begin{bmatrix} \Delta\dot{\mathbf{x}} \\ \mathbf{0} \end{bmatrix} = \begin{bmatrix} \mathbf{J}_{xx} & \mathbf{J}_{xy} \\ \mathbf{J}_{yx} & \mathbf{J}_{yy} \end{bmatrix} \begin{bmatrix} \Delta\mathbf{x} \\ \Delta\mathbf{y} \end{bmatrix} + \begin{bmatrix} \mathbf{B} \\ \mathbf{0} \end{bmatrix} \mathbf{u}$$
$$\mathbf{y} = \begin{bmatrix} \mathbf{c}_x^T & \mathbf{c}_y^T \end{bmatrix} \begin{bmatrix} \Delta\mathbf{x} \\ \Delta\mathbf{y} \end{bmatrix} + \mathbf{Du} \tag{20-3}$$

where $\mathbf{J}_{xx}$, $\mathbf{J}_{xy}$, $\mathbf{J}_{yx}$, and $\mathbf{J}_{yy}$ are Jacobian matrices, and the superscript $T$ denotes a transposition operator. The output vector $\mathbf{y}$ contains input signals for the controllers.

*Differential Equation Power System Models.* Many dynamic phenomena and devices can be represented by linear homogeneous equations of the form

$$\dot{\mathbf{x}} = \mathbf{Ax} + \mathbf{Bu},\ \mathbf{x}(0) = \mathbf{x}_o$$
$$\mathbf{y} = \mathbf{c}^T\mathbf{x} \tag{20-4}$$

where $\mathbf{A}$ and $\mathbf{B}$ are the system matrix and control matrix, respectively. These models can be obtained directly from a time domain simulation program by perturbing the states and inputs, or be obtained from Eqs. (20-1) through (20-3) by eliminating the algebraic equations. In this latter case, the reduced-order matrix $\mathbf{A} = \mathbf{J}_{xx} - \mathbf{J}_{xy}(\mathbf{J}_{yx})^{-1}\mathbf{J}_{yy}$ is called the state matrix of the system, as discussed below.

*Linearization-Based Methods.* These models, used in various software packages for small signal stability analysis can be derived from Eq. (20-1) or constructed directly from a properly modified nonlinear time simulation program. Let the Taylor's theorem be expressed as

$$\mathbf{f}(\mathbf{x}_o + \Delta\mathbf{x}) = \mathbf{f}(\mathbf{x}_o) + \mathbf{f}'(\mathbf{x}_o)\Delta\mathbf{x} + \frac{1}{2}\mathbf{f}''(\mathbf{x}_o)\Delta\mathbf{x}^2 + \cdots + \frac{1}{(r-1)!}\mathbf{f}^{(r-1)}(\mathbf{x}_o)\Delta\mathbf{x}^{(r-1)} + \frac{1}{r!}\mathbf{f}^r(\mathbf{x}_o)\Delta\mathbf{x}^r \tag{20-5}$$

subject to the network constraints $\mathbf{g}(\Delta\mathbf{x}_i, \Delta\mathbf{y}_i) = \mathbf{0}$. Two main approaches have been proposed to construct the state matrix representation: (1) forward-difference approximation and (2) central-difference approximation [11]. The methods are summarized in Table 20-1. Implementation issues are discussed in [12] and are briefly discussed below.

**TABLE 20-1** Finite Difference Approximations to Numerical Derivatives [11, 12]

| Approximation | Expression |
|---|---|
| Forward-difference approximation | $\dfrac{\mathbf{f}(\mathbf{x}_o + \Delta\mathbf{x}) - \mathbf{f}(\mathbf{x}_o)}{\Delta\mathbf{x}} = \mathbf{f}'(\mathbf{x}) + O(\Delta\mathbf{x})$ |
| Central-difference approximation | $\dfrac{\mathbf{f}(\mathbf{x}_o + \Delta\mathbf{x}) - \mathbf{f}(\mathbf{x}_o - \Delta\mathbf{x})}{2\Delta\mathbf{x}} = \mathbf{f}'(\mathbf{x}) + O(\Delta\mathbf{x}^2)$ |

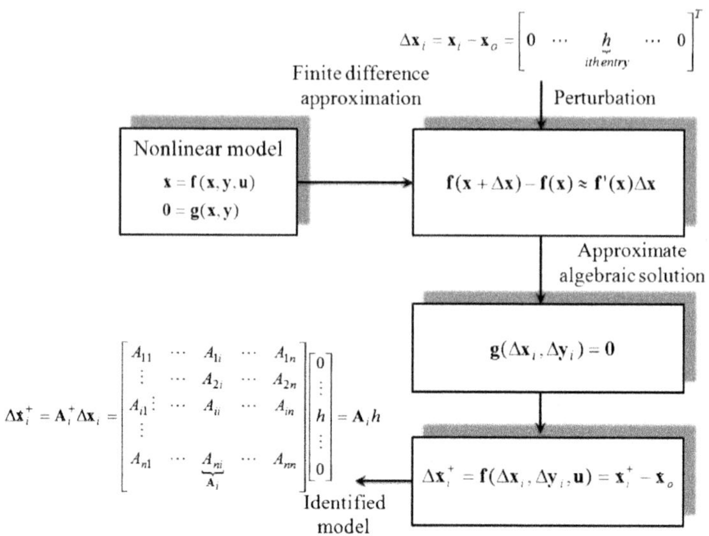


**FIGURE 20-2** Schematic of numerical approximation of the state matrix.

One practical procedure is to start from a given initial state, $x_i$, and then consider a small perturbation $\Delta x_i$. By perturbing each state in turn by a small amount, $h$, the columns $\mathbf{A}_i$, $i = 1, \ldots, n$ of matrix $\mathbf{A}$ can be computed sequentially. The process is summarized in Fig. 20-2. Once the perturbed response is found using Eq. (20-2), the columns of the state matrix are computed as

$$\mathbf{A}_i = \Delta\dot{\mathbf{x}}_i^+/h \tag{20-6}$$

In practice, the accuracy of these approximations depends on the truncation error. Special techniques are needed to deal with the size of perturbations and nonlinear behavior associated with the representation of axis transformations, generator saturation, control limits, and other nonlinearities [10].

*Power System Response to Small Disturbances.* Using eigenvector analysis techniques [13], the general solution of Eq. (20-4) with initial conditions $\mathbf{x(0)} = \mathbf{x}_o$ at $t = 0$ is

$$\mathbf{y}(t) = \sum_{i=1}^{n}\left(\psi_i^T \mathbf{x}_o\right)\phi_i e^{\lambda_i t} + \sum_{i=1}^{n}\int_o^t \left(\psi_i^T u(\tau)\right)\phi_i e^{\lambda_i(t-\tau)}\,d\tau = \mathbf{C}e^{\mathbf{A}t}\mathbf{x}_o + \mathbf{C}\int_o^t e^{\mathbf{A}(t-\tau)}\mathbf{B}\mathbf{u}\,d\tau \tag{20-7}$$

where $\lambda_i$ $(i = 1, \ldots, n)$ are the roots or eigenvalues of the characteristic equation $f(\lambda) = \det[\lambda\mathbf{I} - \mathbf{A}] = 0$, and $\phi_i$, $\psi_i$ are, respectively, the corresponding right and left eigenvectors defined as follows:

$$[\lambda_i\mathbf{I} - \mathbf{A}]\phi_i = \mathbf{0}, \quad \phi_i \neq \mathbf{0}$$
$$\psi_i^T[\lambda_i\mathbf{I} - \mathbf{A}] = \mathbf{0}, \quad \psi_i^T \neq \mathbf{0}$$

where $\phi_i\psi_i^T = 1$ if $i = j, \phi_i\psi_i^T = 0$ if $i \neq j$.

### 20.1.2 Free Oscillations

From a physical point of view, the free system response is obtained when no energy is added or removed from the system and is entirely due to the initial conditions. In terms of the modal parameters, the time evolution of the system states is given by

$$\mathbf{x}(t)=\sum_{i=1}^{n}[(\boldsymbol{\psi}_i^T\mathbf{x}_o)e^{\lambda_i t}]\,\boldsymbol{\phi}_i=\sum_{i=1}^{n}[(\boldsymbol{\phi}_i\boldsymbol{\psi}_i^T)\mathbf{x}_o e^{\lambda_i t}]=\sum_{i=1}^{n}[\mathbf{R}_i\mathbf{x}_o]e^{\lambda_i t} \tag{20-8}$$

where $\mathbf{R}_i=[R_{ij}]=\boldsymbol{\phi}_i\boldsymbol{\psi}_i^T$ is the residue matrix.

Free oscillations are determined by three sets of parameters:

1. The eigenvalues $\lambda_i$, which determine the decay or growth rate of the response
2. The eigenvectors $\boldsymbol{\phi}_i$, which determine the shape of the response
3. The initial conditions $\mathbf{x}_o$, which determine the extent to which each mode will participate in the free system response

This following subsection highlights those aspects of linear system representations that are relevant to the study and control of power system oscillatory behavior.

***The Notion of System Modes.*** The dynamic motion of complex systems with many degrees of freedom can be considered as a superposition of simple oscillatory processes or motions called *modes*. From Eq. (20-7), the system response to small perturbations can be expressed through mode superposition as

$$\mathbf{x}(t)=\sum_{i=1}^{n}(\boldsymbol{\psi}_i^T\mathbf{x}_o)e^{\lambda_i t}\boldsymbol{\phi}_i=\sum_{i=1}^{n}(\alpha_i)e^{\lambda_i t}\boldsymbol{\phi}_i$$

where $\alpha_i=\boldsymbol{\psi}_i^T\mathbf{x}_o$ provides the initial excitation of the $i$th mode. In expanded form,

$$\begin{bmatrix} x_1(t)\\ x_2(t)\\ \cdots\\ x_n(t)\end{bmatrix}=\alpha_1\begin{bmatrix}\phi_{11}(t)\\ \phi_{21}(t)\\ \cdots\\ \phi_{n1}(t)\end{bmatrix}e^{\lambda_1 t}+\alpha_2\begin{bmatrix}\phi_{12}(t)\\ \phi_{22}(t)\\ \cdots\\ \phi_{n2}(t)\end{bmatrix}e^{\lambda_2 t}+\cdots+\alpha_j\underbrace{\begin{bmatrix}\phi_{1j}(t)\\ \phi_{2j}(t)\\ \cdots\\ \phi_{nj}(t)\end{bmatrix}}_{j\text{th mode}}e^{\lambda_j t}+\cdots+\alpha_n\begin{bmatrix}\phi_{1n}(t)\\ \phi_{2n}(t)\\ \cdots\\ \phi_{nn}(t)\end{bmatrix}e^{\lambda_n t} \tag{20-9a}$$

When a fault or small perturbation occurs, multiple modes may be excited and the system response may be affected by linear (nonlinear) interactions between the modes as discussed below.

*Oscillatory Modes of Motion.* Oscillatory modes of motion appear in pairs of complex conjugates $\lambda_i=\sigma_i\pm j\omega_i$. The free system response Eq. (20-9a) now becomes

$$\mathbf{x}(t)=\sum_{i=1}^{n/2}(\boldsymbol{\psi}_i^T\mathbf{x}_o e^{\lambda_i t}\boldsymbol{\phi}_i+\boldsymbol{\psi}_i^{T*}\mathbf{x}_o e^{\lambda_i^* t}\boldsymbol{\phi}_i^*)=\sum_{i=1}^{n/2}e^{\sigma_i t}(\alpha_i\boldsymbol{\phi}_i e^{j\omega_i t}+\alpha_i^*\boldsymbol{\phi}_i^* e^{-j\omega_i t}) \tag{20-9b}$$

with $\alpha_i=\boldsymbol{\psi}_i^T\mathbf{x}_o$, $\alpha_i^*=\boldsymbol{\psi}_i^{T*}\mathbf{x}_o$, $\boldsymbol{\phi}_i=[\phi_{1i}\ \phi_{2i}\ \cdots\ \phi_{ni}]^T$, and $\boldsymbol{\psi}_j=[\psi_{1i}\ \psi_{2i}\ \cdots\ \psi_{ni}]^T$.

Several facts emerge on examination of Eq. (20-9a) and (20-9b) results.

- The system response is the combination of $n/2$ complex conjugate modes. In other words, the oscillations can be represented as a combination of single harmonic motions with appropriate amplitude, frequency and initial phases
- The amount $\alpha_i \phi_{ki}$, provides a measure of the excitation (weight) of each mode in the $k$th state
- The eigenvalues $\lambda_i$ determine system stability: the components of the right eigenvector $\boldsymbol{\phi}_i$ determine the relative activity of each variable in the $i$th mode, while the components of the left eigenvector $\boldsymbol{\psi}_i$ weight the initial conditions in the $i$th mode

***Nonlinear Aspects of System Oscillations.*** The dynamic response of complex interconnected power systems to random perturbations can exhibit strongly nonlinear behavior, especially under conditions of heavy stress or weakening structure of the system [14, 15]. Typically, in such cases, modes are coupled and nonlinear interactions between them have the potential to influence system behavior in key respects.

As the simplest example, the equation of motion of a single-machine infinite-bus (SMIB) system is

$$\Delta\dot{\delta} = \omega_o \Delta\hat{\omega}_r$$
$$2H\Delta\dot{\hat{\omega}}_r = (P_m - P_{\max}\sin\delta - K_D \Delta\hat{\omega}_r)$$

where $\delta$ is the rotor angle position relative to the infinite bus, in radians, $\Delta\hat{\omega}_r = (\omega_r - \omega_o)/\omega_o$ is the pu speed deviation, $P_m$ is the input mechanical power in pu, $P_{\max}$ is the maximum transmitted power in pu, and $K_D$ is the damping coefficient (pu power/pu speed) [6].

Expanding these equations about an equilibrium condition $(\delta_o, \omega_r = \omega_{ro})$ and retaining the first significant nonlinear terms results in the incremental model [14] gives

$$\dot{\mathbf{x}} = \begin{bmatrix} \Delta\dot{\delta} \\ \Delta\dot{\hat{\omega}}_r \end{bmatrix} = \begin{bmatrix} \omega_o \Delta\hat{\omega}_r \\ \underbrace{\left(\frac{-P_{\max}}{2H}\cos\delta_o \Delta\delta - \frac{K_D}{2H}\Delta\hat{\omega}_r\right)}_{\text{Linear terms}} + \underbrace{\frac{1}{2}\frac{P_{\max}}{2H}\sin\delta_o \Delta\delta^2}_{\text{Quadratic terms}} + \underbrace{\frac{1}{6}\frac{P_{\max}}{2H}\sin\delta_o \Delta\delta^3}_{\text{Cubic terms}} + O(4) \end{bmatrix} + \begin{bmatrix} 0 \\ \frac{1}{2H} \end{bmatrix} \Delta P_m$$

or

$$\dot{\mathbf{x}} = \mathbf{Ax} + \underbrace{\mathbf{f}_2(\mathbf{x}) + \mathbf{f}_3(\mathbf{x})}_{\mathbf{f}(\mathbf{x})} + \mathbf{Bu} = \mathbf{Ax} + \begin{bmatrix} 0 \\ \frac{1}{2H} \end{bmatrix} \Delta P_m + \underbrace{\frac{1}{2!}\begin{bmatrix} \mathbf{x}^T\mathbf{H}_2^1\mathbf{x} \\ \mathbf{x}^T\mathbf{H}_2^2\mathbf{x} \end{bmatrix}}_{\mathbf{f}_2(\mathbf{x})} + \underbrace{\frac{1}{3!}\begin{bmatrix} \mathbf{x}^T\mathbf{H}_3^1 \begin{bmatrix} \mathbf{x} & \mathbf{0} \\ \mathbf{0} & \mathbf{x} \end{bmatrix} \mathbf{x} \\ \mathbf{x}^T\mathbf{H}_3^2 \begin{bmatrix} \mathbf{x} & \mathbf{0} \\ \mathbf{0} & \mathbf{x} \end{bmatrix} \mathbf{x} \end{bmatrix}}_{\mathbf{f}_3(\mathbf{x})} + O(4) \qquad (20\text{-}10)$$

in which $\mathbf{H}_2^1, \mathbf{H}_2^2, \mathbf{H}_3^1, \mathbf{H}_3^2$ are real and constant squares matrices of order $n = 2$, defined as follows:

$$\mathbf{H}_2^1 = \begin{bmatrix} 0 & 0 \\ 0 & 0 \end{bmatrix}; \quad \mathbf{H}_2^2 = \begin{bmatrix} \frac{P_{\max}}{2H}\sin\delta_o & 0 \\ 0 & 0 \end{bmatrix} \quad \mathbf{H}_3^1 = \begin{bmatrix} 0 & 0 \\ 0 & 0 \end{bmatrix}; \quad \mathbf{H}_3^2 = \begin{bmatrix} 0 & \frac{P_{\max}}{2H}\sin\delta_o \\ 0 & 0 \end{bmatrix}$$

The system has an equilibrium of interest at $\mathbf{x}_o = [\delta_o \Delta\hat{\omega}_{ro}]^T = [\sin^{-1}(P_m/P_{\max})0]^T$; the Jacobian at the equilibria is

$$\mathbf{J} = \mathbf{A} = \begin{bmatrix} 0 & \omega_o \\ \dfrac{-K_S}{2H} & -\dfrac{K_D}{2H} \end{bmatrix}$$

where $K_S = P_{\max}\cos\delta_o$ and $K_D$ are the *synchronizing and damping power coefficients*, respectively.

Perturbing the initial condition, $\mathbf{x}_o$, will cause the system to oscillate. Two forces influence this behavior: the synchronizing force, $K_S$ which is directly proportional to the angular deviation $\Delta\delta$, and the damping torque, $K_D$, which is proportional to the speed deviation.

Stability of the equilibria is determined by the roots of the characteristic equation $\lambda_{1,2} = -\frac{K_D}{4H} \pm \frac{\omega_o}{4H} j\sqrt{K_D^2/\omega_o^2 - 8K_S/\omega_o} = \sigma \pm j\omega$, in which, without loss of generality it is assumed that $K_D^2/\omega_o^2 < 8K_S/\omega_o$; the eigenvalues are complex conjugates with real part $\sigma = -K_D/4H$ and imaginary part $f(Hz) = \sqrt{K_D^2/\omega_o^2 - 8K_S/\omega_o}$ (undamped behavior). When higher-order terms are of small order, the free system response is a simple damped harmonic movement of the form

$$\begin{bmatrix} \Delta\delta(t) \\ \Delta\omega_r(t) \end{bmatrix} = L^{-1} \begin{bmatrix} \dfrac{s + K_D/2H}{(s-\lambda_1)(s+\lambda_2)}\Delta\delta_o + \dfrac{\omega_o}{(s-\lambda_1)(s+\lambda_2)}\Delta\omega_{ro} \\ \dfrac{-K_S/2H}{(s-\lambda_1)(s+\lambda_2)}\Delta\delta_o + \dfrac{s}{(s-\lambda_1)(s+\lambda_2)}\Delta\omega_{ro} \end{bmatrix}$$

$$= \begin{bmatrix} e^{\sigma t}\left(a_\delta \sin(\omega t + \varphi_\delta)\Delta\delta_o + \dfrac{\omega}{\omega_o} e^{\alpha t}\sin(\omega t)\Delta\omega_{ro}\right) \\ e^{\sigma t}\left(-\dfrac{K_S}{2H\omega}\sin(\omega t)\Delta\delta_o + a_\omega \cos(\omega t - \varphi_\omega)\Delta\omega_{ro}\right) \end{bmatrix}$$

where L is the Laplace operator, and $a_\delta = \sqrt{1 + (K_D/4H)^2}$, $\varphi = tg^{-1}(1/(K_D/4H))$.

Figure 20-3 illustrates the system response to a step change in the initial conditions $\Delta\delta_o$ and its phase plane representation. For a stable case, the phase diagram describes a slowly changing spiral converging to the origin (the stable equilibrium point) as $t \to \infty$. The linear approximation [Eq. (20-10)] is valid for only a limited range of the operating regime. Close to the onset of system instability or under large perturbations and high system stress, the system may exhibit a complex dynamic behavior.

*Modeling of Local Nonlinearities.* Among the several techniques for obtaining approximations to models of the form Eq. (20-10), the method of normal forms (NF) [15] has been used to analyze the system response to small perturbations.

In this approach, a set of nonlinear change of coordinates are used to eliminate or simplify equation nonlinearities. As a first step, a linear transformation $\mathbf{x} = \mathbf{\Phi y}$ is used to transform the system in Eq. (20-10) into its complex Jordan canonical form $\dot{\mathbf{y}} = \Lambda\mathbf{y} + \Phi^{-1}[\sum_{i=2}^{k} \mathbf{f}_i(\Phi\mathbf{y}) + O(k+1)] = \Lambda\mathbf{y} + \mathbf{F}_2(\mathbf{y}) + \sum_{i=2}^{k} \mathbf{F}_i(\mathbf{y}) + \hat{O}(k+1)$, where $\Lambda$ is a diagonal matrix of the eigenvalues, and the nonlinear terms are of the form $\sum_{k=1}^{n}\sum_{l=k}^{n} C_{2_{kl}}^{j} y_k y_l, j = 1, 2, \ldots, n$. This represents a linearly decoupled but nonlinearly coupled system in which internal resonances may exist between modes. Strong interactions may cause performance degradation and adversely affect controller performance.

Once the geometry of the nonlinear system is known, nonlinear coordinate transformations of the form $\mathbf{y} = \mathbf{z} + \mathbf{h}_2(\mathbf{z}), \ldots$ are used to annihilate second- and third-order terms and focus on the

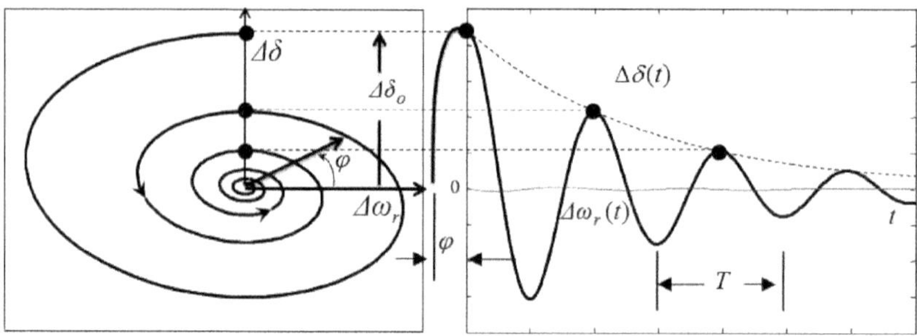


**FIGURE 20-3** Free response and phase plane representation of the swing equation.

governing dynamics. The system response (the NF solution) for the $k$th state is given by, to second order:

$$x_k(t)=\sum_{j=1}^{n}\psi_{kj}z_{jo}e^{\lambda_j t}+\sum_{j=1}^{n}\psi_{kj}\sum_{m=1}^{n}\sum_{p=1}^{n}h_{2mp}^{j}z_{mo}z_{po}e^{(\lambda_m+\lambda_p)t}+\cdots \quad k=1,\ldots,n \tag{20-11}$$

where $z_{jo}, z_{mo}, z_{po}$ represent the initial conditions in the **z** coordinate framework, and the $h_{2mp}^{j}$ are coefficients of the second-order transformation that measure the interaction between modes of the linearized system [15]. Figure 20-4 is an example of a typical simulated system response to a step change in the initial angle $\Delta\delta_o$.

Figure 20-4 highlights three key points: (1) The system response converges to the post-disturbance SEP; (2) higher-order solutions approximate system behavior better for a longer time; and (3) physically interesting results are obtained only when third-order (or higher) terms are retained.

As discussed above, nonlinear modal interaction between system modes contributes to different properties in physical processes and may significantly influence the performance of system controllers. The normalized term $\max_{mp}(|h_{2mp}^{j}z_m^o z_p^o|)/z_j^o$ provides a measure of the nonlinearity present in the initial value and can be used to evaluate the severity of nonlinear interaction.

***Free System Response from Initial Conditions.*** To discuss the basic nature of system electromechanical modes, it is convenient to represent the system by its classical model [16]. The equation of motion of a system with $ng$ classically modeled generators is

$$\begin{bmatrix}\Delta\dot{\delta}_1\\ \Delta\dot{\delta}_2\\ \cdots\\ \Delta\dot{\delta}_{ng}\\ \hline \Delta\dot{\hat{\omega}}_{r1}\\ \Delta\dot{\hat{\omega}}_{r2}\\ \cdots\\ \Delta\dot{\hat{\omega}}_{rng}\end{bmatrix}=\left[\begin{array}{cccc|cccc} & & & & \omega_o & & & \\ & \hat{0} & & & & \omega_o & & \\ & & & & & & \ddots & \\ & & & & & & & \omega_o\\ \hline -A_{g_{11}} & -A_{g_{12}} & \cdots & -A_{g_{1ng}} & -\hat{K}_{D_1} & & & \\ -A_{g_{21}} & -A_{g_2} & \cdots & -A_{g_{2ng}} & & -\hat{K}_{D2} & & \\ \vdots & \vdots & \ddots & \vdots & & & \ddots & \\ -A_{g_{ng1}} & \cdots & \cdots & -A_{g_{ngng}} & & & & -\hat{K}_{Dng}\end{array}\right]\begin{bmatrix}\Delta\delta_1\\ \Delta\delta_2\\ \vdots\\ \Delta\delta_{ng}\\ \hline \Delta\hat{\omega}_{r1}\\ \Delta\hat{\omega}_{r2}\\ \vdots\\ \Delta\hat{\omega}_{rng}\end{bmatrix}+\begin{bmatrix}0\\ 0\\ \vdots\\ 0\\ \hline \dfrac{\Delta P_{m1}}{2H_1}\\ \dfrac{\Delta P_{m2}}{2H_2}\\ \vdots\\ \dfrac{\Delta P_{mng}}{2H_{ng}}\end{bmatrix} \tag{20-12}$$

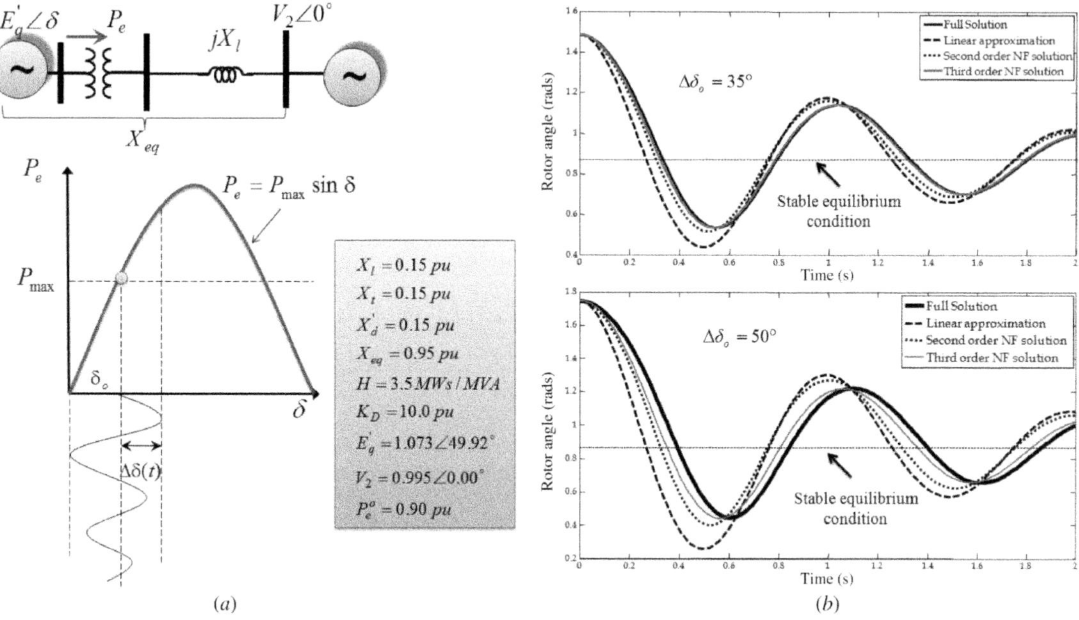


**FIGURE 20-4** System response to a change in $\Delta\delta_o$. (*a*) Test system, (*b*) comparison of linear approximation, $\Delta\delta(t) = a_\delta e^{-\sigma t}\cos(\omega t - \varphi)\Delta\delta_o$, with higher-order approximations.

where

$$A_{g_{ij}} = \frac{1}{2H_i}\frac{\partial P_{e_i}}{\partial \Delta\delta_j}\bigg|_{\delta_i^o,\delta_j^o} = \frac{E'_{qi}E'_{qj}}{2H_i}[G_{ij}\sin(\delta_{io}-\delta_{jo}) - B_{ij}\cos(\delta_{io}-\delta_{jo})]\ ;\ A_{g_{ii}} = -\sum_{\substack{i=1\\ i\neq j}}^{n} A_{g_{ij}}\ ;\ \hat{K}_{Di} = \frac{K_{D_i}}{2H_i}$$

and $G_{ij}$ and $B_{ij}$ are the mutual conductance and susceptance between buses *i* and *j*, respectively.

Equation (20-12) can be rewritten in matrix form as

$$\begin{bmatrix} \Delta\dot{\boldsymbol{\delta}} \\ \Delta\dot{\hat{\boldsymbol{\omega}}}_r \end{bmatrix} = \begin{bmatrix} \mathbf{0} & \omega_o\mathbf{I} \\ -\mathbf{M}^{-1}\mathbf{K}_S & -\mathbf{M}^{-1}\mathbf{K}_D \end{bmatrix}\begin{bmatrix} \Delta\boldsymbol{\delta} \\ \Delta\hat{\boldsymbol{\omega}}_r \end{bmatrix} + \begin{bmatrix} \mathbf{0} \\ \mathbf{M}^{-1} \end{bmatrix}\Delta\mathbf{P}_m\ ;\ \mathbf{y} = \begin{bmatrix} \mathbf{c}_\delta^T & \mathbf{c}_\omega^T \end{bmatrix}\begin{bmatrix} \Delta\boldsymbol{\delta} \\ \Delta\hat{\boldsymbol{\omega}}_r \end{bmatrix} \tag{20-13}$$

where $\Delta\boldsymbol{\delta} = [\Delta\delta_1\ \ \Delta\delta_2\ \cdots\ \Delta\delta_{ng}]^T$, $\Delta\hat{\boldsymbol{\omega}}_r = [\Delta\hat{\omega}_{r1}\ \ \Delta\hat{\omega}_{r2}\ \cdots\ \Delta\hat{\omega}_{rng}]^T$, $\mathbf{M} = \mathrm{diag}[M_1\ \ M_2\ \cdots\ M_n]$, with $M_i = 2H_i$, $\mathbf{K}_D = \mathrm{diag}[K_{D_1}\ \ K_{D_2}\ \cdots\ K_{Dng}]$, $\Delta\mathbf{P}_m = [\Delta P_{m_1}\ \ \Delta P_{m2}\ \cdots\ \Delta P_{mng}]^T$, and $\mathbf{K}_S$ is the synchronizing torques matrix evaluated at the post-disturbance SEP. It follows from this definition that, for a purely inductive network, synchronizing power is small when the synchronous machines are coupled through high reactance, that is, $B_{ij} = 0$.

The eigenvectors of the state model satisfy

$$\begin{bmatrix} \mathbf{0} & \omega_o\mathbf{I} \\ -\mathbf{M}^{-1}\mathbf{K}_S & -\mathbf{M}^{-1}\mathbf{K}_D \end{bmatrix}\begin{bmatrix} \phi_{\delta_i} \\ \phi_{\omega_i} \end{bmatrix} = \lambda_i\begin{bmatrix} \phi_{\delta_i} \\ \phi_{\omega_i} \end{bmatrix}$$

where $\phi_{\delta_i} = \phi_{\delta_{Ri}} + j\phi_{\delta_{Ii}}$ and $\phi_{\omega_i}$ are referred to as the *angle and speed components* of the right eigenvectors, respectively.

**TABLE 20-2** Modal Properties of the State Representation Eq. (20-13)

| Feature | Expression |
|---|---|
| Modal damping | $\sigma_i = -\dfrac{\phi_{\delta_{Ri}}{}^T \mathbf{K}_D \phi_{\delta_{Ri}} + \phi_{\delta_{Ii}}{}^T \mathbf{K}_D \phi_{\delta_{Ii}}}{2(\phi_{\delta_{Ri}\delta_i}{}^T \mathbf{M}\phi_{\delta_{Ri}} + \phi_{\delta_{Ii}}{}^T \mathbf{M}\phi_{\delta_{Ii}})}$ |
| Modal frequency | $\omega_i = \sqrt{\dfrac{\sigma_i^2(\phi_{\delta_{Ri}}{}^T \mathbf{M}\phi_{\delta_{Ri}} + \phi_{\omega_{Ii}}{}^T \mathbf{M}\phi_{\omega_{Ii}}) + \sigma_j \phi_{\delta_{Ri}}{}^T \mathbf{K}_D \phi_{\delta_{Ri}} + \phi_{\omega_{Ii}}{}^T \mathbf{K}_D \phi_{\omega_{Ii}} + \omega_o(\phi_{\delta_{Ri}}{}^T \mathbf{K}_S \phi_{\delta_{Ri}} + \phi_{\omega_{Ii}}{}^T \mathbf{K}_S \phi_{\omega_{Ii}})}{\phi_{\delta_{Ri}}{}^T \mathbf{M}\phi_{\delta_{Ri}} + \phi_{\omega_{Ii}}{}^T \mathbf{M}\phi_{\omega_{Ii}}}}$ |
| Eigenvector relationship | $\phi_{\omega_i} = \dfrac{\lambda_i}{\omega_o}\phi_{\delta_i}$ |

Table 20-2 shows the modal characteristics of the state representation determined using Eq. (20-13). Several key findings should be emphasized: (1) Modal damping is directly proportional to the damping coefficient, $K_D$; (2) modal frequency is determined by the synchronizing torque, $\mathbf{K}_S$, and modal damping $\sigma$. With small damping, frequency is directly proportional to the synchronizing coefficients.

In complex systems, damping may be influenced by control characteristics such as high-gain control systems and speed-governor characteristics.

*Eigenvalues of a Classical Model.* As an example for modal analysis consider a two-area system, four-machine system represented classically, refer to Fig. 20-5. The state matrix $\mathbf{A}$ associated with the classical model in Eq. (20-12) is

$$\mathbf{A} = \left[\begin{array}{cccc|cccc}
0 & 0 & 0 & 0 & 376.9911 & 0 & 0 & 0 \\
0 & 0 & 0 & 0 & 0 & 376.9911 & 0 & 0 \\
0 & 0 & 0 & 0 & 0 & 0 & 376.9911 & 0 \\
0 & 0 & 0 & 0 & 0 & 0 & 0 & 376.9911 \\
\hline
-0.0979 & 0.0936 & 0.0019 & 0.0025 & -0.0769 & 0 & 0 & 0 \\
0.0999 & -0.1114 & 0.0046 & 0.0069 & 0 & -0.0769 & 0 & 0 \\
0.0036 & 0.0072 & -0.1064 & 0.0957 & 0 & 0 & -0.0810 & 0 \\
0.0070 & 0.0145 & 0.1011 & -0.1226 & 0 & 0 & 0 & -0.0810
\end{array}\right]$$

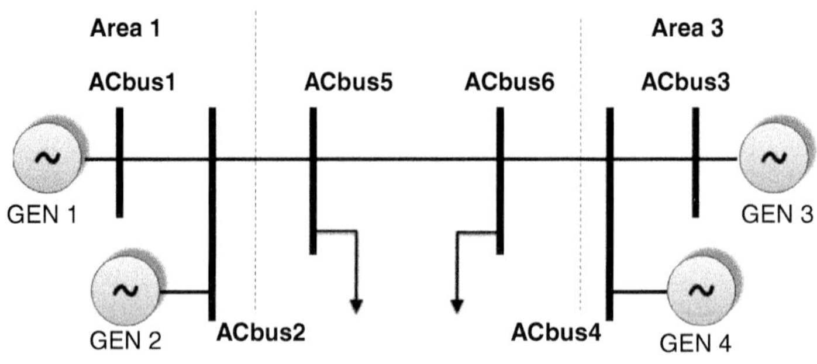


**FIGURE 20-5** Two-area test system.

The eigenvalues of matrix **A** above are $-0.0385 \pm j8.714$, $-0.0404 \pm j8.969, -0.0398 \pm j2.971$; the first two modes represent local oscillations to areas 1 and 2, respectively, while the third mode is an inter-area mode. The eigenvectors associated with the inter-area mode are

$$\phi_5 = \begin{bmatrix} 0.3298\angle 179.92^\circ \\ 0.2921\angle 179.91^\circ \\ 0.6609\angle 0^\circ \\ 0.6075\angle 0.0024^\circ \\ \hdashline 0.0026\angle -89.30^\circ \\ 0.0023\angle -89.31^\circ \\ 0.0052\angle 90.76^\circ \\ 0.0048\angle 90.77^\circ \end{bmatrix} = \begin{bmatrix} \phi_{\delta_5} \\ \hdashline \phi_{\omega_5} \end{bmatrix}; \quad \phi_6 = \begin{bmatrix} 0.3298\angle -179.92^\circ \\ 0.2921\angle -179.91^\circ \\ 0.6609\angle 0^\circ \\ 0.6075\angle -0.0024^\circ \\ \hdashline 0.0026\angle 89.30^\circ \\ 0.0023\angle 89.31^\circ \\ 0.0052\angle -90.76^\circ \\ 0.0048\angle -90.77^\circ \end{bmatrix} = \begin{bmatrix} \phi^*_{\delta 6} \\ \hdashline \phi^*_{\omega 6} \end{bmatrix}$$

The physical meaning of the entries of the right eigenvector is clear: the magnitude of the right eigenvector gives the contribution of the $k$th state to the $i$th mode. From a physical perspective, the right eigenvector entries have large magnitudes at machines having a strong participation in the oscillations, while the phase angle gives the relative oscillation.

From Eq. (20-9), the free system response become

$$\begin{bmatrix} \Delta\delta_1(t) \\ \Delta\delta_2(t) \\ \cdots \\ \Delta\delta_{ng}(t) \\ \hdashline \Delta\omega_{r_1}(t) \\ \Delta\omega_{r_2}(t) \\ \cdots \\ \Delta\omega_{r_{ng}}(t) \end{bmatrix} = \alpha_1 \begin{bmatrix} \phi_{\delta_{11}} \\ \phi_{\delta_{21}} \\ \cdots \\ \phi_{\delta_{ng,1}} \\ \hdashline \phi_{\omega_{1,1}} \\ \phi_{\omega_{2,1}} \\ \cdots \\ \phi_{\omega_{ng,1}} \end{bmatrix} e^{\lambda_1 t} + \alpha_2 \begin{bmatrix} \phi_{\delta_{12}} \\ \phi_{\delta_{22}} \\ \cdots \\ \phi_{\delta_{ng,2}} \\ \hdashline \phi_{\omega_{1,2}} \\ \phi_{\omega_{2,2}} \\ \cdots \\ \phi_{\omega_{ng,2}} \end{bmatrix} e^{\lambda_2 t} + \cdots + \alpha_n \begin{bmatrix} \phi_{\delta_{1n}} \\ \phi_{\delta_{2n}} \\ \cdots \\ \phi_{\delta_{ng,n}} \\ \hdashline \phi_{\omega_{1,n}} \\ \phi_{\omega_{2,n}} \\ \cdots \\ \phi_{\omega_{ng,n}} \end{bmatrix} e^{\lambda_n t} = \begin{bmatrix} \sum_{j=1}^{m} 2a_{1j} e^{\sigma_j t} \cos(\omega_j t + \phi_{1j}) \\ \sum_{j=1}^{m} 2a_{2j} e^{\sigma_j t} \cos(\omega_j t + \phi_{2j}) \\ \cdots \\ \sum_{j=1}^{m} 2a_{ngj} e^{\sigma_j t} \cos(\omega_j t + \phi_{ngj}) \\ \hdashline \sum_{j=1}^{m} 2a_{ng+1j} e^{\sigma_j t} \cos(\omega_j t + \phi_{ng+1j}) \\ \cdots \\ \sum_{j=1}^{m} 2a_{2ngj} e^{\sigma_j t} \cos(\omega_j t + \phi_{2ngj}) \end{bmatrix} \tag{20-14}$$

with $m = n/2$, $a_{kj} = \alpha_j U_{kj} = a_{kj}\angle\phi_{kj}.a^*_{kj} = \alpha^*_j U^*_{kj} = a_{kj}\angle -\phi_{kj}$.

*Mode Shape.* For a given mode $i$, the relative rotor oscillation between machines $k$ and $l$ is given by the phase angle of the right eigenvector $\phi_i$. More formally, two machines swing out of phase if $\phi_{ki} - \phi_{li} \approx 180^\circ$.

*Areas or Regions Connected by Weak Tie Lines.* During a fault or perturbation, the individual generators are accelerated initially according to the electrical proximity of each generator to the disturbance and their individual inertias. The ensuing accelerating energy will initiate both, slow inter-area oscillations between loosely connected geographical areas and fast inter-machine oscillations within the areas.

Assuming that generators within a given area, $A$, are connected through negligible reactances $X_{ij} \approx 0$, the initial angle deviation $\Delta\delta_{ij}$ is the same for all machines, and $\Delta P_A = \Sigma_{j\in A} \frac{\partial P_i}{\partial \delta_j}\Delta\delta_j \approx \left(\frac{\partial P_1}{\partial \delta_1} + \frac{\partial P_2}{\partial \delta_2} + \cdots + \frac{\partial P_{ng}}{\partial \delta_{ng}}\right)\Delta\delta \quad \forall j \in A$ or $\Delta\omega_1 = \Delta\omega_1 = \ldots\Delta\omega_{ng}$. Physically, machines more distant from the fault are expected to swing together in groups with in phase slow motion. The collective behavior of the machines having identical motion is given by

$$\Delta\dot{\hat{\omega}}_{r_A} = \frac{1}{H_{eq}} \sum_{j\in A} \frac{\partial P_{e_i}}{\partial \delta_j} \Delta\delta_j - K_{D_{eq}} \Delta\hat{\omega}_{r_A}$$

Once the machines exhibiting a common behavior are identified, a reduced order model can be extracted. The procedure involves three main stages: (1) coherency identification, (2) generator aggregation, and (3) network reduction.

*Interarea Mode Oscillations.* Large power systems consist of an interconnection of geographical areas or coherent machines that are weakly interconnected. Interarea mode oscillations are electromechanical swings usually associated with groups of machines exchanging kinetic energy across a relatively weak transmission path. Inter-area oscillations typically exhibit frequencies in the range 0.2 to 0.8 Hz and may involve the interaction of two or more modes (multimodal oscillations).

*Local Plant or Inter-Machine Modes of Oscillation.* These oscillations represent more localized system dynamics and may involve a single machine swinging against the rest of the system or the interaction of a small group of closely connected machines.

Figure 20-6 shows a mechanical analogy of a two-area equivalent system for both, local and inter-area modes.

*Modal Representations.* Inspection of the state representation Eq. (20-13) shows that the state equation can be diagonalized using the right eigenvectors associated with matrix $-\mathbf{M}^{-1}\mathbf{K}_s$, that is, $\boldsymbol{\Phi}^{-1}(-\mathbf{M}^{-1}\mathbf{K}_s)\boldsymbol{\Phi} = \boldsymbol{\Lambda} = \mathrm{diag}(\lambda_1 \;\; \lambda_2 \;\; \cdots \;\; \lambda_n)$. Use of the transformation

$$\mathbf{x} = \mathbf{T}\mathbf{x}_m = \begin{bmatrix} \boldsymbol{\Phi} & \\ & \boldsymbol{\Phi} \end{bmatrix} \begin{bmatrix} \Delta\delta_m \\ \Delta\hat{\omega}_{rm} \end{bmatrix}$$

in Eq. (20-13) yields the decoupled system

$$\begin{bmatrix} \Delta\dot{\delta}_m \\ \Delta\dot{\hat{\omega}}_{rm} \end{bmatrix} = \begin{bmatrix} \mathbf{0} & \boldsymbol{\Phi}^{-1}\omega_o\mathbf{I}\boldsymbol{\Phi} \\ -\boldsymbol{\Phi}^{-1}\mathbf{M}^{-1}\mathbf{K}_S\boldsymbol{\Phi} & -\boldsymbol{\Phi}^{-1}\mathbf{M}^{-1}\mathbf{K}_D\boldsymbol{\Phi} \end{bmatrix} \begin{bmatrix} \Delta\delta_m \\ \Delta\hat{\omega}_{rm} \end{bmatrix} = \begin{bmatrix} \mathbf{0} & \omega_o\mathbf{I} \\ \boldsymbol{\Lambda} & -\mathbf{M}^{-1}\mathbf{K}_D \end{bmatrix} \begin{bmatrix} \Delta\delta_m \\ \Delta\hat{\omega}_{rm} \end{bmatrix}$$

The equation above states that the linear transformation $\mathbf{T}$, decomposes the phase-space into uncorrelated patterns of the form, $\Delta\delta_{m_i} = \omega_o\Delta\hat{\omega}_{rm_i}, \Delta\dot{\hat{\omega}}_{rm_i} = \lambda_i\Delta\delta_{m_i} - (K_{D_i}/2H_i)\Delta\hat{\omega}_{rm_i}, i = 1,\ldots, ng$.

***Multimodal Oscillations.*** Transient processes in small signal stability studies exhibit multiscale spatial and temporal patterns, with complex dynamics. Often, modes can interact and result in complex nonlinear interactions. Figure 20-7 shows a typical multimodal oscillation involving the interaction of three major electromechanical modes or oscillatory frequencies. From the perspective of system modeling, three main aspects are of interest:

1. Tracking the temporal behavior the observed oscillations
2. Extracting their spectral content
3. Determining statistical patterns and trends in the oscillations

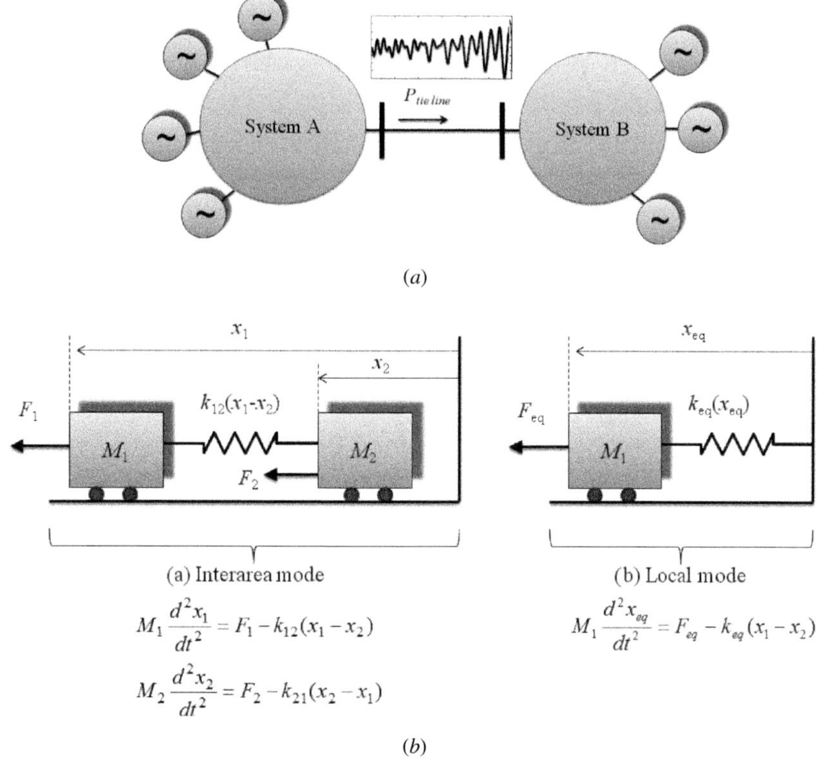


**FIGURE 20-6** Schematic illustrating the notion of ideal local and interarea mode oscillations. (*a*) Conceptual representation; (*b*) mechanical analogy.

*Mode-State Relationships.* In the study of modal interaction it is often desirable to quantify the participation of a particular mode in a given state variable.

*Participation factors (PFs).* The concept of participation factor was developed to measure the degree of participation of a state variable in a mode. From Eq. (20-8), the time evolution of the $k$th state is given by [17]

$$x_k(t) = \sum_{i=1}^{n} (\phi_{ki} \boldsymbol{\Psi}_i^T \mathbf{x}_o) e^{\lambda_i t} = \sum_{i=1}^{n} e^{\lambda_i t} \underbrace{(\phi_{ki} \psi_{ik})}_{\substack{\text{Linear} \\ \text{participation factor}}} x_{ko} + \sum_{i=1}^{n} e^{\lambda_i t} \sum_{\substack{j=1 \\ j \neq k}}^{n} (\phi_{ki} \psi_{ij}) x_{jo} \tag{20-15}$$

where $p_{ki} = \phi_{ki}\psi_{ik}$ is called the linear participation factor, and it is assumed that $x_{jo} = 0$. Eq. (20-15) indicates that the participation factor $p_{ki}$ can be seen as the relative participation of the $i$th mode in the $k$th state at $t = 0$. Then, the participation matrix is defined as $\mathbf{P} = [p_{ki}]$. Participation factors are dimensionless quantities and satisfy the following properties: (1) $\sum_{i=1}^{n} p_{ki} = 1 = \sum_{k=1}^{n} p_{ki} = 1$ and (2) $a_{kk} = \sum_{i=1}^{n} \lambda_i p_{ki}$, where $a_{kk}$ is the $(k, k)$ element of matrix $\mathbf{A}$.

In common practice, PFs are often used to identify the system states that have most influence in producing a particular mode as well as to locate system controllers; the PFs should have large magnitudes at the selected locations or states. It should be observed, however, that participation factors as defined above represent open-loop measures. A similar, but more accurate estimate of the best places

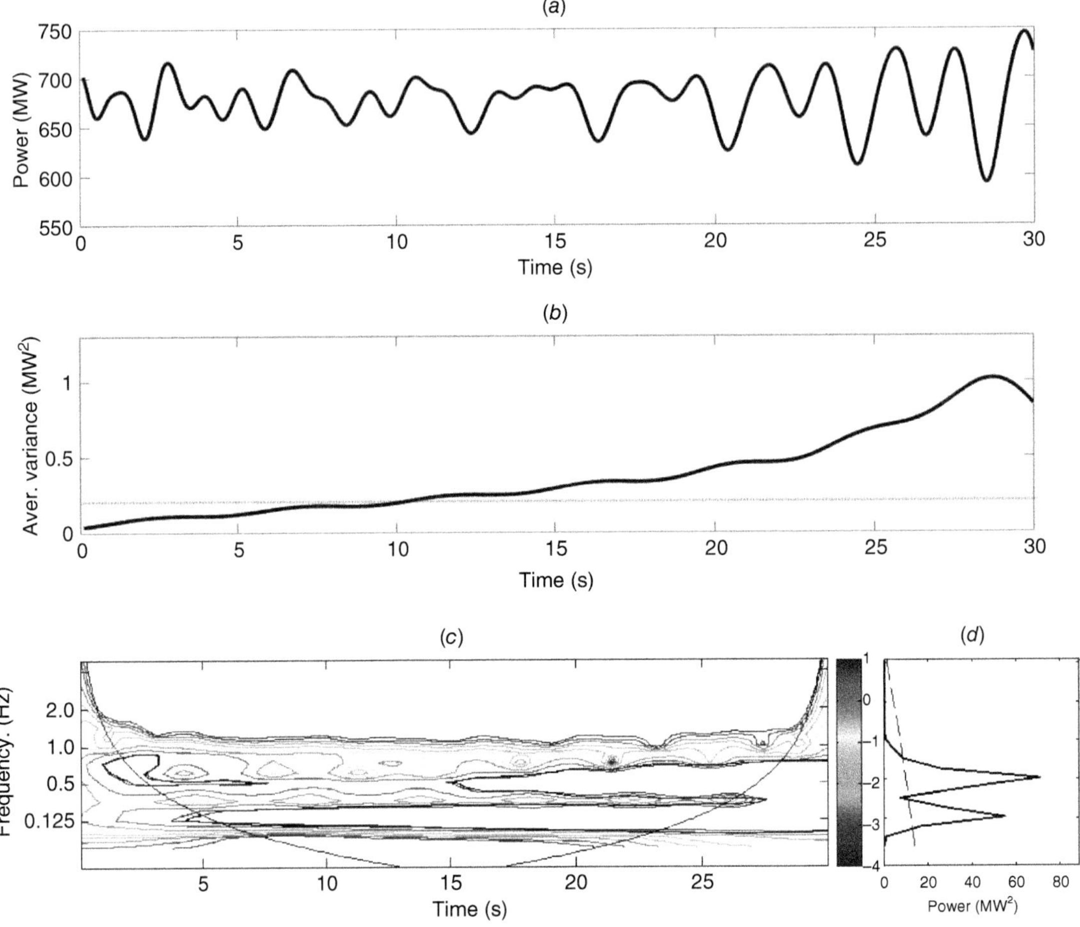


**FIGURE 20-7** Example of multicomponent signal showing the intervals of influence of dominant modes. (*a*) Temporal evolution; (*b*) scaled average significance; (*c*) power spectrum; (*d*) global spectrum.

to locate system controllers can be obtained by using residue analysis. In addition to participation factors, kinetic energy and modal power can also be used as sensitive indicators of coupling between modes and state variables.

Table 20-3 shows energy and power modal expansions based on these notions.

*Modal Power Flow.* Once an efficient method for computing the modal properties of the state representation is available, the modal power flow distribution or energy exchange can be computed. Typically, large power system models represent system behavior by two highly sparse small signal models [6]:

$$\begin{bmatrix} \dot{\mathbf{x}} \\ \Delta\mathbf{i} \end{bmatrix} = \begin{bmatrix} \mathbf{A}_D & \mathbf{B}_D \\ \mathbf{C}_D & -\mathbf{Y}_D \end{bmatrix} \begin{bmatrix} \mathbf{x} \\ \Delta\mathbf{v} \end{bmatrix} + \begin{bmatrix} \mathbf{B}_u \\ \mathbf{0} \end{bmatrix} \mathbf{u} \tag{20-16}$$

$$\Delta\mathbf{i} = \mathbf{Y}_N \Delta\mathbf{v}$$

**TABLE 20-3** Dynamic Energy and Power Relationships

| Feature | Definition | Modal relationship |
|---|---|---|
| Kinetic energy | $KE=\frac{1}{2}\sum_{k=1}^{ng} M_k \omega_{r_k}{}^2=\frac{1}{2}\boldsymbol{\omega}_r^T \mathbf{M}\boldsymbol{\omega}_r$ | $\Delta KE_k=\frac{1}{2}M_k\left\{[\sum_{j=1}^{ng}\alpha_j e^{\lambda_j t}\phi_{\omega r_{kj}}]^2+\sum_{p=1}^{ng}\sum_{q=p+1}^{ng-1}(\phi_{\omega r_{kp}}\phi_{\omega r_{kq}})(\alpha_p\alpha_q)e^{(\lambda_p+\lambda_q)t}\right\}$ |
| System power | $P=\mathbf{x}^T\dot{\mathbf{x}}=(\mathbf{x},\mathbf{A}\,\mathbf{x})$ | $\Delta P_k=\left\{(\alpha_k e^{\lambda_k t})^2\boldsymbol{\phi}_k^T\boldsymbol{\phi}_k+\sum_{j=k+1}^{n-1}2(\lambda_k)(\alpha_k\alpha_j)e^{(\lambda_k+\lambda_j)t}\boldsymbol{\phi}_k^T\boldsymbol{\phi}_j\right\}$ |

where submatrices $\mathbf{A}_D,\mathbf{B}_D,\mathbf{C}_D,\mathbf{Y}_D$ represent sensitivity relations of appropriate dimension and include the effect of FACTS controllers, synchronous machines, wind farms and other dynamic devices, and $\mathbf{Y}_N$ is the network admittance matrix. Models of any degree of complexity can be handled using this approach.

Assuming that $\mathbf{x}(t)=\Sigma_{i=1}^{n}\alpha_i e^{\lambda_i t}\boldsymbol{\phi}_i$, and solving for the bus voltage deviations, $\Delta\mathbf{v}$, in terms of the time evolution of the system states, $\mathbf{x}$, yields

$$\begin{aligned}&\left(\mathbf{Y}_N+\mathbf{Y}_D\right)\Delta\mathbf{v}(\lambda_1,\ldots,\lambda_n)=\mathbf{C}_D\mathbf{x}=\mathbf{C}_D\sum_{j=1}^{n}(\alpha_j)\boldsymbol{\phi}_j e^{\lambda_j t}=\mathbf{I}_{\text{mod}}(\lambda_1,\ldots,\lambda_n)\\ &\Delta\mathbf{i}(\lambda_1,\ldots,\lambda_n)=\mathbf{C}_D\sum_{j=1}^{n}(\alpha_j)\boldsymbol{\phi}_j e^{\lambda_j t}-\mathbf{Y}_D\Delta\mathbf{v}(\lambda_1,\ldots,\lambda_n)\end{aligned}\tag{20-17}$$

where $\mathbf{I}_{\text{mod}}$ essentially represents a modal, current injection vector. Physically, the magnitude of the bus voltage deviations allows the identification of those buses where the swing oscillation has the stronger influence on the bus voltage magnitude. It is equivalent to the notion of bus voltage mode shape.

Of major interest, modal contributions are finally obtained by expressing the incremental variations in power to the corresponding variations in voltage and currents. Noting that, $P_{D_k}=v_{d_k}i_{d_k}+v_{q_k}i_{q_k}, Q_{D_k}=v_{q_k}i_{d_k}-v_{d_k}i_{q_k}$, the modal power can be written as

$$\Delta\mathbf{P}_D(\lambda)=\left|\Delta\mathbf{P}_D\right|\angle\varphi_{P_D}=[\mathbf{v}_D^o]\Delta\mathbf{i}_D(\lambda)+[\mathbf{i}_D^o]\Delta\mathbf{v}_D(\lambda)$$

and

$$\Delta\mathbf{Q}_D(\lambda)=\left|\Delta\mathbf{Q}_D\right|\angle\varphi_{Q_D}=[\hat{\mathbf{i}}_D^o]\Delta\mathbf{v}_D(\lambda)+[\hat{\mathbf{v}}_D^o]\Delta\mathbf{i}_D(\lambda)$$

where $[\mathbf{i}_D^o],[\mathbf{v}_D^o],[\hat{\mathbf{i}}_D^o]$, and $[\hat{\mathbf{v}}_D^o]$ are constant, block-diagonal matrices, whose values are dependent on the initial device and current injections.

Modal power flow analysis is of special interest to analyze the impact of low-to-no-inertia devices such as wind farms and photovoltaic generation, and the exchange of swing oscillation energy in the network.

### 20.1.3 Forced Oscillations

Forced oscillations are associated with a driving force in which energy is continuously supplied to the system. The driving force can take many ways including cyclic load variations, mechanical load changes or stochastic forcing associated with random load variations [18–20].

***Harmonic Excitation.*** In this case, the driving force is usually assumed to be sinusoidal. The equation of motion for a set of $ng$ machines represented classically driven by a simple harmonic force is, in physical space

$$\dot{\mathbf{x}} = \begin{bmatrix} \Delta\dot{\delta} \\ \Delta\dot{\hat{\omega}}_r \end{bmatrix} = \begin{bmatrix} \mathbf{0} & \omega_o \mathbf{I} \\ -\mathbf{M}^{-1}\mathbf{K}_S & -\mathbf{M}^{-1}\mathbf{K}_D \end{bmatrix} \begin{bmatrix} \Delta\delta \\ \Delta\hat{\omega}_r \end{bmatrix} + \underbrace{\begin{bmatrix} \mathbf{0} \\ \mathbf{M}^{-1}\mathbf{f}(e^{j\omega_r t} + e^{-j\omega_r t}) \end{bmatrix}}_{\text{Harmonic forcing term}} \tag{20-18}$$

where $\omega_r$ is the excitation (forcing) frequency, and $\mathbf{f} = [A_1, \ldots, A_{ng}]$ is a vector of amplitudes for each mechanical torque; $A_i$, $i = 1, \ldots, ng$ is the magnitude or level of the excitation.

The total system response is given by

$$\Delta\hat{\omega}_{rm}(s) = \left\{ (s\mathbf{I} - \Lambda)^{-1}\mathbf{x}_o + (s\mathbf{I} - \Lambda)^{-1}\mathbf{T}^{-1}\mathbf{M}^{-1}\mathbf{F}(s) \right\}$$

where $\mathbf{T}$ is the modal transformation matrix, and

$$\mathbf{F}(s) = [2A_1 s/(s - j\omega_r)(s + j\omega_r)2H_1 \quad 2A_2 s/(s - j\omega_r)(s + j\omega_r)2H_2 \quad \cdots \quad 2A_n s/(s - j\omega_r)(s + j\omega_r)2H_n]^T.$$

The modal speed deviations can then be expressed as

$$\Delta\hat{\omega}_{rm_i}(t) = \mathrm{L}^{-1}\left\{ \underbrace{\frac{s\Delta\hat{\omega}_{\mathrm{rm}_i}(0) + \Delta\delta_{m_i}(0)}{(s - \sigma_i - j\omega_i)(s - \sigma_i + j\omega_i)}}_{\text{Free response}} + \sum_{k=1}^{n} \frac{A_k \psi_{ik}}{H_k} \underbrace{\frac{s^2}{(s + j\omega_r)(s - j\omega_r)(s - \sigma_i - j\omega_i)(s - \sigma_i + j\omega_i)}}_{\text{Forced response}} \right\} \tag{20-19}$$

When the $A_i$'s are small (weak excitation), the solution of Eq. (20-18) approximates to the free response. For a stable, well-damped system, the free system response vanishes for large times, and the steady-state modal response due to a sinusoidal excitation is also sinusoidal. A more detailed analysis of Eq. (20-19) shows that the amplitude of the forced oscillation becomes very large when the forcing frequency, $\omega_r$, takes values close to the frequency of weakly damped inter-area (local) modes, and the system is said to be in a state of *resonance*. As an example, Fig. 20-8 shows system responses to resonant frequency, $\omega_r$ for the system shown in Fig. 20-5. The oscillations are, in general, nonperiodic and under some circumstances may not decay with time. References [19, 20] discuss recent experience with the analysis of actual system oscillations.

***Stochastically Driven Linear Dynamical Systems.*** A second case of interest is obtained when the primary source of excitation to the system is random load variations, which result in a low-amplitude stochastic system response. Under these assumptions, it is reasonable to assume that the system dynamic behavior may be approximated by a stochastically driven linear system of the form

$$\begin{aligned} \mathbf{x}(t) &= \mathbf{A}\mathbf{x}(t) + \mathbf{B}\mathbf{u}(t) + \mathbf{F}\xi(t) \\ \mathbf{y}(t) &= \mathbf{C}\mathbf{x}(t) + \upsilon(t) \end{aligned} \tag{20-20}$$

where $\mathbf{x}(t)$ is the vector of states, $\mathbf{A}\mathbf{x}(t)$ is the deterministic part of the model, and $\boldsymbol{\xi}$ is a vector of random vector perturbing the system; the vector $\boldsymbol{\upsilon}$ is measurement noise, and $\mathbf{F}$ is the forcing matrix.

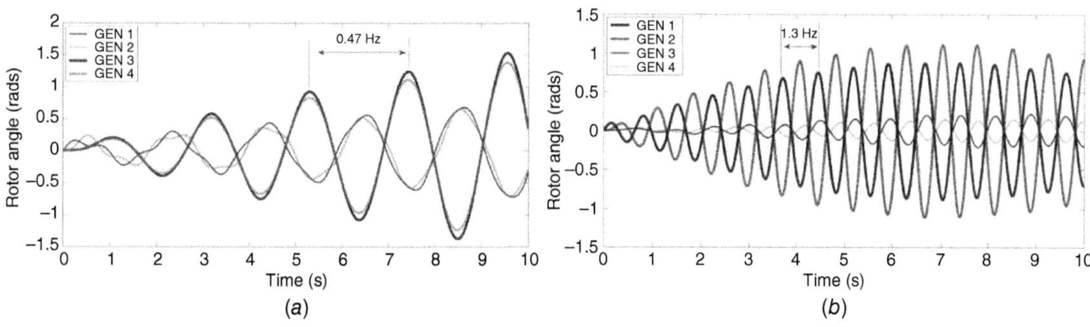


**FIGURE 20-8** Forced system response for the system in Fig. 20-5. (*a*) $f_r = 0.47$ Hz; (*b*) $f_r = 1.32$ Hz.

The output **y** is influenced by the dynamic of the system and results in random variations in measurable signals such as bus voltages, tie-line power and other signals.

The input $\xi(t)$ generally takes the form of a complex Gaussian white-noise process, having zero mean and unit covariance that best represent load characteristics or other effects. Loads are assumed to be of the form $P_{Li} = P_{Li}^o + \xi_{Pi}; Q_{Li} = Q_{Li}^o + \xi_{Pi}$, where $\xi_{Pi}, \xi_{Qi}$ are the random load perturbations associated with the real and reactive components of the load and the superscript *o* denotes initial conditions. Spatial variability can be described by the off-diagonal elements of matrix **F**. Figure 20-9 provides a conceptual representation of the above model.

Similar to the case of forced oscillations, the solution of Eq. (20-20) may be written as

$$\mathbf{x}(t) = \underbrace{e^{\mathbf{A}t}\mathbf{x}_o}_{\text{Deterministic response}} + \underbrace{\int_o^t e^{\mathbf{A}(t-\tau)}\xi(\tau)d\tau}_{\text{Stochastic response}}, \quad \mathbf{y}(t) = \mathbf{C}e^{\mathbf{A}t}\mathbf{x}_o + \mathbf{C}\int_o^t e^{\mathbf{A}(t-\tau)}\xi(\tau)d\tau + \mathbf{D}\xi(\tau) \tag{20-21}$$

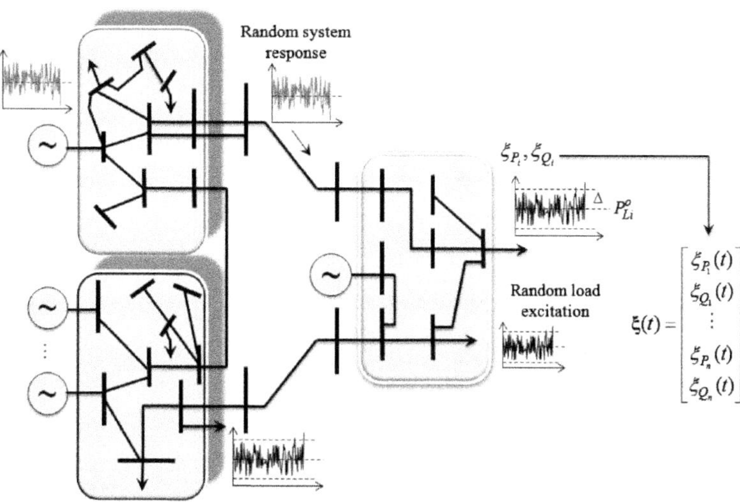


**FIGURE 20-9** Schematic illustrating the nature of system response to random load changes.

In general terms, the idea of this approach is to recover the system properties through modal analysis of the stochastic response when the system has settled to steady-state random behavior. For a stable system, the first part on the *rhs* of Eq. (20-21) (deterministic response) decays to zero and is of no interest for the analysis of the steady-state response; the second term on the *rhs* represents the steady-state stochastic response and can be used to determine the loads associated with dominant system behavior, as well as to assess modal distribution.

Two main parameters define the steady-state stochastic system response:

1. The mean value, and standard deviation, $\Delta$
2. The ensemble average energy of the forcing, $E = \langle \mathbf{x}(t)^T \mathbf{x}(t) \rangle$

Under a statistical steady-state condition, the predominant spatial structure describing the stochastic forcing responses can be obtained from the covariance matrices of the state and measurement vectors.

## 20.1.4 Measurement-Based Analysis Techniques

In recent years, the development of wide-area measurement systems (WAMS) and advanced monitoring devices has added much to the knowledge of system behavior. A major effort has been made to develop methods to extract important patterns (mode shapes) and dynamical variables directly from measured data. A related but separate question is on how to describe the complex system dynamics using a greatly reduced number of measurements.

***Data-Driven Techniques.*** Practical study and analytical experience suggest that the ringdown phase of the observed oscillations following a disturbance is well described by a linear superposition of (complex) oscillation modes. In the noise-free linear case, the set of measurements $y(t)$, $t = 1, \ldots, M$ can be approximated as

$$\hat{y}(t) = \underbrace{b_o}_{\text{offset}} + \underbrace{b_1 t}_{\substack{\text{linear}\\ \text{drift}}} + \underbrace{\sum_{j=1}^{p_c} a_j e^{-\sigma_j t}}_{\substack{\text{Monotonic}\\ \text{components}}} + \underbrace{\sum_{k=1}^{p_o} a_k e^{-\sigma_k t} \cos(\omega_k t + \theta_k)}_{\substack{\text{Oscillatory}\\ \text{components}}} \tag{20-22}$$

where $\sigma_k$ is the damping coefficient in Napiers/s, $\omega_k$ is the modal frequency in rad/s, $\theta_k$ is the phase angle in radians, $a_k$ is the amplitude in *pu* and $N$ is the number of samples; $p_c$ and $p_o$ are the number of significant monotonic and oscillatory components in the approximation.

This model is based on two implicit assumptions: nonlinearities are inconsequential and mode coupling is negligible. In most practical cases, the true dynamics of the observed oscillations may be poorly known due to the difficulties in accounting for various damping effects, nonstationarities, and noise. Further, high levels of noise result in nonstationary signals, which may lead to inefficient performance and inaccurate results. Practical methods are designed to accomplish two major goals: reducing the dimensionality of the data sets by retaining only the dominant modes, and extracting features of interest such as modal parameters. Examples include techniques such as Prony analysis and nonparametric analysis methods [21].

Figure 20-10 illustrates the nonstationary behavior of a measured signal showing phase changes for three different time intervals. Note the changes in the magnitudes and phases of the relative oscillations.

***Oscillation Detection and Analysis.*** Monitoring methods for detection of impending loss of stability or the onset of oscillations have been developed [22]. Given a sufficiently dense set of dynamic recording devices such as phasor measurement units (PMUs), the first step of a wide-area monitoring system is often preprocessing to correct noise, errors, and missing data. Then, an appropriate signal processing identification algorithm is selected to run on the processed data, and produce output patterns.

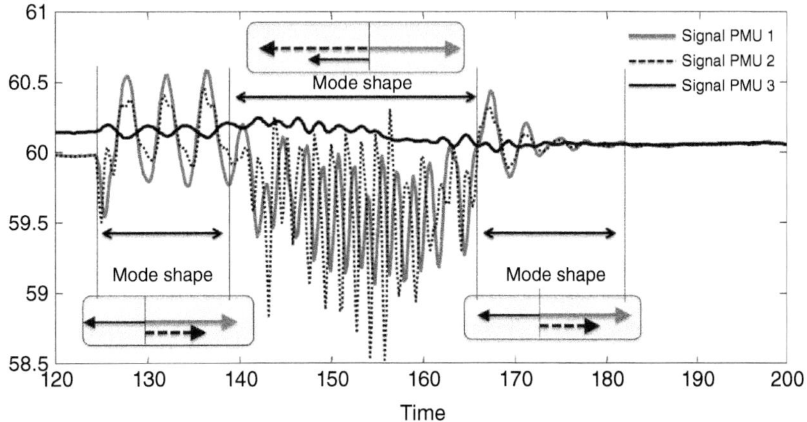


**FIGURE 20-10** Measured signals showing changes in modal parameters. The inset shows the mode shape of the dominant mode.

Typical output patterns include global outliers, mode coupling and clustering, predictive models, as well as change patterns and critical event detection. There also other application domains such as the development of triggering algorithms for PMUs or control and protection systems.

The monitoring system incorporates four basic functions or stages: filtering, detrending and rectification, detection of abnormal activity, and thresholding [22]. In the context of oscillation detection, the definition on an appropriate threshold is a crucial issue in the identification process as it may result in false alarm errors.

Figure 20-11 provides an illustration of the use of time-domain analysis to detect the onset of system oscillations. Other alternatives using filters and user-defined threshold levels are described in [22].

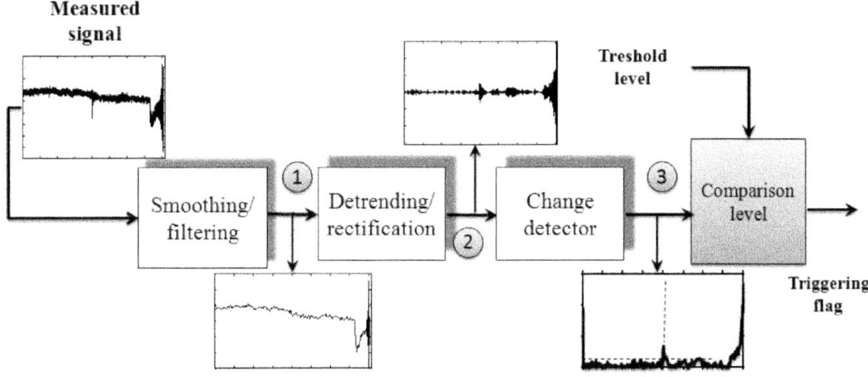


**FIGURE 20-11** Detection of the onset of system oscillations.

## 20.1.5 Mitigation and Control of System Oscillations

***Damping Criteria and Anomaly Detection.*** Grid codes are increasingly specifying criteria for damping of electromechanical modes of oscillation. In practice, a number of well-defined parameters can be used in order to identify the onset and damping of spontaneous oscillations. These include the start time and duration of the oscillatory process, settling times, threshold crossings, rise and decay time, and the envelope of the observed oscillations. With reference to Eq. (20-22), the damping ratio $\xi$, and the associated time constant $\tau$ can be computed as

$$\xi(\%) = \frac{\sigma}{\sqrt{\sigma^2+\omega^2}} x100; \quad \tau = 1/|\sigma| \quad (s) \tag{20-23}$$

Several criteria to define acceptable damping rates for system oscillations have been developed. Figure 20-12 illustrates this notion. Three main time intervals can be considered: (1) ambient, predisturbance operating conditions, (2) increasing oscillations following fault inception, and (3) decaying oscillations.

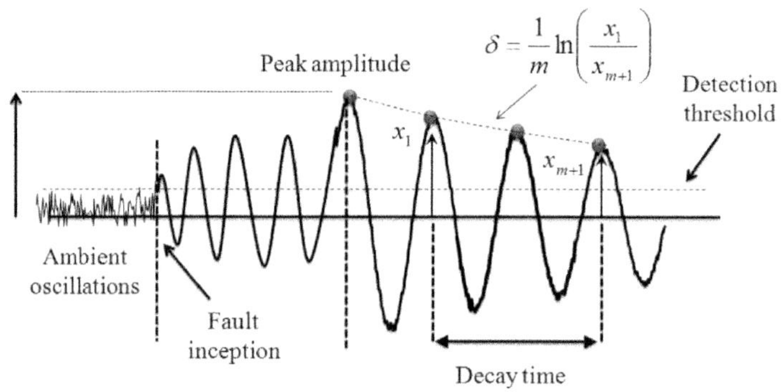


**FIGURE 20-12** Basic parameters for disturbance detection.

*Decay Time.* For a stable system, the oscillations gradually decay with the *envelope* of maximum amplitudes following the dotted curve. Typical time constants for acceptable performance range from 7.0 to 12 s [23] following credible contingency events. According to the above expression, a time constant of 12 s, corresponds to a damping constant with magnitude greater than −0.083 Np/s and a 2% settling time shorter than 47 s.

***Factors Influencing System Damping.*** System damping is influenced by many parameters, including network structure, control actions, and operating conditions. Damping is usually small and can be made negative by the actions of controls, leading to growth of oscillations. Feedback control actions of automatic voltage regulators and speed governors and FACTS devices are generally recognized to have the potential to contribute to negative damping of power swings.

*Damping, Basic Control Idea.* Damping may be affected considerably by control action and control structure. To illustrate how control actions can influence system damping, it suffices to consider a simple two-area system with SVC voltage control at an intermediate bus shown in Fig. 20-13*a* [24].

Under small disturbances, the swing equation of the system can be written as

$$2H\frac{d\Delta\hat{\omega}_r}{dt} \approx \Delta P_m - \frac{\partial P_e}{\partial \delta}\Delta\delta - \frac{\partial P_e}{\partial X_e}\Delta X_e - \frac{\partial P_e}{\partial E'_q}\Delta E'_q - \frac{\partial P_e}{\partial \beta_{\text{svc}}}\Delta\beta_{\text{svc}} - K_D\Delta\hat{\omega}_r \tag{20-24}$$

where $\beta_{\text{svc}}$ is the variable SVC susceptance and the other symbols have the usual meaning. It follows from physical considerations that the transmitted power and transmission bus voltage are given by

$$P_e = \hat{f}_P(\delta,\beta_{\text{svc}}) = \frac{E'_q V_2 \sin\delta}{X_e}\ ; \quad V_m = \hat{f}_v(\delta,\beta_{\text{svc}}) = \frac{1}{X_e}\sqrt{(X_2E'_q\cos\delta + X_1V_2)^2 + (X_2E'_q\sin\delta)^2}$$

where $X_e = (X_1 + X_2 - X_1X_2\beta_{\text{svc}})$.

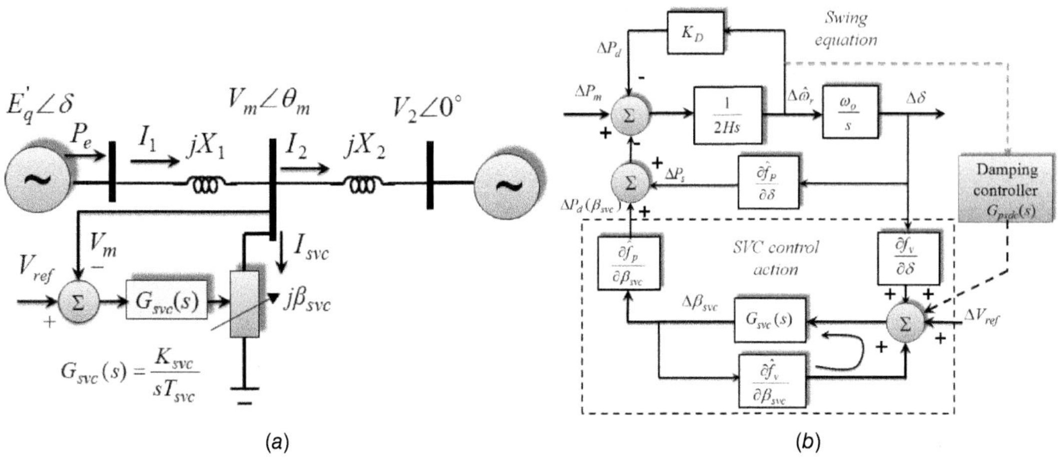


**FIGURE 20-13** Simple test system with SVC voltage support and its block-diagram representation.

Three basic control strategies can be considered: Generator excitation control, constant SVC voltage support, and modulation of the SVC bus voltage. Attention is focused on the effect of SVC voltage support on system damping ($\Delta E_q' = 0$).

*SVC in voltage control mode.* With this control strategy, $V_m = V_{\text{ref}}$. The state representation is

$$\dot{\mathbf{x}} = \begin{bmatrix} \Delta\dot{\delta}(t) \\ \Delta\hat{\omega}_r(t) \end{bmatrix} = \begin{bmatrix} 0 & \omega_o \\ -\dfrac{K_S}{2H} & -\dfrac{K_D}{2H} \end{bmatrix} \begin{bmatrix} \Delta\delta(t) \\ \Delta\hat{\omega}_r(t) \end{bmatrix} + \frac{1}{2H} \begin{bmatrix} 0 \\ \Delta P_m \end{bmatrix}$$

where $K_S = \partial P_e / \partial \delta = E_q' V_2 \sin\delta / X_e$.

*Effect of SVC Voltage Support.* Under closed-loop control, the SVC output can be defined symbolically by $\Delta\beta_{\text{svc}} = G_{\text{svc}}(s)[\Delta V_{\text{ref}}(s) - \Delta V_m(s)]$, where $G_{\text{svc}}(s)$ is the SVC's transfer function, and

$$\Delta V_m = \left.\frac{\partial \hat{f}_v(\delta, \beta_{\text{svc}})}{\partial \delta}\right|_{\substack{\delta=\delta^o \\ \beta_{\text{svc}}=\beta_{\text{svc}}^o}} \Delta\delta + \left.\frac{\partial \hat{f}_v(\delta, \beta_{\text{svc}})}{\partial \beta_{\text{svc}}}\right|_{\substack{\delta=\delta^o \\ \beta_{\text{svc}}=\beta_{\text{svc}}^o}} \Delta\beta_{\text{svc}}$$

In this case, the overall state representation is of the form

$$\begin{bmatrix} \Delta\dot{\delta} \\ \Delta\dot{\hat{\omega}}_r \\ \Delta\dot{\beta}_{\text{svc}} \end{bmatrix} = \begin{bmatrix} 0 & \omega_o & 0 \\ -K_S/(2H) & -K_D/(2H) & (1/2H)\partial\hat{f}_P/\partial\beta_{\text{svc}} \\ -\dfrac{K_{\text{svc}}}{T_{\text{svc}}}\dfrac{\partial\hat{f}_v}{\partial\delta} & 0 & -\dfrac{K_{\text{svc}}}{T_{\text{svc}}}\dfrac{\partial\hat{f}_v}{\partial\beta_{\text{svc}}} \end{bmatrix} \begin{bmatrix} \Delta\delta \\ \Delta\hat{\omega}_r \\ \Delta\beta_{\text{svc}} \end{bmatrix} + \begin{bmatrix} 0 & 0 & 0 \\ 0 & \dfrac{1}{2H} & 0 \\ 0 & 0 & \dfrac{K_{\text{svc}}}{T_{\text{svc}}} \end{bmatrix} \begin{bmatrix} 0 \\ \Delta P_m \\ \Delta V_{\text{ref}} \end{bmatrix}$$

Figure 20-13*b* shows a conceptual representation of the system model. Two control loops can be easily recognized: (a) the electromechanical loop, and (b) the SVC control loop [24, 25]. Several points can be learned from this simple model. (1) With the SVC in voltage control mode, $\Delta\beta_{\text{svc}} = 0$, the SVC provides, essentially, synchronizing power (the SVC contribution is in phase with the rotor angle deviation and damping contribution is rather limited); (2) the SVC control loop introduces a frequency-dependent phase lag; and (3) damping can be increased by introducing a supplementary damping control loop, $G_{psdc}(j\omega)$. The contribution of the SVC to the damping/synchronizing power is given by

$$\Delta P_d(\beta_{svc}) = \frac{\partial \hat{f}_P}{\partial \beta_{svc}} \frac{G_{svc}(s)}{(1 - G_{svc}(s)\frac{\partial \hat{f}_v}{\partial \beta_{svc}})} G_{psdc}(s)\Delta\omega_r$$

Damping contribution in complex systems involves two main mechanisms: (1) A change in the SVC's susceptance changes the terminal voltage $V_t$ and results in changes in power transfers. This in turn, may affect the electric output of the generators and rotor acceleration; (2) suitable modulation of the SVC bus voltage may be result in a torque component in phase with the rotor speed deviations thus indirectly damping system oscillations. In this context, the control synthesis task involves two main subtasks: (1) computing the phase shift introduced by the SVC control action, and (2) designing the PSS damping controller $G_{psdc}$ to cancel the phase shift.

***Controller Design Methodology.*** As discussed above, the design of feedback controllers to damp power system oscillations involves four main subtasks: (1) The identification of oscillatory modes that are most controllable and observable by the controllers and the machines and devices associated with the oscillations; (2) the selection of measurements and controlled variables; (3) the selection of the control structure interconnecting the measured and controlled variable; and (4) the design of system controllers.

*Control Structure.* Study experience shows that wide-area power system controllers may enhance the ability of the power system to control electromechanical oscillations. Control structure, however, can affect control performance and result in adverse interaction problems. The class of control strategies discussed here is divided into three distinct subclasses: *centralized control, partially decentralized control structure,* and *hierarchical control.* The reader is referred to Refs. [26, 27] for more details about a broader class of control structures.

Insight into the problem of loop interaction can be gleaned from the analysis of the *i*th control loop in the linear system representation. From Eqs. (20-16) and (20-17), the dynamic behavior of the *i*th dynamic subsystem or device is given by

$$\dot{\mathbf{x}}_i = \mathbf{A}_{D_i} x_i + \mathbf{B}_{D_i} \Delta \mathbf{v}_i + \mathbf{B}_{u_i} \mathbf{u}_i = \underbrace{\mathbf{A}_{D_{ii}} \mathbf{x}_i}_{\text{Local feedback}} + \underbrace{\sum_{\substack{i=1 \\ i \neq j}}^{n} \mathbf{A}_{D_{ij}} \mathbf{x}_j}_{\text{Interactions with other subsystems}} + \mathbf{B}_{u_i} \mathbf{u} \tag{20-25}$$

The interaction between the *i*th dynamic device and other controllers is given by the *i*th row of the modal relationship Eq. (20-17), $\Delta \mathbf{v}(s) = (\mathbf{Y}_N + \mathbf{Y}_d)^{-1} \mathbf{W}_d \mathbf{x}(s) = \mathbf{L}\mathbf{x}(s)$, and $\Delta \mathbf{v}_i(s) = \sum_j L_{ij} \mathbf{x}_j(s)$. A more physical interpretation is that the output measurement $\Delta \mathbf{v}_i$ will only affect the controlled input via its controlled law if $L_{ij} = 0$ for all $j \neq i$.

In dealing with interacting power system controllers two major aspects must be addressed: the first aspect is the study of the extent of interaction between the control loops, and the second aspect involves assessment of the nature of this interaction. The overall design of decentralized controllers can be broadly divided into two main stages [26]: (1) The selection of feedback signals for the controllers, or pairings of inputs and outputs; and (2) The design of supplemental power system damping controllers.

With the advent of renewable and dispersed generation and the increasing complexity of modern power systems, there has been a surge of interest in utilizing hierarchically structured, distributed control systems for enhancing global behavior.

*Decentralized Control Structures.* In practice, control structures are decentralized or semi-decentralized. Let $\mathbf{G}(s)=[g_{ij}(s)]$ be the open-loop transfer matrix from $\mathbf{u}(s)$ to $\mathbf{y}(s)$ and $\mathbf{G}_c(s)=\text{diag}[g_{c_j}(s)], j = 1, \ldots, k$ be the controller transfer function matrix. A control system is said to be decentralized if the off-diagonal terms of $\mathbf{G}_c(s)$ are zero. Clearly, if, $g_{ij}(s)=0, (i \neq j)$ each controller $g_{c_i}(s)$ can be designed independently for the isolated subsystem $g_{ii}(s)$ without any effect on the other subsystems.

From Eq. (20-4), let the transfer function relating the $i$th output to the $j$th input be expressed as

$$\frac{y_i(s)}{u_j(s)} = g_{ij}(s) = \mathbf{c}_i^T(s\mathbf{I}-\mathbf{A})^{-1}\mathbf{b}_j = \hat{\mathbf{c}}_i^T(s\mathbf{I}-\Lambda)^{-1}\hat{\mathbf{b}}_j = \sum_{k=1}^{n}\frac{\mathbf{c}_i^T\phi_k\psi_k^T\mathbf{b}_j}{s-\lambda_k} = \sum_{k=1}^{n}\frac{R_{ij}^k}{s-\lambda_k} \tag{20-26}$$

Several observations are of interest here

- The quantities $\mathbf{c}_i^T\phi_k$ and $\psi_k^T\mathbf{b}_j$ determine, respectively, the outputs participating in the response of each mode (observability) and the state variables in each mode that are affected by each input (controllability).
- The residue $R_{ij}^k = \mathbf{c}_i^T\phi_k\psi_k^T\mathbf{b}_j$ of $g_{ij}(s)$ associated with the eigenvalue $\lambda_k$ is independent of the scaling of the eigenvector and provides an unambiguous mode-selection criteria.
- Large off-diagonal terms $g_{ij}(s)$ in the transfer function matrix, $\mathbf{G}(s)$ and large terms in the transfer function residue matrix, are associated with strong coupling between control loops. This may result in performance degradation under decentralized control.

Most closed-loop, output-feedback power system controllers use local information to provide system stabilization. In this case, the residue $R_{ij}^k$ can be used for designing system controllers and assessing the effect of critical system modes on system dynamic behavior. Formally, it can be shown that the shift in a selected rotor mode is given by $\Delta\lambda_k = R_{ij}^k g_{c_i}(\lambda_k)\Delta K_{\text{pss}}$, where $g_{c_i}(\lambda_k)$ is the transfer function of the damping controller, and $K_{\text{pss}}$ is the power system stabilizer gain.

As a demonstration of this procedure, Fig. 20-14 compares the reactive power response to a step change in the voltage setpoint of an actual 300 MW hydro unit for the case with no PSS and a designed PSS using the above procedure. From Fig. 20-14 is seen that proper tuning of the PSS can significantly enhance damping of the critical mode of concern.

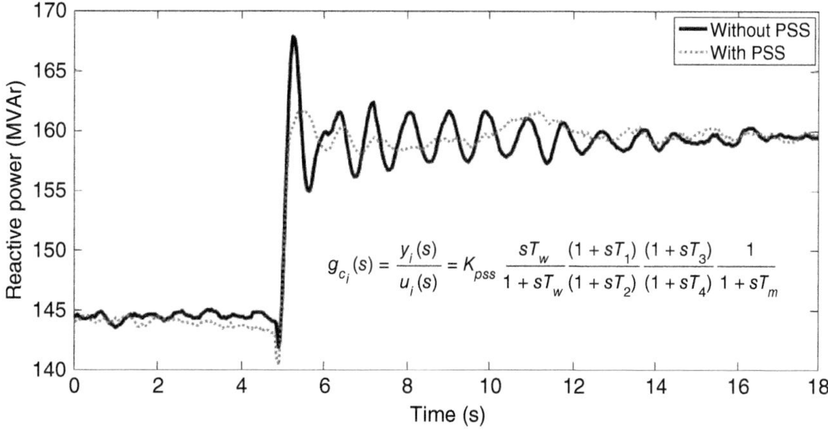


**FIGURE 20-14** Reactive power response to a 5% step change in terminal voltage at a 300-MW hydro generator.

*Hierarchical Control Architectures.* Hierarchical control structures, with multiple levels of control strategies acting at different scales are well suited to achieve control requirements of large interconnected power systems.

Figure 20-15 shows a general conceptual representation of a power system control structure using both, local and hierarchical control structures in which $m$ wide-area damping controllers (WADCs) are used to enhance damping of the system.

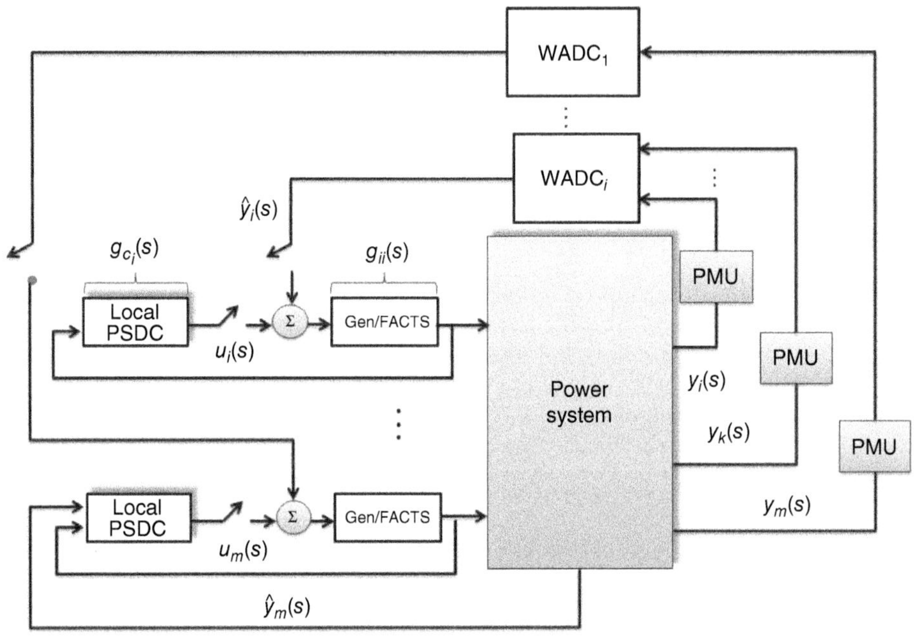


**FIGURE 20-15** General hierarchical control architecture.

At the primary control level, local measurements are used to enhance damping of local modes. At the wide-area level, the input signals to the hierarchical controllers are provided from a WAMS. The outputs of the WADCs are used to modulate the reference signals of excitation systems, FACTS controllers and HVDC systems. Finding suitable pairings from block-decentralized controllers is critical to the whole design procedure due to the interactions that may occur between the input-output variables.

## 20.1.6 Bibliography


1. P. Kundur, J. Paserba, V. Ajjarapu, G. Andersson, A. Bose, C. Canizares, N. Hatziargyriou, et al., (IEEE = CIGRE Joint Task Force on Stability Terms and Definitions), Definition and classification of power system stability, *IEEE Trans. Power Systems*, August 2004.
2. P. Kundur and G. K. Morison, "A Review of Definitions and Classification of Stability Problems in Today's Power Systems," *Panel Session on Stability Terms and Definitions, IEEE PES Meeting*, New York, Feb. 1997.
3. CIGRE Task Force 38.01.07 on Power System Oscillations, Analysis and Control of Power System.
4. "Inter-Area Oscillations in Power Systems," Prepared by Systems Oscillations Working Group, IEEE Technical Publication 95 TP 101.

5. A. A. Andronov, A. A. Vitt, and S. E. Khaikin, *Theory of Oscillators*, Dover Publications Inc., New York, 1966.
6. P. Kundur, *Power Systems Stability and Control*, Electric Power Research Institute, Power System Engineering Series, McGraw-Hill Inc., 1994.
7. IEEE Committee Report, "Terms, Definitions and Symbols for Subsynchronous Oscillations," *IEEE Trans. on Power Apparatus and Systems*, vol. PAS-104, June 1985, pp. 1326–1344.
8. K. Clark, "Overview of Subsynchronous Resonance Related Phenomena," IEEE General Meeting, 2012.
9. CIGRE Task Force 38.02.16, Impact of the Interaction among Power System Controllers, CIGRE Technical Brochure No. 116, 2000, N. Martins, Convenor.
10. E. V. Larsen, and W. W. Price, "Manstab/Possim Power System Dynamic Analysis Programs—A New Approach Combining Nonlinear Simulations and Linearized State-Space/Frequency Domain Analysis," 1977 Power Industry Computer Applications Conference, pp. 350–358.
11. Philip E. Gill, Walter Murray, and Margaret H. Wright, *Practical Optimization*, Academic Press, Inc, San Diego, CA, 1981, IEEE PES Special Publication 95-TP-101, Inter-area Oscillations in Power Systems, 1995.
12. Jonas Persson and Lennart Soder, "Comparison of Three Linearization Methods," 16th Power Systems Computation Conference (PSCC), Glasgow, Scotland, July, 2008.
13. Raymon A. DeCarlo, *Linear Systems—A State Variable Approach with Numerical Implementation*, Prentice Hall, Englewood Cliffs, 1989.
14. Vijay Vittal and A. R. Messina, *Normal Form Analysis of Power Systems*, PowerLearn, 2004.
15. J. Sanchez-Gasca, V. Vittal, M. Gibbard, A. Messina, D. Vowles, S. Liu, and U. Annakkage, "Inclusion of Higher-Order Terms for Small-Signal (Modal) Analysis: Committee Report—Task Force on Assessing the Need to Include Higher-Order Terms for Small-Signal (Modal) Analysis," *IEEE Trans. Power Systems*, vol. 20, no. 4, pp. 1886–1904, Nov. 2005.
16. S. M. Chan, "Modal Controllability and Observability of Power System Models," *Electrical Power and Energy Systems*, vol. 6, no. 2, pp. 83–88, Apr. 1984.
17. J. I. Pérez-Arriaga, G. C. Verghese, and F. C. Schweppe, "Selective Modal Analysis with Applications to Electric Power Systems. Part I: heuristic introduction. Part II: the dynamic stability problem," *IEEE Trans. on Power Apparatus and Systems*, vol. PAS-101, pp. 3117–3134, 1982.
18. M. A. Magdy, and F. Coowar, "Frequency Domain Analysis of Power System Forced Oscillations," *IET Proceedings*, Pt. C, no. 4, pp. 261–268, Jul. 1990.
19. J. E. Van Ness, "Response of Large Power Systems to Cyclic Load Variations," *IEEE Trans. on Power Apparatus and Systems*, vol. PAS-85, no. 7, pp. 723–727, July 1966.
20. Nezam Sarmadi, S. Arash, and Vaithianathan Venkatasubramanian, "Inter-Area Resonance in Power Systems from Forced Oscillations," *IEEE Trans. on Power Systems*, vol. 31, no. 1, pp. 378–386, Jan. 2016.
21. J. J. Sanchez-Gasca and V. Vittal (eds.), "Identification of Electromechanical Modes in Power Systems," Technical Report PES-TR15, IEEE Power & Energy Society, Jun. 2012.
22. J. F. Hauer and F. Vakili, "An Oscillation Detector Used in the BPA Power System Disturbance Monitor," *IEEE Trans. on Power Systems*, vol. 5, no. 1, pp. 7479, 1990.
23. D. J. Vowles, C. Samarasinghe, M. J. Gibbard, and G. Ancell, "Effect of Wind Generation on Small-Signal Stability—A New Zealand Example," *Power and Energy Society General Meeting*, 2008 IEEE, Pittsburgh, US.
24. E. V. Larsen and J. H. Chow, "SVC Control Design Concepts for System Dynamic Performance, Application of Static VAR Systems for System Dynamic Performance," IEEE Special Publication No. 87TH1087-5-PWR on Application of Static Var Systems for System Dynamic Performance, pp. 36–53, 1987.
25. Wei Xuan, Joe H. Choe, and Juan J. Sanchez-Gasca, "On the Sensitivities of Network Variables for FACTS Device Damping Control," *IEEE Power Engineering Society Winter Meeting*, pp. 1188–1193, Jan. 2002.
26. E. N. Reyes, M. A. Messina, and M. A. Pérez, "Design of Wide-Area Damping Controllers Using the Block Relative Gain," Electric Power System Research, 126, pp. 56–67, 2015.
27. Manfred Morari and Evanghelos Zhafiriou, *Robust Process Control*, Prentice Hall, Englewoods Cliff, 1989.

## 20.2 MEASUREMENT-BASED VOLTAGE STABILITY ASSESSMENT TECHNIQUES

*BY KAI SUN*

Growths in electrical energy consumptions and penetration of intermittent renewable resources energy would often make power transmission systems operate closer to their stability limits. Among all stability issues, voltage instability due to the inability of the transmission system to deliver power to loads is one of major concerns in power system operations. Voltage instability usually originates from a local bus but may spread out to a larger area potentially causing a system-wide instability.

At present, electric utilities use model-based online voltage stability assessment (VSA) software tools to assist operators in foreseeing potential voltage instability under a disturbance. Based on the online state estimation of the system, those VSA tools employ power system models to simulate presumed contingencies or load variations. Such a model-based VSA approach has several limitations. First, it needs a convergent system state estimate as the starting point of simulation and stability assessment, which may be hard to obtain under a stressed system condition. Second, the fidelity of its results highly depends on whether the system model used is accurate. Usually, the time-domain simulation on a voltage instability scenario needs to consider a longer period than a transient stability simulation. The reason is that the initial stage of a voltage instability scenario often involves slow dynamics or switches on devices such as shunt compensators and tap changing transformers. If those devices are also modeled and simulated, the simulation will be time consuming for a large system and hence not suitable for online applications.

Many countries deploy GPS-synchronized phasor measurement units (PMUs) on transmission systems to provide wide-area measurements for real-time stability monitoring. That leads to more interests in developing data driven VSA methods that utilize real-time measurements to directly assess voltage stability on selected buses.

### 20.2.1 Concepts on Voltage Stability

***Definitions and Classifications.*** Voltage stability refers to "the ability of a power system to maintain steady voltages at all buses in the system after being subjected to a disturbance from a given initial operating condition" [1]. In contrast to generator rotors in angular stability of a power system, the driving forces for voltage instability are mainly loads and the means by which voltage or reactive power are controlled to support loads.

Occurring more in a heavily stressed power system, voltage instability becomes one of the major concerns in today's system operations and an impacting factor on the power transfer capability of a transmission network. When voltage instability leads to loss of voltage in a significant part of the system, such a process is called *voltage collapse*, in which many bus voltages in the system decay to an unrecoverable level. As a consequence of voltage collapse, the system may experience a power blackout over a large area. A system restoration procedure would then be taken to bring the blackout area back to service.

The literature often classifies voltage stability by the severity of the disturbance for the ease of analyzing causes and phenomena:

1. *Small-disturbance voltage stability* is concerned with a system's ability to control voltages following small perturbations such as gradual load increases. Voltage stability problems of this type can be studied effectively by either steady-state analysis on a power flow model or eigenanalysis on a simplified, linearized system model.
2. *Large-disturbance voltage stability* is concerned with a system's ability to control voltages following large disturbances such as system faults, line outages, loss of load or loss of generation. A study on voltage stability of this type usually requires examination of the dynamic performance of

the system over a period of time that is sufficient to capture the dynamics or switches of voltage control devices such as generator field current limiters, tap changing transformers and shunt compensators. Therefore, nonlinear dynamics of the system need to be analyzed or simulated using a more detailed system model.

Another perspective for classification is the time frame of voltage instability and collapse dynamics, which varies from tens of minutes to a few seconds. Conventionally, electric utilities categorize voltage stability problems as follows:

1. *Long-term voltage stability* involves slower acting equipment such as tap-changing transformers, thermostatically controlled loads and generator field current limiters. Voltage stability problems of this category usually take several minutes to several hours to occur and are usually studied by static analysis techniques with complementary use of dynamic analysis.
2. *Short-term or transient voltage stability* involves dynamics with fast acting load components such as induction motors, electronically controlled loads and HVDC links under contingencies near loads. Voltage stability problems of this category take several seconds or even shorter time to occur.

The above classifications are helpful for simplification of complicated voltage instabilities, identification of the causes and selection of the right tools for voltage stability analysis. However, any realistic voltage instability is essentially a nonlinear dynamic system problem. As a result, the analysis from a perspective of any single category may not be sufficient to understand voltage instability in the real-time operation environment.

***P-V* curve and *V-Q* curve.** As shown in Fig. 20-16, a *P-V* curve is a useful tool for conceptual, steady-state analysis of voltage stability, especially for long-term voltage stability. It is originally developed for a radial system to study how the voltage magnitude $|V|$ of one bus on the system decays with the increase of active power $P$ transferred through the bus. The upper portion of a *P-V* curve can be drawn by power-flow simulation that increases load from an initial light-load operating point until the maximum power is reached. The maximum power point is also called the "nose" or "knee" point of the curve. For a real power system model, power flow simulation usually diverges near the "nose" point, which leads to voltage collapse [2–4].

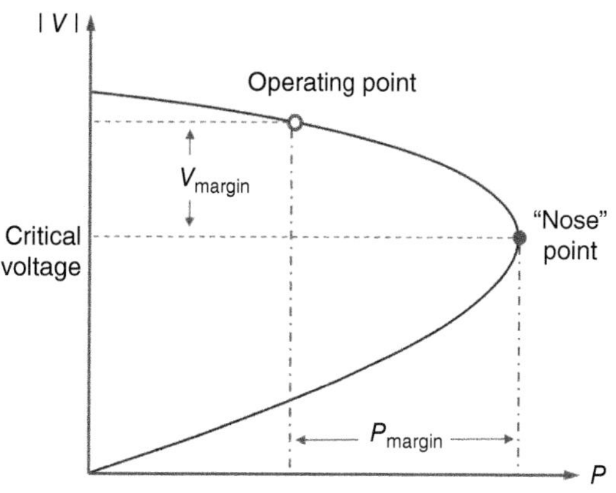


**FIGURE 20-16** *P-V* curve.

In order to obtain the lower portion of the *P-V* curve under the "nose" point, the continuation power flow (CPF) is often used [5]. Projecting the distance between the initial operating point and the "nose" point onto the $|V|$ axis and $P$ axis will give the voltage stability margins respectively in terms of voltage and active power as indicated by $V_{margin}$ and $P_{margin}$ in Fig. 20-16.

For a large meshed network, *P-V* curves are also used by engineers. $P$ could be the total load of a load area, which is assumed to increase in one specific way, for example, all loads of the area being evenly increased, and $|V|$ is often selected to be the voltage of the most critical bus or a representative bus.

Of course, multiple *P-V* curves may be plotted for different buses and for different ways of load increase. However, for a large system, it is difficult to assess voltage stability by *P-V* curves because there are usually a wide variety of possible changes on loads. In addition, associated generators that are supporting the load area must also be realistically redispatched to match any change of the area load in creation of *P-V* curves.

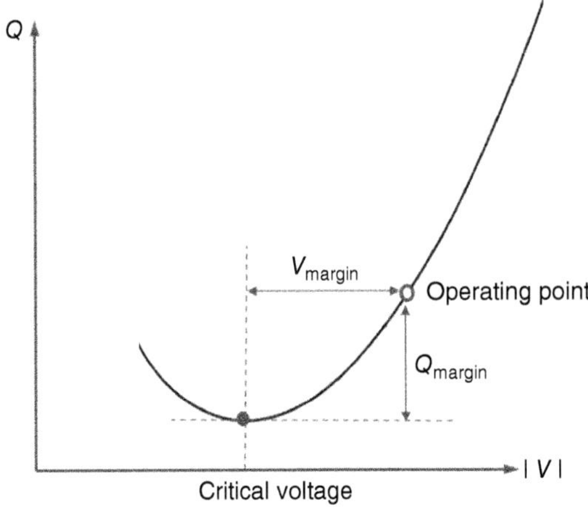


**FIGURE 20-17** *V-Q* curve.

As illustrated in Fig. 20-17, a *V-Q* curve is also often used to plot the voltage magnitude ($|V|$) at a critical bus versus the reactive power injected ($Q$) to the same bus. The bottom point of the curve corresponds to the "nose" point of the *P-V* curve, so a voltage-stable operating point should be on the right portion of the curve with a sufficient margin to the bottom point. Another indicator of voltage stability margin at an operating point is the steepness of the slope $\Delta Q/\Delta|V|$ of the *V-Q* curve. Once that slope decreases to zero, the voltage will dramatically drop to cause voltage instability.

***Voltage Stability Margin.*** Voltage stability margin of an operating point is its distance to the "nose" point on the *P-V* curve. For instance, the stability margin $P_{margin}$ in terms of active power is defined as the distance between the active power of the operating point and the loadability limit; that is, the active power at the "nose" point, as illustrated in Fig. 20-16. Usually, the voltage stability margin refers to the loadability at one load bus. However, it can also be utilized for the interface or tie lines of a load area. If the voltage stability margin is smaller than a certain threshold, a preventive action should be taken to avoid potential voltage instability.

The indices on voltage stability margin can be regarding a pre- or post-contingency condition, which are often called the "$n - 0$" and "$n - 1$" conditions, respectively, by the industry:

1. *Voltage stability margin for the "$n - 0$" condition* considers one specific system topology to be stressed in an anticipated way of load increase together with generation redispatch until the loadability limit is reached, so the margin equals the distance from the operating point of interest on the *P-V* curve to the maximum power point. If applied in the operation environment, this type of margin indices provide the real-time voltage stability margin for the current operating condition.
2. *Voltage stability margin for the "$n - 1$" condition* provides the smallest voltage stability margin on the *P-V* curves about a list of postcontingency conditions such as line outages and generation outages, and is often evaluated by power flow "$n - 1$" simulations.

Depending on whether steady-state analysis or dynamic analysis is employed, model-based VSA techniques adopt either a power flow model or a dynamic model of the system to evaluate voltage stability margin for "$n - 0$" and "$n - 1$" conditions and predict voltage collapse. Many factors influence voltage stability margin for both "$n - 0$" and "$n - 1$" conditions, such as adjustments by transformer tap changers and variations of reactive power outputs from generators and other reactive power sources. Therefore, real-time, accurate voltage stability margin information is critical for system operators to be aware of any potential or already developing voltage instability. However, it is challenging for model-based VSA to provide such real-time margin information especially for a large system. Many generators, busses and transmission lines have to be modeled and also lots of uncertainties need to be addressed such as load variations and intermittent behaviors of renewable generation.

### 20.2.2 Voltage Stability Assessment and Model-Based Techniques

***Modal Analysis.*** For small-disturbance voltage stability, the modal analysis conducts eigenvalue analysis based on the power-flow model of a power system [6]. First, the linearized power-flow equation is calculated as below, which utilizes the Jacobian matrix to give sensitivity of the incremental changes in bus real powers (**ΔP**) and reactive powers (**ΔQ**) to the incremental changes in bus voltage angles (**Δθ**) and voltage magnitudes (**ΔV**).

$$\begin{bmatrix} \Delta \mathbf{P} \\ \Delta \mathbf{Q} \end{bmatrix} = \begin{bmatrix} \mathbf{J}_{P\theta} & \mathbf{J}_{PV} \\ \mathbf{J}_{Q\theta} & \mathbf{J}_{QV} \end{bmatrix} \begin{bmatrix} \Delta \boldsymbol{\theta} \\ \Delta \mathbf{V} \end{bmatrix} \tag{20-27}$$

Let $\Delta \mathbf{P} = 0$, there is

$$\Delta \mathbf{V} = \mathbf{J}_R^{-1} \Delta \mathbf{Q} \qquad \text{where } \mathbf{J}_R = \mathbf{J}_{QV} - \mathbf{J}_{Q\theta} \mathbf{J}_{P\theta}^{-1} \mathbf{J}_{PV} \tag{20-28}$$

$\mathbf{J}_R$ is the reduced Jacobian matrix of the system. It represents the approximate, linearized relationship between the incremental changes in bus voltage magnitudes (**ΔV**) and bus reactive power injections (**ΔQ**). Voltage stability characteristics of the system can be identified by computing the eigenvalues (i.e., modes) and eigenvectors (i.e., mode shapes) of $\mathbf{J}_R$ and following these criteria:

1. Only when all eigenvalues are positive, the system is voltage stable because the modal voltage deviation $\Delta \mathbf{V}$ and modal reactive power variation $\Delta \mathbf{Q}$ regarding each eigenvalue are along the same direction so as to make reactive power control be effective.
2. For a negative eigenvalue, the corresponding modal voltage cannot be improved by adjusting the modal reactive power and hence is unstable.
3. For a zero eigenvalue, $\mathbf{J}_R$ becomes singular and the corresponding modal voltage collapse since any small change in the modal reactive power causes infinite change in the modal voltage.

The *V-Q* curve introduced by Fig. 20-17 plots the characteristics of Eq. (20-28) about a single bus having reactive power injection vary over a range. The eigenvectors of $\mathbf{J}_R$ regarding a mode indicate the bus participation factors, and tell the critical buses and areas associated with that mode. The modal analysis discovers that the voltage collapse is essentially the collapse of the voltage in the critical mode, the mode that corresponds to the eigenvalue having the minimum magnitude. Although that magnitude gives the proximity or margin to voltage instability regarding the critical mode, it is not a good indicator for online applications due to these two reasons: first, it is based on a linearized power-flow model and hence suitable only for analysis on long-term or small-disturbance voltage stability; second, it does not explicitly provide actionable information for system operators.

***Singular Value Decomposition.*** Since the singularity of the reduced Jacobian matrix $\mathbf{J}_R$ indicates occurring voltage instability, a quantitative value on its singularity indicates voltage stability margin. The singular value decomposition (SVD) method is generally used for factorization of a matrix [7]: $\mathbf{M} = \mathbf{U} \times \Sigma \times \mathbf{V}^*$, where **U** and **V** are unitary matrices and Σ is a diagonal matrix with nonnegative real numbers the diagonal, which are called singular values. This method can be applied to $\mathbf{J}_R$ to determine the minimum singular value, which corresponds to the critical mode. If the minimum singular value is equal to zero, the system becomes voltage unstable regarding that mode [8].

***Continuation Power Flow.*** The maximum loadability of a power system can be determined by increasing the loads in a specific way and computing power flows until singularity appears making the power flow unsolvable, where the maximum power point, the "nose" point, is reached from the upper portion of the *P-V* curve. However, the lower portion of the *P-V* curve can hardly be obtained by conventional power flow analysis. Continuation power flow (CPF) is a modification of conventional power flow analysis for finding a static voltage stability margin, $P_{margin}$ or $V_{margin}$, and it overcomes the singularity problem by reformulating the power flow equations so that they can remain well-conditioned even at a loading condition near the "nose" point [5].

As illustrated in Fig. 20-18, starting from an initial power flow solution 1, the general principle of the CPF is to utilize a tangent predictor followed by a vertical or horizontal corrector to determine a series of power flow solutions 3, 5, 7, etc. until the complete *P-V* curve is plotted. Each predictor step

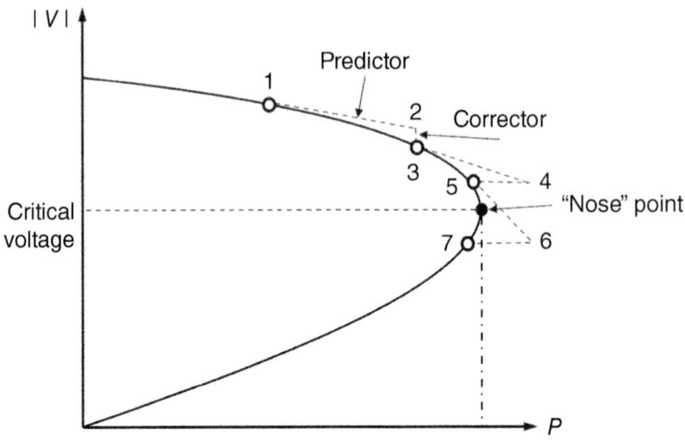


**FIGURE 20-18** Calculations of CPF.

(e.g., 1 → 2, 3 → 4, and 5 → 6) assumes a specific pattern of load increase, following which a vertical corrector step (e.g., 2 → 3) determines the exact solution with the load fixed by conventional power flow analysis. Once the predicted new load (e.g., 4 and 6) is beyond the maximal power, a horizontal corrector step (e.g., 4 → 5 and 6 → 7) with the voltage fixed should be used instead of a vertical corrector to give the exact solution near or below the "nose" node. The CPF is currently supported by some model-based VSA tools for online applications [9]. However, its computation time could be a concern in online applications.

***Dynamic Security Assessment.*** The dynamic security assessment (DSA) performs offline or online time-domain simulation on a list of preselected contingencies to find which may cause transient instability under a specific steady-state power-flow condition. A DSA program can also be used to assess voltage stability if dynamics of the system are concerned especially for large-disturbance, short-term voltage stability problems. The time-domain simulation estimates the system response following a contingency by solving a set of nonlinear differential-algebraic equations that model generators, the transmission network, control devices and static and dynamic loads of the power system. The traditional DSA for rotor angle stability models generators and auxiliary control devices in detail but often represents loads by the ZIP (constant impedance + constant current + constant power) model. Nevertheless, a DSA program for voltage stability studies need to give sufficient considerations to the dynamics on the load side, especially for motor loads and electronically controlled loads, and to controls or switches related to reactive power, for example, the current limiters of generators, transformer tap changers, and reactive power supports. Dynamics such as frequency regulation that are not directly related to voltage stability can be moderately approximated in the simulation models.

A recently increasing operation concern on voltage stability is the FIDVR (fault-induced delayed voltage recovery) issue, which sometime happens in summer to a transmission system with a high percentage of induction motor loads, mainly air conditioners [10]. Those motors are considered "prone to stall" motors and may draw lots of reactive power when they attempt to recover from stalling so as to delay a system voltage recovery and lead to voltage collapse. To simulate FIDVR issues, electricity utilities usually conduct time-domain simulation with detailed models on induction motors and other dynamic loads.

There are increasing needs for online DSA tools to foresee not only rotor angle instability but also voltage instability for predefined "*what-if*" scenarios, critical contingencies influencing the system strongly with high occurrence probability of the voltage collapse. However, the accuracy of such a model-based approach can be influenced by the following factors:

1. The time-domain simulation requires accurately models of all system components related to voltage stability and reactive power control. The use of inaccurate models may influence the creditability of the final VSA results. However, in practice, model validations for both generation and load in a timely manner are not trivial efforts for power system engineers.
2. The simulation is based on a system operating condition, a steady-state power flow solution. Such a solution can be provided by the online state estimator but may become unavailable if the state estimation diverges under, for example, a heavily stressed condition with insufficient voltage stability margin.
3. The high computational burden with online simulation is another concern. A time-domain simulation on voltage stability is usually much longer than a typical 10 to 20 s transient stability simulation because the initial stage of a scenario often involves slow dynamics and the switches on, for example, shunt compensators and tap changing transformers, should be modeled well, which makes the simulation be time consuming for a large system and becomes bottleneck for online applications.

Online assessment of voltage stability margin is critical for system operations to prevent voltage instability and take a timely remedial action, so the margin information is expected to be accurate all the time, even when the system is heavily stressed to approach the edge of voltage stability. However, as discussed above, the model-based approach has many limitations. An alternative approach is to directly using real-time measurements to assess voltage stability margin.

### 20.2.3 Measurement-Based Voltage Stability Assessment for a Load Bus

Unlike model-based methods for VSA, measurement-based methods aim at estimation of voltage stability margin in real time using only measurements. Those methods can be applied to either individual load buses using local voltage and current measurements or a load area using synchronized phasor measurements on boundary buses and tie lines of the area.

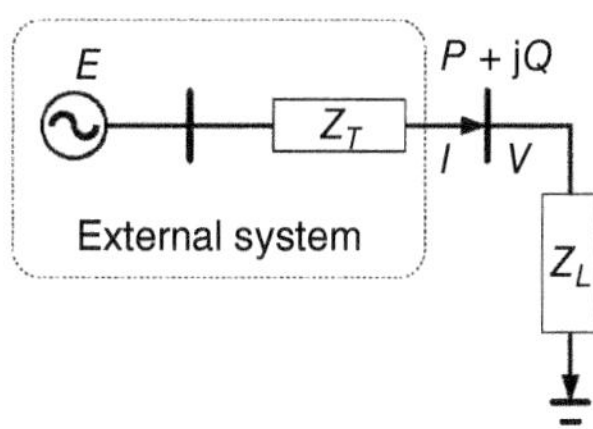


**FIGURE 20-19** Thevenin equivalent for a load bus.

***Thevenin Equivalent-Based Methods.*** A family of measurement-based methods is based on Thevenin's theorem applied to a single load bus [4, 11, 12]. As shown in Fig. 20-19, the external system seen from a load bus can be represented by a Thevenin equivalent (TE), which is a constant voltage source $E$ connected through a Thevenin impedance $Z_T$. The theorem says that the active power $P$ received by the load bus reaches its maximum when the Thevenin impedance has the same magnitude as the load impedance:

$$|Z_L| = |Z_T| \tag{20-29}$$

Typically, the load impedance $Z_L$ has a smaller magnitude than the Thevenin impedance $Z_T$. If no disturbance happens in the external system, $Z_T$ does not change significantly. When the load impedance $Z_L$ decreases its magnitude to increase current as well as active power, the bus voltage magnitude $|V|$ decreases following the characteristics of the $P$-$V$ curve. Once Eq. (20-29) is met, the "nose" point on the $P$-$V$ curve is reached and voltage instability occurs.

The circuit in Fig. 20-19 satisfies Eq. (20-30).

$$E - Z_T I = V \tag{20-30}$$

If Thevenin parameters $E$ and $Z_T$ are assumed to be constant over a time window, they can be estimated from measurements of voltage $V$ and current $I$ over a time window. Let $E = e_r + je_i$, $V = v_r + jv_i$, $I = i_r + ji_i$, and $Z_T = r + jx$. Then Eq. (20-30) becomes Eq. (20-31).

$$\begin{bmatrix} 1 & 0 & -i_r & i_i \\ 0 & 1 & -i_i & -i_r \end{bmatrix} \times \begin{bmatrix} e_r \\ e_i \\ r \\ x \end{bmatrix} = \begin{bmatrix} v_r \\ v_i \end{bmatrix} \tag{20-31}$$

Since $i_r$, $i_i$, $v_r$, and $v_i$ are directly measured, this still leaves four unknowns $e_r$, $e_i$, $r$, and $x$, which require at least two different measurement points from the time window. With $T$ ($T \geq 2$) measurements, the equation is extended to Eq. (20-32).

$$\underbrace{\begin{bmatrix} 1 & 0 & -i_r(1) & i_i(1) \\ 0 & 1 & -i_i(1) & -i_r(1) \\ \vdots & \vdots & \vdots & \vdots \\ 1 & 0 & -i_r(T) & i_i(T) \\ 0 & 1 & -i_i(T) & -i_r(T) \end{bmatrix}}_{\mathbf{H}} \times \underbrace{\begin{bmatrix} e_r \\ e_i \\ r \\ x \end{bmatrix}}_{\mathbf{X}} = \underbrace{\begin{bmatrix} v_r(1) \\ v_i(1) \\ \vdots \\ v_r(T) \\ v_i(T) \end{bmatrix}}_{\mathbf{V}} \tag{20-32}$$

Thus, Thevenin parameters $\mathbf{X} = [e_r, e_i, r, x]T$ can be solved by the method of least squares [4]:

$$\mathbf{X} = (\mathbf{H}^T\mathbf{H})^{-1}\mathbf{H}^T\mathbf{V} \tag{20-33}$$

An alternative approach is the recursive least squares (RLS) method with a forgetting factor [11]. Then, the gap between the estimated $|Z_T|$ and the calculated $|Z_L| = |V/I|$ gives the voltage stability margin in terms of the magnitude of impedance. Consider a two-bus system having the same circuit as in Fig. 20-19, where $E = 1.3\angle 0°$ (p.u.), $Z_T = 0.03\angle 80°$ (p.u.) and the load gradually decreases its impedance magnitude $|Z_L|$ while keeping a constant power factor 0.986. Then, the least squares estimate of $|Z_T|$ gives the lower limit of $|Z_L|$ against voltage instability as shown in Fig. 20-20. Their gap indicates voltage stability margin.

The voltage stability margin indices in terms of the active power $P$ and voltage $V$ are given as follows [12]. Let $Y = 1/Z_T$, that, the Thevenin admittance, and denote the angles of $E$, $V$, and $Z_T$ respectively by $\delta$, $\theta$, and $\beta$. Then, the active power $P$ and reactive power $Q$ transferred from the external system to the load bus are calculated by

$$P = |E \cdot V \cdot Y| \cdot \cos(\theta - \delta + \beta) - |V|^2 \cdot |Y| \cdot \cos\beta \tag{20-34}$$

$$Q = |E \cdot V \cdot Y| \cdot \sin(\theta - \delta + \beta) - |V|^2 \cdot |Y| \cdot \sin\beta \tag{20-35}$$

If the power factor $\phi = \cos[\tan^{-1}(Q/P)]$ of the load bus is constant, $Q$ will change proportionally with $P$, and the maximum of $P$ will be solved as

$$P_{\max} = \frac{|E|^2|Y|\cos\phi}{2[1 + \cos(\phi - \beta)]} \tag{20-36}$$

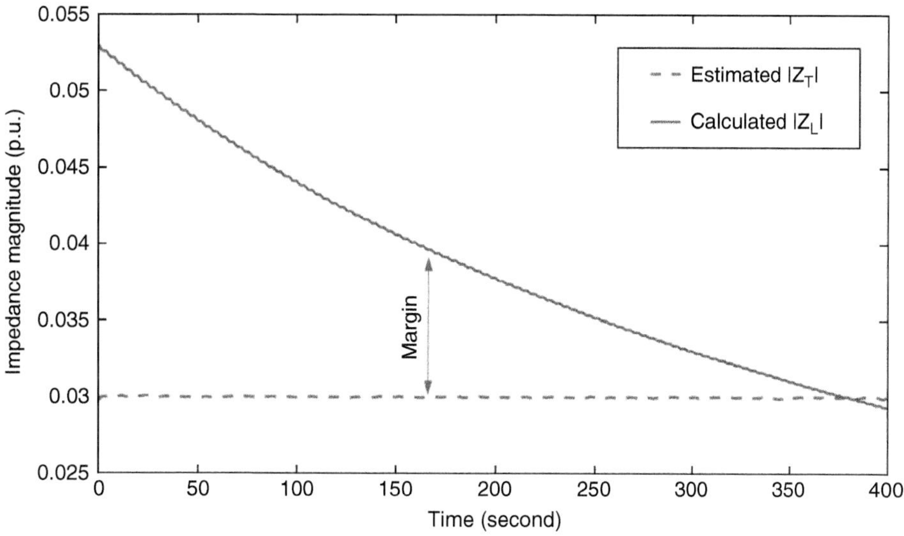


**FIGURE 20-20** Magnitudes of the load impedance and Thevenin impedance.

$P_{\max}$ corresponds to the "nose" point of the $P$-$V$ curve having the critical voltage magnitude

$$V_{\text{critical}} = \frac{|E|}{\sqrt{2[1+\cos(\phi-\beta)]}} \tag{20-37}$$

$V_{\text{critical}}$ is the lowest voltage magnitude the load bus may have before it loses voltage stability. Accordingly, the voltage stability margin indices in terms of the active power and voltage magnitude are

$$P_{\text{margin}} = P_{\max} - P \tag{20-38}$$

$$V_{\text{margin}} = |V| - V_{\text{critical}} \tag{20-39}$$

If the load is represented by a pure resistance and the Thevenin impedance is assumed to be a pure reactance $jX_T$, $P_{\max}$, and $V_{\text{critical}}$ become

$$P_{\max} = \frac{|E|^2}{2X_T} \tag{20-40}$$

$$V_{\text{critical}} = \frac{|E|}{\sqrt{2}} \tag{20-41}$$

For the two-bus system, $P$, $P_{\max}$, and $P_{\text{margin}}$ are shown in Fig. 20-21 and $|V|$, $V_{\text{critical}}$, and $V_{\text{margin}}$ are illustrated in Fig. 20-22.

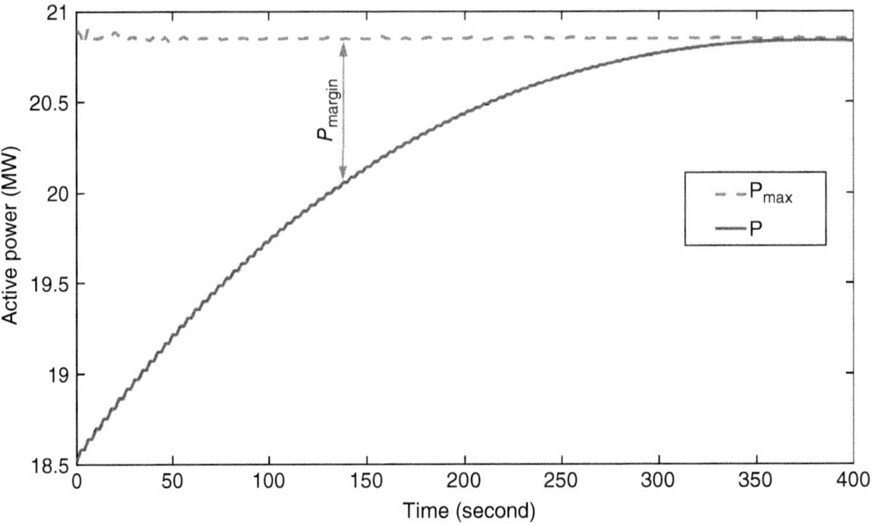


**FIGURE 20-21** The active power and its maximum.

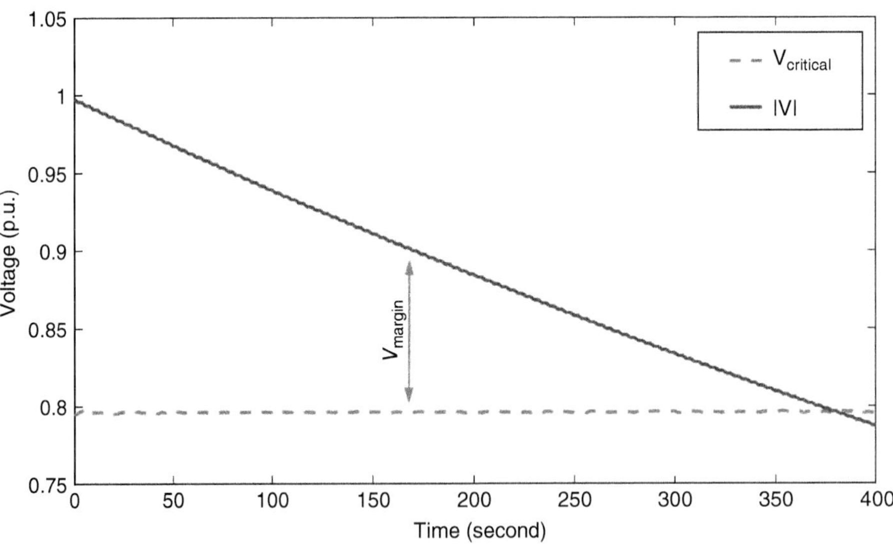


**FIGURE 20-22** The voltage magnitude and critical voltage.

The above method with Eqs. (20-31) to (20-33) assumes $e_r$ and $e_i$ to be two constants for estimation. That is true only when the Thevenin voltage $E$ appears to be a constant phasor over the entire time window for estimation, which may be a too strong assumption in practice. The following algorithm estimates $|E|$ and $Z_T$ using $k$ measurement points of $V$ and $I$ over a time window [13]. The algorithm allows the angle of $E$ and the power factor angle $\phi$ to vary during the window:

1. Let current phasor $I$ be the reference, whose angle at different measurement points is reset to be zero. Then the angle of $V$ relative to $I$ at measurement point $k$ equals the power factor angle $\phi(k)$. Assume $\delta(k)$ to be the angle of $E$ relative to $I$ at measurement point $k$:

$$|E|\angle\delta(k)=|V(k)|\angle\phi(k)+|I(k)|\times Z_T,\quad k=1,\ldots,T \tag{20-42}$$

2. To have a constant $Z_T$ over $T$ measurement points, there must be Eq. (20-43), where $\alpha=\delta(1)-\delta(T)$. Substitute Eq. (20-43) for $Z_T$ in Eq. (20-42), we have Eq. (20-44).

$$Z_T(\alpha)=-\frac{\angle\alpha\times|V(T)|\angle\phi(T)-|V(1)|\angle\phi(1)}{\angle\alpha\times I(T)-I(1)} \tag{20-43}$$

$$E(k)=|V(k)|\angle\phi(k)-|I(1)|\times\frac{\angle\alpha\times|V(T)|\angle\phi(T)-|V(1)|\angle\phi(1)}{\angle\alpha\times I(T)-I(1)} \tag{20-44}$$

3. Find $\alpha$ to minimize the variance of $|E(1)|, \ldots, |E(T)|$ over the time window. Then, $Z_T$ is calculated from Eq. (20-43) and the mean value of $|E(1)|, \ldots, |E(T)|$ gives an estimate of $|E|$.

For the aforementioned two-bus system, if the angle of $E$ has about 5° variation during the period of load increase, the $|Z_L|$ estimated from the least squares method defined by Eqs. (20-31) to (20-33) will contain errors. Figure 20-23 compares that least squares method and the above algorithm given by Eqs. (20-43) and (20-44) regarding the estimated $|Z_T|$ and $P_{max}$. The estimates from the above algorithm are marked by subscript "T2", which is not influenced by the variation in the angle of $E$.

***Voltage Stability Indices.*** As shown in Fig. 20-24, at the maximum power point, the voltage drop $E$-$V$ across the Thevenin impedance has the same magnitude as the voltage phasor at the load bus and hence there is Eq. ( 20-45) [4]:

$$|V|=|E-V|,\qquad |E|=2|V|\cos(\delta-\theta) \tag{20-45}$$

Thus, a voltage stability index (VSI) can be defined to assess the risk of voltage collapse

$$\text{VSI}=\frac{|V|}{|E-V|} \tag{20-46}$$

The proximity to voltage collapse at this load bus is told from how close the VSI is to one. For all load buses, the Thevenin equivalent of the external system can be estimated from measurement data and then such a VSI is calculated in real time. The buses with VSIs closer to one are more vulnerable to voltage instability.

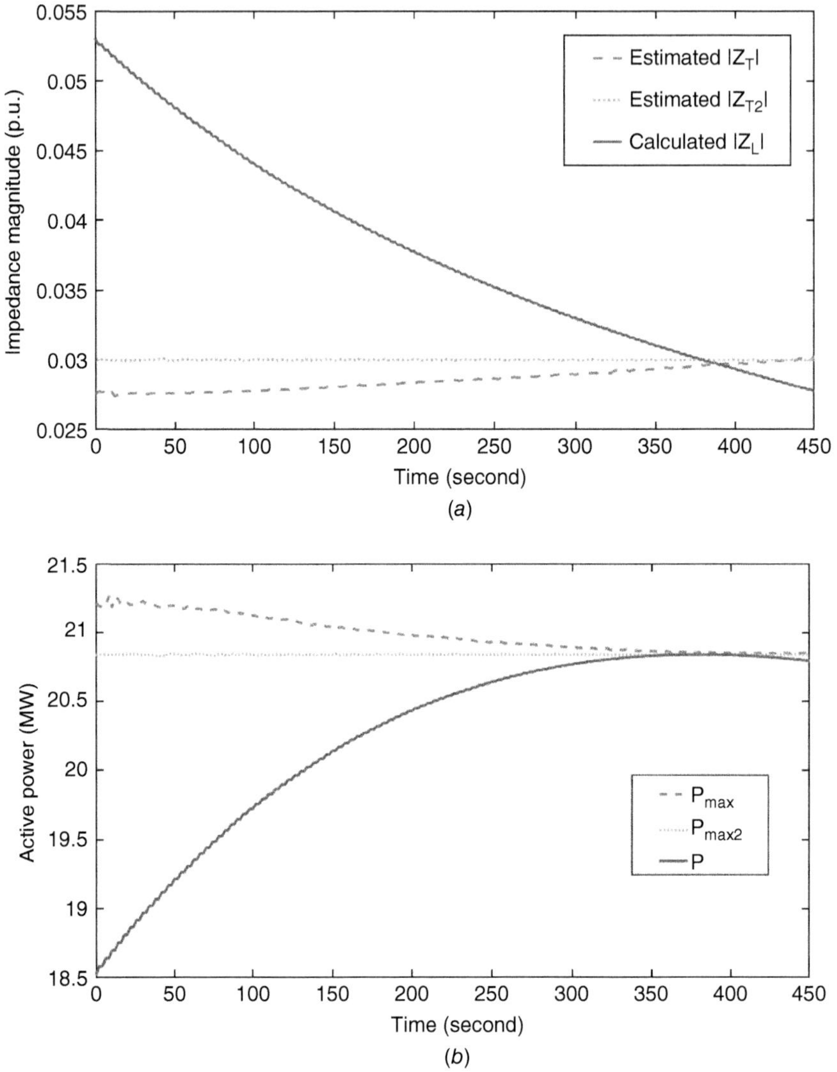

**FIGURE 20-23** Comparison of two methods on (a) estimates of $|Z_L|$ and (b) estimates of $P_{max}$.

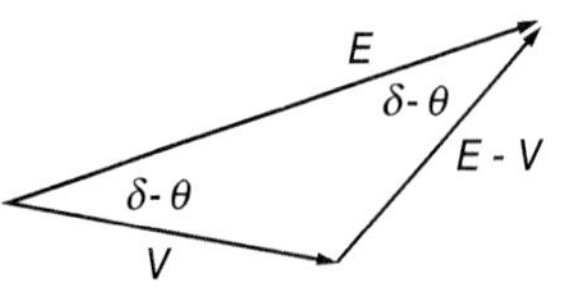


**FIGURE 20-24** Voltage phasors of the Thevenin source and load bus.

Based on measured *V-Q* characteristics at bus *i*, the following *Q-V* sensitivity based voltage stability index $\Gamma_i$ for bus *i* can be defined [14]

$$\Gamma_i = \frac{\Delta Q_i}{\Delta V_i} = \sum_j \frac{\Delta Q_{ij}}{\Delta V_i} \tag{20-47}$$

where $\Delta V_i$ represents the voltage magnitude change during a time window, $\Delta Q_{ij}$ represents the change of the reactive power flow on the line from bus $j$ to bus $i$ and $\Delta Q_i$ is the change of the total reactive power injection to bus $i$. A voltage-stable bus should have positive $\Gamma_i$, or in other words, a positive slope of the $V$-$Q$ curve. When the bus approaches voltage instability, $\Gamma_i$ should decrease to be close to zero. Thus, real-time calculation of this sensitivity index on each bus using the local measurements on the voltage magnitude and reactive power injection can provide an indicator of voltage stability. Thus, real-time applications of this sensitivity index $\Gamma_i$ for voltage stability monitoring needs a credible estimate on the index (mainly its sign) over a window of measurements against measurement noises or data quality issues [15].

### 20.2.4 Measurement-Based Voltage Stability Assessment for a Load Area

***Thevenin Equivalent-Based Voltage Stability Margin.*** This TE-based approach can also be extended to a load pocket area fed by $N$ tie lines from the external system through buses at the boundary of the load area [16, 17], as illustrated in Fig. 20-25. Assume that the $N$ tie lines have currents $I_1$ to $I_N$ and the boundary buses they connect have voltages $V_1$ to $V_N$, whose synchronized phasor measurements can be obtained.

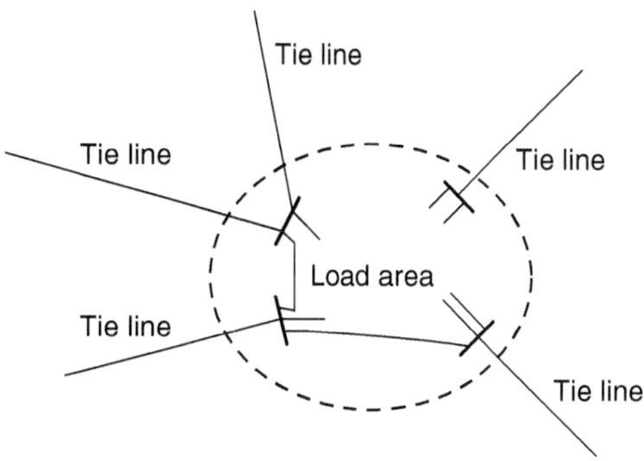


**FIGURE 20-25** Load area.

Let a load pocket area be defined as follows:

$$
\begin{aligned}
S &= P + jQ = \sum_{i=1}^{N} V_i I_i^* \\
I &= \sum_{i=1}^{N} I_i^* \\
V &= S/I^* \\
Z_L &= V/I
\end{aligned}
\tag{20-48}
$$

Then the TE-based approach can still apply to the estimation of Thevenin voltage and impedance for this load area and then calculation of voltage stability margin indices $P_{\text{margin}}$ and $V_{\text{margin}}$. Note: Those margin indices are regarding an average voltage $V$ and the total active power $P$ as defined in Eq. (20-48) rather than any specific bus or tie line.

To summarize, the TE-based VSA method for a load area (a single load bus is a special case with $N = 1$) conducts the following step:

1. For a sliding time window, collect synchronized measurements on voltage and current waveforms at all boundary buses of the load area.
2. Calculate the voltage phasor $V$ and current phasor $I$ by Eq. (20-48), which are simply the measurements if $N = 1$.
3. Estimate the external system's Thevenin parameters $E$ and $Z_T$.
4. Calculate the voltage stability limits and margin indices by Eqs. (20-36) to (20-39).

**N + *1 Buses Equivalent-Based Voltage Stability Margin.*** When applied to a load area fed by $N$ tie lines, the TE-based VSA method merges all tie lines and boundary buses as done in Eq. (20-48) to estimate the voltage stability margin for the entire boundary. However, voltage instability often initiates from one boundary bus or tie line, so it is important to estimate the voltage stability margin

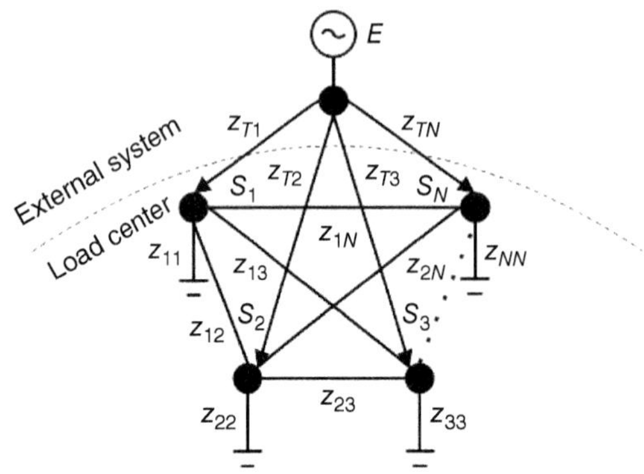


**FIGURE 20-26** $N+1$ buses equivalent.

for each individual boundary bus and tie line. Thus, instead of the TE, a more general $N+1$ buses equivalent can be used [18].

As shown in Fig. 20-26, the $N+1$ buses equivalent system reduces the network details both inside and outside of the load area. Assume that the external system has strong generation coherency and angular stability. Thus, it is represented by a single voltage phasor $E$ like the TE. The source is connected by $N$ branches with impedances $z_{T1}$ to $z_{TN}$ (or admittances $y_{T1}$ to $y_{TN}$) representing $N$ tie lines to $N$ boundary buses, respectively. Each boundary bus is monitored and connects an equivalent load with impedance $z_{ii}$ or admittance $y_{ii}$ modeling the portion of load seen from that bus. Connection between any two boundary buses $i$ and $j$ is modeled by impedance $z_{ij}$ or admittance $y_{ij}$.

The power transfer limit of each tie line is a function of $N(N-1)/2+2N+1$ complex parameters about that equivalent system including:

1. Voltage phasor $E$
2. $N$ tie-line impedances $z_{Ti}$'s
3. $N(N-1)/2$ transfer impedances $z_{ij}$'s
4. $N$ load impedances $z_{ii}$'s

Let $S_i = P_i + jQ_i$ and $I_i$ denote the complex power and current coming to boundary bus $i$ and let $V_i$ denote the bus voltage phasor. Using synchronized measurements on $I_i$ and $V_i$, all parameters of the $N+1$ buses equivalent including external system parameters $E$ and $z_{Ti}$, and load area parameters $z_{ii}$ and $z_{ij}$ can be identified over a sliding time window by, example for, the LS method or other optimization approaches [18].

The admittance matrix of the load area of the equivalent system is given in Eq. (20-49)

$$\mathbf{Y} = \begin{bmatrix} Y_{11} & \cdots & Y_{1i} & \cdots & Y_{1N} \\ \vdots & \ddots & & & \vdots \\ Y_{i1} & & Y_{ii} & & Y_{iN} \\ \vdots & & & \ddots & \vdots \\ Y_{N1} & \cdots & Y_{Ni} & \cdots & Y_{NN} \end{bmatrix}$$

where

$$Y_{ij} = Y_{ji} = -y_{ij} \text{ and } Y_{ii} = \sum_{j=1}^{N} y_{ij} \tag{20-49}$$

Define

$$\mathbf{Y}_{\mathrm{T}} = [y_{T1} \quad y_{T2} \quad \cdots \quad y_{TN}]^T \tag{20-50}$$

$$\mathbf{V} = [V_1 \ V_2 \ \cdots \ V_N]^T \tag{20-51}$$

The vector **I** of injected currents satisfies both equations in Eq. (20-52)

$$\begin{aligned}\mathbf{I} &= \mathbf{V}\mathbf{Y} \\ \mathbf{I} &= E\mathbf{Y}_\mathrm{T} - \mathrm{diag}(\mathbf{Y}_\mathrm{T})\mathbf{V}\end{aligned} \tag{20-52}$$

Then, solve both **V** and **I**:

$$\begin{aligned}\mathbf{V} &= E[(\mathbf{Y} + \mathrm{diag}(\mathbf{Y}_\mathrm{T})]^{-1}\mathbf{Y}_\mathrm{T} \\ \mathbf{I} &= E[(\mathbf{Y} + \mathrm{diag}(\mathbf{Y}_\mathrm{T})]^{-1}\mathbf{Y}_\mathrm{T}\mathbf{Y}\end{aligned} \tag{20-53}$$

The active powers transferred on tie lines are calculated by

$$\mathbf{P} = [P_1 \quad P_2 \quad \cdots \quad P_N]^T = \mathrm{Re}(\mathrm{diag}(\mathbf{I}^*) \times \mathbf{V}) \tag{20-54}$$

which only explicitly depends on $E$, **Y**, and $\mathbf{Y}_\mathrm{T}$, parameters of the equivalent.

If there is no contingency on either the external system or load area, $E$, $y_{Ti}$, and $y_{ij}$ $(i \neq j)$ are constant. Also, if the power factors of all loads are assumed to be constant, over a sliding time window, the active power on each tie line is a function of magnitudes of all load impedances, which is, $P_i = P_i(|y_{11}|, \cdots, |y_{NN}|)$. Its maximum $P_{\max,i,j}$ with respect to the change of the load admittance at bus $j$ is reached if

$$\frac{\partial P_i(|y_{11}|, \cdots, |y_{NN}|)}{\partial |y_{jj}|} = 0 \qquad i, j = 1 \sim N \tag{20-55}$$

$P_{\max,i,j}$ has a long, complex but explicit expression about $E$, $y_{Ti}$, and $y_{ij}$. The expression can be obtained by solving a quadratic equation [18]. $P_{\max,i,j}$ represents the voltage stability limit in terms of the active power on the $i$th tie line if only the load at bus $j$ varies, which can be calculated by plugging the values of $E$, $y_{Ti}$, and $y_{ij}$. For each tie line, there are $N$ such voltage stability limits respectively with respect to the changes of loads at $N$ buses.

Figures 20-27 and 20-28 illustrate the results from this $N + 1$ buses equivalent based method with $N = 3$ for two cases, which are strong connection (Fig. 20-27) and weak connection (Fig. 20-28) between boundary buses. Each tie line $i$ has three limits, $P_{\max,i,1}$, $P_{\max,i,2}$, and $P_{\max,i,3}$, respectively, corresponding to $\partial P_i/\partial|y_{11}| = 0$, $\partial P_i/\partial|y_{22}| = 0$, and $\partial P_i/\partial|y_{33}| = 0$. If only the magnitude of $z_{33}$ decreases, $P_{\max,i,3}$ will be the most reliable power transfer limit among the three. If $z_{12}$, $z_{13}$, and $z_{23}$ are close to zero (indicating strong connection between boundary buses), as shown in Fig. 20-27, all limits as well as the total limit $P_{\Sigma,\max(\mathrm{TE})}$ from the TE-based method are met at the same time, and the $P$-$V$ curves of three boundary buses are very close. However, if $z_{12}$, $z_{13}$, and $z_{23}$ have large magnitudes, as shown in Fig. 20-28, since the three $P$-$V$ curves are very different, the power transfer limits at buses 1, 2, and 3 are met at different times, which are all different from the time when $P_\Sigma^{\mathrm{Max(TE)}}$ is met, indicating the inaccuracy of the TE-based method for this case.

The $N + 1$ buses equivalent based VSA method for a load area conducts the following steps:

1. For a sliding time window, collect synchronized measurements on voltage and current waveforms at $N$ boundary buses of the load area.
2. Estimate the parameters of the $N + 1$ buses equivalent including $E$, $z_{Ei}$'s, $z_{ii}$'s, and $z_{ij}$'s from the measurements.
3. For each tie line $i$, calculate the transfer limits $P_{\max,i,1}, \ldots P_{\max,i,N}$ which give probable limits of the tie line depending on how the actual load increases. For example, if load in this area is found mainly increasing near bus $j$, $P_{\max,i,j}$ will be the most accurate estimation on the transfer limit.

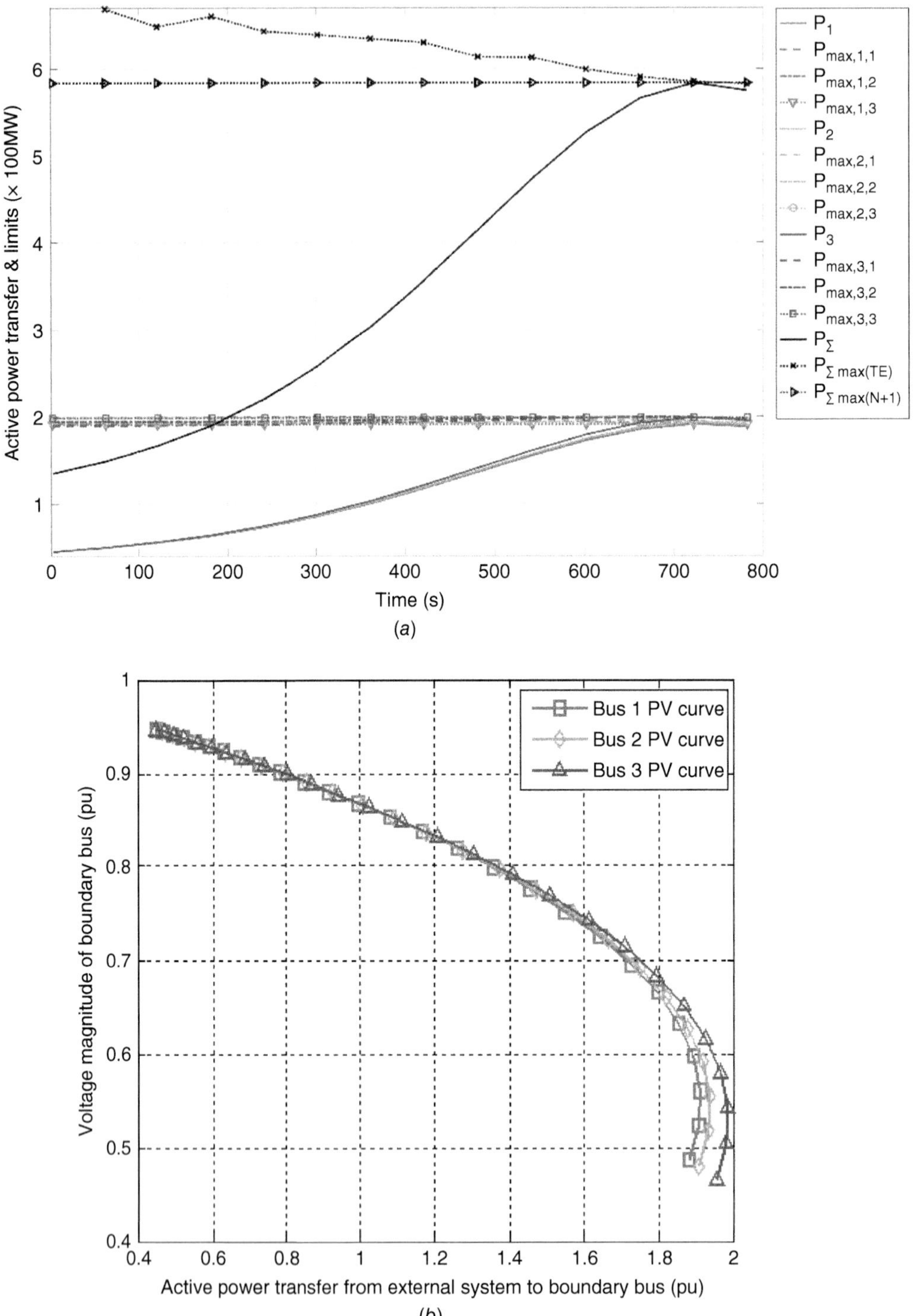


**FIGURE 20-27** 3 + 1 buses system ($z_{12}$, $z_{13}$, and $z_{23}$ are close to zero). (a) Power transfer limits; (b) *P-V* curves.

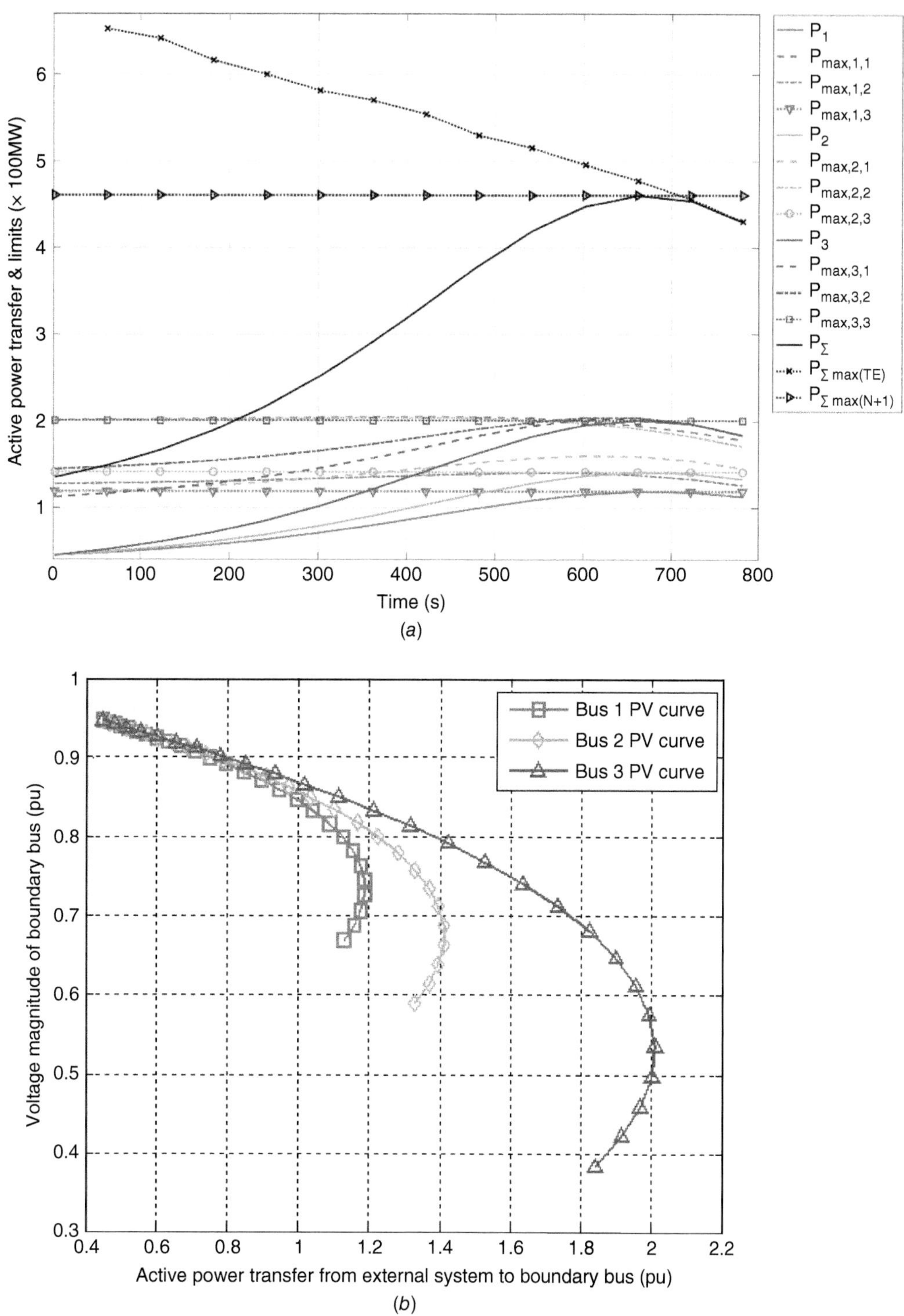


**FIGURE 20-28** 3 + 1 buses system ($z_{12}$, $z_{13}$, and $z_{23}$ have large magnitudes). (a) Power transfer limits; (b) *P-V* curves.

### 20.2.5 Bibliography


1. P. Kundur, J. Paserba, V. Ajjarapu, G. Andersson, A. Bose, C. Canizares, and N. Hatziargyriou, et al., (IEEE = CIGRE Joint Task Force on Stability Terms and Definitions), "Definition and Classification of Power System Stability," *IEEE Trans. Power Systems*, Aug. 2004.
2. I. Dobson and L. Lu, "Voltage Collapse Precipitated by the Immediate Change in Stability When Generator Reactive-Power Limits Are Encountered," *IEEE Trans. Circuits Syst.*, pt. 1, vol. 39, pp. 762–766, Sep. 1992.
3. F. Gubina and B. Strmcnik, "Voltage Collapse Proximity Index Determination Using Voltage Phasor Approach," *IEEE Trans. Power Syst.*, vol. 10, pp. 788–793, May 1995.
4. K. Vu, M. Begovic, D. Novosel, and M. M. Saha, "Use of Local Measurements to Estimate Voltage-stability Margin," *IEEE Transactions on Power Systems*, vol. 14, no. 3, pp. 1029–1035, Aug. 1999.
5. V. Ajjarapu and C. Christy, "The Continuation Power Flow: A Tool for Steady State Voltage Stability Analysis," *IEEE Transactions on Power Systems*, vol. 7, no. 1, pp. 416–423, Feb. 1992.
6. B. Gao, G. K. Morison, and P. Kundur, "Voltage Stability Evaluation Using Modal Analysis," *IEEE Transactions on Power Systems*, vol. 7, no. 4, pp. 1529–1542, Nov. 1992.
7. W. W. Hager, *Applied Numerical Algebra*, Prentice Hall Inc., New Jersey, 1998.
8. P. Löf, T. Smed, G. Andersson, and D. Hill, "Fast Calculation of a Voltage Stability Index," *IEEE Transactions on Power Systems*, vol. 7, no. 1, pp. 54–64, Feb. 1992.
9. C. Cañizares (ed.), *Voltage Stability Assessment: Concepts, Practices, and Tools*, Special publication of the IEEE Power System Stability Subcommittee, 2002. ISBN 0 780 378 695.
10. NERC/WECC Planning Standards, WECC, Apr. 2003.
11. Borka Milosevic and Miroslav Begovic, "Voltage-Stability Protection and Control Using a Wide-Area Network of Phasor Measurements," *IEEE Transactions on Power Systems*, vol. 18, No. 1, pp. 121–127, Feb. 2003.
12. P. Zhang, L. Min, and N. Zhang, "Voltage Instability Load Shedding," EPRI Report No. 1012491, 2006.
13. K. Sun, F. Hu, and N. Bhatt, "A New Approach for Real-Time Voltage Stability Monitoring Using PMUs," *IEEE PES Innovative Smart Grid Technologies Asia (ISGT Asia) Conference*, Kuala Lumpur, Malaysia, May 20–23, 2014.
14. V. Venkatasubramanian, X. Liu, G. Liu, Q. Zhang, and M. Sherwood, "Overview of Wide-Area Stability Monitoring Algorithms in Power Systems Using Synchrophasors," *Proc. IEEE Amer. Control Conf.*, San Francisco, CA, pp. 4172–4176, Jun. 2011.
15. X. Liu, X. Zhang, and V. Venkatasubramanian, "Distributed Voltage Security Monitoring in Large Power Systems Using Synchrophasors," *IEEE Transactions on Smart Grid*, vol. 7, No. 2, pp. 982–991, Mar. 2016.
16. P. Zhang, L. Min, and J. Chen, "Measurement-Based Voltage Stability Monitoring and Control," U.S. Patent 8,126,667, 2012.
17. K. Sun, P. Zhang, and L. Min, "Measurement-Based Voltage Stability Monitoring and Control for Load Centers," EPRI Report No. 1017798, 2009.
18. F. Hu, K. Sun, A. Del Rosso, E. Farantatos, and N. Bhatt, "Measurement-Based Real-Time Voltage Stability Monitoring for Load Areas," *IEEE Transactions on Power Systems*, vol. 31, No. 4, pp. 3189–3201, Jul. 2016.


## 20.3 TRANSIENT STABILITY ASSESSMENT AND CONTROL

*BY DANIEL RUIZ-VEGA*

### 20.3.1 Introduction

Transient stability has always been, and remains, a major concern in power systems. From the initial stages of the electric power industry, where synchronous machines started being operated in parallel, up to late 1950s, most of the stability problems encountered in planning and operating stages were mainly related to transient instabilities. After the interconnection of electric utilities and the resulting formation of large widespread power systems, new forms of instability like small signal angle instabilities, frequency instabilities and, more recently, voltage instabilities, started being also considered as concerns in power systems planning and operation [1].

Being historically the dominant form of instability in most power systems, transient stability is one of the most studied dynamic problems. Transient stability assessment methods evolved along the years in the same way the complexity and size electric power systems did, according to the stability theory of dynamic systems and the developments of the art of computation. However, as was pointed out as early as in 1970, up to now *"no matter how fast computers grew in size and speed, they were never big enough to meet the demands imposed by stability studies"* [2].

This statement is still true today, especially for online transient stability assessment and control. The present trend in power systems is the analysis of huge interconnecting systems that current simulation programs and computer facilities can hardly manage. As an example, some authors claim that rather than 800 individual electric utilities in the United States today, there will be only 50 utilities (others say 4) as a result from deregulation [3, 4]. Analyzing networks comprising 50,000 or more nodes is now a current need for transient stability assessment in existing large interconnected systems in United States and Europe [5]. Even more problematic than size and speed requirements for transient stability assessment, is the design of methods able to extract the most important features of the dynamic phenomena, and to provide suitable control countermeasures.

The next subsections present the basics of the transient stability phenomena, and describe some of the different methods that have been developed to analyze them, culminating with the single machine equivalent method, which provides the dynamic information and the sensitivities required for designing and implementing online and real-time transient security assessment and control techniques, that are suitable to be applied in large scale power systems.

***Definitions.*** Any change in the power system operating variables or parameters can be considered as a disturbance, in the sense that all of them cause a power imbalance. Some disturbances, like the natural change in load consumption, or programmed outages of some power system components for maintenance purposes, can be considered as "normal, or predictable" disturbances. Besides these normal changes, power systems are usually subjected to sudden unpredictable changes in their operating conditions and structure, that are caused by major system component failures, human errors, natural calamities and weather conditions (like lightning, strikes).[a] The consequences of a disturbance can still roughly be considered as mismatches between system generation and load, but are much severer than the ones produced by normal changes. According to these ideas, the following definition is going to be considered in this document:

"A disturbance in a power system is a sudden change or a sequence of changes in one or more of the parameters of the system, or in one or more operating quantities" [6].

In power systems, nearly all electric power is produced by synchronous machines. By its very principle, the parallel operation of these machines requires them to operate in synchronism, that is, in a condition in which the average electrical speed of each machine (product of its rotor angular velocity and its number of pole pairs) equals the angular frequency of the ac network to which they are connected.

Transient stability of a power system is its ability to maintain synchronism when subjected to a large disturbance. This implies that it must always be analyzed using a nonlinear model [6, 7].

***Power System Stability Analysis.*** In order to describe the basics of transient stability it is very useful to use a simplified model of the power system. The simplest power system dynamic representation is known as the "classical model," which is adequate to analyze first swing stability. Modeling considerations of the classical model are [8–10]:

- Machines are represented by an electromagnetic force $\tilde{E}' = E' \angle \delta$ of constant magnitude, where $\delta$ is the rotor angle position, behind the direct axis transient reactance $X'_d$.
- Loads are represented by a constant impedance model.
- The mechanical power $P_m$ is constant and damping is neglected.

[a]A very detailed description of the different causes of possible power system disturbances is given in [4].

- Generator dynamics are described by the electromechanical balance equation (also known as "swing equation"), $w$ is the rotor angular speed, $M$ is the inertia coefficient, $P_e$ is the electrical power and $P_a$ is the accelerating power:

$$M\frac{d^2}{dt^2}(\delta)=M\frac{d}{dt}(\omega)=P_m-P_e=P_a \tag{20-56}$$

In what follows, we illustrate transient stability concepts using a simplified one-machine infinite bus system, (shown in Fig. 20-29),which represents a system in which a single machine is supplying power to a very large external system, modeled by an "infinite" bus (Fig. 20-29*a*).[b] Because of the relative size of the system to which the machine is connected, dynamics of the machine will not cause virtually any change in voltage and frequency of the infinite bus, represented, in this way, by a constant magnitude, phase and frequency voltage source, or by a large synchronous machine having infinite inertia and a zero transient reactance [8, 10]. The equivalent system of the classical model is depicted in Fig. 20-29*b*, where the equivalent reactance $X_{eq}$ is composed by a combination of all the reactances between the infinite bus, and the internal bus of the machine behind the transient reactance. Infinite bus voltage angle is taken as the system angle reference, so $\tilde{V}_{infbus}=V_{infBus}\angle 0$.

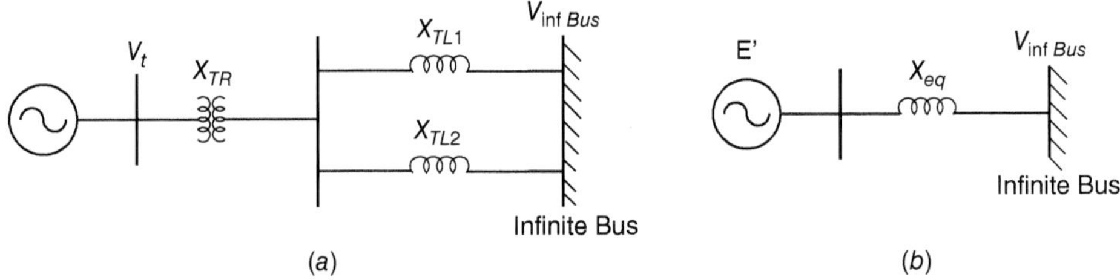


**FIGURE 20-29** One machine-infinite bus system. (*a*) One machine-infinite bus system; (*b*) equivalent system.

Taking into consideration the equivalent one-machine infinite bus system of (Fig. 20-29b), the electrical power of the synchronous generator can be represented by equation Eq. (20-57) [8]:

$$P_e=\frac{E'V_{infBus}}{X_{eq}}\sin(\delta) \tag{20-57}$$

According to the IEEE [7], stability analysis is only valid for one specific operating condition under a given contingency. In this work the term "contingency" is used to denote a disturbance, or a sequence of disturbances, and the corresponding actions performed by the protective devices (protective relays and special discrete controls), in order to correct the disturbance and its effects, and to protect the elements of the power system from being damaged. The sequence of a stability study can be described as follows:

1. The system is initially operating in a predisturbance equilibrium condition. This operating point is usually in steady state, that is, "an operating condition of a power system in which all of the operating quantities that characterize it can be considered to be constant for the purpose of analysis" [6].
2. Next, the specific preselected contingency of interest is applied, and the system enters a faulted or transient state. Usually the contingency is characterized by a specific fault scenario, and its clearing

[b]In [2] an additional simplification of the model is that all generators in the same power plant are represented as a single machine. This restriction had a sense in those times due to the very low computing power digital computers had. In modern analyses, this is not required anymore.

mechanism (e.g., short circuit somewhere in the transmission network, followed by a line disconnection after a pre specified duration of the event—"fault clearing time") [7].

3. After the disturbance is eliminated, the system dynamics is studied in order to determine whether the power system is able to reach a new, acceptable postdisturbance equilibrium condition, the steady postdisturbance operating state, or it becomes unstable.

When the system experiences all the operating conditions described above (prefault, during-fault or disturbed, and postdisturbance states), the effect of the different changes in the system is represented in the one machine-infinite bus system of Fig. 20-29*b*, by a change in the value of the equivalent system reactance $X_{eq}$. In this way, the system model has a different representation depending on the given operating state. This can also be roughly seen as a change in the power transmission capacity of the system, as can be seen in Eq. (20-57); in the faulted and postdisturbance states this capacity usually decreases with respect to the one of the initial, predisturbance state.

***Transient Stability Assessment using the Equal Area Criterion.*** The equal area criterion (EAC) is a powerful graphical technique to assess power system transient stability. It provides a clear explanation of transient stability, based on the physics of this phenomenon, which allows understanding its causes, and possible solutions, in terms of the energy variations of the system during the transients. Its origins are not clear, but the first proposals were made in [8, 11, 12]. This method was initially developed to allow studying the behavior of one-machine infinite bus systems, or two-machine systems, represented by the classical model, without numerical integration. Considering the one machine-infinite bus (OMIB) system of Fig. 20-29*b*, which is governed by the swing equation (Eq. (20-56)), it can be shown that, after solving this equation, we can arrive to the following expression [13]:

$$\frac{1}{2}M(\omega-\omega_0)^2=\int_{\delta_0}^{\delta}P_a\,d\delta \tag{20-58}$$

Equation (20-58) describes the EAC. It states that the kinetic energy of the synchronous machine (the left side term of the equation), equals the integral of the machine's accelerating power $P_a$ during the whole transient period, from an initial steady-state machine angle $\delta_0$, to a final angle $\delta$.

To illustrate the EAC, let us consider a case where the power system of Fig. 20-29*b* is subjected to a three-phase fault at the terminals of the synchronous generator. The system initial steady-state operating point is the intersection of mechanical and electric powers (point 1 in Fig. 20-30). Due to the application of the disturbance (the three phase fault), the electric power transfer falls to zero (point 2) and is smaller than the mechanical power $P_m$ (accelerating power $P_a$ is positive), and the machine accelerates, gaining kinetic energy [see Eq. (20-56)], until fault is cleared at point 3, where the system reaches point 4. Area 1, 2, 3, 4 represents the accelerating area of the system $A_{acc}$.

When the fault is cleared, the system trajectory continues along the post fault electric power curve $P_{e_p}$. In this region, electrical power $P_{e_p}$ is larger than mechanical power $P_m$ and machine decelerates (by dissipating the kinetic energy gained by the machine's rotor in the accelerating area). The EAC states that the system will be stable, if the decelerating area $A_{dec}$ is at least equal to the accelerating area $A_{acc}$.

The EAC provides the more accurate and descriptive assessment of the transient stability phenomena, by evaluating the problem in terms of the energy of the system: in order to be stable, the system should be able to dissipate the kinetic energy that the disturbance produced. The kinetic energy gained by the system during the disturbance application and the potential energy available in the system in the postdisturbance state to dissipate the excess of kinetic energy, can be computed accurately by integrating the accelerating area ($A_{acc}$) and the decelerating area ($A_{dec}$) of Fig. 20-30, respectively.

In addition to permit a better understanding of the physical reasons of transient stability, the graphical description provided in Fig. 20-30 allows to devise means to control the problem: the system

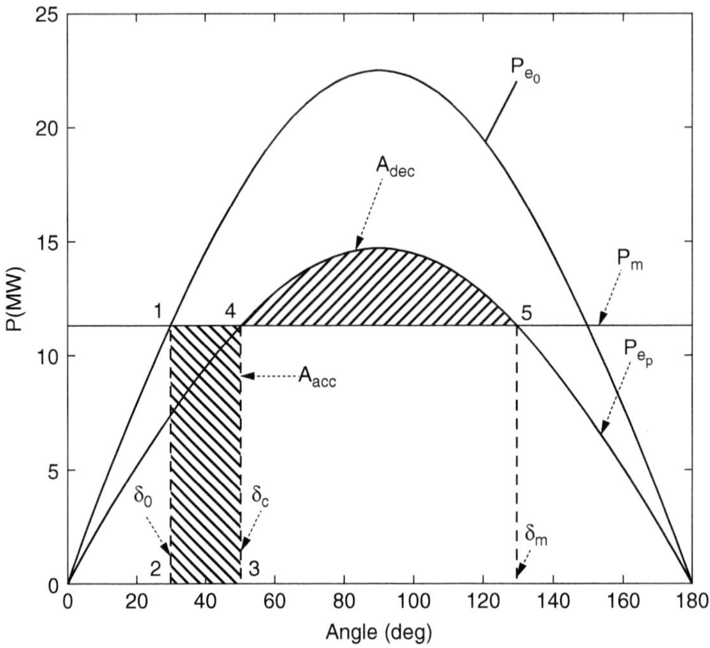


**FIGURE 20-30** EAC of an infinite bus machine equivalent [14].

becomes unstable because of an excess of kinetic energy, so the control measures for transient instability must aim at correcting the energy imbalance by decreasing the accelerating area and increasing the decelerating area. The control measure must, at least, make both areas be equal.

These are the reasons by which, most books in the area of electric power system analysis use the EAC to explain transient stability. However, since its development in the 1930s, this method was relegated to be applied only to very small, academic systems, represented by the classical model. As it is going to be presented in the subsection Nonlinear Aspects of System Oscillations, it was found in mid-1990s that it is possible to apply this method to multi-machine, fully detailed power system models, when combining the EAC with time-domain simulations, giving rise to a new, hybrid transient stability method: SIME [13].

The next subsection describes the transient energy function method, a second direct method that has been proposed to assess transient stability.

***Transient Stability Assessment Using the Transient Energy Function Method.*** The Western technical literature recognizes Magnusson as being the first to use the concept of energy for studying general multi-machine power system transient stability in a pioneering work published in 1947 [15], followed about 10 years later by Aylett [16]. Magnusson and Aylett may therefore be considered as the forerunners of the Lyapunov method application to power system stability. In the 1960s, the research on the application of the Lyapunov method was restarted all around the world, due the appeal of both, the theory as such, and the expected needed practical outcomes (e.g., see [17–19]). From that moment on, more than 50 years of intensive research have been performed in this area, to develop a Lyapunov-based direct method, able to be applied in large-scale realistic power system models. Some of the most important approaches that have been published can be consulted in [20–22, 23–25].

In this document, the transient energy function (TEF) method developed by Athay et al. [26] is considered to explain the basics of the application of Lyapunov-based methods for transient stability assessment. This version of the TEF method is applied to a simplified equivalent one machine–infinite bus power system of Fig. 20-29, represented by the classical model. The analysis of the method is as follows (Fig. 20-31):

- The assessment of transient stability of the TEF method can be understood more clearly by using a mechanical analog in which the system state is a rolling ball, initially resting at the bottom of the potential energy surface of Fig. 20-31*a* (the point labeled as "initial operating condition").
- When the disturbance is applied, the system gains kinetic energy in the transient period, and moves away from its initial condition. In the analogy, the rolling ball starts going up the slope of the potential energy surface (Fig. 20-31*b*).
- When the disturbance is eliminated, the system starts dissipating into potential energy, the kinetic energy that it gained during the transient period; to avoid instability, it must be able to absorb all the kinetic energy. In the rolling ball analogy, this means that if the kinetic energy that the system gained in the transient period is not enough to climb up to the top of the potential energy surface, the ball would go down, oscillate, and regain a new postdisturbance equilibrium (the system would be stable,

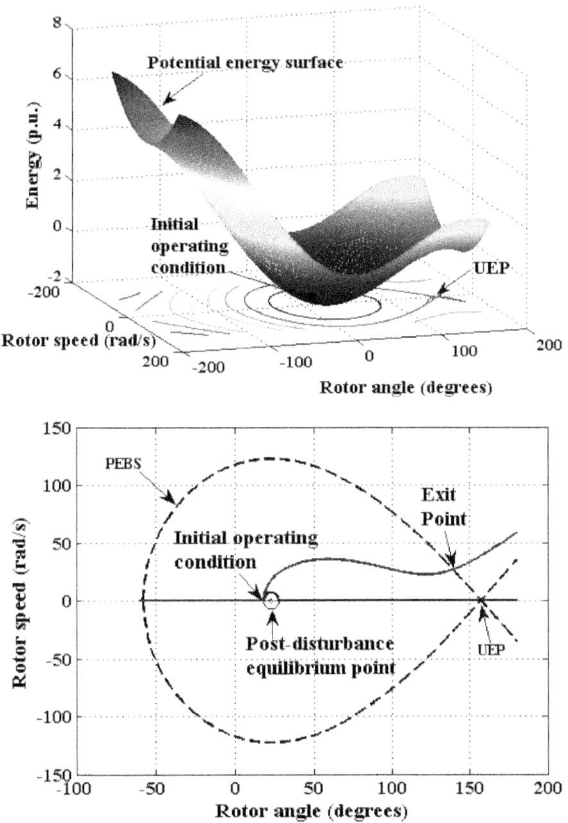


**FIGURE 20-31** Transient energy function method applied to a one machine-infinite bus system. (*a*) Potential energy surface; (*b*) phase plane representation [27].

because all the kinetic energy of the ball was dissipated). On the contrary, if the kinetic energy that the ball acquired during the transient process, is enough to make the ball reach the top of the potential energy surface, the ball would continue its way out of the surface (the system would be unstable).

- The potential energy surface represents the maximum amount of kinetic energy that the system can absorb in the post disturbance period, without loss of synchronism. It depends on both, the system initial operating condition and the specific contingency of interest.

Two main advantages where expected from the direct approach, namely: (1) straightforward computation of transient stability limits, since direct methods reduce the T-D simulations to the sole during-fault period, while avoiding all repetitive runs; (2) suitable definition and computation of margins, which in turn allows sensitivity analysis.

But on the other hand, in order to make the method applicable in practice, one should be able to [1]:

- Construct "good" Lyapunov $V$-functions.
- Assess "good" limit values or, equivalently, good practical stability domain estimates.

Both of the above conditions have been shown to be extremely difficult to meet properly. With reference to condition (i), it appeared quite early that the construction of $V$–functions was only possible with oversimplified (actually unacceptably simplified) system modeling, like the classical model. Among the many interesting attempts to circumvent this difficulty is the "structure preserving modeling" proposed by Bergen and Hill [28]. Another one, which has finally prevailed, led to the use of "pseudo-Lyapunov functions" (the transient energy function) consisting of hybridizing $V$–functions with T-D methods (see below); credit for this may be given to Athay et al. [26].

As for condition (ii), among the large variety of solutions proposed are the methods of Athay et al., [26], of Kakimoto et al. [29], the acceleration approach [30], the BCU or exit point [31], the PEBS [21], and the IPEBS [32] methods.

All these methods aim at identifying the value taken by the Lyapunov function at the boundary of the practical stability domain estimate; but they differ in the way of doing so. Broadly, two different approaches may be distinguished: the one consists of computing $\boldsymbol{x}^{u}$, the "unstable equilibrium point of concern" (UEP point of Fig. 20-31), and hence the energy limit $V(\boldsymbol{x}^{u})$ ; the other relies on criteria able to suggest when the system trajectory comes "close enough" to the boundary of the practical stability domain estimate.

Nevertheless, however inventive, the various solutions were not able to properly overcome the difficulties met. Indeed, the proposed stability domain estimates were found to provide overly conservative stability assessments, with unpredictably varying degree of conservativeness; besides, from a computational point of view, the involved computations were often quite cumbersome, counterbalancing the computer gains expected of direct methods, and even making them much slower than T-D methods.

One very important outcome, of the TEF method developed by Athay et al. [26], is one stability criterion for detecting the moment when the system trajectory has reached the boundary of the stability region, also named as the "potential energy boundary surface" (PEBS in Fig. 20-31*b*). This point is known as "exit point" in the phase plane. As it is shown in the subsection The Dot Product Method, this stability criterion actually works for systems represented by a detailed model, and was used to develop a hybrid stability method, currently known as the "dot product."

## 20.3.2 Transient Stability Assessment Methods for Realistic Power System Models

It was shown in the previous subsection that transient stability is a dynamical phenomenon that is better understood when analyzing the power system energy. Both methods presented above, the

EAC and the TEF method, conduct an energy analysis of the system that provides very valuable information that allows performing a correct stability assessment, and to design some means to control this problem. However, due to different difficulties encountered in its development, the application of both methods was limited to small academic systems (in the case of EAC) or to unacceptably simplified power system models (in the case of the TEF method).

As power system grew in size and interconnections, they could no longer be represented by the OMIB system. With power system interconnection, new stability problems emerged and the modeling requirements to represent the system evolved too mathematical representation of generators and their controls rose in detail and complexity, and new methods to assess stability were employed: the time-domain (T-D) simulation methods.

The advent of the digital computer in late 1960s made possible the development of T-D simulators. Along the years, T-D simulators have shown to have a great flexibility in terms of modeling detail, and of the size of the system that they can analyze. Providing that the parameters of the system are close to the ones of the real power system, T-D simulator is the mean to obtain the most accurate estimation of the system dynamic performance nowadays.

Today, the T-D simulation method using the system nonlinear model, for the reasons mentioned before, has become an indispensable tool for power system stability analysis. However, T-D methods have advantages and disadvantages. Some of them are described below [13]:

***Pros of TD Methods***

- They provide the most accurate dynamic description of the system behavior in the time domain, if parameters are precise enough.
- They can study a power system any degree of modeling detail in system components and contingencies.

***Cons of TD Methods***

- They do not provide tools to identify and discard harmless disturbances from the study.
- They are unable to provide sensitivity variables and stability margins.
- They are helpless in the design of control actions to improve stability.

T-D methods are definitely insufficient, yet indispensable to in-depth transient stability investigations [33]: indispensable, since they are very accurate and effective in handling different models; insufficient, since they are unable to meet the current needs of online transient stability assessment, not to mention control, which is certainly their major weakness ("Achilles' heel"). The inability of T-D methods to meet such needs justified the development of nonconventional methods, aiming at enhancing and complementing T-D program: the hybrid methods [1].

The most successful hybrid transient stability methods combine T-D simulations with direct methods, which analyze the power system energy; in this way, these approaches combine the accuracy and flexibility of T-D simulations with the important information about the power system dynamics that the direct methods provide. In this document, two hybrid direct-time-domain hybrid transient stability assessment methods are described:

- The single machine equivalent (SIME) method, combining the EAC with the T-D method.
- The dot-product method, which applies the exit point instability criterion of the TEF method to the T-D method.

These methods are of interest because they do not require any simplification in modeling detail and can assess power system transient stability under any contingency.

***The Time-Domain Simulation Method.*** T-D methods started being popular with the advent of computers because they allow representing system dynamic behavior closer to its real functioning, making possible the use of detailed models of the elements involved in the system (generators, loads, transmission network elements and some other dynamic devices) and to represent system nonlinearities. T-D methods have two main features [1]:

- Time-domain simulation results are able to accurately predict the dynamic behavior of the system, but it depends on the precision of the parameters employed.
- T-D methods are not capable of providing sensitivity techniques to determine the causes of the stability problem and design adequate control measures.

In consequence most of the research efforts are devoted to improve some aspects of the T-D methods. In these methods, the model of a multi-machine system is divided into two sets of equations:

$$\dot{\mathbf{x}} = \mathbf{f}(\mathbf{x}, \mathbf{y}) \tag{20-59}$$

$$\mathbf{0} = \mathbf{g}(\mathbf{x}, \mathbf{y}) \tag{20-60}$$

where $\mathbf{x}$ is the vector of state variables of power system dynamic elements and $\mathbf{y}$ is the vector of bus voltages. Equation (20-59) is the dynamic differential equations of machines, other dynamic devices and their controls. Equation (20-60) represents the algebraic equations of machine stators, the network, and loads. The sets of Eqs. (20-59) and (20-60) must be coordinately solved. In most modern T-D simulation programs the implicit-simultaneous method is preferred [10]. An implicit numerical integration method (like the trapezoidal rule of integration) is first applied to differential equations [Eq. (20-59)], which become nonlinear algebraic. This new set of algebraic equations is combined with network algebraic equations [Eq. (20-60)] to be solved simultaneously.

Investigation about improvements of the T-D method is still performed nowadays. One interesting problem that has recently appeared is the need for performing the online stability assessment of huge power systems, covering large geographical areas, like the UCPTE system in Western Europe, with systems represented by a model with 50,000 or more buses. In all cases a detailed model representation of all the power system components is considered. One recent example of the improvements proposed for the T-D simulations, modifying the Newton-Raphson method and using parallel programming can be consulted in [5].

***The SIME Method.*** The equivalent system most often considered to study the power system in emergency conditions is a two-machine system comprising two generators with an interconnecting transmission line between them [34]. That system configuration can be achieved by reducing the original multi-machine power system into a one machine-infinite bus (OMIB) equivalent system; this latter is valid since the behavior of a real power system is similar to that of a two-machine system when the system loses synchronism [8]. This fact was known for a long time (almost 80 years) in power system transient stability assessment, but it was considered that the equivalent OMIB could only be properly obtained in two machine systems represented by classical model. One of the most important achievements of SIME is that during its development it was fully demonstrated that the OMIB equivalent could be derived for large-scale multi-machine systems, and considering any modeling level detail in both, system components and contingencies [13].

*Foundations of the SIME Method.* SIME is a hybrid method that combines time-domain simulations with the EAC. It is based on the following two principles [35]:

- *Proposition 1.* However complex, the mechanism of a power system loss of synchronism originates from the irrevocable separation of its machines into two groups. Hence, the multi-machine system transient stability may be inferred from that of an OMIB system properly selected (the critical OMIB).
- *Proposition 2.* The transient stability of an OMIB may be assessed in terms of its transient stability margin $\eta$, defined as the excess of its decelerating over its accelerating energy.

SIME combines the benefits of time-domain and direct methods, while avoiding their disadvantages. Using the time-domain simulation method has the advantage of providing the most accurate system dynamic response available, while the essential advantage of the equal area criterion application is that it provides margins in order to assess the instability severity. Combination of both methods provides a very important "emergent" advantage: the identification of the machines responsible for system loss of synchronism.

SIME's main information: stability margin and identification of the set of critical machines, is importantly complemented with three very useful simplified representations of the multi machine large scale system dynamic response: the OMIB equivalent representation in the time domain, in the $P$-δ plane and in the phase plane, which have been used to derive very important assessment and control techniques that could be implemented either online or in real time.

Depending upon whether the during transients multi-machine temporal data are furnished by a time-domain program or by real-time measurements, SIME takes on two different forms, namely, the *preventive* and the *emergency SIME*.

- *The preventive SIME,* which analyzes the dynamic response of the system provided by a time-domain program, and aims at assessing "what to do" in order to prevent loss of synchronism, that is, to enable the system to face a priori "harmful" contingencies. It is suitable for implementing a transient stability online security function.
- *The emergency SIME,* which uses measurements of the power system transients, and aims at assessing in real-time and closed-loop fashion "what to do" in order to stabilize the system after a "harmful" contingency has actually occurred.

*Overall Formulation of the SIME Method.* SIME concentrates on the postfault configuration of a system after being subjected to a large disturbance that may drive the system to instability. Then, based on the proposition 1 of the method, the two groups of machines (the critical one and the non-critical machines' group) are identified and replaced by a two-machine system and then by an OMIB; transient stability is assessed by means of the EAC in this latter OMIB.

*Identification of critical machines* is done in the following fashion [13]:

- SIME drives the T-D simulation first in the during-fault and then in the post-fault configuration.
- Immediately after the system enters into a post-fault state, SIME begins sorting the machines in decreasing order of their rotor angles at each time step of the T-D simulation.
- SIME identifies the larger rotor angle deviation (gap) between adjacent machines and two groups are conformed: the critical group (those machines whose rotor angle deviation above the largest gap) and the noncritical group (the rest of the machines).
- These two machine groups are first reduced to a two-machine system, then to a candidate OMIB whose parameters are computed as described below, and the procedure is repeated until the candidate OMIB reaches instability conditions (given by the EAC). Then the OMIB is considered the critical OMIB.

*The derivation of OMIB time-varying parameters* is accomplished as follows [13]:

- To apply EAC, the system must be transformed into a two-machine equivalent system, later they are aggregated into their corresponding center of angle (COA) and replaced by an OMIB. All these changes are made using system variables refreshed by the T-D program.
- The following formulas correspond to the pattern which decomposes machines into critical (subscript $C$) and non-critical machines (subscript $NC$). The expressions to calculate the corresponding OMIB parameters are
  - COA of the group of critical machines (CMs) and noncritical machines (NMs)

$$\delta_C(t) \triangleq \frac{1}{M_C}\sum_{k\in C} M_k\delta_k(t) \qquad \delta_{NC}(t) = \frac{1}{M_{NC}}\sum_{j\in NC} M_j\delta_j(t) \tag{20-61}$$

where

$$M_C = \sum_{k \in C} M_k \,; \qquad M_{NC} = \sum_{j \in NC} M_j \tag{20-62}$$

- Rotor angle of the corresponding OMIB

$$\delta(t) \triangleq \delta_C(t) - \delta_{NC}(t) \tag{20-63}$$

- Rotor speed of the corresponding OMIB

$$\omega(t) = \omega_C(t) - \omega_{NC}(t) \tag{20-64}$$

where

$$\omega_C(t) = \frac{1}{M_C} \sum_{k \in C} M_k \omega_k(t); \quad \omega_{NC}(t) = \frac{1}{M_{NC}} \sum_{j \in NC} M_j \omega_j(t) \tag{20-65}$$

- Mechanical power of the corresponding OMIB

$$P_m(t) = M \left( \frac{1}{M_C} \sum_{k \in C} P_{mk}(t) - \frac{1}{M_{NC}} \sum_{j \in NC} P_{mj}(t) \right) \tag{20-66}$$

- Electric power of the corresponding OMIB

$$P_e(t) = M \left( \frac{1}{M_C} \sum_{k \in C} P_{ek}(t) - \frac{1}{M_{NC}} \sum_{j \in NC} P_{ej}(t) \right) \tag{20-67}$$

- Accelerating power of the corresponding OMIB

$$P_a(t) = P_m(t) - P_e(t) \tag{20-68}$$

- Equivalent OMIB inertia coefficient

$$M = \frac{M_C M_{NC}}{M_C + M_{NC}} \tag{20-69}$$

*Transient Stability Assessment.* Transient stability assessment is performed by means of the EAC, described in the subsection Transient Stability Assessment using the EAC. Despite the fact that T-D simulations are very useful methods that can provide a very detailed analysis of the transient stability problem, EAC has demonstrated to be a powerful and unique tool for sensitivity analysis and control issues [13].

EAC states that the stability of a system in faulted conditions may be assessed by means of a stability margin, which is defined as the excess of decelerating area (that represents the maximum potential energy that system can dissipate in post-fault conditions) over the accelerating area (that represents the kinetic energy gained during the fault) of the equivalent OMIB $P$-$\delta$ curve; this margin is written as follows:

$$\eta = A_{\text{dec}} - A_{\text{acc}} \tag{20-70}$$

Expression (20-70) establishes the energy conservation: the energy gained during the fault must be released as potential energy in the post-fault period for the system to be stable, and otherwise the system is unstable [13].

In this context SIME uses EAC to assess transient stability; however, what makes it attractive is that SIME calculates the $P_m$-$\delta$ and the $P_e$-$\delta$ curves using the results of a T-D program (or real-time measurements depending upon the case) only for the period of time EAC requires to assess stability (that is usually quite short). The computation duration is upper bounded by the time the OMIB takes to reach unstable and return angles $\delta_u$ or $\delta_r$. Angle $\delta_u$ is met at the unstable equilibrium point of the $P_e$ and $P_m$ curves at which system becomes unstable. Angle $\delta_r$ represents the maximum angular deviation the system reaches without going unstable (see Fig. 20-32 for a graphical example of both angles).

In general, for during-fault and post-fault scenarios the stability margin can be expressed as:

$$\eta = -\int_{\delta_0}^{\delta_{ch}} P_a\, d\delta - \int_{\delta_{ch}}^{\delta_u} P_a\, d\delta = -\int_{\delta_0}^{\delta_u} P_a\, d\delta \qquad (20\text{-}71)$$

where $\delta_{ch}$ represents the angle where accelerating power changes sign (from positive to negative). The conditions (criteria) for the system to be declared stable or unstable are described below (As in [13]).

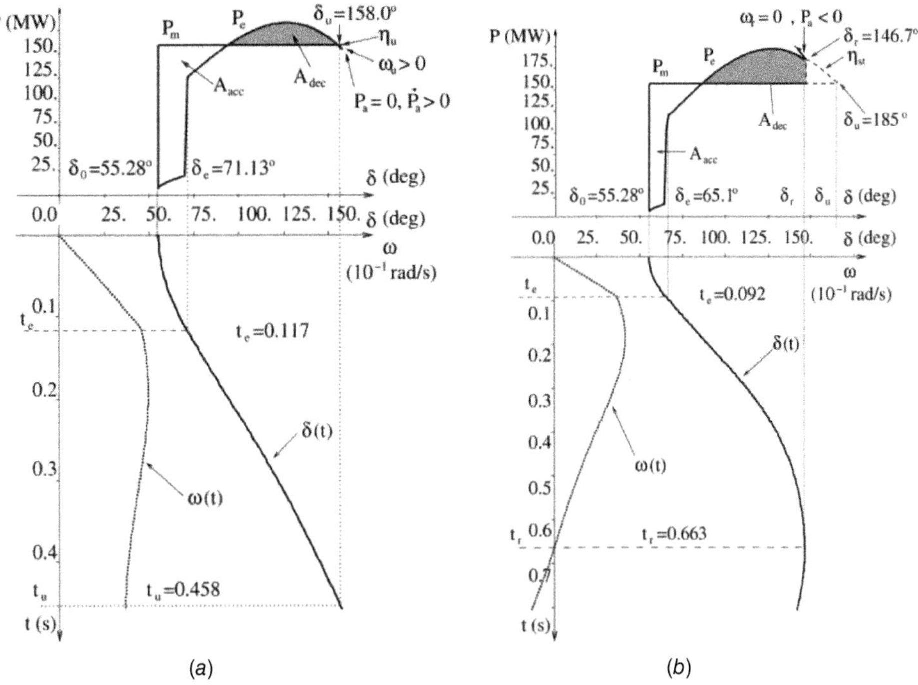


**FIGURE 20-32** SIME stability and instability conditions and computation of their corresponding stability margins. Simulations performed on the three-machine test system. Application of the EAC to the rotor angle-power curve of the OMIB equivalent. Corresponding OMIB rotor angle and speed curves [35]. (*a*) Unstable case illustrating transient instability conditions and negative (unstable) margin; (*b*) stable case illustrating transient stability conditions and positive (stable) margin.

*Unstable Conditions.* In this case, stability margin is negative, $\eta < 0$, what means that the accelerating area is greater than the decelerating area ($A_{dec} < A_{acc}$). The $P_e$ curve crosses $P_m$ and $P_a$ passes by zero and continues increasing, then the OMIB losses synchronism. An OMIB reaches the unstable angle at time $t_u$ when:

$$P_a(t_u) = 0, \quad \overset{\circ}{P}_a(t_u) = \left.\frac{dP_a}{dt}\right|_{t=t_u} > 0 \tag{20-72}$$

where $\omega > 0$ for $t > t_0$. The conditions expressed in Eq. (20-72) are used to determine the critical OMIB of the system.

*Stable Conditions.* In this case, the stability margin is positive, $\eta > 0$, what means that the decelerating area is greater than the accelerating area ($A_{dec} > A_{acc}$). System kinetic energy is less than the potential energy and $P_e$ curve stops at $\delta = \delta_r$ before crossing $P_m$ then the rotor angle starts decreasing. An OMIB is stable when the return angle $\delta_r$ is reached at $t_r$:

$$\omega(t_r) = 0, \quad \text{with } P_a(t_r) < 0 \tag{20-73}$$

*Stability Margins.* The unstable margin is written as follows:

$$\eta_u = -\frac{1}{2} M\omega_u^2 \tag{20-74}$$

Computation of this margin is simple and closed, and avoids accelerating and decelerating areas calculation by numerical integration. The stable margin is defined as follows:

$$\eta_s = \int_{\delta_r}^{\delta_u} |P_a| \, d\delta \tag{20-75}$$

The stable margin can only be approximated because $\delta_u$ and $P_e(\delta)$ ($\delta_u > \delta > \delta_r$) cannot be computed in a direct way, since the $P$-$\delta$ curve returns at $\delta = \delta_r$, before reaching them. To calculate this margin we can only use an approximation.

Figure 20-32 schematically shows stability margins computation from simulations performed in the IEEE 9-bus, three-machine test system [9]. In this work, the approximation of stability margin was performed by means of the least squares technique.

***The Dot Product Method.*** A widely used heuristic criterion for detecting system loss of synchronism using T-D simulations, consists of verifying the evolution in time of rotor angle difference between extreme machines (the most advanced and the most delayed ones), measured with respect to an established reference[c]. If this angle difference is greater than a given maximal (threshold) value (say, 360 electrical degrees), the system is declared unstable. However, the heuristic criterion has the disadvantage of being very subjective; for example, the maximum angle separation values are arbitrarily chosen, and their practical values can be different for different systems [1]. To overcome this disadvantage of the heuristic stability criteria, the dot product method was developed as an objective stability/instability criterion to be used in the T-D simulations.

The dot product method was first proposed in [36], and is a hybrid method, because it applies the transient energy function instability criterion that detects the moment when the post-fault

[c]Rotor angle is the important factor because appreciable increase in machine speeds does not necessarily imply that the synchronism is lost [8]. However, rotor speed variations are sometimes taken into account, generally in combination with rotor angle variations. The choice of the heuristic criterion depends strongly on the power system modeling; the choice of the threshold values (stability/instability boundary) depends upon the very physical system.

system trajectory crosses the PEBS (the exit point), to assess instability of a T-D simulation (see Fig. 20-32) [37–39].

The dot product instability criterion DOT 1 is objective, since it does not depend on any arbitrary or system-dependent parameter, and can be defined as

$$\mathrm{DOT1} = F_k \cdot (\Theta_k - \Theta_{pst}) = \sum_{i=1}^{ng} f_i(\theta_i - \theta_{psti}) \tag{20-76}$$

where all the following variables, are referred to the system center of inertia (COI):

- $F_k$ is the vector of accelerating powers of all the synchronous generators of the system, at the $k$th time-step of the T-D simulation.
- $\Theta_k$ is the vector of load angles of all the synchronous generators of the system, at the $k$th time-step of the T-D simulation.
- $\Theta_{pst}$ is the constant vector of load angles of all the synchronous generators of the system at the postdisturbance stable equilibrium point.
- $ng$ is the number of synchronous machines in the system.

The reason behind the name of the present method can be readily understood, by observing Eq. (20-76): it consists of applying, at each time step of the T-D simulation in the postdisturbance state, a dot product of two vectors of generator variables, to compute DOT1. At the moment DOT1 changes sign, instability is detected; if DOT1 sign remains constant, the system is stable. This makes the method implementation very easy and straightforward. In addition, the accelerating power $f_i$, and the rotor load angle $\delta_i$, of the synchronous generator "$i$" of the system, can be determined as

$$f_i = P_{mi} - P_{ei} - \frac{M_i}{M_T} P_{\mathrm{COI}} \tag{20-77}$$

where $P_{mi}$ and $P_{ei}$ are the mechanical and electrical power of machine "$i$." $M_i$ is the inertia coefficient of machine "$i$," while $M_T$ is the system equivalent inertia coefficient; they are defined as

$$M_i = \frac{2H_i}{\omega_0}, \quad M_T = \sum_{i=1}^{ng} M_i \tag{20-78}$$

In Eq. (20-78), $H_i$ is the per unit inertia constant of generator "$i$," and $\omega_0$ is the base angular synchronous speed of the system. The accelerating power of the COI, $P_{\mathrm{coi}}$, is computed using the individual machine accelerating powers $P_a$ as

$$P_{\mathrm{COI}} = \sum_{i=1}^{ng} P_{ai} \tag{20-79}$$

The load angles of all the synchronous generators of the system, at the $k$th time-step of the T-D simulation and at the stable equilibrium point (denoted by subscript $psti$), referred to the COI are

$$\theta_i = \delta_i - \delta_{\mathrm{COI}} \qquad \theta_{psti} = \delta_{psti} - \delta_{\mathrm{COI}} \tag{20-80}$$

where is $\delta_i$ the rotor load angle of machine "$i$" of the system and the system angular reference $\delta_{\mathrm{COI}}$ is computed as

$$\delta_{\mathrm{COI}} = \frac{1}{M_T} \sum_{i=1}^{ng} M_i \delta_i \tag{20-81}$$

In [38] a stability criterion DOT2, was proposed for the dot product method, to detect first swing stable simulations. This criterion is defined as follows:

$$\mathrm{DOT2} = \Omega_k \cdot (\Theta_k - \Theta_{pst}) = \sum_{i=1}^{ng} \omega_i (\theta_i - \theta_{psti}) \tag{20-82}$$

where all the following variables, are referred to the system COI:

- $\Omega_k$ is the vector of rotor angular speeds of all the synchronous generators of the system, at the $k$th time-step of the T-D simulation.
- $\Theta_k$, $\Theta_{pst}$, and $ng$ are the same variables used in Eq. (20-76) described above.

$\omega_i$ is the rotor angular speed of machine "$i$," referred to $\omega_{\mathrm{COI}}$, the speed of the COI, which are computed as

$$\omega_i = \omega_{Ai} - \omega_{\mathrm{COI}} \tag{20-83}$$

$$\omega_{\mathrm{COI}} = \frac{1}{M} \sum_{i=1}^{n} M_i \, \omega_{Ai} \tag{20-84}$$

For practical reasons, when implementing the stability criteria of Eqs. (20-71) and (20-82), it is recommended in [38] that the vector of initial, pre-fault steady state angles $\Theta_0$ should be used; instead the post fault equilibrium angles $\Theta_{pst}$.

Using the dot products DOT1 and DOT2 presented above [Eqs. (20-71) and (20-72), respectively], two versions of the dot product method can be implemented:

- A first-swing dot product method, using both, DOT1 and DOT2 simultaneously.
- A multi-swing dot product method using DOT1 only.

The first-swing dot product method could be used as an early termination criterion of T-D simulations, limiting the simulation time to the one required for detecting the instability or the stability conditions right after the first-swing of the transient conditions [38]. This early termination criterion of T-D simulations helps saving computational resources and simulation time, and is very useful in the analysis of power systems that do not have poorly damped oscillation problems.

Stability assessment is performed as follows: after the fault is cleared, both dot products, DOT1 and DOT2 are computed at each integration time step of the simulation. The values of these dot products at two successive time steps (the current and the previous one) are compared. If DOT1 (which is usually negative at the beginning of the simulation [26]) changes sign before DOT2, the simulation is classified as unstable, because the post-fault system trajectory has crossed the PEBS; otherwise, if DOT2 changes sign first, the simulation is declared as first-swing stable [38].

The multi-swing dot product method is easily implemented by using DOT1 criterion only: if it changes sign during the simulation, the system is declared to be unstable; if the T-D program reaches its maximum integration time without meeting DOT1 conditions, is classified as multi-swing stable.

## 20.3.3 Transient Stability Control

***Introduction.*** Power system operating states can be classified into five main types [40], that must be born in mind to make an adequate design of control schemes. Figure 20-33 shows a Dy Liacco's diagram or transition states diagram, which shows the operating states and their interactions.

Equality restrictions ($E$) in Fig. 20-33 are related to maintaining the balance between system load and generation. Inequality restrictions ($I$) represent the fact that specific variables (like line power

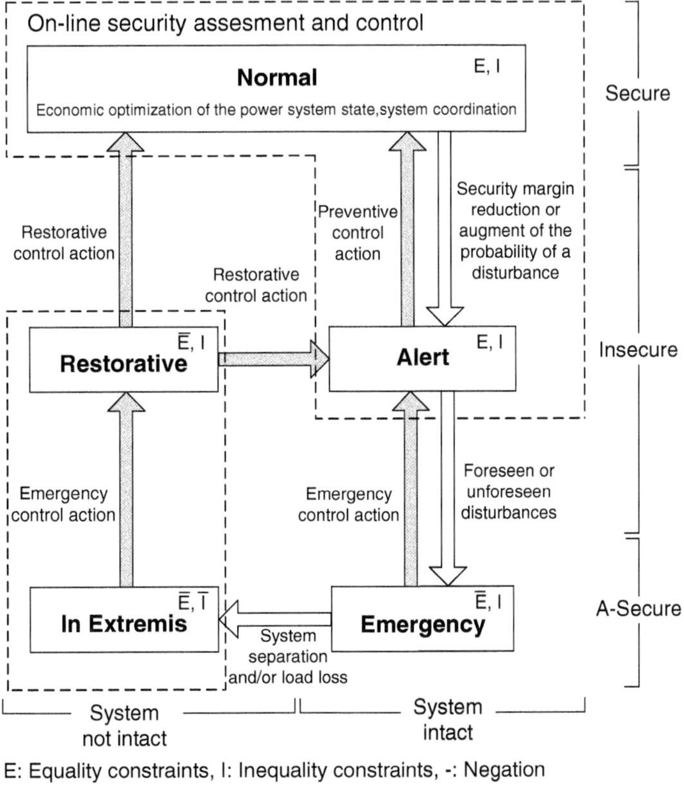


**FIGURE 20-33** Dy Liacco's diagram (adapted from [40]).

and bus voltages) must not exceed or equal their limit values, which represent the equipment physical limits [40].

In the *normal* operating state, all the variables must be within their own limits, which means that all the restrictions are satisfied and the generation is adequate to satisfy the total system load [10, 40]. In this operating state reserve margins must be large enough to maintain an adequate security level respect to the strains, so that the system has to be able to stand the occurrence of a disturbance without violating any restrictions.

If the security level is diminished or the probability of a disturbance augments, then the system enters into an *alert* operating state, in spite of the fact that in this state all the restrictions are satisfied and, within specific margins, the reserve margin would be such that a disturbance would result in a violation of the equality or inequality restrictions. In these conditions, *preventive control actions* must be applied to restore the security level of the system to the normal state.

If the fault actually occurs, the system enters into an *emergency* operating state. Here inequality restrictions are violated. At this point, the system would still remain intact only if *emergency control actions* are performed; otherwise the system enters into an *in extremis* operating state. In this state both, equality and inequality restrictions are violated and the system cannot hold on intact, what would result in the loss of parts of the system (brownouts), unless more drastic emergency control actions ("heroic measures") are carried out to save as many parts of the system as possible, from the total collapse (blackout).

When the collapse has concluded, the system enters into a *restorative* state. In this state, control actions are performed to reconnect the system, and pick up load again. From the restorative state the system could enter into a normal or alert operating state depending on the circumstances.

Table 20-4 displays the control actions that are considered adequate to be applied in each of the operating states.

**TABLE 20-4** Control Actions That Are Suitable for Different Operating States (adapted from [40])

| Operating state | Control action |
|---|---|
| Alert | Economic dispatch or security redispatch, network reconfiguration, voltage reduction, etc. The purpose is to restore the reserve margins of the system. |
| Emergency | Immediate control measures to eliminate the equipment overloaded like: fast fault-clearing (automatic), fast valving (automatic), dynamic braking (automatic), modulation of the excitation system (automatic), capacitors switching (manual), HVDC lines modulation, load modulation, generation tripping (automatic), and all measures for the alert operating state. |
| In extremis | Heroic actions to avoid the system's breakthrough: load shedding (manual), controlled islanding of some areas of the system and all measures mentioned for the alert and emergency operating state. |
| Restorative | Corrective control actions to restore the optimal functioning of the system, such actions are: the units restarting, load restoration, resynchronizing the isolated areas of the system, etc. |

Control systems currently used can be classified into different ways, depending on various factors such as [1]:

- The operating state of the power system.
- The type of control system (open-loop or closed-loop).
- The geographic area covered by the control system (distributed or centralized).

Depending on the operating state of the system, controls can be classified into

- Preventive controls (online controls)
- Corrective controls (real-time or emergency controls)

*Preventive controls* are applicable to the system when the operating state is classified as *normal* or *alert*. When the system is faced to a possible contingency, vulnerability of the system must be determined in order to design and apply a control action before the occurrence of the disturbance and, in consequence, to enlarge the security level of the system. In the *normal* operating state, the control systems commonly employed are those with the main objective of improving the economic efficiency of the system.

Preventive control actions in the *alert* state are mainly performed to increase the system security level. They include security constrained redispatch and online adjustment of system protection schemes.

In order to improve feasibility and acceptation of these control actions, very advanced online dynamic security assessment and control methods have proposed to include economic criteria in the design of these controls. In this way, methods including stability constraints in the optimal power flow problem have been proposed for both, transient stability security redispatch and updating of SPS settings [41, 42].

*Corrective* or *emergency controls* act when the contingency has actually occurred. There are two basic schemes of emergency control [43]:

- Response-based (or measurement-based) emergency controls
- Event-based emergency controls

Response-based emergency controls are based on measured electric variables (such as voltage, frequency, etc.) and initiate their protective actions when the contingency has caused that the measured value hits the trigger level of the corresponding variable [43]. These emergency controls are used to adjust the control action magnitude in accordance to the contingency severity.

Event-based controls are designed to operate upon the recognition of a particular combination of events (like the loss of several lines in a substation). These controls act automatically, applying a previously determined fixed control action, when the occurrence of the event has been detected. The event-based emergency control is faster than the measurement- based control since it does not have to wait for the system response to a specific event [43]. However, these kinds of controls must be designed for all relevant events, while a response-based control operates even for events for which they were not planned.

Applying transient stability controls appropriately, requires the use of a dynamic security assessment and control function that could be online (for preventive controls) or in real time (for corrective controls). The structure of both types of security functions is described below.

***Transient Stability Security Assessment and Control Functions.*** After the 1965 Northeast system blackout in the United States, the concept of security started shifting from that of system robustness designed at the planning stage, to its present sense of risk aversion in the operating stage (firstly known as "adaptive reliability" [44]). Now, power system security is also defined in the system operation framework as "an instantaneous, time-varying condition" reflecting the robustness of the system relative to imminent disturbances [45].

Security analysis should therefore be performed in both, planning and operating environments, with different objectives and requirements. For the planner, the term "security" refers to those aspects of reliability analysis that deal with the ability of the system, as it is expected to be constituted at some future point in time, to withstand unexpected losses of certain system components. What is overlooked in this approach is that even the most reliable systems will inevitably experience periods of severe insecurity from the operator perspective.

This new view of system security (starting after the 1965 blackout) required implementing new strategies and institutions for monitoring, operating and controlling power system online. Some of the most important technical achievements coming from this new vision are the implementation of hierarchical power system control techniques, discrete emergency controls and the creation of a new facility, the control center.

*The Control Center Energy Management System.* In the initial stage of control centers' development, the dispatcher office already had a digital SCADA (for supervisory control and data acquisition) system that collected data, and presented it in a graphical form on a computer monitor or a mimic board. That system also allowed the operator to perform remote control actions on the power system by entering commands or by pushing on correspondingly labeled buttons on the monitor [46].

The main difference between the new control center installed sometime after the 1965 Northeast system blackout and the former dispatcher office in a large power plant is the addition of online security functions, which require the use of an energy management system (EMS). The EMS is a computer system that provides basic support services and a set of applications for performing the operation and control of a power system in a secure and economic fashion. A schematic representation of the EMS software structure is presented in Fig. 20-34.

Power system measurements are transmitted via a digital information system to the central data base. The dispatcher training simulator is an EMS function providing real-time power system simulation under routine and emergency conditions that is required for training the operating personnel. Its importance was highlighted by some blackouts in Europe, and after the 1977 New York Blackout [47], in which operator reactions under emergency situations involving slow phenomena, like network elements overloads or long-term voltage instabilities, were not always correct [48].

One very important component of the EMS is the man-machine interface (MMI). It consists of cathode ray tube (CRT) screens, dynamic wall displays, trend recorder, loggers, and alarms. Its function is to present the measurements collected by the SCADA and the results of the analysis performed

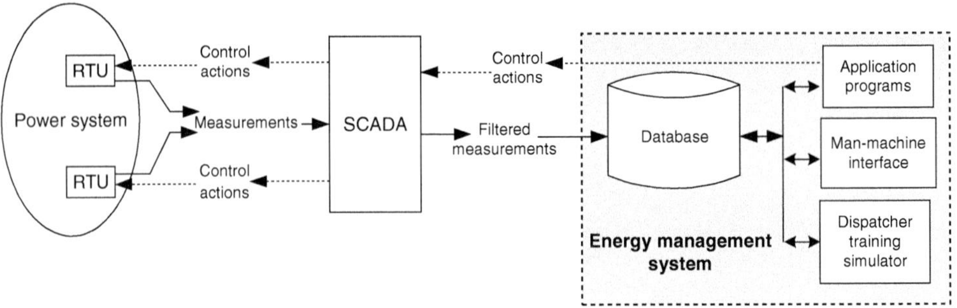


**FIGURE 20-34** Main component functions of a control center EMS (adapted from [1]).

in the application programs, in a way suitable for helping the operator in the decision-making process. EMS application programs can be classified in terms of their functions as [1, 48]:

- Economic functions
- Preprocessing data functions
- Security functions

The scope of this document is limited to security functions related to transient stability, which are classified in the area of dynamic security, which can be defined as [1]:

> ***Dynamic Security Assessment (DSA)*** *are methodologies for evaluating the stability and quality of the transient processes between the precontingency and postcontingency states. In this case, Dynamic Security Assessment and Control (DSA&C) aims at ensuring that the system will be stable after the contingency occurrence and that the transients caused by such a contingency will be well damped, of small amplitude and with little impact on the quality of service* [49, 50].

As mentioned above, power system controls are applied depending on the security level of the power system operating state. For determining this security level, and designing a suitable control action in case it is required, two main types of DSA&C functions have been developed: online security functions for preventive controls and real-time security functions for emergency controls. The structure of both of them is presented in Fig. 20-35, and a description is given below.

*Online Security Assessment and Control Function.* In online analysis, after obtaining the current operating condition, the operator uses security functions to assess the security level of the system. Security assessment of the current power system state is performed with respect to a set of probable contingencies, chosen by the system operator. Sometimes, the "$n - 1$" principle is used for setting up the set of contingencies: this principle implies checking that the loss of any of the transmission system elements does not lead to a power system disruption[d] [48]. Security assessment responds the question "what if next contingency, out of a set of probable contingencies, takes place?," before the contingency of interest actually happens. This concept of analyzing the "next contingency set" is a sound operating strategy in which the operator should always check whether a next contingency would cause a problem, especially when the system has undergone one or more equipment outages due to previous disturbances or operating actions [51].

[d]The "$n - 1$" principle can also be applied in transient stability assessment, but it considers two contingencies for each element of the transmission network: each one of the contingencies applies a short circuit at a different terminal node of the transmission element.

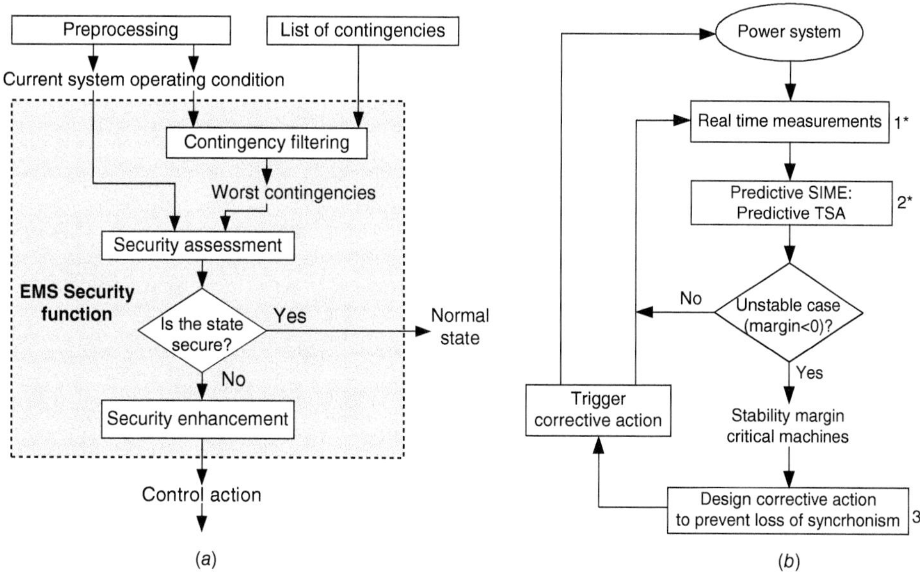


**FIGURE 20-35** Transient stability security assessment and control functions. (*a*) Online security function; (*b*) real-time security function (adapted from [1, 13, 14, 52]).

Determining the security level of the current system operating condition is a very hard task: the security function usually has to analyze a set of probable "next contingencies" in a very short time. The number of contingencies composing the contingency list can be very large. For example, the "$n - 1$" static security assessment of a system containing 1000 buses (with an average of four transmission elements connected to each bus), requires assessing the security of the current operating condition under approximately 4000 possible contingencies.

Due to the huge amount of data and to the complexity of the required power system analysis, all EMS security functions (static and dynamic) share the same general structure, shown in Fig. 20-35*a*.

Major components of the online security function are [45]:

- *Contingency Filtering.* Using the current steady-state operating condition of the power system and a fast approximate screening tool, select the worst contingencies from an initial contingency list.
- *Security Assessment.* Determine whether the system operating state is secure or insecure with respect to each one of the contingencies composing the list.
- *Security Enhancement.* If there is at least one contingency that can cause a potential security problem (system instability, an interruption of the electric service or overloads on transmission network elements) determine a countermeasure able to make the system secure using a control action.

A very important fact of this online security analysis view is that the determination of the vulnerability of the present system conditions does not complete the picture, because *the solution to any existing security problems must also be found*. This implies designing a preventive control action able to stabilize all the harmful contingencies.

To date, practical implementation of online dynamic security analysis functions for a large power system has been feasible, as can be consulted in [53], where many DSA&C online security functions implemented in real power system control centers around the world, are described.

It is important to mention that two main kinds of contingency filtering techniques have been reported to be used in practical DSA functions:

- The contingency filtering-ranking-assessment (FILTRA) technique [13, 54, 55], that is based in the SIME method, and is used in the TSA&C systems of JPM in United States [53], CESI in Italy and GRTN in Greece [53, 56]. One advantage of this technique is that it includes both, the contingency filtering and the security assessment blocks of Fig. 20-7*a* in one single, coordinated process, using the same method.
- Functions that are based in the Lyapunov method and that are used in TEPCO Company in Japan [53, 57] and in the TSA of Northern States Power Company of United States [54, 58, 59]. For these TSA systems, the detailed security assessment is performed using a different technique. In case of TEPCO system, they use T-D simulations for the detailed security assessment [53], while in the Northern States Power system, they use the first-swing version of the dot product method [38].

Regarding the security enhancement functions for transient stability, only the DSA&C functions based in the SIME method, have been implemented in practice. All the companies using SIME-based systems mentioned above, plus the ONS system in Brazil [53], have implemented the design of preventive control actions, mainly security generation redispatch, using a conventional power flow [53, 56]. However, the information that the SIME method provides, and the graphical representation of the transient stability that it allows, permit to extend the scope of the techniques for designing online preventive controls, to more complex functions like:

- Techniques for stabilizing multiple harmful contingencies simultaneously [1, 13, 60]
- Techniques, that allow adapting the settings of discrete emergency controls to the current power system operating conditions (the Open Loop Emergency Control OLEC technique) [1, 13, 42]
- Techniques for the preventive control of poorly damped transient oscillations [1, 35, 61]
- Techniques for the transient stability security redispatching using an optimal power flow [1, 13], [41, 62–64].

*Real-Time Security Assessment and Control Function.* Unlike online preventive control functions, transient stability real-time security control functions have not been practically implemented yet. However, the invention of the synchronized phasor measurements at the beginning of the 1990s [65], and the deployment of this type of measurement systems all around the world since then [66], have provided a platform on which response-based emergency stability controls, could theoretically be developed in a near future, at the interconnected system level [67–69].

The emergency SIME (E-SIME) method, that was first proposed in 1997 [70], could be used in a near future for developing a transient stability real-time security assessment and control function (see Fig. 20-7*b*). In the first publications about this method, it was intended to replace, or improve, event-based primary emergency controls (generation tripping schemes mostly), located in very important generation plants, like the Itaipú Power Plant in Brazil [13, 52, 60, 70, 71]. The performance of E-SIME was tested in transient stability cases, considering the very severe contingency that activated the event-based control, and the method obtained very promising results. However, analyzing this very severe disturbance, strongly limited the time the method has available to design and apply the emergency control actions in real-time [60, 70, 71].

At the beginning of the 2000s, some additional developments of the preventive SIME method, and the possibility of installing wide-area synchronized phasor measurements in different systems all around the world in a near future [66], gave rise to the proposal of a different approach, where E-SIME is applied to control instabilities, at the whole interconnected system level [72].

First, it was proposed that the settings of the event-based emergency controls could be verified and adjusted online, to the current operating conditions of the power system, using the SIME-based dynamic security assessment and control function. This approach is called OLEC, for open loop emergency control [42]. This partially solved the problem, because the uncertainty in system modeling still remains.

From 2001 to 2003, E-SIME was developed in the framework of the EXaMINE European project [72]. The general objective of the project was to design, prototype and test novel monitoring systems supporting the control centers of the European power system in order to adapt and enhance their security control functions. There were two main tasks to be performed [72]: (1) obtain a better knowledge of the network actual or planned status, and (2) design emergency control actions to manage with fast developing phenomenon. Emergency control techniques developed in this project were related to solve the problem of obtaining generator rotor angles from PMU measurements [73] and to apply new techniques like reinforcement learning to improve the performance of emergency control schemes like dynamic braking [74].

In [75] the combination of the open-loop emergency control (OLEC) and the closed-loop emergency control (E-SIME) was proposed, in order to be able to stabilize very severe contingencies in which the instability is met so quickly, that E-SIME alone has not enough time to act. The resulting SIME-based adaptive transient stability control is very powerful, and would represent an efficient solution for online transient stability assessment and control combining both, online analysis and real-time measurements.

In [76] the authors presented the achievements and prospects of E-SIME. They use a generation shedding scheme as a control action to avoid instability that has been tested in various real-life power systems like the South-Southeast Brazilian system, the EPRI test system, the WECC system and Hydro-Québec system using the ST-600 and ETMSP programs as T-D simulations as a base for E-SIME just for want of real-life measurements.

Currently some works have been successfully developed to write a program for applying the E-SIME method at the interconnected system level [14], and to test the predictive stability assessment of E-SIME in a real-time simulator [77]. They have provided very encouraging results toward the development of a practical real-time transient stability assessment and control function, which is still underway.

## 20.3.4 Bibliography


1. D. Ruiz-Vega. "Dynamic Security Assessment and Control: Transient and Small Signal Stability," Ph.D. Dissertation, University of Liège, Belgium, 2002.
2. E. W. Kimbark, "Introduction to Problems of Power System Stability," presented at the in IEEE Committee Tutorial Course No. 70 M 62-PWR. "*Modern Concepts of Power System Dynamics*," 1970.
3. A. M. Wildberger, "Complex Adaptive Systems: Concepts and Power Industry Applications," *IEEE Control Systems*, vol. 17, pp. 77–88, 1997.
4. U. G. Knight, *Power Systems in Emergencies: From Contingency Planning to Crisis Management,* John Wiley & Sons, West Sussex, England, 2001.
5. P. Aristidou, S. Lebeau, and Th. Van Cutsem, "Power System Dynamic Simulations Using a Parallel Two-Level Schur-Complement Decomposition," *IEEE Transactions on Power Systems*, vol. 31, no. 5, Sep. 2016.
6. IEEE Task Force on Terms & Definitions, System Dynamic Performance Subcommittee, Power System Engineering Committee. "Proposed Terms and Definitions for Power System Stability," *IEEE Transactions on Power Apparatus and Systems*, vol. PAS-101, no. 7, pp. 1894–1898, Jul. 1982.
7. IEEE/CIGRE Joint Task Force on Stability Terms and Definitions, "Definition and Classification of Power System," *IEEE Transactions on Power Systems*, vol. 19, pp. 1387–1401, 2004.
8. E. W. Kimbark, *Power System Stability: Vol 1 Elements of Stability Calculations,* John Wiley & Sons, New York, USA, 1948.
9. P. M. Anderson and A. A. Fouad. *Power System Control and Stability*, 2nd ed., Wiley-IEEE Press, New York, USA, 2002.
10. P. Kundur, *Power System Stability and Control,* McGraw Hill, New York, USA, 1994.
11. O. G. C. Dahl, *Electric Power Circuits Theory and Applications Volume II: Power System Stability,* McGraw-Hill, New York, USA, 1938.
12. S. B. Crary, *Power System Stability Volume II-Transient Stability,* John Wiley & Sons, New York, USA, 1947.
13. M. Pavella, D. Ernst, and D. Ruiz-Vega, *Transient Stability of Power Systems: A Unified Approach to Assessment and Control,* Kluwer Academic Publishers, Dordrecht, The Netherlands, 2000.
14. L. L. Juárez-Caltzontzin, "Generation Tripping for Transient Stability Control Using the Emergency Single Machine Equivalent Method," M.Sc. Thesis, Graduate Program in Electrical Engineering, SEPI-ESIME-Zacatenco, IPN Mexico, Jan. 2011.

15. P. C. Magnusson, "The Transient-Energy Method of Calculating Stability," *Transactions of the American Institute of Electrical Engineers*, vol. 66, pp. 747–755, 1947.

16. P. D. Aylett, "The Energy-Integral Criterion of Transient Stability Limits of Power Systems," *IEE Proceedings*, vol. 105, part C, pp. 527–536, 1958.

17. A. A. Gorev, "A Collection of Works on Transient Stability of Power Systems," in *Second Lyapunov's Method and Its Application to Energy Systems*, Naouka Eds, Gosenergoirdat, 1966.

18. A. H. El-abiad and K. Nagappan, "Transient Stability Regions of Multimachine Power Systems," *IEEE Transactions on Power Apparatus and Systems*, vol. PAS-85, pp. 169–179, 1966.

19. G. E. Gless, "Direct Method of Liapunov Applied to Transient Power System Stability," *IEEE Transactions on Power Apparatus and Systems*, vol. PAS-85, pp. 159–168, 1966.

20. M. A. Pai (ed.), *Power System Stability Analysis by the Direct Method of Lyapunov*, North-Holland, Amsterdam, The Netherlands, 1981.

21. M. A. Pai, *Energy Function Analysis for Power System Stability*, Kluwer Academic Publishers, Dordrecht, The Netherlands, 1989.

22. M. Pavella and P. Murthy, *Transient Stability of Power Systems: Theory and Practice*, John Wiley & Sons, 1994.

23. A. A. Fouad and V. Vittal, *Power System Transient Stability Analysis Using the Transient Energy Function Method*, Prentice Hall, New York, USA, 1992.

24. H. D. Chiang, *Direct Methods for Stability Analysis of Electric Power Systems: Theoretical Foundations, BCU Methodologies and Applications*, John Wiley & Sons, New York, USA 2011.

25. K. R. Padiyar, *Structure Preserving Energy Functions in Power Systems: Theory and Applications*, CRC Press, New York, USA, 2013.

26. T. Athay, V. R. Sherkat, R. Podmore, S. Virmani, and C. Puech, *Transient Energy Stability Analysis*, System Engineering for Power. Emergency Operating State Control-Section IV U.S. Department of Energy, Publication No. CONF-790904-PL, 1979.

27. E. L. Belmonte-González, "Transient Stability Studies using the Transient Energy Function Method Incorporating Nonlinear Static Load Models (in Spanish)," M.Sc. Thesis, Graduate Program in Electrical Engineering, SEPI-ESIME-Zacatenco, Mexico City, Mexico, 2004.

28. A. R. Bergen and D. J. Hill, "A Structure Preserving Model for Power System Stability Analysis," *IEEE Transactions on Power Apparatus and Systems*, vol. PAS-100, pp. 25–35, 1981.

29. N. Kakimoto, Y. Ohsawa, and M. Hayashi, "Transient Stability Analysis of Multimachine Power System with Field Flux Decays via Lyapunov's Direct Method," *IEEE Transactions on Power Apparatus and Systems*, vol. PAS-99, pp. 1819–1827, 1980.

30. M. Ribbens-Pavella, P. G. Murthy, and J. L. Howard, "An Acceleration Approach to Practical Stability Domain Estimation in Power Systems," *Proceedings of the 20th IEEE Conference on Decision and Control*, vol. 1, pp. 471–477, San Diego California US, 1981.

31. C. Hsiao-Dong, F. F. Wu, and P. P. Varaiya, "A BCU Method for Direct Analysis of Power System Transient Stability," *IEEE Transactions on Power Systems*, vol. 9, pp. 1194–1208, 1994.

32. L. Fonseca and I. Decker, "Iterative Algorithm for Critical Energy Determination in Transient Stability of Power System," *Proceedings of the IFAC Symposium Planning & Operation in Electric Energy Systems*, Rio de Janeiro, Rio de Janeiro, Brazil, 1985.

33. M. Pavella, "Generalized One-Machine Equivalents in Transient Stability Studies," *IEEE Power Engineering Review*, vol. 18, pp. 50–52, 1998.

34. V. A. Venikov, *Automatic Power System Control*, Mir Publishers, Moscow, USSR, 1982.

35. D. Ruiz-Vega, A. R. Messina, and M. Pavella, "A Novel Approach to the Assessment of Power System Damping," *Proceedings of the Power System Computation Conference*, 2002.

36. G. A. Maria, C. Tang, and J. Kim, "Hybrid Transient Analysis," *IEEE Transactions on Power Systems*, vol. 5, no. 2, May 1990.

37. C. K. Tang, C. E. Graham, M. El-Kady, and R. T. H. Alden, "Transient Stability Index from Conventional Time Domain Simulation," *IEEE Transactions on Power Systems*, vol. 9, no. 3, Aug. 1994.

38. G. C. Ejebe, C. Jing, J. G. Waight, G. Pieper, and F. Jamshidian, "Security Monitor for On-line Dynamic Security Assessment," *Proceedings of the Fourth International Conference on Power System Control and Management*, pp. 58–64, 1996.

39. M. La Scala, G. Lorusso, R. Sbrizzai, and M. Trovato, "A Qualitative Approach to the Transient Stability Analysis," *IEEE Transactions on Power Systems*, vol. 11, no. 4, Nov. 1996.

40. L. H. Fink and K. Carlsen, "Operating under Stress and Strain," *IEEE Spectrum*, Mar. 1978.

41. D. Ruiz-Vega and M. Pavella, "A Comprehensive Approach to Transient Stability Control—Part I: Near-Optimal Preventive Control," *IEEE Transactions on Power Systems*, vol. 18, no. 4, pp. 1446–1453, Nov. 2003.

42. D. Ruiz-Vega and M. Pavella, "A Comprehensive Approach to Transient Stability Control—Part II: Open Loop Emergency Control," *IEEE Transactions on Power Systems*, vol. 18, no. 4, pp. 1454–1460, Nov. 2003.

43. *System Protection Schemes in Power Networks,* CIGRE Task Force 38.02.19, Technical brochure, Jun. 2001.

44. T. E. Dy-Liacco, "The Adaptive Reliability Control System," *IEEE Transactions on Power Apparatus and Systems*, vol. PAS-86, no. 5, pp. 517–531, May 1967.

45. N. Balu, T. Bertram, A. Bose, V. Brandwajn, J. Cauley, D. Curtice, A. A. Fouad, L. Fink, M. G. Lauby, B. F. Wollenberg, and J. N. Wrubel, "On-Line Power System Security Analysis," *Proceedings of the IEEE*, vol. 80, no. 2, pp. 262–280, Feb. 1992.

46. E. Mariani and S. S. Murthy, *Control of Modern Integrated Power Systems,* Springer, London, UK.

47. J. M. Bucciero, J. A. Dodge, R. N. Gillespie, R. O. Hinkel, S. L. Mann, S. M. Martin, R. P. Schulte, and D. Hayward, "Dispatcher Training Simulators: Lessons Learned," *IEEE Transactions on Power Systems*, vol. 6, no. 2, May 1991.

48. E. Handschin and A. Petroianu, *Energy Management Systems,* Springer-Verlag, Berlin, Germany, 1991.

49. *New Trends and Requirements for Dynamic Security Assessment,* CIGRE task force 38.02.13. Technical brochure. Dec. 1997.

50. P. Kundur, "Future Directions in Power Systems," Short Course, *Proceedings of the VII Symposium of Specialists in Electric Operational and Expansion Planning*, SEPOPE 2000. Curitiba, Brazil, May 21, 2000.

51. T. E. Dy-Liacco, "System Security: The Computers Role," *IEEE Spectrum*, pp. 43–50, Jun. 1978.

52. D. Ernst and M. Pavella, "Closed-Loop Transient Stability Emergency Control," *Proceedings of the IEEE Power Engineering Society Winter Meeting*, vol. 1, pp. 58–62, 2000.

53. *Review of On-Line Dynamic Security Assessment Tools and Techniques.* CIGRE Working Group C4.601. Technical brochure. Jun. 2007.

54. *Implementation of Dynamic Security Assessment at Northern States Power*, EPRI, Palo Alto, CA: 2001. 1006218.

55. *Single Machine Equivalent (SIME) Approach to Dynamic Security Assessment (DSA): Contingency Evaluation and Protection Control*, EPRI, Palo Alto, CA, 2000. 1000412.

56. A. Bihain, G. Burt, F. Casamatta, T. Koronides, R. Lopez, S. Massucco, D. Ruiz-Vega, and C. Vournas, "Advanced Perspectives and Implementation of Dynamic Security Assessment in the Open Market Environment," *Proceedings of the CIGRE Session 2002*, Paris, France.

57. H. D. Chiang, C. S. Wang, and H. Li, "Development of BCU Classifiers for On-Line Dynamic Contingency Screening of Electric Power Systems," *IEEE Transactions on Power Systems*, vol. 14, no. 2, May 1999.

58. *Analytical Methods for Contingency Selection and Ranking for Dynamic Security Analysis*, EPRI, Palo Alto, CA: 1994. EPRI TR-104352.

59. G. C. Ejebe, C. Jing, B. Gao, J. G. Waight, G. Pieper, F. Jamshidian, and P. Hirsch, "Online Implementation of Dynamic Security Assessment at Northern States Power Company," *Proceedings 1999 IEEE Power Engineering Society Summer Meeting*, vol. 1, pp. 270–272, Jul. 18–22, 1999.

60. D. Ernst, D. Ruiz-Vega, and M. Pavella, "Preventive and Emergency Transient Stability Control," *Proceedings of the VII Symposium of Specialists in Electric Operational and Expansion Planning (SEPOPE 2000)*, 2000.

61. D. Ruiz-Vega, A. R. Messina, and M. Pavella, "Online Assessment and Control of Transient Oscillations Damping," *IEEE Transactions on Power Systems*, vol. 19, no. 2, May 2004.

62. A. Pizano-Martinez, C. R. Fuerte-Esquivel, and D. Ruiz-Vega, "Global Transient Stability-Constrained Optimal Power Flow Using an OMIB Reference Trajectory," *IEEE Transactions on Power Systems*, vol. 25, no. 1, Feb. 2010.

63. A. Pizano-Martinez, C. R. Fuerte-Esquivel, and D. Ruiz-Vega, "A New Practical Approach to Transient Stability-Constrained Optimal Power Flow," *IEEE Transactions on Power Systems*, vol. 26, no. 3, Aug. 2011.

64. A. Pizano-Martínez, C. R. Fuerte-Esquivel, E. Zamora-Cárdenas, and D. Ruiz-Vega, "Selective Transient Stability-Constrained Optimal Power Flow Using a SIME and Trajectory Sensitivity Unified Analysis," *Electric Power Systems Research*, vol. 109, pp. 32–44, 2014.

65. A. G. Phadke and J. S. Thorp, *Synchronized Phasor Measurements and their Applications*, Springer, New York, USA, 2008.

66. A. G. Phadke, "The Wide World of Wide-Area Measurement," *IEEE Power and Energy Magazine*, Sep./Oct. 2008.

67. C. C. Liu, J. Jung, G. T. Heydt, V. Vittal, and A. G. Phadke, "The Strategic Power Infrastructure Defense (SPID) System: A Conceptual Design," *IEEE Control Systems Magazine*, Aug. 2000.

68. M. Amin, "Toward Self-Healing Energy Infrastructure Systems," *IEEE Computer Applications in Power*, Jan. 2001.

69. E. Santacana, G. Rackliffe, L. Tang, and X. Feng, "Getting Smart," *IEEE Power and Energy Magazine*, Mar./Apr., 2010.

70. Y. Zhang, L. Wehenkel, and M. Pavella, "A Method for Real-Time Transient Stability Emergency Control," *Proceedings of CPSPP'97, IFAC/CIGRE Symp. on Control of Power Systems and Power Plants*, pp. 673–678, Beijing, China, Aug. 1997.

71. D. Ernst, A. Bettiol, Y Zhang, L. Wehenkel, and M. Pavella, "Real-Time Transient Stability Emergency Control of the South-Southeasts Brazilian System," *Proceedings of the V SEPOPE*, Salvador, Brazil, May 1998.

72. A. Diu and L. Wehenkel, "EXaMINE—Experimentation of a Monitoring and Control System for Managing Vulnerabilities of the European Infrastructure for Electrical Power Exchange," *Proceedings of the IEEE PES Summer Meeting*, vol. 3, pp. 1410–1415, Jul. 25, 2002, Chicago, IL.

73. A. Del Angel, M. Glavic, and L. Wehenkel, "Using Artificial Neural Networks to Estimate Rotor Angles and Speeds from Phasor Measurements," *Proceedings of the Intelligent Systems Applications to Power Systems (ISAP) Conference 2003*, Lemnos, Greece, Sep. 2003.

74. M. Glavic, D. Ernst, and L. Wehenkel, "A Reinforcement Learning Based Discrete Supplementary Control for Power System Transient Stability Enhancement," *Proceedings of the Intelligent Systems Applications to Power Systems (ISAP) Conference 2003*, Lemnos, Greece, Sep. 2003.

75. D. Ruiz-Vega, M. Glavic, and D. Ernst, "Transient Stability Emergency Control Approach Combining Open-Loop and Closed-Loop Techniques," *Proceedings of the IEEE PES General Meeting*, Toronto, Canada, Jul. 13–18, 2003.

76. M. Glavic, D. Ernst, D. Ruiz-Vega, L. Wehenkel, and M. Pavella, "E-SIME—A Method for Transient Stability Closed-Loop Emergency Control: Achievements and Prospects," *Proceedings of the Symposium 2007, Bulk Power System Dynamics and Control VII—"Revitalizing Operational Reliability,"* Charleston South Carolina, U.S., Aug. 19–24, 2007.

77. V. M. Gutiérrez-Serrano, :Predicting Electric Power System Transient Stability Using Phasor Measurements (in Spanish), M.Sc. Thesis, Graduate Program in Electrical Engineering, SEPI-ESIME-Zacatenco, IPN Mexico, Jul. 2015.

## 20.4 IMPACT OF WIND GENERATION ON POWER SYSTEM STABILITY

*BY NILANJAN SENROY*

### 20.4.1 Introduction

The penetration of the renewable energy systems into the existing grid has increased significantly since the last one decade and it is expected to dominate the energy mix of different countries worldwide in the near future [1]. Wind is one the leading source of energy added into the system among various renewable energy sources. The huge amount of the wind power generation impacts the complex power system and this has to be addressed in detail in stability studies of the power system. Wind generation at bulk power levels is mainly asynchronous, unlike conventional generation which relies on synchronous alternators. These alternators have inherent capability of inertial response, oscillation damping, reactive power support, and low voltage and fault ride through capabilities. Wind generators behave differently and such network support capabilities may have to be artificially emulated in these systems. Further, the intermittency of wind as a resource adds more uncertainty into the reliable operation of the grid.

Doubly fed induction generator (DFIG) is one of the most suited generators among others like permanent magnet synchronous generator and squirrel-cage induction generator for the variable wind speed and high-power applications. The advantages of using the DFIG includes its maximum power extraction capability, flexible control, lower mechanical stress on the wind turbine, and reduced power rating of the converters employed.

The rest of the discussions will focus primarily on the influence of the high penetration of the DFIG based wind generation into the existing power system, and its impact on the rotor angle stability, small signal stability, voltage stability, and frequency stability.

## 20.4.2 Rotor Angle Stability

The change in the active power flow in an interconnected power system reflects in terms of the change in the rotor angle of the synchronous generators. A significant change of a load or a contingency in the vicinity of a generator leads to a swing of the generator rotor from its steady-state operating speed and this may cause it to fall out of synchronism. Broadly, rotor angle stability is divided in two categories: (1) small disturbance (small signal) rotor angle stability and (2) large disturbance rotor angle disturbance. The small signal stability is associated with the small disturbance in the system and can be observed by the linearization near equilibrium point.

***Small Signal Stability of Wind Penetrated Power System.*** The impact of DFIG-based wind generation integrated with a complex power system on the small signal stability can be analyzed by the linearization of the system at equilibrium point [2, 3]. A small change in the wind speed at a wind farm equivalent to a DFIG wind generator causes a small change in the rotor angle of the other generator in the interconnected system. The small signal stability analysis of a system can be performed by the most commercially available software tools where the DFIG generators along with their control strategy are easily modelled in conjunction with other conventional generators. The various modes of oscillation and the participating dynamic states can be identified for the analysis. The frequency of oscillation and damping due to the addition of the DFIG wind generation can also be determined by the small signal analysis. The formulation of a small signal stability problem could be performed by making use of the system dynamics. An interconnected system consisting wind power generation can be modeled by the following set of equations:

$$\frac{dX}{dt} = f(X,Y,U) \tag{20-85}$$

$$0 = g(X,Y,U) \tag{20-86}$$

where $f$ and $g$ are the set of differential and algebraic equations, respectively; $X$, $Y$, and $U$ are the states, input, and algebraic variable vectors, respectively, which makes a complete model in differential algebraic equations (DAEs) form. In the case of DFIGs, the system states include the speeds of the high-speed and low-speed shaft, the shaft twist angle, stator currents, and transient-induced emfs. The algebraic variables represent the terminal conditions that maintain the active and reactive power exchange between the DFIG and the network. The inputs to the problem are the various controller set-points (in case of closed-loop operation), rotor injected voltage (in case of open-loop operation) and the prevailing wind speed. The linearization of the aforementioned system at an equilibrium operating point $(X_e, Y_e, U_e)$ gives

$$\Delta\dot{X} = \frac{\partial f}{\partial X}\Delta X + \frac{\partial f}{\partial U}\Delta U + \frac{\partial f}{\partial Y}\Delta Y\Big|_{X_e,Y_e,U_e} \tag{20-87}$$

$$0=\frac{\partial g}{\partial X}\Delta X+\frac{\partial g}{\partial U}\Delta U+\frac{\partial g}{\partial Y}\Delta Y\bigg|_{X_e,Y_e,U_e} \tag{20-88}$$

This yields

$$\Delta\dot{X}=A\,\Delta X+B\,\Delta U \tag{20-89}$$

$$A=\left[\frac{\partial f}{\partial X}-\frac{\partial f}{\partial Y}\left(\frac{\partial g}{\partial Y}\right)^{-1}\frac{\partial g}{\partial X}\right] \tag{20-90}$$

$$B=\left[\frac{\partial f}{\partial U}-\frac{\partial f}{\partial Y}\left(\frac{\partial g}{\partial Y}\right)^{-1}\frac{\partial g}{\partial U}\right] \tag{20-91}$$

where, $A$ and $B$ are the system and output matrix, respectively. The eigenvalues of the system matrix indicate the natural modes of the system, determining its small signal stability. The frequency of the oscillation of the system under consideration are in the range of 0.2 to 2 Hz and time frame used for small signal stability is 10 s to minutes.

The placement of the eigenvalues in an *s*-plane depends on the wind generator parameter as well as the operating point (active and reactive power share, terminal voltage, and transmission line reactance) of the system and this determines its stability. The typical modes of a system, consisting a DFIG-based wind generator, connected to grid are shown in Fig. 20-36 [2]. The stator mode depicted in the figure illustrates the damping and frequency of oscillatory modes in which the electrical states such as transient-induced emfs participate. It may be observed from the Fig. 20-36 that at low grid strength as well as at low terminal voltage conditions the machine has a tendency to become unstable, a fact reflected in the eigenvalue crossing over to the right-half of the *s*-plane.

***Large Disturbance Stability of Wind-Penetrated Power System.*** This kind of rotor angle stability is under scrutiny whenever there is a large disturbance in the power system such as a loss of a key

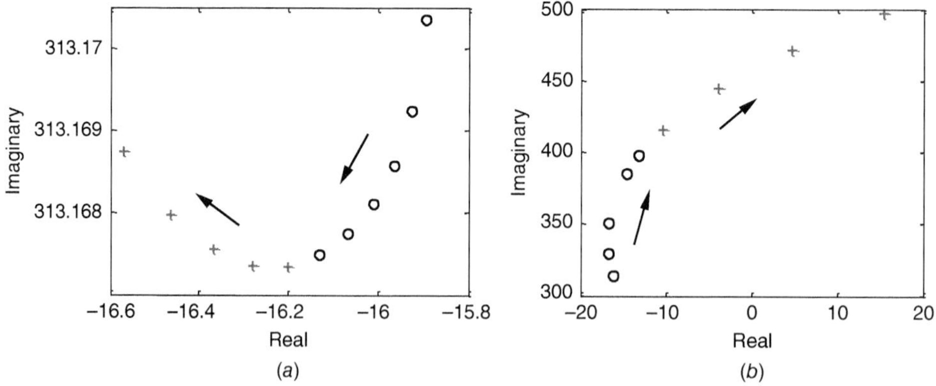


**FIGURE 20-36** Eigenvalue loci of electrical modes of DFIG generator (*a*) for varying terminal voltage (undervoltage to overvoltage), "o" and "+" indicating below and above nominal voltage, (*b*) for grid strength at the point-of-common-coupling (high short circuit MVA to low short circuit MVA), "o" and "+" indicating above and below a selected short circuit MVA of 1.67 pu.

transmission line or unit. Generally it is seen that the presence of DFIG-based wind generation is beneficial to improve the transient stability performance of nearby conventional generators of the power system. The reason is that the DFIG rotor speed being asynchronous and decoupled from the nearby conventional generator speed, the DFIG is able to remain relatively unaffected by the swing in nearby generators rotor speeds.

One of the way suggested in [4, 5] is associated with the reactive power support provided from the wind generators. The terminal voltage control strategy of the variable speed wind generators can be used to support the reactive power to the system. Whenever, the active power flow has been increased, the synchronous generators in the system may be stressed and prone to instability. At such an operating point, if the reactive power support is provided by the wind generators (possible with DFIGs), it stabilizes the system. The reactive power support from the wind generation minimizes the need of reactive power demanded from nearby synchronous generators. Thus, the reactive power output of the synchronous generator remains within the limit. This helps to balance the active power delivered by the synchronous generator and the mechanical power output and improves the rotor angle stability.

### 20.4.3 Voltage Stability

The ability of the power system to regulate the voltage throughout the network when subjected to a disturbance is termed as voltage stability. As the load drawn at a node increases, it increases the line current which further leads to the voltage drop across the lines resulting in a voltage drop at the load bus. If any compensation or an appropriate control is absent to rectify this situation, the voltage may drop to unacceptable levels depending on the nature of the load and there may even be voltage collapse. The factors affecting the voltage stability are the characteristics of loads, generator reactive power controls, the action of the reactive power compensation devices and voltage control devices [6, 7]. A loss of line or a loss of generation may also lead to voltage instability.

At bulk power generation levels, wind turbines employ DFIGs rather than squirrel cage induction generators. Just like conventional synchronous generators, DFIGs also have the capability of independent active and reactive power control. These controls are provided in the power electronic converter employed in the rotor circuit. The DFIG terminal voltage control capability is, however, localized. This means that it is a challenge for the network to rely solely on the connected DFIGs to improve the voltage stability performance, especially during stressed network conditions. The voltage stability of the system with highly integrated wind generation has been investigated using many approaches [8, 9]. It is seen that the reactive power support available from the DFIG is affected by the prevailing wind speed. Therefore, keeping the variability of the wind in mind, the steady-state voltage stability of a system with high penetration of wind power, can be planned by using the historical data to come up with probabilistic scenarios [8, 9].

***Low Voltage Ride-through Capabilities of Wind Generators.*** A large wind farm connected to an interconnected power network through a transmission line is much affected by the change in the voltage at its point of common coupling (PCC) [10]. In the case of prolonged and severe low-voltage conditions existing at the PCC, a wind turbine may be disconnected for safety reasons. The reason behind this intentionally introduced disconnection is that a low voltage on the stator causes a very high current in the rotor network and it may cause the blow-off of the rotor-side converter circuitry. A solution to this is a crowbar protection used with the rotor network which shorts the rotor circuit and bypasses the converter during voltage sage conditions. This improves the low voltage ride-through capability of the generator for a short duration of time, by keeping it online. A typical persisting voltage-dip at the DFIG-connected bus and the rotor current due to it is shown in Fig. 20-37*a*, *b*, respectively. The voltage-dip of to 60% of the nominal stator voltage at $t = 2$ s causes almost double the rotor current.

With the evolution of the technology, it is now expected that the wind generation in the system should support the grid during the low voltage as specified by the grid code in different counties. The wind generator should follow the grid code to maintain the stability and reliability of the power

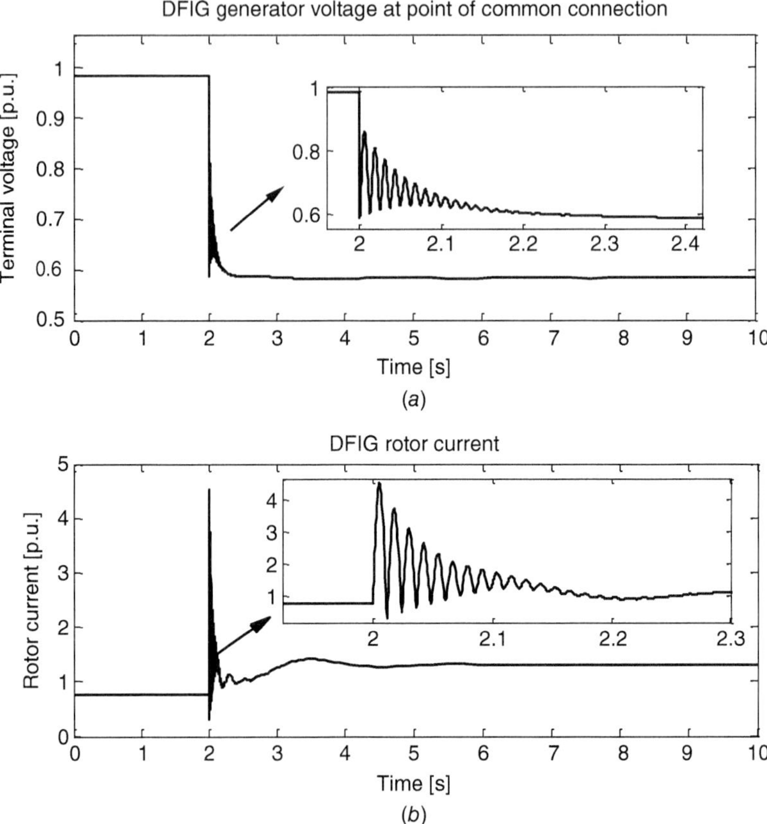


**FIGURE 20-37** (*a*) Typical stator voltage and (*b*) rotor dip of a DFIG connected to an infinite grid subjected to a 40% voltage-dip at infinite bus.

system during stressed conditions. The pictorial view of the grid code for the voltage and fault ride-through of the United States (FERC order 661-A) [11] is as shown in Fig. 20-38. It can be observed form the that a wind generator must not be disconnected from the grid for at least 0.15 s (nine cycles) subjected to a three-phase fault at the high-voltage terminals of the step-up transformer. A voltage-dip below the 90% should be tolerated by the generator for a period of almost 2.00 s. Further, the generator is allowed to disconnect above an overvoltage of more than 120%. Examples of the grid code for other countries can be found in [12] and [13], respectively.

### 20.4.4 Frequency Stability

Understanding the impact of the variations in wind velocities on the grid frequency, requires dynamic models of wind turbine generators (WTGs) relevant for studies at such timescales. The timescales involved range from seconds to tens of minutes. While wind variations may cause minor variations in the reactive power output of a WTG, it is more due to the active power variations that the network frequency is affected.

Variable speed wind turbines connected to DFIGs may be operated under a maximum power extraction strategy, wherein the rotor speed is modulated to enable optimum power generation,

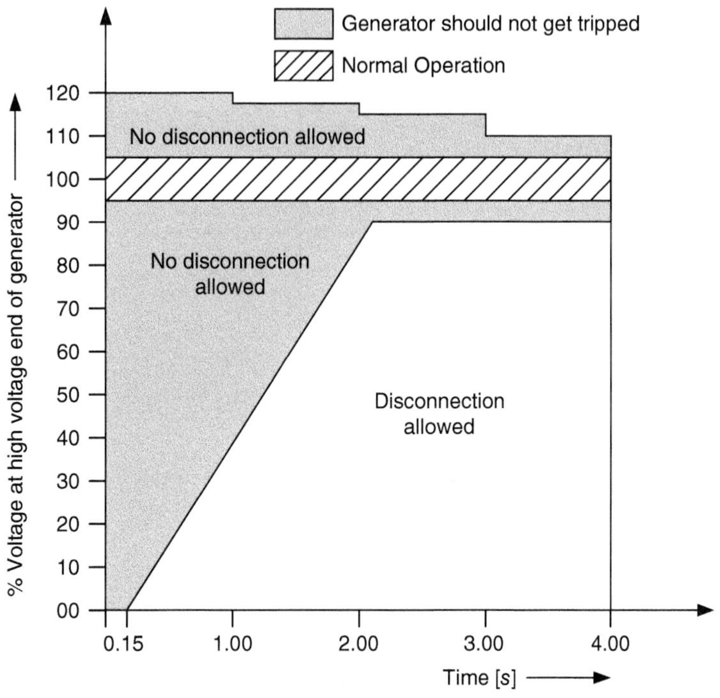


**FIGURE 20-38** Allowable limits of operation (voltage is measured at high-voltage side of the step-up transformer).

at any wind speed. The rotor speed variations are similar to the wind speed variations for such a generator and the output power variations follow a maximum power point tracking characteristics. The input-output dynamics between the variations in wind speed and active power generation resembles that of a first-order system. Due to the increasingly steep nature of the MPPT characteristics, any variation in wind speeds cause greater variations in power output when occurring at higher wind speeds than at lower wind speeds. For instance, a 1% change in wind speed causes a 2.19% change in WTG power output at 12 m/s, while the same variation at 10 m/s causes 1.52% change in the WTG power output. Similarly, the inertia of the rotating shafts within the WTG also has a smoothening impact on the power variations, as also the predisturbance rotor speed. The overall smoothening effect is thus more pronounced at lower wind speeds than at higher wind speeds, as is evident in Fig. 20-39.

Irrespective of the level of penetration of wind generation, its impact on the network frequency response is more pronounced in isolated power systems and microgrids. Wind generation replaces conventional synchronous type generation, by decommitting them during light load conditions. Under heavy load conditions, conventional generation may be brought back into the grid, thus improving the frequency response. In fixed speed WTGs employing squirrel cage induction generators, the shaft speed is locked to the grid frequency, and therefore the kinetic energy stored in the rotating shaft has an automatic stabilizing influence during grid frequency variations. The variable speed WTGs usually connected to DFIG, rely on power electronic interfaces that decouple the shaft speed and the grid frequency. Therefore, the system experiences a loss of inertia when conventional generation is replaced by DFIG-based WTGs. This reduced inertia may significantly degrade the frequency response in isolated grids and microgrids. For instance, a generation trip or a loss of large load is an extremely rare event, when compared to sudden changes

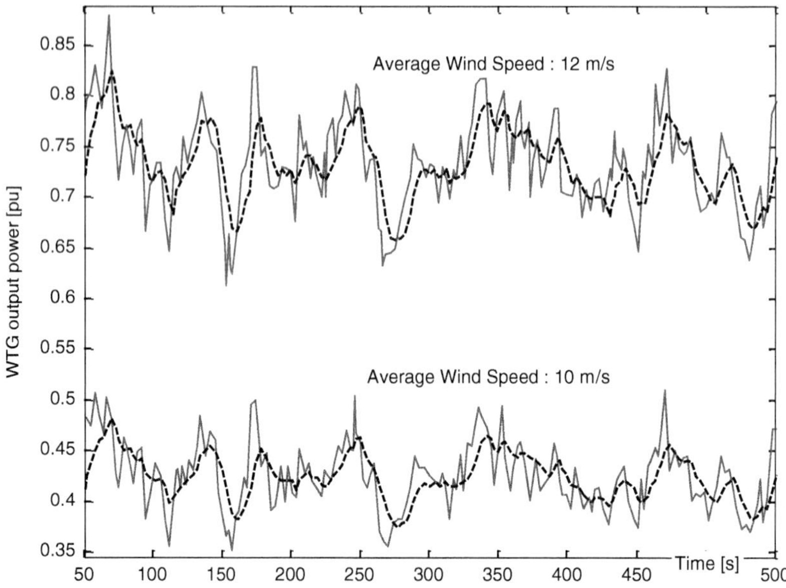


**FIGURE 20-39** Output power variations with no smoothening control applied, due to similar wind variations at two average wind speeds. The solid line indicates the power if no dynamics of the turbine are considered. The dashed line indicates the power output of an actual WTG with an inertia constant of 5 s.

in wind speeds that give rise to equivalent changes in generation. This is particularly true in grids with spatially concentrated wind generation.

In Fig. 20-40, the impact of two instances of insufficient generation on the network frequency response, are compared for a 10-MW microgrid with different levels of penetration of wind power (DFIG type). One event (rare) is the sudden application loss of a 92-kW load, leading to large rate of change of frequency (ROCOF) values, that is then stabilized by the action of secondary governor response from existing conventional generation. During lightly loaded operation, when much of the conventional generation is offline, the reduced inertia of the microgrid would lead to even higher ROCOF values for such an event. The maximum ROCOF as well as the frequency nadir would also increase when the grid has more penetration of wind power, due to lack of inertial response of the grid. Interestingly in comparison, the second event, which is a sudden decrease in wind speed leading to an excess of 92 kW load, causes much lower ROCOF values owing to the inherent smoothening effect of the WTG turbine shaft. While maximum ROCOF values of more than 0.5 Hz/s are cause for concern in any power system, smaller isolated power systems should expect much higher ROCOF values with increasing levels of penetration of wind power. HVDC interconnections, which ultimately replace conventional generation, can further deteriorate these maximum ROCOF values. In order to improve the frequency response of such grids, enough spinning reserve must be planned in order to accommodate unpredicted changes in wind and controllers must be available with WTGs to enable them to support the grid frequency similar to conventional governor response [14, 15]. The situation would improve if the wind generation operating in the grid, comprises of squirrel cage induction generators that offer inertial support to the grid frequency.

Interestingly, the second event shown in Fig. 20-40 is a sudden decrease in the wind leading to a drop in the power injection. This kind of an event is a common event; the ROCOF seen as a result is not so steep as in the case of the load drop event. The reason is the inherent smoothening effect of the

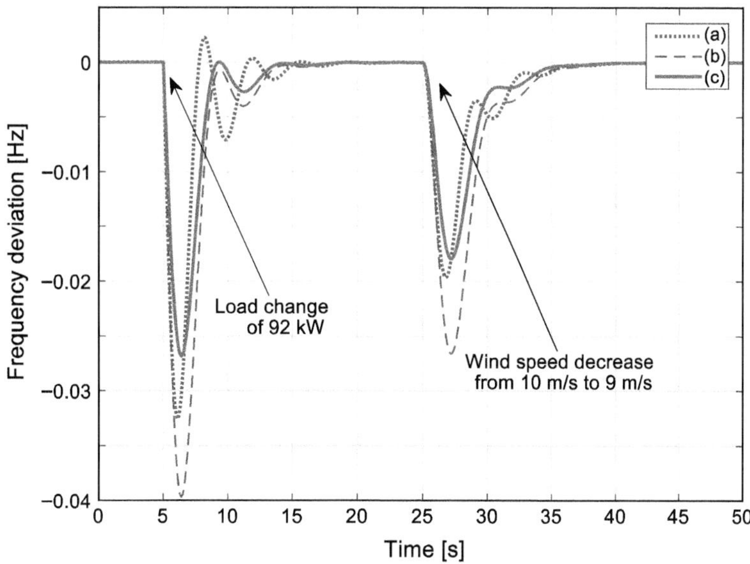


**FIGURE 20-40** Impact of a sudden 92-kW load drop and sudden wind speed change leading to a 92 kW change in wind generation, on a 10-MW microgrid frequency. Secondary frequency controllers present in nearby generators, quickly restore the frequency to nominal. (*a*) Wind penetration 6.6%, (*b*) wind penetration 10%, (*c*) wind penetration 6.6%, spinning reserves 50% more.

WTG, while translating the wind speed variations to grid injections. The frequency nadir, however, significantly deteriorates with increasing wind penetration in the system.

### 20.4.5 Control Techniques to Emulate Inertia

The ability to respond to grid frequency changes, by supplying or drawing active power, must be artificially created in DFIG-based WTGs by additional controllers. The grid frequency changes occur due to load-generation unbalance, caused by unplanned events in the system. For instance, due to a sudden increase in load, the frequency of nearby synchronous generators start to decline, thereby releasing kinetic energy into the grid. This inertial response helps support the additional load initially, until the decline in frequency is detected by various governors, who take control action by increasing the turbine input to arrest the dropping frequency. The governor response, or primary frequency response helps stabilize the grid frequency at an appropriate value that is arrived at dynamically by similar actions from various geographically distributed units. In order to emulate the inertial response in DFIGs, the derivative of the frequency is used to drive the active power control loop of the rotor side converters (RSC). Further, the primary frequency response is emulated by incorporating the actual frequency through a feedback gain (droop control) to drive the RSC.

When the system frequency is used for feedback to the RSC through an appropriate gain, it helps in providing damping support to the network during oscillations caused by disturbances. The electrical torque of the DFIG is modulated in response to the frequency excursions, and the damping relief obtained is highly effective [16]. The DFIG response is available almost immediately, subject to the minor delays in the power electronic converters. For instance, due to a rise in grid frequency, its negative feedback leads to immediate reduction in electrical torque and power. This is however temporary, and the power output soon returns to its predisturbance operating condition, as shown in Fig. 20-41.

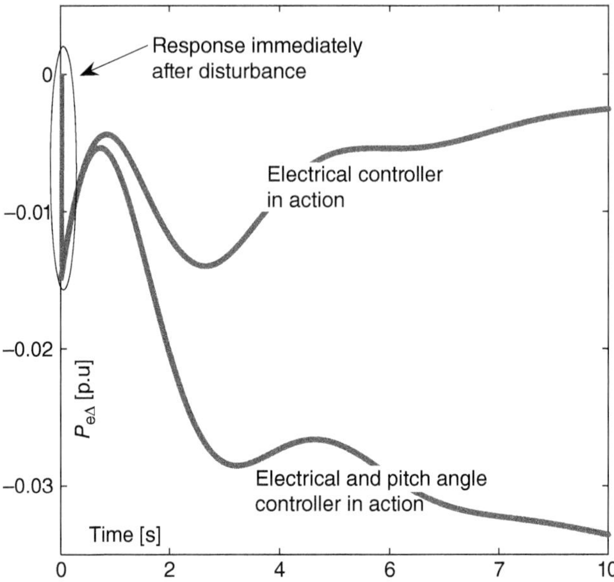


**FIGURE 20-41** Impact of a sudden 92-kW load drop and sudden wind speed change leading to a 92 kW change in wind generation, on a 10-MW microgrid frequency. Secondary frequency controllers present in nearby generators, quickly restore the frequency to nominal.

Sustained primary frequency response from DFIG-WTGs is possible only if there is active power reserve available. Thus, WTGs controlled extract the maximum power from the wind (by operating at zero pitch angle), cannot participate in primary frequency control. Deloading of WTGs is required to create the active power reserve at any wind speed. Increasing the steady-state pitch angle from zero can create a positive active power margin, by wasting some of the wind incident on the turbine. This kind of pitch angle-based deloading results in the deloading percentage being independent of the wind speed. When there is a requirement of the power to support the grid frequency, the pitch angle is modulated accordingly. For instance, when due to a rise in grid frequency, the electrical feedback of the frequency leads to reduction of the electrical torque command to the RSC as well as concomitant increase in the WTG pitch angle, leading to reduction of the mechanical turbine torque, as shown in Fig. 20-41.

The choice of the initial deloading percentage will determine the initial steady-state pitch angle at which the WTG will be operated. Deloading the WTG to the right of the MPPT point will result in a higher predisturbance operating speed, which may stress the RSC converter [17]. However, the benefit is that during depressed grid frequency condition, the pitch angle is reduced and the speed of the WTG also reduces bringing it back toward the MPPT point. The corresponding rise in power injection is aided by the release of the kinetic energy of the shaft slowing down. Deloading to the left of the MPPT point, conversely, leads to lower operating speeds in general and may be preferred to reduce the converter rating requirements [18].

The gain of the electrical feedback look, is referred to as the droop, and is selected carefully to ensure stability of the loop as well as optimum primary frequency response from the DFIG. The primary frequency contribution may be improved by continuously adjusting the droop in response to varying wind velocities. This leads to much more stable operation at low wind speeds and better frequency support at high wind speeds. The annual WTG capacity factor has also been shown to be improved by following a variable droop strategy at locations with relatively higher average wind speeds [18].

### 20.4.6 Bibliography


1. O. Anaya-Lara, N. Jenkins, J. Ekanayake, P. Kartwrite, and M. Hughes, *Wind Energy Generation: Modelling and Control*, John Wiley & Sons, England, 2009.
2. F. Mei and B. Pal, "Modal Analysis of Grid-Connected Doubly Fed Induction Generators," *IEEE Transactions on Energy Conversion*, vol. 22, no. 3, pp. 728–736, Sep. 2007.
3. F. Xu, X. P. Zhang, K. Godfrey, and P. Ju, "Small-Signal Stability Analysis and Optimal Control of a Wind Turbine with Doubly Fed Induction generator," *Proceedings of the 2007 IET Generation, Transmission & Distribution*, pp. 751–760, 2007.
4. D. Gautam, V. Vittal, and T. Harbour, "Impact of Increased Penetration of DFIG-Based Wind Turbine generators on Transient and Small-Signal stability of Power Systems," *IEEE Trans Power Sys*, vol. 24, no. 3, pp. 426–1434, 2009.
5. E. Vittal, M. O'Malley, and A. Keane, "Rotor Angle Stability with High Penetrations of Wind Generation. *IEEE Trans Power Sys*, vol. 27, no. 1, pp. 353–362, 2012.
6. P. Kundur, *Power System Stability and Control*, McGraw-Hill, New York, 1994.
7. E. Muljadi, C. P. Butterfield, B. Parsons, and A. Ellis, "Effect of Variable Speed Wind Turbine Generator on Stability of a Weak Grid," *IEEE Trans Power Sys.*, vol. 22, no. 1, pp. 29–35, 2008.
8. E. Vittal, M. O'Malley, and A. Keane, "A Steady-State Voltage Stability Analysis of Power Systems with High Penetrations of Wind," *IEEE Trans Power Sys.*, vol. 25, no. 1, pp. 433–442, 2010.
9. R. J. Konopinski, P. Vijayan, and V. Ajjarapu, "Extended Reactive Capability of DFIG Wind Parks for Enhanced System Performance," *IEEE Trans Power Sys.*, vol. 24, no. 3, pp. 1346–1355, 2009.
10. M. Kayikci and J. V. Milanovic, "Reactive Power Control Strategies for DFIG-Based Plants," *IEEE Trans Energy Convers*, vol. 22, no. 2, pp. 389–396, 2007.
11. Order 661—Interconnection with Wind Energy, issued by Federal Energy Regulatory Commission (FERC) of United States, Jun. 2, 2005.
12. Grid connection of wind turbines to networks with voltages above 100 kV, Regulation TF 3.2.5, Eltra and Elkraft System.
13. Grid Code, Ver.3.1.3, elaborated by The EirGrid, Apr. 2008.
14. G. Lalor, A. Mullane, and M. O'Malley, "Frequency Control and Wind Turbine Technologies," *IEEE Trans. Power Sys.*, vol. 20, no. 4, Nov. 2005.
15. R. Doherty, A. Mullane, G. Nolan, D. J. Burke, A. Bryson, and M. O'Malley, "An Assessment of the Impact of Wind Generation on System Frequency Control," *IEEE Trans. Power Sys.*, vol. 25, no. 1, Feb. 2010.
16. D. Gautam, L. Goel, R. Ayyanar, V. Vittal, and T. Harbour, "Control Strategy to Mitigate the Impact of Reduced Inertia due to Doubly Fed Induction Generators on Large Power Systems," *IEEE Trans. Power Sys.*, vol. 26, no. 1, Jul. 2010.
17. S. Ghosh and N. Senroy, "Electromechanical Dynamics of Controlled Variable-Speed Wind Turbines," *IEEE Sys. J.*, vol. 9, no. 2, Jun. 2015.
18. K. V. Vidyanandan and N. Senroy, "Primary Frequency Regulation by Deloaded Wind Turbines Using Variable Droop," *IEEE Trans. Power Sys.*, vol. 28, no. 2, May 2013.


## 20.5 IMPACT OF SOLAR GENERATION ON POWER SYSTEM STABILITY

*BY SUKUMAR MISHRA*

### 20.5.1 Introduction

In present day technology, life solar photovoltaic (PV)-based renewable generation has got special focus as it can be integrated at different capacities, from hundreds of MWs to few kWs. These generators can either be operated in grid connected mode or standalone mode configurations based on the location, grid availability and other technical issues.

In grid connected mode, PV generators are usually expected to operate at maximum power point (MPP). As the voltage at the point of common coupling (PCC) is defined by the grid, the PV generation is usually considered as a constant power source with respect to grid. The current supplied by the PV-based power generation is continuously tracked and controlled to maintain the power quality and the stability of the system. There is a limit on the maximum allowable total harmonic distortion (THD) factor in the current flowing to the grid as per the international standards, which can be fulfilled by suitably designing the controller. Hence, the current controller design is pivotal to satisfy standards.

A typical control structure of a grid connected PV system can be classified into two categories:

1. Basic maximum power control scheme
2. Ancillary control scheme

The basic control consists of the maximum power extraction algorithm where PV system is operated at unity power factor with THD being in permissible limits. In addition, islanding detection capability is integrated to the system for protection under abnormal conditions. The control scheme with ancillary services will be able to provide voltage and frequency support along with the above basic functionalities. Ancillary services are considered as optional, as they are selected based on the grid codes amended by the respective government policies.

Figure 20-42 depicts the overall control functionality of the grid connected PV system.

The controller for the voltage source converter is designed to fulfill the basic requirements as given below:

- Control of the active power delivered to the grid, such that maximum power is extracted from the PV
- Control of the reactive power exchange, so as to operate PV system at UPF as per standards

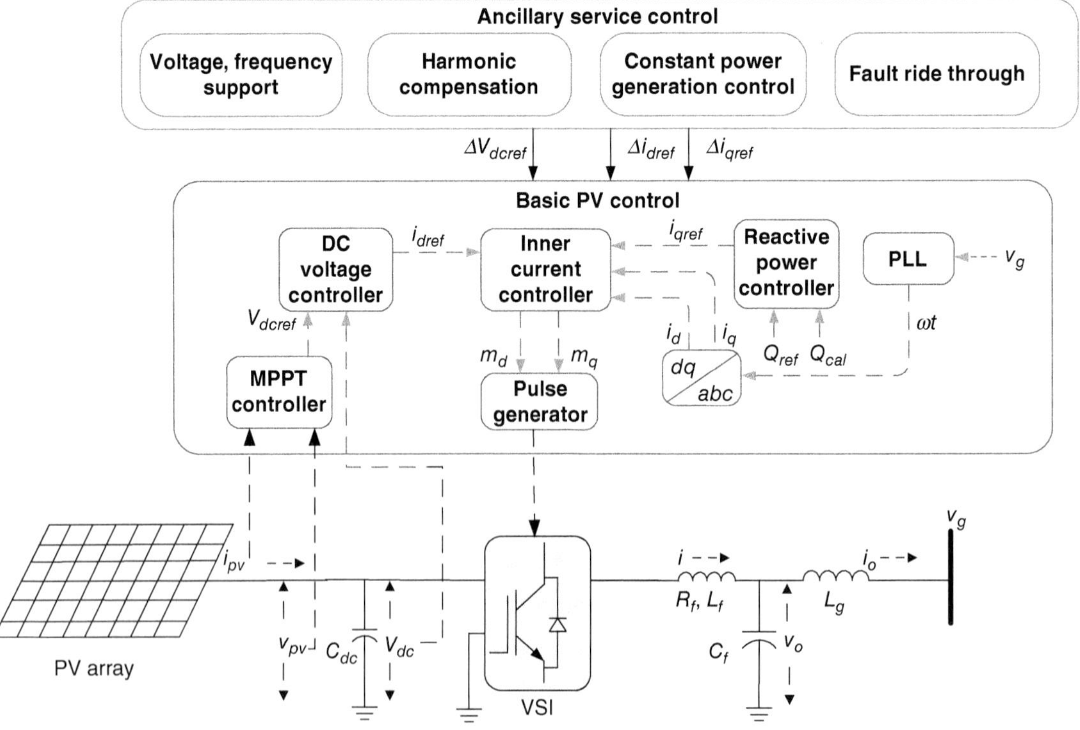


**FIGURE 20-42** Single-line diagram single stage grid connected photovoltaic control.

- High efficiency and high quality of the injected power
- Grid synchronization, and anti-islanding protection

Conventional single-stage grid-connected PV system has a phase locked loop (PLL)-based synchronizing unit along with two cascaded control loops.

***Grid Synchronization.*** To operate the PV system in grid connected mode, the inverter system should be synchronized with the grid. For grid synchronizations the frequency, phase angle, and amplitude of the grid voltage should be identified. Conventionally, synchronous reference frame based on a phase locked loop (PLL) as shown in Fig. 20-43 is used. In this, the $d$- and $q$-axes components $v_d$ and $v_q$ of the grid voltage $v_{\mathrm{abc}}$ are calculated by Park's transformation using the $\theta$ obtained from the PLL scheme. In PLL, the $q$-axis reference voltage component is set to zero such that, PLL angle $\theta$ is made equal to the phase angle of the grid phase voltage $v_{ga}$, as shown in Fig. 20-43.

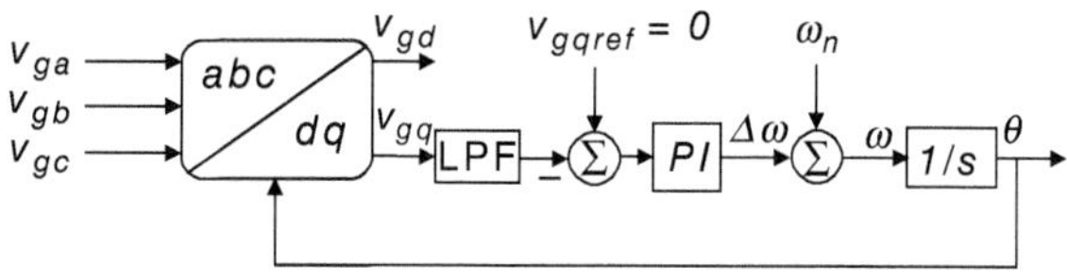


**FIGURE 20-43** Synchronous reference frame-based PLL.

## 20.5.2 Outer Control Loop

***DC Voltage Control.*** The DC voltage controller is used to maintain voltage across the dc link capacitor $C_{dc}$ as per the reference voltage $V_{dcref}$. The $V_{dcref}$ value is obtained from a MPPT algorithm such as perturb and observe (P&O), incremental conductance (INC), and so on. The dc-link voltage is controlled by regulating the real power injection into the grid, which in turn can be regulated by the control of direct axis current flowing through the inverter. The control scheme of the dc voltage controller is shown in Fig. 20-44*a*.

In steady state, the power flow on the either sides of the inverter will be balanced if the inverter losses are ignored, that is, $P_{pv}$ will be equal to inverter ac-side power, which indeed equal to the power $P_{\mathrm{ac}}$ injected into grid assuming a lossless filter component. The dynamic voltage equation across the dc-link capacitor is given by

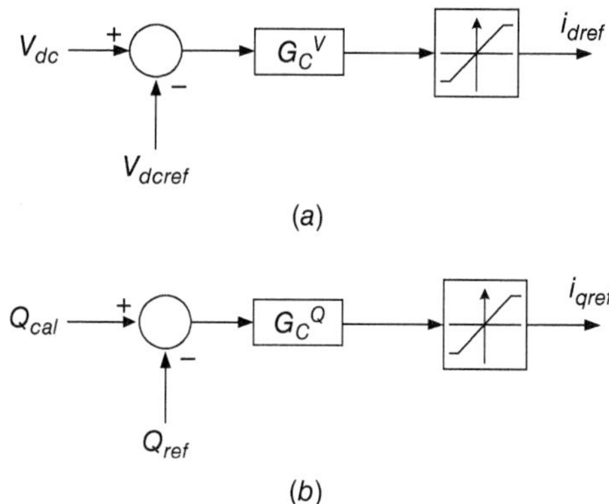


**FIGURE 20-44** (*a*) DC voltage control scheme, (*b*) reactive power control scheme.

$$\frac{d}{dt}\left(\frac{1}{2}C_{\mathrm{dc}}V_{\mathrm{dc}}^2\right) = (P_{pv} - P_{\mathrm{ac}})pu \tag{20-92}$$

This can be simplified as

$$C_{\mathrm{dc}}V_{\mathrm{dc}}\frac{dV_{\mathrm{dc}}}{dt} = (P_{pv} - P_{\mathrm{ac}})pu \tag{20-93}$$

***Reactive Power Control.*** The reactive power injected into grid is given by

$$Q = -\left(\frac{3}{2}\right)v_d i_q \tag{20-94}$$

Similar to the dc voltage control, the reactive power can be controlled by controlling the $q$-axis current component. The reactive power delivered by the shunt branch of the filter, that is, $Q_{cf}$, is compensated by subtracting it from $Q_{ref}$.

***Inner Current Controller.*** The $dq$-frame control of the real-/reactive-power controller is based on current control. The dynamic current equations are

$$L_f \frac{d_{id}}{dt} = L_f \omega i_q + R_f i_d + v_{id} - v_d \tag{20-95}$$

$$L_f \frac{di_q}{dt} = L_f \omega i_d + R_f i_q + v_{iq} - v_q \tag{20-96}$$

where

$$v_{id} = V_{dc} * m_d \quad v_{iq} = V_{dc} * m_q \tag{20-97}$$

To generate the modulating signals $m_d$ and $m_q$ for VSI, the $dq$-axis current control strategy is given by

$$m_d = \frac{u_d - L_f \omega i_q + v_d}{V_{dc}} \tag{20-98}$$

$$m_q = \frac{u_q + L_f \omega i_d + v_q}{V_{dc}} \tag{20-99}$$

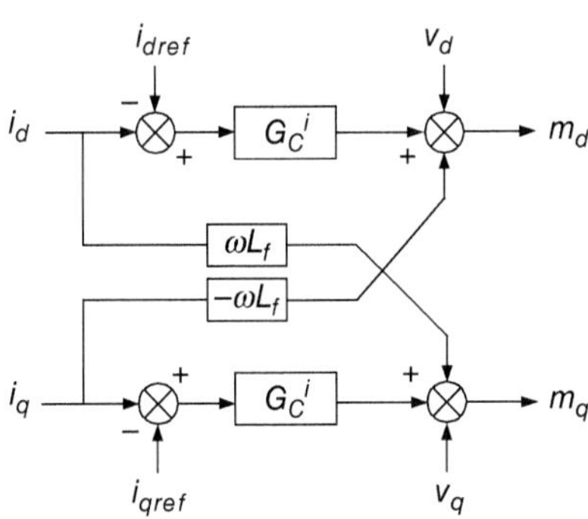


**FIGURE 20-45** Inner current control scheme.

where $u_d$ and $u_q$ are two new control inputs obtained from the compensator output $G_c^i$ as shown in Fig. 20-45. The compensator is designed, such that it compensates the phase margin of the filter dynamics and the settling time of the compensator is ten times that of the filter circuit.

From the current controller, the output modulating signals ($m_d$ and $m_q$) for the inverter are obtained as shown in Fig. 20-45. The inverter dynamics are neglected in modeling of the current controller and it can assumed as a unity gain block since the inverter switching dynamics are faster as compared to that the controller and filter dynamics.

Although grid connected PV system looks effective and promising, when grid fails, the PLL connected to the PV system is unable to provide the system frequency information and hence inverter fails to operate. Further, being in a nascent state and due to the source uncertainties these generating systems are often coupled with diesel generators (DG), where the DG generates voltage at the PCC, is used as the reference signal required for PLL. The combination of such sources is used to serve small regions of the system to improve reliability and to reduce intermittency in the power supply.

However, improper controlling of VSC based PV system, can lead to overall system instability. A general approach toward evaluating stability in converter interfaced microgrids was explained in previous literature. But the conventional synchronous machines have an inherent property of synchronizing torque which contributes in providing first swing system stability. In addition to synchronizing torque, the combination of the machine inertia and damping torques are available in DG which is used to lock the rotor poles along stator magnetic field poles during perturbations. However, in the case of converter-based generation these torques are absent. The small signal model of an isolated PV array delivering to a load, PV generation at the transmission level, have been analyzed. The vital importance of VSC modeling to assess the stability of interconnected systems is stated.

### 20.5.3 Mathematical Modeling of PV-DG System

The PV DG system coupled at common point system as shown in Fig. 20-46. The mathematical model of the PV-DG system can be represented using differential equations.

***Dynamic Model of the PV System.*** The power delivered by the PV system can be expressed as below

$$P_{pv} = f(V_{dc}, S, T) = n_p I_{ph} V_{dc} - n_p I_{rs} V_{dc}\left[\exp\left(\frac{qV_{dc}}{n_s kTA}\right) - 1\right] \tag{20-100}$$

The dynamics of the dc-link voltage are described, based on the principle of power balance is given by Eq. (20-94), which can be written as

$$C_{dc} V_{dc} \frac{dV_{dc}}{dt} = (P_{pv} - v_d i_d - v_q i_q) \tag{20-101}$$

$$\frac{dV_{dc}}{dt} = \frac{\omega}{C_{dc}}\left(\frac{V_{acb}}{V_{dcb}}\right)^2 \left(\frac{P_{pv}}{V_{dc}} - \frac{v_d i_d}{V_{dc}} - \frac{v_q i_q}{V_{dc}}\right) \tag{20-102}$$

The $dq$-components of PV current are dictated by the filter dynamics expressions which are given by

$$\frac{di_d}{dt} = \omega i_q - \frac{\omega v_d}{L_f} + \frac{m_d \omega V_{dc} V_{dcb}}{L_f V_{acb}} \tag{20-103}$$

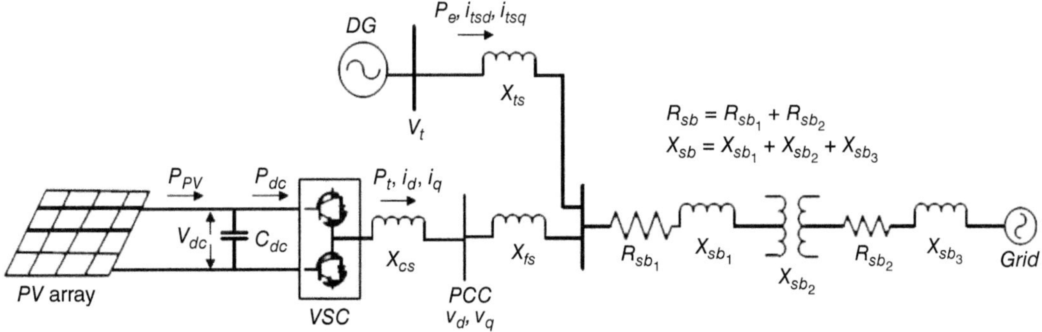


**FIGURE 20-46** Single line diagram of grid connected PV-DG system.

$$\frac{di_q}{dt} = -\omega i_d - \frac{\omega v_q}{L_f} + \frac{m_q \omega V_{dc} V_{dcb}}{L_f V_{acb}} \tag{20-104}$$

***Dynamic Model of the DG System.*** The one-axis dynamic model of synchronous machine that represents the rotor mechanical and electrical dynamics along with the static exciter is given by

$$\frac{d\delta}{dt} = \omega_0(\omega_r - 1) \tag{20-105}$$

$$\frac{d\omega}{dt} = \frac{1}{2H}(P_m - P_e - D(\omega_r - 1)) \tag{20-106}$$

$$\frac{dE'_q}{dt} = \frac{1}{T'_{d0}}(E_{fd} - E'_q - (X_d - X'_d)i_{tsd}) \tag{20-107}$$

$$\frac{dE_{fd}}{dt} = \frac{1}{T_a}(-E_{fd} + K_{avr}(V_{tref} - V_t)) \tag{20-108}$$

The representation for the algebraic variables is as follows:

$$V_t = v_{td} + jv_{tq} \tag{20-109}$$

$$v_{td} = X_d i_{tsq} \tag{20-110}$$

However, to represent the system in the state space form, the algebraic variables have to be represented in terms of the state variables.

$$i_{tsd} = \frac{(X_1 E'_q - R_{sb} v_{bd} - X_1 v_{bq} + (R_{sb} X_{sb} - X_1 R_{sb})\, i_q - (X_1 X_{sb} + R_{sb}^2)\, i_d)}{X_1 X_2 + R_{sb}^2} \tag{20-111}$$

$$i_{tsq} = \frac{R_{sb} E'_q - R_{sb} v_{bq} + X_2 v_{bd} + (R_{sb} X_2 - X_{sb} R_{sb})\, i_d - (X_2 X_{sb} + R_{sb}^2)\, i_q}{X_1 X_2 + R_{sb}^2} \tag{20-112}$$

$$P_e = v_{td} i_{tsd} + v_{tq} i_{tsq} \tag{20-113}$$

$$v_{PCC} = v_d + jv_q \tag{20-114}$$

$$v_d = v_{bd} + i_d R_{sb} - i_q (X_{sb} + X_{fs}) + i_{tsd} R_{sb} - i_{tsq} X_{sb} \tag{20-115}$$

$$v_q = v_{bq} + i_q R_{sb} + i_d\, (X_{sb} + X_{fs}) + i_{tsd} X_{sb} + i_{tsq} R_{sb} \tag{20-116}$$

Using these dynamic equations, the entire system is represented in terms of the state variables. The inputs to the machine state model are the mechanical power $P_m$ and the terminal reference voltage $V_{tref}$ for automatic voltage regulator. The inputs for PV system state model is the dc link reference voltage $V_{dcref}$ to the PV array controller, as $Q_{ref}$ equal to zero at unity power factor as per the existing grid codes. However, based on the requirements the VSC can be made to support reactive power if grid code allows.

The VSC is controlled by linearized feedback technique using a second-order sliding mode controller. Hence, second-order expression is developed from Eq. (20-102) this again divided into two first-order differential equations is developed as Eqs. (20-117), (20-118), using a fictitious state variable "$g$"

$$\frac{dV_{dc}}{dt} = g \tag{20-117}$$

$$\frac{dg}{dt} = \frac{\lambda_1' \omega V_{acb}^2}{C_{dc} V_{dcb}^2} + \frac{-m_d \omega^2 V_{acb} v_d}{L_f C_{dc} V_{dcb}} + \frac{-m_q \omega^2 V_{acb} v_q}{L_f C_{dc} V_{dcb}} \tag{20-118}$$

where

$$\lambda_1' = -\frac{g\lambda_2'}{V_{dc}} - \frac{\omega i_q v_d}{V_{dc}} + \frac{\omega v_d^2}{V_{dc} L_f} + \frac{\omega i_d v_q}{V_{dc}} + \frac{\omega v_q^2}{V_{dc} L_f} \tag{20-119}$$

$$\lambda_2' = \frac{P_{pv}}{V_{dc}} - \frac{v_d i_d}{V_{dc}} - \frac{v_q i_q}{V_{dc}} \tag{20-120}$$

The state variables–based sliding surface is used to obtain the control variables $m_d$ and $m_q$.

The sliding mode controller works on the principle of defining a sliding surface for the error of the state variables. From Eqs. (20-117) and (20-118) it can be observed that $V_{dc}$ and $i_q$ are the only state variables required to develop the controller. Thus, the control variables $m_d$ and $m_q$ are defined by Eqs. (20-121) and (20-122).

$$m_q = \frac{\left(-\rho_2 \tanh(i_q - i_{qref}) + \omega i_d + \frac{\omega v_q}{L_f}\right) L_f V_{acb}}{\omega V_{dc} V_{dcb}} \tag{20-121}$$

$$m_d = \frac{\left(-(\rho_1 \tanh(g + k(V_{dc} - V_{dcref})) + kg) - \frac{\lambda_1' \omega V_{acb}^2}{C_{dc} V_{dcb}^2} + \frac{m_q \omega^2 V_{acb} v_q}{L_f C_{dc} V_{dcb}}\right) L_f V_{acb}}{-\omega^2 v_d V_{acb}} \tag{20-122}$$

The system of equations are linearized about an operating point in order to obtain the state matrix. The operating point is obtained by performing a load flow for a given level of power generation. The resulting state matrix obtained is given by

$$\begin{bmatrix} \Delta\dot{g} \\ \Delta\dot{V}_{dc} \\ \Delta\dot{i}_d \\ \Delta\dot{i}_q \\ \Delta\dot{\delta} \\ \Delta\dot{\omega} \\ \Delta\dot{E}_q' \\ \Delta\dot{E}_{fd} \end{bmatrix} = \begin{bmatrix} A_{21} & A_{22} & A_{50} & A_{48} & A_{49} & 0 & A_{51} & 0 \\ 1 & 0 & 0 & 0 & 0 & 0 & 0 & 0 \\ A_{52} & A_{17} & A_{46} & A_{44} & A_{45} & A_{\omega_1} & A_{53} & 0 \\ 0 & A_{20} & A_{47} & A_{54} & A_{55} & A_{\omega_2} & A_{56} & 0 \\ 0 & 0 & 0 & 0 & 0 & \omega_0 & 0 & 0 \\ 0 & 0 & A_{57} & A_{58} & A_{59} & -D & A_{60} & 0 \\ 0 & 0 & A_{33} & A_{34}' & A_{34} & 0 & A_{32} & A_{31} \\ 0 & 0 & A_{61} & A_{62} & A_{63} & 0 & A_{64} & A_{65} \end{bmatrix} \begin{bmatrix} \Delta g \\ \Delta V_{dc} \\ \Delta i_d \\ \Delta i_q \\ \Delta \delta \\ \Delta \omega \\ \Delta E_q' \\ \Delta E_{fd} \end{bmatrix} \tag{20-123}$$

$$\hat{A}=\left(\frac{\omega}{C_{dc}}\right)\left(\frac{V_{acb}^2}{V_{dcb}^2}\right), X_1=X_q+X_{ts}+X_{sb}; X_2=X_d'+X_{ts}+X_{sb}; D'=X_1X_2+R_{sb}^2$$

$$m_1=\frac{((X_1v_{bd})-(R_{sb}v_{bq}))}{D'}; m_2=\frac{X_1}{D'}; m_3=\frac{(R_{sb}X_{sb}-X_1R_{sb})}{D'}; m_4=\frac{-(X_1X_{sb}+R_{sb}^2)}{D'}$$

$$n_1=\frac{(R_{sb}v_{bd}+X_2v_{bq})}{D'}; n_2=\frac{R_{sb}}{D'}; n_3=\frac{-(R_{sb}^2+X_{sb}X_2)}{D'}; n_4=\frac{(X_2R_{sb}-X_{sb}R_{sb})}{D'}$$

$$A_1=\frac{X_{cs}V_{acb}}{V_{dcb}V_{dc}}; A_2=\frac{V_{acb}}{V_{dcb}V_{dc}}; A_3=-\frac{X_{cs}V_{acb}i_d}{V_{dcb}V_{dc}^2}-\frac{V_{acb}v_q}{V_{dcb}V_{dc}^2}; A_4=-\frac{\rho_2X_{cs}V_{acb}}{\omega V_{dcb}V_{dc}};$$

$$\hat{\hat{A}}=\frac{\left(\frac{\omega}{Cdc}\right)V_{acb}^2}{V_{dcb}^2};$$

$$A_a=\left(-\frac{P_{pv}}{V_{dc}^2}\right)+\left(\frac{v_di_d}{V_{dc}^2}\right)+\left(\frac{v_qi_q}{V_{dc}^2}\right)+\left(\frac{\hat{Y}}{V_{dc}}\right); A_b=\left(-\frac{\omega i_q}{V_{dc}}\right)+\left(\frac{2\omega v_d}{X_{cs}V_{dc}}\right);$$

$$A_c=-\frac{\omega v_d}{V_{dc}}; A_d=\frac{\omega v_q}{V_{dc}}; A_e=\left(\frac{\omega i_d}{V_{dc}}\right)+\left(\frac{2\omega v_q}{X_{cs}V_{dc}}\right);$$

$$A_f=\omega i_qv_d/V_{dc}^2-\omega i_dv_q/V_{dc}^2-\omega v_d^2/X_{cs}V_{dc}^2-\omega v_q^2/X_{cs}V_{dc}^2;$$

$$\hat{\hat{\hat{A}}}=\frac{X_{cs}C_{dc}V_{dcb}}{\omega^2V_{acb}v_d}; A_5=(\rho 1+k)\hat{\hat{A}}+\hat{\hat{\hat{A}}}\hat{\hat{A}}A_a; A_6=\rho 1k\hat{\hat{A}}+\hat{\hat{\hat{A}}}\hat{\hat{A}}A_f;$$

$$A_7=-\frac{v_q}{v_d}; A_8=-\frac{m_q}{v_d}+\hat{\hat{\hat{A}}}\hat{A}A_e;$$

$$B=\left(\frac{1}{V_{dc}}\right)\left(-\omega i_qv_d+\omega i_dv_q+\frac{\omega v_d^2}{X_{cs}}+\frac{\omega v_q^2}{X_{cs}}\right);$$

$$A_9=\frac{\left(-B\hat{A}+\left(\omega^2V_{acb}\frac{v_qm_q}{V_{dc}X_{cs}C_{dc}}\right)\right)X_{cs}C_{dc}V_{dcb}}{\omega^2V_{acb}v_d^2};$$

$$A_9'=\hat{\hat{\hat{A}}}\hat{A}A_b+A_9; A_9''=\hat{\hat{\hat{A}}}\hat{A}A_c; A_9'''=\hat{\hat{\hat{A}}}\hat{A}A_d;$$

$$A_9'''' = \frac{2v_q m_q}{\omega v_d} + \frac{\hat{\hat{\hat{A}}}B\hat{A}}{C_{dc}\omega} + \frac{-B\hat{A} + \left(\dfrac{\omega^2 V_{acb} v_q m_q}{V_{dcb} X_{cs} C_{dc}}\right) 2X_{cs} C_{dc} V_{dcb}}{\omega^3 V_{acb} v_d};$$

$$A_{10} = A_6 + A_7 A_3;\ A_{11} = A_7 A_1 + A_9''';A_{12} = A_8 + A_7 A_2;$$

$$A_{13} = A_7 A_4 + A_9'';$$

$$A' = i_q - \frac{v_d}{X_{cs}} + \frac{m_d V_{dc} V_{dcb}}{X_{cs} V_{acb}};$$

$$A_{\omega 2} = -i_d - \frac{v_q}{X_{cs}} + \frac{m_q V_{dc} V_{dcb}}{X_{cs} V_{acb}};$$

$$A_{14} = \frac{\omega V_{dcb} V_{dc}}{X_{cs} V_{acb}};\ A_{15} = \omega + A_{14} A_{13};\ A_{16} = A_{14} A_9' - \frac{\omega}{X_{cs}};$$

$$A_{17} = \left(\frac{\omega m_d V_{dcb}}{X_{cs} V_{acb}}\right) + A_{14} A_{10};\ A_{18} = A_{14} A_1 - \omega;\ A_{19} = A_{14} A_2 - \frac{\omega}{X_{cs}};$$

$$A_{20} = \left(\frac{\omega m_q V_{dcb}}{X_{cs} V_{acb}}\right) + A_{14} A_3;\ A_{21} = \left(-\frac{\hat{A} v_d A_{14} A_5}{V_{dc}}\right) + \frac{\hat{Y}}{V_{dc}};$$

$$A_{22} = \left(-\frac{\hat{A} v_d A_{17}}{V_{dc}}\right) + \left(-\frac{\hat{A} v_q A_{20}}{V_{dc}}\right);\ A_{23} = \left(-\frac{\hat{A} v_d A_{15}}{V_{dc}}\right) + \left(-\frac{\hat{A} v_q A_{14} A_4}{V_{dc}}\right);$$

$$A_{24} = -\frac{\hat{A} v_d A_{16}}{V_{dc}};\ A_{25} = \left(-\frac{\hat{A} v_d A_{14} A_{11}}{V_{dc}}\right) + \left(-\frac{\hat{A} v_q A_{18}}{V_{dc}}\right);$$

$$A_{26} = \left(-\frac{\hat{A} v_d A_{14} A_{12}}{V_{dc}}\right) + \left(-\frac{\hat{A} v_q A_{19}}{V_{dc}}\right);$$

$$A_{27} = (v_{td} - X_d' i_{tsq}) m_2 + (v_{tq} + X_q i_{tsd}) n_2 + i_{tsq};$$

$$A_{28} = (v_{td} - X_d' i_{tsq}) m_4 + (v_{tq} + X_q i_{tsd}) n_4;$$

$$A_{29} = (v_{td} - X_d' i_{tsq}) m_1 + (v_{tq} + X_q i_{tsd}) n_1;$$

$$A_{30} = (v_{td} - X_d' i_{tsq}) m_3 + (v_{tq} + X_q i_{tsd}) n_3;$$

$$A_{31}=\frac{1}{T'_{d0}};A_{32}=\left(-\frac{1}{T'_{d0}}\right)-\frac{(X_d-X'_d)m_2}{T'_{d0}};$$

$$A_{33}=-\frac{(X_d-X'_d)m_4}{T'_{d0}};A_{34}=-\frac{(X_d-X'_d)m_1}{T'_{d0}};\ A'_{34}=-\frac{(X_d-X'_d)m_3}{T'_{d0}};$$

$$A_{35}=\left(\frac{v_{td}X_q n_1}{v_t}\right)-\left(\frac{v_{tq}X'_d m_1}{v_t}\right);\ A_{36}=\left(\frac{v_{td}X_q n_3}{v_t}\right)-\left(\frac{v_{tq}X'_d m_3}{v_t}\right);$$

$$A_{37}=\left(\frac{v_{td}X_q n_2}{v_t}\right)-\left(\frac{v_{tq}X'_d m_2}{v_t}\right)+\frac{v_{tq}}{v_t};\ A_{38}=\left(\frac{v_{td}X_q n_4}{v_t}\right)-\left(\frac{v_{tq}X'_d m_4}{v_t}\right);$$

$$A_{39}=v_{bq}+R_{sb}m_1-X_{sb}n_1;\ A_{40}=R_{sb}m_3-X_{sb}n_3-X_{sb};A'_{40}=R_{sb}m_2-X_{sb}n_2;$$

$$A''_{40}=R_{sb}m_4-X_{sb}n_4+R_{sb};\ A_{41}=-v_{bd}+X_{sb}m_1+R_{sb}n_1;$$

$$A_{42}=X_{sb}m_2+R_{sb}n_2;A_{43}=X_{sb}m_4+R_{sb}n_4+X_{sb};\ A'_{43}=X_{sb}m_3+R_{sb}n_3+R_{sb};$$

$$A_{44}=A_{15}+A_{16}A_{40}+A_{14}A_{12}A'_{43};\ A_{45}=A_{16}A_{39}+A_{14}A_{12}A_{41};$$

$$A_{46}=A_{14}A_{11}+A_{14}A_{12}A_{43}+A_{16}A''_{40};A_{47}=A_{18}+A_{19}A_{43};$$

$$A_{48}=A_{23}+A_{24}A_{40}+A_{26}A_{43};\ A_{49}=A_{24}A_{39}+A_{26}A_{41};A_{50}=A_{25}+A_{26}A_{43}+A_{24}A'_{40};$$

$$A_{51}=A_{26}A_{42}+A_{24}A_{40};A_{52}=A_{14}A_5;\ A_{53}=A_{14}A_{12}A_{42}+A_{16}A'_{40};$$

$$A_{54}=A_{14}A_4+A_{19}A'_{43};A_{55}=A_{19}A_{41};\ A_{56}=A_{19}A_{42};A_{57}=-\frac{A_{28}}{2H};A_{58}=-\frac{A_{30}}{2H};$$

$$A_{59}=-\frac{A_{29}}{2H};A_{60}=-\frac{A_{27}}{2H};A_{61}=-\frac{K_{avr}A_{38}}{T_a};\ A_{62}=-\frac{K_{avr}A_{36}}{T_a};A_{63}=-\frac{K_{avr}A_{35}}{T_a};$$

$$A_{64}=-\frac{K_{avr}A_{37}}{T_a};A_{65}=-\frac{1}{T_a};A_{\omega 1}=A_{14}A_9''''+\frac{A'}{2}$$

The eigenvalues have been calculated for an operating condition as given below at various ratios of power sharing.

**PV Cell Array:** $N_s=1800, N_p=13, I_{sc}=7.35A, V_{oc}=0.6V$
**Diesel Generator:** $X_q=0.47, X_d=1.3, X_d=0.3, T'_{d0}=5, K_{avr}=10, T_a=0.01, D=0, H=2.4s$
**AC System:** $L_f=X_{cs}=0.3pu,\ X_{ts}=X_{fs}=0.0105pu, X_{sb}=0.18,\ R_{sb}=0.1pu$
**PV Controller parameters:** $\rho_1=3500,\ \rho_2=3500$ and $k=1000$

**TABLE 20-5** Eigenvalues for Varying PV-DG Power Sharing

| $P_{pv}$ (pu) | $P_{DG}$ (pu) | $\lambda_1$ | $\lambda_2$ | $\lambda_3$ | $\lambda_4$ | $\lambda_5$ | $\lambda_6$ | $\lambda_7$ | $\lambda_8$ |
|---|---|---|---|---|---|---|---|---|---|
| 0.1 | 0.9 | −1000.15 | −3499.49 | −3500 | −99.130 | −0.102 + 9.943j | −0.102 − 9.943j | 0 | −1.261 |
| 0.2 | 0.8 | −1000.23 | −3499.18 | −3500 | −99.127 | −0.085 + 9.884j | −0.085 − 9.884j | 0 | −1.297 |
| 0.3 | 0.7 | −1000.32 | −3498.87 | −3500 | −99.126 | −0.068 + 9.841j | −0.068 − 9.841j | 0 | −1.332 |
| 0.4 | 0.6 | −1000.42 | −3498.54 | −3500 | -99.127 | −0.052 + 9.817j | −0.052 − 9.817j | 0 | −1.363 |
| 0.5 | 0.5 | −1000.51 | −3498.21 | −3500 | −99.130 | −0.037 + 9.817j | −0.037 − 9.817j | 0 | −1.389 |
| 0.6 | 0.4 | −1000.61 | −3497.85 | −3500 | −99.135 | −0.025 + 9.844 j | −0.025 − 9.844j | 0 | −1.409 |
| 0.7 | 0.3 | −1000.72 | −3497.49 | −3500 | −99.143 | −0.016 + 9.902j | −0.016 − 9.902j | 0 | −1.420 |
| 0.8 | 0.2 | −1000.83 | −3497.09 | −3500 | −99.152 | −0.010 + 10.003j | −0.010 − 10.003j | 0 | −1.423 |
| 0.9 | 0.1 | −1000.96 | −3496.63 | −3500 | −99.163 | −0.010 + 10.204j | −0.010 − 10.204j | 0 | 20.411 |

The eigenvalues for the system is depicted in Table 20-5. All the eigenvalues $\lambda_1$ to $\lambda_6$ are of negative real components which indicates that the system is stable. There are two eigenvalues $\lambda_5$ and $\lambda_6$ exhibiting distinct electromechanical modes of oscillation at each loading level, as shown in Table 20-5.

From these eigenvalues it can be observed that, although the system is stable, the damping present in the oscillatory modes with the increasing PV power injection is relatively decreasing. The reduction in the damping of the oscillatory mode with increased penetration of PV is apparent.

### 20.5.4 System Mathematical Model with DG Governor and Induction Machine Load

Figure 20-47 shows a diagram of the study system. The following subsections describe the nature of the power system model.

***Governor Model.*** The governor model with dead time constant, actuator time constant, respectively, are $\tau_1$ and $\tau_2$. The actuator, current driver constants and engine torque constants are given by $K_1$, $K_2$, and $K_3$ , respectively. The irrational nonlinear dead time is approximated as a rational function by Padé approximation. A second order approximation obtained from Padé approximation sufficient to satisfy the phase response of the actual dead time for interested range of frequencies as shown in Fig. 20-48.

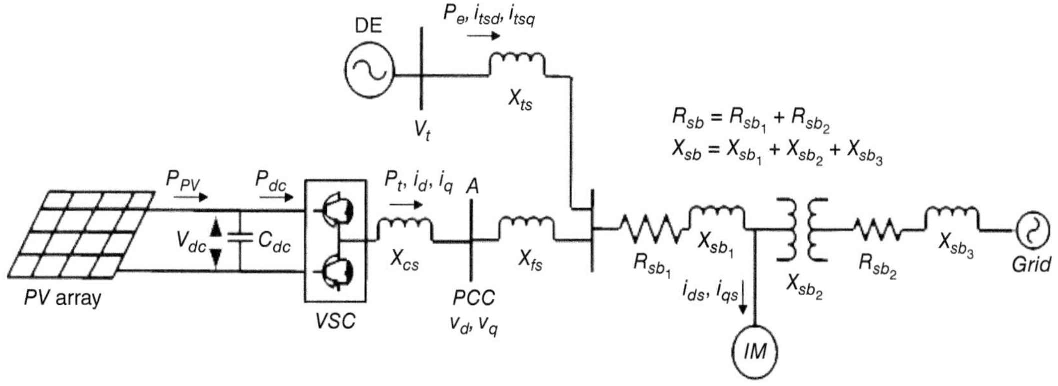


**FIGURE 20-47** Test system.

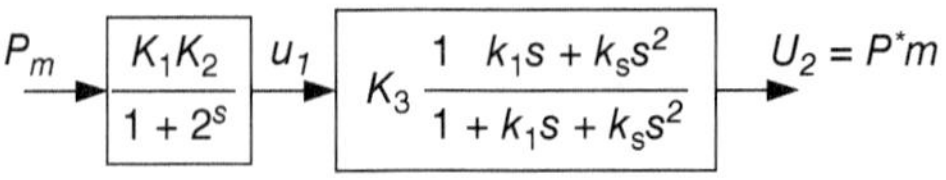

**FIGURE 20-48** Padè approximated governor mode.

The dynamic equations of governor system are

$$\frac{du_1}{dt} = \left(\frac{-1}{\tau_2}\right)u_1 + \left(\frac{K_1K_2}{\tau_2}\right)P_m \tag{20-124}$$

$$\frac{du_2}{dt} = u_3 \tag{20-125}$$

$$\frac{du_3}{dt} = \left(\frac{K_3}{k_2} + \frac{K_3k_1}{\tau_2k_2} + \frac{K_3}{\tau_2^2}\right)u_1 - \frac{u_2}{k_2} - \frac{k_1}{k_2}u_3 - \left(\frac{k_1K_1K_2K_3}{\tau_2k_2} + \frac{K_1K_2K_3}{\tau_2^2}\right)P_m \tag{20-126}$$

The constants are given by $k_1 = \tau_1/2$ and $k_2 = \tau_1^2/12$.

***Induction Motor.*** The induction motor represented using a three equation model is

$$\frac{d\omega_r}{dt} = \frac{1}{2H}(T_e - T_m) \tag{20-127}$$

$$\frac{dv'_d}{dt} = -\frac{1}{T'_0}(v'_d + (X_s - X'_s)i_{qs}) + \omega_r v'_q \tag{20-128}$$

$$\frac{dv_q}{dt} = -\frac{1}{T'_0}(v'_d + (X_s - X'_s)i_{ds}) + \omega_r v'_d \tag{20-129}$$

The entire system state variable set is modified as

$$x = [u_1\ u_2\ u_3\ \delta\ \omega\ E'_q\ E_{fd}\ V_{dc}\ i_d\ i_q\ \omega_r\ v'_d\ v'_q]$$

The comparison of electromechanical eigenmode with and without induction machine is shown in Table 20-6. It can be observed that the system become unstable due to its relatively low damping ratio which varies from $\zeta = 0.002$ at $P_{pv} = 0.1$ pu and $P_{de} = 0.9$ pu to $\zeta = -0.007$ at $P_{pv} = 0.9$ pu and $P_{de} = 0.1$ pu.

In both the cases, the damping ratio of the electromechanical mode becomes less positive with decrease in the DG power injection. Although, DE power decreases, synchronizing torque increases, as its torque angle reduces. Hence the amount of damping torque in system reduces as explained by considering a single-machine infinite bus model. This gives the necessity to provide the auxiliary controller to improve system stability by providing the additional damping to the system.

***PV Auxiliary Controller to Improve System Stability.*** As in power system stabilizer design, an auxiliary signal is developed which is proportional to the frequency deviation. This signal is used to change the PV power levels that makes it operate in variable power mode deviating from MPP point.

To deviate from MPP point the voltage reference signal for the voltage controller is modified as Eq. (20-130)

$$V_{dc\,ref} = V_{dc\,ref\,0} + k_d\Delta\omega \tag{20-130}$$

**TABLE 20-6** Comparison Electromechanical Eigenmode of the System with and Without Induction Motor

| $P_{pv}$ $(pu)$ | $P_{DG}$ $(pu)$ | Electromechanical eigenmode (without induction machine) | Electromechanical eigenmode (with induction machine) |
|---|---|---|---|
| 0.1 | 0.9 | $-0.102 \pm 9.943j$ | $-0.0208 \pm 10.1823j$ |
| 0.2 | 0.8 | $-0.085 \pm 9.884j$ | $-0.0055 \pm 10.1296j$ |
| 0.3 | 0.7 | $-0.068 \pm 9.841j$ | $0.0101 \pm 10.093j$ |
| 0.4 | 0.6 | $-0.052 \pm 9.817j$ | $0.0252 \pm 10.0762j$ |
| 0.5 | 0.5 | $-0.037 \pm 9.817j$ | $0.0393 \pm 10.0819j$ |
| 0.6 | 0.4 | $-0.025 \pm 9.844j$ | $0.0521 \pm 10.1137j$ |
| 0.7 | 0.3 | $-0.016 \pm 9.902j$ | $0.06296 \pm 10.1763j$ |
| 0.8 | 0.2 | $-0.010 \pm 10.003j$ | $0.0716 \pm 10.2820j$ |
| 0.9 | 0.1 | $-0.010 \pm 10.204j$ | $0.07534 \pm 10.4909j$ |

With the addition of the auxiliary signal in the voltage controller a new matrix is defined where input vector is written in terms of the state variables. All elements of the new matrix are zeros except those corresponds to $V_{dcref}$ and $\Delta\omega$.

The electromechanical eigenmodes after addition of this are tabulated in Table 20-7.

**TABLE 20-7** Electromechanical Eigenmode with Induction Machine and Auxiliary Signal Addition

| $P_{pv}$ $(pu)$ | $P_{DG}$ $(pu)$ | Electromechanical eigenmode (with induction machine + auxiliary signal ) |
|---|---|---|
| 0.1 | 0.9 | $-0.092 \pm 10.317j$ |
| 0.2 | 0.8 | $-0.078 \pm 10.293j$ |
| 0.3 | 0.7 | $-0.063 \pm 10.298j$ |
| 0.4 | 0.6 | $-0.05 \pm 10.341j$ |
| 0.5 | 0.5 | $-0.038 \pm 10.44j$ |
| 0.6 | 0.4 | $-0.028 \pm 10.627j$ |
| 0.7 | 0.3 | $-0.023 \pm 10.984j$ |
| 0.8 | 0.2 | $-0.023 \pm 11.834j$ |
| 0.9 | 0.1 | $0.268 \pm 16.894j$ |

From Table 20-7 it can be observed that eigenvalues are shifted to left side of the s-plane with the addition of auxiliary signal controller to the PV system. Hence, even for high penetration of PV power with dynamic loads, it is possible to stabilize the system by contributing some amount of damping with the help of PV power.

# 21

# ELECTRICITY MARKETS

Ross Baldick

*Professor, University of Texas at Austin; Fellow, IEEE*

Resmi Surendran

*Senior Manager, Wholesale Market Operations and Analysis, Electric Reliability Council of Texas, Taylor, Texas*



## 21.1 INTRODUCTION

Electricity markets are artificial constructs created to realize the objective of providing dependable electrical power at the lowest cost of production in the short term and provide incentives for efficient investment in the long term. They have been formed as part of a process to "restructure" electricity systems. Power system coordinators or independent system operators (ISOs) are organizations created to maintain and administer these constructs, typically under authority delegated by a regulatory or governmental agency as part of the restructuring process. Electricity markets are hybrids, incorporating elements of regulatory oversight along with market mechanisms and incentives in an attempt to create an efficient and effective market. While it is true that all other markets also have some form of regulation, including a legal system to enforce contractual agreements, electricity markets have a particularly high degree of regulatory oversight.

Similar to entities in other industries, electricity market participants utilize capital markets to finance equipment and services. Regulatory oversight typically maintains a role of maximizing benefits and equitably allocating costs to market participants and end-use consumers. When designed and operated effectively, these regulations and market controls provide the right incentives for market participants to ensure that their behavior to maximize profit will also maximize reliability and social welfare. A well-designed market will have rules that encourage competition and mitigate the exercise of market power. Similar to other markets, it will have standardized products, ways to manage risk effectively, and will allow open access sufficient to attract participation by traders/investors/speculators, increasing liquidity and bringing stability to the market.

As discussed in an earlier section, ISOs in the United States and analogous entities elsewhere have been established to oversee the operation of the electricity market independent of market participants and asset owners, such as generators, transmission providers, distribution providers, wholesale marketers, retailers, speculators, and customers. The ISO facilitates market operation by enforcing the rules and regulations of the market designed to prevent market manipulation by market participants and incentivize operation that increases system reliability and social welfare.

In the United States, Australia, and several other countries, the ISO (or equivalent entity) is usually responsible for both market and reliability operations. However, in several countries, this responsibility is split into a market operator (MO), which is only responsible for market operation while system reliability is ensured by a reliability operator (RO).

The RO is responsible for long- and short-term transmission planning, long-term generation reserve or capacity planning, outage coordination, ensuring adequate capacity and ancillary service reserves in real time, maintain system reliability/security in real time, performance monitoring and ensuring compliance with the reliability rules. The MO is responsible for operating forward markets, operating real-time markets and sending out price signals, providing methods for hedging against financial risks, identifying and mitigating market power and settling the market. These responsibilities span different time horizons from years ahead to months after the operating day. A sampling of typical tasks for the market and reliability operator are listed below:

- *Year-Ahead Tasks.* Long-term planning (RO), yearly CRR auction (MO)
- *Month-Ahead Tasks.* Monthly CRR auction (MO), outage coordination (RO)
- *Week-Ahead Tasks.* Outage coordination (RO)
- *Day-Ahead Tasks.* Day-ahead energy and ancillary service market (MO), day-ahead reliability unit commitment (RO)
- *Hour-Ahead Tasks.* Hour-ahead reliability unit commitment (RO), ancillary service market and look ahead advisory dispatch (MO)
- *Real Time.* Real-time load balancing market (MO), frequency control (RO), transmission security management (RO)
- *After Real Time—Months Later.* Market settlement (MO), market monitoring (MO), performance monitoring (RO), compliance (RO)

## 21.2 MARKET DEVELOPMENT AND PHILOSOPHY

This subsection examines how the current markets were established, what the basic principles of market economics are, why regulatory rules are needed, what extent of regulation is needed in the market, what role each participant should play in the market and what the market provides to its participants.

### 21.2.1 Establishment of Markets

Most, if not all electric markets have been created in response to dissatisfaction with the level of performance of integrated transmission/generation utilities that were either government-owned or under regulatory governance. Major problems perceived with integrated transmission/generation utilities and regulated/government-owned electricity market model was the insufficient incentive for the operators/providers of the service to optimize investment options or invest in better technology and, in some cases, perceived or actual imprudent investments. Regulatory review suggests that this failure to invest efficiently was, in part, driven by the typical practice of socializing investment costs coupled with the inability of customers to seek an alternative, more efficient supply. Successful investment risks were not rewarded, and unsuccessful investment costs were borne by customers who had no input into the decision making process. This led to large, highly risk averse organizations with no incentive to adopt new technologies.

With the establishment of an efficient market, it is intended that market-based investment decisions will result in better evaluation of investment risks, since those risks will then be borne by the capital investors, which will lead to better investment decisions, and will improve overall efficiency of the system. Hence, the extent of regulation was gradually reduced and markets were established, first in the generation sector and in some cases in the retail sector.

Transmission and distribution assets, however, typically remain as regulated assets, and have not been amenable to imposition of market disciplines. This stems from several issues related to transmission and distribution, including the technical complexity of transmission operations, the economies of scale and lumpiness of transmission assets, contention in the siting of transmission, and the "public good" nature of some of the benefits of transmission capacity. For this reason, the following discussion will primarily be restricted in application to wholesale and retail electrical energy markets, implicitly assuming the existence of an adequate transmission and distribution infrastructure.

### 21.2.2 Principles of Market Economics

Under various assumptions of neoclassical microeconomics, including the lack of economies of scale, perfect competition, and costless entry and exit from the market, the outcomes of markets can be explained by examining the interactions between supply and demand. While these assumptions are not exactly satisfied by any market, and can be violated under various circumstances in electricity markets, perfect competition is a useful paradigm to establish ideal properties of a market.

We consider an abstract, ideal "market" where some product or commodity is being sold by producers to consumers. Perfect competition means that no single actor can unilaterally affect price. This assumption typically holds if all market participants are infinitesimal, and in this case we can aggregate the behavior of the supply side actors together into a "supply function" and aggregate the behavior of the demand side into a "demand function." It is easiest to define the supply and demand functions in terms of their inverse functions. The inverse supply function effectively orders the producers from lowest to highest marginal cost, while the inverse demand function orders the consumers from highest to lowest marginal value of consumption. For any given quantity, the value of the inverse supply function shows the highest marginal cost amongst all of the suppliers that are needed to efficiently produce that quantity, while the value of the inverse demand function shows the lowest marginal valuation amongst all the consumers that should be supplied to maximize the value of that total consumption quantity.

The inverse supply function evaluated at a particular quantity indicates the price needed to infinitesimally increase the production quantity. The inverse demand function evaluated at a particular quantity indicates the price needed to infinitesimally reduce the consumption quantity. In other words, the inverse supply function indicates the price at which it is worthwhile for the producer to produce a given quantity and the inverse demand function indicates the price at which it is worthwhile for the consumer to consume a given quantity of the product. These functions are illustrated in Fig. 21-1.

The intersection of the supply and demand function defines a price $P_E$ and a quantity $Q_E$ such that all supply corresponding to quantities at or below $Q_E$ has marginal cost below or equal to $P_E$, and all

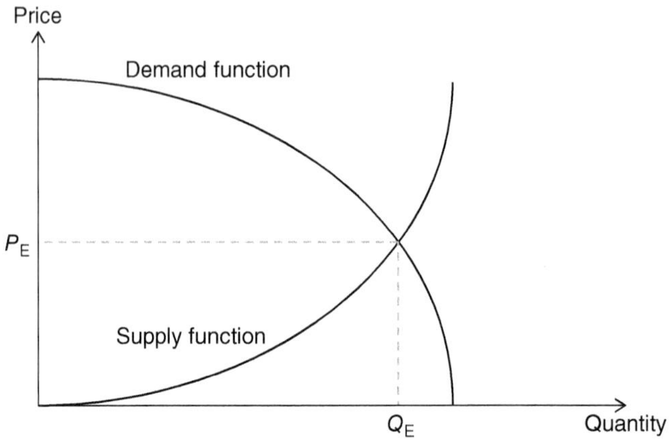


**FIGURE 21-1** Supply and demand functions.

consumption corresponding to quantities at or below $Q_E$ has marginal value above or equal to $P_E$. This quantity $Q_E$ also corresponds to the production and consumption that maximize the short-term welfare of consumption, and the price $P_E$ can be viewed as coordinating this supply and demand:

- All suppliers with marginal cost less than the price $P_E$ will benefit from producing and being paid the price $P_E$, while no suppliers with higher marginal cost would choose to produce given this price.
- All consumers with marginal valuation higher than the price $P_E$ will benefit from consuming and paying the price $P_E$, while no consumers with lower marginal valuation would choose to consume given this price.

If circumstances deviate from this state, there is an incentive for some supplier to alter its production and/or an incentive for some consumer to change its consumption. Under suitable circumstances, the dynamic due to these incentives will drive the outcome of the market to settle on the price $P_E$ for all with consumption equal to $Q_E$. This price and quantity is called the competitive equilibrium of the market.

The area bounded by the price axis and the demand and supply functions up to the equilibrium represents the amount by which the aggregate value of consumption exceeds the aggregate cost of production. This is called the *social welfare* and represents the net benefit of production and consumption. The equilibrium quantity coincides with maximization of social welfare. Because the producer gets paid at the price $P_E$, and the consumer pays at the price $P_E$, the equilibrium price $P_E$ implicitly partitions the total social welfare into the net benefit accruing to the producer and consumer, respectively. In particular, the area bounded by the price axis, the horizontal line at price $P_E$, and the supply function is the "producer surplus" or operating profit. The area bounded by the price axis, the horizontal line at price $P_E$, and the demand function is the "consumer surplus."

To summarize, in a perfectly competitive market, where no single market participant can influence the price, under suitable conditions, the market will converge to an equilibrium at a price $P_E$ and quantity $Q_E$ such that the suppliers in aggregate are willing to produce $Q_E$ at the price $P_E$ and the consumers desire in aggregate to consume $Q_E$ at the price $P_E$. All of the quantity $Q_E$ is sold at the price $P_E$. More information about microeconomic principles can be found in [1].

### 21.2.3 Electricity Markets

In reality, markets, and particularly electricity markets, are not perfectly competitive. Nevertheless, the idea of the equilibrium in a perfectly competitive market is useful as a guide for the design of actual electricity markets in order to emulate the desirable outcomes of an idealized, perfectly competitive market, even though the special characteristics of electricity make the process described above impractical.

A key reason that electricity markets do not literally utilize the process described above is related to the nonstorability of electricity. The dynamic process described above implicitly assumed that production or consumption could adjust from a nonequilibrium condition and that an equilibrium would be reached asymptotically. To be useful as a guide to the outcome of the market, the dynamic process would need to converge fast relative to changes in circumstances that would change the supply or demand functions.

This assumption about the dynamics of price adjustment is most plausible in the context of a storable commodity where we can think of the demand function as representing aggregate behavior of many consumers that is applicable over an extended period such as days. With rapid enough updates, the sequence of prices would approximate the changing equilibria as supply and demand conditions vary. Electricity, however, is not costlessly storable, and adjustment processes cannot feasibly become close to an equilibrium over the short timescales over which supply and demand can change. For this and other reasons, electricity markets are not literally arranged as the ideal market described above, but rather information is solicited from market participants and an equilibrium calculated by the MO, with the goal of reproducing the outcome that would have occurred with an ideal, competitive market. To flesh that out, the following subsection will discuss electricity market participants.

### 21.2.4 Electricity Market Actors and Participants

There are several types of actors and participants in the electricity market. Each plays a significant role in determining the proper functioning of the market. This subsection discusses roles of some significant participants in power markets and what they gain out of participating in these markets.

***Regulators.*** Regulators have a key role in electricity markets to handle the effect of special characteristics of the power system on the market outcome. Regulation is appropriate if done effectively to address market issues or to attain a market equilibrium that is socially more optimal. Although power markets are formed to remove the inefficiencies of the regulatory model, some regulation is found to be still needed, most obviously in setting up the MO process of emulating and implementing the outcome from an idealized market.

Many of the characteristics of the historically vertically integrated electric utility differ significantly from the perfectly competitive market described above. Specifically,

- There is typically no mechanism to update prices as supply and demand conditions change.
- Suppliers do not have the power to set price since the price is regulated.
- Consumers do not have visibility to price.
- Consumers do not have the ability to select between alternate suppliers.
- Consumption is in-elastic (i.e., it does not change in response to price, possibly due to the factors listed above).
- In many environments, the ability to sell electric power is/was strictly regulated and restricted to one provider.
- In many environments other barriers to entry to the market exist(ed) such as
  - Restricted ability to interconnect to the transmission grid
  - Access to use of the transmission grid controlled by competitors
  - Regulated pricing

Regulators desiring a competitive market must take action to change or mitigate these factors if the positive effects of competition are to be realized. In general, regulators attempt to drive a controlled transformation from a fully regulated but inefficient regime, which is often functional and at least perceived as reasonably fair to all involved, to an efficient market infrastructure, while still managing particular characteristics of the regulated regime such as support for low-income consumers and other issues related to access and fairness.

The intent of market-focused regulatory statutes, rules, and structure is to create the infrastructure, environment, and incentive for market participants to willingly seek the maximization of total surplus to society as a natural competitive market outcome of pursuing their own interests. Such regulators must examine the benefits of consuming the product and the opportunity costs of producing the product for society as well as preserving the viability of a particular market segment. They might intervene to adjust the price of the product by introducing subsidies or taxes to attempt to move the market solution closer to a socially optimal solution.

For example, assume that power production from a fuel type has environmental effects that impose indirect costs on society. If these indirect costs are not included in the offer to produce energy then the net cost to the society is not considered when arriving at the market equilibrium. This situation could result in higher than socially optimal production level from the fuel type. In these cases, regulators could introduce rules to put extra tax on the power produced from this fuel type to increase the price and effectively reduce the power production from that fuel type, that is, bring the market equilibrium closer to socially optimal level.

Since external intervention can have significant impact on the market solution, it is critical to evaluate the need for and the effect of any regulatory rule before deciding to use it as a solution to fix market issues. If regulations are imposed without considering the effects on all sectors or are based on incomplete/imperfect information then they may degrade market performance instead of bringing it closer to the socially optimal solution. As an example, in the first implementation of the California market, retailers were discouraged from entering into hedging contracts with generation companies. This had the unintended consequence of exacerbating market power when supply was tight [2].

***Power Producers and Consumers.*** The market participants produce and consume electricity and exchange commodities in wholesale or retail electricity markets. For power producers an efficient electricity market with credit requirements imposed on participants can provide a stable platform for selling power with an assurance of marginal cost recovery without worrying about the credit worthiness of the buyer. Similarly, regulated markets provide a stable platform for retailers to procure power at fair prices consistent with long-term market stability, and be ensured service without multiple contracts with multiple alternate suppliers. Market design should ensure that the most efficient units are dispatched, each load is served at the least possible cost and each entity is paid for the services provided and charged for the energy consumed.

Both producers and consumers are susceptible to uncertainty of prices. These market participants are exposed to price risk due to low elasticity of demand, imbalance of load in real time, uncertainty in real-time load, minimal options to store electricity, potential exercise of market power, congestion due to physical characteristics of the power flow, and uncertainty of congestion, forced outages, or speculative bidding causing high volatility in real-time prices. Different electricity markets create opportunities for these market participants to hedge the risks within bands controllable by the buyer and seller by providing different products like Congestion Revenue Rights and a platform to buy and sell these products.

***Traders and Speculators.*** Several entities in electricity markets are not involved in production or final utilization of an asset, but buy or sell in the expectation of making a profit. Traders make their long-term or short-term strategies based on forecast prices or trends of the prices of these assets. They normally use sophisticated tools to determine the level of risk to be taken and assess the expected future of an asset, and usually have a high level of expertise in the rules and design of the market in which they are participating.

These participants improve the liquidity of products sold within a market, and thus enhance stability. They often play a critical role in price convergence of different markets and their probing into price differentials often exposes flaws in regulated market design. Products that traders can buy/sell in regulated electricity markets are normally restricted by specific market rules. The most common role of traders is dealing with risk management products. Their participation increases the liquidity in the market and helps power producers and consumers hedge risks by assuming these risks if they are found to be possible profit making opportunities.

## 21.3 CHARACTERISTICS OF ELECTRICITY MARKETS

Different regions of North America and internationally are at different stages of market development in the power industry. For example, some regions in the United States do not have any ISO-based markets while others have well-developed markets. There are many different aspects of market designs and their implementations vary in level of complexity and efficiency. The details of design adopted in a regional power industry usually depends upon government policies and typically evolves over time. Markets implemented based on the same basic market design could still be different depending on the specific market rules and government regulations. The power industry with multiple markets could have different market designs for different market or could have a combination of different types for a single market.

In this subsection, we will look into special characteristics of the electricity market that will be fundamental in implementing any market design. Then we will examine characteristics of different market designs and different types of markets in the power industry. References [3–5] compare and contrast the major market designs types and the implementation of different types of markets based on these market designs.

Power markets are different from other commodity markets in many ways. This is due to special characteristics of the power systems:

- The need to obey the physical laws of power transmission
- The need to manage flows below the physical limits of transmission lines, and also such that the flows will be within limits under contingencies
- Minimal options to store electricity causing the need to produce the energy when it needs to be consumed
- The need to maintain the balance between production and consumption to maintain system frequency within limits
- Low elasticity of real-time demand
- Uncertainty of congestion due to forced outages or speculative bidding, etc.

In all existing market designs, it has been concluded that these factors necessitate establishment of a centralized entity to manage the system, which is consistent with the observation that electric utilities prior to restructuring also acted as a centralized entity to manage their systems. As discussed above, electricity markets are designed to reproduce, as closely as possible, the outcome that would have occurred with an ideal, competitive market. This task is accomplished through software models that determine the market outcome (generation commitment and dispatch) by seeking the optimal mix of resources to maximize global welfare. In the next subsection, the basic building blocks of these market designs, including offer-based economic dispatch will be introduced.

## 21.4 BUILDING BLOCKS OF ELECTRICITY MARKET DESIGN

In this subsection, the basic building blocks of electricity market design will be discussed, including

- Offer-based economic dispatch
- Hedging of energy price risk
- Locational marginal pricing
- Hedging of transmission price risk
- Capacity markets

While these basic building blocks together constitute a viable market design in themselves and represent the main features of U.S. markets, various configurations and combinations of these building blocks are used in other markets (and have been used previously in U.S. markets).

### 21.4.1 Offer-Based Economic Dispatch

This subsection describes offer-based economic dispatch, which is sometimes called *bid-based economic dispatch*, and is the underlying "engine" for many market functions. In general, offer-based economic dispatch is used in markets to optimally dispatch resources at minimum system cost to meet the reliability of the system. This is implemented by selecting generation levels of online resources within their ramping limits to match overall generation and load (plus losses) and resolve transmission constraint violation in the most economical way based on the submitted offers from generators and bids from demand side. To illustrate offer-based economic dispatch, we will first consider dispatch and pricing for a simple system with no transmission. A particular type of energy offer, which will be called a *block offer*, will be presented. Then demand specification and demand bids will be described, followed by dispatching and pricing, and the incentives from pricing. Subsequently, variations and additional issues will be discussed. Several simple examples will be introduced to illustrate the development. The rules of the ERCOT nodal market [6] will be referenced, as an example of the implementation of these principles. The other North American markets have broadly similar rules, but differ in many particular details.

***Energy Offers.*** We turn first to the supply side. As a general observation, in a day-ahead or real-time electricity market, generators (or, precisely speaking, their representatives) offer hourly prices to sell into the system for the next day or the next hour at their point of connection. These offers can include a specification of their supply of not only energy, but also reserves and other ancillary services (AS). Initially, we will discuss energy offers only, but then sketch the extension to AS.

Various markets have various rules about the specification of energy offers to the independent system operator (ISO) or market operator (MO). We will focus on a particularly simple example, which we will refer to as a "block offer." (However, it should be understood that "block" is often used in the electricity industry to mean something more restrictive than our definition.) For our purposes, we define a block offer as being specified by three parameters:

1. A quantity, with the understanding that the offer allows for the ISO to select any quantity up to maximum power in the block
2. A price, which is the nominated "offer price" per unit energy
3. A duration, typically an hour or multiple hours

The total offered quantity can be specified in multiple blocks. For example, consider a 150-MW capacity being offered. At the generator's discretion, this could be divided into three blocks each, for example, 50 MW, and with respective offer prices of \$50, \$70, and \$90/MWh. This is illustrated in Fig. 21-2.

Variations on this arrangement are implemented in various markets. For example, in typical U.S. markets, there is a specified minimum production level. In ERCOT, for example, this is called the

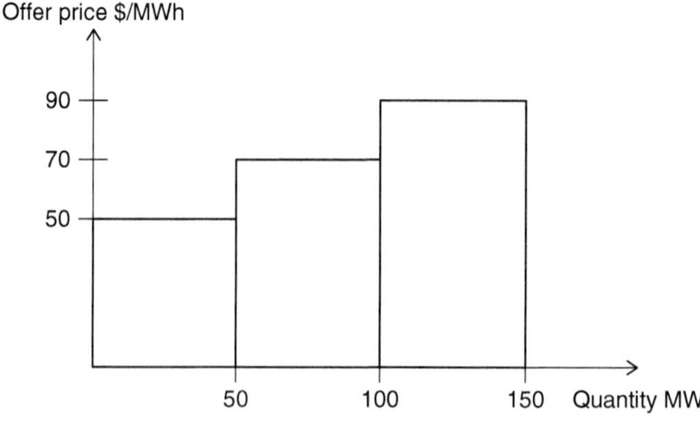


**FIGURE 21-2** Example of offer from generator.

*low sustained level* (LSL), and in the example above this minimum production might be specified as 25 MW. In addition, sometimes blocks are specified as "all-or-nothing" (and, in fact, a "block" is often interpreted in the electricity industry as meaning such a fixed quantity block, although as mentioned above, our definition here is less restrictive.) "Fixed quantity blocks" are specified Sec. 4.4.8.5.1 of the *ERCOT Nodal Protocols* [6].

***Demand and Demand Bids.*** Turning to the demand-side, either the ISO forecasts demand, and/or the representatives of load, which are called *qualified scheduling entities* (QSEs) in ERCOT [6] either

- Bid prices for their "willingness-to-pay" for energy at their point of delivery in each hour
- Effectively specify demand by bidding with a willingness-to-pay equal to the offer cap, which is the highest price that is permitted in an offer or bid

As an example, the total demand could be specified as 220 MW for a particular interval, implicitly meaning that it has been bid at the offer cap.

***Dispatch of Generation.*** Using the generation offers and the demand bids and specifications, the ISO performs a calculation called *offer-based economic dispatch*, which is essentially the same algorithm as in "traditional" economic dispatch as practiced by integrated utilities, except that it uses offer prices from generators instead of generation costs, and uses specified or bid demand. Economic dispatch finds the lowest cost combination of generation offers that will meet the power balance constraint that total generation equals total demand (possibly including an estimate of the line losses.) The results of offer-based economic dispatch are used to select or "dispatch" generation and the selected generation is said to have been "cleared" in the market.

In the simplest case of no transmission constraints, offer-based economic dispatch can be thought of as stacking up offer prices from the cheapest to the most expensive until demand is met, or until the offer price of additional generation would exceed the bid price of additional demand, or until there are no more offers. By design, this process results in the cheapest generators being dispatched to meet demand. It is to be emphasized that this basic description ignores transmission constraints and losses, and ignores unit commitment, start-up costs, and no-load costs.

***Dispatch and Pricing Example I.*** Suppose there are two generators, one of which makes the offers shown in Fig. 21-2 and will be called the *white* generator, and a second generator, the "gray" generator, together with a specification of 220 MW demand. The blocks from Fig. 21-2, together with the offer blocks of the gray generator, are shown in Fig. 21-3 stacked from lowest to higher offer price, together with a vertical line at 220 MW corresponding to the demand specification.

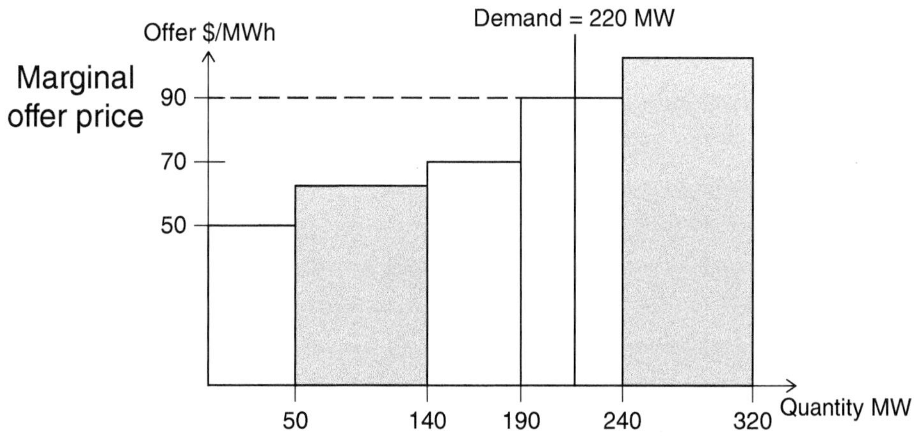


**FIGURE 21-3** Results of offer-based economic dispatch for example I by stacking up the blocks.

As shown in Fig. 21-3, the result of offer-based economic dispatch is to utilize 130 MW of white generation, consisting of all of its first two 50-MW blocks together with 30 MW out of its third 50-MW block and utilize 90 MW of gray generation, consisting of all of its first 90-MW block. The total generation is 220 MW, so that supply equals demand.

***Pricing Rule.*** Some markets, such as for some types of reserves in Europe, use so-called *pay-as-bid* pricing for reserves. Under that pricing rule, each accepted offer is paid at its offer price. This type of pricing rule is not generally used in North American markets, and is not generally used for energy in any market.

The standard pricing rule for energy in offer-based economic dispatch is to set a uniform wholesale price equal to the value of a fictitious extra MW (strictly speaking, an infinitesimal extra quantity) of costless generation. In particular, we imagine an extra MW of generation having zero cost and calculate its value to the system. This is called the *marginal surplus*. The value of this fictitious additional MW is determined by the actual offer costs, demand bids, and "value of lost load" (if involuntary curtailment in the case that there are not enough offers to meet the specified demand level) and is then used as the price for all energy sold.

The simplest case is where the value of the fictitious generation is the savings in reduced offer costs made possible by the additional MW due to reducing the rest of the generation by a MW. In some cases, however, the greatest value would be obtained not by reducing generation of other generators, but rather by increasing the served demand. In this case, the value is equal to the willingness to pay for an extra MW of demand that could be served using the extra MW of generation. Finally, if there is involuntary curtailment due to limited supply, then the value of an additional MW is equal to the "value of lost load" (VOLL) of an extra MW of demand that could be served with the additional MW.

The results of this thought experiment can be simplified in the common case of no demand bids, no curtailment of demand, and the demand level not falling at jump in prices between blocks. In this case, the value of an extra MW of generation is equal to the offer price to serve an additional MW of demand or the offer price that served the last MW of demand, which is called the *marginal offer price*.

It is to be emphasized, however, that when not all demand is served, marginal surplus may instead be equal to the bid price of additional MW of demand, reflecting its "willingness-to-pay," or to the VOLL of curtailed demand. It is often tacitly assumed that the offer cap in the market is a proxy to the VOLL. As will be discussed further below, pricing in practice often deviates from this pricing rule during scarcity.

***Dispatch and Pricing Example II.*** Consider a more elaborate example with three generators, at locations NW, NE, and SE, and 3000 MW demand at SW, as illustrated in Fig. 21-4. There are

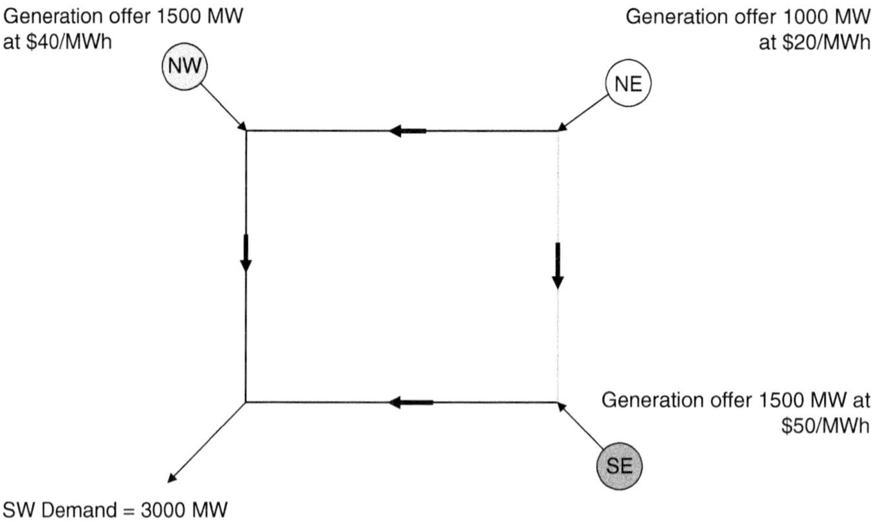


**FIGURE 21-4** Three generator example II.

four transmission lines illustrated in this system, but for now we will assume that the transmission capacity is large enough not to affect or "bind" on the solution. We will revisit this in the context of locational marginal pricing.

Assume that each generator specifies a single block for all of its capacity, with offers as follows:

- NW offer of 1500 MW at $40/MWh
- NE offer of 1000 MW at $20/MWh
- SE offer of 1500 MW at $50/MWh

To find the offer-based economic dispatch, the offer prices are stacked from cheapest to highest until demand is met, as illustrated in Fig. 21-5.

While the graphical construction in Fig. 21-5 is an effective metaphor for illustrating economic dispatch in the absence of transmission constraints, it is not suitable when transmission constraints bind. For later analysis of the transmission-constrained case, we need a more general concept, the "total offer cost," which is the sum over the accepted quantities of the product of the accepted quantity multiplied by the offer price.

The total offer cost measures the area under the selected blocks in Fig. 21-5. Stacking up the blocks is equivalent to choosing quantities of generation, which will be denoted by NW, NE, and SE, so that the total generation meets the demand and also so that the choices of NW, NE, and SE minimize the "total offer costs." The total offer costs are

$$(\text{NW} \times \$40/\text{MWh}) + (\text{NE} \times \$20/\text{MWh}) + (\text{SE} \times \$50/\text{MWh})$$

From Fig. 21-5, or by minimizing the total offer costs, the optimal offer-based dispatch is to utilize 1000 MW of NE, 1500 MW of NW, and 500 MW of SE generation, for which the total offer costs are 105,000 $/h. This is the area under the blocks up to the demand level of 3000 MW. The marginal surplus, equal to the marginal offer price, is $50/MWh, and total payment at this uniform price is 150,000 $/h. The total payment can be visualized as the rectangular area of width 3000 MW and height equal to the marginal offer price, shown by the dashed line. The total payment is larger than the total offer costs.

By design, this process selects the lowest cost combination of generation to meet the demand based on the offer prices. That is, it minimizes the total offer costs, which should be distinguished from the total payment, which is higher than the total offer costs.

***Efficiency of Dispatch.*** An important result relating to operational efficiency is that if offer prices by the market participants are set for each generator to be equal to the slope of its operating costs;

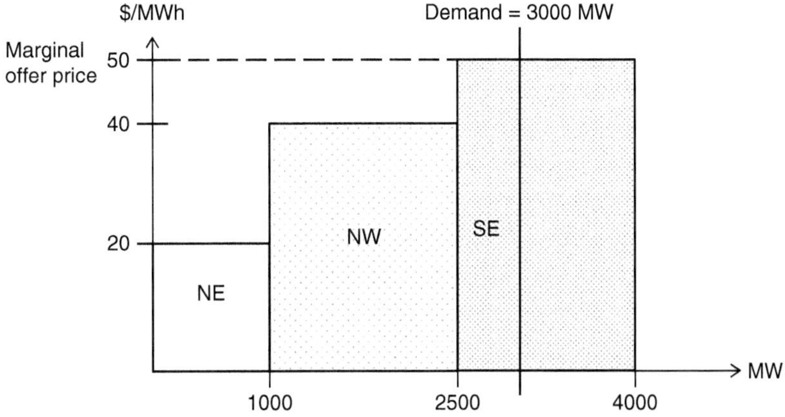


**FIGURE 21-5** Results of offer-based economic dispatch for example I by stacking up the blocks.

that is, equal to the marginal generating costs, if the operating costs are convex, and there are no binding transmission limits then optimal economic dispatch is achieved. That is, demand is satisfied at least generating cost. (Transmission constraints will be considered later and a similar efficiency result applies.)

The efficiency result relies on firms offering at marginal cost, and it is important to establish whether this is a reasonable expectation. A standard argument applies if firms are too small to affect the marginal offer price. In this case, they maximize their profit by offering each generator at its marginal cost. To show that offering at marginal cost maximizes profits, consider the alternatives:

- If a firm offers below the marginal cost then it risks losing money if the marginal offer price is below their marginal cost, which can be expected to occur during times of lower demand.
- If a firm offers above the marginal cost then it risks not being dispatched when it could have made an inframarginal offer, which can be expected to occur during times of higher demand.

To summarize, when firms cannot affect prices, self-interested behavior of offering at marginal cost yields economic dispatch. Under some circumstances, particularly when there is tight supply so that the total generation available can only barely meet demand, individual market participants can affect prices and the result above does not hold. In these cases, it is necessary to impose market power rules to bring the offer prices toward what they would be in the ideal of a perfectly competitive market, and thereby result in efficient dispatch. (Further analysis about the relationship between market power and efficiency can be found in [7].)

As mentioned above, all energy is sold at a uniform price and, therefore, some generators are paid more than their offer price. Offers that are paid more than their offer price are called *inframarginal* offers. A uniform price based on "marginal surplus" allows generators to recover some of their capital costs, start-up costs, and min-load generating costs. (These costs are also sometimes partially recovered by separate explicit payments, to be discussed below.)

The efficiency result can be generalized further to include efficiency of both operations and investment using standard arguments from microeconomics, but the circumstances do not fully apply to electricity markets as currently operating. The result basically says that at an "optimal" level of installed capacity, pricing at marginal surplus covers operating and capital cost of all generators, if we can ignore:

- Start-up and min-load costs of generators when they are committed
- "Lumpiness" of investment and retirement, and economies of scale

The argument is a dynamic one, and observes that if there is less than the "optimal" installed capacity then time average of marginal surplus rises to encourage new construction, while if there is more than the "optimal" level of installed capacity then time average of marginal surplus falls, discouraging new investment. As well as ignoring the commitment and lumpy investment issues, the dynamic depends crucially on the pricing of consumption at its "willingness-to-pay" during times of scarcity when demand is curtailed. As will be discussed further below, this assumption is not satisfied in electricity markets currently because of various reasons, including the historical lack of real-time metering and the historical emphasis on building enough capacity to meet demand under essentially all circumstances.

To summarize, under idealized conditions, if all generating firms are small then pricing all energy at the marginal surplus yields good incentives for operation, and good incentives for new investment. Analogously, pricing demand at the marginal surplus provides good incentives for consumption, since then consumption occurs when willingness to pay exceeds the marginal cost of generation.

***Pricing in Practice and Implications.*** In practice, electricity market rules or procedures may not be able to reproduce the idealized situation described above. For example, ISO practice under tight supply is sometimes to take out-of-market actions to avoid curtailment, and then apply only a (possibly zero) proxy to the cost of ISO actions to avoid curtailment that fall well below the demand's willingness to pay. As an example, the ERCOT market [6, Sec. 6.5.7.1.11] provides for a cost curve for this proxy when regulation ancillary service is used for energy, but this might not fully represent the costs of the reduced provision of regulation. In addition, concerns about the exercise of market power have led to rules in various markets that sometimes result in prices that fall below the competitive levels under tight supply.

To summarize, the proxies used in practice can fall short of the marginal surplus during scarcity. For example, if the VOLL is around $5000/MWh, but the prices in the market reflect values closer to the typical marginal costs of generators even during scarcity conditions, then the difference between the marginal surplus and the price used in practice may be on the order of $5000/MWh. Together with offer caps in markets that are set below the demand willingness to pay, such actions account for what is called the *missing money* in offer-based economic dispatch markets; that is, the deviation between proceeds in an idealized competitive market and the proceeds from pricing rules in practice.

Missing money, or the perception of missing money, has resulted in various actions, including the development of so-called capacity markets in the eastern U.S. markets. Capacity markets provide remuneration to market participants who commit in advance to providing generation capacity to the market. Typically a generator receives a revenue stream in return for some level of guarantee that it will be in-service and offering into the energy market during the contract term of the capacity market.

Another issue that arises in the context of unit commitment (discussed below) is that the marginal cost may not fully represent all the economically significant issues in the cost of dispatch, including start-up and no-load costs. Recently, the Midcontinent ISO has implemented a variation, called *convex hull prices* that aims to reflect start-up and no-load costs into an energy price [8].

***Uplift.*** In various cases, the energy price is increased to include compensation for additional issues. This increase is termed "uplift," and can include a variety of terms to compensate for

- Transmission system losses, if they are not explicitly incorporated into the energy price through the model of the transmission network (to be discussed below)
- The cost of procurement of ancillary services (discussed below), if they are not otherwise charged on the basis of causation
- Start-up and no-load costs incurred by generators that are committed by the ISO (discussed below) and for which the energy price alone does not fully cover the start-up, no-load, and energy costs.

Note that, for example, in the case that the marginal losses are included in the LMP (as in various markets), and to the extent that the costs of start-up and no-load are included in the convex hull prices as in the Midcontinent ISO, then a smaller or even zero amount of losses and start-up and no-load costs are uplifted.

***Energy Offer Curve.*** Our discussion of block offers is convenient for exposition and simple examples, but typical market rules also allow for more complicated forms of offers that can better match the variation of marginal costs of generators with output. In particular, offers can also be specified by an "energy offer curve." In the ERCOT market, an energy offer curve is specified by a sequence of nondecreasing price-quantity pairs. The offer is interpreted as the piecewise linear function that interpolates between these points [6, Sec. 4.4.8.5.1].

***Unit Commitment.*** The discussion of offer-based economic dispatch applies to any given dispatch interval. In a "hour-ahead" or "day-ahead" market context, there are more options than just considering the dispatch. In particular, generators can be started up and shut down. The consideration of start-up and shut-down is called unit commitment. In some markets, such as the Australian market and several European markets, individual generator owners have responsibility for deciding on the unit commitment. In North American markets, individual generator owners typically also have the option to decide their own commitment. However, they can also at their option provide information to the ISO that allows the ISO to decide on unit commitment.

***Process.*** The ISO performs unit commitment hour-ahead or day-ahead by choosing commitment and dispatch of generation so that production matches the forecast, bid, or specified demand for each of the intervals that are modeled in the next day and does so at least cost. Typically, day-ahead unit commitment models each of the 24 h for the next day. Planning in the day-ahead context provides the opportunity to consider start-up of generators and not just the dispatch of the online generators as implicitly assumed in offer-based economic dispatch.

Starting a generator incurs "start-up" costs, while there is typically also a portion of the operating costs that must be incurred independent of operating level. We will initially term the latter "no-load" costs. Additionally, costs are incurred for the production of energy above zero production or above minimum production, as reflected in the energy offer. In this context, we can think of the energy offer as being the offer to generate incrementally above zero or above minimum production. For each hour in the commitment horizon, the ISO would consider the dispatch of the online generators. As well as the costs, the ISO can also consider dynamic limitations on, for example, the minimum number of hours online or offline, or how fast a generator can change its production, that is, its "ramp rate limit."

Analogous to the concept of minimizing the cost of dispatching generation to meet load for a single interval, in unit commitment, the goal is to minimize total offer costs over day, where now the start-up and no-load (or minimum-load) offers and minimum production limits are considered in addition to the incremental energy offers [6, Sec. 4.4.8.2]. The unit commitment process will be generically outlined, and then specialized to the "financial" and the "reliability" unit commitment.

Because both start-up and no-load offers as well as energy offers are considered by the ISO, a higher-priced energy offer by generator might be accepted instead of a lower-priced energy offer by another generator, if start-up and no-load offer of the other generator is too high. That is, the ISO selects the generators to commit, and the dispatch of the committed generators, with a view to minimizing the total costs over the day.

Security constrained unit commitment (SCUC) algorithm is used in markets to optimally commit resources at minimum system cost to meet the reliability criteria of the grid. Typically, this problem is formulated as a mixed-integer linear program. Systems operated by ISOs are typically very large with many resources and transmission network elements, resulting in a large and nonlinear optimization problem. Because of the nonconvex structure of this problem, it can be computationally expensive to solve for large numbers of generators and other operational constraints. In the latest implementations, the SCUC optimization problem is implemented as a master problem and subproblem using benders decomposition. The master problem does unit commitment using mixed-integer programming (MIP) methods and the subproblem identifies network violations and generates benders cuts for the master problem if violations are identified. The master and subproblems are solved iteratively to obtain a commitment and dispatch pattern that meets all constraints at minimum total cost. This methodology effectively allows the programmer to separate the transmission constraint analysis into a separate loop, considering only constraints, which are potentially near the limit of the transmission element while determining the unit commitment. Advances in mixed-integer linear programming software over the last decades have enabled the solution of realistic problems by ISOs. A well-designed SCUC can respect physical flow limits of the transmission lines, resource physical limits (maximum/minimum capacity, ramp rate capabilities, fuel limits, emissions limit, ancillary service limits, etc.), resource temporal constraints (maximum/minimum up/down time, lead time number of start-up/shutdown per day/week, maximum energy per week, etc.), and system limits (regional ancillary service requirement, load balance requirement, etc.).

***Pricing.*** As discussed in the context of offer-based economic dispatch, energy prices are typically set on the basis of the marginal surplus. However, consider a generator that is committed and for which the marginal surplus equals its offer price. The proceeds from energy payments will cover its incremental energy costs, but may not cover its start-up and no-load costs. In this case, the ISO guarantees payment of start-up and no-load costs through an additional payment, if these costs are not covered by inframarginal energy payments. Typically, a daily calculation is made to ensure that the generators (depending on the regional rule either the ISO committed generators or all generators) have nonnegative operating profit on a daily basis [6, Sec. 4.6.2.3]. However, it should be emphasized that for most generators on most days, inframarginal energy payments should cover start-up and no-load.

Since these start-up and no-load costs are additional to the compensation for energy, they must be funded in addition to the energy price. Typically, this cost is "uplifted" to demand.

***Unit Commitment Example.*** Consider dispatch and pricing example I, but suppose that the dispatch conditions applied for every hour of the day. Although this circumstance is unrealistic, it provides a simple illustration of the compensation for start-up and no-load costs.

Suppose energy offers by white generator were as before. We now interpret the energy offer as applying for any generation above zero. Suppose that it also specifies a price of \$2000 for start-up and \$500/h for no-load in its offer, and that before the start of the day it is out of service.

The demand is 220 MW for each hour of day, and we assume that the energy price is the same as in dispatch and pricing example I, namely \$9/MWh. We also assume that offer-based unit commitment results in starting up the white generator for the first hour of the day and maintaining its production constant for all hours of the day.

The ISO energy payment to the white generator over day based on energy price is

$$(\$9/\text{MWh})(130\ \text{MW})(24\ \text{h}) = \$28{,}080$$

However, the energy, start-up, and no-load offer costs are

$$\begin{aligned}(\$5/\text{MWh})(50\ \text{MW})(24\ \text{h}) + (\$7/\text{MWh})(50\ \text{MW})(24\ \text{h}) + (\$9/\text{MWh})(30\ \text{MW})(24\ \text{h})\\ + \$2000 + (\$500/\text{h})(24\ \text{h}) = \$34{,}880\end{aligned}$$

which is more than the payment based on the energy price. The ISO pays difference of \$6800 to the white generator and it is said that the white generator is "made whole" for the day. As mentioned above, the \$6800 would be charged as uplift to either demand that occurs in the real time but was not purchased in the day ahead, or is charged as an uplift generally to all demand, or both. If it is assigned to all demand then it is equivalent to slightly increasing the purchase price for energy so that the total energy sales over the day cover the make whole payment.

Note that if the generator were already in service at the beginning of the day, then the make-whole payment would not include start-up costs and would be \$4800 instead [6, Sec. 4.6.2.3(1)]. Moreover, if the energy payment exceeded the energy, start-up, and no-load offer costs then there would be no make-whole payment.

In several markets, the no-load cost is represented implicitly through the minimum production level. In this case, instead of a no-load offer cost, there is a specification of a minimum-energy offer (\$/MWh), and minimum production level (MW) [6, Sec. 4.4.8.2]. In ERCOT, offers for unit commitment are specified as a "three-part supply offer" consisting of a start-up offer, a minimum-energy offer, and an energy offer curve [6, Sec. 4.4.8.1(1)].

***Day-Ahead Financial Market and Reliability Unit Commitment.*** The ISO typically performs day-ahead unit commitment twice in the "financial" and the "reliability" unit commitments. Financial unit commitment [6, Sec. 4] will be discussed first, then the reliability unit commitment. The subsection following the unit commitment subsection will discuss the real-time market, and finally the interactions.

*Financial Unit Commitment.* The financial market arranges for forward financial commitment of energy prices and quantities based on generation offers, and specified and bid demand, together with so-called *virtual bids*, which as will be discussed below are bids and offers that are not necessarily backed by physical generation. The results of unit commitment using this information determines day-ahead energy prices and quantities, and any make-whole charges. Generally speaking, the results of the financial unit commitment do not impose physical requirements that generators actually start-up; however, the generators are responsible either for physically performing according to the results of the market, or buying out of their financial commitment in the real-time market. That is, although we have described the day-ahead unit commitment as arranging for commitment and dispatch of generation, this should really be interpreted as arranging for responsibility to either commit and dispatch, or effectively to pay another generator to produce. The significance of financial commitment will be further specified in the discussion of the real-time market.

*Reliability Unit Commitment.* The reliability unit commitment [6, Sec. 5] is carried out after the financial unit commitment results are published and is a postprocessing step that is based on the generation actually committed to physically run as nominated to ISO. Various markets have different conventions about how this nomination is specified to the ISO. In ERCOT, it is specified in the current operating plan (COP). The ISO analyzes the nominated committed generation as specified in the COP to see if the capacity is sufficient to reliably meet its forecast of the demand without overloading any transmission elements. The ISO may commit extra generation if generation is not enough to reliably meet forecast demand without overloading any transmission elements represented in the "commercial network model," or if any transmission constraints not represented in the "commercial network model" of the financial market actually require more generation to run at particular locations. The additional commitment costs are either allocated to load that occurs in the real time but was not purchased in the day ahead, or is charged as an uplift generally to all demand, or both.

***Real-Time Markets.*** Real-time markets are typically run every 15 or 5 min and arrange for finer time scale matching of supply and demand, typically based on short-term forecasts. The real-time market is the specialization of offer-based economic dispatch to arrange for the online generators to match the forecast demand level for the end of the upcoming 5 or 15 min dispatch interval. Real-time markets are necessary because demand will vary within each hour and, moreover, predictions of demand levels (and of intermittent renewable production) made day ahead will only be approximate, whereas the supply–demand balance must be maintained essentially continuously.

The solution of the real-time market produces dispatch signals that are sent to generators, which are typically assumed to ramp from current production toward the target levels. Real-time markets arrange for "physical" production of electricity, and (unlike DA financial markets) generators are expected to follow dispatch signals, and are typically penalized if their performance differs significantly.

Some real-time markets also include some representation of the possibility of commitment of fast-start generation, which tends to require a longer computation time than a dispatch-only calculation because of the representation of commitment decisions. In such markets, the computation for a given dispatch interval might then need to begin well before the beginning of the interval, resulting in a compromise between including the possibility of commitment of fast-start units, and the worsening of short-term forecast accuracy. (Additionally, most markets allow for an hour-ahead analysis of the need for additional capacity, which in ERCOT is called the hour-ahead reliability unit commitment, adding an additional layer of unit commitment into the process.)

The generation levels in the real-time market will generally differ at least somewhat from the quantities calculated in the day-ahead market. A generator or load with generation or consumption that differs from the day-ahead quantity will be paid or pay for that deviation at the real-time price. That is, real-time market prices are charged on "deviations" from positions in day-ahead financial market.

This confirms that the day-ahead market is a financially binding short-term forward market. It establishes "ex ante" prices for the quantities "cleared" in the day-ahead market. The real-time market is the closest analogy in electricity to the "spot market," with prices based on real-time generation offers, scheduled generation, and ISO short-term forecast of demand. For historical reasons, although prices are calculated every 5 min (or more often), settlements in real-time markets may be based on the average price over longer intervals. For example, in ERCOT, the settlement is based on a weighted average of prices over 15 min and on the energy produced or consumed in those 15 min [6, Sec. 6.6.1.1]. Some markets are moving toward implementing 5 min settlements so that "demand response" (self-curtailment of demand to real-time price signals send by ISO) gets settled based on the prices to which they responded.

***Interaction between Day-Ahead and Real-Time Markets.*** The interaction between the day-ahead and real-time markets is best explained with reference to an example. Consider a generator that has offered into the day-ahead market, and suppose that the day-ahead market results in 50 MW of generation being cleared at a day-ahead price of \$30/MWh for a particular hour.

The generator also offers into the real-time market. In principle, its energy offer might be different between day ahead and real time, depending on the circumstances. However, in some markets, there are generally restrictions on changing the offers from day ahead to real time and we will assume that the energy offers are the same. This is consistent with the assumption that the fuel and other operating issues for the generator have not changed between the day ahead and real time.

Although the real-time market will generally arrange for varying generation levels for many generators throughout an hour, for simplicity we will assume that the actual generation is constant throughout the hour, but potentially different from the day-ahead quantity. We consider net payment from day-ahead and real-time markets for the particular hour. This net payment depends on day-ahead position, real-time price, and deviation of real-time generation from day-ahead position. Three scenarios will be considered:

1. The generator offers into the real-time market and is dispatched in real time to 60 MW, with real-time price \$60/MWh, and deviation of $60 - 50 = 10$ MW.
2. The generator offers into the real-time market and is dispatched in real time to 40 MW, with real-time price \$20/MWh, and deviation of $40 - 50 = -10$ MW.
3. The generator has a forced outage, with real-time price \$60/MWh, and deviation of $0 - 50 = -50$ MW.

Notice that the first scenario is consistent with demand being higher in real time than forecast day ahead, so that real-time prices are higher than day ahead, while the second scenario represents the opposite situation. The third scenario represents the case that the supply is lower in real time than forecast day ahead because of failure of the generation in question, so that similar to the first scenario, real-time prices are higher than day ahead. For these scenarios, the net payments are

1. $(50 \text{ MW} \times \$30/\text{MWh}) + (10 \text{ MW} \times \$60/\text{MWh}) = \$2100/\text{h}$
2. $(50 \text{ MW} \times \$30/\text{MWh}) + (-10 \text{ MW} \times \$20/\text{MWh}) = \$1300/\text{h}$
3. $(50 \text{ MW} \times \$30/\text{MWh}) + (-50 \text{ MW} \times \$60/\text{MWh}) = -\$1500/\text{h}$

Note that independent of the actual performance in real time, the generator is paid the day-ahead quantity at the day-ahead price. Its actual deviations from day ahead are priced based on the real-time prices. In the third scenario, the generator "buys back" in real time all of the energy that it committed to selling day ahead. An analogous situation applies for the loads.

***Virtual Bids and Offers.*** U.S. markets allow traders and speculators that do not otherwise have generation assets or retail load obligations to participate in the market by taking opposite financial positions in the day-ahead and real-time markets. This is called a virtual bid or a virtual offer. For example, a speculator can offer to sell energy in day ahead without having actual generation available. This is effectively a paired offer to sell energy in day ahead and bid to purchase back that energy in real time. If the real-time price is lower than the day-ahead price then the virtual bidder makes profit.

Conversely, a virtual bidder could buy energy in the day ahead but not match it by actual demand in real time. In this case, the entity profits if the day-ahead price is lower than the real-time price.

This process allows arbitrage between day-ahead and real-time markets. If a virtual bidder profits, then it is acting to bring the day-ahead and real-time prices closer together, which improves the efficiency of overall market by decreasing the dispatch level of higher offer cost generation (from the higher price market) and increasing the dispatch level of lower offer cost generation (from the lower price market). In so doing, the virtual bidder facilitates "price convergence" between the markets.

Various practicalities in the market, including difference in day-ahead/real-time transmission models, the rules for make-whole charges to demand cleared in day-ahead and prudential requirements on virtual bidders, may prevent prices from fully converging. Nevertheless, virtual bidding increases liquidity in the markets and tends to improve market efficiency through bringing day-ahead and real-time prices together.

***Ancillary Services.*** Although the real-time market matches supply to 5 min ahead forecasts of demand, it is necessary to match supply and demand on finer scales in order to keep the system frequency within bounds despite variation in supply and demand and despite failures of generation. To follow the short-term variation of demand, to provide contingency reserves, and to procure other ancillary services (AS), the ISO also sets up further centralized markets for generators (and, in some cases, loads acting as resources) to provide AS. These can be independent of energy market or "co-optimized" with energy.

Some initial AS markets cleared the different types of AS sequentially and without co-optimization with energy. In sequential clearing, AS is procured by clearing one type at a time in the order of the quality of the product. In contrast, simultaneous clearing with AS substitution (substitution of lesser quality product with higher quality product if it is more economical) is more efficient than sequential clearing. The offered capacity available after procuring the highest quality AS product will be considered in the clearing for next higher quality AS product. In simultaneous clearing, participants submit offers for all AS types for the same capacity and the type of AS for which the capacity is procured would depend on combination that maximizes social welfare. Energy and reserves are now considered together (co-optimized) in all U.S. day-ahead markets, including the ERCOT market, because their provision is jointly limited by the capacity of the generator [6, Sec. 4.5].

ERCOT has just a single forward AS market, which is co-optimized with energy in the day-ahead market. However, other U.S. markets have both day-ahead and real-time co-optimization of AS with energy. These markets can be used to meet locational AS reserves and include a reserve demand function to implement scarcity prices when the AS reserve requirement is violated to meet energy demand. In advanced markets, AS is procured at zonal level to meet locational reserves while in simple markets it is procured at system level. The payment for resource awarded in AS market is for the AS capacity procured. The payment for AS energy if deployed in real time is based on the real-time market-clearing price. Loads pay for AS procured by the ISO as a payment for the services needed to serve the load reliably in real time.

Payment for AS in the U.S. is typically based on the ratio of load to the system demand. However, other markets, such as the Australian market, allocate costs to "causers."

The following details the various types of AS

- *Frequency Control/Regulation.* This AS is needed to maintain power balance when there is slight fluctuation in load so that frequency is maintained at an acceptable range to prevent damage/disconnection of generators, loads, and other equipment. This AS is critical for systems that are not interconnected with other systems by ac tie-lines since they would not get the support from the other system during a disturbance event.

  This AS is needed to provide secondary frequency control. Primary frequency control is attained automatically by changing equipment output/consumption within a few seconds based on the change in frequency. In many systems, specific generator droop settings are mandated for reliability from generators participating in the market to help recover the frequency. In some systems, this service can be provided by loads as well. The regulation service is deployed by automatic generation control based on the area control error and frequency characteristics of the system. The droop setting requirements and the amount of regulation needed are generally decided based on the systems characteristics and the level of support from the other systems.

- *Responsive Reserve.* This service is generally provided by online generation reserves that can increase output or loads that can decrease output and can provide the reserve within minutes usually 10 min. This service acts as a backup to the regulation service and is needed mainly to recover from big frequency events due to large generation drop or load increase. The amount of service procured is generally based on the largest single contingency generation loss amount of the system.

- *Operating Reserve.* This service is mainly needed for maintaining long-term load balance during unforeseen events like severe weather, error in load forecast, big generation drop, etc. This is also provided to restore the responsive reserve in case of an event so that the responsive reserves will be available again to protect the system against a secondary event. This service is typically provided by offline resources that need more time to come online, usually more than 10 min and less than 30 min.

This service is also provided by online generating resource that can increase power output by duct firing or load resources that can go offline.

- *Voltage Control.* This ancillary service is needed to maintain voltage at an acceptable range under normal operation and prevent voltage collapse under unforeseen events. Typically, obligations for voltage control are established as requirements for transmission service providers to install reactive power compensators and for generators to provide reactive support if requested by the control centers. Some systems have set up compensation mechanisms for the generating resource for reduction in active power needed to provide the reactive power requested by the control center.
- *Black Start.* This ancillary service is a critical one which is procured to get the ability to restore power after a large black out. This is provided by generators that have the ability to start up and supply load without the support from the power grid.

***Operating Reserve Demand Curve.*** In the discussion of pricing, it was noted that the marginal surplus is equal to the value of lost load if load is curtailed. When supply is tight, operator actions may choose to serve load, but procure less reserves than desired. In this case, prices should be cognizant of the implications of reserve curtailment, effectively pricing the unmet demand for reserves based on what might be called a *value of lost reserves.* Several markets include different demand curves for different reserves, called *operating reserve demand curves,* which set the corresponding reserve prices when the requirements for those reserves are not met. Because of co-optimization of energy and reserves, this also results in pricing of energy to reflect the curtailed reserves. In some markets without co-optimization of energy and reserves, approximations are used to represent the "clearing" of reserve supply against a proxy demand for all reserves together, as in the ERCOT real-time market [6, Sec. 6.5.7.1.11]. There is currently no implementation of this demand curve for reserves in the ERCOT day-ahead market [6, Sec. 4.5.2].

***Intraday Markets.*** So-called intraday markets in Europe are operated during the operating day and consider the remaining hours in the day. They are aimed at "correcting the course" of the market during the day, particularly in markets that do not have a U.S.-style real-time market. While they are not utilized currently in U.S. markets, they are mentioned here for completeness.

### 21.4.2 Hedging Energy Price Risk

***Energy Price Volatility.*** The sequence of prices in successive intervals in an electricity market can vary markedly over time as supply and demand varies. There is both a somewhat predictable daily, weekly, and seasonal fluctuation, together with a superimposed randomness due to generation failures and, increasingly, the effect of variable renewable power production. The resulting variability of prices, strictly speaking its random variation, is called its *volatility.*

Many market participants prefer a steady revenue stream or payment rather than volatile or even periodically varying prices. A "good" price for a generator is "bad" for demand and vice versa, and, as will be discussed, bilateral trades between generation and (the representatives of) demand enable "hedging" of energy price volatility; that is, reduction of the financial exposure to the variability of prices. This hedging happens simultaneously for both the supply and demand by effectively exchanging the good and bad prices.

Bilateral contracting has existed in wholesale electricity markets even well before the existence of restructured markets. In many cases, they took the form of what might be called a *physical bilateral contract,* where a specific generator is contracted to produce power over time for (notionally) a specific load.

In an offer-based economic dispatch market, the short-term goal is always to dispatch the lowest cost generation in preference to higher cost generation. Consequently, pure physical bilateral contracts can potentially constrain the dispatch away from optimality. Instead, so-called financial bilaterals provide similar hedging of exposure to volatility for use in an offer-based economic dispatch market, but without constraining the ISO offer-based economic dispatch process. The clearest example of such a financial bilateral contract is a "contract for differences" (CfD).

***Contract for Differences.*** A CfD hedges a financial position by arranging for a particular "strike price" for an agreed amount of power, resulting in compensation based on the difference between the strike price and an agreed reference price, such as the day-ahead or real-time price. It replaces volatile marginal offer prices with the strike price for agreed quantity. Again, the explanation of the CfD is best made with respect to an example.

Suppose that a generator (G) and demand (D) agree on quantity of 10 MW, a strike price \$30/MWh with reference to the hourly day-ahead price, and a duration consisting of all the hours in a particular day. During each of the hours of the duration of the CfD, D pays to G:

$$10\text{ MW} \times (\$30/\text{MWh} - \text{day-ahead price})$$

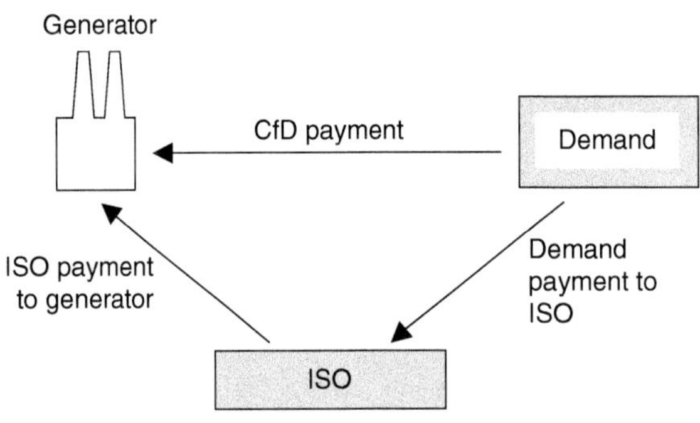


**FIGURE 21-6** Revenue streams from market and contract for differences.

A crucial aspect of the CfD is that it works in conjunction with the generator offering into the market, with the generator being paid for its energy production and the demand paying for its energy consumption as previously discussed. These revenue streams are illustrated in Fig. 21-6.

To be concrete, suppose that the generator offers the same 10 MW into market as the quantity of its CfD. As always under offer-based economic dispatch, the ISO seeks the dispatch of the offers to meet demand at minimum total offer costs. This means that the particular generator G may or may not be dispatched in any given hour depending on whether its offer price is below or above day-ahead market price for that hour. For the purposes of the example, we assume that the demand D is specified as 10 MW and that this quantity is consumed under each of the possible generation offers to be considered.

Table 21-1 illustrates the situation for various possible combinations of generation offer price for the 10 MW offer and day-ahead market prices. If the day-ahead market price is above the generator offer price, then the generator G is cleared in the day-ahead dispatch. If the market price is below the offer price, then the generator is not cleared. If the market price is equal to the offer price, then the generator might be dispatched to any level between 0 and 10 MW; however, for concreteness, it has been assumed that the generator would be fully dispatched in this case. The detailed assumption does not materially affect the net outcome for the generator.

**TABLE 21-1** Payments under Contract for Differences

| Generator | | Day-ahead market | Payments \$/h | | |
|---|---|---|---|---|---|
| Offer \$/MWh | Dispatch MW | price \$/MWh | D→ISO | D→G (CfD) | ISO→G |
| 30 | 10 | 30 | 300 | 0 | 300 |
| 20 | 10 | 25 | 250 | 50 | 250 |
| 20 | 10 | 35 | 350 | −50 | 350 |
| 20 | 0 | 15 | 150 | 150 | 0 |
| 40 | 0 | 35 | 350 | −50 | 0 |
| 40 | 10 | 45 | 450 | −150 | 450 |

The 10 MW of specified demand D is cleared in all cases under the assumption that there is never a need for curtailment. The net payment by the demand, as suggested in Fig. 21-6, is the sum of its payments to the ISO and to the generator under the CfD, as shown in the third-last and second-last columns of Table 21-1. The sum of these terms is always \$300/h, confirming that independent of the market outcome as indicated by the day-ahead market price, D always pays the hedged price of \$30/MWh for its 10 MW consumption.

The net payment to the generator is the sum of the payments from the demand and from the ISO, as shown in the second-last and last columns of Table 21-1. Whenever the generator is actually dispatched, its net payment is \$300/h again reflecting a hedged price of \$30/MWh for its 10 MW production. The generator is dispatched in the first, second, third, and sixth rows of the table.

Whenever it is not dispatched, corresponding to the fourth and fifth rows of the table, the generator is effectively billed for its 10 MW generation obligation at a price equal to the day-ahead market price. Ignoring forced outages, the circumstances where the generator is not dispatched are precisely the circumstances where the market has energy available at a lower offer price than the offer from G. Assuming that G is offering at its marginal cost, it benefits from the ISO finding the lower cost dispatch options in these cases.

Although the day-ahead price varies with demand and supply conditions, the CfD hedges risk of variation in price for the agreed quantity. If the generator or the demand quantities in the offer-based market differ from the CfD quantity, then they are exposed to day-ahead price for quantities that deviate from the agreed quantity. The implication is that the good incentives for generation and demand discussed in the context of offer-based economic dispatch are preserved even in the presence of CfDs, since deviations from the agreed quantity are still priced at the marginal surplus.

***Variations on Contracts for Differences.*** The discussion so far has described a so-called *two-way* CfD, where the payment from demand to generator could be positive or negative. Variations include a "one-way" CfD:

- Demand only pays generator (when marginal offer price is below strike price) not vice versa.
- Generator only pays demand (when marginal offer price is above strike price) not vice versa.

A combination of one-way CfDs at different prices defines a "collar" that brackets payments. More generally, other issues such as weather could be included in the contract.

***Financial Bilateral Summary.*** A key feature of a financial bilateral is that it does not "constrain" ISO dispatch, and so it allows the ISO to achieve economic dispatch. This contrasts with a "pure" physical bilateral, which can prevent economic dispatch if cheaper generation is available.

Conversely, ISO dispatch does not constrain financial contracts. However, it should be understood that price information from offer-based economic dispatch does help with setting strike price for subsequent CfD negotiations.

CfDs enable short-term efficiency of economic dispatch while allowing long-term financial positions to be established through bilateral contracting. The offer-based economic dispatch decisions and day-ahead and real-time prices are determined by ISO, whereas the long-term financial position is established by price of financial bilateral. The bilateral contract hedges exposure to price volatility, and there is considerable flexibility in the form of financial bilateral contracts.

### 21.4.3 Locational Marginal Prices

Transmission systems have various limits on their ability to move power. When flows are at these limits, transmission constraints are binding and congestion is said to occur. In this subsection, we consider congestion and the pricing of energy when congestion occurs. When there are transmission constraints, we cannot simply stack up the generation blocks as in the case described previously of no transmission constraints. However, the goal is still maintained of dispatching to minimize total offer costs, but now the minimization must be subject to satisfying transmission constraints. This will

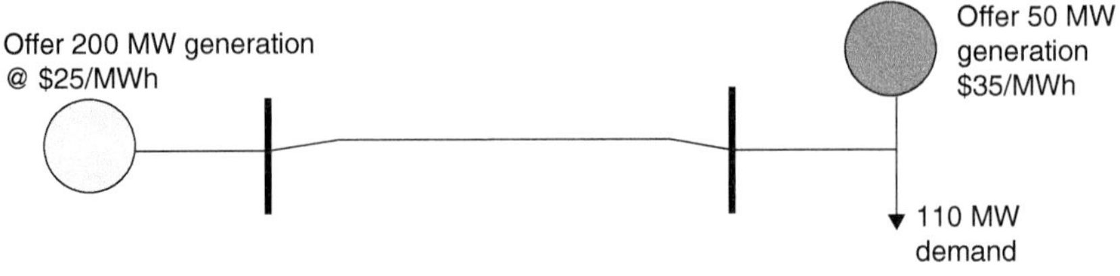


**FIGURE 21-7** Single line, two bus example for dispatch and pricing example III.

first be illustrated with a very simple example and then the dispatch of generation to maintain flows within limits will be discussed more generally, including the pricing of energy when constraints bind.

***Dispatch and Pricing Example III.*** To introduce the process, a very simple one-line example will be used first, as illustrated in Fig. 21-7, which has two buses joined by a single transmission line. The generator at the left offers 200 MW of generation at $25/MWh, while the generator at the right offers 50 MW of generation at $35/MWh. There is also 110 MW of demand at the right.

Before considering transmission constraints, first consider the case if the capability of the line were very large compared to the demand level. This is same type of situation we considered earlier in the context of offer-based economic dispatch. Given 110 MW of local demand at right and no limit on import capability, minimum total offer costs would involve dispatch of

- 110 MW of cheap imports from generator at bus at left
- No expensive local generation at bus at right

For this case, the total offer costs are

$$(110 \text{ MW}) \times (\$25/\text{MWh}) = 2750 \text{ \$/h}$$

The marginal offer price (and the marginal surplus) is $25/MWh, and as in the earlier discussion of offer-based economic dispatch, all generation and all demand is priced at $25/MWh.

***Transmission-Constraints in Example III.*** Now suppose that the transmission line in Fig. 21-7 is limited to transferring at most 100 MW from left to right. Given 110 MW of local demand at right and 100 MW import capability, minimum total offer costs would involve

- 100 MW of cheap imports from left
- 10 MW of expensive local generation at right

Paralleling the observation made in offer-based dispatch, if offer prices are equal to marginal generating costs for each generator, then demand is satisfied at least cost of resources by dispatching according to minimizing the total offer cost subject to the transmission limitations.

***Value of Energy.*** A key difference between transmission-constrained cases and the earlier discussion of offer-based economic dispatch is that the value of energy varies with location when there are transmission constraints at their limits. We can repeat the thought experiment discussed earlier for pricing to value a costless additional MW of generation, but now it will be found that the value differs by location. Locational marginal pricing (LMP) assigns prices based on these locationally differentiated values.

As in the transmission unconstrained case, a simplification of the thought experiment is typically possible by evaluating the cost to serve an additional MW of demand. Applying this to the demand at the right, note that to serve an additional 1 MW of demand at the right, an additional 1 MW of generation at the right would be dispatched, costing $35/MWh. On the other hand, serving an additional 1 MW of demand at the left would require production of an additional 1 MW of generation at the left, costing $25/MWh.

***Locational Marginal Pricing.*** The basic principle in locational marginal pricing (LMP) is to price energy equal to the value of an additional MW of supply, which in simple cases such as this example, is equal to the least offer cost of serving an additional MW of demand based on the offer prices. Ignoring transmission constraints and losses, this price would be uniform throughout system. With "binding" transmission constraints as in the example, the price varies by location of bus or "nodally." In particular, the LMP is \$25/MWh at left and \$35/MWh at right of one line example.

***Congestion Rent and Congestion Cost.*** In this example, the demand pays to ISO (110 MW) × (\$35/MWh) = 3850 \$/h. The generator at left is paid by ISO (100 MW) × (\$25/MWh) = 2500 \$/h. The generator at right is paid by ISO (10 MW) × (\$35/MWh) = 350 \$/h.

Note that the total paid to generators is 2850 \$/h, while the demand pays 3850 \$/h. The difference between the demand and generator payment is called the *congestion rent* of 1000 \$/h.

The congestion rent represents a revenue stream that accrues to ISO, and the disbursement of this revenue stream will be discussed in the context of "financial transmission rights."

It is important to distinguish congestion rent from the related concept of the "congestion cost," which is the difference between

- Minimum total offer costs when satisfying transmission constraints, minus
- Minimum total offer costs ignoring transmission constraints

The congestion cost represents the offer cost of the additional resources that are expended due to limitations of transmission system. (It should be noted that "congestion rent" is sometimes confusingly called *congestion cost* in the literature.)

***Congestion Rent versus Congestion Cost in Example III.*** The congestion rent is the difference between demand payments and payments to generators:

$$(3850\ \$/\text{h}) - (2850\ \$/\text{h}) = 1000\ \$/\text{h}$$

The congestion cost is difference between total offer costs with and without congestion:

$$(2850\ \$/\text{h}) - (2750\ \$/\text{h}) = 100\ \$/\text{h}$$

As in this example, congestion rent can be much larger than congestion cost. Confusion between these two concepts is a principal misconception about LMP.

***Transmission Prices, LMP Differences, and Congestion Rent.*** Consider the case that a new 1-MW generator decided to locate at the left in Fig. 21-7 and wanted to sell its energy to a new 1 MW demand at the right. Note that to accommodate this injection and withdrawal, the generators shown in Fig. 21-7 would need to redispatch, with the generator at the left decreasing production by 1 MW and the generator at the right increasing production by 1 MW. The operating cost of this change to accommodate 1 MW of transmission is equal to the difference in the LMPs between right and left. We can think of this difference between the LMPs at right and left as the transmission price from left to right:

$$\$(35 - 25)/\text{MWh} = \$10/\text{MWh}$$

For the one line case illustrated in Fig. 21-7, the congestion rent is equal to capability of 100 MW times LMP difference of \$10/MWh. However, this radial system is not typical of a real transmission system, which typically has loops.

In looped systems, prices and congestion rent are less "intuitive," and it becomes necessary to use formal optimization techniques and software to calculate offer-based economic dispatch and the LMPs. A slightly more complicated example with a single loop will illustrate the basic issues.

***Dispatch and Pricing Example IV.*** Consider again dispatch and pricing example II, but now assume that there is a 300-MW transmission limit on the right-hand line that joins the NE and SE buses. To determine if a proposed dispatch is feasible with respect to this limit, we must calculate the resulting flow on the right-hand line for the proposed dispatch and compare it to 300 MW.

In general, a power flow model is used to calculate flows, and this depends on the line impedance characteristics, together with

- The values of "injection:" net generation at NW, NE, SE
- The values of "withdrawal:" power delivered to demand at SW

We can combine the two cases by defining the "net injection" to be the injection by the generators minus the withdrawal by demands. An accurate calculation of the flows requires evaluation of the so-called *ac power flow*; however, in most day-ahead markets and some real-time markets a simpler approximation is used based on the "dc power flow" approximation [9]. The dc power flow approximation neglects losses, so that the sum of all net injections is zero. Consequently, we can pick out any particular bus and assume that its net injection is equal to in magnitude and opposite in sign to the sum of the net injections at all other buses. The dc power flow approximation allows the flows to be evaluated in terms of the net injections at all the buses in the system besides the slack bus, given the assumption that the injection at the slack bus is equal and opposite to the sum of the net injections elsewhere. In particular, the flow on a line can be approximated as the sum over all buses besides the slack bus of the net injection at the bus multiplied by a quantity called the *power transfer distribution factor*, or "shift factor" for that line and bus.

***Shift Factors in the Example System.*** The shift factor to a particular line for injection at a particular bus (and, implicitly, withdrawal at the slack bus to satisfy power balance) evaluates by how much the net flow on the line changes for an additional MW of injection at the bus. For this discussion, we will use the SW bus as the slack bus and calculate shift factors to the right-hand line for injections at the NW, NE, and SE buses. We will think of the direction of flow from NE to SE as the "positive" direction, meaning that net flows in this direction will have positive sign, while net flows in the other direction have negative sign. The power balance constraint for the system is

$$\text{NW} + \text{NE} + \text{SE} = 3000 \text{ MW}$$

The shift factors depends on line impedances, and for particular choices of line impedances, the shift factors for injection at NW, NE, and SE, are 0.2, 0.4, and –0.2, respectively. The negative shift factor for injection at SE is because injection at SE and withdrawal at the slack bus SW tends to reduce the flow from NE to SE on the right-hand line.

To summarize, the flow on right-hand line is 0.2 NW + 0.4 NE – 0.2 SE. The flow limit mentioned earlier requires that the flow on right hand is less than or equal to 300 MW. Putting these together results in the transmission limit:

$$0.2 \text{ NW} + 0.4 \text{ NE} - 0.2 \text{ SE} \leq 300 \text{ MW}$$

Again note that increasing net generation at SE acts to decrease net flow on the right-hand line. We say that increasing generation at SE causes "counterflow" on the right-hand line.

***Offer-Based Transmission-Constrained Economic Dispatch.*** We are now ready to seek the least offer cost dispatch to serve 3000 MW demand at SW. Note that if we were to ignore the transmission constraint, then the dispatch would be the same as in dispatch and pricing example II, which involved 1500 MW of NW, 1000 MW of NE, and 500 MW of SE generation. The resulting flow on the right-hand line would be 600 MW, which exceeds the 300 MW limit. This is not a valid solution.

Consider alternative solutions. For example, suppose we were to use the intuition from the case of no transmission constraints and "stack up the blocks" until all demand is met. The cheapest generation is at NE, and if we were to use all 1000 MW of NE and no other generation to meet the first 1000 MW of the demand, then the flow on right-hand line would be 400 MW, which exceeds capacity of right-hand line. Apparently, stacking up the blocks suggests that it will not be possible to fully utilize the cheapest generation. In fact, as will be discussed below, it is possible to utilize all 1000 MW of NE generation, emphasizing that the results of offer-based economic dispatch respecting transmission constraints can be unintuitive. Even though the results may be unintuitive, the calculation of offer-based economic dispatch with transmission constraints is easily accomplished, even for large systems, with linear programming software as will be discussed in the next subsection.

***Linear Programming Solution.*** The choice of NW, NE, and SE generation that meets the 3000 MW of demand, minimizes total offer costs, and keeps flow on right-hand line at or below 300 MW is a linear program [10], which is specified by its objective and constraints:

- The objective is the total offer costs (NW × \$40/MWh) + (NE × \$20/MWh) + (SE × \$50/MWh), which is a linear function of the generations at NW, NE, and SE (in more general formulations, the objective may be quadratic, necessitating the use of quadratic programming software or a piecewise linear approximation).
- The constraint on power flow on the line from NE to SE has been expressed as a linear inequality, $0.2\ \text{NW} + 0.4\ \text{NE} - 0.2\ \text{SE} \leq 300$ MW, through the dc power flow approximation, and considering multiple line flow constraints results in multiple linear inequalities.
- The power balance constraint is a linear equality, $\text{NW} + \text{NE} + \text{SE} = 3000$ MW.
- Other constraints, such as generation ranging between zero and the size of the offer are also linear inequalities, $0 \leq \text{NW} \leq 1500$ MW, $0 \leq \text{NE} \leq 1000$ MW, $0 \leq \text{SE} \leq 1500$ MW.

The solution of the linear program is NW = 750 MW, NE = 1000 MW, SE = 1250 MW, with total offer costs of 112,500 \$/h, the flow on the line from NE to SE is at its limit, and all of the lowest offer price generation from NE is used. The dispatch is illustrated in Fig. 21-8. This solution minimizes the total offer costs subject to the constraints. To see this, note that the low cost NE generation is fully utilized, while decreasing NE generation would not allow for enough transfer from SE to NW to produce net savings, and increasing NW generation would require more SE generation to avoid overloading the line from NE to SE.

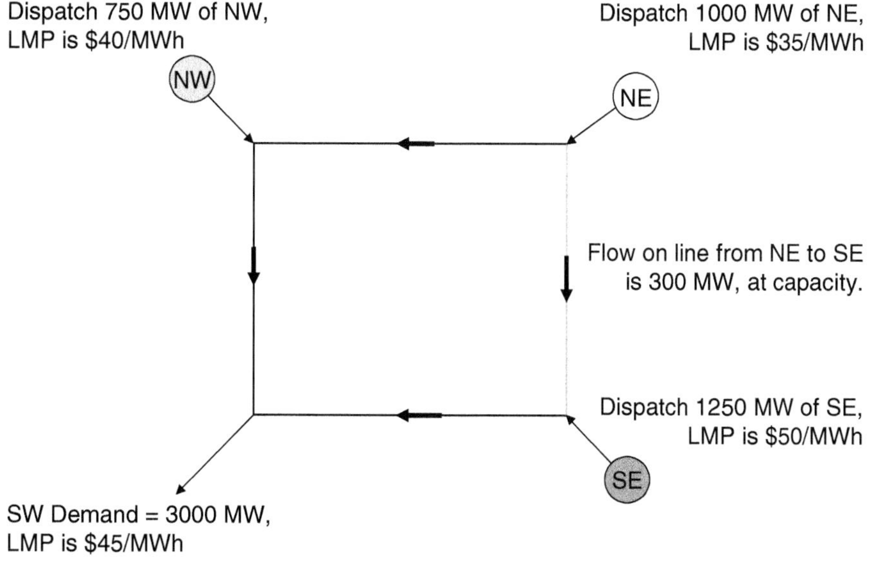


**FIGURE 21-8** Dispatch and prices for four-line, four-bus example IV.

Compared to the solution obtained without enforcing transmission constraints, which had total offer costs of 105,000 \$/h, the total offer costs of 112,500 \$/h is higher in the transmission-constrained problem since the transmission-constrained problem has a smaller set of feasible dispatches. In particular, the transmission-constrained solution will use some more of the high-priced generation from SE and less from NW than in the solution without considering transmission constraints.

Using more of the higher priced generation is sometimes called *out of merit* dispatch, although this phrase is somewhat misleading since the whole point of solving the linear program is to find the most "meritorious" dispatch, while satisfying the transmission constraints. That is, the transmission-constrained solution is more expensive than a counterfactual and infeasible dispatch that did not satisfy the transmission limitations; however, the solution is as inexpensive as allowed by the constraints.

The difference between the total offer costs with and without congestion is the congestion cost. In this case, the congestion cost is 112,500 − 105,000 \$/h = 7500 \$/h and this represents the cost of the increased resources that were required because of the transmission limitation.

***Evaluating the Locational Marginal Prices.*** To calculate the locational marginal prices (LMPs), we consider the marginal surplus at each bus. For this example, this simplifies to evaluating the least offer cost at each bus of serving an extra 1 MW of demand at that bus based on offer prices of available generation. This extra MW of demand is not necessarily most cheaply supplied by local generation, and the least offer cost typically involves combination of changes at several buses. This means that the LMP is not necessarily equal to offer price at bus, but could be the same, higher, or lower.

We will consider each bus in turn and exhibit the least offer cost redispatches to serve an extra MW of demand; however, it should be understood that the LMPs are actually available as a by-product of the solution of the linear programming solution by using the Lagrange multipliers on the constraints [10]. That is, the LMPs do not need to be calculated in practice using the analysis we present. To summarize, the analysis we present serves to rationalize the LMPs, whereas in practice they are calculated directly from the Lagrange multipliers on the constraints.

*Locational Marginal Price at NW.* An additional MW of demand at NW is served at least cost by increasing generation at NW by 1 MW, since NE is at maximum and SE generation is more expensive. If the demand is served by generation at NW, then the resulting net generation at NW stays at 750 MW since generation increases to 751 MW and demand at this bus increases from 0 to 1 MW. Moreover, there is no change in the net generations at other buses, so the flow on the line from NE to SE stays that same, at its capacity. This means that the LMP at NW is \$40/MWh, which is equal to offer price at NW, and NW is a marginal generator. NW is paid at this LMP for its 750 MW of generation, so its revenue is (750 MW) × (\$40/MWh) = 30,000 \$/h.

*Locational Marginal Price at NE.* It turns out that an additional MW of demand at NE is served at least cost by increasing generation at NW by 1.5 MW, and decreasing generation at SE by 0.5 MW. The net injection at NW increases by 1.5 MW, while the net injection at NE decreases by 1 MW due to 1 MW of extra demand, and the net injection at SE decreases by 0.5 MW. Consideration of the shift factors shows that the resulting flow on the line from NE to SE stays the same. The LMP is the net cost of these changes, which equals (1.5 × \$40/MWh) − (0.5 × \$50/MWh) = \$35/MWh. The LMP at NE is higher than the offer price of NE; however, it should be noted that no more NE capacity is available, and so NE is an inframarginal generator. The revenue to NE is (1000 MW) × (\$35/MWh) = 35,000 \$/h.

*Locational Marginal Price at SE.* An additional MW of demand at SE is served at least cost by increasing the generation at SE by 1 MW, since NE is already at capacity, and increasing generation at NW would overload the line from NE to SE. Increasing generation at SE by 1 MW to serve an additional 1 MW of demand at SE results in the flow on the line from NE to SE staying the same. The LMP at SE is 1 × \$50/MWh = \$50/MWh, which is equal to offer price at SE, and so SE is also marginal. The revenue to SE is (1250 MW) × (\$50/MWh) = 62,500 \$/h.

*Locational Marginal Price at SW.* An additional MW of demand at SW is served at least cost by increasing generation at SE by 0.5 MW and increasing generation at NW by 0.5 MW. The net injection at SE increases by 0.5 MW, while the net injection at NW increases by 0.5 MW, and the flow on the line from NE to SE stays the same. The LMP at SW is (0.5 × \$50/MWh) + (0.5 × \$40/MWh) = \$45/MWh, and the demand pays (3000 MW) × (\$45/MWh) = 135,000 \$/h.

*General Calculation.* In practice, as mentioned above, the LMPs are calculated more directly from the Lagrange multipliers from the solution of the linear program. Additionally, the marginal effect of losses can be incorporated into the LMP, typically using an approximation to the dependence of losses on injections. Consideration of the slack bus allows the LMPs to be decomposed into the sum of

- The LMP at the slack bus
- A congestion component due to lines at their limit as adjusted by shift factors
- A loss component

The values of the terms depend on the choice of slack bus, but the sum of the terms is independent of the choice of slack bus.

***Congestion Rent and Congestion Cost in Example IV.*** Recall that the congestion rent is the difference between demand payments (135,000 \$/h) and payments to generators (30,000 + 35,000 + 62,500 \$/h), which equals

$$(135{,}000\ \$/\text{h}) - (127{,}500\ \$/\text{h}) = 7500\ \$/\text{h}$$

On the other hand, congestion cost is difference between total offer costs with and without congestion:

$$(112{,}500\ \$/\text{h}) - (105{,}000\ \$/\text{h}) = 7500\ \$/\text{h}$$

The congestion cost and the congestion rent happen to be the same in this example; however congestion rent can be, and typically is, much larger than congestion cost, as in example III. Since the phrases "congestion cost" and "congestion rent" are similar, while their definitions are different, and their numerical values can be vastly different, it is important to be clear about which definition is intended when dollar values of congestion are discussed.

***Properties of LMPs.*** Examples III and IV point to various properties of LMPs. For example, LMPs are typically different at every bus whenever there is one or more lines operating at its limit. In example IV, the LMPs are \$35/MWh at NE, \$40/MWh at NW, \$45/MWh at SW, and \$50/MWh at SE, which are all different. Early efforts in the United States to find "zones" of similar prices, and avoid the apparent complexity of LMP pricing, all floundered on the fundamental fact that even a single binding constraint results in prices being different throughout an interconnection. While it is true that prices of nearby buses will tend to be similar, geographical variation concerning which constraints are binding at a given time implies that buses cannot typically be grouped into fixed zones such that the LMPs are always similar throughout the zone. Whether the variation of LMPs makes a zonal system unworkable depends on the particulars of the system, including the ability of market participants to arbitrage discrepancies between zonal prices and the underlying LMPs.

*LMPs and Offer Prices at a Bus.* LMPs can be lower than the offer price at a bus if cheaper imports can entirely replace local generation, which occurs in example III if demand were less than 100 MW. The LMP is the same as offer price if a generator is marginal. In example IV, there are two marginal generators. Finally, we can think of the LMP as being higher than the offer price at bus if no more capacity is available at bus.

*Demand versus Supply Prices.* LMPs can be higher or lower at demand than at generation. In most cases, the LMP at demand is higher than at generation, as in example IV in the case of demand at SW and generation at NW and NE. However, in this same example, the LMP at SW is lower than the LMP at SE.

*Direction of Power Flow and LMPs.* While we might normally think that energy should be exported from a low-cost region to a high-cost region, the effect of transmission constraints can be that power can flow from a bus with higher LMP to a bus with lower LMP. In particular, in example IV, the line flows on the line from SE to SW, whereas the LMP is higher at SE than SW. It should be recognized that this flow is actually a "side effect" of the economically significant issue that generation at SW creates counterflow on the line from NE to SE, which allows for cheap generation at NE to be utilized.

*Range of LMPs in System.* LMPs can be higher than any generator offer price, if increasing demand necessitates decreasing generation at cheap generator. For example, if example IV were modified so that SE only offered 1200 MW instead of 1500 MW, then the resulting least offer cost dispatch would involve all 1200 MW of SE generation. In this case, the LMP at SW would be higher, since an extra MW at SW would then require increase at NW by 2 MW and decrease at NE by 1 MW, with LMP at SW of $(2 \times \$40/\text{MWh}) - (1 \times \$20/\text{MWh}) = \$60/\text{MWh}$, which is higher than all the offers. Conversely, under different conditions, the LMP can be lower than any generator offer price, if increasing demand by 1 MW allows for more than 1 MW increase at a cheap generator.

*Comparison to Demand Prices without Transmission Constraints.* If we ignored transmission constraints, then the (hypothetical) price everywhere in the system would be \$50/MWh. Although it is often the case that prices at demand increase when transmission constraints are enforced, example IV shows that the LMP at SW of \$45/MWh can be lower than the hypothetical transmission unconstrained price of \$50/MWh. More typically, the LMP at the demand will be higher than transmission unconstrained price, as in example III.

***Commercial Network Model.*** The small examples have been very simple systems with just a few lines. Actual transmission networks involve thousands of lines, buses (sometimes called *nodes*), and other elements. The dc power flow and other approximations are used to develop the so-called commercial network model that is used to calculate LMPs. Although this model may represent all of the high-voltage lines in the system, it is nevertheless an abstraction from reality that neglects some details. For example, some of the lines may be aggregated together into equivalents. As another example, voltage and reactive power issues are neglected in the dc power flow model.

In the zonal market that was utilized in ERCOT until 2010, the commercial model for the balancing market was a four bus model with five lines. The balancing market that was used for dispatch was effectively a four node, five constrained line LMP market. Each "node" in the model stood for all the buses in a zone, and all generation in the zone was treated as if it was located at one node. This model was so coarse that there was a poor alignment between commercial network model and actual network behavior, which among other reasons, led to ERCOT moving to an LMP market.

Nevertheless, the much more detailed representation of the network in the ERCOT market from December 2010 is still an abstraction, albeit representing thousands of lines and buses using the dc power flow approximation. This more detailed commercial network model results in close alignment of commercial network model with actual network behavior except for voltage and reactive power issues. That is, some details, including voltage and reactive power, are still "hidden" from market, necessitating some out-of-market actions, including reliability unit commitment. Broadly speaking, other market in the United States use similar models, although some have adopted greater degrees of realism, including the explicit representation of losses.

***Charges to Load Serving Entities.*** Wholesale prices charged to retailers and other load serving entities for the consumption of their customers. In some cases, such as ERCOT, these charges are not based on the actual LMPs at the customer buses, but are instead based on zonal demand weighted averages of LMPs. Practices for charging load serving entities vary from ISO to ISO.

***Security-Constrained Economic Dispatch and Unit Commitment.*** Although the transmission-constrained examples above illustrated limits on the flow of power on particular lines under normal or precontingency conditions, many limitations on flow on networks are due to contingency constraints that are designed to limit the flow on lines to within emergency limits even if an outage were to occur. The representation of such contingency constraints in an LMP frameworks is undertaken by so-called *security-constrained economic dispatch* (SCED). It typically also represents a number of other limitations on the system, including ramping constraints, and represents ancillary services. If an unexpected failure results in overloading of lines, SCED will also arrange for redispatch to remove the overload.

Day-ahead unit commitment was introduced without reference to transmission constraints. However, in U.S. markets, unit commitment typically represents both the decisions to commit generators and also the implications of transmission constraints, including contingencies. This combination is called security-constrained unit commitment (SCUC). While it was historically solved approximately by somewhat ad hoc methods, including Lagrangian relaxation, U.S. ISOs have now moved to the use of highly capable mixed-integer linear programming software to solve these problems, including multiple limitations on ramping, provision of ancillary services, and other considerations. Decomposition methods can be used to separate the evaluation of the flows on lines from the commitment and dispatch. Discussion of several algorithms can be found in [11].

### 21.4.4 Hedging Transmission Price Risk

***Transmission Prices.*** The previous examples and discussion has calculated LMPs for several example systems. In this systems, and in realistic systems, the LMP at a given location varies due to variation of demand and supply. As discussed earlier, contracts for differences (CfD) allows the hedging of LMP variation over time at a particular location. In particular, a CfD is sufficient for hedging energy price variation for generation and demand exposed to same prices.

However, LMPs (and differences in LMPs) vary with location. For a generator and demand exposed to two different LMPs, a CfD is insufficient to fully hedge price variation for both LMPs.

Consider again example III. In this example, the LMP is \$25/MWh at the left and \$35/MWh at the right. Contractual terms in a CfD require a reference price, based on, for example, the day-ahead market. In the context of LMPs, this requires the specification of a particular bus for the reference price. Suppose that a generator at left has CfD with a strike price of \$30/MWh and with the LMP of the demand at right as the reference price. We can think of the location of the reference price as the delivery point for the contract. That is, the payment from the demand to the generator is equal to the difference between the strike price and the LMP of the demand. The prices associated with the payoffs from such a CfD under the conditions of example III are as follows:

- The generator at the left is paid \$25/MWh by ISO
- The demand at the right pays ISO \$35/MWh
- The generator at the left pays \$5/MWh to demand for CfD

The net payment to the generator is based on the price \$20/MWh. That is, payment to the generator at left by the ISO minus the payment by the generator for CfD leaves the generator "short" of the \$30/MWh strike price by \$10/MWh. This \$10/MWh amount is precisely the difference in LMPs between the right and left, which is the transmission price in the market to move the power from the generator to the demand.

Similarly, in example IV, if each of NW, NE, and SE have CfDs with the LMP at SW as the reference price, then they are exposed to difference in between the LMP as SW and the LMPs at their respective buses. Again, we can interpret these price differences as the transmission prices:

- Transmission price from NW to SW is (\$45/MWh – \$40/MWh) = \$5/MWh
- Transmission price from NE to SW is (\$45/MWh – \$35/MWh) = \$10/MWh
- Transmission price from SE to SW is (\$45/MWh – \$50/MWh) = –\$5/MWh

***Volatility.*** The price differences between buses can vary significantly with supply and demand, and so if a generator and demand are exposed to different LMPs then a CfD alone is insufficient to hedge variation in LMPs at both generation and demand buses. A generator holding CfD for energy referenced to the LMP at another location is exposed to the risk of variation in LMP differences. An additional hedging instrument is needed to avoid exposure to these volatile price differences.

***Congestion Revenue Rights.*** Recall that the congestion rent is a revenue stream in an LMP market, accruing to the ISO, which reflects the difference between the demand payments minus the generation payments when there are market participants exposed to different LMPs. The basic insight for providing an additional instrument to hedge transmission prices is for ISO to sell the rights to the revenue stream from the congestion rent to enable hedging of the fluctuations in the LMP price differences.

A congestion revenue right (CRR) is the right to a revenue stream of nodal price differences. In particular, the holder of CRR is paid by the ISO:

(Contract quantity) × [difference in LMPs between point of delivery ("sink") and point of injection ("source")]

The CRR hedges against fluctuations in LMP differences between the sink and source due to congestion in network and can be used in conjunction with contracts for differences. The combination of an appropriate CfD and CRR with the same contract quantity perfectly hedges fluctuations in prices at both generator and demand bus. Moreover, CRRs can also be defined between amalgamations of LMPs, such as zones and "hubs" (simple averages of LMPs at several buses [6, Sec. 3.5]). CRRs are also known as "financial transmission rights" (FTRs) and "transmission congestion contracts" (TCCs).

***Example of Revenue Streams with CfD and CRR.*** We continue to assume that in example III the generator (G) has a CfD with the demand (D) using the demand LMP as the reference price. In particular, the CfD is for a quantity of 10 MW at a strike price \$30/MWh with delivery referenced to the LMP at the demand bus, and the contract for all hours in the day. Additionally, we also assume that the generator holds a CRR having the same quantity as the CfD that sources at the generator bus and sinks at the demand bus. As in the earlier discussion of CfDs, we assume that the generator offers 10 MW to ISO at its marginal cost of \$20/MWh, except during any occasions when it is out of service (OS). We also continue to assume that the demand D is specified as 10 MW and that this quantity is consumed under each of the possible generation offers to be considered.

Table 21-2 illustrates various possible combination of LMPs at the generator and demand and shows that the combination of the CfD and CRR hedges risk of variation in LMPs at both generator and demand buses for the agreed quantity.

The net payment to the generator is the sum of the payments from the demand and from the ISO for energy and the CRR, as shown in the third-last, second-last, and last column of Table 21-2. Whenever the generator is actually dispatched, its net payment is \$300/h again reflecting a hedged

**TABLE 21-2** Payments Under Contract for Differences and Congestion Revenue Right

| Generator | | Demand LMP | Payments \$/h | | | |
|---|---|---|---|---|---|---|
| Dispatch MW | LMP \$/MWh | \$/MWh | D→ISO | D→G (CfD) | ISO→G (CRR) | ISO→G |
| 10 | 30 | 30 | 300 | 0 | 0 | 300 |
| 10 | 30 | 50 | 500 | −200 | 200 | 300 |
| 10 | 40 | 50 | 500 | −200 | 100 | 400 |
| 0 | 15 | 25 | 250 | 50 | 100 | 0 |
| (OS)0 | 40 | 50 | 500 | −200 | 100 | 0 |
| (OS)0 | 40 | 20 | 200 | 100 | −200 | 0 |

price of \$30/MWh for its 10 MW production. The generator is dispatched in the first, second, and third rows of the table.

Whenever it is not dispatched, corresponding to the fourth, fifth, and sixth rows of the table, the generator is effectively billed for its 10 MW generation obligation at a price equal to the day-ahead LMP at the generator bus. The fourth row shows the case where the market has energy available at a lower offer price than the offer from G. Assuming that G is offering at its marginal cost, it benefits from the ISO finding the lower cost dispatch options in these cases. For example, in the fourth row it receives \$150/h, reflecting the difference between the \$30/MWh price and the price \$15/MWh of the available energy at the generator bus.

Although the LMPs at the generator and demand vary with demand and supply conditions, the CfD and CRR hedges risk of variation in both LMPs for the agreed quantity. Again, if the generator or the demand quantities in the offer-based market differ from the CfD/CRR quantity, then they are exposed to relevant LMP price for quantities that deviate from the agreed quantity. The implication is that the good incentives for generation and demand discussed in the context of offer-based economic dispatch are again preserved even in the presence of CfDs and CRRs, since deviations from the agreed quantity are still priced at the marginal surplus.

The discussion has presented a so-called point-to-point day-ahead "obligation" CRR, which is the simplest type of CRR to implement, but has the implication that if the LMP is higher at the source than at the sink then CRR holder owes money. This typically necessitates credit requirements to ensure that the CRR holder can cover any amount owed. In addition to obligation rights, other types of rights, including "option" rights and "flowgate" rights can also be defined [6, Sec. 7.3]. With such rights, the payments are always positive. The discussion has focused on CRRs that hedge transmission price risk in day-ahead markets. Real-time CRRs can also be arranged.

Whether the CRRs are obligation or option, and whether they are based on day-ahead or real-time LMPs, it is important to recognize that they are "financial" rights. That is, their definition entails no more than a specific payoff keyed to market prices. There is no notion of "scheduling" to use the transmission system, which contrasts to the "physical" transmission rights model in Eastern and Western Interconnections that involves scheduling to use transmission.

***Acquiring CRRs.*** In the above example, it was tacitly assumed that the generator could acquire a CRR to match the CfD it had with the demand. CfDs are examples of bilateral contracts, and there are various types of bilateral contracts between market participants that span long-term to short-term opportunities. The variety of bilateral contracting options results in a tension between the desire to hedge transmission price risk associated with long-term bilateral contracts, and the need to reconfigure CRRs as contractual positions change.

In some markets, at least some of the CRRs are allocated based on ownership or use of, or rights to use, preexisting "physical" or other transmission rights that predate the advent of the LMP market. For example, in ERCOT, "Preassigned Congestion Revenue Rights" (PCRRs) are allocated to some municipal utilities and co-ops that have longer-term bilateral contracts [6, Sec. 7.4].

Remaining CRRs are typically auctioned [6, Sec. 7.5], with periodic auctions for monthly and longer-term financial instruments available up to several years into the future. To acquire a CRR in the auction, a market participant bids its willingness to pay for a CRR between nominated point of injection and point of withdrawal. The auction is arranged so that the highest priced bids receive CRRs. The winners of the auction pay the clearing price and receive the CRR, that is, the claim to LMP differences over the CRR term, in return.

The auction revenues represent yet an additional revenue stream accruing to the ISO. As a general principle, this revenue stream is used to either

- Defray retail customer obligations to pay for the embedded cost of the transmission system
- In the case that an existing CRR holder sells its CRR in the auction, compensate the current owner of the CRR for giving up the CRR

In ERCOT, the first alternative is accomplished by refunding payments to demand QSEs. In other ISOs, there is typically an assignment of rights to preexisting owners of the transmission system, who then receive the revenues from the auction.

***Revenue Adequacy.*** The holder of the CRR is paid by ISO a revenue stream given by

(Contract quantity) × (difference between LMP at sink minus LMP at source)

The ISO funds the payments to the CRR holders out of the congestion rent and we must face the question as to whether there is enough congestion rent to fund the CRR obligations. For example, in a CRR market based on day-ahead prices, we must determine whether the congestion rent from day-ahead market can fund CRR obligations based on day-ahead prices. Put another way, we need to decide on the limitations on auctioning CRRs so that the congestion rent is adequate to fund the CRR obligations.

The limitations are specified implicitly in terms of the CRRs. To ensure that congestion rent is sufficient to fund the CRRs that are allocated and sold at auction, the dispatch corresponding to collection of CRRs allocated and sold at auction must satisfy the transmission constraints. This is called the *simultaneous feasibility test* (SFT), since it tests the collection of all CRRs simultaneously, and requires a "test system" that is used to represent actual operating conditions [12].

Two examples will serve to illustrate the SFT. First consider example III. In that example, the SFT requires that a total of no more than 100 MW of CRRs from the left to the right be allocated and sold. Note that if more than 100 MW of CRRs were allocated and sold, say 200 MW, then the ISO would be obligated to pay out (200 MW) × (LMP at right minus LMP at left), whereas the congestion rent would only cover (100 MW) × (LMP at right minus LMP at left). This example shows clearly that in a one-line case the total CRR quantity can be no more than the capacity of the system.

Example IV shows that this result is, in fact, much more general. In that example, if all CRRs sink at SW, then revenue from the congestion rent is adequate to fund the CRRs if the CRR quantities satisfy:

$$0.2 \times (\text{NW to SW CRR}) + 0.4 \times (\text{NE to SW CRR}) - 0.2 \times (\text{SE to SW CRR}) \leq 300\ \text{MW}$$

since this condition ensures that the flows resulting from the corresponding injections at the sources and withdrawals at the sink at SW will satisfy the transmission limits of the system.

The implication is that the auction to issue the CRRs should maximize the bid willingness-to-pay for transmission, constrained by transmission limits on the test system. The algorithm to solve the auction is similar to LMP algorithm, with CRR acquisition prices set to "LMP" differences based on transmission bids. It is possible for some CRR acquisition prices to be negative. In particular, the CRR from SE to SW may have negative acquisition price. The interpretation is that the acquirer of a CRR from SE to SW is being paid in the auction to take on the obligation of future payments into the day-ahead market. Typically, the purchaser of such an obligation would need to demonstrate creditworthiness.

***Test System.*** As mentioned, the CRR auction is carried out with reference to a test system, and the revenue adequacy result should be interpreted as saying that if the actual commercial network model is the same as the test system then the ISO will be revenue adequate. However, if there is an outage in the actual system and therefore the commercial network model compared to test system (or a constraint in the commercial network model that is unmodeled in the test system) then the congestion rent may be insufficient to fund the CRR obligations.

Since outages of transmission lines can occur during the contract lifetime, this possibility must be confronted. There are several approaches to dealing with the resulting shortfall. For example,

- The shortfall could be uplifted to demand.
- The transmission owner could be charged for the shortfall.
- A "balancing account" could be set up to average out shortfalls and surpluses.
- The CRR payments could be reduced during outages.
- The CRR capacity could be less than allowed in the test system.

The last three of these approaches are used in ERCOT [6, Secs. 7.6 and 7.5.1].

### 21.4.5 Capacity Markets

Capacity markets are centralized markets for procuring capacity to maintain long-term resource adequacy. These are futures market based on the concept of ensuring the availability of enough capacity reserves to meet future demand. The capacity payments provided to resources are intended to provide an economic incentive for the development of new resources and are in addition to payments received under the scarcity pricing from the real-time market, although capacity markets are intended to significantly reduce the likelihood of scarcity by arranging for capacity to be available to almost always meet demand. Factors that necessitate the implementation of a capacity market are out of market operator actions, regulatory subsidies, price caps, and market power mitigation that prevent real-time prices rising to levels needed to attract adequate new capacity.

The capacity market clearing price should reflect concepts like cost of new entry (CONE), value of lost load (VOLL), loss of load probability (LOLP), etc. In fact, if there is no capacity market then the RT prices should reflect the VOLL during scarcity.

Capacity markets can be used to solve local or system wide capacity problem or quality/performance problems within the currently available resource mixture. The capacity procurement cost is shared across the entities responsible for serving load. The cost is paid by the load entities based on the ratio of their load to the total system demand. By paying this, loads are hedging against future high spot energy and reserve prices.

### 21.4.6 Market Impacts on ISO System Operations

Electricity markets are designed to create incentives that align market participants' profit maximization behavior with actions needed to ensure reliable operation at the lowest cost.

Although the ideal energy market can, in principle, result in adequate generation capacity, issues such as lack of price formation by demand under scarcity and concerns about market power have resulted in the missing money problem as discussed in the subsection Pricing in Practice and Implications. In several regions, capacity markets procure resources to be available several years in the future. This places the responsibility on the ISO to specify planning reserve requirements to reliably serve load in the future years, to some extent undercutting the incentives for market participants to predict future load growth. Capacity markets tend to reduce the likelihood of scarcity because they arrange for new capacity to be built to meet planning reserve requirements. The revenue from capacity market supplements the revenue from energy market and addresses the missing money issue. A poorly designed capacity market could send out price signals for new builds when not needed. Implementing appropriate scarcity pricing mechanism in energy market would ensure that majority of the money needed to sustain needed new builds is obtained from the energy market and hence would reduce the need for a capacity market to provide the missing money. It also ensures better deliverability of the capacity that is procured in the capacity market.

In the shorter term, the day-ahead market ensures that resources will be committed most economically to plan to meet next day forecasted load and by procuring extra reserves so that ISOs can plan to reliably serve load even under unforeseen circumstances. This centralized resource commitment commits enough resources to cope with contingencies, but avoids the typical over-commitment that occurs with decentralized commitment by reducing the inefficiencies introduced by individual entities committing extra to reduce the risk of failing to meet their load obligation.

The reliability or residual unit commitment provides the ISO with the ability to commit extra resources either day-ahead or near real time if needed to provide sufficient capacity for its forecast of load or if transmission constraints not represented in the commercial network model require additional capacity. Generally, it is expected that good market design and appropriate incentives to market participants will result in only occasional commitment by the reliability unit commitment. Significant commitment of resources under reliability unit commitment is an indicator of poor incentives to market participants to reveal expected demand levels in the day ahead or failure to represent significant transmission issues in the day-ahead market.

The real-time market involves the dispatching of resources and (in some markets) re-clearing ancillary services in real time to most economically meet the system load without violating the transmission system and resource operating limits thus ensuring reliable system operations at the least cost. As discussed in the subsection Interaction between Day-Ahead and Real-Time Markets, prices in the real-time market apply to deviation of real-time production or demand from day-ahead financial commitments.

The centralized forecasting of load and renewable and centralized procuring/dispatching of ancillary services in the markets help operations maximize the utilization of renewable resources with minimal reserves by efficiently managing the volatility and uncertainty of these renewable resources.

These markets create a platform for trading at various timescales. The market prices are created based on overall and locational energy/reserve demands. These locational price signals created by these centralized markets gives the right signals for investors to build resources in the locations that minimizes congestion and losses and thus help operations and planning of the power systems. The markets also create very high scarcity prices when the demand is high and reserves are scarce. These high scarcity prices incentivizes generators to maintain their resources so as to be available when the system needs them most. It also incentivizes voluntary price responsive demand reduction and shifting of load from peak time to off-peak time, which helps system operators meet the loads reliability without rolling blackouts. Thus the tools provided by these markets and the pricing incentives created by these markets ensures that the right resources are located at the right location and are dispatched to the appropriate level to help ISO reliably serve the load at the least cost. However, a countervailing issue is that because of the relative difficulty and timing of transmission construction, sometimes the ISO may need to embark on building transmission construction in particular cases in advance of seeing if a response by market participants would avoid the need for construction.

## 21.5 MARKET DESIGN TYPES

The use of LMPs, a day-ahead market based on security-constrained unit commitment, reliability unit commitment, a real-time market based on security-constrained economic dispatch, congestion revenue rights, and bilateral contracting including contracts for differences are basic features of US ISO-based markets. However, these building blocks can be varied and combined in several ways and even combined with preexisting arrangements such as physical bilateral trading. The following variations will be discussed:

- Mandatory pool
- Real-time balancing energy market (imbalance market)
- Physical bilateral trading model
- Power exchanges
- Hybrid model (voluntary pool with bilateral contracts)
- Minimalist model

### 21.5.1 Mandatory Pool

A mandatory pool market (often referred to as a "pool") is a market designed to be run by an ISO and/or market operator (MO) in which all producers are required to offer the full capacity of both online and offline resources to the market. This differs from the typical U.S. market setup in that U.S. markets generally allow for decentralized commitment by asset owners, even though the market design also accommodates the ISO performing commitment for assets that are offered into the day-ahead market. As in other versions, the ISO clears the market and establishes dispatch and prices.

This model can accommodate both a forward market, such as day-ahead, as well as a real-time market. In some day-ahead pool markets, consumers are not required to bid into the market. In such cases, forecasted load is generally used to estimate the amount of offers to be cleared in day ahead. In other cases, consumers bid their estimated load and the market clears it. In mandatory real-time pool markets, all capacity, even the amount cleared in day ahead, are offered and re-cleared to meet real-time load consistent with system reliability.

Producers are required to submit detailed information about their resources to the ISO and are required to submit three-part offers containing start-up cost, no-load or minimum load cost, and incremental energy offer into the market. The ISO considers these offers, together with transmission network congestion, resource parameters, to commit resources if needed to meet reliability criteria and issues dispatch instructions and prices that maximize social welfare while meeting reliability criteria. In forward markets based on this design, transmission network congestion might not be considered while clearing the forward market.

An inherent aspect of an offer-based market, including mandatory pools, is volatility in prices. Market participants can usually hedge against these price volatilities by procuring financial instruments like contracts for differences (CfDs) to enable bilateral trade that are compatible with the pool prices. The initial United Kingdom market design was an example of a pool market.

### 21.5.2 Real-Time Balancing Energy Market (Imbalance Market)

Balancing energy markets are akin to real-time markets but typically used in the case where there is no centralized day-ahead market but where there is instead a day-ahead bilateral scheduling process and/or power exchange such that balanced or roughly balanced combinations of generation and demand are disclosed to the ISO. Balancing energy markets procure differential energy amounts to correct for the real-time energy imbalance caused by energy consumption/production in excess of or deficit to the bilateral schedules or power exchange transactions. The initial wholesale market in ERCOT until December 2010 included only a balancing market without a centralized day-ahead energy market.

Typical balancing markets use incremental and decremental balancing energy offers, which represent a commitment to increase or decrease generation in a traded interval compared to the day-ahead scheduled level of generation. Resources or resource portfolios are dispatched by the operator to make the net real-time imbalance zero so that the system frequency can be maintained at the required set point. The system operator also dispatches incremental and decremental balancing energy offers to resolve congestion, that is, maintain power flows within grid limits. Resources self dispatch for their trades and are redispatched to account for the incremental and decremental balancing energy awards in the balancing market. These redispatches are compensated at the balancing market clearing price. Since balancing energy is offered and transacted near real time and is not transparently linked by virtual bids to a day-ahead price, providers must buy fuel to support the transaction at spot market prices, or arrange storage of fuel. Hence, the balancing energy service is normally considered a premium product and typically clears at a higher price than future markets.

### 21.5.3 Physical Bilateral Trading Model

The earliest form of trade was a negotiated swap between buyer and seller. Only slightly more complicated is a monetized transaction in which cash is exchanged for goods or services. This type of market is often called an over-the-counter market and the transactions called over-the-counter trades. Over-the-counter trades are performed privately without standardized products or centralized coordinating entity to carry out the transaction. Bilateral trades are often longer term and do not consider the many day-to-day changes in the electric system influencing spot prices. The need to match supply to demand essentially continuously typically requires that a real-time market, or other capability to adjust supply, be available in addition to bilateral contracts.

### 21.5.4 Power Exchanges

Power exchanges provide a public platform for financial trading, typically without representation of unit commitment and only coarse representation of transmission issues. The power exchange acts as the counter party for both buyer and seller and hence reduces the risk of default for both buyer and the seller. Power exchanges can coexist with any of the market models.

Some of the exchanges that list electricity futures are CME Group [13] (through its Nymex division), the Intercontinental Exchange [14], Nord Pool [15], Powernext (in France) [16], the European Energy Exchange (in Germany) [17], the Australian Securities Exchange [18], and the Multi Commodity Exchange of India [19].

### 21.5.5 Hybrid Model (Voluntary Pool with Bilateral Contracts)

The hybrid model combines the characteristics of both bilateral trading with RT imbalance market and the pool market. This model provides more flexibility and liquidity in the market allowing wider variety of services and pricing options for customers. There are different flavors of the hybrid model. However, in majority of hybrid models, the day-ahead market is the hybrid one and is supplemented by a real-time market.

The day-ahead market is voluntary. Market participants can schedule their bilateral trades rather than participating in the day-ahead market or they could participate in the day ahead to take advantage of any clearing price lower than their marginal cost of production and obtain trades for the difference between the bilateral trade amount and the updated load forecast. The trades are submitted to ISO and the imbalance with respect to the trades and the imbalance with respect to day-ahead awards are accounted as financial transactions while settling the real-time pool market. The ISO operates the markets and issues prices while ensuring reliability.

With the hybrid model, there is very little incentive to have a physical bilateral trade. During scarcity, if the ISO issues rolling blackouts then even having a physical contract would not ensure that the buyers load will be served if it is controlled by the ISO. Generally, in a system with a day-ahead market based on the hybrid model, participant's bilateral trades are financial in nature and are obtained through a power exchange.

### 21.5.6 Minimalist Model

The minimalist model was the historic way of managing the power system under regulation. This model is followed in areas where power sector is still not deregulated. Vertically integrated single entities own and controls generation, transmission, and consumption in a control area. This single entity commits resources and dispatches them to meet their forecasted load. Different such control areas interconnect with each other and help maintain system reliability.

## 21.6 ERCOT MARKETS

To contextualize the previous discussion, this subsection specifically describes the ERCOT markets, both wholesale and retail. The ERCOT region is one of three North American grid interconnections. The ERCOT grid serves 90% of Texas load. ERCOT peak load in 2016 was roughly 71 GW, and ERCOT has more than 46,500 mi of transmission lines and more than 550 generators. The transmission providers, generators, municipal utilities, and cooperatives own the physical assets. The entity ERCOT, is the independent system operator and reliability coordinator for the ERCOT region.

ERCOT manages the competitive wholesale nodal power market in the ERCOT region and administers customer switching for 7 million Texas customers in competitive choice areas for the retail market. ERCOT both has regions that are open to retail competition and regions that are served by municipal utilities and cooperatives and which did not opt to open their areas to retail competition. All these entities, however, participate in the wholesale market.

The areas of ERCOT with retail competition require that each end-user load be uniquely identified by an electric service identification (ESI ID) that is accessible to competitive retailers (CRs) who vie for access to provide the power across the local distribution utilities' systems. CRs interact directly with ERCOT when submitting switching requests, while customers may choose a new CR. ERCOT processes the switching requests by working with TDSPs to obtain the initial and final meter reads, confirming switches with customers, and confirming the switch with the relevant CRs once the switch is approved. The ESI ID and the standard of data transaction across all distribution utilities in competitive areas are the basis for competitive retail market functions.

The ERCOT wholesale nodal market is built upon ERCOT's function as an ISO and as a facilitator of open electricity markets. The market is based on management of congestion on the transmission grid consisting of more than 10,000 buses, and transactions are priced at LMPs. The market emphasis is on issuing resource specific dispatches. The key components of the wholesale market are the CRRs auction, which provides mechanism to hedge against the energy delivery cost risk, the day-ahead market, which provides mechanism to hedge against the energy cost risk and the real-time market-security constrained economic dispatch (RTM-SCED), which redispatches resources to manage real-time congestion and maintain frequency.

ERCOT markets are run according to the protocols, which were developed by the market participants to implement the rules passed by the PUCT. Any changes to the protocols need to be approved by the ERCOT board, which has representatives from all market participant sectors. Various market participants work closely together and operate according to these standard market rules or protocols to operate the market reliably and efficiently.

### 21.6.1 Market Operations in ERCOT

In ERCOT, the operations related to the market starts with qualified scheduling entities (QSEs) registering their resources with ERCOT. ERCOT conducts tests to qualify each resource for providing different service and monitors performance of these resources to ensure compliance with the market rules. ERCOT does long-term (5 year ahead) and short-term (month/week/day ahead) system planning to ensure adequate transmission is built in time, adequate resource reserve is available for future and maintenance/construction outages are scheduled appropriately to serve load reliably.

ERCOT runs 2 years ahead, yearly and monthly CRRs auctions for selling financial instruments to market participant for them to hedge their energy delivery cost risks. These CRRs give an opportunity for market participants to fix the amount that they are willing to pay for congestion between source and sink and thus hedge against the volatility of the congestion rent due to changes in supply and demand, weather conditions, unanticipated outages or other changes in grid network topology. Day ahead, ERCOT updates forecasts and determine reserve requirement for the next operating day. QSEs update their resource operating-schedule, resource availability information, and submits information about self-arrangement of ancillary services (AS). Day ahead, ERCOT runs a day-ahead market with unit commitment to co-optimize energy, AS, and certain types of CRRs. This ensures energy and ancillary services for the following day and establishes prices. ERCOT runs day-ahead/hourly reliability unit commitment (DRUC/HRUC) to commit resources out of market for reliability if needed. During real-time operations, ERCOT executes load frequency control (LFC) and SCED, issues resource-specific dispatch instructions, and posts market information required by the protocols on the market information system (MIS). ERCOT settles congestion revenue right (CRR) auction, the day-ahead market, and the real-time market (RTM) separately using different settlement timelines. After the operating day, ERCOT gathers meter data, aggregates it and adjusts the real-time settlement based on the actual meter data. Major market components are explained in detail below.

- ERCOT congestion revenue rights (CRR) auction. ERCOT runs annual and monthly auctions to sell CRRs, which are financial instruments that hedge the energy delivery cost risk. CRRs entitle the CRR owner to receive or be charged congestion rents that ERCOT collects. CRRs can be acquired in several ways; through bidding in an auction, by direct allocation, or by trade either within or outside of the ERCOT system. Point-to-point obligations may also be purchased in the day-ahead market, which entitles them to receive congestion rent from real-time market. The monthly or long term auctions (for the next four 6-month blocks) offer a portion of the available network transmission capacity. ERCOT sells up to 60, 45, 30, and 15% of the transmission capacity, respectively, for the next four 6-month blocks in long-term CRR auction. Each 6-month block is run 2 years in advance of the operating day and then offered again 3 more times with more transmission capacity opened up for auction. A month in advance of the operating day, ERCOT runs monthly CRR auction to sell up to 90% of the transmission capacity. CRR account holders, who may be anyone except ERCOT, PUCT, or TDSP staff, can submit bid portfolios for CRRs by specifying the source and sink settlement points, MWs, and duration of the desired CRR. The auction is cleared by running a simultaneous combinatorial feasibility algorithm that uses the nodal network model with planned outages and maximizes revenue with the constraint that all winners pay the same dollar amount per MW per hour for the same product.

- ERCOT day-ahead operations, day-ahead market (DAM), and day-ahead reliability unit commitment (DRUC). ERCOT's day-ahead operations run the day before the operating day. It is primarily a financial market and not a physical market, except that there is no opportunity to trade out of AS obligations in real time, these are somewhat fixed by the DAM. The DAM allows qualified scheduling entities (QSEs) with resources to bid and/or offer energy and to offer ancillary services. Participants have an opportunity to purchase and/or sell energy for speculation even if they do not have resources or loads. QSEs can also bid to purchase point-to-point obligations in the DAM to hedge against delivery risk in the real-time market. Participation in the DAM is voluntary with the exception of election of ancillary service coverage of ERCOT allocation of obligations to QSEs that represent load. ERCOT purchases three types of ancillary services: regulation, responsive reserve, and nonspinning/operational reserve. The amount of ancillary services needed for the next operating day is established by ERCOT day ahead based on historic data. QSEs are responsible to either self-arrange or purchase their load ratio share of the ancillary service requirement for the operating day. The amount of ancillary service that is not self-arranged is procured by ERCOT. DAM co-optimizes the energy offer/bids, ancillary service offers and point-to-point (PTP) bids to maximize revenue while honoring the physical limits of the resources and transmission system.

  Since DAM might not have enough offers to clear energy to cover for the full forecasted load for the next operating day, ERCOT runs the day-ahead reliability unit commitment (DRUC) study after DAM. This process looks at the generation self-commitment plan and ensures that there is sufficient generation capacity committed in the proper locations to reliably serve the forecasted load and forecasted transmission congestion by committing offline resources, if required. Resources started up by ERCOT are given make-whole payments if needed to ensure that they get at least the start-up and operating cost if their revenue based on the market settlement prices will not cover their cost.

- ERCOT adjustments period operations—supplemental ancillary services market (SASM). After day-ahead operations closes, adjustment period (time until operating hour) begins to allow QSEs to submit new information, including energy offers and bids, output schedules and trades.

  Additionally, transmission service providers (TSPs) may submit and/or update information related to forced outages.

  During the adjustment period, ERCOT checks whether QSE has scheduled their resources to provide the ancillary services awarded in DAM or marked as self-arranged in day ahead by the QSE.

  ERCOT also reviews the deliverability of the ancillary services that are scheduled by the QSE. If required, ERCOT then executes supplemental ancillary services market (SASM) to procure additional ancillary services or to procure the ancillary service amount that the QSE is unable to provide or is undeliverable. ERCOT could run multiple SASMs for an operating day if needed. These SASMs are settled as part of the real-time market.

- ERCOT hour-ahead operations—hourly reliability unit commitment (HRUC). QSE proposed adjustment period changes that revoke commitment of resources will be evaluated and accepted or rejected by the hourly reliability unit commitment (HRUC) study, which uses the same algorithm as the DRUC but is run the hour before the operating hour. Each hour, HRUC studies the rest of the operating day and commits resources if needed to account for unforeseen issues like error in day-ahead load forecast and forced outage. During the hour-ahead operations, ERCOT operator reviews the outages, updated wind forecast and updated load forecast for the next hours to prepare action plans for real-time operator to resolve congestion that cannot be resolved by SCED.

- ERCOT real-time operations—real-time market (RTM) and load frequency control. ERCOT reliably serves the load in real-time by running a SCED at least every 5 min, and redispatches resources using offers by individual resources and actual shift factors by each resource on each transmission element. SCED determines the most economical dispatch of individual resources across the grid to resolve constraints and meet the system demand while honoring the physical limits of the resources and the transmissions system. The load frequency control takes these resource specific dispatches and current generation/load level and deploys regulation every 4 s to maintain frequency within limits. The real-time operator deploys responsive reserves and operational reserves if needed to maintain system reliability.

### 21.6.2 Market Dynamics in ERCOT

In the ERCOT market, generation participants serve load reliably in response to economic incentives, that is, maximization of profit and minimization of risk. Each generation participant has assets (generation capability, fuel contracts, etc.), and various opportunities to maximize the revenue from these assets. The participant may sell all, or part of its capacity through bilateral trades, it may sell all, or part of capacity through the day-ahead market, it may sell capacity into the ancillary service markets, or it may retain the capacity for real-time offers. The participant could also offer all its capacity for both energy and ancillary service in day-ahead market and allow the DAM market to select the highest value product. A participant may sell bilaterally either at its point of injection, removing congestion risk, or it may sell remotely, assuming the risk of congestion creating price differentials between the generator site and the purchase site. On any given day, a QSE in ERCOT will normally have a combination of obligations which it is financially obligated to serve, and excess capacity from which it wishes to maximize revenue.

To hedge against volatility in congestion cost, a QSE may procure CRRs in advance if it is selling energy at a point other that its point of injection. However, it may be insufficiently, or over hedged against congestion costs on a given day due to the dynamic nature of load. The QSE representing these assets may protect itself from insufficient hedging by purchasing point-to-point obligations in the day-ahead market. CRR account holders (CRRAHs) are charged in CRR auction for the CRRs awarded in the auction. The CRRs acquired through allocation/auction/trades are settled in DAM to hedge against the delivery risk in DAM by paying the CRRAHs, for each MW of CRR between a source and sink settlement points, the price difference between the source and the sink in DAM.

DAM establishes the market value (congestion rent) of the CRRs and hence indirectly determines the amount that the QSEs are willing to pay for the CRRs in the auction. As the markets stabilize the amount that the QSEs are paying for the CRRs should converge to the amount that they get paid in DAM. QSEs are charged in DAM for any point-to-point obligations awarded to them in DAM. These point-to-point obligations are settled in real-time market to hedge against the delivery risk in real-time market by paying the QSEs, for each MW of PTP obligation between a source and sink settlement points, the real-time price difference between the source and the sink in real-time market, that is, real-time market establishes the market value (congestion rent) of the PTP and hence indirectly determines the amount that the QSEs are willing to pay for the PTPs in the DAM. As the markets stabilize the amount that the QSEs are paying for the PTPs in DAM should converge to the amount that they get paid in real-time market.

To hedge against the volatility in energy cost, QSEs might enter into bilateral contract or purchase energy at power exchanges outside ERCOT markets. In ERCOT markets, QSEs may hedge against volatility in energy cost by offering generation and bidding load in the DAM sufficient to cover its obligation. If the participant hedges its energy obligations by offering of generation and bidding load in DAM, it protects itself by fixing the prices that it will receive for generation and will pay for serving the load. Even if the QSEs have contracts outside ERCOT, offering generation and bidding load in the DAM would enable then to make profit by buying from the market if the energy cost is cheaper in DAM than their contract. QSEs are charged/paid in DAM for the energy that is bought/sold in the DAM. They are charged/paid in real-time market for the imbalance amount, that is, charged in RTM for any amount consumed in excess or produced less than DAM and paid for any amount produced in excess or consumed less than DAM.

Loads pay for the energy and congestion in real-time market for the imbalance amount and in DAM for the amount of energy bought in DAM. Loads also pay the TDSPs for using the transmissions that the TSDPs own. The loads are paid back the revenue from the CRR auction, net amount that remains in DAM (charge for PTPs awards plus charge for loads minus payment to generators minus payment to CRRs), and net amount that remains in real-time market (charge for loads minus payment to generators minus payment to PTPs) at load ratio share of the total ERCOT load. As the markets stabilize the CRR market should converge to DAM which in turn should converge to real-time market, that is, DAM prices should converge to real-time market prices, DAM congestion should converge to real-time market congestion, CRR auction revenue should converge to DAM congestion.

If the convergence happens then loads as a whole will get back the congestion rent paid in real-time market when they get the CRR auction revenue and hence will only be paying for the energy cost alone. If the convergence happens then the congestion rent paid in real-time market by the loads will be negated by the CRR auction revenue and hence they will effectively be just paying the energy cost alone. In short, participants in ERCOT may hedge by buying CRRs in CRR auction, selling energy in DAM and settling in DAM or by purchasing CRRs in CRR auction, purchasing PTPs in DAM, selling energy in real-time market and settling in real-time market.

The ERCOT market provides a high level of market transparency to market participants with directly observable consequences of market behaviors by providing the offer-based marginal cost of serving the next increment of load at more than 10,000 nodes. Because nodal prices rise within congested areas, prices incent development of additional local resources in congested areas, and global energy scarcity increases prices across the grid. Like all markets, however, there are time lags related to the delay between a price signal being applied, and the response of the market. If scarcity develops which requires generator construction, these delays are significant. When developing new generation, a financer will normally insist upon a guaranteed revenue stream sufficient to cover development costs. Thus, new generation will normally have capacity commitments for several years after synchronization.

Load serving entities wish to be protected from scarcity pricing when they commit to serve new customers. In ERCOT, retail customers may switch providers at will, so tension exists between the need to develop additional capacity resources and the risk of losing customers. Load serving entities are incented to develop just enough capacity to serve their customer load, and to provide a high level of service to retain customers. Since it is an energy only market, uncertainty in regulatory rules like cap for the energy offer, pricing rules during scarcity, etc., affect the incentives for generators to build new facilities and incentives for loads to enter into long-term contract with generators. A fine balance between different segments is maintained to ensure resource and transmission adequacies to reliability serve the load.

### 21.6.3 Comparison of Markets in Different Regions

Market systems in many regions are quite different from each other due to the differences in design philosophy. Even market systems based on the same design can be different due to the differences in the market rules and regulations. For example, markets in North America all have LMP-based CRR auction, DAM, and RTM but have differences in features like availability of renewable credits, availability of CRR options, availability of demand response, capacity market, handling of losses, etc.

To understand the variation between similar market systems, two ISOs namely, PJM interconnection (formerly the states of Pennsylvania, New Jersey, and Maryland, but now encompassing a much larger area) [20] and ERCOT [21] are considered. PJM was significant in introducing a nodal pricing model to a market paradigm and ERCOT's nodal market has adopted many of the PJM innovations. PJM and ERCOT, both have a real-time market, day-ahead market, congestion revenue right/financial transmission rights auction, demand response and the three types of ancillary services. However, there are many differences between the two systems. PJM operates a capacity market to ensure construction of adequate generating capacity for the future while ERCOT does not have a capacity market and depends upon energy and ancillary service prices to provide incentives for the construction of sufficient generation. PJM does ancillary service co-optimization in day-ahead market and real-time market and allows substitution between different ancillary services and have regional ancillary services requirements. However, ERCOT performs ancillary service co-optimization only in day-ahead market, and has only system-wide ancillary services requirement. PJM has real-time unit commitment and look ahead real-time market while ERCOT is still in the process of evaluating the need for those. ERCOT has retail market and renewable energy credit tracking which PJM does not have. ERCOT considers the system as lossless in its markets studies while PJM has a marginal loss component considered in its locational marginal prices. PJM does not provide financial instruments in day-ahead market to hedge in real-time market. Even in the components like day-ahead market and congestion revenue right/financial transmission rights auction that are similar between the markets, they have different rules and have different timelines for submission, clearing, and settlements.

Thus, market systems in many regions can be quite different from each other due to the differences in design philosophy. Further, market systems based on the same design can be different due to the differences in the market rules and regulations.

## 21.7 REFERENCES


1. Hal R. Varian, *Microeconomic Analysis*, 3rd ed., W. W. Norton and Company, New York, 1992.
2. Paul Joskow and Edward Kahn, "A Quantitative Analysis of Pricing Behavior In California's Wholesale Electricity Market during Summer 2000," *The Energy Journal*, 23(4):1–35, 2002.
3. Daniel Kirschen and Goran Strbac, *Fundamentals of Power System Economics*, John Wiley and Sons, Inc., Chichester, England, 2010.
4. Chris Harris, *Electricity Markets—Pricing Structure and Economics*, John Wiley and Sons, Inc., United Kingdom, 2006.
5. Steven Stoft, *Power System Economics: Designing Markets for Electricity*, IEEE Press, Piscataway, NJ, 2002.
6. Electric Reliability Council of Texas, "ERCOT Nodal Protocols," 2016. Available from www.ercot.com.
7. Ross Baldick, "Mitigate Market Power to Improve Market Efficiency," In Hong Cheng, *Power Grid Operation in a Market Environment: Economic Efficiency and Risk Mitigation*, Chap. 2, pp. 43–66, IEEE Press, Piscataway, NJ, 2016.
8. Paul Gribik, William Hogan, and Susan Pope, "Market-Clearing Electricity Prices and Energy Uplift," Harvard University, Tech. Rep., 2007. Available from: https://www.hks.harvard.edu/fs/whogan/Gribik_Hogan_Pope_Price_Uplift_123107.pdf.
9. Allen J. Wood and Bruce F. Wollenberg, *Power Generation, Operation, and Control*, Wiley and Sons. New York, NY, 1996.
10. Ross Baldick, *Applied Optimization: Formulation and Algorithms for Engineering Systems*, Cambridge University Press, Cambridge, 2006.
11. Mohammad Shahidehpour, Hatim Yamin, and Zuyi Li, *Market Operations in Electric Power Systems: Forecasting, Scheduling, and Risk Management*, John Wiley and Sons, Inc., New York, NY, 2002.
12. William W. Hogan, "Contract Networks for Electric Power Transmission," *Journal of Regulatory Economics*, 4(3):211–242, 1992.
13. Chicago Mercantile Exchange: http://www.cmegroup.com/.
14. Intercontinental Exchange: https://www.theice.com/homepage.jhtml.

15. Nord Pool: http://www.nordpoolspot.com/.
16. Powernext: http://www.powernext.fr/.
17. European Energy Exchange: http://www.eex.com/en/.
18. Australian Securities Exchange: http://www.asx.com.au/.
19. Multi Commodity Exchange of India Limited, http://www.mcxindia.com/.
20. PJM Interconnection, www.pjm.com.
21. Electricity Reliability Council of Texas, www.ercot.com.

# 22

# POWER QUALITY AND RELIABILITY

Surya Santoso

*Professor, Department of Electrical and Computer Engineering, University of Texas at Austin; Fellow, IEEE*

Mark F. McGranaghan

*Vice President, Electric Power Research Institute, Knoxville, Tennessee; Fellow, IEEE*

Roger C. Dugan

*Senior Technical Executive, Electric Power Research Institute, Knoxville, Tennessee; Fellow, IEEE*

## 22.1 PERSPECTIVE ON POWER QUALITY

### 22.1.1 Introduction

Power quality is about compatibility between the quality of the voltage supplied from the electric power system and the proper operation of end-use equipment. Power quality is also about economics—finding the optimum level of investment in the power system and the end-use equipment to achieve the compatibility. There are two categories of power quality that need to be considered—steady-state (or continuous) power quality and disturbances. Steady-state power quality characteristics include voltage regulation, harmonic distortion, unbalance, and flicker. We can define *compatibility levels* for these characteristics and then the challenge is to maintain performance within these compatibility levels and make sure that equipment can operate with these levels. Power quality disturbances (outages, momentary interruptions, voltage sags, and transients) are much more of a challenge. It is impossible to completely prevent disturbances that may cause equipment disruptions. Therefore, we have to find the best balance between investments to prevent disturbances and investments in equipment and facility protection.

On the technology side, future power quality research will focus on advanced technologies that can be applied at all levels of the system to improve compatibility (both supply-side technologies and end-user technologies) and on the procedures to find the optimum places to make these investments from a system perspective. The result will be guidance regarding expected levels of performance for different types of supply systems that will result in optimum economics if customers also make the associated investments to assure the required equipment performance. Recommendations from the economic analysis will also require regulatory structures to support the implementation of optimum system designs and solutions. Therefore, the research results must be coordinated with development of regulations and market structures for future power systems.

## 22.2 CATEGORIES AND CHARACTERISTICS OF POWER QUALITY DISTURBANCE PHENOMENA

### 22.2.1 General

*Power quality* is a generic term applied to a wide variety of electromagnetic phenomena on the power system. The duration of these phenomena ranges from a few nanoseconds (e.g., lightning strokes) to a few minutes (e.g., feeder voltage regulations) to steady-state disturbances (harmonic distortions and voltage fluctuations). Due to the extensive variety of the phenomena, many power quality terms have sometimes been applied incorrectly and caused confusion among end users, vendors, and service providers in dealing with power quality concerns. For example, a term *power surge* has been used to describe some kind of power disturbances. However, it is ambiguous and in fact has no technical meaning since power surge does not refer to a surge in power. This term has been used to refer to overvoltage transients in voltage. Power is related to the product of voltage and current. Normally, *voltage* is the quantity causing the observed disturbance and the resulting power will not necessarily be directly proportional to the voltage. The solution will generally be to correct or limit the voltage as opposed to addressing the power. Therefore, the use of ambiguous and nonstandard terms is discouraged.

### 22.2.2 General Classes of Power Quality Disturbances

The Institute of Electrical and Electronics Engineers Standards Coordinating Committee 22 (IEEE SCC22) has led the main effort in the United States to coordinate power quality standards. It has the responsibilities across several societies of the IEEE, principally the Industry Applications Society and the Power Engineering Society. It coordinates with international efforts through liaisons with the IEC and CIGRE (International Conference on Large High-Voltage Electric Systems). The IEC classifies electromagnetic phenomena into the groups shown in Table 22-1 [1].

The United States power industry efforts to develop recommended practices for monitoring electric power quality have added a few terms to the IEC terminology [2]. *Sag* is used as a synonym to the IEC term *dip*. The category *short duration variations* is used to refer to voltage dips and short interruptions. The term *swell* is introduced as an inverse to sag (dip). The category *long duration variation* has been added to deal with American National Standards Institute (ANSI) C84.1 limits. The category *noise* has been added to deal with broadband conducted phenomena. The category *waveform distortion* is used as a container category for the IEC *harmonics, interharmonics,* and *dc in ac networks* phenomena, as well as an additional phenomenon from IEEE Std. 519-1992 (Recommended Practices and Requirements for Harmonic Control in Electrical Power Systems) called *notching*. Table 22-2 shows the categorization of electromagnetic phenomena used for the power quality community.

**TABLE 22-1** Principal Phenomena Causing Electromagnetic Disturbances as Classified by the IEC

| |
|---|
| Conducted low-frequency phenomena |
| Harmonics, interharmonics |
| Signal systems (power line carrier) |
| Voltage fluctuations (flicker) |
| Voltage dips and interruptions |
| Voltage imbalance (unbalance) |
| Power-frequency variations |
| Induced low-frequency voltages |
| DC in ac networks |
| Radiated low-frequency phenomena |
| Magnetic fields |
| Electric fields |
| Conducted high-frequency phenomena |
| Induced continuous wave (CW) voltages or currents |
| Unidirectional transients |
| Oscillatory transients |
| Radiated high-frequency phenomena |
| Magnetic fields |
| Electric fields |
| Electromagnetic fields |
| Continuous waves |
| Transients |
| Electrostatic discharge phenomena (ESD) |
| Nuclear electromagnetic pulse (NEMP) |

## 22.2.3 Transient—General

The term *transient* has long been used in the analysis of power system variations to denote an event that is undesirable and momentary in nature. Other definitions in common use are broad in scope and simply state that a transient is "that part of the change in a variable that disappears during transition from one steady-state operating condition to another." Another word in common usage that is often considered synonymous with transient is *surge*. This term should be avoided unless it is qualified with appropriate explanation. In general, transients can be classified into two categories, impulsive and oscillatory. These terms reflect the waveshape of a current or voltage transient.

***Impulsive Transient.*** An *impulsive transient* is a sudden, nonpower frequency change in the steady-state condition of voltage, current, or both, that is unidirectional in polarity (primarily either positive or negative). They are normally characterized by their rise and decay times which can also be revealed by their spectral content. For example, a $1.2 \times 50$ $\mu$s 2000-V impulsive transient nominally rises from zero to its peak value of 2000 V in 1.2 $\mu$s, and then decays to half its peak value in 50 $\mu$s. The most common cause of impulsive transient is lightning. Figure 22-1 illustrates a typical current impulsive transient caused by lightning.

**TABLE 22-2** Categories and Characteristics of Power System Electromagnetic Phenomena

| Categories | Typical spectral content | Typical duration | Typical voltage magnitude |
|---|---|---|---|
| 1.0 Transients | | | |
| 1.1 Impulsive | | | |
| 1.1.1 Nanosecond | 5 ns rise | < 50 ns | |
| 1.1.2 Microsecond | 1 μs rise | 50 ns–1 ms | |
| 1.1.3 Millisecond | 0.1 ms rise | > 1 ms | |
| 1.2 Oscillatory | | | |
| 1.2.1 Low frequency | < 5 kHz | 0.3–50 ms | 0–4 pu* |
| 1.2.2 Medium frequency | 5–500 kHz | 20 μs | 0–8 pu |
| 1.2.3 High frequency | 0.5–5 MHz | 5 μs | 0–4 pu |
| 2.0 Short duration variations | | | |
| 2.1 Instantaneous | | | |
| 2.1.1 Sag (dip) | | 0.5–30 cycle | 0.1–0.9 pu |
| 2.1.2 Swell | | 0.5–30 cycle | 1.1–1.8 pu |
| 2.2 Momentary | | | |
| 2.2.1 Interruption | | 30 cycle–3 s | < 0.1 pu |
| 2.2.2 Sag (dip) | | 30 cycle–3 s | 0.1–0.9 pu |
| 2.2.3 Swell | | 30 cycle–3 s | 1.1–1.4 pu |
| 2.3 Temporary | | | |
| 2.3.1 Interruption | | 3 s–1 min | < 0.1 pu |
| 2.3.2 Sag (dip) | | 3 s–1 min | 0.1–0.9 pu |
| 2.3.3 Swell | | 3 s–1 min | 1.1–1.2 pu |
| 3.0 Long duration variations | | | |
| 3.1 Interruption, sustained | | > 1 min | 0.0 pu |
| 3.2 Undervoltages | | > 1 min | 0.8–0.9 pu |
| 3.3 Overvoltages | | > 1 min | 1.1–1.2 pu |
| 4.0 Voltage unbalance | | steady state | 0.5–2% |
| 5.0 Waveform distortion | | | |
| 5.1 DC offset | | steady state | 0–0.1% |
| 5.2 Harmonics | 0–100 Hz | steady state | 0–20% |
| 5.3 Interharmonics | 0–6 kHz | steady state | 0–2% |
| 5.4 Notching | | steady state | |
| 5.5 Noise | broadband | steady state | 0–1% |
| 6.0 Voltage fluctuations | < 25 Hz | intermittent | 0.1–7%<br>0.2–2 Pst |
| 7.0 Power frequency variations | | < 10 s | |

*pu = per unit.

***Oscillatory Transient.*** An *oscillatory transient* is a sudden, nonpower frequency change in the steady-state condition of voltage, current, or both, that includes both positive and negative polarity values. It consists of a voltage or current whose instantaneous value changes polarity rapidly. It is described by its spectral content (predominate frequency), duration, and magnitude. The spectral content subclasses defined in Table 22-2 are high, medium, and low frequency. The frequency ranges for these classifications are chosen to coincide with common types of power system oscillatory transient phenomena. High- and medium-frequency oscillatory transients are transients with a primary frequency component greater than 500 kHz with a typical duration measured in microseconds, and between 5 and 500 kHz with duration measured in the tens of microseconds, respectively. Figure 22-2 illustrates a medium frequency oscillatory transient event due to back-to-back capacitor energization.

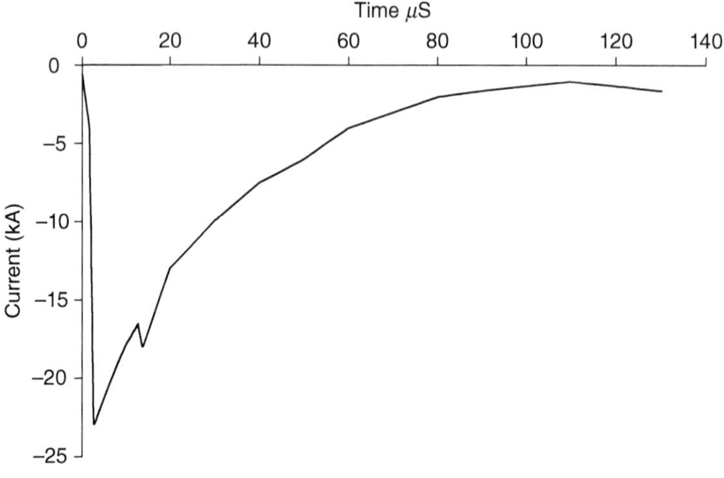


**FIGURE 22-1** Lightning stroke current impulsive transient.

### 22.2.4 Short-Duration Voltage Variations

Short-duration voltage variations are caused by fault conditions, the energization of large loads that require high starting currents, or intermittent loose connections in power wiring. Depending on the fault location and the system conditions, the fault can cause either temporary voltage drops (sags), or voltage rises (swells), or a complete loss of voltage (interruptions). The fault condition can be close to or remote from the point of interest. In either case, the impact on the voltage during the actual fault condition is of short duration variation until protective devices operate to clear the fault.

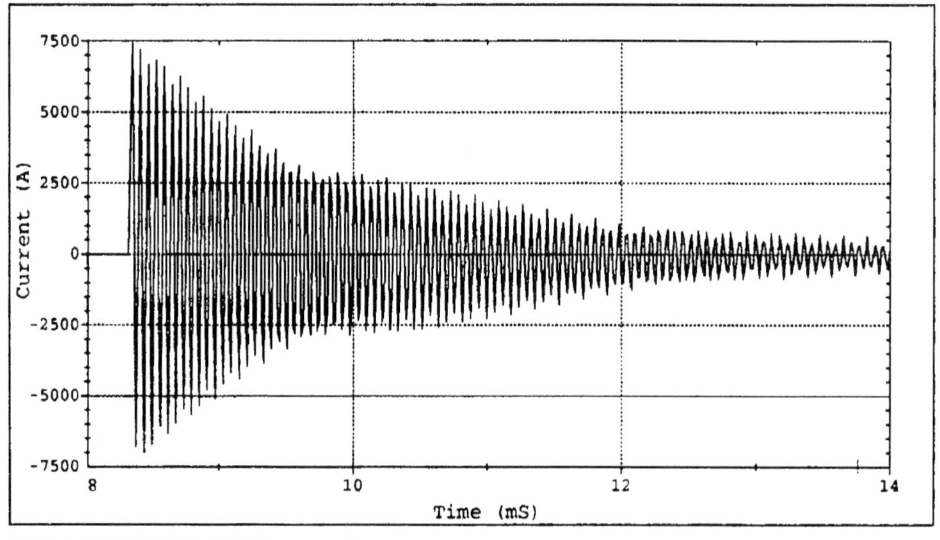


**FIGURE 22-2** Oscillatory transient current caused by back-to-back capacitor switching.

This category encompasses the IEC category of voltage dips and short interruptions. Each type of variation can be designated as instantaneous, momentary, or temporary, depending on its duration as defined in Table 22-2.

***Interruption.*** An interruption occurs when the supply voltage or load current decreases to less than 0.1 pu for a period of time not exceeding 1 min. Interruptions can be the result of power system faults, equipment failures, and control malfunctions. The interruptions are measured by their duration since the voltage magnitude is always less than 10% of nominal. The duration of an interruption due to a fault on the utility system is determined by the operating time of utility protective devices. Instantaneous reclosing generally will limit the interruption caused by a nonpermanent fault to less than 30 cycles. Delayed reclosing of the protective device may cause an instantaneous, momentary, or temporary interruption. The duration of an interruption can be irregular due to equipment malfunction or loose connections. Some interruptions may be preceded by a voltage sag when the interruptions are due to clearing faults on the source system. The voltage sag occurs between the time a fault initiates and the protective device operates. Figure 22-3 shows a plot of the rms voltages for all three phases for such an interruption. The voltage on the faulted phase initially sags to 15% to 25% for 0.6 s while the fault is arcing. A voltage swell occurs on the other two phases at the same time. The breaker then opens, clears the fault, and recloses successfully 0.4 s later. Utility distribution engineers frequently refer to this as an instantaneous reclose.

***Voltage Sags.*** A sag is a decrease to between 0.1 and 0.9 pu in rms voltage or current at the power frequency for durations from 0.5 cycles to 1 min. The IEC definition for this phenomenon is *voltage dip*. The two terms are considered interchangeable, with sag being the preferred synonym in the U.S. power quality community. Figure 22-4 shows a typical voltage sag associated with an SLG fault on another feeder from the same substation. The voltage sags to 60% for about 5 cycles until the substation breaker is able to interrupt the fault current. Typical fault clearing times range from 3 to 30 cycles, depending on the fault current magnitude and the type of overcurrent protection.

***Voltage Swells.*** A *swell* is defined as an increase to between 1.1 and 1.8 pu in rms voltage or current at the power frequency for durations from 0.5 cycle to 1 min. The term *momentary overvoltage* is used by many writers as a synonym for the term swell. As with sags, swells are usually associated with system fault conditions. One way that a swell can occur is from the temporary voltage rise on the unfaulted phases during a single line-to-ground (SLG) fault. An example is shown in Fig. 22-5. Swells can also be caused by switching off a large load or energizing a large capacitor bank.

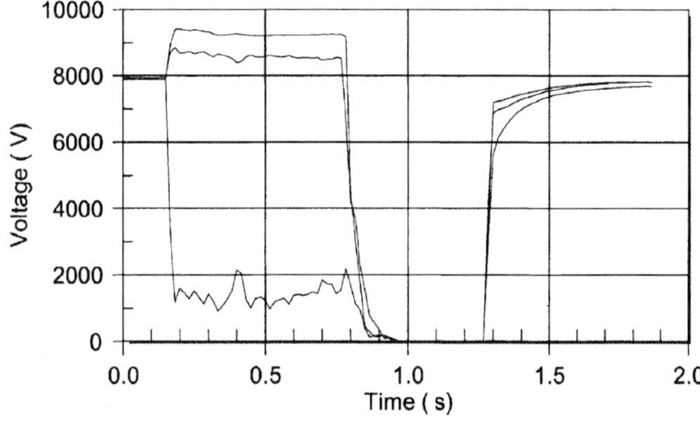


**FIGURE 22-3** An instantaneous interruption due to an SLG fault and subsequent recloser operation.

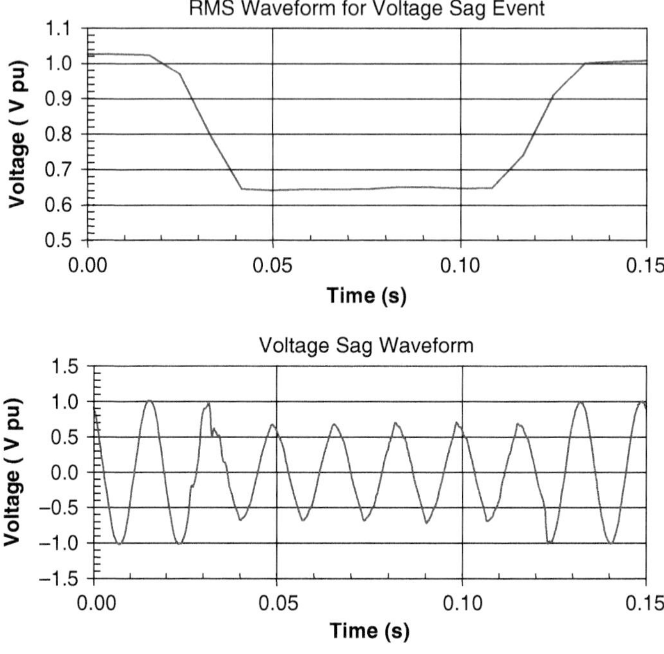


**FIGURE 22-4** Voltage sag caused by an SLG fault.

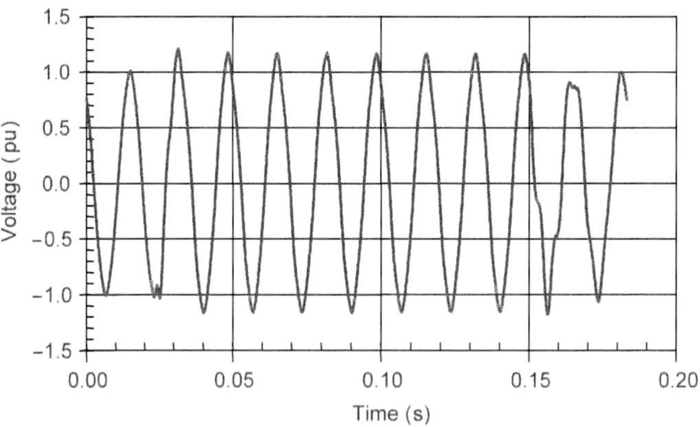


**FIGURE 22-5** An 8-cycle voltage swell caused by an SLG fault.

### 22.2.5 Long-Duration Voltage Variations

Long-duration voltage variations encompass rms deviations at power frequencies for longer than 1 min. ANSI C84.1 specifies the steady-state voltage tolerances expected on a power system. A voltage variation is considered to be long duration when the ANSI limits are exceeded for greater than 1 min. Long-duration variations can be either overvoltages or undervoltages. Overvoltages and undervoltages generally are not the result of system faults, but are caused by load variations on the system and system switching operations. Such variations are typically displayed as plots of rms voltage versus time.

***Overvoltage.*** An *overvoltage* is an increase in the rms ac voltage greater than 110% at the power frequency for a duration longer than 1 min. They are usually the result of load switching (e.g., switching off a large load or energizing a capacitor bank). The overvoltages result because the system is either too weak for the desired voltage regulation or voltage controls are inadequate. Incorrect tap settings on transformers can also result in system overvoltages.

***Undervoltage.*** An *undervoltage* is a decrease in the rms ac voltage to less than 90% at the power frequency for a duration longer than 1 min. They are the result of the events that are the reverse of the events that cause overvoltages. A load switching on or a capacitor bank switching off can cause an undervoltage until voltage regulation equipment on the system can bring the voltage back to within tolerances. Overloaded circuits can result in undervoltages also. The term *brownout* is often used to describe sustained periods of undervoltage initiated as a specific utility dispatch strategy to reduce power demand. Because there is no formal definition for brownout, and it is not as clear as the term undervoltage when trying to characterize a disturbance, the term brownout should be avoided.

### 22.2.6 Sustained Interruption

When the supply voltage has been zero for a period of time in excess of 1 min, the long duration voltage variation is considered a *sustained interruption*. Voltage interruptions longer than 1 min are often permanent and require human intervention to repair the system for restoration. The term sustained interruption refers to specific power system phenomena and, in general, has no relation to the usage of the term *outage*. Utilities use outage or interruption to describe phenomena of similar nature for reliability reporting purposes. However, this causes confusion for end users who think of an outage as any interruption of power that shuts down a process. This could be as little as one-half of a cycle. Outage, as defined in IEEE Std 100 [4], does not refer to a specific phenomenon, but rather to the state of a component in a system that has failed to function as expected. Also, use of the term *interruption* in the context of power quality monitoring has no relation to reliability or other continuity of service statistics. Thus, this term has been defined to be more specific regarding the absence of voltage for long periods.

### 22.2.7 Voltage Imbalance

*Voltage imbalance* (also called voltage unbalance) is sometimes defined as the maximum deviation from the average of the 3-phase voltages or currents, divided by the average of the 3-phase voltages or currents, expressed in percent. Unbalance is more rigorously defined in the standards [3–6] using symmetrical components. The ratio of either the negative or zero sequence component to the positive sequence component can be used to specify the percent unbalance. The most recent standard [5] specifies that the negative sequence method be used. Figure 22-6 shows an example of these two ratios for a 1 week trend of imbalance on a residential feeder.

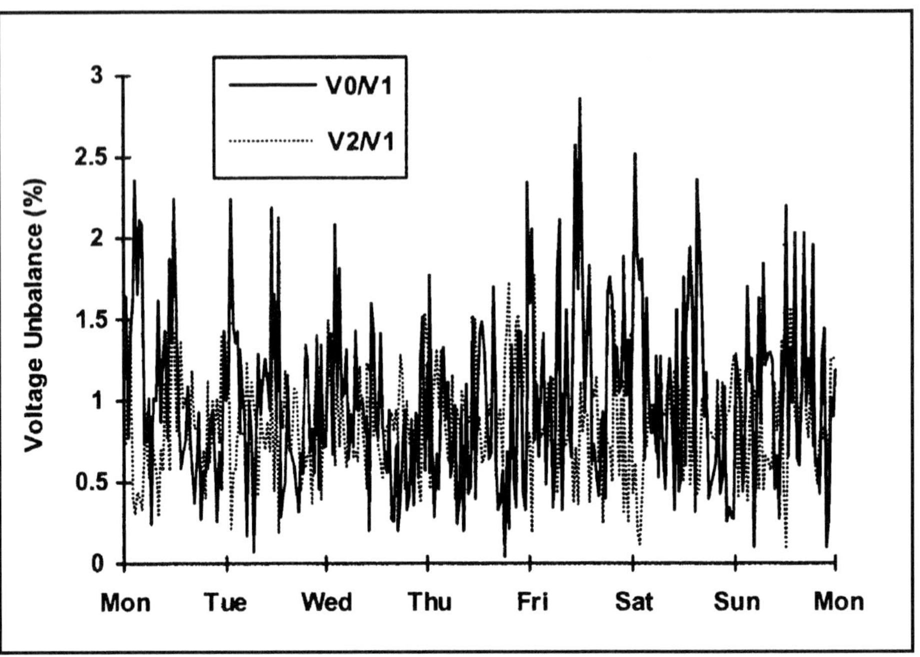


**FIGURE 22-6** Voltage unbalance trend for a residential feeder.

### 22.2.8 Waveform Distortion

*Waveform distortion* is defined as a steady-state deviation from an ideal sine wave of power frequency principally characterized by the spectral content of the deviation. There are five primary types of waveform distortion: dc offset, harmonics, interharmonics, notching, and noise.

***DC Offset.*** The presence of a dc voltage or current in an ac power system is termed *dc offset*. This can occur as the result of a geomagnetic disturbance or asymmetry of electronic power converters. Incandescent lightbulb life extenders, for example, may consist of diodes that reduce the rms voltage supplied to the lightbulb by half-wave rectification. Direct current in alternating current networks can have a detrimental effect by biasing transformer cores so they saturate in normal operation. This causes additional heating and loss of transformer life. DC may also cause the electrolytic erosion of grounding electrodes and other connectors.

***Harmonics and Interharmonics.*** *Harmonics* are sinusoidal voltages or currents having frequencies that are integer multiples of the frequency at which the supply system is designed to operate (termed the fundamental frequency, usually 50 or 60 Hz) [3]. Periodically distorted waveforms can be decomposed into a sum of the fundamental frequency and the harmonics. Harmonic distortion originates in the nonlinear characteristics of devices and loads on the power system. Figure 22-7 illustrates the waveform and harmonic spectrum for a typical adjustable speed drive input current.

Voltages or currents having frequency components that are not integer multiples of the frequency at which the supply system is designed to operate (e.g., 50 or 60 Hz) are called *interharmonics*. They can appear as discrete frequencies or as a wideband spectrum. Interharmonics can be found in networks of all voltage classes. The main sources of interharmonic waveform distortion are static

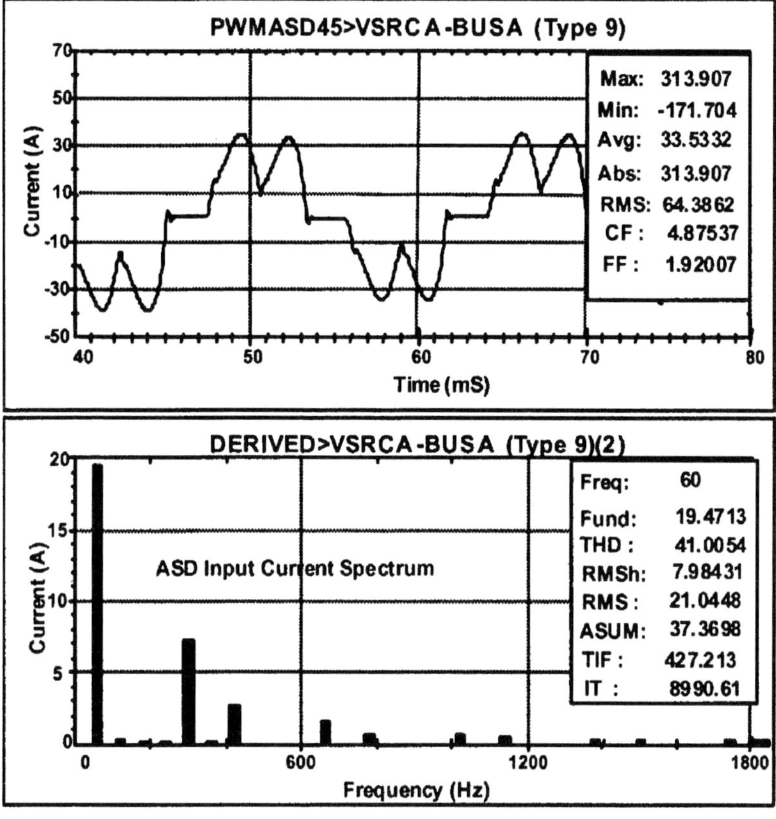


**FIGURE 22-7** Current waveform and harmonic spectrum for an ASD input current.

frequency converters, cycloconverters, induction furnaces, and arcing devices. Power line carrier signals can also be considered as interharmonics.

***Notching.*** *Notching* is a periodic voltage disturbance caused by the normal operation of power electronics devices when current is commutated from one phase to another. Since notching occurs continuously, it can be characterized through the harmonic spectrum of the affected voltage. However, it is generally treated as a special case. The frequency components associated with notching can be quite high and may not be readily characterized with measurement equipment normally used for harmonic analysis. Figure 22-8 shows an example of voltage notching from a 3-phase converter that produces continuous dc current. The notches occur when the current commutates from one phase to another. During this period, there is a momentary short circuit between two phases pulling the voltage as close to zero as permitted by system impedances.

***Noise.*** *Noise* is defined as unwanted electrical signals with broadband spectral content lower than 200 kHz superimposed upon the power system voltage or current in phase conductors, or found on neutral conductors or signal lines. Noise in power systems can be caused by power electronic devices, control circuits, arcing equipment, loads with solid-state rectifiers, and switching power supplies. Noise problems are often exacerbated by improper grounding that fails to conduct noise away from the power system. In principle, noise consists of any unwanted distortion of the power signal that

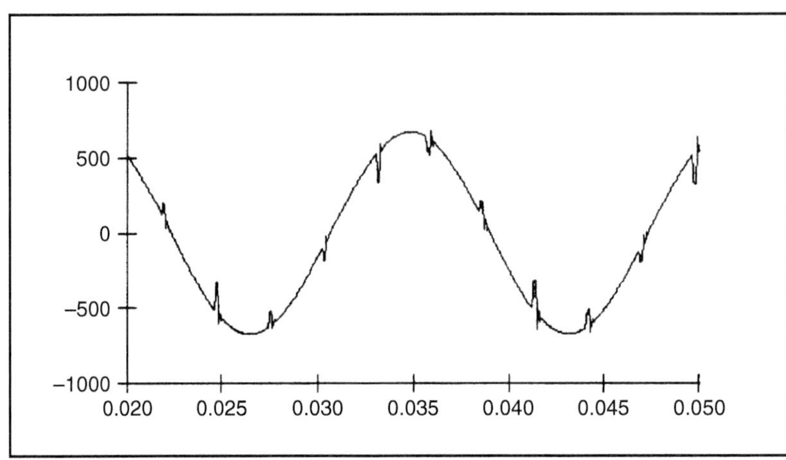


**FIGURE 22-8** Example of voltage notching caused by a 3-phase converter.

cannot be classified as harmonic distortion or transients. Noise disturbs electronic devices such as microcomputer and programmable controllers. The problem can be mitigated by using filters, isolation transformers, and line conditioners.

### 22.2.9 Voltage Fluctuation

*Voltage fluctuations* are systematic variations of the voltage envelope or a series of random voltage changes, the magnitude of which does not normally exceed the voltage ranges specified by ANSI C84.1 of 0.9 to 1.1 pu. IEC 61000-2-1 defines various types of voltage fluctuations. We will restrict our discussion here to IEC 61000-2-1 Type (d) voltage fluctuations, which are characterized as a series of random or continuous voltage fluctuations. Loads that can exhibit continuous, rapid variations in the load current magnitude can cause voltage variations that are often referred to as flicker. The term *flicker* is derived from the impact of the voltage fluctuation on lamps such that they are perceived to flicker by the human eye. To be technically correct, voltage fluctuation is an electromagnetic phenomenon while flicker is an undesirable result of the voltage fluctuation in some loads. However, the two terms are often linked together in standards. Therefore, we will also use the common term voltage flicker to describe such voltage fluctuations. Figure 22-9 illustrates

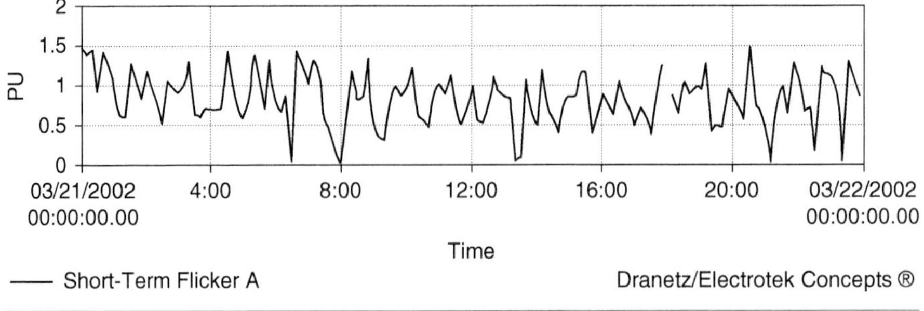


**FIGURE 22-9** Flicker (Pst) at 161-kV substation bus measured according to IEC 61000-4-15.

a voltage waveform which produces flicker. This is caused by an arc furnace, one of the most common causes of voltage fluctuations on utility transmission and distribution systems.

### 22.2.10 Power Frequency Variations

*Power frequency variations* are defined as the deviation of the power system fundamental frequency from its specified nominal value (e.g., 50 or 60 Hz). The power system frequency is directly related to the rotational speed of the generators supplying the system. There are slight variations in frequency as the dynamic balance between load and generation changes. The size of the frequency shift and its duration depends on the load characteristics and the response of the generation control system to load changes. Figure 22-10 illustrates frequency variations for a 24-h period on a typical 13-kV substation bus. Frequency variations that go outside of accepted limits for normal steady-state operation of the power system can be caused by faults on the bulk power transmission system, a large block of load being disconnected, or a large source of generation going offline.

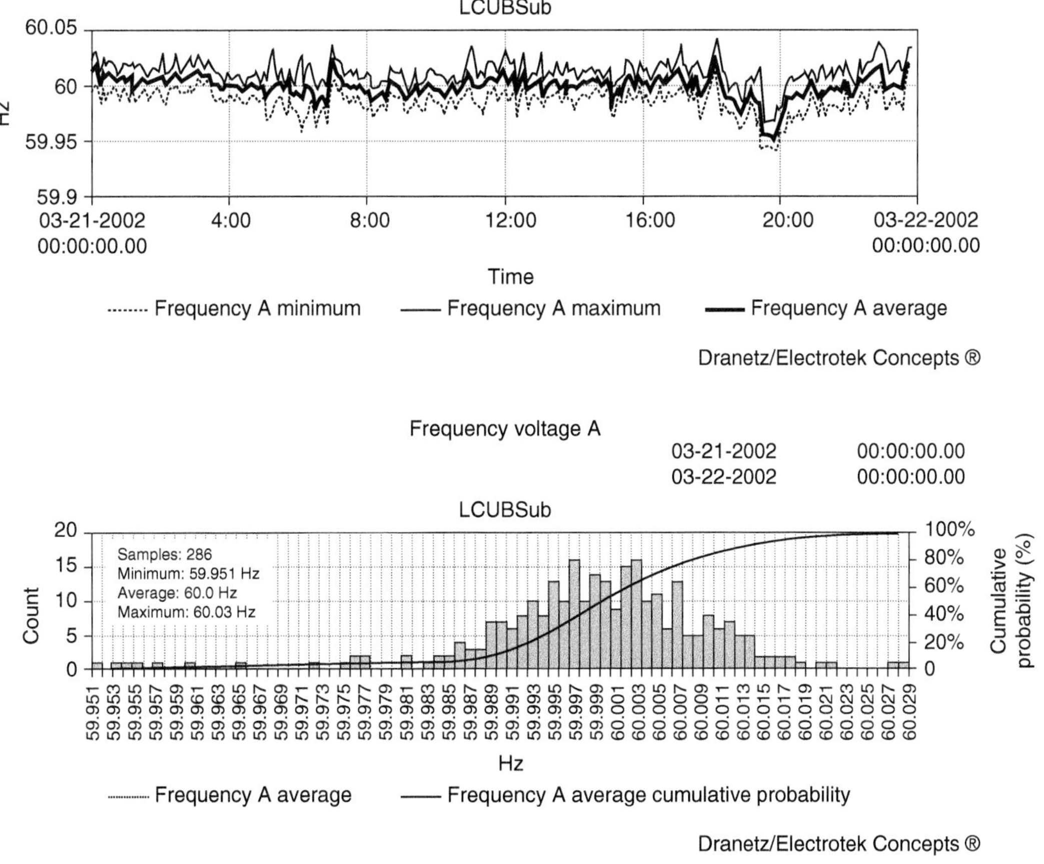


**FIGURE 22-10** Power frequency trend and statistical distribution at 13-kV substation bus.

# 22.3 VOLTAGE SAGS AND INTERRUPTIONS ON POWER SYSTEMS

## 22.3.1 Characteristics

Voltage sags and interruptions are related power quality problems. They are the result of faults in the power system and switching actions to isolate the faulted sections. A voltage sag is characterized by a short duration (typically 0.5 to 30 cycles) reduction in rms voltage caused by faults on the power system and the starting of large loads, such as motors. Momentary interruptions (typically no more than 2 to 5 s) cause a complete loss of voltage and are a common result of the actions taken by utilities to clear transient faults on their systems. Sustained interruptions of longer than 1 min are generally due to permanent faults. Due to the nature of the interconnected power systems and the utility fault-clearing schemes, voltage sags are the most common power quality disturbances.

## 22.3.2 Sources of Sags and Interruptions

Voltage sags and interruptions are generally caused by faults (short circuits) on the utility system and subsequent operations of protective devices in isolating the faults. Transient or temporary faults on the same or parallel feeders can result in voltage sags. Permanent faults usually result in interruptions. It is also possible that voltage sags are the result of starting of large loads, such as large motors. In some rare circumstances, energizing a transformer in a weak power system can also result in voltage sags. The voltage sag and interruption performance is greatly influenced by the utility feeder design and fault-clearing practices.

## 22.3.3 Utility System Fault Clearing

A radial distribution system is designed so that only one fault interrupter must operate to clear a fault. For permanent faults, that same device, or another, operates to *sectionalize* the feeder. That is, the faulted section is isolated so that power may be restored to the rest of the loads served from the sound sections. Orchestrating this process is referred to as the *coordination* of the overcurrent protection devices. While this is simple in concept, some of the behaviors of the devices involved can be quite complex. What is remarkable about this is that nearly all of the process is performed automatically by autonomous devices employing only local intelligence.

Overcurrent protection devices appear in series along a feeder. For permanent fault coordination, the devices operate progressively slower as one moves from the ends of the feeders toward the substation. This helps ensure the proper sectionalizing of the feeder so that only the faulted section is isolated. However, this principle is often violated for temporary faults, particularly if *fuse saving* is employed. The typical hierarchy of overcurrent protection devices on a feeder is

***Feeder Breaker in the Substation.*** This is a circuit breaker capable of interrupting typically 40 kA of current and controlled by separate relays. When the available fault current is less than 20 kA, it is common to find reclosers used in this application.

***Line Reclosers Mounted on Poles at Midfeeder.*** The simplest are self-contained with hydraulically-operated timing, interrupting, and reclosing mechanisms. Others have separate electronic controls.

***Fuses on Many Lateral Taps Off the Main Feeder.*** These protective devices have significant implications on power quality issues.

### 22.3.4 Reclosers

*Reclosers* are special circuit breakers designed to perform interruption and reclosing on temporary faults. They can reclose two or three times if needed in rapid succession. The multiple operations are designed to permit various sectionalizing schemes to operate and to give some more persistent transient faults a second chance to clear. The majority of faults will be cleared on the first operation. These devices can be found in numerous places along distribution feeders and sometimes in substations. They are typically applied at the head of sections subjected to numerous temporary faults. However, they may be applied nearly anywhere a convenient, low-cost primary-side circuit breaker is needed. Figure 22-11 shows a typical pole-mounted line recloser.

In addition to performing interruption and reclosing on temporary faults, reclosers are used for *fuse-saving or fast-tripping* applications. They are some of the fastest mechanical fault interrupters employed on the utility system. While they are typically rated for no faster than 3 to 6 cycles, many examples of interruptions as short as 1.5 cycles have been observed with power quality monitors. This can be beneficial to limiting sag durations. Where fast tripping is not employed, the recloser control will commonly delay operation to more than 6 cycles to allow time for downline fuses to clear.

**FIGURE 22-11** Typical standard 3-phase oil-insulated line recloser with vacuum interrupters. (*Photo courtesy of Cooper Power Systems.*)

### 22.3.5 Reclosing Sequence

Reclosing is quite prevalent in North American utility systems. Utilities in regions of low lightning incidence may reclose only once because they assume that the majority of their faults will be permanent. In lightning-prone regions, it is common to attempt to clear the fault as many as four times. Figure 22-12 illustrates the two most common sequences in use on 4-shot reclosers:

1-fast operation, 3-delayed

2-fast, 2-delayed

Reclosers tend to have uniform reclose intervals between operations. The original hydraulic reclosers were limited to about 1 to 2 s and this setting has been retained by many utilities, although modern electronically controlled reclosers can be set for any value. It is common for the first reclose interval on some types of reclosers to be set for *instantaneous reclose*, which will result in closure in 12 to 30 cycles (0.2 to 0.5 s). This is done to reduce the time of the interruption and improve the power quality. However, there are some conflicts created by this, such as with distributed generation disconnecting times.

Substation circuit breakers often have a different style of reclosing sequence as shown in Fig. 22-13. This stems from a different evolution of relaying technology. Reclosing times are counted from the first tripping signal of the first operation. Thus, the common “0-15-45” operating sequence recloses essentially as fast as possible on the first operation, with approximately 15 and 30 s intervals between the next two operations.

Although the terminology may differ, modern breakers and reclosers can both be set to have the same operating sequences to meet load power quality requirements. Utilities generally choose one technology over the other based on cost or construction standards. It is generally fruitless to automatically reclose in distribution systems that are predominantly underground distribution cable, unless there is a significant portion that is overhead and exposed to trees or lightning.

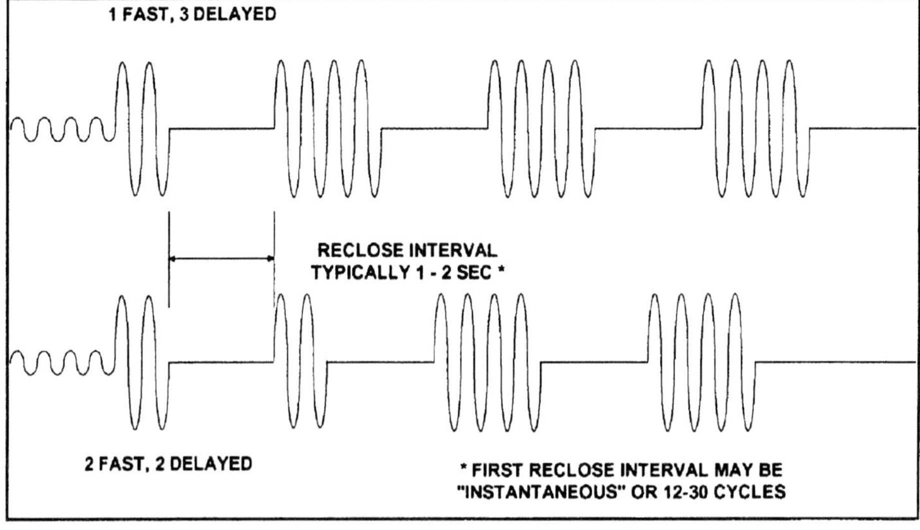


**FIGURE 22-12** Common reclosing sequences for line reclosers in use in the United States.

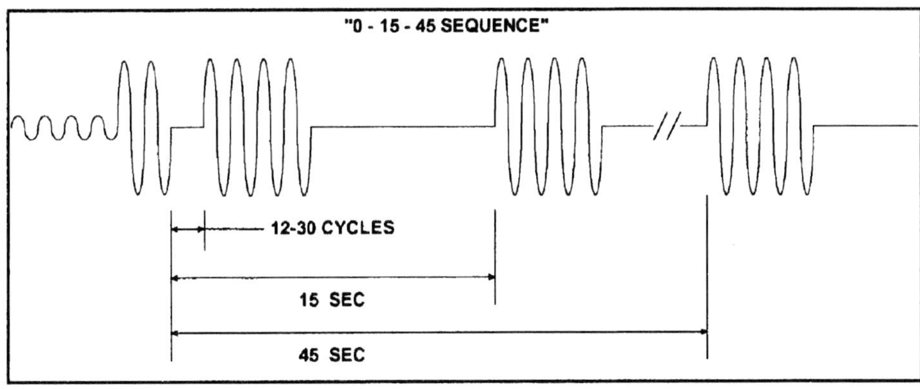


**FIGURE 22-13** A common reclosing sequence for substation breakers in the United States.

### 22.3.6 Fuse Saving or Fast Tripping

Ideally, utility engineers would like to avoid blowing fuses needlessly on transient faults because a line crew must be dispatched to change it. Line reclosers were designed specifically to help save fuses. Substation circuit breakers can use instantaneous ground relaying to accomplish the same objective. The basic idea is to have the mechanical circuit interrupting device operate very quickly on the first operation so that it clears before any fuses downline from it have a chance to melt. When the device closes back in, power is fully restored in the majority of the cases and no human intervention is required. The only inconvenience to the customer is a slight blink. This is called the fast operation of the device, or the instantaneous trip. If the fault is still there upon reclosing, there are two options in common usage:

1. *Switch to a Slow, or Delayed, Tripping Characteristic.* This is frequently the only option for substation circuit breakers; they will operate only one time on the instantaneous trip. This philosophy assumes that the fault is now permanent and switching to a delayed operation will give a downline fuse time to operate and clear the fault by isolating the faulted section.
2. *Try a Second Fast Operation.* This philosophy is used where experience has shown a significant percentage of transient faults need two chances to clear while saving the fuses. Some line constructions and voltage levels have a greater likelihood that a lightning-induced arc may reignite and need a second chance to clear. Also, a certain percentage of tree faults will burn free if given a second shot.

Many utilities have abandoned fuse saving in selected areas due to complaints about power quality. The fast, or instantaneous, trip is eliminated so that breakers and reclosers have only time-delayed operations.

### 22.3.7 Fault-Induced Voltage Sags

The majority of voltage sags are caused by faults on the power systems and the subsequent operations of protective devices. Consider a customer that is supplied from the feeder supplied by circuit breaker no. 1 on the diagram shown in Fig. 22-14. If there is a fault on the same feeder, the customer will experience a voltage sag during the fault followed by an interruption when the breaker opens to clear the fault. If the fault is temporary in nature, a reclosing operation on the breaker should be successful and the interruption will only be temporary. It will usually require about 5 or 6 cycles for the breaker to operate, during which time a voltage sag occurs. The breaker will remain open for typically a minimum of 12 cycles up to 5 s depending on utility reclosing practices. Sensitive equipment will almost surely trip during this interruption.

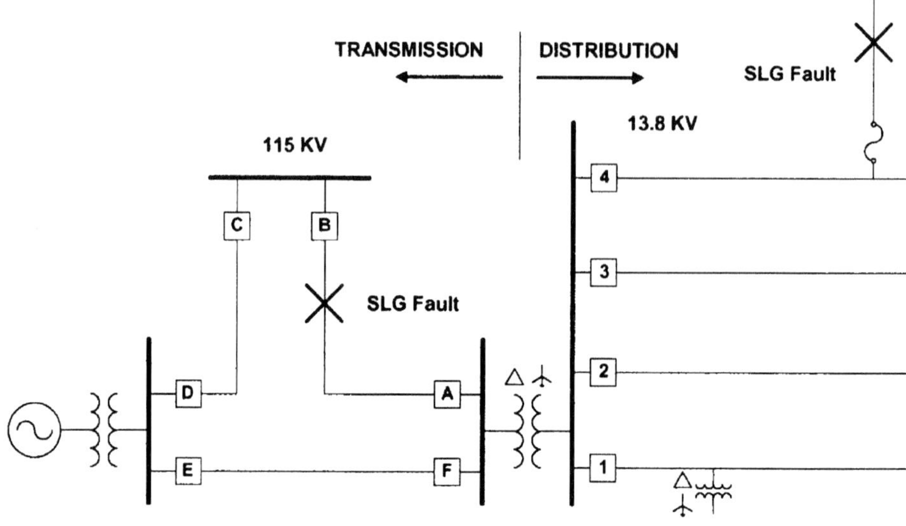


**FIGURE 22-14** Fault locations on the utility power system.

A much more common event would be a fault on one of the other feeders from the substation, that is, a fault is on a parallel feeder, or a fault somewhere on the transmission system (see the fault locations shown on Fig. 22-14). In either of these cases, the customer will experience a voltage sag during the period that the fault is actually on the system. As soon as breakers open to clear the fault, normal voltage will be restored at the customer. Note that to clear the fault shown on the transmission system, both breakers A and B must operate. Transmission breakers will typically clear a fault in 5 or 6 cycles. In this case there are two lines supplying the distribution substation and only one has a fault. Therefore, customers supplied from the substation should expect to see only a sag and not an interruption. The distribution fault on feeder no. 4 may be cleared either by the lateral fuse or the breaker, depending on the utility fuse saving practice. Any of these fault locations can cause equipment to misoperate in customer facilities. The relative importance of faults on the transmission system and the distribution system will depend on the specific characteristics of the systems (underground vs. overhead distribution, lightning flash densities, overhead exposure, etc.) and the sensitivity of the equipment to voltage sags.

***Example of Voltage Sags due to a Fault on a Parallel Feeder.*** This example illustrates voltage sag and momentary interruption events due to a temporary fault on a utility feeder. Figures 22-15 and 22-16 show an interesting utility fault event recorded for an Electric Power Research Institute research project [7, 8] by 8010 PQNode instruments[a] at two locations in the power system. The top chart in each of the figures is the rms voltage variation with time and the bottom chart is the first 175 ms of the actual waveform. Figure 22-15 shows the characteristic measured at a customer location on an unfaulted part of the feeder. Figure 22-16 shows the momentary interruption (actually two separate interruptions) observed downline from the fault. The interrupting device in this case was a line recloser that was able to interrupt the fault very quickly in about 2.5 cycles. This device can have a variety of settings. In this case, it was set for two fast operations and two delayed operations. Figure 22-15 shows only the brief sag to 65% voltage for the first fast operation. There was an identical sag for the second operation. While this is very brief sag that is virtually unnoticeable by observing lighting blinks, many industrial processes would have shut down.

[a]PQNode is a registered trademark of Dranetz, Edison, NJ.

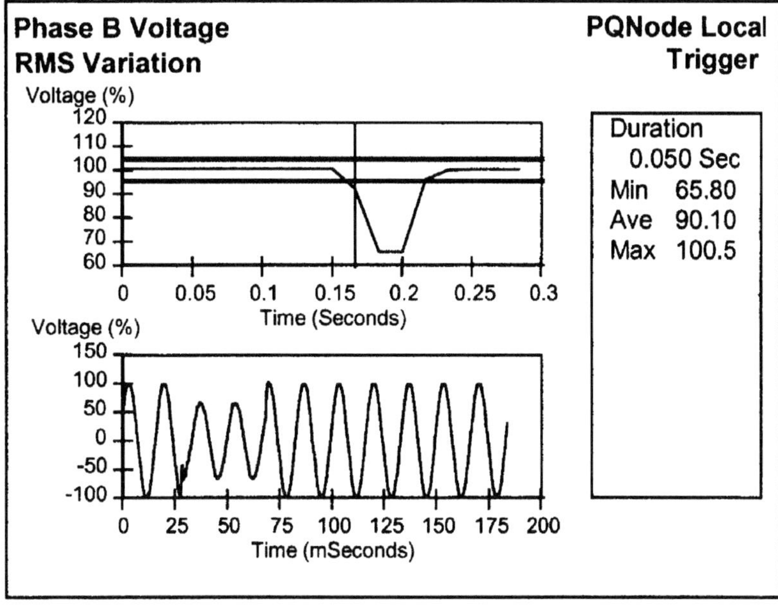


**FIGURE 22-15** Voltage sag due to short-circuit fault on a parallel utility feeder.

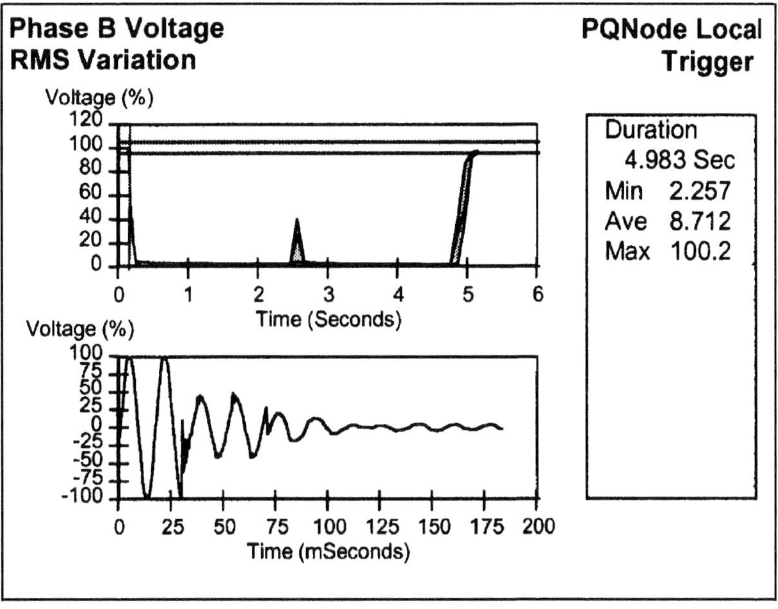


**FIGURE 22-16** Utility short-circuit fault event with two fast trip operations of utility line.

Figure 22-16 clearly shows the voltage sag prior to fault clearing and the subsequent two fast recloser operations. The reclose time (the time the recloser was open) was a little more than 2 s, a very common time for a utility line recloser. Apparently, the fault—perhaps, a tree branch—was not cleared completely by the first operation, forcing a second. The system was restored after the second operation.

### 22.3.8 Motor Starting Sags

Motors have the undesirable effect of drawing several times their full load current while starting. This large current will, by flowing through system impedances, cause a voltage sag which may dim lights, cause contactors to drop out, and disrupt sensitive equipment. The situation is made worse by an extremely poor starting displacement factor—usually in the range of 15% to 30%. The time required for the motor to accelerate to rated speed increases with the magnitude of the sag, and an excessive sag may prevent the motor from starting successfully. Motor starting sags can persist for many seconds, as illustrated in Fig. 22-17.

### 22.3.9 Motor Starting Methods

Energizing the motor in a single step (*full voltage starting*) provides low cost and allows the most rapid acceleration. It is the preferred method unless the resulting voltage sag or mechanical stress is excessive.

*Autotransformer starters* have two autotransformers connected in open delta. Taps provide a motor voltage of 80%, 65%, or 50% of system voltage during start-up. Line current and starting torque vary with the square of the voltage applied to the motor, so the 50% tap will deliver only 25% of the full voltage starting current and torque. The lowest tap which will supply the required starting torque is selected.

*Resistance and reactance starters* initially insert an impedance in series with the motor. After a time delay, this impedance is shorted out. Starting resistors may be shorted out over several steps; starting reactors are shorted out in a single step. Line current and starting torque vary directly with the voltage applied to the motor, so for a given starting voltage, these starters draw more current from

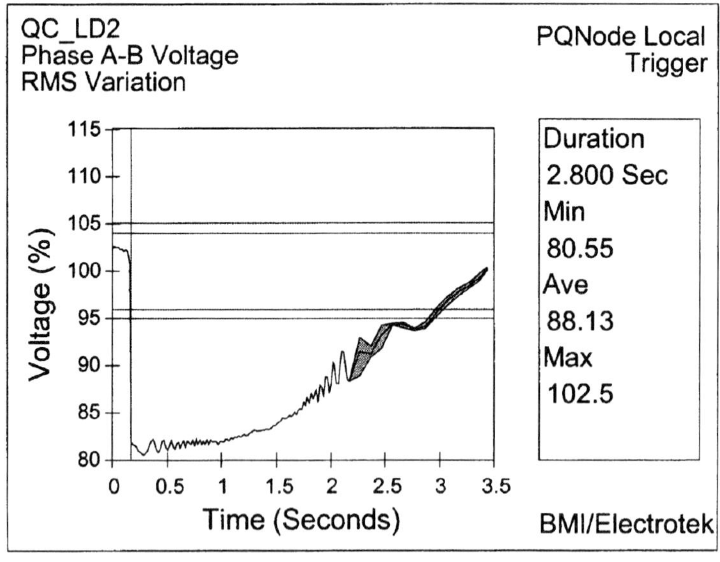


**FIGURE 22-17** Typical motor starting voltage sag.

the line than with autotransformer starters, but provide higher starting torque. Reactors are typically provided with 50%, 45%, and 37.5% taps.

*Part winding starters* are attractive for use with dual-rated motors (220/440 or 230/460 V). The stator of a dual-rated motor consists of two windings connected in parallel at the lower voltage rating, or in series at the higher voltage rating. When operated with a part winding starter at the lower voltage rating, only one winding is energized initially, limiting starting current and starting torque to 50% of the values seen when both windings are energized simultaneously.

*Wye-Delta* starters connect the stator in wye for starting, then after a time delay, reconnect the windings in delta. The wye connecting reduces the starting voltage to 57% of the system line-line voltage; starting current and starting torque are reduced to 33% of their values for full voltage start.

### 22.3.10 Estimating the Sag Severity during Full Voltage Starting

As shown in Fig. 22-17, starting an induction motor results in a steep dip in voltage, followed by a gradual recovery. If full voltage starting is used, the sag voltage, in per unit of nominal system voltage, is

$$V_{\min}(\text{pu}) = \frac{V(\text{pu}) \cdot \text{kVA}_{\text{SC}}}{\text{kVA}_{\text{LR}} + \text{kVA}_{\text{SC}}} \tag{22-1}$$

where $V(\text{pu})$ = actual system voltage, in per unit of nominal
$\text{kVA}_{\text{LR}}$ = motor locked rotor kVA
kVASC = system short circuit kVA at the motor

Figure 22-18 illustrates the results of this computation for sag to 90% of nominal voltage, using typical system impedances and motor characteristics.

If the result is above the minimum allowable steady-state voltage for the effected equipment, then the full voltage starting is acceptable. If not, then the sag magnitude versus duration characteristic must be compared to the voltage tolerance envelope of the effected equipment. The required calculations are fairly complicated, and best left to a motor starting or general transient analysis computer program.

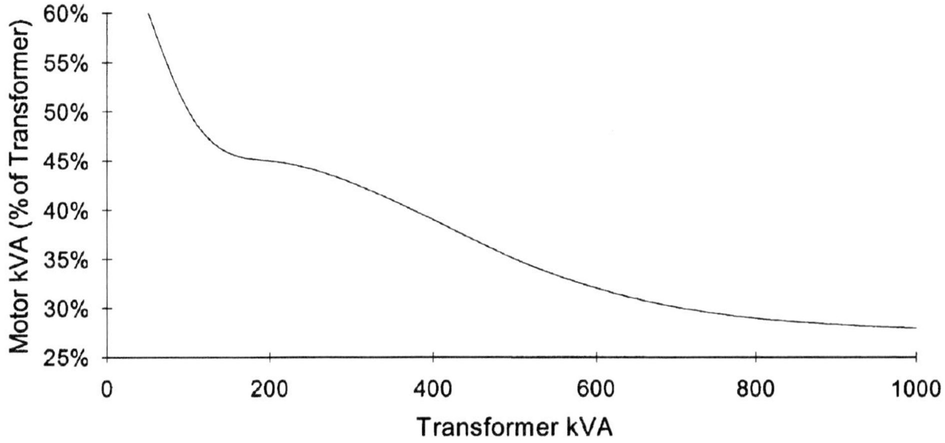


**FIGURE 22-18** Typical motor vs. transformer size for full voltage starting sags of 90%.

## 22.4 ELECTRICAL TRANSIENT PHENOMENA

### 22.4.1 Sources and Characteristics

In principle, electrical transient phenomena can be generated due to natural events such as lightning strokes, and switching operations such as capacitor, load, and transformer energizing, and protective device operations. However, two main sources of transient overvoltages on utility systems are capacitor switching and lightning.

### 22.4.2 Capacitor Switching Transient Overvoltages

*Capacitor switching* is one of the most common switching events on utility systems. Capacitors are used to provide reactive power (vars) to correct the power factor, which reduces losses and supports the voltage on the system. One drawback to capacitors is that they yield oscillatory transients when switched. Some capacitors are energized all the time (a fixed bank) while others are switched according to load levels. Various control means are used to determine when they are switched including time, temperature, voltage, current, and reactive power. It is common for controls to combine two or more of these functions, such as temperature with voltage override.

Figure 22-19 shows the one-line diagram of a typical utility feeder capacitor switching situation. When the switch is closed, a transient similar to the one in Fig. 22-20 may be observed upline from the capacitor at the monitor location. In this particular case, the capacitor switch contacts close at a point near the system voltage peak. This is common for many types of switches because the insulation across the switch contacts tends to break down when the voltage across the switch is at a maximum value. The voltage across the capacitor at this instant is zero. Since the capacitor voltage cannot change instantaneously, the system voltage at the capacitor location is briefly pulled down to zero and rises as the capacitor begins to charge toward the system voltage. Because the power system source is inductive, the capacitor voltage overshoots and rings at the natural frequency of the system. At the monitoring location shown, the initial change in voltage will not go completely to zero because of the impedance between the observation point and the switched capacitor. However, the initial drop and subsequent ringing transient that is indicative of a capacitor switching event will be observable to some degree. The overshoot will generate a transient between 1.0 and 2.0 pu depending on system damping. In this case, the transient observed at the monitoring location is about 1.34 pu. Utility capacitor switching transients are commonly in the 1.3 to 1.4 pu range, but have also been observed near the theoretical maximum.

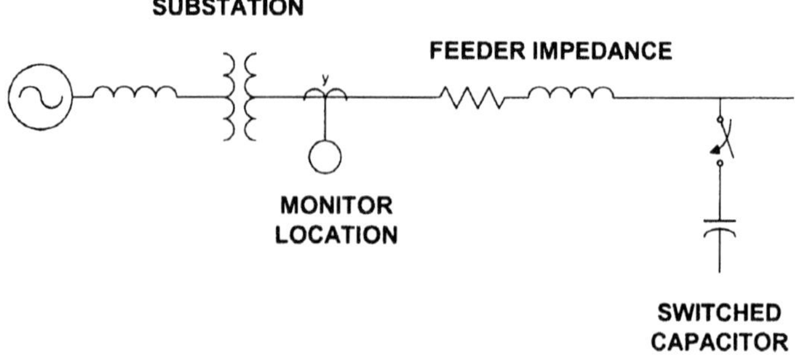


**FIGURE 22-19** One-line diagram of capacitor switching operation.

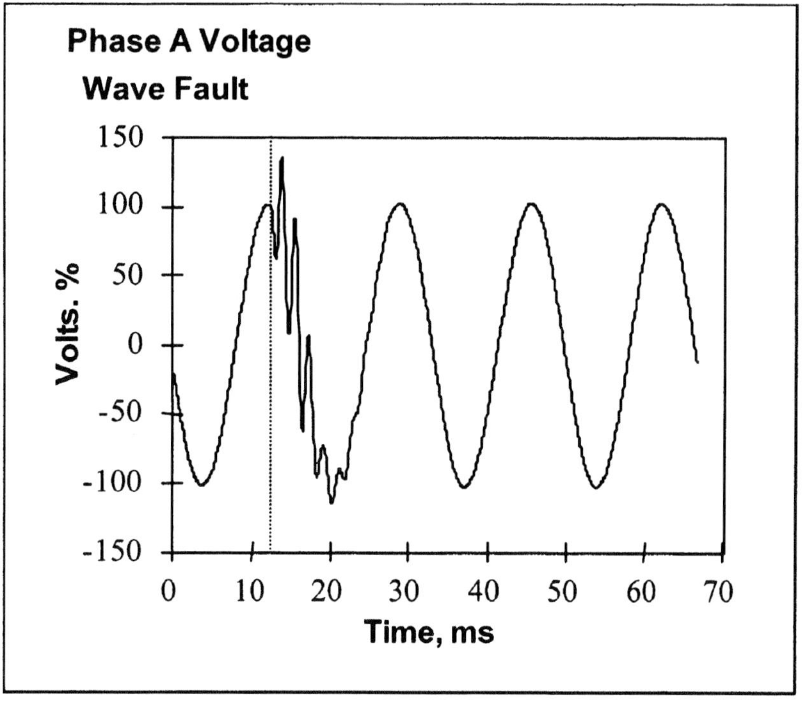


**FIGURE 22-20** Typical utility capacitor switching transient reaching 134% voltage, observed upline from the capacitor.

### 22.4.3 Magnification of Capacitor Switching Transient Overvoltages

Capacitor switching transients can propagate into the local power system and will generally pass through distribution transformers into customer load facilities by nearly the amount related to the turns ratio of the transformer. If there are capacitors on the secondary system, the voltage may actually be magnified on the load side of the transformer if the natural frequencies of the systems are properly aligned. The circuit of concern for this phenomenon is illustrated in Fig. 22-21. Transient overvoltages on the end-user side may reach as high as 3.0 to 4.0 pu on the low-voltage bus under these conditions, with potentially damaging consequences for all types of customer equipment.

### 22.4.4 Options to Limit Magnification

Magnification of utility capacitor switching transients at the end-user location occurs over a wide range of transformer and capacitor sizes. Resizing the customer's power factor correction capacitors or step-down transformer is therefore usually not a practical solution. One solution is to control the transient overvoltage at the utility capacitor. This is sometimes possible using synchronous closing breakers or switches with preinsertion resistors. At the customer location, high-energy surge arresters can be applied to limit the transient voltage magnitude at the customer bus. Energy levels associated with the magnified transient will typically be in the range of 1 kJ. Figure 22-22 shows the expected arrester energy for a range of low-voltage capacitor sizes. High-energy MOV arresters for low-voltage applications can withstand 2 to 4 kJ.

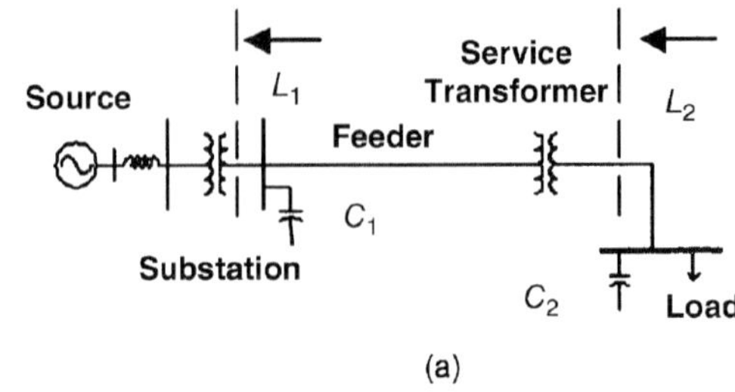


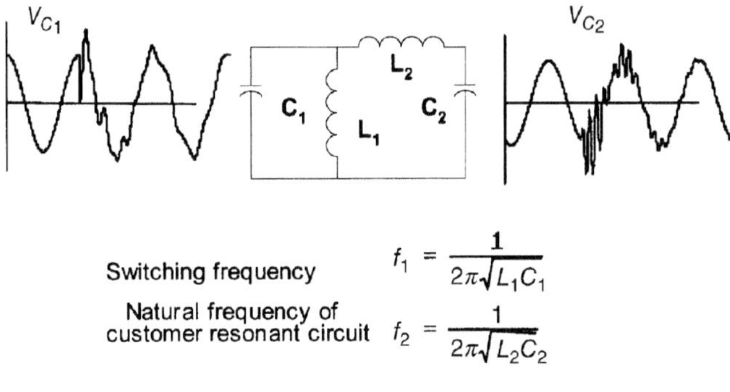


Switching frequency $f_1 = \frac{1}{2\pi\sqrt{L_1C_1}}$

Natural frequency of customer resonant circuit $f_2 = \frac{1}{2\pi\sqrt{L_2C_2}}$

*Voltage magnification* $\Leftrightarrow f_1 \approx f_2$

(b)

**FIGURE 22-21** Voltage magnification of capacitor bank switching. (*a*) Voltage magnification at customer capacitor due to energizing capacitor on utility system; (*b*) equivalent circuit.

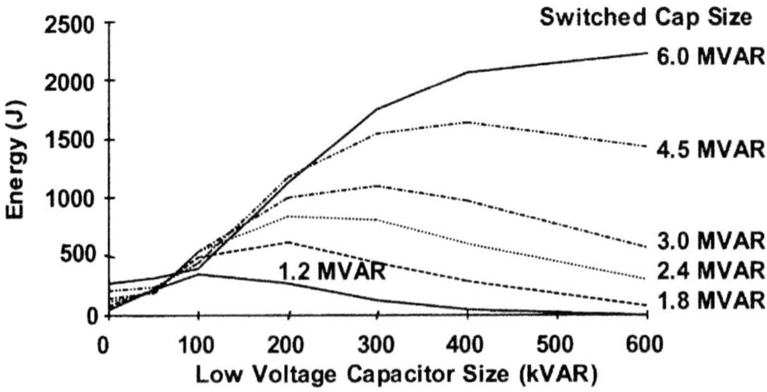


**FIGURE 22-22** Arrester energy duty caused by magnified transient.

While such brief transients up to 2.0 per unit are not generally damaging to the system insulation, it can often cause misoperation of electronic power conversion devices. Controllers may interpret the high voltage as a sign that there is an impending dangerous situation and subsequently disconnect the load to be safe. The transient may also interfere with the gating of thyristors. It is important to note that the arresters can only limit the transient to the arrester protective level. This will typically be approximately 1.8 times the normal peak voltage (1.8 pu).

Another means of limiting the voltage magnification transient is to convert the end-user, power-factor-correction banks to harmonic filters. An inductance in series with the power-factor-correction bank will decrease the transient voltage at the customer bus to acceptable levels. This solution has multiple benefits by providing correction for displacement power factor, controlling harmonic distortion levels within the facility, and limiting the concern for magnified capacitor switching transients.

In many cases, there are only a small number of load devices, such as adjustable-speed motor drives, that are adversely affected by the transient. It is frequently more economical to place line reactors in series with the drives to block the high frequency magnification transient. A 3% reactor is generally effective. While offering only a small impedance to power frequency current, it offers a considerably larger impedance to the transient. Many types of drives have this protection inherently, either through an isolation transformer or a dc bus reactance.

### 22.4.5 Options to Limit Capacitor Switching Transients—Preinsertion

Preinsertion resistors can reduce the capacitor switching transient considerably. The first peak of the transient is usually the most damaging. The idea is to insert a resistor into the circuit briefly so that the first peak is damped significantly. This is old technology, but still quite effective. Figure 22-23 shows one example of a capacitor switch with preinsertion resistors to reduce transients. The preinsertion is accomplished by the movable contacts sliding past the resistor contacts first before mating with the main contacts. This results in a preinsertion time of approximately one-fourth of a cycle at 60 Hz. The effectiveness of the resistors is dependent on capacitor size and available short-circuit current at the capacitor location. Table 22-3 shows expected maximum transient overvoltages upon energization for various conditions, both with and without the preinsertion resistors. These are the maximum values expected; average values are typically 1.3 to 1.4 pu without resistors and 1.1 to 1.2 with resistors.

### 22.4.6 Options to Limit Capacitor Transient Switching—Synchronous Closing

Another popular strategy for reducing transients on capacitor switching is to use a synchronous closing breaker. This is a relatively new technology for controlling capacitor switching transients. Synchronous closing prevents transients by timing the contact closure such that the system voltage closely matches the capacitor voltage at the instant the contacts are made. This avoids the step change in voltage that normally occurs when capacitors are switched, causing the circuit to oscillate. Figure 22-24 shows a vacuum switch made for this purpose. It is applied on 46-kV-class capacitor banks. It consists of three independent poles with separate controls. The timing for synchronous closing is determined by anticipating an upcoming voltage zero. Its success is dependent on the consistent operation of the vacuum switch. The switch reduces capacitor inrush currents by an order of magnitude and voltage transients to about 1.1 pu. A similar switch may also be used at distribution voltages. Each of the switches described here requires a sophisticated microprocessor-based control. Understandably, a synchronous closing system is more expensive than a straightforward capacitor switch. However, it is frequently a cost-effective solution when capacitor switching transients are disrupting end-user loads.

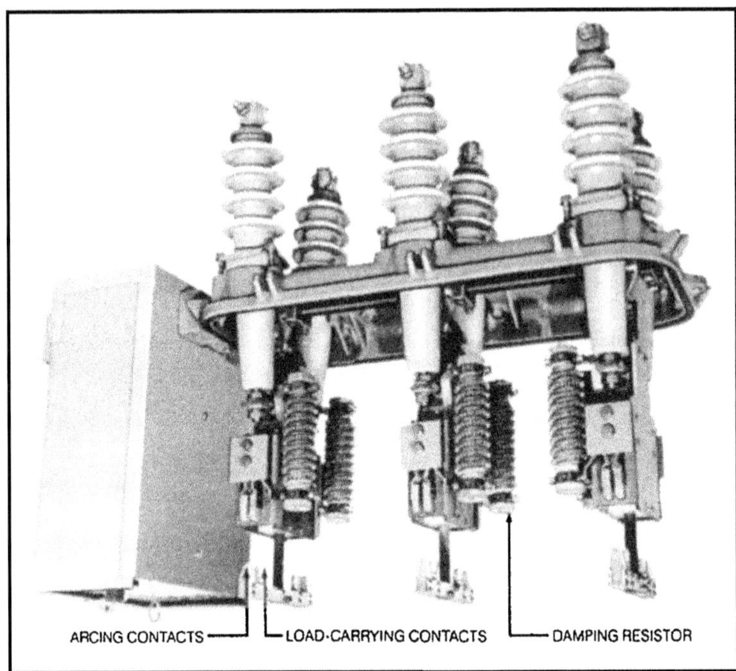


**FIGURE 22-23** Capacitor switch with preinsertion resistors. (*Courtesy of Cooper Power Systems.*)

**TABLE 22-3** Peak Transient Overvoltages Due to Capacitor Switching with and without Preinsertion Resistor

| Size, kvar | Avail. short circuit, kA | Without resistor (pu) | With 6.4 Ω resistor (pu) |
|---|---|---|---|
| 900 | 4 | 1.95 | 1.55 |
| 900 | 9 | 1.97 | 1.45 |
| 900 | 14 | 1.98 | 1.39 |
| 1200 | 4 | 1.94 | 1.50 |
| 1200 | 9 | 1.97 | 1.40 |
| 1200 | 14 | 1.98 | 1.34 |
| 1800 | 4 | 1.92 | 1.42 |
| 1800 | 9 | 1.96 | 1.33 |
| 1800 | 14 | 1.97 | 1.28 |

***Source:*** Cooper Power Systems.

## 22.4.7 Lightning

*Lightning* is a potent source of impulsive transients and can have serious impacts on power system and end-user equipment. Figure 22-25 illustrates some of the places where lightning can strike that results in lightning currents being conducted from the power system into loads. The most obvious conduction path occurs during a direct strike to a phase wire, either on the primary or the secondary side of the transformer. This can generate very high overvoltages, but some analysts question whether this is the most common way that lightning surges enter load facilities and cause damage. Very similar transient overvoltages can be generated by lightning currents flowing along ground conductor paths. Note that there can be numerous paths for lightning currents to enter the grounding system. Common ones, indicated by the dotted lines in Fig. 22-25, include the primary ground, the secondary ground, and the structure of the load facilities. Note also that strokes to the primary phase are conducted to the ground circuits through the arresters on the service transformer. Thus, many more lightning impulses may be observed at loads than one might think. Note that grounds are never perfect conductors, especially for impulses. While most of the surge current may eventually be dissipated into the ground connection closest to the stroke, there will be substantial surge currents flowing in other connected ground conductors in the first few microseconds of the strike.

The chief power quality problems with lightning stroke currents entering the ground system are

**FIGURE 22-24** Synchronous closing capacitor switch. (*Courtesy of Joslyn Hi-Voltage Corporation.*)

- They raise the potential of the local ground above other grounds in the vicinity by several kilovolts. Sensitive electronic equipment that is connected between two ground references, such as a computer connected to the telephone system through a modem, can fail when subjected to the lightning surge voltages.
- They induce high voltages in phase conductors as they pass through cables on the way to a better ground.

## 22.4.8 Low-Side Surges

Some utility and end-user problems with lightning impulses are closely related. One of the most significant ones is called the *low-side surge* problem by many utility engineers. The name was coined

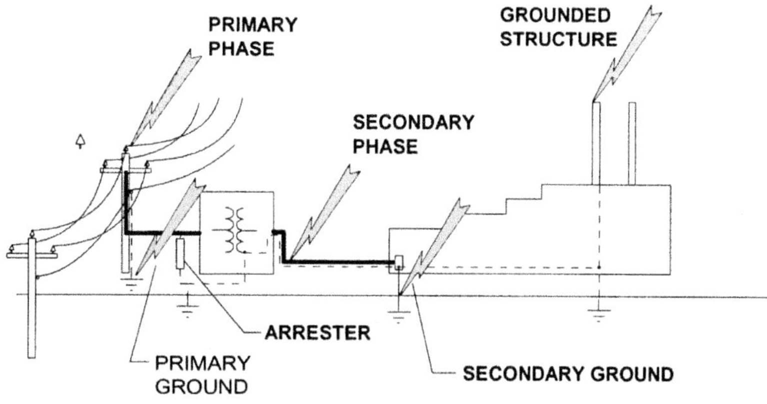


**FIGURE 22-25** Stroke locations for conduction of lightning impulses into load facilities.

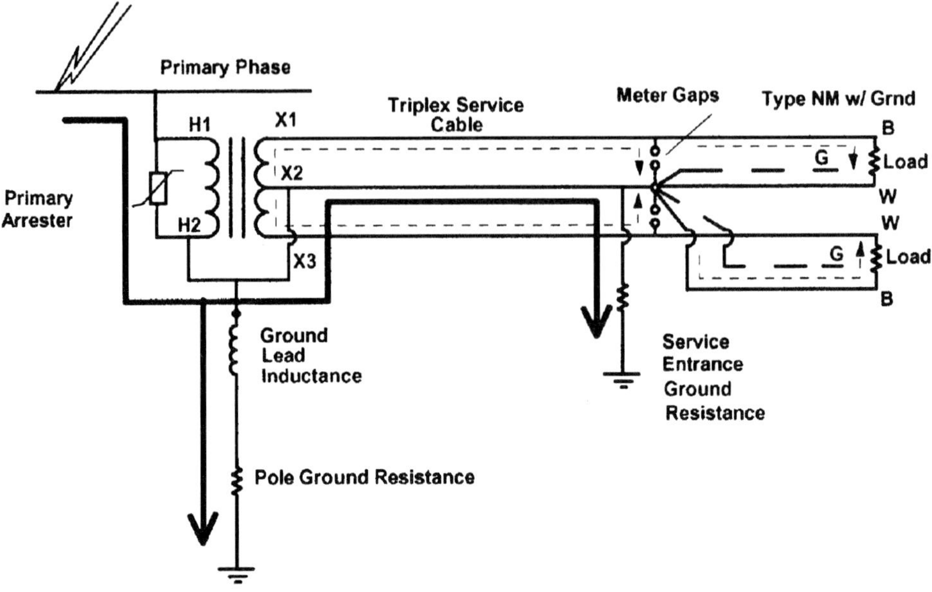


**FIGURE 22-26** Primary arrester discharge current divides between pole and load ground.

by distribution transformer designers because it appears from the transformer's perspective that a current surge is suddenly injected into the low-voltage side terminals. Utilities have not applied secondary arresters at low-voltage levels in great numbers. From the customer's point of view, it appears to be an impulse coming from the utility and is likely to be termed as "secondary surge."

Both problems actually have different side effects of the same surge phenomenon—lightning current flowing from either the utility side or the customer side along the service cable neutral. Figure 22-26 shows one possible scenario. Lightning strikes the primary line and the current is discharged through the primary arrester to the pole ground lead. This lead is also connected to the X2 bushing of the transformer at the top of the pole. Thus, some of the current will flow toward the load ground. The amount of current into the load ground is primarily dependent on the size of the pole ground resistance relative to the load ground. Inductive elements may play a significant role in the current division for the front of the surge, but the ground resistances basically dictate the division of the bulk of the stroke current.

The current that flows through the secondary cables causes a voltage drop in the neutral conductor that is only partially compensated by mutual inductive effects with the phase conductors. Thus, there is a net voltage across the cable, forcing current through the transformer secondary windings and into the load as shown by the dashed lines in the figure. If there is a complete path, substantial surge current will flow. As it flows through the transformer secondary, a surge voltage is induced in the primary, sometimes causing a layer-to-layer insulation failure near the grounded end. If there is not a complete path, the voltage will buildup across the load and may flash over somewhere on the secondary. It is common for the meter gaps to flashover, but not always before there is damage on the secondary because the meter gaps are usually 6 to 8 kV, or higher. The amount of voltage induced in the cable is dependent on the rate-of-rise of the current, which is dependent on other circuit parameters as well as the lightning stroke.

The chief power quality problems this causes are

- The impulse entering the load can cause failure or misoperation of load equipment.
- The utility transformer will fail causing an extended power outage.

- The failing transformer may subject the load to sustained steady-state overvoltages because part of the primary winding is shorted, decreasing the transformer turns ratio. Failure usually occurs in seconds, but has been known to take hours.

The key to this problem is the amount of surge current traveling through the secondary service cable. Keep in mind that the same effect occurs regardless of the direction of the current. All that is required is for the current to get into the ground circuits and for a substantial portion to flow through the cable on its way to another ground. Thus, lightning strikes to either the utility system or the end-user facilities can produce the same symptoms. Transformer protection is more of an issue in residential services, but the secondary transients will appear in industrial systems as well.

### 22.4.9 Low-Side Surges—An Example

Figure 22-27 shows a waveform of the open-circuit voltage measured at an electrical outlet location in a laboratory mock-up of a residential service [9]. For a relatively small stroke to the primary line (2.6 kA), the voltages at the outlet reached nearly 15 kV. In fact, higher current strokes caused random flashovers of the test circuit, which made measurements difficult. This reported experience is indicative of the capacity of these surges to cause overvoltage problems.

The waveform is a very high-frequency, ringing wave riding on the main part of the low-side surge. The ringing is very sensitive to the cable lengths. A small amount of resistive load, such as a lightbulb, would contribute greatly to the damping. The ringing wave differs depending on where the surge was applied while the base low-side surge wave remains about the same; it is more dependent on the waveform of the current through the service cable. One interesting aspect of this wave is that the ringing is so fast that it gets by the spark gaps in the meter base even though the voltage is two times the nominal sparkover value. In the tests, the outlets and lamp sockets could also withstand this kind of wave for about 1 $\mu$s before they flashed over. Thus, it is possible to have some high overvoltages propagating

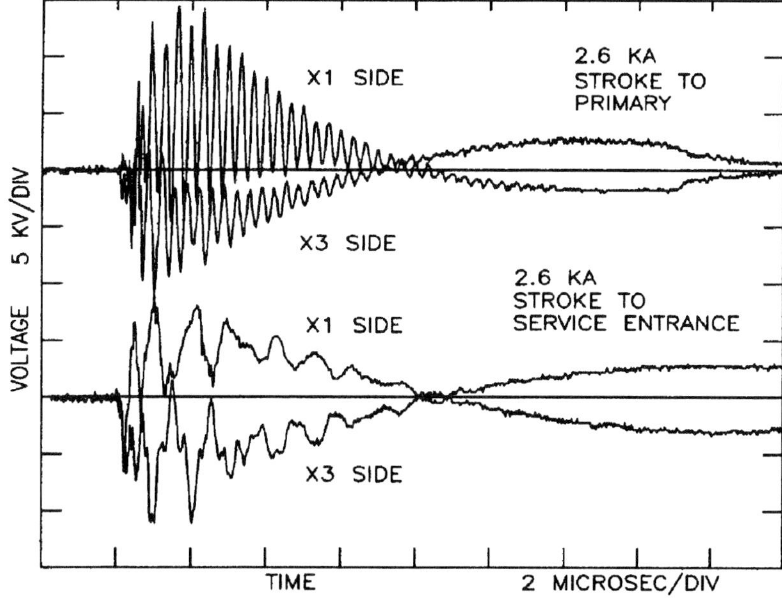


**FIGURE 22-27** Voltage appearing at outlet due to low-side surge phenomena.

throughout the system. The waveform in this figure represents the available open-circuit voltage. In actual practice, a flashover would have occurred somewhere in the circuit after a brief time.

### 22.4.10 Ferroresonance

The term *ferroresonance* refers to a resonance that involves capacitance and iron-core inductance. The most common condition in which it causes disturbances in the power system is when the magnetizing impedance of a transformer is placed in series with a system capacitor due to an open-phase conductor. Under controlled conditions, ferroresonance can be exploited for useful purpose such as in a constant-voltage transformer. In practice, ferroresonance most commonly occurs when unloaded transformers become isolated on underground cables of a certain range of lengths. The capacitance of overhead distribution lines is generally insufficient to yield the appropriate conditions.

The minimum length of cable required to cause ferroresonance varies with system voltage level. The capacitance of cables is nearly the same for all distribution voltage levels, varying from 40 to 100 nF per 1000 ft, depending on conductor size. However, the magnetizing reactance of a 35-kV-class distribution transformer is several times higher (curve is steeper) than a comparably-sized 15-kV-class transformer. Therefore, damaging ferroresonance has been more common at the higher voltages. For delta-connected transformers, ferroresonance can occur for less than 100 ft of cable. For this reason, many utilities avoid this connection on cable-fed transformers. The grounded wye-wye transformer has become the most commonly used connection in underground systems in North America. It is more resistant, but not immune, to ferroresonance because most units use a three-legged or five-legged core design that couples the phases magnetically. It may require a minimum of several hundred feet of cable to provide enough capacitance to create a ferroresonant condition for this connection. The most common events leading to ferroresonance are

- Manual switching of an unloaded, cable-fed, 3-phase transformer where only one phase is closed (Fig. 22-28*a*). Ferroresonance may be noted when the first phase is closed upon energization or before the last phase is opened on de-energization.
- Manual switching of an unloaded, cable-fed, 3-phase transformer where one of the phases is open (Fig. 22-28*b*). Again, this may happen during energization or de-energization.
- One or two riser-pole fuses may blow leaving a transformer with one or two phases open. Single-phase reclosers may also cause this condition. Today, many modern commercial loads will have controls that transfer the load to backup systems when they sense this condition. Unfortunately, this leaves the transformer without any load to damp out the resonance.
- Phase of a cable connected to a wye-connected transformer.

### 22.4.11 Transformer Energizing

Energizing a transformer produces inrush currents that are rich in harmonic components for a period lasting up to 1 s. If the system has a parallel resonance near one of the harmonic frequencies, a dynamic overvoltage condition results that can cause failure of arresters and problems with sensitive equipment. This problem can occur when large transformers are energized simultaneously with large power factor correction capacitor banks in industrial facilities. The equivalent circuit is shown in Fig. 22-29. A dynamic overvoltage waveform caused by a third-harmonic resonance in the circuit is shown in Fig. 22-30. After the expected initial transient, the voltage again swells to nearly 150% for many cycles until the losses and load damp out the oscillations. This can place severe stress on some arresters and has been known to significantly shorten the life of capacitors.

This form of dynamic overvoltage problem can often be eliminated simply by not energizing the capacitor and transformer together. One plant solved the problem by energizing the transformer first and not energizing the capacitor until load was about to be connected to the transformer.

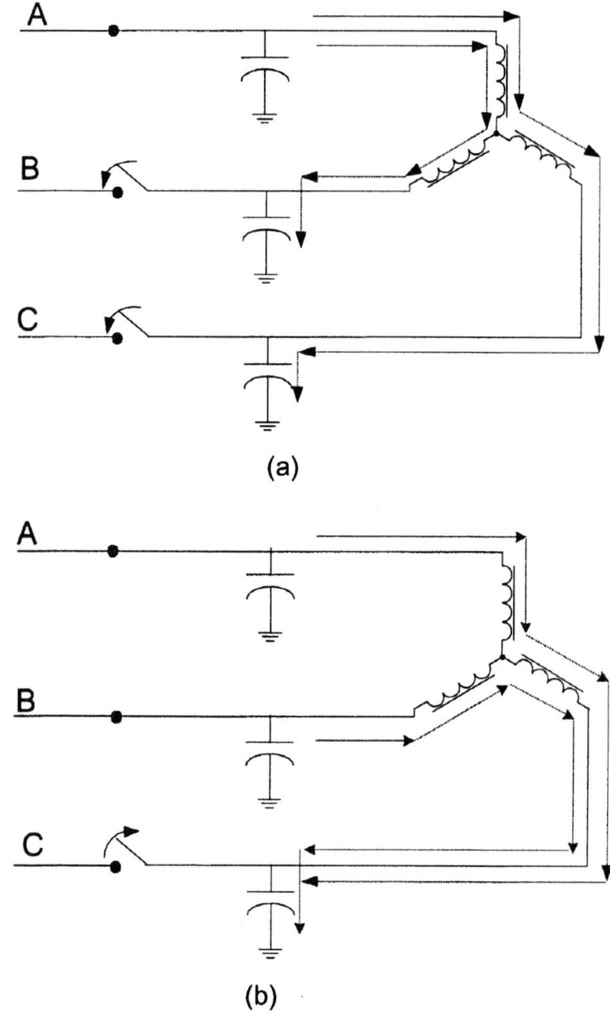


**FIGURE 22-28** Common system conditions where ferroresonance may occur: (*a*) one phase closed, (*b*) one phase open.

## 22.5 POWER SYSTEMS HARMONICS

### 22.5.1 General

Harmonic distortion is not a new phenomenon on power systems. Concern over distortion has ebbed and flowed a number of times during the history of ac electric power systems. Scanning the technical literature of the 1930s and 1940s, one will notice many articles on the subject. Then the primary sources were the transformers and the primary problem was inductive interference with open-wire telephone systems. The forerunners of modern arc lighting were being introduced and were causing quite a stir because of their harmonic content—not unlike the stir caused by electronic power converters in more recent times.

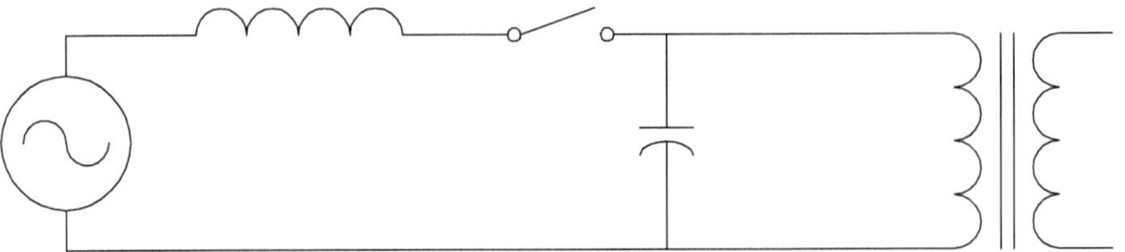

**FIGURE 22-29** Energizing a capacitor and transformer simultaneously can lead to dynamic overvoltages.

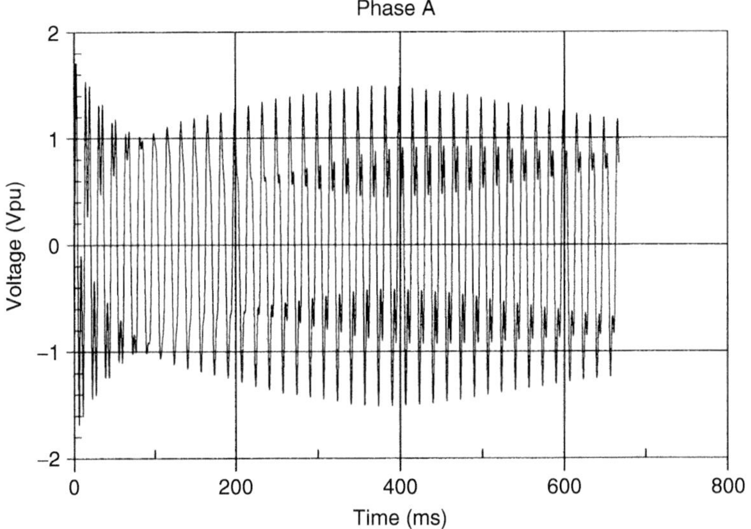


**FIGURE 22-30** Dynamic overvoltages during transformer energizing.

In contrast, voltage sags and interruptions are nearly universal to every feeder and represent the most numerous and significant power quality deviations. The end user sector suffers more from harmonic problems than the utility sector. Industrial users with adjustable speed drives, arc furnaces, induction furnaces, and the like, are much more susceptible to problems stemming from harmonic distortion.

A good assumption for most utilities in the United States is that the sine wave voltage generated in central power stations is very good. In most areas, the voltage found on transmission systems typically has much less than 1% distortion. However, the distortion increases closer to the load. At some loads, the current waveforms barely resemble a sine wave. Electronic power converters can chop the current into seemingly arbitrary waveforms.

### 22.5.2 Harmonic Distortion

Harmonic distortion is caused by nonlinear devices in the power system. A nonlinear device is one in which the current is not proportional to the applied voltage. Figure 22-31 illustrates this concept by the case of a sinusoidal voltage applied to a simple nonlinear resistor in which the voltage and current vary according to the curve shown. While the applied voltage is perfectly sinusoidal, the resulting current is distorted. Increasing the voltage by a few percent may cause the current to double and take on a different waveshape. This is the source of most harmonic distortion in a power system. Figure 22-32 illustrates that any periodic, distorted waveform can be expressed as a

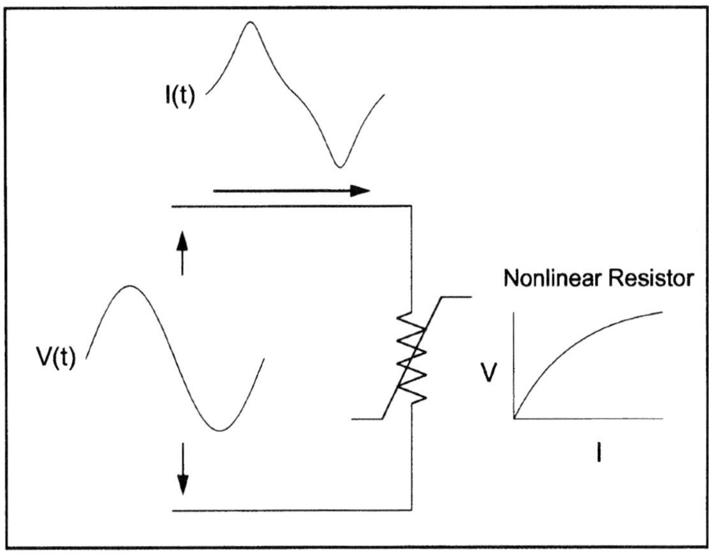


**FIGURE 22-31** Current distortion caused by nonlinear resistance.

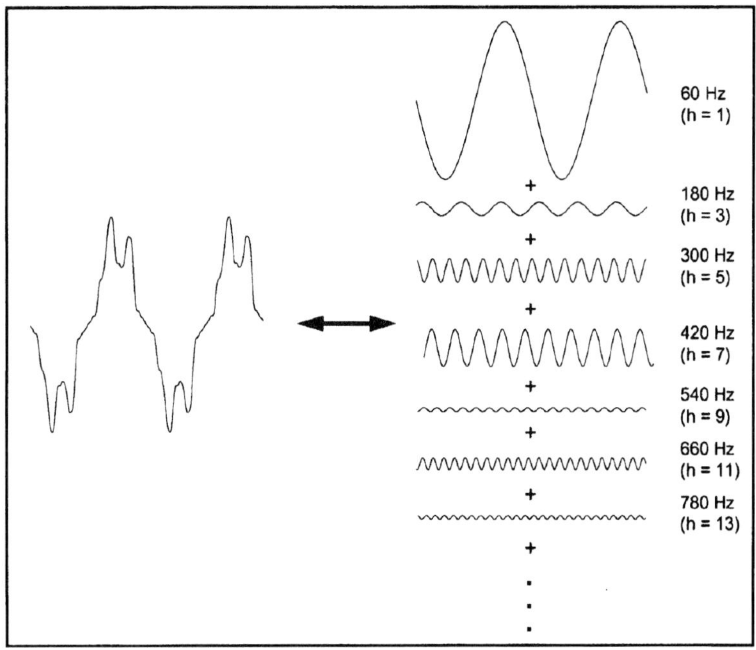


**FIGURE 22-32** Fourier series representation of a distorted waveform.

sum of sinusoids. When a waveform is identical from one cycle to the next, it can be represented as a sum of pure sine waves in which the frequency of each sinusoid is an integer multiple of the fundamental frequency of the distorted wave. This multiple is called a *harmonic* of the fundamental, hence the name of this subject matter. The sum of sinusoids is referred to as a *Fourier series*, named after the great mathematician who discovered the concept.

### 22.5.3 Voltage and Current Distortion

The term "harmonics" is often used by itself without further qualification. Generally, it could mean one of the following three:

1. The harmonic voltages are too great (the voltage too distorted) for the control to properly determine firing angles.
2. The harmonic currents are too great for the capacity of some device in the power supply system such as a transformer and the machine must be operated at a lower than rated power.
3. The harmonic voltages are too great because the harmonic currents produced by the device are too great for the given system condition.

Clearly, there are separate causes and effects for voltages and currents as well as some relationship between them. Thus, the term harmonics by itself is inadequate to definitively describe a problem. Nonlinear loads appear to be sources of harmonic current injecting harmonic currents into the power system. For nearly all analyses, it is sufficient to treat these harmonic-producing loads simply as current sources. There are exceptions to this as described later.

Voltage distortion is the result of distorted currents passing through the linear, series impedance of the power delivery system as illustrated in Fig. 22-33. Although, assuming that the source bus is ultimately a pure sinusoid, there is a nonlinear load that draws a distorted current. The harmonic currents passing through the impedance of the system cause a voltage drop for each harmonic. This results in voltage harmonics appearing at the load bus. The amount of voltage distortion depends on the impedance and the current. Assuming the load bus distortion stays within reasonable limits (e.g., less than 5%), the amount of harmonic current produced by the load is generally constant.

While the load current harmonics ultimately cause the voltage distortion, it should be noted that load has no control over the voltage distortion. The same load put in two different locations on the power system will result in two different voltage distortion values. Recognition of this fact is the basis for the division of responsibilities for harmonic control that are found in standards such as IEEE Std 519-1992.

- The control over the amount of harmonic current injected into the system takes place at the end-use application.
- Assuming the harmonic current injection is within reasonable limits, the control over the voltage distortion is exercised by the entity having control over the system impedance, which is often the utility.

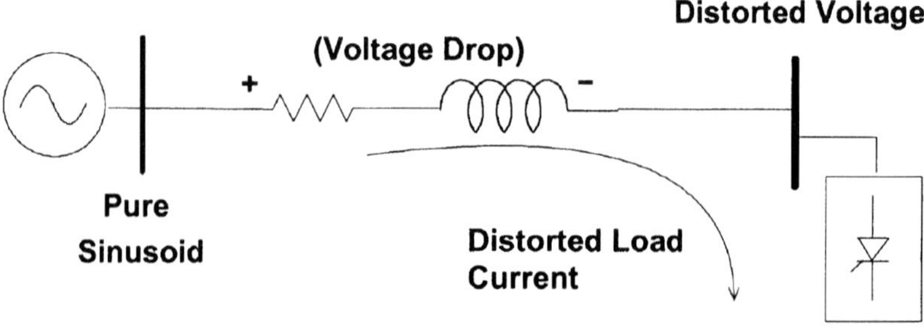


**FIGURE 22-33** Harmonic currents flowing through the system impedance results in harmonic voltages at the load.

One must be careful when describing harmonic phenomena to understand that there are distinct differences between the causes and effects of harmonic voltages and currents. The use of the term harmonics should be qualified accordingly. By popular convention in the power industry, the majority of time the term is used by itself when referring to load apparatus, the speaker is referring to the harmonic currents. When referring to the utility system, the voltages are generally the subject.

### 22.5.4 Power System Quantities under Nonsinusoidal Conditions

Traditional power system quantities such as rms, power (reactive, active, apparent), power factor, and phase sequences are defined for the fundamental frequency context in a pure sinusoidal condition. In the presence of harmonic distortion the power system no longer operates in a sinusoidal condition, and unfortunately many of the simplifications power engineers use for the fundamental frequency analysis do not apply. Therefore, these quantities must be redefined.

### 22.5.5 RMS Values of Voltage and Current

In a sinusoidal condition both the voltage and current waveforms contain only the fundamental frequency component, thus the rms values can be expressed simply as

$$V_{\text{rms}} = \frac{1}{\sqrt{2}}V_1 \quad \text{and} \quad I_{\text{rms}} = \frac{1}{\sqrt{2}}I_1 \tag{22-2}$$

where $V_1$ and $I_1$ are the amplitude of voltage and current waveforms, respectively. The subscript 1 denotes quantities in the fundamental frequency. In a nonsinusoidal condition, a harmonically distorted waveform is made up of sinusoids of harmonic frequencies with different amplitudes as shown in Fig. 22-32. The rms values of the waveforms are computed as the square root of the sum of rms squares of all individual components, that is,

$$V_{\text{rms}} = \sqrt{\sum_{h=1}^{h_{\max}} \left(\frac{1}{\sqrt{2}}V_h\right)^2} = \frac{1}{\sqrt{2}}\sqrt{V_1^2 + V_2^2 + V_3^2 + \cdots + V_{h_{\max}}^2} \tag{22-3}$$

$$I_{\text{rms}} = \sqrt{\sum_{h=1}^{h_{\max}} \left(\frac{1}{\sqrt{2}}I_h\right)^2} = \frac{1}{\sqrt{2}}\sqrt{I_1^2 + I_2^2 + I_3^2 + \cdots + I_{h_{\max}}^2} \tag{22-4}$$

where $V_h$ and $I_h$ are the amplitude of a waveform at the harmonic component $h$. In the sinusoidal condition, harmonic components of $V_h$ and $I_h$ are all zero, and only $V_1$ and $I_1$ remain. Equations (22-3) and (22-4) simplify to Eq. (22-2).

### 22.5.6 Active Power

The *active power*, $P$, is also commonly referred to as the average power, real power, or true power. It represents useful power expended by loads to perform real work, that is, to convert electric energy to other forms of energy. Real work performed by an incandescent light bulb is to convert electric energy into light and heat. In electric power, real work is performed for the portion of the current that is in phase with the voltage. No real work will result, from the portion where the current is not in phase with the voltage. The active power is the rate at which energy is expended, dissipated or consumed by the load, and is measured in units of watts (W). $P$ can be computed by averaging the product of the instantaneous voltage and current, that is,

$$P = \frac{1}{T}\int_0^T v(t)i(t)dt. \tag{22-5}$$

The above equation is valid for both sinusoidal and nonsinusoidal conditions. For sinusoidal condition, $I_{1\text{rms}}$ resolves to the familiar form,

$$P = \frac{V_1 I_1}{2}\cos\theta_1 = V_{1\text{rms}} I_{1\text{rms}} \cos\theta_1 = S\cos\theta_1 \tag{22-6}$$

where $\theta_1$ is the phase angle between voltage and current at the fundamental frequency. Equation 22-6 indicates that the average active power is a function only of the fundamental frequency quantities. In the nonsinusoidal case, the computation of the active power must include contribution from all harmonic components, thus it is the sum of active power at each harmonic. Furthermore, because the voltage distortion is generally very low on power systems (less than 5%), Eq. (22-6) is a good approximation regardless of how distorted the current is. This approximation cannot be applied when computing the apparent and reactive power. These two quantities are greatly influenced by the distortion. The apparent power, $S$, is a measure of the potential impact of the load on the thermal capability of the system. It is proportional to the rms of the distorted current and its computation is straightforward, although slightly more complicated than the sinusoidal case. Also, many current probes can now directly report the true rms value of a distorted waveform.

### 22.5.7 Reactive Power

The *reactive power* is a type of power that does no real work and is generally associated with reactive elements (inductors and capacitors). For example, the inductance of a load such as a motor causes the load current to lag behind the voltage. Power appearing across the inductance sloshes back and forth between the inductance itself and the power system source producing no net work. For this reason it is called imaginary or reactive power since no power is dissipated or expended. It is expressed in units of volt-ampere-reactive or var. In the sinusoidal case, the reactive power is simply defined as

$$Q = S\ \sin\theta_1 = \frac{V_1 I_1}{2}\sin\theta_1 = V_{1\text{rms}} I_{1\text{rms}} \sin\theta_1 \tag{22-7}$$

which is the portion of power in quadrature with the active power shown in Eq. (22-6). Figure 22-34 summarizes the relationship between $P$, $Q$, and $S$ in sinusoidal condition.

There is some disagreement among harmonics analysts on how to define $Q$ in the presence of harmonic distortion. If it were not for the fact that many utilities measure $Q$ and compute demand

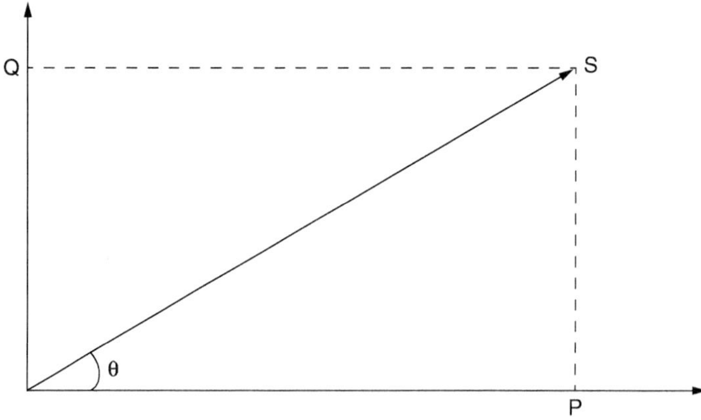


**FIGURE 22-34** Relationship between P, Q, and S in sinusoidal condition.

billing from the power factor computed by $Q$, it might be a moot point. It is more important to determine $P$ and $S$; $P$ defines how much active power is being consumed while $S$ defines the capacity of the power system required to deliver $P$. $Q$ is not actually very useful by itself. However, $Q_1$ the traditional reactive power component at fundamental frequency, may be used to size shunt capacitors.

The reactive power, when distortion is present, has another interesting peculiarity. In fact, it may not be appropriate to call it reactive power. The concept of var flow in the power system is deeply ingrained in the minds of most power engineers. What many do not realize is that this concept is valid only in the sinusoidal steady state. When distortion is present, the component of $S$ that remains after $P$ is taken out, is not conserved—that is, it does not sum to zero at a node. Power quantities are presumed to flow around the system in a conservative manner.

This does not imply that $P$ is not conserved or that current is not conserved because the conservation of energy and Kirchoff's current laws are still applicable for any waveform. The reactive components actually sum in quadrature (square root of the sum of the squares). This has prompted some analysts to propose that $Q$ be used to denote the reactive components that are conserved and introduce a new quantity for the components that are not. Many call this quantity $D$, for *distortion power* or, simply, *distortion voltamperes*. It has units of voltamperes, but it may not be strictly appropriate to refer to this quantity as *power*, because it does not flow through the system as power is assumed to do. In this concept, $Q$ consists of the sum of the traditional reactive power values at each frequency. $D$ represents all cross products of voltage and current at different frequencies, which yield no average power. $P$, $Q$, $D$, and $S$ are related as follows, using the definitions for $S$ and $P$ above as a starting point:

$$S = \sqrt{P^2 + Q^2 + D^2}$$
$$Q = \sum_k V_k I_k \sin\theta_k \tag{22-8}$$

Therefore, $D$ can be determined after $S$, $P$, and $Q$ by

$$D = \sqrt{S^2 - P^2 - Q^2} \tag{22-9}$$

Some prefer to use a three-dimensional vector chart to demonstrate the relationships of the components as shown in Fig. 22-35. $P$ and $Q$ contribute the traditional sinusoidal components to $S$ while $D$ represents the additional contribution to the apparent power by the harmonics.

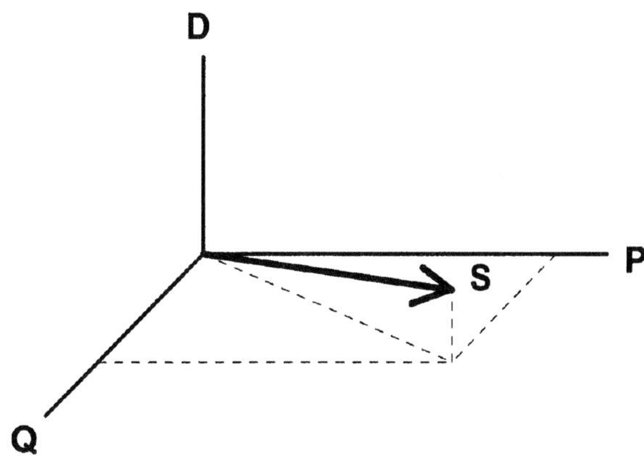


**FIGURE 22-35** Relationship of components of the apparent power.

### 22.5.8 Power Factor

A *power factor* is a ratio of useful power to perform real work (active power) to the power supplied by a utility (apparent power), that is,

$$\mathrm{PF} = \frac{P}{S} \tag{22-10}$$

In other words, the power factor ratio measures the percentage of power expended for its intended use. Power factor ranges from zero to unity. A load with power factor of 0.9 lagging denotes that the load can effectively expend 90% of the apparent power supplied (VA) and convert it to perform useful work ($W$). The term "lagging" denotes that the fundamental current lags behind the fundamental voltage by 25.84°.

In the sinusoidal case there is only one phase angle between the voltage and the current (since only the fundamental frequency is present), the power factor can be computed as the cosine of the phase angle and is commonly referred to as the *displacement power factor*,

$$\mathrm{PF} = \frac{P}{S} = \cos\theta \tag{22-11}$$

In the nonsinusoidal case the power factor cannot be defined as the cosine of the phase angle as in Eq. (22-11). The power factor that takes into account contribution from all active power both fundamental and harmonic frequencies is known as the *true power factor*. The true power factor is simply the ratio of total active power for all frequencies to the apparent power delivered by the utility as shown in Eq. (22-10).

Power quality monitoring instruments now commonly report both displacement and true power factors. Many devices such as switch-mode power supplies and PWM adjustable-speed drives have a near-unity displacement power factor, but the true power factor may be 0.5 to 0.6. An AC-side capacitor will do little to improve the true power factor in this case because $Q_1$ is zero. In fact, if it results in resonance, the distortion may increase, causing the power factor to degrade. The true power factor indicates how large the power delivery system must be built to supply a given load. In this example, using only the displacement power factor would give a false sense of security that all is well.

The bottom line is that distortion results in additional current components flowing in the system that do not yield any net energy except that they cause losses in the power system elements they pass through. This requires the system to be built to a slightly larger capacity to deliver the power to the load than if no distortion were present.

### 22.5.9 Harmonic Phase Sequence

Power engineers have traditionally used symmetrical components to help describe 3-phase system behavior. The 3-phase system is transformed into three single-phase systems that are much simpler to analyze. The method of symmetrical components can be employed for analysis of the system's response to harmonic currents provided care is taken not to violate the fundamental assumptions of the method.

The method allows any unbalanced set of phase currents (or voltages) to be transformed into three balanced sets. The *positive sequence* set contains three sinusoids displaced 120° from each other, with the normal A-B-C phase rotation (e.g., 0°, −120°, 120°). The sinusoids of the *negative-sequence* set are also displaced 120¬∞, but have opposite phase rotation (A-C-B, e.g., 0°, 120°, −120°). The sinusoids of the *zero sequence* are in phase with each other (e.g., 0, 0, 0).

In a perfect balanced 3-phase system, the harmonic phase sequence can be determined by multiplying the harmonic number $h$ with the normal positive sequence phase rotation. For example, for the second harmonic, $h = 2$, produces 2 × (0°, −120°, −120°) or (0°, 120°, −120°) which is the negative sequence. For the third harmonic, $h = 3$, produces 3 × (0°, −120°, −120°) or (0°, 0°, 0°) which is the zero sequence. Phase sequence for all other harmonic orders can be determined in the same fashion. Since a distorted waveform in power systems contains only odd harmonic components (see Sec. 22.5.1), only odd harmonic phase sequence rotations are summarized below:

- Harmonics of order $h = 1, 7, 13, \ldots$ are purely positive sequence.
- Harmonics of order $h = 5, 11, 17, \ldots$ are purely negative sequence.
- Triplens ($h = 3, 9, 15, \ldots$) are purely zero sequence.

## 22.5.10 Triplen Harmonics

*Triplen harmonics* are the odd multiples of the third harmonic ($h = 3, 9, 15, 21, \ldots$). They deserve special consideration because the system response is often considerably different for triplens than for the rest of the harmonics. Triplens become an important issue for grounded-wye systems with current flowing on the neutral. Two typical problems are overloading the neutral and telephone interference. One also hears occasionally of devices that misoperate because the line-to-neutral voltage is badly distorted by the triplen harmonic voltage drop in the neutral conductor.

For the system with perfectly balanced single-phase loads illustrated in Fig. 22-36, an assumption is made that fundamental and third harmonic components are present. Summing the currents at node $N$, the fundamental current components in the neutral are found to be zero, but the third harmonic components are three times the phase currents because they naturally coincide in phase and time.

## 22.5.11 Triplen Harmonics in Transformers

Transformer winding connections have a significant impact on the flow of triplen harmonic currents from single-phase nonlinear loads. Two cases are shown in Fig. 22-37. In the wye-delta transformer ($a$), the triplen harmonic currents are shown entering the wye side. Since they are in phase, they

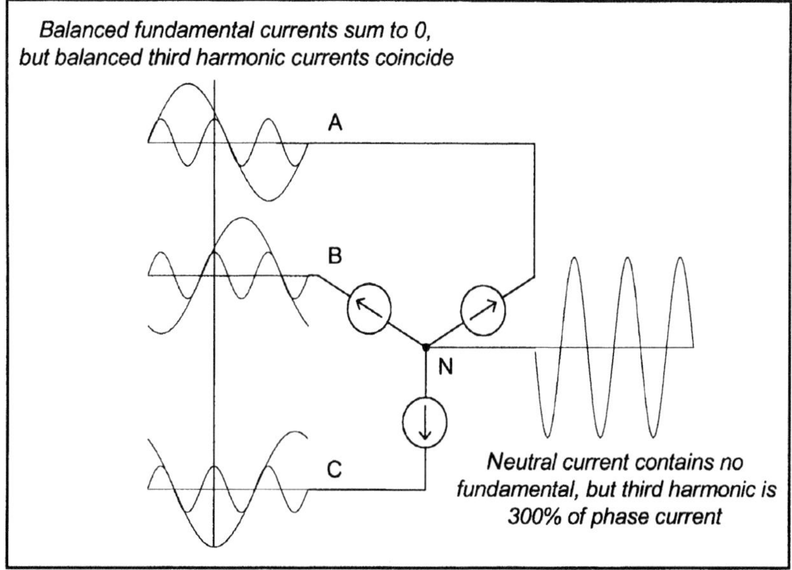


**FIGURE 22-36** High neutral currents in circuits serving single-phase nonlinear loads.

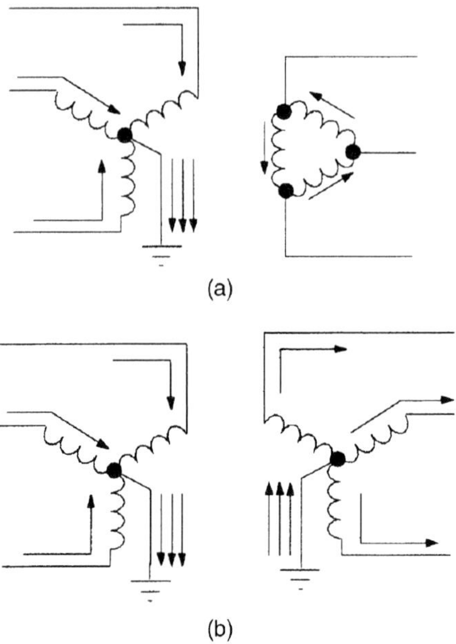


**FIGURE 22-37** Flow of third harmonic current in 3-phase transformers.

add in the neutral. The delta winding provides ampere-turn balance so that they can flow, but they remain trapped in the delta and do not show up in the line currents on the delta side. When the currents are balanced, the triplen harmonic currents behave exactly as zero-sequence currents, which is precisely what they are. This type of transformer connection is the most common employed in utility distribution substations with the delta winding connected to the transmission feed.

Using grounded-wye windings on both sides of the transformer (*b*) allows balanced triplens to flow from the low voltage system to the high voltage system unimpeded. They will be present in equal proportion on both sides. Many loads in the United States are served in this fashion.

Some important implications of this as related to power quality analysis are

1. Transformers, particularly the neutral connections, are susceptible to overheating when serving single phase loads on the wye side that have high third harmonic content.
2. Measuring the current on the delta side of a transformer will not show the triplens and, therefore, not give a true idea of the heating the transformer is being subjected to.

   The flow of triplen harmonic currents can be interrupted by the appropriate isolation transformer connection.
3. Removing the neutral connection in one or both wye windings blocks the flow of triplen harmonic current. There is no place for ampere-turn balance. Likewise, a delta winding blocks the flow from the line. One should note that three-legged core transformers behave as if they have a "phantom" delta tertiary winding. Therefore, a wye-wye with only one neutral point grounded will still be able to conduct the triplen harmonics from that side.

These rules about triplen harmonic current flow in transformers apply only to balanced loading conditions. When the phases are not balanced, currents of normal triplen harmonic frequencies may very well show up where they are not expected. The normal mode for triplen harmonics is to be zero sequence. During imbalances, triplen harmonics may have positive or negative sequence components as well.

One notable case of this is a 3-phase arc furnace. The furnace is nearly always fed by a delta-delta connected transformer to block the flow of the zero sequence currents, as shown in Fig. 22-8. Thinking that third harmonics are synonymous with zero sequence, many engineers are surprised to find substantial third harmonic current present in large magnitudes in the line current. However, during scrap meltdown, the furnace will frequently operate in an unbalanced mode with only two electrodes carrying current. Large third harmonic currents can then freely circulate in these two phases just as a single-phase circuit. However, they are not zero sequence currents. The third harmonic currents are equal amounts of positive and negative sequence currents. But to the extent that the system is mostly balanced, triplens mostly behave in the manner described.

### 22.5.12 Total Harmonic Distortion

The total harmonic distortion (THD) is a measure of the *effective value* of the harmonic components of a distorted waveform. That is, the potential heating value of the harmonics relative to the fundamental. This index can be calculated for either voltage or current:

$$\mathrm{THD} = \frac{\sqrt{\sum_{h>1}^{h_{\max}} M_h^2}}{M_1} \tag{22-12}$$

where $M_h$ is the rms value of harmonic component $h$ of the quantity $M$. The rms value of a distorted waveform is the square root of the sum of the squares as shown in Eqs. (22-3) and (22-4). THD is related to the rms value of the waveform as follows:

$$\mathrm{RMS} = \sqrt{\sum_{h=1}^{h_{\max}} M_h^2} = M_1 \cdot \sqrt{1+\mathrm{THD}^2} \tag{22-13}$$

THD is a very useful quantity for many applications, but its limitations must be realized. It can provide a good idea of how much extra heat will be realized when a distorted voltage is applied across a resistive load. Likewise, it can give an indication of the addition losses caused by the current flowing through a conductor. However, it is not a good indicator of the voltage stress within a capacitor because that is related to the peak value of the voltage wave form, not its heating value. The THD index is most often used to describe voltage harmonic distortion. Harmonic voltages are almost always referenced to the fundamental value of the waveform at the time of the sample. Because voltage varies only a few percent, the voltage THD is nearly always a meaningful number.

### 22.5.13 Total Demand Distortion

Current distortion levels can be characterized by a THD value, as described above, but this can often be misleading. A small current may have a high THD but not be a significant threat to the system. For example, many adjustable speed drives will exhibit high THD values for the input current when they are operating at very light loads. This is not necessarily a significant concern because the magnitude of harmonic current is low, even though its relative current distortion is high.

Some analysts have attempted to avoid this difficulty by referring THD to the fundamental of the peak demand load current rather than the fundamental of the present sample. This is called *total demand distortion* (TDD) and serves as the basis for the guidelines in IEEE STD 519-1992. It is defined as follows:

$$\text{TDD} = \frac{\sqrt{\sum_{h=2}^{h_{\max}} I_h^2}}{I_L} \tag{22-14}$$

where $I_L$ is the peak, or maximum demand load current at the fundamental frequency component measured at the point of common coupling (PCC). There are two ways to measure $I_L$. With a load already in the system, it can be calculated as the average of the maximum demand current for the preceding 12 months. The calculation can simply be done by averaging the 12-month peak demand readings. For a new facility, $I_L$ has to be estimated based on the predicted load profiles.

## 22.5.14 System Response Characteristics

In analyzing harmonic problems, the response of the power system is equally as important as the sources of harmonics. In fact, power systems are quite tolerant of the currents injected by harmonic-producing loads unless there is some adverse interaction with the impedance of the system. Identifying the sources is only half the job of harmonic analysis. The response of the power system at each harmonic frequency determines the true impact of the nonlinear load on harmonic voltage distortion.

There are three primary variables affecting the system response characteristics, that is, the system impedance, the presence of capacitor bank, and the amount of resistive loads in the system.

## 22.5.15 System Impedance

At the fundamental frequency, power systems are primarily inductive, and the equivalent impedance is sometimes called simply the short-circuit reactance. Capacitive effects are frequently neglected on utility distribution systems and industrial power systems. One of most frequently-used quantities in the analysis of harmonics on power systems is the short-circuit impedance to the point on a network at which a capacitor is located. If not directly available, it can be computed from short-circuit study results that give either the short-circuit *MVA* or the short-circuit current as follows:

$$\begin{aligned} Z_{SC} &= R_{SC} + jX_{SC} \\ &= \frac{kV^2}{MVA_{SC}} = \frac{kV}{\sqrt{3}I_{SC}} \end{aligned} \tag{22-15}$$

where $Z_{SC}$ = short-circuit impedance
$R_{SC}$ = short-circuit resistance
$X_{SC}$ = short-circuit reactance
$KV$ = phase-to-phase voltage, kV
$MVA_{SC}$ = 3-phase short-circuit, MVA
$I_{SC}$ = short-circuit current, A

$Z_{SC}$ is a phasor quantity, consisting of both resistance and reactance. However, if the short-circuit data contains no phase information, one is usually constrained to assuming that the impedance is purely reactive. This is a reasonably good assumption for industrial power systems, for buses close to the mains, and for most utility systems. When this is not the case, an effort should be made to determine a more realistic resistance value because that will affect the results once capacitors are considered.

The inductive reactance portion of the impedance changes linearly with frequency. One common error made by novices in harmonic analysis is to forget to adjust the reactance for frequency. The reactance at the *h*-th harmonic is determined from the fundamental-impedance reactance, $X_1$, by

$$X_h = hX_1 \tag{22-16}$$

In most power systems, one can generally assume that the resistance does not change significantly when studying the effects of harmonics less than the ninth. For lines and cables, the resistance varies approximately by the square root of the frequency once skin effect becomes significant in the conductor at a higher frequency. The exception to this rule is with some transformers. Because of stray eddy current losses, the apparent resistance of larger transformers may vary almost proportionately with the frequency. This can have a very beneficial effect on damping of resonance as shown later. In smaller transformers, less than 100 kVA, the resistance of the winding is often so large relative to the other impedances that it swamps out the stray eddy current effects and there is little change in the total apparent resistance until the frequency reaches about 500 Hz. Of course, these smaller transformers may have an $X/R$ ratio of 1.0 to 2.0 at fundamental frequency while large substation transformers might typically be 20 to 30. Therefore, if the bus that is being studied is dominated by transformer impedance rather than line impedance, the system impedance model should be considered more carefully. Neglecting the resistance will generally give a conservatively high prediction of the harmonic distortion.

At utilization voltages, such as industrial power systems, the equivalent system reactance is often dominated by the service transformer impedance. A good approximation for $X_{SC}$ may be based on the impedance of the service entrance transformer only

$$X_{SC} \approx X_{tx} \tag{22-17}$$

While not precise, this is generally at least 90% of the total impedance and is commonly more. This is usually sufficient to evaluate whether or not there will be a significant harmonic resonance problem. Transformer impedance in ohms can be determined from the percent impedance, $Z_{tx}$, found on the nameplate by

$$X_{tx} = \left(\frac{kV^2}{MVA_{3\phi}}\right) \times Z_{tx}(\%) \tag{22-18}$$

where $MVA_3\,\phi$ is the kVA rating of the transformer. This assumes that the impedance is predominantly reactive. For example for a 1500 kVA, 6% transformer, the equivalent impedance on the 480 V side is

$$X_{tx} = \left(\frac{kV^2}{MVA_{3\phi}}\right) \times Z_{tx}(\%) = \left(\frac{0.480^2}{1.5}\right) \times 0.06 = 0.0092\,\Omega$$

## 22.5.16 Capacitor Impedance

Shunt capacitors, either at the customer location for power factor correction, or on the distribution system for voltage control, dramatically alter the system impedance variation with frequency. Capacitors do not create harmonics, but severe harmonic distortion can sometimes be attributed to their presence. While the reactance of inductive components increases proportionately to frequency, capacitive reactance, $X_c$, decreases proportionately:

$$X_C = \frac{1}{2\pi fC} \tag{22-19}$$

where $C$ is the capacitance in farads. This quantity is seldom readily available for power capacitors, which are rated in terms of kvar or Mvar at a given voltage. The equivalent line-to-neutral capacitive reactance at fundamental frequency for a capacitor bank can be determined by

$$X_c = \frac{kV^2}{Mvar} \tag{22-20}$$

For 3-phase banks, use phase-to-phase voltage and the 3-phase reactive power rating. For single-phase units, use the capacitor-can voltage rating and the reactive power rating. For example, for a 3-phase, 1200 kvar, 13.8-kV capacitor bank, the positive-sequence reactance in ohms would be

$$X_c = \frac{kV^2}{Mvar} = \frac{13.8^2}{1.2} = 158.7\ \Omega$$

## 22.5.17 Parallel and Series Resonance

All circuits containing both capacitance and inductance have one or more natural resonant frequencies. When one of these frequencies corresponds to an exciting frequency being produced by nonlinear loads, harmonic resonance can occur. Voltage and current will be dominated by the resonant frequency and can be highly distorted. Thus, the response of the power system at each harmonic frequency determines the true impact of the nonlinear load on harmonic voltage distortion.

Resonance can cause nuisance tripping of sensitive electronic loads and high harmonic currents in feeder capacitor banks. In severe cases, capacitors produce audible noise, and they sometimes bulge. To better understand resonance, consider the simple parallel and series cases shown in the one-line diagrams of Fig. 22-38. Parallel resonance occurs when the power system presents a parallel combination of power system inductance and power factor correction capacitors at the nonlinear load. The product of harmonic impedance and injection current produces high harmonic voltages. Series resonance occurs when the system inductance and capacitors are in series, or nearly in series, with respect to the nonlinear load point. For parallel resonance, the highest voltage distortion is at the nonlinear load. However, for series resonance, the highest voltage distortion is at a remote point, perhaps miles away or on an adjacent feeder served by the same substation transformer. Actual feeders can have five or ten shunt capacitors each, so many parallel and series paths exist, making computer simulations necessary to predict distortion levels throughout the feeder.

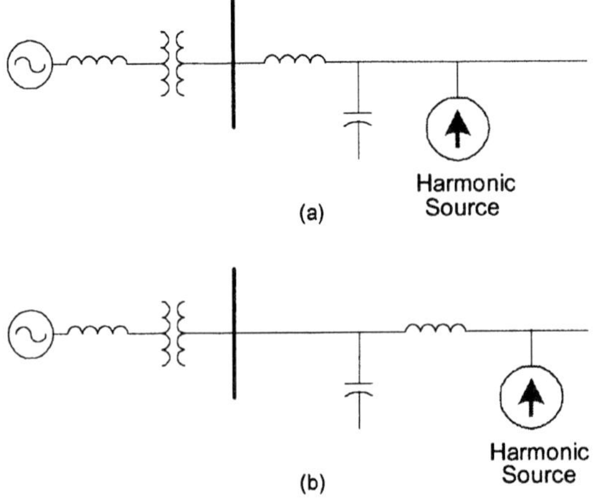


**FIGURE 22-38** Examples of (*a*) parallel and (*b*) series resonance.

In the simplest parallel resonant cases, such as an industrial facility where the system impedance is dominated by the service transformer, shunt capacitors are located inside the facility, and distances are small. In these cases, the simple parallel scenario shown in Fig. 22-38*a* often applies.

## 22.5.18 Effects of Resistance and Resistive Load

Determining that the resonant harmonic aligns with a common harmonic source is not always cause for alarm. The damping provided by resistance in the system is often sufficient to prevent catastrophic voltages and currents. Figure 22-39 shows the parallel resonant circuit impedance characteristic for various amounts of resistive load in parallel with the capacitance. As little as 10% resistive loading can have a significant beneficial impact on peak impedance. Likewise, if there is a significant length of lines or cables between the capacitor bus and the nearest upline transformer, the resonance will be suppressed. Lines and cables can add a significant amount of the resistance to the equivalent circuit.

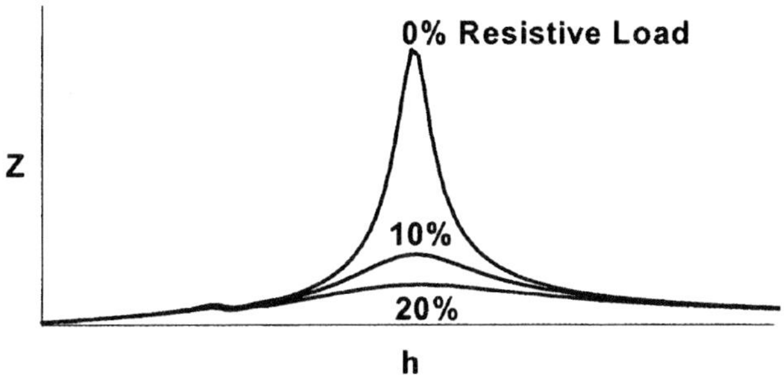


**FIGURE 22-39** Effect of resistive loads on parallel resonance.

Loads and line resistances are the reasons why catastrophic harmonic problems from capacitors on utility distribution feeders are seldom seen. That is not to say that there will not be any harmonic problems due to resonance, but that the problems will generally not cause physical damage to the electrical system components. The most troublesome resonant conditions occur when capacitors are installed on substations buses, either in utility substations or in industrial facilities. In these cases, where the transformer dominates the system impedance and has a high $X/R$ ratio, the relative resistance is low and the corresponding parallel resonant impedance peak is very sharp and high. This is a common cause of capacitor failure, transformer failure, or the failure of load equipment.

It is a misconception that resistive loads damp harmonics because in the absence of resonance, loads of any kind will have little impact on the harmonic currents and resulting voltage distortion. Most of the current will flow back into the power source. However, it is very appropriate to say that resistive loads will damp *resonance*, which will lead to a significant reduction in the harmonic distortion.

## 22.5.19 Harmonic Impacts

Harmonics have a number of undesirable effects on power system components and loads. These fall into two basic categories: short-term and long-term. Short-term effects are usually the most noticeable and are related to excessive voltage distortion. On the other hand, long-term effects often go undetected and are usually related to increased resistive losses or voltage stresses. Short-term effects can cause nuisance tripping of sensitive loads. Some computer-controlled loads are sensitive to voltage distortion. For example, one documented case showed that a voltage distortion of 5.5% regularly shut down computerized lathes at a large pipe company heat treatment operation. While voltage distortions of 5% are not usually a problem, voltage distortions above 10% will almost always cause significant nuisance tripping or transformer overheating.

Harmonics can degrade meter accuracy. This is especially true with common single-phase induction-disk meters. In general, the meter spins 1% to 2% faster when a customer produces harmonic power. However, the greater issue in metering is the question of how active power, and especially reactive power, should be defined and measured when distortion is present. Debate on these definitions continues today.

Blown capacitor fuses and failed capacitor cans are also attributed to harmonics. Harmonic voltages produce excessive harmonic currents in capacitors because of the inverse relationship between capacitor impedance and frequency. Voltage distortions of 5% and 10% can easily increase rms currents by 10% to 50%. Capacitors may also fail because of overvoltage stress on dielectrics. A 10% harmonic voltage for any harmonic above the third increases the peak voltage by approximately 10% because the peak of the harmonic usually coincides, or nearly coincides, with the peak of the fundamental voltage.

Harmonics can also cause transformer overheating. This usually occurs when a dedicated transformer serves only one large nonlinear load. In such a situation, the transformer must be derated accordingly. Derating to 0.80 of nameplate kVA is common.

Overloaded neutrals appear to be the most common problems in commercial buildings. In a 3-phase, four-wire system, the sum of the 3-phase currents returns through the neutral conductor. Positive and negative sequence components add to zero at the neutral point, but zero sequence components are additive at the neutral.

### 22.5.20 Control of Harmonics

The two common causes of harmonic problems are

- Nonlinear loads injecting excessive harmonic currents
- The interaction between harmonic currents and the system frequency response

When harmonics become a problem, commonly-employed solutions are

- Limit harmonic current injection from nonlinear loads. Transformer connections can be employed to reduce harmonics in a 3-phase system by using parallel delta-delta and wye-delta transformers to yield net 12-pulse operation, or delta connected transformers to block triplen harmonics.
- Modify system frequency response to avoid adverse interaction with harmonic currents. This can be done by feeder sectionalizing, adding or removing capacitor banks, changing the size of the capacitor banks, adding shunt filters, or adding reactors to detune the system away from harmful resonances.
- Filter harmonic currents at the load or on the system with shunt filters, or try to block the harmonic currents produced by loads. There are a number of devices to do this. Their selection is largely dependent on the nature of the problems encountered. Solutions can be as simple as an in-line reactor (i.e., a choke) as in PWM-based adjustable speed drive applications, or as complex as an active filter.

## 22.6 ELECTRICAL POWER RELIABILITY AND RECENT BULK POWER OUTAGES

### 22.6.1 Electric Power Distribution Reliability—General

The term *reliability* in the utility context usually refers to the amount of time end users are totally without power for an extended period of time (i.e., a sustained interruption). Definitions of what constitutes a sustained interruption vary among utilities in the range of 1 to 5 min. This is what many utilities refer to as an "outage." Current power quality standards efforts are leaning toward calling any interruption of power for longer than 1 min, a sustained interruption. In any case, reliability is affected by the permanent faults on the system that must be repaired before service can be restored.

### 22.6.2 Electric Power Distribution Reliability Indices

Most commonly used reliability indices for utility distribution systems are defined as follows:

- SAIFI: System Average Interruption Frequency Index

  SAIFI represents the average interruption frequency experienced by customers served in the system over a given period of time. It is computed as follows:

$$\text{SAIFI} = \frac{(\text{no. of customers interrupted})(\text{no. of interruptions})}{\text{total no. of customers}}$$

- SAIDI: System Average Interruption Duration Index

  SAIDI represents the average interruption duration experienced by customers in the system over a given period of time.

$$\text{SAIDI} = \frac{\sum(\text{no. of customers affected})(\text{duration of outage})}{\text{total no. of customers}}$$

- CAIFI: Customer Average Interruption Frequency Index

  CAIFI represents average interruption frequency for affected customers. Customers not experiencing interruption are not included in the calculation.

$$\text{CAIFI} = \frac{\text{total no. of customer interruptions}}{\text{total no. of customers affected}}$$

- CAIDI: Customer Average Interruption Duration Index

  CAIDI represents the average interruption duration for customers experiencing interruptions. In other words, this is the average restoration time for affected customers.

$$\text{CAIDI} = \frac{\sum(\text{customer interruption durations})}{\text{total no. of customer interruptions}}$$

- ASAI: Average System Availability Index

  ASAI represents the average system availability over a given observation period, which is usually a year (or 8760 h). The index is given in percent.

$$\text{ASAI} = \frac{\sum \text{customer hours service availability}}{\text{customer hours service demand}}$$

### 22.6.3 Major Bulk Electric Power Outages

Since the electric power industry was born in the early twentieth century, there have been several notable major bulk power outages. Most common causes of these outages are errors in protective device system design, overgrown vegetation, loss of system awareness due to failure of alarm systems, and a combination of unexpected events, whether they are natural or man-made. Summary of these bulk power outages are compiled from various sources and presented in the next paragraphs.

### 22.6.4 Great Northeast Blackout of 1965

The 1965 power outage started on November 9 at about 5:15 P.M. in Ontario, Canada. It cascaded down through the power system to the majority of New York, Connecticut, Massachusetts, Rhode Island, and some portions of northern Pennsylvania, and New Jersey. There were about 30 million customers out of service for up to 13 h. The power outage left 20 GW of load demand unserved. The outage was triggered by a backup protective relay in opening one of five 230-kW lines delivering power from the Adam Beck Station No. 2 to the Toronto load area. System operators were not aware that the backup relay was set to take the line out of service when the line loading exceeded 375 MW. This relay setting was below the unusually high line loadings of recent months. Higher than normal line loadings was imposed due to higher than normal import from the United States to cover nearby Lakeview power plant (west of Toronto) outage. Upon opening the 230-kV line, the remaining four 230-kV lines were also tripped out successively within 2½ s. Subsequently, two key east–west 345-kV lines between Rochester and Syracuse tripped out due to line instability. Several lower voltage lines tripped open along with 5 of 11 generation units at the St. Lawrence (Massena) Station. Losses of major transmission lines caused 10 generators at Adam Beck Station to shut down due to low governor oil pressure. By 5:30 P.M., the majority of northeast was without power. The service was, however, restored by 4:44 A.M. the next day in Manhattan [11].

### 22.6.5 New York Blackout of 1977

The event started on July 13 at about 8:37 P.M., when a lightning stroke caused a phase B to ground fault on both of a double-circuit 345-kV transmission line between Buchanan South and Millwood West Substations [11,12,14]. The tripping of circuit breakers at Buchanan South Circuit rings isolated Indian Point No. 3 generating unit without a transmission path to any load. The plant tripped off line and shut down causing a generation loss of 883 MW. A coordination error in the protective system played a critical role in the subsequent chain of events in which a transfer trip signal to Ladentown was initiated to open the 345-kV line from Buchanan South to Ladentown. A subsequent lightning stroke also caused a trip out of two more 345-kV lines between Sprain Brook and Buchanan North, and Sprain Brook and Millwood West. The latter was restored to service in about 2 s. However, the Sprain Brook to Buchanan North 345-kV was out of service. Losses of key transmission lines eventually forced the electrical system to separate and collapse. The power outage affected 9 million people. However, it was limited to New York City alone. Unlike the 1965 blackout, the 1977 event was marred by violence and looting [13].

Timeline of key events in the total collapse of the ConEd system are as follows [12,14]:

- At 8:37:17 P.M., July 13, 1997, two 345-kV lines connecting Buchanan South to Millwood West were each subjected to a phase B fault to ground as a result of a severe lightning stroke.
- The tripping of circuit breakers at the Buchanan South ring bus, isolated the Indian Point No. 3 generating unit from any load, and the unit tripped for a loss of 883 MW.
- Loss of the ring bus isolated the 345-kV tie to Ladentown, which had been importing 427 MW, with a total loss now of 1310 MW.
- At 8:55:53 P.M., about 18½ min after the first incident, a severe lightning stroke caused the trip-out of two 345-kV lines, which connect Sprain Brook to Buchanan North, and Sprain Brook to Millwood West. These two 345-kV lines share common towers between Millwood West and Sprain Brook. One line (Millwood West to Sprain Brook) was restored to service in about 2 s. The failure of the other line to reclose isolated the last ConEd interconnection to the northwest.
- The resulting surge of power from the northwest, caused the trip-out of the line between Pleasant Valley and Millwood West (a bent contact on one of the relays at Millwood West caused the improper action).
- At 9:19:11 P.M., a 345-kV line, Leeds Substation to Pleasant Valley, tripped as a result of a phase B fault to ground (fault probably caused by line sag to a tree because of the excessive overload imposed on the line).

- At 9:19:53 P.M., the 345-kV/138-kV transformer at Pleasant Valley, tripped on overcurrent relay and left ConEd with three remaining interconnections.
- At 9:22:11 P.M., the Jamaica/Valley Stream tie was opened manually by the Long Island Lighting Co. system operator after obtaining the approval of the pool dispatcher.
- About 7 min later, the tap-changing mechanism failed on the Goethals phase-angle regulating transformer resulting in the trip of the Linden/Goethals tie to PJM, which was carrying 1150 MW to ConEd.
- The two remaining external 138-kV ties to ConEd tripped on overload isolating the ConEd system.

### 22.6.6 The Northwestern Blackout of July 1996

This outage occurred on July 2 at about 2:24 P.M. when a tree fault tripped a 345-kV taking power from Jim Bridger power plant in southwest Wyoming to southeast Idaho [10]. Protective devices detected the fault and de-energized the line. However, due to a protection coordination error, a parallel 345-kV line was also tripped. The loss of two 345-kV line severely limited power transfers from Jim Bridger plant causing generator protective devices to trip two 500-MW generators to maintain the system stability. With two generators out, frequency in the entire western interconnection began to decline forcing some customers out of service. This move was not successful and the system disintegrated into five islands. About 2 million customers were interrupted for up to several hours with about 11,850 MW of loss of load demand.

### 22.6.7 The Northwestern Blackout of August 1996

This blackout occurred on August 10 when Keeler-Allston 500-kV transmission line sagged into a grove of trees [10]. Prior to the disturbance, the Northwest area was importing about 2300 MW from Canada due to excellent hydroelectric conditions that lead to high electricity transfers. This and other conditions, that is, hot weather, maintenance outage of a transformer that connects a static var compensator to a 500-kV line in Portland, and/or failure to trim trees, led to a cascading outage. A series of tree fault disturbances finally broke the western interconnection area into four islands, interrupting services to 7.5 million customers for up to 9 h.

### 22.6.8 The Great Northeastern Power Blackout of 2003 [15, 16]

This outage on August 14, 2003 is by far the largest and most severe among all major outages. It affected 50 million customers for up to 2 days in two Canadian provinces and eight northeastern U.S. states. The outage caused an estimated $4 to $8 billion in lost economic activity.

The outage was preceded by an abnormal computer software operation, a series of generator tripping, and line outages. This series of events is considered as a precursor to the cascading outage. MISO's (Midwest Independent Service Operator) state estimator software solution did not converge and produced a solution with a high mismatch due to outdated input data in the state estimator. Eastlake No. 5 generating unit tripped along with two other units (Conesville and Greenwood) causing a severe shortage in reactive power supply. Adequate reactive power supply is an important requirement for high-voltage long-distance electric power transmission. At about 2:02 P.M., Stuart–Atlanta 345 kV in southwestern Ohio tripped due to contact with trees causing a short circuit to ground and locked out. This situation was exacerbated by the loss of key alarm functions in FirstEnergy's (Ohio) control room. The controller also lost a series of other important computer functions. Unfortunately, FirstEnergy operators were unaware of computer failures, thus they lost situational awareness of their system.

Precipitating events that lead to the cascading outage began around 3:00 P.M., when three key 345-kV transmission lines into northern Ohio from eastern Ohio tripped out, Harding–Chamberlain (3:05 P.M.), Hanna–Juniper (3:32 P.M.), and Start–South Canton (3:41 P.M.).

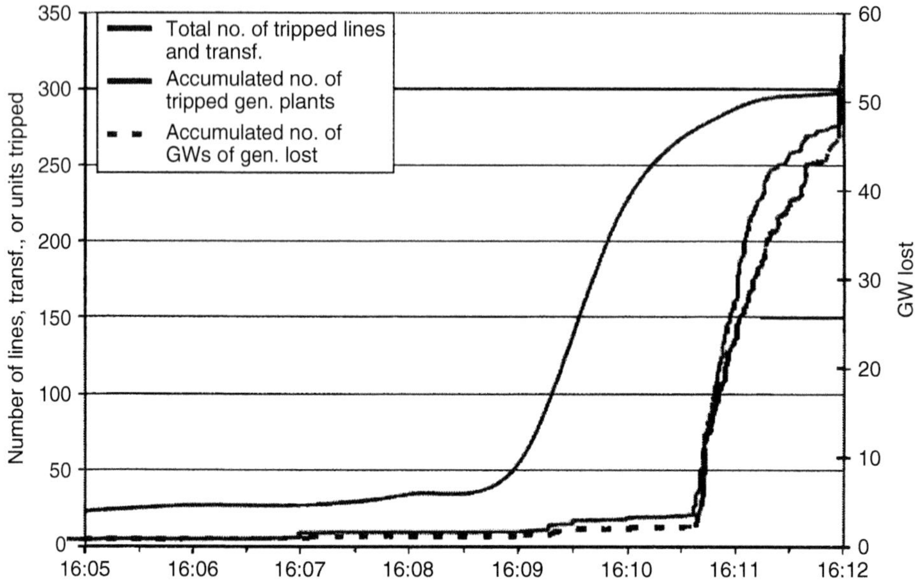


**FIGURE 22-40** Accumulated line and generator trips during the cascade [23].

They were all tripped out due to tree faults caused by overgrown vegetation. With these three 345-kV lines out, power flowed over through other remaining lines including the underlying 138-kV system. This redirection of flow caused severe overloading in 138-kV lines. As a result, between 3:39 P.M. and 3:58 P.M., seven 138-kV lines tripped. At 3:59 P.M., West Akron bus tripped due to breaker failure. This event caused another five 138 kV-lines to trip. A few minutes later, between 4:00 P.M. and 4:08 P.M., another four 138-kV lines tripped. Losses of these transmission lines disconnected northern Ohio from Eastern Ohio. The last 345-kV line between Sammis and Star tripped at 4:06 P.M. The loss of this line left northern Ohio without any 345-kV path to eastern Ohio, and initiated a cascading blackout across the northeast U.S. and Canada. Within 7 min after the loss of 345-kV Sammis–Start line, more than 508 generating units at 265 power plants had been lost, and close to 300 lines and transformers tripped.

## 22.6.9 Power Quality Characteristics in the Great Northeastern Power Blackout of 2003

A major power quality characteristic of the blackout was sagging voltage when transmission lines experienced fault clearing operations (opening and reclosing) due to tree contacts.

An 8-cycle voltage sag was measured at an industrial site in Cleveland at about 3:45 P.M. This was when a series of 138-kV lines experienced tree faults and attempted to isolate them (Fig. 22-41). This particular fault was detected and cleared promptly, but the voltage recovery at the site appears to be slow, suggesting that the system was now much weaker than previously. At least one more line was out of service.

Shortly after 4:00 P.M. there was another instantaneous voltage sag recorded (Fig. 22-42). The voltage drops abruptly and remains at the lower level. Phase unbalance develops, suggesting either the presence of a remote fault or that the system has become very weak due to the tripping of another line.

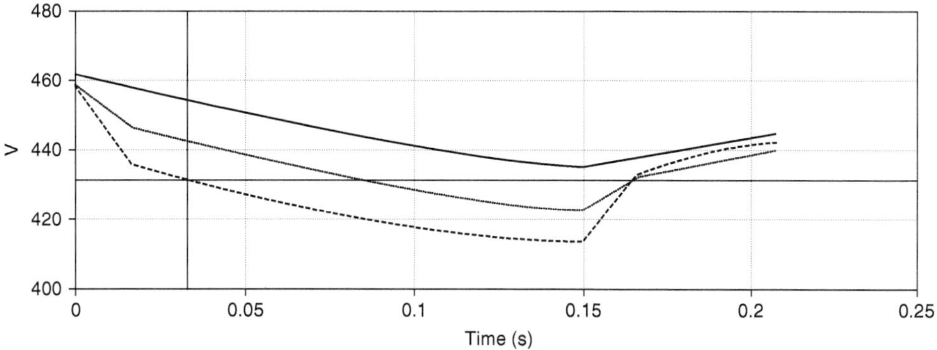


**FIGURE 22-41** Instantaneous sag on phase C captured at service entrance of Cleveland industrial facility at around 3:45 P.M. likely due to a fault on a transmission line. Voltage recovers slowly after fault is cleared, suggesting a weakened system. (*Courtesy of Electrotek Concepts and Dranetz.*)

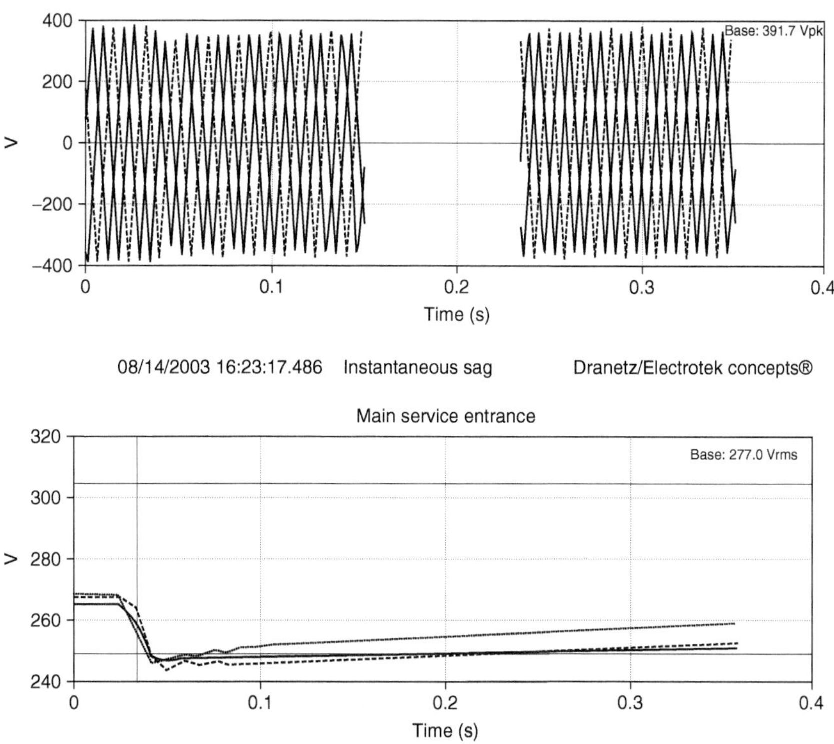


**FIGURE 22-42** 3-Phase instantaneous sag shortly after 4 P.M. on August 14. (*Courtesy of Electrotek Concepts and Dranetz.*)

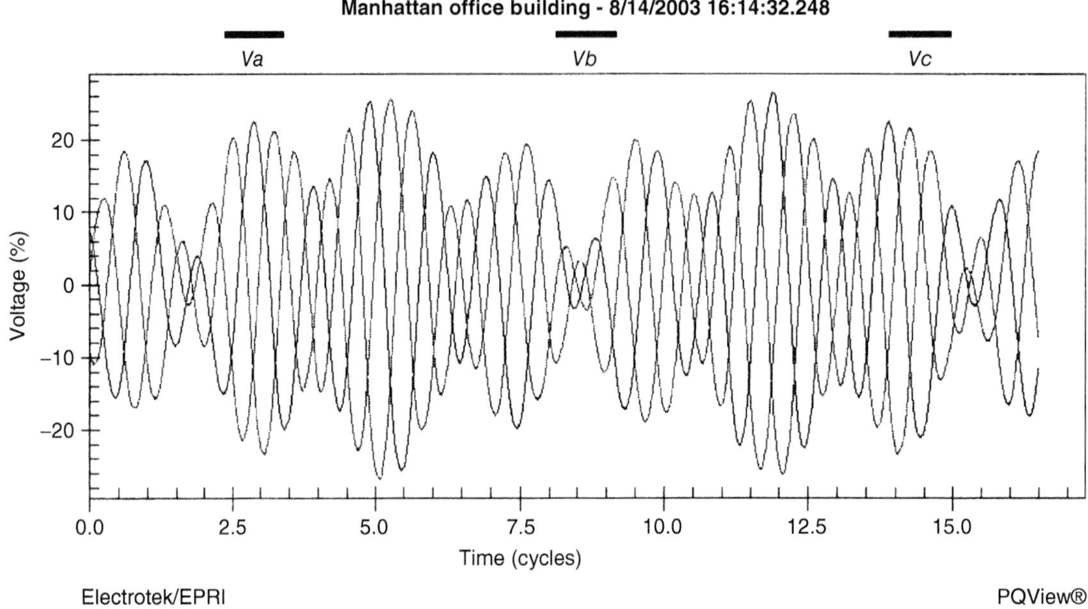


**FIGURE 22-43** Waveform captured as the grid collapsed in New York City showing the instability of the system. (*Courtesy of Electrotek Concepts and Dranetz.*)

A short time later (accuracy of the time stamp is uncertain), the disturbance in the voltage shown in Fig. 22-43 was recorded in an office building in downtown Manhattan. This waveform is consistent with that of a power system that is going unstable.

Waveforms of this type can be observed in system dynamics simulations for buses in the weaker part of the system as it begins to move relative to a more distant part of the system that remains in synchronism. A "beating" frequency develops as the two interconnected systems operate at different frequencies. In this case, it would appear that the system containing this power monitor had drifted by approximately 2 Hz from the rest of the system. Once this occurred, the power system supplying Manhattan detected the instability and immediately shut down the generators and separated from surrounding power systems.

Several of the neighboring systems successfully separated and remained stable throughout the blackout period despite briefly experiencing voltage waveforms like this while they were still interconnected with the part of the system that was collapsing.

Once the massive amount of load in the affected areas was lost, the entire eastern interconnection experienced a jump in frequency of approximately 0.2 Hz. This could be seen over a large geographic area (Fig. 22-44). After a few minutes, generator controls brought the average frequency back to 60 Hz and few energy users outside the affected area realized that anything had happened. While large in terms of system dynamics issues, this frequency change is inconsequential to most loads.

Figure 22-45 shows the complete rms voltage trend for the Manhattan site from the beginning of the blackout shortly after 4:00 P.M. on August 14 until the power was restored at 5:30 A.M. on August 15.

## 22.6.10 Blackouts Due to Natural Disasters

In addition to error in protection coordination and overgrown vegetation, widespread outages can arise from major failures of power apparatus and equipment during natural disasters (earthquakes

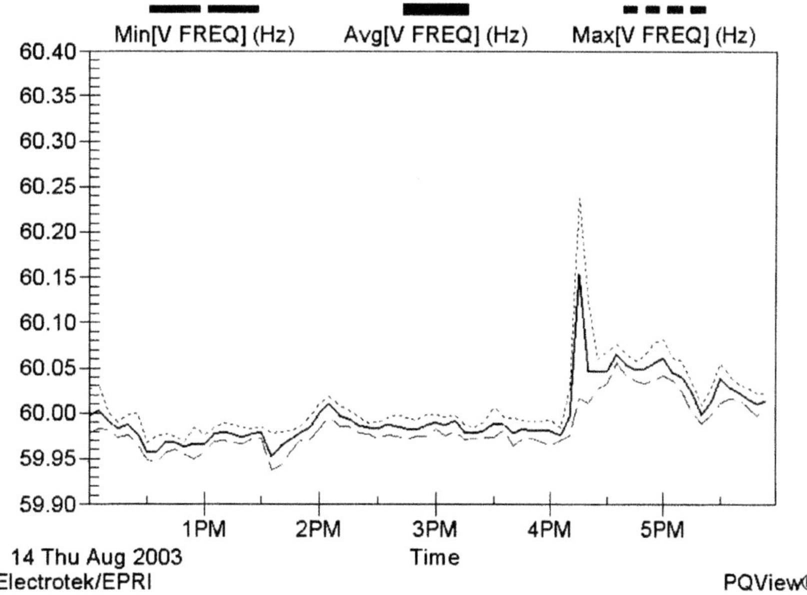


**FIGURE 22-44** System frequency jump when the major systems separated was seen in locations as far away as Knoxville, TN. (*Courtesy of Electrotek Concepts and Dranetz.*)

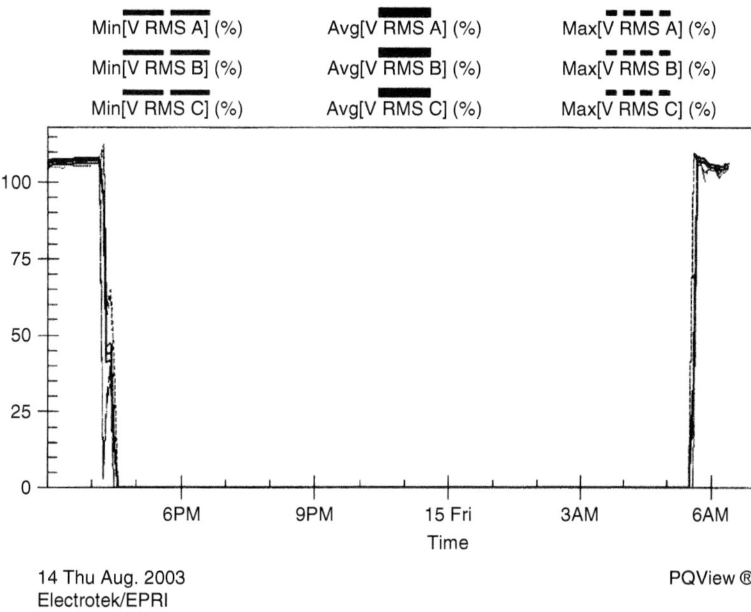


**FIGURE 22-45** Rms voltage trend showing the min/max/avg of all three phases for the entire duration of the blackout at a Manhattan office building. (*Courtesy of Electrotek Concepts and Dranetz.*)

and tsunamis) and extreme weather conditions (ice storms, hurricanes, and flooding). The 9.0 magnitude Tohoku undersea earthquake and the ensuing tsunami on March 11, 2011 in northeast Honshu, Japan resulted in catastrophic failures at the Fukishima nuclear power plant. A total of eleven reactors tripped offline causing a loss of 9.7 GW of generation [17]. Rolling outages of a few hours (up to 5 h) for a few months were employed to avoid a complete power outage.

Another example of rolling outage, but on a much lesser scale, occurred in the ERCOT service area. ERCOT is the Electric Reliability Council of Texas, an independent system operator managing about 85% of Texas' electric load over 40,500 mi of transmission lines with more than 550 generation units for a total of 78 GW capacity. Its service area covers 75% of the land area of the State of Texas.

Significantly colder than usual weather, in the tens of degrees of Fahrenheit, affected much of Texas between February 1 and 4 of 2011. The winter freeze caused a widespread failure of generation units primarily due to improper or non-existent weatherization equipment. In the morning of February 2, 2011, ERCOT experienced more than 7 GW of generation loss. According to the report submitted by Texas Public Utility Commission [18], a total of 152 generators experienced forced outages over the 4-day period. An increase in load demand coupled with the loss of generation triggered rolling outages as long as one to several hours on February 2 and 3. A new all-time record for winter peak demand of 56,334 MW occurred in the evening of February 2. Figure 22-46 shows the locations of customers experiencing outages in the greater Dallas and Fort Worth area in the afternoon of February 3.

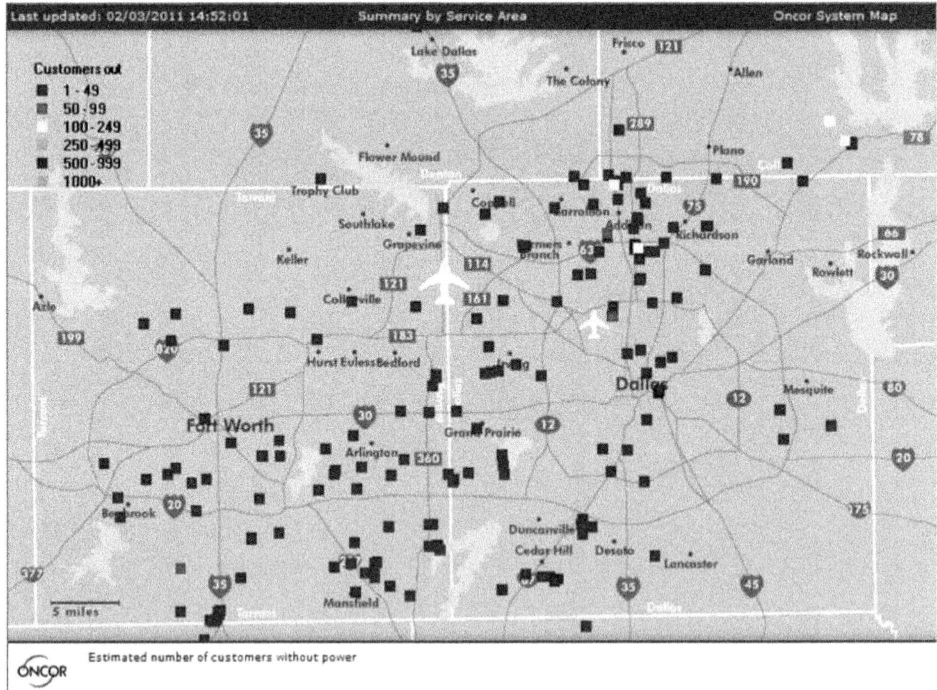


FIGURE 22-46 Locations of customers without power during the rolling outages in the afternoon of February 3, 2011.

## 22.7 REFERENCES

1. TC77WG6 (Secretary) 110-R5, Draft Classification of Electromagnetic Environments, January 1991.
2. IEEE Std 1159-1995, Recommended Practice on Monitoring Electric Power.
3. IEC 61000-2-1(1990-05), "Description of the environment—Electromagnetic environment for low frequency conducted disturbances and signaling in public power supply systems," Electromagnetic Compatibility (EMC)—Part 2 Environment, Section 1, 1990.
4. IEEE Std 100-1992, IEEE Standard Dictionary of Electrical and Electronic Terms.
5. IEC 61000-4-30 77A/356/CDV, Power Quality Measurement Methods.
6. IEC 61000-4-15, Flicker Meter—Functional and Design Specifications.
7. Lamoree, J., Mueller, D., Vinett, P., and Jones, W., "Voltage Sag Analysis Case Studies," 1993 IEEE I&CPS Conference, St. Petersburg, FL.
8. Dugan, R. C., Ray, L. A., Sabin, D. D., et al., "Impact of Fast Tripping of Utility Breakers on Industrial Load Interruptions," Conference Record of the 1994 IEEE/IAS Annual Meeting, Vol III, Denver, October 1994, pp. 2326–2333.
9. Goedde, G. L., Dugan, R. C., and Rowe, L. D., "Full Scale Lightning Surge Tests of Distribution Transformers and Secondary Systems," *Proceedings of the 1991 IEEE PES Transmission and Distribution Conference,* Dallas, September, 1991, pp. 691–97.
10. Consortium for Electric Reliability Technology Solutions, "Review of Recent Reliability Issues and System Events," Grid of the Future White Paper, December, 1999.
11. Website http://blackout.gmu.edu, accessed on February 24, 2005.
12. U.S. Department of Energy, The Con Edison Power Failure of July 13 and 14, 1977, Final Report, June 1978.
13. TIME, "Night of Terror," July 25, 1977, pp. 12–26.
14. Lesson Learned From the 1977 Blackout Case Study 1—Sequence of Events, www.blackout.gmu.edu/archive/pdf/lessons_learned_77.pdf, accessed on February 24, 2005.
15. North American Electric Reliability Council, Technical Analysis of the August 14, 2003, Blackout: What Happened, Why, and What Did We Learn, July 13, 2004.
16. U.S.-Canada Power System Outage Task Force, Final Report on the August 14, 2003, Blackout in the United States and Canada: Causes and Recommendation, April 2004.
17. International Energy Agency, *Impact of Earthquakes and Tsunamis on Energy Sectors in Japan*, http://www.iea.org/files/japanfactsheet.pdf.
18. Reports of the Electric Reliability Council of Texas, Project 27706, to Public Utility Commission of Texas, ERCOT's second supplemental response to Commission's request for information dated February 7, 2011, http://www.ercot.com/content/news/presentations/2011/Forced%20Outage%20List%20-%20PUC%20Filing%2027706.pdf.

# 23 LIGHTNING AND OVERVOLTAGE PROTECTION



## 23.1 INTRODUCTION

Temporary overvoltages in power systems occur from a variety of causes such as faults, switching, and lightning. By far, the most severe overvoltages result from lightning strokes to the power system. Quite often, lightning overvoltages will be very high, resulting in insulation breakdown of power apparatus with destructive results. It is therefore imperative that power systems be designed in such a way that expected overvoltages be below the withstand capability of power apparatus insulation. Many times, this basic requirement is translated into excessive cost. For this reason, one seeks a compromise in which power systems are designed in such a way that the possibility of destructive failure of power apparatus due to lightning overvoltages is minimized. This procedure is based on coordinating the expected overvoltages and the withstand capability of power apparatus. Two steps are typically involved: (1) proper design of the power system to control and minimize the possible overvoltages and (2) application of overvoltage protective devices. Collectively, the two steps are called *overvoltage protection* or *insulation coordination.*

The importance of overvoltage protection cannot be emphasized enough. First it affects system reliability, which translates into economics. Traditionally, overvoltage protection methods were guided by the objective to maximize system reliability with reasonable investment cost. In this sense, transient overvoltages which do not lead to interruptions are acceptable and short-duration interruptions are tolerable. Recently, however, with the introduction of sensitive electronic equipment, such as digital controllers, relays, recorders, power devices with power electronic interfaces such as PV, wind, etc., new concerns have been raised. Lightning overvoltages can damage electronic equipment leading to outages and downtimes. In addition, follow-up power faults (initiated by lightning overvoltages) create power frequency undervoltages and overvoltages (sags and swells), which can also trigger load disruption. Any disturbance, damage, downtime, etc. is characterized as a power quality issue. It is transforming the practices for overvoltage protection. While the application of overvoltage protection devices is pertinent, more and more emphasis is placed on design procedures to minimize the possible overvoltages and control the sources of

Grateful acknowledgment is given to a past contributor to this section: A. P. (Sakis) Meliopoulos.

disturbances. An attempt has been made in this section to provide a balanced treatment of overvoltage protection in view of present-day concerns.

The subject of lightning and overvoltage protection is rather complex. A thorough treatment requires good understanding of many related subjects. First, the mechanisms by which lightning is generated and how its pertinent characteristics are related to power systems must be well understood. Second, the response of power systems to lightning and other causes of overvoltages must be studied. Analysis methods to study the phenomena are indispensable tools, which provide the basis for proper selection of design options. Invariably, overvoltages can be minimized, but they cannot be eliminated. As a result, power systems must be protected against overvoltages using overvoltage protection devices (surge arresters). In recent years, major breakthroughs have occurred in protective device technology. Effective protection requires a deep understanding of the capabilities of present technology as well as its limitations.

## 23.2 BASIC CONCEPTS AND DEFINITIONS

Electric power systems are subjected to external surges (lightning) as well as internally generated surges (switching), which may result in temporary high voltages. To maintain a highly reliable system, protection against these overvoltages is needed. This need is dictated by the fact that the insulation of power equipment (which may be air, oil, $SF_6$, etc.) is subjected to breakdown if sufficiently high voltage is applied. This protection involves a coordinated design of the power system itself and placement of proper protection devices at strategic locations for the purpose of suppressing overvoltages and avoiding or minimizing insulation failures.

Coordinated design involves these items:

- Effective grounding techniques
- Use of shielding conductors
- Preinsertion resistors during switching
- Switching angle control among breaker poles
- Use of surge capacitors
- Use of surge arresters

Protection devices include spark gaps and various designs of surge arresters.

The basic objective of overvoltage protection of power systems is to avoid insulation breakdown and associated outages or damage to equipment. This means that the design objective is to make sure that the actual overvoltages reaching a device should be lower than the breakdown voltage of the insulation. The most common insulators used in power system apparatus and their characteristics are listed in Table 23-1.

**TABLE 23-1** Common Insulators in Power Apparatus

| Insulator | Breakdown, MV/m | Resistivity, $\Omega \cdot$ m | Relative permitivity |
|---|---|---|---|
| Air | 3 | $\rho = \infty$ | $\epsilon_r = 1$ |
| Oil | 10 | $10^4 \times 10^{10}$ | 2.2 |
| $SF_6$ | 15 at 1 atm<br>59 at 5 atm | | |
| Mica | 100 | $10^{11}$–$10^{15}$ | 4.5–7.5 |
| XLPE | 50 MV/m | $10^{11}$–$10^{14}$ | 2.8–3.5 |
| EPR | 60 MV/m | $10^{11}$–$10^{14}$ | 2.5–3.5 |
| Porcelain | 10 | $3 \times 10^{12}$ | 5.7 |
| Glass | ... | $10^{12}$ | 4–7 |

In general, in terms of potential damage to equipment, the insulation of power apparatus can be classified into external and internal as follows:

- External insulation
  Air
  Porcelain
  Glass
- Internal insulation
  Oil
  $SF_6$
  Mica
  XLPE
  EPR

The effects of external insulation breakdown are not as destructive as internal insulation breakdown. The reason is that external insulation is, in general, self-healing (self-restoring) after the cause of breakdown (overvoltage) ceases to exist. On the other hand, internal insulation breakdown generally results in permanent damage to the equipment and possibly catastrophic failure. These facts dictate different approaches for external and internal insulation protection. For external insulation protection, the objective is to minimize the expected number of insulation breakdowns subject to economic constraints. In this sense, many sophisticated approaches have been developed, which balance system reliability (which is mainly related to insulation breakdowns) versus cost. Because many of the exogenous parameters, such as lightning strength and soil parameters are statistical in nature, the methodologies use statistical approaches. For internal insulation protection, deterministic methods are preferred where the objective is to design for zero insulation breakdowns.

The above simplistic characterization of external and internal insulation is not always apparent in power apparatus. Specifically, the insulation of a specific power apparatus may be complex. For example, consider a transformer. The windings of the transformer may be submerged in oil (the dielectric is oil) while the terminals are exposed to air through the bushings (the dielectric is the air). When considering withstand capability of a power apparatus, we are not concerned with which dielectric will break first, although this is part of the design process. But rather we are concerned with the question of at what voltage the insulation (any part) will break down. Because insulation breakdown depends on voltage waveform as well as on some other factors, the following definitions, which have been taken from the ANSI Std C92.1, apply:

*Withstand Voltage.* The voltage that electrical equipment is capable of withstanding without failure or disruptive discharge when tested under specified conditions.

*Insulation Level.* An insulation strength expressed in terms of a withstand voltage (typically 10% less than the withstand voltage).

*Transient Insulation Level (TIL).* An insulation level expressed in terms of the crest value of the withstand voltage for a specified transient wave shape, for example, lightning or a switching impulse.

*Lightning Impulse Insulation Level.* An insulation level expressed in terms of the crest value of a lightning impulse withstand voltage.

*Switching Impulse Insulation Level.* An insulation level expressed in terms of the crest value of a switching impulse withstand voltage.

*Basic Lightning Impulse Insulation Level (BIL).* A specific insulation level expressed in terms of the crest value of a standard lightning impulse.

*Basic Switching Impulse Insulation Level (BSL).* A specific insulation level expressed in terms of the crest value of a standard switching impulse.

Note that two of the most commonly used measures, the basic lightning impulse insulation level and the basic switching impulse insulation level, are the most widely used values to characterize the

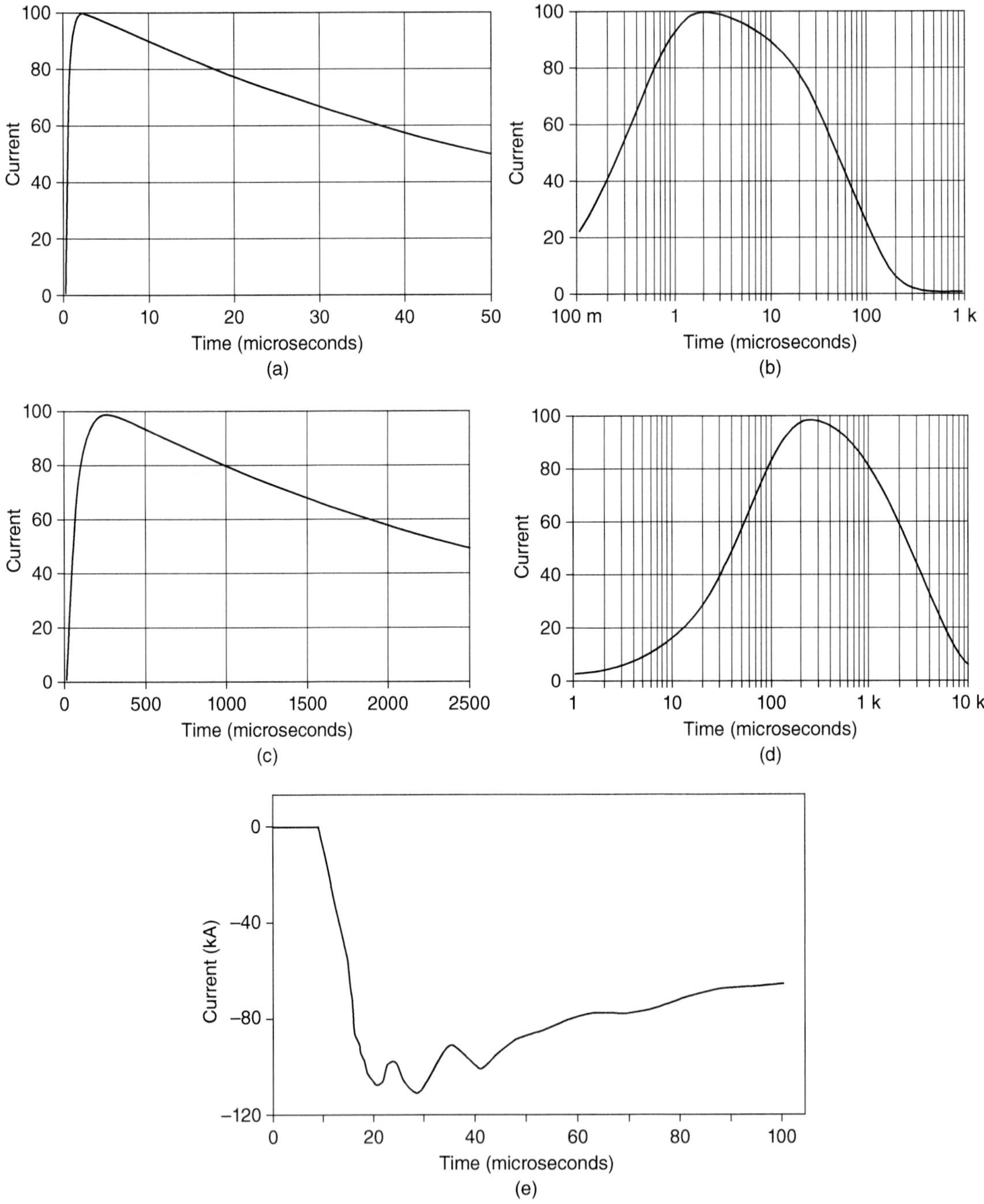


**FIGURE 23-1** Standard waveform: (*a*) standard lightning impulse, linear-linear scale; (*b*) standard lightning impulse, linear-log scale; (*c*) standard switching impulse, linear-linear scale; (*d*) standard switching impulse, linear-log scale; (*e*) example or actual lightning current.

insulation of power apparatus. Note that they are defined in terms of two specific waveforms: (1) the standard lightning impulse and (2) the standard switching impulse. The definitions of these waveforms are

*Standard Lightning Impulse.* A full impulse having a front (rise) time of 1.2 μs and a time to half value of 50 μs. It is described as a 1.2/50 impulse. (See *American National Standard Measurement of Voltage in Dielectric Tests,* C68. 1.)

*Standard Switching Impulse.* A full impulse having a front (rise) time of 250 μs and a time to half value of 2500 μs. It is described as a 250/2500 impulse. (See American National Standard C68.1.)

These waveforms are illustrated in Fig. 23-1.

The standard impulses were introduced because they remotely resemble lightning and switching waveforms, and they can be easily generated in a laboratory via an impulse generator. The basic (single-stage) structure of an impulse generator is illustrated in Fig. 23-2*a*. By stacking many basic structures together (as shown in Fig. 23-2*b*), one can create an impulse generator capable of generating an output impulse many million volts in crest.

The impulse voltage withstand of a power apparatus is strongly dependent on the duration of the impulse voltage. The time dependence is mainly due to the fact that arc generation involves an electron avalanche which takes some time to form. The full development of an arc across an insulator is classified as a breakdown. The time to breakdown is normally quantified with a volt-time characteristic. This characteristic can be determined by applying impulses across an insulator of increasing magnitude and recording the voltage and time at which breakdown occurred. For self-restoring

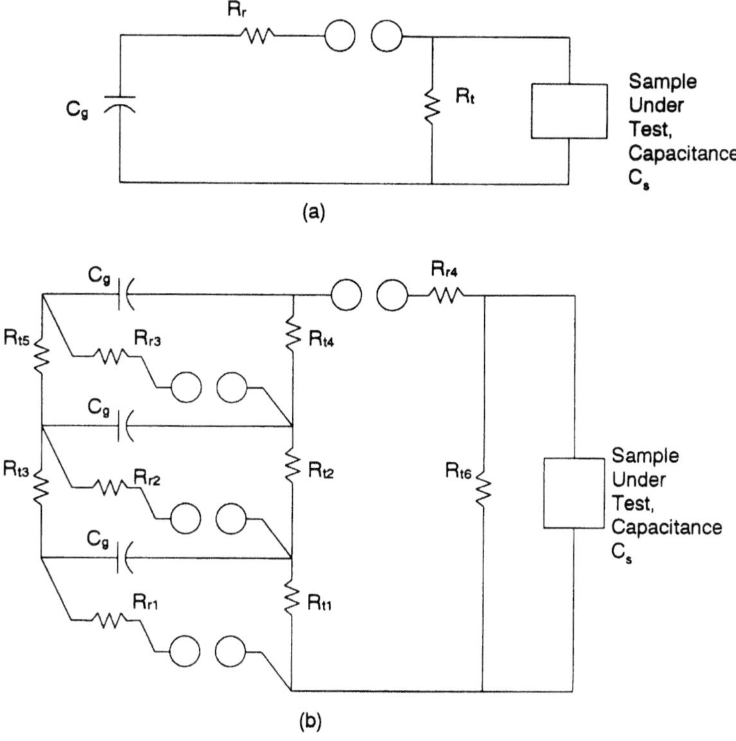


**FIGURE 23-2** Impulse generator: (*a*) single-stage; (*b*) multiple-stage.

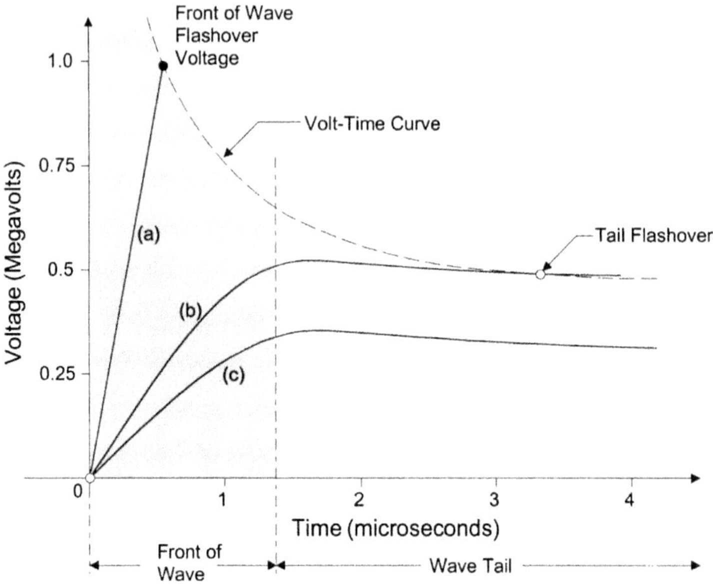


**FIGURE 23-3** Determination of the volt-time curve of insulation breakdown.

insulation, this test is relatively simple and is illustrated in Fig. 23-3. In this way, volt-time curves for all insulators in usage have been determined. Unfortunately, the withstand voltage of non-self-restoring insulation cannot be readily determined without destroying the sample. This means that determining the volt-time curve of non-self-restoring insulation is a practical impossibility. For this reason, the methods for determining withstand voltage for internal insulation are different. Specifically, internal insulation is designed for a specific withstand capability, the *design withstand*. The manufacturer must guarantee a certain withstand at which the insulation, if tested, will not fail. This is the *tested withstand* and it is normally lower than the design withstand. Apparently, the *actual withstand* cannot be known without destroying the sample. The actual withstand is definitely higher than the tested withstand and probably higher than the design withstand.

There is another issue related to the fact that the withstand voltage depends on many other factors that exhibit random variations. Some of them are

- Insulation geometry and smoothness of surfaces
- Insulation contamination
- Atmospheric conditions
- Voltage polarity

It is a practical impossibility to quantify the effects of all variables on voltage withstand. For this reason, voltage withstand is described in statistical terms. In this sense, the following definitions apply with reference to Fig. 23-3:

*Critical flashover* (CFO) is the crest voltage of an applied impulse wave that will cause flashover on the tail of the wave 50% of the time and no flashover the other 50% of the time.

*Critical withstand* is the highest crest voltage insulation can take without flashover under specified conditions—usually less than 1% probability of flashover.

*Rated withstand* is the crest voltage that insulation is required to withstand without flashover when tested by established standards under specified conditions (usually 5% to 10% less than critical withstand).

In summary, in this subsection we reviewed several basic concepts and definitions, which are useful in the process of designing overvoltage protection systems.

## 23.3 MECHANISMS AND CHARACTERISTICS OF LIGHTNING

***Introduction.*** Atmospheric electrical discharges known as *lightning* or *thunderbolts* (from cloud to cloud or cloud to ground) have captured the imagination and fear of the human race since ancient times. The ancient Greeks believed that lightning was Zeus' tool to punish human misbehavior or to demonstrate his anger. It was not until Benjamin Franklin that the first scientific inquiry occurred into the phenomenon of lightning. Since that time, lightning has been extensively studied and many theories have been developed, which reasonably explain the phenomenon. In addition to these theories, there exists an enormous amount of measured data of lightning characteristics. These data are useful for design of protection schemes against lightning.

This subsection presents a brief overview of the theory of thundercloud formation and lightning, the characteristics of lightning, and describes existing relevant data.

***The Electrification of Thunderclouds.*** The cause of lightning is separation and accumulation of electrical charges in clouds via certain microphysical and macrophysical phenomena. This electrification results in electric field intensities high enough to cause air breakdown and subsequent development of lightning. To explain these phenomena, certain theories have been developed. The most useful are the precipitation and convection theories and later improvements, most notably the charge-reversal temperature theory. Understanding of these theories is helpful in the design of protection systems against lightning. A brief description of the cloud electrification theories is provided in this subsection.

The precipitation theory, postulated as early as 1885 by physicists Elster and Geitel, is based on the observation that large water droplets accelerate toward ground because of gravity, while smaller water droplets (mist) remain suspended in air or rise as warmer air moves upward. Collisions between large water droplets and mist of water droplets and possibly ice crystals in the colder altitudes result in transfer of a net negative charge to the large water droplets. As they move toward lower altitudes (by gravity), they cause a net negative charge in the lower part of the cloud. Conservation of charge requires that the upper part of the cloud be positively charged, resulting in a dipole structure in the thundercloud. A simplified illustration of the process is given in Fig. 23-4.

**FIGURE 23-4** Illustration of the precipitation theory of cloud electrification: (*a*) separation of the charge due to collisions; (*b*) cloud electrification due to precipitation of charged water droplets.

The convection theory, which was formulated much later, is based on transfer of charged particles from one location of a cloud to another by the upward and downward drafts in the cloud. The theory suggests that the charged particles are generated by two mechanisms: (1) cosmic rays impinge on air molecules and ionize them, resulting in two ions, one positively charged, the other negatively charged; and (2) high-intensity electric fields around sharp objects on the earth's surface produce corona discharges, which result in positively charged ions. The positive ions are transported to higher altitudes by the upward draft in the cloud. On the other hand, the negative ions attach themselves to water droplets and ice particles, which move to lower altitudes due to gravity or downward drafts. The net result is a dipole structure in the thundercloud. A simplified illustration of the process is given in Fig. 23-5.

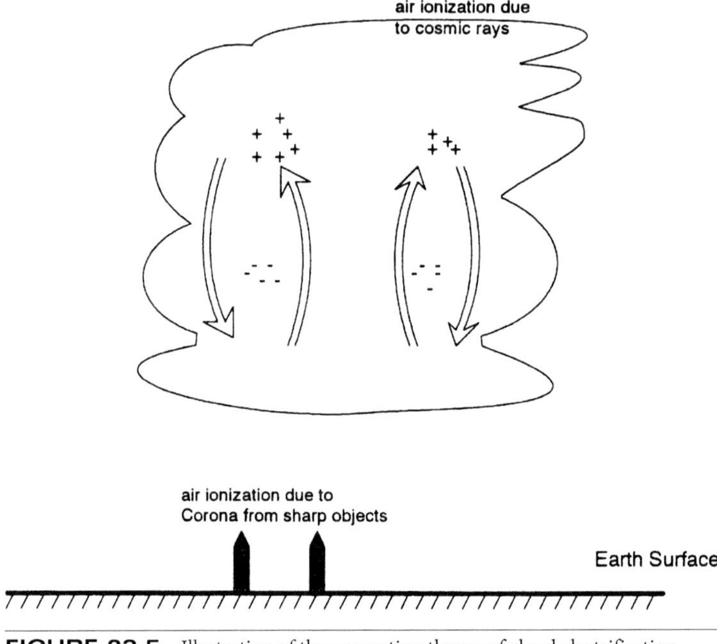


**FIGURE 23-5** Illustration of the convection theory of cloud electrification.

Precipitation and convection occur in a thundercloud simultaneously. Yet the two theories are distinct and independent. Both theories postulate that the thundercloud is a dipole with the negative pole near the earth, that is, negative dipole. Measurements made by Wilson and later by Simpson of the polarity of the dipole resulted in conflicting conclusions which generated debate and further research. Specifically, Wilson's measurements indicate that the thundercloud is a negative dipole (negative charge at the lower part of the cloud) while Simpson's measurements indicated a positive dipole. It took five decades of additional experimentation and measurements to resolve this apparent conflict. Today's most complete theory for lightning phenomena has established the fact that the structure of a thundercloud is *tripolar*, not bipolar. This structure allows the understanding of both Wilson's and Simpson's conclusions. Specifically, an electric tripole of the size of a thundercloud observed from a single specific point will appear as a dipole. Depending on the point of observation one may conclude that it is a negative or positive dipole. Apparently, Wilson and Simpson made their measurements from different observation points. Their measurements were correct but because Wilson made the measurements from a distant point he concluded that the thundercloud is a negative dipole, while Simpson made his measurements from a point underneath the head of the cloud and concluded that the thundercloud is a positive dipole.

Both theories, precipitation and convection, do not completely explain all phenomena occurring in a thundercloud. For example, it has been observed from studies that larger droplets, when they break, acquire positive charge on aggregate. This leads to the hypothesis that the positive charge is due to large droplets that break as they accelerate toward ground. However, this hypothesis is not totally true because it does not explain the fact that precipitation particles below the negative charge carry much greater positive charge than those produced by the droplet fragmentation process. Another hypothesis was based on ice particles accelerating toward ground—as the ice particles reach lower altitudes, they melt and tend to acquire positive charges, which explains the existence of positive charge at altitudes below 4 km. However, this hypothesis still does not explain the existence of positive charges at higher altitudes.

Recent measurements and observations in the past three decades resulted in another hypothesis which explains the tripole nature of a thundercloud. This is the so-called charge-reversal hypothesis, which states that when graupel particles collide with ice crystals, the charge transferred to a graupel particle is dependent on the temperature. At temperatures above a certain value, which is called the *charge-reversal temperature*, the transferred charge is positive. The exact value of the charge-reversal temperature is being debated, but it is believed to be around −15°C. The process is illustrated in Fig. 23-6 in a simplified manner. Considering the fact that the temperature of the atmosphere is −15°C at an approximate altitude of 6 km, this means that due to collisions of graupel particles and ice crystals, the thundercloud will be, on aggregate, negatively charged for altitudes above 6 km and positively charged below 6 km. The situation is illustrated in Fig. 23-7. This hypothesis has been verified in the laboratory and explains the levels of negative and positive charges in a thundercloud. Yet, the exact microphysics of this phenomenon are practically unknown.

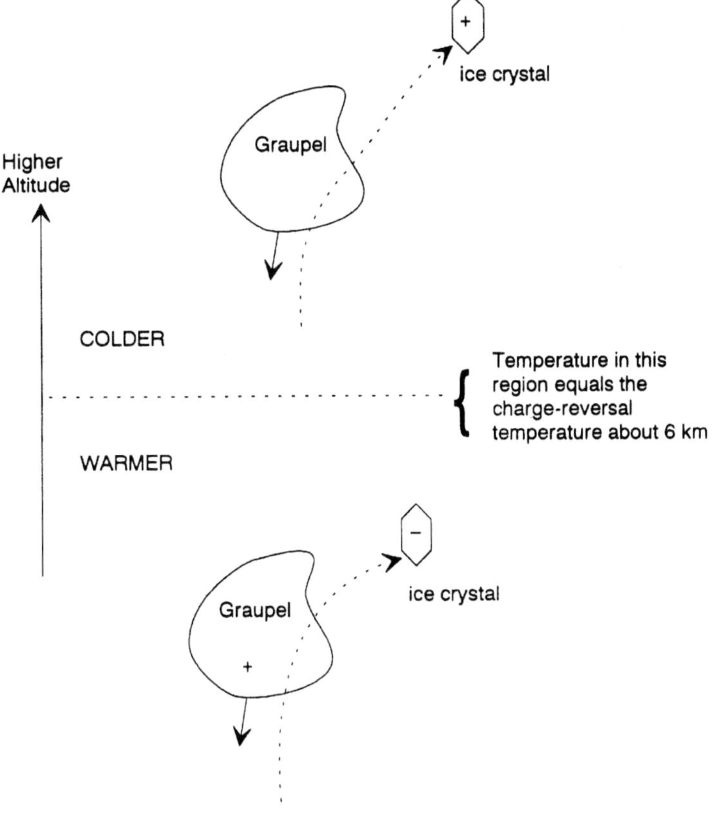


**FIGURE 23-6** Explanation of the charge-reversal temperature theory.

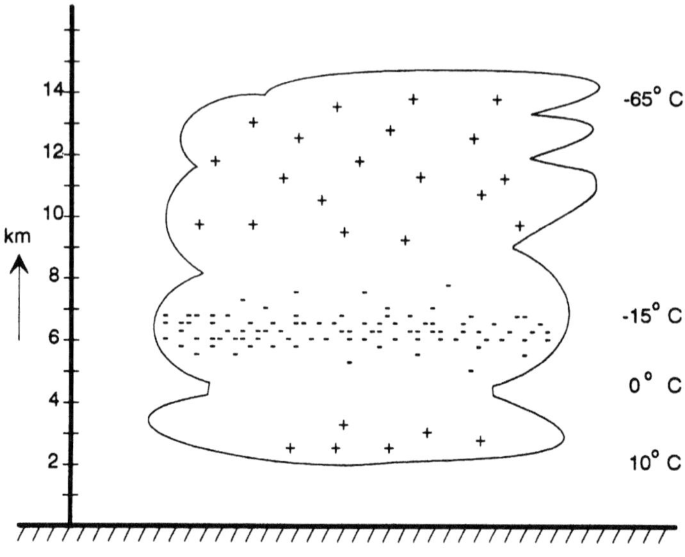


**FIGURE 23-7** An electrified thundercloud is typically tripolar.

In summary, the precipitation model with the graupel—ice crystal interaction and charge-reversal temperature best explains most of the behavior of a thundercloud. Yet this model totally ignores the forceful upward and downward drafts within a thundercloud. The convection model considers these drafts but it is unable to explain certain observed phenomena in a thundercloud. Perhaps one day a theory will be developed, which combines the precipitation and convection models and completely accounts for all phenomena related to the electrification of a thundercloud. What has been verified with measurements are the following facts: a thundercloud can be electrified in such a way that positive charge accumulates at the top of the cloud and negative, at the lower part of cloud. A smaller positive charge may be present at lower altitudes of a thundercloud. These charges are responsible for lightning. The mechanism of lightning is explained next.

***Mechanisms of Lightning.*** Lightning initiates whenever the charge accumulation in a thundercloud is such that the electric field between charge centers inside the cloud or between cloud and earth is very high. For power engineering purposes, only cloud-to-earth lightning strokes (ground flashes) are of importance and will be discussed next. An electrified thundercloud will generate an electric field in the space between the cloud and earth as is illustrated in Fig. 23-8. When the intensity of this field is high enough, a discharge will initiate. Typically, the process involves three phases. In the first phase, the high electric field intensity may generate local ionization and electric discharges, which are known as *pilot streamers*. A pilot streamer is followed by the so-called *stepped leader*. The stepped leader is a sequence of electric discharges, which are luminous; they propagate with a speed approximately 15% to 20% of the speed of light, and they are discrete, progressing approximately 50 m at a time. The time between *steps* is few microseconds to several tens of microseconds. A pictorial view of the stepped leader development is shown in Fig. 23-9. The stepped leader will eventually reach the surface of the earth and will strike an object on the earth. However, where it will strike is not determined until the stepped leader is within a *striking distance* from the object. A model for the striking distance will be described in Sec. 23.7. It is possible that a stepped leader or multiple stepped leaders may also initiate from an object on the surface of the earth. In this case, the two stepped leaders may meet at some point. There is also evidence that the initial stepped leader may originate from a tall structure on the earth and not from the cloud.

The second phase initiates when the stepped leader reaches an object on the earth or meets an upward moving stepped leader. When this occurs, a conductive path between the thundercloud and

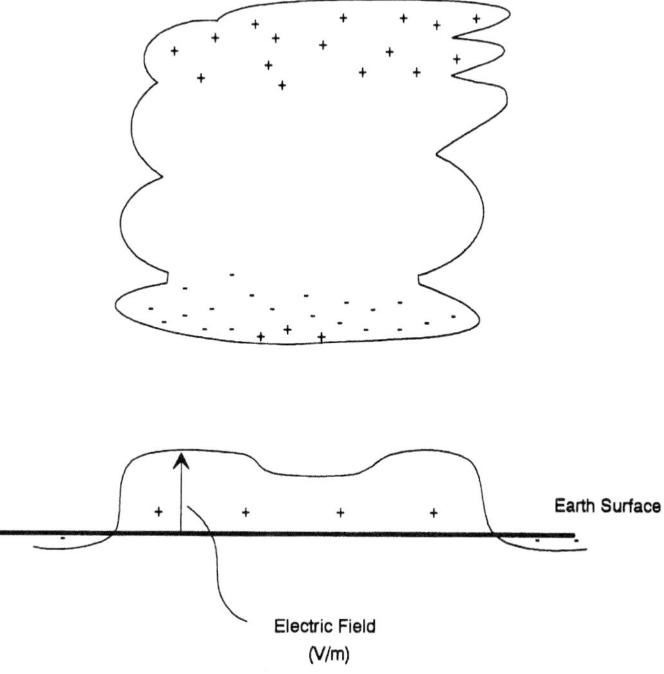


**FIGURE 23-8** Illustration of electric field below an electrified thundercloud.

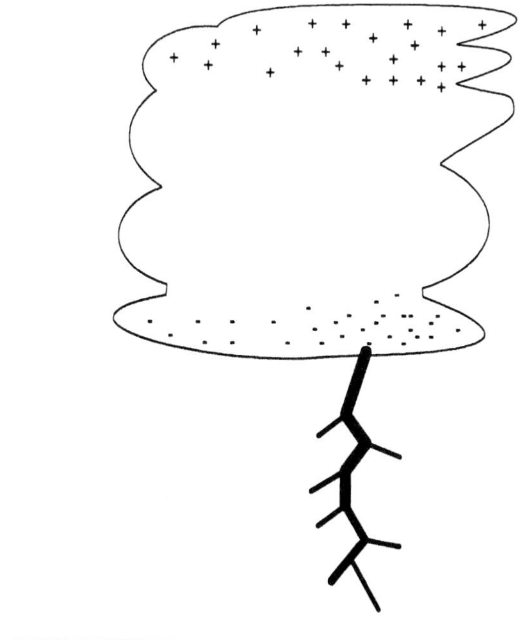

**FIGURE 23-9** Illustration of stepped-leader development.

**FIGURE 23-10** Illustration of return-stroke development.

earth has been established. Then, a high-intensity discharge occurs through the channel established by the stepped leader. This discharge is extremely luminous and therefore visible. It propagates with a speed of about 10% to 50% of the speed of light. The development of the return stroke is illustrated in Fig. 23-10. The return stroke carries an electric current of anywhere from few thousands of amperes to 200 thousands of amperes. The current magnitude rises fast, within 1 to 10 μs, to the peak value and then decreases rapidly. The discharge is known as the *return stroke* or simply the *lightning stroke*. The return stroke transfers a substantial amount of positive charge from the earth to the cloud and specifically to the charge center where the lightning was originated. This transfer results in a significant lowering of the potential of the charge center. This phenomenon initiates the third phase of lightning. In this phase, discharges may occur from other charge centers within the thundercloud to the depleted charge center because of the increased potential difference between them. This discharge will trigger another stroke between cloud and ground through the already established conductive channel with the first stroke. This process may be repeated several times, depending on the electrification status of the thundercloud, resulting in multiple strokes. There is evidence that most lightning to ground involves multiple strokes. For example, an analysis of 1430 strokes to the earth by Anderson (1968) resulted in the following statistics:

| | |
|---|---|
| Single-stroke lightning | 36% |
| Lightning involving six or more strokes | 21% |
| Mean value of multiple strokes | 3 |

It should be mentioned, however, that positive polarity lightning is typically single stroke. Extreme cases have been recorded with a large number of multiple strokes such as 40 or 50 with duration of the entire lightning event approaching 1 s. For example, a 40-stroke lightning event, which lasted 0.624 s is on record. The time interval between successive strokes may be in the order of few milliseconds. However, there is evidence that sometimes the multiple strokes may be so smooth as to appear as a continuous lightning current. This occurs because as long as there is a conductive path between the cloud and ground, electric current will flow until the cloud is neutralized enough for the conductive path to interrupt. The continuous flow of lightning current can be destructive because of its long duration even if the magnitude may be much lower than the crest of a stroke.

***Characteristics of Lightning Strokes.*** The parameters of lightning ground strokes are very important in the design of protection schemes against lightning. The most important parameters are

- Voltage between cloud and earth
- Electric current
- Waveform
- Frequency of occurrence

The voltage between a thundercloud and the earth prior to a ground stroke has been estimated from 10 to 1000 MV. For design work, however, the protection engineer is interested in the voltage appearing on the stricken power apparatus. For reasons explained in the next paragraph, this voltage will be equal to the product of the impedance times the stroke current.

It is generally accepted that the ground-stroke current is independent from the terminating impedance. The reason is that the terminating impedance is much lower than the resistance of the lightning discharge channel, which is on the order of few thousand ohms. Thus, a ground stroke is normally considered as an ideal current source at the point of strike. The crest of the stroke electric current can vary over a wide range: 1 to 200 kA. Data on ground-stroke current magnitudes have been collected by many researchers. Among those, the work of Berger (1967) at Mount San Salvatore in Switzerland has been widely accepted. Statistical representation of these data is shown in Fig. 23-11.

The waveform of the lightning ground stroke current, and especially the rise time, is very important. Again, statistical representation of stroke current rise times data collected by Berger is given in Fig. 23-12.

The frequency of occurrence is also a very important parameter. In order to quantify lightning activity, the crude measure of *thunderstorm day* has been introduced. A thunderstorm day is defined as a 24-h period in which at least one thunder clap has been heard. Collection of historical thunderstorm activity data by the National Weather Service resulted in maps of equi-thunderstorm-day contours.

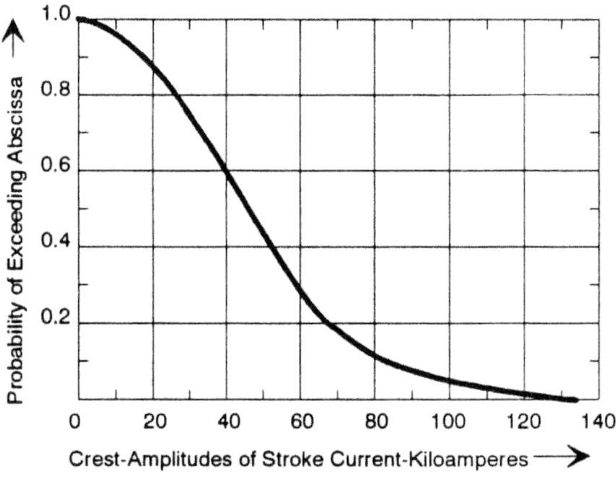


**FIGURE 23-11** Distribution of lightning current magnitudes [recorded by Berger (1967) at Mount San Salvatore].

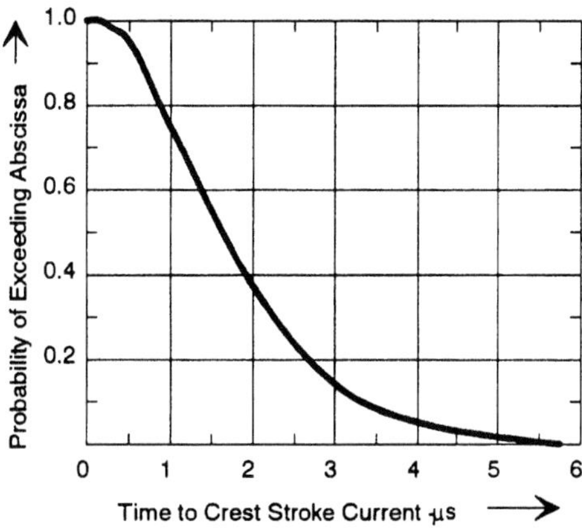


**FIGURE 23-12** Distribution of lightning current rise times [recorded by Berger (1967) at Mount San Salvatore].

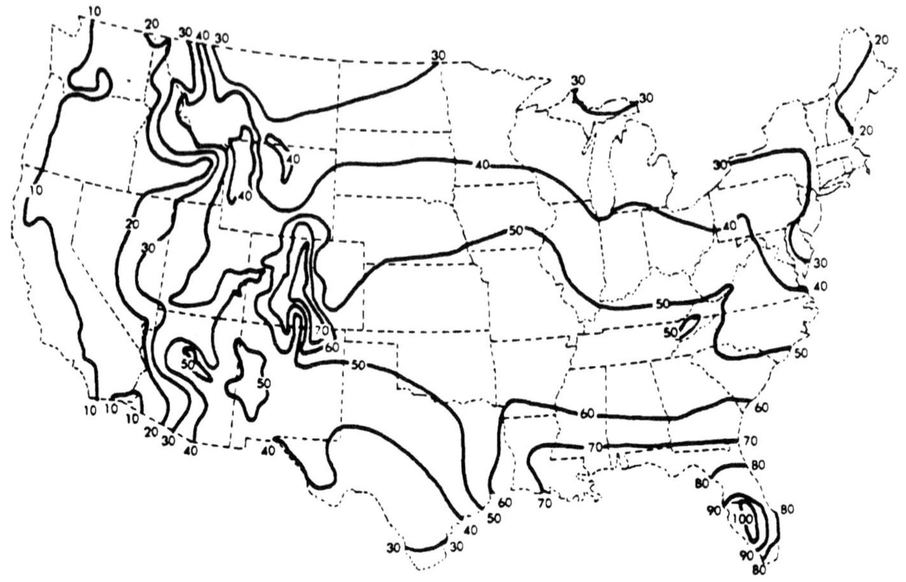

**FIGURE 23-13** Isokeraunic curves for the continental United States.

These maps are known as isokeraunic maps, from the Greek word *keraunos*, lightning. Such a map is illustrated in Fig. 23-13. It is important to note that this map, by definition, provides only a crude measure of lightning activity. Specifically, by definition, a thunderstorm day does not provide any information on the frequency of and total lightning activity. Yet, because of lack of better data before the development of the lightning detection system (see next paragraph), the isokeraunic maps have been used for estimation of lightning activity in an area. There are several models, which provide the approximate number of cloud-to-ground lightning per unit of area as a function of the isokeraunic level. These models will be discussed in detail in Sec. 23.7. As an example, Anderson (1975) has suggested the following

$$N_1 = 0.12T$$

where $N_1$ is the ground flash density per square kilometer per year and $T$ is the number of thunderstorm days (or isokeraunic level).

Early in the 1980s, the Electric Power Research Institute sponsored a project at the State University of New York at Albany (SUNYA) which resulted in the National Lightning Detection Network (NLDN) records. The system integrated two networks and basically records cloud-to-ground lightning discharges. The objective of the project was to collect lightning data over a period of 10 years, which could be used for lightning protection of power systems. Figure 23-14 illustrates average lightning flashes per square kilometer for the state of Florida. The data were collected over a 5-year period (1985 to 1989). Note that these data correlate reasonably well with the isokeraunic maps data of Fig. 23-13 and the model that provides ground flashes from isokeraunic levels. Similar systems have been installed in many countries. Today, the National Lightning Detection Network developed to a system that provides the magnitude (crest) of lightning and the location of lightning with precision in the order of 1 km.

***Summary.*** This subsection has described the mechanism of thundercloud formation and electrification and the initiation, mechanism, and characteristics of lightning. Finally, statistical data on lightning parameters were presented. These data are useful for design work.

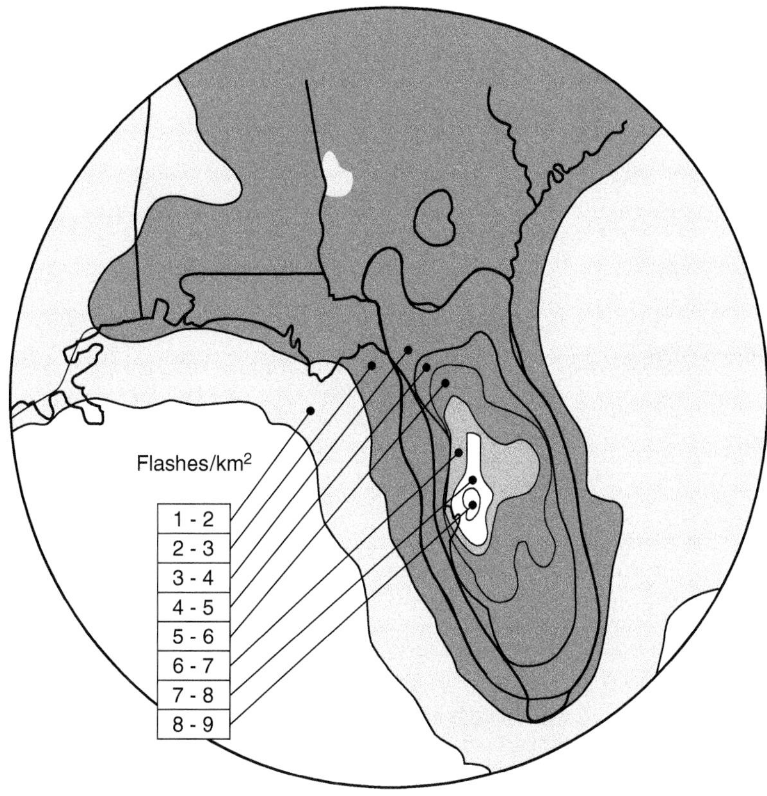


**FIGURE 23-14** Average annual flash density data from 1985 to 1989. (*Electric Power Research Institute.*)

## 23.4 POWER SYSTEM OVERVOLTAGES

The causes of power system overvoltages are numerous and the waveforms are complex. It is customary to classify the transients on the basis of frequency content of the waveforms. In this sense, the following three broad categories are defined:

Power frequency overvoltages

Switching overvoltages

Lightning overvoltages

Table 23-2 provides brief descriptions and typical causes of the most commonly encountered overvoltages in power systems. The relative level of overvoltages due to these causes is illustrated in Fig. 23-15. This figure provides typical and usual levels of overvoltages in a well-designed system and excludes specific cases such as ferroresonance.

In designing a well-protected electric power system, it is extremely important to thoroughly understand the types, frequency, and magnitude of the expected overvoltages on the power system. For this reason, this subsection provides a concise discussion of the nature, generation mechanisms, and characteristics of power frequency, switching, and lightning overvoltages in power systems.

**TABLE 23-2** Power-System Overvoltages

| Category | Description | Causes |
|---|---|---|
| Power frequency overvoltages | Temporary overvoltages dominated by the power frequency component | Electric faults<br>Sudden changes of load<br>Ferroresonance |
| Switching overvoltages | Temporary overvoltages resulting from a switching operation | Energization of lines<br>Deenergization of capacitor banks<br>Fault interruption/TRV<br>High-speed reclosing<br>Energization/deenergization of transformers<br>Other |
| Lightning overvoltages | Temporary overvoltages resulting from a lightning stroke terminating at a phase conductor, shield conductor, any other part of a power system, or a nearby object (tree, etc.) | Lightning—cloud-to-ground flashes |

***Power Frequency Overvoltages.*** The magnitude of power frequency overvoltages is typically low compared to switching or lightning overvoltages. Specifically, for most causes of these types of overvoltage, the magnitude may be few percent to 50% above the nominal operating voltage. In a few cases, power-frequency voltages may reach 300% to 400% of nominal voltages, for example, during ferroresonance. Power-frequency overvoltages play an important role in the application of overvoltage

**Power frequency overvoltages**
**Switching overvoltages**
**Lightning overvoltages**

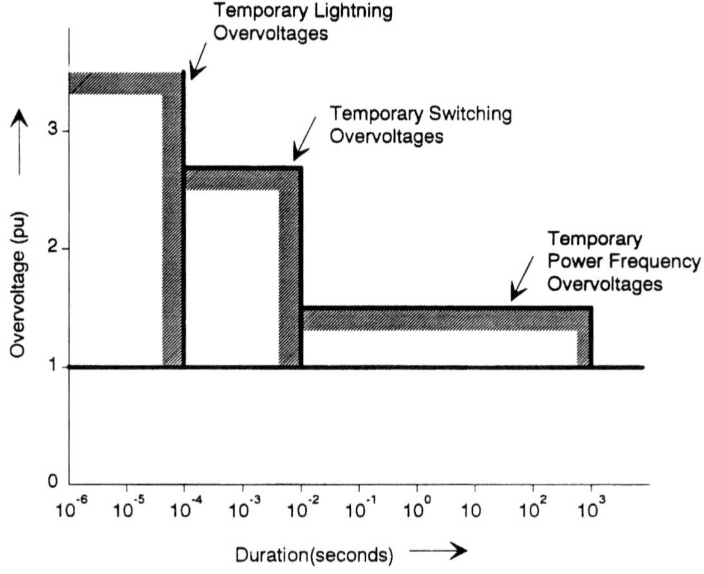


**FIGURE 23-15** Typical range of magnitude and duration of power system temporary overvoltages excluding specific cases such as ferroresonance. [*From Regaller (1980).*]

protection devices. The reason is that modern overvoltage protection devices are not capable of discharging high levels of energy associated with power frequency overvoltages. Thus, it is imperative that protective device ratings be selected in such a way that they do not operate under any foreseeable power frequency overvoltages.

The most common causes of power frequency overvoltages are (1) electric faults, (2) sudden changes of load, and (3) ferroresonance. An electric fault results in voltage collapse for the faulted phase and in a possible overvoltage at the unfaulted phases. The magnitude of the overvoltage depends on the parameters of the circuit, such as positive-, negative-, and zero-sequence impedance, as well as the grounding parameters of the system, such as ground impedance or single- or multiple-grounded system. Figure 23-16 illustrates a typical case of a single-phase-to-ground fault at the end of a 40-mi-long 115-kV transmission line. Because the electric power system is not completely symmetric, the magnitude of the overvoltage on the unfaulted phases may be different; that is, for the case of Fig. 23-16, the overvoltage on phase B is 28.3%, while for phase C the overvoltage is 31.9%. Many studies have been performed over the years to determine simple techniques for determining the power frequency overvoltages. As a first approximation, one can determine the power frequency overvoltage due to a fault from the sequence parameters (positive-, negative-, and zero-sequence impedances) at the fault

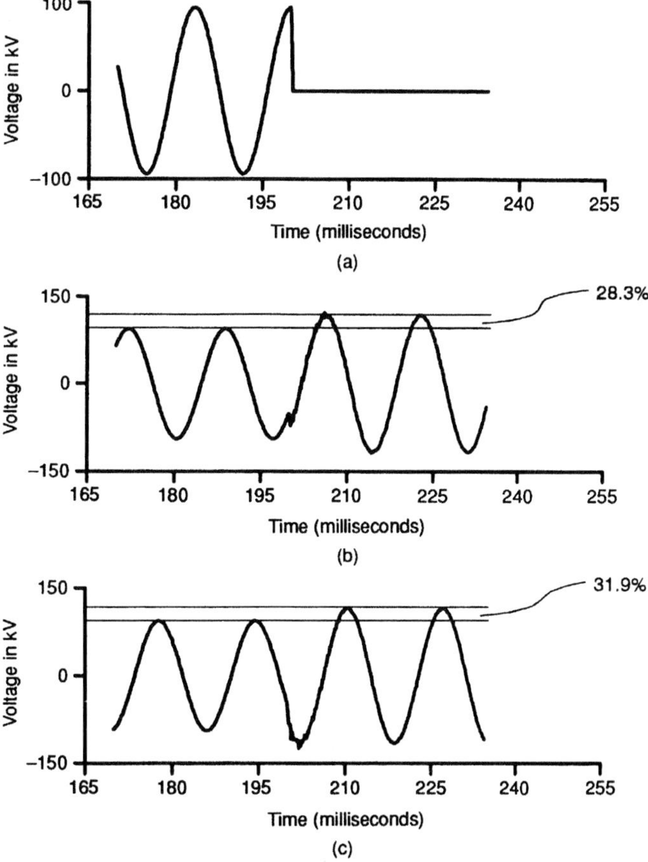


**FIGURE 23-16** Overvoltage due to a single-phase-to-ground fault at the end of a 40-mi-long 115-kV line: (*a*) phase A voltage; (*b*) phase B voltage; (*c*) phase C voltage.

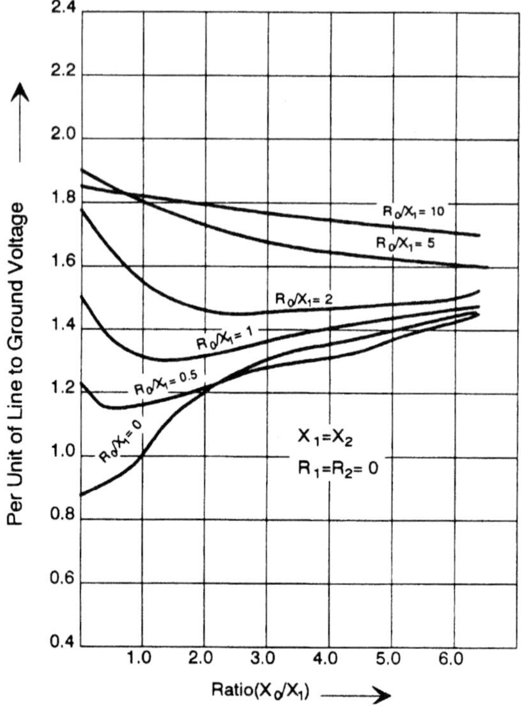


**FIGURE 23-17** Overvoltage on unfaulted phase during single-line-to-ground fault. [*From Johnson (1979).*]

location. Figure 23-17, taken from Johnson (1979), illustrates the power frequency overvoltage at the unfaulted phases due to a ground fault in one phase as a function of the ratios $(X_0/X_1)$ and $(R_0/X_1)$.

Computer models for determining the power frequency overvoltage by taking into consideration all relevant factors have been developed. Using these models, one can determine the exact power frequency overvoltage and the effect of grounding practices. As an example, Fig. 23-18, taken from Mancao et al. (1992), illustrates the maximum line-to-ground overvoltage, per unit (pu), on typical distribution circuits versus fault distance from the feeding substation.

Another source of power frequency overvoltages is the so-called Ferranti effect, which occurs when a load is disconnected at the end of a long transmission line. In this case, the line draws a capacitive current from the source, which generates a voltage gradient along the line of such a phase as to increase the voltage at the open end of the line. An approximate expression of the overvoltage at the open end of the line is given by

$$\text{Overvoltage in pu} = \frac{1.0}{cos(\beta l)}$$

where $\beta$ is the propagation characteristic of the line $(\omega\sqrt{LC})$ and $l$ is the total length of the line. A typical transmission line of 400 mi may experience an overvoltage of 1.30 pu when one end of the line is open.

Ferroresonance is another cause of power frequency overvoltages but is less frequent. Ferroresonance may occur when energizing long transmission lines and unloaded power transformers, in single-phase switching of a 3-phase transformer, and in other cases involving an iron-core magnetic circuit connected to a substantially capacitive circuit. The overvoltages

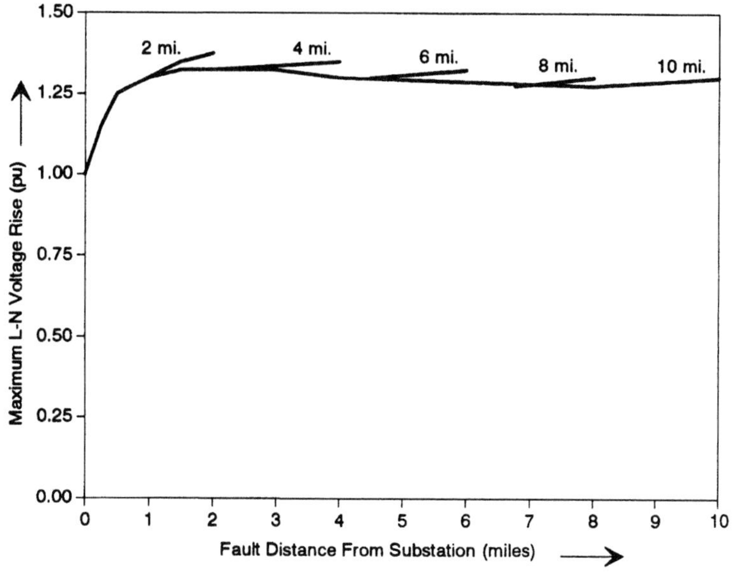


**FIGURE 23-18** Overvoltage on unfaulted phases of a distribution circuit as a function of fault distance and circuit length. [*From Mancao et al. (1992).*]

resulting from ferroresonance can be serious and especially destructive to gapless arresters present in the system. As an example, Fig. 23-19, taken from an IEEE committee report, illustrates the maximum overvoltage due to ferroresonance involving single- or double-phase switching of a 3-phase transformer with an ungrounded primary (delta or wye). The severity of ferroresonance depends on the amount of capacitive reactance present in the system and it can reach a value over 400%.

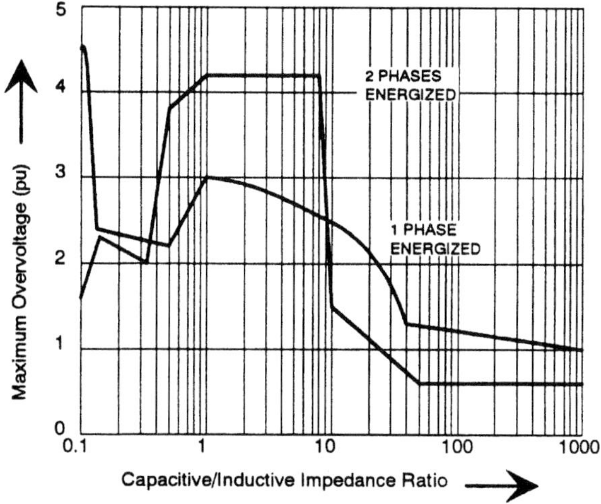


**FIGURE 23-19** Maximum overvoltage due to ferroresonance triggered by single-or double-phase switching of a 3-phase transformer with an ungrounded primary.

Other causes of power frequency overvoltages are (1) generator speedup due to load rejection, (2) generator self-excitation, and (3) malfunction of regulating equipment.

***Switching.*** Switchings in a power system occur frequently. A variety of switchings are performed for routine operations or automatically by control and protection systems. Typical switchings are as follows:

Lines (transmission or distribution)

Cables

Shunt/series capacitors

Shunt reactors

Transformers

Generators/motors

Another class of switching transients are those generated from *insulation flashovers* and breaker *restrikes*. These phenomena are equivalent to the closing of a switch and generate switching surges, which propagate in the system.

Overvoltages resulting from switching operations are typically proportional to the power frequency voltage. For example, energization of a 3-phase line can result in an overvoltage at the open end, which can be as high as 5 pu, depending on the timing of switching with respect to the source. The frequency content of switching transients depends on system parameters. As an example, Fig. 23-20, taken from Johnson (1979), illustrates probability distribution curves of measured line switching overvoltages. Note that there is a substantial probability for overvoltages higher than 5.0 pu.

Switching transients for extra-high-voltage systems, that is, 230 kV and above, can be quite high and must be controlled to avoid the need for higher insulation. There are two methods for controlling the magnitude of switching overvoltages: (1) using breakers with resistor preinsertion and (2) using opening resistors or wound-type potential transformers to discharge trapped charge on lines.

Breakers with resistor preinsertion place a resistor between source and line under energization for a short duration (e.g., 0.8 ms) prior to a direct connection of the source to the line. Proper selection

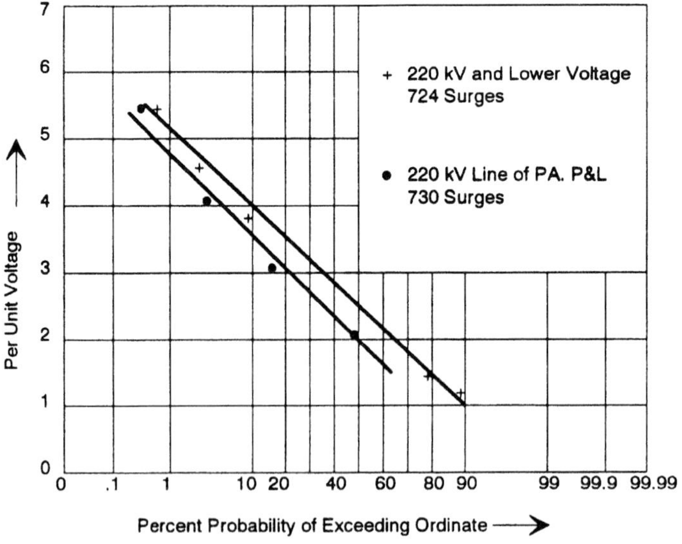


**FIGURE 23-20** Probability distribution curves of measured switching overvoltages. [*From Johnson (1979).*]

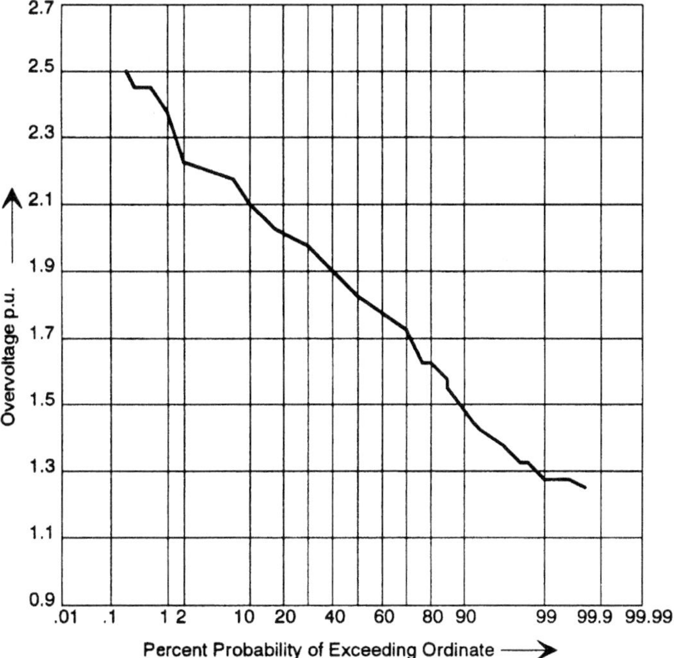


**FIGURE 23-21** Probability distribution curves of computed switching overvoltages (switching with a resistor preinsertion breaker).

of resistor values and insertion time enables effective control of maximum switching overvoltages. As an example, Fig. 23-21 illustrates the probability distribution curve of switching overvoltages when a resistor preinsertion breaker is used. Note that the maximum switching overvoltage is below 2.5 pu.

Trapped charge on a transmission line can cause excessive switching overvoltages (for specific timing of line energization with respect to source phase). In addition, trapped charge can cause breaker restrike because it contributes to overstressing the breaker insulation. Wound-type potential transformers or opening resistors provide a mechanism for quick drainage of trapped charge on lines.

Switchings can cause other undesirable effects such as inrush currents in transformers and ferroresonance, which has already been discussed.

***Lightning Overvoltages.*** Electric power systems are exposed to weather and therefore are subjected to lightning strikes, which result in overvoltages. Lightning overvoltages are generated by direct lightning strikes on a power system apparatus or indirect strikes to nearby objects, from which subsequent overvoltage is transferred to the system via inductive, capacitive, and conductive coupling.

Unlike power frequency overvoltages and switching overvoltages, which are proportional to the system voltage, lightning overvoltages are independent of system voltage but depend on system impedances. For example, a direct lightning hit to a phase conductor of an overhead transmission line will generate an overvoltage proportional to the characteristic impedance of the line and proportional to the current magnitude of the lightning stroke. This overvoltage may be several million volts. It is a practical and economical impossibility to insulate distribution or lower-kilovolt-level transmission lines (i.e., 100 kV and below) to withstand this type of overvoltage. As will be discussed in Sec. 23.7, a coordinated design procedure is applied to minimize the effects of lightning; this procedure involves among other things: (1) shielding of lines and equipment, (2) effective grounding, and (3) application of protective devices (surge arresters). The presence of the shielding system ensures that lightning, which otherwise will terminate to a phase conductor, will terminate on a wire,

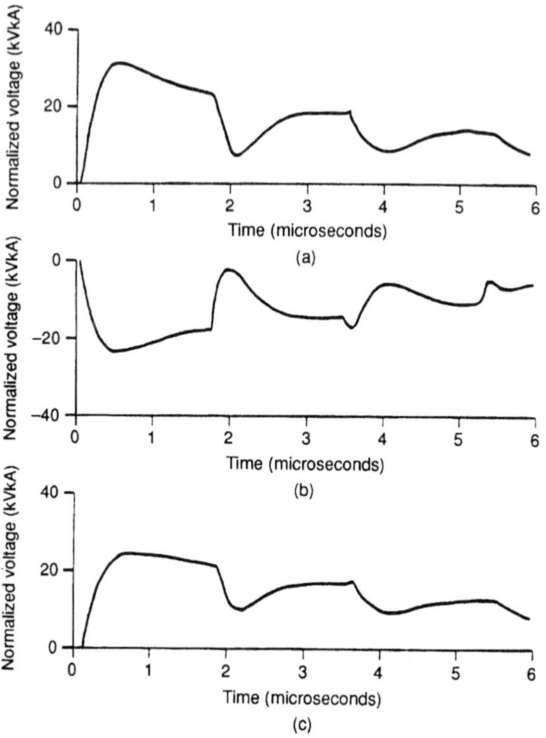


**FIGURE 23-22** Typical lightning overvoltages on a transmission line: (*a*) top of tower voltage; (*b*) voltage across insulator (phase A); (*c*) tower ground potential ruse. [*From Meliopoulos (1988).*]

terminal, etc., which is electrically connected to the grounding system. A well-designed grounding system will divert the majority of the lightning stroke current into the soil and thus will minimize the destructive lightning overvoltages. The subject will be further discussed later. Here, a typical example of lightning overvoltages on a 115-kV shielded line is shown in Fig. 23-22. The figure shows the overvoltage at the top of the tower, voltage across insulator of phase A, and the ground potential rise at the tower base. The voltages are given in kilovolts per kiloampere of lightning stroke current. The case shown is for a relatively short tower with effective grounding. The figure shows the effects of nearby towers, which generate reflections of the lightning surge with the end effect of quickly reducing the lightning overvoltages. It should be apparent that the tower-grounding system plays an important role in determining the magnitude of lightning overvoltages, as illustrated in Fig. 23-23. The topic of grounding will be further discussed later.

Lightning strokes to nearby trees, ground, or other objects can result in voltage surges into the power system through coupling. The coupling can be conductive through the conductive soil and the power system grounding structures, inductive, or capacitive. In a typical situation, all the coupling mechanisms may be present, resulting in a voltage surge to the power system. These voltages are called *induced voltage surges* or *induced lightning overvoltages* and are generally much lower than those occurring after a direct strike. Specifically, they rarely exceed 400 kV. The induced lightning overvoltages are of concern for distribution lines 35 kV or below. Higher-kilovolt-level lines (i.e., 69 kV and above) have sufficient insulation withstand so that induced lightning voltages do not present the risk of flashover.

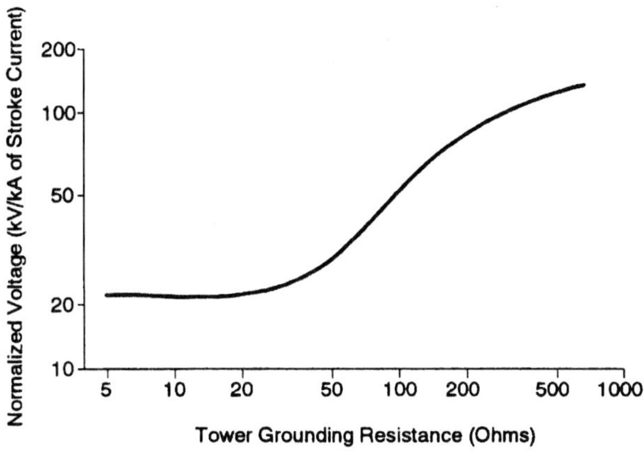


**FIGURE 23-23** Effects of tower footing resistance on a specific 115-kV transmission line (standard lightning wave = 1.2/50 $ms^{-1}$). [*From Meliopoulos (1988).*]

## 23.5 ANALYSIS METHODS

Design of overvoltage protection systems requires a thorough understanding and analysis of transient overvoltages in power systems. Over the years, many analysis methods have been developed for this purpose. All analysis methods require a proper model of the system under study. In this subsection, we shall discuss modeling requirements for transient analysis and various analysis methods.

Modeling is probably the most important task in a study. There are many modeling choices which must be made in a prudent way and in view of the objectives of the study. Modeling choices are affected by

Phenomenon under study

Period of concern

Model selection of individual system components

For overvoltage analysis, typical phenomena under study will be line switching, capacitor bank switching, and lightning. The period of concern may be seconds, milliseconds, or microseconds. Model selection of individual system components should be guided by the expected frequency content of the transient. The selected models should have the proper frequency response required for the study under consideration. There is a large number of components to be modeled. A representative list is given in Table 23-3.

**TABLE 23-3** A Representative List of Power-System Components

| |
|---|
| Transmission lines |
| Single-phase |
| Three-phase |
| Overhead |
| Underground |
| Lumped capacitors |
| Iron-core transformers |
| Generators |
| Surge arresters |
| Grounding |
| Switches, fuses, etc. |

Two other related issues are (1) what to model and (2) how to model. The types of choices to be made regarding the question of what to model may include (1) bus inductance, (2) bus capacitance, (3) transformer winding capacitance, and (4) separation distance between arrester and transformer. The question of how to model is complicated. Typical choices are illustrated in Table 23-4. The task of component model selection is very important. To illustrate the point, consider a 40-mi-long transmission line. The line can be represented as a distributed-parameter

**TABLE 23-4** Typical Model Choices for Components

| Component description | Mathematical model |
|---|---|
| Lumped-parameter model | Ordinary differential equations |
| Resistors | |
| Capacitors | |
| Inductors | |
| Distributed-parameter model | Partial differential equations |
| Lines | |
| Buses | |
| Iron-core transformers | Nonlinear equations |
| Surge arresters | Nonlinear/time-varying equations |
| Circuit-breaker operator | Logical equations |

model or a lumped-parameter model consisting of a set of cascaded pi sections. Figure 23-24 illustrates the switching voltages computed for this line by using the two models. Note that the solutions are different. Of course, the distributed-parameter model provides the correct answer. Examples of simplified lumped-parameter models and distributed-parameter models are illustrated in Fig. 23-25.

Once the model has been selected, analysis can determine the transient overvoltages for a specific event. Analysis methods can be classified into three categories:

Graphical methods

Analytical methods

Numerical methods

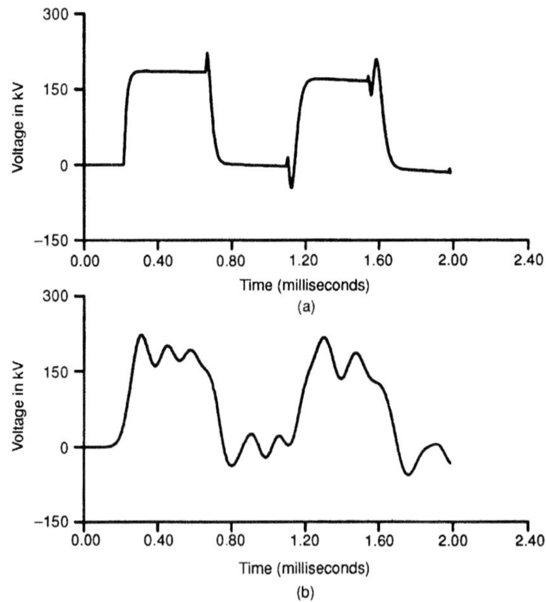


**FIGURE 23-24** Switching overvoltage at the open end of a 40-mi-long 115-kV line: (*a*) distributed-parameter model; (*b*) model with six cascaded pi sections.

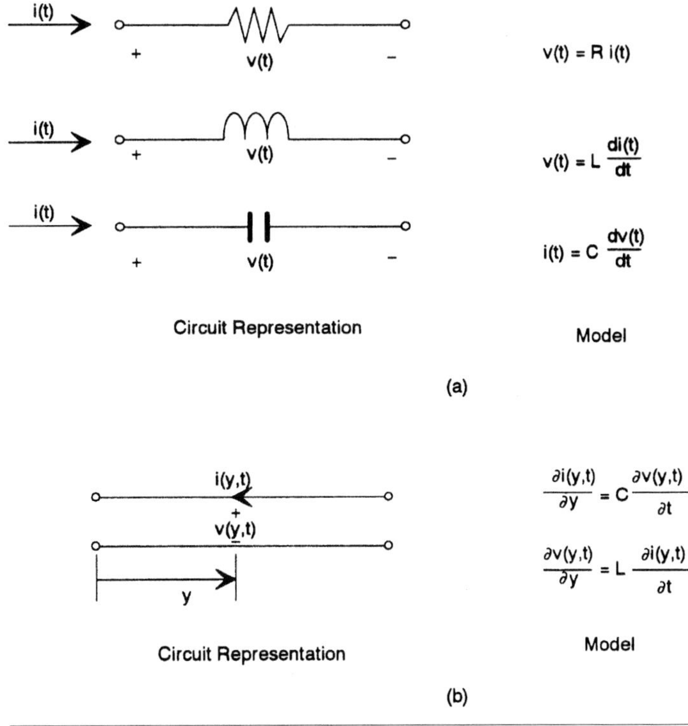


**FIGURE 23-25** Power system models for transient analysis: (*a*) lumped-parameter models; (*b*) distributed-parameter models. [*From Meliopoulos (1988).*]

***Graphical Methods.*** The graphical method is based on the observation that the solution for transient voltages and currents in a transmission line can be represented with traveling waves along the line. Under the assumption of an ideal transmission line (zero losses and constant inductance and capacitance per unit length), the waves travel along a line without distortion. Figure 23-26 illustrates the general solution in this sense. Waves are altered when they reach a discontinuity. Specifically, at discontinuity, a wave will be partially reflected and partially transmitted. If the discontinuity involves

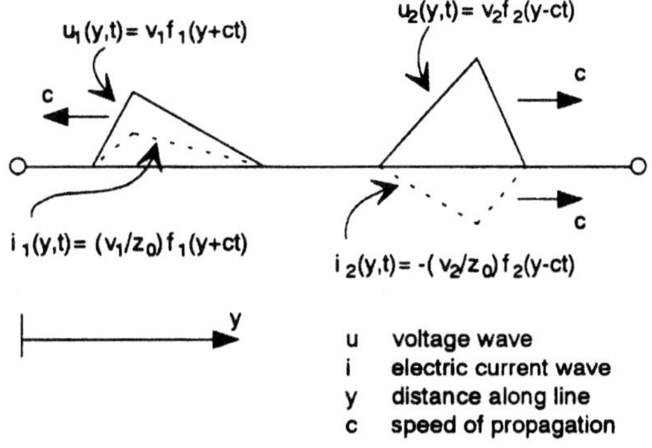


**FIGURE 23-26** General wave solution for an ideal distributed-parameter single-phase transmission line.

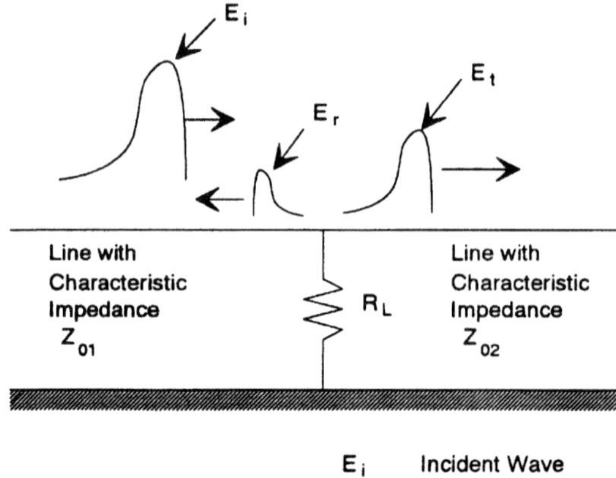


**FIGURE 23-27** Transmission and reflection of waves at discontinuity.

only resistive elements, the coefficients of reflection and transmission will be constant. This situation is illustrated in Fig. 23-27. The basic relationship among incident (subscript *i*), reflected (subscript *r*), and transmitted (subscript *t*) waves are as follows:

*Reflected wave*:

$$E_r = \alpha E_i$$

$$I_r = -\alpha I_i$$

$$\alpha = \frac{Z_{eq} - Z_{01}}{Z_{eq} + Z_{01}} \quad \text{(reflection coefficient)}$$

$$Z_{eq} = \frac{R_L Z_{02}}{R_L + Z_{02}}$$

*Transmitted wave*:

$$E_t = \delta E_i$$

$$I_t = \delta \frac{Z_{01}}{Z_{02}} I_i$$

$$\delta = \frac{2Z_{eq}}{Z_{eq} + Z_{01}} \quad \text{(transmission coefficient)}$$

If the discontinuity involves storage elements, that is, capacitors or inductors, the coefficients are not constant and the analysis becomes much more complex. The graphical method consists of monitoring all traveling waves on a line with the aid of a diagram, known as the *Bewley diagram*. The Bewley diagram provides, for every point in a system, all the waves present and the time at which they arrive. From this information, the actual voltage waveform at a specific point can be constructed as the superposition of all waves at that point. Such a construction is illustrated in Fig. 23-28.

***Analytical Methods.*** Analytical methods are based on systematic algorithms for solution of the differential equations describing a system. A useful method is based on Laplace transforms. This method

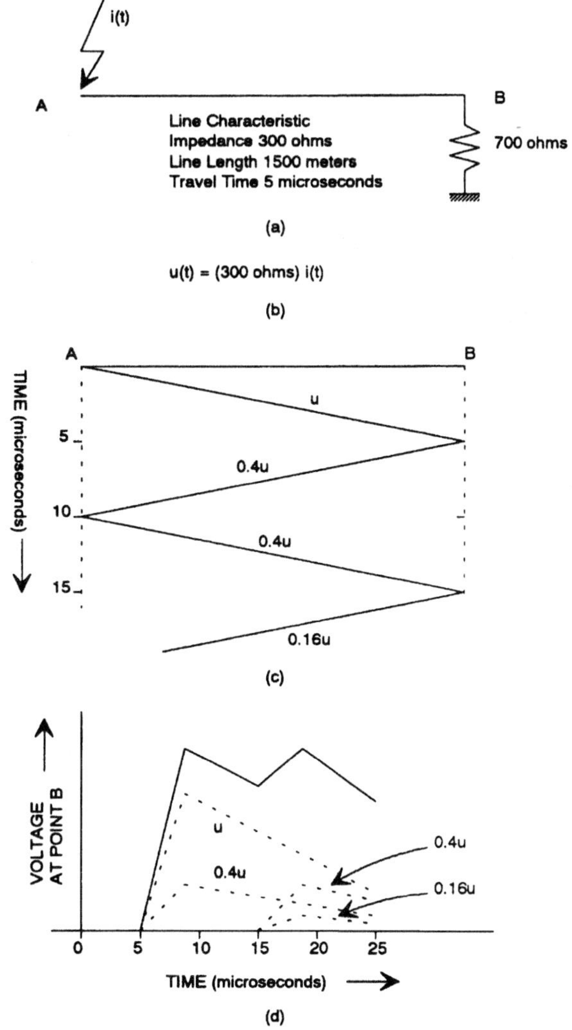


**FIGURE 23-28** Illustration of the graphical method: (*a*) system description; (*b*) voltage surge due to lightning $I(t)$; (*c*) Bewley's diagram; (*d*) construction of voltage at point *B* as the superposition of all surges arriving at point *B*.

transforms the differential equations describing a component into an equivalent circuit. An example follows. Consider the equation describing an inductor

$$v(t) = L\frac{di(t)}{dt}$$

Application of the Laplace transform on this equation yields

$$V(s) = sLI(s) - Li(0)$$

where $V(s)$ is the Laplace transform of $v(t)$ and $I(s)$ is the Laplace transform of $i(t)$.

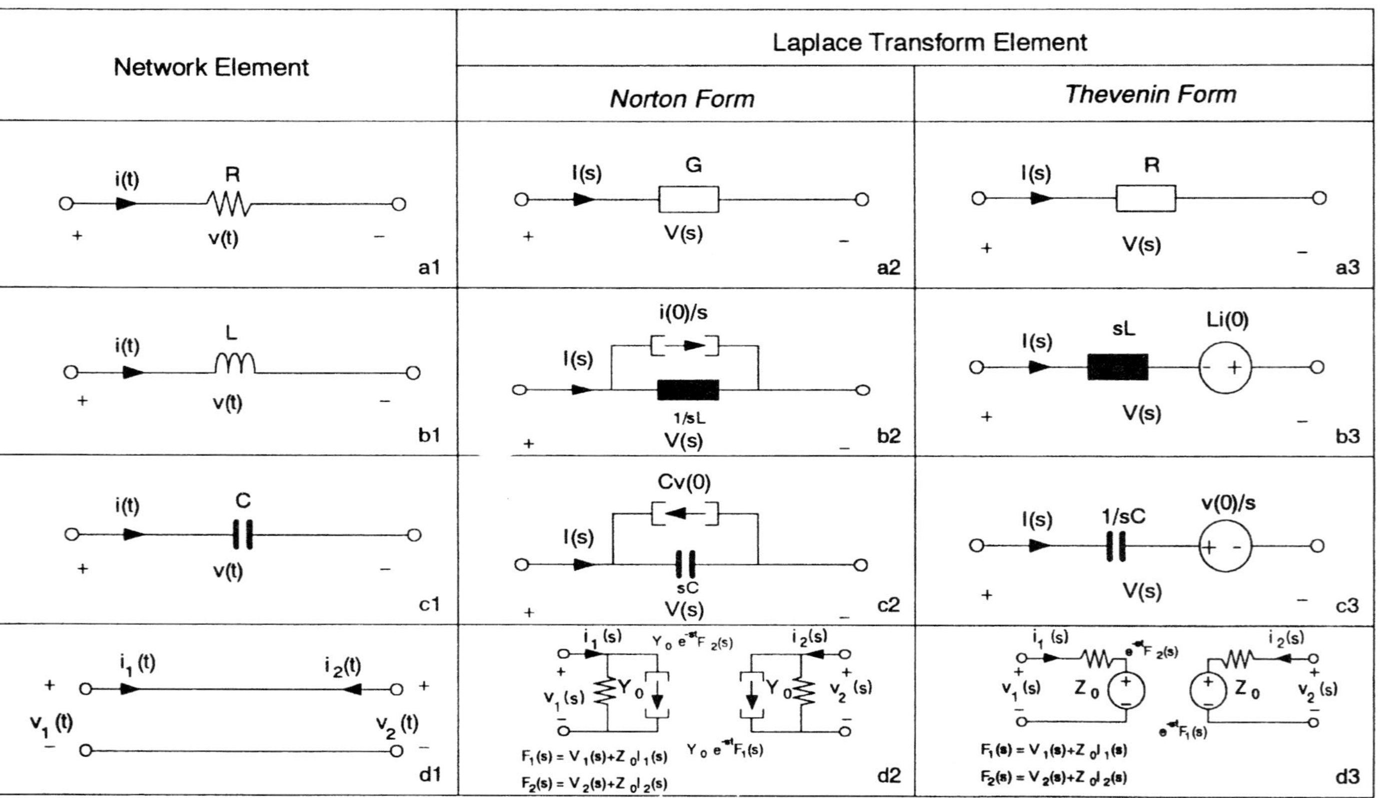


**FIGURE 23-29** Equivalent-circuit representation in the Laplace domain. [*From Meliopoulos (1988).*]

The above equation represents the equivalent circuit of Fig. 23-29 (*b*3). This is known as the Thévenin form of the Laplace domain equivalent circuit. The process can be applied to any power system element described with a set of differential equations. Figure 23-29 illustrates the equivalent circuits in the Laplace domain of typical elements. Note that the equivalent circuits are represented with algebraic equations in complex variables.

Application of the analytical method involves transformation of the differential equations describing individual element with the Laplace transform into an equivalent circuit. Subsequently, nodal analysis (or loop analysis) is applied on the transformed elements to obtain the solution of the voltage at a point of interest as a function of the Laplace variable. Finally, application of the inverse Laplace transform will provide the time waveform of the voltage of interest. Many efficient algorithms were developed during 1975–1995 or so. The details can be found in the literature.

***Numerical Methods.*** Numerical methods are based on transforming the differential equations describing a component into a discrete time equation. This transformation is achieved by proper integration of the differential equations. Many different integration methods can be applied. A very successful method is based on the trapezoidal integration method, because this method is an absolutely stable numerical method. The basic idea is explained as follows. Consider the differential equation

$$\frac{dx(t)}{dt} = ax(t)$$

Integration of this equation in the time interval $(t - h, t)$ yields

$$x(t) - x(t-h) = a\int_{t-h}^{t} x(\tau)d\tau$$

The integral on the right-hand side can be evaluated, *assuming that the function x(t) varies linearly in the time interval (t – h, t)*, yielding

$$\int_{t-h}^{t} x(\tau)d\tau = \frac{h}{2}[x(t) + x(t-h)]$$

This integration is graphically illustrated in Fig. 23-30, which shows that the value of the integral is approximated with the area of the shown trapezoid—thus the name *trapezoidal integration*. Combining above equations and solving for $x(t)$

$$x(t) = \frac{1 + ah/2}{(1 - ah)/2} \cdot x(t-h)$$

If $x(0)$ is known, then the above equation can be applied to obtain the value $x(h)$, then $x(2h)$, $x(3h)$, etc. This is a simple algorithm useful for computing the solution at specified times, $h$, $2h$, $3h$,....

The basic idea described above can be applied to the differential equations of any component. The result will be a set of algebraic equations which can be interpreted as a *resistive companion circuit*. The results for simple elements are illustrated in Fig. 23-31. Application of the method requires transformation of each element of the power system into a resistive companion circuit. The process replaces the actual system with a resistive network. This network includes voltage and current sources that depend on the state of the system at times less than $t$. Application of nodal analysis (or loop analysis) provides the solution for voltages and currents at time $t$. The process

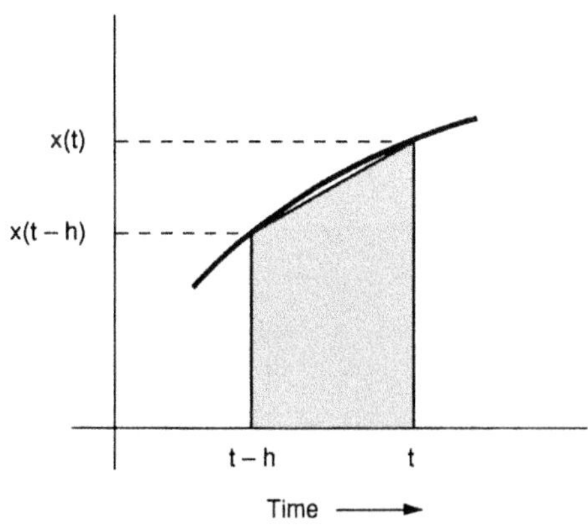


**FIGURE 23-30** Graphical representation of the trapezoidal integration.

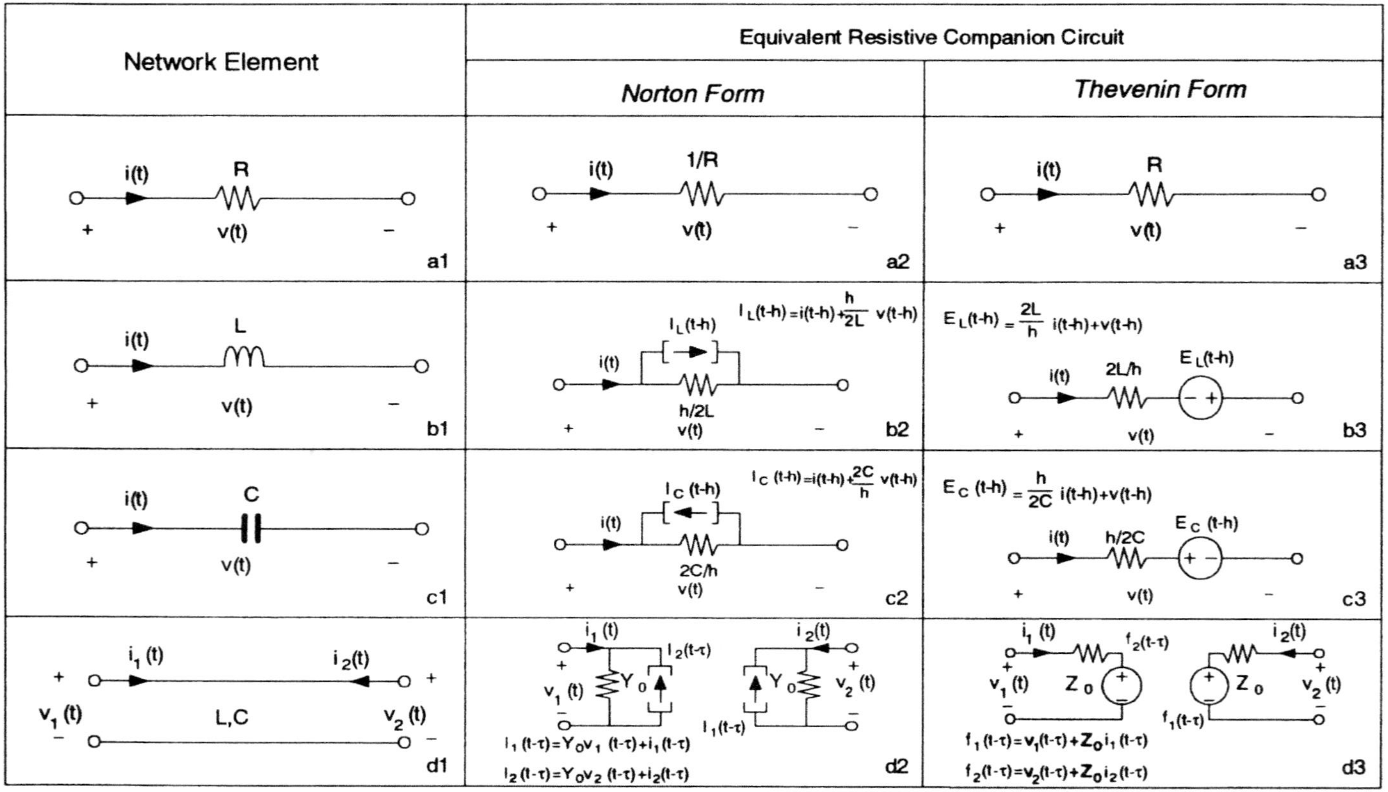

| Network Element | Equivalent Resistive Companion Circuit | |
|---|---|---|
| | *Norton Form* | *Thevenin Form* |
| i(t), R, + v(t) − **a1** | i(t), 1/R, + v(t) − **a2** | i(t), R, + v(t) − **a3** |
| i(t), L, + v(t) − **b1** | $I_L(t-h) = i(t-h) + \frac{h}{2L} v(t-h)$; i(t), $I_L(t-h)$, h/2L, + v(t) − **b2** | $E_L(t-h) = \frac{2L}{h} i(t-h) + v(t-h)$; i(t), 2L/h, $E_L(t-h)$, + v(t) − **b3** |
| i(t), C, + v(t) − **c1** | $I_C(t-h) = i(t-h) + \frac{2C}{h} v(t-h)$; i(t), $I_C(t-h)$, 2C/h, + v(t) − **c2** | $E_C(t-h) = \frac{h}{2C} i(t-h) + v(t-h)$; i(t), h/2C, $E_C(t-h)$, + v(t) − **c3** |
| $i_1(t)$, $i_2(t)$, L,C, $v_1(t)$, $v_2(t)$ **d1** | $i_1(t)$, $I_2(t-\tau)$, $v_1(t)$, $Y_0$, $i_2(t)$, $v_2(t)$, $I_1(t-\tau)$; $I_1(t-\tau) = Y_0 v_1(t-\tau) + i_1(t-\tau)$; $I_2(t-\tau) = Y_0 v_2(t-\tau) + i_2(t-\tau)$ **d2** | $i_1(t)$, $f_2(t-\tau)$, $v_1(t)$, $Z_0$, $i_2(t)$, $v_2(t)$, $f_1(t-\tau)$; $f_1(t-\tau) = v_1(t-\tau) + Z_0 i_1(t-\tau)$; $f_2(t-\tau) = v_2(t-\tau) + Z_0 i_2(t-\tau)$ **d3** |

**FIGURE 23-31** Equivalent resistive comparison circuits. [*From Meliopoulos (1988).*]

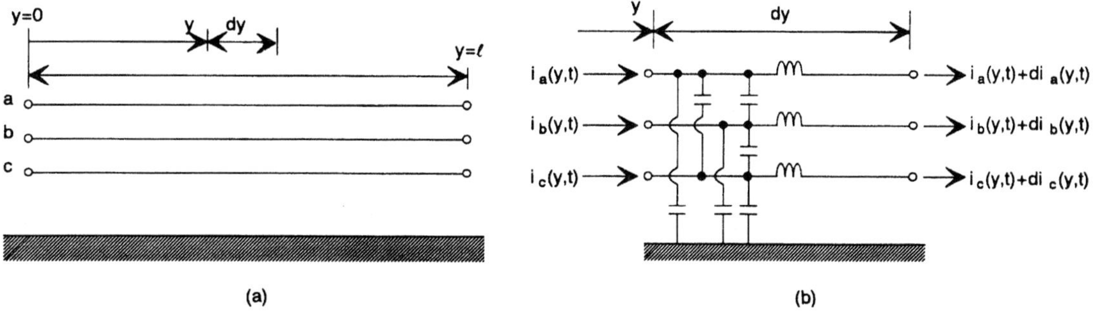


**FIGURE 23-32** 3-Phase line model: (*a*) 3-phase transmission line; (*b*) equivalent circuit of an infinitesimal length (*dy*) of 3-phase line.

can start at time $t = 0$ and be repeated at times $t = h$, $t = 2h$, . . . , yielding the values of voltages and currents at times $t = 0, h, 2h, \ldots$, etc. Further details can be found in Meliopoulos (1988).

***3-Phase Transmission Lines.*** 3-Phase transmission is represented with a distributed-parameter model. This model can be derived by considering an infinitesimal length of a 3-phase line as in Fig. 23-32*b*. Assuming an ideal line (zero losses and constant inductance and capacitance per unit length), the model equations are

$$CL\frac{\partial^2}{\partial t^2}v_{abc}(y,t)=\frac{\partial^2}{\partial y^2}v_{abc}(y,t)$$

$$CL\frac{\partial^2}{\partial t^2}i_{abc}(y,t)=\frac{\partial^2}{\partial y^2}i_{abc}(y,t)$$

$$L\frac{\partial}{\partial t}i_{abc}(y,t)=\frac{\partial}{\partial y}v_{abc}(y,t)$$

where

$$v_{abc}(y,t)=\begin{bmatrix} v_a(y,t) \\ v_b(y,t) \\ v_c(y,t) \end{bmatrix}$$

$$i_{abc}(y,t)=\begin{bmatrix} i_a(y,t) \\ i_b(y,t) \\ i_c(y,t) \end{bmatrix}$$

$$L=\begin{bmatrix} L_{aa} & L_{ab} & L_{ac} \\ L_{ab} & L_{bb} & L_{bc} \\ L_{ac} & L_{bc} & L_{cc} \end{bmatrix}$$

is the inductance matrix per unit length of the line and

$$C=\begin{bmatrix} C_{aa} & C_{ab} & C_{ac} \\ C_{ab} & C_{bb} & C_{bc} \\ C_{ac} & C_{bc} & C_{cc} \end{bmatrix}$$

is the capacitance matrix per unit length of the line, $y$ is distance along line, and $t$ is time.

This is a rather simplified model of a 3-phase line and yet very complex. To make clear the behavior of the 3-phase line, the concept of the *ideal continuously transposed line* will be introduced. This concept assumes a perfectly symmetrical line

$$L_{aa} = L_{bb} = L_{cc} = L_s$$
$$L_{ab} = L_{ac} = L_{bc} = L_m$$
$$C_{aa} = C_{bb} = C_{cc} = C_s$$
$$C_{ab} = C_{ac} = C_{bc} = C_m$$

Next, Karrenbauer's transformation is introduced as follows

$$\upsilon_{abc}(y,t) = K\upsilon_{gll}(y,t)$$
$$i_{abc}(y,t) = Ki_{gll}(y,t)$$

where

$$\upsilon_{gll}(y,t) = \begin{bmatrix} \upsilon_g(y,t) \\ \upsilon_{l1}(y,t) \\ \upsilon_{l2}(y,t) \end{bmatrix}$$

$$i_{gll}(y,t) = \begin{bmatrix} i_g(y,t) \\ i_{l1}(y,t) \\ i_2(y,t) \end{bmatrix}$$

and

$$K = \begin{bmatrix} 1 & 1 & 1 \\ 1 & -2 & 1 \\ 1 & 1 & -2 \end{bmatrix}$$

Replacement of the actual voltages and currents $\upsilon_{abc}(y, t)$ and $i_{abc}(y, t)$ with the voltages and currents $\upsilon_{gll}(y, t)$ and $i_{gll}(y, t)$ through Karrenbauer's transformation yields the following transformed equations for the 3-phase line (Meliopoulos 1988):

*Set 1 (ground-mode equations)*

$$(L_s + 2L_m)(C_s + 2C_m)\frac{\partial^2}{\partial t^2}\upsilon_g(y,t) = \frac{\partial^2}{\partial y^2}\upsilon_g(y,t)$$

$$(L_s + 2L_m)(C_s + 2C_m)\frac{\partial^2}{\partial t^2}i_g(y,t) = \frac{\partial^2}{\partial y^2}i_g(y,t)$$

$$(L_s + 2L_m)\frac{\partial}{\partial t}i_g(y,t) = \frac{\partial}{\partial y}\upsilon_g(y,t)$$

*Set 2 (line-mode 1 equations)*

$$(L_s - L_m)(C_s - C_m)\frac{\partial^2}{\partial t^2}\upsilon_{l1}(y,t) = \frac{\partial^2}{\partial y^2}\upsilon_{l1}(y,t)$$

$$(L_s - L_m)(C_s - C_m)\frac{\partial^2}{\partial t^2}i_{l1}(y,t) = \frac{\partial^2}{\partial y^2}i_{l1}(y,t)$$

$$(L_s - L_m)\frac{\partial}{\partial t}i_{l1}(y,t) = \frac{\partial}{\partial y}\upsilon_{l1}(y,t)$$

*Set 3 (line-mode 2 equations)*

$$(L_s - L_m)(C_s - C_m)\frac{\partial^2}{\partial t^2}v_{l2}(y,t) = \frac{\partial^2}{\partial y^2}v_{l2}(y,t)$$

$$(L_s - L_m)(C_s - C_m)\frac{\partial^2}{\partial t^2}i_{l2}(y,t) = \frac{\partial^2}{\partial y^2}i_{l2}(y,t)$$

$$(L_s - L_m)\frac{\partial}{\partial t}i_{l2}(y,t) = \frac{\partial}{\partial y}v_{l2}(y,t)$$

Note that the complex equations for the 3-phase line have been replaced with three sets of equations, each set representing an ideal single-phase line. The characteristic impedance and speed of propagation of surges for the three ideal single phase lines are

*Ground mode*

$$Z_g = \frac{1}{Y_g} = \sqrt{\frac{L_s + 2L_m}{C_s + 2C_m}}$$

$$c_g = \frac{1}{\sqrt{(L_s + 2L_m)(C_s + 2C_m)}}$$

*Line mode (1 or 2)*

$$Z_l = \frac{1}{Y_l} = \sqrt{\frac{(L_s - L_m)}{(C_s - C_m)}}$$

$$c_l = \frac{1}{\sqrt{(L_s - L_m)(C_s - C_m)}}$$

This model of the line is illustrated in Fig. 23-33. The three sets of equations or three ideal single-phase line models above are known as (1) the ground mode $g$, (2) line mode 1 $l_1$, and (3) the line mode 2 $l_2$. The names become obvious if one considers excitation of the 3-phase line with one mode. For example, consider that the line is excited in such a way that $i_{l1}(y, t) = 0$ and $i_g(y, t) = 0$, $i_{l2}(y, t) = 0$. In this case, the actual phase currents will be

$$i_{abc}(y,t) = Ki_{gll}(y,t)$$

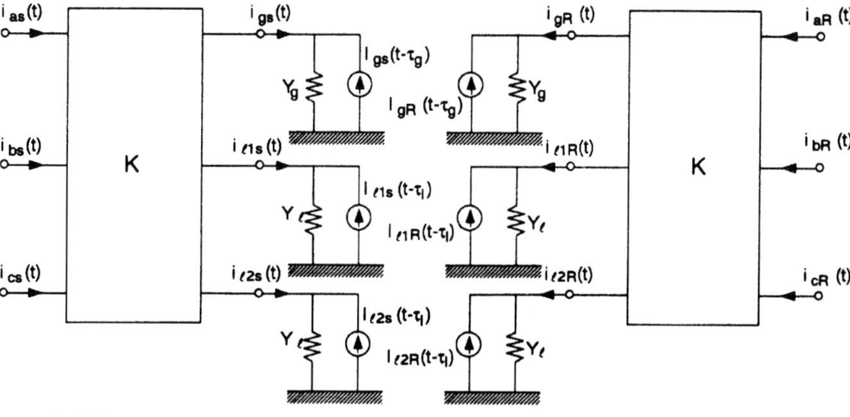


**FIGURE 23-33** Equivalent circuit of a continuously transposed line.

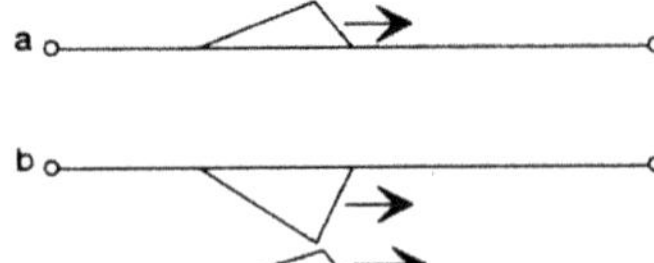

(a)

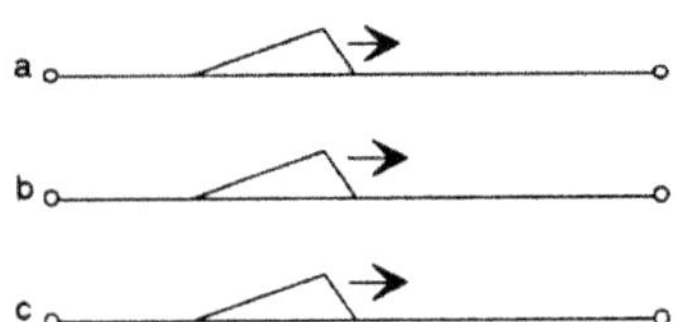


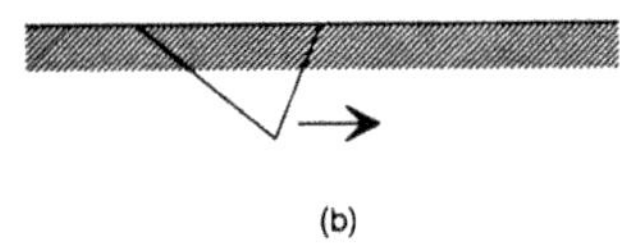

(b)

**FIGURE 23-34** Modes of propagation along a 3-phase line: (*a*) line mode; (*b*) ground mode.

which yields

$$i_a(y,t) = i_{l1}(y,t)$$
$$i_b(y,t) = -2i_{l1}(y,t)$$
$$i_c(y,t) = i_{l1}(y,t)$$

This state of the line is illustrated in Fig. 23-34*a*. Note that there is a positive surge on phases *a* and *c*, which returns through phase *b*. All electric current surges are confined in the line conductors, thus the name line mode.

Consider now excitation with $i_g(y, t) \neq 0$, $i_{l1}(y, t) = 0$, and $i_{l2}(y, t) = 0$. In this case, the equation

$$i_{abc}(y,t) = Ki_{gll}(y,t)$$

yields

$$i_a(y,t) = i_g(y,t)$$
$$i_b(y,t) = i_g(y,t)$$
$$i_c(y,t) = i_g(y,t)$$

This state of the line is illustrated in Fig. 23-34*b*. Note that the surges on all phases are equal. The return current is through the earth, thus the name ground mode. The parameters of the various modes of propagation of surges in a 3-phase line are different. As an example, the following values apply to a 115-kV 3-phase line.

The inductance matrix is

$$L = \begin{bmatrix} 2.2289 & 1.032 & 0.8935 \\ 1.032 & 2.2289 & 1.032 \\ 0.8935 & 1.032 & 2.2289 \end{bmatrix} \mu H/m$$

Therefore, $$L_s = 2.2289 \quad \mu H/m \qquad L_m = 0.9858 \quad \mu H/m$$

The capacitance matrix is

$$C = \begin{bmatrix} 7.6737 & -1.9129 & -1.0065 \\ -1.9129 & 8.0184 & -1.9129 \\ -1.0065 & -1.9129 & 7.6737 \end{bmatrix} \times 10^{-12} \text{ F/m}$$

Therefore, $$C_s = 7.7886 \times 10^{-12} \text{ F/m} \quad C_m = -1.608 \times 10^{-12} \text{ F/m}$$

Line mode (1 or 2):

$$Z_l = \sqrt{\frac{(L_s - L_m)}{(C_s - C_m)}} = 363.7\Omega$$

$$c_l = \frac{1}{\sqrt{(L_s - L_m)(C_s - C_m)}} = 2.926 \times 10^8 \text{ m/s}$$

Ground mode:

$$Z_g = \sqrt{\frac{(L_s + 2L_m)}{(C_s + 2C_m)}} = 959.1\Omega$$

$$c_g = \frac{1}{\sqrt{(L_s + 2L_m)(C_s + 2C_m)}} = 2.283 \times 10^8 \text{ m/s}$$

When lightning hits a 3-phase line, all modes of propagation are excited. The overvoltage of the 3-phase line as a result of the lightning stroke is determined from the parameters of all the modes. Later, examples of these calculations will be provided.

***Frequency-Dependent Models.*** The parameters of power-system elements are frequency dependent. As an example, consider the ground-mode resistance of a 3-phase line. This resistance value may change *several orders of magnitude* in the frequency range 60 Hz (power frequency) to 1 MHz. This increase of resistance is mainly due to skin-effect-type phenomena. The higher resistance values at high frequencies tend to attenuate higher-frequency components of transient overvoltages with the end effect of reducing the maximum overvoltages. Computer models to account for the frequency dependence have been developed for almost all power components. The reader is referred to the literature. Frequency-dependent models are complex, but, on the other hand, they provide a better representation of the real system. As an example, consider Fig. 23-35. The figure illustrates comparison of measured and computed switching transients in a 224.15-mi-long 230-kV line. The single-line diagram of the system is illustrated in Fig. 23-36. The tests were performed by Bonneville Power Administration. The calculated switching transient was performed with a frequency-dependent model (Cokkinides and Meliopoulos 1988). Note that the maximum switching overvoltage is 1.81 pu. It is important to note that simulation of the same system assuming an ideal continuously transposed line results in a maximum switching overvoltage of 2.05 pu. The conclusion is that frequency-dependent models are more realistic.

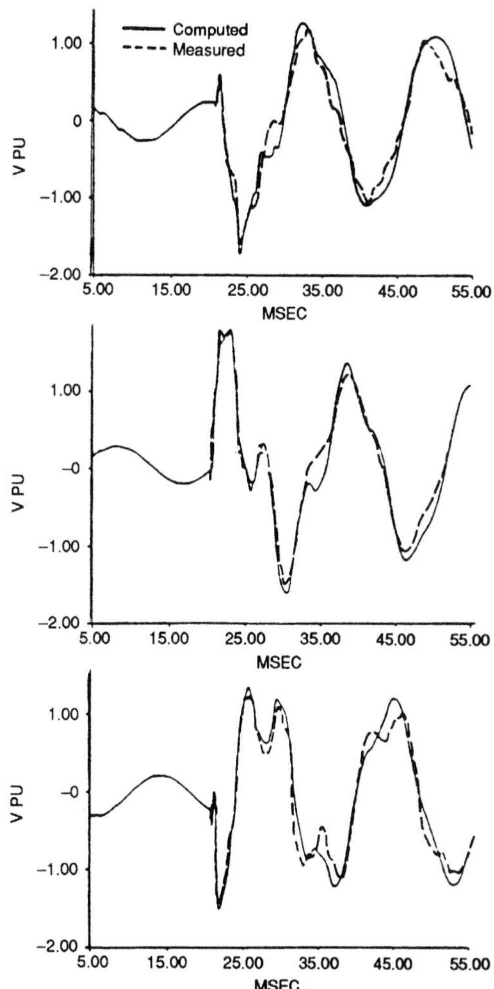


**FIGURE 23-35** Comparison of simulation and test results of switching surges: (*a*) phase A transient voltage; (*b*) phase B transient voltage; (*c*) phase C transient voltage. [*From Cokkinides and Meliopoulos (1988).*]

***Grounding Models.*** Grounding plays an important role in dissipation of lightning strokes and therefore controlling overvoltages resulting from lightning. Yet, grounding has been widely misunderstood and

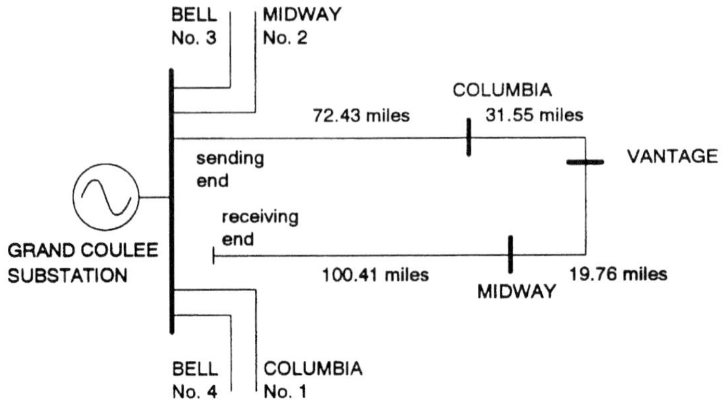


**FIGURE 23-36** Illustration of system for switching surge test. (*Bonneville Power Administration.*)

proper analysis models are scarce. Modeling of grounding systems is a rather complex task, and it is strongly coupled to the overall modeling procedure for power systems. Two distinct approaches apply: (1) grounding models for low frequency generally consider grounds as a pure resistance and thus dc analysis models are utilized, based on the method of moments or relaxation methods; (2) grounding models for higher frequencies require complete electromagnetic analysis. For this purpose, finite element analysis or the method of moments can be utilized. Since the grounds are typically complex systems, simplifications are typically introduced. A rule of thumb for selecting dc models or the more complex models is to compare the largest dimension of a grounding system $l$ to the depth of penetration $\delta = \sqrt{2}/\sqrt{\mu\omega\sigma}$. If $l > 0.1\delta$, then a complete model is necessary. In this expression, $\mu$ is the permeability of the medium, $\omega$ is the frequency in radians per second, and $\sigma$ is the conductivity.

The choice of the grounding model can be crucial. As an example, consider a grounding system consisting of a counterpoise buried in soil of 100 Ω · m. The dc ground resistance of this system is 2.29 Ω. The frequency-dependent impedance is much higher for high frequencies and approaches 2.29 Ω for frequencies below 50 kHz. As a result, a lightning discharge through this ground will require a frequency-dependent ground model to accurately predict the transient overvoltages. As an example, Fig. 23-37 shows the differences between the correct model and the dc model for the standard lightning waveform. The example emphasizes the importance of selecting the proper model for the grounding system.

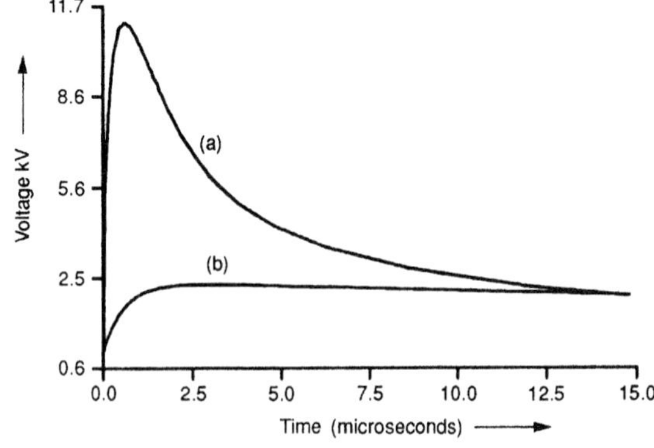


**FIGURE 23-37** Transient voltage on a 300-ft counterpoise from a 1.2- to 50-m/s lightning wave of a 1-kA crest: (*a*) frequency-dependent model; (*b*) frequency-independent model.

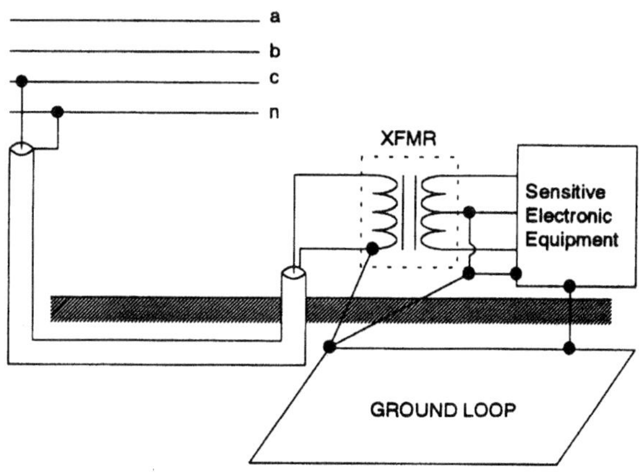


**FIGURE 23-38** Illustration of a typical service entrance to a commercial or industrial facility.

Grounding systems play an important role in propagation of surges from high-voltage power systems to lower-voltage secondary systems. Specifically, transients initiated in the power system can travel through the grounding system and enter the secondary power system of a facility, reaching sensitive electronic equipment. Transformers do not exist in the path of a neutral. Thus, high-frequency transients travel almost unattenuated through the neutral. Attenuation is provided only by the grounds if the neutral is multiply grounded. A proper model of the grounds, neutrals, and the power system can provide a good tool to determine the transient overvoltages reaching a specific piece of equipment. As an example, consider a facility served by a 1-mi (1.6-km) underground distribution cable as in Fig. 23-38. The underground cable is fed from an overhead distribution circuit, which is subject to lightning. At such a service entrance, transients can enter a facility through the cable concentric neutral and the ground conductors.

***Probabilistic Methods.*** The objective of these methods is to provide a probabilistic description of overvoltages in a specific power apparatus. The method takes into consideration the uncertainty in parameters affecting the overvoltages. Some of the uncertain parameters are listed in Table 23-5.

The probabilistic methods consist of the following three components:

1. A probabilistic model for the uncertain parameters
2. An analysis model for the system under study
3. A Monte Carlo simulation method

**TABLE 23-5** Typical Parameters with Uncertainty

| | |
|---|---|
| Lightning | Crest value |
| | Rise time |
| | Frequency of occurrence |
| | Shielding failure |
| | Other |
| Switching | Switch closing time |
| | Time of preinsertion of resistors |
| | Trapped charge |
| | Other |

The result of the method is the probability distribution of the overvoltages at the apparatus of interest. Given the probabilistic model for the uncertain parameters and an analysis model for the system under study, the Monte Carlo simulation consists of the following steps:

*Step 0:* Set count of trial $n$ equal to 0.

*Step 1:* Generate (randomly) a sample of parameters from the probabilistic model of uncertain parameters (crest value of lightning stroke, rise time, fall time, location of incidence, etc.).

*Step 2:* Use the analysis model and the parameter values determined in step 1 to evaluate the maximum overvoltage at points of interest (or other quantities of interest). Store the computed values.

*Step 3:* Repeat steps 1, 2, and 3 until the number of trials = $n$. Then go to step 4.

*Step 4:* Use the stored computed values of overvoltages (or other quantities of interest) to generate histograms and/or probability distribution functions.

The utility of probabilistic methods is quite obvious. They provide the tool to determine not only the expected overvoltages but also the frequency of occurrence. For external insulation protection practices, probabilistic methods provide an indispensable tool for optimal designs.

## 23.6 OVERVOLTAGE PROTECTION DEVICES

With certainty, temporary overvoltages on power-system apparatus will exceed their withstand capability. In this case, if the apparatus is left unprotected, insulation failure will occur. Therefore, it is necessary to protect power system components against overvoltages. The philosophy and objectives of this protection vary depending on the type of overvoltages, frequency, effects of insulation failure, and cost of repair. These issues will be discussed later. In this subsection, we will be concerned with available protection devices, their characteristics, and protection levels. An overvoltage protection device should ideally limit the voltages across the insulation of a power apparatus below a specified value. This specified value is called the *protection level.* The ideal protection device has the voltage-current characteristic indicated in Fig. 23-39. Specifically, if the voltage across the protection device is less than the protection level, then the protection device should have an infinitely large impedance. If the voltage across the protection device is higher than the protection level, then the protection device should allow the flow of electric current through it in such a way that the voltage is clipped to the protection level value. Such a characteristic can be mathematically described with the equation

$$i(t) = i_0 \left\{ \frac{abs[v(t)]}{v_p} \right\}^n sign[v(t)] \tag{23-1}$$

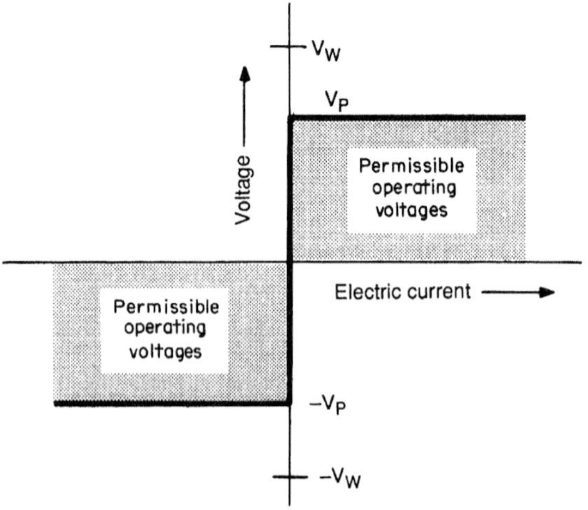


**FIGURE 23-39** Voltage-current characteristics of the ideal overvoltage protection device.

where $i_0$ = a constant
$v_p$ = protection level
abs $(\cdot)$ = absolute value of the argument $(\cdot)$
$v(t)$ = voltage across the protective device
sign $[v(t)]$ = sign of voltage $v(t)$, + or −
$n$ = a very large number

The protection characteristics of Fig. 23-39 correspond to a very large value of $n$.

Of course, a protection device with these characteristics does not exist. Research and development has resulted in protection devices which come close to that of Fig. 23-39 to varying degrees. It is important to note that the research and development of protection devices has been evolutionary; over the years the following major breakthroughs have occurred:

Air gaps
Gapped surge arresters
Expulsion arresters
Gapped valve-type arresters
Gapped silicon carbide (SiC)
Metal oxide varistors
Shunt-gapped metal oxide varistors (MOVs)

The most important application characteristics of a protection device are the protection level and the reseal level. These are defined as follows:

*Protection Level.* The maximum voltage, which will be allowed by the device across its terminals.

*Reseal Level.* The maximum voltage below which the protection device will disallow significant electric current through it.

For best protection, it is expedient that these two parameters be as close as possible. Observe that for the ideal protection device, the protection level is equal to the reseal level. In order to quantify the protection quality of various technologies of protective devices, the following definition is introduced:

*Protection quality index (PQI).* The protection-quality index is defined as the ratio of the reseal level over the protection level:

$$\mathrm{PQI} = \frac{V_r}{V_p}$$

Note that the ideal protective device has a protection-quality index of 1. As should be expected, the protective capabilities of protective devices increase with every breakthrough in the technology. Air gaps provide rudimentary protection, gapped SiC surge arresters provide improved protection, and finally MOVs provide even better protection. Within each class of technology, many variations exist. The next paragraphs will discuss the above technologies.

***Air (Spark) Gaps.*** A spark gap was the first protection device to be applied on a power system. It can be constructed with two electrodes placed at a certain distance. The shape of the electrodes varies depending on the application—for example, two wires, a wire and a planar electrode, or two spheres. When the voltage across the air gap exceeds a certain value, an arc will initiate between the electrodes. The voltage across the arc will depend on the current of the arc, the length of the arc, and time. A typical variation of the voltage across the gap is illustrated in Fig. 23-40. The protection level as well as the reseal level, indicated as maximum permissible operating voltage across the gap, are shown. Note that for good protection, the protection quality index (the ratio $V_r/V_p$) should be maximized. Unfortunately, air gaps exhibit very low protection quality index. In many applications, and in order

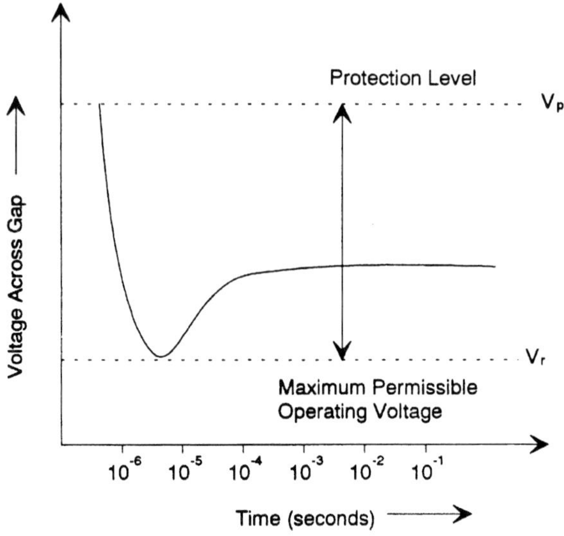


**FIGURE 23-40** Typical voltage-time characteristic of a long air gap.

to improve the protection quality index, the electrodes are shaped in such a way as to force the arc to elongate and therefore increase the voltage reseal value $V_r$.

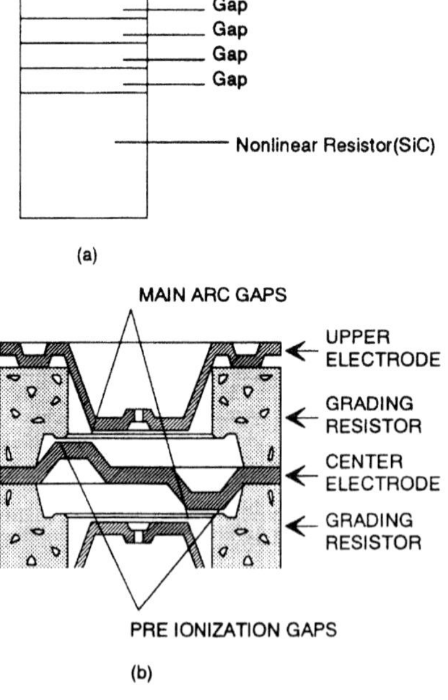


**FIGURE 23-41** Construction of a gapped arrester: (*a*) block diagram; (*b*) detail of a gap.

***Gapped Surge Arresters.*** Gapped surge arresters consist of a series of spark gaps with or without series blocks of nonlinear resistors, which act as current limiters. The construction of such an arrester is illustrated in Fig. 23-41.

The function of the air gap is to isolate the current limiting block of the surge arrester from the power frequency voltage under normal operating conditions. The air gap is necessary because the current limiting block will fail thermally if subjected to the continuous normal power frequency voltage. Elaborate designs of air gaps have been developed, which assure (1) consistent sparkover voltage and (2) arc extinction (resealing) after the overvoltage ceases to exist.

Consistent sparkover voltage is achieved by either a trigger gap or a preionizer. In order to prevent damage of the trigger mechanism, the main current is not allowed to pass through it by use of grading resistors as is illustrated in Fig. 23-41, but through the main arc gaps. To ensure arc extinction and good resealing properties, a mechanism is provided for the control of the arc. Specifically, the main current is passed through a coil. The magnetic field of the coil is established in the space of the arc and is oriented so as to force the arc into a serrated-tooth chamber. The end result of this interaction is that the arc is elongated, cooled, and the voltage

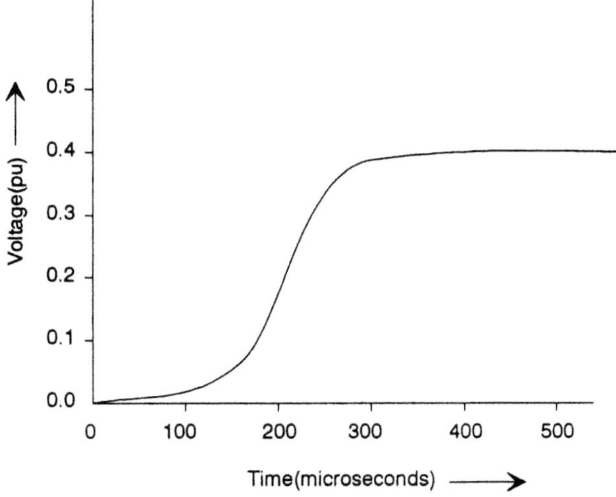


**FIGURE 23-42** Evolution of voltage across the arc of a gapped arrester.

across the arc becomes substantial. Figure 23-42 illustrates the buildup of the voltage across the arc due to this mechanism. An illustration of the serrated-tooth chamber is illustrated in Fig. 23-43. The elongated arc improves the capability of the arrester to reseal when the current through it decreases after the overvoltage ceases to exist. It is important to note that the arc control process takes a relatively long time (hundreds of microseconds) to develop full voltage across the gap. This means that for fast transients, that is lightning, almost the entire voltage appears across the current-limiting block.

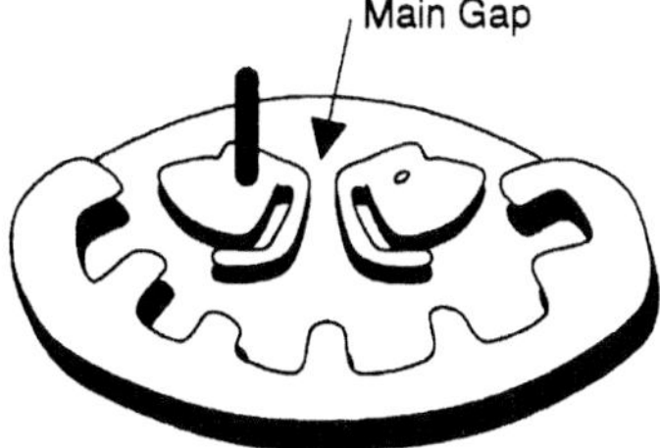


**FIGURE 23-43** Illustration of a serrated-tooth chamber of a gapped arrester.

The current-limiting block in modern gapped arresters is constructed from SiC. The block has nonlinear resistance characteristics that are controlled with the manufacturing process. Specifically, the block is constructed from SiC crystals which are ground into powder form and then are mixed and pressed together with insulating material, forming a block. In this way, an SiC particle partially touches other particles and partially is insulated by the insulating material. The nonlinear properties are due to the resistance-temperature properties of the junction between the SiC crystals. This relationship is illustrated in Fig. 23-44.

The combined operation of the gap and the current-limiting block result in the voltage-current characteristic illustrated in Fig. 23-45. It is important to note that, while the sparkover voltage $V_s$ is typically consistent, the voltage-current curve depends on the waveform of the electric current, since the gap voltage is time dependent.

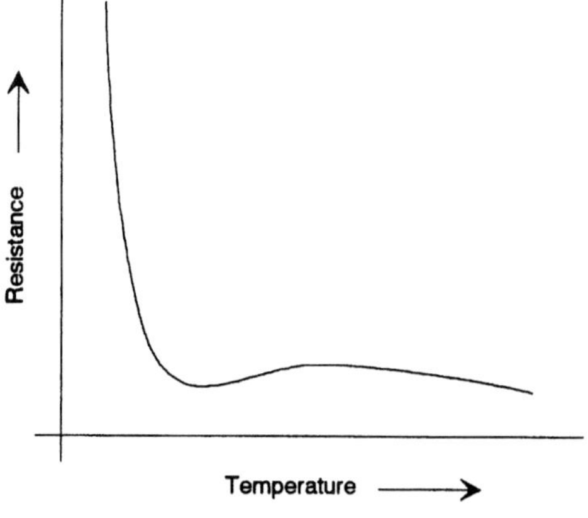


**FIGURE 23-44** Typical resistance versus temperature of junction between SiC crystals.

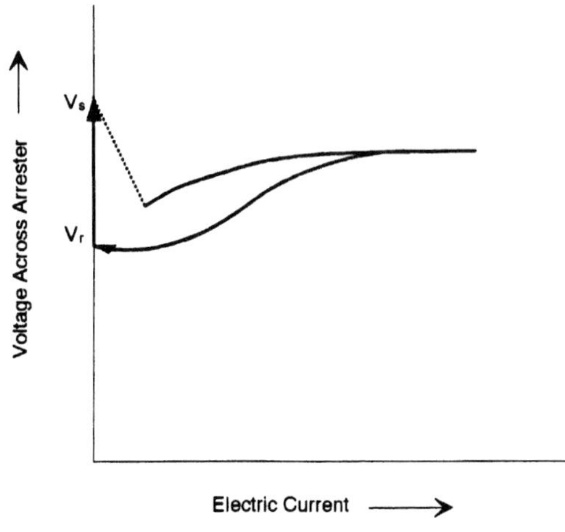


**FIGURE 23-45** Typical voltage-current characteristics of a modern SiC gapped arrester.

Gapped SiC arresters are subject to power-follow current because of the poor nonlinear characteristics of the SiC block. Specifically, after sparkover and after the temporary overvoltage has subsided, if the power voltage across the arrester is substantial, a relatively high current will continue to flow through the arrester. The reason for this behavior can be easily visualized if one considers that the arc in the gap is a low-resistance conductor and that the current is primarily determined by the voltage-current characteristics of the SiC block. The situation is illustrated in Fig. 23-46. In the figure, the power-follow current of a SiC gap arrester is also compared to the power-follow current of an MOV arrester, which is discussed next.

***Metal Oxide Varistor Arresters.*** An MOV is formed from a variety of materials via a manufacturing process, which provides the desired electrical properties to the varistor. These electrical properties are not existent in the raw materials. In other words, the electrical properties of the final product are completely dependent on the manufacturing process. Much research effort has been expended to achieve electrical properties of the varistor close to those of the ideal protective device. The typical structure of a metal oxide varistor consists of highly conductive tiny particles of metal oxide (usually zinc oxide, ZnO) suspended in a semiconducting material. This structure is illustrated in Fig. 23-47. The manufacturing process determines the size of the metal oxide particles as well as the thickness and resistivity of the semiconducting material. The key to a good manufacturing process is that the conductive metal oxide particles do not touch each other but are separated by semiconducting material. This structure gives the varistor the properties of a pair of back-to-back-connected zener diodes. A typical voltage-current characteristic of an MOV is illustrated in Fig. 23-46. It should be understood that this function holds for negative voltages and currents as well.

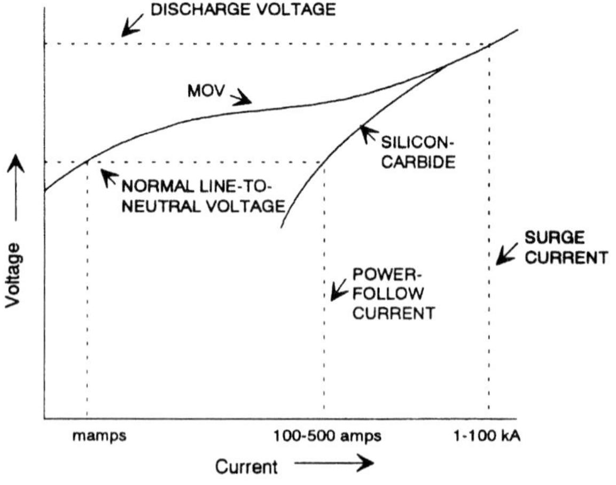


**FIGURE 23-46** Illustration of power-follow current for a SiC gapped arrester and an MOV arrester.

**FIGURE 23-47** Structure of a metal oxide varistor (MOV). Size is exaggerated for illustration purposes.

In addition to this characteristic, it is necessary that the MOV material be capable of absorbing energy losses during conduction. The average energy absorption capability of a conventional ZnO varistor is about 150 to 200 J/cm$^3$. Recent advances in MOV technology under Electric Power Research Institute sponsorship resulted in new formulations and manufacturing processes for MOVs. The new MOVs are capable of absorbing four to five times more energy per volume as above.

The first designs of MOV arresters were subject to problems. The most common were thermal runaway and failures due to moisture ingress causing insulation failures. Thermal runaway can occur because an MOV is typically applied to the energized conductor without a series gap, and a small electric current always flows through the MOV during normal operating conditions. This small current and the associated ohmic losses can potentially raise the temperature of the MOV to the point that its electrical properties deteriorate, allowing more current to flow and further raising the temperature until failure. However, improvements in metal oxide materials and in packaging resulted in robust MOV arresters. The present state of the art and the superior properties of the MOV arresters make them the protection device of choice.

The first commercially available MOV arresters were gapless. As a result, the MOV block is continuously subjected to the power frequency voltage. The performance of the MOV block under this condition is very important for determining the life of the arrester. A deep understanding of these characteristics is essential for the proper application of MOV arresters.

First, observe that the voltage-current characteristic for an MOV arrester in Fig. 23-46 exhibits a knee for small currents (in the milliampere region). The knee is defined to be near the voltage required across the MOV block to cause 1 mA/cm$^2$ of current to flow through the block. It is important to mention that while the voltage-current characteristic of an MOV block in the protection region (high current flow) is insensitive to the temperature of the block, the voltage-current characteristic near the knee is temperature-sensitive. This is illustrated in Fig. 23-48.

Figure 23-49 illustrates the voltage versus current curves for an MOV arrester for different levels of applied voltage (sinusoidal, 60 Hz) expressed in pu of arrester rating. Note that for an applied voltage near nominal (1.05 pu), the electric current through the arrester is mostly capacitive (voltage and current are approximately 90° apart) and of low value on the order of a milliampere. As the voltage increases, the electric current increases much faster. It can be observed from the figure that the increase of the current occurs in the component, which is in phase with the voltage (resistive current) while the capacitive component remains constant.

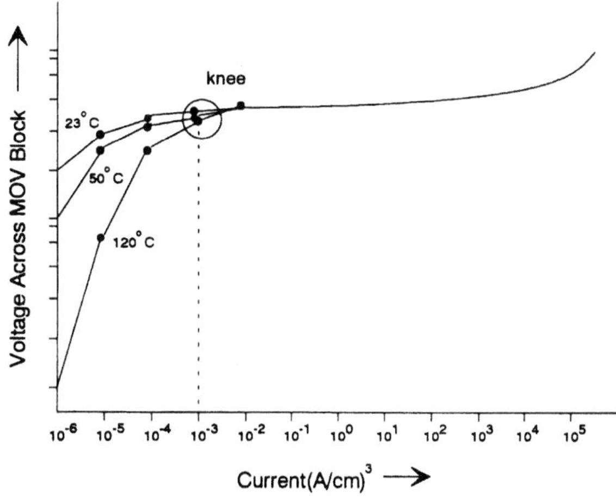


**FIGURE 23-48** Voltage-current characteristic is temperature-sensitive near the knee of the curve.

The current flow through the arrester is responsible for power

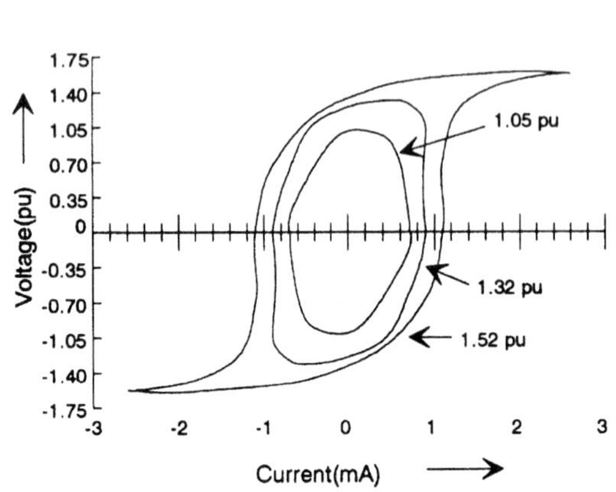


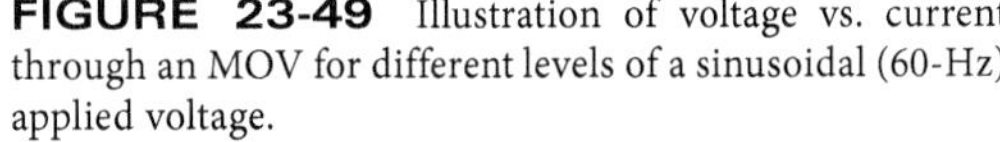
**FIGURE 23-49** Illustration of voltage vs. current through an MOV for different levels of a sinusoidal (60-Hz) applied voltage.

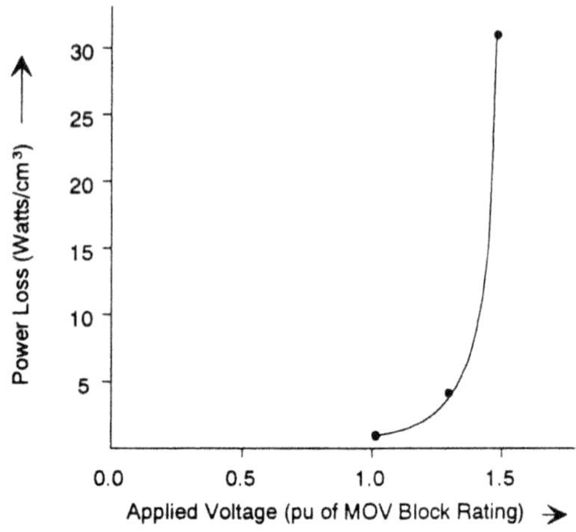


**FIGURE 23-50** Typical power loss of an MOV block vs. applied voltage.

loss within the MOV block, which increases the temperature of the block. A typical power loss—applied voltage function is illustrated in Fig. 23-50. Note that as the voltage increases the power loss increases disproportionately. Increased losses and increased operating temperature have a detrimental effect on the MOV arrester life. Figure 23-51 illustrates the expected MOV arrester life as a function of continuously applied sinusoidal voltage.

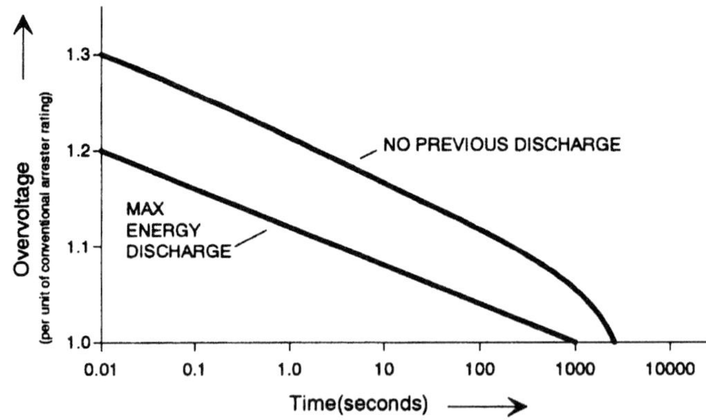


**FIGURE 23-51** Expected MOV arrester life as a function of continuously applied power frequency voltage.

An MOV block, because of its dimensions, exhibits a capacitance and an inductance. Both of them are insignificant at low frequencies. However, at higher frequencies they affect the performance of the arrester. This performance is illustrated in Fig. 23-52, which shows the discharge voltage across the MOV arrester as a function of time for specific values of electric current through the arrester.

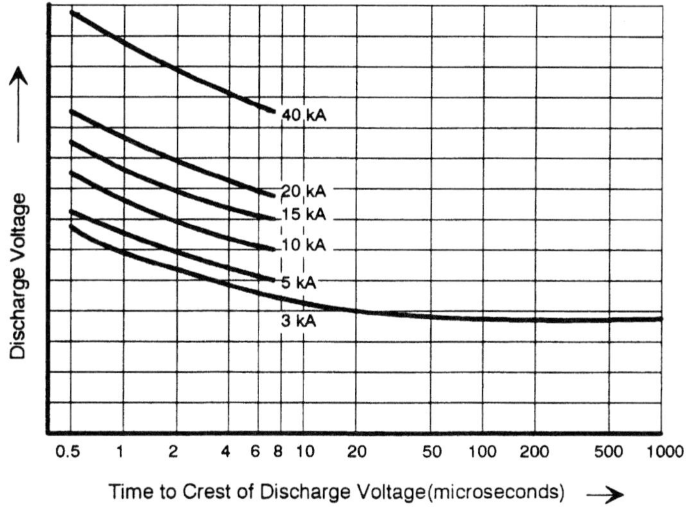


**FIGURE 23-52** Discharge voltage across an MOV arrester vs. time for various current levels through the arrester.

From this figure it is obvious that the protective level of an MOV arrester will depend on the waveform of the overvoltage.

This behavior of the MOV surge arrester can be captured with the model of Fig. 23-53. Note that this model consists of two parallel resistance-inductance (*R-L*) circuits, a parasitic capacitance, and two nonlinear resistors. The nonlinear resistor $RN_0$, represents the nonlinear characteristics of the arrester to fast front end surges, while the nonlinear resistor $RN_1$ represents the nonlinear characteristics of the arrester to slow front-end surges. The parameters of the model are selected to match the measured discharge characteristics of the arrester. Typical parameters are illustrated in Fig. 23-54.

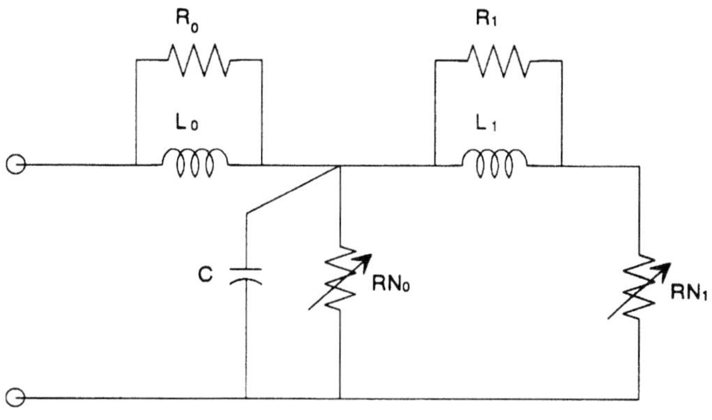


**FIGURE 23-53** Nonlinear MOV arrester model.

It can be observed in Fig. 23-48 that, as the arrester discharge current increases in an MOV, the desirable nonlinear characteristics are deteriorating. This occurs in Fig. 23-48 at discharge

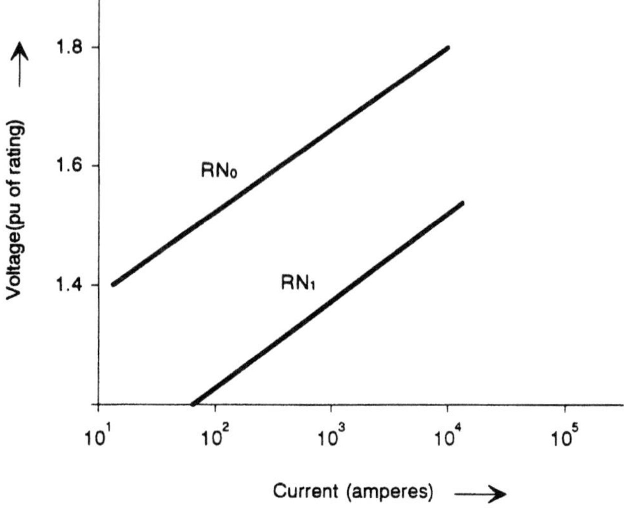


**FIGURE 23-54** Typical nonlinear voltage-current characteristic of the nonlinear resistor $RN_0$ and $RN_1$.

currents above 10 kA. Specifically, the discharge voltage at currents above 10 kA increases, resulting in inferior protection characteristics. In this aspect SiC arresters are superior to MOVs because SiC arresters do not exhibit this abrupt increase of discharge voltage at higher currents. The performance of MOVs at higher discharge currents can be improved by equipping them with a shunt gap (Westrom 1990). Such a construction is illustrated in Fig. 23-55. Specifically, an MOV block in an arrester is equipped with a shunt spark gap. The gap is designed to spark over whenever the discharge current through the arrester exceeds a certain value, for example 10 kA. In this way, when the discharge current increases beyond the value at which the desirable nonlinear characteristics of the MOV block start to deteriorate, the shunt gap sparks over, resulting in reduced discharge voltage and therefore improved protection characteristics.

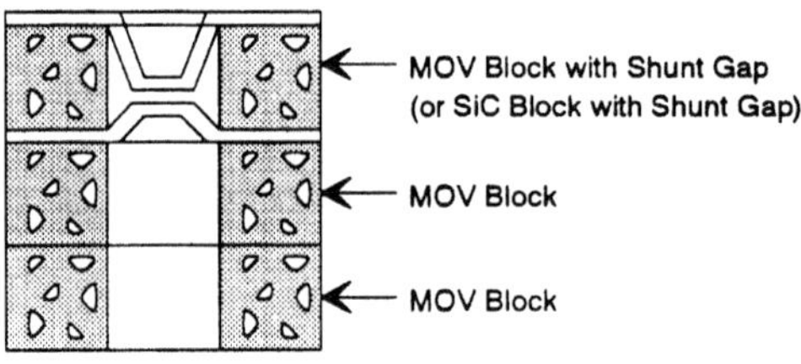


**FIGURE 23-55** Typical construction of a shunt-gap MOV arrester.

***Protective Margins.*** Another important concept is the protection margin (PM). The protective margin is related to the properties of the protective device relative to the withstand capability of the power apparatus under protection. For the ideal situation depicted in Fig. 23-39, the protection margin is defined as

$$\mathrm{PM} = \frac{V_w - V_p}{V_p}$$

where $V_w$ is the withstand capability of the power apparatus under protection and $V_p$ is the protection level.

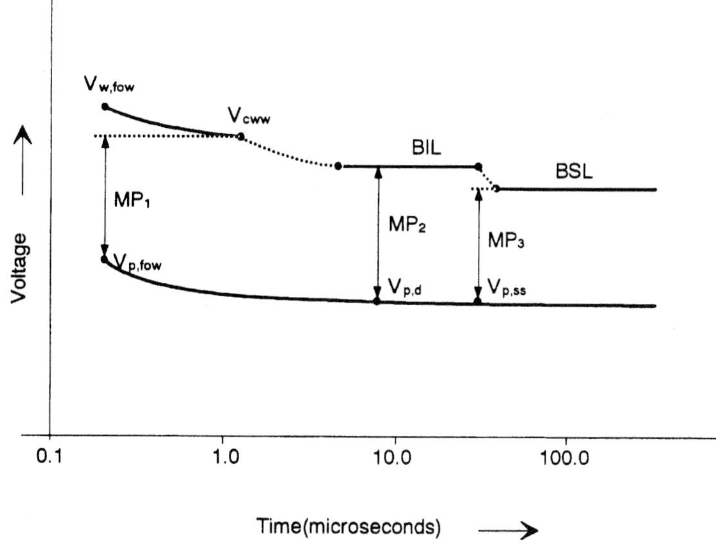


**FIGURE 23-56** Illustration of the protection margin definition.

Because both the protection level of a surge arrester and the withstand capability of a power apparatus depend on the rise time of the transient overvoltage, it is expedient to define protective margins for the standard surge waveforms. In this sense, the following protective margins are defined (see Fig. 23-56).

1. Equivalent front-of-wave protective margin ($PM_{fow}$)

$$PM_{fow} = \frac{V_{cww} - V_{p,fow}}{V_{p,fow}} \times 100$$

where $V_{cww}$ is the equipment chopped-wave withstand (cww) voltage and $V_{p,fow}$ is the arrester protection level for the front of wave (fow).

2. Equipment impulse protection margin ($PM_i$)

$$PM_i = \frac{BIL - V_{p,d}}{BIL} \times 100$$

where BIL is the equipment basic insulation level and $V_{p,d}$ is the arrester protection level for a full impulse wave.

3. Equipment switching protective margin ($PM_s$)

$$PM_s = \frac{BSL - V_{p,ss}}{V_{p,ss}} \times 100$$

where BSL is the equipment basic switching insulation level and $V_{p,ss}$ is the arrester protective level for a switching surge.

The definition of these three protective margins is illustrated in Fig. 23-56. In this definition one should bear in mind that the arrester protective levels may be defined differently for gapped and gapless arresters. Specifically, the definition of the protective levels for gapped and gapless arresters follows.

- $V_{p,\text{fow}}$, arrester protection level for front of wave

  Series-gapped arresters—front-of-wave sparkover: standard test dictates the rate of rise of front of wave to be 5.89 pu V/μs

  Gapless MOV arresters—front-of-wave sparkover: the highest discharge voltage to a surge-current impulse with 0.5 μs rise time and of specified crest (typically 3 to 40 kA)
  Shunt-gapped MOV arresters—front-of-wave sparkover: same as for gapless MOV arresters
- $V_{p,d}$, arrester protection level for a full impulse wave

  Series-gapped arresters—let-through level: the highest full impulse expected to cause sparkover after 8 μs
  Gapless MOV arresters—discharge voltage: the highest discharge voltage to a surge-current impulse (8 × 20 μs) of specified crest (typically 5 to 20 kA)
  Shunt-gapped MOV arresters—discharge voltage: same as for gapless MOV arresters
- $V_{p,ss}$, arrester protection level for a switching impulse

  Series-gapped arresters—switching surge sparkover: the highest full impulse causing sparkover at a time greater than 30 μs
  Gapless MOV arresters—switching surge discharge voltage: the highest discharge voltage to a current switching impulse of specified crest (typically 3 kA)
  Gapped MOV arresters—switching surge discharge voltage: same as for gapless MOV arresters

***Arrester Classification.*** Arresters can be classified into the following groups:

Station arresters

Intermediate arresters

Distribution arresters

Secondary, industrial, and commercial arresters

This classification is based on kilovolt ratings. Historically, station arresters provide the lowest protective levels relative to their rating (i.e., they have the highest protection quality index) and are capable of discharging the highest amount of energy. Intermediate arresters have somewhat higher protective levels relative to their rating and are capable of discharging somewhat less energy. Distribution and secondary, industrial, and commercial arresters have even higher protective levels and are capable of discharging even less energy. This historical classification and description of arresters may not be valid anymore. For example, there are MOV products that use MOV blocks for secondary arresters of the same quality as station class arresters, and therefore exhibit a very high protection quality index. It is important, however, to adopt this classification for consistency with standards.

Specific voltage ratings, protective levels, and other mechanical characteristics of complete arrester units can be found in the ANSI/IEEE standards. As an example, Table 23-6*a*, taken from ANSI/IEEE C62.1-1989, lists the standard voltage ratings for surge arresters. Table 23.6*b* lists the MOV arrester ratings taken from ANSI/IEEE std. C62.11-1999. Another example is Table 23-7, taken from ANSI/IEEE C62.2, which provides several important arrester parameters and application data for station- and intermediate-class arresters. More accurate data for specific products are available from the manufacturers.

There are a variety of commercial transient voltage suppression devices based on MOV technology. The quality of some of these products in terms of (1) their nonlinear characteristics, (2) energy absorption characteristics, and (3) current handling capability is very high. As an example, Table 23-8

**TABLE 23-6a** Surge Arrester Voltage Ratings in Kilovolts

| Secondary arresters | Distribution arresters | Intermediate arresters | Station arresters |
|---|---|---|---|
| 0.175 | | | |
| 0.650 | | | |
| | 1 | | |
| | 3 | 3 | 3 |
| | 6 | 6 | 6 |
| | 9 | 9 | 9 |
| | 10 | | |
| | 12 | 12 | 12 |
| | 15 | 15 | 15 |
| | 18 | | |
| | 21 | 21 | 21 |
| | | 24 | 24 |
| | 25 | | |
| | 27 | | |
| | 30 | 30 | 30 |
| | | 36 | 36 |
| | | 39 | 39 |
| | | 48 | 48 |
| | | 60 | 60 |
| | | 72 | 72 |
| | | 90 | 90 |
| | | 96 | 96 |
| | | 108 | 108 |
| | | 120 | 120 |
| | | | 144 |
| | | | 168 |
| | | | 180 |
| | | | 192 |
| | | | 240 |
| | | | 258 |
| | | | 276 |
| | | | 294 |
| | | | 312 |
| | | | 372 |
| | | | 396 |
| | | | 420 |
| | | | 444 |
| | | | 468 |
| | | | 492 |
| | | | 540 |
| | | | 576 |
| | | | 612 |
| | | | 648 |
| | | | 684 |

**TABLE 23-6b** MOV Arrester Ratings in Kilovolts*

| Duty-cycle voltage (kV rms) | MCOV (kV rms) | Duty-cycle voltage (kV rms) | MCOV (kV rms) |
|---|---|---|---|
| 3 | 2.55 | 144 | 115 |
| 6 | 5.1 | 168 | 131 |
| 9 | 7.65 | 172 | 140 |
| 10 | 8.4 | 180 | 144 |
| 12 | 10.2 | 192 | 152 |
| 15 | 12.7 | 228 | 180 |
| 18 | 15.3 | 240 | 190 |
| 21 | 17 | 258 | 209 |
| 24 | 19.5 | 264 | 212 |
| 27 | 22 | 276 | 220 |
| 30 | 24.4 | 288 | 230 |
| 36 | 29 | 294 | 235 |
| 39 | 31.5 | 312 | 245 |
| 45 | 36.5 | 396 | 318 |
| 48 | 39 | 420 | 335 |
| 54 | 42 | 444 | 353 |
| 60 | 48 | 468 | 372 |
| 72 | 57 | 492 | 392 |
| 90 | 70 | 540 | 428 |
| 96 | 76 | 564 | 448 |
| 108 | 84 | 576 | 462 |
| 120 | 98 | 588 | 470 |
| 132 | 106 | 612 | 485 |

*For ratings not shown, consult with the manufacturer.

illustrates the characteristics of a 480-V rated transient voltage suppression device, HA series. Figure 23-57 also illustrates the voltage-current characteristic of this device. By a simple fitting procedure, the nonlinear characteristic of this device in the range 1 to 1000 A is approximately described by Eq. (23-1), where $n \approx 24$. This is a relatively high exponent, suggesting a rather high protection quality index.

## 23.7 OVERVOLTAGE PROTECTION (INSULATION) COORDINATION

The objective of overvoltage protection coordination, otherwise known as insulation coordination, is to minimize the number of insulation failures and therefore the number of interruptions. This has been a fundamental task of power engineering with a profound effect on the reliability of the system. Insulation coordination methods have evolved from the ad hoc procedures of the past to the highly sophisticated computer methods of today. Yet protection schemes remain an art, and a review of the evolution of methods provides an invaluable insight. By necessity, the methods are based on incomplete data or parameters with substantial uncertainty. As a result, present models are incomplete and interpretation of results is necessary.

The objectives of insulation coordination procedures also depend on the effects of insulation failure. In this respect, power apparatus may be grouped in two broad categories: (1) those that use air as insulation (external insulation) and (2) those that use solid or liquid dielectrics as insulating material (internal insulation). Insulation failure in devices of the first category is not as destructive as insulation failure in devices of the second category. With the exception of secondary effects, failure

**TABLE 23-7** Station and Intermediate Arrester Characteristics

| Ratings, kV rms | Range of application nominal system voltage, kV | Protective levels, per unit crest arrester rating: Front-of-wave sparkover | 1.2/50-μs sparkover | Switching surge sparkover | Discharge voltage, 10 kA, 8/20-μs wave | Durability characteristics: (1) Duty cycle initiating surge, crest amperes | (2) Transmission line discharge, mi | (3) High-current withstand, crest amperes | (4) Pressure relief, rms symmetrical amperes |
|---|---|---|---|---|---|---|---|---|---|
| Station class | | | | | | | | | |
| 3–9 | 2.2–12.47 | 2.24–4.24 | 1.89–3.30 | Test | 1.57–1.77 | 10,000 | 150 | 65,000 | 65,000 25,000 |
| 12–15 | 13.2–18 | 2.12–2.83 | 1.89–2.42 | not | 1.57–1.70 | 10,000 | 150 | 65,000 | 65,000 25,000 |
| 21–48 | 18–46 | 2.09–2.56 | 1.80–2.29 | required | 1.56–1.70 | 10,000 | 150 | 65,000 | 40,000 25,000 |
| 60–120 | 69–138 | 1.99–2.24 | 1.60–1.94 | 1.60–1.80 | 1.56–1.69 | 10,000 | 150 | 65,000 | 40,000 25,000 |
| 144–240 | 161–287 | 1.83–2.22 | 1.57–1.70 | 1.57–1.61 | 1.56–1.79 | 10,000 | 175 | 65,000 | 40,000 25,000 |
| 258–312 | 345 | 2.06–2.17 | 1.56–1.70 | 1.57–1.61 | 1.56–1.58 | 10,000 | 200 | 65,000 | 40,000 25,000 |
| 372 or higher | 500 or higher | 1.94–2.10 | 1.65–1.70 | 1.44–1.58 | 1.54–1.60 | 10,000 | 200 | 65,000 | 40,000 25,000 |
| Intermediate class | | | | | | | | | |
| 3–6 | 2.4–7.2 | 2.47–2.83 | 2.24–2.83 | Test not | 1.77–2.36 | 5000 | 100 | 65,000 | 16,100 |
| 9–48 | 7.2–46 | 2.10–2.59 | 1.78–2.51 | required | 1.77–2.19 | 5000 | 100 | 65,000 | 16,100 |
| 60–120 | 69–138 | 1.76–2.26 | 1.63–1.84 | 2.06–2.43 | 1.77–2.02 | 5000 | 100 | 65,000 | 16,100 |

**TABLE 23-8** Characteristics of a Commercial MOV-Type Device, HA Series, 480 V

| Model number | Size, mm | Continuous rms voltage rating, V | Continuous dc voltage rating, V | Transient energy absorption, J | Peak current 8/20 μs, A | Varistor voltage at 1-mA dc test current: Min., V | Normal, V | Max., V | Maximum clamping voltage at 200 A, 8/20 μs, V | Typical capacitance, pF |
|---|---|---|---|---|---|---|---|---|---|---|
| V481HA32 | 32 | 480 | 640 | 450 | 25,000 | 670 | 750 | 825 | 1290 | 1300 |

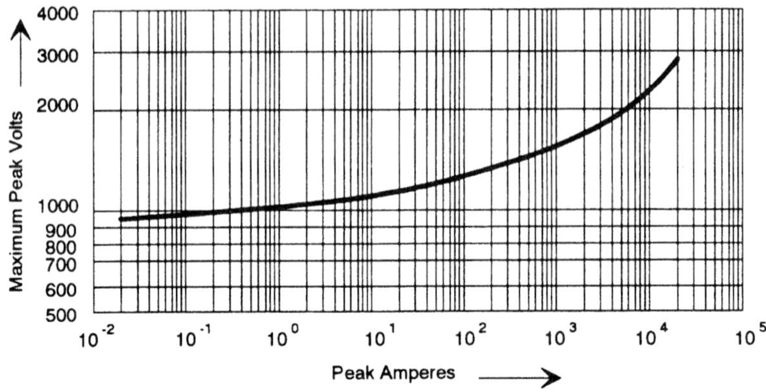


**FIGURE 23-57** Voltage-current characteristic of transient voltage suppressor V481HA32, HA series.

of air insulation is self-healing, since removal of the overvoltage will restore the insulation. For this reason, the objectives of insulation coordination for external insulation are typically the minimization of insulation failure. It is permissible to tolerate a small probability of insulation failure if this will minimize the cost. Probabilistic methods are pertinent for this purpose. On the other hand, the objective of insulation coordination for internal insulation is to disallow any insulation failure from any known causes. Adherence to this objective is dependent on the cost of the apparatus under protection. As an example, this objective is strictly observed for higher MVA-rating power transformers while for distribution transformers the cost of meeting the objective may not be justifiable. Because of this variability in objectives, the process of insulation coordination is examined separately for transmission lines, substations, overhead distribution systems, undergound distribution systems, and industrial/commercial systems.

***Transmission Lines.*** Overhead transmission lines use air or porcelain as the external insulating material. Flashovers across insulator strings or phase conductor to ground wire do not lead to catastrophic failures but rather to momentary outages. In other words, the insulation strength is quickly restored after arc extinction. For this system, the objectives of insulation coordination are to minimize the momentary outages at minimum cost. This is achieved through a procedure by which the cost of momentary outages is balanced against the cost of providing for overvoltage protection. The procedure is statistical, since many driving factors are statistical, for example, lightning strength and rise time. It is illustrated in Fig. 23-58, where the distribution of overvoltages is plotted as well as the distribution of withstand capability. Outages may occur when the overvoltage exceeds the withstand capability. This procedure can be carried one step further, where the performance (reliability) of the system is balanced against cost of designs. This concept is illustrated in Fig. 23-59, where the optimum design is defined at the minimum of the total cost consisting of the investment cost plus the unreliability cost. A drawback of this method is the fact that, while the investment cost can be properly assessed, the cost of unreliability may be elusive and at best subjective. In any case, insulation coordination of overhead transmission lines involves the following basic tasks:

Design of an effective shielding against direct lightning strokes

Design of an adequate insulation strength to withstand power frequency, switching, and lightning overvoltages

Design of an effective grounding system to minimize backflashovers

The design procedure typically consists of selecting a standard design and subsequently estimating the expected number of momentary outages. Design modifications are exercised if objectives are

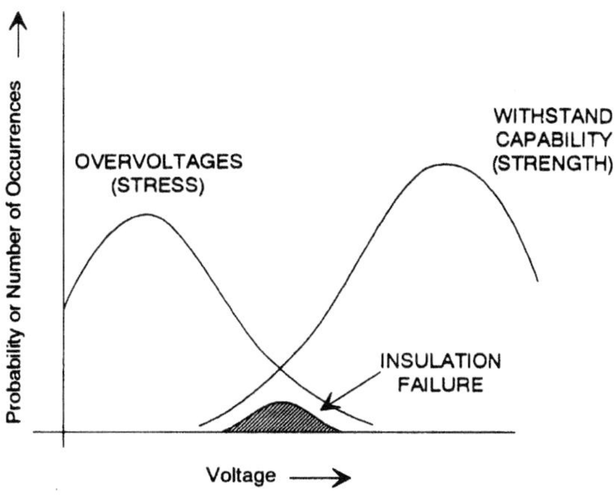


**FIGURE 23-58** Statistical evaluation of insulation failures.

not met. In this subsection, we will be concerned with estimation of the number of momentary outages. This task involves the following procedure:

Estimation of the number of lightning strokes in the area
Estimation of the percentage of strokes reaching the line
Estimation of number of shielding failures
Estimation of probability of backflashover
Effects of corona
Summary of results

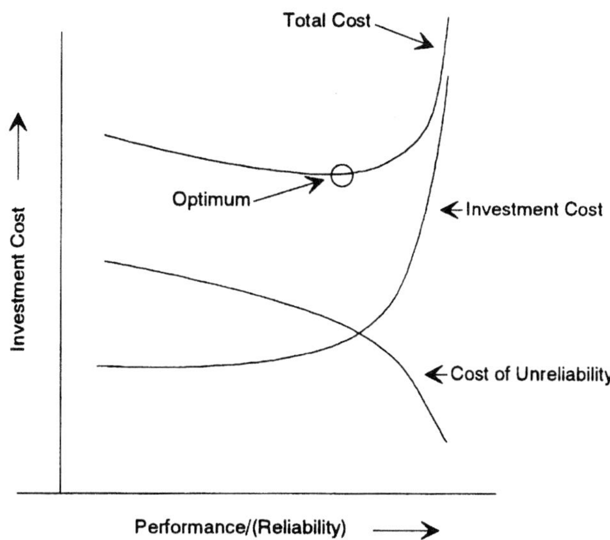


**FIGURE 23-59** Optimal design procedures that provide best trade-off between performance (reliability) and cost.

**TABLE 23-9** Empirical Relationships between Lightning Ground-Flash Density and Annual Thunderstorm Days $T$

| Location | Ground-flash density, number of ground flashes/km$^2$ · year |
|---|---|
| India | $0.1T$ |
| Rhodesia | $0.14T$ |
| South Africa | $0.04T^{1.25}$ |
| Sweden | $0.004T^2$ (approx.) |
| United Kingdom | $aT^b$ ($a = 2.6 \pm 0.2 \times 10^{-3}$; $b = 1.9 \pm 0.1$) |
| United States (north) | $0.11T$ |
| United States (south) | $0.17T$ |
| United States | $0.1T$ |
| United States | $0.15T$ |
| U.S.S.R. (Former) | $0.036T^{1.3}$ |
| World (temperate climate) | $0.19T$ |
| World (temperate climate) | $0.15T$ |
| World (temperate climate) | $0.13T$ |

***Source:*** EPRI (1975).

*Estimation of the Number of Lightning Strokes in the Area.* The ground flash density is defined as the number of lightning strokes (cloud to ground) per unit of area. It is preferable to use actual data on ground flash density if available. If not, then, as a first approximation, the ground flash density is taken to be approximately proportional to the thunderstorm activity, measured in thunderstorm days. Empirical formulas have been developed and are listed in Table 23-9. For the United States, the most commonly used formula is

$$N_1 = 0.12T \qquad \text{or} \qquad N_2 = 0.31T$$

where $N_1$ = ground flash density per square kilometer per year
$N_2$ = ground flash density per square mile per year
$T$ = number of thunderstorm days

It must be emphasized that this formula should be viewed as an average and approximate only, since thunderstorm activity may vary from year to year and the lightning activity in a thunderstorm day may vary.

*Estimation of the Percentage of Strokes Reaching the Line.* The number of strokes reaching the line (shield wire or phase conductor) can be estimated with two basic models:

The geometric model
The electrogeometric model

These basic models have many variations, as many researchers have made improvements to improve their correlation to observed data.

The principle of the geometric model is illustrated in Fig. 23-60. The model postulates that the line casts a shadow around it of width $W$. Any ground flash which would terminate at the shadow of width $W$ if the line was not present will be intercepted by the line. Assuming that the ground flashes are uniformly distributed over the surface of the ground, the following equation applies for the number of strokes to the line:

$$N_{t1} = 0.00012\ WT$$

or

$$N_{t2} = 0.000193\ WT \tag{23-2}$$

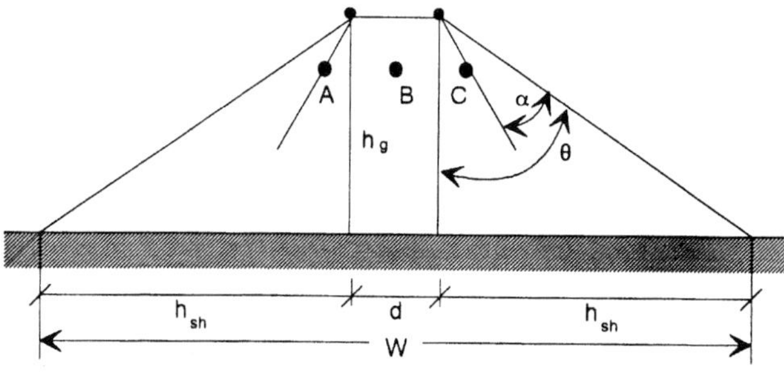


**FIGURE 23-60** Illustration of application of the geometric model.

where $N_{t1}$ = number of strokes to the line per kilometer of line length per year
$N_{t2}$ = number of strokes to the line per mile of line length per year
$W$ = width of the line shadow, m
$T$ = number of thunderstorm days in the area

An expression for the shadow width is

$$W = 4h + d$$

where $d$ = distance between shield wires (zero if only one shield wire)
$h$ = effective height of the line, which can be approximated with $h = h_g + {}^2/_3\,(h_g - h_{gw})$
$h_g$ = shield wire height at the tower or pole
$h_{gw}$ = shield wire height at midspan

The geometric model is very simplistic. It fails to take into account that the striking distance of lightning is determined by the magnitude of the return stroke as well. An improved method is the so-called electrogeometric model. The principle of this model is illustrated in Fig. 23-61. This model is based

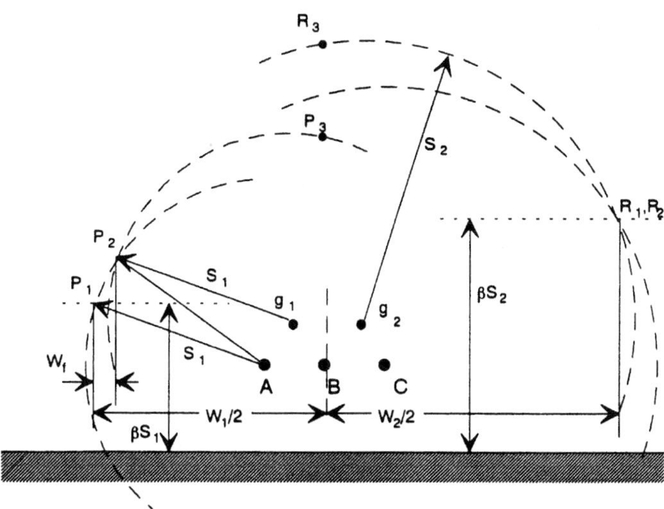


**FIGURE 23-61** Illustration of application of the electrogeometric model. [*Note:* The value of $\beta$ is a function of tower height. Anderson (1975) uses $\beta = 0.67$ for ultra-high-voltage (UHV) lines.]

**TABLE 23-10** Proposed Equations for Determining the Striking Distance
S = striking distance in meters; $I_s$ = first return stroke current, kA

| Proposed formula | Reference |
|---|---|
| $S = 2I_s + 30\,(1 - e^{-0.147I})$ | Darveniza (1979) |
| $S = 10\,I_S^{0.65}$ | Love (1973) |
| $S = 9.4\,I_S^{0.67}$ | Whitehead (1977) |
| $S = 8\,I_S^{0.65}$ | IEEE (1987) |
| $S = 3.3\,I_S^{0.78}$ | Suzuki (1981) |
| $S = 0.67\,h^{0.6}I^{0.74}$ | Erickson (1982) |
| $S = 1.57\,h^{0.45}I^{0.69}$ | Rizk (1994) |
| $S = [3.6 + 1.7\ln(43 - Y_c)]\,I^{0.65}$ for $Y_c < 40$m and $5.5\,I^{0.65}$ for $Y_c >$ (or equal to) 40m (See IEEE Std. 1243) | |

on the observation that the point where the stepped leader of a lightning terminates is not determined until the stepped leader is within a striking distance. The striking distance is defined as the length of the final leg of the stepped leader which establishes contact with a ground object. The striking distance was found to depend on the amplitude of the first return stroke. A number of empirical formulas have been proposed for determining the striking distance. Table 23-10 lists the most common ones.

The equation most commonly used in the United States is the second one listed in Table 23-10. Using a striking distance, which depends on stroke magnitude, generates a model, which suggests that the number of strokes intercepted by the line depends on their magnitude. Figure 23-61 illustrates the application of the method to compute the number of strokes terminating at the line for two different stroke magnitudes, 7 and 12 kA. Specifically, for strokes of 7 kA, the striking distance is $S_1$. For this striking distance, one can construct the width over which the strokes will be intercepted by the line. This width, $W_1$, is shown in the figure. It is constructed by (1) drawing a line above ground at height $\beta S_1$, where $\beta = 0.8$ (from Anderson 1968), (2) drawing a circle of radius $S_1$ with center at the phase conductor, and (3) drawing a circle of radius $S_1$ with center at the shield wire. The construction of the width $W_2$ for strokes of 12 kA is also shown in the figure. The computed widths $W_1$ and $W_2$ can be used in Eq. (23-2) to estimate the number of strokes to the line. It should be understood that the process is repeated for several stroke magnitudes, spanning the distribution of lightning current magnitudes; then all the results are utilized to compute the statistical distribution of the expected strokes to the line. The average value can be extracted from this distribution.

*Estimation of Number of Shielding Failures.* It is important to know how many of the strokes reaching the line will actually terminate on a phase conductor or on a shield wire. The consequences are obvious. A lightning stroke terminating at the phase conductor and of sufficient magnitude will most likely result in a flashover. This is obvious when one considers the magnitude of the overvoltage. When shielding failure occurs and the lighting stroke terminates on the phase conductor, it generates an overvoltage on the stricken phase which is given by

$$v_s(t) = {}^1\!/_6\,(Z_g + 2Z_t)i(t)$$

and an overvoltage on the other two phases which is given by

$$v_n(t) = {}^1\!/_6\,(Z_g - Z_t)i(t)$$

where $i(t)$ = lightning stroke current
$Z_g$ = characteristic impedance for the ground mode of propagation of surges along the line
$Z_t$ = characteristic impedance for the line mode of propagation of surges along the line
$v_s(t)$ = lightning overvoltage on the stricken phase
$v_n(t)$ = lightning overvoltage on any one of the other two phases

For a typical transmission line design

$$Z_g \approx 800\,\Omega$$

$$Z_t = 400\,\Omega$$

In this case, the overvoltage is

$$v_s(t) \approx 300i(t)$$

$$v_n(t) \approx 60i(t)$$

Obviously, a shielding failure for a lightning stroke of 20 kA will generate an overvoltage of about 6 MV, which will with certainty cause a flashover. On the other hand, a lightning stroke terminating on a shield wire will result in much lower overvoltages, if the transmission line is effectively grounded. It is, of course, possible that the overvoltages resulting from a lightning stroke to the shield wire will be of such magnitude as to cause a flashover from the shield wire or tower to the phase conductor. This event is known as a *backflashover*. In any case, the line shielding must be designed in such a way that the number of times lightning reaches a phase conductor is minimized. Occurrences of such incidents are termed *shielding failure*.

The electrogeometric model provides an adequate tool to determine shielding failure. The basic premise of the method is defined in Fig. 23-61. Consider the striking distance $S_1$ for a lightning stroke of magnitude $I_1$. Consider also the line above ground at a height $\beta S_1$, the circle of radius $S_1$ with center at phase A, and the circle of radius $S_1$ with center at shield wire $g_1$. The intersections of these circles define the points $P_1$ and $P_2$. The electrogeometric model states that any stepped leaders, which reach the arc $P_1P_2$ and of magnitude less than $I_1$ will terminate on the phase conductor A, resulting in a shielding failure. Any stepped leader reaching the arc $P_2P_3$ will be attracted by the shield wire. The percentage of lightning strokes of magnitude $I_1$ or lower, which will reach a phase conductor, can be estimated by comparing the length of the arc $P_1P_2$ to the length of the arc $P_2P_3$. This process can be repeated for various stroke magnitudes. Subsequently, and utilizing the statistics of the stroke magnitude distribution, the probability distribution of shielding failures can be computed. The expected number of shielding failures can also be computed.

For lightning strokes of a certain magnitude, the striking distance will be such that the points $P_1$ and $P_2$ may collapse to one point. This condition is illustrated in Fig. 23-61 for a different magnitude of stroke current $I_2$, where the points $R_1$ and $R_2$ coincide. Any stepped leader resulting in stroke current magnitude $I_2$ or higher and reaching the arc $R_2R_3$ will terminate on the shield wire. Any other stepped leaders reaching points to the right of points $R_2$ will terminate to the ground. In this case, the line is shielded for lighting strokes of magnitude greater than $I_2$. Now assume that the line insulation can withstand the overvoltages resulting from a direct hit on a phase conductor of magnitude $I_2$. In this case, a shielding failure for lighting strokes of magnitudes $I_2$ or lower will not result in a flashover. The line is said to be *effectively shielded*.

An extension of this statement results to the concept of the critical striking distance. The critical striking distance is the striking distance for the smallest lightning stroke current magnitude, which, if it terminates on a phase conductor, will cause a flashover. The critical striking distance can be used to define the location of shield wires which will provide an effective shielding.

*Estimation of Probability of Backflashover.* Backflashover can occur when the transmission tower voltage at the location where the phase insulators are suspended (crossarms) becomes so high as to initiate an arc between the crossarm and the phase conductors. The driving force for initiating the arc is the voltage across the insulator which depends on many factors:

Voltage at the crossarm due to the lightning stroke

Induced voltage on the phase conductor due to surges on the shield wire generated by the lightning stroke

Power frequency voltage at the phase conductor

Tower footing impedance

Traveling-wave phenomena along the tower

Lightning strokes to the tower result in the highest overvoltage at the top of the tower as compared to overvoltages from lightning strokes on the shield wire along the span. As an example,

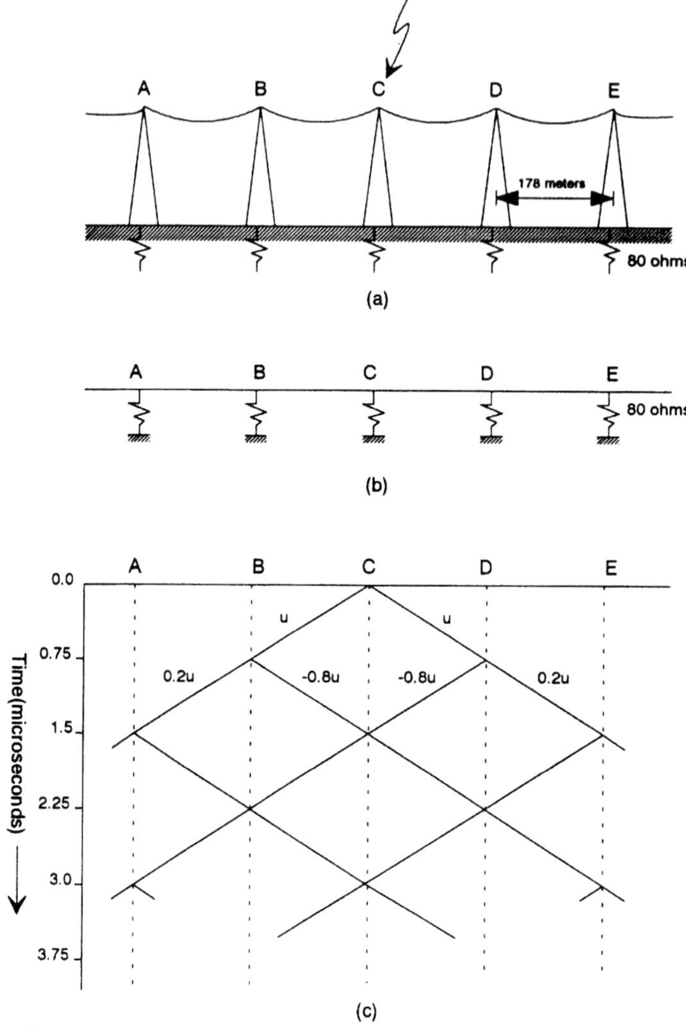


**FIGURE 23-62** Simplified calculations of overvoltages resulting for a lightning stroke to a tower: (*a*) schematic; (*b*) simplified circuit; (*c*) Bewley diagram.

consider a lightning stroke of crest 10 kA and rise time 1 μs terminating on the shield wire of a transmission line. For simplicity, neglect the phase conductors. The system is illustrated in Fig. 23-62*a*. All tower ground resistances are 80 Ω. The parameters of the shield wire, for the example under consideration, are

$$L = 2.7 \mu\text{H/m}$$

and

$$C = 6.58 \times 10^{-12}\,\text{F/m}$$

The characteristic impedance of the shield wire is $Z=\sqrt{L/C}=640\,\Omega$, and the speed of propagation of surges is $u=1\sqrt{LC}=237$ m/$\mu s$. The travel time from one tower to the next is 0.75 μs. To further simplify the computations, traveling-wave phenomena along the tower are neglected, resulting in the simplified circuit of Fig. 23-62*b*. The reflection and transmission coefficients at the tower are $\alpha=-0.80$ and $\delta=0.20$. The Bewley diagram for this circuit is illustrated in Fig. 23-62*c*, where $u(t)$ is computed as

$$u(t)=R_{\mathrm{eq}}i(t)=64i(t)$$

where 64 Ω is the impedance "seen" by the lightning current. 64 Ω resulting from the parallel combination of 80 Ω, 640 Ω to the right of the tower.

Thus, $u(t)$ will have a rise time of 1 μs and crest of 640 kV. By summing up all the waves arriving at point C, the voltage waveform c1 of Fig. 23-63 is obtained. This procedure can be repeated, assuming that the same lightning stroke terminates at the shield wire in the middle of the span. The resulting overvoltage at the top of the tower C is shown as waveform c2 in Fig. 23-63. Note that the crest of this waveform is much lower. In reality, the tower footing impedance is frequency-dependent and traveling-wave phenomena along the tower affect the overvoltages. Figure 23-64 illustrates a typical voltage waveform across the phase insulators using a rather sophisticated model (Cokkinides and Meliopoulos 1988). Note that reflections from adjacent towers quickly reduce the voltage across the phase insulators. Also, the crossarm voltage is in general lower than the top of the tower voltage. As a result, bottom phase insulators, in general, experience the highest overvoltage for backflashover. The tower ground also affects the overvoltage. As an example, Fig. 23-65 illustrates the effect of tower grounding on the lightning voltage at the top of the tower. Note that the voltage at the top of the tower is not affected by tower grounding for the first approximately 0.2 μs after lightning initiation. However, the overall overvoltage waveform is drastically affected by tower grounding. Since the characteristics of arc initiation across the insulator string have a

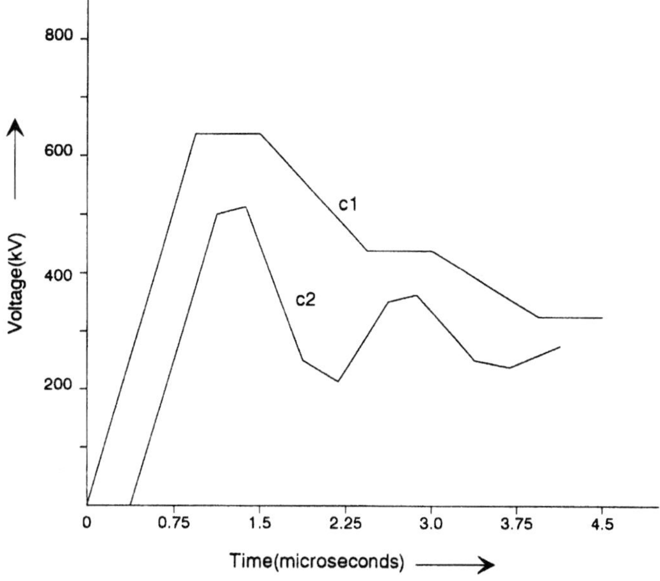


**FIGURE 23-63** Lightning overvoltages at top of tower; in c1 the lightning stroke terminates at tower top, and in c2 the lightning stroke terminates at midspan.

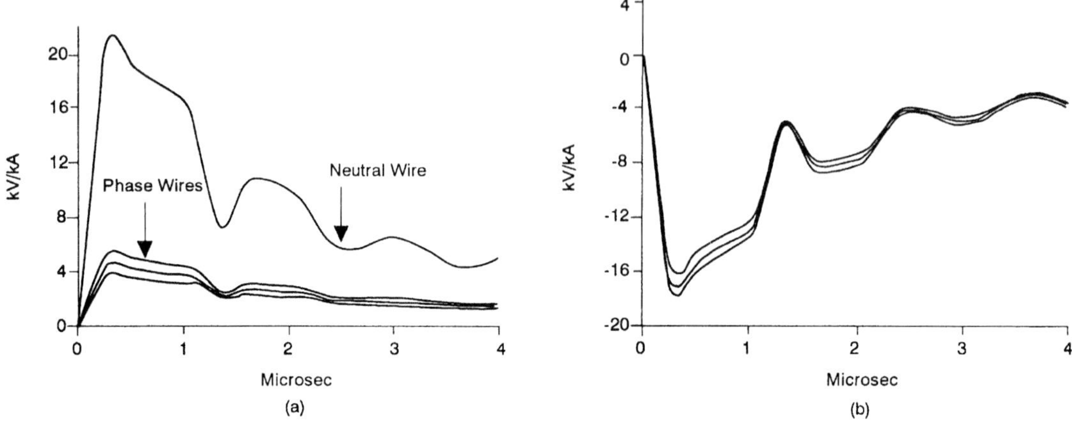


**FIGURE 23-64** Typical voltage waveforms across phase insulators due to lightning: (*a*) transient voltage of phase conductors and shield wire; (*b*) transient voltage across insulators.

volt-time characteristic, it is very important to examine the effect of voltage waveform on backflashover initiation.

CIRGE has accepted the following volt-time curve for line insulator flashover, which is consistent with the work of Darveniza et al. (1975, 1979) and Whitehead (1977)

$$V_{\bar{f}} = 0.4W + \frac{0.71W}{t^{0.75}}$$

where $W$ = line insulator length in meters
$t$ = time to breakdown in microseconds
$V_{\bar{f}}$ = flashover voltage in megavolts for negative surges

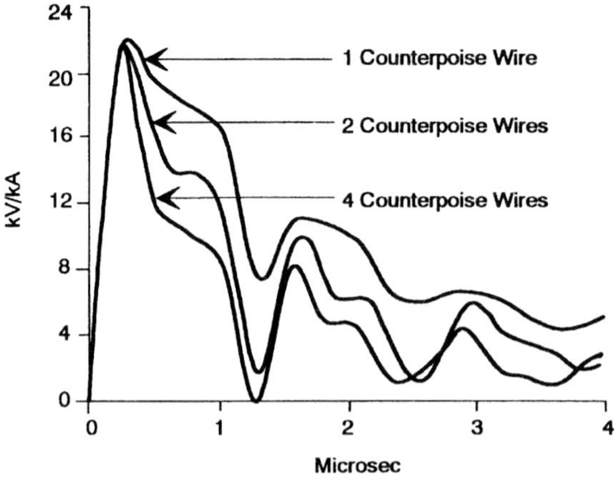


**FIGURE 23-65** Illustration of effects of tower grounding on voltage at top of tower due to a lightning stroke at the tower.

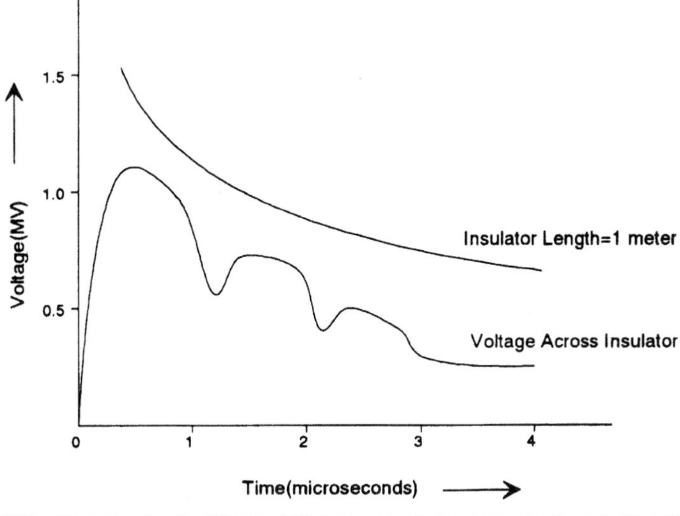


**FIGURE 23-66** Insulator voltage-time curve and insulator voltage.

One should compare the volt-time characteristic to the actual voltage across the phase insulators to determine whether a backflashover will occur. Figure 23-66 illustrates such a construction where the insulator withstand capability versus time of a 1-m-long insulator string is superimposed on the actual voltage experienced by the insulator.

*Effects of Corona.* The overvoltage occurring at a phase conductor or shield wire establishes an extremely high-voltage gradient perpendicular to the conductor. This gradient, wherever it exceeds the withstand capability of air, will generate electric discharges, which will electrify the air surrounding the conductor. This phenomenon is known as corona. Corona increases the apparent radius of the conductor while it does not affect the geometric mean radius of the conductor. As a result, it decreases the characteristic impedance of the line and the level of overvoltages resulting from lightning.

The voltage gradient (electric-field intensity) around an energized conductor is approximately

$$E = \frac{V}{r \ln(2h/a)}$$

where $V$ is the conductor voltage, $r$ is the distance from the conductor centerline, $h$ is the conductor height above the earth, and $a$ is the conductor radius.

The value of 15 kV/cm is accepted as the electric-field intensity required to establish corona. Thus, corona will exist in a radius $a_c$ such that

$$15 \text{ kV/cm} = \frac{V}{a_c \ln(2h/ac)}$$

As an example, a conductor 50 ft above the earth which experiences an overvoltage of 3.5 MV will establish corona at a radius $a_c = 2$ ft.

*Summary of Results.* In summary, methods have been discussed for design of line-shielding systems, estimation of shielding failure, estimation of overvoltage stresses across insulators, and methods to determine the possibility of flashover across an insulator. There is a considerable uncertainty with almost all parameters and models affecting shielding failure, magnitude, and waveform of overvoltage across insulators, etc. This uncertainty can be dealt with by using probabilistic methods. Monte Carlo simulation is utilized for this purpose and is described in Sec. 23.8.

***Substation Lightning Insulation Coordination.*** Substations typically comprise equipment such as transformers, bus work, breakers, reactors, and capacitor banks etc. This equipment must be protected against lightning overvoltages. As a rule, substations are shielded against direct lightning strikes on phase conductors. Thus, lightning overvoltages in a substation may occur for the following reasons:

1. Backflash due to a strike in the shield of the substation
2. Direct lightning hit due to shield failure
3. Lightning surges from lines (most common)
   *a.* Line shielding failure
   *b.* Line backflashover
   *c.* Induced voltage

The overvoltage protection of substations consists of shielding against direct lightning strikes and application of protective devices (surge arresters) to protect specific power apparatus.

Substation shielding methods are similar to those applied for transmission lines. One can use the geometric model or the electrogeometric model to determine the effectiveness of the shielding wires, masts, etc. The application of the electrogeometric model for substation shielding analysis is more complex since the analysis must be performed on a three-dimensional basis as opposed to a two-dimensional basis for a transmission line. A simplified variation of the electrogeometric model for the three-dimensional problems is the so-called rolling-sphere method. The basic idea of the rolling-sphere method is illustrated in Fig. 23-67*a*. Let $S_c$ be the critical striking distance as defined for transmission lines:

$$S_c = 10 I_c^{0.65}$$

where $I_c$ is the critical stroke current, kA, as defined for transmission lines. The rolling-sphere method postulates that by rolling a sphere of radius $S_c$ over the shield system of the substation, the protected equipment are those which are not crossing the path of the sphere as is illustrated in Fig. 23-67. This method provides a simple procedure to design an effective shielding system for the substation. Figure 23-67*b* provides a visualization of the application of this method to an actual substation.

In addition to the shielding system, protection must be provided against overvoltages resulting from (1) lightning strikes to the shielding system and (2) surges entering the substation from the transmission lines (which are the most common). The magnitudes of the overvoltages in (1) above depend on the design of the grounding system. On the other hand, the overvoltages from (2) above depend on many factors, such as distance of point of initiation and source of surge (direct hit, induced). These overvoltages are typically limited by the insulation level of the transmission line. Since the usual case is that the substation equipment will have a BIL below the insulation level of the lines connected to the substation, overvoltage protection for the substation equipment must be provided.

The application of protective devices (surge arresters) for the protection of power transformers is illustrated in Fig. 23-68. The application procedure consists of the following steps:

1. Select arrester rating (preliminary).
2. Determine arrester protection levels
   $V_{p,\,\text{fow}}$ = arrester protection level for a front of wave
   $V_{p,d}$ = arrester protection level for a full-impulse wave
   $V_{p,ss}$ = arrester protection level for a switching impulse
3. Determine transformer withstand capability
   $V_{\text{foww}}$ = front-of-wave withstand
   $V_{\text{cww}}$ = chopped-wave withstand
   BIL = basic insulation level
   BSL = basic switching insulation level
4. Plot arrester protection levels and transformer withstand capability on a common coordinate system, as illustrated in Fig. 23-68.

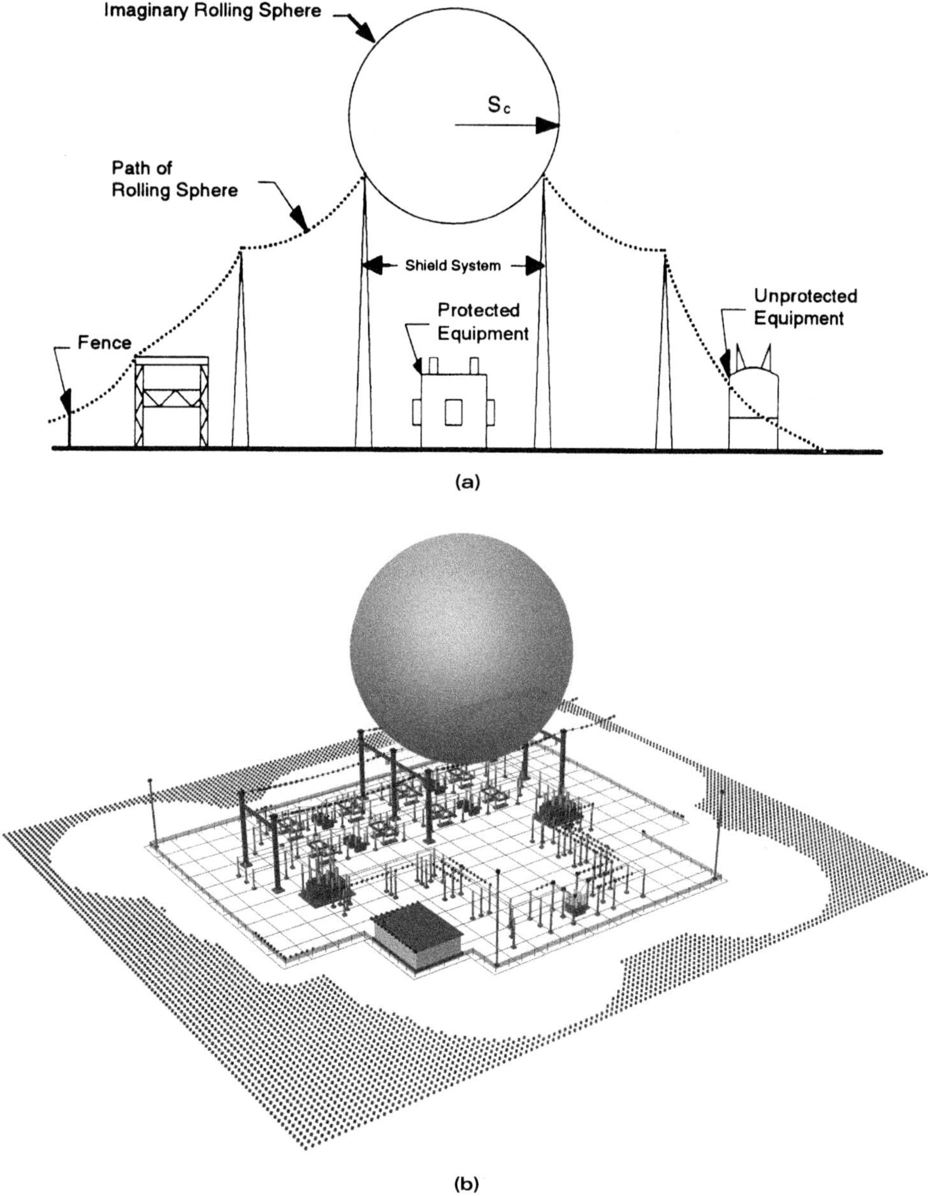


**FIGURE 23-67** Illustration of the basic idea of the rolling-sphere method.

5. Determine protection margins.
6. If protection margins are not higher than the minimum recommended, select the next-lower arrester rating that is compatible with the normal operating voltage of the system, and go to step 2. Otherwise, go to step 7.
7. Determine whether arrester has adequate energy absorption capability for the application.

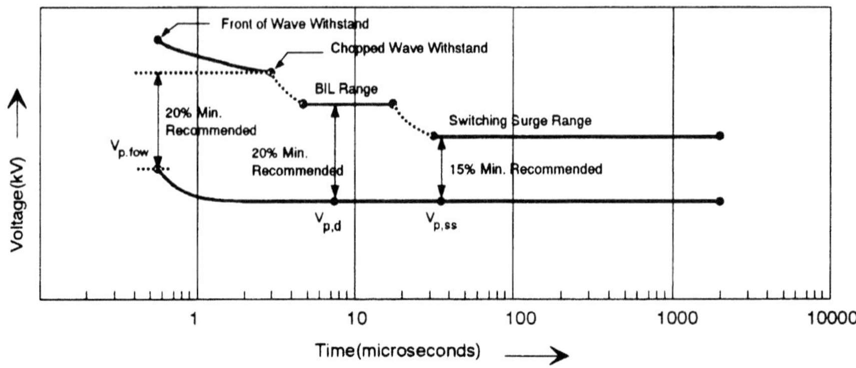


**FIGURE 23-68** Application of surge arresters for protection of power transformers.

This procedure is pertinent for application of surge arresters near the transformer. If there is a separation distance between arrester and transformer, the effects of separation must be accounted for. Specifically, separation results in deterioration of protective margins, as illustrated in Fig. 23-69. The figure shows an application of an ideal arrester 15 m away from the transformer. The ideal arrester is rated 150 kV. Because of the separation distance, the overvoltage reaching the transformer will be 300 kV, double the protective level of the arrester.

***Distribution System Overvoltage Protection.*** Distribution circuits are typically not insulated to withstand direct lightning strokes. As a result, direct strikes will cause a flashover. Direct strikes on distribution lines are not frequent since the poles are not as high and therefore are shielded from trees and structures. On the other hand, distribution lines may be vulnerable to overvoltages resulting from lightning strokes to nearby trees, ground, or other objects. These voltage surges are known as *induced lightning voltages* and are injected into the power system through coupling. The coupling can be conductive through the conductive soil and the power system grounding structures, inductive, or capacitive. In a typical situation, all the coupling mechanisms may be present resulting in a voltage surge to the power system. These voltages are called *induced voltage surges* and are generally much lower than those occurring after a direct strike. Specifically, they rarely exceed 500 kV. The induced lightning overvoltages are of concern for distribution lines 35 kV or below. Higher kilovolt-level lines (i.e., 69 kV and above) have sufficient insulation to withstand induced voltage surges.

The mechanism of induced voltage surges is illustrated in Fig. 23-70. A lightning stroke terminating at a location near a distribution line induces a surge on the line through conductive, inductive, and capacitive coupling. Several models to estimate the level of the induced voltage surges have been reported in the literature (Eriksson et al. 1982; Liew and Mar 1986). A simplified formula suggests the following induced voltage

$$v_p = Z_0 Ih \cdot \left[ \left(\frac{1}{2y}\right) \frac{1+(x+\beta y)}{\sqrt{x^2+2y^2+2\beta xy-\beta^2 y^2}} + \frac{2\beta x+y}{y^2+(2\beta x+y)^2} \cdot \frac{1+2\beta^2 x+\beta y-\beta x}{\sqrt{x^2+2y^2+2\beta xy-\beta^2 y^2}} \right]$$

where $\beta = 0.004I^{0.64} + 0.068$ first stroke
$= 0.004I^{0.86} + 0.18$ subsequent strokes
$Z_0$ = distribution line characteristic impedance
$I$ = stroke current crest

The terms $x$, $y$, and $h$ are as illustrated in Fig. 23-70. In addition, many measurements of induced voltage surges have been performed which practically verify the available models.

Induced voltage surges on distribution lines are quite frequent. It is therefore necessary to insulate distribution lines to withstand these surges. This translates into the requirement of a 300 BIL for

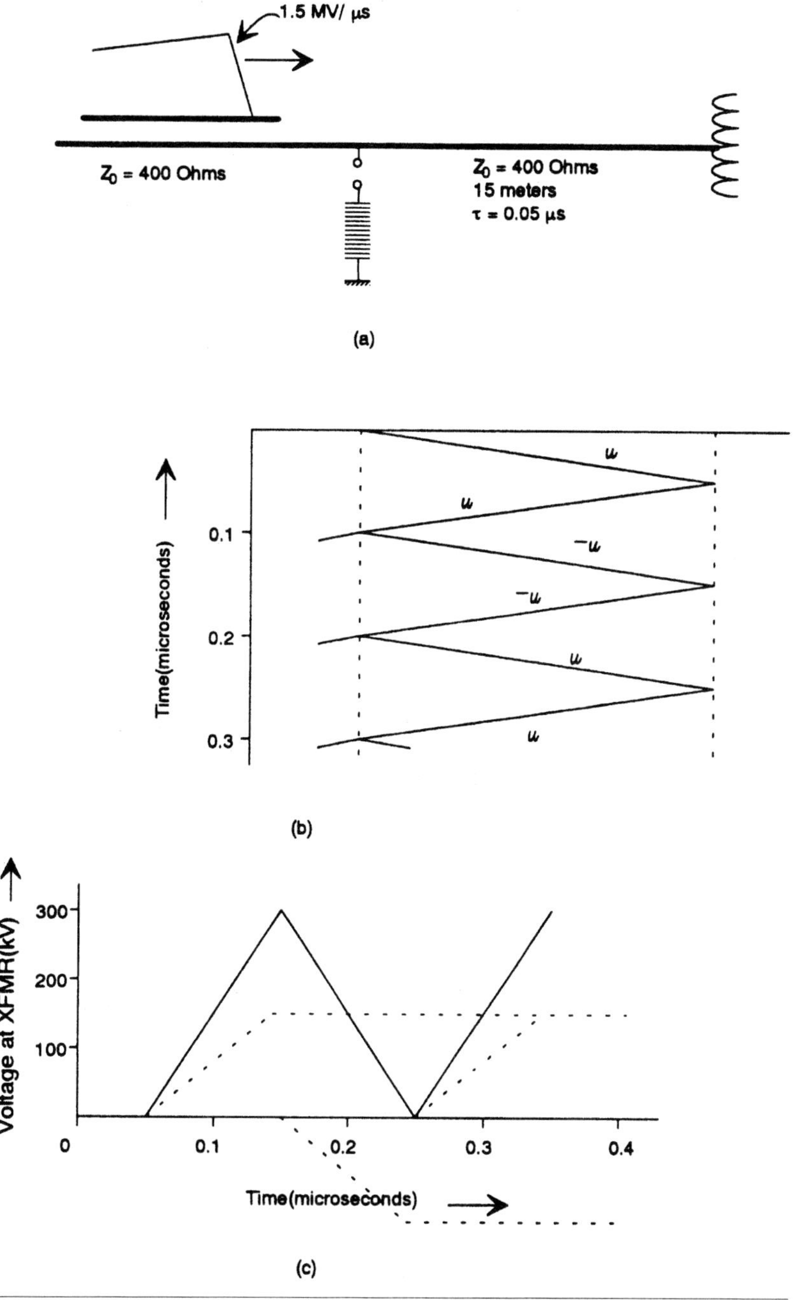


**FIGURE 23-69** Example of arrester-transformer separation effects: (*a*) system configuration; (*b*) Bewley diagram; (*c*) voltage at transformer.

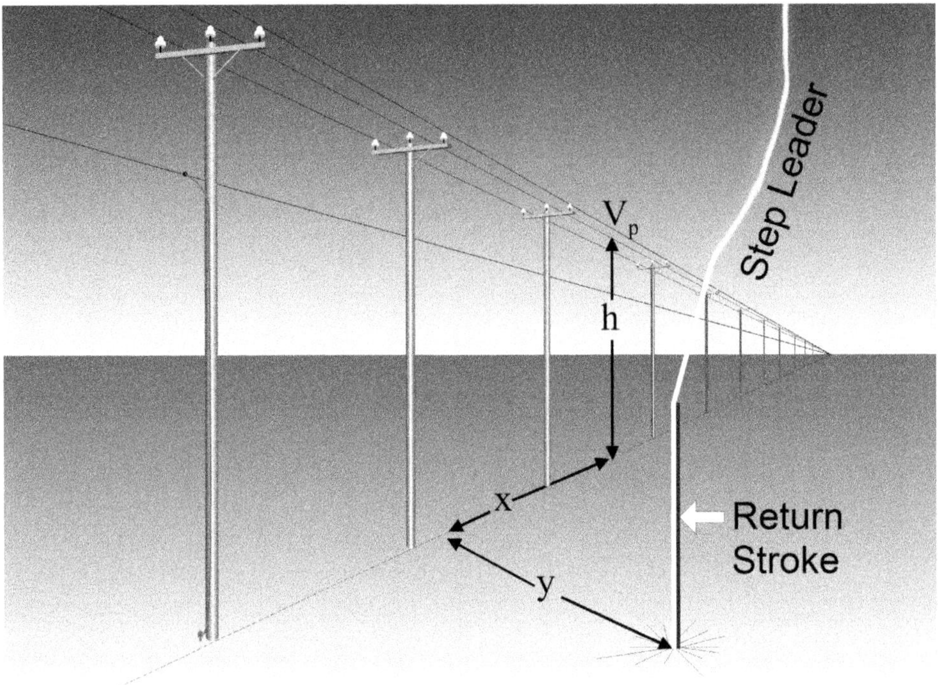


**FIGURE 23-70** Illustration of lightning stroke near a distribution line.

distribution lines. In addition, power apparatus, connected to distribution lines and with BIL lower than the induced voltage surges, such as distribution transformers (typically the BIL of distribution transformers is 100 kV) must be protected. It is practical to protect transformers with surge arresters of appropriate ratings.

***Underground Distribution System Overvoltage Protection.*** A typical configuration of an underground distribution system is shown in Fig. 23-71. Typically a surge arrester will be applied at the riser pole. Underground residential distribution systems present a relatively low characteristic impedance (30 to 50 Ω). When a transient (surge) reaches the URD system through the high-surge-impedance overhead system, and because of the presence of a surge arrester at the riser pole, the surge transmitted to the cable will be of very fast rise time, typically a small fraction of a microsecond. The magnitude of the surge is determined by the characteristics of the arrester. This surge will propagate along the cable and will double when it arrives at an open point or at the end of the cable where a transformer may be present. If this overvoltage is below the BIL level of the URD cable, no additional protection is required. However, the BIL of typical URD cable is relatively low. As an example, Table 23-11 illustrates the typical BIL for the most usual cable classes. In most cases, if a cable is protected with a surge arrester at the riser pole only (of the recommended arrester rating), it is possible that the overvoltage at the open end of the cable exceeds the BIL of the cable with the potential of failure. Use of surge arresters at the open ends, or, better yet, use of elbow type arresters at each transformer, drastically improves cable and transformer protection.

***Overvoltage Protection of Industrial and Commercial Systems.*** Industrial and commercial power systems are subject to overvoltages resulting from lightning or switching operations. By far, lightning overvoltages are the most stressful. Since these systems are interconnected to power systems, disturbances on the power system will be transmitted to them. These systems are also subjected to

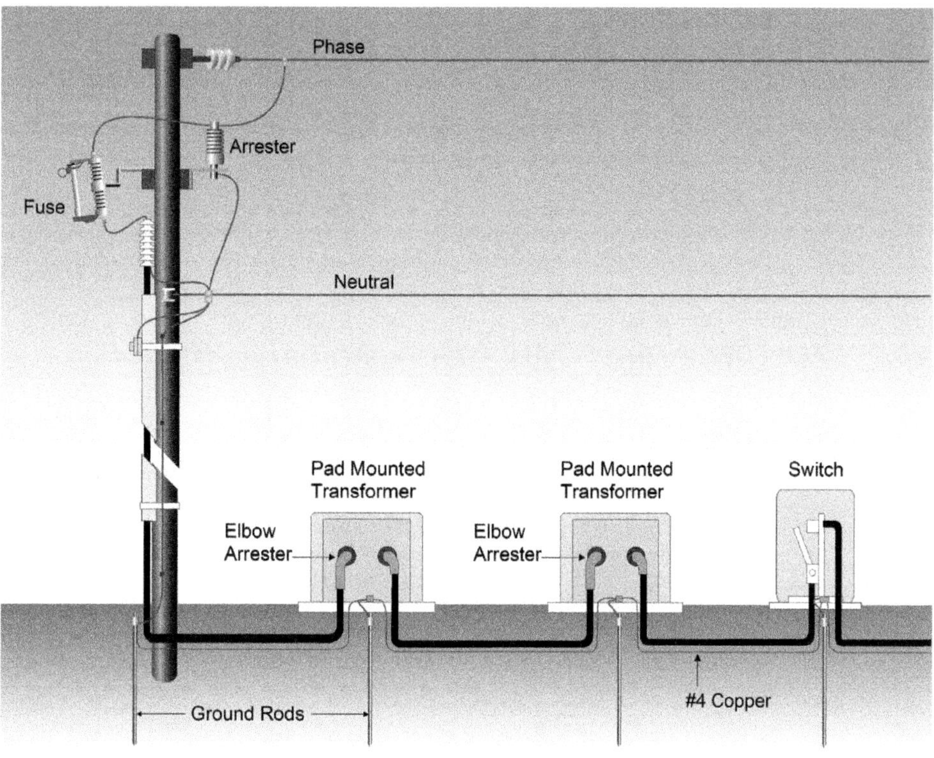


**FIGURE 23-71** Typical overvoltage protection scheme for underground residential distribution (URD) systems.

**TABLE 23-11** Typical Basic Insulation Levels for Cables

| Cable kV class | Typical BIL | Recommended arrester rating |
|---|---|---|
| 15 | 95 | 9 |
| 25 | 125 | 18 |
| 34.5 | 150 | 25 |

direct lightning. Therefore, these systems also require shielding against lightning. Typically, a grounding system will be installed as well as a lightning protection system (shielding) to divert any direct lightning strokes to ground. This grounding system is referred to as *external grounding* to distinguish it from the so-called internal grounding, which refers to the grounding system of various equipment in the facility. The external grounding systems of industrial/commercial power systems are interconnected to the power system, as shown in Fig. 23-72. The lightning protection system is basically a shielding system designed to

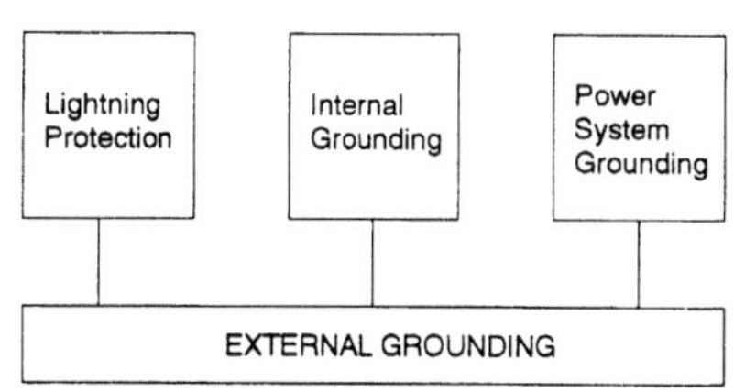


**FIGURE 23-72** Conceptual description of the grounding system of an industrial or commercial facility.

**TABLE 23-12** Lightning Points of Entry

| |
|---|
| Air terminals |
| Communication towers (if present) |
| Power system grounding |
| Fence |

route the lightning surges into the external grounding system for the purpose of minimizing potential differences within the facility. A system like this is subjected to lightning overvoltages which may enter from a number of points. Table 23-12 lists a number of possible points of entry. A coordinated design of the external grounding system, lightning protection system, and internal grounding system can provide a system, which is hardened against lightning and other sources of overvoltages. The effectiveness of the system can be assessed with analysis by using the methods discussed in this subsection.

## 23.8 MONTE CARLO SIMULATION-BASED METHODS

In many parts of the globe, the performance of electrical installations is affected by lightning. In this case, it is important to design the system in such a way that outages and damages from lightning are minimized. Typically, transmission lines, substations, and commercial and industrial buildings are the focus of the design process. Because lightning parameters present substantial variability, the effects of lightning must be evaluated on a statistical basis. Other parameters that affect system performance may also present substantial variability. One important parameter is soil resistivity, which affects ground impedances and therefore lightning overvoltages. Monte Carlo simulation methods are well suited to evaluate the effects of these parameters on system performance. A simplified description of a Monte Carlo simulation, given by Moussa and Wehling (1992), considers two parameters with substantial variability, and is illustrated in the following sequence of steps:

1. Generate a sample of lightning stroke described with respect to (*a*) crest magnitude and (*b*) rise time.
2. Generate a sample of soil resistivity.
3. Generate a sample of power frequency voltages.
4. Compute probability of shielding failure.
5. Perform an experiment and determine the lightning termination point (phase conductor, shield wire, etc.).
6. Compute transient voltages in system for conditions described above.
7. Perform effects (failure) analysis (compare voltage stresses on insulation against withstand capability).
8. If number of trials exceeded allowable, go to step 9. Otherwise go to step 1.
9. Generate histograms of maximum overvoltage and backflashover.

It should be understood that this Monte Carlo simulation requires that a good analysis procedure should be available to reliably compute the overvoltages on the insulation. Such models have been developed and are available. Detailed description of these methods is beyond the scope of this text. Instead, some typical results will be presented and discussed.

A typical application of these methods is to evaluate the performance of a specific power line. Figures 23-73 and 23-74 respectively illustrate examples of shielding analysis for two different samples (trials) of the Monte Carlo simulation. Note that for the same system, when the current is low, there is a probability of shielding failure. As the lightning current increases, shielding failure may be eliminated.

Figures 23-75 and 23-76 respectively illustrate examples of effects (failure) analysis for two different trials of the same system. Note that the conditions illustrated in Fig. 23-75 will lead to insulation flashover, while the system will withstand the conditions illustrated in Fig. 23-76.

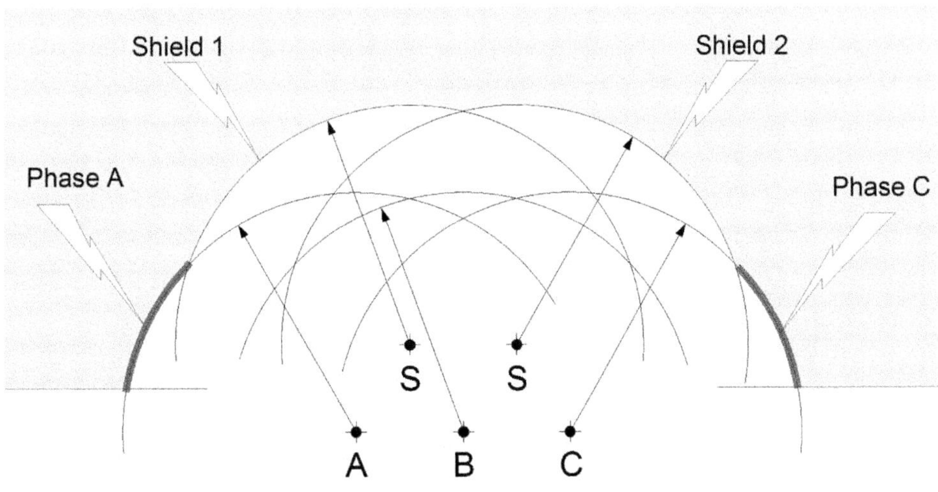


**FIGURE 23-73** Example of shielding analysis. Lightning crest is 1.386 kA. There is a finite probability of direct strike on the phase conductor.

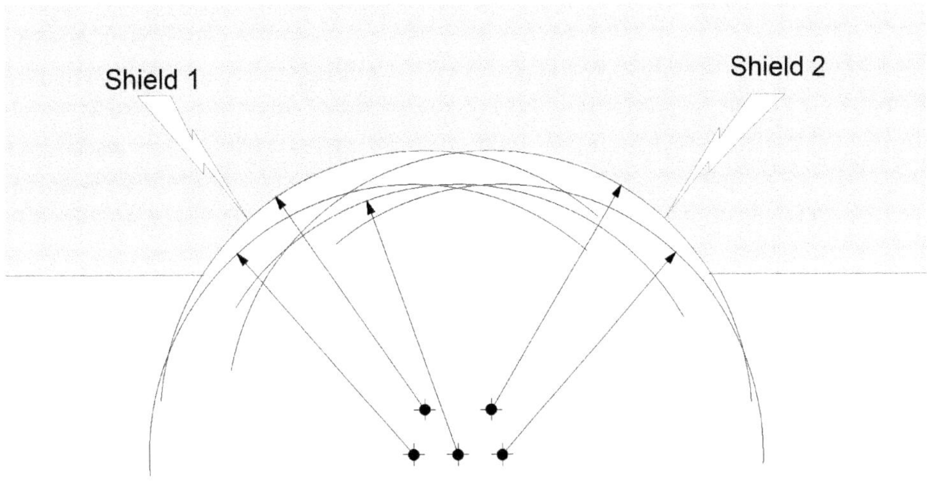


**FIGURE 23-74** Example of shielding analysis. Lightning crest is 10.969 kA. The probability of direct strike on the phase conductor is zero.

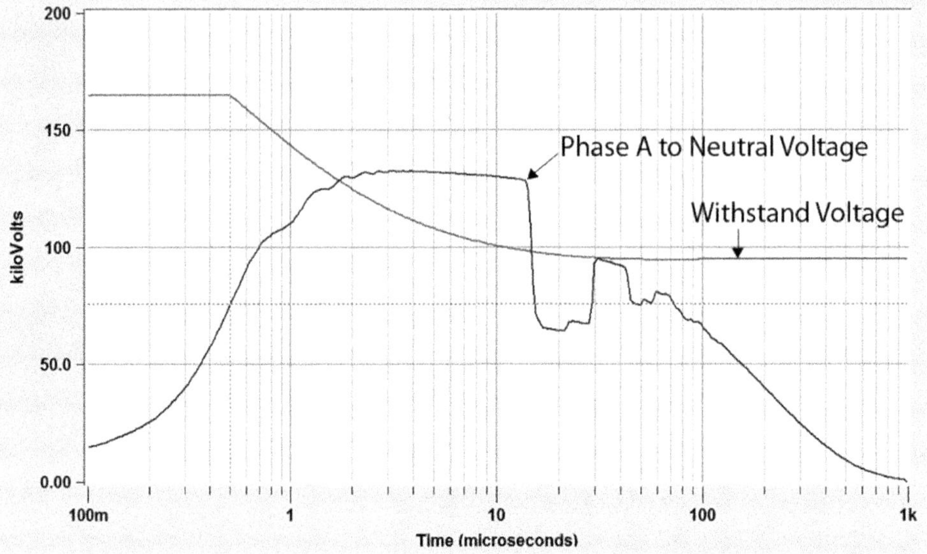


**FIGURE 23-75** Example of effects analysis. Overvoltage across insulator exceeds the volt-time withstand curve of the insulator.

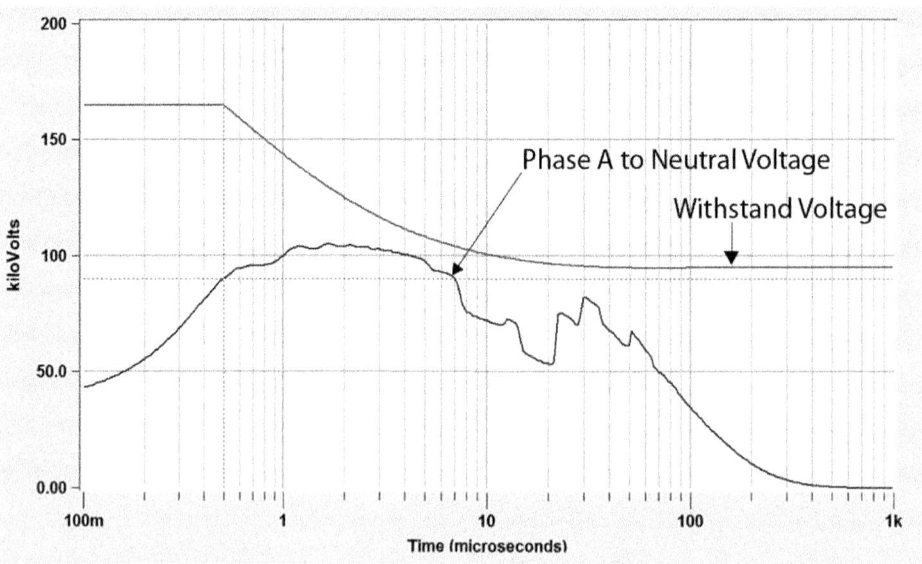


**FIGURE 23-76** Example of effects analysis. Overvoltage across insulator does not exceed the volt-time withstand curve of the insulator.

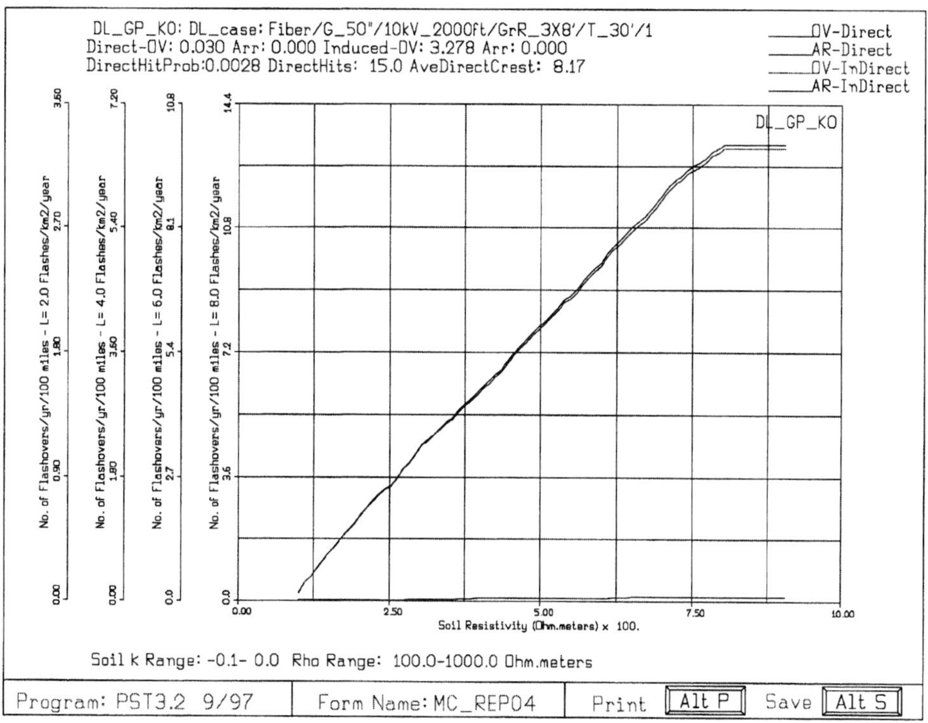


**FIGURE 23-77** Example of Monte Carlo simulation results.

The results of the Monte Carlo simulation are illustrated in Fig. 23-77. Note that for this case, which refers to a transmission line, there is a small probability of direct hit to the phase conductor and that the average crest of direct hit is 8.17 kA s. The results are in terms of expected number of flashovers per 100 mi of the line and per year.

## 23.9 LIGHTNING ELIMINATION DEVICES

It is always desirable to eliminate the destructive effects of lightning on electrical equipment. Benjamin Franklin was the first to invent the lightning rod that provides a controlled path for the flow of the lightning current. Professor Moore (1997) has researched Franklin's investigations and points out that, initially, Franklin believed that his lightning rod would eliminate lightning. Experimental observations have shattered this belief and Franklin accepted the fact that the lightning rod does not eliminate lightning but rather behaves as the control electrode for the termination of lightning so that other structures in the vicinity can be protected. This observation developed into the modern shielding theory we use for design purposes and which was discussed earlier.

The desire to seek a device that prevents lightning did not die. The Czech scientist Prokop Divisch has advocated this goal in 1754. The basic idea in lightning prevention is to provide multiple points of discharge to neutralize an electrified thundercloud. In theory, these charges can reach the electrified cloud and prevent the initiation of lightning. In fact, in 1930 a patent was awarded to J. M. Cage of Los Angeles for a lightning prevention device consisting of *point-bearing* wires suspended from a steel tower for the purpose of protecting petroleum storage tanks. Later, in 1971, lightning prevention systems were commercialized.

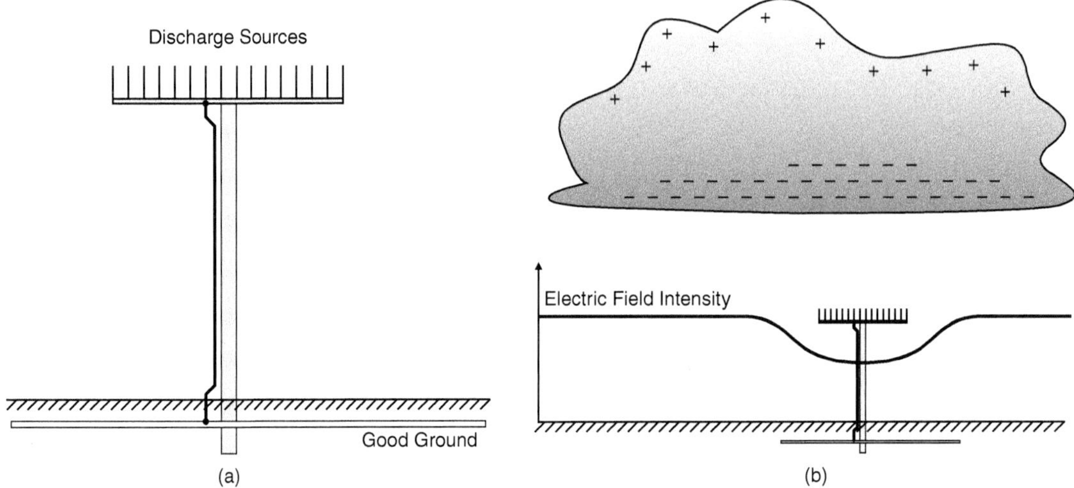


**FIGURE 23-78** Conceptual illustration of a lightning elimination system showing (*a*) a good ground with discharge sources and (*b*) the discharge into the electrified cloud.

There are many technologies of lightning prevention systems: (a) active air terminal that generate ions which blend into the atmosphere to minimize the electric field around the protected facility, (b) terminals with multiple sharp edges that ionize the air around the terminal for the purpose of minimizing the electric field around the protected facility and (c) umbrella terminals, which provide a smooth surface that minimizes the maximum electric field around the protected facility. As an example the theory of operation of technology (b) is discussed. Specifically, the basic idea of this technology is illustrated in Fig. 23-78. The system consists of sharp objects (discharge sources), which are electrically connected to a good grounding system (Fig. 23-78*a*). The theory is that the discharge released by the discharge sources into the atmosphere will reach the electrified cloud (Fig. 23-78*b*) and will partially neutralize it. The idea of using umbrella terminals to provide a smooth surface that minimizes electric field intensity is due to Tesla who has applied and received U.S. Patent 1 266 175 May 14. 1019 on this idea.

The success of lightning prevention systems has not been as advertised. There is indisputable field evidence that the *lightning elimination devices* cannot eliminate lightning. On the other hand, there are cases in which these devices helped reduce the frequency and extent of damage from lightning (Moussa 1998). The present knowledge and theories of lightning are consistent with this field experience.

## 23.10 ACKNOWLEDGMENTS

A large number of people directly or indirectly contributed to the development of this section. Professor Nikolopoulos of the National University of Athens, Greece, introduced me to high voltage engineering principles. Dr. Roger Webb, Director of the School of Electrical Engineering, introduced and developed a course on power system transients at Georgia Tech and subsequently worked with me in teaching this course. Several folks at the Electric Power Research Institute (EPRI) supported and contributed to my work, namely, John Dunlap, Gil Addis, and Mario Rabinowitz, and numerous EPRI project advisors contributed in the development of ideas in many different ways. Art Westrom has been an enthusiastic supporter of this work, reviewed the manuscript, and provided valuable suggestions and inputs. Dr. George Cokkinides developed the transmission-line model for lightning performance evaluation and worked with me on many aspects of lightning shielding and evaluation. Finally, Fan Zhang, a Ph.D. candidate at Georgia Tech, worked with me long hours in typing the manuscript and preparing the drawings.

## 23.11 BIBLIOGRAPHY


Anders, G. J., and M. A. El-Kady. 1992. *Transient Ratings of Buried Power Cables*. Part I: historical perspective and mathematical model, paper 92 WM 044-8 PWRD.

Anderson, J. G. 1968. *EHV Transmission Line Reference Book*. Washington, DC: Edison Electric Institute.

Anderson, J. G. 1975. Lightning Performance of Transmission Lines, in *Transmission Line Reference Book, 345 KV and Above*. Palo Alto, Calif.: Electric Power Research Institute (EPRI).

ANSI Standard C62.2. *Guide for Application of Valve-Type Lightning Arresters for Alternating Current Systems*.

Barker, P. P., et al. 1992. Characteristics of Lightning Surges Measured at Metal Oxide Distribution Arresters, paper 92 WM 255-0 PWRD.

Berger, K. 1967. Novel Observations on Lighting Discharges: Results of Research on Mount San Salvatore. J. Franklin Institute 283(6):478–524.

Boyd, E. L. 1992. *Internal Spark Gap Protection of Distribution Transformers from Low-Side Current Surges*. IEEE Trans. Power Delivery, pp. 1592–1600.

Braunstein, A. 1970. Lightning Strokes to Power Transmission Lines and the Shielding Effect of Ground Wires. *IEEE Trans*. PAS-89(8):1900–1910.

Brown, G. W. 1978. Lightning Performance. Updating Backflash Calculation. *IEEE Trans*. PAS-97:33–38.

Chalmers, J. A. 1967. *Atmospheric Electricity*. New York: Pergamon Press.

Chowdhuri, P., and E. T. B. Gross, 1969. Voltages Induced on Overhead Multiconductor Lines by Lightning Strokes. IEEE Proc. 116:561–564.

Cokkinides, G. J. and A. P. Meliopoulos. 1988. Transmission Line Modeling with Explicit Grounding Representation. Electric Power Syst. Research 14(2):109–119.

Darveniza, M., et al. Nov./Dec. 1979. Modeling for Lightning Performance Calculations. IEEE Trans. Power *Appar. Syst*. PAS-98(6):1900–1908.

Darveniza, M., F. Popolansky, and E. R. Whitehead. July 1975. Lightning Protection of UHV Transmission Lines. Electra (41): 36–39.

Dean, D. S. 1966. Insulation Tests for the Design and Uprating of Wood-pole Transmission Lines. IEEE Trans. PAS-85:1258.

Dommel, H. W. 1969. Digital Computer Solution of Electromagnetic Transients in Single and Multiphase Networks. IEEE Trans. PAS-88:388–399.

Dommel, H. W., and W. S., Meyer, 1974. Computation of Electromagnetic Transients. Proc. IEEE 62:983–993.

Durie, R. C., and C. Pottle, 1992. An Extensible Real-Time Digital Transient Network Analyzer, paper 92 WM 255-0 PWRD.

EPRI. 1984. Report EL-3608, *Single-Phase, Multiphase, and Multiple-Circuit Lightning Flashovers of Transmission Lines*. Palo Alto, Calif.: EPRI.

EPRI. 1990. Report EL-6782, *Characteristics of Lightning Surges on Distribution Lines*. Palo Alto, Calif.: EPRI.

Eriksson, A. J., M. F. Stringefellow, and D. V. Meal 1982. Lightning Induced Overvoltages on Overhead Distribution Lines. IEEE Trans. PAS-101:960–968.

Fink, D. G., and H. Wayne Beaty. 1993. *Standard Handbook for Electrical Engineers*, 13th ed. New York: McGraw-Hill.

Flugum, R. W. 1970. Operation of Lightning Arresters on Abnormal Power Frequency Voltages. IEEE Trans. PAS-89(7):1444–1451.

Furukawa. S. et al. 1989. Development and Application of Lightning Arresters for Transmission Lines. IEEE Trans. Power Delivery 4(4).

Gallagher, T. J., and A. J. Pearmain. 1983. *High Voltage Measurements, Testing and Design*. New York: Wiley.

Goedde, G. L., R. C. Dugan, and L. D. Row. 1991. Full Scale Lightning Surge Tests of Distribution Transformers and Secondary Systems. IEEE Proc. Dallas, Tex. PES T&D Conference pp. 577–581.

Golde, R. H., ed. 1977. *Lightning*. New York: Academic Press.

Hileman, A. R., P. R. Leblanc, and G. W. Brown. 1970. Estimating the Switching Surge Performance of Transmission Lines. IEEE Trans. PAS-89:1455–1466.

IEEE. 1987. *Power System Transient Recovery Voltages. Seminar Report 87THO176-8-PWR*. New York: IEEE.

Ishida, K., et al. 1992. Development of a 500 kV Transmission Line Arrester and Its Characteristics. IEEE Trans. Power Delivery 7(3):1265–1274.

Johnson, I. B., et al. 1979. *Surge Protection in Power System*. IEEE Tutorial Text 79 EH0144-6 PWR.

Kolcio, N., et al. 1992. Transient Overvoltages and Overcurrents on 12.47 kV Distribution Lines; Computer Modeling Results, paper 92 WM 273-3 PWRD.

Kroninger, H. 1974. *Further Analysis of Prof. Berger's San Salvatore Lightning Current Data*, CSIR Spec. Report ELEK 53. Pretoria, South Africa: National Electrical Engineering Research Institute.

Larson, A. 1905. *Annual Report*. Washington, DC: Smithsonian Institute, pp. 119.

Liew, A., and M. Darveniza. 1974. Dynamic Model of Impulse Characteristics of Concentrated Earths. Proc. IEEE 121(2):123–135.

Liew, A. C., and S. C. Mar. 1986. Extension of Chowdhuyri-Gross Model for Lightning Induced Voltage on Overhead Lines. *IEEE Trans. Power Delivery* PWRD-1(4):1073–1081.

Love, R. R., 1973. Improvements on Lightning Stroke Modeling and Applications to the Design of EHV and UHV Transmission Lines, M.Sc. thesis, University of Colorado.

McComb, T. R., et al. 1992. Digital Techniques in HV Tests: Summary of 1989 Panel Session, paper 92 WM 175-0 PWRS.

Mak, S. T. 1992. Propagation of Transients in a Distribution Network, paper 2/92 WM 272-5 PWRD.

Mancao, R. T., J. J. Burke, and A. Myers 1992. The Effect of Distribution System Grounding on MOV Selection, paper 92 WM 223-7 PWRD.

Meliopoulos, A. P. 1972. Problems and Concise Theory of High Voltage Structures (in Greek). Athens, Greece: S. B. H. Sellountos Publishing Co.

Meliopoulos, A. P. 1988. *Power System Grounding and Transients: An Introduction*. New York: Marcel Dekker.

Meliopoulos, A. P., and M. G. Moharam. 1983. Transient Analysis of Grounding Systems. IEEE Trans. PAS-102(2): 389–397.

Meliopoulos, A. P., Sakis, W. Adams, and R. Casey. May 1997. An Integrated Backflashover Model for Insulation Coordination of Overhead Transmission Lines. *Electrical Power Energy Syst.* 19(4):229–234.

Melvold, D. J., S. A. Miske, Jr., and E. C. Sakshung May 15, 1974. Switching Cables Can Stress Arresters. New York: *Electrical World* magazine.

Moore, C. B. Jan. 1997. Study of Behavior of Sharp and Blunt Lightning Rods in Strong Electric Fields: Review of Lightning Protection Technology for Tall Structures. Publication AD-A075 449, pp. 96–107.

Mousa, A. M. Oct. 1998. The Applicability of Lightning Elimination Devices to Substation and Power Lines. *IEEE Trans. Power Delivery* 13(4):1120–1127.

Mousa, A. M., and R. J. Wehling 1992. A Survey of Industry Practices Regarding Shielding of Substation against Direct Lightning Strokes, paper 92 WM 224-6 PWRD.

Ogbuobiri, E. C., W. F. Tinney, and J. W. Walker 1970. Sparsity-Directed Decomposition for Gaussian Elimination on Matrices. *IEEE Trans. Power Appar. Syst.* PAS-89:141–150.

Papalexopoulos, A. D., and A. P. Meliopoulos 1987. Frequency Dependent Characteristics of Grounding Systems. *IEEE Trans. Power Delivery* PWRD-2(4):1073–1081.

Powell R. W. 1967. Lightning Protection of Underground Residential Distribution Circuits. IEEE Trans. PAS-86: 1052–1056.

Regaller, K. ed. 1980. *Surges in High-Voltage Networks*. New York: Plenum Press.

Rizk, F. A. M. 1994. *Modeling of Lightning Incidence to Tall Structures, Part II: Application*, IEEE Transactions on Power Delivery, pp. 172–193.

Sakshung, E. C., T. J. Carpenter, and E. W. Stetson. 1965. Duty Cycle Testing of Current-Limiting Station and Intermediate Lightning Arresters. *IEEE Trans.* PAS-84(3):422–425.

Sakshaug, E. C., J. S. Kresge, and S. A. Miske, Jr. 1977. A New Concept in Station Arrester Design. *IEEE Trans.* PAS-96(2):647–656.

Sargent, M. A. 1972. Monte Carlo Simulation of the Lightning Performance of Overhead Shielding Networks of High Voltage Stations. IEEE Trans. PAS(pt. III)-91:1651–1656.

Sargent, M. A., and M. Darveniza 1969. Tower Surge Impedance. *IEEE Trans.* PAS-88:680–687.

Semlyen, A., and M. R. Iravani 1992. Frequency Domain Modeling of External Systems in an Electromagnetic Transient Program, paper 92 WM 304-6 PWRS.

Shindo, T., et al. 1992. Model Experiments of Laser-Triggered Lightning, paper 92 WM 25804 PWRD.

Smith, S. B., and R. B. Standler 1991. The Effects of Surges on Electronic Appliances, paper 91 SM 384-8 PWRD, IEEE summer meeting.

Stone, G. C., R. G. van Heeswijk, and R. Bartnikas. Investigation of the Effect of Repetitive Voltage Surges on Epoxy Insulation, paper 92 WM 067-9 EC.

Suzuki, T., K. Miyake, and T. Shindo 1981. *Discharge Path Model in Model Test of Lightning Strokes to Tall Mast* IEEE Transactions on Power Apparatus and Systems, vol. PAS-100, No. 7, pp. 3553–3562.

Thomas, H. 1968. Transport Time-Delay Simulation for Transmission Line Representation. *IEEE Trans. Computers* C-17:204–214.

Tominager, S. 1980. Stability and Long-Term Degradation of Metal Oxide Surge Arresters. *IEEE Trans.* PAS-99(4): 1548–1556.

*Transmission Line Reference Book, 345 kV and Above.* Palo Alto, Calif.: EPRI.

Udo, T. 1965. Switching Surge and Impulse Sparkover Characteristics of Large Gap Spacing and Long Insulator Strings. IEEE Trans. PAS-84:304.

Uman, M. A. 1969. *Lightning.* New York: McGraw-Hill.

Uman, M. A. 1987. *The Lightning Discharge.* New York: Academic Press.

Walsh, G. W. 1977. A New Technology Station Class Arrester for Industrial and Commercial Power Systems, Conf. Rec. IEEE Industrial and Commercial Power System Technical Conference, catalog no. 77 CH1198-11A.

Westrom, A. C. 1990. Surge Arrester with Shunt Gap. U.S. Patent 4,908,730.

Whitehead, E. R. 1977. Protection of Transmission Lines, in *Lightning*, vol. 2, R. H. Golde, ed. New York: Academic Press, pp. 697–745.

Williams, E. R. Nov. 1988. The Electrification of Thunderstorms. Scientific American pp. 88–99.

# 24

# COMPUTER APPLICATIONS IN THE ELECTRIC POWER INDUSTRY

Juan A. Martinez-Velasco

*Professor, Universitat Politècnica de Catalunya, Spain*

## 24.1 INTRODUCTION

### 24.1.1 A Vision of the Power Industry

The power industry is engaged in the generation, transmission, and distribution of electrical energy which is obtained by conversion from other forms of energy such as coal, gas, oil, nuclear, water, or other renewable energy. These activities often include mining, rail transport, shipping, slurry pipelines, and storage of energy in many forms.

In the first 90 years of its history, the industry expanded at a pace nearly twice that of the overall economy, doubling roughly every 10 years. During that period, real prices per kilowatt hour (kWh) decreased steadily because of technical improvements, productivity increases, and stable fuel prices. Throughout the 1970s, increased fuel costs, limits in economies of scale, diminishing returns in technology improvement, and increased regulation costs led to increased kilowatt hour costs and reduced demand growth.

The political and economic response to increasing costs has been a movement to smaller generator sizes, minimization of capital investment, and attempts to control costs by fostering competition in generation supply. Incentives were also established to reduce demands and increase load factors. Today power supply is diversifying away from large central station technologies and toward increased use and availability of the transmission and distribution systems.

The industry's purpose is to provide adequate, reliable, environmentally compatible electricity at reasonable cost with the ultimate goal of improving its productivity and net earnings. In spite of the differences between publicly and privately owned utilities, this goal is applicable to each, in different form.

Several issues have recently emerged, whose impacts will increase in future years:

- Large blackouts have raised concerns that the current control structure for the power grid is or will be unable to cope with an increasing demand for reliable electric power. Improving the reliability of power systems will require experience with uncommon system behaviors, broader situational awareness at control centers, fast and reliable automatic controls, and active participation by power consumers in the control process.
- There are currently many legislative efforts to increase the percentage of power derived from low-carbon sources and reduce carbon-intensive generation. Renewable power generation depends largely on local conditions (e.g., wind or sunlight availability). The best locations for production are often not where the greatest power demands are. Power transmission, then, could be a limiting factor in the growth of distributed, renewable energy generation. In addition, without a proper design, a too high penetration of intermittent power sources, such as wind or solar, can result in grid instability.
- The difficulties and barriers to developing new transmission network capacity are leading many transmission and distribution parts to an operation closer to the thermal, mechanical, and electrical limits. This reflects the need to manage the congestions on transmission and distribution networks and identify solutions that could defer investments.
- The operation of the power system will need to address an increase in plug-in hybrid electric vehicles, which will significantly alter the demand pattern.

- The increase of sensitive semiconductor-based electronic equipment has increased concern about power line disturbances. In addition, interruption costs for industrial facilities may be too high. All this calls for a better quality and reliability.
- The connection of distributed energy resources (DERs) to the distribution system raises some technical concerns. For instance, although distributed generation (DG) units may successfully operate in island when there is a balance between load and generation, islanded operation is not generally allowed because power quality cannot be always maintained within an acceptable level by the islanded generators, and restoring the outaged circuits is complicated when having islanded generators since automatic reclosing is generally used. In addition, DG can cause miscoordination of protective devices. The design and installation of *microgrids* is foreseen as a solution to some of these problems.
- The increasing use of sensors (e.g., phasor measurement units—PMUs) to monitor the network state in both transmission and distribution levels, and smart meters at costumer facilities to remotely read consumption or store electricity price will require an advanced two-way communication system.
- The network of physical objects (e.g., home appliances, electrical vehicles, buildings) connected to the Internet framework and that form the Internet of Things (IoT) can be seen as a new step of the smart grid development in which devices are embedded with both intelligence and sensors that provide them with decision-making capabilities.
- The proliferation of any type of intelligent electronic devices (IEDs) and the deployment of the IoT raises concerns about cybersecurity. Energy management systems (EMSs) are critical to the flow of power through the grids, and a malicious manipulation of one or several IEDs can cause a major grid failure. On the other hand, the increase of information exchange within the utility and between the utility and costumers raises a concern about privacy and confidentiality.

For more details about some topics covered in this section, readers are referred to other sections of this handbook (e.g., Secs. 8 and 17 through 23).

### 24.1.2 The Transition to the Smart Grid

The state of the current power grid in developed nations is considered mature. Although its reliability is very high, its structure still has some significant limitations, such as inefficiency at managing peak load, too high cost of power outages and power quality variations, or relatively slow data transmission rates. The power supply has to change according to the power demand continuously and the power grid will also need to maintain a buffer of excess supply, which results in lower efficiency and higher costs. In addition, environmental constraints push power providers for lower greenhouse gas (GHG) emissions. On the other hand, advances in computers, controls, communications, alternative energy sources, and other components may enable methods of using the existing grid structure more efficiently and help to moderate electrical demand at critical times, resulting in less pollution and lower costs. A smarter grid will use new technologies to reduce the environmental impact of power grid and increase efficiency as well as renewable energy utilization.

Although there is no coincidence with an accurate definition for the term "smart grid," the features it should have or the goals it should achieve, the smart grid may be seen as an upgrade of the present power system in which the current and new functionalities will monitor, protect, and automatically optimize the operation of its interconnected elements to maintain a reliable and secure environment.

The new functionalities represent challenges that depend on each utility's context (e.g., legacy systems). The smart grid offers better management of the energy consumption by the usage of dual-way advanced metering infrastructure and real-time communication, improved power reliability and quality, enhanced security by reducing outages and cascading problems, better integration of DERs and new loads (e.g., plug-in hybrid electric vehicles), reduction in greenhouse gases and other pollutants, or financial benefits for suppliers and consumers. Although the smart grid will build upon the basic design of the current power grid, it will have features essential to its operation that will involve monitors, sensors, switching devices and sophisticated two-way communication systems that will allow it to be a highly automated power delivery system. By delivering real-time information over a distributed computing and communications infrastructure, the smart grid will be able to provide a near-instantaneous balance between electricity supply and electricity demand.

According to the model proposed by the U.S. National Institute of Standards and Technology (NIST) [1], a smart grid may be considered an interconnected set of domains: *markets* (operators and participants in electricity markets), *operations* (managers of the movement of electricity), *service providers* (organizations providing service to electrical customers and utilities), *bulk generation* (generators of electricity in bulk quantities), *transmission* (carriers of bulk electricity over long distances), *distribution* (distributors of electricity to and from customers), and *customers* (end users of electricity). The actors in each domain may be organizations, devices, computer systems or software applications that make decisions and exchange information. Each domain will have secure communications connections with the other actors, whereas the electrical connections will be between bulk generation, transmission, distribution, and customer, although the customer may also be an electrical supplier [1]. The European version of the smart grid reference architecture was presented in a CENELEC report released in November 2012 [2].

The drivers behind smart grid solutions are needs for more energy, sustainability, increased usage of renewable energy resources, competitive energy prices, or security of supply. To achieve smart grid goals, utilities have to face several challenges, such as high power system loading, increased use of DERs and intermittent renewables, new loads (hybrid/e-cars), cost pressure, utility unbundling, increased energy trading, transparent consumption and pricing for the consumer, or significant regulatory push. Obviously, the priority of drivers and challenges will differ from place to place.

Although the smart grid does not yet exist as such, certain aspects of what the smart grid will become are already evident. Moreover, specific devices or techniques may already exist commercially and/or may be part of pilot projects currently underway in various geographic regions. The list of these technologies include power electronics technologies for transmission and distribution (e.g., HVDC, FACTS, and custom power), sensing, measurement and metering technologies (e.g., smart meters, meter reading systems and phasor measurement units), DERs, advanced telecommunications technologies, demand-side technologies, or an information management system (i.e., a system including functions such as collection and processing, analysis, integration, improved interfaces, and information security).

The implementation of the smart grid involves new technologies and regulations. Existing equipment that functions adequately in the present grid may not fulfill the demands of the smart grid, the technologies required to realize the smart grid in all of its functionality may not yet exist along with the skilled workforce to implement and maintain it, and some regulatory agencies may not have yet experienced with the smart grid concept. The smart grid will greatly change the way residential customers have always used electrical power as they would have to consider the actual cost of power at time of use. In addition, the access to and information available from customers through the telecommunication system will also cause concern for data privacy and cybersecurity. The list of barriers to the adoption of the smart grid may thus include regulatory uncertainty, economic disincentives, reluctance to change the existing system, data privacy and cybersecurity, available technology and required skills.

Lack of technology standards is being another major obstacle to smart grid deployment. The utilities will deploy many new products and will be greatly affected by adopted standards, particularly those regarding information transfer, interoperability, cybersecurity, and data privacy. The importance of these standards will vary in their relation to smart grid applications. Deployment of various smart grid elements, including smart sensors on distribution lines, smart meters in homes, and widely dispersed energy resources, is already underway and without standards there is the potential for technologies developed or implemented to become obsolete prematurely or to be implemented without measures necessary to ensure security. Lack of standards may also impede future innovation and the realization of promising applications.

The International Electrotechnical Commission (IEC) Smart Grid Standardization Roadmap recognizes that a very high level of interoperability of an increased number of intelligent devices, solutions, and organizations is one major common requirement for most of the smart grid applications [3].

Interoperability has different aspects which will be present in most of the new applications. Syntactic interoperability is the ability of two or more systems to communicate and exchange data; it is mainly done through standardized data formats and protocols. Another aspect is the ability of two or more systems to automatically interpret the exchanged data. This is called semantic interoperability. To achieve this, one must accept a common information exchange reference model. Both aspects are typical domains for standardization. A detailed list of standards was provided in a CENELEC document [4].

Major steps to be made toward achieving a Smart Grid Interoperability Framework are proposed in the NIST Framework and Roadmap for Smart Grid Interoperability Standards [1], which identifies applicable smart grid interoperability standards, discusses the gaps present in currently available standards, and lays out priorities for near-term development of smart grid standards. The report describes a high-level conceptual reference model for the smart grid, identifies existing standards that are applicable (or likely to be applicable) to the ongoing development of the smart grid, specifies high-priority gaps and harmonization issues (in addition to cybersecurity) for which new or revised standards and requirements are needed, documents action plans with timelines by which designated standards-setting organizations will address these gaps, and describes the strategy to establish requirements and standards to help ensure smart grid cybersecurity. The smart grid will require hundreds of standards, specifications, and requirements. According to NIST, the priority areas are wide-area situational awareness (WASA), demand response, electric storage, electric transportation, advanced metering infrastructure, distribution grid management, and cybersecurity.

### 24.1.3 Computer Applications

***Utility Information Systems.*** In scheduling its day-to-day operation, and in planning for its future growth, the industry has made extensive use of analytical tools and mathematical models which, through optimization and simulation, help in the decision-making process. The digital computer is applied in a variety of fields within a utility, whether to control its operations in real-time or to perform complex calculations. Nearly every piece of equipment installed in modern power system has been designed with the help of a software application. The power industry has long been one of the largest users of computers and among the most sophisticated in applying computational techniques. This use is quite understandable when one considers the high cost of power system equipment, the complexity of power systems, and the severe operational, reliability, and environmental requirements on the electricity supply.

Computer applications have assisted the industry in achieving its objectives: reducing the cost of energy delivered to consumers, improving the quality of service, enhancing the quality of the environment, and extending the life of existing equipment. These objectives have been achieved as follows:

1. Since the industry is one in which capital investment is usually high, unit costs have been reduced by operating facilities closer to their design limits, allowing better utilization of equipment.
2. Unit costs also have been reduced by automation, allowing operation with fewer personnel, and by optimization, lowering fuel consumption per kilowatt hour delivered.
3. Although there have been significant advances in energy storage technologies, electricity cannot be stored in large quantities; therefore, production and consumption must be simultaneous. Hence enough capacity is required to meet the maximum coincident demand or peak load of all customers. Interconnections between power systems provide important economies arising from different time patterns or diversity of use of the component systems in the network. They allow higher power system reliability at lower capital cost.
4. Quality of service has been improved by reducing the number, extent, and duration of service interruptions, thus providing a more reliable service.
5. Quality of environment has been maintained by operating facilities within acceptable bounds of emission, thermal discharge, waste disposal, and more effective land use. It is foreseen a significant improvement by increasing the percentage of renewable energy in the total energy production.

The electric power industry has been taking advantage of the tremendous power provided by computer and microprocessor-based technology. Protection and control equipment, SCADA (supervisory control and data acquisition), remote control and monitoring, and many other applications are routinely implemented with this technology. Recent experience has shown that security-related issues must be addressed by the power industry. Government regulations are legislating for proactive measures to be taken in terms of securing the computer network infrastructure within the power grid.

Today the industry has reached a stage where computer systems are no longer merely an engineering tool. The effectiveness of computer applications is one of the key elements in achieving the basic functions associated with the planning, designing, construction, operation, and maintenance of the power system. In fact, engineering and computers have been integrated. This integration may be viewed as tending toward the construction of a utility industry information system. Such a system is shown in Fig. 24-1. It depicts a typical information system which may be viewed as a combination and integration of several functional information systems.

Such an information system can extend the company capabilities by making relevant and current information accessible to both technical and management personnel. Designs can be refined by using measured data or operations experience, projects can be monitored, revenue requirements can be

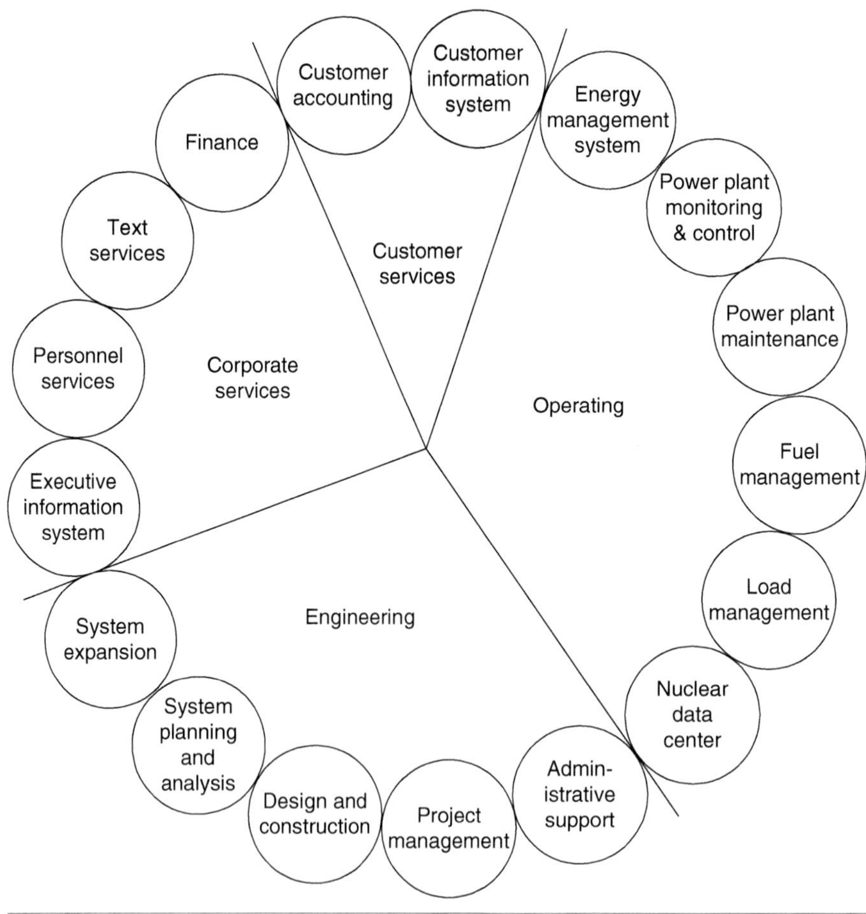


**FIGURE 24-1** Electric utility information systems.

predicted more closely, and the experience of operations can be reflected in the methods and criteria used in planning and engineering. The information system thus can provide meaningful data at proper times and locations to make decisions and concentrate resources in the most effective manner.

***Computing Support.*** Engineering users require several categories of computing support:

1. *Performance Requirements.* Less than 0.4-s response time is required for human-intensive work at a terminal. Longer response times reduce concentration and increase frustration. A predictable response time is required for machine-intensive work. Consistent timings for load flow are expected when operating under deadlines. High-bandwidth terminals are required to support graphics user interface (GUI) color graphics applications, which are used to speed work flow and to reduce volumes of data to a presentation of results in a form that can be readily assimilated.
2. *Technical Support Requirements.* The engineering user requires a minimum of system programming knowledge to use the computing system. Included in this are user-friendly selection panels, menus, aids, and help functions; a library of programming tools, languages, and application packages, all of which are maintained by the central data processing department; and training facilities and technical support to assist engineers with the use of computing resources.
3. *Operational Support Requirements.* Security features and operational procedures are required to protect the user's data and access to the computing system. Included are security features such as password protection of user data files and sign-on's and operational procedures to provide backup of user data and prevent its destruction or loss.

***Computer Configurations.*** Computer configurations such as shown in Figs. 24-2 and 24-3 are used to support engineering functions. Although the centralized computing approach has traditionally been used, the trend during the last years has been to a distributed approach in which a separate engineering information center is assigned strictly to engineering work.

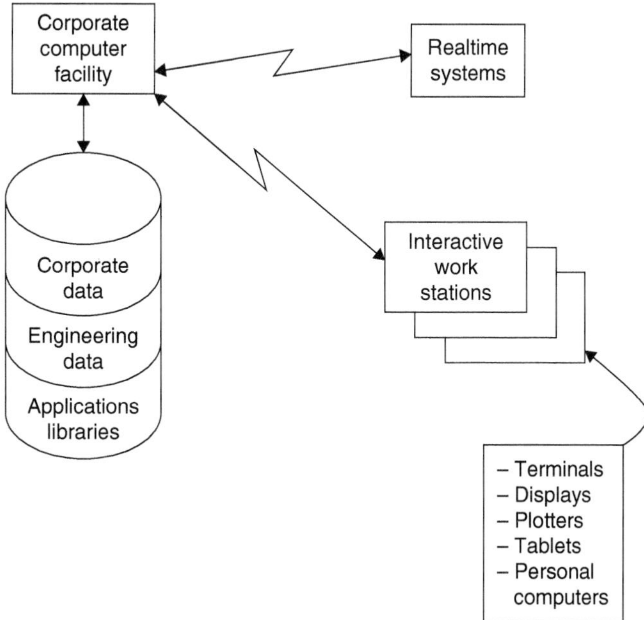


**FIGURE 24-2** Power engineering centralized architecture.

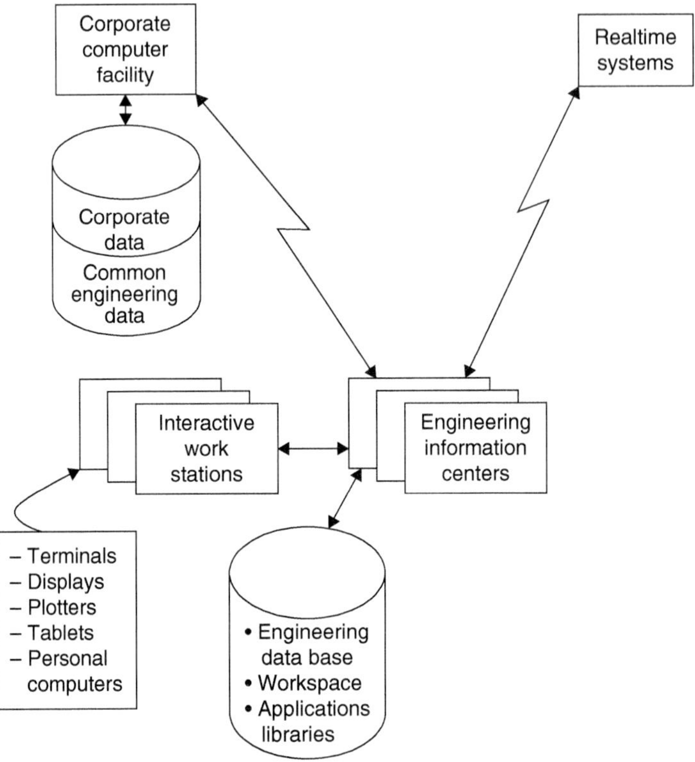


**FIGURE 24-3** Power engineering distributed architecture.

## 24.1.4 Spectrum of Computer Usage

A review of engineering and operating computer applications indicates that they fall within several broad categories, as shown in Fig. 24-4 and as described below.

***System Expansion.*** These applications are related to 20-, 10-, and 5-year construction programs and cover planning, design, and construction of new facilities. These functions are performed at least once a year and use long-range load forecasts and other predictions as input data. Competitive pressures and complexity of expansion options demand that engineers have sophisticated interactive computer tools, decision-support and communication systems, and report-generating mechanisms.

***System Planning and Analysis.*** These applications deal with 3- and 1-year construction of new facilities and the economic and reliable operation of these additions in conjunction with other interconnected power systems. Nuclear fuel management, annual hydrothermal coordination, and coordination for firm transmission and generation planning are among these functions. Because these programs are more frequently called on, they normally reside on disk storage devices. Thus only changes in data and programs need be entered when using specific programs.

***Long-Range Scheduling (Operating).*** These applications are related to annual, monthly, and daily operation of the power system. In this category are transmission and generation maintenance scheduling, unit commitment and withdrawal, and other functions dealing with both reliability and economy of operation. Electric power systems are more complex and stressed than ever before.

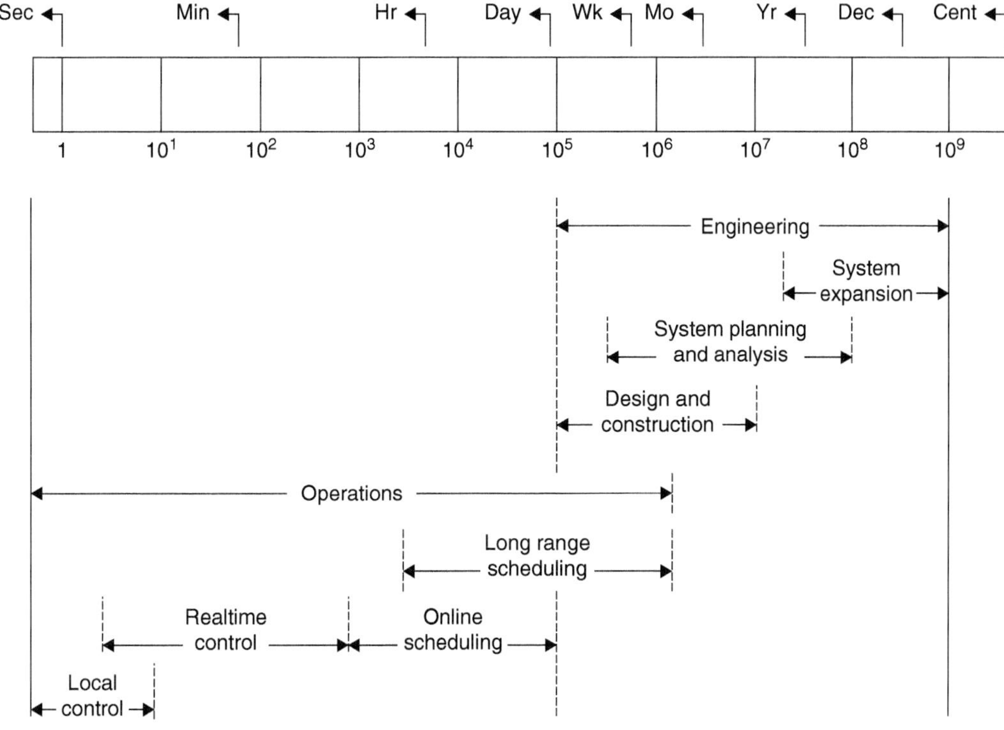


**FIGURE 24-4** Spectrum of computer use.

Maintenance of reliability and cost reduction require fast interactive computation in order to evaluate contingencies, operating options, and limits.

***Online Scheduling.*** These applications are related to security monitoring and determination of reserve indexes and hourly data recording. These schedules are performed at least once an hour, although some applications such as pumped storage scheduling are performed weekly and daily. They are based on historical data but also need current power system data such as facilities in and out of operation, generation outputs, and line flows. Therefore, they require direct data flow into the computer. The results, however, are presented to the user for consideration and execution. Because of the scheduling nature of these applications, very fast data acquisition is not a prerequisite; however, accuracy and timeliness of schedules are related to the extent that they include direct data acquisition.

***Real-Time Control.*** These regulating functions are carried out to meet the changing demands on the power system. Power system monitoring, security assessment, and display, rescheduling, and control of system frequency, tie-line flows, voltage conditions, and transmission flows are examples of this category. Other examples are closed-loop automatic control of generating units and interchange scheduling with neighboring companies and pool areas. These functions are performed in a time range of a few seconds to several minutes and therefore not only require direct data flows into the central computer but, in addition, require signals from the computer to the various remote controllers and actuators.

***Local Control.*** These applications require a response speed beyond the capability of central computer control and related communication. Most of these functions are initiated immediately after a fault develops or a variable exceeds certain limits. Their objective is to react quickly and correct the situation or to isolate and contain a disturbance. These functions are performed in the few milliseconds to several seconds range and can best be handled by local computers: (1) by directly sensing variables and controlling through actuators (e.g., direct digital control of boilers or digital relaying of the substations) and (2) by superimposing the computer on the local controllers or protective relays in order to reset their operating positions. The latter applications are in the 1- to 10-s range.

The computational requirements shown in Fig. 24-4 cover both engineering and operating functions. These areas of computer activities are interrelated. From the preceding discussion it is clear that the power system operating functions do not have to be performed necessarily in real time. In addition, both engineering analysis and power equipment operation may be related to other functions whose time scale is much shorter than 1 s.

***Power Quality.*** The increased use of equipment sensitive to voltage deviations has been raising concerns about power quality. Power-line disturbances, such as voltage sags and momentary interruptions, cause electric utility customers high economic losses every year. The operation of modern power quality conditioners is initiated as soon as a voltage deviation is detected and may be required to response within half a cycle of the power frequency.

***Overcurrent Calculation and Protection.*** Overcurrents can be caused by faults and short circuits. They can cause not only thermal overload but power system instability too. A fast isolation of the fault location can be then vital for equipment survival and power system security. Protective systems aimed at detecting and isolating have a different design depending whether they are installed at transmission or distribution level. Fault detection and isolation must be faster at transmission and the required response can be in the range of half a cycle of the power frequency.

***Overvoltage Calculation and Protection.*** Power system overvoltages can be caused by transients whose duration may as short as a few nanoseconds. Power system overvoltages can induce power system faults as a result of damage to power equipment insulation. Several mitigation techniques can be used to prevent or limit overvoltages. One of the most useful means for limiting power system overvoltages is the surge arrester which is installed in parallel to the protected equipment and whose response time is in the order of one microsecond.

### 24.1.5 Organization of the Section

The organization of the section is as follows. Sections 24.2 and 24.3 describe the applications used by two of the functional information systems shown in Fig. 24-1, namely engineering and operating systems. The tools required for the design, analysis, and optimization of the smart grid are discussed in Sec. 24.4; this covers some topics of crucial importance for the future grid such as the combined simulation of power and communication systems, the fields for which *high-performance computing* (HPC) capabilities are required, the main features of real-time simulation platforms, the benefits that big data and analytics will bring to the smart grid operation, and the application of cloud computing to smart grid analysis and operation.

Section 24.5 summarizes the main features, the present state and the harmonization progress of the common information model (CIM), a solution implemented for solving the problem of integrating software applications from different vendors with different data formats.

The IoT is aimed at achieving real-time interaction between the grid and the user, and enhancing power grid services. The deployment of advances networking technologies and more reliable communications, as well as the development of more powerful data processing tools, will increase the intelligence level of electricity services, enhance user interaction with the grid, and provide more secure, convenient, and environmentally friendly electricity service. Section 24.6 provides as short introduction to the elements and architectures of the IoT, and the applications of the IoT in the future Smart Grid.

Issues related to computer security are addressed in Sec. 24.7. Cybersecurity is the term used to cover issues that involve automation and communications that can affect the power grid. This subsection describes cybersecurity threats and vulnerabilities, as well as the measures recommended for preventing and detecting computer intrusions, and those measures aimed at recovering from a cyber attack.

## 24.2 ENGINEERING APPLICATIONS

As the electric utility industry has grown in size and complexity, modifications, and additions to existing electric power networks have become increasingly costly. Therefore, it is vital that different design possibilities for additions and modifications to the network be studied in detail to determine their effect on the network, their effect during abnormal operating conditions, and their applicability as a flexible solution to current and future power demands.

The design and construction of planned facilities involves the efforts of a sizable engineering staff and a substantial investment in facilities. To provide support in these activities, computer programs have been developed for analysis of specified designs. The application of these programs contributes to the installation of reliable and economic facilities. The major engineering applications are shown in Fig. 24-5 and summarized below.

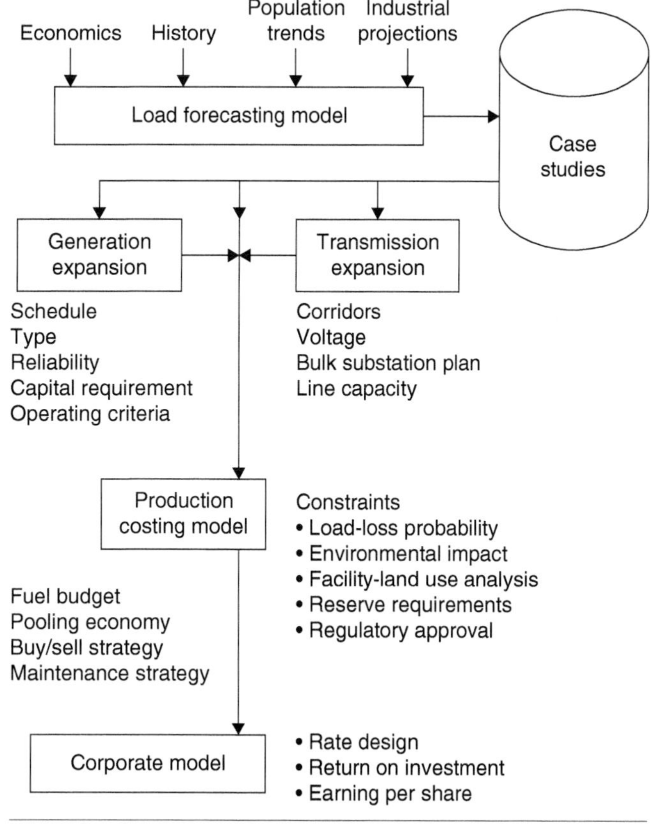


**FIGURE 24-5** System expansion applications.

### 24.2.1 System Expansion

The system-expansion applications (see Fig. 24-5) support the long-term (5 to 20 years) planning function for generation and transmission of power. The system-expansion application area represents the typical decision support environment in that many cases are produced and a variety of options and strategies are considered in the planning process [5]. This area controls large common data sets from multiple sources. Lengthy reports are produced for internal documentation and regulation approval. In the past, most of the processing was batch-oriented because of the length of computation. Today online dialog with the applications is feasible and essential for evaluation of alternatives.

Currently, the emphasis in the industry is to develop more efficient use of existing facilities rather than new construction. Load forecasting and production costing are becoming the most significant items in system expansion to predict load requirements and operating costs. Trade-offs between expansion and new facilities are increasingly important. The applications in this category are as follows.

***Load Forecasting.*** This application is the basis for all planning functions; see [6–8]. It utilizes historical data, trends, economic factors, and residential and industrial projections by geographic area to produce load requirements and load duration plots by area. The effects of demand-side management are incorporated to evaluate the most cost-effective options. It also predicts the load factor.

***Generation Mix Analysis.*** This plans the optimal mix of peaking, base-loaded, or independent power producer units; fuel type; and location of units to meet the future load requirements. It also provides a buy-and-sell analysis and accounts for reliability of generation. In a liberalized environment, generation is not planned, it is instead proposed by generation companies based on foreseen market demand. Generation can be also proposed to secure long-term commitments for power generation that could be disconnected. In recent years, the use of renewable energy resources has increased due to environmental concerns and to a significant reduction in costs. This raises some challenges and requires a different modeling approach that should account for intermittent power generation.

***Production Costing.*** This stimulates the operation of the existing and planned generation facilities for several years in order to predict the fuel budget. It meets the load forecast and accounts for the generation availability and the hour-by-hour dispatch of generation. New techniques use statistical approaches versus detailed models.

***Loss of Load Probability.*** This accounts for unit availability and the reserve requirement to produce a probability of loss of specific loads.

***Voltage Level Analysis.*** This application is a tool to plan voltage levels of existing and planned transmission facilities. It provides trade-offs of network losses versus capital requirements.

***Environmental and Facility Land-Use Analysis.*** This set of applications assists the planner in locating plants, substations, transmission towers, and lines. Trade-offs considered are expansion versus new facilities, right-of-way utilization, and environmental impact of planned facilities. Tighter environmental controls are increasingly affecting expansion and operating plans.

### 24.2.2 Transmission Planning and Analysis

Planning is aimed at identifying major capital investment projects needed to maintain reliable and secure system performance [9–11]. *Reliability* refers to the probability of its satisfactory operation over the long run, while *security* refers to the degree of risk in its ability to survive imminent disturbances (contingencies) without interruption of customer service. This application area supports the short-term planning process and provides tools for analyzing incremental transmission expansion. System planning and analysis applications are high in floating-point content and represent a significant computational requirement. Many cases are analyzed, and there is a rapid turnaround requirement. While there is online dialog with the application and it is common to provide online display and edit of results, the trend is to use interactive graphics in all phases of this decision support process. The specific applications are shown in Fig. 24-6 and described below.

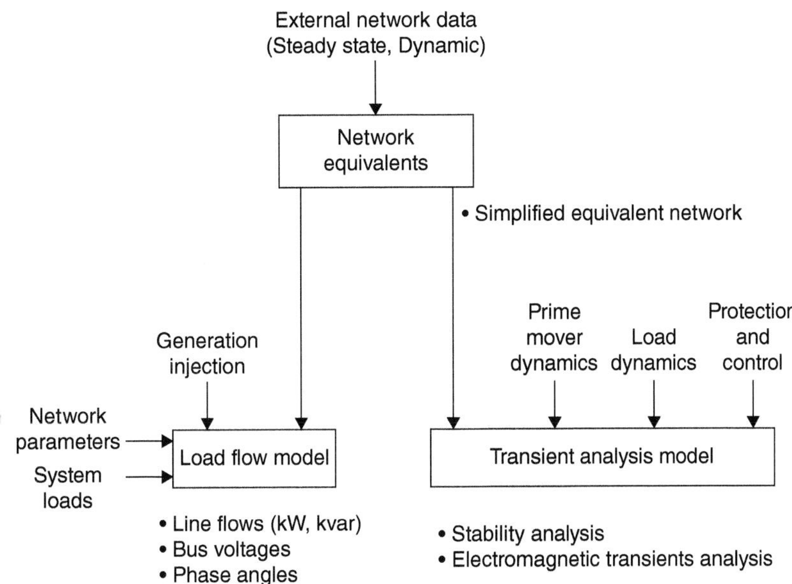


**FIGURE 24-6** System planning and analysis applications.

***Load Flow.*** The load-flow program is one of the major tools of system planning and is utilized extensively. Important to the system planner are that input data errors be minimized and that there be an easy and rapid turnaround for answers when the frequency of program use is high. To accomplish this, interactive capability is provided with the ability to store base cases or numerous power system models on the computer's disks. The storage capability provides many different cases that the planning engineer can access for studying or varying a particular system condition. Load flow enables the power system planning engineer to simulate and solve various power system expansion alternatives in an interactive mode. It utilizes a graphics color terminal specified with a special set of graphic characters that presents results in the form of system online diagrams. The multicolor feature of the terminal is used to indicate heavily loaded lines, bus voltages outside normal limits, and open circuit breakers. The engineer working at such a terminal may, with a mouse and alphanumeric keyboard, remove, add, or change elements of the system being studied and request a solution from the host computer. The simulation programs associated with the load flow program are a system of linked programs that have the following capabilities [12–14]:

1. Basic programs involving calculation of voltages, power flows, angles, and interchanges between areas of a power system.
2. A network reduction program to represent large networks as equivalents in conjunction with the specified area to be studied.
3. A distribution-factors program that indicates the sensitivity of response of the various circuits to outages of specified transmission lines, used to predict thermal limits with linear (superposition) techniques.
4. The series of programs associated with the stored load flow files which permit accessing a particular case, deleting a case, adding a case, and changing a case.
5. The var allocation program which selects the minimum amount of kilovars of compensation necessary to maintain bus voltages within specified limits under normal and/or emergency conditions.

***Optimal Power Flow (OPF).*** It is a versatile alternative for var planning, economic dispatch, and performance improvement [14, 15]. Optimal power flow applications take the load flow solutions and analyze them according to user-defined *objective functions*, such as least cost or minimization of transmission loading. Where the ordinary load flow solution provides only engineering information (e.g., voltage, power, and phase angle values), OPF applications assist operators in ranking alternatives according to economic and other criteria.

***Stability Analysis.*** Transient stability is the ability of a power system to maintain synchronism when subjected to a severe disturbance (e.g., the loss of a generator or a transmission line, a fault, or a sudden variation in demand) [16, 17]. An accurate system response requires an adequate modeling of the nonlinear behavior of the system; these studies use detailed modeling of generator characteristics including excitation systems, control systems, inertia, and governor response. Subsets of stability studies are done to determine the allowable time intervals for protective relays and circuit breakers to sense and isolate faults without causing instability on the system, or to address imbalances between generation and load when electrical islands are formed due to major disturbances. The transient stability application will usually include tools for analyzing other stability aspects:

1. *Small Signal Stability.* Small Signal Stability is the ability of the system to maintain stability under small disturbances.
2. *Mid-Term Stability.* It is essentially an extension of transient stability to cover a longer period of time up to 2 min. This program is used for study of such a phenomenon.
3. *Long-Term Stability.* It is used for the study of very severe disturbances lasting over 2 min. This application has extended modeling capabilities, such as boiler and nuclear reactors, which are not required for the other dynamic programs.
4. *Voltage Stability.* The program is used for study of the voltage collapse phenomenon [18, 19].
5. *Network Equivalent.* A dynamic reduction application is used to obtain a reduced model of a power system, maintaining its dynamic characteristics, so that the computational burden for a study is reduced.

***Short Circuit Calculations.*** Short-circuit calculations are needed to compute fault currents for various types of short circuits (i.e., three-phase, phase-to-phase, phase-to-ground) [14, 20]. Short circuit current calculations are used to determine the required specifications for protection equipment such as circuit breakers and relays, and to determine the proper settings for relays to clear faults, or to assess the ability of circuit breakers to interrupt fault currents and the ability of all equipment to withstand the mechanical forces associated with fault currents.

***Protective System Design.*** This application automates the work of the relay engineer in designing the protective system. It includes relay-coordination studies and calculations of complex relay settings [21, 22].

***Electromagnetic Transient Studies.*** They are used to compute the magnitude of transient voltages and current spikes caused by lightning or circuit switching [23, 24]. The solutions are used to specify the insulation requirements for power equipment (e.g., lines, cables, transformers), to design grounding schemes, or to obtain surge arrester specifications. These types of studies are conducted to ensure that planned expansions are designed in a manner that will not impose transient stresses beyond the capability of equipment on their system, either existing or new. Scenarios studied include energizing and deenergizing, fault clearing under normal and stuck breaker conditions, backfeed conditions, potential resonance conditions, ferroresonance, and lightning overvoltages. These studies may be also conducted to address unusual or unexpected electrical phenomena observed on the power system in real-time operation.

***Contingency Analysis.*** This is the offline analysis of predetermined outages and network contingencies [10, 13]. These studies are run with a load flow application and used by planning in design studies. Results are used as input to the operating area for guidance during problem or alert conditions. A standard reliability requirement is that utilities meet the $n - 1$ criterion, aimed at assessing whether the system will be able to continue to supply all loads despite the loss of a large generator or the outage of large transmission line. If the solution indicates a problem, it must be addressed by adding new generation and/or transmission capacity, or by changing operational procedures. When combined with OPF, this is referred to as *security constrained optimization*. Extreme contingency scenarios that stress the transmission system beyond its design criteria must be also are assessed. Recent studies have shown that the $n - 1$ criterion may not be adequate to assess the vulnerability of cascading failures, so $n - 2$ and even higher order (i.e., $n - x$) contingency events need to be considered [25].

### 24.2.3 Distribution Planning and Analysis

Distribution system analysis has been usually applied to radially connected systems with simple power flow methods. However, the actual distribution system is more complex than transmission systems. In addition, distribution systems may contain a mixture of three-phase, two-phase, and single-phase lines and transformers interconnected in nearly every imaginable way. Distribution system analysis is becoming much more sophisticated, and distribution system modeling is now commonly done with full three-phase models [26–29]. The list of capabilities implemented in modern packages for load flow calculations could include full three-phase analysis, simulations over periods of time with variable time step, meshed network analysis, end use load modeling made generally with time-invariant ZIP models.

Distribution system analysis is a fast-changing field, being models and solution techniques regularly updated and improved. This subsection discusses not only the present status, but also the current limitations and future needs for simulation tools. See also Sec. 24.4.

***Load Flow.*** The primary needs for distribution system load-flow software are to assess voltage profile, power flows, and losses [29]. However, it must also capture the behavior of DER devices and voltage-sensitive loads, and be capable of reproducing unbalance. Transmission system methods are generally based on positive-sequence models while distribution system analysis requires full multiphase analysis. Some distribution programs accommodate only one-, two-, or three-phase sections while others will accommodate an arbitrary number of phases (and other conductors on a pole). The connection of DERs adds special requirements, and when low-voltage distribution networks are modeled detailed models of single-phase DERs may be required.

***Short Circuit Calculation.*** This is essential for determining overcurrent protective device settings, and is a component of reliability analysis. Traditional radial circuit methods assume that there is only one source, but DER addition increases the time varying nature of the fault current, and its contribution to short-circuit currents may be important. The list of capabilities of a short-circuit simulator for distribution systems must include a broad array of features: single- and three-phase analyses; dc analysis; balanced and unbalanced networks; minimum and maximum faults; derating of breakers; arcing fault contributions; accurate and flexible DG models; a full range of transformer connections; interface with protection and reliability software; fault current flow under numerous switching states; overvoltage estimation during faults.

***Reliability Analysis.*** Reliability evaluations are a major part of distribution planning [30, 31]. There are many widely varying approaches for managing reliability that make reliability evaluations for planning purposes very difficult. In general, a reliability tool uses equipment outage frequency and repair time statistics to calculate standard industry customer and system reliability indices [32]. The results can be used to evaluate the performance of a network configuration or a protection scheme. Capabilities that should be available in distribution reliability planning tools include full three-phase representations; integration with advanced metering systems and information for characterizing load profiles and for forecasting; built-in equipment reliability databases; addition of risk assessment methods, economics of reliability, and economics of different maintenance and operation approaches for improving reliability; automatic reconfiguration algorithms; calculation of voltage sag and momentary interruption indices.

***Power Quality.*** Very different tools can be used for analysis of harmonics, flicker, voltage sags, and any type of current and voltage waveform analysis [33, 34]. Time-domain tools are a common approach in power quality studies, since they can accurately represent almost any scenario. But for some cases (e.g., harmonic studies), a frequency-domain approach can be faster and accurate enough. Some programs only model balanced three-phase harmonics; however, for analyzing voltage/load unbalance or single-phase DER applications, three-phase modeling is important. Flicker analysis is important for generation with fluctuating output, since it may be the limiting factor for certain types of generators. Modeling multiple generators is another challenge, since the flicker generated by some units may be totally independent, but others, such as PV, may show a high correlation since they will be located close together geographically. Voltage sag analysis can be also performed by means of simulators with capabilities for short-circuit calculations.

***Transient Studies.*** Several types of transient overvoltages can occur in a distribution system (ground-fault overvoltage, load-rejection overvoltage, ferroresonance); under some conditions, these overvoltages can be severe enough to damage equipment and customer loads. Overvoltages are generally simulated by means of a time-domain solution technique [35].

***Feasibility Analysis of Distributed Energy Resources.*** A DER installation is by default connected to a nearby load and may consist of any combination of electrical generation and energy storage technologies. Current simulation tools for economic operation and design of distributed microgeneration plants vary in terms of capabilities, structure, scale of application, and computing code/platform; in general they have been designed as decision support tools that can be used to select the optimal technology and size, and allow users to analyze different technologies and sizes from among the available alternatives to adequately address the trade-offs between economics, financial risks, and environmental impacts [28, 36].

***Planning.*** A distribution planning package is a set of tools that can be grouped into three distinct categories [6]: (1) electrical performance simulators, (2) analytical tools for reliability analysis, and (3) decision support methods to assist in evaluating and selecting from the possible alternatives. Present planning tools can be used to assess the trade-offs between deploying small DER units and building new or upgrading existing networks, or building new conventional central power plants. The integration of DER devices must take into account multiple factors, such as the existing resources, costs, or the environmental impact. Geographic information systems (GIS) may solve these problems, since they can handle information of very diverse origins and formats (maps, photographs, satellite images, tables, records, or historical time series), and offer a variety of structured data models suitable for the storage, manipulation, and analysis of the information needed in DG planning [37]. GIS tools can perform calculations aimed at determining the optimal location for DG facilities with a given technology (i.e., photovoltaic or wind systems), or be used in applications of spatial load forecasting that allow users to identify areas with a future increase in demand. See also the subsection Distribution Facilities Information System.

### 24.2.4 Design and Construction

This function includes all software applications associated with the design and construction of power plants, transmission facilities, substations, service centers, and distribution facilities. Although the primary objective of electric utility companies is the usual production and sale of a product, they must be concerned with a large capital investment program. Efficient planning, scheduling, and control of labor and material resources are necessary if customer demand is to be met and at the same time a fair rate of return is to be provided on the investment.

***Some Major Programs.*** Design and construction is a multidiscipline activity that employs computer applications dealing with electrical, mechanical, and civil engineering functions. A partial listing of specific applications is provided below.

The *tower analysis* program provides a summary of the maximum tension and maximum compression for each member of a three-dimensional structure over the entire load range specified. This program also spots structure locations, plots a profile of the transmission line, and calculates sag, insulator swing, and ground clearance.

*Line sag* calculates sag and tension of conductors under a given situation.

*Branch circuit design* uses the load, distance, number of cables in a raceway, wire temperature rating, motor starting, and full-load amps to compute the voltage drops and sizes of breakers, cable, and conduit in a circuit.

*Structural design* programs are used to design concrete and steel structures using as input the structure configuration and loads such as floor, roof, and impact.

The *structural steel framing* program is used to design the beams, columns, girders, and baseplates of power plant structures.

The *foundation-slab analysis* program is used to design large, complex foundation mats. The results permit evaluation of various slab thicknesses, soil bearing pressures, shears and bending moments, and reinforcement areas.

The *concrete stack analysis* program is used to analyze proposed stacks by determining loadings, resulting stresses, and required steel reinforcement. This program is used extensively in the design of very tall concrete stacks selected for new power plants.

*Piping* programs are used to perform stress analysis of piping systems and determine hanger design information. The *power plant piping* program analyzes the flexibility of a piping system under the influence of temperature.

The *cable routing* program provides the shortest cable route between nodes, percent raceway fill, and number of cables in a tray or raceway.

*Interference analysis* resolves the interference between pipe, cable tray, and structures occupying the same space.

The *heating, ventilating, and air-conditioning design* program uses thermal loads and the building configuration to calculate the size of refrigeration equipment and ductwork required.

*Fluid dynamics analysis* analyzes piping systems for pressure drop, flow distribution, and power requirements.

*Hydrologic analysis* is used to determine seepage flow networks, underground flow, and rainfall and runoff drainage for culvert and bridge size and design.

*Earthwork design* is used to design embankments and roadways, and perform settlement and embankment stability analysis.

*Geotechnical evaluation* is used to evaluate soil testing results and determine the strength and swell of soil for dam and foundation design.

*Foundation design* programs are used to design foundation pile, pier, mat, and spread footing.

A *statistical analysis of equipment failures* uses data collected on the frequency and cause of equipment failure to determine the likelihood of similar failures in the future based on various changes. The purpose of establishing an equipment operation database is to record and summarize the specific causes of service interruption to generation, transmission, distribution, and communication system equipment as well as to customers. The data provide the basis for designing new systems to specific reliability levels, monitoring equipment and manufacturer adherence to desired availability standards, and carrying out maintenance scheduling activities. The database consists of a main file for each major equipment category and supplementary files which supply input to a family of programs designed to provide the engineers with periodic statistical reports. Engineers also have the ability to retrieve from this database any combination of data of their own choosing.

*Transformer load management* consists of a series of programs to process manufacturing performance data for distribution transformers and derive an economic evaluation based on unit cost and expected loss contributions over the expected lifetime. Since distribution transformers represent such a substantial proportion of system investment, it is imperative that they be utilized to their fullest economic capability. A large percentage of distribution transformers are nominally underloaded; that is, they are oversized for the load being served and hence waste money through overinvestment and excessive core losses. Overloaded transformers also waste money in terms of losses, loss of life, fuse, and transformer burnouts and replacements.

*Drafting* includes engineering sketches and standard symbols used to lay out a drawing on a terminal and the results are printed or plotted. Included in this application are computer-aided design and drafting packages.

*Economic analysis* is used to make economic decisions between alternative sets of system designs or equipment.

Most of the preceding applications require common data. The trend is to treat the data as a corporate resource and to capture and maintain them in a common database. This provides for consistency of data, avoids duplication, and minimizes errors due to entry of the same data in different programs. This common database can then be used by many of the design and construction programs.

Resource-management subsystems that support the design and construction applications are briefly described below. The purpose of resource management is to help utilities in more effective utilization, control, and management of their basic resource (people, equipment, and facilities) in the distribution system.

***Distribution Construction Information System.*** This application supports the management of new investment and maintenance in the distribution area. It provides information for planning, scheduling, and controlling equipment, labor, and material resources and becomes a tool to assist public utilities in providing consistent service and meeting customer demands while realizing fair rates of return on capital investment.

The term "distribution construction" refers to the entire process of work requesting, design, scheduling, reporting, and closing of that portion of the facilities closest in service to customers. New distribution work stems from three types of activity: system maintenance and improvement requirements, customer requests, and inspections or surveys.

There exists a continuing concern with the process by which facilities are constructed and maintained and by which costs are transferred to property accounts or charged to expense appropriations. This concern is often focused on improving the distribution work process to support the planning, design, scheduling, controlling, and tracking of jobs. Such improvements are undertaken to obtain a more effective and efficient work process leading to an earlier plant-in-service and to improving the utilization of the many resources devoted to distribution construction, maintenance, and operating tasks.

Computerizing the distribution work process is desirable because of the significant capital investment and expense components of utility costs and because the work process has characteristics that lend themselves to a high degree of computerization and to improved productivity and control.

***Distribution Facilities Information System.*** This application provides the information required to plan, control, maintain, locate, account for, and manage the distribution facilities of an electric utility.

It is also referred to as an *automated mapping and facilities management system*. When combined with terrain and other landmark information, it is referred to as a *geographic information system* (GIS). It is composed of a graphics system and a database system and has been already mentioned as a fundamental tool in Sec. 24.2.3.

The graphics system provides the interactive functions required to capture information needed to maintain a database for facilities by locations. The user is required to define to the graphics system the facilities and the data elements which are associated with them. This includes its data fields, the pictures that are displayed on a map to represent it, and the connectivity requirements, if it is a network facility.

The graphics system employs a graphics workstation composed of at least one high-resolution display, an alphanumeric display, a keyboard, a mouse, and various hard-copy devices. This workstation is used to enter geographically related data, making it subsequently possible to display the data pictorially (maps), interact with the display (zoom in, window, edit, etc.), display facilities data and alter them, or to make additions or revisions. These are functions that formerly involved manual drafting and filing methods.

Maps or data generated at the workstation may be stored in a common facilities database accessible to many users. The manner in which these data are stored varies. Some systems are able to retrieve only the map facets that were entered; the user must establish the relationship between adjoining facets. In other systems the data exist as a continuous network, and the user requests only the portion and type of data needed. Storage techniques based on the common corporate database management system are becoming prevalent because of the common requirements for facilities data by many departments throughout the company.

The production of maps, diagrams, and pictures is a by-product of this system. Of far greater importance is the network relationship of the facilities data. This allows such applications as feeder analysis, transformer load management, fuse coordination, branch circuit design, and fault current calculations to be executed using the common facilities database.

***Material-Management Information System.*** This system is used to plan for and control the flow of materials in and out of the company. A material items database may be accessed by many departments for multiple purposes. The main subsystems are stores operations, materials planning, and purchasing.

Stores operations relates to all day-to-day activities within the warehouse location. Included in these are functions such as stock inquiry handling, recording of stores transactions (receipts, issues), item location management, order and requisition initiation, and material reservation and allocation control.

*Materials planning* refers to the control and management of an inventory, both repairable and expendable parts. The functions under this application are acquisition analysis, item forecasting, materials requirements planning, reporting, and stock taking control.

The purchasing area includes the functions of ordering material from the suppliers and transferring to the inventory on receipt of the material. The functions within this are purchase order writing, quotation preparation, receiving, returns, implementation, quality assurance, vendor performance analysis, and invoice matching.

***Asset Management.*** Asset Management is a systematic process of operating, maintaining, and upgrading assets cost-effectively, and refers to any system where things that are of value are monitored and maintained. An asset management tool is a composite financial/engineering method, automated by a set of software modules that provides decision makers improved quantitative information regarding investments, supports strategic long-term planning and life cycle management, portfolio evaluation, and risk assessment and management at the control area, corporate, or even regional levels. Decisions affecting the design, operation, and maintenance of components and systems will not only impact reliability and power quality, but will also affect the economic performance of the asset through their impact on expected revenues, costs, and profitability. For a review of asset management tools and a proposal for implementation of an asset management toolkit, see reference [38].

### 24.2.5 Project Management

The objective of project management is to control project costs and schedules in the maintenance and construction of power system facilities. Project management in the engineering departments includes project control, project scheduling, and resource optimization tools. These tools are required to manage small procurement projects. However, they also could be major, long-term projects such as construction of a facility, installation of a major program, or daily tracking of activities within a department. Automated techniques, taking into consideration the control of time, resources, and costs, allow more productive utilization of project management personnel and stricter control of projects than manual methods. There are three major components of a project management system:

1. *Critical-Path Method.* A network represents a project which consists of a mixture of serial and parallel activities and employs a combination of personnel resources, materials, and facilities. When time is associated with each activity within a network, critical-path methodology can be used to analyze the network and determine the longest time path to completion of the project. All other time paths through the network will then have some slack in terms of the critical path.
2. *Resource Management.* Project management and scheduling provides the means to plan and control a variety of projects. These systems permit tasks to be scheduled, resources assigned, costs allocated, and progress reported. Using this process, management can address identified problem areas and adjust its plans accordingly.
3. *Project Costing and Estimating.* Cost control techniques involve the ability to estimate and assign costs for labor, material, facilities, test equipment, and other resources to all activities comprising the execution of all phases of a project. In addition, some application programs permit extending rates; accommodate matrix and other organization structures; compute general, administrative, and overhead expenses; and summarize project costs over selected parts of projects as well as multiproject groups.

### 24.2.6 Administrative Support

Administrative functions have been automated to serve the various requirements of engineering departments. Because these requirements are common throughout the company, integrated or common solutions are often used. Administrative support typically falls into the following categories.

*Text processing* is the preparation, output, and data entry and editing of text using an interactive host-based or stand-alone computer system. This service may be used by a secretary to compose a letter or modify an existing memo. It may also be used directly by an engineer for notes, lists, progress reports, and general documentation. Text processing in a power company is used to prepare and maintain operating standards and procedures, standard material lists, nuclear records, training manuals, maintenance and safety procedures, regulatory reports, and specifications.

*Administrative processing* allows the user to manage electronic document images. It includes activities such as copying and reproducing, document distribution, records file processing, mailing, and office correspondence. Documents may be filed on disk, searched for, and retrieved. Document search may be by name or by complex search parameters as in nuclear records.

Other administrative services provide a convenient means for writing notes, reminders, messages, and appointment records. Typical functions include calendaring, tickler file (diary), meeting schedule, phone list, and to-do lists.

*Text and data integration* applications provide the ability to include data created outside of text applications to form reports and letters. These may be used for the creation of manuals which include specification data to reduce redundant keystrokes and increase accuracy.

*Communications* applications provide an informal and unstructured method of communication within an organization. This provides the means of handling messages that might otherwise require a phone call or memo. It also allows for distribution to multiple locations and receipt acknowledgment.

*Personal computing* provides engineers with the tools, packages, and techniques that allow them to enter and manipulate data and accomplish in hours what might otherwise take weeks. The intelligent workstation provides local processing, user-friendly interfaces, and access to host applications. Many times, all applications programs can be executed on a personal computer.

*Presentation graphics* applications are used to present data in a pictorial form. The graphics may be displayed on a terminal or converted to a hard copy using a printer, plotter, terminal copier, or an attached camera device. Typical uses include the presentation of engineering or statistical data as line, bar, or pie charts, preparation of foils for a presentation, or drawing sketches or diagrams for inclusion in a publication.

## 24.2.7 Power Market Computer Simulation

***Power Market Overview.*** The electric power industry has changed worldwide from vertically integrated monopolies to multiple independent companies. This process has replaced the centralized cost-based market to supply- and demand-based competition. The major goals of this reform have been to promote energy conservation, increase the use of more efficient generation technologies, and reduce energy consumption. Over time, the number of business transactions has increased and new market products have emerged. These products range from long-term contracts, to participation in ancillary services or demand response programs. The future energy system will apply the ever increasing information technology (IT) capabilities to coordinate DERs with bulk generation resources, enhance system performance and reduce the impact of component failure. To accomplish this transformation, the traditional paradigm of meeting all demand at a fixed cost at all times is giving way to mechanisms that recognize the value of an array of energy services to market participants [39].

Economic modeling forms the foundation for power market settlement and simulation. Under vertically integrated utilities, all decisions across the power system are integrated and balanced by the company. In a deregulated and competitive environment, firms interact through market mechanisms, so the nature of the model becomes that of an economic equilibrium, wherein agents make buy and sell decisions. Several equilibrium models have been applied (e.g., the Competitive Equilibrium, the Nash Equilibrium, and the Supply Function Equilibrium) [40]. Although, all of these equilibrium models rely on simplifying assumptions that may be unrealistic, they represent important benchmarks because they are well understood and have shed insights on the detailed behaviors and interactions that make up markets of many different forms.

Models for analyzing the economics of power markets range from stochastic models that correlate trades, prices, weather conditions and energy resources, to optimization models that aim at maximizing the benefit of market agents. *Game theory* models are knowledge-based models that attempt to simulate the behavior of market agents, and can be used for developing/testing trading strategies or detecting/monitoring unusual trades [41].

***Power Market Simulation.*** A power market simulator is a software tool that mimics the operation of electricity markets and can help users to simulate hedging strategies in electricity markets before they are put into practice. In a broader sense, a power market simulator can be seen as a tool for predicting the economic and physical behavior of a power system. The capabilities of a simulator will depend on the goals for which it was developed or the vendor product. Some power market simulators include only generation and transmission, while other tools add also distribution representation. The most advanced tools may be used for

- system contingency analysis to develop hedging strategies for system contingencies;
- market contingency analysis to develop hedging strategies for market contingencies, such as changes in price caps, bidding mechanisms, and new entrants;
- congestion management to develop hedging strategies for congestion charges;
- planning and expansion to demonstrate improved efficiency, reliability, and service from planned generation expansion in specific network locations;
- inter-regional coordination to simulate seams issues and help test strategies for coordinating production across multiple markets.

The input may include information about generation (e.g., size, cost, availability), power delivery system (e.g., system data, constraints) and load (e.g., location, hourly variation), while the output may provide information about generator utilization (dispatch, production cost, revenues, hours on marginal), clearing prices for energy and ancillary services, transmission loading and congestion (e.g., hours of congestion), or reliability indices (e.g., energy not supplied).

***Agent-based Simulation of Electricity Markets.*** Agent-based simulation (ABS) is a powerful approach for the development of power market simulators. ABS techniques have received increasing attention due to their advantages in modeling large-scale complex systems. The increasing popularity of ABS among power market modelers can be explained by the additional opportunities that this modeling paradigm offers for the analysis of economic systems, as compared to other models. Aspects like the interaction structure, the individual behavior of agents, learning effects in repeated interactions, asymmetric information or imperfect competition can be included in a more realistic way in agent-based models. Increasingly, powerful computational resources as well as the development of toolkits that facilitate the implementation of ABS models in object-oriented programming languages have further pushed their applications. Two different reviews of the application of ABS to power market simulation have been presented in [42] and [43].

The works on power market simulation using ABS can be classified into three main categories [42]: analysis of market power and design, modeling agent decisions, and coupling of long-term and short-term decisions. Starting from simple models with aggregated demand curves and few supply bidders on one market, the development of agent-based power market simulation has progressed to concepts of large scale simulation platforms capable of dealing with multiple markets and time scales. However, the development of these tools is a demanding task and raises several challenges. A first issue is the development of an adequate architecture and the use of suitable learning algorithms. Another important issue is the provision of data for agent decisions. For this purpose, the use of conventional optimization models as a source of data for agent decisions has been successfully applied. Since the concept of ABS is applied to large scale simulation models, the requirements on data and architecture are considerable, ranging from detailed electricity market and load data to future power plant options. A common challenge to all ABS models is the validation of simulation results. In the reviewed literature, outcomes are compared to real-world market data (e.g., with respect to market prices). Another issue is the validation of the behavior of single agents. The analysis of agent decisions, such as consumer contract choice or investment decisions, may require considerable efforts to improve the empirical data basis. The concept of ABS as a test bed for the electricity sector can provide additional insights for market and policy design.

Figure 24-7 illustrates a power market model based on the ABS framework proposed in [43]. The figure shows the interactions among market participants and their roles in this framework. The common design issues are the physical transmission system configuration, the power market model, the definition of agents, their roles and their interactions, and the decision making and adaptation in each agent. The physical model includes only a transmission system and several generators and customers located at each local region. Any transaction between two regions is conducted by the physical transmission system, which links generators from one region to customers in another region.

The market model includes two components: a day-ahead pool market and a bilateral contact market. The participants in the power market model include generators, generation companies, customers, customer companies, transmission system, and the independent system operator (ISO). Each participant is modeled as an agent, is responsible for some functionalities, and has a specific decision making framework with possible adaptation mechanisms. Generation companies interact with the ISO by submitting bids to the pool market, receiving bidding results, and conducting dispatch specified by the ISO. Although the generation companies could directly negotiate with the customer companies for long-term bilateral contracts, these contracts should be approved by the ISO. The interactions between customer companies and the ISO are similar. The objective of customer companies is to reduce the amount of payment to the generation companies and transmission companies. Finally, the transmission companies deliver transmission information to the market information service (MIS) and implement the transmission schedules determined by the ISO.

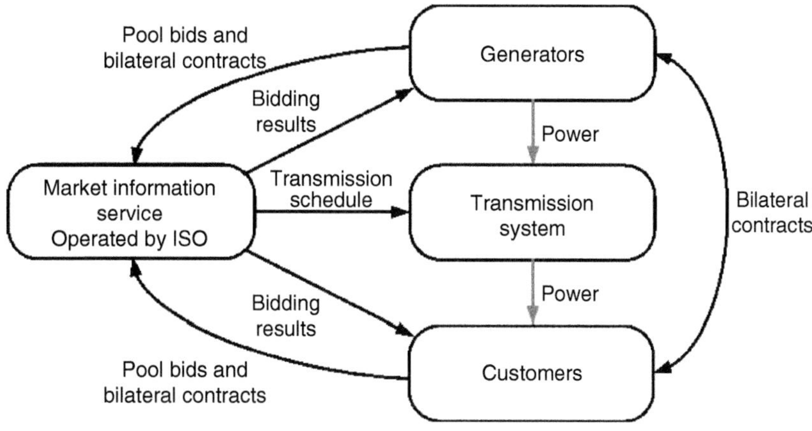


**FIGURE 24-7** Framework for an agent-based power market simulator.

At the physical system level, transmission lines transmit electricity directly from generators to customers. Participants with strategic decision making are the ISO, generation companies, and customer companies. The decision-making procedures include two types of decisions: short-term operational decisions and long-term planning. An *adaptation mechanism* is applied to the bidding process of generation companies and/or customer companies; there is no adaptation in the ISO's decision making.

## 24.3 OPERATING APPLICATIONS

The prime concern of the electric utility industry is to meet the consumer's power demand at all times and under all conditions. Electric utilities are continually seeking and have been most receptive to every available technique which would reduce the capital investment per kilowatt of installation, reduce the operating expenditures per kilowatt hour of energy delivered, and improve the quality of service to the consumer.

Computer systems can be found in use at all levels of operation in power systems. At the generating-plant level, for example, they are used to control and monitor unit startup and operating conditions. In bulk power substations they serve such functions as monitoring, event recording, and switching. And at the system operating center level, they help to improve the economy of operation, improve the quality of service, and simplify system operation. The major computer applications for the operating area of a utility are briefly described below. See also Fig. 24-1.

### 24.3.1 Supervisory Control and Data Acquisition System

Most electric utilities have means to monitor their power system activity and control substation equipment from a central location that would be classified as a SCADA (supervisory control and data acquisition) system [44]. This system connects two distinctly different environments: the substation, where it measures, monitors, and digitizes, and the operations center, where it collects, stores, displays, and processes substation data. A communications pathway connects the two environments. Interfaces to substation equipment and conversations and communications resources complete the system. The substation terminus for traditional SCADA system is the *remote terminal unit* (RTU) where the communications and substation interface interconnect.

RTUs collect measurements of power system parameters and transport them to an operations center where the SCADA master presents them to system operators. Predominantly, there are real and reactive power flow (watts and vars), voltages and currents, but other measurements like tank levels/pressures and tap positions are common to SCADA systems. These belong to the class of measurements termed analogs. Almost anything that can be viewed as a continuous variable over a range fits this category. Analog data is refreshed periodically so that the operator can be assured that data on this screen is relevant. The refresh rate is often dependent on the characteristics of the data being viewed and the communications resources available.

SCADA master stations monitor the incoming stream of analog variables and flag values that are outside prescribed limits with warnings and alarms to alert the system operator to potential problems. Data are screened for "bad" (i.e., out of reasonability limits) data as well.

SCADA systems also collect the state of power equipment such as circuit breakers and switches. These data are presented usually on graphical displays, to give the operator a view of the connectivity of the power system at any given moment. Various state change-reporting techniques have been used to report changes for the system operator. These include flagging momentary changes, counting changes and time tagging them with varying degrees of resolution (sometimes as short as 1 ms).

SCADA systems almost always provide a means for the system operator to control power equipment. This includes circuit breakers, switches, tap changers, and generators. It may include some peripheral equipment in the substation as well.

In the operation center, a SCADA system has at least one computer, communicating to substations and/or generating stations collecting data, issuing control commands, and storing the incoming data. The system operator views data and messages through a set of displays on "view stations." The displays allow the operator to control power equipment and make system changes through a screen dialog. Besides these basic functions, the operations center computer archives data and displays selected data sets, such as trends and logs in special ways for the operators. More modern systems provide data to other areas of the utility enterprise in any number of different forms and services.

## 24.3.2 Energy Management System

***EMS/SCADA Technologies.*** As the central nervous system of the power network, the control center, along with its energy management system (EMS), is a critical component of the power system operational reliability picture [44]; see Fig. 24-8.

The following technologies have been or are being implemented in the EMS/SCADA solutions:

- *Visualization Control.* Control-room visualization today is still limited primarily to one-line diagrams, which are insufficient when it comes to today's needs to understand the availability of electricity at any given time and location and in understanding load, voltage levels, real and reactive power flow, phase angles, the impact of transmission-line loading relief measures on existing and proposed transactions, and network overloads. Three-dimensional, geo-spatial, and other visualization software will become increasingly indispensable as electricity transactions continue to increase in number and complexity and as power data, historically relevant to a contained group of entities, is increasingly communicated more widely to the ISOs and regional transmission organizations (RTOs) charged with managing an open grid. Not only do visualization capabilities enable all parties to display much larger volumes of data as more readily understandable computer-generated images, but they also provide the ability to immediately comprehend rapidly changing situations and react almost instantaneously. Visualization is an invaluable tool for using calculated values to graphically depict reactive power output, impacts of enforcing transmission line constraints, line loadings, and voltages magnitudes, making large volumes of data with complex relationships easily understood.
- *Advanced Metering Technology.* Automated meter reading (AMR) has set new standards by which the energy market can more closely match energy supply and demand through more precise load forecasting and management, along with programs like demand-side management and

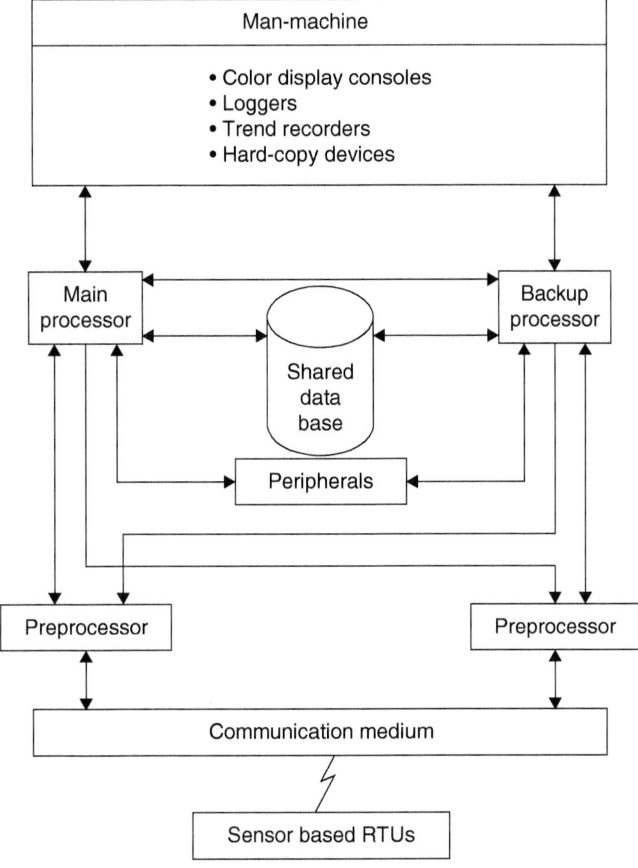


**FIGURE 24-8** EMS architecture.

time-of-use rate structures. Beyond AMR, however, a host of real-time energy management capabilities are now on the market, which, through wireless communication with commercial, residential, or industrial meters, enables utilities to read meters and collect load data as frequently as once every minute. This enables utilities to better cope with dynamic market changes through real-time access to the critical load forecasting and consumption information needed to optimize decision support. The convergence of demand-response technologies and real-time pricing, wireless communications, and the need for more reliable and timely settlement processes are all drivers for enhanced metering capabilities. This, in turn, will create a demand for EMS solutions capable of handling much larger volumes of data and the analytical tools to manage this data. See Sec. 24.4.5.

- *More Stringent Alarm Performance.* The 2003 blackout drew attention to what has become a potentially overwhelming problem: SCADA/EMS has little ability to suppress the bombardment of alarms that can overwhelm control room personnel during a rapidly escalating event. In a matter of minutes, thousands of warnings can flood the screens of dispatchers facing an outage situation, causing them to ignore the very system that has been put in place to help them. Although distribution SCADA has been able to take advantage of straightforward priority and filtering schemes to reduce the alarm overload, the transmission operations systems have not. This is because

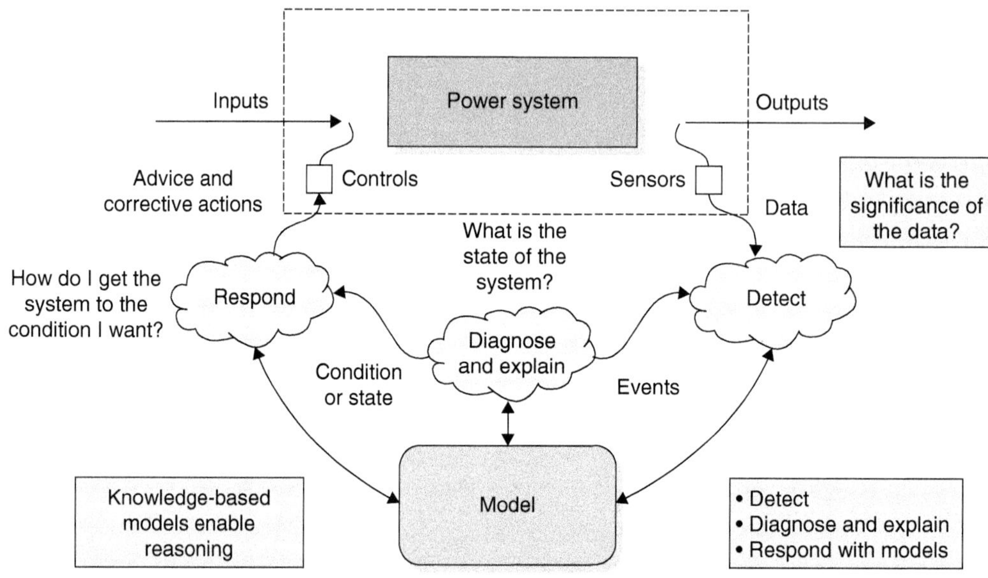


**FIGURE 24-9** Real-time event management. (Courtesy of Gensym Corp.)

transmission systems are networked, and it is more difficult to analyze the alarms to determine what needs to be shown to help the operator reach a conclusion. Also, reaction time is not an issue in distribution, and there is more value in taking the time to locate the fault before taking action; short outages can be tolerated. New analytical tools are needed in the EMS to enable operators to manage and respond to abnormal events and conditions. See Fig. 24-9.

- *Data Warehousing.* For many years, utilities have been archiving the operational (real-time) and nonoperational (historic) information captured by energy management systems. A current trend is to focus on how this archived operational and nonoperational data can be combined with emerging analytic functionality to meet a host of business needs; for example, to more readily identify parts of the network that are at the greatest risk of potential failure. If integrated properly, heads-up information stored by these systems can also aid utilities in proactive replacement or reinforcement of weak links, thus reducing the probability of unplanned events. A data mart is a repository of the measurement and event data recorded by automated systems. This data might be stored in an enterprise-wide database, data warehouse, or specialized database. In practice, the terms data mart and data warehouse are sometimes used interchangeably; however, a data mart tends to start from the analysis of user needs, while a data warehouse starts from an analysis of what data already exist and how it can be collected in such a way.
- *Communication Protocols.* EMS systems must have the capacity to talk to legacy (i.e., preexisting RTUs). Consequently, they are severely handicapped today in that many still rely on serial RTU protocols that evolved in an era of very limited bandwidth. As a result, many EMS solutions in use today are unable to exploit breakthroughs in communications, in particular, secure communications such as encryption and validation. This will need to change. Eventually, the need for encrypted, secure communications to the RTU, combined with adoption of substation automation and substation computers, may lead to the end of RTU protocols as they are known today.
- *Enterprise Architectures.* To achieve the benefits offered by the technologies described here, EMS solutions need to be able to take advantage of modern enterprise architectures (EAs). EMS systems are typically not included as part of utility EA initiatives, but as their importance becomes readily apparent, this will change. Though EA definitions vary, they share the notion of a comprehensive

blueprint for an organization's business processes and IT investments. The scope is the entire enterprise, including the control room, and, increasingly, the utility's partners, vendors, and customers. Though no industry-standard technical/technology reference model exists for defining an EA, it is clear that component-based software standards, such as Web services, as well as popular data-exchange standards, such as the Extensible Markup Language (XML), are preferred, as are systems that are interoperable, scalable, and secure, such as Sun Microsystem's Java 2, Enterprise Edition (J2EE) platform, or Microsoft's.Net framework. By using shared, reusable business models (not just objects) on an enterprise wide scale, the EA provides tremendous benefits through the combination of improved organizational, operational, and technological effectiveness for the entire enterprise.

- *Web Services Architecture.* There are no EMS deployments today that take full advantage of modern Web services architecture, although the architecture is providing tremendous benefits to businesses around the world and holds big promise for control room operations. Like object-oriented design, Web services encompass fundamental concepts like encapsulation, message passing, dynamic binding, and service description and querying. With Web services architecture, everything is a "service," encapsulating behavior and providing the behavior through an API that can be invoked for use by other services on the network. Systems built with these principles are more likely to dominate the next generation of e-business systems, with flexibility being the overriding characteristic of their success. As utilities move more of their applications to the Internet, a Web services architecture will enable them to take strong advantage of e-portals and to leverage standards, such as Universal Description, Discovery and Integration (UDDI); Simple Object Access Protocol (SOAP); Web Services Definition Language (WSDL); Web Services Flow Language (WSFL); J2EE; and Microsoft.NET.

***EMS Functions.*** The dispatcher's interface to the EMS system is through color display terminals used to display online diagrams and tabular displays. These displays are updated cyclically or whenever a change is detected. Various data and operating conditions are highlighted with color. The dispatcher uses the displays to interact with the system such as acknowledging an alarm, changing operating limits, controlling devices, determining system status, or executing a program.

The major EMS functions perform tasks that include periodic work to be completed within a specific cycle, random work that results from alarms and system events, real-time or demand work with specific response-time requirements, and interactive work which is dispatcher-initiated and also has response-time requirements.

- *Network Surveillance and Control.* This consists of continuous scanning of remote sensor-based units, acquiring all key network and generation data on a 2-s cycle, checking the data for problems, and presenting the status of the network to the dispatcher through displays. The control function allows the dispatcher to take selective control action on power system devices such as circuit breakers. These functions are those performed by SCADA systems, described above.

  *Network data acquisition* programs acquire power system data through periodic scanning of local and remote sensors attached to RTUs. The raw data are checked for missing or invalid data before being sent to the central processors by the communications front end. Retransmissions are requested for missing status data.

  *Data conversion and limit checking* programs are used to process the raw data. The data are scaled and converted to engineering units and stored in a database for use by application programs. Reasonability checks are made on the data and values not found to be within limits cause alarms.

  *Alarm processing* programs generate alarms for scanned data and calculated results as changes of network status are discovered, limits are exceeded, or other invalid conditions are encountered.

  *Logging and reporting* programs store selected data in historical log files, where they are available for analysis through displays and reports.

  *Power device control* functions provide for the control of power system devices and for the placement or removal of device tags. Control actions made by the dispatcher to activate power devices are verified, then forwarded through the network to the proper destination.

*Load shedding* provides rapid access to data for guiding the action of the dispatcher during abnormal operating conditions. Typical of the data that are displayed is the amount of load relief that can be obtained through emergency output of all generating sources, generation requirements, interruptible loads, load curtailment, voltage reduction, and emergency backup through tie lines.

- *Generation Commitment and Control.* This consists of a set of programs used to optimize the production and delivery of power for fuel savings. See also Sec. 24.2.1.

*Automatic generation control* (AGC) regulates generator output to match the load in order to maintain the frequency in an area and the sum of all active tie-line power exchanges with neighboring power systems. This program executes cyclically every 2 s.

*Economic dispatch* minimizes the cost of meeting the energy requirements of the system over a period of time and in a manner consistent with reliable service. Desired generator settings are computed and fed to the automatic generation control program. This cyclic program executes every 5 min and also on demand.

*Load forecasting* computes the total system hourly load for a specified number of days. It provides an adaptive forecasting system based on observed values of demand and estimated weather conditions. The program generally consists of three mathematical models. A load forecasting model uses past load data to compute hourly load forecasts; see [6–8]. A weather forecasting model computes hourly weather forecasts based on past weather history. A weather correction model uses telemetered values of load and weather conditions to correct the forecast. The dispatcher may optionally enter these values through a display. The historical load and weather data are stored and maintained in files which also contain special events that affect load such as holidays or unusual weather phenomena.

*Renewable energy forecasting* has been incorporated into some EMSs to estimate the expected power production from intermittent renewable energy resources (i.e., wind and sun). The increasing penetration of renewable energy resources raises some concerns about the effects that intermittent energy can have of the power grid performance. Power forecast models are computer programs that can estimate power production with different time scales, which may range from 1 min to several days, depending on the energy resource and the user's need. In general, operational planners and ISOs use 1-h ahead and 1-day ahead forecasts for system operating, power scheduling, or power trading. Several approaches have been developed for wind and sun power forecasting, and some commercial products are already available. For a review of models and products see [45] and [46].

*Unit commitment* determines a schedule for optimal startup and shutdown of thermal units which minimizes unit startup costs subject to generation objectives, predicted area requirements (load forecast), security (spinning reserve requirements and off-system capacity), and operational constraints (unit minimum up and down times, limits, ramp rates, maintenance and derating schedule). It minimizes the operating costs of the dispatchable generating units. The total dispatchable generation is the sum of the load forecast and net scheduled interchange minus the total nondispatchable generation. The operating cost is defined as the sum of production, startup, shutdown, and maintenance costs. Production cost is calculated by the use of input and output curves adjusted by fuel prices. Transmission losses are also considered using one or more sets of penalty factors. Transmission limits are considered with simple approximations. The output of unit commitment is the hourly unit schedule. Several sets of schedules (or strategies) may be output for operator review and selection. The output is stored and made available to other applications such as transaction evaluation.

*Reserve monitor* calculates the available operating reserve necessary to meet company operating policies. Typically, a spinning reserve, 30-min reserve, and 2-h reserve are computed. This program executes cyclically every 4 min or on demand. The dispatcher is alerted when there is inadequate reserve.

*Maintenance scheduling* assists operating personnel in scheduling generator maintenance in an optimum manner. It generates a maintenance schedule while taking into account constraints and limitations on available resources required to perform maintenance work. The maintenance schedule indicates when particular generating units will be out of service (unit outages). Maintenance is performed routinely at desired intervals or may be required because of unexpected forced outages. During the maintenance of a given generator, the outage of this unit is compensated for by other capacities in the power system. This is done by maintaining a level of system reserves. The scheduling of generator outages is performed in a manner which maintains a flat megawatt reserve level or a consistent level of probabilistic level of risk (i.e., probability of emergency procedures).

- *Interchange Management.* Formal and informal interchange agreements may exist between neighboring companies. Operating the power system as an interconnected system has economic and security advantages. While many power pools have control centers to manage the interconnection, most companies prefer to manage, or at least monitor, their own interchange of power. The interchange management programs provide the dispatcher with the ability to make good deals with the neighbors. These interactive routines allow the dispatcher to evaluate a buy or sell transaction before being committed to it. Once initiated, the interchange scheduling program automatically schedules the transaction through the automatic generation control program.

  *Production costing* provides the capability to compare a change in interchange with the current schedule. The program always starts by accessing a study file. The calculations are based on current or forecast system load and a proposed interchange schedule entered by the operator.

  *Transaction evaluation* calculates the costs and savings associated with the sale and purchase of power with a selected interconnected company. The program can be used to evaluate past, present, and future operation costs. A production cost calculation is generally included as a part of this program or may be an independent routine. For evaluating transactions, the transaction evaluation program computes the costs of savings of a proposed transaction by comparing the production costs computed from two economic dispatch calculations: a base economic dispatch calculation for the operating conditions with and without the proposed transaction. The transaction evaluation program usually can be executed in two modes which specify the starting point for all calculations. These two modes are whether the unit commitment program will be called or not. Essentially, the two options determine how the generation schedule is to be determined. Mode A (also called *economy A*) is executed without unit commitment. Current system load, interchange schedules in study file, and existing unit commitment form the starting conditions. Mode B (also called *economy B*) is executed with unit commitment and is used for transactions in the future.

  *Interchange scheduling* is used to process and display each scheduled transaction with interconnected utilities. It computes the total net scheduled interchange as a function of time for use by the automatic generation control program. This calculation considers the start and stop times, generator ramp rate, magnitude, and direction of each active interchange transaction. Interchange schedules and cost information may be logged and displayed.

  *Interchange accounting* provides the capability to account for electric power in the system. This power, in the form of measured, calculated, or scheduled megawatts, includes megawatts which are generated, consumed, lost, passed through, sold, and purchased. It includes functions for logging, displaying, reporting, and updating data recorded by the real-time system.

- *System Security.* These programs help to reduce the chance of a major outage or blackout condition [47, 48]. They operate on a study database which has been cleansed of missing data and metering errors by the network status and state estimation programs.

  *Network status* determines the current configuration of the network based on the status of circuit breakers and disconnectors obtained from the real-time data and from manually entered status information for devices not telemetered. The program is executed periodically, whenever a status change is detected, or on dispatcher request. In addition, the program develops the corresponding mathematical model of the network using the impedance data for the current base load flow or state estimation case. The model will be used in the subsequent calculation of real-time system conditions using the load flow or state estimation programs.

  *State estimation* determines the current state of the power system, including voltage levels and power flows, and calculates loss factors for use by the economic dispatch program for generation scheduling [49]. It filters real-time measurements to detect and eliminate known errors; estimates expected values of the next real-time measurements; and determines the current network configuration. State estimation corrects the mathematical model of the current network configuration, so this program acts as a filter between the raw real-time data and the real-time data requirements of other security applications. This task is executed whenever the network status program detects a change, periodically, or on dispatcher request.

*Contingency analysis* computes the potential effect of contingencies involving the loss of generation and transmission facilities. A specific set of predefined contingencies is analyzed on a cyclic basis. It simulates a contingency and calculates the changes in bus voltages and power flows resulting from the contingency. The base conditions for this calculation are the bus voltages or power flows obtained from the load flow program.

*Interactive load flow* (*ILF*) allows the dispatcher to perform load flow studies for a scheduled outage or analyze corrective actions after an unexpected outage. ILF uses real-time data to project bus loads in a network. A bus is a connection where power lines change direction or voltage. A bus load is the load at this point. The ILF program stores mathematical models of the present and planned networks. A color display terminal is used to display one-line diagrams of these models. Colors are used to differentiate voltages, heavily loaded lines, bus voltages outside of limits, and open circuit lines. The dispatcher can retrieve previous load flow cases, make modifications, create displays, execute load flow cases, and store them.

*Training simulator* uses a model of the power network to produce realistic reactions to a dispatcher in training. This function is usually run on the backup central processing unit (CPU). The program can operate in either of two modes: monitor or simulation mode. In monitor or playback mode, the simulator reflects the changing status of the power system during the time period when a playback tape was written. The trainee can view displays, thus monitoring power system changes, but cannot take any control action. In simulator mode, the trainer can input system changes, and the trainee is allowed to perform control actions. The simulator software will reflect the state of the power system based upon the changes input and control actions taken. In this manner the trainee can operate the control system without affecting the true state of the power system.

*Customer response systems control* centers are a critical link in responding to customer complaints and reacting to outages. Customer information systems, used for many business and commercial purposes, are increasingly being combined with the engineering and operations information in order to speed repair crew dispatch and reduce outage times. Work orders, repairs, and inspections are increasingly being merged with inventory, warehouse, and engineering records to reduce costs, speed reaction times, and improve technical performance.

***Wide-Area Measurement System (WAMS).*** The restructuring of the electric power industry has changed operational planning tasks, which are moving toward more close interaction with operating activities [50]. The increased need for accuracy is moving operational planning toward real-time operations (e.g., voltage and transient stability could be handled in real time), although other aspects (e.g., transaction scheduling), by their nature, are not likely to converge to real time. There are several limitations and constraints for developing accurate simulations. In some cases, the data does not match real-time data gathered by an EMS; in other cases, data gathered from a SCADA system is not sufficiently accurate. For example, to determine steady state and dynamic operating margins input data come from a state estimator, whose output consists of estimated values which limit the accuracy of some data. As a result, some studies (e.g., stability margin) produce conservative estimates. Moreover, gathered data might not cover a sufficiently large geographical area to meet the needs of some operational planning studies. Another limitation is the lack of modeling of the details of circuit breakers in operational planning studies; that is, the level of detail that is needed to ensure accuracy. The term wide-area measurement system (WAMS) denotes a more complete and accurate system of measuring a wide range of data on power systems that can be used to improve the quality of the input data and that of power system simulations [51–53]. This system can gather a wide range of time-stamped data in the form of real-time synchronized phasor measurements (i.e., synchrophasors), across broad geographical areas, and make it rapidly available via satellite to receiving stations.

***Substation Automation.*** A substation automation (SA) system uses microprocessor-based IEDs integrated by a communication technology for the purpose of monitoring, controlling, and configuring the substation. IEDs provide inputs and outputs to the system while performing some primary control or processing service [54–56]. Common IEDs are protective relays, load survey and/or operator indicating meters, revenue meters, programmable logic controllers, and power

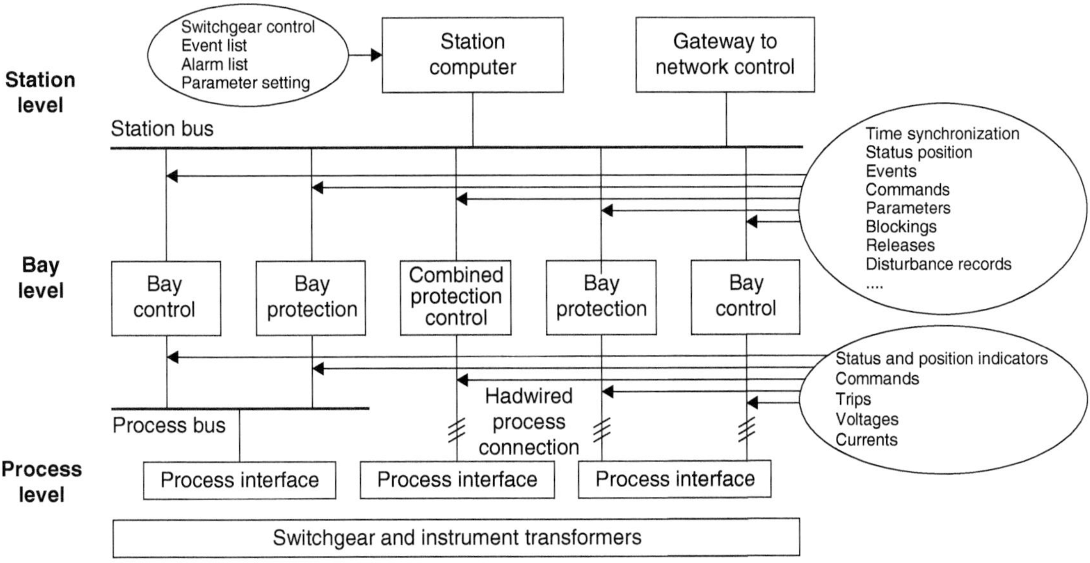


**FIGURE 24-10** Configuration of substation automation system and data exchange (adapted from Ref. [58]).

equipment controllers of various descriptions. Other devices dedicated to specific functions for the SA system may also be present. These may include transducers, position sensors, and clusters of interposing relays. Common communications connections include utility operations centers, maintenance offices, and/or engineering centers. Most SA systems connect to a traditional SCADA system master station serving the real-time needs for operating the utility network from one or more operations center. SA systems may also incorporate a variation of SCADA RTU for this purpose or the RTU function may appear in a SA controller or substation host computer.

SA systems may be broken down into three hierarchical levels, which are found in most implementations as physical levels also [57, 58], see Fig. 24-10: (1) the *process level*, which refers to the power system equipment in the substation represented by the process interface, (2) the *bay level*, which consists of bay protection and control IEDs hosting the related functions, and (3) the *station level*, which refers to tasks for the complete substation and consists typically of the substation computer with central functions and human-machine interface and of the gateway to the network control center. Station and bay level are connected by the *station bus*. Present bay and process levels are connected by a lot of parallel copper wires, the future substation will use the *process bus*. This architecture has some data paths from the substation to the utility enterprise, the SCADA system and the data warehouse [55].

For modern SA systems or for legacy SA systems that need to be updated, IEC 61850 is becoming the general choice for standard protocols [59–61]. IEC 61850 is also a comprehensive approach for the conception of substation automation and protection with serial communication.

SA design involves a series of steps from the specification to the commissioning of the project system. The first step of the engineering of a SA system is to define the functional specification according to the selected protection, automation and control design. The list of elements that are part of the specification may include the single line diagram of the substation, the geographical layout (extension, buildings, cable channels, etc.), protection functions required for each primary distribution substation or system component, measurements and status information needed, controls to be used, reporting requirements, monitoring and recording requirements, redundancy requirements, communications architecture, distribution substation level functions, and others [57].

Substation configuration description language (SCL) is the language proposed by IEC 61850 for describing the configuration of substation devices. The SCL is based on the XML, and its goal is to provide a formal description of the SA on engineering level (i.e., files that can be exchanged between proprietary tools of different suppliers). The formal SCL description will allow exchanges between compliant tools independent from the supplier and result in a machine-readable documentation of the data and communication structure of the SA system.

***Distribution Automation.*** The concept distribution automation (DA) and its implementation vary from one utility system to another. The traditional DA function has been the remote control of switches to locate, isolate the fault and restore the service, when a fault occurs in the power distribution line [62]. A more advanced DA system involves utilization of communications infrastructure and information technologies to enable remote control of switching devices including substation breakers, reclosers, and field switches as well as capacitor banks and line regulators. This category of DA is commonly referred to as feeder automation (FA). With the application of remote data collection and control through FA, switching tasks are accomplished in an automated fashion giving rise to shorter restoration times. DA has evolved since then and turned into an established concept.

DA has gained renewed attention with the transition toward the smart grid and the need for more reliable and efficient distribution systems. The DA system to be implemented in the smart grid will be based on new emergent technologies and concerned with complete automation of all the controllable equipment and functions in the distribution system. A broad definition of DA includes any automation which is used in the planning, engineering, construction, operation, and maintenance of the distribution power system, including interactions with the transmission system, interconnected DERs, and automated interfaces with end users. A smarter DA system will have to address enhancements in efficiency as well as reliability and quality of power distribution, reduce operation and maintenance costs, enable new customer services, or defer capacity expansion projects. The future DA will cover the complete range of functions from existing SCADA systems to ever-increasing deployment of *advanced metering infrastructure* (AMI) technologies at the customer level in which local automation, remote control, and central decision making are brought together to deliver a cost-effective, flexible, and cohesive operating architecture [63].

Feeder switching and protection systems utilizing powerful IEDs, sophisticated algorithms, various technologies of sensing devices, and all connected by increasingly fast and secure data communications, enable the implementation of distributed intelligence, which is fundamental to implementation of the future smart grid. Centralized control of the distribution system will move to distributed computing, which may effectively eliminate communication bottlenecks and time delays associated with centrally controlled SCADA systems, and may be sustainable even if single computing nodes do not function. Properly designed systems based on distributed intelligence offer a completely scalable advanced FA system that can improve reliability, or grow to deliver system-wide automation functionality and improved asset utilization.

A challenging application of the smart grid DA system will be the extension to the consumer location, since there will be a large number of installation points, and automation at the consumer's location will include the ability to remotely read meters, program time-of-use meters, connect/disconnect services, and control consumer loads.

DA functions that have been implemented by utilities throughout the world vary greatly in nature and so does their communication requirement. To date most DA functions have been implemented by using proprietary protocols. The natural shift from proprietary communication protocols to standardized protocols is being accelerated and directed toward more advanced solutions that will provide an interoperable environment. The IEC 61850 standard is also becoming the common communication protocol for DA.

Although IEC 61850 standard can be potentially used as the communication protocol for FA applications or communication with the control centers, it does not standardize the representation of combinatorial, sequential, rule-based (or any other form of) power system control and automation logic (e.g., the interlocking logic for determining whether a control operation can be performed or not). To play this role, reference [64] proposes the application of the IEC 61499 standard to define the algorithms for control and automation functions.

### 24.3.3 Power Plant Monitoring and Control

This application approaches the traditional process control application. Several applications such as fuel monitoring and performance calculations increase the requirement for the data processing resource. The three major computer systems for fossil power plant monitoring and control are described below.

1. *Process Control System.* This is a closed-loop control system which takes its direction from EMS and automatically collects plant data by reading instruments. Physical and electrical parameters associated with the boiler, turbine, and generator are monitored on a continuous cyclic basis. Alarms and events are logged, and control of pumps, valves, and switches for routine functions and for startup and shutdown are provided.
2. *Plant Monitoring System.* This is strictly a data-collection system for fuel monitoring, performance calculations, and balance-of-plant calculations; no control actions are performed. Data are stored and retrieved as required to prepare reports and perform analysis. These reports include those required by the plant management, load dispatchers, and planning and engineering groups. Periodic reports are prepared to reflect plant operation. Unit incremental generation cost is determined periodically by collecting data that continuously reflect actual operating conditions. This information is transmitted to the dispatcher for use in load dispatching.
3. *Operational Monitoring System.* This is used by plant operators to enter manually collected operational data for record keeping, report writing, and engineering analysis. In addition to these systems, the power plant also may use mini- or microcomputers for security systems, environmental systems, controlled access systems, and chemical analysis systems.

### 24.3.4 Power Plant Maintenance

Power plant maintenance stores pertinent plant maintenance information for analysis of maintenance costs and evaluation of equipment performance. The interactive portion of the system provides power plant personnel with the capability to enter problem data, planning data, and work execution data. Approval and verification functions are at each step of a work order's progress from problem description through work completion and commitment to history. Interactive functions also are provided for entry and maintenance of an equipment database and for access to equipment history. The batch portion of the system provides for moving completed and rejected work from the active work database to the equipment history database. Batch functions are also used for work backlogs, scheduled work, and other reports. Because of its varied data requirements, the maintenance information system also has interfaces to other computer systems in the power company. These are the materials-information system for equipment stocking levels, the personnel information system for labor resources, and the general accounting system for cost tracking.

### 24.3.5 Fuel Management

Operating support personnel must plan for both short-term (1 year) and long-term (5 years) fuel availability. Thus control and improvement of fuel cost represent the most substantial contribution to the overall economy of power production. The system is used to administer the procurement of coal and oil. It facilitates the monitoring, reconciliation, and performance of fuel contracts. Accounting functions are used for fuel purchases, transportation costs, fuel usage, and inventory value. Also included are short- and long-term cash flow projections and managerial and regulatory reporting capabilities. The overall fuel management application is an integrated model of the load forecast, generation scheduling and dispatch, and fuel allocation with an objective of optimizing fuel contracts. This application is used routinely with growth and on demand as system perturbations occur.

### 24.3.6 Load Management

The objective of the load management application is to improve the load factor (ratio of average load to peak load) and to be able to shed selected load during emergency conditions. Most utilities have pilot tested some form of load management. The pilot solutions range from consumer guidance (voluntary) to complete control and metering of the consumer's load.

The benefits of load management have been shown to be enhanced by combining two-way utility to customer communications and control into a distribution automation system. An automated distribution system can have valuable additional functions such as automated meter reading, performance monitoring, and quality-of-service improvement.

***Load Curtailment System.*** This scheme employs one master computer at the power company and several computers at major industrial and commercial load centers. In this scheme there is a limited sensor-based data-acquisition requirement and no closed-loop control. Reduction of load will be manual.

***Automatic Meter Reading and Load Management.*** This hierarchical approach employs the corporate customer information system and the EMS system to direct and receive the meter readings, an intermediate level to serve as a communication concentrator, and an intelligent data control unit to scan remote transponders at the meter locations.

## 24.4 TOOLS FOR THE SMART GRID

### 24.4.1 Introduction

The smart grid uses a wide variety of technologies and performs functions ranging from wide-area monitoring at transmission levels to home automation at low-voltage distribution levels. The design, analysis, and optimization of the smart grid have to be carried out taking into account, among others, the following aspects:

- Power system planning, operation, and control require large amount of computational work due to large mathematical models and, in certain applications, fast time response.
- Traditional computational resources do not meet the emerging requirements of a grid model that is evolving into a more dynamic, probabilistic, and complex representation. New computing techniques and solutions should enable faster and more comprehensive analysis. In addition, the focus is moving to real-time simulation platforms.
- In the foreseen smart grid scenario, the manipulation of large amount of data collected from smart meters and sensors will make necessary the application of new computing capabilities, like those provided by graphics processing units (GPU), and geographically distributed processing techniques, like cloud computing.
- The study of the smart grid requires a combined simulation of power systems and information and communications technology (ICT) infrastructures. Since the operation of the power system increasingly depends on communication and data networks [65], it is crucial to understand the impact of the ICT infrastructures on the operation of the power grid.
- The operation of the smart grid also involves issues such as safety and security (including protection against potential cyber attacks) [65].

To deal with those aspects and achieve the goals mentioned above, new analytical methods and simulation tools are required. Simulation has always been an important tool for the design of power systems; in a smart grid context it can be useful to reduce the costs associated with upgrades to both power and communication infrastructures, analyze the potential loss of service that can occur as a consequence of a failure or a cyber attack, or enable the design and evaluation of different solutions before deploying them in the field [65]. Simulation faster than real time can be crucial for the development and implementation of smart grids [66].

Some features of the tools required for the design, analysis and optimization of the smart grid are discussed below; the rest of Sec. 24.4 covers with some detail the combined simulation of energy and communication systems, the fields in which HPC has been applied to date, the development and capabilities of real-time simulation platforms, the advantages that big data and analytics will bring to the smart grid operation, and the application of cloud computing to smart grid analysis and operation.

***Simulation of Very Large-Scale Systems.*** Several large-scale grid concerns and threats cannot be adequately modeled using present capabilities. Among these concerns are wide-area disruptive events, including natural events, cascading accidents, coordinated cyber and physical attacks, interdependencies of the power grid system and critical infrastructures, or scenarios including wide-scale deployment of intermittent distributed generation. For instance, understanding the interdependencies of the electric power grids with other critical infrastructures is a serious need, since disruptions in one infrastructure (e.g., the electric grid system) can have severe consequences for other infrastructures (e.g., natural gas and water supply systems). New modeling approaches could span different applications (operations, planning, training, and policymaking) and concerns (security, reliability, economics, resilience, and environmental impact) on a set of spatial and temporal scales wider than those now available. To fulfill this role, new simulation tools are needed. Such tools could be built through a combination of existing distributed and new capabilities, and take either the form of a single, centralized facility or a virtual, integrated environment. Several issues would have to be addressed, however, in their development and implementation. In particular, acquisition of and access to validated electric infrastructure data, physical and administrative protection of controlled information, including protection of sensitive information generated as model output, opportunity for data sharing, data verification and validation, identification of data use, and an environment that simplifies the integration of diverse system models. For a discussion of this topic see reference [67]. However, the main challenges will not be in the size of the systems to be analyzed but on the effectiveness of the analytical methods and the accuracy of the implemented models. Some experience is already available in the implementation of tools for simulating huge power systems at a reasonable time; see for instance [68] and [69].

***Multidomain Simulation Tools.*** An accurate modeling of some generation and energy storage technologies may require the application of simulation tools capable of connecting and interfacing applications from different types of physical systems (mechanical, thermal, chemical, electrical, electronics). Several packages offer a flexible and adequate environment for these purposes. They can be used to develop custom-made models not implemented in specialized packages. Open connectivity for coupling to other tools, a programming language for development of custom-made models and a powerful graphical interface are capabilities available in some circuit-oriented tools that can be used for expanding their own applications and for developing more sophisticated models. These tools can be applied for the development and testing of highly detailed and accurate device models, or linked to other tools to expand their modeling capabilities [70].

***Interfacing Techniques.*** Power system studies are numerous and each has its own modeling requirements and solution techniques. Over time, several software applications have been developed to meet the individual needs of each study. However, a need is also recognized to exploit the complementary strengths by interfacing two or more applications for model validation or to exchange data between different tools. Examples are the interface of time- and frequency-domain tools, the interface of a circuit-oriented tool and a tool based on the finite element method (FEM), or the interface for analysis of interactions between communication and power systems, each system being represented by a specialized tool [70–76]. Although there have been significant improvements in tools that can simultaneously reproduce different physical systems (e.g., multidomain simulation tools), the interfacing will become an increasing necessity for the simulation of complex systems whose modeling requirements cannot be met by a single tool.

***Agent-Based Simulation.*** Agent-based simulation (ABS) is a powerful approach for simulation of real-world systems with a group of interacting autonomous agents modeled as computer programs that interact with each other. The design of an agent-based simulator involves communication protocols and languages, negotiation strategies, software architecture, and formalisms. ABS can conveniently model the complex behavior of system participants and is particularly suitable for large-scale systems involving various types of interacting system participants with distinct roles, functionalities, behavior, and decisions. ABS is a suitable approach for some type of simulations: power market simulation [42, 43], system vulnerability assessment [77], or co-simulation of power and communication systems [78]. ABS has been applied in power system studies for many years and will increase its application mainly due to the increasing demand for large-scale system simulators.

### 24.4.2 Combined Modeling of Power Grids and Communication Systems

Communication technology will have a prominent role in the future smart grid; therefore, an accurate prediction of the power grid behavior will require accurate models of the communication infrastructure and the integrated simulation of generators, transmission and distribution systems, control processes, loads, and data networks. Current approaches to hybrid system simulation focus on integrating available software packages. However, the results often have limited application or make significant sacrifices of precision and accuracy; the analysis of integrated power and ICT infrastructures requires integrated simulation frameworks [65]. The scenarios to be considered for the electric grid require the systematic construction of nontrivial hybrid models, which should include complex continuous and discrete dynamics [78–80].

***The Role of Communication Networks in Smart Grids.*** Communication networks already play an important role in the power system, and they will play an even more crucial role in the future smart grid. The smart grid communications layer can be seen as consisting of several types of networks, each having a distinct scale and range. For instance, *wide-area networks* (WAN) are bandwidth communication networks that operate at the scale of the medium voltage network and beyond, handle long-distance data transmission and provide communication between the electric utility and substations; *AMIs* interconnect WANs and end-user networks, and provide communication for low voltage power distribution areas; *home area networks* (HAN) provide communication between electrical appliances and smart meters within the home, building or industrial complex. *Power line communication* (PLC) is an option that uses the existing power wires for data communication (i.e., the power grid itself becomes the communication network); narrowband PLC technologies that operate over distribution systems are also used for monitoring (e.g., AMI) or grid control. However, the power grid was not designed for communication purposes; from a communication perspective, existing power grid networks suffer from several drawbacks [65]: fragmented architectures, lack of adequate bandwidth for two-way communications, lack of interoperability between system components, or inability to handle increasing amount of data from smart devices.

***Continuous Time and Discrete Event Simulation Models.*** Power system and communication network simulators use different modeling approaches and solution techniques:

- Dynamic power system simulation uses continuous time where variables are described as continuous functions of time, but since some discrete dynamics are to be introduced, a time stepped approach is used (numerical algorithms with discrete time slots are applied).
- Communication networks are packet switching networks, which can be adequately modeled as discrete events that occur unevenly distributed in time. This is different approach from that used for power system dynamic simulation, based on a fixed interval between events.

Figure 24-11 illustrates the difference between the two types of simulators: it is evident that synchronizing the time of different components is a crucial aspect when combining several tools.

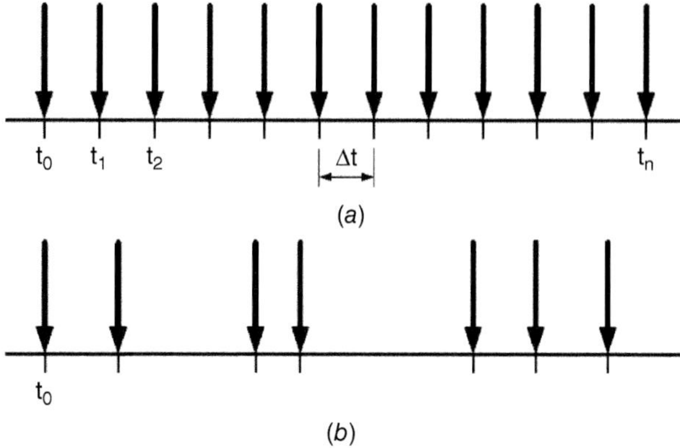


**FIGURE 24-11** Continuous time versus discrete event simulation. (*a*) Continuous time simulation—evenly distributed time steps; (*b*) discrete event simulation—unevenly distributed time steps.

An option to deal with both approaches is the use of predefined synchronization points: each simulator pauses when their simulation clock reaches a synchronization point; after each simulator is paused, information is exchanged. However, this can have some drawback as messages that need to be exchanged between both simulators are delayed if they occur between synchronization points. A solution to this problem is to reduce the time step between synchronization points, although this will degrade performance. Consequently, the co-simulation needs to achieve balance between accuracy and simulation speed, and take into account that not all time instants at which communication between the different simulators must occur are known a priori. See [65] and [81].

***Combined Simulation of Power and Communication Systems.*** The combined simulation of the power systems and communication networks can be achieved by two approaches [65].

- *Co-simulation.* An approach that combines existing specialized simulators can be a faster and cheaper solution that constructing a new environment that combines power and communication systems in a single tool. On the other hand, using existing models and algorithms that have already been implemented and validated reduces the risk of errors. In the smart grid context, a co-simulator would consist of a specialized communication network simulator and a specialized power system simulator. When multiple simulators are interfaced, the main challenge is to connect, handle and synchronize data between simulators using their respective interfaces: time management between two simulators can be challenging if each simulator manages its time individually, and the necessary synchronization between simulators running separately implies penalties, such as start-up times, reading and processing input data. A significant effort on this field has been carried out to date; see [82–88].
- *Integrated Simulation.* An approach in which the power system and the communication network are simulated in a single environment simplifies the interface between tasks and allows sharing the management of time, data, and power/communication system interactions among the simulator parts.

Several tutorials on the combined simulation of power systems and communication networks have been presented in [65, 89, 90].

### 24.4.3 High Performance Computing

***Definition.*** Simulation, optimization, and control of the smart grid can be included in the category of highly intensive computer problems. Smart grid studies require very complex mathematical models due to the increasing use of power electronic based control devices, the implementation of deregulation policies, or the co-simulation of energy and communications systems. All these facts are increasing the computer requirements for power system applications. The concept *high performance computing* (HPC), usually associated to supercomputing, is now used to denote a multidisciplinary field that combines powerful computer architectures (e.g., computer clusters) with powerful computational techniques (e.g., algorithms and software) [91–93]. The availability of affordable medium-size computer clusters and multicore processors, together with the increased complexity of power system studies, is promoting the application of HPC [94], which may involve grid computing [95] or multicore computing.

Early attempts focused in two categories: naturally parallel applications, like Monte Carlo simulation and contingency analysis, and coupled problems, like the simulation of large-scale electromechanical transients [96, 97]. Parallel computing is a type of computation in which calculations are carried out simultaneously: a large problem is divided into smaller ones, which are then solved at the same time. Parallelism has been employed for many years but interest in it has grown lately, and parallel computing is becoming a dominant paradigm in computer architecture, mainly in the form of multicore processors. Parallel computers can be roughly classified according to the level at which the hardware supports parallelism, with multicore computers having multiple processing elements within a single machine, while clusters and grids use multiple computers to work on the same task [98–102]. The availability of desktop computers with parallel computing capabilities raises new challenges in software development, since these capabilities can be exploited for developing tools that could support the smart grid simulation requirements. A first option is to adapt current methods, since some current algorithms can be easily adapted to parallel computing environments; however, the real challenge is the development of new solution methods specially adapted to a parallel processing environment.

***Application to Power System Analysis.*** A first classification of power system computations may distinguish between steady-state and dynamic analysis: steady-state analysis determines a snapshot of the power grid, without considering the transition from a snapshot to the next; dynamic analysis solves the set of differential-algebraic equations that represent the power system to determine its evolving path. Current power system operation is based on steady-state analysis. Central to power grid operations is state estimation. State estimation typically receives telemetered data from the SCADA system every few seconds and extrapolates a full set of conditions based on the grid configuration and a theoretically based power flow model. State estimation provides the current power grid status and drives other key functions such as contingency analysis, economic dispatch, OPF, and AGC.

A computational effort aimed at obtaining adequate simulators must account for the following aspects [91–93]: (1) power grid operation functions are built on complex mathematical algorithms, which require significant time to solve; (2) a low computational efficiency in grid operations can lead to inability to respond to adverse situations; (3) a power grid can become unstable and collapse within seconds; (4) studies are usually conducted for areas within individual utility boundaries and examine an incomplete set of contingencies prioritized from experience as the most significant; (5) dynamic analysis still remains as an offline application, but grid operation margins determined by offline analysis do not always reflect real-time operating conditions. Consequently, power system operations demand the application of HPC technologies to transform grid operations with improved computational efficiency and dynamic analysis capabilities.

Potential areas of HPC application in power systems are dynamic simulation, real-time assessment, planning and optimization, or probabilistic assessment. For a review of HPC development and applications in power system analysis see references [92, 93].

### 24.4.4 Real-Time Simulation

***Introduction.*** Continuous advances of hardware and software have led to the replacement of expensive analog simulators by fully digital real-time simulators (RTSs) [103–107]. Due to advances in digital processors, parallel processing, and communication technology, RTSs are becoming increasingly popular for a variety of applications [108–110]. Field programmable gate arrays (FPGAs) are also making significant inroads into real-time simulators: this technology can offer high-speed high-precision simulations in standalone configurations, and work as accelerator components in PC-cluster simulators [111, 112].

Real-time simulators can be used for testing equipment in a hardware-in-the-loop (HIL) configuration or for rapid control prototyping (RCP), where a model-based controller interacts in real time with the actual hardware; see Fig. 24-12 [35, 103]. To achieve their goals, RTSs must have the ability to simulate all phenomena within a specified time step and maintain its real-time performance so all signals must be *exactly* updated at the specified time step; a failure to update outputs and inputs at the specified time step can cause distortion that will affect the simulation or the performance of the equipment under test. The rest of this subsection summarizes the main features of a real-time simulator and their applications.

***Main Features of Real-Time Simulation.*** The implementation and development of real-time simulation platforms are based on the features detailed below [106].

*Real-Time Constraint.* A real-time computation must be fast enough to keep up with real time; this constraint must be respected at any time step. In general, the differential-algebraic equations (DAEs) of the system under study are discretized and computed along a sequence of equal time-spaced points. Then, all these points must be computed and completed within the specified time step. If a time step is not completed in time, there is an overrun. Overruns create distortion of the waveforms injected into the equipment under test, and can lead to equipment misoperation.

*Bandwidth.* The time step selected for a simulation must be compatible with the frequency range of the phenomena to be simulated. Electric systems are difficult to simulate because their bandwidth is high. Mechanical systems with slow dynamics generally require a simulation time-step between 1 and 10 ms, although a smaller time-step may be required to maintain numerical stability in stiff systems. A common practice is to use a simulation time-step below 50 μs to provide acceptable results for transients up to 2 kHz, a time-step of approximately 10 μs for phenomena with frequency content up to 10 kHz, and time steps shorter than 1 μs for simulating fast-switching power electronic devices used in transmission and distribution systems, or used to interface distributed generators. Power electronic converters with a higher PWM carrier frequency in the range of 10 kHz may require time-steps of less than 0.5 μs.

*Parallel Processing.* Real-time constraints require the use of highly optimized solvers that can take advantage of parallel processing. The application of parallel processing can be facilitated by using distributed-parameters transmission lines to make the admittance matrix block-diagonal (i.e., creating subsystems that can be solved independently from each other). This approach allows dividing the network into subsystems with smaller admittance matrices; however, an effective implementation of this technique assumes that the processing time for data exchange between processors is much smaller than the time used to simulate each subsystem. Some software packages have been specifically developed to take advantage of parallel processing, with great improvements in simulation speed. Linkage is performed to generate different executable modules that can be assigned to different processor cores and increase the simulation speed according to the number of assigned processor cores. Using hundreds of cores to simulate very large grids requires the use of efficient and automatic processor allocation software to facilitate to use of such powerful parallel computers.

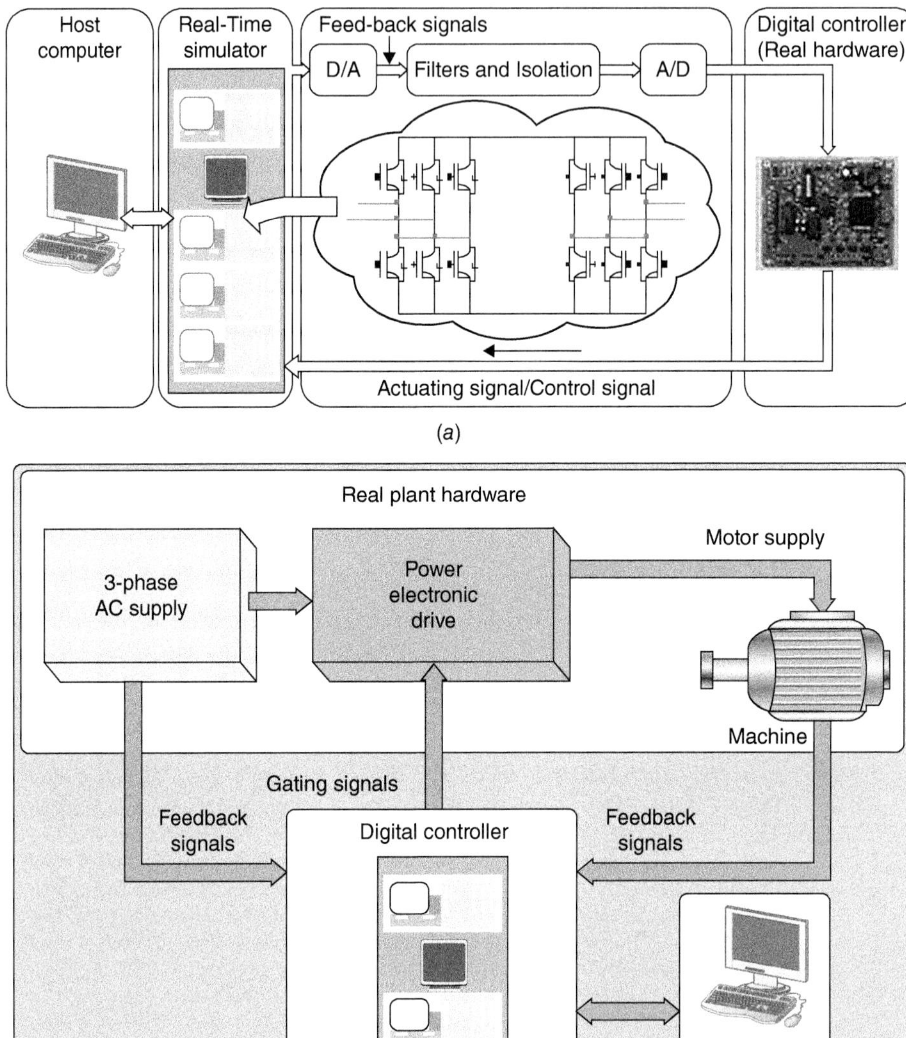


**FIGURE 24-12** Applications of real-time digital simulators. (*a*) Hardware-in-the-loop simulation; (*b*) controller prototyping [103]. (©IEEE 2011.)

*Latency.* In power electronic converter simulation it can be defined as the time elapsed between the semiconductor firing pulses sent by the controller under test and the reception by the controller of voltage and current signals sent back by the simulator. The use of voltage source converters with PWM frequency higher than 10 kHz as well as modular multilevel converters (MMC) with a very large number of levels may require time step values below 1 μs to achieve a total latency below 2 μs. Reaching such a low latency requires the use of FPGA chips. The difficulties in implementing low-latency inter-processor communications capable of fast data transfer without overloading processors is one of the main impediments for the development of very high performance simulators.

*Solvers.* Two important aspects to be implemented in real-time solvers are parallelization and stiffness. Parallel solvers for real-time applications are generally based on the presence of lines or cables that split the system into smaller subsystems, each of which can be simulated within the specified time step using only one processor. If the computational time for one of these subsystems becomes too long because of its size, the common practice is to add artificial delays to obtain a new reduction of size and computational time. An artificial delay is usually implemented with a stub line, which is a line with a length adjusted to obtain a propagation time of one time step. Large capacitors and inductors can also be used to split a large system in several smaller subsystems to take advantage of parallel processing; however, the addition of these artificial delays may be problematic as parasitic Ls and Cs could be large compared to actual component values. Consequently, circuit solvers capable of simulating large circuits in parallel without adding parasitic elements can be very useful to increase simulation speed and accuracy. This is the key feature of the state-space-nodal algorithm [113], a nodal admittance based solver that minimizes the number of nodes (and thus the size of the nodal admittance matrix) to achieve a faster simulation. Another important issue is the stiffness exhibited by the equations of electric circuits: the circuit has a spread of natural frequencies. Non-stiff solvers (e.g., explicit Runge-Kutta) will be unstable due to the high-frequency components of the DAEs unless a very short-time step corresponding to the highest frequency is used; under such circumstance the simulation becomes extremely long. Some stiff solvers are able to cut through the high-frequency components and be less influenced by frequency components higher than sampling frequency. For switching transient studies, the concern is usually for components below 2 kHz, but the equations may have eigenvalues in the MHz range that cannot easily be eliminated. This is the reason why most solvers are based on the A-stable order-2 trapezoidal rule of integration, although it can exhibit numerical oscillations and in many cases users must use snubber circuits across the switches to avoid such oscillations [114]. Other more stable and accurate rules, such as the order-5 L-stable discretization rule, enable the use of larger time step values [106].

*Input/Output Requirements.* Real-time simulators are built around multicore PCs with extensive input/output (I/O) capabilities. FPGA chips mounted on electronic boards can provide a direct and rapid interface and be used to implement models and solvers of moderate complexity, as well as fast control systems and signal processing. I/O requirements for real-time simulation are increasing mainly due to the increasing complexity of current power electronics systems. Real-time simulators must be able to physically interface with communication protocols (e.g., DNP3 protocol used for power system relays and substations) with an appropriate driver. The user can also program the FPGA board to interface to the desired protocol. The simulator must also provide proper signal conditioning for all I/Os such as filtering and isolation.

***Applications of Real-Time Simulators.*** The applications of real-time simulators can be classified into the following categories [106]:

1. *Rapid Control Prototyping (RCP).* A real-time simulator is used to quickly implement a controller prototype that is connected to either the real or a simulated plant.
2. *Hardware-in-the-Loop (HIL).* This approach acts in an opposite manner; its main purpose is to test actual hardware (e.g., a controller) connected to a simulated plant. See also [115].
3. *Software-in-the-Loop (SIL).* It is applied when controller object code can be embedded in the simulator to analyze the global system performance and to perform tests prior to the use of actual controller hardware (HIL).
4. *Power-Hardware-in-the-Loop (PHIL).* This option consists of using an actual power component in the loop with the simulator.
5. *Rapid Simulation (RS).* It takes advantage of parallel processing to accelerate simulation in massive batch run tests (e.g., a Monte-Carlo simulation). RS is very useful for reducing the simulation time of large systems using detailed models.

The list of fields in which RTSs have been applied include among others FACTS and HVDC applications, Monte Carlo simulations, and protection system studies [106]. Research is also done to develop real-time and faster-than-real-time transient stability simulators. Although real-time simulation of power system transient is well established, the need for offline real-time phasor-domain simulation has been recognized for a long time; real-time phasor-type simulators can be useful for testing the functionality of controllers and protective devices in large-scale power systems, for training purposes, as operator training tool in energy management centers, for online prediction of system instability, or for implementing corrective actions to prevent system collapse. For a more detailed list of applications of real-time simulation, see [110].

The future generation of real-time simulation platforms will be capable of simulating long-term phenomena simultaneously with very short transients and fast switching events requiring sub-microseconds time-steps, performing multidomain and multirate simulation (i.e., capable of simulating the dynamic response of all aspects and components affecting the system performance and security assessment), and integrating high-end general purpose processors with reconfigurable processor technologies, such as FPGAs, to achieve the best performance at a low cost [116].

## 24.4.5 Big Data and Analytics

***Introduction.*** Big data is a term coined to denote large and complex data sets for which traditional data processing applications are inadequate. The term often refers to the use of advanced methods to collect, manipulate and extract value from huge collections of data. To achieve such goal it is important to account for the following aspects:

- Big data manipulation includes capture, search, storage, transfer, visualization, analysis, curation, sharing, updating, and information privacy. The list of challenges includes how to characterize the uncertainty in large and complex data sets, reconcile information from multiple sources, and extract information from large volumes of data.
- Data can be complex in many ways (volume, noise, heterogeneity, multisource, collected over a range of temporal and spatial scales). Deducing information from such complexity can be useful to optimize performance, improve design, or create predictive models.
- The data are usually generated as random information from an unknown probability distribution. Understanding data means to deduce the probabilistic model from which they have been generated. The information to be gained is the relationship between the variables of the model.
- Accuracy in big data may lead to more confident decision making, and better decisions can result in greater operational efficiency, as well as reduction of cost and risk. However, there are always sources of error (e.g., noise in the measurements, lossy data compression, mistakes in model assumptions, unknown failures in algorithm executions) in the process from data generation to inference.

For more details on big data and data engineering, see references [117]–[119].

Data is increasing at exponential rates, which requires new framework for modeling uncertainty and predicting the change of the uncertainty. Powerful techniques are needed to efficiently process very large volume of data within limited time. The concept *analytics* is generally used to denote the discovery, interpretation, and communication of meaningful patterns in data. Big data analytics refers to the set of technologies, such as statistics, data mining, machine learning, signal processing, pattern recognition, optimization and visualization methods, which can be used to capture, organize, and analyze massive quantity of information as it is produced, and obtain meaningful insight to provide a better service, at a lesser cost, and taking advantage of multiple data sources. The main challenges faced by analytics software (i.e., massive and complex data sets in a constant state of change) are leading to the development of tools that will be useful for its application in the smart grid.

Two concepts closely related to big data are data mining and machine learning [120–122]. Data mining is the process of discovering patterns in large data sets: it is aimed at extracting information from data and transforms it into knowledge for further use. The goal is the extraction of patterns and knowledge from large amounts of data, not the extraction (mining) of data itself; the actual task of data mining is the analysis of large quantities of data to extract previously unknown patterns such as groups of data records (cluster analysis), unusual records (anomaly detection), and dependencies (association rule mining). These patterns can then be seen as a kind of summary of the input data, and may be used in further analyses. Machine learning explores the study and construction of algorithms that can learn from and make predictions on data: such algorithms build a model from inputs in order to make predictions or decisions expressed as outputs. Machine learning is employed where designing and programming explicit algorithms is infeasible. Within the field of data analytics, machine learning is used to devise complex models and algorithms that lend themselves to prediction. These models allow users to produce reliable decisions and uncover hidden insights through learning from relationships and trends in the data.

***The Role of Big Data in Smart Grid Operation [123–130].*** Utilities are collecting an increasing amount of real-time information from homes, factories, power plants, and transmission/distribution infrastructures. Big data is generated in the grid from various sources [130]: (1) synchrophasor-measured data; (2) condition-based measurements acquired by IEDs; (3) data from smart meters and other customer interaction channels (e.g., voice, Internet, or mobile); (4) data from home automation and intelligent home devices; (5) data from new technologies requiring additional monitoring (e.g., electric vehicles, wind generation, photovoltaic panels, microgrids); (6) energy market pricing and bidding data; (7) offline entered nameplate and maintenance data; (8) data on customers, service connection and assets; and (9) management, control, and maintenance data in the power generation, transmission, and distribution networks acquired by IEDs. Some data widely used in decision making, such as weather and GIS information, are not directly obtained through grid measurements. Utilities can also collect large amount of data from computer simulations based on reliable power system models and powerful software tools. Figure 24-13 shows the main sources of big data in the smart grid.

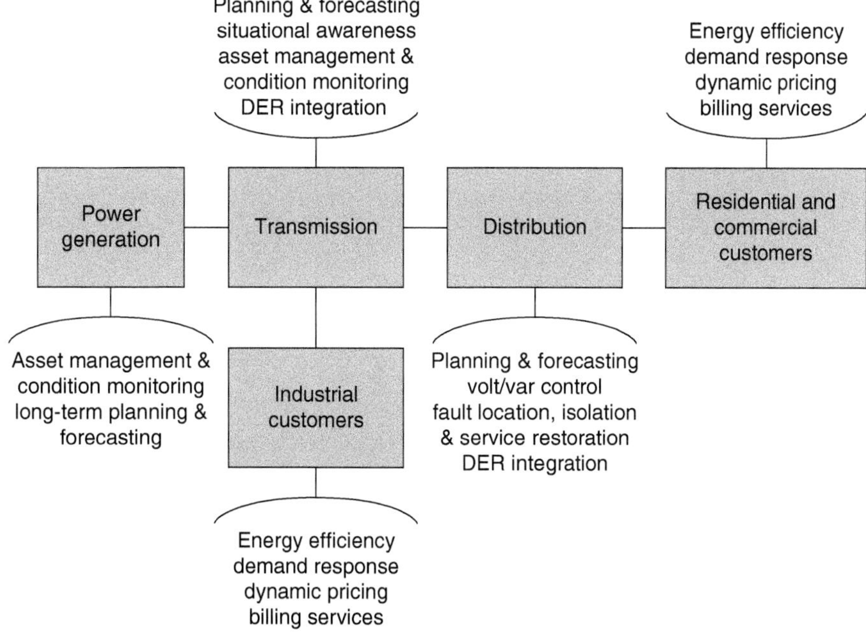


**FIGURE 24-13** Sources of big data in the smart grid.

Key characteristics of data from power grids are volume (e.g., immense amounts of data), velocity (i.e., some information is made available in real or near real time), and variety (i.e., either structured or unstructured data can come from multiple data sources, such as network monitors, meters, or even images and videos).

The data from AMI generates information on how individual customers respond to requests of consumption reduction. Both the supply of power and the demand for it can be managed more efficiently if utilities and consumers get accurate information about power use (e.g., consumers could be guided to move some electricity consumption to off-peak hours and reduce the need for nonrenewable power plants to be activated at peak hours). This information can also be used to improve grid reliability, outage response, reducing the cost of distribution operations, or measuring the impact of a demand response program.

For efficient asset management and outage prevention, real-time operation can be crucial, often requiring fast processing of large volumes of data. To unlock the full value of this information, utilities need complex event-processing engines that are still under development. However, many utilities do not have the capabilities to transmit, store and manipulate such information. In addition, focusing on high volumes of data rather than on the data management can lead to nonadequate decisions; although volume is a significant challenge in managing big data, data information variety and velocity must be focused on as well.

A challenge for utilities is that they run their central control operations on a mix of legacy computer systems, many of which cannot communicate with each other. Past experience, particularly from large blackouts, has shown the need for better situational awareness about network disturbances such as faults and dynamic events, sudden changes of intermittent power from renewable resources such as wind generation, outage management tasks such as fault location and restoration, and monitoring of system operating conditions such as voltage stability. These and other tasks have been handled reasonably well by existing solutions, but improvements in decision making are highly desirable in order to produce more cost-effective and timely decisions, facilitating more efficient and secure grid operation.

Appropriate big data analytics capabilities will deliver value to (1) anticipate failures of distribution assets, such as transformers, to organize maintenance operations; (2) improve balance of generation and demand through better forecasting and demand response management; (3) improve energy planning forecasts to decrease energy costs; (4) improve customer service quality, identifying power cuts more accurately at the moment they occur for faster restoration; (5) enable optimization and control of delocalized generation to facilitate the connection of a large number of electrical vehicles and renewable energy sources; (6) support distribution network operations, such as voltage optimization or outage management; (7) structure energy saving policies; (8) improve customer service with error-free metering and billing services; (9) allow utilities to identify and reduce losses.

The incorporation of big data analytics into utilities is finding some evident barriers, such as the heterogeneity of data types and formats, security, confidentiality and privacy issues [131], the ownership of private data, or the unavailability of adequate tools.

Some emerging technologies, such as the IoT and cloud computing, are closely related to big data in the smart grid [132]. The IoT represents an enormous amount of networking sensors which collect various kinds of data (e.g., environmental, geographical, operation and customer data). These new technologies offer new opportunities, as they can contribute to create the value that big data will generate to smart grid through efficiently integrating, managing and mining large quantities of complex.

## 24.4.6 Cloud Computing

***Definition.*** Cloud computing is a model for enabling ubiquitous, on-demand access to a pool of configurable computing resources (e.g., networks, servers, storage, applications, and services) [133]. Cloud computing provides users with capabilities to process data in third-party data centers. The current availability of high-capacity networks, low-cost computers and storage devices, as well as the widespread adoption of hardware virtualization, service-oriented architecture, and autonomic and utility computing, have led to a growth in cloud computing, which is becoming a highly demanded service due to high computing power, low cost of services, high performance, scalability, accessibility, and availability. For more details on cloud computing see [134–139].

Cloud computing is often compared to the following technologies with which it shares certain aspects [134]:

- *Virtualization* abstracts away the details of physical hardware and provides virtualized resources for high-level applications. Virtualization forms the foundation of cloud computing, as it provides the capability of pooling computing resources from clusters of servers and dynamically assigning or reassigning virtual resources to applications on demand.
- *Grid computing* coordinates networked resources to achieve a common computational objective. Cloud computing and grid computing employ distributed resources to achieve application-level objectives; however, cloud computing takes one step further by leveraging virtualization technologies at multiple levels to realize resource sharing and dynamic resource provisioning.
- *Utility computing* provides resources on demand and charges customers based on usage rather than a flat rate. Cloud computing can be perceived as a realization of utility computing, and adopts a utility-based pricing scheme for economic reasons. With on-demand resource provisioning and utility-based pricing, service providers can maximize resource utilization and minimize their operating costs.
- *Autonomic computing* aims at building computing systems capable of self-management (i.e., they react without human intervention). Its goal is to overcome the management complexity of current computer systems. Although cloud computing exhibits certain autonomic features, such as automatic resource provisioning, its objective is to reduce cost rather than complexity.

***Deployment Models.*** Several models of cloud computing, with variations in physical location and distribution, have been adopted. Basically, cloud computing can be grouped into four subcategories: public, private, community, or hybrid [133, 137]. A public cloud is a cloud made available to the general public in a pay-as-you-go manner while a private cloud is as internal data center not made available to the general public. In most cases, establishing a private cloud means restructuring an existing infrastructure by adding virtualization and cloud-like interfaces to allow users to interact with the local data center while experiencing the same advantages of public clouds, most notably self-service interface, privileged access to virtual servers, and per-usage metering and billing. A community cloud is shared by several organizations with common concerns (e.g., goal, security requirements, policy, compliance considerations) [137]. A hybrid cloud takes shape when a private cloud is supplemented with computing capacity from public clouds.

***Types of Clouds.*** From the perspective of service model, cloud computing is divided into three classes, depending on the abstraction level of the capability provided and the service model of providers [133–135] (see Fig. 24-14):

- *Infrastructure as a Service (IaaS)*. This option offers virtualized resources (i.e., computation, storage, and communication) on demand. A cloud infrastructure provisions servers running several choices of operating systems and a customized software stack. Users are given privileges to perform numerous activities to the server (e.g., starting and stopping it, customizing it by installing software packages, attaching virtual disks to it, and configuring access permissions and firewalls rules).
- *Platform as a Service (PaaS)*. This option offers a higher level of abstraction to make a cloud easily programmable: an environment on which developers create and deploy applications and do not necessarily need to know how many processors or how much memory applications will be using. In addition, multiple programming models and specialized services (e.g., data access, authentication, and payments) are offered as building blocks to new applications.
- *Software as a Service (SaaS)*. Services provided by this option can be accessed by end users through Web portals. Consumers are increasingly shifting from locally installed computer programs to

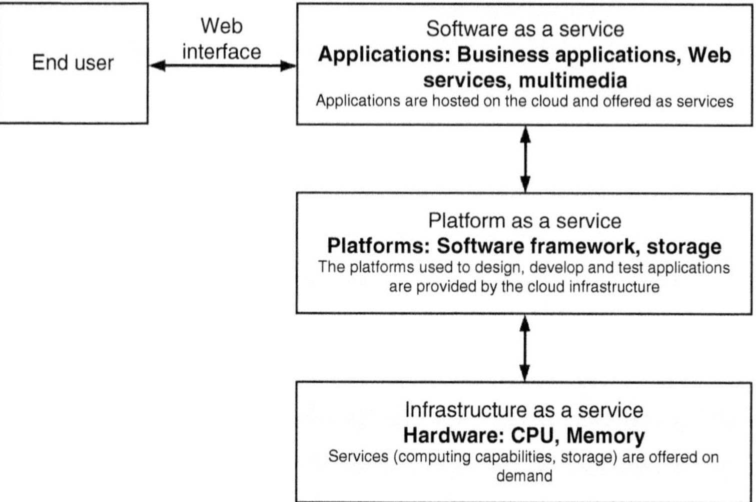


**FIGURE 24-14** Cloud computing classes.

online software services that offer the same functionally; traditional desktop applications such as word processing and spreadsheet can now be accessed as a service in the Web. This model alleviates the burden of software maintenance for customers and simplifies development and testing for providers.

***A Practical Experience.*** Utilities will have to provide new business capabilities in the face of rapidly changing technologies. Cloud computing solutions promise great technological capabilities and benefits; however, not much experience is yet available on its application by utilities. The experience of ISO-NE presented in [140] is a good example from which some important conclusions were derived.

The list of challenges associated with migrating power system simulations to the cloud-computing platform includes topics related to development, software license, cost management, and security. The main findings provided in reference [140] can be summarized as follows:

- A platform developed on the basis of an open architecture should not only accommodate different resource management and job balancing tools, but also support different power system simulation programs.
- The cloud-computing platform can comply with current cybersecurity and data privacy requirements, comparable to existing on-premise infrastructures.
- Deploying power system applications to the cloud can result in significant cost savings and performance improvement without the compromise of system security and quality of service.
- Utilities need to develop a strong business case when making the decisions on which data and applications will be deployed in the cloud so as to take advantage of all the benefits.
- The way cloud computing functions is significantly different from the traditional on-premise infrastructure, so understanding all the implications is important to equip utilities with the information needed to make the appropriate assessment.

Cloud computing can achieve significant cost saving by avoiding unnecessary over-expenditure on IT infrastructure and eliminating capital investment expenditures. Public cloud computing can be a robust solution far beyond the economic reach of many organizations. The high resiliency and availability of services provided by cloud computing are critical to utilities as their daily operations become increasingly dependent on IT services and makes cloud computing an attractive solution to meet the growing internal computational needs; in addition, migrating power system applications to public cloud services can comply with current cybersecurity and data privacy requirements.

The potential cloud computing applications in power system analysis and operations have been analyzed in several works; see [141–145].

## 24.5 THE COMMON INFORMATION MODEL

### 24.5.1 Introduction

Computer engineering and operation applications require digitally stored data and their exchange between these applications. Large-scale EMSs and asset-management systems use database schemes for defining the structure of the data storage, often custom-written to reflect the specific requirement of the operator. They communicate with each other, generally using a vendor-custom format. In the past this required the user to purchase each piece of software from the same vendor to ensure compatibility. Offline applications for performing calculations use proprietary formats that represent the data required by each application. When subsequent versions of the program require additional details, the file format is changed, resulting in multiple formats for a single application. Changing the file format for each new software version has been common practice within the software industry, although its impact has been mitigated since the new software contained import facilities to convert previous versions of the file format into the new format.

In addition, power companies have long needed to share system modeling information with one another in order to construct simulation environments for power system economics and security analysis. The major motivation has been to support system planning functions including transmission planning, maintenance scheduling, and operations planning. Typically, a utility must also model significant portions of neighboring systems, even for proper analysis. Finally, the deregulation of the power industry has resulted in multiple utilities running software from a number of different vendors and having to exchange large data sets on a regular basis. ISOs must maintain operational power system models that span multiple service areas for reliable operation of the transmission grid; to accomplish this, they must regularly exchange detailed models with their member utilities.

These situations may become unmanageable when the software bridges are built by different vendors using diverse technologies and the number of interfaces is too large because of the so many areas of integration. The situation is further aggravated when systems are upgraded or replaced and/or when processes change. The large number of proprietary formats used by these applications requires a myriad of translators to import and export the data between multiple systems. As a result of this situation, new vendors face high market entry barriers, because it is necessary to interface their products with large and customized data systems at each potential utility customer.

A way to reduce the effort and complexity of this task is for each system to map its data elements to a common data model. Instead of mapping the data elements from each system directly to the data elements of each other connected system, the mapping is performed once from each system to this common model. In cases where both the number of systems and the number of interconnects are large, this method can significantly reduce the number of mappings required. Consider the example shown in Fig. 24-15 with six different systems. Within this configuration, the SCADA is used to monitor the current status of the power system; data about the system topology is stored within a GIS and must be transferred into the SCADA, which is connected to the EMS in order to track and control the power flow. Since there are six foreign formats to support, then six interface modules must be developed; however, without a common object model, each pair-wise interface might have to be implemented separately, and that would mean up to 15 different modules to develop. Using a common language, only six adaptors are required to connect the six systems. The number of systems within an electric utility is higher than six, although there can be systems from the same vendor having the same formats and not every system is linked to all others.

Another important issue that requires using commonly accepted and compatible data exchange formats is the integration of renewable energy sources (RES), which is affecting the electric grid

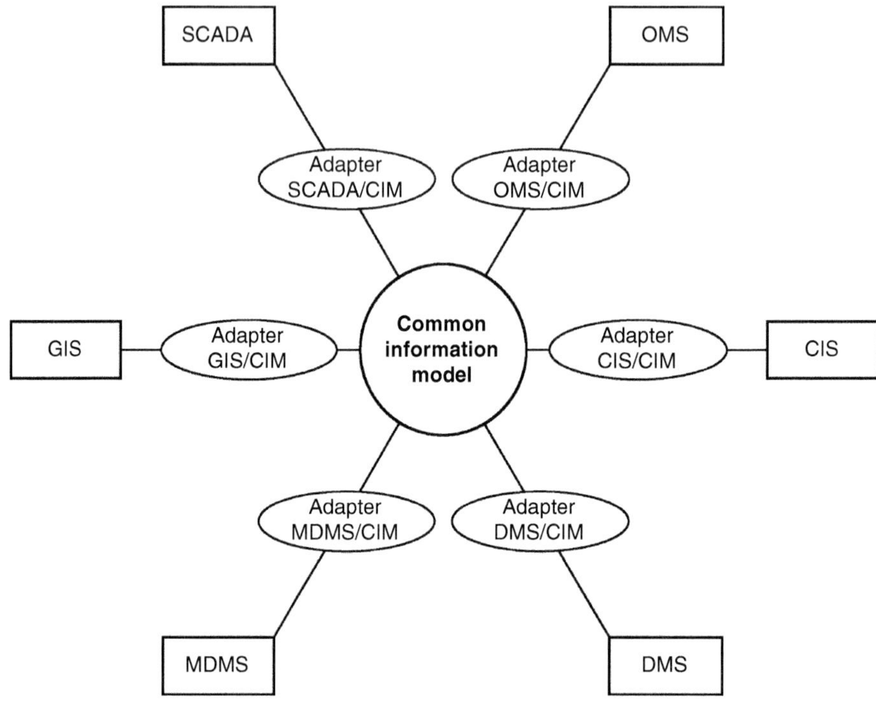


**FIGURE 24-15** Adapters between CIM and different systems. (SCADA: supervisory control and data acquisition; OMS: outage management system; GIS: geographic information system; CIS: customer information system; MDMS: meter data management system; DMS: distribution management system.)

infrastructure and the functioning of the electricity market, and requires for their intermittency to be balanced. To achieve this, an efficient information exchange is necessary between an increasing number of companies (i.e., generators, TSOs, DSOs). Such information exchanges have become indispensable in network planning (e.g., network development, interconnection development to tackle congestions), power system operation (e.g., real-time information, balancing control), and market operation (e.g., generation schedules, trades, balancing resource management).

A technique for further simplifying this task is to utilize an industry standard information model, so that the common model does not have to be created, it is instead adopted. A common language used within EMSs is the standard *Common Information Model* (CIM) [146, 147].

### 24.5.2 Background

***CIM History.*** According to [148], the beginning of CIM can be traced back to the efforts made by the IEEE PES Energy Management System Architecture Task Force, whose purpose was to think about and make recommendations on future EMS architectures. The task force met from 1982 to 1989, when it was dismantled after writing a final report.

The next efforts were headed by EPRI in early 1990s. To integrate different systems from different vendors, EPRI sponsored the Control Center Application Programming Interface (CCAPI), an attempt to produce a set of common Application Programming Interfaces (APIs) that could be provided and used by vendors to communicate information between applications, potentially provided and used by different vendors. Over time, it was evident that having a common definition of the data between applications was the fundamental problem that needed to be solved. It was also becoming apparent that APIs are typically tied to specific technologies. Thus, it would be very difficult to get vendors and users who had made significant investments in specific technologies to agree upon a set of APIs. For these reasons, the focus shifted toward defining what is called an information model.

As the research progressed and started to produce a concrete information model, the next step was to promote its use among the vendors. One way to do this was to turn the information model into an international standard since vendors would build products using the standard. For this reason, the development of CIM standards was started by the IEC. The committee chartered with turning the information model into a standard was Technical Committee 57 (TC57): *Power Systems Management and Associated Information Exchange.*

A new working group, WG13, was appointed to start this work primarily in the transmission operations domain, although the application to distribution operations and market communications quickly followed with the establishment of WG14, Distribution Management Systems Interfaces, and WG16, Market Communications. Other working groups followed later to incorporate the CIM semantic model into their work as well. The TC57 has expanded to include many working groups that are based on the CIM. Today, the CIM is composed of a series of standards comprising a single, unified semantic model and multiple profiles developed in these different working groups defining specific information exchanges between applications/systems/devices in support of business processes in transmission, distribution, and markets. See Secs. 24.5.5 and 24.5.6.

The two most relevant CIM standards are the IEC 61970 and IEC 61968 series [149–152]. The IEC 61970-301 standard is the core of CIM and includes information associated with control center applications, such as EMS, SCADA, and network planning. The IEC 61968 standard extends CIM to include distribution management system (DMS) functions. These two standards are known as the CIM for power systems and have two primary uses: to facilitate the exchange of power system network data between companies and allow the exchange of data between applications within a company.

Because IEC TC 57 is focused on defining and revising the official standard and is formally part of the IEC, another forum, the CIM Users Group (CIMug), was created in 2005 to provide a venue for utilities, vendors, consultants, and integrators who use the CIM. The CIMug also provides a channel to the standard organization for users to make suggestions for changes and extensions to the current standard.

In 2006, EPRI initiated the CIM for planning models with the objective of developing a common power system network model that both operations and planning groups can use as a basis for information exchange. EPRI launched in 2008 the CIM for dynamic models [153].

In 2008, the European Union for the Coordination of Transmission of Electricity (UCTE) decided to use the CIM-based data exchange format and joined the international efforts of EPRI projects and CIMug. These activities were expanded with the creation of the European Network of Transmission System Operators for Electricity (ENTSO-E) in 2009. ENTSO-E has established liaisons with IEC TC 57/WG13 (the working group dealing with CIM for transmission) and IEC TC57/WG16 (the working group responsible for energy markets).

The establishment of an electricity energy market requires information to be exchanged among utilities but also from a number of participants from various sectors (e.g., traders, power exchanges, aggregators of information, meter data collectors). A harmonized approach is necessary to define data interchanges to reduce the complexity and cost of IT systems while ensuring that ongoing quality and usability is maintained. With this objective in mind, the IEC 62325 series of standards was developed to facilitate efficient interactions among market participants, market operator, and system operators to provide a unified approach to conducting market operations [154]. Two market models are being developed: one for the North American–style market (nodal market) and the other one for the European-style market (zonal market). The initial focus was on European-style markets, which resulted in the series of IEC 62325-451-n for dedicated business processes such as acknowledgment, scheduling transmission capacity allocation and nomination, settlement and reconciliation, status request and problem statement, and transparency data publication [155]. IEC 62325-450 provides the modeling rules necessary to ensure that contextual models derived from the CIM are in conformity with the CIM model, ensures modeling consistency and avoids ambiguity between objects by providing a clear understanding on what they are based within the CIM [156]. The work for North American–style markets will be in the form of a series of standards dealing with day ahead and intraday markets. These standards will describe the exchange of supplemental market definition data and run-time market data (bids, offers, schedules of award, and locational marginal prices).

For a good introduction to the CIM foundations, see the EPRI CIM Primer [157]. A tutorial introduction to CIM was provided in a special issue of the IEEE Power & Energy Magazine [158]. The articles included in the issue aim to explain the high-level challenges of power systems and related use cases, demonstrate where the CIM can be applied to help resolve problems that power system engineers are facing, and bring forward current information on approaches that can resolve important issues in a key domain such as data exchanges in a complex and increasingly challenging environment.

### 24.5.3 Main CIM Features

***The Unified Modeling Language.*** An information model is an abstract and formal representation of objects, their attributes, their associations to other objects, and the behavior and operations that can be performed on them. The primary purpose of the information model is to formally describe a domain without constraining how that description is implemented in a software application. The information model is formally described in a particular, well-defined fashion. The Unified Modeling Language (UML) was chosen as the language to represent the CIM semantics used in electric utilities: the UML is a formal descriptive language that unifies several methodologies commonly used by software engineers to model systems [159, 160]. A UML model can be directly used in electronic form by software tools to generate artifacts (e.g., power system network models) that vendors can download and incorporate in their products. A complete model includes not only diagrams but also supporting written documents; UML diagrams are used to provide functional requirements, static structure and dynamic behavior of a model. At implementation level, the IEC has proposed the simplified CIM XML exchange format [161]. The XML is a markup language developed by the World Wide Web Consortium (W3C). A markup language is a way to encode both information and meta-information so that a clear, unambiguous communication of the information can be exchanged between computer applications and systems. HyperText Markup Language (HTML) is an example of well-known markup language. A pragmatic reason for using XML is the availability of an extensive technology infrastructure covering functions such as transformation, presentation, query, schema and exchange protocols. The adoption of XML has other benefits; for instance, it is compatible to different industry standards.

***Resource Description Framework.*** The IEC proposes a resource description framework (RDF) schema as a proper way to exchange data in a common format. RDF is a method of defining information models and is based upon the idea of making statements in a subject-predicate-object expression. That is, each expression in RDF terminology has a subject, defined by naming a *resource* (i.e., anything that can be identified), an object, which denotes traits or attributes associated with the subject, and a predicate, which expresses the relationship between the subject and the object. The object in a statement can be a literal, such as a string, or another resource. The CIM XML language is an application of RDF to CIM. It is defined by a confluence of the CIM, RDF schema, and RDF syntax specifications. Since RDF is general enough to describe UML concepts, the conversion is straightforward. The standard RDF schema vocabulary is extended to represent additional UML association concepts such as inverse roles and multiplicity. This enables the CIM to be translated from UML to RDF schema with sufficient fidelity. The final result is a concrete schema, encoded in RDF syntax and employing RDF concepts that software tools can readily interpret. CIM RDF is both machine and human readable and self-describing. Standard mechanisms to convert the UML model into an RDF scheme are defined in IEC 61970-501 [150].

### 24.5.4 CIM Packages and Profiles

The basic CIM data model consists of several packages that describe a unique subdomain of the information model. Figure 24-16 shows an incomplete list of packages. Each package contains a set of classes along with their inheritance structure, their attributes, and their associations. For instance,

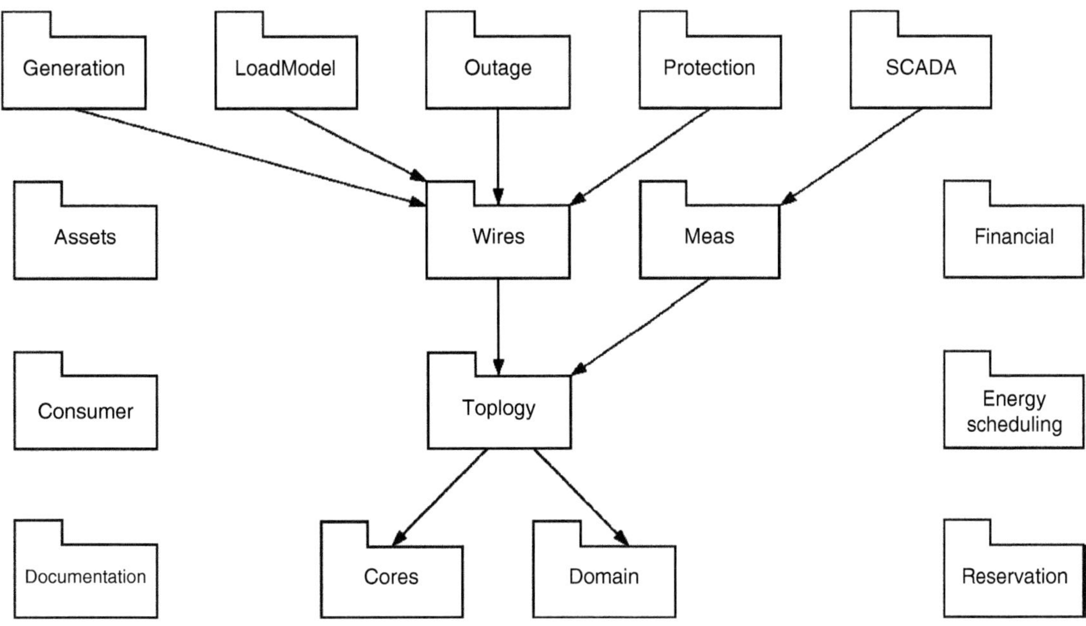


**FIGURE 24-16** CIM packages.

the *core* package contains definitions of classes that are parent classes to many of the more specific classes in other packages, including classes defined in both IEC 61970 and 61968 series. The *wires* package is a relatively large package that can be broken down into three parts (lines, transformers, and switches), each one having classes that define the corresponding devices. The *EnergyScheduling* package is used in applications that support the exchange of electricity, provides the capability to schedule and account for transactions for the exchange of electricity between energy utility companies, and includes transactions for power which is generated, consumed, lost, passed through, sold and purchased. The *assets* package defines the attributes and relationships that are commonly part of an asset management program; it includes a large number of attributes, such as manufacturer, serial number, manufacture date, or purchase date.

A *profile* is defined as the set of classes, attributes, and relationships that are a subset of the classes, attributes, and relationships found in another schema. Thus, a given profile is a subset of a parent schema. Profiles are used to define domain models and can be realized in a variety of forms (e.g., text documents or HTML). Examples of CIM profiles are the *common power system model* (CPSM), which defines the subset of classes, attributes, and associations necessary to execute the EMS applications of power flow and state estimation; the *common distribution power System Model* (CDPSM), which defines the subset of classes, attributes, and associations necessary to execute several DMS applications; the *planning profile*, which is required for defining models required in performing transmission planning power flow and other related applications; or the *dynamics profile*, which is required to handle dynamic stability packages.

## 24.5.5 CIM Interoperability Tests

Interoperability (IOP) tests provide a mechanism to ensure compliance of applications with the IEC standards and integration architectures. The main goal of an IOP test is to check whether applications can exchange models using the CIM standards. The general objectives are to [162]: (1) demonstrate interoperability between different products based on the latest CIM version; (2) verify compliance with the CIM for those CIM classes/attributes involved in the information exchanges supported by the tests; (3) demonstrate the exchange of power system models using the CIM with an RDF Schema and XML representation of the model data; (4) check that the profile documentation is correct, complete and ready to be implemented by the tests; (5) validate the correctness and completeness of IEC draft standards; and (6) validate the test models, the UML, and the RDF files created for the test. A result of IOP tests is the development of tools that can be used to validate compliance with the standards and provide testing artifacts to enable IOP testing [162].

ENTSO-E runs tests on a yearly basis to demonstrate the interoperability of CIM standards and to support CIM development. These tests are tailored to ensure adequate representation of TSO requirements and designed to allow vendors to verify the correctness of the interpretation of the CIM standards [163]. The last ENTSO-E IOP test, held in July 2016, was aimed at (1) validating the proposal for CIM extensions to be included in the next Common Grid Model Exchange Standard (CGMES) version of IEC standards that will be used to support the data exchanges required by the European legislation; (2) demonstrating exchange of dynamic models following the implementation effort performed by a group of vendors; (3) demonstrating exchange based on efficient XML interexchange defined by the W3C. ENTSO-E has also conducted IOP tests on the IEC 62325 series for the electricity market; these tests aim to demonstrate that the IEC 62325-451-n series satisfy the information requirements for the corresponding business processes in the European-style market profile [164].

The results of IOP tests have proved that the work carried out on the IEC standards is in line with the business requirements, there are no major deficiencies in the standards, and the standards include all the capabilities needed to support the exchanges already implemented.

### 24.5.6 CIM Harmonization

The CIM standards are continuously revised to correct errors, incorporate improvements, include extensions desirable to standardize, and meet the changing requirements for data exchange, which are increasing in both frequency and type, with higher RES integration and the introduction of smart grids. To unambiguously specify the CIM XML version, a uniform resource identifier (URI) is used for the CIM namespace. A CIM XML document declares the CIM namespace with the version being used in a statement. Software reading the CIM XML document can then detect the CIM XML version and respond accordingly. CIM XML documents can be extended to model vendor or utility special needs. The CIM RDF schema can be extended with new classes and attributes by providing a separate namespace. Because a separate namespace is used, the customized CIM XML documents can clearly delineate what is standard and what is custom. Several different custom extensions can exist and be clearly identified within the same XML document. When these customized documents are imported to information systems that know nothing about the extensions, the elements with the unknown tags can be simply ignored.

Although the IEC 61970 and IEC 61850 standards were developed within the IEC TC57, they were independently developed and historically have not shared a common model or modeling approach. A project to harmonize both standards, following the approach for interoperability recommended by the NIST Smart Grid Roadmap, suggests the development of a common semantic model to be used for the smart grid [165]. The goal is define a IEC 61850 profile that could be used to generate IEC 61850-6 SCL files in a fashion already used to create CIM XML files and message payloads for the IEC 61968/70 CIM standards. A unified 61850/CIM model offers the possibility to create an environment that can be used to automatically configure SCADA communication interfaces and to exchange the communication related information to other systems.

IEC CIM is a sophisticated approach to data exchange still under development. A significant activity on CIM applications has been carried out during the last years; for more details about the scope and the applications of this language; see [166–175].

## 24.6 THE IoT

### 24.6.1 Introduction

The Internet of Things (IoT) can be defined as a network of physical objects (e.g., home appliances, electrical vehicles, buildings) connected to the Internet framework, able to identify themselves to other devices, and enabled to collect and exchange data. The IoT can be seen as an extension of Internet and communication infrastructures in which devices are embedded with both intelligence and sensors that provide them with decision-making capabilities. Each object, uniquely identifiable through its embedded computing system, is able to interoperate within the Internet infrastructure: objects can sense, analyze and process data, exchange information with other objects to achieve real-time status of the physical world, and take intelligent decisions.

The IoT is enabled by the latest developments in radio frequency identification (RFID), smart sensors, communication technologies and Internet protocols, and will enable new applications by connecting physical objects together in support of intelligent decision making. For a thorough overview of the IoT, see [176–181].

The IoT can also be seen as a step of the smart grid development: the electric system, protected by means of relays, breakers, and disconnectors, uses a centralized management (i.e., SCADA and EMSs) that allows control centers to remotely sense and reconfigure grids; a further step is an interaction between devices and the electrical grid, in which power plants maintain constant communication with consumers, and connected objects (e.g., home appliances) can self-adjust to consume less power or even turn off. From the data collected from all the devices, an accurate energy demand can be estimated, so generation can be automatically shifted according to a predicted demand variation. The interconnection of devices can promote automation in nearly all fields and enable advanced smart grid applications. Since IoT is expected to generate large amounts of data from diverse locations, an aggregation of the data will be required, as well as an increasing need to store and process such data.

### 24.6.2 Drivers, Barriers, and Challenges

***Drivers.*** In addition to the advances in sensors, communications, Internet infrastructures, and computing capabilities, the IoT implementation and development will also be supported by other drivers [182]:

1. The price of solar photovoltaic panels is declining, and as the adoption of solar increases, there will be an adoption and usage of devices and applications that should allow users to monitor their system.
2. The increasing availability of onsite backup power will make solar useful at any time, while the expansion of ancillary services in energy markets will enable cost-effective energy storage solutions (e.g., battery-based technologies).
3. It is expected a rise in demand response (DR) programs that could be connected via the cloud. Newly designed programs will use DR bidding and market-based measurement and verification to measure success. The market will see DR direct participation in energy markets.
4. Consumers will utilize smart meters for measurement and verification of participation in demand response.
5. Utilities are already increasing the use of data analytics to optimize grid operations, including outage management and volt/var control; massive data collected from the deployment of the IoT will require new and more powerful analytics tools, and at the same time will promote the development of these tools.

***Barriers.*** The connection of billions of devices through the Internet infrastructure will raise some concerns about the generated data [183]:

- *Privacy and Confidentiality.* The data gathered by billions of devices create concerns in privacy, confidentiality, and integrity. Providers of IoT-enabled services will have to develop new techniques to collect and use data, provide transparency into what data are used and how they are used, and ensure that the data are appropriately protected.
- *Security.* Organizations that gather data from devices will need to deal with new categories of risk that the IoT can introduce: extending IT systems to new devices creates many more opportunities for potential breaches. If the IoT is used to control physical assets, the consequences associated with a breach in security extend beyond the unauthorized release of information since they could potentially cause physical harm.
- *Intellectual Property.* A common understanding of ownership rights to data produced by various connected devices will be required to unlock the full potential of IoT. Clarification about the rights to the data from a sensor manufactured by one company and part of a solution deployed by another in a setting owned by a third party.

***Challenges.*** The deployment of the IoT is raising the need for

- *Standards.* The implementation of the IoT should be based on standards that facilitated and simplified the tasks of programmers and service providers. Different groups worldwide are presently involved in the development of IoT protocols. Reference [181] classifies the IoT protocols into four categories: application protocols, service discovery protocols, infrastructure protocols, other influential protocols. However, not all of these protocols have to be bundled together to deliver a given IoT application, and some of them may not be required to be supported in an application. As with the smart grid, the variety of standards and technologies required for the deployment of the IoT and the way IoT objects should interoperate are some of the main challenges that can slow down the development of IoT applications.
- *Software Tools.* The connection of billions of devices to power and Internet/communication networks will demand for specialized tools with capabilities not available in any present software package. The simulation of an IoT-type system will require not only complex co-simulation capabilities for simultaneously reproduce power systems, Internet protocols, and decision-making algorithms, but also skills for analyzing cybersecurity issues and capabilities for simulating very large systems.

### 24.6.3 IoT Applications in the Smart Grid

A smart grid is an energy delivery system that moves from a centrally controlled system to a more consumer driven one, relying on bidirectional communications to constantly adapt and tune the delivery of energy: a smart grid includes a broad range of sophisticated sensors that are constantly assessing the state of the grid, the availability of power flowing into the grid, and the demand on the grid. The sensors must collect a vast amount of information that can be used to estimate the changes required to optimize energy delivery. Through adaptation and tuning, the energy can be delivered at the right time and at the best price. Since the IoT brings together traditionally separate systems (e.g., lightning, electric vehicle, security, and IT) into one platform that increases energy management requirements, it can be assumed the IoT will significantly impact the smart grid performance. Although the design of the IoT framework for energy management capabilities requires a significant commitment in resources, a smart grid in which all devices are in communication can provide improved energy efficiency. IoT technology can

- Support the smart grid by improving the grid reliability, providing users with more intelligent and diversified services, and improving the power supply service (including user energy information, two-way interactive services, home energy management services, intelligent home control, or distributed energy electric car charging for electric auxiliary management).
- Help utilities to optimize and control assets, increase safety, control the grid, and keep the lights on.

Two already-implemented examples of IoT usage in the electric power industry are SCADA and AMI [184]. PMUs collect a vast amount of information on the status of the grid and smart meters can allow energy pricing to be based on the cost to deliver at a very granular level, offering a wider range of options on how and when energy should be consumed.

Some IoT functionalities will enable reaching balanced resources that previously were only possible through less accurate central control: (1) with a unique identification, devices will not only provide their information for monitoring but will be capable of dynamically discovering nearby devices to collaborate with, so peer-to-peer interactions will emerge with advanced capabilities to interact with networked-based services hosted in enterprise systems or somewhere on the Internet; (2) devices will be able to enhance their own functionality in dynamic ways through emerging services; available price signals will also be used to affect device behavior; and (3) consumers can behave as an energy storage system by agreeing to control mechanisms that run home appliances when the system demand is low (e.g., in the middle of the night).

The interaction between the IoT and the smart grid as well as the impact that the IoT could have on the smart grid has been the subject of a number of works; see [185–189].

## 24.7 CYBERSECURITY

### 24.7.1 Introduction

Traditionally, the concerns of electric utilities about the security of their assets have centered on protecting them from physical threats (i.e., electromagnetic pulse, geomagnetic storm, inclement weather). However, the nature and the magnitude of the threats have now changed because the equipment to monitor and control devices is frequently connected by communication lines to WANs potentially accessible by the general public; as a consequence, the potential threat can be very far from the target and impact multiple devices simultaneously. The smart grid is introducing new security risks related to communications, system automation, and data collection technologies. In addition, the smart grid has to support legacy systems, which typically do not implement security features that newer systems will have, and new smart grid technologies exhibit security vulnerabilities that can be exploited by intruders. Approaches to secure these new technologies and to protect privacy must be designed and implemented in the design stage. The smart grid security must be addressed in a manner that balances protection with the need to provide affordable energy to consumers [190].

Cybersecurity is the term used for security with respect to the area of computers. Computers in power industry can be subject to malicious attacks and inadvertent errors, and vulnerable not only to direct and focused attacks, but also to indirect attacks (e.g., from viruses). Impacts by cyber incidents in electric power can be ranged from trivial to significant, and include, equipment damage, environmental damage, and even deaths. For a summary of the difference between traditional and contemporary threats to utility substation assets, see [191]. In a broad sense, cybersecurity for the power industry must cover all issues involving automation and communications that affect the operation of electric power systems, and the functioning of the utilities that manage them, as well as the business processes that support the customer base [192, 193].

As the grid is increasingly networked, it is becoming increasingly vulnerable to intrusions. With the increased use of smart grid and IoT technologies, there will be a greater number of access points, requiring increased security awareness by utilities, device manufacturers, and the general public. The interconnectivity between grid components and the increasing reliance on its cyber system makes the grid more vulnerable to a multitude of cyber-physical attacks (CPAs) that aim at compromising its functionalities. The security of the smart grid is one of the most critical technical challenges facing its deployment.

Since the attacks against computers are increasing in frequency, intensity, and variety and not all cyber intrusions and/or cyber attacks are the same, understanding the purpose of the attack can be crucial for the security response. Many of the viruses that have caused substantial damage to critical facilitates across the world were specifically designed to compromise SCADA systems (e.g., the Stuxnet virus intercepts commands sent to high-speed frequency converters, which are devices that control functions, such as motors). A list of recent attacks and their targets is presented in [190].

The objectives of the smart grid cybersecurity are different from most of the other industries: any security countermeasure implemented in the smart grid should not impede power availability, which is the most important security objective. The electricity in the power grid not only needs to always be available, but it must also have quality, and this quality depends on the quality of the state estimation. The integrity of data collected by various sensors and agents, used to monitor the current state of the electrical power system, is very important; inadvertent or unauthorized modification of the data can cause failures or damage in the power system. The final security objective is confidentiality. Although the loss of data confidentiality has a lower risk than loss of availability or integrity, there are certain areas where confidentiality is more important. The privacy of customer information, general corporation information, and electric market information are some examples.

For an introduction to cybersecurity of the smart grid, see the special issue of the IEEE Power & Energy Magazine [194]. See also [195–202]. For an interesting discussion about misconceptions related to cybersecurity threats; see also [203].

### 24.7.2 Cybersecurity Threats and Vulnerabilities

***Threats.*** Although most blackouts have been caused by a combination of natural events and human errors, it is, however, not difficult to imagine scenarios in which a deliberate action could cause a power outage. In addition, power system cybersecurity can be attacked from an unintentional threat; for instance, substation control can be impacted by inappropriate testing or procedures [191]. In general, threats are directed toward information held by the utility, but the target of the threat may be an entity other than the utility, such as a customer or a supplier. Although a physical attack can damage or destroy a single device, it can compromise a highly networked device and exploit electronic vulnerabilities to cause outages in one or several remote devices. Attackers cannot only exploit electronic vulnerabilities from publicly accessible networks, but they may also launch an automated attack on several target sites from a single location geographically separated from the target. If a physical attack can be effective on a limited basis, an automated electronic attack can multiply the force of the attack because it enables just one or few individuals to simultaneously attack many targets. Utilities can properly design SCADA systems against physical threats such as radio-frequency interference, electromagnetic pulse, and voltage transients. However, there are a variety of threats that can compromise securely installed SCADA systems: unauthorized individuals who gained access to the facility could install a key logger to expose the system, and a cyber attack could be used to damage the entire network. Cyber threats include denial of service (DoS), data interception, unauthorized user access, data alteration, and/or data-retransmission. In addition, these threats could interrupt the ability for other critical infrastructures to function. Securing against such challenges not only requires the development of improved security protocols for utility employees, but also improved behavioral analysis aimed at preventing or detecting unauthorized access or software installation, inadvertent or malicious.

***Vulnerabilities.*** Traditional IT business systems have been the object of a wide variety of cyber attacks, which include exploitation of programming errors in operating systems and application software, cracking user passwords, use of ports left open to attack or use improperly configured firewalls. In addition to these vulnerabilities, a highly networked power system may have a number of special cyber vulnerabilities (e.g., insecure communications media, open protocols, lack of authentication, or lack of centralized system administration) [191]. Vulnerabilities related to protection can be organized into several groups and categorized by source and attack methods, see [204, 205].

### 24.7.3 Cybersecurity Measures

***Goals and Objectives.*** Cybersecurity measures must be aimed at achieving three primary goals [194, 206]: (1) *confidentiality* to ensure that only authorized entities have access to sensitive information (e.g., electricity market data and transaction information should only be accessible to authorized market agents); (2) *integrity* to ensure that any unauthorized modifications to data and information are detected (e.g., an adversary should not be able to modify sensor data without detection); (3) *availability* to ensure that critical systems and information must be available when needed (e.g., communication networks supporting WAMSs must be available to deliver synchrophasor measurements even in the presence of malicious activity such as a DOS attack). For an infrastructure such as the smart grid, availability and integrity are considered to be more important than confidentiality.

Other security properties include nonrepudiation and privacy [194]: nonrepudiation ensures that a particular message was actually sent as the receiving entity claims, while privacy refers to adequate protection of personally identifiable information accessible only to authorized entities (e.g., consumer energy consumption data need to be kept private).

A common approach to achieving these properties is to design, develop, and deploy technologies for protection, detection, and response. Protection systems devise security components such as key management, authentication and authorization, and perimeter defense that help ensure the above goals against a range of attacks (e.g., encryption tools help provide confidentiality, cryptographic message authentication tools help provide integrity, and redundancy helps provide availability). Detection tools observe network and system behavior to identify malicious activities and attacks (e.g., intrusion-detection systems may look for malware signatures on the network). Response tools enable administrators to deal with detected attacks (e.g., they may allow dynamic changes in firewall policies in order to limit information flow to and from adversaries to contain an attack).

There are regulatory mandates (e.g., NERC CIP 002-009 [207]), as well as business reasons for establishing a cybersecurity program, which should be correlated with a program for ensuring the physical security of the utility assets. The main components of a cybersecurity strategy can be classified into four types of actions: prevention, detection, response, and recovery [208]. The following paragraphs summarize the actions that may be considered for each activity.

***Preventing Cyber Intrusion.*** The power system operations environment is very different from the Internet environment and poses many security challenges that are different from most other industries. Security services and technologies developed primarily for industries do not have many of the requirements that are needed by power system operations. For instance, many communication channels used in the power industry are narrowband, thus not permitting some of the overhead needed for certain security measures, such as encryption and key exchanges. However, IEC 61850 is designed to operate efficiently over 10 Mbps switched, 100 Mbps shared, or switched Ethernet, so a prerequisite for shifting IEC 61850 is the installation of high-speed communications. The impact of implementing cybersecurity requirements related to the use of IEC 61850 was analyzed by a CIGRE working group and presented in [209].

Potential barriers to the success of any electronic attack may be based on making difficult: (1) the *access to the communications channel* (i.e., the attacker must be in a position to read data from the communications channel and/or write data to the communications channel); (2) *bypassing link-security mechanisms* (i.e., the attacker must defeat any defensive technologies protecting access to the communications channel), or (3) *bypassing electronic access-control mechanisms* (i.e., the attacker must defeat any electronic access-control mechanisms on the target device itself) [205].

Cybersecurity measures can consider electronic access control and link security. The distinction between these two terms lies in the fact that many potential target devices in the electric power infrastructure do not have integrated electronic access-control mechanisms. SCADA connections are automated data retrieval links that do not contain security provisions such as requirements for username and password entry; to secure such connections, security devices, like cryptographic modules, can be implemented. In contrast, engineering access connections are often protected by electronic access-control mechanisms, such as password requirements.

The list of *electronic access control* alternatives that can be considered include strong password, multilevel password, or time-outs and channel disconnects [205]. Link-security technologies provide a very effective means of limiting access to the communications media itself and protecting the contents of the data that travels over the media. Many devices provide strong cryptographic security in the form of *encryption*, which ensures that data cannot be read by unauthorized individuals, and *authentication*, which ensures that data is sent by an authorized individual.

Encryption modifies a file or message so it cannot be read without reversing the modifications using another piece of information called an encryption key. The key provides data needed for controlling modifications, which are performed according to an encryption algorithm. Encrypting all data transmitted on a communications link protects sensitive data like passwords, device settings, and system status information from being intercepted and read by unauthorized individuals. A strong password is extremely hard to guess with automated attack tools, but even the strongest password will fail if an attacker can intercept it and read it directly from the communications channel. Strong link encryption prevents password interception by scrambling the data prior to transmission. The scrambling function can only be reversed by an authorized individual with knowledge of the secret encryption key.

Authentication is the process of determining that the user is authentic. This is done by receiving information about the user and comparing the received information to a stored version of the information for the authentic user; for instance, something the user knows, such as a password. There are numerous ways in which an authentication system can be attacked and compromised; for instance, capturing a password as it moves in the system. There are also ways in which an authentication system can be bypassed, essentially involving attacks on the security of the overall system. Link-authentication mechanisms are independent of any electronic access controls (e.g., passwords) in the protected devices themselves. Because of this, they provide a means of locking out unauthorized access to otherwise unprotected communications access points such as SCADA protocol ports.

***Detecting Cyber Intrusion.*** Although measures to prevent intrusion can make difficult a successful attack, they do not make it impossible. It is extremely important to prevent intrusions, but it is also essential that intrusions be detected if they do occur. It can be very difficult to detect software modifications when the intrusion does not produce any damage. But even when the intrusion causes damage, it cannot be always said that the damage was caused due to an intrusion rather than some other failure (e.g., a software bug). For these reasons, it is important to detect intrusions when they occur [191]. Time is an important concern. If attackers are given unlimited time to check system vulnerabilities, they may then eventually succeed. A way to combat this is to install technologies that allow the utility to monitor electronic connections for suspicious activity and receive timely notification of a possible attack. Measures to be taken to quickly detect and react to electronic attacks include decreasing alarm response time, securing status visibility, or consolidating and monitoring electronic access. Some manufacturers have developed *intrusion detection systems* (IDSs), which are designed to recognize intrusions when communications are attempted from unauthorized/unusual addresses and when there is an unusual pattern of activity. They generate logs of suspicious events, which must be inspected to distinguish between a true intrusion and a false alarm. However, unusual activities cannot be easily defined, and more research is necessary to investigate what would constitute unusual activity in a computer-based environment [191]. With the implementation of IDSs, operators have the ability to use firewalls to prevent the attack from spreading, thus mitigating the overall damage the cyber threat could cause.

To secure SCADA systems from cyber threats, utilities can integrate software management and documentation systems into SCADA to assist operators with an application restoration following a catastrophic event, and to develop security solutions to prevent unwanted access to a network. Through the use of a virtual private network (VPN) utilities can ensure the proper authentication and authorization of data transactions between different networks: VPN gives a private use of a public network through the development of an encrypted tunnel between the server and client. To fully secure the VPN from unauthorized access, a high level of authentication must be implemented in all networked devices.

Securing systems will also require an increased emphasis on behavioral analysis, background investigations, psychological profiling, and analysis of individual motives; all these factors are key tools for detecting potential threats and revealing how potential adversaries may seek to target individuals who can provide access to vital networks.

***Responding and Recovering from Cyber Intrusion.*** Even if an intrusion is neither detected nor recorded when it occurs, it is essential to develop a procedure for the restoration of service after a cyber attack. After discovering a cyber attack, it is critical to make a full assessment of the situation as quickly as possible since it might not be an isolated incident and more attacks might occur if left unmitigated. Reference [204] recommends that the recovery and remediation procedure determines five things regarding the attack: who, what, where, when, and why. Depending on the security features of the device and administrative procedures, it may not be possible to determine all of these parameters; in such case, consideration should be given to upgrading technology and installation/maintenance procedures to provide a better analysis of the attack.

The utility has to maintain backups of the software of all programmable substation units and documentation regarding the standard parameters and settings of all IEDs. When a programmable device has been compromised, the secure backup software should be reloaded, and if the settings have been changed, the original settings must be restored. However, unless the nature of the security breach is known and can be repaired, the utility must consider taking the device offline or making it inaccessible to prevent a future attack [191].

Theoretically, it is desirable to record all data communications into and out of all IEDs. After a successful attack, the recordings can be used to determine what intrusion technique was used, in order to modify the system and close that particular vulnerability, and also to identify the intruder. However, it may be impractical to record all communications. So this option will be deferred until fast, voluminous, and inexpensive storage media are developed or IDSs with capabilities to filter out the nonsuspicious usual traffic and record only the deviant patterns are available [191].

### 24.7.4 Cybersecurity Standards

Standardization is a crucial aspect for the development of a secure smart grid. All institutions involved in the development of standards for the smart grid are also concerned with cybersecurity standards. IEEE has developed several standards that affect smart grid and substation cybersecurity, see [210–212]. The IEC TC57 is responsible for developing international standards for power system information exchanges (e.g., CIM standards). In 1999 the TC57 WG15 was formed to undertake the development of standards for security of the communication protocols, standards and technical reports on end-to-end security issues. The result is the IEC 62351 series, some of which are still under development [213].

Of special importance are NERC Critical Infrastructure Protection (CIP) standards [207], which require compliance by North American entities such as generator owners, generator operators, transmission owners, transmission operators, or regional reliability organizations. The purpose of CIP standards is to ensure that the bulk electric system in North America is reliable, adequate, and secure, so each interconnected entity should maintain an independent level of security, as well as manage the self-certification and compliance of partners with whom the utility interconnects. The suite of standards documents the requirements to demonstrate due diligence in creating and executing a cybersecurity plan.

The three-volume NIST report, *NISTIR* 7628, *Guidelines for Smart Grid Cybersecurity*, presents an analytical framework for effective cybersecurity strategies tailored to the particular combinations of smart grid-related characteristics, risks, and vulnerabilities [206]. The report recognizes that the electric grid is changing from a relatively closed system to a complex, highly interconnected environment, so cybersecurity requirements of each organization should evolve as technology advances and threats to grid security multiply and diversify. The *Guidelines* report is a companion document to the *NIST Framework and Roadmap for Smart Grid Interoperability Standards, Release 3.0* [1]; it describes a high-level conceptual reference model for the smart grid, identifies standards that are applicable (or likely to be applicable) to the ongoing development of an interoperable smart grid, and specifies a set of high-priority standards-related gaps and issues. Cybersecurity is recognized as a critical, cross-cutting issue that must be addressed in all standards developed for smart grid applications. The *Framework* document is the first instalment in an ongoing standards and harmonization process. Ultimately, this process will deliver the hundreds of communication protocols, standard interfaces, and other technical specifications necessary to build an advanced, secure electric power grid with two-way communication and control capabilities.

Reference [214] presents a tutorial on how to deploy IEDs and communications and security technology to satisfy the NERC CIP and reduce the chances of electronic intrusion.

## 24.7.5 Ongoing Efforts

Technologies are not yet mature for ensuring the cybersecurity of the future smart grid. Utilities must evaluate what assets have the highest value and deserve the greatest effort at protecting. Some of the ongoing efforts and activities aimed at achieving a better security are summarized below.

***New Tools.*** Simulation can be a powerful tool to test and design cybersecurity measures. Software manufacturers have begun to develop big data analytics; their application as well as that of grid monitoring software can be useful to manage risks, increase efficiency and reliability. The patterns of grid usage and data access can provide a better foundation for detecting malicious activity: a rapid identification of grid behavior can allow grid operators to quickly isolate affected systems, the ideal system would respond with an instantaneous and automated isolation of an affected system and would subsequently reroute electric power to minimize disruption. Physical hardware under development allow for automatic adjustment of grid voltage and load; technologies that have been developed to automatically balance the input of solar and wind generation can also work in case of a cyberattack to balance electric power loads: these systems can be designed to work without computerized input, so that in the event of a cyberattack, there could still be an automated, hardware-based response.

***Development of Test Beds.*** A cybersecurity test bed is a model that captures the relevant complexity of a test system from a perspective of cybersecurity concerns and can be used to determine the vulnerabilities of the system by taking into account hardware and software interactions between system components: cybersecurity test beds are a powerful mean for studying grid vulnerabilities and identifying security enhancements. Several efforts have been made to develop test beds for cybersecurity assessment [215]. A test bed was also presented in the Special Issue on Cybersecurity [194]: it is based on the IEEE 39-bus system and can be used to analyze intrusions originated from remote access connections to a substation communication network; IEC 61850-based communication is used between IEDs and the user interface, which is able to acquire monitored data generated by power system simulation tools; there are remote access points using dial-up, VPN, or wireless technology, which can serve as intrusion paths; a dispatcher training simulator is used for training of operators and simulation of system operation, control, and restoration scenarios.

***Development of Maturity Models.*** A maturity model is a management tool that provides a common language and framework for defining key elements of smart grid transformation and helping utilities develop a programmatic approach and track their progress. Two examples of maturity models are the Smart Grid Maturity Model (SGMM), an initiative of the Global Intelligent Utility Coalition currently administrated by the DOE [216, 217], and the Electricity Subsector Cybersecurity Capability Maturity Model (ES-C2M2), an initiative of the DoE in partnership with the U.S. Department of Homeland Security (DHS) and in collaboration with industry, private-sector, and public-sector experts [218]. In January 2015, the DoE released the Cybersecurity Framework Implementation Guidance [219]; ES-C2M2 is the recommended implementation tool to help organizations evaluate, prioritize, and improve cybersecurity capabilities. The C2M2 model does not measure cybersecurity performance; rather it should be used to assess the maturity of a cybersecurity program.

***Grid Security Exercises.*** The NERC conducts biennial grid security and emergency response exercises. The third one (GridEx III) was organized in November 2015 and provided an opportunity for participants to respond to simulated cyber and physical attacks affecting the reliable operation of the grid [220]. The exercise was aimed at practicing response and recovery, improving communication, and identifying the learned lessons. Results can be used to propose recommendations that can enhance grid security and reliability (e.g., coordinated response and communication; reporting mechanisms; active participation of personnel; introduction of new exercise tools).

## 24.8 ACKNOWLEDGMENT

This section covers many topics. For some of them only a short tutorial introduction is provided. This means that for some topics the material presented here mostly comes from review or state-of-the-art publications. The author wants to thank all those who produced such publications, which have been extremely useful for writing this section.

## 24.9 REFERENCES

1. National Institute of Standards and Technology, "NIST Framework and Roadmap for Smart Grid Interoperability Standards," NIST Special Publication 1108r3, Release 3.0, September 2014.
2. CEN-CENELEC-ETSI Smart Grid Coordination Group, "Smart Grid Reference Architecture," November 2012.
3. IEC SMB Smart Grid Strategic Group (SG3), "IEC Smart Grid Standardization Roadmap," edition 1.0, June 2010.
4. CEN-CENELEC-ETSI Smart Grid Coordination Group, "SGCG/M490/G-Smart Grid Set of Standards," Version 3.1, October 2014.
5. F. Elkarmi and N. Abu Shikhah, *Power System Planning Technologies and Applications. Concepts, Solutions and Management*, IGI Global, 2012.
6. H. Lee Willis, *Power Distribution Planning Reference Book*, 2nd ed., Marcel Dekker,New York, NY, USA, 2004.
7. H. Lee Willis, *Spatial Electric Load Forecasting*, Marcel Dekker, New York, NY, USA, 1996.
8. S. A. Soliman and A. M. Al-Kandari, *Electrical Load Forecasting: Modeling and Model Construction*, Butterworth-Heinemann, Elsevier, Burlington, MA, USA, 2010.
9. EPRI, "Scoping Study for Identifying the Need for New Tools for the Planning of Transmission and Distribution Systems: Converting the Legacy Infrastructure into an Intelligent Grid," Report 1015285, 2007.
10. W. Li, *Probabilistic Transmission System Planning*, John Wiley, Hoboken, NJ, USA, 2011.
11. H. Seifi and M.S. Sepasian, *Electric Power System Planning. Issues, Algorithms and Solutions*, Springer, Berlin, Germany, 2011.
12. G. Kusic, *Computer-Aided Power Systems Analysis*, 2nd ed., CRC Press, Boca Raton, FL, USA, 2009.
13. A. Gómez-Expósito, A. Conejo, and C. Cañizares (eds.), *Electric Energy Systems: Analysis and Operation*, CRC Press, Boca Raton, FL, USA, 2009.
14. J. C. Das, *Power System Analysis*, 2nd ed., CRC Press, Boca Raton, FL, USA, 2012.
15. J. A. Momoh, *Electric Power System Applications of Optimization*, 2nd ed., CRC Press, Boca Raton, FL, USA, 2009.
16. P. Kundur, *Power System Stability and Control*, McGraw-Hill, New York, NY, USA, 1994.
17. P. M. Anderson and A. A. Fouad, *Power System Control and Stability*, John Wiley—IEEE Press, Piscataway, NJ, USA, 2002.
18. C. W. Taylor, *Power System Voltage Stability*, McGraw-Hill, New York, NY, USA, 1994.
19. V. Ajjarapu, *Computational Techniques for Voltage Stability Assessment and Control*, Springer, New York, NY, USA, 2006.
20. P. M. Anderson, *Analysis of Faulted Power Systems*, John Wiley—IEEE Press, New York, NY, USA, 1995.
21. P. M. Anderson, *Power System Protection*, McGraw-Hill—IEEE Press, Piscataway, NJ, USA, 1999.
22. J. Gers and T. Holmes, *Protection of Electricity Distribution Networks*, 3rd ed., IET, Power and Energy Series 63, Stevenage, UK, 2011.
23. N. Watson and J. Arrillaga, *Power System Electromagnetic Transients Simulation*, IET, Power and Energy Series 39, Stevenage, UK, 2003.
24. J. A. Martinez-Velasco (ed.), *Transient Analysis of Power Systems. Solution Techniques, Tools and Applications*, John Wiley—IEEE Press, Chichester, UK, 2015.

25. IEEE TF on Understanding, Prediction, Mitigation and Restoration of Cascading Failures, "Survey of Tools for Risk Assessment of Cascading Outages," *IEEE PES General Meeting*, Detroit, MI, July 2011.

26. W. H. Kersting and R. C. Dugan, "Recommended Practices for Distribution System Analysis," *IEEE PES Power Systems Conference and Exposition*, Atlanta, November 2006.

27. R. C. Dugan, R. F. Arritt, T. E. McDermott, S. M. Brahma, and K. Schneider, "Distribution System Analysis to Support the Smart Grid," *IEEE PES General Meeting*, Minneapolis, MN, July 2010.

28. J. A. Martinez, F. de León, and V. Dinavahi, "Tools for Analysis of Distribution Systems with Distributed Resources: Present and future trends," *IEEE PES General Meeting*, Minneapolis, MN, July 2010.

29. W. H. Kersting, *Distribution System Modeling and Analysis*, CRC Press, 3rd ed., 2012.

30. R. E. Brown, *Electric Power Reliability*, CRC Press, 2nd ed., Boca Raton, FL, USA, 2009.

31. A. Chowdhury and D. Koval, *Power Distribution System Reliability: Practical Methods and Applications*, John Wiley, Hoboken, NJ, USA, 2009.

32. IEEE Std. 1366-2003, "IEEE Trial Use Guide for Electric Power Distribution Reliability Indices," 2003.

33. R. C. Dugan, M. F. McGranaghan, S. Santoso, and H. Wayne Beaty, *Electrical Power Systems Quality*, 3rd ed., McGraw-Hill, New York, NY, USA, 2012.

34. M. H. Bollen, *Understanding Power Quality Problems: Voltage Sags and Interruptions*, John Wiley—IEEE Press, Piscataway, NJ, USA, 1999.

35. J. Mahseredjian, V. Dinavahi, and J. A. Martinez, "Simulation Tools for Electromagnetic Transients in Power Systems: Overview and Challenges," *IEEE Trans. on Power Delivery*, vol. 24, no. 3, pp. 1657–1669, July 2009.

36. T. Lambert, P. Gilman, and P. Lilienthal, "Micropower System Modeling with HOMER," Chap. 15 of *Integration of Alternative Sources of Energy*, by F. A. Farret and M. Godoy Simões, John Wiley, Hoboken, NJ, USA, 2006.

37. I. J. Ramirez-Rosado, L.A. Fernandez-Jimenez, C. Monteiro, V. Miranda, E. Garcia-Garrido, and P. J. Zorzano-Santamaria, "Powerful planning tools," *IEEE Power & Energy Magazine*, vol. 3, no. 2, pp. 56–63, March/April 2005.

38. EPRI, "Asset Management Toolkit Modules: An Approach for Risk-Informed, Performance-Focused Asset Management in the Power Delivery Industry," Report 1011365, 2005.

39. S. E. Widergren, J. M. Roop, R. T. Guttromson, and Z. Huang, "Simulating the Dynamic Coupling of Market and Physical System Operations," *IEEE PES General Meeting*, Denver, CO, June 2004.

40. EPRI and New York ISO, "Survey of Electricity Market Simulation," Report 1010703, Palo Alto, CA, 2005.

41. H. Singh (ed.), "Game Theory Applications in Electricity Power Markets," IEEE Special Publication, TP-136-0, 1999.

42. F. Sensuß, M. Genoese, M. Ragwitz, and D. Möst, "Agent-Based Simulation of Electricity Markets: A Literature Review," *Energy Studies Review*, vol. 15, no. 2, Article 2, 2007.

43. Z. Zhou, W. K. Chan, and J. H. Chow, "Agent-Based Simulation of Electricity Markets: A Survey of Tools," *Artificial Intelligence Review*, vol. 28, pp. 305–342, 2007.

44. M. S. Thomas and J. D. McDonald, *Power System SCADA and Smart Grids*, CRC Press, Boca Raton, FL, USA, 2015.

45. C. Monteiro, R. Bessa, V. Miranda, A. Botterud, J. Wang, and G. Conzelmann, "Wind Power Forecasting: State-of-the-Art 2009," Report ANL/DIS-10-1, Argonne National Laboratory, Argonne, Illinois, November 6, 2009.

46. W. Glassley, J. Kleissl, C. P. van Dam, H. Shiu, J. Huang, G. Braun, and R. Holland, "California Renewable Energy Forecasting, Resource Data and Mapping," California Institute for Energy and Environment, 2010.

47. G. N. Sorebo and M. C. Echols, *Smart Grid Security. An End-to-End View of Security in the New Electrical Grid*, CRC Press, Boca Raton, FL, USA, 2012.

48. P. Venkatesh, B. V. Manikandan, S. Charles Raja, and A. Srinivasan, *Electrical Power Systems: Analysis, Security and Deregulation*, PHI Learning, New Delhi, India, 2010.

49. A. Abur and A. Gómez Expósito, *Power System State Estimation. Theory and Implementation*, Marcel Dekker, New York, NY, USA, 2004.

50. EPRI, "Scoping Study for Identifying the Need for New Tools for the Planning of Transmission and Distribution Systems: Converting the Legacy Infrastructure into an Intelligent Grid," Report 1015285, 2007.

51. A. Monti, C. Muscas, and F. Ponci, *Phasor Measurement Units and Wide Area Monitoring Systems*, Elsevier, London, UK, 2016.

52. A. Román Mesina, "Wide Area Monitoring of Interconnected Power Systems," IET, Power and Energy Systems 77, 2015.

53. Y. Li, D. Yang, F. Liu, Y. Cao, and C. Rehtanz, *Interconnected Power Systems. Wide-Area Dynamic Monitoring and Control Applications*, Springer, Berlin, Germany, 2016.

54. J. W. Evans, "Interface between Automation and the Substation," Section 6 of *Electric Power Substations Engineering* (J. D. McDonald, Ed.), 2nd ed., CRC Press, Boca Raton, FL, 2007.

55. J. D. McDonald, "Substation Integration and Automation," Sec. 7 of *Electric Power Substations Engineering* (J. D. McDonald, ed.), 2nd ed., CRC Press, Boca Raton, FL, 2007.

56. E. Padilla, *Substation Automation Systems: Design and Implementation*, John Wiley, Chichester, UK, 2015.

57. K. P. Brand, Ch. Brunner, and I. De Mesmaeker, "How to Use IEC 61850 in Protection and Automation," *Electra*, Issue 222, pp. 11–21, October 2005.

58. K. P. Brand, Ch. Brunner, and I. De Mesmaeker, "How to Complete a Substation Automation System with an IEC 61850 Process Bus," *Electra*, Issue 255, pp. 12–24, April 2011.

59. IEC TR 61850-1, "Communication Networks and Systems in Substations. Part 1: Introduction and Overview," 2003.

60. IEC 61850-5, "Communication Networks and Systems in Substations—Part 5: Communication Requirements for Functions and Device Models," 2003.

61. CIGRE Working Group B5.11, "The Introduction of IEC 61850 and Its Impact on Protection and Automation within Substations," CIGRE Brochure 326, August 2007.

62. J. Northcote-Green and R. Wilson, *Control and Automation of Electrical Power Distribution Systems*, CRC Press, Boca Raton, FL, USA, 2007.

63. EPRI, "Technical and System Requirements for Advanced Distribution Automation," Report 1010915, Palo Alto, CA, June 2004.

64. N. Higgins, V. Vyatkin, N. K. C. Nair, and K. Schwarz, "Distributed Power System Automation with IEC 61850, IEC 61499, and Intelligent Control," *IEEE Trans. on Systems, Man, and Cybernetics-Part C: Applications and Reviews*, vol. 41, no. 1, pp. 81–92, January 2011.

65. K. Mets, J. A. Ojea, and C. Develder, "Combining Power and Communication Network Simulation for Cost-Effective Smart Grid Analysis," *IEEE Communications Surveys & Tutorials*, vol. 16, no. 3, pp. 1771–1796, Third Quarter 2014.

66. R. Podmore and M. R. Robinson, "The Role of Simulators for Smart Grid Development," *IEEE Trans. on Smart Grid*, vol. 1, no. 2, pp. 205–212, September 2010.

67. "National Power Grid Simulation Capability: Needs and Issues," *National Power Grid Simulator Workshop*, U.S. Dept. of Homeland Security, Science and Technology Directorate, Argonne, Illinois, December 2008.

68. R. Idema, D. J. P. Lahaye, C. Vuik, and L. van der Sluis, "Scalable Newton-Krylov Solver for Very Large Power Flow Problems," *IEEE Trans. on Power Systems*, vol. 27, no. 1, pp. 390–396, February 2012.

69. R. Idema, G. Papaefthymiou, D. Lahaye, C. Vuik, and L. van der Sluis, "Towards Faster Solution of Large Power Flow Problems," *IEEE Trans. on Power Systems*, vol. 28, no. 4, pp. 4918–4925, November 2013.

70. M. O. Faruque, V. Dinavahi, M. Steurer, A. Monti, K. Strunz, J. A. Martinez, G. W. Chang, J. Jatskevich, R. Iravani, and A. Davoudi, "Interfacing Issues in Multi-Domain Simulation Tools," *IEEE Trans. on Power Delivery*, vol. 27, no. 1, pp. 439–448, January 2012.

71. S. Filizadeh, M. Heidari, A. Mehrizi-Sani, J. Jatskevich, and J. A. Martinez, "Techniques for Interfacing Electromagnetic Transient Simulation Programs with General Mathematical Tools," *IEEE Trans. on Power Delivery*, vol. 23, no. 4, pp. 2610–2622, October 2008.

72. B. Asghari, V. Dinavahi, M. Rioual, J. A. Martinez, and R. Iravani, "Interfacing Techniques for Electromagnetic Field and Circuit Simulation Programs," *IEEE Trans. on Power Delivery*, vol. 24, no. 2, pp. 939–950, April 2009.

73. V. Jalili-Marandi, V. Dinavahi, K. Strunz, J. A. Martinez, and A. Ramirez, "Interfacing Techniques for Transient Stability and Electromagnetic Transient Programs," *IEEE Trans. on Power Delivery*, vol. 24, no. 4, pp. 2385–2395, October 2009.

74. W. Ren, M. Sloderbeck, M. Steurer, V. Dinavahi, T. Noda, S. Filizadeh, A. R. Chevrefils, et al., "Interfacing Issues in Real-Time Digital Simulators," *IEEE Trans. on Power Delivery*, vol. 26, no. 2, pp. 1221–1230, April 2011.

75. X. Wang, P. Zhang, Z. Wang, V. Dinavahi, G. Chang, J. A. Martinez, A. Davoudi, A. Mehrizi-Sani, and S. Abhyankar, "Interfacing Issues in Multiagent Simulation for Smart Grid Applications," *IEEE Trans. on Power Delivery*, vol. 28, no. 3, pp. 1918–1927, July 2013.

76. S. C. Mueller, H. Georg, J. J. Nutaro, E. Widl, Y. Deng, P. Palensky, M. Usman Awais, et al., "Interfacing Power System and ICT Simulators: Challenges, State-of-the-Art, and Case Studies," to be published in *IEEE Trans. on Smart Grid.*

77. EPRI, "Development of Analytical and Computational Methods for the Strategic Power Infrastructure Defense (SPID) System: EPRI/DoD Complex Interactive Networks/Systems Initiative: Second Annual Report," Report 1006089, 2001.

78. K. Hopkinson, X. Wang, R. Giovanini, J. Thorp, K. Birman, and D. Coury, "EPOCHS: A Platform for Agent-Based Electric Power and Communication Simulation Built from Commercial Off-the-Shelf Components," *IEEE Trans. on Power Systems,* vol. 21, pp. 548–558, May 2006.

79. K. Tomsovic, D. E. Bakken, V. Venkatasubramanian, and A. Bose, "Designing the Next Generation of Real-Time Control, Communication, and Computations for Large Power Systems," *Proc. of IEEE*, vol. 93, no. 5, pp. 965–979, May 2005.

80. J. Nutaro, P. T. Kuruganti, M. Shankar, L. Miller, and S. Mullen, "Integrated Modeling of the Electric Grid, Communications, and Control," *Int. Journal of Energy Sector Management*, vol. 2, no. 3, pp. 420–438, 2008.

81. S. Ciraci, J. Daily, K. Agarwal, J. Fuller, L. Marinovici, and A. Fisher, "Synchronization Algorithms for Co-simulation of Power Grid and Communication Networks," *22nd Int. Symp. on Modelling, Analysis & Simulation of Computer and Telecommunication Systems*, Paris, September 2014.

82. T. Godfrey, S. Mullen, R. C. Dugan, C. Rodine, D. W. Griffith, and N. Golmie, "Modeling Smart Grid Applications with Co-simulation," *1st IEEE Int. Conf. on Smart Grid Communications (SmartGridComm)*, Gaithersburg, MD, October 2010.

83. H. Lin, S. Sambamoorthy, S. Shukla, J. Thorp, and L. Mili, "Power System and Communication Network Co-simulation for Smart Grid Applications," *IEEE PES Innovative Smart Grid Technologies (ISGT)*, Anaheim, CA, January 2011.

84. M. Lévesque, D. Qian Xu, M. Maier, and G. Joós, "Communications and Power Distribution Network Co-simulation for Multidisciplinary Smart Grid Experimentations," *45th Annual Simulation Symposium*, Orlando, FL, March 2012.

85. H. Lin, S. S. Veda, S. S. Shukla, L. Mili, and J. Thorp, "GECO: Global Event-Driven Co-simulation Framework for Interconnected Power System and Communication Network," *IEEE Trans. on Smart Grid*, vol. 3, no. 3, pp. 1444–1456, September 2012.

86. S. C. Muller, H. Georg, C. Rehtanz, and C. Wietfeld, "Hybrid Simulation of Power Systems and ICT for Real-Time Applications," *3rd IEEE PES Innovative Smart Grid Technologies Europe (ISGT Europe)*, Berlin, October 2012.

87. R. Majumder, G. Bag, G. Velotto, and A. Marinopoulos, "Closed Loop Simulation of Communication and Power Network in a Zone Based System," *Electric Power Systems Research*, vol. 95, pp. 247–256, 2013.

88. S. Ciraci, J. Daily, J. Fuller, A. Fisher, L. Marinovici, and K. Agarwal, "FNCS: A Framework for Power System and Communication Networks Co-simulation," *Symp. on Theory of Modeling & Simulation*, San Diego, CA, April 2014.

89. E. Moradi-Pari, N. Nasiriani, Y. P. Fallah, P. Famouri, S. Bossart, and K. Dodrill, "Design, Modeling, and Simulation of On-Demand Communication Mechanisms for Cyber-Physical Energy Systems," *IEEE Trans. on Industrial Informatics*, vol. 10, no. 4, pp. 2330–2339, November 2014.

90. W. Li, M. Ferdowsi, M. Stevic, A. Monti, and F. Ponci, "Cosimulation for Smart Grid Communications," *IEEE Trans. on Industrial Informatics*, vol. 10, no. 4, pp. 2374–2384, November 2014.

91. M. Collette, B. Corey, and J. Johnson, "High Performance Tools and Technologies," Report UCRL-TR-209289, Lawrence Livermore National Laboratory, U.S. Department of Energy, 2004.

92. R. C. Green II, L. Wang, and M. Alam, "High Performance Computing for Electric Power Systems: Applications and Trends," *IEEE PES General Meeting*, Detroit, CO, July 2011, 2007.

93. S. Kumar Khaitan and A. Gupta (eds.), *High Performance Computing in Power and Energy Systems*, Springer, Berlin, Germany, 2013.

94. L. Zhang, A. Hoyt, and B. Calson, "A Review of High Performance Technical Computing in Midwest ISO," *IEEE PES General Meeting*, Minneapolis, MN, July 2010.

95. M. Irving, G. Taylor, and P. Hobson, "Plug in to Grid Computing," *IEEE Power and Energy Magazine*, vol. 2, no. 2, pp. 40–44, March 2004.

96. D. J. Tylavsky, A. Bose, F. Alvarado, R. Betancourt, K. Clements, G. T. Heydt, G. Huang, et al., "Parallel Processing in Power Systems Computation," *IEEE Trans. on Power Systems*, vol. 7, no. 2, pp. 629–638, May 1992.

97. D. M. Falcao, "High Performance Computing in Power System Applications," in *Vector and Parallel Processing—VECPAR'96*, Volume 1215 of the series Lecture Notes in Computer Science, J. M. L. M. Palma and J. Dongarra (eds.), Springer, Heidelberg, Germany, 1997.

98. C. Dufour, V. Jalili-Marandi, J. Bélanger, and L. Snider, "Power System Simulation Algorithms for Parallel Computer Architectures," *IEEE PES General Meeting*, San Diego, CA, July 2012.

99. K. Asanovíc, R. Bodik, B. Catanzaro, J. Gebis, P. Husbands, K. Keutzer, D. Patterson, W. Plishker, J. Shalf, S. Williams, and K. Yelick, "The Landscape of Parallel Computing Research: A View from Berkeley," Technical Report No. UCB/EECS-2006-183, December 18, 2006.

100. K. Asanovic, R. Bodik, J. Demmel, T. Keaveny, K. Keutzer, J. Kubiatowicz, N. Morgan, et al., "A View of the Parallel Computing Landscape," *Communications of the ACM*, vol. 52, no. 10, pp. 56–67, October 2009.

101. K. H. Hoffmann and A. Meyer (eds.), *Parallel Algorithms and Cluster Computing. Implementations, Algorithms and Applications*, Springer, Berlin, Germany, 2006.

102. R. Trobec, M. Vajtersic, and P. Zinterhof (eds.), *Parallel Computing: Numerics, Applications, and Trends*, Springer, London, UK, 2009.

103. J. A. Martinez, V. Dinavahi, M. H. Nehrir, and X. Guillaud, "Tools for Analysis and Design of Distributed Resources. Part IV: Future Trends," *IEEE Trans. on Power Delivery*, vol. 26, no. 2, pp. 1663–1670, July 2011.

104. C. Dufour, T. Ould Bachir, L.-A., Grégoire, and J. Bélanger, "Real-Time Simulation of Power Electronic Systems and Devices," Chap. 15 of *Dynamics and Control of Switched Electronic Systems: Advanced Perspectives for Modeling, Simulation and Control of Power Converters*, F. Vasca and L. Iannelli (eds.), Springer Series on Advances in Industrial Control, Springer, London, UK, 2012.

105. K. Popovici and P. J. Mosterman (eds.), *Real-Time Simulation Technologies: Principles, Methodologies, and Applications*, CRC Press, Boca Raton, FL, USA, 2013.

106. C. Dufour and J. Bélanger, "Real-Time Simulation Technologies in Engineering," Chap. 4 of *Transient Analysis of Power Systems: Solution Techniques, Tools and Applications*, J. A. Martinez-Velasco (ed.), John Wiley—IEEE Press, Chichester, UK, 2015.

107. M. O. Faruque, T. Strasser, G. Lauss, V. Jalili-Marandi, P. Forsyth, C. Dufour, V. Dinavahi, et al., "Real-Time Simulation Technologies for Power Systems Design, Testing, and Analysis," *IEEE Power and Energy Technology Systems Journal*, vol. 2, no. 2, pp. 63–73, June 2015.

108. P. G. McLaren, R. Kuffel, R. Wierckx, J. Giesbrecht, and L. Arendt, "A Real Time Digital Simulator for Testing Relays," *IEEE Trans. on Power Delivery*, vol. 7, no. 1, pp. 207–213, January 1992.

109. G. G. Parma and V. Dinavahi, "Real-Time Digital Hardware Simulation of Power Electronics and Drives," *IEEE Trans. on Power Delivery*, vol. 22, no. 2, pp. 1235–1246, April 2007.

110. X. Guillaud, M. O. Faruque, A. Teninge, A. Hasan Hariri, L. Vanfretti, M. Paolone, V. Dinavahi, et al., "Applications of Real-Time Simulation Technologies in Power and Energy Systems," *IEEE Power and Energy Technology Systems Journal*, vol. 2, no. 3, pp. 103–115, September 2015.

111. L.-F. Pak, M. O. Faruque, X. Nie, and V. Dinavahi, "A Versatile Cluster-Based Real-Time Digital Simulator for Power Engineering Research," *IEEE Trans. on Power Systems*, vol. 21, no. 2, pp. 455–465, May 2006.

112. Y. Chen and V. Dinavahi, "FPGA-Based Real-Time EMTP," *IEEE Trans. on Power Delivery*, vol. 24, no. 2, pp. 892–902, April 2009.

113. C. Dufour, J. Mahseredjian, and J. Bélanger, "A Combined State-Space Nodal Method for the Simulation of Power System Transients," *IEEE Trans. on Power Delivery*, vol. 26, no. 2, pp. 928–935, April 2011.

114. H. W. Dommel, *Electromagnetic Transients Program Reference Manual* (*EMTP Theory Book*), Bonneville Power Administration, Portland, OR, August 1986.

115. G. F. Lauss, M. O. Faruque, K. Schoder, C. Dufour, A. Viehweider, and J. Langston, "Characteristics and Design of Power Hardware-in-the-Loop Simulations for Electrical Power Systems," *IEEE Trans. on Industrial Electronics*, vol. 63, no. 1, pp. 406–417, January 2016.

116. J. Bélanger, S. Abourida, and C. Dufour, "Real-Time Digital Simulation and Control Laboratory for Distributed Power Electronic Generation and Distribution," *Huntsville Simulation Conference*, Huntsville, AL, USA, 2005.

117. J. Manyika, M. Chui, B. Brown, J. Bughin, R. Dobbs, C. Roxburgh, and A. Hung Byers, "Big Data: The Next Frontier for Innovation, Competition, and Productivity," McKinsey Global Institute, May 2011.

118. C. L. Stimmel, *Big Data Analytics Strategies for the Smart Grid*, CRC Press, Boca Raton, FL, USA, 2014.

119. X. W. Chen and X. Lin, "Big Data Deep Learning: Challenges and Perspectives," *IEEE Access*, vol. 2, pp. 514–525, 2014.

120. M. J. Zaki and W. Meira Jr., *Data Mining and Analysis. Fundamental Concepts and Algorithms*, Cambridge University Press, New York, 2014.

121. X. Wu, X. Zhu, G. Q. Wu, and W. Ding, "Data Mining with Big Data," *IEEE Trans. on Knowledge and Data Engineering*, vol. 26, no. 1, pp. 97–107, January 2014.

122. P. Flach, *Machine Learning: The Art and Science of Algorithms that Make Sense of Data*, Cambridge University Press, Cambridge, UK, 2012.

123. M. Kezunovic, L. Xie, and S. Grijalva, "The Role of Big Data in Improving Power System Operation and Protection," *Symposium-Bulk Power System Dynamics and Control (IREP)*, Rethymnon, Greece, August 2013.

124. A. Ukil and R. Zivanovic, "Automated Analysis of Power Systems Disturbance Records: Smart Grid Big Data Perspective," *IEEE Innovative Smart Grid Technologies—Asia (ISGT ASIA)*, Kuala Lumpur, Malaysia, May 2014.

125. P. C. Chen, T. Dokic, and M. Kezunovic, "The Use of Big Data for Outage Management in Distribution Systems," *CIRED*, Paper 0406, Rome, June 2014.

126. R. Meier, E. Cotilla-Sanchez, B. McCamish, D. Chiu, M. Histand, J. Landford, and R. B. Bass, "Power System Data Management and Analysis Using Synchrophasor Data," *IEEE Conf. on Technologies for Sustainability (SusTech)*, Portland, OR, July 2014.

127. N. Yu, S. Shah, R. Johnson, R. Sherick, M. Hong, and K. Loparo, "Big Data Analytics in Power Distribution Systems," *Innovative Smart Grid Technologies Conference (ISGT)*, Washington DC, February 2015.

128. G. C. Zweigle, "A Wide-Area, Wide-Spectrum Big Data System," *IEEE PES General Meeting*, Denver, CO, July 2015.

129. M. Kezunovic, L. Xie, S. Grijalva, P. Chau, T. Dokic, Y. Gu, A. Anand, and S. T. Cheng, "Systematic Integration of Large Data Sets for Improved Decision-Making," PSERC Publication 15-05, September 2015.

130. J. Deign and C. Márquez Salazar, "Data Management and Analytics for Utilities," Smart Grid Update, 2013.

131. L. Xu, C. Jiang, J. Wang, J. Yuan, and Y. Ren, "Information Security in Big Data: Privacy and Data Mining," *IEEE Access*, vol. 2, pp. 1149–1176, 2014.

132. Y. Sun, H. Song, A. J. Jara, and R. Bie, "Internet of Things and Big Data Analytics for Smart and Connected Communities," *IEEE Access*, vol. 4, pp. 766–773, 2016.

133. P. Mell and T. Grance, "The NIST Definition of Cloud Computing," NIST Special Publication 800-145, September 2011.

134. Q. Zhang, L. Cheng, and R. Boutaba, "Cloud Computing: State-of-the-Art and Research Challenges," *J. Internet Serv. Appl.*, vol. 1, pp. 7–18, 2010.

135. C. N. Höfer and G. Karagiannis, "Cloud Computing Services: Taxonomy and Comparison," *J. Internet Serv. Appl.*, vol. 2, pp. 81–94, 2011.

136. L. Qian, Z. Luo, Y. Du, and L. Guo, "Cloud Computing: An Overview," in *Cloud Computing*, M. G. Jaatun, G. Zhao, and C. Rong (eds.), Springer, Berlin, Germany, 2009.

137. N. Antonopoulos and L. Gillam (eds.), *Cloud Computing: Principles, Systems and Applications*, Springer, London, UK, 2010.

138. R. Buyya, J. Broberg, and A. M. Goscinski (eds.), *Cloud Computing: Principles and Paradigms*, John Wiley, Hoboken, NJ, USA, 2010.

139. T. Erl, R. Puttini, and Z. Mahmood, *Cloud Computing: Concepts, Technology & Architecture*, Prentice Hall, 2014.

140. F. Ma, X. Luo, and E. Litvinov, "Cloud Computing for Power System Simulations at ISO New England—Experiences and Challenges," to be published in *IEEE Trans. on Smart Grids*.

141. D. S. Markovic, D. Zivkovic, I. Branovic, R. Popovic, and D. Cvetkovic, "Smart Power Grid and Cloud Computing," *Renewable and Sustainable Energy Reviews*, vol. 24, pp. 566–577, 2013.

142. C. Xu, F. Zhao, Z. Wang, X. Lin, S. He, and C. Shao, "Design of Cloud Computing Architecture for Power System Analysis," *IEEE Region 10 Conference (TENCON)*, Xian, China, October 2013.

143. Z. Liang and L. Xiuqing, "The Core of Constructing the Future Power Systems Computation Platform is Cloud Computing," *Int. Conf. on Mechatronic Science, Electric Engineering and Computer (MEC)*, Jilin, China, August 2011.

144. Q. Huang, M. Zhou, Y. Zhang, and Z. Wu, "Exploiting Cloud Computing for Power System Analysis," *Int. Conf. on Power System Technology (POWERCON)*, Hangzhou, China, October 2010.

145. S. Rusitschka, K. Eger, and C. Gerdes, "Smart Grid Data Cloud: A Model for Utilizing Cloud Computing in the Smart Grid Domain," *1st IEEE Int. Conf. on Smart Grid Communications (SmartGridComm)*, Gaithersburg, MD, October 2010.

146. EPRI, "Common Information Model (CIM). CIM 10 Version," Report 1001976, Palo Alto, CA, 2001.

147. M. Uslar, M. Specht, S. Rohjans, J. Trefke, and J. M. Vasquez González, *The Common Information Model CIM*, Springer, Berlin, Germany, 2012.

148. B. Wollenberg, J. Britton, E. Dobrowolski, R. Podmore, J. Resek, J. Scheidt, J. Russell, T. Saxton, and C. Ivanov, "A Brief History. The Common Information Model," *IEEE Energy & Power Magazine*, vol. 14, no. 1, January/February 2016.

149. IEC 61970-301, "Energy Management System Application Program Interface (EMS-API)—Part 301: Common Information Model (CIM) base," edition 5.0, 2013.

150. IEC 61970-501, "Energy Management System Application Program Interface (EMS-API)—Part 501: Common Information Model Resource Description Framework (CIM RDF) schema," 2006.

151. IEC 61968-11, "Application Integration at Electric Utilities—System Interfaces for Distribution Management—Part 11: Common Information Model (CIM) Extensions for Distribution," 2010.

152. IEC 61968-13, "Application Integration at Electric Utilities—System Interfaces for Distribution Management—Part 13: CIM RDF Model Exchange Format for Distribution," 2008.

153. EPRI, "Reference Manual for Exchanging Standard Power System Dynamic Models: Based on the IEC 61970 Common Information Model (CIM)," Report 1020200, Palo Alto, CA, 2009.

154. IEC 62325-301, "Framework for Energy Market Communications—Part 301: Common Information Model (CIM) Extensions for Markets," 2014.

155. IEC 62325-451-1, "Framework for Energy Market Communications—Part 451-1: Acknowledgement Business Process and Contextual Model for CIM European Market," 2013.

156. IEC 62325-450, "Framework for Energy Market Communications—Part 450: Profile and Context Modelling Rules," 2013.

157. EPRI, "Common Information Model Primer," 3rd ed., Technical Report 3002006001, Palo Alto, CA, 2015.

158. Special Issue on The Common Information Model, "The Root of All Data Exchange," *IEEE Energy & Power Magazine*, vol. 14, no. 1, pp. 30–104, January/February 2016.

159. G. Booch, J. Rumbaugh, and I. Jacobson, *The Unified Modeling Language User Guide*, 2nd ed., Addison-Wesley, Upper Saddle River, NJ, USA, 2005.

160. J. Rumbaugh, I. Jacobson, and G. Booch, *Unified Modeling Language Reference Manual*, Addison-Wesley, Upper Saddle River, NJ, USA, 1999.

161. IEC 61970-552, "Energy Management System Application Program Interface (EMS-API)—Part 552: CIMXML Model Exchange Format," 2013.

162. EPRI, "Major 2010 Common Information Model Interoperability Test of Power System Model Revisions: CIM Standards Revisions for Major Energy Management System Vendors," Report 1017804, Palo Alto, CA, 2010.

163. ENTSO-E, "Interoperability Test—CIM for System Development and Operations," Final Report, August 2013.

164. ENTSO-E, "ENTSO-E Interoperability Tests for the Market," June 2015.

165. EPRI, "Harmonizing the International Electrotechnical Commission Common Information Model (CIM) and 61850 Standards via a Unified Model: Key to Achieve Smart Grid Interoperability Objectives," Report 1020098, Palo Alto, CA, 2010.

166. X. Wang, N. N. Schulz, and S. Neumann, "CIM Extensions to Electrical Distribution and CIM XML for the IEEE Radial Test Feeders," *IEEE Trans. on Power Systems*, vol. 18, no. 3, pp. 1021–1028, August 2003.

167. A. W. McMorran, G. W. Ault, I. M. Elders, C. E. T. Foote, G. M. Burt, and J. R. McDonald, "Translating CIM XML Power System Data to a Proprietary Format for System Simulation," *IEEE Trans. on Power Systems*, vol. 19, no. 1, pp. 229–235, February 2004.

168. J. P. Britton and A. N. deVos, "CIM-Based Standards and CIM Evolution," *IEEE Trans. on Power Systems*, vol. 20, no. 2, pp. 758–764, May 2005.

169. A. W. McMorran, G. W. Ault, C. Morgan, I. M. Elders, and J. R. McDonald, "A Common Information Model (CIM) Toolkit Framework Implemented in Java," *IEEE Trans. on Power Systems*, vol. 21, no. 1, pp. 194–201, February 2006.

170. D. S. Popovic, E. Varga, and Z. Perlic, "Extension of the Common Information Model with a Catalog of Topologies," *IEEE Trans. on Power Systems*, vol. 22, no. 2, pp. 770–777, May 2007.

171. G. Ravikumar, Y. Pradeep, and S. A. Khaparde, "Graphics Model for Power Systems Using Layouts and Relative Coordinates in CIM Framework," *IEEE Trans. on Power Systems*, vol. 28, no. 4, pp. 3906–3915, November 2013.

172. R. Santodomingo, S. Rohjans, M. Uslar, J. A. Rodríguez-Mondèjar, and M. A. Sanz-Bobi, "Facilitating the Automatic Mapping of IEC 61850 Signals and CIM Measurements," *IEEE Trans. on Power Systems*, vol. 28, no. 4, pp. 4348–4355, November 2013.

173. L. Sabari Chandramohan, G. Ravikumar, S. Doolla, and S. A. Khaparde, "Business Process Model for Deriving CIM Profile: A Case Study for Indian Utility," *IEEE Trans. on Power Systems*, vol. 30, no. 1, pp. 132–141, January 2015.

174. B. Lee, D. K. Kim, H. Yang, H. Jang, D. Hong, and H. Falk, "Unifying Data Types of IEC 61850 and CIM," *IEEE Trans. on Power Systems*, vol. 30, no. 1, pp. 448–456, January 2015.

175. J. D. Moseley, N. V. Mago, N. D. R. Sarma, W. Mack Grady, and S. Santoso, "Extending CIM Standards to Support Exchange of Ratings on Dynamically Rated Equipment," *IEEE Trans. on Power Systems*, vol. 31, no. 1, pp. 296–303, January 2016.

176. A. McEwen and H. Cassimally, *Designing the Internet of Things*, John Wiley, Chichester, UK, 2013.

177. S. Greengard, *The Internet of Things*, The MIT Press, Essential Knowledge Series, Cambridge, MA, USA, 2015.

178. L. Atzori, A. Iera, and G. Morabito, The Internet of Things: A Survey, *Computer Networks*, vol. 54, pp. 2787–2805, 2010.

179. J. A. Stankovic, "Research Directions for the Internet of Things," *IEEE Internet of Things Journal*, vol. 1, no. 1, pp. 3–9, February 2014.

180. L. Da Xu, W. He, and S. Li, "Internet of Things in industries: A Survey," *IEEE Trans. on Industrial Informatics*, vol. 10, no. 4, pp. 2233–2242, November 2014.

181. A. Al-Fuqaha, M. Guizani, M. Mohammadi, M. Aledhari, and M. Ayyash, "Internet of Things: A Survey on Enabling Technologies, Protocols, and Applications," *IEEE Communication Surveys & Tutorials*, vol. 17, no. 4, pp. 2347–2376, Fourth Quarter 2015.

182. Parks Associates Whitepaper, "Top 2016 Trends for the Consumer IoT," 2016.

183. McKinsey Global Institute, "The Internet of Things: Mapping the Value Beyond the Hype," 2015.

184. C. Grant, J. McCue, and R. Young, "The Power Is On. How IoT Technology Is Driving Energy Innovation," Report, Deloitte Center for Energy Solutions, 2015.

185. S. Karnouskos, "The Cooperative Internet of Things Enabled Smart Grid," 14*th IEEE Int. Symp. on Consumer Electronics*, 2010.

186. IEC White Paper, "Internet of Things: Wireless Sensor Networks," 2014.

187. E. Spanò, L. Niccolini, S. Di Pascoli, and G. Iannaccone, "Last-Meter Smart Grid Embedded in an Internet-of-Things Platform," *IEEE Trans. on Smart Grid*, vol. 6, no. 1, pp. 468–476, January 2015.

188. G. Bedi, G. Kumar Venayagamoorthy, and R. Singh, "Navigating the Challenges of Internet of Things (IoT) for Power and Energy Systems," *Clemson University Power Systems Conference (PSC)*, Clemson, South Carolina, March 2016.

189. S. Ciavarella, J. Y. Joo, and S. Silvestri, "Managing Contingencies in Smart Grids via the Internet of Things," *IEEE Trans. on Smart Grid*, vol. 7, no. 4, pp. 2134–2141, July 2016.

190. Center for the Study of the Presidency and Congress, "Securing the U.S. Electrical Grid," 2014.
191. J. Weiss and M. Delson, "Cyber Security of Substation Control and Diagnostic Systems," Chap. 17 of *Electric Power Substations Engineering* (J. D. McDonald, ed.), 2nd ed., CRC Press, Boca Raton, FL 2007.
192. E. D. Knapp and R. Samani, *Applied Cyber Security and the Smart Grid*, Syngress-Elsevier, Waltham, MA, USA, 2013.
193. Y. Xiao (ed.), *Security and Privacy in Smart Grids*, CRC Press, Boca Raton, FL, USA, 2013.
194. Special Issue on Cybersecurity for Electric Systems, "Keeping the Smart Grid Safe," *IEEE Power & Energy Magazine*, vol. 10, no. 1, pp. 18–73, January/February 2012.
195. T. Baumeister, "Literature Review on Smart Grid Cyber Security," Dept. of Information and Computer Sciences, University of Hawaii, December 2010.
196. S. Sridhar, A. Hahn, and M. Govindarasu, "Cyber-Physical System Security for the Electric Power Grid," *Proc. of the IEEE*, vol. 100, no. 1, pp. 210–224, January 2012.
197. K. S. Wilson and M. Ayse Kiy, "Some Fundamental Cybersecurity Concepts," *IEEE Access*, vol. 2, pp. 116–124, February 2014.
198. Y. Yan, Y. Qian, H. Sharif, and D. Tipper, "A Survey on Cyber Security for Smart Grid Communications," *IEEE Communications Surveys & Tutorials*, vol. 14, no. 4, pp. 998–1010, Fourth Quarter 2012.
199. W. Wang and Z. Lu, "Cyber Security in the Smart Grid: Survey and Challenges," *Computer Networks*, vol. 57, pp. 1344–1371, 2013.
200. P. Jokar, N. Arianpoo, and V. C. M. Leung, "A Survey on Security Issues in Smart Grids," *Security and Communication Networks*, vol. 9, no. 3, pp. 262–273, February 2016.
201. A. Sanjab, W. Saad, I. Guvenc, A. Sarwat, and S. Biswas, "Smart Grid Security: Threats, Challenges, and Solutions," Available at http://arxiv.org/pdf/1606.06992v1.pdf.
202. I. L. G. Pearson, "Smart Grid Cyber Security for Europe," *Energy Policy*, vol. 39, pp. 5211–5218, 2011.
203. L. Piètre-Cambacédès, M. Tritschler, and G. N. Ericsson, "Cybersecurity Myths on Power Control Systems: 21 Misconceptions and False Beliefs," *IEEE Trans. on Power Delivery*, vol. 26, no. 1, pp. 161–172, January 2011.
204. IEEE PES Power System Relaying Committee Working Group C1, "Cyber Security Issues for Protective Relays," *IEEE PES General Meeting*, Tampa, FL, July 2007. (There is an expanded version available at http://www.pes-psrc.org/kb/published/reports.html).
205. A. Risley and K. Carson, "Low- or No-Cost Cybersecurity Solutions for Defending the Electric Power System Against Electronic Intrusions," Schweitzer Engineering Laboratories, Inc.
206. The Smart Grid Interoperability Panel—Cyber Security Working Group, "NISTIR 7628 Rev. 1- Guidelines for Smart Grid Cyber Security: Vol. 1, Smart Grid Cyber Security Strategy, Architecture, and High-Level Requirements; Vol. 2, Privacy and the Smart Grid; Vol. 3, Supportive Analyses and References," NIST, U.S. Dept. of Commerce, September 2014.
207. NERC Standards CIP-002 through CIP-009—Cyber Security, January 2011.
208. The Smart Grid Interoperability Panel—Cyber Security Working Group, "Introduction to NISTIR 7628 Guidelines for Smart Grid Cyber Security," September 2010.
209. CIGRE Working Group B5.38, "The Impact of Implementing Cyber Security Requirements Using IEC 61850," CIGRE Brochure 427, August 2010.
210. IEEE 1686-2007, "Standard for Substation Intelligent Electronic Devices (IEDs) Cyber Security Capabilities," 2007.
211. IEEE-1711-2010, "IEEE Trial-Use Standard for a Cryptographic Protocol for Cyber Security of Substation Serial Links," 2010.
212. IEEE Std 2030, "IEEE Guide for Smart Grid Interoperability of Energy Technology and Information Technology Operation with the Electric Power System (EPS), End-Use Applications, and Loads," 2011.
213. IEC TC57 WG15 (F. Cleveland, Convenor), "IEC 62351 Security Standards for the Power System Information Infrastructure," 2012.
214. T. Tibbals and D. Dolezilek, "Communications Technologies and Practices to Satisfy NERC Critical Infrastructure Protection (CIP)," *5th Annual Clemson University Power Systems Conference*, Clemson, South Carolina, March 2006.

215. A. Hahn, A. Ashok, S. Sridhar, and M. Govindarasu, "Cyber-Physical Security Testbeds: Architecture, Application, and Evaluation for Smart Grid," *IEEE Trans. on Smart Grid*, vol. 4, no. 2, pp. 847–855, June 2013.

216. Software Engineering Institute, "Smart Grid Maturity Model," Update, October 2010.

217. R. Caralli, M. Knight, and A. Montgomery, "Maturity Models 101: A Primer for Applying Maturity Models to Smart Grid Security, Resilience, and Interoperability," Carnegie Mellon University, November 2012.

218. Smart Grid Cybersecurity Committee, "Implementing Cybersecurity Frameworks: Utility Lessons Learned," SGIP White Paper, April 25, 2016.

219. National Institute of Standards and Technology (NIST), "Framework for Improving Critical Infrastructure Cybersecurity," Version 1.0, February 12, 2014.

220. NERC, "Grid Security Exercise. GridEx III Report," March 2016.

## 24.10 ACRONYMS

| | |
|---|---|
| ABS | Agent-based simulation |
| AGC | Automatic generation control |
| AMI | Advanced metering infrastructure |
| AMR | Automated meter reading |
| API | Application programming interface |
| CCAPI | Control center application programming interface |
| CDPSM | Common distribution power system model |
| CENELEC | Comité Européen de Normalisation Electrotechnique—European Committee for Electrotechnical Standardization |
| CGMES | Common Grid Model Exchange Standard |
| CIGRE | Conseil International des Grands Réseaux Electriques—Council on Large Electric Systems |
| CIM | Common information model |
| CIMug | CIM users group |
| CIP | Critical infrastructure protection |
| CIS | Component interface specification |
| CPA | Cyber-physical attack |
| CPSM | Common power system model |
| CPU | Central processing unit |
| DA | Distribution automation |
| DAE | Differential-algebraic equation |
| DER | Distributed energy resource |
| DG | Distributed generation |
| DHS | Department of Homeland Security |
| DMS | Distribution management system |
| DoE | Department of Energy |
| DoS | Denial of service |
| DR | Demand response |
| DSO | Distribution system operator |

| | |
|---|---|
| EA | Enterprise architecture |
| EMS | Energy management system |
| EMTP | Electromagnetic transients program |
| ENTSO-E | European network of transmission system operators for electricity |
| EPRI | Electric Power Research Institute |
| FA | Feeder automation |
| FACTS | Flexible AC transmission system |
| FEM | Finite element method |
| FPGA | Field programmable gate array |
| GHG | Greenhouse gas |
| GIS | Geographic information system |
| GPU | Graphics processing unit |
| GUI | Graphics user interface |
| HAN | Home area network |
| HIL | Hardware-in-the-loop |
| HPC | High-performance computing |
| HTML | HyperText Markup Language |
| HVDC | High-voltage direct current |
| ICCP | Inter-Control Center Communications Protocol |
| ICT | Information and communications technology |
| ID | Identification |
| IDS | Intrusion detection systems |
| IEC | International Electrotechnical Commission |
| IED | Intelligent electronic device |
| IEEE | Institute of Electrical and Electronics Engineers |
| ILF | Interactive load flow |
| I/O | Input/output |
| IOP | Interoperability |
| IoT | Internet of Things |
| ISO | Independent system operator |
| IT | Information technology |
| MIS | Market information service |
| MMC | Modular multilevel converter |
| NERC | North American Electric Reliability Corporation |
| NIST | National Institute of Standards and Technology |
| NRECA | National Rural Electric Cooperative |
| OPF | Optimal power flow |
| OWL | Ontology Web language |
| PHIL | Power-hardware-in-the-loop |
| PLC | Power line communication |
| PMU | Phasor measurement unit |
| PWM | Pulse width modulation |

| | |
|---|---|
| RCP | Rapid control prototyping |
| RDF | Resource description framework |
| RES | Renewable energy source |
| RFID | Radio frequency identification |
| RS | Rapid simulation |
| RTO | Regional transmission organization |
| RTS | Real-time simulator |
| RTU | Remote terminal unit |
| SA | Substation automation |
| SCADA | Supervisory control and data acquisition |
| SCL | Substation configuration description language |
| SGMM | Smart grid maturity model |
| SIL | Software-in-the-loop |
| TC | Technical committee |
| TSO | Transmission system operator |
| XML | eXtensible Markup Language |
| UCTE | Union for the co-ordination of transmission of electricity |
| UML | Unified Modeling Language |
| URI | Uniform resource identifier |
| VPN | Virtual private network |
| W3C | World Wide Web Consortium |
| WAMS | Wide-area measurement system |
| WAN | Wide-area network |
| WASA | Wide-area situational awareness |
| WG | Working group |

# 25

# STANDARDS IN ELECTROTECHNOLOGY, TELECOMMUNICATIONS, AND INFORMATION TECHNOLOGY

Marco W. Migliaro

*President and CEO, IEEE Industry Standards and Technology Organization (IEEE-ISTO); Fellow, IEEE*

Adam C. Newman

*Senior Director, Business Development and Alliance Management, IEEE Standards Association*



## 25.1 INTRODUCTION

*Standards* are technical agreements published in documents on which agreement has been reached, normally by consensus, that contain specifications or criteria to be used to ensure that materials, products, processes, tests, or services are suitable for their intended purpose. Those documents have gone from being published on paper to being published in electronic formats such as Adobe's Portable Document Format (PDF) and are now often being published in hypertext markup language (html) or eXtensible markup language (xml) for online and electronic uses.

Standards apply to virtually everything in the world today. The average person is not even aware of their existence, but life would not be the same without them. Engineers, computer scientists, and other scientists, however, are acutely aware of standards and their impact on the work they perform. Development of many of the original standards associated with electrotechnology was a slow process. The products, tests, or specifications being standardized were often in use in industry and had, in reality, become de facto standards before the standards that referred to them were written and approved. In general, most of the standards written were not mandatory, and it was voluntary for users to apply them. In some instances, these voluntary standards became part of government regulations, were adopted by government agencies, or were mandated by companies in specifications. When these types of events occurred, compliance with a specific standard became mandatory. Early standards also became regional to varying degrees. For example, the United States had its electric power standards and European countries had their own electric power standards. Although the two sets of standards had many similarities, there were significant differences between them. Overall, although there were some complaints, the standards development processes remained slow and the differences between standards continued to exist.

The 1980s brought some dramatic changes. The deregulation of the telephone industry, the privatization of government-run electric and telecommunication organizations, the birth of the information age, and the realization of a global market created a huge demand for standards in the fields of telecommunications and information technology. The Internet has, by itself, created the need for standards that govern its use, domain registration, and so forth. Many new players have entered the standards arena and found the traditional methods for developing standards unacceptable, primarily because they were too slow and bureaucratic. New standards in telecommunications and information technology were needed immediately, and the affected industries were moving so fast that many of the standards developed using traditional methods were obsolete before they were issued. Standards developers responded by streamlining processes and adopting fast-track systems. Even these proved too slow for some, and standards saw the birth of consortia whose charters included standards development. The global economy also caused people to take a hard look at existing regional standards. The differences between the standards were viewed as potential "barriers to trade," and harmonization efforts began to make these standards more widely accepted. For example, the United States and Canada are now harmonizing their electrical codes and standards. The global market brought about another significant change in standards. In addition to the traditional standards that dealt with products, tests, and specifications, the protection of intellectual property (e.g., trademarks, inventions, and copyrights) became an important issue. In addition, certification and conformity assessment has, in many instances, become a formal follow-on activity to standards development. Standards activity in electrotechnology, telecommunications, and information technology is greater today than ever before as standards developing organizations have focused on providing products for the full life-cycle of a standard. One example of increased standards activity is in the area of "smart grid." The first interoperability cross-discipline guide on smart grid was issued by IEEE in August 2011.[a] The guide includes a knowledge base that addresses terminology, characteristics, functional performance and evaluation criteria, and the application of engineering principles for smart grid interoperability of the electric power system with end use applications and loads. It is intended for use by the power engineering, communications, and information technology industries worldwide as a foundational, cross-discipline guide for smart grid interoperability. Many standards developers have already begun to make their standards available on the Internet for a fee, although grassroots initiatives have continued to allow free access to standards via the Internet.

Many consortia make their standards and specifications available at no charge. A few standards development organizations have also initiated programs that allow access to a limited number of standards without fee (e.g., the IEEE's "Get 802" program).

## 25.2 HISTORY OF ELECTRICAL STANDARDS

***Early History.*** The early history of electrical standards stems from activities dominated by the American Institute of Electrical Engineers (AIEE).[b] In 1884, the institute actively began to develop standard specifications for the growing electrical industry. In 1890, it proposed that the practical unit

[a]IEEE 2030-2011, IEEE Guide for Smart Grid Interoperability of Energy Technology and Information Technology Operation with the Electric Power System (EPS), and End-Use Applications and Loads.

[b]In 1963, the AIEE merged with the Institute of Radio Engineers to form the IEEE.

of self-induction be named the henry. At the same time, the institute appointed its first committee on standardization—the Committee on Units and Standards. The members of this committee were A. E. Kennelly, chairman, F. B. Crocker, W. E. Geyer, G. A. Hamilton, and G. B. Prescott, Jr. The institute also appointed a "Standard Wiring Table Committee" under the chairmanship of E. B. Crocker, to assign linear resistance of standard-conductivity copper wire and at standard temperatures.

A committee was also appointed to prepare a program for the delegates to the International Electrical Congress, held in Chicago in 1893, in regard to units, standards, and nomenclature. As a result of the congress, there were adopted units for magnetomotive force (gilbert), flux (weber), reluctance (oersted), and flux density (gauss). Subsequently, as a result of correspondence with engineering organizations in England, France, and Germany, the term *inductance* was adopted to represent the coefficient of induction (with the symbol $L$) and the present definition of the term *reactance* was proposed by Steinmetz and adopted.

***First Electrical Standards.*** In 1896, a "National Conference of Standard Electrical Rules" was held. The institute's delegate, Professor F. B. Crocker, was made its president, and in cooperation with other national organizations, the conference promulgated the "Underwriters' Rules," which finally resulted in the *National Electrical Code*® (NEC®).[c]

In 1897, the Units and Standards Committee recommended adoption of the standard of luminous intensity, or candlepower, as the output of the amylacetate Hefner-Alteneck lamp. It also recommended that the Lummer-Brodhun photometer screen be adopted for measuring the mean horizontal intensity of incandescent lamps.

At the beginning of 1898, a discussion was organized on the subject of "standardization of generators, motors, and transformers." This resulted in the formation of the first AIEE product standards committee, which in 1899 published the first electrical standard under the unique title *Report of the Committee on Standardization.*

***National Institute of Standards and Technology.*** The AIEE was a prime mover in the endorsement of a bill before the U.S. Congress, in 1901, for establishing a national standardizing bureau in Washington, DC, "for the construction, custody, and comparison of standards used in scientific and technical work." This bureau became known as the National Bureau of Standards (NBS) and has had a marked influence on the growth of U.S. technology. In 1988, the mission of the NBS was broadened by The Omnibus Trade and Competitiveness Act and other legislation, to help enhance competitiveness of U.S. industry and speed up the commercialization of new technology. At that time, the NBS was renamed the National Institute of Standards and Technology (NIST).

***International Electrical Standards.*** In 1904, an International Electrical Congress was held in St. Louis which set a precedent for international congresses related to electrical units and standards. The congress unanimously recommended the establishment of two committees. Committee 1 consisted of government representatives and was responsible for legal maintenance of units and standards. This committee evolved into the International Conference on Weights and Measures (GPMU). Committee 2, of which Lord Kelvin was elected president, was responsible for standards related to commercial products in the electrical industry and became the International Electrotechnical Commission (IEC) in 1906.

Another international body, the International Commission on Illumination [Commission International de l'Eclairage (CIE)], had its first meeting in 1913. The CIE establishes international units, standards, and nomenclature, in the science and technology of light and illumination.

***International Telecommunications Standards.*** In 1865 the first International Telegraph Convention was signed by 20 countries. This marked the formation of the International Telegraph Union (ITU). After the invention of the telephone in 1876 and wireless telegraphy (the first type of radiocommunication) in 1896, the scope of the ITU was broadened to include these new technologies. In 1906, the first International Radiotelegraph Convention was signed. The International Telephone

[c]*National Electrical Code* and NEC are registered trademarks of the *National Fire Protection Association (NFPA).*

Consultative Committee (CCIF) was formed in 1924, and the International Telegraph Consultative Committee (CCIT) was formed in 1925. In 1920 sound broadcasting began, and in 1927 the International Radio Consultative Committee (CCIR) was formed. At the Madrid Conference in 1932, the previous conventions were combined into the International Telecommunication Convention. The ITU changed its name in 1934, to the International Telecommunication Union. After World War II, the ITU became a specialized agency of the United Nations in October 1947. In 1956, the CCIF and CCIT merged to form the International Telephone and Telegraph Consultative Committee (CCITT). The year 1963 saw the first telecommunications satellite and the ITU set up a study group on space communications. Most recently, the Plenipotentiary Conference held in 1992 has remodeled the ITU to meet the challenges of the future. Today, work in the ITU covers all sectors of information and communication technologies including digital broadcasting, the Internet, mobile devices, and 3D television.

***International Standards Outside the Electrical Field.*** The original standards work in the fields outside electrotechnology was performed under the International Federation of the National Standardizing Associations (ISA), which was formed in 1926. ISA's activities ended in 1942 as a result of World War II. In 1947, the International Organization for Standardization (ISO) was established, as a result of a meeting of delegates from 25 countries that was held in London in 1946. Like the IEC, the ISO is a nongovernmental organization that promotes the development of international standardization and related activities. Its areas of responsibility are fields outside electrotechnology, light, and telecommunications.

***International Information Technology Standards.*** When the need for international standards in the field of information technology arose, it was clear that both the IEC and the ISO needed to be involved. In 1987, an agreement between the IEC and the ISO created the Joint Technical Committee on Information Technology (JTC-1). The ITU provides input to JTC-1 as an official liasion.

Another organization, The Internet Society (ISOC), was formed in 1992. Its formation came as a result of the INET Conference held in Copenhagen in 1991, where it was decided that a neutral and internationally recognized body devoted to the support of Internet administrative infrastructure was needed.

***National Standardization.*** Although an international standards organization for electrotechnology existed, representation on the IEC was by national committees from its member countries. Many of these countries had their own national standards organizations responsible for their national standards program, endorsement of national standards, participation in international standards development, and so forth. A number of these national organizations later became founding members of the ISO.

In the United States, five professional engineering societies and three government agencies spearheaded by the AIEE organized the American Engineering Standards Committee (AESC) in 1918. The AESC has been aptly described as a "national clearinghouse for industrial standardization." In its early years, this body was organized with 12 divisions, each based on its own area of technology. Few of these became active. The electrical engineering division actually became the strongest, even to the point of having its own bylaws. Today, the AESC is known as the American National Standards Institute (ANSI); however, at times during its history it was also known as the American Standards Association (ASA) and the United States of America Standards Institute (USASI). In 1926, under the auspices of the ASA, engineering abbreviations and symbols were standardized. The AIEE, in cooperation with ASA, sponsored in 1928 the development of a glossary of terms used in electrical engineering. This work was coordinated with the IEC.[d] Over the years, ANSI (and its predecessors) has had many responsibilities in the standards arena, including development of standards. Although many people still believe ANSI develops standards, it has not done so for many years. Standards that become American National Standards (ANSs) are written by 1 of more than 270 standards developers that can submit their completed standards to ANSI for acceptance as ANSs. It is interesting to

[d]The 6th edition of the IEC Multilingual Dictionary was published in 2005, with over 19,400 definitions in English and French. Equivalent terms wherever available are included in up to 11 additional languages, including Arabic, Chinese, Dutch, German, Italian, Japanese, Polish, Portuguese, Russian, Spanish, and Swedish.

note that, in the electrical industry, basic standardization was first in order of development, dating back before 1890. Technical standardization came next, with the formation of the Standards Committee of the AIEE in 1898. Manufacturing standardization came only as a result of World War I and did not take effect until 1920.

In Canada, the Canadian Standards Association (CSA) was formed in 1919. It is a not-for-profit organization supported by its members and develops standards in many fields. In 1970, a new organization called the Standards Council of Canada (SCC) was established by an act of parliament to coordinate voluntary standardization in Canada. The SCC is a federal Crown Corporation. The CSA represents Canada on a number of ISO Committees on behalf of the SCC. Standards may be submitted by accredited standards developing organizations to the SCC for approval as a National Standard of Canada.

***Regional Standardization.*** Once standards began to be developed, it did not take long for regional organizations such as the Organization of American States (OAS), the Pan American Standards Commission (COPANT), or the Pacific Area Standards Congress (PASC), and alliances such as the North American Treaty Organization (NATO) to see the value in having common (or harmonized) standards. For example, in May 1923, the OAS (then known as American States of the Pan-American Union) established the Inter-American Electrical Communication Commission (now known as the Inter-American Telecommunication Commission). However, it was not until the formation of three regional standards organizations by the European Economic Community that the world really took notice. In the area of electrotechnology, the European Committee for Electrotechnical Standardization (CENELEC) was formed in 1973. Telecommunications standardization is the responsibility of the European Telecommunications Standard Institute (ETSI). All other standardization is the responsibility of the European Committee for Standardization (CEN).

***Associations.*** Many associations have come to exist for various reasons, which may include standards. One of the earliest was the Association of Edison Illuminating Companies (AEIC). It was founded in 1885 to provide guidance to the Edison Illuminating companies that were being formed around the United States. The AEIC became a place where problems facing the growing electric utility industry could be solved by pooling the knowledge and experience of managers, engineers, and operators. Most of the work of the AEIC was technical in nature until 1948; however, in 1948 a committee was formed to deal with load forecasting and end-use management. The AEIC today continues to produce standards for equipment, such as cable, used by electric utilities.

Another example of an early association was Aeronautical Radio, Inc. (ARINC), formed in 1929 by four major airlines. ARINC was incorporated to serve as "the single licensee and coordinator of radio communication outside of the [U.S.] government." Once ARINC was organized, the Federal Radio Commission (predecessor of the Federal Communications Commission) transferred responsibility for all aeronautical ground radio stations to ARINC. ARINC, now part of Rockwell Collins, continues to provide services today to the airlines, aviation-related companies, and government agencies.

In the field of telecommunications, the Exchange Carriers Association was formed in 1983 as part of the breakup of the Bell System (i.e., AT&T) in the United States. It later changed its name to the Alliance for Telecommunications Industry Solutions (ATIS). Its membership is open to those involved in telecommunications in North America and the Caribbean. Committee T1 was formed in 1984 to give exchange carriers a voice in the creation of telecommunication standards, which had previously been developed, de facto, by AT&T. ATIS eventually became the secretariat for committee T1; however, committee T1 was retired in 2004 and its standards work was assumed by ATIS.

***Standards in Current Times.*** The information age and global economy have increased the demand for new standards that are internationally acceptable. The completion of the Uruguay Round of negotiations (1986–1994) of the General Agreement on Tariffs and Trade (GATT) led to the establishment of the World Trade Organization (WTO) in 1995 and a new set of agreements covering goods, services, and intellectual property. It also established a new dispute settlement mechanism. The WTO is the only international agency overseeing the rules of international trade with 164 member nations (as of July 2016). In 2001, the WTO began to host a new round of negotiations under the

Doha Development Agency. The formation of the European Commission (EC) along with its regional entities and requirements for compliance with European Norms has had a significant impact on standards. For example, in 1990 the European Organization for Certification and Testing (EOTC) was created under a memorandum of understanding between CEN, CENELEC, and the European Free Trade Agreement (EFTA) countries. EOTC was formed to promote the mutual recognition of test results, certification procedures, and so forth throughout the EC and EFTA countries. The North American Free Trade Agreement (NAFTA) and EFTA have had similar effects on the standards community.

Programs intended to harmonize standards to make them more internationally acceptable were instituted by many standards developers. Certification (or registration) began to take on additional importance to those organizations that wanted to compete in the global market. As a result, conformity assessment (which includes both registration and certification) programs began to expand, and in a number of instances certification organizations in one country expanded into other countries or partnered with a certification organization in another country. In 1979, ISO established a technical committee to harmonize the increasing international activity in quality management and quality assurance standards. One product of this committee was the ISO 9000 series of standards, which are internationally accepted and can provide a company that uses them with a route to the world markets. After ISO 9000, the ISO 14000 series of standards on environmental management tools and systems was developed. These standards address a company's system for managing its day-to-day operations as they impact the environment. ISO now has a number of series standards that focus on various topics including risk management, food safety management systems, and social responsibility.

Other changes that have occurred in the standards development arena are that the standards developers themselves are changing. Many have been renamed to reflect a more international flavor, and most have reengineered their processes to provide standards in a more timely manner or have begun to introduce new products such as emerging technology standards. For example, the IEC introduced the Industry Technical Agreement (ITA) as a new product in 1997 "in its drive to remain relevant in the field of electrotechnology."An ITA is a document that addresses specifications for areas of rapidly developing technology that can be created by any group of interested parties, including a single or large company or a consortium. The stakeholders themselves decide who participates in the creation of an ITA and what its content is. ITAs are not produced within the traditional IEC committee structure, nor are they consensus documents; however, they can be produced in months rather than years. Later, the Technology Trend Assessment (TTA) was introduced. A TTA presents the state-of-the-art or trend in a field of emerging technology that might become an area for standardization in the near-to-medium-term. TTAs are typically the result of research or prestandardization work. Although these programs have met with some degree of success, those in the information technology and telecommunications market want standards in place before, not after, their products are created. These industries are more interested in the anticipatory information provided by the standards process than they are in the final standard. They are also interested in new and more flexible forms of standards development.

As a result, these industries have turned to the formation of trade associations and consortia to develop their standards. In the United States, the number of standards produced in this manner will be greater than those produced by traditional SDOs. In contrast to traditional standards which are typically produced by volunteers, trade associations and consortia use paid professionals and provide them with budgets for expenses, research, legal advice, and so forth. Additionally, international standards organizations such as IEC and ISO have recognized a number of these standards as Publicly Available Specifications (PASs). A PAS is a publication that responds to an urgent market need, representing consensus in an organization (including consortia) outside the IEC or experts within a working group. Although PASs do not conflict with international standards, it is possible to have more than one PAS on the same subject. One example of an industry consortium is the World Wide Web Consortium (W3C). It was founded in 1994 and its mission is "to lead the World Wide Web to its full potential by developing protocols and guidelines that ensure long-term growth of the web."

Other examples are the Unicode Consortium, established in 1991, to bring together leading software corporations and researchers at the leading edge of standardizing international character encoding; the Open Group, established to answer questions in IT that corporate IT users need answers to by aiding in the development and implementation of a secure and reliable IT infrastructure; the

DSDM Consortium, established in 1994 to develop and promote a public-domain rapid application development method; and the North American Electric Reliability Corporation (NERC), founded in 1968 after the Northeast Blackout to promote reliability of the electrical supply for North America.

## 25.3 STANDARDS AND THE LAW

***Voluntary Standards System.*** The word *standard* has a number of meanings, but in the context of trade or engineering, it refers to voluntary technical standards that are normally developed by a consensus of experts. Many standards set safety or performance requirements for products or services, for example, standards for workshoes for those involved in electrical work to be "nonconductive with a reinforced toe" for safety purposes. These standards are not normally developed by lawmakers, and because they are outside of the mandatory scope of laws, they are sometimes referred to as *voluntary standards*. In general, standards are developed under a voluntary system. To the extent that their adoption is also voluntary, there is less vulnerability to legal liability. However, many standards are made mandatory, either through reference in purchase specifications and contracts or through adoption by government bodies as regulatory documents. For example, in certain states it may be illegal for a person to drive (or ride as a passenger) a motorcycle without a helmet that meets the requirements of a specific standard. Under such circumstances, compliance ceases to be voluntary and the effect of the document is to disqualify or limit the acceptability of certain products or services.

The ability of standards to limit acceptable suppliers is a potential danger of standards and the processes under which they are developed must minimize the possibility of discrimination against specific companies. One can begin to see that standards could be developed that contain absurd requirements that could act as a barrier to trade from a foreign nation, or within the same nation violate antitrust laws. Additionally, once a standard is written into the law, if the law simply states that compliance with standard *xyz* is required, then any revision to the standard (as occurs periodically) has the effect of amending the law.

***Legality of Standards.*** The legality of standards activities is primarily affected by laws related to the fixing of prices, conspiracy in restraint of trade, and intellectual property. Throughout history, standards have been well-known barriers to trade as countries hide protectionism in the veil of an absurd standard. An example would be a standard written by a country that requires the use of a specific material available only in that country for a particular part of the product. Today, however, standards are covered by the GATT, the WTO, and other trade agreements such as NAFTA. The WTO supports the use of international standards developed under the auspices of international standards organizations such as IEC and ISO. One reason for this is the belief that it is felt that the international development process will identify and exclude any documents that contain hidden trade barriers. From the users' perspective, certification to international standards should result in greater international acceptance for their product or service.

In the United States, the two key governmental agencies involved are the Federal Trade Commission (FTC) and the Department of Justice. In 1975, the FTC and the Justice Department held hearings on a number of abuses of standards development and certification activities. These abuses involved individuals involved in the standards process who attempted to use standards for market advantage or to deny competitors entrance into an established market. One conclusion of the hearings was that there needed to be some fundamental guidelines and practices that would ensure that the activities related to standards development would be "fair." Those fundamentals were due process, openness, balance, public notice, and the right to participate and appeal. These are the same basic principles that govern standards development around the world; however, they have, in many instances, been interpreted and reinterpreted to the point where they slow down the process. This, as stated earlier, has driven many away from the traditional standards organizations, especially in the areas of new, emerging, or rapidly evolving technologies.

Since standards activities involve meetings in which representatives of competing organizations make agreements that affect engineering and industrial practices (both of which have economic implications), such meetings must take place under conditions that are subject to carefully regulated procedures. Failing this, participants could be subject to charges of violation of antitrust or conspiracy statutes. Trade associations and consortia are particularly vulnerable in this respect, as meetings restricted to their membership involve participants who tend to be exclusively competitive manufacturers, whereas meetings of committees of professional societies involve technical personnel who are more apt to be representative of the total industry (both manufacturers and users), independent consultants, government personnel, educators, and scientists. Similarly, international standards committees are populated by experts from the national committees from the member countries. These persons individually represent the consensus of experts in their country when developing or voting on standards. However, the degree of liability of participants in standards development activities is virtually negligible when these activities are conducted under the auspices of, and under the strict rules of, an organization experienced in standards development; that is, an organization whose procedures are designed to promote fair and unprejudiced participation by all eligible parties.

***Certification.*** The certification of a product provides additional assurance that a product is reasonably safe and reasonably suited for its intended function. Certification is particularly important for products that are purchased by the general public (i.e., consumers). Legal action against certifying organizations is rare; however, negligence in the certification process could cause a certifying organization to become exposed to a claim for liability.

***Patents.*** The issue of patents as related to their use (or specification) in standards has become something that most standards developers have had to deal with, particularly for new technology standards. Most standards development organizations (SDOs) have patent policies that require disclosure of patents at the time a proposal for development of a standard is submitted. They further require disclosure at any time in the standards development process that it becomes known that a patent is applicable to a standard. Some consortia/associations that are developing standards/requirements may not require patent disclosure.

***Personal Liability.*** An area of legal concern for participants in the standards generation or approval process is the question of legal liability. A typical situation deals with the case where an accident occurs under circumstances where potentially negligent parties demonstrate that they faithfully complied with the provisions of the applicable safety standards. The question here is one of the extent of liability of those who participated in the generation or adoption of the standard. A somewhat equivalent situation arises in product liability cases. Any such claim in a legal action turns on allegations of negligence in writing the standard. The general conclusion held by counsel is that members of voluntary standards committees operating under procedures that embody the fundamental principles of due process, openness, balance, public notice, and the right to participate and appeal are not likely to incur significant legal risks.

Some standards developers indemnify those persons who are members of the organization, provided the processes of the organization have been followed during the standards development process. Other standards developers require those participating in its standards writing groups to sign a statement attesting to the fact that they will follow the organization's procedures when participating in standards activities for the organization.

## 25.4 THE VOLUNTARY STANDARDS PROCESS

***Voluntary Development of Standards.*** There are literally tens of thousands of experts in the fields of electrotechnology, telecommunications, and information technology who participate in standards development worldwide. It should be clear that it is the process by which the standards are developed that is of importance. The process should embody the fundamental principles; however, there are almost as many processes as there are standards developers. Although it would be impossible to describe them all in this subsection, the information age has made access for interested parties easier than ever before. Today, the World Wide Web allows direct access to a wealth of standards information. A listing of organizations (including their acronyms) and their websites, is provided at the end of this section to enable an interested party to begin exploration of the "world of standards." Many of the websites contain the full text of the policies and procedures followed by the standards developer. These are typically available for download.

Although the development processes are designed to enable the broadest possible input to standards, duplicate standards, conflicts between standards, and other problems may exist. The resolution of these problems may be by the standards developers themselves, or by a national or international organization. For example, in the field of electric power cable, conflicts have arisen between standards developed by ASTM and IEEE that had to be resolved by the developers, or by ANSI where resolution was not achieved by the developers. In the case of duplicate standards in a voluntary system, the standard of choice will usually be decided by the market (i.e., the standard referenced by users will become the standard used). Complementary standards for a product are sometimes developed by two or more standards organizations, with each organization responsible for preparing standards within its area of expertise. For example, for power switchgear standards in the United States, IEEE develops those standards related to specifications and tests, while NEMA is responsible for those standards related to ratings. Occasionally, two or more organizations may develop a standard jointly; however, issues related to copyright and reproduction need to be agreed to by each organization before work begins. An example of this is standard IEEE/ASTM SI10, which was jointly developed by IEEE and ASTM International and first published in 1976. This standard is intended to give authoritative information on use of the SI system of units and guidance on its application to U.S. citizens and industry. Its use became more important in 1988 when the U.S. Metric Conversion Act was amended to designate "the metric system of measurement as the preferred system of weights and measures for U.S. trade and commerce." Many U.S.-based standards developers today have policies requiring the use of SI units as the primary unit in its standards, with English units shown in parentheses or in an annex.

Approval of standards is normally by consensus, with the definition of what constitutes consensus defined by the developing organization's procedures. What this means is that a standard may be made available to users once consensus is reached by the sponsor. A problem here is that some sponsors apply a more rigid interpretation of consensus than intended (some even try to attain unanimity), which ultimately delays the standard. In these instances, the sponsor is doing a disservice to the users that need the standard. Although the sponsor is responsible for the technical content of the standard, the standards developer often has some sort of authority (e.g., a standards board) that is responsible for ensuring the organization's process has been followed. Approval by that authority is the final step in the process before publication. Here, too, interpretation of the process by the authority can cause standards to be returned to the sponsor, resulting in an unnecessary delay of the standard.

Another part of the process that has not been discussed is the maintenance of a standard once it is developed. All standards require periodic review to ensure that the information contained in them is current. Once reviewed, the standard may be reaffirmed (or confirmed), revised, or withdrawn. The standards developers also have the ability to administratively withdraw a standard if it has not been reviewed by its sponsor within some maximum period of time. For example, in the United States, the maximum time a standard can exist without being revised or reaffirmed is 10 years. The IEEE requires the majority of its standards to be reviewed every 5 years (a stabilized standard that has little likelihood of revision is reviewed every 10 years). The standard is administratively withdrawn by IEEE after 5 (or 10 for stabilized standards) years if maintenance of the standard by its sponsor has not been initiated. Internationally, both ISO and IEC require a review of their standards by the responsible Technical Committee (or Subcommittee) periodically. The review periods can vary based upon an assessment of the maturity of the standard and foreseen changes to associated standards. Based on this, a year is targeted for completion of maintenance (e.g., revision) of the current standard. This year is called the stability date and is included on the website www.webstore.iec.ch. The process for maintenance of standards developed by some consortia is an area that traditional standards developers are targeting for new work (i.e., to provide standards maintenance services to consortia).

***Intellectual Property Rights and the Voluntary Standards Process.*** Another important aspect to the standardization process is the consideration of intellectual property (IP) issues. In many cases today companies have IP in the technology they seek to include in voluntary or mandatory standards. In some cases those IP rights may include patent rights in one or more jurisdictions. SDOs of all types must have IP rights (IPR) policies that address the contribution and inclusion of such technologies to their standardization work. Over the last two decades patents have been granted to system and software design innovations that have significantly impacted the work of standards developers, and SDOs have been working to understand and in some cases update their IPR policies in light of these events. As noted above, most SDOs require the declaration and disclosure of such IPR when it is contributed to a standards project. SDOs generally have one of three regimes for acceptance of IP: Reasonable and Non-Discriminatory (RAND), Fair Reasonable and Non-Discriminatory (FRAND), and Free RAND where access to the IP is expected to be free reasonable and nondiscriminatory. Determination of fair and reasonable can be subject to significant differences between parties involved. Additionally, as part of the standards process in most SDOs the patent holder is normally required to provide a letter stating that (1) the patentee will not enforce any of its present or future patent(s) whose use would be required to implement the proposed standard against any person or entity using the patent(s) to comply with the standard or (2) a license will be made available to all applicants without compensation or under reasonable rates, with reasonable terms and conditions that are demonstrably free of any unfair discrimination. If these conditions are met, the patent may be included in the standard; however, the standards developer normally publishes a disclaimer in the standard making no claims as to the validity of the patent or the reasonableness of rates and/or terms and conditions of the license. In the case of SDOs that have individual membership as participants rather than entities, there is some allowance often made for the individual to declare or disclose to the best of their knowledge but not require a complete patent search within their organization. Some large entities that have large IP portfolios and seek monetization via licensing rather than sales of an end product (or in addition to such sales) may look to choosing to contribute to an SDO or consortia/alliance based in significant part on the IPR policy in place and not necessarily on the traditional scope of that SDO. This is part of the marketplace that has lead to a burgeoning number of alliances/consortia, especially in the design and software technology areas.

## 25.5 TERMINOLOGY IN STANDARDS

***Standards Terms.*** The following comprises a partial list of terms used by participants in standards activities. Many of these terms have unique and specialized meaning when used in the context of standardization, and a brief definition is given for each as applied in this context. Some of the terms are specific to the processes of the IEC and are identified by an (IEC) at the end of the explanation of the term.

**accreditation** A process in which certification of authority, competency, or creditability is verified typically by an external (or third-party) review or assessment.

**approval stage** The point after the enquiry stage, at which the final-draft international standard (FDIS) is circulated to the National Committees for a 2-month voting period. If the FDIS is approved it is published, and if it is not approved it is referred back to committee for reconsideration (IEC).

**balance** The characteristic of a standards approving unit (committee, subcommittee, or working group) which assures that all classifications of interests are represented and that no single classification has a representation sufficiently large to enable it to unduly influence the resulting output.

**balanced committee** A committee so constituted as to maintain a balance among its members. Many committees are balanced among manufacturers, users, and general-interest classifications.

**basic standard** A standard common to all disciplines, or to an overall technology.

**canvass** A method used for approval of standards which is dependent on circulation of a draft document to a list of concerned organizations for review and ballot.

**certification** An attestation to the effect that a particular product or service meets the requirements of a relevant standard.

**certification mark** A special kind of trademark that appears *only* on products that have been certified against a standard.

**classification of membership** The classification assigned to a participant or member of a standards developing unit which identifies the member's functional relationship or interest in the subject to be standardized. Thus, a participant may be a manufacturer of a product being standardized, a user or purchaser of the product, a technically qualified expert with no well-defined functional relationship (classified as general-interest), a labor or insurance representative (in the case of safety standards), or a constructor (one who installs the product for use by others). A variety of other classifications is possible as dictated by the scope of the standards activity.

**code** (*a*) A body of recommendations of good practice to be followed during design, manufacture, construction, installation, operation, and maintenance to satisfy considerations of safety, quality, economy, or performance in a given application. (*b*) A particular form of identification marking or reference which serves the dual purpose of establishing in a systematic manner the complete identity of an individual product and of identifying its similarity with other products. It may consist of a brief, systematic combination of letters, numerals, and symbols.

**committee stage** The point at which the document is submitted as a committee draft (CD) to the National Committees for comment (IEC).

**conformity assessment** An activity or set of activities that determines directly or indirectly whether relevant local product requirements have been fulfilled. Typical forms of conformity assessment include testing, inspection, assessment, auditing, registration, and certification.

**consensus** A substantial agreement of those concerned. It implies that no important interested parties are strongly opposed on substantive grounds, or alternatively, that any opposition is in a small minority and the changes required to effect agreement by this minority would lead to substantive disagreement by the majority. Consensus implies that all disagreements have been given careful consideration and all reasonable attempts have been made for their resolution.

**designation** A definite and distinguishing name or symbol given to a product or to a group of functionally similar products or to an abstract matter. It emphasizes the group similarity but does not bring out the differences among the various members of the group.

**dimensional interchangeability** A condition in which the dimensions of two or more products are such that one can physically replace another in a given application.

**dimensional standard** A standard whose main content is dimensions and sizes of a product or group of products.

**e.ballot** A "letter ballot" or equivalent that is conducted electronically (e.g., via e-mail).

**enquiry stage** A point before the approval stage where the bilingual Committee Draft for Vote (CDV) is submitted to all national committees for a 5-month voting period. It is the last stage at which technical comments can be taken into consideration (IEC).

**functional interchangeability** A condition where the characteristics of two or more products are such that they are able to perform the same functions.

**guide** A standards document that provides alternative information which comprises good engineering practice. Guides may contain application information for use of products and may be tutorial in nature. The user should be cautioned that the use of the word "guide" in the title of a document does not guarantee that the document is in fact nonmandatory. There are many governmental regulatory guides which in fact set forth mandatory requirements. Conversely, many documents that are differently titled are in fact guides.

**harmonization** The act of coordinating requirements from multiple standards (e.g., multiple countries or multiple SDO's standards) and copublishing the resulting document.

**harmonization committee (or task force)** A group of individuals responsible for technically developing the proposed draft of the harmonized standard. The group typically consists of a representative of each involved country (or SDO), the secretariat, and the chair.

**interface standard** A standard whose main purpose is to ensure coordination between systems.

**international standard** A standard that has been adopted by a recognized international standards body (such as IEC or ISO).

**joint publication** A standard that has been submitted through the standards development process of two or more SDOs and is published separately by all involved SDOs. A joint publication may also be referred to as a harmonized or copublished standard, or if the SDOs represent different countries, as a binational or trinational standard.

**letter ballot** A ballot used in standards development to determine agreement on a draft standard, or to generate comments that will be instrumental in developing a document on which consensus agreement can be achieved. Such ballots provide for affirmative and negative votes. Negative votes, however, must be accompanied by reasons in sufficient detail to enable the writers of the document to determine what steps need to be taken in revision to change the vote from negative to affirmative. The primary advantage of a letter ballot is that it provides adequate time for the recipients to review thoroughly the document that is subject to ballot.

**marking** The action and the result of stamping, inscribing, printing, or labeling marks, symbols, letters, or numerals on a product or its package for the purposes of identifying the product.

**may** An operative verb used in a standards document which identifies a possible means for satisfying a requirement. For example, several alternative procedures may be indicated for measuring a particular characteristic or phenomenon, and the selection of the most suitable procedure is left to the user of the document.

**national standard** A standard that has been adopted by a recognized national standards body (such as ANSI), or a standard that is in effect recognized and used nationally in preference to other documents.

**O-member** An observer member of a technical committee who has the right, but not the obligation, to submit comments and attend meetings (IEC).

**performance characteristic** A characteristic of a product which determines the product's suitability for a specific application.

**P-member** A participating member of a technical committee who is obliged to contribute to meetings and vote at all stages (IEC).

**preliminary stage** Projects envisaged for the future but not yet ripe for immediate development, or preliminary work, such as better definition of a project for new work, data collection, or round-robin tests necessary to develop standards, which is not part of the standardization process (IEC).

**preparatory stage** The phase during which a working draft (WD) of a document is prepared (IEC).

**product standard** A standard containing requirements to be met by a product or group of products, usually including, directly or by reference to other standards, all or some of the following elements: dimensions, performance characteristics, other characteristics, and test methods.

**proposal stage** A proposal for new work originated from industry via a National Committee, communicated to the members of a technical committee or subcommittee with a form. A simple majority vote takes place within 3 months; if the result is positive and the minimum number of experts (4 or 5 depending on committee size) are nominated by the P-members that approved the proposal, it is included in the work program (IEC).

**publication stage** The period after approval of an FDIS that ends with the publication of the international standard, normally within 1.5 months. During this stage, minor editorial changes can be made to the final text (IEC).

**rating** A characteristic of a product which is determined in an arbitrary, yet consistent, manner, based on the intended function of the product.

**recommended practice** A standards document that provides information on good engineering practice. Such documents may contain application information for use of products.

**registration authority** A body given the authority for maintaining lists of codes under standards and issuing new codes to those wishing to register them.

**safety standard** A standard whose primary purpose is to ensure the safety of people and property.

**SDO** Standards Developing Organization.

**secretariat** An organization that assumes the responsibility for providing administrative oversight of a standards committee's activities and assures compliance with all applicable procedures.

**self-certification** An attestation by a manufacturer or supplier of a product or service that it meets the requirements of a relevant standard.

**shall** An operative verb used in a standards document which indicates a mandatory requirement that must be specifically complied with for conformance to the document.

**should** An operative verb used in a standards document which indicates a problem area that must be resolved and specifies a requirement, compliance with which resolves the problem. In this sense, the verb "should" can be read as "shall." Alternatively, it is allowable under the document to use some other method which can be proved to resolve adequately the condition or problem area addressed. In some cases, it is also possible to demonstrate clearly that the condition or problem area addressed does not in fact exist, or apply to the product or circumstance in a specific instance.

**simplification** A form of standardization consisting of the reduction of the number of types of products within a definite range to that number which is adequate to meet prevailing needs at a given time.

**specification** A standards document that specifies all the characteristics and conditions to be met by a product or service to be supplied to the purchaser. Such a document may refer to other standards, selecting among the specific allowable options. A specification is intended to be a complete purchasing document.

**sponsor** The group (e.g., a technical committee) that assumes responsibility for the development and/or maintenance of a standard.

**standard** A documented agreement containing technical specifications or other precise criteria to be used consistently as rules, guidelines, or definitions of characteristics to ensure that materials, products, processes, and services are fit for their purpose.

**standardization** An activity aimed at an increase of order, giving solutions for recurring problems in the spheres of scientific, technological, and economic activity. Generally it consists of the processes of formulating, issuing, and implementing standards.

**technical specification** Similar to an international standard in that it is normative in nature and developed according to consensus procedures, except that the final vote is taken at the draft technical specification (DTS) stage immediately following the committee draft (CD) stage (IEC).

**terminology standard** A standard containing exclusively terms and their definitions.

**test standard** A standard containing test methods which may be combined with other requirements related to testing, such as sampling, use of statistical methods, and sequence of tests.

**third-party certification** An attestation by a recognized, technically qualified, independent organization that a product or service supplied by others meets the requirements of a relevant standard. Such certification may be based on inspections and tests conducted by the certifying organization, or on supervision, monitoring, or auditing by the organization of such tests which may be conducted by others. The tests may be performed by the manufacturer or supplier of the service or product while being witnessed or audited by the certifier.

**trial-use** A publication (standard, recommended practice, or guide) that is effective for not more than 2 years, during which time comments and criticisms from a broad constituency are sought. In the absence of comments during the trial-use period, the document is subject to automatic approval on recommendation of the sponsor.

**unification** A form of standardization in which two or more specifications are combined into one in such manner that the products obtained are interchangeable in use.

## 25.6 ISO 9000 AND ISO 14000 STANDARDS

***ISO 9000 and ISO 14000 Overview.*** Unlike the vast majority of ISO standards which are highly specific, ISO 9000 and ISO 14000 are generic families of standards and guidelines relating to management systems. When introduced in 1987 and 1997, ISO 9000 and ISO 14000, respectively, introduced standards to a much broader business community (i.e., beyond organizations in the field of engineering and science only) than did any standards published previously. ISO 9000 is concerned primarily with quality management systems, while ISO 14000 is concerned mostly with environmental management systems. In the context of the standards, *quality management* means what an organization does to conform to its customers' requirements; *environmental management* means what an organization does to minimize harmful effects on the environment caused by its activities. Neither standard is a product standard, and organizations such as law firms, consulting engineers, or standards developers can become ISO 9000–certified. The reason for this is that both ISO 9000 and ISO 14000 are concerned with the way an organization goes about its work (i.e., the process), rather than the direct result of the work (e.g., a product or a service). (For additional information visit the ANSI, ASQ, ISO, and NIST websites listed at the end of this section.)

***ISO 9000 History.*** In 1959, the U.S. Department of Defense (DoD) established a quality management program designated as MIL-Q-9858, which was later revised to MIL-Q-9858A. NATO essentially adopted the provisions of MIL-Q-9858A in 1968, and published them in Allied Quality Assurance Publication 1 (AQAP-1). In 1970, the U.K. Ministry of Defense adopted the provisions of AQAP-1 when it published its Management Programs Defence Standard DEF/STAN 05-8. In 1979, the British Standards Institution (BSI) developed the first commercial quality management system standard, BS 5750. It was from all of these documents, and BS 5750, in particular, that ISO created the ISO 9000 standards family of documents (the base standards of which are shown next).

***When Countries Adopt ISO 9000.*** By 1992, the European Economic Community (EEC) and 56 countries had adopted ISO 9000. The EEC and other countries assigned numbers to the adopted ISO 9000 standards according to their own national standards numbering system. In the United States, the ISO 9000 series was adopted as ANSI/ASQ Q9000 in 1987. The ANSI/ASQ Q9000 series is essentially identical to the ISO 9000 series with the exception that the text incorporates customary American English language and spelling. Some other examples of adoption of the ISO 9000 series are the EEC as the European Norm (EN) 29000 series, the United Kingdom as BS 5750 Parts 0 to 3, Pakistan as PS 3000-3004 series, Tanzania as TZS 500-504, and China as GB/T 10300.1–10300.5.

***Certification.*** ISO does not itself certify conformity to ISO 9000. This is done by independent certification bodies in different countries. There is also no "official" database of entities certified to ISO 9000.

***ISO 9000.*** This standard, entitled *Quality Management Systems—Fundamentals and Vocabulary*, explains fundamental quality concepts. Additionally, it defines terms and provides guidance on selecting, using, and tailoring the standards in the series.

***ISO 9001.*** This standard, entitled *Quality Management Systems—Requirements*, is the most comprehensive standard in the series. It addresses all the elements in design, development, and so on and provides requirements for quality planning.

***ISO 9002.*** Superseded by the 2000 Edition of ISO 9001.

***ISO 9003.*** Superseded by the 2000 Edition of ISO 9001.

***ISO 9004.*** This standard, entitled *Quality Management Systems—Guidelines for Performance Improvements*, provides guidance for developing and implementing an internal quality system.

Since their introduction, the number of standards in the ISO 9000 and ISO 14000 families has grown to more than 40 documents. In 2004, ISO 90003 was introduced which addresses guidelines for the application of ISO 9001 to computer software.

## 25.7 INTERNATIONAL ORGANIZATIONS

***The International Electrotechnical Commission.*** The IEC, a nongovernmental body located in Geneva, Switzerland, is the world organization that develops and publishes international standards for electrotechnology and related technology. Its membership is limited to countries. Today, the IEC membership consists of 82 participating countries, including all the world's major trading countries (see Table 25-1). Full membership (presently at 60) in the IEC allows the country to participate in

**TABLE 25-1** Member Countries of the IEC

| | | |
|---|---|---|
| Albania* | Hungary | Oman |
| Algeria | Iceland* | Pakistan |
| Argentina | India | Republic of the Philippines |
| Australia | Indonesia | Poland |
| Austria | Iran | Portugal |
| Bahrain* | Iraq | Qatar |
| Belarus | Ireland | Romania |
| Belgium | Israel | Russian Federation |
| Bosnia and Herzegovina* | Italy | Saudi Arabia |
| Brazil | Japan | Serbia |
| Bulgaria | Jordan* | Singapore |
| Canada | Kazakhstan* | Slovakia |
| Chile | Kenya* | Slovenia |
| China | Republic of Korea | South Africa |
| Colombia | Latvia* | Spain |
| Croatia | Libya | Sri Lanka* |
| Cuba* | Libian Arab Jamahiriya | Sweden |
| Cyprus* | Lithuania* | Switzerland |
| Czech Republic | Luxembourg | Thailand |
| Democratic People's | Macedonia* (former Yugoslav Rep. of) | Tunisia* |
| Republic of Korea* | Malaysia | Turkey |
| Denmark | Malta* | Ukraine |
| Egypt | Mexico | United Arab Emirates |
| Estonia* | Montenegro* | United Kingdom |
| Finland | Morocco* | United States of America |
| France | Netherlands | Vietnam* |
| Georgia* | New Zealand | |
| Germany | Nigeria* | |
| Greece | Norway | |

*Associate member.

all international standardization activities. Participation of a country is by a national committee. Each national committee agrees to open access and balanced representation from all electrotechnical interests in its country (i.e., public and private). A country may also become an associate member (presently 22), which allows for limited participation in the IEC. Associate members have observer status at all IEC meetings, but have no voting rights.

The mission of the IEC is to promote, through its members, international cooperation on all questions of electrotechnical standardization and related matters, such as the assessment of conformity to standards, in the fields of electricity, electronics, and related technologies, including magnetics and electromagnetics, electroacoustics, telecommunication, energy production and distribution, terminology and symbols, measurement and performance, dependability, design and development, safety, and the environment.

The work of the IEC is carried out by more than 10,000 experts worldwide who participate on more than 175 technical committees and subcommittees, and more than 1100 working groups, project teams, and maintenance teams. All IEC publications are bilingual (English and French). Certain documents have also been translated into Spanish. The Russian Federation National Committee develops Russian-language editions of IEC documents. English, French, and Russian are the three official languages of the IEC; however, in all bodies of the IEC other than the Council, discussions may be held in English and/or French, following agreement of the delegates. Standards developed by the IEC follow the procedures in the ISO/IEC Directives, published jointly by both organizations and administered by the Joint Technical Program Committee (JTPC). These directives are published in two parts. Part 1 covers the procedures for the technical work. Part 2 covers drafting and presentation of international standards. These common procedures were adopted by ISO/IEC in recognition of the need to develop timely and cost-effective international standards. The joint ISO/IEC Committee JCT 1, Information Technology develops and maintains its own procedures. All of these procedures may be accessed on each organization's website.

Figure 25-1 is an organization chart for the IEC. The Council, which is a "general assembly" of committees, is the supreme authority of the IEC and sets policy, financial objectives, and strategy.

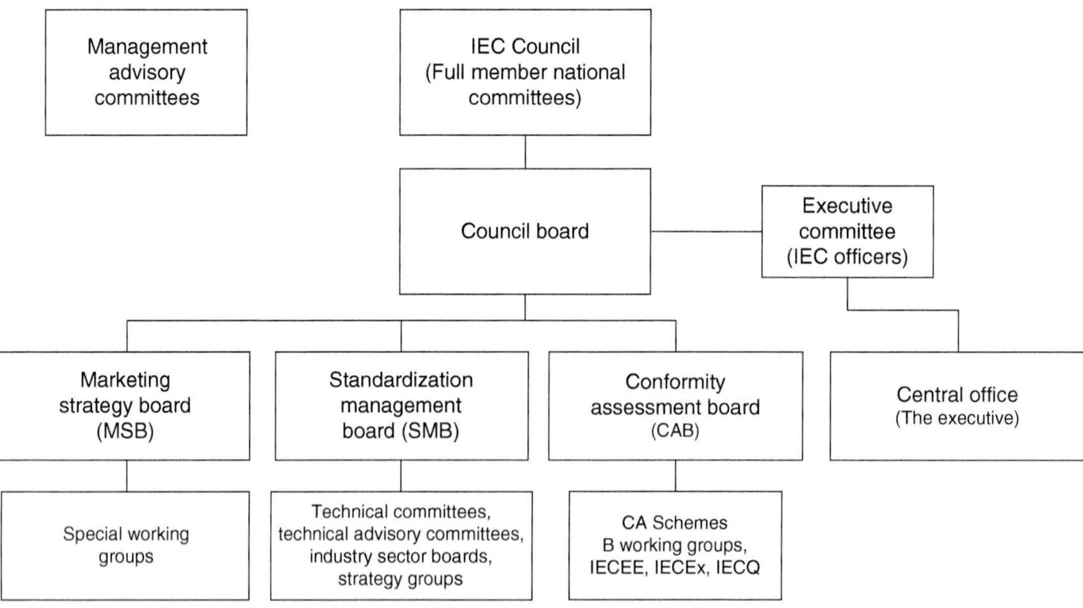


**FIGURE 25-1** IEC structure and management.

The Council delegates the management of all IEC work to the Council Board. The responsibility for standards and conformity assessment is assumed by the Standardization Management Board and Conformity Assessment Board, respectively. In the areas of standards and conformity assessment, the IEC works closely with other international organizations such as the ISO, WTO, and ITU. It also has relationships with governmental agencies and regional standardization organizations. For example, an agreement between IEC and CENELEC ratified in 1996, known as the Dresden Agreement, addresses common planning of new work and parallel IEC/CENELEC voting. Other standards organizations may interface with the IEC as liaisons. For example, the IEEE is recognized as a Class D liaison to the IEC.

***International Organization for Standardization.*** Like the IEC, ISO is a nongovernmental body, located in Geneva, Switzerland, that is a worldwide federation of national standard bodies. ISO is the world's largest developer of standards, publishing approximately 18,000 documents. Its membership is limited to countries and, at present, 163 countries are members of ISO. The member body of a country to ISO is the national body "most representative of standardization in its country," and only one body in each country can be admitted to membership in ISO. So, for example, ANSI is the ISO member from the United States, Standards Council of Canada is the ISO member from Canada, Standards Australia is the ISO member from Australia, and DIN is the ISO member from Germany. ISO has two other categories of membership for countries. The first is correspondent member, for countries which do not yet have a fully developed national standards activity; the second is subscriber member, for countries with very small economies.

ISO likes to think of its documents as international agreements that are published as international standards. Some might note an apparent inconsistency between the short form of the organization's name, specifically, ISO and its official title. That is because the short form of the name is not an acronym as many believe, but rather a word ISO derived from the Greek *isos* meaning equal. On this basis, the connection between "standard" and "equal" is easy to understand.

The mission of ISO is essentially to promote the development and standards-related activities in the world with a view to facilitating the international exchange of goods and services, and to developing cooperation in the spheres of intellectual, scientific, technology, and economic activity. The scope of ISO's work is not limited to any particular field, except that electrotechnology is the responsibility of the IEC and information technology is carried out by JTC 1 (a joint committee of ISO/IEC).

Among the documents that affect those in electrotechnology that ISO is responsible for is the universal system of measurements, known as SI (Système International d'Unités) units, which are described in a series of 14 international standards. The work of ISO is carried out by thousands of experts worldwide who participate on more than 1100 technical committees and subcommittees, and more than 2400 working groups. An organization chart for ISO is shown in Fig. 25-2. ISO cooperates with other international bodies such as the IEC and ITU, with regional standardization organizations and is building a strategic partnership with the WTO. ISO also has liaisons with over 700 entities worldwide that are interested in specific aspects of its standardization work. All ISO publications are bilingual (English and French) and are developed following the ISO/IEC directives.

***The International Telecommunication Union.*** The ITU, unlike the IEC and ISO, is part of the United Nations (UN), and its members are governments. It is located in Geneva, Switzerland and has 12 regional and area offices around the world. Today the ITU has 193 member states (i.e., countries) and approximately 700 private-sector entities. The ITU is an organization "within which governments and the private sector coordinate global telecom networks and services." It is the leading publisher of telecommunication and information technology, regulatory, and standards information. The governing bodies of the ITU are shown in Fig. 25-3. The Plenipotentiary Conference, held every 4 years (most recently in 2014, in the Republic of Korea), has the responsibility to determine the structural and operating changes necessary in the ITU to effectively influence and affect the development of information and communication technologies (ICTs) worldwide.

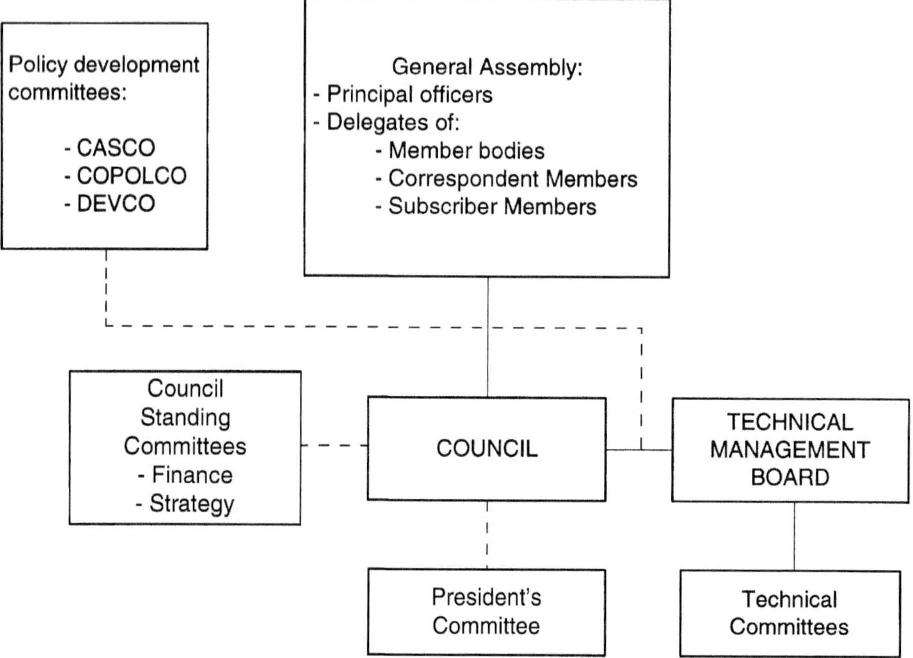


**FIGURE 25-2** The ISO organization. (*Legend:* CASCO—Committee on Conformity Assessment; COPOLCO—Committee on Consumer Policy; DEVCO—Committee on Developing Country Matter.) Solid lines indicate Reporting/responsibility and dashed lines indicate Advisory.

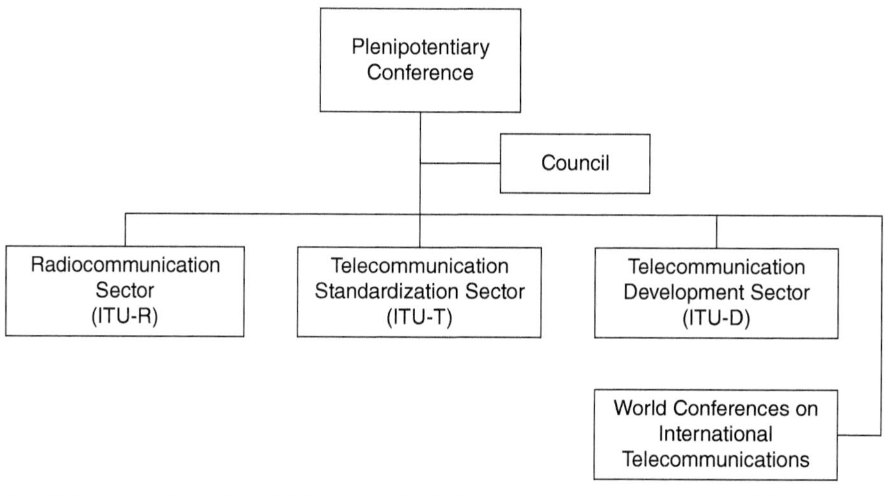


**FIGURE 25-3** Governing bodies of the ITU.

***Joint Technical Committee 1 on Information Technology.*** In recognition of the broad scope and applicability of international standards in the field of information technology, ISO and the IEC approved an agreement in 1987 that formed a joint committee, known as JTC 1 to carry out this work. Membership in JTC 1, like ISO and IEC, is limited to countries; its publications are bilingual (English and French). JTC 1 has 32 member countries (P-members) and 63 observer countries (O-members). It has 20 subcommittees and 4 working groups and an Advisory Group that accomplish its work, and even though it is part of the ISO/IEC, its website is contained on the ISO website.

***The International Committee on Illumination.*** The International Committee on Illumination (CIE), located in Wien (Vienna), Austria, is "an organization devoted to worldwide cooperation and exchange of information on all matters relating to the science and art of light and lighting, color and vision, photobiology and image technology." Its short form, CIE, is an acronym from its French title, Commission Internationale de l'Eclairage. The subjects covered by CIE in the fields of light and lighting include vision, photometry, colorimetry, application of light both indoors and out, environmental effects, aesthetic effects, and means for the production and control of light and radiation. The spectrum of light covered is both natural and synthetic over the ultraviolet, visible, and infrared regions. CIE also addresses the optical, visual, and metrological aspects of the processing and reproduction of images using all types of analogous and digital imaging devices, storage media, and imaging media. Participation in CIE is like that of other international organizations, by national committee, with 37 countries represented at present. In addition to its own publications, the CIE has published standards jointly with IEC (e.g., the IEC/CIE International Lighting Vocabulary) and ISO.

***The Internet Society.*** The ISOC, located in Reston, Va. (U.S.A.), is a nonprofit, nongovernmental, international organization. It was formed to provide an institution home for the Internet Standards process, as well as to provide financial support for the process. It provides the leadership for addressing issues that affect the future of the Internet and it is the organizational home for the groups responsible for Internet infrastructure standards. The ISOC also acts as a global clearinghouse for Internet information and education and facilitates and coordinates Internet-related efforts around the world. The primary organization supported is the Internet Engineering Task Force (IETF), which since its inception had been supported primarily from research supporting agencies of the U.S. government. The combined efforts of the ISOC and IETF have formed and maintained the foundation for all modern networks and Internet products and services. The ISOC also charters the Internet Assigned Numbers Authority (IANA) as the central clearing house to assign and coordinate unique parameter values for Internet protocols. Today, the ISOC has more than 150 organizational members in more than 180 countries. The ISOC also encourages individual memberships.

***The World Trade Organization.*** The WTO, located in Geneva, Switzerland, is an intergovernmental organization and is the only international organization involved in the rules of trade between nations. The WTO is a place where member governments can go to try to resolve trade problems that are between them. The main decision-making bodies are councils and committees consisting of the WTO's entire membership. The authority of the WTO comes from agreements, which are the legal ground rules for international trade and commerce policy. These agreements have three main objectives: "to help trade flow as freely as possible; to achieve further liberalization gradually through negotiation; and to set up an impartial means for settling disputes." Prior to the establishment of the WTO, the GATT had governed international trade and commerce since 1948. The GATT itself was initially an agreement that was provisional. It later became an international organization created to support the agreement but was not recognized in law as an international organization; however, the GATT (i.e., the agreement) has been incorporated into the WTO agreements. Unlike the GATT, which only dealt with trade in goods, the WTO agreements deal with other issues such as copyright, trademarks, patents, industrial designs, and trade secrets. Members of the WTO agree to use international standards (or harmonized standards) to minimize the risk of barriers to trade being introduced by standards. The WTO rules, however, sometimes support maintaining trade barriers when necessary such as in cases where there is a need to protect consumers or to prevent the spread of disease.

***The World Intellectual Property Organization.*** The World Intellectual Property Organization (WIPO), located in Geneva, Switzerland, is one of the United Nation's specialized agencies. As such, it is an intergovernmental agency, dedicated to "the promotion of innovation and creativity for the economic, social, and cultural development of all countries through a balanced and effective international IP system." It is responsible for "the promotion and protection of intellectual property throughout the world through cooperation among States, and for the administration of various multilateral treaties dealing with the legal and administrative aspects of intellectual property." The WIPO currently has 189 member states and administers 26 international treaties. Although the WIPO officially became part of the UN in 1974, its "roots" go back to 1883 when the Paris Convention for the Protection of Industrial Property was the first treaty designed to protect intellectual creations through patents and trademarks. Protection of copyrights was added in 1886 with the Bern Convention for the Protection of Literacy and Artistic Works. These groups united in 1893 to form what became the WIPO in 1970.

## 25.8 REGIONAL ORGANIZATIONS[e]

***European Committee for Electrotechnical Standardization.*** CENELEC, located in Brussels, Belgium, is a not-for-profit organization and has been officially recognized as the European Standards Organization in its field by the EC in Directive 83/189 EEC. Its membership is made up of national committees from 33 countries. Sixteen additional national committees from Eastern Europe, the Balkans, Northern Africa, and the Middle-East work in CENELEC as affiliates. CENELEC has entered into agreements with European associations and federations (currently three) that give them a status of "cooperating partners" and also offers a special partnership status PSB (partner standardization body) to countries outside of Europe. In 2009, it introduced the Technical Liaison Partnership to organizations active in rapidly evolving and innovative market segments. Thousands of technical experts are involved in the development of its standards in more than 300 technical bodies, which are identified as European Norms (ENs) or Harmonization Documents (HDs). At the end of 2010, CENELEC had more than 5800 active standards. ENs in the field of information technology are developed and published jointly with the European Commission for Standardization (CEN).

In developing standards, it is the policy of CENELEC to use an IEC standard if it exists (approximately 70% of CENELEC standards).

Voting in CENELEC is weighted according to the size of the country. Larger countries such as Germany, Italy, France, and the United Kingdom have 29 votes. The smaller countries have three weighted votes depending on size. An affirmative vote must meet two conditions. First, a majority of the national committees must vote affirmative and second, at least 71% of the weighted votes must be affirmative. Once an EN or HD is approved, it is mandatory that it be adopted as a national standard by CENELEC members (this is unlike most standards processes, e.g., those of the IEC, which allow for voluntary adoption). Each CENELEC member is also obligated to withdraw any of its national standards that conflict with an approved EN. On this basis, it is acceptable for a national committee to vote one way in CENELEC (e.g., negative) and another way in IEC (e.g., affirmative). All documents are published in three official languages, English, French, and German; however, any national committee may translate documents into their own language.

[e]*Author's note:* It is not possible, nor practical, to list all the regional and national organizations, and other standards organizations, that are involved with electrotechnology, telecommunications, and information technology standards. For those readers having additional interest, the World Wide Web provides one with an excellent means of researching these organizations. The reader is also referred to the listing of websites at the end of this section.

***European Telecommunications Standards Institute.*** European Telecommunications Standards Institute (ETSI), located in Sophia Antipolis, France, is a nonprofit organization whose mission is "to produce telecommunications standards for use today and in the future." It is an open forum (i.e., to European organizations) that brings together over 800 members from 66 countries, representing governments, network operators, manufacturers, service providers, and users across 5 continents. ETSI, unlike CEN and CENELEC, produces voluntary standards, some of which may be adopted by the EC as the technical base for directives or regulations. Promotion of international standardization is favored by ETSI, and as such, it coordinates its activities with international bodies (primarily with the ITU).

Voting on standards within ETSI is done according to its Rules of Procedure depending on the issue, with only those that are full and associate members allowed to vote. The voting method is weighted individual voting.

## 25.9 NATIONAL ORGANIZATIONS

***The American National Standards Institute.*** ANSI, headquartered in Washington, DC with its operation center located in New York City, is a federation founded in 1918 that has been the administrator and coordinator of the U.S. private-sector voluntary standardization for over 90 years. It is a private, not-for-profit organization supported by a number of organizations (i.e., its membership) in the public and private sectors. Today ANSI has more than 1100 members from businesses, professional societies, trade associations, standards developers, consumer and labor organizations, and government agencies and institutions.

ANSI's primary goal is "the enhancement of global competitiveness of U.S. business and the American quality of life by promoting and facilitating voluntary consensus standards and ensuring their integrity." ANSI also "promotes the use of U.S. standards internationally, advocates U.S. policy and technical positions in international and regional standards organizations, and encourages the adoption of international standards as national standards where these needs meet the needs of the user country." ANSI is the U.S. member of ISO and, via the United States National Committee (USNC), the IEC.

ANSI does not itself develop standards but functions, rather, as a coordinating body for the purpose of encouraging development and adoption of worthwhile standards as American National Standards. It looks to its organizational members, as well as to other concerned organizations, for accomplishing the task of standards development. These development activities may be performed wholly within one of these organizations, or in accredited committees organized and administered by one of these organizations and operating under a set of rules meeting the basic procedures of ANSI.

A large number of standards (more than 11,000 in 2015) that are processed for adoption by ANSI are developed, approved, and published by any of more than 240 standards development organizations accredited by ANSI. Organizations wishing to become ANSI accredited must develop and maintain procedures (subject to periodic audit by ANSI) that consistently adhere to a set of ANSI requirements or procedures entitled "ANSI Essential Requirements: Due Process Requirements for American National Standards,"[f] that govern the consensus development process. Three of the more commonly used methods are described below:

1. *Accredited Organization Method.* The organization method is most often used by associations and societies that have, among other activities, an interest in developing standards. Although participation on the consensus body is open to all interested parties, members of the consensus body often participate as members in the association or society. The organization's procedures must meet the general requirements of the ANSI procedures. By choosing to use this method, flexibility is provided, allowing the standards developer to utilize a system that accommodates its particular structure and practices.

[f]Available for download from the ANSI website.

2. *Accredited Standards Committee Method.* Accredited standards committees are standing committees of directly and materially affected interests created for the purpose of developing a document and establishing consensus in support of this document for submittal to ANSI. The committee method is most often used when a standard affects a broad range of diverse interests or where multiple associations or societies with similar interests exist. The committee serves as a forum where many different interests, without a common membership in an organization or society, can be represented. Accredited standards committees are administered by a secretariat. An accredited standards committee develops and maintains its own operating procedures consistent with the requirements of the ANSI procedures.
3. *Accredited Canvass Method.* A standards developer using the canvass method identifies, to the extent possible, those who are directly and materially affected by the activity in question and conducts a letter ballot or canvass of those interests to determine consensus on a document. Although canvass developers provide ANSI with internal procedures used in the development of the draft American National Standard, the due process used to determine consensus begins after the draft standard has been developed. Standards developers using the canvass method must use procedures consistent with the requirements of the ANSI procedures.

Figure 25-4 is an organization chart for ANSI. Standards work is under the Executive Committee (EC). The EC has responsibilities for American National Standards and U.S. participation in those international standards activities in which ANSI participates. EC is also responsible for maintaining ANSI procedures for standards development, development and coordination of U.S. position in international standards activities, and establishing and supervising such groups as needed to carry out these responsibilities. The approval and withdrawal of American National Standards has been delegated by the EC to the Board of Standards Review (BSR), one of the boards that report to the National Policy Committee.

***Chinese Standardization Organizations.*** China has become a driver on international standardization over the last decade and as one of the largest emerging markets its national standard development activities are of interest to those wishing to enter and succeed in that market. China is very active in most major international and regional standardization organizations. The standards system in China is administered by the General Administration of Quality Supervision, Inspection and Quarantine and headed primarily by the Standardization Administration of the People's Republic of China. Other governmental and private sector organizations provide a significant amount of additional input and participate in the standards process. Within China there are several organizations officially working on standardization that are outlined below. In addition, China has recently adopted new policies encouraging the formation of consortia or alliances to allow for the rapid introduction of new technologies to be brought to the marketplace. These consortia seem to be still subject to more governmental oversight than consortia from North America or Europe but are a beginning for China and Asia in general to be more nimble in the development of voluntary standards and technologies.

***China Communications Standards Association.*** A major research institute in information and communications technology (ICT) standardization in China. **China Communications Standards Association (CCSA)** carries out standardization activities throughout China. The organization is composed of a General Assembly and a Council that oversees various committees focusing on standardization in ICT fields. The Ministry of Information Industry is the government body that oversees CCSA activities. The scope of activities for CCSA include standards drafting, interoperability testing, study and surveys on communication standards, and organizing domestic and international exchanges in ICT and standardization.

***China Electronics Standardization Institute.*** A major research institute in ICT standardization in China. **China Electronics Standardization Institute (CESI)** was founded in 1963 and is an organization involved in standardization in the IT and electronics fields under the Ministry of Industry and Information Technology. Some significant steps that CESI has taken include, taking charge of

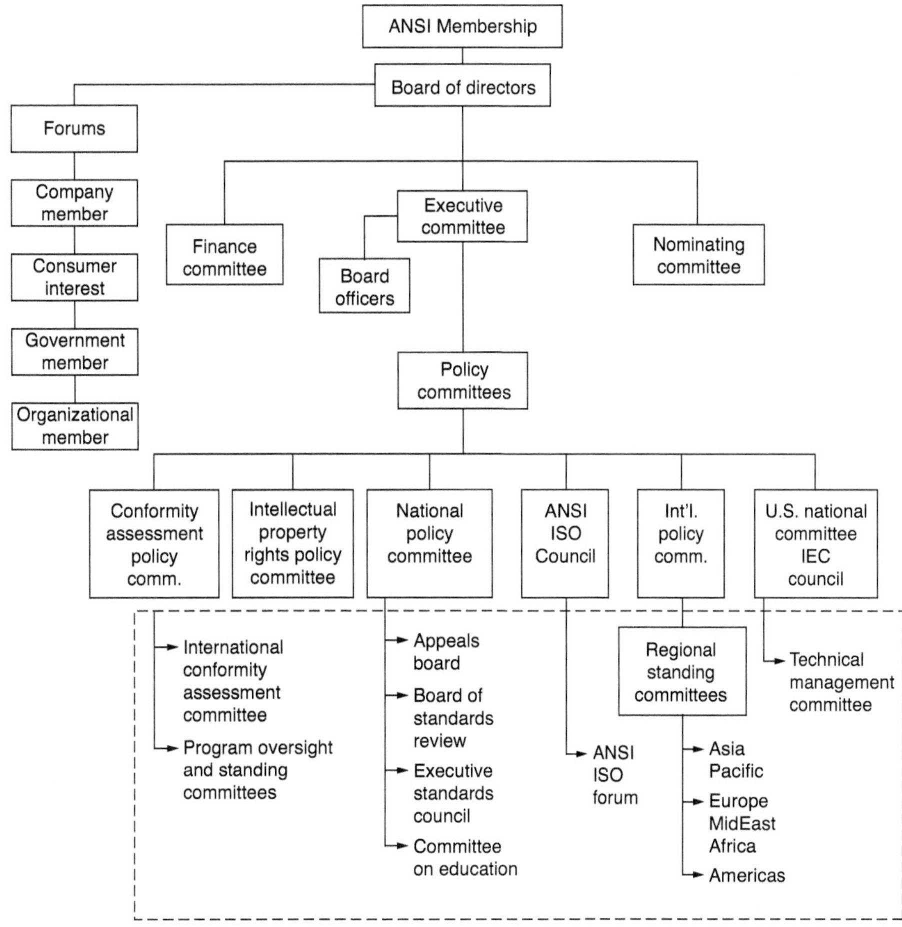


**FIGURE 25-4** ANSI organization chart.

centralized management for 51 national technical mirrors to IEC TC/SCs and ISO/IEC JTC 1 SCs. With 11 secretariats of national standardization technical committees, CESI has played a role in disseminating and applying standards in the field of electronics and information technology.

***The Association for Electrical, Electronic and Information Technologies.*** The Association for Electrical, Electronic and Information Technologies (VDE), located in Frankfurt am Main, Germany, is a nonprofit, nongovernmental organizations founded in 1893 and is one of the largest technical and scientific organizations in Europe with more than 34,000 members. In the area of standards in electrotechnology, VDE participates with The German Institute for Standardization (DIN) on the German Commission for Electrical, Electronic and Information Technologies of DIN and VDE (DKE), in both the IEC and CENELEC. The DKE is a joint organization of VDE and DIN; however, VDE is responsible for the administration of DKE. The focus of standardization has shifted in the recent past toward the development of European Standards (based on international standards where possible). Pure national standards activity has dropped to approximately 5% of all work done.

VDE also performs certification and testing through the VDE Testing and Certification Institute, which was originally established in 1920 as the VDE Test Center. The VDE Testing and Certification Institute works together with other organizations, such as **Underwriters' Laboratories, Inc. (UL)** in the United States and CSA in Canada.

***Standards Australia.*** SA, located in Sydney, Australia, is an independent not-for-profit organization whose primary role is "to meet Australia's need for contemporary, internationally aligned standards and related services." SA was originally founded in 1922, as the Commonwealth Engineering Standards Association. In 1929, it became the Standards Association of Australia and was granted a Royal Charter in 1950. Its name officially changed to Standards Australia in 1988. SA represents Australia on both the ISO and the IEC. It is also involved in regional standardization activities and was a founding member of the Pacific Area Standards Congress (PASC). It maintains close ties with Standards New Zealand (SNZ) and has a formal agreement with SNZ for developing and publishing joint standards.

At present, approximately 8000 volunteer experts serve on the many committees that are responsible for the development and maintenance of more than 7000 Australian Standards. SA is no longer involved in system and product certification services.

***Standards Council of Canada.*** The SCC, located in Ottawa, Canada, is a federal crown corporation with a mandate to "promote efficient and effective standardization in Canada." The SCC reports to Parliament through the Minister of Industry. The Governing Council of the SCC consists of 13 members, appointed by the federal government, representing a broad spectrum of stakeholder interests. The council is responsible for setting the strategic direction of the organization, ensuring the fulfillment of the SCC's mandate, and providing direction on governance matters. Canadian participation in the ISO and the IEC operates under a program under the SCC. The SCC acts as the secretariat for the Canadian National Committee of the ISO (CNC/ISO) and the Canadian National Committee of the IEC (CNC/IEC).

There are more than 15,000 Canadian volunteers involved in standardization activities and some 400 organizations have been accredited by the SCC, as SDOs, certification organizations, testing and calibration laboratories, registration organizations, or auditor certifiers and trainers. Of these organizations, four have been accredited as SDOs. These are Bureau de normalization du Québec (BNQ), Canadian General Standards Board (CGSB), Canadian Standards Association (CSA), and Underwriters' Laboratories of Canada (ULC). These accredited SDOs may submit standards they develop to the SCC for approval as National Standards of Canada (NSC). To become an NSC, a standard must be developed by a process incorporating the fundamental principles, be subject to public scrutiny, be consistent with or incorporate appropriate international and national standards, and be available in the English and French languages. A NSC cannot in any way be presented in such a manner that it will act as a barrier to trade.

## 25.10 OTHER STANDARDS DEVELOPERS

***The Institute of Electrical and Electronics Engineers—Standards Association.*** The Institute of Electrical and Electronics Engineers—Standards Association (IEEE-SA), located in Piscataway, N.J. (U.S.A.), has been responsible for all matters related to standards since its formation in 1998. Actually, the concept of the IEEE-SA was approved by the IEEE Board of Directors in 1996 as part of an overall restructuring of IEEE as it moved into the twenty-first century. Prior to 1998, the IEEE standards program was administered by a Standards Board that reported to the Board of Directors. Under the IEEE-SA, the Standards Board still exists; however, it reports to the IEEE-SA Board of Governors. Standards development in the IEEE, which is the world's largest professional society, with more than 400,000 members in 160 countries, goes back to 1884, when the AIEE began to develop standards. Even today, the IEEE publishes 25% of all of the world's literature in electrotechnology.

The IEEE-SA was formed "to provide a major entity that would offer increased responsiveness to the standards interests of IEEE societies and their represented industries." The IEEE-SA provides services and innovative standards development processes that will keep pace with the needs of users. New classes of membership in the IEEE-SA have been approved, such as corporate membership, which will assist the IEEE-SA in realizing its strategic objectives. At present, the IEEE-SA has more than 30,000 volunteers and more than 200 corporate members participating in its standards activities. The IEEE-SA develops standards in such diverse subjects as broadcasting and communications, electrical practices for large industry (mining, textiles, shipbuilding, transportation, cement plants, and others), instrumentation and measurement, insulators and insulation, magnetics, motors and generators, nuclear power, power apparatus and systems, recording, symbols and units, electrical transmission and distribution, medical devices, electrical practices for commercial buildings and hospitals, emergency power systems, stationary batteries, and information technology. The IEEE-SA also has dual logo agreements and joint standards development agreements for certain standards with the IEC, ISO, and other standards bodies. The IEEE-SA also serves as secretariat for a number of accredited standards committees (ASCs), an example of which is ASC C2, the *National Electrical Safety Code®* (NESC®).[g]

***The National Electrical Manufacturers Association.*** National Electrical Manufacturers Association (NEMA), located in Rosslyn, Va. (U.S.A.), is responsible for the development and maintenance of over 500 standards. NEMA is the largest trade organization for manufacturers of electrical products in the United States, and its more than 475 member companies are domestic firms varying in size from small companies to large diversified companies. It develops standards in the technical committees of its eight divisions covering products in such fields as building equipment, power electronics, industrial electrical equipment, insulation, lighting, power equipment, wire and cable, radiation imaging products, and industrial automation.

NEMA technical committees comprise engineers designated to represent member companies who are manufacturers of electrical equipment. Since manufacturers are most knowledgeable in the technology associated with their respective products, NEMA committees are highly competent in developing product standards that realistically take into consideration the economic tradeoffs that are essential to practical standardization. Standards are adopted by consensus with final approval given by the NEMA Codes and Standards Committee.

NEMA Standards are generated in five classifications:

1. NEMA Standard—defines a commercially standardized product subject to repetitive manufacture.
2. Suggested Standard for Future Design—suggests an approach to future product improvement or development.
3. Authorized Engineering Information—included as part of other NEMA standards to explain data or information.
4. Application guides—provides information for the layperson, inspector, installers, and employees working with a product.
5. White papers—intended for a broad audience, they contain practical tips on subjects like phantom voltages and hazards of working with "hot" electrical equipment.

NEMA is a member of and actively participates in ANSI. It administers the work of a number of ASCs, acting as the secretariat.

***The National Fire Protection Association.***[h] The NFPA, located in Quincy, Mass. (U.S.A.), has been active in standards development since its founding in 1896. Working under the direction of

[g]*National Electrical Safety Code®* is a registered IEEE trademark, and its acronym, NESC®, is a registered IEEE service mark of the IEEE.

[h]*National Electrical Code®*, *Life Safety Code®*, and NEC® are registered trademarks of NFPA.

its Standards Council, the technical committees of NFPA, composed of organization representatives, personal members, and liaison members from other technical committees, develop standards documents that are then subject to public review as a result of advance publication, and are finally approved at a semiannual meeting of the entire NFPA membership. Although dedicated to fire prevention, NFPA is responsible for a series of electrical standards, the most noted of which are the *National Electrical Code* (NFPA 70) and the *Life Safety Code* (NFPA 101).

The NEC covers "the installation of electric conductors and equipment in public and private buildings or other structures (including mobile homes, recreational vehicles, and floating buildings), industrial substations, and other premises (such as yards, carnivals, and parking lots). The NEC also covers installation of optical fiber cable." The NEC is adopted and enforced in all the 50 U.S. states. It is also the basis for electrical codes in several other countries. The first edition of the NEC was published in 1897. The NFPA took over responsibility for the NEC in 1911. The NEC is produced and maintained by volunteers on 19 codemaking panels and a correlating committee that oversees the work of the panels.

The *Life Safety Code* "provides minimum requirements for the design, operation, and maintenance of buildings and other structures for safety to life from fire and similar emergencies." The *Life Safety Code* has been adopted by many states. Additionally, all health-care facilities in the United States receiving Medicare or Medicaid funding must comply with the *Life Safety Code*. The groundwork for the *Life Safety Code* began in 1913, when the NFPA appointed its first "Committee on Safety to Life." Its first pamphlet was on exit drills in factories, schools, department stores, and theaters. By 1921, the committee had produced a more comprehensive *Building Exit Code*. The *Life Safety Code* is produced and maintained by members of 15 technical committees.

***Underwriters' Laboratories, Inc. (UL).***[i] UL, located in Northbrook, Ill., (U.S.A.) is the leading third-party certification organization in the United States and the largest in North America. It is an independent, not-for-profit organization, founded in 1894, whose mark is recognized throughout the world as a symbol of safety. It has expanded its business to focus on five areas, namely, product safety, environmental, life and health, university, and verification services.

UL maintains and operates laboratories for testing devices, systems, and materials with relation to public safety. Products so tested and meeting its requirements are eligible for UL "listing." UL maintains an inspection and follow-up program in factories where UL-listed devices are manufactured. UL representatives conduct in-factory and in-the-field inspections of manufacturers' procedures for assuring production compliance with UL requirements. Such requirements appear in appropriate UL Standards for Safety, which are developed by UL under procedures that involve consultation with industry and government experts and consumers, among others.

The majority of insurance underwriters in the United States, and many federal, state, and municipal authorities, either accept or require listing or classification by UL as a condition of their recognition of devices, systems, and materials having a bearing upon life and fire hazards.

UL is divided into several engineering departments, each dealing with distinct and separate subjects as follows: electrical heating, air conditioning, and refrigeration; casualty and chemical hazards; burglary protection and signaling; fire protection; and marine. Each department has prepared standards for systems, materials, and appliances.

UL publishes (lists) the names of companies who have demonstrated the ability to provide products conforming to its requirements. Listing authorizes the manufacturer to use the laboratories' listing mark (classification marking, recognition marking, or certificate) on the listed products. UL submits its standards to ANSI for adoption as American National Standards.

UL is active in international standards development and product certification, and is a registrar for quality systems for various systems, including ISO 9000. UL acquired the Danish National Testing and Certification Organization (DEMKO) in July 1996 from the kingdom of Denmark, making it a wholly owned subsidiary of UL. This acquisition has enabled direct European certification of products to customers worldwide. In 2011 UL acquired the quality assurance business of STR Holdings,

[i]Underwriters' Laboratories, Inc.® and its acronym, UL®, are registered trademarks of Underwriters' Laboratories.

Inc., allowing UL to complement its existing expertise and to provide a complete suite of QA services focused on consumer products.

UL is responsible for the development and maintenance of more than 1000 safety standards, the evaluation of more than 19,000 products, and more than 5700 registrations to a management system standard. In 2004, UL estimated that approximately 19 billion products enter the marketplace each year bearing a UL mark.

## 25.11 U.S. GOVERNMENT REGULATORY STANDARDS BODIES

***U.S. Government Policy on Voluntary Standardization.*** In 1995 the National Technology Transfer and Advancement Act (NTTAA) codified pre-existing polices on the development and use of standards in Circular A-110 and assigned to NIST a coordinating role in conformity assessment activities. In January 2016, the U.S. Office of Management and Budget published an new circular OMB Circular A-119 "Federal Participation and Use of Voluntary Consensus Standards and in Conformity Assessment Activities." The revision to the Circular was in part to provide guidance for Federal agencies to meet the U.S. strategic objectives for Federal engagement in standards. The Circular applies to all agencies and agency representatives who use standards or conformity assessment and/or participate in the development of standards. The NTTAA and A-119 note that all Federal agencies "must use voluntary consensus standards in lieu of government unique standards in their procurement and regulatory activities except where inconsistent with law or otherwise impractical." The revision also stressed the importance of coordination and harmonization with international standardization efforts. The comments in the proceeding for this revision, and the text of the circular itself demonstrate the continued value of standardization to the U.S. government and in the U.S. market. Below is a brief discussion of some of the major U.S. regulators

***The U.S. Consumer Products Safety Commission.*** The Consumer Products Safety Commission (CPSC) is an independent federal regulatory agency that was created by Congress in 1972, in the Consumer Products Safety Act. The mission of the CPSC is to "protect the public against unreasonable risks of injury and death from thousands of types of consumer products under the agency's jurisdiction." The CPSC has jurisdiction over approximately tens of thousands of consumer products. There are some exceptions such as motor vehicles, trucks, motorcycles, tires, and car seats, which are covered by the Department of Transportation; foods, drugs, radiation, medical devices, veterinary medicines, and cosmetics, which are covered by the Food and Drug Administration (FDA); pesticides and fungicides, which are covered by the Environment Protection Agency; radioactive materials, which are covered under the Nuclear Regulatory Commission (NRC); and alcohol, tobacco, ammunition, and firearms, which are covered by the Department of the Treasury. Jurisdiction of amusement rides is governed by the states.

The CPSC works to reduce the risk of death and injury to consumers by developing voluntary standards with industry, and issuing and enforcing mandatory standards. If no feasible standard would adequately protect the public, the CPSC can ban a product. It can also require a recall of products to arrange for their repair. The CPSC also conducts research on potential product hazards and sponsors consumer educational activities (e.g., through the media). All work related to standards or regulations are published in the *Federal Register*. Initially, notice of the proposed standards or regulation is published for comment. Final regulations are also published in the *Federal Register*.

***The U.S. Environmental Protection Agency.*** The Environmental Protection Agency (EPA) is an independent federal agency created by Congress in 1970. Its mission is "to protect human health and the environment." The EPA develops regulations and has additional responsibilities for enforcement of these regulations.

The EPA, like other U.S. government agencies, follows a prescribed rulemaking process. The U.S. Congress establishes these requirements to support the development of quality rulemaking and protect the rights of those affected by the rules. Examples of these requirements include the Administrative Procedure Act, the Regulatory Flexibility Act, the Unfunded Mandates Reform Act, the Paperwork Reduction Act, and the National Technology Transfer and Advancement Act. The five stages of rulemaking are

1. *Prerulemaking*—actions to determine whether the agency should initiate rulemaking
2. *Proposed rules*
3. *Final rules*
4. *Long-term actions*—prerulemaking, proposed rules, and final rules expected to be published beyond the next 12 months
5. *Completed actions*—actions that are promulgated and published or actions that are no longer being considered

Comments from interested parties are encouraged during the rulemaking process. All rulemaking activities are published in the *Federal Register*.

***The U.S. Federal Communications Commission.*** The Federal Communications Commission (FCC) is an independent U.S. government agency established in 1934 by Congress in the Communications Act of 1934. It "regulates interstate and international communications by radio, television, wire, satellite and cable in all 50 states, the District of Columbia and all U.S. territories." The FCC is organized in bureaus and offices, and the official statement of the FCC actions is called an *order*. Although there are 14 bureaus and offices, most of the FCC's documents are issued by 7 major regulatory bureaus or offices. When the FCC considers a change to its regulations, it issues a Notice of Proposed Rule Making. Essentially all documents issued by the FCC since 1994 can be found online through its website.

***The U.S. Federal Trade Commission.*** The FTC deals with issues that touch the economic life of every American in that it enforces federal antitrust and consumer protection laws and seeks to ensure that the nation's markets function competitively, and are vigorous, efficient, and free of undue restrictions. It is the only federal agency with both consumer protection and competition jurisdiction in broad sectors of the economy. The FTC has enforcement and administrative responsibilities under numerous acts. Its rules are published in Title 16 of the Code of Federal Regulations.

***The U.S. Food and Drug Administration.*** The FDA is one of the oldest federal agencies involved with the protection of consumers. In the area of electrotechnology standards, it is involved with setting standards for medical devices and radiation-emitting products. This work is done under the FDA's Center for Devices and Radiological Health (CDRH). The CDRH is responsible for "ensuring the safety and effectiveness of medical devices and eliminating unnecessary human exposure to man-made radiation from medical, occupational, and consumer products." Products covered in the CDRH's scope include thousands of medical devices (e.g., pacemakers), videodisplay terminals, microwave ovens, medical x-ray machines, and medical ultrasound devices.

***National Institute of Standards and Technology.*** NIST, which is under the U.S. Department of Commerce, has been involved in the development of standards since its founding in 1901 (then known as the National Bureau of Standards). The documents it produced played a great role in the industrial development of the United States in a broad range of industries, including steel manufacturing, railroads, electric power, and telephone communications. Under NTTAA NIST coordinates federal, state, and local technical standards and conformity assessment activities, as well as coordinating with those in the private sector. NIST has also developed an National Computer Security

Resource Center to coordinate efforts in this important area. In quality systems, it sponsors the Baldridge Performance Excellence Program and the National Voluntary Laboratory Accreditation Program.

NIST also develops standards and guidelines for federal computer systems. These documents, known as *Federal Information Processing Standards* (FIPSs), are approved by the Secretary of Commerce. FIPSs are generally developed only when there are no acceptable industry standards available or there are compelling federal requirements (e.g., security). There are also Federal Standards (FED-STDS) developed for telecommunications; however, these are developed by the National Communications System (NCS) and are approved by the General Services Administration (GSA).

The major focus of NIST is on information technologies although NIST has been actively involved in smart grid initiatives in recent years. NIST cooperates with national and international standards organizations, trade associations, consortia, user groups, and so forth to have needed standards developed.

***The U.S. Nuclear Regulatory Commission.*** The NRC is an independent federal agency established by Congress under the Energy Reorganization Act of 1974. Its mission is to regulate the nation's civilian use of byproduct, source, and special nuclear materials to ensure adequate protection of the public health and safety, the common defense and security, and to protect the environment. Its predecessor was the U.S. Atomic Energy Commission. NRC regulations are issued under the U.S. *Code of Federal Regulations* (CFR) Title 10, Chapter 1. Typically, rulemaking is initiated by the NRC's staff; however, any member of the public may petition the NRC to develop, modify, or rescind any regulation. During the rulemaking process, the documents are published in the *Federal Register* and interested parties are allowed at least one opportunity to comment. In some instances, the NRC will hold meetings and workshops before a proposed rule is drafted to obtain a broad range of input from interested parties. Most often in the area of electrotechnology, the NRC will endorse existing industry standards by issuing a regulatory guide (RG). The RG will describe methods that are acceptable to the NRC for applying the standard. The NRC may also use the RG to make recommendations or guides presented in a standard's mandatory requirements. The NRC also issues NUREG reports to provide information and expertise to support the NRC's decision making and to assess potential technical issues. In the area of electrical standards there has been renewed activity as a result of recent applications to build and license new nuclear power generating stations. There has also been significant activity in the area of nuclear generating station security.

***The U.S. Occupational Safety and Health Administration.*** Occupational Safety and Health Administration (OSHA), which is under the U.S. Department of Labor, was established by Congress in 1970 under the Occupational Safety and Health Act to "ensure safe and healthful working conditions for working men and women by setting and enforcing standards and by providing training, outreach, education, and assistance." OSHA has responsibilities for developing standards and regulations, and enforcement. OSHA may set standards on its own initiative or in response to petitions from state and local governments, employers, labor representatives, standards organizations, or any other interested party. If it is decided to develop a standard, any one of several advisory committees may be tasked with the development. The process is very similar to that for other U.S. government agencies, including notice and comment period for interested parties.

## 25.12 CONTACTING STANDARDS ORGANIZATIONS

There are literally thousands of organizations around the world that are involved in standards development. The World Wide Web (WWW) makes contacting these organizations easier than ever. The following listing is just a small sample of what is out on the Web. Many of the websites listed contain links to the websites of other standards developers.

## International

**IEC (International Electrotechnical Commission)**
3, rue de Varembé
PO Box 131
CH-1211 Geneva 20
Switzerland
http://www.iec.ch

**ISO (International Organization for Standardization)** ISO
Central Secretariat
Chemin de Blandonnet 8
CP 401
1214 Vernier, Geneva
Switzerland
http://www.iso.org

**JTC1 (Joint Technical Committee 1 Information Technology)**[j]
http://www.iso.org/iso/jtc1_home.html

**ITU (International Telecommunication Union)**
Place des Nations
CH-1211 Geneva 20
Switzerland
http://www.itu.ch

**ISOC (Internet Society)**
1755 Wiehle Ave.
Suite 201
Reston VA 20190-5108
U.S.A.
http://www.isoc.org
The ISOC also has an office located in Geneva, Switzerland and the address can be found on the website.
(At this website the information on the following related groups can be found.)

**IETF (Internet Engineering Task Force)**

**IESG (Internet Engineering Steering Group)**

**IAB (Internet Architecture Board)**

**IRTF (Internet Research Task Force)**

**ICANN (The Internet Corporation for Assigned Names and Numbers) and IANA (Internet Assigned Numbers Authority)**
http://www.iana.org

**NRO (The Numbers Resource Organization) and RIRs (Regional Internet Registries)**

**CIE (International Commission on Illumination)**
Babenbergerstraße 9/9A
1010 Vienna
Austria
http://www.cie.co.at

## Regional Organizations

**CEN (European Committee for Standardization)**
http://www.cen.eu/cen

**ETSI (European Telecommunications Standards Institute)**
650, route des
Lucioles
Sophia-Antipolis 06560 Valbonne France
http://www.etsi.org

**CENELEC (European Committee for Electrotechnical Standardization)**
[In 2010, CEN and CENELEC created a joint CEN-CENELEC Management Centre (CCMC) that replaced the CEN Management Centre and the CENELEC Central Secretariat.]

**CCMC**
4th Floor
Avenue Marnix 17
B-1000 Brussels
Belgium
http://www.cenelec.eu
[At this website a considerable number of contacts are provided for other regional organizations such as the European Computer Manufacturers Association (ECMA), as well as the national standards bodies for member countries such as the National Standards Authority of Ireland (NSAI), and affiliate countries.]

[j]JTC1 is a joint committee of both the ISO and the IEC. The ITU has an official liaison to the committee.

## Selected National Organizations

*Australia*
**SA (Standards Australia)**
http://www.standards.org.au

*Canada*
**CSA (Canadian Standards Association)**
178 Rexdale Blvd.
Toronto, ON Canada M9W 1R3
http://www.csagroup.org/
**SCC (Standards Council of Canada)**
600-55 Metcalfe Street
Ottawa, ON K1P 6L5
Canada
**CNC/IEC (Canadian National Committee of the IEC)**
**CNC/ISO (Canadian National Committee of the ISO)**
http://www.scc.ca

*France*
**UTE (Union Technique de l'Electricité)**
http://www.afnor.org
**AFNOR (L'Association FranÇaise de Normalisation)**
http://www.afnor.org

*Germany*
**DIN (German Institute for Standardization)**
http://www.din.de
**VDE (The Association for Electrical, Electronic and Information Technologies)**
**DKE (German Commission for Electrical, Electronic and Information Technologies of DIN and VDE)**
http://www.vde.com/de

*Italy*
**CEI (Comitato Elettrotecnico Italiano)**
http://www.ceiweb.it
(Click on English flag for the website in English)

*United Kingdom*
**BSI (British Standards Institution)**
**BEC (British Electrotechnical Committee)**
http://www.bsigroup.com

*United States*

**ANSI (American National Standards Institute)**

**USNC (U.S. National Committee of the IEC)**

25 West 43rd Street, 4th Floor

New York, NY 10036

U.S.A.

http://www.ansi.org

(At this website a considerable number of links are provided to websites for standards-developing organizations located in the United States.)

**ASC T1 (Accredited Standards Committee T1—Telecommunications)**

(T1 was retired in 2004 and its work assumed by ATIS.)

http://www.atis.org

**INCITS (InterNational Committee for Information Technology Standards)**

(Formerly Accredited Standards Committee X3 and NCITS.)

http://www.incits.org

## Selected Standards-Developing Organizations

**AAMI (Association for the Advancement of Medical Instrumentation)**

http://www.aami.org

**ABMA (American Bearing Manufacturers Association)**

http://www.americanbearings.org

**AEIC (Association of Edison Illuminating Companies)**

http://www.aeic.org

**AHAM (Association of Home Appliance Manufacturers)**

http://www.aham.org

**AIIM**

**(Formerly Association for Information and Image Management-The EMC Association)**

http://www.aiim.org

**ANS (American Nuclear Society)**

http://www.ans.org

**API (American Petroleum Institute)**

http://api.org

**ARINC (Aeronautical Radio, Inc.)**

**ARINC is now part of Rockwell Collins**

http://www.rockwellcollins.com/

**ASA (Acoustical Society of America)**

http://acousticalsociety.org

**ASQ (American Society for Quality)**

http://www.asq.org

**ASTM International**

http://www.astm.org

**ATIS (Alliance for Telecommunications Industry Solutions)**

http://www.atis.org

**EEI (Edison Electric Institute)**

http://www.eei.org

**ECIA (Electronic Components Industry Association)**

[Formed by a merger of the Electronic Components Association (ECA) and the National Electronic Distributors Association (NEDA); standards produced by the former Electronic Industries Alliance (EIA) are maintained by ECIA]

www.eciaonline.org

**TechAmerica**

[Formed by a merger of the Government Electronics & Information Technology Association (GEIA), American Electronics Association (AeA), Cyber Security Industry Alliance (CSIA), and Information Technology Association of America (ITAA)]

www.geia.org

**JEDEC**

http://www.jedec.org

**TIA (Telecommunications Industry Association)**

http://www.tiaonline.org

**IS Alliance (Internet Security Alliance)**

http://www.isalliance.org/

**Electro (Electro-Federation of Canada)**
**CAMA (Canadian Appliance Manufacturers Association)**
**CEMC (Consumer Electronics Marketers of Canada)**
**EC (Electrical Council)**
**IMR (Installation, Maintenance and Repair Sector Council and Trade Association)**
http://www.electrofed.com
**IEEE (Institute of Electrical and Electronics Engineers)**
http://www.standards.ieee.org
**IESNA (Illuminating Engineering Society of North America)**
http://www.iesna.org
**ISA (The International Society of Automation)**
http://www.isa.org
**IMAPS (International Microelectronics and Packaging Society)**
http://www.imaps.org
**ITI (Information Technology Industry Council)**
http://www.itic.org
**NACE (NACE International)**
http://www.nace.org
**NEMA (The Association of Electrical Equipment and Medical Imaging Manufacturers)**
(Includes standards of the Insulated Cable Engineers Association, ICEA)
http://www.nema.org
**NETA (InterNational Electrical Testing Association)**
http://www.netaworld.org
**NFPA (National Fire Protection Association)**
http://www.nfpa.org
**SAE (SAE International)**
http://www.sae.org
**SEMI (Semiconductor Equipment and Materials International)**
http://www.semi.org/en/
**SMPTE (Society of Motion Picture & Television Engineers)**
http://www.smpte.org
**UL (Underwriters' Laboratories, Inc.)**
http://www.ul.com
**VITA (VMEbus International Trade Association)**
http://www.vita.com

## U.S. Government

**CPSC (Consumer Products Safety Commission)**
http://www.cpsc.gov
**EPA (Environmental Protection Agency)**
http://www.epa.gov
**FCC (Federal Communications Commission)**
http://www.fcc.gov
**FDA (Food and Drug Administration)**
http://www.fda.gov
**FTC (Federal Trade Commission)**
http://www.ftc.gov
**NIST (National Institute of Standards and Technology)**
http://www.nist.gov
**NRC (Nuclear Regulatory Commission)**
http://www.nrc.gov
**NCS (National Communications System)**
http://www.hhs.gov/ocio/ea/National%20Communication%20System/
**NTIA (National Telecommunications & Information Administration)**
http://www.ntia.doc.gov
**OSHA (Occupational Safety and Health Administration)**
http://www.osha.gov

## Other Organizations

**CITEL (Inter-American Telecommunication Commission)**
CITEL Organization of American States
1889 F Street NW, 6th Floor
Washington, DC 20006
U.S.A.
http://www.citel.oas.org

**WIPO (World Intellectual Property Organization)**
34, chemin des Colombettes
CH-1211 Geneva 20
Switzerland
http://www.wipo.int

**WTO (World Trade Organization)**
World Trade Organization
Centre William Rappard
Rue de Lausanne 154
CH-1211 Geneva 21
Switzerland
http://www.wto.org

**The World Wide Web Consortium (W3C)**
In the Americas, contact W3C at:
WC3/MIT
32 Vasser Street, Room 32-G515
Cambridge, MA 02139
U.S.A.
http://www.w3.org

## Webpages of Interest to Those Searching for Standards-Related Information

**NSSN (National Standards System Network)**
(A free online information service providing bibliographic information for more than 225,000 approved standards.)
http://www.nssn.org

**New Approach Standardisation in the Internal Market**
(A website sponsored by CEN, CENELEC, ETSI, the European Commission, and EFTA.)
http://www.newapproach.org

**University of Waterloo—Canada: standards and specifications written by scholarly societies**
(Provides links to the Internet sites of scholarly societies around the world many of which include standards organizations.)
http://www.lib.uwaterloo.ca/society/

**U.S. General Services Administration—Index of U.S. Federal Specifications, Standards and Commercial Items**
(An alpha numerical listing of federal specifications and standards, including a separate listing of all canceled or superseded federal specifications and standards.)
http://www.gsa.gov/portal/content/100847

**FDSys—Federal Digital System**
(Provides online access to official federal government publications, allowing a user to search, browse, and download documents in various formats.)
http://www.gpo.gov/fdsys/

**World Wide Legal Information Alliance**
(Provides information on product standards and the law.)
http://www.wwlia.org

# INDEX